OpenStax
Rice University
6100 Main Street MS-375
Houston, Texas 77005

To learn more about OpenStax, visit https://openstax.org.
Individual print copies and bulk orders can be purchased through our website.

HARDCOVER BOOK ISBN-13	**978-1-947172-51-7**
PAPERBACK BOOK ISBN-13	**978-1-947172-95-1**
B&W PAPERBACK BOOK ISBN-13	**978-1-50669-985-1**
DIGITAL VERSION ISBN-13	**978-1-947172-52-4**
ENHANCED TEXTBOOK ISBN-13	**978-1-947172-53-1**
ORIGINAL PUBLICATION YEAR	**2018**

9 8 7 6 5 4 3 2 1

Printed by
XanEdu
4750 Venture Drive, Suite 400
Ann Arbor, MI 48108
800-562-2147
www.xanedu.com

OpenStax

OpenStax provides free, peer-reviewed, openly licensed textbooks for introductory college and Advanced Placement® courses and low-cost, personalized courseware that helps students learn. A nonprofit ed tech initiative based at Rice University, we're committed to helping students access the tools they need to complete their courses and meet their educational goals.

Rice University

OpenStax, OpenStax CNX, and OpenStax Tutor are initiatives of Rice University. As a leading research university with a distinctive commitment to undergraduate education, Rice University aspires to path-breaking research, unsurpassed teaching, and contributions to the betterment of our world. It seeks to fulfill this mission by cultivating a diverse community of learning and discovery that produces leaders across the spectrum of human endeavor.

RICE

Philanthropic Support

OpenStax is grateful for our generous philanthropic partners, who support our vision to improve educational opportunities for all learners.

Laura and John Arnold Foundation

Arthur and Carlyse Ciocca Charitable Foundation

Ann and John Doerr

Bill & Melinda Gates Foundation

Girard Foundation

Google Inc.

The William and Flora Hewlett Foundation

Rusty and John Jaggers

The Calvin K. Kazanjian Economics Foundation

Charles Koch Foundation

Leon Lowenstein Foundation, Inc.

The Maxfield Foundation

Burt and Deedee McMurtry

Michelson 20MM Foundation

National Science Foundation

The Open Society Foundations

Jumee Yhu and David E. Park III

Brian D. Patterson USA-International Foundation

The Bill and Stephanie Sick Fund

Robin and Sandy Stuart Foundation

The Stuart Family Foundation

Tammy and Guillermo Treviño

Contents

UNIT 2 THE CELL

UNIT 3 GENETICS

UNIT 4 EVOLUTIONARY PROCESSES

UNIT 5 BIOLOGICAL DIVERSITY

CHAPTER 32
Plant Reproduction

UNIT 7 ANIMAL STRUCTURE AND FUNCTION

CHAPTER 33
The Animal Body: Basic Form and Function

CHAPTER 34
Animal Nutrition and the Digestive System

CHAPTER 35
The Nervous System

PREFACE

Welcome to *Biology 2e* (2nd edition), an OpenStax resource. This textbook was written to increase student access to high-quality learning materials, maintaining highest standards of academic rigor at little to no cost.

About OpenStax

OpenStax is a nonprofit based at Rice University, and it's our mission to improve student access to education. Our first openly licensed college textbook was published in 2012, and our library has since scaled to over 25 books for college and AP® courses used by hundreds of thousands of students. OpenStax Tutor, our low-cost personalized learning tool, is being used in college courses throughout the country. Through our partnerships with philanthropic foundations and our alliance with other educational resource organizations, OpenStax is breaking down the most common barriers to learning and empowering students and instructors to succeed.

About OpenStax resources

Customization

Biology 2e is licensed under a Creative Commons Attribution 4.0 International (CC BY) license, which means that you can distribute, remix, and build upon the content, as long as you provide attribution to OpenStax and its content contributors.

Because our books are openly licensed, you are free to use the entire book or pick and choose the sections that are most relevant to the needs of your course. Feel free to remix the content by assigning your students certain chapters and sections in your syllabus, in the order that you prefer. You can even provide a direct link in your syllabus to the sections in the web view of your book.

Instructors also have the option of creating a customized version of their OpenStax book. The custom version can be made available to students in low-cost print or digital form through their campus bookstore. Visit the Instructor Resources section of your book page on OpenStax.org for more information.

Errata

All OpenStax textbooks undergo a rigorous review process. However, like any professional-grade textbook, errors sometimes occur. Since our books are web based, we can make updates periodically when deemed pedagogically necessary. If you have a correction to suggest, submit it through the link on your book page on OpenStax.org. Subject matter experts review all errata suggestions. OpenStax is committed to remaining transparent about all updates, so you will also find a list of past errata changes on your book page on OpenStax.org.

Format

You can access this textbook for free in web view or PDF through OpenStax.org, and for a low cost in print.

About *Biology 2e*

Biology 2e (2nd edition) is designed to cover the scope and sequence requirements of a typical two-semester biology course for science majors. The text provides comprehensive coverage of foundational research and core biology concepts through an evolutionary lens. *Biology* includes rich features that engage students in scientific inquiry, highlight careers in the biological sciences, and offer everyday applications. The book also includes various types of practice and homework questions that help students understand — and apply — key concepts.

The 2nd edition has been revised to incorporate clearer, more current, and more dynamic explanations, while maintaining the same organization as the first edition. Art and illustrations have been substantially improved, and the textbook features additional assessments and related resources.

Coverage and scope

Biology was one of the first textbooks published by OpenStax and has been used by hundreds of faculty and thousands of students since 2012. We mined our adopters' extensive and helpful feedback to identify the most significant revision needs while maintaining the organization that many instructors had incorporated into their courses. Specific surveys, focus groups, and pre-revision reviews, as well as data from our OpenStax Tutor users, all aided in planning the revision.

The result is a book that thoroughly treats biology's foundational concepts while adding current and meaningful coverage in specific areas. *Biology 2e* retains its manageable scope and contains ample features to draw learners into the discipline.

Structurally, the textbook remains similar to the first edition, with no chapter reorganization and very targeted changes at the section level (mostly in biodiversity).

- **Unit 1: The Chemistry of Life.** Our opening unit introduces students to the sciences, including the scientific method and the fundamental concepts of chemistry and physics that provide a framework within which learners comprehend biological processes.

- **Unit 2: The Cell.** Students will gain solid understanding of the structures, functions, and processes of the most basic unit of life: the cell.
- **Unit 3: Genetics.** Our comprehensive genetics unit takes learners from the earliest experiments that revealed the basis of genetics through the intricacies of DNA to current applications in the emerging studies of biotechnology and genomics.
- **Unit 4: Evolutionary Processes.** The core concepts of evolution are discussed in this unit with examples illustrating evolutionary processes. Additionally, the evolutionary basis of biology reappears throughout the textbook in general discussion and is reinforced through special call-out features highlighting specific evolution-based topics.
- **Unit 5: Biological Diversity.** The diversity of life is explored with detailed study of various organisms and discussion of emerging phylogenetic relationships. This unit moves from viruses to living organisms like bacteria, discusses the organisms formerly grouped as protists, and devotes multiple chapters to plant and animal life.
- **Unit 6: Plant Structure and Function.** Our plant unit thoroughly covers the fundamental knowledge of plant life essential to an introductory biology course.
- **Unit 7: Animal Structure and Function.** An introduction to the form and function of the animal body is followed by chapters on specific body systems and processes. This unit touches on the biology of all organisms while maintaining an engaging focus on human anatomy and physiology that helps students connect to the topics.
- **Unit 8: Ecology.** Ecological concepts are broadly covered in this unit, with features highlighting localized, real-world issues of conservation and biodiversity.

Changes to the Second Edition

OpenStax only undertakes second editions when significant modifications to the text are necessary. In the case of *Biology 2e*, user feedback indicated that we needed to focus on a few key areas, which we have done in the following ways:

- **Content revisions for clarity, accuracy, and currency.** The revision plan varied by chapter based on need. About twenty chapters were wholly revised with significant updates to conceptual coverage, research-informed data, and clearer language. In about fifteen other chapters, the revisions focused mostly on readability and clearer language with fewer conceptual and factual changes.
- **Additional end-of-chapter questions.** The authors added new assessments to nearly every chapter, including both review and critical thinking questions. The additions total over 350 new items.
- **Art and illustrations.** Under the guidance of the authors and expert scientific illustrators, especially those well versed in creating accessible art, the OpenStax team made changes to most of the art in *Biology*. You will find examples in the section below. The revisions fall into the following categories:
    - Revisions for accuracy
    - Redesigns for greater understanding and impact
    - Recoloring art for overall consistency
- **Accessibility improvements.** As with all OpenStax books, the first edition of *Biology* was created with a focus on accessibility. We have emphasized and improved that approach in the second edition.
    - To accommodate users of specific assistive technologies, all alternative text was reviewed and revised for comprehensiveness and clarity.
    - Many illustrations were revised to improve the color contrast, which is important for some visually impaired students.
    - Overall, the OpenStax platform has been continually upgraded to improve accessibility.

A transition guide will be available on OpenStax.org to highlight the specific chapter-level changes to the second edition.

Pedagogical foundation

The pedagogical choices, chapter arrangements, and learning objective fulfillment were developed and vetted with the feedback of another one hundred reviewers, who thoroughly read the material and offered detailed critical commentary.

- **Evolution Connection** features uphold the importance of evolution to all biological study through discussions like "The Evolution of Metabolic Pathways" and "Algae and Evolutionary Paths to Photosynthesis."
- **Scientific Method Connection** call-outs walk students through actual or thought experiments that elucidate the steps of the scientific process as applied to the topic. Features include "Determining the Time Spent in Cell Cycle Stages" and "Testing the Hypothesis of Independent Assortment."
- **Career Connection** features present information on a variety of careers in the biological sciences, introducing students to the educational requirements and day-to-day work life of a variety of professions, such as microbiologist, ecologist, neurologist, and forensic scientist.
- **Everyday Connection** features tie biological concepts to emerging issues and discuss science in terms of everyday life. Topics include "Chesapeake Bay" and "Can Snail Venom Be Used as a Pharmacological Pain Killer?"

Art and animations that engage

Our art program takes a straightforward approach designed

to help students learn the concepts of biology through simple, effective illustrations, photos, and micrographs. Biology 2e also incorporates links to relevant animations and interactive exercises that help bring biology to life for students.

- **Visual Connection** features call out core figures in each chapter for student study. Questions about key figures, including clicker questions that can be used in the classroom, engage students' critical thinking and analytical abilities to ensure their genuine understanding.
- **Link to Learning** features direct students to online interactive exercises and animations to add a fuller context and examples to core content.

Below are a few examples of the revised art for *Biology 2e*:

Desmosome

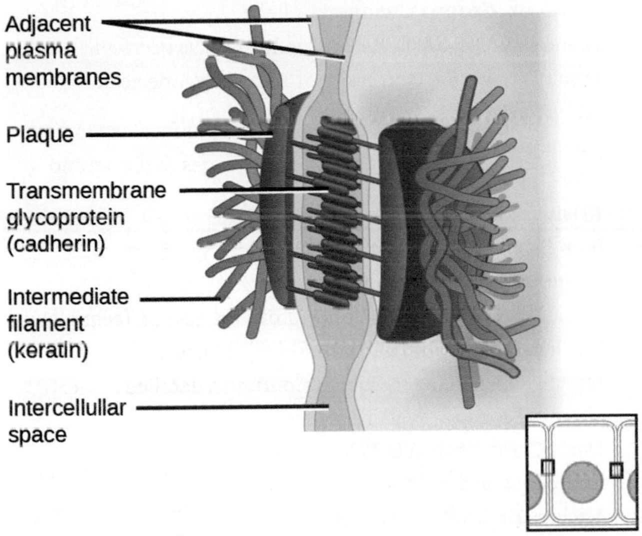

Phagocytosis

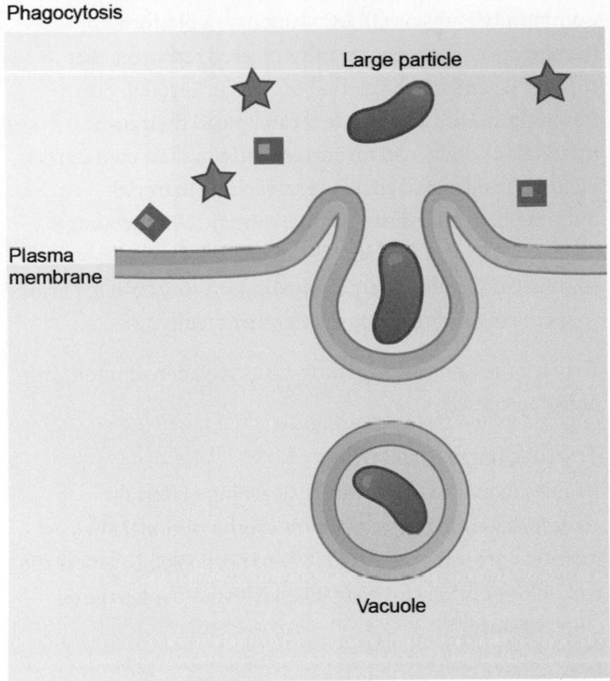

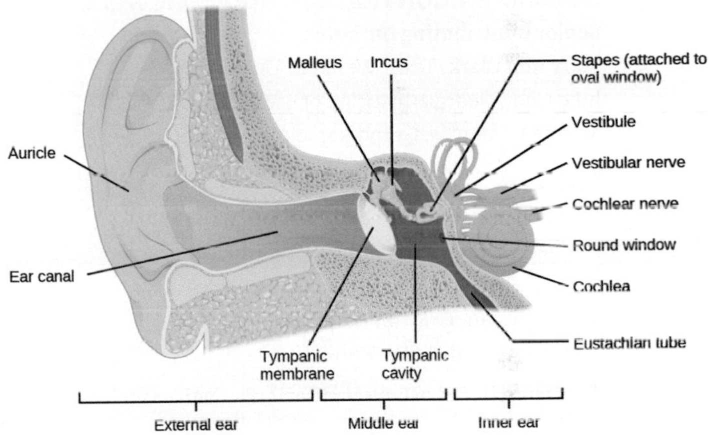

Additional resources

Student and instructor resources

We've compiled additional resources for both students and instructors, including Getting Started Guides, an instructor solution guide, and PowerPoint lecture slides. Instructor resources require a verified instructor account, which you can apply for when you log in or create your account on OpenStax.org. Take advantage of these resources to supplement your OpenStax book.

Community Hubs

OpenStax partners with the Institute for the Study of Knowledge Management in Education (ISKME) to offer

Community Hubs on OER Commons – a platform for instructors to share community-created resources that support OpenStax books, free of charge. Through our Community Hubs, instructors can upload their own materials or download resources to use in their own courses, including additional ancillaries, teaching material, multimedia, and relevant course content. We encourage instructors to join the hubs for the subjects most relevant to your teaching and research as an opportunity both to enrich your courses and to engage with other faculty.

To reach the Community Hubs, visit www.oercommons.org/hubs/OpenStax.

Technology partners

As allies in making high-quality learning materials accessible, our technology partners offer optional low-cost tools that are integrated with OpenStax books. To access the technology options for your text, visit your book page on OpenStax.org.

About the authors

Second edition authors and reviewers

Senior Contributing Authors

Mary Ann Clark, Texas Wesleyan University
Jung Choi, Georgia Institute of Technology
Matthew Douglas, Grand Rapids Community College

Reviewers

Kathleen Berlyn, Baltimore City Community College
Bridgett Brinton, Armstrong State University
Jennifer Chase, Northwest Nazarene University
Amy Hoffman, Grayson County College
Olga Kopp, Utah Valley University
Jennifer Larson, Capital University
Jason Locklin, Austin Community College
Hongmei Ma, American University
Melissa Masse, Tulsa Community College
Shannon McDermott, Central Virginia Community College
Bryan Monesson-Olson, University of Massachusetts Amherst
Amber Reece, California State University Fresno
Monique Reed, Texas A&M University
Jeffrey Roberts, American River College
Matthew Smith, North Dakota State University
Dawn Wankowski, Cardinal Stritch University

First edition authors and reviewers

Senior Contributing Authors

Yael Avissar (Cell Biology), Rhode Island College
Jung Choi (Genetics), Georgia Institute of Technology
Jean DeSaix (Evolution), University of North Carolina at Chapel Hill
Vladimir Jurukovski (Animal Physiology), Suffolk County Community College
Robert Wise (Plant Biology), University of Wisconsin, Oshkosh
Connie Rye (General Content Lead), East Mississippi Community College

Contributing Authors and Reviewers

Julie Adams, Aurora University
Summer Allen, Brown University
James Bader, Case Western Reserve University
David Bailey, St. Norbert College
Mark Belk, Brigham Young University
Nancy Boury, Iowa State University
Lisa Bonneau, Metropolitan Community College – Blue River
Graciela Brelles-Marino, California State University Pomona
Mark Browning, Purdue University
Sue Chaplin, University of St. Thomas
George Cline, Jacksonville State University
Deb Cook, Georgia Gwinnett College
Diane Day, Clayton State University
Frank Dirrigl, The University of Texas Pan American
Waneene Dorsey, Grambling State University
Nick Downey, University of Wisconsin La Crosse
Rick Duhrkopf, Baylor University
Kristy Duran, Adams State University
Stan Eisen, Christian Brothers University
Brent Ewers, University of Wyoming
Myriam Feldman, Lake Washington Institute of Technology
Michael Fine, Virginia Commonwealth University
Linda Flora, Delaware County Community College
Thomas Freeland, Walsh University
David Grisé, Texas A & M University – Corpus Christi
Andrea Hazard, SUNY Cortland
Michael Hedrick, University of North Texas
Linda Hensel, Mercer University
Mark Kopeny, University of Virginia
Norman Johnson, University of Massachusetts Amherst
Grace Lasker, Lake Washington Institute of Technology; Walden University
Sandy Latourelle, SUNY Plattsburgh
Theo Light, Shippensburg University
Clark Lindgren, Grinnell College
James Malcolm, University of Redlands
Mark Meade, Jacksonville State University
Richard Merritt, Houston Community College
James Mickle, North Carolina State University
Jasleen Mishra, Houston Community College
Dudley Moon, Albany College of Pharmacy and Health Sciences
Shobhana Natarajan, Brookhaven College
Jonas Okeagu, Fayetteville State University
Diana Oliveras, University of Colorado Boulder
John Peters, College of Charleston
Joel Piperberg, Millersville University

Johanna Porter-Kelley, Winston-Salem State University
Robyn Puffenbarger, Bridgewater College
Dennis Revie, California Lutheran University
Ann Rushing, Baylor University
Sangha Saha, City College of Chicago
Edward Saiff, Ramapo College of New Jersey
Brian Shmaefsky, Lone Star College System
Robert Sizemore, Alcorn State University
Marc Smith, Sinclair Community College
Frederick Spiegel, University of Arkansas
Frederick Sproull, La Roche College
Bob Sullivan, Marist College
Mark Sutherland, Hendrix College
Toure Thompson, Alabama A&M University
Scott Thomson, University of Wisconsin – Parkside
Allison van de Meene, University of Melbourne
Mary White, Southeastern Louisiana University
Steven Wilt, Bellarmine University
James Wise, Hampton University
Renna Wolfe
Virginia Young, Mercer University
Leslie Zeman, University of Washington
Daniel Zurek, Pittsburg State University
Shobhana Natarajan, Alcon Laboratories, Inc.

CHAPTER 1
The Study of Life

Figure 1.1 This NASA image is a composite of several satellite-based views of Earth. To make the whole-Earth image, NASA scientists combine observations of different parts of the planet. (credit: NASA/GSFC/NOAA/USGS)

INTRODUCTION Viewed from space, Earth offers no clues about the diversity of life forms that reside there. Scientists believe that the first forms of life on Earth were microorganisms that existed for billions of years in the ocean before plants and animals appeared. The mammals, birds, and flowers so familiar to us are all relatively recent, originating 130 to 250 million years ago. The earliest representatives of the genus *Homo*, to which we belong, have inhabited this planet for only the last 2.5 million years, and only in the last 300,000 years have humans started looking like we do today.

1.1 The Science of Biology

By the end of this section, you will be able to do the following:

- Identify the shared characteristics of the natural sciences
- Summarize the steps of the scientific method
- Compare inductive reasoning with deductive reasoning
- Describe the goals of basic science and applied science

Figure 1.2 Formerly called blue-green algae, these (a) cyanobacteria, magnified 300x under a light microscope, are some of Earth's oldest life forms. These (b) stromatolites along the shores of Lake Thetis in Western Australia are ancient structures formed by layering cyanobacteria in shallow waters. (credit a: modification of work by NASA; credit b: modification of work by Ruth Ellison; scale-bar data from Matt Russell)

What is biology? In simple terms, **biology** is the study of life. This is a very broad definition because the scope of biology is vast. Biologists may study anything from the microscopic or submicroscopic view of a cell to ecosystems and the whole living planet (Figure 1.2). Listening to the daily news, you will quickly realize how many aspects of biology we discuss every day. For example, recent news topics include *Escherichia coli* (Figure 1.3) outbreaks in spinach and *Salmonella* contamination in peanut butter. Other subjects include efforts toward finding a cure for AIDS, Alzheimer's disease, and cancer. On a global scale, many researchers are committed to finding ways to protect the planet, solve environmental issues, and reduce the effects of climate change. All of these diverse endeavors are related to different facets of the discipline of biology.

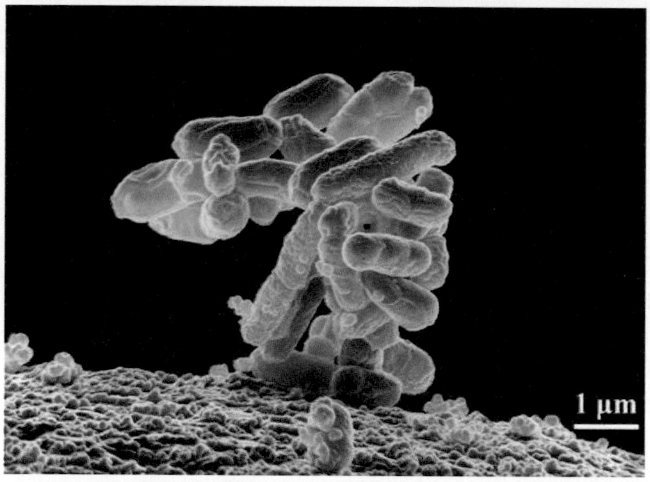

Figure 1.3 *Escherichia coli* (*E. coli*) bacteria, in this scanning electron micrograph, are normal residents of our digestive tracts that aid in absorbing vitamin K and other nutrients. However, virulent strains are sometimes responsible for disease outbreaks. (credit: Eric Erbe, digital colorization by Christopher Pooley, both of USDA, ARS, EMU)

The Process of Science

Biology is a science, but what exactly is science? What does the study of biology share with other scientific disciplines? We can define **science** (from the Latin *scientia*, meaning "knowledge") as knowledge that covers general truths or the operation of general laws, especially when acquired and tested by the scientific method. It becomes clear from this definition that applying scientific method plays a major role in science. The **scientific method** is a method of research with defined steps that include experiments and careful observation.

We will examine scientific method steps in detail later, but one of the most important aspects of this method is the testing of hypotheses by means of repeatable experiments. A **hypothesis** is a suggested explanation for an event, which one can test. Although using the scientific method is inherent to science, it is inadequate in determining what science is. This is because it is relatively easy to apply the scientific method to disciplines such as physics and chemistry, but when it comes to disciplines like archaeology, psychology, and geology, the scientific method becomes less applicable as repeating experiments becomes more difficult.

These areas of study are still sciences, however. Consider archaeology—even though one cannot perform repeatable experiments, hypotheses may still be supported. For instance, an archaeologist can hypothesize that an ancient culture existed based on finding a piece of pottery. He or she could make further hypotheses about various characteristics of this culture, which could be correct or false through continued support or contradictions from other findings. A hypothesis may become a verified theory. A **theory** is a tested and confirmed explanation for observations or phenomena. Therefore, we may be better off to define science as fields of study that attempt to comprehend the nature of the universe.

Natural Sciences

What would you expect to see in a museum of natural sciences? Frogs? Plants? Dinosaur skeletons? Exhibits about how the brain functions? A planetarium? Gems and minerals? Maybe all of the above? Science includes such diverse fields as astronomy, biology, computer sciences, geology, logic, physics, chemistry, and mathematics (Figure 1.4). However, scientists consider those fields of science related to the physical world and its phenomena and processes **natural sciences**. Thus, a museum of natural sciences might contain any of the items listed above.

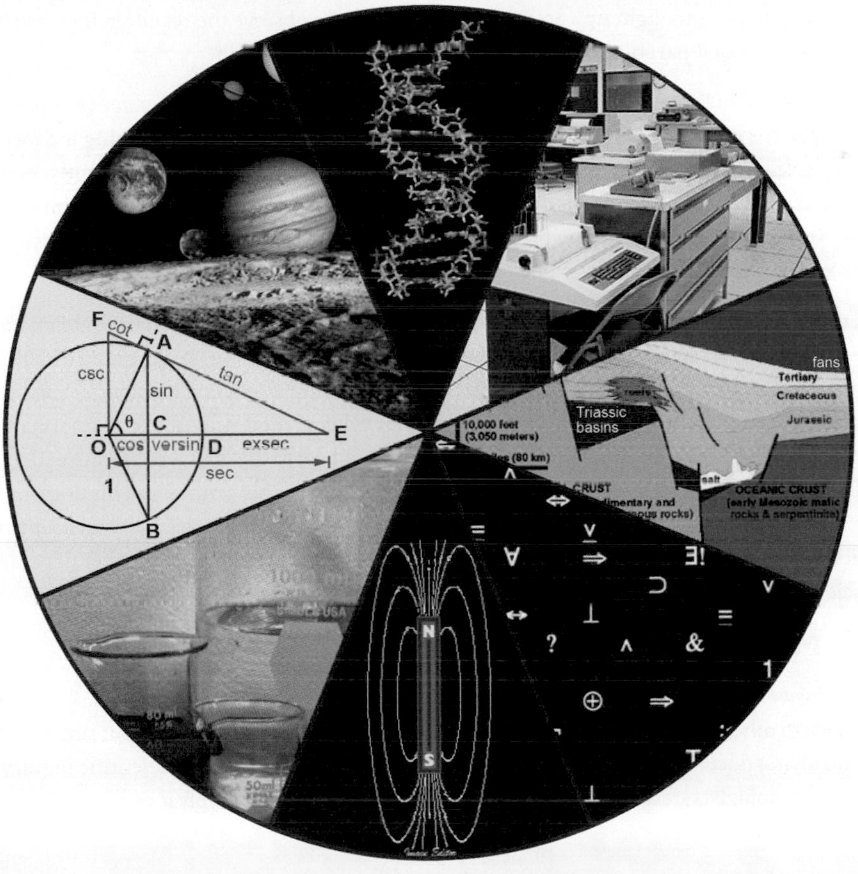

Figure 1.4 The diversity of scientific fields includes astronomy, biology, computer science, geology, logic, physics, chemistry, mathematics, and many other fields. (credit: "Image Editor"/Flickr)

There is no complete agreement when it comes to defining what the natural sciences include, however. For some experts, the natural sciences are astronomy, biology, chemistry, earth science, and physics. Other scholars choose to divide natural sciences into **life sciences**, which study living things and include biology, and **physical sciences**, which study nonliving matter and include astronomy, geology, physics, and chemistry. Some disciplines such as biophysics and biochemistry build on both life and physical sciences and are interdisciplinary. Some refer to natural sciences as "hard science" because they rely on the use of quantitative data. Social sciences that study society and human behavior are more likely to use qualitative assessments to drive investigations and findings.

Not surprisingly, the natural science of biology has many branches or subdisciplines. Cell biologists study cell structure and function, while biologists who study anatomy investigate the structure of an entire organism. Those biologists studying physiology, however, focus on the internal functioning of an organism. Some areas of biology focus on only particular types of living things. For example, botanists explore plants, while zoologists specialize in animals.

Scientific Reasoning

One thing is common to all forms of science: an ultimate goal "to know." Curiosity and inquiry are the driving forces for the development of science. Scientists seek to understand the world and the way it operates. To do this, they use two methods of logical thinking: inductive reasoning and deductive reasoning.

Inductive reasoning is a form of logical thinking that uses related observations to arrive at a general conclusion. This type of reasoning is common in descriptive science. A life scientist such as a biologist makes observations and records them. These data can be qualitative or quantitative, and one can supplement the raw data with drawings, pictures, photos, or videos. From many observations, the scientist can infer conclusions (inductions) based on evidence. Inductive reasoning involves formulating generalizations inferred from careful observation and analyzing a large amount of data. Brain studies provide an example. In this type of research, scientists observe many live brains while people are engaged in a specific activity, such as viewing images of food. The scientist then predicts the part of the brain that "lights up" during this activity to be the part controlling the response to the selected stimulus, in this case, images of food. Excess absorption of radioactive sugar derivatives by active areas of the brain causes the various areas to "light up". Scientists use a scanner to observe the resultant increase in radioactivity. Then, researchers can stimulate that part of the brain to see if similar responses result.

Deductive reasoning or deduction is the type of logic used in hypothesis-based science. In deductive reason, the pattern of thinking moves in the opposite direction as compared to inductive reasoning. **Deductive reasoning** is a form of logical thinking that uses a general principle or law to forecast specific results. From those general principles, a scientist can extrapolate and predict the specific results that would be valid as long as the general principles are valid. Studies in climate change can illustrate this type of reasoning. For example, scientists may predict that if the climate becomes warmer in a particular region, then the distribution of plants and animals should change.

Both types of logical thinking are related to the two main pathways of scientific study: descriptive science and hypothesis-based science. **Descriptive (or discovery) science**, which is usually inductive, aims to observe, explore, and discover, while **hypothesis-based science**, which is usually deductive, begins with a specific question or problem and a potential answer or solution that one can test. The boundary between these two forms of study is often blurred, and most scientific endeavors combine both approaches. The fuzzy boundary becomes apparent when thinking about how easily observation can lead to specific questions. For example, a gentleman in the 1940s observed that the burr seeds that stuck to his clothes and his dog's fur had a tiny hook structure. On closer inspection, he discovered that the burrs' gripping device was more reliable than a zipper. He eventually experimented to find the best material that acted similar, and produced the hook-and-loop fastener popularly known today as Velcro. Descriptive science and hypothesis-based science are in continuous dialogue.

The Scientific Method

Biologists study the living world by posing questions about it and seeking science-based responses. Known as scientific method, this approach is common to other sciences as well. The scientific method was used even in ancient times, but England's Sir Francis Bacon (1561–1626) first documented it (Figure 1.5). He set up inductive methods for scientific inquiry. The scientific method is not used only by biologists; researchers from almost all fields of study can apply it as a logical, rational problem-solving method.

Figure 1.5 Historians credit Sir Francis Bacon (1561–1626) as the first to define the scientific method. (credit: Paul van Somer)

The scientific process typically starts with an observation (often a problem to solve) that leads to a question. Let's think about a simple problem that starts with an observation and apply the scientific method to solve the problem. One Monday morning, a student arrives at class and quickly discovers that the classroom is too warm. That is an observation that also describes a problem: the classroom is too warm. The student then asks a question: "Why is the classroom so warm?"

Proposing a Hypothesis

Recall that a hypothesis is a suggested explanation that one can test. To solve a problem, one can propose several hypotheses. For example, one hypothesis might be, "The classroom is warm because no one turned on the air conditioning." However, there could be other responses to the question, and therefore one may propose other hypotheses. A second hypothesis might be, "The classroom is warm because there is a power failure, and so the air conditioning doesn't work."

Once one has selected a hypothesis, the student can make a prediction. A prediction is similar to a hypothesis but it typically has the format "If . . . then . . . ." For example, the prediction for the first hypothesis might be, "*If* the student turns on the air conditioning, *then* the classroom will no longer be too warm."

Testing a Hypothesis

A valid hypothesis must be testable. It should also be **falsifiable**, meaning that experimental results can disprove it. Importantly, science does not claim to "prove" anything because scientific understandings are always subject to modification with further information. This step—openness to disproving ideas—is what distinguishes sciences from non-sciences. The presence of the supernatural, for instance, is neither testable nor falsifiable. To test a hypothesis, a researcher will conduct one or more experiments designed to eliminate one or more of the hypotheses. Each experiment will have one or more variables and one or more controls. A **variable** is any part of the experiment that can vary or change during the experiment. The **control group** contains every feature of the experimental group except it is not given the manipulation that the researcher hypothesizes. Therefore, if the experimental group's results differ from the control group, the difference must be due to the hypothesized manipulation, rather than some outside factor. Look for the variables and controls in the examples that follow. To test the first hypothesis, the student would find out if the air conditioning is on. If the air conditioning is turned on but does not work, there should be another reason, and the student should reject this hypothesis. To test the second hypothesis, the student could check if the lights in the classroom are functional. If so, there is no power failure and the student should reject this hypothesis. The students should test each hypothesis by carrying out appropriate experiments. Be aware that rejecting one hypothesis does not determine whether or not one can accept the other hypotheses. It simply eliminates one hypothesis that is not valid (Figure 1.6). Using the scientific method, the student rejects the hypotheses that are inconsistent with experimental data.

While this "warm classroom" example is based on observational results, other hypotheses and experiments might have clearer controls. For instance, a student might attend class on Monday and realize she had difficulty concentrating on the lecture. One observation to explain this occurrence might be, "When I eat breakfast before class, I am better able to pay attention." The student could then design an experiment with a control to test this hypothesis.

In hypothesis-based science, researchers predict specific results from a general premise. We call this type of reasoning deductive reasoning: deduction proceeds from the general to the particular. However, the reverse of the process is also possible: sometimes, scientists reach a general conclusion from a number of specific observations. We call this type of reasoning inductive reasoning, and it proceeds from the particular to the general. Researchers often use inductive and deductive reasoning in tandem to advance scientific knowledge (Figure 1.7).

VISUAL CONNECTION

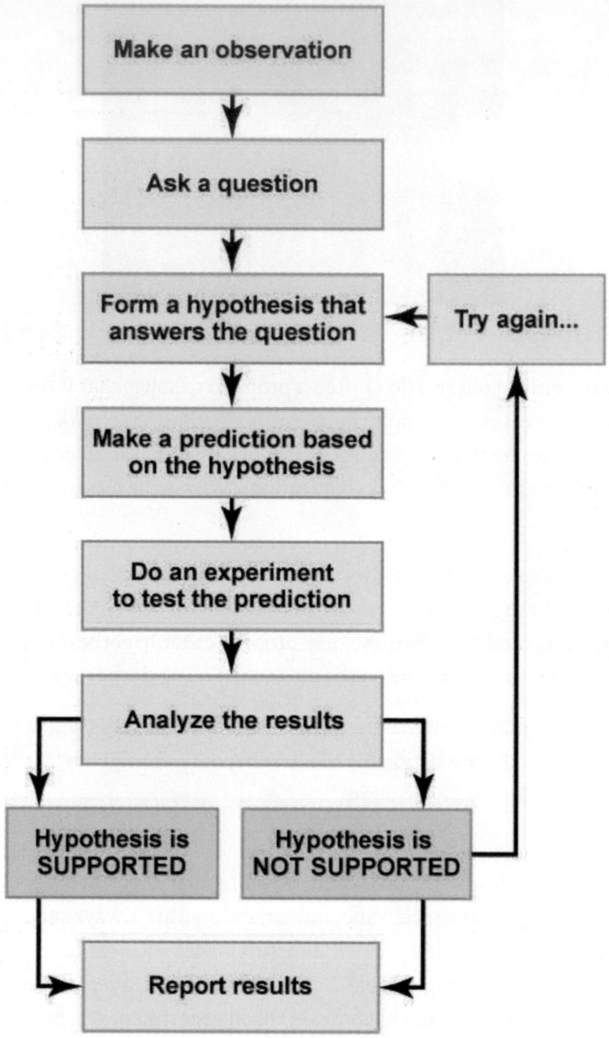

Figure 1.6 The scientific method consists of a series of well-defined steps. If a hypothesis is not supported by experimental data, one can propose a new hypothesis.

In the example below, the scientific method is used to solve an everyday problem. Match the scientific method steps (numbered items) with the process of solving the everyday problem (lettered items). Based on the results of the experiment, is the hypothesis correct? If it is incorrect, propose some alternative hypotheses.

1. Observation	a. There is something wrong with the electrical outlet.
2. Question	b. If something is wrong with the outlet, my coffeemaker also won't work when plugged into it.

3. Hypothesis (answer) c. My toaster doesn't toast my bread.

4. Prediction d. I plug my coffee maker into the outlet.

5. Experiment e. My coffeemaker works.

6. Result f. Why doesn't my toaster work?

VISUAL CONNECTION

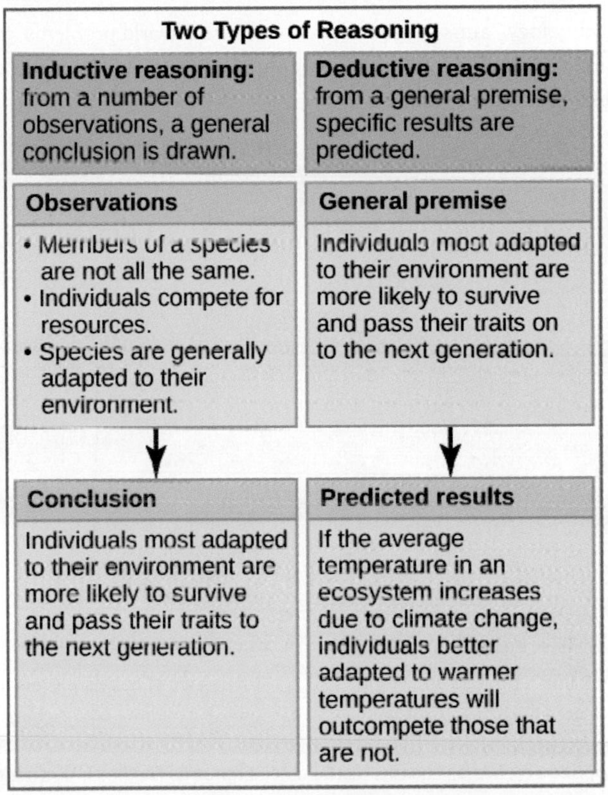

Figure 1.7 Scientists use two types of reasoning, inductive and deductive reasoning, to advance scientific knowledge. As is the case in this example, the conclusion from inductive reasoning can often become the premise for deductive reasoning.

Decide if each of the following is an example of inductive or deductive reasoning.

1. All flying birds and insects have wings. Birds and insects flap their wings as they move through the air. Therefore, wings enable flight.
2. Insects generally survive mild winters better than harsh ones. Therefore, insect pests will become more problematic if global temperatures increase.
3. Chromosomes, the carriers of DNA, are distributed evenly between the daughter cells during cell division. Therefore, each daughter cell will have the same chromosome set as the mother cell.
4. Animals as diverse as humans, insects, and wolves all exhibit social behavior. Therefore, social behavior must have an evolutionary advantage.

The scientific method may seem too rigid and structured. It is important to keep in mind that, although scientists often follow this sequence, there is flexibility. Sometimes an experiment leads to conclusions that favor a change in approach. Often, an experiment brings entirely new scientific questions to the puzzle. Many times, science does not operate in a linear fashion. Instead, scientists continually draw inferences and make generalizations, finding patterns as their research proceeds. Scientific reasoning is more complex than the scientific method alone suggests. Notice, too, that we can apply the scientific method to solving problems that aren't necessarily scientific in nature.

Two Types of Science: Basic Science and Applied Science

The scientific community has been debating for the last few decades about the value of different types of science. Is it valuable to pursue science for the sake of simply gaining knowledge, or does scientific knowledge only have worth if we can apply it to solving a specific problem or to bettering our lives? This question focuses on the differences between two types of science: basic science and applied science.

Basic science or "pure" science seeks to expand knowledge regardless of the short-term application of that knowledge. It is not focused on developing a product or a service of immediate public or commercial value. The immediate goal of basic science is knowledge for knowledge's sake, although this does not mean that, in the end, it may not result in a practical application.

In contrast, **applied science** or "technology," aims to use science to solve real-world problems, making it possible, for example, to improve a crop yield, find a cure for a particular disease, or save animals threatened by a natural disaster (Figure 1.8). In applied science, the problem is usually defined for the researcher.

Figure 1.8 After Hurricane Irma struck the Caribbean and Florida in 2017, thousands of baby squirrels like this one were thrown from their nests. Thanks to applied science, scientists knew how to rehabilitate the squirrel. (credit: audreyjm529, Flickr)

Some individuals may perceive applied science as "useful" and basic science as "useless." A question these people might pose to a scientist advocating knowledge acquisition would be, "What for?" However, a careful look at the history of science reveals that basic knowledge has resulted in many remarkable applications of great value. Many scientists think that a basic understanding of science is necessary before researchers develop an application therefore, applied science relies on the results that researchers generate through basic science. Other scientists think that it is time to move on from basic science in order to find solutions to actual problems. Both approaches are valid. It is true that there are problems that demand immediate attention; however, scientists would find few solutions without the help of the wide knowledge foundation that basic science generates.

One example of how basic and applied science can work together to solve practical problems occurred after the discovery of DNA structure led to an understanding of the molecular mechanisms governing DNA replication. DNA strands, unique in every human, are in our cells, where they provide the instructions necessary for life. When DNA replicates, it produces new copies of itself, shortly before a cell divides. Understanding DNA replication mechanisms enabled scientists to develop laboratory techniques that researchers now use to identify genetic diseases, pinpoint individuals who were at a crime scene, and determine paternity. Without basic science, it is unlikely that applied science would exist.

Another example of the link between basic and applied research is the Human Genome Project, a study in which researchers analyzed and mapped each human chromosome to determine the precise sequence of DNA subunits and each gene's exact location. (The gene is the basic unit of heredity represented by a specific DNA segment that codes for a functional molecule. An individual's complete collection of genes is his or her genome.) Researchers have studied other less complex organisms as part of this project in order to gain a better understanding of human chromosomes. The Human Genome Project (Figure 1.9) relied on basic research with simple organisms and, later, with the human genome. An important end goal eventually became using the data for applied research, seeking cures and early diagnoses for genetically related diseases.

Figure 1.9 The Human Genome Project was a 13-year collaborative effort among researchers working in several different science fields. Researchers completed the project, which sequenced the entire human genome, in 2003. (credit: the U.S. Department of Energy Genome Programs (http://genomics.energy.gov (http://openstax.org/l/genomics_gov))

While scientists usually carefully plan research efforts in both basic science and applied science, note that some discoveries are made by **serendipity**, that is, by means of a fortunate accident or a lucky surprise. Scottish biologist Alexander Fleming discovered penicillin when he accidentally left a petri dish of *Staphylococcus* bacteria open. An unwanted mold grew on the dish, killing the bacteria. Fleming's curiosity to investigate the reason behind the bacterial death, followed by his experiments, led to the discovery of the antibiotic penicillin, which is produced by the fungus *Penicillium*. Even in the highly organized world of science, luck—when combined with an observant, curious mind—can lead to unexpected breakthroughs.

Reporting Scientific Work

Whether scientific research is basic science or applied science, scientists must share their findings in order for other researchers to expand and build upon their discoveries. Collaboration with other scientists—when planning, conducting, and analyzing results—are all important for scientific research. For this reason, important aspects of a scientist's work are communicating with peers and disseminating results to peers. Scientists can share results by presenting them at a scientific meeting or conference, but this approach can reach only the select few who are present. Instead, most scientists present their results in peer-reviewed manuscripts that are published in scientific journals. **Peer-reviewed manuscripts** are scientific papers that a scientist's colleagues or peers review. These colleagues are qualified individuals, often experts in the same research area, who judge whether or not the scientist's work is suitable for publication. The process of peer review helps to ensure that the research in a scientific paper or grant proposal is original, significant, logical, and thorough. Grant proposals, which are requests for research funding, are also subject to peer review. Scientists publish their work so other scientists can reproduce their experiments under similar or different conditions to expand on the findings. The experimental results must be consistent with the findings of other scientists.

A scientific paper is very different from creative writing. Although creativity is required to design experiments, there are fixed guidelines when it comes to presenting scientific results. First, scientific writing must be brief, concise, and accurate. A scientific paper needs to be succinct but detailed enough to allow peers to reproduce the experiments.

The scientific paper consists of several specific sections—introduction, materials and methods, results, and discussion. This structure is sometimes called the "IMRaD" format. There are usually acknowledgment and reference sections as well as an **abstract** (a concise summary) at the beginning of the paper. There might be additional sections depending on the type of paper and the journal where it will be published. For example, some review papers require an outline.

The **introduction** starts with brief, but broad, background information about what is known in the field. A good introduction also gives the rationale of the work. It justifies the work carried out and also briefly mentions the end of the paper, where the researcher will present the hypothesis or research question driving the research. The introduction refers to the published scientific work of others and therefore requires citations following the style of the journal. Using the work or ideas of others without proper citation is **plagiarism**.

The **materials and methods** section includes a complete and accurate description of the substances the researchers use, and the method and techniques they use to gather data. The description should be thorough enough to allow another researcher to repeat the experiment and obtain similar results, but it does not have to be verbose. This section will also include information on how the researchers made measurements and the types of calculations and statistical analyses they used to examine raw data. Although the materials and methods section gives an accurate description of the experiments, it does not discuss them.

Some journals require a results section followed by a discussion section, but it is more common to combine both. If the journal does not allow combining both sections, the **results** section simply narrates the findings without any further interpretation. The researchers present results with tables or graphs, but they do not present duplicate information. In the **discussion** section, the researchers will interpret the results, describe how variables may be related, and attempt to explain the observations. It is indispensable to conduct an extensive literature search to put the results in the context of previously published scientific research. Therefore, researchers include proper citations in this section as well.

Finally, the **conclusion** section summarizes the importance of the experimental findings. While the scientific paper almost certainly answers one or more scientific questions that the researchers stated, any good research should lead to more questions. Therefore, a well-done scientific paper allows the researchers and others to continue and expand on the findings.

Review articles do not follow the IMRAD format because they do not present original scientific findings, or primary literature. Instead, they summarize and comment on findings that were published as primary literature and typically include extensive reference sections.

1.2 Themes and Concepts of Biology

By the end of this section, you will be able to do the following:

- Identify and describe the properties of life
- Describe the levels of organization among living things
- Recognize and interpret a phylogenetic tree
- List examples of different subdisciplines in biology

Biology is the science that studies life, but what exactly is life? This may sound like a silly question with an obvious response, but it is not always easy to define life. For example, a branch of biology called virology studies viruses, which exhibit some of the characteristics of living entities but lack others. Although viruses can attack living organisms, cause diseases, and even reproduce, they do not meet the criteria that biologists use to define life. Consequently, virologists are not biologists, strictly speaking. Similarly, some biologists study the early molecular evolution that gave rise to life. Since the events that preceded life are not biological events, these scientists are also excluded from biology in the strict sense of the term.

From its earliest beginnings, biology has wrestled with three questions: What are the shared properties that make something "alive"? Once we know something is alive, how do we find meaningful levels of organization in its structure? Finally, when faced with the remarkable diversity of life, how do we organize the different kinds of organisms so that we can better understand them? As scientists discover new organisms every day, biologists continue to seek answers to these and other questions.

Properties of Life

All living organisms share several key characteristics or functions: order, sensitivity or response to the environment, reproduction, adaptation, growth and development, regulation/homeostasis, energy processing, and evolution. When viewed together, these eight characteristics serve to define life.

Order

Figure 1.10 A toad represents a highly organized structure consisting of cells, tissues, organs, and organ systems. (credit: "Ivengo"/Wikimedia Commons)

Organisms are highly organized, coordinated structures that consist of one or more cells. Even very simple, single-celled organisms are remarkably complex: inside each cell, atoms comprise molecules. These in turn comprise cell organelles and other cellular inclusions. In multicellular organisms (Figure 1.10), similar cells form tissues. Tissues, in turn, collaborate to create organs (body structures with a distinct function). Organs work together to form organ systems.

Sensitivity or Response to Stimuli

Figure 1.11 The leaves of this sensitive plant (*Mimosa pudica*) will instantly droop and fold when touched. After a few minutes, the plant returns to normal. (credit: Alex Lomas)

Organisms respond to diverse stimuli. For example, plants can bend toward a source of light, climb on fences and walls, or respond to touch (Figure 1.11). Even tiny bacteria can move toward or away from chemicals (a process called *chemotaxis*) or light (*phototaxis*). Movement toward a stimulus is a positive response, while movement away from a stimulus is a negative response.

LINK TO LEARNING

Watch this video (http://openstax.org/l/movement_plants) to see how plants respond to a stimulus—from opening to light, to wrapping a tendril around a branch, to capturing prey.

Reproduction

Single-celled organisms reproduce by first duplicating their DNA, and then dividing it equally as the cell prepares to divide to form two new cells. Multicellular organisms often produce specialized reproductive germline, gamete, oocyte, and sperm cells. After fertilization (the fusion of an oocyte and a sperm cell), a new individual develops. When reproduction occurs, DNA containing genes are passed along to an organism's offspring. These genes ensure that the offspring will belong to the same species and will have similar characteristics, such as size and shape.

Adaptation

All living organisms exhibit a "fit" to their environment. Biologists refer to this fit as adaptation, and it is a consequence of evolution by natural selection, which operates in every lineage of reproducing organisms. Examples of adaptations are diverse and unique, from heat-resistant Archaea that live in boiling hotsprings to the tongue length of a nectar-feeding moth that matches the size of the flower from which it feeds. All adaptations enhance the reproductive potential of the individuals exhibiting them, including their ability to survive to reproduce. Adaptations are not constant. As an environment changes, natural selection causes the characteristics of the individuals in a population to track those changes.

Growth and Development

Organisms grow and develop as a result of genes providing specific instructions that will direct cellular growth and development. This ensures that a species' young (Figure 1.12) will grow up to exhibit many of the same characteristics as its parents.

Figure 1.12 Although no two look alike, these kittens have inherited genes from both parents and share many of the same characteristics. (credit: Rocky Mountain Feline Rescue)

Regulation/Homeostasis

Even the smallest organisms are complex and require multiple regulatory mechanisms to coordinate internal functions, respond to stimuli, and cope with environmental stresses. Two examples of internal functions regulated in an organism are nutrient transport and blood flow. Organs (groups of tissues working together) perform specific functions, such as carrying oxygen throughout the body, removing wastes, delivering nutrients to every cell, and cooling the body.

Figure 1.13 Polar bears (*Ursus maritimus*) and other mammals living in ice-covered regions maintain their body temperature by generating heat and reducing heat loss through thick fur and a dense layer of fat under their skin. (credit: "longhorndave"/Flickr)

In order to function properly, cells require appropriate conditions such as proper temperature, pH, and appropriate concentration of diverse chemicals. These conditions may, however, change from one moment to the next. Organisms are able to maintain internal conditions within a narrow range almost constantly, despite environmental changes, through **homeostasis** (literally, "steady state"). For example, an organism needs to regulate body temperature through the thermoregulation process. Organisms that live in cold climates, such as the polar bear (Figure 1.13), have body structures that help them withstand low temperatures and conserve body heat. Structures that aid in this type of insulation include fur, feathers, blubber, and fat. In hot climates, organisms have methods (such as perspiration in humans or panting in dogs) that help them to shed excess body heat.

Energy Processing

Figure 1.14 The California condor (*Gymnogyps californianus*) uses chemical energy derived from food to power flight. California condors are an endangered species. This bird has a wing tag that helps biologists identify the individual. (credit: Pacific Southwest Region U.S. Fish and Wildlife Service)

All organisms use a source of energy for their metabolic activities. Some organisms capture energy from the sun and convert it into chemical energy in food. Others use chemical energy in molecules they take in as food (Figure 1.14).

Evolution

The diversity of life on Earth is a result of mutations, or random changes in hereditary material over time. These mutations allow the possibility for organisms to adapt to a changing environment. An organism that evolves characteristics fit for the environment will have greater reproductive success, subject to the forces of natural selection.

Levels of Organization of Living Things

Living things are highly organized and structured, following a hierarchy that we can examine on a scale from small to large. The **atom** is the smallest and most fundamental unit of matter. It consists of a nucleus surrounded by electrons. Atoms form molecules. A **molecule** is a chemical structure consisting of at least two atoms held together by one or more chemical bonds. Many molecules that are biologically important are **macromolecules**, large molecules that are typically formed by polymerization (a polymer is a large molecule that is made by combining smaller units called monomers, which are simpler than macromolecules). An example of a macromolecule is deoxyribonucleic acid (DNA) (Figure 1.15), which contains the instructions for the structure and functioning of all living organisms.

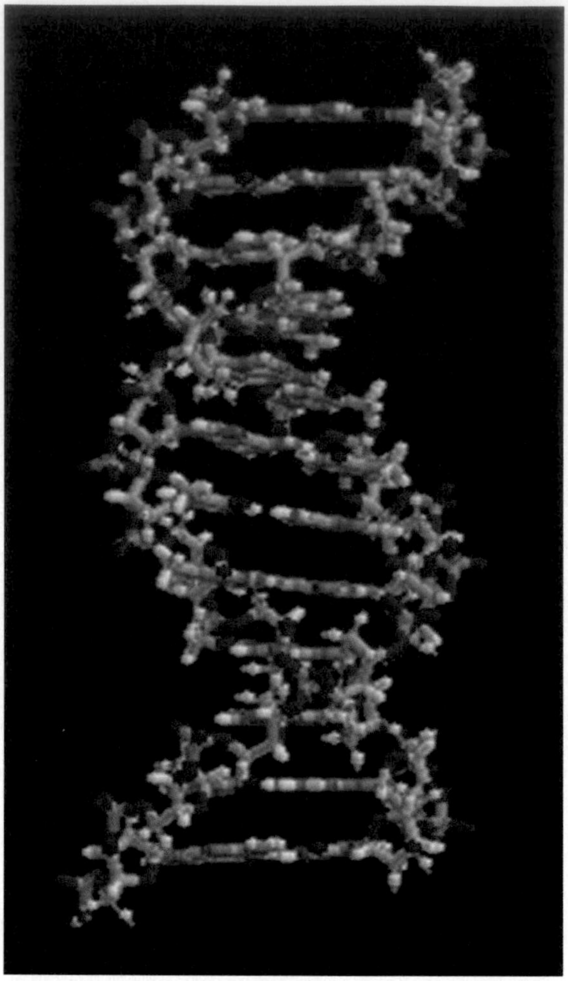

Figure 1.15 All molecules, including this DNA molecule, are comprised of atoms. (credit: “brian0918”/Wikimedia Commons)

LINK TO LEARNING

Watch this video (http://openstax.org/l/rotating_DNA) that animates the three-dimensional structure of the DNA molecule in Figure 1.15.

Some cells contain aggregates of macromolecules surrounded by membranes. We call these **organelles**. Organelles are small structures that exist within cells. Examples of organelles include mitochondria and chloroplasts, which carry out indispensable functions: mitochondria produce energy to power the cell, while chloroplasts enable green plants to utilize the energy in sunlight to make sugars. All living things are made of cells. The **cell** itself is the smallest fundamental unit of structure and function in living organisms. (This requirement is why scientists do not consider viruses living: they are not made of cells. To make new viruses, they have to invade and hijack the reproductive mechanism of a living cell. Only then can they obtain the materials they need to reproduce.) Some organisms consist of a single cell and others are multicellular. Scientists classify cells as prokaryotic or eukaryotic. **Prokaryotes** are single-celled or colonial organisms that do not have membrane-bound nuclei. In contrast, the cells of **eukaryotes** do have membrane-bound organelles and a membrane-bound nucleus.

In larger organisms, cells combine to make **tissues**, which are groups of similar cells carrying out similar or related functions. **Organs** are collections of tissues grouped together performing a common function. Organs are present not only in animals but also in plants. An **organ system** is a higher level of organization that consists of functionally related organs. Mammals have many organ systems. For instance, the circulatory system transports blood through the body and to and from the lungs. It includes organs such as the heart and blood vessels. **Organisms** are individual living entities. For example, each tree in a forest is an organism. Single-celled prokaryotes and single-celled eukaryotes are also organisms, which biologists typically call microorganisms.

Biologists collectively call all the individuals of a species living within a specific area a **population**. For example, a forest may include many pine trees, which represent the population of pine trees in this forest. Different populations may live in the same specific area. For example, the forest with the pine trees includes populations of flowering plants, insects, and microbial populations. A **community** is the sum of populations inhabiting a particular area. For instance, all of the trees, flowers, insects, and other populations in a forest form the forest's community. The forest itself is an ecosystem. An **ecosystem** consists of all the living things in a particular area together with the abiotic, nonliving parts of that environment such as nitrogen in the soil or rain water. At the highest level of organization (Figure 1.16), the **biosphere** is the collection of all ecosystems, and it represents the zones of life on Earth. It includes land, water, and even the atmosphere to a certain extent.

VISUAL CONNECTION

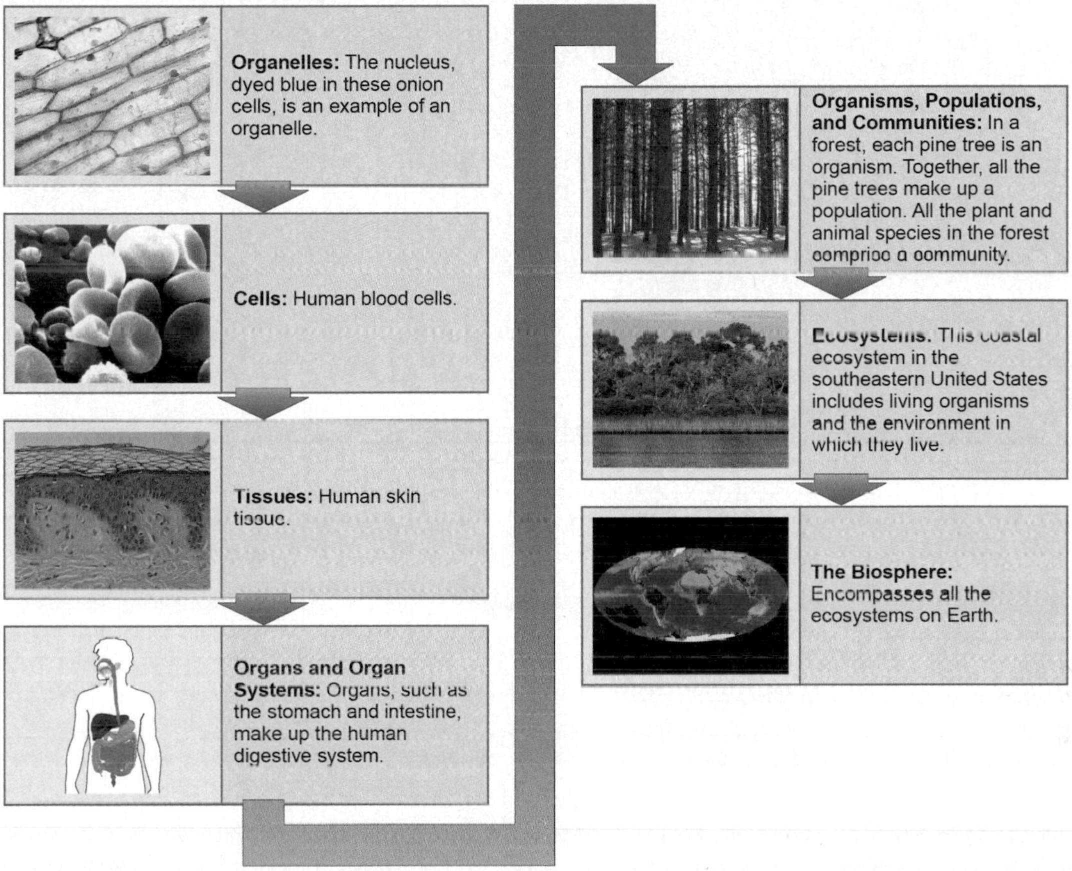

Figure 1.16 shows the biological levels of organization of living things. From a single organelle to the entire biosphere, living organisms are parts of a highly structured hierarchy. (credit "organelles": modification of work by Umberto Salvagnin; credit "cells": modification of work by Bruce Wetzel, Harry Schaefer/ National Cancer Institute; credit "tissues": modification of work by Kilbad; Fama Clamosa; Mikael Häggström; credit "organs": modification of work by Mariana Ruiz Villareal; credit "organisms": modification of work by "Crystal"/Flickr; credit "ecosystems": modification of work by US Fish and Wildlife Service Headquarters; credit "biosphere": modification of work by NASA)

Which of the following statements is false?

a. Tissues exist within organs which exist within organ systems.
b. Communities exist within populations which exist within ecosystems.
c. Organelles exist within cells which exist within tissues.
d. Communities exist within ecosystems which exist in the biosphere.

The Diversity of Life

The fact that biology, as a science, has such a broad scope has to do with the tremendous diversity of life on earth. The source of this diversity is **evolution**, the process of gradual change in a population or species over time. Evolutionary biologists study the evolution of living things in everything from the microscopic world to ecosystems.

A phylogenetic tree (Figure 1.17) can summarize the evolution of various life forms on Earth. It is a diagram showing the evolutionary relationships among biological species based on similarities and differences in genetic or physical traits or both. Nodes and branches comprise a phylogenetic tree. The internal nodes represent ancestors and are points in evolution when, based on scientific evidence, researchers believe an ancestor has diverged to form two new species. The length of each branch is proportional to the time elapsed since the split.

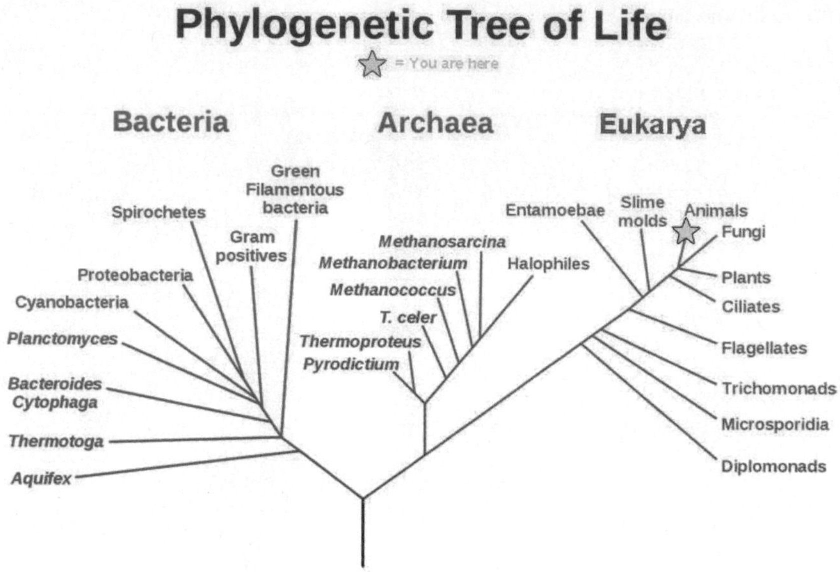

Figure 1.17 Microbiologist Carl Woese constructed this phylogenetic tree using data that he obtained from sequencing ribosomal RNA genes. The tree shows the separation of living organisms into three domains: Bacteria, Archaea, and Eukarya. Bacteria and Archaea are prokaryotes, single-celled organisms lacking intracellular organelles. (credit: Eric Gaba; NASA Astrobiology Institute)

EVOLUTION CONNECTION

Carl Woese and the Phylogenetic Tree

In the past, biologists grouped living organisms into five kingdoms: animals, plants, fungi, protists, and bacteria. They based the organizational scheme mainly on physical features, as opposed to physiology, biochemistry, or molecular biology, all of which modern systematics use. American microbiologist Carl Woese's pioneering work in the early 1970s has shown, however, that life on Earth has evolved along three lineages, now called domains—Bacteria, Archaea, and Eukarya. The first two are prokaryotic cells with microbes that lack membrane-enclosed nuclei and organelles. The third domain contains the eukaryotes and includes unicellular microorganisms (protists), together with the three remaining kingdoms (fungi, plants, and animals). Woese defined Archaea as a new domain, and this resulted in a new taxonomic tree (Figure 1.17). Many organisms belonging to the Archaea domain live under extreme conditions and are called extremophiles. To construct his tree, Woese used genetic relationships rather than similarities based on morphology (shape).

Woese constructed his tree from universally distributed comparative gene sequencing that are present in every organism, and conserved (meaning that these genes have remained essentially unchanged throughout evolution). Woese's approach was revolutionary because comparing physical features are insufficient to differentiate between the prokaryotes that appear fairly similar in spite of their tremendous biochemical diversity and genetic variability (Figure 1.18). Comparing homologous DNA and RNA sequences provided Woese with a sensitive device that revealed the extensive variability of prokaryotes, and which justified separating the prokaryotes into two domains: bacteria and archaea.

Figure 1.18 These images represent different domains. The (a) bacteria in this micrograph belong to Domain Bacteria, while the (b) extremophiles (not visible) living in this hot vent belong to Domain Archaea. Both the (c) sunflower and (d) lion are part of Domain Eukarya. (credit a: modification of work by Drew March; credit b: modification of work by Steve Jurvetson; credit c: modification of work by Michael Arrighi; credit d: modification of work by Leszek Leszcynski)

Branches of Biological Study

The scope of biology is broad and therefore contains many branches and subdisciplines. Biologists may pursue one of those subdisciplines and work in a more focused field. For instance, **molecular biology** and **biochemistry** study biological processes at the molecular and chemical level, including interactions among molecules such as DNA, RNA, and proteins, as well as the way they are regulated. **Microbiology**, the study of microorganisms, is the study of the structure and function of single-celled organisms. It is quite a broad branch itself, and depending on the subject of study, there are also microbial physiologists, ecologists, and geneticists, among others.

CAREER CONNECTION

Forensic Scientist

Forensic science is the application of science to answer questions related to the law. Biologists as well as chemists and biochemists can be forensic scientists. Forensic scientists provide scientific evidence for use in courts, and their job involves examining trace materials associated with crimes. Interest in forensic science has increased in the last few years, possibly because of popular television shows that feature forensic scientists on the job. Also, developing molecular techniques and establishing DNA databases have expanded the types of work that forensic scientists can do. Their job activities are primarily related to crimes against people such as murder, rape, and assault. Their work involves analyzing samples such as hair, blood, and other body fluids and also processing DNA (Figure 1.19) found in many different environments and materials. Forensic scientists also analyze other biological evidence left at crime scenes, such as insect larvae or pollen grains. Students who want to pursue careers in forensic science will most likely have to take chemistry and biology courses as well as some intensive math courses.

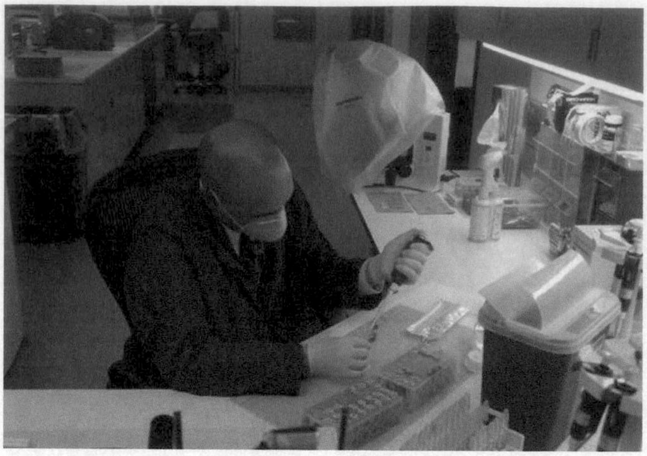

Figure 1.19 This forensic scientist works in a DNA extraction room at the U.S. Army Criminal Investigation Laboratory at Fort Gillem, GA. (credit: United States Army CID Command Public Affairs)

Another field of biological study, **neurobiology**, studies the biology of the nervous system, and although it is a branch of biology, it is also an interdisciplinary field of study known as neuroscience. Because of its interdisciplinary nature, this subdiscipline studies different nervous system functions using molecular, cellular, developmental, medical, and computational approaches.

Figure 1.20 Researchers work on excavating dinosaur fossils at a site in Castellón, Spain. (credit: Mario Modesto)

Paleontology, another branch of biology, uses fossils to study life's history (Figure 1.20). **Zoology** and **botany** are the study of animals and plants, respectively. Biologists can also specialize as biotechnologists, ecologists, or physiologists, to name just a few areas. This is just a small sample of the many fields that biologists can pursue.

Biology is the culmination of the achievements of the natural sciences from their inception to today. Excitingly, it is the cradle of emerging sciences, such as the biology of brain activity, genetic engineering of custom organisms, and the biology of evolution that uses the laboratory tools of molecular biology to retrace the earliest stages of life on Earth. A scan of news headlines—whether reporting on immunizations, a newly discovered species, sports doping, or a genetically-modified food—demonstrates the way biology is active in and important to our everyday world.

KEY TERMS

abstract opening section of a scientific paper that summarizes the research and conclusions

applied science form of science that aims to solve real-world problems

atom smallest and most fundamental unit of matter

basic science science that seeks to expand knowledge and understanding regardless of the short-term application of that knowledge

biochemistry study of the chemistry of biological organisms

biology the study of life

biosphere collection of all the ecosystems on Earth

botany study of plants

cell smallest fundamental unit of structure and function in living things

community set of populations inhabiting a particular area

conclusion section of a scientific paper that summarizes the importance of the experimental findings

control part of an experiment that does not change during the experiment

deductive reasoning form of logical thinking that uses a general inclusive statement to forecast specific results

descriptive science (also, discovery science) form of science that aims to observe, explore, and investigate

discussion section of a scientific paper in which the author interprets experimental results, describes how variables may be related, and attempts to explain the phenomenon in question

ecosystem all the living things in a particular area together with the abiotic, nonliving parts of that environment

eukaryote organism with cells that have nuclei and membrane-bound organelles

evolution the process of gradual change in a population or species over time

falsifiable able to be disproven by experimental results

homeostasis ability of an organism to maintain constant internal conditions

hypothesis suggested explanation for an observation, which one can test

hypothesis-based science form of science that begins with a specific question and potential testable answers

inductive reasoning form of logical thinking that uses related observations to arrive at a general conclusion

introduction opening section of a scientific paper, which provides background information about what was known in the field prior to the research reported in the paper

life science field of science, such as biology, that studies living things

macromolecule large molecule, typically formed by the joining of smaller molecules

materials and methods section of a scientific paper that includes a complete description of the substances, methods, and techniques that the researchers used to gather data

microbiology study of the structure and function of microorganisms

molecular biology study of biological processes and their regulation at the molecular level, including interactions among molecules such as DNA, RNA, and proteins

molecule chemical structure consisting of at least two atoms held together by one or more chemical bonds

natural science field of science that is related to the physical world and its phenomena and processes

neurobiology study of the biology of the nervous system

organ collection of related tissues grouped together performing a common function

organ system level of organization that consists of functionally related interacting organs

organelle small structures that exist within cells and carry out cellular functions

organism individual living entity

paleontology study of life's history by means of fossils

peer-reviewed manuscript scientific paper that a scientist's colleagues review who are experts in the field of study

phylogenetic tree diagram showing the evolutionary relationships among various biological species based on similarities and differences in genetic or physical traits or both; in essence, a hypothesis concerning evolutionary connections

physical science field of science, such as geology, astronomy, physics, and chemistry, that studies nonliving matter

plagiarism using other people's work or ideas without proper citation, creating the false impression that those are the author's original ideas

population all of the individuals of a species living within a specific area

prokaryote single-celled organism that lacks organelles and does not have nuclei surrounded by a nuclear membrane

results section of a scientific paper in which the author narrates the experimental findings and presents relevant figures, pictures, diagrams, graphs, and tables, without any further interpretation

review article paper that summarizes and comments on findings that were published as primary literature

science knowledge that covers general truths or the operation of general laws, especially when acquired and tested by the scientific method

scientific method method of research with defined steps that include observation, formulation of a hypothesis, testing, and confirming or falsifying the hypothesis

serendipity fortunate accident or a lucky surprise

theory tested and confirmed explanation for observations or phenomena

tissue group of similar cells carrying out related functions

variable part of an experiment that the experimenter can vary or change

zoology study of animals

CHAPTER SUMMARY

1.1 The Science of Biology

Biology is the science that studies living organisms and their interactions with one another and their environments. Science attempts to describe and understand the nature of the universe in whole or in part by rational means. Science has many fields. Those fields related to the physical world and its phenomena are natural sciences.

Science can be basic or applied. The main goal of basic science is to expand knowledge without any expectation of short-term practical application of that knowledge. The primary goal of applied research, however, is to solve practical problems.

Science uses two types of logical reasoning. Inductive reasoning uses particular results to produce general scientific principles. Deductive reasoning is a form of logical thinking that predicts results by applying general principles. The common thread throughout scientific research is using the scientific method, a step-based process that consists of making observations, defining a problem, posing hypotheses, testing these hypotheses, and drawing one or more conclusions. The testing uses proper controls. Scientists present their results in peer-reviewed scientific papers published in scientific journals. A scientific research paper consists of several well-defined sections: introduction, materials and methods, results, and, finally, a concluding discussion. Review papers summarize the conducted research in a particular field over a period of time.

1.2 Themes and Concepts of Biology

Biology is the science of life. All living organisms share several key properties such as order, sensitivity or response to stimuli, reproduction, growth and development, regulation, homeostasis, and energy processing. Living things are highly organized parts of a hierarchy that includes atoms, molecules, organelles, cells, tissues, organs, and organ systems. In turn, biologists group organisms as populations, communities, ecosystems, and the biosphere. The great diversity of life today evolved from less-diverse ancestral organisms over billions of years. We can use a phylogenetic tree to show evolutionary relationships among organisms.

Biology is very broad and includes many branches and subdisciplines. Examples include molecular biology, microbiology, neurobiology, zoology, and botany, among others.

VISUAL CONNECTION QUESTIONS

1. Figure 1.6 In the example below, the scientific method is used to solve an everyday problem. Match the scientific method steps (numbered items) with the process of solving the everyday problem (lettered items). Based on the results of the experiment, is the hypothesis correct? If it is incorrect, propose some alternative hypotheses.

1. Observation	a. There is something wrong with the electrical outlet.
2. Question	b. If something is wrong with the outlet, my coffeemaker also won't work when plugged into it.
3. Hypothesis (answer)	c. My toaster doesn't toast my bread.
4. Prediction	d. I plug my coffee maker into the outlet.
5. Experiment	e. My coffeemaker works.
6. Result	f. Why doesn't my toaster work?

2. Figure 1.7 Decide if each of the following is an example of inductive or deductive reasoning.

1. All flying birds and insects have wings. Birds and insects flap their wings as they move through the air. Therefore, wings enable flight.
2. Insects generally survive mild winters better than harsh ones. Therefore, insect pests will become more problematic if global temperatures increase.
3. Chromosomes, the carriers of DNA, separate into daughter cells during cell division. Therefore, each daughter cell will have the same chromosome set as the mother cell.
4. Animals as diverse as humans, insects, and wolves all exhibit social behavior. Therefore, social behavior must have an evolutionary advantage.

3. Figure 1.16 Which of the following statements is false?

a. Tissues exist within organs which exist within organ systems.
b. Communities exist within populations which exist within ecosystems.
c. Organelles exist within cells which exist within tissues.
d. Communities exist within ecosystems which exist in the biosphere.

REVIEW QUESTIONS

4. The first forms of life on Earth were _______.

a. plants
b. microorganisms
c. birds
d. dinosaurs

5. A suggested and testable explanation for an event is called a _______.

a. hypothesis
b. variable
c. theory
d. control

6. Which of the following sciences is not considered a natural science?

a. biology
b. astronomy
c. physics
d. computer science

7. The type of logical thinking that uses related observations to arrive at a general conclusion is called _______.

a. deductive reasoning
b. the scientific method
c. hypothesis-based science
d. inductive reasoning

8. The process of _______ helps to ensure that a scientist's research is original, significant, logical, and thorough.

a. publication
b. public speaking
c. peer review
d. the scientific method

9. A person notices that her houseplants that are regularly exposed to music seem to grow more quickly than those in rooms with no music. As a result, she determines that plants grow better when exposed to music. This example most closely resembles which type of reasoning?
   a. inductive reasoning
   b. deductive reasoning
   c. neither, because no hypothesis was made
   d. both inductive and deductive reasoning

10. The smallest unit of biological structure that meets the functional requirements of "living" is the _______.
   a. organ
   b. organelle
   c. cell
   d. macromolecule

11. Viruses are not considered living because they _______.
   a. are not made of cells
   b. lack cell nuclei
   c. do not contain DNA or RNA
   d. cannot reproduce

12. The presence of a membrane-enclosed nucleus is a characteristic of _______.
   a. prokaryotic cells
   b. eukaryotic cells
   c. living organisms
   d. bacteria

13. A group of individuals of the same species living in the same area is called a(n) _______.
   a. family
   b. community
   c. population
   d. ecosystem

14. Which of the following sequences represents the hierarchy of biological organization from the most inclusive to the least complex level?
   a. organelle, tissue, biosphere, ecosystem, population
   b. organ, organism, tissue, organelle, molecule
   c. organism, community, biosphere, molecule, tissue, organ
   d. biosphere, ecosystem, community, population, organism

15. Where in a phylogenetic tree would you expect to find the organism that had evolved most recently?
   a. at the base
   b. within the branches
   c. at the nodes
   d. at the branch tips

CRITICAL THINKING QUESTIONS

16. Although the scientific method is used by most of the sciences, it can also be applied to everyday situations. Think about a problem that you may have at home, at school, or with your car, and apply the scientific method to solve it.
17. Give an example of how applied science has had a direct effect on your daily life.
18. Name two topics that are likely to be studied by biologists, and two areas of scientific study that would fall outside the realm of biology.
19. Thinking about the topic of cancer, write a basic science question and an applied science question that a researcher interested in this topic might ask.
20. Select two items that biologists agree are necessary in order to consider an organism "alive." For each, give an example of a nonliving object that otherwise fits the definition of "alive."
21. Consider the levels of organization of the biological world, and place each of these items in order from smallest level of organization to most encompassing: skin cell, elephant, water molecule, planet Earth, tropical rainforest, hydrogen atom, wolf pack, liver.
22. You go for a long walk on a hot day. Give an example of a way in which homeostasis keeps your body healthy.
23. Using examples, explain how biology can be studied from a microscopic approach to a global approach.

CHAPTER 2
The Chemical Foundation of Life

Figure 2.1 Atoms are the building blocks of molecules in the universe—air, soil, water, rocks . . . and also the cells of all living organisms. In this model of an organic molecule, the atoms of carbon (black), hydrogen (white), nitrogen (blue), oxygen (red), and sulfur (yellow) are in proportional atomic size. The silver rods indicate chemical bonds. (credit: modification of work by Christian Guthier)

Chapter Outline

INTRODUCTION Elements in various combinations comprise all matter, including living things. Some of the most abundant elements in living organisms include carbon, hydrogen, nitrogen, oxygen, sulfur, and phosphorus. These form the nucleic acids, proteins, carbohydrates, and lipids that are the fundamental components of living matter. Biologists must understand these important building blocks and the unique structures of the atoms that comprise molecules, allowing for cells, tissues, organ systems, and entire organisms to form.

All biological processes follow the laws of physics and chemistry, so in order to understand how biological systems work, it is important to understand the underlying physics and chemistry. For example, the flow of blood within the circulatory system follows the laws of physics that regulate the modes of fluid flow. The breakdown of the large, complex molecules of food into smaller molecules—and the conversion of these to release energy to be stored in adenosine triphosphate (ATP)—is a series of chemical reactions that follow chemical laws. The properties of water and the formation of hydrogen bonds are key to understanding living processes. Recognizing the properties of acids and bases is important, for example, to our understanding of the digestive process. Therefore, the fundamentals of physics and chemistry are important for gaining insight into biological processes.

2.1 Atoms, Isotopes, Ions, and Molecules: The Building Blocks

By the end of this section, you will be able to do the following:

- Define matter and elements
- Describe the interrelationship between protons, neutrons, and electrons
- Compare the ways in which electrons can be donated or shared between atoms
- Explain the ways in which naturally occurring elements combine to create molecules, cells, tissues, organ systems, and organisms

At its most fundamental level, life is made up of matter. **Matter** is any substance that occupies space and has mass. **Elements** are unique forms of matter with specific chemical and physical properties that cannot break down into smaller substances by ordinary chemical reactions. There are 118 elements, but only 98 occur naturally. The remaining elements are unstable and require scientists to synthesize them in laboratories.

Each element is designated by its chemical symbol, which is a single capital letter or, when the first letter is already "taken" by another element, a combination of two letters. Some elements follow the English term for the element, such as C for carbon and Ca for calcium. Other elements' chemical symbols derive from their Latin names. For example, the symbol for sodium is Na, referring to *natrium*, the Latin word for sodium.

The four elements common to all living organisms are oxygen (O), carbon (C), hydrogen (H), and nitrogen (N). In the nonliving world, elements are found in different proportions, and some elements common to living organisms are relatively rare on the earth as a whole, as Table 2.1 shows. For example, the atmosphere is rich in nitrogen and oxygen but contains little carbon and hydrogen, while the earth's crust, although it contains oxygen and a small amount of hydrogen, has little nitrogen and carbon. In spite of their differences in abundance, all elements and the chemical reactions between them obey the same chemical and physical laws regardless of whether they are a part of the living or nonliving world.

Approximate Percentage of Elements in Living Organisms (Humans) Compared to the Nonliving World

Element	Life (Humans)	Atmosphere	Earth's Crust
Oxygen (O)	65%	21%	46%
Carbon (C)	18%	trace	trace
Hydrogen (H)	10%	trace	0.1%
Nitrogen (N)	3%	78%	trace

Table 2.1

The Structure of the Atom

To understand how elements come together, we must first discuss the element's smallest component or building block, the atom. An **atom** is the smallest unit of matter that retains all of the element's chemical properties. For example, one gold atom has all of the properties of gold in that it is a solid metal at room temperature. A gold coin is simply a very large number of gold atoms molded into the shape of a coin and contains small amounts of other elements known as impurities. We cannot break down gold atoms into anything smaller while still retaining the properties of gold.

An atom is composed of two regions: the **nucleus**, which is in the atom's center and contains protons and neutrons. The atom's outermost region holds its electrons in orbit around the nucleus, as Figure 2.2 illustrates. Atoms contain protons, electrons, and neutrons, among other subatomic particles. The only exception is hydrogen (H), which is made of one proton and one electron with no neutrons.

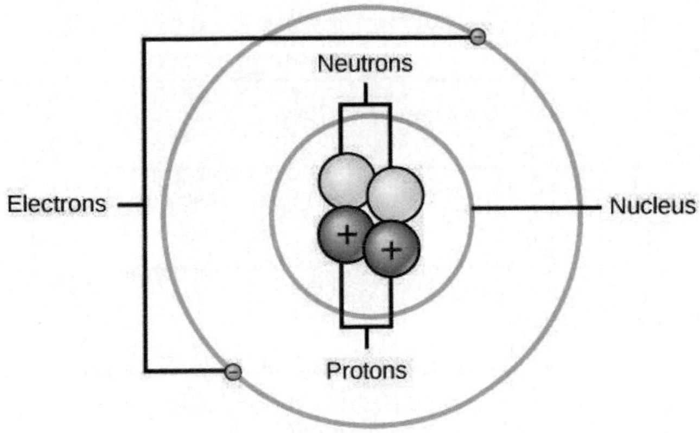

Figure 2.2 Elements, such as helium, depicted here, are made up of atoms. Atoms are made up of protons and neutrons located within the nucleus, with electrons in orbitals surrounding the nucleus.

Protons and neutrons have approximately the same mass, about 1.67×10^{-24} grams. Scientists arbitrarily define this amount of mass as one atomic mass unit (amu) or one Dalton, as Table 2.2 shows. Although similar in mass, protons and neutrons differ in their electric charge. A **proton** is positively charged; whereas, a **neutron** is uncharged. Therefore, the number of neutrons in an atom contributes significantly to its mass, but not to its charge. **Electrons** are much smaller in mass than protons, weighing only 9.11×10^{-28} grams, or about 1/1800 of an atomic mass unit. Hence, they do not contribute much to an element's overall atomic mass. Therefore, when considering atomic mass, it is customary to ignore the mass of any electrons and calculate the atom's mass based on the number of protons and neutrons alone. Although not significant contributors to mass, electrons do contribute greatly to the atom's charge, as each electron has a negative charge equal to the proton's positive charge. In uncharged, neutral atoms, the number of electrons orbiting the nucleus is equal to the number of protons inside the nucleus. In these atoms, the positive and negative charges cancel each other out, leading to an atom with no net charge.

Accounting for the sizes of protons, neutrons, and electrons, most of the atom's volume—greater than 99 percent—is empty space. With all this empty space, one might ask why so-called solid objects do not just pass through one another. The reason they do not is that the electrons that surround all atoms are negatively charged and negative charges repel each other.

Protons, Neutrons, and Electrons

	Charge	Mass (amu)	Location
Proton	+1	1	nucleus
Neutron	0	1	nucleus
Electron	–1	0	orbitals

Table 2.2

Atomic Number and Mass

Atoms of each element contain a characteristic number of protons and electrons. The number of protons determines an element's **atomic number**, which scientists use to distinguish one element from another. The number of neutrons is variable, resulting in isotopes, which are different forms of the same atom that vary only in the number of neutrons they possess. Together, the number of protons and neutrons determine an element's **mass number**, as Figure 2.3 illustrates. Note that we disregard the small contribution of mass from electrons in calculating the mass number. We can use this approximation of mass to easily calculate how many neutrons an element has by simply subtracting the number of protons from the mass number. Since an element's isotopes will have slightly different mass numbers, scientists also determine the **atomic mass**, which is the calculated mean of the mass number for its naturally occurring isotopes. Often, the resulting number contains a fraction. For example, the atomic mass of chlorine (Cl) is 35.45 because chlorine is composed of several isotopes, some (the majority) with atomic mass 35 (17 protons and 18 neutrons) and some with atomic mass 37 (17 protons and 20 neutrons).

VISUAL CONNECTION

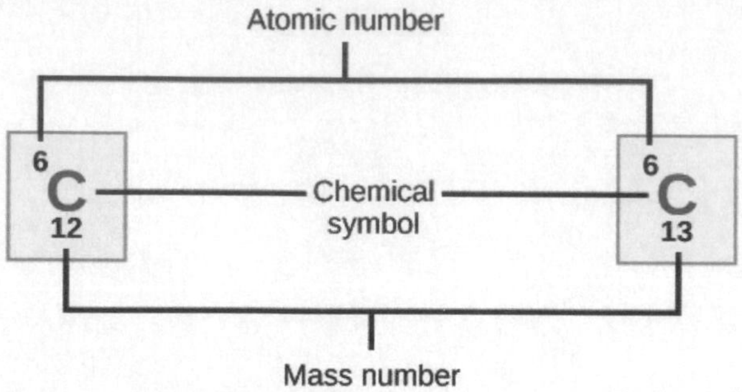

Figure 2.3 Carbon has an atomic number of six, and two stable isotopes with mass numbers of twelve and thirteen, respectively. Its relative atomic mass is 12.011

How many neutrons do carbon-12 and carbon-13 have, respectively?

Isotopes

Isotopes are different forms of an element that have the same number of protons but a different number of neutrons. Some elements—such as carbon, potassium, and uranium—have naturally occurring isotopes. Carbon-12 contains six protons, six neutrons, and six electrons; therefore, it has a mass number of 12 (six protons and six neutrons). Carbon-14 contains six protons, eight neutrons, and six electrons; its atomic mass is 14 (six protons and eight neutrons). These two alternate forms of carbon are isotopes. Some isotopes may emit neutrons, protons, and electrons, and attain a more stable atomic configuration (lower level of potential energy); these are radioactive isotopes, or **radioisotopes**. Radioactive decay (carbon-14 decaying to eventually become nitrogen-14) describes the energy loss that occurs when an unstable atom's nucleus releases radiation.

EVOLUTION CONNECTION

Carbon Dating

Carbon is normally present in the atmosphere in the form of gaseous compounds like carbon dioxide and methane. Carbon-14 (^{14}C) is a naturally occurring radioisotope that is created in the atmosphere from atmospheric ^{14}N (nitrogen) by the addition of a neutron and the loss of a proton because of cosmic rays. This is a continuous process, so more ^{14}C is always being created. As a living organism incorporates ^{14}C initially as carbon dioxide fixed in the process of photosynthesis, the relative amount of ^{14}C in its body is equal to the concentration of ^{14}C in the atmosphere. When an organism dies, it is no longer ingesting ^{14}C, so the ratio between ^{14}C and ^{12}C will decline as ^{14}C decays gradually to ^{14}N by a process called beta decay—electrons or positrons emission. This decay emits energy in a slow process.

After approximately 5,730 years, half of the starting concentration of ^{14}C will convert back to ^{14}N. We call the time it takes for half of the original concentration of an isotope to decay back to its more stable form its half-life. Because the half-life of ^{14}C is long, scientists use it to date formerly living objects such as old bones or wood. Comparing the ratio of the ^{14}C concentration in an object to the amount of ^{14}C in the atmosphere, scientists can determine the amount of the isotope that has not yet decayed. On the basis of this amount, Figure 2.4 shows that we can calculate the age of the material, such as the pygmy mammoth, with accuracy if it is not much older than about 50,000 years. Other elements have isotopes with different half lives. For example, ^{40}K (potassium-40) has a half-life of 1.25 billion years, and ^{235}U (Uranium 235) has a half-life of about 700 million years. Through the use of radiometric dating, scientists can study the age of fossils or other remains of extinct organisms to understand how organisms have evolved from earlier species.

Figure 2.4 Scientists can determine the age of carbon-containing remains less than about 50,000 years old, such as this pygmy mammoth, using carbon dating. (credit: Bill Faulkner, NPS)

LINK TO LEARNING

To learn more about atoms, isotopes, and how to tell one isotope from another, run the simulation.

Click to view content (https://openstax.org/l/atoms_isotopes)

The Periodic Table

The **periodic table** organizes and displays different elements. Devised by Russian chemist Dmitri Mendeleev (1834–1907) in 1869, the table groups elements that, although unique, share certain chemical properties with other elements. The properties of elements are responsible for their physical state at room temperature: they may be gases, solids, or liquids. Elements also have specific **chemical reactivity**, the ability to combine and to chemically bond with each other.

In the periodic table in Figure 2.5, the elements are organized and displayed according to their atomic number and are arranged in a series of rows and columns based on shared chemical and physical properties. In addition to providing the atomic number for each element, the periodic table also displays the element's atomic mass. Looking at carbon, for example, its symbol (C) and name appear, as well as its atomic number of six (in the upper left-hand corner) and its atomic mass of 12.01.

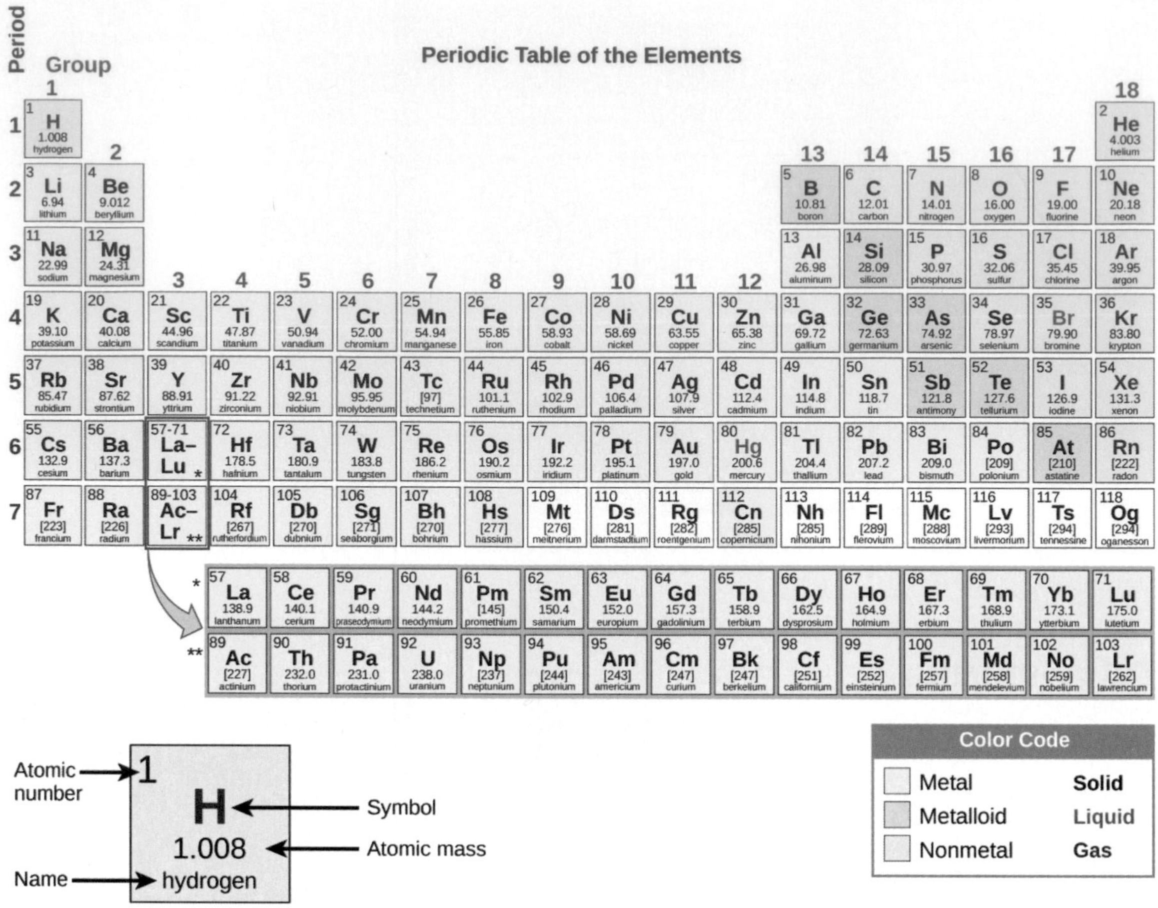

Figure 2.5 The periodic table shows each element's atomic mass and atomic number. The atomic number appears above the symbol for the element and the approximate atomic mass appears below it.

The periodic table groups elements according to chemical properties. Scientists base the differences in chemical reactivity between the elements on the number and spatial distribution of an atom's electrons. Atoms that chemically react and bond to each other form molecules. **Molecules** are simply two or more atoms chemically bonded together. Logically, when two atoms chemically bond to form a molecule, their electrons, which form the outermost region of each atom, come together first as the atoms form a chemical bond.

Electron Shells and the Bohr Model

Note that there is a connection between the number of protons in an element, the atomic number that distinguishes one element from another, and the number of electrons it has. In all electrically neutral atoms, the number of electrons is the same as the number of protons. Thus, each element, at least when electrically neutral, has a characteristic number of electrons equal to its atomic number.

In 1913, Danish scientist Niels Bohr (1885–1962) developed an early model of the atom. The Bohr model shows the atom as a central nucleus containing protons and neutrons, with the electrons in circular **orbitals** at specific distances from the nucleus, as Figure 2.6 illustrates. These orbits form electron shells or energy levels, which are a way of visualizing the number of electrons in the outermost shells. These energy levels are designated by a number and the symbol "n." For example, 1n represents the first energy level located closest to the nucleus.

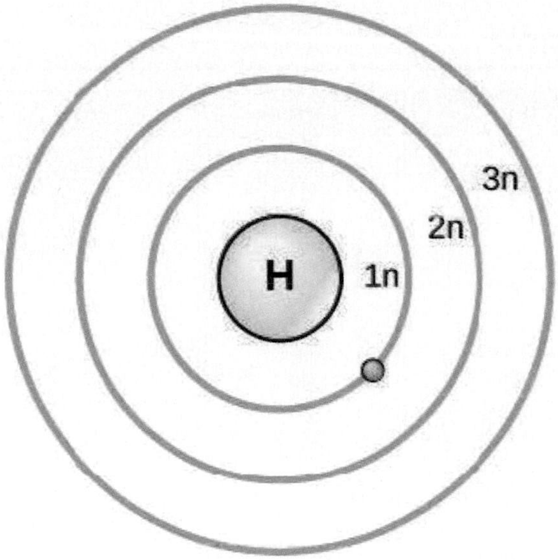

Figure 2.6 In 1913, Niels Bohrs developed the Bohr model in which electrons exist within principal shells. An electron normally exists in the lowest energy shell available, which is the one closest to the nucleus. Energy from a photon of light can bump it up to a higher energy shell, but this situation is unstable, and the electron quickly decays back to the ground state. In the process, it releases a photon of light.

Electrons fill orbitals in a consistent order: they first fill the orbitals closest to the nucleus, then they continue to fill orbitals of increasing energy further from the nucleus. If there are multiple orbitals of equal energy, they fill with one electron in each energy level before adding a second electron. The electrons of the outermost energy level determine the atom's energetic stability and its tendency to form chemical bonds with other atoms to form molecules.

Under standard conditions, atoms fill the inner shells first, often resulting in a variable number of electrons in the outermost shell. The innermost shell has a maximum of two electrons but the next two electron shells can each have a maximum of eight electrons. This is known as the **octet rule**, which states, with the exception of the innermost shell, that atoms are more stable energetically when they have eight electrons in their **valence shell**, the outermost electron shell. Figure 2.7 shows examples of some neutral atoms and their electron configurations. Notice that in Figure 2.7, helium has a complete outer electron shell, with two electrons filling its first and only shell. Similarly, neon has a complete outer 2n shell containing eight electrons. In contrast, chlorine and sodium have seven and one in their outer shells, respectively, but theoretically they would be more energetically stable if they followed the octet rule and had eight.

VISUAL CONNECTION

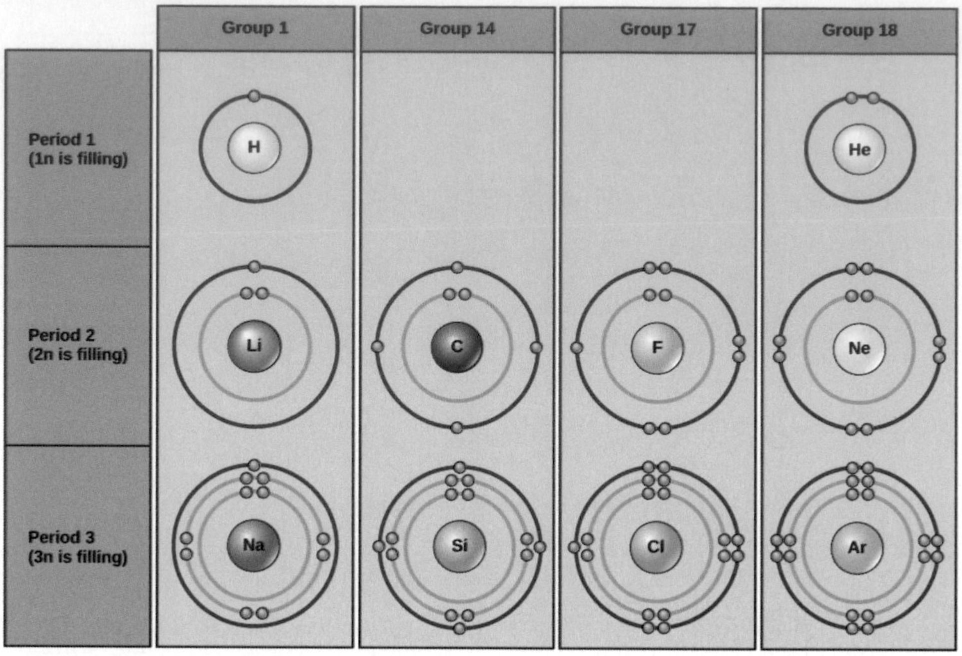

Figure 2.7 Bohr diagrams indicate how many electrons fill each principal shell. Group 18 elements (helium, neon, and argon) have a full outer, or valence, shell. A full valence shell is the most stable electron configuration. Elements in other groups have partially filled valence shells and gain or lose electrons to achieve a stable electron configuration.

An atom may give, take, or share electrons with another atom to achieve a full valence shell, the most stable electron configuration. Looking at this figure, how many electrons do elements in group 1 need to lose in order to achieve a stable electron configuration? How many electrons do elements in groups 14 and 17 need to gain to achieve a stable configuration?

Understanding that the periodic table's organization is based on the total number of protons (and electrons) helps us know how electrons distribute themselves among the shells. The periodic table is arranged in columns and rows based on the number of electrons and their location. Examine more closely some of the elements in the table's far right column in Figure 2.5. The group 18 atoms helium (He), neon (Ne), and argon (Ar) all have filled outer electron shells, making it unnecessary for them to share electrons with other atoms to attain stability. They are highly stable as single atoms. Because they are non reactive, scientists coin them **inert** (or **noble gases**). Compare this to the group 1 elements in the left-hand column. These elements, including hydrogen (H), lithium (Li), and sodium (Na), all have one electron in their outermost shells. That means that they can achieve a stable configuration and a filled outer shell by donating or sharing one electron with another atom or a molecule such as water. Hydrogen will donate or share its electron to achieve this configuration, while lithium and sodium will donate their electron to become stable. As a result of losing a negatively charged electron, they become positively charged **ions**. Group 17 elements, including fluorine and chlorine, have seven electrons in their outmost shells, so they tend to fill this shell with an electron from other atoms or molecules, making them negatively charged ions. Group 14 elements, of which carbon is the most important to living systems, have four electrons in their outer shell allowing them to make several covalent bonds (discussed below) with other atoms. Thus, the periodic table's columns represent the potential shared state of these elements' outer electron shells that is responsible for their similar chemical characteristics.

Electron Orbitals

Although useful to explain the reactivity and chemical bonding of certain elements, the Bohr model does not accurately reflect how electrons spatially distribute themselves around the nucleus. They do not circle the nucleus like the earth orbits the sun, but we find them in **electron orbitals**. These relatively complex shapes result from the fact that electrons behave not just like particles, but also like waves. Mathematical equations from quantum mechanics, which scientists call wave functions, can predict within a certain level of probability where an electron might be at any given time. Scientists call the area where an electron is most likely to be found its orbital.

Recall that the Bohr model depicts an atom's electron shell configuration. Within each electron shell are subshells, and each subshell has a specified number of orbitals containing electrons. While it is impossible to calculate exactly an electron's location, scientists know that it is most probably located within its orbital path. The letter s, *p*, *d*, and *f* designate the subshells. The *s* subshell is spherical in shape and has one orbital. Principal shell 1n has only a single *s* orbital, which can hold two electrons. Principal shell 2n has one *s* and one *p* subshell, and can hold a total of eight electrons. The *p* subshell has three dumbbell-shaped orbitals, as Figure 2.8 illustrates. Subshells *d* and *f* have more complex shapes and contain five and seven orbitals, respectively. We do not show these in the illustration. Principal shell 3n has *s*, *p*, and *d* subshells and can hold 18 electrons. Principal shell 4n has *s*, *p*, *d* and *f* orbitals and can hold 32 electrons. Moving away from the nucleus, the number of electrons and orbitals in the energy levels increases. Progressing from one atom to the next in the periodic table, we can determine the electron structure by fitting an extra electron into the next available orbital.

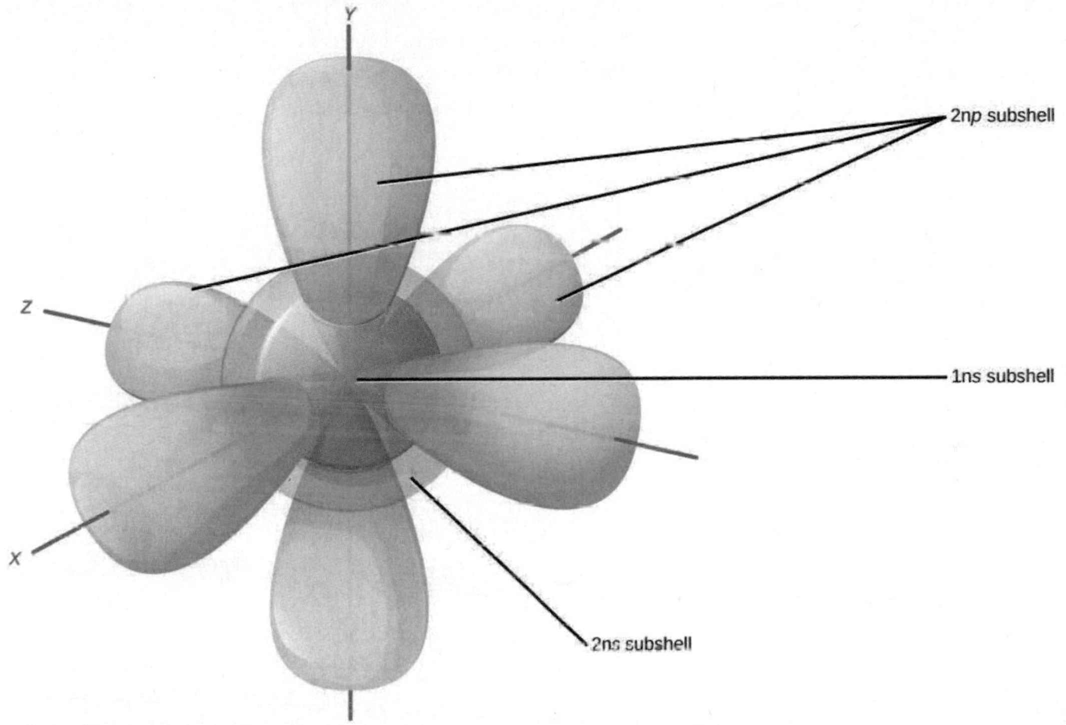

Figure 2.8 The *s* subshells are shaped like spheres. Both the 1n and 2n principal shells have an *s* orbital, but the size of the sphere is larger in the 2n orbital. Each sphere is a single orbital. Three dumbbell-shaped orbitals comprise *p* subshells. Principal shell 2n has a *p* subshell, but shell 1 does not.

The closest orbital to the nucleus, the 1*s* orbital, can hold up to two electrons. This orbital is equivalent to the Bohr model's innermost electron shell. Scientists call it the 1*s* orbital because it is spherical around the nucleus. The 1*s* orbital is the closest orbital to the nucleus, and it is always filled first, before any other orbital fills. Hydrogen has one electron; therefore, it occupies only one spot within the 1*s* orbital. We designate this as $1s^1$, where the superscripted 1 refers to the one electron within the 1*s* orbital. Helium has two electrons; therefore, it can completely fill the 1*s* orbital with its two electrons. We designate this as $1s^2$, referring to the two electrons of helium in the 1*s* orbital. On the periodic table Figure 2.5, hydrogen and helium are the only two elements in the first row (period). This is because they only have electrons in their first shell, the 1*s* orbital. Hydrogen and helium are the only two elements that have the 1*s* and no other electron orbitals in the electrically neutral state.

The second electron shell may contain eight electrons. This shell contains another spherical *s* orbital and three "dumbbell" shaped *p* orbitals, each of which can hold two electrons, as Figure 2.8 shows. After the 1*s* orbital fills, the second electron shell fills, first filling its 2*s* orbital and then its three *p* orbitals. When filling the *p* orbitals, each takes a single electron. Once each *p* orbital has an electron, it may add a second. Lithium (Li) contains three electrons that occupy the first and second shells. Two electrons fill the 1*s* orbital, and the third electron then fills the 2*s* orbital. Its **electron configuration** is $1s^2 2s^1$. Neon (Ne), alternatively, has a total of ten electrons: two are in its innermost 1*s* orbital and eight fill its second shell (two each in the 2*s* and three *p* orbitals). Thus it is an inert gas and energetically stable as a single atom that will rarely form a chemical bond with other atoms. Larger elements have additional orbitals, comprising the third electron shell. While the concepts of electron shells and orbitals are closely related, orbitals provide a more accurate depiction of an atom's electron configuration because the orbital model specifies the different shapes and special orientations of all the places that electrons may occupy.

LINK TO LEARNING

Watch this visual animation to see the spatial arrangement of the p and s orbitals.

Click to view content (https://www.openstax.org/l/orbitals)

Chemical Reactions and Molecules

All elements are most stable when their outermost shell is filled with electrons according to the octet rule. This is because it is energetically favorable for atoms to be in that configuration and it makes them stable. However, since not all elements have enough electrons to fill their outermost shells, atoms form **chemical bonds** with other atoms thereby obtaining the electrons they need to attain a stable electron configuration. When two or more atoms chemically bond with each other, the resultant chemical structure is a molecule. The familiar water molecule, H_2O, consists of two hydrogen atoms and one oxygen atom. These bond together to form water, as Figure 2.9 illustrates. Atoms can form molecules by donating, accepting, or sharing electrons to fill their outer shells.

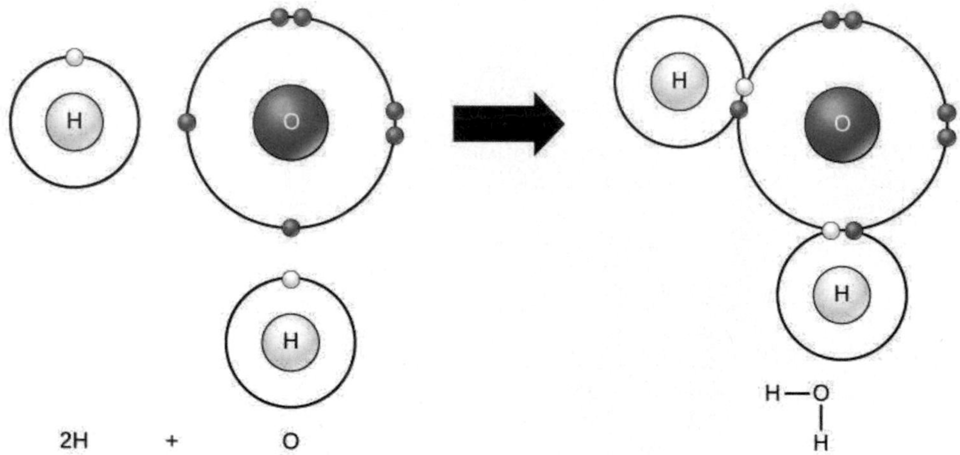

Figure 2.9 Two or more atoms may bond with each other to form a molecule. When two hydrogens and an oxygen share electrons via covalent bonds it forms a water molecule.

Chemical reactions occur when two or more atoms bond together to form molecules or when bonded atoms break apart. Scientists call the substances used in the beginning of a chemical reaction **reactants** (usually on the left side of a chemical equation), and we call the substances at the end of the reaction **products** (usually on the right side of a chemical equation). We typically draw an arrow between the reactants and products to indicate the chemical reaction's direction. This direction is not always a "one-way street." To create the water molecule above, the chemical equation would be:

$$2H + O \rightarrow H_2O$$

An example of a simple chemical reaction is breaking down hydrogen peroxide molecules, each of which consists of two hydrogen atoms bonded to two oxygen atoms (H_2O_2). The reactant hydrogen peroxide breaks down into water, containing one oxygen atom bound to two hydrogen atoms (H_2O), and oxygen, which consists of two bonded oxygen atoms (O_2). In the equation below, the reaction includes two hydrogen peroxide molecules and two water molecules. This is an example of a **balanced chemical equation**, wherein each element's number of atoms is the same on each side of the equation. According to the law of conservation of matter, the number of atoms before and after a chemical reaction should be equal, such that no atoms are, under normal circumstances, created or destroyed.

$$2H_2O_2 \text{ (hydrogen peroxide)} \rightarrow 2H_2O \text{ (water)} + O_2 \text{ (oxygen)}$$

Even though all of the reactants and products of this reaction are molecules (each atom remains bonded to at least one other atom), in this reaction only hydrogen peroxide and water are representatives of **compounds**: they contain atoms of more than one type of element. Molecular oxygen, alternatively, as Figure 2.10 shows, consists of two doubly bonded oxygen atoms and is not classified as a compound but as a homonuclear molecule.

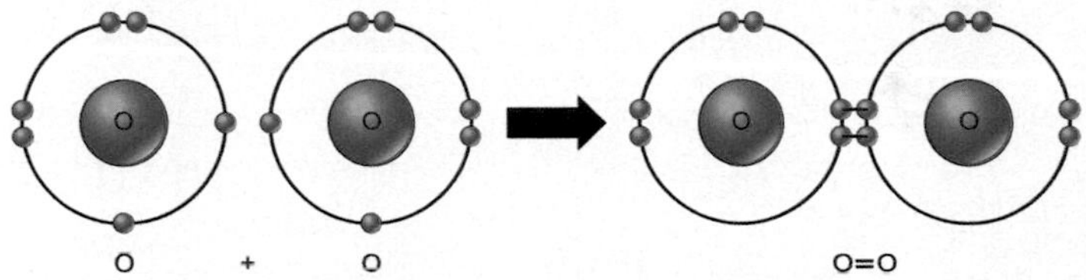

Figure 2.10 A double bond joins the oxygen atoms in an O_2 molecule.

Some chemical reactions, such as the one above, can proceed in one direction until they expend all the reactants. The equations that describe these reactions contain a unidirectional arrow and are **irreversible**. **Reversible reactions** are those that can go in either direction. In reversible reactions, reactants turn into products, but when the product's concentration goes beyond a certain threshold (characteristic of the particular reaction), some of these products convert back into reactants. At this point, product and reactant designations reverse. This back and forth continues until a certain relative balance between reactants and products occurs—a state called **equilibrium**. A chemical equation with a double headed arrow pointing towards both the reactants and products often denote these reversible reaction situations.

For example, in human blood, excess hydrogen ions (H^+) bind to bicarbonate ions (HCO_3^-) forming an equilibrium state with carbonic acid (H_2CO_3). If we added carbonic acid to this system, some of it would convert to bicarbonate and hydrogen ions.

$$HCO_3^- + H^+ \leftrightarrow H_2CO_3$$

However, biological reactions rarely obtain equilibrium because the concentrations of the reactants or products or both are constantly changing, often with one reaction's product a reactant for another. To return to the example of excess hydrogen ions in the blood, forming carbonic acid will be the reaction's major direction. However, the carbonic acid can also leave the body as carbon dioxide gas (via exhalation) instead of converting back to bicarbonate ion, thus driving the reaction to the right by the **law of mass action**. These reactions are important for maintaining homeostasis in our blood.

$$HCO_3^- + H^+ \leftrightarrow H_2CO_3 \leftrightarrow CO_2 + H_2O$$

Ions and Ionic Bonds

Some atoms are more stable when they gain or lose an electron (or possibly two) and form ions. This fills their outermost electron shell and makes them energetically more stable. Because the number of electrons does not equal the number of protons, each ion has a net charge. **Cations** are positive ions that form by losing electrons. Negative ions form by gaining electrons, which we call anions. We designate **anions** by their elemental name and change the ending to "-ide", thus the anion of chlorine is chloride, and the anion of sulfur is sulfide.

Scientists refer to this movement of electrons from one element to another as **electron transfer**. As Figure 2.11 illustrates, sodium (Na) only has one electron in its outer electron shell. It takes less energy for sodium to donate that one electron than it does to accept seven more electrons to fill the outer shell. If sodium loses an electron, it now has 11 protons, 11 neutrons, and only 10 electrons, leaving it with an overall charge of +1. We now refer to it as a sodium ion. Chlorine (Cl) in its lowest energy state (called the ground state) has seven electrons in its outer shell. Again, it is more energy-efficient for chlorine to gain one electron than to lose seven. Therefore, it tends to gain an electron to create an ion with 17 protons, 17 neutrons, and 18 electrons, giving it a net negative (–1) charge. We now refer to it as a chloride ion. In this example, sodium will donate its one electron to empty its shell, and chlorine will accept that electron to fill its shell. Both ions now satisfy the octet rule and have complete outermost shells. Because the number of electrons is no longer equal to the number of protons, each is now an ion and has a +1 (sodium cation) or –1 (chloride anion) charge. Note that these transactions can normally only take place simultaneously: in order for a sodium atom to lose an electron, it must be in the presence of a suitable recipient like a chlorine atom.

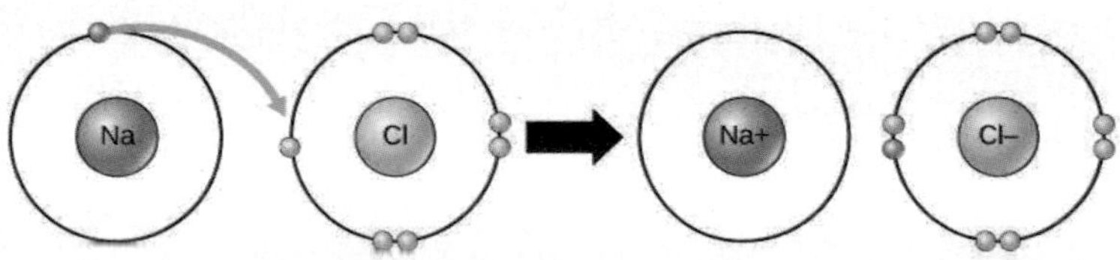

Figure 2.11 In the formation of an ionic compound, metals lose electrons and nonmetals gain electrons to achieve an octet.

Ionic bonds form between ions with opposite charges. For instance, positively charged sodium ions and negatively charged chloride ions bond together to make crystals of sodium chloride, or table salt, creating a crystalline molecule with zero net charge.

Physiologists refer to certain salts as **electrolytes** (including sodium, potassium, and calcium), ions necessary for nerve impulse conduction, muscle contractions, and water balance. Many sports drinks and dietary supplements provide these ions to replace those lost from the body via sweating during exercise.

Covalent Bonds and Other Bonds and Interactions

Another way to satisfy the octet rule is by sharing electrons between atoms to form **covalent bonds**. These bonds are stronger and much more common than ionic bonds in the molecules of living organisms. We commonly find covalent bonds in carbon-based organic molecules, such as our DNA and proteins. We also find covalent bonds in inorganic molecules like H_2O, CO_2, and O_2. The bonds may share one, two, or three pairs of electrons, making single, double, and triple bonds, respectively. The more covalent bonds between two atoms, the stronger their connection. Thus, triple bonds are the strongest.

The strength of different levels of covalent bonding is one of the main reasons living organisms have a difficult time in acquiring nitrogen for use in constructing their molecules, even though molecular nitrogen, N_2, is the most abundant gas in the atmosphere. Molecular nitrogen consists of two nitrogen atoms triple bonded to each other and, as with all molecules, sharing these three pairs of electrons between the two nitrogen atoms allows for filling their outer electron shells, making the molecule more stable than the individual nitrogen atoms. This strong triple bond makes it difficult for living systems to break apart this nitrogen in order to use it as constituents of proteins and DNA.

Forming water molecules provides an example of covalent bonding. Covalent bonds bind the hydrogen and oxygen atoms that combine to form water molecules as Figure 2.9 shows. The electron from the hydrogen splits its time between the hydrogen atoms' incomplete outer shell and the oxygen atoms' incomplete outer shell. To completely fill the oxygen's outer shell, which has six electrons but which would be more stable with eight, two electrons (one from each hydrogen atom) are needed: hence, the well-known formula H_2O. The two elements share the electrons to fill the outer shell of each, making both elements more stable.

LINK TO LEARNING

View this short video to see an animation of ionic and covalent bonding.

Click to view content (https://www.openstax.org/l/ionic_covalent)

Polar Covalent Bonds

There are two types of covalent bonds: polar and nonpolar. In a **polar covalent bond**, Figure 2.12 shows atoms unequally share the electrons and are attracted more to one nucleus than the other. Because of the unequal electron distribution between the atoms of different elements, a slightly positive (δ+) or slightly negative (δ–) charge develops. This partial charge is an important property of water and accounts for many of its characteristics.

Water is a polar molecule, with the hydrogen atoms acquiring a partial positive charge and the oxygen a partial negative charge. This occurs because the oxygen atom's nucleus is more attractive to the hydrogen atoms' electrons than the hydrogen nucleus is to the oxygen's electrons. Thus, oxygen has a higher **electronegativity** than hydrogen and the shared electrons spend more time near the oxygen nucleus than the hydrogen atoms' nucleus, giving the oxygen and hydrogen atoms slightly negative and positive charges, respectively. Another way of stating this is that the probability of finding a shared electron near an oxygen nucleus is more likely than finding it near a hydrogen nucleus. Either way, the atom's relative electronegativity contributes to developing partial charges whenever one element is significantly more electronegative than the other, and the charges that these polar bonds generate may then be used to form hydrogen bonds based on the attraction of opposite partial charges. (Hydrogen bonds, which we discuss in detail below, are weak bonds between slightly positively charged hydrogen atoms to slightly negatively charged atoms in other molecules.) Since macromolecules often have atoms within them that differ in electronegativity, polar bonds are often present in organic molecules.

Nonpolar Covalent Bonds

Nonpolar covalent bonds form between two atoms of the same element or between different elements that share electrons equally. For example, molecular oxygen (O_2) is nonpolar because the electrons distribute equally between the two oxygen atoms.

Figure 2.12 also shows another example of a nonpolar covalent bond—methane (CH_4). Carbon has four electrons in its outermost shell and needs four more to fill it. It obtains these four from four hydrogen atoms, each atom providing one, making a stable outer shell of eight electrons. Carbon and hydrogen do not have the same electronegativity but are similar; thus, nonpolar bonds form. The hydrogen atoms each need one electron for their outermost shell, which is filled when it contains two electrons. These elements share the electrons equally among the carbons and the hydrogen atoms, creating a nonpolar covalent molecule.

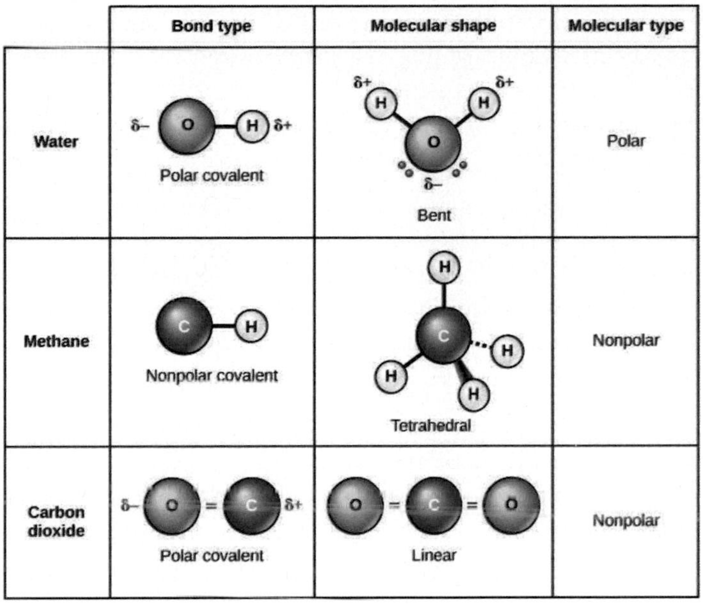

Figure 2.12 Whether a molecule is polar or nonpolar depends both on bond type and molecular shape. Both water and carbon dioxide have polar covalent bonds, but carbon dioxide is linear, so the partial charges on the molecule cancel each other out.

Hydrogen Bonds and Van Der Waals Interactions

Ionic and covalent bonds between elements require energy to break. Ionic bonds are not as strong as covalent, which determines their behavior in biological systems. However, not all bonds are ionic or covalent bonds. Weaker bonds can also form between molecules. Two weak bonds that occur frequently are hydrogen bonds and van der Waals interactions. Without these two types of bonds, life as we know it would not exist. Hydrogen bonds provide many of the critical, life-sustaining properties of water and also stabilize the structures of proteins and DNA, the building block of cells.

When polar covalent bonds containing hydrogen form, the hydrogen in that bond has a slightly positive charge because hydrogen's electron is pulled more strongly toward the other element and away from the hydrogen. Because the hydrogen is slightly positive, it will be attracted to neighboring negative charges. When this happens, a weak interaction occurs between the hydrogen's δ^{+} from one molecule and the molecule's δ^{-} charge on another molecule with the more electronegative atoms, usually oxygen. Scientists call this interaction a **hydrogen bond**. This type of bond is common and occurs regularly between water molecules. Individual hydrogen bonds are weak and easily broken; however, they occur in very large numbers in water and in organic polymers, creating a major force in combination. Hydrogen bonds are also responsible for zipping together the DNA double helix.

Like hydrogen bonds, **van der Waals interactions** are weak attractions or interactions between molecules. Van der Waals attractions can occur between any two or more molecules and are dependent on slight fluctuations of the electron densities, which are not always symmetrical around an atom. For these attractions to happen, the molecules need to be very close to one another. These bonds—along with ionic, covalent, and hydrogen bonds—contribute to the proteins' three-dimensional structure in our cells that is necessary for their proper function.

CAREER CONNECTION

Pharmaceutical Chemist

Pharmaceutical chemists are responsible for developing new drugs and trying to determine the mode of action of both old and new drugs. They are involved in every step of the drug development process. We can find drugs in the natural environment or we can synthesize them in the laboratory. In many cases, chemists chemically change potential drugs from nature chemically in the laboratory to make them safer and more effective, and sometimes synthetic versions of drugs substitute for the version we find in nature.

After a drug's initial discovery or synthesis, the chemist then develops the drug, perhaps chemically altering it, testing it to see if it is toxic, and then designing methods for efficient large-scale production. Then, the process of approving the drug for human use begins. In the United States, the Food and Drug Administration (FDA) handles drug approval. This involves a series of large-scale experiments using human subjects to ensure the drug is not harmful and effectively treats the condition for which it is intended. This process often takes several years and requires the participation of physicians and scientists, in addition to chemists, to complete testing and gain approval.

An example of a drug that was originally discovered in a living organism is Paclitaxel (Taxol), an anti-cancer drug used to treat breast cancer. This drug was discovered in the bark of the pacific yew tree. Another example is aspirin, originally isolated from willow tree bark. Finding drugs often means testing hundreds of samples of plants, fungi, and other forms of life to see if they contain any biologically active compounds. Sometimes, traditional medicine can give modern medicine clues as to where to find an active compound. For example, mankind has used willow bark to make medicine for thousands of years, dating back to ancient Egypt. However, it was not until the late 1800s that scientists and pharmaceutical companies purified and marketed the aspirin molecule, acetylsalicylic acid, for human use.

Occasionally, drugs developed for one use have unforeseen effects that allow usage in other, unrelated ways. For example, scientists originally developed the drug minoxidil (Rogaine) to treat high blood pressure. When tested on humans, researchers noticed that individuals taking the drug would grow new hair. Eventually the pharmaceutical company marketed the drug to men and women with baldness to restore lost hair.

A pharmaceutical chemist's career may involve detective work, experimentation, and drug development, all with the goal of making human beings healthier.

2.2 Water

By the end of this section, you will be able to do the following:

- Describe the properties of water that are critical to maintaining life
- Explain why water is an excellent solvent
- Provide examples of water's cohesive and adhesive properties
- Discuss the role of acids, bases, and buffers in homeostasis

Why do scientists spend time looking for water on other planets? Why is water so important? It is because water is essential to life as we know it. Water is one of the more abundant molecules and the one most critical to life on Earth. Water comprises approximately 60–70 percent of the human body. Without it, life as we know it simply would not exist.

The polarity of the water molecule and its resulting hydrogen bonding make water a unique substance with special properties that are intimately tied to the processes of life. Life originally evolved in a watery environment, and most of an organism's cellular chemistry and metabolism occur inside the watery contents of the cell's cytoplasm. Special properties of water are its high heat capacity and heat of vaporization, its ability to dissolve polar molecules, its cohesive and adhesive properties, and its dissociation into ions that leads to generating pH. Understanding these characteristics of water helps to elucidate its importance in maintaining life.

Water's Polarity

One of water's important properties is that it is composed of polar molecules: the hydrogen and oxygen within water molecules (H_2O) form polar covalent bonds. While there is no net charge to a water molecule, water's polarity creates a slightly positive charge on hydrogen and a slightly negative charge on oxygen, contributing to water's properties of attraction. Water generates charges because oxygen is more electronegative than hydrogen, making it more likely that a shared electron would be near the oxygen nucleus than the hydrogen nucleus, thus generating the partial negative charge near the **oxygen**.

As a result of water's polarity, each water molecule attracts other water molecules because of the opposite charges between water molecules, forming hydrogen bonds. Water also attracts or is attracted to other polar molecules and ions. We call a polar substance that interacts readily with or dissolves in water **hydrophilic** (hydro- = "water"; -philic = "loving"). In contrast, nonpolar molecules such as oils and fats do not interact well with water, as Figure 2.13 shows. A good example of this is vinegar and oil salad dressing (an acidic water solution). We call such nonpolar compounds **hydrophobic** (hydro- = "water"; -phobic = "fearing").

Figure 2.13 Oil and water do not mix. As this macro image of oil and water shows, oil does not dissolve in water but forms droplets instead. This is because it is a nonpolar compound. (credit: Gautam Dogra).

Water's States: Gas, Liquid, and Solid

The formation of hydrogen bonds is an important quality of the liquid water that is crucial to life as we know it. As water molecules make hydrogen bonds with each other, water takes on some unique chemical characteristics compared to other liquids and, since living things have a high water content, understanding these chemical features is key to understanding life. In liquid water, hydrogen bonds constantly form and break as the water molecules slide past each other. The water molecules' motion (kinetic energy) causes the bonds to break due to the heat contained in the system. When the heat rises as water boils, the water molecules' higher kinetic energy causes the hydrogen bonds to break completely and allows water molecules to escape into the air as gas (steam or water vapor). Alternatively, when water temperature reduces and water freezes, the water molecules form a crystalline structure maintained by hydrogen bonding (there is not enough energy to break the hydrogen bonds) that makes ice less dense than liquid water, a phenomenon that we do not see when other liquids solidify.

Water's lower density in its solid form is due to the way hydrogen bonds orient as they freeze: the water molecules push farther apart compared to liquid water. With most other liquids, solidification when the temperature drops includes lowering kinetic energy between molecules, allowing them to pack even more tightly than in liquid form and giving the solid a greater density than the liquid.

The lower density of ice, as Figure 2.14 depicts, an anomaly causes it to float at the surface of liquid water, such as in an iceberg or ice cubes in a glass of water. In lakes and ponds, ice will form on the water's surface creating an insulating barrier that protects the animals and plant life in the pond from freezing. Without this insulating ice layer, plants and animals living in the pond would freeze in the solid block of ice and could not survive. The expansion of ice relative to liquid water causes the detrimental effect of freezing on living organisms. The ice crystals that form upon freezing rupture the delicate membranes essential for living cells to function, irreversibly damaging them. Cells can only survive freezing if another liquid like glycerol temporarily replaces the water in them.

Figure 2.14 Hydrogen bonding makes ice less dense than liquid water. The (a) lattice structure of ice makes it less dense than the liquid water's freely flowing molecules, enabling it to (b) float on water. (credit a: modification of work by Jane Whitney, image created using Visual Molecular Dynamics (VMD) software[1]; credit b: modification of work by Carlos Ponte)

LINK TO LEARNING

Click here (http://openstax.org/l/ice_lattice2) to see a 3-D animation of an ice lattice structure.

Water's High Heat Capacity

Water's high heat capacity is a property that hydrogen bonding among water molecules causes. Water has the highest **specific heat capacity** of any liquids. We define specific heat as the amount of heat one gram of a substance must absorb or lose to change its temperature by one degree Celsius. For water, this amount is one **calorie**. It therefore takes water a long time to heat and a long time to cool. In fact, water's specific heat capacity is about five times more than that of sand. This explains why the land cools faster than the sea. Due to its high heat capacity, warm blooded animals use water to more evenly disperse heat in their bodies: it acts in a similar manner to a car's cooling system, transporting heat from warm places to cool places, causing the body to maintain a more even temperature.

Water's Heat of Vaporization

Water also has a high **heat of vaporization**, the amount of energy required to change one gram of a liquid substance to a gas. A considerable amount of heat energy (586 cal) is required to accomplish this change in water. This process occurs on the water's surface. As liquid water heats up, hydrogen bonding makes it difficult to separate the liquid water molecules from each other, which is required for it to enter its gaseous phase (steam). As a result, water acts as a heat sink or heat reservoir and requires much more heat to boil than does a liquid such as ethanol (grain alcohol), whose hydrogen bonding with other ethanol molecules is weaker than water's hydrogen bonding. Eventually, as water reaches its boiling point of 100° Celsius (212° Fahrenheit), the heat is able to break the hydrogen bonds between the water molecules, and the kinetic energy (motion) between the water molecules allows them to escape from the liquid as a gas. Even when below its boiling point, water's individual molecules acquire enough energy from other water molecules such that some surface water molecules can escape and vaporize: we call this process **evaporation**.

The fact that hydrogen bonds need to be broken for water to evaporate means that bonds use a substantial amount of energy in the process. As the water evaporates, energy is taken up by the process, cooling the environment where the evaporation is taking place. In many living organisms, including in humans, the evaporation of sweat, which is 90 percent water, allows the organism to cool so that it can maintain homeostasis of body temperature.

Water's Solvent Properties

Since water is a polar molecule with slightly positive and slightly negative charges, ions and polar molecules can readily dissolve in it. Therefore, we refer to water as a **solvent**, a substance capable of dissolving other polar molecules and ionic compounds. The charges associated with these molecules will form hydrogen bonds with water, surrounding the particle with water molecules. We refer to this as a **sphere of hydration**, or a hydration shell, as Figure 2.15 illustrates and serves to keep the particles separated or dispersed in the water.

1W. Humphrey W., A. Dalke, and K. Schulten, "VMD—Visual Molecular Dynamics," *Journal of Molecular Graphics* 14 (1996): 33-38.

When we add ionic compounds to water, the individual ions react with the water molecules' polar regions and their ionic bonds are disrupted in the process of **dissociation**. Dissociation occurs when atoms or groups of atoms break off from molecules and form ions. Consider table salt (NaCl, or sodium chloride): when we add NaCl crystals to water, the NaCl molecules dissociate into Na^+ and Cl^- ions, and spheres of hydration form around the ions, as Figure 2.15 illustrates. The partially negative charge of the water molecule's oxygen surrounds the positively charged sodium ion. The hydrogen's partially positive charge on the water molecule surrounds the negatively charged chloride ion.

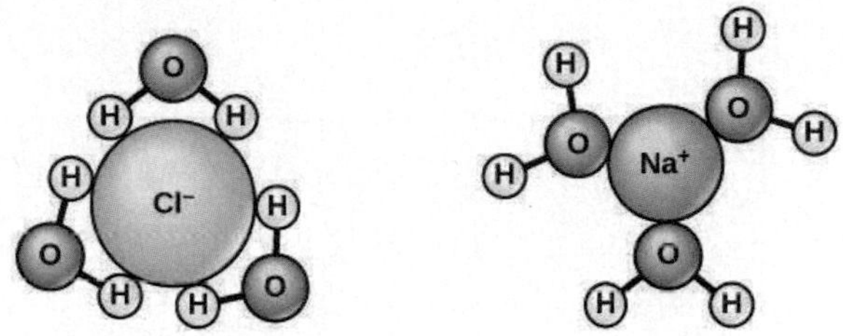

Figure 2.15 When we mix table salt (NaCl) in water, it forms spheres of hydration around the ions.

Water's Cohesive and Adhesive Properties

Have you ever filled a glass of water to the very top and then slowly added a few more drops? Before it overflows, the water forms a dome-like shape above the rim of the glass. This water can stay above the glass because of the property of **cohesion**. In cohesion, water molecules are attracted to each other (because of hydrogen bonding), keeping the molecules together at the liquid-gas (water-air) interface, although there is no more room in the glass.

Cohesion allows for **surface tension**, the capacity of a substance to withstand rupturing when placed under tension or stress. This is also why water forms droplets when on a dry surface rather than flattening by gravity. When we place a small scrap of paper onto a water droplet, the paper floats on top even though paper is denser (heavier) than the water. Cohesion and surface tension keep the water molecules' hydrogen bonds intact and support the item floating on the top. It's even possible to "float" a needle on top of a glass of water if you place it gently without breaking the surface tension, as Figure 2.16 shows.

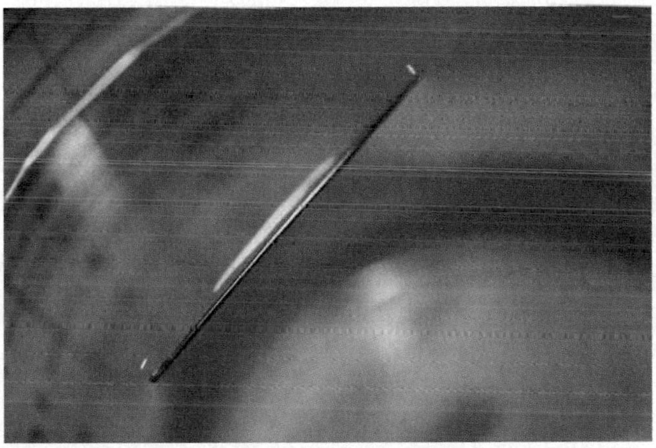

Figure 2.16 A needle's weight pulls the surface downward. At the same time, the surface tension pulls it up, suspending it on the water's surface preventing it from sinking. Notice the indentation in the water around the needle. (credit: Cory Zanker)

These cohesive forces are related to water's property of **adhesion**, or the attraction between water molecules and other molecules. This attraction is sometimes stronger than water's cohesive forces, especially when the water is exposed to charged surfaces such as those on the inside of thin glass tubes known as capillary tubes. We observe adhesion when water "climbs" up the tube placed in a glass of water: notice that the water appears to be higher on the tube's sides than in the middle. This is because the water molecules are attracted to the capillary's charged glass walls more than they are to each other and therefore adhere to it. We call this type of adhesion **capillary action**, as Figure 2.17 illustrates.

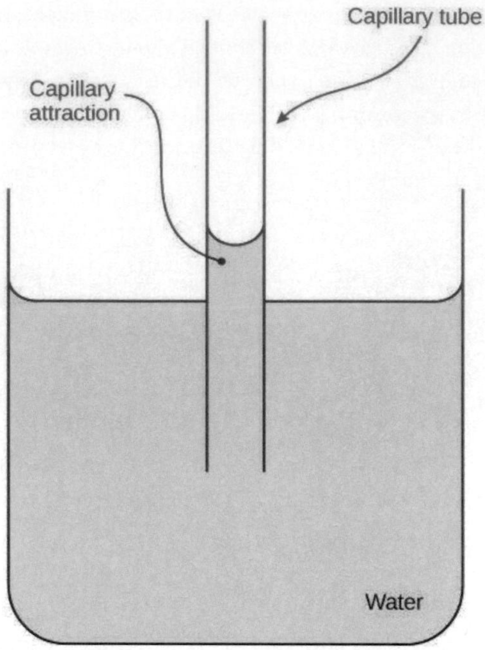

Figure 2.17 The adhesive forces exerted by the glass' internal surface exceeding the cohesive forces between the water molecules themselves causes capillary action in a glass tube. (credit: modification of work by Pearson-Scott Foresman, donated to the Wikimedia Foundation)

Why are cohesive and adhesive forces important for life? Cohesive and adhesive forces are important for transporting water from the roots to the leaves in plants. These forces create a "pull" on the water column. This pull results from the tendency of water molecules evaporating on the plant's surface to stay connected to water molecules below them, and so they are pulled along. Plants use this natural phenomenon to help transport water from their roots to their leaves. Without these properties of water, plants would be unable to receive the water and the dissolved minerals they require. In another example, insects such as the water strider, as Figure 2.18 shows, use the water's surface tension to stay afloat on the water's surface layer and even mate there.

Figure 2.18 Water's cohesive and adhesive properties allow this water strider (*Gerris* sp.) to stay afloat. (credit: Tim Vickers)

pH, Buffers, Acids, and Bases

The pH of a solution indicates its acidity or basicity.

$$H_2O(I) \leftrightarrow H^+(aq) + OH^-(aq)$$

You may have used **litmus** or pH paper, filter paper treated with a natural water-soluble dye for use as a pH indicator, tests how much acid (acidity) or base (basicity) exists in a solution. You might have even used some to test whether the water in a swimming pool is properly treated. In both cases, the pH test measures hydrogen ions' concentration in a given solution.

Hydrogen ions spontaneously generate in pure water by the dissociation (ionization) of a small percentage of water molecules into equal numbers of hydrogen (H^+) ions and hydroxide (OH^-) ions. While the hydroxide ions are kept in solution by their hydrogen bonding with other water molecules, the hydrogen ions, consisting of naked protons, immediately attract to un-ionized water molecules, forming hydronium ions (H_3O^+). Still, by convention, scientists refer to hydrogen ions and their concentration as if they were free in this state in liquid water.

The concentration of hydrogen ions dissociating from pure water is 1×10^{-7} moles H^+ ions per liter of water. Moles (mol) are a way to express the amount of a substance (which can be atoms, molecules, ions, etc.). One mole represents the atomic weight of a substance, expressed in grams, which equals the amount of the substance containing as many units as there are atoms in 12 grams of ^{12}C. Mathematically, one mole is equal to 6.02×10^{23} particles of the substance. Therefore, 1 mole of water is equal to 6.02×10^{23} water molecules. We calculate the pH as the negative of the base 10 logarithm of this concentration. The log10 of 1×10^{-7} is -7.0, and the negative of this number (indicated by the "p" of "pH") yields a pH of 7.0, which is also a neutral pH. The pH inside of human cells and blood are examples of two body areas where near-neutral pH is maintained.

Non-neutral pH readings result from dissolving acids or bases in water. Using the negative logarithm to generate positive integers, high concentrations of hydrogen ions yield a low pH number; whereas, low levels of hydrogen ions result in a high pH. An **acid** is a substance that increases hydrogen ions' (H^+) concentration in a solution, usually by having one of its hydrogen atoms dissociate. A **base** provides either hydroxide ions (OH^-) or other negatively charged ions that combine with hydrogen ions, reducing their concentration in the solution and thereby raising the pH. In cases where the base releases hydroxide ions, these ions bind to free hydrogen ions, generating new water molecules.

The stronger the acid, the more readily it donates H^+. For example, hydrochloric acid (HCl) completely dissociates into hydrogen and chloride ions and is highly acidic; whereas the acids in tomato juice or vinegar do not completely dissociate and are weak acids. Conversely, strong bases are those substances that readily donate OH^- or take up hydrogen ions. Sodium hydroxide (NaOH) and many household cleaners are highly alkaline and give up OH^- rapidly when we place them in water, thereby raising the pH. An example of a weak basic solution is seawater, which has a pH near 8.0 This is close enough to a neutral pH that marine organisms have adapted in order to live and thrive in a saline environment.

The **pH scale** is, as we previously mentioned, an inverse logarithm and ranges from 0 to 14 (Figure 2.19). Anything below 7.0 (ranging from 0.0 to 6.9) is acidic, and anything above 7.0 (from 7.1 to 14.0) is alkaline. Extremes in pH in either direction from 7.0 are usually inhospitable to life. The pH inside cells (6.8) and the pH in the blood (7.4) are both very close to neutral. However, the environment in the stomach is highly acidic, with a pH of 1 to 2. As a result, how do stomach cells survive in such an acidic environment? How do they homeostatically maintain the near neutral pH inside them? The answer is that they cannot do it and are constantly dying. The stomach constantly produces new cells to replace dead ones, which stomach acids digest. Scientists estimate that the human body completely replaces the stomach lining every seven to ten days.

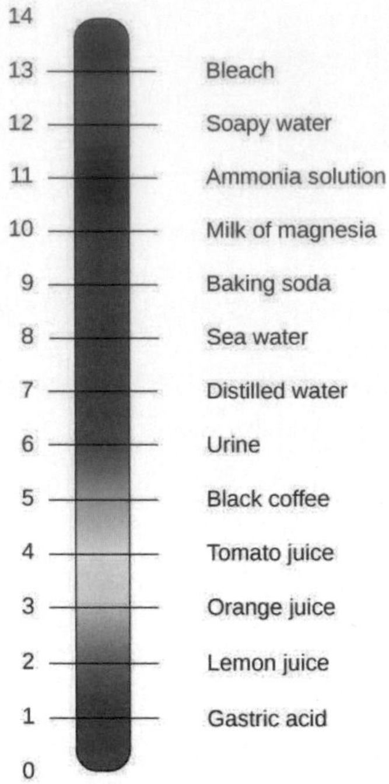

Figure 2.19 The pH scale measures hydrogen ions' (H^+) concentration in a solution. (credit: modification of work by Edward Stevens)

LINK TO LEARNING

Watch this video for a straightforward explanation of pH and its logarithmic scale.

Click to view content (https://www.openstax.org/l/pH_scale)

How can organisms whose bodies require a near-neutral pH ingest acidic and basic substances (a human drinking orange juice, for example) and survive? Buffers are the key. **Buffers** readily absorb excess H^+ or OH^-, keeping the body's pH carefully maintained in the narrow range required for survival. Maintaining a constant blood pH is critical to a person's well-being. The buffer maintaining the pH of human blood involves carbonic acid (H_2CO_3), bicarbonate ion (HCO_3^-), and carbon dioxide (CO_2). When bicarbonate ions combine with free hydrogen ions and become carbonic acid, it removes hydrogen ions and moderates pH changes. Similarly, as Figure 2.20 shows, excess carbonic acid can convert to carbon dioxide gas which we exhale through the lungs. This prevents too many free hydrogen ions from building up in the blood and dangerously reducing the blood's pH. Likewise, if too much OH^- enters into the system, carbonic acid will combine with it to create bicarbonate, lowering the pH. Without this buffer system, the body's pH would fluctuate enough to put survival in jeopardy.

Figure 2.20 This diagram shows the body's buffering of blood pH levels. The blue arrows show the process of raising pH as more CO_2 is made. The purple arrows indicate the reverse process: the lowering of pH as more bicarbonate is created.

Other examples of buffers are antacids that some people use to combat excess stomach acid. Many of these over-the-counter medications work in the same way as blood buffers, usually with at least one ion capable of absorbing hydrogen and moderating pH, bringing relief to those who suffer "heartburn" after eating. Water's unique properties that contribute to this capacity to balance pH—as well as water's other characteristics—are essential to sustaining life on Earth.

LINK TO LEARNING

To learn more about water, visit the U.S. Geological Survey Water Science for Schools (http://openstax.org/l/all_about_water) All About Water! website.

2.3 Carbon

By the end of this section, you will be able to do the following:

- Explain why carbon is important for life
- Describe the role of functional groups in biological molecules

Many complex molecules called macromolecules, such as proteins, nucleic acids (RNA and DNA), carbohydrates, and lipids comprise cells. The macromolecules are a subset of **organic molecules** (any carbon-containing liquid, solid, or gas) that are especially important for life. The fundamental component for all of these macromolecules is carbon. The carbon atom has unique properties that allow it to form covalent bonds to as many as four different atoms, making this versatile element ideal to serve as the basic structural component, or "backbone," of the macromolecules.

Individual carbon atoms have an incomplete outermost electron shell. With an atomic number of 6 (six electrons and six protons), the first two electrons fill the inner shell, leaving four in the second shell. Therefore, carbon atoms can form up to four covalent bonds with other atoms to satisfy the octet rule. The methane molecule provides an example: it has the chemical formula CH_4. Each of its four hydrogen atoms forms a single covalent bond with the carbon atom by sharing a pair of electrons. This results in a filled outermost shell.

Hydrocarbons

Hydrocarbons are organic molecules consisting entirely of carbon and hydrogen, such as methane (CH_4) described above. We often use hydrocarbons in our daily lives as fuels—like the propane in a gas grill or the butane in a lighter. The many covalent bonds between the atoms in hydrocarbons store a great amount of energy, which releases when these molecules burn (oxidize) Methane, an excellent fuel, is the simplest hydrocarbon molecule, with a central carbon atom bonded to four different hydrogen atoms, as Figure 2.21 illustrates. The shape of its electron orbitals determines the shape of the methane molecule's geometry, where the atoms reside in three dimensions. The carbons and the four hydrogen atoms form a tetrahedron, with four triangular faces. For this reason, we describe methane as having tetrahedral geometry.

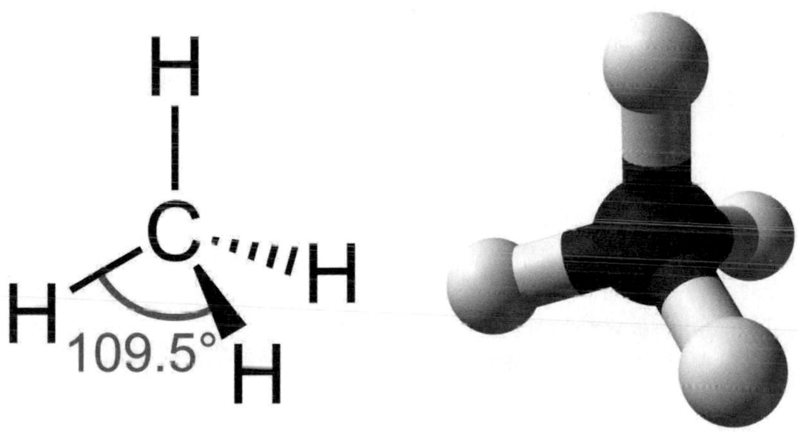

Figure 2.21 Methane has a tetrahedral geometry, with each of the four hydrogen atoms spaced 109.5° apart.

As the backbone of the large molecules of living things, hydrocarbons may exist as linear carbon chains, carbon rings, or combinations of both. Furthermore, individual carbon-to-carbon bonds may be single, double, or triple covalent bonds, and each type of bond affects the molecule's geometry in a specific way. This three-dimensional shape or conformation of the large molecules of life (macromolecules) is critical to how they function.

Hydrocarbon Chains

Successive bonds between carbon atoms form hydrocarbon chains. These may be branched or unbranched. Furthermore, a molecule's different geometries of single, double, and triple covalent bonds alter the overall molecule's geometry as Figure 2.22 illustrates. The hydrocarbons ethane, ethene, and ethyne serve as examples of how different carbon-to-carbon bonds affect the molecule's geometry. The names of all three molecules start with the prefix "eth-," which is the prefix for two carbon hydrocarbons. The suffixes "-ane," "-ene," and "-yne" refer to the presence of single, double, or triple carbon-carbon bonds, respectively. Thus, propane, propene, and propyne follow the same pattern with three carbon molecules, butane, butene, and butyne for four carbon molecules, and so on. Double and triple bonds change the molecule's geometry: single bonds allow rotation along the bond's axis; whereas, double bonds lead to a planar configuration and triple bonds to a linear one. These geometries have a significant impact on the shape a particular molecule can assume.

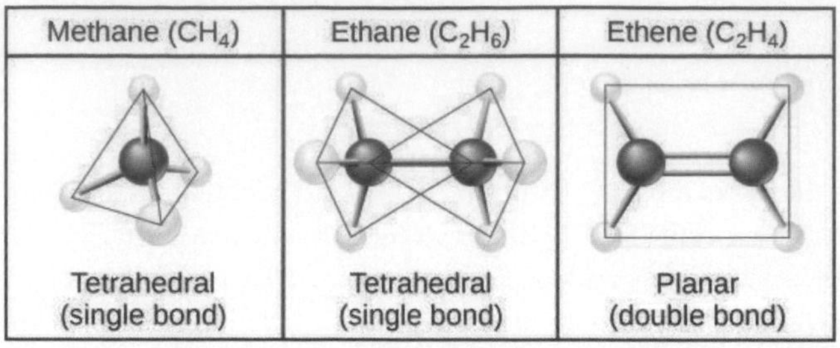

Figure 2.22 When carbon forms single bonds with other atoms, the shape is tetrahedral. When two carbon atoms form a double bond, the shape is planar, or flat. Single bonds, like those in ethane, are able to rotate. Double bonds, like those in ethene, cannot rotate, so the atoms on either side are locked in place.

Hydrocarbon Rings

So far, the hydrocarbons we have discussed have been **aliphatic hydrocarbons**, which consist of linear chains of carbon atoms. Another type of hydrocarbon, **aromatic hydrocarbons**, consists of closed rings of carbon atoms with alternating single and double bonds. We find ring structures in aliphatic hydrocarbons, sometimes with the presence of double bonds, which we can see by comparing cyclohexane's structure to benzene in Figure 2.23. Examples of biological molecules that incorporate the benzene ring include some amino acids and cholesterol and its derivatives, including the hormones estrogen and testosterone. We also find the benzene ring in the herbicide 2,4-D. Benzene is a natural component of crude oil and has been classified as a carcinogen. Some hydrocarbons have both aliphatic and aromatic portions. Beta-carotene is an example of such a hydrocarbon.

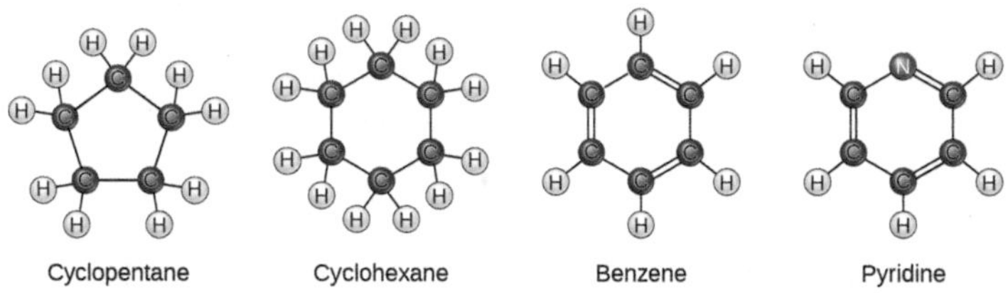

Figure 2.23 Carbon can form five- and six-membered rings. Single or double bonds may connect the carbons in the ring, and nitrogen may be substituted for carbon.

Isomers

The three-dimensional placement of atoms and chemical bonds within organic molecules is central to understanding their chemistry. We call molecules that share the same chemical formula but differ in the placement (structure) of their atoms and/or chemical bonds **isomers**. **Structural isomers** (like butane and isobutane in Figure 2.24a) differ in the placement of their covalent bonds: both molecules have four carbons and ten hydrogens (C_4H_{10}), but the different atom arrangement within the molecules leads to differences in their chemical properties. For example, butane is suited for use as a fuel for cigarette lighters and torches; whereas, isobutane is suited for use as a refrigerant and a propellant in spray cans.

Geometric isomers, alternatively have similar placements of their covalent bonds but differ in how these bonds are made to the surrounding atoms, especially in carbon-to-carbon double bonds. In the simple molecule butene (C_4H_8), the two methyl groups (CH_3) can be on either side of the double covalent bond central to the molecule, as Figure 2.24b illustrates. When the carbons are bound on the same side of the double bond, this is the *cis* configuration. If they are on opposite sides of the double bond, it is a *trans* configuration. In the *trans* configuration, the carbons form a more or less linear structure; whereas, the carbons in the *cis* configuration make a bend (change in direction) of the carbon backbone.

VISUAL CONNECTION

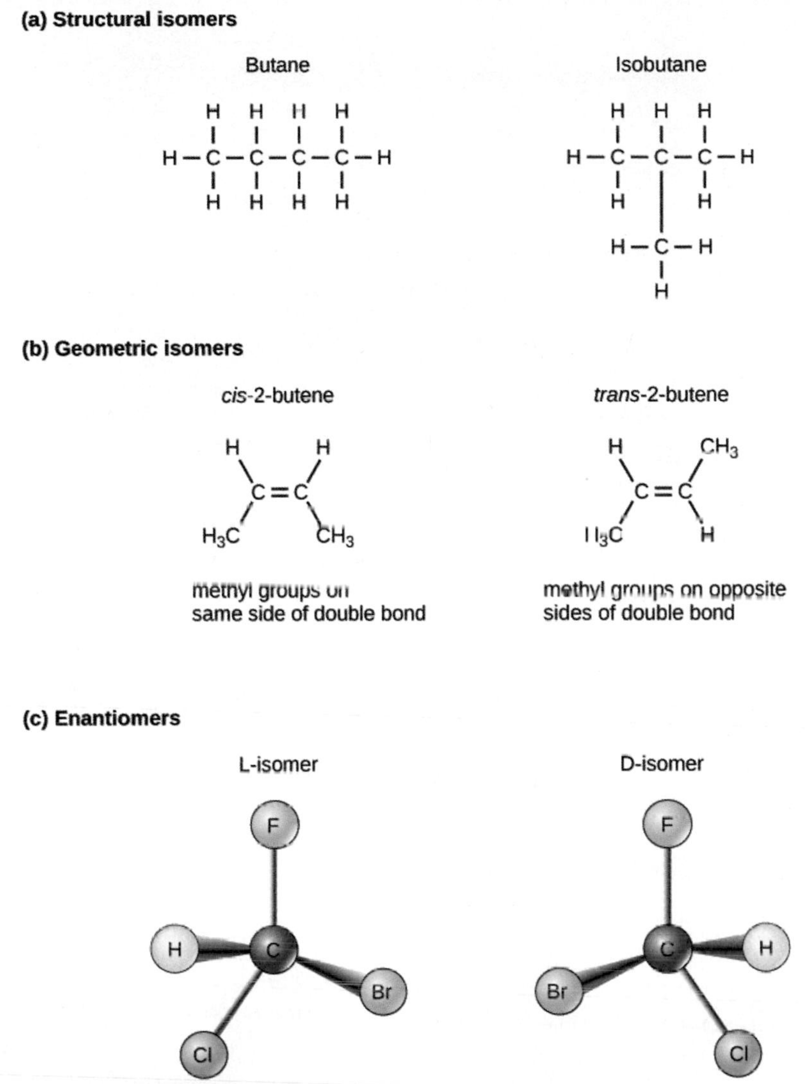

Figure 2.24 We call molecules that have the same number and type of atoms arranged differently isomers. (a) Structural isomers have a different covalent arrangement of atoms. (b) Geometric isomers have a different arrangement of atoms around a double bond. (c) Enantiomers are mirror images of each other.

Which of the following statements is false?

a. Molecules with the formulas CH_3CH_2COOH and $C_3H_6O_2$ could be structural isomers.
b. Molecules must have a double bond to be *cis-trans* isomers.
c. To be enantiomers, a molecule must have at least three different atoms or groups connected to a central carbon.
d. To be enantiomers, a molecule must have at least four different atoms or groups connected to a central carbon.

In triglycerides (fats and oils), long carbon chains known as fatty acids may contain double bonds, which can be in either the *cis* or *trans* configuration, as Figure 2.25 illustrates. Fats with at least one double bond between carbon atoms are unsaturated fats. When some of these bonds are in the *cis* configuration, the resulting bend in the chain's carbon backbone means that triglyceride molecules cannot pack tightly, so they remain liquid (oil) at room temperature. Alternatively, triglycerides with *trans* double bonds (popularly called trans fats), have relatively linear fatty acids that are able to pack tightly together at room temperature and form solid fats. In the human diet, trans fats are linked to an increased risk of cardiovascular disease, so many food manufacturers have reduced or eliminated their use in recent years. In contrast to unsaturated fats, we call triglycerides without double bonds between carbon atoms saturated fats, meaning that they contain all the hydrogen atoms available. Saturated fats are a solid at room temperature and usually of animal origin.

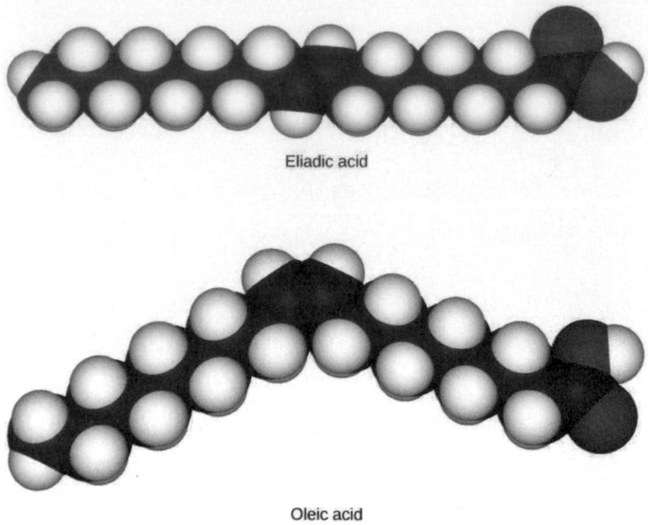

Figure 2.25 These space-filling models show a *cis* (oleic acid) and a *trans* (eliadic acid) fatty acid. Notice the bend in the molecule caused by the *cis* configuration.

Enantiomers

Enantiomers are molecules that share the same chemical structure and chemical bonds but differ in the three-dimensional placement of atoms so that they are non-superimposable mirror images. Figure 2.26 shows an amino acid alanine example, where the two structures are nonsuperimposable. In nature, the L-forms of amino acids are predominant in proteins. Some D forms of amino acids are seen in the cell walls of bacteria and polypeptides in other organisms. Similarly, the D-form of glucose is the main product of photosynthesis and we rarely see the molecule's L-form in nature.

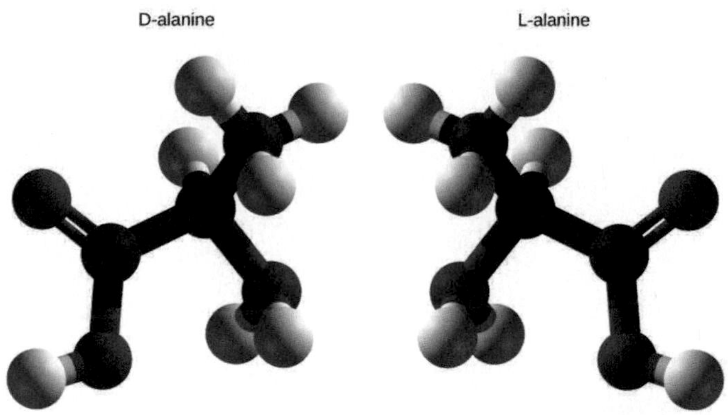

Figure 2.26 D-alanine and L-alanine are examples of enantiomers or mirror images. L-forms of amino acids are predominant in proteins.

Functional Groups

Functional groups are groups of atoms that occur within molecules and confer specific chemical properties to those molecules. We find them along the "carbon backbone" of macromolecules. Chains and/or rings of carbon atoms with the occasional substitution of an element such as nitrogen or oxygen form this carbon backbone. Molecules with other elements in their carbon backbone are **substituted hydrocarbons**.

The functional groups in a macromolecule are usually attached to the carbon backbone at one or several different places along its chain and/or ring structure. Each of the four types of macromolecules—proteins, lipids, carbohydrates, and nucleic acids—has its own characteristic set of functional groups that contributes greatly to its differing chemical properties and its function in living organisms.

A functional group can participate in specific chemical reactions. Figure 2.27 shows some of the important functional groups in biological molecules. They include: hydroxyl, methyl, carbonyl, carboxyl, amino, phosphate, and sulfhydryl. These groups play an important role in forming molecules like DNA, proteins, carbohydrates, and lipids. We usually classify functional groups as hydrophobic or hydrophilic depending on their charge or polarity characteristics. An example of a hydrophobic group is the nonpolar methyl molecule. Among the hydrophilic functional groups is the carboxyl group in amino acids, some amino acid side chains, and the fatty acids that form triglycerides and phospholipids. This carboxyl group ionizes to release hydrogen ions (H^+) from the COOH group resulting in the negatively charged COO^- group. This contributes to the hydrophilic nature of whatever molecule on which it is found. Other functional groups, such as the carbonyl group, have a partially negatively charged oxygen atom that may form hydrogen bonds with water molecules, again making the molecule more hydrophilic.

Functional Group	Structure	Properties
Hydroxyl	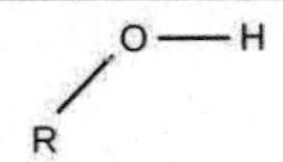	Polar
Methyl		Nonpolar
Carbonyl		Polar
Carboxyl	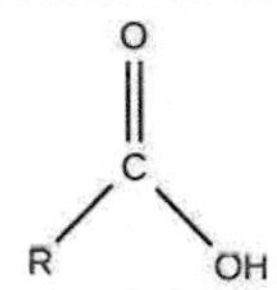	Charged, ionizes to release H^+. Since carboxyl groups can release H^+ ions into solution, they are considered acidic.
Amino	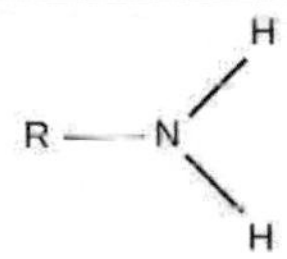	Charged, accepts H^+ to form NH_3^+. Since amino groups can remove H^+ from solution, they are considered basic.
Phosphate	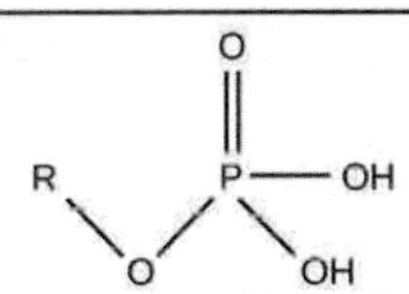	Charged, ionizes to release H^+. Since phosphate groups can release H^+ ions into solution, they are considered acidic.
Sulfhydryl	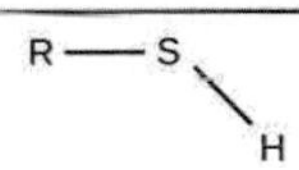	Polar

Figure 2.27 These functional groups are in many different biological molecules. R, also known as R-group, is an abbreviation for any group in which a carbon or hydrogen atom is attached to the rest of the molecule.

Hydrogen bonds between functional groups (within the same molecule or between different molecules) are important to the function of many macromolecules and help them to fold properly into and maintain the appropriate shape for functioning. Hydrogen bonds are also involved in various recognition processes, such as DNA complementary base pairing and the binding of an enzyme to its substrate, as Figure 2.28 illustrates.

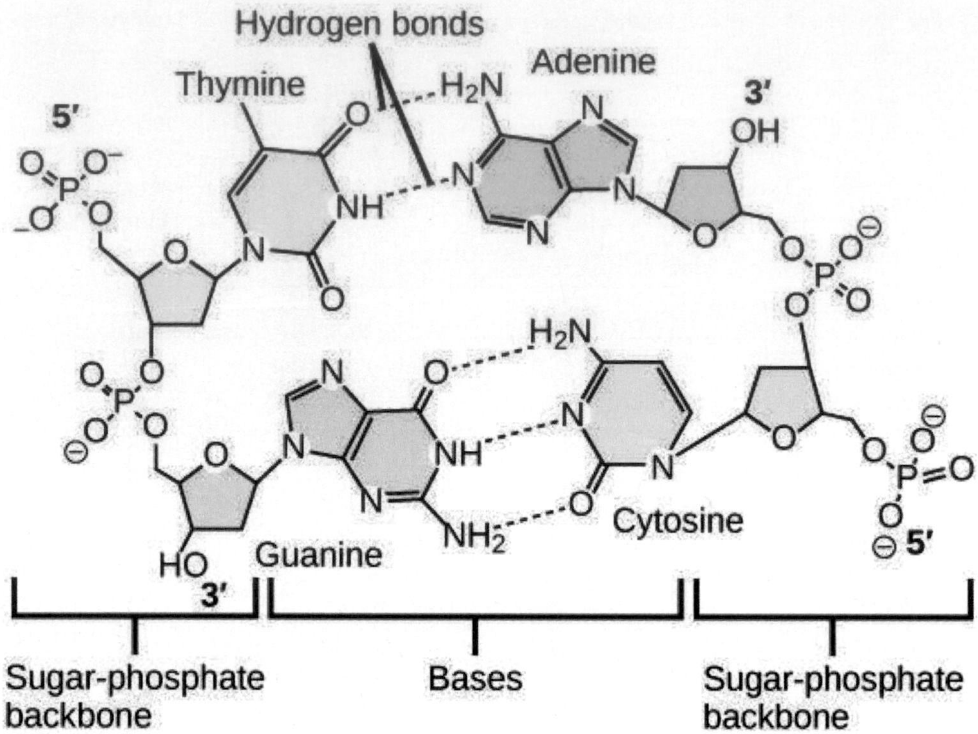

Figure 2.28 Hydrogen bonds connect two strands of DNA together to create the double-helix structure.

KEY TERMS

acid molecule that donates hydrogen ions and increases the concentration of hydrogen ions in a solution

adhesion attraction between water molecules and other molecules

aliphatic hydrocarbon hydrocarbon consisting of a linear chain of carbon atoms

anion negative ion that is formed by an atom gaining one or more electrons

aromatic hydrocarbon hydrocarbon consisting of closed rings of carbon atoms

atom the smallest unit of matter that retains all of the chemical properties of an element

atomic mass calculated mean of the mass number for an element's isotopes

atomic number total number of protons in an atom

balanced chemical equation statement of a chemical reaction with the number of each type of atom equalized for both the products and reactants

base molecule that donates hydroxide ions or otherwise binds excess hydrogen ions and decreases the hydrogen ions' concentration in a solution

buffer substance that resists a change in pH by absorbing or releasing hydrogen or hydroxide ions

calorie amount of heat required to change the temperature of one gram of water by one degree Celsius

capillary action occurs because water molecules are attracted to charges on the inner surfaces of narrow tubular structures such as glass tubes, drawing the water molecules to the tubes' sides

cation positive ion that is formed by an atom losing one or more electrons

chemical bond interaction between two or more of the same or different atoms that results in forming molecules

chemical reaction process leading to rearranging atoms in molecules

chemical reactivity the ability to combine and to chemically bond with each other

cohesion intermolecular forces between water molecules caused by the polar nature of water; responsible for surface tension

compound substance composed of molecules consisting of atoms of at least two different elements

covalent bond type of strong bond formed between two atoms of the same or different elements; forms when electrons are shared between atoms

dissociation release of an ion from a molecule such that the original molecule now consists of an ion and the charged remains of the original, such as when water dissociates into H^+ and OH^-

electrolyte ion necessary for nerve impulse conduction, muscle contractions, and water balance

electron negatively charged subatomic particle that resides outside of the nucleus in the electron orbital; lacks functional mass and has a negative charge of –1 unit

electron configuration arrangement of electrons in an atom's electron shell (for example, $1s^2 2s^2 2p^6$)

electron orbital how electrons are spatially distributed surrounding the nucleus; the area where we are most likely to find an electron

electron transfer movement of electrons from one element to another; important in creating ionic bonds

electronegativity ability of some elements to attract electrons (often of hydrogen atoms), acquiring partial negative charges in molecules and creating partial positive charges on the hydrogen atoms

element one of 118 unique substances that cannot break down into smaller substances; each element has unique properties and a specified number of protons

enantiomers molecules that share overall structure and bonding patterns, but differ in how the atoms are three dimensionally placed such that they are mirror images of each other

equilibrium steady state of relative reactant and product concentration in reversible chemical reactions in a closed system

evaporation change from liquid to gaseous state at a body of water's surface, plant leaves, or an organism's skin

functional group group of atoms that provides or imparts a specific function to a carbon skeleton

geometric isomer isomer with similar bonding patterns differing in the placement of atoms alongside a double covalent bond

heat of vaporization of water high amount of energy required for liquid water to turn into water vapor

hydrocarbon molecule that consists only of carbon and hydrogen

hydrogen bond weak bond between slightly positively charged hydrogen atoms and slightly negatively charged atoms in other molecules

hydrophilic describes ions or polar molecules that interact well with other polar molecules such as water

hydrophobic describes uncharged nonpolar molecules that do not interact well with polar molecules such as water

inert gas (also, noble gas) element with filled outer electron shell that is unreactive with other atoms

ion atom or chemical group that does not contain equal numbers of protons and electrons

ionic bond chemical bond that forms between ions with opposite charges (cations and anions)

irreversible chemical reaction chemical reaction where reactants proceed unidirectionally to form products

isomers molecules that differ from one another even though they share the same chemical formula

isotope one or more forms of an element that have

different numbers of neutrons

law of mass action chemical law stating that the rate of a reaction is proportional to the concentration of the reacting substances

litmus paper (also, pH paper) filter paper treated with a natural water-soluble dye that changes its color as the pH of the environment changes in order to use it as a pH indicator

mass number total number of protons and neutrons in an atom

matter anything that has mass and occupies space

molecule two or more atoms chemically bonded together

neutron uncharged particle that resides in an atom's nucleus; has a mass of one amu

noble gas see inert gas

nonpolar covalent bond type of covalent bond that forms between atoms when electrons are shared equally between them

nucleus core of an atom; contains protons and neutrons

octet rule rule that atoms are most stable when they hold eight electrons in their outermost shells

orbital region surrounding the nucleus; contains electrons

organic molecule any molecule containing carbon (except carbon dioxide)

periodic table organizational chart of elements indicating each element's atomic number and atomic mass; provides key information about the elements' properties

pH paper see litmus paper

pH scale scale ranging from zero to 14 that is inversely proportional to the hydrogen ions' concentration in a solution

polar covalent bond type of covalent bond that forms as a result of unequal electron sharing, resulting in creating slightly positive and negative charged molecule regions

product molecule that is result of chemical reaction

proton positively charged particle that resides in the atom's nucleus; has a mass of one amu and a charge of +1

radioisotope isotope that emits radiation comprised of subatomic particles to form more stable elements

reactant molecule that takes part in a chemical reaction

reversible chemical reaction chemical reaction that functions bidirectionally, where products may turn into reactants if their concentration is great enough

solvent substance capable of dissolving another substance

specific heat capacity the amount of heat one gram of a substance must absorb or lose to change its temperature by one degree Celsius

sphere of hydration when a polar water molecule surrounds charged or polar molecules thus keeping them dissolved and in solution

structural isomers molecules that share a chemical formula but differ in the placement of their chemical bonds

substituted hydrocarbon hydrocarbon chain or ring containing an atom of another element in place of one of the backbone carbons

surface tension tension at the surface of a body of liquid that prevents the molecules from separating; created by the attractive cohesive forces between the liquid's molecules

valence shell outermost shell of an atom

van der Waals interaction very weak interaction between molecules due to temporary charges attracting atoms that are very close together

CHAPTER SUMMARY

2.1 Atoms, Isotopes, Ions, and Molecules: The Building Blocks

Matter is anything that occupies space and has mass. It is comprised of elements. All of the 98 elements that occur naturally have unique qualities that allow them to combine in various ways to create molecules, which in turn combine to form cells, tissues, organ systems, and organisms. Atoms, which consist of protons, neutrons, and electrons, are the smallest units of an element that retain all of the properties of that element. Electrons can transfer, share, or cause charge disparities between atoms to create bonds, including ionic, covalent, and hydrogen bonds, as well as van der Waals interactions.

2.2 Water

Water has many properties that are critical to maintaining life. It is a polar molecule, allowing for forming hydrogen bonds. Hydrogen bonds allow ions and other polar molecules to dissolve in water. Therefore, water is an excellent solvent. The hydrogen bonds between water molecules cause the water to have a high heat capacity, meaning it takes considerable added heat to raise its temperature. As the temperature rises, the hydrogen bonds between water continually break and form anew. This allows for the overall temperature to remain stable, although energy is added to the system. Water also exhibits a high heat of vaporization, which is key to how organisms cool themselves by evaporating sweat. Water's cohesive forces allow for the property of surface tension; whereas, we see its adhesive properties as water rises inside capillary tubes. The pH value is a measure of hydrogen ion concentration in a solution and is one of many chemical characteristics that is highly regulated in living organisms through homeostasis. Acids and bases can change pH values, but buffers tend to moderate the changes they cause. These properties of water are intimately connected to the biochemical and physical processes performed by living organisms, and life would be very different if these properties were altered, if it could exist at all.

2.3 Carbon

The unique properties of carbon make it a central part of biological molecules. Carbon binds to oxygen, hydrogen, and nitrogen covalently to form the many molecules important for cellular function. Carbon has four electrons in its outermost shell and can form four bonds. Carbon and hydrogen can form hydrocarbon chains or rings. Functional groups are groups of atoms that confer specific properties to hydrocarbon (or substituted hydrocarbon) chains or rings that define their overall chemical characteristics and function.

VISUAL CONNECTION QUESTIONS

1. Figure 2.3 How many neutrons do carbon-12 and carbon-13 have, respectively?
2. Figure 2.7 An atom may give, take, or share electrons with another atom to achieve a full valence shell, the most stable electron configuration. Looking at this figure, how many electrons do elements in group 1 need to lose in order to achieve a stable electron configuration? How many electrons do elements in groups 14 and 17 need to gain to achieve a stable configuration?
3. Figure 2.24 Which of the following statements is false?
   a. Molecules with the formulas CH_3CH_2COOH and $C_3H_6O_2$ could be structural isomers.
   b. Molecules must have a double bond to be *cis-trans* isomers.
   c. To be enantiomers, a molecule must have at least three different atoms or groups connected to a central carbon.
   d. To be enantiomers, a molecule must have at least four different atoms or groups connected to a central carbon.

REVIEW QUESTIONS

4. If xenon has an atomic number of 54 and a mass number of 108, how many neutrons does it have?
   a. 54
   b. 27
   c. 100
   d. 108
5. Atoms that vary in the number of neutrons found in their nuclei are called ________.
   a. ions
   b. neutrons
   c. neutral atoms
   d. isotopes
6. Potassium has an atomic number of 19. What is its electron configuration?
   a. shells 1 and 2 are full, and shell 3 has nine electrons
   b. shells 1, 2 and 3 are full and shell 4 has three electrons
   c. shells 1, 2 and 3 are full and shell 4 has one electron
   d. shells 1, 2 and 3 are full and no other electrons are present
7. Which type of bond represents a weak chemical bond?
   a. hydrogen bond
   b. atomic bond
   c. covalent bond
   d. nonpolar covalent bond
8. Which of the following statements is not true?
   a. Water is polar.
   b. Water stabilizes temperature.
   c. Water is essential for life.
   d. Water is the most abundant molecule in the Earth's atmosphere.
9. When acids are added to a solution, the pH should ________.
   a. decrease
   b. increase
   c. stay the same
   d. cannot tell without testing
10. We call a molecule that binds up excess hydrogen ions in a solution a(n) ________.
    a. acid
    b. isotope
    c. base
    d. donator
11. Which of the following statements is true?
    a. Acids and bases cannot mix together.
    b. Acids and bases will neutralize each other.
    c. Acids, but not bases, can change the pH of a solution.
    d. Acids donate hydroxide ions (OH^-); bases donate hydrogen ions (H^+).

12. Each carbon molecule can bond with as many as________ other atom(s) or molecule(s).
    a. one
    b. two
    c. six
    d. four

13. Which of the following is not a functional group that can bond with carbon?
    a. sodium
    b. hydroxyl
    c. phosphate
    d. carbonyl

CRITICAL THINKING QUESTIONS

14. What makes ionic bonds different from covalent bonds?
15. Why are hydrogen bonds and van der Waals interactions necessary for cells?
16. Discuss how buffers help prevent drastic swings in pH.
17. Why can some insects walk on water?
18. What property of carbon makes it essential for organic life?
19. Compare and contrast saturated and unsaturated triglycerides.

CHAPTER 3
Biological Macromolecules

Figure 3.1 Foods such as bread, fruit, and cheese are rich sources of biological macromolecules. (credit: modification of work by Bengt Nyman)

INTRODUCTION Food provides the body with the nutrients it needs to survive. Many of these critical nutrients are biological macromolecules, or large molecules, necessary for life. Different smaller organic molecule (monomer) combinations build these macromolecules (polymers). What specific biological macromolecules do living things require? How do these molecules form? What functions do they serve? We explore these questions in this chapter.

3.1 Synthesis of Biological Macromolecules

By the end of this section, you will be able to do the following:

- Understand macromolecule synthesis
- Explain dehydration (or condensation) and hydrolysis reactions

As you've learned, **biological macromolecules** are large molecules, necessary for life, that are built from smaller organic molecules. There are four major biological macromolecule classes (carbohydrates, lipids, proteins, and nucleic acids). Each is an important cell component and performs a wide array of functions.

Combined, these molecules make up the majority of a cell's dry mass (recall that water makes up the majority of its complete mass). Biological macromolecules are organic, meaning they contain carbon. In addition, they may contain hydrogen, oxygen, nitrogen, and additional minor elements.

Dehydration Synthesis

Most macromolecules are made from single subunits, or building blocks, called **monomers**. The monomers combine with each other using covalent bonds to form larger molecules known as **polymers**. In doing so, monomers release water molecules as byproducts. This type of reaction is **dehydration synthesis**, which means "to put together while losing water."

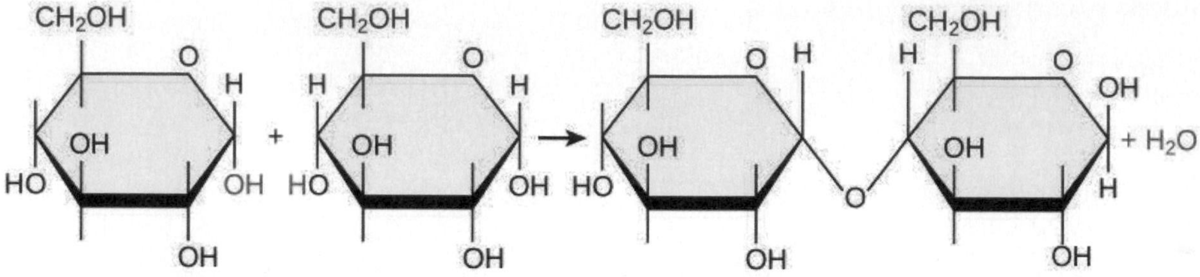

Figure 3.2 In the dehydration synthesis reaction above, two glucose molecules link to form the disaccharide maltose. In the process, it forms a water molecule.

In a dehydration synthesis reaction (Figure 3.2), the hydrogen of one monomer combines with the hydroxyl group of another monomer, releasing a water molecule. At the same time, the monomers share electrons and form covalent bonds. As additional monomers join, this chain of repeating monomers forms a polymer. Different monomer types can combine in many configurations, giving rise to a diverse group of macromolecules. Even one kind of monomer can combine in a variety of ways to form several different polymers. For example, glucose monomers are the constituents of starch, glycogen, and cellulose.

Hydrolysis

Polymers break down into monomers during hydrolysis. A chemical reaction occurs when inserting a water molecule across the bond. Breaking a covalent bond with this water molecule in the compound achieves this (Figure 3.3). During these reactions, the polymer breaks into two components: one part gains a hydrogen atom (H+) and the other gains a hydroxyl molecule (OH−) from a split water molecule.

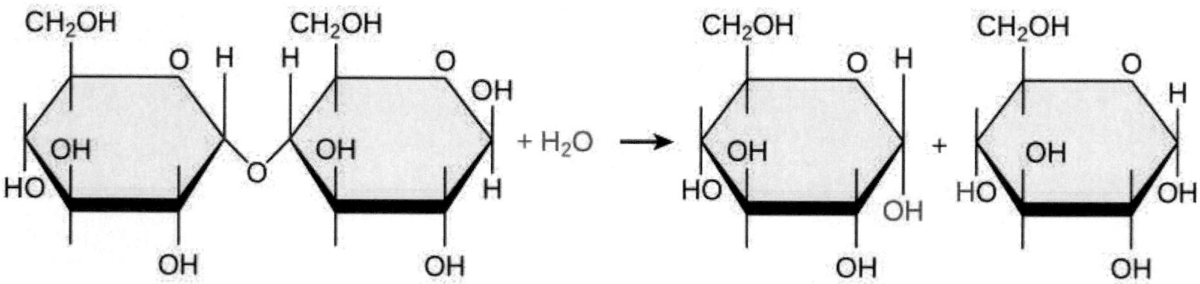

Figure 3.3 In the hydrolysis reaction here, the disaccharide maltose breaks down to form two glucose monomers by adding a water molecule. Note that this reaction is the reverse of the synthesis reaction in Figure 3.2.

Dehydration and **hydrolysis reactions** are catalyzed, or "sped up," by specific enzymes; dehydration reactions involve the formation of new bonds, requiring energy, while hydrolysis reactions break bonds and release energy. These reactions are similar for most macromolecules, but each monomer and polymer reaction is specific for its class. For example, catalytic enzymes in the digestive system hydrolyze or break down the food we ingest into smaller molecules. This allows cells in our body to easily absorb nutrients in the intestine. A specific enzyme breaks down each macromolecule. For instance, amylase, sucrase, lactase, or maltase break down carbohydrates. Enzymes called proteases, such as pepsin and peptidase, and hydrochloric acid break down proteins. Lipases break down lipids. These broken down macromolecules provide energy for cellular activities.

LINK TO LEARNING

Visit this site (http://openstax.org/l/hydrolysis) to see visual representations of dehydration synthesis and hydrolysis.

3.2 Carbohydrates

By the end of this section, you will be able to do the following:

- Discuss the role of carbohydrates in cells and in the extracellular materials of animals and plants
- Explain carbohydrate classifications
- List common monosaccharides, disaccharides, and polysaccharides

Most people are familiar with carbohydrates, one type of macromolecule, especially when it comes to what we eat. To lose weight, some individuals adhere to "low-carb" diets. Athletes, in contrast, often "carb-load" before important competitions to ensure that they have enough energy to compete at a high level. Carbohydrates are, in fact, an essential part of our diet. Grains, fruits, and vegetables are all natural carbohydrate sources that provide energy to the body, particularly through glucose, a simple sugar that is a component of **starch** and an ingredient in many staple foods. Carbohydrates also have other important functions in humans, animals, and plants.

Molecular Structures

The stoichiometric formula $(CH_2O)_n$, where *n* is the number of carbons in the molecule represents **carbohydrates**. In other words, the ratio of carbon to hydrogen to oxygen is 1:2:1 in carbohydrate molecules. This formula also explains the origin of the term "carbohydrate": the components are carbon ("carbo") and the components of water (hence, "hydrate"). Scientists classify carbohydrates into three subtypes: monosaccharides, disaccharides, and polysaccharides.

Monosaccharides

Monosaccharides (mono- = "one"; sacchar- = "sweet") are simple sugars, the most common of which is glucose. In monosaccharides, the number of carbons usually ranges from three to seven. Most monosaccharide names end with the suffix -ose. If the sugar has an aldehyde group (the functional group with the structure R-CHO), it is an aldose, and if it has a ketone group (the functional group with the structure RC(=O)R'), it is a ketose. Depending on the number of carbons in the sugar, they can be trioses (three carbons), pentoses (five carbons), and/or hexoses (six carbons). Figure 3.4 illustrates monosaccharides.

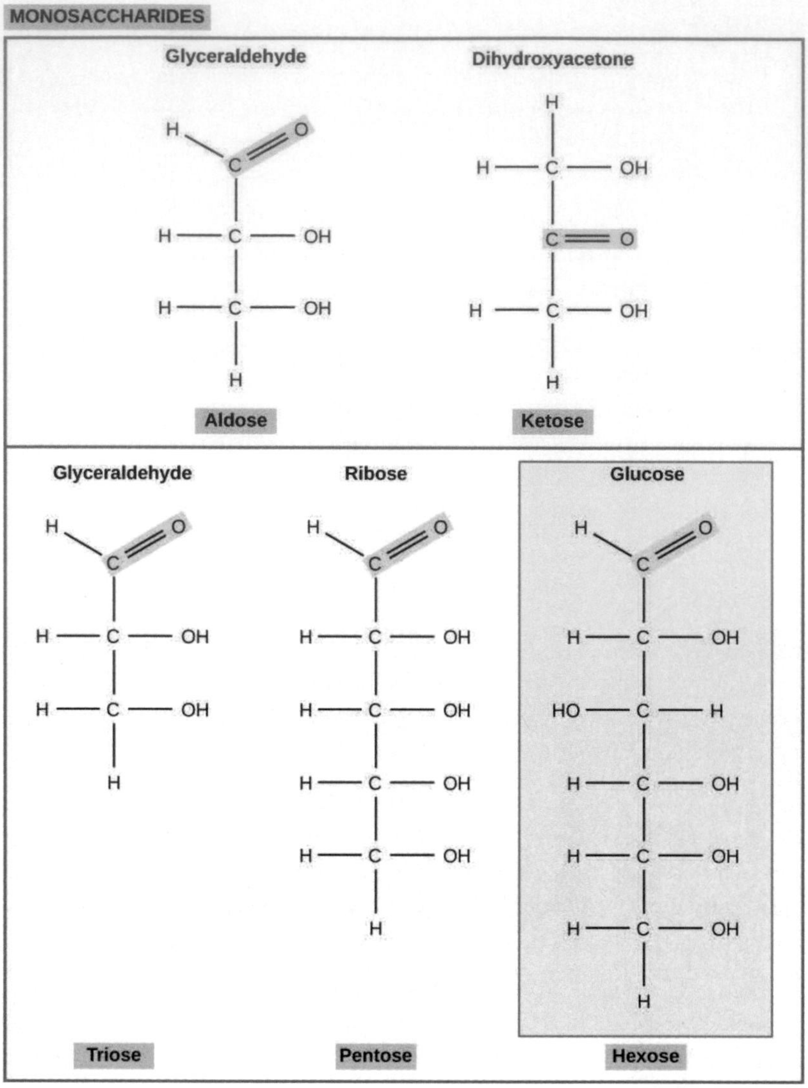

Figure 3.4 Scientists classify monosaccharides based on the position of their carbonyl group and the number of carbons in the backbone. Aldoses have a carbonyl group (indicated in green) at the end of the carbon chain, and ketoses have a carbonyl group in the middle of the carbon chain. Trioses, pentoses, and hexoses have three-, five-, and six- carbon backbones, respectively.

The chemical formula for glucose is $C_6H_{12}O_6$. In humans, glucose is an important source of energy. During cellular respiration, energy releases from glucose, and that energy helps make adenosine triphosphate (ATP). Plants synthesize glucose using carbon dioxide and water, and glucose in turn provides energy requirements for the plant. Humans and other animals that feed on plants often store excess glucose as catabolized (cell breakdown of larger molecules) starch.

Galactose (part of lactose, or milk sugar) and fructose (found in sucrose, in fruit) are other common monosaccharides. Although glucose, galactose, and fructose all have the same chemical formula ($C_6H_{12}O_6$), they differ structurally and chemically (and are isomers) because of the different arrangement of functional groups around the asymmetric carbon. All these monosaccharides have more than one asymmetric carbon (Figure 3.5).

VISUAL CONNECTION

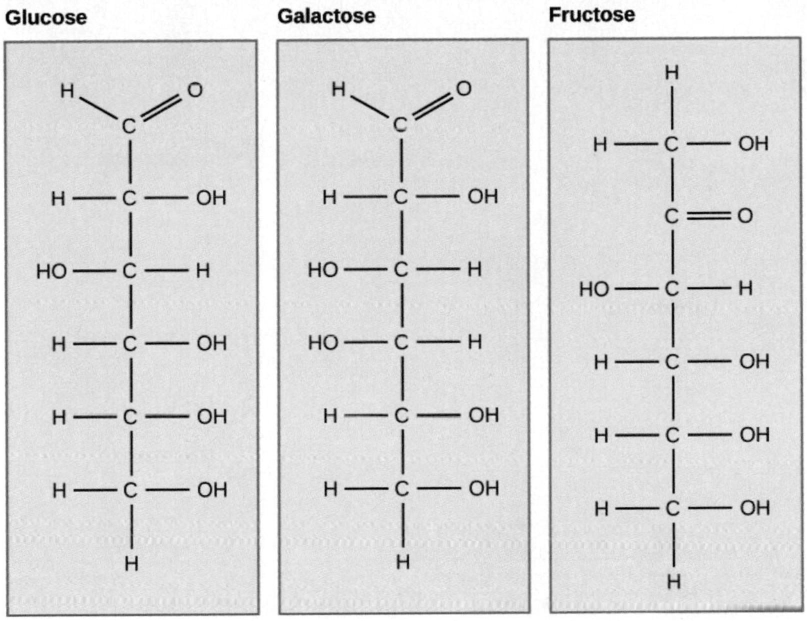

Figure 3.5 Glucose, galactose, and fructose are all hexoses. They are structural isomers, meaning they have the same chemical formula ($C_6H_{12}O_6$) but a different atom arrangement.

What kind of sugars are these, aldose or ketose?

Glucose, galactose, and fructose are isomeric monosaccharides (hexoses), meaning they have the same chemical formula but have slightly different structures. Glucose and galactose are aldoses, and fructose is a ketose.

Monosaccharides can exist as a linear chain or as ring-shaped molecules. In aqueous solutions they are usually in ring forms (Figure 3.6). Glucose in a ring form can have two different hydroxyl group arrangements (OH) around the anomeric carbon (carbon 1 that becomes asymmetric in the ring formation process). If the hydroxyl group is below carbon number 1 in the sugar, it is in the alpha (α) position, and if it is above the plane, it is in the beta (β) position.

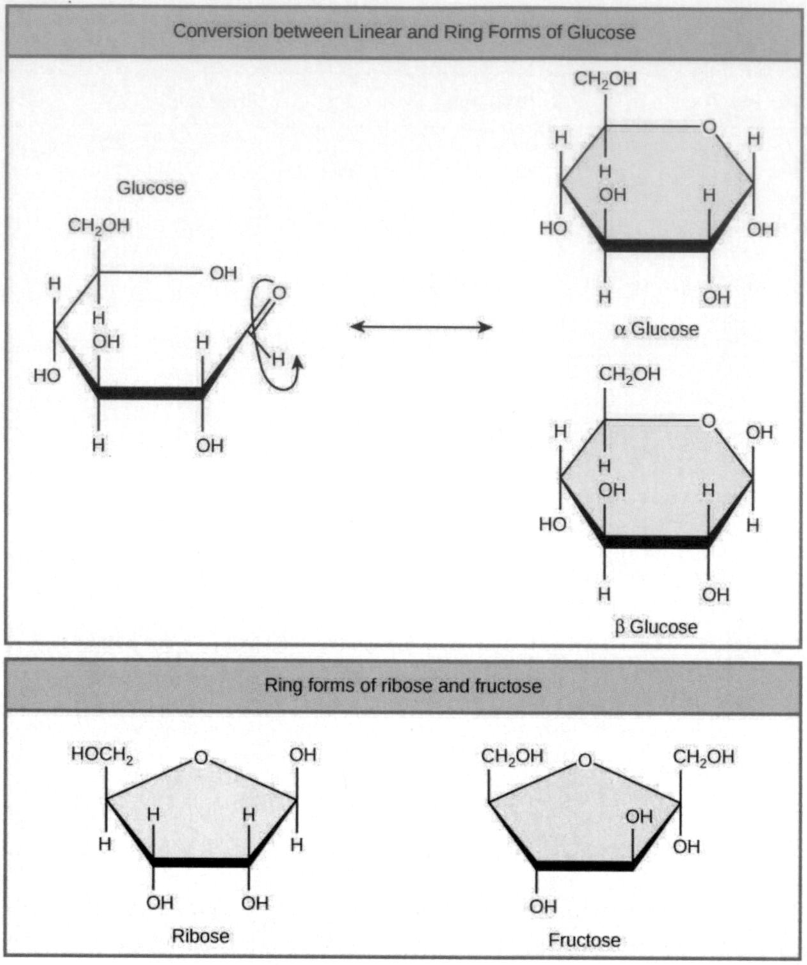

Figure 3.6 Five and six carbon monosaccharides exist in equilibrium between linear and ring forms. When the ring forms, the side chain it closes on locks into an *α* or *β* position. Fructose and ribose also form rings, although they form five-membered rings as opposed to the six-membered ring of glucose.

Disaccharides

Disaccharides (di- = "two") form when two monosaccharides undergo a dehydration reaction (or a condensation reaction or dehydration synthesis). During this process, one monosaccharide's hydroxyl group combines with another monosaccharide's hydrogen, releasing a water molecule and forming a covalent bond. A covalent bond forms between a carbohydrate molecule and another molecule (in this case, between two monosaccharides). Scientists call this a **glycosidic bond** (Figure 3.7). Glycosidic bonds (or glycosidic linkages) can be an alpha or beta type. An alpha bond is formed when the OH group on the carbon-1 of the first glucose is below the ring plane, and a beta bond is formed when the OH group on the carbon-1 is above the ring plane.

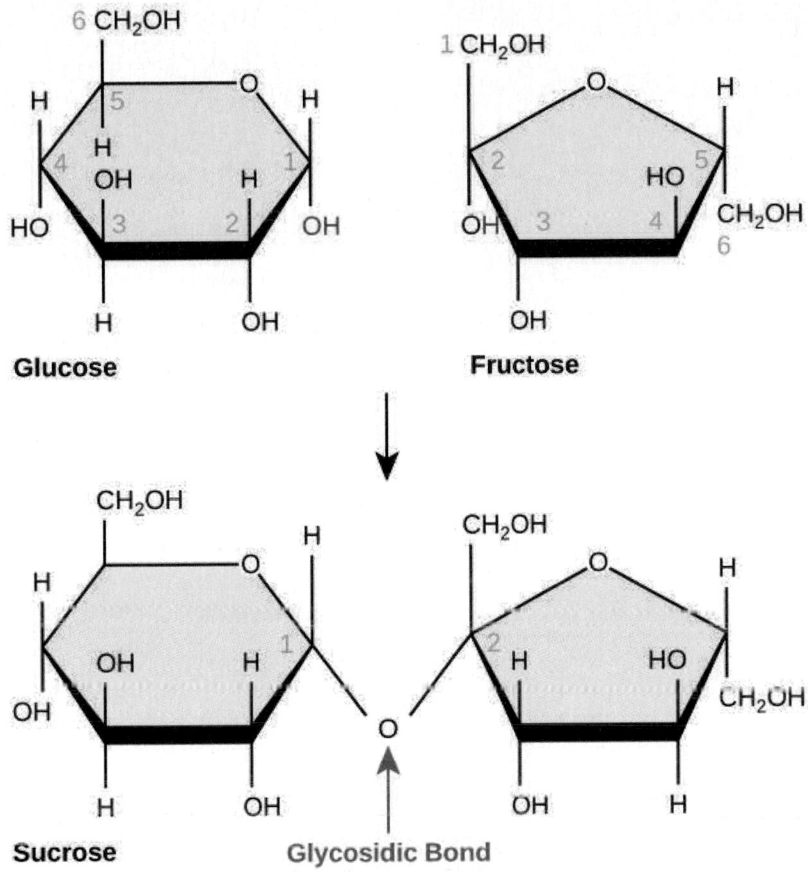

Figure 3.7 Sucrose forms when a glucose monomer and a fructose monomer join in a dehydration reaction to form a glycosidic bond. In the process, a water molecule is lost. By convention, the carbon atoms in a monosaccharide are numbered from the terminal carbon closest to the carbonyl group. In sucrose, a glycosidic linkage forms between carbon 1 in glucose and carbon 2 in fructose.

Common disaccharides include lactose, maltose, and sucrose (Figure 3.8). Lactose is a disaccharide consisting of the monomers glucose and galactose. It is naturally in milk. Maltose, or malt sugar, is a disaccharide formed by a dehydration reaction between two glucose molecules. The most common disaccharide is sucrose, or table sugar, which is comprised of glucose and fructose monomers.

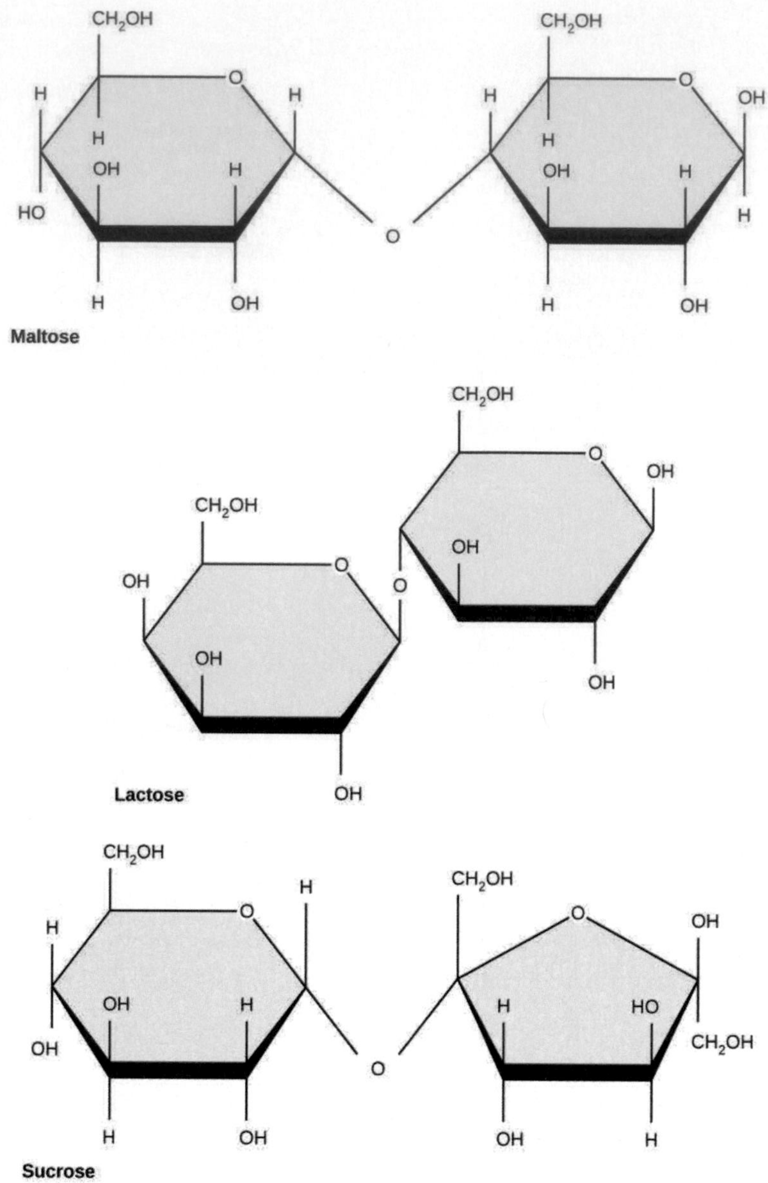

Figure 3.8 Common disaccharides include maltose (grain sugar), lactose (milk sugar), and sucrose (table sugar).

Polysaccharides

A long chain of monosaccharides linked by glycosidic bonds is a **polysaccharide** (poly- = "many"). The chain may be branched or unbranched, and it may contain different types of monosaccharides. The molecular weight may be 100,000 daltons or more depending on the number of joined monomers. Starch, glycogen, cellulose, and chitin are primary examples of polysaccharides.

Plants store starch in the form of sugars. In plants, an amylose and amylopectic mixture (both glucose polymers) comprise these sugars. Plants are able to synthesize glucose, and they store the excess glucose, beyond their immediate energy needs, as starch in different plant parts, including roots and seeds. The starch in the seeds provides food for the embryo as it germinates and can also act as a food source for humans and animals. Enzymes break down the starch that humans consume. For example, an amylase present in saliva catalyzes, or breaks down this starch into smaller molecules, such as maltose and glucose. The cells can then absorb the glucose.

Glucose starch comprises monomers that are joined by α 1-4 or α 1-6 glycosidic bonds. The numbers 1-4 and 1-6 refer to the carbon number of the two residues that have joined to form the bond. As Figure 3.9 illustrates, unbranched glucose monomer chains (only α 1-4 linkages) form the starch; whereas, amylopectin is a branched polysaccharide (α 1-6 linkages at the branch points).

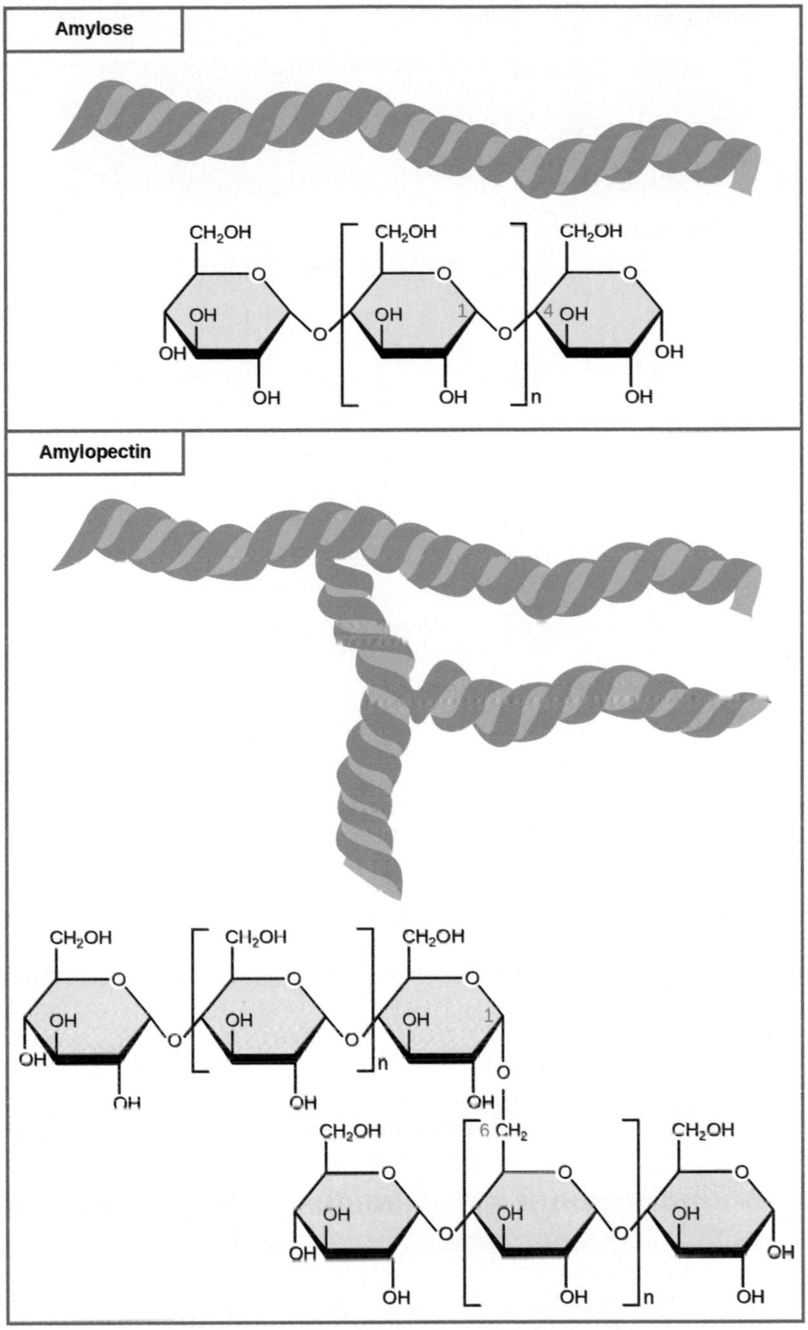

Figure 3.9 Amylose and amylopectin are two different starch forms. Unbranched glucose monomer chains comprise amylose by α 1-4 glycosidic linkages. Unbranched glucose monomer chains comprise amylopectin by α 1-4 and α 1-6 glycosidic linkages. Because of the way the subunits are joined, the glucose chains have a helical structure. Glycogen (not shown) is similar in structure to amylopectin but more highly branched.

Glycogen is the storage form of glucose in humans and other vertebrates and is comprised of monomers of glucose. Glycogen is the animal equivalent of starch and is a highly branched molecule usually stored in liver and muscle cells. Whenever blood glucose levels decrease, glycogen breaks down to release glucose in a process scientists call glycogenolysis.

Cellulose is the most abundant natural biopolymer. Cellulose mostly comprises a plant's cell wall. This provides the cell structural support. Wood and paper are mostly cellulosic in nature. Glucose monomers comprise cellulose that β 1-4 glycosidic bonds link (Figure 3.10).

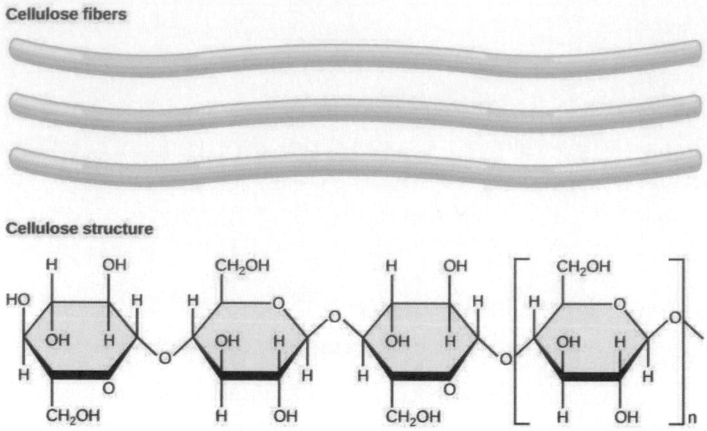

Figure 3.10 In cellulose, glucose monomers are linked in unbranched chains by β 1-4 glycosidic linkages. Because of the way the glucose subunits are joined, every glucose monomer is flipped relative to the next one resulting in a linear, fibrous structure.

As Figure 3.10 shows, every other glucose monomer in cellulose is flipped over, and the monomers are packed tightly as extended long chains. This gives cellulose its rigidity and high tensile strength—which is so important to plant cells. While human digestive enzymes cannot break down the β 1-4 linkage, herbivores such as cows, koalas, and buffalos are able, with the help of the specialized flora in their stomach, to digest plant material that is rich in cellulose and use it as a food source. In some of these animals, certain species of bacteria and protists reside in the rumen (part of the herbivore's digestive system) and secrete the enzyme cellulase. The appendix of grazing animals also contains bacteria that digest cellulose, giving it an important role in ruminants' digestive systems. Cellulases can break down cellulose into glucose monomers that animals use as an energy source. Termites are also able to break down cellulose because of the presence of other organisms in their bodies that secrete cellulases.

Carbohydrates serve various functions in different animals. Arthropods (insects, crustaceans, and others) have an outer skeleton, the exoskeleton, which protects their internal body parts (as we see in the bee in Figure 3.11). This exoskeleton is made of the biological macromolecule **chitin**, which is a polysaccharide-containing nitrogen. It is made of repeating N-acetyl-β-d-glucosamine units, which are a modified sugar. Chitin is also a major component of fungal cell walls. Fungi are neither animals nor plants and form a kingdom of their own in the domain Eukarya.

Figure 3.11 Insects have a hard outer exoskeleton made of chitin, a type of polysaccharide. (credit: Louise Docker)

CAREER CONNECTION

Registered Dietitian

Obesity is a worldwide health concern, and many diseases such as diabetes and heart disease are becoming more prevalent because of obesity. This is one of the reasons why people increasingly seek out registered dietitians for advice. Registered dietitians help plan nutrition programs for individuals in various settings. They often work with patients in health care facilities, designing nutrition plans to treat and prevent diseases. For example, dietitians may teach a patient with diabetes how to manage blood sugar levels by eating the correct types and amounts of carbohydrates. Dietitians may also work in nursing homes, schools, and private practices.

To become a registered dietitian, one needs to earn at least a bachelor's degree in dietetics, nutrition, food technology, or a related field. In addition, registered dietitians must complete a supervised internship program and pass a national exam. Those who pursue careers in dietetics take courses in nutrition, chemistry, biochemistry, biology, microbiology, and human physiology. Dietitians must become experts in the chemistry and physiology (biological functions) of food (proteins, carbohydrates, and fats).

Benefits of Carbohydrates

Are carbohydrates good for you? Some people believe that carbohydrates are bad and they should avoid them. Some diets completely forbid carbohydrate consumption, claiming that a low-carbohydrate diet helps people to lose weight faster. However, carbohydrates have been an important part of the human diet for thousands of years. Artifacts from ancient civilizations show the presence of wheat, rice, and corn in our ancestors' storage areas.

As part of a well balanced diet, we should supplement carbohydrates with proteins, vitamins, and fats. Calorie-wise, a gram of carbohydrate provides 4.3 Kcal. For comparison, fats provide 9 Kcal/g, a less desirable ratio. Carbohydrates contain soluble and insoluble elements. The insoluble part, fiber, is mostly cellulose. Fiber has many uses. It promotes regular bowel movement by adding bulk, and it regulates the blood glucose consumption rate. Fiber also helps to remove excess cholesterol from the body. Fiber binds to the cholesterol in the small intestine, then attaches to the cholesterol and prevents the cholesterol particles from entering the bloodstream. Cholesterol then exits the body via the feces. Fiber-rich diets also have a protective role in reducing the occurrence of colon cancer. In addition, a meal containing whole grains and vegetables gives a feeling of fullness. As an immediate source of energy, glucose breaks down during the cellular respiration process, which produces ATP, the cell's energy currency. Without consuming carbohydrates, we reduce the availability of "instant energy". Eliminating carbohydrates from the diet may be necessary for some people, but such a step may not be healthy for everyone.

LINK TO LEARNING

For an additional perspective on carbohydrates, explore "Biomolecules: the Carbohydrates" through this interactive animation (http://openstax.org/l/carbohydrates) .

3.3 Lipids

By the end of this section, you will be able to do the following:

- Describe the four major types of lipids
- Explain the role of fats in storing energy
- Differentiate between saturated and unsaturated fatty acids
- Describe phospholipids and their role in cells
- Define the basic structure of a steroid and some steroid functions
- Explain how cholesterol helps maintain the plasma membrane's fluid nature

Lipids include a diverse group of compounds that are largely nonpolar in nature. This is because they are hydrocarbons that include mostly nonpolar carbon–carbon or carbon–hydrogen bonds. Non-polar molecules are hydrophobic ("water fearing"), or insoluble in water. Lipids perform many different functions in a cell. Cells store energy for long-term use in the form of fats. Lipids also provide insulation from the environment for plants and animals (Figure 3.12). For example, they help keep aquatic birds and mammals dry when forming a protective layer over fur or feathers because of their water-repellant hydrophobic nature. Lipids are also the building blocks of many hormones and are an important constituent of all cellular membranes. Lipids include fats, oils, waxes, phospholipids, and steroids.

Figure 3.12 Hydrophobic lipids in aquatic mammals' fur, such as this river otter, protect them from the elements. (credit: Ken Bosma)

Fats and Oils

A fat molecule consists of two main components—glycerol and fatty acids. Glycerol is an organic compound (alcohol) with three carbons, five hydrogens, and three hydroxyl (OH) groups. Fatty acids have a long chain of hydrocarbons to which a carboxyl group is attached, hence the name "fatty acid." The number of carbons in the fatty acid may range from 4 to 36. The most common are those containing 12–18 carbons. In a fat molecule, the fatty acids attach to each of the glycerol molecule's three carbons with an ester bond through an oxygen atom (Figure 3.13).

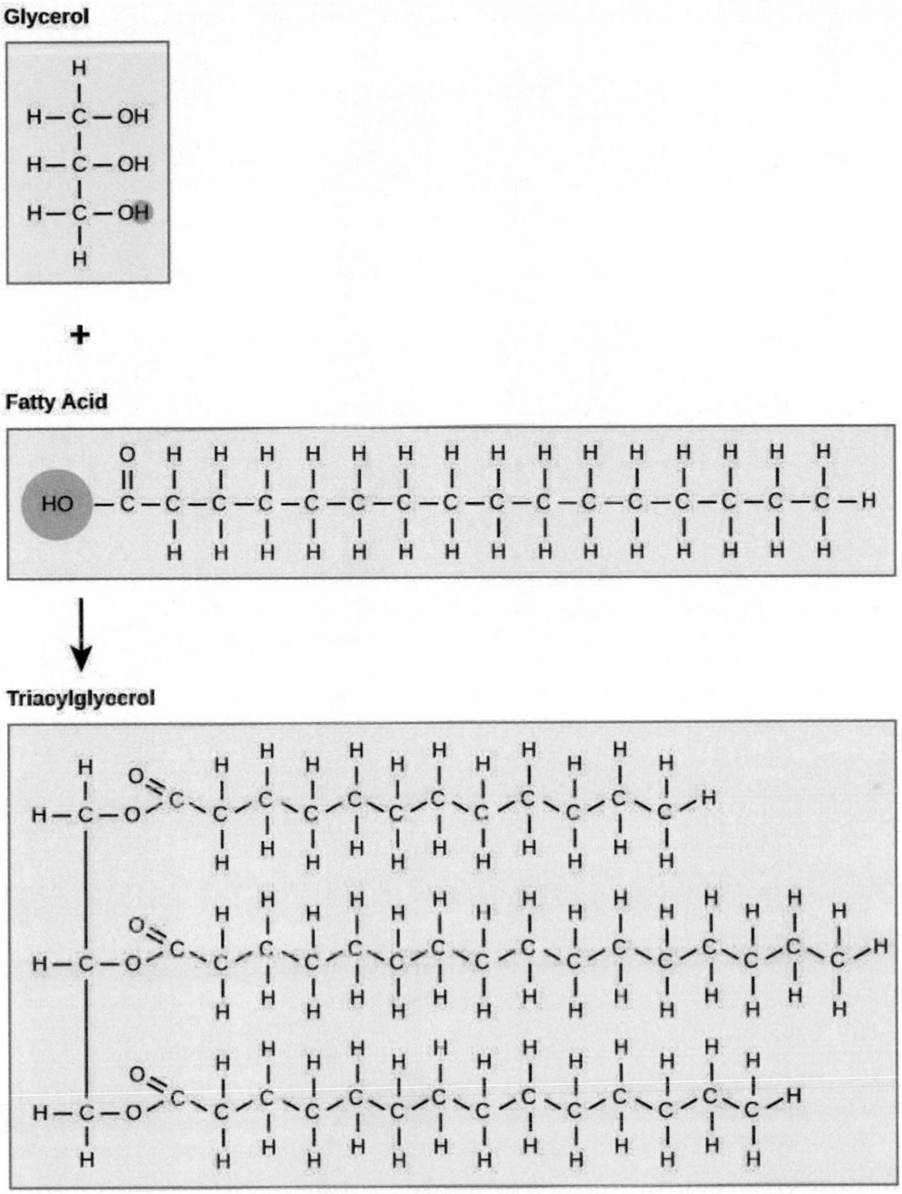

Figure 3.13 Joining three fatty acids to a glycerol backbone in a dehydration reaction forms triacylglycerol. Three water molecules release in the process.

During this ester bond formation, three water molecules are released. The three fatty acids in the triacylglycerol may be similar or dissimilar. We also call fats **triacylglycerols** or **triglycerides** because of their chemical structure. Some fatty acids have common names that specify their origin. For example, palmitic acid, a **saturated fatty acid**, is derived from the palm tree. Arachidic acid is derived from *Arachis hypogea,* the scientific name for groundnuts or peanuts.

Fatty acids may be saturated or unsaturated. In a fatty acid chain, if there are only single bonds between neighboring carbons in the hydrocarbon chain, the fatty acid is saturated. Saturated fatty acids are saturated with hydrogen. In other words, the number of hydrogen atoms attached to the carbon skeleton is maximized. Stearic acid is an example of a saturated fatty acid (Figure 3.14).

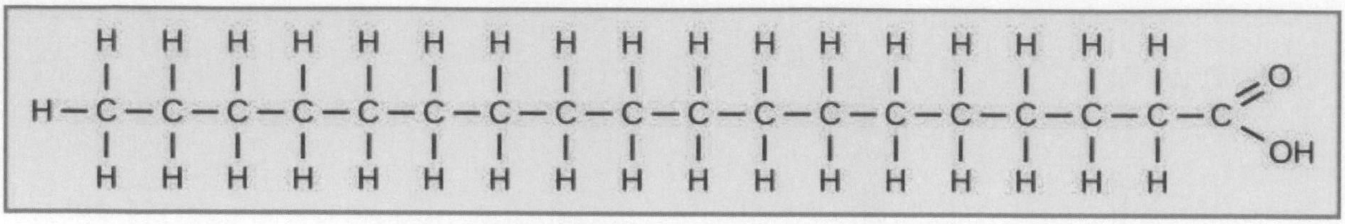

Figure 3.14 Stearic acid is a common saturated fatty acid.

When the hydrocarbon chain contains a double bond, the fatty acid is **unsaturated**. Oleic acid is an example of an unsaturated fatty acid (Figure 3.15).

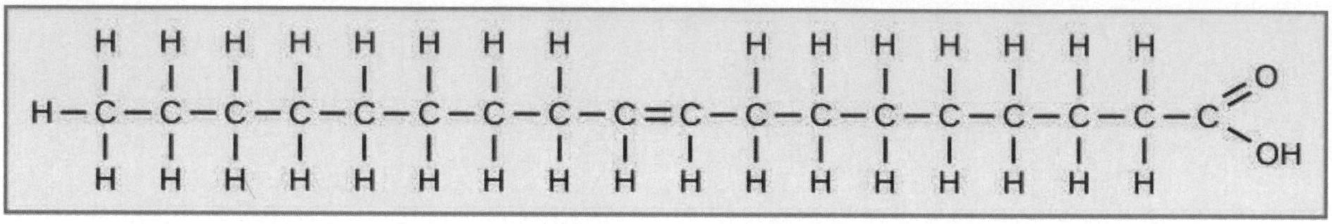

Figure 3.15 Oleic acid is a common unsaturated fatty acid.

Most unsaturated fats are liquid at room temperature. We call these oils. If there is one double bond in the molecule, then it is a monounsaturated fat (e.g., olive oil), and if there is more than one double bond, then it is a polyunsaturated fat (e.g., canola oil).

When a fatty acid has no double bonds, it is a saturated fatty acid because it is not possible to add more hydrogen to the chain's carbon atoms. A fat may contain similar or different fatty acids attached to glycerol. Long straight fatty acids with single bonds generally pack tightly and are solid at room temperature. Animal fats with stearic acid and palmitic acid (common in meat) and the fat with butyric acid (common in butter) are examples of saturated fats. Mammals store fats in specialized cells, or adipocytes, where fat globules occupy most of the cell's volume. Plants store fat or oil in many seeds and use them as a source of energy during seedling development. Unsaturated fats or oils are usually of plant origin and contain *cis* unsaturated fatty acids. *Cis* and *trans* indicate the configuration of the molecule around the double bond. If hydrogens are present in the same plane, it is a cis fat. If the hydrogen atoms are on two different planes, it is a **trans fat**. The *cis* double bond causes a bend or a "kink" that prevents the fatty acids from packing tightly, keeping them liquid at room temperature (Figure 3.16). Olive oil, corn oil, canola oil, and cod liver oil are examples of unsaturated fats. Unsaturated fats help to lower blood cholesterol levels; whereas, saturated fats contribute to plaque formation in the arteries.

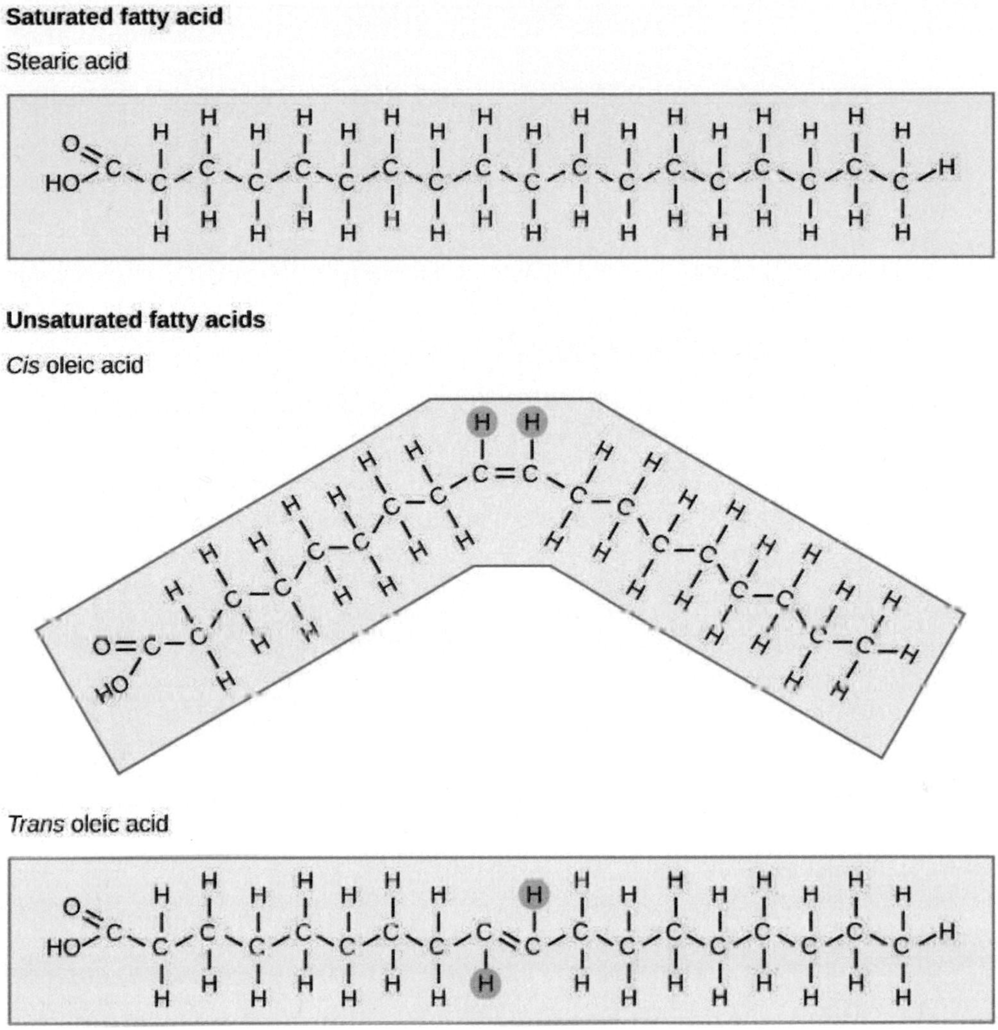

Figure 3.16 Saturated fatty acids have hydrocarbon chains connected by single bonds only. Unsaturated fatty acids have one or more double bonds. Each double bond may be in a *cis* or *trans* configuration. In the *cis* configuration, both hydrogens are on the same side of the hydrocarbon chain. In the *trans* configuration, the hydrogens are on opposite sides. A *cis* double bond causes a kink in the chain.

Trans Fats

The food industry artificially hydrogenates oils to make them semi-solid and of a consistency desirable for many processed food products. Simply speaking, hydrogen gas is bubbled through oils to solidify them. During this hydrogenation process, double bonds of the *cis*- conformation in the hydrocarbon chain may convert to double bonds in the *trans*- conformation.

Margarine, some types of peanut butter, and shortening are examples of artificially hydrogenated trans fats. Recent studies have shown that an increase in trans fats in the human diet may lead to higher levels of low-density lipoproteins (LDL), or "bad" cholesterol, which in turn may lead to plaque deposition in the arteries, resulting in heart disease. Many fast food restaurants have recently banned using trans fats, and food labels are required to display the trans fat content.

Omega Fatty Acids

Essential fatty acids are those that the human body requires but does not synthesize. Consequently, they have to be supplemented through ingestion via the diet. **Omega**-3 fatty acids (like those in Figure 3.17) fall into this category and are one of only two known for humans (the other is omega-6 fatty acid). These are polyunsaturated fatty acids and are omega-3 because a double bond connects the third carbon from the hydrocarbon chain's end to its neighboring carbon.

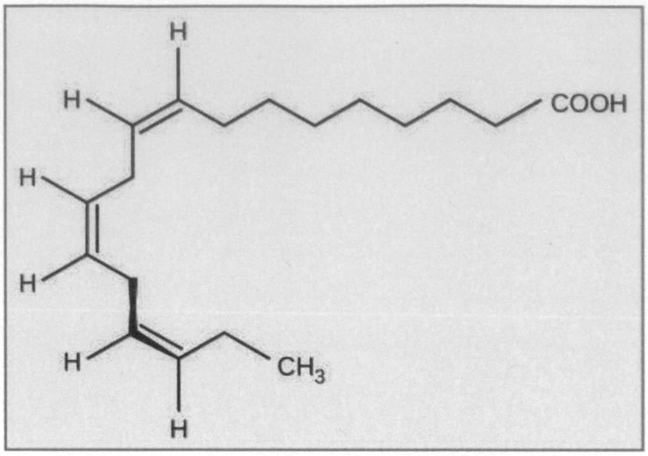

Figure 3.17 Alpha-linolenic acid is an example of an omega-3 fatty acid. It has three *cis* double bonds and, as a result, a curved shape. For clarity, the diagram does not show the carbons. Each singly bonded carbon has two hydrogens associated with it, which the diagram also does not show.

The farthest carbon away from the carboxyl group is numbered as the omega (ω) carbon, and if the double bond is between the third and fourth carbon from that end, it is an omega-3 fatty acid. Nutritionally important because the body does not make them, omega-3 fatty acids include alpha-linoleic acid (ALA), eicosapentaenoic acid (EPA), and docosahexaenoic acid (DHA), all of which are polyunsaturated. Salmon, trout, and tuna are good sources of omega-3 fatty acids. Research indicates that omega-3 fatty acids reduce the risk of sudden death from heart attacks, lower triglycerides in the blood, decrease blood pressure, and prevent thrombosis by inhibiting blood clotting. They also reduce inflammation, and may help lower the risk of some cancers in animals.

Like carbohydrates, fats have received considerable bad publicity. It is true that eating an excess of fried foods and other "fatty" foods leads to weight gain. However, fats do have important functions. Many vitamins are fat soluble, and fats serve as a long-term storage form of fatty acids: a source of energy. They also provide insulation for the body. Therefore, we should consume "healthy" fats in moderate amounts on a regular basis.

Waxes

Wax covers some aquatic birds' feathers and some plants' leaf surfaces. Because of waxes' hydrophobic nature, they prevent water from sticking on the surface (Figure 3.18). Long fatty acid chains esterified to long-chain alcohols comprise waxes.

Figure 3.18 Lipids comprise waxy coverings on some leaves. (credit: Roger Griffith)

Phospholipids

Phospholipids are major plasma membrane constituents that comprise cells' outermost layer. Like fats, they are comprised of fatty acid chains attached to a glycerol or sphingosine backbone. However, instead of three fatty acids attached as in triglycerides, there are two fatty acids forming diacylglycerol, and a modified phosphate group occupies the glycerol backbone's third carbon (Figure 3.19). A phosphate group alone attached to a diacylglycerol does not qualify as a phospholipid. It is phosphatidate (diacylglycerol 3-phosphate), the precursor of phospholipids. An alcohol modifies the phosphate group. Phosphatidylcholine and phosphatidylserine are two important phospholipids that are in plasma membranes.

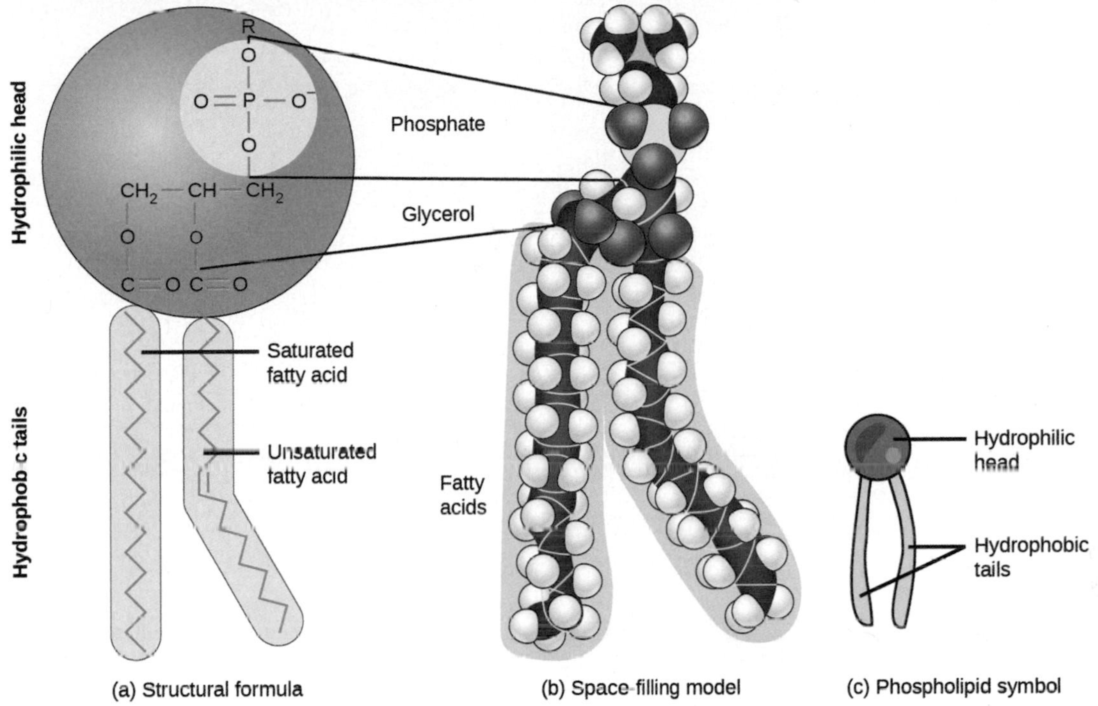

Figure 3.19 A phospholipid is a molecule with two fatty acids and a modified phosphate group attached to a glycerol backbone. Adding a charged or polar chemical group may modify the phosphate.

A phospholipid is an amphipathic molecule, meaning it has a hydrophobic and a hydrophilic part. The fatty acid chains are hydrophobic and cannot interact with water; whereas, the phosphate-containing group is hydrophilic and interacts with water (Figure 3.20).

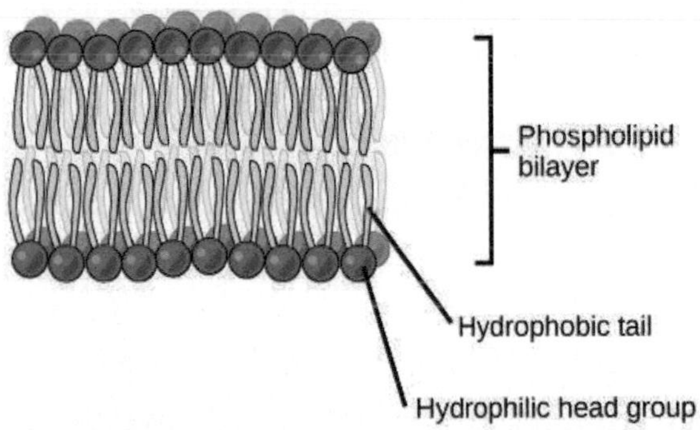

Figure 3.20 The phospholipid bilayer is the major component of all cellular membranes. The hydrophilic head groups of the phospholipids face the aqueous solution. The hydrophobic tails are sequestered in the middle of the bilayer.

The head is the hydrophilic part, and the tail contains the hydrophobic fatty acids. In a membrane, a bilayer of phospholipids forms the structure's matrix, phospholipids' fatty acid tails face inside, away from water; whereas, the phosphate group faces the outside, aqueous side (Figure 3.20).

Phospholipids are responsible for the plasma membrane's dynamic nature. If a drop of phospholipids is placed in water, it spontaneously forms a structure that scientists call a micelle, where the hydrophilic phosphate heads face the outside and the fatty acids face the structure's interior.

Steroids

Unlike the phospholipids and fats that we discussed earlier, **steroids** have a fused ring structure. Although they do not resemble the other lipids, scientists group them with them because they are also hydrophobic and insoluble in water. All steroids have four linked carbon rings and several of them, like cholesterol, have a short tail (Figure 3.21). Many steroids also have the –OH functional group, which puts them in the alcohol classification (sterols).

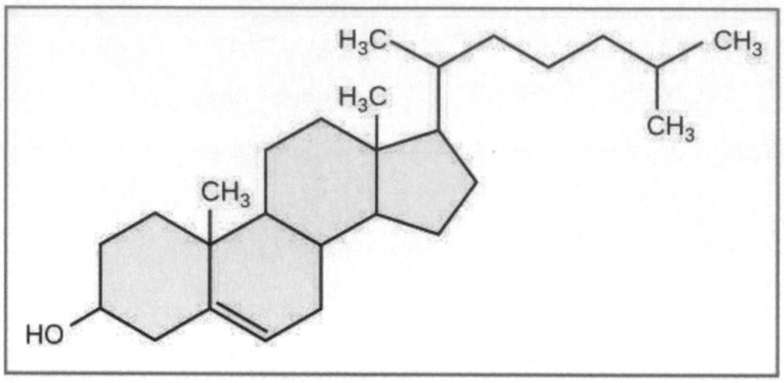

Cholesterol

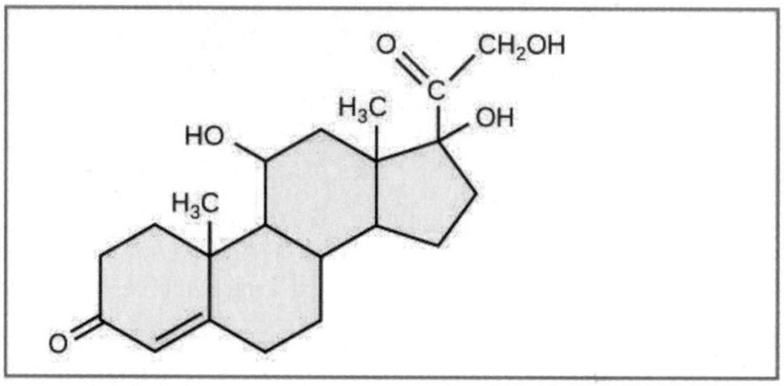

Cortisol

Figure 3.21 Four fused hydrocarbon rings comprise steroids such as cholesterol and cortisol.

Cholesterol is the most common steroid. The liver synthesizes cholesterol and is the precursor to many steroid hormones such as testosterone and estradiol, which gonads and endocrine glands secrete. It is also the precursor to Vitamin D. Cholesterol is also the precursor of bile salts, which help emulsifying fats and their subsequent absorption by cells. Although lay people often speak negatively about cholesterol, it is necessary for the body's proper functioning. Sterols (cholesterol in animal cells, phytosterol in plants) are components of the plasma membrane of cells and are found within the phospholipid bilayer.

LINK TO LEARNING

For an additional perspective on lipids, explore the interactive animation "Biomolecules: The Lipids" (http://openstax.org/l/lipids) .

3.4 Proteins

By the end of this section, you will be able to do the following:

- Describe the functions proteins perform in the cell and in tissues
- Discuss the relationship between amino acids and proteins
- Explain the four levels of protein organization
- Describe the ways in which protein shape and function are linked

Proteins are one of the most abundant organic molecules in living systems and have the most diverse range of functions of all macromolecules. Proteins may be structural, regulatory, contractile, or protective. They may serve in transport, storage, or membranes; or they may be toxins or enzymes. Each cell in a living system may contain thousands of proteins, each with a unique function. Their structures, like their functions, vary greatly. They are all, however, amino acid polymers arranged in a linear sequence.

Types and Functions of Proteins

Enzymes, which living cells produce, are catalysts in biochemical reactions (like digestion) and are usually complex or conjugated proteins. Each enzyme is specific for the substrate (a reactant that binds to an enzyme) upon which it acts. The enzyme may help in breakdown, rearrangement, or synthesis reactions. We call enzymes that break down their substrates catabolic enzymes. Those that build more complex molecules from their substrates are anabolic enzymes, and enzymes that affect the rate of reaction are catalytic enzymes. Note that all enzymes increase the reaction rate and, therefore, are organic catalysts. An example of an enzyme is salivary amylase, which hydrolyzes its substrate amylose, a component of starch.

Hormones are chemical-signaling molecules, usually small proteins or steroids, secreted by endocrine cells that act to control or regulate specific physiological processes, including growth, development, metabolism, and reproduction. For example, insulin is a protein hormone that helps regulate the blood glucose level. Table 3.1 lists the primary types and functions of proteins.

Protein Types and Functions

Type	Examples	Functions
Digestive Enzymes	Amylase, lipase, pepsin, trypsin	Help in food by catabolizing nutrients into monomeric units
Transport	Hemoglobin, albumin	Carry substances in the blood or lymph throughout the body
Structural	Actin, tubulin, keratin	Construct different structures, like the cytoskeleton
Hormones	Insulin, thyroxine	Coordinate different body systems' activity
Defense	Immunoglobulins	Protect the body from foreign pathogens
Contractile	Actin, myosin	Effect muscle contraction
Storage	Legume storage proteins, egg white (albumin)	Provide nourishment in early embryo development and the seedling

Table 3.1

Proteins have different shapes and molecular weights. Some proteins are globular in shape; whereas, others are fibrous in nature. For example, hemoglobin is a globular protein, but collagen, located in our skin, is a fibrous protein. Protein shape is critical to its function, and many different types of chemical bonds maintain this shape. Changes in temperature, pH, and exposure to chemicals may lead to permanent changes in the protein's shape, leading to loss of function, or **denaturation**. Different arrangements of the same 20 types of amino acids comprise all proteins. Two rare new amino acids were discovered recently (selenocystein and pirrolysine), and additional new discoveries may be added to the list.

Amino Acids

Amino acids are the monomers that comprise proteins. Each amino acid has the same fundamental structure, which consists of a central carbon atom, or the alpha (α) carbon, bonded to an amino group (NH_2), a carboxyl group (COOH), and to a hydrogen atom. Every amino acid also has another atom or group of atoms bonded to the central atom known as the R group (Figure 3.22).

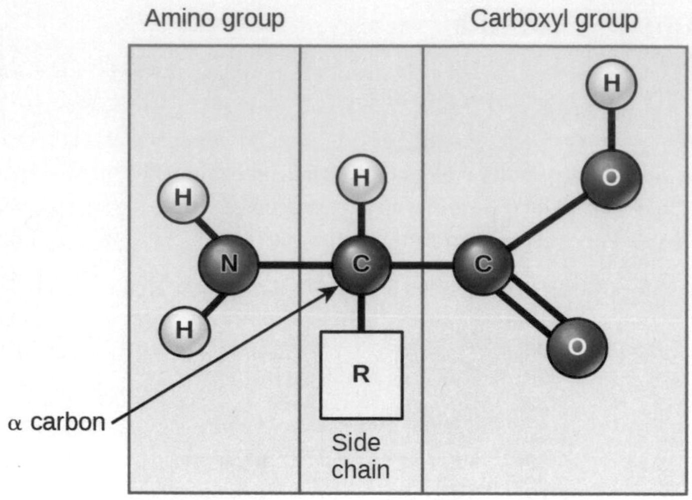

Figure 3.22 Amino acids have a central asymmetric carbon to which an amino group, a carboxyl group, a hydrogen atom, and a side chain (R group) are attached.

Scientists use the name "amino acid" because these acids contain both amino group and carboxyl-acid-group in their basic structure. As we mentioned, there are 20 common amino acids present in proteins. Nine of these are essential amino acids in humans because the human body cannot produce them and we obtain them from our diet. For each amino acid, the R group (or side chain) is different (Figure 3.23).

VISUAL CONNECTION

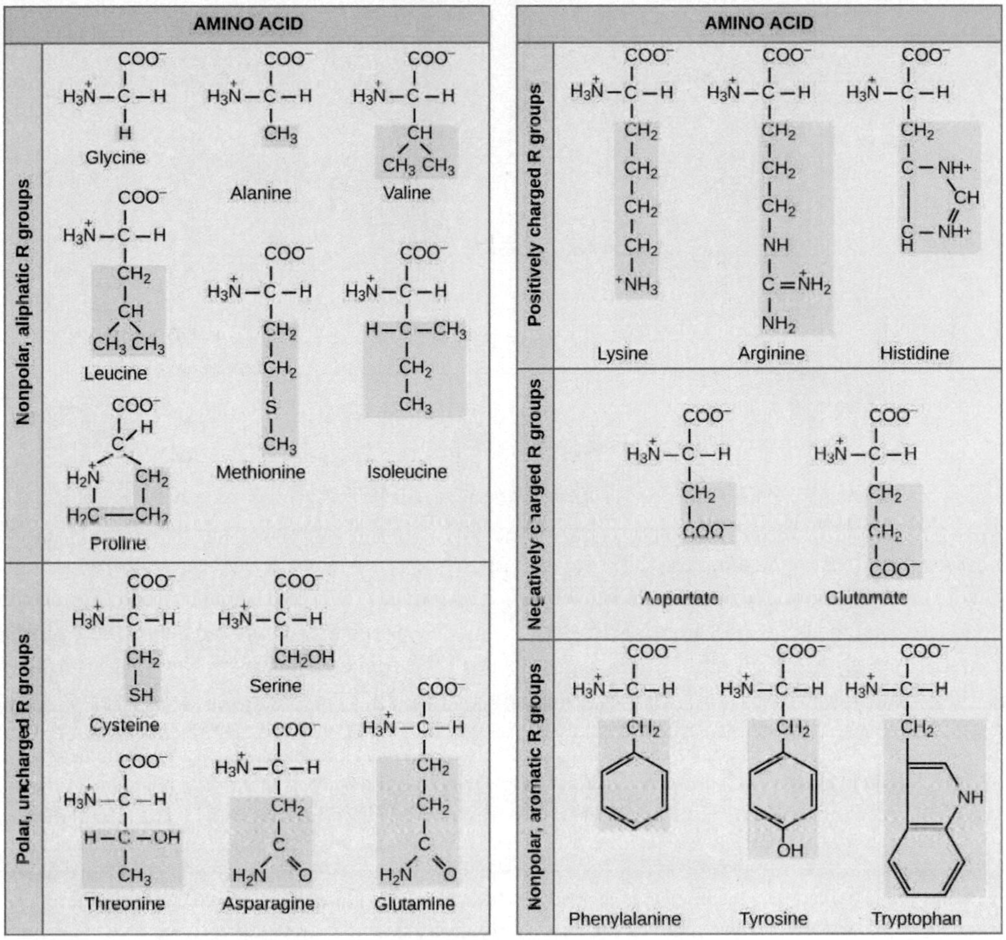

Figure 3.23 There are 20 common amino acids commonly found in proteins, each with a different R group (variant group) that determines its chemical nature.

Which categories of amino acid would you expect to find on a soluble protein's surface and which would you expect to find in the interior? What distribution of amino acids would you expect to find in a protein embedded in a lipid bilayer?

The chemical nature of the side chain determines the amino acid's nature (that is, whether it is acidic, basic, polar, or nonpolar). For example, the amino acid glycine has a hydrogen atom as the R group. Amino acids such as valine, methionine, and alanine are nonpolar or hydrophobic in nature, while amino acids such as serine, threonine, and cysteine are polar and have hydrophilic side chains. The side chains of lysine and arginine are positively charged, and therefore these amino acids are also basic amino acids. Proline has an R group that is linked to the amino group, forming a ring-like structure. Proline is an exception to the amino acid's standard structure since its amino group is not separate from the side chain (Figure 3.23).

A single upper case letter or a three-letter abbreviation represents amino acids. For example, the letter V or the three-letter symbol val represent valine. Just as some fatty acids are essential to a diet, some amino acids also are necessary. These essential amino acids in humans include isoleucine, leucine, and cysteine. Essential amino acids refer to those necessary to build proteins in the body, but not those that the body produces. Which amino acids are essential varies from organism to organism.

The sequence and the number of amino acids ultimately determine the protein's shape, size, and function. A covalent bond, or **peptide bond**, attaches to each amino acid, which a dehydration reaction forms. One amino acid's carboxyl group and the incoming amino acid's amino group combine, releasing a water molecule. The resulting bond is the peptide bond (Figure 3.24).

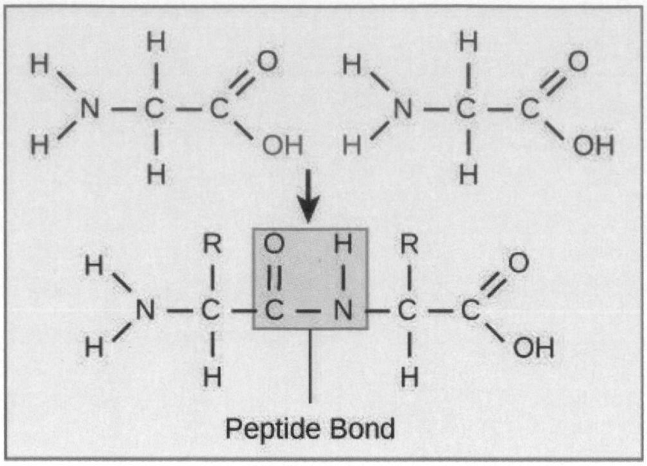

Figure 3.24 Peptide bond formation is a dehydration synthesis reaction. The carboxyl group of one amino acid is linked to the incoming amino acid's amino group. In the process, it releases a water molecule.

The products that such linkages form are peptides. As more amino acids join to this growing chain, the resulting chain is a polypeptide. Each polypeptide has a free amino group at one end. This end the N terminal, or the amino terminal, and the other end has a free carboxyl group, also the C or carboxyl terminal. While the terms polypeptide and protein are sometimes used interchangeably, a polypeptide is technically a polymer of amino acids, whereas the term protein is used for a polypeptide or polypeptides that have combined together, often have bound non-peptide prosthetic groups, have a distinct shape, and have a unique function. After protein synthesis (translation), most proteins are modified. These are known as post-translational modifications. They may undergo cleavage, phosphorylation, or may require adding other chemical groups. Only after these modifications is the protein completely functional.

LINK TO LEARNING

Click through the steps of protein synthesis in this interactive tutorial (http://openstax.org/l/protein_synth) .

EVOLUTION CONNECTION

The Evolutionary Significance of Cytochrome c

Cytochrome c is an important component of the electron transport chain, a part of cellular respiration, and it is normally located in the cellular organelle, the mitochondrion. This protein has a heme prosthetic group, and the heme's central ion alternately reduces and oxidizes during electron transfer. Because this essential protein's role in producing cellular energy is crucial, it has changed very little over millions of years. Protein sequencing has shown that there is a considerable amount of cytochrome c amino acid sequence homology among different species. In other words, we can assess evolutionary kinship by measuring the similarities or differences among various species' DNA or protein sequences.

Scientists have determined that human cytochrome c contains 104 amino acids. For each cytochrome c molecule from different organisms that scientists have sequenced to date, 37 of these amino acids appear in the same position in all cytochrome c samples. This indicates that there may have been a common ancestor. On comparing the human and chimpanzee protein sequences, scientists did not find a sequence difference. When researchers compared human and rhesus monkey sequences, the single difference was in one amino acid. In another comparison, human to yeast sequencing shows a difference in the 44th position.

Protein Structure

As we discussed earlier, a protein's shape is critical to its function. For example, an enzyme can bind to a specific substrate at an active site. If this active site is altered because of local changes or changes in overall protein structure, the enzyme may be unable to bind to the substrate. To understand how the protein gets its final shape or conformation, we need to understand the four levels of protein structure: primary, secondary, tertiary, and quaternary.

Primary Structure

Amino acids' unique sequence in a polypeptide chain is its **primary structure**. For example, the pancreatic hormone insulin has two polypeptide chains, A and B, and they are linked together by disulfide bonds. The N terminal amino acid of the A chain is glycine; whereas, the C terminal amino acid is asparagine (Figure 3.25). The amino acid sequences in the A and B chains are unique to insulin.

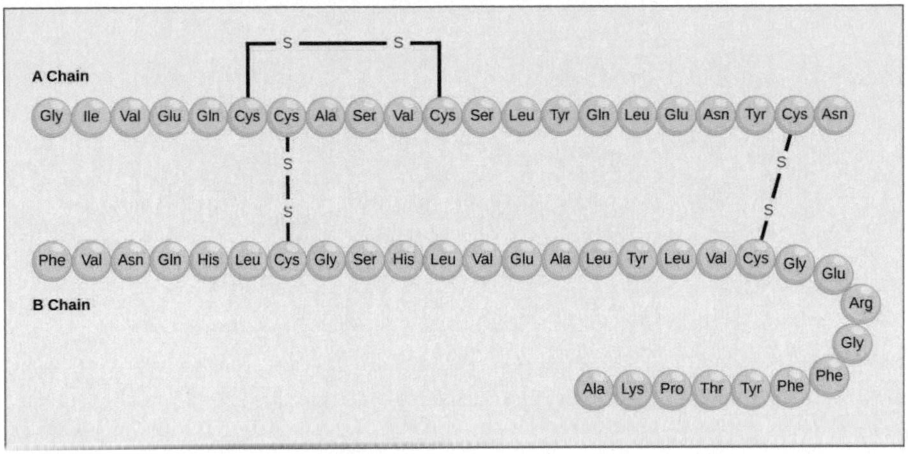

Figure 3.25 Bovine serum insulin is a protein hormone comprised of two peptide chains, A (21 amino acids long) and B (30 amino acids long). In each chain, three-letter abbreviations that represent the amino acids' names in the order they are present indicate primary structure. The amino acid cysteine (cys) has a sulfhydryl (SH) group as a side chain. Two sulfhydryl groups can react in the presence of oxygen to form a disulfide (S-S) bond. Two disulfide bonds connect the A and B chains together, and a third helps the A chain fold into the correct shape. Note that all disulfide bonds are the same length, but we have drawn them different sizes for clarity.

The gene encoding the protein ultimately determines the unique sequence for every protein. A change in nucleotide sequence of the gene's coding region may lead to adding a different amino acid to the growing polypeptide chain, causing a change in protein structure and function. In sickle cell anemia, the hemoglobin β chain (a small portion of which we show in Figure 3.26) has a single amino acid substitution, causing a change in protein structure and function. Specifically, valine in the β chain substitutes the amino acid glutamic. What is most remarkable to consider is that a hemoglobin molecule is comprised of two alpha and two beta chains that each consist of about 150 amino acids. The molecule, therefore, has about 600 amino acids. The structural difference between a normal hemoglobin molecule and a sickle cell molecule—which dramatically decreases life expectancy—is a single amino acid of the 600. What is even more remarkable is that three nucleotides each encode those 600 amino acids, and a single base change (point mutation), 1 in 1800 bases causes the mutation.

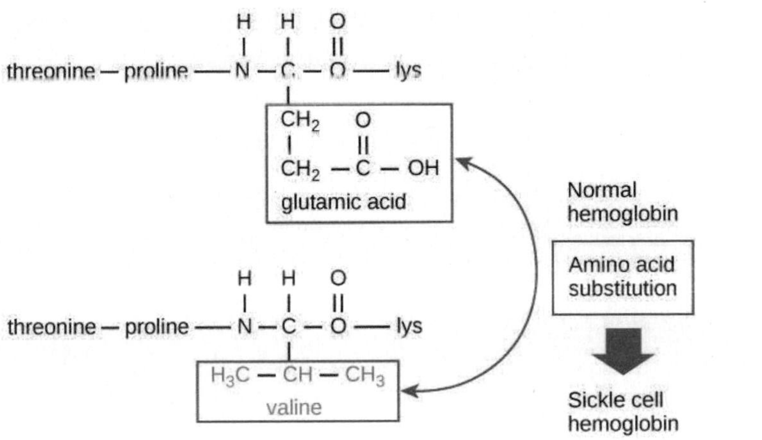

Figure 3.26 The beta chain of hemoglobin is 147 residues in length, yet a single amino acid substitution leads to sickle cell anemia. In normal hemoglobin, the amino acid at position seven is glutamate. In sickle cell hemoglobin, a valine replaces glutamate.

Because of this change of one amino acid in the chain, hemoglobin molecules form long fibers that distort the biconcave, or disc-shaped, red blood cells and causes them to assume a crescent or “sickle” shape, which clogs blood vessels (Figure 3.27). This can lead to myriad serious health problems such as breathlessness, dizziness, headaches, and abdominal pain for those affected by this disease.

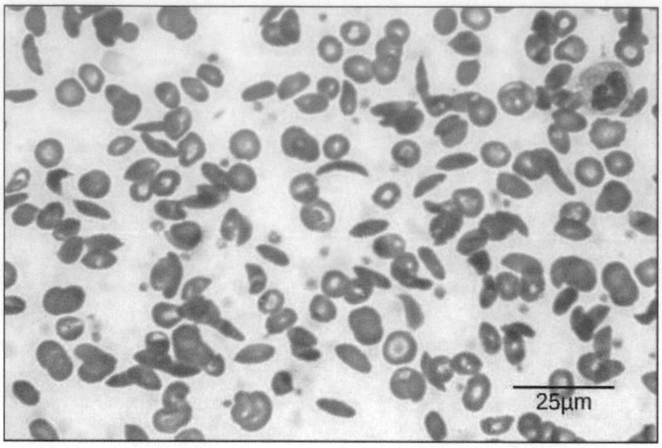

Figure 3.27 In this blood smear, visualized at 535x magnification using bright field microscopy, sickle cells are crescent shaped, while normal cells are disc-shaped. (credit: modification of work by Ed Uthman; scale-bar data from Matt Russell)

Secondary Structure

The local folding of the polypeptide in some regions gives rise to the **secondary structure** of the protein. The most common are the **α-helix** and **β-pleated sheet** structures (Figure 3.28). Both structures are held in shape by hydrogen bonds. The hydrogen bonds form between the oxygen atom in the carbonyl group in one amino acid and another amino acid that is four amino acids farther along the chain.

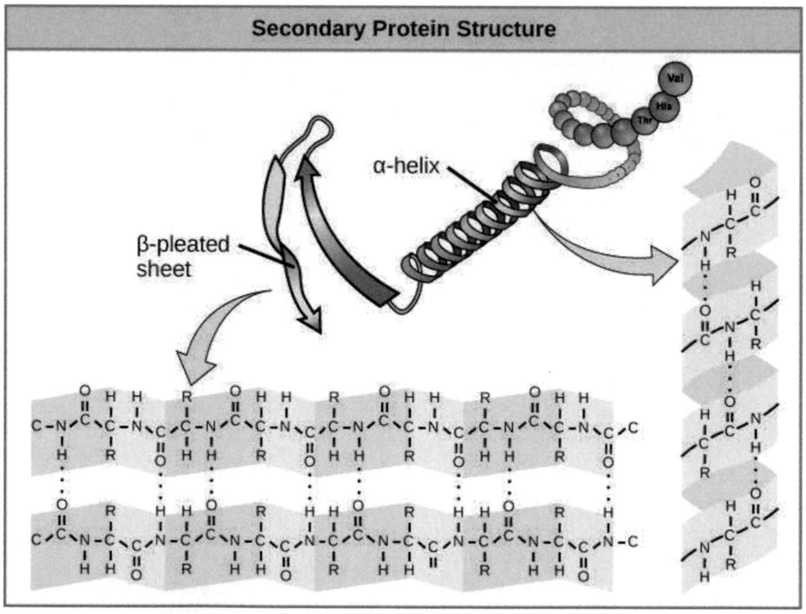

Figure 3.28 The α-helix and β-pleated sheet are secondary structures of proteins that form because of hydrogen bonding between carbonyl and amino groups in the peptide backbone. Certain amino acids have a propensity to form an α-helix, while others have a propensity to form a β-pleated sheet.

Every helical turn in an alpha helix has 3.6 amino acid residues. The polypeptide's R groups (the variant groups) protrude out from the α-helix chain. In the β-pleated sheet, hydrogen bonding between atoms on the polypeptide chain's backbone form the "pleats". The R groups are attached to the carbons and extend above and below the pleat's folds. The pleated segments align parallel or antiparallel to each other, and hydrogen bonds form between the partially positive nitrogen atom in the amino group and the partially negative oxygen atom in the peptide backbone's carbonyl group. The α-helix and β-pleated sheet structures are in most globular and fibrous proteins and they play an important structural role.

Tertiary Structure

The polypeptide's unique three-dimensional structure is its **tertiary structure** (Figure 3.29). This structure is in part due to chemical interactions at work on the polypeptide chain. Primarily, the interactions among R groups create the protein's complex three-dimensional tertiary structure. The nature of the R groups in the amino acids involved can counteract forming the hydrogen bonds we described for standard secondary structures. For example, R groups with like charges repel each other and those with unlike charges are attracted to each other (ionic bonds). When protein folding takes place, the nonpolar amino acids' hydrophobic R groups lie in the protein's interior; whereas, the hydrophilic R groups lie on the outside. Scientists also call the former interaction types hydrophobic interactions. Interaction between cysteine side chains forms disulfide linkages in the presence of oxygen, the only covalent bond that forms during protein folding.

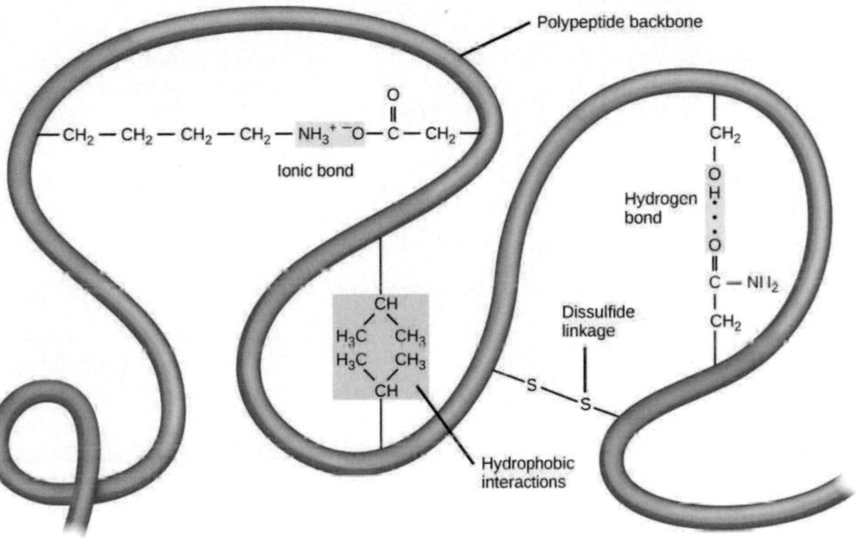

Figure 3.29 A variety of chemical interactions determine the proteins' tertiary structure. These include hydrophobic interactions, ionic bonding, hydrogen bonding, and disulfide linkages.

All of these interactions, weak and strong, determine the protein's final three-dimensional shape. When a protein loses its three-dimensional shape, it may no longer be functional.

Quaternary Structure

In nature, some proteins form from several polypeptides, or subunits, and the interaction of these subunits forms the **quaternary structure**. Weak interactions between the subunits help to stabilize the overall structure. For example, insulin (a globular protein) has a combination of hydrogen and disulfide bonds that cause it to mostly clump into a ball shape. Insulin starts out as a single polypeptide and loses some internal sequences in the presence of post translational modification after forming the disulfide linkages that hold the remaining chains together. Silk (a fibrous protein), however, has a β-pleated sheet structure that is the result of hydrogen bonding between different chains.

Figure 3.30 illustrates the four levels of protein structure (primary, secondary, tertiary, and quaternary).

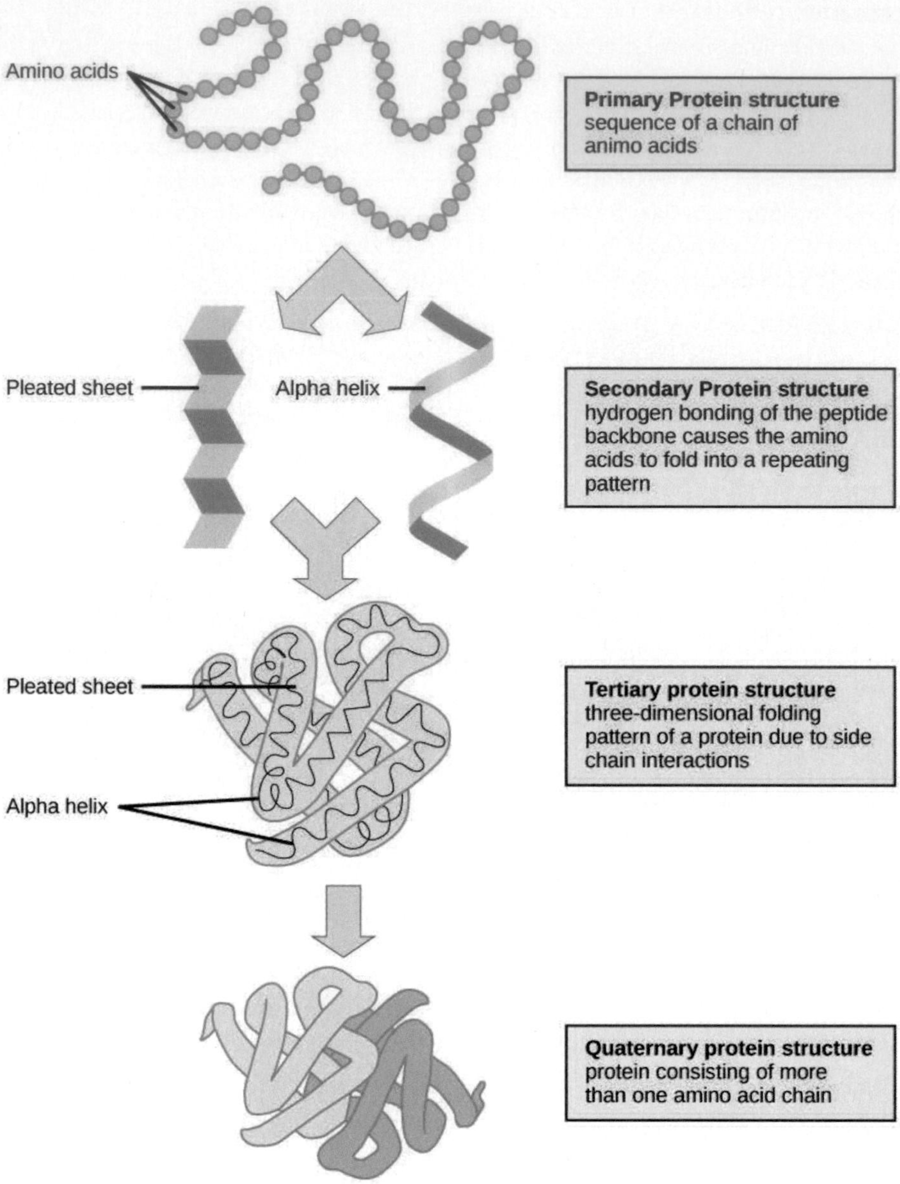

Figure 3.30 Observe the four levels of protein structure in these illustrations. (credit: modification of work by National Human Genome Research Institute)

Denaturation and Protein Folding

Each protein has its own unique sequence and shape that chemical interactions hold together. If the protein is subject to changes in temperature, pH, or exposure to chemicals, the protein structure may change, losing its shape without losing its primary sequence in what scientists call denaturation. Denaturation is often reversible because the polypeptide's primary structure is conserved in the process if the denaturing agent is removed, allowing the protein to resume its function. Sometimes denaturation is irreversible, leading to loss of function. One example of irreversible protein denaturation is frying an egg. The albumin protein in the liquid egg white denatures when placed in a hot pan. Not all proteins denature at high temperatures. For instance, bacteria that survive in hot springs have proteins that function at temperatures close to boiling. The stomach is also very acidic, has a low pH, and denatures proteins as part of the digestion process; however, the stomach's digestive enzymes retain their activity under these conditions.

Protein folding is critical to its function. Scientists originally thought that the proteins themselves were responsible for the folding process. Only recently researchers discovered that often they receive assistance in the folding process from protein helpers, or **chaperones** (or chaperonins) that associate with the target protein during the folding process. They act by preventing polypeptide aggregation that comprise the complete protein structure, and they disassociate from the protein once the target protein is folded.

LINK TO LEARNING

For an additional perspective on proteins, view this animation (http://openstax.org/l/proteins) called "Biomolecules: The Proteins."

3.5 Nucleic Acids

By the end of this section, you will be able to do the following:

- Describe nucleic acids' structure and define the two types of nucleic acids
- Explain DNA's structure and role
- Explain RNA's structure and roles

Nucleic acids are the most important macromolecules for the continuity of life. They carry the cell's genetic blueprint and carry instructions for its functioning.

DNA and RNA

The two main types of nucleic acids are **deoxyribonucleic acid (DNA)** and **ribonucleic acid (RNA)**. DNA is the genetic material in all living organisms, ranging from single-celled bacteria to multicellular mammals. It is in the nucleus of eukaryotes and in the organelles, chloroplasts, and mitochondria. In prokaryotes, the DNA is not enclosed in a membranous envelope.

The cell's entire genetic content is its genome, and the study of genomes is genomics. In eukaryotic cells but not in prokaryotes, DNA forms a complex with histone proteins to form chromatin, the substance of eukaryotic chromosomes. A chromosome may contain tens of thousands of genes. Many genes contain the information to make protein products. Other genes code for RNA products. DNA controls all of the cellular activities by turning the genes "on" or "off."

The other type of nucleic acid, RNA, is mostly involved in protein synthesis. The DNA molecules never leave the nucleus but instead use an intermediary to communicate with the rest of the cell. This intermediary is the **messenger RNA (mRNA)**. Other types of RNA—like rRNA, tRNA, and microRNA—are involved in protein synthesis and its regulation.

DNA and RNA are comprised of monomers that scientists call **nucleotides**. The nucleotides combine with each other to form a **polynucleotide**, DNA or RNA. Three components comprise each nucleotide: a nitrogenous base, a pentose (five-carbon) sugar, and a phosphate group (Figure 3.31). Each nitrogenous base in a nucleotide is attached to a sugar molecule, which is attached to one or more phosphate groups.

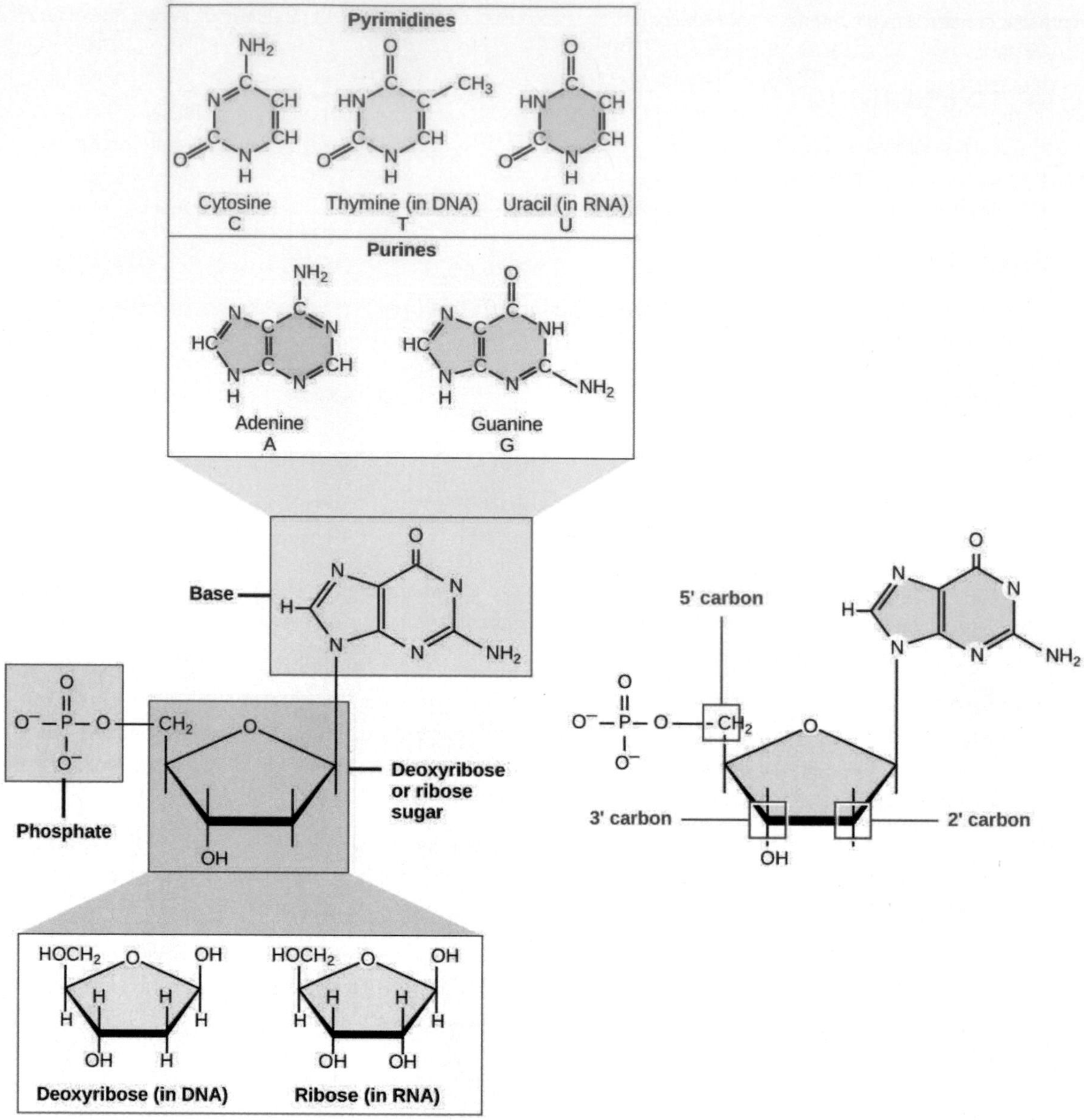

Figure 3.31 Three components comprise a nucleotide: a nitrogenous base, a pentose sugar, and one or more phosphate groups. Carbon residues in the pentose are numbered 1′ through 5′ (the prime distinguishes these residues from those in the base, which are numbered without using a prime notation). The base is attached to the ribose's 1′ position, and the phosphate is attached to the 5′ position. When a polynucleotide forms, the incoming nucleotide's 5′ phosphate attaches to the 3′ hydroxyl group at the end of the growing chain. Two types of pentose are in nucleotides, deoxyribose (found in DNA) and ribose (found in RNA). Deoxyribose is similar in structure to ribose, but it has an H instead of an OH at the 2′ position. We can divide bases into two categories: purines and pyrimidines. Purines have a double ring structure, and pyrimidines have a single ring.

The nitrogenous bases, important components of nucleotides, are organic molecules and are so named because they contain carbon and nitrogen. They are bases because they contain an amino group that has the potential of binding an extra hydrogen, and thus decreasing the hydrogen ion concentration in its environment, making it more basic. Each nucleotide in DNA contains one of four possible nitrogenous bases: adenine (A), guanine (G) cytosine (C), and thymine (T).

Scientists classify adenine and guanine as **purines**. The purine's primary structure is two carbon-nitrogen rings. Scientists classify cytosine, thymine, and uracil as **pyrimidines** which have a single carbon-nitrogen ring as their primary structure (Figure 3.31). Each of these basic carbon-nitrogen rings has different functional groups attached to it. In molecular biology shorthand, we know the nitrogenous bases by their symbols A, T, G, C, and U. DNA contains A, T, G, and C; whereas, RNA contains A, U, G, and C.

The pentose sugar in DNA is deoxyribose, and in RNA, the sugar is ribose (Figure 3.31). The difference between the sugars is the presence of the hydroxyl group on the ribose's second carbon and hydrogen on the deoxyribose's second carbon. The carbon atoms of the sugar molecule are numbered as 1', 2', 3', 4', and 5' (1' is read as "one prime"). The phosphate residue attaches to the hydroxyl group of the 5' carbon of one sugar and the hydroxyl group of the 3' carbon of the sugar of the next nucleotide, which forms a 5'–3' **phosphodiester** linkage. A simple dehydration reaction like the other linkages connecting monomers in macromolecules does not form the phosphodiester linkage. Its formation involves removing two phosphate groups. A polynucleotide may have thousands of such phosphodiester linkages.

DNA Double-Helix Structure

DNA has a double-helix structure (Figure 3.32). The sugar and phosphate lie on the outside of the helix, forming the DNA's backbone. The nitrogenous bases are stacked in the interior, like a pair of staircase steps. Hydrogen bonds bind the pairs to each other. Every base pair in the double helix is separated from the next base pair by 0.34 nm. The helix's two strands run in opposite directions, meaning that the 5' carbon end of one strand will face the 3' carbon end of its matching strand. (Scientists call this an antiparallel orientation and is important to DNA replication and in many nucleic acid interactions.)

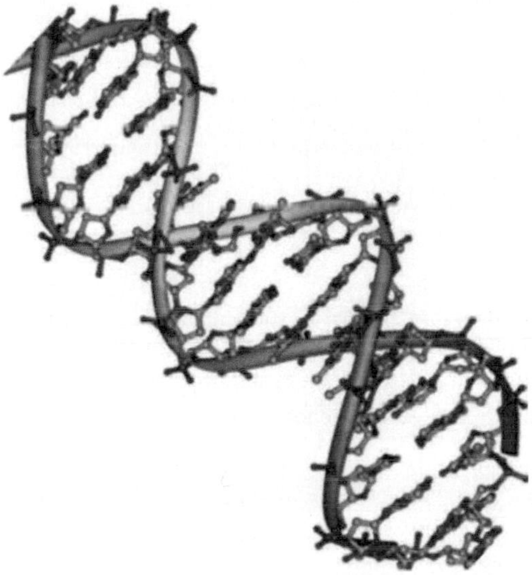

Figure 3.32 Native DNA is an antiparallel double helix. The phosphate backbone (indicated by the curvy lines) is on the outside, and the bases are on the inside. Each base from one strand interacts via hydrogen bonding with a base from the opposing strand. (credit: Jerome Walker/Dennis Myts)

Only certain types of base pairing are allowed. For example, a certain purine can only pair with a certain pyrimidine. This means A can pair with T, and G can pair with C, as Figure 3.33 shows. This is the base complementary rule. In other words, the DNA strands are complementary to each other. If the sequence of one strand is AATTGGCC, the complementary strand would have the sequence TTAACCGG. During DNA replication, each strand copies itself, resulting in a daughter DNA double helix containing one parental DNA strand and a newly synthesized strand.

VISUAL CONNECTION

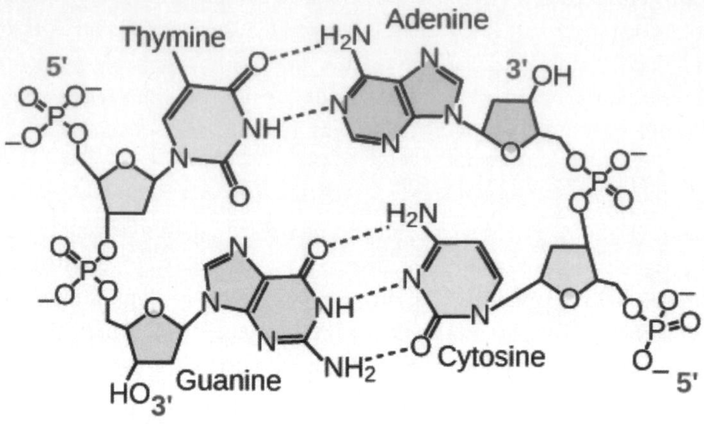

Figure 3.33 In a double stranded DNA molecule, the two strands run antiparallel to one another so that one strand runs 5′ to 3′ and the other 3′ to 5′. The phosphate backbone is located on the outside, and the bases are in the middle. Adenine forms hydrogen bonds (or base pairs) with thymine, and guanine base pairs with cytosine.

A mutation occurs, and adenine replaces cytosine. What impact do you think this will have on the DNA structure?

RNA

Ribonucleic acid, or RNA, is mainly involved in the process of protein synthesis under the direction of DNA. RNA is usually single-stranded and is comprised of ribonucleotides that are linked by phosphodiester bonds. A ribonucleotide in the RNA chain contains ribose (the pentose sugar), one of the four nitrogenous bases (A, U, G, and C), and the phosphate group.

There are four major types of RNA: messenger RNA (mRNA), ribosomal RNA (rRNA), transfer RNA (tRNA), and microRNA (miRNA). The first, mRNA, carries the message from DNA, which controls all of the cellular activities in a cell. If a cell requires synthesizing a certain protein, the gene for this product turns "on" and the messenger RNA synthesizes in the nucleus. The RNA base sequence is complementary to the DNA's coding sequence from which it has been copied. However, in RNA, the base T is absent and U is present instead. If the DNA strand has a sequence AATTGCGC, the sequence of the complementary RNA is UUAACGCG. In the cytoplasm, the mRNA interacts with ribosomes and other cellular machinery (Figure 3.34).

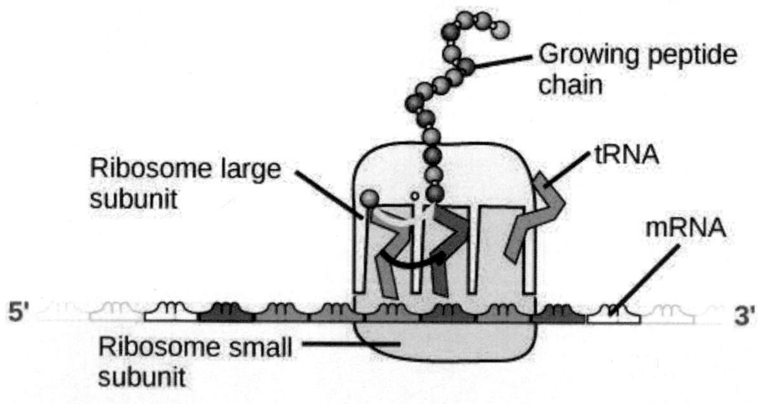

Figure 3.34 A ribosome has two parts: a large subunit and a small subunit. The mRNA sits in between the two subunits. A tRNA molecule recognizes a codon on the mRNA, binds to it by complementary base pairing, and adds the correct amino acid to the growing peptide chain.

The mRNA is read in sets of three bases known as codons. Each codon codes for a single amino acid. In this way, the mRNA is read and the protein product is made. **Ribosomal RNA (rRNA)** is a major constituent of ribosomes on which the mRNA binds. The rRNA ensures the proper alignment of the mRNA and the Ribosomes. The ribosome's rRNA also has an enzymatic activity (peptidyl transferase) and catalyzes peptide bond formation between two aligned amino acids. **Transfer RNA (tRNA)** is one of the smallest of the four types of RNA, usually 70–90 nucleotides long. It carries the correct amino acid to the protein synthesis site. It is the base pairing between the tRNA and mRNA that allows for the correct amino acid to insert itself in the polypeptide chain. MicroRNAs are the smallest RNA molecules and their role involves regulating gene expression by interfering with the expression of certain mRNA messages. Table 3.2 summarizes DNA and RNA features.

DNA and RNA Features

	DNA	RNA
Function	Carries genetic information	Involved in protein synthesis
Location	Remains in the nucleus	Leaves the nucleus
Structure	Double helix	Usually single-stranded
Sugar	Deoxyribose	Ribose
Pyrimidines	Cytosine, thymine	Cytosine, uracil
Purines	Adenine, guanine	Adenine, guanine

Table 3.2

Even though the RNA is single stranded, most RNA types show extensive intramolecular base pairing between complementary sequences, creating a predictable three-dimensional structure essential for their function.

As you have learned, information flow in an organism takes place from DNA to RNA to protein. DNA dictates the structure of mRNA in a process scientists call **transcription**, and RNA dictates the protein's structure in a process scientists call **translation**. This is the Central Dogma of Life, which holds true for all organisms; however, exceptions to the rule occur in connection with viral infections.

LINK TO LEARNING

To learn more about DNA, explore the Howard Hughes Medical Institute BioInteractive animations (http://openstax.org/l/DNA) on the topic of DNA.

KEY TERMS

alpha-helix structure (α-helix) type of secondary protein structure formed by folding the polypeptide into a helix shape with hydrogen bonds stabilizing the structure

amino acid a protein's monomer; has a central carbon or alpha carbon to which an amino group, a carboxyl group, a hydrogen, and an R group or side chain is attached; the R group is different for all 20 common amino acids

beta-pleated sheet (β-pleated) secondary structure in proteins in which hydrogen bonding forms "pleats" between atoms on the polypeptide chain's backbone

biological macromolecule large molecule necessary for life that is built from smaller organic molecules

carbohydrate biological macromolecule in which the ratio of carbon to hydrogen and to oxygen is 1:2:1; carbohydrates serve as energy sources and structural support in cells and form arthropods' cellular exoskeleton

cellulose polysaccharide that comprises the plants' cell wall; provides structural support to the cell

chaperone (also, chaperonin) protein that helps nascent protein in the folding process

chitin type of carbohydrate that forms the outer skeleton of all arthropods that include crustaceans and insects; it also forms fungi cell walls

dehydration synthesis (also, condensation) reaction that links monomer molecules, releasing a water molecule for each bond formed

denaturation loss of shape in a protein as a result of changes in temperature, pH, or chemical exposure

deoxyribonucleic acid (DNA) double-helical molecule that carries the cell's hereditary information

disaccharide two sugar monomers that a glycosidic bond links

enzyme catalyst in a biochemical reaction that is usually a complex or conjugated protein

glycogen storage carbohydrate in animals

glycosidic bond bond formed by a dehydration reaction between two monosaccharides with eliminating a water molecule

hormone chemical signaling molecule, usually protein or steroid, secreted by endocrine cells that act to control or regulate specific physiological processes

hydrolysis reaction that causes breakdown of larger molecules into smaller molecules by utilizing water

lipid macromolecule that is nonpolar and insoluble in water

messenger RNA (mRNA) RNA that carries information from DNA to ribosomes during protein synthesis

monomer smallest unit of larger molecules that are polymers

monosaccharide single unit or monomer of carbohydrates

nucleic acid biological macromolecule that carries the cell's genetic blueprint and carries instructions for the cell's functioning

nucleotide monomer of nucleic acids; contains a pentose sugar, one or more phosphate groups, and a nitrogenous base

omega fat type of polyunsaturated fat that the body requires; numbering the carbon omega starts from the methyl end or the end that is farthest from the carboxylic end

peptide bond bond formed between two amino acids by a dehydration reaction

phosphodiester linkage covalent chemical bond that holds together the polynucleotide chains with a phosphate group linking neighboring nucleotides' two pentose sugars

phospholipid membranes' major constituent; comprised of two fatty acids and a phosphate-containing group attached to a glycerol backbone

polymer chain of monomer residues that covalent bonds link; polymerization is the process of polymer formation from monomers by condensation

polynucleotide long chain of nucleotides

polypeptide long chain of amino acids that peptide bonds link

polysaccharide long chain of monosaccharides; may be branched or unbranched

primary structure linear sequence of amino acids in a protein

protein biological macromolecule comprised of one or more amino acid chains

purine type of nitrogenous base in DNA and RNA; adenine and guanine are purines

pyrimidine type of nitrogenous base in DNA and RNA; cytosine, thymine, and uracil are pyrimidines

quaternary structure association of discrete polypeptide subunits in a protein

ribonucleic acid (RNA) single-stranded, often internally base paired, molecule that is involved in protein synthesis

ribosomal RNA (rRNA) RNA that ensures the proper alignment of the mRNA and the ribosomes during protein synthesis and catalyzes forming the peptide linkage

saturated fatty acid long-chain hydrocarbon with single covalent bonds in the carbon chain; the number of hydrogen atoms attached to the carbon skeleton is maximized

secondary structure regular structure that proteins form by intramolecular hydrogen bonding between the oxygen atom of one amino acid residue and the hydrogen attached to the nitrogen atom of another amino acid residue

starch storage carbohydrate in plants

steroid type of lipid comprised of four fused hydrocarbon rings forming a planar structure

tertiary structure a protein's three-dimensional conformation, including interactions between secondary structural elements; formed from interactions between amino acid side chains

trans fat fat formed artificially by hydrogenating oils, leading to a different arrangement of double bond(s) than those in naturally occurring lipids

transcription process through which messenger RNA forms on a template of DNA

transfer RNA (tRNA) RNA that carries activated amino acids to the site of protein synthesis on the ribosome

translation process through which RNA directs the protein's formation

triacylglycerol (also, triglyceride) fat molecule; consists of three fatty acids linked to a glycerol molecule

unsaturated fatty acid long-chain hydrocarbon that has one or more double bonds in the hydrocarbon chain

wax lipid comprised of a long-chain fatty acid that is esterified to a long-chain alcohol; serves as a protective coating on some feathers, aquatic mammal fur, and leaves

CHAPTER SUMMARY

3.1 Synthesis of Biological Macromolecules

Proteins, carbohydrates, nucleic acids, and lipids are the four major classes of biological macromolecules—large molecules necessary for life that are built from smaller organic molecules. Macromolecules are comprised of single units scientists call monomers that are joined by covalent bonds to form larger polymers. The polymer is more than the sum of its parts: it acquires new characteristics, and leads to an osmotic pressure that is much lower than that formed by its ingredients. This is an important advantage in maintaining cellular osmotic conditions. A monomer joins with another monomer with water molecule release, leading to a covalent bond forming. Scientists call these dehydration or condensation reactions. When polymers break down into smaller units (monomers), they use a water molecule for each bond broken by these reactions. Such reactions are hydrolysis reactions. Dehydration and hydrolysis reactions are similar for all macromolecules, but each monomer and polymer reaction is specific to its class. Dehydration reactions typically require an investment of energy for new bond formation, while hydrolysis reactions typically release energy by breaking bonds.

3.2 Carbohydrates

Carbohydrates are a group of macromolecules that are a vital energy source for the cell and provide structural support to plant cells, fungi, and all of the arthropods that include lobsters, crabs, shrimp, insects, and spiders. Scientists classify carbohydrates as monosaccharides, disaccharides, and polysaccharides depending on the number of monomers in the molecule. Monosaccharides are linked by glycosidic bonds that form as a result of dehydration reactions, forming disaccharides and polysaccharides with eliminating a water molecule for each bond formed. Glucose, galactose, and fructose are common monosaccharides; whereas, common disaccharides include lactose, maltose, and sucrose. Starch and glycogen, examples of polysaccharides, are the storage forms of glucose in plants and animals, respectively. The long polysaccharide chains may be branched or unbranched. Cellulose is an example of an unbranched polysaccharide; whereas, amylopectin, a constituent of starch, is a highly branched molecule. Glucose storage, in the form of polymers like starch of glycogen, makes it slightly less accessible for metabolism; however, this prevents it from leaking out of the cell or creating a high osmotic pressure that could cause the cell to uptake excessive water.

3.3 Lipids

Lipids are a class of macromolecules that are nonpolar and hydrophobic in nature. Major types include fats and oils, waxes, phospholipids, and steroids. Fats are a stored form of energy and are also known as triacylglycerols or triglycerides. Fats are comprised of fatty acids and either glycerol or sphingosine. Fatty acids may be unsaturated or saturated, depending on the presence or absence of double bonds in the hydrocarbon chain. If only single bonds are present, they are saturated fatty acids. Unsaturated fatty acids may have one or more double bonds in the hydrocarbon chain. Phospholipids comprise the membrane's matrix. They have a glycerol or sphingosine backbone to which two fatty acid chains and a phosphate-containing group are attached. Steroids are another class of lipids. Their basic structure has four fused carbon rings. Cholesterol is a type of steroid and is an important constituent of the plasma membrane, where it helps to maintain the membrane's fluid nature. It is also the precursor of steroid hormones such as testosterone.

3.4 Proteins

Proteins are a class of macromolecules that perform a diverse range of functions for the cell. They help in metabolism by acting as enzymes, carriers, or hormones, and provide structural support. The building blocks of proteins (monomers) are amino acids. Each amino acid has a central carbon that bonds to an amino group, a carboxyl group, a hydrogen atom, and an R group or side chain. There are 20 commonly occurring amino acids, each of which differs in the R group. A peptide bond links each amino acid to its neighbors. A long amino acid chain is a polypeptide.

Proteins are organized at four levels: primary, secondary, tertiary, and (optional) quaternary. The primary structure is the amino acids' unique sequence. The polypeptide's local folding to form structures such as the α-helix and β-pleated sheet constitutes the secondary structure. The overall three-dimensional structure is the tertiary structure. When two or more polypeptides combine to form the complete protein structure, the configuration is the protein's quaternary structure. Protein shape and function are intricately linked. Any change in shape caused by changes in temperature or pH may lead to protein denaturation and a loss in function.

3.5 Nucleic Acids

Nucleic acids are molecules comprised of nucleotides that direct cellular activities such as cell division and protein synthesis. Pentose sugar, a nitrogenous base, and a phosphate group comprise each nucleotide. There are two types of nucleic acids: DNA and RNA. DNA carries the cell's genetic blueprint and passes it on from parents to offspring (in the form of chromosomes). It has a double-helical structure with the two strands running in opposite directions, connected by hydrogen bonds, and complementary to each other. RNA is a single-stranded polymer composed of linked nucleotides made up of a pentose sugar (ribose), a nitrogenous base, and a phosphate group. RNA is involved in protein synthesis and its regulation. Messenger RNA (mRNA) copies from the DNA, exports itself from the nucleus to the cytoplasm, and contains information for constructing proteins. Ribosomal RNA (rRNA) is a part of the ribosomes at the site of protein synthesis; whereas, transfer RNA (tRNA) carries the amino acid to the site of protein synthesis. The microRNA regulates using mRNA for protein synthesis.

VISUAL CONNECTION QUESTIONS

1. Figure 3.5 What kind of sugars are these, aldose or ketose?

2. Figure 3.23 Which categories of amino acid would you expect to find on the surface of a soluble protein, and which would you expect to find in the interior? What distribution of amino acids would you expect to find in a protein embedded in a lipid bilayer?

3. Figure 3.33 A mutation occurs, and cytosine is replaced with adenine. What impact do you think this will have on the DNA structure?

REVIEW QUESTIONS

4. Dehydration synthesis leads to formation of
   a. monomers
   b. polymers
   c. water and polymers
   d. none of the above

5. During the breakdown of polymers, which of the following reactions takes place?
   a. hydrolysis
   b. dehydration
   c. condensation
   d. covalent bond

6. The following chemical reactants produce the ester ethyl ethanoate ($C_4H_8O_2$):
   $C_2H_6O + CH_3COOH$
   What type of reaction occurs to make ethyl ethanoate?
   a. condensation
   b. hydrolysis
   c. combustion
   d. acid-base reaction

7. An example of a monosaccharide is ________.
   a. fructose
   b. glucose
   c. galactose
   d. all of the above

8. Cellulose and starch are examples of:
   a. monosaccharides
   b. disaccharides
   c. lipids
   d. polysaccharides

9. Plant cell walls contain which of the following in abundance?
   a. starch
   b. cellulose
   c. glycogen
   d. lactose

10. Lactose is a disaccharide formed by the formation of a ________ bond between glucose and ________.
   a. glycosidic; lactose
   b. glycosidic; galactose
   c. hydrogen; sucrose
   d. hydrogen; fructose

11. Which of the following is not an extracellular matrix role of carbohydrates?
   a. protect an insect's internal organs from external trauma
   b. prevent plant cells from lysing after the plant is watered
   c. maintain the shape of a fungal spore
   d. provide energy for muscle movement

12. Saturated fats have all of the following characteristics except:
   a. they are solid at room temperature
   b. they have single bonds within the carbon chain
   c. they are usually obtained from animal sources
   d. they tend to dissolve in water easily

13. Phospholipids are important components of ________.
   a. the plasma membrane of cells
   b. the ring structure of steroids
   c. the waxy covering on leaves
   d. the double bond in hydrocarbon chains

14. Cholesterol is an integral part of plasma membranes. Based on its structure, where is it found in the membrane?
   a. on the extracellular surface
   b. embedded with the phospholipid heads
   c. within the tail bilayer
   d. attached to the intracellular surface

15. The monomers that make up proteins are called ________.
   a. nucleotides
   b. disaccharides
   c. amino acids
   d. chaperones

16. The α-helix and the β-pleated sheet are part of which protein structure?
   a. primary
   b. secondary
   c. tertiary
   d. quaternary

17. Mad cow disease is an infectious disease where one misfolded protein causes all other copies of the protein to begin misfolding. This is an example of a disease impacting ____ structure.
   a. primary
   b. secondary
   c. tertiary
   d. quaternary

18. A nucleotide of DNA may contain ________.
   a. ribose, uracil, and a phosphate group
   b. deoxyribose, uracil, and a phosphate group
   c. deoxyribose, thymine, and a phosphate group
   d. ribose, thymine, and a phosphate group

19. The building blocks of nucleic acids are ________.
   a. sugars
   b. nitrogenous bases
   c. peptides
   d. nucleotides

20. How does the double helix structure of DNA support its role in encoding the genome?
   a. The sugar-phosphate backbone provides a template for DNA replication.
   b. tRNA pairing with the template strand creates proteins encoded by the genome.
   c. Complementary base pairing creates a very stable structure.
   d. Complementary base pairing allows for easy editing of both strands of DNA.

CRITICAL THINKING QUESTIONS

21. Why are biological macromolecules considered organic?
22. What role do electrons play in dehydration synthesis and hydrolysis?

23. Amino acids have the generic structure seen below, where R represents different carbon-based side chains.

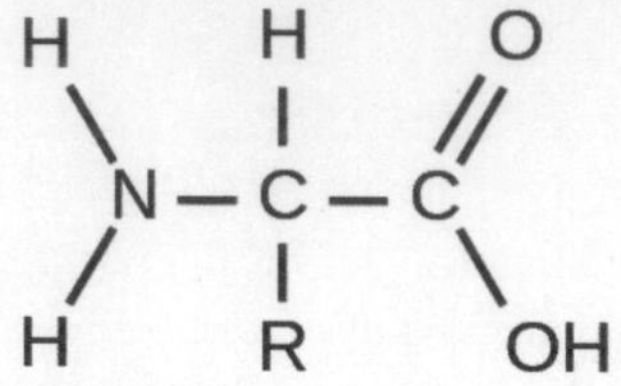

Describe how the structure of amino acids allows them to be linked into long peptide chains to form proteins.

24. Describe the similarities and differences between glycogen and starch.
25. Why is it impossible for humans to digest food that contains cellulose?
26. Draw the ketose and aldose forms of a monosaccharide with the chemical formula $C_3H_6O_3$. How is the structure of the monosaccharide changed from one form to the other in the human body?
27. Explain at least three functions that lipids serve in plants and/or animals.
28. Why have trans fats been banned from some restaurants? How are they created?
29. Why are fatty acids better than glycogen for storing large amounts of chemical energy?
30. Part of cortisol's role in the body involves passing through the plasma membrane to initiate signaling inside a cell. Describe how the structures of cortisol and the plasma membrane allow this to occur.
31. Explain what happens if even one amino acid is substituted for another in a polypeptide chain. Provide a specific example.
32. Describe the differences in the four protein structures.
33. Aquaporins are proteins embedded in the plasma membrane that allow water molecules to move between the extracellular matrix and the intracellular space. Based on its function and location, describe the key features of the protein's shape and the chemical characteristics of its amino acids.
34. What are the structural differences between RNA and DNA?
35. What are the four types of RNA and how do they function?

CHAPTER 4
Cell Structure

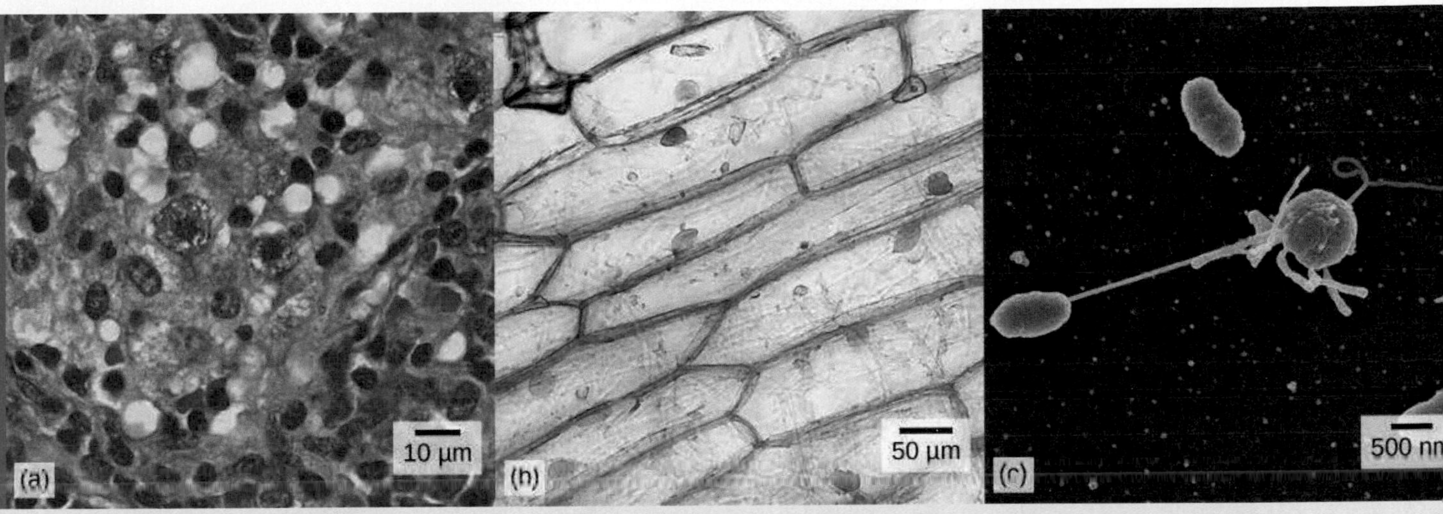

Figure 4.1 (a) Nasal sinus cells (viewed with a light microscope), (b) onion cells (viewed with a light microscope), and (c) *Vibrio tasmaniensis* bacterial cells (seen through a scanning electron microscope) are from very different organisms, yet all share certain basic cell structure characteristics. (credit a: modification of work by Ed Uthman, MD; credit b: modification of work by Umberto Salvagnin; credit c: modification of work by Anthony D'Onofrio, William H. Fowle, Eric J. Stewart, and Kim Lewis of the Lewis Lab at Northeastern University; scale-bar data from Matt Russell)

INTRODUCTION Close your eyes and picture a brick wall. What is the wall's basic building block? It is a single brick. Like a brick wall, cells are the building blocks that make up your body.

Your body has many kinds of cells, each specialized for a specific purpose. Just as we use a variety of materials to build a home, the human body is constructed from many cell types. For example, epithelial cells protect the body's surface and cover the organs and body cavities within. Bone cells help to support and protect the body. Immune system cells fight invading bacteria. Additionally, blood and blood cells carry nutrients and oxygen throughout the body while removing carbon dioxide. Each of these cell types plays a vital role during the body's growth, development, and day-to-day maintenance. In spite of their enormous variety, however, cells from all organisms—even ones as diverse as bacteria, onion, and human—share certain fundamental characteristics.

Chapter Outline

4.1 Studying Cells

By the end of this section, you will be able to do the following:

- Describe the role of cells in organisms
- Compare and contrast light microscopy and electron microscopy
- Summarize cell theory

A cell is the smallest unit of a living thing. Whether comprised of one cell (like bacteria) or many cells (like a human), we call it an organism. Thus, cells are the basic building blocks of all organisms.

Several cells of one kind that interconnect with each other and perform a shared function form tissues. These tissues combine to form an organ (your stomach, heart, or brain), and several organs comprise an organ system (such as the digestive system, circulatory system, or nervous system). Several systems that function together form an organism (like a human being). Here, we will examine the structure and function of cells.

There are many types of cells, which scientists group into one of two broad categories: prokaryotic and eukaryotic. For example, we classify both animal and plant cells as eukaryotic cells; whereas, we classify bacterial cells as prokaryotic. Before discussing the criteria for determining whether a cell is prokaryotic or eukaryotic, we will first examine how biologists study cells.

Microscopy

Cells vary in size. With few exceptions, we cannot see individual cells with the naked eye, so scientists use microscopes (micro- = "small"; -scope = "to look at") to study them. A **microscope** is an instrument that magnifies an object. We photograph most cells with a microscope, so we can call these images micrographs.

The optics of a microscope's lenses change the image orientation that the user sees. A specimen that is right-side up and facing right on the microscope slide will appear upside-down and facing left when one views through a microscope, and vice versa. Similarly, if one moves the slide left while looking through the microscope, it will appear to move right, and if one moves it down, it will seem to move up. This occurs because microscopes use two sets of lenses to magnify the image. Because of the manner by which light travels through the lenses, this two lens system produces an inverted image (binocular, or dissecting microscopes, work in a similar manner, but include an additional magnification system that makes the final image appear to be upright).

Light Microscopes

To give you a sense of cell size, a typical human red blood cell is about eight millionths of a meter or eight micrometers (abbreviated as eight µm) in diameter. A pin head is about two thousandths of a meter (two mm) in diameter. That means about 250 red blood cells could fit on a pinhead.

Most student microscopes are **light microscopes** (Figure 4.2a). Visible light passes and bends through the lens system to enable the user to see the specimen. Light microscopes are advantageous for viewing living organisms, but since individual cells are generally transparent, their components are not distinguishable unless they are colored with special stains. Staining, however, usually kills the cells.

Light microscopes that undergraduates commonly use in the laboratory magnify up to approximately 400 times. Two parameters that are important in microscopy are magnification and resolving power. Magnification is the process of enlarging an object in appearance. Resolving power is the microscope's ability to distinguish two adjacent structures as separate: the higher the resolution, the better the image's clarity and detail. When one uses oil immersion lenses to study small objects, magnification usually increases to 1,000 times. In order to gain a better understanding of cellular structure and function, scientists typically use electron microscopes.

Figure 4.2 (a) Most light microscopes in a college biology lab can magnify cells up to approximately 400 times and have a resolution of about 200 nanometers. (b) Electron microscopes provide a much higher magnification, 100,000x, and a have a resolution of 50 picometers. (credit a: modification of work by "GcG"/Wikimedia Commons; credit b: modification of work by Evan Bench)

Electron Microscopes

In contrast to light microscopes, **electron microscopes** (Figure 4.2b) use a beam of electrons instead of a beam of light. Not only does this allow for higher magnification and, thus, more detail (Figure 4.3), it also provides higher resolving power. The method to prepare the specimen for viewing with an electron microscope kills the specimen. Electrons have short wavelengths (shorter than photons) that move best in a vacuum, so we cannot view living cells with an electron microscope.

In a scanning electron microscope, a beam of electrons moves back and forth across a cell's surface, creating details of cell surface characteristics. In a transmission electron microscope, the electron beam penetrates the cell and provides details of a cell's internal structures. As you might imagine, electron microscopes are significantly more bulky and expensive than light microscopes.

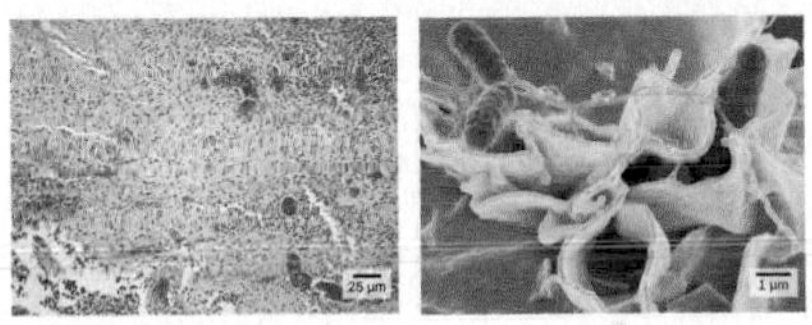

Figure 4.3 (a) These *Salmonella* bacteria appear as tiny purple dots when viewed with a light microscope. (b) This scanning electron microscope micrograph shows *Salmonella* bacteria (in red) invading human cells (yellow). Even though subfigure (b) shows a different *Salmonella* specimen than subfigure (a), you can still observe the comparative increase in magnification and detail. (credit a: modification of work by CDC/Armed Forces Institute of Pathology, Charles N. Farmer, Rocky Mountain Laboratories; credit b: modification of work by NIAID, NIH; scale-bar data from Matt Russell)

LINK TO LEARNING

For another perspective on cell size, try the HowBig interactive at this site (http://openstax.org/l/cell_sizes).

Cell Theory

The microscopes we use today are far more complex than those that Dutch shopkeeper Antony van Leeuwenhoek, used in the 1600s. Skilled in crafting lenses, van Leeuwenhoek observed the movements of single-celled organisms, which he collectively termed "animalcules."

In the 1665 publication *Micrographia*, experimental scientist Robert Hooke coined the term "cell" for the box-like structures he observed when viewing cork tissue through a lens. In the 1670s, van Leeuwenhoek discovered bacteria and protozoa. Later advances in lenses, microscope construction, and staining techniques enabled other scientists to see some components inside cells.

By the late 1830s, botanist Matthias Schleiden and zoologist Theodor Schwann were studying tissues and proposed the **unified cell theory**, which states that one or more cells comprise all living things, the cell is the basic unit of life, and new cells arise from existing cells. Rudolf Virchow later made important contributions to this theory.

CAREER CONNECTION

Cytotechnologist

Have you ever heard of a medical test called a Pap smear (Figure 4.4)? In this test, a doctor takes a small sample of cells from the patient's uterine cervix and sends it to a medical lab where a cytotechnologist stains the cells and examines them for any changes that could indicate cervical cancer or a microbial infection.

Cytotechnologists (cyto- = "cell") are professionals who study cells via microscopic examinations and other laboratory tests. They are trained to determine which cellular changes are within normal limits and which are abnormal. Their focus is not limited to cervical cells. They study cellular specimens that come from all organs. When they notice abnormalities, they consult a pathologist, a medical doctor who interprets and diagnoses changes that disease in body tissue and fluids cause.

Cytotechnologists play a vital role in saving people's lives. When doctors discover abnormalities early, a patient's treatment can begin sooner, which usually increases the chances of a successful outcome.

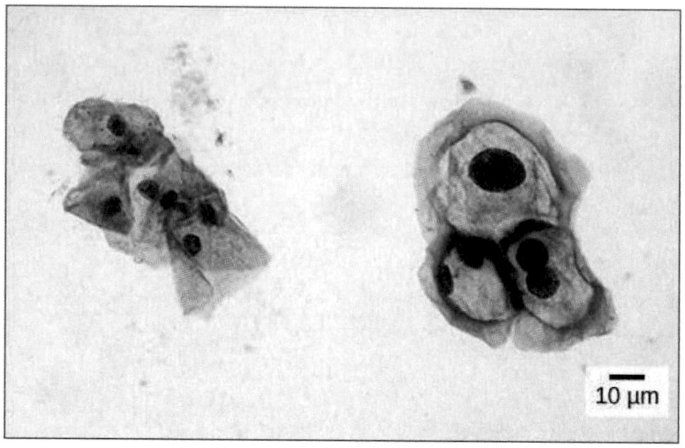

Figure 4.4 These uterine cervix cells, viewed through a light microscope, are from a Pap smear. Normal cells are on the left. The cells on the right are infected with human papillomavirus (HPV). Notice that the infected cells are larger. Also, two of these cells each have two nuclei instead of one, the normal number. (credit: modification of work by Ed Uthman, MD; scale-bar data from Matt Russell)

4.2 Prokaryotic Cells

By the end of this section, you will be able to do the following:

- Name examples of prokaryotic and eukaryotic organisms
- Compare and contrast prokaryotic and eukaryotic cells
- Describe the relative sizes of different cells
- Explain why cells must be small

Cells fall into one of two broad categories: prokaryotic and eukaryotic. We classify only the predominantly single-celled organisms Bacteria and Archaea as prokaryotes (pro- = "before"; -kary- = "nucleus"). Animal cells, plants, fungi, and protists are all eukaryotes (eu- = "true").

Components of Prokaryotic Cells

All cells share four common components: 1) a plasma membrane, an outer covering that separates the cell's interior from its surrounding environment; 2) cytoplasm, consisting of a jelly-like cytosol within the cell in which there are other cellular components; 3) DNA, the cell's genetic material; and 4) ribosomes, which synthesize proteins. However, prokaryotes differ from eukaryotic cells in several ways.

A **prokaryote** is a simple, mostly single-celled (unicellular) organism that lacks a nucleus, or any other membrane-bound organelle. We will shortly come to see that this is significantly different in eukaryotes. Prokaryotic DNA is in the cell's central part: the **nucleoid** (Figure 4.5).

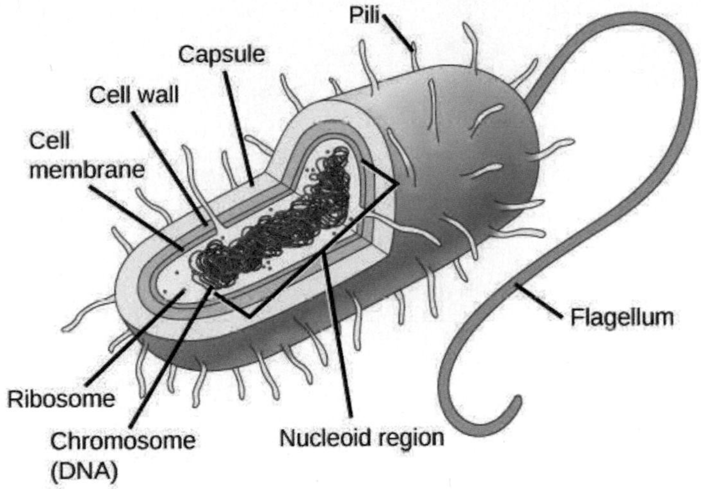

Figure 4.5 This figure shows the generalized structure of a prokaryotic cell. All prokaryotes have chromosomal DNA localized in a nucleoid, ribosomes, a cell membrane, and a cell wall. The other structures shown are present in some, but not all, bacteria.

Most prokaryotes have a peptidoglycan cell wall and many have a polysaccharide capsule (Figure 4.5). The cell wall acts as an extra layer of protection, helps the cell maintain its shape, and prevents dehydration. The capsule enables the cell to attach to surfaces in its environment. Some prokaryotes have flagella, pili, or fimbriae. Flagella are used for locomotion. Pili exchange genetic material during conjugation, the process by which one bacterium transfers genetic material to another through direct contact. Bacteria use fimbriae to attach to a host cell.

CAREER CONNECTION

Microbiologist

The most effective action anyone can take to prevent the spread of contagious illnesses is to wash his or her hands. Why? Because microbes (organisms so tiny that they can only be seen with microscopes) are ubiquitous. They live on doorknobs, money, your hands, and many other surfaces. If someone sneezes into his hand and touches a doorknob, and afterwards you touch that same doorknob, the microbes from the sneezer's mucus are now on your hands. If you touch your hands to your mouth, nose, or eyes, those microbes can enter your body and could make you sick.

However, not all microbes (also called microorganisms) cause disease; most are actually beneficial. You have microbes in your gut that make vitamin K. Other microorganisms are used to ferment beer and wine.

Microbiologists are scientists who study microbes. Microbiologists can pursue a number of careers. Not only do they work in the food industry, they are also employed in the veterinary and medical fields. They can work in the pharmaceutical sector, serving key roles in research and development by identifying new antibiotic sources that can treat bacterial infections.

Environmental microbiologists may look for new ways to use specially selected or genetically engineered microbes to remove pollutants from soil or groundwater, as well as hazardous elements from contaminated sites. We call using these microbes bioremediation technologies. Microbiologists can also work in the bioinformatics field, providing specialized knowledge and insight for designing, developing, and specificity of computer models of, for example, bacterial epidemics.

Cell Size

At 0.1 to 5.0 µm in diameter, prokaryotic cells are significantly smaller than eukaryotic cells, which have diameters ranging from 10 to 100 µm (Figure 4.6). The prokaryotes' small size allows ions and organic molecules that enter them to quickly diffuse to other parts of the cell. Similarly, any wastes produced within a prokaryotic cell can quickly diffuse. This is not the case in eukaryotic cells, which have developed different structural adaptations to enhance intracellular transport.

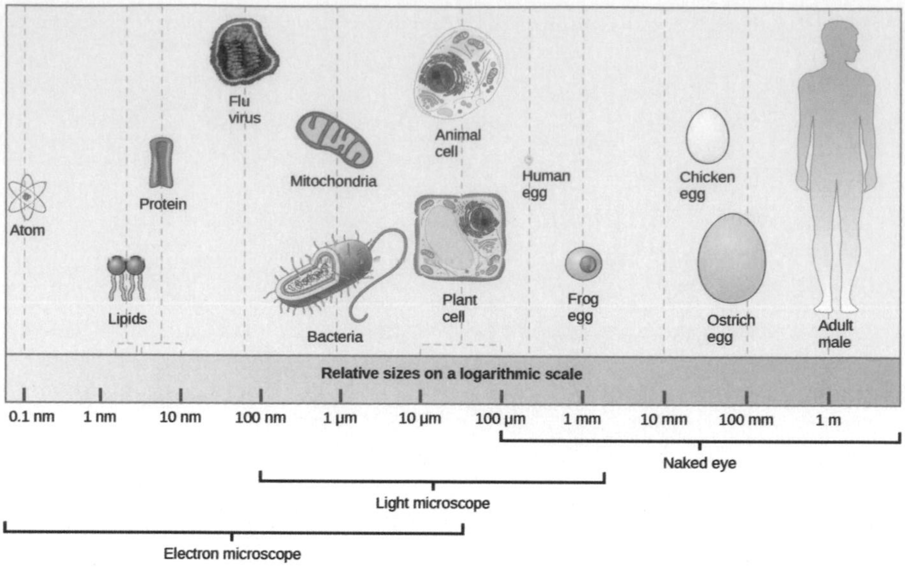

Figure 4.6 This figure shows relative sizes of microbes on a logarithmic scale (recall that each unit of increase in a logarithmic scale represents a 10-fold increase in the quantity measured).

Small size, in general, is necessary for all cells, whether prokaryotic or eukaryotic. Let's examine why that is so. First, we'll consider the area and volume of a typical cell. Not all cells are spherical in shape, but most tend to approximate a sphere. You may remember from your high school geometry course that the formula for the surface area of a sphere is $4\pi r^2$, while the formula for its volume is $4\pi r^3/3$. Thus, as the radius of a cell increases, its surface area increases as the square of its radius, but its volume increases as the cube of its radius (much more rapidly). Therefore, as a cell increases in size, its surface area-to-volume ratio decreases. This same principle would apply if the cell had a cube shape (Figure 4.7). If the cell grows too large, the plasma membrane will not have sufficient surface area to support the rate of diffusion required for the increased volume. In other words, as a cell grows, it becomes less efficient. One way to become more efficient is to divide. Other ways are to increase surface area by foldings of the cell membrane, become flat or thin and elongated, or develop organelles that perform specific tasks. These adaptations lead to developing more sophisticated cells, which we call eukaryotic cells.

VISUAL CONNECTION

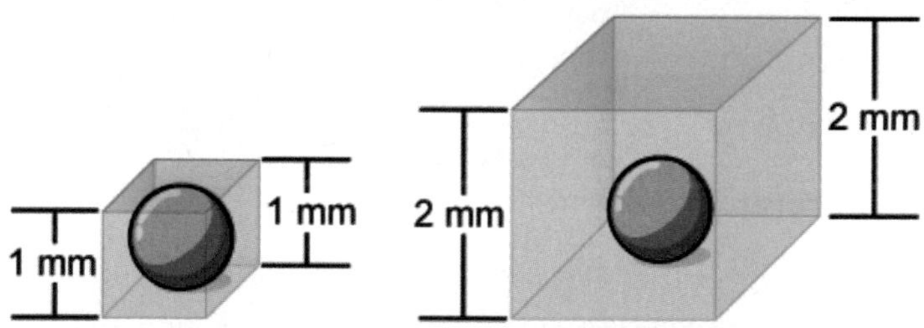

Figure 4.7 Notice that as a cell increases in size, its surface area-to-volume ratio decreases. When there is insufficient surface area to support a cell's increasing volume, a cell will either divide or die. The cell on the left has a volume of 1 mm^3 and a surface area of 6 mm^2, with a surface area-to-volume ratio of 6 to 1; whereas, the cell on the right has a volume of 8 mm^3 and a surface area of 24 mm^2, with a surface area-to-volume ratio of 3 to 1.

Prokaryotic cells are much smaller than eukaryotic cells. What advantages might small cell size confer on a cell? What advantages might large cell size have?

4.3 Eukaryotic Cells

By the end of this section, you will be able to do the following:

- Describe the structure of eukaryotic cells
- Compare animal cells with plant cells
- State the role of the plasma membrane
- Summarize the functions of the major cell organelles

Have you ever heard the phrase "form follows function?" It's a philosophy that many industries follow. In architecture, this means that buildings should be constructed to support the activities that will be carried out inside them. For example, a skyscraper should include several elevator banks. A hospital should have its emergency room easily accessible.

Our natural world also utilizes the principle of form following function, especially in cell biology, and this will become clear as we explore eukaryotic cells (Figure 4.8). Unlike prokaryotic cells, **eukaryotic cells** have: 1) a membrane-bound nucleus; 2) numerous membrane-bound **organelles** such as the endoplasmic reticulum, Golgi apparatus, chloroplasts, mitochondria, and others; and 3) several, rod-shaped chromosomes. Because a membrane surrounds eukaryotic cell's nucleus, it has a "true nucleus." The word "organelle" means "little organ," and, as we already mentioned, organelles have specialized cellular functions, just as your body's organs have specialized functions.

At this point, it should be clear to you that eukaryotic cells have a more complex structure than prokaryotic cells. Organelles allow different functions to be compartmentalized in different areas of the cell. Before turning to organelles, let's first examine two important components of the cell: the plasma membrane and the cytoplasm

VISUAL CONNECTION

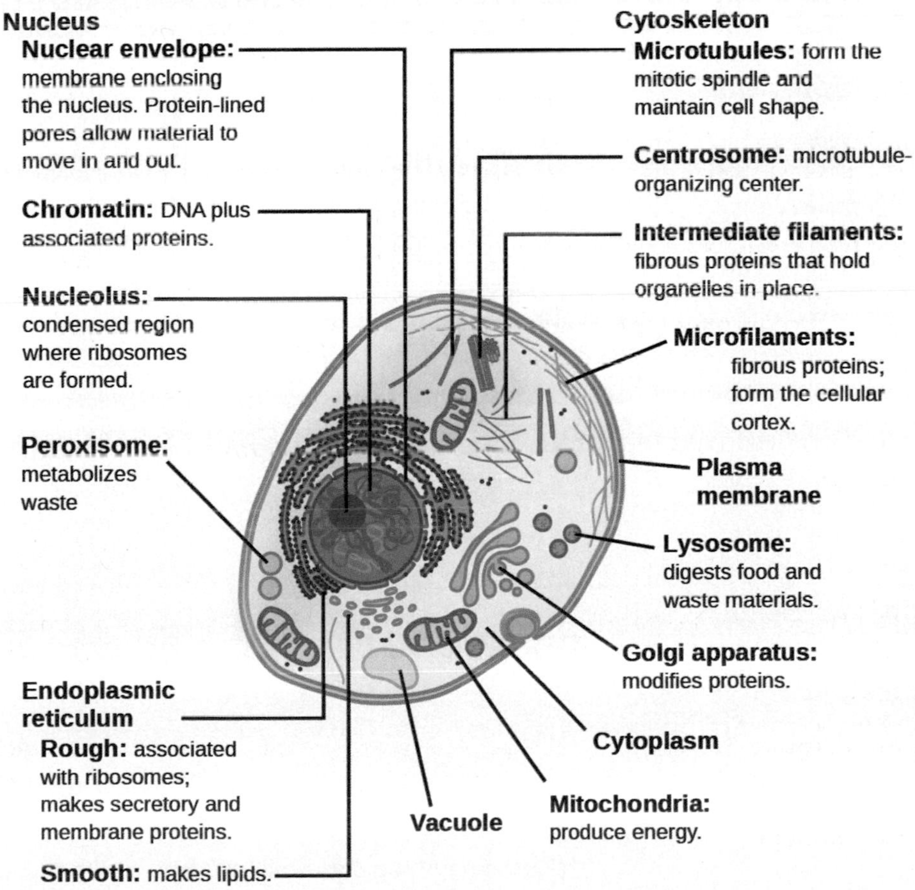

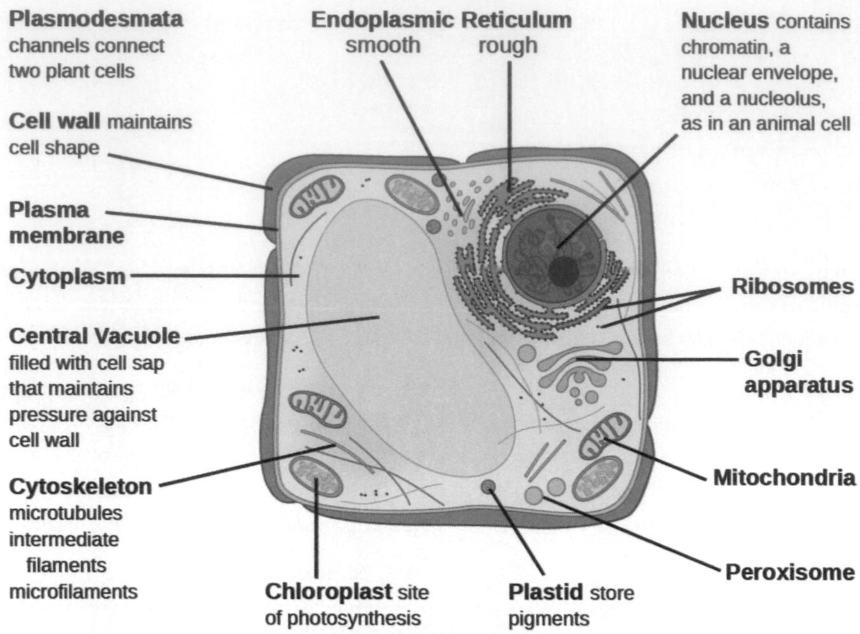

Figure 4.8 These figures show the major organelles and other cell components of (a) a typical animal cell and (b) a typical eukaryotic plant cell. The plant cell has a cell wall, chloroplasts, plastids, and a central vacuole—structures not in animal cells. Most cells do not have lysosomes or centrosomes.

If the nucleolus were not able to carry out its function, what other cellular organelles would be affected?

The Plasma Membrane

Like prokaryotes, eukaryotic cells have a **plasma membrane** (Figure 4.9), a phospholipid bilayer with embedded proteins that separates the internal contents of the cell from its surrounding environment. A phospholipid is a lipid molecule with two fatty acid chains and a phosphate-containing group. The plasma membrane controls the passage of organic molecules, ions, water, and oxygen into and out of the cell. Wastes (such as carbon dioxide and ammonia) also leave the cell by passing through the plasma membrane.

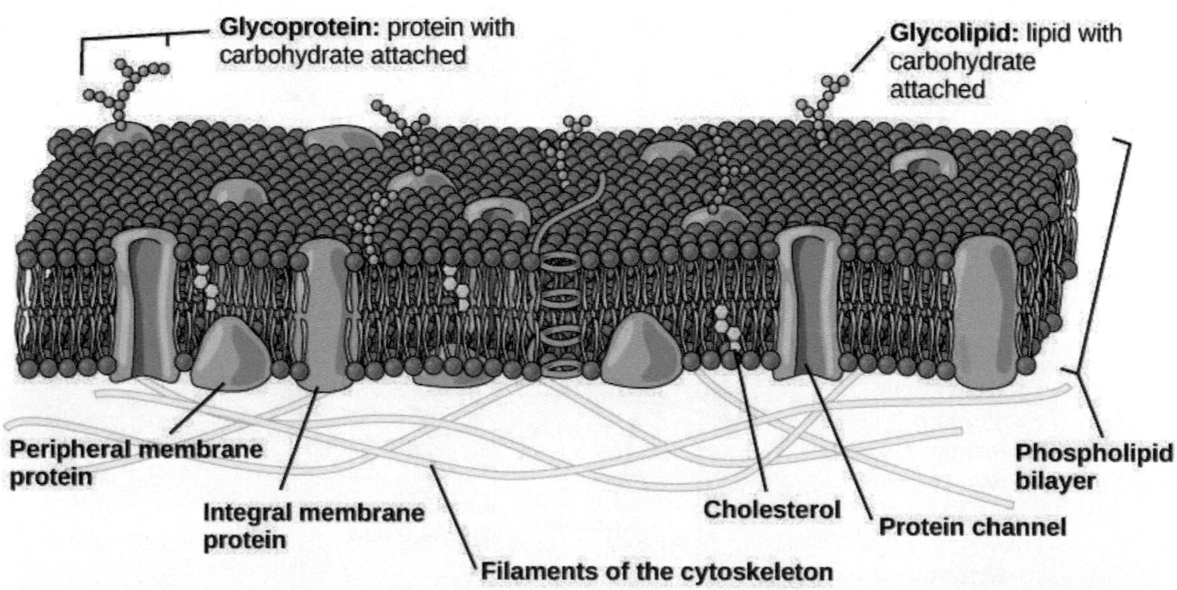

Figure 4.9 The eukaryotic plasma membrane is a phospholipid bilayer with proteins and cholesterol embedded in it.

The plasma membranes of cells that specialize in absorption fold into fingerlike projections that we call microvilli (singular =

microvillus); (Figure 4.10). Such cells typically line the small intestine, the organ that absorbs nutrients from digested food. This is an excellent example of form following function. People with celiac disease have an immune response to gluten, which is a protein in wheat, barley, and rye. The immune response damages microvilli, and thus, afflicted individuals cannot absorb nutrients. This leads to malnutrition, cramping, and diarrhea. Patients suffering from celiac disease must follow a gluten-free diet.

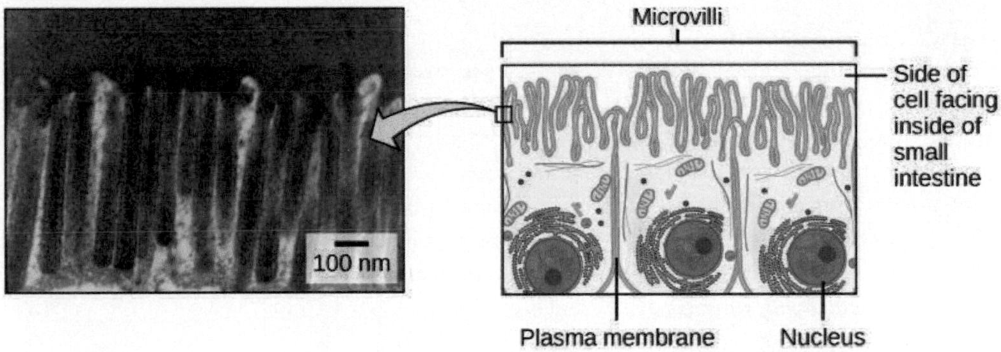

Figure 4.10 Microvilli, as they appear on cells lining the small intestine, increase the surface area available for absorption. These microvilli are only on the area of the plasma membrane that faces the cavity from which substances will be absorbed. (credit "micrograph": modification of work by Louisa Howard)

The Cytoplasm

The **cytoplasm** is the cell's entire region between the plasma membrane and the nuclear envelope (a structure we will discuss shortly). It is comprised of organelles suspended in the gel-like **cytosol**, the cytoskeleton, and various chemicals (Figure 4.8). Even though the cytoplasm consists of 70 to 80 percent water, it has a semi-solid consistency, which comes from the proteins within it. However, proteins are not the only organic molecules in the cytoplasm. Glucose and other simple sugars, polysaccharides, amino acids, nucleic acids, fatty acids, and derivatives of glycerol are also there. Ions of sodium, potassium, calcium, and many other elements also dissolve in the cytoplasm. Many metabolic reactions, including protein synthesis, take place in the cytoplasm.

The Nucleus

Typically, the nucleus is the most prominent organelle in a cell (Figure 4.8). The **nucleus** (plural = nuclei) houses the cell's DNA and directs the synthesis of ribosomes and proteins. Let's look at it in more detail (Figure 4.11).

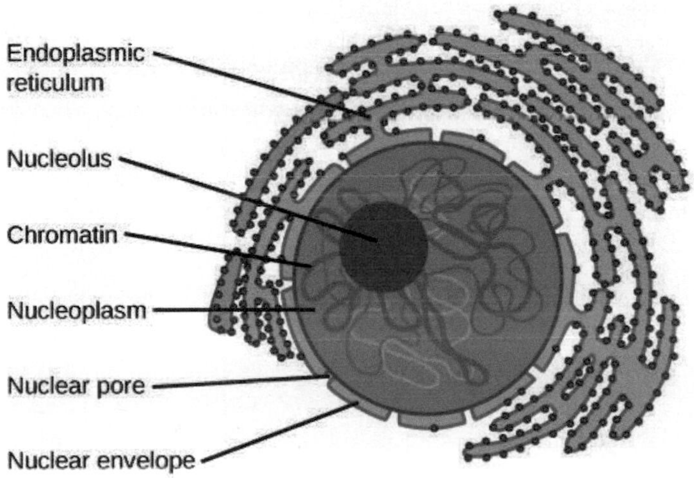

Figure 4.11 The nucleus stores chromatin (DNA plus proteins) in a gel-like substance called the nucleoplasm. The nucleolus is a condensed chromatin region where ribosome synthesis occurs. We call the nucleus' boundary the nuclear envelope. It consists of two phospholipid bilayers: an outer and an inner membrane. The nuclear membrane is continuous with the endoplasmic reticulum. Nuclear pores allow substances to enter and exit the nucleus.

The Nuclear Envelope

The **nuclear envelope** is a double-membrane structure that constitutes the nucleus' outermost portion (Figure 4.11). Both the nuclear envelope's inner and outer membranes are phospholipid bilayers.

The nuclear envelope is punctuated with pores that control the passage of ions, molecules, and RNA between the nucleoplasm and cytoplasm. The **nucleoplasm** is the semi-solid fluid inside the nucleus, where we find the chromatin and the nucleolus.

Chromatin and Chromosomes

To understand chromatin, it is helpful to first explore **chromosomes**, structures within the nucleus that are made up of DNA, the hereditary material. You may remember that in prokaryotes, DNA is organized into a single circular chromosome. In eukaryotes, chromosomes are linear structures. Every eukaryotic species has a specific number of chromosomes in the nucleus of each cell. For example, in humans, the chromosome number is 46, while in fruit flies, it is eight. Chromosomes are only visible and distinguishable from one another when the cell is getting ready to divide. When the cell is in the growth and maintenance phases of its life cycle, proteins attach to chromosomes, and they resemble an unwound, jumbled bunch of threads. We call these unwound protein-chromosome complexes **chromatin** (Figure 4.12). Chromatin describes the material that makes up the chromosomes both when condensed and decondensed.

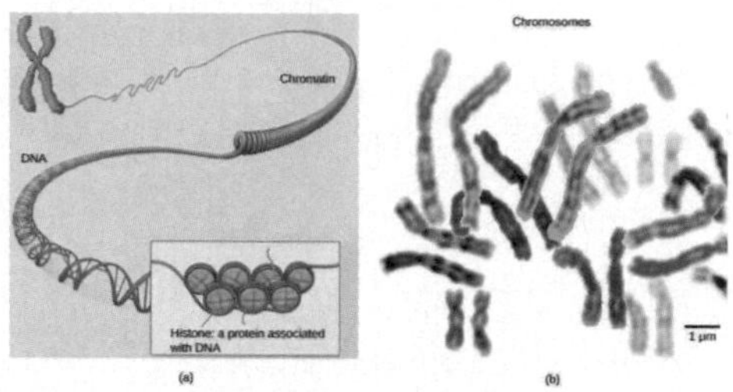

Figure 4.12 (a) This image shows various levels of chromatin's organization (DNA and protein). (b) This image shows paired chromosomes. (credit b: modification of work by NIH; scale-bar data from Matt Russell)

The Nucleolus

We already know that the nucleus directs the synthesis of ribosomes, but how does it do this? Some chromosomes have sections of DNA that encode ribosomal RNA. A darkly staining area within the nucleus called the **nucleolus** (plural = nucleoli) aggregates the ribosomal RNA with associated proteins to assemble the ribosomal subunits that are then transported out through the pores in the nuclear envelope to the cytoplasm.

Ribosomes

Ribosomes are the cellular structures responsible for protein synthesis. When we view them through an electron microscope, ribosomes appear either as clusters (polyribosomes) or single, tiny dots that float freely in the cytoplasm. They may be attached to the plasma membrane's cytoplasmic side or the endoplasmic reticulum's cytoplasmic side and the nuclear envelope's outer membrane (Figure 4.8). Electron microscopy shows us that ribosomes, which are large protein and RNA complexes, consist of two subunits, large and small (Figure 4.13). Ribosomes receive their "orders" for protein synthesis from the nucleus where the DNA transcribes into messenger RNA (mRNA). The mRNA travels to the ribosomes, which translate the code provided by the sequence of the nitrogenous bases in the mRNA into a specific order of amino acids in a protein. Amino acids are the building blocks of proteins.

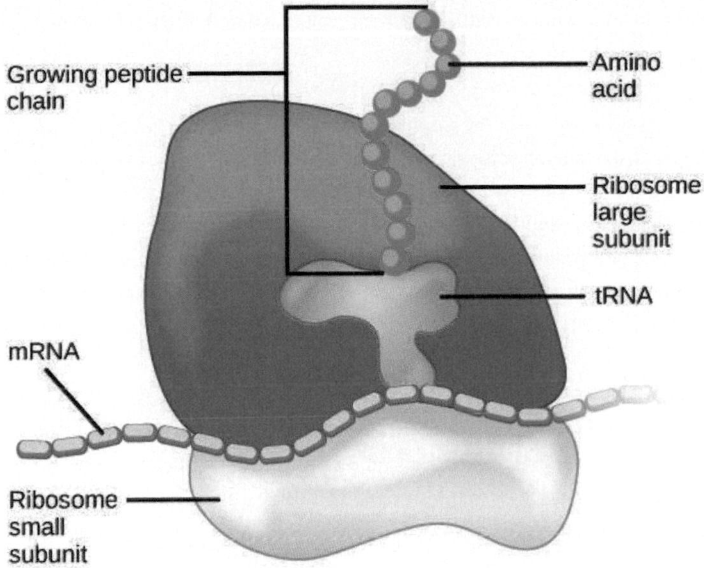

Figure 4.13 A large subunit (top) and a small subunit (bottom) comprise ribosomes. During protein synthesis, ribosomes assemble amino acids into proteins.

Because protein synthesis is an essential function of all cells (including enzymes, hormones, antibodies, pigments, structural components, and surface receptors), there are ribosomes in practically every cell. Ribosomes are particularly abundant in cells that synthesize large amounts of protein. For example, the pancreas is responsible for creating several digestive enzymes and the cells that produce these enzymes contain many ribosomes. Thus, we see another example of form following function.

Mitochondria

Scientists often call **mitochondria** (singular = mitochondrion) "powerhouses" or "energy factories" of both plant and animal cells because they are responsible for making adenosine triphosphate (ATP), the cell's main energy-carrying molecule. ATP represents the cell's short-term stored energy. Cellular respiration is the process of making ATP using the chemical energy in glucose and other nutrients. In mitochondria, this process uses oxygen and produces carbon dioxide as a waste product. In fact, the carbon dioxide that you exhale with every breath comes from the cellular reactions that produce carbon dioxide as a byproduct.

In keeping with our theme of form following function, it is important to point out that muscle cells have a very high concentration of mitochondria that produce ATP. Your muscle cells need considerable energy to keep your body moving. When your cells don't get enough oxygen, they do not make much ATP. Instead, producing lactic acid accompanies the small amount of ATP they make in the absence of oxygen.

Mitochondria are oval-shaped, double membrane organelles (Figure 4.14) that have their own ribosomes and DNA. Each membrane is a phospholipid bilayer embedded with proteins. The inner layer has folds called cristae. We call the area surrounded by the folds the mitochondrial matrix. The cristae and the matrix have different roles in cellular respiration.

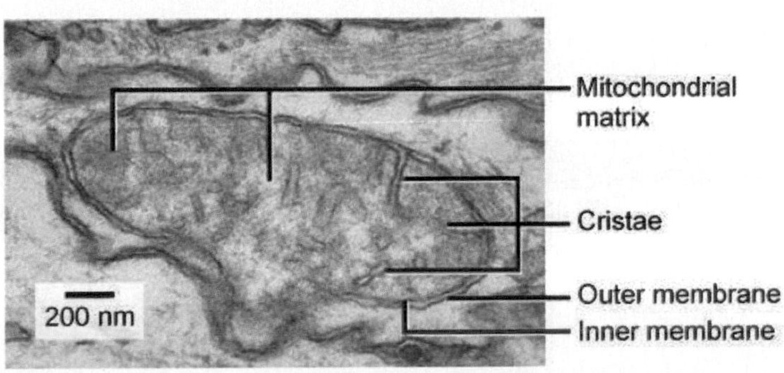

Figure 4.14 This electron micrograph shows a mitochondrion through an electron microscope. This organelle has an outer membrane and an inner membrane. The inner membrane contains folds, called cristae, which increase its surface area. We call the space between the two membranes the intermembrane space, and the space inside the inner membrane the mitochondrial matrix. ATP synthesis takes place on

the inner membrane. (credit: modification of work by Matthew Britton; scale-bar data from Matt Russell)

Peroxisomes

Peroxisomes are small, round organelles enclosed by single membranes. They carry out oxidation reactions that break down fatty acids and amino acids. They also detoxify many poisons that may enter the body. (Many of these oxidation reactions release hydrogen peroxide, H_2O_2, which would be damaging to cells; however, when these reactions are confined to peroxisomes, enzymes safely break down the H_2O_2 into oxygen and water.) For example, peroxisomes in liver cells detoxify alcohol. Glyoxysomes, which are specialized peroxisomes in plants, are responsible for converting stored fats into sugars. Plant cells contain many different types of peroxisomes that play a role in metabolism, pathogene defense, and stress response, to mention a few.

Vesicles and Vacuoles

Vesicles and **vacuoles** are membrane-bound sacs that function in storage and transport. Other than the fact that vacuoles are somewhat larger than vesicles, there is a very subtle distinction between them. Vesicle membranes can fuse with either the plasma membrane or other membrane systems within the cell. Additionally, some agents such as enzymes within plant vacuoles break down macromolecules. The vacuole's membrane does not fuse with the membranes of other cellular components.

Animal Cells versus Plant Cells

At this point, you know that each eukaryotic cell has a plasma membrane, cytoplasm, a nucleus, ribosomes, mitochondria, peroxisomes, and in some, vacuoles, but there are some striking differences between animal and plant cells. While both animal and plant cells have microtubule organizing centers (MTOCs), animal cells also have centrioles associated with the MTOC: a complex we call the centrosome. Animal cells each have a centrosome and lysosomes; whereas, most plant cells do not. Plant cells have a cell wall, chloroplasts and other specialized plastids, and a large central vacuole; whereas, animal cells do not.

The Centrosome

The **centrosome** is a microtubule-organizing center found near the nuclei of animal cells. It contains a pair of centrioles, two structures that lie perpendicular to each other (Figure 4.15). Each centriole is a cylinder of nine triplets of microtubules.

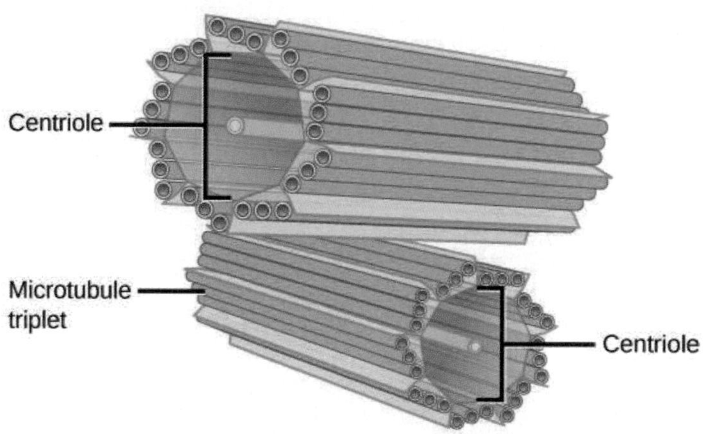

Figure 4.15 The centrosome consists of two centrioles that lie at right angles to each other. Each centriole is a cylinder comprised of nine triplets of microtubules. Nontubulin proteins (indicated by the green lines) hold the microtubule triplets together.

The centrosome (the organelle where all microtubules originate) replicates itself before a cell divides, and the centrioles appear to have some role in pulling the duplicated chromosomes to opposite ends of the dividing cell. However, the centriole's exact function in cell division isn't clear, because cells that have had the centrosome removed can still divide, and plant cells, which lack centrosomes, are capable of cell division.

Lysosomes

Animal cells have another set of organelles that most plant cells do not: lysosomes. The **lysosomes** are the cell's "garbage disposal." In plant cells, the digestive processes take place in vacuoles. Enzymes within the lysosomes aid in breaking down proteins, polysaccharides, lipids, nucleic acids, and even worn-out organelles. These enzymes are active at a much lower pH than the cytoplasm's. Therefore, the pH within lysosomes is more acidic than the cytoplasm's pH. Many reactions that take place in the cytoplasm could not occur at a low pH, so again, the advantage of compartmentalizing the eukaryotic cell into organelles is apparent.

The Cell Wall

If you examine Figure 4.8, the plant cell diagram, you will see a structure external to the plasma membrane. This is the **cell wall**, a rigid covering that protects the cell, provides structural support, and gives shape to the cell. Fungal and some protistan cells also have cell walls. While the prokaryotic cell walls' chief component is peptidoglycan, the major organic molecule in the plant (and some protists') cell wall is cellulose (Figure 4.16), a polysaccharide comprised of glucose units. Have you ever noticed that when you bite into a raw vegetable, like celery, it crunches? That's because you are tearing the celery cells' rigid cell walls with your teeth.

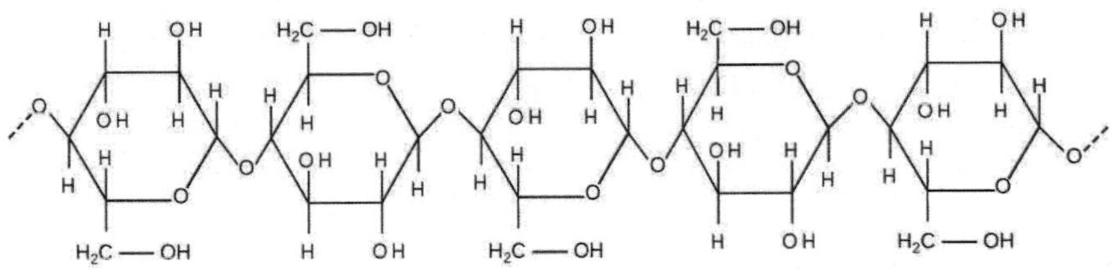

Figure 4.16 Cellulose is a long chain of β-glucose molecules connected by a 1-4 linkage. The dashed lines at each end of the figure indicate a series of many more glucose units. The size of the page makes it impossible to portray an entire cellulose molecule.

Chloroplasts

Like the mitochondria, chloroplasts have their own DNA and ribosomes, but chloroplasts have an entirely different function. **Chloroplasts** are plant cell organelles that carry out photosynthesis. Photosynthesis is the series of reactions that use carbon dioxide, water, and light energy to make glucose and oxygen. This is a major difference between plants and animals. Plants (autotrophs) are able to make their own food, like sugars used in cellular respiration to provide ATP energy generated in the plant mitochondria. Animals (heterotrophs) must ingest their food.

Like mitochondria, chloroplasts have outer and inner membranes, but within the space enclosed by a chloroplast's inner membrane is a set of interconnected and stacked fluid-filled membrane sacs we call thylakoids (Figure 4.17). Each thylakoid stack is a granum (plural = grana). We call the fluid enclosed by the inner membrane that surrounds the grana the stroma.

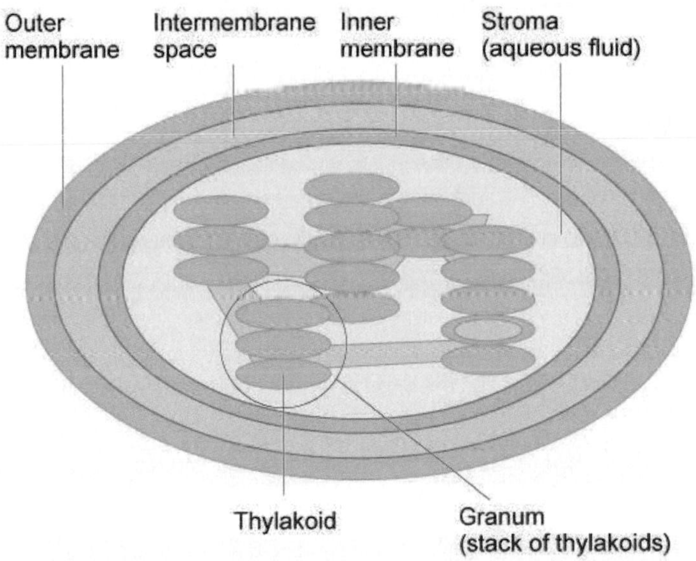

Figure 4.17 The chloroplast has an outer membrane, an inner membrane, and membrane structures - thylakoids that are stacked into grana. We call the space inside the thylakoid membranes the thylakoid space. The light harvesting reactions take place in the thylakoid membranes, and sugar synthesis takes place in the fluid inside the inner membrane, which we call the stroma. Chloroplasts also have their own genome, which is contained on a single circular chromosome.

The chloroplasts contain a green pigment, **chlorophyll**, which captures the light energy that drives the reactions of photosynthesis. Like plant cells, photosynthetic protists also have chloroplasts. Some bacteria perform photosynthesis, but their chlorophyll is not relegated to an organelle.

EVOLUTION CONNECTION

Endosymbiosis

We have mentioned that both mitochondria and chloroplasts contain DNA and ribosomes. Have you wondered why? Strong evidence points to endosymbiosis as the explanation.

Symbiosis is a relationship in which organisms from two separate species depend on each other for their survival. Endosymbiosis (endo- = "within") is a mutually beneficial relationship in which one organism lives inside the other. Endosymbiotic relationships abound in nature. We have already mentioned that microbes that produce vitamin K live inside the human gut. This relationship is beneficial for us because we are unable to synthesize vitamin K. It is also beneficial for the microbes because they are protected from other organisms and from drying out, and they receive abundant food from the environment of the large intestine.

Scientists have long noticed that bacteria, mitochondria, and chloroplasts are similar in size. We also know that bacteria have DNA and ribosomes, just like mitochondria and chloroplasts. Scientists believe that host cells and bacteria formed an endosymbiotic relationship when the host cells ingested both aerobic and autotrophic bacteria (cyanobacteria) but did not destroy them. Through many millions of years of evolution, these ingested bacteria became more specialized in their functions, with the aerobic bacteria becoming mitochondria and the autotrophic bacteria becoming chloroplasts.

The Central Vacuole

Previously, we mentioned vacuoles as essential components of plant cells. If you look at Figure 4.8b, you will see that plant cells each have a large central vacuole that occupies most of the cell's area. The **central vacuole** plays a key role in regulating the cell's concentration of water in changing environmental conditions. Have you ever noticed that if you forget to water a plant for a few days, it wilts? That's because as the water concentration in the soil becomes lower than the water concentration in the plant, water moves out of the central vacuoles and cytoplasm. As the central vacuole shrinks, it leaves the cell wall unsupported. This loss of support to the plant's cell walls results in the wilted appearance.

The central vacuole also supports the cell's expansion. When the central vacuole holds more water, the cell becomes larger without having to invest considerable energy in synthesizing new cytoplasm.

4.4 The Endomembrane System and Proteins

By the end of this section, you will be able to do the following:

- List the components of the endomembrane system
- Recognize the relationship between the endomembrane system and its functions

The endomembrane system (endo = "within") is a group of membranes and organelles (Figure 4.18) in eukaryotic cells that works together to modify, package, and transport lipids and proteins. It includes the nuclear envelope, lysosomes, and vesicles, which we have already mentioned, and the endoplasmic reticulum and Golgi apparatus, which we will cover shortly. Although not technically *within* the cell, the plasma membrane is included in the endomembrane system because, as you will see, it interacts with the other endomembranous organelles. The endomembrane system does not include either mitochondria or chloroplast membranes.

VISUAL CONNECTION

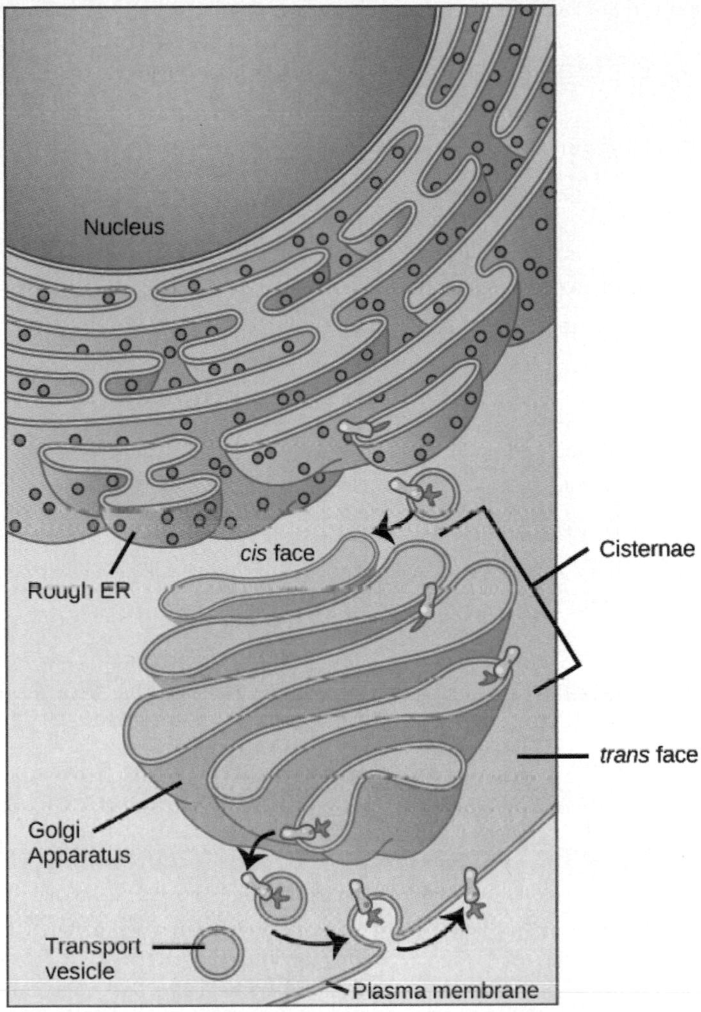

Figure 4.18 Membrane and secretory proteins are synthesized in the rough endoplasmic reticulum (RER). The RER also sometimes modifies proteins. In this illustration, an attachment of a (purple) carbohydrate modifies a (green) integral membrane protein in the ER. Vesicles with the integral protein bud from the ER and fuse with the Golgi apparatus' cis face. As the protein passes along the Golgi's cisternae, the addition of more carbohydrates further modifies it. After its synthesis is complete, it exits as an integral membrane protein of the vesicle that buds from the Golgi's **trans** face. When the vesicle fuses with the cell membrane, the protein becomes an integral portion of that cell membrane. (credit: modification of work by Magnus Manske)

If a peripheral membrane protein were synthesized in the lumen (inside) of the ER, would it end up on the inside or outside of the plasma membrane?

The Endoplasmic Reticulum

The **endoplasmic reticulum (ER)** (Figure 4.18) is a series of interconnected membranous sacs and tubules that collectively modifies proteins and synthesizes lipids. However, these two functions take place in separate areas of the ER: the rough ER and the smooth ER, respectively.

We call the ER tubules' hollow portion the lumen or cisternal space. The ER's membrane, which is a phospholipid bilayer embedded with proteins, is continuous with the nuclear envelope.

Rough ER

Scientists have named the **rough endoplasmic reticulum (RER)** as such because the ribosomes attached to its cytoplasmic surface give it a studded appearance when viewing it through an electron microscope (Figure 4.19).

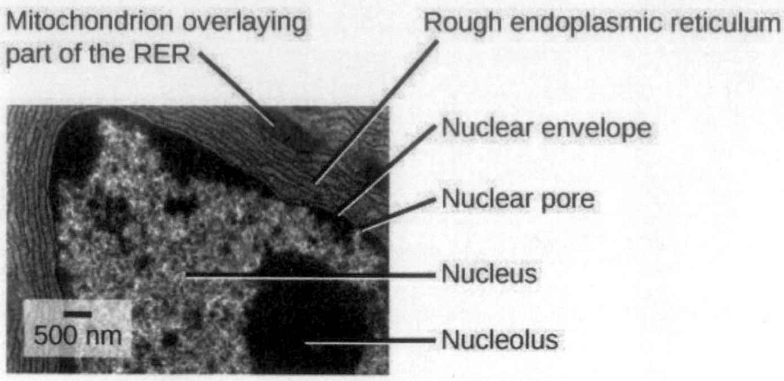

Figure 4.19 This transmission electron micrograph shows the rough endoplasmic reticulum and other organelles in a pancreatic cell. (credit: modification of work by Louisa Howard)

Ribosomes transfer their newly synthesized proteins into the RER's lumen where they undergo structural modifications, such as folding or acquiring side chains. These modified proteins incorporate into cellular membranes—the ER or the ER's or other organelles' membranes. The proteins can also secrete from the cell (such as protein hormones, enzymes). The RER also makes phospholipids for cellular membranes.

If the phospholipids or modified proteins are not destined to stay in the RER, they will reach their destinations via transport vesicles that bud from the RER's membrane (Figure 4.18).

Since the RER is engaged in modifying proteins (such as enzymes, for example) that secrete from the cell, you would be correct in assuming that the RER is abundant in cells that secrete proteins. This is the case with liver cells, for example.

Smooth ER

The **smooth endoplasmic reticulum (SER)** is continuous with the RER but has few or no ribosomes on its cytoplasmic surface (Figure 4.18). SER functions include synthesis of carbohydrates, lipids, and steroid hormones; detoxification of medications and poisons; and storing calcium ions.

In muscle cells, a specialized SER, the sarcoplasmic reticulum, is responsible for storing calcium ions that are needed to trigger the muscle cells' coordinated contractions.

LINK TO LEARNING

You can watch an excellent animation of the endomembrane system here (http://openstax.org/l/insane_in_the_endomembrane) . At the end of the animation, there is a short self-assessment.

CAREER CONNECTION

Cardiologist

Heart disease is the leading cause of death in the United States. This is primarily due to our sedentary lifestyle and our high trans-fat diets.

Heart failure is just one of many disabling heart conditions. Heart failure does not mean that the heart has stopped working. Rather, it means that the heart can't pump with sufficient force to transport oxygenated blood to all the vital organs. Left untreated, heart failure can lead to kidney failure and other organ failure.

Cardiac muscle tissue comprises the heart's wall. Heart failure occurs when cardiac muscle cells' endoplasmic reticula do not function properly. As a result, an insufficient number of calcium ions are available to trigger a sufficient contractile force.

Cardiologists (cardi- = "heart"; -ologist = "one who studies") are doctors who specialize in treating heart diseases, including heart failure. Cardiologists can diagnose heart failure via a physical examination, results from an electrocardiogram (ECG, a test that measures the heart's electrical activity), a chest X-ray to see whether the heart is enlarged, and other tests. If the cardiologist diagnoses heart failure, he or she will typically prescribe appropriate medications and recommend a reduced table salt intake and a supervised exercise program.

The Golgi Apparatus

We have already mentioned that vesicles can bud from the ER and transport their contents elsewhere, but where do the vesicles go? Before reaching their final destination, the lipids or proteins within the transport vesicles still need sorting, packaging, and tagging so that they end up in the right place. Sorting, tagging, packaging, and distributing lipids and proteins takes place in the **Golgi apparatus** (also called the Golgi body), a series of flattened membranes (Figure 4.20).

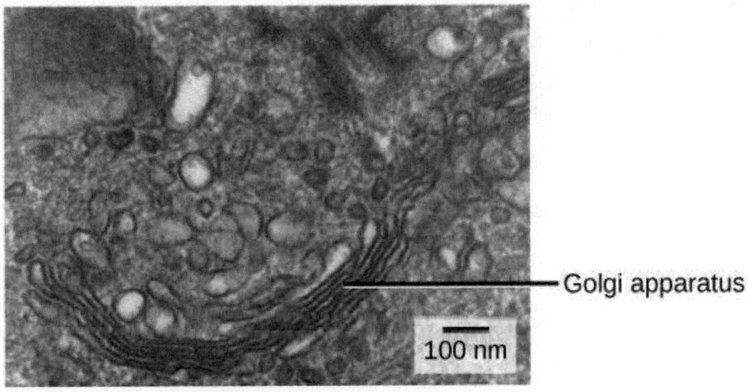

Figure 4.20 The Golgi apparatus in this white blood cell is visible as a stack of semicircular, flattened rings in the lower portion of the image. You can see several vesicles near the Golgi apparatus. (credit: modification of work by Louisa Howard)

We call the Golgi apparatus' the *cis* face. The opposite side is the *trans* face. The transport vesicles that formed from the ER travel to the *cis* face, fuse with it, and empty their contents into the Golgi apparatus' lumen. As the proteins and lipids travel through the Golgi, they undergo further modifications that allow them to be sorted. The most frequent modification is adding short sugar molecule chains. These newly modified proteins and lipids then tag with phosphate groups or other small molecules in order to travel to their proper destinations.

Finally, the modified and tagged proteins are packaged into secretory vesicles that bud from the Golgi's *trans* face. While some of these vesicles deposit their contents into other cell parts where they will be used, other secretory vesicles fuse with the plasma membrane and release their contents outside the cell.

In another example of form following function, cells that engage in a great deal of secretory activity (such as salivary gland cells that secrete digestive enzymes or immune system cells that secrete antibodies) have an abundance of Golgi.

In plant cells, the Golgi apparatus has the additional role of synthesizing polysaccharides, some of which are incorporated into the cell wall and some of which other cell parts use.

CAREER CONNECTION

Geneticist

Many diseases arise from genetic mutations that prevent synthesizing critical proteins. One such disease is Lowe disease (or oculocerebrorenal syndrome, because it affects the eyes, brain, and kidneys). In Lowe disease, there is a deficiency in an enzyme localized to the Golgi apparatus. Children with Lowe disease are born with cataracts, typically develop kidney disease after the first year of life, and may have impaired mental abilities.

A mutation on the X chromosome causes Lowe disease. The X chromosome is one of the two human sex chromosomes, as these chromosomes determine a person's sex. Females possess two X chromosomes while males possess one X and one Y chromosome. In females, the genes on only one of the two X chromosomes are expressed. Females who carry the Lowe disease gene on one of their X chromosomes are carriers and do not show symptoms of the disease. However, males only have one X chromosome and the genes on this chromosome are always expressed. Therefore, males will always have Lowe disease if their X chromosome carries the Lowe disease gene. Geneticists have identified the mutated gene's location, as well as many other mutation locations that cause genetic diseases. Through prenatal testing, a woman can find out if the fetus she is carrying may be afflicted with one of several genetic diseases.

Geneticists analyze prenatal genetic test results and may counsel pregnant women on available options. They may also conduct genetic research that leads to new drugs or foods, or perform DNA analyses for forensic investigations.

Lysosomes

In addition to their role as the digestive component and organelle-recycling facility of animal cells, lysosomes are part of the endomembrane system. Lysosomes also use their hydrolytic enzymes to destroy pathogens (disease-causing organisms) that might enter the cell. A good example of this occurs in macrophages, a group of white blood cells which are part of your body's immune system. In a process that scientists call phagocytosis or endocytosis, a section of the macrophage's plasma membrane invaginates (folds in) and engulfs a pathogen. The invaginated section, with the pathogen inside, then pinches itself off from the plasma membrane and becomes a vesicle. The vesicle fuses with a lysosome. The lysosome's hydrolytic enzymes then destroy the pathogen (Figure 4.21).

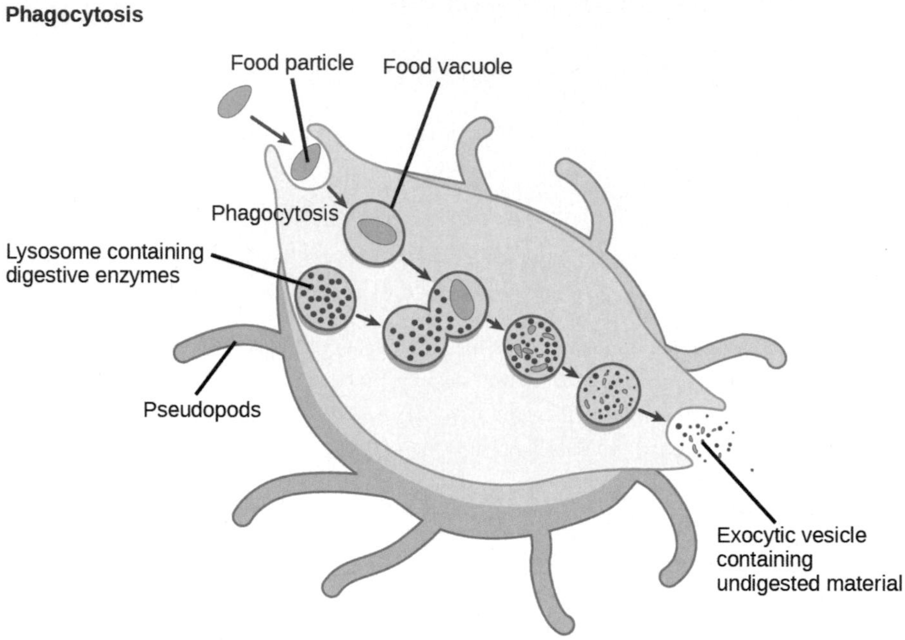

Figure 4.21 A macrophage has engulfed (phagocytized) a potentially pathogenic bacterium and then fuses with lysosomes within the cell to destroy the pathogen. Other organelles are present in the cell but for simplicity we do not show them.

4.5 The Cytoskeleton

By the end of this section, you will be able to do the following:

- Describe the cytoskeleton
- Compare the roles of microfilaments, intermediate filaments, and microtubules
- Compare and contrast cilia and flagella
- Summarize the differences among the components of prokaryotic cells, animal cells, and plant cells

If you were to remove all the organelles from a cell, would the plasma membrane and the cytoplasm be the only components left? No. Within the cytoplasm, there would still be ions and organic molecules, plus a network of protein fibers that help maintain the cell's shape, secure some organelles in specific positions, allow cytoplasm and vesicles to move within the cell, and enable cells within multicellular organisms to move. Collectively, scientists call this network of protein fibers the **cytoskeleton**. There are three types of fibers within the cytoskeleton: microfilaments, intermediate filaments, and microtubules (Figure 4.22). Here, we will examine each.

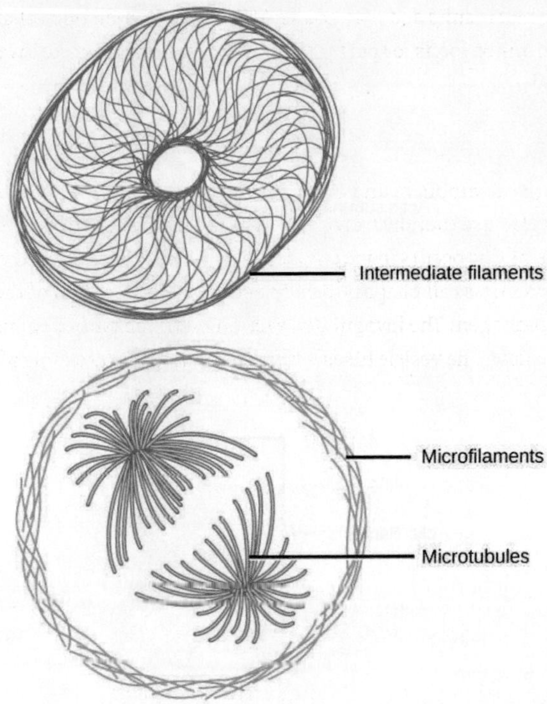

Figure 4.22 Microfilaments thicken the cortex around the cell's inner edge. Like rubber bands, they resist tension. There are microtubules in the cell's interior where they maintain their shape by resisting compressive forces. There are intermediate filaments throughout the cell that hold organelles in place.

Microfilaments

Of the three types of protein fibers in the cytoskeleton, **microfilaments** are the narrowest. They function in cellular movement, have a diameter of about 7 nm, and are comprised of two globular protein intertwined strands, which we call actin (Figure 4.23). For this reason, we also call microfilaments actin filaments.

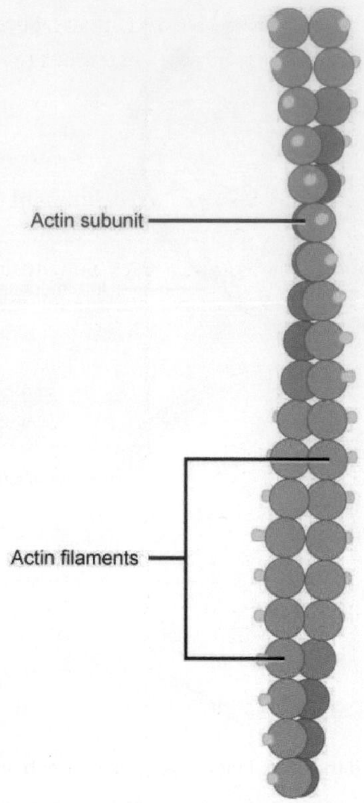

Figure 4.23 Two intertwined actin strands comprise microfilaments.

ATP powers actin to assemble its filamentous form, which serves as a track for the movement of a motor protein we call myosin. This enables actin to engage in cellular events requiring motion, such as cell division in eukaryotic cells and cytoplasmic streaming, which is the cell cytoplasm's circular movement in plant cells. Actin and myosin are plentiful in muscle cells. When your actin and myosin filaments slide past each other, your muscles contract.

Microfilaments also provide some rigidity and shape to the cell. They can depolymerize (disassemble) and reform quickly, thus enabling a cell to change its shape and move. White blood cells (your body's infection-fighting cells) make good use of this ability. They can move to an infection site and phagocytize the pathogen.

LINK TO LEARNING

To see an example of a white blood cell in action, watch a short time-lapse video of the cell capturing two bacteria. It engulfs one and then moves on to the other.

Click to view content (https://www.openstax.org/l/chasing_bcteria)

Intermediate Filaments

Several strands of fibrous proteins that are wound together comprise intermediate filaments (Figure 4.24). Cytoskeleton elements get their name from the fact that their diameter, 8 to 10 nm, is between those of microfilaments and microtubules.

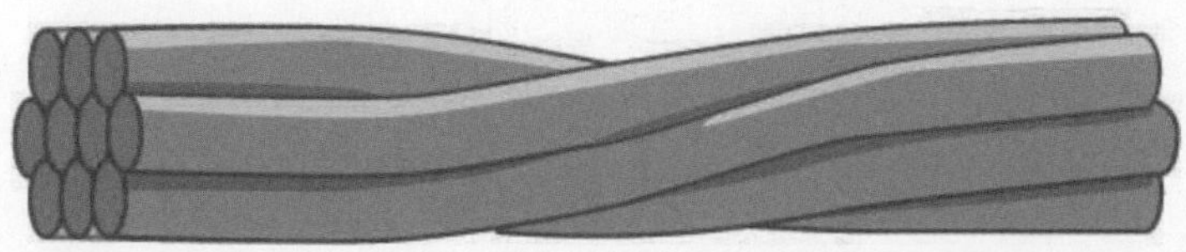

Figure 4.24 Intermediate filaments consist of several intertwined strands of fibrous proteins.

Intermediate filaments have no role in cell movement. Their function is purely structural. They bear tension, thus maintaining the cell's shape, and anchor the nucleus and other organelles in place. Figure 4.22 shows how intermediate filaments create a supportive scaffolding inside the cell.

The intermediate filaments are the most diverse group of cytoskeletal elements. Several fibrous protein types are in the intermediate filaments. You are probably most familiar with keratin, the fibrous protein that strengthens your hair, nails, and the skin's epidermis.

Microtubules

As their name implies, microtubules are small hollow tubes. Polymerized dimers of α-tubulin and β-tubulin, two globular proteins, comprise the microtubule's walls (Figure 4.25). With a diameter of about 25 nm, **microtubules** are cytoskeletons' widest components. They help the cell resist compression, provide a track along which vesicles move through the cell, and pull replicated chromosomes to opposite ends of a dividing cell. Like microfilaments, microtubules can disassemble and reform quickly.

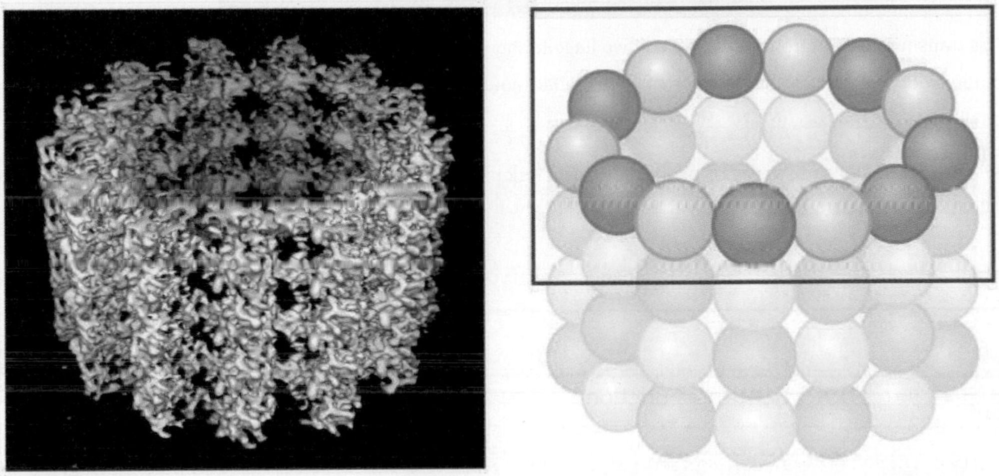

Figure 4.25 Microtubules are hollow. Their walls consist of 13 polymerized dimers of α-tubulin and β-tubulin (right image). The left image shows the tube's molecular structure.

Microtubules are also the structural elements of flagella, cilia, and centrioles (the latter are the centrosome's two perpendicular bodies). In animal cells, the centrosome is the microtubule-organizing center. In eukaryotic cells, flagella and cilia are quite different structurally from their counterparts in prokaryotes, as we discuss below.

Flagella and Cilia

The **flagella** (singular = flagellum) are long, hair-like structures that extend from the plasma membrane and enable an entire cell to move (for example, sperm, *Euglena*, and some prokaryotes). When present, the cell has just one flagellum or a few flagella. However, when **cilia** (singular = cilium) are present, many of them extend along the plasma membrane's entire surface. They are short, hair-like structures that move entire cells (such as paramecia) or substances along the cell's outer surface (for example, the cilia of cells lining the Fallopian tubes that move the ovum toward the uterus, or cilia lining the cells of the respiratory tract that trap particulate matter and move it toward your nostrils.)

Despite their differences in length and number, flagella and cilia share a common structural arrangement of microtubules called a "9 + 2 array." This is an appropriate name because a single flagellum or cilium is made of a ring of nine microtubule doublets, surrounding a single microtubule doublet in the center (Figure 4.26).

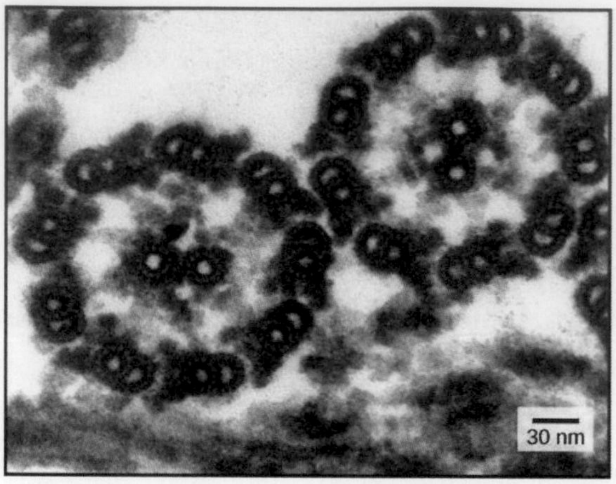

Figure 4.26 This transmission electron micrograph of two flagella shows the microtubules' 9 + 2 array: nine microtubule doublets surround a single microtubule doublet. (credit: modification of work by Dartmouth Electron Microscope Facility, Dartmouth College; scale-bar data from Matt Russell)

You have now completed a broad survey of prokaryotic and eukaryotic cell components. For a summary of cellular components in prokaryotic and eukaryotic cells, see Table 4.1.

Components of Prokaryotic and Eukaryotic Cells

Cell Component	Function	Present in Prokaryotes?	Present in Animal Cells?	Present in Plant Cells?
Plasma membrane	Separates cell from external environment; controls passage of organic molecules, ions, water, oxygen, and wastes into and out of cell	Yes	Yes	Yes
Cytoplasm	Provides turgor pressure to plant cells as fluid inside the central vacuole; site of many metabolic reactions; medium in which organelles are found	Yes	Yes	Yes
Nucleolus	Darkened area within the nucleus where ribosomal subunits are synthesized.	No	Yes	Yes
Nucleus	Cell organelle that houses DNA and directs synthesis of ribosomes and proteins	No	Yes	Yes
Ribosomes	Protein synthesis	Yes	Yes	Yes
Mitochondria	ATP production/cellular respiration	No	Yes	Yes
Peroxisomes	Oxidize and thus break down fatty acids and amino acids, and detoxify poisons	No	Yes	Yes
Vesicles and vacuoles	Storage and transport; digestive function in plant cells	No	Yes	Yes

Table 4.1

Cell Component	Function	Present in Prokaryotes?	Present in Animal Cells?	Present in Plant Cells?
Centrosome	Unspecified role in cell division in animal cells; microtubule source in animal cells	No	Yes	No
Lysosomes	Digestion of macromolecules; recycling of worn-out organelles	No	Yes	Some
Cell wall	Protection, structural support, and maintenance of cell shape	Yes, primarily peptidoglycan	No	Yes, primarily cellulose
Chloroplasts	Photosynthesis	No	No	Yes
Endoplasmic reticulum	Modifies proteins and synthesizes lipids	No	Yes	Yes
Golgi apparatus	Modifies, sorts, tags, packages, and distributes lipids and proteins	No	Yes	Yes
Cytoskeleton	Maintains cell's shape, secures organelles in specific positions, allows cytoplasm and vesicles to move within cell, and enables unicellular organisms to move independently	Yes	Yes	Yes
Flagella	Cellular locomotion	Some	Some	No, except for some plant sperm cells
Cilia	Cellular locomotion, movement of particles along plasma membrane's extracellular surface, and filtration	Some	Some	No

Table 4.1

4.6 Connections between Cells and Cellular Activities

By the end of this section, you will be able to do the following:

- Describe the extracellular matrix
- List examples of the ways that plant cells and animal cells communicate with adjacent cells
- Summarize the roles of tight junctions, desmosomes, gap junctions, and plasmodesmata

You already know that tissue is a group of similar cells working together. As you might expect, if cells are to work together, they must communicate with each other, just as you need to communicate with others if you work on a group project. Let's take a look at how cells communicate with each other.

Extracellular Matrix of Animal Cells

While cells in most multicellular organisms release materials into the extracellular space, animal cells will be discussed as an example. The primary components of these materials are proteins, and the most abundant protein is collagen. Collagen fibers are interwoven with proteoglycans, which are carbohydrate-containing protein molecules. Collectively, we call these materials the **extracellular matrix** (Figure 4.27). Not only does the extracellular matrix hold the cells together to form a tissue, but it also allows the cells within the tissue to communicate with each other. How can this happen?

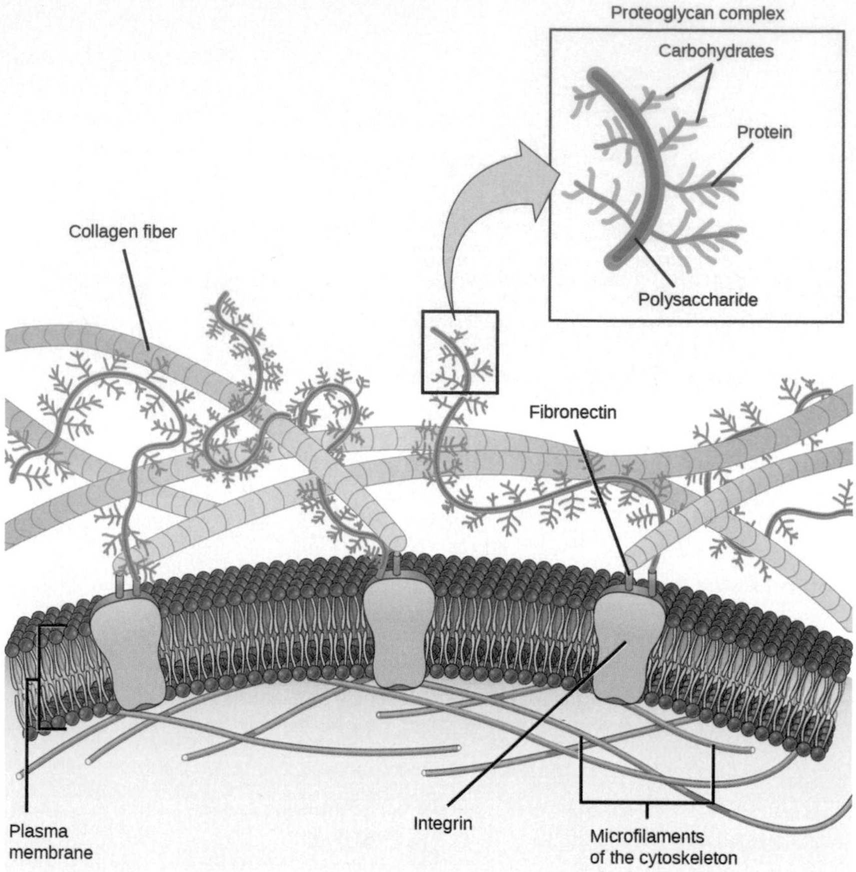

Figure 4.27 The extracellular matrix consists of a network of proteins and carbohydrates.

Cells have protein receptors on their plasma membranes' extracellular surfaces. When a molecule within the matrix binds to the receptor, it changes the receptor's molecular structure. The receptor, in turn, changes the microfilaments' conformation positioned just inside the plasma membrane. These conformational changes induce chemical signals inside the cell that reach the nucleus and turn "on" or "off" the transcription of specific DNA sections, which affects the associated protein production, thus changing the activities within the cell.

Blood clotting provides an example of the extracellular matrix's role in cell communication. When the cells lining a blood vessel are damaged, they display a protein receptor, which we call tissue factor. When tissue factor binds with another factor in the extracellular matrix, it causes platelets to adhere to the damaged blood vessel's wall, stimulates the adjacent smooth muscle cells in the blood vessel to contract (thus constricting the blood vessel), and initiates a series of steps that stimulate the platelets to produce clotting factors.

Intercellular Junctions

Cells can also communicate with each other via direct contact, or intercellular junctions. There are differences in the ways that plant and animal and fungal cells communicate. Plasmodesmata are junctions between plant cells; whereas, animal cell contacts include tight junctions, gap junctions, and desmosomes.

Plasmodesmata

In general, long stretches of the plasma membranes of neighboring plant cells cannot touch one another because the cell wall that surrounds each cell separates them (Figure 4.8). How then, can a plant transfer water and other soil nutrients from its roots, through its stems, and to its leaves? Such transport uses the vascular tissues (xylem and phloem) primarily. There also exist structural modifications, which we call **plasmodesmata** (singular = plasmodesma). Numerous channels that pass between adjacent plant cells' cell walls connect their cytoplasm, and enable transport of materials from cell to cell, and thus throughout the plant (Figure 4.28).

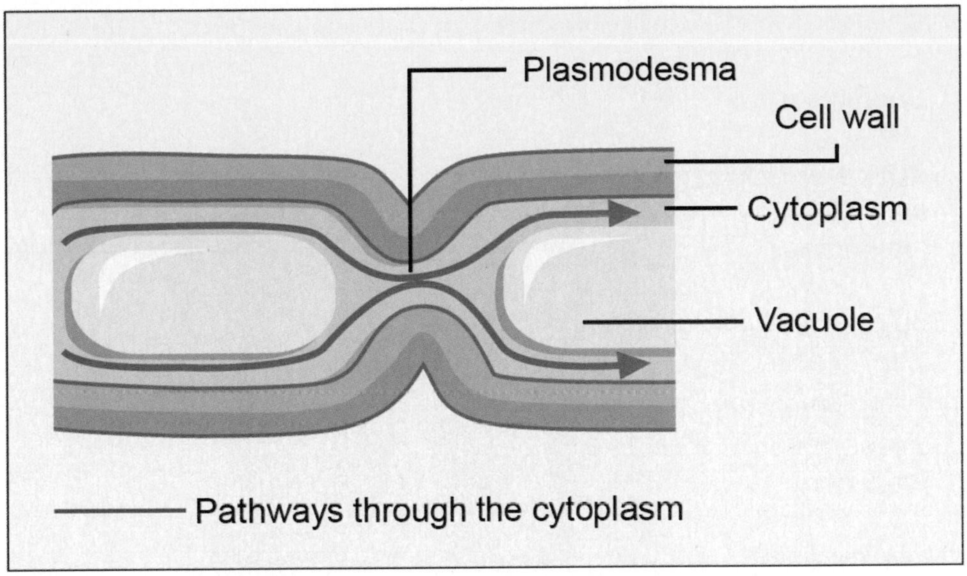

Figure 4.28 A plasmodesma is a channel between two adjacent plant cells' cell walls. Plasmodesmata allow materials to pass from one plant cell's cytoplasm to an adjacent cell's cytoplasm.

Tight Junctions

A **tight junction** is a watertight seal between two adjacent animal cells (Figure 4.29). Proteins (predominantly two proteins called claudins and occludins) tightly hold the cells against each other.

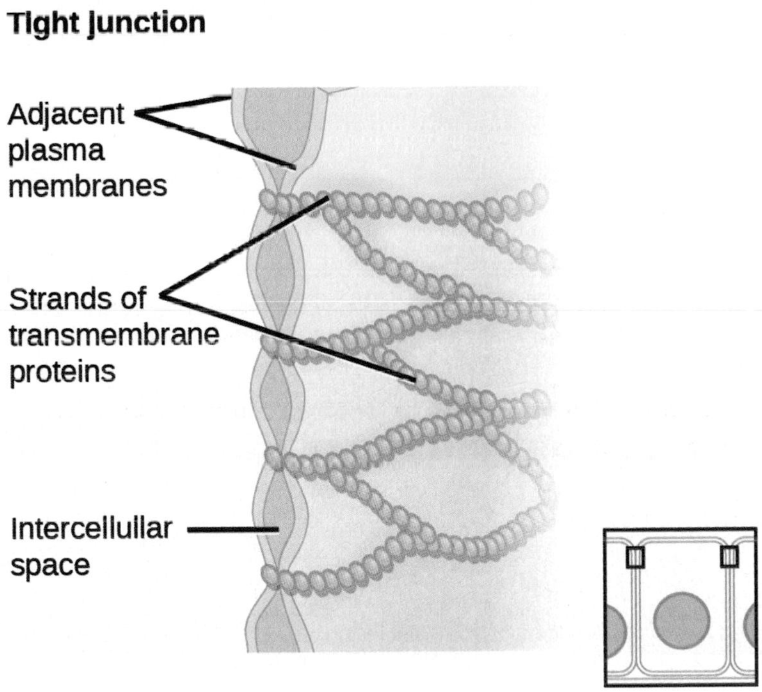

Figure 4.29 Tight junctions form watertight connections between adjacent animal cells. Proteins create tight junction adherence. (credit: modification of work by Mariana Ruiz Villareal)

This tight adherence prevents materials from leaking between the cells; tight junctions are typically found in epithelial tissues that line internal organs and cavities, and comprise most of the skin. For example, the tight junctions of the epithelial cells lining your urinary bladder prevent urine from leaking out into the extracellular space.

Desmosomes

Also only in animal cells are **desmosomes**, which act like spot welds between adjacent epithelial cells (Figure 4.30). Cadherins, short proteins in the plasma membrane connect to intermediate filaments to create desmosomes. The cadherins connect two adjacent cells and maintain the cells in a sheet-like formation in organs and tissues that stretch, like the skin, heart, and muscles.

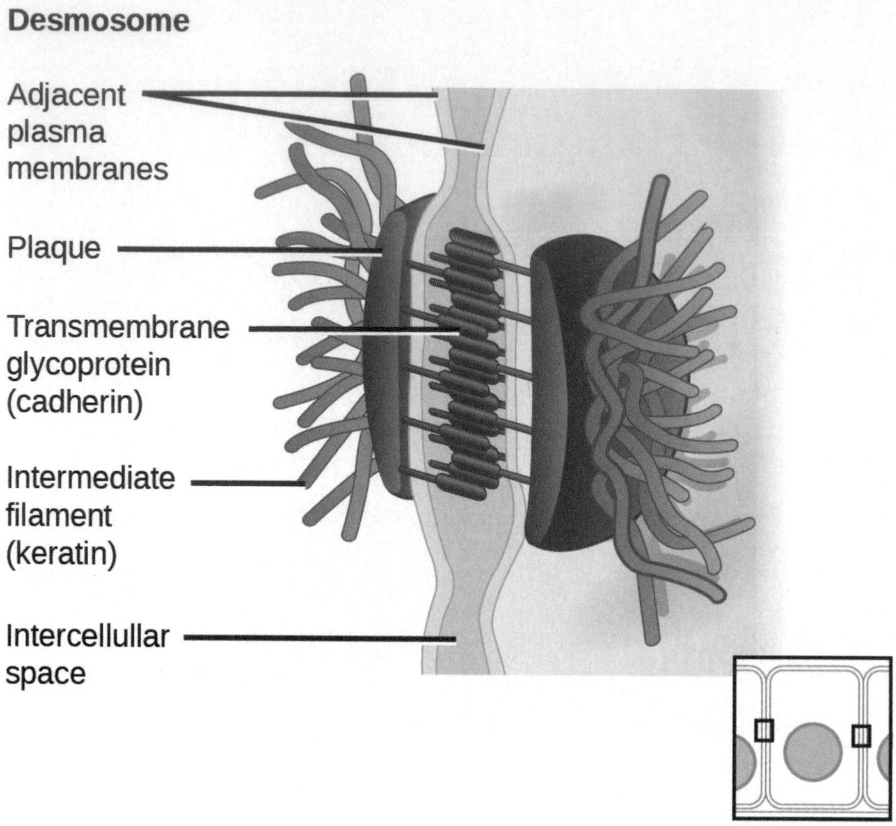

Figure 4.30 A desmosome forms a very strong spot weld between cells. Linking cadherins and intermediate filaments create it. (credit: modification of work by Mariana Ruiz Villareal)

Gap Junctions

Gap junctions in animal cells are like plasmodesmata in plant cells in that they are channels between adjacent cells that allow for transporting ions, nutrients, and other substances that enable cells to communicate (Figure 4.31). Structurally, however, gap junctions and plasmodesmata differ.

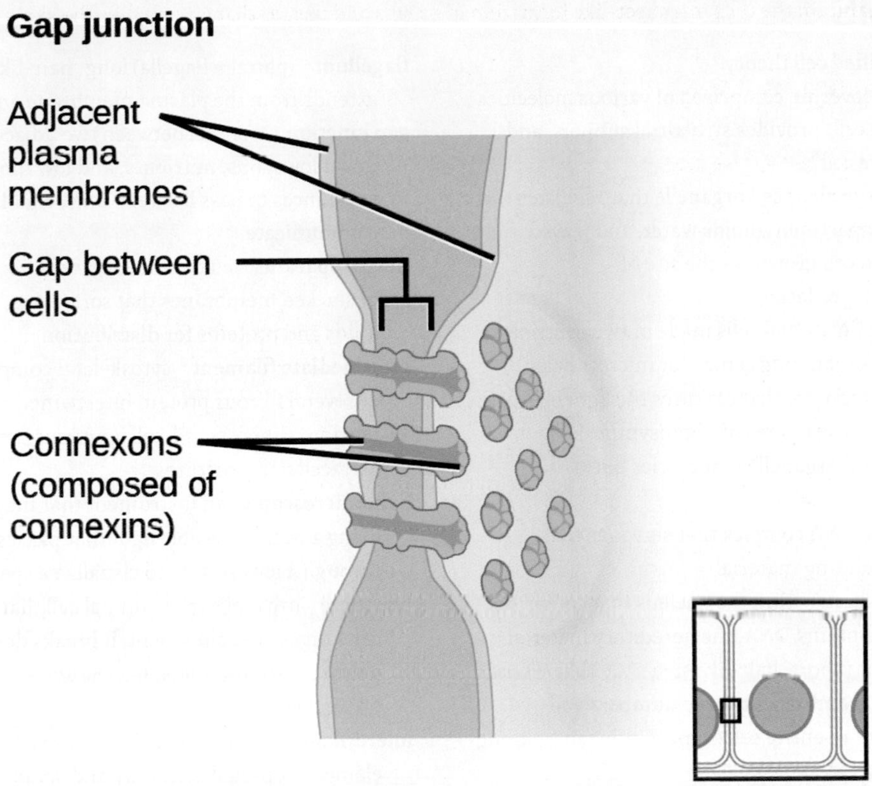

Figure 4.31 A gap junction is a protein-lined pore that allows water and small molecules to pass between adjacent animal cells. (credit: modification of work by Mariana Ruiz Villareal)

Gap junctions develop when a set of six proteins (connexins) in the plasma membrane arrange themselves in an elongated donut-like configuration - a connexon. When the connexon's pores ("doughnut holes") in adjacent animal cells align, a channel between the two cells forms. Gap junctions are particularly important in cardiac muscle. The electrical signal for the muscle to contract passes efficiently through gap junctions, allowing the heart muscle cells to contract in tandem.

LINK TO LEARNING

To conduct a virtual microscopy lab and review the parts of a cell, work through the steps of this interactive assignment (http://openstax.org/l/microscopy_lab) .

KEY TERMS

cell theory see unified cell theory

cell wall rigid cell covering comprised of various molecules that protects the cell, provides structural support, and gives shape to the cell

central vacuole large plant cell organelle that regulates the cell's storage compartment, holds water, and plays a significant role in cell growth as the site of macromolecule degradation

centrosome region in animal cells made of two centrioles that serves as an organizing center for microtubules

chlorophyll green pigment that captures the light energy that drives the light reactions of photosynthesis

chloroplast plant cell organelle that carries out photosynthesis

chromatin protein-DNA complex that serves as the chromosomes' building material

chromosome structure within the nucleus that comprises chromatin that contains DNA, the hereditary material

cilium (plural = cilia) short, hair-like structure that extends from the plasma membrane in large numbers and functions to move an entire cell or move substances along the cell's outer surface

cytoplasm entire region between the plasma membrane and the nuclear envelope, consisting of organelles suspended in the gel-like cytosol, the cytoskeleton, and various chemicals

cytoskeleton protein fiber network that collectively maintains the cell's shape, secures some organelles in specific positions, allows cytoplasm and vesicles to move within the cell, and enables unicellular organisms to move independently

cytosol the cytoplasm's gel-like material in which cell structures are suspended

desmosome linkages between adjacent epithelial cells that form when cadherins in the plasma membrane attach to intermediate filaments

electron microscope an instrument that magnifies an object using an electron beam that passes and bends through a lens system to visualize a specimen

endomembrane system group of organelles and membranes in eukaryotic cells that work together modifying, packaging, and transporting lipids and proteins

endoplasmic reticulum (ER) series of interconnected membranous structures within eukaryotic cells that collectively modify proteins and synthesize lipids

eukaryotic cell cell that has a membrane-bound nucleus and several other membrane-bound compartments or sacs

extracellular matrix material secreted from animal or fungal cells that provides mechanical protection and anchoring for the cells in the tissue

flagellum (plural = flagella) long, hair-like structure that extends from the plasma membrane and moves the cell

gap junction channel between two adjacent animal cells that allows ions, nutrients, and low molecular weight substances to pass between cells, enabling the cells to communicate

Golgi apparatus eukaryotic organelle comprised of a series of stacked membranes that sorts, tags, and packages lipids and proteins for distribution

intermediate filament cytoskeletal component, comprised of several fibrous protein intertwined strands, that bears tension, supports cell-cell junctions, and anchors cells to extracellular structures

light microscope an instrument that magnifies an object using a beam of visible light that passes and bends through a lens system to visualize a specimen

lysosome organelle in an animal cell that functions as the cell's digestive component; it breaks down proteins, polysaccharides, lipids, nucleic acids, and even worn-out organelles

microfilament the cytoskeleton system's narrowest element; it provides rigidity and shape to the cell and enables cellular movements

microscope an instrument that magnifies an object

microtubule the cytoskeleton system's widest element; it helps the cell resist compression, provides a track along which vesicles move through the cell, pulls replicated chromosomes to opposite ends of a dividing cell, and is the structural element of centrioles, flagella, and cilia

mitochondria (singular = mitochondrion) cellular organelles responsible for carrying out cellular respiration, resulting in producing ATP, the cell's main energy-carrying molecule

nuclear envelope double-membrane structure that constitutes the nucleus' outermost portion

nucleoid central part of a prokaryotic cell's central part where the chromosome is located

nucleolus darkly staining body within the nucleus that is responsible for assembling ribosome subunits

nucleoplasm semi-solid fluid inside the nucleus that contains the chromatin and nucleolus

nucleus cell organelle that houses the cell's DNA and directs ribosome and protein synthesis

organelle compartment or sac within a cell

peroxisome small, round organelle that contains hydrogen peroxide, oxidizes fatty acids and amino acids, and detoxifies many poisons

plasma membrane phospholipid bilayer with embedded (integral) or attached (peripheral) proteins, and separates the cell's internal content from its surrounding environment

plasmodesma (plural = plasmodesmata) channel that

passes between adjacent plant cells' cell walls, connects their cytoplasm, and allows transporting of materials from cell to cell

prokaryote unicellular organism that lacks a nucleus or any other membrane-bound organelle

ribosome cellular structure that carries out protein synthesis

rough endoplasmic reticulum (RER) region of the endoplasmic reticulum that is studded with ribosomes and engages in protein modification and phospholipid synthesis

smooth endoplasmic reticulum (SER) region of the endoplasmic reticulum that has few or no ribosomes on its cytoplasmic surface and synthesizes carbohydrates, lipids, and steroid hormones; detoxifies certain chemicals (like pesticides, preservatives, medications, and environmental pollutants), and stores calcium ions

tight junction protein adherence that creates a firm seal between two adjacent animal cells

unified cell theory a biological concept that states that one or more cells comprise all organisms; the cell is the basic unit of life; and new cells arise from existing cells

vacuole membrane-bound sac, somewhat larger than a vesicle, which functions in cellular storage and transport

vesicle small, membrane-bound sac that functions in cellular storage and transport; its membrane is capable of fusing with the plasma membrane and the membranes of the endoplasmic reticulum and Golgi apparatus

CHAPTER SUMMARY

4.1 Studying Cells

A cell is the smallest unit of life. Most cells are so tiny that we cannot see them with the naked eye. Therefore, scientists use microscopes to study cells. Electron microscopes provide higher magnification, higher resolution, and more detail than light microscopes. The unified cell theory states that one or more cells comprise all organisms, the cell is the basic unit of life, and new cells arise from existing cells.

4.2 Prokaryotic Cells

Prokaryotes are single-celled organisms of the domains Bacteria and Archaea. All prokaryotes have plasma membranes, cytoplasm, ribosomes, and DNA that is not membrane-bound. Most have peptidoglycan cell walls and many have polysaccharide capsules. Prokaryotic cells range in diameter from 0.1 to 5.0 μm.

As a cell increases in size, its surface area-to-volume ratio decreases. If the cell grows too large, the plasma membrane will not have sufficient surface area to support the rate of diffusion required for the increased volume.

4.3 Eukaryotic Cells

Like a prokaryotic cell, a eukaryotic cell has a plasma membrane, cytoplasm, and ribosomes, but a eukaryotic cell is typically larger than a prokaryotic cell, has a true nucleus (meaning a membrane surrounds its DNA), and has other membrane-bound organelles that allow for compartmentalizing functions. The plasma membrane is a phospholipid bilayer embedded with proteins. The nucleus's nucleolus is the site of ribosome assembly. We find ribosomes either in the cytoplasm or attached to the cytoplasmic side of the plasma membrane or endoplasmic reticulum. They perform protein synthesis. Mitochondria participate in cellular respiration. They are responsible for the majority of ATP produced in the cell. Peroxisomes hydrolyze fatty acids, amino acids, and some toxins. Vesicles and vacuoles are storage and transport compartments. In plant cells, vacuoles also help break down macromolecules.

Animal cells also have a centrosome and lysosomes. The centrosome has two bodies perpendicular to each other, the centrioles, and has an unknown purpose in cell division. Lysosomes are the digestive organelles of animal cells.

Plant cells and plant-like cells each have a cell wall, chloroplasts, and a central vacuole. The plant cell wall, whose primary component is cellulose, protects the cell, provides structural support, and gives the cell shape. Photosynthesis takes place in chloroplasts. The central vacuole can expand without having to produce more cytoplasm.

4.4 The Endomembrane System and Proteins

The endomembrane system includes the nuclear envelope, lysosomes, vesicles, the ER, and Golgi apparatus, as well as the plasma membrane. These cellular components work together to modify, package, tag, and transport proteins and lipids that form the membranes.

The RER modifies proteins and synthesizes phospholipids in cell membranes. The SER synthesizes carbohydrates, lipids, and steroid hormones; engages in the detoxification of medications and poisons; and stores calcium ions. Sorting, tagging, packaging, and distributing lipids and proteins take place in the Golgi apparatus. Budding RER and Golgi membranes create lysosomes. Lysosomes digest macromolecules, recycle worn-out organelles, and destroy pathogens.

4.5 The Cytoskeleton

The cytoskeleton has three different protein element types. From narrowest to widest, they are the microfilaments (actin filaments), intermediate filaments, and microtubules.

Biologists often associate microfilaments with myosin. They provide rigidity and shape to the cell and facilitate cellular movements. Intermediate filaments bear tension and anchor the nucleus and other organelles in place. Microtubules help the cell resist compression, serve as tracks for motor proteins that move vesicles through the cell, and pull replicated chromosomes to opposite ends of a dividing cell. They are also the structural element of centrioles, flagella, and cilia.

4.6 Connections between Cells and Cellular Activities

Animal cells communicate via their extracellular matrices and are connected to each other via tight junctions, desmosomes, and gap junctions. Plant cells are connected and communicate with each other via plasmodesmata.

When protein receptors on the plasma membrane's surface of an animal cell bind to a substance in the extracellular matrix, a chain of reactions begins that changes activities taking place within the cell. Plasmodesmata are channels between adjacent plant cells, while gap junctions are channels between adjacent animal cells. However, their structures are quite different. A tight junction is a watertight seal between two adjacent cells, while a desmosome acts like a spot weld.

VISUAL CONNECTION QUESTIONS

1. Figure 4.7 Prokaryotic cells are much smaller than eukaryotic cells. What advantages might small cell size confer on a cell? What advantages might large cell size have?

2. Figure 4.8 If the nucleolus were not able to carry out its function, what other cellular organelles would be affected?

3. Figure 4.18 If a peripheral membrane protein were synthesized in the lumen (inside) of the ER, would it end up on the inside or outside of the plasma membrane?

REVIEW QUESTIONS

4. When viewing a specimen through a light microscope, scientists use _______ to distinguish the individual components of cells.
   a. a beam of electrons
   b. radioactive isotopes
   c. special stains
   d. high temperatures

5. The _______ is the basic unit of life.
   a. organism
   b. cell
   c. tissue
   d. organ

6. Prokaryotes depend on _______ to obtain some materials and to get rid of wastes.
   a. ribosomes
   b. flagella
   c. cell division
   d. diffusion

7. Bacteria that lack fimbriae are less likely to _______.
   a. adhere to cell surfaces
   b. swim through bodily fluids
   c. synthesize proteins
   d. retain the ability to divide

8. Which of the following organisms is a prokaryote?
   a. amoeba
   b. influenza A virus
   c. charophyte algae
   d. *E. coli*

9. Which of the following is surrounded by two phospholipid bilayers?
   a. the ribosomes
   b. the vesicles
   c. the cytoplasm
   d. the nucleoplasm

10. Peroxisomes got their name because hydrogen peroxide is:
    a. used in their detoxification reactions
    b. produced during their oxidation reactions
    c. incorporated into their membranes
    d. a cofactor for the organelles' enzymes

11. In plant cells, the function of the lysosomes is carried out by _______.
    a. vacuoles
    b. peroxisomes
    c. ribosomes
    d. nuclei

12. Which of the following is both in eukaryotic and prokaryotic cells?
    a. nucleus
    b. mitochondrion
    c. vacuole
    d. ribosomes

13. Tay-Sachs disease is a genetic disorder that results in the destruction of neurons due to a buildup of sphingolipids in the cells. Which organelle is malfunctioning in Tay-Sachs?
    a. lysosome
    b. endoplasmic reticulum
    c. peroxisome
    d. mitochondria

14. Which of the following is not a component of the endomembrane system?
    a. mitochondrion
    b. Golgi apparatus
    c. endoplasmic reticulum
    d. lysosome

15. The process by which a cell engulfs a foreign particle is known as:
    a. endosymbiosis
    b. phagocytosis
    c. hydrolysis
    d. membrane synthesis

16. Which of the following is most likely to have the greatest concentration of smooth endoplasmic reticulum?
    a. a cell that secretes enzymes
    b. a cell that destroys pathogens
    c. a cell that makes steroid hormones
    d. a cell that engages in photosynthesis

17. Which of the following sequences correctly lists in order the steps involved in the incorporation of a proteinaceous molecule within a cell?
    a. protein synthesis of the protein on the ribosome; modification in the Golgi apparatus; packaging in the endoplasmic reticulum; tagging in the vesicle
    b. synthesis of the protein on the lysosome; tagging in the Golgi; packaging in the vesicle; distribution in the endoplasmic reticulum
    c. synthesis of the protein on the ribosome; modification in the endoplasmic reticulum; tagging in the Golgi; distribution via the vesicle
    d. synthesis of the protein on the lysosome; packaging in the vesicle; distribution via the Golgi; tagging in the endoplasmic reticulum

18. Congenital disorders of glycosylation are a growing class of rare diseases. Which organelle would be most commonly involved in the glycoprotein disorder portion of the group?
    a. RER
    b. ribosomes
    c. endosomes
    d. Golgi apparatus

19. Which of the following have the ability to disassemble and reform quickly?
    a. microfilaments and intermediate filaments
    b. microfilaments and microtubules
    c. intermediate filaments and microtubules
    d. only intermediate filaments

20. Which of the following do not play a role in intracellular movement?
    a. microfilaments and intermediate filaments
    b. microfilaments and microtubules
    c. intermediate filaments and microtubules
    d. only intermediate filaments

21. In humans, _____ are used to move a cell within its environment while _____ are used to move the environment relative to the cell.
    a. cilia, pseudopodia
    b. flagella; cilia
    c. microtubules; flagella
    d. microfilaments; microtubules

22. Which of the following are only in plant cells?
    a. gap junctions
    b. desmosomes
    c. plasmodesmata
    d. tight junctions

23. The key components of desmosomes are cadherins and __________.
    a. actin
    b. microfilaments
    c. intermediate filaments
    d. microtubules

24. Diseased animal cells may produce molecules that activate death cascades to kill the cells in a controlled manner. Why would neighboring healthy cells also die?
    a. The death molecule is passed through desmosomes.
    b. The death molecule is passed through plasmodesmata.
    c. The death molecule disrupts the extracellular matrix.
    d. The death molecule passes through gap junctions.

CRITICAL THINKING QUESTIONS

25. In your everyday life, you have probably noticed that certain instruments are ideal for certain situations. For example, you would use a spoon rather than a fork to eat soup because a spoon is shaped for scooping, while soup would slip between the tines of a fork. The use of ideal instruments also applies in science. In what situation(s) would the use of a light microscope be ideal, and why?
26. In what situation(s) would the use of a scanning electron microscope be ideal, and why?
27. In what situation(s) would a transmission electron microscope be ideal, and why?
28. What are the advantages and disadvantages of each of these types of microscopes?
29. Explain how the formation of an adult human follows the cell theory.
30. Antibiotics are medicines that are used to fight bacterial infections. These medicines kill prokaryotic cells without harming human cells. What part or parts of the bacterial cell do you think antibiotics target? Why?
31. Explain why not all microbes are harmful.
32. You already know that ribosomes are abundant in red blood cells. In what other cells of the body would you find them in great abundance? Why?
33. What are the structural and functional similarities and differences between mitochondria and chloroplasts?
34. Why are plasma membranes arranged as a bilayer rather than a monolayer?
35. In the context of cell biology, what do we mean by form follows function? What are at least two examples of this concept?
36. In your opinion, is the nuclear membrane part of the endomembrane system? Why or why not? Defend your answer.
37. What are the similarities and differences between the structures of centrioles and flagella?
38. How do cilia and flagella differ?
39. Describe how microfilaments and microtubules are involved in the phagocytosis and destruction of a pathogen by a macrophage.
40. Compare and contrast the boundaries that plant, animal, and bacteria cells use to separate themselves from their surrounding environment.
41. How does the structure of a plasmodesma differ from that of a gap junction?
42. Explain how the extracellular matrix functions.
43. Pathogenic *E. coli* have recently been shown to degrade tight junction proteins during infection. How would this provide an advantage to the bacteria?

CHAPTER 5
Structure and Function of Plasma Membranes

Figure 5.1 Despite its seeming hustle and bustle, Grand Central Station functions with a high level of organization: People and objects move from one location to another, they cross or are contained within certain boundaries, and they provide a constant flow as part of larger activity. Analogously, a plasma membrane's functions involve movement within the cell and across boundaries' activities. (credit: modification of work by Randy Le'Moine)

INTRODUCTION The plasma membrane, the cell membrane, has many functions, but the most basic one is to define the cell's borders and keep the cell functional. The plasma membrane is selectively permeable. This means that the membrane allows some materials to freely enter or leave the cell, while other materials cannot move freely, but require a specialized structure, and occasionally, even energy investment for crossing.

Chapter Outline

5.1 Components and Structure

By the end of this section, you will be able to do the following:

- Understand the cell membrane fluid mosaic model
- Describe phospholipid, protein, and carbohydrate functions in membranes
- Discuss membrane fluidity

A cell's plasma membrane defines the cell, outlines its borders, and determines the nature of its interaction with its environment (see Table 5.1 for a summary). Cells exclude some substances, take in others, and excrete still others, all in controlled quantities. The plasma membrane must be very flexible to allow certain cells, such as red and white blood cells, to change shape as they pass through narrow capillaries. These are the more obvious plasma membrane functions. In addition, the plasma membrane's surface carries markers that allow cells to recognize one another, which is vital for tissue and organ formation during early development, and which later plays a role in the immune response's "self" versus "non-self" distinction.

Among the most sophisticated plasma membrane functions is the ability for complex, integral proteins, receptors to transmit signals. These proteins act both as extracellular input receivers and as intracellular processing activators. These membrane receptors provide extracellular attachment sites for effectors like hormones and growth factors, and they activate intracellular response cascades when their effectors are bound. Occasionally, viruses hijack receptors (HIV, human immunodeficiency virus, is one example) that use them to gain entry into cells, and at times, the genes encoding receptors become mutated, causing the signal transduction process to malfunction with disastrous consequences.

Fluid Mosaic Model

Scientists identified the plasma membrane in the 1890s, and its chemical components in 1915. The principal components they identified were lipids and proteins. In 1935, Hugh Davson and James Danielli proposed the plasma membrane's structure. This was the first model that others in the scientific community widely accepted. It was based on the plasma membrane's "railroad track" appearance in early electron micrographs. Davson and Danielli theorized that the plasma membrane's structure resembles a sandwich. They made the analogy of proteins to bread, and lipids to the filling. In the 1950s, advances in microscopy, notably transmission electron microscopy (TEM), allowed researchers to see that the plasma membrane's core consisted of a double, rather than a single, layer. In 1972, S.J. Singer and Garth L. Nicolson proposed a new model that provides microscopic observations and better explains plasma membrane function.

The explanation, the **fluid mosaic model**, has evolved somewhat over time, but it still best accounts for plasma membrane structure and function as we now understand them. The fluid mosaic model describes the plasma membrane structure as a mosaic of components—including phospholipids, cholesterol, proteins, and carbohydrates—that gives the membrane a fluid character. Plasma membranes range from 5 to 10 nm in thickness. For comparison, human red blood cells, visible via light microscopy, are approximately 8 µm wide, or approximately 1,000 times wider than a plasma membrane. The membrane does look a bit like a sandwich (Figure 5.2).

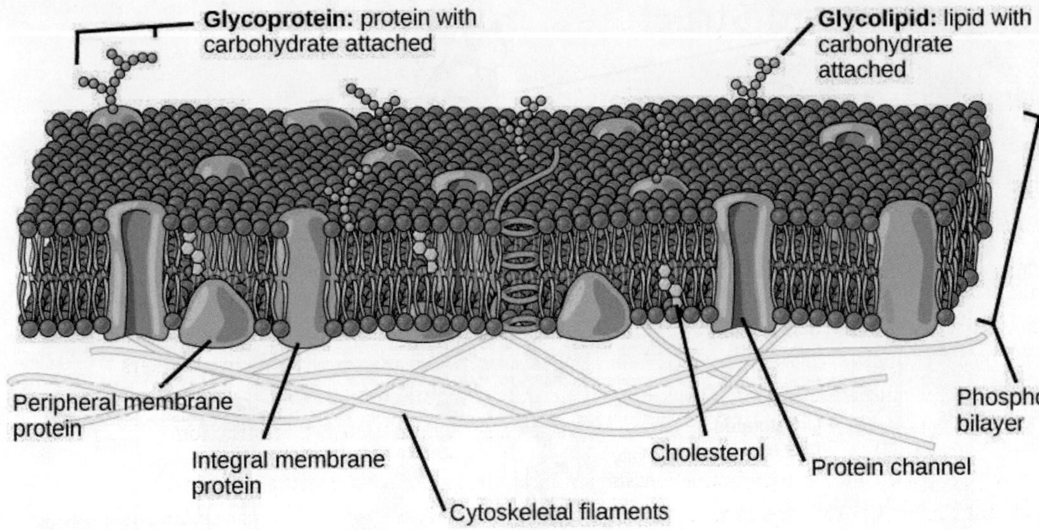

Figure 5.2 The plasma membrane fluid mosaic model describes the plasma membrane as a fluid combination of phospholipids, cholesterol, and proteins. Carbohydrates attached to lipids (glycolipids) and to proteins (glycoproteins) extend from the membrane's outward-facing surface.

A plasma membrane's principal components are lipids (phospholipids and cholesterol), proteins, and carbohydrates attached to some of the lipids and proteins. A phospholipid is a molecule consisting of glycerol, two fatty acids, and a phosphate-linked head group. Cholesterol, another lipid comprised of four fused carbon rings, is situated alongside the phospholipids in the membrane's core. The protein, lipid, and carbohydrate proportions in the plasma membrane vary with cell type, but for a typical human cell, protein accounts for about 50 percent of the composition by mass, lipids (of all types) account for about 40 percent, and carbohydrates comprise the remaining 10 percent. However, protein and lipid concentration varies with different cell membranes. For example, myelin, an outgrowth of specialized cells' membrane that insulates the peripheral nerves' axons, contains only 18 percent protein and 76 percent lipid. The mitochondrial inner membrane contains 76 percent protein and only 24 percent lipid. The plasma membrane of human red blood cells is 30 percent lipid. Carbohydrates are present only on the plasma membrane's exterior surface and are attached to proteins, forming **glycoproteins**, or attached to lipids, forming **glycolipids**.

Phospholipids

The membrane's main fabric comprises amphiphilic, phospholipid molecules. The **hydrophilic** or "water-loving" areas of these molecules (which look like a collection of balls in an artist's rendition of the model) (Figure 5.2) are in contact with the aqueous fluid both inside and outside the cell. **Hydrophobic**, or water-hating molecules, tend to be non-polar. They interact with other non-polar molecules in chemical reactions, but generally do not interact with polar molecules. When placed in water, hydrophobic molecules tend to form a ball or cluster. The phospholipids' hydrophilic regions form hydrogen bonds with water and other polar molecules on both the cell's exterior and interior. Thus, the membrane surfaces that face the cell's interior and exterior are hydrophilic. In contrast, the cell membrane's interior is hydrophobic and will not interact with water. Therefore, phospholipids form an excellent two-layer cell membrane that separates fluid within the cell from the fluid outside the cell.

A phospholipid molecule (Figure 5.3) consists of a three-carbon glycerol backbone with two fatty acid molecules attached to carbons 1 and 2, and a phosphate-containing group attached to the third carbon. This arrangement gives the overall molecule a head area (the phosphate-containing group), which has a polar character or negative charge, and a tail area (the fatty acids), which has no charge. The head can form hydrogen bonds, but the tail cannot. Scientists call a molecule with a positively or negatively charged area and an uncharged, or non-polar, area **amphiphilic** or "dual-loving."

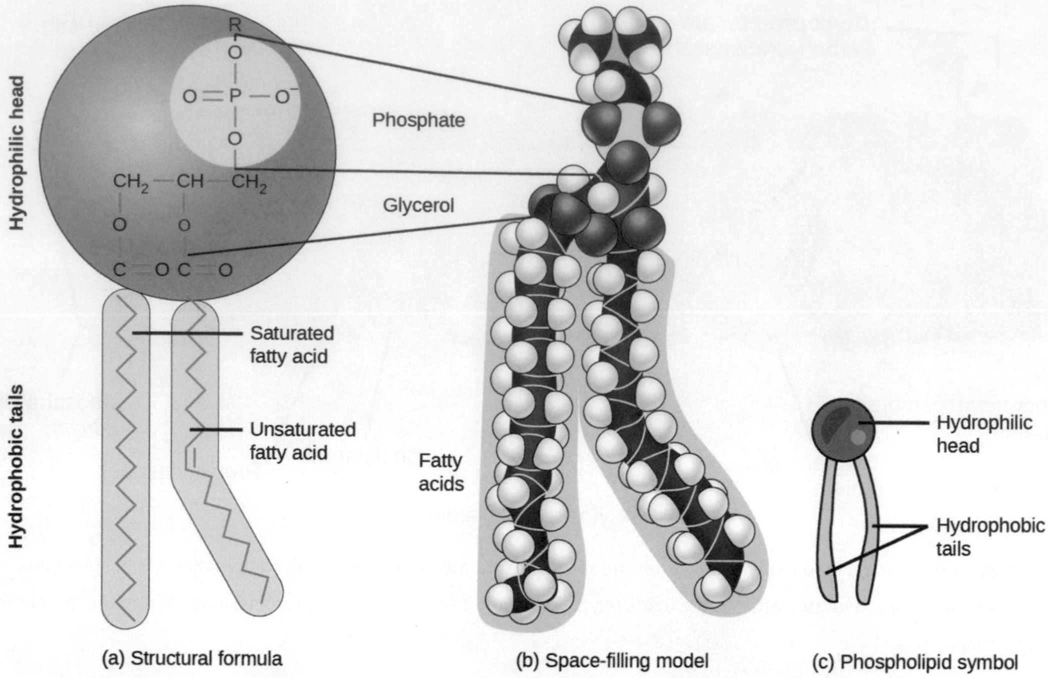

Figure 5.3 A hydrophilic head and two hydrophobic tails comprise this phospholipid molecule. The hydrophilic head group consists of a phosphate-containing group attached to a glycerol molecule. The hydrophobic tails, each containing either a saturated or an unsaturated fatty acid, are long hydrocarbon chains.

This characteristic is vital to the plasma membrane's structure because, in water, phospholipids arrange themselves with their hydrophobic tails facing each other and their hydrophilic heads facing out. In this way, they form a lipid bilayer—a double layered phospholipid barrier that separates the water and other materials on one side from the water and other materials on the other side. Phosopholipids heated in an aqueous solution usually spontaneously form small spheres or droplets (micelles or liposomes), with their hydrophilic heads forming the exterior and their hydrophobic tails on the inside (Figure 5.4).

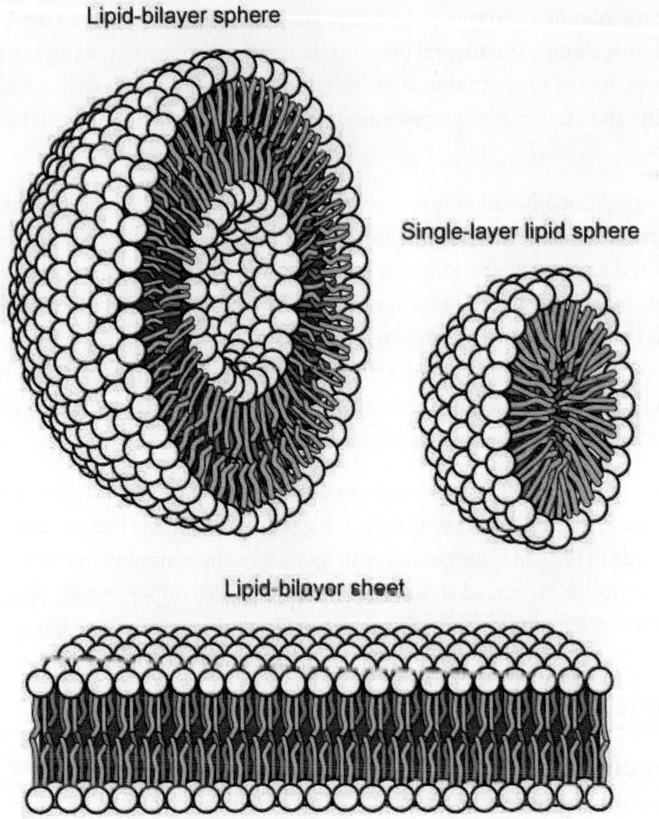

Figure 5.4 In an aqueous solution, phospholipids usually arrange themselves with their polar heads facing outward and their hydrophobic tails facing inward. (credit: modification of work by Mariana Ruiz Villareal)

Proteins

Proteins comprise the plasma membranes' second major component. **Integral proteins**, or integrins, as their name suggests, integrate completely into the membrane structure, and their hydrophobic membrane-spanning regions interact with the phospholipid bilayer's hydrophobic region (Figure 5.2). Single-pass integral membrane proteins usually have a hydrophobic transmembrane segment that consists of 20–25 amino acids. Some span only part of the membrane—associating with a single layer—while others stretch from one side to the other, and are exposed on either side. Up to 12 single protein segments comprise some complex proteins, which are extensively folded and embedded in the membrane (Figure 5.5). This protein type has a hydrophilic region or regions, and one or several mildly hydrophobic regions. This arrangement of protein regions orients the protein alongside the phospholipids, with the protein's hydrophobic region adjacent to the phosopholipids' tails and the protein's hydrophilic region or regions protruding from the membrane and in contact with the cytosol or extracellular fluid.

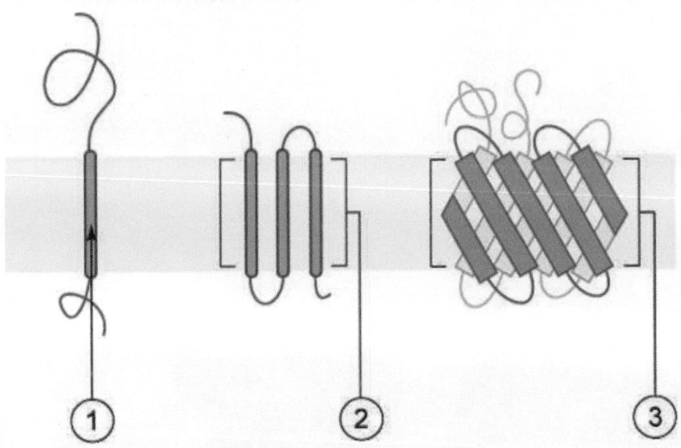

Figure 5.5 Integral membrane proteins may have one or more alpha-helices that span the membrane (examples 1 and 2), or they may have beta-sheets that span the membrane (example 3). (credit: "Foobar"/Wikimedia Commons)

Peripheral proteins are on the membranes' exterior and interior surfaces, attached either to integral proteins or to phospholipids. Peripheral proteins, along with integral proteins, may serve as enzymes, as structural attachments for the cytoskeleton's fibers, or as part of the cell's recognition sites. Scientists sometimes refer to these as "cell-specific" proteins. The body recognizes its own proteins and attacks foreign proteins associated with invasive pathogens.

Carbohydrates

Carbohydrates are the third major plasma membrane component. They are always on the cells' exterior surface and are bound either to proteins (forming glycoproteins) or to lipids (forming glycolipids) (Figure 5.2). These carbohydrate chains may consist of 2–60 monosaccharide units and can be either straight or branched. Along with peripheral proteins, carbohydrates form specialized sites on the cell surface that allow cells to recognize each other. These sites have unique patterns that allow for cell recognition, much the way that the facial features unique to each person allow individuals to recognize him or her. This recognition function is very important to cells, as it allows the immune system to differentiate between body cells ("self") and foreign cells or tissues ("non-self"). Similar glycoprotein and glycolipid types are on the surfaces of viruses and may change frequently, preventing immune cells from recognizing and attacking them.

We collectively refer to these carbohydrates on the cell's exterior surface—the carbohydrate components of both glycoproteins and glycolipids—as the glycocalyx (meaning "sugar coating"). The glycocalyx is highly hydrophilic and attracts large amounts of water to the cell's surface. This aids in the cell's interaction with its watery environment and in the cell's ability to obtain substances dissolved in the water. As we discussed above, the glycocalyx is also important for cell identification, self/non-self determination, and embryonic development, and is used in cell to cell attachments to form tissues.

EVOLUTION CONNECTION

How Viruses Infect Specific Organs

Glycoprotein and glycolipid patterns on the cells' surfaces give many viruses an opportunity for infection. HIV and hepatitis viruses infect only specific organs or cells in the human body. HIV is able to penetrate the plasma membranes of a subtype of lymphocytes called T-helper cells, as well as some monocytes and central nervous system cells. The hepatitis virus attacks liver cells.

These viruses are able to invade these cells, because the cells have binding sites on their surfaces that are specific to and compatible with certain viruses (Figure 5.6). Other recognition sites on the virus's surface interact with the human immune system, prompting the body to produce antibodies. Antibodies are made in response to the antigens or proteins associated with invasive pathogens, or in response to foreign cells, such as might occur with an organ transplant. These same sites serve as places for antibodies to attach and either destroy or inhibit the virus' activity. Unfortunately, these recognition sites on HIV change at a rapid rate because of mutations, making an effective vaccine against the virus very difficult, as the virus evolves and adapts. A person infected with HIV will quickly develop different populations, or variants, of the virus that differences in these recognition sites distinguish. This rapid change of surface markers decreases the effectiveness of the person's immune system in attacking the virus, because the antibodies will not recognize the surface patterns' new variations. In the case of HIV, the problem is compounded because the virus specifically infects and destroys cells involved in the immune response, further incapacitating the host.

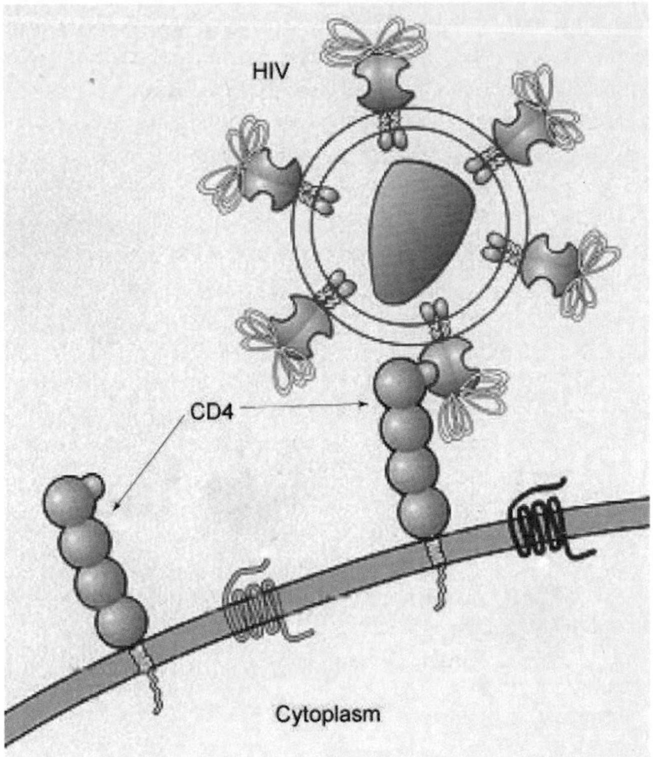

Figure 5.6 HIV binds to the CD4 receptor, a glycoprotein on T cell surfaces. (credit: modification of work by NIH, NIAID)

Membrane Fluidity

The membrane's mosaic characteristic helps to illustrate its nature. The integral proteins and lipids exist in the membrane as separate but loosely attached molecules. These resemble the separate, multicolored tiles of a mosaic picture, and they float, moving somewhat with respect to one another. The membrane is not like a balloon, however, that can expand and contract; rather, it is fairly rigid and can burst if penetrated or if a cell takes in too much water. However, because of its mosaic nature, a very fine needle can easily penetrate a plasma membrane without causing it to burst, and the membrane will flow and self-seal when one extracts the needle.

The membrane's mosaic characteristics explain some but not all of its fluidity. There are two other factors that help maintain this fluid characteristic. One factor is the nature of the phospholipids themselves. In their saturated form, the fatty acids in phospholipid tails are saturated with bound hydrogen atoms. There are no double bonds between adjacent carbon atoms. This results in tails that are relatively straight. In contrast, unsaturated fatty acids do not contain a maximal number of hydrogen atoms, but they do contain some double bonds between adjacent carbon atoms. A double bond results in a bend in the carbon string of approximately 30 degrees (Figure 5.3).

Thus, if decreasing temperatures compress saturated fatty acids with their straight tails, they press in on each other, making a dense and fairly rigid membrane. If unsaturated fatty acids are compressed, the "kinks" in their tails elbow adjacent phospholipid molecules away, maintaining some space between the phospholipid molecules. This "elbow room" helps to maintain fluidity in the membrane at temperatures at which membranes with saturated fatty acid tails in their phospholipids would "freeze" or solidify. The membrane's relative fluidity is particularly important in a cold environment. A cold environment usually compresses membranes comprised largely of saturated fatty acids, making them less fluid and more susceptible to rupturing. Many organisms (fish are one example) are capable of adapting to cold environments by changing the proportion of unsaturated fatty acids in their membranes in response to lower temperature.

LINK TO LEARNING

Visit this site (http://openstax.org/l/biological_memb) to see animations of the membranes' fluidity and mosaic quality.

Animals have an additional membrane constituent that assists in maintaining fluidity. Cholesterol, which lies alongside the phospholipids in the membrane, tends to dampen temperature effects on the membrane. Thus, this lipid functions as a buffer, preventing lower temperatures from inhibiting fluidity and preventing increased temperatures from increasing fluidity too much. Thus, cholesterol extends, in both directions, the temperature range in which the membrane is appropriately fluid and consequently functional. Cholesterol also serves other functions, such as organizing clusters of transmembrane proteins into lipid rafts.

Plasma Membrane Components and Functions

Component	Location
Phospholipid	Main membrane fabric
Cholesterol	Attached between phospholipids and between the two phospholipid layers
Integral proteins (for example, integrins)	Embedded within the phospholipid layer(s); may or may not penetrate through both layers
Peripheral proteins	On the phospholipid bilayer's inner or outer surface; not embedded within the phospholipids
Carbohydrates (components of glycoproteins and glycolipids)	Generally attached to proteins on the outside membrane layer

Table 5.1

CAREER CONNECTION

Immunologist

The variations in peripheral proteins and carbohydrates that affect a cell's recognition sites are of prime interest in immunology. In developing vaccines, researchers have been able to conquer many infectious diseases, such as smallpox, polio, diphtheria, and tetanus.

Immunologists are the physicians and scientists who research and develop vaccines, as well as treat and study allergies or other immune problems. Some immunologists study and treat autoimmune problems (diseases in which a person's immune system attacks his or her own cells or tissues, such as lupus) and immunodeficiencies, whether acquired (such as acquired immunodeficiency syndrome, or AIDS) or hereditary (such as severe combined immunodeficiency, or SCID). Immunologists also help treat organ transplantation patients, who must have their immune systems suppressed so that their bodies will not reject a transplanted organ. Some immunologists work to understand natural immunity and the effects of a person's environment on it. Others work on questions about how the immune system affects diseases such as cancer. In the past, researchers did not understand the importance of having a healthy immune system in preventing cancer.

To work as an immunologist, one must have a PhD or MD. In addition, immunologists undertake at least two to three years of training in an accredited program and must pass the American Board of Allergy and Immunology exam. Immunologists must possess knowledge of the human body's function as they relate to issues beyond immunization, and knowledge of pharmacology and medical technology, such as medications, therapies, test materials, and surgical procedures.

5.2 Passive Transport

By the end of this section, you will be able to do the following:

- Explain why and how passive transport occurs
- Understand the osmosis and diffusion processes
- Define tonicity and its relevance to passive transport

Plasma membranes must allow certain substances to enter and leave a cell, and prevent some harmful materials from entering and some essential materials from leaving. In other words, plasma membranes are **selectively permeable**—they allow some substances to pass through, but not others. If they were to lose this selectivity, the cell would no longer be able to sustain itself, and it would be destroyed. Some cells require larger amounts of specific substances. They must have a way of obtaining these materials from extracellular fluids. This may happen passively, as certain materials move back and forth, or the cell may have special mechanisms that facilitate transport. Some materials are so important to a cell that it spends some of its energy, hydrolyzing adenosine triphosphate (ATP), to obtain these materials. Red blood cells use some of their energy doing just that. Most cells spend the majority of their energy to maintain an imbalance of sodium and potassium ions between the cell's interior and exterior, as well as on protein synthesis.

The most direct forms of membrane transport are passive. **Passive transport** is a naturally occurring phenomenon and does not require the cell to exert any of its energy to accomplish the movement. In passive transport, substances move from an area of higher concentration to an area of lower concentration. A physical space in which there is a single substance concentration range has a **concentration gradient**.

Selective Permeability

Plasma membranes are asymmetric: the membrane's interior is not identical to its exterior. There is a considerable difference between the array of phospholipids and proteins between the two leaflets that form a membrane. On the membrane's interior, some proteins serve to anchor the membrane to cytoskeleton's fibers. There are peripheral proteins on the membrane's exterior that bind extracellular matrix elements. Carbohydrates, attached to lipids or proteins, are also on the plasma membrane's exterior surface. These carbohydrate complexes help the cell bind required substances in the extracellular fluid. This adds considerably to plasma membrane's selective nature (Figure 5.7).

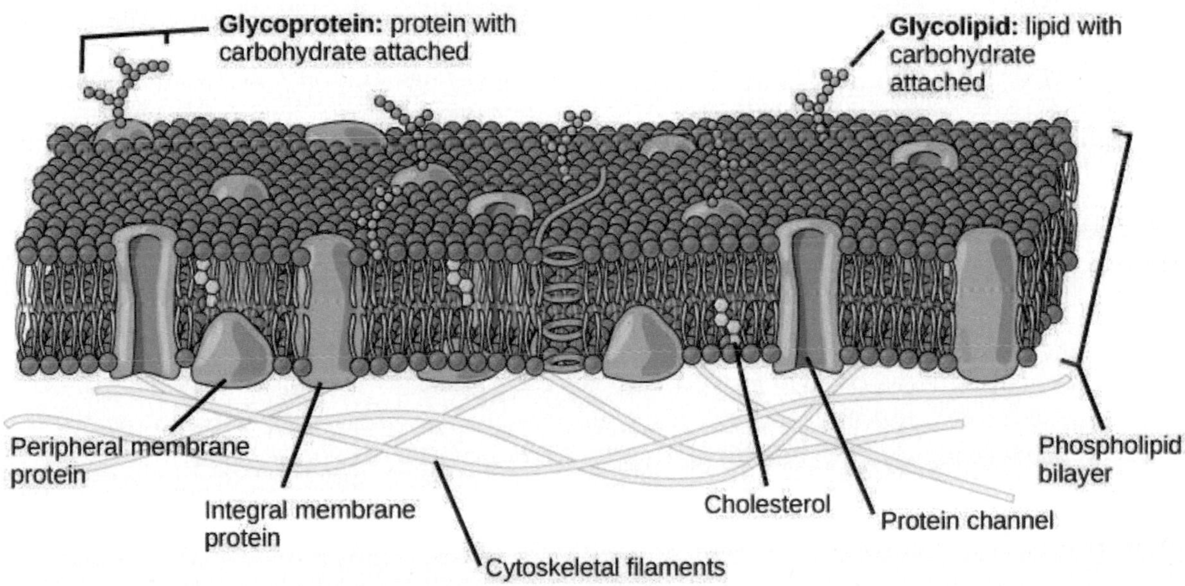

Figure 5.7 The plasma membrane's exterior surface is not identical to its interior surface.

Recall that plasma membranes are amphiphilic: They have hydrophilic and hydrophobic regions. This characteristic helps move some materials through the membrane and hinders the movement of others. Non-polar and lipid-soluble material with a low molecular weight can easily slip through the membrane's hydrophobic lipid core. Substances such as the fat-soluble vitamins A, D, E, and K readily pass through the plasma membranes in the digestive tract and other tissues. Fat-soluble drugs and hormones also gain easy entry into cells and readily transport themselves into the body's tissues and organs. Oxygen and carbon dioxide molecules have no charge and pass through membranes by simple diffusion.

Polar substances present problems for the membrane. While some polar molecules connect easily with the cell's outside, they cannot readily pass through the plasma membrane's lipid core. Additionally, while small ions could easily slip through the spaces in the membrane's mosaic, their charge prevents them from doing so. Ions such as sodium, potassium, calcium, and chloride must have special means of penetrating plasma membranes. Simple sugars and amino acids also need the help of various transmembrane proteins (channels) to transport themselves across plasma membranes.

Diffusion

Diffusion is a passive process of transport. A single substance moves from a high concentration to a low concentration area until the concentration is equal across a space. You are familiar with diffusion of substances through the air. For example, think about someone opening a bottle of ammonia in a room filled with people. The ammonia gas is at its highest concentration in the bottle. Its lowest concentration is at the room's edges. The ammonia vapor will diffuse, or spread away, from the bottle, and gradually, increasingly more people will smell the ammonia as it spreads. Materials move within the cell's cytosol by diffusion, and certain materials move through the plasma membrane by diffusion (Figure 5.8). Diffusion expends no energy. On the contrary, concentration gradients are a form of potential energy, which dissipates as the gradient is eliminated.

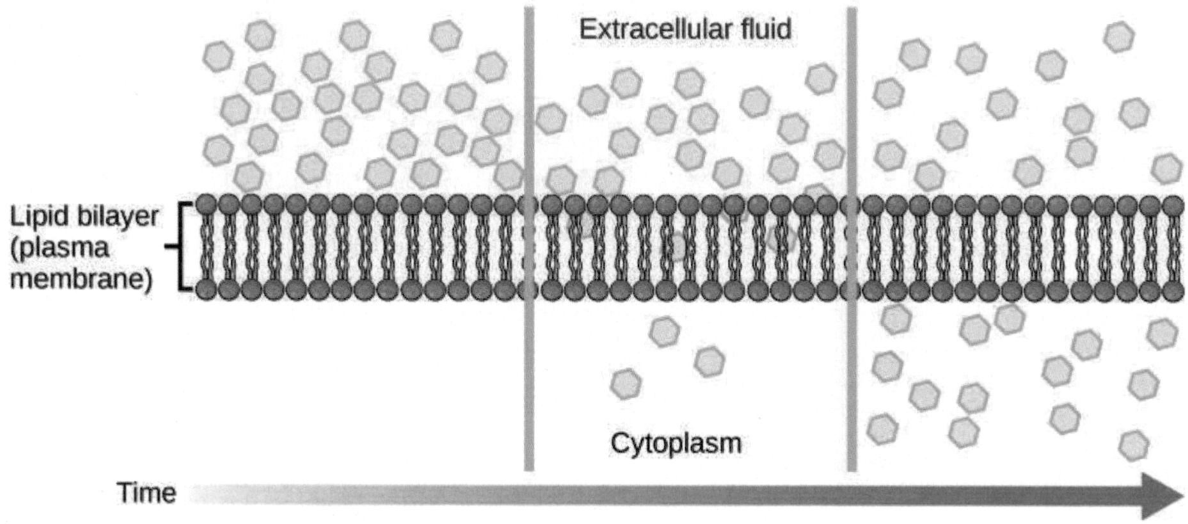

Figure 5.8 Diffusion through a permeable membrane moves a substance from a high concentration area (extracellular fluid, in this case) down its concentration gradient (into the cytoplasm). (credit: modification of work by Mariana Ruiz Villareal)

Each separate substance in a medium, such as the extracellular fluid, has its own concentration gradient, independent of other materials' concentration gradients. In addition, each substance will diffuse according to that gradient. Within a system, there will be different diffusion rates of various substances in the medium.

Factors That Affect Diffusion

Molecules move constantly in a random manner, at a rate that depends on their mass, their environment, and the amount of thermal energy they possess, which in turn is a function of temperature. This movement accounts for molecule diffusion through whatever medium in which they are localized. A substance moves into any space available to it until it evenly distributes itself throughout. After a substance has diffused completely through a space, removing its concentration gradient, molecules will still move around in the space, but there will be no *net* movement of the number of molecules from one area to another. We call this lack of a concentration gradient in which the substance has no net movement dynamic equilibrium. While diffusion will go forward in the presence of a substance's concentration gradient, several factors affect the diffusion rate.

- Extent of the concentration gradient: The greater the difference in concentration, the more rapid the diffusion. The closer the distribution of the material gets to equilibrium, the slower the diffusion rate.
- Mass of the molecules diffusing: Heavier molecules move more slowly; therefore, they diffuse more slowly. The reverse is true for lighter molecules.
- Temperature: Higher temperatures increase the energy and therefore the molecules' movement, increasing the diffusion rate. Lower temperatures decrease the molecules' energy, thus decreasing the diffusion rate.
- Solvent density: As the density of a solvent increases, the diffusion rate decreases. The molecules slow down because they

have a more difficult time passing through the denser medium. If the medium is less dense, diffusion increases. Because cells primarily use diffusion to move materials within the cytoplasm, any increase in the cytoplasm's density will inhibit the movement of the materials. An example of this is a person experiencing dehydration. As the body's cells lose water, the diffusion rate decreases in the cytoplasm, and the cells' functions deteriorate. Neurons tend to be very sensitive to this effect. Dehydration frequently leads to unconsciousness and possibly coma because of the decrease in diffusion rate within the cells.

- Solubility: As we discussed earlier, nonpolar or lipid-soluble materials pass through plasma membranes more easily than polar materials, allowing a faster diffusion rate.
- Surface area and plasma membrane thickness: Increased surface area increases the diffusion rate; whereas, a thicker membrane reduces it.
- Distance travelled: The greater the distance that a substance must travel, the slower the diffusion rate. This places an upper limitation on cell size. A large, spherical cell will die because nutrients or waste cannot reach or leave the cell's center, respectively. Therefore, cells must either be small in size, as in the case of many prokaryotes, or be flattened, as with many single-celled eukaryotes.

A variation of diffusion is the process of filtration. In filtration, material moves according to its concentration gradient through a membrane. Sometimes pressure enhances the diffusion rate, causing the substances to filter more rapidly. This occurs in the kidney, where blood pressure forces large amounts of water and accompanying dissolved substances, or **solutes**, out of the blood and into the renal tubules. The diffusion rate in this instance is almost totally dependent on pressure. One of the effects of high blood pressure is the appearance of protein in the urine, which abnormally high pressure "squeezes through".

Facilitated transport

In **facilitated transport**, or facilitated diffusion, materials diffuse across the plasma membrane with the help of membrane proteins. A concentration gradient exists that would allow these materials to diffuse into the cell without expending cellular energy. However, these materials are polar molecule ions that the cell membrane's hydrophobic parts repel. Facilitated transport proteins shield these materials from the membrane's repulsive force, allowing them to diffuse into the cell.

The transported material first attaches to protein or glycoprotein receptors on the plasma membrane's exterior surface. This allows removal of material from the extracellular fluid that the cell needs. The substances then pass to specific integral proteins that facilitate their passage. Some of these integral proteins are collections of beta-pleated sheets that form a pore or channel through the phospholipid bilayer. Others are carrier proteins which bind with the substance and aid its diffusion through the membrane.

Channels

The integral proteins involved in facilitated transport are **transport proteins**, and they function as either channels for the material or carriers. In both cases, they are transmembrane proteins. Channels are specific for the transported substance. **Channel proteins** have hydrophilic domains exposed to the intracellular and extracellular fluids. In addition, they have a hydrophilic channel through their core that provides a hydrated opening through the membrane layers (Figure 5.9). Passage through the channel allows polar compounds to avoid the plasma membrane's nonpolar central layer that would otherwise slow or prevent their entry into the cell. **Aquaporins** are channel proteins that allow water to pass through the membrane at a very high rate.

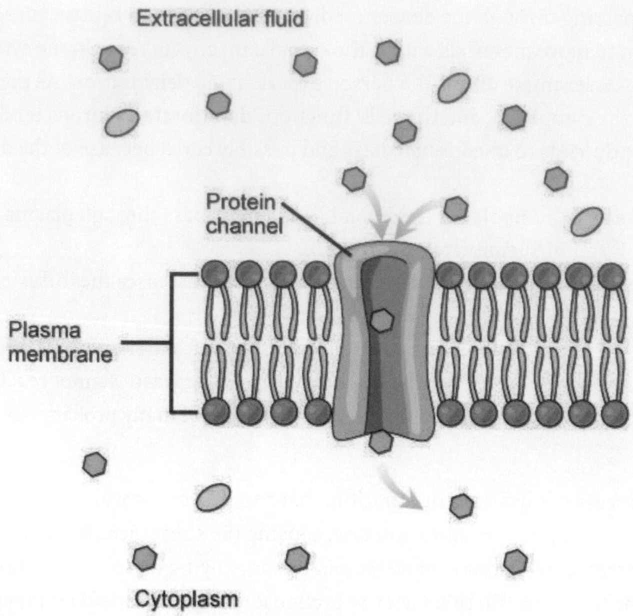

Figure 5.9 Facilitated transport moves substances down their concentration gradients. They may cross the plasma membrane with the aid of channel proteins. (credit: modification of work by Mariana Ruiz Villareal)

Channel proteins are either open at all times or they are "gated," which controls the channel's opening. When a particular ion attaches to the channel protein it may control the opening, or other mechanisms or substances may be involved. In some tissues, sodium and chloride ions pass freely through open channels; whereas, in other tissues a gate must open to allow passage. An example of this occurs in the kidney, where there are both channel forms in different parts of the renal tubules. Cells involved in transmitting electrical impulses, such as nerve and muscle cells, have gated channels for sodium, potassium, and calcium in their membranes. Opening and closing these channels changes the relative concentrations on opposing sides of the membrane of these ions, resulting in facilitating electrical transmission along membranes (in the case of nerve cells) or in muscle contraction (in the case of muscle cells).

Carrier Proteins

Another type of protein embedded in the plasma membrane is a **carrier protein**. This aptly named protein binds a substance and, thus triggers a change of its own shape, moving the bound molecule from the cell's outside to its interior (Figure 5.10). Depending on the gradient, the material may move in the opposite direction. Carrier proteins are typically specific for a single substance. This selectivity adds to the plasma membrane's overall selectivity. Scientists poorly understand the exact mechanism for the change of shape. Proteins can change shape when their hydrogen bonds are affected, but this may not fully explain this mechanism. Each carrier protein is specific to one substance, and there are a finite number of these proteins in any membrane. This can cause problems in transporting enough material for the cell to function properly. When all of the proteins are bound to their ligands, they are saturated and the rate of transport is at its maximum. Increasing the concentration gradient at this point will not result in an increased transport rate.

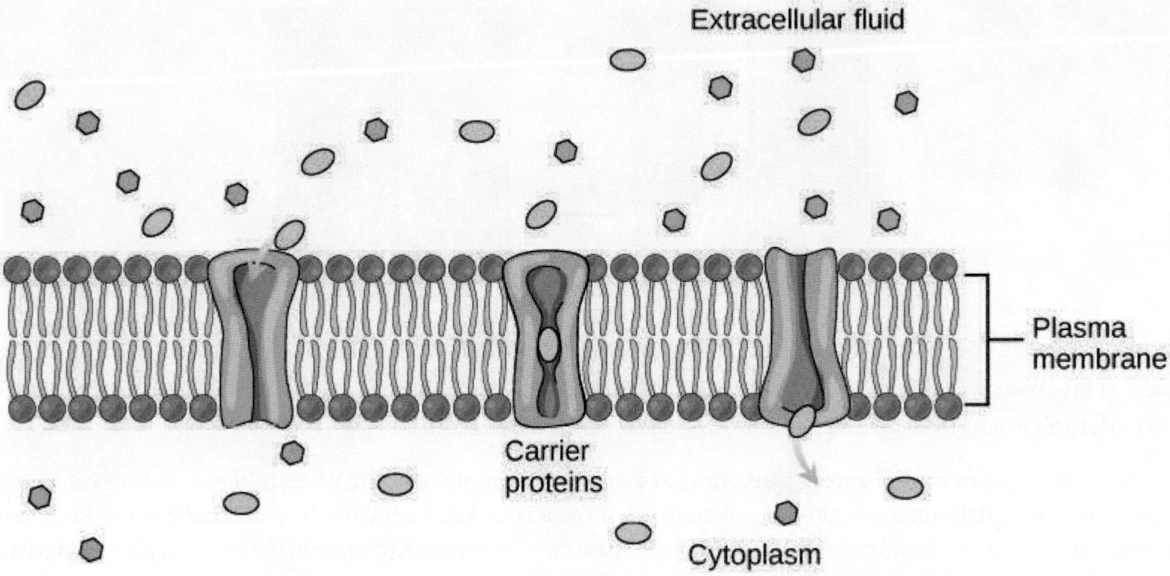

Figure 5.10 Some substances are able to move down their concentration gradient across the plasma membrane with the aid of carrier proteins. Carrier proteins change shape as they move molecules across the membrane. (credit: modification of work by Mariana Ruiz Villareal)

An example of this process occurs in the kidney. In one part, the kidney filters glucose, water, salts, ions, and amino acids that the body requires. This filtrate, which includes glucose, then reabsorbs in another part of the kidney. Because there are only a finite number of carrier proteins for glucose, if more glucose is present than the proteins can handle, the excess is not transported and the body excretes this through urine. In a diabetic individual, the term is "spilling glucose into the urine." A different group of carrier proteins, glucose transport proteins, or GLUTs, are involved in transporting glucose and other hexose sugars through plasma membranes within the body.

Channel and carrier proteins transport material at different rates. Channel proteins transport much more quickly than carrier proteins. Channel proteins facilitate diffusion at a rate of tens of millions of molecules per second; whereas, carrier proteins work at a rate of a thousand to a million molecules per second.

Osmosis

Osmosis is the movement of water through a semipermeable membrane according to the water's concentration gradient across the membrane, which is inversely proportional to the solutes' concentration. While diffusion transports material across membranes and within cells, osmosis transports *only water* across a membrane and the membrane limits the solutes' diffusion in the water. Not surprisingly, the aquaporins that facilitate water movement play a large role in osmosis, most prominently in red blood cells and the membranes of kidney tubules.

Mechanism

Osmosis is a special case of diffusion. Water, like other substances, moves from an area of high concentration to one of low concentration. An obvious question is what makes water move at all? Imagine a beaker with a semipermeable membrane separating the two sides or halves (Figure 5.11). On both sides of the membrane the water level is the same, but there are different dissolved substance concentrations, or **solute**, that cannot cross the membrane (otherwise the solute crossing the membrane would balance concentrations on each side). If the solution's volume on both sides of the membrane is the same, but the solute's concentrations are different, then there are different amounts of water, the solvent, on either side of the membrane.

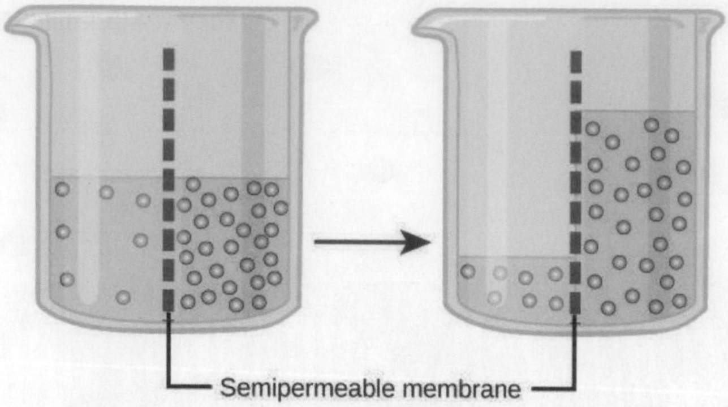

Figure 5.11 In osmosis, water always moves from an area of higher water concentration to one of lower concentration. In the diagram, the solute cannot pass through the selectively permeable membrane, but the water can.

To illustrate this, imagine two full water glasses. One has a single teaspoon of sugar in it; whereas, the second one contains one-quarter cup of sugar. If the total volume of the solutions in both cups is the same, which cup contains more water? Because the large sugar amount in the second cup takes up much more space than the teaspoon of sugar in the first cup, the first cup has more water in it.

Returning to the beaker example, recall that it has a solute mixture on either side of the membrane. A principle of diffusion is that the molecules move around and will spread evenly throughout the medium if they can. However, only the material capable of getting through the membrane will diffuse through it. In this example, the solute cannot diffuse through the membrane, but the water can. Water has a concentration gradient in this system. Thus, water will diffuse down its concentration gradient, crossing the membrane to the side where it is less concentrated. This diffusion of water through the membrane—osmosis—will continue until the water's concentration gradient goes to zero or until the water's hydrostatic pressure balances the osmotic pressure. Osmosis proceeds constantly in living systems.

Tonicity

Tonicity describes how an extracellular solution can change a cell's volume by affecting osmosis. A solution's tonicity often directly correlates with the solution's osmolarity. **Osmolarity** describes the solution's total solute concentration. A solution with low osmolarity has a greater number of water molecules relative to the number of solute particles. A solution with high osmolarity has fewer water molecules with respect to solute particles. In a situation in which a membrane permeable to water, though not to the solute separates two different osmolarities, water will move from the membrane's side with lower osmolarity (and more water) to the side with higher osmolarity (and less water). This effect makes sense if you remember that the solute cannot move across the membrane, and thus the only component in the system that can move—the water—moves along its own concentration gradient. An important distinction that concerns living systems is that osmolarity measures the number of particles (which may be molecules) in a solution. Therefore, a solution that is cloudy with cells may have a lower osmolarity than a solution that is clear, if the second solution contains more dissolved molecules than there are cells.

Hypotonic Solutions

Scientists use three terms—hypotonic, isotonic, and hypertonic—to relate the cell's osmolarity to the extracellular fluid's osmolarity that contains the cells. In a **hypotonic** situation, the extracellular fluid has lower osmolarity than the fluid inside the cell, and water enters the cell. (In living systems, the point of reference is always the cytoplasm, so the prefix *hypo-* means that the extracellular fluid has a lower solute concentration, or a lower osmolarity, than the cell cytoplasm.) It also means that the extracellular fluid has a higher water concentration in the solution than does the cell. In this situation, water will follow its concentration gradient and enter the cell.

Hypertonic Solutions

As for a **hypertonic** solution, the prefix *hyper-* refers to the extracellular fluid having a higher osmolarity than the cell's cytoplasm; therefore, the fluid contains less water than the cell does. Because the cell has a relatively higher water concentration, water will leave the cell.

Isotonic Solutions

In an **isotonic** solution, the extracellular fluid has the same osmolarity as the cell. If the cell's osmolarity matches that of the extracellular fluid, there will be no net movement of water into or out of the cell, although water will still move in and out. Blood cells and plant cells in hypertonic, isotonic, and hypotonic solutions take on characteristic appearances (Figure 5.12).

VISUAL CONNECTION

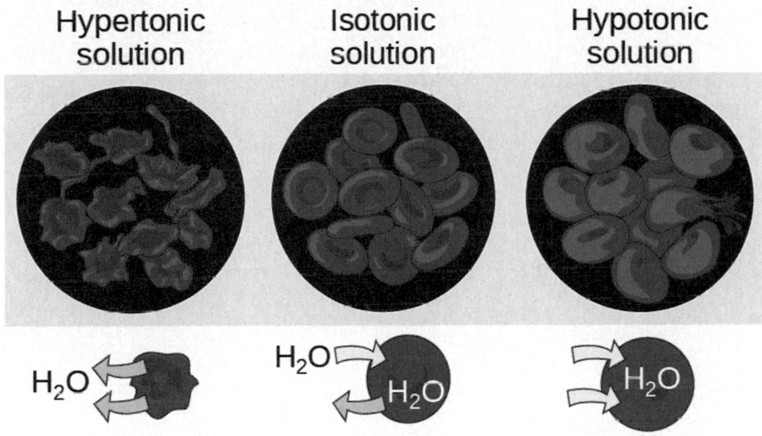

Figure 5.12 Osmotic pressure changes red blood cells' shape in hypertonic, isotonic, and hypotonic solutions. (credit: Mariana Ruiz Villareal)

A doctor injects a patient with what the doctor thinks is an isotonic saline solution. The patient dies, and an autopsy reveals that many red blood cells have been destroyed. Do you think the solution the doctor injected was really isotonic?

LINK TO LEARNING

For a video illustrating the diffusion process in solutions, visit this site (http://openstax.org/l/dispersion) .

Tonicity in Living Systems

In a hypotonic environment, water enters a cell, and the cell swells. In an isotonic condition, the relative solute and solvent concentrations are equal on both membrane sides. There is no net water movement; therefore, there is no change in the cell's size. In a hypertonic solution, water leaves a cell and the cell shrinks. If either the hypo- or hyper- condition goes to excess, the cell's functions become compromised, and the cell may be destroyed.

A red blood cell will burst, or lyse, when it swells beyond the plasma membrane's capability to expand. Remember, the membrane resembles a mosaic, with discrete spaces between the molecules comprising it. If the cell swells, and the spaces between the lipids and proteins become too large, the cell will break apart.

In contrast, when excessive water amounts leave a red blood cell, the cell shrinks, or crenates. This has the effect of concentrating the solutes left in the cell, making the cytosol denser and interfering with diffusion within the cell. The cell's ability to function will be compromised and may also result in the cell's death.

Various living things have ways of controlling the effects of osmosis—a mechanism we call osmoregulation. Some organisms, such as plants, fungi, bacteria, and some protists, have cell walls that surround the plasma membrane and prevent cell lysis in a hypotonic solution. The plasma membrane can only expand to the cell wall's limit, so the cell will not lyse. The cytoplasm in plants is always slightly hypertonic to the cellular environment, and water will always enter a cell if water is available. This water inflow produces turgor pressure, which stiffens the plant's cell walls (Figure 5.13). In nonwoody plants, turgor pressure supports the plant. Conversly, if you do not water the plant, the extracellular fluid will become hypertonic, causing water to leave the cell. In this condition, the cell does not shrink because the cell wall is not flexible. However, the cell membrane detaches from the wall and constricts the cytoplasm. We call this **plasmolysis**. Plants lose turgor pressure in this condition and wilt (Figure 5.14).

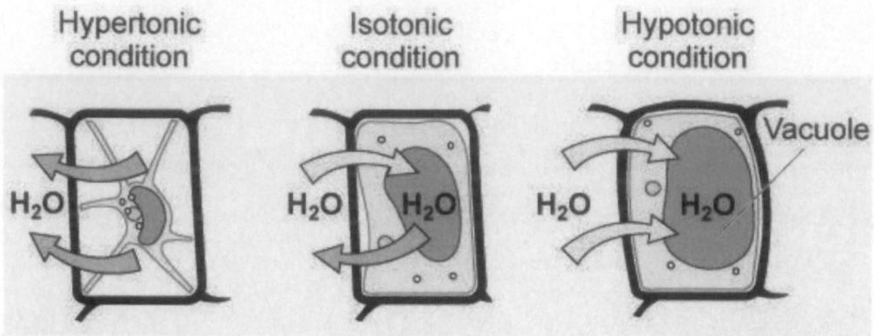

Figure 5.13 The turgor pressure within a plant cell depends on the solution's tonicity in which it is bathed. (credit: modification of work by Mariana Ruiz Villareal)

Figure 5.14 Without adequate water, the plant on the left has lost turgor pressure, visible in its wilting. Watering the plant (right) will restore the turgor pressure. (credit: Victor M. Vicente Selvas)

Tonicity is a concern for all living things. For example, paramecia and amoebas, which are protists that lack cell walls, have contractile vacuoles. This vesicle collects excess water from the cell and pumps it out, keeping the cell from lysing as it takes on water from its environment (Figure 5.15).

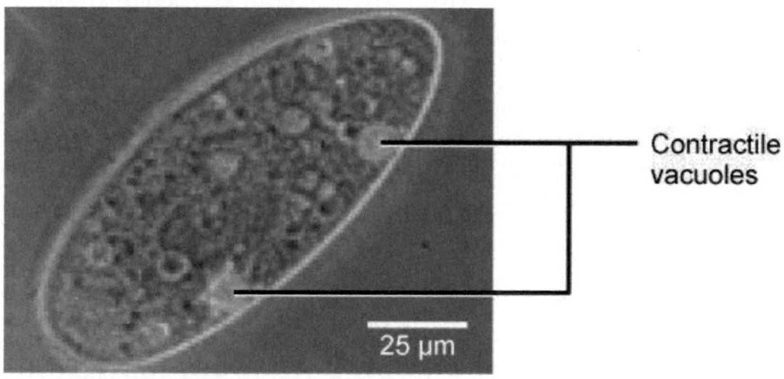

Figure 5.15 A paramecium's contractile vacuole, here visualized using bright field light microscopy at 480x magnification, continuously pumps water out of the organism's body to keep it from bursting in a hypotonic medium. (credit: modification of work by NIH; scale-bar data from Matt Russell)

Many marine invertebrates have internal salt levels matched to their environments, making them isotonic with the water in which they live. Fish, however, must spend approximately five percent of their metabolic energy maintaining osmotic homeostasis. Freshwater fish live in an environment that is hypotonic to their cells. These fish actively take in salt through their gills and excrete diluted urine to rid themselves of excess water. Saltwater fish live in the reverse environment, which is hypertonic to their cells, and they secrete salt through their gills and excrete highly concentrated urine.

In vertebrates, the kidneys regulate the water amount in the body. Osmoreceptors are specialized cells in the brain that monitor solute concentration in the blood. If the solute levels increase beyond a certain range, a hormone releases that slows water loss through the kidney and dilutes the blood to safer levels. Animals also have high albumin concentrations, which the liver produces, in their blood. This protein is too large to pass easily through plasma membranes and is a major factor in controlling the osmotic pressures applied to tissues.

5.3 Active Transport

By the end of this section, you will be able to do the following:

- Understand how electrochemical gradients affect ions
- Distinguish between primary active transport and secondary active transport

Active transport mechanisms require the cell's energy, usually in the form of adenosine triphosphate (ATP). If a substance must move into the cell against its concentration gradient—that is, if the substance's concentration inside the cell is greater than its concentration in the extracellular fluid (and vice versa)—the cell must use energy to move the substance. Some active transport mechanisms move small-molecular weight materials, such as ions, through the membrane. Other mechanisms transport much larger molecules.

Electrochemical Gradient

We have discussed simple concentration gradients—a substance's differential concentrations across a space or a membrane—but in living systems, gradients are more complex. Because ions move into and out of cells and because cells contain proteins that do not move across the membrane and are mostly negatively charged, there is also an electrical gradient, a difference of charge, across the plasma membrane. The interior of living cells is electrically negative with respect to the extracellular fluid in which they are bathed, and at the same time, cells have higher concentrations of potassium (K^+) and lower concentrations of sodium (Na^+) than the extracellular fluid. Thus in a living cell, the concentration gradient of Na^+ tends to drive it into the cell, and its electrical gradient (a positive ion) also drives it inward to the negatively charged interior. However, the situation is more complex for other elements such as potassium. The electrical gradient of K^+, a positive ion, also drives it into the cell, but the concentration gradient of K^+ drives K^+ *out* of the cell (Figure 5.16). We call the combined concentration gradient and electrical charge that affects an ion its **electrochemical gradient**.

VISUAL CONNECTION

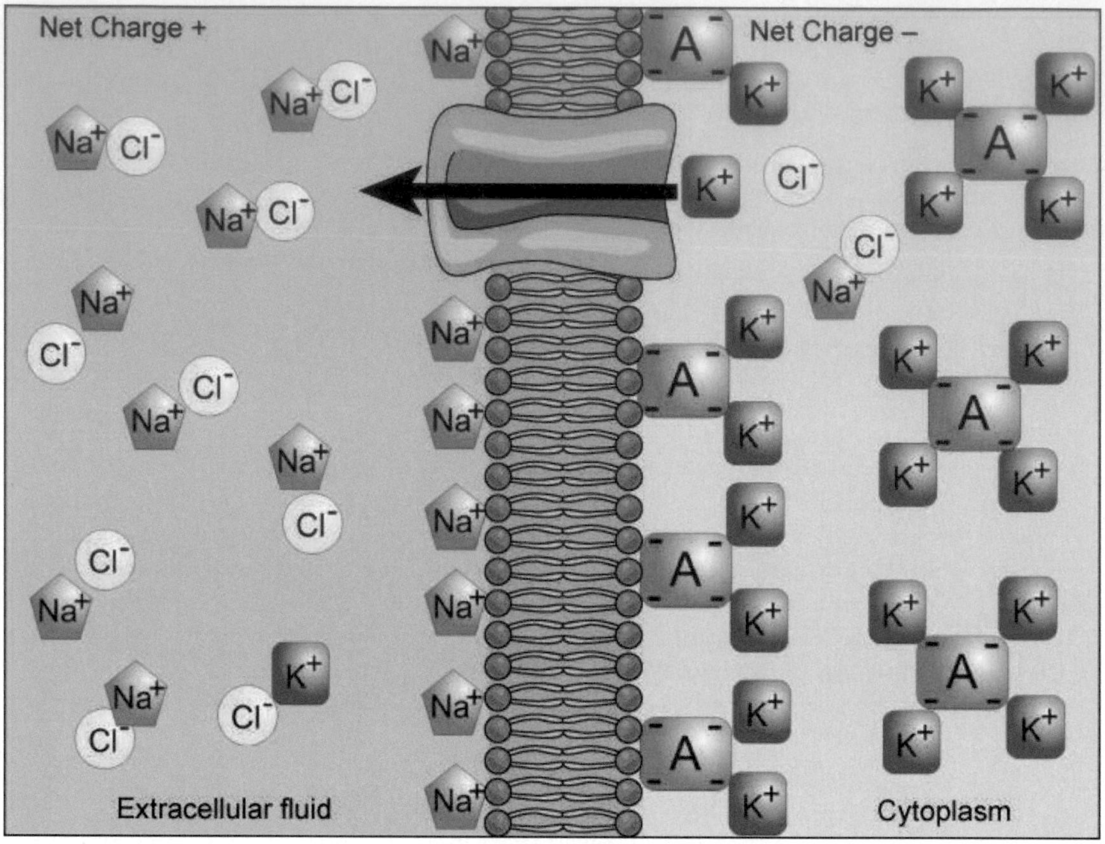

Figure 5.16 Electrochemical gradients arise from the combined effects of concentration gradients and electrical gradients. Structures labeled A represent proteins. (credit: “Synaptitude”/Wikimedia Commons)

Injecting a potassium solution into a person’s blood is lethal. This is how capital punishment and euthanasia subjects die. Why do you think a potassium solution injection is lethal?

Moving Against a Gradient

To move substances against a concentration or electrochemical gradient, the cell must use energy. This energy comes from ATP generated through the cell’s metabolism. Active transport mechanisms, or **pumps**, work against electrochemical gradients. Small substances constantly pass through plasma membranes. Active transport maintains concentrations of ions and other substances that living cells require in the face of these passive movements. A cell may spend much of its metabolic energy supply maintaining these processes. (A red blood cell uses most of its metabolic energy to maintain the imbalance between exterior and interior sodium and potassium levels that the cell requires.) Because active transport mechanisms depend on a cell’s metabolism for energy, they are sensitive to many metabolic poisons that interfere with the ATP supply.

Two mechanisms exist for transporting small-molecular weight material and small molecules. **Primary active transport** moves ions across a membrane and creates a difference in charge across that membrane, which is directly dependent on ATP. **Secondary active transport** does not directly require ATP: instead, it is the movement of material due to the electrochemical gradient established by primary active transport.

Carrier Proteins for Active Transport

An important membrane adaption for active transport is the presence of specific carrier proteins or pumps to facilitate movement: there are three protein types or **transporters** (Figure 5.17). A **uniporter** carries one specific ion or molecule. A **symporter** carries two different ions or molecules, both in the same direction. An **antiporter** also carries two different ions or molecules, but in different directions. All of these transporters can also transport small, uncharged organic molecules like glucose. These three types of carrier proteins are also in facilitated diffusion, but they do not require ATP to work in that process. Some examples of pumps for active transport are Na^+-K^+ ATPase, which carries sodium and potassium ions, and H^+-K^+ ATPase, which carries hydrogen and potassium ions. Both of these are antiporter carrier proteins. Two other carrier proteins are Ca^{2+} ATPase and H^+ ATPase, which carry only calcium and only hydrogen ions, respectively. Both are pumps.

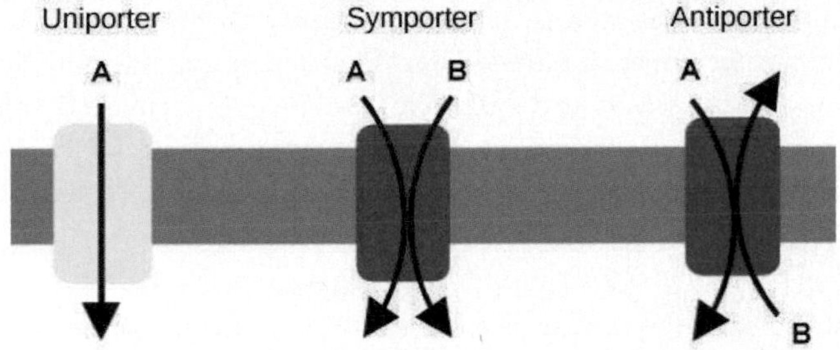

Figure 5.17 A uniporter carries one molecule or ion. A symporter carries two different molecules or ions, both in the same direction. An antiporter also carries two different molecules or ions, but in different directions. (credit: modification of work by "Lupask"/Wikimedia Commons)

Primary Active Transport

The primary active transport that functions with the active transport of sodium and potassium allows secondary active transport to occur. The second transport method is still active because it depends on using energy as does primary transport (Figure 5.18).

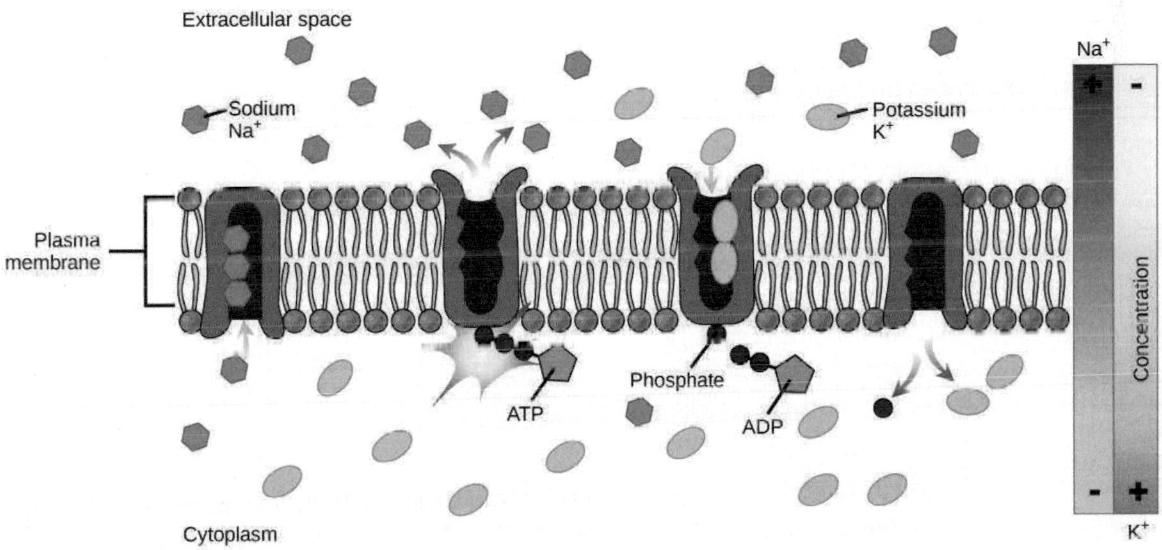

Figure 5.18 Primary active transport moves ions across a membrane, creating an electrochemical gradient (electrogenic transport). (credit: modification of work by Mariana Ruiz Villareal)

One of the most important pumps in animal cells is the sodium-potassium pump (Na^+-K^+ ATPase), which maintains the electrochemical gradient (and the correct concentrations of Na^+ and K^+) in living cells. The sodium-potassium pump moves K^+ into the cell while moving Na^+ out at the same time, at a ratio of three Na^+ for every two K^+ ions moved in. The Na^+-K^+ ATPase exists in two forms, depending on its orientation to the cell's interior or exterior and its affinity for either sodium or potassium ions. The process consists of the following six steps.

1. With the enzyme oriented towards the cell's interior, the carrier has a high affinity for sodium ions. Three ions bind to the protein.
2. The protein carrier hydrolyzes ATP and a low-energy phosphate group attaches to it.
3. As a result, the carrier changes shape and reorients itself towards the membrane's exterior. The protein's affinity for sodium decreases and the three sodium ions leave the carrier.
4. The shape change increases the carrier's affinity for potassium ions, and two such ions attach to the protein. Subsequently, the low-energy phosphate group detaches from the carrier.
5. With the phosphate group removed and potassium ions attached, the carrier protein repositions itself towards the cell's

interior.

6. The carrier protein, in its new configuration, has a decreased affinity for potassium, and the two ions moves into the cytoplasm. The protein now has a higher affinity for sodium ions, and the process starts again.

Several things have happened as a result of this process. At this point, there are more sodium ions outside the cell than inside and more potassium ions inside than out. For every three sodium ions that move out, two potassium ions move in. This results in the interior being slightly more negative relative to the exterior. This difference in charge is important in creating the conditions necessary for the secondary process. The sodium-potassium pump is, therefore, an **electrogenic pump** (a pump that creates a charge imbalance), creating an electrical imbalance across the membrane and contributing to the membrane potential.

LINK TO LEARNING

Watch this video (https://openstax.org/l/Na_K_ATPase) to see an active transport simulation in a sodium-potassium ATPase.

Secondary Active Transport (Co-transport)

Secondary active transport brings sodium ions, and possibly other compounds, into the cell. As sodium ion concentrations build outside of the plasma membrane because of the primary active transport process, this creates an electrochemical gradient. If a channel protein exists and is open, the sodium ions will pull through the membrane. This movement transports other substances that can attach themselves to the transport protein through the membrane (Figure 5.19). Many amino acids, as well as glucose, enter a cell this way. This secondary process also stores high-energy hydrogen ions in the mitochondria of plant and animal cells in order to produce ATP. The potential energy that accumulates in the stored hydrogen ions translates into kinetic energy as the ions surge through the channel protein ATP synthase, and that energy then converts ADP into ATP.

VISUAL CONNECTION

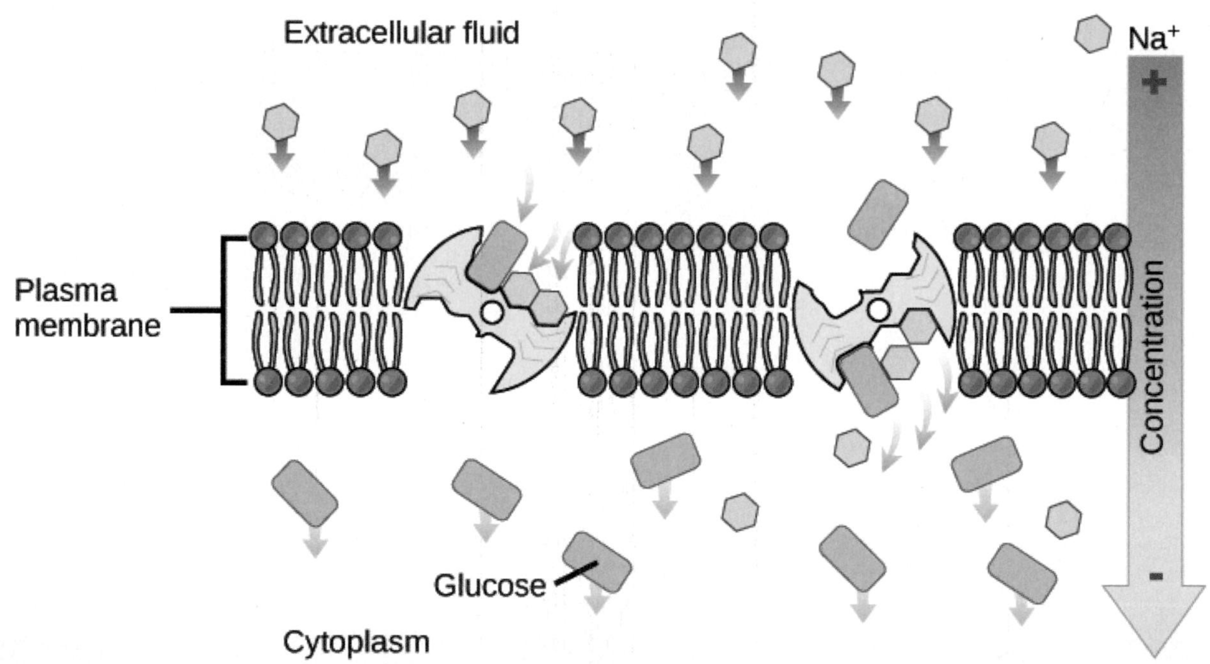

Figure 5.19 An electrochemical gradient, which primary active transport creates, can move other substances against their concentration gradients, a process scientists call co-transport or secondary active transport. (credit: modification of work by Mariana Ruiz Villareal)

If the pH outside the cell decreases, would you expect the amount of amino acids transported into the cell to increase or decrease?

5.4 Bulk Transport

By the end of this section, you will be able to do the following:

- Describe endocytosis, including phagocytosis, pinocytosis, and receptor-mediated endocytosis
- Understand the process of exocytosis

In addition to moving small ions and molecules through the membrane, cells also need to remove and take in larger molecules and particles (see Table 5.2 for examples). Some cells are even capable of engulfing entire unicellular microorganisms. You might have correctly hypothesized that when a cell uptakes and releases large particles, it requires energy. A large particle, however, cannot pass through the membrane, even with energy that the cell supplies.

Endocytosis

Endocytosis is a type of active transport that moves particles, such as large molecules, parts of cells, and even whole cells, into a cell. There are different endocytosis variations, but all share a common characteristic: the cell's plasma membrane invaginates, forming a pocket around the target particle. The pocket pinches off, resulting in the particle containing itself in a newly created intracellular vesicle formed from the plasma membrane.

Phagocytosis

Phagocytosis (the condition of "cell eating") is the process by which a cell takes in large particles, such as other cells or relatively large particles. For example, when microorganisms invade the human body, a type of white blood cell, a neutrophil, will remove the invaders through this process, surrounding and engulfing the microorganism, which the neutrophil then destroys (Figure 5.20).

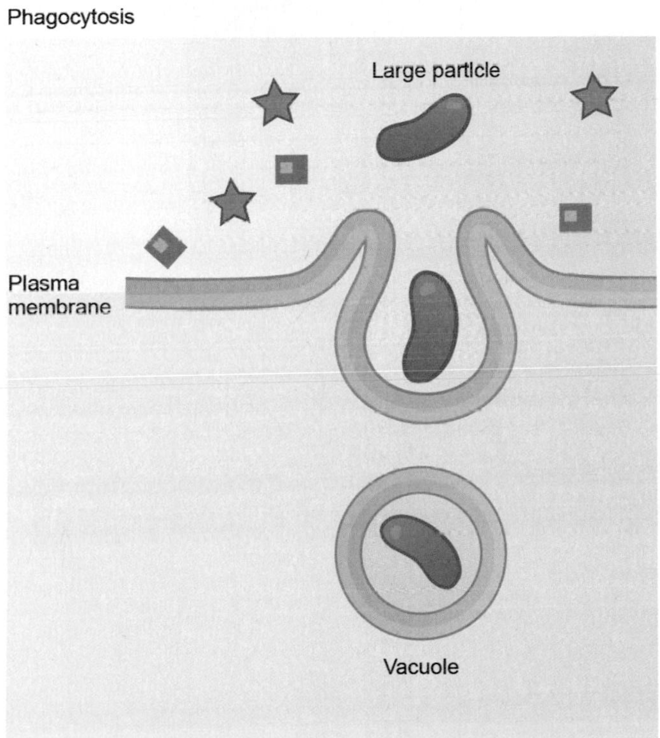

Figure 5.20 In phagocytosis, the cell membrane surrounds the particle and engulfs it. (credit: modification of work by Mariana Ruiz Villareal)

In preparation for phagocytosis, a portion of the plasma membrane's inward-facing surface becomes coated with the protein **clathrin**, which stabilizes this membrane's section. The membrane's coated portion then extends from the cell's body and surrounds the particle, eventually enclosing it. Once the vesicle containing the particle is enclosed within the cell, the clathrin disengages from the membrane and the vesicle merges with a lysosome for breaking down the material in the newly formed compartment (endosome). When accessible nutrients from the vesicular contents' degradation have been extracted, the newly formed endosome merges with the plasma membrane and releases its contents into the extracellular fluid. The endosomal membrane again becomes part of the plasma membrane.

Pinocytosis

A variation of endocytosis is **pinocytosis**. This literally means "cell drinking". Discovered by Warren Lewis in 1929, this American embryologist and cell biologist described a process whereby he assumed that the cell was purposefully taking in extracellular fluid. In reality, this is a process that takes in molecules, including water, which the cell needs from the extracellular fluid. Pinocytosis results in a much smaller vesicle than does phagocytosis, and the vesicle does not need to merge with a lysosome (Figure 5.21).

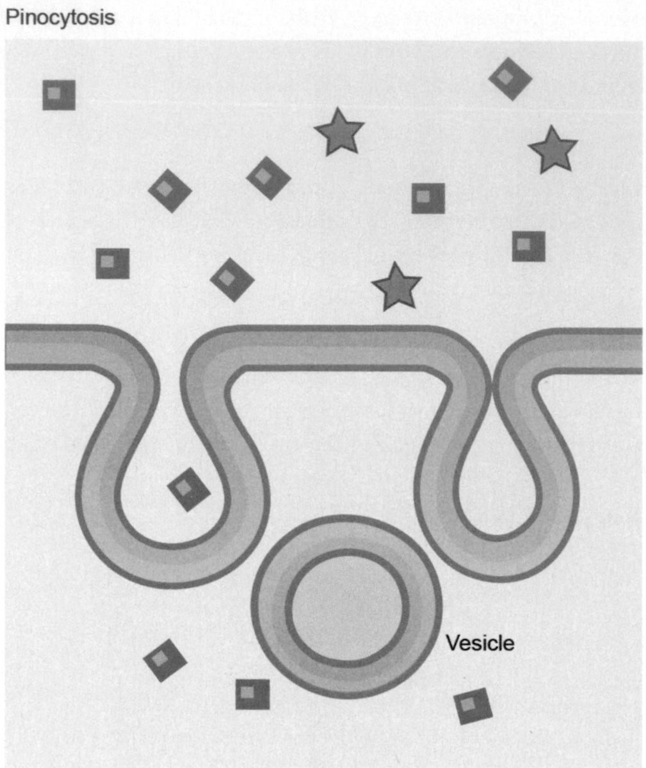

Figure 5.21 In pinocytosis, the cell membrane invaginates, surrounds a small volume of fluid, and pinches off. (credit: modification of work by Mariana Ruiz Villareal)

A variation of pinocytosis is **potocytosis**. This process uses a coating protein, **caveolin**, on the plasma membrane's cytoplasmic side, which performs a similar function to clathrin. The cavities in the plasma membrane that form the vacuoles have membrane receptors and lipid rafts in addition to caveolin. The vacuoles or vesicles formed in caveolae (singular caveola) are smaller than those in pinocytosis. Potocytosis brings small molecules into the cell and transports them through the cell for their release on the other side, a process we call transcytosis.

Receptor-mediated Endocytosis

A targeted variation of endocytosis employs receptor proteins in the plasma membrane that have a specific binding affinity for certain substances (Figure 5.22).

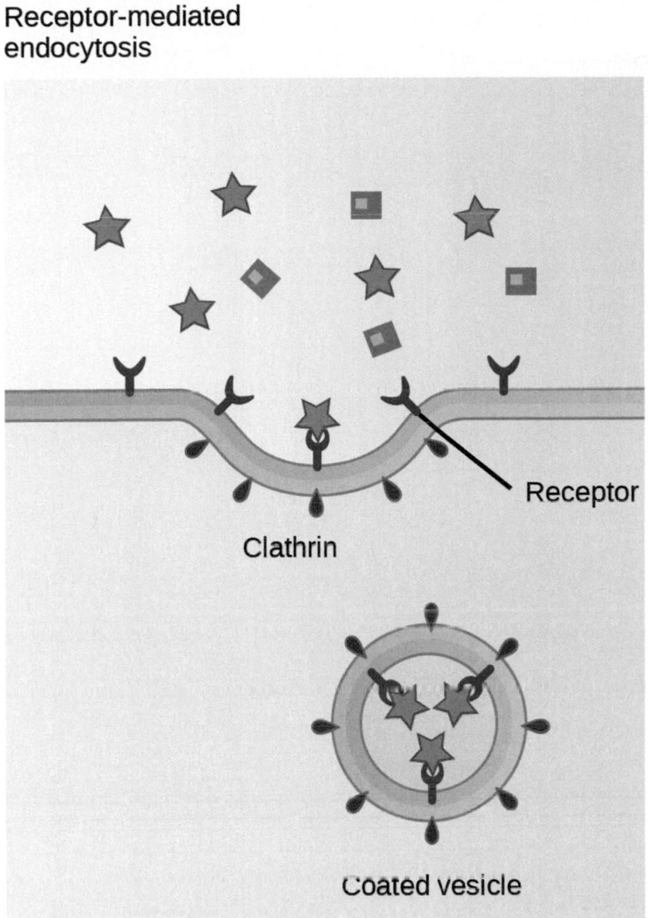

Figure 5.22 In receptor-mediated endocytosis, the cell's uptake of substances targets a single type of substance that binds to the receptor on the cell membrane's external surface. (credit: modification of work by Mariana Ruiz Villareal)

In **receptor-mediated endocytosis**, as in phagocytosis, clathrin attaches to the plasma membrane's cytoplasmic side. If a compound's uptake is dependent on receptor-mediated endocytosis and the process is ineffective, the material will not be removed from the tissue fluids or blood. Instead, it will stay in those fluids and increase in concentration. The failure of receptor-mediated endocytosis causes some human diseases. For example, receptor mediated endocytosis removes low density lipoprotein or LDL (or "bad" cholesterol) from the blood. In the human genetic disease familial hypercholesterolemia, the LDL receptors are defective or missing entirely. People with this condition have life-threatening levels of cholesterol in their blood, because their cells cannot clear LDL particles.

Although receptor-mediated endocytosis is designed to bring specific substances that are normally in the extracellular fluid into the cell, other substances may gain entry into the cell at the same site. Flu viruses, diphtheria, and cholera toxin all have sites that cross-react with normal receptor-binding sites and gain entry into cells.

LINK TO LEARNING

See receptor-mediated endocytosis in action, and click on different parts (http://openstax.org/l/endocytosis) for a focused animation.

Exocytosis

The reverse process of moving material into a cell is the process of exocytosis. **Exocytosis** is the opposite of the processes we discussed above in that its purpose is to expel material from the cell into the extracellular fluid. Waste material is enveloped in a membrane and fuses with the plasma membrane's interior. This fusion opens the membranous envelope on the cell's exterior, and the waste material expels into the extracellular space (Figure 5.23). Other examples of cells releasing molecules via exocytosis include extracellular matrix protein secretion and neurotransmitter secretion into the synaptic cleft by synaptic vesicles.

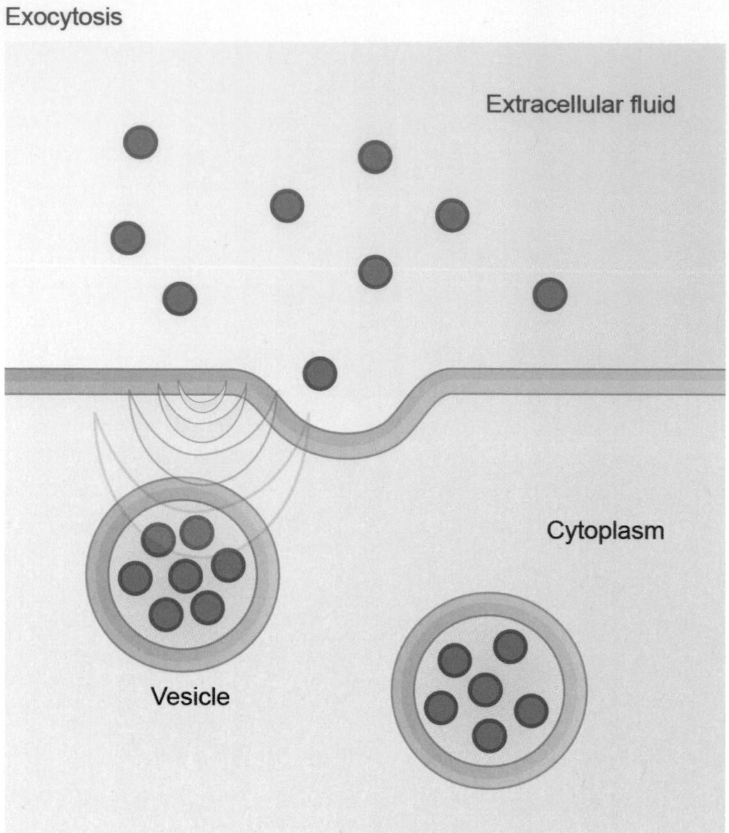

Figure 5.23 In exocytosis, vesicles containing substances fuse with the plasma membrane. The contents then release to the cell's exterior. (credit: modification of work by Mariana Ruiz Villareal)

Methods of Transport, Energy Requirements, and Types of Transported Material

Transport Method	Active/Passive	Material Transported
Diffusion	Passive	Small-molecular weight material
Osmosis	Passive	Water
Facilitated transport/diffusion	Passive	Sodium, potassium, calcium, glucose
Primary active transport	Active	Sodium, potassium, calcium
Secondary active transport	Active	Amino acids, lactose
Phagocytosis	Active	Large macromolecules, whole cells, or cellular structures
Pinocytosis and potocytosis	Active	Small molecules (liquids/water)
Receptor-mediated endocytosis	Active	Large quantities of macromolecules

Table 5.2

KEY TERMS

active transport method of transporting material that requires energy

amphiphilic molecule possessing a polar or charged area and a nonpolar or uncharged area capable of interacting with both hydrophilic and hydrophobic environments

antiporter transporter that carries two ions or small molecules in different directions

aquaporin channel protein that allows water through the membrane at a very high rate

carrier protein membrane protein that moves a substance across the plasma membrane by changing its own shape

caveolin protein that coats the plasma membrane's cytoplasmic side and participates in the liquid uptake process by potocytosis

channel protein membrane protein that allows a substance to pass through its hollow core across the plasma membrane

clathrin protein that coats the plasma membrane's inward-facing surface and assists in forming specialized structures, like coated pits, for phagocytosis

concentration gradient area of high concentration adjacent to an area of low concentration

diffusion passive transport process of low-molecular weight material according to its concentration gradient

electrochemical gradient a combined electrical and chemical force that produces a gradient

electrogenic pump pump that creates a charge imbalance

endocytosis type of active transport that moves substances, including fluids and particles, into a cell

exocytosis process of passing bulk material out of a cell

facilitated transport process by which material moves down a concentration gradient (from high to low concentration) using integral membrane proteins

fluid mosaic model describes the plasma membrane's structure as a mosaic of components including phospholipids, cholesterol, proteins, glycoproteins, and glycolipids (sugar chains attached to proteins or lipids, respectively), resulting in a fluid character (fluidity)

glycolipid combination of carbohydrates and lipids

glycoprotein combination of carbohydrates and proteins

hydrophilic molecule with the ability to bond with water; "water-loving"

hydrophobic molecule that does not have the ability to bond with water; "water-hating"

hypertonic situation in which extracellular fluid has a higher osmolarity than the fluid inside the cell, resulting in water moving out of the cell

hypotonic situation in which extracellular fluid has a lower osmolarity than the fluid inside the cell, resulting in water moving into the cell

integral protein protein integrated into the membrane structure that interacts extensively with the membrane lipids' hydrocarbon chains and often spans the membrane

isotonic situation in which the extracellular fluid has the same osmolarity as the fluid inside the cell, resulting in no net water movement into or out of the cell

osmolarity total amount of substances dissolved in a specific amount of solution

osmosis transport of water through a semipermeable membrane according to the water's concentration gradient across the membrane that results from the presence of solute that cannot pass through the membrane

passive transport method of transporting material through a membrane that does not require energy

peripheral protein protein at the plasma membrane's surface either on its exterior or interior side

pinocytosis a variation of endocytosis that imports macromolecules that the cell needs from the extracellular fluid

plasmolysis detaching the cell membrane from the cell wall and constricting the cell membrane when a plant cell is in a hypertonic solution

potocytosis variation of pinocytosis that uses a different coating protein (caveolin) on the plasma membrane's cytoplasmic side

primary active transport active transport that moves ions or small molecules across a membrane and may create a difference in charge across that membrane

pump active transport mechanism that works against electrochemical gradients

receptor-mediated endocytosis variation of endocytosis that involves using specific binding proteins in the plasma membrane for specific molecules or particles, and clathrin-coated pits that become clathrin-coated vesicles

secondary active transport movement of material that results from primary active transport to the electrochemical gradient

selectively permeable membrane characteristic that allows some substances through

solute substance dissolved in a liquid to form a solution

symporter transporter that carries two different ions or small molecules, both in the same direction

tonicity amount of solute in a solution

transport protein membrane protein that facilitates a substance's passage across a membrane by binding it

transporter specific carrier proteins or pumps that facilitate movement

uniporter transporter that carries one specific ion or molecule

CHAPTER SUMMARY

5.1 Components and Structure

Modern scientists refer to the plasma membrane as the fluid mosaic model. A phospholipid bilayer comprises the plasma membrane, with hydrophobic, fatty acid tails in contact with each other. The membrane's landscape is studded with proteins, some which span the membrane. Some of these proteins serve to transport materials into or out of the cell. Carbohydrates are attached to some of the proteins and lipids on the membrane's outward-facing surface, forming complexes that function to identify the cell to other cells. The membrane's fluid nature is due to temperature, fatty acid tail configuration (some kinked by double bonds), cholesterol presence embedded in the membrane, and the mosaic nature of the proteins and protein-carbohydrate combinations, which are not firmly fixed in place. Plasma membranes enclose and define the cells' borders. Not static, they are dynamic and constantly in flux.

5.2 Passive Transport

The passive transport forms, diffusion and osmosis, move materials of small molecular weight across membranes. Substances diffuse from high to lower concentration areas, and this process continues until the substance evenly distributes itself in a system. In solutions containing more than one substance, each molecule type diffuses according to its own concentration gradient, independent of other substances diffusing. Many factors can affect the diffusion rate, such as concentration gradient, diffusing, particle sizes, and the system's temperature.

In living systems, the plasma membrane mediates substances diffusing in and out of cells. Some materials diffuse readily through the membrane, but others are hindered and only can pass through due to specialized proteins such as channels and transporters. The chemistry of living things occurs in aqueous solutions, and balancing the concentrations of those solutions is an ongoing problem. In living systems, diffusing some substances would be slow or difficult without membrane proteins that facilitate transport.

5.3 Active Transport

The combined gradient that affects an ion includes its concentration gradient and its electrical gradient. A positive ion, for example, might diffuse into a new area, down its concentration gradient, but if it is diffusing into an area of net positive charge, its electrical gradient hampers its diffusion. When dealing with ions in aqueous solutions, one must consider electrochemical and concentration gradient combinations, rather than just the concentration gradient alone. Living cells need certain substances that exist inside the cell in concentrations greater than they exist in the extracellular space. Moving substances up their electrochemical gradients requires energy from the cell. Active transport uses energy stored in ATP to fuel this transport. Active transport of small molecular-sized materials uses integral proteins in the cell membrane to move the materials. These proteins are analogous to pumps. Some pumps, which carry out primary active transport, couple directly with ATP to drive their action. In co-transport (or secondary active transport), energy from primary transport can move another substance into the cell and up its concentration gradient.

5.4 Bulk Transport

Active transport methods require directly using ATP to fuel the transport. In a process scientists call phagocytosis, other cells can engulf large particles, such as macromolecules, cell parts, or whole cells. In phagocytosis, a portion of the membrane invaginates and flows around the particle, eventually pinching off and leaving the particle entirely enclosed by a plasma membrane's envelope. The cell breaks down vesicle contents, with the particles either used as food or dispatched. Pinocytosis is a similar process on a smaller scale. The plasma membrane invaginates and pinches off, producing a small envelope of fluid from outside the cell. Pinocytosis imports substances that the cell needs from the extracellular fluid. The cell expels waste in a similar but reverse manner. It pushes a membranous vacuole to the plasma membrane, allowing the vacuole to fuse with the membrane and incorporate itself into the membrane structure, releasing its contents to the exterior.

VISUAL CONNECTION QUESTIONS

1. Figure 5.12 A doctor injects a patient with what the doctor thinks is an isotonic saline solution. The patient dies, and an autopsy reveals that many red blood cells have been destroyed. Do you think the solution the doctor injected was really isotonic?
2. Figure 5.16 Injecting a potassium solution into a person's blood is lethal. Capital punishment and euthanasia utilize this method in their subjects. Why do you think a potassium solution injection is lethal?
3. Figure 5.19 If the pH outside the cell decreases, would you expect the amount of amino acids transported into the cell to increase or decrease?

REVIEW QUESTIONS

4. Which plasma membrane component can be either found on its surface or embedded in the membrane structure?
   a. protein
   b. cholesterol
   c. carbohydrate
   d. phospholipid

5. Which characteristic of a phospholipid contributes to the fluidity of the membrane?
   a. its head
   b. cholesterol
   c. a saturated fatty acid tail
   d. double bonds in the fatty acid tail

6. What is the primary function of carbohydrates attached to the exterior of cell membranes?
   a. identification of the cell
   b. flexibility of the membrane
   c. strengthening the membrane
   d. channels through membrane

7. A scientist compares the plasma membrane composition of an animal from the Mediterranean coast with one from the Mojave Desert. Which hypothesis is most likely to be correct?
   a. The cells from the Mediterranean coast animal will have more fluid plasma membranes.
   b. The cells from the Mojave Desert animal will have a higher cholesterol concentration in the plasma membranes.
   c. The cells' plasma membranes will be indistinguishable.
   d. The cells from the Mediterranean coast animal will have a higher glycoprotein content, while the cells from the Mojave Desert animal will have a higher lipoprotein content.

8. Water moves via osmosis _________.
   a. throughout the cytoplasm
   b. from an area with a high concentration of other solutes to a lower one
   c. from an area with a high concentration of water to one of lower concentration
   d. from an area with a low concentration of water to higher concentration

9. The principal force driving movement in diffusion is the _________.
   a. temperature
   b. particle size
   c. concentration gradient
   d. membrane surface area

10. What problem is faced by organisms that live in fresh water?
   a. Their bodies tend to take in too much water.
   b. They have no way of controlling their tonicity.
   c. Only salt water poses problems for animals that live in it.
   d. Their bodies tend to lose too much water to their environment.

11. In which situation would passive transport **not** use a transport protein for entry into a cell?
   a. water flowing into a hypertonic environment
   b. glucose being absorbed from the blood
   c. an ion flowing into a nerve cell to create an electrical potential
   d. oxygen moving into a cell after oxygen deprivation

12. Active transport must function continuously because _________.
   a. plasma membranes wear out
   b. not all membranes are amphiphilic
   c. facilitated transport opposes active transport
   d. diffusion is constantly moving solutes in opposite directions

13. How does the sodium-potassium pump make the interior of the cell negatively charged?
   a. by expelling anions
   b. by pulling in anions
   c. by expelling more cations than are taken in
   d. by taking in and expelling an equal number of cations

14. What is the combination of an electrical gradient and a concentration gradient called?
   a. potential gradient
   b. electrical potential
   c. concentration potential
   d. electrochemical gradient

15. What happens to the membrane of a vesicle after exocytosis?
   a. It leaves the cell.
   b. It is disassembled by the cell.
   c. It fuses with and becomes part of the plasma membrane.
   d. It is used again in another exocytosis event.

16. Which transport mechanism can bring whole cells into a cell?
    a. pinocytosis
    b. phagocytosis
    c. facilitated transport
    d. primary active transport
17. In what important way does receptor-mediated endocytosis differ from phagocytosis?
    a. It transports only small amounts of fluid.
    b. It does not involve the pinching off of membrane.
    c. It brings in only a specifically targeted substance.
    d. It brings substances into the cell, while phagocytosis removes substances.
18. Many viruses enter host cells through receptor-mediated endocytosis. What is an advantage of this entry strategy?
    a. The virus directly enters the cytoplasm of the cell.
    b. The virus is protected from recognition by white blood cells.
    c. The virus only enters its target host cell type.
    d. The virus can directly inject its genome into the cell's nucleus.
19. Which of the following organelles relies on exocytosis to complete its function?
    a. Golgi apparatus
    b. vacuole
    c. mitochondria
    d. endoplasmic reticulum
20. Imagine a cell can perform exocytosis, but only minimal endocytosis. What would happen to the cell?
    a. The cell would secrete all its intracellular proteins.
    b. The plasma membrane would increase in size over time.
    c. The cell would stop expressing integral receptor proteins in its plasma membrane.
    d. The cell would lyse.

CRITICAL THINKING QUESTIONS

21. Why is it advantageous for the cell membrane to be fluid in nature?
22. Why do phospholipids tend to spontaneously orient themselves into something resembling a membrane?
23. How can a cell use an extracellular peripheral protein as the receptor to transmit a signal into the cell?
24. Discuss why the following affect the rate of diffusion: molecular size, temperature, solution density, and the distance that must be traveled.
25. Why does water move through a membrane?
26. Both of the regular intravenous solutions administered in medicine, normal saline and lactated Ringer's solution, are isotonic. Why is this important?
27. Describe two ways that decreasing temperature would affect the rate of diffusion of molecules across a cell's plasma membrane.
28. A cell develops a mutation in its potassium channels that prevents the ions from leaving the cell. If the cell's aquaporins are still active, what will happen to the cell? Be sure to describe the tonicity and osmolarity of the cell.
29. Where does the cell get energy for active transport processes?
30. How does the sodium-potassium pump contribute to the net negative charge of the interior of the cell?
31. Glucose from digested food enters intestinal epithelial cells by active transport. Why would intestinal cells use active transport when most body cells use facilitated diffusion?
32. The sodium/calcium exchanger (NCX) transports sodium into and calcium out of cardiac muscle cells. Describe why this transporter is classified as secondary active transport.
33. Why is it important that there are different types of proteins in plasma membranes for the transport of materials into and out of a cell?
34. Why do ions have a difficult time getting through plasma membranes despite their small size?

CHAPTER 6 Metabolism

Figure 6.1 A hummingbird needs energy to maintain prolonged periods of flight. The bird obtains its energy from taking in food and transforming the nutrients into energy through a series of biochemical reactions. The flight muscles in birds are extremely efficient in energy production. (credit: modification of work by Cory Zanker)

INTRODUCTION Virtually every task performed by living organisms requires energy. Organisms require energy to perform heavy labor and exercise, but humans also use considerable energy while thinking, and even during sleep. Every organism's living cells constantly use energy. Organisms import nutrients and other molecules. They metabolize (break down) and possibly synthesize into new molecules. If necessary, molecules modify, move around the cell and may distribute themselves to the entire organism. For example, the large proteins that make up muscles are actively built from smaller molecules. Complex carbohydrates break down into simple sugars that the cell uses for energy. Just as energy is required to both build and demolish a building, energy is required to synthesize and break down molecules. Additionally, signaling molecules such as hormones and neurotransmitters transport between cells. Cells ingest and break down bacteria and viruses. Cells must also export waste and toxins to stay healthy, and many cells must swim or move surrounding materials via the beating motion of cellular appendages like cilia and flagella.

The cellular processes that we listed above require a steady supply of energy. From where, and in what form, does this energy come? How do living cells obtain energy, and how do they use it? This chapter will discuss different forms of energy and the physical laws that govern energy transfer. This chapter will also describe how cells use energy and replenish it, and how chemical reactions in the cell perform with great efficiency.

Chapter Outline

6.1 Energy and Metabolism

By the end of this section, you will be able to do the following:

- Explain metabolic pathways and describe the two major types
- Discuss how chemical reactions play a role in energy transfer

Scientists use the term **bioenergetics** to discuss the concept of energy flow (Figure 6.2) through living systems, such as cells. Cellular processes such as building and breaking down complex molecules occur through stepwise chemical reactions. Some of these chemical reactions are spontaneous and release energy; whereas, others require energy to proceed. Just as living things must continually consume food to replenish what they have used, cells must continually obtain more energy to replenish that which the many energy-requiring chemical reactions that constantly take place use. All of the chemical reactions that transpire inside cells, including those that use and release energy, are the cell's **metabolism**.

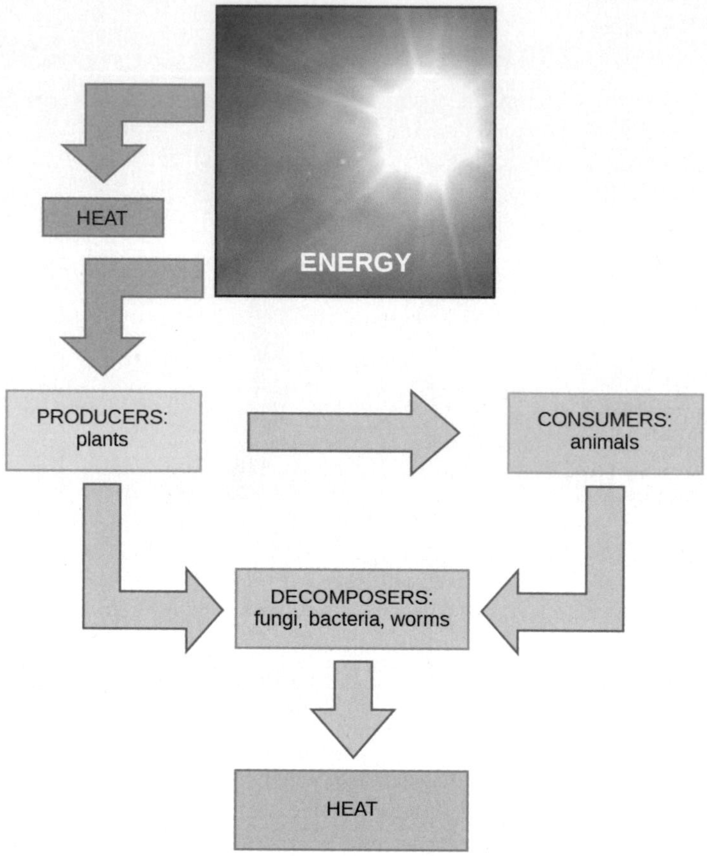

Figure 6.2 Most life forms on earth obtain their energy from the sun. Plants use photosynthesis to capture sunlight, and herbivores eat those plants to obtain energy. Carnivores eat the herbivores, and decomposers digest plant and animal matter.

Carbohydrate Metabolism

Sugar (a simple carbohydrate) metabolism (chemical reactions) is a classic example of the many cellular processes that use and produce energy. Living things consume sugar as a major energy source, because sugar molecules have considerable energy stored within their bonds. The following equation describes the breakdown of glucose, a simple sugar:

$$C_6H_{12}O_6 + 6O_2 \rightarrow 6CO_2 + 6H_2O + \text{energy}$$

Consumed carbohydrates have their origins in photosynthesizing organisms like plants (Figure 6.3). During photosynthesis, plants use the energy of sunlight to convert carbon dioxide gas (CO_2) into sugar molecules, like glucose ($C_6H_{12}O_6$). Because this process involves synthesizing a larger, energy-storing molecule, it requires an energy input to proceed. The following equation (notice that it is the reverse of the previous equation) describes the synthesis of glucose:

$$6CO_2 + 6H_2O + \text{energy} \rightarrow C_6H_{12}O_6 + 6O_2$$

During photosynthesis chemical reactions, energy is in the form of a very high-energy molecule scientists call ATP, or adenosine triphosphate. This is the primary energy currency of all cells. Just as the dollar is the currency we use to buy goods, cells use ATP molecules as energy currency to perform immediate work. The sugar (glucose) is stored as starch or glycogen. Energy-storing polymers like these break down into glucose to supply ATP molecules.

Solar energy is required to synthesize a glucose molecule during the photosynthesis reactions. In photosynthesis, light energy from the sun initially transforms into chemical energy that temporally stores itself in the energy carrier molecules ATP and NADPH (nicotinamide adenine dinucleotide phosphate). Photosynthesis later uses the stored energy in ATP and NADPH to build one glucose molecule from six molecules of CO_2. This process is analogous to eating breakfast in the morning to acquire energy for your body that you can use later in the day. Under ideal conditions, energy from 18 molecules of ATP is required to synthesize one glucose molecule during photosynthesis reactions. Glucose molecules can also combine with and convert into other sugar types. When an organism consumes sugars, glucose molecules eventually make their way into each organism's living cell. Inside the cell, each sugar molecule breaks down through a complex series of chemical reactions. The goal of these reactions is to harvest the energy stored inside the sugar molecules. The harvested energy makes high-energy ATP molecules, which perform work, powering many chemical reactions in the cell. The amount of energy needed to make one glucose molecule from six carbon dioxide molecules is 18 ATP molecules and 12 NADPH molecules (each one of which is energetically equivalent to three ATP molecules), or a total of 54 molecule equivalents required for synthesizing one glucose molecule. This process is a fundamental and efficient way for cells to generate the molecular energy that they require.

Figure 6.3 Plants, like this oak tree and acorn, use energy from sunlight to make sugar and other organic molecules. Both plants and animals (like this squirrel) use cellular respiration to derive energy from the organic molecules that plants originally produced. (credit "acorn": modification of work by Noel Reynolds; credit "squirrel": modification of work by Dawn Huczek)

Metabolic Pathways

The processes of making and breaking down sugar molecules illustrate two types of metabolic pathways. A metabolic pathway is a series of interconnected biochemical reactions that convert a substrate molecule or molecules, step-by-step, through a series of metabolic intermediates, eventually yielding a final product or products. In the case of sugar metabolism, the first metabolic pathway synthesized sugar from smaller molecules, and the other pathway broke sugar down into smaller molecules. Scientists call these two opposite processes—the first requiring energy and the second producing energy—anabolic (building) and catabolic (breaking down) pathways, respectively. Consequently, building (anabolism) and degradation (catabolism) comprise metabolism.

EVOLUTION CONNECTION

Evolution of Metabolic Pathways

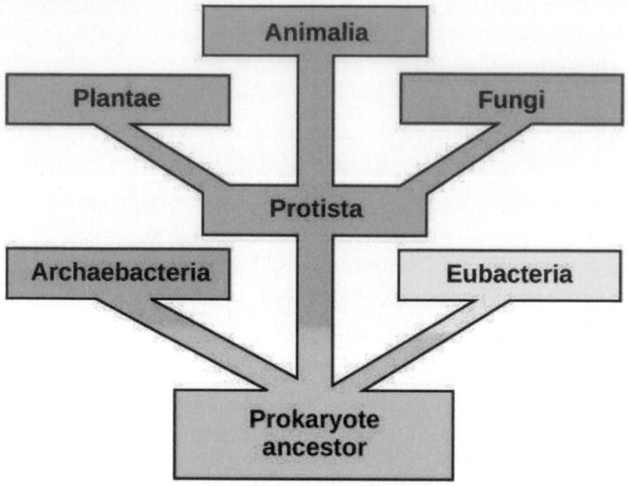

Figure 6.4 This tree shows the evolution of the various branches of life. The vertical dimension is time. Early life forms, in blue, used anaerobic metabolism to obtain energy from their surroundings.

There is more to the complexity of metabolism than understanding the metabolic pathways alone. Metabolic complexity varies from organism to organism. Photosynthesis is the primary pathway in which photosynthetic organisms like plants (planktonic algae perform the majority of global photosynthesis) harvest the sun's energy and convert it into carbohydrates. The by-product of photosynthesis is oxygen, which some cells require to carry out cellular respiration. During cellular respiration, oxygen aids in the catabolic breakdown of carbon compounds, like carbohydrates. Among the products are CO_2 and ATP. In addition, some eukaryotes perform catabolic processes without oxygen (fermentation); that is, they perform or use anaerobic metabolism.

Organisms probably evolved anaerobic metabolism to survive (living organisms came into existence about 3.8 billion years ago, when the atmosphere lacked oxygen). Despite the differences between organisms and the complexity of metabolism, researchers have found that all branches of life share some of the same metabolic pathways, suggesting that all organisms evolved from the same ancient common ancestor (Figure 6.4). Evidence indicates that over time, the pathways diverged, adding specialized enzymes to allow organisms to better adapt to their environment, thus increasing their chance to survive. However, the underlying principle remains that all organisms must harvest energy from their environment and convert it to ATP to carry out cellular functions.

Anabolic and Catabolic Pathways

Anabolic pathways require an input of energy to synthesize complex molecules from simpler ones. Synthesizing sugar from CO_2 is one example. Other examples are synthesizing large proteins from amino acid building blocks, and synthesizing new DNA strands from nucleic acid building blocks. These biosynthetic processes are critical to the cell's life, take place constantly, and demand energy that ATP and other high-energy molecules like NADH (nicotinamide adenine dinucleotide) and NADPH provide (Figure 6.5).

ATP is an important molecule for cells to have in sufficient supply at all times. The breakdown of sugars illustrates how a single glucose molecule can store enough energy to make a great deal of ATP, 36 to 38 molecules. This is a **catabolic** pathway. Catabolic pathways involve degrading (or breaking down) complex molecules into simpler ones. Molecular energy stored in complex molecule bonds release in catabolic pathways and harvest in such a way that it can produce ATP. Other energy-storing molecules, such as fats, also break down through similar catabolic reactions to release energy and make ATP (Figure 6.5).

It is important to know that metabolic pathway chemical reactions do not take place spontaneously. A protein called an enzyme facilitates or catalyzes each reaction step. Enzymes are important for catalyzing all types of biological reactions—those that require energy as well as those that release energy.

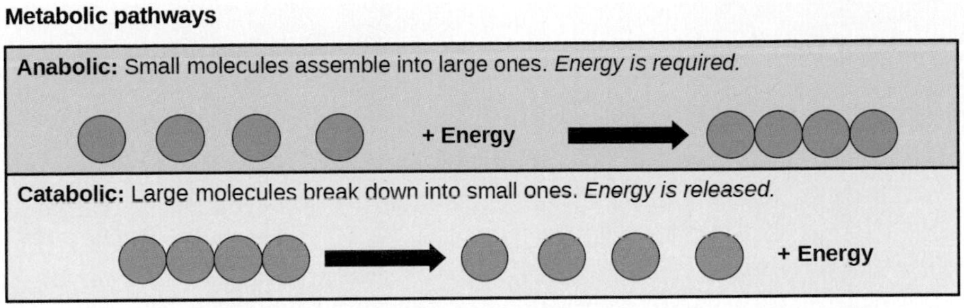

Figure 6.5 Anabolic pathways are those that require energy to synthesize larger molecules. Catabolic pathways are those that generate energy by breaking down larger molecules. Both types of pathways are required for maintaining the cell's energy balance.

6.2 Potential, Kinetic, Free, and Activation Energy

By the end of this section, you will be able to do the following:

- Define "energy"
- Explain the difference between kinetic and potential energy
- Discuss the concepts of free energy and activation energy
- Describe endergonic and exergonic reactions

We define energy as the ability to do work. As you've learned, energy exists in different forms. For example, electrical energy, light energy, and heat energy are all different energy types. While these are all familiar energy types that one can see or feel, there is another energy type that is much less tangible. Scientists associate this energy with something as simple as an object above the ground. In order to appreciate the way energy flows into and out of biological systems, it is important to understand more about the different energy types that exist in the physical world.

Energy Types

When an object is in motion, there is energy. For example, an airplane in flight produces considerable energy. This is because moving objects are capable of enacting a change, or doing work. Think of a wrecking ball. Even a slow-moving wrecking ball can do considerable damage to other objects. However, a wrecking ball that is not in motion is incapable of performing work. Energy with objects in motion is **kinetic energy**. A speeding bullet, a walking person, rapid molecule movement in the air (which produces heat), and electromagnetic radiation like light all have kinetic energy.

What if we lift that same motionless wrecking ball two stories above a car with a crane? If the suspended wrecking ball is unmoving, can we associate energy with it? The answer is yes. The suspended wrecking ball has associated energy that is fundamentally different from the kinetic energy of objects in motion. This energy form results from the *potential* for the wrecking ball to do work. If we release the ball it would do work. Because this energy type refers to the potential to do work, we call it **potential energy**. Objects transfer their energy between kinetic and potential in the following way: As the wrecking ball hangs motionless, it has 0 kinetic and 100 percent potential energy. Once it releases, its kinetic energy begins to increase because it builds speed due to gravity. Simultaneously, as it nears the ground, it loses potential energy. Somewhere mid-fall it has 50 percent kinetic and 50 percent potential energy. Just before it hits the ground, the ball has nearly lost its potential energy and has near-maximal kinetic energy. Other examples of potential energy include water's energy held behind a dam (Figure 6.6), or a person about to skydive from an airplane.

Figure 6.6 Water behind a dam has potential energy. Moving water, such as in a waterfall or a rapidly flowing river, has kinetic energy. (credit "dam": modification of work by "Pascal"/Flickr; credit "waterfall": modification of work by Frank Gualtieri)

We associate potential energy not only with the matter's location (such as a child sitting on a tree branch), but also with the matter's structure. A spring on the ground has potential energy if it is compressed; so does a tautly pulled rubber band. The very existence of living cells relies heavily on structural potential energy. On a chemical level, the bonds that hold the molecules' atoms together have potential energy. Remember that anabolic cellular pathways require energy to synthesize complex molecules from simpler ones, and catabolic pathways release energy when complex molecules break down. That certain chemical bonds' breakdown can release energy implies that those bonds have potential energy. In fact, there is potential energy stored within the bonds of all the food molecules we eat, which we eventually harness for use. This is because these bonds can release energy when broken. Scientists call the potential energy type that exists within chemical bonds that releases when those bonds break **chemical energy** (Figure 6.7). Chemical energy is responsible for providing living cells with energy from food. Breaking the molecular bonds within fuel molecules brings about the energy's release.

Figure 6.7 The molecules in gasoline contain chemical energy within the chemical bonds. This energy transforms into kinetic energy that allows a car to race on a racetrack. (credit "car": modification of work by Russell Trow)

LINK TO LEARNING

Visit this site (http://openstax.org/l/simple_pendulum) and select "A simple pendulum" on the menu (under "Harmonic Motion") to see the shifting kinetic (K) and potential energy (U) of a pendulum in motion.

Free Energy

After learning that chemical reactions release energy when energy-storing bonds break, an important next question is how do we quantify and express the chemical reactions with the associated energy? How can we compare the energy that releases from one reaction to that of another reaction? We use a measurement of **free energy** to quantitate these energy transfers. Scientists call this free energy Gibbs free energy (abbreviated with the letter G) after Josiah Willard Gibbs, the scientist who developed the measurement. Recall that according to the second law of thermodynamics, all energy transfers involve losing some energy in an unusable form such as heat, resulting in entropy. Gibbs free energy specifically refers to the energy that takes place with a chemical reaction that is available after we account for entropy. In other words, Gibbs free energy is usable energy, or energy that is available to do work.

Every chemical reaction involves a change in free energy, called delta G (ΔG). We can calculate the change in free energy for any system that undergoes such a change, such as a chemical reaction. To calculate ΔG, subtract the amount of energy lost to entropy (denoted as ΔS) from the system's total energy change. Scientists call this total energy change in the system **enthalpy** and we denote it as ΔH. The formula for calculating ΔG is as follows, where the symbol T refers to absolute temperature in Kelvin (degrees Celsius + 273):

$$\Delta G = \Delta H - T\Delta S$$

We express a chemical reaction's standard free energy change as an amount of energy per mole of the reaction product (either in kilojoules or kilocalories, kJ/mol or kcal/mol; 1 kJ = 0.239 kcal) under standard pH, temperature, and pressure conditions. We generally calculate standard pH, temperature, and pressure conditions at pH 7.0 in biological systems, 25 degrees Celsius, and 100 kilopascals (1 atm pressure), respectively. Note that cellular conditions vary considerably from these standard conditions, and so standard calculated ΔG values for biological reactions will be different inside the cell.

Endergonic Reactions and Exergonic Reactions

If energy releases during a chemical reaction, then the resulting value from the above equation will be a negative number. In other words, reactions that release energy have a $\Delta G < 0$. A negative ΔG also means that the reaction's products have less free energy than the reactants, because they gave off some free energy during the reaction. Scientists call reactions that have a negative ΔG and consequently release free energy **exergonic reactions**. Think: *ex*ergonic means energy is *ex*iting the system. We also refer to these reactions as spontaneous reactions, because they can occur without adding energy into the system. Understanding which chemical reactions are spontaneous and release free energy is extremely useful for biologists, because these reactions can be harnessed to perform work inside the cell. We must draw an important distinction between the term spontaneous and the idea of a chemical reaction that occurs immediately. Contrary to the everyday use of the term, a spontaneous reaction is not one that suddenly or quickly occurs. Rusting iron is an example of a spontaneous reaction that occurs slowly, little by little, over time.

If a chemical reaction requires an energy input rather than releasing energy, then the ΔG for that reaction will be a positive value. In this case, the products have more free energy than the reactants. Thus, we can think of the reactions' products as energy-storing molecules. We call these chemical reactions **endergonic reactions**, and they are non-spontaneous. An endergonic reaction will not take place on its own without adding free energy.

Let's revisit the example of the synthesis and breakdown of the food molecule, glucose. Remember that building complex molecules, such as sugars, from simpler ones is an anabolic process and requires energy. Therefore, the chemical reactions involved in anabolic processes are endergonic reactions. Alternatively the catabolic process of breaking sugar down into simpler molecules releases energy in a series of exergonic reactions. Like the rust example above, the sugar breakdown involves spontaneous reactions, but these reactions do not occur instantaneously. Figure 6.8 shows some other examples of endergonic and exergonic reactions. Later sections will provide more information about what else is required to make even spontaneous reactions happen more efficiently.

VISUAL CONNECTION

Figure 6.8 This figure shows some examples of endergonic processes (ones that require energy) and exergonic processes (ones that release energy). These include (a) a compost pile decomposing, (b) a chick developing from a fertilized egg, (c) sand art destruction, and (d) a ball rolling down a hill. (credit a: modification of work by Natalie Maynor; credit b: modification of work by USDA; credit c: modification of work by "Athlex"/Flickr; credit d: modification of work by Harry Malsch)

Look at each of the processes, and decide if it is endergonic or exergonic. In each case, does enthalpy increase or decrease, and does entropy increase or decrease?

An important concept in studying metabolism and energy is that of chemical equilibrium. Most chemical reactions are reversible. They can proceed in both directions, releasing energy into their environment in one direction, and absorbing it from the environment in the other direction (Figure 6.9). The same is true for the chemical reactions involved in cell metabolism, such as the breaking down and building up of proteins into and from individual amino acids, respectively. Reactants within a closed system will undergo chemical reactions in both directions until they reach a state of equilibrium, which is one of the lowest possible free energy and a state of maximal entropy. To push the reactants and products away from a state of equilibrium requires energy. Either reactants or products must be added, removed, or changed. If a cell were a closed system, its chemical reactions would reach equilibrium, and it would die because there would be insufficient free energy left to perform the necessary work to maintain life. In a living cell, chemical reactions are constantly moving towards equilibrium, but never reach it. This is because a living cell is an open system. Materials pass in and out, the cell recycles the products of certain chemical reactions into other reactions, and there is never chemical equilibrium. In this way, living organisms are in a constant energy-requiring, uphill battle against equilibrium and entropy. This constant energy supply ultimately comes from sunlight, which produces nutrients in the photosynthesis process.

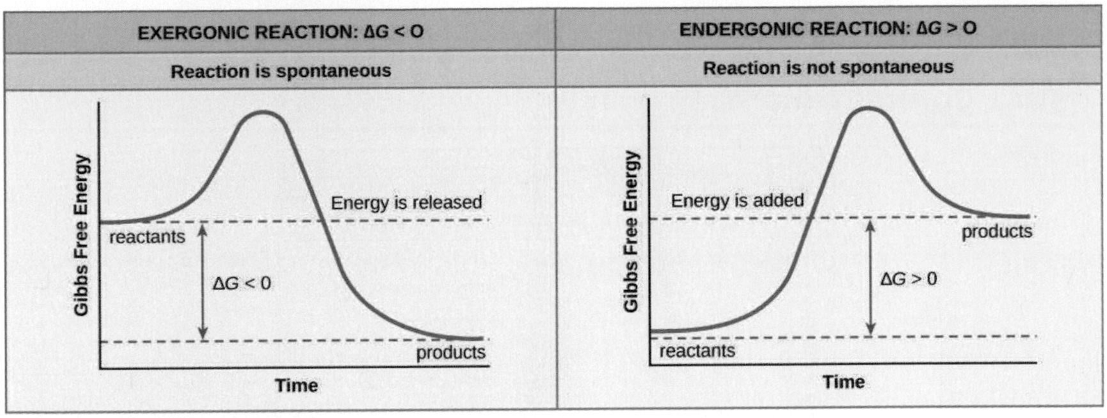

Figure 6.9 Exergonic and endergonic reactions result in changes in Gibbs free energy. Exergonic reactions release energy. Endergonic reactions require energy to proceed.

Activation Energy

There is another important concept that we must consider regarding endergonic and exergonic reactions. Even exergonic reactions require a small amount of energy input before they can proceed with their energy-releasing steps. These reactions have a net release of energy, but still require some initial energy. Scientists call this small amount of energy input necessary for all chemical reactions to occur the **activation energy** (or free energy of activation) abbreviated as E_A (Figure 6.10).

Why would an energy-releasing, negative ΔG reaction actually require some energy to proceed? The reason lies in the steps that take place during a chemical reaction. During chemical reactions, certain chemical bonds break and new ones form. For example, when a glucose molecule breaks down, bonds between the molecule's carbon atoms break. Since these are energy-storing bonds, they release energy when broken. However, to get them into a state that allows the bonds to break, the molecule must be somewhat contorted. A small energy input is required to achieve this contorted state. This contorted state is the **transition state**, and it is a high-energy, unstable state. For this reason, reactant molecules do not last long in their transition state, but very quickly proceed to the chemical reaction's next steps. Free energy diagrams illustrate the energy profiles for a given reaction. Whether the reaction is exergonic or endergonic determines whether the products in the diagram will exist at a lower or higher energy state than both the reactants and the products. However, regardless of this measure, the transition state of the reaction exists at a higher energy state than the reactants, and thus, E_A is always positive.

LINK TO LEARNING

Watch an animation of the move from free energy to transition state at this (http://openstax.org/l/energy_reaction) site.

From where does the activation energy that chemical reactants require come? The activation energy's required source to push reactions forward is typically heat energy from the surroundings. **Heat energy** (the total bond energy of reactants or products in a chemical reaction) speeds up the molecule's motion, increasing the frequency and force with which they collide. It also moves atoms and bonds within the molecule slightly, helping them reach their transition state. For this reason, heating a system will cause chemical reactants within that system to react more frequently. Increasing the pressure on a system has the same effect. Once reactants have absorbed enough heat energy from their surroundings to reach the transition state, the reaction will proceed.

The activation energy of a particular reaction determines the rate at which it will proceed. The higher the activation energy, the slower the chemical reaction. The example of iron rusting illustrates an inherently slow reaction. This reaction occurs slowly over time because of its high E_A. Additionally, burning many fuels, which is strongly exergonic, will take place at a negligible rate unless sufficient heat from a spark overcomes their activation energy. However, once they begin to burn, the chemical reactions release enough heat to continue the burning process, supplying the activation energy for surrounding fuel molecules. Like these reactions outside of cells, the activation energy for most cellular reactions is too high for heat energy to overcome at efficient rates. In other words, in order for important cellular reactions to occur at appreciable rates (number of reactions per unit time), their activation energies must be lowered (Figure 6.10). Scientist refer to this as catalysis. This is a very good thing as far as living cells are concerned. Important macromolecules, such as proteins, DNA, and RNA, store considerable energy, and their breakdown is exergonic. If cellular temperatures alone provided enough heat energy for these exergonic reactions to overcome their activation barriers, the cell's essential components would disintegrate.

VISUAL CONNECTION

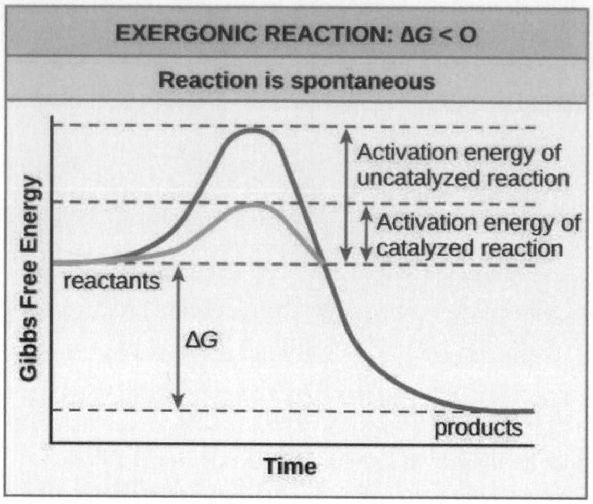

Figure 6.10 Activation energy is the energy required for a reaction to proceed, and it is lower if the reaction is catalyzed. This diagram's horizontal axis describes the sequence of events in time.

If no activation energy were required to break down sucrose (table sugar), would you be able to store it in a sugar bowl?

6.3 The Laws of Thermodynamics

By the end of this section, you will be able to do the following:

- Discuss the concept of entropy
- Explain the first and second laws of thermodynamics

Thermodynamics refers to the study of energy and energy transfer involving physical matter. The matter and its environment relevant to a particular case of energy transfer are classified as a system, and everything outside that system is the surroundings. For instance, when heating a pot of water on the stove, the system includes the stove, the pot, and the water. Energy transfers within the system (between the stove, pot, and water). There are two types of systems: open and closed. An open system is one in which energy can transfer between the system and its surroundings. The stovetop system is open because it can lose heat into the air. A closed system is one that cannot transfer energy to its surroundings.

Biological organisms are open systems. Energy exchanges between them and their surroundings, as they consume energy-storing molecules and release energy to the environment by doing work. Like all things in the physical world, energy is subject to the laws of physics. The laws of thermodynamics govern the transfer of energy in and among all systems in the universe.

The First Law of Thermodynamics

The first law of thermodynamics deals with the total amount of energy in the universe. It states that this total amount of energy is constant. In other words, there has always been, and always will be, exactly the same amount of energy in the universe. Energy exists in many different forms. According to the first law of thermodynamics, energy may transfer from place to place or transform into different forms, but it cannot be created or destroyed. The transfers and transformations of energy take place around us all the time. Light bulbs transform electrical energy into light energy. Gas stoves transform chemical energy from natural gas into heat energy. Plants perform one of the most biologically useful energy transformations on earth: that of converting sunlight energy into the chemical energy stored within organic molecules (Figure 6.2). Figure 6.11 examples of energy transformations.

The challenge for all living organisms is to obtain energy from their surroundings in forms that they can transfer or transform into usable energy to do work. Living cells have evolved to meet this challenge very well. Chemical energy stored within organic molecules such as sugars and fats transforms through a series of cellular chemical reactions into energy within ATP molecules. Energy in ATP molecules is easily accessible to do work. Examples of the types of work that cells need to do include building complex molecules, transporting materials, powering the beating motion of cilia or flagella, contracting muscle fibers to create movement, and reproduction.

Figure 6.11 Here are two examples of energy transferring from one system to another and transformed from one form to another. Humans can convert the chemical energy in food, like this ice cream cone, into kinetic energy (the energy of movement to ride a bicycle). Plants can convert electromagnetic radiation (light energy) from the sun into chemical energy. (credit "ice cream": modification of work by D. Sharon Pruitt; credit "kids on bikes": modification of work by Michelle Riggen-Ransom; credit "leaf": modification of work by Cory Zanker)

The Second Law of Thermodynamics

A living cell's primary tasks of obtaining, transforming, and using energy to do work may seem simple. However, the second law of thermodynamics explains why these tasks are harder than they appear. None of the energy transfers that we have discussed, along with all energy transfers and transformations in the universe, is completely efficient. In every energy transfer, some amount of energy is lost in a form that is unusable. In most cases, this form is heat energy. Thermodynamically, scientists define **heat energy** as energy that transfers from one system to another that is not doing work. For example, when an airplane flies through the air, it loses some of its energy as heat energy due to friction with the surrounding air. This friction actually heats the air by temporarily increasing air molecule speed. Likewise, some energy is lost as heat energy during cellular metabolic reactions. This is good for warm-blooded creatures like us, because heat energy helps to maintain our body temperature. Strictly speaking, no energy transfer is completely efficient, because some energy is lost in an unusable form.

An important concept in physical systems is that of order and disorder (or randomness). The more energy that a system loses to its surroundings, the less ordered and more random the system. Scientists refer to the measure of randomness or disorder within a system as **entropy**. High entropy means high disorder and low energy (Figure 6.12). To better understand entropy, think of a student's bedroom. If no energy or work were put into it, the room would quickly become messy. It would exist in a very disordered state, one of high entropy. Energy must be put into the system, in the form of the student doing work and putting everything away, in order to bring the room back to a state of cleanliness and order. This state is one of low entropy. Similarly, a car or house must be constantly maintained with work in order to keep it in an ordered state. Left alone, a house's or car's entropy gradually increases through rust and degradation. Molecules and chemical reactions have varying amounts of entropy as well. For example, as chemical reactions reach a state of equilibrium, entropy increases, and as molecules at a high concentration in one place diffuse and spread out, entropy also increases.

SCIENTIFIC METHOD CONNECTION

Transfer of Energy and the Resulting Entropy

Set up a simple experiment to understand how energy transfers and how a change in entropy results.

1. Take a block of ice. This is water in solid form, so it has a high structural order. This means that the molecules cannot move very much and are in a fixed position. The ice's temperature is 0°C. As a result, the system's entropy is low.
2. Allow the ice to melt at room temperature. What is the state of molecules in the liquid water now? How did the energy transfer take place? Is the system's entropy higher or lower? Why?
3. Heat the water to its boiling point. What happens to the system's entropy when the water is heated?

Think of all physical systems of in this way: Living things are highly ordered, requiring constant energy input to maintain themselves in a state of low entropy. As living systems take in energy-storing molecules and transform them through chemical reactions, they lose some amount of usable energy in the process, because no reaction is completely efficient. They also produce waste and by-products that are not useful energy sources. This process increases the entropy of the system's surroundings. Since all energy transfers result in losing some usable energy, the second law of thermodynamics states that every energy transfer or transformation increases the universe's entropy. Even though living things are highly ordered and maintain a state of low entropy, the universe's entropy in total is constantly increasing due to losing usable energy with each energy transfer that occurs. Essentially, living things are in a continuous uphill battle against this constant increase in universal entropy.

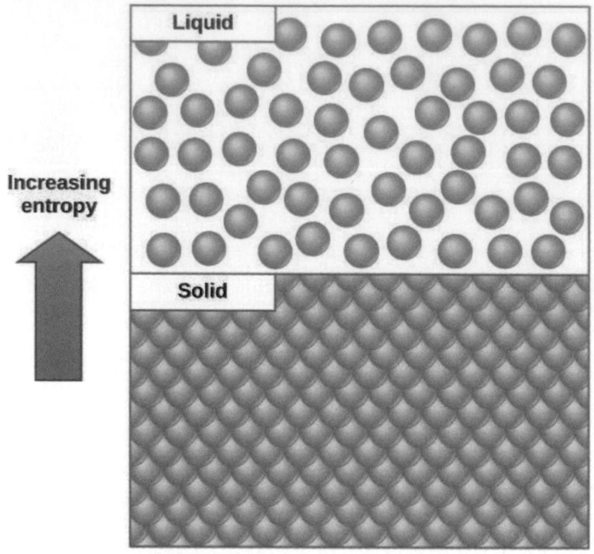

Figure 6.12 Entropy is a measure of randomness or disorder in a system. Gases have higher entropy than liquids, and liquids have higher entropy than solids.

6.4 ATP: Adenosine Triphosphate

By the end of this section, you will be able to do the following:

- Explain ATP's role as the cellular energy currency
- Describe how energy releases through ATP hydrolysis

Even exergonic, energy-releasing reactions require a small amount of activation energy in order to proceed. However, consider endergonic reactions, which require much more energy input, because their products have more free energy than their reactants. Within the cell, from where does energy to power such reactions come? The answer lies with an energy-supplying molecule scientists call **adenosine triphosphate**, or **ATP**. This is a small, relatively simple molecule (Figure 6.13), but within some of its bonds, it contains the potential for a quick burst of energy that can be harnessed to perform cellular work. Think of this molecule as the cells' primary energy currency in much the same way that money is the currency that people exchange for things they need. ATP powers the majority of energy-requiring cellular reactions.

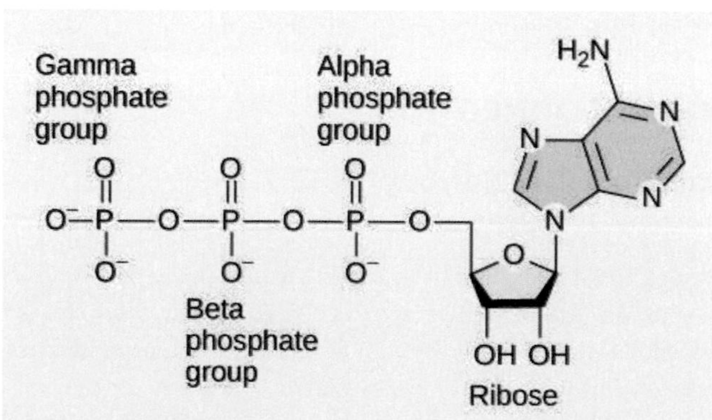

Figure 6.13 ATP is the cell's primary energy currency. It has an adenosine backbone with three phosphate groups attached.

As its name suggests, adenosine triphosphate is comprised of adenosine bound to three phosphate groups (Figure 6.13). Adenosine is a nucleoside consisting of the nitrogenous base adenine and a five-carbon sugar, ribose. The three phosphate groups, in order of closest to furthest from the ribose sugar, are alpha, beta, and gamma. Together, these chemical groups constitute an energy powerhouse. However, not all bonds within this molecule exist in a particularly high-energy state. Both bonds that link the phosphates are equally high-energy bonds (**phosphoanhydride bonds**) that, when broken, release sufficient energy to power a variety of cellular reactions and processes. These high-energy bonds are the bonds between the second and third (or beta and gamma) phosphate groups and between the first and second phosphate groups. These bonds are "high-energy" because the products of such bond breaking—adenosine diphosphate (ADP) and one inorganic phosphate group (P_i)—have considerably lower free energy than the reactants: ATP and a water molecule. Because this reaction takes place using a water molecule, it is a hydrolysis reaction. In other words, ATP hydrolyzes into ADP in the following reaction:

$$\text{ATP} + \text{H}_2\text{O} \rightarrow \text{ADP} + \text{P}_\text{i} + \text{free energy}$$

Like most chemical reactions, ATP to ADP hydrolysis is reversible. The reverse reaction regenerates ATP from ADP + P_i. Cells rely on ATP regeneration just as people rely on regenerating spent money through some sort of income. Since ATP hydrolysis releases energy, ATP regeneration must require an input of free energy. This equation expresses ATP formation:

$$\text{ADP} + \text{P}_\text{i} + \text{free energy} \rightarrow \text{ATP} + \text{H}_2\text{O}$$

Two prominent questions remain with regard to using ATP as an energy source. Exactly how much free energy releases with ATP hydrolysis, and how does that free energy do cellular work? The calculated ∆G for the hydrolysis of one ATP mole into ADP and P_i is –7.3 kcal/mole (–30.5 kJ/mol). Since this calculation is true under standard conditions, one would expect a different value exists under cellular conditions. In fact, the ∆G for one ATP mole's hydrolysis in a living cell is almost double the value at standard conditions: –14 kcal/mol (–57 kJ/mol).

ATP is a highly unstable molecule. Unless quickly used to perform work, ATP spontaneously dissociates into ADP + P_i, and the free energy released during this process is lost as heat. The second question we posed above discusses how ATP hydrolysis energy release performs work inside the cell. This depends on a strategy scientists call energy coupling. Cells couple the ATP hydrolysis' exergonic reaction allowing them to proceed. One example of energy coupling using ATP involves a transmembrane ion pump that is extremely important for cellular function. This sodium-potassium pump (Na^+/K^+ pump) drives sodium out of the cell and potassium into the cell (Figure 6.14). A large percentage of a cell's ATP powers this pump, because cellular processes bring considerable sodium into the cell and potassium out of it. The pump works constantly to stabilize cellular concentrations of sodium and potassium. In order for the pump to turn one cycle (exporting three Na+ ions and importing two K^+ ions), one ATP molecule must hydrolyze. When ATP hydrolyzes, its gamma phosphate does not simply float away, but it actually transfers onto the pump protein. Scientists call this process of a phosphate group binding to a molecule phosphorylation. As with most ATP hydrolysis cases, a phosphate from ATP transfers onto another molecule. In a phosphorylated state, the Na^+/K^+ pump has more free energy and is triggered to undergo a conformational change. This change allows it to release Na^+ to the cell's outside. It then binds extracellular K^+, which, through another conformational change, causes the phosphate to detach from the pump. This phosphate release triggers the K^+ to release to the cell's inside. Essentially, the energy released from the ATP hydrolysis couples with the energy required to power the pump and transport Na^+ and K^+ ions. ATP performs cellular work using this basic form of energy coupling through phosphorylation.

VISUAL CONNECTION

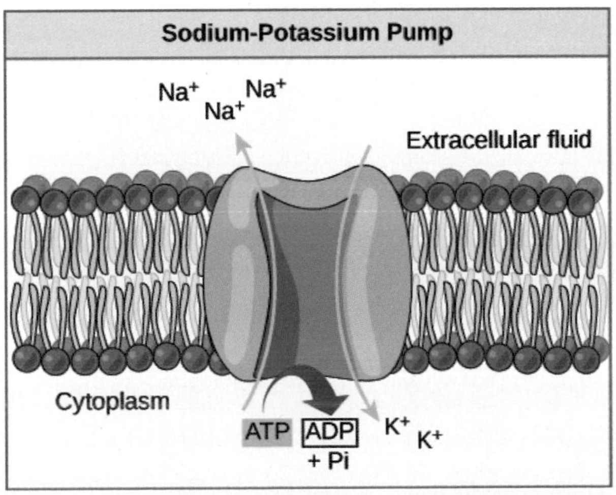

Figure 6.14 The sodium-potassium pump is an example of energy coupling. The energy derived from exergonic ATP hydrolysis pumps

sodium and potassium ions across the cell membrane.

One ATP molecule's hydrolysis releases 7.3 kcal/mol of energy ($\Delta G = -7.3$ kcal/mol of energy). If it takes 2.1 kcal/mol of energy to move one Na^+ across the membrane ($\Delta G = +2.1$ kcal/mol of energy), how many sodium ions could one ATP molecule's hydrolysis move?

Often during cellular metabolic reactions, such as nutrient synthesis and breakdown, certain molecules must alter slightly in their conformation to become substrates for the next step in the reaction series. One example is during the very first steps of cellular respiration, when a sugar glucose molecule breaks down in the process of glycolysis. In the first step, ATP is required to phosphorylze glucose, creating a high-energy but unstable intermediate. This phosphorylation reaction powers a conformational change that allows the phosphorylated glucose molecule to convert to the phosphorylated sugar fructose. Fructose is a necessary intermediate for glycolysis to move forward. Here, ATP hydrolysis' exergonic reaction couples with the endergonic reaction of converting glucose into a phosphorylated intermediate in the pathway. Once again, the energy released by breaking a phosphate bond within ATP was used for phosphorylyzing another molecule, creating an unstable intermediate and powering an important conformational change.

LINK TO LEARNING

See an animation of the ATP-producing glycolysis process at this site (http://openstax.org/l/glycolysis_stgs) .

6.5 Enzymes

By the end of this section, you will be able to do the following:

- Describe the role of enzymes in metabolic pathways
- Explain how enzymes function as molecular catalysts
- Discuss enzyme regulation by various factors

A substance that helps a chemical reaction to occur is a catalyst, and the special molecules that catalyze biochemical reactions are enzymes. Almost all enzymes are proteins, comprised of amino acid chains, and they perform the critical task of lowering the activation energies of chemical reactions inside the cell. Enzymes do this by binding to the reactant molecules, and holding them in such a way as to make the chemical bond-breaking and bond-forming processes take place more readily. It is important to remember that enzymes do not change the reaction's ΔG. In other words, they do not change whether a reaction is exergonic (spontaneous) or endergonic. This is because they do not change the reactants' or products' free energy. They only reduce the activation energy required to reach the transition state (Figure 6.15).

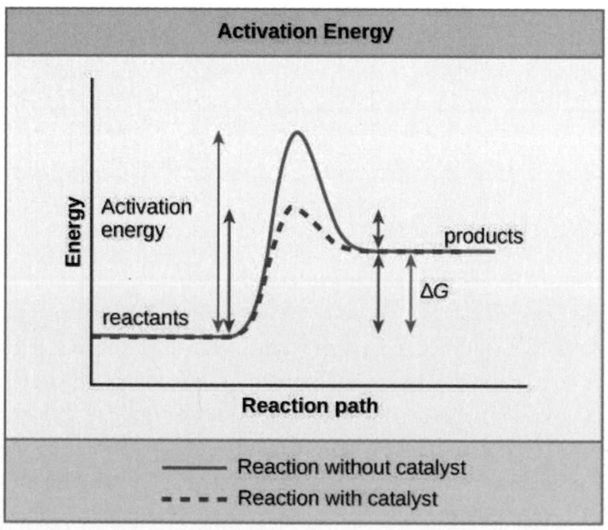

Figure 6.15 Enzymes lower the reaction's activation energy but do not change the reaction's free energy.

Enzyme Active Site and Substrate Specificity

The chemical reactants to which an enzyme binds are the enzyme's **substrates**. There may be one or more substrates, depending on the particular chemical reaction. In some reactions, a single-reactant substrate breaks down into multiple products. In others, two substrates may come together to create one larger molecule. Two reactants might also enter a reaction, both become modified, and leave the reaction as two products. The location within the enzyme where the substrate binds is the enzyme's **active site**. This is where the "action" happens. Since enzymes are proteins, there is a unique combination of amino acid residues (also side chains, or R groups) within the active site. Different properties characterize each residue. These can be large or small, weakly acidic or basic, hydrophilic or hydrophobic, positively or negatively charged, or neutral. The unique combination of amino acid residues, their positions, sequences, structures, and properties, creates a very specific chemical environment within the active site. This specific environment is suited to bind, albeit briefly, to a specific chemical substrate (or substrates). Due to this jigsaw puzzle-like match between an enzyme and its substrates (which adapts to find the best fit between the transition state and the active site), enzymes are known for their specificity. The "best fit" results from the shape and the amino acid functional group's attraction to the substrate. There is a specifically matched enzyme for each substrate and, thus, for each chemical reaction; however, there is flexibility as well.

The fact that active sites are so perfectly suited to provide specific environmental conditions also means that they are subject to local enviromental influences. It is true that increasing the environmental temperature generally increases reaction rates, enzyme-catalyzed or otherwise. However, increasing or decreasing the temperature outside of an optimal range can affect chemical bonds within the active site in such a way that they are less well suited to bind substrates. High temperatures will eventually cause enzymes, like other biological molecules, to **denature**, a process that changes the substance's natural properties. Likewise, the local environment's pH can also affect enzyme function. Active site amino acid residues have their own acidic or basic properties that are optimal for catalysis. These residues are sensitive to changes in pH that can impair the way substrate molecules bind. Enzymes are suited to function best within a certain pH range, and, as with temperature, extreme environmental pH values (acidic or basic) can cause enzymes to denature.

Induced Fit and Enzyme Function

For many years, scientists thought that enzyme-substrate binding took place in a simple "lock-and-key" fashion. This model asserted that the enzyme and substrate fit together perfectly in one instantaneous step. However, current research supports a more refined view scientists call **induced fit** (Figure 6.16). This model expands upon the lock-and-key model by describing a more dynamic interaction between enzyme and substrate. As the enzyme and substrate come together, their interaction causes a mild shift in the enzyme's structure that confirms an ideal binding arrangement between the enzyme and the substrate's transition state. This ideal binding maximizes the enzyme's ability to catalyze its reaction.

LINK TO LEARNING

View an induced fit animation at this website (http://openstax.org/l/hexokinase) .

When an enzyme binds its substrate, it forms an enzyme-substrate complex. This complex lowers the reaction's activation energy and promotes its rapid progression in one of many ways. On a basic level, enzymes promote chemical reactions that involve more than one substrate by bringing the substrates together in an optimal orientation. The appropriate region (atoms and bonds) of one molecule is juxtaposed to the other molecule's appropriate region with which it must react. Another way in which enzymes promote substrate reaction is by creating an optimal environment within the active site for the reaction to occur. Certain chemical reactions might proceed best in a slightly acidic or non-polar environment. The chemical properties that emerge from the particular arrangement of amino acid residues within an active site create the perfect environment for an enzyme's specific substrates to react.

You have learned that the activation energy required for many reactions includes the energy involved in manipulating or slightly contorting chemical bonds so that they can easily break and allow others to reform. Enzymatic action can aid this process. The enzyme-substrate complex can lower the activation energy by contorting substrate molecules in such a way as to facilitate bond-breaking, helping to reach the transition state. Finally, enzymes can also lower activation energies by taking part in the chemical reaction itself. The amino acid residues can provide certain ions or chemical groups that actually form covalent bonds with substrate molecules as a necessary step of the reaction process. In these cases, it is important to remember that the enzyme will always return to its original state at the reaction's completion. One of enzymes' hallmark properties is that they remain ultimately unchanged by the reactions they catalyze. After an enzyme catalyzes a reaction, it releases its product(s).

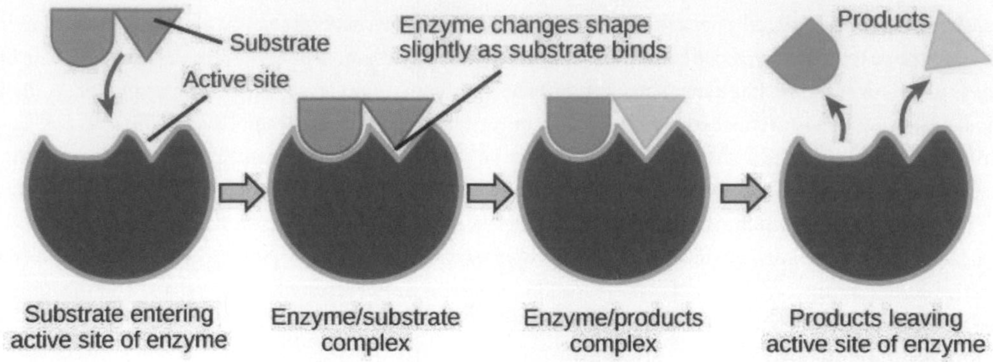

Figure 6.16 According to the induced-fit model, both enzyme and substrate undergo dynamic conformational changes upon binding. The enzyme contorts the substrate into its transition state, thereby increasing the reaction's rate.

Metabolism Control Through Enzyme Regulation

It would seem ideal to have a scenario in which all the encoded enzymes in an organism's genome existed in abundant supply and functioned optimally under all cellular conditions, in all cells, at all times. In reality, this is far from the case. A variety of mechanisms ensure that this does not happen. Cellular needs and conditions vary from cell to cell, and change within individual cells over time. The required enzymes and energetic demands of stomach cells are different from those of fat storage cells, skin cells, blood cells, and nerve cells. Furthermore, a digestive cell works much harder to process and break down nutrients during the time that closely follows a meal compared with many hours after a meal. As these cellular demands and conditions vary, so do the amounts and functionality of different enzymes.

Since the rates of biochemical reactions are controlled by activation energy, and enzymes lower and determine activation energies for chemical reactions, the relative amounts and functioning of the variety of enzymes within a cell ultimately determine which reactions will proceed and at which rates. This determination is tightly controlled. In certain cellular environments, environmental factors like pH and temperature partly control enzyme activity. There are other mechanisms through which cells control enzyme activity and determine the rates at which various biochemical reactions will occur.

Molecular Regulation of Enzymes

Enzymes can be regulated in ways that either promote or reduce their activity. There are many different kinds of molecules that inhibit or promote enzyme function, and various mechanisms exist for doing so. For example, in some cases of enzyme inhibition, an inhibitor molecule is similar enough to a substrate that it can bind to the active site and simply block the substrate from binding. When this happens, the enzyme is inhibited through **competitive inhibition**, because an inhibitor molecule competes with the substrate for active site binding (Figure 6.17). Alternatively, in noncompetitive inhibition, an inhibitor molecule binds to the enzyme at an allosteric site, a binding site away from the active site, and still manages to block substrate binding to the active site.

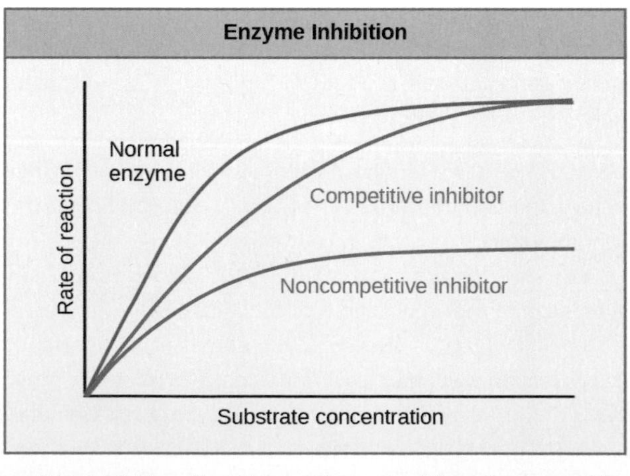

Figure 6.17 Competitive and noncompetitive inhibition affect the reaction's rate differently. Competitive inhibitors affect the initial rate but do not affect the maximal rate; whereas, noncompetitive inhibitors affect the maximal rate.

Some inhibitor molecules bind to enzymes in a location where their binding induces a conformational change that reduces the enzyme's affinity for its substrate. This type of inhibition is an **allosteric inhibition** (Figure 6.18). More than one polypeptide comprise most allosterically regulated enzymes, meaning that they have more than one protein subunit. When an allosteric inhibitor binds to an enzyme, all active sites on the protein subunits change slightly such that they bind their substrates with less efficiency. There are allosteric activators as well as inhibitors. Allosteric activators bind to locations on an enzyme away from the active site, inducing a conformational change that increases the affinity of the enzyme's active site(s) for its substrate(s).

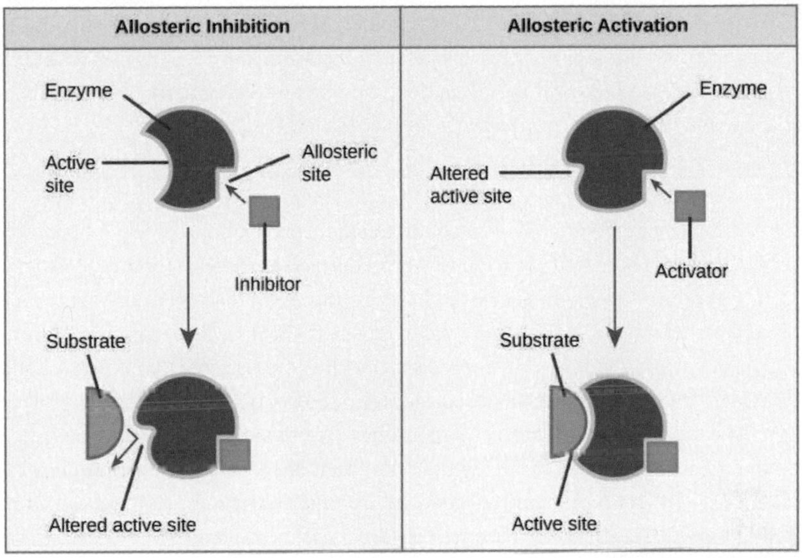

Figure 6.18 Allosteric inhibitors modify the enzyme's active site so that substrate binding is reduced or prevented. In contrast, allosteric activators modify the enzyme's active site so that the affinity for the substrate increases.

Everyday Connection

Drug Discovery by Looking for Inhibitors of Key Enzymes in Specific Pathways

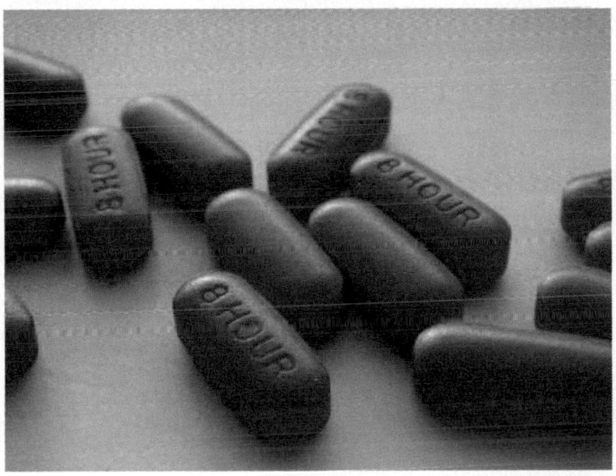

Figure 6.19 Have you ever wondered how pharmaceutical drugs are developed? (credit: Deborah Austin)

Enzymes are key components of metabolic pathways. Understanding how enzymes work and how they can be regulated is a key principle behind developing many pharmaceutical drugs (Figure 6.19) on the market today. Biologists working in this field collaborate with other scientists, usually chemists, to design drugs.

Consider statins for example—which is a class of drugs that reduces cholesterol levels. These compounds are essentially inhibitors of the enzyme HMG-CoA reductase. HMG-CoA reductase is the enzyme that synthesizes cholesterol from lipids in the body. By inhibiting this enzyme, the drug reduces cholesterol levels synthesized in the body. Similarly, acetaminophen, popularly marketed under the brand name Tylenol, is an inhibitor of the enzyme cyclooxygenase. While it is effective in providing relief from fever and inflammation (pain), scientists still do not completely understand its mechanism of action.

How are drugs developed? One of the first challenges in drug development is identifying the specific molecule that the drug is intended to target. In the case of statins, HMG-CoA reductase is the drug target. Researchers identify targets through painstaking research in the laboratory. Identifying the target alone is not sufficient. Scientists also need to know how the target acts inside the cell and which reactions go awry in the case of disease. Once researchers identify the target and the pathway, then the actual drug design process begins. During this stage, chemists and biologists work together to design and synthesize molecules that can either block or activate a particular reaction. However, this is only the beginning: both if and when a drug prototype is successful in performing its function, then it must undergo many tests from *in vitro* experiments to clinical trials before it can obtain FDA approval to be on the market.

Many enzymes don't work optimally, or even at all, unless bound to other specific non-protein helper molecules, either temporarily through ionic or hydrogen bonds or permanently through stronger covalent bonds. Two types of helper molecules are **cofactors** and **coenzymes**. Binding to these molecules promotes optimal conformation and function for their respective enzymes. Cofactors are inorganic ions such as iron (Fe++) and magnesium (Mg++). One example of an enzyme that requires a metal ion as a cofactor is the enzyme that builds DNA molecules, DNA polymerase, which requires a bound zinc ion (Zn++) to function. Coenzymes are organic helper molecules, with a basic atomic structure comprised of carbon and hydrogen, which are required for enzyme action. The most common sources of coenzymes are dietary vitamins (Figure 6.20). Some vitamins are precursors to coenzymes and others act directly as coenzymes. Vitamin C is a coenzyme for multiple enzymes that take part in building the important connective tissue component, collagen. An important step in breaking down glucose to yield energy is catalysis by a multi-enzyme complex scientists call pyruvate dehydrogenase. Pyruvate dehydrogenase is a complex of several enzymes that actually requires one cofactor (a magnesium ion) and five different organic coenzymes to catalyze its specific chemical reaction. Therefore, enzyme function is, in part, regulated by an abundance of various cofactors and coenzymes, which the diets of most organisms supply.

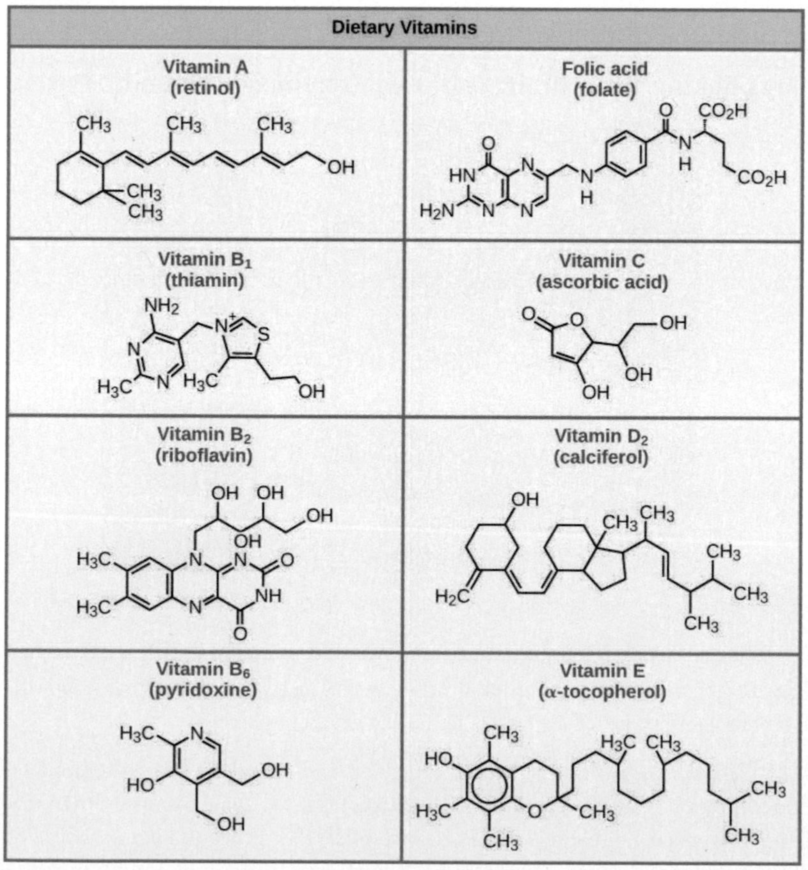

Figure 6.20 Vitamins are important coenzymes or precursors of coenzymes, and are required for enzymes to function properly. Multivitamin

capsules usually contain mixtures of all the vitamins at different percentages.

Enzyme Compartmentalization

In eukaryotic cells, molecules such as enzymes are usually compartmentalized into different organelles. This allows for yet another level of regulation of enzyme activity. Enzymes required only for certain cellular processes are sometimes housed separately along with their substrates, allowing for more efficient chemical reactions. Examples of this sort of enzyme regulation based on location and proximity include the enzymes involved in the latter stages of cellular respiration, which take place exclusively in the mitochondria, and the enzymes involved in digesting cellular debris and foreign materials, located within lysosomes.

Feedback Inhibition in Metabolic Pathways

Molecules can regulate enzyme function in many ways. However, a major question remains: What are these molecules and from where do they come? Some are cofactors and coenzymes, ions, and organic molecules, as you have learned. What other molecules in the cell provide enzymatic regulation, such as allosteric modulation, and competitive and noncompetitive inhibition? The answer is that a wide variety of molecules can perform these roles. Some include pharmaceutical and non-pharmaceutical drugs, toxins, and poisons from the environment. Perhaps the most relevant sources of enzyme regulatory molecules, with respect to cellular metabolism, are cellular metabolic reaction products themselves. In a most efficient and elegant way, cells have evolved to use their own reactions' products for feedback inhibition of enzyme activity. **Feedback inhibition** involves using a reaction product to regulate its own further production (Figure 6.21). The cell responds to the abundance of specific products by slowing down production during anabolic or catabolic reactions. Such reaction products may inhibit the enzymes that catalyzed their production through the mechanisms that we described above.

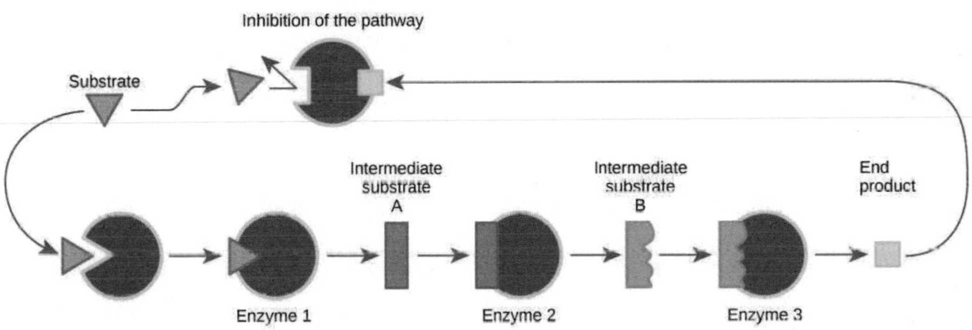

Figure 6.21 Metabolic pathways are a series of reactions that multiple enzymes catalyze. Feedback inhibition, where the pathway's end product inhibits an upstream step, is an important regulatory mechanism in cells.

Producing both amino acids and nucleotides is controlled through feedback inhibition. Additionally, ATP is an allosteric regulator of some of the enzymes involved in sugar's catabolic breakdown, the process that produces ATP. In this way, when ATP is abundant, the cell can prevent its further production. Remember that ATP is an unstable molecule that can spontaneously dissociate into ADP. If too much ATP were present in a cell, much of it would go to waste. Alternatively, ADP serves as a positive allosteric regulator (an allosteric activator) for some of the same enzymes that ATP inhibits. Thus, when relative ADP levels are high compared to ATP, the cell is triggered to produce more ATP through sugar catabolism.

KEY TERMS

activation energy energy necessary for reactions to occur

active site enzyme's specific region to which the substrate binds

allosteric inhibition inhibition by a binding event at a site different from the active site, which induces a conformational change and reduces the enzyme's affinity for its substrate

anabolic (also, anabolism) pathways that require an energy input to synthesize complex molecules from simpler ones

ATP adenosine triphosphate, the cell's energy currency

bioenergetics study of energy flowing through living systems

catabolic (also, catabolism) pathways in which complex molecules break down into simpler ones

chemical energy potential energy in chemical bonds that releases when those bonds are broken

coenzyme small organic molecule, such as a vitamin or its derivative, which is required to enhance an enzyme's activity

cofactor inorganic ion, such as iron and magnesium ions, required for optimal enzyme activity regulation

competitive inhibition type of inhibition in which the inhibitor competes with the substrate molecule by binding to the enzyme's active site

denature process that changes a subtance's natural properties

endergonic describes chemical reactions that require energy input

enthalpy a system's total energy

entropy (S) measure of randomness or disorder within a system

exergonic describes chemical reactions that release free energy

feedback inhibition a product's effect of a reaction sequence to decrease its further production by inhibiting the first enzyme's activity in the pathway that produces it

free energy Gibbs free energy is the usable energy, or energy that is available to do work

heat energy transferred from one system to another that is not work (energy of the molecules' motion or particles)

heat energy total bond energy of reactants or products in a chemical reaction

induced fit dynamic fit between the enzyme and its substrate, in which both components modify their structures to allow for ideal binding

kinetic energy energy type that takes place with objects or particles in motion

metabolism all the chemical reactions that take place inside cells, including anabolism and catabolism

phosphoanhydride bond bond that connects phosphates in an ATP molecule

potential energy energy type that has the potential to do work; stored energy

substrate molecule on which the enzyme acts

thermodynamics study of energy and energy transfer involving physical matter

transition state high-energy, unstable state (an intermediate form between the substrate and the product) occurring during a chemical reaction

CHAPTER SUMMARY

6.1 Energy and Metabolism

Cells perform the functions of life through various chemical reactions. A cell's metabolism refers to the chemical reactions that take place within it. There are metabolic reactions that involve breaking down complex chemicals into simpler ones, such as breaking down large macromolecules. Scientists refer to this process as catabolism, and we associate such reactions an energy release. On the other end of the spectrum, anabolism refers to metabolic processes that build complex molecules out of simpler ones, such as macromolecule synthesis. Anabolic processes require energy. Glucose synthesis and glucose breakdown are examples of anabolic and catabolic pathways, respectively.

6.2 Potential, Kinetic, Free, and Activation Energy

Energy comes in many different forms. Objects in motion do physical work, and kinetic energy is the energy of objects in motion. Objects that are not in motion may have the potential to do work, and thus, have potential energy. Molecules also have potential energy because breaking molecular bonds has the potential to release energy. Living cells depend on harvesting potential energy from molecular bonds to perform work. Free energy is a measure of energy that is available to do work. A system's free energy changes during energy transfers such as chemical reactions, and scientists refer to this change as ΔG.

A reaction's ΔG can be negative or positive, meaning that the reaction releases energy or consumes energy, respectively. A reaction with a negative ΔG that gives off energy is an exergonic reaction. One with a positive ΔG that requires energy input is an endergonic reaction. Exergonic reactions are spontaneous because their products have less energy than their reactants. Endergonic reactions' products have a higher energy state than the reactants, and so these are nonspontaneous reactions. However, all reactions (including spontaneous -ΔG reactions) require an initial energy input in order to reach the transition state, at which they will proceed. This initial input of energy is the activation energy.

6.3 The Laws of Thermodynamics

In studying energy, scientists use the term "system" to refer to the matter and its environment involved in energy transfers. Everything outside of the system is the surroundings. Single cells are biological systems. We can think of systems as having a certain amount of order. It takes energy to make a system more ordered. The more ordered a system, the lower its entropy. Entropy is a measure of a system's disorder. As a system becomes more disordered, the lower its energy and the higher its entropy.

The laws of thermodynamics are a series of laws that describe the properties and processes of energy transfer. The first law states that the total amount of energy in the universe is constant. This means that energy cannot be created or destroyed, only transferred or transformed. The second law of thermodynamics states that every energy transfer involves some loss of energy in an unusable form, such as heat energy, resulting in a more disordered system. In other words, no energy transfer is completely efficient, and all transfers trend toward disorder.

6.4 ATP: Adenosine Triphosphate

ATP is the primary energy-supplying molecule for living cells. ATP is comprised of a nucleotide, a five-carbon sugar, and three phosphate groups. The bonds that connect the phosphates (phosphoanhydride bonds) have high-energy content. The energy released from ATP hydrolysis into ADP + P_i performs cellular work. Cells use ATP to perform work by coupling ATP hydrolysis' exergonic reaction with endergonic reactions. ATP donates its phosphate group to another molecule via phosphorylation. The phosphorylated molecule is at a higher-energy state and is less stable than its unphosphorylated form, and this added energy from phosphate allows the molecule to undergo its endergonic reaction.

6.5 Enzymes

Enzymes are chemical catalysts that accelerate chemical reactions at physiological temperatures by lowering their activation energy. Enzymes are usually proteins consisting of one or more polypeptide chains. Enzymes have an active site that provides a unique chemical environment, comprised of certain amino acid R groups (residues). This unique environment is perfectly suited to convert particular chemical reactants for that enzyme, scientists call substrates, into unstable intermediates that they call transition states. Enzymes and substrates bind with an induced fit, which means that enzymes undergo slight conformational adjustments upon substrate contact, leading to full, optimal binding. Enzymes bind to substrates and catalyze reactions in four different ways: bringing substrates together in an optimal orientation, compromising the bond structures of substrates so that bonds can break down more easily, providing optimal environmental conditions for a reaction to occur, or participating directly in their chemical reaction by forming transient covalent bonds with the substrates.

Enzyme action must be regulated so that in a given cell at a given time, the desired reactions catalyze and the undesired reactions are not. Enzymes are regulated by cellular conditions, such as temperature and pH. They are also regulated through their location within a cell, sometimes compartmentalized so that they can only catalyze reactions under certain circumstances. Enzyme inhibition and activation via other molecules are other important ways that enzymes are regulated. Inhibitors can act competitively, noncompetitively, or allosterically. Noncompetitive inhibitors are usually allosteric. Activators can also enhance enzyme function allosterically. The most common method by which cells regulate the enzymes in metabolic pathways is through feedback inhibition. During feedback inhibition, metabolic pathway products serve as inhibitors (usually allosteric) of one or more of the enzymes (usually the first committed enzyme of the pathway) involved in the pathway that produces them.

VISUAL CONNECTION QUESTIONS

1. Figure 6.8 Look at each of the processes, and decide if it is endergonic or exergonic. In each case, does enthalpy increase or decrease, and does entropy increase or decrease?
2. Figure 6.10 If no activation energy were required to break down sucrose (table sugar), would you be able to store it in a sugar bowl?
3. Figure 6.14 One ATP molecule's hydrolysis releases 7.3 kcal/mol of energy ($\Delta G = -7.3$ kcal/mol of energy). If it takes 2.1 kcal/mol of energy to move one Na^+ across the membrane ($\Delta G = +2.1$ kcal/mol of energy), how many sodium ions could one ATP molecule's hydrolysis move?

REVIEW QUESTIONS

4. Energy is stored long-term in the bonds of_____ and used short-term to perform work from a(n) _____ molecule.
   a. ATP : glucose
   b. an anabolic molecule : catabolic molecule
   c. glucose : ATP
   d. a catabolic molecule : anabolic molecule

5. DNA replication involves unwinding two strands of parent DNA, copying each strand to synthesize complementary strands, and releasing the parent and daughter DNA. Which of the following accurately describes this process?
   a. This is an anabolic process.
   b. This is a catabolic process.
   c. This is both anabolic and catabolic.
   d. This is a metabolic process but is neither anabolic nor catabolic.

6. Consider a pendulum swinging. Which type(s) of energy is/are associated with the pendulum in the following instances: i. the moment at which it completes one cycle, just before it begins to fall back towards the other end, ii. the moment that it is in the middle between the two ends, and iii. just before it reaches the end of one cycle (just before instant i.).
   a. i. potential and kinetic, ii. potential and kinetic, iii. kinetic
   b. i. potential, ii. potential and kinetic, iii. potential and kinetic
   c. i. potential, ii. kinetic, iii. potential and kinetic
   d. i. potential and kinetic, ii. kinetic iii. kinetic

7. Which of the following comparisons or contrasts between endergonic and exergonic reactions is false?
   a. Endergonic reactions have a positive ΔG and exergonic reactions have a negative ΔG.
   b. Endergonic reactions consume energy and exergonic reactions release energy.
   c. Both endergonic and exergonic reactions require a small amount of energy to overcome an activation barrier.
   d. Endergonic reactions take place slowly and exergonic reactions take place quickly.

8. Which of the following is the best way to judge the relative activation energies between two given chemical reactions?
   a. Compare the ΔG values between the two reactions.
   b. Compare their reaction rates.
   c. Compare their ideal environmental conditions.
   d. Compare the spontaneity between the two reactions.

9. Which of the following is not an example of an energy transformation?
   a. turning on a light switch
   b. solar panels at work
   c. formation of static electricity
   d. none of the above

10. In each of the three systems, determine the state of entropy (low or high) when comparing the first and second: i. the instant that a perfume bottle is sprayed compared with 30 seconds later, ii. an old 1950s car compared with a brand new car, and iii. a living cell compared with a dead cell.
   a. i. low, ii. high, iii. low
   b. i. low, ii. high, iii. high
   c. i. high, ii. low, iii. high
   d. i. high, ii. low, iii. low

11. The energy released by the hydrolysis of ATP is____
   a. primarily stored between the alpha and beta phosphates
   b. equal to –57 kcal/mol
   c. harnessed as heat energy by the cell to perform work
   d. providing energy to coupled reactions

12. Which of the following molecules is likely to have the most potential energy?
   a. sucrose
   b. ATP
   c. glucose
   d. ADP

13. Which of the following is not true about enzymes:
   a. They increase ΔG of reactions.
   b. They are usually made of amino acids.
   c. They lower the activation energy of chemical reactions.
   d. Each one is specific to the particular substrate(s) to which it binds.

14. An allosteric inhibitor does which of the following?
   a. Binds to an enzyme away from the active site and changes the conformation of the active site, increasing its affinity for substrate binding.
   b. Binds to the active site and blocks it from binding substrate.
   c. Binds to an enzyme away from the active site and changes the conformation of the active site, decreasing its affinity for the substrate.
   d. Binds directly to the active site and mimics the substrate.

15. Which of the following analogies best describes the induced-fit model of enzyme-substrate binding?
    a. a hug between two people
    b. a key fitting into a lock
    c. a square peg fitting through the square hole and a round peg fitting through the round hole of a children's toy
    d. the fitting together of two jigsaw puzzle pieces

CRITICAL THINKING QUESTIONS

16. Does physical exercise involve anabolic and/or catabolic processes? Give evidence for your answer.
17. Name two different cellular functions that require energy that parallel human energy-requiring functions.
18. Explain in your own words the difference between a spontaneous reaction and one that occurs instantaneously, and what causes this difference.
19. Describe the position of the transition state on a vertical energy scale, from low to high, relative to the position of the reactants and products, for both endergonic and exergonic reactions.
20. Imagine an elaborate ant farm with tunnels and passageways through the sand where ants live in a large community. Now imagine that an earthquake shook the ground and demolished the ant farm. In which of these two scenarios, before or after the earthquake, was the ant farm system in a state of higher or lower entropy?
21. Energy transfers take place constantly in everyday activities. Think of two scenarios: cooking on a stove and driving. Explain how the second law of thermodynamics applies to these two scenarios.
22. Do you think that the E_A for ATP hydrolysis is relatively low or high? Explain your reasoning.
23. With regard to enzymes, why are vitamins necessary for good health? Give examples.
24. Explain in your own words how enzyme feedback inhibition benefits a cell.

CHAPTER 7
Cellular Respiration

Figure 7.1 This geothermal energy plant transforms thermal energy from deep in the ground into electrical energy, which can be easily used. (credit: modification of work by the U.S. Department of Defense)

INTRODUCTION The electrical energy plant in Figure 7.1 converts energy from one form to another form that can be more easily used. This type of generating plant starts with underground thermal energy (heat) and transforms it into electrical energy that will be transported to homes and factories. Like a generating plant, plants and animals also must take in energy from the environment and convert it into a form that their cells can use. Mass and its stored energy enter an organism's body in one form and are converted into another form that can fuel the organism's life functions. In the process of photosynthesis, plants and other photosynthetic producers take in energy in the form of light (solar energy) and convert it into chemical energy in the form of glucose, which stores this energy in its chemical bonds. Then, a series of metabolic pathways, collectively called cellular respiration, extracts the energy from the bonds in glucose and converts it into a form that all living things can use.

Chapter Outline

7.1 Energy in Living Systems

By the end of this section, you will be able to do the following:

- Discuss the importance of electrons in the transfer of energy in living systems
- Explain how ATP is used by cells as an energy source

Energy production within a cell involves many coordinated chemical pathways. Most of these pathways are combinations of oxidation and reduction reactions, which occur at the same time. An oxidation reaction strips an electron from an atom in a compound, and the addition of this electron to another compound is a reduction reaction. Because oxidation and reduction usually occur together, these pairs of reactions are called oxidation reduction reactions, or **redox reactions**.

Electrons and Energy

The removal of an electron from a molecule (oxidizing it), results in a decrease in potential energy in the oxidized compound. The electron (sometimes as part of a hydrogen atom) does not remain unbonded, however, in the cytoplasm of a cell. Rather, the electron is shifted to a second compound, reducing the second compound. *The shift of an electron from one compound to another removes some potential energy from the first compound (the oxidized compound) and increases the potential energy of the second compound (the reduced compound).* The transfer of electrons between molecules is important because most of the energy stored in atoms and used to fuel cell functions is in the form of high-energy electrons. The transfer of energy in the form of high-energy electrons allows the cell to transfer and use energy in an incremental fashion—in small packages rather than in a single, destructive burst. This chapter focuses on the extraction of energy from food; you will see that as you track the path of the transfers, you are tracking the path of electrons moving through metabolic pathways.

Electron Carriers

In living systems, a small class of compounds functions as electron shuttles: they bind and carry high-energy electrons between compounds in biochemical pathways. The principal electron carriers we will consider are derived from the B vitamin group and are derivatives of nucleotides. These compounds can be easily reduced (that is, they accept electrons) or oxidized (they lose electrons). Nicotinamide adenine dinucleotide (NAD) (Figure 7.2) is derived from vitamin B_3, niacin. NAD^+ is the oxidized form of the molecule; NADH is the reduced form of the molecule after it has accepted two electrons and a proton (which together are the equivalent of a hydrogen atom with an extra electron). Note that if a compound has an "H" on it, it is generally reduced (e.g., NADH is the reduced form of NAD).

NAD^+ can accept electrons from an organic molecule according to the general equation:

$$\underset{\substack{\text{Reducing}\\\text{agent}}}{\text{RH}} + \underset{\substack{\text{Oxidizing}\\\text{agent}}}{\text{NAD}^+} \rightarrow \underset{\text{Reduced}}{\text{NADH}} + \underset{\text{Oxidized}}{\text{R}}$$

When electrons are added to a compound, *it is reduced.* A compound that reduces another is called a reducing agent. In the above equation, RH is a reducing agent, and NAD^+ is reduced to NADH. When electrons are removed from a compound, *it is oxidized.* A compound that oxidizes another is called an oxidizing agent. In the above equation, NAD^+ is an oxidizing agent, and RH is oxidized to R.

Similarly, flavin adenine dinucleotide (FAD^+) is derived from vitamin B_2, also called riboflavin. Its reduced form is $FADH_2$. A second variation of NAD, NADP, contains an extra phosphate group. Both NAD^+ and FAD^+ are extensively used in energy extraction from sugars, and NADP plays an important role in anabolic reactions and photosynthesis in plants.

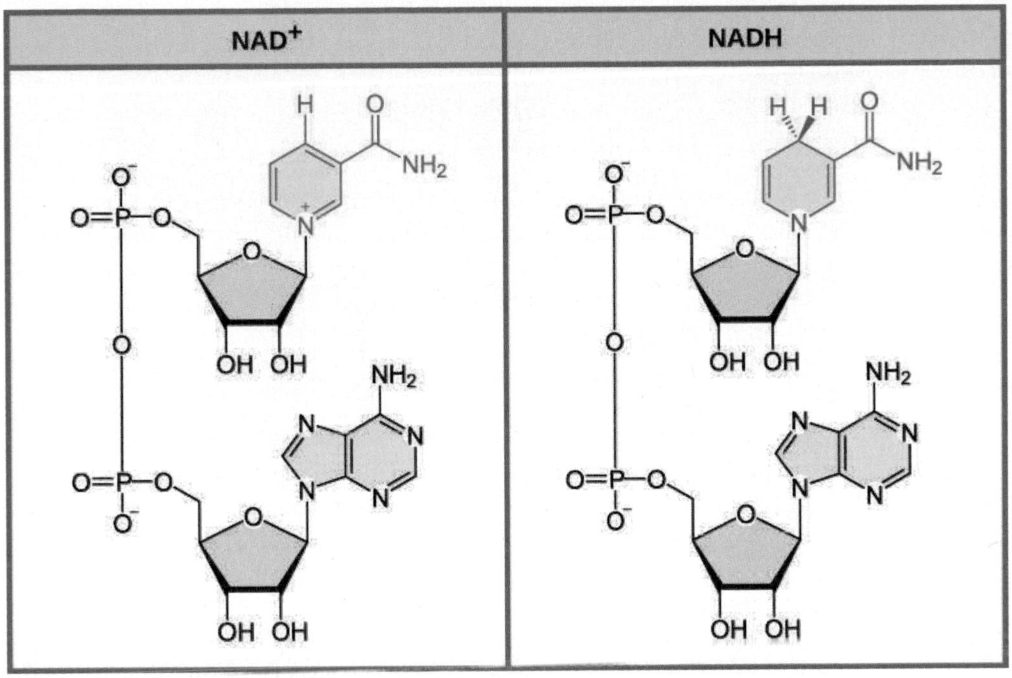

Figure 7.2 The oxidized form of the electron carrier (NAD^+) is shown on the left, and the reduced form (NADH) is shown on the right. The nitrogenous base in NADH has one more hydrogen ion and two more electrons than in NAD^+.

ATP in Living Systems

A living cell cannot store significant amounts of free energy. Excess free energy would result in an increase of heat in the cell, which would result in excessive thermal motion that could damage and then destroy the cell. Rather, a cell must be able to handle that energy in a way that enables the cell to store energy safely and release it for use only as needed. Living cells accomplish this by using the compound adenosine triphosphate (ATP). ATP is often called the "energy currency" of the cell, and, like currency, this versatile compound can be used to fill any energy need of the cell. How? It functions similarly to a rechargeable battery.

When ATP is broken down, usually by the removal of its terminal phosphate group, energy is released. The energy is used to do work by the cell, usually when the released phosphate binds to another molecule, thereby activating it. For example, in the mechanical work of muscle contraction, ATP supplies the energy to move the contractile muscle proteins. Recall the active transport work of the sodium-potassium pump in cell membranes. ATP alters the structure of the integral protein that functions as the pump, changing its affinity for sodium and potassium. In this way, the cell performs work, pumping ions against their electrochemical gradients.

ATP Structure and Function

At the heart of ATP is a molecule of adenosine monophosphate (AMP), which is composed of an adenine molecule bonded to a ribose molecule and to a single phosphate group (Figure 7.3). Ribose is a five-carbon sugar found in RNA, and AMP is one of the nucleotides in RNA. The addition of a second phosphate group to this core molecule results in the formation of adenosine diphosphate (ADP); the addition of a third phosphate group forms adenosine triphosphate (ATP).

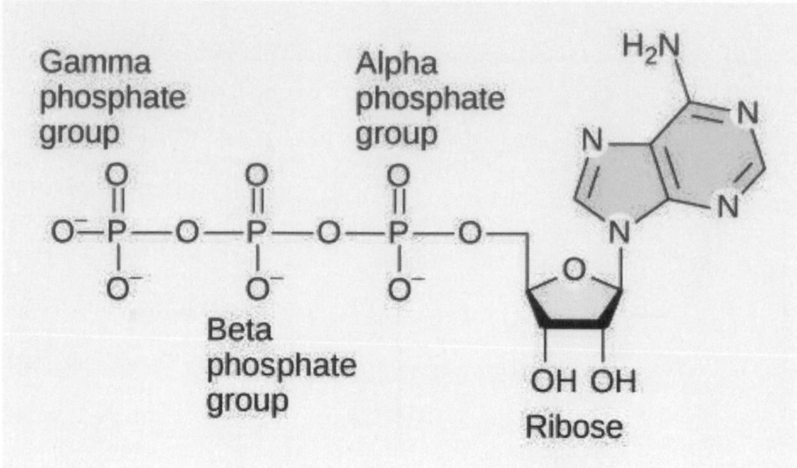

Figure 7.3 ATP (adenosine triphosphate) has three phosphate groups that can be removed by hydrolysis (addition of H_2O) to form ADP (adenosine diphosphate) or AMP (adenosine monophosphate). The negative charges on the phosphate group naturally repel each other, requiring energy to bond them together and releasing energy when these bonds are broken.

The addition of a phosphate group to a molecule requires energy. Phosphate groups are negatively charged and thus repel one another when they are arranged in series, as they are in ADP and ATP. This repulsion makes the ADP and ATP molecules inherently unstable. The release of one or two phosphate groups from ATP, a process called **dephosphorylation**, releases energy.

Energy from ATP

Hydrolysis is the process of breaking complex macromolecules apart. During hydrolysis, water is split, or lysed, and the resulting hydrogen atom (H^+) and a hydroxyl group (OH^-), or *hydroxide,* are added to the larger molecule. The hydrolysis of ATP produces ADP, together with an inorganic phosphate ion (P_i), and the release of free energy. To carry out life processes, ATP is continuously broken down into ADP, and like a rechargeable battery, ADP is continuously regenerated into ATP by the reattachment of a third phosphate group. Water, which was broken down into its hydrogen atom and hydroxyl group (hydroxide) during ATP hydrolysis, is regenerated when a third phosphate is added to the ADP molecule, reforming ATP.

Obviously, energy must be infused into the system to regenerate ATP. Where does this energy come from? In nearly every living thing on Earth, the energy comes from the metabolism of glucose, fructose, or galactose, all isomers with the chemical formula $C_6H_{12}O_6$ but different molecular configurations. In this way, ATP is a direct link between the limited set of exergonic pathways of glucose catabolism and the multitude of endergonic pathways that power living cells.

Phosphorylation

Recall that, in some chemical reactions, enzymes may bind to several substrates that react with each other on the enzyme, forming an intermediate complex. An intermediate complex is a temporary structure, and it allows one of the substrates (such as ATP) and reactants to more readily react with each other; in reactions involving ATP, ATP is one of the substrates and ADP is a product. During an endergonic chemical reaction, ATP forms an intermediate complex with the substrate and enzyme in the reaction. This intermediate complex allows the ATP to transfer its third phosphate group, with its energy, to the substrate, a process called phosphorylation. **Phosphorylation** refers to the addition of the phosphate (~P). This is illustrated by the following generic reaction, in which A and B represent two different substrates:

$$\text{A} + \text{enzyme} + \text{ATP} \rightarrow \left[\text{A} - \text{enzyme} - \sim\text{P}\right] \rightarrow \text{B} + \text{enzyme} + \text{ADP} + \text{phosphate ion}$$

When the intermediate complex breaks apart, the energy is used to modify the substrate and convert it into a product of the reaction. The ADP molecule and a free phosphate ion are released into the medium and are available for recycling through cell metabolism.

Substrate Phosphorylation

ATP is generated through two mechanisms during the breakdown of glucose. A few ATP molecules are generated (that is, regenerated from ADP) as a direct result of the chemical reactions that occur in the catabolic pathways. A phosphate group is removed from an intermediate reactant in the pathway, and the free energy of the reaction is used to add the third phosphate to an available ADP molecule, producing ATP (Figure 7.4). This very direct method of phosphorylation is called **substrate-level phosphorylation**.

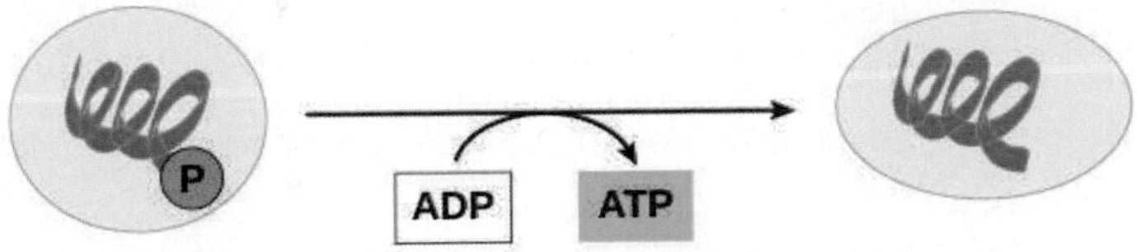

Figure 7.4 In phosphorylation reactions, the gamma (third) phosphate of ATP is attached to a protein.

Oxidative Phosphorylation

Most of the ATP generated during glucose catabolism, however, is derived from a much more complex process, chemiosmosis, which takes place in mitochondria (Figure 7.5) within a eukaryotic cell or the plasma membrane of a prokaryotic cell. **Chemiosmosis**, a process of ATP production in cellular metabolism, is used to generate 90 percent of the ATP made during glucose catabolism and is also the method used in the light reactions of photosynthesis to harness the energy of sunlight. The production of ATP using the process of chemiosmosis is called **oxidative phosphorylation** because of the involvement of oxygen in the process.

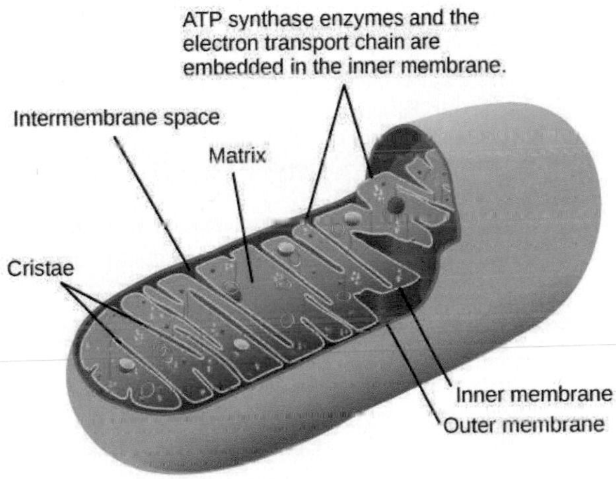

Figure 7.5 In eukaryotes, oxidative phosphorylation takes place in mitochondria. In prokaryotes, this process takes place in the plasma membrane. (Credit: modification of work by Mariana Ruiz Villareal)

CAREER CONNECTION

Mitochondrial Disease Physician

What happens when the critical reactions of cellular respiration do not proceed correctly? This may happen in mitochondrial diseases, which are genetic disorders of metabolism. Mitochondrial disorders can arise from mutations in nuclear or mitochondrial DNA, and they result in the production of less energy than is normal in body cells. In type 2 diabetes, for instance, the oxidation efficiency of NADH is reduced, impacting oxidative phosphorylation but not the other steps of respiration. Symptoms of mitochondrial diseases can include muscle weakness, lack of coordination, stroke-like episodes, and loss of vision and hearing. Most affected people are diagnosed in childhood, although there are some adult-onset diseases. Identifying and treating mitochondrial disorders is a specialized medical field. The educational preparation for this profession requires a college education, followed by medical school with a specialization in medical genetics. Medical geneticists can be board certified by the American Board of Medical Genetics and go on to become associated with professional organizations devoted to the study of mitochondrial diseases, such as the Mitochondrial Medicine Society and the Society for Inherited Metabolic Disorders.

7.2 Glycolysis

By the end of this section, you will be able to do the following:

- Describe the overall result in terms of molecules produced during the chemical breakdown of glucose by glycolysis
- Compare the output of glycolysis in terms of ATP molecules and NADH molecules produced

As you have read, nearly all of the energy used by living cells comes to them in the bonds of the sugar glucose. **Glycolysis** is the first step in the breakdown of glucose to extract energy for cellular metabolism. In fact, nearly all living organisms carry out glycolysis as part of their metabolism. The process does not use oxygen directly and therefore is termed **anaerobic**. Glycolysis takes place in the cytoplasm of both prokaryotic and eukaryotic cells. Glucose enters heterotrophic cells in two ways. One method is through secondary active transport in which the transport takes place against the glucose concentration gradient. The other mechanism uses a group of integral proteins called **GLUT proteins**, also known as glucose transporter proteins. These transporters assist in the facilitated diffusion of glucose.

Glycolysis begins with the six-carbon ring-shaped structure of a single glucose molecule and ends with two molecules of a three-carbon sugar called **pyruvate**. Glycolysis consists of two distinct phases. The first part of the glycolysis pathway traps the glucose molecule in the cell and uses energy to modify it so that the six-carbon sugar molecule can be split evenly into the two three-carbon molecules. The second part of glycolysis extracts energy from the molecules and stores it in the form of ATP and NADH—remember: this is the reduced form of NAD.

First Half of Glycolysis (Energy-Requiring Steps)

Step 1. The first step in glycolysis (Figure 7.6) is catalyzed by hexokinase, an enzyme with broad specificity that catalyzes the phosphorylation of six-carbon sugars. Hexokinase phosphorylates glucose using ATP as the source of the phosphate, producing glucose-6-phosphate, a more reactive form of glucose. This reaction prevents the phosphorylated glucose molecule from continuing to interact with the GLUT proteins, and it can no longer leave the cell because the negatively charged phosphate will not allow it to cross the hydrophobic interior of the plasma membrane.

Step 2. In the second step of glycolysis, an isomerase converts glucose-6-phosphate into one of its isomers, fructose-6-phosphate (this isomer has a phosphate attached at the location of the sixth carbon of the ring). An **isomerase** is an enzyme that catalyzes the conversion of a molecule into one of its isomers. (This change from phosphoglucose to phosphofructose allows the eventual split of the sugar into two three-carbon molecules.)

Step 3. The third step is the phosphorylation of fructose-6-phosphate, catalyzed by the enzyme phosphofructokinase. A second ATP molecule donates a high-energy phosphate to fructose-6-phosphate, producing fructose-1,6-bisphosphate. In this pathway, phosphofructokinase is a rate-limiting enzyme. It is active when the concentration of ADP is high; it is less active when ADP levels are low and the concentration of ATP is high. Thus, if there is "sufficient" ATP in the system, the pathway slows down. This is a type of end product inhibition, since ATP is the end product of glucose catabolism.

Step 4. The newly added high-energy phosphates further destabilize fructose-1,6-bisphosphate. The fourth step in glycolysis employs an enzyme, aldolase, to cleave fructose-1,6-bisphosphate into two three-carbon isomers: dihydroxyacetone phosphate and glyceraldehyde-3-phosphate.

Step 5. In the fifth step, an isomerase transforms the dihydroxyacetone-phosphate into its isomer, glyceraldehyde-3-phosphate. Thus, the pathway will continue with two molecules of a glyceraldehyde-3-phosphate. At this point in the pathway, there is a net investment of energy from two ATP molecules in the breakdown of one glucose molecule.

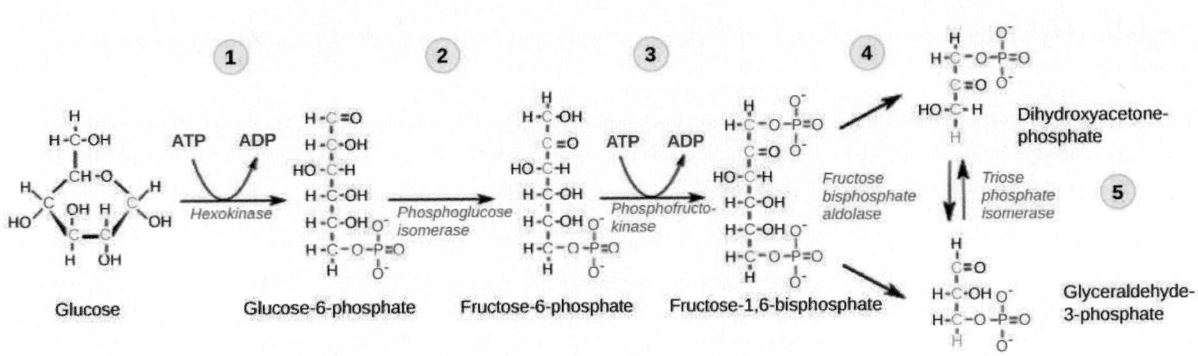

Figure 7.6 The first half of glycolysis uses two ATP molecules in the phosphorylation of glucose, which is then split into two three-carbon molecules.

Second Half of Glycolysis (Energy-Releasing Steps)

So far, glycolysis has cost the cell two ATP molecules and produced two small, three-carbon sugar molecules. Both of these molecules will proceed through the second half of the pathway, and sufficient energy will be extracted to pay back the two ATP molecules used as an initial investment and produce a profit for the cell of two additional ATP molecules and two even higher-energy NADH molecules.

Step 6. The sixth step in glycolysis (Figure 7.7) oxidizes the sugar (glyceraldehyde-3-phosphate), extracting high-energy electrons, which are picked up by the electron carrier NAD^+, producing NADH. The sugar is then phosphorylated by the addition of a second phosphate group, producing 1,3-bisphosphoglycerate. Note that the second phosphate group does not require another ATP molecule.

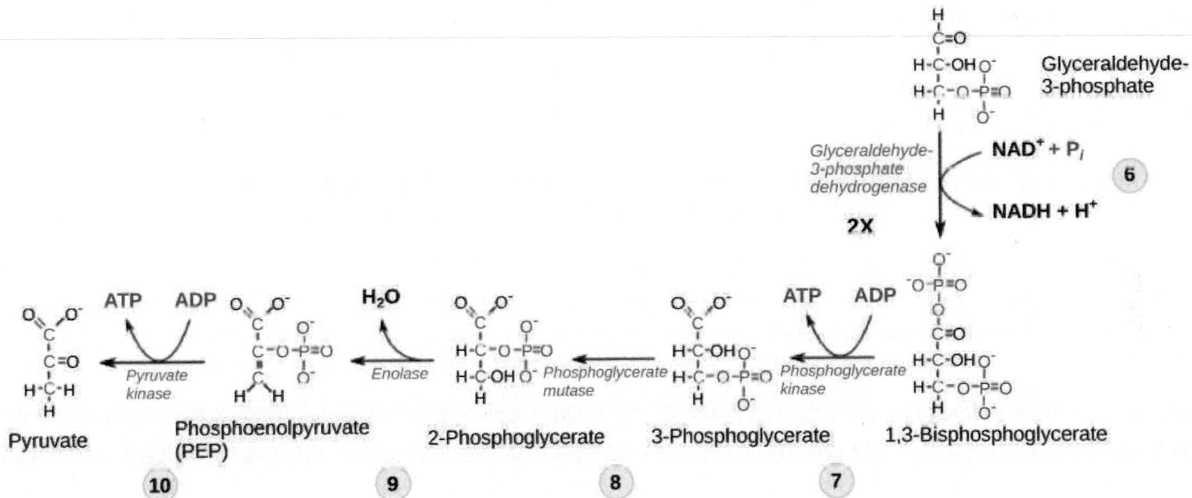

Figure 7.7 The second half of glycolysis involves phosphorylation without ATP investment (step 6) and produces two NADH and four ATP molecules per glucose.

Here again is a potential limiting factor for this pathway. The continuation of the reaction depends upon the availability of the oxidized form of the electron carrier, NAD^+. Thus, NADH must be continuously oxidized back into NAD^+ in order to keep this step going. If NAD^+ is not available, the second half of glycolysis slows down or stops. If oxygen is available in the system, the NADH will be oxidized readily, though indirectly, and the high-energy electrons from the hydrogen released in this process will be used to produce ATP. In an environment without oxygen, an alternate pathway (fermentation) can provide the oxidation of NADH to NAD^+.

Step 7. In the seventh step, catalyzed by phosphoglycerate kinase (an enzyme named for the reverse reaction), 1,3-bisphosphoglycerate donates a high-energy phosphate to ADP, forming one molecule of ATP. (This is an example of substrate-level phosphorylation.) A carbonyl group on the 1,3-bisphosphoglycerate is oxidized to a carboxyl group, and 3-phosphoglycerate is formed.

Step 8. In the eighth step, the remaining phosphate group in 3-phosphoglycerate moves from the third carbon to the second carbon, producing 2-phosphoglycerate (an isomer of 3-phosphoglycerate). The enzyme catalyzing this step is a mutase (isomerase).

Step 9. Enolase catalyzes the ninth step. This enzyme causes 2-phosphoglycerate to lose water from its structure; this is a dehydration reaction, resulting in the formation of a double bond that increases the potential energy in the remaining phosphate bond and produces phosphoenolpyruvate (PEP).

Step 10. The last step in glycolysis is catalyzed by the enzyme pyruvate kinase (the enzyme in this case is named for the reverse reaction of pyruvate's conversion into PEP) and results in the production of a second ATP molecule by substrate-level phosphorylation and the compound pyruvic acid (or its salt form, pyruvate). Many enzymes in enzymatic pathways are named for the reverse reactions, since the enzyme can catalyze both forward and reverse reactions (these may have been described initially by the reverse reaction that takes place in vitro, under nonphysiological conditions).

LINK TO LEARNING

Gain a better understanding of the breakdown of glucose by glycolysis by visiting this site (http://openstax.org/l/glycolysis) to see the process in action.

Outcomes of Glycolysis

Glycolysis begins with glucose and produces two pyruvate molecules, four new ATP molecules, and two molecules of NADH. (Note: two ATP molecules are used in the first half of the pathway to prepare the six-carbon ring for cleavage, so the cell has a *net gain of two ATP molecules* and two NADH molecules for its use). If the cell cannot catabolize the pyruvate molecules further, it will harvest only two ATP molecules from one molecule of glucose. Mature mammalian red blood cells do not have mitochondria and thus are not capable of **aerobic respiration**—the process in which organisms convert energy in the presence of oxygen—and glycolysis is their sole source of ATP. If glycolysis is interrupted, these cells lose their ability to maintain their sodium-potassium pumps, and eventually, they die.

The last step in glycolysis will not occur if pyruvate kinase, the enzyme that catalyzes the formation of pyruvate, is not available in sufficient quantities. In this situation, the entire glycolysis pathway will proceed, but only two ATP molecules will be made in the second half. Thus, pyruvate kinase is a rate-limiting enzyme for glycolysis.

7.3 Oxidation of Pyruvate and the Citric Acid Cycle

By the end of this section, you will be able to do the following:

- Explain how a circular pathway, such as the citric acid cycle, fundamentally differs from a linear biochemical pathway, such as glycolysis
- Describe how pyruvate, the product of glycolysis, is prepared for entry into the citric acid cycle

If oxygen is available, aerobic respiration will go forward. In eukaryotic cells, the pyruvate molecules produced at the end of glycolysis are transported into the mitochondria, which are the sites of cellular respiration. There, pyruvate is transformed into an acetyl group that will be picked up and activated by a carrier compound called coenzyme A (CoA). The resulting compound is called **acetyl CoA**. CoA is derived from vitamin B5, pantothenic acid. Acetyl CoA can be used in a variety of ways by the cell, but its major function is to deliver the acetyl group derived from pyruvate to the next stage of the pathway in glucose catabolism.

Breakdown of Pyruvate

In order for pyruvate, the product of glycolysis, to enter the next pathway, it must undergo several changes. The conversion is a three-step process (Figure 7.8).

Step 1. A carboxyl group is removed from pyruvate, releasing a molecule of carbon dioxide into the surrounding medium. This reaction creates a two-carbon hydroxyethyl group bound to the enzyme (pyruvate dehydrogenase). We should note that this is the first of the six carbons from the original glucose molecule to be removed. (This step proceeds twice because there are *two* pyruvate molecules produced at the end of glycolsis for every molecule of glucose metabolized anaerobically; thus, two of the six carbons will have been removed at the end of both steps.)

Step 2. The hydroxyethyl group is oxidized to an acetyl group, and the electrons are picked up by NAD^+, forming NADH. The high-energy electrons from NADH will be used later to generate ATP.

Step 3. The enzyme-bound acetyl group is transferred to CoA, producing a molecule of acetyl CoA.

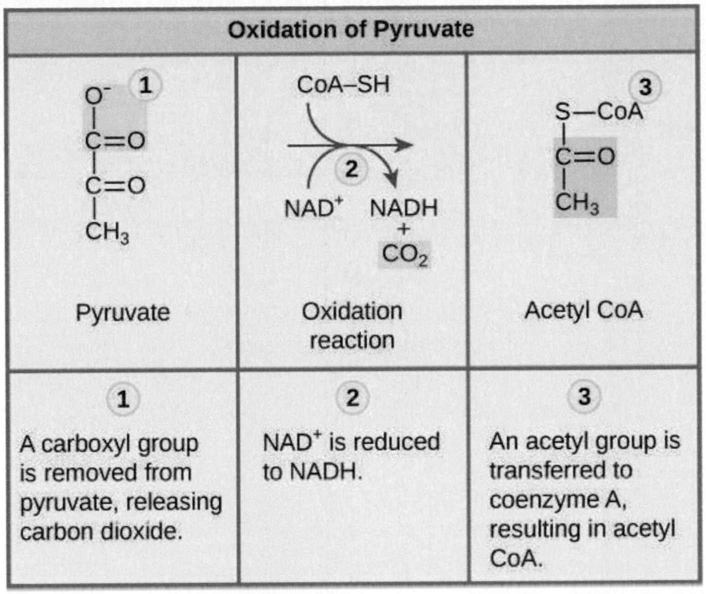

Figure 7.8 Upon entering the mitochondrial matrix, a multienzyme complex converts pyruvate into acetyl CoA. In the process, carbon dioxide is released, and one molecule of NADH is formed.

Note that during the second stage of glucose metabolism, whenever a carbon atom is removed, it is bound to two oxygen atoms, producing carbon dioxide, one of the major end products of cellular respiration.

Acetyl CoA to CO_2

In the presence of oxygen, acetyl CoA delivers its acetyl (2C) group to a four-carbon molecule, oxaloacetate, to form citrate, a six-carbon molecule with three carboxyl groups; this pathway will harvest the remainder of the extractable energy from what began as a glucose molecule and release the remaining four CO_2 molecules. This single pathway is called by different names: the **citric acid cycle** (for the first intermediate formed—citric acid, or citrate—when acetate joins to the oxaloacetate), the **TCA cycle** (because citric acid or citrate and isocitrate are tricarboxylic acids), and the **Krebs cycle**, after Hans Krebs, who first identified the steps in the pathway in the 1930s in pigeon flight muscles.

Citric Acid Cycle

Like the conversion of pyruvate to acetyl CoA, the citric acid cycle takes place in the matrix of mitochondria. Almost all of the enzymes of the citric acid cycle are soluble, with the single exception of the enzyme succinate dehydrogenase, which is embedded in the inner membrane of the mitochondrion. Unlike glycolysis, the citric acid cycle is a closed loop: the last part of the pathway regenerates the compound used in the first step. The eight steps of the cycle are a series of redox, dehydration, hydration, and decarboxylation reactions that produce two carbon dioxide molecules, one GTP/ATP, and the reduced carriers NADH and $FADH_2$ (Figure 7.9). *This is considered an aerobic pathway because the NADH and $FADH_2$ produced must transfer their electrons to the next pathway in the system, which will use oxygen.* If this transfer does not occur, the oxidation steps of the citric acid cycle also do not occur. Note that the citric acid cycle produces very little ATP directly and does not directly consume oxygen.

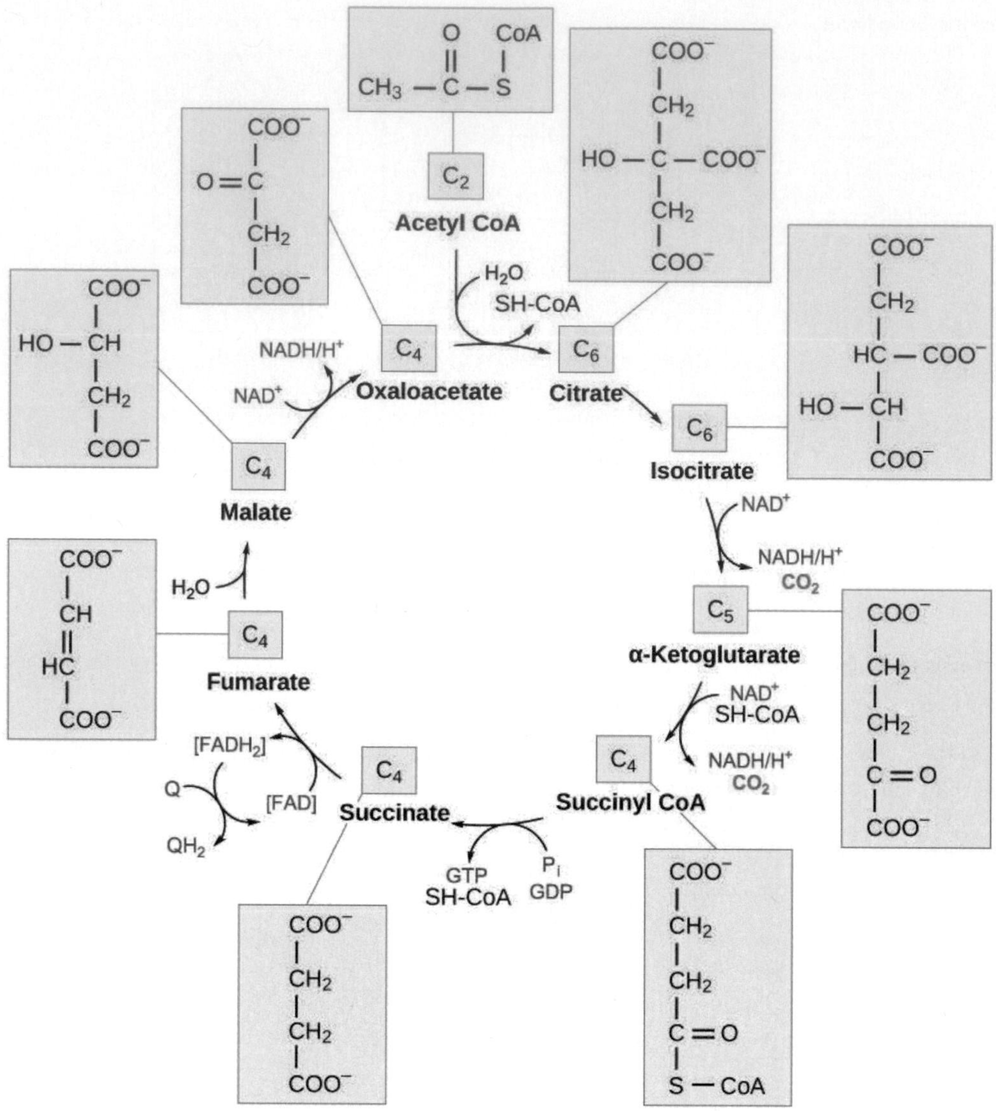

Figure 7.9 In the citric acid cycle, the acetyl group from acetyl CoA is attached to a four-carbon oxaloacetate molecule to form a six-carbon citrate molecule. Through a series of steps, citrate is oxidized, releasing two carbon dioxide molecules for each acetyl group fed into the cycle. In the process, three NAD^+ molecules are reduced to NADH, one FAD molecule is reduced to $FADH_2$, and one ATP or GTP (depending on the cell type) is produced (by substrate-level phosphorylation). Because the final product of the citric acid cycle is also the first reactant, the cycle runs continuously in the presence of sufficient reactants. (credit: modification of work by "Yikrazuul"/Wikimedia Commons)

Steps in the Citric Acid Cycle

Step 1. Prior to the first step, a transitional phase occurs during which pyruvic acid is converted to acetyl CoA. Then, the first step of the cycle begins: This condensation step combines the two-carbon acetyl group with a four-carbon oxaloacetate molecule to form a six-carbon molecule of citrate. CoA is bound to a sulfhydryl group (-SH) and diffuses away to eventually combine with another acetyl group. This step is irreversible because it is highly exergonic. The rate of this reaction is controlled by negative feedback and the amount of ATP available. If ATP levels increase, the rate of this reaction decreases. If ATP is in short supply, the rate increases.

Step 2. In step two, citrate loses one water molecule and gains another as citrate is converted into its isomer, isocitrate.

Step 3. In step three, isocitrate is oxidized, producing a five-carbon molecule, α-ketoglutarate, along with a molecule of CO_2 and two electrons, which reduce NAD^+ to NADH. This step is also regulated by negative feedback from ATP and NADH and a positive effect of ADP.

Step 4. Steps three and four are both oxidation and decarboxylation steps, which as we have seen, release electrons that reduce

NAD^+ to NADH and release carboxyl groups that form CO_2 molecules. Alpha-ketoglutarate is the product of step three, and a succinyl group is the product of step four. CoA binds with the succinyl group to form succinyl CoA. The enzyme that catalyzes step four is regulated by feedback inhibition of ATP, succinyl CoA, and NADH.

Step 5. In step five, a phosphate group is substituted for coenzyme A, and a high-energy bond is formed. This energy is used in substrate-level phosphorylation (during the conversion of the succinyl group to succinate) to form either guanine triphosphate (GTP) or ATP. There are two forms of the enzyme, called isoenzymes, for this step, depending upon the type of animal tissue in which they are found. One form is found in tissues that use large amounts of ATP, such as heart and skeletal muscle. This form produces ATP. The second form of the enzyme is found in tissues that have a high number of anabolic pathways, such as liver. This form produces GTP. GTP is energetically equivalent to ATP; however, its use is more restricted. In particular, protein synthesis primarily uses GTP.

Step 6. Step six is a dehydration process that converts succinate into fumarate. Two hydrogen atoms are transferred to FAD, reducing it to $FADH_2$. (Note: the energy contained in the electrons of these hydrogens is insufficient to reduce NAD^+ but adequate to reduce FAD.) Unlike NADH, this carrier remains attached to the enzyme and transfers the electrons to the electron transport chain directly. This process is made possible by the localization of the enzyme catalyzing this step inside the inner membrane of the mitochondrion.

Step 7. Water is added by hydrolysis to fumarate during step seven, and malate is produced. The last step in the citric acid cycle regenerates oxaloacetate by oxidizing malate. Another molecule of NADH is then produced in the process.

LINK TO LEARNING

View an animation of the citric acid cycle here (http://openstax.org/l/krebs_cycle) .

Products of the Citric Acid Cycle

Two carbon atoms come into the citric acid cycle from each acetyl group, representing four out of the six carbons of one glucose molecule. Two carbon dioxide molecules are released on each turn of the cycle; however, these do not necessarily contain the most recently added carbon atoms. The two acetyl carbon atoms will eventually be released on later turns of the cycle; thus, all six carbon atoms from the original glucose molecule are eventually incorporated into carbon dioxide. Each turn of the cycle forms three NADH molecules and one $FADH_2$ molecule. These carriers will connect with the last portion of aerobic respiration, the electron transport chain, to produce ATP molecules. One GTP or ATP is also made in each cycle. Several of the intermediate compounds in the citric acid cycle can be used in synthesizing nonessential amino acids; therefore, the cycle is **amphibolic** (both catabolic and anabolic).

7.4 Oxidative Phosphorylation

By the end of this section, you will be able to do the following:

- Describe how electrons move through the electron transport chain and explain what happens to their energy levels during this process
- Explain how a proton (H^+) gradient is established and maintained by the electron transport chain

You have just read about two pathways in glucose catabolism—glycolysis and the citric acid cycle—that generate ATP. Most of the ATP generated during the aerobic catabolism of glucose, however, is not generated directly from these pathways. Instead, it is derived from a process that begins by moving electrons through a series of electron carriers that undergo redox reactions. This process causes hydrogen ions to accumulate within the intermembranous space. Therefore, a concentration gradient forms in which hydrogen ions diffuse out of the intermembranous space into the mitochondrial matrix by passing through ATP synthase. The current of hydrogen ions powers the catalytic action of ATP synthase, which phosphorylates ADP, producing ATP.

Electron Transport Chain

The electron transport chain (Figure 7.10) is the last component of aerobic respiration and is the only part of glucose metabolism that uses atmospheric oxygen. Oxygen continuously diffuses into plant tissues (typically through stomata), as well as into fungi and bacteria; however, in animals, oxygen enters the body through a variety of respiratory systems. Electron transport is a series of redox reactions that resembles a relay race or bucket brigade in that electrons are passed rapidly from one component to the next, to the endpoint of the chain where the electrons reduce molecular oxygen and, along with associated protons, produces water. There are four complexes composed of proteins, labeled I through IV in Figure 7.10, and the aggregation of these four complexes, together with associated mobile, accessory electron carriers, is called the **electron transport chain**. The electron transport chain is present with multiple copies in the inner mitochondrial membrane of eukaryotes and within the plasma membrane of prokaryotes.

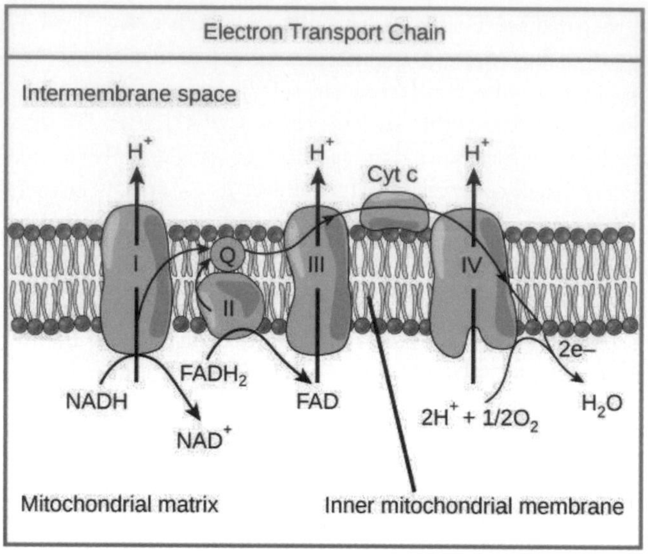

Figure 7.10 The electron transport chain is a series of electron transporters embedded in the inner mitochondrial membrane that shuttles electrons from NADH and $FADH_2$ to molecular oxygen. In the process, protons are pumped from the mitochondrial matrix to the intermembrane space, and oxygen is reduced to form water.

Complex I

First, two electrons are carried to the first complex via NADH. This complex, labeled **I**, is composed of flavin mononucleotide (FMN) and an iron-sulfur (Fe-S)-containing protein. FMN, which is derived from vitamin B_2 (also called riboflavin), is one of several prosthetic groups or cofactors in the electron transport chain. A **prosthetic group** is a nonprotein molecule required for the activity of a protein. Prosthetic groups are organic or inorganic, nonpeptide molecules bound to a protein that facilitate its function. Prosthetic groups include coenzymes, which are the prosthetic groups of enzymes. The enzyme in complex I is NADH dehydrogenase and is a very large protein, containing 45 amino acid chains. Complex I can pump four hydrogen ions across the membrane from the matrix into the intermembrane space, and it is in this way that the hydrogen ion gradient is established and maintained between the two compartments separated by the inner mitochondrial membrane.

Q and Complex II

Complex II directly receives $FADH_2$—which does not pass through complex I. The compound connecting the first and second complexes to the third is **ubiquinone** B. The Q molecule is lipid soluble and freely moves through the hydrophobic core of the membrane. Once it is reduced (QH_2), ubiquinone delivers its electrons to the next complex in the electron transport chain. Q receives the electrons derived from NADH from complex I, and the electrons derived from $FADH_2$ from complex II. This enzyme and $FADH_2$ form a small complex that delivers electrons directly to the electron transport chain, bypassing the first complex. Since these electrons bypass and thus do not energize the proton pump in the first complex, fewer ATP molecules are made from the $FADH_2$ electrons. *The number of ATP molecules ultimately obtained is directly proportional to the number of protons pumped across the inner mitochondrial membrane.*

Complex III

The third complex is composed of cytochrome b—another Fe-S protein, a Rieske center (2Fe-2S center), and cytochrome c proteins. This complex is also called cytochrome oxidoreductase. Cytochrome proteins have a prosthetic group of heme. The heme molecule is similar to the heme in hemoglobin, but it carries electrons, not oxygen. As a result, the iron ion at its core is reduced and oxidized as it passes the electrons, fluctuating between different oxidation states: Fe^{++} (reduced) and Fe^{+++} (oxidized). The heme molecules in the cytochromes have slightly different characteristics due to the effects of the different proteins binding to them, giving slightly different characteristics to each complex. Complex III pumps protons through the membrane and passes its electrons to cytochrome c for transport to the fourth complex of proteins and enzymes. (Cytochrome c receives electrons from Q; however, whereas Q carries pairs of electrons, cytochrome c can accept only one at a time.)

Complex IV

The fourth complex is composed of cytochrome proteins c, a, and a_3. This complex contains two heme groups (one in each of the two cytochromes, a, and a_3) and three copper ions (a pair of Cu_A and one Cu_B in cytochrome a_3). The cytochromes hold an oxygen molecule very tightly between the iron and copper ions until the oxygen is completely reduced by the gain of two electrons. The reduced oxygen then picks up two hydrogen ions from the surrounding medium to make water (H_2O). The removal of the hydrogen ions from the system contributes to the ion gradient that forms the foundation for the process of chemiosmosis.

Chemiosmosis

In chemiosmosis, the free energy from the series of redox reactions just described is used to pump hydrogen ions (protons) across the mitochondrial membrane. The uneven distribution of H^+ ions across the membrane establishes both concentration and electrical gradients (thus, an electrochemical gradient), owing to the hydrogen ions' positive charge and their aggregation on one side of the membrane.

If the membrane were continuously open to simple diffusion by the hydrogen ions, the ions would tend to diffuse back across into the matrix, driven by the concentrations producing their electrochemical gradient. Recall that many ions cannot diffuse through the nonpolar regions of phospholipid membranes without the aid of ion channels. Similarly, hydrogen ions in the matrix space can only pass through the inner mitochondrial membrane by an integral membrane protein called ATP synthase (Figure 7.11). This complex protein acts as a tiny generator, turned by the force of the hydrogen ions diffusing through it, down their electrochemical gradient. The turning of parts of this molecular machine facilitates the addition of a phosphate to ADP, forming ATP, using the potential energy of the hydrogen ion gradient.

VISUAL CONNECTION

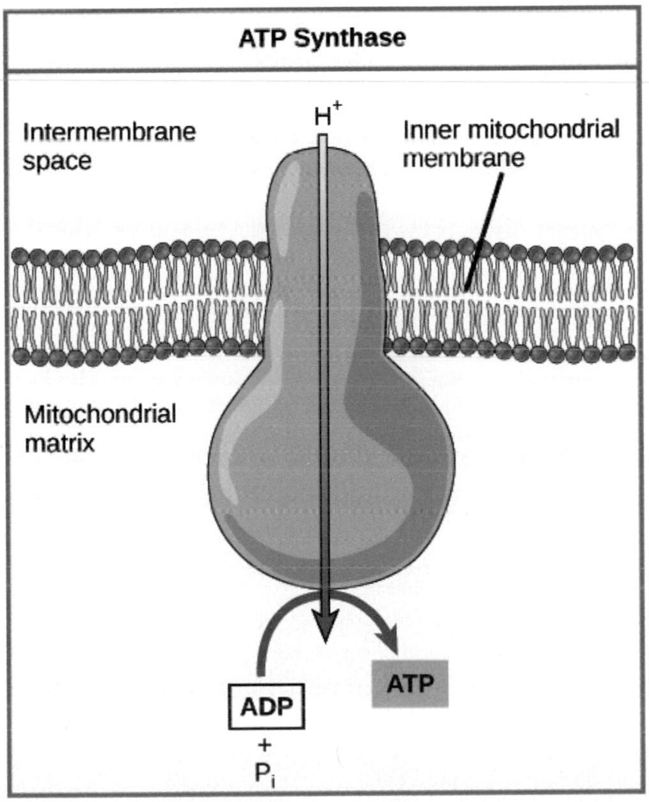

Figure 7.11 ATP synthase is a complex, molecular machine that uses a proton (H^+) gradient to form ATP from ADP and inorganic phosphate (Pi). (Credit: modification of work by Klaus Hoffmeier)

Dinitrophenol (DNP) is an "uncoupler" that makes the inner mitochondrial membrane "leaky" to protons. It was used until 1938 as a weight-loss drug. What effect would you expect DNP to have on the change in pH across the inner mitochondrial membrane? Why do you think this might be an effective weight-loss drug?

Chemiosmosis (Figure 7.12) is used to generate 90 percent of the ATP made during aerobic glucose catabolism; it is also the method used in the light reactions of photosynthesis to harness the energy of sunlight in the process of photophosphorylation. Recall that the production of ATP using the process of chemiosmosis in mitochondria is called oxidative phosphorylation. The overall result of these reactions is the production of ATP from the energy of the electrons removed from hydrogen atoms. These atoms were originally part of a glucose molecule. At the end of the pathway, the electrons are used to reduce an oxygen molecule to oxygen ions. The extra electrons on the oxygen attract hydrogen ions (protons) from the surrounding medium, and water is formed. Thus, oxygen is the final electron acceptor in the electron transport chain.

VISUAL CONNECTION

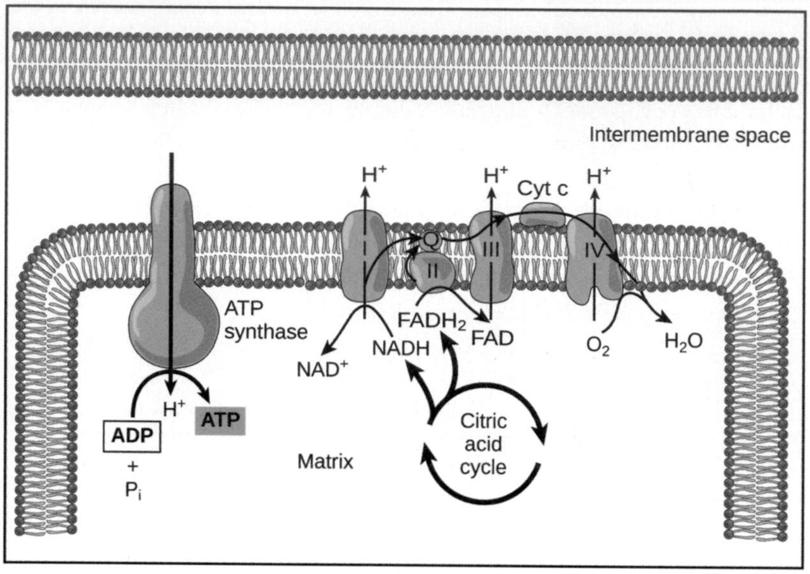

Figure 7.12 In oxidative phosphorylation, the pH gradient formed by the electron transport chain is used by ATP synthase to form ATP.

Cyanide inhibits cytochrome c oxidase, a component of the electron transport chain. If cyanide poisoning occurs, would you expect the pH of the intermembrane space to increase or decrease? What effect would cyanide have on ATP synthesis?

ATP Yield

The number of ATP molecules generated from the catabolism of glucose varies. For example, the number of hydrogen ions that the electron transport chain complexes can pump through the membrane varies between species. Another source of variance stems from the shuttle of electrons across the membranes of the mitochondria. (The NADH generated from glycolysis cannot easily enter mitochondria.) Thus, electrons are picked up on the inside of mitochondria by either NAD^+ or FAD^+. As you have learned earlier, these FAD^+ molecules can transport fewer ions; consequently, fewer ATP molecules are generated when FAD^+ acts as a carrier. NAD^+ is used as the electron transporter in the liver and FAD^+ acts in the brain.

Another factor that affects the yield of ATP molecules generated from glucose is the fact that intermediate compounds in these pathways are also used for other purposes. Glucose catabolism connects with the pathways that build or break down all other biochemical compounds in cells, and the result is somewhat messier than the ideal situations described thus far. For example, sugars other than glucose are fed into the glycolytic pathway for energy extraction. In addition, the five-carbon sugars that form nucleic acids are made from intermediates in glycolysis. Certain nonessential amino acids can be made from intermediates of both glycolysis and the citric acid cycle. Lipids, such as cholesterol and triglycerides, are also made from intermediates in these pathways, and both amino acids and triglycerides are broken down for energy through these pathways. Overall, in living systems, these pathways of glucose catabolism extract about 34 percent of the energy contained in glucose, with the remainder being released as heat.

7.5 Metabolism without Oxygen

By the end of this section, you will be able to do the following:

- Discuss the fundamental difference between anaerobic cellular respiration and fermentation
- Describe the type of fermentation that readily occurs in animal cells and the conditions that initiate that fermentation

In aerobic respiration, the final electron acceptor is an oxygen molecule, O_2. If aerobic respiration occurs, then ATP will be produced using the energy of high-energy electrons carried by NADH or $FADH_2$ to the electron transport chain. If aerobic respiration does not occur, NADH must be reoxidized to NAD^+ for reuse as an electron carrier for the glycolytic pathway to continue. How is this done? Some living systems use an organic molecule as the final electron acceptor. Processes that use an organic molecule to regenerate NAD^+ from NADH are collectively referred to as **fermentation**. In contrast, some living systems use an inorganic molecule as a final electron acceptor. Both methods are called **anaerobic cellular respiration**, in which organisms convert energy for their use in the absence of oxygen.

Anaerobic Cellular Respiration

Certain prokaryotes, including some species in the domains Bacteria and Archaea, use anaerobic respiration. For example, a group of archaeans called methanogens reduces carbon dioxide to methane to oxidize NADH. These microorganisms are found in soil and in the digestive tracts of ruminants, such as cows and sheep. Similarly, sulfate-reducing bacteria, most of which are anaerobic (Figure 7.13), reduce sulfate to hydrogen sulfide to regenerate NAD^+ from NADH.

Figure 7.13 The green color seen in these coastal waters is from an eruption of hydrogen sulfide–producing bacteria. These anaerobic, sulfate-reducing bacteria release hydrogen sulfide gas as they decompose algae in the water. (credit: modification of work by NASA/Jeff Schmaltz, MODIS Land Rapid Response Team at NASA GSFC, Visible Earth Catalog of NASA images)

LINK TO LEARNING

Visit this site (http://openstax.org/l/fermentation) to see anaerobic cellular respiration in action.

Lactic Acid Fermentation

The fermentation method used by animals and certain bacteria, such as those in yogurt, is lactic acid fermentation (Figure 7.14). This type of fermentation is used routinely in mammalian red blood cells, which do not have mitochondria, and in skeletal muscle that has an insufficient oxygen supply to allow aerobic respiration to continue (that is, in muscles used to the point of fatigue). In muscles, lactic acid accumulation must be removed by the blood circulation, and when the lactic acid loses a hydrogen, the resulting lactate is brought to the liver for further metabolism. The chemical reactions of lactic acid fermentation are the following:

$$\text{Pyruvic acid} + \text{NADH} \leftrightarrow \text{lactic acid} + \text{NAD}^+$$

The enzyme used in this reaction is lactate dehydrogenase (LDH). The reaction can proceed in either direction, but the reaction from left to right is inhibited by acidic conditions. Such lactic acid accumulation was once believed to cause muscle stiffness, fatigue, and soreness, although more recent research disputes this hypothesis. Once the lactic acid has been removed from the muscle and circulated to the liver, it can be reconverted into pyruvic acid and further catabolized for energy.

VISUAL CONNECTION

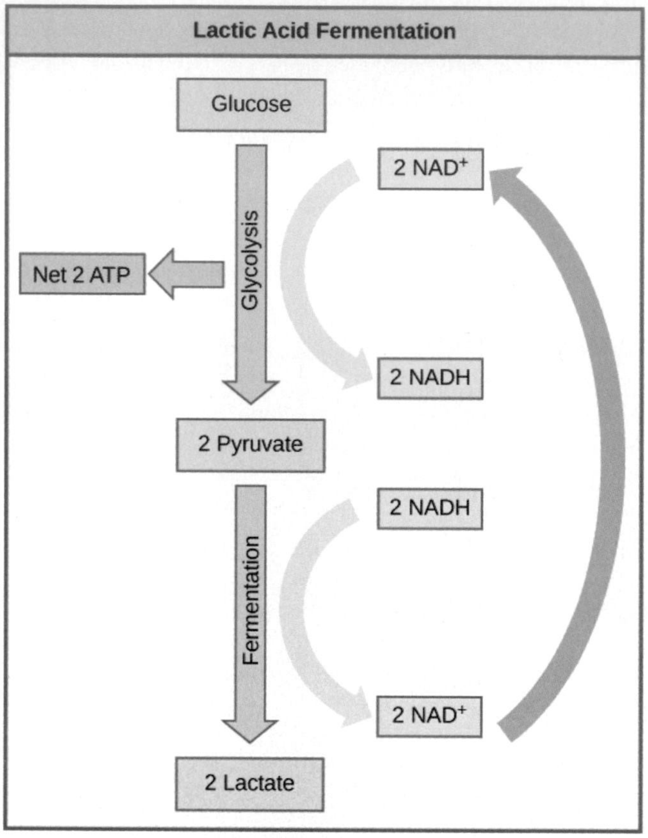

Figure 7.14 Lactic acid fermentation is common in muscle cells that have run out of oxygen.

Tremetol, a metabolic poison found in the white snakeroot plant, prevents the metabolism of lactate. When cows eat this plant, tremetol is concentrated in the milk they produce. Humans who consume the milk can become seriously ill. Symptoms of this disease, which include vomiting, abdominal pain, and tremors, become worse after exercise. Why do you think this is the case?

Alcohol Fermentation

Another familiar fermentation process is alcohol fermentation (Figure 7.15), which produces ethanol. The first chemical reaction of alcohol fermentation is the following (CO_2 does not participate in the second reaction):

$$\text{pyruvic acid} + H^+ \rightarrow CO_2 + \text{acetaldehyde} + \text{NADH} + H^+ \rightarrow \text{ethanol} + \text{NAD}^+$$

The first reaction is catalyzed by pyruvate decarboxylase, a cytoplasmic enzyme, with a coenzyme of thiamine pyrophosphate (TPP, derived from vitamin B_1 and also called thiamine). A carboxyl group is removed from pyruvic acid, releasing carbon dioxide as a gas. The loss of carbon dioxide reduces the size of the molecule by one carbon, producing acetaldehyde. The second reaction is catalyzed by alcohol dehydrogenase to oxidize NADH to NAD$^+$ and reduce acetaldehyde to ethanol. The fermentation of pyruvic acid by yeast produces the ethanol found in alcoholic beverages. Ethanol tolerance of yeast is variable, ranging from about 5 percent to 21 percent, depending on the yeast strain and environmental conditions.

Figure 7.15 Fermentation of grape juice into wine produces CO_2 as a byproduct. Fermentation tanks have valves so that the pressure inside the tanks created by the carbon dioxide produced can be released.

Other Types of Fermentation

Other fermentation methods take place in bacteria. We should note that many prokaryotes are *facultatively* anaerobic. This means that they can switch between aerobic respiration and fermentation, depending on the availability of free oxygen. Certain prokaryotes, such as *Clostridia*, are obligate anaerobes. Obligate anaerobes live and grow in the absence of molecular oxygen. Oxygen is a poison to these microorganisms and kills them on exposure. We should also note that all forms of fermentation, except lactic acid fermentation, produce gas. The production of particular types of gas is used as an indicator of the fermentation of specific carbohydrates, which plays a role in the laboratory identification of the bacteria. Various methods of fermentation are used by assorted organisms to ensure an adequate supply of NAD^+ for the sixth step in glycolysis. Without these pathways, this step would not occur, and ATP could not be harvested from the breakdown of glucose.

7.6 Connections of Carbohydrate, Protein, and Lipid Metabolic Pathways

By the end of this section, you will be able to do the following:

- Discuss the ways in which carbohydrate metabolic pathways, glycolysis, and the citric acid cycle interrelate with protein and lipid metabolic pathways
- Explain why metabolic pathways are not considered closed systems

You have learned about the catabolism of glucose, which provides energy to living cells. But living things consume organic compounds other than glucose for food. How does a turkey sandwich end up as ATP in your cells? This happens because all of the catabolic pathways for carbohydrates, proteins, and lipids eventually connect into glycolysis and the citric acid cycle pathways (see Figure 7.17). Metabolic pathways should be thought of as porous and interconnecting—that is, substances enter from other pathways, and intermediates leave for other pathways. These pathways are not closed systems! Many of the substrates, intermediates, and products in a particular pathway are reactants in other pathways.

Connections of Other Sugars to Glucose Metabolism

Glycogen, a polymer of glucose, is an energy storage molecule in animals. When there is adequate ATP present, excess glucose is stored as glycogen in both liver and muscle cells. The glycogen will be hydrolyzed into glucose 1-phosphate monomers (G-1-P) if blood sugar levels drop. The presence of glycogen as a source of glucose allows ATP to be produced for a longer period of time during exercise. Glycogen is broken down into glucose-1-phosphate (G-1-P) and converted into glucose-6-phosphate (G-6-P) in both muscle and liver cells, and this product enters the glycolytic pathway.

Sucrose is a disaccharide with a molecule of glucose and a molecule of fructose bonded together with a glycosidic linkage. Fructose is one of the three "dietary" monosaccharides, along with glucose and galactose (part of the milk sugar dissacharide lactose), which are absorbed directly into the bloodstream during digestion. The catabolism of both fructose and galactose produces the same number of ATP molecules as glucose.

Connections of Proteins to Glucose Metabolism

Proteins are hydrolyzed by a variety of enzymes in cells. Most of the time, the amino acids are recycled into the synthesis of new proteins. If there are excess amino acids, however, or if the body is in a state of starvation, some amino acids will be shunted into the pathways of glucose catabolism (Figure 7.16). It is very important to note that each amino acid must have its amino group removed prior to entry into these pathways. The amino group is converted into ammonia. In mammals, the liver synthesizes urea from two ammonia molecules and a carbon dioxide molecule. Thus, urea is the principal waste product in mammals, produced from the nitrogen originating in amino acids, and it leaves the body in urine. It should be noted that amino acids can be synthesized from the intermediates and reactants in the cellular respiration cycle.

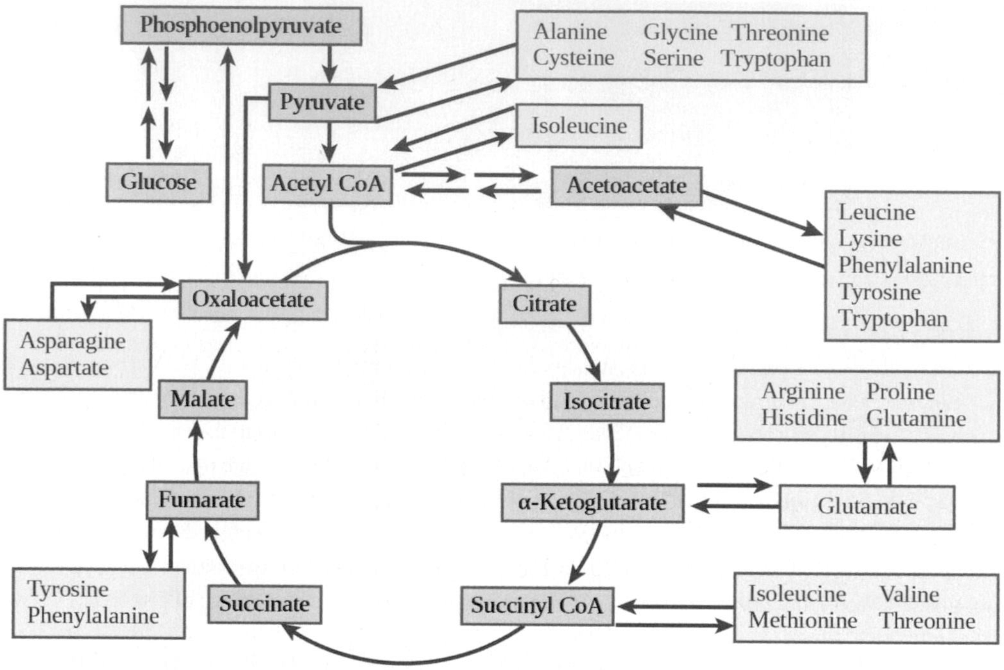

Figure 7.16 The carbon skeletons of certain amino acids (indicated in boxes) derived from proteins can feed into the citric acid cycle. (credit: modification of work by Mikael Häggström)

Connections of Lipid and Glucose Metabolisms

The lipids connected to the glucose pathway include cholesterol and triglycerides. Cholesterol is a lipid that contributes to cell membrane flexibility and is a precursor of steroid hormones. The synthesis of cholesterol starts with acetyl groups and proceeds in only one direction. The process cannot be reversed.

Triglycerides—made from the bonding of glycerol and three fatty acids—are a form of long-term energy storage in animals. Animals can make most of the fatty acids they need. Triglycerides can be both made and broken down through parts of the glucose catabolism pathways. Glycerol can be phosphorylated to glycerol-3-phosphate, which continues through glycolysis. Fatty acids are catabolized in a process called beta-oxidation, which takes place in the matrix of the mitochondria and converts their fatty acid chains into two-carbon units of acetyl groups. The acetyl groups are picked up by CoA to form acetyl CoA that proceeds into the citric acid cycle.

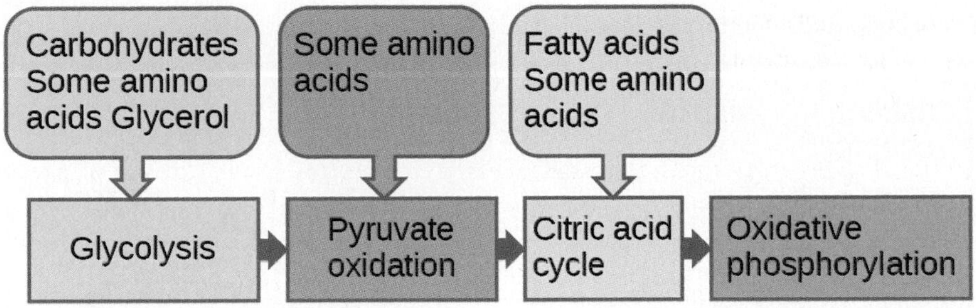

Figure 7.17 Glycogen from the liver and muscles, as well as other carbohydrates, hydrolyzed into glucose-1-phosphate, together with fats and proteins, can feed into the catabolic pathways for carbohydrates.

EVOLUTION CONNECTION

Pathways of Photosynthesis and Cellular Metabolism

The processes of photosynthesis and cellular metabolism consist of several very complex pathways. It is generally thought that the first cells arose in an aqueous environment—a "soup" of nutrients—possibly on the surface of some porous clays, perhaps in warm marine environments. If these cells reproduced successfully and their numbers climbed steadily, it follows that the cells would begin to deplete the nutrients from the medium in which they lived as they shifted the nutrients into the components of their own bodies. This hypothetical situation would have resulted in natural selection favoring those organisms that could exist by using the nutrients that remained in their environment and by manipulating these nutrients into materials upon which they could survive. Selection would favor those organisms that could extract maximal value from the nutrients to which they had access.

An early form of photosynthesis developed that harnessed the sun's energy using water as a source of hydrogen atoms, but this pathway did not produce free oxygen (anoxygenic photosynthesis). (Another type of anoxygenic photosynthesis did not produce free oxygen because it did not use water as the source of hydrogen ions; instead, it used materials such as hydrogen sulfide and consequently produced sulfur). It is thought that glycolysis developed at this time and could take advantage of the simple sugars being produced but that these reactions were unable to fully extract the energy stored in the carbohydrates. The development of glycolysis probably predated the evolution of photosynthesis, as it was well suited to extract energy from materials spontaneously accumulating in the "primeval soup." A later form of photosynthesis used water as a source of electrons and hydrogen and generated free oxygen. Over time, the atmosphere became oxygenated, but not before the oxygen released oxidized metals in the ocean and created a "rust" layer in the sediment, permitting the dating of the rise of the first oxygenic photosynthesizers. Living things adapted to exploit this new atmosphere that allowed aerobic respiration as we know it to evolve. When the full process of oxygenic photosynthesis developed and the atmosphere became oxygenated, cells were finally able to use the oxygen expelled by photosynthesis to extract considerably more energy from the sugar molecules using the citric acid cycle and oxidative phosphorylation.

7.7 Regulation of Cellular Respiration

By the end of this section, you will be able to do the following:

- Describe how feedback inhibition would affect the production of an intermediate or product in a pathway
- Identify the mechanism that controls the rate of the transport of electrons through the electron transport chain

Cellular respiration must be regulated in order to provide balanced amounts of energy in the form of ATP. The cell also must generate a number of intermediate compounds that are used in the anabolism and catabolism of macromolecules. Without controls, metabolic reactions would quickly come to a standstill as the forward and backward reactions reached a state of equilibrium. Resources would be used inappropriately. A cell does not need the maximum amount of ATP that it can make all the time: At times, the cell needs to shunt some of the intermediates to pathways for amino acid, protein, glycogen, lipid, and nucleic acid production. In short, the cell needs to control its metabolism.

Regulatory Mechanisms

A variety of mechanisms is used to control cellular respiration. Some type of control exists at each stage of glucose metabolism. Access of glucose to the cell can be regulated using the **GLUT (glucose transporter) proteins** that transport glucose (Figure 7.18).

Different forms of the GLUT protein control passage of glucose into the cells of specific tissues.

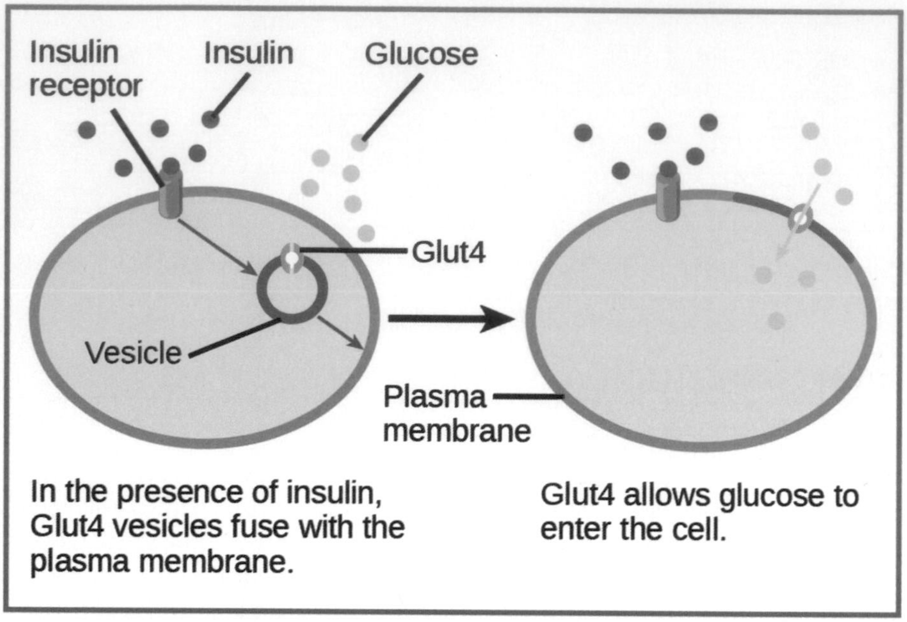

Figure 7.18 GLUT4 is a glucose transporter that is stored in vesicles. A cascade of events that occurs upon insulin binding to a receptor in the plasma membrane causes GLUT4-containing vesicles to fuse with the plasma membrane so that glucose may be transported into the cell.

Some reactions are controlled by having two different enzymes—one each for the two directions of a reversible reaction. Reactions that are catalyzed by only one enzyme can go to equilibrium, stalling the reaction. In contrast, if two different enzymes (each specific for a given direction) are necessary for a reversible reaction, the opportunity to control the rate of the reaction increases, and equilibrium is not reached.

A number of enzymes involved in each of the pathways—in particular, the enzyme catalyzing the first committed reaction of the pathway—are controlled by attachment of a molecule to an allosteric site on the protein. The molecules most commonly used in this capacity are the nucleotides ATP, ADP, AMP, NAD^+, and NADH. These regulators—allosteric effectors—may increase or decrease enzyme activity, depending on the prevailing conditions. The allosteric effector alters the steric structure of the enzyme, usually affecting the configuration of the active site. This alteration of the protein's (the enzyme's) structure either increases or decreases its affinity for its substrate, with the effect of increasing or decreasing the rate of the reaction. The attachment signals to the enzyme. This binding can increase or decrease the enzyme's activity, providing a feedback mechanism. This feedback type of control is effective as long as the chemical affecting it is attached to the enzyme. Once the overall concentration of the chemical decreases, it will diffuse away from the protein, and the control is relaxed.

Control of Catabolic Pathways

Enzymes, proteins, electron carriers, and pumps that play roles in glycolysis, the citric acid cycle, and the electron transport chain tend to catalyze nonreversible reactions. In other words, if the initial reaction takes place, the pathway is committed to proceeding with the remaining reactions. Whether a particular enzyme activity is released depends upon the energy needs of the cell (as reflected by the levels of ATP, ADP, and AMP).

Glycolysis

The control of glycolysis begins with the first enzyme in the pathway, hexokinase (Figure 7.19). This enzyme catalyzes the phosphorylation of glucose, which helps to prepare the compound for cleavage in a later step. The presence of the negatively charged phosphate in the molecule also prevents the sugar from leaving the cell. When hexokinase is inhibited, glucose diffuses out of the cell and does not become a substrate for the respiration pathways in that tissue. The product of the hexokinase reaction is glucose-6-phosphate, which accumulates when a later enzyme, phosphofructokinase, is inhibited.

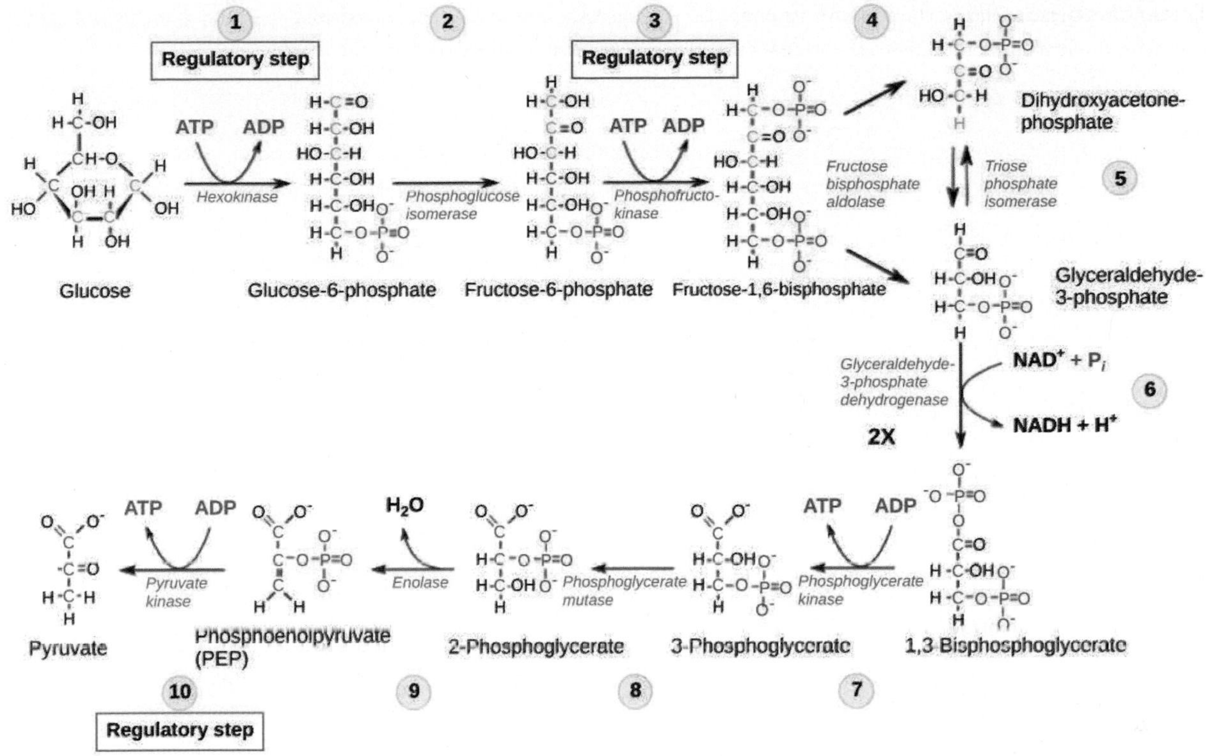

Figure 7.19 The glycolysis pathway is primarily regulated at the three key enzymatic steps (1, 3, and 10) as indicated. Note that the first two steps that are regulated occur early in the pathway and involve hydrolysis of ATP.

Phosphofructokinase is the main enzyme controlled in glycolysis. High levels of ATP or citrate or a lower, more acidic pH decreases the enzyme's activity. An increase in citrate concentration can occur because of a blockage in the citric acid cycle. Fermentation, with its production of organic acids such as lactic acid, frequently accounts for the increased acidity in a cell; however, the products of fermentation do not typically accumulate in cells.

The last step in glycolysis is catalyzed by pyruvate kinase. The pyruvate produced can proceed to be catabolized or converted into the amino acid alanine. If no more energy is needed and alanine is in adequate supply, the enzyme is inhibited. The enzyme's activity is increased when fructose-1,6-bisphosphate levels increase. (Recall that fructose-1,6-bisphosphate is an intermediate in the first half of glycolysis.) The regulation of pyruvate kinase involves phosphorylation by a kinase (pyruvate kinase), resulting in a less-active enzyme. Dephosphorylation by a phosphatase reactivates it. Pyruvate kinase is also regulated by ATP (a negative allosteric effect).

If more energy is needed, more pyruvate will be converted into acetyl CoA through the action of pyruvate dehydrogenase. If either acetyl groups or NADH accumulates, there is less need for the reaction, and the rate decreases. Pyruvate dehydrogenase is also regulated by phosphorylation: a kinase phosphorylates it to form an inactive enzyme, and a phosphatase reactivates it. The kinase and the phosphatase are also regulated.

Citric Acid Cycle

The citric acid cycle is controlled through the enzymes that catalyze the reactions that make the first two molecules of NADH (Figure 7.9). These enzymes are isocitrate dehydrogenase and α-ketoglutarate dehydrogenase. When adequate ATP and NADH levels are available, the rates of these reactions decrease. When more ATP is needed, as reflected in rising ADP levels, the rate increases. Alpha-ketoglutarate dehydrogenase will also be affected by the levels of succinyl CoA—a subsequent intermediate in the cycle—causing a decrease in activity. A decrease in the rate of operation of the pathway at this point is not necessarily negative, as the increased levels of the α-ketoglutarate not used by the citric acid cycle can be used by the cell for amino acid (glutamate) synthesis.

Electron Transport Chain

Specific enzymes of the electron transport chain are unaffected by feedback inhibition, but the rate of electron transport through the pathway is affected by the levels of ADP and ATP. Greater ATP consumption by a cell is indicated by a buildup of

ADP. As ATP usage decreases, the concentration of ADP decreases, and now, ATP begins to build up in the cell. This change in the relative concentration of ADP to ATP triggers the cell to slow down the electron transport chain.

LINK TO LEARNING

Visit this site (http://openstax.org/l/electron_transp) to see an animation of the electron transport chain and ATP synthesis.

For a summary of feedback controls in cellular respiration, see Table 7.1.

Summary of Feedback Controls in Cellular Respiration

Pathway	Enzyme affected	Elevated levels of effector	Effect on pathway activity
glycolysis	hexokinase	glucose-6-phosphate	decrease
	phosphofructokinase	low-energy charge (ATP, AMP), fructose-6-phosphate via fructose-2,6-bisphosphate	increase
		high-energy charge (ATP, AMP), citrate, acidic pH	decrease
	pyruvate kinase	fructose-1,6-bisphosphate	increase
		high-energy charge (ATP, AMP), alanine	decrease
pyruvate to acetyl CoA conversion	pyruvate dehydrogenase	ADP, pyruvate	increase
		acetyl CoA, ATP, NADH	decrease
citric acid cycle	isocitrate dehydrogenase	ADP	increase
		ATP, NADH	decrease
	α-ketoglutarate dehydrogenase	calcium ions, ADP	increase
		ATP, NADH, succinyl CoA	decrease
electron transport chain		ADP	increase
		ATP	decrease

Table 7.1

KEY TERMS

acetyl CoA combination of an acetyl group derived from pyruvic acid and coenzyme A, which is made from pantothenic acid (a B-group vitamin)

aerobic respiration process in which organisms convert energy in the presence of oxygen

anaerobic process that does not use oxygen

anaerobic cellular respiration process in which organisms convert energy for their use in the absence of oxygen

ATP synthase (also F1Fo ATP synthase) membrane-embedded protein complex that adds a phosphate to ADP with energy from protons diffusing through it

chemiosmosis process in which there is a production of adenosine triphosphate (ATP) in cellular metabolism by the involvement of a proton gradient across a membrane

citric acid cycle (also Krebs cycle) series of enzyme-catalyzed chemical reactions of central importance in all living cells for extraction of energy from carbohydrates

dephosphorylation removal of a phosphate group from a molecule

fermentation process of regenerating NAD^+ with either an inorganic or organic compound serving as the final electron acceptor; occurs in the absence of oxygen

GLUT protein integral membrane protein that transports glucose

glycolysis process of breaking glucose into two three-carbon molecules with the production of ATP and NADH

isomerase enzyme that converts a molecule into its isomer

Krebs cycle (also citric acid cycle) alternate name for the citric acid cycle, named after Hans Krebs, who first identified the steps in the pathway in the 1930s in pigeon flight muscles; see citric acid cycle

oxidative phosphorylation production of ATP using the process of chemiosmosis in the presence of oxygen

phosphorylation addition of a high-energy phosphate to a compound, usually a metabolic intermediate, a protein, or ADP

prosthetic group (also prosthetic cofactor) molecule bound to a protein that facilitates the function of the protein

pyruvate three-carbon sugar that can be decarboxylated and oxidized to make acetyl CoA, which enters the citric acid cycle under aerobic conditions; the end product of glycolysis

redox reaction chemical reaction that consists of the coupling of an oxidation reaction and a reduction reaction

substrate-level phosphorylation production of ATP from ADP using the excess energy from a chemical reaction and a phosphate group from a reactant

TCA cycle (also citric acid cycle) alternate name for the citric acid cycle, named after the group name for citric acid, tricarboxylic acid (TCA); see citric acid cycle

ubiquinone soluble electron transporter in the electron transport chain that connects the first or second complex to the third

CHAPTER SUMMARY

7.1 Energy in Living Systems

ATP functions as the energy currency for cells. It allows the cell to store energy briefly and transport it within the cell to support endergonic chemical reactions. The structure of ATP is that of an RNA nucleotide with three phosphates attached. As ATP is used for energy, a phosphate group or two are detached, and either ADP or AMP is produced. Energy derived from glucose catabolism is used to convert ADP into ATP. When ATP is used in a reaction, the third phosphate is temporarily attached to a substrate in a process called phosphorylation. The two processes of ATP regeneration that are used in conjunction with glucose catabolism are substrate-level phosphorylation and oxidative phosphorylation through the process of chemiosmosis.

7.2 Glycolysis

Glycolysis is the first pathway within the cytoplasm used in the breakdown of glucose to extract energy. It was probably one of the earliest metabolic pathways to evolve and is used by nearly all of the organisms on Earth. Glycolysis consists of two parts: The first part prepares the six-carbon ring of glucose for cleavage into two three-carbon sugars. ATP is invested in the process during this half to energize the separation. The second half of glycolysis extracts ATP and high-energy electrons from hydrogen atoms and attaches them to NAD^+. Two ATP molecules are invested in the first half and four ATP molecules are formed by substrate phosphorylation during the second half. This produces a net gain of two ATP and two NADH molecules for the cell.

7.3 Oxidation of Pyruvate and the Citric Acid Cycle

In the presence of oxygen, pyruvate is transformed into an acetyl group attached to a carrier molecule of coenzyme A. The resulting acetyl CoA can enter several pathways, but most often, the acetyl group is delivered to the citric acid cycle for further catabolism. During the conversion of pyruvate into the acetyl group, a molecule of carbon dioxide and two high-energy electrons are removed. The carbon dioxide accounts for two (conversion of two pyruvate molecules) of the six carbons of the original glucose molecule. The electrons are picked up by NAD^+, and the NADH carries the electrons to a later pathway for ATP production. At this point, the glucose molecule that originally entered cellular respiration has been completely oxidized. Chemical potential energy stored within the glucose molecule has been transferred to electron carriers or has been used to synthesize a few ATPs.

The citric acid cycle is a series of redox and decarboxylation reactions that removes high-energy electrons and carbon dioxide. The electrons, temporarily stored in molecules of NADH and $FADH_2$, are used to generate ATP in a subsequent pathway. One molecule of either GTP or ATP is produced by substrate-level phosphorylation on each turn of the cycle. There is no comparison of the cyclic pathway with a linear one.

7.4 Oxidative Phosphorylation

The electron transport chain is the portion of aerobic respiration that uses free oxygen as the final electron acceptor of the electrons removed from the intermediate compounds in glucose catabolism. The electron transport chain is composed of four large, multiprotein complexes embedded in the inner mitochondrial membrane and two small diffusible electron carriers shuttling electrons between them. The electrons are passed through a series of redox reactions, with a small amount of free energy used at three points to transport hydrogen ions across a membrane. This process contributes to the gradient used in chemiosmosis. The electrons passing through the electron transport chain gradually lose energy. High-energy electrons donated to the chain by either NADH or $FADH_2$ complete the chain, as low-energy electrons reduce oxygen molecules and form water. The level of free energy of the electrons drops from about 60 kcal/mol in NADH or 45 kcal/mol in $FADH_2$ to about 0 kcal/mol in water. The end products of the electron transport chain are water and ATP. A number of intermediate compounds of the citric acid cycle can be diverted into the anabolism of other biochemical molecules, such as nonessential amino acids, sugars, and lipids. These same molecules can serve as energy sources for the glucose pathways.

7.5 Metabolism without Oxygen

If NADH cannot be oxidized through aerobic respiration, another electron acceptor is used. Most organisms will use some form of fermentation to accomplish the regeneration of NAD^+, ensuring the continuation of glycolysis. The regeneration of NAD^+ in fermentation is not accompanied by ATP production; therefore, the potential of NADH to produce ATP using an electron transport chain is not utilized.

7.6 Connections of Carbohydrate, Protein, and Lipid Metabolic Pathways

The breakdown and synthesis of carbohydrates, proteins, and lipids connect with the pathways of glucose catabolism. The simple sugars are galactose, fructose, glycogen, and pentose. These are catabolized during glycolysis. The amino acids from proteins connect with glucose catabolism through pyruvate, acetyl CoA, and components of the citric acid cycle. Cholesterol synthesis starts with acetyl groups, and the components of triglycerides come from glycerol-3-phosphate from glycolysis and acetyl groups produced in the mitochondria from pyruvate.

7.7 Regulation of Cellular Respiration

Cellular respiration is controlled by a variety of means. The entry of glucose into a cell is controlled by the transport proteins that aid glucose passage through the cell membrane. Most of the control of the respiration processes is accomplished through the control of specific enzymes in the pathways. This is a type of negative feedback mechanism, turning the enzymes off. The enzymes respond most often to the levels of the available nucleosides ATP, ADP, AMP, NAD^+, and FAD. Other intermediates of the pathway also affect certain enzymes in the systems.

VISUAL CONNECTION QUESTIONS

1. Figure 7.11 Dinitrophenol (DNP) is an "uncoupler" that makes the inner mitochondrial membrane "leaky" to protons. It was used until 1938 as a weight-loss drug. What effect would you expect DNP to have on the change in pH across the inner mitochondrial membrane? Why do you think this might be an effective weight-loss drug?
2. Figure 7.12 Cyanide inhibits cytochrome c oxidase, a component of the electron transport chain. If cyanide poisoning occurs, would you expect the pH of the intermembrane space to increase or decrease? What effect would cyanide have on ATP synthesis?
3. Figure 7.14 Tremetol, a metabolic poison found in the white snake root plant, prevents the metabolism of lactate. When cows eat this plant, tremetol is concentrated in the milk they produce. Humans who consume the milk can become seriously ill. Symptoms of this disease, which include vomiting, abdominal pain, and tremors, become worse after exercise. Why do you think this is the case?

REVIEW QUESTIONS

4. The energy currency used by cells is ________.
   a. ATP
   b. ADP
   c. AMP
   d. adenosine

5. A reducing chemical reaction ________.
   a. reduces the compound to a simpler form
   b. adds an electron to the substrate
   c. removes a hydrogen atom from the substrate
   d. is a catabolic reaction

6. During the second half of glycolysis, what occurs?
   a. ATP is used up.
   b. Fructose is split in two.
   c. ATP is made.
   d. Glucose becomes fructose.

7. What is removed from pyruvate during its conversion into an acetyl group?
   a. oxygen
   b. ATP
   c. B vitamin
   d. carbon dioxide

8. What do the electrons added to NAD^+ do?
   a. They become part of a fermentation pathway.
   b. They go to another pathway for ATP production.
   c. They energize the entry of the acetyl group into the citric acid cycle.
   d. They are converted to NADP.

9. GTP or ATP is produced during the conversion of ________.
   a. isocitrate into α-ketoglutarate
   b. succinyl CoA into succinate
   c. fumarate into malate
   d. malate into oxaloacetate

10. How many NADH molecules are produced on each turn of the citric acid cycle?
    a. one
    b. two
    c. three
    d. four

11. What compound receives electrons from NADH?
    a. FMN
    b. ubiquinone
    c. cytochrome c_1
    d. oxygen

12. Chemiosmosis involves ________.
    a. the movement of electrons across the cell membrane
    b. the movement of hydrogen atoms across a mitochondrial membrane
    c. the movement of hydrogen ions across a mitochondrial membrane
    d. the movement of glucose through the cell membrane

13. Which of the following fermentation methods can occur in animal skeletal muscles?
    a. lactic acid fermentation
    b. alcohol fermentation
    c. mixed acid fermentation
    d. propionic fermentation

14. A major connection for sugars in glycolysis is ________.
    a. glucose-6-phosphate
    b. fructose-1,6-bisphosphate
    c. dihydroxyacetone phosphate
    d. phosphoenolpyruvate

15. Beta-oxidation is ________.
    a. the breakdown of sugars
    b. the assembly of sugars
    c. the breakdown of fatty acids
    d. the removal of amino groups from amino acids

16. The effect of high levels of ADP is to ________ in cellular respiration.
    a. increase the activity of specific enzymes
    b. decrease the activity of specific enzymes
    c. have no effect on the activity of specific enzymes
    d. slow down the pathway

17. The control of which enzyme exerts the most control on glycolysis?
    a. hexokinase
    b. phosphofructokinase
    c. glucose-6-phosphatase
    d. aldolase

CRITICAL THINKING QUESTIONS

18. Why is it beneficial for cells to use ATP rather than energy directly from the bonds of carbohydrates? What are the greatest drawbacks to harnessing energy directly from the bonds of several different compounds?

19. Nearly all organisms on Earth carry out some form of glycolysis. How does this fact support or not support the assertion that glycolysis is one of the oldest metabolic pathways?

20. Because they lose their mitochondria during development, red blood cells cannot perform aerobic respiration; however, they do perform glycolysis in the cytoplasm. Why do all cells need an energy source, and what would happen if glycolysis were blocked in a red blood cell?

21. What is the primary difference between a circular pathway and a linear pathway?

22. How do the roles of ubiquinone and cytochrome c differ from the roles of the other components of the electron transport chain?

23. What accounts for the different number of ATP molecules that are formed through cellular respiration?

24. What is the primary difference between fermentation and anaerobic respiration?

25. Would you describe metabolic pathways as inherently wasteful or inherently economical? Why?

26. How does citrate from the citric acid cycle affect glycolysis?

27. Why might negative feedback mechanisms be more common than positive feedback mechanisms in living cells?

CHAPTER 8
Photosynthesis

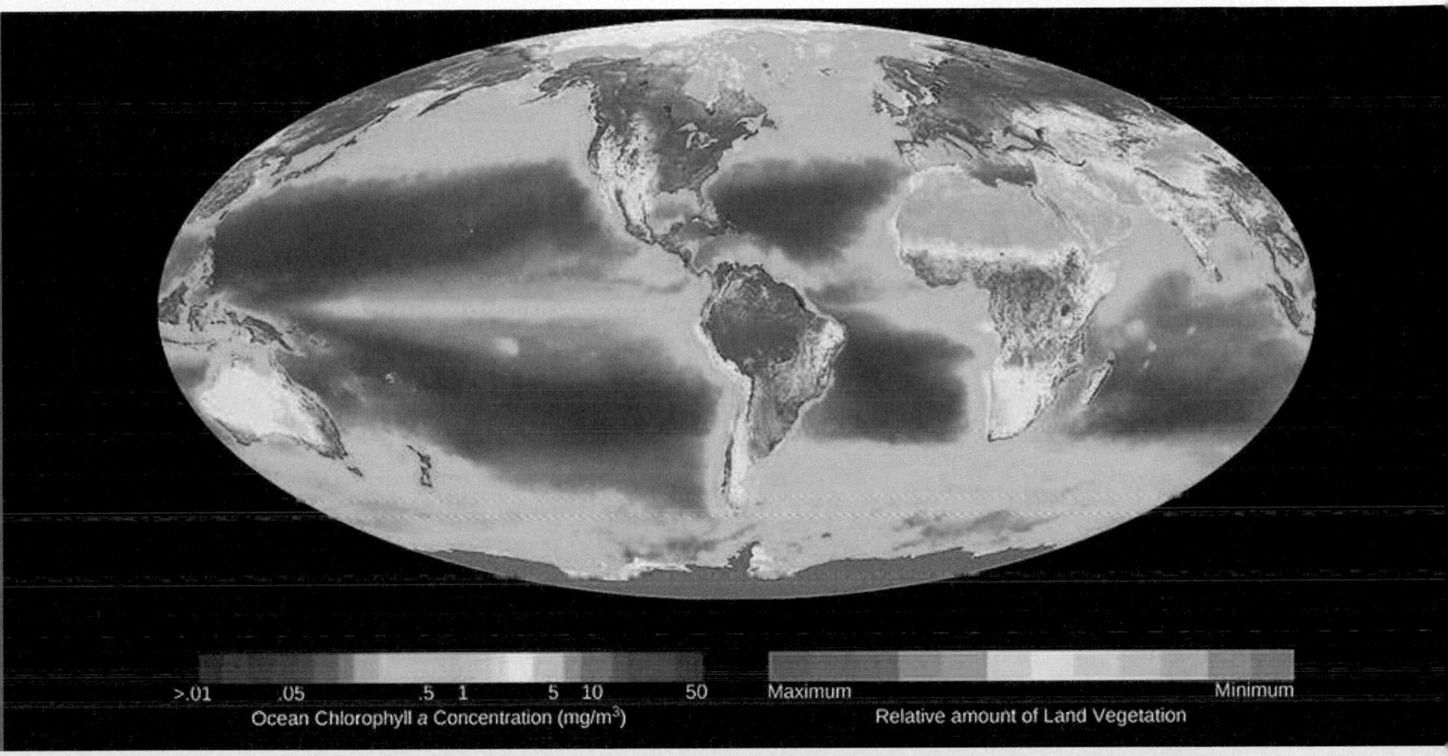

Figure 8.1 This world map shows Earth's distribution of photosynthetic activity determined by chlorophyll *a* concentrations. On land, chlorophyll is evident from terrestrial plants, and within oceanic zones, from chlorophyll from phytoplankton. (credit: modification of work by SeaWiFS Project, NASA/Goddard Space Flight Center and ORBIMAGE)

INTRODUCTION The metabolic processes in all organisms—from bacteria to humans—require energy. To get this energy, many organisms access stored energy by eating, that is, by ingesting other organisms. But where does the stored energy in food originate? All of this energy can be traced back to photosynthesis.

Chapter Outline

8.1 Overview of Photosynthesis

By the end of this section, you will be able to do the following:

- Explain the significance of photosynthesis to other living organisms
- Describe the main structures involved in photosynthesis
- Identify the substrates and products of photosynthesis

Photosynthesis is essential to all life on earth; both plants and animals depend on it. It is the only biological process that can capture energy that originates from sunlight and converts it into chemical compounds (carbohydrates) that every organism uses to power its metabolism. It is also a source of oxygen necessary for many living organisms. In brief, the energy of sunlight is "captured" to energize electrons, whose energy is then stored in the covalent bonds of sugar molecules. How long lasting and stable are those covalent bonds? The energy extracted today by the burning of coal and petroleum products represents sunlight energy captured and stored by photosynthesis 350 to 200 million years ago during the Carboniferous Period.

Plants, algae, and a group of bacteria called cyanobacteria are the only organisms capable of performing photosynthesis (Figure 8.2). Because they use light to manufacture their own food, they are called **photoautotrophs** (literally, "self-feeders using light"). Other organisms, such as animals, fungi, and most other bacteria, are termed **heterotrophs** ("other feeders"), because they must rely on the sugars produced by photosynthetic organisms for their energy needs. A third very interesting group of bacteria synthesize sugars, not by using sunlight's energy, but by extracting energy from inorganic chemical compounds. For this reason, they are referred to as **chemoautotrophs**.

Figure 8.2 Photoautotrophs including (a) plants, (b) algae, and (c) cyanobacteria synthesize their organic compounds via photosynthesis using sunlight as an energy source. Cyanobacteria and planktonic algae can grow over enormous areas in water, at times completely covering the surface. In a (d) deep sea vent, chemoautotrophs, such as these (e) thermophilic bacteria, capture energy from inorganic compounds to produce organic compounds. The ecosystem surrounding the vents has a diverse array of animals, such as tubeworms, crustaceans, and octopuses that derive energy from the bacteria. (credit a: modification of work by Steve Hillebrand, U.S. Fish and Wildlife Service; credit b: modification of work by "eutrophication&hypoxia"/Flickr; credit c: modification of work by NASA; credit d: University of Washington, NOAA; credit e: modification of work by Mark Amend, West Coast and Polar Regions Undersea Research Center, UAF, NOAA)

The importance of photosynthesis is not just that it can capture sunlight's energy. After all, a lizard sunning itself on a cold day can use the sun's energy to warm up in a process called *behavioral thermoregulation*. In contrast, photosynthesis is vital because it evolved as a way to *store the energy from solar radiation (the "photo-" part) to energy in the carbon-carbon bonds of carbohydrate molecules (the "-synthesis" part).* Those carbohydrates are the energy source that heterotrophs use to power the synthesis of ATP via respiration. Therefore, photosynthesis powers 99 percent of Earth's ecosystems. When a top predator, such as a wolf, preys on a deer (Figure 8.3), the wolf is at the end of an energy path that went from nuclear reactions on the surface of the sun, to visible light, to photosynthesis, to vegetation, to deer, and finally to the wolf.

Figure 8.3 The energy stored in carbohydrate molecules from photosynthesis passes through the food chain. The predator that eats these deer receives a portion of the energy that originated in the photosynthetic vegetation that the deer consumed. (credit: modification of work by Steve VanRiper, U.S. Fish and Wildlife Service)

Main Structures and Summary of Photosynthesis

Photosynthesis is a multi-step process that requires specific wavelengths of visible sunlight, carbon dioxide (which is low in energy), and water as substrates (Figure 8.4). After the process is complete, it releases oxygen and produces glyceraldehyde-3-phosphate (G3P), as well as simple carbohydrate molecules (high in energy) that can then be converted into glucose, sucrose, or any of dozens of other sugar molecules. These sugar molecules contain energy and the energized carbon that all living things need to survive.

Figure 8.4 Photosynthesis uses solar energy, carbon dioxide, and water to produce energy-storing carbohydrates. Oxygen is generated as a waste product of photosynthesis.

The following is the chemical equation for photosynthesis (Figure 8.5):

Photosynthesis Equation				
Carbon dioxide	+ Water	SUNLIGHT →	Sugar	+ Oxygen
$6CO_2$	$6H_2O$		$C_6H_{12}O_6$	$6O_2$

Figure 8.5 The basic equation for photosynthesis is deceptively simple. In reality, the process takes place in many steps involving intermediate reactants and products. Glucose, the primary energy source in cells, is made from two three-carbon G3Ps.

Although the equation looks simple, the many steps that take place during photosynthesis are actually quite complex. Before learning the details of how photoautotrophs turn sunlight into food, it is important to become familiar with the structures involved.

Basic Photosynthetic Structures

In plants, photosynthesis generally takes place in leaves, which consist of several layers of cells. The process of photosynthesis occurs in a middle layer called the **mesophyll**. The gas exchange of carbon dioxide and oxygen occurs through small, regulated openings called **stomata** (singular: stoma), which also play roles in the regulation of gas exchange and water balance. The stomata are typically located on the underside of the leaf, which helps to minimize water loss due to high temperatures on the upper surface of the leaf. Each stoma is flanked by guard cells that regulate the opening and closing of the stomata by swelling or shrinking in response to osmotic changes.

In all autotrophic eukaryotes, photosynthesis takes place inside an organelle called a **chloroplast**. For plants, chloroplast-containing cells exist mostly in the mesophyll. Chloroplasts have a double membrane envelope (composed of an outer membrane and an inner membrane), and are ancestrally derived from ancient free-living cyanobacteria. Within the chloroplast are stacked, disc-shaped structures called **thylakoids**. Embedded in the thylakoid membrane is chlorophyll, a **pigment** (molecule that absorbs light) responsible for the initial interaction between light and plant material, and numerous proteins that make up the electron transport chain. The thylakoid membrane encloses an internal space called the **thylakoid lumen**. As shown in Figure 8.6, a stack of thylakoids is called a **granum**, and the liquid-filled space surrounding the granum is called **stroma** or "bed" (not to be confused with stoma or "mouth," an opening on the leaf epidermis).

VISUAL CONNECTION

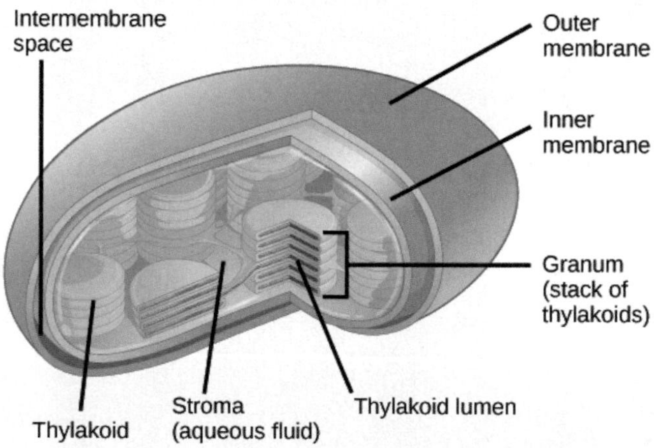

Figure 8.6 Photosynthesis takes place in chloroplasts, which have an outer membrane and an inner membrane. Stacks of thylakoids called grana form a third membrane layer.

On a hot, dry day, the guard cells of plants close their stomata to conserve water. What impact will this have on photosynthesis?

The Two Parts of Photosynthesis

Photosynthesis takes place in two sequential stages: the light-dependent reactions and the light-independent reactions. In the **light-dependent reactions**, energy from sunlight is absorbed by chlorophyll and that energy is converted into stored chemical energy. In the **light-independent reactions**, the chemical energy harvested during the light-dependent reactions drives the assembly of sugar molecules from carbon dioxide. Therefore, although the light-independent reactions do not use light as a reactant, they require the products of the light-dependent reactions to function. In addition, however, several enzymes of the light-independent reactions are activated by light. The light-dependent reactions utilize certain molecules to temporarily store the energy: These are referred to as *energy carriers*. The energy carriers that move energy from light-dependent reactions to light-independent reactions can be thought of as "full" because they are rich in energy. After the energy is released, the "empty" energy carriers return to the light-dependent reaction to obtain more energy. Figure 8.7 illustrates the components inside the chloroplast where the light-dependent and light-independent reactions take place.

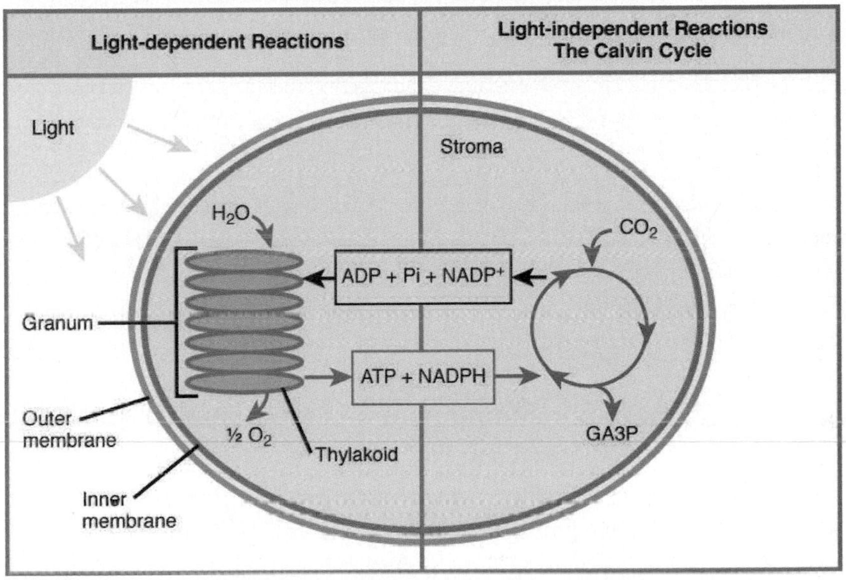

Figure 8.7 Photosynthesis takes place in two stages: light-dependent reactions and the Calvin cycle. Light-dependent reactions, which take place in the thylakoid membrane, use light energy to make ATP and NADPH. The Calvin cycle, which takes place in the stroma, uses energy derived from these compounds to make G3P from CO_2.

LINK TO LEARNING

Click the link (http://openstax.org/l/photosynthesis) to learn more about photosynthesis.

Everyday Connection

Photosynthesis at the Grocery Store

Figure 8.8 Foods that humans consume originate from photosynthesis. (credit: Associação Brasileira de Supermercados)

Major grocery stores in the United States are organized into departments, such as dairy, meats, produce, bread, cereals, and so forth. Each aisle (Figure 8.8) contains hundreds, if not thousands, of different products for customers to buy and consume.

Although there is a large variety, each item ultimately can be linked back to photosynthesis. Meats and dairy link, because the animals were fed plant-based foods. The breads, cereals, and pastas come largely from starchy grains, which are the seeds of photosynthesis-dependent plants. What about desserts and drinks? All of these products contain sugar—sucrose is a plant product, a disaccharide, a carbohydrate molecule, which is built directly from photosynthesis. Moreover, many items are less obviously derived from plants: For instance, paper goods are generally plant products, and many plastics (abundant as products and packaging) are derived from "algae" (unicellular plant-like organisms, and cyanobacteria). Virtually every spice and flavoring in the spice aisle was produced by a plant as a leaf, root, bark, flower, fruit, or stem. Ultimately, photosynthesis connects to every meal and every food a person consumes.

8.2 The Light-Dependent Reactions of Photosynthesis

By the end of this section, you will be able to do the following:

- Explain how plants absorb energy from sunlight
- Describe short and long wavelengths of light
- Describe how and where photosynthesis takes place within a plant

How can light energy be used to make food? When a person turns on a lamp, electrical energy becomes light energy. Like all other forms of kinetic energy, light can travel, change form, and be harnessed to do work. In the case of photosynthesis, light energy is converted into chemical energy, which photoautotrophs use to build basic carbohydrate molecules (Figure 8.9). However, autotrophs only use a few specific wavelengths of sunlight.

Figure 8.9 Photoautotrophs can capture visible light energy in specific wavelengths from the sun, converting it into the chemical energy used to build food molecules. (credit: Gerry Atwell)

What Is Light Energy?

The sun emits an enormous amount of electromagnetic radiation (solar energy in a spectrum from very short gamma rays to very long radio waves). Humans can see only a tiny fraction of this energy, which we refer to as "visible light." The manner in which solar energy travels is described as waves. Scientists can determine the amount of energy of a wave by measuring its **wavelength** (shorter wavelengths are more powerful than longer wavelengths)—the distance between consecutive crest points of a wave. Therefore, a single wave is measured from two consecutive points, such as from crest to crest or from trough to trough (Figure 8.10).

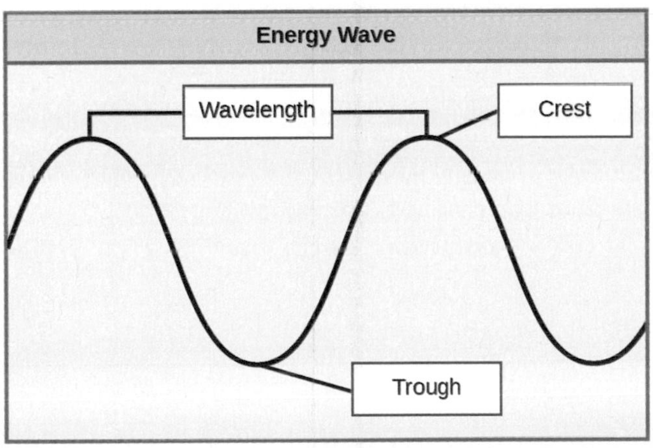

Figure 8.10 The wavelength of a single wave is the distance between two consecutive points of similar position (two crests or two troughs) along the wave.

Visible light constitutes only one of many types of electromagnetic radiation emitted from the sun and other stars. Scientists differentiate the various types of radiant energy from the sun within the electromagnetic spectrum. The **electromagnetic spectrum** is the range of all possible frequencies of radiation (Figure 8.11). The difference between wavelengths relates to the amount of energy carried by them.

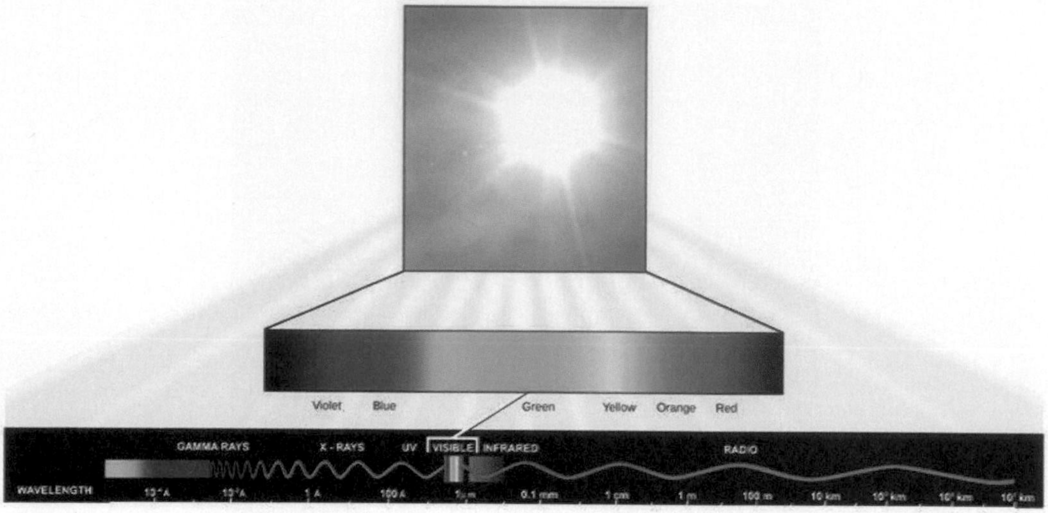

Figure 8.11 The sun emits energy in the form of electromagnetic radiation. This radiation exists at different wavelengths, each of which has its own characteristic energy. All electromagnetic radiation, including visible light, is characterized by its wavelength.

Each type of electromagnetic radiation travels at a particular wavelength. The longer the wavelength, the less energy it carries. Short, tight waves carry the most energy. This may seem illogical, but think of it in terms of a piece of moving heavy rope. It takes little effort by a person to move a rope in long, wide waves. To make a rope move in short, tight waves, a person would need to apply significantly more energy.

The electromagnetic spectrum (Figure 8.11) shows several types of electromagnetic radiation originating from the sun, including X-rays and ultraviolet (UV) rays. The higher-energy waves can penetrate tissues and damage cells and DNA, which explains why both X-rays and UV rays can be harmful to living organisms.

Absorption of Light

Light energy initiates the process of photosynthesis when pigments absorb specific wavelengths of visible light. Organic pigments, whether in the human retina or the chloroplast thylakoid, have a narrow range of energy levels that they can absorb. Energy levels lower than those represented by red light are insufficient to raise an orbital electron to a excited (quantum) state. Energy levels higher than those in blue light will physically tear the molecules apart, in a process called bleaching. Our retinal pigments can only "see" (absorb) wavelengths between 700 nm and 400 nm of light, a spectrum that is therefore called visible light. For the same reasons, plants, pigment molecules absorb only light in the wavelength range of 700 nm to 400 nm; plant physiologists refer to this range for plants as photosynthetically active radiation.

The visible light seen by humans as white light actually exists in a rainbow of colors. Certain objects, such as a prism or a drop of water, disperse white light to reveal the colors to the human eye. The visible light portion of the electromagnetic spectrum shows the rainbow of colors, with violet and blue having shorter wavelengths, and therefore higher energy. At the other end of the spectrum toward red, the wavelengths are longer and have lower energy (Figure 8.12).

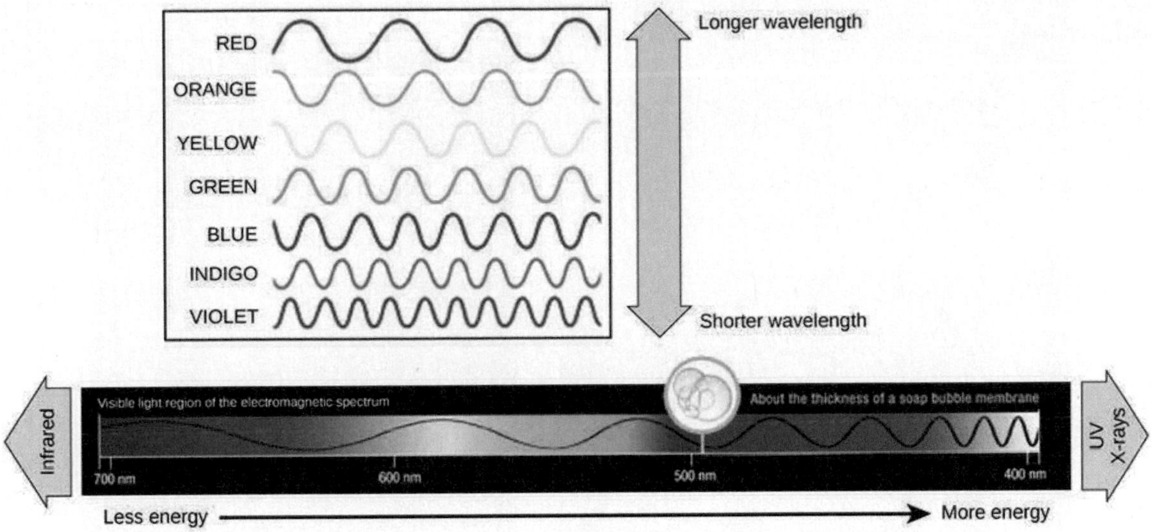

Figure 8.12 The colors of visible light do not carry the same amount of energy. Violet has the shortest wavelength and therefore carries the most energy, whereas red has the longest wavelength and carries the least amount of energy. (credit: modification of work by NASA)

Understanding Pigments

Different kinds of pigments exist, and each absorbs only specific wavelengths (colors) of visible light. Pigments reflect or transmit the wavelengths they cannot absorb, making them appear a mixture of the reflected or transmitted light colors.

Chlorophylls and carotenoids are the two major classes of photosynthetic pigments found in plants and algae; each class has multiple types of pigment molecules. There are five major chlorophylls: *a*, *b*, *c* and *d* and a related molecule found in prokaryotes called *bacteriochlorophyll*. **Chlorophyll *a*** and **chlorophyll *b*** are found in higher plant chloroplasts and will be the focus of the following discussion.

With dozens of different forms, carotenoids are a much larger group of pigments. The carotenoids found in fruit—such as the red of tomato (lycopene), the yellow of corn seeds (zeaxanthin), or the orange of an orange peel (β-carotene)—are used as advertisements to attract seed dispersers. In photosynthesis, **carotenoids** function as photosynthetic pigments that are very efficient molecules for the disposal of excess energy. When a leaf is exposed to full sun, the light-dependent reactions are required to process an enormous amount of energy; if that energy is not handled properly, it can do significant damage. Therefore, many carotenoids reside in the thylakoid membrane, absorb excess energy, and safely dissipate that energy as heat.

Each type of pigment can be identified by the specific pattern of wavelengths it absorbs from visible light: This is termed the **absorption spectrum**. The graph in Figure 8.13 shows the absorption spectra for chlorophyll *a*, chlorophyll *b*, and a type of carotenoid pigment called β-carotene (which absorbs blue and green light). Notice how each pigment has a distinct set of peaks and troughs, revealing a highly specific pattern of absorption. Chlorophyll *a* absorbs wavelengths from either end of the visible spectrum (blue and red), but not green. Because green is reflected or transmitted, chlorophyll appears green. Carotenoids absorb in the short-wavelength blue region, and reflect the longer yellow, red, and orange wavelengths.

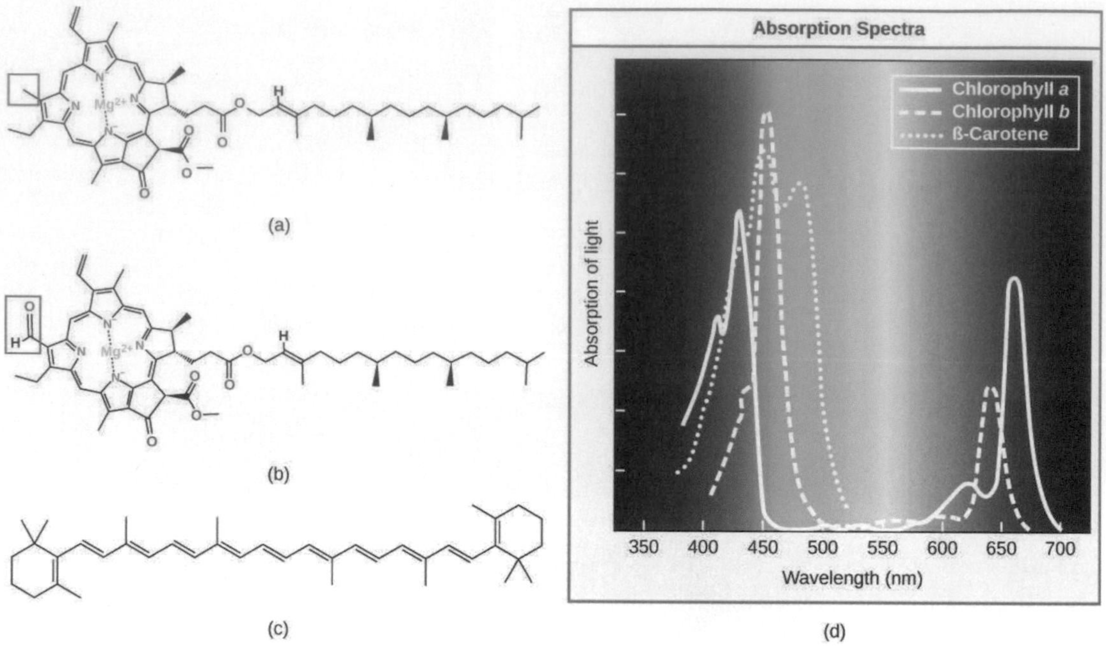

Figure 8.13 (a) Chlorophyll *a*, (b) chlorophyll *b*, and (c) *β*-carotene are hydrophobic organic pigments found in the thylakoid membrane. Chlorophyll *a* and *b*, which are identical except for the part indicated in the red box, are responsible for the green color of leaves. *β*-carotene is responsible for the orange color in carrots. Each pigment has (d) a unique absorbance spectrum.

Many photosynthetic organisms have a mixture of pigments, and by using these pigments, the organism can absorb energy from a wider range of wavelengths. Not all photosynthetic organisms have full access to sunlight. Some organisms grow underwater where light intensity and quality decrease and change with depth. Other organisms grow in competition for light. Plants on the rainforest floor must be able to absorb any bit of light that comes through, because the taller trees absorb most of the sunlight and scatter the remaining solar radiation (Figure 8.14).

Figure 8.14 Plants that commonly grow in the shade have adapted to low levels of light by changing the relative concentrations of their chlorophyll pigments. (credit: Jason Hollinger)

When studying a photosynthetic organism, scientists can determine the types of pigments present by generating absorption spectra. An instrument called a **spectrophotometer** can differentiate which wavelengths of light a substance can absorb. Spectrophotometers measure transmitted light and compute from it the absorption. By extracting pigments from leaves and placing these samples into a spectrophotometer, scientists can identify which wavelengths of light an organism can absorb. Additional methods for the identification of plant pigments include various types of chromatography that separate the pigments by their relative affinities to solid and mobile phases.

How Light-Dependent Reactions Work

The overall function of light-dependent reactions is to convert solar energy into chemical energy in the form of NADPH and ATP. This chemical energy supports the light-independent reactions and fuels the assembly of sugar molecules. The light-dependent reactions are depicted in Figure 8.15. Protein complexes and pigment molecules work together to produce NADPH and ATP. The numbering of the photosystems is derived from the order in which they were discovered, not in the order of the transfer of electrons.

(a) Photosystem II (P680)

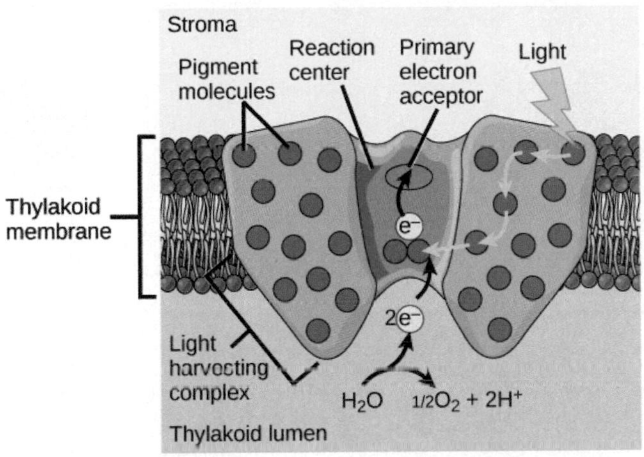

(b) Photosystem I (P700)

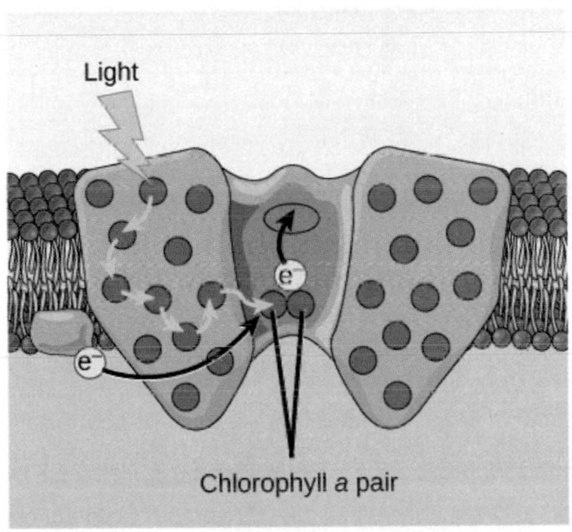

Figure 8.15 A photosystem consists of 1) a light-harvesting complex and 2) a reaction center. Pigments in the light-harvesting complex pass light energy to two special chlorophyll *a* molecules in the reaction center. The light excites an electron from the chlorophyll *a* pair, which passes to the primary electron acceptor. The excited electron must then be replaced. In (a) photosystem II, the electron comes from the splitting of water, which releases oxygen as a waste product. In (b) photosystem I, the electron comes from the chloroplast electron transport chain discussed below.

The actual step that converts light energy into chemical energy takes place in a multiprotein complex called a **photosystem**, two types of which are found embedded in the thylakoid membrane: **photosystem II** (PSII) and **photosystem I** (PSI) (Figure 8.16). The two complexes differ on the basis of what they oxidize (that is, the source of the low-energy electron supply) and what they reduce (the place to which they deliver their energized electrons).

Both photosystems have the same basic structure; a number of **antenna proteins** to which the chlorophyll molecules are bound surround the **reaction center** where the photochemistry takes place. Each photosystem is serviced by the **light-harvesting complex**, which passes energy from sunlight to the reaction center; it consists of multiple antenna proteins that contain a mixture of 300 to 400 chlorophyll *a* and *b* molecules as well as other pigments like carotenoids. The absorption of a single **photon** or distinct quantity or "packet" of light by any of the chlorophylls pushes that molecule into an excited state. In short, the light energy has now been captured by biological molecules but is not stored in any useful form yet. The energy is transferred from chlorophyll to chlorophyll until eventually (after about a millionth of a second), it is delivered to the reaction center. Up to this point, only energy has been transferred between molecules, not electrons.

VISUAL CONNECTION

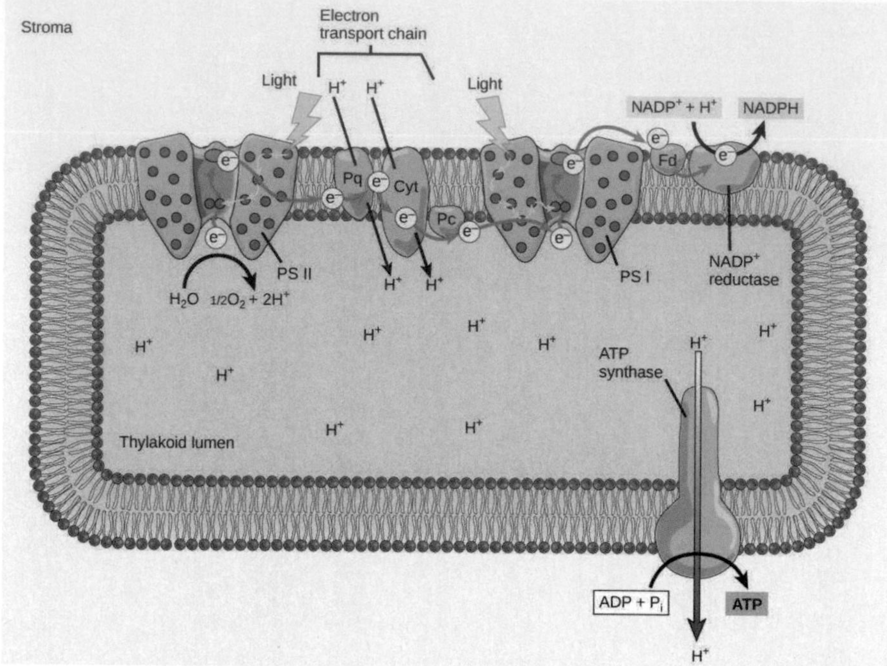

Figure 8.16 In the photosystem II (PSII) reaction center, energy from sunlight is used to extract electrons from water. The electrons travel through the chloroplast electron transport chain to photosystem I (PSI), which reduces $NADP^+$ to NADPH. The electron transport chain moves protons across the thylakoid membrane into the lumen. At the same time, splitting of water adds protons to the lumen, and reduction of NADPH removes protons from the stroma. The net result is a low pH in the thylakoid lumen, and a high pH in the stroma. ATP synthase uses this electrochemical gradient to make ATP.

What is the initial source of electrons for the chloroplast electron transport chain?

a. water
b. oxygen
c. carbon dioxide
d. NADPH

The reaction center contains a pair of chlorophyll *a* molecules with a special property. Those two chlorophylls can undergo oxidation upon excitation; they can actually give up an electron in a process called a **photoact**. It is at this step in the reaction center during photosynthesis that light energy is converted into an excited electron. All of the subsequent steps involve getting that electron onto the energy carrier NADPH for delivery to the Calvin cycle where the electron is deposited onto carbon for long-term storage in the form of a carbohydrate. PSII and PSI are two major components of the photosynthetic **electron transport chain**, which also includes the *cytochrome complex*. The cytochrome complex, an enzyme composed of two protein complexes, transfers the electrons from the carrier molecule plastoquinone (Pq) to the protein plastocyanin (Pc), thus enabling both the transfer of protons across the thylakoid membrane and the transfer of electrons from PSII to PSI.

The reaction center of PSII (called **P680**) delivers its high-energy electrons, one at the time, to the **primary electron acceptor**, and through the electron transport chain (Pq to cytochrome complex to plastocyanine) to PSI. P680's missing electron is replaced by extracting a low-energy electron from water; thus, water is "split" during this stage of photosynthesis, and PSII is re-reduced after every photoact. Splitting one H_2O molecule releases two electrons, two hydrogen atoms, and one atom of oxygen. However, splitting two molecules is required to form one molecule of diatomic O_2 gas. About 10 percent of the oxygen is used by mitochondria in the leaf to support oxidative phosphorylation. The remainder escapes to the atmosphere where it is used by aerobic organisms to support respiration.

As electrons move through the proteins that reside between PSII and PSI, they lose energy. This energy is used to move hydrogen atoms from the stromal side of the membrane to the thylakoid lumen. Those hydrogen atoms, plus the ones produced by splitting water, accumulate in the thylakoid lumen and will be used synthesize ATP in a later step. Because the electrons have lost energy prior to their arrival at PSI, they must be re-energized by PSI, hence, another photon is absorbed by the PSI antenna. That energy is relayed to the PSI reaction center (called **P700**). P700 is oxidized and sends a high-energy electron to $NADP^+$ to form NADPH. Thus, PSII captures the energy to create proton gradients to make ATP, and PSI captures the energy to reduce $NADP^+$ into NADPH. The two photosystems work in concert, in part, to guarantee that the production of NADPH will roughly equal the production of ATP. Other mechanisms exist to fine-tune that ratio to exactly match the chloroplast's constantly changing energy needs.

Generating an Energy Carrier: ATP

As in the intermembrane space of the mitochondria during cellular respiration, the buildup of hydrogen ions inside the thylakoid lumen creates a *concentration gradient*. The passive diffusion of hydrogen ions from high concentration (in the thylakoid lumen) to low concentration (in the stroma) is harnessed to create ATP, just as in the electron transport chain of cellular respiration. The ions build up energy because of diffusion and because they all have the same electrical charge, repelling each other.

To release this energy, hydrogen ions will rush through any opening, similar to water jetting through a hole in a dam. In the thylakoid, that opening is a passage through a specialized protein channel called the ATP synthase. The energy released by the hydrogen ion stream allows ATP synthase to attach a third phosphate group to ADP, which forms a molecule of ATP (Figure 8.16). The flow of hydrogen ions through ATP synthase is called chemiosmosis because the ions move from an area of high to an area of low concentration through a semi-permeable structure of the thylakoid.

LINK TO LEARNING

Visit this site (http://openstax.org/l/light_reactions) and click through the animation to view the process of photosynthesis within a leaf.

8.3 Using Light Energy to Make Organic Molecules

By the end of this section, you will be able to do the following:

- Describe the Calvin cycle
- Define carbon fixation
- Explain how photosynthesis works in the energy cycle of all living organisms

After the energy from the sun is converted into chemical energy and temporarily stored in ATP and NADPH molecules, the cell has the fuel needed to build carbohydrate molecules for long-term energy storage. The products of the light-dependent reactions, ATP and NADPH, have lifespans in the range of millionths of seconds, whereas the products of the light-independent reactions (carbohydrates and other forms of reduced carbon) can survive almost indefinitely. The carbohydrate molecules made will have a backbone of carbon atoms. But where does the carbon come from? It comes from carbon dioxide—the gas that is a waste product of respiration in microbes, fungi, plants, and animals.

The Calvin Cycle

In plants, carbon dioxide (CO_2) enters the leaves through stomata, where it diffuses over short distances through intercellular spaces until it reaches the mesophyll cells. Once in the mesophyll cells, CO_2 diffuses into the stroma of the chloroplast—the site of light-independent reactions of photosynthesis. These reactions actually have several names associated with them. Another term, the **Calvin cycle**, is named for the man who discovered it, and because these reactions function as a cycle. Others call it the Calvin-Benson cycle to include the name of another scientist involved in its discovery. The most outdated name is "dark reaction," because light is not directly required (Figure 8.17). However, the term dark reaction can be misleading because it implies incorrectly that the reaction only occurs at night or is independent of light, which is why most scientists and instructors no longer use it.

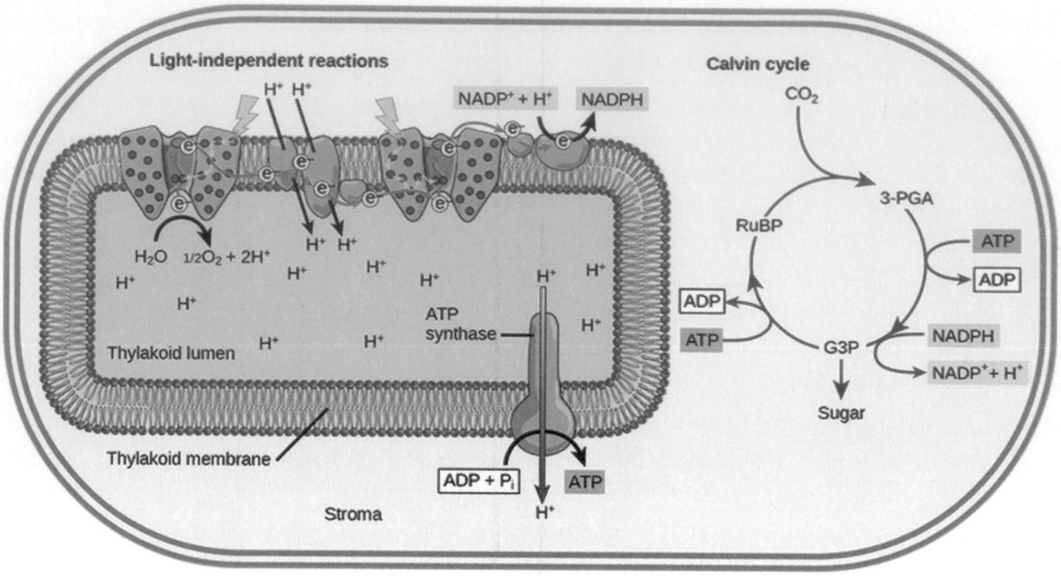

Figure 8.17 Light reactions harness energy from the sun to produce chemical bonds, ATP, and NADPH. These energy-carrying molecules are made in the stroma where carbon fixation takes place.

The light-independent reactions of the Calvin cycle can be organized into three basic stages: *fixation, reduction, and regeneration.*

Stage 1: Fixation

In the stroma, in addition to CO_2, two other components are present to initiate the light-independent reactions: an enzyme called ribulose-1,5-bisphosphate carboxylase/oxygenase (RuBisCO), and three molecules of ribulose bisphosphate (RuBP), as shown in Figure 8.18. RuBP has five atoms of carbon, flanked by two phosphates.

VISUAL CONNECTION

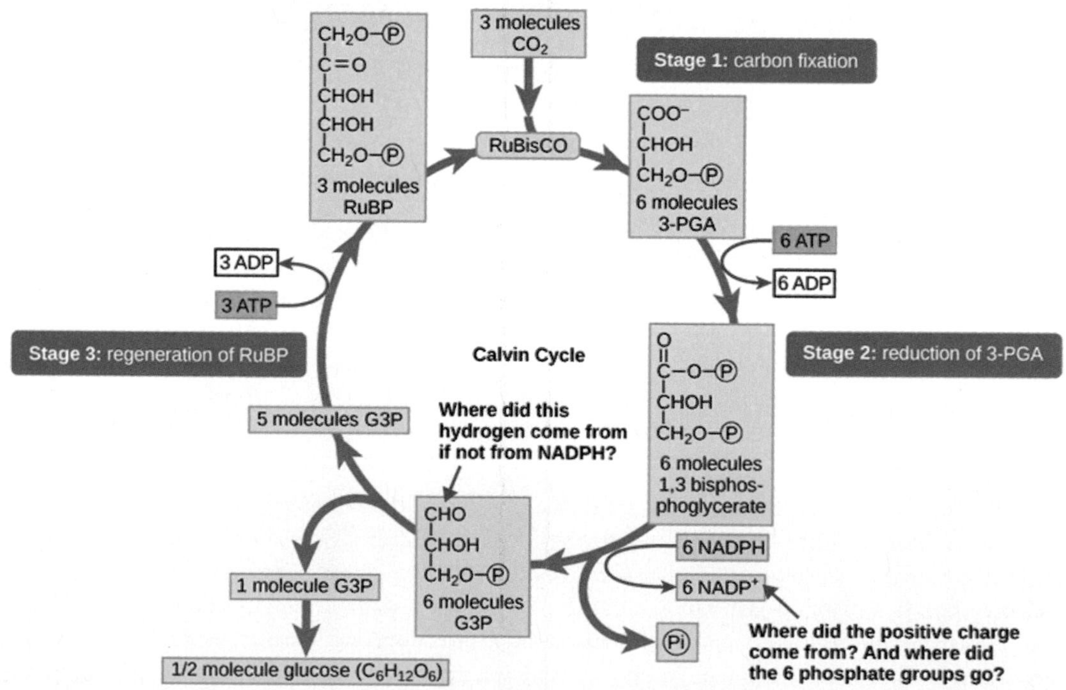

Figure 8.18 The Calvin cycle has three stages. In stage 1, the enzyme RuBisCO incorporates carbon dioxide into an organic molecule, 3-PGA. In stage 2, the organic molecule is reduced using electrons supplied by NADPH. In stage 3, RuBP, the molecule that starts the cycle,

is regenerated so that the cycle can continue. Only one carbon dioxide molecule is incorporated at a time, so the cycle must be completed three times to produce a single three-carbon G3P molecule, and six times to produce a six-carbon glucose molecule.

Which of the following statements is true?

a. In photosynthesis, oxygen, carbon dioxide, ATP, and NADPH are reactants. G3P and water are products.
b. In photosynthesis, chlorophyll, water, and carbon dioxide are reactants. G3P and oxygen are products.
c. In photosynthesis, water, carbon dioxide, ATP, and NADPH are reactants. RuBP and oxygen are products.
d. In photosynthesis, water and carbon dioxide are reactants. G3P and oxygen are products.

RuBisCO catalyzes a reaction between CO_2 and RuBP. For each CO_2 molecule that reacts with one RuBP, two molecules of another compound 3-phospho glyceric acid (3-PGA) form. PGA has three carbons and one phosphate. Each turn of the cycle involves only one RuBP and one carbon dioxide and forms two molecules of 3-PGA. The number of carbon atoms remains the same, as the atoms move to form new bonds during the reactions (3 C atoms from $3CO_2$ + 15 C atoms from 3RuBP = 18 C atoms in 6 molecules of 3-PGA). This process is called **carbon fixation**, because CO_2 is "fixed" from an inorganic form into organic molecules.

Stage 2: Reduction

ATP and NADPH are used to convert the six molecules of 3-PGA into six molecules of a chemical called glyceraldehyde 3-phosphate (G3P). That is a reduction reaction because it involves the gain of electrons by 3-PGA. (Recall that a **reduction** is the gain of an electron by an atom or molecule.) Six molecules of both ATP and NADPH are used. For ATP, energy is released with the loss of the terminal phosphate atom, converting it into ADP; for NADPH, both energy and a hydrogen atom are lost, converting it into $NADP^+$. Both of these molecules return to the nearby light-dependent reactions to be reused and re-energized.

Stage 3: Regeneration

Interestingly, at this point, only one of the G3P molecules leaves the Calvin cycle and is sent to the cytoplasm to contribute to the formation of other compounds needed by the plant. Because the G3P exported from the chloroplast has three carbon atoms, it takes three "turns" of the Calvin cycle to fix enough net carbon to export one G3P. But each turn makes two G3Ps, thus three turns make six G3Ps. One is exported while the remaining five G3P molecules remain in the cycle and are used to regenerate RuBP, which enables the system to prepare for more CO_2 to be fixed. Three more molecules of ATP are used in these regeneration reactions.

LINK TO LEARNING

This link (http://openstax.org/l/calvin_cycle) leads to an animation of photosynthesis and the Calvin cycle.

EVOLUTION CONNECTION

Photosynthesis

During the evolution of photosynthesis, a major shift occurred from the bacterial type of photosynthesis that involves only one photosystem and is typically anoxygenic (does not generate oxygen) into modern oxygenic (does generate oxygen) photosynthesis, employing two photosystems. This modern oxygenic photosynthesis is used by many organisms—from giant tropical leaves in the rainforest to tiny cyanobacterial cells—and the process and components of this photosynthesis remain largely the same. Photosystems absorb light and use electron transport chains to convert energy into the chemical energy of ATP and NADH. The subsequent light-independent reactions then assemble carbohydrate molecules with this energy.

In the harsh dry heat of the desert, plants must conserve every drop of water must be used to survive. Because stomata must open to allow for the uptake of CO_2, water escapes from the leaf during active photosynthesis. Desert plants have evolved processes to conserve water and deal with harsh conditions. Mechanisms to capture and store CO_2 allows plants to adapt to living with less water. Some plants such as cacti (Figure 8.19) can prepare materials for photosynthesis during the night by a temporary carbon fixation/storage process, because opening the stomata at this time conserves water due to cooler temperatures. During the day cacti use the captured CO_2 for photosynthesis, and keep their stomata closed.

Figure 8.19 The harsh conditions of the desert have led plants like these cacti to evolve variations of the light-independent reactions of photosynthesis. These variations increase the efficiency of water usage, helping to conserve water and energy. (credit: Piotr Wojtkowski)

The Energy Cycle

Whether the organism is a bacterium, plant, or animal, all living things access energy by breaking down carbohydrate and other carbon-rich organic molecules. But if plants make carbohydrate molecules, why would they need to break them down, especially when it has been shown that the gas organisms release as a "waste product" (CO_2) acts as a substrate for the formation of more food in photosynthesis? Remember, living things need energy to perform life functions. In addition, an organism can either make its own food or eat another organism—either way, the food still needs to be broken down. Finally, in the process of breaking down food, called cellular respiration, heterotrophs release needed energy and produce "waste" in the form of CO_2 gas.

However, in nature, there is no such thing as "waste." Every single atom of matter and energy is conserved, recycled over and over infinitely. Substances change form or move from one type of molecule to another, but their constituent atoms never disappear (Figure 8.20).

In reality, CO_2 is no more a form of waste than oxygen is wasteful to photosynthesis. Both are byproducts of reactions that move on to other reactions. Photosynthesis absorbs light energy to build carbohydrates in chloroplasts, and aerobic cellular respiration releases energy by using oxygen to metabolize carbohydrates in the cytoplasm and mitochondria. Both processes use electron transport chains to capture the energy necessary to drive other reactions. These two powerhouse processes, photosynthesis and cellular respiration, function in biological, cyclical harmony to allow organisms to access life-sustaining energy that originates millions of miles away in a burning star humans call the sun.

Figure 8.20 Photosynthesis consumes carbon dioxide and produces oxygen. Aerobic respiration consumes oxygen and produces carbon dioxide. These two processes play an important role in the carbon cycle. (credit: modification of work by Stuart Bassil)

KEY TERMS

absorption spectrum range of wavelengths of electromagnetic radiation absorbed by a given substance

antenna protein pigment molecule that directly absorbs light and transfers the energy absorbed to other pigment molecules

Calvin cycle light-independent reactions of photosynthesis that convert carbon dioxide from the atmosphere into carbohydrates using the energy and reducing power of ATP and NADPH

carbon fixation process of converting inorganic CO_2 gas into organic compounds

carotenoid photosynthetic pigment (yellow-orange-red) that functions to dispose of excess energy

chemoautotroph organism that can build organic molecules using energy derived from inorganic chemicals instead of sunlight

chlorophyll *a* form of chlorophyll that absorbs violet-blue and red light and consequently has a bluish-green color; the only pigment molecule that performs the photochemistry by getting excited and losing an electron to the electron transport chain

chlorophyll *b* accessory pigment that absorbs blue and red-orange light and consequently has a yellowish-green tint

chloroplast organelle in which photosynthesis takes place

cytochrome complex group of reversibly oxidizable and reducible proteins that forms part of the electron transport chain between photosystem II and photosystem I

electromagnetic spectrum range of all possible frequencies of radiation

electron transport chain group of proteins between PSII and PSI that pass energized electrons and use the energy released by the electrons to move hydrogen ions against their concentration gradient into the thylakoid lumen

granum stack of thylakoids located inside a chloroplast

heterotroph organism that consumes organic substances or other organisms for food

light harvesting complex complex that passes energy from sunlight to the reaction center in each photosystem; it consists of multiple antenna proteins that contain a mixture of 300 to 400 chlorophyll *a* and *b* molecules as well as other pigments like carotenoids

light-dependent reaction first stage of photosynthesis where certain wavelengths of the visible light are absorbed to form two energy-carrying molecules (ATP and NADPH)

light-independent reaction second stage of photosynthesis, through which carbon dioxide is used to build carbohydrate molecules using energy from ATP and NADPH

mesophyll middle layer of chlorophyll-rich cells in a leaf

P680 reaction center of photosystem II

P700 reaction center of photosystem I

photoact ejection of an electron from a reaction center using the energy of an absorbed photon

photoautotroph organism capable of producing its own organic compounds from sunlight

photon distinct quantity or "packet" of light energy

photosystem group of proteins, chlorophyll, and other pigments that are used in the light-dependent reactions of photosynthesis to absorb light energy and convert it into chemical energy

photosystem I integral pigment and protein complex in thylakoid membranes that uses light energy to transport electrons from plastocyanin to $NADP^+$ (which becomes reduced to NADPH in the process)

photosystem II integral protein and pigment complex in thylakoid membranes that transports electrons from water to the electron transport chain; oxygen is a product of PSII

pigment molecule that is capable of absorbing certain wavelengths of light and reflecting others (which accounts for its color)

primary electron acceptor pigment or other organic molecule in the reaction center that accepts an energized electron from the reaction center

reaction center complex of chlorophyll molecules and other organic molecules that is assembled around a special pair of chlorophyll molecules and a primary electron acceptor; capable of undergoing oxidation and reduction

reduction gain of electron(s) by an atom or molecule

spectrophotometer instrument that can measure transmitted light and compute the absorption

stoma opening that regulates gas exchange and water evaporation between leaves and the environment, typically situated on the underside of leaves

stroma fluid-filled space surrounding the grana inside a chloroplast where the light-independent reactions of photosynthesis take place

thylakoid disc-shaped, membrane-bound structure inside a chloroplast where the light-dependent reactions of photosynthesis take place; stacks of thylakoids are called grana

thylakoid lumen aqueous space bound by a thylakoid membrane where protons accumulate during light-driven electron transport

wavelength distance between consecutive points of equal position (two crests or two troughs) of a wave in a graphic representation; inversely proportional to the energy of the radiation

CHAPTER SUMMARY

8.1 Overview of Photosynthesis

The process of photosynthesis transformed life on Earth. By harnessing energy from the sun, the evolution of photosynthesis allowed living things access to enormous amounts of energy. Because of photosynthesis, living things gained access to sufficient energy that allowed them to build new structures and achieve the biodiversity evident today.

Only certain organisms (photoautotrophs), can perform photosynthesis; they require the presence of chlorophyll, a specialized pigment that absorbs certain wavelengths of the visible spectrum and can capture energy from sunlight. Photosynthesis uses carbon dioxide and water to assemble carbohydrate molecules and release oxygen as a byproduct into the atmosphere. Eukaryotic autotrophs, such as plants and algae, have organelles called chloroplasts in which photosynthesis takes place, and starch accumulates. In prokaryotes, such as cyanobacteria, the process is less localized and occurs within folded membranes, extensions of the plasma membrane, and in the cytoplasm.

8.2 The Light-Dependent Reactions of Photosynthesis

The pigments of the first part of photosynthesis, the light-dependent reactions, absorb energy from sunlight. A photon strikes the antenna pigments of photosystem II to initiate photosynthesis. The energy travels to the reaction center that contains chlorophyll *a* and then to the electron transport chain, which pumps hydrogen ions into the thylakoid interior. This action builds up a high concentration of hydrogen ions. The hydrogen ions flow through ATP synthase during chemiosmosis to form molecules of ATP, which are used for the formation of sugar molecules in the second stage of photosynthesis. Photosystem I absorbs a second photon, which results in the formation of an NADPH molecule, another energy and reducing carrier for the light-independent reactions.

8.3 Using Light Energy to Make Organic Molecules

Using the energy carriers formed in the first steps of photosynthesis, the light-independent reactions, or the Calvin cycle, take in CO_2 from the atmosphere. An enzyme, RuBisCO, catalyzes a reaction with CO_2 and another organic compound, RuBP. After three cycles, a three-carbon molecule of G3P leaves the cycle to become part of a carbohydrate molecule. The remaining G3P molecules stay in the cycle to be regenerated into RuBP, which is then ready to react with more CO_2. Photosynthesis forms an energy cycle with the process of cellular respiration. Because plants contain both chloroplasts and mitochondria, they rely upon both photosynthesis and respiration for their ability to function in both the light and dark, and to be able to interconvert essential metabolites.

VISUAL CONNECTION QUESTIONS

1. Figure 8.6 On a hot, dry day, the guard cells of plants close their stomata to conserve water. What impact will this have on photosynthesis?

2. Figure 8.16 What is the initial source of electrons for the chloroplast electron transport chain?
   a. Water
   b. Oxygen
   c. Carbon dioxide
   d. NADPH

3. Figure 8.18 Which of the following statements is true?
   a. In photosynthesis, oxygen, carbon dioxide, ATP, and NADPH are reactants. G3P and water are products.
   b. In photosynthesis, chlorophyll, water, and carbon dioxide are reactants. G3P and oxygen are products.
   c. In photosynthesis, water, carbon dioxide, ATP, and NADPH are reactants. RuBP and oxygen are products.
   d. In photosynthesis, water and carbon dioxide are reactants. G3P and oxygen are products.

REVIEW QUESTIONS

4. Which of the following components is *not* used by both plants and cyanobacteria to carry out photosynthesis?
   a. chloroplasts
   b. chlorophyll
   c. carbon dioxide
   d. water

5. What two main products result from photosynthesis?
   a. oxygen and carbon dioxide
   b. chlorophyll and oxygen
   c. sugars/carbohydrates and oxygen
   d. sugars/carbohydrates and carbon dioxide

6. In which compartment of the plant cell do the light-independent reactions of photosynthesis take place?
   a. thylakoid
   b. stroma
   c. outer membrane
   d. mesophyll

7. Which statement about thylakoids in eukaryotes is *not* correct?
   a. Thylakoids are assembled into stacks.
   b. Thylakoids exist as a maze of folded membranes.
   c. The space surrounding thylakoids is called stroma.
   d. Thylakoids contain chlorophyll.

8. Predict the end result if a chloroplast's light-independent enzymes developed a mutation that prevented them from activating in response to light.
   a. G3P accumulation
   b. ATP and NADPH accumulation
   c. Water accumulation
   d. Carbon dioxide depletion

9. How are the NADPH and G3P molecules made during photosynthesis similar?
   a. They are both end products of photosynthesis.
   b. They are both substrates for photosynthesis.
   c. They are both produced from carbon dioxide.
   d. They both store energy in chemical bonds.

10. Which of the following structures is *not* a component of a photosystem?
    a. ATP synthase
    b. antenna molecule
    c. reaction center
    d. primary electron acceptor

11. How many photons does it take to fully reduce one molecule of $NADP^+$ to NADPH?
    a. 1
    b. 2
    c. 4
    d. 8

12. Which complex is *not* involved in the establishment of conditions for ATP synthesis?
    a. photosystem I
    b. ATP synthase
    c. photosystem II
    d. cytochrome complex

13. From which component of the light-dependent reactions does NADPH form most directly?
    a. photosystem II
    b. photosystem I
    c. cytochrome complex
    d. ATP synthase

14. Three of the same species of plant are each grown under a different colored light for the same amount of time. Plant A is grown under blue light, Plant B is grown under green light, and Plant C is grown under orange light. Assuming the plants use only chlorophyll *a* and chlorophyll *b* for photosynthesis, what would be the predicted order of the plants from most growth to least growth?
    a. A, C, B
    b. A, B, C
    c. C, A, B
    d. B, A, C

15. Plants containing only chlorophyll *b* are exposed to radiation with the following wavelengths: 10nm (x-rays), 450nm (blue light), 670nm (red light), and 800nm (infrared light). Which plants harness the most energy for photosynthesis?
    a. X-ray irradiated plants
    b. Blue light irradiated plants
    c. Red light irradiated plants
    d. Infrared irradiated plants

16. Which molecule must enter the Calvin cycle continually for the light-independent reactions to take place?
    a. RuBisCO
    b. RuBP
    c. 3-PGA
    d. CO_2

17. Which order of molecular conversions is correct for the Calvin cycle?
    a. $\text{RuBP} + \text{G3P} \rightarrow \text{3-PGA} \rightarrow \text{sugar}$
    b. $\text{RuBisCO} \rightarrow CO_2 \rightarrow \text{RuBP} \rightarrow \text{G3P}$
    c. $\text{RuBP} + CO_2 \rightarrow [\text{RuBisCO}]\ \text{3-PGA} \rightarrow \text{G3P}$
    d. $CO_2 \rightarrow \text{3-PGA} \rightarrow \text{RuBP} \rightarrow \text{G3P}$

18. Where in eukaryotic cells does the Calvin cycle take place?
    a. thylakoid membrane
    b. thylakoid lumen
    c. chloroplast stroma
    d. granum

19. Which statement correctly describes carbon fixation?
    a. the conversion of CO_2 into an organic compound
    b. the use of RuBisCO to form 3-PGA
    c. the production of carbohydrate molecules from G3P
    d. the formation of RuBP from G3P molecules
    e. the use of ATP and NADPH to reduce CO_2

20. If four molecules of carbon dioxide enter the Calvin cycle (four "turns" of the cycle), how many G3P molecules are produced and how many are exported?
    a. 4 G3P made, 1 G3P exported
    b. 4 G3P made, 2 G3P exported
    c. 8 G3P made, 1 G3P exported
    d. 8 G3P made, 4 G3P exported

CRITICAL THINKING QUESTIONS

21. What is the overall outcome of the light reactions in photosynthesis?
22. Why are carnivores, such as lions, dependent on photosynthesis to survive?
23. Why are energy carriers thought of as either "full" or "empty"?
24. Describe how the grey wolf population would be impacted by a volcanic eruption that spewed a dense ash cloud that blocked sunlight in a section of Yellowstone National Park.
25. How does the closing of the stomata limit photosynthesis?
26. Describe the pathway of electron transfer from photosystem II to photosystem I in light-dependent reactions.
27. What are the roles of ATP and NADPH in photosynthesis?
28. How and why would the end products of photosynthesis be changed if a plant had a mutation that eliminated its photosystem II complex?
29. Why is the third stage of the Calvin cycle called the regeneration stage?
30. Which part of the light-independent reactions would be affected if a cell could not produce the enzyme RuBisCO?
31. Why does it take three turns of the Calvin cycle to produce G3P, the initial product of photosynthesis?
32. Imagine a sealed terrarium containing a plant and a beetle. How does each organism provide resources for the other? Could each organism survive if it was the only living thing in the terrarium? Why or why not?
33. Compare the flow of energy with the flow of nutrients in a closed, sunny ecosystem consisting of a giraffe and a tree.

CHAPTER 9
Cell Communication

Figure 9.1 Have you ever become separated from a friend while in a crowd? If so, you know the challenge of searching for someone when surrounded by thousands of other people. If you and your friend have cell phones, your chances of finding each other are good. Cell phone networks use various methods of encoding to ensure that the signals reach their intended recipients without interference. Similarly, cells must communicate using specific signals and receptors to ensure that messages are clear. (credit: modification of work by Vincent and Bella Productions)

Chapter Outline

INTRODUCTION Imagine what life would be like if you and the people around you could not communicate. You would not be able to express your wishes to others, nor could you ask questions about your location. Social organization is dependent on communication between the individuals that comprise that society; without communication, society would fall apart.

As with people, it is vital for individual cells to be able to interact with their environment. This is true for both a one-celled organism growing in a puddle and a large animal living on a savanna. In order to properly respond to external stimuli, cells have developed complex mechanisms of communication that can receive a message, transfer the information across the plasma membrane, and then produce changes within the cell in response to the message.

In multicellular organisms, cells send and receive chemical messages constantly to coordinate the actions of distant organs, tissues, and cells. The ability to send messages quickly and efficiently enables cells to coordinate and fine-tune their functions.

While the necessity for cellular communication in larger organisms seems obvious, even single-celled organisms communicate with each other. Yeast cells signal each other to aid in finding other yeast cells for reproduction. Some forms of bacteria coordinate their actions in order to form large complexes called biofilms or to organize the production of toxins to remove competing organisms. The ability of cells to communicate through chemical signals originated in single cells and was essential for the evolution of multicellular organisms. The efficient and relatively error-free function of communication systems is vital for all life as we know it.

9.1 Signaling Molecules and Cellular Receptors

By the end of this section, you will be able to do the following:

- Describe four types of signaling mechanisms found in multicellular organisms
- Compare internal receptors with cell-surface receptors
- Recognize the relationship between a ligand's structure and its mechanism of action

There are two kinds of communication in the world of living cells. Communication between cells is called **intercellular signaling**, and communication within a cell is called **intracellular signaling**. An easy way to remember the distinction is by understanding the Latin origin of the prefixes: *inter-* means "between" (for example, intersecting lines are those that cross each other) and *intra-* means "inside" (as in intravenous).

Chemical signals are released by **signaling cells** in the form of small, usually volatile or soluble molecules called ligands. A **ligand** is a molecule that binds another specific molecule, in some cases, delivering a signal in the process. Ligands can thus be thought of as signaling molecules. Ligands interact with proteins in **target cells**, which are cells that are affected by chemical signals; these proteins are also called **receptors**. Ligands and receptors exist in several varieties; however, a specific ligand will have a specific receptor that typically binds only that ligand.

Forms of Signaling

There are four categories of chemical signaling found in multicellular organisms: paracrine signaling, endocrine signaling, autocrine signaling, and direct signaling across gap junctions (Figure 9.2). The main difference between the different categories of signaling is the distance that the signal travels through the organism to reach the target cell. We should note here that not all cells are affected by the same signals.

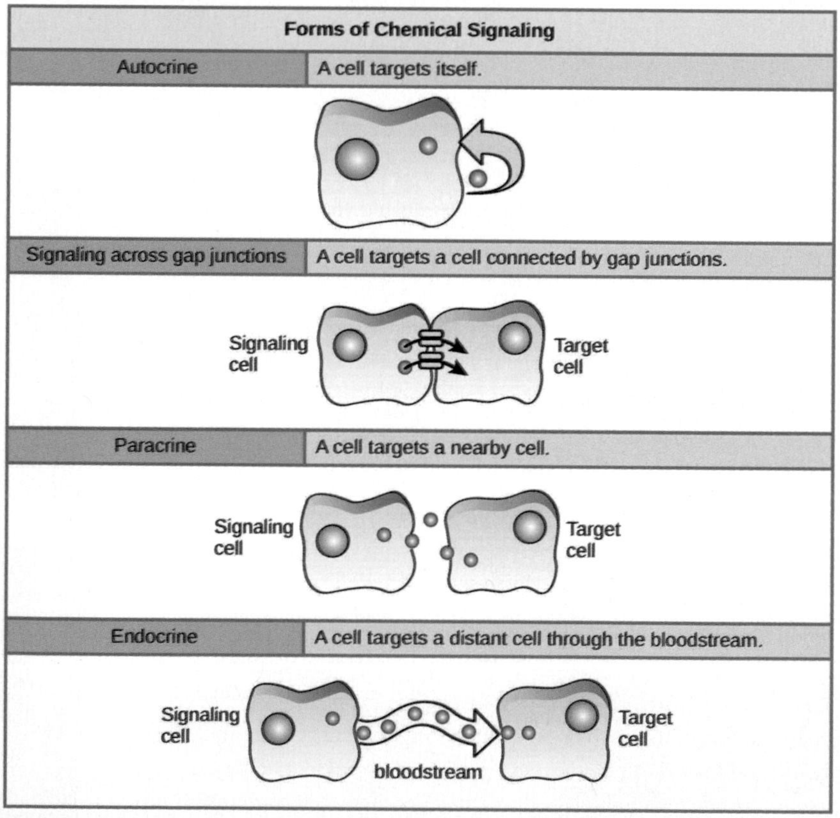

Figure 9.2 In chemical signaling, a cell may target itself (autocrine signaling), a cell connected by gap junctions, a nearby cell (paracrine signaling), or a distant cell (endocrine signaling). Paracrine signaling acts on nearby cells, endocrine signaling uses the circulatory system to transport ligands, and autocrine signaling acts on the signaling cell. Signaling via gap junctions involves signaling molecules moving directly between adjacent cells.

Paracrine Signaling

Signals that act locally between cells that are close together are called **paracrine signals**. Paracrine signals move by diffusion through the extracellular matrix. These types of signals usually elicit quick responses that last only a short period of time. In order to keep the response localized, paracrine ligand molecules are normally quickly degraded by enzymes or removed by neighboring cells. Removing the signals will reestablish the concentration gradient for the signal, allowing them to quickly diffuse through the intracellular space if released again.

One example of paracrine signaling is the transfer of signals across synapses between nerve cells. A nerve cell consists of a cell body, several short, branched extensions called dendrites that receive stimuli, and a long extension called an axon, which transmits signals to other nerve cells or muscle cells. The junction between nerve cells where signal transmission occurs is called a synapse. A **synaptic signal** is a chemical signal that travels between nerve cells. Signals within the nerve cells are propagated by fast-moving electrical impulses. When these impulses reach the end of the axon, the signal continues on to a dendrite of the next cell by the release of chemical ligands called **neurotransmitters** from the presynaptic cell (the cell emitting the signal). The neurotransmitters are transported across the very small distances (20–40 nanometers) between nerve cells, which are called **chemical synapses** (Figure 9.3). The small distance between nerve cells allows the signal to travel quickly; this enables an immediate response, such as, "Take your hand off the stove!"

When the neurotransmitter binds the receptor on the surface of the postsynaptic cell, the electrochemical potential of the target cell changes, and the next electrical impulse is launched. The neurotransmitters that are released into the chemical synapse are degraded quickly or get reabsorbed by the presynaptic cell so that the recipient nerve cell can recover quickly and be prepared to respond rapidly to the next synaptic signal.

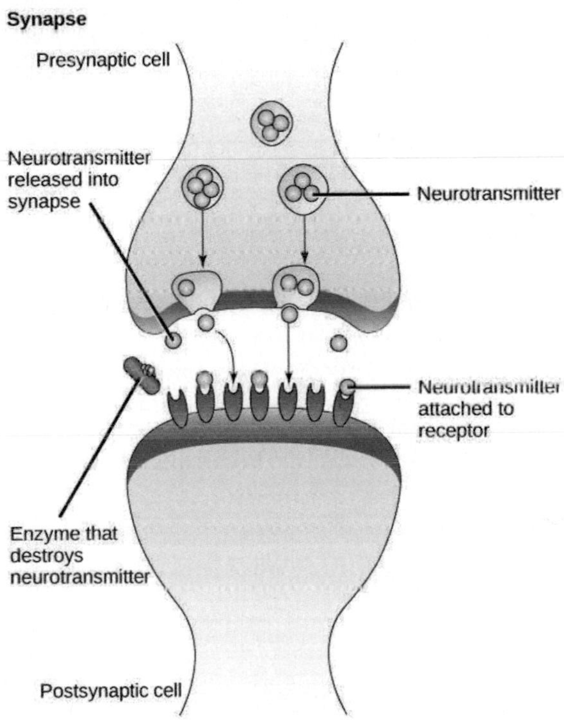

Figure 9.3 The distance between the presynaptic cell and the postsynaptic cell—called the synaptic gap—is very small and allows for rapid diffusion of the neurotransmitter. Enzymes in the synapatic gap degrade some types of neurotransmitters to terminate the signal.

Endocrine Signaling

Signals from distant cells are called **endocrine signals**, and they originate from **endocrine cells**. (In the body, many endocrine cells are located in endocrine glands, such as the thyroid gland, the hypothalamus, and the pituitary gland.) These types of signals usually produce a slower response but have a longer-lasting effect. The ligands released in endocrine signaling are called hormones, signaling molecules that are produced in one part of the body but affect other body regions some distance away.

Hormones travel the large distances between endocrine cells and their target cells via the bloodstream, which is a relatively slow way to move throughout the body. Because of their form of transport, hormones become diluted and are present in low concentrations when they act on their target cells. This is different from paracrine signaling, in which local concentrations of ligands can be very high.

Autocrine Signaling

Autocrine signals are produced by signaling cells that can also bind to the ligand that is released. This means the signaling cell and the target cell can be the same or a similar cell (the prefix *auto-* means self, a reminder that the signaling cell sends a signal to itself). This type of signaling often occurs during the early development of an organism to ensure that cells develop into the correct tissues and take on the proper function. Autocrine signaling also regulates pain sensation and inflammatory responses. Further, if a cell is infected with a virus, the cell can signal itself to undergo programmed cell death, killing the virus in the process. In some cases, neighboring cells of the same type are also influenced by the released ligand. In embryological development, this process of stimulating a group of neighboring cells may help to direct the differentiation of identical cells into the same cell type, thus ensuring the proper developmental outcome.

Direct Signaling Across Gap Junctions

Gap junctions in animals and *plasmodesmata* in plants are connections between the plasma membranes of neighboring cells. These fluid-filled channels allow small signaling molecules, called **intracellular mediators**, to diffuse between the two cells. Small molecules or ions, such as calcium ions (Ca^{2+}), are able to move between cells, but large molecules like proteins and DNA cannot fit through the channels. The specificity of the channels ensures that the cells remain independent but can quickly and easily transmit signals. The transfer of signaling molecules communicates the current state of the cell that is directly next to the target cell; this allows a group of cells to coordinate their response to a signal that only one of them may have received. In plants, *plasmodesmata* are ubiquitous, making the entire plant into a giant communication network.

Types of Receptors

Receptors are protein molecules in the target cell or on its surface that bind ligand. There are two types of receptors, internal receptors and cell-surface receptors.

Internal receptors

Internal receptors, also known as intracellular or cytoplasmic receptors, are found in the cytoplasm of the cell and respond to hydrophobic ligand molecules that are able to travel across the plasma membrane. Once inside the cell, many of these molecules bind to proteins that act as regulators of mRNA synthesis (transcription) to mediate gene expression. Gene expression is the cellular process of transforming the information in a cell's DNA into a sequence of amino acids, which ultimately forms a protein. When the ligand binds to the internal receptor, a conformational change is triggered that exposes a DNA-binding site on the protein. The ligand-receptor complex moves into the nucleus, then binds to specific regulatory regions of the chromosomal DNA and promotes the initiation of transcription (Figure 9.4). Transcription is the process of copying the information in a cell's DNA into a special form of RNA called messenger RNA (mRNA); the cell uses information in the mRNA (which moves out into the cytoplasm and associates with ribosomes) to link specific amino acids in the correct order, producing a protein. Internal receptors can directly influence gene expression without having to pass the signal on to other receptors or messengers.

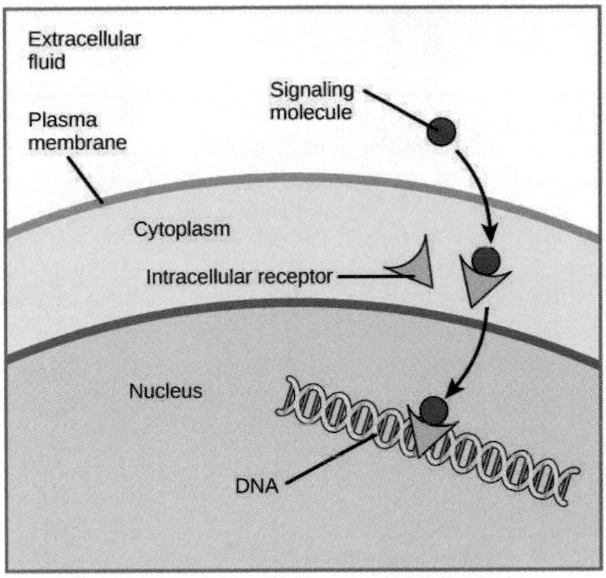

Figure 9.4 Hydrophobic signaling molecules typically diffuse across the plasma membrane and interact with intracellular receptors in the cytoplasm. Many intracellular receptors are transcription factors that interact with DNA in the nucleus and regulate gene expression.

Cell-Surface Receptors

Cell-surface receptors, also known as transmembrane receptors, are cell surface, membrane-anchored (integral) proteins that bind to external ligand molecules. This type of receptor spans the plasma membrane and performs signal transduction, through which an extracellular signal is converted into an intracellular signal. Ligands that interact with cell-surface receptors do not have to enter the cell that they affect. Cell-surface receptors are also called cell-specific proteins or markers because they are specific to individual cell types.

Because cell-surface receptor proteins are fundamental to normal cell functioning, it should come as no surprise that a malfunction in any one of these proteins could have severe consequences. Errors in the protein structures of certain receptor molecules have been shown to play a role in hypertension (high blood pressure), asthma, heart disease, and cancer.

Each cell-surface receptor has three main components: an external ligand-binding domain called the **extracellular domain**, a hydrophobic membrane-spanning region called a transmembrane domain, and an intracellular domain inside the cell. The size and extent of each of these domains vary widely, depending on the type of receptor.

EVOLUTION CONNECTION

How Viruses Recognize a Host

Unlike living cells, many viruses do not have a plasma membrane or any of the structures necessary to sustain metabolic life. Some viruses are simply composed of an inert protein shell enclosing DNA or RNA. To reproduce, viruses must invade a living cell, which serves as a host, and then take over the hosts cellular apparatus. But how does a virus recognize its host?

Viruses often bind to cell-surface receptors on the host cell. For example, the virus that causes human influenza (flu) binds specifically to receptors on membranes of cells of the respiratory system. Chemical differences in the cell-surface receptors among hosts mean that a virus that infects a specific species (for example, humans) often cannot infect another species (for example, chickens).

However, viruses have very small amounts of DNA or RNA compared to humans, and, as a result, viral reproduction can occur rapidly. Viral reproduction invariably produces errors that can lead to changes in newly produced viruses; these changes mean that the viral proteins that interact with cell-surface receptors may evolve in such a way that they can bind to receptors in a new host. Such changes happen randomly and quite often in the reproductive cycle of a virus, but the changes only matter if a virus with new binding properties comes into contact with a suitable host. In the case of influenza, this situation can occur in settings where animals and people are in close contact, such as poultry and swine farms.[1] Once a virus jumps the former "species barrier" to a new host, it can spread quickly. Scientists watch newly appearing viruses (called emerging viruses) closely in the hope that

1A. B. Sigalov, The School of Nature. IV. Learning from Viruses, *Self/Nonself* 1, no. 4 (2010): 282-298. Y. Cao, X. Koh, L. Dong, X. Du, A. Wu, X. Ding, H. Deng, Y. Shu, J. Chen, T. Jiang, Rapid Estimation of Binding Activity of Influenza Virus Hemagglutinin to Human and Avian Receptors, *PLoS One* 6, no. 4 (2011): e18664.

such monitoring can reduce the likelihood of global viral epidemics.

Cell-surface receptors are involved in most of the signaling in multicellular organisms. There are three general categories of cell-surface receptors: ion channel-linked receptors, G-protein-linked receptors, and enzyme-linked receptors.

Ion channel-linked receptors bind a ligand and open a channel through the membrane that allows specific ions to pass through. To form a channel, this type of cell-surface receptor has an extensive membrane-spanning region. In order to interact with the double layer of phospholipid fatty acid tails that form the center of the plasma membrane, many of the amino acids in the membrane-spanning region are hydrophobic in nature. Conversely, the amino acids that line the inside of the channel are hydrophilic to allow for the passage of water or ions. When a ligand binds to the extracellular region of the channel, there is a conformational change in the protein's structure that allows ions such as sodium, calcium, magnesium, and hydrogen to pass through (Figure 9.5).

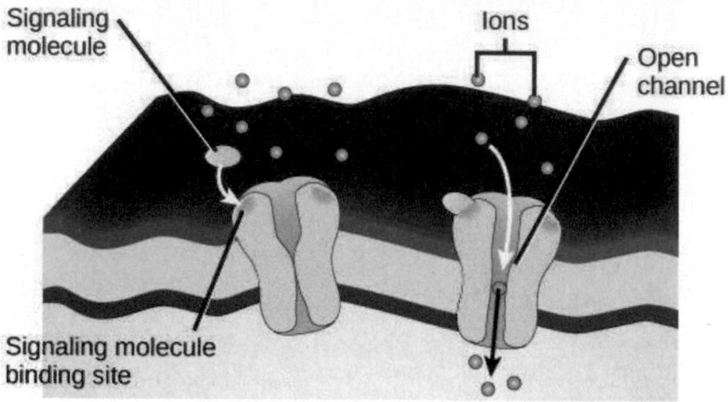

Figure 9.5 Gated ion channels form a pore through the plasma membrane that opens when the signaling molecule binds. The open pore then allows ions to flow into or out of the cell.

G-protein-linked receptors bind a ligand and activate a membrane protein called a G-protein. The activated G-protein then interacts with either an ion channel or an enzyme in the membrane (Figure 9.6). All G-protein-linked receptors have seven transmembrane domains, but each receptor has its own specific extracellular domain and G-protein-binding site.

Cell signaling using G-protein-linked receptors occurs as a cyclic series of events. Before the ligand binds, the inactive G-protein can bind to a newly revealed site on the receptor specific for its binding. Once the G-protein binds to the receptor, the resulting change in shape activates the G-protein, which releases guanosine diposphate (GDP) and picks up guanosine 3-phosphate (GTP). The subunits of the G-protein then split into the α subunit and the $\beta\gamma$ subunit. One or both of these G-protein fragments may be able to activate other proteins as a result. After awhile, the GTP on the active α subunit of the G-protein is hydrolyzed to GDP and the $\beta\gamma$ subunit is deactivated. The subunits reassociate to form the inactive G-protein and the cycle begins anew.

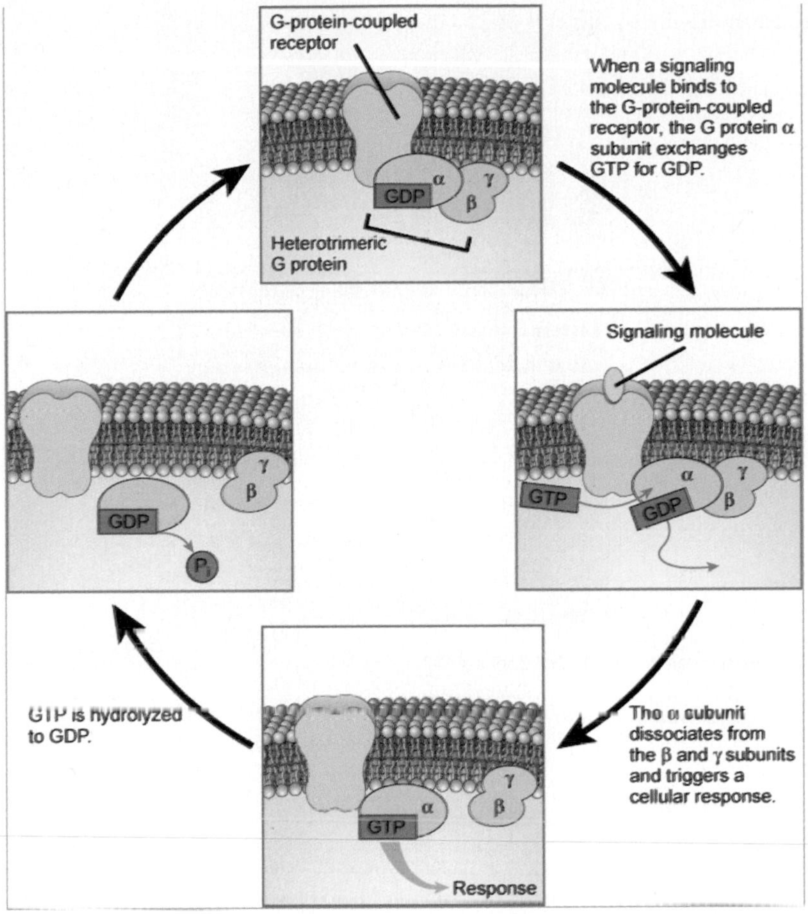

Figure 9.6 Heterotrimeric G-proteins have three subunits: α, β, and γ. When a signaling molecule binds to a G-protein-coupled receptor in the plasma membrane, a GDP molecule associated with the α subunit is exchanged for GTP. The β and γ subunits dissociate from the α subunit, and a cellular response is triggered either by the α subunit or the dissociated $\beta\gamma$ pair. Hydrolysis of GTP to GDP terminates the signal.

G-protein-linked receptors have been extensively studied and much has been learned about their roles in maintaining health. Bacteria that are pathogenic to humans can release poisons that interrupt specific G-protein-linked receptor function, leading to illnesses such as pertussis, botulism, and cholera. In cholera (Figure 9.7), for example, the water-borne bacterium *Vibrio cholerae* produces a toxin, choleragen, that binds to cells lining the small intestine. The toxin then enters these intestinal cells, where it modifies a G-protein that controls the opening of a chloride channel and causes it to remain continuously active, resulting in large losses of fluids from the body and potentially fatal dehydration as a result.

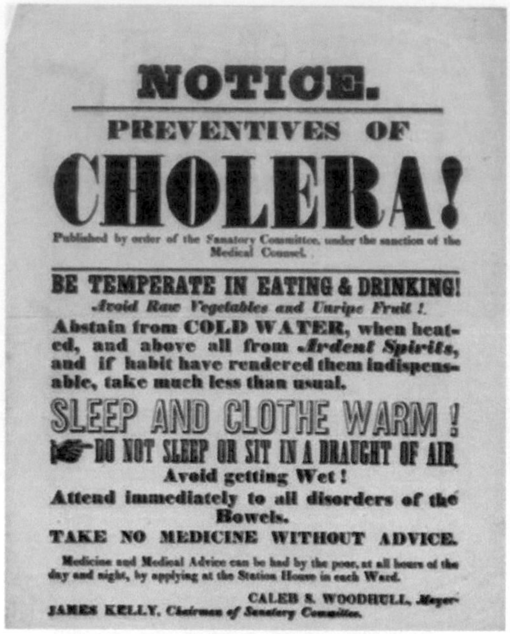

Figure 9.7 Transmitted primarily through contaminated drinking water, cholera is a major cause of death in the developing world and in areas where natural disasters interrupt the availability of clean water. The cholera bacterium, *Vibrio cholerae*, creates a toxin that modifies G-protein-mediated cell signaling pathways in the intestines. Modern sanitation eliminates the threat of cholera outbreaks, such as the one that swept through New York City in 1866. This poster from that era shows how, at that time, the way that the disease was transmitted was not understood. (credit: New York City Sanitary Commission)

Enzyme-linked receptors are cell-surface receptors with intracellular domains that are associated with an enzyme. In some cases, the intracellular domain of the receptor itself is an enzyme. Other enzyme-linked receptors have a small intracellular domain that interacts directly with an enzyme. The enzyme-linked receptors normally have large extracellular and intracellular domains, but the membrane-spanning region consists of a single alpha-helical region of the peptide strand. When a ligand binds to the extracellular domain, a signal is transferred through the membrane, activating the enzyme. Activation of the enzyme sets off a chain of events within the cell that eventually leads to a response. One example of this type of enzyme-linked receptor is the tyrosine kinase receptor (Figure 9.8). A kinase is an enzyme that transfers phosphate groups from ATP to another protein. The tyrosine kinase receptor transfers phosphate groups to tyrosine molecules (tyrosine residues). First, signaling molecules bind to the extracellular domain of two nearby tyrosine kinase receptors. The two neighboring receptors then bond together, or dimerize. Phosphates are then added to tyrosine residues on the intracellular domain of the receptors (phosphorylation). The phosphorylated residues can then transmit the signal to the next messenger within the cytoplasm.

VISUAL CONNECTION

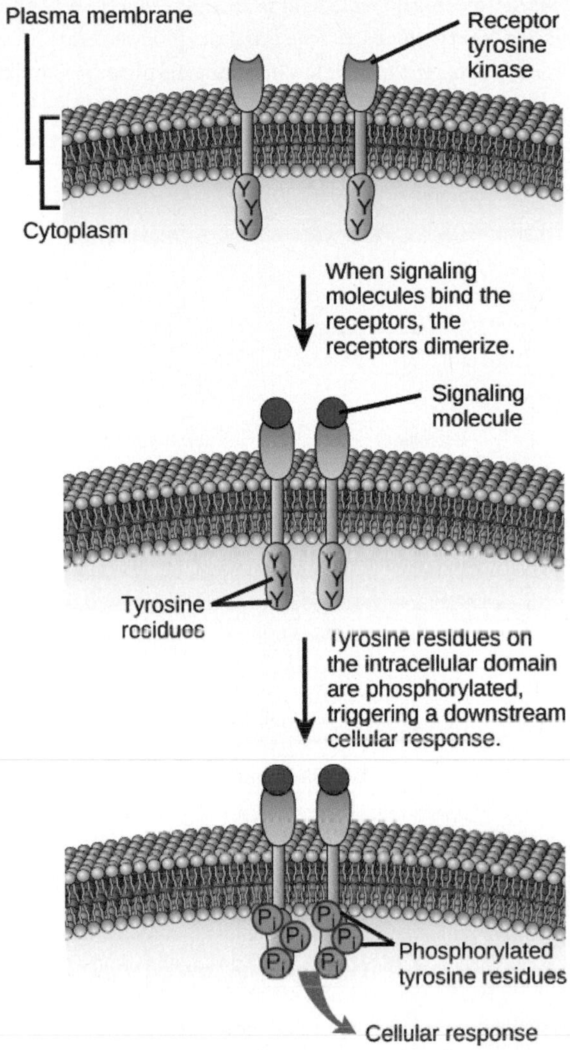

Figure 9.8 A receptor tyrosine kinase is an enzyme-linked receptor with a single helical transmembrane region, and extracellular and intracellular domains. Binding of a signaling molecule to the extracellular domain causes the receptor to dimerize. Tyrosine residues on the intracellular domain are then autophosphorylated, triggering a downstream cellular response. The signal is terminated by a phosphatase that removes the phosphates from the phosphotyrosine residues.

HER2 is a receptor tyrosine kinase. In 30 percent of human breast cancers, HER2 is permanently activated, resulting in unregulated cell division. Lapatinib, a drug used to treat breast cancer, inhibits HER2 receptor tyrosine kinase autophosphorylation (the process by which the receptor adds phosphates onto itself), thus reducing tumor growth by 50 percent. Besides autophosphorylation, which of the following steps would be inhibited by Lapatinib?

a. Signaling molecule binding, dimerization, and the downstream cellular response
b. Dimerization, and the downstream cellular response
c. The downstream cellular response
d. Phosphatase activity, dimerization, and the downsteam cellular response

Signaling Molecules

Produced by signaling cells and the subsequent binding to receptors in target cells, ligands act as chemical signals that travel to the target cells to coordinate responses. The types of molecules that serve as ligands are incredibly varied and range from small proteins to small ions like calcium (Ca^{2+}).

Small Hydrophobic Ligands

Small hydrophobic ligands can directly diffuse through the plasma membrane and interact with internal receptors. Important members of this class of ligands are the steroid hormones. Steroids are lipids that have a hydrocarbon skeleton with four fused rings; different steroids have different functional groups attached to the carbon skeleton. Steroid hormones include the female sex hormone, estradiol, which is a type of estrogen; the male sex hormone, testosterone; and cholesterol, which is an important structural component of biological membranes and a precursor of steriod hormones (Figure 9.9). Other hydrophobic hormones include thyroid hormones and vitamin D. In order to be soluble in blood, hydrophobic ligands must bind to carrier proteins while they are being transported through the bloodstream.

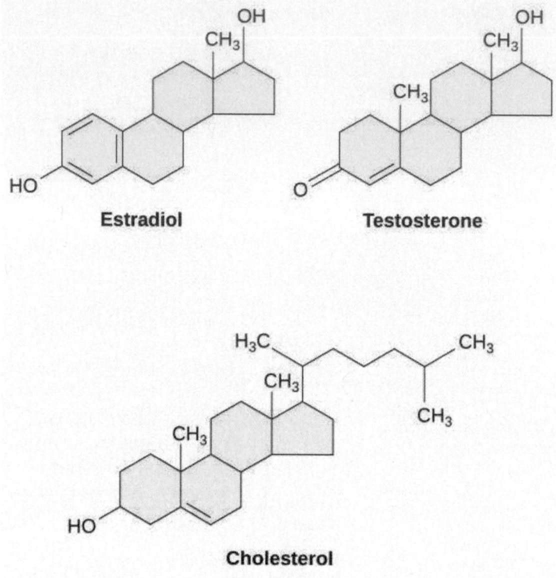

Figure 9.9 Steroid hormones have similar chemical structures to their precursor, cholesterol. Because these molecules are small and hydrophobic, they can diffuse directly across the plasma membrane into the cell, where they interact with internal receptors.

Water-Soluble Ligands

Water-soluble ligands are polar and, therefore, cannot pass through the plasma membrane unaided; sometimes, they are too large to pass through the membrane at all. Instead, most water-soluble ligands bind to the extracellular domain of cell-surface receptors. This group of ligands is quite diverse and includes small molecules, peptides, and proteins.

Other Ligands

Nitric oxide (NO) is a gas that also acts as a ligand. It is able to diffuse directly across the plasma membrane, and one of its roles is to interact with receptors in smooth muscle and induce relaxation of the tissue. NO has a very short half-life and, therefore, only functions over short distances. Nitroglycerin, a treatment for heart disease, acts by triggering the release of NO, which causes blood vessels to dilate (expand), thus restoring blood flow to the heart. NO has become better known recently because the pathway that it affects is targeted by prescription medications for erectile dysfunction, such as Viagra (erection involves dilated blood vessels).

9.2 Propagation of the Signal

By the end of this section, you will be able to do the following:

- Explain how the binding of a ligand initiates signal transduction throughout a cell
- Recognize the role of phosphorylation in the transmission of intracellular signals
- Evaluate the role of second messengers in signal transmission

Once a ligand binds to a receptor, the signal is transmitted through the membrane and into the cytoplasm. Continuation of a signal in this manner is called **signal transduction**. Signal transduction only occurs with cell-surface receptors, which cannot interact with most components of the cell such as DNA. Only internal receptors are able to interact directly with DNA in the nucleus to initiate protein synthesis.

When a ligand binds to its receptor, conformational changes occur that affect the receptor's intracellular domain.

Conformational changes of the extracellular domain upon ligand binding can propagate through the membrane region of the receptor and lead to activation of the intracellular domain or its associated proteins. In some cases, binding of the ligand causes **dimerization** of the receptor, which means that two receptors bind to each other to form a stable complex called a dimer. A **dimer** is a chemical compound formed when two molecules (often identical) join together. The binding of the receptors in this manner enables their intracellular domains to come into close contact and activate each other.

Binding Initiates a Signaling Pathway

After the ligand binds to the cell-surface receptor, the activation of the receptor's intracellular components sets off a chain of events that is called a **signaling pathway**, sometimes called a signaling cascade. In a signaling pathway, second messengers–enzymes–and activated proteins interact with specific proteins, which are in turn activated in a chain reaction that eventually leads to a change in the cell's environment (Figure 9.10), such as an increase in metabolism or specific gene expression. The events in the cascade occur in a series, much like a current flows in a river. Interactions that occur before a certain point are defined as upstream events, and events after that point are called downstream events.

VISUAL CONNECTION

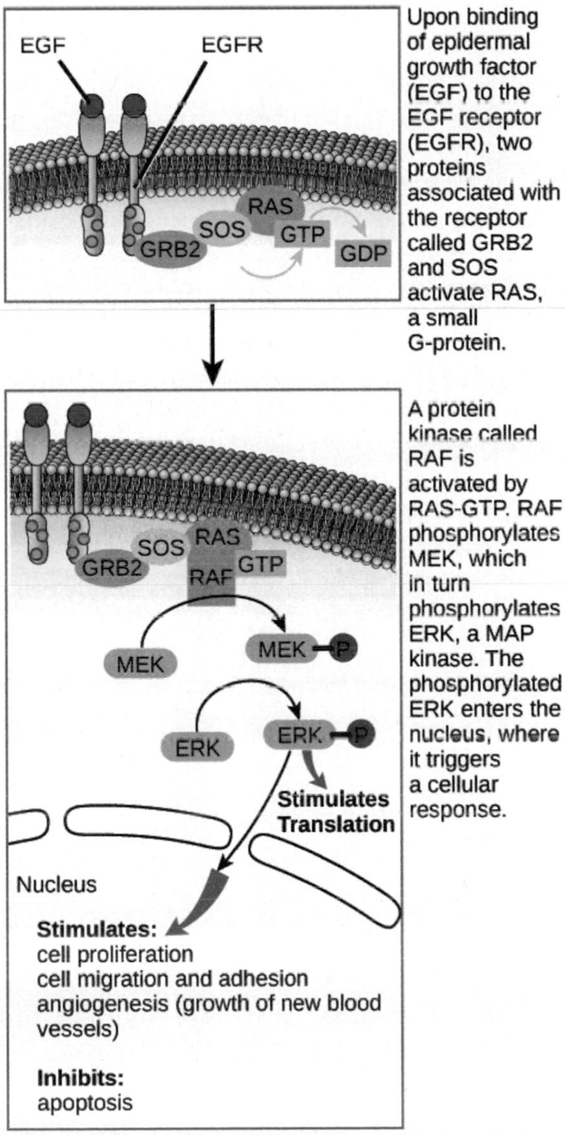

Figure 9.10 The epidermal growth factor (EGF) receptor (EGFR) is a receptor tyrosine kinase involved in the regulation of cell growth, wound healing, and tissue repair. When EGF binds to the EGFR, a cascade of downstream events causes the cell to grow and divide. If EGFR

is activated at inappropriate times, uncontrolled cell growth (cancer) may occur.

In certain cancers, the GTPase activity of the RAS G-protein is inhibited. This means that the RAS protein can no longer hydrolyze GTP into GDP. What effect would this have on downstream cellular events?

You can see that signaling pathways can get very complicated very quickly because most cellular proteins can affect different downstream events, depending on the conditions within the cell. A single pathway can branch off toward different endpoints based on the interplay between two or more signaling pathways, and the same ligands are often used to initiate different signals in different cell types. This variation in response is due to differences in protein expression in different cell types. Another complicating element is **signal integration** of the pathways, in which signals from two or more different cell-surface receptors merge to activate the same response in the cell. This process can ensure that multiple external requirements are met before a cell commits to a specific response.

The effects of extracellular signals can also be amplified by enzymatic cascades. At the initiation of the signal, a single ligand binds to a single receptor. However, activation of a receptor-linked enzyme can activate many copies of a component of the signaling cascade, which amplifies the signal.

LINK TO LEARNING

Observe an animation of cell signaling at this site (http://openstax.org/l/cell_signals) .

Methods of Intracellular Signaling

The induction of a signaling pathway depends on the modification of a cellular component by an enzyme. There are numerous enzymatic modifications that can occur, and they are recognized in turn by the next component downstream. The following are some of the more common events in intracellular signaling.

Phosphorylation

One of the most common chemical modifications that occurs in signaling pathways is the addition of a phosphate group (PO_4^{-3}) to a molecule such as a protein in a process called phosphorylation. The phosphate can be added to a nucleotide such as GMP to form GDP or GTP. Phosphates are also often added to serine, threonine, and tyrosine residues of proteins, where they replace the hydroxyl group of the amino acid (Figure 9.11). The transfer of the phosphate is catalyzed by an enzyme called a **kinase**. Various kinases are named for the substrate they phosphorylate. Phosphorylation of serine and threonine residues often activates enzymes. Phosphorylation of tyrosine residues can either affect the activity of an enzyme or create a binding site that interacts with downstream components in the signaling cascade. Phosphorylation may activate or inactivate enzymes, and the reversal of phosphorylation, dephosphorylation by a phosphatase, will reverse the effect.

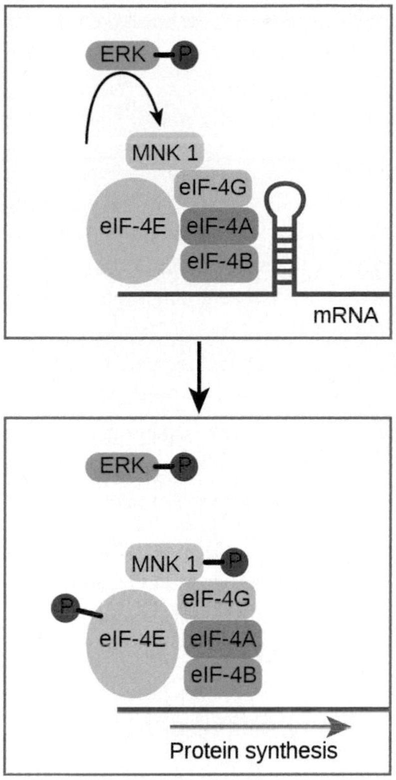

Figure 9.14 ERK is a MAP kinase that activates translation when it is phosphorylated. ERK phosphorylates MNK1, which in turn phosphorylates eIF-4E, an elongation initiation factor that, with other initiation factors, is associated with mRNA. When eIF-4E becomes phosphorylated, the mRNA unfolds, allowing protein synthesis in the nucleus to begin. (See Figure 9.10 for the phosphorylation pathway that activates ERK.)

Another mechanism of gene regulation involves PKC, which can interact with is a protein that acts as an inhibitor. An **inhibitor** is a molecule that binds to a protein and prevents it from functioning or reduces its function. In this case, the inhibitor is a protein called Iκ-B, which binds to the regulatory protein NF-κB. (The symbol κ represents the Greek letter kappa.) When Iκ-B is bound to NF-κB, the complex cannot enter the nucleus of the cell, but when Iκ-B is phosphorylated by PKC, it can no longer bind NF-κB, and NF-κB (a transcription factor) can enter the nucleus and initiate RNA transcription. In this case, the effect of phosphorylation is to inactivate an inhibitor and thereby activate the process of transcription.

Increase in Cellular Metabolism

The result of another signaling pathway affects muscle cells. The activation of β-adrenergic receptors in muscle cells by adrenaline leads to an increase in cyclic AMP (cAMP) inside the cell. Also known as epinephrine, adrenaline is a hormone (produced by the adrenal gland located on top of the kidney) that readies the body for short-term emergencies. Cyclic AMP activates PKA (protein kinase A), which in turn phosphorylates two enzymes. The first enzyme promotes the degradation of glycogen by activating intermediate glycogen phosphorylase kinase (GPK) that in turn activates glycogen phosphorylase (GP) that catabolizes glycogen into its constituent glucose monomers. (Recall that your body converts excess glucose to glycogen for short-term storage. When energy is needed, glycogen is quickly reconverted to glucose.) Phosphorylation of the second enzyme, glycogen synthase (GS), inhibits its ability to form glycogen from glucose. In this manner, a muscle cell obtains a ready pool of glucose by activating its formation via glycogen degradation and by inhibiting the use of glucose to form glycogen, thus preventing a futile cycle of glycogen degradation and synthesis. The glucose is then available for use by the muscle cell in response to a sudden surge of adrenaline—the "fight or flight" reflex.

Cell Growth

Cell signaling pathways also play a major role in cell division. Cells do not normally divide unless they are stimulated by signals from other cells. The ligands that promote cell growth are called **growth factors**. Most growth factors bind to cell-surface receptors that are linked to tyrosine kinases. These cell-surface receptors are called receptor tyrosine kinases (RTKs). Activation of RTKs initiates a signaling pathway that includes a G-protein called RAS, which activates the MAP kinase pathway described earlier. The enzyme MAP kinase then stimulates the expression of proteins that interact with other cellular components to initiate cell division.

CAREER CONNECTION

Cancer Biologist

Cancer biologists study the molecular origins of cancer with the goal of developing new prevention methods and treatment strategies that will inhibit the growth of tumors without harming the normal cells of the body. As mentioned earlier, signaling pathways control cell growth. These signaling pathways are controlled by signaling proteins, which are, in turn, expressed by genes. Mutations in these genes can result in malfunctioning signaling proteins. This prevents the cell from regulating its cell cycle, triggering unrestricted cell division and perhaps cancer. The genes that regulate the signaling proteins are one type of oncogene, which is a gene that has the potential to cause cancer. The gene encoding RAS is an oncogene that was originally discovered when mutations in the RAS protein were linked to cancer. Further studies have indicated that 30 percent of cancer cells have a mutation in the RAS gene that leads to uncontrolled growth. If left unchecked, uncontrolled cell division can lead to tumor formation and metastasis, the growth of cancer cells in new locations in the body.

Cancer biologists have been able to identify many other oncogenes that contribute to the development of cancer. For example, HER2 is a cell-surface receptor that is present in excessive amounts in 20 percent of human breast cancers. Cancer biologists realized that gene duplication led to HER2 overexpression in 25 percent of breast cancer patients and developed a drug called Herceptin (trastuzumab). Herceptin is a monoclonal antibody that targets HER2 for removal by the immune system. Herceptin therapy helps to control signaling through HER2. The use of Herceptin in combination with chemotherapy has helped to increase the overall survival rate of patients with metastatic breast cancer.

More information on cancer biology research can be found at the National Cancer Institute website (https://www.cancer.gov/research/areas/biology (http://openstax.org/l/cancer_research)).

Cell Death

When a cell is damaged, superfluous, or potentially dangerous to an organism, a cell can initiate a mechanism to trigger programmed cell death, or **apoptosis**. Apoptosis allows a cell to die in a controlled manner that prevents the release of potentially damaging molecules from inside the cell. There are many internal checkpoints that monitor a cell's health; if abnormalities are observed, a cell can spontaneously initiate the process of apoptosis. However, in some cases, such as a viral infection or uncontrolled cell division due to cancer, the cell's normal checks and balances fail. External signaling can also initiate apoptosis. For example, most normal animal cells have receptors that interact with the extracellular matrix, a network of glycoproteins that provides structural support for cells in an organism. The binding of cellular receptors to the extracellular matrix initiates a signaling cascade within the cell. However, if the cell moves away from the extracellular matrix, the signaling ceases, and the cell undergoes apoptosis. This system keeps cells from traveling through the body and proliferating out of control, as happens with tumor cells that metastasize.

Another example of external signaling that leads to apoptosis occurs in T-cell development. T-cells are immune cells that bind to foreign macromolecules and particles, and target them for destruction by the immune system. Normally, T-cells do not target "self" proteins (those of their own organism), a process that can lead to autoimmune diseases. In order to develop the ability to discriminate between self and non-self, immature T-cells undergo screening to determine whether they bind to so-called self proteins. If the T-cell receptor binds to self proteins, the cell initiates apoptosis to remove the potentially dangerous cell.

Apoptosis is also essential for normal embryological development. In vertebrates, for example, early stages of development include the formation of web-like tissue between individual fingers and toes (Figure 9.15). During the course of normal development, these unneeded cells must be eliminated, enabling fully separated fingers and toes to form. A cell signaling mechanism triggers apoptosis, which destroys the cells between the developing digits.

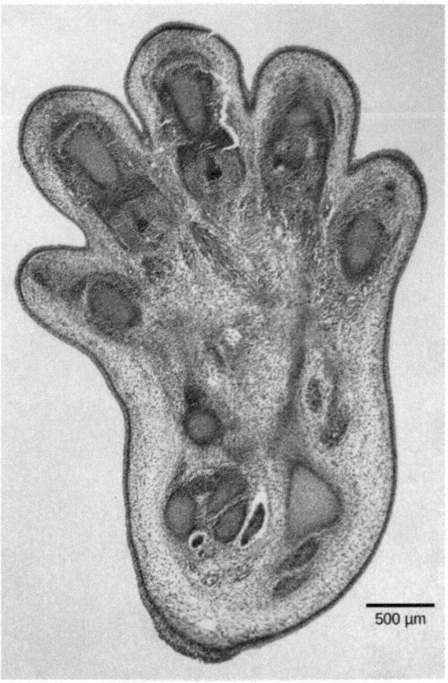

Figure 9.15 The histological section of a foot of a 15-day-old mouse embryo, visualized using light microscopy, reveals areas of tissue between the toes, which apoptosis will eliminate before the mouse reaches its full gestational age at 27 days. (credit: modification of work by Michal Mañas)

Termination of the Signal Cascade

The aberrant signaling often seen in tumor cells is proof that the termination of a signal at the appropriate time can be just as important as the initiation of a signal. One method of stopping a specific signal is to degrade the ligand or remove it so that it can no longer access its receptor. One reason that hydrophobic hormones like estrogen and testosterone trigger long-lasting events is because they bind carrier proteins. These proteins allow the insoluble molecules to be soluble in blood, but they also protect the hormones from degradation by circulating enzymes.

Inside the cell, many different enzymes reverse the cellular modifications that result from signaling cascades. For example, **phosphatases** are enzymes that remove the phosphate group attached to proteins by kinases in a process called dephosphorylation. Cyclic AMP (cAMP) is degraded into AMP by **phosphodiesterase**, and the release of calcium stores is reversed by the Ca^{2+} pumps that are located in the external and internal membranes of the cell.

9.4 Signaling in Single-Celled Organisms

By the end of this section, you will be able to do the following:

- Describe how single-celled yeasts use cell signaling to communicate with one another
- Relate the role of quorum sensing to the ability of some bacteria to form biofilms

Within-cell signaling allows bacteria to respond to environmental cues, such as nutrient levels. Some single-celled organisms also release molecules to signal to each other.

Signaling in Yeast

Yeasts are eukaryotes (fungi), and the components and processes found in yeast signals are similar to those of cell-surface receptor signals in multicellular organisms. Budding yeasts (Figure 9.16) are able to participate in a process that is similar to sexual reproduction that entails two haploid cells (cells with one-half the normal number of chromosomes) combining to form a diploid cell (a cell with two sets of each chromosome, which is what normal body cells contain). In order to find another haploid yeast cell that is prepared to mate, budding yeasts secrete a signaling molecule called **mating factor**. When mating factor binds to cell-surface receptors in other yeast cells that are nearby, they stop their normal growth cycles and initiate a cell signaling cascade that includes protein kinases and GTP-binding proteins that are similar to G-proteins.

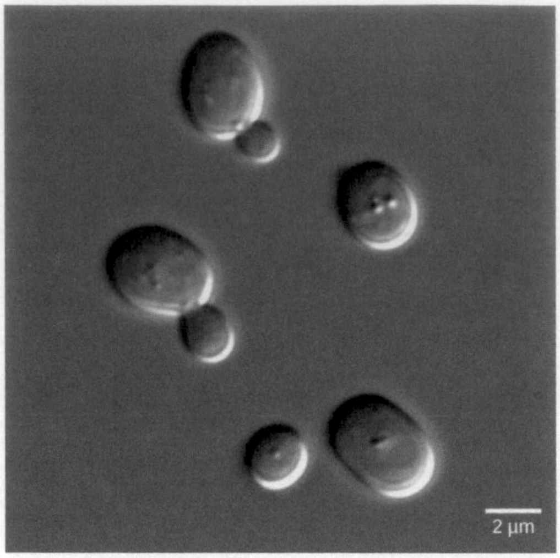

Figure 9.16 Budding *Saccharomyces cerevisiae* yeast cells can communicate by releasing a signaling molecule called mating factor. In this micrograph, they are visualized using differential interference contrast microscopy, a light microscopy technique that enhances the contrast of the sample.

Signaling in Bacteria

Signaling in bacteria enables bacteria to monitor extracellular conditions, ensure that there are sufficient amounts of nutrients, and ensure that hazardous situations are avoided. There are circumstances, however, when bacteria communicate with each other.

The first evidence of bacterial communication was observed in a bacterium that has a symbiotic relationship with Hawaiian bobtail squid. When the population density of the bacteria reaches a certain level, specific gene expression is initiated, and the bacteria produce bioluminescent proteins that emit light. Because the number of cells present in the environment (cell density) is the determining factor for signaling, bacterial signaling was named **quorum sensing**. In politics and business, a quorum is the minimum number of members required to be present to vote on an issue.

Quorum sensing uses autoinducers as signaling molecules. **Autoinducers** are signaling molecules secreted by bacteria to communicate with other bacteria of the same kind. The secreted autoinducers can be small, hydrophobic molecules, such as acyl-homoserine lactone (AHL), or larger peptide-based molecules; each type of molecule has a different mode of action. When AHL enters target bacteria, it binds to transcription factors, which then switch gene expression on or off. When the number of bacteria increases so does the concentration of the autoinducer, triggering increased expression of certain genes including autoinducers, which results in a self-amplifying cycle, also known as a positive feedback loop (Figure 9.17). The peptide autoinducers stimulate more complicated signaling pathways that include bacterial kinases. The changes in bacteria following exposure to autoinducers can be quite extensive. The pathogenic bacterium *Pseudomonas aeruginosa* has 616 different genes that respond to autoinducers.

Some species of bacteria that use quorum sensing form biofilms, complex colonies of bacteria (often containing several species) that exchange chemical signals to coordinate the release of toxins that will attack the host. Bacterial biofilms (Figure 9.18) can sometimes be found on medical equipment; when biofilms invade implants such as hip or knee replacements or heart pacemakers, they can cause life-threatening infections.

VISUAL CONNECTION

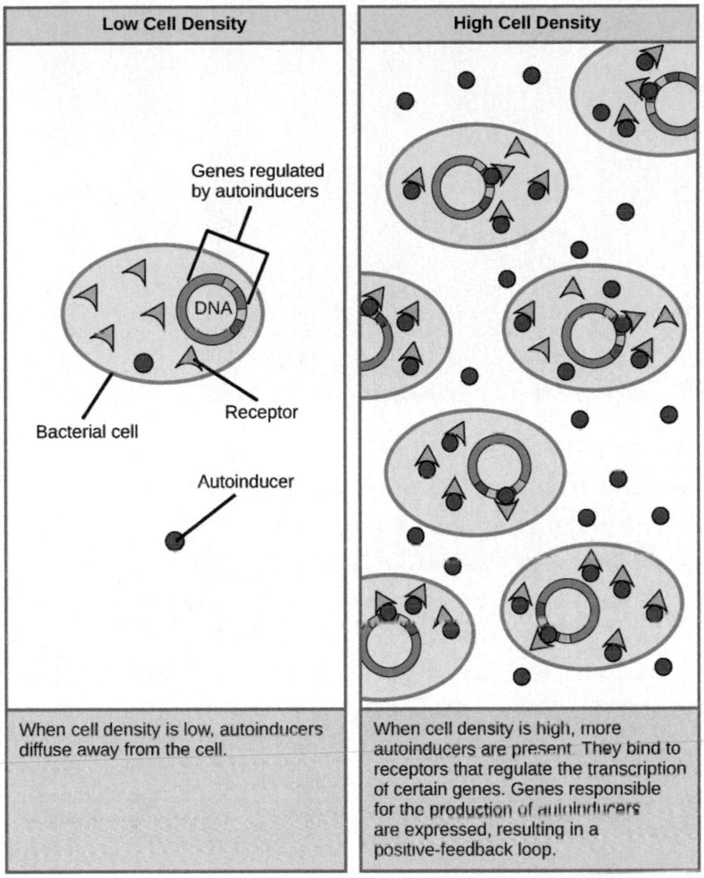

Figure 9.17 Autoinducers are small molecules or proteins produced by bacteria that regulate gene expression.

Which of the following statements about quorum sensing is false?

a. Autoinducer must bind to receptor to turn on transcription of genes responsible for the production of more autoinducer.
b. The receptor stays in the bacterial cell, but the autoinducer diffuses out.
c. Autoinducer can only act on a different cell: it cannot act on the cell in which it is made.
d. Autoinducer turns on genes that enable the bacteria to form a biofilm.

VISUAL CONNECTION

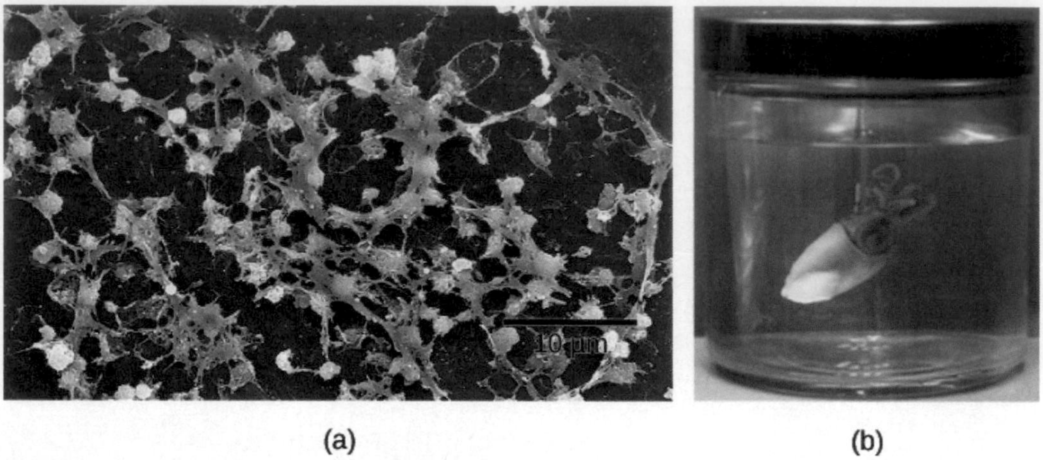

Figure 9.18 Cell-cell communication enables these (a) *Staphylococcus aureus* bacteria to work together to form a biofilm inside a hospital patient's catheter, seen here via scanning electron microscopy. *S. aureus* is the main cause of hospital-acquired infections. (b) Hawaiian bobtail squid have a symbiotic relationship with the bioluminescent bacteria *Vibrio fischeri*. The luminescence makes it difficult to see the squid from below because it effectively eliminates its shadow. In return for camouflage, the squid provides food for the bacteria. Free-living *V. fischeri* do not produce luciferase, the enzyme responsible for luminescence, but *V. fischeri* living in a symbiotic relationship with the squid do. Quorum sensing determines whether the bacteria should produce the luciferase enzyme. (credit a: modifications of work by CDC/ Janice Carr; credit b: modifications of work by Cliff1066/Flickr)

What advantage might biofilm production confer on the *S. aureus* inside the catheter?

Research on the details of quorum sensing has led to advances in growing bacteria for industrial purposes. Recent discoveries suggest that it may be possible to exploit bacterial signaling pathways to control bacterial growth; this process could replace or supplement antibiotics that are no longer effective in certain situations.

LINK TO LEARNING

Watch geneticist Bonnie Bassler discuss her discovery (http://openstax.org/l/bacteria_talk) of quorum sensing in biofilm bacteria in squid.

EVOLUTION CONNECTION

Cellular Communication in Yeasts

The first cellular form of life on our planet likely consisted of single-celled prokaryotic organisms that had limited interaction with each other. While some external signaling occurs between different species of single-celled organisms, the majority of signaling within bacteria and yeasts concerns only other members of the same species. The evolution of cellular communication is an absolute necessity for the development of multicellular organisms, and this innovation is thought to have required approximately 2 billion years to appear in early life forms.

Yeasts are single-celled eukaryotes and, therefore, have a nucleus and organelles characteristic of more complex life forms. Comparisons of the genomes of yeasts, nematode worms, fruit flies, and humans illustrate the evolution of increasingly complex signaling systems that allow for the efficient inner workings that keep humans and other complex life forms functioning correctly.

Kinases are a major component of cellular communication, and studies of these enzymes illustrate the evolutionary connectivity of different species. Yeasts have 130 types of kinases. More complex organisms such as nematode worms and fruit flies have 454 and 239 kinases, respectively. Of the 130 kinase types in yeast, 97 belong to the 55 subfamilies of kinases that are found in other eukaryotic organisms. The only obvious deficiency seen in yeasts is the complete absence of tyrosine kinases. It is hypothesized that phosphorylation of tyrosine residues is needed to control the more sophisticated functions of development, differentiation, and cellular communication used in multicellular organisms.

Because yeasts contain many of the same classes of signaling proteins as humans, these organisms are ideal for studying signaling cascades. Yeasts multiply quickly and are much simpler organisms than humans or other multicellular animals. Therefore, the signaling cascades are also simpler and easier to study, although they contain similar counterparts to human signaling.[2]

LINK TO LEARNING

Watch this collection of interview clips with biofilm researchers in "What Are Bacterial Biofilms?"

Click to view content (https://openstax.org/l/bacteria_biofilm)

2G. Manning, G.D. Plowman, T. Hunter, S. Sudarsanam, "Evolution of Protein Kinase Signaling from Yeast to Man," *Trends in Biochemical Sciences* 27, no. 10 (2002). 514–520.

KEY TERMS

apoptosis programmed cell death

autocrine signal signal that is sent and received by the same or similar nearby cells

autoinducer signaling molecule secreted by bacteria to communicate with other bacteria of its kind

cell-surface receptor cell-surface protein that transmits a signal from the exterior of the cell to the interior, even though the ligand does not enter the cell

chemical synapse small space between axon terminals and dendrites of nerve cells where neurotransmitters function

cyclic AMP (cAMP) second messenger that is derived from ATP

cyclic AMP-dependent kinase (also, protein kinase A, or PKA) kinase that is activated by binding to cAMP

diacylglycerol (DAG) cleavage product of PIP_2 that is used for signaling within the plasma membrane

dimer chemical compound formed when two molecules join together

dimerization (of receptor proteins) interaction of two receptor proteins to form a functional complex called a dimer

endocrine cell cell that releases ligands involved in endocrine signaling (hormones)

endocrine signal long-distance signal that is delivered by ligands (hormones) traveling through an organism's circulatory system from the signaling cell to the target cell

enzyme-linked receptor cell-surface receptor with intracellular domains that are associated with membrane-bound enzymes

extracellular domain region of a cell-surface receptor that is located on the cell surface

G-protein-linked receptor cell-surface receptor that activates membrane-bound G-proteins to transmit a signal from the receptor to nearby membrane components

growth factor ligand that binds to cell-surface receptors and stimulates cell growth

inhibitor molecule that binds to a protein (usually an enzyme) and keeps it from functioning

inositol phospholipid lipid present at small concentrations in the plasma membrane that is converted into a second messenger; it has inositol (a carbohydrate) as its hydrophilic head group

inositol triphosphate (IP_3) cleavage product of PIP_2 that is used for signaling within the cell

intercellular signaling communication between a cell

internal receptor (also, intracellular receptor) receptor protein that is located in the cytosol of a cell and binds to ligands that pass through the plasma membrane

intracellular mediator (also, second messenger) small molecule that transmits signals within a cell

intracellular signaling communication within cells

ion channel-linked receptor cell-surface receptor that forms a plasma membrane channel, which opens when a ligand binds to the extracellular domain (ligand-gated channels)

kinase enzyme that catalyzes the transfer of a phosphate group from ATP to another molecule

ligand molecule produced by a signaling cell that binds with a specific receptor, delivering a signal in the process

mating factor signaling molecule secreted by yeast cells to communicate to nearby yeast cells that they are available to mate

neurotransmitter chemical ligand that carries a signal from one nerve cell to the next

paracrine signal signal between nearby cells that is delivered by ligands traveling in the liquid medium in the space between the cells

phosphatase enzyme that removes the phosphate group from a molecule that has been previously phosphorylated

phosphodiesterase enzyme that degrades cAMP, producing AMP, to terminate signaling

quorum sensing method of cellular communication used by bacteria that informs them of the abundance of similar (or different) bacteria in the environment

receptor protein in or on a target cell that bind to ligands

second messenger small, non-protein molecule that propagates a signal within the cell after activation of a receptor causes its release

signal integration interaction of signals from two or more different cell-surface receptors that merge to activate the same response in the cell

signal transduction propagation of the signal through the cytoplasm (and sometimes also the nucleus) of the cell

signaling cell cell that releases signal molecules that allow communication with another cell

signaling pathway (also signaling cascade) chain of events that occurs in the cytoplasm of the cell to propagate the signal from the plasma membrane to produce a response

synaptic signal chemical signal (neurotransmitter) that travels between nerve cells

target cell cell that has a receptor for a signal or ligand from a signaling cell

CHAPTER SUMMARY

9.1 Signaling Molecules and Cellular Receptors

Cells communicate by both inter- and intracellular signaling. Signaling cells secrete ligands that bind to target cells and initiate a chain of events within the target cell. The four categories of signaling in multicellular organisms are paracrine signaling, endocrine signaling, autocrine signaling, and direct signaling across gap junctions. Paracrine signaling takes place over short distances. Endocrine signals are carried long distances through the bloodstream by hormones, and autocrine signals are received by the same cell that sent the signal or other nearby cells of the same kind. Gap junctions allow small molecules, including signaling molecules, to flow between neighboring cells.

Internal receptors are found in the cell cytoplasm. Here, they bind ligand molecules that cross the plasma membrane; these receptor-ligand complexes move to the nucleus and interact directly with cellular DNA. Cell surface receptors transmit a signal from outside the cell to the cytoplasm. Ion channel-linked receptors, when bound to their ligands, form a pore through the plasma membrane through which certain ions can pass. G-protein-linked receptors interact with a G-protein on the cytoplasmic side of the plasma membrane, promoting the exchange of bound GDP for GTP and interacting with other enzymes or ion channels to transmit a signal. Enzyme-linked receptors transmit a signal from outside the cell to an intracellular domain of a membrane-bound enzyme. Ligand binding causes activation of the enzyme. Small hydrophobic ligands (like steroids) are able to penetrate the plasma membrane and bind to internal receptors. Water-soluble hydrophilic ligands are unable to pass through the membrane; instead, they bind to cell-surface receptors, which transmit the signal to the inside of the cell.

9.2 Propagation of the Signal

Ligand binding to the receptor allows for signal transduction through the cell. The chain of events that conveys the signal through the cell is called a signaling pathway or cascade. Signaling pathways are often very complex because of the interplay between different proteins. A major component of cell signaling cascades is the phosphorylation of molecules by enzymes known as kinases. Phosphorylation adds a phosphate group to serine, threonine, and tyrosine residues in a protein, changing their shapes, and activating or inactivating the protein. Small molecules like nucleotides can also be phosphorylated. Second messengers are small, non-protein molecules that are used to transmit a signal within a cell. Some examples of second messengers are calcium ions (Ca^{2+}), cyclic AMP (cAMP), diacylglycerol (DAG), and inositol triphosphate (IP_3).

9.3 Response to the Signal

The initiation of a signaling pathway is a response to external stimuli. This response can take many different forms, including protein synthesis, a change in the cell's metabolism, cell growth, or even cell death. Many pathways influence the cell by initiating gene expression, and the methods utilized are quite numerous. Some pathways activate enzymes that interact with DNA transcription factors. Others modify proteins and induce them to change their location in the cell. Depending on the status of the organism, cells can respond by storing energy as glycogen or fat, or making it available in the form of glucose. A signal transduction pathway allows muscle cells to respond to immediate requirements for energy in the form of glucose. Cell growth is almost always stimulated by external signals called growth factors. Uncontrolled cell growth leads to cancer, and mutations in the genes encoding protein components of signaling pathways are often found in tumor cells. Programmed cell death, or apoptosis, is important for removing damaged or unnecessary cells. The use of cellular signaling to organize the dismantling of a cell ensures that harmful molecules from the cytoplasm are not released into the spaces between cells, as they are in uncontrolled death, necrosis. Apoptosis also ensures the efficient recycling of the components of the dead cell. Termination of the cellular signaling cascade is very important so that the response to a signal is appropriate in both timing and intensity. Degradation of signaling molecules and dephosphorylation of phosphorylated intermediates of the pathway by phosphatases are two ways to terminate signals within the cell.

9.4 Signaling in Single-Celled Organisms

Yeasts and multicellular organisms have similar signaling mechanisms. Yeasts use cell-surface receptors and signaling cascades to communicate information on mating with other yeast cells. The signaling molecule secreted by yeasts is called mating factor.

Bacterial signaling is called quorum sensing. Bacteria secrete signaling molecules called autoinducers that are either small, hydrophobic molecules or peptide-based signals. The hydrophobic autoinducers, such as AHL, bind transcription factors and directly affect gene expression. The peptide-based molecules bind kinases and initiate signaling cascades in the cells.

VISUAL CONNECTION QUESTIONS

1. Figure 9.8 HER2 is a receptor tyrosine kinase. In 30 percent of human breast cancers, HER2 is permanently activated, resulting in unregulated cell division. Lapatinib, a drug used to treat breast cancer, inhibits HER2 receptor tyrosine kinase autophosphorylation (the process by which the receptor adds phosphates onto itself), thus reducing tumor growth by 50 percent. Besides autophosphorylation, which of the following steps would be inhibited by Lapatinib?
   a. Signaling molecule binding, dimerization, and the downstream cellular response.
   b. Dimerization, and the downstream cellular response.
   c. The downstream cellular response.
   d. Phosphatase activity, dimerization, and the downsteam cellular response.

2. Figure 9.10 In certain cancers, the GTPase activity of the RAS G-protein is inhibited. This means that the RAS protein can no longer hydrolyze GTP into GDP. What effect would this have on downstream cellular events?

3. Figure 9.17 Which of the following statements about quorum sensing is false?
   a. Autoinducer must bind to receptor to turn on transcription of genes responsible for the production of more autoinducer.
   b. The receptor stays in the bacterial cell, but the autoinducer diffuses out.
   c. Autoinducer can only act on a different cell: it cannot act on the cell in which it is made.
   d. Autoinducer turns on genes that enable the bacteria to form a biofilm.

4. Figure 9.18 What advantage might biofilm production confer on the *S. aureus* inside the catheter?

REVIEW QUESTIONS

5. What property prevents the ligands of cell-surface receptors from entering the cell?
   a. The molecules bind to the extracellular domain.
   b. The molecules are hydrophilic and cannot penetrate the hydrophobic interior of the plasma membrane.
   c. The molecules are attached to transport proteins that deliver them through the bloodstream to target cells.
   d. The ligands are able to penetrate the membrane and directly influence gene expression upon receptor binding.

6. The secretion of hormones by the pituitary gland is an example of ______________.
   a. autocrine signaling
   b. paracrine signaling
   c. endocrine signaling
   d. direct signaling across gap junctions

7. Why are ion channels necessary to transport ions into or out of a cell?
   a. Ions are too large to diffuse through the membrane.
   b. Ions are charged particles and cannot diffuse through the hydrophobic interior of the membrane.
   c. Ions do not need ion channels to move through the membrane.
   d. Ions bind to carrier proteins in the bloodstream, which must be removed before transport into the cell.

8. Endocrine signals are transmitted more slowly than paracrine signals because __________.
   a. the ligands are transported through the bloodstream and travel greater distances
   b. the target and signaling cells are close together
   c. the ligands are degraded rapidly
   d. the ligands don't bind to carrier proteins during transport

9. A scientist notices that when she adds a small, water-soluble molecule to a dish of cells, the cells turn off transcription of a gene. She hypothesizes that the ligand she added binds to a(n) ______ receptor.
   a. Intracellular
   b. Hormone
   c. Enzyme-linked
   d. Gated ion channel-linked

10. Where do DAG and IP_3 originate?
   a. They are formed by phosphorylation of cAMP.
   b. They are ligands expressed by signaling cells.
   c. They are hormones that diffuse through the plasma membrane to stimulate protein production.
   d. They are the cleavage products of the inositol phospholipid, PIP_2.

11. What property enables the residues of the amino acids serine, threonine, and tyrosine to be phosphorylated?
    a. They are polar.
    b. They are non-polar.
    c. They contain a hydroxyl group.
    d. They occur more frequently in the amino acid sequence of signaling proteins.

12. Histamine binds to the H1 G-protein-linked receptor to initiate the itchiness and airway constriction associated with an allergic response. If a mutation in the associated G-protein's alpha subunit prevented the hydrolysis of GTP how would the allergic response change?
    a. More severe allergic response compared to normal G-protein signaling.
    b. Less severe allergic response compared to normal G-protein signaling.
    c. No allergic response.
    d. No change compared to normal G-protein signaling.

13. A scientist observes a mutation in the transmembrane region of EGFR that eliminates its ability to be stabilized by binding interactions during dimerization after ligand binding. Which hypothesis regarding the effect of this mutation on EGF signaling is most likely to be correct?
    a. EGF signaling cascades would be active for longer in the cell.
    b. EGF signaling cascades would be active for a shorter period of time in the cell.
    c. EGF signaling cascades would not occur.
    d. EGF signaling would be unaffected.

14. What is the function of a phosphatase?
    a. A phosphatase removes phosphorylated amino acids from proteins.
    b. A phosphatase removes the phosphate group from phosphorylated amino acid residues in a protein.
    c. A phosphatase phosphorylates serine, threonine, and tyrosine residues.
    d. A phosphatase degrades second messengers in the cell.

15. How does NF-κB induce gene expression?
    a. A small, hydrophobic ligand binds to NF-κB, activating it.
    b. Phosphorylation of the inhibitor Iκ-B dissociates the complex between it and NF-κB, and allows NF-κB to enter the nucleus and stimulate transcription.
    c. NF-κB is phosphorylated and is then free to enter the nucleus and bind DNA.
    d. NF-κB is a kinase that phosphorylates a transcription factor that binds DNA and promotes protein production.

16. Apoptosis can occur in a cell when the cell is ________.
    a. damaged
    b. no longer needed
    c. infected by a virus
    d. all of the above

17. What is the effect of an inhibitor binding an enzyme?
    a. The enzyme is degraded.
    b. The enzyme is activated.
    c. The enzyme is inactivated.
    d. The complex is transported out of the cell.

18. How does PKC's signaling role change in response to growth factor signaling versus an immune response?
    a. PKC interacts directly with signaling molecules in both cascades, but only exhibits kinase activity during growth factor signaling.
    b. PKC interacts directly with signaling molecules in growth factor cascades, but interacts with signaling inhibitors during immune signaling.
    c. PKC amplifies growth factor cascades, but turns off immune cascades.
    d. PKC is activated during growth factor cascades, but is inactivated during immune response cascades.

19. A scientist notices that a cancer cell line fails to die when he adds an inducer of apoptosis to his culture of cells. Which hypothesis could explain why the cells fail to die?
    a. The cells have a mutation that prevents the initiation of apoptosis signaling.
    b. The cells have lost expression of the receptor for the apoptosis-inducing ligand.
    c. The cells overexpress a growth factor pathway that inhibits apoptosis.
    d. All of the above.

20. Which type of molecule acts as a signaling molecule in yeasts?
    a. steroid
    b. autoinducer
    c. mating factor
    d. second messenger
21. Quorum sensing is triggered to begin when ________.
    a. treatment with antibiotics occurs
    b. bacteria release growth hormones
    c. bacterial protein expression is switched on
    d. a sufficient number of bacteria are present
22. A doctor is researching new ways to treat biofilms on artificial joints. Which approach would best help prevent bacterial colonization of the medical implants?
    a. Increase antibiotic dosing
    b. Create implants with rougher surfaces
    c. Vaccinate patients against all pathogenic bacteria
    d. Inhibit quorum sensing

CRITICAL THINKING QUESTIONS

23. What is the difference between intracellular signaling and intercellular signaling?
24. How are the effects of paracrine signaling limited to an area near the signaling cells?
25. What are the differences between internal receptors and cell-surface receptors?
26. Cells grown in the laboratory are mixed with a dye molecule that is unable to pass through the plasma membrane. If a ligand is added to the cells, observations show that the dye enters the cells. What type of receptor did the ligand bind to on the cell surface?
27. Insulin is a hormone that regulates blood sugar by binding to its receptor, insulin receptor tyrosine kinase. How does insulin's behavior differ from steroid hormone signaling, and what can you infer about its structure?
28. The same second messengers are used in many different cells, but the response to second messengers is different in each cell. How is this possible?
29. What would happen if the intracellular domain of a cell-surface receptor was switched with the domain from another receptor?
30. If a cell developed a mutation in its *MAP2K1* gene (encodes the MEK protein) that prevented MEK from being recognized by phosphatases, how would the EGFR signaling cascade and the cell's behavior change?
31. What is a possible result of a mutation in a kinase that controls a pathway that stimulates cell growth?
32. How does the extracellular matrix control the growth of cells?
33. A scientist notices that a cancer cell line shows high levels of phosphorylated ERK in the absence of EGF. What are two possible explanations for the increase in phosphorylated ERK? Be specific in which proteins are involved.
34. What characteristics make yeasts a good model for learning about signaling in humans?
35. Why is signaling in multicellular organisms more complicated than signaling in single-celled organisms?
36. *Pseudomonas* infections are very common in hospital settings. Why would it be important for doctors to determine the bacterial load before treating an infected patient?

CHAPTER 10
Cell Reproduction

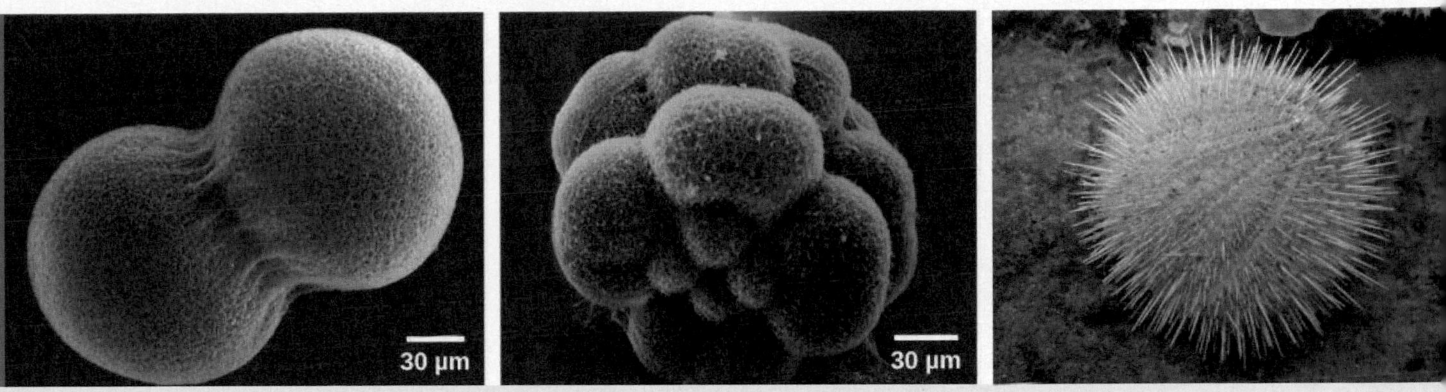

Figure 10.1 A sea urchin begins life as a single diploid cell (zygote) that (a) divides through cell division to form two genetically identical daughter cells, visible here through scanning electron microscopy (SEM). After four rounds of cell division, (b) there are 16 cells, as seen in this SEM image. After many rounds of cell division, the individual develops into a complex, multicellular organism, as seen in this (c) mature sea urchin. (credit a: modification of work by Evelyn Spiegel, Louisa Howard; credit b: modification of work by Evelyn Spiegel, Louisa Howard; credit c: modification of work by Marco Busdraghi; scale-bar data from Matt Russell)

INTRODUCTION A human, like every sexually reproducing organism, begins life as a fertilized egg (embryo) or **zygote**. In our species, billions of cell divisions subsequently must occur in a controlled manner in order to produce a complex, multicellular human comprising trillions of cells. Thus, the original single-celled zygote is literally the ancestor of all cells in the body. However, once a human is fully grown, cell reproduction is still necessary to repair and regenerate tissues, and sometimes to increase our size! In fact, all multicellular organisms use cell division for growth and the maintenance and repair of cells and tissues. Cell division is closely regulated, and the occasional failure of this regulation can have life-threatening consequences. Single-celled organisms may also use cell division as their method of reproduction.

Chapter Outline

10.1 Cell Division

By the end of this section, you will be able to do the following:

- Describe the structure of prokaryotic and eukaryotic genomes
- Distinguish between chromosomes, genes, and traits
- Describe the mechanisms of chromosome compaction

The continuity of life from one cell to another has its foundation in the reproduction of cells by way of the cell cycle. The cell cycle is an orderly sequence of events that describes the stages of a cell's life from the division of a single parent cell to the production of two new genetically identical daughter cells.

Genomic DNA

Before discussing the steps a cell must undertake to replicate and divide its DNA, a deeper understanding of the structure and function of a cell's genetic information is necessary. A cell's DNA, packaged as a double-stranded DNA molecule, is called its **genome**. In prokaryotes, the genome is composed of a single, double-stranded DNA molecule in the form of a loop or circle (Figure 10.2). The region in the cell containing this genetic material is called a nucleoid. Some prokaryotes also have smaller loops of DNA called plasmids that are not essential for normal growth. Bacteria can exchange these plasmids with other bacteria, sometimes receiving beneficial new genes that the recipient can add to their chromosomal DNA. *Antibiotic resistance* is one trait that often spreads through a bacterial colony through plasmid exchange from resistant donors to recipient cells.

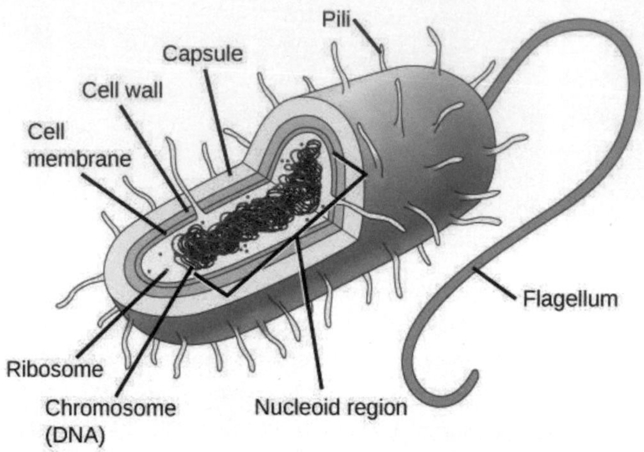

Figure 10.2 Prokaryotes, including both Bacteria and Archaea, have a single, circular chromosome located in a central region called the nucleoid.

In eukaryotes, the genome consists of several double-stranded linear DNA molecules (Figure 10.3). Each species of eukaryotes has a characteristic number of chromosomes in the nuclei of its cells. Human body (somatic) cells have 46 chromosomes, while human **gametes** (sperm or eggs) have 23 chromosomes each. A typical body cell contains two matched or homologous sets of chromosomes (one set from each biological parent)—a configuration known as **diploid**. (Note: The letter *n* is used to represent a single set of chromosomes; therefore, a diploid organism is designated 2*n*.) Human cells that contain one set of chromosomes are called gametes, or sex cells; these are eggs and sperm, and are designated 1*n*, or **haploid**.

Upon fertilization, each gamete contributes one set of chromosomes, creating a diploid cell containing matched pairs of chromosomes called **homologous** ("same knowledge") **chromosomes**. Homologous chromosomes are the same length and have specific nucleotide segments called **genes** in exactly the same location, or **locus**. Genes, the functional units of chromosomes, determine specific characteristics by coding for specific proteins. Traits are the variations of those characteristics. For example, hair color is a characteristic with traits that are blonde, brown, or black, and many colors in between.

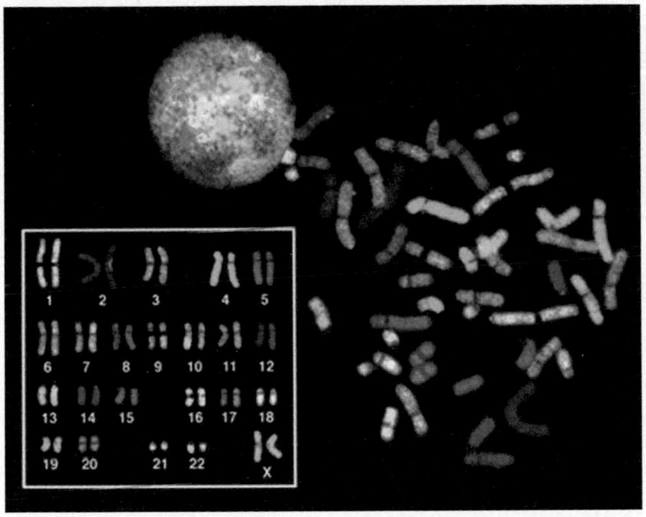

Figure 10.3 There are 23 pairs of homologous chromosomes in a female human somatic cell. The condensed chromosomes are viewed within the nucleus (top), removed from a cell during mitosis (also called karyokinesis or nuclear division) and spread out on a slide (right), and artificially arranged according to length (left); an arrangement like this is called a karyotype. In this image, the chromosomes were exposed to fluorescent stains for differentiation of the different chromosomes. A method of staining called *"chromosome painting"* employs fluorescent dyes that highlight chromosomes in different colors. (credit: National Human Genome Project/NIH)

Each copy of a homologous pair of chromosomes originates from a different parent; therefore, the different genes (alleles) themselves are not identical, although they code for the same traits such as "hair color." The variation of individuals within a species is due to the specific combination of the genes inherited from both parents. Even a slightly altered sequence of nucleotides within a gene can result in an alternative trait. For example, there are three possible gene sequences on the human chromosome that code for blood type: sequence A, sequence B, and sequence O. Because all diploid human cells have two copies of the chromosome that determines blood type, the blood type (the trait) is determined by the two alleles of the marker gene that are inherited. It is possible to have two copies of the same gene sequence on both homologous chromosomes, with one on each (for example, AA, BB, or OO), or two different sequences, such as AB, AO, or BO.

Apparently minor variations of traits, such as blood type, eye color, and handedness, contribute to the natural variation found within a species, but even though they seem minor, these traits may be connected with the expression of other traits as of yet unknown. However, if the entire DNA sequence from any pair of human homologous chromosomes is compared, the difference is much less than one percent. The sex chromosomes, X and Y, are the single exception to the rule of homologous chromosome uniformity: Other than a small amount of homology that is necessary to accurately produce gametes, the genes found on the X and Y chromosomes are different.

Eukaryotic Chromosomal Structure and Compaction

If the DNA from all 46 chromosomes in a human cell nucleus were laid out end-to-end, it would measure approximately two meters; however, its diameter would be only 2 nm! Considering that the size of a typical human cell is about 10 µm (100,000 cells lined up to equal one meter), DNA must be *tightly packaged* to fit in the cell's nucleus. At the same time, it must also be readily accessible for the genes to be expressed. For this reason, the long strands of DNA are condensed into compact chromosomes during certain stages of the cell cycle. There are a number of ways that chromosomes are compacted.

In the first level of compaction, short stretches of the DNA double helix wrap around a core of eight **histone proteins** at regular intervals along the entire length of the chromosome (Figure 10.4). The DNA-histone complex is called chromatin. The beadlike, histone DNA complex is called a **nucleosome**, and DNA connecting the nucleosomes is called linker DNA. A DNA molecule in this form is about seven times shorter than the double helix without the histones, and the beads are about 10 nm in diameter, in contrast with the 2-nm diameter of a DNA double helix.

The second level of compaction occurs as the nucleosomes and the linker DNA between them coil into a 30-nm chromatin fiber. This coiling further *condenses* the chromosome so that it is now about 50 times shorter than the extended form.

In the third level of compaction, a variety of *fibrous proteins* is used to "pack the chromatin." These fibrous proteins also ensure that each chromosome in a non-dividing cell occupies a particular area of the nucleus that does not overlap with that of any other chromosome (see the top image in Figure 10.3).

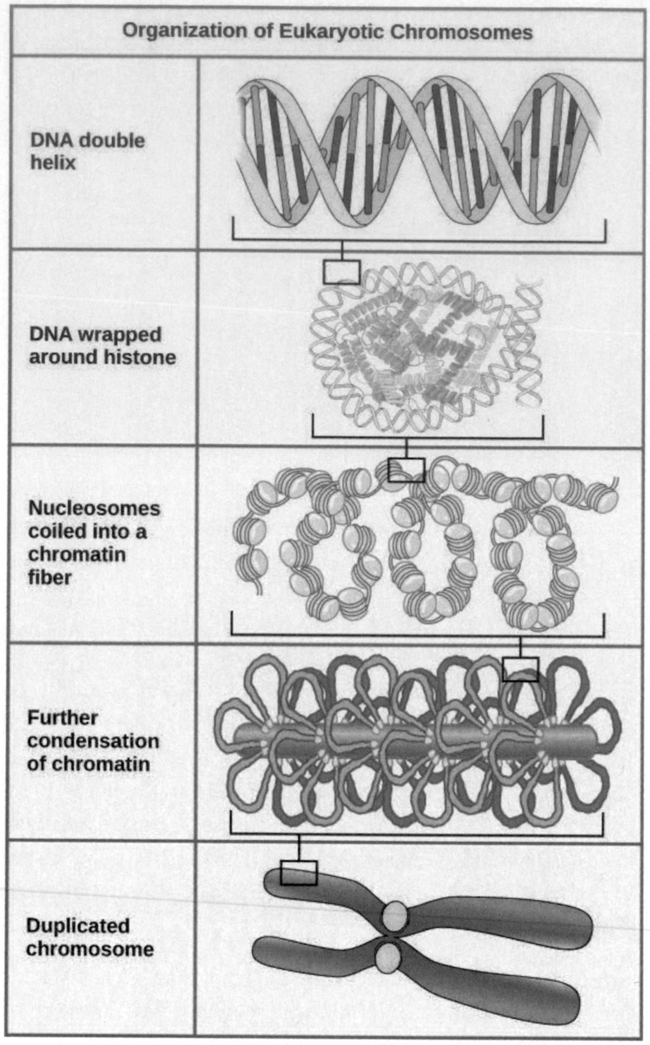

Figure 10.4 Double-stranded DNA wraps around histone proteins to form nucleosomes that create the appearance of "beads on a string." The nucleosomes are coiled into a 30-nm chromatin fiber. When a cell undergoes mitosis, the chromosomes condense even further.

DNA replicates in the S phase of interphase, which technically is not a part of mitosis, but must always precede it. After replication, the chromosomes are composed of two linked sister **chromatids**. When fully compact, the pairs of identically packed chromosomes are bound to each other by cohesin proteins. The connection between the sister chromatids is closest in a region called the **centromere**. The conjoined sister chromatids, with a diameter of about 1 µm, are visible under a light microscope. The centromeric region is highly condensed and thus will appear as a constricted area.

LINK TO LEARNING

This animation illustrates the different levels of chromosome packing.

Click to view content (https://www.openstax.org/l/Packaged_DNA)

10.2 The Cell Cycle

By the end of this section, you will be able to do the following:

- Describe the three stages of interphase
- Discuss the behavior of chromosomes during karyokinesis/mitosis
- Explain how the cytoplasmic content is divided during cytokinesis
- Define the quiescent G_0 phase

The **cell cycle** is an ordered series of events involving cell growth and cell division that produces two new daughter cells. Cells on the path to cell division proceed through a series of precisely timed and carefully regulated stages of growth, DNA replication, and nuclear and cytoplasmic division that ultimately produces two identical (clone) cells. The cell cycle has two major phases: interphase and the mitotic phase (Figure 10.5). During **interphase**, the cell grows and DNA is replicated. During the **mitotic phase**, the replicated DNA and cytoplasmic contents are separated, and the cell cytoplasm is typically partitioned by a third process of the cell cycle called **cytokinesis**. We should note, however, that interphase and mitosis (karyokinesis) may take place without cytokinesis, in which case cells with multiple nuclei (multinucleate cells) are produced.

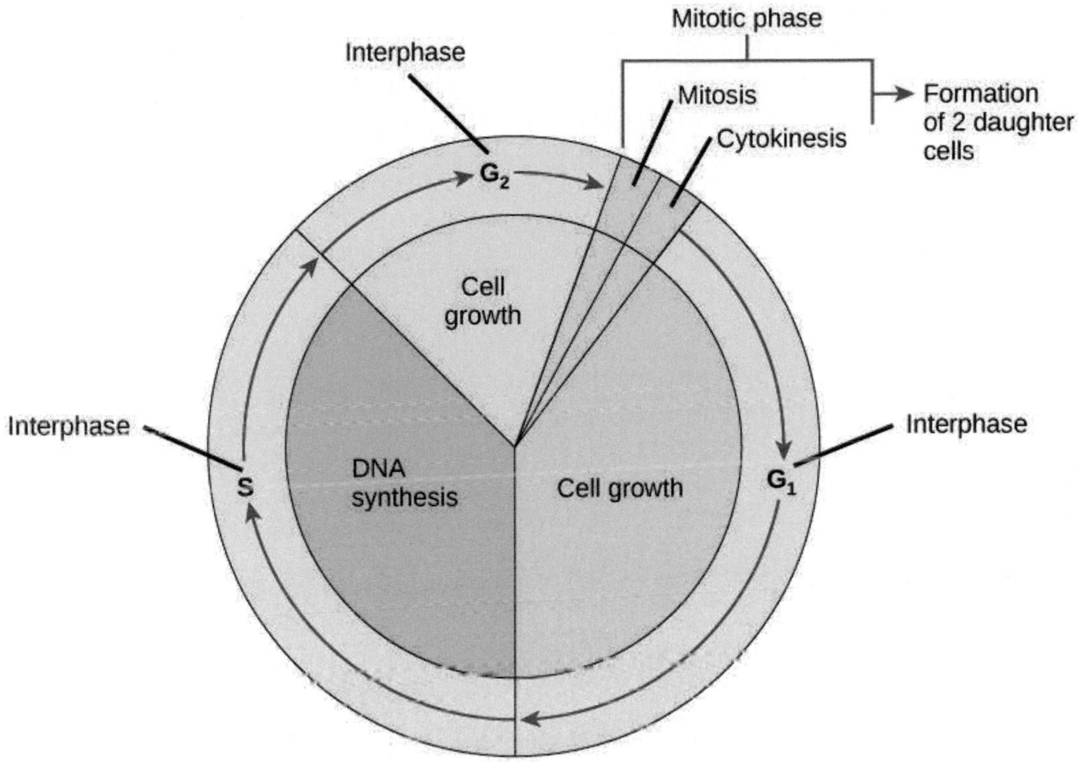

Figure 10.5 The cell cycle in multicellular organisms consists of interphase and the mitotic phase. During interphase, the cell grows and the nuclear DNA is duplicated. Interphase is followed by the mitotic phase. During the mitotic phase, the duplicated chromosomes are segregated and distributed into daughter nuclei. Following mitosis, the cytoplasm is usually divided as well by cytokinesis, resulting in two genetically identical daughter cells.

Interphase

During interphase, the cell undergoes normal growth processes while also preparing for cell division. In order for a cell to move from interphase into the mitotic phase, many internal and external conditions must be met. The three stages of interphase are called *G_1, S, and G_2*.

G_1 Phase (First Gap)

The first stage of interphase is called the **G_1 phase** (first gap) because, from a microscopic point of view, little change is visible. However, during the G_1 stage, the cell is quite active at the biochemical level. The cell is accumulating the building blocks of chromosomal DNA and the associated proteins as well as accumulating sufficient energy reserves to complete the task of replicating each chromosome in the nucleus.

S Phase (Synthesis of DNA)

Throughout interphase, nuclear DNA remains in a semi-condensed chromatin configuration. In the **S phase**, DNA replication can proceed through the mechanisms that result in the formation of identical pairs of DNA molecules—sister chromatids—that are firmly attached to the centromeric region. The centrosome is also duplicated during the S phase. The two centrosomes of homologous chromosomes will give rise to the **mitotic spindle**, the apparatus that orchestrates the movement of chromosomes during mitosis. For example, roughly at the center of each animal cell, the centrosomes are associated with a pair of rod-like objects, the **centrioles**, which are positioned at right angles to each other. Centrioles help organize cell division. We should note, however, that centrioles are not present in the centrosomes of other eukaryotic organisms, such as plants and most fungi.

G_2 Phase (Second Gap)

In the **G_2 phase**, the cell replenishes its energy stores and synthesizes proteins necessary for chromosome manipulation and movement. Some cell organelles are duplicated, and the cytoskeleton is dismantled to provide resources for the mitotic phase. There may be additional cell growth during G_2. The final preparations for the mitotic phase must be completed before the cell is able to enter the first stage of mitosis.

The Mitotic Phase

The mitotic phase is a multistep process during which the duplicated chromosomes are aligned, separated, and move into two new, identical daughter cells. The first portion of the mitotic phase is called **karyokinesis**, or nuclear division. As we have just seen, the second portion of the mitotic phase (and often viewed as a process separate from and following mitosis) is called cytokinesis—the physical separation of the cytoplasmic components into the two daughter cells.

LINK TO LEARNING

Revisit the stages of mitosis at this site (http://openstax.org/l/Cell_cycle_mito) .

Karyokinesis (Mitosis)

Karyokinesis, also known as **mitosis**, is divided into a series of phases—prophase, prometaphase, metaphase, anaphase, and telophase—that result in the division of the cell nucleus (Figure 10.6).

VISUAL CONNECTION

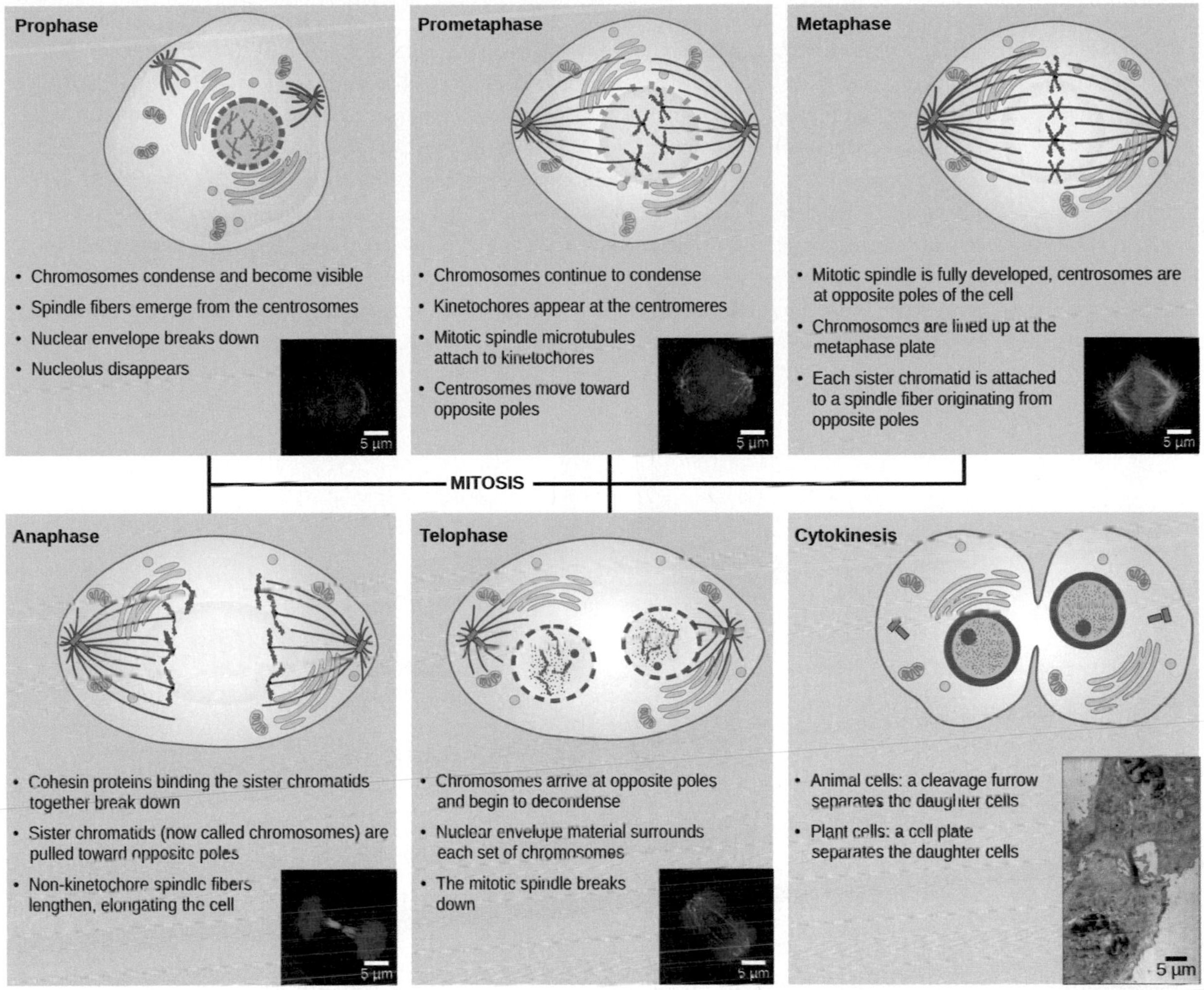

Figure 10.6 Karyokinesis (or mitosis) is divided into five stages—prophase, prometaphase, metaphase, anaphase, and telophase. The pictures at the bottom were taken by fluorescence microscopy (hence, the black background) of cells artificially stained by fluorescent dyes: blue fluorescence indicates DNA (chromosomes) and green fluorescence indicates microtubules (spindle apparatus). (credit "mitosis drawings": modification of work by Mariana Ruiz Villareal; credit "micrographs": modification of work by Roy van Heesbeen; credit "cytokinesis micrograph": Wadsworth Center/New York State Department of Health; scale-bar data from Matt Russell)

Which of the following is the correct order of events in mitosis?

a. Sister chromatids line up at the metaphase plate. The kinetochore becomes attached to the mitotic spindle. The nucleus reforms and the cell divides. Cohesin proteins break down and the sister chromatids separate.
b. The kinetochore becomes attached to the mitotic spindle. Cohesin proteins break down and the sister chromatids separate. Sister chromatids line up at the metaphase plate. The nucleus reforms and the cell divides.
c. The kinetochore becomes attached to the cohesin proteins. Sister chromatids line up at the metaphase plate. The kinetochore breaks down and the sister chromatids separate. The nucleus reforms and the cell divides.
d. The kinetochore becomes attached to the mitotic spindle. Sister chromatids line up at the metaphase plate. Cohesin proteins break down and the sister chromatids separate. The nucleus reforms and the cell divides.

Prophase (the "first phase"): the nuclear envelope starts to dissociate into small vesicles, and the membranous organelles (such as the Golgi complex [Golgi apparatus] and the endoplasmic reticulum), fragment and disperse toward the periphery of the cell. The nucleolus disappears (disperses) as well, and the centrosomes begin to move to opposite poles of the cell. Microtubules that will form the *mitotic spindle* extend between the centrosomes, *pushing them farther apart* as the microtubule fibers lengthen. The sister chromatids begin to coil more tightly with the aid of **condensin proteins** and now become visible under a light microscope.

Prometaphase (the "first change phase"): Many processes that began in prophase continue to advance. The remnants of the nuclear envelope fragment further, and the mitotic spindle continues to develop as more microtubules assemble and stretch across the length of the former nuclear area. Chromosomes become even more condensed and discrete. Each sister chromatid develops a protein structure called a **kinetochore** in its centromeric region (Figure 10.7). The proteins of the kinetochore attract and bind to the mitotic spindle microtubules. As the spindle microtubules extend from the centrosomes, some of these microtubules come into contact with and firmly bind to the kinetochores. Once a mitotic fiber attaches to a chromosome, the chromosome will be oriented until the kinetochores of sister chromatids face the *opposite poles*. Eventually, all the sister chromatids will be attached via their kinetochores to microtubules from opposing poles. Spindle microtubules that do not engage the chromosomes are called **polar microtubules**. These microtubules overlap each other midway between the two poles and contribute to *cell elongation*. Astral microtubules are located near the poles, aid in spindle orientation, and are required for the regulation of mitosis.

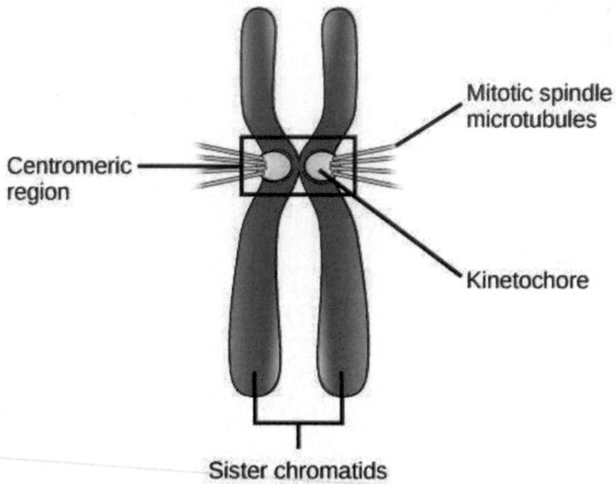

Figure 10.7 During prometaphase, mitotic spindle microtubules from opposite poles attach to each sister chromatid at the kinetochore. In anaphase, the connection between the sister chromatids breaks down, and the microtubules pull the chromosomes toward opposite poles.

Metaphase (the "change phase"): All the chromosomes are aligned in a plane called the **metaphase plate**, or the equatorial plane, roughly midway between the two poles of the cell. The sister chromatids are still tightly attached to each other by cohesin proteins. At this time, the chromosomes are maximally condensed.

Anaphase ("upward phase"): The cohesin proteins degrade, and the sister chromatids separate at the centromere. Each chromatid, now called a single chromosome, is pulled rapidly toward the centrosome to which its microtubule is attached. The cell becomes visibly elongated (oval shaped) as the polar microtubules slide against each other at the metaphase plate where they overlap.

Telophase (the "distance phase"): the chromosomes reach the opposite poles and begin to *decondense* (unravel), relaxing once again into a stretched-out chromatin configuration. The mitotic spindles are depolymerized into tubulin monomers that will be used to assemble cytoskeletal components for each daughter cell. Nuclear envelopes form around the chromosomes, and nucleosomes appear within the nuclear area.

Cytokinesis

Cytokinesis, or "cell motion," is sometimes viewed as the second main stage of the mitotic phase, during which cell division is completed via the physical separation of the cytoplasmic components into two daughter cells However, as we have seen earlier, cytokinesis can also be viewed as a separate phase, which may or may not take place following mitosis. If cytokinesis does take place, cell division is not complete until the cell components have been apportioned and completely separated into the two daughter cells. Although the stages of mitosis are similar for most eukaryotes, the process of cytokinesis is quite different for eukaryotes that have cell walls, such as plant cells.

In animal cells, cytokinesis typically starts during late anaphase. A contractile ring composed of actin filaments forms just inside the plasma membrane at the former metaphase plate. The actin filaments pull the equator of the cell inward, forming a fissure. This fissure is called the **cleavage furrow**. The furrow deepens as the actin ring contracts, and eventually the membrane is cleaved in two (Figure 10.8).

In plant cells, a new cell wall must form between the daughter cells. During interphase, the Golgi apparatus accumulates enzymes, structural proteins, and glucose molecules prior to breaking into vesicles and dispersing throughout the dividing cell. During telophase, these Golgi vesicles are transported on microtubules to form a *phragmoplast* (a vesicular structure) at the metaphase plate. There, the vesicles fuse and coalesce from the center toward the cell walls; this structure is called a **cell plate**. As more vesicles fuse, the cell plate enlarges until it merges with the cell walls at the periphery of the cell. Enzymes use the glucose that has accumulated between the membrane layers to build a new cell wall. The Golgi membranes become parts of the plasma membrane on either side of the new cell wall (Figure 10.8).

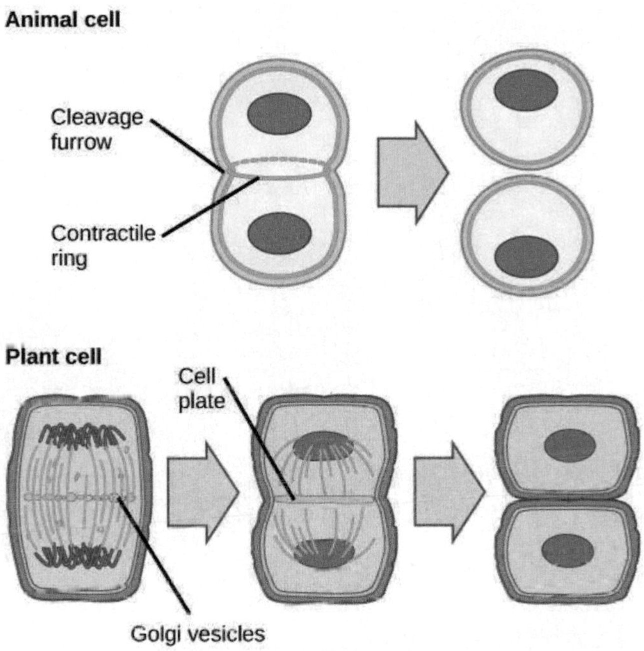

Figure 10.8 During cytokinesis in animal cells, a ring of actin filaments forms at the metaphase plate. The ring contracts, forming a cleavage furrow, which divides the cell in two. In plant cells, Golgi vesicles coalesce at the former metaphase plate, forming a phragmoplast. A cell plate formed by the fusion of the vesicles of the phragmoplast grows from the center toward the cell walls, and the membranes of the vesicles fuse to form a plasma membrane that divides the cell in two.

G_0 Phase

Not all cells adhere to the classic cell-cycle pattern in which a newly formed daughter cell immediately enters the preparatory phases of interphase, closely followed by the mitotic phase, and cytokinesis. Cells in **G_0 phase** are not actively preparing to divide. The cell is in a **quiescent** (inactive) stage that occurs when cells exit the cell cycle. Some cells enter G_0 temporarily due to environmental conditions such as availability of nutrients, or stimulation by growth factors. The cell will remain in this phase until conditions improve or until an external signal triggers the onset of G_1. Other cells that never or rarely divide, such as mature cardiac muscle and nerve cells, remain in G_0 permanently.

SCIENTIFIC METHOD CONNECTION

Determine the Time Spent in Cell-Cycle Stages

Problem: How long does a cell spend in interphase compared to each stage of mitosis?

Background: A prepared microscope slide of whitefish blastula cross-sections will show cells arrested in various stages of the cell cycle. (Note: It is not visually possible to separate the stages of interphase from each other, but the mitotic stages are readily identifiable.) If 100 cells are examined, the number of cells in each identifiable cell-cycle stage will give an estimate of the time it takes for the cell to complete that stage.

Problem Statement: Given the events included in all of interphase and those that take place in each stage of mitosis, estimate the length of each stage based on a 24-hour cell cycle. Before proceeding, state your hypothesis.

Test your hypothesis: Test your hypothesis by doing the following:

1. Place a fixed and stained microscope slide of whitefish blastula cross-sections under the scanning objective of a light microscope.
2. Locate and focus on one of the sections using the low-power objective of your microscope. Notice that the section is a circle composed of dozens of closely packed individual cells.
3. Switch to the medium-power objective and refocus. With this objective, individual cells are clearly visible, but the chromosomes will still be very small.
4. Switch to the high-power objective and slowly move the slide left to right, and up and down to view all the cells in the section (Figure 10.9). As you scan, you will notice that most of the cells are not undergoing mitosis but are in the interphase period of the cell cycle.

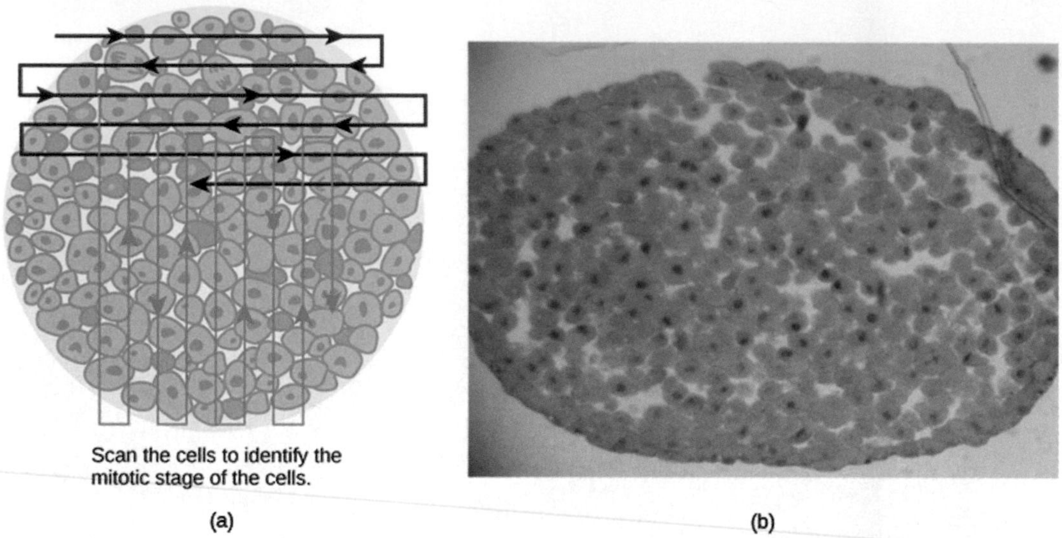

Figure 10.9 Slowly scan whitefish blastula cells with the high-power objective as illustrated in image (a) to identify their mitotic stage. (b) A microscopic image of the scanned cells is shown. (credit "micrograph": modification of work by Linda Flora; scale-bar data from Matt Russell)

5. Practice identifying the various stages of the cell cycle, using the drawings of the stages as a guide (Figure 10.6).
6. Once you are confident about your identification, begin to record the stage of each cell you encounter as you scan left to right, and top to bottom across the blastula section.
7. Keep a tally of your observations and stop when you reach 100 cells identified.
8. The larger the sample size (total number of cells counted), the more accurate the results. If possible, gather and record group data prior to calculating percentages and making estimates.

Record your observations: Make a table similar to Table 10.1 within which to record your observations.

Results of Cell Stage Identification

Phase or Stage	Individual Totals	Group Totals	Percent
Interphase			
Prophase			

Table 10.1

Phase or Stage	Individual Totals	Group Totals	Percent
Metaphase			
Anaphase			
Telophase			
Cytokinesis			
Totals	100	100	100 percent

Table 10.1

Analyze your data/report your results: To find the length of time whitefish blastula cells spend in each stage, multiply the percent (recorded as a decimal) by 24 hours. Make a table similar to Table 10.2 to illustrate your data.

Estimate of Cell Stage Length

Phase or Stage	Percent	Time in Hours
Interphase		
Prophase		
Metaphase		
Anaphase		
Telophase		
Cytokinesis		

Table 10.2

Draw a conclusion: Did your results support your estimated times? Were any of the outcomes unexpected? If so, discuss those events in that stage that may have contributed to the calculated time.

10.3 Control of the Cell Cycle

By the end of this section, you will be able to do the following:

- Understand how the cell cycle is controlled by mechanisms that are both internal and external to the cell
- Explain how the three internal “control checkpoints” occur at the end of G_1, at the G_2/M transition, and during metaphase
- Describe the molecules that control the cell cycle through positive and negative regulation

The length of the cell cycle is highly variable, even within the cells of a single organism. In humans, the frequency of cell turnover ranges from a few hours in early embryonic development, to an average of two to five days for epithelial cells, and to an entire human lifetime spent in G_0 by specialized cells, such as cortical neurons or cardiac muscle cells.

There is also variation in the time that a cell spends in each phase of the cell cycle. When rapidly dividing mammalian cells are grown in a culture (outside the body under optimal growing conditions), the length of the cell cycle is about 24 hours. In rapidly dividing human cells with a 24-hour cell cycle, the G_1 phase lasts approximately nine hours, the S phase lasts 10 hours, the G_2

phase lasts about four and one-half hours, and the M phase lasts approximately one-half hour. By comparison, in fertilized eggs (and early embryos) of fruit flies, the cell cycle is completed in about eight minutes. This is because the nucleus of the fertilized egg divides many times by mitosis but does not go through cytokinesis until a multinucleate "zygote" has been produced, with many nuclei located along the periphery of the cell membrane, thereby shortening the time of the cell division cycle. The timing of events in the cell cycle of both "invertebrates" and "vertebrates" is controlled by mechanisms that are both internal and external to the cell.

Regulation of the Cell Cycle by External Events

Both the initiation and inhibition of cell division are triggered by events external to the cell when it is about to begin the replication process. An event may be as simple as the death of nearby cells or as sweeping as the release of growth-promoting hormones, such as **human growth hormone (HGH or hGH)**. A lack of HGH can *inhibit* cell division, resulting in dwarfism, whereas too much HGH can result in gigantism. Crowding of cells can also inhibit cell division. In contrast, a factor that can initiate cell division is the size of the cell: As a cell grows, it becomes physiologically inefficient due to its decreasing surface-to-volume ratio. The solution to this problem is to divide.

Whatever the source of the message, the cell receives the signal, and a series of events within the cell allows it to proceed into interphase. Moving forward from this initiation point, every parameter required during each cell cycle phase must be met or the cycle cannot progress.

Regulation at Internal Checkpoints

It is essential that the daughter cells produced be exact duplicates of the parent cell. Mistakes in the duplication or distribution of the chromosomes lead to mutations that may be passed forward to every new cell produced from an abnormal cell. To prevent a compromised cell from continuing to divide, there are internal control mechanisms that operate at three main **cell-cycle checkpoints**: A checkpoint is one of several points in the eukaryotic cell cycle at which the progression of a cell to the next stage in the cycle can be halted until conditions are favorable. These checkpoints occur near the end of G_1, at the G_2/M transition, and during metaphase (Figure 10.10).

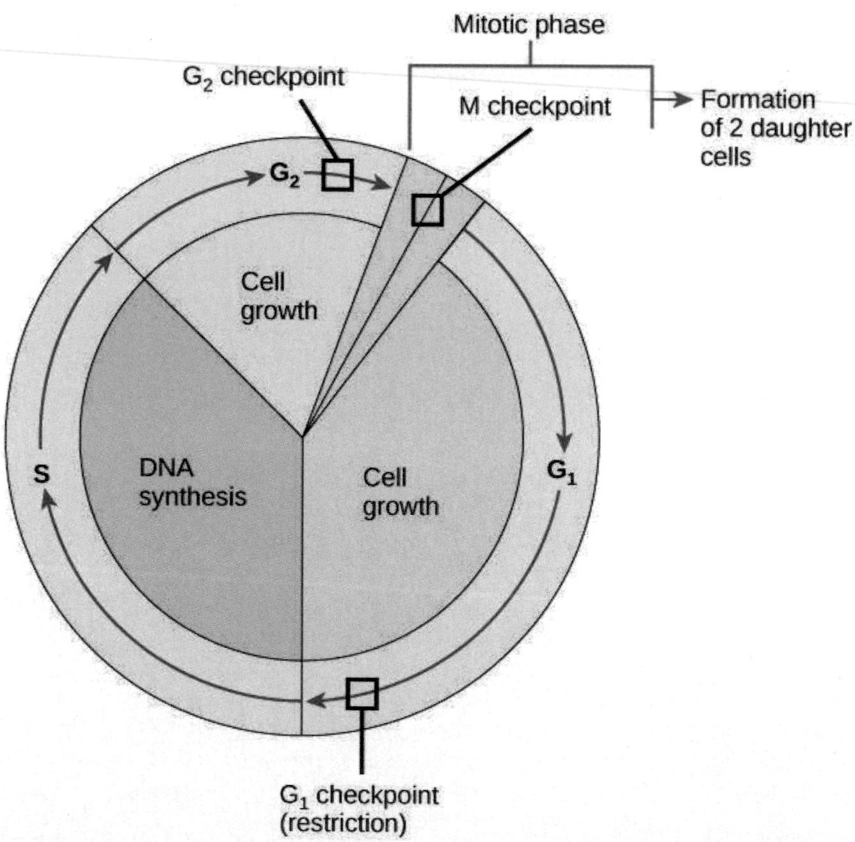

Figure 10.10 The cell cycle is controlled at three checkpoints. The integrity of the DNA is assessed at the G_1 checkpoint. Proper chromosome duplication is assessed at the G_2 checkpoint. Attachment of each kinetochore to a spindle fiber is assessed at the M

checkpoint.

The G_1 Checkpoint

The **G_1 checkpoint** determines whether all conditions are favorable for cell division to proceed. The G_1 checkpoint, also called the restriction point (in yeast), is a point at which the cell irreversibly commits to the cell division process. External influences, such as growth factors, play a large role in carrying the cell past the G_1 checkpoint. In addition to adequate reserves and cell size, there is a check for genomic DNA damage at the G_1 checkpoint. A cell that does not meet all the requirements will not be allowed to progress into the S phase. The cell can halt the cycle and attempt to remedy the problematic condition, or the cell can advance into G_0 and await further signals when conditions improve.

The G_2 Checkpoint

The G_2 checkpoint bars entry into the mitotic phase if certain conditions are not met. As at the G_1 checkpoint, cell size and protein reserves are assessed. However, the most important role of the G_2 checkpoint is to ensure that all of the chromosomes have been replicated and that the replicated DNA is not damaged. If the checkpoint mechanisms detect problems with the DNA, the cell cycle is halted, and the cell attempts to either complete DNA replication or repair the damaged DNA.

The M Checkpoint

The M checkpoint occurs near the end of the metaphase stage of karyokinesis. The M checkpoint is also known as the spindle checkpoint, because it determines whether all the sister chromatids are correctly attached to the spindle microtubules. Because the separation of the sister chromatids during anaphase is an irreversible step, the cycle will not proceed until the kinetochores of each pair of sister chromatids are firmly anchored to at least two spindle fibers arising from opposite poles of the cell.

LINK TO LEARNING

Watch what occurs at the G_1, G_2, and M checkpoints by visiting this website (http://openstax.org/l/cell_checkpnts) to see an animation of the cell cycle.

Regulator Molecules of the Cell Cycle

In addition to the internally controlled checkpoints, there are two groups of intracellular molecules that regulate the cell cycle. These regulatory molecules either promote progress of the cell to the next phase (positive regulation) or halt the cycle (negative regulation). Regulator molecules may act individually, or they can influence the activity or production of other regulatory proteins. Therefore, the failure of a single regulator may have almost no effect on the cell cycle, especially if more than one mechanism controls the same event. However, the effect of a deficient or non-functioning regulator can be wide-ranging and possibly fatal to the cell if multiple processes are affected.

Positive Regulation of the Cell Cycle

Two groups of proteins, called **cyclins** and **cyclin-dependent kinases** (Cdks), are termed positive regulators. They are responsible for the progress of the cell through the various checkpoints. The levels of the four cyclin proteins fluctuate throughout the cell cycle in a predictable pattern (Figure 10.11). Increases in the concentration of cyclin proteins are triggered by both external and internal signals. After the cell moves to the next stage of the cell cycle, the cyclins that were active in the previous stage are degraded by cytoplasmic enzymes, as shown in Figure 10.11 below.

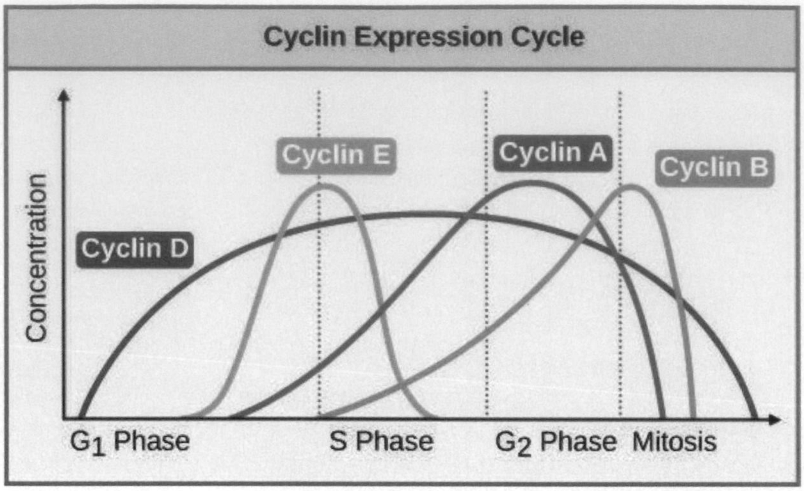

Figure 10.11 The concentrations of cyclin proteins change throughout the cell cycle. There is a direct correlation between cyclin accumulation and the three major cell-cycle checkpoints. Also note the sharp decline of cyclin levels following each checkpoint (the transition between phases of the cell cycle), as cyclin is degraded by cytoplasmic enzymes. (credit: modification of work by "WikiMiMa"/Wikimedia Commons)

Cyclins regulate the cell cycle only when they are tightly bound to Cdks. To be fully active, the Cdk/cyclin complex must also be phosphorylated in specific locations to activate the complex. Like all kinases, Cdks are enzymes (*kinases*) that in turn phosphorylate other proteins. Phosphorylation activates the protein by changing its shape. The proteins phosphorylated by Cdks are involved in advancing the cell to the next phase. (Figure 10.12). The levels of Cdk proteins are relatively stable throughout the cell cycle; however, the concentrations of cyclin fluctuate and determine when Cdk/cyclin complexes form. The different cyclins and Cdks bind at specific points in the cell cycle and thus regulate different checkpoints.

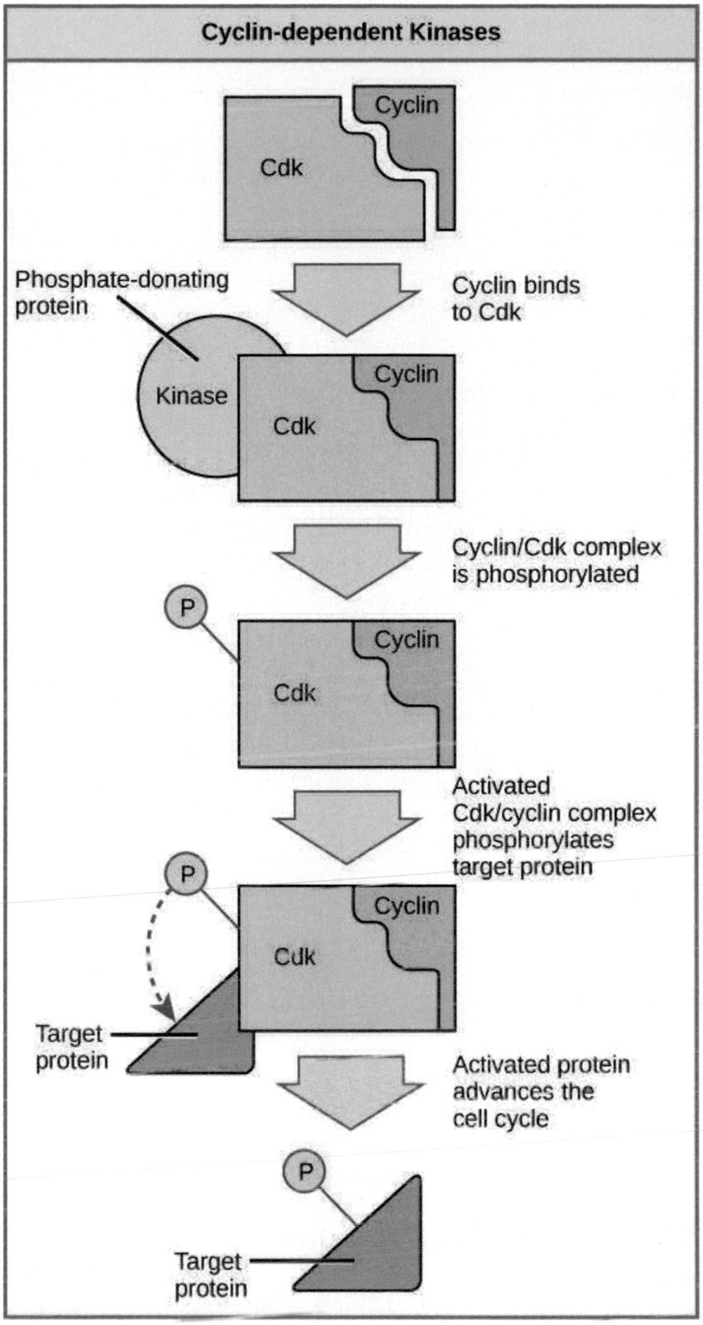

Figure 10.12 *Cyclin-dependent kinases (Cdks)* are protein kinases that, when fully activated, can phosphorylate and thus activate other proteins that advance the cell cycle past a checkpoint. To become fully activated, a Cdk must bind to a cyclin protein and then be phosphorylated by another kinase.

Because the cyclic fluctuations of cyclin levels are largely based on the *timing of the cell cycle* and not on specific events, regulation of the cell cycle usually occurs by either the Cdk molecules alone or the Cdk/cyclin complexes. Without a specific concentration of fully activated cyclin/Cdk complexes, the cell cycle cannot proceed through the checkpoints.

Although the cyclins are the main regulatory molecules that determine the forward momentum of the cell cycle, there are several other mechanisms that fine-tune the progress of the cycle with negative, rather than positive, effects. These mechanisms essentially block the progression of the cell cycle until problematic conditions are resolved. Molecules that prevent the full activation of Cdks are called Cdk inhibitors. Many of these inhibitor molecules directly or indirectly monitor a particular cell-cycle event. The block placed on Cdks by inhibitor molecules will not be removed until the specific event that the inhibitor monitors is completed.

Negative Regulation of the Cell Cycle

The second group of cell-cycle regulatory molecules are *negative regulators*, which stop the cell cycle. Remember that in positive regulation, active molecules cause the cycle to progress.

The best understood negative regulatory molecules are **retinoblastoma protein (Rb)**, **p53**, and **p21**. Retinoblastoma proteins are a group of *tumor-suppressor proteins* common in many cells. We should note here that the 53 and 21 designations refer to the functional molecular masses of the proteins (p) in kilodaltons (a dalton is equal to an *atomic mass unit*, which is equal to one proton or one neutron or 1 g/mol). Much of what is known about cell-cycle regulation comes from research conducted with cells that have *lost regulatory control*. All three of these regulatory proteins were discovered to be damaged or non-functional in cells that had begun to replicate uncontrollably (i.e., became cancerous). In each case, the main cause of the unchecked progress through the cell cycle was a faulty copy of the regulatory protein.

Rb, p53, and p21 act primarily at the G_1 checkpoint. p53 is a multi-functional protein that has a major impact on the commitment of a cell to division because it acts when there is damaged DNA in cells that are undergoing the preparatory processes during G_1. If damaged DNA is detected, p53 halts the cell cycle and then recruits specific enzymes to repair the DNA. If the DNA cannot be repaired, p53 can trigger apoptosis, or cell suicide, to prevent the duplication of damaged chromosomes. As p53 levels rise, the production of p21 is triggered. p21 enforces the halt in the cycle dictated by p53 by binding to and inhibiting the activity of the Cdk/cyclin complexes. As a cell is exposed to more stress, higher levels of p53 and p21 accumulate, making it less likely that the cell will move into the S phase.

Rb, which largely monitors cell size, exerts its regulatory influence on other positive regulator proteins. In the *active*, dephosphorylated state, Rb binds to proteins called *transcription factors*, most commonly, E2F (Figure 10.13). Transcription factors "turn on" specific genes, allowing the production of proteins encoded by that gene. When Rb is bound to E2F, production of proteins necessary for the G_1/S transition is blocked. As the cell increases in size, Rb is slowly phosphorylated until it becomes *inactivated*. Rb releases E2F, which can now turn on the gene that produces the transition protein, and this particular block is removed. For the cell to move past each of the checkpoints, all positive regulators must be "turned on," and all negative regulators must be "turned off."

VISUAL CONNECTION

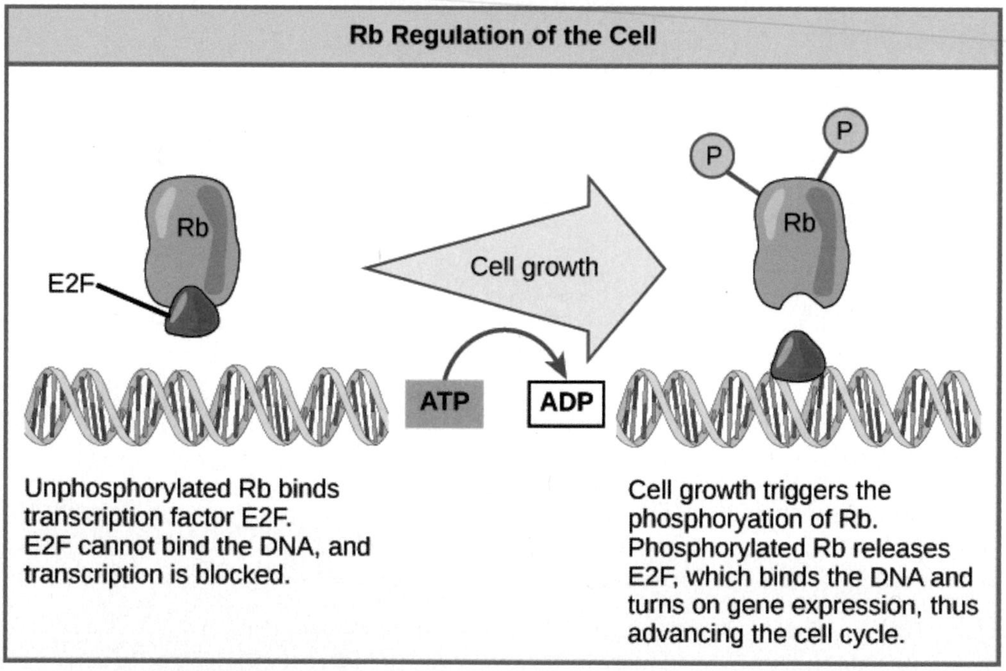

Figure 10.13 Rb halts the cell cycle and releases its hold in response to cell growth.

Rb and other proteins that negatively regulate the cell cycle are sometimes called tumor suppressors. Why do you think the name tumor suppressor might be appropriate for these proteins?

10.4 Cancer and the Cell Cycle

By the end of this section, you will be able to do the following:

- Describe how cancer is caused by uncontrolled cell growth
- Understand how proto-oncogenes are normal cell genes that, when mutated, become oncogenes
- Describe how tumor suppressors function
- Explain how mutant tumor suppressors cause cancer

Cancer comprises many different diseases caused by a common mechanism: uncontrolled cell growth. Despite the redundancy and overlapping levels of cell cycle control, errors do occur. One of the critical processes monitored by the cell-cycle checkpoint surveillance mechanism is the proper replication of DNA during the S phase. Even when all of the cell-cycle controls are fully functional, a small percentage of replication errors (mutations) will be passed on to the daughter cells. If changes to the DNA nucleotide sequence occur within a coding portion of a gene and are not corrected, a gene mutation results. All cancers start when a gene mutation gives rise to a faulty protein that plays a key role in cell reproduction.

The change in the cell that results from the malformed protein may be minor: perhaps a slight delay in the binding of Cdk to cyclin or an Rb protein that detaches from its target DNA while still phosphorylated. Even minor mistakes, however, may allow subsequent mistakes to occur more readily. Over and over, small uncorrected errors are passed from the parent cell to the daughter cells and amplified as each generation produces more non-functional proteins from uncorrected DNA damage. Eventually, the pace of the cell cycle speeds up as the effectiveness of the control and repair mechanisms decreases. Uncontrolled growth of the mutated cells outpaces the growth of normal cells in the area, and a tumor ("-oma") can result.

Proto-oncogenes

The genes that code for the positive cell-cycle regulators are called **proto-oncogenes**. Proto-oncogenes are normal genes that, when mutated in certain ways, become **oncogenes**—genes that cause a cell to become cancerous. Consider what might happen to the cell cycle in a cell with a recently acquired oncogene. In most instances, the alteration of the DNA sequence will result in a less functional (or non-functional) protein. The result is detrimental to the cell and will likely prevent the cell from completing the cell cycle; however, the organism is not harmed because the mutation will not be carried forward. If a cell cannot reproduce, the mutation is not propagated and the damage is minimal. Occasionally, however, a gene mutation causes a change that increases the activity of a positive regulator. For example, a mutation that allows Cdk to be activated without being partnered with cyclin could push the cell cycle past a checkpoint before all of the required conditions are met. If the resulting daughter cells are too damaged to undergo further cell divisions, the mutation would not be propagated and no harm would come to the organism. However, if the atypical daughter cells are able to undergo further cell divisions, subsequent generations of cells may accumulate even more mutations, some possibly in additional genes that regulate the cell cycle.

The Cdk gene in the above example is only one of many genes that are considered proto-oncogenes. In addition to the cell-cycle regulatory proteins, any protein that influences the cycle can be altered in such a way as to override cell-cycle checkpoints. An oncogene is any gene that, when altered, leads to an increase in the rate of cell-cycle progression.

Tumor Suppressor Genes

Like proto-oncogenes, many of the negative cell-cycle regulatory proteins were discovered in cells that had become cancerous. **Tumor suppressor genes** are segments of DNA that code for negative regulator proteins, the type of regulators that, when activated, can prevent the cell from undergoing uncontrolled division. The collective function of the best-understood tumor suppressor gene proteins, Rb, p53, and p21, is to put up a roadblock to cell-cycle progression until certain events are completed. A cell that carries a mutated form of a negative regulator might not be able to halt the cell cycle if there is a problem. Tumor suppressors are similar to brakes in a vehicle: Malfunctioning brakes can contribute to a car crash!

Mutated p53 genes have been identified in more than 50 percent of all human tumor cells. This discovery is not surprising in light of the multiple roles that the p53 protein plays at the G_1 checkpoint. A cell with a faulty p53 may fail to detect errors present in the genomic DNA (Figure 10.14). Even if a partially functional p53 does identify the mutations, it may no longer be able to signal the necessary DNA repair enzymes. Either way, damaged DNA will remain uncorrected. At this point, a functional p53 will deem the cell unsalvageable and trigger programmed cell death (apoptosis). The damaged version of p53 found in cancer cells, however, cannot trigger apoptosis.

VISUAL CONNECTION

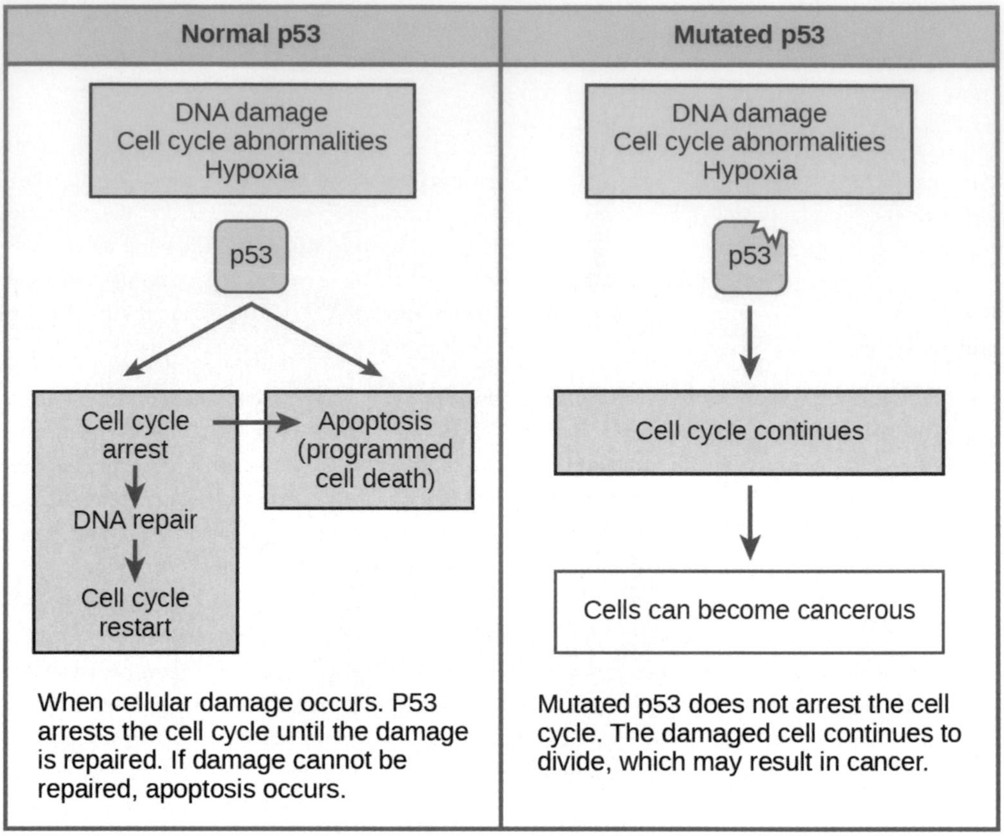

Figure 10.14 The role of normal p53 is to monitor DNA and the supply of oxygen (hypoxia is a condition of reduced oxygen supply). If damage is detected, p53 triggers repair mechanisms. If repairs are unsuccessful, p53 signals apoptosis. A cell with an abnormal p53 protein cannot repair damaged DNA and thus cannot signal apoptosis. Cells with abnormal p53 can become cancerous. (credit: modification of work by Thierry Soussi)

Human papillomavirus can cause cervical cancer. The virus encodes E6, a protein that binds p53. Based on this fact and what you know about p53, what effect do you think E6 binding has on p53 activity?

a. E6 activates p53
b. E6 inactivates p53
c. E6 mutates p53
d. E6 binding marks p53 for degradation

The loss of p53 function has other repercussions for the cell cycle. Mutated p53 might lose its ability to trigger p21 production. Without adequate levels of p21, there is no effective block on Cdk activation. Essentially, without a fully functional p53, the G_1 checkpoint is severely compromised and the cell proceeds directly from G_1 to S regardless of internal and external conditions. At the completion of this shortened cell cycle, two daughter cells are produced that have inherited the mutated p53 gene. Given the non-optimal conditions under which the parent cell reproduced, it is likely that the daughter cells will have acquired other mutations in addition to the faulty tumor-suppressor gene. Cells such as these daughter cells quickly accumulate both oncogenes and non-functional tumor-suppressor genes. Again, the result is tumor growth.

LINK TO LEARNING

Watch an animation of how cancer results from errors in the cell cycle.

Click to view content (https://www.openstax.org/l/cancer)

10.5 Prokaryotic Cell Division

By the end of this section, you will be able to do the following:

- Describe the process of binary fission in prokaryotes
- Explain how FtsZ and tubulin proteins are examples of homology

Prokaryotes, such as bacteria, produce daughter cells by binary fission. For unicellular organisms, cell division is the only method to produce new individuals. In both prokaryotic and eukaryotic cells, the outcome of cell reproduction is a pair of daughter cells that are genetically identical to the parent cell. In unicellular organisms, daughter cells are individuals.

To achieve the outcome of cloned offspring, certain steps are essential. The genomic DNA must be replicated and then allocated into the daughter cells; the cytoplasmic contents must also be divided to give both new cells the cellular machinery to sustain life. As we've seen with bacterial cells, the genome consists of a single, circular DNA chromosome; therefore, the process of cell division is simplified. Karyokinesis is unnecessary because there is no true nucleus and thus no need to direct one copy of the multiple chromosomes into each daughter cell. This type of cell division is called **binary (prokaryotic) fission**.

Binary Fission

Due to the relative simplicity of the prokaryotes, the cell division process is a less complicated and much more rapid process than cell division in eukaryotes. As a review of the general information on cell division we discussed at the beginning of this chapter, recall that the single, circular DNA chromosome of bacteria occupies a specific location, the nucleoid region, within the cell (Figure 10.2). Although the DNA of the nucleoid is associated with proteins that aid in packaging the molecule into a compact size, there are no histone proteins and thus no nucleosomes in prokaryotes. The packing proteins of bacteria are, however, related to the cohesin and condensin proteins involved in the chromosome compaction of eukaryotes.

The bacterial chromosome is attached to the plasma membrane at about the midpoint of the cell. The starting point of replication, the **origin**, is close to the binding site of the chromosome to the plasma membrane (Figure 10.15). Replication of the DNA is bidirectional, moving away from the origin on both strands of the loop simultaneously. As the new double strands are formed, each origin point moves away from the cell wall attachment toward the opposite ends of the cell. As the cell elongates, the growing membrane aids in the transport of the chromosomes. After the chromosomes have cleared the midpoint of the elongated cell, cytoplasmic separation begins. The formation of a ring composed of repeating units of a protein called **FtsZ** (short for "filamenting temperature-sensitive mutant Z") directs the partition between the nucleoids. Formation of the FtsZ ring triggers the accumulation of other proteins that work together to recruit new membrane and cell wall materials to the site. A **septum** is formed between the daughter nucleoids, extending gradually from the periphery toward the center of the cell. When the new cell walls are in place, the daughter cells separate.

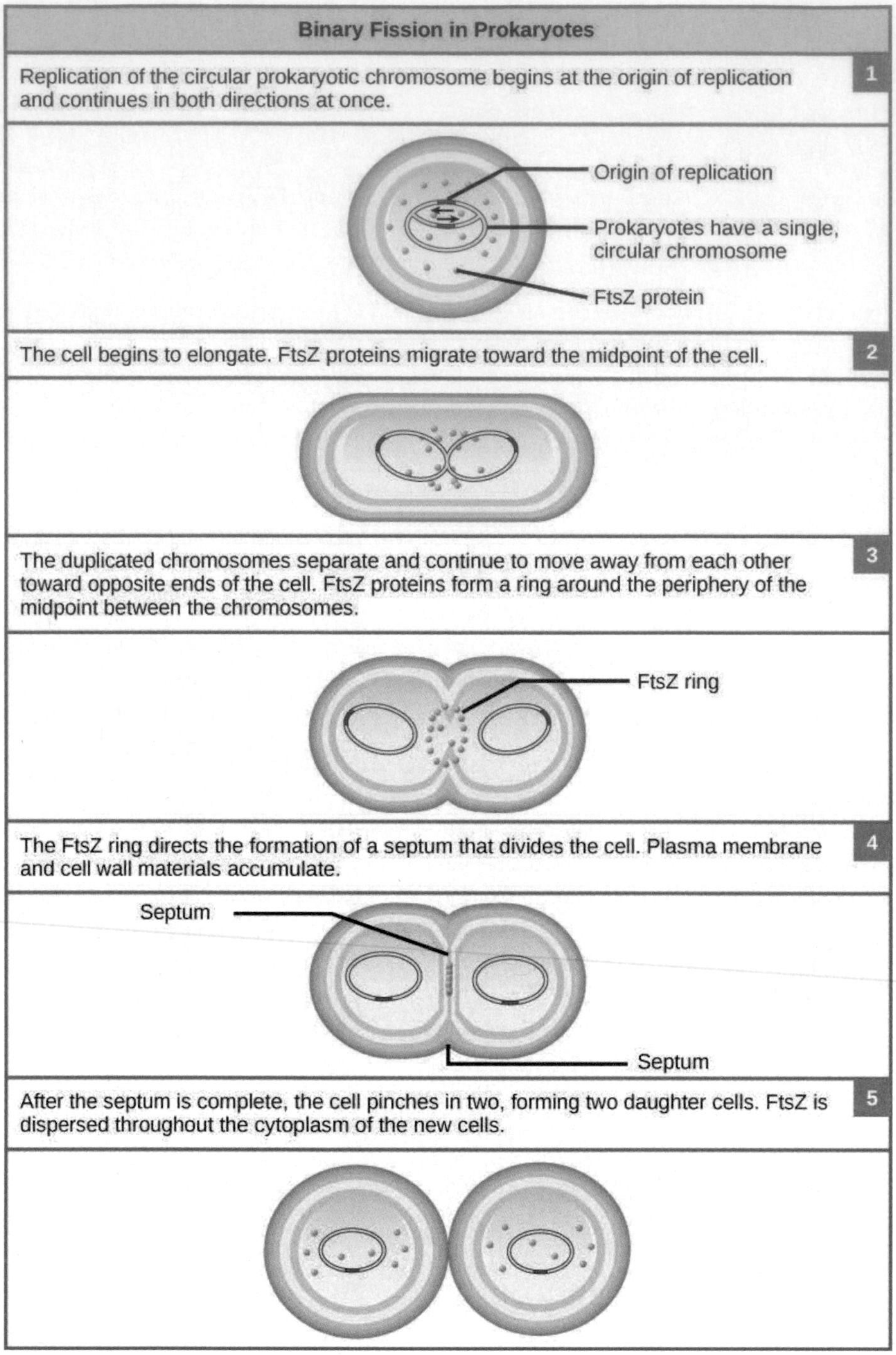

Figure 10.15 These images show the steps of binary fission in prokaryotes. (credit: modification of work by “Mcstrother”/Wikimedia Commons)

EVOLUTION CONNECTION

Mitotic Spindle Apparatus

The precise timing and formation of the mitotic spindle is critical to the success of eukaryotic cell division. Prokaryotic cells, on the other hand, do not undergo karyokinesis and therefore have no need for a mitotic spindle. However, the FtsZ protein that plays such a vital role in prokaryotic cytokinesis is structurally and functionally very similar to tubulin, the building block of the microtubules which make up the mitotic spindle fibers that are necessary for eukaryotic nuclear division. FtsZ proteins can form filaments, rings, and other three-dimensional structures that resemble the way tubulin forms microtubules, centrioles, and various cytoskeletal components. In addition, both FtsZ and tubulin employ the same energy source, GTP (guanosine triphosphate), to rapidly assemble and disassemble complex structures.

FtsZ and tubulin are considered to be homologous structures derived from common evolutionary origins. In this example, FtsZ is the ancestor protein to tubulin (an evolutionarily derived protein). While both proteins are found in extant organisms, tubulin function has evolved and diversified tremendously since evolving from its FtsZ prokaryotic origin. A survey of mitotic assembly components found in present-day unicellular eukaryotes reveals crucial intermediary steps to the complex membrane-enclosed genomes of multicellular eukaryotes (Table 10.3).

Cell Division Apparatus among Various Organisms

	Structure of genetic material	Division of nuclear material	Separation of daughter cells
Prokaryotes	There is no nucleus. The single, circular chromosome exists in a region of cytoplasm called the nucleoid.	Occurs through binary fission. As the chromosome is replicated, the two copies move to opposite ends of the cell by an unknown mechanism.	FtsZ proteins assemble into a ring that pinches the cell in two.
Some protists	Linear chromosomes exist in the nucleus.	Chromosomes attach to the nuclear envelope, which remains intact. The mitotic spindle passes through the envelope and elongates the cell. No centrioles exist.	Microfilaments form a cleavage furrow that pinches the cell in two.
Other protists	Linear chromosomes wrapped around histones exist in the nucleus.	A mitotic spindle forms from the centrioles and passes through the nuclear membrane, which remains intact. Chromosomes attach to the mitotic spindle, which separates the chromosomes and elongates the cell.	Microfilaments form a cleavage furrow that pinches the cell in two.
Animal cells	Linear chromosomes exist in the nucleus.	A mitotic spindle forms from the centrosomes. The nuclear envelope dissolves. Chromosomes attach to the mitotic spindle, which separates the chromosomes and elongates the cell.	Microfilaments form a cleavage furrow that pinches the cell in two.

Table 10.3

KEY TERMS

anaphase stage of mitosis during which sister chromatids are separated from each other

binary fission prokaryotic cell division process

cell cycle ordered series of events involving cell growth and cell division that produces two new daughter cells

cell plate structure formed during plant cell cytokinesis by Golgi vesicles, forming a temporary structure (phragmoplast) and fusing at the metaphase plate; ultimately leads to the formation of cell walls that separate the two daughter cells

cell-cycle checkpoint mechanism that monitors the preparedness of a eukaryotic cell to advance through the various cell-cycle stages

centriole rod-like structure constructed of microtubules at the center of each animal cell centrosome

centromere region at which sister chromatids are bound together; a constricted area in condensed chromosomes

chromatid single DNA molecule of two strands of duplicated DNA and associated proteins held together at the centromere

cleavage furrow constriction formed by an actin ring during cytokinesis in animal cells that leads to cytoplasmic division

condensin proteins that help sister chromatids coil during prophase

cyclin one of a group of proteins that act in conjunction with cyclin-dependent kinases to help regulate the cell cycle by phosphorylating key proteins; the concentrations of cyclins fluctuate throughout the cell cycle

cyclin-dependent kinase (Cdk) one of a group of protein kinases that helps to regulate the cell cycle when bound to cyclin; it functions to phosphorylate other proteins that are either activated or inactivated by phosphorylation

cytokinesis division of the cytoplasm following mitosis that forms two daughter cells.

diploid cell, nucleus, or organism containing two sets of chromosomes ($2n$)

FtsZ tubulin-like protein component of the prokaryotic cytoskeleton that is important in prokaryotic cytokinesis (name origin: **F**ilamenting **t**emperature-**s**ensitive mutant **Z**)

G_0 phase distinct from the G_1 phase of interphase; a cell in G_0 is not preparing to divide

G_1 phase (also, first gap) first phase of interphase centered on cell growth during mitosis

G_2 phase (also, second gap) third phase of interphase during which the cell undergoes final preparations for mitosis

gamete haploid reproductive cell or sex cell (sperm, pollen grain, or egg)

gene physical and functional unit of heredity, a sequence of DNA that codes for a protein.

genome total genetic information of a cell or organism

haploid cell, nucleus, or organism containing one set of chromosomes (n)

histone one of several similar, highly conserved, low molecular weight, basic proteins found in the chromatin of all eukaryotic cells; associates with DNA to form nucleosomes

homologous chromosomes chromosomes of the same morphology with genes in the same location; diploid organisms have pairs of homologous chromosomes (homologs), with each homolog derived from a different parent

interphase period of the cell cycle leading up to mitosis; includes G_1, S, and G_2 phases (the interim period between two consecutive cell divisions)

karyokinesis mitotic nuclear division

kinetochore protein structure associated with the centromere of each sister chromatid that attracts and binds spindle microtubules during prometaphase

locus position of a gene on a chromosome

metaphase stage of mitosis during which chromosomes are aligned at the metaphase plate

metaphase plate equatorial plane midway between the two poles of a cell where the chromosomes align during metaphase

mitosis (also, karyokinesis) period of the cell cycle during which the duplicated chromosomes are separated into identical nuclei; includes prophase, prometaphase, metaphase, anaphase, and telophase

mitotic phase period of the cell cycle during which duplicated chromosomes are distributed into two nuclei and cytoplasmic contents are divided; includes karyokinesis (mitosis) and cytokinesis

mitotic spindle apparatus composed of microtubules that orchestrates the movement of chromosomes during mitosis

nucleosome subunit of chromatin composed of a short length of DNA wrapped around a core of histone proteins

oncogene mutated version of a normal gene involved in the positive regulation of the cell cycle

origin (also, ORI) region of the prokaryotic chromosome where replication begins (origin of replication)

p21 cell-cycle regulatory protein that inhibits the cell cycle; its levels are controlled by p53

p53 cell-cycle regulatory protein that regulates cell growth and monitors DNA damage; it halts the progression of the cell cycle in cases of DNA damage and may induce apoptosis

prometaphase stage of mitosis during which the nuclear membrane breaks down and mitotic spindle fibers attach to kinetochores

prophase stage of mitosis during which chromosomes

condense and the mitotic spindle begins to form

proto-oncogene normal gene that when mutated becomes an oncogene

quiescent refers to a cell that is performing normal cell functions and has not initiated preparations for cell division

retinoblastoma protein (Rb) regulatory molecule that exhibits negative effects on the cell cycle by interacting with a transcription factor (E2F)

S phase second, or synthesis, stage of interphase during which DNA replication occurs

septum structure formed in a bacterial cell as a precursor to the separation of the cell into two daughter cells

telophase stage of mitosis during which chromosomes arrive at opposite poles, decondense, and are surrounded by a new nuclear envelope

tumor suppressor gene segment of DNA that codes for regulator proteins that prevent the cell from undergoing uncontrolled division

CHAPTER SUMMARY

10.1 Cell Division

Prokaryotes have a single circular chromosome composed of double-stranded DNA, whereas eukaryotes have multiple, linear chromosomes composed of chromatin wrapped around histones, all of which are surrounded by a nuclear membrane. The 46 chromosomes of human somatic cells are composed of 22 pairs of autosomes (matched pairs) and a pair of sex chromosomes, which may or may not be matched. This is the 2*n* or diploid state. Human gametes have 23 chromosomes, or one complete set of chromosomes; a set of chromosomes is complete with either one of the sex chromosomes, X or Y. This is the *n* or haploid state. Genes are segments of DNA that code for a specific functional molecule (a protein or RNA). An organism's traits are determined by the genes inherited from each parent. Duplicated chromosomes are composed of two sister chromatids. Chromosomes are compacted using a variety of mechanisms during certain stages of the cell cycle. Several classes of protein are involved in the organization and packing of the chromosomal DNA into a highly condensed structure. The condensing complex compacts chromosomes, and the resulting condensed structure is necessary for chromosomal segregation during mitosis.

10.2 The Cell Cycle

The cell cycle is an orderly sequence of events. Cells on the path to cell division proceed through a series of precisely timed and carefully regulated stages. In eukaryotes, the cell cycle consists of a long preparatory period, called interphase, during which chromosomes are replicated. Interphase is divided into G_1, S, and G_2 phases. The mitotic phase begins with karyokinesis (mitosis), which consists of five stages: prophase, prometaphase, metaphase, anaphase, and telophase. The final stage of the cell division process, and sometimes viewed as the final stage of the mitotic phase, is cytokinesis, during which the cytoplasmic components of the daughter cells are separated either by an actin ring (animal cells) or by cell plate formation (plant cells).

10.3 Control of the Cell Cycle

Each step of the cell cycle is monitored by internal controls called checkpoints. There are three major checkpoints in the cell cycle: one near the end of G_1, a second at the G_2/M transition, and the third during metaphase. Positive regulator molecules allow the cell cycle to advance to the next stage of cell division. Negative regulator molecules monitor cellular conditions and can halt the cycle until specific requirements are met.

10.4 Cancer and the Cell Cycle

Cancer is the result of unchecked cell division caused by a breakdown of the mechanisms that regulate the cell cycle. The loss of control begins with a change in the DNA sequence of a gene that codes for one of the regulatory molecules. Faulty instructions lead to a protein that does not function as it should. Any disruption of the monitoring system can allow other mistakes to be passed on to the daughter cells. Each successive cell division will give rise to daughter cells with even more accumulated damage. Eventually, all checkpoints become nonfunctional, and rapidly reproducing cells crowd out normal cells, resulting in a tumor or leukemia (blood cancer).

10.5 Prokaryotic Cell Division

In both prokaryotic and eukaryotic cell division, the genomic DNA is replicated and then each copy is allocated into a daughter cell. In addition, the cytoplasmic contents are divided evenly and distributed to the new cells. However, there are many differences between prokaryotic and eukaryotic cell division. Bacteria have a single, circular DNA chromosome but no nucleus. Therefore, mitosis (karyokinesis) is not necessary in bacterial cell division. Bacterial cytokinesis is directed by a ring composed of a protein called FtsZ. Ingrowth of membrane and cell wall material from the periphery of the cells results in the formation of a septum that eventually constructs the separate cell walls of the daughter cells.

VISUAL CONNECTION QUESTIONS

1. Figure 10.13 Rb and other proteins that negatively regulate the cell cycle are sometimes called tumor suppressors. Why do you think the name tumor suppressor might be appropriate for these proteins?

2. Figure 10.14 Human papillomavirus can cause cervical cancer. The virus encodes E6, a protein that binds p53. Based on this fact and what you know about p53, what effect do you think E6 binding has on p53 activity?
   a. E6 activates p53
   b. E6 inactivates p53
   c. E6 mutates p53
   d. E6 binding marks p53 for degradation

REVIEW QUESTIONS

3. A diploid cell has_______ the number of chromosomes as a haploid cell.
   a. one-fourth
   b. half
   c. twice
   d. four times

4. An organism's traits are determined by the specific combination of inherited _____.
   a. cells.
   b. genes.
   c. proteins.
   d. chromatids.

5. The first level of DNA organization in a eukaryotic cell is maintained by which molecule?
   a. cohesin
   b. condensin
   c. chromatin
   d. histone

6. Identical copies of chromatin held together by cohesin at the centromere are called _____.
   a. histones.
   b. nucleosomes.
   c. chromatin.
   d. sister chromatids.

7. Chromosomes are duplicated during what stage of the cell cycle?
   a. G_1 phase
   b. S phase
   c. prophase
   d. prometaphase

8. Which of the following events does not occur during some stages of interphase?
   a. DNA duplication
   b. organelle duplication
   c. increase in cell size
   d. separation of sister chromatids

9. The mitotic spindles arise from which cell structure?
   a. centromere
   b. centrosome
   c. kinetochore
   d. cleavage furrow

10. Attachment of the mitotic spindle fibers to the kinetochores is a characteristic of which stage of mitosis?
    a. prophase
    b. prometaphase
    c. metaphase
    d. anaphase

11. Unpacking of chromosomes and the formation of a new nuclear envelope is a characteristic of which stage of mitosis?
    a. prometaphase
    b. metaphase
    c. anaphase
    d. telophase

12. Separation of the sister chromatids is a characteristic of which stage of mitosis?
    a. prometaphase
    b. metaphase
    c. anaphase
    d. telophase

13. The chromosomes become visible under a light microscope during which stage of mitosis?
    a. prophase
    b. prometaphase
    c. metaphase
    d. anaphase

14. The fusing of Golgi vesicles at the metaphase plate of dividing plant cells forms what structure?
    a. cell plate
    b. actin ring
    c. cleavage furrow
    d. mitotic spindle

15. Figure 10.6 Which of the following is the correct order of events in mitosis?
    a. Sister chromatids line up at the metaphase plate. The kinetochore becomes attached to the mitotic spindle. The nucleus reforms and the cell divides. Cohesin proteins break down and the sister chromatids separate.
    b. The kinetochore becomes attached to the mitotic spindle. Cohesin proteins break down and the sister chromatids separate. Sister chromatids line up at the metaphase plate. The nucleus reforms and the cell divides.
    c. The kinetochore becomes attached to the cohesin proteins. Sister chromatids line up at the metaphase plate. The kinetochore breaks down and the sister chromatids separate. The nucleus reforms and the cell divides.
    d. The kinetochore becomes attached to the mitotic spindle. Sister chromatids line up at the metaphase plate. Cohesin proteins break down and the sister chromatids separate. The nucleus reforms and the cell divides.

16. At which of the cell-cycle checkpoints do external forces have the greatest influence?
    a. G_1 checkpoint
    b. G_2 checkpoint
    c. M checkpoint
    d. G_0 checkpoint

17. What is the main prerequisite for clearance at the G_2 checkpoint?
    a. cell has reached a sufficient size
    b. an adequate stockpile of nucleotides
    c. accurate and complete DNA replication
    d. proper attachment of mitotic spindle fibers to kinetochores

18. If the M checkpoint is not cleared, what stage of mitosis will be blocked?
    a. prophase
    b. prometaphase
    c. metaphase
    d. anaphase

19. Which protein is a positive regulator that phosphorylates other proteins when activated?
    a. p53
    b. retinoblastoma protein (Rb)
    c. cyclin
    d. cyclin-dependent kinase (Cdk)

20. Many of the negative regulator proteins of the cell cycle were discovered in what type of cells?
    a. gametes
    b. cells in G_0
    c. cancer cells
    d. stem cells

21. Which negative regulatory molecule can trigger cell suicide (apoptosis) if vital cell cycle events do not occur?
    a. p53
    b. p21
    c. retinoblastoma protein (Rb)
    d. cyclin-dependent kinase (Cdk)

22. __________ are changes to the order of nucleotides in a segment of DNA that codes for a protein.
    a. Proto-oncogenes
    b. Tumor suppressor genes
    c. Gene mutations
    d. Negative regulators

23. A gene that codes for a positive cell-cycle regulator is called a(n) _____.
    a. kinase inhibitor.
    b. tumor suppressor gene.
    c. proto-oncogene.
    d. oncogene.

24. A mutated gene that codes for an altered version of Cdk that is active in the absence of cyclin is a(n) _____.
    a. kinase inhibitor.
    b. tumor suppressor gene.
    c. proto-oncogene.
    d. oncogene.

25. Which molecule is a Cdk inhibitor that is controlled by p53?
    a. cyclin
    b. anti-kinase
    c. Rb
    d. p21

26. Which eukaryotic cell-cycle event is missing in binary fission?
    a. cell growth
    b. DNA duplication
    c. karyokinesis
    d. cytokinesis

27. FtsZ proteins direct the formation of a _______ that will eventually form the new cell walls of the daughter cells.
    a. contractile ring
    b. cell plate
    c. cytoskeleton
    d. septum

CRITICAL THINKING QUESTIONS

28. Compare and contrast a human somatic cell to a human gamete.

29. What is the relationship between a genome, chromosomes, and genes?

30. Eukaryotic chromosomes are thousands of times longer than a typical cell. Explain how chromosomes can fit inside a eukaryotic nucleus.

31. Briefly describe the events that occur in each phase of interphase.

32. Chemotherapy drugs such as *vincristine* (derived from Madagascar periwinkle plants) and *colchicine* (derived from autumn crocus plants) disrupt mitosis by binding to tubulin (the subunit of microtubules) and interfering with microtubule assembly and disassembly. Exactly what mitotic structure is targeted by these drugs and what effect would that have on cell division?

33. Describe the similarities and differences between the cytokinesis mechanisms found in animal cells versus those in plant cells.

34. List some reasons why a cell that has just completed cytokinesis might enter the G_0 phase instead of the G_1 phase.

35. What cell-cycle events will be affected in a cell that produces mutated (non-functional) cohesin protein?

36. Describe the general conditions that must be met at each of the three main cell-cycle checkpoints.

37. Compare and contrast the roles of the positive cell-cycle regulators negative regulators.

38. What steps are necessary for Cdk to become fully active?

39. Rb is a negative regulator that blocks the cell cycle at the G_1 checkpoint until the cell achieves a requisite size. What molecular mechanism does Rb employ to halt the cell cycle?

40. Outline the steps that lead to a cell becoming cancerous.

41. Explain the difference between a proto-oncogene and a tumor-suppressor gene.

42. List the regulatory mechanisms that might be lost in a cell producing faulty p53.

43. p53 can trigger apoptosis if certain cell-cycle events fail. How does this regulatory outcome benefit a multicellular organism?

44. Name the common components of eukaryotic cell division and binary fission.

45. Describe how the duplicated bacterial chromosomes are distributed into new daughter cells without the direction of the mitotic spindle.

CHAPTER 11
Meiosis and Sexual Reproduction

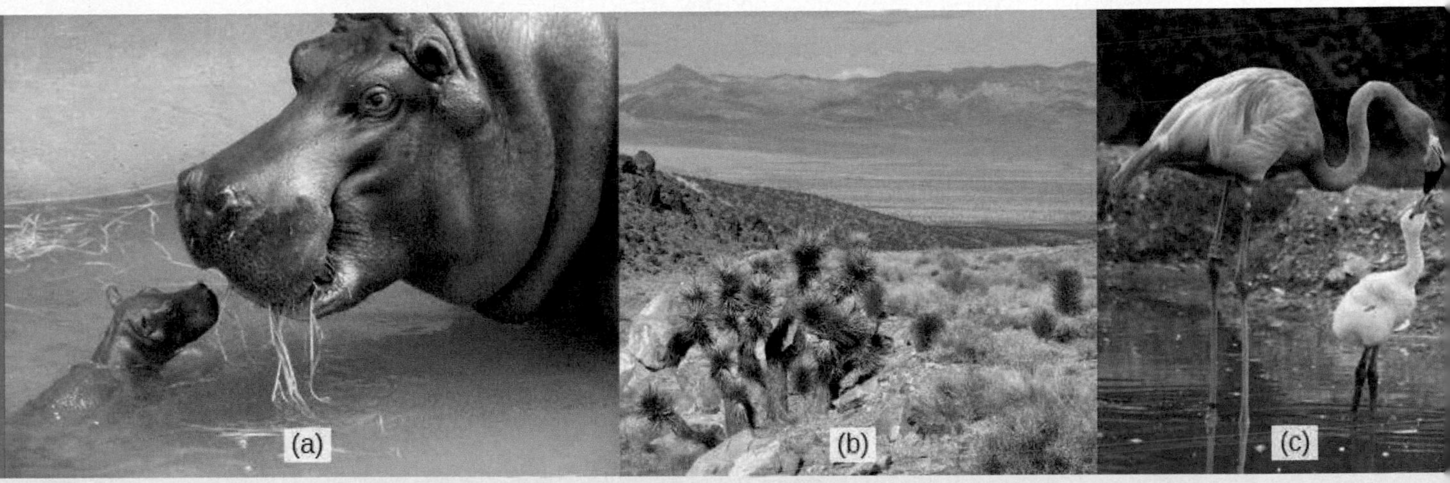

Figure 11.1 Each of us, like the organisms shown above, begins life as a fertilized egg (zygote). After trillions of cell divisions, each of us develops into a complex, multicellular organism. (credit a: modification of work by Frank Wouters; credit b: modification of work by Ken Cole, USGS; credit c: modification of work by Martin Pettitt)

INTRODUCTION The ability to reproduce is a basic characteristic of all organisms: Hippopotamuses give birth to hippopotamus calves; Joshua trees produce seeds from which Joshua tree seedlings emerge; and adult flamingos lay eggs that hatch into flamingo chicks. However, unlike the organisms shown above, offspring may or may not resemble their parents. For example, in the case of most insects such as butterflies (with a complete metamorphosis), the larval forms rarely resemble the adult forms.

Although many unicellular organisms and a few multicellular organisms can produce genetically identical clones of themselves through ***asexual reproduction***, many single-celled organisms and most multicellular organisms reproduce regularly using another method—***sexual reproduction***. This highly evolved method involves the production by parents of two haploid cells and the fusion of two haploid cells to form a single, genetically recombined diploid cell—a genetically unique organism. Haploid cells that are part of the sexual reproductive cycle are produced by a type of cell division called **meiosis**. Sexual reproduction, involving both meiosis and fertilization, introduces variation into offspring that may account for the evolutionary success of sexual reproduction. The vast majority of eukaryotic organisms, both multicellular and unicellular, can or must employ some form of meiosis and fertilization to reproduce.

In most plants and animals, through thousands of rounds of mitotic cell division, diploid cells (whether produced by asexual or sexual reproduction) will develop into an adult organism.

Chapter Outline

11.1 The Process of Meiosis

By the end of this section, you will be able to do the following:

- Describe the behavior of chromosomes during meiosis, and the differences between the first and second meiotic divisions
- Describe the cellular events that take place during meiosis
- Explain the differences between meiosis and mitosis
- Explain the mechanisms within the meiotic process that produce genetic variation among the haploid gametes

Sexual reproduction requires the union of two specialized cells, called **gametes**, each of which contains one set of chromosomes. When gametes unite, they form a **zygote**, or fertilized egg that contains two sets of chromosomes. (Note: Cells that contain one set of chromosomes are called **haploid**; cells containing two sets of chromosomes are called **diploid**.) If the reproductive cycle is to continue for any sexually reproducing species, then the diploid cell must somehow reduce its number of chromosome sets to produce haploid gametes; otherwise, the number of chromosome sets will double with every future round of fertilization. Therefore, sexual reproduction requires a nuclear division that reduces the number of chromosome sets by half.

Most animals and plants and many unicellular organisms are diploid and therefore have two sets of chromosomes. In each **somatic cell** of the organism (all cells of a multicellular organism except the gametes or reproductive cells), the nucleus contains two copies of each chromosome, called **homologous chromosomes**. Homologous chromosomes are matched pairs containing the same genes in identical locations along their lengths. Diploid organisms inherit one copy of each homologous chromosome from each parent.

Meiosis is the *nuclear division* that forms haploid cells from diploid cells, and it employs many of the same cellular mechanisms as mitosis. However, as you have learned, **mitosis** produces daughter cells whose nuclei are genetically identical to the original parent nucleus. In mitosis, both the parent and the daughter nuclei are at the same "ploidy level"—diploid in the case of most multicellular most animals. Plants use mitosis to grow as sporophytes, and to grow and produce eggs and sperm as gametophytes; so they use mitosis for both haploid and diploid cells (as well as for all other ploidies). In meiosis, the starting nucleus is always diploid and the daughter nuclei that result are haploid. To achieve this reduction in chromosome number, meiosis consists of one round of chromosome replication followed by two rounds of nuclear division. Because the events that occur during each of the division stages are analogous to the events of mitosis, the same stage names are assigned. However, because there are two rounds of division, the major process and the stages are designated with a "I" or a "II." Thus, **meiosis I** is the first round of meiotic division and consists of prophase I, prometaphase I, and so on. Likewise, **Meiosis II** (during which the second round of meiotic division takes place) includes prophase II, prometaphase II, and so on.

Meiosis I

Meiosis is preceded by an interphase consisting of G_1, S, and G_2 phases, which are nearly identical to the phases preceding mitosis. The G_1 phase (the "first gap phase") is focused on cell growth. During the S phase—the second phase of interphase—the cell copies or *replicates* the DNA of the chromosomes. Finally, in the G_2 phase (the "second gap phase") the cell undergoes the final preparations for meiosis.

During DNA duplication in the S phase, each chromosome is replicated to produce two identical copies—*sister chromatids* that are held together at the centromere by **cohesin** proteins, which hold the chromatids together until anaphase II.

Prophase I

Early in prophase I, before the chromosomes can be seen clearly with a microscope, the homologous chromosomes are attached at their tips to the nuclear envelope by proteins. As the nuclear envelope begins to break down, the proteins associated with homologous chromosomes bring the pair closer together. Recall that in mitosis, homologous chromosomes do not pair together. The **synaptonemal complex**, a lattice of proteins between the homologous chromosomes, first forms at specific locations and then spreads outward to cover the entire length of the chromosomes. The tight pairing of the homologous chromosomes is called *synapsis*. In **synapsis**, the genes on the chromatids of the homologous chromosomes are aligned precisely with each other. The synaptonemal complex supports the exchange of chromosomal segments between homologous nonsister chromatids—a process called **crossing over**. Crossing over can be observed visually after the exchange as *chiasmata* (singular = chiasma) (Figure 11.2).

In humans, even though the X and Y sex chromosomes are not completely homologous (that is, most of their genes differ), there is a small region of homology that allows the X and Y chromosomes to pair up during prophase I. A partial synaptonemal complex develops only between the regions of homology.

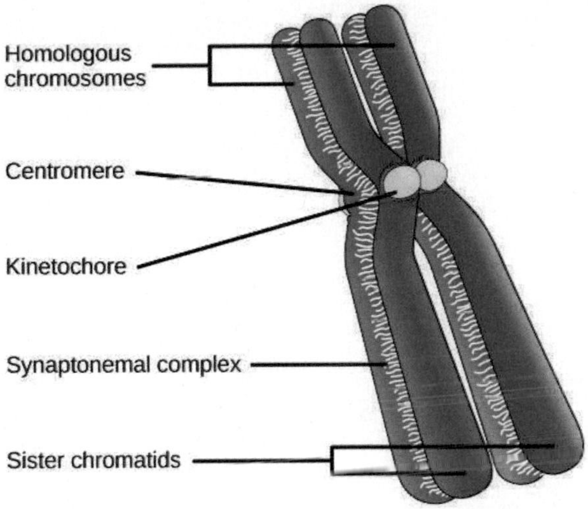

Figure 11.2 Early in prophase I, homologous chromosomes come together to form a synapse. The chromosomes are bound tightly together and in perfect alignment by a protein lattice called a synaptonemal complex and by cohesin proteins at the centromere.

Located at intervals along the synaptonemal complex are large protein assemblies called **recombination nodules**. These assemblies mark the points of later chiasmata and mediate the multistep process of **crossover**—or genetic recombination—between the nonsister chromatids. Near the recombination nodule, the double-stranded DNA of each chromatid is cleaved, the cut ends are modified, and a new connection is made between the nonsister chromatids. As prophase I progresses, the synaptonemal complex begins to break down and the chromosomes begin to condense. When the synaptonemal complex is gone, the homologous chromosomes remain attached to each other at the centromere and at chiasmata. The chiasmata remain until anaphase I. The number of chiasmata varies according to the species and the length of the chromosome. There must be at least one chiasma per chromosome for proper separation of homologous chromosomes during meiosis I, but there may be as many as 25. Following crossover, the synaptonemal complex breaks down and the cohesin connection between homologous pairs is removed. At the end of prophase I, the pairs are held together only at the chiasmata (Figure 11.3). These pairs are called **tetrads** because the four sister chromatids of each pair of homologous chromosomes are now visible.

The crossover events are the first source of genetic variation in the nuclei produced by meiosis. A single crossover event between homologous nonsister chromatids leads to a reciprocal exchange of equivalent DNA between a maternal chromosome and a paternal chromosome. When a recombinant sister chromatid is moved into a gamete cell it will carry some DNA from one parent and some DNA from the other parent. The recombinant chromatid has a combination of maternal and paternal genes that did not exist before the crossover. Crossover events can occur almost anywhere along the length of the synapsed chromosomes. Different cells undergoing meiosis will therefore produce different recombinant chromatids, with varying combinations of maternal and parental genes. Multiple crossovers in an arm of the chromosome have the same effect, exchanging segments of DNA to produce genetically recombined chromosomes.

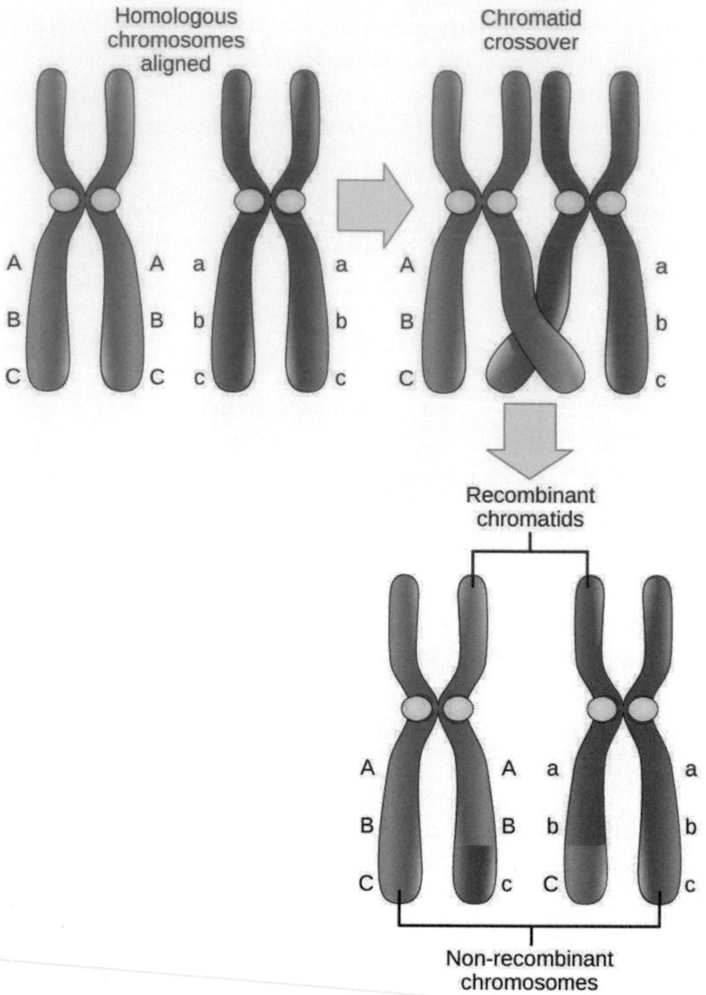

Figure 11.3 Crossover occurs between *nonsister chromatids of homologous chromosomes.* The result is an exchange of genetic material between homologous chromosomes.

Prometaphase I

The key event in prometaphase I is the attachment of the spindle fiber microtubules to the kinetochore proteins at the centromeres. Kinetochore proteins are multiprotein complexes that bind the centromeres of a chromosome to the microtubules of the mitotic spindle. Microtubules grow from microtubule-organizing centers (MTOCs). In animal cells, MTOCs are centrosomes located at opposite poles of the cell. The microtubules from each pole move toward the middle of the cell and attach to one of the kinetochores of the two fused homologous chromosomes. Each member of the homologous pair attaches to a microtubule extending from opposite poles of the cell so that in the next phase, the microtubules can pull the homologous pair apart. A spindle fiber that has attached to a kinetochore is called a *kinetochore microtubule.* At the end of prometaphase I, each tetrad is attached to microtubules from both poles, with one homologous chromosome facing each pole. The homologous chromosomes are still held together at the chiasmata. In addition, the nuclear membrane has broken down entirely.

Metaphase I

During metaphase I, the homologous chromosomes are arranged at the **metaphase plate**—roughly in the midline of the cell, with the kinetochores facing opposite poles. The homologous pairs orient themselves randomly at the equator. For example, if the two homologous members of chromosome 1 are labeled *a* and *b*, then the chromosomes could line up a-b or b-a. This is important in determining the genes carried by a gamete, as each will only receive one of the two homologous chromosomes. (Recall that after crossing over takes place, homologous chromosomes are not identical. They contain slight differences in their genetic information, causing each gamete to have a unique genetic makeup.)

The randomness in the alignment of recombined chromosomes at the metaphase plate, coupled with the crossing over events between nonsister chromatids, are responsible for much of the genetic variation in the offspring. To clarify this further, remember that the homologous chromosomes of a sexually reproducing organism are originally inherited as two separate sets, one from each parent. Using humans as an example, one set of 23 chromosomes is present in the egg donated by the mother. The father provides the other set of 23 chromosomes in the sperm that fertilizes the egg. Every cell of the multicellular offspring has copies of the original two sets of homologous chromosomes. In prophase I of meiosis, the homologous chromosomes form the tetrads. In metaphase I, these pairs line up at the midway point between the two poles of the cell to form the metaphase plate. Because there is an equal chance that a microtubule fiber will encounter a maternally or paternally inherited chromosome, the arrangement of the tetrads at the metaphase plate is random. Thus, any maternally inherited chromosome may face either pole. Likewise, any paternally inherited chromosome may also face either pole. *The orientation of each tetrad is independent of the orientation of the other 22 tetrads.*

This event—the *random* (or *independent*) assortment of homologous chromosomes at the metaphase plate—is the second mechanism that introduces variation into the gametes or spores. In each cell that undergoes meiosis, the arrangement of the tetrads is different. The number of variations is dependent on the number of chromosomes making up a set. There are two possibilities for orientation at the metaphase plate; the possible number of alignments therefore equals 2^n in a diploid cell, where n is the number of chromosomes per haploid set. Humans have 23 chromosome pairs, which results in over eight million (2^{23}) possible genetically-distinct gametes just from the random alignment of chromosomes at the metaphase plate. This number does not include the variability that was previously produced by crossing over between the nonsister chromatids. Given these two mechanisms, it is highly unlikely that any two haploid cells resulting from meiosis will have the same genetic composition (Figure 11.4).

To summarize, meiosis I creates genetically diverse gametes in two ways. First, during prophase I, crossover events between the nonsister chromatids of each homologous pair of chromosomes generate recombinant chromatids with new combinations of maternal and paternal genes. Second, the random assortment of tetrads on the metaphase plate produces unique combinations of maternal and paternal chromosomes that will make their way into the gametes.

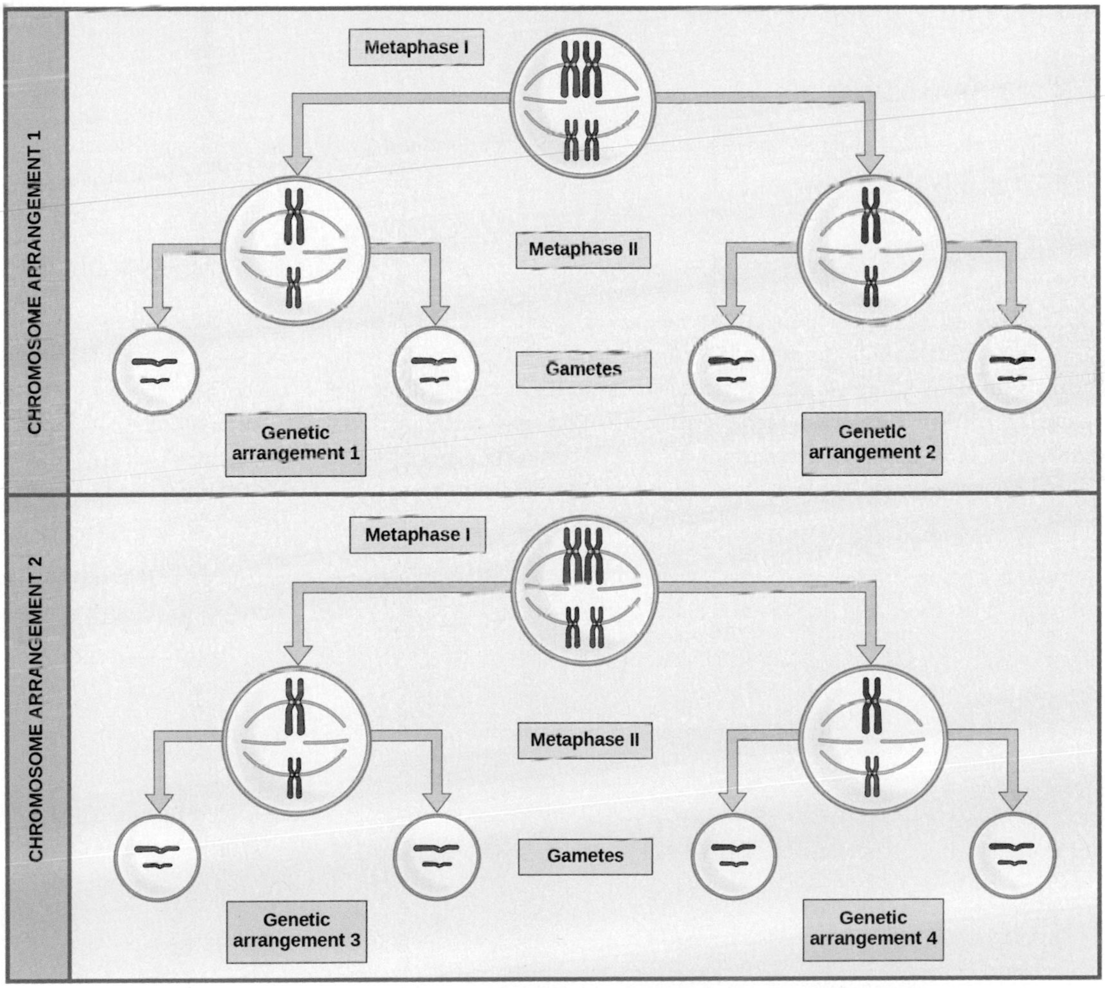

Figure 11.4 Random, independent assortment during metaphase I can be demonstrated by considering a cell with a set of two

chromosomes ($n = 2$). In this case, there are two possible arrangements at the equatorial plane in metaphase I. The total possible number of different gametes is 2^n, where n equals the number of chromosomes in a set. In this example, there are four possible genetic combinations for the gametes. With $n = 23$ in human cells, there are over eight million possible combinations of paternal and maternal chromosomes.

Anaphase I

In anaphase I, the microtubules pull the linked chromosomes apart. The sister chromatids remain tightly bound together at the centromere. The chiasmata are broken in anaphase I as the microtubules attached to the fused kinetochores pull the homologous chromosomes apart (Figure 11.5).

Telophase I and Cytokinesis

In telophase, the separated chromosomes arrive at opposite poles. The remainder of the typical telophase events may or may not occur, depending on the species. In some organisms, the chromosomes "decondense" and nuclear envelopes form around the separated sets of chromatids produced during telophase I. In other organisms, **cytokinesis**—the physical separation of the cytoplasmic components into two daughter cells—occurs without reformation of the nuclei. In nearly all species of animals and some fungi, cytokinesis separates the cell contents via a *cleavage furrow* (constriction of the actin ring that leads to cytoplasmic division). In plants, a *cell plate* is formed during cell cytokinesis by Golgi vesicles fusing at the metaphase plate. This cell plate will ultimately lead to the formation of cell walls that separate the two daughter cells.

Two haploid cells are the result of the first meiotic division of a diploid cell. The cells are haploid because at each pole, there is just one of each pair of the homologous chromosomes. Therefore, only one full set of the chromosomes is present. This is why the cells are considered haploid—there is only one chromosome set, *even though each chromosome still consists of two sister chromatids.* Recall that sister chromatids are merely duplicates of one of the two homologous chromosomes (except for changes that occurred during crossing over). In meiosis II, these two sister chromatids will separate, creating four haploid daughter cells.

LINK TO LEARNING

Review the process of meiosis, observing how chromosomes align and migrate, at Meiosis: An Interactive Animation (http://openstax.org/l/animal_meiosis) .

Meiosis II

In some species, cells enter a brief interphase, or **interkinesis**, before entering meiosis II. Interkinesis lacks an S phase, so chromosomes are not duplicated. The two cells produced in meiosis I go through the events of meiosis II in synchrony. During meiosis II, the sister chromatids within the two daughter cells separate, forming four new haploid gametes. The mechanics of meiosis II are similar to mitosis, except that each dividing cell has only one set of homologous chromosomes, each with two chromatids. Therefore, each cell has half the number of sister chromatids to separate out as a diploid cell undergoing mitosis. In terms of chromosomal content, cells at the start of meiosis II are similar to haploid cells in G_2, preparing to undergo mitosis.

Prophase II

If the chromosomes decondensed in telophase I, they condense again. If nuclear envelopes were formed, they fragment into vesicles. The MTOCs that were duplicated during interkinesis move away from each other toward opposite poles, and new spindles are formed.

Prometaphase II

The nuclear envelopes are completely broken down, and the spindle is fully formed. Each sister chromatid forms an individual kinetochore that attaches to microtubules from opposite poles.

Metaphase II

The sister chromatids are maximally condensed and aligned at the equator of the cell.

Anaphase II

The sister chromatids are pulled apart by the kinetochore microtubules and move toward opposite poles. Nonkinetochore microtubules elongate the cell.

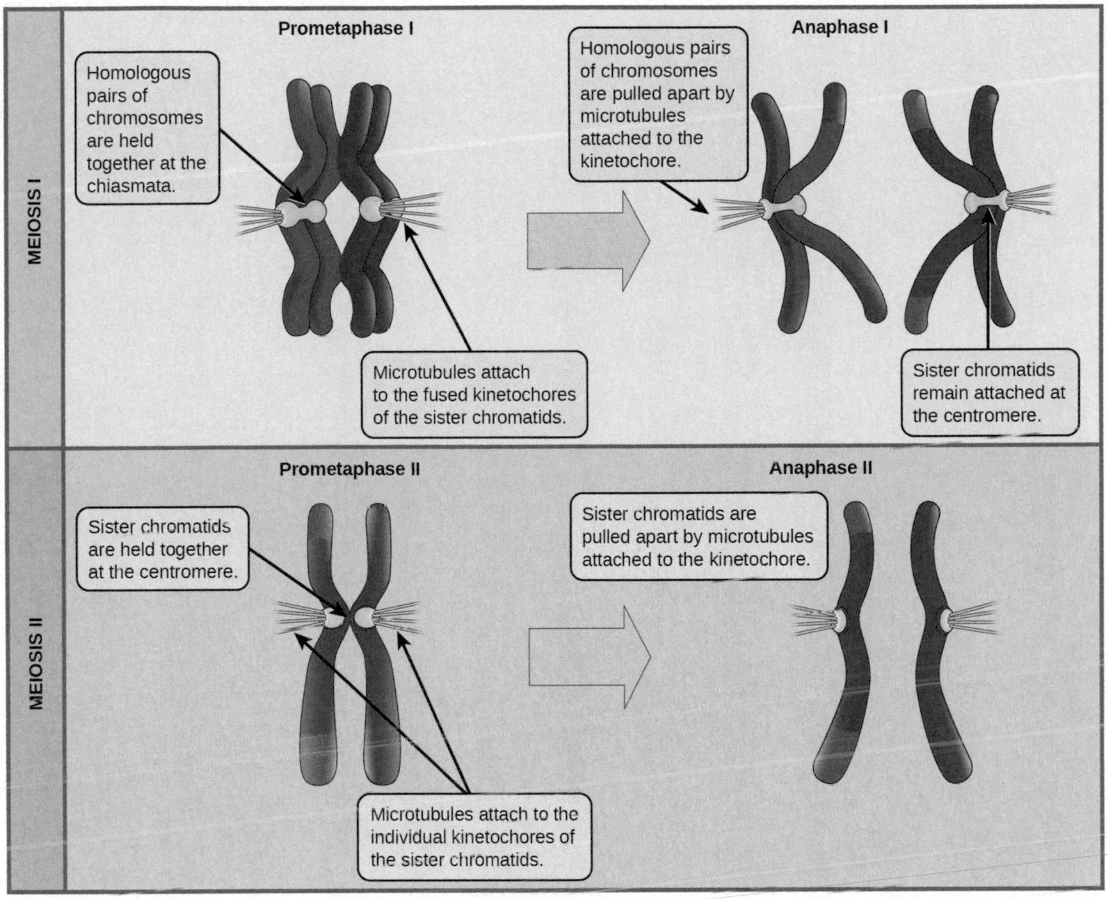

Figure 11.5 The process of chromosome alignment differs between meiosis I and meiosis II. In prometaphase I, microtubules attach to the fused kinetochores of homologous chromosomes, and the homologous chromosomes are arranged at the midline of the cell (the metaphase plate) in metaphase I. In anaphase I, the homologous chromosomes separate. In prometaphase II, microtubules attach to the kinetochores of sister chromatids, and the sister chromatids are arranged at the midpoint of the cells in metaphase II. In anaphase II, the sister chromatids separate.

Telophase II and Cytokinesis

The chromosomes arrive at opposite poles and begin to decondense. Nuclear envelopes form around the chromosomes. If the parent cell was diploid, as is most commonly the case, then cytokinesis now separates the two cells into four unique haploid cells. The cells produced are genetically unique because of the random assortment of paternal and maternal homologs and because of the recombination of maternal and paternal segments of chromosomes (with their sets of genes) that occurs during crossover. The entire process of meiosis is outlined in Figure 11.6.

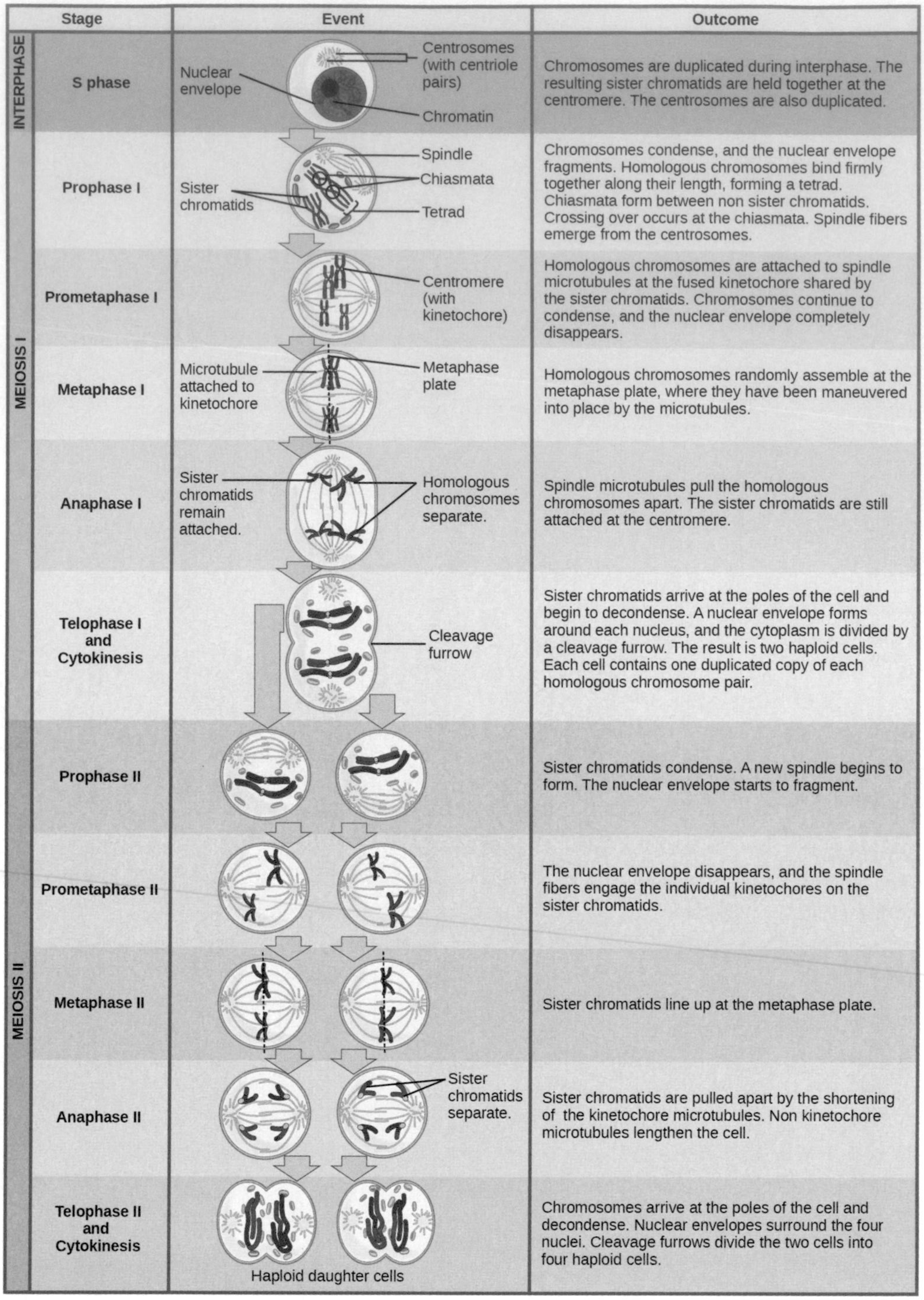

Figure 11.6 An animal cell with a diploid number of four ($2n = 4$) proceeds through the stages of meiosis to form four haploid daughter cells.

Comparing Meiosis and Mitosis

Mitosis and meiosis are both forms of division of the nucleus in eukaryotic cells. They share some similarities, but also exhibit a number of important and distinct differences that lead to very different outcomes (Figure 11.7). Mitosis is a single nuclear division that results in two nuclei that are usually partitioned into two new cells. The nuclei resulting from a mitotic division are genetically identical to the original nucleus. They have the same number of sets of chromosomes: one set in the case of haploid cells and two sets in the case of diploid cells. In contrast, meiosis consists of two nuclear divisions resulting in four nuclei that are usually partitioned into four new, genetically distinct cells. The four nuclei produced during meiosis are not genetically identical, and they contain one chromosome set only. This is half the number of chromosome sets in the original cell, which is diploid.

The main differences between mitosis and meiosis occur in meiosis I, which is a very different nuclear division than mitosis. In meiosis I, the homologous chromosome pairs physically meet and are bound together with the synaptonemal complex. Following this, the chromosomes develop chiasmata and undergo crossover between nonsister chromatids. In the end, the chromosomes line up along the metaphase plate as tetrads—with kinetochore fibers from opposite spindle poles attached to each kinetochore of a homolog to form a tetrad. *All of these events occur only in meiosis I.*

When the chiasmata resolve and the tetrad is broken up with the homologs moving to one pole or another, the ploidy level—the number of sets of chromosomes in each future nucleus—has been reduced from two to one. For this reason, meiosis I is referred to as a **reductional division**. There is no such reduction in ploidy level during mitosis.

Meiosis II is analogous to a mitotic division. In this case, the duplicated chromosomes (only one set of them) line up on the metaphase plate with divided kinetochores attached to kinetochore fibers from opposite poles. During anaphase II, as in mitotic anaphase, the kinetochores divide and one sister chromatid—now referred to as a chromosome—is pulled to one pole while the other sister chromatid is pulled to the other pole. If it were not for the fact that there had been crossover, the two products of each individual meiosis II division would be identical (as in mitosis). Instead, they are different because there has always been at least one crossover per chromosome. Meiosis II is not a reduction division because although there are fewer copies of the genome in the resulting cells, there is still one set of chromosomes, as there was at the end of meiosis I.

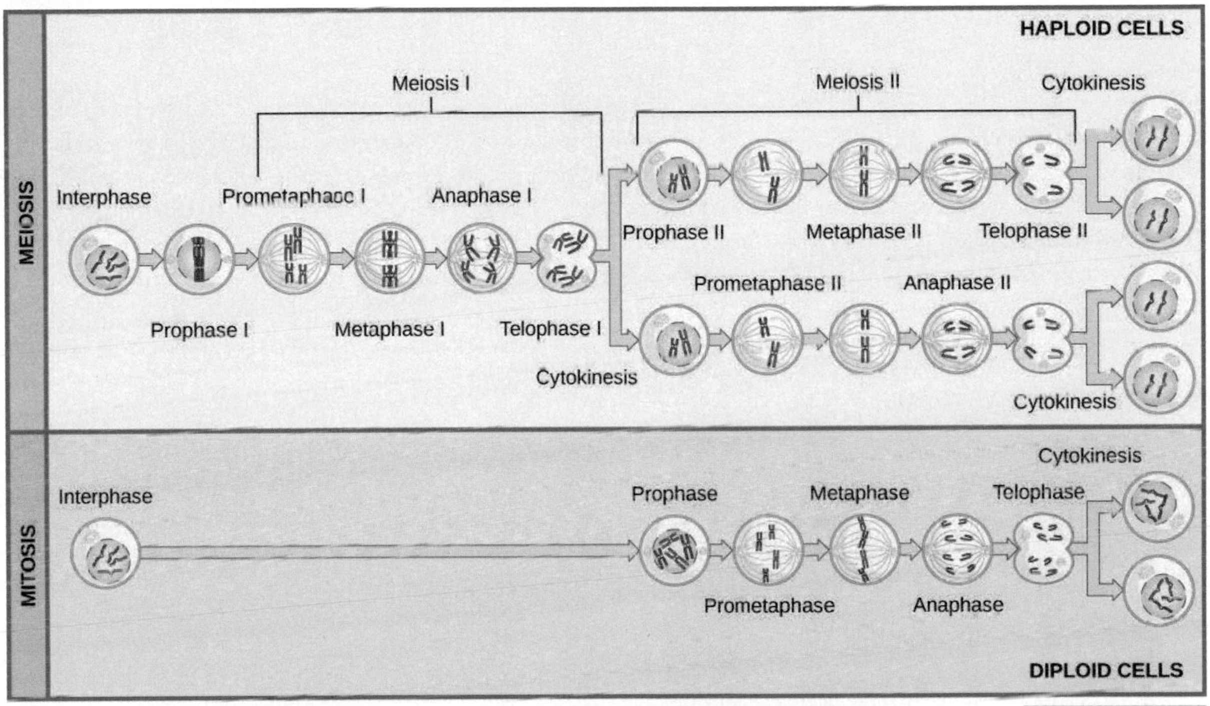

						OUTCOME
PROCESS	DNA synthesis	Synapsis of homologous chromosomes	Crossover	Homologous chromosomes line up at metaphase plate	Sister chromatids line up at metaphase plate	Number and genetic composition of daughter cells
MEIOSIS	Occurs in S phase of interphase	During prophase I	During prophase I	During metaphase I	During metaphase II	Four haploid cells at the end of meiosis II
MITOSIS	Occurs in S phase of interphase	Does not occur in mitosis	Does not occur in mitosis	Does not occur in mitosis	During metaphase	Two diploid cells at the end of mitosis

Figure 11.7 Meiosis and mitosis are both preceded by one cycle of DNA replication; however, meiosis includes two nuclear divisions. The four daughter cells resulting from meiosis are haploid and genetically distinct. The daughter cells resulting from mitosis are diploid and

identical to the parent cell.

EVOLUTION CONNECTION

The Mystery of the Evolution of Meiosis

Some characteristics of organisms are so widespread and fundamental that it is sometimes difficult to remember that they evolved like other simple traits. Meiosis is such an extraordinarily complex series of cellular events that biologists have had trouble testing hypotheses concerning how it may have evolved. Although meiosis is inextricably entwined with sexual reproduction and its advantages and disadvantages, it is important to separate the questions of the evolution of meiosis and the evolution of sex, because early meiosis may have been advantageous for different reasons than it is now. Thinking outside the box and imagining what the early benefits from meiosis might have been is one approach to uncovering how it may have evolved.

Meiosis and mitosis share obvious cellular processes, and it makes sense that meiosis evolved from mitosis. The difficulty lies in the clear differences between meiosis I and mitosis. Adam Wilkins and Robin Holliday[1] summarized the unique events that needed to occur for the evolution of meiosis from mitosis. These steps are homologous chromosome pairing and synapsis, crossover exchanges, sister chromatids remaining attached during anaphase, and suppression of DNA replication in interphase. They argue that the first step is the hardest and most important and that understanding how it evolved would make the evolutionary process clearer. They suggest genetic experiments that might shed light on the evolution of synapsis.

There are other approaches to understanding the evolution of meiosis in progress. Different forms of meiosis exist in single-celled protists. Some appear to be simpler or more "primitive" forms of meiosis. Comparing the meiotic divisions of different protists may shed light on the evolution of meiosis. Marilee Ramesh and colleagues[2] compared the genes involved in meiosis in protists to understand when and where meiosis might have evolved. Although research is still ongoing, recent scholarship into meiosis in protists suggests that some aspects of meiosis may have evolved later than others. This kind of genetic comparison can tell us what aspects of meiosis are the oldest and what cellular processes they may have borrowed from in earlier cells.

LINK TO LEARNING

Click through the steps of this interactive animation to compare the meiotic process of cell division to that of mitosis at How Cells Divide (http://openstax.org/l/how_cells_dvide) .

11.2 Sexual Reproduction

By the end of this section, you will be able to do the following:

- Explain that meiosis and sexual reproduction are highly evolved traits
- Identify variation among offspring as a potential evolutionary advantage of sexual reproduction
- Describe the three different life-cycle types among sexually reproducing multicellular organisms.

Sexual reproduction was likely an early evolutionary innovation after the appearance of eukaryotic cells. It appears to have been very successful because most eukaryotes are able to reproduce sexually and, in many animals, it is the only mode of reproduction. And yet, scientists also recognize some real disadvantages to sexual reproduction. On the surface, creating offspring that are genetic clones of the parent appears to be a better system. If the parent organism is successfully occupying a habitat, offspring with the same traits should be similarly successful. There is also the obvious benefit to an organism that can produce offspring whenever circumstances are favorable by asexual budding, fragmentation, or by producing eggs asexually. These methods of reproduction do not require another organism of the opposite sex. Indeed, some organisms that lead a solitary lifestyle have retained the ability to reproduce asexually. In addition, in asexual populations, every individual is capable of reproduction. In sexual populations, the males are not producing the offspring themselves, so hypothetically an asexual

1Adam S. Wilkins and Robin Holliday, "The Evolution of Meiosis from Mitosis," *Genetics* 181 (2009): 3–12.

2Marilee A. Ramesh, Shehre-Banoo Malik and John M. Logsdon, Jr, "A Phylogenetic Inventory of Meiotic Genes: Evidence for Sex in *Giardia* and an Early Eukaryotic Origin of Meiosis," *Current Biology* 15 (2005):185–91.

population could grow twice as fast.

However, multicellular organisms that exclusively depend on asexual reproduction are exceedingly rare. Why are meiosis and sexual reproductive strategies so common? These are important (and as yet unanswered) questions in biology, even though they have been the focus of much research beginning in the latter half of the 20th century. There are several possible explanations, one of which is that the variation that sexual reproduction creates among offspring is very important to the survival and reproduction of the population. Thus, on average, a sexually reproducing population will leave more descendants than an otherwise similar asexually reproducing population. The only source of variation in asexual organisms is mutation. Mutations that take place during the formation of germ cell lines are also the ultimate source of variation in sexually reproducing organisms. However, in contrast to mutation during asexual reproduction, the mutations during sexual reproduction can be continually reshuffled from one generation to the next when different parents combine their unique genomes and the genes are mixed into different combinations by crossovers during prophase I and random assortment at metaphase I.

EVOLUTION CONNECTION

The Red Queen Hypothesis

Genetic variation is the outcome of sexual reproduction, but why are ongoing variations necessary, even under seemingly stable environmental conditions? Enter the Red Queen hypothesis, first proposed by Leigh Van Valen in 1973.[3] The concept was named in reference to the Red Queen's race in Lewis Carroll's book, *Through the Looking-Glass*.

All species **coevolve** (evolve together) with other organisms. For example, predators evolve with their prey, and parasites evolve with their hosts. Each tiny advantage gained by favorable variation gives a species a reproductive edge over close competitors, predators, parasites, or even prey. However, survival of any given genotype or phenotype in a population is dependent on the reproductive fitness of other genotypes or phenotypes within a given species. The only method that will allow a coevolving species to maintain its own share of the resources is to also *continually improve its* **fitness** (the capacity of the members to produce more reproductively viable offspring relative to others within a species). As one species gains an advantage, this increases selection on the other species; they must also develop an advantage or they will be outcompeted. No single species progresses too far ahead because genetic variation among the progeny of sexual reproduction provides all species with a mechanism to improve rapidly. Species that cannot keep up become extinct. The Red Queen's catchphrase was, "It takes all the running you can do to stay in the same place." This is an apt description of coevolution between competing species.

Life Cycles of Sexually Reproducing Organisms

Fertilization and meiosis alternate in sexual **life cycles**. What happens between these two events depends on the organism's "reproductive strategy." The process of meiosis reduces the chromosome number by half. Fertilization, the joining of two haploid gametes, restores the diploid condition. Some organisms have a multicellular diploid stage that is most obvious and only produce haploid reproductive cells. Animals, including humans, have this type of life cycle. Other organisms, such as fungi, have a multicellular haploid stage that is most obvious. Plants and some algae have alternation of generations, in which they have multicellular diploid and haploid life stages that are apparent to different degrees depending on the group.

Nearly all animals employ a diploid-dominant life-cycle strategy in which the only haploid cells produced by the organism are the gametes. Early in the development of the embryo, specialized diploid cells, called **germ cells**, are produced within the gonads (such as the testes and ovaries). Germ cells are capable of mitosis to perpetuate the germ cell line and meiosis to produce haploid gametes. Once the haploid gametes are formed, they lose the ability to divide again. There is no multicellular haploid life stage. Fertilization occurs with the fusion of two gametes, usually from different individuals, restoring the diploid state (Figure 11.8).

3Leigh Van Valen, "A New Evolutionary Law," *Evolutionary Theory* 1 (1973): 1–30

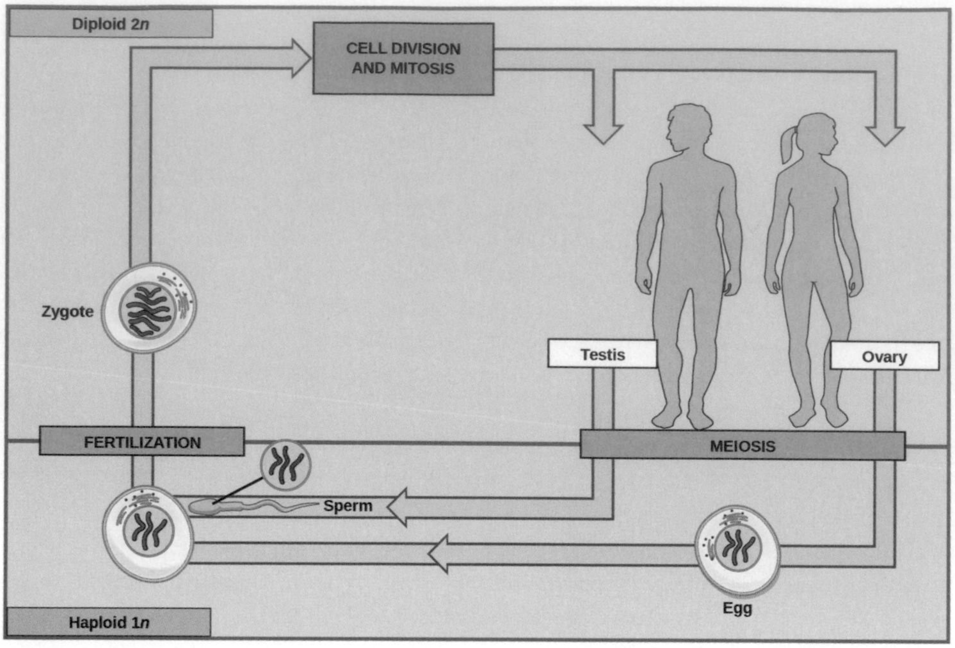

Figure 11.8 In animals, sexually reproducing adults form haploid gametes from diploid germ cells. Fusion of the gametes gives rise to a fertilized egg cell, or zygote. The zygote will undergo multiple rounds of mitosis to produce a multicellular offspring. The germ cells are generated early in the development of the zygote.

Most fungi and algae employ a life-cycle type in which the "body" of the organism—the ecologically important part of the life cycle—is haploid. The haploid cells that make up the tissues of the dominant multicellular stage are formed by mitosis. During sexual reproduction, specialized haploid cells from two individuals—designated the (+) and (–) mating types—join to form a diploid zygote. The zygote immediately undergoes meiosis to form four haploid cells called *spores*. Although these spores are haploid like the "parents," they contain a new genetic combination from two parents. The spores can remain dormant for various time periods. Eventually, when conditions are favorable, the spores form multicellular haploid structures through many rounds of mitosis (Figure 11.9).

VISUAL CONNECTION

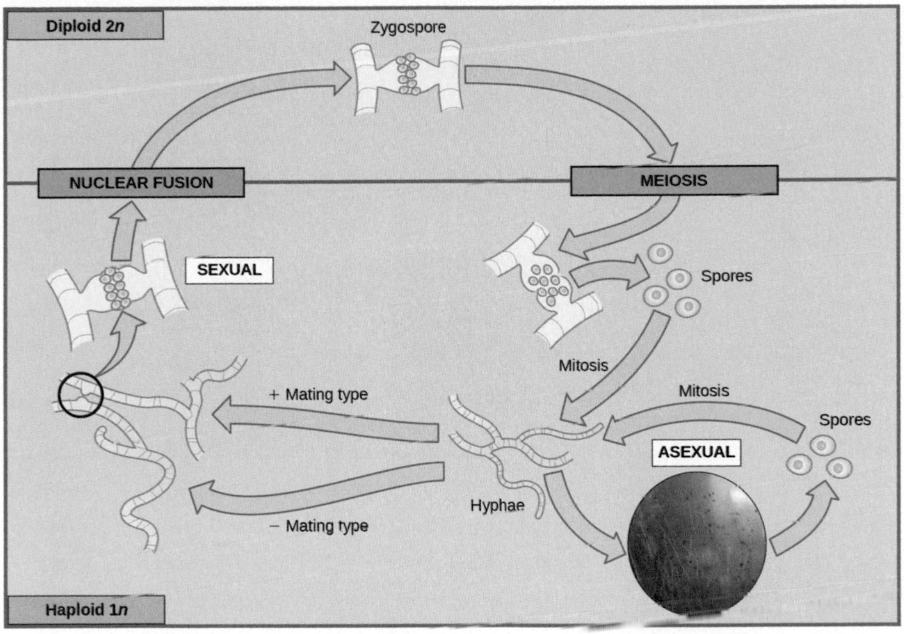

Figure 11.9 Fungi, such as black bread mold (*Rhizopus nigricans*), have a haploid multicellular stage that produces specialized haploid cells by mitosis that fuse to form a diploid zygote. The zygote undergoes meiosis to produce haploid spores. Each spore gives rise to a multicellular haploid organism by mitosis. Above, different mating hyphae types (denoted as + and –) join to form a zygospore through nuclear fusion. (credit “zygomycota” micrograph: modification of work by “Fanaberka”/Wikimedia Commons)

If a mutation occurs so that a fungus is no longer able to produce a minus mating type, will it still be able to reproduce?

The third life-cycle type, employed by some algae and all plants, is a blend of the haploid-dominant and diploid-dominant extremes. Species with alternation of generations have both haploid and diploid multicellular organisms as part of their life cycle. The haploid multicellular plants are called **gametophytes**, because they produce gametes from specialized cells. Meiosis is not directly involved in the production of gametes in this case, because the organism that produces the gametes is already haploid. Fertilization between the gametes forms a diploid zygote. The zygote will undergo many rounds of mitosis and give rise to a diploid multicellular plant called a **sporophyte**. Specialized cells of the sporophyte will undergo meiosis and produce haploid spores. The spores will subsequently develop into the gametophytes (Figure 11.10).

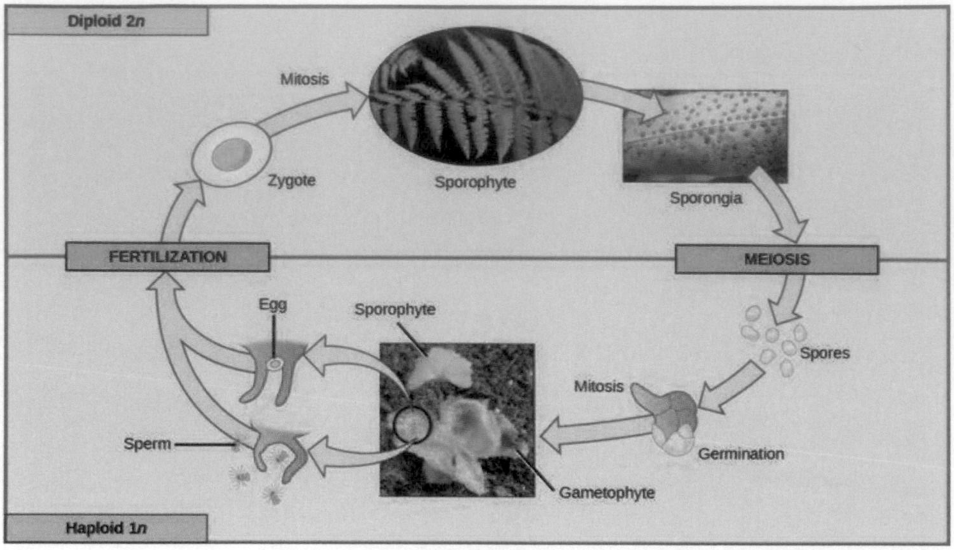

Figure 11.10 Plants have a life cycle that alternates between a multicellular haploid organism and a multicellular diploid organism. In some plants, such as ferns, both the haploid and diploid plant stages are free-living. The diploid plant is called a sporophyte because it produces haploid spores by meiosis. The spores develop into multicellular, haploid plants that are called *gametophytes* because they produce gametes. The gametes of two individuals will fuse to form a diploid zygote that becomes the sporophyte. (credit "fern": modification of work by Cory Zanker; credit "sporangia": modification of work by "Obsidian Soul"/Wikimedia Commons; credit "gametophyte and sporophyte": modification of work by "Vlmastra"/Wikimedia Commons)

Although all plants utilize some version of the alternation of generations, the relative size of the sporophyte and the gametophyte and the relationship between them vary greatly. In plants such as moss, the gametophyte organism is the free-living plant and the sporophyte is physically dependent on the gametophyte. In other plants, such as ferns, both the gametophyte and sporophyte plants are free-living; however, the sporophyte is much larger. In seed plants, such as magnolia trees and daisies, the gametophyte is composed of only a few cells and, in the case of the female gametophyte, is completely retained within the sporophyte.

Sexual reproduction takes many forms in multicellular organisms. The fact that nearly every multicellular organism on Earth employs sexual reproduction is strong evidence for the benefits of producing offspring with unique gene combinations, though there are other possible benefits as well.

KEY TERMS

alternation of generations life-cycle type in which the diploid and haploid stages alternate

chiasmata (singular, *chiasma*) the structure that forms at the crossover points after genetic material is exchanged

cohesin proteins that form a complex that seals sister chromatids together at their centromeres until anaphase II of meiosis

crossover exchange of genetic material between nonsister chromatids resulting in chromosomes that incorporate genes from both parents of the organism

fertilization union of two haploid cells from two individual organisms

gametophyte a multicellular haploid life-cycle stage that produces gametes

germ cells specialized cell line that produces gametes, such as eggs or sperm

interkinesis (also, *interphase II*) brief period of rest between meiosis I and meiosis II

life cycle the sequence of events in the development of an organism and the production of cells that produce offspring

meiosis a nuclear division process that results in four haploid cells

meiosis I first round of meiotic cell division; referred to as *reduction division* because the ploidy level is reduced from diploid to haploid

meiosis II second round of meiotic cell division following meiosis I; sister chromatids are separated into individual chromosomes, and the result is four unique haploid cells

recombination nodules protein assemblies formed on the synaptonemal complex that mark the points of crossover events and mediate the multistep process of genetic recombination between nonsister chromatids

reduction division nuclear division that produces daughter nuclei each having one-half as many chromosome sets as the parental nucleus; meiosis I is a reduction division

somatic cell all the cells of a multicellular organism except the gametes or reproductive cells

spore haploid cell that can produce a haploid multicellular organism or can fuse with another spore to form a diploid cell

sporophyte a multicellular diploid life-cycle stage that produces haploid spores by meiosis

synapsis formation of a close association between homologous chromosomes during prophase I

synaptonemal complex protein lattice that forms between homologous chromosomes during prophase I, supporting crossover

tetrad two duplicated homologous chromosomes (four chromatids) bound together by chiasmata during prophase I

CHAPTER SUMMARY

11.1 The Process of Meiosis

Sexual reproduction requires that organisms produce cells that can fuse during fertilization to produce offspring. In most animals, meiosis is used to produce haploid eggs and sperm from diploid parent cells so that the fusion of an egg and sperm produces a diploid zygote. As with mitosis, DNA replication occurs prior to meiosis during the S-phase of the cell cycle so that each chromosome becomes a pair of sister chromatids. In meiosis, there are two rounds of nuclear division resulting in four nuclei and usually four daughter cells, each with half the number of chromosomes as the parent cell. The first division separates homologs, and the second—like mitosis—separates chromatids into individual chromosomes. Meiosis generates variation in the daughter nuclei during crossover in prophase I as well as during the random alignment of tetrads at metaphase I. The cells that are produced by meiosis are genetically unique.

Meiosis and mitosis share similar processes, but have distinct outcomes. Mitotic divisions are single nuclear divisions that produce genetically identical daughter nuclei (i.e., each daughter nucleus has the same number of chromosome sets as the original cell). In contrast, meiotic divisions include two nuclear divisions that ultimately produce four genetically different daughter nuclei that have only one chromosome set (instead of the two sets of chromosomes in the parent cell). The main differences between the two nuclear division processes take place during the first division of meiosis: homologous chromosomes pair, crossover, and exchange homologous nonsister chromatid segments. The homologous chromosomes separate into different nuclei during meiosis I, causing a *reduction of ploidy level in the first division*. The second division of meiosis is similar to a mitotic division, except that the daughter cells do not contain identical genomes because of crossover and chromosome recombination in prophase I.

11.2 Sexual Reproduction

Nearly all eukaryotes undergo sexual reproduction. The variation introduced into the reproductive cells by meiosis provides an important advantage that has made sexual reproduction evolutionarily successful. Meiosis and fertilization alternate in sexual life cycles. The process of meiosis produces unique reproductive cells called gametes, which have half the number of chromosomes as the parent cell. When two haploid gametes fuse, this restores the diploid condition in the new zygote. Thus, most sexually reproducing organisms alternate between haploid and diploid stages. However, the ways in which reproductive cells are produced and the timing between meiosis and fertilization vary greatly.

VISUAL CONNECTION QUESTIONS

1. Figure 11.9 If a mutation occurs so that a fungus is no longer able to produce a minus mating type, will it still be able to reproduce?

REVIEW QUESTIONS

2. Meiosis usually produces ________ daughter cells.
   a. two haploid
   b. two diploid
   c. four haploid
   d. four diploid

3. What structure is most important in forming the tetrads?
   a. centromere
   b. synaptonemal complex
   c. chiasma
   d. kinetochore

4. At which stage of meiosis are sister chromatids separated from each other?
   a. prophase I
   b. prophase II
   c. anaphase I
   d. anaphase II

5. At metaphase I, homologous chromosomes are connected only at what structures?
   a. chiasmata
   b. recombination nodules
   c. microtubules
   d. kinetochores

6. Which of the following is *not* true in regard to crossover?
   a. Spindle microtubules guide the transfer of DNA across the synaptonemal complex.
   b. Nonsister chromatids exchange genetic material.
   c. Chiasmata are formed.
   d. Recombination nodules mark the crossover point.

7. What phase of mitotic interphase is missing from meiotic interkinesis?
   a. G_0 phase
   b. G_1 phase
   c. S phase
   d. G_2 phase

8. The part of meiosis that is similar to mitosis is ________.
   a. meiosis I
   b. anaphase I
   c. meiosis II
   d. interkinesis

9. If a muscle cell of a typical organism has 32 chromosomes, how many chromosomes will be in a gamete of that same organism?
   a. 8
   b. 16
   c. 32
   d. 64

10. Which statement best describes the genetic content of the two daughter cells in prophase II of meiosis?
    a. haploid with one copy of each gene
    b. haploid with two copies of each gene
    c. diploid with two copies of each gene
    d. diploid with four copies of each gene

11. The pea plants used in Mendel's genetic inheritance studies were diploid, with 14 chromosomes in somatic cells. Assuming no crossing over events occur, how many unique gametes could one pea plant produce?
    a. 28
    b. 128
    c. 196
    d. 16,384

12. How do telophase I and telophase II differ during meiosis in animal cells?
    a. Cells remain diploid at the end of telophase I, but are haploid at the end of telophase II.
    b. Daughter cells form a cell plate to divide during telophase I, but divide by cytokinesis during telophase II.
    c. Cells enter interphase after telophase I, but not after telophase II.
    d. Chromosomes can remain condensed at the end of telophase I, but decondense after telophase II.

13. What is a likely evolutionary advantage of sexual reproduction over asexual reproduction?
    a. Sexual reproduction involves fewer steps.
    b. There is a lower chance of using up the resources in a given environment.
    c. Sexual reproduction results in variation in the offspring.
    d. Sexual reproduction is more cost-effective.

14. Which type of life cycle has both a haploid and diploid multicellular stage?
    a. asexual life cycles
    b. most animal life cycles
    c. most fungal life cycles
    d. alternation of generations

15. What is the ploidy of the most conspicuous form of most fungi?
    a. diploid
    b. haploid
    c. alternation of generations
    d. asexual

16. A diploid, multicellular life-cycle stage that gives rise to haploid cells by meiosis is called a ________.
    a. sporophyte
    b. gametophyte
    c. spore
    d. gamete

17. Hydras and jellyfish both live in a freshwater lake that is slowly being acidified by the runoff from a chemical plant built upstream. Which population is predicted to be better able to cope with the changing environment?
    a. jellyfish
    b. hydra
    c. The populations will be equally able to cope.
    d. Both populations will die.

18. Many farmers are worried about the decreasing genetic diversity of plants associated with generations of artificial selection and inbreeding. Why is limiting random sexual reproduction of food crops concerning?
    a. Mutations during asexual reproduction decrease plant fitness.
    b. Consumers do not trust identical-appearing produce.
    c. Larger portions of the plant populations are susceptible to the same diseases.
    d. Spores are not viable in an agricultural setting.

CRITICAL THINKING QUESTIONS

19. Describe the process that results in the formation of a tetrad.
20. Explain how the random alignment of homologous chromosomes during metaphase I contributes to the variation in gametes produced by meiosis.
21. What is the function of the fused kinetochore found on sister chromatids in prometaphase I?
22. In a comparison of the stages of meiosis to the stages of mitosis, which stages are unique to meiosis and which stages have the same events in both meiosis and mitosis?
23. Why would an individual with a mutation that prevented the formation of recombination nodules be considered less fit than other members of its species?
24. Does crossing over occur during prophase II? From an evolutionary perspective, why is this advantageous?
25. List and briefly describe the three processes that lead to variation in offspring with the same parents.
26. Animals and plants both have diploid and haploid cells. How does the animal life cycle differ from the alternation of generations exhibited by plants?
27. Explain why sexual reproduction is beneficial to a population but can be detrimental to an individual offspring.
28. How does the role of meiosis in gamete production differ between organisms with a diploid-dominant life cycle and organisms with an alternation of generations life cycle?
29. How do organisms with haploid-dominant life cycles ensure continued genetic diversification in offspring without using a meiotic process to make gametes?

CHAPTER 12
Mendel's Experiments and Heredity

Figure 12.1 Experimenting with thousands of garden peas, Mendel uncovered the fundamentals of genetics. (credit: modification of work by Jerry Kirkhart)

INTRODUCTION Genetics is the study of heredity. Johann Gregor Mendel set the framework for genetics long before chromosomes or genes had been identified, at a time when meiosis was not well understood. Mendel selected a simple biological system and conducted methodical, quantitative analyses using large sample sizes. Because of Mendel's work, the fundamental principles of heredity were revealed. We now know that genes, carried on chromosomes, are the basic functional units of heredity with the capability to be replicated, expressed, or mutated. Today, the postulates put forth by Mendel form the basis of classical, or Mendelian, genetics. Not all genes are transmitted from parents to offspring according to Mendelian genetics, but Mendel's experiments serve as an excellent starting point for thinking about inheritance.

Chapter Outline

12.1 Mendel's Experiments and the Laws of Probability

By the end of this section, you will be able to do the following:

- Describe the scientific reasons for the success of Mendel's experimental work
- Describe the expected outcomes of monohybrid crosses involving dominant and recessive alleles
- Apply the sum and product rules to calculate probabilities

Figure 12.2 Johann Gregor Mendel is considered the father of genetics.

Johann Gregor Mendel (1822–1884) (Figure 12.2) was a lifelong learner, teacher, scientist, and man of faith. As a young adult, he joined the Augustinian Abbey of St. Thomas in Brno in what is now the Czech Republic. Supported by the monastery, he taught physics, botany, and natural science courses at the secondary and university levels. In 1856, he began a decade-long research pursuit involving inheritance patterns in honeybees and plants, ultimately settling on pea plants as his primary **model system** (a system with convenient characteristics used to study a specific biological phenomenon to be applied to other systems). In 1865, Mendel presented the results of his experiments with nearly 30,000 pea plants to the local Natural History Society. He demonstrated that traits are transmitted from parents to offspring independently of other traits and in dominant and recessive patterns. In 1866, he published his work, *Experiments in Plant Hybridization,*[1] in the proceedings of the Natural History Society of Brünn.

Mendel's work went virtually unnoticed by the scientific community, which believed, incorrectly, that the process of inheritance involved a blending of parental traits that produced an intermediate physical appearance in offspring. The **blending theory of inheritance** asserted that the original parental traits were lost or absorbed by the blending in the offspring, but we now know that this is not the case. This hypothetical process appeared to be correct because of what we know now as continuous variation. **Continuous variation** results from the action of many genes to determine a characteristic like human height. Offspring appear to be a "blend" of their parents' traits.

Instead of continuous characteristics, Mendel worked with traits that were inherited in distinct classes (specifically, violet versus white flowers); this is referred to as **discontinuous variation**. Mendel's choice of these kinds of traits allowed him to see experimentally that the traits were not blended in the offspring, nor

1Johann Gregor Mendel, *Versuche über Pflanzenhybriden Verhandlungen des naturforschenden Vereines in Brünn, Bd. IV für das Jahr,* 1865 Abhandlungen, 3–47. [for English translation see http://www.mendelweb.org/Mendel.plain.html (http://openstax.org/l/mendel_experiments)]

were they absorbed, but rather that they kept their distinctness and could be passed on. In 1868, Mendel became abbot of the monastery and exchanged his scientific pursuits for his pastoral duties. He was not recognized for his extraordinary scientific contributions during his lifetime. In fact, it was not until 1900 that his work was rediscovered, reproduced, and revitalized by scientists on the brink of discovering the chromosomal basis of heredity.

Mendel's Model System

Mendel's seminal work was accomplished using the garden pea, *Pisum sativum*, to study inheritance. This species naturally self-fertilizes, such that pollen encounters ova within individual flowers. The flower petals remain sealed tightly until after pollination, preventing pollination from other plants. The result is highly inbred, or "true-breeding," pea plants. These are plants that always produce offspring that look like the parent. By experimenting with true-breeding pea plants, Mendel avoided the appearance of unexpected traits in offspring that might occur if the plants were not true breeding. The garden pea also grows to maturity within one season, meaning that several generations could be evaluated over a relatively short time. Finally, large quantities of garden peas could be cultivated simultaneously, allowing Mendel to conclude that his results did not come about simply by chance.

Mendelian Crosses

Mendel performed **hybridizations**, which involve mating two true-breeding individuals that have different traits. In the pea, which is naturally self-pollinating, this is done by manually transferring pollen from the anther of a mature pea plant of one variety to the stigma of a separate mature pea plant of the second variety. In plants, pollen carries the male gametes (sperm) to the stigma, a sticky organ that traps pollen and allows the sperm to move down the pistil to the female gametes (ova) below. To prevent the pea plant that was receiving pollen from self-fertilizing and confounding his results, Mendel painstakingly removed all of the anthers from the plant's flowers before they had a chance to mature.

Plants used in first-generation crosses were called $\mathbf{P_0}$, or parental generation one (Figure 12.3). After each cross, Mendel collected the seeds belonging to the P_0 plants and grew them the following season. These offspring were called the $\mathbf{F_1}$, or the first filial (*filial* = offspring, daughter or son) generation. Once Mendel examined the characteristics in the F_1 generation of plants, he allowed them to self-fertilize naturally. He then collected and grew the seeds from the F_1 plants to produce the $\mathbf{F_2}$, or second filial, generation. Mendel's experiments extended beyond the F_2 generation to the F_3 and F_4 generations, and so on, but it was the ratio of characteristics in the P_0–F_1–F_2 generations that were the most intriguing and became the basis for Mendel's postulates.

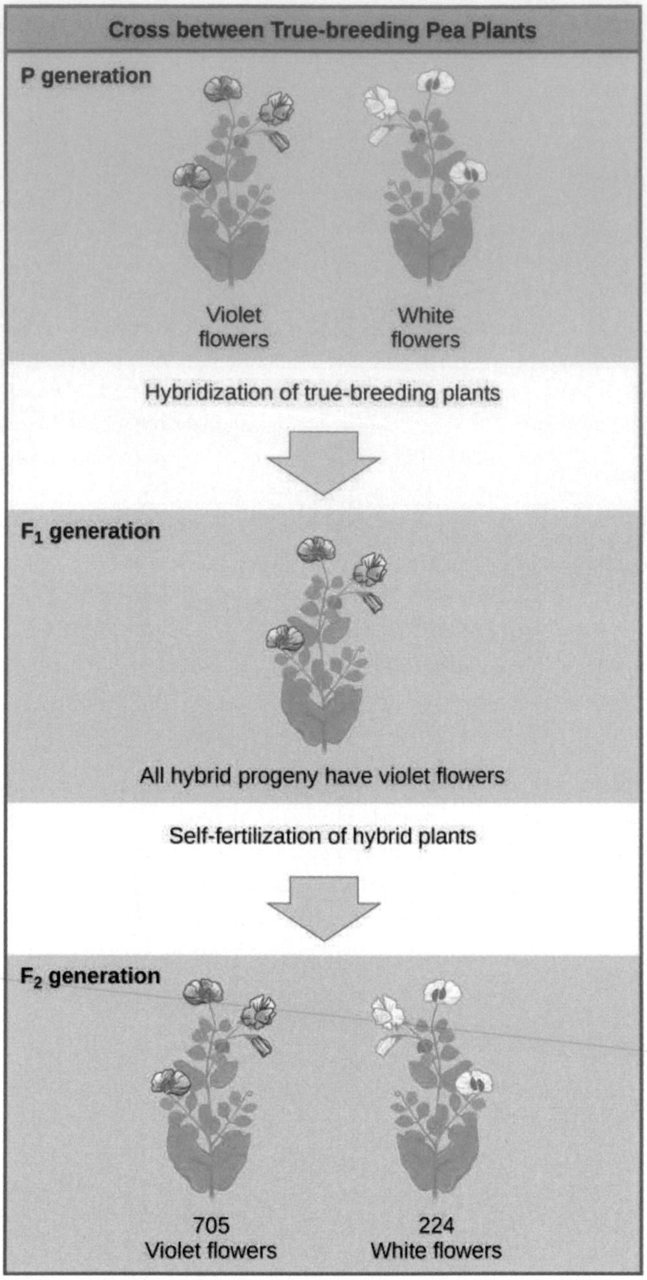

Figure 12.3 In one of his experiments on inheritance patterns, Mendel crossed plants that were true-breeding for violet flower color with plants true-breeding for white flower color (the P generation). The resulting hybrids in the F_1 generation all had violet flowers. In the F_2 generation, approximately three quarters of the plants had violet flowers, and one quarter had white flowers.

Garden Pea Characteristics Revealed the Basics of Heredity

In his 1865 publication, Mendel reported the results of his crosses involving seven different characteristics, each with two contrasting traits. A **trait** is defined as a variation in the physical appearance of a heritable characteristic. The characteristics included plant height, seed texture, seed color, flower color, pea pod size, pea pod color, and flower position. For the characteristic of flower color, for example, the two contrasting traits were white versus violet. To fully examine each characteristic, Mendel generated large numbers of F_1 and F_2 plants, reporting results from 19,959 F_2 plants alone. His findings were consistent.

What results did Mendel find in his crosses for flower color? First, Mendel confirmed that he had plants that bred true for white or violet flower color. Regardless of how many generations Mendel examined, all self-crossed offspring of parents with white flowers had white flowers, and all self-crossed offspring of parents with violet flowers had violet flowers. In addition, Mendel confirmed that, other than flower color, the pea plants were physically identical.

Once these validations were complete, Mendel applied the pollen from a plant with violet flowers to the stigma of a plant with white flowers. After gathering and sowing the seeds that resulted from this cross, *Mendel found that 100 percent of the F_1 hybrid generation had violet flowers.* Conventional wisdom at that time (the blending theory) would have predicted the hybrid flowers to be pale violet or for hybrid plants to have equal numbers of white and violet flowers. In other words, the contrasting parental traits were expected to blend in the offspring. Instead, Mendel's results demonstrated that the white flower trait in the F_1 generation had completely disappeared.

Importantly, Mendel did not stop his experimentation there. He allowed the F_1 plants to self-fertilize and found that, of F_2-generation plants, 705 had violet flowers and 224 had white flowers. This was a ratio of 3.15 violet flowers per one white flower, or approximately 3:1. When Mendel transferred pollen from a plant with violet flowers to the stigma of a plant with white flowers and vice versa, he obtained about the same ratio regardless of which parent, male or female, contributed which trait. This is called a **reciprocal cross**—a paired cross in which the respective traits of the male and female in one cross become the respective traits of the female and male in the other cross. For the other six characteristics Mendel examined, the F_1 and F_2 generations behaved in the same way as they had for flower color. One of the two traits would disappear completely from the F_1 generation only to reappear in the F_2 generation at a ratio of approximately 3:1 (Table 12.1).

The Results of Mendel's Garden Pea Hybridizations

Characteristic	Contrasting P_0 Traits	F_1 Offspring Traits	F_2 Offspring Traits	F_2 Trait Ratios
Flower color	Violet vs. white	100 percent violet	705 violet 224 white	3.15:1
Flower position	Axial vs. terminal	100 percent axial	651 axial 207 terminal	3.14:1
Plant height	Tall vs. dwarf	100 percent tall	787 tall 277 dwarf	2.84:1
Seed texture	Round vs. wrinkled	100 percent round	5,474 round 1,850 wrinkled	2.96:1
Seed color	Yellow vs. green	100 percent yellow	6,022 yellow 2,001 green	3.01:1
Pea pod texture	Inflated vs. constricted	100 percent inflated	882 inflated 299 constricted	2.95:1
Pea pod color	Green vs. yellow	100 percent green	428 green 152 yellow	2.82:1

Table 12.1

Upon compiling his results for many thousands of plants, Mendel concluded that the characteristics could be divided into expressed and latent traits. He called these, respectively, dominant and recessive traits. **Dominant traits** are those that are inherited unchanged in a hybridization. **Recessive traits** become latent, or disappear, in the offspring of a hybridization. The recessive trait does, however, reappear in the progeny of the hybrid offspring. An example of a dominant trait is the violet-flower trait. For this same characteristic (flower color), white-colored flowers are a recessive trait. The fact that the recessive trait reappeared in the F_2 generation meant that the traits remained separate (not blended) in the plants of the F_1 generation. Mendel also proposed that plants possessed two copies of the trait for the flower-color characteristic, and that each parent transmitted one of its two copies to its offspring, where they came together. Moreover, the physical observation of a dominant trait could mean that the genetic composition of the organism included two dominant versions of the characteristic or that it included one dominant and one recessive version. Conversely, the observation of a recessive trait meant that the organism lacked any dominant versions of this characteristic.

So why did Mendel repeatedly obtain 3:1 ratios in his crosses? To understand how Mendel deduced the basic mechanisms of inheritance that lead to such ratios, we must first review the laws of probability.

Probability Basics

Probabilities are mathematical measures of likelihood. The empirical probability of an event is calculated by dividing the number of times the event occurs by the total number of opportunities for the event to occur. It is also possible to calculate theoretical probabilities by dividing the number of times that an event is *expected* to occur by the number of times that it could occur. Empirical probabilities come from observations, like those of Mendel. Theoretical probabilities, on the other hand, come from knowing how the events are produced and assuming that the probabilities of individual outcomes are equal. A probability of one for some event indicates that it is guaranteed to occur, whereas a probability of zero indicates that it is guaranteed not to occur. An example of a genetic event is a round seed produced by a pea plant.

In one experiment, Mendel demonstrated that the probability of the event "round seed" occurring was one in the F_1 offspring of true-breeding parents, one of which has round seeds and one of which has wrinkled seeds. When the F_1 plants were subsequently self-crossed, the probability of any given F_2 offspring having round seeds was now three out of four. In other words, in a large population of F_2 offspring chosen at random, 75 percent were expected to have round seeds, whereas 25 percent were expected to have wrinkled seeds. Using large numbers of crosses, Mendel was able to calculate probabilities and use these to predict the outcomes of other crosses.

The Product Rule and Sum Rule

Mendel demonstrated that pea plants transmit characteristics as discrete units from parent to offspring. As will be discussed, Mendel also determined that different characteristics, like seed color and seed texture, were transmitted independently of one another and could be considered in separate probability analyses. For instance, performing a cross between a plant with green, wrinkled seeds and a plant with yellow, round seeds still produced offspring that had a 3:1 ratio of yellow:green seeds (ignoring seed texture) and a 3:1 ratio of wrinkled:round seeds (ignoring seed color). The characteristics of color and texture did not influence each other.

The **product rule** of probability can be applied to this phenomenon of the independent transmission of characteristics. The product rule states that the probability of two independent events occurring together can be calculated by multiplying the individual probabilities of each event occurring alone. To demonstrate the product rule, imagine that you are rolling a six-sided die (D) and flipping a penny (P) at the same time. The die may roll any number from 1–6 ($D_\#$), whereas the penny may turn up heads (P_H) or tails (P_T). The outcome of rolling the die has no effect on the outcome of flipping the penny and vice versa. There are 12 possible outcomes of this action (Table 12.2), and each event is expected to occur with equal probability.

Twelve Equally Likely Outcomes of Rolling a Die and Flipping a Penny

Rolling Die	Flipping Penny
D_1	P_H
D_1	P_T
D_2	P_H
D_2	P_T

Table 12.2

Rolling Die	Flipping Penny
D_3	P_H
D_3	P_T
D_4	P_H
D_4	P_T
D_5	P_H
D_5	P_T
D_6	P_H
D_6	P_T

Table 12.2

Of the 12 possible outcomes, the die has a 2/12 (or 1/6) probability of rolling a two, and the penny has a 6/12 (or 1/2) probability of coming up heads. By the product rule, the probability that you will obtain the combined outcome 2 and heads is: $(D_2) \times (P_H) = (1/6) \times (1/2)$ or 1/12 (Table 12.3). Notice the word "and" in the description of the probability. The "and" is a signal to apply the product rule. For example, consider how the product rule is applied to the dihybrid cross: the probability of having both dominant traits (for example, yellow and round) in the F_2 progeny is the product of the probabilities of having the dominant trait for each characteristic, as shown here:

$$^3/_4 \times {}^3/_4 = {}^9/_{16}$$

On the other hand, the **sum rule** of probability is applied when considering two mutually exclusive outcomes that can come about by more than one pathway. The sum rule states that the probability of the occurrence of one event or the other event, of two mutually exclusive events, is the sum of their individual probabilities. Notice the word "or" in the description of the probability. The "or" indicates that you should apply the sum rule. In this case, let's imagine you are flipping a penny (P) and a quarter (Q). What is the probability of one coin coming up heads and one coin coming up tails? This outcome can be achieved by two cases: the penny may be heads (P_H) and the quarter may be tails (Q_T), or the quarter may be heads (Q_H) and the penny may be tails (P_T). Either case fulfills the outcome. By the sum rule, we calculate the probability of obtaining one head and one tail as $[(P_H) \times (Q_T)] + [(Q_H) \times (P_T)] = [(1/2) \times (1/2)] + [(1/2) \times (1/2)] = 1/2$ (Table 12.3). You should also notice that we used the product rule to calculate the probability of P_H and Q_T, and also the probability of P_T and Q_H, before we summed them. Again, the sum rule can be applied to show the probability of having at least one dominant trait in the F_2 generation of a dihybrid cross:

$$\left({}^1/_4 \times {}^3/_4\right) + \left({}^3/_4 \times {}^1/_4\right) = {}^3/_{16} + {}^3/_{16} = {}^6/_{16} = {}^3/_8$$

The Product Rule and Sum Rule

Product Rule	Sum Rule
For independent events A and B, the probability (P) of them both occurring (A *and* B) is $(P_A \times P_B)$	For mutually exclusive events A and B, the probability (P) that at least one occurs (A *or* B) is $(P_A + P_B)$

Table 12.3

To use probability laws in practice, we must work with large sample sizes because small sample sizes are prone to deviations caused by chance. The large quantities of pea plants that Mendel examined allowed him calculate the probabilities of the traits appearing in his F_2 generation. As you will learn, this discovery meant that when parental traits were known, the offspring's traits could be predicted accurately even before fertilization.

12.2 Characteristics and Traits

By the end of this section, you will be able to do the following:

- Explain the relationship between genotypes and phenotypes in dominant and recessive gene systems
- Develop a Punnett square to calculate the expected proportions of genotypes and phenotypes in a monohybrid cross
- Explain the purpose and methods of a test cross
- Identify non-Mendelian inheritance patterns such as incomplete dominance, codominance, recessive lethals, multiple alleles, and sex linkage

Physical characteristics are expressed through genes carried on chromosomes. The genetic makeup of peas consists of two similar, or homologous, copies of each chromosome, one from each parent. Each pair of homologous chromosomes has the same linear order of genes. In other words, peas are diploid organisms in that they have two copies of each chromosome. The same is true for many other plants and for virtually all animals. Diploid organisms produce haploid gametes, which contain one copy of each homologous chromosome that unite at fertilization to create a diploid zygote.

For cases in which a single gene controls a single characteristic, a diploid organism has two genetic copies that may or may not encode the same version of that characteristic. Gene variants that arise by mutation and exist at the same relative locations on homologous chromosomes are called **alleles**. Mendel examined the inheritance of genes with just two allele forms, but it is common to encounter more than two alleles for any given gene in a natural population.

Phenotypes and Genotypes

Two alleles for a given gene in a diploid organism are expressed and interact to produce physical characteristics. The observable traits expressed by an organism are referred to as its **phenotype**. An organism's underlying genetic makeup, consisting of both physically visible and non-expressed alleles, is called its **genotype**. Mendel's hybridization experiments demonstrate the difference between phenotype and genotype. When true-breeding plants in which one parent had yellow pods and one had green pods were cross-fertilized, all of the F_1 hybrid offspring had yellow pods. That is, the hybrid offspring were phenotypically identical to the true-breeding parent with yellow pods. However, we know that the allele donated by the parent with green pods was not simply lost because it reappeared in some of the F_2 offspring. Therefore, the F_1 plants must have been genotypically different from the parent with yellow pods.

The P_1 plants that Mendel used in his experiments were each homozygous for the trait he was studying. Diploid organisms that are **homozygous** at a given gene, or locus, have two identical alleles for that gene on their homologous chromosomes. Mendel's parental pea plants always bred true because both of the gametes produced carried the same trait. When P_1 plants with contrasting traits were cross-fertilized, all of the offspring were **heterozygous** for the contrasting trait, meaning that their genotype reflected that they had different alleles for the gene being examined.

Dominant and Recessive Alleles

Our discussion of homozygous and heterozygous organisms brings us to why the F_1 heterozygous offspring were identical to one of the parents, rather than expressing both alleles. In all seven pea-plant characteristics, one of the two contrasting alleles was dominant, and the other was recessive. Mendel called the dominant allele the expressed unit factor; the recessive allele was referred to as the latent unit factor. We now know that these so-called unit factors are actually genes on homologous chromosome pairs. For a gene that is expressed in a dominant and recessive pattern, homozygous dominant and heterozygous organisms will look identical (that is, they will have different genotypes but the same phenotype). The recessive allele will only be observed in homozygous recessive individuals (Table 12.4).

Human Inheritance in Dominant and Recessive Patterns

Dominant Traits	Recessive Traits
Achondroplasia	Albinism
Brachydactyly	Cystic fibrosis
Huntington's disease	Duchenne muscular dystrophy
Marfan syndrome	Galactosemia
Neurofibromatosis	Phenylketonuria
Widow's peak	Sickle-cell anemia
Wooly hair	Tay-Sachs disease

Table 12.4

Several conventions exist for referring to genes and alleles. For the purposes of this chapter, we will abbreviate genes using the first letter of the gene's corresponding dominant trait. For example, violet is the dominant trait for a pea plant's flower color, so the flower-color gene would be abbreviated as *V* (note that it is customary to italicize gene designations). Furthermore, we will use uppercase and lowercase letters to represent dominant and recessive alleles, respectively. Therefore, we would refer to the genotype of a homozygous dominant pea plant with violet flowers as *VV*, a homozygous recessive pea plant with white flowers as *vv*, and a heterozygous pea plant with violet flowers as *Vv*.

The Punnett Square Approach for a Monohybrid Cross

When fertilization occurs between two true-breeding parents that differ in only one characteristic, the process is called a **monohybrid** cross, and the resulting offspring are monohybrids. Mendel performed seven monohybrid crosses involving contrasting traits for each characteristic. On the basis of his results in F_1 and F_2 generations, Mendel postulated that each parent in the monohybrid cross contributed one of two paired unit factors to each offspring, and every possible combination of unit factors was equally likely.

To demonstrate a monohybrid cross, consider the case of true-breeding pea plants with yellow versus green pea seeds. The dominant seed color is yellow; therefore, the parental genotypes were *YY* for the plants with yellow seeds and *yy* for the plants with green seeds, respectively. A **Punnett square**, devised by the British geneticist Reginald Punnett, can be drawn that applies the rules of probability to predict the possible outcomes of a genetic cross or mating and their expected frequencies. To prepare a Punnett square, all possible combinations of the parental alleles are listed along the top (for one parent) and side (for the other parent) of a grid, representing their meiotic segregation into haploid gametes. Then the combinations of egg and sperm are made in the boxes in the table to show which alleles are combining. Each box then represents the diploid genotype of a zygote, or fertilized egg, that could result from this mating. Because each possibility is equally likely, genotypic ratios can be determined from a Punnett square. If the pattern of inheritance (dominant or recessive) is known, the phenotypic ratios can be inferred as well. For a monohybrid cross of two true-breeding parents, each parent contributes one type of allele. In this case, only one genotype is possible. All offspring are *Yy* and have yellow seeds (Figure 12.4).

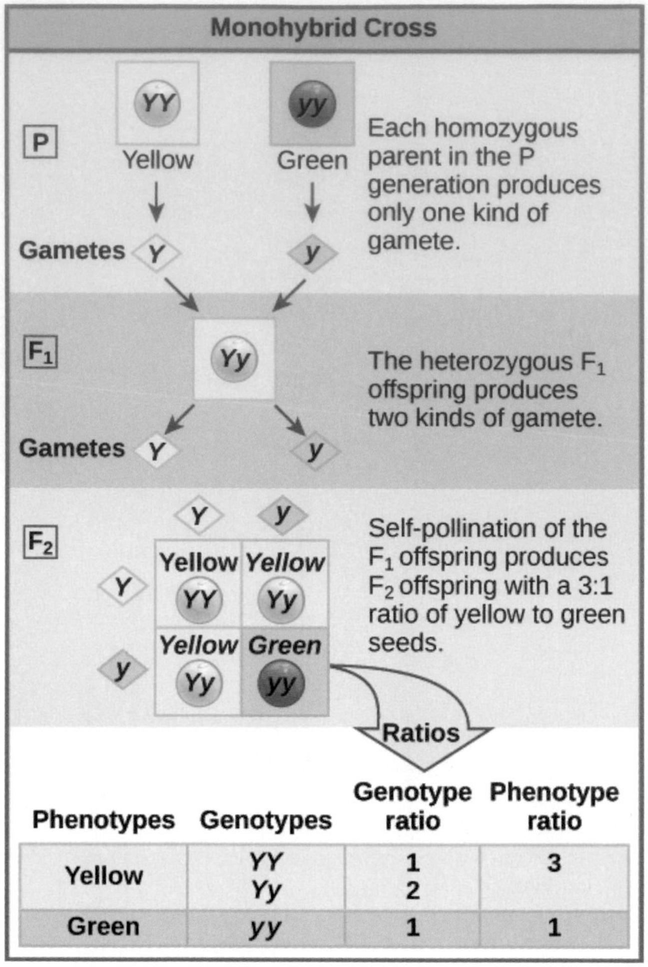

Figure 12.4 In the P generation, pea plants that are true-breeding for the dominant yellow phenotype are crossed with plants with the recessive green phenotype. This cross produces F_1 heterozygotes with a yellow phenotype. Punnett square analysis can be used to predict the genotypes of the F_2 generation.

A self-cross of one of the *Yy* heterozygous offspring can be represented in a 2 × 2 Punnett square because each parent can donate one of two different alleles. Therefore, the offspring can potentially have one of four allele combinations: *YY*, *Yy*, *yY*, or *yy* (Figure 12.4). Notice that there are two ways to obtain the *Yy* genotype: a *Y* from the egg and a *y* from the sperm, or a *y* from the egg and a *Y* from the sperm. Both of these possibilities must be counted. Recall that Mendel's pea-plant characteristics behaved in the same way in reciprocal crosses. Therefore, the two possible heterozygous combinations produce offspring that are genotypically and phenotypically identical despite their dominant and recessive alleles deriving from different parents. They are grouped together. Because fertilization is a random event, we expect each combination to be equally likely and for the offspring to exhibit a ratio of *YY*:*Yy*:*yy* genotypes of 1:2:1 (Figure 12.4). Furthermore, because the *YY* and *Yy* offspring have yellow seeds and are phenotypically identical, applying the sum rule of probability, we expect the offspring to exhibit a phenotypic ratio of 3 yellow:1 green. Indeed, working with large sample sizes, Mendel observed approximately this ratio in every F_2 generation resulting from crosses for individual traits.

Mendel validated these results by performing an F_3 cross in which he self-crossed the dominant- and recessive-expressing F_2 plants. When he self-crossed the plants expressing green seeds, all of the offspring had green seeds, confirming that all green seeds had homozygous genotypes of *yy*. When he self-crossed the F_2 plants expressing yellow seeds, he found that one-third of the plants bred true, and two-thirds of the plants segregated at a 3:1 ratio of yellow:green seeds. In this case, the true-breeding plants had homozygous (*YY*) genotypes, whereas the segregating plants corresponded to the heterozygous (*Yy*) genotype. When these plants self-fertilized, the outcome was just like the F_1 self-fertilizing cross.

The Test Cross Distinguishes the Dominant Phenotype

Beyond predicting the offspring of a cross between known homozygous or heterozygous parents, Mendel also developed a way to determine whether an organism that expressed a dominant trait was a heterozygote or a homozygote. Called the **test cross**, this technique is still used by plant and animal breeders. In a test cross, the dominant-expressing organism is crossed with an organism that is homozygous recessive for the same characteristic. If the dominant-expressing organism is a homozygote, then all F_1 offspring will be heterozygotes expressing the dominant trait (Figure 12.5). Alternatively, if the dominant expressing organism is a heterozygote, the F_1 offspring will exhibit a 1:1 ratio of heterozygotes and recessive homozygotes (Figure 12.5). The test cross further validates Mendel's postulate that pairs of unit factors segregate equally.

VISUAL CONNECTION

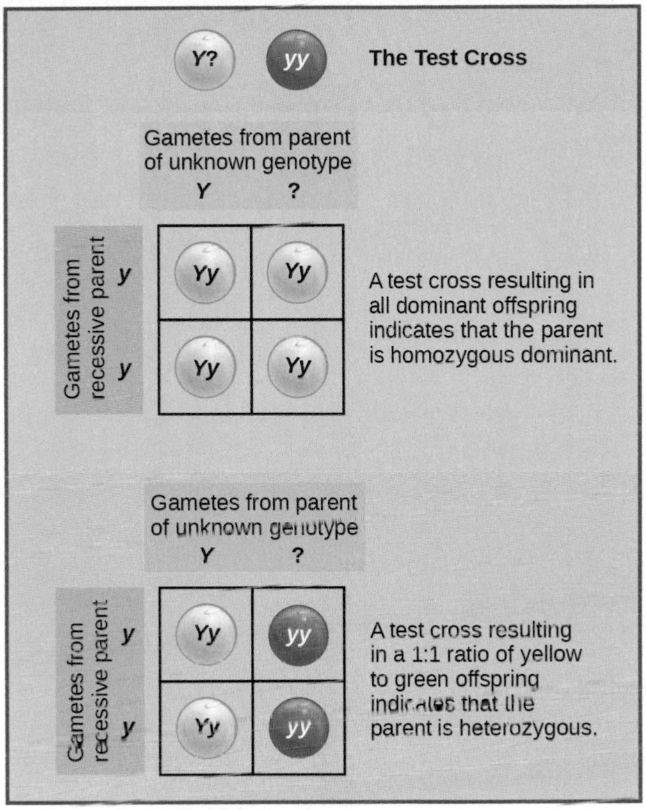

Figure 12.5 A test cross can be performed to determine whether an organism expressing a dominant trait is a homozygote or a heterozygote.

In pea plants, round peas (*R*) are dominant to wrinkled peas (*r*). You do a test cross between a pea plant with wrinkled peas (genotype *rr*) and a plant of unknown genotype that has round peas. You end up with three plants, all which have round peas. From this data, can you tell if the round pea parent plant is homozygous dominant or heterozygous? If the round pea parent plant is heterozygous, what is the probability that a random sample of 3 progeny peas will all be round?

Many human diseases are genetically inherited. A healthy person in a family in which some members suffer from a recessive genetic disorder may want to know if he or she has the disease-causing gene and what risk exists of passing the disorder on to his or her offspring. Of course, doing a test cross in humans is unethical and impractical. Instead, geneticists use **pedigree analysis** to study the inheritance pattern of human genetic diseases (Figure 12.6).

VISUAL CONNECTION

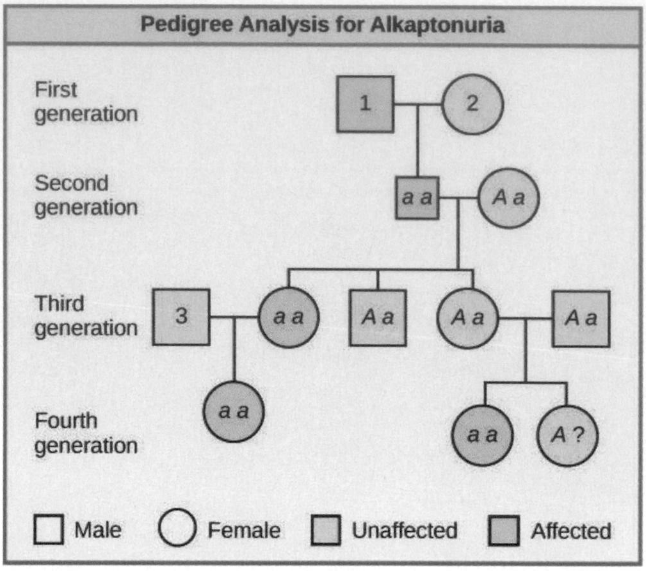

Figure 12.6 Alkaptonuria is a recessive genetic disorder in which two amino acids, phenylalanine and tyrosine, are not properly metabolized. Affected individuals may have darkened skin and brown urine, and may suffer joint damage and other complications. In this pedigree, individuals with the disorder are indicated in blue and have the genotype *aa*. Unaffected individuals are indicated in yellow and have the genotype *AA* or *Aa*. Note that it is often possible to determine a person's genotype from the genotype of their offspring. For example, if neither parent has the disorder but their child does, they must be heterozygous. Two individuals on the pedigree have an unaffected phenotype but unknown genotype. Because they do not have the disorder, they must have at least one normal allele, so their genotype gets the "*A?*" designation.

What are the genotypes of the individuals labeled 1, 2, and 3?

Alternatives to Dominance and Recessiveness

Mendel's experiments with pea plants suggested that: (1) two "units" or alleles exist for every gene; (2) alleles maintain their integrity in each generation (no blending); and (3) in the presence of the dominant allele, the recessive allele is hidden and makes no contribution to the phenotype. Therefore, recessive alleles can be "carried" and not expressed by individuals. Such heterozygous individuals are sometimes referred to as "carriers." Further genetic studies in other plants and animals have shown that much more complexity exists, but that the fundamental principles of Mendelian genetics still hold true. In the sections to follow, we consider some of the extensions of Mendelism. If Mendel had chosen an experimental system that exhibited these genetic complexities, it's possible that he would not have understood what his results meant.

Incomplete Dominance

Mendel's results, that traits are inherited as dominant and recessive pairs, contradicted the view at that time that offspring exhibited a blend of their parents' traits. However, the heterozygote phenotype occasionally does appear to be intermediate between the two parents. For example, in the snapdragon, *Antirrhinum majus* (Figure 12.7), a cross between a homozygous parent with white flowers (C^WC^W) and a homozygous parent with red flowers (C^RC^R) will produce offspring with pink flowers (C^RC^W). (Note that different genotypic abbreviations are used for Mendelian extensions to distinguish these patterns from simple dominance and recessiveness.) This pattern of inheritance is described as **incomplete dominance**, denoting the expression of two contrasting alleles such that the individual displays an intermediate phenotype. The allele for red flowers is incompletely dominant over the allele for white flowers. However, the results of a heterozygote self-cross can still be predicted, just as with Mendelian dominant and recessive crosses. In this case, the genotypic ratio would be 1 C^RC^R:2 C^RC^W:1 C^WC^W, and the phenotypic ratio would be 1:2:1 for red:pink:white.

Figure 12.7 These pink flowers of a heterozygote snapdragon result from incomplete dominance. (credit: "storebukkebruse"/Flickr)

Codominance

A variation on incomplete dominance is **codominance**, in which both alleles for the same characteristic are simultaneously expressed in the heterozygote. An example of codominance is the MN blood groups of humans. The M and N alleles are expressed in the form of an M or N antigen present on the surface of red blood cells. Homozygotes (L^ML^M and L^NL^N) express either the M or the N allele, and heterozygotes (L^ML^N) express both alleles equally. In a self-cross between heterozygotes expressing a codominant trait, the three possible offspring genotypes are phenotypically distinct. However, the 1:2:1 genotypic ratio characteristic of a Mendelian monohybrid cross still applies.

Multiple Alleles

Mendel implied that only two alleles, one dominant and one recessive, could exist for a given gene. We now know that this is an oversimplification. Although individual humans (and all diploid organisms) can only have two alleles for a given gene, multiple alleles may exist at the population level such that many combinations of two alleles are observed. Note that when many alleles exist for the same gene, the convention is to denote the most common phenotype or genotype among wild animals as the **wild type** (often abbreviated "+"); this is considered the standard or norm. All other phenotypes or genotypes are considered **variants** of this standard, meaning that they deviate from the wild type. The variant may be recessive or dominant to the wild-type allele.

An example of multiple alleles is coat color in rabbits (Figure 12.8). Here, four alleles exist for the *c* gene. The wild-type version, C^+C^+, is expressed as brown fur. The chinchilla phenotype, $c^{ch}c^{ch}$, is expressed as black-tipped white fur. The Himalayan phenotype, c^hc^h, has black fur on the extremities and white fur elsewhere. Finally, the albino, or "colorless" phenotype, *cc*, is expressed as white fur. In cases of multiple alleles, dominance hierarchies can exist. In this case, the wild-type allele is dominant over all the others, chinchilla is incompletely dominant over Himalayan and albino, and Himalayan is dominant over albino. This hierarchy, or allelic series, was revealed by observing the phenotypes of each possible heterozygote offspring.

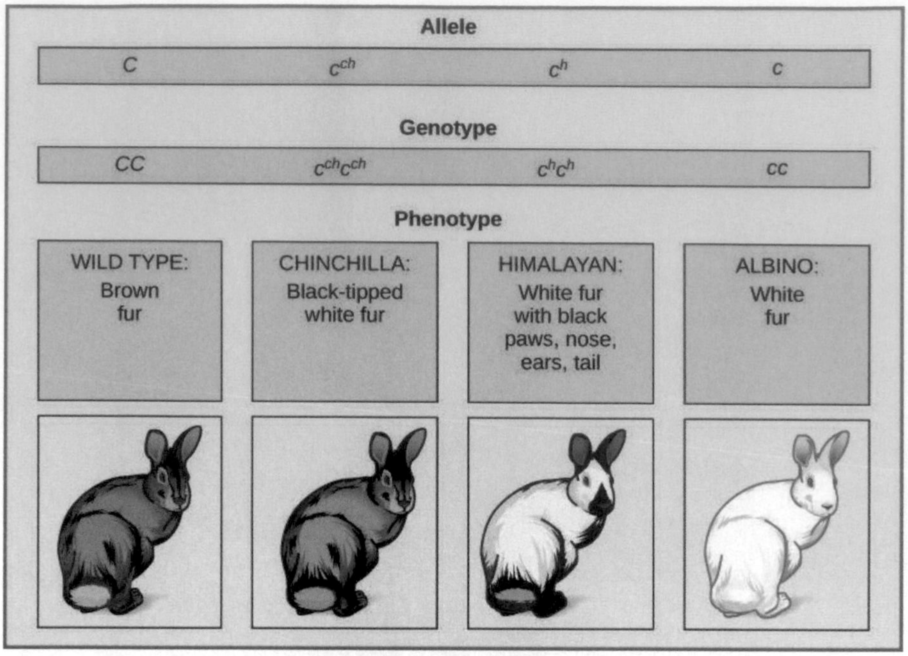

Figure 12.8 Four different alleles exist for the rabbit coat color (*C*) gene.

The complete dominance of a wild-type phenotype over all other mutants often occurs as an effect of "dosage" of a specific gene product, such that the wild-type allele supplies the correct amount of gene product whereas the mutant alleles cannot. For the allelic series in rabbits, the wild-type allele may supply a given dosage of fur pigment, whereas the mutants supply a lesser dosage or none at all. Interestingly, the Himalayan phenotype is the result of an allele that produces a temperature-sensitive gene product that only produces pigment in the cooler extremities of the rabbit's body.

Alternatively, one mutant allele can be dominant over all other phenotypes, including the wild type. This may occur when the mutant allele somehow interferes with the genetic message so that even a heterozygote with one wild-type allele copy expresses the mutant phenotype. One way in which the mutant allele can interfere is by enhancing the function of the wild-type gene product or changing its distribution in the body. One example of this is the *Antennapedia* mutation in *Drosophila* (Figure 12.9). In this case, the mutant allele expands the distribution of the gene product, and as a result, the *Antennapedia* heterozygote develops legs on its head where its antennae should be.

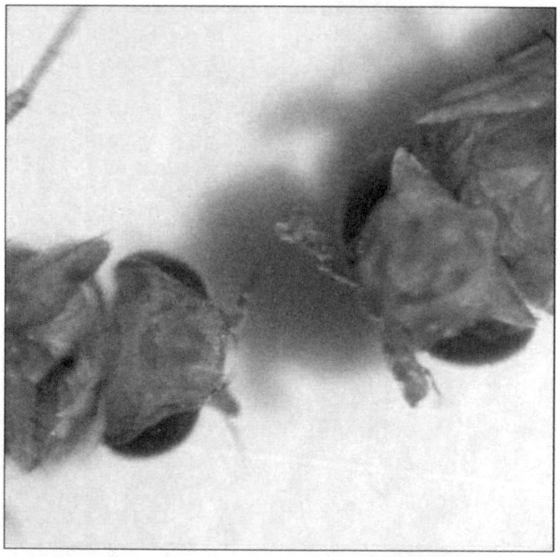

Figure 12.9 As seen in comparing the wild-type *Drosophila* (left) and the *Antennapedia* mutant (right), the *Antennapedia* mutant has legs on its head in place of antennae.

EVOLUTION CONNECTION

Multiple Alleles Confer Drug Resistance in the Malaria Parasite

Malaria is a parasitic disease in humans that is transmitted by infected female mosquitoes, including *Anopheles gambiae* (Figure 12.10a), and is characterized by cyclic high fevers, chills, flu-like symptoms, and severe anemia. *Plasmodium falciparum* and *P. vivax* are the most common causative agents of malaria, and *P. falciparum* is the most deadly (Figure 12.10b). When promptly and correctly treated, *P. falciparum* malaria has a mortality rate of 0.1 percent. However, in some parts of the world, the parasite has evolved resistance to commonly used malaria treatments, so the most effective malarial treatments can vary by geographic region.

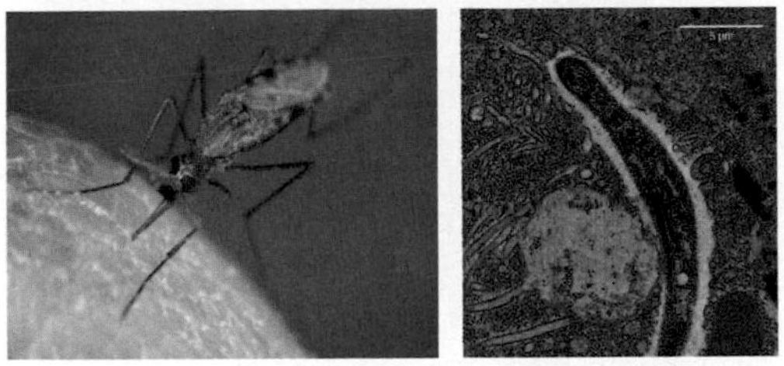

Figure 12.10 The (a) *Anopheles gambiae*, or African malaria mosquito, acts as a vector in the transmission to humans of the malaria-causing parasite (b) *Plasmodium falciparum*, here visualized using false-color transmission electron microscopy. (credit a: James D. Gathany; credit b: Ute Frevert; false color by Margaret Shear; scale-bar data from Matt Russell)

In Southeast Asia, Africa, and South America, *P. falciparum* has developed resistance to the anti-malarial drugs chloroquine, mefloquine, and sulfadoxine-pyrimethamine. *P. falciparum*, which is haploid during the life stage in which it is infectious to humans, has evolved multiple drug-resistant mutant alleles of the *dhps* gene. Varying degrees of sulfadoxine resistance are associated with each of these alleles. Being haploid, *P. falciparum* needs only one drug-resistant allele to express this trait.

In Southeast Asia, different sulfadoxine-resistant alleles of the *dhps* gene are localized to different geographic regions. This is a common evolutionary phenomenon that occurs because drug-resistant mutants arise in a population and interbreed with other *P. falciparum* isolates in close proximity. Sulfadoxine-resistant parasites cause considerable human hardship in regions where this drug is widely used as an over-the-counter malaria remedy. As is common with pathogens that multiply to large numbers within an infection cycle, *P. falciparum* evolves relatively rapidly (over a decade or so) in response to the selective pressure of commonly used anti-malarial drugs. For this reason, scientists must constantly work to develop new drugs or drug combinations to combat the worldwide malaria burden.[2]

X-Linked Traits

In humans, as well as in many other animals and some plants, the sex of the individual is determined by sex chromosomes. The sex chromosomes are one pair of non-homologous chromosomes. Until now, we have only considered inheritance patterns among non-sex chromosomes, or **autosomes**. In addition to 22 homologous pairs of autosomes, human females have a homologous pair of X chromosomes, whereas human males have an XY chromosome pair. Although the Y chromosome contains a small region of similarity to the X chromosome so that they can pair during meiosis, the Y chromosome is much shorter and contains many fewer genes. When a gene being examined is present on the X chromosome, but not on the Y chromosome, it is said to be **X-linked**.

Eye color in *Drosophila* was one of the first X-linked traits to be identified. Thomas Hunt Morgan mapped this trait to the X chromosome in 1910. Like humans, *Drosophila* males have an XY chromosome pair, and females are XX. In flies, the wild-type eye color is red (X^W) and it is dominant to white eye color (X^w) (Figure 12.11). Because of the location of the eye-color gene, reciprocal crosses do not produce the same offspring ratios. Males are said to be **hemizygous**, because they have only one allele for any X-linked characteristic. Hemizygosity makes the descriptions of dominance and recessiveness irrelevant for XY males. *Drosophila* males lack a second allele copy on the Y chromosome; that is, their genotype can only be X^WY or X^wY. In contrast, females have two allele copies of this gene and can be X^WX^W, X^WX^w, or X^wX^w.

2Sumiti Vinayak, et al., "Origin and Evolution of Sulfadoxine Resistant *Plasmodium falciparum*," *Public Library of Science Pathogens* 6, no. 3 (2010): e1000830, doi:10.1371/journal.ppat.1000830.

Figure 12.11 In *Drosophila*, several genes determine eye color. The genes for white and vermilion eye colors are located on the X chromosome. Others are located on the autosomes. Clockwise from top left are brown, cinnabar, sepia, vermilion, white, and red. Red eye color is wild-type and is dominant to white eye color.

In an X-linked cross, the genotypes of F_1 and F_2 offspring depend on whether the recessive trait was expressed by the male or the female in the P_1 generation. With regard to *Drosophila* eye color, when the P_1 male expresses the white-eye phenotype and the female is homozygous red-eyed, all members of the F_1 generation exhibit red eyes (Figure 12.12). The F_1 females are heterozygous (X^WX^w), and the males are all X^WY, having received their X chromosome from the homozygous dominant P_1 female and their Y chromosome from the P_1 male. A subsequent cross between the X^WX^w female and the X^WY male would produce only red-eyed females (with X^WX^W or X^WX^w genotypes) and both red- and white-eyed males (with X^WY or X^wY genotypes). Now, consider a cross between a homozygous white-eyed female and a male with red eyes. The F_1 generation would exhibit only heterozygous red-eyed females (X^WX^w) and only white-eyed males (X^wY). Half of the F_2 females would be red-eyed (X^WX^w) and half would be white-eyed (X^wX^w). Similarly, half of the F_2 males would be red-eyed (X^WY) and half would be white-eyed (X^wY).

VISUAL CONNECTION

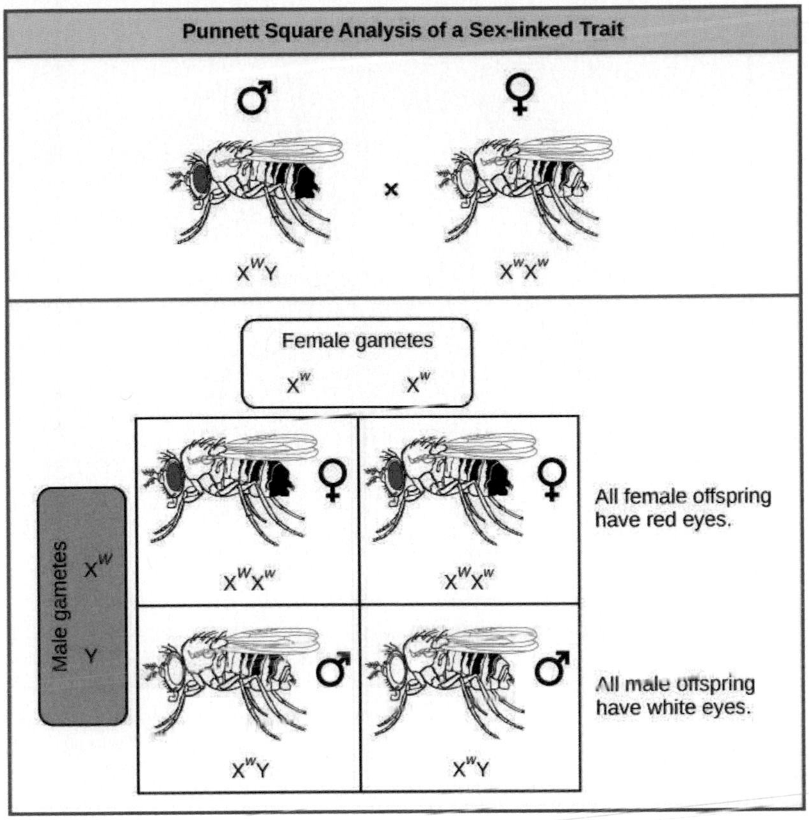

Figure 12.12 Punnett square analysis is used to determine the ratio of offspring from a cross between a red-eyed male fruit fly and a white eyed female fruit fly.

What ratio of offspring would result from a cross between a white-eyed male and a female that is heterozygous for red eye color?

Discoveries in fruit fly genetics can be applied to human genetics. When a female parent is homozygous for a recessive X-linked trait, she will pass the trait on to 100 percent of her offspring. Her male offspring are, therefore, destined to express the trait, as they will inherit their father's Y chromosome. In humans, the alleles for certain conditions (some forms of color blindness, hemophilia, and muscular dystrophy) are X-linked. Females who are heterozygous for these diseases are said to be carriers and may not exhibit any phenotypic effects. These females will pass the disease to half of their sons and will pass carrier status to half of their daughters; therefore, recessive X-linked traits appear more frequently in males than females.

In some groups of organisms with sex chromosomes, the sex with the non-homologous sex chromosomes is the female rather than the male. This is the case for all birds. In this case, sex-linked traits will be more likely to appear in the female, in which they are hemizygous.

Human Sex-linked Disorders

Sex-linkage studies in Morgan's laboratory provided the fundamentals for understanding X-linked recessive disorders in humans, which include red-green color blindness, and Types A and B hemophilia. Because human males need to inherit only one recessive mutant X allele to be affected, X-linked disorders are disproportionately observed in males. Females must inherit recessive X-linked alleles from both of their parents in order to express the trait. When they inherit one recessive X-linked mutant allele and one dominant X-linked wild-type allele, they are carriers of the trait and are typically unaffected. Carrier females can manifest mild forms of the trait due to the inactivation of the dominant allele located on one of the X chromosomes. However, female carriers can contribute the trait to their sons, resulting in the son exhibiting the trait, or they can contribute the recessive allele to their daughters, resulting in the daughters being carriers of the trait (Figure 12.13). Although some Y-linked recessive disorders exist, typically they are associated with infertility in males and are therefore not transmitted to subsequent generations.

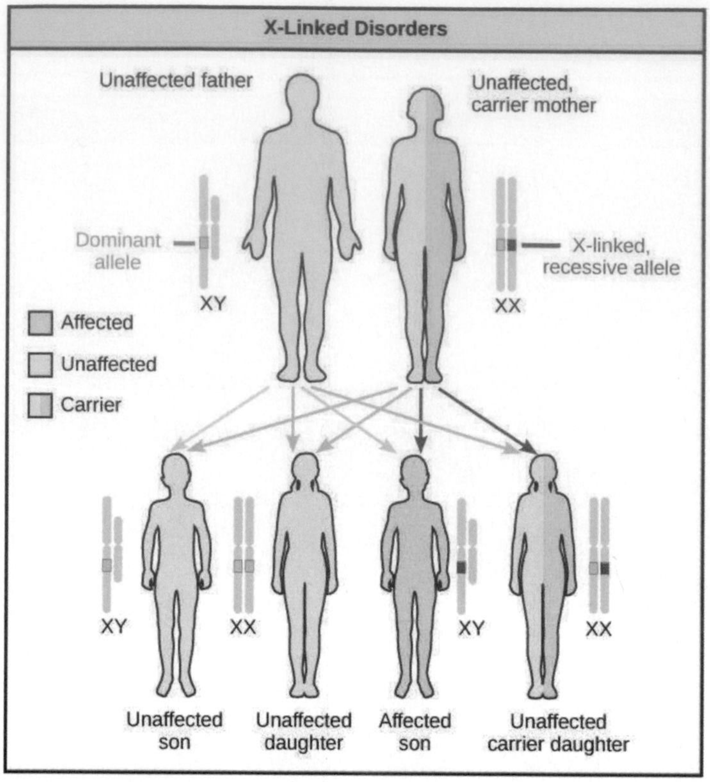

Figure 12.13 The son of a woman who is a carrier of a recessive X-linked disorder will have a 50 percent chance of being affected. A daughter will not be affected, but she will have a 50 percent chance of being a carrier like her mother.

LINK TO LEARNING

Watch this video to learn more about sex-linked traits.

Click to view content (https://www.openstax.org/l/sex-linked_trts)

Lethality

A large proportion of genes in an individual's genome are essential for survival. Occasionally, a nonfunctional allele for an essential gene can arise by mutation and be transmitted in a population as long as individuals with this allele also have a wild-type, functional copy. The wild-type allele functions at a capacity sufficient to sustain life and is therefore considered to be dominant over the nonfunctional allele. However, consider two heterozygous parents that have a genotype of wild-type/nonfunctional mutant for a hypothetical essential gene. In one quarter of their offspring, we would expect to observe individuals that are homozygous recessive for the nonfunctional allele. Because the gene is essential, these individuals might fail to develop past fertilization, die *in utero*, or die later in life, depending on what life stage requires this gene. An inheritance pattern in which an allele is only lethal in the homozygous form and in which the heterozygote may be normal or have some altered nonlethal phenotype is referred to as **recessive lethal**.

For crosses between heterozygous individuals with a recessive lethal allele that causes death before birth when homozygous, only wild-type homozygotes and heterozygotes would be observed. The genotypic ratio would therefore be 2:1. In other instances, the recessive lethal allele might also exhibit a dominant (but not lethal) phenotype in the heterozygote. For instance, the recessive lethal *Curly* allele in *Drosophila* affects wing shape in the heterozygote form but is lethal in the homozygote.

A single copy of the wild-type allele is not always sufficient for normal functioning or even survival. The **dominant lethal** inheritance pattern is one in which an allele is lethal both in the homozygote and the heterozygote; this allele can only be transmitted if the lethality phenotype occurs after reproductive age. Individuals with mutations that result in dominant lethal alleles fail to survive even in the heterozygote form. Dominant lethal alleles are very rare because, as you might expect, the allele only lasts one generation and is not transmitted. However, just as the recessive lethal allele might not immediately manifest the phenotype of death, dominant lethal alleles also might not be expressed until adulthood. Once the individual reaches reproductive age, the allele may be unknowingly passed on, resulting in a delayed death in both generations. An example of this in humans is Huntington's disease, in which the nervous system gradually wastes away (Figure 12.14). People who are heterozygous for the dominant Huntington allele (*Hh*) will inevitably develop the fatal disease. However, the onset of Huntington's disease may not occur until age 40, at which point the afflicted persons may have already passed the allele to 50 percent of their offspring.

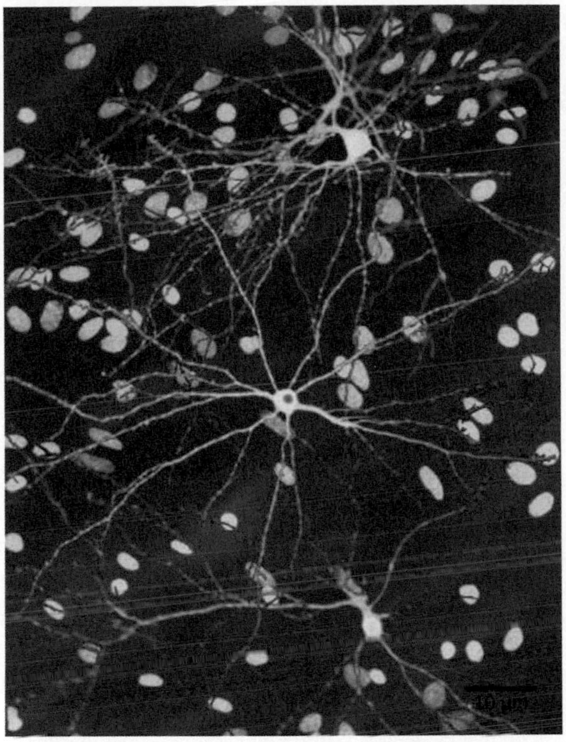

Figure 12.14 The neuron in the center of this micrograph (yellow) has nuclear inclusions characteristic of Huntington's disease (orange area in the center of the neuron). Huntington's disease occurs when an abnormal dominant allele for the Huntington gene is present. (credit: Dr. Steven Finkbeiner, Gladstone Institute of Neurological Disease, The Taube-Koret Center for Huntington's Disease Research, and the University of California San Francisco/Wikimedia)

12.3 Laws of Inheritance

By the end of this section, you will be able to do the following:

- Explain Mendel's law of segregation and independent assortment in terms of genetics and the events of meiosis
- Use the forked-line method and the probability rules to calculate the probability of genotypes and phenotypes from multiple gene crosses
- Explain the effect of linkage and recombination on gamete genotypes
- Explain the phenotypic outcomes of epistatic effects between genes

Mendel generalized the results of his pea-plant experiments into four postulates, some of which are sometimes called "laws," that describe the basis of dominant and recessive inheritance in diploid organisms. As you have learned, more complex extensions of Mendelism exist that do not exhibit the same F_2 phenotypic ratios (3:1). Nevertheless, these laws summarize the basics of classical genetics.

Pairs of Unit Factors, or Genes

Mendel proposed first that paired unit factors of heredity were transmitted faithfully from generation to generation by the dissociation and reassociation of paired factors during gametogenesis and fertilization, respectively. After he crossed peas with contrasting traits and found that the recessive trait resurfaced in the F_2 generation, Mendel deduced that hereditary factors must be inherited as discrete units. This finding contradicted the belief at that time that parental traits were blended in the offspring.

Alleles Can Be Dominant or Recessive

Mendel's **law of dominance** states that in a heterozygote, one trait will conceal the presence of another trait for the same characteristic. Rather than both alleles contributing to a phenotype, the dominant allele will be expressed exclusively. The recessive allele will remain "latent" but will be transmitted to offspring by the same manner in which the dominant allele is transmitted. The recessive trait will only be expressed by offspring that have two copies of this allele (Figure 12.15), and these offspring will breed true when self-crossed.

Since Mendel's experiments with pea plants, researchers have found that the law of dominance does not always hold true. Instead, several different patterns of inheritance have been found to exist.

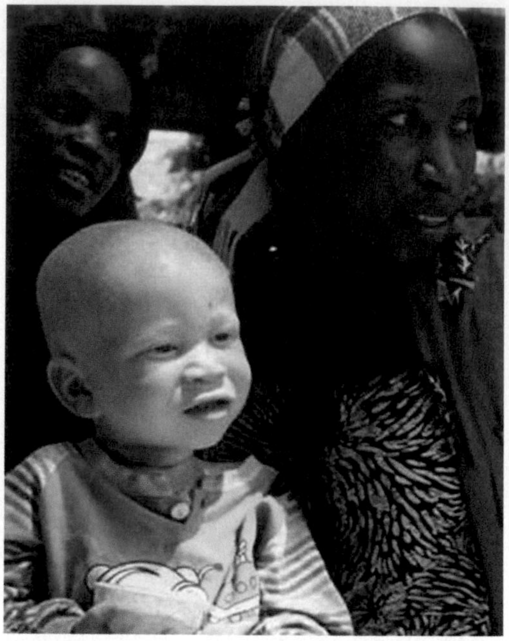

Figure 12.15 The child in the photo expresses albinism, a recessive trait.

Equal Segregation of Alleles

Observing that true-breeding pea plants with contrasting traits gave rise to F_1 generations that all expressed the dominant trait and F_2 generations that expressed the dominant and recessive traits in a 3:1 ratio, Mendel proposed the **law of segregation**. This law states that paired unit factors (genes) must segregate equally into gametes such that offspring have an equal likelihood of inheriting either factor. For the F_2 generation of a monohybrid cross, the following three possible combinations of genotypes could result: homozygous dominant, heterozygous, or homozygous recessive. Because heterozygotes could arise from two different pathways (receiving one dominant and one recessive allele from either parent), and because heterozygotes and homozygous dominant individuals are phenotypically identical, the law supports Mendel's observed 3:1 phenotypic ratio. The equal segregation of alleles is the reason we can apply the Punnett square to accurately predict the offspring of parents with known genotypes. The physical basis of Mendel's law of segregation is the first division of meiosis, in which the homologous chromosomes with their different versions of each gene are segregated into daughter nuclei. The role of the meiotic segregation of chromosomes in sexual reproduction was not understood by the scientific community during Mendel's lifetime.

Independent Assortment

Mendel's **law of independent assortment** states that genes do not influence each other with regard to the sorting of alleles into gametes, and every possible combination of alleles for every gene is equally likely to occur. The independent assortment of genes can be illustrated by the **dihybrid** cross, a cross between two true-breeding parents that express different traits for two characteristics. Consider the characteristics of seed color and seed texture for two pea plants, one that has green, wrinkled seeds (*yyrr*) and another that has yellow, round seeds (*YYRR*). Because each parent is homozygous, the law of segregation indicates that the gametes for the green/wrinkled plant all are *yr*, and the gametes for the yellow/round plant are all *YR*. Therefore, the F_1 generation of offspring all are *YyRr* (Figure 12.16).

VISUAL CONNECTION

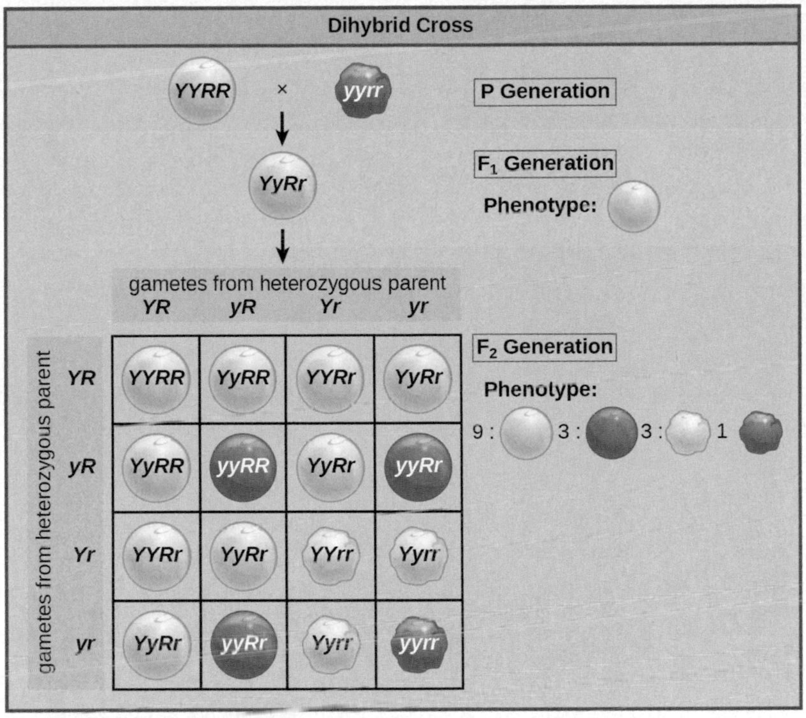

Figure 12.16 This dihybrid cross of pea plants involves the genes for seed color and texture.

In pea plants, purple flowers (P) are dominant to white flowers (p) and yellow peas (Y) are dominant to green peas (y). What are the possible genotypes and phenotypes for a cross between PpYY and ppYy pea plants? How many squares do you need to do a Punnett square analysis of this cross?

For the F2 generation, the law of segregation requires that each gamete receive either an *R* allele or an *r* allele along with either a *Y* allele or a *y* allele. The law of independent assortment states that a gamete into which an *r* allele sorted would be equally likely to contain either a *Y* allele or a *y* allele. Thus, there are four equally likely gametes that can be formed when the *YyRr* heterozygote is self-crossed, as follows: *YR*, *Yr*, *yR*, and *yr*. Arranging these gametes along the top and left of a 4 × 4 Punnett square (Figure 12.16) gives us 16 equally likely genotypic combinations. From these genotypes, we infer a phenotypic ratio of 9 round/yellow:3 round/green:3 wrinkled/yellow:1 wrinkled/green (Figure 12.16). These are the offspring ratios we would expect, assuming we performed the crosses with a large enough sample size.

Because of independent assortment and dominance, the 9:3:3:1 dihybrid phenotypic ratio can be collapsed into two 3:1 ratios, characteristic of any monohybrid cross that follows a dominant and recessive pattern. Ignoring seed color and considering only seed texture in the above dihybrid cross, we would expect that three quarters of the F_2 generation offspring would be round, and one quarter would be wrinkled. Similarly, isolating only seed color, we would assume that three quarters of the F_2 offspring would be yellow and one quarter would be green. The sorting of alleles for texture and color are independent events, so we can apply the product rule. Therefore, the proportion of round and yellow F_2 offspring is expected to be (3/4) × (3/4) = 9/16, and the proportion of wrinkled and green offspring is expected to be (1/4) × (1/4) = 1/16. These proportions are identical to those obtained using a Punnett square. Round, green and wrinkled, yellow offspring can also be calculated using the product rule, as each of these genotypes includes one dominant and one recessive phenotype. Therefore, the proportion of each is calculated as (3/4) × (1/4) = 3/16.

The law of independent assortment also indicates that a cross between yellow, wrinkled (*YYrr*) and green, round (*yyRR*) parents would yield the same F_1 and F_2 offspring as in the *YYRR* x *yyrr* cross.

The physical basis for the law of independent assortment also lies in meiosis I, in which the different homologous pairs line up in random orientations. Each gamete can contain any combination of paternal and maternal chromosomes (and therefore the genes on them) because the orientation of tetrads on the metaphase plane is random.

Forked-Line Method

When more than two genes are being considered, the Punnett-square method becomes unwieldy. For instance, examining a cross involving four genes would require a 16 × 16 grid containing 256 boxes. It would be extremely cumbersome to manually enter each genotype. For more complex crosses, the forked-line and probability methods are preferred.

To prepare a forked-line diagram for a cross between F_1 heterozygotes resulting from a cross between *AABBCC* and *aabbcc* parents, we first create rows equal to the number of genes being considered, and then segregate the alleles in each row on forked lines according to the probabilities for individual monohybrid crosses (Figure 12.17). We then multiply the values along each forked path to obtain the F_2 offspring probabilities. Note that this process is a diagrammatic version of the product rule. The values along each forked pathway can be multiplied because each gene assorts independently. For a trihybrid cross, the F_2 phenotypic ratio is 27:9:9:9:3:3:3:1.

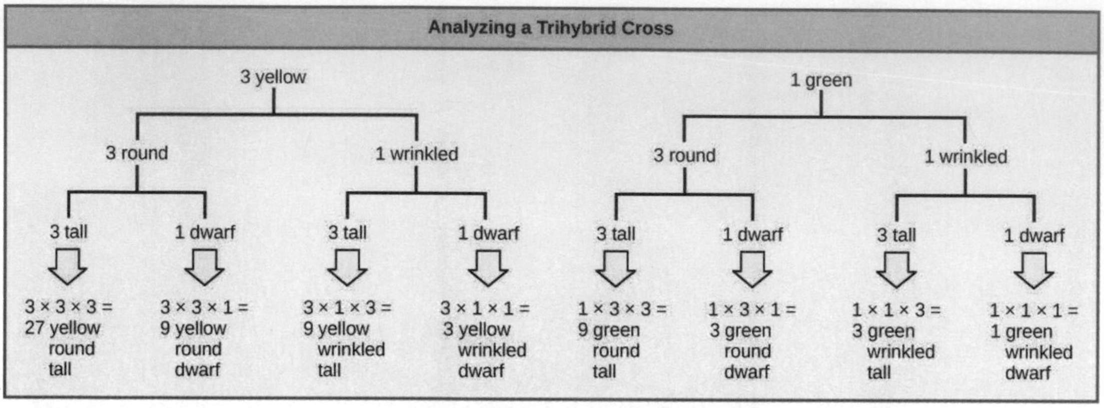

Figure 12.17 The forked-line method can be used to analyze a trihybrid cross. Here, the probability for color in the F_2 generation occupies the top row (3 yellow:1 green). The probability for shape occupies the second row (3 round: 1 wrinkled), and the probability for height occupies the third row (3 tall:1 dwarf). The probability for each possible combination of traits is calculated by multiplying the probability for each individual trait. Thus, the probability of F_2 offspring having yellow, round, and tall traits is 3 × 3 × 3, or 27.

Probability Method

While the forked-line method is a diagrammatic approach to keeping track of probabilities in a cross, the probability method gives the proportions of offspring expected to exhibit each phenotype (or genotype) without the added visual assistance. Both methods make use of the product rule and consider the alleles for each gene separately. Earlier, we examined the phenotypic proportions for a trihybrid cross using the forked-line method; now we will use the probability method to examine the genotypic proportions for a cross with even more genes.

For a trihybrid cross, writing out the forked-line method is tedious, albeit not as tedious as using the Punnett-square method. To fully demonstrate the power of the probability method, however, we can consider specific genetic calculations. For instance, for a tetrahybrid cross between individuals that are heterozygotes for all four genes, and in which all four genes are sorting independently and in a dominant and recessive pattern, what proportion of the offspring will be expected to be homozygous recessive for all four alleles? Rather than writing out every possible genotype, we can use the probability method. We know that for each gene, the fraction of homozygous recessive offspring will be 1/4. Therefore, multiplying this fraction for each of the four genes, (1/4) × (1/4) × (1/4) × (1/4), we determine that 1/256 of the offspring will be quadruply homozygous recessive.

For the same tetrahybrid cross, what is the expected proportion of offspring that have the dominant phenotype at all four loci? We can answer this question using phenotypic proportions, but let's do it the hard way—using genotypic proportions. The question asks for the proportion of offspring that are 1) homozygous dominant at *A* or heterozygous at *A*, and 2) homozygous at *B* or heterozygous at *B*, and so on. Noting the "or" and "and" in each circumstance makes clear where to apply the sum and product rules. The probability of a homozygous dominant at *A* is 1/4 and the probability of a heterozygote at *A* is 1/2. The probability of the homozygote or the heterozygote is 1/4 + 1/2 = 3/4 using the sum rule. The same probability can be obtained in the same way for each of the other genes, so that the probability of a dominant phenotype at *A* and *B* and *C* and *D* is, using the product rule, equal to 3/4 × 3/4 × 3/4 × 3/4, or 81/256. If you are ever unsure about how to combine probabilities, returning to the forked-line method should make it clear.

Rules for Multihybrid Fertilization

Predicting the genotypes and phenotypes of offspring from given crosses is the best way to test your knowledge of Mendelian genetics. Given a multihybrid cross that obeys independent assortment and follows a dominant and recessive pattern, several generalized rules exist; you can use these rules to check your results as you work through genetics calculations (Table 12.5). To apply these rules, first you must determine *n*, the number of heterozygous gene pairs (the number of genes segregating two alleles each). For example, a cross between *AaBb* and *AaBb* heterozygotes has an *n* of 2. In contrast, a cross between *AABb* and *AABb* has an *n* of 1 because *A* is not heterozygous.

General Rules for Multihybrid Crosses

General Rule	Number of Heterozygous Gene Pairs
Number of different F_1 gametes	2^n
Number of different F_2 genotypes	3^n
Given dominant and recessive inheritance, the number of different F_2 phenotypes	2^n

Table 12.5

Linked Genes Violate the Law of Independent Assortment

Although all of Mendel's pea characteristics behaved according to the law of independent assortment, we now know that some allele combinations are not inherited independently of each other. Genes that are located on separate non-homologous chromosomes will always sort independently. However, each chromosome contains hundreds or thousands of genes, organized linearly on chromosomes like beads on a string. The segregation of alleles into gametes can be influenced by **linkage**, in which genes that are located physically close to each other on the same chromosome are more likely to be inherited as a pair. However, because of the process of recombination, or "crossover," it is possible for two genes on the same chromosome to behave independently, or as if they are not linked. To understand this, let's consider the biological basis of gene linkage and recombination.

Homologous chromosomes possess the same genes in the same linear order. The alleles may differ on homologous chromosome pairs, but the genes to which they correspond do not. In preparation for the first division of meiosis, homologous chromosomes replicate and synapse. Like genes on the homologs align with each other. At this stage, segments of homologous chromosomes exchange linear segments of genetic material (Figure 12.18). This process is called *recombination*, or crossover, and it is a common genetic process. Because the genes are aligned during recombination, the gene order is not altered. Instead, the result of recombination is that maternal and paternal alleles are combined onto the same chromosome. Across a given chromosome, several recombination events may occur, causing extensive shuffling of alleles.

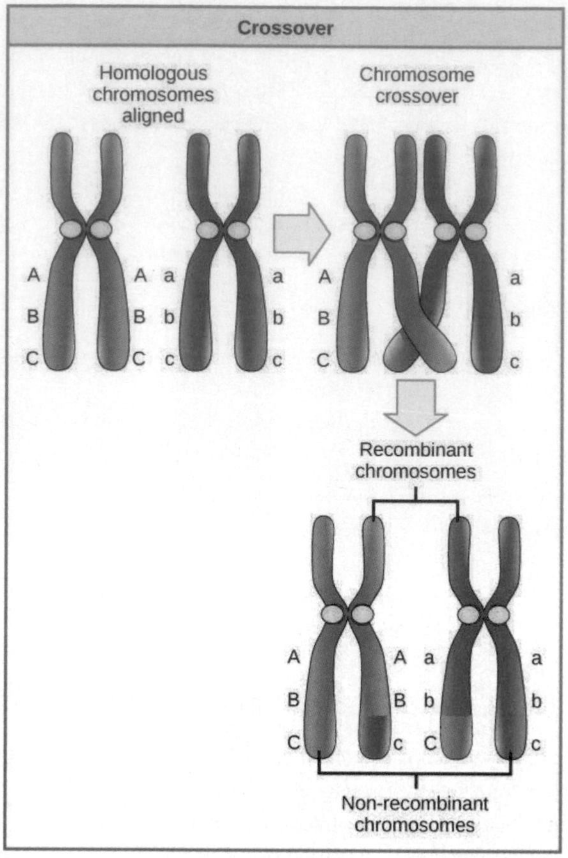

Figure 12.18 The process of crossover, or recombination, occurs when two homologous chromosomes align during meiosis and exchange a segment of genetic material. Here, the alleles for gene C were exchanged. The result is two recombinant and two non-recombinant chromosomes.

When two genes are located in close proximity on the same chromosome, they are considered linked, and their alleles tend to be transmitted through meiosis together. To exemplify this, imagine a dihybrid cross involving flower color and plant height in which the genes are next to each other on the chromosome. If one homologous chromosome has alleles for tall plants and red flowers, and the other chromosome has genes for short plants and yellow flowers, then when the gametes are formed, the tall and red alleles will go together into a gamete and the short and yellow alleles will go into other gametes. These are called the parental genotypes because they have been inherited intact from the parents of the individual producing gametes. But unlike if the genes were on different chromosomes, there will be no gametes with tall and yellow alleles and no gametes with short and red alleles. If you create the Punnett square with these gametes, you will see that the classical Mendelian prediction of a 9:3:3:1 outcome of a dihybrid cross would not apply. As the distance between two genes increases, the probability of one or more crossovers between them increases, and the genes behave more like they are on separate chromosomes. Geneticists have used the proportion of recombinant gametes (the ones not like the parents) as a measure of how far apart genes are on a chromosome. Using this information, they have constructed elaborate maps of genes on chromosomes for well-studied organisms, including humans.

Mendel's seminal publication makes no mention of linkage, and many researchers have questioned whether he encountered linkage but chose not to publish those crosses out of concern that they would invalidate his independent assortment postulate. The garden pea has seven chromosomes, and some have suggested that his choice of seven characteristics was not a coincidence. However, even if the genes he examined were not located on separate chromosomes, it is possible that he simply did not observe linkage because of the extensive shuffling effects of recombination.

SCIENTIFIC METHOD CONNECTION

Testing the Hypothesis of Independent Assortment

To better appreciate the amount of labor and ingenuity that went into Mendel's experiments, proceed through one of Mendel's dihybrid crosses.

Question: What will be the offspring of a dihybrid cross?

Background: Consider that pea plants mature in one growing season, and you have access to a large garden in which you can cultivate thousands of pea plants. There are several true-breeding plants with the following pairs of traits: tall plants with inflated pods, and dwarf plants with constricted pods. Before the plants have matured, you remove the pollen-producing organs from the tall/inflated plants in your crosses to prevent self-fertilization. Upon plant maturation, the plants are manually crossed by transferring pollen from the dwarf/constricted plants to the stigmata of the tall/inflated plants.

Hypothesis: Both trait pairs will sort independently according to Mendelian laws. When the true-breeding parents are crossed, all of the F_1 offspring are tall and have inflated pods, which indicates that the tall and inflated traits are dominant over the dwarf and constricted traits, respectively. A self-cross of the F_1 heterozygotes results in 2,000 F_2 progeny.

Test the hypothesis: Because each trait pair sorts independently, the ratios of tall:dwarf and inflated:constricted are each expected to be 3:1. The tall/dwarf trait pair is called *T/t*, and the inflated/constricted trait pair is designated *I/i*. Each member of the F_1 generation therefore has a genotype of *TtIi*. Construct a grid analogous to Figure 12.16, in which you cross two *TtIi* individuals. Each individual can donate four combinations of two traits: *TI*, *Ti*, *tI*, or *ti*, meaning that there are 16 possibilities of offspring genotypes. Because the *T* and *I* alleles are dominant, any individual having one or two of those alleles will express the tall or inflated phenotypes, respectively, regardless if they also have a *t* or *i* allele. Only individuals that are *tt* or *ii* will express the dwarf and constricted alleles, respectively. As shown in Figure 12.19, you predict that you will observe the following offspring proportions: tall/inflated:tall/constricted:dwarf/inflated:dwarf/constricted in a 9:3:3:1 ratio. Notice from the grid that when considering the tall/dwarf and inflated/constricted trait pairs in isolation, they are each inherited in 3:1 ratios.

TtIi × *TtIi*	*TI*	*Ti*	*tI*	*ti*
TI	*TTII*	*TTIi*	*TtII*	*TtIi*
Ti	*TTIi*	*TTii*	*TtIi*	*Ttii*
tI	*TtII*	*TtIi*	*ttII*	*ttIi*
ti	*TtIi*	*Ttii*	*ttIi*	*ttii*

Figure 12.19 This figure shows all possible combinations of offspring resulting from a dihybrid cross of pea plants that are heterozygous for the tall/dwarf and inflated/constricted alleles.

Test the hypothesis: You cross the dwarf and tall plants and then self-cross the offspring. For best results, this is repeated with hundreds or even thousands of pea plants. What special precautions should be taken in the crosses and in growing the plants?

Analyze your data: You observe the following plant phenotypes in the F_2 generation: 2706 tall/inflated, 930 tall/constricted, 888 dwarf/inflated, and 300 dwarf/constricted. Reduce these findings to a ratio and determine if they are consistent with Mendelian laws.

Form a conclusion: Were the results close to the expected 9:3:3:1 phenotypic ratio? Do the results support the prediction? What might be observed if far fewer plants were used, given that alleles segregate randomly into gametes? Try to imagine growing that many pea plants, and consider the potential for experimental error. For instance, what would happen if it was extremely windy one day?

Epistasis

Mendel's studies in pea plants implied that the sum of an individual's phenotype was controlled by genes (or as he called them, unit factors), such that every characteristic was distinctly and completely controlled by a single gene. In fact, single observable characteristics are almost always under the influence of multiple genes (each with two or more alleles) acting in unison. For example, at least eight genes contribute to eye color in humans.

LINK TO LEARNING

Eye color in humans is determined by multiple genes. Use the Eye Color Calculator (http://openstax.org/l/eye_color_calc) to predict the eye color of children from parental eye color.

In some cases, several genes can contribute to aspects of a common phenotype without their gene products ever directly interacting. In the case of organ development, for instance, genes may be expressed sequentially, with each gene adding to the complexity and specificity of the organ. Genes may function in complementary or synergistic fashions, such that two or more genes need to be expressed simultaneously to affect a phenotype. Genes may also oppose each other, with one gene modifying the expression of another.

In **epistasis**, the interaction between genes is antagonistic, such that one gene masks or interferes with the expression of another. "Epistasis" is a word composed of Greek roots that mean "standing upon." The alleles that are being masked or silenced are said to be hypostatic to the epistatic alleles that are doing the masking. Often the biochemical basis of epistasis is a gene pathway in which the expression of one gene is dependent on the function of a gene that precedes or follows it in the pathway.

An example of epistasis is pigmentation in mice. The wild-type coat color, agouti (*AA*), is dominant to solid-colored fur (*aa*). However, a separate gene (*C*) is necessary for pigment production. A mouse with a recessive *c* allele at this locus is unable to produce pigment and is albino regardless of the allele present at locus *A* (Figure 12.20). Therefore, the genotypes *AAcc*, *Aacc*, and *aacc* all produce the same albino phenotype. A cross between heterozygotes for both genes (*AaCc* x *AaCc*) would generate offspring with a phenotypic ratio of 9 agouti:3 solid color:4 albino (Figure 12.20). In this case, the *C* gene is epistatic to the *A* gene.

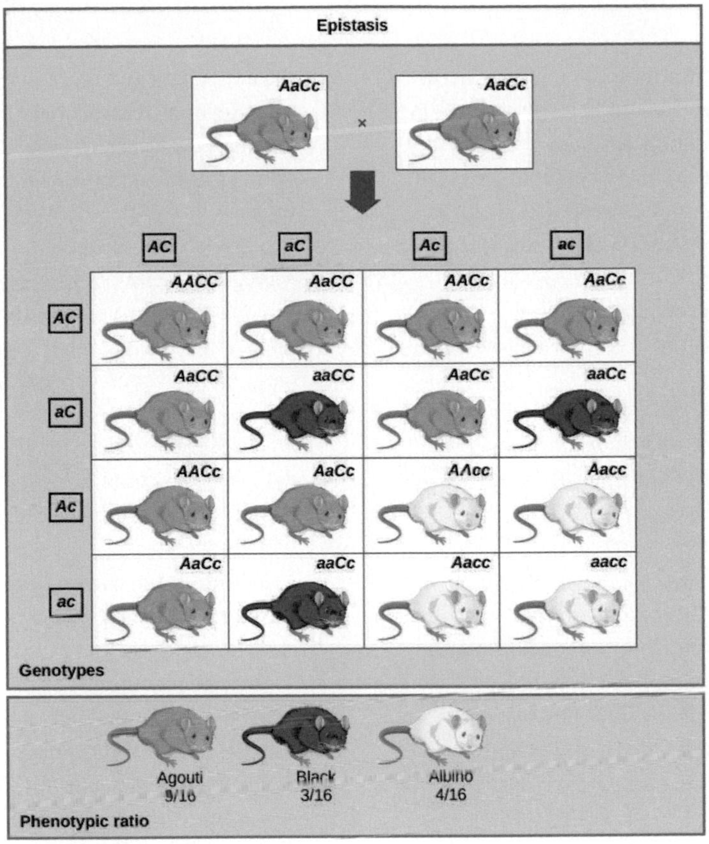

Figure 12.20 In mice, the mottled agouti coat color (*A*) is dominant to a solid coloration, such as black or gray. A gene at a separate locus (*C*) is responsible for pigment production. The recessive *c* allele does not produce pigment, and a mouse with the homozygous recessive *cc* genotype is albino regardless of the allele present at the *A* locus. Thus, the *C* gene is epistatic to the *A* gene.

Epistasis can also occur when a dominant allele masks expression at a separate gene. Fruit color in summer squash is expressed in this way. Homozygous recessive expression of the *W* gene (*ww*) coupled with homozygous dominant or heterozygous expression of the *Y* gene (*YY* or *Yy*) generates yellow fruit, and the *wwyy* genotype produces green fruit. However, if a dominant copy of the *W* gene is present in the homozygous or heterozygous form, the summer squash will produce white fruit regardless of the *Y* alleles. A cross between white heterozygotes for both genes (*WwYy* × *WwYy*) would produce offspring with a phenotypic ratio of 12 white:3 yellow:1 green.

Finally, epistasis can be reciprocal such that either gene, when present in the dominant (or recessive) form, expresses the same phenotype. In the shepherd's purse plant (*Capsella bursa-pastoris*), the characteristic of seed shape is controlled by two genes in a dominant epistatic relationship. When the genes *A* and *B* are both homozygous recessive (*aabb*), the seeds are ovoid. If the dominant allele for either of these genes is present, the result is triangular seeds. That is, every possible genotype other than *aabb* results in triangular seeds, and a cross between heterozygotes for both genes (*AaBb* x *AaBb*) would yield offspring with a phenotypic ratio of 15 triangular:1 ovoid.

As you work through genetics problems, keep in mind that any single characteristic that results in a phenotypic ratio that totals 16 is typical of a two-gene interaction. Recall the phenotypic inheritance pattern for Mendel's dihybrid cross, which considered two noninteracting genes—9:3:3:1. Similarly, we would expect interacting gene pairs to also exhibit ratios expressed as 16 parts. Note that we are assuming the interacting genes are not linked; they are still assorting independently into gametes.

LINK TO LEARNING

For an excellent review of Mendel's experiments and to perform your own crosses and identify patterns of inheritance, visit the Mendel's Peas (http://openstax.org/l/mendels_peas) web lab.

KEY TERMS

allele gene variations that arise by mutation and exist at the same relative locations on homologous chromosomes

autosomes any of the non-sex chromosomes

blending theory of inheritance hypothetical inheritance pattern in which parental traits are blended together in the offspring to produce an intermediate physical appearance

codominance in a heterozygote, complete and simultaneous expression of both alleles for the same characteristic

continuous variation inheritance pattern in which a character shows a range of trait values with small gradations rather than large gaps between them

dihybrid result of a cross between two true-breeding parents that express different traits for two characteristics

discontinuous variation inheritance pattern in which traits are distinct and are transmitted independently of one another

dominant trait which confers the same physical appearance whether an individual has two copies of the trait or one copy of the dominant trait and one copy of the recessive trait

dominant lethal inheritance pattern in which an allele is lethal both in the homozygote and the heterozygote; this allele can only be transmitted if the lethality phenotype occurs after reproductive age

epistasis antagonistic interaction between genes such that one gene masks or interferes with the expression of another

F_1 first filial generation in a cross; the offspring of the parental generation

F_2 second filial generation produced when F_1 individuals are self-crossed or fertilized with each other

genotype underlying genetic makeup, consisting of both physically visible and non-expressed alleles, of an organism

hemizygous presence of only one allele for a characteristic, as in X-linkage; hemizygosity makes descriptions of dominance and recessiveness irrelevant

heterozygous having two different alleles for a given gene on the homologous chromosome

homozygous having two identical alleles for a given gene on the homologous chromosome

hybridization process of mating two individuals that differ with the goal of achieving a certain characteristic in their offspring

incomplete dominance in a heterozygote, expression of two contrasting alleles such that the individual displays an intermediate phenotype

law of dominance in a heterozygote, one trait will conceal the presence of another trait for the same characteristic

law of independent assortment genes do not influence each other with regard to sorting of alleles into gametes; every possible combination of alleles is equally likely to occur

law of segregation paired unit factors (i.e., genes) segregate equally into gametes such that offspring have an equal likelihood of inheriting any combination of factors

linkage phenomenon in which alleles that are located in close proximity to each other on the same chromosome are more likely to be inherited together

model system species or biological system used to study a specific biological phenomenon to be applied to other different species

monohybrid result of a cross between two true-breeding parents that express different traits for only one characteristic

P_0 parental generation in a cross

phenotype observable traits expressed by an organism

product rule probability of two independent events occurring simultaneously can be calculated by multiplying the individual probabilities of each event occurring alone

Punnett square visual representation of a cross between two individuals in which the gametes of each individual are denoted along the top and side of a grid, respectively, and the possible zygotic genotypes are recombined at each box in the grid

recessive trait that appears "latent" or non-expressed when the individual also carries a dominant trait for that same characteristic; when present as two identical copies, the recessive trait is expressed

recessive lethal inheritance pattern in which an allele is only lethal in the homozygous form; the heterozygote may be normal or have some altered, nonlethal phenotype

reciprocal cross paired cross in which the respective traits of the male and female in one cross become the respective traits of the female and male in the other cross

sex-linked any gene on a sex chromosome

sum rule probability of the occurrence of at least one of two mutually exclusive events is the sum of their individual probabilities

test cross cross between a dominant expressing individual with an unknown genotype and a homozygous recessive individual; the offspring phenotypes indicate whether the unknown parent is heterozygous or homozygous for the dominant trait

trait variation in the physical appearance of a heritable characteristic

X-linked gene present on the X, but not the Y chromosome

CHAPTER SUMMARY

12.1 Mendel's Experiments and the Laws of Probability

Working with garden pea plants, Mendel found that crosses between parents that differed by one trait produced F_1 offspring that all expressed the traits of one parent. Observable traits are referred to as dominant, and non-expressed traits are described as recessive. When the offspring in Mendel's experiment were self-crossed, the F_2 offspring exhibited the dominant trait or the recessive trait in a 3:1 ratio, confirming that the recessive trait had been transmitted faithfully from the original P_0 parent. Reciprocal crosses generated identical F_1 and F_2 offspring ratios. By examining sample sizes, Mendel showed that his crosses behaved reproducibly according to the laws of probability, and that the traits were inherited as independent events.

Two rules in probability can be used to find the expected proportions of offspring of different traits from different crosses. To find the probability of two or more independent events occurring together, apply the product rule and multiply the probabilities of the individual events. The use of the word "and" suggests the appropriate application of the product rule. To find the probability of two or more events occurring in combination, apply the sum rule and add their individual probabilities together. The use of the word "or" suggests the appropriate application of the sum rule.

12.2 Characteristics and Traits

When true-breeding or homozygous individuals that differ for a certain trait are crossed, all of the offspring will be heterozygotes for that trait. If the traits are inherited as dominant and recessive, the F_1 offspring will all exhibit the same phenotype as the parent homozygous for the dominant trait. If these heterozygous offspring are self-crossed, the resulting F_2 offspring will be equally likely to inherit gametes carrying the dominant or recessive trait, giving rise to offspring of which one quarter are homozygous dominant, half are heterozygous, and one quarter are homozygous recessive. Because homozygous dominant and heterozygous individuals are phenotypically identical, the observed traits in the F_2 offspring will exhibit a ratio of three dominant to one recessive.

Alleles do not always behave in dominant and recessive patterns. Incomplete dominance describes situations in which the heterozygote exhibits a phenotype that is intermediate between the homozygous phenotypes. Codominance describes the simultaneous expression of both of the alleles in the heterozygote. Although diploid organisms can only have two alleles for any given gene, it is common for more than two alleles of a gene to exist in a population. In humans, as in many animals and some plants, females have two X chromosomes and males have one X and one Y chromosome. Genes that are present on the X but not the Y chromosome are said to be X-linked, such that males only inherit one allele for the gene, and females inherit two. Finally, some alleles can be lethal. Recessive lethal alleles are only lethal in homozygotes, but dominant lethal alleles are fatal in heterozygotes as well.

12.3 Laws of Inheritance

Mendel postulated that genes (characteristics) are inherited as pairs of alleles (traits) that behave in a dominant and recessive pattern. Alleles segregate into gametes such that each gamete is equally likely to receive either one of the two alleles present in a diploid individual. In addition, genes are assorted into gametes independently of one another. That is, alleles are generally not more likely to segregate into a gamete with a particular allele of another gene. A dihybrid cross demonstrates independent assortment when the genes in question are on different chromosomes or distant from each other on the same chromosome. For crosses involving more than two genes, use the forked line or probability methods to predict offspring genotypes and phenotypes rather than a Punnett square.

Although chromosomes sort independently into gametes during meiosis, Mendel's law of independent assortment refers to genes, not chromosomes, and a single chromosome may carry more than 1,000 genes. When genes are located in close proximity on the same chromosome, their alleles tend to be inherited together. This results in offspring ratios that violate Mendel's law of independent assortment. However, recombination serves to exchange genetic material on homologous chromosomes such that maternal and paternal alleles may be recombined on the same chromosome. This is why alleles on a given chromosome are not always inherited together. Recombination is a random event occurring anywhere on a chromosome. Therefore, genes that are far apart on the same chromosome are likely to still assort independently because of recombination events that occurred in the intervening chromosomal space.

Whether or not they are sorting independently, genes may interact at the level of gene products such that the expression of an allele for one gene masks or modifies the expression of an allele for a different gene. This is called epistasis.

VISUAL CONNECTION QUESTIONS

1. Figure 12.5 In pea plants, round peas (*R*) are dominant to wrinkled peas (*r*). You do a test cross between a pea plant with wrinkled peas (genotype *rr*) and a plant of unknown genotype that has round peas. You end up with three plants, all which have round peas. From this data, can you tell if the round pea parent plant is homozygous dominant or heterozygous? If the round pea parent plant is heterozygous, what is the probability that a random sample of 3 progeny peas will all be round?
2. Figure 12.6 What are the genotypes of the individuals labeled 1, 2, and 3?
3. Figure 12.12 What ratio of offspring would result from a cross between a white-eyed male and a female that is heterozygous for red eye color?
4. Figure 12.16 In pea plants, purple flowers (P) are dominant to white flowers (p) and yellow peas (Y) are dominant to green peas (y). What are the possible genotypes and phenotypes for a cross between PpYY and ppYy pea plants? How many squares do you need to do a Punnett square analysis of this cross?

REVIEW QUESTIONS

5. Mendel performed hybridizations by transferring pollen from the _______ of the male plant to the female ova.
   a. anther
   b. pistil
   c. stigma
   d. seed
6. Which is one of the seven characteristics that Mendel observed in pea plants?
   a. flower size
   b. seed texture
   c. leaf shape
   d. stem color
7. Imagine you are performing a cross involving seed color in garden pea plants. What F_1 offspring would you expect if you cross true-breeding parents with green seeds and yellow seeds? Yellow seed color is dominant over green.
   a. 100 percent yellow-green seeds
   b. 100 percent yellow seeds
   c. 50 percent yellow, 50 percent green seeds
   d. 25 percent green, 75 percent yellow seeds
8. Consider a cross to investigate the pea pod texture trait, involving constricted or inflated pods. Mendel found that the traits behave according to a dominant/recessive pattern in which inflated pods were dominant. If you performed this cross and obtained 650 inflated-pod plants in the F_2 generation, approximately how many constricted-pod plants would you expect to have?
   a. 600
   b. 165
   c. 217
   d. 468
9. A scientist pollinates a true-breeding pea plant with violet, terminal flowers with pollen from a true-breeding pea plant with white, axial flowers. Which of the following observations would most accurately describe the F_2 generation?
   a. 75% violet flowers; 75% terminal flowers
   b. 75% white flowers in a terminal position
   c. 75% violet flowers; 75% axial flowers
   d. 75% violet flowers in an axial position
10. The observable traits expressed by an organism are described as its ________.
   a. phenotype
   b. genotype
   c. alleles
   d. zygote
11. A recessive trait will be observed in individuals that are ________ for that trait.
   a. heterozygous
   b. homozygous or heterozygous
   c. homozygous
   d. diploid
12. If black and white true-breeding mice are mated and the result is all gray offspring, what inheritance pattern would this be indicative of?
   a. dominance
   b. codominance
   c. multiple alleles
   d. incomplete dominance

13. The ABO blood groups in humans are expressed as the I^A, I^B, and i alleles. The I^A allele encodes the A blood group antigen, I^B encodes B, and i encodes O. Both A and B are dominant to O. If a heterozygous blood type A parent (I^Ai) and a heterozygous blood type B parent (I^Bi) mate, one quarter of their offspring will have AB blood type (I^AI^B) in which both antigens are expressed equally. Therefore, ABO blood groups are an example of:
    a. multiple alleles and incomplete dominance
    b. codominance and incomplete dominance
    c. incomplete dominance only
    d. multiple alleles and codominance

14. In a mating between two individuals that are heterozygous for a recessive lethal allele that is expressed *in utero*, what genotypic ratio (homozygous dominant:heterozygous:homozygous recessive) would you expect to observe in the offspring?
    a. 1:2:1
    b. 3:1:1
    c. 1:2:0
    d. 0:2:1

15. If the allele encoding polydactyly (six fingers) is dominant why do most people have five fingers?
    a. Genetic elements suppress the polydactyl gene.
    b. Polydactyly is embryonic lethal.
    c. The sixth finger is removed at birth.
    d. The polydactyl allele is very rare in the human population.

16. A farmer raises black and white chickens. To his surprise, when the first generation of eggs hatch all the chickens are black with white speckles throughout their feathers. What should the farmer expect when the eggs laid after interbreeding the speckled chickens hatch?
    a. All the offspring will be speckled.
    b. 75% of the offspring will be speckled, and 25% will be black.
    c. 50% of the offspring will be speckled, 25% will be black, and 25% will be white.
    d. 50% of the offspring will be black and 50% of the offspring will be white.

17. Assuming no gene linkage, in a dihybrid cross of *AABB* x *aabb* with *AaBb* F_1 heterozygotes, what is the ratio of the F_1 gametes (*AB*, *aB*, *Ab*, *ab*) that will give rise to the F_2 offspring?
    a. 1:1:1:1
    b. 1:3:3:1
    c. 1:2:2:1
    d. 4:3:2:1

18. The forked line and probability methods make use of what probability rule?
    a. test cross
    b. product rule
    c. monohybrid rule
    d. sum rule

19. How many different offspring genotypes are expected in a trihybrid cross between parents heterozygous for all three traits when the traits behave in a dominant and recessive pattern? How many phenotypes?
    a. 64 genotypes; 16 phenotypes
    b. 16 genotypes; 64 phenotypes
    c. 8 genotypes; 27 phenotypes
    d. 27 genotypes; 8 phenotypes

20. Labrador retrievers' fur color is controlled by two alleles, E and B. Any dog with the ee_ genotype develops into a yellow lab, while B_E_ dogs become black labs and bbE_ dogs become chocolate labs. This is an example of ____.
    a. epistasis
    b. codominance
    c. incomplete dominance
    d. linkage

21. Which of the following situations does **not** follow the Law of Independent Assortment?
    a. A blond man and a brunette woman produce three offspring over time, all of who have blond hair.
    b. A white cow crossed with a brown bull produces roan cattle.
    c. Mating a hog with a sow produces six female piglets.
    d. Men are more likely to experience hemophilia than women.

CRITICAL THINKING QUESTIONS

22. Describe one of the reasons why the garden pea was an excellent choice of model system for studying inheritance.
23. How would you perform a reciprocal cross for the characteristic of stem height in the garden pea?
24. Mendel performs a cross using a true-breeding pea plant with round, yellow seeds and a true-breeding pea plant with green, wrinkled seeds. What is the probability that offspring will have green, round seeds? Calculate the probability for the F_1 and F_2 generations.
25. Calculate the probability of selecting a heart or a face card from a standard deck of cards. Is this outcome more or less likely than selecting a heart suit face card?

26. The gene for flower position in pea plants exists as axial or terminal alleles. Given that axial is dominant to terminal, list all of the possible F_1 and F_2 genotypes and phenotypes from a cross involving parents that are homozygous for each trait. Express genotypes with conventional genetic abbreviations.

27. Use a Punnett square to predict the offspring in a cross between a dwarf pea plant (homozygous recessive) and a tall pea plant (heterozygous). What is the phenotypic ratio of the offspring?

28. Can a human male be a carrier of red-green color blindness?

29. Why is it more efficient to perform a test cross with a homozygous recessive donor than a homozygous dominant donor? How could the same information still be found with a homozygous dominant donor?

30. Use the probability method to calculate the genotypes and genotypic proportions of a cross between *AABBCc* and *Aabbcc* parents.

31. Explain epistasis in terms of its Greek-language roots "standing upon."

32. In Section 12.3, "Laws of Inheritance," an example of epistasis was given for the summer squash. Cross white *WwYy* heterozygotes to prove the phenotypic ratio of 12 white:3 yellow:1 green that was given in the text.

33. People with trisomy 21 develop Down's syndrome. What law of Mendelian inheritance is violated in this disease? What is the most likely way this occurs?

34. A heterozygous pea plant produces violet flowers and yellow, round seeds. Describe the expected genotypes of the gametes produced by Mendelian inheritance. If all three genes are found on the same arm of one chromosome should a scientist predict that inheritance patterns will follow Mendelian genetics?

CHAPTER 13
Modern Understandings of Inheritance

Figure 13.1 Chromosomes are threadlike nuclear structures consisting of DNA and proteins that serve as the repositories for genetic information. The chromosomes depicted here were isolated from a fruit fly's salivary gland, stained with dye, and visualized under a microscope. Akin to miniature bar codes, chromosomes absorb different dyes to produce characteristic banding patterns, which allows for their routine identification. (credit: modification of work by "LPLT"/Wikimedia Commons; scale-bar data from Matt Russell)

INTRODUCTION The gene is the physical unit of inheritance, and genes are arranged in a linear order on chromosomes. Chromosome behavior and interaction during meiosis explain, at a cellular level, inheritance patterns that we observe in populations. Genetic disorders involving alterations in chromosome number or structure may have dramatic effects and can prevent a fertilized egg from developing.

Chapter Outline

13.1 Chromosomal Theory and Genetic Linkage

By the end of this section, you will be able to do the following:

- Discuss Sutton's Chromosomal Theory of Inheritance
- Describe genetic linkage
- Explain the process of homologous recombination, or crossing over
- Describe chromosome creation
- Calculate the distances between three genes on a chromosome using a three-point test cross

Long before scientists visualized chromosomes under a microscope, the father of modern genetics, Gregor Mendel, began studying heredity in 1843. With improved microscopic techniques during the late 1800s, cell biologists could stain and visualize subcellular structures with dyes and observe their actions during cell division and meiosis. With each mitotic division, chromosomes replicated, condensed from an amorphous (no constant shape) nuclear mass into distinct X-shaped bodies (pairs of identical sister chromatids), and migrated to separate cellular poles.

Chromosomal Theory of Inheritance

The speculation that chromosomes might be the key to understanding heredity led several scientists to examine Mendel's publications and reevaluate his model in terms of chromosome behavior during mitosis and meiosis. In 1902, Theodor Boveri observed that proper sea urchin embryonic development does not occur unless chromosomes are present. That same year, Walter Sutton observed chromosome separation into daughter cells during meiosis (Figure 13.2). Together, these observations led to the **Chromosomal Theory of Inheritance**, which identified chromosomes as the genetic material responsible for Mendelian inheritance.

Figure 13.2 (a) Walter Sutton and (b) Theodor Boveri developed the Chromosomal Theory of Inheritance, which states that chromosomes carry the unit of heredity (genes).

The Chromosomal Theory of Inheritance was consistent with Mendel's laws, which the following observations supported:

- During meiosis, homologous chromosome pairs migrate as discrete structures that are independent of other chromosome pairs.
- Chromosome sorting from each homologous pair into pre-gametes appears to be random.
- Each parent synthesizes gametes that contain only half their chromosomal complement.
- Even though male and female gametes (sperm and egg) differ in size and morphology, they have the same number of chromosomes, suggesting equal genetic contributions from each parent.
- The gametic chromosomes combine during fertilization to produce offspring with the same chromosome number as their parents.

Despite compelling correlations between chromosome behavior during meiosis and Mendel's abstract laws, scientists proposed the Chromosomal Theory of Inheritance long before there was any direct evidence that chromosomes carried traits. Critics pointed out that individuals had far more independently segregating traits than they had chromosomes. It was only after several years of carrying out crosses with the fruit fly, *Drosophila melanogaster*, that Thomas Hunt Morgan provided experimental evidence to support the Chromosomal Theory of Inheritance.

Genetic Linkage and Distances

Mendel's work suggested that traits are inherited independently of each other. Morgan identified a 1:1 correspondence between a segregating trait and the X chromosome, suggesting that random chromosome segregation was the physical basis of Mendel's model. This also demonstrated that linked genes disrupt Mendel's predicted outcomes. That each chromosome can carry many linked genes explains how individuals can have many more traits than they have chromosomes. However, researchers in Morgan's laboratory suggested that alleles positioned on the same chromosome were not always inherited together. During meiosis, linked genes somehow became unlinked.

Homologous Recombination

In 1909, Frans Janssen observed chiasmata—the point at which chromatids are in contact with each other and may exchange segments—prior to the first meiosis division. He suggested that alleles become unlinked and chromosomes physically exchange segments. As chromosomes condensed and paired with their homologs, they appeared to interact at distinct points. Janssen suggested that these points corresponded to regions in which chromosome segments exchanged. We now know that the pairing and interaction between homologous chromosomes, or synapsis, does more than simply organize the homologs for migration to separate daughter cells. When synapsed, homologous chromosomes undergo reciprocal physical exchanges at their arms in **homologous recombination**, or more simply, "crossing over."

To better understand the type of experimental results that researchers were obtaining at this time, consider a heterozygous individual that inherited dominant maternal alleles for two genes on the same chromosome (such as *AB*) and two recessive paternal alleles for those same genes (such as *ab*). If the genes are linked, one would expect this individual to produce gametes that are either *AB* or *ab* with a 1:1 ratio. If the genes are unlinked, the individual should produce *AB*, *Ab*, *aB*, and *ab* gametes with equal frequencies, according to the Mendelian concept of independent assortment. Because they correspond to new allele combinations, the genotypes Ab and aB are **nonparental types** that result from homologous recombination during meiosis. **Parental types** are progeny that exhibit the same allelic combination as their parents. Morgan and his colleagues, however, found that when they test crossed such heterozygous individuals to a homozygous recessive parent (*AaBb* × *aabb*), both parental and nonparental cases occurred. For example, 950 offspring might be recovered that were either *AaBb* or *aabb*, but 50 offspring would also result that were either *Aabb* or *aaBb*. These results suggested that linkage occurred most often, but a significant minority of offspring were the products of recombination.

VISUAL CONNECTION

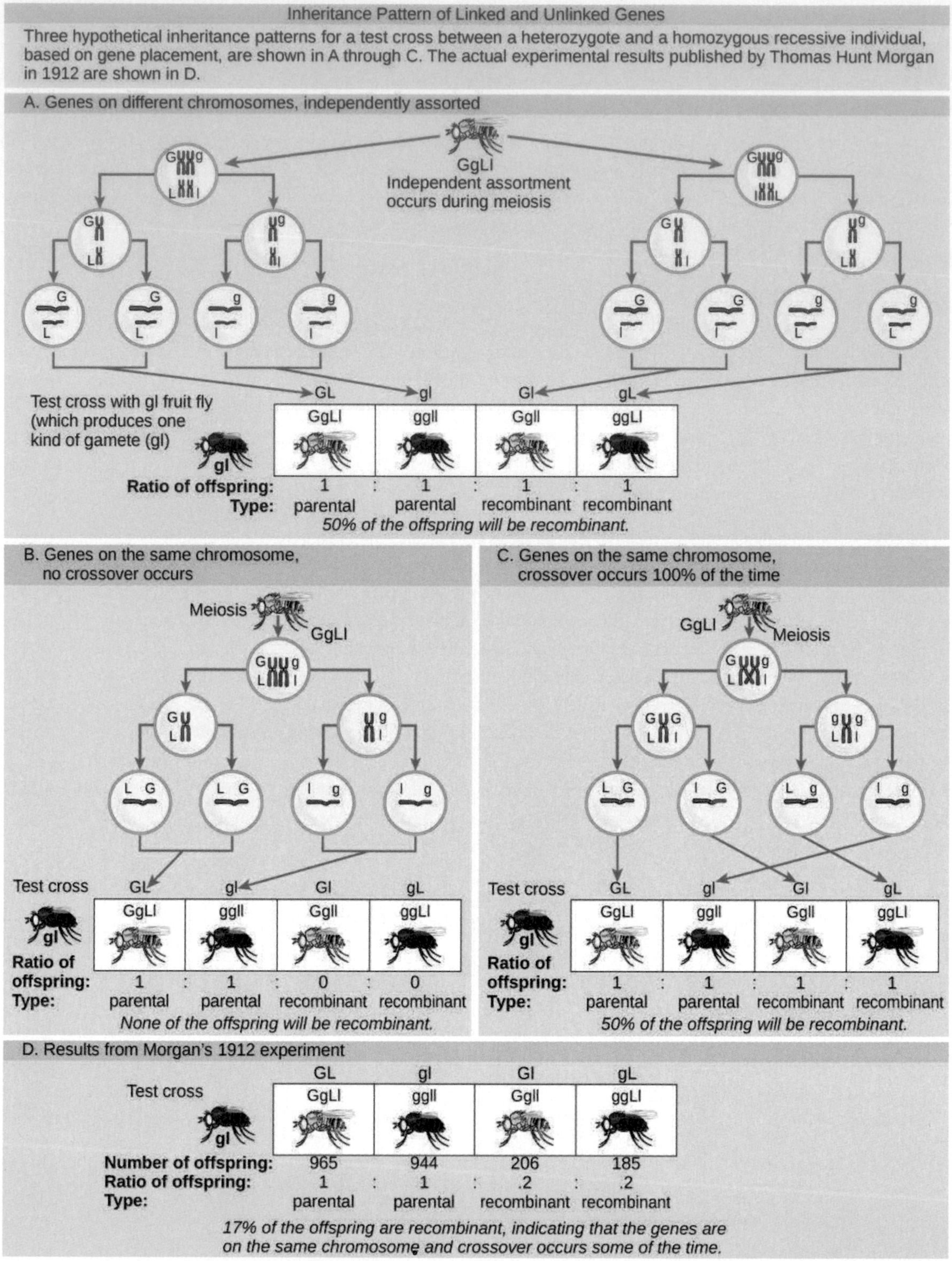

Figure 13.3 This figure shows unlinked and linked gene inheritance patterns. In (a), two genes are located on different chromosomes so independent assortment occurs during meiosis. The offspring have an equal chance of being the parental type (inheriting the same combination of traits as the parents) or a nonparental type (inheriting a different combination of traits than the parents). In (b), two genes are very close together on the same chromosome so that no crossing over occurs between them. Therefore, the genes are always inherited together and all the offspring are the parental type. In (c), two genes are far apart on the chromosome such that crossing over occurs during

every meiotic event. The recombination frequency will be the same as if the genes were on separate chromosomes. (d) The actual recombination frequency of fruit fly wing length and body color that Thomas Morgan observed in 1912 was 17 percent. A crossover frequency between 0 percent and 50 percent indicates that the genes are on the same chromosome and crossover sometimes occurs.

In a test cross for two characteristics such as the one here, can the recombinant offspring's predicted frequency be 60 percent? Why or why not?

Genetic Maps

Janssen did not have the technology to demonstrate crossing over so it remained an abstract idea that scientists did not widely believe. Scientists thought chiasmata were a variation on synapsis and could not understand how chromosomes could break and rejoin. Yet, the data were clear that linkage did not always occur. Ultimately, it took a young undergraduate student and an "all-nighter" to mathematically elucidate the linkage and recombination problem.

In 1913, Alfred Sturtevant, a student in Morgan's laboratory, gathered results from researchers in the laboratory, and took them home one night to mull them over. By the next morning, he had created the first "chromosome map," a linear representation of gene order and relative distance on a chromosome (Figure 13.4).

VISUAL CONNECTION

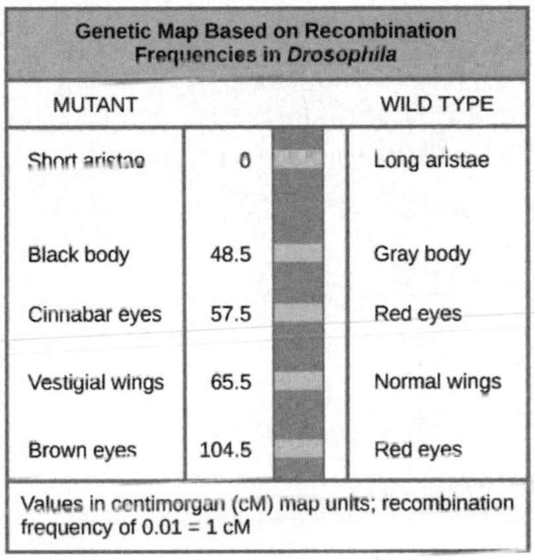

Figure 13.4 This genetic map orders *Drosophila* genes on the basis of recombination frequency.

Which of the following statements is true?

a. Recombination of the body color and red/cinnabar eye alleles will occur more frequently than recombination of the alleles for wing length and aristae length.
b. Recombination of the body color and aristae length alleles will occur more frequently than recombination of red/brown eye alleles and the aristae length alleles.
c. Recombination of the gray/black body color and long/short aristae alleles will not occur.
d. Recombination of the red/brown eye and long/short aristae alleles will occur more frequently than recombination of the alleles for wing length and body color.

As Figure 13.4 shows, by using recombination frequency to predict genetic distance, we can infer the relative gene order on chromosome 2. The values represent map distances in centimorgans (cM), which correspond to recombination frequencies (in percent). Therefore, the genes for body color and wing size were 65.5 − 48.5 = 17 cM apart, indicating that the maternal and paternal alleles for these genes recombine in 17 percent of offspring, on average.

To construct a chromosome map, Sturtevant assumed that genes were ordered serially on threadlike chromosomes. He also assumed that the incidence of recombination between two homologous chromosomes could occur with equal likelihood anywhere along the chromosome's length. Operating under these assumptions, Sturtevant postulated that alleles that were far apart on a chromosome were more likely to dissociate during meiosis simply because there was a larger region over which recombination could occur. Conversely, alleles that were close to each other on the chromosome were likely to be inherited together. The average number of crossovers between two alleles—that is, their **recombination frequency**—correlated with their genetic distance from each other, relative to the locations of other genes on that chromosome. Considering the example cross between *AaBb* and *aabb* above, we could calculate the recombination's frequency as 50/1000 = 0.05. That is, the likelihood of a crossover between genes *A/a* and *B/b* was 0.05, or 5 percent. Such a result would indicate that the genes were definitively linked, but that they were far enough apart for crossovers to occasionally occur. Sturtevant divided his genetic map into map units, or **centimorgans (cM)**, in which a 0,01 recombination frequency corresponds to 1 cM.

By representing alleles in a linear map, Sturtevant suggested that genes can range from linking perfectly (recombination frequency = 0) to unlinking perfectly (recombination frequency = 0.5) when genes are on different chromosomes or genes separate very far apart on the same chromosome. Perfectly unlinked genes correspond to the frequencies Mendel predicted to assort independently in a dihybrid cross. A 0.5 recombination frequency indicates that 50 percent of offspring are recombinants and the other 50 percent are parental types. That is, every type of allele combination is represented with equal frequency. This representation allowed Sturtevant to additively calculate distances between several genes on the same chromosome. However, as the genetic distances approached 0.50, his predictions became less accurate because it was not clear whether the genes were very far apart on the same or on different chromosomes.

In 1931, Barbara McClintock and Harriet Creighton demonstrated the crossover of homologous chromosomes in corn plants. Weeks later, Curt Stern demonstrated microscopically homologous recombination in *Drosophila*. Stern observed several X-linked phenotypes that were associated with a structurally unusual and dissimilar X chromosome pair in which one X was missing a small terminal segment, and the other X was fused to a piece of the Y chromosome. By crossing flies, observing their offspring, and then visualizing the offspring's chromosomes, Stern demonstrated that every time the offspring allele combination deviated from either of the parental combinations, there was a corresponding exchange of an X chromosome segment. Using mutant flies with structurally distinct X chromosomes was the key to observing the products of recombination because DNA sequencing and other molecular tools were not yet available. We now know that homologous chromosomes regularly exchange segments in meiosis by reciprocally breaking and rejoining their DNA at precise locations.

LINK TO LEARNING

Review Sturtevant's process to create a genetic map on the basis of recombination frequencies here (http://openstax.org/l/gene_crossover) .

Mendel's Mapped Traits

Homologous recombination is a common genetic process, yet Mendel never observed it. Had he investigated both linked and unlinked genes, it would have been much more difficult for him to create a unified model of his data on the basis of probabilistic calculations. Researchers who have since mapped the seven traits that Mendel investigated onto a pea plant genome's seven chromosomes have confirmed that all the genes he examined are either on separate chromosomes or are sufficiently far apart as to be statistically unlinked. Some have suggested that Mendel was enormously lucky to select only unlinked genes; whereas, others question whether Mendel discarded any data suggesting linkage. In any case, Mendel consistently observed independent assortment because he examined genes that were effectively unlinked.

13.2 Chromosomal Basis of Inherited Disorders

By the end of this section, you will be able to do the following:

- Describe how a karyogram is created
- Explain how nondisjunction leads to disorders in chromosome number
- Compare disorders that aneuploidy causes
- Describe how errors in chromosome structure occur through inversions and translocations

Inherited disorders can arise when chromosomes behave abnormally during meiosis. We can divide chromosome disorders into two categories: abnormalities in chromosome number and chromosomal structural rearrangements. Because even small chromosome segments can span many genes, chromosomal disorders are characteristically dramatic and often fatal.

Chromosome Identification

Chromosome isolation and microscopic observation forms the basis of cytogenetics and is the primary method by which clinicians detect chromosomal abnormalities in humans. A **karyotype** is the number and appearance of chromosomes, and includes their length, banding pattern, and centromere position. To obtain a view of an individual's karyotype, cytologists photograph the chromosomes and then cut and paste each chromosome into a chart, or **karyogram**. Another name is an ideogram (Figure 13.5).

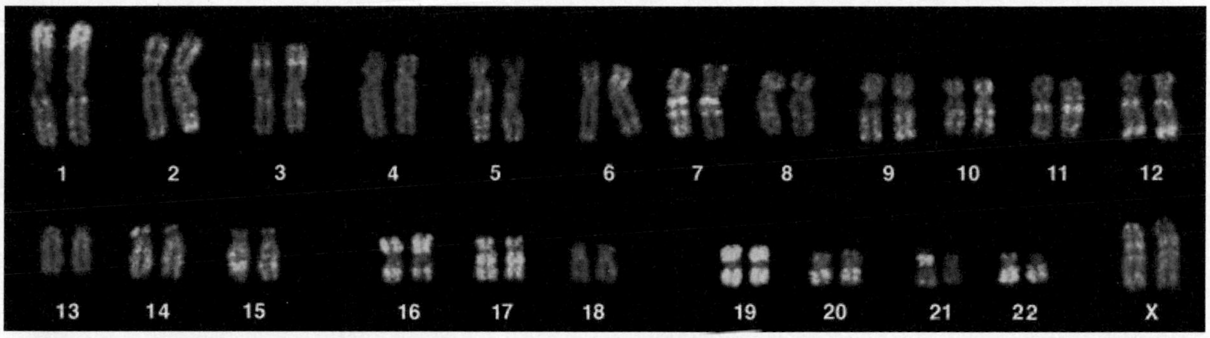

Figure 13.5 This karyotype is of a female human. Notice that homologous chromosomes are the same size, and have the same centromere positions and banding patterns. A human male would have an XY chromosome pair instead of the XX pair. (credit: Andreas Blozer et al)

In a given species, we can identify chromosomes by their number, size, centromere position, and banding pattern. In a human karyotype, **autosomes** or "body chromosomes" (all of the non–sex chromosomes) are generally organized in approximate order of size from largest (chromosome 1) to smallest (chromosome 22). The X and Y chromosomes are not autosomes. However, chromosome 21 is actually shorter than chromosome 22. Researchers discovered this after naming Down syndrome as trisomy 21, reflecting how this disease results from possessing one extra chromosome 21 (three total). Not wanting to change the name of this important disease, scientists retained the numbering of chromosome 21 despite describing it having the shortest set of chromosomes. We may designate the chromosome "arms" projecting from either end of the centromere as short or long, depending on their relative lengths. We abbreviate the short arm *p* (for "petite"); whereas, we abbreviate the long arm *q* (because it follows "p" alphabetically). Numbers further subdivide and denote each arm. Using this naming system, we can describe chromosome locations consistently in the scientific literature.

CAREER CONNECTION

Geneticists Use Karyograms to Identify Chromosomal Aberrations

Although we refer to Mendel as the "father of modern genetics," he performed his experiments with none of the tools that the geneticists of today routinely employ. One such powerful cytological technique is karyotyping, a method in which geneticists can identify traits characterized by chromosomal abnormalities from a single cell. To observe an individual's karyotype, a geneticist first collects a person's cells (like white blood cells) from a blood sample or other tissue. In the laboratory, he or she stimulates the isolated cells to begin actively dividing. The geneticist then applies the chemical colchicine to cells to arrest condensed chromosomes in metaphase. The geneticist then induces swelling in the cells using a hypotonic solution so the chromosomes spread apart. Finally, the geneticist preserves the sample in a fixative and applies it to a slide.

The geneticist then stains chromosomes with one of several dyes to better visualize each chromosome pair's distinct and reproducible banding patterns. Following staining, the geneticist views the chromosomes using bright-field microscopy. A common stain choice is the Giemsa stain. Giemsa staining results in approximately 400–800 bands (of tightly coiled DNA and condensed proteins) arranged along all 23 chromosome pairs. An experienced geneticist can identify each band. In addition to the banding patterns, geneticists further identify chromosomes on the basis of size and centromere location. To obtain the classic depiction of the karyotype in which homologous chromosome pairs align in numerical order from longest to shortest, the geneticist obtains a digital image, identifies each chromosome, and manually arranges the chromosomes into this pattern (Figure 13.5).

At its most basic, the karyogram may reveal genetic abnormalities in which an individual has too many or too few chromosomes per cell. Examples of this are Down Syndrome, which one identifies by a third copy of chromosome 21, and Turner Syndrome, which has the presence of only one X chromosome in women instead of the normal two characterizes. Geneticists can also identify large DNA deletions or insertions. For instance, geneticists can identify Jacobsen Syndrome—which involves distinctive facial features as well as heart and bleeding defects—by a deletion on chromosome 11. Finally, the karyotype can pinpoint **translocations**, which occur when a segment of genetic material breaks from one chromosome and reattaches to another chromosome or to a different part of the same chromosome. Translocations are implicated in certain cancers, including chronic myelogenous leukemia.

During Mendel's lifetime, inheritance was an abstract concept that one could only infer by performing crosses and observing the traits that offspring expressed. By observing a karyogram, today's geneticists can actually visualize an individual's chromosomal composition to confirm or predict genetic abnormalities in offspring, even before birth.

Chromosome Number Disorders

Of all of the chromosomal disorders, chromosome number abnormalities are the most obviously identifiable from a karyogram. Chromosome number disorders include duplicating or losing entire chromosomes, as well as changes in the number of complete sets of chromosomes. They are caused by **nondisjunction**, which occurs when homologous chromosome pairs or sister chromatids fail to separate during meiosis. Misaligned or incomplete synapsis, or a spindle apparatus dysfunction that facilitates chromosome migration, can cause nondisjunction. The risk of nondisjunction occurring increases with the parents' age.

Nondisjunction can occur during either meiosis I or II, with differing results (Figure 13.6). If homologous chromosomes fail to separate during meiosis I, the result is two gametes that lack that particular chromosome and two gametes with two chromosome copies. If sister chromatids fail to separate during meiosis II, the result is one gamete that lacks that chromosome, two normal gametes with one chromosome copy, and one gamete with two chromosome copies.

VISUAL CONNECTION

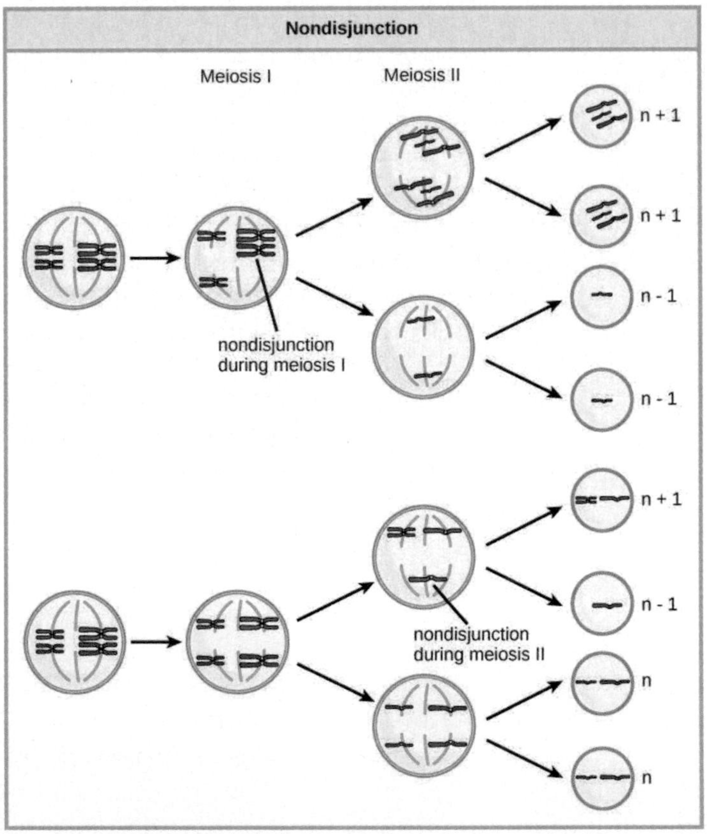

Figure 13.6 Nondisjunction occurs when homologous chromosomes or sister chromatids fail to separate during meiosis, resulting in an abnormal chromosome number. Nondisjunction may occur during meiosis I or meiosis II.

Which of the following statements about nondisjunction is true?

a. Nondisjunction only results in gametes with n+1 or n−1 chromosomes.

b. Nondisjunction occurring during meiosis II results in 50 percent normal gametes.
c. Nondisjunction during meiosis I results in 50 percent normal gametes.
d. Nondisjunction always results in four different kinds of gametes.

Aneuploidy

Scientists call an individual with the appropriate number of chromosomes for their species **euploid**. In humans, euploidy corresponds to 22 pairs of autosomes and one pair of sex chromosomes. An individual with an error in chromosome number is described as **aneuploid**, a term that includes **monosomy** (losing one chromosome) or **trisomy** (gaining an extraneous chromosome). Monosomic human zygotes missing any one copy of an autosome invariably fail to develop to birth because they lack essential genes. This underscores the importance of "gene dosage" in humans. Most autosomal trisomies also fail to develop to birth; however, duplications of some smaller chromosomes (13, 15, 18, 21, or 22) can result in offspring that survive for several weeks to many years. Trisomic individuals suffer from a different type of genetic imbalance: an excess in gene dose. Individuals with an extra chromosome may synthesize an abundance of the gene products, which that chromosome encodes. This extra dose (150 percent) of specific genes can lead to a number of functional challenges and often precludes development. The most common trisomy among viable births is that of chromosome 21, which corresponds to Down Syndrome. Short stature and stunted digits, facial distinctions that include a broad skull and large tongue, and significant developmental delays characterize individuals with this inherited disorder. We can correlate the incidence of Down syndrome with maternal age. Older women are more likely to become pregnant with fetuses carrying the trisomy 21 genotype (Figure 13.7).

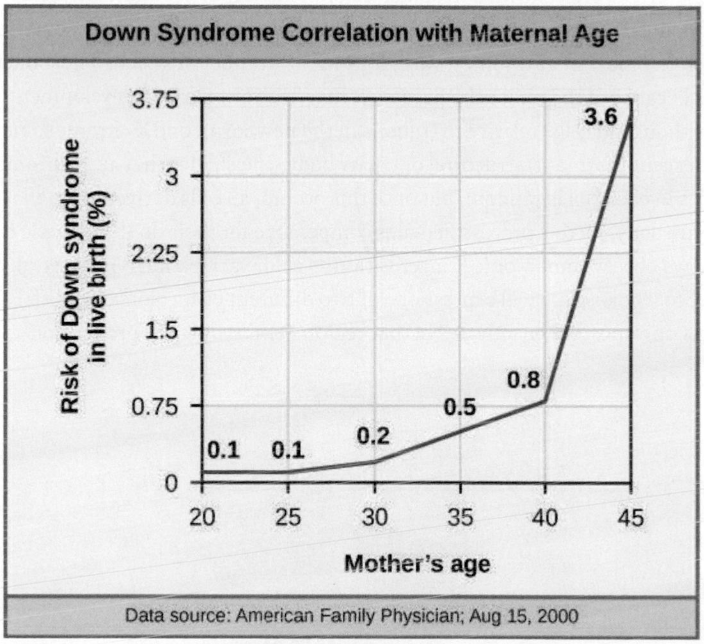

Figure 13.7 The incidence of having a fetus with trisomy 21 increases dramatically with maternal age.

LINK TO LEARNING

Visualize adding a chromosome that leads to Down syndrome in this video simulation (http://openstax.org/l/down_syndrome) .

Polyploidy

We call an individual with more than the correct number of chromosome sets (two for diploid species) **polyploid**. For instance, fertilizing an abnormal diploid egg with a normal haploid sperm would yield a triploid zygote. Polyploid animals are extremely rare, with only a few examples among the flatworms, crustaceans, amphibians, fish, and lizards. Polyploid animals are sterile because meiosis cannot proceed normally and instead produces mostly aneuploid daughter cells that cannot yield viable zygotes. Rarely, polyploid animals can reproduce asexually by haplodiploidy, in which an unfertilized egg divides mitotically to produce offspring. In contrast, polyploidy is very common in the plant kingdom, and polyploid plants tend to be larger and more robust than euploids of their species (Figure 13.8).

Figure 13.8 As with many polyploid plants, this triploid orange daylily (*Hemerocallis fulva*) is particularly large and robust, and grows flowers with triple the number of petals of its diploid counterparts. (credit: Steve Karg)

Sex Chromosome Nondisjunction in Humans

Humans display dramatic deleterious effects with autosomal trisomies and monosomies. Therefore, it may seem counterintuitive that human females and males can function normally, despite carrying different numbers of the X chromosome. Rather than a gain or loss of autosomes, variations in the number of sex chromosomes occur with relatively mild effects. In part, this happens because of the molecular process **X inactivation**. Early in development, when female mammalian embryos consist of just a few thousand cells (relative to trillions in the newborn), one X chromosome in each cell inactivates by tightly condensing into a quiescent (dormant) structure, or a Barr body. The chance that an X chromosome (maternally or paternally derived) inactivates in each cell is random, but once this occurs, all cells derived from that one will have the same inactive X chromosome or Barr body. By this process, females compensate for their double genetic dose of X chromosome. In so-called "tortoiseshell" cats, we observe embryonic X inactivation as color variegation (Figure 13.9). Females that are heterozygous for an X-linked coat color gene will express one of two different coat colors over different regions of their body, corresponding to whichever X chromosome inactivates in that region's embryonic cell progenitor.

Figure 13.9 In cats, the gene for coat color is located on the X chromosome. In female cats' embryonic development, one of the two X chromosomes randomly inactivates in each cell, resulting in a tortoiseshell pattern if the cat has two different alleles for coat color. Male cats, having only one X chromosome, never exhibit a tortoiseshell coat color. (credit: Michael Bodega)

An individual carrying an abnormal number of X chromosomes will inactivate all but one X chromosome in each of her cells. However, even inactivated X chromosomes continue to express a few genes, and X chromosomes must reactivate for the proper maturation of female ovaries. As a result, X-chromosomal abnormalities typically occur with mild mental and physical defects, as well as sterility. If the X chromosome is absent altogether, the individual will not develop in utero.

Scientists have identified and characterized several errors in sex chromosome number. Individuals with three X chromosomes, triplo-X, are phenotypically female but express developmental delays and reduced fertility. The XXY genotype, corresponding to one type of Klinefelter syndrome, corresponds to phenotypically male individuals with small testes, enlarged breasts, and reduced body hair. More complex types of Klinefelter syndrome exist in which the individual has as many as five X chromosomes. In all types, every X chromosome except one undergoes inactivation to compensate for the excess genetic dosage. We see this as several Barr bodies in each cell nucleus. Turner syndrome, characterized as an Xo genotype (i.e., only a single sex chromosome), corresponds to a phenotypically female individual with short stature, webbed skin in the neck region, hearing and cardiac impairments, and sterility.

Duplications and Deletions

In addition to losing or gaining an entire chromosome, a chromosomal segment may duplicate or lose itself. Duplications and deletions often produce offspring that survive but exhibit physical and mental abnormalities. Duplicated chromosomal segments may fuse to existing chromosomes or may be free in the nucleus. Cri-du-chat (from the French for "cry of the cat") is a syndrome that occurs with nervous system abnormalities and identifiable physical features that result from a deletion of most 5p (the small arm of chromosome 5) (Figure 13.10). Infants with this genotype emit a characteristic high-pitched cry on which the disorder's name is based.

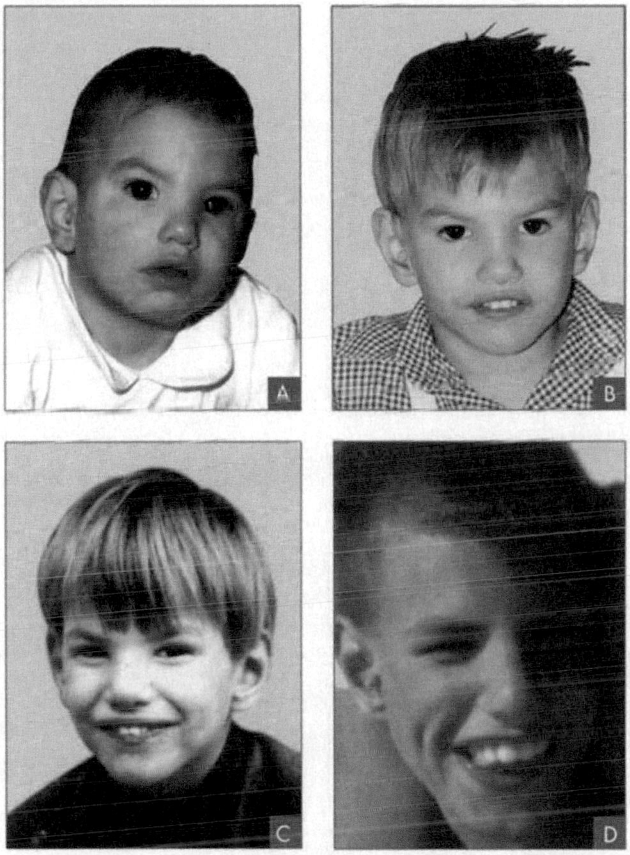

Figure 13.10 This figure shows an individual with cri-du-chat syndrome at two, four, nine, and 12 years of age. (credit: Paola Cerruti Mainardi)

Chromosomal Structural Rearrangements

Cytologists have characterized numerous structural rearrangements in chromosomes, but chromosome inversions and translocations are the most common. We can identify both during meiosis by the adaptive pairing of rearranged chromosomes with their former homologs to maintain appropriate gene alignment. If the genes on two homologs are not oriented correctly, a recombination event could result in losing genes from one chromosome and gaining genes on the other. This would produce aneuploid gametes.

Chromosome Inversions

A **chromosome inversion** is the detachment, 180° rotation, and reinsertion of part of a chromosome. Inversions may occur in nature as a result of mechanical shear, or from transposable elements' action (special DNA sequences capable of facilitating rearranging chromosome segments with the help of enzymes that cut and paste DNA sequences). Unless they disrupt a gene sequence, inversions only change gene orientation and are likely to have more mild effects than aneuploid errors. However, altered gene orientation can result in functional changes because regulators of gene expression could move out of position with respect to their targets, causing aberrant levels of gene products.

An inversion can be **pericentric** and include the centromere, or **paracentric** and occur outside the centromere (Figure 13.11). A pericentric inversion that is asymmetric about the centromere can change the chromosome arms' relative lengths, making these inversions easily identifiable.

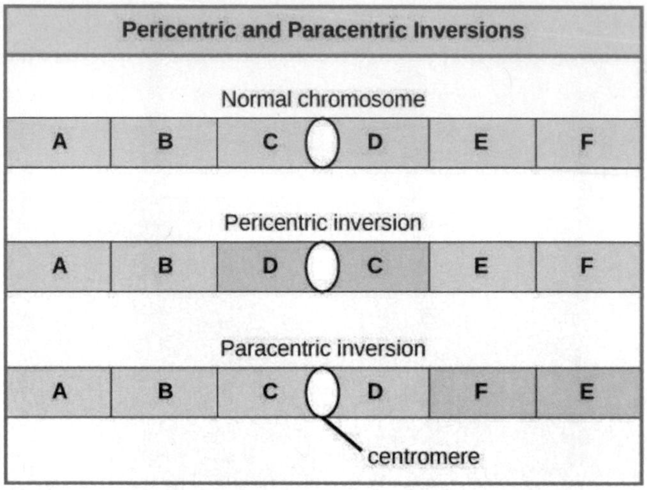

Figure 13.11 Pericentric inversions include the centromere, and paracentric inversions do not. A pericentric inversion can change the chromosome arms' relative lengths. A paracentric inversion cannot.

When one homologous chromosome undergoes an inversion but the other does not, the individual is an inversion heterozygote. To maintain point-for-point synapsis during meiosis, one homolog must form a loop, and the other homolog must mold around it. Although this topology can ensure that the genes correctly align, it also forces the homologs to stretch and can occur with imprecise synapsis regions (Figure 13.12).

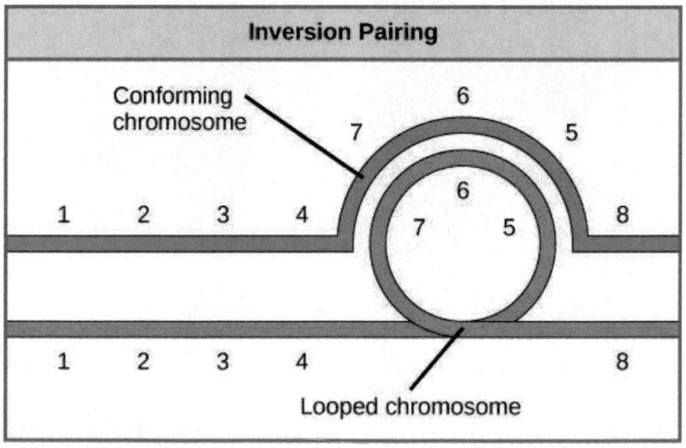

Figure 13.12 When one chromosome undergoes an inversion but the other does not, one chromosome must form an inverted loop to retain point-for-point interaction during synapsis. This inversion pairing is essential to maintaining gene alignment during meiosis and to allow for recombination.

EVOLUTION CONNECTION

The Chromosome 18 Inversion

Not all chromosomes' structural rearrangements produce nonviable, impaired, or infertile individuals. In rare instances, such a change can result in new species evolving. In fact, a pericentric inversion in chromosome 18 appears to have contributed to human evolution. This inversion is not present in our closest genetic relatives, the chimpanzees. Humans and chimpanzees differ cytogenetically by pericentric inversions on several chromosomes and by the fusion of two separate chromosomes in chimpanzees that correspond to chromosome two in humans.

Scientists believe the pericentric chromosome 18 inversion occurred in early humans following their divergence from a common ancestor with chimpanzees approximately five million years ago. Researchers characterizing this inversion have suggested that approximately 19,000 nucleotide bases were duplicated on 18p, and the duplicated region inverted and reinserted on chromosome 18 of an ancestral human.

A comparison of human and chimpanzee genes in the region of this inversion indicates that two genes—*ROCK1* and *USP14*— that are adjacent on chimpanzee chromosome 17 (which corresponds to human chromosome 18) are more distantly positioned on human chromosome 18. This suggests that one of the inversion breakpoints occurred between these two genes. Interestingly, humans and chimpanzees express *USP14* at distinct levels in specific cell types, including cortical cells and fibroblasts. Perhaps the chromosome 18 inversion in an ancestral human repositioned specific genes and reset their expression levels in a useful way. Because both *ROCK1* and *USP14* encode cellular enzymes, a change in their expression could alter cellular function. We do not know how this inversion contributed to hominid evolution, but it appears to be a significant factor in the divergence of humans from other primates.[1]

Translocations

A **translocation** occurs when a chromosome segment dissociates and reattaches to a different, nonhomologous chromosome. Translocations can be benign or have devastating effects depending on how the positions of genes are altered with respect to regulatory sequences. Notably, specific translocations have occurred with several cancers and with schizophrenia. Reciprocal translocations result from exchanging chromosome segments between two nonhomologous chromosomes such that there is no genetic information gain or loss (Figure 13.13).

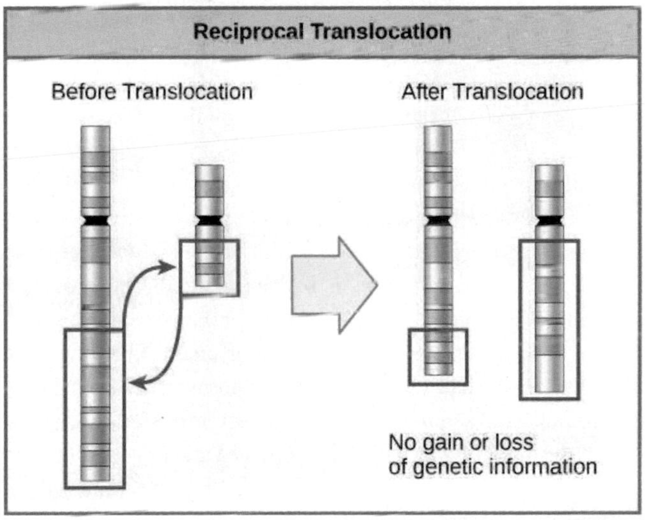

Figure 13.13 A reciprocal translocation occurs when a DNA segment transfers from one chromosome to another, nonhomologous chromosome. (credit: modification of work by National Human Genome Research/USA)

1Violaine Goidts et al., "Segmental duplication associated with the human-specific inversion of chromosome 18: a further example of the impact of segmental duplications on karyotype and genome evolution in primates," *Human Genetics*. 115 (2004):116-122

KEY TERMS

aneuploid individual with an error in chromosome number; includes chromosome segment deletions and duplications

autosome any of the non-sex chromosomes

centimorgan (cM) (also, map unit) relative distance that corresponds to a 0,01 recombination frequency

Chromosomal Theory of Inheritance theory proposing that chromosomes are the genes' vehicles and that their behavior during meiosis is the physical basis of the inheritance patterns that Mendel observed

chromosome inversion detachment, 180° rotation, and chromosome arm reinsertion

euploid individual with the appropriate number of chromosomes for their species

homologous recombination process by which homologous chromosomes undergo reciprocal physical exchanges at their arms, also crossing over

karyogram a karyotype's photographic image

karyotype an individual's chromosome number and appearance; includes the size, banding patterns, and centromere position

monosomy otherwise diploid genotype in which one chromosome is missing

nondisjunction failure of synapsed homologs to completely separate and migrate to separate poles during the meiosis' first cell division

nonparental (recombinant) type progeny resulting from homologous recombination that exhibits a different allele combination compared with its parents

paracentric inversion that occurs outside the centromere

parental types progeny that exhibits the same allelic combination as its parents

pericentric inversion that involves the centromere

polyploid individual with an incorrect number of chromosome sets

recombination frequency average number of crossovers between two alleles; observed as the number of nonparental types in a progeny's population

translocation process by which one chromosome segment dissociates and reattaches to a different, nonhomologous chromosome

trisomy otherwise diploid genotype in which one entire chromosome duplicates

X inactivation condensing X chromosomes into Barr bodies during embryonic development in females to compensate for the double genetic dose

CHAPTER SUMMARY

13.1 Chromosomal Theory and Genetic Linkage

Sutton and Boveri's Chromosomal Theory of Inheritance states that chromosomes are the vehicles of genetic heredity. Neither Mendelian genetics nor gene linkage is perfectly accurate. Instead, chromosome behavior involves segregation, independent assortment, and occasionally, linkage. Sturtevant devised a method to assess recombination frequency and infer linked genes' relative positions and distances on a chromosome on the basis of the average number of crossovers in the intervening region between the genes. Sturtevant correctly presumed that genes are arranged in serial order on chromosomes and that recombination between homologs can occur anywhere on a chromosome with equal likelihood. Whereas linkage causes alleles on the same chromosome to be inherited together, homologous recombination biases alleles toward an independent inheritance pattern.

13.2 Chromosomal Basis of Inherited Disorders

The number, size, shape, and banding pattern of chromosomes make them easily identifiable in a karyogram and allows for the assessment of many chromosomal abnormalities. Disorders in chromosome number, or aneuploidies, are typically lethal to the embryo, although a few trisomic genotypes are viable. Because of X inactivation, aberrations in sex chromosomes typically have milder phenotypic effects. Aneuploidies also include instances in which a chromosome's segments duplicate or delete themselves. Inversion or translocation also may rearrange chromosome structures. Both of these aberrations can result in problematic phenotypic effects. Because they force chromosomes to assume unnatural topologies during meiosis, inversions and translocations often occur with reduced fertility because of the likelihood of nondisjunction.

VISUAL CONNECTION QUESTIONS

1. Figure 13.3 In a test cross for two characteristics such as the one shown here, can the predicted frequency of recombinant offspring be 60 percent? Why or why not?

2. Figure 13.4 Which of the following statements is true?
   a. Recombination of the body color and red/cinnabar eye alleles will occur more frequently than recombination of the alleles for wing length and aristae length.
   b. Recombination of the body color and aristae length alleles will occur more frequently than recombination of red/brown eye alleles and the aristae length alleles.
   c. Recombination of the gray/black body color and long/short aristae alleles will not occur.
   d. Recombination of the red/brown eye and long/short aristae alleles will occur more frequently than recombination of the alleles for wing length and body color.

3. Figure 13.6 Which of the following statements about nondisjunction is true?
   a. Nondisjunction only results in gametes with n+1 or n−1 chromosomes.
   b. Nondisjunction occurring during meiosis II results in 50 percent normal gametes.
   c. Nondisjunction during meiosis I results in 50 percent normal gametes.
   d. Nondisjunction always results in four different kinds of gametes.

REVIEW QUESTIONS

4. X-linked recessive traits in humans (or in *Drosophila*) are observed ________.
   a. in more males than females
   b. in more females than males
   c. in males and females equally
   d. in different distributions depending on the trait

5. The first suggestion that chromosomes may physically exchange segments came from the microscopic identification of ________.
   a. synapsis
   b. sister chromatids
   c. chiasmata
   d. alleles

6. Which recombination frequency corresponds to independent assortment and the absence of linkage?
   a. 0
   b. 0.25
   c. 0.50
   d. 0.75

7. Which recombination frequency corresponds to perfect linkage and violates the law of independent assortment?
   a. 0
   b. 0.25
   c. 0.50
   d. 0.75

8. Which of the following codes describes position 12 on the long arm of chromosome 13?
   a. 13p12
   b. 13q12
   c. 12p13
   d. 12q13

9. In agriculture, polyploid crops (like coffee, strawberries, or bananas) tend to produce ________.
   a. more uniformity
   b. more variety
   c. larger yields
   d. smaller yields

10. Assume a pericentric inversion occurred in one of two homologs prior to meiosis. The other homolog remains normal. During meiosis, what structure—if any—would these homologs assume in order to pair accurately along their lengths?
    a. V formation
    b. cruciform
    c. loop
    d. pairing would not be possible

11. The genotype XXY corresponds to
    a. Klinefelter syndrome
    b. Turner syndrome
    c. Triplo-X
    d. Jacob syndrome

12. Abnormalities in the number of X chromosomes tends to have milder phenotypic effects than the same abnormalities in autosomes because of ________.
    a. deletions
    b. nonhomologous recombination
    c. synapsis
    d. X inactivation

13. By definition, a pericentric inversion includes the ________.
    a. centromere
    b. chiasma
    c. telomere
    d. synapse

CRITICAL THINKING QUESTIONS

14. Explain how the Chromosomal Theory of Inheritance helped to advance our understanding of genetics.

15. Using diagrams, illustrate how nondisjunction can result in an aneuploid zygote.

CHAPTER 14
DNA Structure and Function

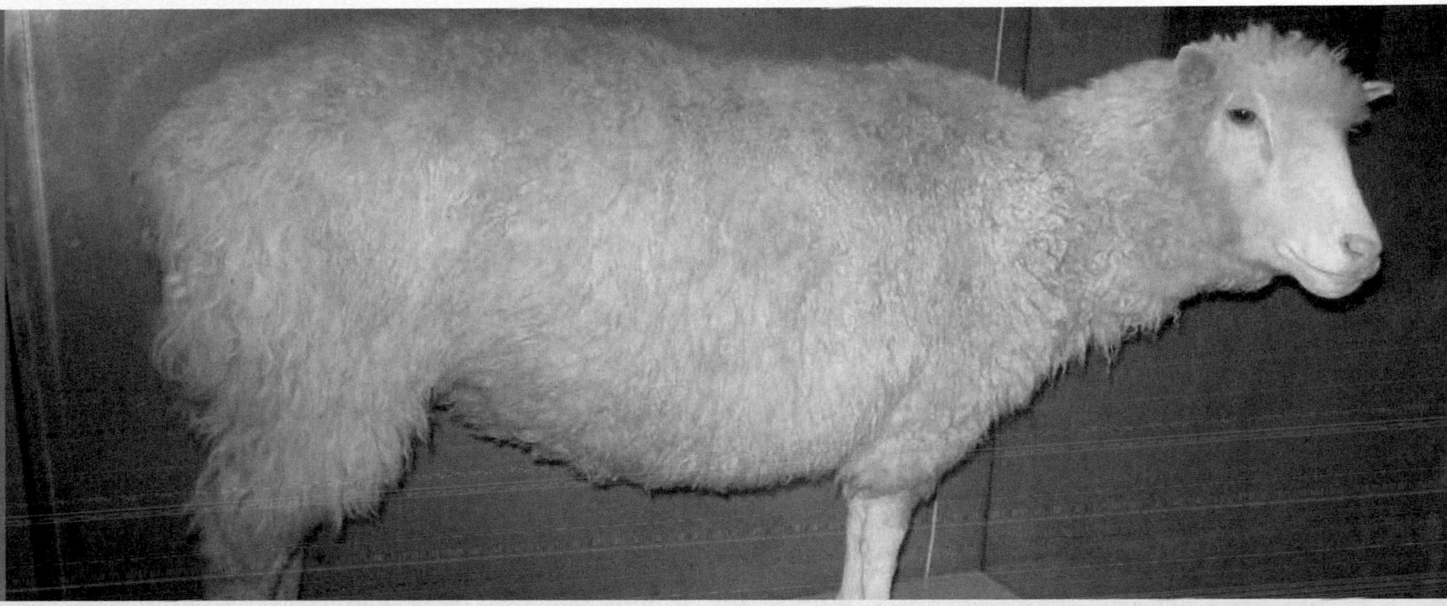

Figure 14.1 Dolly the sheep was the first large mammal to be cloned.

INTRODUCTION The three letters "DNA" have now become synonymous with crime solving and genetic testing. DNA can be retrieved from hair, blood, or saliva. Each person's DNA is unique, and it is possible to detect differences between individuals within a species on the basis of these unique features.

DNA analysis has many practical applications beyond forensics. In humans, DNA testing is applied to numerous uses: determining paternity, tracing genealogy, identifying pathogens, archeological research, tracing disease outbreaks, and studying human migration patterns. In the medical field, DNA is used in diagnostics, new vaccine development, and cancer therapy. It is now possible to determine predisposition to diseases by looking at genes.

Each human cell has 23 pairs of chromosomes: one set of chromosomes is inherited from the mother and the other set is inherited from the father. There is also a mitochondrial genome, inherited exclusively from the mother, which can be involved in inherited genetic disorders. On each chromosome, there are thousands of genes that are responsible for determining the genotype and phenotype of the individual. A gene is defined as a sequence of DNA that codes for a functional product. The human haploid genome contains 3 billion base pairs and has between 20,000 and 25,000 functional genes.

Chapter Outline

14.1 Historical Basis of Modern Understanding

By the end of this section, you will be able to do the following:

- Explain transformation of DNA
- Describe the key experiments that helped identify that DNA is the genetic material
- State and explain Chargaff's rules

Our current understanding of DNA began with the discovery of nucleic acids followed by the development of the double-helix model. In the 1860s, Friedrich Miescher (Figure 14.2), a physician by profession, isolated phosphate-rich chemicals from white blood cells (leukocytes). He named these chemicals (which would eventually be known as DNA) *nuclein* because they were isolated from the nuclei of the cells.

Figure 14.2 Friedrich Miescher (1844–1895) discovered nucleic acids.

LINK TO LEARNING

To see Miescher conduct his experiment that led to his discovery of DNA and associated proteins in the nucleus, click through this review (http://openstax.org/l/miescher_levene) .

A half century later, in 1928, British bacteriologist Frederick Griffith reported the first demonstration of bacterial **transformation**—a process in which external DNA is taken up by a cell, thereby changing its morphology and physiology. Griffith conducted his experiments with *Streptococcus pneumoniae,* a bacterium that causes pneumonia. Griffith worked with two strains of this bacterium called rough (R) and smooth (S). (The two cell types were called "rough" and "smooth" after the appearance of their colonies grown on a nutrient agar plate.)

The R strain is non-pathogenic (does not cause disease). The S strain is pathogenic (disease-causing), and has a capsule outside its cell wall. The capsule allows the cell to escape the immune responses of the host mouse.

When Griffith injected the living S strain into mice, they died from pneumonia. In contrast, when Griffith injected the live R strain into mice, they survived. In another experiment, when he injected mice with the heat-killed S strain, they also survived. This experiment showed that the capsule alone was not the cause of death. In a third set of experiments, a mixture of live R strain and heat-killed S strain were injected into mice, and—to his surprise—the mice died. Upon isolating the live bacteria from the dead mouse, only the S strain of bacteria was recovered. When this isolated S strain was injected into fresh mice, the mice died. Griffith concluded that something had passed from the heat-killed S strain into the live R strain and transformed it into the pathogenic S strain. He called this the *transforming principle* (Figure 14.3). These experiments are now known as Griffith's transformation experiments.

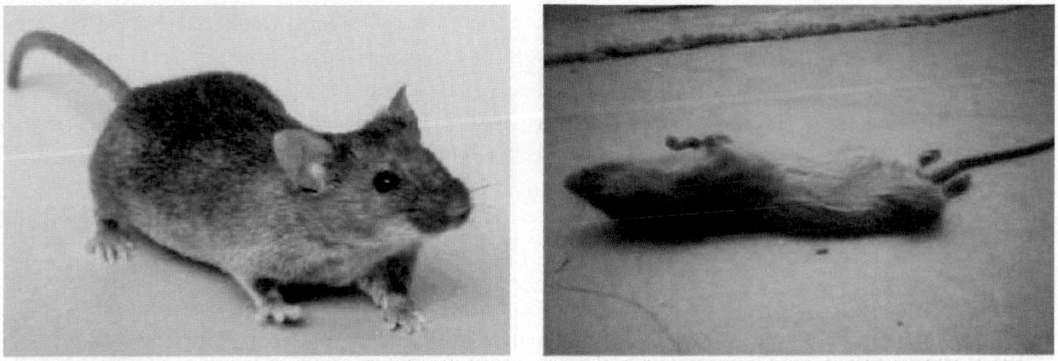

Figure 14.3 Two strains of *S. pneumoniae* were used in Griffith's transformation experiments. The R strain is non-pathogenic, whereas the S strain is pathogenic and causes death. When Griffith injected a mouse with the heat-killed S strain and a live R strain, the mouse died. The S strain was recovered from the dead mouse. Griffith concluded that something had passed from the heat-killed S strain to the R strain, transforming the R strain into the S strain in the process. (credit "living mouse": modification of work by NIH; credit "dead mouse": modification of work by Sarah Marriage)

Scientists Oswald Avery, Colin MacLeod, and Maclyn McCarty (1944) were interested in exploring this transforming principle further. They isolated the S strain from the dead mice and isolated the proteins and nucleic acids (RNA and DNA) as these were possible candidates for the molecule of heredity. They used enzymes that specifically degraded each component and then used each mixture separately to transform the R strain. They found that when DNA was degraded, the resulting mixture was no longer able to transform the bacteria, whereas all of the other combinations were able to transform the bacteria. This led them to conclude that DNA was the transforming principle.

CAREER CONNECTION

Forensic Scientist

Forensic Scientists used DNA analysis evidence for the first time to solve an immigration case. The story started with a teenage boy returning to London from Ghana to be with his mother. Immigration authorities at the airport were suspicious of him, thinking that he was traveling on a forged passport. After much persuasion, he was allowed to go live with his mother, but the immigration authorities did not drop the case against him. All types of evidence, including photographs, were provided to the authorities, but deportation proceedings were started nevertheless. Around the same time, Dr. Alec Jeffreys of Leicester University in the United Kingdom had invented a technique known as **DNA fingerprinting**. The immigration authorities approached Dr. Jeffreys for help. He took DNA samples from the mother and three of her children, as well as an unrelated mother, and compared the samples with the boy's DNA. Because the biological father was not in the picture, DNA from the three children was compared with the boy's DNA. He found a match in the boy's DNA for both the mother and his three siblings. He concluded that the boy was indeed the mother's son.

Forensic scientists analyze many items, including documents, handwriting, firearms, and biological samples. They analyze the DNA content of hair, semen, saliva, and blood, and compare it with a database of DNA profiles of known criminals. Analysis includes DNA isolation, sequencing, and sequence analysis. Forensic scientists are expected to appear at court hearings to present their findings. They are usually employed in crime labs of city and state government agencies. Geneticists experimenting with DNA techniques also work for scientific and research organizations, pharmaceutical industries, and college and university labs. Students wishing to pursue a career as a forensic scientist should have at least a bachelor's degree in chemistry, biology, or physics, and preferably some experience working in a laboratory.

Although the experiments of Avery, McCarty and McLeod had demonstrated that DNA was the informational component transferred during transformation, DNA was still considered to be too simple a molecule to carry biological information. Proteins, with their 20 different amino acids, were regarded as more likely candidates. The decisive experiment, conducted by Martha Chase and Alfred Hershey in 1952, provided confirmatory evidence that DNA was indeed the genetic material and not proteins. Chase and Hershey were studying a **bacteriophage**—a virus that infects bacteria. Viruses typically have a simple structure: a protein coat, called the capsid, and a nucleic acid core that contains the genetic material (either DNA or RNA). The bacteriophage infects the host bacterial cell by attaching to its surface, and then it injects its nucleic acids inside the cell. The phage DNA makes multiple copies of itself using the host machinery, and eventually the host cell bursts, releasing a large number of bacteriophages. Hershey and Chase selected radioactive elements that would specifically distinguish the protein from the DNA in infected cells. They labeled one batch of phage with radioactive sulfur, ^{35}S, to label the protein coat. Another batch of phage were labeled with radioactive phosphorus, ^{32}P. Because phosphorous is found in DNA, but not protein, the DNA and not the protein would be tagged with radioactive phosphorus. Likewise, sulfur is absent from DNA, but present in several amino acids such as methionine and cysteine.

Each batch of phage was allowed to infect the cells separately. After infection, the phage bacterial suspension was put in a blender, which caused the phage coat to detach from the host cell. Cells exposed long enough for infection to occur were then examined to see which of the two radioactive molecules had entered the cell. The phage and bacterial suspension was spun down in a centrifuge. The heavier bacterial cells settled down and formed a pellet, whereas the lighter phage particles stayed in the supernatant. In the tube that contained phage labeled with ^{35}S, the supernatant contained the radioactively labeled phage, whereas no radioactivity was detected in the pellet. In the tube that contained the phage labeled with ^{32}P, the radioactivity was detected in the pellet that contained the heavier bacterial cells, and no radioactivity was detected in the supernatant. Hershey and Chase concluded that it was the phage DNA that was injected into the cell and carried information to produce more phage particles, thus providing evidence that DNA was the genetic material and not proteins (Figure 14.4).

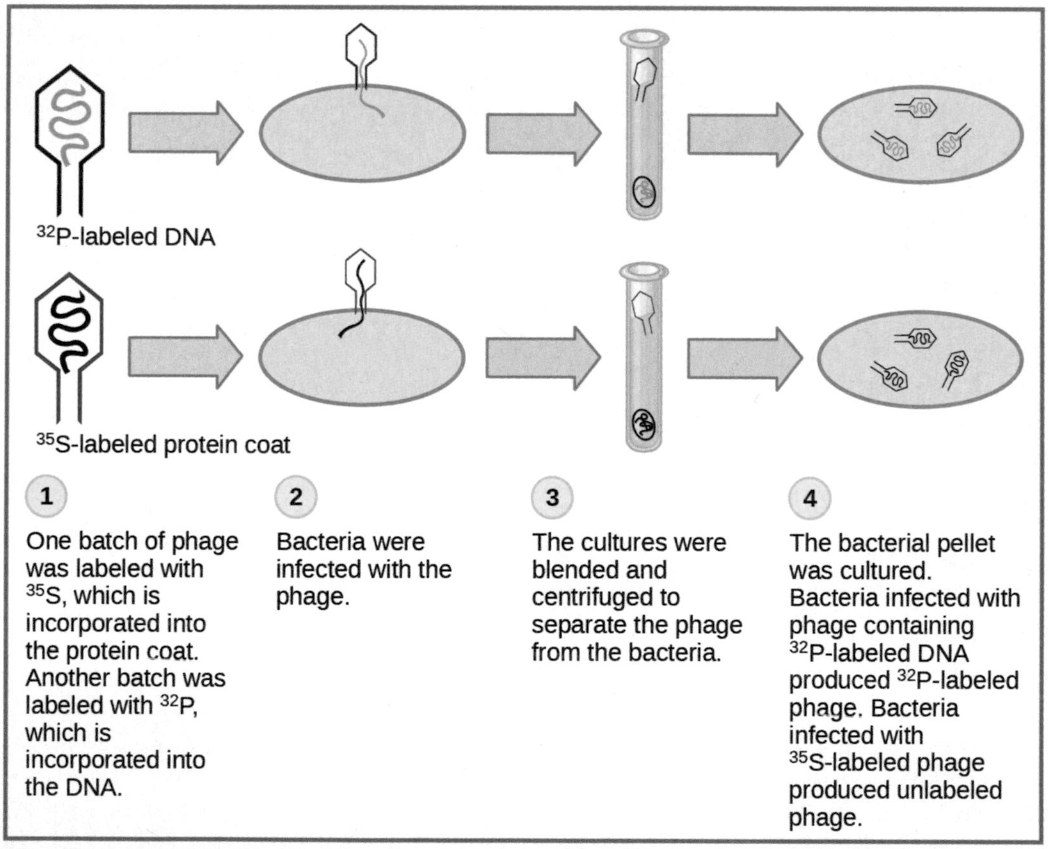

Figure 14.4 In Hershey and Chase's experiments, bacteria were infected with phage radiolabeled with either ^{35}S, which labels protein, or ^{32}P, which labels DNA. Only ^{32}P entered the bacterial cells, indicating that DNA is the genetic material.

Around this same time, Austrian biochemist Erwin Chargaff examined the content of DNA in different species and found that the amounts of adenine, thymine, guanine, and cytosine were not found in equal quantities, and that relative concentrations of the four nucleotide bases varied from species to species, but not within tissues of the same individual or between individuals of the same species. He also discovered something unexpected: That the amount of adenine equaled the amount of thymine, and the amount of cytosine equaled the amount of guanine (that is, A = T and G = C). Different species had equal amounts of *purines*

(A+G) and **pyrimidines** (T + C), but different ratios of A+T to G+C. These observations became known as **Chargaff's rules**. Chargaff's findings proved immensely useful when Watson and Crick were getting ready to propose their DNA double helix model! You can see after reading the past few pages how science builds upon previous discoveries, sometimes in a slow and laborious process.

14.2 DNA Structure and Sequencing

By the end of this section, you will be able to do the following:

- Describe the structure of DNA
- Explain the Sanger method of DNA sequencing
- Discuss the similarities and differences between eukaryotic and prokaryotic DNA

The building blocks of DNA are nucleotides. The important components of the nucleotide are a nitrogenous (nitrogen-bearing) base, a 5-carbon sugar (pentose), and a phosphate group (Figure 14.5). The nucleotide is named depending on the nitrogenous base. The nitrogenous base can be a purine such as adenine (A) and guanine (G), or a pyrimidine such as cytosine (C) and thymine (T).

VISUAL CONNECTION

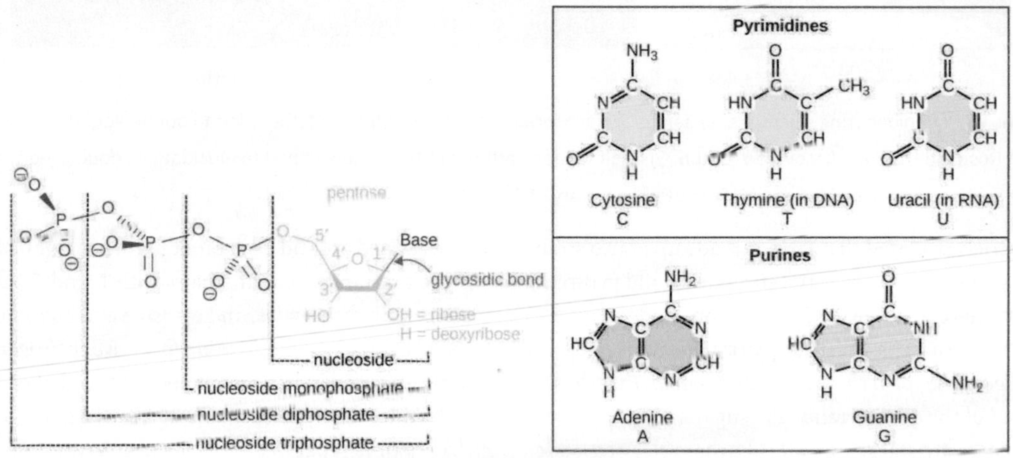

Figure 14.5 The purines have a double ring structure with a six-membered ring fused to a five-membered ring. Pyrimidines are smaller in size; they have a single six-membered ring structure.

The images above illustrate the five bases of DNA and RNA. Examine the images and explain why these are called "nitrogenous bases." How are the purines different from the pyrimidines? How is one purine or pyrimidine different from another, e.g., adenine from guanine? How is a nucleoside different from a nucleotide?

The purines have a double ring structure with a six-membered ring fused to a five-membered ring. Pyrimidines are smaller in size; they have a single six-membered ring structure.

The sugar is deoxyribose in DNA and ribose in RNA. The carbon atoms of the five-carbon sugar are numbered 1', 2', 3', 4', and 5' (1' is read as "one prime"). The phosphate, which makes DNA and RNA acidic, is connected to the 5' carbon of the sugar by the formation of an ester linkage between phosphoric acid and the 5'-OH group (an ester is an acid + an alcohol). In DNA nucleotides, the 3' carbon of the sugar deoxyribose is attached to a hydroxyl (OH) group. In RNA nucleotides, the 2' carbon of the sugar ribose also contains a hydroxyl group. The base is attached to the 1'carbon of the sugar.

The nucleotides combine with each other to produce phosphodiester bonds. The phosphate residue attached to the 5' carbon of the sugar of one nucleotide forms a second ester linkage with the hydroxyl group of the 3' carbon of the sugar of the next nucleotide, thereby forming a 5'-3' phosphodiester bond. In a polynucleotide, one end of the chain has a free 5' phosphate, and the other end has a free 3'-OH. These are called the 5' and 3' ends of the chain.

In the 1950s, Francis Crick and James Watson worked together to determine the structure of DNA at the University of Cambridge, England. Other scientists like Linus Pauling and Maurice Wilkins were also actively exploring this field. Pauling previously had discovered the secondary structure of proteins using X-ray crystallography. In Wilkins' lab, researcher Rosalind Franklin was using X-ray diffraction methods to understand the structure of DNA. Watson and Crick were able to piece together the puzzle of the DNA molecule on the basis of Franklin's data because Crick had also studied X-ray diffraction (Figure 14.6). In 1962, James Watson, Francis Crick, and Maurice Wilkins were awarded the Nobel Prize in Medicine. Unfortunately, by then Franklin had died, and Nobel prizes are not awarded posthumously.

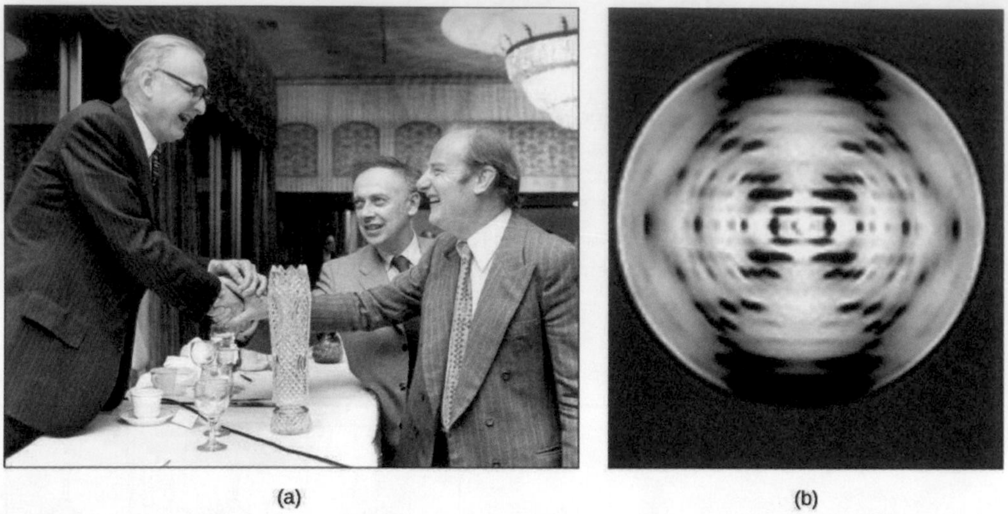

Figure 14.6 The work of pioneering scientists (a) James Watson, Francis Crick, and Maclyn McCarty led to our present day understanding of DNA. Scientist Rosalind Franklin discovered (b) the X-ray diffraction pattern of DNA, which helped to elucidate its double-helix structure. (credit a: modification of work by Marjorie McCarty, Public Library of Science)

Watson and Crick proposed that DNA is made up of two strands that are twisted around each other to form a right-handed helix. Base pairing takes place between a purine and pyrimidine on opposite strands, so that A pairs with T, and G pairs with C (suggested by Chargaff's Rules). Thus, adenine and thymine are complementary base pairs, and cytosine and guanine are also complementary base pairs. The base pairs are stabilized by hydrogen bonds: *adenine and thymine form two hydrogen bonds and cytosine and guanine form three hydrogen bonds.* The two strands are anti-parallel in nature; that is, the 3' end of one strand faces the 5' end of the other strand. The sugar and phosphate of the nucleotides form the backbone of the structure, whereas the nitrogenous bases are stacked inside, like the rungs of a ladder. Each base pair is separated from the next base pair by a distance of 0.34 nm, and each turn of the helix measures 3.4 nm. Therefore, 10 base pairs are present per turn of the helix. The diameter of the DNA double-helix is 2 nm, and it is uniform throughout. Only the pairing between a purine and pyrimidine and the antiparallel orientation of the two DNA strands can explain the uniform diameter. The twisting of the two strands around each other results in the formation of uniformly spaced major and minor grooves (Figure 14.7).

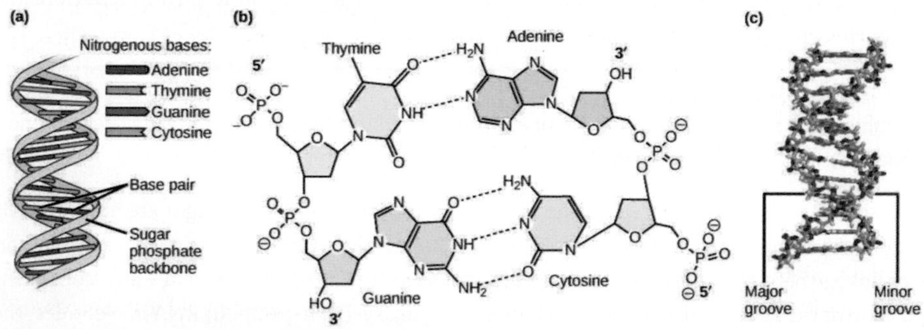

Figure 14.7 DNA has (a) a double helix structure and (b) phosphodiester bonds; the dotted lines between Thymine and Adenine and Guanine and Cytosine represent hydrogen bonds. The (c) major and minor grooves are binding sites for DNA binding proteins during processes such as transcription (the copying of RNA from DNA) and replication.

DNA Sequencing Techniques

Until the 1990s, the sequencing of DNA (reading the sequence of DNA) was a relatively expensive and long process. Using radiolabeled nucleotides also compounded the problem through safety concerns. With currently available technology and automated machines, the process is cheaper, safer, and can be completed in a matter of hours. Fred Sanger developed the sequencing method used for the human genome sequencing project, which is widely used today (Figure 14.8).

LINK TO LEARNING

Visit this site (http://openstax.org/l/DNA_sequencing) to watch a video explaining the DNA sequence-reading technique that resulted from Sanger's work.

The sequencing method is known as the dideoxy chain termination method. The method is based on the use of chain terminators, the dideoxynucleotides (ddNTPs). The ddNTPSs differ from the deoxynucleotides by the lack of a free 3' OH group on the five-carbon sugar. If a ddNTP is added to a growing DNA strand, the chain cannot be extended any further because the free 3' OH group needed to add another nucleotide is not available. By using a predetermined ratio of deoxynucleotides to dideoxynucleotides, it is possible to generate DNA fragments of different sizes.

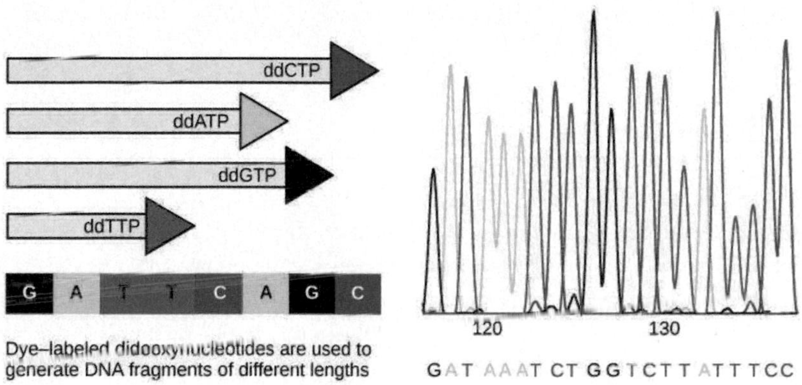

Figure 14.8 In Frederick Sanger's dideoxy chain termination method, dye-labeled dideoxynucleotides are used to generate DNA fragments that terminate at different points. The DNA is separated by capillary electrophoresis (not defined) on the basis of size, and from the order of fragments formed, the DNA sequence can be read. The DNA sequence readout is shown on an electropherogram (not defined) that is generated by a laser scanner.

The DNA sample to be sequenced is denatured (separated into two strands by heating it to high temperatures). The DNA is divided into four tubes in which a primer, DNA polymerase, and all four nucleoside triphosphates (A, T, G, and C) are added. In addition, limited quantities of one of the four dideoxynucleoside triphosphates (ddCTP, ddATP, ddGTP, and ddTTP) are added to each tube respectively. The tubes are labeled as A, T, G, and C according to the ddNTP added. For detection purposes, each of the four dideoxynucleotides carries a different fluorescent label. Chain elongation continues until a fluorescent dideoxy nucleotide is incorporated, after which no further elongation takes place. After the reaction is over, electrophoresis is performed. Even a difference in length of a single base can be detected. The sequence is read from a laser scanner that detects the fluorescent marker of each fragment. For his work on DNA sequencing, Sanger received a Nobel Prize in Chemistry in 1980.

LINK TO LEARNING

Sanger's genome sequencing has led to a race to sequence human genomes at rapid speed and low cost. Learn more by viewing the animation here (https://openstax.org/l/DNA_and_genomes) .

Gel electrophoresis is a technique used to separate DNA fragments of different sizes. Usually the gel is made of a chemical called *agarose* (a polysaccharide polymer extracted from seaweed that is high in galactose residues). Agarose powder is added to a buffer and heated. After cooling, the gel solution is poured into a casting tray. Once the gel has solidified, the DNA is loaded on the gel and electric current is applied. The DNA has a net negative charge and moves from the negative electrode toward the positive electrode. The electric current is applied for sufficient time to let the DNA separate according to size; the smallest fragments will be farthest from the well (where the DNA was loaded), and the heavier molecular weight fragments will be closest to the well. Once the DNA is separated, the gel is stained with a DNA-specific dye for viewing it (Figure 14.9).

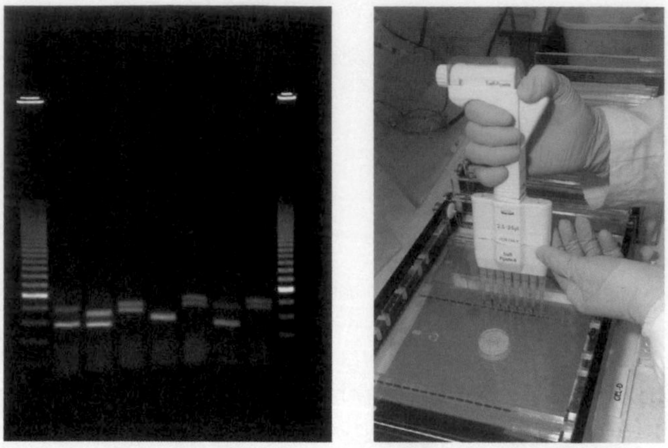

Figure 14.9 DNA can be separated on the basis of size using gel electrophoresis. (credit: James Jacob, Tompkins Cortland Community College)

EVOLUTION CONNECTION

Neanderthal Genome: How Are We Related?

The first draft sequence of the Neanderthal genome was recently published by Richard E. Green et al. in 2010.[1] Neanderthals are the closest ancestors of present-day humans. They were known to have lived in Europe and Western Asia (and now, perhaps, in Northern Africa) before they disappeared from fossil records approximately 30,000 years ago. Green's team studied almost 40,000-year-old fossil remains that were selected from sites across the world. Extremely sophisticated means of sample preparation and DNA sequencing were employed because of the fragile nature of the bones and heavy microbial contamination. In their study, the scientists were able to sequence some four billion base pairs. The Neanderthal sequence was compared with that of present-day humans from across the world. After comparing the sequences, the researchers found that the Neanderthal genome had 2 to 3 percent greater similarity to people living outside Africa than to people in Africa. While current theories have suggested that all present-day humans can be traced to a small ancestral population in Africa, the data from the Neanderthal genome suggest some interbreeding between Neanderthals and early modern humans.

Green and his colleagues also discovered DNA segments among people in Europe and Asia that are more similar to Neanderthal sequences than to other contemporary human sequences. Another interesting observation was that Neanderthals are as closely related to people from Papua New Guinea as to those from China or France. This is surprising because Neanderthal fossil remains have been located only in Europe and West Asia. Most likely, genetic exchange took place between Neanderthals and modern humans as modern humans emerged out of Africa, before the divergence of Europeans, East Asians, and Papua New Guineans.

Several genes seem to have undergone changes from Neanderthals during the evolution of present-day humans. These genes are involved in cranial structure, metabolism, skin morphology, and cognitive development. One of the genes that is of particular interest is *RUNX2*, which is different in modern day humans and Neanderthals. This gene is responsible for the prominent frontal bone, bell-shaped rib cage, and dental differences seen in Neanderthals. It is speculated that an evolutionary change in *RUNX2* was important in the origin of modern-day humans, and this affected the cranium and the upper body.

LINK TO LEARNING

Watch Svante Pääbo's talk (http://openstax.org/l/neanderthal) explaining the Neanderthal genome research at the 2011 annual TED (Technology, Entertainment, Design) conference.

DNA Packaging in Cells

Prokaryotes are much simpler than eukaryotes in many of their features (Figure 14.10). Most prokaryotes contain a single, circular chromosome that is found in an area of the cytoplasm called the *nucleoid region*.

1Richard E. Green et al., "A Draft Sequence of the Neandertal Genome," *Science* 328 (2010): 710-22.

VISUAL CONNECTION

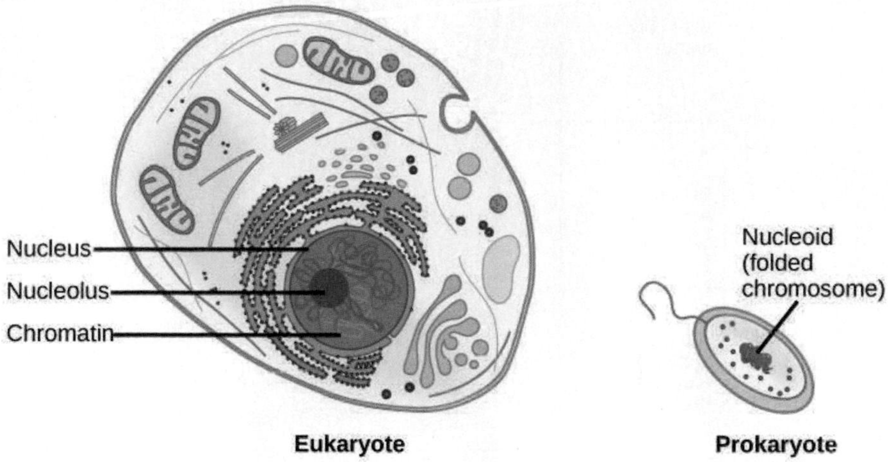

Figure 14.10 A eukaryote contains a well-defined nucleus, whereas in prokaryotes, the chromosome lies in the cytoplasm in an area called the nucleoid.

In eukaryotic cells, DNA and RNA synthesis occur in a separate compartment from protein synthesis. In prokaryotic cells, both processes occur together. What advantages might there be to separating the processes? What advantages might there be to having them occur together?

The size of the genome in one of the most well-studied prokaryotes, *E.coli,* is 4.6 million base pairs (approximately 1.1 mm, if cut and stretched out). So how does this fit inside a small bacterial cell? The DNA is twisted by what is known as supercoiling. Supercoiling suggests that DNA is either "under-wound" (less than one turn of the helix per 10 base pairs) or "over-wound" (more than 1 turn per 10 base pairs) from its normal relaxed state. Some proteins are known to be involved in the supercoiling; other proteins and enzymes such as DNA gyrase help in maintaining the supercoiled structure.

Eukaryotes, whose chromosomes each consist of a linear DNA molecule, employ a different type of packing strategy to fit their DNA inside the nucleus (Figure 14.11). At the most basic level, DNA is wrapped around proteins known as **histones** to form structures called nucleosomes. The histones are evolutionarily conserved proteins that are rich in basic amino acids and form an octamer composed of two molecules of each of four different histones. The DNA (remember, it is negatively charged because of the phosphate groups) is wrapped tightly around the histone core. This nucleosome is linked to the next one with the help of a *linker DNA*. This is also known as the "beads on a string" structure. With the help of a fifth histone, a string of nucleosomes is further compacted into a 30-nm fiber, which is the diameter of the structure. Metaphase chromosomes are even further condensed by association with scaffolding proteins. At the metaphase stage, the chromosomes are at their most compact, approximately 700 nm in width.

In interphase, eukaryotic chromosomes have two distinct regions that can be distinguished by staining. The tightly packaged region is known as heterochromatin, and the less dense region is known as euchromatin. Heterochromatin usually contains genes that are not expressed, and is found in the regions of the centromere and telomeres. The euchromatin usually contains genes that are transcribed, with DNA packaged around nucleosomes but not further compacted.

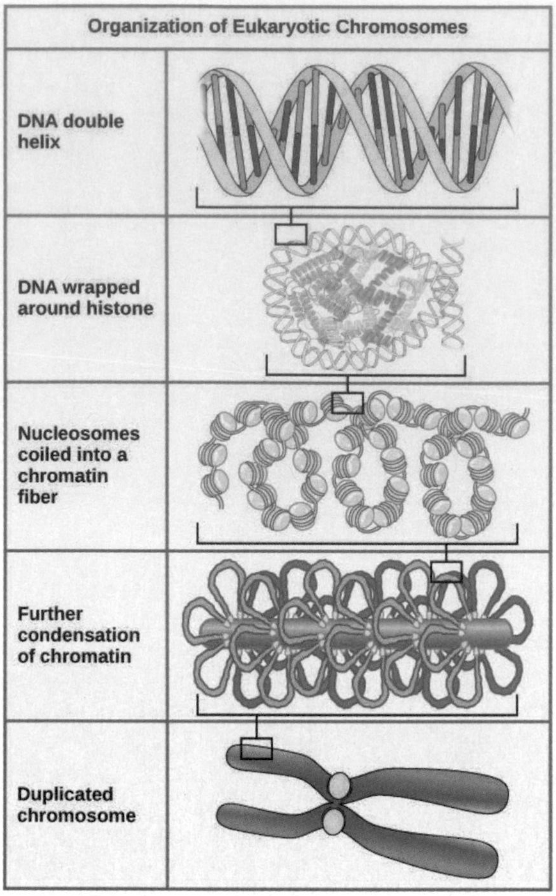

Figure 14.11 These figures illustrate the compaction of the eukaryotic chromosome.

14.3 Basics of DNA Replication

By the end of this section, you will be able to do the following:

- Explain how the structure of DNA reveals the replication process
- Describe the Meselson and Stahl experiments

The elucidation of the structure of the double helix provided a hint as to how DNA divides and makes copies of itself. In their 1953 paper, Watson and Crick penned an incredible understatement: "It has not escaped our notice that the specific pairing we have postulated immediately suggests a possible copying mechanism for the genetic material." With specific base pairs, the sequence of one DNA strand can be predicted from its complement. The double-helix model suggests that the two strands of the double helix separate during replication, and each strand serves as a template from which the new complementary strand is copied. What was not clear was how the replication took place. There were three models suggested (Figure 14.12): *conservative, semi-conservative, and dispersive.*

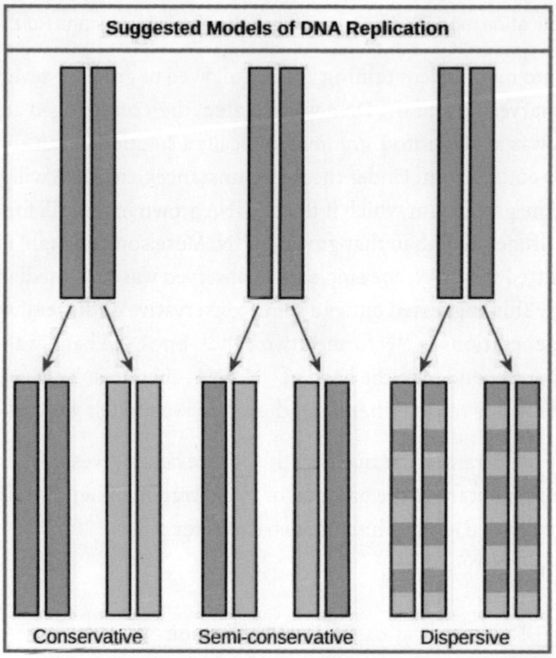

Figure 14.12 The three suggested models of DNA replication. Gray indicates the original DNA strands, and blue indicates newly synthesized DNA.

In conservative replication, the parental DNA remains together, and the newly formed daughter strands are together. The semi-conservative method suggests that each of the two parental DNA strands acts as a template for new DNA to be synthesized; after replication, each double-stranded DNA includes one parental or "old" strand and one "new" strand. In the dispersive model, both copies of DNA have double-stranded segments of parental DNA and newly synthesized DNA interspersed.

Meselson and Stahl were interested in understanding how DNA replicates. They grew *E. coli* for several generations in a medium containing a "heavy" isotope of nitrogen (^{15}N), which gets incorporated into nitrogenous bases, and eventually into the DNA (Figure 14.13).

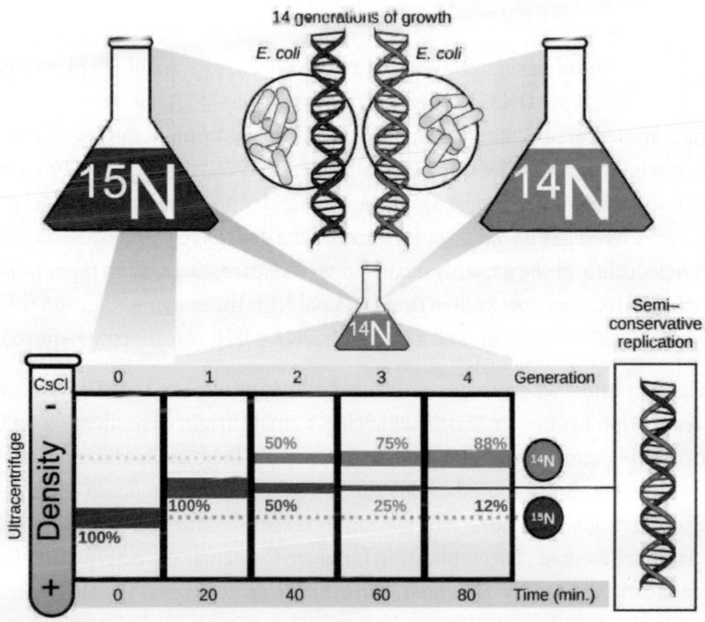

Figure 14.13 Meselson and Stahl experimented with *E. coli* grown first in heavy nitrogen (^{15}N) then in ^{14}N. DNA grown in ^{15}N (red band) is heavier than DNA grown in ^{14}N (orange band), and sediments to a lower level in cesium chloride solution in an ultracentrifuge. When DNA grown in ^{15}N is switched to media containing ^{14}N, after one round of cell division the DNA sediments halfway between the ^{15}N and ^{14}N levels, indicating that it now contains fifty percent ^{14}N. In subsequent cell divisions, an increasing amount of DNA contains ^{14}N only. These

data support the semi-conservative replication model. (credit: modification of work by Mariana Ruiz Villareal)

The *E. coli* culture was then placed into medium containing ^{14}N and allowed to grow for several generations. After each of the first few generations, the cells were harvested and the DNA was isolated, then centrifuged at high speeds in an ultracentrifuge. During the centrifugation, the DNA was loaded into a *gradient* (typically a solution of salt such as cesium chloride or sucrose) and spun at high speeds of 50,000 to 60,000 rpm. Under these circumstances, the DNA will form a band according to its *buoyant density*: the density within the gradient at which it floats. DNA grown in ^{15}N will form a band at a higher density position (i.e., farther down the centrifuge tube) than that grown in ^{14}N. Meselson and Stahl noted that after one generation of growth in ^{14}N after they had been shifted from ^{15}N, the single band observed was intermediate in position in between DNA of cells grown exclusively in ^{15}N and ^{14}N. This suggested either a semi-conservative or dispersive mode of replication. The DNA harvested from cells grown for two generations in ^{14}N formed two bands: one DNA band was at the intermediate position between ^{15}N and ^{14}N, and the other corresponded to the band of ^{14}N DNA. These results could only be explained if DNA replicates in a semi-conservative manner. And for this reason, therefore, the other two models were ruled out.

During DNA replication, each of the two strands that make up the double helix serves as a template from which new strands are copied. The new strands will be complementary to the parental or "old" strands. When two daughter DNA copies are formed, they have the same sequence and are divided equally into the two daughter cells.

LINK TO LEARNING

View this video (http://openstax.org/l/DNA_replicatio2) on DNA replication.

14.4 DNA Replication in Prokaryotes

By the end of this section, you will be able to do the following:

- Explain the process of DNA replication in prokaryotes
- Discuss the role of different enzymes and proteins in supporting this process

DNA replication has been well studied in prokaryotes primarily because of the small size of the genome and because of the large variety of mutants that are available. *E. coli* has 4.6 million base pairs in a single circular chromosome and all of it gets replicated in approximately 42 minutes, starting from a single site along the chromosome and proceeding around the circle in both directions. This means that approximately 1000 nucleotides are added per second. Thus, the process is quite rapid and occurs without many mistakes.

DNA replication employs a large number of structural proteins and enzymes, each of which plays a critical role during the process. One of the key players is the enzyme **DNA polymerase**, also known as DNA pol, which adds nucleotides one-by-one to the growing DNA chain that is complementary to the template strand. The addition of nucleotides requires energy; this energy is obtained from the nucleoside triphosphates ATP, GTP, TTP and CTP. Like ATP, the other **NTPs** (nucleoside triphosphates) are high-energy molecules that can serve both as the source of DNA nucleotides and the source of energy to drive the polymerization. When the bond between the phosphates is "broken," the energy released is used to form the phosphodiester bond between the incoming nucleotide and the growing chain. In prokaryotes, three main types of polymerases are known: DNA pol I, DNA pol II, and DNA pol III. It is now known that DNA pol III is the enzyme required for DNA synthesis; DNA pol I is an important accessory enzyme in DNA replication, and along with DNA pol II, is primarily required for repair.

How does the replication machinery know where to begin? It turns out that there are specific nucleotide sequences called *origins of replication* where replication begins. In *E. coli,* which has a single origin of replication on its one chromosome (as do most prokaryotes), this origin of replication is approximately 245 base pairs long and is rich in AT sequences. The origin of replication is recognized by certain proteins that bind to this site. An enzyme called *helicase* unwinds the DNA by breaking the hydrogen bonds between the nitrogenous base pairs. ATP hydrolysis is required for this process. As the DNA opens up, Y-shaped structures called *replication forks* are formed. Two replication forks are formed at the origin of replication and these get extended bi-directionally as replication proceeds. **Single-strand binding proteins** coat the single strands of DNA near the replication fork to prevent the single-stranded DNA from winding back into a double helix.

DNA polymerase has two important restrictions: it is able to add nucleotides only in the 5' to 3' direction (a new DNA strand can be only extended in this direction). It also requires a free 3'-OH group to which it can add nucleotides by forming a phosphodiester bond between the 3'-OH end and the 5' phosphate of the next nucleotide. This essentially means that it cannot add nucleotides if a free 3'-OH group is not available. Then how does it add the first nucleotide? The problem is solved with the help of a primer that provides the free 3'-OH end. Another enzyme, RNA **primase**, synthesizes an RNA segment that is about five to ten nucleotides long and complementary to the template DNA. Because this sequence primes the DNA synthesis, it is appropriately called the **primer**. DNA polymerase can now extend this RNA primer, adding nucleotides one-by-one that are complementary to the template strand (Figure 14.14).

VISUAL CONNECTION

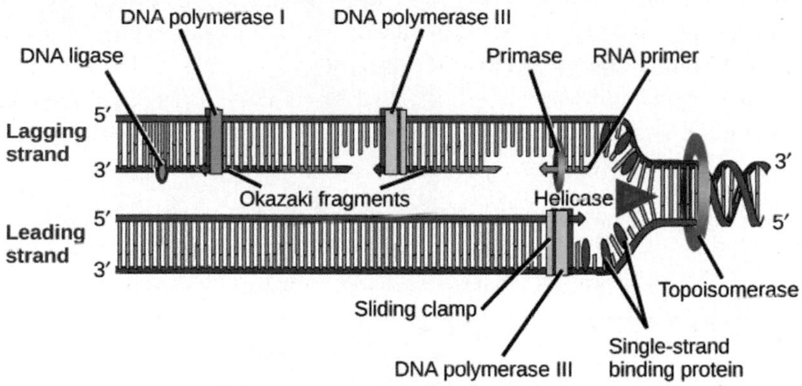

Figure 14.14 A replication fork is formed when helicase separates the DNA strands at the origin of replication. The DNA tends to become more highly coiled ahead of the replication fork. Topoisomerase breaks and reforms DNA's phosphate backbone ahead of the replication fork, thereby relieving the pressure that results from this "supercoiling." Single-strand binding proteins bind to the single-stranded DNA to prevent the helix from re-forming. Primase synthesizes an RNA primer. DNA polymerase III uses this primer to synthesize the daughter DNA strand. On the leading strand, DNA is synthesized continuously, whereas on the lagging strand, DNA is synthesized in short stretches called Okazaki fragments. DNA polymerase I replaces the RNA primer with DNA. DNA ligase seals the gaps between the Okazaki fragments, joining the fragments into a single DNA molecule. (credit: modification of work by Mariana Ruiz Villareal)

Question: You isolate a cell strain in which the joining of Okazaki fragments is impaired and suspect that a mutation has occurred in an enzyme found at the replication fork. Which enzyme is most likely to be mutated?

The replication fork moves at the rate of 1000 nucleotides per second. Topoisomerase prevents the over-winding of the DNA double helix ahead of the replication fork as the DNA is opening up; it does so by causing temporary nicks in the DNA helix and then resealing it. Because DNA polymerase can only extend in the 5' to 3' direction, and because the DNA double helix is *antiparallel*, there is a slight problem at the replication fork. *The two template DNA strands have opposing orientations:* one strand is in the 5' to 3' direction and the other is oriented in the 3' to 5' direction. Only one new DNA strand, the one that is complementary to the 3' to 5' parental DNA strand, can be synthesized continuously towards the replication fork. This continuously synthesized strand is known as the **leading strand**. The other strand, complementary to the 5' to 3' parental DNA, is extended away from the replication fork, in small fragments known as **Okazaki fragments**, each requiring a primer to start the synthesis. New primer segments are laid down in the direction of the replication fork, but each pointing away from it. (Okazaki fragments are named after the Japanese scientist who first discovered them. The strand with the Okazaki fragments is known as the **lagging strand**.)

The leading strand can be extended from a single primer, whereas the lagging strand needs a new primer for each of the short Okazaki fragments. The overall direction of the lagging strand will be 3' to 5', and that of the leading strand 5' to 3'. A protein called the **sliding clamp** holds the DNA polymerase in place as it continues to add nucleotides. The sliding clamp is a ring-shaped protein that binds to the DNA and holds the polymerase in place. As synthesis proceeds, the RNA primers are replaced by DNA. The primers are removed by the exonuclease activity of DNA pol I, which uses DNA behind the RNA as its own primer and fills in the gaps left by removal of the RNA nucleotides by the addition of DNA nucleotides. The nicks that remain between the newly synthesized DNA (that replaced the RNA primer) and the previously synthesized DNA are sealed by the enzyme DNA **ligase**, which catalyzes the formation of phosphodiester linkages between the 3'-OH end of one nucleotide and the 5' phosphate end of the other fragment.

Once the chromosome has been completely replicated, the two DNA copies move into two different cells during cell division.

The process of DNA replication can be summarized as follows:

1. DNA unwinds at the origin of replication.
2. Helicase opens up the DNA-forming replication forks; these are extended bidirectionally.
3. Single-strand binding proteins coat the DNA around the replication fork to prevent rewinding of the DNA.
4. Topoisomerase binds at the region ahead of the replication fork to prevent supercoiling.
5. Primase synthesizes RNA primers complementary to the DNA strand.
6. DNA polymerase III starts adding nucleotides to the 3'-OH end of the primer.
7. Elongation of both the lagging and the leading strand continues.
8. RNA primers are removed by exonuclease activity.
9. Gaps are filled by DNA pol I by adding dNTPs.
10. The gap between the two DNA fragments is sealed by DNA ligase, which helps in the formation of phosphodiester bonds.

Table 14.1 summarizes the enzymes involved in prokaryotic DNA replication and the functions of each.

Prokaryotic DNA Replication: Enzymes and Their Function

Enzyme/protein	Specific Function
DNA pol I	Removes RNA primer and replaces it with newly synthesized DNA
DNA pol III	Main enzyme that adds nucleotides in the 5'-3' direction
Helicase	Opens the DNA helix by breaking hydrogen bonds between the nitrogenous bases
Ligase	Seals the gaps between the Okazaki fragments to create one continuous DNA strand
Primase	Synthesizes RNA primers needed to start replication
Sliding Clamp	Helps to hold the DNA polymerase in place when nucleotides are being added
Topoisomerase	Helps relieve the strain on DNA when unwinding by causing breaks, and then resealing the DNA
Single-strand binding proteins (SSB)	Binds to single-stranded DNA to prevent DNA from rewinding back.

Table 14.1

LINK TO LEARNING

Review the full process of DNA replication here (http://openstax.org/l/replication_DNA) .

14.5 DNA Replication in Eukaryotes

By the end of this section, you will be able to do the following:

- Discuss the similarities and differences between DNA replication in eukaryotes and prokaryotes
- State the role of telomerase in DNA replication

Eukaryotic genomes are much more complex and larger in size than prokaryotic genomes. Eukaryotes also have a number of different linear chromosomes. The human genome has 3 billion base pairs per haploid set of chromosomes, and 6 billion base pairs are replicated during the S phase of the cell cycle. There are multiple origins of replication on each eukaryotic chromosome; humans can have up to 100,000 origins of replication across the genome. The rate of replication is approximately

100 nucleotides per second, much slower than prokaryotic replication. In yeast, which is a eukaryote, special sequences known as autonomously replicating sequences (ARS) are found on the chromosomes. These are equivalent to the origin of replication in *E. coli*.

The number of DNA polymerases in eukaryotes is much more than in prokaryotes: 14 are known, of which five are known to have major roles during replication and have been well studied. They are known as pol α, pol β, pol γ, pol δ, and pol ε.

The essential steps of replication are the same as in prokaryotes. Before replication can start, the DNA has to be made available as a template. Eukaryotic DNA is bound to basic proteins known as histones to form structures called nucleosomes. Histones must be removed and then replaced during the replication process, which helps to account for the lower replication rate in eukaryotes. The chromatin (the complex between DNA and proteins) may undergo some chemical modifications, so that the DNA may be able to slide off the proteins or be accessible to the enzymes of the DNA replication machinery. At the origin of replication, a pre-replication complex is made with other initiator proteins. Helicase and other proteins are then recruited to start the replication process (Table 14.2).

Difference between Prokaryotic and Eukaryotic Replication

Property	Prokaryotes	Eukaryotes
Origin of replication	Single	Multiple
Rate of replication	1000 nucleotides/s	50 to 100 nucleotides/s
DNA polymerase types	5	14
Telomerase	Not present	Present
RNA primer removal	DNA pol I	RNase H
Strand elongation	DNA pol III	Pol α, pol δ, pol ε
Sliding clamp	Sliding clamp	PCNA

Table 14.2

A helicase using the energy from ATP hydrolysis opens up the DNA helix. Replication forks are formed at each replication origin as the DNA unwinds. The opening of the double helix causes over-winding, or supercoiling, in the DNA ahead of the replication fork. These are resolved with the action of topoisomerases. Primers are formed by the enzyme primase, and using the primer, DNA pol can start synthesis. Three major DNA polymerases are then involved: α, δ and ε. DNA pol α adds a short (20 to 30 nucleotides) DNA fragment to the RNA primer on both strands, and then hands off to a second polymerase. While the leading strand is continuously synthesized by the enzyme pol ε, the lagging strand is synthesized by pol δ. A sliding clamp protein known as PCNA (proliferating cell nuclear antigen) holds the DNA pol in place so that it does not slide off the DNA. As pol δ runs into the primer RNA on the lagging strand, it displaces it from the DNA template. The displaced primer RNA is then removed by RNase H (AKA flap endonuclease) and replaced with DNA nucleotides. The Okazaki fragments in the lagging strand are joined after the replacement of the RNA primers with DNA. The gaps that remain are sealed by DNA ligase, which forms the phosphodiester bond.

Telomere replication

Unlike prokaryotic chromosomes, eukaryotic chromosomes are linear. As you've learned, the enzyme DNA pol can add nucleotides only in the 5' to 3' direction. In the leading strand, synthesis continues until the end of the chromosome is reached. On the lagging strand, DNA is synthesized in short stretches, each of which is initiated by a separate primer. When the replication fork reaches the end of the linear chromosome, there is no way to replace the primer on the 5' end of the lagging strand. The DNA at the ends of the chromosome thus remains unpaired, and over time these ends, called **telomeres,** may get progressively shorter as cells continue to divide.

Telomeres comprise repetitive sequences that code for no particular gene. In humans, a six-base-pair sequence, TTAGGG, is repeated 100 to 1000 times in the telomere regions. In a way, these telomeres protect the genes from getting deleted as cells continue to divide. The telomeres are added to the ends of chromosomes by a separate enzyme, telomerase (Figure 14.16), whose discovery helped in the understanding of how these repetitive chromosome ends are maintained. The **telomerase** enzyme contains a catalytic part and a built-in RNA template. It attaches to the end of the chromosome, and DNA nucleotides complementary to the RNA template are added on the 3' end of the DNA strand. Once the 3' end of the lagging strand template is sufficiently elongated, DNA polymerase can add the nucleotides complementary to the ends of the chromosomes. Thus, the ends of the chromosomes are replicated.

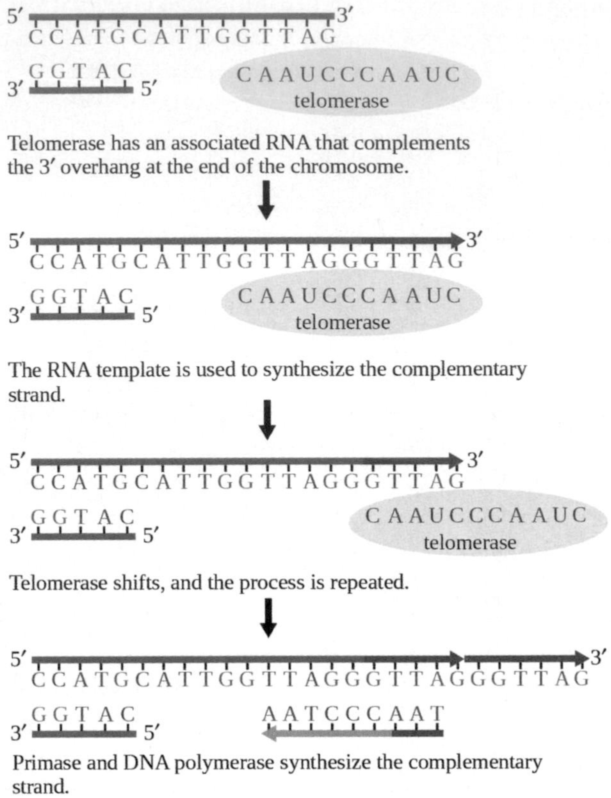

Figure 14.15 The ends of linear chromosomes are maintained by the action of the telomerase enzyme.

Telomerase is typically active in germ cells and adult stem cells. It is not active in adult somatic cells. For their discovery of telomerase and its action, Elizabeth Blackburn, Carol W. Greider, and Jack W. Szostak (Figure 14.16) received the Nobel Prize for Medicine and Physiology in 2009.

Figure 14.16 Elizabeth Blackburn, 2009 Nobel Laureate, is one of the scientists who discovered how telomerase works. (credit: US

Embassy Sweden)

Telomerase and Aging

Cells that undergo cell division continue to have their telomeres shortened because most somatic cells do not make telomerase. This essentially means that telomere shortening is associated with aging. With the advent of modern medicine, preventative health care, and healthier lifestyles, the human life span has increased, and there is an increasing demand for people to look younger and have a better quality of life as they grow older.

In 2010, scientists found that telomerase can reverse some age-related conditions in mice. This may have potential in regenerative medicine.[2] Telomerase-deficient mice were used in these studies; these mice have tissue atrophy, stem cell depletion, organ system failure, and impaired tissue injury responses. Telomerase reactivation in these mice caused extension of telomeres, reduced DNA damage, reversed neurodegeneration, and improved the function of the testes, spleen, and intestines. Thus, telomere reactivation may have potential for treating age-related diseases in humans.

Cancer is characterized by uncontrolled cell division of abnormal cells. The cells accumulate mutations, proliferate uncontrollably, and can migrate to different parts of the body through a process called metastasis. Scientists have observed that cancerous cells have considerably shortened telomeres and that telomerase is active in these cells. Interestingly, only after the telomeres were shortened in the cancer cells did the telomerase become active. If the action of telomerase in these cells can be inhibited by drugs during cancer therapy, then the cancerous cells could potentially be stopped from further division.

14.6 DNA Repair

By the end of this section, you will be able to do the following:

- Discuss the different types of mutations in DNA
- Explain DNA repair mechanisms

DNA replication is a highly accurate process, but mistakes can occasionally occur, such as a DNA polymerase inserting a wrong base. Uncorrected mistakes may sometimes lead to serious consequences, such as cancer. Repair mechanisms correct the mistakes. In rare cases, mistakes are not corrected, leading to mutations; in other cases, repair enzymes are themselves mutated or defective.

Most of the mistakes during DNA replication are promptly corrected by the proofreading ability of DNA polymerase itself. (Figure 14.17). In **proofreading**, the DNA pol reads the newly added base before adding the next one, so a correction can be made. The polymerase checks whether the newly added base has paired correctly with the base in the template strand. If it is the right base, the next nucleotide is added. If an incorrect base has been added, the enzyme makes a cut at the phosphodiester bond and releases the wrong nucleotide. This is performed by the 3' exonuclease action of DNA pol. Once the incorrect nucleotide has been removed, it can be replaced by the correct one.

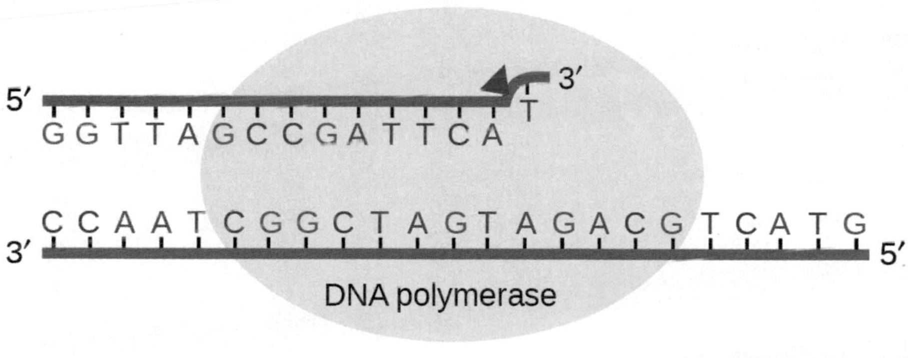

Figure 14.17 Proofreading by DNA polymerase corrects errors during replication.

Some errors are not corrected during replication, but are instead corrected after replication is completed; this type of repair is known as **mismatch repair** (Figure 14.18). Specific repair enzymes recognize the mispaired nucleotide and excise part of the strand that contains it; the excised region is then resynthesized. If the mismatch remains uncorrected, it may lead to more permanent damage when the mismatched DNA is replicated. How do mismatch repair enzymes recognize which of the two bases is the incorrect one? In *E. coli*, after replication, the nitrogenous base adenine acquires a methyl group; the parental DNA

2Jaskelioff et al., "Telomerase reactivation reverses tissue degeneration in aged telomerase-deficient mice," *Nature* 469 (2011): 102-7.

strand will have methyl groups, whereas the newly synthesized strand lacks them. Thus, DNA polymerase is able to remove the wrongly incorporated bases from the newly synthesized, non-methylated strand. In eukaryotes, the mechanism is not very well understood, but it is believed to involve recognition of unsealed nicks in the new strand, as well as a short-term continuing association of some of the replication proteins with the new daughter strand after replication has completed.

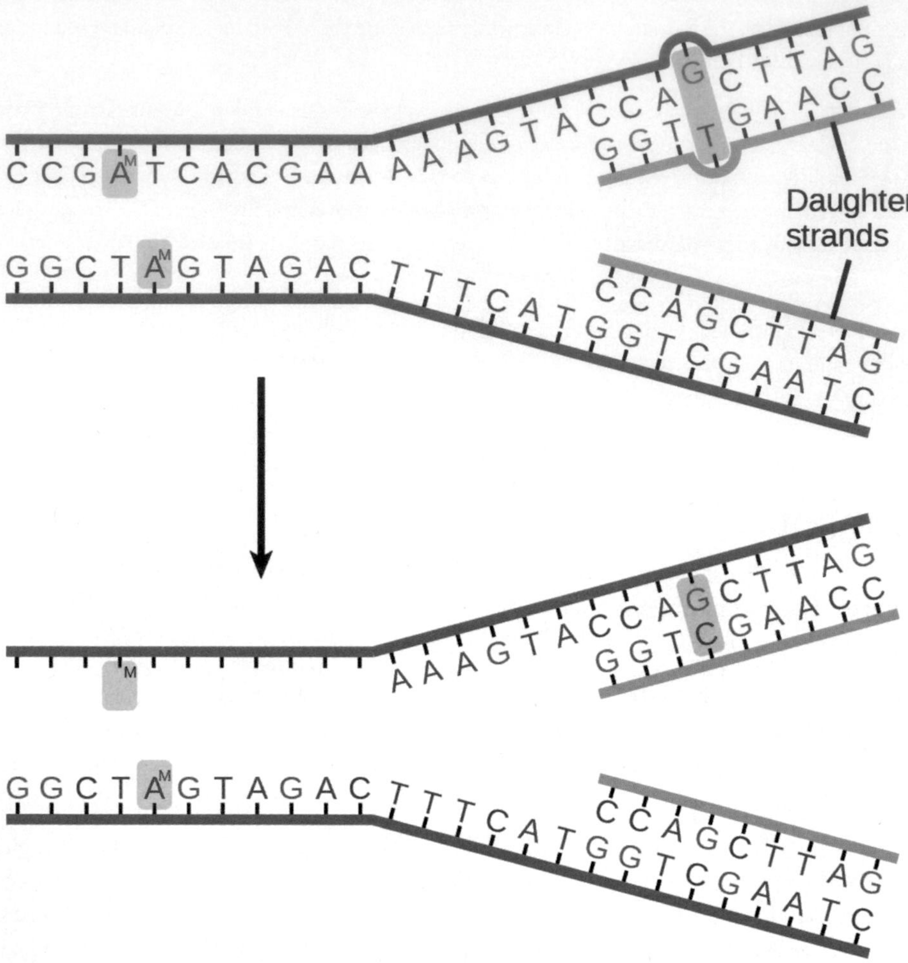

Figure 14.18 In mismatch repair, the incorrectly added base is detected after replication. The mismatch repair proteins detect this base and remove it from the newly synthesized strand by nuclease action. The gap is now filled with the correctly paired base.

Another type of repair mechanism, **nucleotide excision repair**, is similar to mismatch repair, except that it is used to remove damaged bases rather than mismatched ones. The repair enzymes replace abnormal bases by making a cut on both the 3' and 5' ends of the damaged base (Figure 14.19). The segment of DNA is removed and replaced with the correctly paired nucleotides by the action of DNA pol. Once the bases are filled in, the remaining gap is sealed with a phosphodiester linkage catalyzed by DNA ligase. This repair mechanism is often employed when UV exposure causes the formation of pyrimidine dimers.

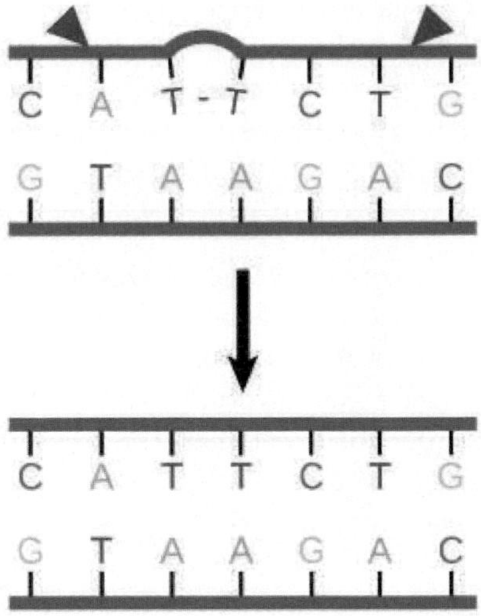

Figure 14.19 Nucleotide excision repairs thymine dimers. When exposed to UV light, thymines lying adjacent to each other can form thymine dimers. In normal cells, they are excised and replaced.

A well-studied example of mistakes not being corrected is seen in people suffering from xeroderma pigmentosa (Figure 14.20). Affected individuals have skin that is highly sensitive to UV rays from the sun. When individuals are exposed to UV light, pyrimidine dimers, especially those of thymine, are formed; people with xeroderma pigmentosa are not able to repair the damage. These are not repaired because of a defect in the nucleotide excision repair enzymes, whereas in normal individuals, the thymine dimers are excised and the defect is corrected. The thymine dimers distort the structure of the DNA double helix, and this may cause problems during DNA replication. People with xeroderma pigmentosa may have a higher risk of contracting skin cancer than those who don't have the condition.

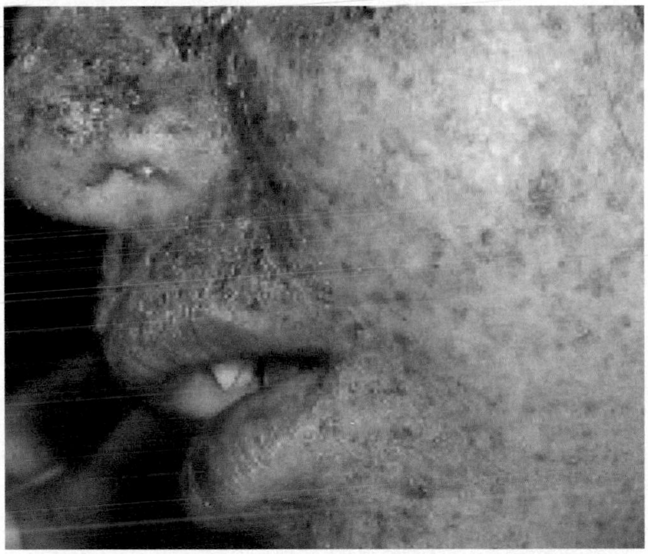

Figure 14.20 Xeroderma pigmentosa is a condition in which thymine dimerization from exposure to UV light is not repaired. Exposure to sunlight results in skin lesions. (credit: James Halpern et al.)

Errors during DNA replication are not the only reason why mutations arise in DNA. **Mutations**, variations in the nucleotide sequence of a genome, can also occur because of damage to DNA. Such mutations may be of two types: induced or spontaneous. **Induced mutations** are those that result from an exposure to chemicals, UV rays, x-rays, or some other environmental agent. **Spontaneous mutations** occur without any exposure to any environmental agent; they are a result of natural reactions taking place within the body.

Mutations may have a wide range of effects. Point mutations are those mutations that affect a single base pair. The most common nucleotide mutations are substitutions, in which one base is replaced by another. These substitutions can be of two types, either transitions or transversions. **Transition substitution** refers to a purine or pyrimidine being replaced by a base of the same kind; for example, a purine such as adenine may be replaced by the purine guanine. **Transversion substitution** refers to a purine being replaced by a pyrimidine, or vice versa; for example, cytosine, a pyrimidine, is replaced by adenine, a purine. Some point mutations are not expressed; these are known as silent mutations. Silent mutations are usually due to a substitution in the third base of a codon, which often represents the same amino acid as the original codon. Other point mutations can result in the replacement of one amino acid by another, which may alter the function of the protein. Point mutations that generate a stop codon can terminate a protein early.

Some mutations can result in an increased number of copies of the same codon. These are called trinucleotide repeat expansions and result in repeated regions of the same amino acid. Mutations can also be the result of the addition of a base, known as an insertion, or the removal of a base, also known as deletion. If an insertion or deletion results in the alteration of the translational reading frame (a frameshift mutation), the resultant protein is usually nonfunctional. Sometimes a piece of DNA from one chromosome may get translocated to another chromosome or to another region of the same chromosome; this is also known as translocation. These mutation types are shown in Figure 14.21.

VISUAL CONNECTION

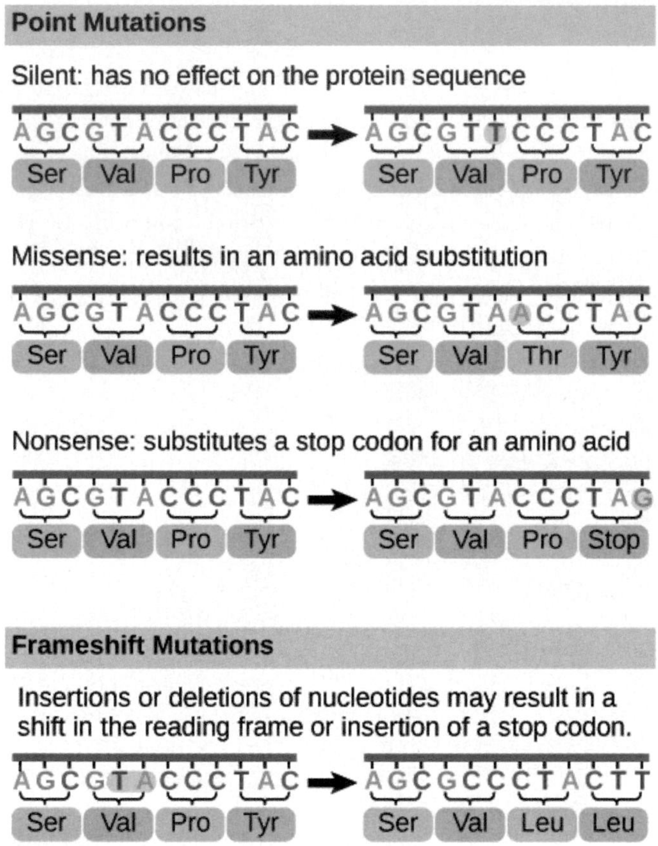

Figure 14.21 Mutations can lead to changes in the protein sequence encoded by the DNA.

A frameshift mutation that results in the insertion of three nucleotides is often less deleterious than a mutation that results in the insertion of one nucleotide. Why?

Mutations in repair genes have been known to cause cancer. Many mutated repair genes have been implicated in certain forms of pancreatic cancer, colon cancer, and colorectal cancer. Mutations can affect either somatic cells or germ cells. If many mutations accumulate in a somatic cell, they may lead to problems such as the uncontrolled cell division observed in cancer. If a mutation takes place in germ cells, the mutation will be passed on to the next generation, as in the case of hemophilia and xeroderma pigmentosa.

KEY TERMS

electrophoresis technique used to separate DNA fragments according to size

helicase during replication, this enzyme helps to open up the DNA helix by breaking the hydrogen bonds

induced mutation mutation that results from exposure to chemicals or environmental agents

lagging strand during replication, the strand that is replicated in short fragments and away from the replication fork

leading strand strand that is synthesized continuously in the 5'-3' direction, which is synthesized in the direction of the replication fork

ligase enzyme that catalyzes the formation of a phosphodiester linkage between the 3' OH and 5' phosphate ends of the DNA

mismatch repair type of repair mechanism in which mismatched bases are removed after replication

mutation variation in the nucleotide sequence of a genome

nucleotide excision repair type of DNA repair mechanism in which the wrong base, along with a few nucleotides upstream or downstream, are removed

Okazaki fragment DNA fragment that is synthesized in short stretches on the lagging strand

point mutation mutation that affects a single base

primase enzyme that synthesizes the RNA primer; the primer is needed for DNA pol to start synthesis of a new DNA strand

primer short stretch of nucleotides that is required to initiate replication; in the case of replication, the primer has RNA nucleotides

proofreading function of DNA pol in which it reads the newly added base before adding the next one

replication fork Y-shaped structure formed during initiation of replication

silent mutation mutation that is not expressed

single-strand binding protein during replication, protein that binds to the single-stranded DNA; this helps in keeping the two strands of DNA apart so that they may serve as templates

sliding clamp ring-shaped protein that holds the DNA pol on the DNA strand

spontaneous mutation mutation that takes place in the cells as a result of chemical reactions taking place naturally without exposure to any external agent

telomerase enzyme that contains a catalytic part and an inbuilt RNA template; it functions to maintain telomeres at chromosome ends

telomere DNA at the end of linear chromosomes

topoisomerase enzyme that causes underwinding or overwinding of DNA when DNA replication is taking place

transformation process in which external DNA is taken up by a cell

transition substitution when a purine is replaced with a purine or a pyrimidine is replaced with another pyrimidine

transversion substitution when a purine is replaced by a pyrimidine or a pyrimidine is replaced by a purine

CHAPTER SUMMARY

14.1 Historical Basis of Modern Understanding

DNA was first isolated from white blood cells by Friedrich Miescher, who called it nuclein because it was isolated from nuclei. Frederick Griffith's experiments with strains of *Streptococcus pneumoniae* provided the first hint that DNA may be the transforming principle. Avery, MacLeod, and McCarty showed that DNA is required for the transformation of bacteria. Later experiments by Hershey and Chase using bacteriophage T2 proved that DNA is the genetic material. Chargaff found that the ratio of A = T and C = G, and that the percentage content of A, T, G, and C is different for different species.

14.2 DNA Structure and Sequencing

The currently accepted model of the double-helix structure of DNA was proposed by Watson and Crick. Some of the salient features are that the two strands that make up the double helix have complementary base sequences and anti-parallel orientations. Alternating deoxyribose sugars and phosphates form the backbone of the structure, and the nitrogenous bases are stacked like rungs inside. The diameter of the double helix, 2 nm, is uniform throughout. A purine always pairs with a pyrimidine; A pairs with T, and G pairs with C. One turn of the helix has 10 base pairs. Prokaryotes are much simpler than eukaryotes in many of their features. Most prokaryotes contain a single, circular chromosome. In general, eukaryotic chromosomes contain a linear DNA molecule packaged into nucleosomes, and have two distinct regions that can be distinguished by staining, reflecting different states of packaging and compaction.

14.3 Basics of DNA Replication

During cell division, each daughter cell receives a copy of each molecule of DNA by a process known as DNA replication. The single chromosome of a prokaryote or each chromosome of a eukaryote consists of a single continuous double helix. The model for DNA replication suggests that the two strands of the double helix separate during replication, and each strand serves as a template from which the new complementary strand is copied. In the conservative model of replication, the parental DNA is conserved, and the daughter DNA is newly synthesized. The semi-conservative model suggests that each of the two parental DNA strands acts as template for new DNA to be synthesized; after replication, each double-stranded DNA retains the parental or "old" strand and one "new" strand. The dispersive model suggested that the two copies of the DNA would have segments of parental DNA and newly synthesized DNA. The Meselson and Stahl experiment supported the semi-conservative model of replication, in which an entire replicated chromosome consists of one parental strand and one newly synthesized strand of DNA.

14.4 DNA Replication in Prokaryotes

Replication in prokaryotes starts from a sequence found on the chromosome called the origin of replication—the point at which the DNA opens up. Helicase opens up the DNA double helix, resulting in the formation of the replication fork. Single-strand binding proteins bind to the single-stranded DNA near the replication fork to keep the fork open. Primase synthesizes an RNA primer to initiate synthesis by DNA polymerase, which can add nucleotides only to the 3' end of a previously synthesized primer strand. Both new DNA strands grow according to their respective 5'-3' directions. One strand is synthesized continuously in the direction of the replication fork; this is called the leading strand. The other strand is synthesized in a direction away from the replication fork, in short stretches of DNA known as Okazaki fragments. This strand is known as the lagging strand. Once replication is completed, the RNA primers are replaced by DNA nucleotides and the DNA is sealed with DNA ligase, which creates phosphodiester bonds between the 3'-OH of one end and the 5' phosphate of the other strand.

14.5 DNA Replication in Eukaryotes

Replication in eukaryotes starts at multiple origins of replication. The mechanism is quite similar to that in prokaryotes. A primer is required to initiate synthesis, which is then extended by DNA polymerase as it adds nucleotides one by one to the growing chain. The leading strand is synthesized continuously, whereas the lagging strand is synthesized in short stretches called Okazaki fragments. The RNA primers are replaced with DNA nucleotides; the DNA Okazaki fragments are linked into one continuous strand by DNA ligase. The ends of the chromosomes pose a problem as the primer RNA at the 5' ends of the DNA cannot be replaced with DNA, and the chromosome is progressively shortened. Telomerase, an enzyme with an inbuilt RNA template, extends the ends by copying the RNA template and extending one strand of the chromosome. DNA polymerase can then fill in the complementary DNA strand using the regular replication enzymes. In this way, the ends of the chromosomes are protected.

14.6 DNA Repair

DNA polymerase can make mistakes while adding nucleotides. It edits the DNA by proofreading every newly added base. Incorrect bases are removed and replaced by the correct base before proceeding with elongation. Most mistakes are corrected during replication, although when this does not happen, the mismatch repair mechanism is employed. Mismatch repair enzymes recognize the wrongly incorporated base and excise it from the DNA, replacing it with the correct base. In yet another type of repair, nucleotide excision repair, a damaged base is removed along with a few bases on the 5' and 3' end, and these are replaced by copying the template with the help of DNA polymerase. The ends of the newly synthesized fragment are attached to the rest of the DNA using DNA ligase, which creates a phosphodiester bond.

Most mistakes are corrected, and if they are not, they may result in a mutation, defined as a permanent change in the DNA sequence. Mutations can be of many types, such as substitution, deletion, insertion, and trinucleotide repeat expansions. Mutations in repair genes may lead to serious consequences such as cancer. Mutations can be induced or may occur spontaneously.

VISUAL CONNECTION QUESTIONS

1. Figure 14.10 In eukaryotic cells, DNA and RNA synthesis occur in a separate compartment from protein synthesis. In prokaryotic cells, both processes occur together. What advantages might there be to separating the processes? What advantages might there be to having them occur together?

2. Figure 14.14 You isolate a cell strain in which the joining of Okazaki fragments is impaired and suspect that a mutation has occurred in an enzyme found at the replication fork. Which enzyme is most likely to be mutated?

3. Figure 14.21 A frameshift mutation that results in the insertion of three nucleotides is often less deleterious than a mutation that results in the insertion of one nucleotide. Why?

REVIEW QUESTIONS

4. If DNA of a particular species was analyzed and it was found that it contains 27 percent A, what would be the percentage of C?
   a. 27 percent
   b. 30 percent
   c. 23 percent
   d. 54 percent

5. The experiments by Hershey and Chase helped confirm that DNA was the hereditary material on the basis of the finding that:
   a. radioactive phage were found in the pellet
   b. radioactive cells were found in the supernatant
   c. radioactive sulfur was found inside the cell
   d. radioactive phosphorus was found in the cell

6. Bacterial transformation is a major concern in many medical settings. Why might health care providers be concerned?
   a. Pathogenic bacteria could introduce disease-causing genes in non-pathogenic bacteria.
   b. Antibiotic resistance genes could be introduced to new bacteria to create "superbugs."
   c. Bacteriophages could spread DNA encoding toxins to new bacteria.
   d. All of the above.

7. DNA double helix does not have which of the following?
   a. antiparallel configuration
   b. complementary base pairing
   c. major and minor grooves
   d. uracil

8. In eukaryotes, what is the DNA wrapped around?
   a. single-stranded binding proteins
   b. sliding clamp
   c. polymerase
   d. histones

9. Meselson and Stahl's experiments proved that DNA replicates by which mode?
   a. conservative
   b. semi-conservative
   c. dispersive
   d. none of the above

10. If the sequence of the 5'-3' strand is AATGCTAC, then the complementary sequence has the following sequence:
    a. 3'-AATGCTAC-5'
    b. 3'-CATCGTAA-5'
    c. 3'-TTACGATG-5'
    d. 3'-GTAGCATT-5'

11. How did Meselson and Stahl support Watson and Crick's double-helix model?
    a. They demonstrated that each strand serves as a template for synthesizing a new strand of DNA.
    b. They showed that the DNA strands break and recombine without losing genetic material.
    c. They proved that DNA maintains a double-helix structure while undergoing semi-conservative replication.
    d. They demonstrated that conservative replication maintains the complementary base pairing of each DNA helix.

12. Which of the following components is not involved during the formation of the replication fork?
    a. single-strand binding proteins
    b. helicase
    c. origin of replication
    d. ligase

13. Which of the following does the enzyme primase synthesize?
    a. DNA primer
    b. RNA primer
    c. Okazaki fragments
    d. phosphodiester linkage

14. In which direction does DNA replication take place?
    a. 5'-3'
    b. 3'-5'
    c. 5'
    d. 3'

15. A scientist randomly mutates the DNA of a bacterium. She then sequences the bacterium's daughter cells, and finds that the daughters have many errors in their replicated DNA. The parent bacterium likely acquired a mutation in which enzyme?
    a. DNA ligase
    b. DNA pol II
    c. Primase
    d. DNA pol I

16. The ends of the linear chromosomes are maintained by
    a. helicase
    b. primase
    c. DNA pol
    d. telomerase

17. Which of the following is not a true statement comparing prokaryotic and eukaryotic DNA replication?
    a. Both eukaryotic and prokaryotic DNA polymerases build off RNA primers made by primase.
    b. Eukaryotic DNA replication requires multiple replication forks, while prokaryotic replication uses a single origin to rapidly replicate the entire genome.
    c. DNA replication always occurs in the nucleus.
    d. Eukaryotic DNA replication involves more polymerases than prokaryotic replication.

18. During proofreading, which of the following enzymes reads the DNA?
    a. primase
    b. topoisomerase
    c. DNA pol
    d. helicase

19. The initial mechanism for repairing nucleotide errors in DNA is ________.
    a. mismatch repair
    b. DNA polymerase proofreading
    c. nucleotide excision repair
    d. thymine dimers

20. A scientist creates fruit fly larvae with a mutation that eliminates the exonuclease function of DNA pol III. Which prediction about the mutational load in the adult fruit flies is most likely to be correct?
    a. The adults with the DNA pol III mutation will have significantly more mutations than average.
    b. The adults with the DNA pol III mutation will have slightly more mutations than average.
    c. The adults with the DNA pol III mutation will have the same number of mutations as average.
    d. The adults with the DNA pol III mutation will have fewer mutations than average.

CRITICAL THINKING QUESTIONS

21. Explain Griffith's transformation experiments. What did he conclude from them?
22. Why were radioactive sulfur and phosphorous used to label bacteriophage in Hershey and Chase's experiments?
23. When Chargaff was performing his experiments, the tetranucleotide hypothesis, which stated that DNA was composed of GACT nucleotide repeats, was the most widely accepted view of DNA's composition. How did Chargaff disprove this hypothesis?
24. Provide a brief summary of the Sanger sequencing method.
25. Describe the structure and complementary base pairing of DNA.
26. Prokaryotes have a single circular chromosome while eukaryotes have linear chromosomes. Describe one advantage and one disadvantage to the eukaryotic genome packaging compared to the prokaryotcs.
27. How did the scientific community learn that DNA replication takes place in a semi-conservative fashion?
28. Imagine the Meselson and Stahl experiments had supported conservative replication instead of semi-conservative replication. What results would you predict to observe after two rounds of replication? Be specific regarding percent distributions of DNA incorporating ^{15}N and ^{14}N in the gradient.
29. DNA replication is bidirectional and discontinuous; explain your understanding of those concepts.
30. What are Okazaki fragments and how they are formed?
31. If the rate of replication in a particular prokaryote is 900 nucleotides per second, how long would it take 1.2 million base pair genomes to make two copies?
32. Explain the events taking place at the replication fork. If the gene for helicase is mutated, what part of replication will be affected?
33. What is the role of a primer in DNA replication? What would happen if you forgot to add a primer in a tube containing the reaction mix for a DNA sequencing reaction?
34. Quinolone antibiotics treat bacterial infections by blocking the activity of topoisomerase. Why does this treatment work? Explain what occurs at the molecular level.
35. How do the linear chromosomes in eukaryotes ensure that its ends are replicated completely?
36. What is the consequence of mutation of a mismatch repair enzyme? How will this affect the function of a gene?
37. An adult with a history of tanning has his genome sequenced. The beginning of a protein-coding region of his DNA reads ATGGGGATATGGCAT. If the protein-coding region of a healthy adult reads ATGGGGATATGAGCAT, identify the site and type of mutation.

CHAPTER 15
Genes and Proteins

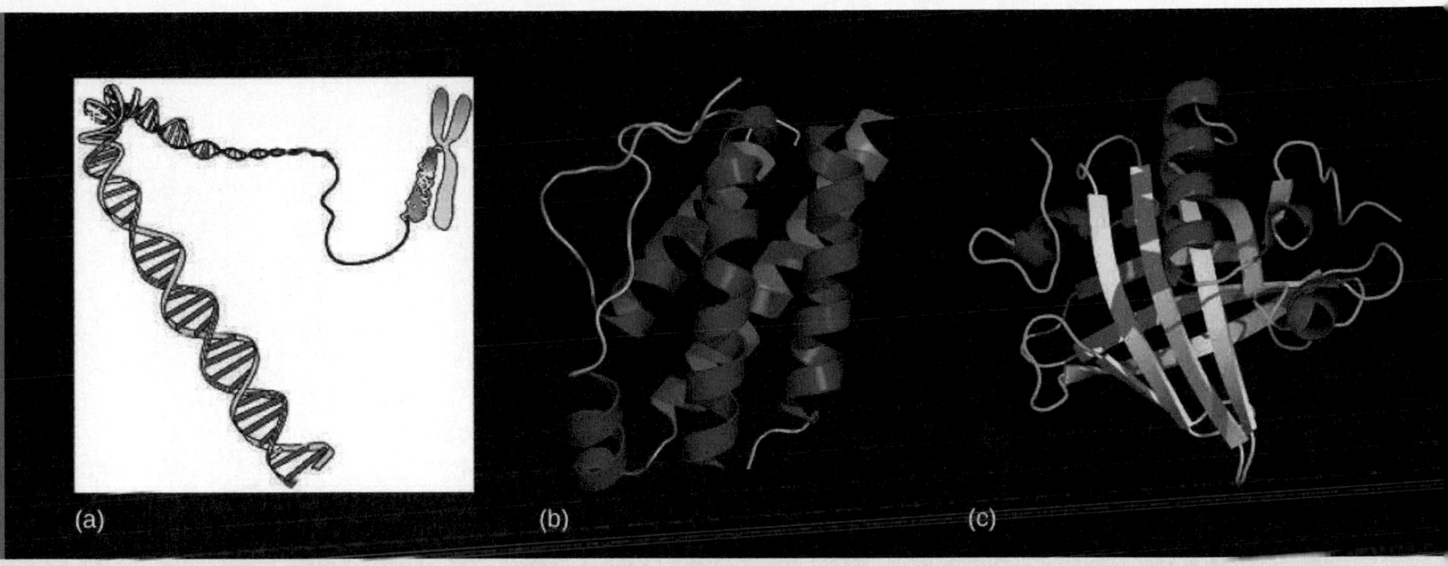

Figure 15.1 Genes, which are carried on (a) chromosomes, are linearly organized instructions for making the RNA and protein molecules that are necessary for all of the processes of life. The (b) interleukin-2 protein and (c) alpha-2u-globulin protein are just two examples of the array of different molecular structures that are encoded by genes. (credit "chromosome: National Human Genome Research Institute; credit "interleukin-2": Ramin Herati/Created from PDB 1M47 and rendered with Pymol; credit "alpha-2u-globulin": Darren Logan/rendered with AISMIG)

INTRODUCTION Since the rediscovery of Mendel's work in 1900, the definition of the gene has progressed from an abstract unit of heredity to a tangible molecular entity capable of replication, expression, and mutation (Figure 15.1). Genes are composed of DNA and are linearly arranged on chromosomes. Genes specify the sequences of amino acids, which are the building blocks of proteins. In turn, proteins are responsible for orchestrating nearly every function of the cell. Both genes and the proteins they encode are absolutely essential to life as we know it.

Chapter Outline

15.1 The Genetic Code

By the end of this section, you will be able to do the following:

- Explain the "central dogma" of DNA-protein synthesis
- Describe the genetic code and how the nucleotide sequence prescribes the amino acid and the protein sequence

The cellular process of transcription generates messenger RNA (mRNA), a mobile molecular copy of one or more genes with an alphabet of A, C, G, and uracil (U). Translation of the mRNA template on ribosomes converts nucleotide-based genetic information into a protein product. That is the central dogma of DNA-protein synthesis. Protein sequences consist of 20 commonly occurring amino acids; therefore, it can be said that the protein alphabet consists of 20 "letters" (Figure 15.2). Different amino acids have different chemistries (such as acidic versus basic, or polar and nonpolar) and different structural constraints. Variation in amino acid sequence is responsible for the enormous variation in protein structure and function.

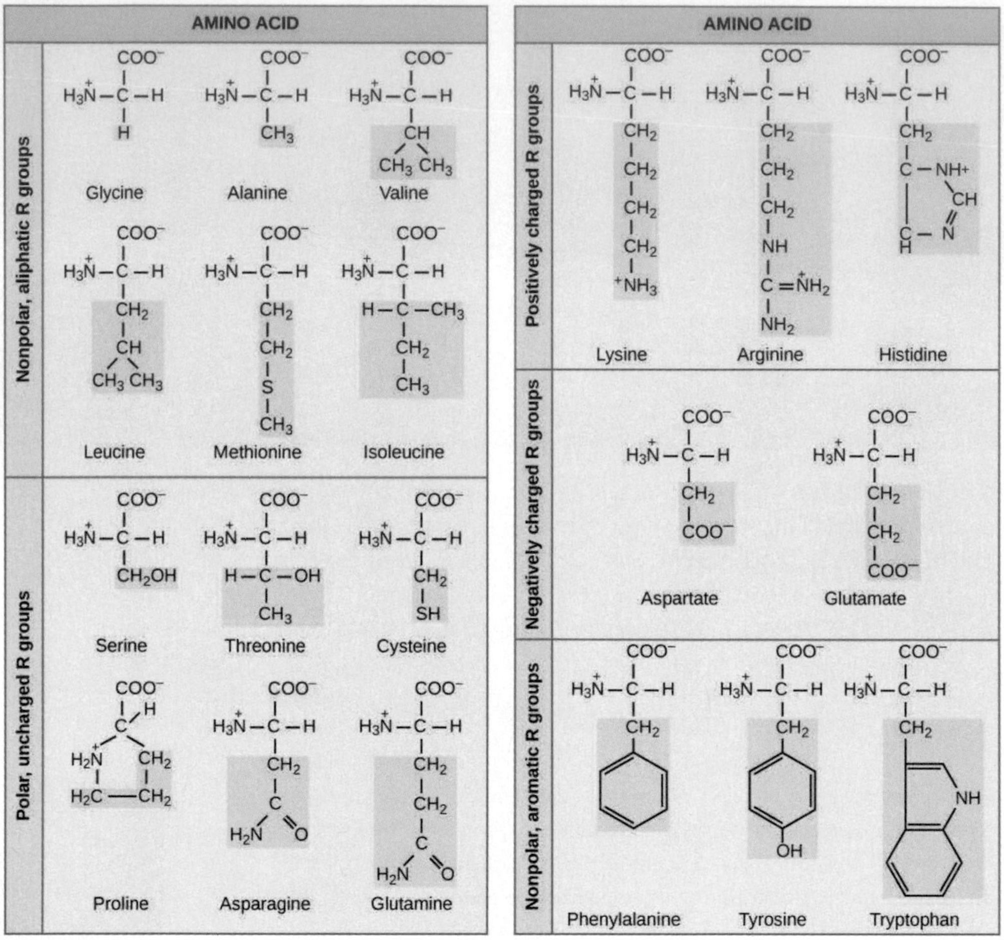

Figure 15.2 Structures of the 20 amino acids found in proteins are shown. Each amino acid is composed of an amino group (NH_3^+), a carboxyl group (COO^-), and a side chain (blue). The side chain may be nonpolar, polar, or charged, as well as large or small. It is the variety of amino acid side chains that gives rise to the incredible variation of protein structure and function.

The Central Dogma: DNA Encodes RNA; RNA Encodes Protein

The flow of genetic information in cells from DNA to mRNA to protein is described by the **central dogma** (Figure 15.3), which states that genes specify the sequence of mRNAs, which in turn specify the sequence of amino acids making up all proteins. The decoding of one molecule to another is performed by specific proteins and RNAs. Because the information stored in DNA is so central to cellular function, it makes intuitive sense that the cell would make mRNA copies of this information for protein synthesis, while keeping the DNA itself intact and protected. The copying of DNA to RNA is relatively straightforward, with one nucleotide being added to the mRNA strand for every nucleotide read in the DNA strand. The translation to protein is a bit more complex because three mRNA nucleotides correspond to one amino acid in the polypeptide sequence. However, the translation to protein is still systematic and **colinear**, such that nucleotides 1 to 3 correspond to amino acid 1, nucleotides 4 to 6 correspond to amino acid 2, and so on.

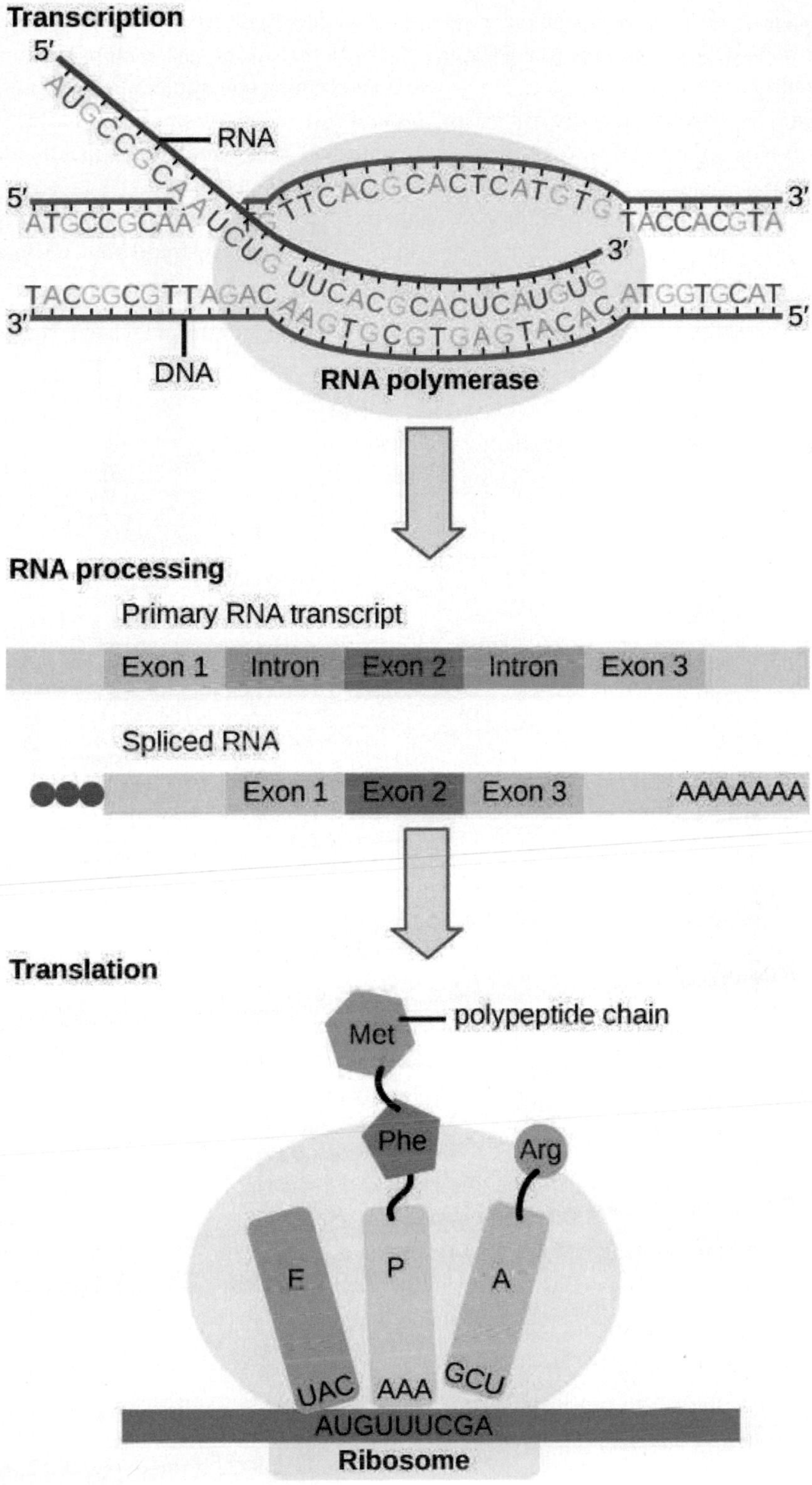

Figure 15.3 Instructions on DNA are transcribed onto messenger RNA. Ribosomes are able to read the genetic information inscribed on a strand of messenger RNA and use this information to string amino acids together into a protein.

The Genetic Code Is Degenerate and Universal

Each amino acid is defined by a three-nucleotide sequence called the triplet codon. Given the different numbers of "letters" in the mRNA and protein "alphabets," scientists theorized that single amino acids must be represented by combinations of nucleotides. Nucleotide doublets would not be sufficient to specify every amino acid because there are only 16 possible two-nucleotide combinations (4^2). In contrast, there are 64 possible nucleotide triplets (4^3), which is far more than the number of amino acids. Scientists theorized that amino acids were encoded by nucleotide triplets and that the genetic code was **"degenerate."** In other words, a given amino acid could be encoded by more than one nucleotide triplet. This was later confirmed experimentally: Francis Crick and Sydney Brenner used the chemical mutagen proflavin to insert one, two, or three nucleotides into the gene of a virus. When one or two nucleotides were inserted, the normal proteins were not produced. When three nucleotides were inserted, the protein was synthesized and functional. This demonstrated that the amino acids must be specified by groups of three nucleotides. These nucleotide triplets are called **codons**. The insertion of one or two nucleotides completely changed the triplet **reading frame**, thereby altering the message for every subsequent amino acid (Figure 15.5). Though insertion of three nucleotides caused an extra amino acid to be inserted during translation, the integrity of the rest of the protein was maintained.

Scientists painstakingly solved the genetic code by translating synthetic mRNAs in vitro and sequencing the proteins they specified (Figure 15.4).

Second letter

First letter	U	C	A	G	Third letter
U	UUU Phe UUC Phe UUA Leu UUG Leu	UCU Ser UCC Ser UCA Ser UCG Ser	UAU Tyr UAC Tyr **UAA Stop** **UAG Stop**	UGU Cys UGC Cys **UGA Stop** UGG Trp	U C A G
C	CUU Leu CUC Leu CUA Leu CUG Leu	CCU Pro CCC Pro CCA Pro CCG Pro	CAU His CAC His CAA Gln CAG Gln	CGU Arg CGC Arg CGA Arg CGG Arg	U C A G
A	AUU Ile AUC Ile AUA Ile **AUG Met**	ACU Thr ACC Thr ACA Thr ACG Thr	AAU Asn AAC Asn AAA Lys AAG Lys	AGU Ser AGC Ser AGA Arg AGG Arg	U C A G
G	GUU Val GUC Val GUA Val GUG Val	GCU Ala GCC Ala GCA Ala GCG Ala	GAU Asp GAC Asp GAA Glu GAG Glu	GGU Gly GGC Gly GGA Gly GGG Gly	U C A G

Figure 15.4 This figure shows the genetic code for translating each nucleotide triplet in mRNA into an amino acid or a termination signal in a protein. (credit: modification of work by NIH)

In addition to codons that instruct the addition of a specific amino acid to a polypeptide chain, three of the 64 codons terminate protein synthesis and release the polypeptide from the translation machinery. These triplets are called **nonsense codons**, or *stop codons*. Another codon, AUG, also has a special function. In addition to specifying the amino acid methionine, it also serves as the start codon to initiate translation. The **reading frame** for translation is set by the AUG start codon near the 5' end of the mRNA. Following the start codon, the mRNA is read in groups of three until a stop codon is encountered.

The arrangement of the coding table reveals the structure of the code. There are sixteen "blocks" of codons, each specified by the first and second nucleotides of the codons within the block, e.g., the "AC*" block that corresponds to the amino acid threonine (Thr). Some blocks are divided into a pyrimidine half, in which the codon ends with U or C, and a purine half, in which the codon ends with A or G. Some amino acids get a whole block of four codons, like alanine (Ala), threonine (Thr) and proline (Pro). Some get the pyrimidine half of their block, like histidine (His) and asparagine (Asn). Others get the purine half of their block, like glutamate (Glu) and lysine (Lys). Note that some amino acids get a block and a half-block for a total of six codons.

The specification of a single amino acid by multiple similar codons is called "degeneracy." Degeneracy is believed to be a cellular mechanism to reduce the negative impact of random mutations. Codons that specify the same amino acid typically only differ by one nucleotide. In addition, amino acids with chemically similar side chains are encoded by similar codons. For example, aspartate (Asp) and glutamate (Glu), which occupy the GA* block, are both negatively charged. This nuance of the genetic code ensures that a single-nucleotide substitution mutation might specify the same amino acid but have no effect or specify a similar amino acid, preventing the protein from being rendered completely nonfunctional.

The genetic code is nearly universal. With a few minor exceptions, virtually all species use the same genetic code for protein synthesis. Conservation of codons means that a purified mRNA encoding the globin protein in horses could be transferred to a tulip cell, and the tulip would synthesize horse globin. That there is only one genetic code is powerful evidence that all of life on Earth shares a common origin, especially considering that there are about 10^{84} possible combinations of 20 amino acids and 64 triplet codons.

LINK TO LEARNING

Transcribe a gene and translate it to protein using complementary pairing and the genetic code at this site (http://openstax.org/l/create_protein) .

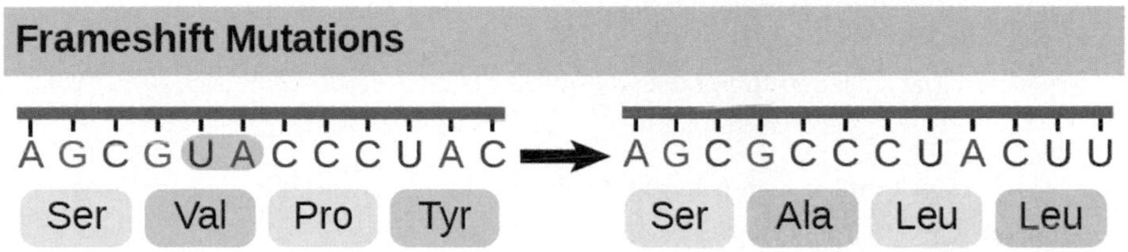

Figure 15.5 The deletion of two nucleotides shifts the reading frame of an mRNA and changes the entire protein message, creating a nonfunctional protein or terminating protein synthesis altogether.

SCIENTIFIC METHOD CONNECTION

Which Has More DNA: A Kiwi or a Strawberry?

Figure 15.6 Do you think that a kiwi or a strawberry has more DNA per fruit? (credit “kiwi”: "Kelbv"/Flickr; credit: “strawberry”: Alisdair McDiarmid)

Question: Would a kiwi and strawberry that are approximately the same size (Figure 15.6) also have approximately the same amount of DNA?

Background: Genes are carried on chromosomes and are made of DNA. All mammals are diploid, meaning they have two copies of each chromosome. However, not all plants are diploid. The common strawberry is octoploid (8*n*) and the cultivated kiwi is hexaploid (6*n*). Research the total number of chromosomes in the cells of each of these fruits and think about how this might correspond to the amount of DNA in these fruits’ cell nuclei. What other factors might contribute to the total amount of DNA in a single fruit? Read about the technique of DNA isolation to understand how each step in the isolation protocol helps liberate and precipitate DNA.

Hypothesis: Hypothesize whether you would be able to detect a difference in DNA quantity from similarly sized strawberries and kiwis. Which fruit do you think would yield more DNA?

Test your hypothesis: Isolate the DNA from a strawberry and a kiwi that are similarly sized. Perform the experiment in at least triplicate for each fruit

1. Prepare a bottle of DNA extraction buffer from 900 mL water, 50 mL dish detergent, and two teaspoons of table salt. Mix by

inversion (cap it and turn it upside down a few times).
2. Grind a strawberry and a kiwi by hand in a plastic bag, or using a mortar and pestle, or with a metal bowl and the end of a blunt instrument. Grind for at least two minutes per fruit.
3. Add 10 mL of the DNA extraction buffer to each fruit, and mix well for at least one minute.
4. Remove cellular debris by filtering each fruit mixture through cheesecloth or porous cloth and into a funnel placed in a test tube or an appropriate container.
5. Pour ice-cold ethanol or isopropanol (rubbing alcohol) into the test tube. You should observe white, precipitated DNA.
6. Gather the DNA from each fruit by winding it around separate glass rods.

Record your observations: Because you are not quantitatively measuring DNA volume, you can record for each trial whether the two fruits produced the same or different amounts of DNA as observed by eye. If one or the other fruit produced noticeably more DNA, record this as well. Determine whether your observations are consistent with several pieces of each fruit.

Analyze your data: Did you notice an obvious difference in the amount of DNA produced by each fruit? Were your results reproducible?

Draw a conclusion: Given what you know about the number of chromosomes in each fruit, can you conclude that chromosome number necessarily correlates to DNA amount? Can you identify any drawbacks to this procedure? If you had access to a laboratory, how could you standardize your comparison and make it more quantitative?

15.2 Prokaryotic Transcription

By the end of this section, you will be able to do the following:
- List the different steps in prokaryotic transcription
- Discuss the role of promoters in prokaryotic transcription
- Describe how and when transcription is terminated

The prokaryotes, which include Bacteria and Archaea, are mostly single-celled organisms that, by definition, lack membrane-bound nuclei and other organelles. A bacterial chromosome is a closed circle that, unlike eukaryotic chromosomes, is not organized around histone proteins. The central region of the cell in which prokaryotic DNA resides is called the nucleoid region. In addition, prokaryotes often have abundant **plasmids**, which are shorter, circular DNA molecules that may only contain one or a few genes. Plasmids can be transferred independently of the bacterial chromosome during cell division and often carry traits such as those involved with antibiotic resistance.

Transcription in prokaryotes (and in eukaryotes) requires the DNA double helix to partially unwind in the region of mRNA synthesis. The region of unwinding is called a **transcription bubble.** Transcription always proceeds from the same DNA strand for each gene, which is called the **template strand**. The mRNA product is complementary to the template strand and is almost identical to the other DNA strand, called the **nontemplate strand**, or the coding strand. The only nucleotide difference is that in mRNA, all of the T nucleotides are replaced with U nucleotides (Figure 15.7). In an RNA double helix, A can bind U via two hydrogen bonds, just as in A–T pairing in a DNA double helix.

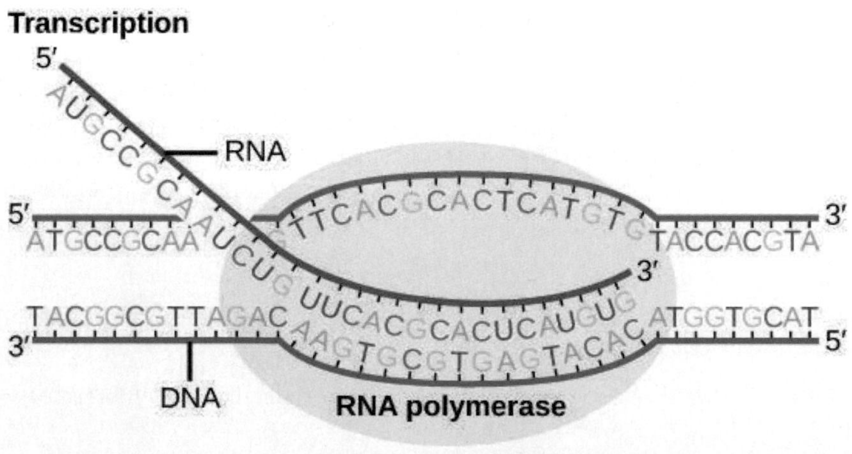

Figure 15.7 Messenger RNA is a copy of protein-coding information in the coding strand of DNA, with the substitution of U in the RNA for T

in the coding sequence. However, new RNA nucleotides base pair with the nucleotides of the template strand. RNA is synthesized in its 5'-3' direction, using the enzyme RNA polymerase. As the template is read, the DNA unwinds ahead of the polymerase and then rewinds behind it.

The nucleotide pair in the DNA double helix that corresponds to the site from which the first 5' mRNA nucleotide is transcribed is called the +1 site, or the **initiation site**. Nucleotides preceding the initiation site are denoted with a "-" and are designated *upstream nucleotides*. Conversely, nucleotides following the initiation site are denoted with "+" numbering and are called *downstream nucleotides*.

Initiation of Transcription in Prokaryotes

Prokaryotes do not have membrane-enclosed nuclei. Therefore, the processes of transcription, translation, and mRNA degradation can all occur simultaneously. The intracellular level of a bacterial protein can quickly be amplified by multiple transcription and translation events that occur concurrently on the same DNA template. Prokaryotic genomes are very compact, and prokaryotic transcripts often cover more than one gene or cistron (a coding sequence for a single protein). Polycistronic mRNAs are then translated to produce more than one kind of protein.

Our discussion here will exemplify transcription by describing this process in *Escherichia coli*, a well-studied eubacterial species. Although some differences exist between transcription in *E. coli* and transcription in archaea, an understanding of *E. coli* transcription can be applied to virtually all bacterial species.

Prokaryotic RNA Polymerase

Prokaryotes use the same RNA polymerase to transcribe all of their genes. In *E. coli*, the polymerase is composed of five polypeptide subunits, two of which are identical. Four of these subunits, denoted α, α, β, and β', comprise the polymerase **core enzyme**. These subunits assemble every time a gene is transcribed, and they disassemble once transcription is complete. Each subunit has a unique role; the two α-subunits are necessary to assemble the polymerase on the DNA; the β-subunit binds to the ribonucleoside triphosphate that will become part of the nascent mRNA molecule; and the β' subunit binds the DNA template strand. The fifth subunit, σ, is involved only in transcription initiation. It confers transcriptional specificity such that the polymerase begins to synthesize mRNA from an appropriate initiation site. Without σ, the core enzyme would transcribe from random sites and would produce mRNA molecules that specified protein gibberish. The polymerase comprised of all five subunits is called the **holoenzyme**.

Prokaryotic Promoters

A **promoter** is a DNA sequence onto which the transcription machinery, including RNA polymerase, binds and initiates transcription. In most cases, promoters exist upstream of the genes they regulate. The specific sequence of a promoter is very important because it determines whether the corresponding gene is transcribed all the time, some of the time, or infrequently. Although promoters vary among prokaryotic genomes, a few elements are evolutionarily conserved in many species. At the -10 and -35 regions upstream of the initiation site, there are two *promoter consensus sequences*, or regions that are similar across all promoters and across various bacterial species (Figure 15.8). The -10 sequence, called the -10 region, has the consensus sequence TATAAT. The -35 sequence has the consensus sequence TTGACA. These consensus sequences are recognized and bound by σ. Once this interaction is made, the subunits of the core enzyme bind to the site. The A–T-rich -10 region facilitates unwinding of the DNA template, and several phosphodiester bonds are made. The transcription initiation phase ends with the production of abortive transcripts, which are polymers of approximately 10 nucleotides that are made and released.

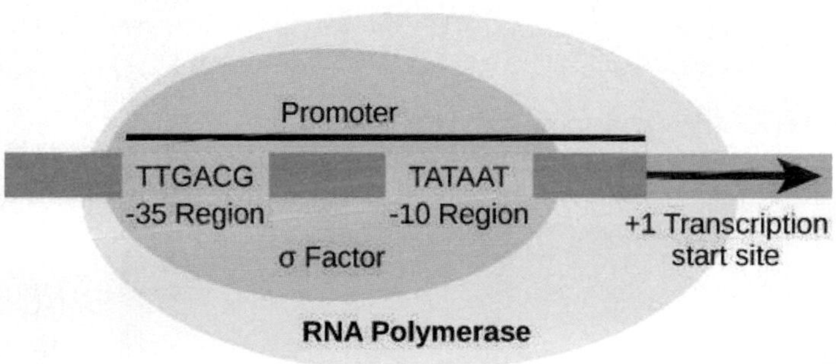

Figure 15.8 The σ subunit of prokaryotic RNA polymerase recognizes consensus sequences found in the promoter region upstream of the transcription start site. The σ subunit dissociates from the polymerase after transcription has been initiated.

LINK TO LEARNING

View this MolecularMovies animation (http://openstax.org/l/transcription) to see the first part of transcription and the base sequence repetition of the TATA box.

Elongation and Termination in Prokaryotes

The transcription elongation phase begins with the release of the σ subunit from the polymerase. The dissociation of σ allows the core enzyme to proceed along the DNA template, synthesizing mRNA in the 5' to 3' direction at a rate of approximately 40 nucleotides per second. As elongation proceeds, the DNA is continuously unwound ahead of the core enzyme and rewound behind it. The base pairing between DNA and RNA is not stable enough to maintain the stability of the mRNA synthesis components. Instead, the RNA polymerase acts as a stable linker between the DNA template and the nascent RNA strands to ensure that elongation is not interrupted prematurely.

Prokaryotic Termination Signals

Once a gene is transcribed, the prokaryotic polymerase needs to be instructed to dissociate from the DNA template and liberate the newly made mRNA. Depending on the gene being transcribed, there are two kinds of termination signals. One is protein-based and the other is RNA-based. **Rho-dependent termination** is controlled by the rho protein, which tracks along behind the polymerase on the growing mRNA chain. Near the end of the gene, the polymerase encounters a run of G nucleotides on the DNA template and it stalls. As a result, the rho protein collides with the polymerase. The interaction with rho releases the mRNA from the transcription bubble.

Rho-independent termination is controlled by specific sequences in the DNA template strand. As the polymerase nears the end of the gene being transcribed, it encounters a region rich in C–G nucleotides. The mRNA folds back on itself, and the complementary C–G nucleotides bind together. The result is a **stable hairpin** that causes the polymerase to stall as soon as it begins to transcribe a region rich in A–T nucleotides. The complementary U–A region of the mRNA transcript forms only a weak interaction with the template DNA. This, coupled with the stalled polymerase, induces enough instability for the core enzyme to break away and liberate the new mRNA transcript.

Upon termination, the process of transcription is complete. By the time termination occurs, the prokaryotic transcript would already have been used to begin synthesis of numerous copies of the encoded protein because these processes can occur concurrently. The unification of transcription, translation, and even mRNA degradation is possible because all of these processes occur in the same 5' to 3' direction, and because there is no membranous compartmentalization in the prokaryotic cell (Figure 15.9). In contrast, the presence of a nucleus in eukaryotic cells precludes simultaneous transcription and translation.

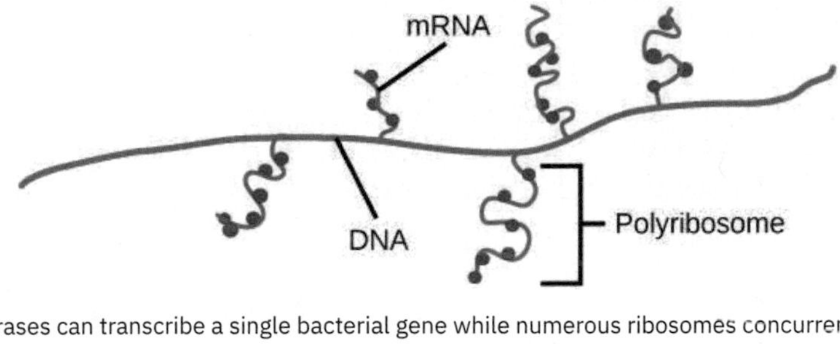

Figure 15.9 Multiple polymerases can transcribe a single bacterial gene while numerous ribosomes concurrently translate the mRNA transcripts into polypeptides. In this way, a specific protein can rapidly reach a high concentration in the bacterial cell.

LINK TO LEARNING

Visit this BioStudio animation (http://openstax.org/l/transcription2) to see the process of prokaryotic transcription.

15.3 Eukaryotic Transcription

By the end of this section, you will be able to do the following:

- List the steps in eukaryotic transcription
- Discuss the role of RNA polymerases in transcription
- Compare and contrast the three RNA polymerases
- Explain the significance of transcription factors

Prokaryotes and eukaryotes perform fundamentally the same process of transcription, with a few key differences. The most important difference between prokaryote and eukaryote transcription is due to the latter's membrane-bound nucleus and organelles. With the genes bound in a nucleus, the eukaryotic cell must be able to transport its mRNA to the cytoplasm and must protect its mRNA from degrading before it is translated. Eukaryotes also employ three different polymerases that each transcribe a different subset of genes. Eukaryotic mRNAs are usually *monogenic,* meaning that they specify a single protein.

Initiation of Transcription in Eukaryotes

Unlike the prokaryotic polymerase that can bind to a DNA template on its own, eukaryotes require several other proteins, called transcription factors, to first bind to the promoter region and then to help recruit the appropriate polymerase.

The Three Eukaryotic RNA Polymerases

The features of eukaryotic mRNA synthesis are markedly more complex than those of prokaryotes. Instead of a single polymerase comprising five subunits, the eukaryotes have three polymerases that are each made up of 10 subunits or more. Each eukaryotic polymerase also requires a distinct set of transcription factors to bring it to the DNA template.

RNA polymerase I is located in the nucleolus, a specialized nuclear substructure in which ribosomal RNA (rRNA) is transcribed, processed, and assembled into ribosomes (Table 15.1). The rRNA molecules are considered structural RNAs because they have a cellular role but are not translated into protein. The rRNAs are components of the ribosome and are essential to the process of translation. RNA polymerase I synthesizes all of the rRNAs from the tandemly duplicated set of 18S, 5.8S, and 28S ribosomal genes. (Note that the "S" designation applies to "Svedberg" units, a nonadditive value that characterizes the speed at which a particle sediments during centrifugation.)

Locations, Products, and Sensitivities of the Three Eukaryotic RNA Polymerases

RNA Polymerase	Cellular Compartment	Product of Transcription	α-Amanitin Sensitivity
I	Nucleolus	All rRNAs except 5S rRNA	Insensitive
II	Nucleus	All protein-coding nuclear pre-mRNAs	Extremely sensitive
III	Nucleus	5S rRNA, tRNAs, and small nuclear RNAs	Moderately sensitive

Table 15.1

RNA polymerase II is located in the nucleus and synthesizes all protein-coding nuclear pre-mRNAs. Eukaryotic pre-mRNAs undergo extensive processing after transcription but before translation. For clarity, this module's discussion of transcription and translation in eukaryotes will use the term "mRNAs" to describe only the mature, processed molecules that are ready to be translated. RNA polymerase II is responsible for transcribing the overwhelming majority of eukaryotic genes.

RNA polymerase III is also located in the nucleus. This polymerase transcribes a variety of structural RNAs that includes the 5S pre-rRNA, transfer pre-RNAs (pre-tRNAs), and **small nuclear** pre-**RNAs.** The tRNAs have a critical role in translation; they serve as the "adaptor molecules" between the mRNA template and the growing polypeptide chain. Small nuclear RNAs have a variety of functions, including "splicing" pre-mRNAs and regulating transcription factors.

A scientist characterizing a new gene can determine which polymerase transcribes it by testing whether the gene is expressed in the presence of α-amanitin, an oligopeptide toxin produced by the fly agaric toadstool mushroom and other species of *Amanita.* Interestingly, the α-amanitin affects the three polymerases very differently (Table 15.1). RNA polymerase I is completely insensitive to α-amanitin, meaning that the polymerase can transcribe DNA in vitro in the presence of this poison. RNA polymerase III is moderately sensitive to the toxin. In contrast, RNA polymerase II is extremely sensitive to α-amanitin. The toxin prevents the enzyme from progressing down the DNA, and thus inhibits transcription. Knowing the transcribing polymerase can provide clues as to the general function of the gene being studied. Because RNA polymerase II transcribes the vast majority of genes, we will focus on this polymerase in our subsequent discussions about eukaryotic transcription factors and promoters.

RNA Polymerase II Promoters and Transcription Factors

Eukaryotic promoters are much larger and more intricate than prokaryotic promoters. However, both have a sequence similar to the -10 sequence of prokaryotes. In eukaryotes, this sequence is called the TATA box, and has the consensus sequence TATAAA on the coding strand. It is located at -25 to -35 bases relative to the initiation (+1) site (Figure 15.10). This sequence is not identical to the *E. coli* -10 box, but it conserves the A–T rich element. The thermostability of A–T bonds is low and this helps the DNA template to locally unwind in preparation for transcription.

Instead of the simple σ factor that helps bind the prokaryotic RNA polymerase to its promoter, eukaryotes assemble a complex of transcription factors required to recruit RNA polymerase II to a protein coding gene. Transcription factors that bind to the promoter are called *basal transcription factors*. These basal factors are all called TFII (for Transcription Factor/polymerase II) plus an additional letter (A-J). The core complex is TFIID, which includes a TATA-binding protein (TBP). The other transcription factors systematically fall into place on the DNA template, with each one further stabilizing the pre-initiation complex and contributing to the recruitment of RNA polymerase II.

VISUAL CONNECTION

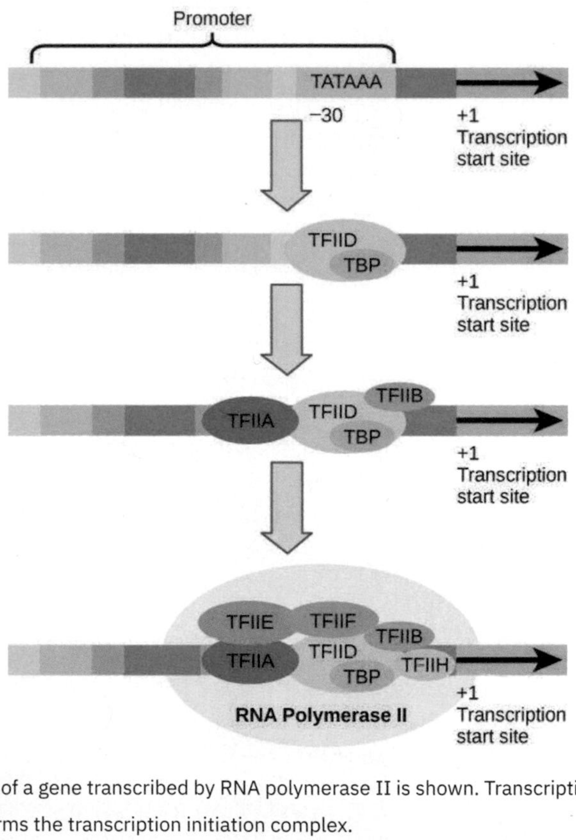

Figure 15.10 A generalized promoter of a gene transcribed by RNA polymerase II is shown. Transcription factors recognize the promoter. RNA polymerase II then binds and forms the transcription initiation complex.

A scientist splices a eukaryotic promoter in front of a bacterial gene and inserts the gene in a bacterial chromosome. Would you expect the bacteria to transcribe the gene?

Some eukaryotic promoters also have a conserved **CAAT box** (GGCCAATCT) at approximately -80. Further upstream of the TATA box, eukaryotic promoters may also contain one or more **GC-rich boxes** (GGCG) or **octamer boxes** (ATTTGCAT). These elements bind cellular factors that increase the efficiency of transcription initiation and are often identified in more "active" genes that are constantly being expressed by the cell.

Basal transcription factors are crucial in the formation of a preinitiation complex on the DNA template that subsequently recruits RNA polymerase II for transcription initiation. The complexity of eukaryotic transcription does not end with the polymerases and promoters. An army of other transcription factors, which bind to upstream enhancers and silencers, also help to regulate the frequency with which pre-mRNA is synthesized from a gene. Enhancers and silencers affect the efficiency of transcription but are not necessary for transcription to proceed.

Promoter Structures for RNA Polymerases I and III

The processes of bringing RNA polymerases I and III to the DNA template involve slightly less complex collections of transcription factors, but the general theme is the same.

The conserved promoter elements for genes transcribed by polymerases I and III differ from those transcribed by RNA polymerase II. RNA polymerase I transcribes genes that have two GC-rich promoter sequences in the -45 to +20 region. These sequences alone are sufficient for transcription initiation to occur, but promoters with additional sequences in the region from -180 to -105 upstream of the initiation site will further enhance initiation. Genes that are transcribed by RNA polymerase III have upstream promoters or promoters that occur within the genes themselves.

Eukaryotic transcription is a tightly regulated process that requires a variety of proteins to interact with each other and with the DNA strand. Although the process of transcription in eukaryotes involves a greater metabolic investment than in prokaryotes, it ensures that the cell transcribes precisely the pre-mRNAs that it needs for protein synthesis.

EVOLUTION CONNECTION

The Evolution of Promoters

The evolution of genes may be a familiar concept. Mutations can occur in genes during DNA replication, and the result may or may not be beneficial to the cell. By altering an enzyme, structural protein, or some other factor, the process of mutation can transform functions or physical features. However, eukaryotic promoters and other gene regulatory sequences may evolve as well. For instance, consider a gene that, over many generations, becomes more valuable to the cell. Maybe the gene encodes a structural protein that the cell needs to synthesize in abundance for a certain function. If this is the case, it would be beneficial to the cell for that gene's promoter to recruit transcription factors more efficiently and increase gene expression.

Scientists examining the evolution of promoter sequences have reported varying results. In part, this is because it is difficult to infer exactly where a eukaryotic promoter begins and ends. Some promoters occur within genes; others are located very far upstream, or even downstream, of the genes they are regulating. However, when researchers limited their examination to human core promoter sequences that were defined experimentally as sequences that bind the preinitiation complex, they found that promoters evolve even faster than protein-coding genes.

It is still unclear how promoter evolution might correspond to the evolution of humans or other complex organisms. However, the evolution of a promoter to effectively make more or less of a given gene product is an intriguing alternative to the evolution of the genes themselves.[1]

Eukaryotic Elongation and Termination

Following the formation of the preinitiation complex, the polymerase is released from the other transcription factors, and elongation is allowed to proceed as it does in prokaryotes with the polymerase synthesizing pre-mRNA in the 5' to 3' direction. As discussed previously, RNA polymerase II transcribes the major share of eukaryotic genes, so in this section we will focus on how this polymerase accomplishes elongation and termination.

Although the enzymatic process of elongation is essentially the same in eukaryotes and prokaryotes, the DNA template is considerably more complex. When eukaryotic cells are not dividing, their genes exist as a diffuse mass of DNA and proteins called chromatin. The DNA is tightly packaged around charged histone proteins at repeated intervals. These *DNA–histone complexes*, collectively called nucleosomes, are regularly spaced and include 146 nucleotides of DNA wound around eight

1H Liang et al., "Fast evolution of core promoters in primate genomes," *Molecular Biology and Evolution* 25 (2008): 1239–44.

histones like thread around a spool.

For polynucleotide synthesis to occur, the transcription machinery needs to move histones out of the way every time it encounters a nucleosome. This is accomplished by a special protein complex called **FACT**, which stands for "*facilitates chromatin transcription*." This complex pulls histones away from the DNA template as the polymerase moves along it. Once the pre-mRNA is synthesized, the FACT complex replaces the histones to recreate the nucleosomes.

The termination of transcription is different for the different polymerases. Unlike in prokaryotes, elongation by RNA polymerase II in eukaryotes takes place 1,000 to 2,000 nucleotides beyond the end of the gene being transcribed. This pre-mRNA tail is subsequently removed by cleavage during mRNA processing. On the other hand, RNA polymerases I and III require termination signals. Genes transcribed by RNA polymerase I contain a specific 18-nucleotide sequence that is recognized by a termination protein. The process of termination in RNA polymerase III involves an mRNA hairpin similar to rho-independent termination of transcription in prokaryotes.

15.4 RNA Processing in Eukaryotes

By the end of this section, you will be able to do the following:

- Describe the different steps in RNA processing
- Understand the significance of exons, introns, and splicing for mRNAs
- Explain how tRNAs and rRNAs are processed

After transcription, eukaryotic pre-mRNAs must undergo several processing steps before they can be translated. Eukaryotic (and prokaryotic) tRNAs and rRNAs also undergo processing before they can function as components in the protein-synthesis machinery.

mRNA Processing

The eukaryotic pre-mRNA undergoes extensive processing before it is ready to be translated. Eukaryotic protein-coding sequences are not continuous, as they are in prokaryotes. The coding sequences (exons) are interrupted by noncoding introns, which must be removed to make a translatable mRNA. The additional steps involved in eukaryotic mRNA maturation also create a molecule with a much longer half-life than a prokaryotic mRNA. Eukaryotic mRNAs last for several hours, whereas the typical *E. coli* mRNA lasts no more than five seconds.

Pre-mRNAs are first coated in RNA-stabilizing proteins; these protect the pre-mRNA from degradation while it is processed and exported out of the nucleus. The three most important steps of pre-mRNA processing are the addition of stabilizing and signaling factors at the 5' and 3' ends of the molecule, and the removal of the introns (Figure 15.11). In rare cases, the mRNA transcript can be "edited" after it is transcribed.

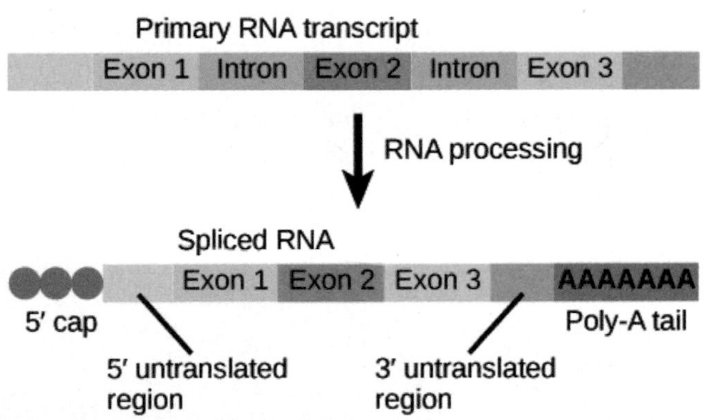

Figure 15.11 Eukaryotic mRNA contains introns that must be spliced out. A 5' cap and 3' poly-A tail are also added.

EVOLUTION CONNECTION

RNA Editing in Trypanosomes

The trypanosomes are a group of protozoa that include the pathogen *Trypanosoma brucei*, which causes nagana in cattle and sleeping sickness in humans throughout great areas of Africa (Figure 15.12). The trypanosome is carried by biting flies in the genus *Glossina* (commonly called tsetse flies). Trypanosomes, and virtually all other eukaryotes, have organelles called mitochondria that supply the cell with chemical energy. Mitochondria are organelles that express their own DNA and are believed to be the remnants of a symbiotic relationship between a eukaryote and an engulfed prokaryote. The mitochondrial DNA of trypanosomes exhibit an interesting exception to the central dogma: their pre-mRNAs do not have the correct information to specify a functional protein. Usually, this is because the mRNA is missing several U nucleotides. The cell performs an additional RNA processing step called **RNA editing** to remedy this.

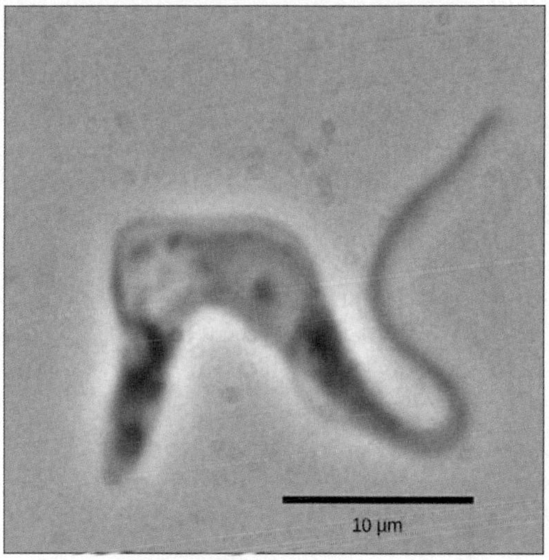

Figure 15.12 *Trypanosoma brucei* is the causative agent of sleeping sickness in humans. The mRNAs of this pathogen must be modified by the addition of nucleotides before protein synthesis can occur. (credit: modification of work by Torsten Ochsenreiter)

Other genes in the mitochondrial genome encode 40- to 80-nucleotide guide RNAs. One or more of these molecules interacts by complementary base pairing with some of the nucleotides in the pre-mRNA transcript. However, the *guide RNA* has more A nucleotides than the pre-mRNA has U nucleotides with which to bind. In these regions, the guide RNA loops out. The 3' ends of guide RNAs have a long poly-U tail, and these U bases are inserted in regions of the pre-mRNA transcript at which the guide RNAs are looped. This process is entirely mediated by RNA molecules. That is, guide RNAs—rather than proteins—serve as the catalysts in RNA editing.

RNA editing is not just a phenomenon of trypanosomes. In the mitochondria of some plants, almost all pre-mRNAs are edited. RNA editing has also been identified in mammals such as rats, rabbits, and even humans. What could be the evolutionary reason for this additional step in pre-mRNA processing? One possibility is that the mitochondria, being remnants of ancient prokaryotes, have an equally ancient RNA-based method for regulating gene expression. In support of this hypothesis, edits made to pre-mRNAs differ depending on cellular conditions. Although speculative, the process of RNA editing may be a holdover from a primordial time when RNA molecules, instead of proteins, were responsible for catalyzing reactions.

5' Capping

While the pre-mRNA is still being synthesized, a **7-methylguanosine cap** is added to the 5' end of the growing transcript by a phosphate linkage. This functional group protects the nascent mRNA from degradation. In addition, factors involved in protein synthesis recognize the cap to help initiate translation by ribosomes.

3' Poly-A Tail

Once elongation is complete, the pre-mRNA is cleaved by an endonuclease between an AAUAAA consensus sequence and a GU-rich sequence, leaving the AAUAAA sequence on the pre-mRNA. An enzyme called poly-A polymerase then adds a string of approximately 200 A residues, called the **poly-A tail**. This modification further protects the pre-mRNA from degradation and is also the binding site for a protein necessary for exporting the processed mRNA to the cytoplasm.

Pre-mRNA Splicing

Eukaryotic genes are composed of **exons**, which correspond to protein-coding sequences (*ex*-on signifies that they are expressed), and *int*ervening sequences called **introns** (*int*-ron denotes their *int*ervening role), which may be involved in gene regulation but are removed from the pre-mRNA during processing. Intron sequences in mRNA do not encode functional proteins.

The discovery of introns came as a surprise to researchers in the 1970s who expected that pre-mRNAs would specify protein sequences without further processing, as they had observed in prokaryotes. The genes of higher eukaryotes very often contain one or more introns. These regions may correspond to regulatory sequences; however, the biological significance of having many introns or having very long introns in a gene is unclear. It is possible that introns slow down gene expression because it takes longer to transcribe pre-mRNAs with lots of introns. Alternatively, introns may be nonfunctional sequence remnants left over from the fusion of ancient genes throughout the course of evolution. This is supported by the fact that separate exons often encode separate protein subunits or domains. For the most part, the sequences of introns can be mutated without ultimately affecting the protein product.

All of a pre-mRNA's introns must be completely and precisely removed before protein synthesis. If the process errs by even a single nucleotide, the reading frame of the rejoined exons would shift, and the resulting protein would be dysfunctional. The process of removing introns and reconnecting exons is called **splicing** (Figure 15.13). Introns are removed and degraded while the pre-mRNA is still in the nucleus. Splicing occurs by a sequence-specific mechanism that ensures introns will be removed and exons rejoined with the accuracy and precision of a single nucleotide. Although the intron itself is noncoding, the beginning and end of each intron is marked with specific nucleotides: GU at the 5' end and AG at the 3' end of the intron. The splicing of pre-mRNAs is conducted by complexes of proteins and RNA molecules called spliceosomes.

VISUAL CONNECTION

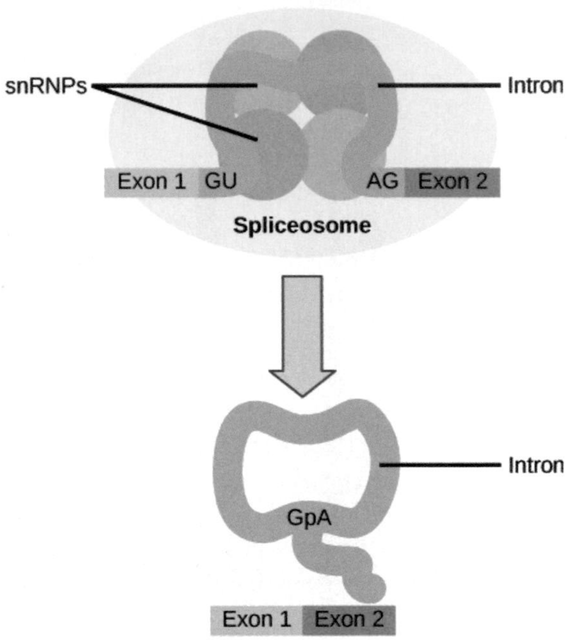

Figure 15.13 Pre-mRNA splicing involves the precise removal of introns from the primary RNA transcript. The splicing process is catalyzed by protein complexes called spliceosomes that are composed of proteins and RNA molecules called small nuclear RNAs (snRNAs). Spliceosomes recognize sequences at the 5' and 3' end of the intron.

Errors in splicing are implicated in cancers and other human diseases. What kinds of mutations might lead to splicing errors? Think of different possible outcomes if splicing errors occur.

Note that more than 70 individual introns can be present, and each has to undergo the process of splicing—in addition to 5' capping and the addition of a poly-A tail—just to generate a single, translatable mRNA molecule.

LINK TO LEARNING

See how introns are removed during RNA splicing at this website (http://openstax.org/l/RNA_splicing) .

Processing of tRNAs and rRNAs

The tRNAs and rRNAs are structural molecules that have roles in protein synthesis; however, these RNAs are not themselves translated. Pre-rRNAs are transcribed, processed, and assembled into ribosomes in the nucleolus. Pre-tRNAs are transcribed and processed in the nucleus and then released into the cytoplasm where they are linked to free amino acids for protein synthesis.

Most of the tRNAs and rRNAs in eukaryotes and prokaryotes are first transcribed as a long precursor molecule that spans multiple rRNAs or tRNAs. Enzymes then cleave the precursors into subunits corresponding to each structural RNA. Some of the bases of pre-rRNAs are *methylated*; that is, a $-CH_3$ methyl functional group is added for stability. Pre-tRNA molecules also undergo methylation. As with pre-mRNAs, subunit excision occurs in eukaryotic pre-RNAs destined to become tRNAs or rRNAs.

Mature rRNAs make up approximately 50 percent of each ribosome. Some of a ribosome's RNA molecules are purely structural, whereas others have catalytic or binding activities. Mature tRNAs take on a three-dimensional structure through local regions of base pairing stabilized by intramolecular hydrogen bonding. The tRNA folds to position the amino acid binding site at one end and the **anticodon** at the other end (Figure 15.14). The anticodon is a three-nucleotide sequence in a tRNA that interacts with an mRNA codon through complementary base pairing.

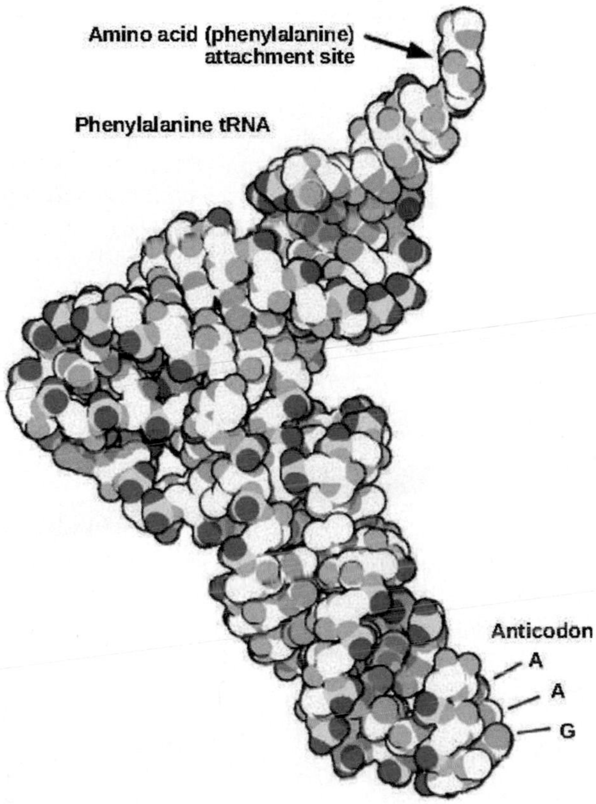

Figure 15.14 This is a space filling model of a tRNA molecule that adds the amino acid phenylalanine to a growing polypeptide chain. The anticodon AAG binds the Codon UUC on the mRNA. The amino acid phenylalanine is attached to the other end of the tRNA.

15.5 Ribosomes and Protein Synthesis

By the end of this section, you will be able to do the following:

- Describe the different steps in protein synthesis
- Discuss the role of ribosomes in protein synthesis

The synthesis of proteins consumes more of a cell's energy than any other metabolic process. In turn, proteins account for more mass than any other component of living organisms (with the exception of water), and proteins perform virtually every function of a cell. The process of translation, or protein synthesis, involves the decoding of an mRNA message into a polypeptide product. Amino acids are covalently strung together by interlinking peptide bonds in lengths ranging from approximately 50 to more than 1000 amino acid residues. Each individual amino acid has an amino group (NH_2) and a carboxyl (COOH) group. Polypeptides are formed when the amino group of one amino acid forms an amide (i.e., peptide) bond with the carboxyl group of another amino acid (Figure 15.15). This reaction is catalyzed by ribosomes and generates one water molecule.

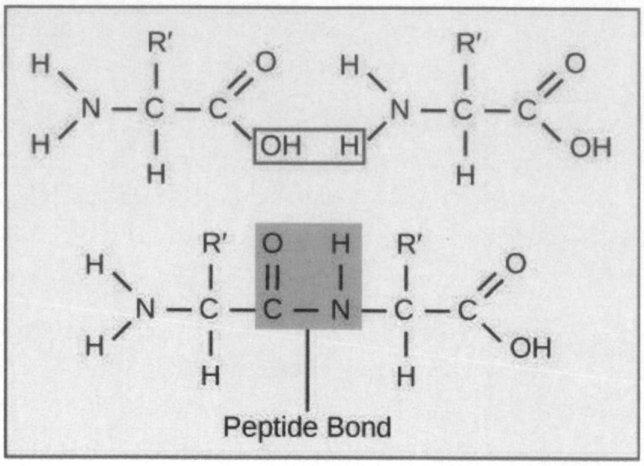

Figure 15.15 A peptide bond links the carboxyl end of one amino acid with the amino end of another, producing one water molecule during the process. For simplicity in this image, only the functional groups involved in the peptide bond are shown. The R and R' designations refer to the rest of each amino acid structure.

The Protein Synthesis Machinery

In addition to the mRNA template, many molecules and macromolecules contribute to the process of translation. The composition of each component may vary across species; for example, ribosomes may consist of different numbers of rRNAs and polypeptides depending on the organism. However, the general structures and functions of the protein synthesis machinery are comparable from bacteria to human cells. Translation requires the input of an mRNA template, ribosomes, tRNAs, and various enzymatic factors. (Note: A ribosome can be thought of as an enzyme whose amino acid binding sites are specified by mRNA.)

LINK TO LEARNING

Click through the steps of this PBS interactive (http://openstax.org/l/prokary_protein) to see protein synthesis in action.

Ribosomes

Even before an mRNA is translated, a cell must invest energy to build each of its ribosomes. In *E. coli*, there are between 10,000 and 70,000 ribosomes present in each cell at any given time. A **ribosome** is a complex macromolecule composed of structural and catalytic rRNAs, and many distinct polypeptides. In eukaryotes, the nucleolus is completely specialized for the synthesis and assembly of rRNAs.

Ribosomes exist in the cytoplasm of prokaryotes and in the cytoplasm and rough endoplasmic reticulum of eukaryotes. Mitochondria and chloroplasts also have their own ribosomes in the matrix and stroma, which look more similar to prokaryotic ribosomes (and have similar drug sensitivities) than the ribosomes just outside their outer membranes in the cytoplasm. Ribosomes dissociate into large and small subunits when they are not synthesizing proteins and reassociate during the initiation of translation. *In E. coli, the small subunit is described as 30S, and the large subunit is 50S, for a total of 70S (recall that Svedberg units are not additive).* Mammalian ribosomes have a small 40S subunit and a large 60S subunit, for a total of 80S. The small subunit is responsible for binding the mRNA template, whereas the large subunit sequentially binds tRNAs. Each mRNA molecule is simultaneously translated by many ribosomes, all synthesizing protein in the same direction: reading the mRNA from 5' to 3' and synthesizing the polypeptide from the N terminus to the C terminus. The complete mRNA/poly-ribosome structure is called a **polysome**.

tRNAs

The tRNAs are structural RNA molecules that were transcribed from genes by RNA polymerase III. Depending on the species, 40 to 60 types of tRNAs exist in the cytoplasm. Transfer RNAs serve as adaptor molecules. Each tRNA carries a specific amino acid and recognizes one or more of the mRNA codons that define the order of amino acids in a protein. Aminoacyl-tRNAs bind to the ribosome and add the corresponding amino acid to the polypeptide chain. Therefore, tRNAs are the molecules that actually "translate" the language of RNA into the language of proteins.

Of the 64 possible mRNA codons—or triplet combinations of A, U, G, and C—three specify the termination of protein synthesis and 61 specify the addition of amino acids to the polypeptide chain. Of these 61, one codon (AUG) also encodes the initiation of translation. Each tRNA anticodon can base pair with one or more of the mRNA codons for its amino acid. For instance, if the sequence CUA occurred on an mRNA template in the proper reading frame, it would bind a leucine tRNA expressing the complementary sequence, GAU. The ability of some tRNAs to match more than one codon is what gives the genetic code its blocky structure.

As the adaptor molecules of translation, it is surprising that tRNAs can fit so much specificity into such a small package. Consider that tRNAs need to interact with three factors: 1) they must be recognized by the correct aminoacyl synthetase (see below); 2) they must be recognized by ribosomes; and 3) they must bind to the correct sequence in mRNA.

Aminoacyl tRNA Synthetases

The process of pre-tRNA synthesis by RNA polymerase III only creates the RNA portion of the adaptor molecule. The corresponding amino acid must be added later, once the tRNA is processed and exported to the cytoplasm. Through the process of tRNA "charging," each tRNA molecule is linked to its correct amino acid by one of a group of enzymes called **aminoacyl tRNA synthetases**. At least one type of aminoacyl tRNA synthetase exists for each of the 20 amino acids; the exact number of aminoacyl tRNA synthetases varies by species. These enzymes first bind and hydrolyze ATP to catalyze a high-energy bond between an amino acid and adenosine monophosphate (AMP); a pyrophosphate molecule is expelled in this reaction. The activated amino acid is then transferred to the tRNA, and AMP is released. The term "charging" is appropriate, since the high-energy bond that attaches an amino acid to its tRNA is later used to drive the formation of the peptide bond. Each tRNA is named for its amino acid.

The Mechanism of Protein Synthesis

As with mRNA synthesis, protein synthesis can be divided into three phases: *initiation, elongation, and termination*. The process of translation is similar in prokaryotes and eukaryotes. Here we'll explore how translation occurs in *E. coli*, a representative prokaryote, and specify any differences between prokaryotic and eukaryotic translation.

Initiation of Translation

Protein synthesis begins with the formation of an **initiation complex**. In *E. coli*, this complex involves the small 30S ribosome, the mRNA template, three initiation factors (IFs; IF-1, IF-2, and IF-3), and a special **initiator tRNA**, called $tRNA^{Metf}$.

In *E. coli* mRNA, a sequence upstream of the first AUG codon, called the Shine-Dalgarno sequence (AGGAGG), interacts with the rRNA molecules that compose the ribosome. This interaction anchors the 30S ribosomal subunit at the correct location on the mRNA template. Guanosine triphosphate (GTP), which is a purine nucleotide triphosphate, acts as an energy source during translation—both at the start of elongation and during the ribosome's translocation. Binding of the mRNA to the 30S ribosome also requires IF-III.

The initiator tRNA then interacts with the **start codon** AUG (or rarely, GUG). This tRNA carries the amino acid methionine, which is formylated after its attachment to the tRNA. The formylation creates a "faux" peptide bond between the formyl carboxyl group and the amino group of the methionine. Binding of the fMet-$tRNA^{Metf}$ is mediated by the initiation factor IF-2. The fMet begins every polypeptide chain synthesized by *E. coli*, but it is usually removed after translation is complete. When an in-frame AUG is encountered during translation elongation, a non-formylated methionine is inserted by a regular Met-$tRNA^{Met}$. After the formation of the initiation complex, the 30S ribosomal subunit is joined by the 50S subunit to form the translation complex. In eukaryotes, a similar initiation complex forms, comprising mRNA, the 40S small ribosomal subunit, eukaryotic IFs, and nucleoside triphosphates (GTP and ATP). The methionine on the charged initiator tRNA, called Met-$tRNA_i$, is not formylated. However, Met-$tRNA_i$ is distinct from other Met-tRNAs in that it can bind IFs.

Instead of depositing at the Shine-Dalgarno sequence, the eukaryotic initiation complex recognizes the 7-methylguanosine cap at the 5' end of the mRNA. A cap-binding protein (CBP) and several other IFs assist the movement of the ribosome to the 5' cap. Once at the cap, the initiation complex tracks along the mRNA in the 5' to 3' direction, searching for the AUG start codon. Many eukaryotic mRNAs are translated from the first AUG, but this is not always the case. According to **Kozak's rules**, the nucleotides around the AUG indicate whether it is the correct start codon. Kozak's rules state that the following consensus sequence must appear around the AUG of vertebrate genes: 5'-gccRccAUGG-3'. The R (for purine) indicates a site that can be either A or G, but cannot be C or U. Essentially, the closer the sequence is to this consensus, the higher the efficiency of translation.

Once the appropriate AUG is identified, the other proteins and CBP dissociate, and the 60S subunit binds to the complex of Met-$tRNA_i$, mRNA, and the 40S subunit. This step completes the initiation of translation in eukaryotes.

Translation, Elongation, and Termination

In prokaryotes and eukaryotes, the basics of elongation are the same, so we will review elongation from the perspective of *E. coli*. When the translation complex is formed, the tRNA binding region of the ribosome consists of three compartments. The A (aminoacyl) site binds incoming charged aminoacyl tRNAs. The P (peptidyl) site binds charged tRNAs carrying amino acids that have formed peptide bonds with the growing polypeptide chain but have not yet dissociated from their corresponding tRNA. The E (exit) site releases dissociated tRNAs so that they can be recharged with free amino acids. The initiating methionyl-tRNA, however, occupies the P site at the beginning of the elongation phase of translation in both prokaryotes and eukaryotes.

During translation elongation, the mRNA template provides tRNA binding specificity. As the ribosome moves along the mRNA, each mRNA codon comes into register, and specific binding with the corresponding charged tRNA anticodon is ensured. If mRNA were not present in the elongation complex, the ribosome would bind tRNAs nonspecifically and randomly (?).

Elongation proceeds with charged tRNAs sequentially entering and leaving the ribosome as each new amino acid is added to the polypeptide chain. Movement of a tRNA from A to P to E site is induced by conformational changes that advance the ribosome by three bases in the 3' direction. The energy for each step along the ribosome is donated by elongation factors that hydrolyze GTP. GTP energy is required both for the binding of a new aminoacyl-tRNA to the A site and for its translocation to the P site after formation of the peptide bond. Peptide bonds form between the amino group of the amino acid attached to the A-site tRNA and the carboxyl group of the amino acid attached to the P-site tRNA. The formation of each peptide bond is catalyzed by **peptidyl transferase**, an RNA-based enzyme that is integrated into the 50S ribosomal subunit. The energy for each peptide bond formation is derived from the high-energy bond linking each amino acid to its tRNA. After peptide bond formation, the A-site tRNA that now holds the growing peptide chain moves to the P site, and the P-site tRNA that is now empty moves to the E site and is expelled from the ribosome (Figure 15.16). Amazingly, the *E. coli* translation apparatus takes only 0.05 seconds to add each amino acid, meaning that a 200-amino-acid protein can be translated in just 10 seconds.

VISUAL CONNECTION

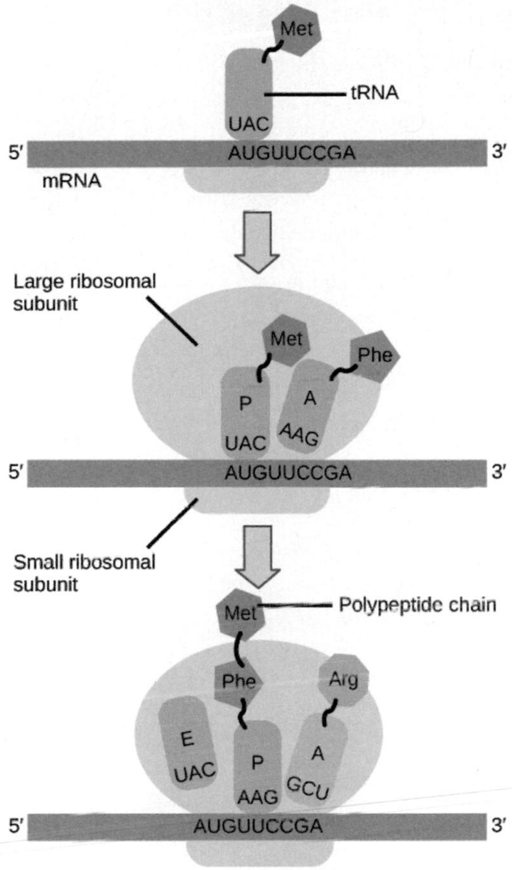

Figure 15.16 Translation begins when an initiator tRNA anticodon recognizes a start codon on mRNA bound to a small ribosomal subunit. The large ribosomal subunit joins the small subunit, and a second tRNA is recruited. As the mRNA moves relative to the ribosome, successive tRNAs move through the ribosome and the polypeptide chain is formed. Entry of a release factor into the A site terminates translation and the components dissociate.

Many antibiotics inhibit bacterial protein synthesis. For example, tetracycline blocks the A site on the bacterial ribosome, and chloramphenicol blocks peptidyl transfer. What specific effect would you expect each of these antibiotics to have on protein synthesis?

Tetracycline would directly affect:

a. tRNA binding to the ribosome
b. ribosome assembly
c. growth of the protein chain

Chloramphenicol would directly affect:

a. tRNA binding to the ribosome
b. ribosome assembly
c. growth of the protein chain

Termination of translation occurs when a nonsense codon (UAA, UAG, or UGA) is encountered. Upon aligning with the A site, these nonsense codons are recognized by *protein release factors* that resemble tRNAs. The releasing factors in both prokaryotes and eukaryotes instruct peptidyl transferase to add a water molecule to the carboxyl end of the P-site amino acid. This reaction forces the P-site amino acid to detach from its tRNA, and the newly made protein is released. The small and large ribosomal subunits dissociate from the mRNA and from each other; they are recruited almost immediately into another translation initiation complex. After many ribosomes have completed translation, the mRNA is degraded so the nucleotides can be reused in another transcription reaction.

Protein Folding, Modification, and Targeting

During and after translation, individual amino acids may be chemically modified, signal sequences appended, and the new protein "folded" into a distinct three-dimensional structure as a result of intramolecular interactions. A **signal sequence** is a short sequence at the amino end of a protein that directs it to a specific cellular compartment. These sequences can be thought of as the protein's "train ticket" to its ultimate destination, and are recognized by signal-recognition proteins that act as conductors. For instance, a specific signal sequence terminus will direct a protein to the mitochondria or chloroplasts (in plants). Once the protein reaches its cellular destination, the signal sequence is usually clipped off.

Many proteins fold spontaneously, but some proteins require helper molecules, called *chaperones*, to prevent them from aggregating during the complicated process of folding. Even if a protein is properly specified by its corresponding mRNA, it could take on a completely dysfunctional shape if abnormal temperature or pH conditions prevent it from folding correctly.

KEY TERMS

7-methylguanosine cap modification added to the 5' end of pre-mRNAs to protect mRNA from degradation and assist translation

aminoacyl tRNA synthetase enzyme that "charges" tRNA molecules by catalyzing a bond between the tRNA and a corresponding amino acid

anticodon three-nucleotide sequence in a tRNA molecule that corresponds to an mRNA codon

CAAT box (GGCCAATCT) essential eukaryotic promoter sequence involved in binding transcription factors

central dogma states that genes specify the sequence of mRNAs, which in turn specify the sequence of proteins

codon three consecutive nucleotides in mRNA that specify the insertion of an amino acid or the release of a polypeptide chain during translation

colinear in terms of RNA and protein, three "units" of RNA (nucleotides) specify one "unit" of protein (amino acid) in a consecutive fashion

consensus DNA sequence that is used by many species to perform the same or similar functions

core enzyme prokaryotic RNA polymerase consisting of α, α, β, and β' but missing σ; this complex performs elongation

degeneracy (of the genetic code) describes that a given amino acid can be encoded by more than one nucleotide triplet; the code is degenerate, but not ambiguous

downstream nucleotides following the initiation site in the direction of mRNA transcription; in general, sequences that are toward the 3' end relative to a site on the mRNA

exon sequence present in protein-coding mRNA after completion of pre-mRNA splicing

FACT complex that "facilitates chromatin transcription" by disassembling nucleosomes ahead of a transcribing RNA polymerase II and reassembling them after the polymerase passes by

GC-rich box (GGCG) nonessential eukaryotic promoter sequence that binds cellular factors to increase the efficiency of transcription; may be present several times in a promoter

hairpin structure of RNA when it folds back on itself and forms intramolecular hydrogen bonds between complementary nucleotides

holoenzyme prokaryotic RNA polymerase consisting of α, α, β, β', and σ; this complex is responsible for transcription initiation

initiation site nucleotide from which mRNA synthesis proceeds in the 5' to 3' direction; denoted with a "+1"

initiator tRNA in prokaryotes, called $tRNA_f^{Met}$; in eukaryotes, called $tRNA_i$; a tRNA that interacts with a start codon, binds directly to the ribosome P site, and links to a special methionine to begin a polypeptide chain

intron non–protein-coding intervening sequences that are spliced from mRNA during processing

Kozak's rules determines the correct initiation AUG in a eukaryotic mRNA; the following consensus sequence must appear around the AUG: 5'-GCC(**purine**)CCAUG**G**-3'; the bolded bases are most important

nonsense codon one of the three mRNA codons that specifies termination of translation

nontemplate strand strand of DNA that is not used to transcribe mRNA; this strand is identical to the mRNA except that T nucleotides in the DNA are replaced by U nucleotides in the mRNA

Octamer box (ATTTGCAT) nonessential eukaryotic promoter sequence that binds cellular factors to increase the efficiency of transcription; may be present several times in a promoter

peptidyl transferase RNA-based enzyme that is integrated into the 50S ribosomal subunit and catalyzes the formation of peptide bonds

plasmid extrachromosomal, covalently closed, circular DNA molecule that may only contain one or a few genes; common in prokaryotes

poly-A tail modification added to the 3' end of pre-mRNAs to protect mRNA from degradation and assist mRNA export from the nucleus

polysome mRNA molecule simultaneously being translated by many ribosomes all going in the same direction

preinitiation complex cluster of transcription factors and other proteins that recruit RNA polymerase II for transcription of a DNA template

promoter DNA sequence to which RNA polymerase and associated factors bind and initiate transcription

reading frame sequence of triplet codons in mRNA that specify a particular protein; a ribosome shift of one or two nucleotides in either direction completely abolishes synthesis of that protein

rho-dependent termination in prokaryotes, termination of transcription by an interaction between RNA polymerase and the rho protein at a run of G nucleotides on the DNA template

rho-independent termination sequence-dependent termination of prokaryotic mRNA synthesis; caused by hairpin formation in the mRNA that stalls the polymerase

RNA editing direct alteration of one or more nucleotides in an mRNA that has already been synthesized

Shine-Dalgarno sequence (AGGAGG); initiates prokaryotic translation by interacting with rRNA molecules comprising the 30S ribosome

signal sequence short tail of amino acids that directs a protein to a specific cellular compartment

small nuclear RNA molecules synthesized by RNA

polymerase III that have a variety of functions, including splicing pre-mRNAs and regulating transcription factors

splicing process of removing introns and reconnecting exons in a pre-mRNA

start codon AUG (or rarely, GUG) on an mRNA from which translation begins; always specifies methionine

TATA box conserved promoter sequence in eukaryotes and prokaryotes that helps to establish the initiation site for transcription

template strand strand of DNA that specifies the complementary mRNA molecule

transcription bubble region of locally unwound DNA that allows for transcription of mRNA

upstream nucleotides preceding the initiation site; in general, sequences toward the 5' end relative to a site on the mRNA

CHAPTER SUMMARY

15.1 The Genetic Code

The genetic code refers to the DNA alphabet (A, T, C, G), the RNA alphabet (A, U, C, G), and the polypeptide alphabet (20 amino acids). The central dogma describes the flow of genetic information in the cell from genes to mRNA to proteins. Genes are used to make mRNA by the process of transcription; mRNA is used to synthesize proteins by the process of translation. The genetic code is degenerate because 64 triplet codons in mRNA specify only 20 amino acids and three nonsense codons. Most amino acids have several similar codons. Almost every species on the planet uses the same genetic code.

15.2 Prokaryotic Transcription

In prokaryotes, mRNA synthesis is initiated at a promoter sequence on the DNA template comprising two consensus sequences that recruit RNA polymerase. The prokaryotic polymerase consists of a core enzyme of four protein subunits and a σ protein that assists only with initiation. Elongation synthesizes mRNA in the 5' to 3' direction at a rate of 40 nucleotides per second. Termination liberates the mRNA and occurs either by rho protein interaction or by the formation of an mRNA hairpin.

15.3 Eukaryotic Transcription

Transcription in eukaryotes involves one of three types of polymerases, depending on the gene being transcribed. RNA polymerase II transcribes all of the protein-coding genes, whereas RNA polymerase I transcribes the tandemly duplicated rRNA genes, and RNA polymerase III transcribes various small RNAs, like the 5S rRNA, tRNA, and small nuclear RNA genes. The initiation of transcription in eukaryotes involves the binding of several transcription factors to complex promoter sequences that are usually located upstream of the gene being copied. The mRNA is synthesized in the 5' to 3' direction, and the FACT complex moves and reassembles nucleosomes as the polymerase passes by. Whereas RNA polymerases I and III terminate transcription by protein- or RNA hairpin-dependent methods, RNA polymerase II transcribes for 1,000 or more nucleotides beyond the gene template and cleaves the excess during pre-mRNA processing.

15.4 RNA Processing in Eukaryotes

Eukaryotic pre-mRNAs are modified with a 5' methylguanosine cap and a poly-A tail. These structures protect the mature mRNA from degradation and help export it from the nucleus. Pre-mRNAs also undergo splicing, in which introns are removed and exons are reconnected with single-nucleotide accuracy. Only finished mRNAs that have undergone 5' capping, 3' polyadenylation, and intron splicing are exported from the nucleus to the cytoplasm. Pre-rRNAs and pre-tRNAs may be processed by intramolecular cleavage, splicing, methylation, and chemical conversion of nucleotides. Rarely, RNA editing is also performed to insert missing bases after an mRNA has been synthesized.

15.5 Ribosomes and Protein Synthesis

The players in translation include the mRNA template, ribosomes, tRNAs, and various enzymatic factors. The small ribosomal subunit binds to the mRNA template either at the Shine-Dalgarno sequence (prokaryotes) or the 5' cap (eukaryotes). Translation begins at the initiating AUG on the mRNA, specifying methionine. The formation of peptide bonds occurs between sequential amino acids matched to the mRNA template by their tRNAs according to the genetic code. Charged tRNAs enter the ribosomal A site, and their amino acid bonds with the amino acid at the P site. The entire mRNA is translated in three-nucleotide "steps" of the ribosome. When a nonsense codon is encountered, a release factor binds and dissociates the components and frees the new protein. Folding of the protein occurs during and after translation.

VISUAL CONNECTION QUESTIONS

1. Figure 15.11 A scientist splices a eukaryotic promoter in front of a bacterial gene and inserts the gene in a bacterial chromosome. Would you expect the bacteria to transcribe the gene?

2. Figure 15.13 Errors in splicing are implicated in cancers and other human diseases. What kinds of mutations might lead to splicing errors? Think of different possible outcomes if splicing errors occur.

3. Figure 15.16 Many antibiotics inhibit bacterial protein synthesis. For example, tetracycline blocks the A site on the bacterial ribosome, and chloramphenicol blocks peptidyl transfer. What specific effect would you expect each of these antibiotics to have on protein synthesis?
Tetracycline would directly affect:
   a. tRNA binding to the ribosome
   b. ribosome assembly
   c. growth of the protein chain

   Chloramphenicol would directly affect
   a. tRNA binding to the ribosome
   b. ribosome assembly
   c. growth of the protein chain

REVIEW QUESTIONS

4. The AUC and AUA codons in mRNA both specify isoleucine. What feature of the genetic code explains this?
   a. complementarity
   b. nonsense codons
   c. universality
   d. degeneracy

5. How many nucleotides are in 12 mRNA codons?
   a. 12
   b. 24
   c. 36
   d. 48

6. Which event contradicts the central dogma of molecular biology?
   a. Poly-A polymerase enzymes process mRNA in the nucleus.
   b. Endonuclease enzymes splice out and repair damaged DNA.
   c. Scientists use reverse transcriptase enzymes to make DNA from RNA.
   d. Codons specifying amino acids are degenerate and universal.

7. Which subunit of the *E. coli* polymerase confers specificity to transcription?
   a. α
   b. β
   c. β'
   d. σ

8. The -10 and -35 regions of prokaryotic promoters are called consensus sequences because _______.
   a. they are identical in all bacterial species
   b. they are similar in all bacterial species
   c. they exist in all organisms
   d. they have the same function in all organisms

9. Three different bacteria species have the following consensus sequences upstream of a conserved gene.

	Species A	Species B	Species C
-10	TAATAAT	TTTAAT	TATATT
-35	TTGACA	TTGGCC	TTGAAA

Table 15.2

Order the bacteria from most to least efficient initiation of gene transcription.
   a. A > B > C
   b. B > C > A
   c. C > B > A
   d. A > C > B

10. Which feature of promoters can be found in both prokaryotes and eukaryotes?
   a. GC box
   b. TATA box
   c. octamer box
   d. -10 and -35 sequences

11. What transcripts will be most affected by low levels of α-amanitin?
    a. 18S and 28S rRNAs
    b. pre-mRNAs
    c. 5S rRNAs and tRNAs
    d. other small nuclear RNAs

12. How do enhancers and promoters differ?
    a. Enhancers bind transcription factors to silence gene expression, while promoters activate transcription.
    b. Enhancers increase the efficiency of gene expression, but are not essential for transcription. Promoter recognition is essential to transcription initiation.
    c. Promoters bind transcription factors to increase the efficiency of transcription. Enhancers bind RNA polymerases to initiate transcription.
    d. There is no difference. Both are transcription factor-binding sequences in DNA.

13. Which pre-mRNA processing step is important for initiating translation?
    a. poly-A tail
    b. RNA editing
    c. splicing
    d. 7-methylguanosine cap

14. What processing step enhances the stability of pre-tRNAs and pre-rRNAs?
    a. methylation
    b. nucleotide modification
    c. cleavage
    d. splicing

15. A scientist identifies a pre-mRNA with the following structure.

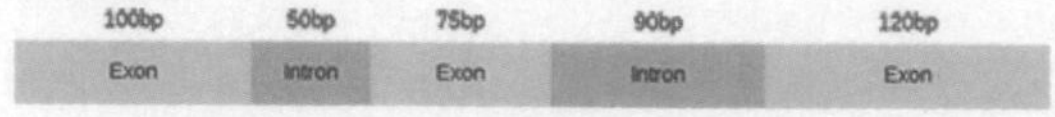

What is the predicted size of the corresponding mature mRNA in base pairs (bp), excluding the 5' cap and 3' poly-A tail?
    a. 220bp
    b. 295bp
    c. 140bp
    d. 435bp

16. The RNA components of ribosomes are synthesized in the ________.
    a. cytoplasm
    b. nucleus
    c. nucleolus
    d. endoplasmic reticulum

17. In any given species, there are at least how many types of aminoacyl tRNA synthetases?
    a. 20
    b. 40
    c. 100
    d. 200

18. A scientist introduces a mutation that makes the 60S ribosomal subunit nonfunctional in a human cell line. What would be the predicted effect on translation?
    a. Translation stalls after the initiation AUG codon is identified.
    b. The ribosome cannot catalyze the formation of peptide bonds between the tRNAs in the A and P sites.
    c. The ribosome cannot interact with mRNAs.
    d. tRNAs cannot exit the E site of the ribosome.

CRITICAL THINKING QUESTIONS

19. Imagine if there were 200 commonly occurring amino acids instead of 20. Given what you know about the genetic code, what would be the shortest possible codon length? Explain.
20. Discuss how degeneracy of the genetic code makes cells more robust to mutations.
21. A scientist sequencing mRNA identifies the following strand: CUAUGUGUCGUAACAGCCGAUGACCCG
    What is the sequence of the amino acid chain this mRNA makes when it is translated?

22. If mRNA is complementary to the DNA template strand and the DNA template strand is complementary to the DNA nontemplate strand, then why are base sequences of mRNA and the DNA nontemplate strand not identical? Could they ever be?

23. In your own words, describe the difference between rho-dependent and rho-independent termination of transcription in prokaryotes.
24. A fragment of bacterial DNA reads:
    3' –TACCTATAATCTCAATTGATAGAAGCACTCTAC– 5'
    Assuming that this fragment is the template strand, what is the sequence of mRNA that would be transcribed? (Hint: Be sure to identify the initiation site.)

25. A scientist observes that a cell has an RNA polymerase deficiency that prevents it from making proteins. Describe three additional observations that would together support the conclusion that a defect in RNA polymerase I activity, and not problems with the other polymerases, causes the defect.

26. Chronic lymphocytic leukemia patients often harbor nonsense mutations in their spliceosome machinery. Describe how this mutation of the spliceosome would change the final location and sequence of a pre-mRNA.

27. Transcribe and translate the following DNA sequence (nontemplate strand): 5'-ATGGCCGGTTATTAAGCA-3'

28. Explain how single nucleotide changes can have vastly different effects on protein function.

29. A normal mRNA that reads 5' – UGCCAUGGUAAUAACACAUGAGGCCUGAAC– 3' has an insertion mutation that changes the sequence to 5' -UGCCAUGG**U**UAAUAACACAUGAGGCCUGAAC– 3'. Translate the original mRNA and the mutated mRNA, and explain how insertion mutations can have dramatic effects on proteins. (Hint: Be sure to find the initiation site.)

CHAPTER 16
Gene Expression

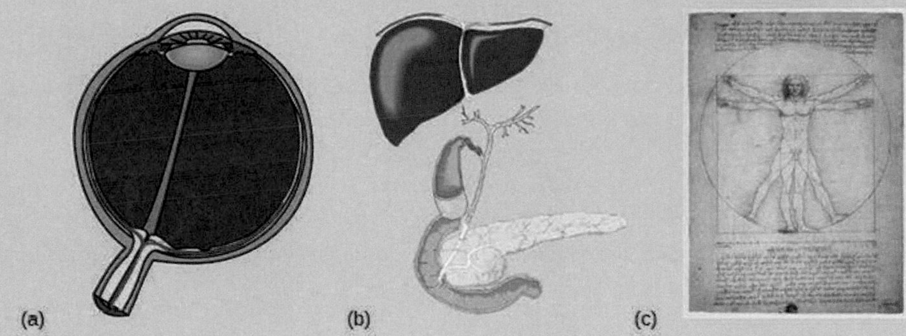

Figure 16.1 The genetic content of each somatic cell in an organism is the same, but not all genes are expressed in every cell. The control of which genes are expressed dictates whether a cell is, for example, (a) an eye cell or (b) a liver cell. It is the differential gene expression patterns that arise in different cells that give rise to (c) a complete organism.

Chapter Outline

INTRODUCTION Each somatic cell in the body generally contains the same DNA. A few exceptions include red blood cells, which contain no DNA in their mature state, and some immune system cells that rearrange their DNA while producing antibodies. In general, however, the genes that determine whether you have green eyes, brown hair, and how fast you metabolize food are the same in the cells in your eyes and your liver, even though these organs function quite differently. If each cell has the same DNA, how is it that cells or organs are different? Why do cells in the eye differ so dramatically from cells in the liver?

Whereas each cell shares the same genome and DNA sequence, each cell does not turn on, or express, the same set of genes. Each cell type needs a different set of proteins to perform its function. Therefore, only a small subset of proteins is expressed in a cell. For the proteins to be expressed, the DNA must be transcribed into RNA and the RNA must be translated into protein. In a given cell type, not all genes encoded in the DNA are transcribed into RNA or translated into protein because specific cells in our body have specific functions. Specialized proteins that make up the eye (iris, lens, and cornea) are only expressed in the eye, whereas the specialized proteins in the heart (pacemaker cells, heart muscle, and valves) are only expressed in the heart. At any given time, only a subset of all of the genes encoded by our DNA are expressed and translated into proteins. The expression of specific genes is a highly regulated process with many levels and stages of control. This complexity ensures the proper expression in the proper cell at the proper time.

16.1 Regulation of Gene Expression

By the end of this section, you will be able to do the following:

- Discuss why every cell does not express all of its genes all of the time
- Describe how prokaryotic gene regulation occurs at the transcriptional level
- Discuss how eukaryotic gene regulation occurs at the epigenetic, transcriptional, post-transcriptional, translational, and post-translational levels

For a cell to function properly, necessary proteins must be synthesized at the proper time and place. All cells control or regulate the synthesis of proteins from information encoded in their DNA. The process of turning on a gene to produce RNA and protein is called **gene expression**. Whether in a simple unicellular organism or a complex multi-cellular organism, each cell controls when and how its genes are expressed. For this to occur, there must be internal chemical mechanisms that control when a gene is expressed to make RNA and protein, how much of the protein is made, and when it is time to stop making that protein because it is no longer needed.

The regulation of gene expression conserves energy and space. It would require a significant amount of energy for an organism to express every gene at all times, so it is more energy efficient to turn on the genes only when they are required. In addition, only expressing a subset of genes in each cell saves space because DNA must be unwound from its tightly coiled structure to transcribe and translate the DNA. Cells would have to be enormous if every protein were expressed in every cell all the time.

The control of gene expression is extremely complex. Malfunctions in this process are detrimental to the cell and can lead to the development of many diseases, including cancer.

Prokaryotic versus Eukaryotic Gene Expression

To understand how gene expression is regulated, we must first understand how a gene codes for a functional protein in a cell. The process occurs in both prokaryotic and eukaryotic cells, just in slightly different manners.

Prokaryotic organisms are single-celled organisms that lack a cell nucleus, and their DNA therefore floats freely in the cell cytoplasm. To synthesize a protein, the processes of transcription and translation occur almost simultaneously. When the resulting protein is no longer needed, transcription stops. As a result, the primary method to control what type of protein and how much of each protein is expressed in a prokaryotic cell is the regulation of DNA transcription. All of the subsequent steps occur automatically. When more protein is required, more transcription occurs. Therefore, in prokaryotic cells, the control of gene expression is mostly at the transcriptional level.

Eukaryotic cells, in contrast, have intracellular organelles that add to their complexity. In eukaryotic cells, the DNA is contained inside the cell's nucleus and there it is transcribed into RNA. The newly synthesized RNA is then transported out of the nucleus into the cytoplasm, where ribosomes translate the RNA into protein. The processes of transcription and translation are *physically separated* by the nuclear membrane; transcription occurs only within the nucleus, and translation occurs only outside the nucleus in the cytoplasm. The regulation of gene expression can occur at all stages of the process (Figure 16.2). Regulation may occur when the DNA is uncoiled and loosened from nucleosomes to bind transcription factors (**epigenetic** level), when the RNA is transcribed (**transcriptional** level), when the RNA is processed and exported to the cytoplasm after it is transcribed (**post-transcriptional** level), when the RNA is translated into protein (**translational** level), or after the protein has been made (**post-translational** level).

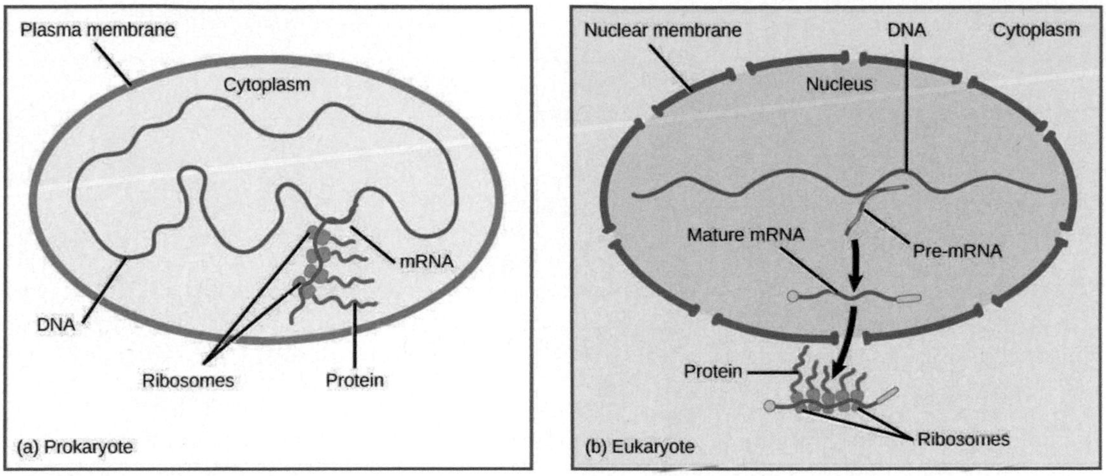

Figure 16.2 Regulation in prokaryotes and eukaryotes. Prokaryotic transcription and translation occur simultaneously in the cytoplasm, and regulation occurs at the transcriptional level. Eukaryotic gene expression is regulated during transcription and RNA processing, which take place in the nucleus, and during protein translation, which takes place in the cytoplasm. Further regulation may occur through post-translational modifications of proteins.

The differences in the regulation of gene expression between prokaryotes and eukaryotes are summarized in Table 16.1. The regulation of gene expression is discussed in detail in subsequent modules.

Differences in the Regulation of Gene Expression of Prokaryotic and Eukaryotic Organisms

Prokaryotic organisms	Eukaryotic organisms
Lack a membrane-bound nucleus	Contain nucleus
DNA is found in the cytoplasm	DNA is confined to the nuclear compartment
RNA transcription and protein formation occur almost simultaneously	RNA transcription occurs prior to protein formation, and it takes place in the nucleus. Translation of RNA to protein occurs in the cytoplasm.
Gene expression is regulated primarily at the transcriptional level	Gene expression is regulated at many levels (epigenetic, transcriptional, nuclear shuttling, post-transcriptional, translational, and post-translational)

Table 16.1

EVOLUTION CONNECTION

Evolution of Gene Regulation

Prokaryotic cells can only regulate gene expression by controlling the amount of transcription. As eukaryotic cells evolved, the complexity of the control of gene expression increased. For example, with the evolution of eukaryotic cells came compartmentalization of important cellular components and cellular processes. A nuclear region that contains the DNA was formed. Transcription and translation were physically separated into two different cellular compartments. It therefore became possible to control gene expression by regulating transcription in the nucleus, and also by controlling the RNA levels and protein translation present outside the nucleus.

Most gene regulation is done to conserve cell resources. However, other regulatory processes may be defensive. Cellular processes such as developed to protect the cell from viral or parasitic infections. If the cell could quickly shut off gene expression for a short period of time, it would be able to survive an infection when other organisms could not. Therefore, the organism evolved a new process that helped it survive, and it was able to pass this new development to offspring.

16.2 Prokaryotic Gene Regulation

By the end of this section, you will be able to do the following:

- Describe the steps involved in prokaryotic gene regulation
- Explain the roles of activators, inducers, and repressors in gene regulation

The DNA of prokaryotes is organized into a circular chromosome, supercoiled within the nucleoid region of the cell cytoplasm. Proteins that are needed for a specific function, or that are involved in the same biochemical pathway, are encoded together in blocks called **operons**. For example, all of the genes needed to use lactose as an energy source are coded next to each other in the lactose (or *lac*) operon, and transcribed into a single mRNA.

In prokaryotic cells, there are three types of regulatory molecules that can affect the expression of operons: repressors, activators, and inducers. Repressors and activators are proteins produced in the cell. Both repressors and activators regulate gene expression by binding to specific DNA sites *adjacent* to the genes they control. *In general, activators bind to the promoter site, while repressors bind to operator regions.* **Repressors** prevent transcription of a gene in response to an external stimulus, whereas **activators** increase the transcription of a gene in response to an external stimulus. Inducers are small molecules that may be produced by the cell or that are in the cell's environment. Inducers either activate or repress transcription depending on the needs of the cell and the availability of substrate.

The *trp* Operon: A Repressible Operon

Bacteria such as *Escherichia coli* need amino acids to survive, and are able to synthesize many of them. **Tryptophan** is one such amino acid that *E. coli* can either ingest from the environment or synthesize using enzymes that are encoded by five genes. These five genes are next to each other in what is called the **tryptophan (*trp*) operon** (Figure 16.3). The genes are transcribed into a single mRNA, which is then translated to produce all five enzymes. If tryptophan is present in the environment, then *E. coli* does not need to synthesize it and the *trp* operon is switched off. However, when tryptophan availability is low, the switch controlling the operon is turned on, the mRNA is transcribed, the enzyme proteins are translated, and tryptophan is synthesized.

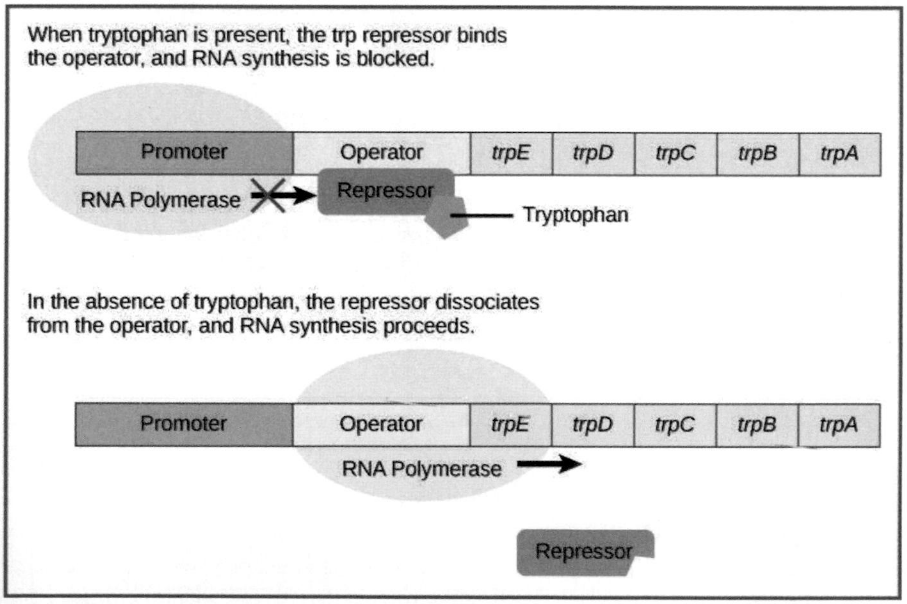

Figure 16.3 The tryptophan operon. The five genes that are needed to synthesize tryptophan in *E. coli* are located next to each other in the *trp* operon. When tryptophan is plentiful, two tryptophan molecules bind the repressor protein at the operator sequence. This physically blocks the RNA polymerase from transcribing the tryptophan genes. When tryptophan is absent, the repressor protein does not bind to the operator and the genes are transcribed.

The *trp* operon includes three important regions: the coding region, the *trp* operator and the *trp* promoter. The coding region includes the genes for the five tryptophan biosynthesis enzymes. Just before the coding region is the **transcriptional start site**. The promoter sequence, to which RNA polymerase binds to initiate transcription, is before or "upstream" of the transcriptional start site. Between the promoter and the transcriptional start site is the operator region.

The *trp* **operator** contains the DNA code to which the *trp* repressor protein can bind. However, the repressor alone cannot bind to the operator. When tryptophan is present in the cell, two tryptophan molecules bind to the *trp* repressor, which changes the shape of the repressor protein to a form that can bind to the *trp* operator. Binding of the tryptophan–repressor complex at the operator physically prevents the RNA polymerase from binding to the promoter and transcribing the downstream genes.

When tryptophan is not present in the cell, the repressor by itself does not bind to the operator, the polymerase can transcribe the enzyme genes, and tryptophan is synthesized. Because the repressor protein actively binds to the operator to keep the genes turned off, the *trp* operon is said to be *negatively regulated* and the proteins that bind to the operator to silence *trp* expression are **negative regulators**.

LINK TO LEARNING

Watch this video to learn more about the *trp* operon.

Click to view content (https://www.openstax.org/l/trp_operon)

Catabolite Activator Protein (CAP): A Transcriptional Activator

Just as the *trp* operon is negatively regulated by tryptophan molecules, there are proteins that bind to the promoter sequences that act as **positive regulators** to turn genes on and activate them. For example, when glucose is scarce, *E. coli* bacteria can turn to other sugar sources for fuel. To do this, new genes to process these alternate sugars must be transcribed. When glucose levels drop, cyclic AMP (cAMP) begins to accumulate in the cell. The cAMP molecule is a signaling molecule that is involved in glucose and energy metabolism in *E. coli*. Accumulating cAMP binds to the positive regulator **catabolite activator protein (CAP)**, a protein that binds to the promoters of operons which control the processing of alternative sugars. When cAMP binds to CAP, the complex then binds to the promoter region of the genes that are needed to use the alternate sugar sources (Figure 16.4). In these operons, a CAP-binding site is located upstream of the RNA-polymerase-binding site in the promoter. CAP binding stabilizes the binding of RNA polymerase to the promoter region and increases transcription of the associated protein-coding genes.

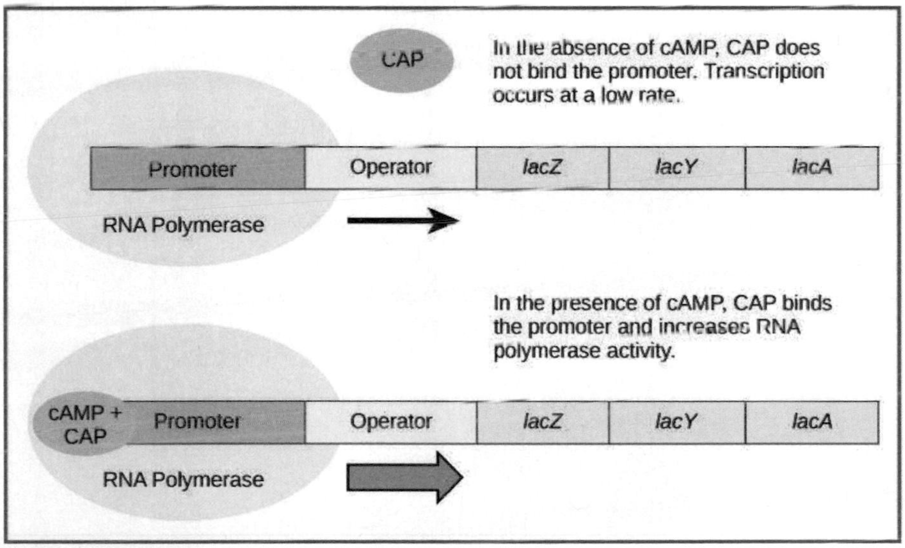

Figure 16.4 Transcriptional activation by the CAP protein. When glucose levels fall, *E. coli* may use other sugars for fuel but must transcribe new genes to do so. As glucose supplies become limited, cAMP levels increase. This cAMP binds to the CAP protein, a positive regulator that binds to a promoter region upstream of the genes required to use other sugar sources.

The *lac* Operon: An Inducible Operon

The third type of gene regulation in prokaryotic cells occurs through *inducible operons*, which have proteins that bind to activate or repress transcription depending on the local environment and the needs of the cell. The *lac* operon is a typical inducible operon. As mentioned previously, *E. coli* is able to use other sugars as energy sources when glucose concentrations are low. One such sugar source is lactose. The ***lac* operon** encodes the genes necessary to acquire and process the lactose from the local environment. The Z gene of the *lac* operon encodes beta-galactosidase, which breaks lactose down to glucose and galactose.

However, for the *lac* operon to be activated, two conditions must be met. First, the level of glucose must be very low or non-existent. Second, lactose must be present. Only when glucose is absent and lactose is present will the *lac* operon be transcribed (Figure 16.5). In the absence of glucose, the binding of the CAP protein makes transcription of the *lac* operon more effective. When lactose is present, its metabolite, allolactose, binds to the *lac* repressor and changes its shape so that it cannot bind to the *lac* operator to prevent transcription. This combination of conditions makes sense for the cell, because it would be energetically wasteful to synthesize the enzymes to process lactose if glucose was plentiful or lactose was not available. It should be mentioned that the lac operon is transcribed at a very low rate even when glucose is present and lactose absent.

VISUAL CONNECTION

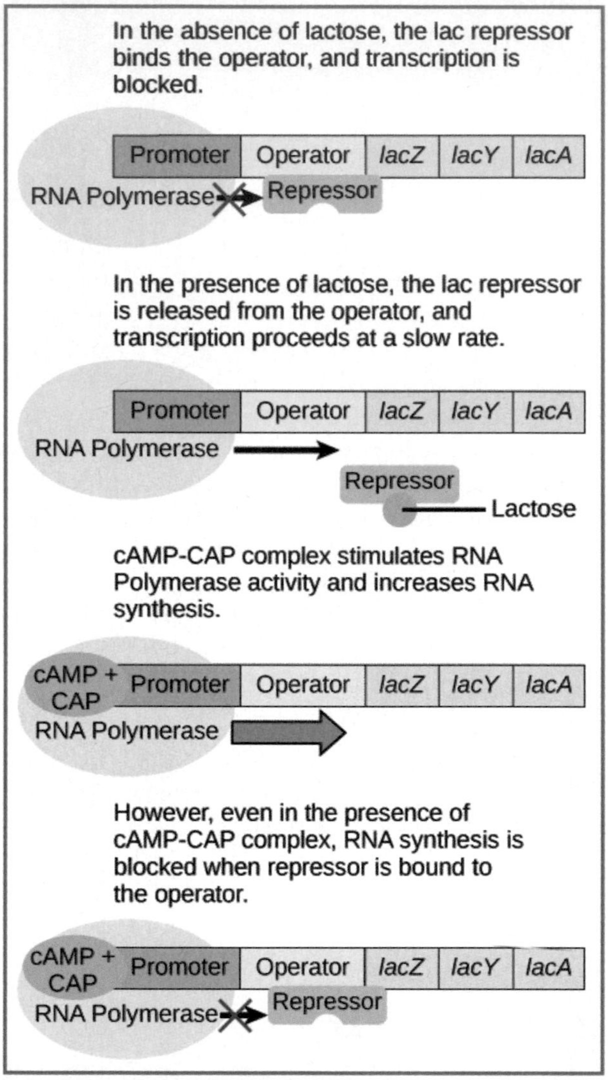

Figure 16.5 Regulation of the *lac* operon. Transcription of the *lac* operon is carefully regulated so that its expression only occurs when glucose is limited and lactose is present to serve as an alternative fuel source.

In *E. coli*, the *trp* operon is on by default, while the *lac* operon is off. Why do you think this is the case?

If glucose is present, then CAP fails to bind to the promoter sequence to activate transcription. If lactose is absent, then the repressor binds to the operator to prevent transcription. If either of these conditions is met, then transcription remains off. Only when glucose is absent and lactose is present is the *lac* operon transcribed (Table 16.2).

Signals that Induce or Repress Transcription of the *lac* Operon

Glucose	CAP binds	Lactose	Repressor binds	Transcription
+	-	-	+	No
+	-	+	-	Some
-	+	-	+	No
-	+	+	-	Yes

Table 16.2

LINK TO LEARNING

Watch an animated tutorial about the workings of *lac* operon here.

Click to view content (https://www.openstax.org/l/lac_operon)

16.3 Eukaryotic Epigenetic Gene Regulation

By the end of this section, you will be able to do the following:

- Explain how chromatin remodeling controls transcriptional access
- Describe how access to DNA is controlled by histone modification
- Describe how DNA methylation is related to epigenetic gene changes

Eukaryotic gene expression is more complex than prokaryotic gene expression because the processes of transcription and translation are physically separated. Unlike prokaryotic cells, eukaryotic cells can regulate gene expression at many different levels. Epigenetic changes are inheritable changes in gene expression that do not result from changes in the DNA sequence. Eukaryotic gene expression begins with control of access to the DNA. Transcriptional access to the DNA can be controlled in two general ways: chromatin remodeling and DNA methylation. Chromatin remodeling changes the way that DNA is associated with chromosomal histones. DNA methylation is associated with developmental changes and gene silencing.

Epigenetic Control: Regulating Access to Genes within the Chromosome

The human genome encodes over 20,000 genes, with hundreds to thousands of genes on each of the 23 human chromosomes. The DNA in the nucleus is precisely wound, folded, and compacted into chromosomes so that it will fit into the nucleus. It is also organized so that specific segments can be accessed as needed by a specific cell type.

The first level of organization, or **packing**, is the winding of DNA strands around histone proteins. Histones package and order DNA into structural units called nucleosome complexes, which can control the access of proteins to the DNA regions (Figure 16.6a). Under the electron microscope, this winding of DNA around histone proteins to form nucleosomes looks like small beads on a string (Figure 16.6b).

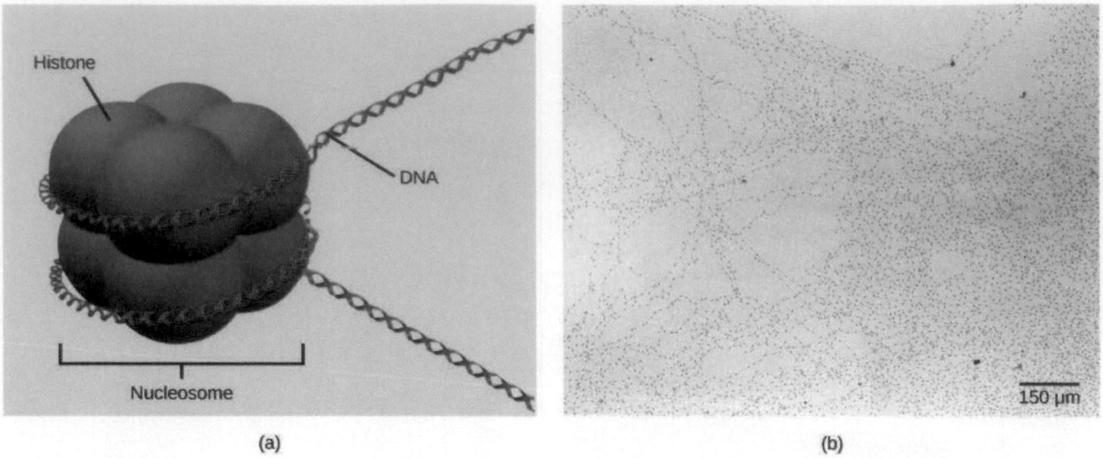

Figure 16.6 DNA is folded around histone proteins to create (a) nucleosome complexes. These nucleosomes control the access of proteins to the underlying DNA. When viewed through an electron microscope (b), the nucleosomes look like beads on a string. (credit "micrograph": modification of work by Chris Woodcock)

These beads (histone proteins) can move along the string (DNA) to expose different sections of the molecule. If DNA encoding a specific gene is to be transcribed into RNA, the nucleosomes surrounding that region of DNA can slide down the DNA to open that specific chromosomal region and allow for the transcriptional machinery (RNA polymerase) to initiate transcription (Figure 16.7).

VISUAL CONNECTION

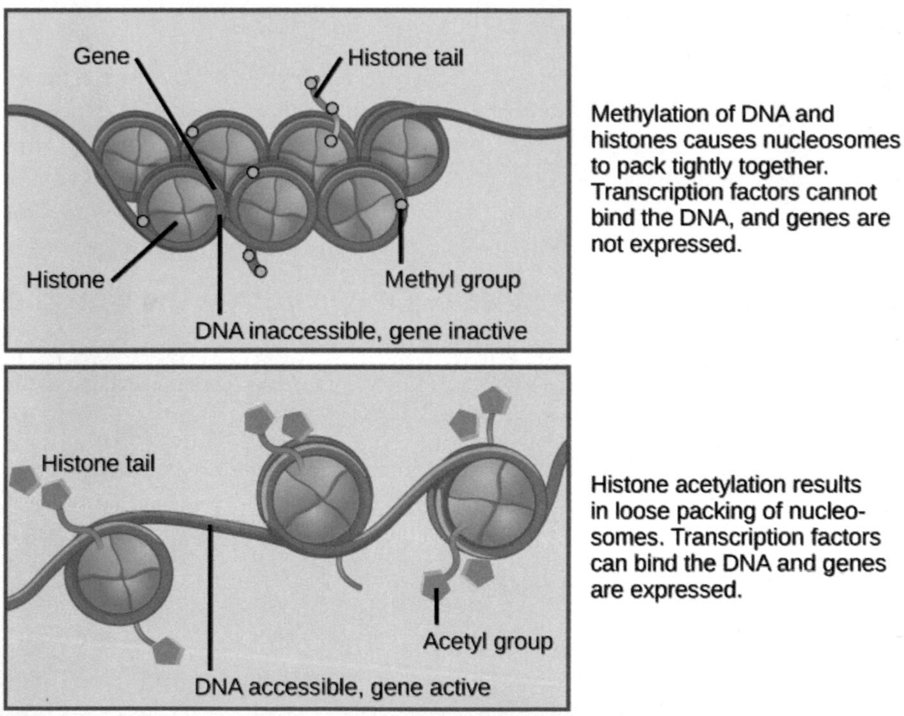

Figure 16.7 Nucleosomes can slide along DNA. When nucleosomes are spaced closely together (top), transcription factors cannot bind and gene expression is turned off. When the nucleosomes are spaced far apart (bottom), the DNA is exposed. Transcription factors can bind, allowing gene expression to occur. Modifications to the histones and DNA affect nucleosome spacing.

In females, one of the two X chromosomes is inactivated during embryonic development because of epigenetic changes to the chromatin. What impact do you think these changes would have on nucleosome packing?

How closely the histone proteins associate with the DNA is regulated by signals found on both the histone proteins and on the DNA. These signals are functional groups added to histone proteins or to DNA and determine whether a chromosomal region should be open or closed (Figure 16.8 depicts modifications to histone proteins and DNA). These tags are not permanent, but may be added or removed as needed. Some chemical groups (phosphate, methyl, or acetyl groups) are attached to specific amino acids in histone "tails" at the N-terminus of the protein. These groups do not alter the DNA base sequence, but they do alter how tightly wound the DNA is around the histone proteins. DNA is a negatively charged molecule and unmodified histones are positively charged; therefore, changes in the charge of the histone will change how tightly wound the DNA molecule will be. By adding chemical modifications like acetyl groups, the charge becomes less positive, and the binding of DNA to the histones is relaxed. Altering the location of nucleosomes and the tightness of histone binding opens some regions of chromatin to transcription and closes others.

The DNA molecule itself can also be modified by methylation. DNA methylation occurs within very specific regions called CpG islands. These are stretches with a high frequency of cytosine and guanine dinucleotide DNA pairs (CG) found in the promoter regions of genes. The cytosine member of the CG pair can be methylated (a methyl group is added). Methylated genes are usually silenced, although methylation may have other regulatory effects. In some cases, genes that are silenced during the development of the gametes of one parent are transmitted in their silenced condition to the offspring. Such genes are said to be imprinted. Parental diet or other environmental conditions may also affect the methylation patterns of genes, which in turn modifies gene expression. Changes in chromatin organization interact with DNA methylation. DNA methyltransferases appear to be attracted to chromatin regions with specific histone modifications. Highly methylated (*hypermethylated*) DNA regions with deacetylated histones are tightly coiled and transcriptionally inactive.

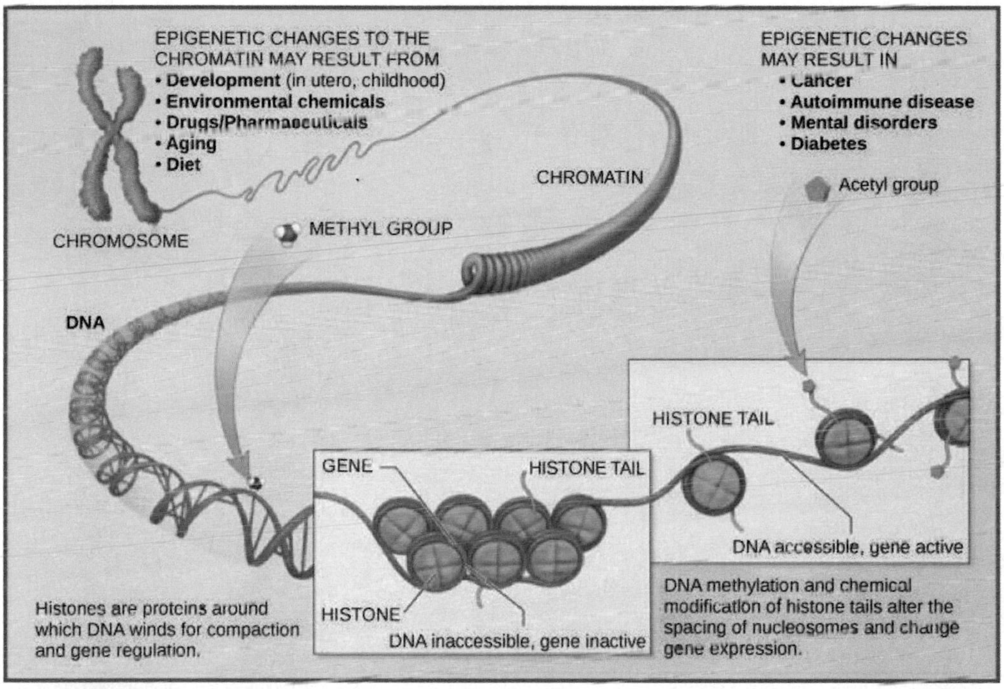

Figure 16.8 Histone proteins and DNA nucleotides can be modified chemically. Modifications affect nucleosome spacing and gene expression. (credit: modification of work by NIH)

Epigenetic changes are not permanent, although they often persist through multiple rounds of cell division and may even cross generational lines. Chromatin remodeling alters the chromosomal structure (open or closed) as needed. If a gene is to be transcribed, the histone proteins and DNA in the chromosomal region encoding that gene are modified in a way that opens the promoter region to allow RNA polymerase and other proteins, called **transcription factors**, to bind and initiate transcription. If a gene is to remain turned off, or silenced, the histone proteins and DNA have different modifications that signal a closed chromosomal configuration. In this closed configuration, the RNA polymerase and transcription factors do not have access to the DNA and transcription cannot occur (Figure 16.8).

LINK TO LEARNING

View this video that describes how epigenetic regulation controls gene expression.

Click to view content (https://www.openstax.org/l/epigenetic_reg)

16.4 Eukaryotic Transcription Gene Regulation

By the end of this section, you will be able to do the following:

- Discuss the role of transcription factors in gene regulation
- Explain how enhancers and repressors regulate gene expression

Like prokaryotic cells, the transcription of genes in eukaryotes requires the action of an RNA polymerase to bind to a DNA sequence upstream of a gene in order to initiate transcription. However, unlike prokaryotic cells, the eukaryotic RNA polymerase requires other proteins, or transcription factors, to facilitate transcription initiation. RNA polymerase by itself cannot initiate transcription in eukaryotic cells. There are two types of transcription factors that regulate eukaryotic transcription: *General (or basal) transcription factors* bind to the core promoter region to assist with the binding of RNA polymerase. *Specific transcription factors* bind to various regions outside of the core promoter region and interact with the proteins at the core promoter to enhance or repress the activity of the polymerase.

LINK TO LEARNING

View the process of transcription—the making of RNA from a DNA template.

Click to view content (https://www.openstax.org/l/transcript_RNA)

The Promoter and the Transcription Machinery

Genes are organized to make the control of gene expression easier. The promoter region is immediately upstream of the coding sequence. This region can be short (only a few nucleotides in length) or quite long (hundreds of nucleotides long). The longer the promoter, the more available space for proteins to bind. This also adds more control to the transcription process. The length of the promoter is gene-specific and can differ dramatically between genes. Consequently, the level of control of gene expression can also differ quite dramatically between genes. The purpose of the **promoter** is to bind transcription factors that control the initiation of transcription.

Within the core promoter region, 25 to 35 bases upstream of the transcriptional start site, resides the TATA box. The TATA box has the consensus sequence of 5'-TATAAA-3'. The TATA box is the binding site for a protein complex called TFIID, which contains a TATA-binding protein. Binding of TFIID recruits other transcription factors, including TFIIB, TFIIE, TFIIF, and TFIIH. Some of these transcription factors help to bind the RNA polymerase to the promoter, and others help to activate the transcription initiation complex.

In addition to the TATA box, other binding sites are found in some promoters. Some biologists prefer to restrict the range of the eukaryotic promoter to the core promoter, or polymerase binding site, and refer to these additional sites as promoter-proximal elements, because they are usually found within a few hundred base pairs upstream of the transcriptional start site. Examples of these elements are the CAAT box, with the consensus sequence 5'-CCAAT-3' and the GC box, with the consensus sequence 5'-GGGCGG-3'. Specific transcription factors can bind to these promoter-proximal elements to regulate gene transcription. A given gene may have its own combination of these specific transcription-factor binding sites. There are hundreds of transcription factors in a cell, each of which binds specifically to a particular DNA sequence motif. When transcription factors bind to the promoter just upstream of the encoded gene, it is referred to as a **cis-acting element**, because it is on the same chromosome just next to the gene. Transcription factors respond to environmental stimuli that cause the proteins to find their binding sites and initiate transcription of the gene that is needed.

Enhancers and Transcription

In some eukaryotic genes, there are additional regions that help increase or enhance transcription. These regions, called **enhancers**, are not necessarily close to the genes they enhance. They can be located upstream of a gene, within the coding region of the gene, downstream of a gene, or may be thousands of nucleotides away.

Enhancer regions are binding sequences, or sites, for specific transcription factors. When a protein transcription factor binds to its enhancer sequence, the shape of the protein changes, allowing it to interact with proteins at the promoter site. However, since the enhancer region may be distant from the promoter, the DNA must bend to allow the proteins at the two sites to come into contact. DNA bending proteins help to bend the DNA and bring the enhancer and promoter regions together (Figure 16.9). This shape change allows for the interaction of the specific activator proteins bound to the enhancers with the general transcription factors bound to the promoter region and the RNA polymerase.

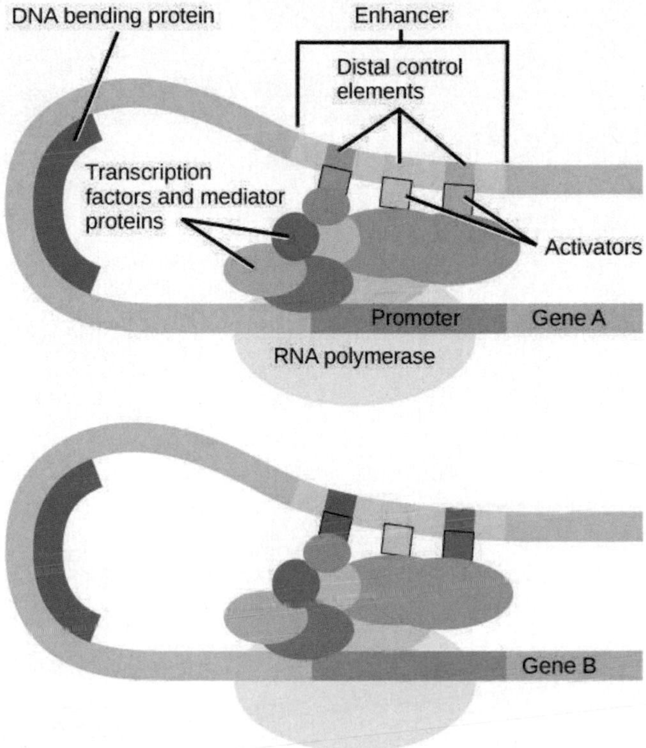

Figure 16.9 Interaction between proteins at the promoter and enhancer sites. An enhancer is a DNA sequence that promotes transcription. Each enhancer is made up of short DNA sequences called distal control elements. Activators bound to the distal control elements interact with mediator proteins and transcription factors. Two different genes may have the same promoter but different distal control elements, enabling differential gene expression.

Turning Genes Off: Transcriptional Repressors

Like prokaryotic cells, eukaryotic cells also have mechanisms to prevent transcription. Transcriptional repressors can bind to promoter or enhancer regions and block transcription. Like the transcriptional activators, repressors respond to external stimuli to prevent the binding of activating transcription factors.

16.5 Eukaryotic Post-transcriptional Gene Regulation

By the end of this section, you will be able to do the following:

- Understand RNA splicing and explain its role in regulating gene expression
- Describe the importance of RNA stability in gene regulation

RNA is transcribed, but must be processed into a mature form before translation can begin. This processing that takes place after an RNA molecule has been transcribed, but before it is translated into a protein, is called *post-transcriptional modification*. As with the epigenetic and transcriptional stages of processing, this post-transcriptional step can also be regulated to control gene expression in the cell. If the RNA is not processed, shuttled, or translated, then no protein will be synthesized.

RNA Splicing, the First Stage of Post-transcriptional Control

In eukaryotic cells, the RNA transcript often contains regions, called introns, that are removed prior to translation. The regions of RNA that code for protein are called **exons**. (Figure 16.10). After an RNA molecule has been transcribed, but prior to its departure from the nucleus to be translated, the RNA is processed and the introns are removed by splicing. Splicing is done by spliceosomes, ribonucleoprotein complexes that can recognize the two ends of the intron, cut the transcript at those two points, and bring the exons together for ligation.

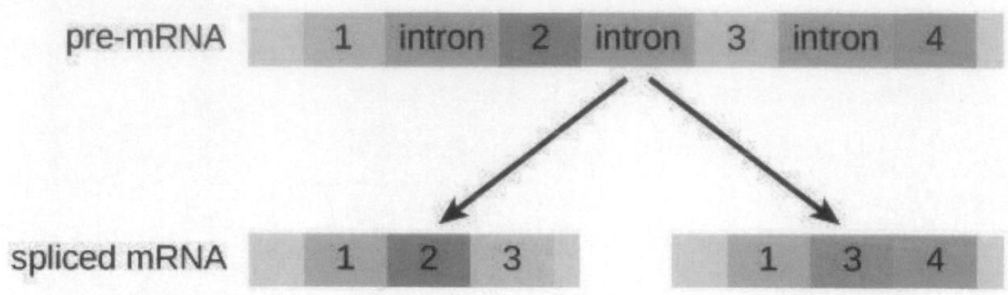

Figure 16.10 Pre-mRNA can be alternatively spliced to create different proteins.

EVOLUTION CONNECTION

Alternative RNA Splicing

In the 1970s, genes were first observed that exhibited alternative RNA splicing. Alternative RNA splicing is a mechanism that allows different protein products to be produced from one gene when different combinations of exons are combined to form the mRNA (Figure 16.11). This alternative splicing can be haphazard, but more often it is controlled and acts as a mechanism of *gene regulation*, with the frequency of different splicing alternatives controlled by the cell as a way to control the production of different protein products in different cells or at different stages of development. Alternative splicing is now understood to be a common mechanism of gene regulation in eukaryotes; according to one estimate, 70 percent of genes in humans are expressed as multiple proteins through alternative splicing. Although there are multiple ways to alternatively splice RNA transcripts, the original 5'-3' order of the exons is *always conserved*. That is, a transcript with exons 1 2 3 4 5 6 7 might be spliced 1 2 4 5 6 7 or 1 2 3 6 7, but never 1 2 5 4 3 6 7.

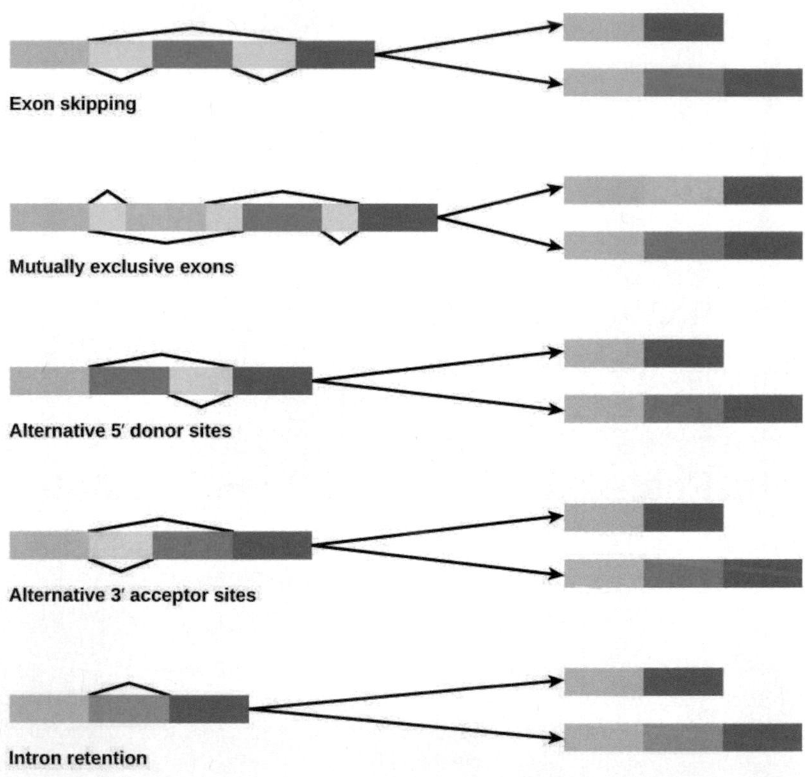

Figure 16.11 There are five basic modes of alternative splicing.

How could alternative splicing evolve? Introns have a beginning- and ending-recognition sequence; it is easy to imagine the failure of the splicing mechanism to identify the end of an intron and instead find the end of the next intron, thus removing two introns and the intervening exon. In fact, there are mechanisms in place to prevent such intron skipping, but mutations are likely to lead to their failure. Such "mistakes" would more than likely produce a nonfunctional protein. Indeed, the cause of many genetic diseases is abnormal splicing rather than mutations in a coding sequence. However, alternative splicing could possibly create a protein variant without the loss of the original protein, opening up possibilities for adaptation of the new variant to new functions. Gene duplication has played an important role in the evolution of new functions in a similar way by providing genes that may evolve without eliminating the original, functional protein.

Question: In the corn snake *Pantherophis guttatus*, there are several different color variants, including amelanistic snakes whose skin patterns display only red and yellow pigments. The cause of amelanism in these snakes was recently identified as the insertion of a transposable element into an intron in the OCA2 (oculocutaneous albinism) gene. How might the insertion of extra genetic material into an intron lead to a nonfunctional protein?

LINK TO LEARNING

Visualize how mRNA splicing happens by watching the process in action in this video.

Click to view content (https://www.openstax.org/l/mRNA_splicing)

Control of RNA Stability

Before the mRNA leaves the nucleus, it is given two protective "caps" that prevent the ends of the strand from degrading during its journey. 5' and 3' exonucleases can degrade unprotected RNAs. The **5' cap**, which is placed on the 5' end of the mRNA, is usually composed of a methylated guanosine triphosphate molecule (GTP). The GTP is placed "backward" on the 5' end of the mRNA, so that the 5' carbons of the GTP and the terminal nucleotide are linked through three phosphates. The **poly-A tail**, which is attached to the 3' end, is usually composed of a long chain of adenine nucleotides. These changes protect the two ends of the RNA from exonuclease attack.

Once the RNA is transported to the cytoplasm, the length of time that the RNA resides there can be controlled. Each RNA molecule has a defined lifespan and decays at a specific rate. This rate of decay can influence how much protein is in the cell. If the decay rate is increased, the RNA will not exist in the cytoplasm as long, shortening the time available for translation of the mRNA to occur. Conversely, if the rate of decay is decreased, the mRNA molecule will reside in the cytoplasm longer and more protein can be translated. This rate of decay is referred to as the RNA stability. If the RNA is stable, it will be detected for longer periods of time in the cytoplasm.

Binding of proteins to the RNA can also influence its stability. Proteins called **RNA-binding proteins**, or RBPs, can bind to the regions of the mRNA just upstream or downstream of the protein-coding region. These regions in the RNA that are not translated into protein are called the **untranslated regions**, or UTRs. They are not introns (those have been removed in the nucleus). Rather, these are regions that regulate mRNA localization, stability, and protein translation. The region just before the protein-coding region is called the **5' UTR**, whereas the region after the coding region is called the **3' UTR** (Figure 16.12). The binding of RBPs to these regions can increase or decrease the stability of an RNA molecule, depending on the specific RBP that binds.

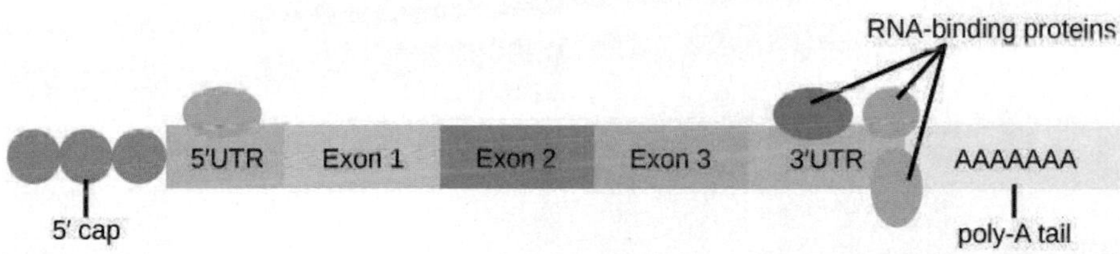

Figure 16.12 RNA-binding proteins. The protein-coding region of this processed mRNA is flanked by 5' and 3' untranslated regions (UTRs). The presence of RNA-binding proteins at the 5' or 3' UTR influences the stability of the RNA molecule.

RNA Stability and microRNAs

In addition to RBPs that bind to and control (increase or decrease) RNA stability, other elements called microRNAs can bind to the RNA molecule. These **microRNAs**, or miRNAs, are short RNA molecules that are only 21 to 24 nucleotides in length. The miRNAs are made in the nucleus as longer pre-miRNAs. These pre-miRNAs are chopped into mature miRNAs by a protein called ***Dicer***. Like transcription factors and RBPs, mature miRNAs recognize a specific sequence and bind to the RNA; however, miRNAs also associate with a ribonucleoprotein complex called the **RNA-induced silencing complex (RISC)**. The RNA component of the RISC base-pairs with complementary sequences on an mRNA and either impede translation of the message or lead to the degradation of the mRNA.

16.6 Eukaryotic Translational and Post-translational Gene Regulation

By the end of this section, you will be able to do the following:

- Understand the process of translation and discuss its key factors
- Describe how the initiation complex controls translation
- Explain the different ways in which the post-translational control of gene expression takes place

After RNA has been transported to the cytoplasm, it is translated into protein. Control of this process is largely dependent on the RNA molecule. As previously discussed, the stability of the RNA will have a large impact on its translation into a protein. As the stability changes, the amount of time that it is available for translation also changes.

The Initiation Complex and Translation Rate

Like transcription, translation is controlled by proteins that bind and initiate the process. In translation, the complex that assembles to start the process is referred to as the translation **initiation complex**. In eukaryotes, translation is initiated by binding the initiating met-tRNAi to the 40S ribosome. This tRNA is brought to the 40S ribosome by a protein initiation factor, **eukaryotic initiation factor-2 (eIF-2)**. The eIF-2 protein binds to the high-energy molecule **guanosine triphosphate (GTP)**. The tRNA-eIF2-GTP complex then binds to the 40S ribosome. A second complex forms on the mRNA. Several different initiation factors recognize the 5' cap of the mRNA and proteins bound to the poly-A tail of the same mRNA, forming the mRNA into a loop. The cap-binding protein eIF4F brings the mRNA complex together with the 40S ribosome complex. The ribosome then scans along the mRNA until it finds a start codon AUG. When the anticodon of the initiator tRNA and the start codon are aligned, the GTP is hydrolyzed, the initiation factors are released, and the large **60S ribosomal subunit** binds to form the translation complex. The binding of eIF-2 to the RNA is controlled by phosphorylation. If eIF-2 is phosphorylated, it undergoes a conformational change and cannot bind to GTP. Therefore, the initiation complex cannot form properly and translation is impeded (Figure 16.13). When eIF-2 remains unphosphorylated, the initiation complex can form normally and translation can proceed.

VISUAL CONNECTION

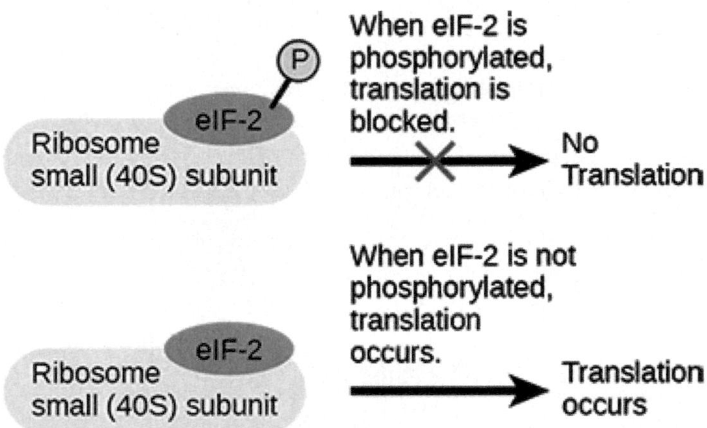

Figure 16.13 Gene expression can be controlled by factors that bind the translation initiation complex.

An increase in phosphorylation levels of eIF-2 has been observed in patients with neurodegenerative diseases such as Alzheimer's, Parkinson's, and Huntington's. What impact do you think this might have on protein synthesis?

Chemical Modifications, Protein Activity, and Longevity

Proteins can be chemically modified with the addition of groups including methyl, phosphate, acetyl, and ubiquitin groups. The addition or removal of these groups from proteins regulates their activity or the length of time they exist in the cell. Sometimes these modifications can regulate where a protein is found in the cell—for example, in the nucleus, in the cytoplasm, or attached to the plasma membrane.

Chemical modifications occur in response to external stimuli such as stress, the lack of nutrients, heat, or ultraviolet light exposure. These changes can alter epigenetic accessibility, transcription, mRNA stability, or translation—all resulting in changes in expression of various genes. This is an efficient way for the cell to rapidly change the levels of specific proteins in response to the environment. Because proteins are involved in every stage of gene regulation, the phosphorylation of a protein (depending on the protein that is modified) can alter accessibility to the chromosome, can alter translation (by altering transcription factor binding or function), can change nuclear shuttling (by influencing modifications to the nuclear pore complex), can alter RNA stability (by binding or not binding to the RNA to regulate its stability), can modify translation (increase or decrease), or can change post-translational modifications (add or remove phosphates or other chemical modifications).

The addition of an ubiquitin group to a protein marks that protein for degradation. Ubiquitin acts like a flag indicating that the protein lifespan is complete. These proteins are moved to the **proteasome**, an organelle that functions to remove proteins, to be degraded (Figure 16.14). One way to control gene expression, therefore, is to alter the longevity of the protein.

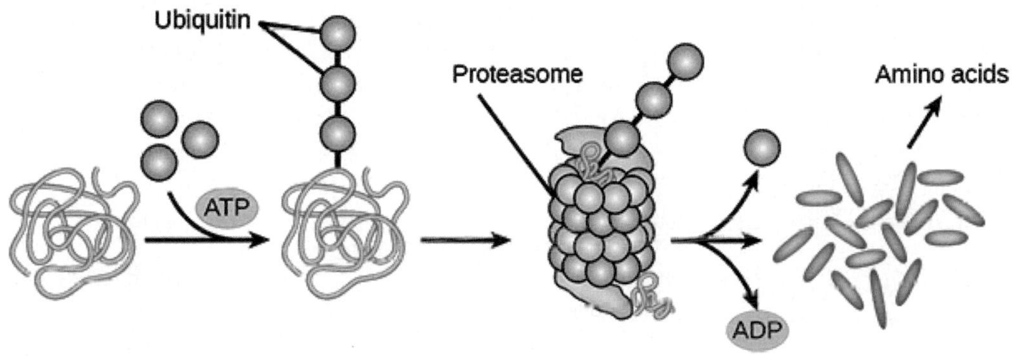

Figure 16.14 Proteins with ubiquitin tags are marked for degradation within the proteasome.

16.7 Cancer and Gene Regulation

By the end of this section, you will be able to do the following:

- Describe how changes to gene expression can cause cancer
- Explain how changes to gene expression at different levels can disrupt the cell cycle
- Discuss how understanding regulation of gene expression can lead to better drug design

Cancer is not a single disease but includes many different diseases. In cancer cells, mutations modify cell-cycle control and cells don't stop growing as they normally would. Mutations can also alter the growth rate or the progression of the cell through the cell cycle. One example of a gene modification that alters the growth rate is increased phosphorylation of cyclin B, a protein that controls the progression of a cell through the cell cycle and serves as a cell-cycle checkpoint protein.

For cells to move through each phase of the cell cycle, the cell must pass through checkpoints. This ensures that the cell has properly completed the step and has not encountered any mutation that will alter its function. Many proteins, including cyclin B, control these checkpoints. The phosphorylation of cyclin B, a post-translational event, alters its function. As a result, cells can progress through the cell cycle unimpeded, even if mutations exist in the cell and its growth should be terminated. This post-translational change of cyclin B prevents it from controlling the cell cycle and contributes to the development of cancer.

Cancer: Disease of Altered Gene Expression

Cancer can be described as a disease of altered gene expression. There are many proteins that are turned on or off (gene activation or gene silencing) that dramatically alter the overall activity of the cell. A gene that is not normally expressed in that cell can be switched on and expressed at high levels. This can be the result of gene mutation or changes in gene regulation (epigenetic, transcription, post-transcription, translation, or post-translation).

Changes in epigenetic regulation, transcription, RNA stability, protein translation, and post-translational control can be detected in cancer. While these changes don't occur simultaneously in one cancer, changes at each of these levels can be detected when observing cancer at different sites in different individuals. Therefore, changes in **histone acetylation** (epigenetic modification that leads to gene silencing), activation of transcription factors by phosphorylation, increased RNA stability, increased translational control, and protein modification can all be detected at some point in various cancer cells. Scientists are working to understand the common changes that give rise to certain types of cancer or how a modification might be exploited to destroy a tumor cell.

Tumor Suppressor Genes, Oncogenes, and Cancer

In normal cells, some genes function to prevent excess, inappropriate cell growth. These are tumor-suppressor genes, which are active in normal cells to prevent uncontrolled cell growth. There are many tumor-suppressor genes in cells. The most studied tumor-suppressor gene is p53, which is mutated in over 50 percent of all cancer types. The p53 protein itself functions as a transcription factor. It can bind to sites in the promoters of genes to initiate transcription. Therefore, the mutation of p53 in cancer will dramatically alter the transcriptional activity of its target genes.

LINK TO LEARNING

Watch this animation (http://openstax.org/l/p53_cancer) to learn more about the use of p53 in fighting cancer.

Proto-oncogenes are positive cell-cycle regulators. When mutated, proto-oncogenes can become oncogenes and cause cancer. Overexpression of the oncogene can lead to uncontrolled cell growth. This is because oncogenes can alter transcriptional activity, stability, or protein translation of another gene that directly or indirectly controls cell growth. An example of an oncogene involved in cancer is a protein called myc. **Myc** is a transcription factor that is aberrantly activated in Burkett's Lymphoma, a cancer of the lymph system. Overexpression of myc transforms normal B cells into cancerous cells that continue to grow uncontrollably. High B-cell numbers can result in tumors that can interfere with normal bodily function. Patients with Burkett's lymphoma can develop tumors on their jaw or in their mouth that interfere with the ability to eat.

Cancer and Epigenetic Alterations

Silencing genes through epigenetic mechanisms is also very common in cancer cells. There are characteristic modifications to histone proteins and DNA that are associated with silenced genes. In cancer cells, the DNA in the promoter region of silenced genes is methylated on cytosine DNA residues in CpG islands. Histone proteins that surround that region lack the acetylation modification that is present when the genes are expressed in normal cells. This combination of DNA methylation and histone deacetylation (epigenetic modifications that lead to gene silencing) is commonly found in cancer. When these modifications occur, the gene present in that chromosomal region is silenced. Increasingly, scientists understand how epigenetic changes are altered in cancer. Because these changes are temporary and can be reversed—for example, by preventing the action of the histone deacetylase protein that removes acetyl groups, or by DNA methyl transferase enzymes that add methyl groups to cytosines in DNA—it is possible to design new drugs and new therapies to take advantage of the reversible nature of these processes. Indeed, many researchers are testing how a silenced gene can be switched back on in a cancer cell to help re-establish normal growth patterns.

Genes involved in the development of many other illnesses, ranging from allergies to inflammation to autism, are thought to be regulated by epigenetic mechanisms. As our knowledge of how genes are controlled deepens, new ways to treat diseases like cancer will emerge.

Cancer and Transcriptional Control

Alterations in cells that give rise to cancer can affect the transcriptional control of gene expression. Mutations that activate transcription factors, such as increased phosphorylation, can increase the binding of a transcription factor to its binding site in a promoter. This could lead to increased transcriptional activation of that gene that results in modified cell growth. Alternatively, a mutation in the DNA of a promoter or enhancer region can increase the binding ability of a transcription factor. This could also lead to the increased transcription and aberrant gene expression that is seen in cancer cells.

Researchers have been investigating how to control the transcriptional activation of gene expression in cancer. Identifying how a transcription factor binds, or a pathway that activates where a gene can be turned off, has led to new drugs and new ways to treat cancer. In breast cancer, for example, many proteins are overexpressed. This can lead to increased phosphorylation of key transcription factors that increase transcription. One such example is the overexpression of the epidermal growth-factor receptor (EGFR) in a subset of breast cancers. The EGFR pathway activates many protein kinases that, in turn, activate many transcription factors which control genes involved in cell growth. New drugs that prevent the activation of EGFR have been developed and are used to treat these cancers.

Cancer and Post-transcriptional Control

Changes in the post-transcriptional control of a gene can also result in cancer. Recently, several groups of researchers have shown that specific cancers have altered expression of miRNAs. Because miRNAs bind to the 3' UTR of RNA molecules to degrade them, overexpression of these miRNAs could be detrimental to normal cellular activity. Too many miRNAs could dramatically decrease the RNA population, leading to a decrease in protein expression. Several studies have demonstrated a change in the miRNA population in specific cancer types. It appears that the subset of miRNAs expressed in breast cancer cells is quite different from the subset expressed in lung cancer cells or even from normal breast cells. This suggests that alterations in miRNA activity can contribute to the growth of breast cancer cells. These types of studies also suggest that if some miRNAs are specifically expressed only in cancer cells, they could be potential drug targets. It would, therefore, be conceivable that new drugs that turn off miRNA expression in cancer could be an effective method to treat cancer.

Cancer and Translational/Post-translational Control

There are many examples of how translational or post-translational modifications of proteins arise in cancer. Modifications are found in cancer cells from the increased translation of a protein to changes in protein phosphorylation to alternative splice variants of a protein. An example of how the expression of an alternative form of a protein can have dramatically different outcomes is seen in colon cancer cells. The c-Flip protein, a protein involved in mediating the cell-death pathway, comes in two forms: long (c-FLIPL) and short (c-FLIPS). Both forms appear to be involved in initiating controlled cell-death mechanisms in normal cells. However, in colon cancer cells, expression of the long form results in increased cell growth instead of cell death. Clearly, the expression of the wrong protein dramatically alters cell function and contributes to the development of cancer.

New Drugs to Combat Cancer: Targeted Therapies

Scientists are using what is known about the regulation of gene expression in disease states, including cancer, to develop new ways to treat and prevent disease development. Many scientists are designing drugs on the basis of the gene expression patterns within individual tumors. This idea, that therapy and medicines can be tailored to an individual, has given rise to the field of personalized medicine. With an increased understanding of gene regulation and gene function, medicines can be designed to specifically target diseased cells without harming healthy cells. Some new medicines, called targeted therapies, have exploited the overexpression of a specific protein or the mutation of a gene to develop a new medication to treat disease. One such example is the use of anti-EGF receptor medications to treat the subset of breast cancer tumors that have very high levels of the EGF protein. Undoubtedly, more targeted therapies will be developed as scientists learn more about how gene expression changes can cause cancer.

CAREER CONNECTION

Clinical Trial Coordinator

A clinical trial coordinator is the person managing the proceedings of the clinical trial. This job includes coordinating patient schedules and appointments, maintaining detailed notes, building the database to track patients (especially for long-term follow-up studies), ensuring proper documentation has been acquired and accepted, and working with the nurses and doctors to facilitate the trial and publication of the results. A clinical trial coordinator may have a science background, like a nursing degree, or other certification. People who have worked in science labs or in clinical offices are also qualified to become a clinical trial coordinator. These jobs are generally in hospitals; however, some clinics and doctor's offices also conduct clinical trials and may hire a coordinator.

KEY TERMS

3' UTR 3' untranslated region; region just downstream of the protein-coding region in an RNA molecule that is not translated

5' cap a methylated guanosine triphosphate (GTP) molecule that is attached to the 5' end of a messenger RNA to protect the end from degradation

5' UTR 5' untranslated region; region just upstream of the protein-coding region in an RNA molecule that is not translated

activator protein that binds to prokaryotic operators to increase transcription

catabolite activator protein (CAP) protein that complexes with cAMP to bind to the promoter sequences of operons which control sugar processing when glucose is not available

***cis*-acting element** transcription factor binding sites within the promoter that regulate the transcription of a gene adjacent to it

Dicer enzyme that chops the pre-miRNA into the mature form of the miRNA

DNA methylation epigenetic modification that leads to gene silencing; a process involving adding a methyl group to the DNA molecule

enhancer segment of DNA that is upstream, downstream, perhaps thousands of nucleotides away, or on another chromosome that influence the transcription of a specific gene

epigenetic heritable changes that do not involve changes in the DNA sequence

eukaryotic initiation factor-2 (eIF-2) protein that binds first to an mRNA to initiate translation

gene expression processes that control the turning on or turning off of a gene

guanine diphosphate (GDP) molecule that is left after the energy is used to start translation

guanine triphosphate (GTP) energy-providing molecule that binds to eIF-2 and is needed for translation

histone acetylation epigenetic modification that leads to gene silencing; a process involving adding or removing an acetyl functional group

inducible operon operon that can be activated or repressed depending on cellular needs and the surrounding environment

initiation complex protein complex containing eIF-2 that starts translation

***lac* operon** operon in prokaryotic cells that encodes genes required for processing and intake of lactose

large 60S ribosomal subunit second, larger ribosomal subunit that binds to the RNA to translate it into protein

microRNA (miRNA) small RNA molecules (approximately 21 nucleotides in length) that bind to RNA molecules to degrade them

myc oncogene that causes cancer in many cancer cells

negative regulator protein that prevents transcription

operator region of DNA outside of the promoter region that binds activators or repressors that control gene expression in prokaryotic cells

operon collection of genes involved in a pathway that are transcribed together as a single mRNA in prokaryotic cells

poly-A tail a series of adenine nucleotides that are attached to the 3' end of an mRNA to protect the end from degradation

positive regulator protein that increases transcription

post-transcriptional control of gene expression after the RNA molecule has been created but before it is translated into protein

post-translational control of gene expression after a protein has been created

proteasome organelle that degrades proteins

repressor protein that binds to the operator of prokaryotic genes to prevent transcription

RISC protein complex that binds along with the miRNA to the RNA to degrade it

RNA stability how long an RNA molecule will remain intact in the cytoplasm

RNA-binding protein (RBP) protein that binds to the 3' or 5' UTR to increase or decrease the RNA stability

small 40S ribosomal subunit ribosomal subunit that binds to the RNA to translate it into protein

***trans*-acting element** transcription factor binding site found outside the promoter or on another chromosome that influences the transcription of a particular gene

transcription factor protein that binds to the DNA at the promoter or enhancer region and that influences transcription of a gene

transcription factor binding site sequence of DNA to which a transcription factor binds

transcriptional start site site at which transcription begins

***trp* operon** series of genes necessary to synthesize tryptophan in prokaryotic cells

tryptophan amino acid that can be synthesized by prokaryotic cells when necessary

untranslated region segment of the RNA molecule that is not translated into protein. These regions lie before (upstream or 5') and after (downstream or 3') the protein-coding region

CHAPTER SUMMARY

16.1 Regulation of Gene Expression

While all somatic cells within an organism contain the same DNA, not all cells within that organism express the same proteins. Prokaryotic organisms express most of their genes most of the time. However, some genes are expressed only when they are needed. Eukaryotic organisms, on the other hand, express only a subset of their genes in any given cell. To express a protein, the DNA is first transcribed into RNA, which is then translated into proteins, which are then targeted to specific cellular locations. In prokaryotic cells, transcription and translation occur almost simultaneously. In eukaryotic cells, transcription occurs in the nucleus and is separate from the translation that occurs in the cytoplasm. Gene expression in prokaryotes is mostly regulated at the transcriptional level (some epigenetic and post-translational regulation is also present), whereas in eukaryotic cells, gene expression is regulated at the epigenetic, transcriptional, post-transcriptional, translational, and post-translational levels.

16.2 Prokaryotic Gene Regulation

The regulation of gene expression in prokaryotic cells occurs at the transcriptional level. There are two majors kinds of proteins that control prokaryotic transcription: repressors and activators. Repressors bind to an operator region to block the action of RNA polymerase. Activators bind to the promoter to enhance the binding of RNA polymerase. Inducer molecules can increase transcription either by inactivating repressors or by activating activator proteins. In the *trp* operon, the *trp* repressor is itself activated by binding to tryptophan. Therefore, if tryptophan is not needed, the repressor is bound to the operator and transcription remains off. The *lac* operon is activated by the CAP (catabolite activator protein), which binds to the promoter to stabilize RNA polymerase binding. CAP is itself activated by cAMP, whose concentration rises as the concentration of glucose falls. However, the *lac* operon also requires the presence of lactose for transcription to occur. Lactose inactivates the *lac* repressor, and prevents the repressor protein from binding to the *lac* operator. With the repressor inactivated, transcription may proceed. Therefore glucose must be absent and lactose must be present for effective transcription of the *lac* operon.

16.3 Eukaryotic Epigenetic Gene Regulation

In eukaryotic cells, the first stage of gene-expression control occurs at the epigenetic level. Epigenetic mechanisms control access to the chromosomal region to allow genes to be turned on or off. Chromatin remodeling controls how DNA is packed into the nucleus by regulating how tightly the DNA is wound around histone proteins. The DNA itself may be methylated to selectively silence genes. The addition or removal of chemical modifications (or flags) to histone proteins or DNA signals the cell to open or close a chromosomal region. Therefore, eukaryotic cells can control whether a gene is expressed by controlling accessibility to the binding of RNA polymerase and its transcription factors.

16.4 Eukaryotic Transcription Gene Regulation

To start transcription, general transcription factors, such as TFIID, TFIIB, and others, must first bind to the TATA box and recruit RNA polymerase to that location. Additional transcription factors may also bind to other regulatory elements at the promoter to increase or prevent transcription. In addition to promoter sequences, enhancer regions help augment transcription. Enhancers can be upstream, downstream, within a gene itself, or on other chromosomes. Specific transcription factors bound to enhancer regions may either increase or prevent transcription.

16.5 Eukaryotic Post-transcriptional Gene Regulation

Post-transcriptional control can occur at any stage after transcription, including RNA splicing and RNA stability. Once RNA is transcribed, it must be processed to create a mature RNA that is ready to be translated. This involves the removal of introns that do not code for protein. Spliceosomes bind to the signals that mark the exon/intron border to remove the introns and ligate the exons together. Once this occurs, the RNA is mature and can be translated. Alternative splicing can produce more than one mRNA from a given transcript. Different splicing variants may be produced under different conditions.

RNA is created and spliced in the nucleus, but needs to be transported to the cytoplasm to be translated. RNA is transported to the cytoplasm through the nuclear pore complex. Once the RNA is in the cytoplasm, the length of time it resides there before being degraded, called RNA stability, can also be altered to control the overall amount of protein that is synthesized. The RNA stability can be increased, leading to longer residency time in the cytoplasm, or decreased, leading to shortened time and less protein synthesis. RNA stability is controlled by RNA-binding proteins (RPBs) and microRNAs (miRNAs). These RPBs and miRNAs bind to the 5' UTR or the 3' UTR of the RNA to increase or decrease RNA stability. MicroRNAs associated with RISC complexes may repress translation or lead to mRNA breakdown.

16.6 Eukaryotic Translational and Post-translational Gene Regulation

Changing the status of the RNA or the protein itself can affect the amount of protein, the function of the protein, or how long it is found in the cell. To translate the protein, a protein initiator complex must assemble on the RNA. Modifications (such as phosphorylation) of proteins in this complex can prevent proper translation from occurring. Once a protein has been synthesized, it can be modified (phosphorylated, acetylated, methylated, or ubiquitinated). These post-translational modifications can greatly impact the stability, degradation, or function of the protein.

16.7 Cancer and Gene Regulation

Cancer can be described as a disease of altered gene expression. Changes at every level of eukaryotic gene expression can be detected in some form of cancer at some point in time. In order to understand how changes to gene expression can cause cancer, it is critical to understand how each stage of gene regulation works in normal cells. By understanding the mechanisms of control in normal, non-diseased cells, it will be easier for scientists to understand what goes wrong in disease states including complex ones like cancer.

VISUAL CONNECTION QUESTIONS

1. Figure 16.5 In *E. coli*, the *trp* operon is on by default, while the *lac* operon is off. Why do you think that this is the case?

2. Figure 16.7 In females, one of the two X chromosomes is inactivated during embryonic development because of epigenetic changes to the chromatin. What impact do you think these changes would have on nucleosome packing?

3. Figure 16.13 An increase in phosphorylation levels of eIF-2 has been observed in patients with neurodegenerative diseases such as Alzheimer's, Parkinson's, and Huntington's. What impact do you think this might have on protein synthesis?

REVIEW QUESTIONS

4. Control of gene expression in eukaryotic cells occurs at which level(s)?
   a. only the transcriptional level
   b. epigenetic and transcriptional levels
   c. epigenetic, transcriptional, and translational levels
   d. epigenetic, transcriptional, post-transcriptional, translational, and post-translational levels

5. Post-translational control refers to:
   a. regulation of gene expression after transcription
   b. regulation of gene expression after translation
   c. control of epigenetic activation
   d. period between transcription and translation

6. How does the regulation of gene expression support continued evolution of more complex organisms?
   a. Cells can become specialized within a multicellular organism.
   b. Organisms can conserve energy and resources.
   c. Cells grow larger to accommodate protein production.
   d. Both A and B.

7. If glucose is absent, but so is lactose, the *lac* operon will be _______.
   a. activated
   b. repressed
   c. activated, but only partially
   d. mutated

8. Prokaryotic cells lack a nucleus. Therefore, the genes in prokaryotic cells are:
   a. all expressed, all of the time
   b. transcribed and translated almost simultaneously
   c. transcriptionally controlled because translation begins before transcription ends
   d. b and c are both true

9. The *ara* operon is an inducible operon that controls the production of the sugar arabinose. When arabinose is present in a bacterium it binds to the protein AraC, and the complex binds to the initiator site to promote transcription. In this scenario, AraC is a(n) _______.
   a. activator
   b. inducer
   c. repressor
   d. operator

10. What are epigenetic modifications?
 a. the addition of reversible changes to histone proteins and DNA
 b. the removal of nucleosomes from the DNA
 c. the addition of more nucleosomes to the DNA
 d. mutation of the DNA sequence

11. Which of the following are true of epigenetic changes?
 a. allow DNA to be transcribed
 b. move histones to open or close a chromosomal region
 c. are temporary
 d. all of the above

12. The binding of _______ is required for transcription to start.
 a. a protein
 b. DNA polymerase
 c. RNA polymerase
 d. a transcription factor

13. What will result from the binding of a transcription factor to an enhancer region?
 a. decreased transcription of an adjacent gene
 b. increased transcription of a distant gene
 c. alteration of the translation of an adjacent gene
 d. initiation of the recruitment of RNA polymerase

14. A scientist compares the promoter regions of two genes. Gene A's core promoter plus proximal promoter elements encompasses 70bp. Gene B's core promoter plus proximal promoter elements encompasses 250bp. Which of the scientist's hypotheses is most likely to be correct?
 a. More transcripts will be made from Gene B.
 b. Transcription of Gene A involves fewer transcription factors.
 c. Enhancers control Gene B's transcription.
 d. Transcription of Gene A is more controlled than transcription of Gene B.

15. Which of the following are involved in post-transcriptional control?
 a. control of RNA splicing
 b. control of RNA shuttling
 c. control of RNA stability
 d. all of the above

16. Binding of an RNA binding protein will _______ the stability of the RNA molecule.
 a. increase
 b. decrease
 c. neither increase nor decrease
 d. either increase or decrease

17. An unprocessed pre-mRNA has the following structure.

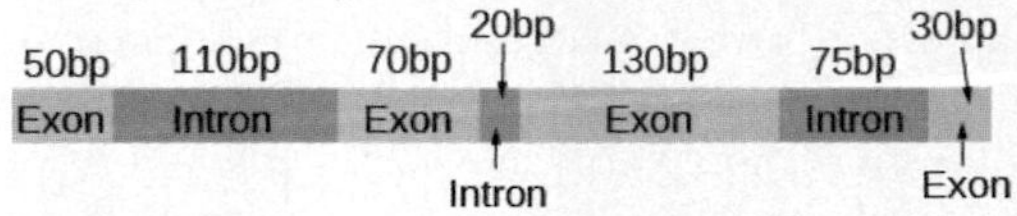

Which of the following is not a possible size (in bp) of the mature mRNA?
 a. 205bp
 b. 180bp
 c. 150bp
 d. 100bp

18. Alternative splicing has been estimated to occur in more than 95% of multi-exon genes. Which of the following is not an evolutionary advantage of alternative splicing?
 a. Alternative splicing increases diversity without increasing genome size.
 b. Different gene isoforms can be expressed in different tissues.
 c. Alternative splicing creates shorter mRNA transcripts
 d. Different gene isoforms can be expressed during different stages of development.

19. Post-translational modifications of proteins can affect which of the following?
 a. protein function
 b. transcriptional regulation
 c. chromatin modification
 d. all of the above

20. A scientist mutates eIF-2 to eliminate its GTP hydrolysis capability. How would this mutated form of eIF-2 alter translation?
 a. Initiation factors would not be able to bind to mRNA.
 b. The large ribosomal subunit would not be able to interact with mRNA transcripts.
 c. tRNAi-Met would not scan mRNA transcripts for the start codon.
 d. eIF-2 would not be able to interact with the small ribosomal subunit.

21. Cancer causing genes are called _______.
 a. transformation genes
 b. tumor suppressor genes
 c. oncogenes
 d. mutated genes

22. Targeted therapies are used in patients with a set gene expression pattern. A targeted therapy that prevents the activation of the estrogen receptor in breast cancer would be beneficial to which type of patient?
    a. patients who express the EGFR receptor in normal cells
    b. patients with a mutation that inactivates the estrogen receptor
    c. patients with lots of the estrogen receptor expressed in their tumor
    d. patients that have no estrogen receptor expressed in their tumor

CRITICAL THINKING QUESTIONS

23. Name two differences between prokaryotic and eukaryotic cells and how these differences benefit multicellular organisms.
24. Describe how controlling gene expression will alter the overall protein levels in the cell.
25. Describe how transcription in prokaryotic cells can be altered by external stimulation such as excess lactose in the environment.
26. What is the difference between a repressible and an inducible operon?
27. In cancer cells, alteration to epigenetic modifications turns off genes that are normally expressed. Hypothetically, how could you reverse this process to turn these genes back on?
28. A scientific study demonstrated that rat mothering behavior impacts the stress response in their pups. Rats that were born and grew up with attentive mothers showed low activation of stress-response genes later in life, while rats with inattentive mothers had high activation of stress-response genes in the same situation. An additional study that swapped the pups at birth (i.e., rats born to inattentive mothers grew up with attentive mothers and vice versa) showed the same positive effect of attentive mothering. How do genetics and/or epigenetics explain the results of this study?
29. Some autoimmune diseases show a positive correlation with dramatically decreased expression of histone deacetylase 9 (HDAC9, an enzyme that removes acetyl groups from histones). Why would the decreased expression of HDAC9 cause immune cells to produce inflammatory genes at inappropriate times?
30. A mutation within the promoter region can alter transcription of a gene. Describe how this can happen.
31. What could happen if a cell had too much of an activating transcription factor present?
32. A scientist identifies a potential transcription regulation site 300bp downstream of a gene and hypothesizes that it is a repressor. What experiment (with results) could he perform to support this hypothesis?
33. Describe how RBPs can prevent miRNAs from degrading an RNA molecule.
34. How can external stimuli alter post-transcriptional control of gene expression?
35. Protein modification can alter gene expression in many ways. Describe how phosphorylation of proteins can alter gene expression.
36. Alternative forms of a protein can be beneficial or harmful to a cell. What do you think would happen if too much of an alternative protein bound to the 3' UTR of an RNA and caused it to degrade?
37. Changes in epigenetic modifications alter the accessibility and transcription of DNA. Describe how environmental stimuli, such as ultraviolet light exposure, could modify gene expression.
38. A scientist discovers a virus encoding a Protein X that degrades a subunit of the eIF4F complex. Knowing that this virus transcribes its own mRNAs in the cytoplasm of human cells, why would Protein X be an effective virulence factor?
39. New drugs are being developed that decrease DNA methylation and prevent the removal of acetyl groups from histone proteins. Explain how these drugs could affect gene expression to help kill tumor cells.
40. How can understanding the gene expression pattern in a cancer cell tell you something about that specific form of cancer?

CHAPTER 17
Biotechnology and Genomics

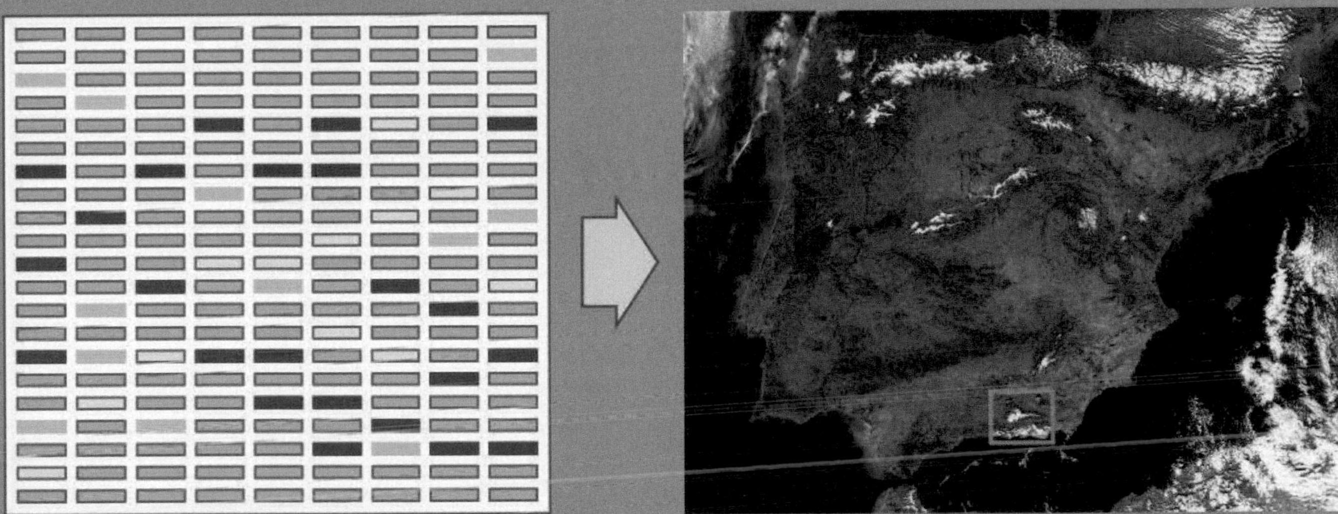

Figure 17.1 Genomics compares the DNA of different organisms, enabling scientists to create maps with which to navigate different organisims' DNA. (credit "map": modification of photo by NASA)

INTRODUCTION The study of nucleic acids began with the discovery of DNA, progressed to the study of genes and small fragments, and has now exploded to the field of genomics. Genomics is the study of entire genomes, including the complete set of genes, their nucleotide sequence and organization, and their interactions within a species and with other species. DNA sequencing technology has contributed to advances in genomics. Just as information technology has led to Google maps that enable people to obtain detailed information about locations around the globe, researchers use genomic information to create similar DNA maps of different organisms. These findings have helped anthropologists to better understand human migration and have aided the medical field through mapping human genetic diseases. Genomic information can contribute to scientific understanding in various ways and knowledge in the field is quickly growing.

Chapter Outline

17.1 Biotechnology

By the end of this section, you will be able to do the following:

- Describe gel electrophoresis
- Explain molecular and reproductive cloning
- Describe biotechnology uses in medicine and agriculture

Biotechnology is the use of biological agents for technological advancement. Biotechnology was used for breeding livestock and crops long before people understood the scientific basis of these techniques. Since the discovery of the structure of DNA in 1953, the biotechnology field has grown rapidly through both academic research and private companies. The primary applications of this technology are in medicine (vaccine and antibiotic production) and agriculture (crop genetic modification in order to increase yields). Biotechnology also has many industrial applications, such as fermentation, treating oil spills, and producing biofuels (Figure 17.2).

Figure 17.2 Fungi, bacteria, and other organisms that have antimicrobial properties produce antibiotics. The first antibiotic discovered was penicillin. Pharmaceutical companies now commercially produce and test antibiotics for their potential to inhibit bacterial growth. (credit "advertisement": modification of work by NIH; credit "test plate": modification of work by Don Stalons/CDC; scale-bar data from Matt Russell)

Basic Techniques to Manipulate Genetic Material (DNA and RNA)

To understand the basic techniques used to work with nucleic acids, remember that nucleic acids are macromolecules made of nucleotides (a sugar, a phosphate, and a nitrogenous base) linked by phosphodiester bonds. The phosphate groups on these molecules each have a net negative charge. An entire set of DNA molecules in the nucleus is called the genome. DNA has two complementary strands linked by hydrogen bonds between the paired bases. Exposure to high temperatures (DNA denaturation) can separate the two strands and cooling can reanneal them. The DNA polymerase enzyme can replicate the DNA. Unlike DNA, which is located in the eukaryotic cells' nucleus, RNA molecules leave the nucleus. The most common type of RNA that researchers analyze is the messenger RNA (mRNA) because it represents the protein-coding genes that are actively expressed. However, RNA molecules present some other challenges to analysis, as they are often less stable than DNA.

DNA and RNA Extraction

To study or manipulate nucleic acids, one must first isolate or extract the DNA or RNA from the cells. Researchers use various techniques to extract different types of DNA (Figure 17.3). Most nucleic acid extraction techniques involve steps to break open the cell and use enzymatic reactions to destroy all macromolecules that are not desired (such as unwanted molecule degradation and separation from the DNA sample). A **lysis buffer** (a solution which is mostly a detergent) breaks cells. Note that lysis means "to split". These enzymes break apart lipid molecules in the cell membranes and nuclear membranes. Enzymes such as **proteases** that break down proteins inactivate macromolecules, and **ribonucleases** (RNAses) that break down RNA. Using alcohol precipitates the DNA. Human genomic DNA is usually visible as a gelatinous, white mass. One can store the DNA samples frozen at –80°C for several years.

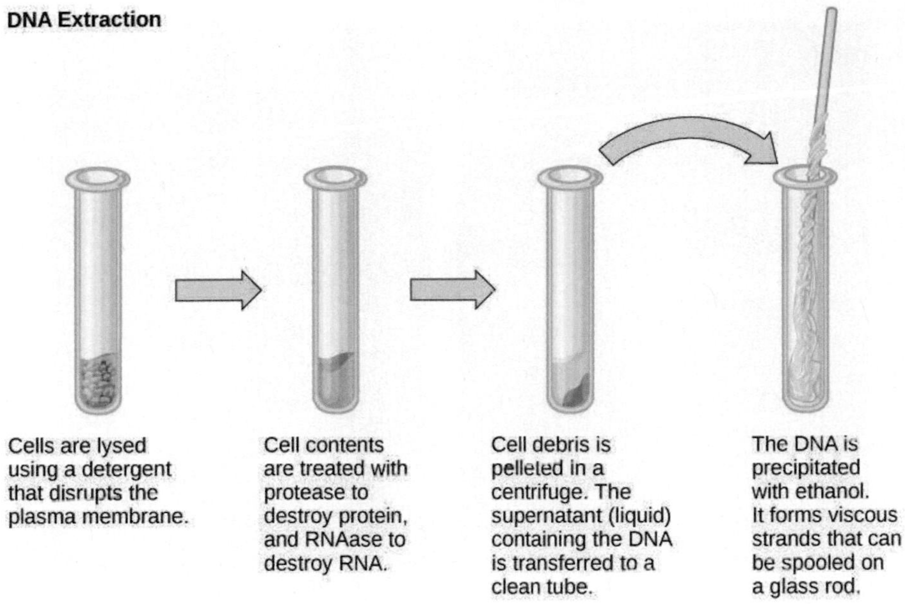

Figure 17.3 This diagram shows the basic method of DNA extraction.

Scientists perform RNA analysis to study gene expression patterns in cells. RNA is naturally very unstable because RNAses are commonly present in nature and very difficult to inactivate. Similar to DNA, RNA extraction involves using various buffers and enzymes to inactivate macromolecules and preserve the RNA.

Gel Electrophoresis

Because nucleic acids are negatively charged ions at neutral or basic pH in an aqueous environment, an electric field can mobilize them. **Gel electrophoresis** is a technique that scientists use to separate molecules on the basis of size, using this charge. One can separate the nucleic acids as whole chromosomes or fragments. The nucleic acids load into a slot near the semisolid, porous gel matrix's negative electrode, and pulled toward the positive electrode at the gel's opposite end. Smaller molecules move through the gel's pores faster than larger molecules. This difference in the migration rate separates the fragments on the basis of size. There are molecular weight standard samples that researchers can run alongside the molecules to provide a size comparison. We can observe nucleic acids in a gel matrix using various fluorescent or colored dyes. Distinct nucleic acid fragments appear as bands at specific distances from the gel's top (the negative electrode end) on the basis of their size (Figure 17.4). A mixture of genomic DNA fragments of varying sizes appear as a long smear; whereas, uncut genomic DNA is usually too large to run through the gel and forms a single large band at the gel's top.

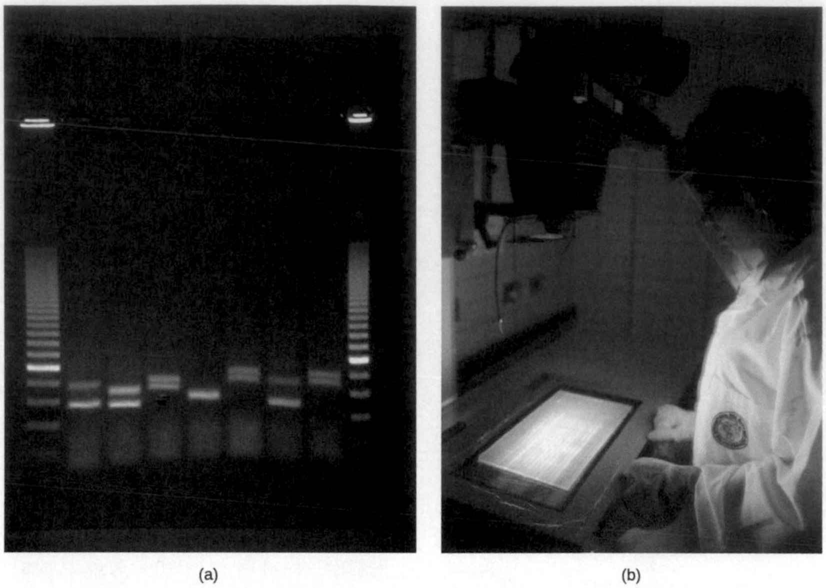

Figure 17.4 a) Shown are DNA fragments from seven samples run on a gel, stained with a fluorescent dye, and viewed under UV light; and b) a researcher from International Rice Research Institute, reviewing DNA profiles using UV light. (credit: a: James Jacob, Tompkins Cortland Community College b: International Rice Research Institute)

Nucleic Acid Fragment Amplification by Polymerase Chain Reaction

Although genomic DNA is visible to the naked eye when it is extracted in bulk, DNA analysis often requires focusing on one or more specific genome regions. **Polymerase chain reaction** (**PCR**) is a technique that scientists use to amplify specific DNA regions for further analysis (Figure 17.5). Researchers use PCR for many purposes in laboratories, such as cloning gene fragments to analyze genetic diseases, identifying contaminant foreign DNA in a sample, and amplifying DNA for sequencing. More practical applications include determining paternity and detecting genetic diseases.

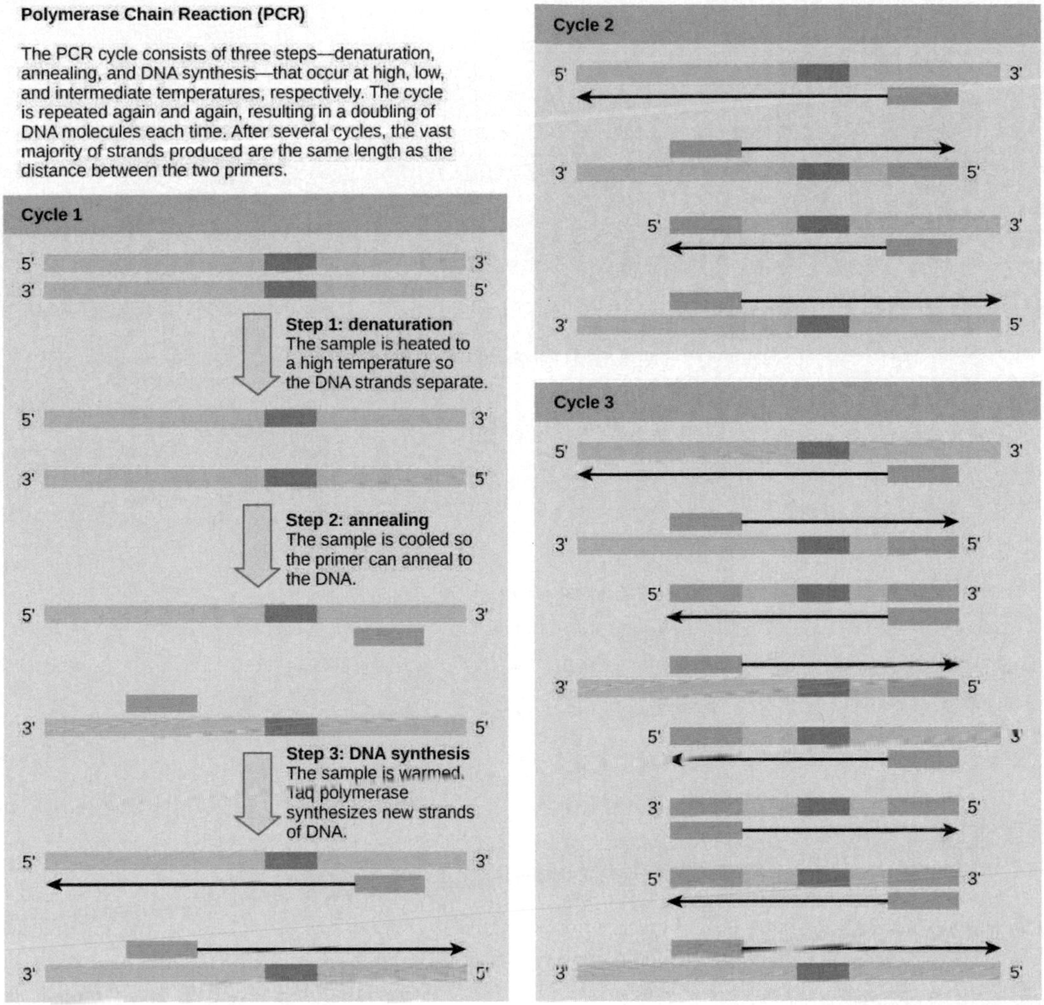

Figure 17.5 Scientists use polymerase chain reaction, or PCR, to amplify a specific DNA sequence. Primers—short pieces of DNA complementary to each end of the target sequence combine with genomic DNA, Taq polymerase, and deoxynucleotides. Taq polymerase is a DNA polymerase isolated from the thermostable bacterium *Thermus aquaticus* that is able to withstand the high temperatures that scientists use in PCR. *Thermus aquaticus* grows in the Lower Geyser Basin of Yellowstone National Park. Reverse transcriptase PCR (RT-PCR) is similar to PCR, but cDNA is made from an RNA template before PCR begins.

DNA fragments can also be amplified from an RNA template in a process called **reverse transcriptase PCR (RT-PCR)**. The first step is to recreate the original DNA template strand (called cDNA) by applying DNA nucleotides to the mRNA. This process is called reverse transcription. This requires the presence of an enzyme called reverse transcriptase. After the cDNA is made, regular PCR can be used to amplify it.

LINK TO LEARNING

Deepen your understanding of the polymerase chain reaction by watching this video (http://openstax.org/l/PCR) .

Hybridization, Southern Blotting, and Northern Blotting

Scientists can probe nucleic acid samples, such as fragmented genomic DNA and RNA extracts, for the presence of certain sequences. Scientists design and label short DNA fragments, or **probes** with radioactive or fluorescent dyes to aid detection. Gel electrophoresis separates the nucleic acid fragments according to their size. Scientists then transfer the fragments in the gel onto a nylon membrane in a procedure we call **blotting** (Figure 17.6). Scientists can then probe the nucleic acid fragments that are bound to the membrane's surface with specific radioactively or fluorescently labeled probe sequences. When scientists transfer DNA to a nylon membrane, they refer to the technique as **Southern blotting**. When they transfer the RNA to a nylon membrane, they call it **Northern blotting**. Scientists use Southern blots to detect the presence of certain DNA sequences in a given genome, and Northern blots to detect gene expression.

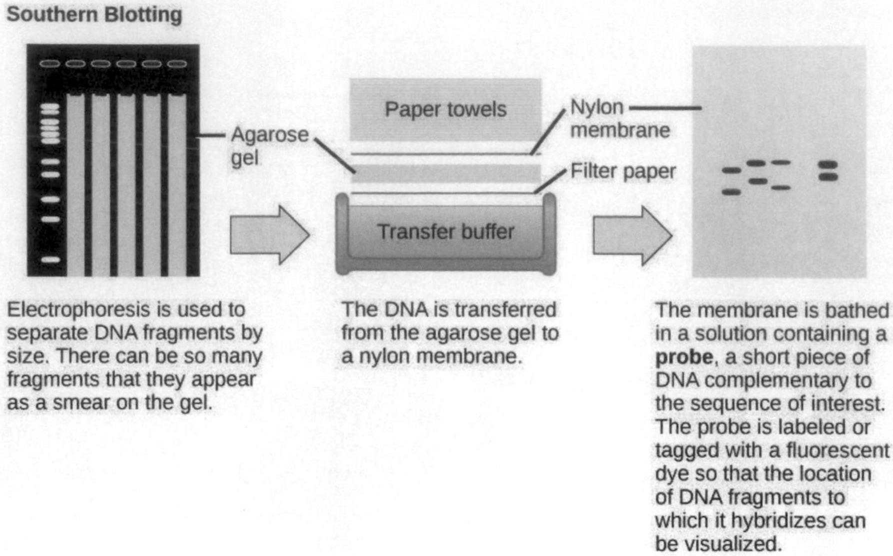

Figure 17.6 Scientists use Southern blotting to find a particular sequence in a DNA sample. Scientists separate DNA fragments on a gel, transfer them to a nylon membrane, and incubate them with a DNA probe complementary to the sequence of interest. Northern blotting is similar to Southern blotting, but scientists run RNA on the gel instead of DNA. In Western blotting, scientists run proteins on a gel and detect them using antibodies.

Molecular Cloning

In general, the word "cloning" means the creation of a perfect replica; however, in biology, the re-creation of a whole organism is referred to as "reproductive cloning." Long before attempts were made to clone an entire organism, researchers learned how to reproduce desired regions or fragments of the genome, a process that is referred to as molecular cloning.

Cloning small genome fragments allows researchers to manipulate and study specific genes (and their protein products), or noncoding regions in isolation. A plasmid, or vector, is a small circular DNA molecule that replicates independently of the chromosomal DNA. In cloning, scientists can use the plasmid molecules to provide a "folder" in which to insert a desired DNA fragment. Plasmids are usually introduced into a bacterial host for proliferation. In the bacterial context, scientists call the DNA fragment from the human genome (or the genome of another studied organism) **foreign DNA**, or a transgene, to differentiate it from the bacterium's DNA, or the **host DNA**.

Plasmids occur naturally in bacterial populations (such as *Escherichia coli*) and have genes that can contribute favorable traits to the organism, such as **antibiotic resistance** (the ability to be unaffected by antibiotics). Scientists have repurposed and engineered plasmids as vectors for molecular cloning and the large-scale production of important reagents, such as insulin and human growth hormone. An important feature of plasmid vectors is the ease with which scientists can introduce a foreign DNA fragment via the **multiple cloning site (MCS)**. The MCS is a short DNA sequence containing multiple sites that different commonly available restriction endonucleases can cut. **Restriction endonucleases** recognize specific DNA sequences and cut them in a predictable manner. They are naturally produced by bacteria as a defense mechanism against foreign DNA. Many restriction endonucleases make staggered cuts in the two DNA strands, such that the cut ends have a 2- or 4-base single-stranded overhang. Because these overhangs are capable of annealing with complementary overhangs, we call them "sticky ends." Adding the enzyme DNA ligase permanently joins the DNA fragments via phosphodiester bonds. In this way, scientists can splice any DNA fragment generated by restriction endonuclease cleavage between the plasmid DNA's two ends that has been cut with the same restriction endonuclease (Figure 17.7).

Recombinant DNA Molecules

Plasmids with foreign DNA inserted into them are called **recombinant DNA** molecules because they are created artificially and do not occur in nature. They are also called chimeric molecules because the origin of different molecule parts of the molecules can be traced back to different species of biological organisms or even to chemical synthesis. We call proteins that are expressed from recombinant DNA molecules **recombinant proteins**. Not all recombinant plasmids are capable of expressing genes. The recombinant DNA may need to move into a different vector (or host) that is better designed for gene expression. Scientists may also engineer plasmids to express proteins only when certain environmental factors stimulate them, so they can control the recombinant proteins' expression.

VISUAL CONNECTION

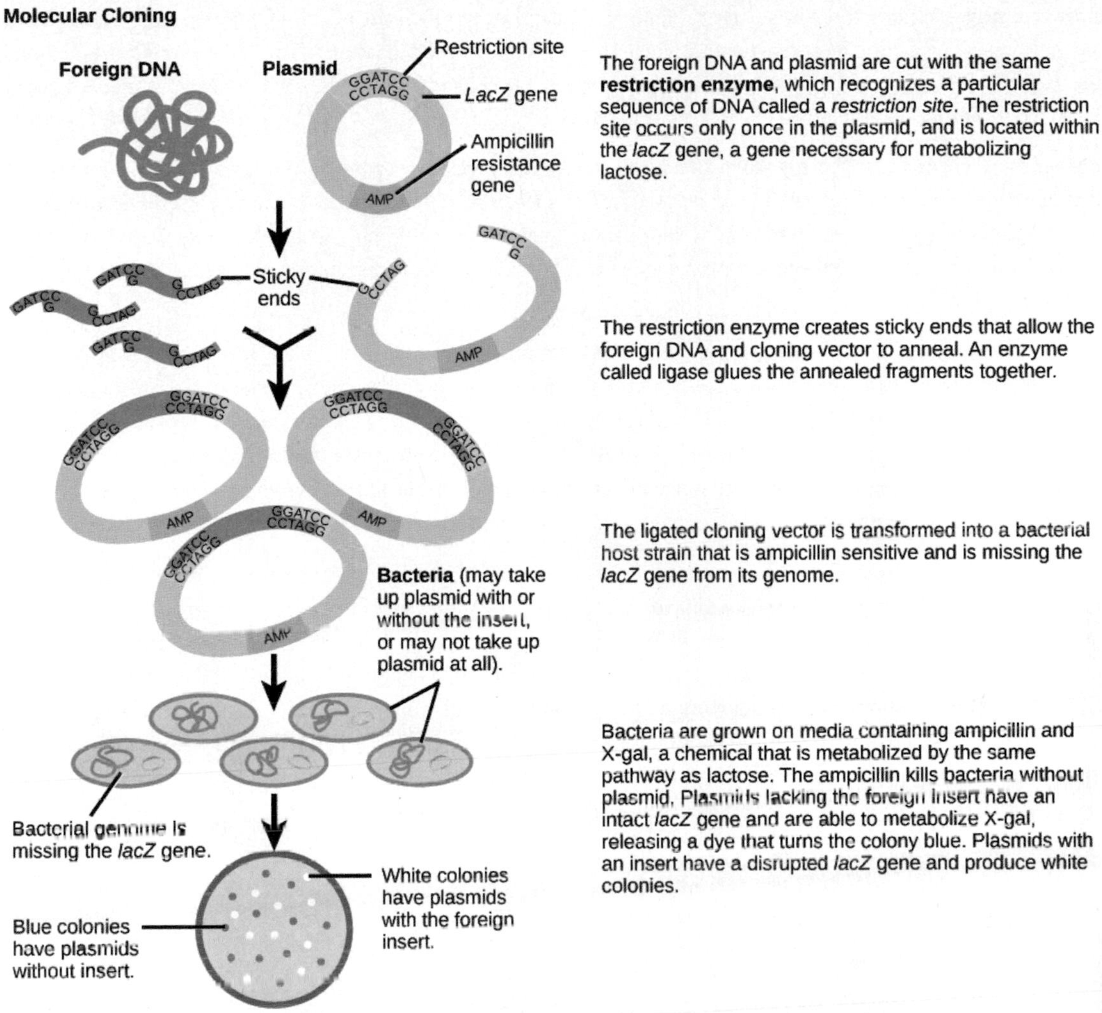

Figure 17.7 This diagram shows the steps involved in molecular cloning.

You are working in a molecular biology lab and, unbeknownst to you, your lab partner left the foreign genomic DNA that you are planning to clone on the lab bench overnight instead of storing it in the freezer. As a result, it was degraded by nucleases, but still used in the experiment. The plasmid, on the other hand, is fine. What results would you expect from your molecular cloning experiment?

a. There will be no colonies on the bacterial plate.
b. There will be blue colonies only.
c. There will be blue and white colonies.
d. The will be white colonies only.

LINK TO LEARNING

View an animation of recombination in cloning (http://openstax.org/l/recombination) from the DNA Learning Center.

Cellular Cloning

Unicellular organisms, such as bacteria and yeast, naturally produce clones of themselves when they replicate asexually by binary fission; this is known as **cellular cloning**. The nuclear DNA duplicates by the process of mitosis, which creates an exact replica of the genetic material.

Reproductive Cloning

Reproductive cloning is a method scientists use to clone or identically copy an entire multicellular organism. Most multicellular organisms undergo reproduction by sexual means, which involves genetic hybridization of two individuals (parents), making it impossible to generate an identical copy or a clone of either parent. Recent advances in biotechnology have made it possible to artificially induce mammal asexual reproduction in the laboratory.

Parthenogenesis, or "virgin birth," occurs when an embryo grows and develops without egg fertilization. This is a form of asexual reproduction. An example of parthenogenesis occurs in species in which the female lays an egg and if the egg is fertilized, it is a diploid egg and the individual develops into a female. If the egg is not fertilized, it remains a haploid egg and develops into a male. The unfertilized egg is a parthenogenic, or virgin egg. Some insects and reptiles lay parthenogenic eggs that can develop into adults.

Sexual reproduction requires two cells. When the haploid egg and sperm cells fuse, a diploid zygote results. The zygote nucleus contains the genetic information to produce a new individual. However, early embryonic development requires the cytoplasmic material contained in the egg cell. This idea forms the basis for reproductive cloning. Therefore, if we replace the egg cell's haploid nucleus with a diploid nucleus from the cell of any individual of the same species (a donor), it will become a zygote that is genetically identical to the donor. Somatic cell nuclear transfer is the technique of transferring a diploid nucleus into an enucleated egg. Scientists can use it for either therapeutic cloning or reproductive cloning.

The first cloned animal was Dolly, a sheep born in 1996. The reproductive cloning success rate at the time was very low. Dolly lived for seven years and died of respiratory complications (Figure 17.8). There is speculation that because the cell DNA belongs to an older individual, DNA's age may affect a cloned individual's life expectancy. Since Dolly, scientists have cloned successfully several animals such as horses, bulls, and goats, although these animals often exhibit facial, limb, and cardiac abnormalities. There have been attempts at producing cloned human embryos as sources of embryonic stem cells for therapeutic purposes. Therapeutic cloning produces stem cells in the attempt to remedy detrimental diseases or defects (unlike reproductive cloning, which aims to reproduce an organism). Still, some have met therapeutic cloning efforts with resistance because of bioethical considerations.

VISUAL CONNECTION

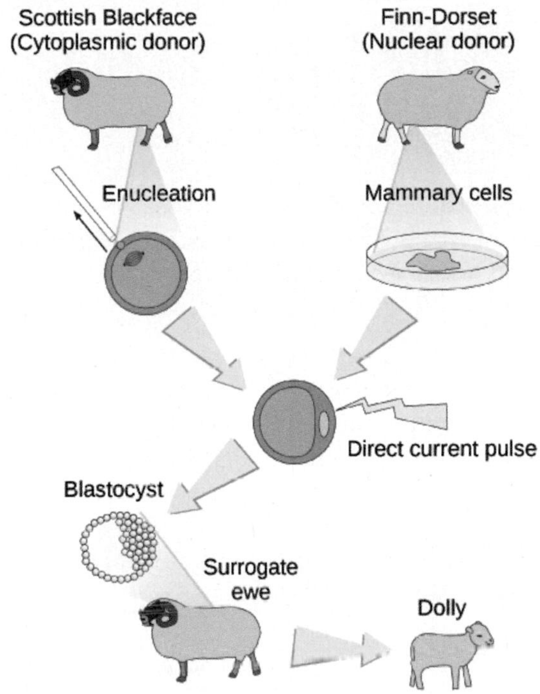

Figure 17.8 Dolly the sheep was the first mammal to be cloned. To create Dolly, they removed the nucleus from a donor egg cell. They then introduced the nucleus from a second sheep into the cell, which divided to the blastocyst stage before they implanted it in a surrogate mother. (credit: modification of work by "Squidonius"/Wikimedia Commons)

Do you think Dolly was a Finn-Dorset or a Scottish Blackface sheep?

Genetic Engineering

Genetic engineering is the alteration of an organism's genotype using recombinant DNA technology to modify an organism's DNA to achieve desirable traits. The addition of foreign DNA in the form of recombinant DNA vectors generated by molecular cloning is the most common method of genetic engineering. The organism that receives the recombinant DNA is a **genetically modified organism** (GMO). If the foreign DNA comes from a different species, the host organism is **transgenic**. Scientists have genetically modified bacteria, plants, and animals since the early 1970s for academic, medical, agricultural, and industrial purposes. In the US, GMOs such as Roundup-ready soybeans and borer-resistant corn are part of many common processed foods.

Gene Targeting

Although classical methods of studying gene function began with a given phenotype and determined the genetic basis of that phenotype, modern techniques allow researchers to start at the DNA sequence level and ask: "What does this gene or DNA element do?" This technique, reverse genetics, has resulted in reversing the classic genetic methodology. This method would be similar to damaging a body part to determine its function. An insect that loses a wing cannot fly, which means that the wing's function is flight. The classical genetic method would compare insects that cannot fly with insects that can fly, and observe that the non-flying insects have lost wings. Similarly, mutating or deleting genes provides researchers with clues about gene function. We collectively call the methods they use to disable gene function gene targeting. **Gene targeting** is the use of recombinant DNA vectors to alter a particular gene's expression, either by introducing mutations in a gene, or by eliminating a certain gene's expression by deleting a part or all of the gene sequence from the organism's genome.

Biotechnology in Medicine and Agriculture

It is easy to see how biotechnology can be used for medicinal purposes. Knowledge of the genetic makeup of our species, the genetic basis of heritable diseases, and the invention of technology to manipulate and fix mutant genes provides methods to treat the disease. Biotechnology in agriculture can enhance resistance to disease, pest, and environmental stress, and improve both crop yield and quality.

Genetic Diagnosis and Gene Therapy

Scientists call the process of testing for suspected genetic defects before administering treatment **genetic diagnosis** by **genetic testing**. Depending on the inheritance patterns of a disease-causing gene, family members are advised to undergo genetic testing. For example, doctors usually advise women diagnosed with breast cancer to have a biopsy so that the medical team can determine the genetic basis of cancer development. Doctors base treatment plans on genetic test findings that determine the type of cancer. If inherited gene mutations cause the cancer, doctors also advise other female relatives to undergo genetic testing and periodic screening for breast cancer. Doctors also offer genetic testing for fetuses (or embryos with in vitro fertilization) to determine the presence or absence of disease-causing genes in families with specific debilitating diseases.

Gene therapy is a genetic engineering technique used to cure disease. In its simplest form, it involves the introduction of a good gene at a random location in the genome to aid the cure of a disease that is caused by a mutated gene. The good gene is usually introduced into diseased cells as part of a vector transmitted by a virus that can infect the host cell and deliver the foreign DNA (Figure 17.9). More advanced forms of gene therapy try to correct the mutation at the original site in the genome, such as is the case with treatment of severe combined immunodeficiency (SCID).

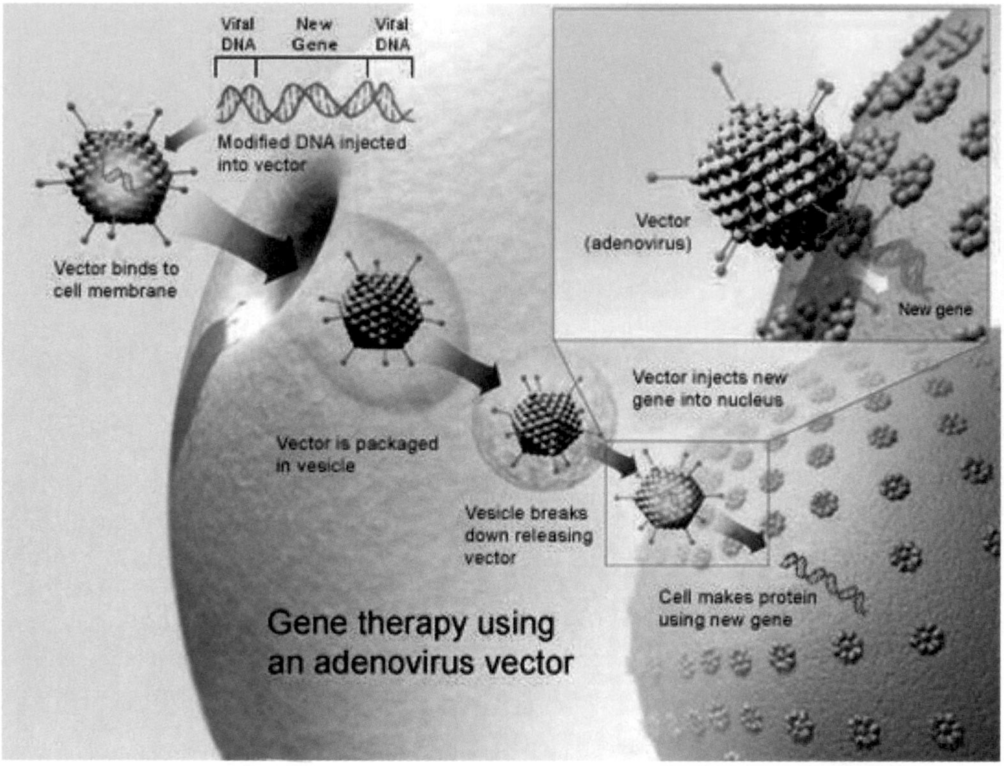

Figure 17.9 Gene therapy using an adenovirus vector can be used to cure certain genetic diseases in which a person has a defective gene. (credit: NIH)

Production of Vaccines, Antibiotics, and Hormones

Traditional vaccination strategies use weakened or inactive forms of microorganisms to mount the initial immune response. Modern techniques use the genes of microorganisms cloned into vectors to mass produce the desired antigen. Doctors then introduce the antigen into the body to stimulate the primary immune response and trigger immune memory. The medical field has used genes cloned from the influenza virus to combat the constantly changing strains of this virus.

Antibiotics are a biotechnological product. Microorganisms, such as fungi, naturally produce them to attain an advantage over bacterial populations. Cultivating and manipulating fungal cells produces antibodies.

Scientists used recombinant DNA technology to produce large-scale quantities of human insulin in *E. coli* as early as 1978. Previously, it was only possible to treat diabetes with pig insulin, which caused allergic reactions in humans because of differences in the gene product. In addition, doctors use human growth hormone (HGH) to treat growth disorders in children. Researchers cloned the HGH gene from a cDNA library and inserted it into *E. coli* cells by cloning it into a bacterial vector.

Transgenic Animals

Although several recombinant proteins in medicine are successfully produced in bacteria, some proteins require a eukaryotic animal host for proper processing. For this reason, the desired genes are cloned and expressed in animals, such as sheep, goats, chickens, and mice. We call animals that have been modified to express recombinant DNA transgenic animals. Several human proteins are expressed in transgenic sheep and goat milk, and some are expressed in chicken eggs. Scientists have used mice extensively for expressing and studying recombinant gene and mutation effects.

Transgenic Plants

Manipulating the DNA of plants (i.e., creating GMOs) has helped to create desirable traits, such as disease resistance, herbicide and pesticide resistance, better nutritional value, and better shelf-life (Figure 17.10). Plants are the most important source of food for the human population. Farmers developed ways to select for plant varieties with desirable traits long before modern-day biotechnology practices were established.

Figure 17.10 Corn, a major agricultural crop used to create products for a variety of industries, is often modified through plant biotechnology. (credit: Keith Weller, USDA)

We call plants that have received recombinant DNA from other species transgenic plants. Because they are not natural, government agencies closely monitor transgenic plants and other GMOs to ensure that they are fit for human consumption and do not endanger other plant and animal life. Because foreign genes can spread to other species in the environment, extensive testing is required to ensure ecological stability. Staples like corn, potatoes, and tomatoes were the first crop plants that scientists genetically engineered.

Transformation of Plants Using *Agrobacterium tumefaciens*

Gene transfer occurs naturally between species in microbial populations. Many viruses that cause human diseases, such as cancer, act by incorporating their DNA into the human genome. In plants, tumors caused by the bacterium *Agrobacterium tumefaciens* occur by DNA transfer from the bacterium to the plant. Although the tumors do not kill the plants, they stunt the plants and they become more susceptible to harsh environmental conditions. *A. tumefaciens* affects many plants such as walnuts, grapes, nut trees, and beets. Artificially introducing DNA into plant cells is more challenging than in animal cells because of the thick plant cell wall.

Researchers used the natural transfer of DNA from *Agrobacterium* to a plant host to introduce DNA fragments of their choice into plant hosts. In nature, the disease-causing *A. tumefaciens* have a set of plasmids, **Ti plasmids** (tumor-inducing plasmids), that contain genes to produce tumors in plants. DNA from the Ti plasmid integrates into the infected plant cell's genome. Researchers manipulate the Ti plasmids to remove the tumor-causing genes and insert the desired DNA fragment for transfer into the plant genome. The Ti plasmids carry antibiotic resistance genes to aid selection and researchers can propagate them in *E. coli* cells as well.

The Organic Insecticide *Bacillus thuringiensis*

Bacillus thuringiensis (Bt) is a bacterium that produces protein crystals during sporulation that are toxic to many insect species that affect plants. Insects need to ingest Bt toxin in order to activate the toxin. Insects that have eaten Bt toxin stop feeding on the plants within a few hours. After the toxin activates in the insects' intestines, they die within a couple of days. Modern biotechnology has allowed plants to encode their own crystal Bt toxin that acts against insects. Scientists have cloned the crystal toxin genes from Bt and introduced them into plants. Bt toxin is safe for the environment, nontoxic to humans and other mammals, and organic farmers have approved it as a natural insecticide.

Flavr Savr Tomato

The first GM crop on the market was the Flavr Savr Tomato in 1994. Scientists used antisense RNA technology to slow the softening and rotting process caused by fungal infections, which led to increased shelf life of the GM tomatoes. Additional genetic modification improved the tomato's flavor. The Flavr Savr tomato did not successfully stay in the market because of problems maintaining and shipping the crop.

17.2 Mapping Genomes

By the end of this section, you will be able to do the following:

- Define genomics
- Describe genetic and physical maps
- Describe genomic mapping methods

Genomics is the study of entire genomes, including the complete set of genes, their nucleotide sequence and organization, and their interactions within a species and with other species. **Genome mapping** is the process of finding the locations of genes on each chromosome. The maps that genome mapping create are comparable to the maps that we use to navigate streets. A **genetic map** is an illustration that lists genes and their location on a chromosome. Genetic maps provide the big picture (similar to an interstate highway map) and use genetic markers (similar to landmarks). A **genetic marker** is a gene or sequence on a chromosome that co-segregates (shows genetic linkage) with a specific trait. Early geneticists called this linkage analysis. Physical maps present the intimate details of smaller chromosome regions (similar to a detailed road map). A **physical map** is a representation of the physical distance, in nucleotides, between genes or genetic markers. Both genetic linkage maps and physical maps are required to build a genome's complete picture. Having a complete genome map of the genome makes it easier for researchers to study individual genes. Human genome maps help researchers in their efforts to identify human disease-causing genes related to illnesses like cancer, heart disease, and cystic fibrosis. We can use genome mapping in a variety of other applications, such as using live microbes to clean up pollutants or even prevent pollution. Research involving plant genome mapping may lead to producing higher crop yields or developing plants that better adapt to climate change.

Genetic Maps

The study of genetic maps begins with **linkage analysis**, a procedure that analyzes the recombination frequency between genes to determine if they are linked or show independent assortment. Scientists used the term *linkage* before the discovery of DNA. Early geneticists relied on observing phenotypic changes to understand an organism's genotype. Shortly after Gregor Mendel (the father of modern genetics) proposed that traits were determined by what we now call genes, other researchers observed that different traits were often inherited together, and thereby deduced that the genes were physically linked by their location on the same chromosome. Gene mapping relative to each other based on linkage analysis led to developing the first genetic maps.

Observations that certain traits were always linked and certain others were not linked came from studying the offspring of crosses between parents with different traits. For example, in garden pea experiments, researchers discovered, that the flower's color and plant pollen's shape were linked traits, and therefore the genes encoding these traits were in close proximity on the same chromosome. We call exchanging DNA between homologous chromosome pairs **genetic recombination**, which occurs by crossing over DNA between homologous DNA strands, such as nonsister chromatids. Linkage analysis involves studying the recombination frequency between any two genes. The greater the distance between two genes, the higher the chance that a recombination event will occur between them, and the higher the recombination frequency between them. Figure 17.11 shows two possibilities for recombination between two nonsister chromatids during meiosis. If the recombination frequency between two genes is less than 50 percent, they are linked.

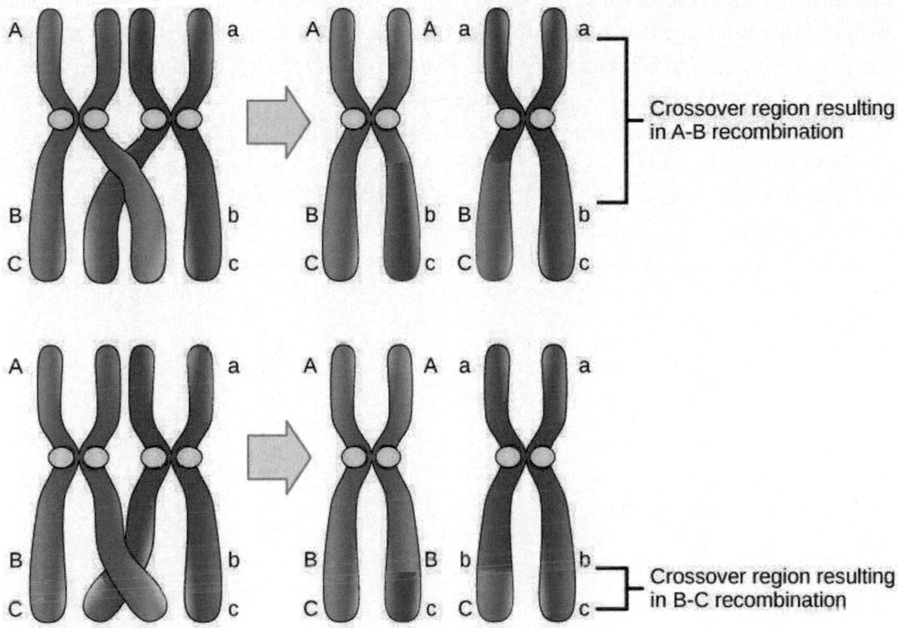

Figure 17.11 Crossover may occur at different locations on the chromosome. Recombination between genes *A* and *B* is more frequent than recombination between genes *B* and *C* because genes *A* and *B* are farther apart. Therefore, a crossover is more likely to occur between them.

The generation of genetic maps requires markers, just as a road map requires landmarks (such as rivers and mountains). Scientists based early genetic maps on using known genes as markers. Scientists now use more sophisticated markers, including those based on non-coding DNA, to compare individuals' genomes in a population. Although individuals of a given species are genetically similar, they are not identical. Every individual has a unique set of traits. These minor differences in the genome between individuals in a population are useful for genetic mapping purposes. In general, a good genetic marker is a region on the chromosome that shows variability or polymorphism (multiple forms) in the population.

Some genetic markers that scientists use in generating genetic maps are **restriction fragment length polymorphisms** (RFLP), variable number of tandem repeats (VNTRs), **microsatellite polymorphisms**, and the **single nucleotide polymorphisms** (SNPs). We can detect RFLPs (sometimes pronounced "rif-lips") when the DNA of an individual is cut with a restriction endonuclease that recognizes specific sequences in the DNA to generate a series of DNA fragments, which we can then analyze using gel electrophoresis. Every individual's DNA will give rise to a unique pattern of bands when cut with a particular set of restriction endonucleases. Scientists sometimes refer to this as an individual's DNA "fingerprint." Certain chromosome regions that are subject to polymorphism will lead to generating the unique banding pattern. VNTRs are repeated sets of nucleotides present in DNA's non-coding regions. Non-coding, or "junk," DNA has no known biological function; however, research shows that much of this DNA is actually transcribed. While its function is uncertain, it is certainly active, and it may be involved in regulating coding genes. The number of repeats may vary in a population's individual organisms. Microsatellite polymorphisms are similar to VNTRs, but the repeat unit is very small. SNPs are variations in a single nucleotide.

Because genetic maps rely completely on the natural process of recombination, natural increases or decreases in the recombination level given genome area affects mapping. Some parts of the genome are recombination hotspots; whereas, others do not show a propensity for recombination. For this reason, it is important to look at mapping information developed by multiple methods.

Physical Maps

A physical map provides detail of the actual physical distance between genetic markers, as well as the number of nucleotides. There are three methods scientists use to create a physical map: cytogenetic mapping, radiation hybrid mapping, and sequence mapping. **Cytogenetic mapping** uses information from microscopic analysis of stained chromosome sections (Figure 17.12). It is possible to determine the approximate distance between genetic markers using cytogenetic mapping, but not the exact distance (number of base pairs). **Radiation hybrid mapping** uses radiation, such as x-rays, to break the DNA into fragments. We can adjust the radiation amount to create smaller or larger fragments. This technique overcomes the limitation of genetic mapping, and we can adjust the radiation so that increased or decreased recombination frequency does not affect it. **Sequence mapping** resulted from DNA sequencing technology that allowed for creating detailed physical maps with distances measured in terms of the number of base pairs. Creating **genomic libraries** and **complementary DNA (cDNA) libraries** (collections of cloned sequences or all DNA from a genome) has sped the physical mapping process. A genetic site that scientists use to generate a physical map with sequencing technology (a sequence-tagged site, or STS) is a unique sequence in the genome with a known exact chromosomal location. An **expressed sequence tag (EST)** and a single sequence length polymorphism (SSLP) are common STSs. An EST is a short STS that we can identify with cDNA libraries, while we obtain SSLPs from known genetic markers, which provide a link between genetic and physical maps.

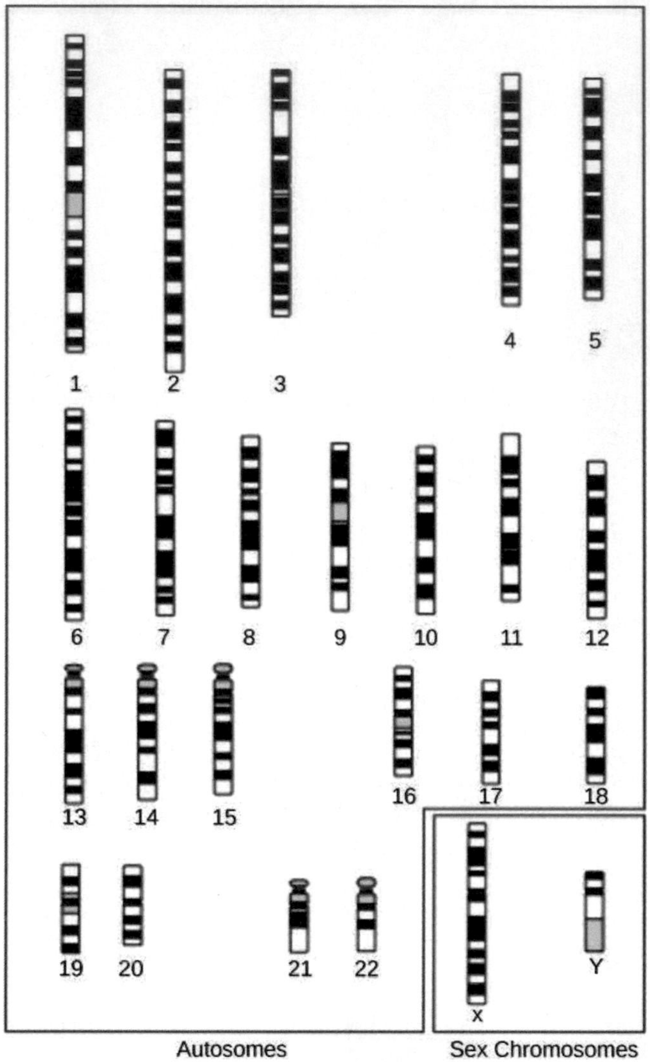

Figure 17.12 A cytogenetic map shows the appearance of a chromosome after scientists stain and exam it under a microscope. (credit: National Human Genome Research Institute)

Genetic and Physical Maps Integration

Genetic maps provide the outline and physical maps provide the details. It is easy to understand why both genome mapping technique types are important to show the big picture. Scientists use information from each technique in combination to study the genome. Scientists are using genomic mapping with different model organisms for research. Genome mapping is still an ongoing process, and as researchers develop more advanced techniques, they expect more breakthroughs. Genome mapping is similar to completing a complicated puzzle using every piece of available data. Mapping information generated in laboratories all over the world goes into central databases, such as GenBank at the National Center for Biotechnology Information (NCBI).

Researchers are making efforts for the information to be more easily accessible to other researchers and the general public. Just as we use global positioning systems instead of paper maps to navigate through roadways, NCBI has created a genome viewer tool to simplify the data-mining process.

SCIENTIFIC METHOD CONNECTION

How to Use a Genome Map Viewer

Problem statement: Do the human, macaque, and mouse genomes contain common DNA sequences?

Develop a hypothesis.

Go to this website (https://openstax.org/l/GeneINSR) to test the hypothesis.

The web page displays the comparison of the gene sequences of many organisms to the Human Insulin Receptor gene. Explore the type of information provided, select the groups of organisms needed for testing of the hypothesis from the top portion of the displayed data. Focus the attention to the bottom part, the Selected Orthologues. Explore which columns are relevant to the needed information.

On the same page, there are other options to explore, not all are necessary for the task, however it might give more insight to the value of genome/gene comparisons.

LINK TO LEARNING

Online Mendelian Inheritance in Man (OMIM) is a searchable online catalog of human genes and genetic disorders. This website shows genome mapping information, and also details the history and research of each trait and disorder. Click this link (http://openstax.org/l/OMIM) to search for traits (such as handedness) and genetic disorders (such as diabetes).

17.3 Whole-Genome Sequencing

By the end of this section, you will be able to do the following:

- Describe three types of sequencing
- Define whole-genome sequencing

Although there have been significant advances in the medical sciences in recent years, doctors are still confounded by some diseases, and they are using whole-genome sequencing to discover the root of the problem. **Whole-genome sequencing** is a process that determines an entire genome's DNA sequence. Whole-genome sequencing is a brute-force approach to problem solving when there is a genetic basis at the core of a disease. Several laboratories now provide services to sequence, analyze, and interpret entire genomes.

For example, whole-exome sequencing is a lower-cost alternative to whole genome sequencing. In exome sequencing, the doctor sequences only the DNA's coding, exon-producing regions. In 2010, doctors used whole-exome sequencing to save a young boy whose intestines had multiple mysterious abscesses. The child had several colon operations with no relief. Finally, they performed whole-exome sequencing, which revealed a defect in a pathway that controls apoptosis (programmed cell death). The doctors used a bone-marrow transplant to overcome this genetic disorder, leading to a cure for the boy. He was the first person to receive successful treatment based on a whole-exome sequencing diagnosis. Today, human genome sequencing is more readily available and results are available within two days for about $1000.

Strategies Used in Sequencing Projects

The basic sequencing technique used in all modern day sequencing projects is the chain termination method (also known as the dideoxy method), which Fred Sanger developed in the 1970s. The chain termination method involves DNA replication of a single-stranded template by using a primer and a regular **deoxynucleotide** (dNTP), which is a monomer, or a single DNA unit. The primer and dNTP mix with a small proportion of fluorescently labeled **dideoxynucleotides** (ddNTPs). The ddNTPs are monomers that are missing a hydroxyl group (–OH) at the site at which another nucleotide usually attaches to form a chain (Figure 17.13). Scientists label each ddNTP with a different color of fluorophore. Every time a ddNTP incorporates in the growing complementary strand, it terminates the DNA replication process, which results in multiple short strands of replicated DNA that each terminate at a different point during replication. When gel electrophoresis processes the reaction mixture after separating into single strands, the multiple newly replicated DNA strands form a ladder because of the differing sizes. Because the ddNTPs are fluorescently labeled, each band on the gel reflects the DNA strand's size and the ddNTP that terminated the reaction. The different colors of the fluorophore-labeled ddNTPs help identify the ddNTP incorporated at that position. Reading the gel on the basis of each band's color on the ladder produces the template strand's sequence (Figure 17.14).

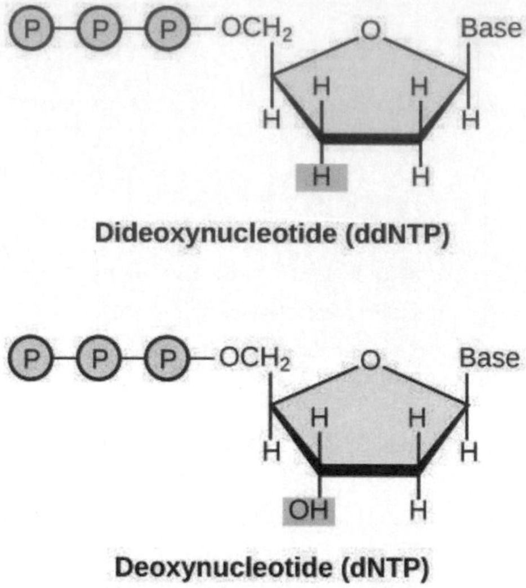

Figure 17.13 A dideoxynucleotide is similar in structure to a deoxynucleotide, but is missing the 3' hydroxyl group (indicated by the box). When a dideoxynucleotide is incorporated into a DNA strand, DNA synthesis stops.

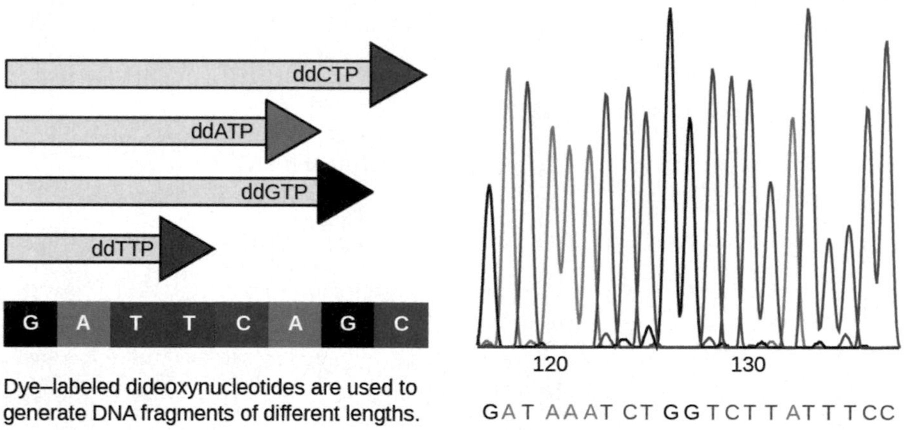

Figure 17.14 This figure illustrates Frederick Sanger's dideoxy chain termination method. Using dideoxynucleotides, the DNA fragment can terminate at different points. The DNA separates on the basis of size, and we can read these bands based on the fragments' size.

Early Strategies: Shotgun Sequencing and Pair-Wise End Sequencing

In **shotgun sequencing** method, several DNA fragment copies cut randomly into many smaller pieces (somewhat like what happens to a round shot cartridge when fired from a shotgun). All of the segments sequence using the chain-sequencing method. Then, with sequence computer assistance, scientists can analyze the fragments to see where their sequences overlap. By matching overlapping sequences at each fragment's end, scientists can reform the entire DNA sequence. A larger sequence that is assembled from overlapping shorter sequences is called a **contig**. As an analogy, consider that someone has four copies of a landscape photograph that you have never seen before and know nothing about how it should appear. The person then rips up each photograph with their hands, so that different size pieces are present from each copy. The person then mixes all of the pieces together and asks you to reconstruct the photograph. In one of the smaller pieces you see a mountain. In a larger piece, you see that the same mountain is behind a lake. A third fragment shows only the lake, but it reveals that there is a cabin on the shore of the lake. Therefore, from looking at the overlapping information in these three fragments, you know that the picture contains a mountain behind a lake that has a cabin on its shore. This is the principle behind reconstructing entire DNA sequences using shotgun sequencing.

Originally, shotgun sequencing only analyzed one end of each fragment for overlaps. This was sufficient for sequencing small genomes. However, the desire to sequence larger genomes, such as that of a human, led to developing double-barrel shotgun sequencing, or **pairwise-end sequencing**. In pairwise-end sequencing, scientists analyze each fragment's end for overlap. Pairwise-end sequencing is, therefore, more cumbersome than shotgun sequencing, but it is easier to reconstruct the sequence because there is more available information.

Next-generation Sequencing

Since 2005, automated sequencing techniques used by laboratories are under the umbrella of **next-generation sequencing**, which is a group of automated techniques used for rapid DNA sequencing. These automated low-cost sequencers can generate sequences of hundreds of thousands or millions of short fragments (25 to 500 base pairs) in the span of one day. These sequencers use sophisticated software to get through the cumbersome process of putting all the fragments in order.

EVOLUTION CONNECTION

Comparing Sequences

A sequence alignment is an arrangement of proteins, DNA, or RNA. Scientists use it to identify similar regions between cell types or species, which may indicate function or structure conservation. We can use sequence alignments to construct phylogenetic trees. The following website uses a software program called BLAST (basic local alignment search tool) (http://blast.ncbi.nlm.nih.gov/Blast.cgi) .

Under "Basic Blast," click "Nucleotide Blast." Input the following sequence into the large "query sequence" box: ATTGCTTCGATTGCA. Below the box, locate the "Species" field and type "human" or "Homo sapiens". Then click "BLAST" to compare the inputted sequence against the human genome's known sequences. The result is that this sequence occurs in over a hundred places in the human genome. Scroll down below the graphic with the horizontal bars and you will see a short description of each of the matching hits. Pick one of the hits near the top of the list and click on "Graphics". This will bring you to a page that shows the sequence's location within the entire human genome. You can move the slider that looks like a green flag back and forth to view the sequences immediately around the selected gene. You can then return to your selected sequence by clicking the "ATG" button.

Use of Whole-Genome Sequences of Model Organisms

British biochemist and Nobel Prize winner Fred Sanger used a bacterial virus, the bacteriophage fx174 (5368 base pairs), to completely sequence the first genome. Other scientists later sequenced several other organelle and viral genomes. American biotechnologist, biochemist, geneticist, and businessman Craig Venter sequenced the bacterium *Haemophilus influenzae* in the 1980s. Approximately 74 different laboratories collaborated on sequencing the genome of the yeast *Saccharomyces cerevisiae*, which began in 1989 and was completed in 1996, because it was 60 times bigger than any other genome sequencing. By 1997, the genome sequences of two important model organisms were available: the bacterium *Escherichia coli* K12 and the yeast *Saccharomyces cerevisiae*. We now know the genomes of other model organisms, such as the mouse *Mus musculus*, the fruit fly *Drosophila melanogaster*, the nematode *Caenorhabditis. elegans*, and humans *Homo sapiens*. Researchers perform extensive basic research in model organisms because they can apply the information to genetically similar organisms. A **model organism** is a species that researchers use as a model to understand the biological processes in other species that the model organism represents. Having entire genomes sequenced helps with the research efforts in these model organisms. The process of attaching biological information to gene sequences is **genome annotation**. Annotating gene sequences helps with basic experiments in molecular biology, such as designing PCR primers and RNA targets.

LINK TO LEARNING

Click through each genome sequencing step at this site (http://openstax.org/l/DNA_sequence) .

Genome Sequence Uses

DNA microarrays are methods that scientists use to detect gene expression by analyzing different DNA fragments that are fixed to a glass slide or a silicon chip to identify active genes and sequences. We can discover almost one million genotypic abnormalities using microarrays; whereas, whole-genome sequencing can provide information about all six billion base pairs in the human genome. Although studying genome sequencing medical applications is interesting, this discipline dwells on abnormal gene function. Knowing about the entire genome will allow researchers to discover future onset diseases and other genetic disorders early. This will allow for more informed decisions about lifestyle, medication, and having children. Genomics is still in its infancy, although someday it may become routine to use whole-genome sequencing to screen every newborn to detect genetic abnormalities.

In addition to disease and medicine, genomics can contribute to developing novel enzymes that convert biomass to biofuel, which results in higher crop and fuel production, and lower consumer cost. This knowledge should allow better methods of control over the microbes that industry uses to produce biofuels. Genomics could also improve monitoring methods that measure the impact of pollutants on ecosystems and help clean up environmental contaminants. Genomics has aided in developing agrochemicals and pharmaceuticals that could benefit medical science and agriculture.

It sounds great to have all the knowledge we can get from whole-genome sequencing; however, humans have a responsibility to use this knowledge wisely. Otherwise, it could be easy to misuse the power of such knowledge, leading to discrimination based on a person's genetics, human genetic engineering, and other ethical concerns. This information could also lead to legal issues regarding health and privacy.

17.4 Applying Genomics

By the end of this section, you will be able to do the following:

- Explain pharmacogenomics
- Define polygenic

Introducing DNA sequencing and whole genome sequencing projects, particularly the Human Genome project, has expanded the applicability of DNA sequence information. Many fields, such as metagenomics, pharmacogenomics, and mitochondrial genomics are using genomics. Understanding and finding cures for diseases is the most common application of genomics.

Predicting Disease Risk at the Individual Level

Predicting disease risk involves screening currently healthy individuals by genome analysis at the individual level. Health care professionals can recommend intervention with lifestyle changes and drugs before disease onset. However, this approach is most applicable when the problem resides within a single gene defect. Such defects only account for approximately 5 percent of diseases in developed countries. Most of the common diseases, such as heart disease, are multi-factored or **polygenic**, which is a phenotypic characteristic that involves two or more genes, and also involve environmental factors such as diet. In April 2010, scientists at Stanford University published the genome analysis of a healthy individual (Stephen Quake, a scientist at Stanford University, who had his genome sequenced. The analysis predicted his propensity to acquire various diseases. The medical team performed a risk assessment to analyze Quake's percentage of risk for 55 different medical conditions. The team found a rare genetic mutation, which showed him to be at risk for sudden heart attack. The results also predicted that Quake had a 23 percent risk of developing prostate cancer and a 1.4 percent risk of developing Alzheimer's. The scientists used databases and several publications to analyze the genomic data. Even though genomic sequencing is becoming more affordable and analytical tools are becoming more reliable, researchers still must address ethical issues surrounding genomic analysis at a population level.

VISUAL CONNECTION

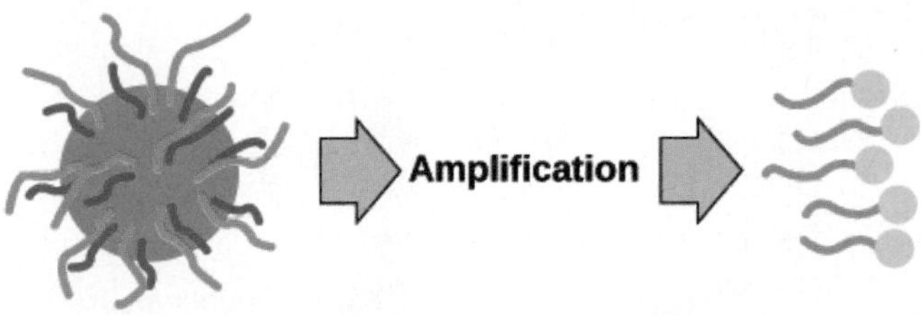

Figure 17.15 *PCA3* is a gene that is expressed in prostate epithelial cells and overexpressed in cancerous cells. A high *PCA3* concentration in urine is indicative of prostate cancer. The *PCA3* test is a better indicator of cancer than the more well known PSA test, which measures the level of PSA (prostate-specific antigen) in the blood.

In 2011, the United States Preventative Services Task Force recommended against using the PSA test to screen healthy men for prostate cancer. Their recommendation is based on evidence that screening does not reduce the risk of death from prostate cancer. Prostate cancer often develops very slowly and does not cause problems, while the cancer treatment can have severe side effects. The *PCA3* test is more accurate, but screening may still result in men who would not have been harmed by the cancer itself suffering side effects from treatment. What do you think? Should all healthy men receive prostate cancer screenings using the *PCA3* or PSA test? Should people in general receive screenings to find out if they have a genetic risk for cancer or other diseases?

Pharmacogenomics and Toxicogenomics

Pharmacogenomics, or toxicogenomics, involves evaluating drug effectiveness and safety on the basis of information from an individual's genomic sequence. We can study genomic responses to drugs using experimental animals (such as laboratory rats or mice) or live cells in the laboratory before embarking on studies with humans. Studying changes in gene expression could provide information about the transcription profile in the drug's presence, which we can use as an early indicator of the potential for toxic effects. For example, genes involved in cellular growth and controlled cell death, when disturbed, could lead to cancerous cell growth. Genome-wide studies can also help to find new genes involved in drug toxicity. Medical professionals can use personal genome sequence information to prescribe medications that will be most effective and least toxic on the basis of the individual patient's genotype. The gene signatures may not be completely accurate, but medical professionals can test them further before pathologic symptoms arise.

Microbial Genomics: Metagenomics

Traditionally, scholars have taught microbiology with the view that it is best to study microorganisms under **pure culture** conditions. This involves isolating a single cell type and culturing it in the laboratory. Because microorganisms can go through several generations in a matter of hours, their gene expression profiles adapt to the new laboratory environment very quickly. In addition, the vast majority of bacterial species resist culturing in isolation. Most microorganisms do not live as isolated entities, but in microbial communities or biofilms. For all of these reasons, pure culture is not always the best way to study microorganisms. **Metagenomics** is the study of the collective genomes of multiple species that grow and interact in an environmental niche. Metagenomics can be used to identify new species more rapidly and to analyze the effect of pollutants on the environment (Figure 17.16).

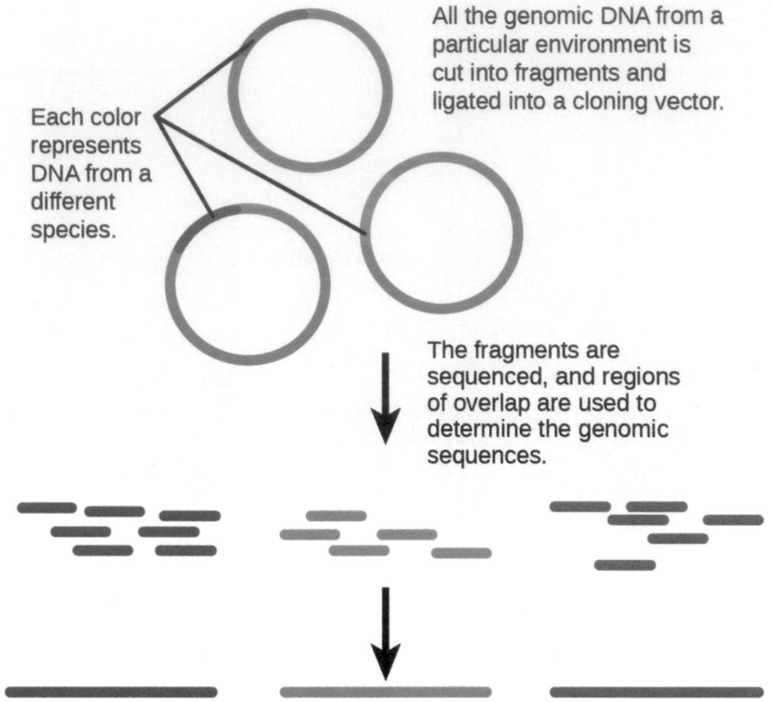

Figure 17.16 Metagenomics involves isolating DNA from multiple species within an environmental niche.

Microbial Genomics: Creation of New Biofuels

Knowledge of the genomics of microorganisms is being used to find better ways to harness biofuels from algae and cyanobacteria. The primary sources of fuel today are coal, oil, wood, and other plant products, such as ethanol. Although plants are renewable resources, there is still a need to find more alternative renewable sources of energy to meet our population's energy demands. The microbial world is one of the largest resources for genes that encode new enzymes and produce new organic compounds, and it remains largely untapped. Microorganisms are used to create products, such as enzymes that are used in research, antibiotics, and other antimicrobial mechanisms. Microbial genomics is helping to develop diagnostic tools, improved vaccines, new disease treatments, and advanced environmental cleanup techniques.

Mitochondrial Genomics

Mitochondria are intracellular organelles that contain their own DNA. Mitochondrial DNA mutates at a rapid rate and scientists often use it to study evolutionary relationships. Another feature that makes studying the mitochondrial genome interesting is that the mitochondrial DNA in most multicellular organisms passes from the mother during the fertilization process. For this reason, scientists often use mitochondrial genomics to trace genealogy.

Experts have used information and clues from DNA samples at crime scenes as evidence in court cases, and they have used genetic markers in forensic analysis. Genomic analysis has also become useful in this field. The first publication showcasing the first use of genomics in forensics came out in 2001. It was a collaborative attempt between academic research institutions and the FBI to solve the mysterious cases of anthrax communicated via the US Postal Service. Using microbial genomics, researchers determined that the culprit used a specific anthrax strain in all the mailings.

Genomics in Agriculture

Genomics can reduce the trials and failures involved in scientific research to a certain extent, which could improve agricultural crop yield quality and quantity. Linking traits to genes or gene signatures helps improve crop breeding to generate hybrids with the most desirable qualities. Scientists use genomic data to identify desirable traits, and then transfer those traits to a different organism. Researchers are discovering how genomics can improve agricultural production's quality and quantity. For example, scientists could use desirable traits to create a useful product or enhance an existing product, such as making a drought-sensitive crop more tolerant of the dry season.

17.5 Genomics and Proteomics

By the end of this section, you will be able to do the following:

- Explain systems biology
- Describe a proteome
- Define protein signature

Proteins are the final products of genes, which help perform the function that the gene encodes. Amino acids comprise proteins and play important roles in the cell. All enzymes (except ribozymes) are proteins that act as catalysts to affect the rate of reactions. Proteins are also regulatory molecules, and some are hormones. Transport proteins, such as hemoglobin, help transport oxygen to various organs. Antibodies that defend against foreign particles are also proteins. In the diseased state, protein function can be impaired because of changes at the genetic level or because of direct impact on a specific protein.

A **proteome** is the entire set of proteins that a cell type produces. We can study proteoms using the knowledge of genomes because genes code for mRNAs, and the mRNAs encode proteins. Although mRNA analysis is a step in the right direction, not all mRNAs are translated into proteins. **Proteomics** is the study of proteomes' function. Proteomics complements genomics and is useful when scientists want to test their hypotheses that they based on genes. Even though all multicellular organisms' cells have the same set of genes, the set of proteins produced in different tissues is different and dependent on gene expression. Thus, the genome is constant, but the proteome varies and is dynamic within an organism. In addition, RNAs can be alternately spliced (cut and pasted to create novel combinations and novel proteins) and many proteins modify themselves after translation by processes such as proteolytic cleavage, phosphorylation, glycosylation, and ubiquitination. There are also protein-protein interactions, which complicate studying proteomes. Although the genome provides a blueprint, the final architecture depends on several factors that can change the progression of events that generate the proteome.

Metabolomics is related to genomics and proteomics. **Metabolomics** involves studying small molecule metabolites in an organism. The **metabolome** is the complete set of metabolites that are related to an organism's genetic makeup. Metabolomics offers an opportunity to compare genetic makeup and physical characteristics, as well as genetic makeup and environmental factors. The goal of metabolome research is to identify, quantify, and catalogue all the metabolites in living organisms' tissues and fluids.

Basic Techniques in Protein Analysis

The ultimate goal of proteomics is to identify or compare the proteins expressed from a given genome under specific conditions, study the interactions between the proteins, and use the information to predict cell behavior or develop drug targets. Just as scientists analyze the genome using the basic DNA sequencing technique, proteomics requires techniques for protein analysis. The basic technique for protein analysis, analogous to DNA sequencing, is mass spectrometry. Mass spectrometry identifies and determines a molecule's characteristics. Advances in spectrometry have allowed researchers to analyze very small protein samples. X-ray crystallography, for example, enables scientists to determine a protein crystal's three-dimensional structure at atomic resolution. Another protein imaging technique, nuclear magnetic resonance (NMR), uses atoms' magnetic properties to determine the protein's three-dimensional structure in aqueous solution. Scientists have also used protein microarrays to study protein interactions. Large-scale adaptations of the basic two-hybrid screen (Figure 17.17) have provided the basis for protein microarrays. Scientists use computer software to analyze the vast amount of data for proteomic analysis.

Genomic- and proteomic-scale analyses are part of **systems biology**, which is the study of whole biological systems (genomes and proteomes) based on interactions within the system. The European Bioinformatics Institute and the Human Proteome Organization (HUPO) are developing and establishing effective tools to sort through the enormous pile of systems biology data. Because proteins are the direct products of genes and reflect activity at the genomic level, it is natural to use proteomes to compare the protein profiles of different cells to identify proteins and genes involved in disease processes. Most pharmaceutical drug trials target proteins. Researchers use information that they obtain from proteomics to identify novel drugs and to understand their mechanisms of action.

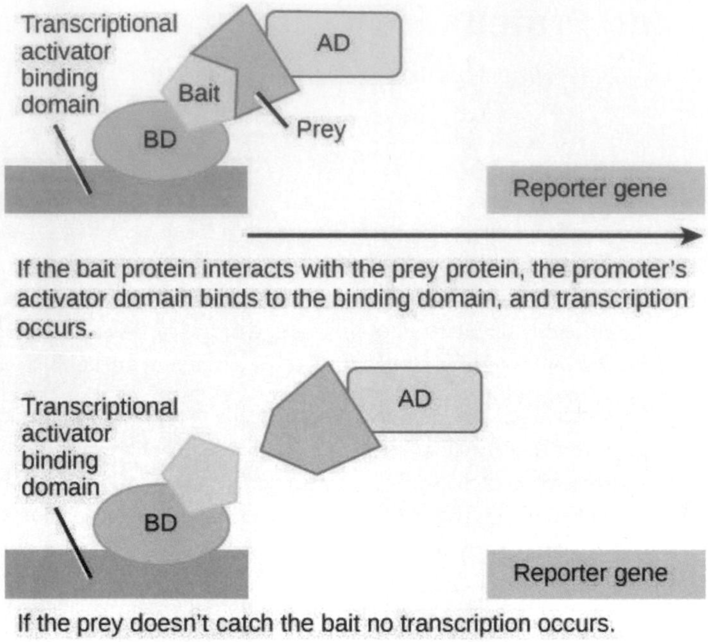

Figure 17.17 Scientists use two-hybrid screening to determine whether two proteins interact. In this method, a transcription factor splits into a DNA-binding domain (BD) and an activator domain (AD). The binding domain is able to bind the promoter in the activator domain's absence, but it does not turn on transcription. The bait protein attaches to the BD, and the prey protein attaches to the AD. Transcription occurs only if the prey "catches" the bait.

Scientists are challenged when implementing proteomic analysis because it is difficult to detect small protein quantities. Although mass spectrometry is good for detecting small protein amounts, variations in protein expression in diseased states can be difficult to discern. Proteins are naturally unstable molecules, which makes proteomic analysis much more difficult than genomic analysis.

Cancer Proteomics

Researchers are studying patients' genomes and proteomes to understand the genetic basis of diseases. The most prominent disease researchers are studying with proteomic approaches is cancer. These approaches improve screening and early cancer detection. Researchers are able to identify proteins whose expression indicates the disease process. An individual protein is a **biomarker**; whereas, a set of proteins with altered expression levels is a **protein signature**. For a biomarker or protein signature to be useful as a candidate for early cancer screening and detection, they must secrete in body fluids, such as sweat, blood, or urine, such that health professionals can perform large-scale screenings in a noninvasive fashion. The current problem with using biomarkers for early cancer detection is the high rate of false-negative results. A **false negative** is an incorrect test result that should have been positive. In other words, many cancer cases go undetected, which makes biomarkers unreliable. Some examples of protein biomarkers in cancer detection are CA-125 for ovarian cancer and PSA for prostate cancer. Protein signatures may be more reliable than biomarkers to detect cancer cells. Researchers are also using proteomics to develop individualized treatment plans, which involves predicting whether or not an individual will respond to specific drugs and the side effects that the individual may experience. Researchers also use proteomics to predict the possibility of disease recurrence.

The National Cancer Institute has developed programs to improve cancer detection and treatment. The Clinical Proteomic Technologies for Cancer and the Early Detection Research Network are efforts to identify protein signatures specific to different cancer types. The Biomedical Proteomics Program identifies protein signatures and designs effective therapies for cancer patients.

KEY TERMS

antibiotic resistance ability of an organism to be unaffected by an antibiotic's actions

biomarker individual protein that is uniquely produced in a diseased state

biotechnology use of biological agents for technological advancement

cDNA library collection of cloned cDNA sequences

cellular cloning production of identical cell populations by binary fission

chain termination method method of DNA sequencing using labeled dideoxynucleotides to terminate DNA replication; it is also called the dideoxy method or the Sanger method

clone exact replica

contig larger sequence of DNA assembled from overlapping shorter sequences

cytogenetic mapping technique that uses a microscope to create a map from stained chromosomes

deoxynucleotide individual DNA monomer (single unit)

dideoxynucleotide individual DNA monomer that is missing a hydroxyl group (–OH)

DNA microarray method to detect gene expression by analyzing many DNA fragments that are fixed to a glass slide or a silicon chip to identify active genes and identify sequences

expressed sequence tag (EST) short STS that is identified with cDNA

false negative incorrect test result that should have been positive

foreign DNA DNA that belongs to a different species or DNA that is artificially synthesized

gel electrophoresis technique used to separate molecules on the basis of size using electric charge

gene targeting method for altering the sequence of a specific gene by introducing the modified version on a vector

gene therapy technique used to cure inheritable diseases by replacing mutant genes with good genes

genetic diagnosis diagnosis of the potential for disease development by analyzing disease-causing genes

genetic engineering alteration of the genetic makeup of an organism

genetic map outline of genes and their location on a chromosome

genetic marker gene or sequence on a chromosome with a known location that is associated with a specific trait

genetic recombination DNA exchange between homologous chromosome pairs

genetic testing process of testing for the presence of disease-causing genes

genetically modified organism (GMO) organism whose genome has been artificially changed

genome annotation process of attaching biological information to gene sequences

genome mapping process of finding the location of genes on each chromosome

genomic library collection of cloned DNA which represents all of the sequences and fragments from a genome

genomics study of entire genomes including the complete set of genes, their nucleotide sequence and organization, and their interactions within a species and with other species

host DNA DNA that is present in the genome of the organism of interest

linkage analysis procedure that analyzes recombining genes to determine if they are linked

lysis buffer solution to break the cell membrane and release cell contents

metabolome complete set of metabolites which are related to an organism's genetic makeup

metabolomics study of small molecule metabolites in an organism

metagenomics study of multiple species' collective genomes that grow and interact in an environmental niche

microsatellite polymorphism variation between individuals in the sequence and number of microsatellite DNA repeats

model organism species that researchers study and use as a model to understand the biological processes in other species represented by the model organism

molecular cloning cloning of DNA fragments

multiple cloning site (MCS) site that multiple restriction endonucleases can recognize

next-generation sequencing group of automated techniques for rapid DNA sequencing

Northern blotting transfer of RNA from a gel to a nylon membrane

pharmacogenomics study of drug interactions with the genome or proteome; also called toxicogenomics

physical map representation of the physical distance between genes or genetic markers

polygenic phenotypic characteristic caused by two or more genes

polymerase chain reaction (PCR) technique to amplify DNA

probe small DNA fragment to determine if the complementary sequence is present in a DNA sample

protease enzyme that breaks down proteins

protein signature set of uniquely expressed proteins in the diseased state

proteome entire set of proteins that cell type produces

proteomics study of proteomes' function

pure culture growth of a single cell type in the laboratory

radiation hybrid mapping information obtained by fragmenting the chromosome with x-rays
recombinant DNA combining DNA fragments that molecular cloning generates that do not exist in nature; also a chimeric molecule
recombinant protein a gene's protein product derived by molecular cloning
reproductive cloning entire organism cloning
restriction endonuclease enzyme that can recognize and cleave specific DNA sequences
restriction fragment length polymorphism (RFLP) variation between individuals in the length of DNA fragments, which restriction endonucleases generate
reverse genetics method of determining the gene's function by starting with the gene itself instead of starting with the gene product
reverse transcriptase PCR (RT-PCR) PCR technique that involves converting RNA to DNA by reverse transcriptase
ribonuclease enzyme that breaks down RNA
sequence mapping mapping information obtained after DNA sequencing
shotgun sequencing method used to sequence multiple DNA fragments to generate the sequence of a large piece of DNA
single nucleotide polymorphism (SNP) variation between individuals in a single nucleotide
Southern blotting DNA transfer from a gel to a nylon membrane
systems biology study of whole biological systems (genomes and proteomes) based on interactions within the system
Ti plasmid plasmid system derived from *Agrobacterium tumifaciens* that scientists have used to introduce foreign DNA into plant cells
transgenic organism that receives DNA from a different species
variable number of tandem repeats (VNTRs) variation in the number of tandem repeats between individuals in the population
whole-genome sequencing process that determines an entire genome's DNA sequence

CHAPTER SUMMARY

17.1 Biotechnology

Nucleic acids can be isolated from cells for the purposes of further analysis by breaking open the cells and enzymatically destroying all other major macromolecules. Fragmented or whole chromosomes can separate on the basis of size by gel electrophoresis. PCR can amplify short DNA or RNA stretches. Researchers can use Southern and Northern blotting to detect the presence of specific short sequences in a DNA or RNA sample. The term "cloning" may refer to cloning small DNA fragments (molecular cloning), cloning cell populations (cellular cloning), or cloning entire organisms (reproductive cloning). Medical professionals perform genetic testing to identify disease-causing genes, and use gene therapy to cure an inheritable disease.

Transgenic organisms possess DNA from a different species, usually generated by molecular cloning techniques. Vaccines, antibiotics, and hormones are examples of products obtained by recombinant DNA technology. Scientists usually create transgenic plants to improve crop plant characteristics.

17.2 Mapping Genomes

Genome mapping is similar to solving a big, complicated puzzle with pieces of information coming from laboratories all over the world. Genetic maps provide an outline for locating genes within a genome, and they estimate the distance between genes and genetic markers on the basis of recombination frequencies during meiosis. Physical maps provide detailed information about the physical distance between the genes. The most detailed information is available through sequence mapping. Researchers combine information from all mapping and sequencing sources to study an entire genome.

17.3 Whole-Genome Sequencing

Whole-genome sequencing is the latest available resource to treat genetic diseases. Some doctors are using whole-genome sequencing to save lives. Genomics has many industrial applications including biofuel development, agriculture, pharmaceuticals, and pollution control. The basic principle of all modern-day sequencing strategies involves the chain termination method of sequencing.

Although the human genome sequences provide key insights to medical professionals, researchers use whole-genome sequences of model organisms to better understand the species' genome. Automation and the decreased cost of whole-genome sequencing may lead to personalized medicine in the future.

17.4 Applying Genomics

Imagination is the only barrier to the applicability of genomics. Researchers are applying genomics to most fields of biology. They use it for personalized medicine, prediction of disease risks at an individual level, studying drug interactions before conducting clinical trials, and studying microorganisms in the environment as opposed to the laboratory. They are also applying it to developments such as generating new biofuels, genealogical assessment using mitochondria, advances in forensic science, and improvements in agriculture.

17.5 Genomics and Proteomics

Proteomics is the study of the entire set of proteins expressed by a given type of cell under certain environmental conditions. In a multicellular organism, different cell types will have different proteomes, and these will vary with environmental changes. Unlike a genome, a proteome is dynamic and in constant flux, which makes it both more complicated and more useful than the knowledge of genomes alone.

Proteomics approaches rely on protein analysis. Researchers are constantly upgrading these techniques. Researchers have used proteomics to study different cancer types. Medical professionals are using different biomarkers and protein signatures to analyze each cancer type. The future goal is to have a personalized treatment plan for each individual.

VISUAL CONNECTION QUESTIONS

1. Figure 17.7 You are working in a molecular biology lab and, unbeknownst to you, your lab partner left the foreign genomic DNA that you are planning to clone on the lab bench overnight instead of storing it in the freezer. As a result, it was degraded by nucleases, but still used in the experiment. The plasmid, on the other hand, is fine. What results would you expect from your molecular cloning experiment?
   a. There will be no colonies on the bacterial plate.
   b. There will be blue colonies only.
   c. There will be blue and white colonies.
   d. The will be white colonies only.
2. Figure 17.8 Do you think Dolly was a Finn-Dorset or a Scottish Blackface sheep?
3. Figure 17.15 In 2011, the United States Preventative Services Task Force recommended against using the PSA test to screen healthy men for prostate cancer. Their recommendation is based on evidence that screening does not reduce the risk of death from prostate cancer. Prostate cancer often develops very slowly and does not cause problems, while the cancer treatment can have severe side effects. The *PCA3* test is more accurate, but screening may still result in men who would not have been harmed by the cancer itself suffering side effects from treatment. What do you think? Should all healthy men be screened for prostate cancer using the *PCA3* or PSA test? Should people in general be screened to find out if they have a genetic risk for cancer or other diseases?

REVIEW QUESTIONS

4. GMOs are created by _______.
   a. generating genomic DNA fragments with restriction endonucleases
   b. introducing recombinant DNA into an organism by any means
   c. overexpressing proteins in *E. coli*
   d. all of the above
5. Gene therapy can be used to introduce foreign DNA into cells _______.
   a. for molecular cloning
   b. by PCR
   c. of tissues to cure inheritable disease
   d. all of the above
6. Insulin produced by molecular cloning:
   a. is of pig origin
   b. is a recombinant protein
   c. is made by the human pancreas
   d. is recombinant DNA
7. Bt toxin is considered to be _______.
   a. a gene for modifying insect DNA
   b. an organic insecticide produced by bacteria
   c. useful for humans to fight against insects
   d. a recombinant protein
8. The Flavr Savr Tomato:
   a. is a variety of vine-ripened tomato in the supermarket
   b. was created to have better flavor and shelf-life
   c. does not undergo soft rot
   d. all of the above
9. ESTs are _______.
   a. generated after a cDNA library is made
   b. unique sequences in the genome
   c. useful for mapping using sequence information
   d. all of the above
10. Linkage analysis _______.
   a. is used to create a physical map
   b. is based on the natural recombination process
   c. requires radiation hybrid mapping
   d. involves breaking and rejoining of DNA artificially

11. Genetic recombination occurs by which process?
    a. independent assortment
    b. crossing over
    c. chromosome segregation
    d. sister chromatids
12. Individual genetic maps in a given species are:
    a. genetically similar
    b. genetically identical
    c. genetically dissimilar
    d. not useful in species analysis
13. Information obtained by microscopic analysis of stained chromosomes is used in:
    a. radiation hybrid mapping
    b. sequence mapping
    c. RFLP mapping
    d. cytogenetic mapping
14. The chain termination method of sequencing:
    a. uses labeled ddNTPs
    b. uses only dideoxynucleotides
    c. uses only deoxynucleotides
    d. uses labeled dNTPs
15. Whole-genome sequencing can be used for advances in:
    a. the medical field
    b. agriculture
    c. biofuels
    d. all of the above
16. Sequencing an individual person's genome
    a. is currently possible
    b. could lead to legal issues regarding discrimination and privacy
    c. could help make informed choices about medical treatment
    d. all of the above
17. What is the most challenging issue facing genome sequencing?
    a. the inability to develop fast and accurate sequencing techniques
    b. the ethics of using information from genomes at the individual level
    c. the availability and stability of DNA
    d. all of the above
18. Genomics can be used in agriculture to:
    a. generate new hybrid strains
    b. improve disease resistance
    c. improve yield
    d. all of the above
19. Genomics can be used on a personal level to:
    a. decrease transplant rejection
    b. predict genetic diseases that a person may have inherited
    c. determine the risks of genetic diseases for an individual's children
    d. all of the above
20. What is a biomarker?
    a. the color coding of different genes
    b. a protein that is uniquely produced in a diseased state
    c. a molecule in the genome or proteome
    d. a marker that is genetically inherited
21. A protein signature is:
    a. the path followed by a protein after it is synthesized in the nucleus
    b. the path followed by a protein in the cytoplasm
    c. a protein expressed on the cell surface
    d. a unique set of proteins present in a diseased state

CRITICAL THINKING QUESTIONS

22. Describe the process of Southern blotting.
23. A researcher wants to study cancer cells from a patient with breast cancer. Is cloning the cancer cells an option?
24. How would a scientist introduce a gene for herbicide resistance into a plant?
25. If you had a chance to get your genome sequenced, what are some questions you might be able to have answered about yourself?
26. Why is so much effort being poured into genome mapping applications?
27. How could a genetic map of the human genome help find a cure for cancer?
28. Explain why metagenomics is probably the most revolutionary application of genomics.
29. How can genomics be used to predict disease risk and treatment options?
30. How has proteomics been used in cancer detection and treatment?
31. What is personalized medicine?

CHAPTER 18
Evolution and the Origin of Species

Figure 18.1 All organisms are products of evolution adapted to their environment. (a) Saguaro (*Carnegiea gigantea*) can soak up 750 liters of water in a single rain storm, enabling these cacti to survive the dry conditions of the Sonora desert in Mexico and the Southwestern United States. (b) The Andean semiaquatic lizard (*Potamites montanicola*) discovered in Peru in 2010 lives between 1,570 to 2,100 meters in elevation, and, unlike most lizards, is nocturnal and swims. Scientists still do not know how these cold-blood animals are able to move in the cold (10 to 15°C) temperatures of the Andean night. (credit a: modification of work by Gentry George, U.S. Fish and Wildlife Service; credit b: modification of work by Germán Chávez and Diego Vásquez, *ZooKeys*)

Chapter Outline

INTRODUCTION All living organisms, from bacteria to baboons to blueberries, evolved at some point from a different species. Although it may seem that living things today stay much the same, that is not the case—evolution is an ongoing process.

The theory of evolution is the unifying theory of biology, meaning it is the framework within which biologists ask questions about the living world. Its power is that it provides direction for predictions about living things that are borne out in ongoing experiments. The Ukrainian-born American geneticist Theodosius Dobzhansky famously wrote that "nothing makes sense in biology except in the light of evolution."[1] He meant that the tenet that all life has evolved and diversified from a common ancestor is the foundation from which we approach all questions in biology.

1Theodosius Dobzhansky. "Biology, Molecular and Organismic." *American Zoologist* 4, no. 4 (1964): 449.

18.1 Understanding Evolution

By the end of this section, you will be able to do the following:

- Describe how scientists developed the present-day theory of evolution
- Define adaptation
- Explain convergent and divergent evolution
- Describe homologous and vestigial structures
- Discuss misconceptions about the theory of evolution

Evolution by natural selection describes a mechanism for how species change over time. Scientists, philosophers, researchers, and others had made suggestions and debated this topic well before Darwin began to explore this idea. Classical Greek philosopher Plato emphasized in his writings that species were static and unchanging, yet there were also ancient Greeks who expressed evolutionary ideas. In the eighteenth century, naturalist Georges-Louis Leclerc Comte de Buffon reintroduced ideas about the evolution of animals and observed that various geographic regions have different plant and animal populations, even when the environments are similar. Some at this time also accepted that there were extinct species.

Also during the eighteenth century, James Hutton, a Scottish geologist and naturalist, proposed that geological change occurred gradually by accumulating small changes from processes operating like they are today over long periods of time. This contrasted with the predominant view that the planet's geology was a consequence of catastrophic events occurring during a relatively brief past. Nineteenth century geologist Charles Lyell popularized Hutton's view. A friend to Darwin. Lyell's ideas were influential on Darwin's thinking: Lyell's notion of the greater age of Earth gave more time for gradual change in species, and the process of change provided an analogy for this change. In the early nineteenth century, Jean-Baptiste Lamarck published a book that detailed a mechanism for evolutionary change. We now refer to this mechanism as an inheritance of acquired characteristics by which the environment causes modifications in an individual, or offspring could use or disuse of a structure during its lifetime, and thus bring about change in a species. While many discredited this mechanism for evolutionary change, Lamarck's ideas were an important influence on evolutionary thought.

Charles Darwin and Natural Selection

In the mid-nineteenth century, two naturalists, Charles Darwin and Alfred Russel Wallace, independently conceived and described the actual mechanism for evolution. Importantly, each naturalist spent time exploring the natural world on expeditions to the tropics. From 1831 to 1836, Darwin traveled around the world on *H.M.S. Beagle*, including stops in South America, Australia, and the southern tip of Africa. Wallace traveled to Brazil to collect insects in the Amazon rainforest from 1848 to 1852 and to the Malay Archipelago from 1854 to 1862. Darwin's journey, like Wallace's later journeys to the Malay Archipelago, included stops at several island chains, the last being the Galápagos Islands west of Ecuador. On these islands, Darwin observed species of organisms on different islands that were clearly similar, yet had distinct differences. For example, the ground finches inhabiting the Galápagos Islands comprised several species with a unique beak shape (Figure 18.2). The species on the islands had a graded series of beak sizes and shapes with very small differences between the most similar. He observed that these finches closely resembled another finch species on the South American mainland. Darwin imagined that the island species might be species modified from one of the original mainland species. Upon further study, he realized that each finch's varied beaks helped the birds acquire a specific type of food. For example, seed-eating finches had stronger, thicker beaks for breaking seeds, and insect-eating finches had spear-like beaks for stabbing their prey.

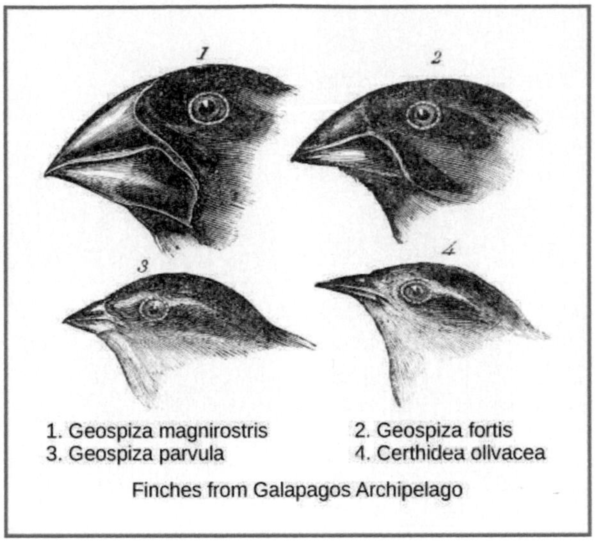

Figure 18.2 Darwin observed that beak shape varies among finch species. He postulated that ancestral species' beaks had adapted over time to equip the finches to acquire different food sources.

Wallace and Darwin both observed similar patterns in other organisms and they independently developed the same explanation for how and why such changes could take place. Darwin called this mechanism natural selection. **Natural selection**, or "survival of the fittest," is the more prolific reproduction of individuals with favorable traits that survive environmental change because of those traits. This leads to evolutionary change.

For example, Darwin observed a population of giant tortoises in the Galápagos Archipelago to have longer necks than those that lived on other islands with dry lowlands. These tortoises were "selected" because they could reach more leaves and access more food than those with short necks. In times of drought when fewer leaves would be available, those that could reach more leaves had a better chance to eat and survive than those that couldn't reach the food source. Consequently, long-necked tortoises would be more likely to be reproductively successful and pass the long-necked trait to their offspring. Over time, only long-necked tortoises would be present in the population.

Natural selection, Darwin argued, was an inevitable outcome of three principles that operated in nature. First, most characteristics of organisms are inherited, or passed from parent to offspring. Although no one, including Darwin and Wallace, knew how this happened at the time, it was a common understanding. Second, more offspring are produced than are able to survive, so resources for survival and reproduction are limited. The capacity for reproduction in all organisms outstrips the availability of resources to support their numbers. Thus, there is competition for those resources in each generation. Both Darwin and Wallace's understanding of this principle came from reading economist Thomas Malthus' essay that explained this principle in relation to human populations. Third, offspring vary among each other in regard to their characteristics and those variations are inherited. Darwin and Wallace reasoned that offspring with inherited characteristics which allow them to best compete for limited resources will survive and have more offspring than those individuals with variations that are less able to compete. Because characteristics are inherited, these traits will be better represented in the next generation. This will lead to change in populations over generations in a process that Darwin called descent with modification. Ultimately, natural selection leads to greater adaptation of the population to its local environment. It is the only mechanism known for adaptive evolution.

In 1858, Darwin and Wallace (Figure 18.3) presented papers at the Linnean Society in London that discussed the idea of natural selection. The following year Darwin's book, *On the Origin of Species*, was published. His book outlined in considerable detail his arguments for evolution by natural selection.

(a) (b)

Figure 18.3 Both (a) Charles Darwin and (b) Alfred Wallace wrote scientific papers on natural selection that they presented together at the Linnean Society in 1858.

It is difficult and time-consuming to document and present examples of evolution by natural selection. The Galápagos finches are an excellent example. Peter and Rosemary Grant and their colleagues have studied Galápagos finch populations every year since 1976 and have provided important evidence of natural selection. The Grants found changes from one generation to the next in beak shape distribution with the medium ground finch on the Galápagos island of Daphne Major. The birds have inherited a variation in their bill shape with some having wide deep bills and others having thinner bills. During a period in which rainfall was higher than normal because of an El Niño, there was a lack of large hard seeds of which the large-billed birds ate; however, there was an abundance of the small soft seeds which the small-billed birds ate. Therefore, the small-billed birds were able to survive and reproduce. In the years following this El Niño, the Grants measured beak sizes in the population and found that the average bill size was smaller. Since bill size is an inherited trait, parents with smaller bills had more offspring and the bill evolved into a much smaller size. As conditions improved in 1987 and larger seeds became more available, the trend toward smaller average bill size ceased.

CAREER CONNECTION

Field Biologist

Many people hike, explore caves, scuba dive, or climb mountains for recreation. People often participate in these activities hoping to see wildlife. Experiencing the outdoors can be incredibly enjoyable and invigorating. What if your job entailed working in the wilderness? Field biologists by definition work outdoors in the "field." The term field in this case refers to any location outdoors, even under water. A field biologist typically focuses research on a certain species, group of organisms, or a single habitat (Figure 18.4).

Figure 18.4 A field biologist tranquilizes a polar bear for study. (credit: Karen Rhode)

One objective of many field biologists includes discovering new, unrecorded species. Not only do such findings expand our understanding of the natural world, but they also lead to important innovations in fields such as medicine and agriculture. Plant and microbial species, in particular, can reveal new medicinal and nutritive knowledge. Other organisms can play key roles in ecosystems or if rare require protection. When discovered, researchers can use these important species as evidence for environmental regulations and laws.

Processes and Patterns of Evolution

Natural selection can only take place if there is **variation**, or differences, among individuals in a population. Importantly, these differences must have some genetic basis; otherwise, the selection will not lead to change in the next generation. This is critical because nongenetic reasons can cause variation among individuals such as an individual's height because of better nutrition rather than different genes.

Genetic diversity in a population comes from two main mechanisms: mutation and sexual reproduction. Mutation, a change in DNA, is the ultimate source of new alleles, or new genetic variation in any population. The genetic changes that mutation causes can have one of three outcomes on the phenotype. A mutation affects the organism's phenotype in a way that gives it reduced fitness—lower likelihood of survival or fewer offspring. A mutation may produce a phenotype with a beneficial effect on fitness. Many mutations will also have no effect on the phenotype's fitness. We call these neutral mutations. Mutations may also have a whole range of effect sizes on the organism's fitness that expresses them in their phenotype, from a small effect to a great effect. Sexual reproduction also leads to genetic diversity: when two parents reproduce, unique combinations of alleles assemble to produce the unique genotypes and thus phenotypes in each offspring.

We call a heritable trait that helps an organism's survival and reproduction in its present environment an **adaptation**. Scientists describe groups of organisms adapting to their environment when a genetic variation occurs over time that increases or maintains the population's "fit" to its environment. A platypus's webbed feet are an adaptation for swimming. A snow leopard's thick fur is an adaptation for living in the cold. A cheetah's fast speed is an adaptation for catching prey.

Whether or not a trait is favorable depends on the current environmental conditions. The same traits are not always selected because environmental conditions can change. For example, consider a plant species that grew in a moist climate and did not need to conserve water. Large leaves were selected because they allowed the plant to obtain more energy from the sun. Large leaves require more water to maintain than small leaves, and the moist environment provided favorable conditions to support large leaves. After thousands of years, the climate changed, and the area no longer had excess water. The direction of natural selection shifted so that plants with small leaves were selected because those populations were able to conserve water to survive the new environmental conditions.

The evolution of species has resulted in enormous variation in form and function. Sometimes, evolution gives rise to groups of organisms that become tremendously different from each other. We call two species that evolve in diverse directions from a common point **divergent evolution**. We can see such divergent evolution in the forms of the reproductive organs of flowering plants which share the same basic anatomies; however, they can look very different as a result of selection in different physical environments and adaptation to different kinds of pollinators (Figure 18.5).

(a) (b)

Figure 18.5 Flowering plants evolved from a common ancestor. Notice that the (a) dense blazing star (*Liatrus spicata*) and the (b) purple coneflower (*Echinacea purpurea*) vary in appearance, yet both share a similar basic morphology. (credit a: modification of work by Drew Avery; credit b: modification of work by Cory Zanker)

In other cases, similar phenotypes evolve independently in distantly related species. For example, flight has evolved in both bats and insects, and they both have structures we refer to as wings, which are adaptations to flight. However, bat and insect wings have evolved from very different original structures. We call this phenomenon **convergent evolution**, where similar traits evolve independently in species that do not share a common ancestry. The two species came to the same function, flying, but did so separately from each other.

These physical changes occur over enormous time spans and help explain how evolution occurs. Natural selection acts on individual organisms, which can then shape an entire species. Although natural selection may work in a single generation on an individual, it can take thousands or even millions of years for an entire species' genotype to evolve. It is over these large time spans that life on earth has changed and continues to change.

Evidence of Evolution

The evidence for evolution is compelling and extensive. Looking at every level of organization in living systems, biologists see the signature of past and present evolution. Darwin dedicated a large portion of his book, *On the Origin of Species*, to identifying patterns in nature that were consistent with evolution, and since Darwin, our understanding has become clearer and broader.

Fossils

Fossils provide solid evidence that organisms from the past are not the same as those today, and fossils show a progression of evolution. Scientists determine the age of fossils and categorize them from all over the world to determine when the organisms lived relative to each other. The resulting fossil record tells the story of the past and shows the evolution of form over millions of years (Figure 18.6). For example, scientists have recovered highly detailed records showing the evolution of humans and horses (Figure 18.6). The whale flipper shares a similar morphology to bird and mammal appendages (Figure 18.7) indicating that these species share a common ancestor.

Figure 18.6 In this (a) display, fossil hominids are arranged from oldest (bottom) to newest (top). As hominids evolved, the skull's shape changed. An artist's rendition of (b) extinct species of the genus *Equus* reveals that these ancient species resembled the modern horse (*Equus ferus*) but varied in size.

Anatomy and Embryology

Another type of evidence for evolution is the presence of structures in organisms that share the same basic form. For example, the bones in human, dog, bird, and whale appendages all share the same overall construction (Figure 18.7) resulting from their origin in a common ancestor's appendages. Over time, evolution led to changes in the bones' shapes and sizes different species, but they have maintained the same overall layout. Scientists call these synonymous parts **homologous structures**.

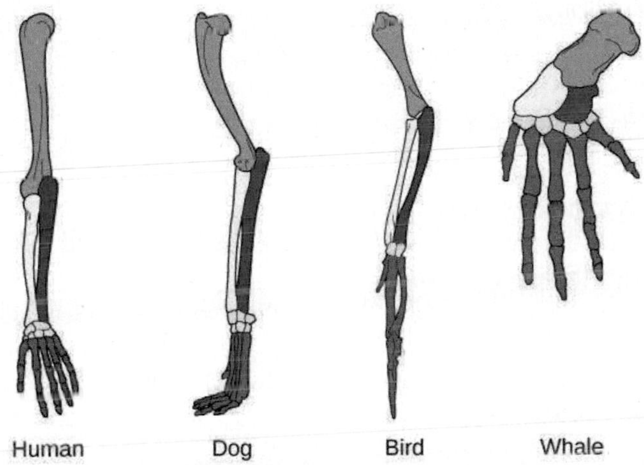

Figure 18.7 The similar construction of these appendages indicates that these organisms share a common ancestor.

Some structures exist in organisms that have no apparent function at all, and appear to be residual parts from a past common ancestor. We call these unused structures without function **vestigial structures**. Other examples of vestigial structures are wings on flightless birds, leaves on some cacti, and hind leg bones in whales. Not all similarities represent homologous structures. As explained in Determining Evolutionary Relationships, when similar characteristics occur because of environmental constraints and not due to a close evolutionary relationship, it is an analogy or homoplasy. For example, insects use wings to fly like bats and birds, but the wing structure and embryonic origin are completely different. These are analogous structures (Figure 20.8).

LINK TO LEARNING

Watch this video (http://openstax.org/l/bone_structures) exploring the bones in the human body.

Another evidence of evolution is the convergence of form in organisms that share similar environments. For example, species of unrelated animals, such as the arctic fox and ptarmigan, living in the arctic region have been selected for seasonal white phenotypes during winter to blend with the snow and ice (Figure 18.8). These similarities occur not because of common ancestry, but because of similar selection pressures—the benefits of predators not seeing them.

(a) (b)

Figure 18.8 The white winter coat of the (a) arctic fox and the (b) ptarmigan's plumage are adaptations to their environments. (credit a: modification of work by Keith Morehouse)

Embryology, the study of the anatomy of an organism's development to its adult form, also provides evidence of relatedness between now widely divergent groups of organisms. Mutational tweaking in the embryo can have such magnified consequences in the adult that tends to conserve embryo formation. As a result, structures that are absent in some groups often appear in their embryonic forms and disappear when they reach the adult or juvenile form. For example, all vertebrate embryos, including humans, exhibit gill slits and tails at some point in their early development. These disappear in the adults of terrestrial groups but adult forms of aquatic groups such as fish and some amphibians maintain them. Great ape embryos, including humans, have a tail structure during their development that they lose when they are born.

Biogeography

The geographic distribution of organisms on the planet follows patterns that we can explain best by evolution in conjunction with tectonic plate movement over geological time. Broad groups that evolved before the supercontinent Pangaea broke up (about 200 million years ago) are distributed worldwide. Groups that evolved since the breakup appear uniquely in regions of the planet, such as the unique flora and fauna of northern continents that formed from the supercontinent Laurasia and of the southern continents that formed from the supercontinent Gondwana. The presence of members of the plant family Proteaceae in Australia, southern Africa, and South America was most predominant prior to the southern supercontinent Gondwana breaking up.

Marsupial diversification in Australia and the absence of other mammals reflect Australia's long isolation. Australia has an abundance of endemic species—species found nowhere else—which is typical of islands whose isolation by expanses of water prevents species to migrate. Over time, these species diverge evolutionarily into new species that look very different from their ancestors that may exist on the mainland. Australia's marsupials, the Galápagos' finches, and many species on the Hawaiian Islands are all unique to their one point of origin, yet they display distant relationships to ancestral species on mainlands.

Molecular Biology

Like anatomical structures, the molecular structures of life reflect descent with modification. DNA's universality reflects evidence of a common ancestor for all of life. Fundamental divisions in life between the genetic code, DNA replication, and expression are reflected in major structural differences in otherwise conservative structures such as ribosome components and membrane structures. In general, the relatedness of groups of organisms is reflected in the similarity of their DNA sequences—exactly the pattern that we would expect from descent and diversification from a common ancestor.

DNA sequences have also shed light on some of the mechanisms of evolution. For example, it is clear that the evolution of new functions for proteins commonly occurs after gene duplication events that allow freely modifying one copy by mutation, selection, or drift (changes in a population's gene pool resulting from chance), while the second copy continues to produce a functional protein.

Misconceptions of Evolution

Although the theory of evolution generated some controversy when Darwin first proposed it, biologists almost universally accepted it, particularly younger biologists, within 20 years after publication of *On the Origin of Species*. Nevertheless, the theory of evolution is a difficult concept and misconceptions about how it works abound.

LINK TO LEARNING

This site (http://openstax.org/l/misconceptions) addresses some of the main misconceptions associated with the theory of evolution.

Evolution Is Just a Theory

Critics of the theory of evolution dismiss its importance by purposefully confounding the everyday usage of the word "theory" with the way scientists use the word. In science, we understand a "theory" to be a body of thoroughly tested and verified explanations for a set of observations of the natural world. Scientists have a theory of the atom, a theory of gravity, and the theory of relativity, each which describes understood facts about the world. In the same way, the theory of evolution describes facts about the living world. As such, a theory in science has survived significant efforts to discredit it by scientists. In contrast, a "theory" in common vernacular is a word meaning a guess or suggested explanation. This meaning is more akin to the scientific concept of "hypothesis." When critics of evolution say it is "just a theory," they are implying that there is little evidence supporting it and that it is still in the process of rigorous testing. This is a mischaracterization.

Individuals Evolve

Evolution is the change in a population's genetic composition over time, specifically over generations, resulting from differential reproduction of individuals with certain alleles. Individuals do change over their lifetime, obviously, but this is development and involves changes programmed by the set of genes the individual acquired at birth in coordination with the individual's environment. When thinking about the evolution of a characteristic, it is probably best to think about the change of the average value of the characteristic in the population over time. For example, when natural selection leads to bill-size change in medium ground finches in the Galápagos, this does not mean that individual bills on the finches are changing. If one measures the average bill size among all individuals in the population at one time and then measures them in the population several years later, this average value will be different as a result of evolution. Although some individuals may survive from the first time to the second, they will still have the same bill size; however, there will be many new individuals who contribute to the shift in average bill size.

Evolution Explains the Origin of Life

It is a common misunderstanding that evolution includes an explanation of life's origins. Conversely, some of the theory's critics believe that it cannot explain the origin of life. The theory does not try to explain the origin of life. The theory of evolution explains how populations change over time and how life diversifies the origin of species. It does not shed light on the beginnings of life including the origins of the first cells, which define life. Importantly, biologists believe that the presence of life on Earth precludes the possibility that the events that led to life on Earth can repeat themselves because the intermediate stages would immediately become food for existing living things.

However, once a mechanism of inheritance was in place in the form of a molecule like DNA either within a cell or pre-cell, these entities would be subject to the principle of natural selection. More effective reproducers would increase in frequency at the expense of inefficient reproducers. While evolution does not explain the origin of life, it may have something to say about some of the processes operating once pre-living entities acquired certain properties.

Organisms Evolve on Purpose

Statements such as "organisms evolve in response to a change in an environment" are quite common, but such statements can lead to two types of misunderstandings. First, do not interpret the statement to mean that individual organisms evolve. The statement is shorthand for "a population evolves in response to a changing environment." However, a second misunderstanding may arise by interpreting the statement to mean that the evolution is somehow intentional. A changed environment results in some individuals in the population, those with particular phenotypes, benefiting and therefore producing proportionately more offspring than other phenotypes. This results in change in the population if the characteristics are genetically determined.

It is also important to understand that the variation that natural selection works on is already in a population and does not arise in response to an environmental change. For example, applying antibiotics to a population of bacteria will, over time, select a population of bacteria that are resistant to antibiotics. The resistance, which a gene causes, did not arise by mutation because of applying the antibiotic. The gene for resistance was already present in the bacteria's gene pool, likely at a low frequency. The antibiotic, which kills the bacterial cells without the resistance gene, strongly selects individuals that are resistant, since these would be the only ones that survived and divided. Experiments have demonstrated that mutations for antibiotic resistance do not arise as a result of antibiotic.

In a larger sense, evolution is not goal directed. Species do not become "better" over time. They simply track their changing environment with adaptations that maximize their reproduction in a particular environment at a particular time. Evolution has no goal of making faster, bigger, more complex, or even smarter species, despite the commonness of this kind of language in popular discourse. What characteristics evolve in a species are a function of the variation present and the environment, both of which are constantly changing in a nondirectional way. A trait that fits in one environment at one time may well be fatal at some point in the future. This holds equally well for insect and human species.

18.2 Formation of New Species

By the end of this section, you will be able to do the following:

- Define species and describe how scientists identify species as different
- Describe genetic variables that lead to speciation
- Identify prezygotic and postzygotic reproductive barriers
- Explain allopatric and sympatric speciation
- Describe adaptive radiation

Although all life on earth shares various genetic similarities, only certain organisms combine genetic information by sexual reproduction and have offspring that can then successfully reproduce. Scientists call such organisms members of the same biological species.

Species and the Ability to Reproduce

A **species** is a group of individual organisms that interbreed and produce fertile, viable offspring. According to this definition, one species is distinguished from another when, in nature, it is not possible for matings between individuals from each species to produce fertile offspring.

Members of the same species share both external and internal characteristics, which develop from their DNA. The closer relationship two organisms share, the more DNA they have in common, just like people and their families. People's DNA is likely to be more like their father or mother's DNA than their cousin or grandparent's DNA. Organisms of the same species have the highest level of DNA alignment and therefore share characteristics and behaviors that lead to successful reproduction.

Species' appearance can be misleading in suggesting an ability or inability to mate. For example, even though domestic dogs (*Canis lupus familiaris*) display phenotypic differences, such as size, build, and coat, most dogs can interbreed and produce viable puppies that can mature and sexually reproduce (Figure 18.9).

Figure 18.9 The (a) poodle and (b) cocker spaniel can reproduce to produce a breed known as (c) the cockapoo. (credit a: modification of work by Sally Eller, Tom Reese; credit b: modification of work by Jeremy McWilliams; credit c: modification of work by Kathleen Conklin)

In other cases, individuals may appear similar although they are not members of the same species. For example, even though bald eagles (*Haliaeetus leucocephalus*) and African fish eagles (*Haliaeetus vocifer*) are both birds and eagles, each belongs to a separate species group (Figure 18.10). If humans were to artificially intervene and fertilize a bald eagle's egg with an African fish eagle's sperm and a chick did hatch, that offspring, called a **hybrid** (a cross between two species), would probably be infertile—unable to successfully reproduce after it reached maturity. Different species may have different genes that are active in development; therefore, it may not be possible to develop a viable offspring with two different sets of directions. Thus, even though hybridization may take place, the two species still remain separate.

(a) (b)

Figure 18.10 The (a) African fish eagle is similar in appearance to the (b) bald eagle, but the two birds are members of different species. (credit a: modification of work by Nigel Wedge; credit b: modification of work by U.S. Fish and Wildlife Service)

Populations of species share a gene pool: a collection of all the gene variants in the species. Again, the basis to any changes in a group or population of organisms must be genetic for this is the only way to share and pass on traits. When variations occur within a species, they can only pass to the next generation along two main pathways: asexual reproduction or sexual reproduction. The change will pass on asexually simply if the reproducing cell possesses the changed trait. For the changed trait to pass on by sexual reproduction, a gamete, such as a sperm or egg cell, must possess the changed trait. In other words, sexually-reproducing organisms can experience several genetic changes in their body cells, but if these changes do not occur in a sperm or egg cell, the changed trait will never reach the next generation. Only heritable traits can evolve. Therefore, reproduction plays a paramount role for genetic change to take root in a population or species. In short, organisms must be able to reproduce with each other to pass new traits to offspring.

Speciation

The biological definition of species, which works for sexually reproducing organisms, is a group of actual or potential interbreeding individuals. There are exceptions to this rule. Many species are similar enough that hybrid offspring are possible and may often occur in nature, but for the majority of species this rule generally holds. The presence in nature of hybrids between similar species suggests that they may have descended from a single interbreeding species, and the speciation process may not yet be completed.

Given the extraordinary diversity of life on the planet there must be mechanisms for **speciation**: the formation of two species from one original species. Darwin envisioned this process as a branching event and diagrammed the process in the only illustration in *On the Origin of Species* (Figure 18.11a). Compare this illustration to the diagram of elephant evolution (Figure 18.11), which shows that as one species changes over time, it branches to form more than one new species, repeatedly, as long as the population survives or until the organism becomes extinct.

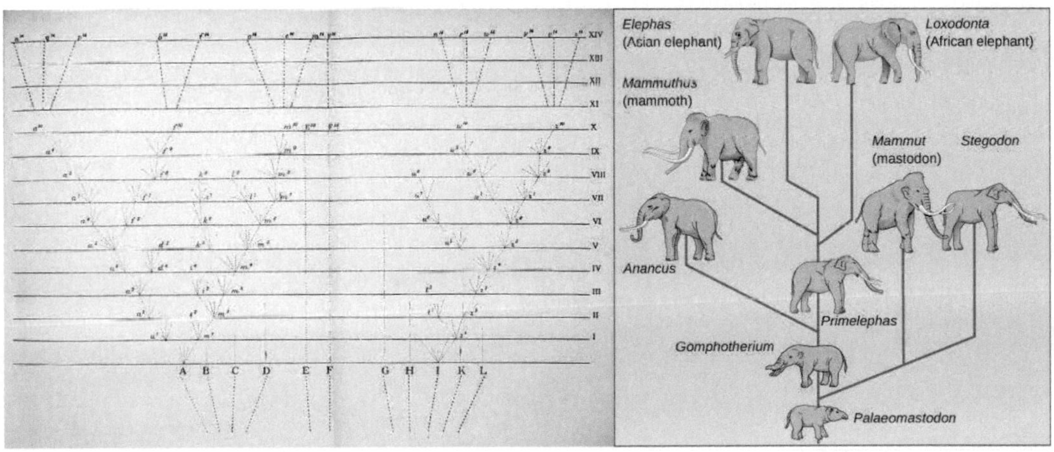

(a) (b)

Figure 18.11 The only illustration in Darwin's *On the Origin of Species* is (a) a diagram showing speciation events leading to biological diversity. The diagram shows similarities to phylogenetic charts that today illustrate the relationships of species. (b) Modern elephants evolved from the *Palaeomastodon*, a species that lived in Egypt 35–50 million years ago.

For speciation to occur, two new populations must form from one original population and they must evolve in such a way that it becomes impossible for individuals from the two new populations to interbreed. Biologists have proposed mechanisms by which this could occur that fall into two broad categories. **Allopatric speciation** (allo- = "other"; -patric = "homeland") involves geographic separation of populations from a parent species and subsequent evolution. **Sympatric speciation** (sym- = "same"; -patric = "homeland") involves speciation occurring within a parent species remaining in one location.

Biologists think of speciation events as the splitting of one ancestral species into two descendant species. There is no reason why more than two species might not form at one time except that it is less likely and we can conceptualize multiple events as single splits occurring close in time.

Allopatric Speciation

A geographically continuous population has a gene pool that is relatively homogeneous. Gene flow, the movement of alleles across a species' range, is relatively free because individuals can move and then mate with individuals in their new location. Thus, an allele's frequency at one end of a distribution will be similar to the allele's frequency at the other end. When populations become geographically discontinuous, it prevents alleles' free-flow. When that separation lasts for a period of time, the two populations are able to evolve along different trajectories. Thus, their allele frequencies at numerous genetic loci gradually become increasingly different as new alleles independently arise by mutation in each population. Typically, environmental conditions, such as climate, resources, predators, and competitors for the two populations will differ causing natural selection to favor divergent adaptations in each group.

Isolation of populations leading to allopatric speciation can occur in a variety of ways: a river forming a new branch, erosion creating a new valley, a group of organisms traveling to a new location without the ability to return, or seeds floating over the ocean to an island. The nature of the geographic separation necessary to isolate populations depends entirely on the organism's biology and its potential for dispersal. If two flying insect populations took up residence in separate nearby valleys, chances are, individuals from each population would fly back and forth continuing gene flow. However, if a new lake divided two rodent populations continued gene flow would be unlikely; therefore, speciation would be more likely.

Biologists group allopatric processes into two categories: dispersal and vicariance. **Dispersal** is when a few members of a species move to a new geographical area, and **vicariance** is when a natural situation arises to physically divide organisms.

Scientists have documented numerous cases of allopatric speciation taking place. For example, along the west coast of the United States, two separate spotted owl subspecies exist. The northern spotted owl has genetic and phenotypic differences from its close relative: the Mexican spotted owl, which lives in the south (Figure 18.12).

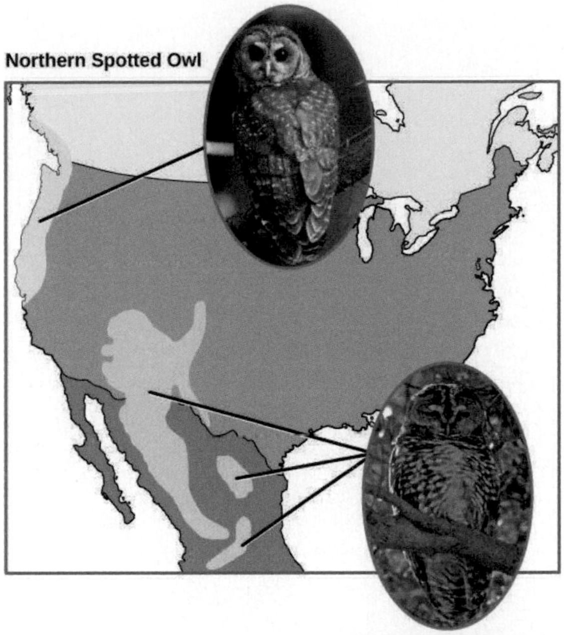

Figure 18.12 The northern spotted owl and the Mexican spotted owl inhabit geographically separate locations with different climates and ecosystems. The owl is an example of allopatric speciation. (credit "northern spotted owl": modification of work by John and Karen Hollingsworth; credit "Mexican spotted owl": modification of work by Bill Radke)

Additionally, scientists have found that the further the distance between two groups that once were the same species, the more likely it is that speciation will occur. This seems logical because as the distance increases, the various environmental factors would likely have less in common than locations in close proximity. Consider the two owls: in the north, the climate is cooler than in the south. The types of organisms in each ecosystem differ, as do their behaviors and habits. Also, the hunting habits and prey choices of the southern owls vary from the northern owls. These variances can lead to evolved differences in the owls, and speciation likely will occur.

Adaptive Radiation

In some cases, a population of one species disperses throughout an area, and each finds a distinct niche or isolated habitat. Over time, the varied demands of their new lifestyles lead to multiple speciation events originating from a single species. We call this **adaptive radiation** because many adaptations evolve from a single point of origin; thus, causing the species to radiate into several new ones. Island archipelagos like the Hawaiian Islands provide an ideal context for adaptive radiation events because water surrounds each island which leads to geographical isolation for many organisms. The Hawaiian honeycreeper illustrates one example of adaptive radiation. From a single species, the founder species, numerous species have evolved, including the six in Figure 18.13.

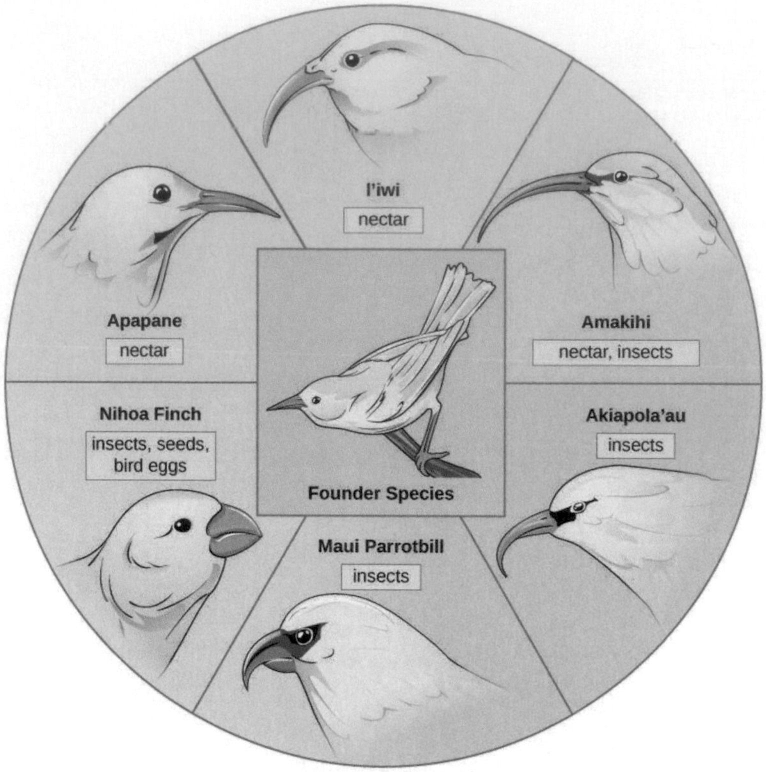

Figure 18.13 The honeycreeper birds illustrate adaptive radiation. From one original species of bird, multiple others evolved, each with its own distinctive characteristics.

Notice the differences in the species' beaks in Figure 18.13. Evolution in response to natural selection based on specific food sources in each new habitat led to evolution of a different beak suited to the specific food source. The seed-eating bird has a thicker, stronger beak which is suited to break hard nuts. The nectar-eating birds have long beaks to dip into flowers to reach the nectar. The insect-eating birds have beaks like swords, appropriate for stabbing and impaling insects. Darwin's finches are another example of adaptive radiation in an archipelago.

LINK TO LEARNING

Watch this video (http://openstax.org/l/bird_evolution) to see how scientists use evidence to understand how birds evolved.

Sympatric Speciation

Can divergence occur if no physical barriers are in place to separate individuals who continue to live and reproduce in the same habitat? The answer is yes. We call the process of speciation within the same space sympatric. The prefix "sym" means same, so "sympatric" means "same homeland" in contrast to "allopatric" meaning "other homeland." Scientists have proposed and studied many mechanisms.

One form of sympatric speciation can begin with a serious chromosomal error during cell division. In a normal cell division event chromosomes replicate, pair up, and then separate so that each new cell has the same number of chromosomes. However, sometimes the pairs separate and the end cell product has too many or too few individual chromosomes in a condition that we call **aneuploidy** (Figure 18.14).

VISUAL CONNECTION

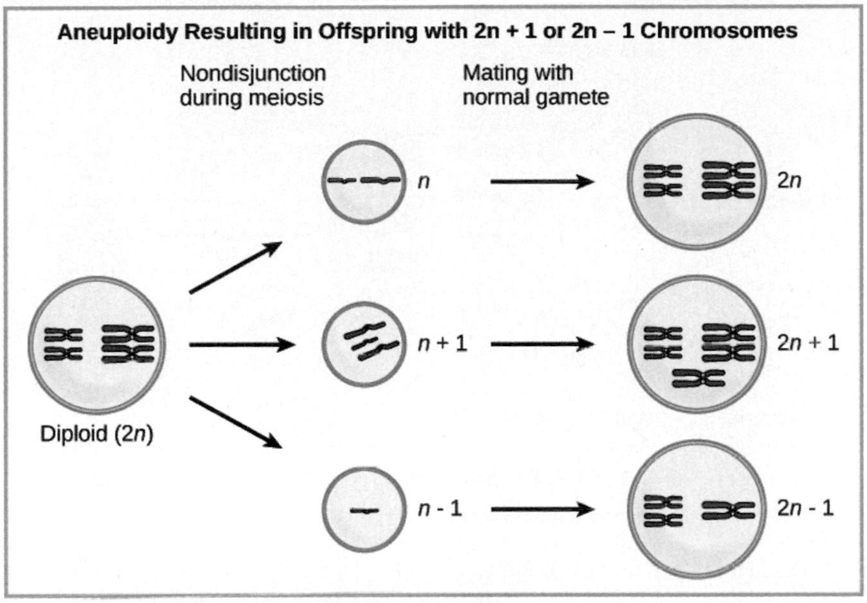

Figure 18.14 Aneuploidy results when the gametes have too many or too few chromosomes due to nondisjunction during meiosis. In this example, the resulting offspring will have $2n+1$ or $2n-1$ chromosomes.

Which is most likely to survive, offspring with $2n+1$ chromosomes or offspring with $2n-1$ chromosomes?

Polyploidy is a condition in which a cell or organism has an extra set, or sets, of chromosomes. Scientists have identified two main types of polyploidy that can lead to reproductive isolation of an individual in the polyploidy state. Reproductive isolation is the inability to interbreed. In some cases, a polyploid individual will have two or more complete sets of chromosomes from its own species in a condition that we call **autopolyploidy** (Figure 18.15). The prefix "auto-" means "self," so the term means multiple chromosomes from one's own species. Polyploidy results from an error in meiosis in which all of the chromosomes move into one cell instead of separating.

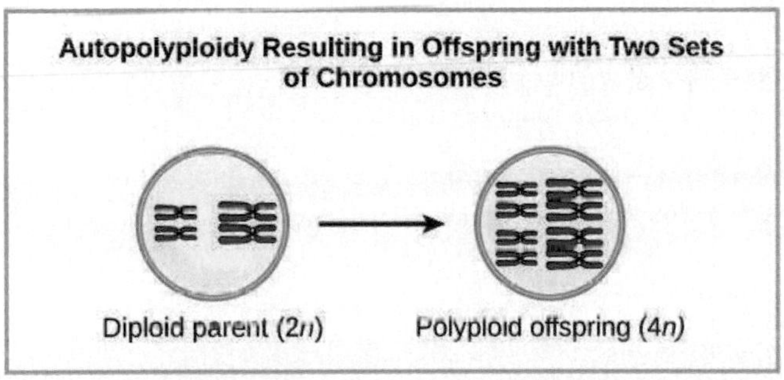

Figure 18.15 Autopolyploidy results when mitosis is not followed by cytokinesis.

For example, if a plant species with $2n = 6$ produces autopolyploid gametes that are also diploid ($2n = 6$, when they should be $n = 3$), the gametes now have twice as many chromosomes as they should have. These new gametes will be incompatible with the normal gametes that this plant species produces. However, they could either self-pollinate or reproduce with other autopolyploid plants with gametes having the same diploid number. In this way, sympatric speciation can occur quickly by forming offspring with $4n$ that we call a tetraploid. These individuals would immediately be able to reproduce only with those of this new kind and not those of the ancestral species.

The other form of polyploidy occurs when individuals of two different species reproduce to form a viable offspring that we call an **allopolyploid**. The prefix "allo-" means "other" (recall from allopatric): therefore, an allopolyploid occurs when gametes from two different species combine. Figure 18.16 illustrates one possible way an allopolyploid can form. Notice how it takes two generations, or two reproductive acts, before the viable fertile hybrid results.

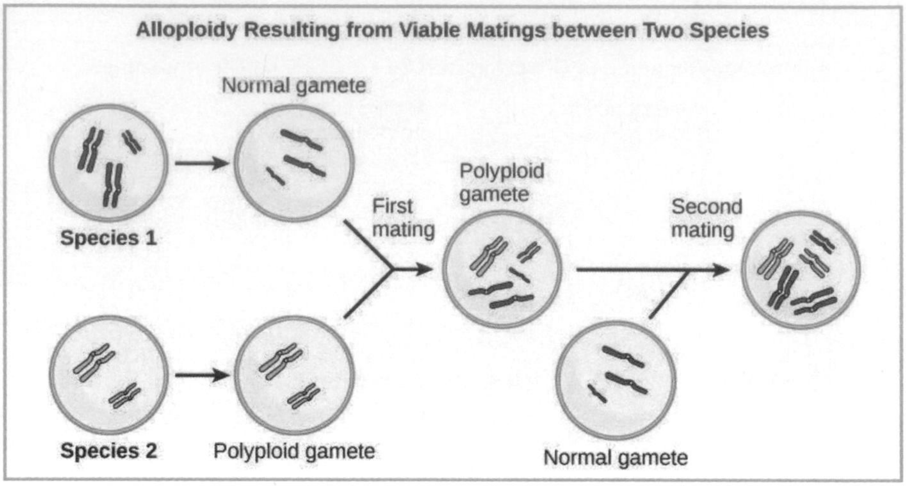

Figure 18.16 Alloploidy results when two species mate to produce viable offspring. In this example, a normal gamete from one species fuses with a polyploidy gamete from another. Two matings are necessary to produce viable offspring.

The cultivated forms of wheat, cotton, and tobacco plants are all allopolyploids. Although polyploidy occurs occasionally in animals, it takes place most commonly in plants. (Animals with any of the types of chromosomal aberrations that we describe here are unlikely to survive and produce normal offspring.) Scientists have discovered more than half of all plant species studied relate back to a species evolved through polyploidy. With such a high rate of polyploidy in plants, some scientists hypothesize that this mechanism takes place more as an adaptation than as an error.

Reproductive Isolation

Given enough time, the genetic and phenotypic divergence between populations will affect characters that influence reproduction: if individuals of the two populations were brought together, mating would be less likely, but if mating occurred, offspring would be nonviable or infertile. Many types of diverging characters may affect the **reproductive isolation**, the ability to interbreed, of the two populations.

Reproductive isolation can take place in a variety of ways. Scientists organize them into two groups: prezygotic barriers and postzygotic barriers. Recall that a zygote is a fertilized egg: the first cell of an organism's development that reproduces sexually. Therefore, a **prezygotic barrier** is a mechanism that blocks reproduction from taking place. This includes barriers that prevent fertilization when organisms attempt reproduction. A **postzygotic barrier** occurs after zygote formation. This includes organisms that don't survive the embryonic stage and those that are born sterile.

Some types of prezygotic barriers prevent reproduction entirely. Many organisms only reproduce at certain times of the year, often just annually. Differences in breeding schedules, which we call **temporal isolation**, can act as a form of reproductive isolation. For example, two frog species inhabit the same area, but one reproduces from January to March; whereas, the other reproduces from March to May (Figure 18.17).

(a) (b)

Figure 18.17 These two related frog species exhibit temporal reproductive isolation. (a) *Rana aurora* breeds earlier in the year than (b) *Rana boylii*. (credit a: modification of work by Mark R. Jennings, USFWS; credit b: modification of work by Alessandro Catenazzi)

In some cases, populations of a species move or are moved to a new habitat and take up residence in a place that no longer overlaps with the same species' other populations. We call this situation **habitat isolation**. Reproduction with the parent species ceases, and a new group exists that is now reproductively and genetically independent. For example, a cricket population that was divided after a flood could no longer interact with each other. Over time, natural selection forces, mutation, and genetic drift will likely result in the two groups diverging (Figure 18.18).

Figure 18.18 Speciation can occur when two populations occupy different habitats. The habitats need not be far apart. The cricket (a) *Gryllus pennsylvanicus* prefers sandy soil, and the cricket (b) *Gryllus firmus* prefers loamy soil. The two species can live in close proximity, but because of their different soil preferences, they became genetically isolated.

Behavioral isolation occurs when the presence or absence of a specific behavior prevents reproduction. For example, male fireflies use specific light patterns to attract females. Various firefly species display their lights differently. If a male of one species tried to attract the female of another, she would not recognize the light pattern and would not mate with the male.

Other prezygotic barriers work when differences in their gamete cells (eggs and sperm) prevent fertilization from taking place. We call this a **gametic barrier**. Similarly, in some cases closely related organisms try to mate, but their reproductive structures simply do not fit together. For example, damselfly males of different species have differently shaped reproductive organs. If one species tries to mate with the female of another, their body parts simply do not fit together. (Figure 18.19).

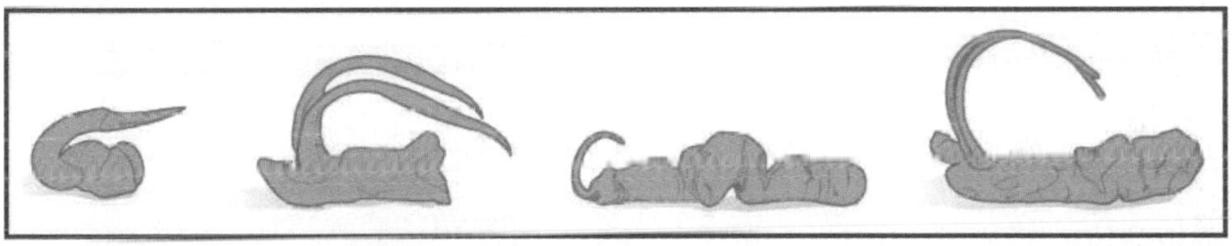

Figure 18.19 The shape of the male reproductive organ varies among male damselfly species, and is only compatible with the female of that species. Reproductive organ incompatibility keeps the species reproductively isolated.

In plants, certain structures aimed to attract one type of pollinator simultaneously prevent a different pollinator from accessing the pollen. The tunnel through which an animal must access nectar can vary widely in length and diameter, which prevents the plant from cross-pollinating with a different species (Figure 18.20).

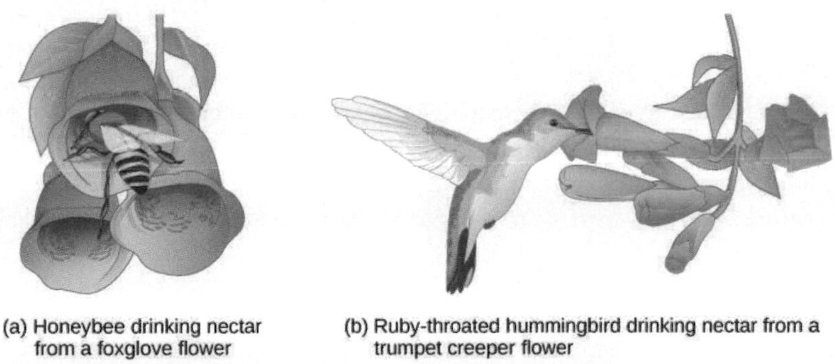

Figure 18.20 Some flowers have evolved to attract certain pollinators. The (a) wide foxglove flower is adapted for pollination by bees, while the (b) long, tube-shaped trumpet creeper flower is adapted for pollination by hummingbirds.

When fertilization takes place and a zygote forms, postzygotic barriers can prevent reproduction. Hybrid individuals in many cases cannot form normally in the womb and simply do not survive past the embryonic stages. We call this **hybrid inviability** because the hybrid organisms simply are not viable. In another postzygotic situation, reproduction leads to hybrid birth and growth that is sterile. Therefore, the organisms are unable to reproduce offspring of their own. We call this hybrid sterility.

Habitat Influence on Speciation

Sympatric speciation may also take place in ways other than polyploidy. For example, consider a fish species that lives in a lake. As the population grows, competition for food increases. Under pressure to find food, suppose that a group of these fish had the genetic flexibility to discover and feed off another resource that other fish did not use. What if this new food source was located at a different depth of the lake? Over time, those feeding on the second food source would interact more with each other than the other fish; therefore, they would breed together as well. Offspring of these fish would likely behave as their parents: feeding and living in the same area and keeping separate from the original population. If this group of fish continued to remain separate from the first population, eventually sympatric speciation might occur as more genetic differences accumulated between them.

This scenario does play out in nature, as do others that lead to reproductive isolation. One such place is Lake Victoria in Africa, famous for its sympatric speciation of cichlid fish. Researchers have found hundreds of sympatric speciation events in these fish, which have not only happened in great number, but also over a short period of time. Figure 18.21 shows this type of speciation among a cichlid fish population in Nicaragua. In this locale, two types of cichlids live in the same geographic location but have come to have different morphologies that allow them to eat various food sources.

Figure 18.21 Cichlid fish from Lake Apoyeque, Nicaragua, show evidence of sympatric speciation. Lake Apoyeque, a crater lake, is 1800 years old, but genetic evidence indicates that a single population of cichlid fish populated the lake only 100 years ago. Nevertheless, two populations with distinct morphologies and diets now exist in the lake, and scientists believe these populations may be in an early stage of speciation.

18.3 Reconnection and Speciation Rates

By the end of this section, you will be able to do the following:

- Describe pathways of species evolution in hybrid zones
- Explain the two major theories on rates of speciation

Speciation occurs over a span of evolutionary time, so when a new species arises, there is a transition period during which the closely related species continue to interact.

Reconnection

After speciation, two species may recombine or even continue interacting indefinitely. Individual organisms will mate with any nearby individual with whom they are capable of breeding. We call an area where two closely related species continue to interact and reproduce, forming hybrids a **hybrid zone**. Over time, the hybrid zone may change depending on the fitness of the hybrids and the reproductive barriers (Figure 18.22). If the hybrids are less fit than the parents, speciation reinforcement occurs, and the species continue to diverge until they can no longer mate and produce viable offspring. If reproductive barriers weaken, fusion occurs and the two species become one. Barriers remain the same if hybrids are fit and reproductive: stability may occur and hybridization continues.

VISUAL CONNECTION

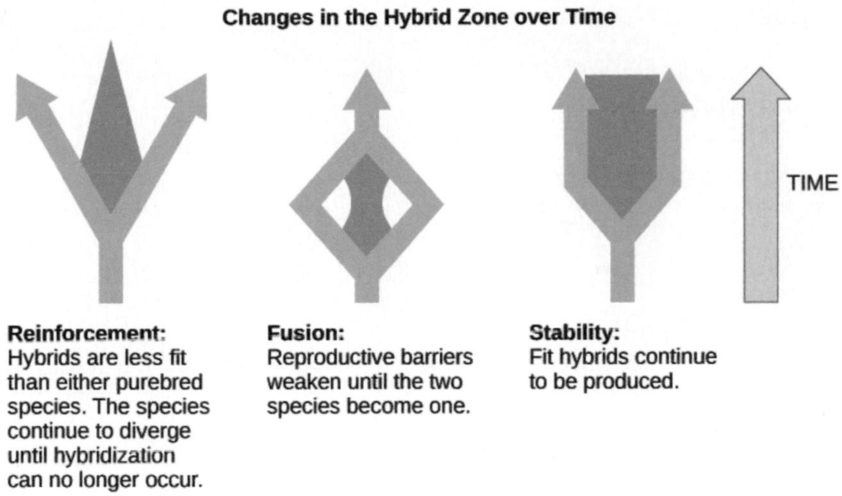

Figure 18.22 After speciation has occurred, the two separate but closely related species may continue to produce offspring in an area called the hybrid zone. Reinforcement, fusion, or stability may result, depending on reproductive barriers and the relative fitness of the hybrids.

If two species eat a different diet but one of the food sources is eliminated and both species are forced to eat the same foods, what change in the hybrid zone is most likely to occur?

Hybrids can be either less fit than the parents, more fit, or about the same. Usually hybrids tend to be less fit; therefore, such reproduction diminishes over time, nudging the two species to diverge further in a process we call **reinforcement**. Scientists use this term because the hybrids' low success reinforces the original speciation. If the hybrids are as fit or more fit than the parents, the two species may fuse back into one species (Figure 18.23). Scientists have also observed that sometimes two species will remain separate but also continue to interact to produce some individuals. Scientists classify this as stability because no real net change is taking place.

Varying Rates of Speciation

Scientists around the world study speciation, documenting observations both of living organisms and those found in the fossil record. As their ideas take shape and as research reveals new details about how life evolves, they develop models to help explain speciation rates. In terms of how quickly speciation occurs, we can observe two current patterns: gradual speciation model and punctuated equilibrium model.

In the **gradual speciation model**, species diverge gradually over time in small steps. In the **punctuated equilibrium** model, a new species undergoes changes quickly from the parent species, and then remains largely unchanged for long periods of time afterward (Figure 18.23). We call this early change model punctuated equilibrium, because it begins with a punctuated or periodic change and then remains in balance afterward. While punctuated equilibrium suggests a faster tempo, it does not necessarily exclude gradualism.

VISUAL CONNECTION

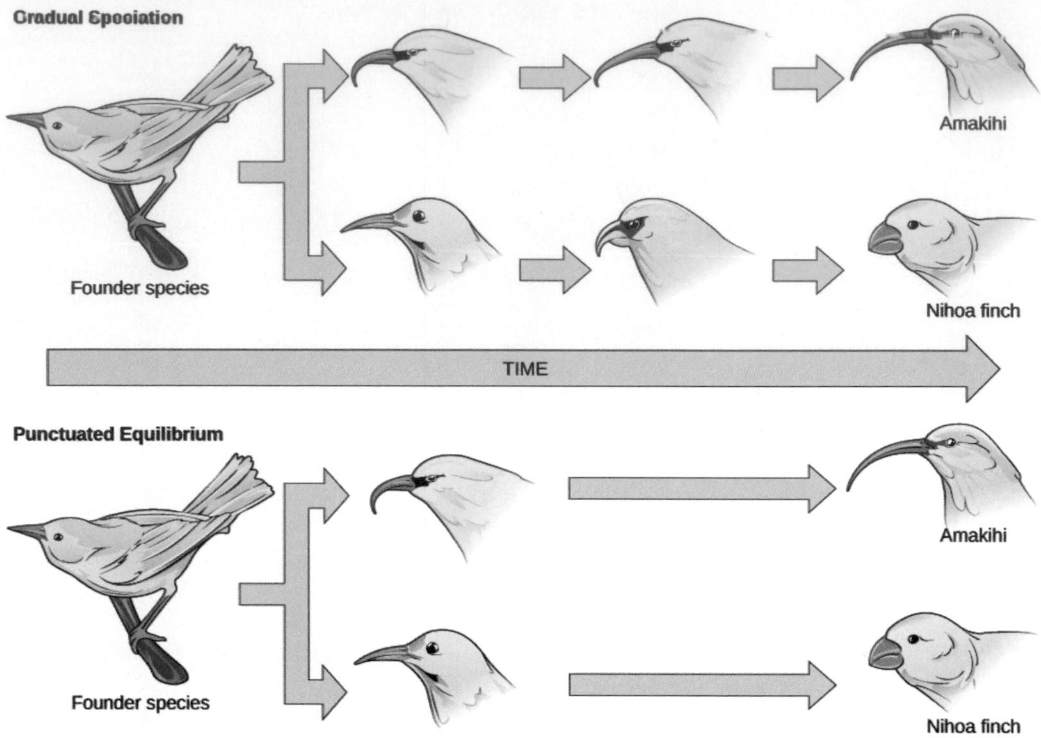

Figure 18.23 In (a) gradual speciation, species diverge at a slow, steady pace as traits change incrementally. In (b) punctuated equilibrium, species diverge quickly and then remain unchanged for long periods of time.

Which of the following statements is false?

a. Punctuated equilibrium is most likely to occur in a small population that experiences a rapid change in its environment.
b. Punctuated equilibrium is most likely to occur in a large population that lives in a stable climate.
c. Gradual speciation is most likely to occur in species that live in a stable climate.
d. Gradual speciation and punctuated equilibrium both result in the divergence of species.

The primary influencing factor on changes in speciation rate is environmental conditions. Under some conditions, selection occurs quickly or radically. Consider a species of snails that had been living with the same basic form for many thousands of years. Layers of their fossils would appear similar for a long time. When a change in the environment takes place—such as a drop in the water level—a small number of organisms are separated from the rest in a brief period of time, essentially forming one large and one tiny population. The tiny population faces new environmental conditions. Because its gene pool quickly became so small, any variation that surfaces and that aids in surviving the new conditions becomes the predominant form.

LINK TO LEARNING

Visit this website (http://openstax.org/l/snails) to continue the speciation story of the snails.

KEY TERMS

adaptation heritable trait or behavior in an organism that aids in its survival and reproduction in its present environment

adaptive radiation speciation when one species radiates to form several other species

allopatric speciation speciation that occurs via geographic separation

allopolyploid polyploidy formed between two related, but separate species

aneuploidy condition of a cell having an extra chromosome or missing a chromosome for its species

autopolyploid polyploidy formed within a single species

behavioral isolation type of reproductive isolation that occurs when a specific behavior or lack of one prevents reproduction from taking place

convergent evolution process by which groups of organisms independently evolve to similar forms

dispersal allopatric speciation that occurs when a few members of a species move to a new geographical area

divergent evolution process by which groups of organisms evolve in diverse directions from a common point

gametic barrier prezygotic barrier occurring when closely related individuals of different species mate, but differences in their gamete cells (eggs and sperm) prevent fertilization from taking place

gradual speciation model model that shows how species diverge gradually over time in small steps

habitat isolation reproductive isolation resulting when species' populations move or are moved to a new habitat, taking up residence in a place that no longer overlaps with the same species' other populations

homologous structures parallel structures in diverse organisms that have a common ancestor

hybrid offspring of two closely related individuals, not of the same species

hybrid zone area where two closely related species continue to interact and reproduce, forming hybrids

natural selection reproduction of individuals with favorable genetic traits that survive environmental change because of those traits, leading to evolutionary change

postzygotic barrier reproductive isolation mechanism that occurs after zygote formation

prezygotic barrier reproductive isolation mechanism that occurs before zygote formation

punctuated equilibrium model for rapid speciation that can occur when an event causes a small portion of a population to be cut off from the rest of the population

reinforcement continued speciation divergence between two related species due to low fitness of hybrids between them

reproductive isolation situation that occurs when a species is reproductively independent from other species; behavior, location, or reproductive barriers may cause this to happen

speciation formation of a new species

species group of populations that interbreed and produce fertile offspring

sympatric speciation speciation that occurs in the same geographic space

temporal isolation differences in breeding schedules that can act as a form of prezygotic barrier leading to reproductive isolation

variation genetic differences among individuals in a population

vestigial structure physical structure present in an organism but that has no apparent function and appears to be from a functional structure in a distant ancestor

vicariance allopatric speciation that occurs when something in the environment separates organisms of the same species into separate groups

CHAPTER SUMMARY

18.1 Understanding Evolution

Evolution is the process of adaptation through mutation which allows more desirable characteristics to pass to the next generation. Over time, organisms evolve more characteristics that are beneficial to their survival. For living organisms to adapt and change to environmental pressures, genetic variation must be present. With genetic variation, individuals have differences in form and function that allow some to survive certain conditions better than others. These organisms pass their favorable traits to their offspring. Eventually, environments change, and what was once a desirable, advantageous trait may become an undesirable trait and organisms may further evolve. Evolution may be convergent with similar traits evolving in multiple species or divergent with diverse traits evolving in multiple species that came from a common ancestor. We can observe evidence of evolution by means of DNA code and the fossil record, and also by the existence of homologous and vestigial structures.

18.2 Formation of New Species

Speciation occurs along two main pathways: geographic separation (allopatric speciation) and through mechanisms that occur within a shared habitat (sympatric speciation). Both pathways isolate a population reproductively in some form. Mechanisms of reproductive isolation act as barriers between closely related species, enabling them to diverge and exist as genetically independent species. Prezygotic barriers block reproduction prior to formation of a zygote; whereas, postzygotic barriers block reproduction after fertilization occurs. For a new species to develop, something must introduce a reproductive barrier. Sympatric speciation can occur through errors in meiosis that form gametes with extra chromosomes (polyploidy). Autopolyploidy occurs within a single species; whereas, allopolyploidy occurs between closely related species.

18.3 Reconnection and Speciation Rates

Speciation is not a precise division: overlap between closely related species can occur in areas called hybrid zones. Organisms reproduce with other similar organisms. The fitness of these hybrid offspring can affect the two species' evolutionary path. Scientists propose two models for the rate of speciation: one model illustrates how a species can change slowly over time. The other model demonstrates how change can occur quickly from a parent generation to a new species. Both models continue to follow natural selection patterns.

VISUAL CONNECTION QUESTIONS

1. Figure 18.14 Which is most likely to survive, offspring with $2n+1$ chromosomes or offspring with $2n-1$ chromosomes?

2. Figure 18.22 If two species eat a different diet but one of the food sources is eliminated and both species are forced to eat the same foods, what change in the hybrid zone is most likely to occur?

3. Figure 18.23 Which of the following statements is false?
   a. Punctuated equilibrium is most likely to occur in a small population that experiences a rapid change in its environment.
   b. Punctuated equilibrium is most likely to occur in a large population that lives in a stable climate.
   c. Gradual speciation is most likely to occur in species that live in a stable climate.
   d. Gradual speciation and punctuated equilibrium both result in the evolution of new species.

REVIEW QUESTIONS

4. Which scientific concept did Charles Darwin and Alfred Wallace independently discover?
   a. mutation
   b. natural selection
   c. overbreeding
   d. sexual reproduction

5. Which of the following situations will lead to natural selection?
   a. The seeds of two plants land near each other and one grows larger than the other.
   b. Two types of fish eat the same kind of food, and one is better able to gather food than the other.
   c. Male lions compete for the right to mate with females, with only one possible winner.
   d. all of the above

6. Which description is an example of a phenotype?
   a. A certain duck has a blue beak.
   b. A mutation occurred to a flower.
   c. Most cheetahs live solitary lives.
   d. both a and c

7. Which situation is most likely an example of convergent evolution?
   a. Squid and humans have eyes similar in structure.
   b. Worms and snakes both move without legs.
   c. Some bats and birds have wings that allow them to fly.
   d. all of the above

8. Which situation would most likely lead to allopatric speciation?
   a. Flood causes the formation of a new lake.
   b. A storm causes several large trees to fall down.
   c. A mutation causes a new trait to develop.
   d. An injury causes an organism to seek out a new food source.

9. What is the main difference between dispersal and vicariance?
   a. One leads to allopatric speciation, whereas the other leads to sympatric speciation.
   b. One involves the movement of the organism, and the other involves a change in the environment.
   c. One depends on a genetic mutation occurring, and the other does not.
   d. One involves closely related organisms, and the other involves only individuals of the same species.

10. Which variable increases the likelihood of allopatric speciation taking place more quickly?
    a. lower rate of mutation
    b. longer distance between divided groups
    c. increased instances of hybrid formation
    d. equivalent numbers of individuals in each population

11. What is the main difference between autopolyploid and allopolyploid?
    a. the number of chromosomes
    b. the functionality of the chromosomes
    c. the source of the extra chromosomes
    d. the number of mutations in the extra chromosomes

12. Which reproductive combination produces hybrids?
    a. when individuals of the same species in different geographical areas reproduce
    b. when any two individuals sharing the same habitat reproduce
    c. when members of closely related species reproduce
    d. when offspring of the same parents reproduce

13. Which condition is the basis for a species to be reproductively isolated from other members?
    a. It does not share its habitat with related species.
    b. It does not exist out of a single habitat.
    c. It does not exchange genetic information with other species.
    d. It does not undergo evolutionary changes for a significant period of time.

14. Which situation is *not* an example of a prezygotic barrier?
    a. Two species of turtles breed at different times of the year.
    b. Two species of flowers attract different pollinators.
    c. Two species of birds display different mating dances.
    d. Two species of insects produce infertile offspring.

15. Which term is used to describe the continued divergence of species based on the low fitness of hybrid offspring?
    a. reinforcement
    b. fusion
    c. stability
    d. punctuated equilibrium

16. Which components of speciation would be least likely to be a part of punctuated equilibrium?
    a. a division of populations
    b. a change in environmental conditions
    c. ongoing gene flow among all individuals
    d. a large number of mutations taking place at once

CRITICAL THINKING QUESTIONS

17. If a person scatters a handful of garden pea plant seeds in one area, how would natural selection work in this situation?
18. Why do scientists consider vestigial structures evidence for evolution?
19. How does the scientific meaning of "theory" differ from the common vernacular meaning?
20. Explain why the statement that a monkey is more evolved than a mouse is incorrect.
21. Why do island chains provide ideal conditions for adaptive radiation to occur?
22. Two species of fish had recently undergone sympatric speciation. The males of each species had a different coloring through which the females could identify and choose a partner from her own species. After some time, pollution made the lake so cloudy that it was hard for females to distinguish colors. What might take place in this situation?
23. Why can polyploidy individuals lead to speciation fairly quickly?
24. What do both rate of speciation models have in common?
25. Describe a situation where hybrid reproduction would cause two species to fuse into one.

CHAPTER 19

The Evolution of Populations

Figure 19.1 Living things may be single-celled or complex, multicellular organisms. They may be plants, animals, fungi, bacteria, or archaea. This diversity results from evolution. (credit "wolf": modification of work by Gary Kramer; credit "coral": modification of work by William Harrigan, NOAA; credit "river": modification of work by Vojtěch Dostál; credit "fish" modification of work by Christian Mehlführer; credit "mushroom": modification of work by Cory Zanker; credit "tree": modification of work by Joseph Kranak; credit "bee": modification of work by Cory Zanker)

Chapter Outline

INTRODUCTION All life on Earth is related. Evolutionary theory states that humans, beetles, plants, and bacteria all share a common ancestor, but that millions of years of evolution have shaped each of these organisms into the forms we see today. Scientists consider evolution a key concept to understanding life. It is one of the most dominant evolutionary forces. Natural selection acts to promote traits and behaviors that increase an organism's chances of survival and reproduction, while eliminating those traits and behaviors that are detrimental to the organism. However, natural selection can only, as its name implies, select—it cannot create. We can attribute novel traits and behaviors to another evolutionary force—mutation. Mutation and other sources of variation among individuals, as well as the evolutionary forces that act upon them, alter populations and species. This combination of processes has led to the world of life we see today.

19.1 Population Evolution

By the end of this section, you will be able to do the following:

- Define population genetics and describe how scientists use population genetics in studying population evolution
- Define the Hardy-Weinberg principle and discuss its importance

People did not understand the mechanisms of inheritance, or genetics, at the time Charles Darwin and

Alfred Russel Wallace were developing their idea of natural selection. This lack of knowledge was a stumbling block to understanding many aspects of evolution. The predominant (and incorrect) genetic theory of the time, blending inheritance, made it difficult to understand how natural selection might operate. Darwin and Wallace were unaware of the Austrian monk Gregor Mendel's 1866 publication "Experiments in Plant Hybridization", which came out not long after Darwin's book, *On the Origin of Species*. Scholars rediscovered Mendel's work in the early twentieth century at which time geneticists were rapidly coming to an understanding of the basics of inheritance. Initially, the newly discovered particulate nature of genes made it difficult for biologists to understand how gradual evolution could occur. However, over the next few decades scientists integrated genetics and evolution in what became known as the **modern synthesis**—the coherent understanding of the relationship between natural selection and genetics that took shape by the 1940s. Generally, this concept is generally accepted today. In short, the modern synthesis describes how evolutionary processes, such as natural selection, can affect a population's genetic makeup, and, in turn, how this can result in the gradual evolution of populations and species. The theory also connects population change over time **(microevolution)**, with the processes that gave rise to new species and higher taxonomic groups with widely divergent characters, called **(macroevolution)**.

Everyday Connection

Evolution and Flu Vaccines

Every fall, the media starts reporting on flu vaccinations and potential outbreaks. Scientists, health experts, and institutions determine recommendations for different parts of the population, predict optimal production and inoculation schedules, create vaccines, and set up clinics to provide inoculations. You may think of the annual flu shot as media hype, an important health protection, or just a briefly uncomfortable prick in your arm. However, do you think of it in terms of evolution?

The media hype of annual flu shots is scientifically grounded in our understanding of evolution. Each year, scientists across the globe strive to predict the flu strains that they anticipate as most widespread and harmful in the coming year. They base this knowledge on how flu strains have evolved over time and over the past few flu seasons. Scientists then work to create the most effective vaccine to combat those selected strains. Pharmaceutical companies produce hundreds of millions of doses in a short period in order to provide vaccinations to key populations at the optimal time.

Because viruses, like the flu, evolve very quickly (especially in evolutionary time), this poses quite a challenge. Viruses mutate and replicate at a fast rate, so the vaccine developed to protect against last year's flu strain may not provide the protection one needs against the coming year's strain. Evolution of these viruses means continued adaptions to ensure survival, including adaptations to survive previous vaccines.

Population Genetics

Recall that a gene for a particular character may have several alleles, or variants, that code for different traits associated with that character. For example, in the ABO blood type system in humans, three alleles determine the particular blood-type carbohydrate on the surface of red blood cells. Each individual in a population of diploid organisms can only carry two alleles for a particular gene, but more than two may be present in the individuals that comprise the population. Mendel followed alleles as they were inherited from parent to offspring. In the early twentieth century, biologists in the area of **population genetics** began to study how selective forces change a population through changes in allele and genotypic frequencies.

The **allele frequency** (or gene frequency) is the rate at which a specific allele appears within a population. Until now we have discussed evolution as a change in the characteristics of a population of organisms, but behind that phenotypic change is genetic change. In population genetics, scientists define the term evolution as a change in the allele's frequency in a population. Using the ABO blood type system as an example, the frequency of one of the alleles, I^A, is the number of copies of that allele divided by all the copies of the ABO gene in the population. For example, a study in Jordan[1] found a frequency of I^A to be 26.1 percent.

1Sahar S. Hanania, Dhia S. Hassawi, and Nidal M. Irshaid, "Allele Frequency and Molecular Genotypes of ABO Blood Group System in a Jordanian Population," *Journal of Medical Sciences* 7 (2007): 51-58, doi:10.3923/jms.2007.51.58.

The I^B and I^O alleles comprise 13.4 percent and 60.5 percent of the alleles respectively, and all of the frequencies added up to 100 percent. A change in this frequency over time would constitute evolution in the population.

The allele frequency within a given population can change depending on environmental factors; therefore, certain alleles become more widespread than others during the natural selection process. Natural selection can alter the population's genetic makeup. An example is if a given allele confers a phenotype that allows an individual to better survive or have more offspring. Because many of those offspring will also carry the beneficial allele, and often the corresponding phenotype, they will have more offspring of their own that also carry the allele, thus, perpetuating the cycle. Over time, the allele will spread throughout the population. Some alleles will quickly become fixed in this way, meaning that every individual of the population will carry the allele, while detrimental mutations may be swiftly eliminated if derived from a dominant allele from the gene pool. The **gene pool** is the sum of all the alleles in a population.

Sometimes, allele frequencies within a population change randomly with no advantage to the population over existing allele frequencies. We call this phenomenon genetic drift. Natural selection and genetic drift usually occur simultaneously in populations and are not isolated events. It is hard to determine which process dominates because it is often nearly impossible to determine the cause of change in allele frequencies at each occurrence. We call an event that initiates an allele frequency change in an isolated part of the population, which is not typical of the original population, the **founder effect**. Natural selection, random drift, and founder effects can lead to significant changes in a population's genome.

Hardy-Weinberg Principle of Equilibrium

In the early twentieth century, English mathematician Godfrey Hardy and German physician Wilhelm Weinberg stated the principle of equilibrium to describe the population's genetic makeup. The theory, which later became known as the Hardy-Weinberg principle of equilibrium, states that a population's allele and genotype frequencies are inherently stable— unless some kind of evolutionary force is acting upon the population, neither the allele nor the genotypic frequencies would change. The Hardy-Weinberg principle assumes conditions with no mutations, migration, emigration, or selective pressure for or against genotype, plus an infinite population. While no population can satisfy those conditions, the principle offers a useful model against which to compare real population changes.

Working under this theory, population geneticists represent different alleles as different variables in their mathematical models. The variable p, for example, often represents the frequency of a particular allele, say Y for the trait of yellow in Mendel's peas, while the variable q represents the frequency of y alleles that confer the color green. If these are the only two possible alleles for a given locus in the population, $p + q = 1$. In other words, all the p alleles and all the q alleles comprise all of the alleles for that locus in the population.

However, what ultimately interests most biologists is not the frequencies of different alleles, but the frequencies of the resulting genotypes, known as the population's **genetic structure**, from which scientists can surmise phenotype distribution. If we observe the phenotype, we can know only the homozygous recessive allele's genotype. The calculations provide an estimate of the remaining genotypes. Since each individual carries two alleles per gene, if we know the allele frequencies (p and q), predicting the genotypes' frequencies is a simple mathematical calculation to determine the probability of obtaining these genotypes if we draw two alleles at random from the gene pool. In the above scenario, an individual pea plant could be pp (YY), and thus produce yellow peas; pq (Yy), also yellow; or qq (yy), and thus produce green peas (Figure 19.2). In other words, the frequency of pp individuals is simply p^2; the frequency of pq individuals is $2pq$; and the frequency of qq individuals is q^2. Again, if p and q are the only two possible alleles for a given trait in the population, these genotypes frequencies will sum to one: $p^2 + 2pq + q^2 = 1$.

VISUAL CONNECTION

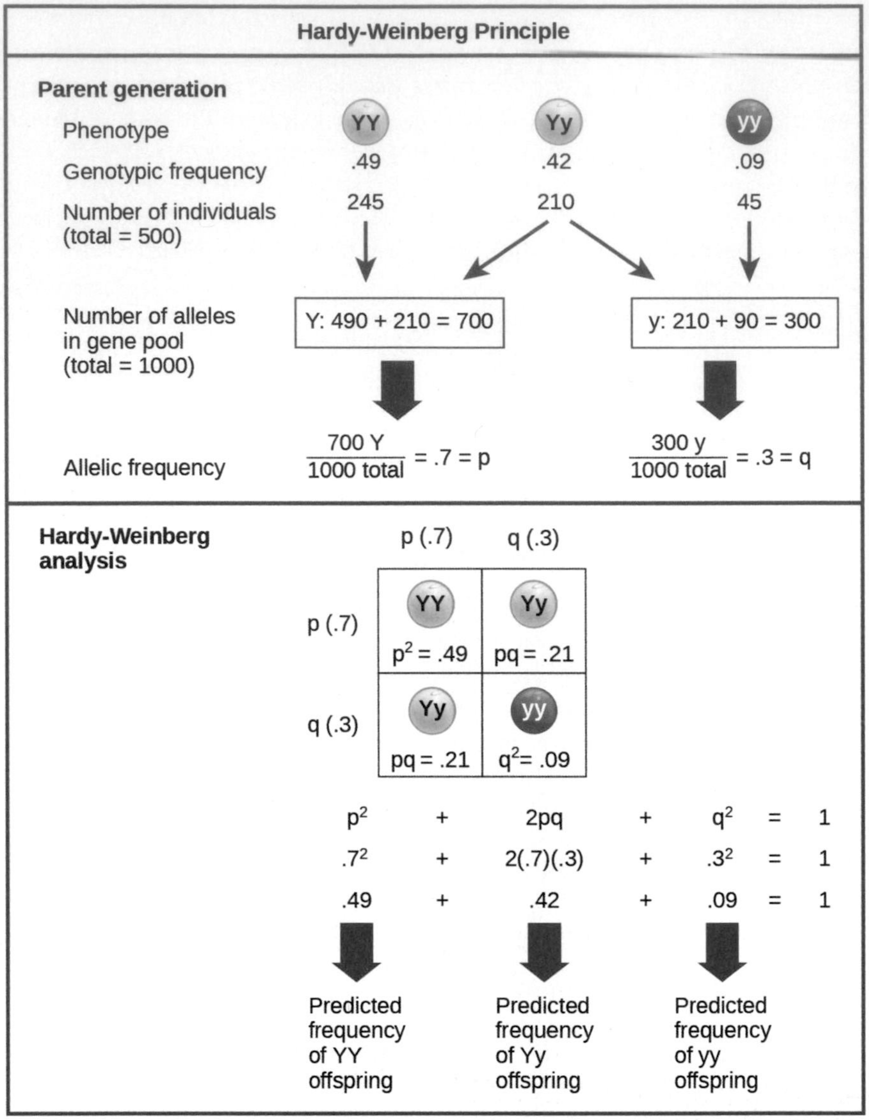

Figure 19.2 When populations are in the Hardy-Weinberg equilibrium, the allelic frequency is stable from generation to generation and we can determine the allele distribution from the Hardy-Weinberg equation. If the allelic frequency measured in the field differs from the predicted value, scientists can make inferences about what evolutionary forces are at play.

In plants, violet flower color (V) is dominant over white (v). If $p = 0.8$ and $q = 0.2$ in a population of 500 plants, how many individuals would you expect to be homozygous dominant (VV), heterozygous (Vv), and homozygous recessive (vv)? How many plants would you expect to have violet flowers, and how many would have white flowers?

In theory, if a population is at equilibrium—that is, there are no evolutionary forces acting upon it—generation after generation would have the same gene pool and genetic structure, and these equations would all hold true all of the time. Of course, even Hardy and Weinberg recognized that no natural population is immune to evolution. Populations in nature are constantly changing in genetic makeup due to drift, mutation, possibly migration, and selection. As a result, the only way to determine the exact distribution of phenotypes in a population is to go out and count them. However, the Hardy-Weinberg principle gives scientists a mathematical baseline of a non-evolving population to which they can compare evolving populations and thereby infer what evolutionary forces might be at play. If the frequencies of alleles or genotypes deviate from the value expected from the Hardy-Weinberg equation, then the population is evolving.

LINK TO LEARNING

Use this online calculator (http://openstax.org/l/hardy-weinberg) to determine a population's genetic structure.

19.2 Population Genetics

By the end of this section, you will be able to do the following:

- Describe the different types of variation in a population
- Explain why only natural selection can act upon heritable variation
- Describe genetic drift and the bottleneck effect
- Explain how each evolutionary force can influence a population's allele frequencies

A population's individuals often display different phenotypes, or express different alleles of a particular gene, which scientists refer to as polymorphisms. We call populations with two or more variations of particular characteristics polymorphic. A number of factors, including the population's genetic structure and the environment (Figure 19.3) influence **population variation**, the distribution of phenotypes among individuals. Understanding phenotypic variation sources in a population is important for determining how a population will evolve in response to different evolutionary pressures.

Figure 19.3 The distribution of phenotypes in this litter of kittens illustrates population variation. (credit: Pieter Lanser)

Genetic Variance

Natural selection and some of the other evolutionary forces can only act on heritable traits, namely an organism's genetic code. Because alleles are passed from parent to offspring, those that confer beneficial traits or behaviors may be selected, while deleterious alleles may not. Acquired traits, for the most part, are not heritable. For example, if an athlete works out in the gym every day, building up muscle strength, the athlete's offspring will not necessarily grow up to be a body builder. If there is a genetic basis for the ability to run fast, on the other hand, a parent may pass this to a child.

LINK TO LEARNING

Before Darwinian evolution became the prevailing theory of the field, French naturalist Jean-Baptiste Lamarck theorized that organisms could inherit acquired traits. While the majority of scientists have not supported this hypothesis, some have recently begun to realize that Lamarck was not completely wrong. Visit this site (http://openstax.org/l/epigenetic) to learn more.

Heritability is the fraction of phenotype variation that we can attribute to genetic differences, or genetic variance, among individuals in a population. The greater the heritability of a population's phenotypic variation, the more susceptible it is to the evolutionary forces that act on heritable variation.

We call the diversity of alleles and genotypes within a population **genetic variance**. When scientists are involved in the breeding of a species, such as with animals in zoos and nature preserves, they try to increase a population's genetic variance to preserve as much of the phenotypic diversity as possible. This also helps reduce associated risks of **inbreeding**, the mating of closely related individuals, which can have the undesirable effect of bringing together deleterious recessive mutations that can cause abnormalities and susceptibility to disease. For example, a disease that is caused by a rare, recessive allele might exist in a population, but it will only manifest itself when an individual carries two copies of the allele. Because the allele is rare in a normal, healthy population with unrestricted habitat, the chance that two carriers will mate is low, and even then, only 25 percent of their offspring will inherit the disease allele from both parents. While it is likely to happen at some point, it will not happen frequently enough for natural selection to be able to swiftly eliminate the allele from the population, and as a result, the allele maintains itself at low levels in the gene pool. However, if a family of carriers begins to interbreed with each other, this will dramatically increase the likelihood of two carriers mating and eventually producing diseased offspring, a phenomenon that scientists call **inbreeding depression**.

Changes in allele frequencies that we identify in a population can shed light on how it is evolving. In addition to natural selection, there are other evolutionary forces that could be in play: genetic drift, gene flow, mutation, nonrandom mating, and environmental variances.

Genetic Drift

The theory of natural selection stems from the observation that some individuals in a population are more likely to survive longer and have more offspring than others; thus, they will pass on more of their genes to the next generation. A big, powerful male gorilla, for example, is much more likely than a smaller, weaker one to become the population's silverback, the pack's leader who mates far more than the other males of the group. The pack leader will father more offspring, who share half of his genes, and are likely to also grow bigger and stronger like their father. Over time, the genes for bigger size will increase in frequency in the population, and the population will, as a result, grow larger on average. That is, this would occur if this particular **selection pressure**, or driving selective force, were the only one acting on the population. In other examples, better camouflage or a stronger resistance to drought might pose a selection pressure.

Another way a population's allele and genotype frequencies can change is **genetic drift** (Figure 19.4), which is simply the effect of chance. By chance, some individuals will have more offspring than others—not due to an advantage conferred by some genetically-encoded trait, but just because one male happened to be in the right place at the right time (when the receptive female walked by) or because the other one happened to be in the wrong place at the wrong time (when a fox was hunting).

VISUAL CONNECTION

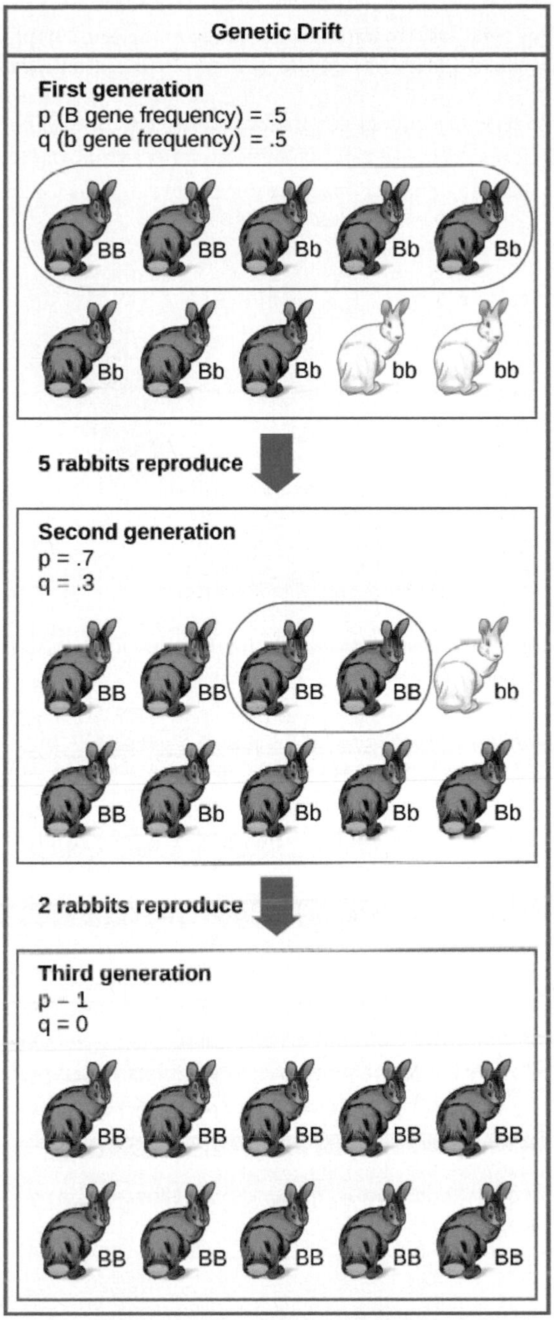

Figure 19.4 Genetic drift in a population can lead to eliminating an allele from a population by chance. In this example, rabbits with the brown coat color allele (*B*) are dominant over rabbits with the white coat color allele (*b*). In the first generation, the two alleles occur with equal frequency in the population, resulting in p and q values of .5. Only half of the individuals reproduce, resulting in a second generation with p and q values of .7 and .3, respectively. Only two individuals in the second generation reproduce, and by chance these individuals are homozygous dominant for brown coat color. As a result, in the third generation the recessive *b* allele is lost.

Do you think genetic drift would happen more quickly on an island or on the mainland?

Small populations are more susceptible to the forces of genetic drift. Large populations, alternatively, are buffered against the effects of chance. If one individual of a population of 10 individuals happens to die at a young age before it leaves any offspring to the next generation, all of its genes—1/10 of the population's gene pool—will be suddenly lost. In a population of 100, that's only 1 percent of the overall gene pool; therefore, it is much less impactful on the population's genetic structure.

LINK TO LEARNING

Go to this site (http://openstax.org/l/genetic_drift) to watch an animation of random sampling and genetic drift in action.

Natural events, such as an earthquake disaster that kills—at random—a large portion of the population, can magnify genetic drift. Known as the **bottleneck effect**, it results in suddenly wiping out a large portion of the genome (Figure 19.5). At once, the survivors' genetic structure becomes the entire population's genetic structure, which may be very different from the pre-disaster population.

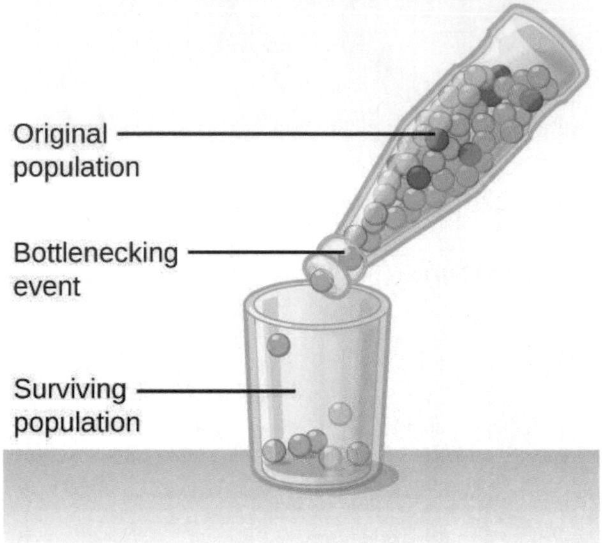

Figure 19.5 A chance event or catastrophe can reduce the genetic variability within a population.

Another scenario in which populations might experience a strong influence of genetic drift is if some portion of the population leaves to start a new population in a new location or if a physical barrier divides a population. In this situation, those individuals are an unlikely representation of the entire population, which results in the founder effect. The founder effect occurs when the genetic structure changes to match that of the new population's founding fathers and mothers. Researchers believe that the founder effect was a key factor in the genetic history of the Afrikaner population of Dutch settlers in South Africa, as evidenced by mutations that are common in Afrikaners but rare in most other populations. This is probably because a higher-than-normal proportion of the founding colonists carried these mutations. As a result, the population expresses unusually high incidences of Huntington's disease (HD) and Fanconi anemia (FA), a genetic disorder known to cause blood marrow and congenital abnormalities—even cancer.[2]

LINK TO LEARNING

Watch this short video to learn more about the founder and bottleneck effects.

Click to view content (https://www.openstax.org/l/founder_bottle)

SCIENTIFIC METHOD CONNECTION

Testing the Bottleneck Effect

Question: How do natural disasters affect a population's genetic structure?

2A. J. Tipping et al., "Molecular and Genealogical Evidence for a Founder Effect in Fanconi Anemia Families of the Afrikaner Population of South Africa," *PNAS* 98, no. 10 (2001): 5734-5739, doi: 10.1073/pnas.091402398.

Background: When an earthquake or hurricane suddenly wipes out much of a population, the surviving individuals are usually a random sampling of the original group. As a result, the population's genetic makeup can change dramatically. We call this phenomenon the bottleneck effect.

Hypothesis: Repeated natural disasters will yield different population genetic structures; therefore, each time one runs this experiment the results will vary.

Test the hypothesis: Count out the original population using different colored beads. For example, red, blue, and yellow beads might represent red, blue, and yellow individuals. After recording the number of each individual in the original population, place them all in a bottle with a narrow neck that will only allow a few beads out at a time. Then, pour 1/3 of the bottle's contents into a bowl. This represents the surviving individuals after a natural disaster kills a majority of the population. Count the number of the different colored beads in the bowl, and record it. Then, place all of the beads back in the bottle and repeat the experiment four more times.

Analyze the data: Compare the five populations that resulted from the experiment. Do the populations all contain the same number of different colored beads, or do they vary? Remember, these populations all came from the same exact parent population.

Form a conclusion: Most likely, the five resulting populations will differ quite dramatically. This is because natural disasters are not selective—they kill and spare individuals at random. Now think about how this might affect a real population. What happens when a hurricane hits the Mississippi Gulf Coast? How do the seabirds that live on the beach fare?

Gene Flow

Another important evolutionary force is **gene flow**: the flow of alleles in and out of a population due to the migration of individuals or gametes (Figure 19.6). While some populations are fairly stable, others experience more flux. Many plants, for example, send their pollen far and wide, by wind or by bird, to pollinate other populations of the same species some distance away. Even a population that may initially appear to be stable, such as a pride of lions, can experience its fair share of immigration and emigration as developing males leave their mothers to seek out a new pride with genetically unrelated females. This variable flow of individuals in and out of the group not only changes the population's gene structure, but it can also introduce new genetic variation to populations in different geological locations and habitats.

Figure 19.6 Gene flow can occur when an individual travels from one geographic location to another.

Mutation

Mutations are changes to an organism's DNA and are an important driver of diversity in populations. Species evolve because of mutations accumulating over time. The appearance of new mutations is the most common way to introduce novel genotypic and phenotypic variance. Some mutations are unfavorable or harmful and are quickly eliminated from the population by natural selection. Others are beneficial and will spread through the population. Whether or not a mutation is beneficial or harmful is determined by whether it helps an organism survive to sexual maturity and reproduce. Some mutations do not do anything and can linger, unaffected by natural selection, in the genome. Some can have a dramatic effect on a gene and the resulting phenotype.

Nonrandom Mating

If individuals nonrandomly mate with their peers, the result can be a changing population. There are many reasons **nonrandom mating** occurs. One reason is simple mate choice. For example, female peahens may prefer peacocks with bigger, brighter tails. Natural selection picks traits that lead to more mating selections for an individual. One common form of mate choice, called **assortative mating**, is an individual's preference to mate with partners who are phenotypically similar to themselves.

Another cause of nonrandom mating is physical location. This is especially true in large populations spread over vast geographic distances where not all individuals will have equal access to one another. Some might be miles apart through woods or over rough terrain, while others might live immediately nearby.

Environmental Variance

Genes are not the only players involved in determining population variation. Other factors, such as the environment (Figure 19.7) also influence phenotypes. A beachgoer is likely to have darker skin than a city dweller, for example, due to regular exposure to the sun, an environmental factor. For some species, the environment determines some major characteristics, such as gender. For example, some turtles and other reptiles have temperature-dependent sex determination (TSD). TSD means that individuals develop into males if their eggs are incubated within a certain temperature range, or females at a different temperature range.

Figure 19.7 The temperature at which the eggs are incubated determine the American alligator's (*Alligator mississippiensis*) sex. Eggs incubated at 30°C produce females, and eggs incubated at 33°C produce males. (credit: Steve Hillebrand, USFWS)

Geographic separation between populations can lead to differences in the phenotypic variation between those populations. We see such **geographical variation** between most populations and it can be significant. We can observe one type of geographic variation, a **cline**, as given species' populations vary gradually across an ecological gradient. Species of warm-blooded animals, for example, tend to have larger bodies in the cooler climates closer to the earth's poles, allowing them to better conserve heat. This is a latitudinal cline. Alternatively, flowering plants tend to bloom at different times depending on where they are along a mountain slope. This is an altitudinal cline.

If there is gene flow between the populations, the individuals will likely show gradual differences in phenotype along the cline. Restricted gene flow, alternatively can lead to abrupt differences, even speciation.

19.3 Adaptive Evolution

By the end of this section, you will be able to do the following:

- Explain the different ways natural selection can shape populations
- Describe how these different forces can lead to different outcomes in terms of the population variation

Natural selection acts on the population's heritable traits: selecting for beneficial alleles that allow for environmental adaptation, and thus increasing their frequency in the population, while selecting against deleterious alleles and thereby decreasing their frequency. Scientists call this process **adaptive evolution**. Natural selection acts on entire organisms, not on an individual allele within the organism. An individual may carry a very beneficial genotype with a resulting phenotype that, for example, increases the ability to reproduce (fecundity), but if that same individual also carries an allele that results in a fatal childhood disease, that fecundity phenotype will not pass to the next generation because the individual will not live to reach reproductive age. Natural selection acts at the individual's level. It selects for individuals with greater contributions to the gene pool of the next generation. Scientists call this an organism's **evolutionary (Darwinian) fitness**.

Fitness is often quantifiable and is measured by scientists in the field. However, it is not an individual's absolute fitness that counts, but rather how it compares to the other organisms in the population. Scientists call this concept **relative fitness**, which allows researchers to determine which individuals are contributing additional offspring to the next generation, and thus, how the population might evolve.

There are several ways selection can affect population variation: stabilizing selection, directional selection, diversifying selection, frequency-dependent selection, and sexual selection. As natural selection influences the allele frequencies in a population, individuals can either become more or less genetically similar and the phenotypes can become more similar or more disparate.

Stabilizing Selection

If natural selection favors an average phenotype, selecting against extreme variation, the population will undergo **stabilizing selection** (Figure 19.8). In a mouse population that live in the woods, for example, natural selection is likely to favor mice that best blend in with the forest floor and are less likely for predators to spot. Assuming the ground is a fairly consistent shade of brown, those mice whose fur is most closely matched to that color will be most likely to survive and reproduce, passing on their genes for their brown coat. Mice that carry alleles that make them a bit lighter or a bit darker will stand out against the ground and be more likely to fall victim to predation. As a result of this selection, the population's genetic variance will decrease.

Directional Selection

When the environment changes, populations will often undergo **directional selection** (Figure 19.8), which selects for phenotypes at one end of the spectrum of existing variation. A classic example of this type of selection is the evolution of the peppered moth in eighteenth- and nineteenth-century England. Prior to the Industrial Revolution, the moths were predominately light in color, which allowed them to blend in with the light-colored trees and lichens in their environment. However, as soot began spewing from factories, the trees darkened, and the light-colored moths became easier for predatory birds to spot. Over time, the frequency of the moth's melanic form increased because they had a higher survival rate in habitats affected by air pollution because their darker coloration blended with the sooty trees. Similarly, the hypothetical mouse population may evolve to take on a different coloration if something were to cause the forest floor where they live to change color. The result of this type of selection is a shift in the population's genetic variance toward the new, fit phenotype.

LINK TO LEARNING

In science, we sometimes believe some things are true, and then new information becomes available that changes our understanding. The peppered moth story is an example: some scientists recently have questioned the facts behind the selection toward darker moths. Read this article (http://openstax.org/l/peppered_moths) to learn more.

Diversifying Selection

Sometimes two or more distinct phenotypes can each have their advantages for natural selection, while the intermediate phenotypes are, on average, less fit. Scientists call this **diversifying selection** (Figure 19.8) We see this in many animal populations that have multiple male forms. Large, dominant alpha males use brute force to obtain mates, while small males can sneak in for furtive copulations with the females in an alpha male's territory. In this case, both the alpha males and the "sneaking" males will be selected for, but medium-sized males, who can't overtake the alpha males and are too big to sneak copulations, are selected against. Diversifying selection can also occur when environmental changes favor individuals on either end of the phenotypic spectrum. Imagine a mouse population living at the beach where there is light-colored sand interspersed with patches of tall grass. In this scenario, light-colored mice that blend in with the sand would be favored, as well as dark-colored mice that can hide in the grass. Medium-colored mice, alternatively would not blend in with either the grass or the sand, and thus predators would most likely eat them. The result of this type of selection is increased genetic variance as the population becomes more diverse.

VISUAL CONNECTION

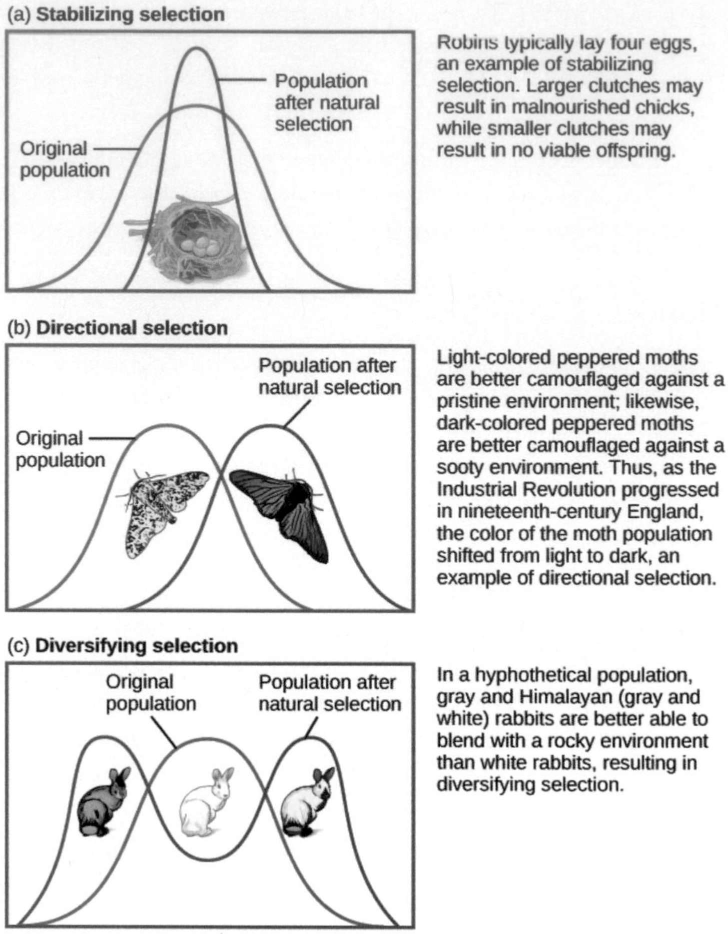

Figure 19.8 Different types of natural selection can impact the distribution of phenotypes within a population. In (a) stabilizing selection, an average phenotype is favored. In (b) directional selection, a change in the environment shifts the spectrum of observed phenotypes. In (c) diversifying selection, two or more extreme phenotypes are selected for, while the average phenotype is selected against.

In recent years, factories have become cleaner, and release less soot into the environment. What impact do you think this has had on the distribution of moth color in the population?

Frequency-Dependent Selection

Another type of selection, **frequency-dependent selection**, favors phenotypes that are either common (positive frequency-dependent selection) or rare (negative frequency-dependent selection). We can observe an interesting example of this type of selection in a unique group of Pacific Northwest lizards. Male common side-blotched lizards come in three throat-color patterns: orange, blue, and yellow. Each of these forms has a different reproductive strategy: orange males are the strongest and can fight other males for access to their females. Blue males are medium-sized and form strong pair bonds with their mates. Yellow males (Figure 19.9) are the smallest, and look a bit like females, which allows them to sneak copulations. Like a game of rock-paper-scissors, orange beats blue, blue beats yellow, and yellow beats orange in the competition for females. That is, the big, strong orange males can fight off the blue males to mate with the blue's pair-bonded females, the blue males are successful at guarding their mates against yellow sneaker males, and the yellow males can sneak copulations from the potential mates of the large, polygynous orange males.

Figure 19.9 A yellow-throated side-blotched lizard is smaller than either the blue-throated or orange-throated males and appears a bit like the females of the species, allowing it to sneak copulations. (credit: "tinyfroglet"/Flickr)

In this scenario, natural selection favors orange males when blue males dominate the population. Blue males will thrive when the population is mostly yellow males, and yellow males will be selected for when orange males are the most populous. As a result, populations of side-blotched lizards cycle in the distribution of these phenotypes—in one generation, orange might predominate, and then yellow males will begin to rise in frequency. Once yellow males comprise a majority of the population, blue males will be selected. Finally, when blue males become common, orange males once again will be favored.

Negative frequency-dependent selection serves to increase the population's genetic variance by selecting for rare phenotypes; whereas, positive frequency-dependent selection usually decreases genetic variance by selecting for common phenotypes.

Sexual Selection

Males and females of certain species are often quite different from one another in ways beyond the reproductive organs. Males are often larger, for example, and display many elaborate colors and adornments, like the peacock's tail, while females tend to be smaller and duller in decoration. We call such differences **sexual dimorphisms** (Figure 19.10), which arise in many populations, particularly animal populations, where there is more variance in the male's reproductive success than that of the females. That is, some males—often the bigger, stronger, or more decorated males—obtain the vast majority of the total matings, while others receive none. This can occur because the males are better at fighting off other males, or because females will choose to mate with the bigger or more decorated males. In either case, this variation in reproductive success generates a strong selection pressure among males to obtain those matings, resulting in the evolution of bigger body size and elaborate ornaments to attract the females' attention. Females, however, tend to achieve a handful of selected matings; therefore, they are more likely to select more desirable males.

Sexual dimorphism varies widely among species, and some species are even sex-role reversed. In such cases, females tend to have a greater variance in their reproductive success than males and are correspondingly selected for the bigger body size and elaborate traits usually characteristic of males.

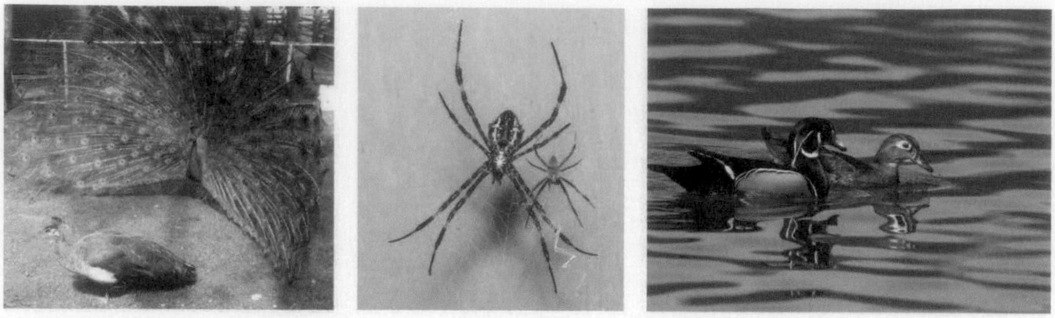

Figure 19.10 Sexual dimorphism in (a) peacocks and peahens, (b) *Argiope appensa* spiders (the female spider is the large one), and in (c) wood ducks. (credit "spiders": modification of work by "Sanba38"/Wikimedia Commons; credit "duck": modification of work by Kevin Cole)

We call the selection pressures on males and females to obtain matings sexual selection. It can result in developing secondary sexual characteristics that do not benefit the individual's likelihood of survival but help to maximize its reproductive success. Sexual selection can be so strong that it selects traits that are actually detrimental to the individual's survival. Think, once again, about the peacock's tail. While it is beautiful and the male with the largest, most colorful tail is more likely to win the female, it is not the most practical appendage. In addition to greater visibility to predators, it makes the males slower in their attempted escapes. There is some evidence that this risk is why females like the big tails in the first place. The speculation is that large tails carry risk, and only the best males survive that risk: the bigger the tail, the more fit the male. We call this the **handicap principle**.

The **good genes hypothesis** states that males develop these impressive ornaments to show off their efficient metabolism or their ability to fight disease. Females then choose males with the most impressive traits because it signals their genetic superiority, which they will then pass on to their offspring. Although one may argue that females should not be picky because it will likely reduce their number of offspring, if better males father more fit offspring, it may be beneficial. Fewer, healthier offspring may increase the chances of survival more than many, weaker offspring.

LINK TO LEARNING

In 1915, biologist Ronald Fisher proposed another model of sexual selection: the Fisherian runaway model (http://openstax.org/l/sexual_select) , which suggests that selection of certain traits is a result of sexual preference.

In both the handicap principle and the good genes hypothesis, the trait is an **honest signal** of the males' quality, thus giving females a way to find the fittest mates— males that will pass the best genes to their offspring.

No Perfect Organism

Natural selection is a driving force in evolution and can generate populations that are better adapted to survive and successfully reproduce in their environments. However, natural selection cannot produce the perfect organism. Natural selection can only select on existing variation in the population. It does not create anything from scratch. Thus, it is limited by a population's existing genetic variance and whatever new alleles arise through mutation and gene flow.

Natural selection is also limited because it works at the individual, not allele level, and some alleles are linked due to their physical proximity in the genome, making them more likely to pass on together (linkage disequilibrium). Any given individual may carry some beneficial and some unfavorable alleles. It is the alleles' net effect, or the organism's fitness, upon which natural selection can act. As a result, good alleles can be lost if individuals who carry them also have several overwhelmingly bad alleles. Likewise, bad alleles can be kept if individuals who have enough good alleles to result in an overall fitness benefit carry them.

Furthermore, natural selection can be constrained by the relationships between different polymorphisms. One morph may confer a higher fitness than another, but may not increase in frequency because going from the less beneficial to the more beneficial trait would require going through a less beneficial phenotype. Think back to the mice that live at the beach. Some are light-colored and blend in with the sand, while others are dark and blend in with the patches of grass. The dark-colored mice may be, overall, more fit than the light-colored mice, and at first glance, one might expect the light-colored mice to be selected for a darker coloration. However, remember that the intermediate phenotype, a medium-colored coat, is very bad for the mice—they cannot blend in with either the sand or the grass and predators are more likely to eat them. As a result, the light-colored mice would not be selected for a dark coloration because those individuals who began moving in that direction (began selection for a darker coat) would be less fit than those that stayed light.

Finally, it is important to understand that not all evolution is adaptive. While natural selection selects the fittest individuals and often results in a more fit population overall, other forces of evolution, including genetic drift and gene flow, often do the opposite: introducing deleterious alleles to the population's gene pool. Evolution has no purpose—it is not changing a population into a preconceived ideal. It is simply the sum of the various forces that we have described in this chapter and how they influence the population's genetic and phenotypic variance.

KEY TERMS

adaptive evolution increase in frequency of beneficial alleles and decrease in deleterious alleles due to selection

allele frequency (also, gene frequency) rate at which a specific allele appears within a population

assortative mating when individuals tend to mate with those who are phenotypically similar to themselves

bottleneck effect magnification of genetic drift as a result of natural events or catastrophes

cline gradual geographic variation across an ecological gradient

directional selection selection that favors phenotypes at one end of the spectrum of existing variation

diversifying selection selection that favors two or more distinct phenotypes

evolutionary fitness (also, Darwinian fitness) individual's ability to survive and reproduce

founder effect event that initiates an allele frequency change in part of the population, which is not typical of the original population

frequency-dependent selection selection that favors phenotypes that are either common (positive frequency-dependent selection) or rare (negative frequency-dependent selection)

gene flow flow of alleles in and out of a population due to the individual or gamete migration

gene pool all the alleles that the individuals in the population carry

genetic drift effect of chance on a population's gene pool

genetic structure distribution of the different possible genotypes in a population

genetic variance diversity of alleles and genotypes in a population

geographical variation differences in the phenotypic variation between populations that are separated geographically

good genes hypothesis theory of sexual selection that argues individuals develop impressive ornaments to show off their efficient metabolism or ability to fight disease

handicap principle theory of sexual selection that argues only the fittest individuals can afford costly traits

heritability fraction of population variation that can be attributed to its genetic variance

honest signal trait that gives a truthful impression of an individual's fitness

inbreeding mating of closely related individuals

inbreeding depression increase in abnormalities and disease in inbreeding populations

macroevolution broader scale evolutionary changes that scientists see over paleontological time

microevolution changes in a population's genetic structure

modern synthesis overarching evolutionary paradigm that took shape by the 1940s and scientists generally accept today

nonrandom mating changes in a population's gene pool due to mate choice or other forces that cause individuals to mate with certain phenotypes more than others

population genetics study of how selective forces change the allele frequencies in a population over time

population variation distribution of phenotypes in a population

relative fitness individual's ability to survive and reproduce relative to the rest of the population

selective pressure environmental factor that causes one phenotype to be better than another

sexual dimorphism phenotypic difference between a population's males and females

stabilizing selection selection that favors average phenotypes

CHAPTER SUMMARY

19.1 Population Evolution

The modern synthesis of evolutionary theory grew out of the cohesion of Darwin's, Wallace's, and Mendel's thoughts on evolution and heredity, along with the more modern study of population genetics. It describes the evolution of populations and species, from small-scale changes among individuals to large-scale changes over paleontological time periods. To understand how organisms evolve, scientists can track populations' allele frequencies over time. If they differ from generation to generation, scientists can conclude that the population is not in Hardy-Weinberg equilibrium, and is thus evolving.

19.2 Population Genetics

Both genetic and environmental factors can cause phenotypic variation in a population. Different alleles can confer different phenotypes, and different environments can also cause individuals to look or act differently. Only those differences encoded in an individual's genes, however, can pass to its offspring and, thus, be a target of natural selection. Natural selection works by selecting for alleles that confer beneficial traits or behaviors, while selecting against those for deleterious qualities. Genetic drift stems from the chance occurrence that some individuals in the gene line have more offspring than others. When individuals leave or join the population, allele frequencies can change as a result of gene flow. Mutations to an individual's DNA may introduce new variation into a population. Allele frequencies can also alter when individuals do not randomly mate with others in the group.

19.3 Adaptive Evolution

Because natural selection acts to increase the frequency of beneficial alleles and traits while decreasing the frequency of deleterious qualities, it is adaptive evolution. Natural selection acts at the individual level, selecting for those that have a higher overall fitness compared to the rest of the population. If the fit phenotypes are those that are similar, natural selection will result in stabilizing selection, and an overall decrease in the population's variation. Directional selection works to shift a population's variance toward a new, fit phenotype, as environmental conditions change. In contrast, diversifying selection results in increased genetic variance by selecting for two or more distinct phenotypes.

Other types of selection include frequency-dependent selection, in which individuals with either common (positive frequency-dependent selection) or rare (negative frequency-dependent selection) are selected. Finally, sexual selection results from one sex having more variance in the reproductive success than the other. As a result, males and females experience different selective pressures, which can often lead to the evolution of phenotypic differences, or sexual dimorphisms, between the two.

VISUAL CONNECTION QUESTIONS

1. Figure 19.2 In plants, violet flower color (V) is dominant over white (v). If $p = .8$ and $q = 0.2$ in a population of 500 plants, how many individuals would you expect to be homozygous dominant (VV), heterozygous (Vv), and homozygous recessive (vv)? How many plants would you expect to have violet flowers, and how many would have white flowers?

2. Figure 19.4 Do you think genetic drift would happen more quickly on an island or on the mainland?

3. Figure 19.8 In recent years, factories have become cleaner, and less soot is released into the environment. What impact do you think this has had on the distribution of moth color in the population?

REVIEW QUESTIONS

4. What is the difference between micro- and macroevolution?
   a. Microevolution describes the evolution of small organisms, such as insects, while macroevolution describes the evolution of large organisms, like people and elephants.
   b. Microevolution describes the evolution of microscopic entities, such as molecules and proteins, while macroevolution describes the evolution of whole organisms.
   c. Microevolution describes the evolution of organisms in populations, while macroevolution describes the evolution of species over long periods of time.
   d. Microevolution describes the evolution of organisms over their lifetimes, while macroevolution describes the evolution of organisms over multiple generations.

5. Population genetics is the study of:
   a. how selective forces change the allele frequencies in a population over time
   b. the genetic basis of population-wide traits
   c. whether traits have a genetic basis
   d. the degree of inbreeding in a population

6. Which of the following populations is not in Hardy-Weinberg equilibrium?
   a. a population with 12 homozygous recessive individuals (yy), 8 homozygous dominant individuals (YY), and 4 heterozygous individuals (Yy)
   b. a population in which the allele frequencies do not change over time
   c. $p^2 + 2pq + q^2 = 1$
   d. a population undergoing natural selection

7. One of the original Amish colonies rose from a ship of colonists that came from Europe. The ship's captain, who had polydactyly, a rare dominant trait, was one of the original colonists. Today, we see a much higher frequency of polydactyly in the Amish population. This is an example of:
   a. natural selection
   b. genetic drift
   c. founder effect
   d. b and c

8. When male lions reach sexual maturity, they leave their group in search of a new pride. This can alter the allele frequencies of the population through which of the following mechanisms?
   a. natural selection
   b. genetic drift
   c. gene flow
   d. random mating

9. Which of the following evolutionary forces can introduce new genetic variation into a population?
   a. natural selection and genetic drift
   b. mutation and gene flow
   c. natural selection and nonrandom mating
   d. mutation and genetic drift

10. What is assortative mating?
   a. when individuals mate with those who are similar to themselves
   b. when individuals mate with those who are dissimilar to themselves
   c. when individuals mate with those who are the most fit in the population
   d. when individuals mate with those who are least fit in the population

11. When closely related individuals mate with each other, or inbreed, the offspring are often not as fit as the offspring of two unrelated individuals. Why?
   a. Close relatives are genetically incompatible.
   b. The DNA of close relatives reacts negatively in the offspring.
   c. Inbreeding can bring together rare, deleterious mutations that lead to harmful phenotypes.
   d. Inbreeding causes normally silent alleles to be expressed.

12. What is a cline?
   a. the slope of a mountain where a population lives
   b. the degree to which a mutation helps an individual survive
   c. the number of individuals in the population
   d. gradual geographic variation across an ecological gradient

13. Which type of selection results in greater genetic variance in a population?
   a. stabilizing selection
   b. directional selection
   c. diversifying selection
   d. positive frequency-dependent selection

14. When males and females of a population look or act differently, it is referred to as _______.
   a. sexual dimorphism
   b. sexual selection
   c. diversifying selection
   d. a cline

15. The good genes hypothesis is a theory that explains what?
   a. why more fit individuals are more likely to have more offspring
   b. why alleles that confer beneficial traits or behaviors are selected for by natural selection
   c. why some deleterious mutations are maintained in the population
   d. why individuals of one sex develop impressive ornamental traits

CRITICAL THINKING QUESTIONS

16. Solve for the genetic structure of a population with 12 homozygous recessive individuals (yy), 8 homozygous dominant individuals (YY), and 4 heterozygous individuals (Yy).
17. Explain the Hardy-Weinberg principle of equilibrium theory.
18. Imagine you are trying to test whether a population of flowers is undergoing evolution. You suspect there is selection pressure on the color of the flower: bees seem to cluster around the red flowers more often than the blue flowers. In a separate experiment, you discover blue flower color is dominant to red flower color. In a field, you count 600 blue flowers and 200 red flowers. What would you expect the genetic structure of the flowers to be?
19. Describe a situation in which a population would undergo the bottleneck effect and explain what impact that would have on the population's gene pool.
20. Describe natural selection and give an example of natural selection at work in a population.
21. Explain what a cline is and provide examples.
22. Give an example of a trait that may have evolved as a result of the handicap principle and explain your reasoning.
23. List the ways in which evolution can affect population variation and describe how they influence allele frequencies.

CHAPTER 20
Phylogenies and the History of Life

Figure 20.1 A bee's life is very different from a flower's, but the two organisms are related. Both are members of the domain Eukarya and have cells containing many similar organelles, genes, and proteins. (credit: modification of work by John Beetham)

INTRODUCTION This bee and *Echinacea* flower (Figure 20.1) could not look more different, yet they are related, as are all living organisms on Earth. By following pathways of similarities and changes—both visible and genetic—scientists seek to map the evolutionary past of how life developed from single-celled organisms to the tremendous collection of creatures that have germinated, crawled, floated, swum, flown, and walked on this planet.

Chapter Outline

20.1 Organizing Life on Earth

By the end of this section, you will be able to do the following:

- Discuss the need for a comprehensive classification system
- List the different levels of the taxonomic classification system
- Describe how systematics and taxonomy relate to phylogeny
- Discuss a phylogenetic tree's components and purpose

In scientific terms, **phylogeny** is the evolutionary history and relationship of an organism or group of organisms. A phylogeny describes the organisim's relationships, such as from which organisms it may have evolved, or to which species it is most closely related. Phylogenetic relationships provide information on shared ancestry but not necessarily on how organisms are similar or different.

Phylogenetic Trees

Scientists use a tool called a phylogenetic tree to show the evolutionary pathways and connections among organisms. A **phylogenetic tree** is a diagram used to reflect evolutionary relationships among organisms or groups of organisms. Scientists consider phylogenetic trees to be a hypothesis of the evolutionary past since one cannot go back to confirm the proposed relationships. In other words, we can construct a "tree of life" to illustrate when different organisms evolved and to show the relationships among different organisms (Figure 20.2).

Unlike a taxonomic classification diagram, we can read a phylogenetic tree like a map of evolutionary history. Many phylogenetic trees have a single lineage at the base representing a common ancestor. Scientists call such trees **rooted**, which means there is a single ancestral lineage (typically drawn from the bottom or left) to which all organisms represented in the diagram relate. Notice in the rooted phylogenetic tree that the three domains— Bacteria, Archaea, and Eukarya—diverge from a single point and branch off. The small branch that plants and animals (including humans) occupy in this diagram shows how recent and miniscule these groups are compared with other organisms. Unrooted trees do not show a common ancestor but do show relationships among species.

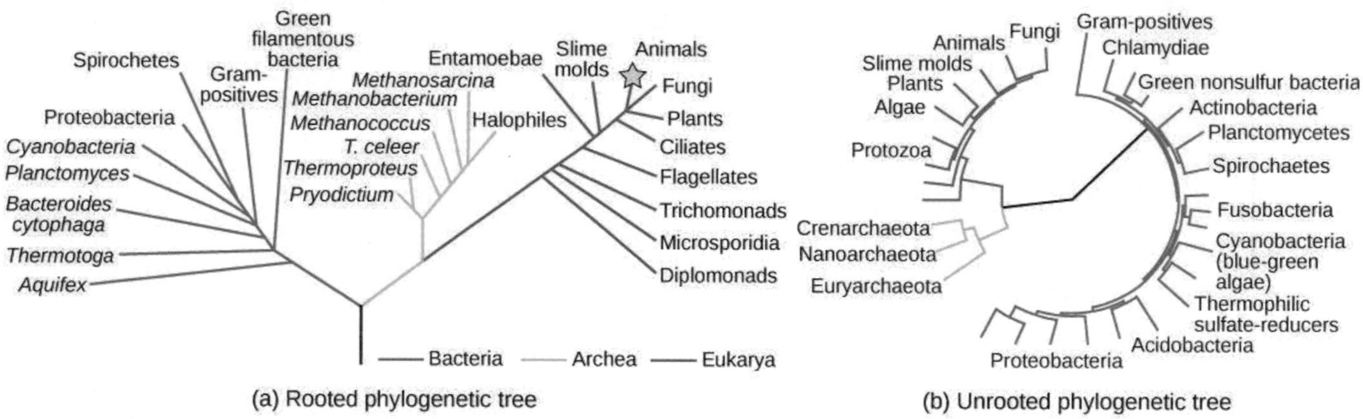

Figure 20.2 Both of these phylogenetic trees show the relationship of the three domains of life—Bacteria, Archaea, and Eukarya—but the (a) rooted tree attempts to identify when various species diverged from a common ancestor while the (b) unrooted tree does not. (credit a: modification of work by Eric Gaba)

In a rooted tree, the branching indicates evolutionary relationships (Figure 20.3). The point where a split occurs, a **branch point**, represents where a single lineage evolved into a distinct new one. We call a lineage that evolved early from the root that remains unbranched a **basal taxon**. We call two lineages stemming from the same branch point **sister taxa**. A branch with more than two lineages is a **polytomy** and serves to illustrate where scientists have not definitively determined all of the relationships. Note that although sister taxa and polytomy do share an ancestor, it does not mean that the groups of organisms split or evolved from each other. Organisms in two taxa may have split at a specific branch point, but neither taxon gave rise to the other.

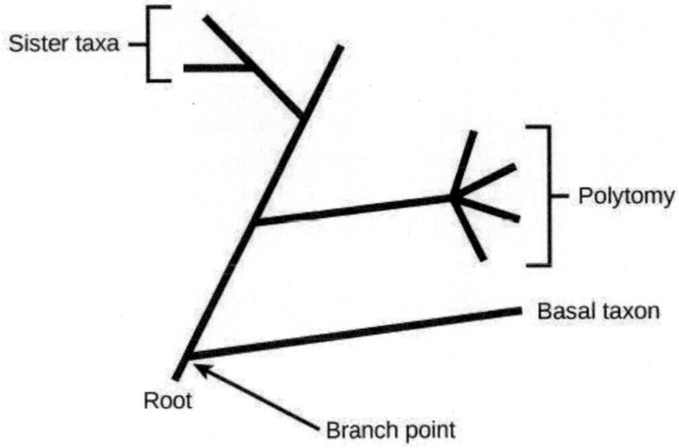

Figure 20.3 A phylogenetic tree's root indicates that an ancestral lineage gave rise to all organisms on the tree. A branch point indicates where two lineages diverged. A lineage that evolved early and remains unbranched is a basal taxon. When two lineages stem from the same branch point, they are sister taxa. A branch with more than two lineages is a polytomy.

The diagrams above can serve as a pathway to understanding evolutionary history. We can trace the pathway from the origin of life to any individual species by navigating through the evolutionary branches between the two points. Also, by starting with a single species and tracing back towards the "trunk" of the tree, one can discover species' ancestors, as well as where lineages share a common ancestry. In addition, we can use the tree to study entire groups of organisms.

Another point to mention on phylogenetic tree structure is that rotation at branch points does not change the information. For example, if a branch point rotated and the taxon order changed, this would not alter the information because each taxon's evolution from the branch point was independent of the other.

Many disciplines within the study of biology contribute to understanding how past and present life evolved over time; these disciplines together contribute to building, updating, and maintaining the "tree of life." **Systematics** is the field that scientists use to organize and classify organisms based on evolutionary relationships. Researchers may use data from fossils, from studying the body part structures, or molecules that an organism uses, and DNA analysis. By combining data from many sources, scientists can construct an organism's phylogeny Since phylogenetic trees are hypotheses, they will continue to change as researchers discover new types of life and learn new information.

Limitations of Phylogenetic Trees

It may be easy to assume that more closely related organisms look more alike, and while this is often the case, it is not always true. If two closely related lineages evolved under significantly varied surroundings, it is possible for the two groups to appear more different than other groups that are not as closely related. For example, the phylogenetic tree in Figure 20.4 shows that lizards and rabbits both have amniotic eggs; whereas, frogs do not. Yet lizards and frogs appear more similar than lizards and rabbits.

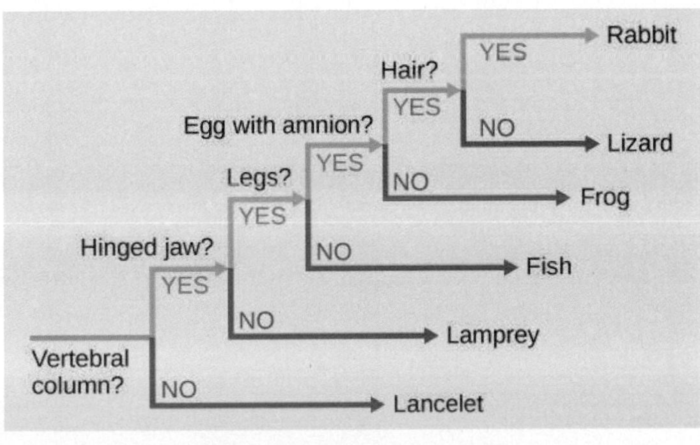

Figure 20.4 An organism that lacked a vertebral column roots this ladder-like phylogenetic tree of vertebrates. At each branch point, scientists place organisms with different characters in different groups based on shared characteristics.

Another aspect of phylogenetic trees is that, unless otherwise indicated, the branches do not account for length of time, only the evolutionary order. In other words, a branch's length does not typically mean more time passed, nor does a short branch mean less time passed— unless specified on the diagram. For example, in Figure 20.4, the tree does not indicate how much time passed between the evolution of amniotic eggs and hair. What the tree does show is the order in which things took place. Again using Figure 20.4, the tree shows that the oldest trait is the vertebral column, followed by hinged jaws, and so forth. Remember that any phylogenetic tree is a part of the greater whole, and like a real tree, it does not grow in only one direction after a new branch develops. Thus, for the organisms in Figure 20.4, just because a vertebral column evolved does not mean that invertebrate evolution ceased. It only means that a new branch formed. Also, groups that are not closely related, but evolve under similar conditions, may appear more phenotypically similar to each other than to a close relative.

LINK TO LEARNING

Head to this website (http://openstax.org/l/tree_of_life) to see interactive exercises that allow you to explore the evolutionary relationships among species.

Classification Levels

Taxonomy (which literally means "arrangement law") is the science of classifying organisms to construct internationally shared classification systems with each organism placed into increasingly more inclusive groupings. Think about a grocery store's organization. One large space is divided into departments, such as produce, dairy, and meats. Then each department further divides into aisles, then each aisle into categories and brands, and then finally a single product. We call this organization from larger to smaller, more specific categories a hierarchical system.

The taxonomic classification system (also called the Linnaean system after its inventor, Carl Linnaeus, a Swedish botanist, zoologist, and physician) uses a hierarchical model. Moving from the point of origin, the groups become more specific, until one branch ends as a single species. For example, after the common beginning of all life, scientists divide organisms into three large categories called domains: Bacteria, Archaea, and Eukarya. Within each domain is a second category called a **kingdom**. After kingdoms, the subsequent categories of increasing specificity are: **phylum**, **class**, **order**, **family**, **genus**, and **species** (Figure 20.5).

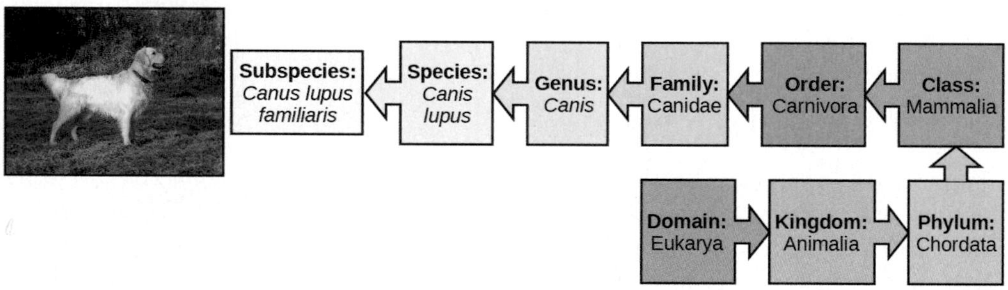

Figure 20.5 The taxonomic classification system uses a hierarchical model to organize living organisms into increasingly specific categories. The common dog, *Canis lupus familiaris*, is a subspecies of *Canis lupus*, which also includes the wolf and dingo. (credit "dog": modification of work by Janneke Vreugdenhil)

The kingdom Animalia stems from the Eukarya domain. Figure 20.5 above shows the classification for the common dog. Therefore, the full name of an organism technically has eight terms. For the dog it is: Eukarya, Animalia, Chordata, Mammalia, Carnivora, Canidae, *Canis,* and *lupus.* Notice that each name is capitalized except for species, and the genus and species names are italicized. Scientists generally refer to an organism only by its genus and species, which is its two-word scientific name, or **binomial nomenclature**. Therefore, the scientific name of the dog is *Canis lupus.* The name at each level is also a **taxon**. In other words, dogs are in order Carnivora. Carnivora is the name of the taxon at the order level; Canidae is the taxon at the family level, and so forth. Organisms also have a common name that people typically use, in this case, dog. Note that the dog is additionally a subspecies: the "*familiaris*" in *Canis lupus familiaris.* Subspecies are members of the same species that are capable of mating and reproducing viable offspring, but they are separate subspecies due to geographic or behavioral isolation or other factors.

Figure 20.6 shows how the levels move toward specificity with other organisms. Notice how the dog shares a domain with the widest diversity of organisms, including plants and butterflies. At each sublevel, the organisms become more similar because they are more closely related. Historically, scientists classified organisms using characteristics, but as DNA technology developed, they have determined more precise phylogenies.

VISUAL CONNECTION

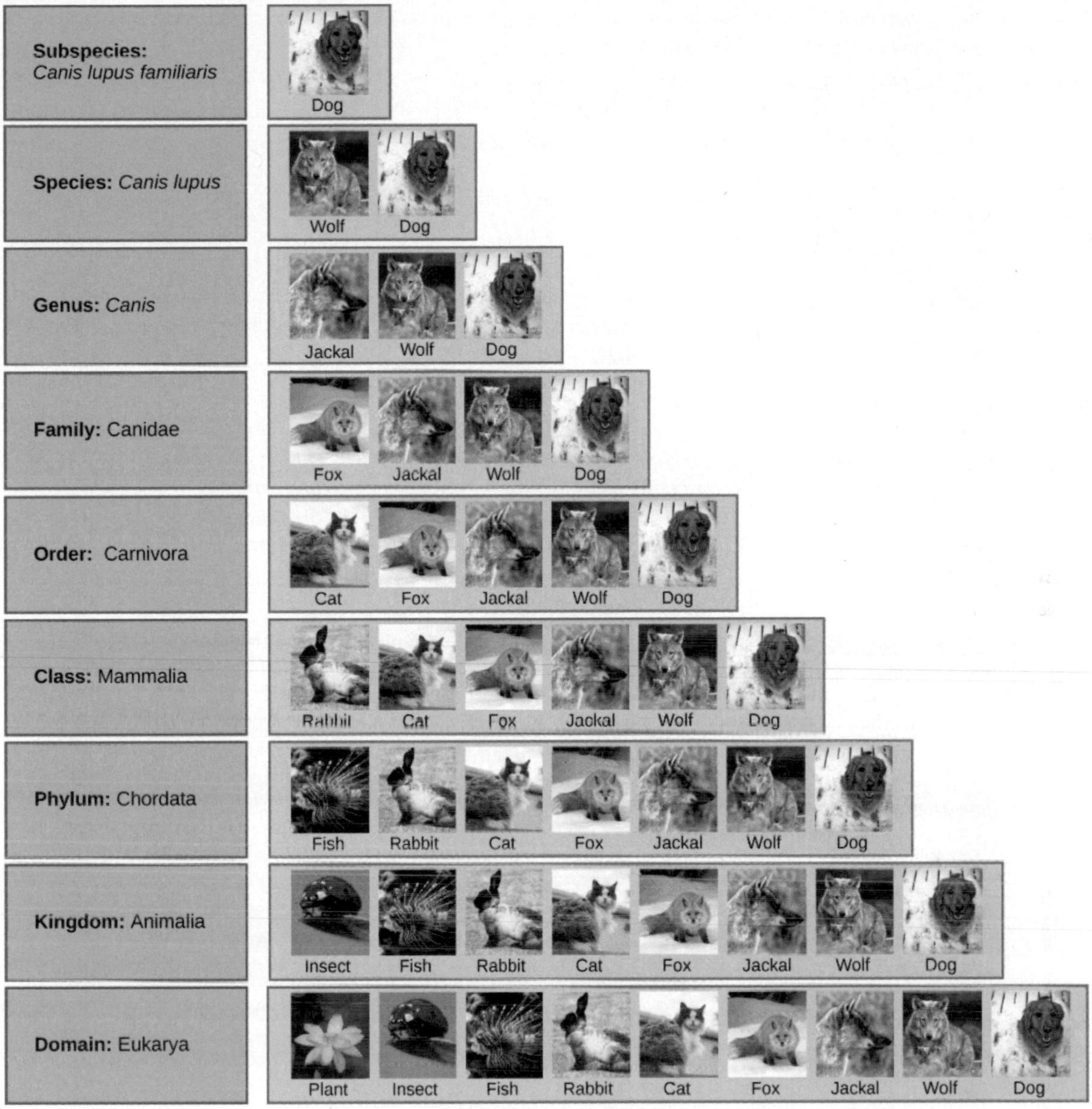

Figure 20.6 At each sublevel in the taxonomic classification system, organisms become more similar. Dogs and wolves are the same species because they can breed and produce viable offspring, but they are different enough to be classified as different subspecies. (credit "plant": modification of work by "berduchwal"/Flickr; credit "insect": modification of work by Jon Sullivan; credit "fish": modification of work by Christian Mehlführer; credit "rabbit": modification of work by Aidan Wojtas; credit "cat": modification of work by Jonathan Lidbeck; credit "fox": modification of work by Kevin Bacher, NPS; credit "jackal": modification of work by Thomas A. Hermann, NBII, USGS; credit "wolf": modification of work by Robert Dewar; credit "dog": modification of work by "digital_image_fan"/Flickr)

At what levels are cats and dogs part of the same group?

LINK TO LEARNING

Visit this website (http://openstax.org/l/classify_life) to explore the classifications of thousands of organisms. This reference site contains about 10% of the described species on the planet.

Recent genetic analysis and other advancements have found that some earlier phylogenetic classifications do not align with the evolutionary past; therefore, researchers must make changes and updates as new discoveries occur. Recall that phylogenetic trees are hypotheses and are modified as data becomes available. In addition, classification historically has focused on grouping organisms mainly by shared characteristics and does not necessarily illustrate how the various groups relate to each other from an evolutionary perspective. For example, despite the fact that a hippopotamus resembles a pig more than a whale, the hippopotamus may be the whale's closest living relative.

20.2 Determining Evolutionary Relationships

By the end of this section, you will be able to do the following:

- Compare homologous and analogous traits
- Discuss the purpose of cladistics
- Describe maximum parsimony

Scientists must collect accurate information that allows them to make evolutionary connections among organisms. Similar to detective work, scientists must use evidence to uncover the facts. In the case of phylogeny, evolutionary investigations focus on two types of evidence: morphologic (form and function) and genetic.

Two Options for Similarities

In general, organisms that share similar physical features and genomes are more closely related than those that do not. We refer to such features that overlap both morphologically (in form) and genetically as homologous structures. They stem from developmental similarities that are based on evolution. For example, the bones in bat and bird wings have homologous structures (Figure 20.7).

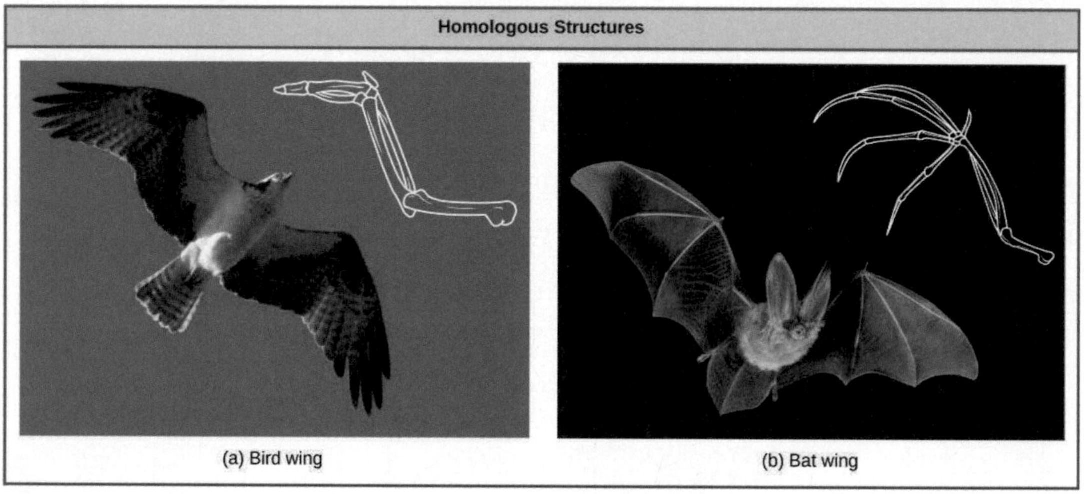

Figure 20.7 Bat and bird wings are homologous structures, indicating that bats and birds share a common evolutionary past. (credit a: modification of work by Steve Hillebrand, USFWS; credit b: modification of work by U.S. DOI BLM)

Notice it is not simply a single bone, but rather a grouping of several bones arranged in a similar way. The more complex the feature, the more likely any kind of overlap is due to a common evolutionary past. Imagine two people from different countries both inventing a car with all the same parts and in exactly the same arrangement without any previous or shared knowledge. That outcome would be highly improbable. However, if two people both invented a hammer, we can reasonably conclude that both could have the original idea without the help of the other. The same relationship between complexity and shared evolutionary history is true for homologous structures in organisms.

Misleading Appearances

Some organisms may be very closely related, even though a minor genetic change caused a major morphological difference to make them look quite different. Similarly, unrelated organisms may be distantly related, but appear very much alike. This usually happens because both organisms were in common adaptations that evolved within similar environmental conditions. When similar characteristics occur because of environmental constraints and not due to a close evolutionary relationship, it is an **analogy** or homoplasy. For example, insects use wings to fly like bats and birds, but the wing structure and embryonic origin is completely different. These are analogous structures (Figure 20.8).

Similar traits can be either homologous or analogous. Homologous structures share a similar embryonic origin. Analogous organs have a similar function. For example, the bones in a whale's front flipper are homologous to the bones in the human arm. These structures are not analogous. A butterfly or bird's wings are analogous but not homologous. Some structures are both analogous and homologous: bird and bat wings are both homologous and analogous. Scientists must determine which type of similarity a feature exhibits to decipher the organisms' phylogeny.

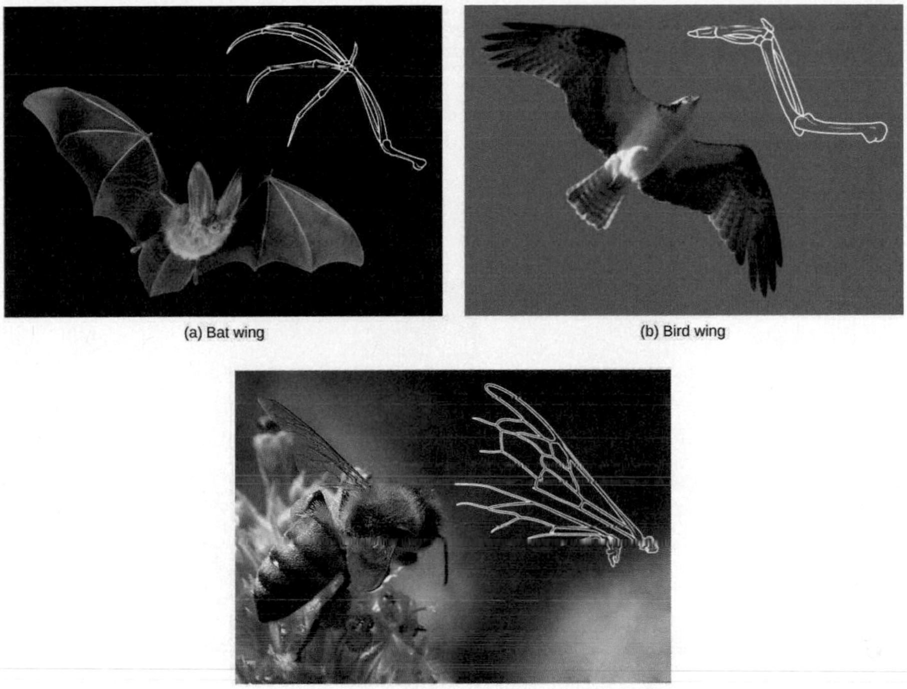

(a) Bat wing

(b) Bird wing

(c) Insect wing

Figure 20.8 The (c) wing of a honeybee is similar in shape to a (b) bird wing and (a) bat wing, and it serves the same function. However, the honeybee wing is not composed of bones and has a distinctly different structure and embryonic origin. These wing types (insect versus bat and bird) illustrate an analogy—similar structures that do not share an evolutionary history. (credit a: modification of work by U.S. DOI BLM; credit b: modification of work by Steve Hillebrand, USFWS; credit c: modification of work by Jon Sullivan)

LINK TO LEARNING

This website (http://openstax.org/l/relationships) has several examples to show how appearances can be misleading in understanding organisms' phylogenetic relationships.

Molecular Comparisons

The advancement of DNA technology has given rise to **molecular systematics**, which is use of molecular data in taxonomy and biological geography (biogeography). New computer programs not only confirm many earlier classified organisms, but also uncover previously made errors. As with physical characteristics, even the DNA sequence can be tricky to read in some cases. For some situations, two very closely related organisms can appear unrelated if a mutation occurred that caused a shift in the genetic code. Inserting or deleting a mutation would move each nucleotide base over one place, causing two similar codes to appear unrelated.

Sometimes two segments of DNA code in distantly related organisms randomly share a high percentage of bases in the same locations, causing these organisms to appear closely related when they are not. For both of these situations, computer technologies help identify the actual relationships, and, ultimately, the coupled use of both morphologic and molecular information is more effective in determining phylogeny.

EVOLUTION CONNECTION

Why Does Phylogeny Matter?

Evolutionary biologists could list many reasons why understanding phylogeny is important to everyday life in human society. For botanists, phylogeny acts as a guide to discovering new plants that can be used to benefit people. Think of all the ways humans use plants—food, medicine, and clothing are a few examples. If a plant contains a compound that is effective in treating cancer, scientists might want to examine all of the compounds for other useful drugs.

A research team in China identified a DNA segment that they thought to be common to some medicinal plants in the family Fabaceae (the legume family). They worked to identify which species had this segment (Figure 20.9). After testing plant species in this family, the team found a DNA marker (a known location on a chromosome that enabled them to identify the species) present. Then, using the DNA to uncover phylogenetic relationships, the team could identify whether a newly discovered plant was in this family and assess its potential medicinal properties.

Figure 20.9 *Dalbergia sissoo (D. sissoo)* is in the Fabaceae, or legume family. Scientists found that *D. sissoo* shares a DNA marker with species within the Fabaceae family that have antifungal properties. Subsequently, researchers found that *D. sissoo* had fungicidal activity, supporting the idea that DNA markers are useful to screen plants with potential medicinal properties.

Building Phylogenetic Trees

How do scientists construct phylogenetic trees? After they sort the homologous and analogous traits, scientists often organize the homologous traits using **cladistics**. This system sorts organisms into clades: groups of organisms that descended from a single ancestor. For example, in Figure 20.10, all the organisms in the orange region evolved from a single ancestor that had amniotic eggs. Consequently, these organisms also have amniotic eggs and make a single clade, or a **monophyletic group**. Clades must include all descendants from a branch point.

VISUAL CONNECTION

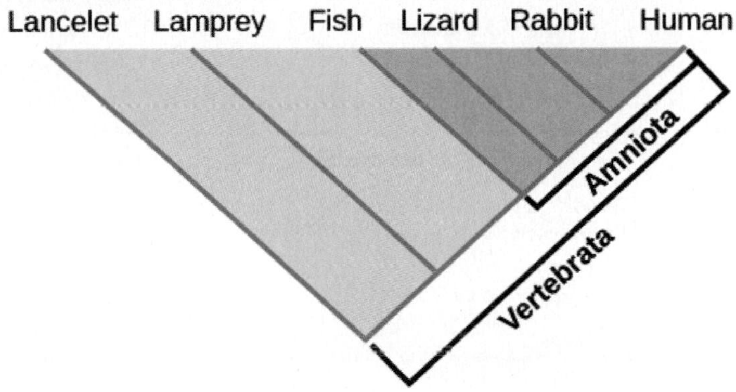

Figure 20.10 Lizards, rabbits, and humans all descend from a common ancestor that had an amniotic egg. Thus, lizards, rabbits, and humans all belong to the clade Amniota. Vertebrata is a larger clade that also includes fish and lamprey.

Which animals in this figure belong to a clade that includes animals with hair? Which evolved first, hair or the amniotic egg?

Clades can vary in size depending on which branch point one references. The important factor is that all organisms in the clade or monophyletic group stem from a single point on the tree. You can remember this because monophyletic breaks down into "mono," meaning one, and "phyletic," meaning evolutionary relationship. Figure 20.11 shows various clade examples. Notice how each clade comes from a single point; whereas, the non-clade groups show branches that do not share a single point.

VISUAL CONNECTION

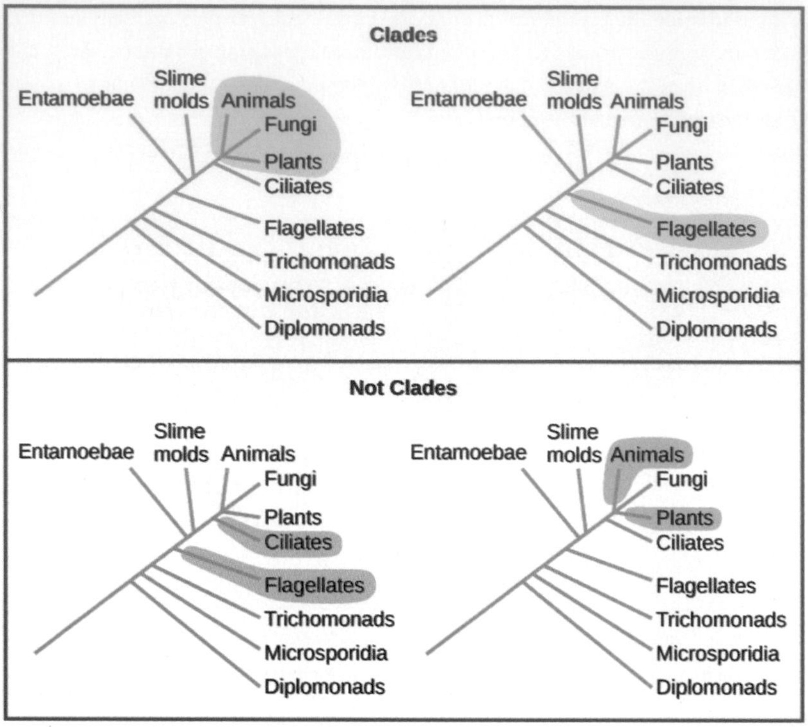

Figure 20.11 All the organisms within a clade stem from a single point on the tree. A clade may contain multiple groups, as in the case of animals, fungi and plants, or a single group, as in the case of flagellates. Groups that diverge at a different branch point, or that do not include all groups in a single branch point, are not clades.

What is the largest clade in this diagram?

Shared Characteristics

Organisms evolve from common ancestors and then diversify. Scientists use the phrase "descent with modification" because even though related organisms have many of the same characteristics and genetic codes, changes occur. This pattern repeats as one goes through the phylogenetic tree of life:

1. A change in an organism's genetic makeup leads to a new trait which becomes prevalent in the group.
2. Many organisms descend from this point and have this trait.
3. New variations continue to arise: some are adaptive and persist, leading to new traits.
4. With new traits, a new branch point is determined (go back to step 1 and repeat).

If a characteristic is found in the ancestor of a group, it is considered a **shared ancestral character** because all of the organisms in the taxon or clade have that trait. The vertebrate in Figure 20.10 is a shared ancestral character. Now consider the amniotic egg characteristic in the same figure. Only some of the organisms in Figure 20.10 have this trait, and to those that do, it is called a **shared derived character** because this trait derived at some point but does not include all of the ancestors in the tree.

The tricky aspect to shared ancestral and shared derived characters is that these terms are relative. We can consider the same trait one or the other depending on the particular diagram that we use. Returning to Figure 20.10, note that the amniotic egg is a shared ancestral character for the Amniota clade, while having hair is a shared derived character for some organisms in this group. These terms help scientists distinguish between clades in building phylogenetic trees.

Choosing the Right Relationships

Imagine being the person responsible for organizing all department store items properly—an overwhelming task. Organizing the evolutionary relationships of all life on Earth proves much more difficult: scientists must span enormous blocks of time and work with information from long-extinct organisms. Trying to decipher the proper connections, especially given the presence of homologies and analogies, makes the task of building an accurate tree of life extraordinarily difficult. Add to that advancing DNA technology, which now provides large quantities of genetic sequences for researchers to use and analzye. Taxonomy is a subjective discipline: many organisms have more than one connection to each other, so each taxonomist will decide the order of connections.

To aid in the tremendous task of describing phylogenies accurately, scientists often use the concept of **maximum parsimony**, which means that events occurred in the simplest, most obvious way. For example, if a group of people entered a forest preserve to hike, based on the principle of maximum parsimony, one could predict that most would hike on established trails rather than forge new ones.

For scientists deciphering evolutionary pathways, the same idea is used: the pathway of evolution probably includes the fewest major events that coincide with the evidence at hand. Starting with all of the homologous traits in a group of organisms, scientists look for the most obvious and simple order of evolutionary events that led to the occurrence of those traits.

LINK TO LEARNING

Head to this website (http://openstax.org/l/using_parsimony) to learn how researchers use maximum parsimony to create phylogenetic trees.

These tools and concepts are only a few strategies scientists use to tackle the task of revealing the evolutionary history of life on Earth. Recently, newer technologies have uncovered surprising discoveries with unexpected relationships, such as the fact that people seem to be more closely related to fungi than fungi are to plants. Sound unbelievable? As the information about DNA sequences grows, scientists will become closer to mapping the evolutionary history of all life on Earth.

20.3 Perspectives on the Phylogenetic Tree

By the end of this section, you will be able to do the following:

- Describe horizontal gene transfer
- Illustrate how prokaryotes and eukaryotes transfer genes horizontally
- Identify the web and ring models of phylogenetic relationships and describe how they differ from the original phylogenetic tree concept

Phylogenetic modeling concepts are constantly changing. It is one of the most dynamic fields of study in all biology. Over the last several decades, new research has challenged scientists' ideas about how organisms are related. The scientific community has proposed new models of these relationships.

Many phylogenetic trees are models of the evolutionary relationship among species. Phylogenetic trees originated with Charles Darwin, who sketched the first phylogenetic tree in 1837 (Figure 20.12a). This served as a prototype for subsequent studies for more than a century. The phylogenetic tree concept with a single trunk representing a common ancestor, with the branches representing the divergence of species from this ancestor, fits well with the structure of many common trees, such as the oak (Figure 20.12b). However, evidence from modern DNA sequence analysis and newly developed computer algorithms has caused skepticism about the standard tree model's validity in the scientific community.

(a) (b)

Figure 20.12 The (a) concept of the “tree of life” dates to an 1837 Charles Darwin sketch. Like an (b) oak tree, the “tree of life” has a single trunk and many branches. (credit b: modification of work by "Amada44"/Wikimedia Commons)

Limitations to the Classic Model

Classical thinking about prokaryotic evolution, included in the classic tree model, is that species evolve clonally. That is, they produce offspring themselves with only random mutations causing the descent into the variety of modern-day and extinct species known to science. This view is somewhat complicated in eukaryotes that reproduce sexually, but the laws of Mendelian genetics explain the variation in offspring, again, to be a result of a mutation within the species. Scientists did not consider the concept of genes transferring between unrelated species as a possibility until relatively recently. Horizontal gene transfer (HGT), or lateral gene transfer, is the transfer of genes between unrelated species. HGT is an ever-present phenomenon, with many evolutionists postulating a major role for this process in evolution, thus complicating the simple tree model. Genes pass between species which are only distantly related using standard phylogeny, thus adding a layer of complexity to understanding phylogenetic relationships.

The various ways that HGT occurs in prokaryotes is important to understanding phylogenies. Although at present some do not view HGT as important to eukaryotic evolution, HGT does occur in this domain as well. Finally, as an example of the ultimate gene transfer, some scientists have proposed genome fusion theories between symbiotic or endosymbiotic organisms to explain an event of great importance—the evolution of the first eukaryotic cell, without which humans could not have come into existence.

Horizontal Gene Transfer

Horizontal gene transfer (HGT) is the introduction of genetic material from one species to another species by mechanisms other than the vertical transmission from parent(s) to offspring. These transfers allow even distantly related species to share genes, influencing their phenotypes. Scientists believe that HGT is more prevalent in prokaryotes, but that this process transfers only about 2% of the prokaryotic genome. Some researchers believe such estimates are premature: we must view the actual importance of HGT to evolutionary processes as a work in progress. As scientists investigate this phenomenon more thoroughly, they may reveal more HGT transfer. Many scientists believe that HGT and mutation are (especially in prokaryotes) a significant source of genetic variation, which is the raw material in the natural selection process. These transfers may occur between any two species that share an intimate relationship (Table 20.1).

Prokaryotic and Eukaryotic HGT Mechanisms Summary

	Mechanism	Mode of Transmission	Example
Prokaryotes	transformation	DNA uptake	many prokaryotes

Table 20.1

	Mechanism	Mode of Transmission	Example
	transduction	bacteriophage (virus)	bacteria
	conjugation	pilus	many prokaryotes
	gene transfer agents	phage-like particles	purple non-sulfur bacteria
Eukaryotes	from food organisms	unknown	aphid
	jumping genes	transposons	rice and millet plants
	epiphytes/parasites	unknown	yew tree fungi
	from viral infections		

Table 20.1

HGT in Prokaryotes

HGT mechanisms are quite common in the Bacteria and Archaea domains, thus significantly changing the way scientists view their evolution. The majority of evolutionary models, such as in the Endosymbiont Theory, propose that eukaryotes descended from multiple prokaryotes, which makes HGT all the more important to understanding the phylogenetic relationships of all extant and extinct species. The Endosymbiont Theory purports that the eukaryotes' mitochondria and the green plants' chloroplasts and flagellates originated as free-living prokaryotes that invaded primitive eukaryotic cells and become established as permanent symbionts in the cytoplasm.

Microbiology students are well aware that genes transfer among common bacteria. These gene transfers between species are the major mechanism whereby bacteria acquire resistance to antibiotics. Classically, scientists believe that three different mechanisms drive such transfers.

1. Transformation: bacteria takes up naked DNA
2. Transduction: a virus transfers the genes
3. Conjugation: a hollow tube, or pilus transfers genes between organisms

More recently, scientists have discovered a fourth gene transfer mechanism between prokaryotes. Small, virus-like particles, or **gene transfer agents (GTAs)** transfer random genomic segments from one prokaryote species to another. GTAs are responsible for genetic changes, sometimes at a very high frequency compared to other evolutionary processes. Scientists characterized the first GTA in 1974 using purple, non-sulfur bacteria. These GTAs, which are most likely derived from bacteriophage DNA inserted in a prokaryote that lost the ability to produce new bacteriophages, carry random DNA pieces from one organism to another. Controlled studies using marine bacteria have demonstrated GTAs' ability to act with high frequency. Scientists have estimated gene transfer events in marine prokaryotes, either by GTAs or by viruses, to be as high as 10^{13} per year in the Mediterranean Sea alone. GTAs and viruses are efficient HGT vehicles with a major impact on prokaryotic evolution.

As a consequence of this modern DNA analysis, the idea that eukaryotes evolved directly from Archaea has fallen out of favor. While eukaryotes share many features that are absent in bacteria, such as the TATA box (located in many genes' promoter region), the discovery that some eukaryotic genes were more homologous with bacterial DNA than Archaea DNA made this idea less tenable. Furthermore, scientists have proposed genome fusion from Archaea and Bacteria by endosymbiosis as the ultimate event in eukaryotic evolution.

HGT in Eukaryotes

Although it is easy to see how prokaryotes exchange genetic material by HGT, scientists initially thought that this process was absent in eukaryotes. After all, prokaryotes are but single cells exposed directly to their environment; whereas, the multicellular organisms' sex cells are usually sequestered in protected parts of the body. It follows from this idea that the gene transfers between multicellular eukaryotes should be more difficult. Scientists believe this process is rarer in eukaryotes and has a much smaller evolutionary impact than in prokaryotes. In spite of this, HGT between distantly related organisms is evident in several eukaryotic species, and it is possible that scientists will discover more examples in the future.

In plants, researchers have observed gene transfer in species that cannot cross-pollinate by normal means. Transposons or "jumping genes" have shown a transfer between rice and millet plant species. Furthermore, fungal species feeding on yew trees, from which the anti-cancer drug TAXOL® is derived from the bark, have acquired the ability to make taxol themselves, a clear example of gene transfer.

In animals, a particularly interesting example of HGT occurs within the aphid species (Figure 20.13). Aphids are insects that vary in color based on carotenoid content. Carotenoids are pigments that a variety of plants, fungi, and microbes produce, and they serve a variety of functions in animals, who obtain these chemicals from their food. Humans require carotenoids to synthesize vitamin A, and we obtain them by eating orange fruits and vegetables: carrots, apricots, mangoes, and sweet potatoes. Alternatively, aphids have acquired the ability to make the carotenoids on their own. According to DNA analysis, this ability is due to fungal genes transferring into the insect by HGT, presumably as the insect consumed fungi for food. A carotenoid enzyme, or desaturase, is responsible for the red coloration in certain aphids, and when mutation of this gene leads to formation of inactive enzyme, the aphids revert to their more common green color (Figure 20.13).

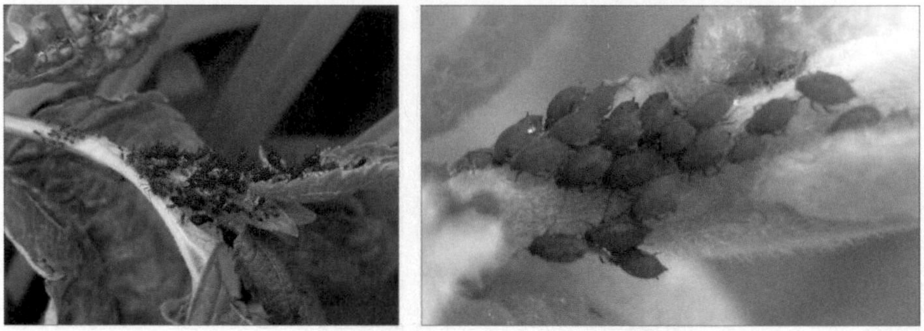

Figure 20.13 (a) Red aphids get their color from red carotenoid pigment. Genes necessary to make this pigment are present in certain fungi, and scientists speculate that aphids acquired these genes through HGT after consuming fungi for food. If mutation inactivates the genes for making carotenoids, the aphids revert back to (b) their green color. Red coloration makes the aphids considerably more conspicuous to predators, but evidence suggests that red aphids are more resistant to insecticides than green ones. Thus, red aphids may be more fit to survive in some environments than green ones. (credit a: modification of work by Benny Mazur; credit b: modification of work by Mick Talbot)

Genome Fusion and Eukaryote Evolution

Scientists believe the ultimate in HGT occurs through **genome fusion** between different prokaryote species when two symbiotic organisms become endosymbiotic. This occurs when one species is taken inside another species' cytoplasm, which ultimately results in a genome consisting of genes from both the endosymbiont and the host. This mechanism is an aspect of the Endosymbiont Theory, which most biologists accept as the mechanism whereby eukaryotic cells obtained their mitochondria and chloroplasts. However, the role of endosymbiosis in developing the nucleus is more controversial. Scientists believe that nuclear and mitochondrial DNA have different (separate) evolutionary origins, with the mitochondrial DNA derived from the bacteria's circular genomes ancient prokaryotic cells engulfed. We can regard mitochondrial DNA as the smallest chromosome. Interestingly enough, mitochondrial DNA is inherited only from the mother. The mitochondrial DNA degrades in sperm when the sperm degrades in the fertilized egg or in other instances when the mitochondria located in the sperm's flagellum fails to enter the egg.

Within the past decade, James Lake of the UCLA/NASA Astrobiology Institute proposed that the genome fusion process is responsible for the evolution of the first eukaryotic cells (Figure 20.14a). Using DNA analysis and a new mathematical algorithm, conditioned reconstruction (CR), his laboratory proposed that eukaryotic cells developed from an endosymbiotic gene fusion between two species, one an Archaea and the other a Bacteria. As mentioned, some eukaryotic genes resemble those of Archaea; whereas, others resemble those from Bacteria. An endosymbiotic fusion event, such as Lake has proposed, would clearly explain this observation. Alternatively, this work is new and the CR algorithm is relatively unsubstantiated, which causes many scientists to resist this hypothesis.

Lake's more recent work (Figure 20.14b) proposes that gram-negative bacteria, which are unique within their domain in that they contain two lipid bilayer membranes, resulted from an endosymbiotic fusion of archaeal and bacterial species. The double membrane would be a direct result of the endosymbiosis, with the endosymbiont picking up the second membrane from the host as it was internalized. Scientists have also used this mechanism to explain the double membranes in mitochondria and chloroplasts. Lake's work is not without skepticism, and the biological science community still debates his ideas. In addition to Lake's hypothesis, there are several other competing theories as to the origin of eukaryotes. How did the eukaryotic nucleus evolve? One theory is that the prokaryotic cells produced an additional membrane that surrounded the bacterial chromosome. Some bacteria have the DNA enclosed by two membranes; however, there is no evidence of a nucleolus or nuclear pores. Other proteobacteria also have membrane-bound chromosomes. If the eukaryotic nucleus evolved this way, we would expect one of the two types of prokaryotes to be more closely related to eukaryotes.

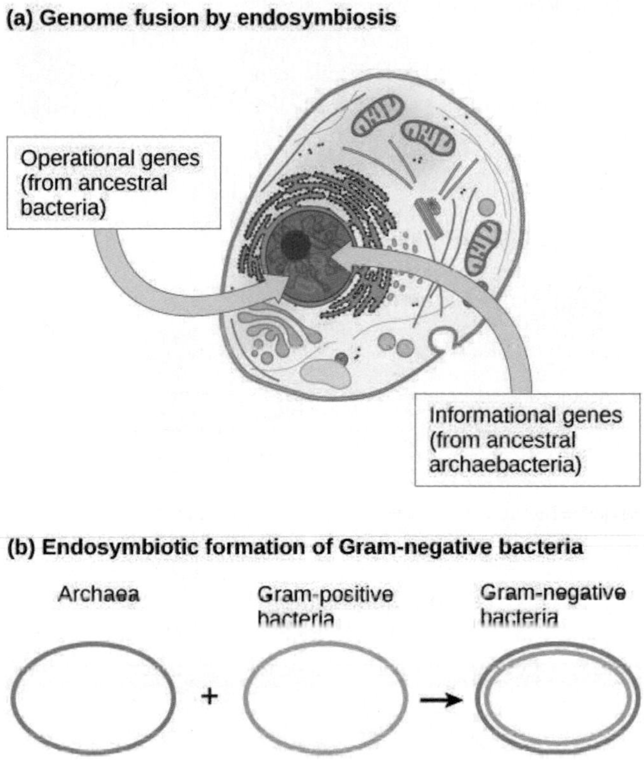

Figure 20.14 The scientific community now widely accepts the theory that mitochondria and chloroplasts are endosymbiotic in origin. More controversial is the proposal that (a) the eukaryotic nucleus resulted from fusing archaeal and bacterial genomes, and that (b) Gram-negative bacteria, which have two membranes, resulted from fusing Archaea and Gram-positive bacteria, each of which has a single membrane.

The **nucleus-first** hypothesis proposes that the nucleus evolved in prokaryotes first (Figure 20.15a), followed by a later fusion of the new eukaryote with bacteria that became mitochondria. The **mitochondria-first** hypothesis proposes that mitochondria were first established in a prokaryotic host (Figure 20.15b), which subsequently acquired a nucleus, by fusion or other mechanisms, to become the first eukaryotic cell. Most interestingly, the **eukaryote-first** hypothesis proposes that prokaryotes actually evolved from eukaryotes by losing genes and complexity (Figure 20.15c). All of these hypotheses are testable. Only time and more experimentation will determine which hypothesis data best supports.

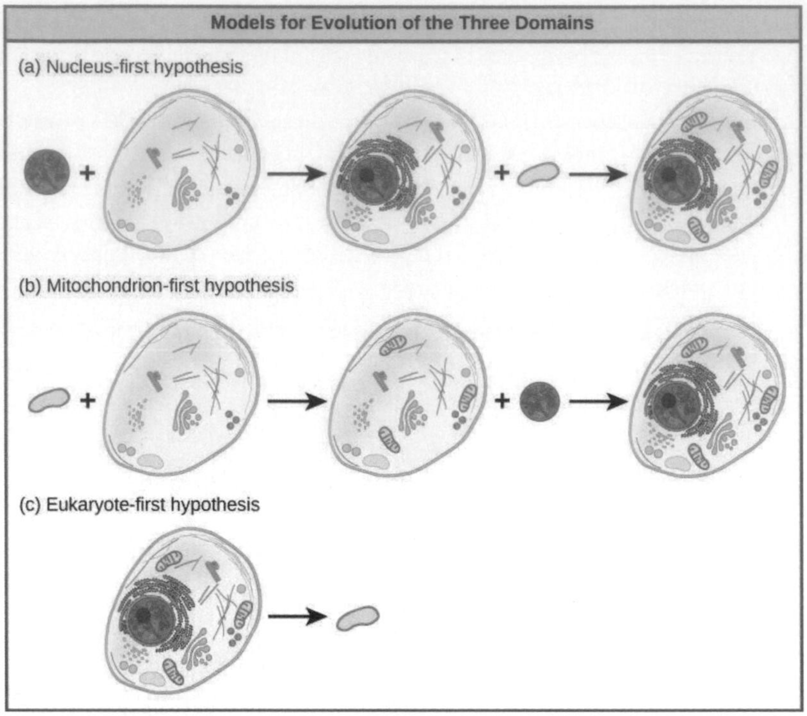

Figure 20.15 Three alternate hypotheses of eukaryotic and prokaryotic evolution are (a) the nucleus-first hypothesis, (b) the mitochondrion-first hypothesis, and (c) the eukaryote-first hypothesis.

Web and Network Models

Recognizing the importance of HGT, especially in prokaryote evolution, has caused some to propose abandoning the classic "tree of life" model. In 1999, W. Ford Doolittle proposed a phylogenetic model that resembles a web or a network more than a tree. The hypothesis is that eukaryotes evolved not from a single prokaryotic ancestor, but from a pool of many species that were sharing genes by HGT mechanisms. As Figure 20.16a shows, some individual prokaryotes were responsible for transferring the bacteria that caused mitochondrial development to the new eukaryotes; whereas, other species transferred the bacteria that gave rise to chloroplasts. Scientists often call this model the "**web of life**." In an effort to save the tree analogy, some have proposed using the *Ficus* tree (Figure 20.16b) with its multiple trunks as a phylogenetic way to represent a diminished evolutionary role for HGT.

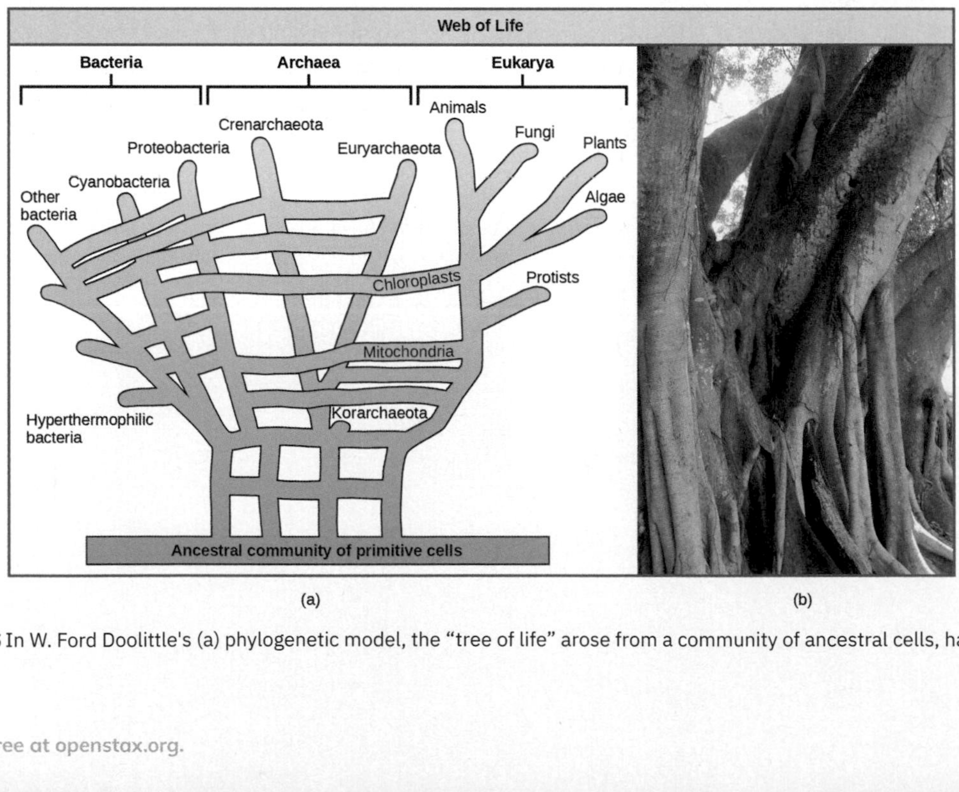

Figure 20.16 In W. Ford Doolittle's (a) phylogenetic model, the "tree of life" arose from a community of ancestral cells, has multiple trunks,

and has connections between branches where horizontal gene transfer has occurred. Visually, this concept is better represented by (b) the multi-trunked **Ficus** than by an oak's single trunk similar to Darwin's tree in Figure 20.12. (credit b: modification of work by "psyberartist"/Flickr)

Ring of Life Models

Others have proposed abandoning any tree-like model of phylogeny in favor of a ring structure, the so-called "**ring of life**" (Figure 20.17). This is a phylogenetic model where all three domains of life evolved from a pool of primitive prokaryotes. Lake, again using the conditioned reconstruction algorithm, proposes a ring-like model in which species of all three domains—Archaea, Bacteria, and Eukarya—evolved from a single pool of gene-swapping prokaryotes. His laboratory proposes that this structure is the best fit for data from extensive DNA analyses performed in his laboratory, and that the ring model is the only one that adequately takes HGT and genomic fusion into account. However, other phylogeneticists remain highly skeptical of this model.

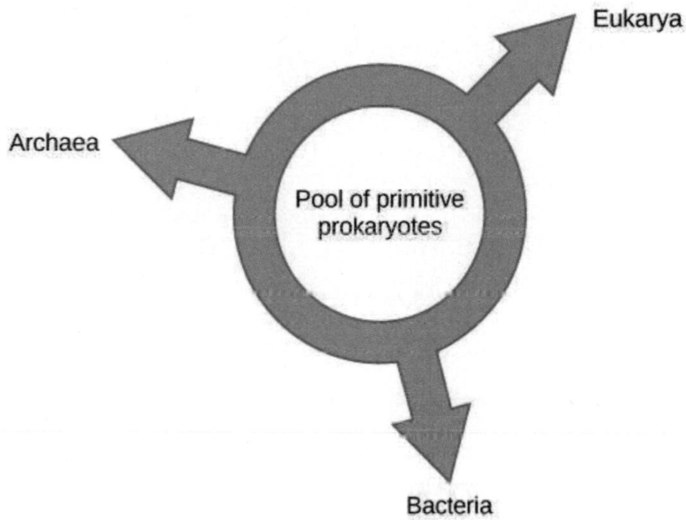

Figure 20.17 According to the "ring of life" phylogenetic model, the three domains of life evolved from a pool of primitive prokaryotes.

In summary, we must modify Darwin's "tree of life" model to include HGT. Does this mean abandoning the tree model completely? Even Lake argues that scientists should attempt to modify the tree model to allow it to accurately fit his data, and only the inability to do so will sway people toward his ring proposal.

This doesn't mean a tree, web, or a ring will correlate completely to an accurate description of phylogenetic relationships of life. A consequence of the new thinking about phylogenetic models is the idea that Darwin's original phylogenetic tree concept is too simple, but made sense based on what scientists knew at the time. However, the search for a more useful model moves on: each model serves as hypotheses to test with the possibility of developing new models. This is how science advances. Researchers use these models as visualizations to help construct hypothetical evolutionary relationships and understand the massive amount of data that requires analysis.

KEY TERMS

analogy (also, homoplasy) characteristic that is similar between organisms by convergent evolution, not due to the same evolutionary path

basal taxon branch on a phylogenetic tree that has not diverged significantly from the root ancestor

binomial nomenclature system of two-part scientific names for an organism, which includes genus and species names

branch point node on a phylogenetic tree where a single lineage splits into distinct new ones

cladistics system to organize homologous traits to describe phylogenies

class division of phylum in the taxonomic classification system

eukaryote-first hypothesis proposal that prokaryotes evolved from eukaryotes

family division of order in the taxonomic classification system

gene transfer agent (GTA) bacteriophage-like particle that transfers random genomic segments from one species of prokaryote to another

genome fusion fusion of two prokaryotic genomes, presumably by endosymbiosis

genus division of family in the taxonomic classification system; the first part of the binomial scientific name

horizontal gene transfer (HGT) (also, lateral gene transfer) transfer of genes between unrelated species

kingdom domain division in the taxonomic classification system

maximum parsimony applying the simplest, most obvious way with the least number of steps

mitochondria-first hypothesis proposal that prokaryotes acquired a mitochondrion first, followed by nuclear development

molecular systematics technique using molecular evidence to identify phylogenetic relationships

monophyletic group (also, clade) organisms that share a single ancestor

nucleus-first hypothesis proposal that prokaryotes acquired a nucleus first, and then the mitochondrion

order class division in the taxonomic classification system

phylogenetic tree diagram that reflects the evolutionary relationships among organisms or groups of organisms

phylogeny evolutionary history and relationship of an organism or group of organisms

phylum (plural: phyla) kingdom division in the taxonomic classification system

polytomy branch on a phylogenetic tree with more than two groups or taxa

ring of life phylogenetic model where all three domains of life evolved from a pool of primitive prokaryotes

rooted single ancestral lineage on a phylogenetic tree to which all organisms represented in the diagram relate

shared ancestral character describes a characteristic on a phylogenetic tree that all organisms on the tree share

shared derived character describes a characteristic on a phylogenetic tree that only a certain clade of organisms share

sister taxa two lineages that diverged from the same branch point

systematics field of organizing and classifying organisms based on evolutionary relationships

taxon (plural: taxa) single level in the taxonomic classification system

taxonomy science of classifying organisms

web of life phylogenetic model that attempts to incorporate the effects of horizontal gene transfer on evolution

CHAPTER SUMMARY

20.1 Organizing Life on Earth

Scientists continually gain new information that helps understand the evolutionary history of life on Earth. Each group of organisms went through its own evolutionary journey, or its phylogeny. Each organism shares relatedness with others, and based on morphologic and genetic evidence, scientists attempt to map the evolutionary pathways of all life on Earth. Historically, scientists organized organisms into a taxonomic classification system. However, today many scientists build phylogenetic trees to illustrate evolutionary relationships.

20.2 Determining Evolutionary Relationships

To build phylogenetic trees, scientists must collect accurate information that allows them to make evolutionary connections between organisms. Using morphologic and molecular data, scientists work to identify homologous characteristics and genes. Similarities between organisms can stem either from shared evolutionary history (homologies) or from separate evolutionary paths (analogies). Scientists can use newer technologies to help distinguish homologies from analogies. After identifying homologous information, scientists use cladistics to organize these events as a means to determine an evolutionary timeline. They then apply the concept of maximum parsimony, which states that the order of events probably occurred in the most obvious and simple way with the least amount of steps. For evolutionary events, this would be the path with the least number of major divergences that correlate with the evidence.

20.3 Perspectives on the Phylogenetic Tree

The phylogenetic tree, which Darwin first used, is the classic "tree of life" model describing phylogenetic relationships among species, and the most common model that scientists use today. New ideas about HGT and genome fusion have caused some to suggest revising the model to resemble webs or rings.

VISUAL CONNECTION QUESTIONS

1. Figure 20.6 At what levels are cats and dogs part of the same group?
2. Figure 20.10 Which animals in this figure belong to a clade that includes animals with hair? Which evolved first, hair or the amniotic egg?
3. Figure 20.11 What is the largest clade in this diagram?

REVIEW QUESTIONS

4. What is used to determine phylogeny?
   a. mutations
   b. DNA
   c. evolutionary history
   d. organisms on earth
5. What do scientists in the field of systematics accomplish?
   a. discover new fossil sites
   b. organize and classify organisms
   c. name new species
   d. communicate among field biologists
6. Which statement about the taxonomic classification system is correct?
   a. There are more domains than kingdoms.
   b. Kingdoms are the top category of classification.
   c. Classes are divisions of orders.
   d. Subspecies are the most specific category of classification.
7. On a phylogenetic tree, which term refers to lineages that diverged from the same place?
   a. sister taxa
   b. basal taxa
   c. rooted taxa
   d. dichotomous taxa
8. Which statement about analogies is correct?
   a. They occur only as errors.
   b. They are synonymous with homologous traits.
   c. They are derived by similar environmental constraints.
   d. They are a form of mutation.
9. What do scientists use to apply cladistics?
   a. homologous traits
   b. homoplasies
   c. analogous traits
   d. monophyletic groups
10. What is true about organisms that are a part of the same clade?
    a. They all share the same basic characteristics.
    b. They evolved from a shared ancestor.
    c. They usually fall into the same classification taxa.
    d. They have identical phylogenies.
11. Why do scientists apply the concept of maximum parsimony?
    a. to decipher accurate phylogenies
    b. to eliminate analogous traits
    c. to identify mutations in DNA codes
    d. to locate homoplasies
12. The transfer of genes by a mechanism not involving asexual reproduction is called:
    a. meiosis
    b. web of life
    c. horizontal gene transfer
    d. gene fusion
13. Particles that transfer genetic material from one species to another, especially in marine prokaryotes:
    a. horizontal gene transfer
    b. lateral gene transfer
    c. genome fusion device
    d. gene transfer agents

14. What does the trunk of the classic phylogenetic tree represent?
    a. single common ancestor
    b. pool of ancestral organisms
    c. new species
    d. old species

15. Which phylogenetic model proposes that all three domains of life evolved from a pool of primitive prokaryotes?
    a. tree of life
    b. web of life
    c. ring of life
    d. network model

CRITICAL THINKING QUESTIONS

16. How does a phylogenetic tree relate to the passing of time?
17. Some organisms that appear very closely related on a phylogenetic tree may not actually be closely related. Why is this?
18. List the different levels of the taxonomic classification system.
19. Dolphins and fish have similar body shapes. Is this feature more likely a homologous or analogous trait?
20. Why is it so important for scientists to distinguish between homologous and analogous characteristics before building phylogenetic trees?
21. Describe maximum parsimony.
22. Compare three different ways that eukaryotic cells may have evolved.
23. Describe how aphids acquired the ability to change color.

CHAPTER 21
Viruses

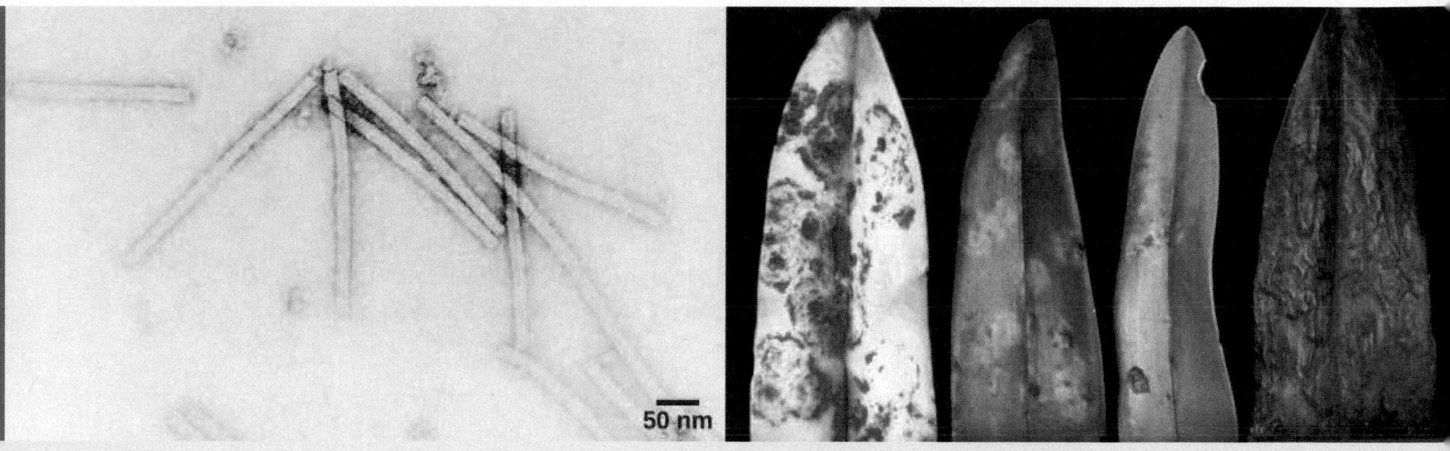

Figure 21.1 The tobacco mosaic virus, seen here by transmission electron microscopy (left), was the first virus to be discovered. The virus causes disease in tobacco and other plants, such as the orchid (right). (credit a: USDA ARS; credit b: modification of work by USDA Forest Service, Department of Plant Pathology Archive North Carolina State University; scale-bar data from Matt Russell)

INTRODUCTION Viruses are noncellular parasitic entities that cannot be classified within any kingdom. They can infect organisms as diverse as bacteria, plants, and animals. In fact, viruses exist in a sort of netherworld between a living organism and a nonliving entity. Living things grow, metabolize, and reproduce. In contrast, viruses are not cellular, do not have a metabolism or grow, and cannot divide by cell division. Viruses *can* copy, or replicate themselves; however, they are entirely dependent on resources derived from their host cells to produce progeny viruses which are assembled in their mature form. No one knows exactly when or how viruses evolved or from what ancestral source because viruses have not left a fossil record. Some virologists contend that modern viruses are a mosaic of bits and pieces of nucleic acids picked up from various sources along their respective evolutionary paths.

Chapter Outline

21.1 Viral Evolution, Morphology, and Classification

By the end of this section, you will be able to do the following:

- Describe how viruses were first discovered and how they are detected
- Discuss three hypotheses about how viruses evolved
- Describe the general structure of a virus
- Recognize the basic shapes of viruses
- Understand past and emerging classification systems for viruses
- Describe the basis for the Baltimore classification system

Viruses are diverse entities: They vary in structure, methods of replication, and the hosts they infect. Nearly all forms of life—from prokaryotic bacteria and archaeans, to eukaryotes such as plants, animals, and fungi—have viruses that infect them. While most biological diversity can be understood through evolutionary history (such as how species have adapted to changing environmental conditions and how different species are related to one another through common descent), much about virus origins and evolution remains unknown.

Discovery and Detection

Viruses were first discovered after the development of a porcelain filter—the Chamberland-Pasteur filter—that could remove all bacteria visible in the microscope from any liquid sample. In 1886, Adolph Meyer demonstrated that a disease of tobacco plants—**tobacco mosaic disease**—could be transferred from a diseased plant to a healthy one via liquid plant extracts. In 1892, Dmitri Ivanowski showed that this disease could be transmitted in this way even after the Chamberland-Pasteur filter had removed all viable bacteria from the extract. Still, it was many years before it was proved that these "filterable" infectious agents were not simply very small bacteria but were a new type of very small, disease-causing particle.

Most **virions**, or single virus particles, are very small, about 20 to 250 nanometers in diameter. However, some recently discovered viruses from amoebae range up to 1000 nm in diameter. With the exception of large virions, like the poxvirus and other large DNA viruses, viruses cannot be seen with a light microscope. It was not until the development of the electron microscope in the late 1930s that scientists got their first good view of the structure of the tobacco mosaic virus (TMV) (Figure 21.1), discussed above, and other viruses (Figure 21.2). The surface structure of virions can be observed by both scanning and transmission electron microscopy, whereas the internal structures of the virus can only be observed in images from a transmission electron microscope. The use of electron microscopy and other technologies has allowed for the discovery of many viruses of all types of living organisms.

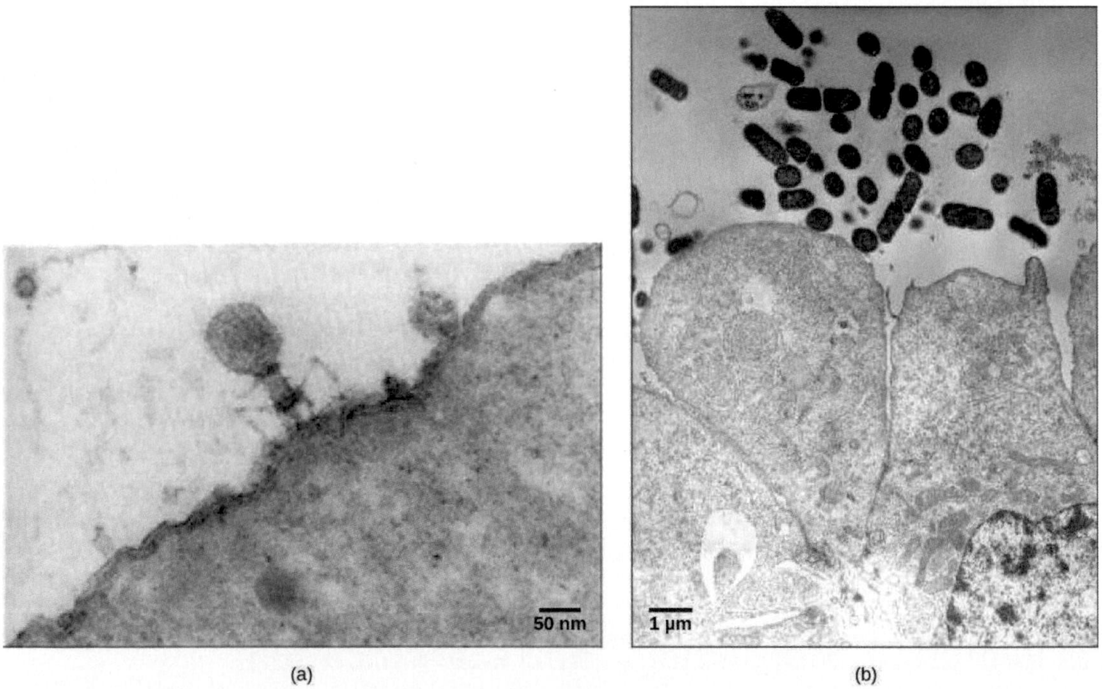

Figure 21.2 Most virus particles are visible only by electron microscopy. In these transmission electron micrographs, (a) a virus is as dwarfed by the bacterial cell it infects, as (b) these *E. coli* cells are dwarfed by cultured colon cells. (credit a: modification of work by U.S. Dept. of Energy, Office of Science, LBL, PBD; credit b: modification of work by J.P. Nataro and S. Sears, unpub. data, CDC; scale-bar data from Matt Russell)

Evolution of Viruses

Although biologists have a significant amount of knowledge about how present-day viruses mutate and adapt, much less is known about how viruses originated in the first place. When exploring the evolutionary history of most organisms, scientists can look at fossil records and similar historic evidence. However, viruses do not fossilize, as far as we know, so researchers must extrapolate from investigations of how today's viruses evolve and by using biochemical and genetic information to create speculative virus histories.

Most scholars agree that viruses don't have a single common ancestor, nor is there a single reasonable hypothesis about virus origins. There are current evolutionary scenarios that may explain the origin of viruses. One such hypothesis, the "devolution" or the *regressive hypothesis*, suggests that viruses evolved from free-living cells, or from intracellular prokaryotic parasites. However, many components of how this process might have occurred remain a mystery. A second hypothesis, the escapist or the *progressive hypothesis*, suggests that viruses originated from RNA and DNA molecules, or self-replicating entities similar to transposons or other mobile genetic elements, that escaped from a host cell with the ability to enter another. A third hypothesis, the *virus first hypothesis,* suggests that viruses may have been the first self-replicating entities before the first cells. In all cases, viruses are probably continuing to evolve along with the cells on which they rely on as hosts.

As technology advances, scientists may develop and refine additional hypotheses to explain the origins of viruses. The emerging field called virus molecular systematics attempts to do just that through comparisons of sequenced genetic material. These researchers hope one day to better understand the origin of viruses—a discovery that could lead to advances in the treatments for the ailments they produce.

Viral Morphology

Viruses are **noncellular**, meaning they are biological entities that do not have a cellular structure. They therefore lack most of the components of cells, such as organelles, ribosomes, and the plasma membrane. A virion consists of a nucleic acid core, an outer protein coating or **capsid**, and sometimes an outer **envelope** made of protein and phospholipid membranes derived from the host cell. Viruses may also contain additional proteins, such as enzymes, within the capsid or attached to the viral genome. The most obvious difference between members of different viral families is the variation in their morphology, which is quite diverse. An interesting feature of viral complexity is that the complexity of the host does not necessarily correlate with the complexity of the virion. In fact, some of the most complex virion structures are found in the **bacteriophages**—viruses that infect the simplest living organisms, bacteria.

Morphology

Viruses come in many shapes and sizes, but these features are consistent for each viral family. As we have seen, all virions have a nucleic acid genome covered by a protective capsid. The proteins of the capsid are encoded in the viral genome, and are called ***capsomeres***. Some viral capsids are simple helices or polyhedral "spheres," whereas others are quite complex in structure (Figure 21.3).

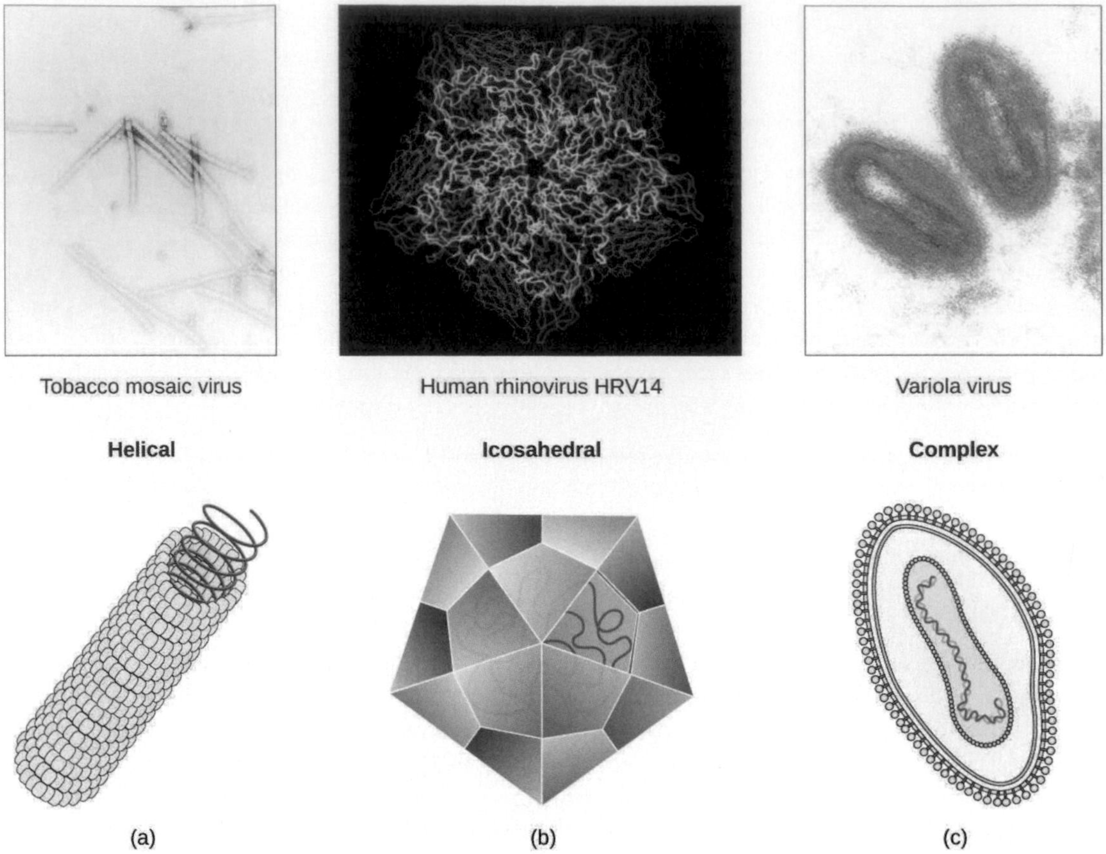

Figure 21.3 Viral capsids can be (a) helical, (b) polyhedral, or (c) have a complex shape. (credit a "micrograph": modification of work by USDA ARS; credit b "micrograph": modification of work by U.S. Department of Energy)

In general, the capsids of viruses are classified into four groups: helical, icosahedral, enveloped, and head-and-tail. *Helical capsids* are long and cylindrical. Many plant viruses are helical, including TMV. *Icosahedral viruses* have shapes that are roughly spherical, such as those of poliovirus or herpesviruses. *Enveloped viruses* have membranes derived from the host cell that surrounds the capsids. Animal viruses, such as HIV, are frequently enveloped. *Head-and-tail viruses* infect bacteria and have a head that is similar to icosahedral viruses and a tail shaped like helical viruses.

Many viruses use some sort of *glycoprotein* to attach to their host cells via molecules on the cell called ***viral receptors***. For these viruses, attachment is required for later penetration of the cell membrane; only after penetration takes place can the virus complete its replication inside the cell. The receptors that viruses use are molecules that are normally found on cell surfaces and have their own physiological functions. It appears that viruses have simply evolved to make use of these molecules for their own replication. For example, HIV uses the CD4 molecule on T lymphocytes as one of its receptors (Figure 21.4). CD4 is a type of molecule called a *cell adhesion molecule*, which functions to keep different types of immune cells in close proximity to each other during the generation of a T lymphocyte immune response.

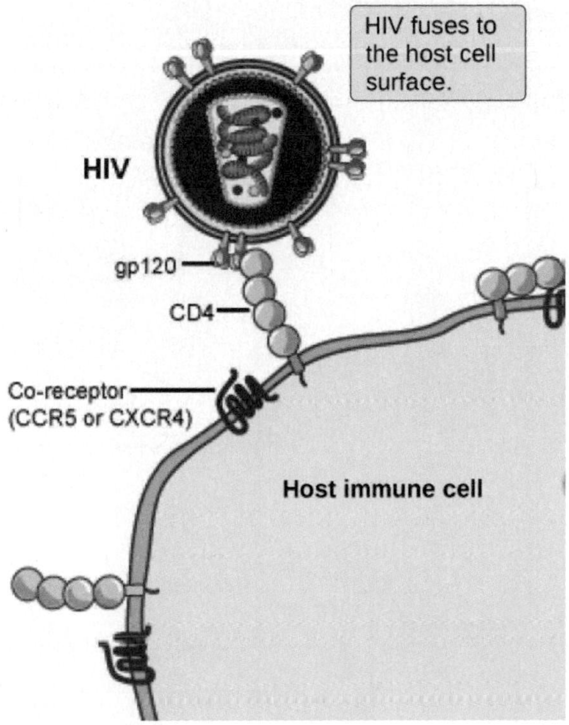

Figure 21.4 A virus and its host receptor protein. The HIV virus binds the CD4 receptor on the surface of human cells. CD4 receptors help white blood cells to communicate with other cells of the immune system when producing an immune response. (credit: modification of work by NIAID, NIH)

One of the most complex virions known, the T4 bacteriophage (which infects the *Escherichia coli*) bacterium, has a tail structure that the virus uses to attach to host cells and a head structure that houses its DNA.

Adenovirus, a non-enveloped animal virus that causes respiratory illnesses in humans, uses glycoprotein spikes protruding from its capsomeres to attach to host cells. Non-enveloped viruses also include those that cause polio (poliovirus), plantar warts (papillomavirus), and hepatitis A (hepatitis A virus).

Enveloped virions, such as the influenza virus, consist of nucleic acid (RNA in the case of influenza) and capsid proteins surrounded by a phospholipid bilayer envelope that contains virus-encoded proteins. Glycoproteins embedded in the viral envelope are used to attach to host cells. Other envelope proteins are the **matrix proteins** that stabilize the envelope and often play a role in the assembly of progeny virions. Chicken pox, HIV, and mumps are other examples of diseases caused by viruses with envelopes. Because of the fragility of the envelope, non-enveloped viruses are more resistant to changes in temperature, pH, and some disinfectants than enveloped viruses.

Overall, the shape of the virion and the presence or absence of an envelope tell us little about what disease the virus may cause or what species it might infect, but they are still useful means to begin viral classification (Figure 21.5).

VISUAL CONNECTION

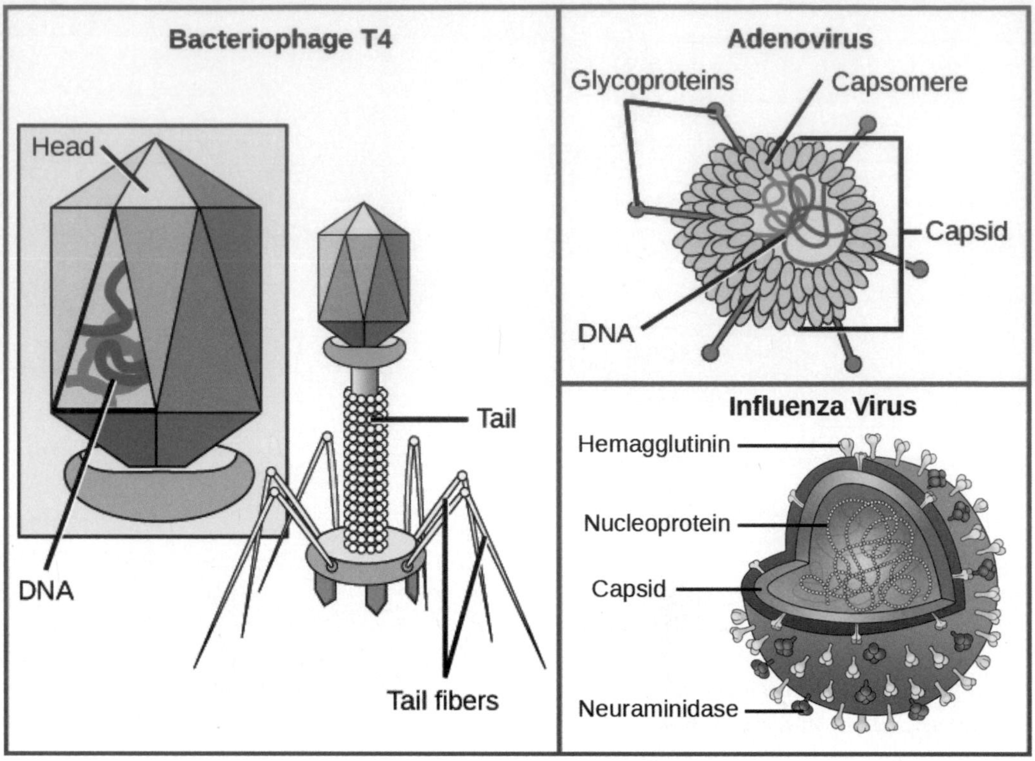

Figure 21.5 Complex Viruses. Viruses can be either complex or relatively simple in shape. This figure shows three relatively complex virions: the bacteriophage T4, with its DNA-containing head group and tail fibers that attach to host cells; adenovirus, which uses spikes from its capsid to bind to host cells; and the influenza virus, which uses glycoproteins embedded in its envelope to bind to host cells. The influenza virus also has matrix proteins, internal to the envelope, which help stabilize the virion's shape. (credit "bacteriophage, adenovirus": modification of work by NCBI, NIH; credit "influenza virus": modification of work by Dan Higgins, Centers for Disease Control and Prevention)

Which of the following statements about virus structure is true?

a. All viruses are encased in a viral membrane.
b. The capsomere is made up of small protein subunits called capsids.
c. DNA is the genetic material in all viruses.
d. Glycoproteins help the virus attach to the host cell.

Types of Nucleic Acid

Unlike nearly all living organisms that use DNA as their genetic material, viruses may use either DNA or RNA. The **virus core** contains the genome—the total genetic content of the virus. Viral genomes tend to be small, containing only those genes that encode proteins which the virus cannot get from the host cell. This genetic material may be single- or double-stranded. It may also be linear or circular. While most viruses contain a single nucleic acid, others have genomes divided into several segments. The RNA genome of the influenza virus is segmented, which contributes to its variability and continuous evolution, and explains why it is difficult to develop a vaccine against it.

In DNA viruses, the viral DNA directs the host cell's replication proteins to synthesize new copies of the viral genome and to transcribe and translate that genome into viral proteins. Human diseases caused by DNA viruses include chickenpox, hepatitis B, and adenoviruses. Sexually transmitted DNA viruses include the herpes virus and the human papilloma virus (HPV), which has been associated with cervical cancer and genital warts.

RNA viruses contain only RNA as their genetic material. To replicate their genomes in the host cell, the RNA viruses must encode their own enzymes that can replicate RNA into RNA or, in the retroviruses, into DNA. These *RNA polymerase enzymes* are more likely to make copying errors than DNA polymerases, and therefore often make mistakes during transcription. For this reason, mutations in RNA viruses occur more frequently than in DNA viruses. This causes them to change and adapt more rapidly to their host. Human diseases caused by RNA viruses include influenza, hepatitis C, measles, and rabies. The HIV virus, which is sexually transmitted, is an RNA retrovirus.

The Challenge of Virus Classification

Because most viruses probably evolved from different ancestors, the systematic methods that scientists have used to classify prokaryotic and eukaryotic cells are not very useful. If viruses represent "remnants" of different organisms, then even genomic or protein analysis is not useful. Why?, *Because viruses have no common genomic sequence that they all share.* For example, the 16S rRNA sequence so useful for constructing prokaryote phylogenies is of no use for a creature with no ribosomes! Biologists have used several classification systems in the past. Viruses were initially grouped by shared morphology. Later, groups of viruses were classified by the type of nucleic acid they contained, DNA or RNA, and whether their nucleic acid was single- or double-stranded. However, these earlier classification methods grouped viruses differently, because they were based on different sets of characters of the virus. The most commonly used classification method today is called the Baltimore classification scheme, and is based on how messenger RNA (mRNA) is generated in each particular type of virus.

Past Systems of Classification

Viruses contain only a few elements by which they can be classified: the viral genome, the type of capsid, and the envelope structure for the enveloped viruses. All of these elements have been used in the past for viral classification (Table 21.1 and Figure 21.6). Viral genomes may vary in the type of genetic material (DNA or RNA) and its organization (single- or double-stranded, linear or circular, and segmented or non-segmented). In some viruses, additional proteins needed for replication are associated directly with the genome or contained within the viral capsid.

Virus Classification by Genome Structure

Genome Structure	Examples
RNA DNA	Rabies virus, retroviruses Herpesviruses, smallpox virus
Single-stranded Double-stranded	Rabies virus, retroviruses Herpesviruses, smallpox virus
Linear Circular	Rabies virus, retroviruses, herpesviruses, smallpox virus Papillomaviruses, many bacteriophages
Non-segmented: genome consists of a single segment of genetic material Segmented: genome is divided into multiple segments	Parainfluenza viruses Influenza viruses

Table 21.1

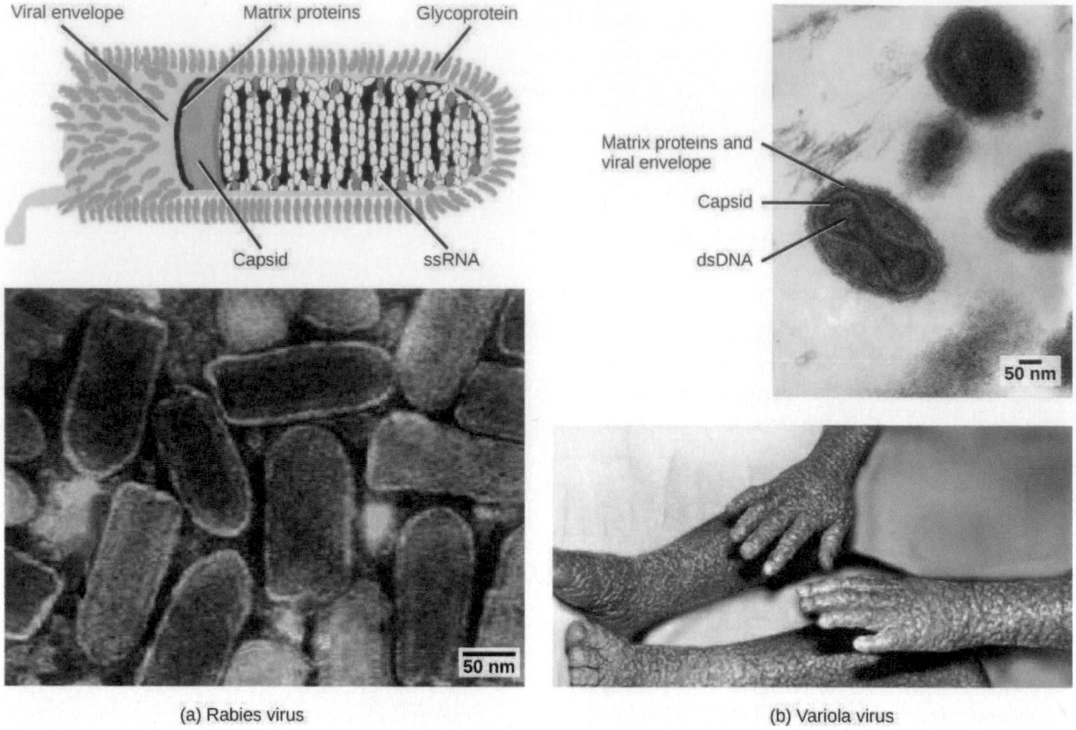

Figure 21.6 Viruses can be classified according to their core genetic material and capsid design. (a) Rabies virus has a single-stranded RNA (ssRNA) core and an enveloped helical capsid, whereas (b) variola virus, the causative agent of smallpox, has a double-stranded DNA (dsDNA) core and a complex capsid. Rabies transmission occurs when saliva from an infected mammal enters a wound. The virus travels through neurons in the peripheral nervous system to the central nervous system, where it impairs brain function, and then travels to other tissues. The virus can infect any mammal, and most die within weeks of infection. Smallpox is a human virus transmitted by inhalation of the variola virus, localized in the skin, mouth, and throat, which causes a characteristic rash. Before its eradication in 1979, infection resulted in a 30 to 35 percent mortality rate. (credit "rabies diagram": modification of work by CDC; "rabies micrograph": modification of work by Dr. Fred Murphy, CDC; credit "small pox micrograph": modification of work by Dr. Fred Murphy, Sylvia Whitfield, CDC; credit "smallpox photo": modification of work by CDC; scale-bar data from Matt Russell)

Viruses can also be classified by the design of their capsids (Table 21.2 and Figure 21.7). Capsids are classified as naked icosahedral, enveloped icosahedral, enveloped helical, naked helical, and complex. The type of genetic material (DNA or RNA) and its structure (single- or double-stranded, linear or circular, and segmented or non-segmented) are used to classify the virus core structures (Table 21.2).

Virus Classification by Capsid Structure

Capsid Classification	Examples
Naked icosahedral	Hepatitis A virus, polioviruses
Enveloped icosahedral	Epstein-Barr virus, herpes simplex virus, rubella virus, yellow fever virus, HIV-1
Enveloped helical	Influenza viruses, mumps virus, measles virus, rabies virus
Naked helical	Tobacco mosaic virus

Table 21.2

Capsid Classification	Examples
Complex with many proteins; some have combinations of icosahedral and helical capsid structures	Herpesviruses, smallpox virus, hepatitis B virus, T4 bacteriophage

Table 21.2

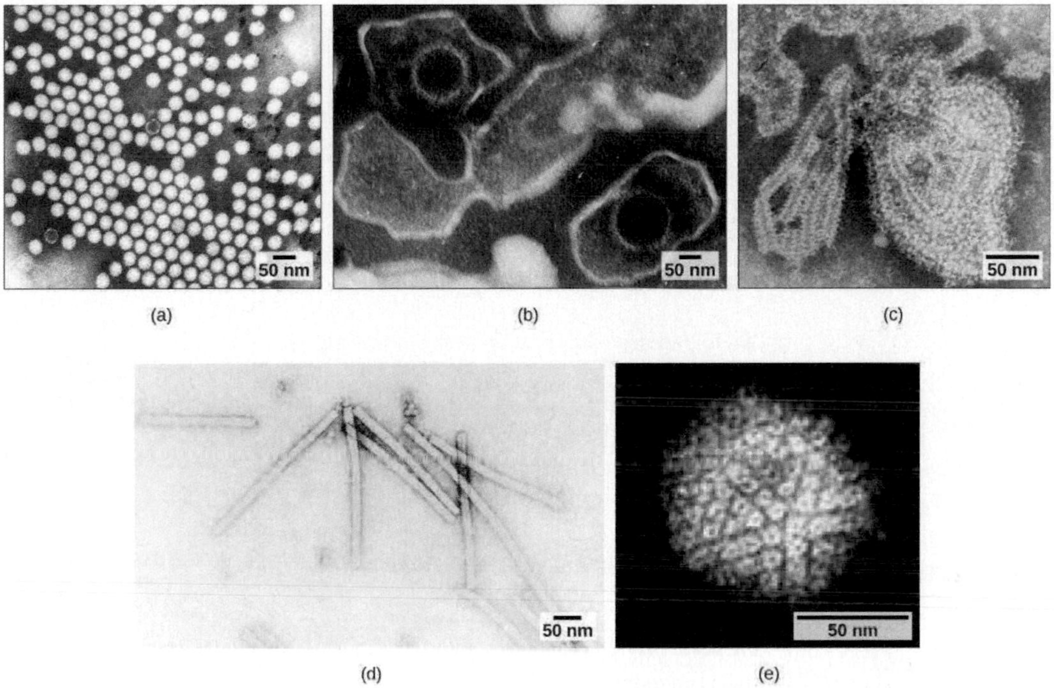

Figure 21.7 Transmission electron micrographs of various viruses show their capsid structures. The capsid of the (a) polio virus is naked icosahedral; (b) the Epstein-Barr virus capsid is enveloped icosahedral; (c) the mumps virus capsid is an enveloped helix; (d) the tobacco mosaic virus capsid is naked helical; and (e) the herpesvirus capsid is complex. (credit a: modification of work by Dr. Fred Murphy, Sylvia Whitfield; credit b: modification of work by Liza Gross; credit c: modification of work by Dr. F. A. Murphy, CDC; credit d: modification of work by USDA ARS; credit e: modification of work by Linda Stannard, Department of Medical Microbiology, University of Cape Town, South Africa, NASA; scale-bar data from Matt Russell)

Baltimore Classification

The most commonly and currently used system of virus classification was first developed by Nobel Prize-winning biologist David Baltimore in the early 1970s. In addition to the differences in morphology and genetics mentioned above, the Baltimore classification scheme groups viruses according to how the mRNA is produced during the replicative cycle of the virus.

Group I viruses contain double-stranded DNA (dsDNA) as their genome. Their mRNA is produced by transcription in much the same way as with cellular DNA, using the enzymes of the host cell.

Group II viruses have single-stranded DNA (ssDNA) as their genome. They convert their single-stranded genomes into a dsDNA intermediate before transcription to mRNA can occur.

Group III viruses use dsRNA as their genome. The strands separate, and one of them is used as a template for the generation of mRNA using the RNA-dependent RNA polymerase encoded by the virus.

Group IV viruses have ssRNA as their genome with a *positive polarity*, which means that the genomic RNA can serve directly as mRNA. Intermediates of dsRNA, called ***replicative intermediates***, are made in the process of copying the genomic RNA. Multiple, full-length RNA strands of *negative polarity* (complementary to the positive-stranded genomic RNA) are formed from these intermediates, which may then serve as templates for the production of RNA with positive polarity, including both full-length genomic RNA and shorter viral mRNAs.

Group V viruses contain ssRNA genomes with a **negative polarity**, meaning that their sequence is complementary to the mRNA. As with Group IV viruses, dsRNA intermediates are used to make copies of the genome and produce mRNA. In this case, the negative-stranded genome can be converted directly to mRNA. Additionally, full-length positive RNA strands are made to serve as templates for the production of the negative-stranded genome.

Group VI viruses have diploid (two copies) ssRNA genomes that must be converted, using the enzyme **reverse transcriptase**, to dsDNA; the dsDNA is then transported to the nucleus of the host cell and inserted into the host genome. Then, mRNA can be produced by transcription of the viral DNA that was integrated into the host genome.

Group VII viruses have partial dsDNA genomes and make ssRNA intermediates that act as mRNA, but are also converted back into dsDNA genomes by reverse transcriptase, necessary for genome replication.

The characteristics of each group in the Baltimore classification are summarized in Table 21.3 with examples of each group.

Baltimore Classification

Group	Characteristics	Mode of mRNA Production	Example
I	Double-stranded DNA	mRNA is transcribed directly from the DNA template	Herpes simplex (herpesvirus)
II	Single-stranded DNA	DNA is converted to double-stranded form before RNA is transcribed	Canine parvovirus (parvovirus)
III	Double-stranded RNA	mRNA is transcribed from the RNA genome	Childhood gastroenteritis (rotavirus)
IV	Single stranded RNA (+)	Genome functions as mRNA	Common cold (picornavirus)
V	Single stranded RNA (-)	mRNA is transcribed from the RNA genome	Rabies (rhabdovirus)
VI	Single stranded RNA viruses with reverse transcriptase	Reverse transcriptase makes DNA from the RNA genome; DNA is then incorporated in the host genome; mRNA is transcribed from the incorporated DNA	Human immunodeficiency virus (HIV)
VII	Double stranded DNA viruses with reverse transcriptase	The viral genome is double-stranded DNA, but viral DNA is replicated through an RNA intermediate; the RNA may serve directly as mRNA or as a template to make mRNA	Hepatitis B virus (hepadnavirus)

Table 21.3

21.2 Virus Infections and Hosts

By the end of this section, you will be able to do the following:

- List the steps of replication and explain what occurs at each step
- Describe the lytic and lysogenic cycles of virus replication
- Explain the transmission of plant and animal viruses
- Discuss some of the diseases caused by plant and animal viruses
- Discuss the economic impact of plant and animal viruses

Viruses are obligate, intracellular parasites. A virus must first recognize and attach to a specific living cell prior to entering it. After penetration, the invading virus must copy its genome and manufacture its own proteins. Finally, the progeny virions must escape the host cell so that they can infect other cells. Viruses can infect only certain species of hosts and only certain cells within that host. Specific host cells that a virus must occupy and use to replicate are called **permissive**. In most cases, the molecular basis for this specificity is due to a particular surface molecule known as the *viral receptor* on the host cell surface. A specific viral receptor is required for the virus to attach. In addition, differences in metabolism and host-cell immune responses (based on differential gene expression) are a likely factor in determining which cells a virus may target for replication.

Steps of Virus Infections

A virus must use its host-cell processes to replicate. The viral replication cycle can produce dramatic biochemical and structural changes in the host cell, which may cause cell damage. These changes, called **cytopathic effects**, can change cell functions or even destroy the cell. Some infected cells, such as those infected by the common cold virus known as rhinovirus, die through **lysis** (bursting) or **apoptosis** (programmed cell death or "cell suicide"), releasing all progeny virions at once. The symptoms of viral diseases result both from such cell damage caused by the virus and from the immune response to the virus, which attempts to control and eliminate the virus from the body.

Many animal viruses, such as **HIV (human immunodeficiency virus)**, leave the infected cells of the immune system by a process known as ***budding***, where virions leave the cell individually. During the budding process, the cell does not undergo lysis and is not immediately killed. However, the damage to the cells that the virus infects may make it impossible for the cells to function normally, even though the cells remain alive for a period of time. Most productive viral infections follow similar steps in the virus replication cycle: *attachment, penetration, uncoating, replication, assembly, and release* (Figure 21.8).

Attachment

A virus attaches to a specific receptor site on the host cell membrane through attachment proteins in the capsid or via glycoproteins embedded in the viral envelope. The specificity of this interaction determines the host—and the cells within the host—that can be infected by a particular virus. This can be illustrated by thinking of several keys and several locks, where each key will fit only one specific lock.

LINK TO LEARNING

This video (http://openstax.org/l/influenza) explains how influenza attacks the body.

Entry

Viruses may enter a host cell either with or without the viral capsid. The nucleic acid of bacteriophages enters the host cell "naked," leaving the capsid outside the cell. Plant and animal viruses can enter through *endocytosis* (as you may recall, the cell membrane surrounds and engulfs the entire virus). Some enveloped viruses enter the cell when the viral envelope fuses directly with the cell membrane. Once inside the cell, the viral capsid degrades, and then the viral nucleic acid is released and becomes available for replication and transcription.

Replication and Assembly

The replication mechanism depends on the viral genome. DNA viruses usually use host-cell proteins and enzymes to replicate the viral DNA and to transcribe viral mRNA, which is then used to direct viral protein synthesis. RNA viruses usually use the RNA core as a template for synthesis of viral genomic RNA and mRNA. The viral mRNA directs the host cell to synthesize viral enzymes and capsid proteins, and assemble new virions.

Of course, there are exceptions to this pattern. If a host cell does not provide the enzymes necessary for viral replication, viral genes supply the information to direct synthesis of the missing proteins. Retroviruses, such as HIV (group VI of the Baltimore classification scheme), have an RNA genome that must be reverse transcribed into DNA, which then is incorporated into the host cell genome. To convert RNA into DNA, retroviruses must contain genes that encode the virus-specific enzyme reverse transcriptase that transcribes an RNA template to DNA. Reverse transcription never occurs in uninfected host cells—the enzyme reverse transcriptase is only derived from the expression of viral genes within the infected host cells. The fact that HIV produces some of its own enzymes not found in the host has allowed researchers to develop drugs that inhibit these enzymes without affecting the host's metabolism.

This approach has led to the development of a variety of drugs used to treat HIV and has been effective at reducing the number of infectious virions (copies of viral RNA) in the blood to non-detectable levels in many HIV-infected individuals.

Egress

The last stage of viral replication is the release of the new virions produced in the host organism, where they are able to infect adjacent cells and repeat the replication cycle. As you've learned, some viruses are released when the host cell dies, and other viruses can leave infected cells by budding through the membrane without directly killing the cell.

VISUAL CONNECTION

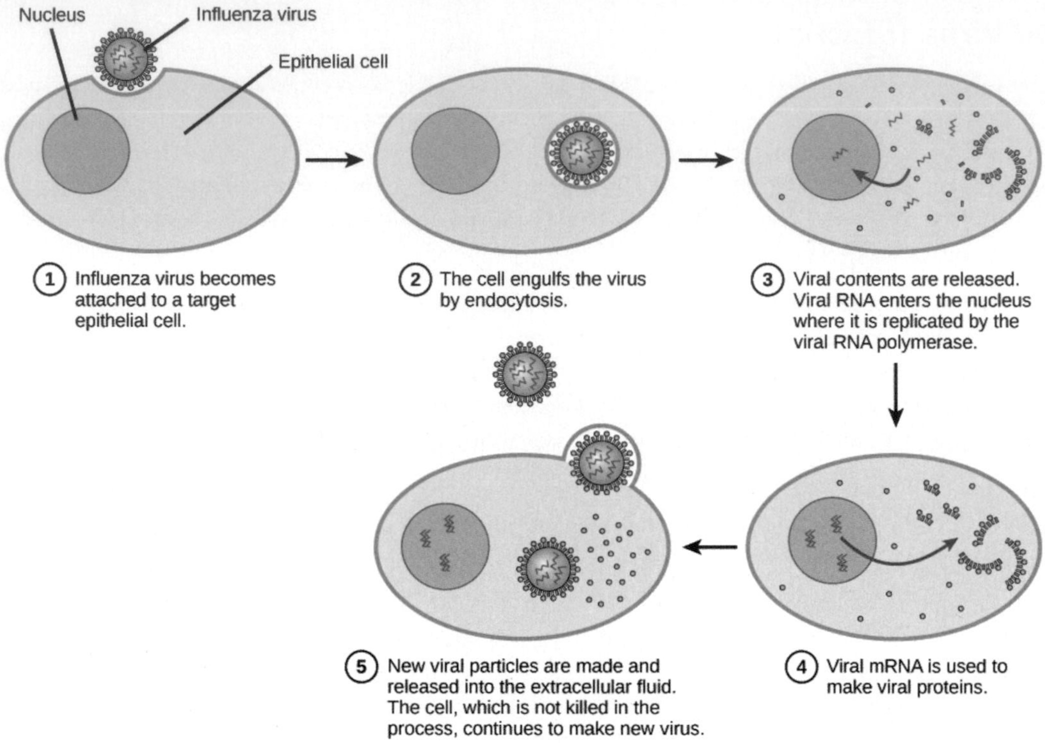

Figure 21.8 The influenza reproductive cycle. In influenza virus infection, glycoproteins on the capsid attach to a host epithelial cell. Following this, the virus is engulfed. RNA and proteins are then made and assembled into new virions.

Influenza virus is packaged in a viral envelope that fuses with the plasma membrane. This way, the virus can exit the host cell without killing it. What advantage does the virus gain by keeping the host cell alive?

LINK TO LEARNING

Watch a video (https://openstax.org/l/viruses) on viruses, identifying structures, modes of transmission, replication, and more.

Different Hosts and Their Viruses

As you've learned, viruses often infect very specific hosts, as well as specific cells within the host. This feature of a virus makes it specific to one or a few species of life on Earth. On the other hand, so many different types of viruses exist on Earth that nearly every living organism has its own set of *viruses* trying to infect its cells. Even prokaryotes, the smallest and simplest of cells, may be attacked by specific types of viruses. In the following section, we will look at some of the features of viral infection of prokaryotic cells. As we have learned, viruses that infect bacteria are called *bacteriophages* (Figure 21.9). Archaea have their own similar viruses.

Bacteriophages

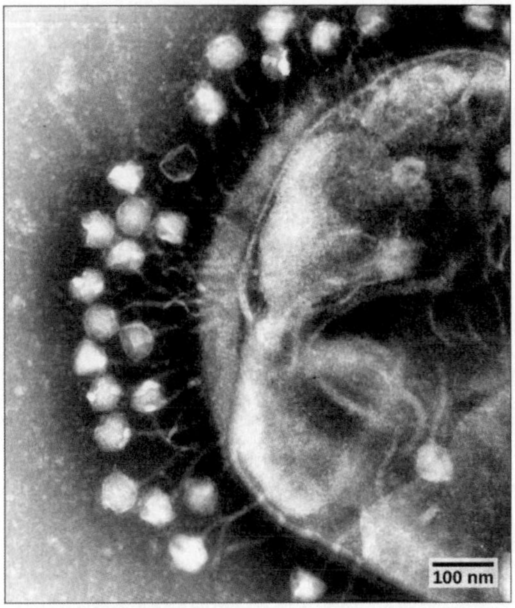

Figure 21.9 Bacteriophages attached to a host cell (transmission electron micrograph). In bacteriophage with tails, like the one shown here, the tails serve as a passageway for transmission of the phage genome. (credit: modification of work by Dr. Graham Beards; scale-bar data from Matt Russell)

Most bacteriophages are dsDNA viruses, which use host enzymes for DNA replication and RNA transcription. Phage particles must bind to specific surface receptors and actively insert the genome into the host cell. (The complex tail structures seen in many bacteriophages are actively involved in getting the viral genome across the prokaryotic cell wall.) When infection of a cell by a bacteriophage results in the production of new virions, the infection is said to be ***productive***. If the virions are released by bursting the cell, the virus replicates by means of a ***lytic cycle*** (Figure 21.10). An example of a lytic bacteriophage is T4, which infects *Escherichia coli* found in the human intestinal tract. Sometimes, however, a virus can remain within the cell without being released. For example, when a temperate bacteriophage infects a bacterial cell, it replicates by means of a ***lysogenic cycle*** (Figure 21.10), and the viral genome is incorporated into the genome of the host cell. When the phage DNA is incorporated into the host-cell genome, it is called a **prophage**. An example of a lysogenic bacteriophage is the λ (lambda) virus, which also infects the *E. coli* bacterium. Viruses that infect plant or animal cells may sometimes undergo infections where they are not producing virions for long periods. An example is the animal *herpesviruses*, including herpes simplex viruses, the cause of oral and genital herpes in humans. In a process called **latency**, these viruses can exist in nervous tissue for long periods of time without producing new virions, only to leave latency periodically and cause lesions in the skin where the virus replicates. Even though there are similarities between lysogeny and latency, the term lysogenic cycle is usually reserved to describe bacteriophages. Latency will be described in more detail in the next section.

VISUAL CONNECTION

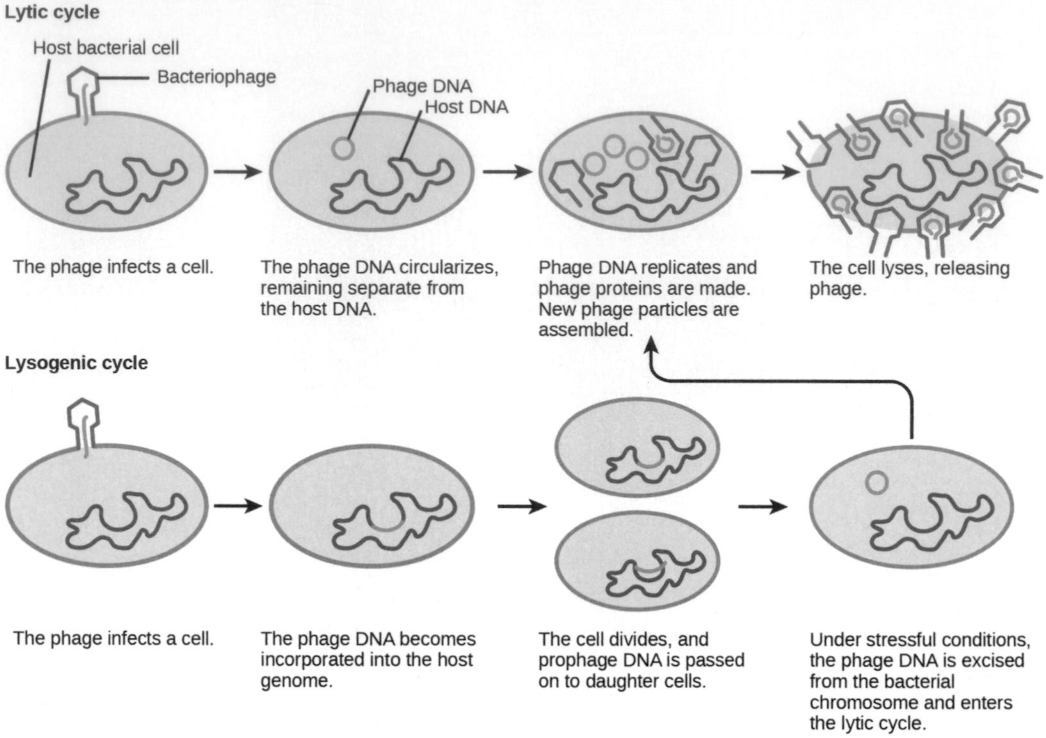

Figure 21.10 A temperate bacteriophage has both lytic and lysogenic cycles. In the lytic cycle, the phage replicates and lyses the host cell. In the lysogenic cycle, phage DNA is incorporated into the host genome, where it is passed on to subsequent generations. Environmental stressors such as starvation or exposure to toxic chemicals may cause the prophage to excise and enter the lytic cycle.

Which of the following statements is false?

a. In the lytic cycle, new phages are produced and released into the environment.
b. In the lysogenic cycle, phage DNA is incorporated into the host genome.
c. An environmental stressor can cause the phage to initiate the lysogenic cycle.
d. Cell lysis only occurs in the lytic cycle.

Plant Viruses

Most plant viruses, like the tobacco mosaic virus, have single-stranded (+) RNA genomes. However, there are also plant viruses in most other virus categories. Unlike bacteriophages, plant viruses do not have active mechanisms for delivering the viral genome across the protective cell wall. For a plant virus to enter a new host plant, some type of mechanical damage must occur. This damage is often caused by weather, insects, animals, fire, or human activities like farming or landscaping. Movement from cell to cell within a plant can be facilitated by viral modification of plasmodesmata (cytoplasmic threads that pass from one plant cell to the next). Additionally, plant offspring may inherit viral diseases from parent plants. Plant viruses can be transmitted by a variety of vectors, through contact with an infected plant's sap, by living organisms such as insects and nematodes, and through pollen. The transfer of a virus from one plant to another is known as **horizontal transmission**, whereas the inheritance of a virus from a parent is called **vertical transmission**.

Symptoms of viral diseases vary according to the virus and its host (Table 21.4). One common symptom is ***hyperplasia***, the abnormal proliferation of cells that causes the appearance of plant tumors known as **galls**. Other viruses induce ***hypoplasia***, or decreased cell growth, in the leaves of plants, causing thin, yellow areas to appear. Still other viruses affect the plant by directly killing plant cells, a process known as ***cell necrosis***. Other symptoms of plant viruses include malformed leaves; black streaks on the stems of the plants; altered growth of stems, leaves, or fruits; and ring spots, which are circular or linear areas of discoloration found in a leaf.

Some Common Symptoms of Plant Viral Diseases

Symptom	Appears as
Hyperplasia	Galls (tumors)
Hypoplasia	Thinned, yellow splotches on leaves
Cell necrosis	Dead, blackened stems, leaves, or fruit
Abnormal growth patterns	Malformed stems, leaves, or fruit
Discoloration	Yellow, red, or black lines, or rings in stems, leaves, or fruit

Table 21.4

Plant viruses can seriously disrupt crop growth and development, significantly affecting our food supply. They are responsible for poor crop quality and quantity globally, and can bring about huge economic losses annually. Others viruses may damage plants used in landscaping. Some viruses that infect agricultural food plants include the name of the plant they infect, such as tomato spotted wilt virus, bean common mosaic virus, and cucumber mosaic virus. In plants used for landscaping, two of the most common viruses are peony ring spot and rose mosaic virus. There are far too many plant viruses to discuss each in detail, but symptoms of bean common mosaic virus result in lowered bean production and stunted, unproductive plants. In the ornamental rose, the rose mosaic disease causes wavy yellow lines and colored splotches on the leaves of the plant.

Animal Viruses

Animal viruses, unlike the viruses of plants and bacteria, do not have to penetrate a cell wall to gain access to the host cell. The virus may even induce the host cell to cooperate in the infection process. Non-enveloped or "naked" animal viruses may enter cells in two different ways. As a protein in the viral capsid binds to its receptor on the host cell, the virus may be taken inside the cell via a vesicle during the normal cell process of *receptor-mediated endocytosis*. An alternative method of cell penetration used by non-enveloped viruses is for capsid proteins to undergo shape changes after binding to the receptor, creating channels in the host cell membrane. The viral genome is then "injected" into the host cell through these channels in a manner analogous to that used by many bacteriophages.

Enveloped viruses also have two ways of entering cells after binding to their receptors: receptor-mediated endocytosis, or **fusion**. Many enveloped viruses enter the cell by receptor-mediated endocytosis in a fashion similar to that seen in some non-enveloped viruses. On the other hand, fusion only occurs with enveloped virions. These viruses, which include HIV among others, use special fusion proteins in their envelopes to cause the envelope to fuse with the plasma membrane of the cell, thus releasing the genome and capsid of the virus into the cell cytoplasm.

After making their proteins and copying their genomes, animal viruses complete the assembly of new virions and exit the cell. As we have already discussed using the example the influenza virus, enveloped animal viruses may bud from the cell membrane as they assemble themselves, taking a piece of the cell's plasma membrane in the process. On the other hand, non-enveloped viral progeny, such as rhinoviruses, accumulate in infected cells until there is a signal for lysis or apoptosis, and all virions are released together.

As you will learn in the next module, animal viruses are associated with a variety of human diseases. Some of them follow the classic pattern of **acute disease**, where symptoms get increasingly worse for a short period followed by the elimination of the virus from the body by the immune system and eventual recovery from the infection. Examples of acute viral diseases are the common cold and influenza. Other viruses cause long-term **chronic infections**, such as the virus causing hepatitis C, whereas others, like herpes simplex virus, only cause **intermittent** symptoms. Still other viruses, such as human herpesviruses 6 and 7, which in some cases can cause the minor childhood disease roseola, often successfully cause productive infections without causing any symptoms at all in the host, and thus we say these patients have an **asymptomatic infection**.

In hepatitis C infections, the virus grows and reproduces in liver cells, causing low levels of liver damage. The damage is so low that infected individuals are often unaware that they are infected, and many infections are detected only by routine blood work on patients with risk factors such as intravenous drug use. On the other hand, since many of the symptoms of viral diseases are caused by immune responses, a lack of symptoms is an indication of a weak immune response to the virus. This allows the virus to escape elimination by the immune system and persist in individuals for years, all the while producing low levels of progeny virions in what is known as a chronic viral disease. Chronic infection of the liver by this virus leads to a much greater chance of developing liver cancer, sometimes as much as 30 years after the initial infection.

As already discussed, herpes simplex virus can remain in a state of latency in nervous tissue for months, even years. As the virus "hides" in the tissue and makes few if any viral proteins, there is nothing for the immune response to act against, and immunity to the virus slowly declines. Under certain conditions, including various types of physical and psychological stress, the latent herpes simplex virus may be reactivated and undergo a lytic replication cycle in the skin, causing the lesions associated with the disease. Once virions are produced in the skin and viral proteins are synthesized, the immune response is again stimulated and resolves the skin lesions in a few days or weeks by destroying viruses in the skin. As a result of this type of replicative cycle, appearances of cold sores and genital herpes outbreaks only occur intermittently, even though the viruses remain in the nervous tissue for life. Latent infections are common with other herpesviruses as well, including the varicella-zoster virus that causes chickenpox. After having a chickenpox infection in childhood, the varicella-zoster virus can remain latent for many years and reactivate in adults to cause the painful condition known as "shingles" (Figure 21.11).

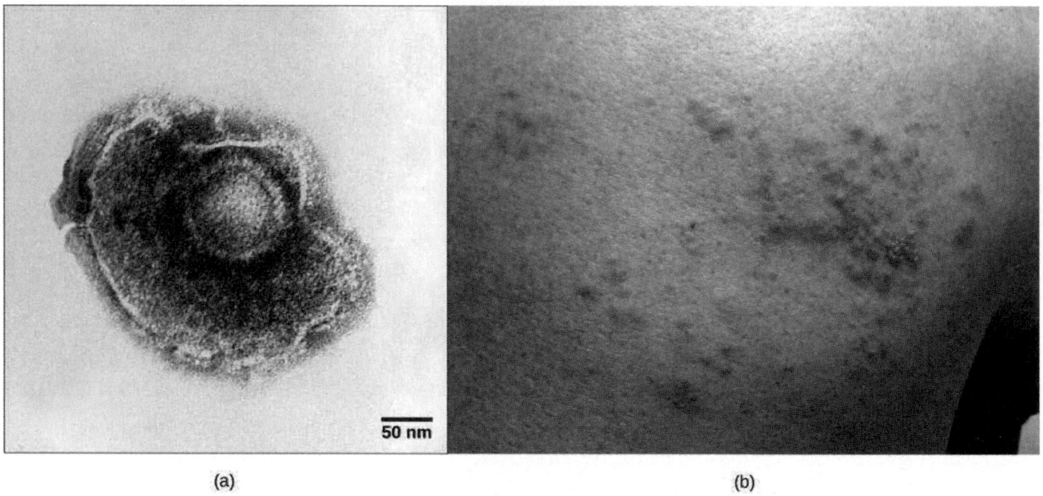

Figure 21.11 A latent virus infection. (a) Varicella-zoster, the virus that causes chickenpox, has an enveloped icosahedral capsid visible in this transmission electron micrograph. Its double-stranded DNA genome becomes incorporated in the host DNA and can reactivate after latency in the form of (b) shingles, often exhibiting a rash. (credit a: modification of work by Dr. Erskine Palmer, B. G. Martin, CDC; credit b: modification of work by "rosmary"/Flickr; scale-bar data from Matt Russell)

Some animal-infecting viruses, including the hepatitis C virus discussed above, are known as **oncogenic viruses**: They have the ability to cause cancer. These viruses interfere with the normal regulation of the host cell cycle either by introducing genes that stimulate unregulated cell growth (oncogenes) or by interfering with the expression of genes that inhibit cell growth. Oncogenic viruses can be either DNA or RNA viruses. Cancers known to be associated with viral infections include cervical cancer, caused by human papillomavirus (HPV) (Figure 21.12), liver cancer caused by hepatitis B virus, T-cell leukemia, and several types of lymphoma.

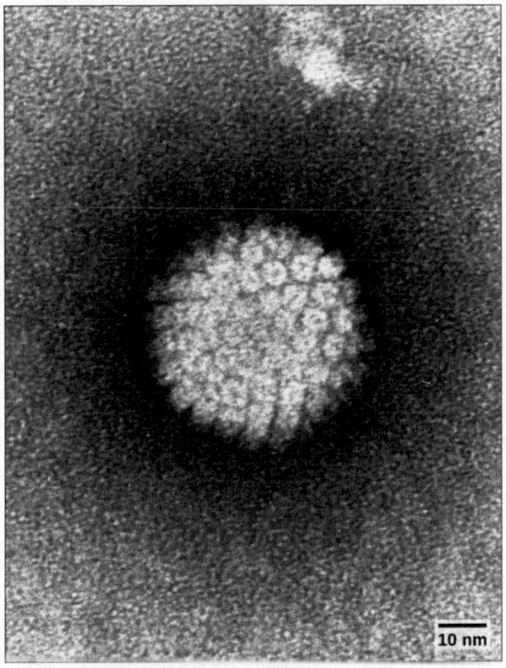

Figure 21.12 HPV, or human papillomavirus, has a naked icosahedral capsid visible in this transmission electron micrograph and a double-stranded DNA genome that is incorporated into the host DNA. The virus, which is sexually transmitted, is oncogenic and can lead to cervical cancer. (credit: modification of work by NCI, NIH; scale-bar data from Matt Russell)

LINK TO LEARNING

This video (http://openstax.org/l/animal_viruses) shows the life cycle of a virus.

21.3 Prevention and Treatment of Viral Infections

By the end of this section, you will be able to do the following:

- Identify major viral illnesses that affect humans
- Compare vaccinations and anti-viral drugs as medical approaches to viruses

Viruses cause a variety of diseases in animals, including humans, ranging from the common cold to potentially fatal illnesses like meningitis (Figure 21.13). These diseases can be treated by antiviral drugs or by vaccines; however, some viruses, such as HIV, are capable both of avoiding the immune response and of mutating within the host organism to become resistant to antiviral drugs.

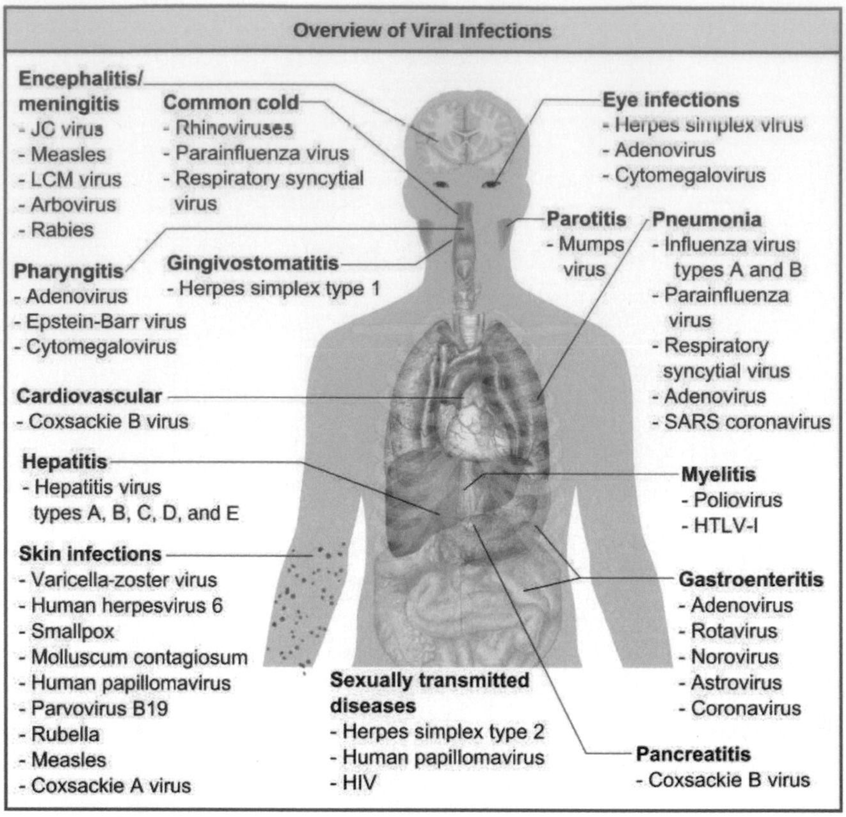

Figure 21.13 A sampling of human viruses. Viruses can cause dozens of ailments in humans, ranging from mild illnesses to serious diseases. (credit: modification of work by Mikael Häggström)

Vaccines for Prevention

The primary method of controlling viral disease is by **vaccination**, which is intended to prevent outbreaks by building immunity to a virus or virus family (Figure 21.14). **Vaccines** may be prepared using live viruses, killed viruses, or molecular subunits of the virus. Note that the killed viral vaccines and subunit viruses are both incapable of causing disease, nor is there any valid evidence that vaccinations contribute to autism.

Live viral vaccines are designed in the laboratory to cause few symptoms in recipients while giving them protective immunity against future infections. Polio was one disease that represented a milestone in the use of vaccines. Mass immunization campaigns in the 1950s (killed vaccine) and 1960s (live vaccine) significantly reduced the incidence of the disease, which caused muscle paralysis in children and generated a great amount of fear in the general population when regional epidemics occurred. The success of the polio vaccine paved the way for the routine dispensation of childhood vaccines against measles, mumps, rubella, chickenpox, and other diseases.

The issue with using live vaccines (which are usually more effective than killed vaccines), is the low but significant danger that these viruses will revert to their disease-causing form by **back mutations**. Live vaccines are usually made by **attenuating** (weakening) the "wild-type" (disease-causing) virus by growing it in the laboratory in tissues or at temperatures different from what the virus is accustomed to in the host. Adaptations to these new cells or temperatures induce mutations in the genomes of the virus, allowing it to grow better in the laboratory while inhibiting its ability to cause disease when reintroduced into conditions found in the host. These attenuated viruses thus still cause infection, but they do not grow very well, allowing the immune response to develop in time to prevent major disease. Back mutations occur when the vaccine undergoes mutations in the host such that it readapts to the host and can again cause disease, which can then be spread to other humans in an epidemic. This type of scenario happened as recently as 2007 in Nigeria where mutations in a polio vaccine led to an epidemic of polio in that country.

Some vaccines are in continuous development because certain viruses, such as influenza and HIV, have a high mutation rate compared to that of other viruses and normal host cells. With influenza, mutations in the surface molecules of the virus help the organism evade the protective immunity that may have been obtained in a previous influenza season, making it necessary for individuals to get vaccinated every year. Other viruses, such as those that cause the childhood diseases measles, mumps, and rubella, mutate so infrequently that the same vaccine is used year after year.

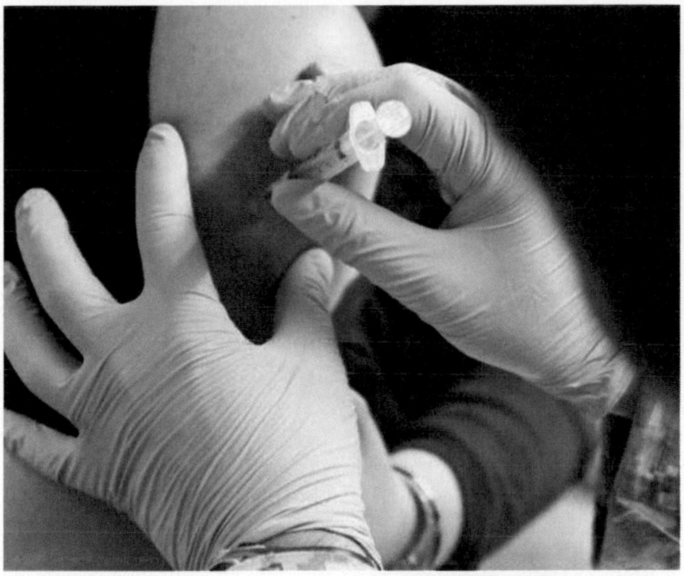

Figure 21.14 Vaccinations are designed to boost immunity to a virus to prevent infection. (credit: USACE Europe District)

LINK TO LEARNING

Watch this NOVA video (http://openstax.org/l/1918_flu) to learn how microbiologists are attempting to replicate the deadly 1918 Spanish influenza virus so they can understand more about virology.

Vaccines and Antiviral Drugs for Treatment

In some cases, vaccines can be used to treat an active viral infection. The concept behind this is that by giving the vaccine, immunity is boosted without adding more disease-causing virus. In the case of *rabies*, a fatal neurological disease transmitted via the saliva of rabies virus-infected animals, the progression of the disease from the time of the animal bite to the time it enters the central nervous system may be two weeks or longer. This is enough time to vaccinate individuals who suspect that they have been bitten by a rabid animal, and their boosted immune response is sufficient to prevent the virus from entering nervous tissue. Thus, the potentially fatal neurological consequences of the disease are averted, and the individual only has to recover from the infected bite. This approach is also being used for the treatment of Ebola, one of the fastest and most deadly viruses on Earth. Transmitted by bats and great apes, this disease can cause death in 70 to 90 percent of infected humans within two weeks. Using newly developed vaccines that *boost the immune response* in this way, there is hope that affected individuals will be better able to control the virus, potentially saving a greater percentage of infected persons from a rapid and very painful death.

Another way of treating viral infections is the use of antiviral drugs. Because viruses use the resources of the host cell for replication and the production of new virus proteins, it is difficult to block their activities without damaging the host. However, we do have some effective antiviral drugs, such as those used to treat HIV and influenza. Some antiviral drugs are specific for a particular virus and others have been used to control and reduce symptoms for a wide variety of viral diseases. For most viruses, these drugs can inhibit the virus by blocking the actions of one or more of its proteins. *It is important to note that the targeted proteins be encoded by viral genes and that these molecules are not present in a healthy host cell.* In this way, viral growth is inhibited without damaging the host.

Antivirals have been developed to treat genital herpes (herpes simplex II) and influenza. For genital herpes, drugs such as acyclovir can reduce the number and duration of episodes of active viral disease, during which patients develop viral lesions in their skin cells. As the virus remains latent in nervous tissue of the body for life, this drug is not curative but can make the symptoms of the disease more manageable. For influenza, drugs like Tamiflu (oseltamivir) (Figure 21.15) can reduce the duration of "flu" symptoms by one or two days, but the drug does not prevent symptoms entirely. Tamiflu works by inhibiting an enzyme (viral neuraminidase) that allows new virions to leave their infected cells. Thus, Tamiflu inhibits the spread of virus from infected to uninfected cells. Other antiviral drugs, such as Ribavirin, have been used to treat a variety of viral infections, although its mechanism of action against certain viruses remains unclear.

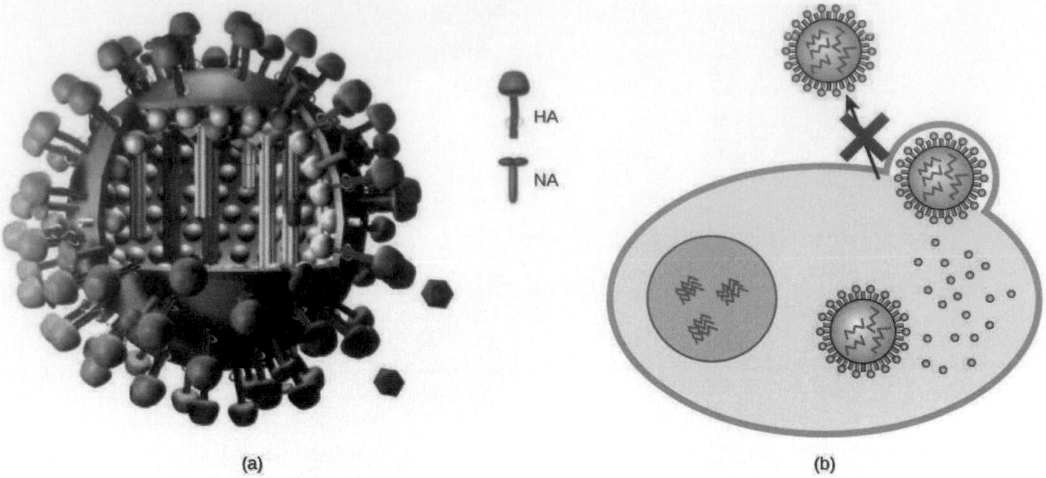

Figure 21.15 Action of an antiviral drug. (a) Tamiflu inhibits a viral enzyme called neuraminidase (NA) found in the influenza viral envelope. (b) Neuraminidase cleaves the connection between viral hemagglutinin (HA), also found in the viral envelope, and glycoproteins on the host cell surface. Inhibition of neuraminidase prevents the virus from detaching from the host cell, thereby blocking further infection. (credit a: modification of work by M. Eickmann)

By far, the most successful use of antivirals has been in the treatment of the retrovirus HIV, which causes a disease that, if untreated, is usually fatal within 10 to 12 years after infection. Anti-HIV drugs have been able to control viral replication to the point that individuals receiving these drugs survive for a significantly longer time than the untreated.

Anti-HIV drugs inhibit viral replication at many different phases of the HIV replicative cycle (Figure 21.16). Drugs have been developed that inhibit the fusion of the HIV viral envelope with the plasma membrane of the host cell (fusion inhibitors), the conversion of its RNA genome into double-stranded DNA (reverse transcriptase inhibitors, like AZT), the integration of the viral DNA into the host genome (integrase inhibitors), and the processing of viral proteins (protease inhibitors).

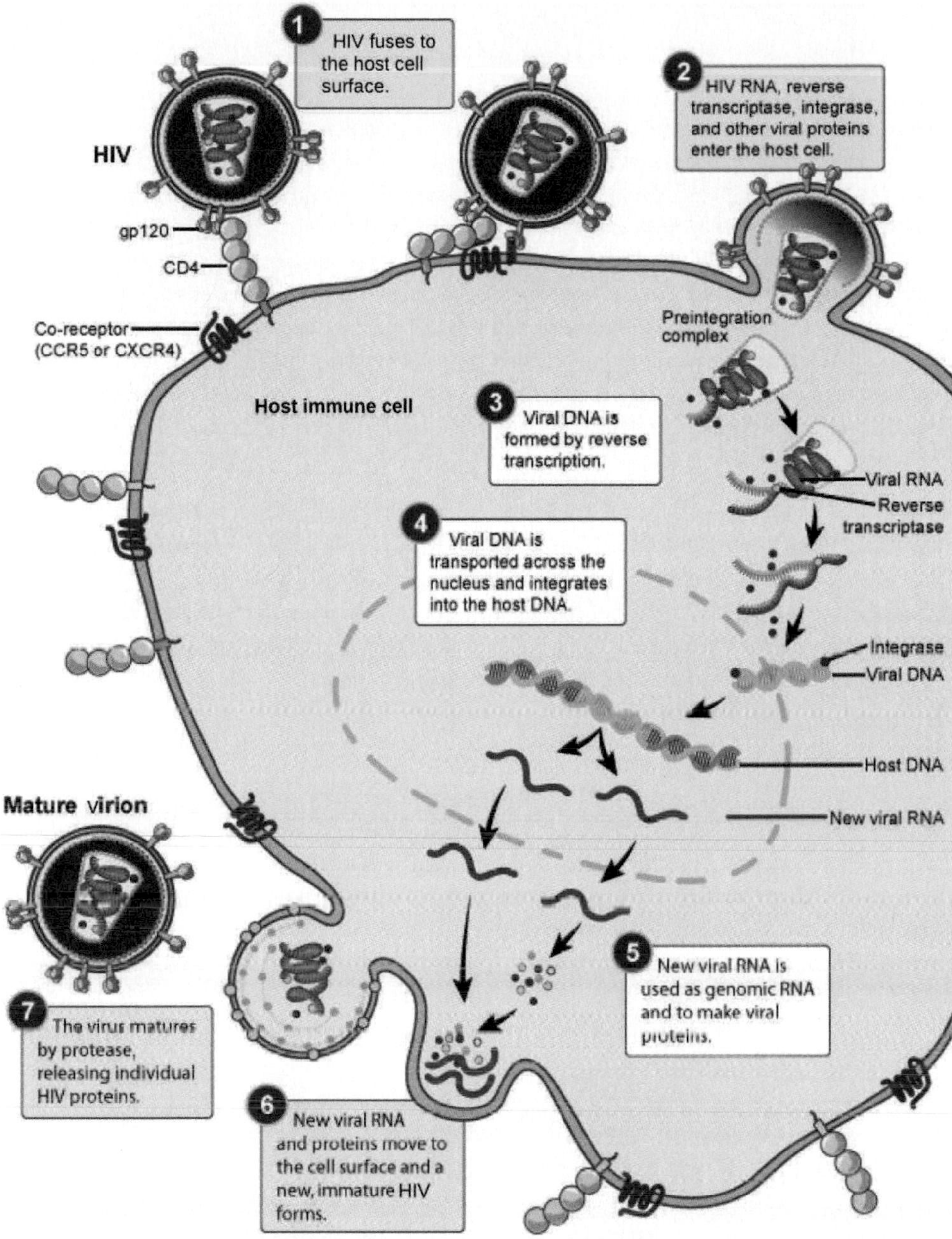

Figure 21.16 Life cycle of HIV. HIV, an enveloped, icosahedral virus, attaches to the CD4 receptor of an immune cell and fuses with the cell membrane. Viral contents are released into the cell, where viral enzymes convert the single-stranded RNA genome into DNA and incorporate it into the host genome. (credit: NIAID, NIH)

Unfortunately, when any of these drugs are used individually, the high mutation rate of the virus allows it to easily and rapidly develop resistance to the drug, limiting the drug's effectiveness. The breakthrough in the treatment of HIV was the development of HAART, *highly active anti-retroviral therapy*, which involves a mixture of different drugs, sometimes called a drug "cocktail." By attacking the virus at different stages of its replicative cycle, it is much more difficult for the virus to develop resistance to multiple drugs at the same time. Still, even with the use of combination HAART therapy, there is concern that, over time, the virus will develop resistance to this therapy. Thus, new anti-HIV drugs are constantly being developed with the hope of continuing the battle against this highly fatal virus.

Everyday Connection

Applied Virology:

The study of viruses has led to the development of a variety of new ways to treat non-viral diseases. Viruses have been used in **gene therapy**. Gene therapy is used to treat genetic diseases such as severe combined immunodeficiency (SCID), a heritable, recessive disease in which children are born with severely compromised immune systems. One common type of SCID is due to the lack of an enzyme, adenosine deaminase (ADA), which breaks down purine bases. To treat this disease by gene therapy, bone marrow cells are taken from a SCID patient and the ADA gene is inserted. This is where viruses come in, and their use relies on their ability to penetrate living cells and bring genes in with them. Viruses such as adenovirus, an upper-respiratory human virus, are modified by the addition of the ADA gene, and the virus then transports this gene into the cell. The modified cells, now capable of making ADA, are then given back to the patients in the hope of curing them. Gene therapy using viruses as carriers of genes (viral vectors), although still experimental, holds promise for the treatment of many genetic diseases. Still, many technological problems need to be solved for this approach to be a viable method for treating genetic disease.

Another medical use for viruses relies on their specificity and ability to kill the cells they infect. **Oncolytic viruses** are engineered in the laboratory specifically to attack and kill cancer cells. A genetically modified adenovirus known as H101 has been used since 2005 in clinical trials in China to treat head and neck cancers. The results have been promising, with a greater short-term response rate to the combination of chemotherapy and viral therapy than to chemotherapy treatment alone. This ongoing research may herald the beginning of a new age of cancer therapy, where viruses are engineered to find and specifically kill cancer cells, regardless of where in the body they may have spread.

A third use of viruses in medicine relies on their specificity and involves using bacteriophages in the treatment of bacterial infections. Bacterial diseases have been treated with antibiotics since the 1940s. However, over time, many bacteria have evolved resistance to antibiotics. A good example is methicillin-resistant *Staphylococcus aureus* (MRSA, pronounced "mersa"), an infection commonly acquired in hospitals. This bacterium is resistant to a variety of antibiotics, making it difficult to treat. The use of bacteriophages specific for such bacteria would bypass their resistance to antibiotics and specifically kill them. Although **phage therapy** is in use in the Republic of Georgia to treat antibiotic-resistant bacteria, its use to treat human diseases has not been approved in most countries. However, the safety of the treatment was confirmed in the United States when the U.S. Food and Drug Administration approved spraying meats with bacteriophages to destroy the food pathogen *Listeria*. As more and more antibiotic-resistant strains of bacteria evolve, the use of bacteriophages might be a potential solution to the problem, and the development of phage therapy is of much interest to researchers worldwide.

21.4 Other Acellular Entities: Prions and Viroids

By the end of this section, you will be able to do the following:

- Describe prions and their basic properties
- Define viroids and their targets of infection

Prions and viroids are **pathogens** (agents with the ability to cause disease) that have simpler structures than viruses but, in the case of prions, still can produce deadly diseases.

Prions

Prions, so-called because they are proteinaceous, are infectious particles—smaller than viruses—that contain no nucleic acids (neither DNA nor RNA). Historically, the idea of an infectious agent that did not use nucleic acids was considered impossible, but pioneering work by Nobel Prize-winning biologist Stanley Prusiner has convinced the majority of biologists that such agents do indeed exist.

Fatal neurodegenerative diseases, such as kuru in humans and bovine spongiform encephalopathy (BSE) in cattle (commonly known as "mad cow disease") were shown to be transmitted by prions. The disease was spread by the consumption of meat, nervous tissue, or internal organs between members of the same species. Kuru, native to humans in Papua New Guinea, was spread from human to human via ritualistic cannibalism. BSE, originally detected in the United Kingdom, was spread between cattle by the practice of including cattle nervous tissue in feed for other cattle. Individuals with kuru and BSE show symptoms of loss of motor control and unusual behaviors, such as uncontrolled bursts of laughter with kuru, followed by death. Kuru was controlled by inducing the population to abandon its ritualistic cannibalism.

On the other hand, BSE was initially thought to only affect cattle. Cattle dying of the disease were shown to have developed lesions or "holes" in the brain, causing the brain tissue to resemble a sponge. Later on in the outbreak, however, it was shown that a similar encephalopathy in humans, known as variant Creutzfeldt-Jakob disease (CJD), could be acquired from eating beef from animals infected with BSE, sparking bans by various countries on the importation of British beef and causing considerable economic damage to the British beef industry (Figure 21.17). BSE still exists in various areas, and although a rare disease, individuals that acquire CJD are difficult to treat. The disease can be spread from human to human by blood, so many countries have banned blood donation from regions associated with BSE.

The cause of spongiform encephalopathies, such as kuru and BSE, is an infectious structural variant of a normal cellular protein called PrP (prion protein). It is this variant that constitutes the prion particle. PrP exists in two forms, **PrP^c**, the normal form of the protein, and **PrP^{sc}**, the infectious form. Once introduced into the body, the PrP^{sc} contained within the prion binds to PrP^c and converts it to PrP^{sc}. This leads to an exponential increase of the PrP^{sc} protein, which aggregates. PrP^{sc} is folded abnormally, and the resulting conformation (shape) is directly responsible for the lesions seen in the brains of infected cattle. Thus, although not without some detractors among scientists, the prion seems likely to be an entirely new form of infectious agent, the first one found whose transmission is not reliant upon genes made of DNA or RNA.

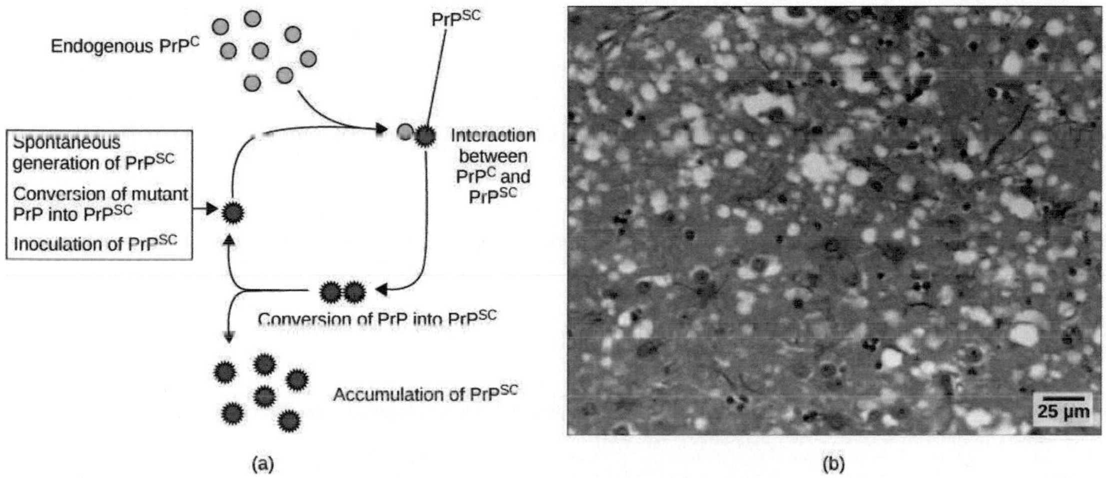

Figure 21.17 Mad Cow Disease in humans. (a) Endogenous normal prion protein (PrP^c) is converted into the disease-causing form (PrP^{sc}) when it encounters this variant form of the protein. PrP^{sc} may arise spontaneously in brain tissue, especially if a mutant form of the protein is present, or it may occur via the spread of misfolded prions consumed in food into brain tissue. (b) This prion-infected brain tissue, visualized using light microscopy, shows the vacuoles that give it a spongy texture, typical of transmissible spongiform encephalopathies. (credit b: modification of work by Dr. Al Jenny, USDA APHIS; scale-bar data from Matt Russell)

Viroids

Viroids are plant pathogens: small, single-stranded, circular RNA particles that are much simpler than a virus. They do not have a capsid or outer envelope, but like viruses can reproduce only within a host cell. Viroids do not, however, manufacture any proteins, and they only produce a single, specific RNA molecule. Human diseases caused by viroids have yet to be identified.

Viroids are known to infect plants (Figure 21.18) and are responsible for crop failures and the loss of millions of dollars in agricultural revenue each year. Some of the plants they infect include potatoes, cucumbers, tomatoes, chrysanthemums, avocados, and coconut palms.

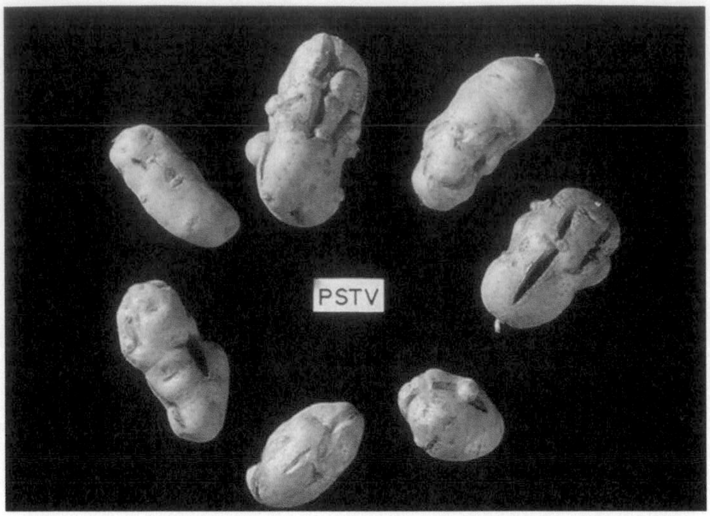

Figure 21.18 These potatoes have been infected by the potato spindle tuber viroid (PSTV), which is typically spread when infected knives are used to cut healthy potatoes, which are then planted. (credit: Pamela Roberts, University of Florida Institute of Food and Agricultural Sciences, USDA ARS)

CAREER CONNECTION

Virologist

Virology is the study of viruses, and a virologist is an individual trained in this discipline. Training in virology can lead to many different career paths. Virologists are actively involved in academic research and teaching in colleges and medical schools. Some virologists treat patients or are involved in the generation and production of vaccines. They might participate in epidemiologic studies (Figure 21.19) or become science writers, to name just a few possible careers.

Figure 21.19 This virologist is engaged in fieldwork, sampling eggs from this nest for avian influenza. (credit: Don Becker, USGS EROS, U.S. Fish and Wildlife Service)

If you think you may be interested in a career in virology, find a mentor in the field. Many large medical centers have departments of virology, and smaller hospitals usually have virology labs within their microbiology departments. Volunteer in a virology lab for a semester or work in one over the summer. Discussing the profession and getting a first-hand look at the work will help you decide whether a career in virology is right for you. The American Society of Virology's website (http://openstax.org/l/asv) is a good resource for information regarding training and careers in virology.

KEY TERMS

acellular lacking cells
acute disease disease where the symptoms rise and fall within a short period of time
asymptomatic disease disease where there are no symptoms and the individual is unaware of being infected unless lab tests are performed
attenuation weakening of a virus during vaccine development
AZT anti-HIV drug that inhibits the viral enzyme reverse transcriptase
back mutation when a live virus vaccine reverts back to it disease-causing phenotype
bacteriophage virus that infects bacteria
budding method of exit from the cell used in certain animal viruses, where virions leave the cell individually by capturing a piece of the host plasma membrane
capsid protein coating of the viral core
capsomere protein subunit that makes up the capsid
cell necrosis cell death
chronic infection describes when the virus persists in the body for a long period of time
cytopathic causing cell damage
envelope lipid bilayer that encircles some viruses
fusion method of entry by some enveloped viruses, where the viral envelope fuses with the plasma membrane of the host cell
gall appearance of a plant tumor
gene therapy treatment of genetic disease by adding genes, using viruses to carry the new genes inside the cell
group I virus virus with a dsDNA genome
group II virus virus with an ssDNA genome
group III virus virus with a dsRNA genome
group IV virus virus with an ssRNA genome with positive polarity
group V virus virus with an ssRNA genome with negative polarity
group VI virus virus with an ssRNA genome converted into dsDNA by reverse transcriptase
group VII virus virus with a single-stranded mRNA converted into dsDNA for genome replication
horizontal transmission transmission of a disease between unrelated individuals
hyperplasia abnormally high cell growth and division
hypoplasia abnormally low cell growth and division
intermittent symptom symptom that occurs periodically
latency virus that remains in the body for a long period of time but only causes intermittent symptoms
lysis bursting of a cell
lysogenic cycle type of virus replication in which the viral genome is incorporated into the genome of the host cell
lytic cycle type of virus replication in which virions are released through lysis, or bursting, of the cell
matrix protein envelope protein that stabilizes the envelope and often plays a role in the assembly of progeny virions
negative polarity ssRNA viruses with genomes complementary to their mRNA
oncogenic virus virus that has the ability to cause cancer
oncolytic virus virus engineered to specifically infect and kill cancer cells
pathogen agent with the ability to cause disease
permissive cell type that is able to support productive replication of a virus
phage therapy treatment of bacterial diseases using bacteriophages specific to a particular bacterium
positive polarity ssRNA virus with a genome that contains the same base sequences and codons found in their mRNA
prion infectious particle that consists of proteins that replicate without DNA or RNA
productive viral infection that leads to the production of new virions
prophage phage DNA that is incorporated into the host cell genome
PrP^c normal prion protein
PrP^{sc} infectious form of a prion protein
replicative intermediate dsRNA intermediate made in the process of copying genomic RNA
reverse transcriptase enzyme found in Baltimore groups VI and VII that converts single-stranded RNA into double-stranded DNA
vaccine weakened solution of virus components, viruses, or other agents that produce an immune response
vertical transmission transmission of disease from parent to offspring
viral receptor glycoprotein used to attach a virus to host cells via molecules on the cell
virion individual virus particle outside a host cell
viroid plant pathogen that produces only a single, specific RNA
virus core contains the virus genome

CHAPTER SUMMARY

21.1 Viral Evolution, Morphology, and Classification

Viruses are tiny, noncellular entities that usually can be seen only with an electron microscope. Their genomes contain either DNA or RNA—never both—and they replicate either by using the replication proteins of a host cell or by using proteins encoded in the viral genome. Viruses are diverse, infecting archaea, bacteria, fungi, plants, and animals. Viruses consist of a nucleic acid core surrounded by a protein capsid with or without an outer lipid envelope. The capsid shape, presence of an envelope, and core composition dictate some elements of the classification of viruses. The most commonly used classification method, the Baltimore classification, categorizes viruses based on how they produce their mRNA.

21.2 Virus Infections and Hosts

Plant viruses may be transmitted either vertically from parent reproductive cells or horizontally through damaged plant tissues. Viruses of plants are responsible for significant economic damage in both crop plants and plants used for ornamentation. Animal viruses enter their hosts through several types of virus-host cell interactions and cause a variety of infections. Viral infections can be either acute, with a brief period of infection terminated by host immune responses, or chronic, in which the infection persists. Persistent infections may cause chronic symptoms (hepatitis C), intermittent symptoms (latent viruses such a herpes simplex virus 1), or even be effectively asymptomatic (human herpesviruses 6 and 7). Oncogenic viruses in animals have the ability to cause cancer by interfering with the regulation of the host cell cycle.

21.3 Prevention and Treatment of Viral Infections

Viruses cause a variety of diseases in humans. Many of these diseases can be prevented by the use of viral vaccines, which stimulate protective immunity against the virus without causing major disease. Viral vaccines may also be used in active viral infections, boosting the ability of the immune system to control or destroy the virus. A series of antiviral drugs that target enzymes and other protein products of viral genes have been developed and used with mixed success. Combinations of anti-HIV drugs have been used to effectively control the virus, extending the lifespans of infected individuals. Viruses have many uses in medicines, such as in the treatment of genetic disorders, cancer, and bacterial infections.

21.4 Other Acellular Entities: Prions and Viroids

Prions are infectious agents that consist of protein, but no DNA or RNA, and seem to produce their deadly effects by duplicating their shapes and accumulating in tissues. They are thought to contribute to several progressive brain disorders, including mad cow disease and Creutzfeldt-Jakob disease. Viroids are single-stranded RNA pathogens that infect plants. Their presence can have a severe impact on the agriculture industry.

VISUAL CONNECTION QUESTIONS

1. Figure 21.5 Which of the following statements about virus structure is true?
   a. All viruses are encased in a viral membrane.
   b. The capsomere is made up of small protein subunits called capsids.
   c. DNA is the genetic material in all viruses.
   d. Glycoproteins help the virus attach to the host cell.

2. Figure 21.8 Influenza virus is packaged in a viral envelope that fuses with the plasma membrane. This way, the virus can exit the host cell without killing it. What advantage does the virus gain by keeping the host cell alive?

3. Figure 21.10 Which of the following statements is false?
   a. In the lytic cycle, new phages are produced and released into the environment.
   b. In the lysogenic cycle, phage DNA is incorporated into the host genome.
   c. An environmental stressor can cause the phage to initiate the lysogenic cycle.
   d. Cell lysis only occurs in the lytic cycle.

REVIEW QUESTIONS

4. Which statement is true?
   a. A virion contains DNA and RNA.
   b. Viruses are acellular.
   c. Viruses replicate outside of the cell.
   d. Most viruses are easily visualized with a light microscope.

5. The viral _______ play(s) a role in attaching a virion to the host cell.
   a. core
   b. capsid
   c. envelope
   d. both b and c

6. Viruses______.
   a. all have a round shape
   b. cannot have a long shape
   c. do not maintain any shape
   d. vary in shape

7. The observation that the bacteria genus *Chlamydia* contains species that can only survive as intracellular parasites supports which viral origin hypothesis?
   a. Progressive
   b. Regressive
   c. Self-replicating
   d. Virus-first

8. A scientist discovers a new virus with a linear, RNA genome surrounded by a helical capsid. The virus is most likely a member of which family based on structure classification?
   a. Rabies virus
   b. Herpesviruses
   c. Retroviruses
   d. Influenza viruses

9. Which statement is *not* true of viral replication?
   a. A lysogenic cycle kills the host cell.
   b. There are six basic steps in the viral replication cycle.
   c. Viral replication does not affect host cell function.
   d. Newly released virions can infect adjacent cells.

10. Which statement is true of viral replication?
    a. In the process of apoptosis, the cell survives.
    b. During attachment, the virus attaches at specific sites on the cell surface.
    c. The viral capsid helps the host cell produce more copies of the viral genome.
    d. mRNA works outside of the host cell to produce enzymes and proteins.

11. Which statement is true of reverse transcriptase?
    a. It is a nucleic acid.
    b. It infects cells.
    c. It transcribes RNA to make DNA.
    d. It is a lipid.

12. Oncogenic virus cores can be_______.
    a. RNA
    b. DNA
    c. neither RNA nor DNA
    d. either RNA or DNA

13. Which is true of DNA viruses?
    a. They use the host cell's machinery to produce new copies of their genome.
    b. They all have envelopes.
    c. They are the only kind of viruses that can cause cancer.
    d. They are not important plant pathogens.

14. A bacteriophage can infect _______.
    a. the lungs
    b. viruses
    c. prions
    d. bacteria

15. People with the CCR5Δ32 mutation of a T-cell surface protein can be exposed to some strains of HIV-1 without becoming sick. What step of the virus life cycle is likely to be inhibited with this mutation?
    a. Release
    b. Reverse transcription
    c. Uncoating
    d. Attachment

16. An apple grower notices that several of his apple trees with fungi growing on their trunks have developed necrotic ring spots, while other trees in the orchard that lack fungi appear healthy. What is the most likely conclusion the farmer can make about the virus infecting his apple trees?
    a. The apple trees were infected by horizontal transmission.
    b. The fungi carry disease.
    c. The fungi attract disease-carrying insects.
    d. The apple trees were infected by vertical transmission.

17. Which of the following is NOT used to treat active viral disease?
    a. Vaccines
    b. Antiviral drugs
    c. Antibiotics
    d. Phage therapy

18. Vaccines_______.
    a. are similar to viroids
    b. are only needed once
    c. kill viruses
    d. stimulate an immune response
19. A patient presents at the clinic with an acute viral infection. Assays that analyze the viral life cycle classify the virus into Group V with a segmented genome. Which virus is the most likely diagnosis for the patient?
    a. Rabies virus
    b. Picornavirus
    c. HIV-1
    d. Influenza A virus
20. Which of the following is not associated with prions?
    a. Replicating shapes
    b. Mad cow disease
    c. DNA
    d. Toxic proteins
21. Which statement is true of viroids?
    a. They are single-stranded RNA particles.
    b. They reproduce only outside of the cell.
    c. They produce proteins.
    d. They affect both plants and animals.

CRITICAL THINKING QUESTIONS

22. The first electron micrograph of a virus (tobacco mosaic virus) was produced in 1939. Before that time, how did scientists know that viruses existed if they could not see them? (Hint: Early scientists called viruses "filterable agents.")
23. Varicella-zoster virus is a double-stranded DNA virus that causes chickenpox. How does its genome structure provide an evolutionary advantage over a single-stranded DNA virus?
24. Classify the Rabies virus (a rhabdovirus family member) and HIV-1 with both the Baltimore and genomic structure systems. Compare your results. What conclusions can be made about these two different methods?
25. Why can't dogs catch the measles?
26. One of the first and most important targets for drugs to fight infection with HIV (a retrovirus) is the reverse transcriptase enzyme. Why?
27. In this section, you were introduced to different types of viruses and viral diseases. Briefly discuss the most interesting or surprising thing you learned about viruses.
28. Although plant viruses cannot infect humans, what are some of the ways in which they affect humans?
29. A bacteriophage with a lytic life cycle develops a mutation that allows it to now also go through the lysogenic cycle. How would this provide an evolutionary advantage over the other bacteriophages that can only spread through lytic cycles?
30. Why is immunization after being bitten by a rabid animal so effective and why aren't people vaccinated for rabies like dogs and cats are?
31. The vaccine Gardasil that targets human papilloma virus (HPV), the etiological agent of genital warts, was developed after the anti-HPV medication podofilox. Why would doctors still want a vaccine created after anti-viral medications were available?
32. Prions are responsible for variant Creutzfeldt-Jakob Disease, which has resulted in over 100 human deaths in Great Britain during the last 10 years. How do humans contract this disease?
33. How are viroids like viruses?
34. A botanist notices that a tomato plant looks diseased. How could the botanist confirm that the agent causing disease is a viroid, and not a virus?

CHAPTER 22
Prokaryotes: Bacteria and Archaea

Figure 22.1 Certain prokaryotes can live in extreme environments such as the Morning Glory pool, a hot spring in Yellowstone National Park. The spring's vivid blue color is from the prokaryotes that thrive in its very hot waters. (credit: modification of work by Jon Sullivan)

Chapter Outline

INTRODUCTION In the recent past, scientists grouped living things into five kingdoms— animals, plants, fungi, protists, and prokaryotes—based on several criteria, such as the absence or presence of a nucleus and other membrane-bound organelles, the absence or presence of cell walls, multicellularity, and so on. In the late 20th century, the pioneering work of Carl Woese and others compared sequences of small-subunit ribosomal RNA (SSU rRNA), which resulted in a more fundamental way to group organisms on Earth. Based on differences in the structure of cell membranes and in rRNA, Woese and his colleagues proposed that all life on Earth evolved along three lineages, called domains. The domain Bacteria comprises all organisms in the kingdom Bacteria, the domain Archaea comprises the rest of the prokaryotes, and the domain Eukarya comprises all eukaryotes—including organisms in the kingdoms Animalia, Plantae, Fungi, and Protista.

Two of the three domains—Bacteria and Archaea—are prokaryotic. Prokaryotes were the first inhabitants on Earth, appearing 3.5 to 3.8 billion years ago. These organisms are abundant and ubiquitous; that is, they are present everywhere. In addition to inhabiting moderate environments, they are found in extreme conditions: from boiling springs to permanently frozen environments in Antarctica; from salty environments like the Dead Sea to environments under tremendous pressure, such as the depths of the ocean; and from areas without oxygen, such as a waste management plant, to radioactively contaminated regions, such as Chernobyl. Prokaryotes reside in the human digestive system and on the skin, are responsible for certain illnesses, and serve an important role in the preparation of many foods.

22.1 Prokaryotic Diversity

By the end of this section, you will be able to do the following:

- Describe the evolutionary history of prokaryotes
- Discuss the distinguishing features of extremophiles
- Explain why it is difficult to culture prokaryotes

Prokaryotes are ubiquitous. They cover every imaginable surface where there is sufficient moisture, and they also live on and inside virtually all other living things. In the typical human body, prokaryotic cells outnumber human body cells by about ten to one. They comprise the majority of living things in all ecosystems. Some prokaryotes thrive in environments that are inhospitable for most living things. Prokaryotes recycle **nutrients**—essential substances (such as carbon and nitrogen)—and they drive the evolution of new ecosystems, some of which are natural and others man-made. Prokaryotes have been on Earth since long before multicellular life appeared. Indeed, eukaryotic cells are thought to be the descendants of ancient prokaryotic communities.

Prokaryotes, the First Inhabitants of Earth

When and where did cellular life begin? What were the conditions on Earth when life began? We now know that prokaryotes were likely the first forms of cellular life on Earth, and they existed for billions of years before plants and animals appeared. The Earth and its moon are dated at about 4.54 billion years in age. This estimate is based on evidence from radiometric dating of meteorite material together with other substrate material from Earth and the moon. Early Earth had a very different atmosphere (contained less molecular oxygen) than it does today and was subjected to strong solar radiation; thus, the first organisms probably would have flourished where they were more protected, such as in the deep ocean or far beneath the surface of the Earth. Strong volcanic activity was common on Earth at this time, so it is likely that these first organisms—the first prokaryotes—were adapted to very high temperatures. Because early Earth was prone to geological upheaval and volcanic eruption, and was subject to bombardment by mutagenic radiation from the sun, the first organisms were prokaryotes that must have withstood these harsh conditions.

Microbial Mats

Microbial mats or large biofilms may represent the earliest forms of prokaryotic life on Earth; there is fossil evidence of their presence starting about 3.5 billion years ago. It is remarkable that cellular life appeared on Earth only a billion years after the Earth itself formed, suggesting that pre-cellular "life" that could replicate itself had evolved much earlier. A **microbial mat** is a multi-layered sheet of prokaryotes (Figure 22.2) that includes mostly bacteria, but also archaeans. Microbial mats are only a few centimeters thick, and they typically grow where different types of materials interface, mostly on moist surfaces. The various types of prokaryotes that comprise them carry out different metabolic pathways, and that is the reason for their various colors. Prokaryotes in a microbial mat are held together by a glue-like sticky substance that they secrete called *extracellular matrix*.

The first microbial mats likely obtained their energy from chemicals found near hydrothermal vents. A ***hydrothermal vent*** is a breakage or fissure in the Earth's surface that releases geothermally heated water. With the evolution of photosynthesis about three billion years ago, some prokaryotes in microbial mats came to use a more widely available energy source—sunlight—whereas others were still dependent on chemicals from hydrothermal vents for energy and food.

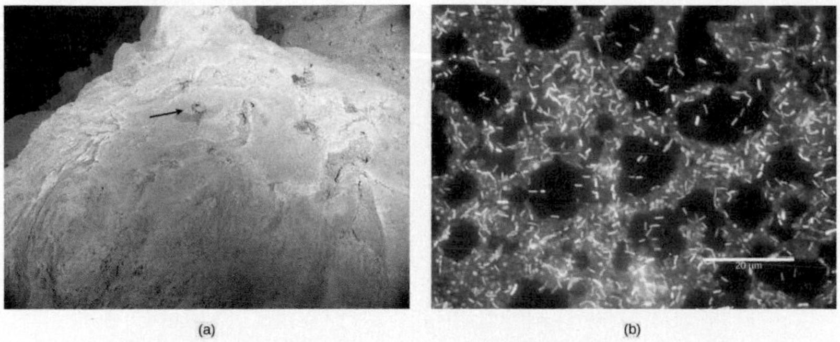

Figure 22.2 A microbial mat. (a) This microbial mat, about one meter in diameter, is growing over a hydrothermal vent in the Pacific Ocean in a region known as the "Pacific Ring of Fire." The mat's colony of bacteria helps retain microbial nutrients. Chimneys such as the one indicated by the arrow allow gases to escape. (b) In this micrograph, bacteria are visualized using fluorescence microscopy. (credit a: modification of work by Dr. Bob Embley, NOAA PMEL, Chief Scientist; credit b: modification of work by Ricardo Murga, Rodney Donlan, CDC; scale-bar data from Matt Russell)

Stromatolites

Fossilized microbial mats represent the earliest record of life on Earth. A **stromatolite** is a sedimentary structure formed when minerals are precipitated out of water by prokaryotes in a microbial mat (Figure 22.3). Stromatolites form layered rocks made of carbonate or silicate. Although most stromatolites are artifacts from the past, there are places on Earth where stromatolites are still forming. For example, growing stromatolites have been found in the Anza-Borrego Desert State Park in San Diego County, California.

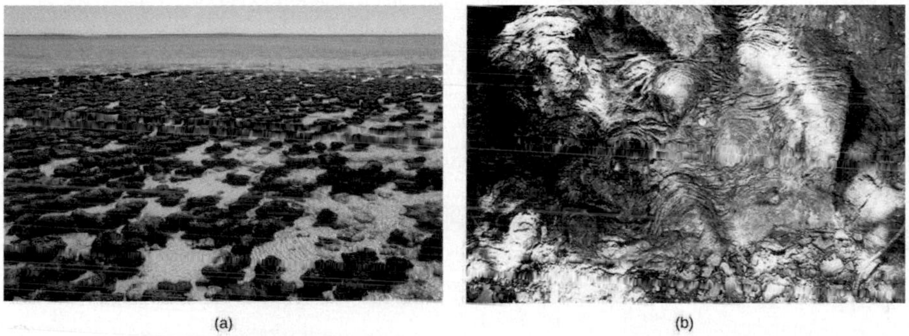

Figure 22.3 Stromatolites. (a) These living stromatolites are located in Shark Bay, Australia. (b) These fossilized stromatolites, found in Glacier National Park, Montana, are nearly 1.5 billion years old. (credit a: Robert Young; credit b: P. Carrara, NPS)

The Ancient Atmosphere

Evidence indicates that during the first two billion years of Earth's existence, the atmosphere was **anoxic**, meaning that there was no molecular oxygen. Therefore, only those organisms that can grow without oxygen—*anaerobic organisms*—were able to live. Autotrophic organisms that convert solar energy into chemical energy are called **phototrophs**, and they appeared within one billion years of the formation of Earth. Then, **cyanobacteria**, also known as "blue-green algae," evolved from these simple phototrophs at least one billion years later. It was the ancestral cyanobacteria (Figure 22.4) that began the "oxygenation" of the atmosphere: Increased atmospheric oxygen allowed the evolution of more efficient O_2-utilizing catabolic pathways. It also opened up the land to increased colonization, because some O_2 is converted into O_3 (ozone) and ozone effectively absorbs the ultraviolet light that could have otherwise caused lethal mutations in DNA. The current evidence suggests that the increase in O_2 concentrations allowed the evolution of other life forms.

Figure 22.4 Cyanobacteria. This hot spring in Yellowstone National Park flows toward the foreground. Cyanobacteria in the spring are green, and as water flows down the gradient, the intensity of the color increases as cell density increases. The water is cooler at the edges of the stream than in the center, causing the edges to appear greener. (credit: Graciela Brelles-Mariño)

Microbes Are Adaptable: Life in Moderate and Extreme Environments

Some organisms have developed strategies that allow them to survive harsh conditions. Almost all prokaryotes have a cell wall, a protective structure that allows them to survive in both hypertonic and hypotonic aqueous conditions. Some soil bacteria are able to form *endospores* that resist heat and drought, thereby allowing the organism to survive until favorable conditions recur. These adaptations, along with others, allow bacteria to remain the most abundant life form in all terrestrial and aquatic ecosystems.

Prokaryotes thrive in a vast array of environments: Some grow in conditions that would seem very normal to us, whereas others are able to thrive and grow under conditions that would kill a plant or an animal. Bacteria and archaea that are adapted to grow under extreme conditions are called **extremophiles**, meaning "lovers of extremes." Extremophiles have been found in all kinds of environments: the depths of the oceans, hot springs, the Arctic and the Antarctic, in very dry places, deep inside Earth, in harsh chemical environments, and in high radiation environments (Figure 22.5), just to mention a few. Because they have specialized adaptations that allow them to live in extreme conditions, many extremophiles cannot survive in moderate environments. There are many different groups of extremophiles: They are identified based on the conditions in which they grow best, and several habitats are extreme in multiple ways. For example, a soda lake is both salty and alkaline, so organisms that live in a soda lake must be both alkaliphiles and halophiles (Table 22.1). Other extremophiles, like **radioresistant** organisms, do not prefer an extreme environment (in this case, one with high levels of radiation), but have adapted to survive in it (Figure 22.5). Organisms like these give us a better understanding of prokaryotic diversity and open up the possibility of finding new prokaryotic species that may lead to the discovery of new therapeutic drugs or have industrial applications.

Extremophiles and Their Preferred Conditions

Extremophile	Conditions for Optimal Growth
Acidophiles	pH 3 or below
Alkaliphiles	pH 9 or above
Thermophiles	Temperature 60–80 °C (140–176 °F)
Hyperthermophiles	Temperature 80–122 °C (176–250 °F)
Psychrophiles	Temperature of -15-10 °C (5-50 °F) or lower

Table 22.1

Extremophile	Conditions for Optimal Growth
Halophiles	Salt concentration of at least 0.2 M
Osmophiles	High sugar concentration

Table 22.1

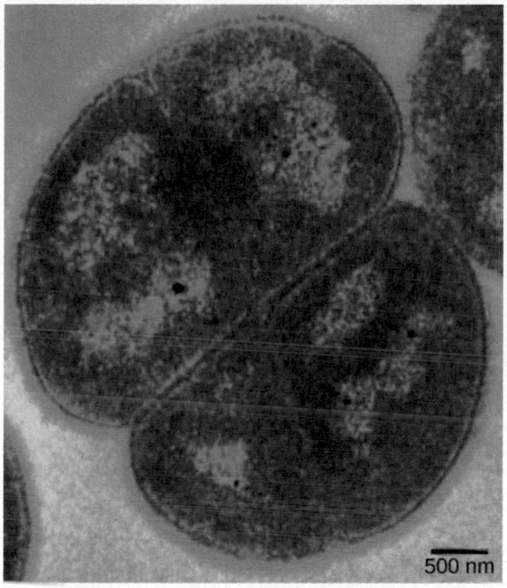

Figure 22.5 Radiation-tolerant prokaryotes. *Deinococcus radiodurans*, visualized in this false color transmission electron micrograph, is a prokaryote that can tolerate very high doses of ionizing radiation. It has developed DNA repair mechanisms that allow it to reconstruct its chromosome even if it has been broken into hundreds of pieces by radiation or heat. (credit: modification of work by Michael Daly; scale-bar data from Matt Russell)

Prokaryotes in the Dead Sea

One example of a very harsh environment is the Dead Sea, a hypersaline basin that is located between Jordan and Israel. Hypersaline environments are essentially concentrated seawater. In the Dead Sea, the sodium concentration is 10 times higher than that of seawater, and the water contains high levels of magnesium (about 40 times higher than in seawater) that would be toxic to most living things. Iron, calcium, and magnesium, elements that form divalent ions (Fe^{2+}, Ca^{2+}, and Mg^{2+}), produce what is commonly referred to as "hard" water. Taken together, the high concentration of divalent cations, the acidic pH (6.0), and the intense solar radiation flux make the Dead Sea a unique, and uniquely hostile, ecosystem[1] (Figure 22.6).

What sort of prokaryotes do we find in the Dead Sea? The extremely salt-tolerant bacterial mats include *Halobacterium*, *Haloferax volcanii* (which is found in other locations, not only the Dead Sea), *Halorubrum sodomense*, and *Halobaculum gomorrense*, and the archaean *Haloarcula marismortui*, among others.

1Bodaker, I, Itai, S, Suzuki, MT, Feingersch, R, Rosenberg, M, Maguire, ME, Shimshon, B, and others. Comparative community genomics in the Dead Sea: An increasingly extreme environment. *The ISME Journal* 4 (2010): 399–407, doi:10.1038/ismej.2009.141. published online 24 December 2009.

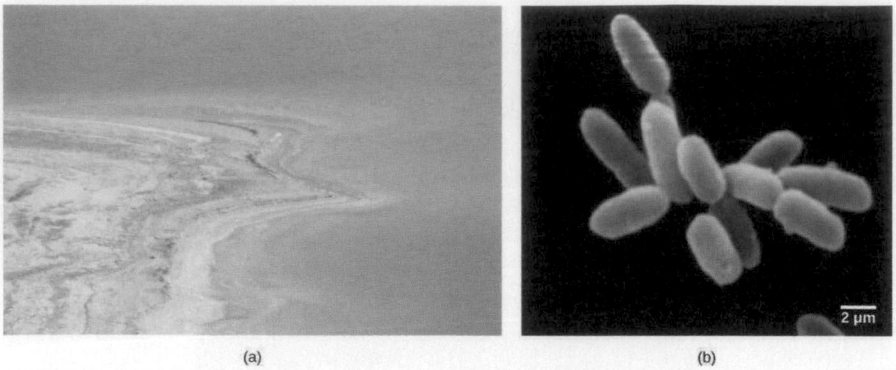

Figure 22.6 Halophilic prokaryotes. (a) The Dead Sea is hypersaline. Nevertheless, salt-tolerant bacteria thrive in this sea. (b) These halobacteria cells can form salt-tolerant bacterial mats. (credit a: Julien Menichini; credit b: NASA; scale-bar data from Matt Russell)

Unculturable Prokaryotes and the Viable-but-Non-Culturable State

The process of culturing bacteria is complex and is one of the greatest discoveries of modern science. German physician Robert Koch is credited with discovering the techniques for pure culture, including staining and using growth media. Microbiologists typically grow prokaryotes in the laboratory using an appropriate culture medium containing all the nutrients needed by the target organism. The medium can be liquid, broth, or solid. After an incubation time at the right temperature, there should be evidence of microbial growth (Figure 22.7). Koch's assistant Julius Petri invented the Petri dish, whose use persists in today's laboratories. Koch worked primarily with the *Mycobacterium tuberculosis* bacterium that causes tuberculosis and developed guidelines, called **Koch's postulates**, to identify the organisms responsible for specific diseases. Koch's postulates continue to be widely used in the medical community. Koch's postulates include that an organism can be identified as the cause of disease when it is present in all infected samples and absent in all healthy samples, and it is able to reproduce the infection after being cultured multiple times. Today, cultures remain a primary diagnostic tool in medicine and other areas of molecular biology.

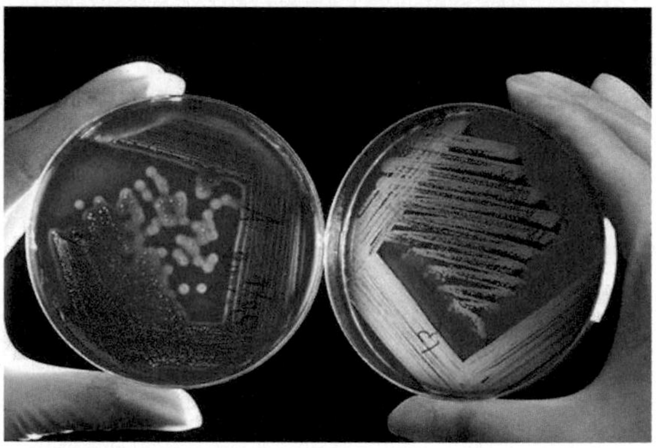

Figure 22.7 Bacteria growing on blood agar plates. In these agar plates, the growth medium is supplemented with red blood cells. Blood agar becomes transparent in the presence of hemolytic *Streptococcus*, which destroys red blood cells and is used to diagnose *Streptococcus* infections. The plate on the left is inoculated with non-hemolytic *Staphylococcus* (large white colonies), and the plate on the right is inoculated with hemolytic *Streptococcus* (tiny clear colonies). If you look closely at the right plate, you can see that the agar surrounding the bacteria has turned clear. (credit: Bill Branson, NCI)

Koch's postulates can be fully applied only to organisms that can be isolated and cultured. Some prokaryotes, however, cannot grow in a laboratory setting. In fact, over 99 percent of bacteria and archaea are *unculturable*. For the most part, this is due to a lack of knowledge as to what to feed these organisms and how to grow them; they may have special requirements for growth that remain unknown to scientists, such as needing specific micronutrients, pH, temperature, pressure, co-factors, or co-metabolites. Some bacteria cannot be cultured because they are obligate intracellular parasites and cannot be grown outside a host cell.

In other cases, *culturable organisms* become unculturable under stressful conditions, even though the same organism could be cultured previously. Those organisms that cannot be cultured but are not dead are in a **viable-but-non-culturable** (VBNC) state. The VBNC state occurs when prokaryotes respond to environmental stressors by entering a dormant state that allows their survival. The criteria for entering into the VBNC state are not completely understood. In a process called **resuscitation**, the prokaryote can go back to "normal" life when environmental conditions improve.

Is the VBNC state an unusual way of living for prokaryotes? In fact, most of the prokaryotes living in the soil or in oceanic waters are non-culturable. It has been said that only a small fraction, perhaps one percent, of prokaryotes can be cultured under laboratory conditions. If these organisms are non-culturable, then how is it known whether they are present and alive? Microbiologists use molecular techniques, such as the polymerase chain reaction (PCR), to amplify selected portions of DNA of prokaryotes, e.g., 16S rRNA genes, demonstrating their existence. (Recall that PCR can make billions of copies of a DNA segment in a process called amplification.)

The Ecology of Biofilms

Some prokaryotes may be unculturable because they require the presence of other prokaryotic species. Until a couple of decades ago, microbiologists used to think of prokaryotes as isolated entities living apart. This model, however, does not reflect the true ecology of prokaryotes, most of which prefer to live in communities where they can interact. As we have seen, a **biofilm** is a microbial community (Figure 22.8) held together in a gummy-textured matrix that consists primarily of polysaccharides secreted by the organisms, together with some proteins and nucleic acids. Biofilms typically grow attached to surfaces. Some of the best-studied biofilms are composed of prokaryotes, although fungal biofilms have also been described, as well as some composed of a mixture of fungi and bacteria.

Biofilms are present almost everywhere: they can cause the clogging of pipes and readily colonize surfaces in industrial settings. In recent, large-scale outbreaks of bacterial contamination of food, biofilms have played a major role. They also colonize household surfaces, such as kitchen counters, cutting boards, sinks, and toilets, as well as places on the human body, such as the surfaces of our teeth.

Interactions among the organisms that populate a biofilm, together with their protective *exopolysaccharidic (EPS)* environment, make these communities more robust than free-living, or planktonic, prokaryotes. The sticky substance that holds bacteria together also excludes most antibiotics and disinfectants, making biofilm bacteria hardier than their planktonic counterparts. Overall, biofilms are very difficult to destroy because they are resistant to many common forms of sterilization.

VISUAL CONNECTION

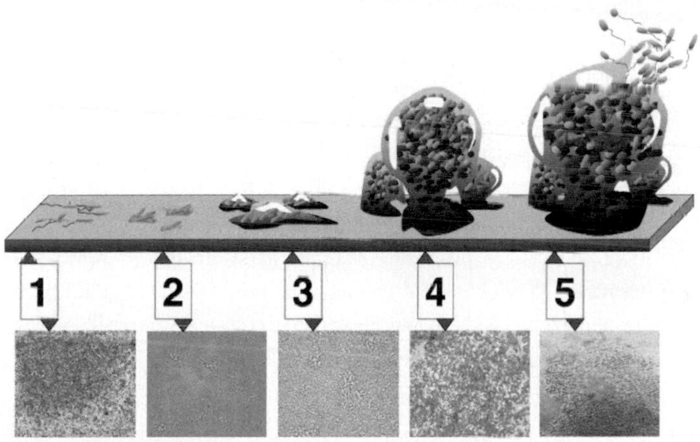

Figure 22.8 Development of a biofilm. Five stages of biofilm development are shown. During stage 1, initial attachment, bacteria adhere to a solid surface via weak *van der Waals interactions* (forces produced by induced electrical interactions between atoms). During stage 2, irreversible attachment, hairlike appendages called *pili* permanently anchor the bacteria to the surface. During stage 3, maturation I, the biofilm grows through cell division and recruitment of other bacteria. An extracellular matrix composed primarily of polysaccharides holds the biofilm together. During stage 4, maturation II, the biofilm continues to grow and takes on a more complex shape. During stage 5, dispersal, the biofilm matrix is partly broken down, allowing some bacteria to escape and colonize another surface. Micrographs of a *Pseudomonas aeruginosa* biofilm in each of the stages of development are shown. (credit: D. Davis, Don Monroe, PLoS)

Compared to free-floating bacteria, bacteria in biofilms often show increased resistance to antibiotics and detergents. Why do you think this might be the case?

22.2 Structure of Prokaryotes: Bacteria and Archaea

By the end of this section, you will be able to do the following:

- Describe the basic structure of a typical prokaryote
- Describe important differences in structure between Archaea and Bacteria

There are many differences between prokaryotic and eukaryotic cells. The name "prokaryote" suggests that prokaryotes are defined by exclusion—they are not eukaryotes, or organisms whose cells contain a nucleus and other internal membrane-bound organelles. However, all cells have four common structures: the plasma membrane, which functions as a barrier for the cell and separates the cell from its environment; the cytoplasm, a complex solution of organic molecules and salts inside the cell; a double-stranded DNA genome, the informational archive of the cell; and ribosomes, where protein synthesis takes place. Prokaryotes come in various shapes, but many fall into three categories: *cocci* (spherical), *bacilli* (rod-shaped), and *spirilli* (spiral-shaped) (Figure 22.9).

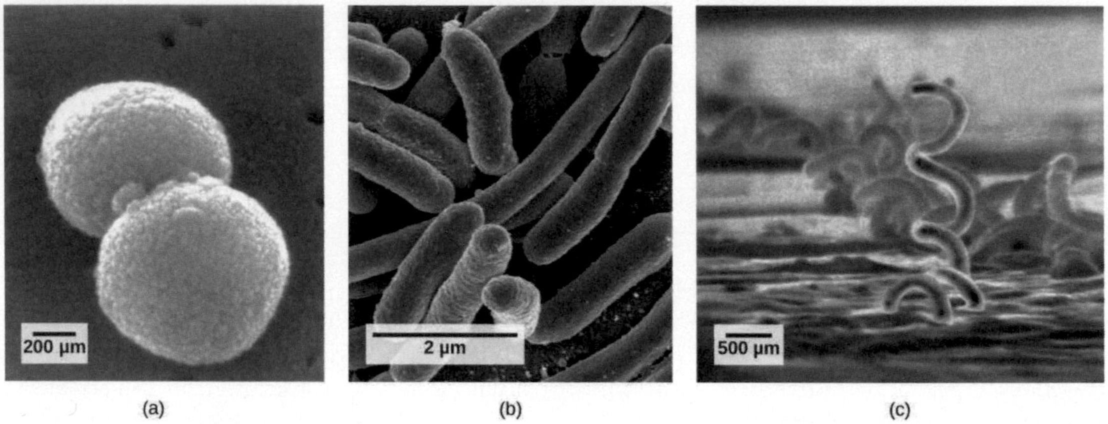

Figure 22.9 Common prokaryotic cell types. Prokaryotes fall into three basic categories based on their shape, visualized here using scanning electron microscopy: (a) cocci, or spherical (a pair is shown); (b) bacilli, or rod-shaped; and (c) spirilli, or spiral-shaped. (credit a: modification of work by Janice Haney Carr, Dr. Richard Facklam, CDC; credit c: modification of work by Dr. David Cox; scale-bar data from Matt Russell)

The Prokaryotic Cell

Recall that prokaryotes are unicellular organisms that lack membrane-bound organelles or other internal membrane-bound structures (Figure 22.10). Their chromosome—usually single—consists of a piece of circular, double-stranded DNA located in an area of the cell called the **nucleoid**. Most prokaryotes have a cell wall outside the plasma membrane. The **cell wall** functions as a protective layer, and it is responsible for the organism's shape. Some bacterial species have a capsule outside the cell wall. The capsule enables the organism to attach to surfaces, protects it from dehydration and attack by phagocytic cells, and makes pathogens more resistant to our immune responses. Some species also have flagella (singular, flagellum) used for locomotion, and pili (singular, pilus) used for attachment to surfaces including the surfaces of other cells. Plasmids, which consist of extra-chromosomal DNA, are also present in many species of bacteria and archaea.

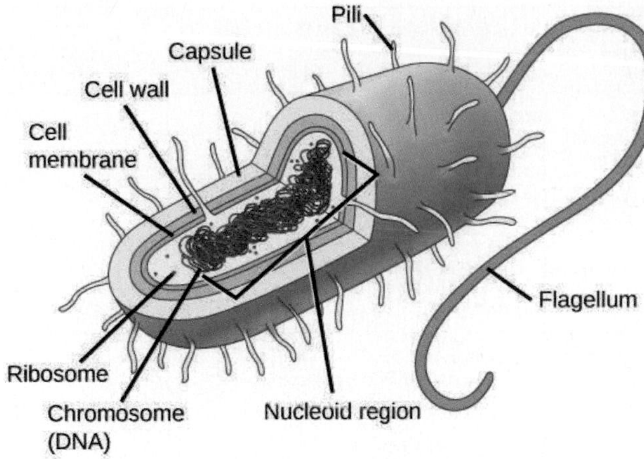

Figure 22.10 The features of a typical prokaryotic cell. Flagella, capsules, and pili are not found in all prokaryotes.

Recall that prokaryotes are divided into two different domains, Bacteria and Archaea, which together with Eukarya, comprise the three domains of life (Figure 22.11).

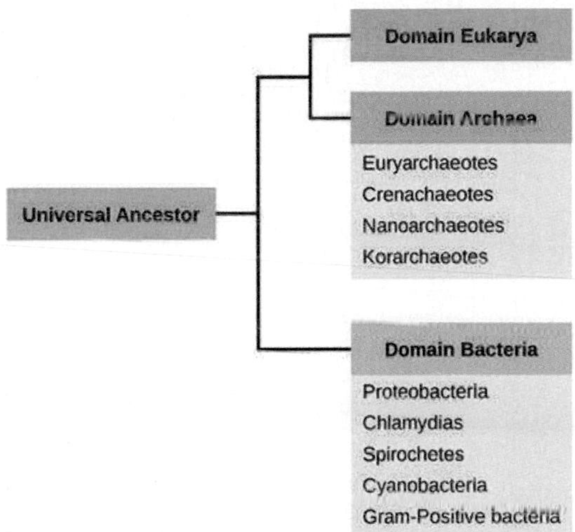

Figure 22.11 The three domains of living organisms. Bacteria and Archaea are both prokaryotes but differ enough to be placed in separate domains. An ancestor of modern Archaea is believed to have given rise to Eukarya, the third domain of life. Major groups of Archaea and Bacteria are shown.

Characteristics of bacterial phyla are described in Figure 22.12 and Figure 22.13. Major bacterial phyla include the Proteobacteria, the Chlamydias, the Spirochaetes, the photosynthetic Cyanobacteria, and the Gram-positive bacteria. The Proteobacteria are in turn subdivided into several classes, from the Alpha- to the Epsilon proteobacteria. Eukaryotic mitochondria are thought to be the descendants of alphaproteobacteria, while eukaryotic chloroplasts are derived from cyanobacteria. Archaeal phyla are described in Figure 22.14.

Bacteria of Phylum Proteobacteria		
Class	**Representative organisms**	**Representative micrograph**
Alpha Proteobacteria Some species are photoautotrophic but some are symbionts of plants and animals and others are pathogens. Eukaryotic mitochondria are thought be derived from bacteria in this group.	*Rhizobium* Nitrogen-fixing endosymbiont associated with the roots of legumes *Rickettsia* Obligate intracellular parasite that causes typhus and Rocky Mountain Spotted Fever (but not rickets, which is caused by Vitamin C deficiency)	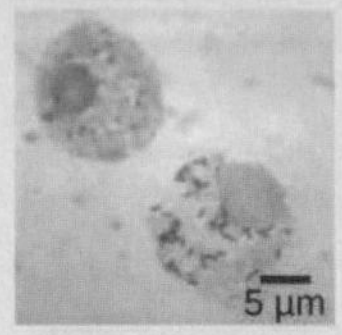 *Rickettsia rickettsia*, stained red, grow inside a host cell.
Beta Proteobacteria This group of bacteria is diverse. Some species play an important role in the nitrogen cycle.	*Nitrosomas* Species from this group oxidize ammonia into nitrite. *Spirillum minus* Causes rat-bite fever	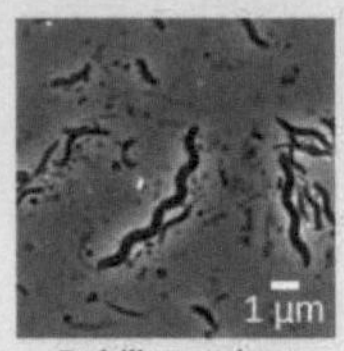 *Spirillum minus*
Gamma Proteobacteria Many are beneficial symbionts that populate the human gut, but others are familiar human pathogens. Some species from this subgroup oxidize sulfur compounds.	*Escherichia coli* Normally beneficial microbe of the human gut, but some strains cause disease *Salmonella* Certain strains cause food poisoning or typhoid fever *Yersinia pestis* Causative agent of Bubonic plague *Psuedomonas aeruginosa* Causes lung infections *Vibrio cholera* Causative agent of cholera *Chromatium* Sulfur-producing bacteria that oxidize sulfur, producing H_2S	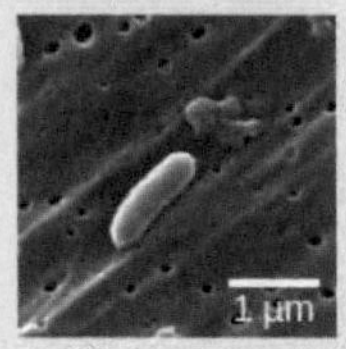 *Vibrio cholera*
Delta Proteobacteria Some species generate a spore-forming fruiting body in adverse conditions. Others reduce sulfate and sulfur.	*Myxobacteria* Generate spore-forming fruiting bodies in adverse conditions *Desulfovibrio vulgaris* Aneorobic, sulfate-reducing bacterium	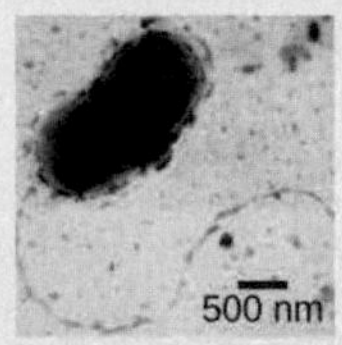 *Desulfovibrio vulgaris*
Epsilon Proteobacteria Many species inhabit the digestive tract of animals as symbionts or pathogens. Bacteria from this group have been found in deep-sea hydrothermal vents and cold seep habitats.	*Campylobacter* Causes blood poisoning and intestinal inflammation *Heliobacter pylori* Causes stomach ulcers	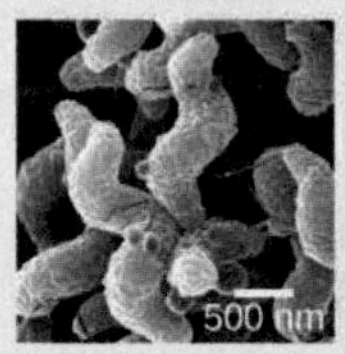 *Campylobacter*

Figure 22.12 The Proteobacteria. Phylum *Proteobacteria* is one of up to 52 bacteria phyla. *Proteobacteria* is further subdivided into five classes, Alpha through Epsilon. (credit "Rickettsia rickettsia": modification of work by CDC; credit "Spirillum minus": modification of work by Wolframm Adlassnig; credit "Vibrio cholera": modification of work by Janice Haney Carr, CDC; credit "Desulfovibrio vulgaris": modification of work by Graham Bradley; credit "Campylobacter": modification of work by De Wood, Pooley, USDA, ARS, EMU; scale-bar data from Matt Russell)

Bacteria: Chlamydia, Spirochaetae, Cyanobacteria, and Gram-positive		
Phylum	**Representative organisms**	**Representative micrograph**
Chlamydias All members of this group are obligate intracellular parasites of animal cells. Cells walls lack peptidoglycan.	*Chlamydia trachomatis* Common sexually transmitted disease that can lead to blindness	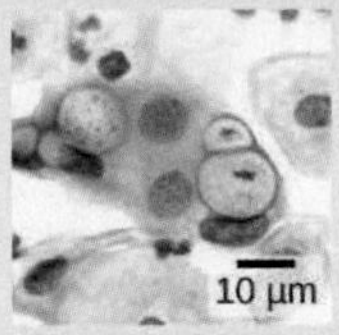 In this pap smear, *Chlamydia trachomatis* appear as pink inclusions inside cells.
Spirochetes Most members of this species, which has spiral-shaped cells, are free-living aneaerobes, but some are pathogenic. Flagella run lengthwise in the periplasmic space between the inner and outer membrane.	*Treponema pallidum* Causative agent of syphilis *Borrelia burgdorferi* Causative agent of Lyme disease	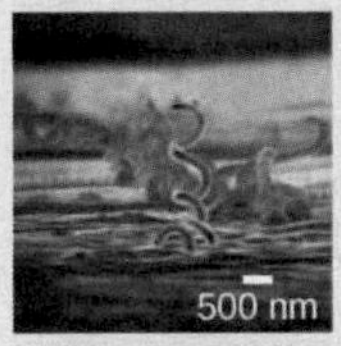 *Treponema pallidum*
Cyanobacteria Also known as blue-green algae, these bacteria obtain their energy through photosynthesis. They are ubiquitous, found in terrestrial, marine, and freshwater environments. Eukaryotic chloroplasts are thought be derived from bacteria in this group.	*Prochlorococcus* Believed to be the most abundant photosynthetic organism on earth; responsible for generating half the world's oxygen	 *Phormidium*
Gram-positive Bacteria Soil-dwelling members of this subgroup decompose organic matter. Some species cause disease. They have a thick cell wall and lack an outer membrane.	*Bacillus anthracis* Causes anthrax *Clostridium botulinum* Causes Botulism *Clostridium difficile* Causes diarrhea during antibiotic therapy *Streptomyces* Many antibiotics, including streptomyocin, are derived from these bacteria. *Mycoplasmas* These tiny bacteria, the smallest known, lack a cell wall. Some are free-living, and some are pathogenic.	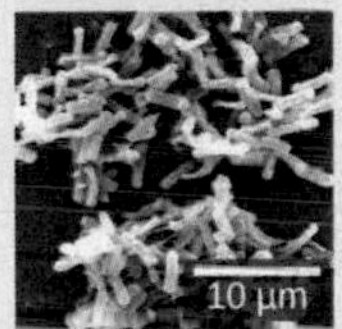 *Clostridium difficile*

Figure 22.13 Other bacterial phyla. Chlamydia, Spirochetes, Cyanobacteria, and Gram-positive bacteria are described in this table. Note that bacterial shape is not phylum-dependent; bacteria within a phylum may be cocci, rod-shaped, or spiral. (credit *"Chlamydia trachomatis"*: modification of work by Dr. Lance Liotta Laboratory, NCI; credit *"Treponema pallidum"*: modification of work by Dr. David Cox, CDC; credit *"Phormidium"*: modification of work by USGS; credit *"Clostridium difficile"*: modification of work by Lois S. Wiggs, CDC; scale-bar data from Matt Russell)

Archaea		
Phylum	**Representative organisms**	**Representative micrograph**
Euryarchaeota This phylum includes methanogens, which produce methane as a metabolic waste product, and halobacteria, which live in an extreme saline environment.	*Methanogens* Methane production causes flatulence in humans and other animals. *Halobacteria* Large blooms of this salt-loving archaea appear reddish due to the presence of bacterirhodopsin in the membrane. Bacteriorhodopsin is related to the retinal pigment rhodopsin.	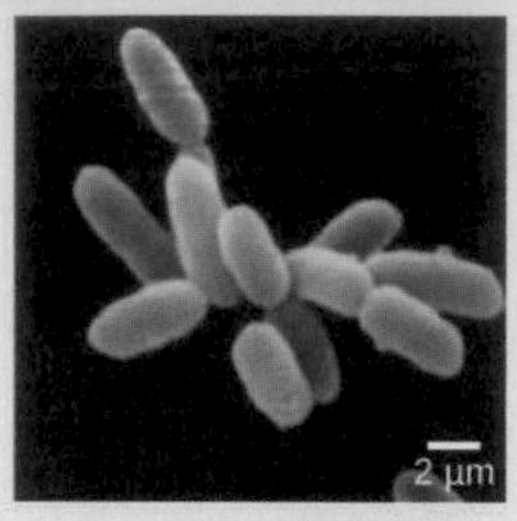 *Halobacterium* strain NRC-1
Crenarchaeota Members of the ubiquitous phylum play an important role in the fixation of carbon. Many members of this group are sulfur-dependent extremophiles. Some are thermophilic or hyperthermophilic.	*Sulfolobus* Members of this genus grow in volcanic springs at temperatures between 75° and 80°C and at a pH between 2 and 3.	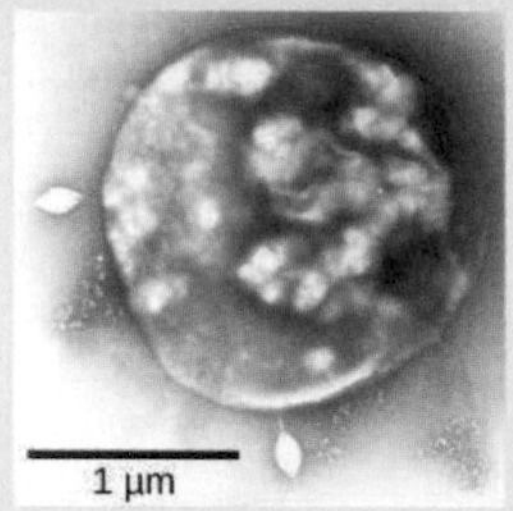 *Sulfolobus* being infected by bacteriophage
Nanoarchaeota This group currently contains only one species, *Nanoarchaeum equitans*.	*Nanoarchaeum equitans* This species was isolated from the bottom of the Atlantic Ocean and from a hydrothermal vent at Yellowstone National Park. It is an obligate symbiont with *Ignicoccus*, another species of archaea.	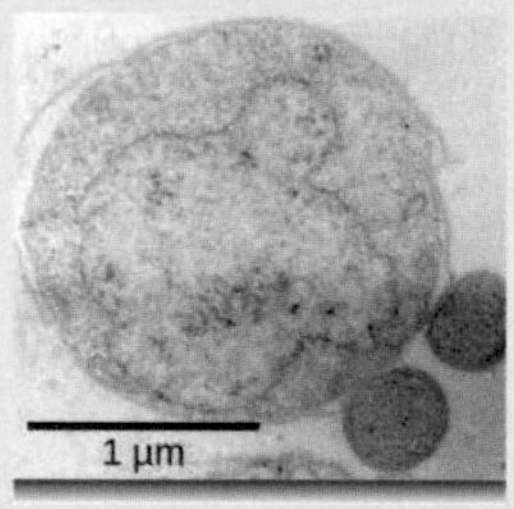 *Nanoarchaeum equitans* (small dark spheres) are in contact with their larger host, *Ignicoccus*.
Korarchaeota Members of this phylum, considered to be one of the most primitive forms of life, have only been found in the Obsidian Pool, a hot spring at Yellowstone National Park.	No members of this species have been cultivated.	 This image shows a variety of korarchaeota species from the Obsidian Pool at Yellowstone National Park.

Figure 22.14 Archaeal phyla. Archaea are separated into four phyla: the Korarchaeota, Euryarchaeota, Crenarchaeota, and Nanoarchaeota. (credit *"Halobacterium"*: modification of work by NASA; credit *"Nanoarchaeotum equitans"*: modification of work by Karl O. Stetter; credit "Korarchaeota": modification of work by Office of Science of the U.S. Dept. of Energy; scale-bar data from Matt Russell)

The Plasma Membrane of Prokaryotes

The prokaryotic plasma membrane is a thin lipid bilayer (6 to 8 nanometers) that completely surrounds the cell and separates the inside from the outside. Its selectively permeable nature keeps ions, proteins, and other molecules within the cell and prevents them from diffusing into the extracellular environment, while other molecules may move through the membrane. Recall that the general structure of a cell membrane is a phospholipid bilayer composed of two layers of lipid molecules. In archaeal cell membranes, *isoprene (phytanyl) chains* linked to glycerol replace the fatty acids linked to glycerol in bacterial membranes. Some archaeal membranes are lipid monolayers instead of bilayers (Figure 22.15).

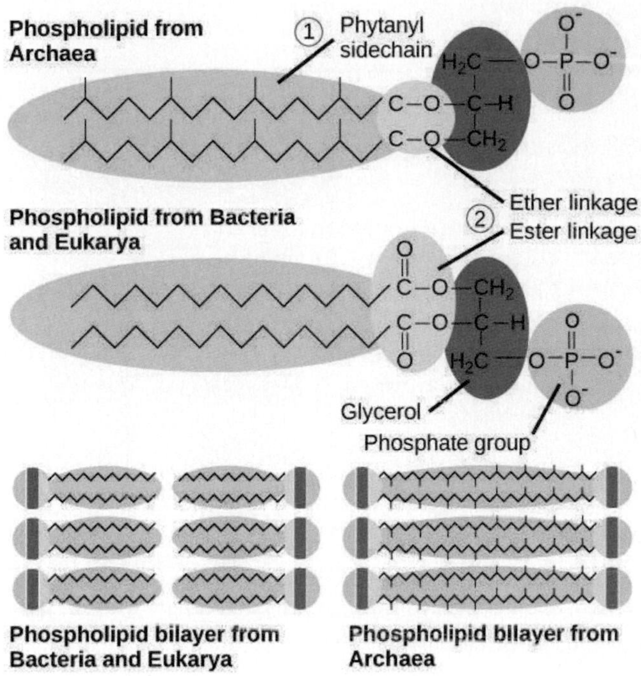

Figure 22.15 Bacterial and archaeal phospholipids. Archaeal phospholipids differ from those found in Bacteria and Eukarya in two ways. First, they have branched phytanyl sidechains instead of linear ones. Second, an ether bond instead of an ester bond connects the lipid to the glycerol.

The Cell Wall of Prokaryotes

The cytoplasm of prokaryotic cells has a high concentration of dissolved solutes. Therefore, the osmotic pressure within the cell is relatively high. The cell wall is a protective layer that surrounds some cells and gives them shape and rigidity. It is located outside the cell membrane and prevents *osmotic lysis* (bursting due to increasing volume). The chemical composition of the cell wall varies between Archaea and Bacteria, and also varies between bacterial species.

Bacterial cell walls contain **peptidoglycan**, composed of polysaccharide chains that are cross-linked by unusual peptides containing both L- and D-amino acids including D-glutamic acid and D-alanine. (Proteins normally have only L-amino acids; as a consequence, many of our antibiotics work by mimicking D-amino acids and therefore have specific effects on bacterial cell-wall development.) There are more than 100 different forms of peptidoglycan. *S-layer (surface layer) proteins* are also present on the outside of cell walls of both Archaea and Bacteria.

Bacteria are divided into two major groups: ***Gram positive*** and ***Gram negative***, based on their reaction to Gram staining. Note that all Gram-positive bacteria belong to one phylum; bacteria in the other phyla (Proteobacteria, Chlamydias, Spirochetes, Cyanobacteria, and others) are Gram-negative. The Gram staining method is named after its inventor, Danish scientist Hans Christian Gram (1853–1938). The different bacterial responses to the staining procedure are ultimately due to cell wall structure. *Gram-positive organisms typically lack the outer membrane found in Gram-negative organisms* (Figure 22.16). Up to 90 percent of the cell-wall in Gram-positive bacteria is composed of peptidoglycan, and most of the rest is composed of acidic substances called *teichoic acids*. Teichoic acids may be covalently linked to lipids in the plasma membrane to form *lipoteichoic acids*. Lipoteichoic acids anchor the cell wall to the cell membrane. Gram-negative bacteria have a relatively thin cell wall composed of a few layers of peptidoglycan (only 10 percent of the total cell wall), surrounded by an outer envelope containing lipopolysaccharides (LPS) and lipoproteins. This outer envelope is sometimes referred to as a second lipid bilayer. The chemistry of this outer envelope is very different, however, from that of the typical lipid bilayer that forms plasma membranes.

VISUAL CONNECTION

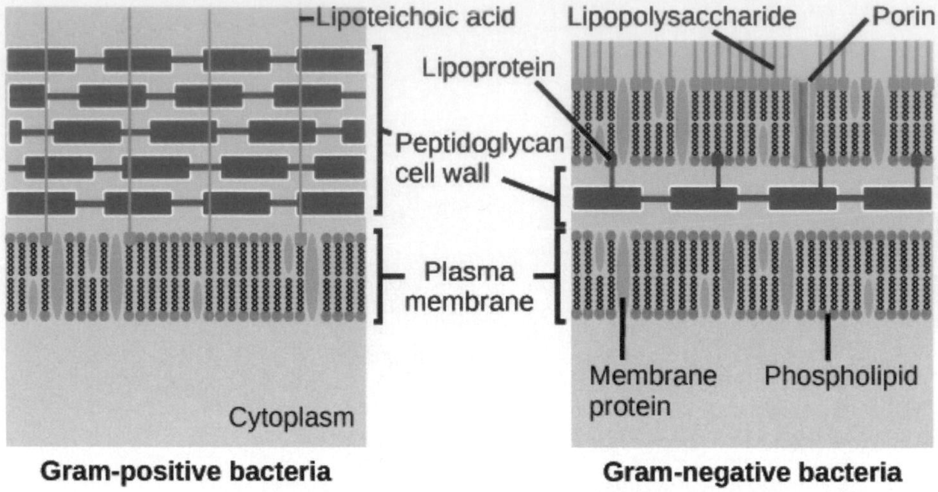

Figure 22.16 Cell walls in Gram-positive and Gram-negative bacteria. Bacteria are divided into two major groups: Gram positive and Gram negative. Both groups have a cell wall composed of peptidoglycan: in Gram-positive bacteria, the wall is thick, whereas in Gram-negative bacteria, the wall is thin. In Gram-negative bacteria, the cell wall is surrounded by an outer membrane that contains lipopolysaccharides and lipoproteins. Porins are proteins in this cell membrane that allow substances to pass through the outer membrane of Gram-negative bacteria. In Gram-positive bacteria, lipoteichoic acid anchors the cell wall to the cell membrane. (credit: modification of work by "Franciscosp2"/Wikimedia Commons)

Which of the following statements is true?

a. Gram-positive bacteria have a single cell wall anchored to the cell membrane by lipoteichoic acid.
b. Porins allow entry of substances into both Gram-positive and Gram-negative bacteria.
c. The cell wall of Gram-negative bacteria is thick, and the cell wall of Gram-positive bacteria is thin.
d. Gram-negative bacteria have a cell wall made of peptidoglycan, whereas Gram-positive bacteria have a cell wall made of lipoteichoic acid.

Archaean cell walls do not have peptidoglycan. There are four different types of archaean cell walls. One type is composed of **pseudopeptidoglycan**, which is similar to peptidoglycan in morphology but contains different sugars in the polysaccharide chain. The other three types of cell walls are composed of polysaccharides, glycoproteins, or pure protein. Other differences between Bacteria and Archaea are seen in Table 22.2. Note that features related to DNA replication, transcription and translation in Archaea are similar to those seen in eukaryotes.

Differences and Similarities between Bacteria and Archaea

Structural Characteristic	Bacteria	Archaea
Cell type	Prokaryotic	Prokaryotic
Cell morphology	Variable	Variable
Cell wall	Contains peptidoglycan	Does not contain peptidoglycan
Cell membrane type	Lipid bilayer	Lipid bilayer or lipid monolayer

Table 22.2

Structural Characteristic	Bacteria	Archaea
Plasma membrane lipids	Fatty acids-glycerol ester	Phytanyl-glycerol ethers
Chromosome	Typically circular	Typically circular
Replication origins	Single	Multiple
RNA polymerase	Single	Multiple
Initiator tRNA	Formyl-methionine	Methionine
Streptomycin inhibition	Sensitive	Resistant
Calvin cycle	Yes	No

Table 22.2

Reproduction

Reproduction in prokaryotes is *asexual* and usually takes place by binary fission. (Recall that the DNA of a prokaryote is a single, circular chromosome.) Prokaryotes do not undergo mitosis; instead, the chromosome is replicated and the two resulting copies separate from one another, due to the growth of the cell. The prokaryote, now enlarged, is pinched inward at its equator and the two resulting cells, which are *clones*, separate. Binary fission does not provide an opportunity for genetic recombination or genetic diversity, but prokaryotes can share genes by three other mechanisms.

In **transformation**, the prokaryote takes in DNA shed by other prokaryotes into its environment. If a nonpathogenic bacterium takes up DNA for a toxin gene from a pathogen and incorporates the new DNA into its own chromosome, it too may become pathogenic. In **transduction**, bacteriophages, the viruses that infect bacteria, may move short pieces of chromosomal DNA from one bacterium to another. Transduction results in a *recombinant organism*. Archaea also have viruses that may translocate genetic material from one individual to another. In **conjugation**, DNA is transferred from one prokaryote to another by means of a *pilus*, which brings the organisms into contact with one another, and provides a channel for transfer of DNA. The DNA transferred can be in the form of a plasmid or as a composite molecule, containing both plasmid and chromosomal DNA. These three processes of DNA exchange are shown in Figure 22.17.

Reproduction can be very rapid: a few minutes for some species. This short generation time coupled with mechanisms of genetic recombination and high rates of mutation result in the rapid evolution of prokaryotes, allowing them to respond to environmental changes (such as the introduction of an antibiotic) very quickly.

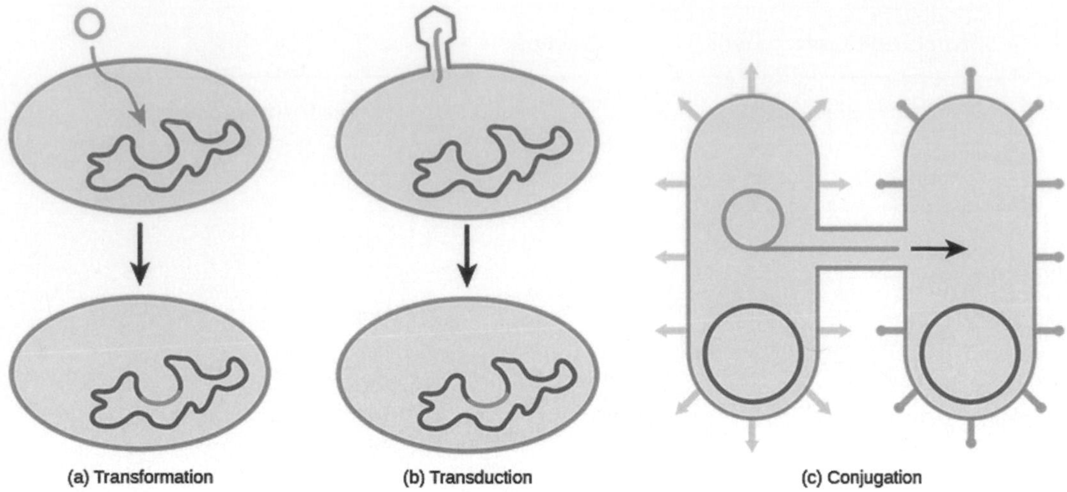

Figure 22.17 Gene transfer mechanisms in prokaryotes. There are three mechanisms by which prokaryotes can exchange DNA. In (a) transformation, the cell takes up prokaryotic DNA directly from the environment. The DNA may remain separate as plasmid DNA or be incorporated into the host genome. In (b) transduction, a bacteriophage injects DNA into the cell that contains a small fragment of DNA from a different prokaryote. In (c) conjugation, DNA is transferred from one cell to another via a mating bridge, or pilus, that connects the two cells after the sex pilus draws the two bacteria close enough to form the bridge.

EVOLUTION CONNECTION

The Evolution of Prokaryotes

How do scientists answer questions about the evolution of prokaryotes? Unlike with animals, artifacts in the fossil record of prokaryotes offer very little information. Fossils of ancient prokaryotes look like tiny bubbles in rock. Some scientists turn to genetics and to the principle of the molecular clock, which holds that the more recently two species have diverged, the more similar their genes (and thus proteins) will be. Conversely, species that diverged long ago will have more genes that are dissimilar.

Scientists at the NASA Astrobiology Institute and at the European Molecular Biology Laboratory collaborated to analyze the molecular evolution of 32 specific proteins common to 72 species of prokaryotes.[2] The model they derived from their data indicates that three important groups of bacteria—Actinobacteria, *Deinococcus*, and Cyanobacteria (collectively called *Terrabacteria* by the authors)—were the first to colonize land. Actinobacteria are a group of very common Gram-positive bacteria that produce branched structures like fungal mycelia, and include species important in decomposition of organic wastes. You will recall that *Deinococcus* is a genus of bacterium that is highly resistant to ionizing radiation. It has a thick peptidoglycan layer in addition to a second external membrane, so it has features of both Gram-positive and Gram-negative bacteria.

Cyanobacteria are photosynthesizers, and were probably responsible for the production of oxygen on the ancient earth. The timelines of divergence suggest that bacteria (members of the domain Bacteria) diverged from common ancestral species between 2.5 and 3.2 billion years ago, whereas the Archaea diverged earlier: between 3.1 and 4.1 billion years ago. Eukarya later diverged from the archaean line. The work further suggests that stromatolites that formed prior to the advent of cyanobacteria (about 2.6 billion years ago) photosynthesized in an anoxic environment and that because of the modifications of the Terrabacteria for land (resistance to drying and the possession of compounds that protect the organism from excess light), photosynthesis using oxygen may be closely linked to adaptations to survive on land.

2Battistuzzi, FU, Feijao, A, and Hedges, SB. A genomic timescale of prokaryote evolution: Insights into the origin of methanogenesis, phototrophy, and the colonization of land. *BioMed Central: Evolutionary Biology* 4 (2004): 44, doi:10.1186/1471-2148-4-44.

22.3 Prokaryotic Metabolism

By the end of this section, you will be able to do the following:

- Identify the macronutrients needed by prokaryotes, and explain their importance
- Describe the ways in which prokaryotes get energy and carbon for life processes
- Describe the roles of prokaryotes in the carbon and nitrogen cycles

Prokaryotes are metabolically diverse organisms. In many cases, a prokaryote may be placed into a species clade by its defining metabolic features: Can it metabolize lactose? Can it grow on citrate? Does it produce H_2S? Does it ferment carbohydrates to produce acid and gas? Can it grow under anaerobic conditions? Since metabolism and metabolites are the product of enzyme pathways, and enzymes are encoded in genes, the metabolic capabilities of a prokaryote are a reflection of its genome. There are many different environments on Earth with various energy and carbon sources, and variable conditions to which prokaryotes may be able to adapt. Prokaryotes have been able to live in every environment from deep-water volcanic vents to Antarctic ice by using whatever energy and carbon sources are available. Prokaryotes fill many niches on Earth, including involvement in nitrogen and carbon cycles, photosynthetic production of oxygen, decomposition of dead organisms, and thriving as parasitic, commensal, or mutualistic organisms inside multicellular organisms, including humans. The very broad range of environments that prokaryotes occupy is possible because they have diverse metabolic processes.

Needs of Prokaryotes

The diverse environments and ecosystems on Earth have a wide range of conditions in terms of temperature, available nutrients, acidity, salinity, oxygen availability, and energy sources. Prokaryotes are very well equipped to make their living out of a vast array of nutrients and environmental conditions. To live, prokaryotes need a source of energy, a source of carbon, and some additional nutrients.

Macronutrients

Cells are essentially a well-organized assemblage of macromolecules and water. Recall that macromolecules are produced by the polymerization of smaller units called monomers. For cells to build all of the molecules required to sustain life, they need certain substances, collectively called **nutrients**. When prokaryotes grow in nature, they must obtain their nutrients from the environment. Nutrients that are required in large amounts are called *macronutrients*, whereas those required in smaller or trace amounts are called *micronutrients*. Just a handful of elements are considered macronutrients—carbon, hydrogen, oxygen, nitrogen, phosphorus, and sulfur. (A mnemonic for remembering these elements is the acronym *CHONPS*.)

Why are these macronutrients needed in large amounts? They are the components of organic compounds in cells, including water. Carbon is the major element in all macromolecules: carbohydrates, proteins, nucleic acids, lipids, and many other compounds. Carbon accounts for about 50 percent of the composition of the cell. In contrast, nitrogen represents only 12 percent of the total dry weight of a typical cell. Nitrogen is a component of proteins, nucleic acids, and other cell constituents. Most of the nitrogen available in nature is either atmospheric nitrogen (N_2) or another inorganic form. Diatomic (N_2) nitrogen, however, can be converted into an organic form only by certain microorganisms, called nitrogen-fixing organisms. Both hydrogen and oxygen are part of many organic compounds and of water. Phosphorus is required by all organisms for the synthesis of nucleotides and phospholipids. Sulfur is part of the structure of some amino acids such as cysteine and methionine, and is also present in several vitamins and coenzymes. Other important macronutrients are potassium (K), magnesium (Mg), calcium (Ca), and sodium (Na). Although these elements are required in smaller amounts, they are very important for the structure and function of the prokaryotic cell.

Micronutrients

In addition to these macronutrients, prokaryotes require various metallic elements in small amounts. These are referred to as micronutrients or trace elements. For example, iron is necessary for the function of the cytochromes involved in electron-transport reactions. Some prokaryotes require other elements—such as boron (B), chromium (Cr), and manganese (Mn)—primarily as enzyme cofactors.

The Ways in Which Prokaryotes Obtain Energy

Prokaryotes are classified both by the way they obtain energy, and by the carbon source they use for producing organic molecules. These categories are summarized in Table 22.3. Prokaryotes can use different sources of energy to generate the ATP needed for biosynthesis and other cellular activities. **Phototrophs** (or phototrophic organisms) obtain their energy from sunlight. Phototrophs trap the energy of light using chlorophylls, or in a few cases, bacterial rhodopsin. (Rhodopsin-using phototrophs, oddly, are phototrophic, but not photosynthetic, since they do not fix carbon.) **Chemotrophs** (or chemosynthetic organisms) obtain their energy from chemical compounds. Chemotrophs that can use organic compounds as energy sources are called chemoorganotrophs. Those that can use inorganic compounds, like sulfur or iron compounds, as energy sources are called chemolithotrophs.

Energy-producing pathways may be either **aerobic**, using oxygen as the terminal electron acceptor, or anaerobic, using either simple inorganic compounds or organic molecules as the *terminal electron acceptor*. Since prokaryotes lived on Earth for nearly a billion years before photosynthesis produced significant amounts of oxygen for aerobic respiration, many species of both Bacteria and Archaea are anaerobic and their metabolic activities are important in the carbon and nitrogen cycles discussed below.

The Ways in Which Prokaryotes Obtain Carbon

Prokaryotes not only can use different sources of energy, but also different sources of carbon compounds. Autotrophic prokaryotes synthesize organic molecules from carbon dioxide. In contrast, heterotrophic prokaryotes obtain carbon from organic compounds. To make the picture more complex, the terms that describe how prokaryotes obtain energy and carbon can be combined. Thus, photoautotrophs use energy from sunlight, and carbon from carbon dioxide and water, whereas chemoheterotrophs obtain both energy and carbon from an organic chemical source. Chemolithoautotrophs obtain their energy from inorganic compounds, and they build their complex molecules from carbon dioxide. Finally, prokaryotes that get their energy from light, but their carbon from organic compounds, are photoheterotrophs. The table below (Table 22.3) summarizes carbon and energy sources in prokaryotes.

Carbon and Energy Sources in Prokaryotes

Energy Source	Electron Source	Carbon Source	Nutritional Type
Light (phototroph)	Organic material (organotroph)	Organic material (heterotroph)	Photoorganoheterotroph
		Carbon dioxide (autotroph)	
	Inorganic material (lithotroph)	Organic material (heterotroph)	
		Carbon dioxide (autotroph)	Photolithoautotroph
Chemicals (chemotroph)	Organic material (organotroph)	Organic material (heterotroph)	Chemoorganoheterotroph
		Carbon dioxide (autotroph)	
	Inorganic material (lithotroph)	Organic material (heterotroph)	Chemolithoheterotroph
		Carbon dioxide (autotroph)	Chemolithoautotroph

Table 22.3

Role of Prokaryotes in Ecosystems

Prokaryotes are ubiquitous: There is no niche or ecosystem in which they are not present. Prokaryotes play many roles in the environments they occupy. The roles they play in the carbon and nitrogen cycles are vital to life on Earth. In addition, the current scientific consensus suggests that metabolically interactive prokaryotic communities may have been the basis for the emergence of eukaryotic cells.

Prokaryotes and the Carbon Cycle

Carbon is one of the most important macronutrients, and prokaryotes play an important role in the carbon cycle (Figure 22.18). The carbon cycle traces the movement of carbon from inorganic to organic compounds and back again. Carbon is cycled through Earth's major reservoirs: land, the atmosphere, aquatic environments, sediments and rocks, and biomass. In a way, the carbon cycle echoes the role of the "four elements" first proposed by the ancient Greek philosopher, Empedocles: fire, water, earth, and air. Carbon dioxide is removed from the atmosphere by land plants and marine prokaryotes, and is returned to the atmosphere via the respiration of chemoorganotrophic organisms, including prokaryotes, fungi, and animals. Although the largest carbon reservoir in terrestrial ecosystems is in rocks and sediments, that carbon is not readily available.

Participants in the carbon cycle are roughly divided among producers, consumers, and decomposers of organic carbon compounds. The *primary producers* of organic carbon compounds from CO_2 are land plants and photosynthetic bacteria. A large amount of available carbon is found in living land plants. A related source of carbon compounds is *humus*, which is a mixture of organic materials from dead plants and prokaryotes that have resisted decomposition. (The term "humus," by the way, is the root of the word "human.") Consumers such as animals and other heterotrophs use organic compounds generated by producers and release carbon dioxide to the atmosphere. Other bacteria and fungi, collectively called ***decomposers***, carry out the breakdown (decomposition) of plants and animals and their organic compounds. Most carbon dioxide in the atmosphere is derived from the respiration of microorganisms that decompose dead animals, plants, and humus.

In aqueous environments and their anoxic sediments, there is another carbon cycle taking place. In this case, the cycle is based on one-carbon compounds. In anoxic sediments, prokaryotes, mostly archaea, produce methane (CH_4). This methane moves into the zone above the sediment, which is richer in oxygen and supports bacteria called *methane oxidizers* that oxidize methane to carbon dioxide, which then returns to the atmosphere.

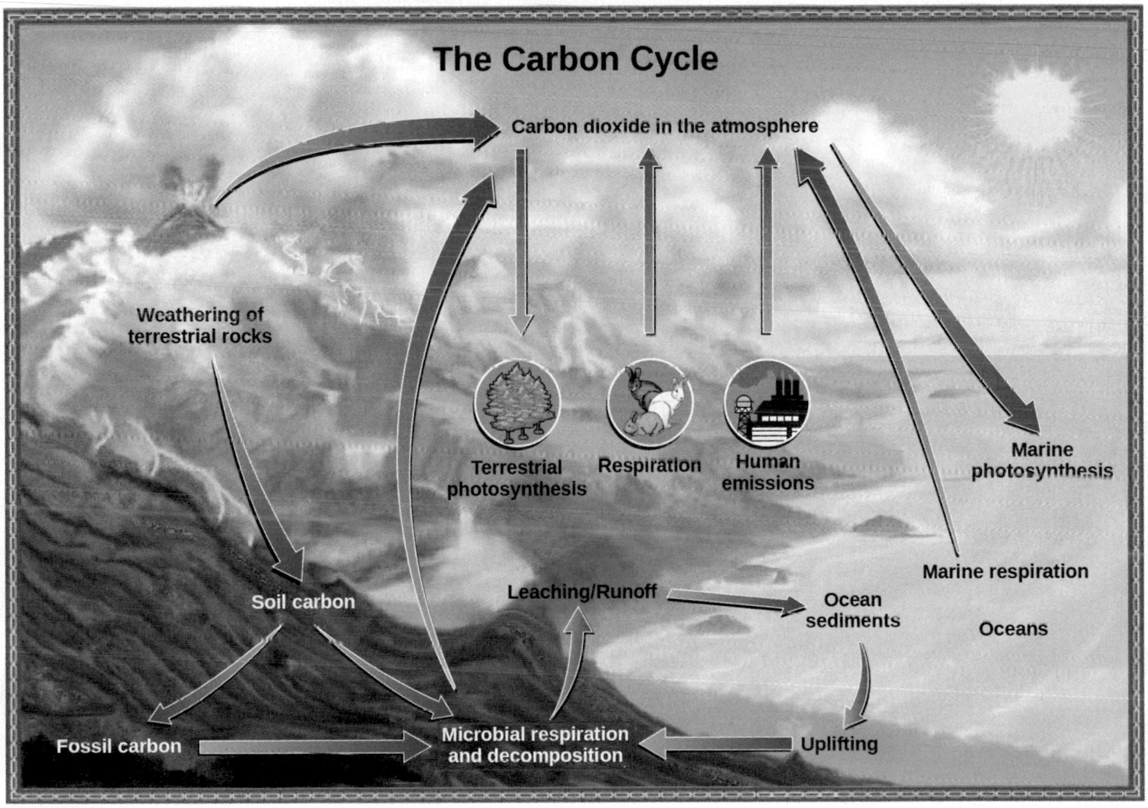

Figure 22.18 The carbon cycle. Prokaryotes play a significant role in continuously moving carbon through the biosphere. (credit: modification of work by John M. Evans and Howard Perlman, USGS)

Prokaryotes and the Nitrogen Cycle

Nitrogen is a very important element for life because it is a major constituent of proteins and nucleic acids. It is a macronutrient, and in nature, it is recycled from organic compounds to ammonia, ammonium ions, nitrate, nitrite, and nitrogen gas by many processes, many of which are carried out only by prokaryotes. As illustrated in Figure 22.19, prokaryotes are key to the nitrogen cycle. The largest pool of nitrogen available in the terrestrial ecosystem is *gaseous nitrogen* (N_2) from the air, but this nitrogen is not usable by plants, which are primary producers. Gaseous nitrogen is transformed, or "fixed" into more readily available forms, such as ammonia (NH_3), through the process of **nitrogen fixation**. Nitrogen-fixing bacteria include *Azotobacter* in soil and the ubiquitous photosynthetic cyanobacteria. Some nitrogen fixing bacteria, like *Rhizobium*, live in symbiotic relationships in the roots of legumes. Another source of ammonia is **ammonification**, the process by which ammonia is released during the decomposition of nitrogen-containing organic compounds. The ammonium ion is progressively oxidized by different species of bacteria in a process called nitrification. The nitrification process begins with the conversion of ammonium to nitrite (NO_2^-), and continues with the conversion of nitrite to nitrate. Nitrification in soils is carried out by bacteria belonging to the genera *Nitrosomas*, *Nitrobacter*, and *Nitrospira*. Most nitrogen in soil is in the form of ammonium (NH_4^+) or nitrate (NO_3^-). Ammonia and nitrate can be used by plants or converted to other forms.

Ammonia released into the atmosphere, however, represents only 15 percent of the total nitrogen released; the rest is as N_2 and N_2O (nitrous oxide). Ammonia is catabolized anaerobically by some prokaryotes, yielding N_2 as the final product. Denitrifying bacteria reverse the process of nitrification, reducing the nitrate from soils to gaseous compounds such as N_2O, NO, and N_2.

VISUAL CONNECTION

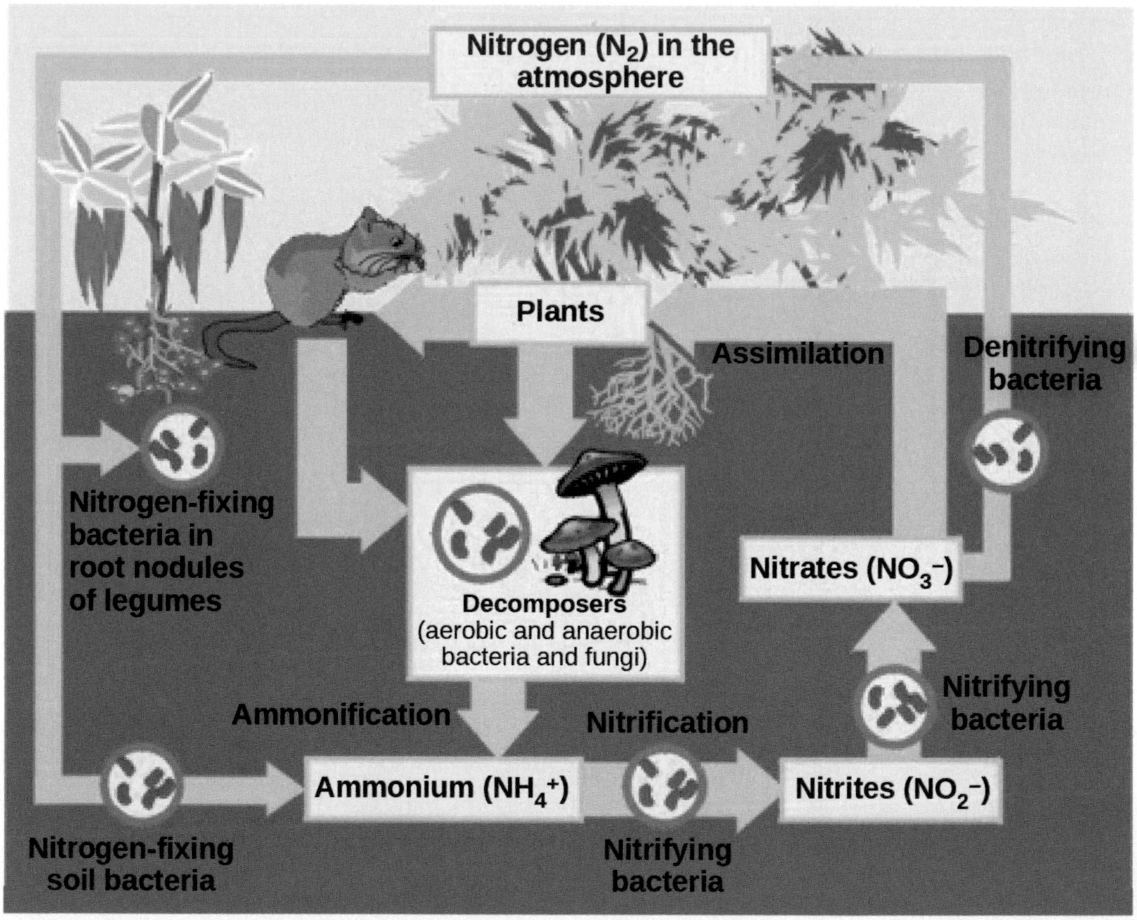

Figure 22.19 The nitrogen cycle. Prokaryotes play a key role in the nitrogen cycle. (credit: Environmental Protection Agency)

Which of the following statements about the nitrogen cycle is false?

a. Nitrogen-fixing bacteria exist on the root nodules of legumes and in the soil.
b. Denitrifying bacteria convert nitrates (NO_3^-) into nitrogen gas (N_2).
c. Ammonification is the process by which ammonium ion (NH_4^+) is released from decomposing organic compounds.
d. Nitrification is the process by which nitrites (NO_2^-) are converted to ammonium ion (NH_4^+).

22.4 Bacterial Diseases in Humans

By the end of this section, you will be able to do the following:

- Identify bacterial diseases that caused historically important plagues and epidemics
- Describe the link between biofilms and foodborne diseases
- Explain how overuse of antibiotics may be creating "super bugs"
- Explain the importance of MRSA with respect to the problems of antibiotic resistance

To a prokaryote, humans may be just another housing opportunity. Unfortunately, the tenancy of some species can have harmful effects and cause disease. Bacteria or other infectious agents that cause harm to their human hosts are called **pathogens**. Devastating pathogen-borne diseases and plagues, both viral and bacterial in nature, have affected humans and their ancestors for millions of years. The true cause of these diseases was not understood until modern scientific thought developed, and many people thought that diseases were a "spiritual punishment." Only within the past several centuries have people understood that staying away from afflicted persons, disposing of the corpses and personal belongings of victims of illness, and sanitation practices reduced their own chances of getting sick.

Epidemiologists study how diseases are transmitted and how they affect a population. Often, they must following the course of an **epidemic**—a disease that occurs in an unusually high number of individuals in a population at the same time. In contrast, a **pandemic** is a widespread, and usually worldwide, epidemic. An **endemic disease** is a disease that is always present, usually at low incidence, in a population.

Long History of Bacterial Disease

There are records about infectious diseases as far back as 3000 B.C. A number of significant pandemics caused by bacteria have been documented over several hundred years. Some of the most memorable pandemics led to the decline of cities and entire nations.

In the 21st century, infectious diseases remain among the leading causes of death worldwide, despite advances made in medical research and treatments in recent decades. A disease *spreads* when the pathogen that causes it is passed from one person to another. For a pathogen to cause disease, it must be able to reproduce in the host's body and damage the host in some way.

The Plague of Athens

In 430 B.C., the Plague of Athens killed one-quarter of the Athenian troops who were fighting in the great Peloponnesian War and weakened Athens's dominance and power. The plague impacted people living in overcrowded Athens as well as troops aboard ships that had to return to Athens. The source of the plague may have been identified recently when researchers from the University of Athens were able to use DNA from teeth recovered from a mass grave. The scientists identified nucleotide sequences from a pathogenic bacterium, *Salmonella enterica* serovar Typhi (Figure 22.20), which causes *typhoid fever*.[3] This disease is commonly seen in overcrowded areas and has caused epidemics throughout recorded history.

3Papagrigorakis MJ, Synodinos PN, and Yapijakis C. Ancient typhoid epidemic reveals possible ancestral strain of *Salmonella enterica* serovar Typhi. *Infect Genet Evol* 7 (2007): 126–7, Epub 2006 Jun.

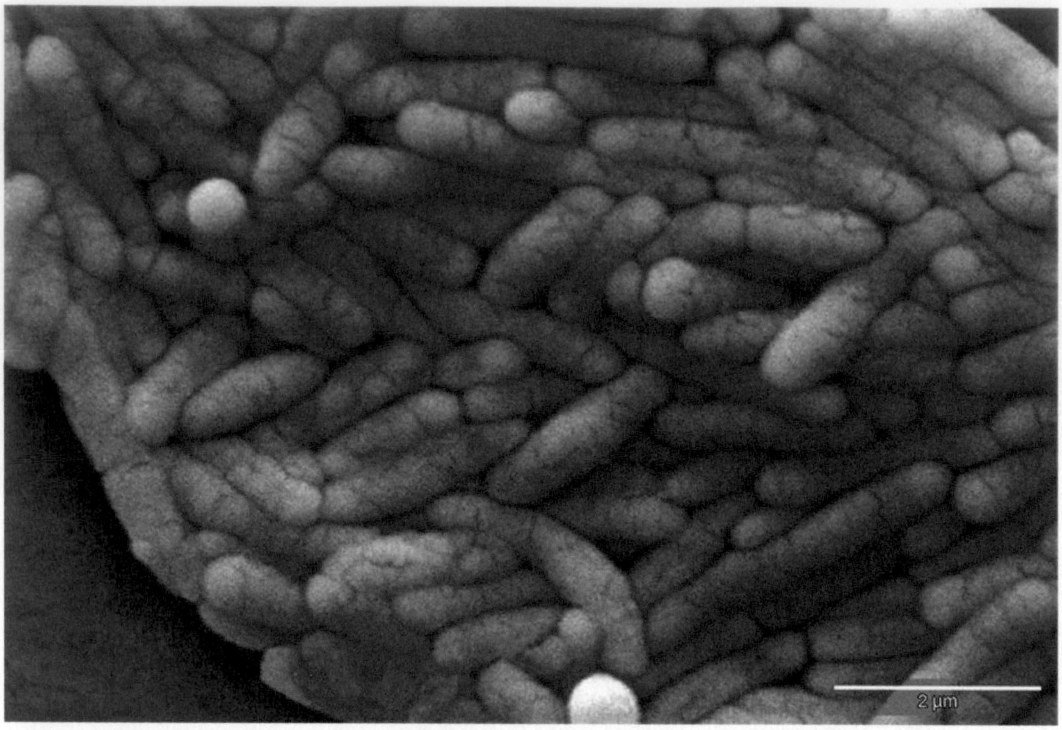

Figure 22.20 *Salmonella enterica. Salmonella enterica* serovar Typhi, the causative agent of Typhoid fever, is a Gram-negative, rod-shaped gamma proteobacterium. Typhoid fever, which is spread through feces, causes intestinal hemorrhage, high fever, delirium, and dehydration. Today, between 16 and 33 million cases of this re-emerging disease occur annually, resulting in over 200,000 deaths. Carriers of the disease can be asymptomatic. In a famous case in the early 1900s, a cook named Mary Mallon ("Typhoid Mary") unknowingly spread the disease to over fifty people, three of whom died. Other serotypes of *Salmonella* cause food poisoning. (credit: modification of work by NCI, CDC)

Bubonic Plagues

From 541 to 750, the Plague of Justinian, an outbreak of what was likely *bubonic plague*, eliminated one-quarter to one-half of the human population in the eastern Mediterranean region. The population in Europe dropped by 50 percent during this outbreak. Astoundingly, bubonic plague would strike Europe more than once!

Bubonic plague is caused by the bacterium *Yersinia pestis*. One of the most devastating pandemics attributed to bubonic plague was the **Black Death** (1346 to 1361). It is thought to have originated in China and spread along the Silk Road, a network of land and sea trade routes, to the Mediterranean region and Europe, carried by fleas living on black rats that were always present on ships. The Black Death was probably named for the tissue necrosis (Figure 22.21c) that can be one of the symptoms. The "buboes" of bubonic plague were painfully swollen areas of lymphatic tissue. A *pneumonic form* of the plague, spread by the coughing and sneezing of infected individuals, spreads directly from human to human and can cause death within a week. The pneumonic form was responsible for the rapid spread of the Black Death in Europe. The Black Death reduced the world's population from an estimated 450 million to about 350 to 375 million. Bubonic plague struck London yet again in the mid-1600s (Figure 22.21). In modern times, approximately 1,000 to 3,000 cases of plague arise globally each year, and a "sylvatic" form of plague, carried by fleas living on rodents such as prairie dogs and black footed ferrets, infects 10 to 20 people annually in the American Southwest. Although contracting bubonic plague before antibiotics meant almost certain death, the bacterium responds to several types of modern antibiotics, and mortality rates from plague are now very low.

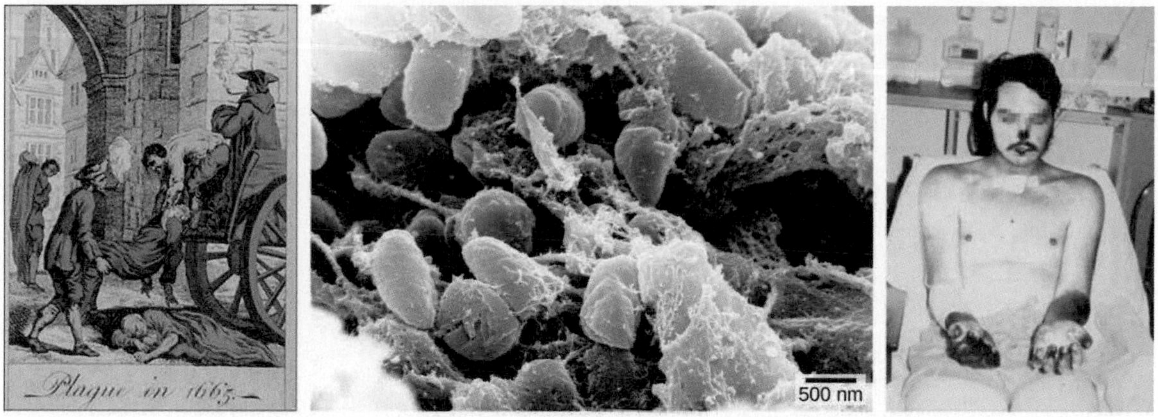

Figure 22.21 The Black Death. The (a) Great Plague of London killed an estimated 200,000 people, or about 20 percent of the city's population. The causative agent, the (b) bacterium *Yersinia pestis*, is a Gram-negative, rod-shaped bacterium from the class Gammaproteobacteria. The disease is transmitted through the bite of an infected flea, which is carried on a rodent. Symptoms include swollen lymph nodes, fever, seizure, vomiting of blood, and (c) gangrene. (credit b: Rocky Mountain Laboratories, NIAID, NIH; scale-bar data from Matt Russell; credit c: Textbook of Military Medicine, Washington, D.C., U.S. Dept. of the Army, Office of the Surgeon General, Borden Institute)

LINK TO LEARNING

Watch a video (http://openstax.org/l/black_death) on the modern understanding of the Black Death—bubonic plague in Europe during the 14th century.

Migration of Diseases to New Populations

One of the negative consequences of human exploration was the accidental "biological warfare" that resulted from the transport of a pathogen into a population that had not previously been exposed to it. Over the centuries, Europeans tended to develop genetic immunity to endemic infectious diseases, but when European conquerors reached the western hemisphere, they brought with them disease-causing bacteria and viruses, which triggered epidemics that completely devastated many diverse populations of Native Americans, who had no natural resistance to many European diseases. It has been estimated that up to 90 percent of Native Americans died from infectious diseases after the arrival of Europeans, making conquest of the New World a foregone conclusion.

Emerging and Re-emerging Diseases

The distribution of a particular disease is *dynamic*. Changes in the environment, the pathogen, or the host population can dramatically impact the spread of a disease. According to the World Health Organization (WHO), an **emerging disease** (Figure 22.22) is one that has appeared in a population for the first time, or that may have existed previously but is rapidly increasing in incidence or geographic range. This definition also includes *re-emerging diseases* that were previously under control. Approximately 75 percent of recently emerging infectious diseases affecting humans are zoonotic diseases. **Zoonoses** are diseases that primarily infect animals but can be transmitted to humans; some are of viral origin and some are of bacterial origin. Brucellosis is an example of a prokaryotic zoonosis that is re-emerging in some regions, and *necrotizing fasciitis* (commonly known as flesh-eating bacteria) has been increasing in virulence for the last 80 years for unknown reasons.

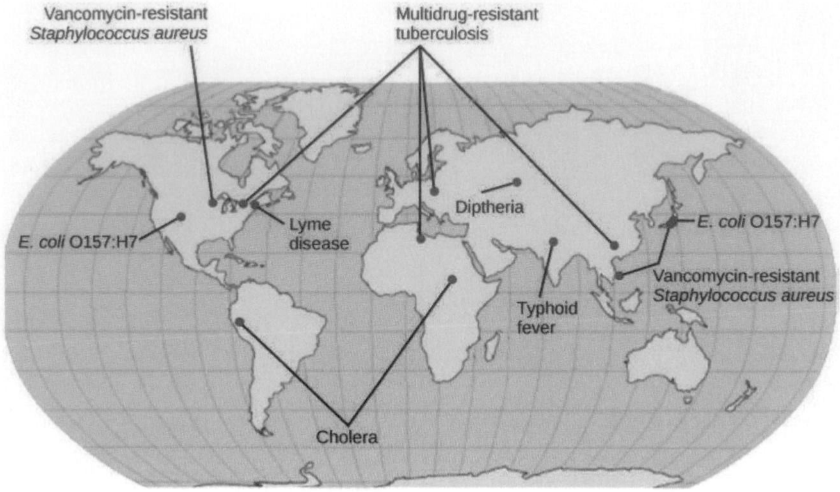

Figure 22.22 Emerging diseases. The map shows regions where bacterial diseases are emerging or re-emerging. (credit: modification of work by NIH)

Some of the present emerging diseases are not actually new, but are diseases that were catastrophic in the past (Figure 22.23). They devastated populations and became dormant for a while, just to come back, sometimes more virulent than before, as was the case with bubonic plague. Other diseases, like tuberculosis, were never eradicated but were under control in some regions of the world until coming back, mostly in urban centers with high concentrations of immunocompromised people. WHO has identified certain diseases whose worldwide re-emergence should be monitored. Among these are three viral diseases (dengue fever, yellow fever, and zika), and three bacterial diseases (diphtheria, cholera, and bubonic plague). The war against infectious diseases has no foreseeable end.

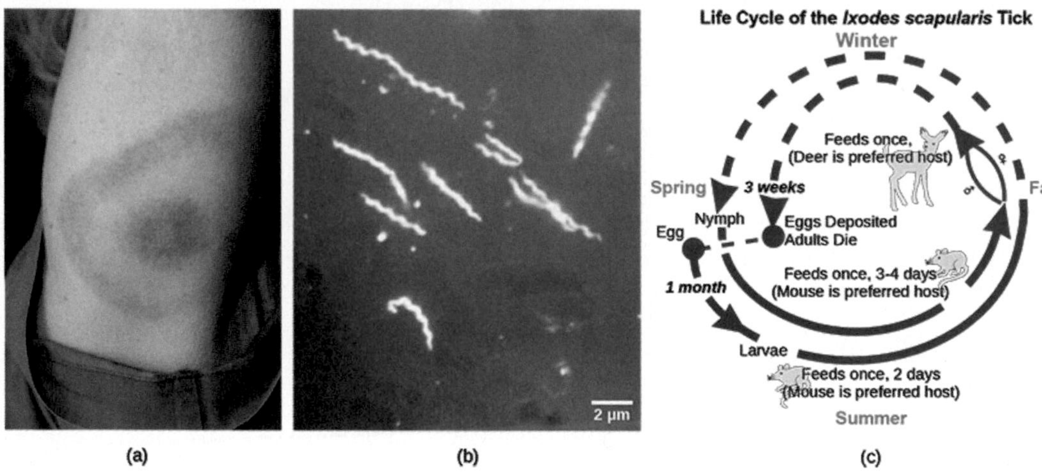

Figure 22.23 Lyme Disease. Lyme disease often, but not always, results in (a) a characteristic bullseye rash. The disease is caused by a (b) Gram-negative spirochete bacterium of the genus *Borrelia*. The bacteria (c) infect ticks, which in turn infect mice. Deer are the preferred secondary host, but the ticks also may feed on humans. Untreated, the disease causes chronic disorders in the nervous system, eyes, joints, and heart. The disease is named after Lyme, Connecticut, where an outbreak occurred in 1995 and has subsequently spread. The disease is not new, however. Genetic evidence suggests that Ötzi the Iceman, a 5,300-year-old mummy found in the Alps, was infected with *Borrelia*. (credit a: James Gathany, CDC; credit b: CDC; scale-bar data from Matt Russell)

Foodborne Diseases

Prokaryotes are everywhere: They readily colonize the surface of any type of material, and food is not an exception. Most of the time, prokaryotes colonize food and food-processing equipment in the form of a *biofilm*, as we have discussed earlier. Outbreaks of bacterial infection related to food consumption are common. A foodborne disease (commonly called "food poisoning") is an illness resulting from the consumption the pathogenic bacteria, viruses, or other parasites that contaminate food. Although the United States has one of the safest food supplies in the world, the U.S. Centers for Disease Control and Prevention (CDC) has reported that "76 million people get sick, more than 300,000 are hospitalized, and 5,000 Americans die each year from foodborne illness."

The characteristics of foodborne illnesses have changed over time. In the past, it was relatively common to hear about sporadic cases of botulism, the potentially fatal disease produced by a toxin from the anaerobic bacterium *Clostridium botulinum*. Some of the most common sources for this bacterium were non-acidic canned foods, homemade pickles, and processed meat and sausages. The can, jar, or package created a suitable anaerobic environment where *Clostridium* could grow. Proper sterilization and canning procedures have reduced the incidence of this disease.

While people may tend to think of foodborne illnesses as associated with animal-based foods, most cases are now linked to produce. There have been serious, produce-related outbreaks associated with raw spinach in the United States and with vegetable sprouts in Germany, and these types of outbreaks have become more common. The raw spinach outbreak in 2006 was produced by the bacterium *E. coli* serotype O157:H7. A serotype is a strain of bacteria that carries a set of similar antigens on its cell surface, and there are often many different serotypes of a bacterial species. Most *E. coli* are not particularly dangerous to humans, but serotype O157:H7 can cause bloody diarrhea and is potentially fatal.

All types of food can potentially be contaminated with bacteria. Recent outbreaks of *Salmonella* reported by the CDC occurred in foods as diverse as peanut butter, alfalfa sprouts, and eggs. A deadly outbreak in Germany in 2010 was caused by *E. coli* contamination of vegetable sprouts (Figure 22.24). The strain that caused the outbreak was found to be a new serotype not previously involved in other outbreaks, which indicates that *E. coli* is continuously evolving. Outbreaks of listeriosis, due to contamination of meats, raw cheeses, and frozen or fresh vegetables with *Listeria monocytogenes*, are becoming more frequent.

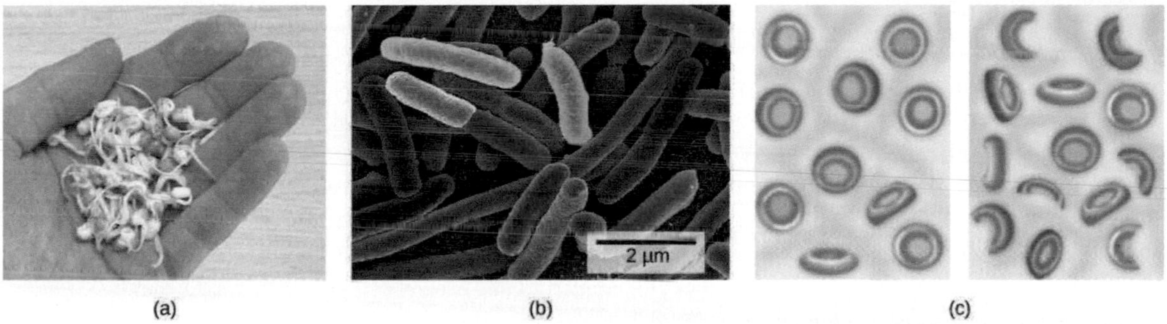

Figure 22.24 Foodborne pathogens. (a) Vegetable sprouts grown at an organic farm were the cause of an (b) *E. coli* outbreak that killed 32 people and sickened 3,800 in Germany in 2011. The strain responsible, *E. coli* O104:H4, produces Shiga toxin, a substance that inhibits protein synthesis in the host cell. The toxin (c) destroys red blood cells resulting in bloody diarrhea. Deformed red blood cells clog the capillaries of the kidney, which can lead to kidney failure, as happened to 845 patients in the 2011 outbreak. Kidney failure is usually reversible, but some patients experience kidney problems years later. (credit c: NIDDK, NIH)

Biofilms and Disease

Recall that biofilms are microbial communities that are very difficult to destroy. They are responsible for diseases such as Legionnaires' disease, otitis media (ear infections), and various infections in patients with cystic fibrosis. They produce dental plaque and colonize catheters, prostheses, transcutaneous and orthopedic devices, contact lenses, and internal devices such as pacemakers. They also form in open wounds and burned tissue. In healthcare environments, biofilms grow on hemodialysis machines, mechanical ventilators, shunts, and other medical equipment. In fact, 65 percent of all infections acquired in the hospital (nosocomial infections) are attributed to biofilms. Biofilms are also related to diseases contracted from food because they colonize the surfaces of vegetable leaves and meat, as well as food-processing equipment that isn't adequately cleaned.

Biofilm infections develop gradually and may not cause immediate symptoms. They are rarely resolved by host defense mechanisms. Once an infection by a biofilm is established, it is very difficult to eradicate, because biofilms tend to be resistant to most methods used to control microbial growth, including antibiotics. The matrix that attaches the cells to a substrate and to other another protects the cells from antibiotics or drugs. In addition, since biofilms grow slowly, they are less responsive to agents that interfere with cell growth. It has been reported that biofilms can resist up to 1,000 times the antibiotic concentrations used to kill the same bacteria when they are free-living or planktonic. An antibiotic dose that large would harm the patient; therefore, scientists are working on new ways to get rid of biofilms.

Antibiotics: Are We Facing a Crisis?

The word *antibiotic* comes from the Greek *anti* meaning "against" and *bios* meaning "life." An **antibiotic** is a chemical, produced either by microbes or synthetically, that is hostile to or prevents the growth of other organisms. Today's media often address concerns about an antibiotic crisis. Are the antibiotics that easily treated bacterial infections in the past becoming obsolete? Are there new "superbugs"—bacteria that have evolved to become more resistant to our arsenal of antibiotics? Is this the beginning of the end of antibiotics? All these questions challenge the healthcare community.

One of the main causes of antibiotic resistance in bacteria is overexposure to antibiotics. The imprudent and excessive use of antibiotics has resulted in the natural selection of resistant forms of bacteria. The antibiotic kills most of the infecting bacteria, and therefore only the resistant forms remain. These resistant forms reproduce, resulting in an increase in the proportion of resistant forms over non-resistant ones. In addition to transmission of resistance genes to progeny, lateral transfer of resistance genes on plasmids can rapidly spread these genes through a bacterial population. A major misuse of antibiotics is in patients with viral infections like colds or the flu, against which antibiotics are useless. Another problem is the excessive use of antibiotics in livestock. The routine use of antibiotics in animal feed promotes bacterial resistance as well. In the United States, 70 percent of the antibiotics produced are fed to animals. These antibiotics are given to livestock in low doses, which maximize the probability of resistance developing, and these resistant bacteria are readily transferred to humans.

LINK TO LEARNING

Watch a recent news report (http://openstax.org/l/antibiotics) on the problem of routine antibiotic administration to livestock and antibiotic-resistant bacteria.

One of the Superbugs: MRSA

The imprudent use of antibiotics has paved the way for the expansion of resistant bacterial populations. For example, *Staphylococcus aureus*, often called "staph," is a common bacterium that can live in the human body and is usually easily treated with antibiotics. However, a very dangerous strain, **methicillin-resistant *Staphylococcus aureus* (MRSA)** has made the news over the past few years (Figure 22.25). This strain is resistant to many commonly used antibiotics, including methicillin, amoxicillin, penicillin, and oxacillin. MRSA can cause infections of the skin, but it can also infect the bloodstream, lungs, urinary tract, or sites of injury. While MRSA infections are common among people in healthcare facilities, they have also appeared in healthy people who haven't been hospitalized, but who live or work in tight populations (like military personnel and prisoners). Researchers have expressed concern about the way this latter source of MRSA targets a much younger population than those residing in care facilities. *The Journal of the American Medical Association* reported that, among MRSA-afflicted persons in healthcare facilities, the average age is 68, whereas people with "community-associated MRSA" (**CA-MRSA**) have an average age of 23.[4]

4Naimi, TS, LeDell, KH, Como-Sabetti, K, et al. Comparison of community- and health care-associated methicillin-resistant *Staphylococcus aureus* infection. *JAMA* 290 (2003): 2976–84, doi: 10.1001/jama.290.22.2976.

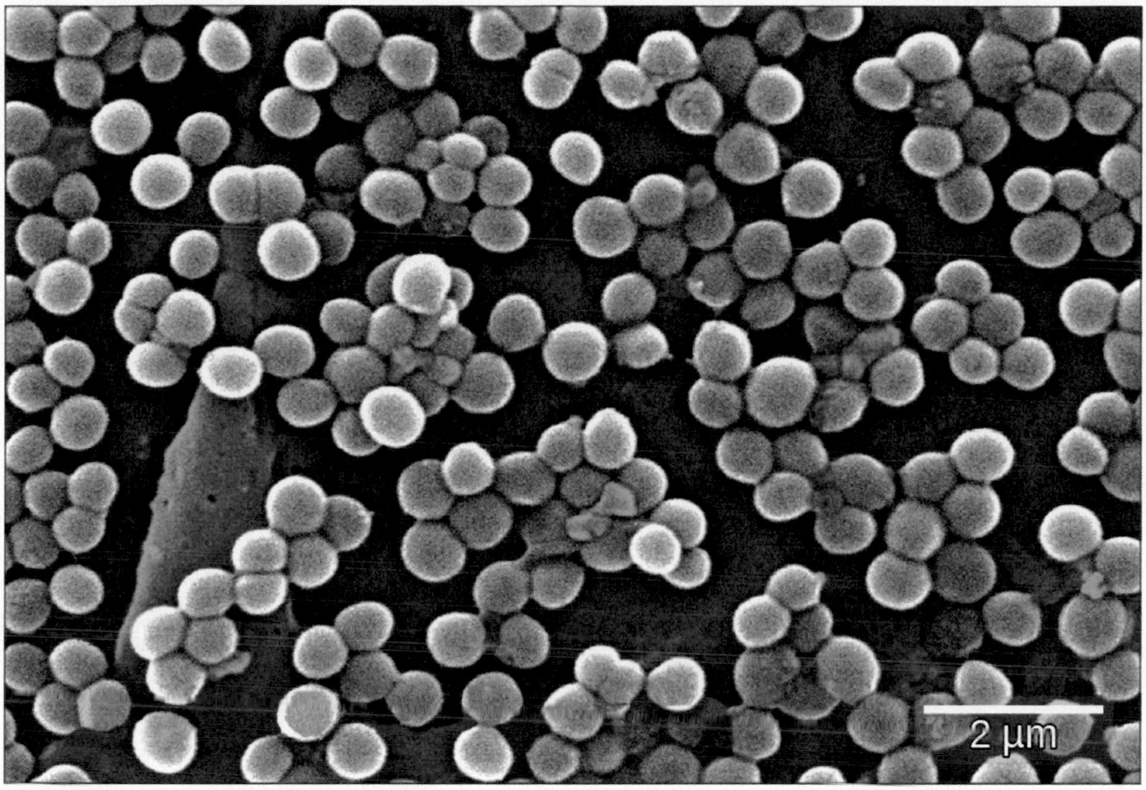

Figure 22.25 MRSA. This scanning electron micrograph shows methicillin-resistant *Staphylococcus aureus* bacteria, commonly known as MRSA. *S. aureus* is not always pathogenic, but can cause diseases such as food poisoning and skin and respiratory infections. (credit: modification of work by Janice Haney Carr; scale-bar data from Matt Russell)

In summary, the medical community is facing an antibiotic crisis. Some scientists believe that after years of being protected from bacterial infections by antibiotics, we may be returning to a time in which a simple bacterial infection could again devastate the human population. Researchers are developing new antibiotics, but it takes many years of research and clinical trials, plus financial investments in the millions of dollars, to generate an effective and approved drug.

CAREER CONNECTION

Epidemiologist

Epidemiology is the study of the occurrence, distribution, and determinants of health and disease in a population. It is, therefore, part of public health. An epidemiologist studies the frequency and distribution of diseases within human populations and environments.

Epidemiologists collect data about a particular disease and track its spread to identify the original mode of transmission. They sometimes work in close collaboration with historians to try to understand the way a disease evolved geographically and over time, tracking the natural history of pathogens. They gather information from clinical records, patient interviews, surveillance, and any other available means. That information is used to develop strategies, such as vaccinations (Figure 22.26), and design public health policies to reduce the incidence of a disease or to prevent its spread. Epidemiologists also conduct rapid investigations in case of an outbreak to recommend immediate measures to control it.

An epidemiologist has a bachelor's degree, plus a master's degree in public health (MPH). Many epidemiologists are also physicians (and have an M.D. or D.O degree), or they have a Ph.D. in an associated field, such as biology or microbiology.

Figure 22.26 Vaccination. Vaccinations can slow the spread of communicable diseases. (credit: modification of work by Daniel Paquet)

22.5 Beneficial Prokaryotes

By the end of this section, you will be able to do the following:

- Explain the need for nitrogen fixation and how it is accomplished
- Describe the beneficial effects of bacteria that colonize our skin and digestive tracts
- Identify prokaryotes used during the processing of food
- Describe the use of prokaryotes in bioremediation

Fortunately, only a few species of prokaryotes are pathogenic! Prokaryotes also interact with humans and other organisms in a number of ways that are beneficial. For example, prokaryotes are major participants in the carbon and nitrogen cycles. They produce or process nutrients in the digestive tracts of humans and other animals. Prokaryotes are used in the production of some human foods, and also have been recruited for the degradation of hazardous materials. In fact, our life would not be possible without prokaryotes!

Cooperation between Bacteria and Eukaryotes: Nitrogen Fixation

Nitrogen is a very important element to living things, because it is part of nucleotides and amino acids that are the building blocks of nucleic acids and proteins, respectively. Nitrogen is usually the most limiting element in terrestrial ecosystems, with atmospheric nitrogen, N_2, providing the largest pool of available nitrogen. However, eukaryotes cannot use atmospheric, gaseous nitrogen to synthesize macromolecules. Fortunately, nitrogen can be “fixed,” meaning it is converted into a more accessible form—ammonia (NH_3)—either biologically or abiotically.

Abiotic nitrogen fixation occurs as a result of physical processes such as lightning or by industrial processes. **Biological nitrogen fixation** (BNF) is exclusively carried out by prokaryotes: soil bacteria, cyanobacteria, and *Frankia* spp. (filamentous bacteria interacting with actinorhizal plants such as alder, bayberry, and sweet fern). After photosynthesis, BNF is the most important biological process on Earth. The overall nitrogen fixation equation below represents a series of *redox reactions* (Pi stands for inorganic phosphate).

$$N_2 + 16ATP + 8e^- + 8H^+ \rightarrow 2NH_3 + 16ADP + 16Pi + H_2$$

The total fixed nitrogen through BNF is about 100 to 180 million metric tons per year, which contributes about 65 percent of the nitrogen used in agriculture.

Cyanobacteria are the most important nitrogen fixers in aquatic environments. In soil, members of the genera *Clostridium* and *Azotobacter* are examples of free-living, nitrogen-fixing bacteria. Other bacteria live symbiotically with legume plants, providing the most important source of fixed nitrogen. Symbionts may fix more nitrogen in soils than free-living organisms by a factor of 10. Soil bacteria, collectively called rhizobia, are able to symbiotically interact with legumes to form **nodules**, specialized structures where nitrogen fixation occurs (Figure 22.27). *Nitrogenase*, the enzyme that fixes nitrogen, is inactivated by oxygen, so the nodule provides an oxygen-free area for nitrogen fixation to take place. The oxygen is sequestered by a form of plant hemoglobin called *leghemoglobin*, which protects the *nitrogenase*, but releases enough oxygen to support respiratory activity.

Symbiotic nitrogen fixation provides a natural and inexpensive plant fertilizer: It reduces atmospheric nitrogen to ammonia, which is easily usable by plants. The use of legumes is an excellent alternative to chemical fertilization and is of special interest to *sustainable agriculture*, which seeks to minimize the use of chemicals and conserve natural resources. Through symbiotic nitrogen fixation, the plant benefits from using an endless source of nitrogen: the atmosphere. The bacteria benefit from using photosynthates (carbohydrates produced during photosynthesis) from the plant and having a protected niche. In addition, the soil benefits from being naturally fertilized. Therefore, the use of rhizobia as biofertilizers is a sustainable practice.

Why are legumes so important? Some, like soybeans, are key sources of agricultural protein. Some of the most important legumes consumed by humans are soybeans, peanuts, peas, chickpeas, and beans. Other legumes, such as alfalfa, are used to feed cattle.

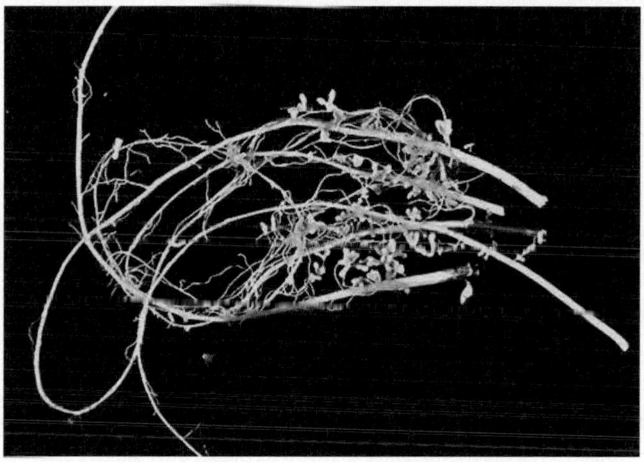

Figure 22.27 Nitrogen-fixation nodules on soybean roots. Soybean (*Glycine max*) is a legume that interacts symbiotically with the soil bacterium *Bradyrhizobium japonicum* to form specialized structures on the roots called *nodules* where nitrogen fixation occurs. (credit: USDA)

Everyday Connection

Microbes on the Human Body

The commensal bacteria that inhabit our skin and gastrointestinal tract do a host of good things for us. They protect us from pathogens, help us digest our food, and produce some of our vitamins and other nutrients. These activities have been known for a long time. More recently, scientists have gathered evidence that these bacteria may also help regulate our moods, influence our activity levels, and even help control weight by affecting our food choices and absorption patterns. The Human Microbiome Project has begun the process of cataloging our normal bacteria (and archaea) so we can better understand these functions.

A particularly fascinating example of our normal flora relates to our digestive systems. People who take high doses of antibiotics tend to lose many of their normal gut bacteria, allowing a naturally antibiotic-resistant species called *Clostridium difficile* to overgrow and cause severe gastric problems, especially chronic diarrhea (Figure 22.28). Obviously, trying to treat this problem with antibiotics only makes it worse. However, it has been successfully treated by giving the patients fecal transplants from healthy donors to reestablish the normal intestinal microbial community. Clinical trials are underway to ensure the safety and effectiveness of this technique.

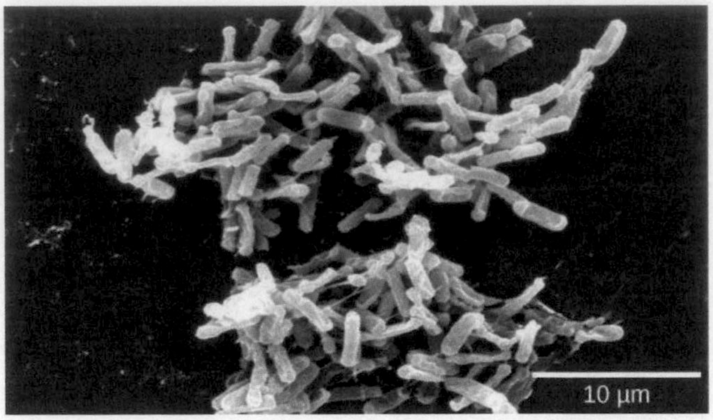

Figure 22.28 *Clostridium difficile*. This scanning electron micrograph shows *Clostridium difficile*, a Gram-positive, rod-shaped bacterium that causes severe diarrhea. Infection commonly occurs after the normal gut fauna are eradicated by antibiotics, and in the hospital can be deadly to seriously ill patients. (credit: modification of work by CDC, HHS; scale-bar data from Matt Russell)

Scientists are also discovering that the absence of certain key microbes from our intestinal tract may set us up for a variety of problems. This seems to be particularly true regarding the appropriate functioning of the immune system. There are intriguing findings that suggest that the absence of these microbes is an important contributor to the development of allergies and some autoimmune disorders. Research is currently underway to test whether adding certain microbes to our internal ecosystem may help in the treatment of these problems, as well as in treating some forms of autism.

Early Biotechnology: Cheese, Bread, Wine, Beer, and Yogurt

According to the United Nations Convention on Biological Diversity, **biotechnology** is “any technological application that uses biological systems, living organisms, or derivatives thereof, to make or modify products or processes for specific use."[5] The concept of “specific use” involves some sort of commercial application. Genetic engineering, artificial selection, antibiotic production, and cell culture are current topics of study in biotechnology and will be described in later chapters. However, humans were using prokaryotes before the term biotechnology was even coined. Some of the products of this early biotechnology are as familiar as cheese, bread, wine, beer, and yogurt, which employ both bacteria and other microbes, such as yeast, a fungus (Figure 22.29).

5http://www.cbd.int/convention/articles/?a=cbd-02 (http://openstax.org/l/UN_convention) , United Nations Convention on Biological Diversity: Article 2: Use of Terms.

Figure 22.29 Some foods produced by microorganisms. Some of the products derived from the use of prokaryotes in early biotechnology include (a) cheese, (b) wine, (c) beer and bread, and (d) yogurt. (credit bread: modification of work by F. Rodrigo/Wikimedia Commons; credit wine: modification of work by Jon Sullivan; credit beer and bread: modification of work by Kris Miller; credit yogurt: modification of work by Jon Sullivan)

Cheese production began around 4,000 to 7,000 years ago when humans began to breed animals and process their milk. Fermentation in this case preserves nutrients: Milk will spoil relatively quickly, but when processed as cheese, it is more stable. As for beer, the oldest records of brewing are about 6,000 years old and were an integral part of the Sumerian culture. Evidence indicates that the Sumerians discovered fermentation by chance. Wine has been produced for about 4,500 years, and evidence suggests that cultured milk products, like yogurt, have existed for at least 4,000 years.

Using Prokaryotes to Clean up Our Planet: Bioremediation

Microbial **bioremediation** is the use of prokaryotes (or microbial metabolism) to remove pollutants. Bioremediation has been used to remove agricultural chemicals (e.g., pesticides, fertilizers) that leach from soil into groundwater and the subsurface. Certain toxic metals and oxides, such as selenium and arsenic compounds, can also be removed from water by bioremediation. The reduction of SeO_4^{-2} to SeO_3^{-2} and to Se^{0} (metallic selenium) is a method used to remove selenium ions from water. Mercury (Hg) is an example of a toxic metal that can be removed from an environment by bioremediation. As an active ingredient of some pesticides, mercury is used in industry and is also a by-product of certain processes, such as battery production. Methyl mercury is usually present in very low concentrations in natural environments, but it is highly toxic because it accumulates in living tissues. Several species of bacteria can carry out the biotransformation of toxic mercury into nontoxic forms. These bacteria, such as *Pseudomonas aeruginosa*, can convert Hg^{+2} into Hg^{0}, which is nontoxic to humans.

One of the most useful and interesting examples of the use of prokaryotes for bioremediation purposes is the cleanup of oil spills. The significance of prokaryotes to petroleum bioremediation has been demonstrated in several oil spills in recent years, such as the Exxon Valdez spill in Alaska (1989) (Figure 22.30), the Prestige oil spill in Spain (2002), the spill into the Mediterranean from a Lebanon power plant (2006), and more recently, the BP oil spill in the Gulf of Mexico (2010). In the case of oil spills in the ocean, ongoing natural bioremediation tends to occur, since there are oil-consuming bacteria in the ocean prior to the spill. In addition to these naturally occurring oil-degrading bacteria, humans select and engineer bacteria that possess the same capability with increased efficacy and spectrum of hydrocarbon compounds that can be processed. Bioremediation is enhanced by the addition of inorganic nutrients that help bacteria to grow.

Some hydrocarbon-degrading bacteria feed on hydrocarbons in the oil droplet, breaking down the hydrocarbons into smaller subunits. Some species, such as *Alcanivorax borkumensis*, produce surfactants that *solubilize* the oil (making it soluble in water), whereas other bacteria degrade the oil into carbon dioxide. Under ideal conditions, it has been reported that up to 80 percent of the nonvolatile components in oil can be degraded within one year of the spill. Other oil fractions containing aromatic and highly branched hydrocarbon chains are more difficult to remove and remain in the environment for longer periods of time.

Figure 22.30 Prokaryotes and bioremediation. (a) Cleaning up oil after the Exxon Valdez spill in Alaska, workers hosed oil from beaches and then used a floating boom to corral the oil, which was finally skimmed from the water surface. Some species of bacteria are able to solubilize and degrade the oil. (b) One of the most catastrophic consequences of oil spills is the damage to fauna. (credit a: modification of work by NOAA; credit b: modification of work by GOLUBENKOV, NGO: Saving Taman)

KEY TERMS

acidophile organism with optimal growth pH of three or below

alkaliphile organism with optimal growth pH of nine or above

ammonification process by which ammonia is released during the decomposition of nitrogen-containing organic compounds

anaerobic refers to organisms that grow without oxygen

anoxic without oxygen

antibiotic biological substance that, in low concentration, is antagonistic to the growth of prokaryotes

biofilm microbial community that is held together by a gummy-textured matrix

biological nitrogen fixation conversion of atmospheric nitrogen into ammonia exclusively carried out by prokaryotes

bioremediation use of microbial metabolism to remove pollutants

biotechnology any technological application that uses living organisms, biological systems, or their derivatives to produce or modify other products

Black Death devastating pandemic that is believed to have been an outbreak of bubonic plague caused by the bacterium *Yersinia pestis*

botulism disease produced by the toxin of the anaerobic bacterium *Clostridium botulinum*

CA-MRSA MRSA acquired in the community rather than in a hospital

capsule external structure that enables a prokaryote to attach to surfaces and protects it from dehydration

chemotroph organism that obtains energy from chemical compounds

conjugation process by which prokaryotes move DNA from one individual to another using a pilus

cyanobacteria bacteria that evolved from early phototrophs and oxygenated the atmosphere; also known as blue-green algae

decomposer organism that carries out the decomposition of dead organisms

denitrification transformation of nitrate from soil to gaseous nitrogen compounds such as N_2O, NO, and N_2

emerging disease disease making an initial appearance in a population or that is increasing in incidence or geographic range

endemic disease disease that is constantly present, usually at low incidence, in a population

epidemic disease that occurs in an unusually high number of individuals in a population at the same time

extremophile organism that grows under extreme or harsh conditions

foodborne disease any illness resulting from the consumption of contaminated food, or of the pathogenic bacteria, viruses, or other parasites that contaminate food

Gram negative bacterium whose cell wall contains little peptidoglycan but has an outer membrane

Gram positive bacterium that contains mainly peptidoglycan in its cell walls

halophile organism that require a salt concentration of at least 0.2 M

hydrothermal vent fissure in Earth's surface that releases geothermally heated water

hyperthermophile organism that grows at temperatures between 80–122 °C

microbial mat multi-layered sheet of prokaryotes that may include bacteria and archaea

MRSA (methicillin-resistant *Staphylococcus aureus*) very dangerous *Staphylococcus aureus* strain resistant to multiple antibiotics

nitrification conversion of ammonium into nitrite and nitrate in soils

nitrogen fixation process by which gaseous nitrogen is transformed, or "fixed" into more readily available forms such as ammonia

nodule novel structure on the roots of certain plants (legumes) that results from the symbiotic interaction between the plant and soil bacteria, and is the site of nitrogen fixation

nutrient essential substances for growth, such as carbon and nitrogen

osmophile organism that grows in a high sugar concentration

pandemic widespread, usually worldwide, epidemic disease

peptidoglycan material composed of polysaccharide chains cross-linked to unusual peptides

phototroph organism that is able to make its own food by converting solar energy to chemical energy

pilus surface appendage of some prokaryotes used for attachment to surfaces including other prokaryotes

pseudopeptidoglycan component of archaea cell walls that is similar to peptidoglycan in morphology but contains different sugars

psychrophile organism that grows at temperatures of -15 °C or lower

radioresistant organism that grows in high levels of radiation

resuscitation process by which prokaryotes that are in the VBNC state return to viability

S-layer surface-layer protein present on the outside of cell walls of archaea and bacteria

serotype strain of bacterium that carries a set of similar antigens on its cell surface, often many in a bacterial species

stromatolite layered sedimentary structure formed by precipitation of minerals by prokaryotes in microbial mats

teichoic acid polymer associated with the cell wall of Gram-positive bacteria

thermophile organism that lives at temperatures between 60–80 °C

transduction process by which a bacteriophage moves DNA from one prokaryote to another

transformation process by which a prokaryote takes in DNA found in its environment that is shed by other prokaryotes

viable-but-non-culturable (VBNC) state survival mechanism of bacteria facing environmental stress conditions

zoonosis disease that primarily infects animals that is transmitted to humans

CHAPTER SUMMARY

22.1 Prokaryotic Diversity

Prokaryotes existed for billions of years before plants and animals appeared. Hot springs and hydrothermal vents may have been the environments in which life began. Microbial mats are thought to represent the earliest forms of life on Earth. A microbial mat is a multi-layered sheet of prokaryotes that grows at interfaces between different types of material, mostly on moist surfaces. Fossilized microbial mats are called stromatolites and consist of laminated organo-sedimentary structures formed by precipitation of minerals by prokaryotes. They represent the earliest fossil record of life on Earth.

During the first two billion years, the atmosphere was anoxic and only anaerobic organisms were able to live. Cyanobacteria evolved from early phototrophs and began the oxygenation o the atmosphere. The increase in oxygen concentration allowed the evolution of other life forms.

Bacteria and archaea grow in virtually every environment. Those that survive under extreme conditions are called extremophiles (extreme lovers). Some prokaryotes cannot grow in a laboratory setting, but they are not dead. They are in the viable-but-non-culturable (VBNC) state. The VBNC state occurs when prokaryotes enter a dormant state in response to environmental stressors. Most prokaryotes are colonial and prefer to live in communities where interactions take place. A biofilm is a microbial community held together in a gummy-textured matrix.

22.2 Structure of Prokaryotes: Bacteria and Archaea

Prokaryotes (domains Archaea and Bacteria) are single-celled organisms that lack a nucleus. They have a single piece of circular DNA in the nucleoid area of the cell. Most prokaryotes have a cell wall that lies outside the boundary of the plasma membrane. Some prokaryotes may have additional structures such as a capsule, flagella, and pili. Bacteria and Archaea differ in the lipid composition of their cell membranes and the characteristics of the cell wall. In archaeal membranes, phytanyl units, rather than fatty acids, are linked to glycerol. Some archaeal membranes are lipid monolayers instead of bilayers.

The cell wall is located outside the cell membrane and prevents osmotic lysis. The chemical composition of cell walls varies between species. Bacterial cell walls contain peptidoglycan. Archaean cell walls do not have peptidoglycan, but they may have pseudopeptidoglycan, polysaccharides, glycoproteins, or protein-based cell walls. Bacteria can be divided into two major groups: Gram positive and Gram negative, based on the Gram stain reaction. Gram-positive organisms have a thick peptidoglycan layer fortified with teichoic acids. Gram-negative organisms have a thin cell wall and an outer envelope containing lipopolysaccharides and lipoproteins.

Prokaryotes can transfer DNA from one cell to another by three mechanisms: transformation (uptake of environmental DNA), transduction (transfer of genomic DNA via viruses), and conjugation (transfer of DNA by direct cell contact).

22.3 Prokaryotic Metabolism

As the oldest living inhabitants of Earth, prokaryotes are also the most metabolically diverse; they flourish in many different environments with various energy and carbon sources, variable temperature, pH, pressure, oxygen and water availability. Nutrients required in large amounts are called macronutrients, whereas those required in trace amounts are called micronutrients or trace elements. Macronutrients include C, H, O, N, P, S, K, Mg, Ca, and Na. In addition to these macronutrients, prokaryotes require various metallic elements for growth and enzyme function. Prokaryotes use different sources of energy to assemble macromolecules from smaller molecules. Phototrophs obtain their energy from sunlight, whereas chemotrophs obtain energy from chemical compounds. Energy-producing pathways may be either aerobic or anaerobic.

Prokaryotes play roles in the carbon and nitrogen cycles. Producers capture carbon dioxide from the atmosphere and convert it to organic compounds. Consumers (animals and other chemoorganotrophic organisms) use organic compounds generated by producers and release carbon dioxide into the atmosphere by respiration. Carbon dioxide is also returned to the atmosphere by the microbial decomposers of dead organisms. Nitrogen also cycles in and out of living organisms, from organic compounds to ammonia, ammonium ions, nitrite, nitrate, and nitrogen gas. Prokaryotes are essential for most of these conversions. Gaseous nitrogen is transformed into ammonia through nitrogen fixation. Ammonia is anaerobically catabolized by some prokaryotes, yielding N_2 as the final product. Nitrification is the conversion of ammonium into nitrite. Nitrification in soils is carried out by bacteria. Denitrification is also performed by bacteria and transforms nitrate from soils into gaseous nitrogen compounds, such as N_2O, NO, and N_2.

22.4 Bacterial Diseases in Humans

Some prokaryotes are human pathogens. Devastating diseases and plagues have been among us since early times and remain among the leading causes of death worldwide. Emerging diseases are those rapidly increasing in incidence or geographic range. They can be new or re-emerging diseases (previously under control). Many emerging diseases affecting humans originate in animals (zoonoses), such as brucellosis. A group of re-emerging bacterial diseases recently identified by WHO for monitoring include bubonic plague, diphtheria, and cholera. Foodborne diseases result from the consumption of food contaminated with food, pathogenic bacteria, viruses, or parasites.

Some bacterial infections have been associated with biofilms: Legionnaires' disease, otitis media, and infection of patients with cystic fibrosis. Biofilms can grow on human tissues, like dental plaque; colonize medical devices; and cause infection or produce foodborne disease by growing on the surfaces of food and food-processing equipment. Biofilms are resistant to most of the methods used to control microbial growth. The excessive use of antibiotics has resulted in a major global problem, since resistant forms of bacteria have been selected over time. A very dangerous strain, methicillin-resistant *Staphylococcus aureus* (MRSA), has wreaked havoc recently across the world.

22.5 Beneficial Prokaryotes

Pathogens are only a small percentage of all prokaryotes. In fact, prokaryotes provide essential services to humans and other organisms. Nitrogen, which is not usable by eukaryotes in its plentiful atmospheric form, can be "fixed," or converted into ammonia (NH_3) either biologically or abiotically. Biological nitrogen fixation (BNF) is exclusively carried out by prokaryotes, and constitutes the second most important biological process on Earth. Although some terrestrial nitrogen is fixed by free-living bacteria, most BNF comes from the symbiotic interaction between soil rhizobia and the roots of legume plants.

Human life is only possible due to the action of microbes, both those in the environment and those species that call us home. Internally, they help us digest our food, produce vital nutrients for us, protect us from pathogenic microbes, and help train our immune systems to function properly.

Microbial bioremediation is the use of microbial metabolism to remove pollutants. Bioremediation has been used to remove agricultural chemicals that leach from soil into groundwater and the subsurface. Toxic metals and oxides, such as selenium and arsenic compounds, can also be removed by bioremediation. Probably one of the most useful and interesting examples of the use of prokaryotes for bioremediation purposes is the cleanup of oil spills.

VISUAL CONNECTION QUESTIONS

1. Figure 22.8 Compared to free-floating bacteria, bacteria in biofilms often show increased resistance to antibiotics and detergents. Why do you think this might be the case?

2. Figure 22.16 Which of the following statements is true?
   a. Gram-positive bacteria have a single cell wall anchored to the cell membrane by lipoteichoic acid.
   b. Porins allow entry of substances into both Gram-positive and Gram-negative bacteria.
   c. The cell wall of Gram-negative bacteria is thick, and the cell wall of Gram-positive bacteria is thin.
   d. Gram-negative bacteria have a cell wall made of peptidoglycan, whereas Gram-positive bacteria have a cell wall made of lipoteichoic acid.

3. Figure 22.19 Which of the following statements about the nitrogen cycle is false?
   a. Nitrogen fixing bacteria exist on the root nodules of legumes and in the soil.
   b. Denitrifying bacteria convert nitrates (NO_3^-) into nitrogen gas (N_2).
   c. Ammonification is the process by which ammonium ion (NH_4^+) is released from decomposing organic compounds.
   d. Nitrification is the process by which nitrites (NO_2^-) are converted to ammonium ion (NH_4^+).

REVIEW QUESTIONS

4. The first forms of life on Earth were thought to be________.
   a. single-celled plants
   b. prokaryotes
   c. insects
   d. large animals such as dinosaurs

5. Microbial mats _________.
   a. are the earliest forms of life on Earth
   b. obtained their energy and food from hydrothermal vents
   c. are multi-layered sheets of prokaryotes including mostly bacteria but also archaea
   d. all of the above

6. The first organisms that oxygenated the atmosphere were
   a. cyanobacteria
   b. phototrophic organisms
   c. anaerobic organisms
   d. all of the above

7. Halophiles are organisms that require________.
   a. a salt concentration of at least 0.2 M
   b. high sugar concentration
   c. the addition of halogens
   d. all of the above

8. Many of the first prokaryotes to be cultured in a scientific lab were human or animal pathogens. Why would these species be more readily cultured than non-pathogenic prokaryotes?
   a. Pathogenic prokaryotes are hardier than non-pathogenic prokaryotes.
   b. Non-pathogenic prokaryotes require more supplements in their growth media.
   c. Most of the necessary culture conditions could be inferred for pathogenic prokaryotes.
   d. Pathogenic bacteria can grow as free bacteria, but non-pathogenic bacteria only grow as parts of large colonies.

9. The presence of a membrane-enclosed nucleus is a characteristic of_______.
   a. prokaryotic cells
   b. eukaryotic cells
   c. all cells
   d. viruses

10. Which of the following consist of prokaryotic cells?
    a. bacteria and fungi
    b. archaea and fungi
    c. protists and animals
    d. bacteria and archaea

11. The cell wall is _______.
    a. interior to the cell membrane
    b. exterior to the cell membrane
    c. a part of the cell membrane
    d. interior or exterior, depending on the particular cell

12. Organisms most likely to be found in extreme environments are _______.
    a. fungi
    b. bacteria
    c. viruses
    d. archaea

13. Prokaryotes stain as Gram-positive or Gram-negative because of differences in the cell ______.
    a. wall
    b. cytoplasm
    c. nucleus
    d. chromosome

14. Pseudopeptidoglycan is a characteristic of the walls of _______.
    a. eukaryotic cells
    b. bacterial prokaryotic cells
    c. archaean prokaryotic cells
    d. bacterial and archaean prokaryotic cells

15. The lipopolysaccharide layer (LPS) is a characteristic of the wall of_______.
    a. archaean cells
    b. Gram-negative bacteria
    c. bacterial prokaryotic cells
    d. eukaryotic cells

16. Which of the following elements is *not* a micronutrient?
    a. boron
    b. calcium
    c. chromium
    d. manganese

17. Prokaryotes that obtain their energy from chemical compounds are called ____.
    a. phototrophs
    b. auxotrophs
    c. chemotrophs
    d. lithotrophs

18. Ammonification is the process by which _____.
    a. ammonia is released during the decomposition of nitrogen-containing organic compounds
    b. ammonium is converted to nitrite and nitrate in soils
    c. nitrate from soil is transformed to gaseous nitrogen compounds such as NO, N_2O, and N_2
    d. gaseous nitrogen is fixed to yield ammonia

19. Plants use carbon dioxide from the air and are therefore called _____.
    a. consumers
    b. producers
    c. decomposer
    d. carbon fixers

20. Cyanobacteria harness energy from the sun through photosynthesis, and oxidize water to provide electrons for energy generation. Thus, we classify cyanobacteria as ________.
    a. photolithotrophs
    b. photoautotrophs
    c. chemolithoautotrophs
    d. chemo-organotrophs

21. A disease that is constantly present in a population is called _____.
    a. pandemic
    b. epidemic
    c. endemic
    d. re-emerging

22. Which of the statements about biofilms is incorrect?
    a. Biofilms are considered responsible for diseases such as cystic fibrosis.
    b. Biofilms produce dental plaque, and colonize catheters and prostheses.
    c. Biofilms colonize open wounds and burned tissue.
    d. All statements are incorrect.

23. Which of these statements is true?
    a. An antibiotic is any substance produced by a organism that is antagonistic to the growth of prokaryotes.
    b. An antibiotic is any substance produced by a prokaryote that is antagonistic to the growth of other viruses.
    c. An antibiotic is any substance produced by a prokaryote that is antagonistic to the growth of eukaryotic cells.
    d. An antibiotic is any substance produced by a prokaryote that prevents growth of the same prokaryote.

24. A person in England arrives at a medical clinic with a fever and swollen lymph nodes shortly after returning from a visit to New Mexico. For which bacteria should the doctor test the patient?
    a. *Salmonella enterica*
    b. *Borrelia burgdorferi*
    c. *Clostridium botulinum*
    d. *Yersinia pestis*

25. MRSA has emerged as a serious infectious disease, with the first case of methicillin-resistant S. *aureus* being detected in 1961. Why are medical professionals so concerned when antibiotics exist that can kill MRSA?
    a. MRSA can transfer methicillin-resistance to other bacteria.
    b. Patients are not treated with correct antibiotics rapidly enough to prevent serious illness.
    c. MRSA could acquire additional antibiotic resistance genes from other bacteria to become a "super bug."
    d. All of the above.

26. Which of these occurs through symbiotic nitrogen fixation?
    a. The plant benefits from using an endless source of nitrogen.
    b. The soil benefits from being naturally fertilized.
    c. Bacteria benefit from using photosynthates from the plant.
    d. All of the above occur.

27. Synthetic compounds found in an organism but not normally produced or expected to be present in that organism are called _____.
    a. pesticides
    b. bioremediators
    c. recalcitrant compounds
    d. xenobiotics

28. Bioremediation includes _____.
    a. the use of prokaryotes that can fix nitrogen
    b. the use of prokaryotes to clean up pollutants
    c. the use of prokaryotes as natural fertilizers
    d. All of the above

29. In addition to providing yogurt with its unique flavor and texture, lactic acid-producing bacteria also provide which additional benefit during food production?
    a. Providing xenobiotics
    b. Lowering the pH to kill pathogenic bacteria
    c. Pasteurizing milk products
    d. Breaking down lactose for lactose-intolerant individuals

CRITICAL THINKING QUESTIONS

30. Describe briefly how you would detect the presence of a non-culturable prokaryote in an environmental sample.
31. Why do scientists believe that the first organisms on Earth were extremophiles?
32. A new bacterial species is discovered and classified as an endolith, an extremophile that lives inside rock. If the bacteria were discovered in the permafrost of Antarctica, describe two extremophile features the bacteria must possess.
33. Mention three differences between bacteria and archaea.
34. Explain the statement that both types, bacteria and archaea, have the same basic structures, but built from different chemical components.
35. A scientist isolates a new species of prokaryote. He notes that the specimen is a bacillus with a lipid bilayer and cell wall that stains positive for peptidoglycan. Its circular chromosome replicates from a single origin of replication. Is the specimen most likely an Archaea, a Gram-positive bacterium, or a Gram-negative bacterium? How do you know?
36. Think about the conditions (temperature, light, pressure, and organic and inorganic materials) that you may find in a deep-sea hydrothermal vent. What type of prokaryotes, in terms of their metabolic needs (autotrophs, phototrophs, chemotrophs, etc.), would you expect to find there?
37. Farmers continually rotate the crops grown in different fields to maintain nutrients in the soil. How would planting soybeans in a field the year after the field was used to grow carrots help maintain nitrogen in the soil?
38. Imagine a region of soil became contaminated, killing bacteria that decompose dead plants and animals. How would this effect the carbon cycle in the area? Be specific in stating where carbon would accumulate in the cycle.
39. Explain the reason why the imprudent and excessive use of antibiotics has resulted in a major global problem.
40. Researchers have discovered that washing spinach with water several times does not prevent foodborne diseases due to *E. coli*. How can you explain this fact?
41. Your friend believes that prokaryotes are always detrimental and pathogenic. How would you explain to them that they are wrong?
42. Many people use antimicrobial soap to kill bacteria on their hands. However, overuse may actually increase the risk of infection. How could this occur?

CHAPTER 23
Protists

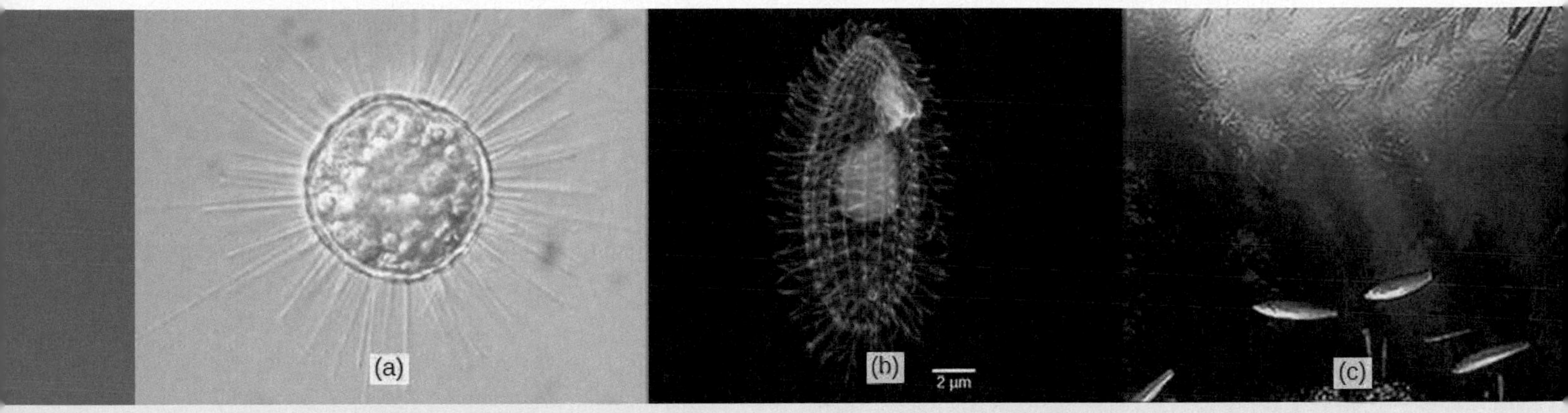

Figure 23.1 Protists range from the microscopic, single-celled (a) *Acanthocystis turfacea* and the (b) ciliate *Tetrahymena thermophila*, both visualized here using light microscopy, to the enormous, multicellular (c) kelps (Chromalveolata) that extend for hundreds of feet in underwater "forests." (credit a: modification of work by Yuluji Tsukii; credit b: modification of work by Richard Robinson, Public Library of Science; credit c: modification of work by Kip Evans, NOAA; scale-bar data from Matt Russell)

Chapter Outline

INTRODUCTION Humans have been familiar with macroscopic organisms (organisms big enough to see with the unaided eye) since before there was a written history, and it is likely that most cultures distinguished between animals and land plants, and most probably included the macroscopic fungi as plants. Therefore, it became an interesting challenge to deal with the world of microorganisms once microscopes were developed a few centuries ago. Many different naming schemes were used over the last couple of centuries, but it has become the most common practice to refer to eukaryotes that are not land plants, animals, or fungi as protists.

This name was first suggested by Ernst Haeckel in the late nineteenth century. It has been applied in many contexts and has been formally used to represent a kingdom-level taxon called Protista. However, many modern systematists (biologists who study the relationships among organisms) are beginning to shy away from the idea of formal ranks such as kingdom and phylum. Instead, they are naming taxa as groups of organisms thought to include all the descendants of a last common ancestor (monophyletic group). During the past two decades, the field of molecular genetics has demonstrated that some protists are more related to animals, plants, or fungi than they are to other protists. Therefore, not including animals, plants, and fungi make the kingdom Protista a paraphyletic group, or one that does not include all descendents of its common ancestor. For this reason, protist lineages originally classified into the kingdom Protista continue to be examined and debated. In the meantime, the term "protist" still is used informally to describe this tremendously diverse group of eukaryotes.

Most protists are microscopic, unicellular organisms that are abundant in soil, freshwater, brackish, and marine environments. They are also common in the digestive tracts of animals and in the vascular tissues of plants. Others invade the cells of other protists, animals, and plants. Not all protists are microscopic. Some have huge, macroscopic cells, such as the plasmodia (giant amoebae) of myxomycete slime molds or the marine green alga *Caulerpa*, which can have single cells that can be several meters in size. Some protists are multicellular, such as the red, green, and brown seaweeds. It is among the protists that one finds the wealth of ways that organisms can grow.

23.1 Eukaryotic Origins

By the end of this section, you will be able to do the following:

- List the unifying characteristics of eukaryotes
- Describe what scientists know about the origins of eukaryotes based on the last common ancestor
- Explain the endosymbiotic theory

Organisms are classified into three domains: Archaea, Bacteria, and Eukarya. The first two lineages comprise all prokaryotic cells, and the third contains all eukaryotes. A very sparse fossil record prevents us from determining what the first members of each of these lineages looked like, so it is possible that all the events that led to the last common ancestor of extant eukaryotes will remain unknown. However, comparative biology of extant (living) organisms and the limited fossil record provide some insight into the evolution of Eukarya.

The earliest fossils found appear to be those of domain Bacteria, most likely cyanobacteria. They are about 3.5 to 3.8 billion years old and are recognizable because of their relatively complex structure and, for prokaryotes, relatively large cells. Most other prokaryotes have small cells, 1 or 2 µm in size, and would be difficult to pick out as fossils. Fossil stromatolites suggest that at least some prokaryotes lived in interactive communities, and evidence from the structure of living eukaryotic cells suggests that it was similar ancestral interactions that gave rise to the eukaryotes. Most living eukaryotes have cells measuring 10 µm or greater. Structures this size, which might be fossilized remains of early eukaryotes, appear in the geological record in deposits dating to about 2.1 billion years ago.

Characteristics of Eukaryotes

Data from these fossils, as well as from the study of living genomes, have led comparative biologists to conclude that living eukaryotes are all descendants of a single common ancestor. Mapping the characteristics found in all major groups of eukaryotes reveals that the following characteristics are present in at least some of the members of each major lineage, or during some part of their life cycle, and therefore must have been present in the *last common ancestor.*

1. **Cells with nuclei surrounded by a nuclear envelope with nuclear pores**: This is the single characteristic that is both necessary and sufficient to define an organism as a eukaryote. All extant eukaryotes have cells with nuclei.
2. **Mitochondria**: Most extant eukaryotes have "typical" mitochondria, although some eukaryotes have very reduced mitochondrial "remnants" and a few lack detectable mitochondria.
3. **Cytoskeleton of microtubules and microfilaments**: Eukaryotic cells possess the structural and motility components called *actin microfilaments and microtubules.* All extant eukaryotes have these cytoskeletal elements.
4. **Flagella and cilia**: Organelles associated with cell motility. Some extant eukaryotes lack flagella and/or cilia, but their presence in related lineages suggests that they are descended from ancestors that possessed these organelles.
5. **Chromosomes organized by histones**: Each eukaryotic chromosome consists of a linear DNA molecule coiled around basic (alkaline) proteins called histones. The few eukaryotes with chromosomes lacking histones clearly evolved from ancestors that had them.
6. **Mitosis**: A process of nuclear division in which replicated chromosomes are divided and separated using elements of the cytoskeleton. Mitosis is universally present in eukaryotes.
7. **Sexual reproduction**: A meiotic process of nuclear division and genetic recombination unique to eukaryotes. During this process, diploid nuclei at one stage of the life cycle undergo meiosis to yield haploid nuclei, which subsequently fuse together (karyogamy) to create a diploid zygote nucleus.
8. **Cell walls**: It might be reasonable to conclude that the last common ancestor could make cell walls during some stage of its life cycle, simple because cell walls were present in their prokaryote precursors. However, not enough is known about eukaryotes' cell walls and their development to know how much homology exists between those of prokaryotes and eukaryotes. If the last common ancestor could make cell walls, it is clear that this ability must have been lost in many groups.

Endosymbiosis and the Evolution of Eukaryotes

Before we discuss the origins of eukaryotes, it is first important to understand that all extant eukaryotes are likely the descendants of a chimera-like organism that was a composite of a host cell and the cell(s) of an alpha-proteobacterium that "took up residence" inside it. This major theme in the origin of eukaryotes is known as **endosymbiosis**, one cell engulfing another such that the engulfed cell survives and both cells benefit. Over many generations, a symbiotic relationship can result in two organisms that depend on each other so completely that neither could survive on its own. Endosymbiotic events likely contributed to the origin of the last common ancestor of today's eukaryotes and to later diversification in certain lineages of eukaryotes (Figure 23.5). Similar endosymbiotic associations are not uncommon in living eukaryotes. Before explaining this further, it is necessary to consider metabolism in prokaryotes.

Prokaryotic Metabolism

Many important metabolic processes arose in prokaryotes; however, some of these processes, such as nitrogen fixation, are never found in eukaryotes. The process of aerobic respiration is found in all major lineages of eukaryotes, and it is localized in the mitochondria. Aerobic respiration is also found in many lineages of prokaryotes, but it is not present in all of them, and a great deal of evidence suggests that such anaerobic prokaryotes never carried out aerobic respiration nor did their ancestors.

While today's atmosphere is about 20 percent molecular oxygen (O_2), geological evidence shows that it originally lacked O_2. Without oxygen, aerobic respiration would not be expected, and living things would have relied on anaerobic respiration or the process of fermentation instead. At some point before about 3.8 billion years ago, some prokaryotes began using energy from sunlight to power anabolic processes that reduce carbon dioxide to form organic compounds. That is, they evolved the ability to photosynthesize. Hydrogen, derived from various sources, was "captured" using light-powered reactions to reduce fixed carbon dioxide in the Calvin cycle. The group of Gram-negative bacteria that gave rise to cyanobacteria used water as the hydrogen source and released O_2 as a "waste" product.

Eventually, the amount of photosynthetic oxygen built up in some environments to levels that posed a risk to living organisms, since it can damage many organic compounds. Various metabolic processes evolved that protected organisms from oxygen, one of which, aerobic respiration, also generated high levels of ATP. It became widely present among prokaryotes, including in a free-living group we now call alpha-proteobacteria. Organisms that did not acquire aerobic respiration had to remain in oxygen-free environments. Originally, oxygen-rich environments were likely localized around places where cyanobacteria were abundant and active, but by about 2 billion years ago, geological evidence shows that oxygen was building up to higher concentrations in the atmosphere. Oxygen levels similar to today's levels only arose within the last 700 million years.

Recall that the first fossils that we believe to be eukaryotes date to about 2 billion years old, so they seemed to have evolved and diversified rapidly as oxygen levels were increasing. Also, recall that all extant eukaryotes descended from an ancestor with mitochondria. These organelles were first observed by light microscopists in the late 1800s, where they appeared to be somewhat worm-shaped structures that seemed to be moving around in the cell. Some early observers suggested that they might be bacteria living inside host cells, but these hypotheses remained unknown or rejected in most scientific communities.

Endosymbiotic Theory

As cell biology developed in the twentieth century, it became clear that mitochondria were the organelles responsible for producing ATP using aerobic respiration, in which oxygen was the final electron acceptor. In the 1960s, American biologist Lynn Margulis of Boston University developed the **endosymbiotic theory**, which states that eukaryotes may have been a product of one cell engulfing another, one living within another, and coevolving over time until the separate cells were no longer recognizable as such and shared genetic control of a mutualistic metabolic pathway to produce ATP. In 1967, Margulis introduced new data to support her work on the theory and substantiated her findings through microbiological evidence. Although Margulis's work initially was met with resistance, this basic component of this once-revolutionary hypothesis is now widely accepted, with work progressing on uncovering the steps involved in this evolutionary process and the key players involved.

While the metabolic organelles and genes responsible for many energy-harvesting processes appear to have had their origins in bacteria, our nuclear genes and the molecular machinery responsible for replication and expression appear to be more closely related to those found in the Archaea. Much remains to be clarified about how this relationship occurred; this continues to be an exciting field of discovery in biology. For instance, it is *not* known whether the endosymbiotic event that led to mitochondria occurred before or after the host cell had a nucleus. Such organisms would be among the extinct precursors of the last common ancestor of eukaryotes.

Mitochondria

One of the major features distinguishing prokaryotes from eukaryotes is the presence of mitochondria, or their reduced derivatives, in virtually all eukaryotic cells. Eukaryotic cells may contain anywhere from one to several thousand mitochondria, depending on the cell's level of energy consumption, in humans being most abundant in the liver and skeletal muscles. Each mitochondrion measures 1 to 10 or greater micrometers in length and exists in the cell as an organelle that can be ovoid to worm-shaped to intricately branched (Figure 23.2). However, although they may have originated as free-living aerobic organisms, mitochondria can no longer survive and reproduce outside the cell.

Mitochondria have several features that suggest their relationship to alpha-proteobacteria (Figure 23.5). Alpha-proteobacteria are a large group of bacteria that includes species symbiotic with plants, disease organisms that can infect humans via ticks, and many free-living species that use light for energy. Mitochondria have their own genomes, with a circular chromosome stabilized by attachments to the inner membrane. Mitochondria also have special ribosomes and transfer RNAs that resemble these same components in prokaryotes. An intriguing feature of mitochondria is that many of them exhibit minor differences from the universal genetic code. However, many of the genes for respiratory proteins are now relocated in the nucleus. When these genes are compared to those of other organisms, they appear to be of alpha-proteobacterial origin. In some eukaryotic groups, such genes are found in the mitochondria, whereas in other groups, they are found in the nucleus. This has been interpreted as evidence that over evolutionary time, genes have been transferred from the endosymbiont chromosome to those of the host genome. This apparent "loss" of genes by the endosymbiont is probably one explanation why mitochondria cannot live without a host.

Another line of evidence supporting the idea that mitochondria were derived by endosymbiosis comes from the structure of the mitochondrian itself. Most mitochondria are shaped like alpha-proteobacteria and are surrounded by two membranes; the inner membrane is bacterial in nature whereas the outer membrane is eukaryotic in nature. This is exactly what one would expect if one membrane-bound organism was engulfed into a vacuole by another membrane-bound organism. The outer mitochondrial membrane was derived by the enclosing vesicle, while the inner membrane was derived from the plasma membrane of the endosymbiont. The mitochondrial inner membrane is extensive and involves substantial infoldings called **cristae** that resemble the textured, outer surface of alpha-proteobacteria. The matrix and inner membrane are rich with the enzymes necessary for aerobic respiration.

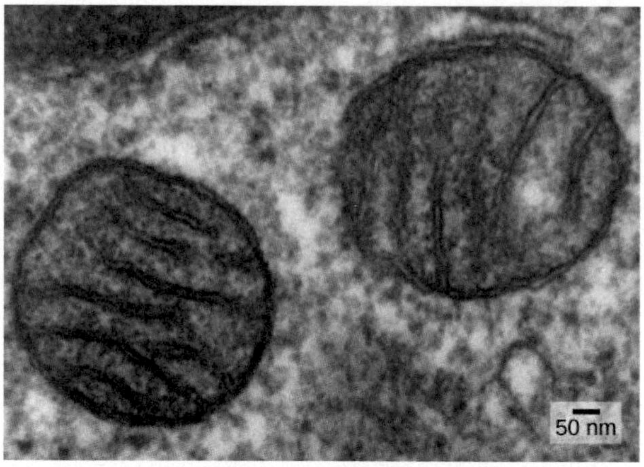

Figure 23.2 Mitochondria. In this transmission electron micrograph of mitochondria in a mammalian lung cell, the cristae, infoldings of the mitochondrial inner membrane, can be seen in cross-section. (credit: Louise Howard)

The third line of evidence comes from the production of new mitochondria. Mitochondria divide independently by a process that resembles binary fission in prokaryotes. Mitochondria arise only from previous mitochondria; they are not formed from scratch (de novo) by the eukaryotic cell. Mitochondria may fuse together; and they may be moved around inside the cell by interactions with the cytoskeleton. They reproduce within their enclosing cell and are distributed with the cytoplasm when a cell divides or two cells fuse. Therefore, although these organelles are highly integrated into the eukaryotic cell, they still reproduce as if they were independent organisms within the cell. However, their reproduction is synchronized with the activity and division of the cell. These features all support the theory that mitochondria were once free-living prokaryotes.

Some living eukaryotes are anaerobic and cannot survive in the presence of too much oxygen. However, a few appear to lack organelles that could be recognized as mitochondria. In the 1970s and on into the early 1990s, many biologists suggested that some of these eukaryotes were descended from ancestors whose lineages had diverged from the lineage of mitochondrion-containing eukaryotes before endosymbiosis occurred. Later findings suggest that *reduced organelles* are found in most, if not all, anaerobic eukaryotes, and that virtually all eukaryotes appear to carry some genes in their nuclei that are of mitochondrial origin.

In addition to the aerobic generation of ATP, mitochondria have several other metabolic functions. One of these functions is to generate clusters of iron and sulfur that are important cofactors of many enzymes. Such functions are often associated with the reduced mitochondrion-derived organelles of anaerobic eukaryotes. The protist *Monocercomonoides,* an inhabitant of vertebrate digestive tracts, appears to be an exception; it has no mitochondria and its genome contains neither genes derived from mitochondria nor nuclear genes related to mitochondrial maintenance. However, it is related to other protists with reduced mitochondria and probably represents an end-point in mitochondrial reduction. Although most biologists accept that the last common ancestor of eukaryotes had mitochondria, it appears that the complex relationship between mitochondria and their host cell continues to evolve.

Plastids

Some groups of eukaryotes are photosynthetic. Their cells contain, in addition to the standard eukaryotic organelles, another kind of organelle called a **plastid**. When such cells are carrying out photosynthesis, their plastids are rich in the pigment chlorophyll *a* and a range of other pigments, called *accessory pigments*, which are involved in harvesting energy from light. Photosynthetic plastids are called chloroplasts (Figure 23.3).

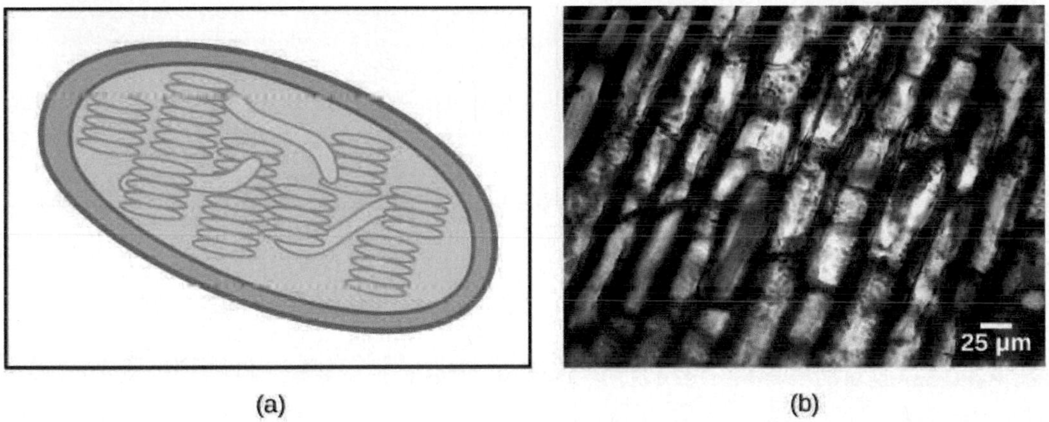

Figure 23.3 Chloroplasts. (a) This chloroplast cross-section illustrates its elaborate inner membrane organization. Stacks of thylakoid membranes compartmentalize photosynthetic enzymes and provide scaffolding for chloroplast DNA. (b) In this micrograph of *Elodea* sp., the chloroplasts can be seen as small green spheres. (credit b: modification of work by Brandon Zierer; scale-bar data from Matt Russell)

Like mitochondria, plastids appear to have an endosymbiotic origin. This hypothesis was also proposed and championed with the first direct evidence by Lynn Margulis. We now know that plastids are derived from cyanobacteria that lived inside the cells of an ancestral, aerobic, heterotrophic eukaryote. This is called primary endosymbiosis, and plastids of primary origin are surrounded by two membranes. However, the best evidence is that the acquisition of cyanobacterial endosymbionts has happened *twice* in the history of eukaryotes. In one case, the common ancestor of the major lineage/supergroup Archaeplastida took on a cyanobacterial endosymbiont; in the other, the ancestor of the small amoeboid rhizarian taxon, *Paulinella,* took on a different cyanobacterial endosymbiont. Almost all photosynthetic eukaryotes are descended from the first event, and only a couple of species are derived from the other, which in evolutionary terms, appears to be more recent.

Cyanobacteria are a group of Gram-negative bacteria with all the conventional structures of the group. However, unlike most prokaryotes, they have extensive, internal membrane-bound sacs called thylakoids. Chlorophyll is a component of these membranes, as are many of the proteins of the light reactions of photosynthesis. Cyanobacteria also have the peptidoglycan wall and lipopolysaccharide layer associated with Gram-negative bacteria.

Chloroplasts of primary endosymbiotic origin have thylakoids, a circular DNA chromosome, and ribosomes similar to those of cyanobacteria. As in mitochondria, each chloroplast is surrounded by two membranes. The outer membrane is thought to be derived from the enclosing vacuole of the host, and the inner membrane is thought to be derived from the plasma membrane of the cyanobacterial endosymbiont. In the group of Archaeplastida called the glaucophytes and in the rhizarian *Paulinella,* a thin peptidoglycan layer is still present between the outer and inner plastid membranes. All other plastids lack this relict of the cyanobacterial wall.

There is also, as with the case of mitochondria, strong evidence that many of the genes of the endosymbiont were transferred to the nucleus. Plastids, like mitochondria, cannot live independently outside the host. In addition, like mitochondria, plastids are derived from the division of other plastids and never built from scratch. Researchers have suggested that the endosymbiotic event that led to Archaeplastida occurred 1 to 1.5 billion years ago, at least 5 hundred million years after the fossil record suggests that eukaryotes were present.

Not all plastids in eukaryotes are derived directly from primary endosymbiosis. Some of the major groups of algae became photosynthetic by secondary endosymbiosis, that is, by taking in either green algae or red algae (both from Archaeplastida) as endosymbionts (Figure 23.4). Numerous microscopic and genetic studies have supported this conclusion. Secondary plastids are surrounded by three or more membranes, and some secondary plastids even have clear remnants of the nucleus (nucleomorphs) of endosymbiotic algae. There are even cases where tertiary or higher-order endosymbiotic events are the best explanations for the features of some eukaryotic plastids.

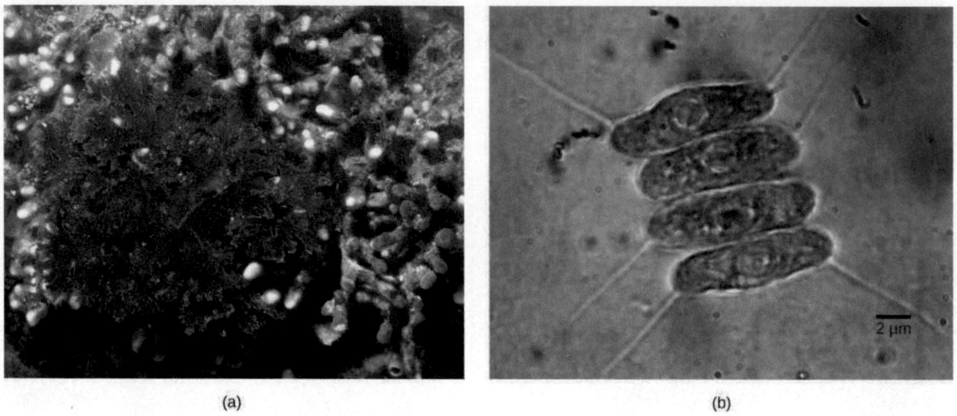

Figure 23.4 Algae. (a) Red algae and (b) green algae (seen here by light microscopy) share similar DNA sequences with photosynthetic cyanobacteria. Scientists speculate that, in a process called endosymbiosis, an ancestral prokaryote engulfed a photosynthetic cyanobacterium that evolved into modern-day chloroplasts. (credit a: modification of work by Ed Bierman; credit b: modification of work by G. Fahnenstiel, NOAA; scale-bar data from Matt Russell)

VISUAL CONNECTION

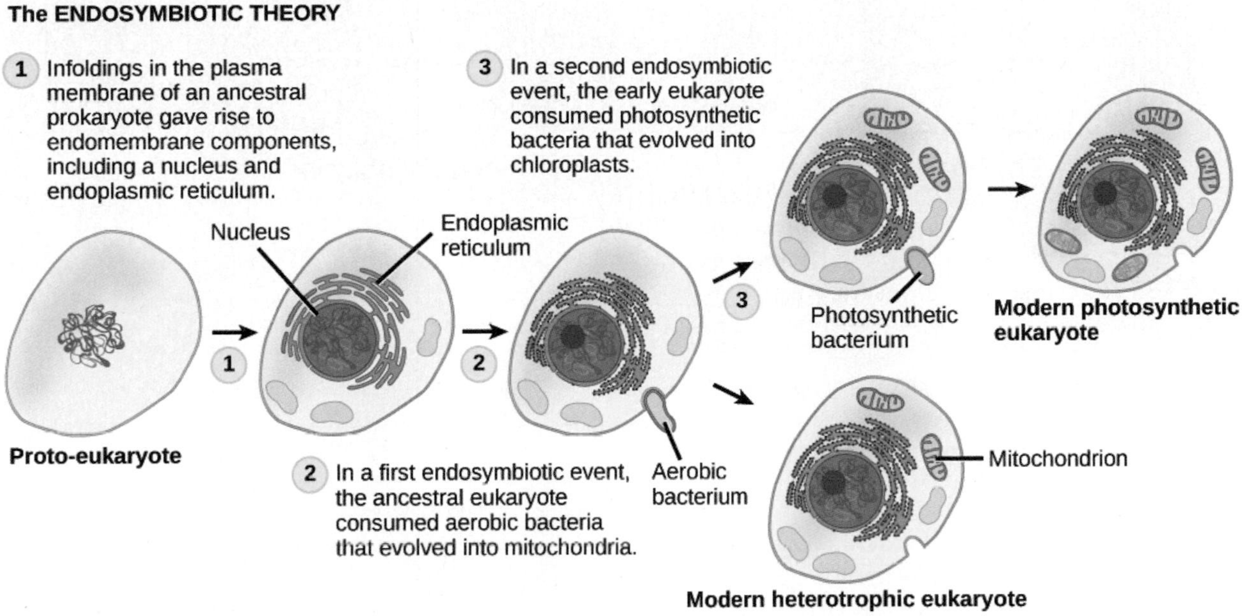

Figure 23.5 The Endosymbiotic Theory. The first eukaryote may have originated from an ancestral prokaryote that had undergone membrane proliferation, compartmentalization of cellular function (into a nucleus, lysosomes, and an endoplasmic reticulum), and the establishment of endosymbiotic relationships with an aerobic prokaryote, and, in some cases, a photosynthetic prokaryote, to form mitochondria and chloroplasts, respectively.

What evidence is there that mitochondria were incorporated into the ancestral eukaryotic cell before chloroplasts?

EVOLUTION CONNECTION

Secondary Endosymbiosis in Chlorarachniophytes

Endosymbiosis involves one cell engulfing another to produce, over time, a coevolved relationship in which neither cell could survive alone. The chloroplasts of red and green algae, for instance, are derived from the engulfment of a photosynthetic cyanobacterium by an ancestral prokaryote.

This evidence suggests the possibility that an ancestral cell (already containing a photosynthetic endosymbiont) was engulfed by another eukaryote cell, resulting in a secondary endosymbiosis. Molecular and morphological evidence suggest that the *chlorarachniophyte protists* are derived from a secondary endosymbiotic event. Chlorarachniophytes are rare algae indigenous to tropical seas and sand. They are classified into the Rhizarian supergroup. Chlorarachniophytes are reticulose amoebae, extending thin cytoplasmic strands that interconnect them with other chlorarachniophytes in a cytoplasmic network. These protists are thought to have originated when a eukaryote engulfed a green alga, the latter of which had previously established an endosymbiotic relationship with a photosynthetic cyanobacterium (Figure 23.6).

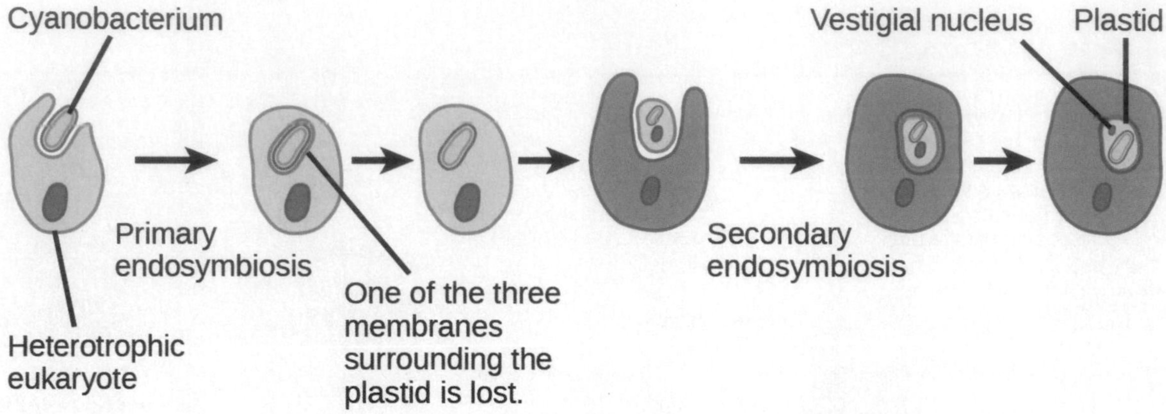

Figure 23.6 Secondary endosymbiosis. The hypothesized process of several endosymbiotic events leading to the evolution of chlorarachniophytes is shown. In a primary endosymbiotic event, a heterotrophic eukaryote consumed a cyanobacterium. In a secondary endosymbiotic event, the cell resulting from primary endosymbiosis was consumed by a second cell. The resulting organelle became a plastid in modern chlorarachniophytes.

Several lines of evidence support that chlorarachniophytes evolved from secondary endosymbiosis. The chloroplasts contained within the green algal endosymbionts still are capable of photosynthesis, making chlorarachniophytes photosynthetic. The green algal endosymbiont also exhibits a vestigial nucleus. In fact, it appears that chlorarachniophytes are the products of an evolutionarily recent secondary endosymbiotic event. The plastids of chlorarachniophytes are surrounded by *four membranes*: The first two correspond to the inner and outer membranes of the photosynthetic cyanobacterium, the third corresponds to plasma membrane of the green alga, and the fourth corresponds to the vacuole that surrounded the green alga when it was engulfed by the chlorarachniophyte ancestor. In other lineages that involved secondary endosymbiosis, only *three membranes* can be identified around plastids. This is currently interpreted as a sequential loss of a membrane during the course of evolution.

The process of secondary endosymbiosis is not unique to chlorarachniophytes. Secondary plastids are also found in the Excavates and the Chromalveolates. In the Excavates, secondary endosymbiosis of green algae led to euglenid protists, while in the Chromalveolates, secondary endosymbiosis of red algae led to the evolution of plastids in dinoflagellates, apicomplexans, and stramenopiles.

23.2 Characteristics of Protists

By the end of this section, you will be able to do the following:

- Describe the cell structure characteristics of protists
- Describe the metabolic diversity of protists
- Describe the life cycle diversity of protists

There are over 100,000 described living species of protists, and it is unclear how many undescribed species may exist. Since many protists live as commensals or parasites in other organisms and these relationships are often species-specific, there is a huge potential for protist diversity that matches the diversity of their hosts. Because the name "protist" serves as a catchall term for eukaryotic organisms that are not animal, plant, or fungi, it is not surprising that very few characteristics are common to all protists. On the other hand, familiar characteristics of plants and animals are foreshadowed in various protists.

Cell Structure

The cells of protists are among the most elaborate of all cells. Multicellular plants, animals, and fungi are embedded among the protists in eukaryotic phylogeny. In most plants and animals and some fungi, complexity arises out of *multicellularity, tissue specialization,* and subsequent interaction because of these features. Although a rudimentary form of multicellularity exists among some of the organisms labelled as "protists," those that have remained unicellular show how complexity can evolve in the absence of true multicellularity, with the differentiation of cellular morphology and function. A few protists live as colonies that behave in some ways as a group of free-living cells and in other ways as a multicellular organism. Some protists are composed of enormous, multinucleate, single cells that look like amorphous blobs of slime, or in other cases, like ferns. In some species of protists, the nuclei are different sizes and have distinct roles in protist cell function.

Single protist cells range in size from less than a micrometer to three meters in length to hectares! Protist cells may be enveloped by animal-like cell membranes or plant-like cell walls. Others are encased in glassy silica-based shells or wound with **pellicles** of interlocking protein strips. The pellicle functions like a flexible coat of armor, preventing the protist from being torn or pierced without compromising its range of motion.

Metabolism

Protists exhibit many forms of nutrition and may be aerobic or anaerobic. Those that store energy by photosynthesis belong to a group of *photoautotrophs* and are characterized by the presence of chloroplasts. Other protists are *heterotrophic* and consume organic materials (such as other organisms) to obtain nutrition. Amoebas and some other heterotrophic protist species ingest particles by a process called *phagocytosis*, in which the cell membrane engulfs a food particle and brings it inward, pinching off an intracellular membranous sac, or vesicle, called a food vacuole (Figure 23.7). In some protists, food vacuoles can be formed anywhere on the body surface, whereas in others, they may be restricted to the base of a specialized feeding structure. The vesicle containing the ingested particle, the phagosome, then fuses with a lysosome containing hydrolytic enzymes to produce a **phagolysosome**, and the food particle is broken down into small molecules that can diffuse into the cytoplasm and be used in cellular metabolism. Undigested remains ultimately are expelled from the cell via *exocytosis*.

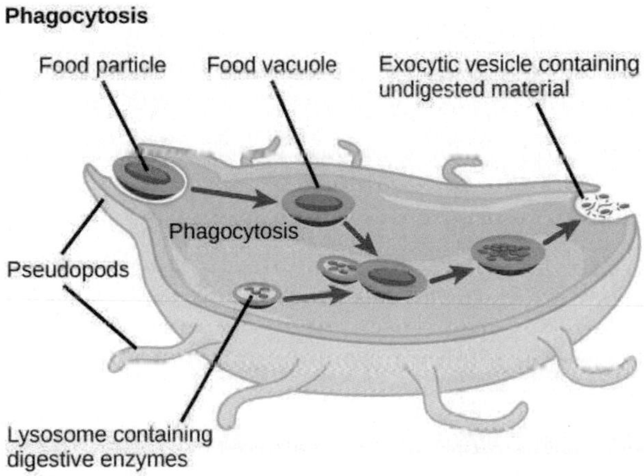

Figure 23.7 Phagocytosis. The stages of phagocytosis include the engulfment of a food particle, the digestion of the particle using hydrolytic enzymes contained within a lysosome, and the expulsion of undigested materials from the cell.

Subtypes of heterotrophs, called saprobes, absorb nutrients from dead organisms or their organic wastes. Some protists can function as **mixotrophs**, obtaining nutrition by photoautotrophic or heterotrophic routes, depending on whether sunlight or organic nutrients are available.

Motility

The majority of protists are motile, but different types of protists have evolved varied modes of movement (Figure 23.8). Some protists have one or more flagella, which they rotate or whip. Others are covered in rows or tufts of tiny cilia that they beat in a coordinated manner to swim. Still others form cytoplasmic extensions called *pseudopodia* anywhere on the cell, anchor the pseudopodia to a substrate, and pull themselves forward. Some protists can move toward or away from a stimulus, a movement referred to as taxis. For example, movement toward light, termed phototaxis, is accomplished by coupling their locomotion strategy with a light-sensing organ.

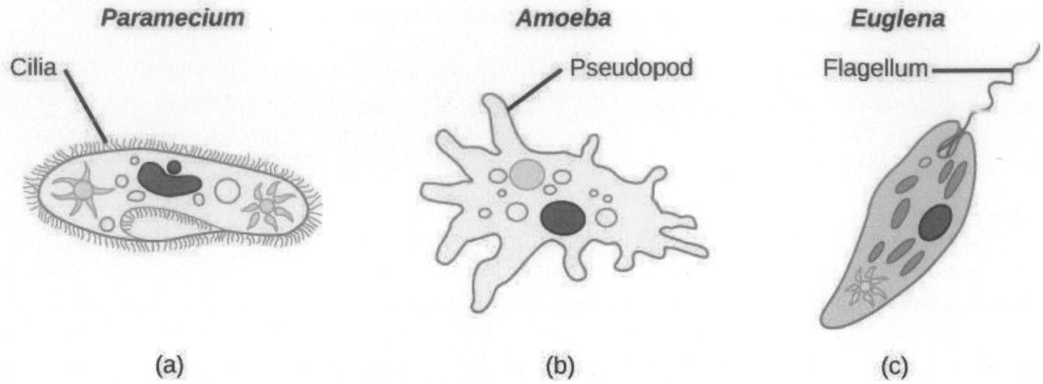

Figure 23.8 Locomotor organelles in protists. Protists use various methods for transportation. (a) *Paramecium* waves hair-like appendages called cilia to propel itself. (b) *Amoeba* uses lobe-like pseudopodia to anchor itself to a solid surface and pull itself forward. (c) *Euglena* uses a whip-like tail called a flagellum to propel itself.

Life Cycles

Protists reproduce by a variety of mechanisms. Most undergo some form of *asexual reproduction*, such as binary fission, to produce two daughter cells. In protists, binary fission can be divided into transverse or longitudinal, depending on the axis of orientation; sometimes *Paramecium* exhibits this method. Some protists such as the true slime molds exhibit multiple fission and simultaneously divide into many daughter cells. Others produce tiny buds that go on to divide and grow to the size of the parental protist.

Sexual reproduction, involving meiosis and fertilization, is common among protists, and many protist species can switch from asexual to sexual reproduction when necessary. Sexual reproduction is often associated with periods when nutrients are depleted or environmental changes occur. Sexual reproduction may allow the protist to recombine genes and produce new variations of progeny, some of which may be better suited to surviving changes in a new or changing environment. However, sexual reproduction is often associated with resistant **cysts** that are a protective, resting stage. Depending on habitat of the species, the cysts may be particularly resistant to temperature extremes, desiccation, or low pH. This strategy allows certain protists to "wait out" stressors until their environment becomes more favorable for survival or until they are carried (such as by wind, water, or transport on a larger organism) to a different environment, because cysts exhibit virtually no cellular metabolism.

Protist life cycles range from simple to extremely elaborate. Certain parasitic protists have complicated life cycles and must infect different host species at different developmental stages to complete their life cycle. Some protists are unicellular in the haploid form and multicellular in the diploid form, a strategy employed by animals. Other protists have multicellular stages in both haploid and diploid forms, a strategy called alternation of generations, analogous to that used by plants.

Habitats

Nearly all protists exist in some type of aquatic environment, including freshwater and marine environments, damp soil, and even snow. Several protist species are parasites that infect animals or plants. A few protist species live on dead organisms or their wastes, and contribute to their decay.

23.3 Groups of Protists

By the end of this section, you will be able to do the following:

- Describe representative protist organisms from each of the six presently recognized supergroups of eukaryotes
- Identify the evolutionary relationships of plants, animals, and fungi within the six presently recognized supergroups of eukaryotes
- Identify defining features of protists in each of the six supergroups of eukaryotes.

In the span of several decades, the Kingdom Protista has been disassembled because sequence analyses have revealed new genetic (and therefore evolutionary) relationships among these eukaryotes. Moreover, protists that exhibit similar morphological features may have evolved analogous structures because of similar selective pressures—rather than because of recent common ancestry. This phenomenon, called convergent evolution, is one reason why protist classification is so challenging. The emerging classification scheme groups the entire domain Eukarya into six "supergroups" that contain all of the protists as well as animals, plants, and fungi that evolved from a common ancestor (Figure 23.9). Each of the supergroups is believed to be monophyletic, meaning that all organisms within each supergroup are believed to have evolved from a single common ancestor, and thus all members are most closely related to each other than to organisms outside that group. There is still evidence lacking for the monophyly of some groups. Each supergroup can be viewed as representing one of many variants on eukaryotic cell structure. In each group, one or more of the defining characters of the eukaryotic cell—the nucleus, the cytoskeleton, and the endosymbiotic organelles—may have diverged from the "typical" pattern.

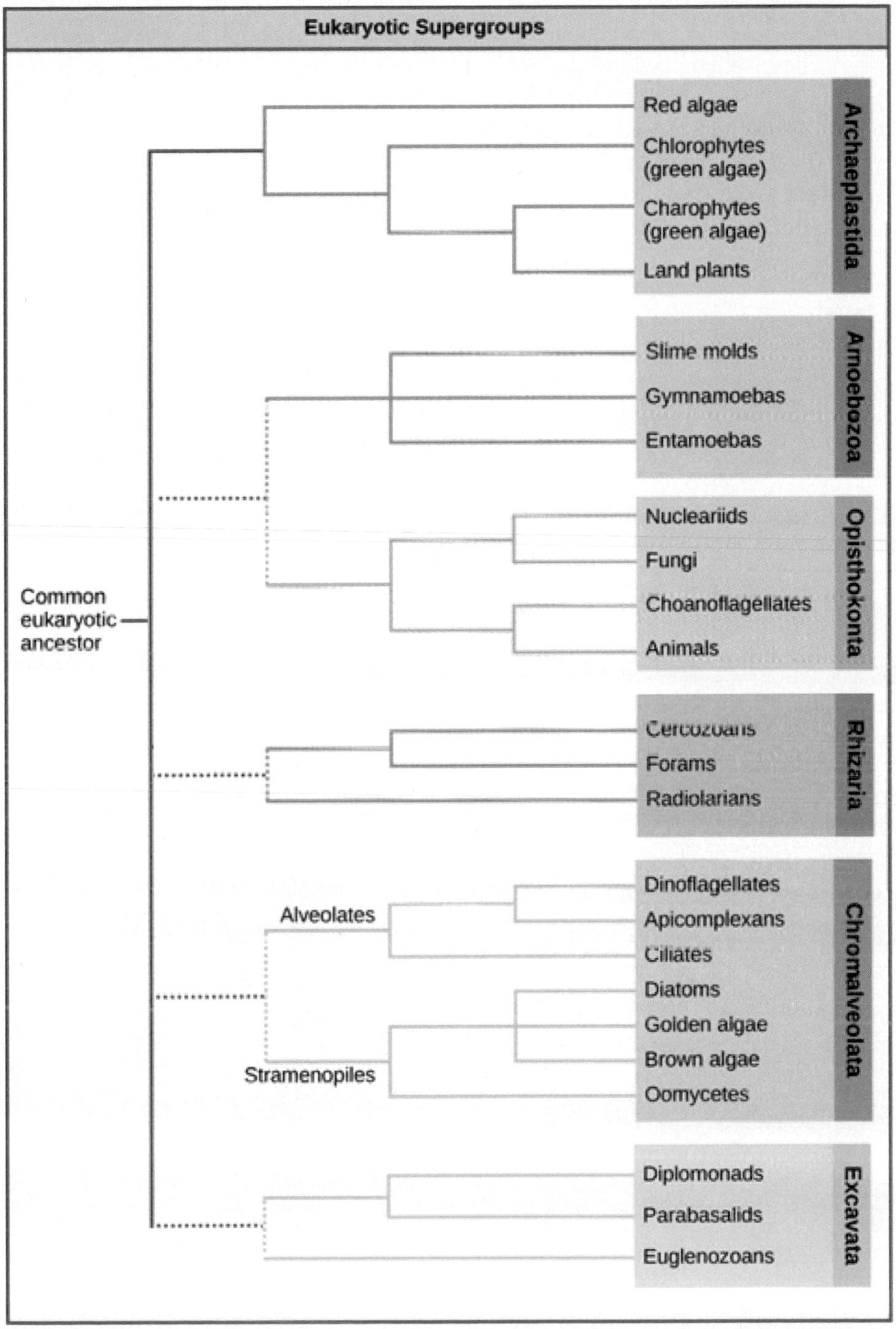

Figure 23.9 Eukaryotic supergroups. This diagram shows a proposed classification of the domain Eukarya. Currently, the domain Eukarya is divided into six supergroups. Within each supergroup are multiple kingdoms. Although each supergroup is believed to be monophyletic, the dotted lines suggest evolutionary relationships among the supergroups that continue to be debated.

Keep in mind that the classification scheme presented here represents just one of several hypotheses, and the true evolutionary relationships are still to be determined. The six supergroups may be modified or replaced by a more appropriate hierarchy as genetic, morphological, and ecological data accumulate. When learning about protists, it is helpful to focus less on the nomenclature and more on the commonalities and differences that illustrate how each group has exploited the possibilities of eukaryotic life.

Archaeplastida

Molecular evidence supports the hypothesis that all Archaeplastida are descendents of an endosymbiotic relationship between a heterotrophic protist and a cyanobacterium. The protist members of the group include the red algae and green algae. It was from a common ancestor of these protists that the land plants evolved, since their closest relatives are found in this group. The red and green algae include unicellular, multicellular, and colonial forms. A variety of algal life cycles exists, but the most complex is alternation of generations, in which both haploid and diploid stages are multicellular. A diploid sporophyte contains cells that undergo meiosis to produce haploid spores. The spores germinate and grow into a haploid gametophyte, which then makes gametes by mitosis. The gametes fuse to form a zygote that grows into a diploid sporophyte. Alternation of generations is seen in some species of Archaeplastid algae, as well as some species of Stramenopiles (Figure 23.10). In some species, the gametophyte and sporophyte look quite different, while in others they are nearly indistinguishable.

Glaucophytes

Glaucophytes are a small group of Archaeplastida interesting because their chloroplasts retain remnants of the peptidoglycan cell wall of the ancestral cyanobacterial endosymbiont (Figure 23.10).

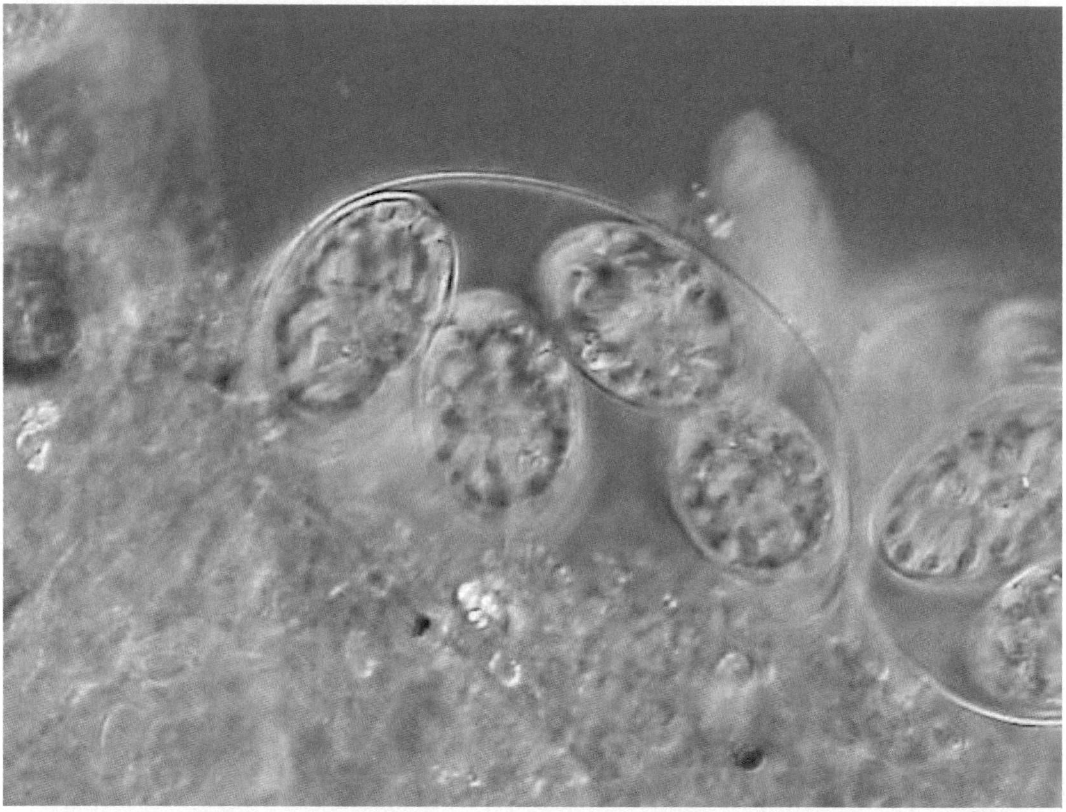

Figure 23.10 Glaucocystis. (credit: By ja:User:NEON / commons:User:NEON_ja - Own work, CC BY-SA 2.5, https://commons.wikimedia.org/w/index.php?curid=1706641 (http://openstax.org/l/Glaucocystis))

Red Algae

Red algae, or rhodophytes lack flagella, and are primarily multicellular, although they range in size from microscopic, unicellular protists to large, multicellular forms grouped into the informal seaweed category. Red algae have a second cell wall outside an inner cellulose cell wall. Carbohydrates in this wall are the source of agarose used for electrophoresis gels and agar for solidifying bacterial media. The "red" in the red algae comes from phycoerythrins, accessory photopigments that are red in color and obscure the green tint of chlorophyll in some species. Other protists classified as red algae lack phycoerythrins and are parasites. Both the red algae and the glaucophytes store carbohydrates in the cytoplasm rather than in the plastid. Red algae are common in tropical waters where they have been detected at depths of 260 meters. Other red algae exist in terrestrial or freshwater environments. The red algae life cycle is an unusual alternation of generations that includes two sporophyte phases, with meiosis occurring only in the second sporophyte.

Green Algae: Chlorophytes and Charophytes

The most abundant group of algae is the green algae. The green algae exhibit features similar to those of the land plants, particularly in terms of chloroplast structure. In both green algae and plants, carbohydrates are stored in the plastid. That this group of protists shared a relatively recent common ancestor with land plants is well supported. The green algae are subdivided into the chlorophytes and the charophytes. The charophytes are the closest living relatives to land plants and resemble them in morphology and reproductive strategies. The familiar *Spirogyra* is a charophyte. Charophytes are common in wet habitats, and their presence often signals a healthy ecosystem.

The chlorophytes exhibit great diversity of form and function. Chlorophytes primarily inhabit freshwater and damp soil, and are a common component of plankton. *Chlamydomonas* is a simple, unicellular chlorophyte with a pear-shaped morphology and two opposing, anterior flagella that guide this protist toward light sensed by its eyespot. More complex chlorophyte species exhibit haploid gametes and spores that resemble *Chlamydomonas*.

The chlorophyte *Volvox* is one of only a few examples of a colonial organism, which behaves in some ways like a collection of individual cells, but in other ways like the specialized cells of a multicellular organism (Figure 23.11). *Volvox* colonies contain 500 to 60,000 cells, each with two flagella, contained within a hollow, spherical matrix composed of a gelatinous glycoprotein secretion. Individual cells in a *Volvox* colony move in a coordinated fashion and are interconnected by cytoplasmic bridges. Only a few of the cells reproduce to create daughter colonies, an example of basic cell specialization in this organism. Daughter colonies are produced with their flagella on the inside and have to evert as they are released.

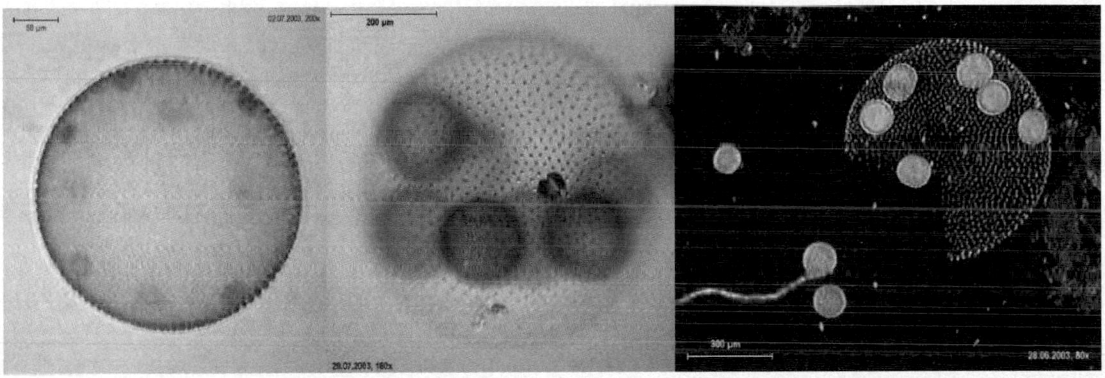

Figure 23.11 Volvox. *Volvox aureus* is a green alga in the supergroup Archaeplastida. This species exists as a colony, consisting of cells immersed in a gel-like matrix and intertwined with each other via hair-like cytoplasmic extensions. (credit: Dr. Ralf Wagner)

True multicellular organisms, such as the sea lettuce, *Ulva*, are also represented among the chlorophytes. In addition, some chlorophytes exist as large, multinucleate, single cells. Species in the genus *Caulerpa* exhibit flattened fern-like foliage and can reach lengths of 3 meters (Figure 23.12). *Caulerpa* species undergo nuclear division, but their cells do not complete cytokinesis, remaining instead as massive and elaborate single cells.

Figure 23.12 A multinucleate alga. *Caulerpa taxifolia* is a chlorophyte consisting of a single cell containing potentially thousands of nuclei. (credit: NOAA). An interesting question is how a single cell can produce such complex shapes.

LINK TO LEARNING

Take a look at this video to see cytoplasmic streaming in a green alga.

Click to view content (https://www.openstax.org/l/chara_corallina)

Amoebozoa

Like the Archaeplastida, the Amoebozoa include species with single cells, species with large multinucleated cells, and species that have multicellular phases. Amoebozoan cells characteristically exhibit pseudopodia that extend like tubes or flat lobes. These pseudopods project outward from anywhere on the cell surface and can anchor to a substrate. The protist then transports its cytoplasm into the pseudopod, thereby moving the entire cell. This type of motion is similar to the cytoplasmic streaming used to move organelles in the Archaeplastida, and is also used by other protists as a means of locomotion or as a method to distribute nutrients and oxygen. The Amoebozoa include both free-living and parasitic species.

Gymnomoebae

The Gymnamoeba or lobose amoebae include both naked amoebae like the familiar *Amoeba proteus* and shelled amoebae, whose bodies protrude like snails from their protective tests. *Amoeba proteus* is a large amoeba about 500 µm in diameter but is dwarfed by the multinucleate amoebae *Pelomyxa*, which can be 10 times its size. Although *Pelomyxa* may have hundreds of nuclei, it has lost its mitochondria, but replaced them with bacterial endosymbionts. The secondary loss or modification of mitochondria is a feature also seen in other protist groups.

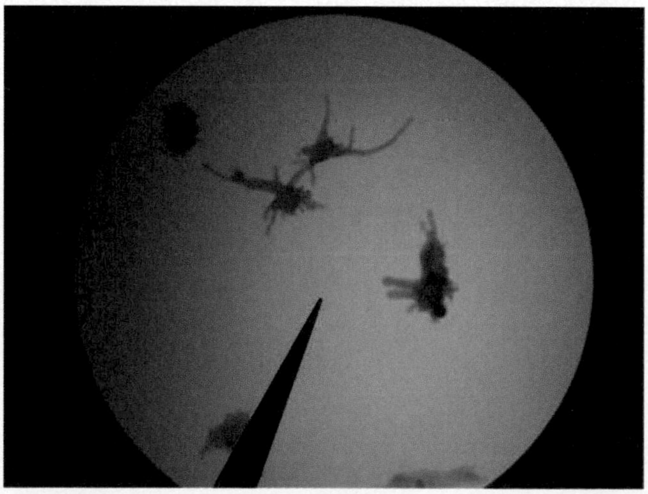

Figure 23.13 Amoeba. Amoebae with tubular and lobe-shaped pseudopodia are seen under a microscope. These isolates would be morphologically classified as amoebozoans.

Slime Molds

A subset of the amoebozoans, the slime molds, has several morphological similarities to fungi that are thought to be the result of convergent evolution. For instance, during times of stress, some slime molds develop into spore-generating fruiting bodies, much like fungi.

The slime molds are categorized on the basis of their life cycles into plasmodial or cellular types. Plasmodial slime molds are composed of large, multinucleate cells and move along surfaces like an amorphous blob of slime during their feeding stage (Figure 23.14). Food particles are lifted and engulfed into the slime mold as it glides along. The "dog vomit" slime mold seen in Figure 23.14 is a particularly colorful specimen and its ability to creep about might well trigger suspicion of alien invasion. Upon maturation, the plasmodium takes on a net-like appearance with the ability to form fruiting bodies, or sporangia, during times of stress. Haploid spores are produced by meiosis within the sporangia, and spores can be disseminated through the air or water to potentially land in more favorable environments. If this occurs, the spores germinate to form ameboid or flagellate haploid cells that can combine with each other and produce a diploid zygotic slime mold to complete the life cycle.

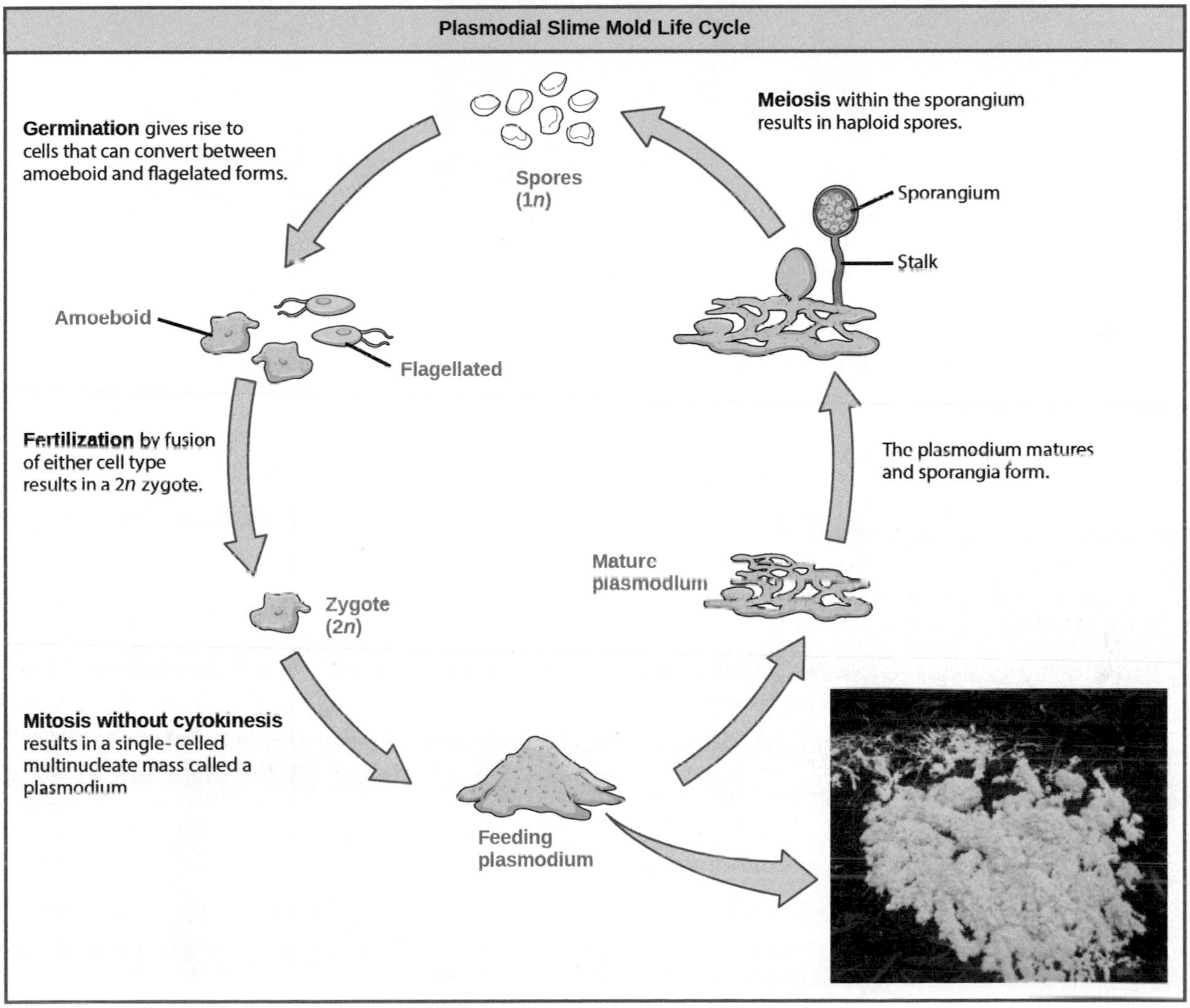

Figure 23.14 Plasmodial slime molds. The life cycle of the plasmodial slime mold is shown. The brightly colored plasmodium in the inset photo is a single-celled, multinucleate mass. (credit: modification of work by Dr. Jonatha Gott and the Center for RNA Molecular Biology, Case Western Reserve University)

The cellular slime molds function as independent amoeboid cells when nutrients are abundant. When food is depleted, cellular slime molds aggregate into a mass of cells that behaves as a single unit, called a slug. Some cells in the slug contribute to a 2–3-millimeter stalk, drying up and dying in the process. Cells atop the stalk form an asexual fruiting body that contains haploid spores (Figure 23.15). As with plasmodial slime molds, the spores are disseminated and can germinate if they land in a moist environment. One representative genus of the cellular slime molds is *Dictyostelium*, which commonly exists in the damp soil of forests.

Figure 23.15 Cellular Slime Mold. The image shows several stages in the life cycle of *Dictyostelium discoideum*, including aggregated cells, mobile slugs and their transformation into fruiting bodies with a cluster of spores supported by a stalk. (credit: By Usman Bashir (Own work) [CC BY-SA 4.0 (http://creativecommons.org/licenses/by-sa/4.0 (http://openstax.org/l/CCBY))], via Wikimedia Commons)

LINK TO LEARNING

View this video to see the formation of a fruiting body by a cellular slime mold.

Click to view content (https://www.openstax.org/l/slime_mold)

Opisthokonta

The Opisthokonts are named for the single posterior flagellum seen in flagellated cells of the group. The flagella of other protists are anterior and their movement pulls the cells along, while the opisthokonts are pushed. Protist members of the opisthokonts include the animal-like choanoflagellates, which are believed to resemble the common ancestor of sponges and perhaps, all animals. Choanoflagellates include unicellular and colonial forms (Figure 23.16), and number about 244 described species. In these organisms, the single, apical flagellum is surrounded by a contractile collar composed of microvilli. The collar is used to filter and collect bacteria for ingestion by the protist. A similar feeding mechanism is seen in the collar cells of sponges, which suggests a possible connection between choanoflagellates and animals.

The Mesomycetozoa form a small group of parasites, primarily of fish, and at least one form that can parasitize humans. Their life cycles are poorly understood. These organisms are of special interest, because they appear to be so closely related to animals. In the past, they were grouped with fungi and other protists based on their morphology.

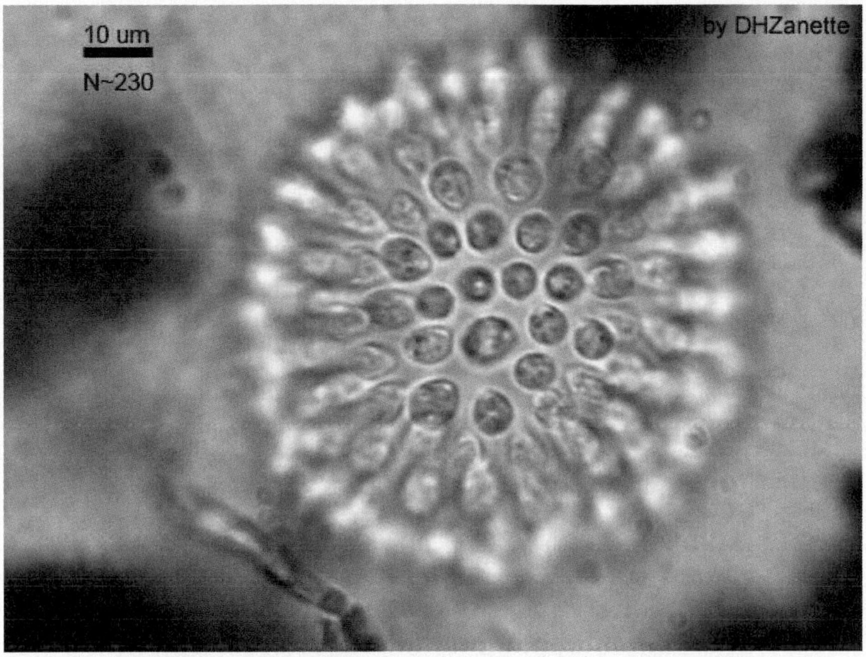

Figure 23.16 A Colonial Choanoflagellate. (credit: By Dhzanette (http://en.wikipedia.org/wiki/Choanoflagellate (http://openstax.org/l/choano)) [Public domain], via Wikimedia Commons)

The previous supergroups are all the products of primary endosymbiontic events and their organelles—nucleus, mitochondria, and chloroplasts—are what would be considered "typical," i.e., matching the diagrams you would find in an introductory biology book. The next three supergroups all contain at least some photosynthetic members whose chloroplasts were derived by secondary endosymbiosis. They also show some interesting variations in nuclear structure, and modification of mitochondria or chloroplasts.

Rhizaria

The Rhizaria supergroup includes many of the amoebas with thin threadlike, needle-like or root-like pseudopodia (Figure 23.17), rather than the broader lobed pseudopodia of the Amoebozoa. Many rhizarians make elaborate and beautiful tests—armor-like coverings for the body of the cell—composed of calcium carbonate, silicon, or strontium salts. Rhizarians have important roles in both carbon and nitrogen cycles. When rhizarians die, and their tests sink into deep water, the carbonates are out of reach of most decomposers, locking carbon dioxide away from the atmosphere. In general, this process by which carbon is transported deep into the ocean is described as the biological carbon pump, because carbon is "pumped" to the ocean depths where it is inaccessible to the atmosphere as carbon dioxide. The biological carbon pump is a crucial component of the carbon cycle that maintains lower atmospheric carbon dioxide levels. Foraminiferans are unusual in that they are the only eukaryotes known to participate in the nitrogen cycle by denitrification, an activity usually served only by prokaryotes.

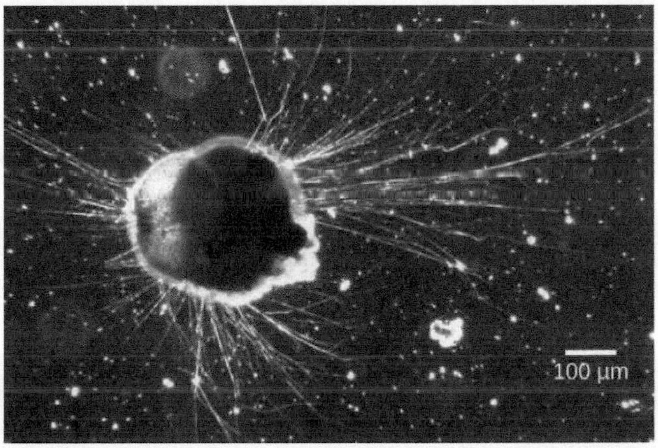

Figure 23.17 Rhizaria. *Ammonia tepida*, a Rhizaria species viewed here using phase contrast light microscopy, exhibits many threadlike

pseudopodia. It also has a chambered calcium carbonate shell or test. (credit: modification of work by Scott Fay, UC Berkeley; scale-bar data from Matt Russell)

Foraminiferans

Foraminiferans, or forams, are unicellular heterotrophic protists, ranging from approximately 20 micrometers to several centimeters in length, and occasionally resembling tiny snails (Figure 23.18). As a group, the forams exhibit porous shells, called **tests** that are built from various organic materials and typically hardened with calcium carbonate. The tests may house photosynthetic algae, which the forams can harvest for nutrition. Foram pseudopodia extend through the pores and allow the forams to move, feed, and gather additional building materials. Typically, forams are associated with sand or other particles in marine or freshwater habitats. Foraminiferans are also useful as indicators of pollution and changes in global weather patterns.

Figure 23.18 Foraminiferan Tests. These shells from foraminifera sank to the sea floor. (credit: Deep East 2001, NOAA/OER)

Radiolarians

A second subtype of Rhizaria, the radiolarians, exhibit intricate exteriors of glassy silica with radial or bilateral symmetry (Figure 23.19). Needle-like pseudopods supported by microtubules radiate outward from the cell bodies of these protists and function to catch food particles. The shells of dead radiolarians sink to the ocean floor, where they may accumulate in 100 meter-thick depths. Preserved, sedimented radiolarians are very common in the fossil record.

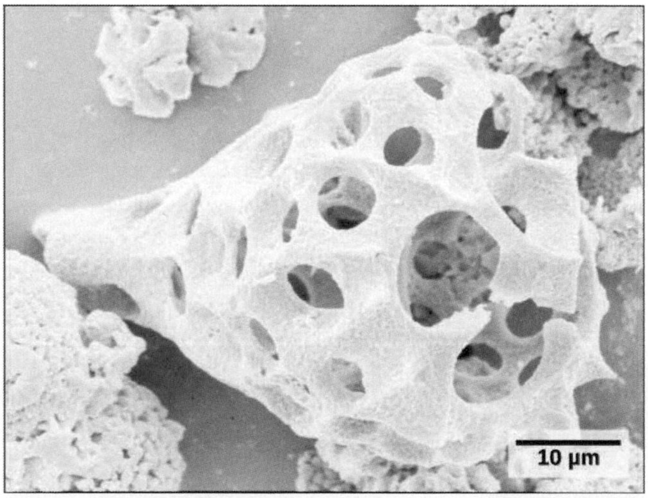

Figure 23.19 Radiolarian shell. This fossilized radiolarian shell was imaged using a scanning electron microscope. (credit: modification of work by Hannes Grobe, Alfred Wegener Institute; scale-bar data from Matt Russell)

Cercozoa

The Cercozoa are both morphologically and metabolically diverse, and include both naked and shelled forms. The Chlorarachniophytes (Figure 23.20) are photosynthetic, having acquired chloroplasts by secondary endosymbiosis. The chloroplast contains a remnant of the chlorophyte endosymbiont nucleus, sandwiched between the two sets of chloroplast membranes. Vampyrellids or "vampire amoebae," as their name suggests, obtain their nutrients by thrusting a pseudopod into the interior of other cells and sucking out their contents.

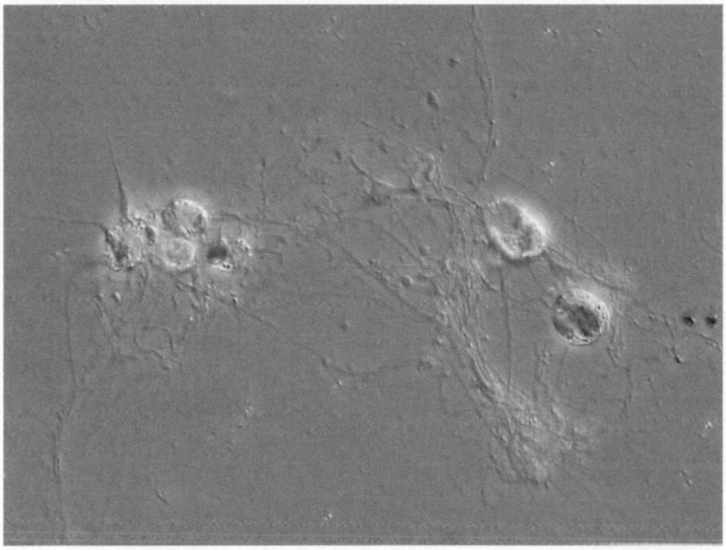

Figure 23.20 A Chlorarachniophyte. This rhizarian is mixotrophic, and can obtain nutrients both by photosynthesis and by trapping various microorganisms with its network of pseudopodia. (credit: By ja:User:NEON / commons:User:NEON_ja (Own work) [CC BY-SA 2.5 (http://creativecommons.org/licenses/by-sa/2.5 (http://openstax.org/l/CCBY_25)) or CC BY-SA 2.5 (http://creativecommons.org/licenses/by-sa/2.5 (http://openstax.org/l/CCBY_25))], via Wikimedia Commons)

Chromalveolata

Current evidence suggests that species classified as chromalveolates are derived from a common ancestor that engulfed a photosynthetic red algal cell, which itself had already evolved chloroplasts from an endosymbiotic relationship with a photosynthetic prokaryote. Therefore, the ancestor of chromalveolates is believed to have resulted from a secondary endosymbiotic event. However, some chromalveolates appear to have lost red alga-derived plastid organelles or lack plastid genes altogether. Therefore, this supergroup should be considered a hypothesis-based working group that is subject to change. Chromalveolates include very important photosynthetic organisms, such as diatoms, brown algae, and significant disease agents in animals and plants. The chromalveolates can be subdivided into alveolates and stramenopiles.

Alveolates: Dinoflagellates, Apicomplexians, and Ciliates

A large body of data supports that the alveolates are derived from a shared common ancestor. The alveolates are named for the presence of an alveolus, or membrane-enclosed sac, beneath the cell membrane. The exact function of the alveolus is unknown, but it may be involved in osmoregulation. The alveolates are further categorized into some of the better-known protists: the dinoflagellates, the apicomplexans, and the ciliates.

Dinoflagellates exhibit extensive morphological diversity and can be photosynthetic, heterotrophic, or mixotrophic. The chloroplast of photosynthetic dinoflagellates was derived by secondary endosymbiosis of a red alga. Many dinoflagellates are encased in interlocking plates of cellulose. Two perpendicular flagella fit into the grooves between the cellulose plates, with one flagellum extending longitudinally and a second encircling the dinoflagellate (Figure 23.21). Together, the flagella contribute to the characteristic spinning motion of dinoflagellates. These protists exist in freshwater and marine habitats, and are a component of **plankton**, the typically microscopic organisms that drift through the water and serve as a crucial food source for larger aquatic organisms.

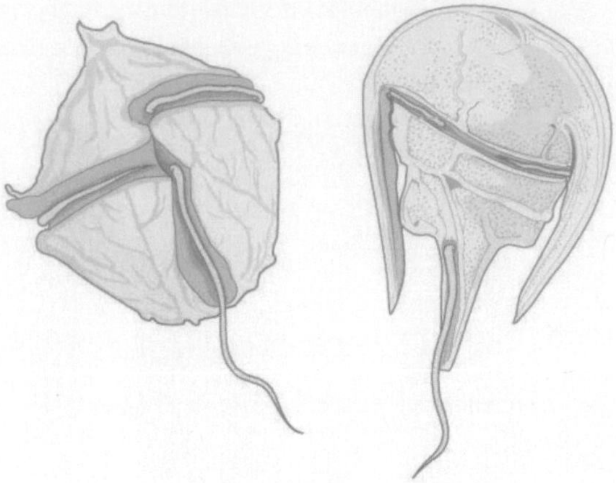

Figure 23.21 Dinoflagellates. The dinoflagellates exhibit great diversity in shape. Many are encased in cellulose armor and have two flagella that fit in grooves between the plates. Movement of these two perpendicular flagella causes a spinning motion.

Dinoflagellates have a nuclear variant called a dinokaryon. The chromosomes in the dinokaryon are highly condensed throughout the cell cycle and do not have typical histones. Mitosis in dinoflagellates is closed, that is, the spindle separates the chromosomes from outside of the nucleus without breakdown of the nuclear envelope.

Some dinoflagellates generate light, called **bioluminescence**, when they are jarred or stressed. Large numbers of marine dinoflagellates (billions or trillions of cells per wave) can emit light and cause an entire breaking wave to twinkle or take on a brilliant blue color (Figure 23.22). For approximately 20 species of marine dinoflagellates, population explosions (also called blooms) during the summer months can tint the ocean with a muddy red color. This phenomenon is called a red tide, and it results from the abundant red pigments present in dinoflagellate plastids. In large quantities, these dinoflagellate species secrete an asphyxiating toxin that can kill fish, birds, and marine mammals. Red tides can be massively detrimental to commercial fisheries, and humans who consume these protists may become poisoned.

Figure 23.22 Dinoflagellate bioluminescence. Bioluminescence is emitted from dinoflagellates in a breaking wave, as seen from the New Jersey coast. (credit: “catalano82”/Flickr)

The apicomplexan protists are named for a structure called an apical complex (Figure 23.23), which appears to be a highly modified secondary chloroplast. The apicoplast genome is similar to those of dinoflagellate chloroplasts. The apical complex is specialized for entry and infection of host cells. Indeed, all apicomplexans are parasitic. This group includes the genus *Plasmodium*, which causes malaria in humans. Apicomplexan life cycles are complex, involving multiple hosts and stages of sexual and asexual reproduction.

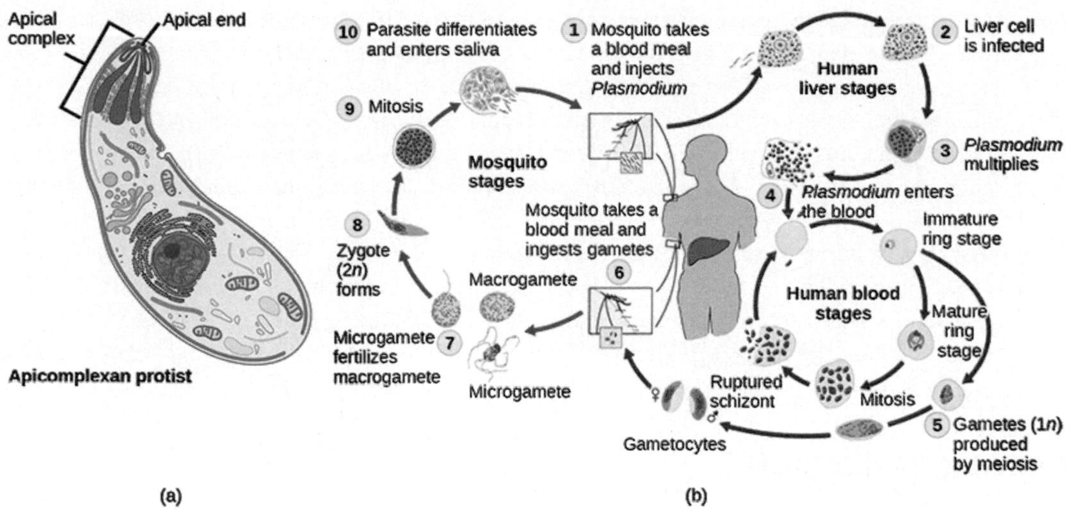

Figure 23.23 Apicomplexa. (a) Apicomplexans are parasitic protists. They have a characteristic apical complex that enables them to infect host cells. (b) *Plasmodium*, the causative agent of malaria, has a complex life cycle typical of apicomplexans. (credit b: modification of work by CDC)

The ciliates, which include *Paramecium* and *Tetrahymena*, are a group of protists 10 to 3,000 micrometers in length that are covered in rows, tufts, or spirals of tiny cilia. By beating their cilia synchronously or in waves, ciliates can coordinate directed movements and ingest food particles. Certain ciliates have fused cilia-based structures that function like paddles, funnels, or fins. Ciliates also are surrounded by a pellicle, providing protection without compromising agility. The genus *Paramecium* includes protists that have organized their cilia into a plate-like primitive mouth, called an oral groove, which is used to capture and digest bacteria (Figure 23.24). Food captured in the oral groove enters a food vacuole, where it combines with digestive enzymes. Waste particles are expelled by an exocytic vesicle that fuses at a specific region on the cell membrane, called the anal pore. In addition to a vacuole-based digestive system, *Paramecium* also uses **contractile vacuoles**, which are osmoregulatory vesicles that fill with water as it enters the cell by osmosis and then contract to squeeze water from the cell. Ciliates therefore exhibit considerable structural complexity without having achieved multicellularity.

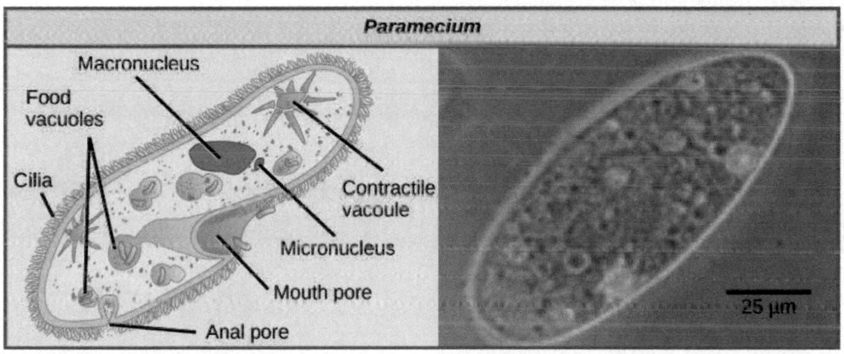

Figure 23.24 Paramecium. *Paramecium* has a primitive mouth (called an oral groove) to ingest food, and an anal pore to eliminate waste. Contractile vacuoles allow the organism to excrete excess water. Cilia enable the organism to move. (credit "paramecium micrograph": modification of work by NIH; scale-bar data from Matt Russell)

LINK TO LEARNING

Watch the video of the contractile vacuole of *Paramecium* expelling water to keep the cell osmotically balanced.

Click to view content (https://www.openstax.org/l/paramecium)

Paramecium has two nuclei, a macronucleus and a micronucleus, in each cell. The micronucleus is essential for sexual reproduction, and is in many ways a typical eukaryotic nucleus, except that its genes are not transcribed. The transcribed nucleus is the macronucleus, which directs asexual binary fission and all other biological functions. The macronucleus is a multiploid nucleus constructed from the micronucleus during sexual reproduction. Periodic reconstruction of the macronucleus is necessary because the macronucleus divides amitotically, and thus becomes genetically unbalanced over a period of successive cell replications. *Paramecium* and most other ciliates reproduce sexually by conjugation. This process begins when two different mating types of *Paramecium* make physical contact and join with a cytoplasmic bridge (Figure 23.25). The diploid micronucleus in each cell then undergoes meiosis to produce four haploid micronuclei. Three of these degenerate in each cell, leaving one micronucleus that then undergoes mitosis, generating two haploid micronuclei. The cells each exchange one of these haploid nuclei and move away from each other. Fusion of the haploid micronuclei generates a completely novel diploid pre-micronucleus in each conjugative cell. This pre-micronucleus undergoes three rounds of mitosis to produce eight copies, and the original macronucleus disintegrates. Four of the eight pre-micronuclei become full-fledged micronuclei, whereas the other four perform multiple rounds of DNA replication. The copies of the micronuclear chromosomes are severely edited to form hundreds of smaller chromosomes that contain only the protein coding genes. Each of these smaller chromosomes gets new telomeres as the macronucleus differentiates. Two cycles of cell division then yield four new *Paramecia* from each original conjugative cell.

VISUAL CONNECTION

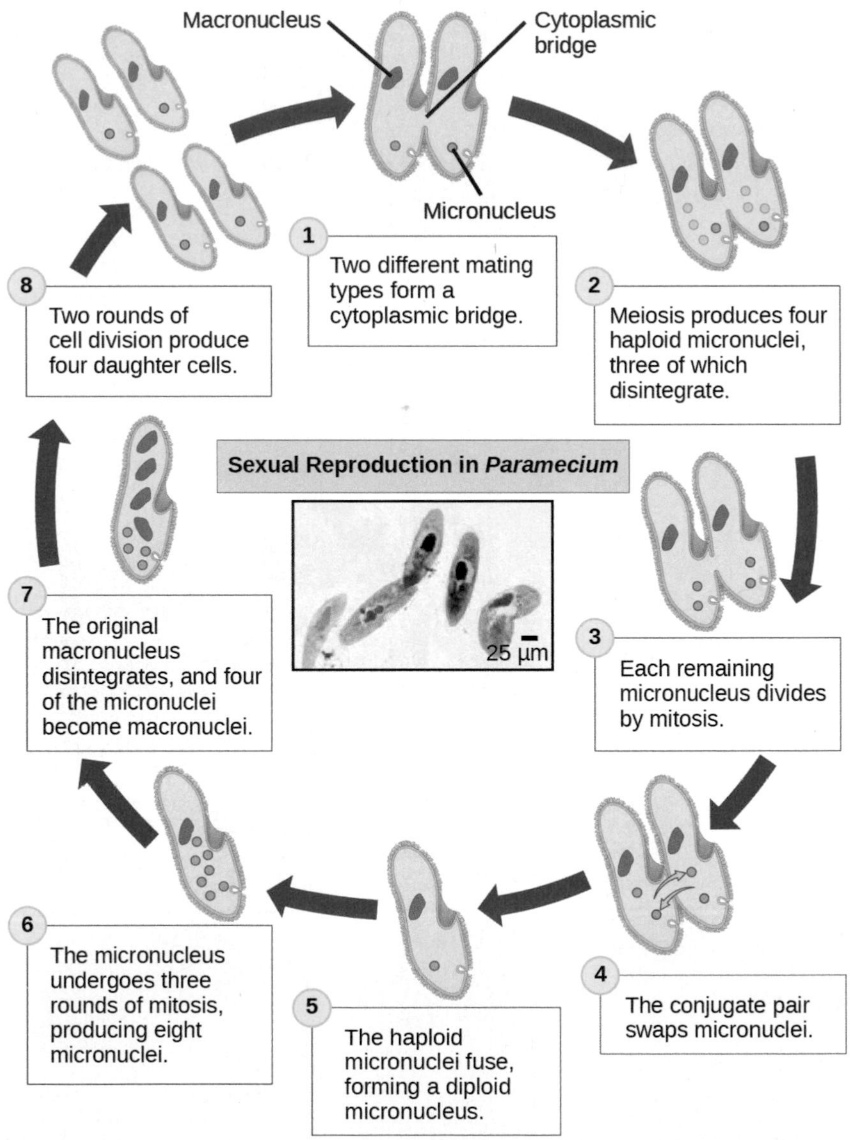

Figure 23.25 Conjugation in Paramecium. The complex process of sexual reproduction in *Paramecium* creates eight daughter cells from two original cells. Each cell has a macronucleus and a micronucleus. During sexual reproduction, the macronucleus dissolves and is

replaced by a micronucleus. (credit "micrograph": modification of work by Ian Sutton; scale-bar data from Matt Russell)

Which of the following statements about *Paramecium* sexual reproduction is false?

a. The macronuclei are derived from micronuclei.
b. Both mitosis and meiosis occur during sexual reproduction.
c. The conjugate pair swaps macronucleii.
d. Each parent produces four daughter cells.

Stramenopiles: Diatoms, Brown Algae, Golden Algae and Oomycetes

The other subgroup of chromalveolates, the stramenopiles, includes photosynthetic marine algae and heterotrophic protists. The chloroplast of these algae is derived from red alga. The identifying feature of this group is the presence of a textured, or "hairy," flagellum. Many stramenopiles also have an additional flagellum that lacks hair-like projections (Figure 23.26). Members of this subgroup range in size from single-celled diatoms to the massive and multicellular kelp.

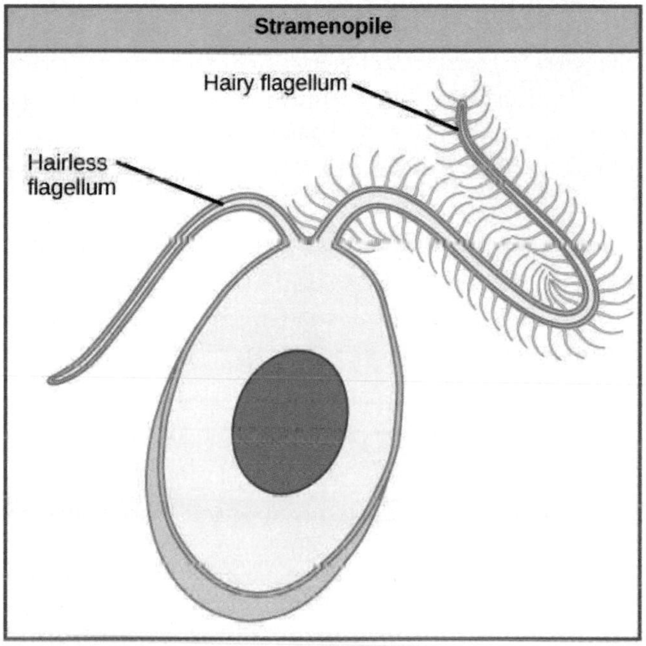

Figure 23.26 Stramenopile flagella. This stramenopile cell has a single hairy flagellum and a secondary smooth flagellum.

The diatoms are unicellular photosynthetic protists that encase themselves in intricately patterned, glassy cell walls composed of silicon dioxide in a matrix of organic particles (Figure 23.27). These protists are a component of freshwater and marine plankton. Most species of diatoms reproduce asexually, although some instances of sexual reproduction and sporulation also exist. Some diatoms exhibit a slit in their silica shell, called a **raphe**. By expelling a stream of mucopolysaccharides from the raphe, the diatom can attach to surfaces or propel itself in one direction.

Figure 23.27 Diatoms. Assorted diatoms, visualized here using light microscopy, live among annual sea ice in McMurdo Sound, Antarctica. Diatoms range in size from 2 to 200 μm. (credit: Prof. Gordon T. Taylor, Stony Brook University, NSF, NOAA)

During periods of nutrient availability, diatom populations bloom to numbers greater than can be consumed by aquatic organisms. The excess diatoms die and sink to the sea floor where they are not easily reached by saprobes that feed on dead organisms. As a result, the carbon dioxide that the diatoms had consumed and incorporated into their cells during photosynthesis is not returned to the atmosphere. Along with rhizarians and other shelled protists, diatoms help to maintain a balanced carbon cycle.

Like diatoms, golden algae are largely unicellular, although some species can form large colonies. Their characteristic gold color results from their extensive use of carotenoids, a group of photosynthetic pigments that are generally yellow or orange in color. Golden algae are found in both freshwater and marine environments, where they form a major part of the plankton community.

The brown algae are primarily marine, multicellular organisms that are known colloquially as seaweeds. Giant kelps are a type of brown alga. Some brown algae have evolved specialized tissues that resemble terrestrial plants, with root-like holdfasts, stem-like stipes, and leaf-like blades that are capable of photosynthesis. The stipes of giant kelps are enormous, extending in some cases for 60 meters. Like the green algae, brown algae have a variety of life cycles, including alternation of generations. In the brown algae genus *Laminaria*, haploid spores develop into multicellular gametophytes, which produce haploid gametes that combine to produce diploid organisms that then become multicellular organisms with a different structure from the haploid form (Figure 23.28).

VISUAL CONNECTION

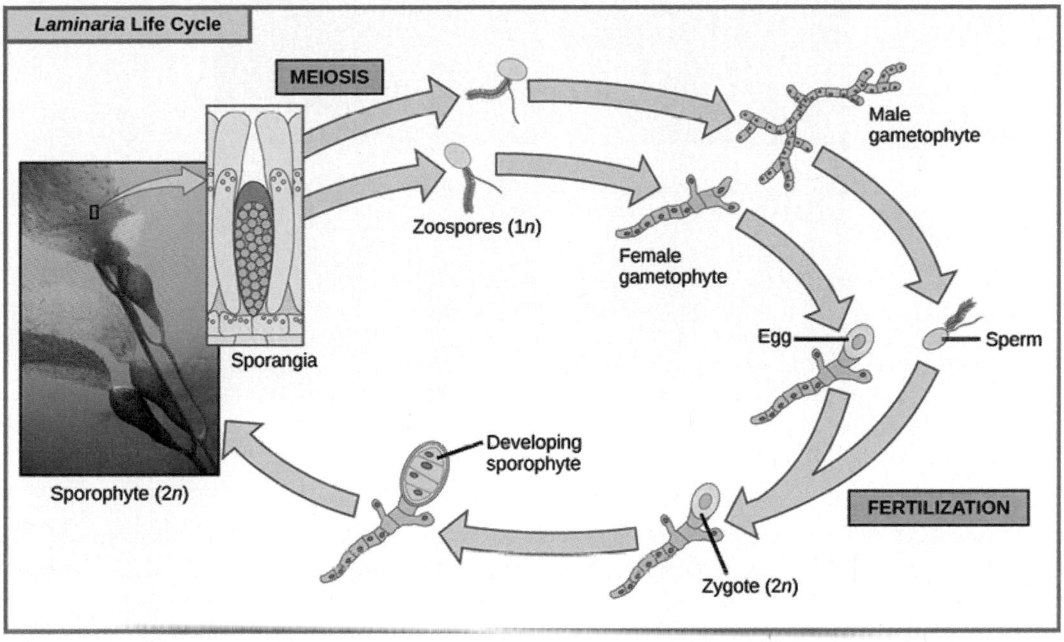

Figure 23.28 Alternation of generations in a brown alga. Several species of brown algae, such as the *Laminaria* shown here, have evolved life cycles in which both the haploid (gametophyte) and diploid (sporophyte) forms are multicellular. The gametophyte is different in structure than the sporophyte. (credit "laminaria photograph": modification of work by Claire Fackler, CINMS, NOAA Photo Library)

Which of the following statements about the *Laminaria* life cycle is false?

a. 1*n* zoospores form in the sporangia.
b. The sporophyte is the 2*n* plant.
c. The gametophyte is diploid.
d. Both the gametophyte and sporophyte stages are multicellular.

The water molds, oomycetes ("egg fungus"), were so-named based on their fungus-like morphology, but molecular data have shown that the water molds are not closely related to fungi. The oomycetes are characterized by a cellulose-based cell wall and an extensive network of filaments that allow for nutrient uptake. As diploid spores, many oomycetes have two oppositely directed flagella (one hairy and one smooth) for locomotion. The oomycetes are nonphotosynthetic and include many saprobes and parasites. The saprobes appear as white fluffy growths on dead organisms (Figure 23.29). Most oomycetes are aquatic, but some parasitize terrestrial plants. One plant pathogen is *Phytophthora infestans*, the causative agent of late blight of potatoes, such as occurred in the nineteenth century Irish potato famine.

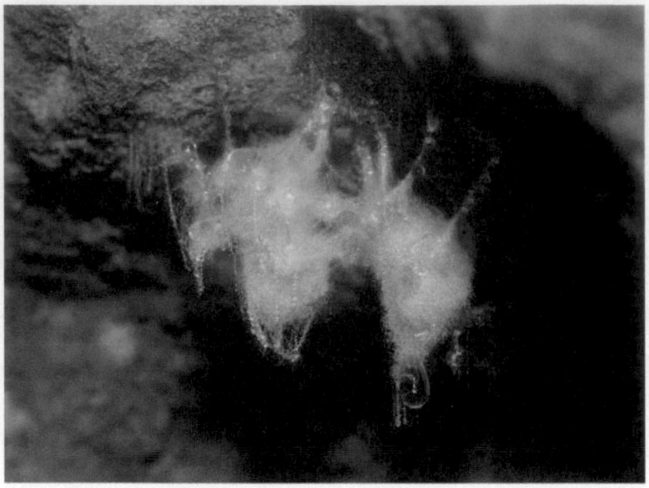

Figure 23.29 Oomycetes. A saprobic oomycete engulfs a dead insect. (credit: modification of work by Thomas Bresson)

Excavata

Many of the protist species classified into the supergroup Excavata are asymmetrical, single-celled organisms with a feeding groove "excavated" from one side. This supergroup includes heterotrophic predators, photosynthetic species, and parasites. Its subgroups are the diplomonads, parabasalids, and euglenozoans. The group includes a variety of modified mitochondria, as well as chloroplasts derived from green algae by secondary endosymbiosis. Many of the euglenozoans are free-living, but most diplomonads and parabasalids are symbionts or parasites.

Diplomonads

Among the Excavata are the diplomonads, which include the intestinal parasite, *Giardia lamblia* (Figure 23.30). Until recently, these protists were believed to lack mitochondria. Mitochondrial remnant organelles, called mitosomes, have since been identified in diplomonads, but although these mitosomes are essentially nonfunctional as respiratory organelles, they do function in iron and sulfur metabolism. Diplomonads exist in anaerobic environments and use alternative pathways, such as glycolysis, to generate energy. Each diplomonad cell has two similar, but not identical haploid nuclei. Diplomonads have four pairs of locomotor flagella that are fairly deeply rooted in basal bodies that lie between the two nuclei.

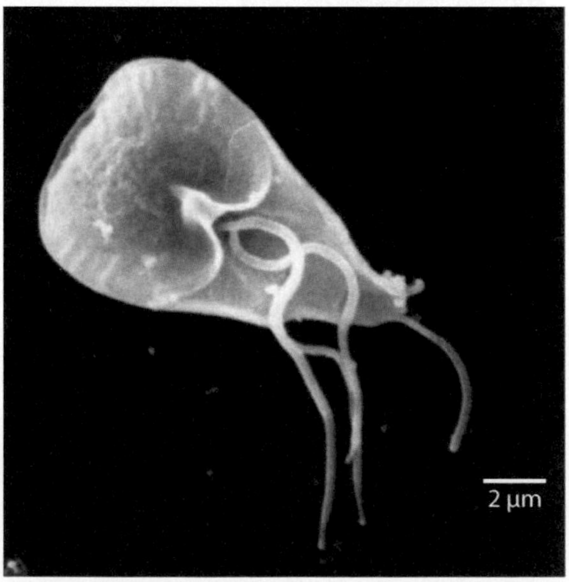

Figure 23.30 Giardia. The mammalian intestinal parasite *Giardia lamblia*, visualized here using scanning electron microscopy, is a waterborne protist that causes severe diarrhea when ingested. (credit: modification of work by Janice Carr, CDC; scale-bar data from Matt Russell)

Parabasalids

A second Excavata subgroup, the parabasalids, are named for the parabasal apparatus, which consists of a Golgi complex associated with cytoskeletal fibers. Other cytoskeletal features include an axostyle, a bundle of fibers that runs the length of the cell and may even extend beyond it. Parabasalids move with flagella and membrane rippling, and these and other cytoskeletal modifications may assist locomotion. Like the diplomonads, the parabasalids exhibit modified mitochondria. In parabasalids these structures function anaerobically and are called hydrogenosomes because they produce hydrogen gas as a byproduct.

The parabasalid *Trichomonas vaginalis* causes trichomoniasis, a sexually transmitted disease in humans, which appears in an estimated 180 million cases worldwide each year. Whereas men rarely exhibit symptoms during an infection with this protist, infected women may become more susceptible to secondary infection with human immunodeficiency virus (HIV) and may be more likely to develop cervical cancer. Pregnant women infected with *T. vaginalis* are at an increased risk of serious complications, such as pre-term delivery.

Some of the most complex of the parabasalids are those that colonize the rumen of ruminant animals and the guts of termites. These organisms can digest cellulose, a metabolic talent that is unusual among eukaryotic cells. They have multiple flagella arranged in complex patterns and some additionally recruit spirochetes that attach to their surface to act as accessory locomotor structures.

LINK TO LEARNING

Termite gut endosymbionts

Click to view content (https://www.youtube.com/embed/HOx7SDdIqyU)

Euglenozoans

Euglenozoans includes parasites, heterotrophs, autotrophs, and mixotrophs, ranging in size from 10 to 500 µm. Euglenoids move through their aquatic habitats using two long flagella that guide them toward light sources sensed by a primitive ocular organ called an eyespot. The familiar genus, *Euglena*, encompasses some mixotrophic species that display a photosynthetic capability only when light is present. The chloroplast of *Euglena* descends from a green alga by secondary endosymbiosis. In the dark, the chloroplasts of Euglena shrink up and temporarily cease functioning, and the cells instead take up organic nutrients from their environment. Euglena has a tough pellicle composed of bands of protein attached to the cytoskeleton. The bands spiral around the cell and give Euglena its exceptional flexibility.

The human parasite, *Trypanosoma brucei*, belongs to a different subgroup of Euglenozoa, the kinetoplastids. The kinetoplastid subgroup is named after the kinetoplast, a large modified mitochondrion carrying multiple circular DNAs. This subgroup includes several parasites, collectively called trypanosomes, which cause devastating human diseases and infect an insect species during a portion of their life cycle. *T. brucei* develops in the gut of the tsetse fly after the fly bites an infected human or other mammalian host. The parasite then travels to the insect salivary glands to be transmitted to another human or other mammal when the infected tsetse fly consumes another blood meal. *T. brucei* is common in central Africa and is the causative agent of African sleeping sickness, a disease associated with severe chronic fatigue, coma, and can be fatal if left untreated since it leads to progressive decline of the function of the central nervous system.

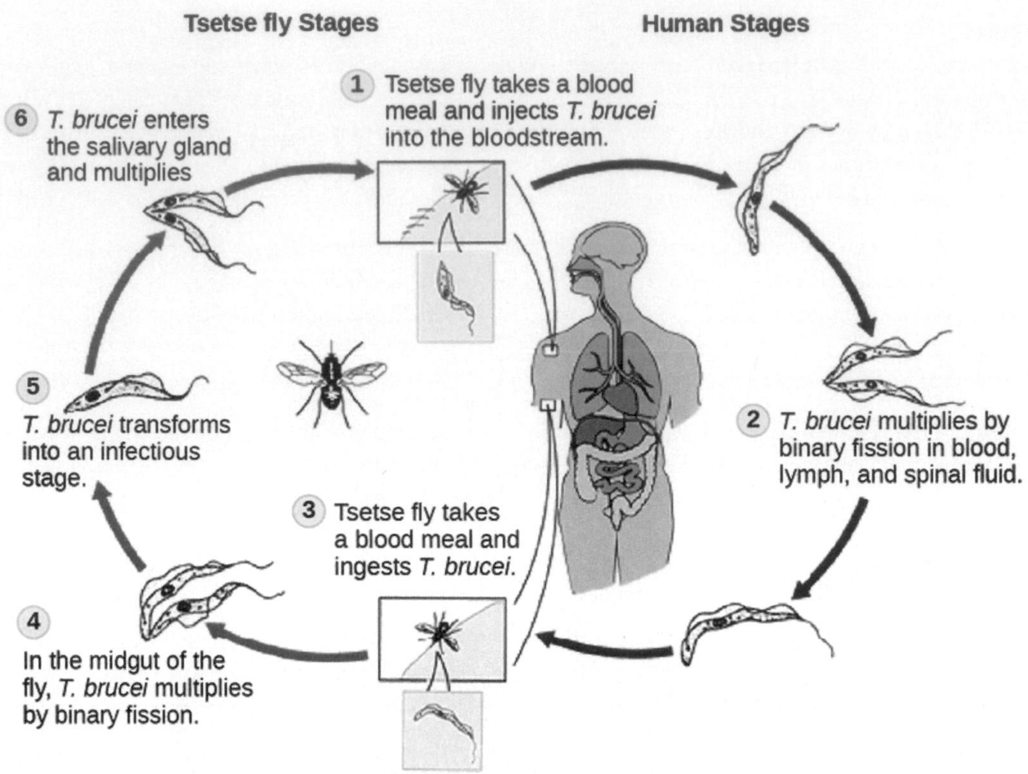

Figure 23.31 Sleeping sickness. *Trypanosoma brucei*, the causative agent of sleeping sickness, spends part of its life cycle in the tsetse fly and part in humans. (credit: modification of work by CDC)

LINK TO LEARNING

Watch this video to see *T. brucei* swimming.

Click to view content (https://www.openstax.org/l/T_brucei)

23.4 Ecology of Protists

By the end of this section, you will be able to do the following:

- Describe the role that protists play in the ecosystem
- Describe important pathogenic species of protists

Protists function in various ecological niches. Whereas some protist species are essential components of the food chain and generators of biomass, others function in the decomposition of organic materials. Still other protists are dangerous human pathogens or causative agents of devastating plant diseases.

Primary Producers/Food Sources

Protists are essential sources of food and provide nutrition for many other organisms. In some cases, as with zooplankton, protists are consumed directly. Alternatively, photosynthetic protists serve as producers of nutrition for other organisms. *Paramecium bursaria* and several other species of ciliates are *mixotrophic* due to a symbiotic relationship with green algae. This is a temporary version of the secondarily endosymbiotic chloroplast found in *Euglena*. But these symbiotic associations are not restricted to protists. For instance, photosynthetic dinoflagellates called *zooxanthellae* provide nutrients for the coral polyps (Figure 23.32) that house them, giving corals a boost of energy to secrete their calcium carbonate skeleton. In turn, the corals provide the protist with a protected environment and the compounds needed for photosynthesis. This type of symbiotic relationship is important in nutrient-poor environments. Without dinoflagellate symbionts, corals lose algal pigments in a process called *coral bleaching*, and they eventually die. This explains why reef-building corals typically do not reside in waters deeper than 20 meters: insufficient light reaches those depths for dinoflagellates to photosynthesize.

Figure 23.32 Coral with symbiotic dinoflagellates. Coral polyps obtain nutrition through a symbiotic relationship with dinoflagellates.

The protists and their products of photosynthesis are essential—directly or indirectly—to the survival of organisms ranging from bacteria to mammals (Figure 23.33). As primary producers, protists feed a large proportion of the world's aquatic species. (On land, terrestrial plants serve as primary producers.) In fact, approximately 25 percent of the world's photosynthesis is conducted by photosynthetic protists, particularly dinoflagellates, diatoms, and multicellular algae.

Figure 23.33 Protists contribute to the food chain. Virtually all aquatic organisms depend directly or indirectly on protists for food. (credit "mollusks": modification of work by Craig Stihler, USFWS; credit "crab": modification of work by David Berkowitz; credit "dolphin": modification of work by Mike Baird; credit "fish": modification of work by Tim Sheerman-Chase; credit "penguin": modification of work by Aaron Logan)

Protists do not create food sources only for sea-dwelling organisms. Recall that certain anaerobic parabasalid species exist in the digestive tracts of termites and wood-eating cockroaches, where they contribute an essential step in the digestion of cellulose ingested by these insects as they consume wood.

Human Pathogens

As we have seen, a pathogen is anything that causes disease. Parasitic organisms live in or on a host organism and harm the organism. A small number of protists are serious pathogenic parasites that must infect other organisms to survive and propagate. For example, protist parasites include the causative agents of malaria, African sleeping sickness, amoebic encephalitis, and waterborne gastroenteritis in humans. Other protist pathogens prey on plants, effecting massive destruction of food crops.

Plasmodium Species

In 2015 WHO reported over 200 million cases of malaria, mostly in Africa, South America, and southern Asia. However, it is not well known that malaria was also a prevalent and debilitating disease of the North Central region of the United States, particularly Michigan, with its thousands of lakes and numerous swamps. Prior to the civil war, and the drainage of many swamps, virtually everyone who immigrated to Michigan picked up malaria (*ague* as it was called in the late 1800s), and the pale, sallow, bloated faces of that period were the rule. The only healthy faces were worn by those immigrants who had just arrived. In fact, there were more deaths due to malaria in Michigan than those from the Civil War.

We now know that malaria is caused by several species of the apicomplexan protist genus *Plasmodium*. Members of *Plasmodium* must sequentially require both a mosquito and a vertebrate to complete their life cycle. In vertebrates, the parasite develops in liver cells (the exoerythrocytic stage) and goes on to infect red blood cells (the erythrocytic stage), bursting from and destroying the blood cells with each asexual replication cycle (Figure 23.34). Of the four *Plasmodium* species known to infect humans, *P. falciparum* accounts for 50 percent of all malaria cases and is the primary (and deadliest) cause of disease-related fatalities in tropical regions of the world. In 2015, it was estimated that malaria caused over 400,000 deaths, mostly in African children. During the course of malaria, *P. falciparum* can infect and destroy more than one-half of a human's circulating blood cells, leading to severe anemia. In response to waste products released as the parasites burst from infected blood cells, the host immune system mounts a massive inflammatory response with episodes of delirium-inducing fever (paroxysms) as parasites lyse red blood cells, spilling parasite waste into the bloodstream. *P. falciparum* is transmitted to humans by the African mosquito, *Anopheles gambiae*. Techniques to kill, sterilize, or avoid exposure to this highly aggressive mosquito species are crucial to malaria control. Ironically, a type of genetic control has arisen in parts of the world where malaria is endemic. Possession of one copy of the HbS beta globin allele results in malaria resistance. Unfortunately, this allele also has an unfortunate second effect; when homozygous it causes sickle cell disease.

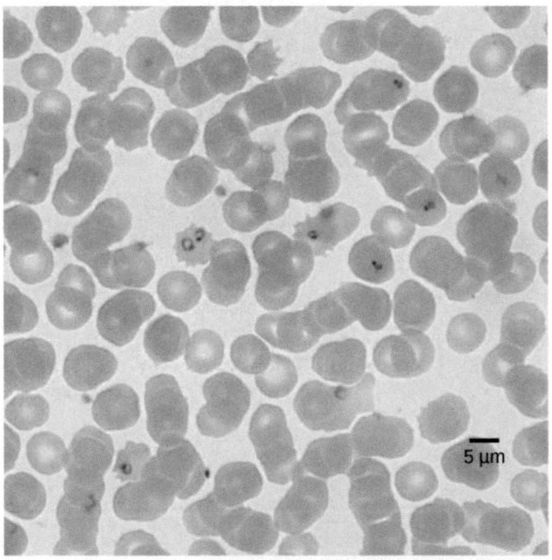

Figure 23.34 Malaria parasite. Red blood cells are shown to be infected with *P. falciparum*, the causative agent of malaria. In this light microscopic image taken using a 100× oil immersion lens, the ring-shaped *P. falciparum* stains purple. (credit: modification of work by Michael Zahniser; scale-bar data from Matt Russell)

LINK TO LEARNING

This movie (http://openstax.org/l/malaria) depicts the pathogenesis of *Plasmodium falciparum*, the causative agent of malaria.

Trypanosomes

Trypanosoma brucei (Figure 23.35), transmitted by tsetse flies (*Glossina* spp) in Africa, and related flies in South America, is an flagellated endoparasite responsible for the deadly disease nagana in cattle and horses, and for African sleeping sickness in humans. This trypanosome confounds the human immune system by changing its thick layer of surface glycoproteins with each infectious cycle. (The glycoproteins are identified by the immune system as foreign antigens, and a specific antibody defense is mounted against the parasite.) However, *T. brucei* has thousands of possible antigens, and with each subsequent generation, the protist switches to a glycoprotein coating with a different molecular structure. In this way, *T. brucei* is capable of replicating continuously without the immune system ever succeeding in clearing the parasite. Without treatment, *T. brucei* attacks red blood cells, causing the patient to lapse into a coma and eventually die. During epidemic periods, mortality from the disease can be high. Greater surveillance and control measures lead to a reduction in reported cases; some of the lowest numbers reported in 50 years (fewer than 10,000 cases in all of sub-Saharan Africa) have happened since 2009.

LINK TO LEARNING

This movie (http://openstax.org/l/African_sleep) discusses the pathogenesis of *Trypanosoma brucei*, the causative agent of African sleeping sickness.

In Latin America, another species of trypanosome, *T. cruzi*, is responsible for Chagas disease. *T. cruzi* infections are mainly caused by a blood-sucking "kissing bug" in the genus *Triatoma*. These "true bugs" bite the host during the night and then defecate on the wound, transmitting the trypanosome to the victim. The victim scratches the wound, further inoculating the site with trypanosomes at the location of the bite. After about 10 weeks, individuals enter the chronic phase but most never develop further symptoms. In about 30 percent of cases, however, the trypanosome causes further damage, especially to the heart and digestive system tissues in the chronic phase of infection, leading to malnutrition and heart failure due to abnormal heart rhythms. An estimated 10 million people are infected with Chagas disease, and it caused 10,000 deaths in 2008.

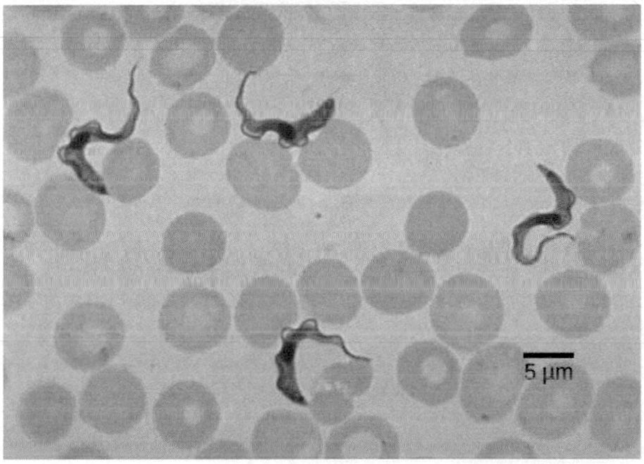

Figure 23.35 Trypanosomes. Trypanosomes are shown among red blood cells. (credit: modification of work by Dr. Myron G. Shultz; scale-bar data from Matt Russell)

Plant Parasites

Protist parasites of terrestrial plants include agents that destroy food crops. The oomycete *Plasmopara viticola* parasitizes grape plants, causing a disease called *downy mildew* (Figure 23.36). Grape plants infected with *P. viticola* appear stunted and have discolored, withered leaves. The spread of downy mildew nearly collapsed the French wine industry in the nineteenth century

Figure 23.36 Protist plant infections. Both downy and powdery mildews on this grape leaf are caused by an infection of *P. viticola*. (credit: modification of work by USDA)

Phytophthora infestans is an oomycete responsible for potato late blight, which causes potato stalks and stems to decay into black slime (Figure 23.37). Widespread potato blight caused by *P. infestans* precipitated the well-known Irish potato famine in the nineteenth century that claimed the lives of approximately 1 million people and led to the emigration of at least 1 million more from Ireland. Late blight continues to plague potato crops in certain parts of the United States and Russia, wiping out as much as 70 percent of crops when no pesticides are applied.

Figure 23.37 Potato blight. These unappetizing remnants result from an infection with *P. infestans*, the causative agent of potato late blight. (credit: USDA)

Protist Decomposers

The fungus-like protist *saprobes* are specialized to absorb nutrients from nonliving organic matter, such as dead organisms or their wastes. For instance, many types of oomycetes grow on dead animals or algae. Saprobic protists have the essential function of returning inorganic nutrients to the soil and water. This process allows for new plant growth, which in turn generates sustenance for other organisms along the food chain. Indeed, without saprobe species, such as protists, fungi, and bacteria, life would cease to exist as all organic carbon became "tied up" in dead organisms.

KEY TERMS

biological carbon pump process by which inorganic carbon is fixed by photosynthetic species that then die and fall to the sea floor where they cannot be reached by saprobes and their carbon dioxide consumption cannot be returned to the atmosphere

bioluminescence generation and emission of light by an organism, as in dinoflagellates

contractile vacuole vesicle that fills with water (as it enters the cell by osmosis) and then contracts to squeeze water from the cell; an osmoregulatory vesicle

cytoplasmic streaming movement of cytoplasm into an extended pseudopod such that the entire cell is transported to the site of the pseudopod

endosymbiosis engulfment of one cell within another such that the engulfed cell survives, and both cells benefit; the process responsible for the evolution of mitochondria and chloroplasts in eukaryotes

endosymbiotic theory theory that states that eukaryotes may have been a product of one cell engulfing another, one living within another, and evolving over time until the separate cells were no longer recognizable as such

hydrogenosome organelle carried by parabasalids (Excavata) that functions anaerobically and outputs hydrogen gas as a byproduct; likely evolved from mitochondria

kinetoplast mass of DNA carried within the single, oversized mitochondrion, characteristic of kinetoplastids (phylum: Euglenozoa)

mitosome nonfunctional organelle carried in the cells of diplomonads (Excavata) that likely evolved from a mitochondrion

mixotroph organism that can obtain nutrition by autotrophic or heterotrophic means, usually facultatively

pellicle outer cell covering composed of interlocking protein strips that function like a flexible coat of armor, preventing cells from being torn or pierced without compromising their range of motion

phagolysosome cellular body formed by the union of a phagosome containing the ingested particle with a lysosome that contains hydrolytic enzymes

plankton diverse group of mostly microscopic organisms that drift in marine and freshwater systems and serve as a food source for larger aquatic organisms

plastid one of a group of related organelles in plant cells that are involved in the storage of starches, fats, proteins, and pigments

raphe slit in the silica shell of diatoms through which the protist secretes a stream of mucopolysaccharides for locomotion and attachment to substrates

test porous shell of a foram that is built from various organic materials and typically hardened with calcium carbonate

CHAPTER SUMMARY

23.1 Eukaryotic Origins

The oldest fossil evidence of eukaryotes is about 2 billion years old. Fossils older than this all appear to be prokaryotes. It is probable that today's eukaryotes are descended from an ancestor that had a prokaryotic organization. The last common ancestor of today's Eukarya had several characteristics, including cells with nuclei and an endomembrane system (which includes the nuclear envelope). Its chromosomes were linear and contained DNA associated with histones. The nuclear genome seems to be descended from an archaean ancestor. This ancestor would have had a cytoskeleton and divided its chromosomes mitotically.

The ancestral cytoskeletal system included the ability to make cilia/flagella during at least part of its life cycle. It was aerobic because it had mitochondria derived from an aerobic alpha-proteobacterium that lived inside a host cell. Whether this host had a nucleus at the time of the initial symbiosis remains unknown. The last common ancestor may have had a cell wall for at least part of its life cycle, but more data are needed to confirm this hypothesis. Today's eukaryotes are very diverse in their shapes, organization, life cycles, and number of cells per individual.

23.2 Characteristics of Protists

Protists are extremely diverse in terms of their biological and ecological characteristics, partly because they are an artificial assemblage of phylogenetically unrelated groups. Protists display highly varied cell structures, several types of reproductive strategies, virtually every possible type of nutrition, and varied habitats. Most single-celled protists are motile, but these organisms use diverse structures for transportation.

23.3 Groups of Protists

The process of classifying protists into meaningful groups is ongoing, but genetic data in the past 20 years have clarified many relationships that were previously unclear or mistaken. The majority view at present is to order all eukaryotes into six supergroups: Archaeplastida, Amoebozoa, Opisthokonta, Rhizaria, Chromalveolata, and Excavata. The goal of this classification scheme is to create clusters of species that all are derived from a common ancestor. At present, the monophyly of some of the supergroups are better supported by genetic data than others. Although tremendous variation exists within the supergroups, commonalities at the morphological, physiological, and ecological levels can be identified.

23.4 Ecology of Protists

Protists function at several levels of the ecological food web: as primary producers, as direct food sources, and as decomposers. In addition, many protists are parasites of plants and animals and can cause deadly human diseases or destroy valuable crops.

VISUAL CONNECTION QUESTIONS

1. Figure 23.5 What evidence is there that mitochondria were incorporated into the ancestral eukaryotic cell before chloroplasts?

2. Figure 23.25 Which of the following statements about *Paramecium* sexual reproduction is false?
   a. The macronuclei are derived from micronuclei.
   b. Both mitosis and meiosis occur during sexual reproduction.
   c. The conjugate pair swaps macronuclei.
   d. Each parent produces four daughter cells.

3. Figure 23.28 Which of the following statements about the *Laminaria* life cycle is false?
   a. 1*n* zoospores form in the sporangia.
   b. The sporophyte is the 2*n* plant.
   c. The gametophyte is diploid.
   d. Both the gametophyte and sporophyte stages are multicellular.

REVIEW QUESTIONS

4. What event is thought to have contributed to the evolution of eukaryotes?
   a. global warming
   b. glaciation
   c. volcanic activity
   d. oxygenation of the atmosphere

5. Which characteristic is shared by prokaryotes and eukaryotes?
   a. cytoskeleton
   b. nuclear envelope
   c. DNA-based genome
   d. mitochondria

6. Mitochondria most likely evolved by ___________.
   a. a photosynthetic cyanobacterium
   b. cytoskeletal elements
   c. endosymbiosis
   d. membrane proliferation

7. Which of these protists is believed to have evolved following a secondary endosymbiosis?
   a. green algae
   b. cyanobacteria
   c. red algae
   d. chlorarachniophytes

8. In 2016, scientists published the genome of *Monocercomonoides*, and demonstrated that this organism has no detectable mitochondrial genes. However, its genome was arranged in linear chromosomes wrapped around histones which are contained within the nucleus. *Monocercomonoides* is therefore a(n) ________.
   a. Bacteria
   b. Archea
   c. Eukaryote
   d. Endosymbiont

9. Which of the following observations about a bacterium would distinguish it from the last eukaryotic common ancestor?
   a. A double-stranded DNA genome
   b. Lack of a membrane-bound structure surrounding the genome
   c. Fatty acids in the lipid bilayer of the plasma membrane
   d. Enclosed by a cell wall

10. Protists that have a pellicle are surrounded by ___________.
   a. silica dioxide
   b. calcium carbonate
   c. carbohydrates
   d. proteins

11. Protists with the capabilities to perform photosynthesis and to absorb nutrients from dead organisms are called ____________.
   a. photoautotrophs
   b. mixotrophs
   c. saprobes
   d. heterotrophs

12. Which of these locomotor organs would likely be the shortest?
   a. a flagellum
   b. a cilium
   c. an extended pseudopod
   d. a pellicle

13. Alternation of generations describes which of the following?
   a. The haploid form can be multicellular; the diploid form is unicellular.
   b. The haploid form is unicellular; the diploid form can be multicellular.
   c. Both the haploid and diploid forms can be multicellular.
   d. Neither the haploid nor the diploid forms can be multicellular.

14. The amoeba *E. histolytica* is a pathogen that forms liver abscesses in infected individuals. Its metabolic classification is most likely ______.
   a. Anaerobic heterotroph
   b. Mixotroph
   c. Aerobic phototroph
   d. Phagocytic autotroph

15. Which protist group exhibits mitochondrial remnants with reduced functionality?
   a. slime molds
   b. diatoms
   c. parabasalids
   d. dinoflagellates

16. Conjugation between two *Paramecia* produces _______ total daughter cells.
   a. 2
   b. 4
   c. 8
   d. 16

17. What is the function of the raphe in diatoms?
   a. locomotion
   b. defense
   c. capturing food
   d. photosynthesis

18. What genus of protists appears to contradict the statement that unicellularity restricts cell size?
   a. *Dictyostelium*
   b. *Ulva*
   c. *Plasmodium*
   d. *Caulerpa*

19. A marine biologist analyzing water samples notices a protist with a calcium carbonate shell that moves by pseudopodia extension. The protist is likely to be closely related to which species?
   a. *Fuligo septica* (Dog Vomit slime mold)
   b. *Circogonia icosahedra* (Radiolarian)
   c. *Euglena viridis*
   d. *Ammonia tepida*

20. An example of carbon fixation is ____________.
   a. photosynthesis
   b. decomposition
   c. phagocytosis
   d. parasitism

21. Which parasitic protist evades the host immune system by altering its surface proteins with each generation?
   a. *Paramecium caudatum*
   b. *Trypanosoma brucei*
   c. *Plasmodium falciparum*
   d. *Phytophthora infestans*

22. Which of the following is **not** a way that protists contribute to the food web?
   a. They fix carbon into organic molecules.
   b. They occupy the apex producer niche.
   c. They enter symbiotic relationships with animals.
   d. They recycle nutrients back into the carbon and nitrogen cycles.

CRITICAL THINKING QUESTIONS

23. Describe the hypothesized steps in the origin of eukaryotic cells.

24. Some aspects of eukaryotes are more similar to Archaea, while other aspects of eukaryotic cell composition appear more closely related to Bacteria. Explain how endosymbiosis could resolve this paradox.

25. Explain in your own words why sexual reproduction can be useful if a protist's environment changes.

26. *Giardia lamblia* is a cyst-forming protist parasite that causes diarrhea if ingested. Given this information, against what type(s) of environments might *G. lamblia* cysts be particularly resistant?

27. Explain how the definition of protists ensures that the kingdom Protista includes a wide diversity of cellular structures. Provide an example of two different structures that perform the same function for their respective protist.

28. The chlorophyte (green algae) genera *Ulva* and *Caulerpa* both have macroscopic leaf-like and stem-like structures, but only *Ulva* species are considered truly multicellular. Explain why.

29. Why might a light-sensing eyespot be ineffective for an obligate saprobe? Suggest an alternative organ for a saprobic protist.

30. Opisthokonta includes animals and fungi, as well as protists. Describe the key feature of this phylum, and an example of how an organism in each kingdom uses this feature.

31. Describe two ways in which paramecium differs from the projected traits of the last eukaryotic common ancestor.

32. How does killing *Anopheles* mosquitoes affect the *Plasmodium* protists?

33. Without treatment, why does African sleeping sickness invariably lead to death?

34. Describe how increasing stress to the ocean would affect a food chain containing zooxanthellae, corals, parrotfish, and sharks.

CHAPTER 24
Fungi

Figure 24.1 Many species of fungus produce the familiar mushroom (a) which is a reproductive structure. This (b) coral fungus displays brightly colored fruiting bodies. This electron micrograph shows (c) the spore-bearing structures of *Aspergillus*, a type of toxic fungus found mostly in soil and plants. (credit "mushroom": modification of work by Chris Wee; credit "coral fungus": modification of work by Cory Zanker; credit "*Aspergillus*": modification of work by Janice Haney Carr, Robert Simmons, CDC; scale-bar data from Matt Russell)

INTRODUCTION The word *fungus* comes from the Latin word for mushrooms. Indeed, the familiar mushroom is a reproductive structure used by many types of fungi. However, there are also many fungus species that don't produce mushrooms at all. Being eukaryotes, a typical fungal cell contains a true nucleus and many membrane-bound organelles. The kingdom Fungi includes an enormous variety of living organisms collectively referred to as Eumycota, or true Fungi. While scientists have identified about 100,000 species of fungi, this is only a fraction of the 1.5 million species of fungus likely present on Earth. Edible mushrooms, yeasts, black mold, and the producer of the antibiotic penicillin, *Penicillium notatum*, are all members of the kingdom Fungi, which belongs to the domain Eukarya.

Chapter Outline

24.1 Characteristics of Fungi

By the end of this section, you will be able to do the following:

- List the characteristics of fungi
- Describe the composition of the mycelium
- Describe the mode of nutrition of fungi
- Explain sexual and asexual reproduction in fungi

Fungi, once considered plant-like organisms, are more closely related to animals than plants. Fungi are not capable of photosynthesis: they are heterotrophic because they use complex organic compounds as sources of energy and carbon. Fungi share a few other traits with animals. Their cell walls are composed of chitin, which is found in the exoskeletons of arthropods. Fungi produce a number of pigments, including melanin,also found in the hair and skin of animals. Like animals, fungi also store carbohydrates as glycogen. However, like bacteria, fungi absorb nutrients across the cell surface and act as decomposers, helping to recycle nutrients by breaking down organic materials to simple molecules.

Some fungal organisms multiply only asexually, whereas others undergo both asexual reproduction and sexual reproduction with alternation of generations. Most fungi produce a large number of **spores**, which are haploid cells that can undergo mitosis to form multicellular, haploid individuals.

Fungi often interact with other organisms, forming beneficial or mutualistic associations. For example, most terrestrial plants form symbiotic relationships with fungi. The roots of the plant connect with the underground parts of the fungus, which form **mycorrhizae**. Through mycorrhizae, the fungus and plant exchange nutrients and water, greatly aiding the survival of both species. Alternatively, lichens are an association between a fungus and its photosynthetic partner (usually an alga).

Fungi also cause serious infections in plants and animals. For example, Dutch elm disease, which is caused by the fungus *Ophiostoma ulmi*, is a particularly devastating type of fungal infestation that destroys many native species of elm (*Ulmus* sp.) by infecting the tree's vascular system. The elm bark beetle acts as a vector, transmitting the disease from tree to tree. Accidentally introduced in the 1900s, the fungus decimated elm trees across the continent. Many European and Asiatic elms are less susceptible to Dutch elm disease than American elms.

In humans, fungal infections are generally considered challenging to treat. Unlike bacteria, fungi do not respond to traditional antibiotic therapy, since they are eukaryotes. Fungal infections may prove deadly for individuals with compromised immune systems.

Fungi have many commercial applications. The food industry uses yeasts in baking, brewing, and cheese and wine making. Many industrial compounds are byproducts of fungal fermentation. Fungi are the source of many commercial enzymes and antibiotics.

Although humans have used yeasts and mushrooms since prehistoric times, until recently, the biology of fungi was poorly understood. In fact, up until the mid-20th century, many scientists classified fungi as plants! Fungi, like plants, are mostly sessile and seemingly rooted in place. They possess a stem-like structure similar to plants, as well as having a root-like fungal mycelium in the soil. In addition, their mode of nutrition was poorly understood. Progress in the field of fungal biology was the result of **mycology**: the scientific study of fungi. Based on fossil evidence, fungi appeared in the pre-Cambrian era, about 450 million years ago. Molecular biology analysis of the fungal genome demonstrates that fungi are more closely related to animals than plants. Under some current systematic phylogenies, they continue to be a *polyphyletic group* of organisms that share characteristics, rather than sharing a single common ancestor.

CAREER CONNECTION

Mycologist

Mycologists are biologists who study fungi. Historically, mycology was a branch of microbiology, and many mycologists start their careers with a degree in microbiology. To become a mycologist, a bachelor's degree in a biological science (preferably majoring in microbiology) and a master's degree in mycology are minimally necessary. Mycologists can specialize in taxonomy and fungal genomics, molecular and cellular biology, plant pathology, biotechnology, or biochemistry. Some medical microbiologists concentrate on the study of infectious diseases caused by fungi, called mycoses. Mycologists collaborate with zoologists and plant pathologists to identify and control difficult fungal infections, such as the devastating chestnut blight, the mysterious decline in frog populations in many areas of the world, or the deadly epidemic called white nose syndrome, which is decimating bats in the Eastern United States.

Government agencies hire mycologists as research scientists and technicians to monitor the health of crops, national parks, and national forests. Mycologists are also employed in the private sector by companies that develop chemical and biological control products or new agricultural products, and by companies that provide disease control services. Because of the key role played by fungi in the fermentation of alcohol and the preparation of many important foods, scientists with a good understanding of fungal physiology routinely work in the food technology industry. Oenology, the science of wine making, relies not only on the knowledge of grape varietals and soil composition, but also on a solid understanding of the characteristics of the wild yeasts that thrive in different wine-making regions. It is possible to purchase yeast strains isolated from specific grape-growing regions. The great French chemist and microbiologist, Louis Pasteur, made many of his essential discoveries working on the humble brewer's yeast, thus discovering the process of *fermentation*.

Cell Structure and Function

Fungi are eukaryotes, and as such, have a complex cellular organization. As eukaryotes, fungal cells contain a membrane-bound nucleus. The DNA in the nucleus is wrapped around histone proteins, as is observed in other eukaryotic cells. A few types of fungi have accessory genomic structures comparable to bacterial plasmids (loops of DNA), however, the horizontal transfer of genetic information that occurs between one bacterium and another rarely occurs in fungi. Fungal cells also contain mitochondria and a complex system of internal membranes, including the endoplasmic reticulum and Golgi apparatus.

Unlike plant cells, fungal cells do not have chloroplasts or chlorophyll. Many fungi display bright colors arising from other cellular pigments, ranging from red to green to black. The poisonous *Amanita muscaria* (fly agaric) is recognizable by its bright red cap with white patches (Figure 24.2). Pigments in fungi are associated with the cell wall and play a protective role against ultraviolet radiation. Some fungal pigments are toxic to humans.

Figure 24.2 Amanita. The poisonous *Amanita muscaria* is native to temperate and boreal regions of North America. (credit: Christine Majul)

Like plant cells, fungal cells have a thick cell wall. The rigid layers of fungal cell walls contain complex polysaccharides called *chitin* and *glucans*. Chitin **(N-acetyl-D-glucosamine)**, also found in the exoskeleton of arthropods such as insects, gives structural strength to the cell walls of fungi. The wall protects the cell from desiccation and some predators. Fungi have plasma membranes similar to those of other eukaryotes, except that the structure is stabilized by *ergosterol*: a steroid molecule that replaces the cholesterol found in animal cell membranes. Most members of the kingdom Fungi are nonmotile. However, flagella are produced by the spores and gametes in the primitive Phylum Chytridiomycota.

Growth

The vegetative body of a fungus is a unicellular or multicellular *thallus*. Unicellular fungi are called yeasts. Multicellular fungi produce threadlike *hyphae* (singular hypha). Dimorphic fungi can change from the unicellular to multicellular state depending on environmental conditions. *Saccharomyces cerevisiae* (baker's yeast) and *Candida* species (the agents of thrush, a common fungal infection) are examples of unicellular fungi (Figure 24.3).

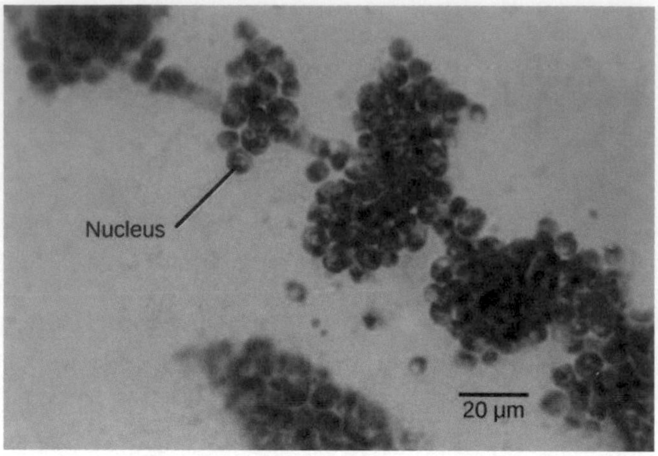

Figure 24.3 Candida albicans. *Candida albicans* is a yeast cell and the agent of *candidiasis* and *thrush*. This organism has a similar morphology to coccus bacteria; however, yeast is a eukaryotic organism (note the nucleus). (credit: modification of work by Dr. Godon Roberstad, CDC; scale-bar data from Matt Russell)

Most fungi are multicellular organisms. They display two distinct morphological stages: the vegetative and reproductive. The vegetative stage consists of a tangle of hyphae, whereas the reproductive stage can be more conspicuous. The mass of hyphae is a **mycelium** (Figure 24.4). It can grow on a surface, in soil or decaying material, in a liquid, or even on living tissue. Although individual hyphae must be observed under a microscope, the mycelium of a fungus can be very large, with some species truly being "the fungus humongous." The giant *Armillaria solidipes* (honey mushroom) is considered the largest organism on Earth, spreading across more than 2,000 acres of underground soil in eastern Oregon; it is estimated to be at least 2,400 years old.

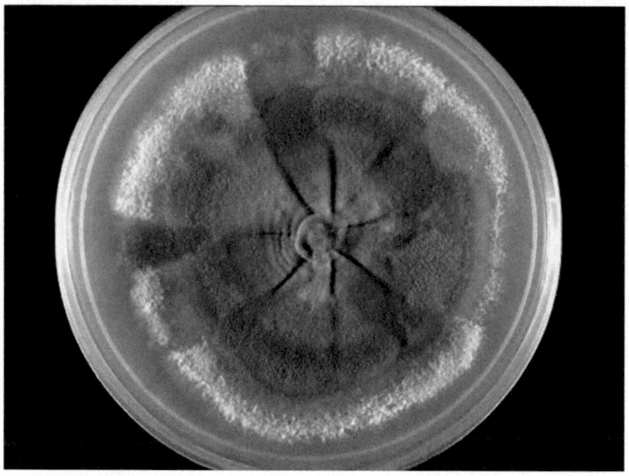

Figure 24.4 A fungal mycelium. The mycelium of the fungus *Neotestudina rosati* can be pathogenic to humans. The fungus enters through a cut or scrape and develops a *mycetoma*, a chronic subcutaneous infection. (credit: CDC)

Most fungal hyphae are divided into separate cells by *endwalls* called **septa** (singular, **septum**) (Figure 24.5a, c). In most phyla of fungi, tiny holes in the septa allow for the rapid flow of nutrients and small molecules from cell to cell along the hypha. They are described as *perforated septa*. The hyphae in bread molds (which belong to the Phylum Zygomycota) are not separated by septa. Instead, they are formed by large cells containing many nuclei (multinucleate), an arrangement described as *coenocytic hyphae* (Figure 24.5b).

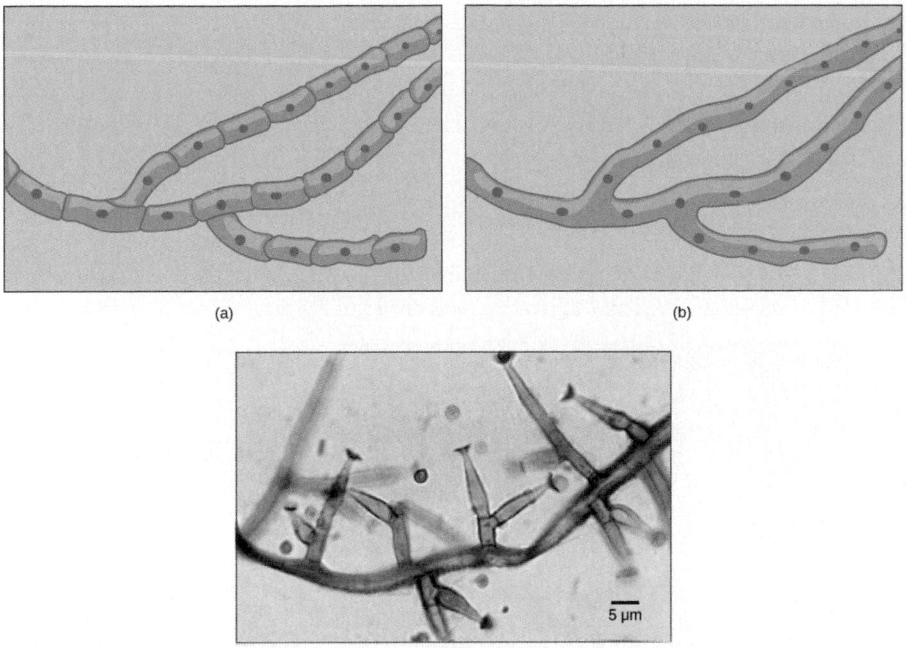

Figure 24.5 Fungal hyphae. Fungal hyphae may be (a) septated or (b) coenocytic (coeno- = "common"; -cytic = "cell") with many nuclei present in a single hypha. A bright field light micrograph of (c) *Phialophora richardsiae* shows septa that divide the hyphae. (credit c: modification of work by Dr. Lucille Georg, CDC; scale-bar data from Matt Russell)

Fungi thrive in environments that are moist and slightly acidic, and can grow with or without light. They vary in their oxygen requirement. Most fungi are **obligate aerobes**, requiring oxygen to survive. Other species, such as members of the Chytridiomycota that reside in the rumen of cattle, are **obligate anaerobes**, in that they only use anaerobic respiration because oxygen will disrupt their metabolism or kill them. Yeasts are intermediate, being **facultative anaerobes**. This means that they grow best in the presence of oxygen using aerobic respiration, but can survive using anaerobic respiration when oxygen is not available. The alcohol produced from yeast fermentation is used in wine and beer production.

Nutrition

Like animals, fungi are heterotrophs; they use complex organic compounds as a source of carbon, rather than fix carbon dioxide from the atmosphere as do some bacteria and most plants. In addition, fungi do not fix nitrogen from the atmosphere. Like animals, they must obtain it from their diet. However, unlike most animals, which ingest food and then digest it internally in specialized organs, fungi perform these steps in the reverse order; digestion precedes ingestion. First, *exoenzymes* are transported out of the hyphae, where they process nutrients in the environment. Then, the smaller molecules produced by this *external digestion* are absorbed through the large surface area of the mycelium. As with animal cells, the polysaccharide of storage is *glycogen*, a branched polysaccaride, rather than amylopectin, a less densely branched polysaccharide, and amylose, a linear polysaccharide, as found in plants.

Fungi are mostly **saprobes** (saprophyte is an equivalent term): organisms that derive nutrients from decaying organic matter. They obtain their nutrients from dead or decomposing organic material derived mainly from plants. Fungal exoenzymes are able to break down insoluble compounds, such as the cellulose and lignin of dead wood, into readily absorbable glucose molecules. The carbon, nitrogen, and other elements are thus released into the environment. Because of their varied metabolic pathways, fungi fulfill an important ecological role and are being investigated as potential tools in *bioremediation* of chemically damaged ecosystems. For example, some species of fungi can be used to break down diesel oil and polycyclic aromatic hydrocarbons (PAHs). Other species take up heavy metals, such as cadmium and lead.

Some fungi are parasitic, infecting either plants or animals. Smut and Dutch elm disease affect plants, whereas athlete's foot and candidiasis (thrush) are medically important fungal infections in humans. In environments poor in nitrogen, some fungi resort to predation of nematodes (small non-segmented roundworms). In fact, species of *Arthrobotrys* fungi have a number of mechanisms to trap nematodes: One mechanism involves constricting rings within the network of hyphae. The rings swell when they touch the nematode, gripping it in a tight hold. The fungus then penetrates the tissue of the worm by extending specialized hyphae called **haustoria**. Many parasitic fungi possess haustoria, as these structures penetrate the tissues of the host, release digestive enzymes within the host's body, and absorb the digested nutrients.

Reproduction

Fungi reproduce sexually and/or asexually. Perfect fungi reproduce both sexually and asexually, while the so-called imperfect fungi reproduce only asexually (by mitosis).

In both sexual and asexual reproduction, fungi produce spores that disperse from the parent organism by either floating on the wind or hitching a ride on an animal. Fungal spores are smaller and lighter than plant seeds. For example, the giant puffball mushroom bursts open and releases trillions of spores in a massive cloud of what looks like finely particulate dust. The huge number of spores released increases the likelihood of landing in an environment that will support growth (Figure 24.6).

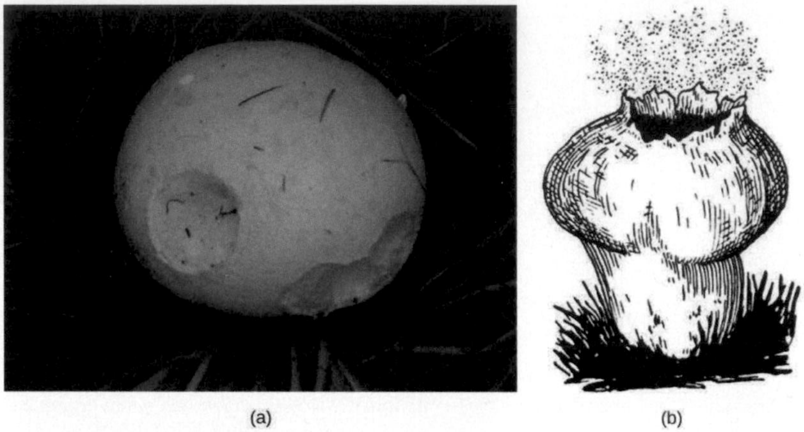

Figure 24.6 Puffball and spores. The (a) giant puffball mushroom releases (b) a cloud of spores when it reaches maturity. (credit a: modification of work by Roger Griffith; credit b: modification of work by Pearson Scott Foresman, donated to the Wikimedia Foundation)

Asexual Reproduction

Fungi reproduce asexually by *fragmentation, budding*, or *producing spores*. Fragments of hyphae can grow new colonies. Somatic cells in yeast form buds. During budding (an expanded type of cytokinesis), a bulge forms on the side of the cell, the nucleus divides mitotically, and the bud ultimately detaches itself from the mother cell (Figure 24.7).

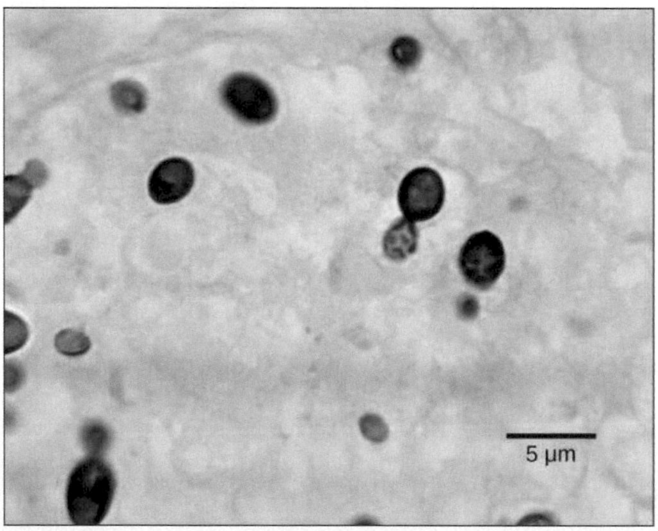

Figure 24.7 Budding in *Histoplasma*. The dark cells in this bright field light micrograph are the pathogenic yeast *Histoplasma capsulatum,* seen against a backdrop of light blue tissue. Histoplasma primarily infects lungs but can spread to other tissues, causing histoplasmosis, a potentially fatal disease. (credit: modification of work by Dr. Libero Ajello, CDC; scale-bar data from Matt Russell)

The most common mode of asexual reproduction is through the formation of asexual spores, which are produced by a single individual *thallus* (through mitosis) and are genetically identical to the parent thallus (Figure 24.8). Spores allow fungi to expand their distribution and colonize new environments. They may be released from the parent thallus either outside or within a special reproductive sac called a **sporangium**.

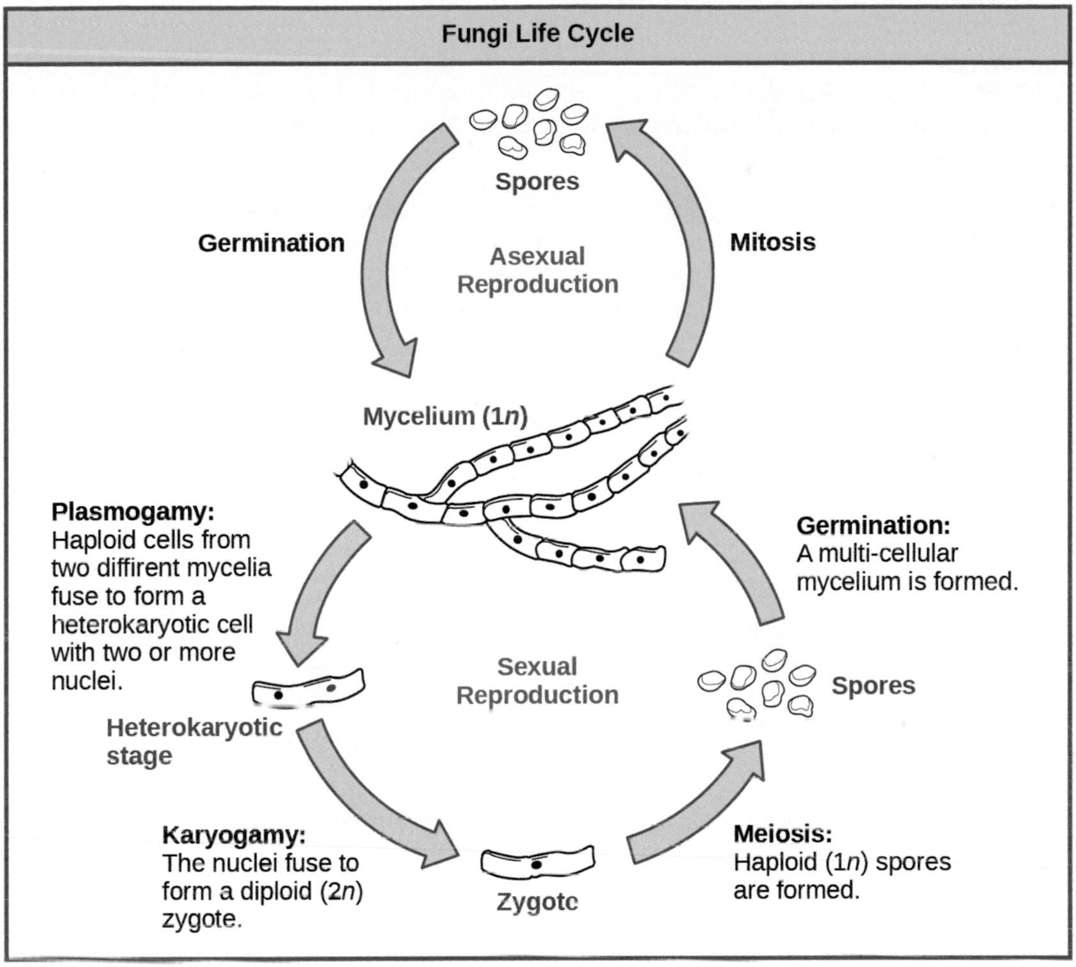

Figure 24.8 Generalized fungal life cycle. Fungi may have both asexual and sexual stages of reproduction.

There are many types of asexual spores. **Conidiospores** are unicellular or multicellular spores that are released directly from the tip or side of the hypha. Other asexual spores originate in the fragmentation of a hypha to form single cells that are released as spores; some of these have a thick wall surrounding the fragment. Yet others bud off the vegetative parent cell. In contrast to conidiospores, sporangiospores are produced directly from a sporangium (Figure 24.9).

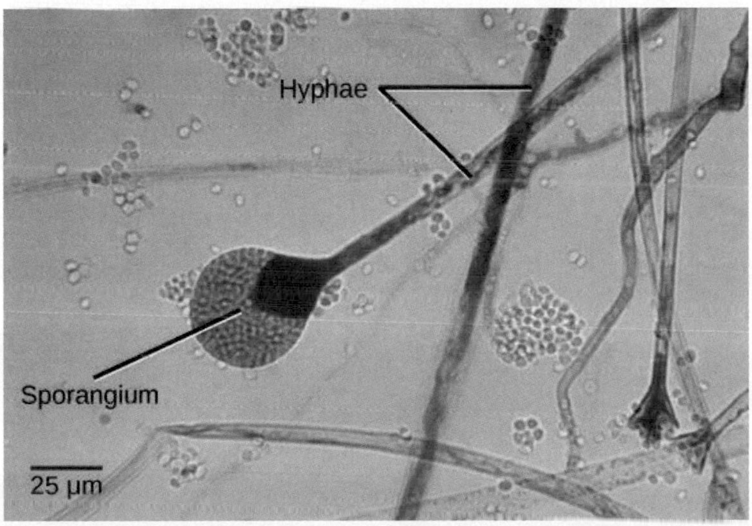

Figure 24.9 Sporangiospores. This bright field light micrograph shows the release of spores from a sporangium at the end of a hypha called a sporangiophore. The organism is a *Mucor* sp. fungus, a mold often found indoors. (credit: modification of work by Dr. Lucille Georg, CDC; scale-bar data from Matt Russell)

Sexual Reproduction

Sexual reproduction introduces genetic variation into a population of fungi. In fungi, *sexual reproduction* often occurs in response to adverse environmental conditions. During sexual reproduction, two *mating types* are produced. When both mating types are present in the same mycelium, it is called **homothallic**, or self-fertile. **Heterothallic** mycelia require two different, but compatible, mycelia to reproduce sexually.

Although there are many variations in fungal sexual reproduction, all include the following three stages (Figure 24.8). First, during **plasmogamy** (literally, "marriage or union of cytoplasm"), two haploid cells fuse, leading to a dikaryotic stage where two haploid nuclei coexist in a single cell. During **karyogamy** ("nuclear marriage"), the haploid nuclei fuse to form a diploid zygote nucleus. Finally, meiosis takes place in the gametangia (singular, gametangium) organs, in which gametes of different mating types are generated. At this stage, spores are disseminated into the environment.

LINK TO LEARNING

Review the characteristics of fungi by visiting this interactive site (http://openstax.org/l/fungi_kingdom) from Wisconsin-online.

24.2 Classifications of Fungi

By the end of this section, you will be able to do the following:

- Identify fungi and place them into the five major phyla according to current classification
- Describe each phylum in terms of major representative species and patterns of reproduction

The kingdom Fungi contains five major phyla that were established according to their mode of sexual reproduction or using molecular data. Polyphyletic, unrelated fungi that reproduce without a sexual cycle, were once placed for convenience in a sixth group, the Deuteromycota, called a "form phylum," because superficially they appeared to be similar. However, most mycologists have discontinued this practice. Rapid advances in molecular biology and the sequencing of 18S rRNA (ribosomal RNA) continue to show new and different relationships among the various categories of fungi.

The five true phyla of fungi are the Chytridiomycota (Chytrids), the Zygomycota (conjugated fungi), the Ascomycota (sac fungi), the Basidiomycota (club fungi) and the recently described Phylum Glomeromycota (Figure 24.10).

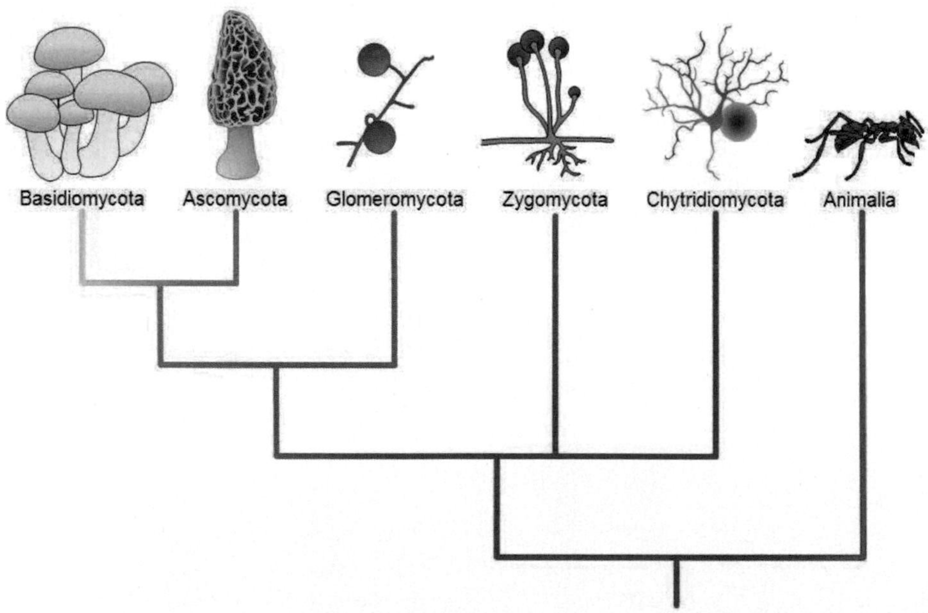

Figure 24.10 Fungal phyla. Note: "-mycota" is used to designate a phylum while "-mycetes" formally denotes a class or is used informally to refer to all members of the phylum.

Chytridiomycota: The Chytrids

The only class in the Phylum Chytridiomycota is the **Chytridiomycetes**. The chytrids are the simplest and most primitive

Eumycota, or true fungi. The evolutionary record shows that the first recognizable chytrids appeared during the late pre-Cambrian period, more than 500 million years ago. Like all fungi, chytrids have chitin in their cell walls, but one group of chytrids has both cellulose and chitin in the cell wall. Most chytrids are unicellular; however, a few form multicellular organisms and hyphae, which have no septa between cells (coenocytic). The Chytrids are the only fungi that have retained flagella. They produce both gametes and diploid zoospores that swim with the help of a single flagellum. An unusual feature of the chytrids is that both male and female gametes are flagellated.

The ecological habitat and cell structure of chytrids have much in common with protists. Chytrids usually live in aquatic environments, although some species live on land. Some species thrive as parasites on plants, insects, or amphibians (Figure 24.11), while others are saprobes. The chytrid species *Allomyces* is well characterized as an experimental organism. Its reproductive cycle includes both asexual and sexual phases. *Allomyces* produces diploid or haploid flagellated zoospores in a sporangium.

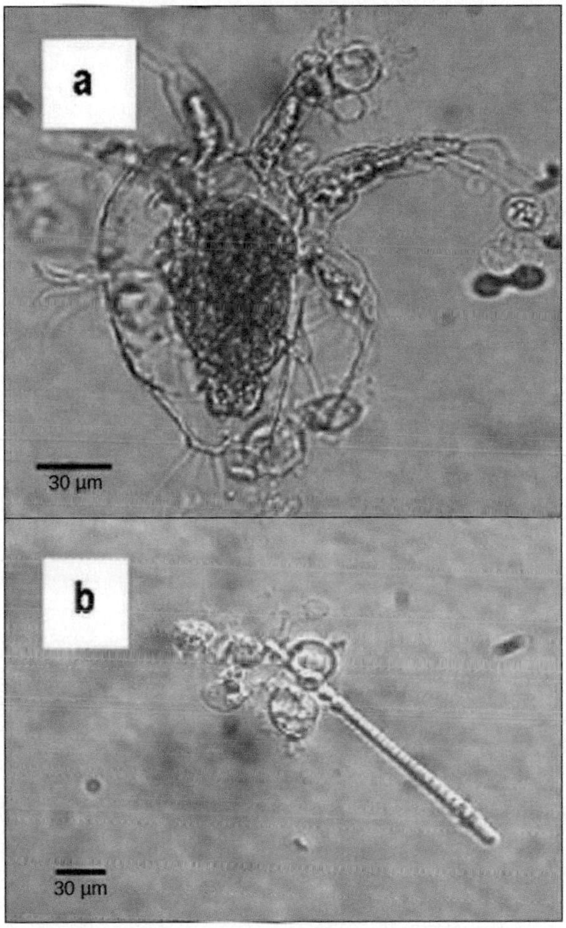

Figure 24.11 Chytrids. The chytrid *Batrachochytrium dendrobatidis* is seen in these light micrographs as transparent spheres growing on (a) a freshwater arthropod (water mite) and (b) algae. This chytrid causes skin diseases in many species of amphibians, resulting in species decline and extinction. (credit: modification of work by Johnson ML, Speare R., CDC)

Zygomycota: The Conjugated Fungi

The zygomycetes are a relatively small group of fungi belonging to the Phylum **Zygomycota**. They include the familiar bread mold, *Rhizopus stolonifer*, which rapidly propagates on the surfaces of breads, fruits, and vegetables. Most species are saprobes, living off decaying organic material; a few are parasites, particularly of insects. Zygomycetes play a considerable commercial role. For example, the metabolic products of some species of *Rhizopus* are intermediates in the synthesis of semi-synthetic steroid hormones.

Zygomycetes have a thallus of coenocytic hyphae in which the nuclei are haploid when the organism is in the vegetative stage. The fungi usually reproduce asexually by producing sporangiospores (Figure 24.12). The black tips of bread mold are the swollen sporangia packed with black spores (Figure 24.13). When spores land on a suitable substrate, they germinate and produce a new mycelium. Sexual reproduction starts when environmental conditions become unfavorable. Two opposing mating strains (type + and type –) must be in close proximity for gametangia from the hyphae to be produced and fuse, leading to karyogamy. Each zygospore can contain several diploid nuclei. The developing diploid **zygospores** have thick coats that protect them from desiccation and other hazards. They may remain dormant until environmental conditions are favorable. When the zygospore germinates, it undergoes meiosis and produces haploid spores, which will, in turn, grow into a new organism. This form of sexual reproduction in fungi is called conjugation (although it differs markedly from conjugation in bacteria and protists), giving rise to the name "conjugated fungi".

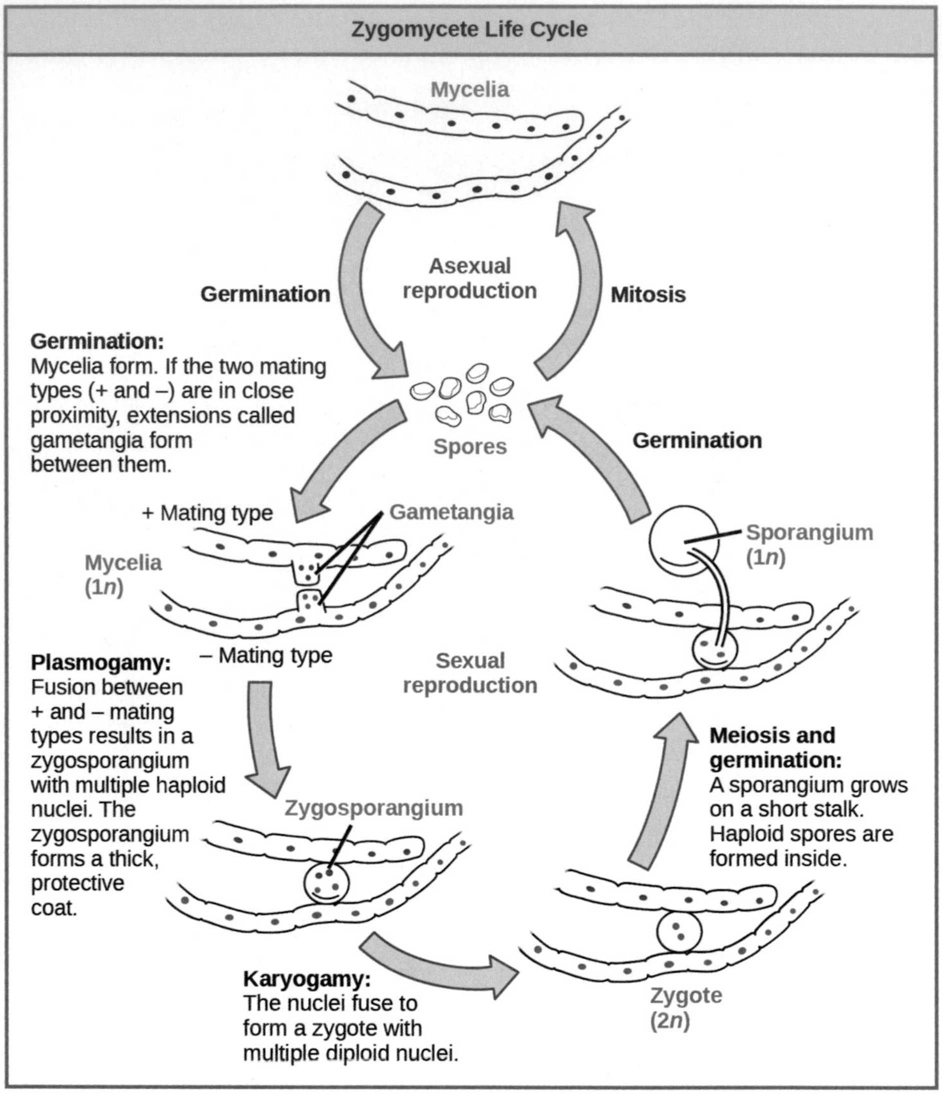

Figure 24.12 Zygomycete life cycle. Zygomycetes have asexual and sexual phases in their life cycles. In the asexual phase, spores are produced from haploid sporangia by mitosis (not shown). In the sexual phase, plus and minus haploid mating types conjugate to form a heterokaryotic zygosporangium. Karyogamy then produces a diploid zygote. Diploid cells in the zygote undergo meiosis and germinate to form a haploid sporangium, which releases the next generation of haploid spores.

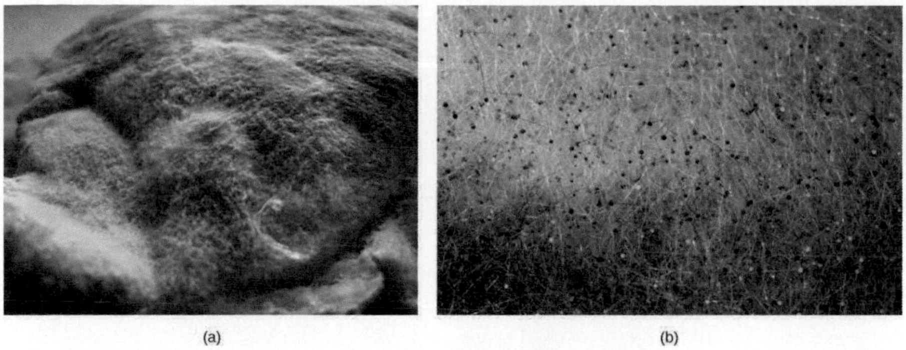

Figure 24.13 Rhizopus spores. Asexual sporangia grow at the end of stalks, which appear as (a) white fuzz seen on this bread mold, *Rhizopus stolonifer*. The black tips (b) of bread mold are the spore-containing sporangia. (credit b: modification of work by "polandeze"/Flickr)

Ascomycota: The Sac Fungi

The majority of known fungi belong to the Phylum **Ascomycota**, which is characterized by the formation of an **ascus** (plural, asci), a sac-like structure that contains haploid *ascospores*. Filamentous ascomycetes produce hyphae divided by perforated septa, allowing streaming of cytoplasm from one cell to another. Conidia and asci, which are used respectively for asexual and sexual reproduction, are usually separated from the vegetative hyphae by blocked (non-perforated) septa. Many ascomycetes are of commercial importance. Some play a beneficial role for humanity, such as the yeasts used in baking, brewing, and wine fermentation, and directly as food delicacies such as truffles and morels. *Aspergillus oryzae* is used in the fermentation of rice to produce sake. Other ascomycetes parasitize plants and animals, including humans. For example, fungal pneumonia poses a significant threat to AIDS patients who have a compromised immune system. Ascomycetes not only infest and destroy crops directly; they also produce poisonous secondary metabolites that make crops unfit for consumption.

Asexual reproduction is frequent and involves the production of conidiophores that release haploid *conidiospores* (Figure 24.14). Sexual reproduction starts with the development of special hyphae from either one of two types of mating strains (Figure 24.14). The "male" strain produces an antheridium and the "female" strain develops an ascogonium. At fertilization, the antheridium and the ascogonium combine in plasmogamy, without nuclear fusion. Special dikaryotic ascogenous (ascus-producing) hyphae arise from this **dikaryon**, in which each cell has pairs of nuclei: one from the "male" strain and one from the "female" strain. In each ascus, two haploid nuclei fuse in karyogamy. Thousands of asci fill a fruiting body called the **ascocarp**. The diploid nucleus in each ascus gives rise to haploid nuclei by meiosis, and spore walls form around each nucleus. The spores in each ascus contain the meiotic products of a single diploid nucleus. The ascospores are then released, germinate, and form hyphae that are disseminated in the environment and start new mycelia (Figure 24.15).

VISUAL CONNECTION

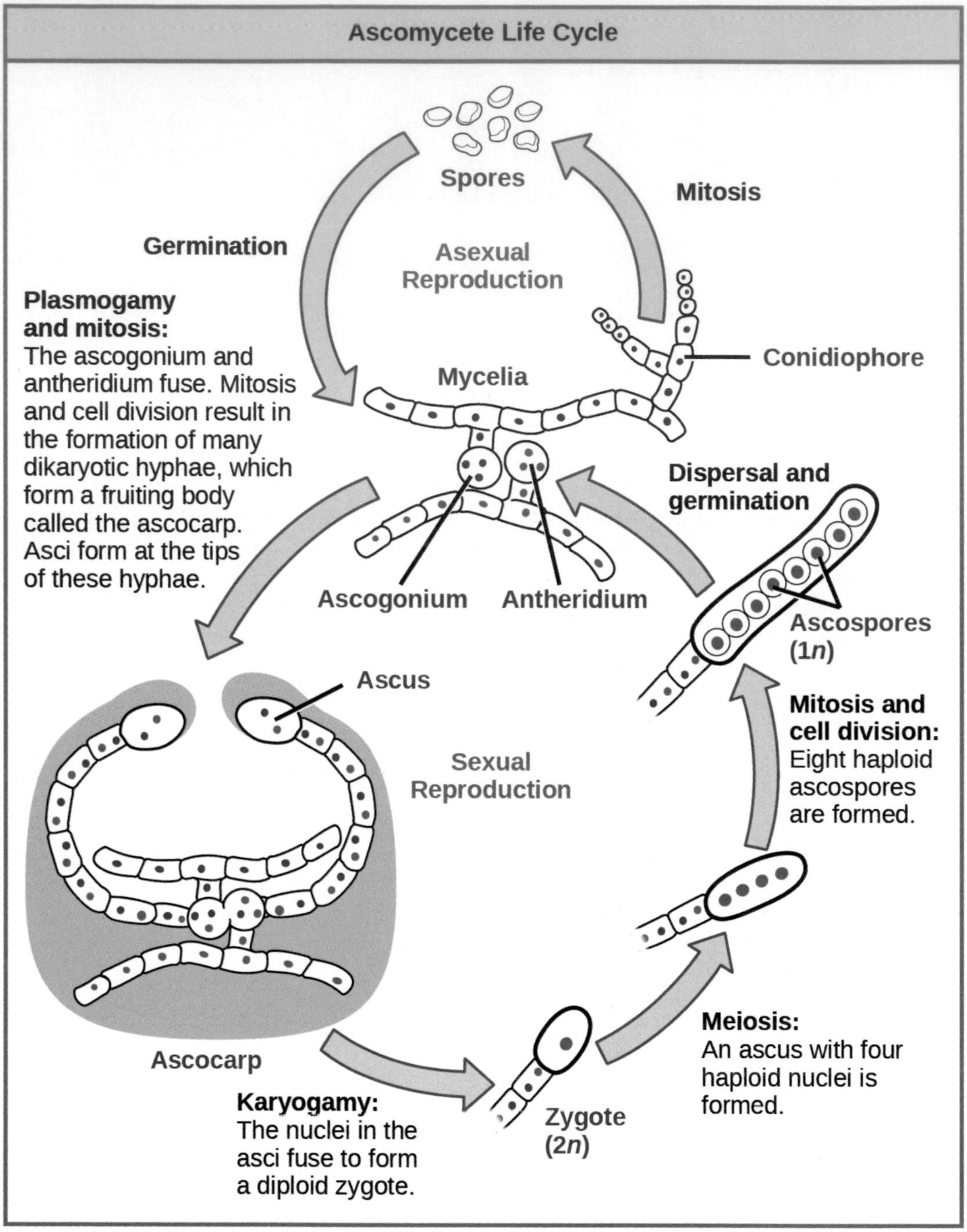

Figure 24.14 Ascomycete life cycle. The lifecycle of an ascomycete is characterized by the production of asci during the sexual phase. In each ascus, the four nuclei produced by meiosis divide once mitotically for a total of eight haploid ascospores. The haploid phase is the predominant phase of the life cycle in Ascomycetes.

Which of the following statements is true?

a. A dikaryotic ascus that forms in the ascocarp undergoes karyogamy, meiosis, and mitosis to form eight ascospores.
b. A diploid ascus that forms in the ascocarp undergoes karyogamy, meiosis, and mitosis to form eight ascospores.
c. A haploid zygote that forms in the ascocarp undergoes karyogamy, meiosis, and mitosis to form eight ascospores.
d. A dikaryotic ascus that forms in the ascocarp undergoes plasmogamy, meiosis, and mitosis to form eight ascospores.

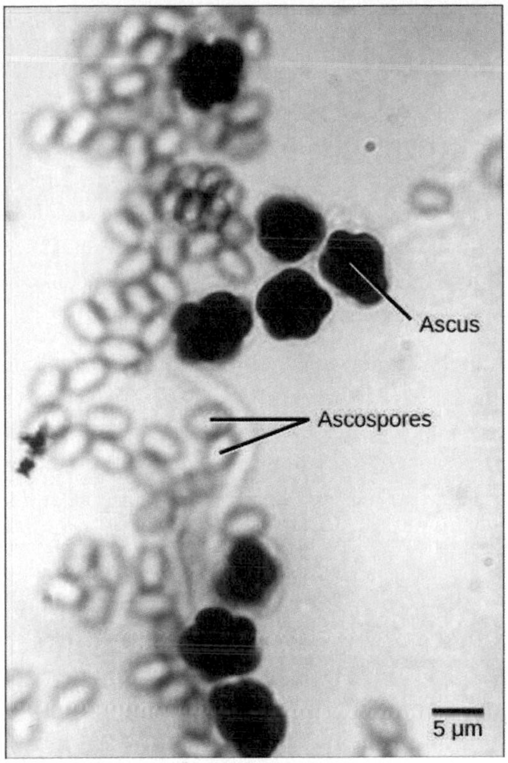

Figure 24.15 Ascospores. The bright field light micrograph shows ascospores being released from asci in the fungus *Talaromyces flavus* var. *flavus*. (credit: modification of work by Dr. Lucille Georg, CDC; scale-bar data from Matt Russell)

Basidiomycota: The Club Fungi

The fungi in the Phylum **Basidiomycota** are easily recognizable under a light microscope by their club-shaped fruiting bodies called **basidia** (singular, **basidium**), which are the swollen terminal cells of hyphae. The basidia, which are the reproductive organs of these fungi, are often contained within the familiar mushroom, commonly seen in fields after rain, on the supermarket shelves, and growing on your lawn (Figure 24.16). These mushroom-producing basidiomycetes are sometimes referred to as *"gill fungi"* because of the presence of gill-like structures on the underside of the cap. The gills are actually *compacted hyphae* on which the basidia are borne. This group also includes shelf fungi, which cling to the bark of trees like small shelves. In addition, the basidiomycota include smuts and rusts, which are important plant pathogens. Most edible fungi belong to the Phylum Basidiomycota; however, some basidiomycota are inedible and produce deadly toxins. For example, *Cryptococcus neoformans* causes severe respiratory illness. The infamous death cap mushroom (*Amanita phalloides*) is related to the fly agaric seen at the beginning of the previous section.

Figure 24.16 Fairy ring. The fruiting bodies of a basidiomycete form a ring in a meadow, commonly called "fairy ring." The best-known fairy ring fungus has the scientific name *Marasmius oreades*. The body of this fungus, its mycelium, is underground and grows outward in a

circle. As it grows, the mycelium depletes the soil of nitrogen, causing the mycelia to grow away from the center and leading to the "fairy ring" of fruiting bodies where there is adequate soil nitrogen. (Credit: "Cropcircles"/Wikipedia Commons)]

The lifecycle of basidiomycetes includes *alternation of generations* (Figure 24.17). Most fungi are haploid through most of their life cycles, but the basidiomycetes produce *both* haploid and dikaryotic mycelia, with the dikaryotic phase being dominant. (Note: The dikaryotic phase is technically not diploid, since the nuclei remain unfused until shortly before spore production.) In the basidiomycetes, sexual spores are more common than asexual spores. The sexual spores form in the club-shaped basidium and are called basidiospores. In the basidium, nuclei of two different mating strains fuse (karyogamy), giving rise to a diploid zygote that then undergoes meiosis. The haploid nuclei migrate into four different chambers appended to the basidium, and then become basidiospores.

Each basidiospore germinates and generates *monokaryotic haploid hyphae.* The mycelium that results is called a primary mycelium. Mycelia of different mating strains can combine and produce a secondary mycelium that contains haploid nuclei of two different mating strains. This is the dominant dikaryotic stage of the basidiomycete life cycle. Thus, each cell in this mycelium has two haploid nuclei, which will not fuse until formation of the basidium. Eventually, the secondary mycelium generates a **basidiocarp**, a fruiting body that protrudes from the ground—this is what we think of as a mushroom. The basidiocarp bears the developing basidia on the gills under its cap.

VISUAL CONNECTION

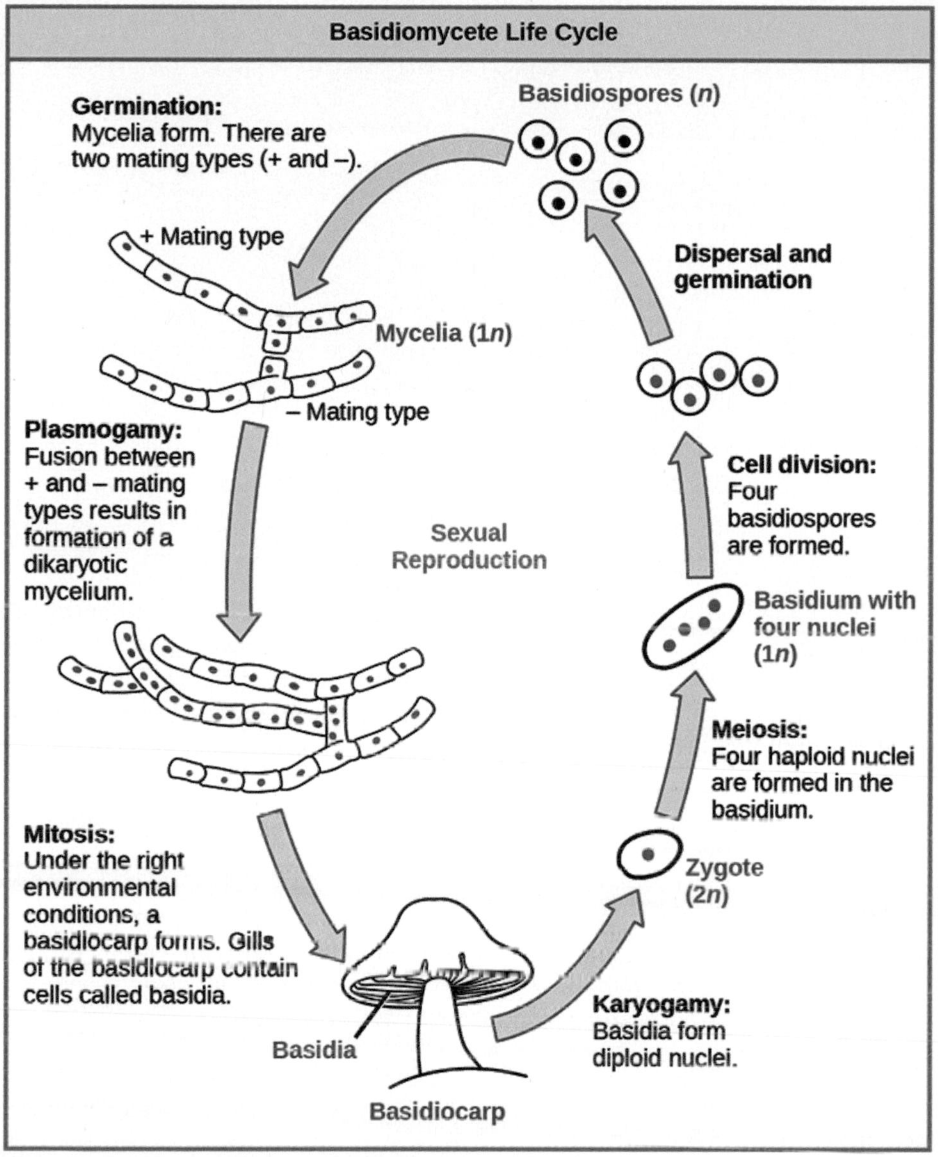

Figure 24.17 Basidiomycete life cycle. The lifecycle of a basidiomycete has alternate generations with haploid and dikaryotic mycelia. Haploid primary mycelia fuse to form a dikaryotic secondary mycelium, which is the dominant stage of the life cycle, and produces the basidiocarp.

Which of the following statements is true?

a. A basidium is the fruiting body of a mushroom-producing fungus, and it forms four basidiocarps.
b. The result of the plasmogamy step is four basidiospores.
c. Karyogamy results directly in the formation of mycelia.
d. A basidiocarp is the fruiting body of a mushroom-producing fungus.

Asexual Ascomycota and Basidiomycota

Imperfect fungi—those that do not display a sexual phase—were formerly classified in the form phylum **Deuteromycota**, an invalid taxon no longer used in the present, ever-developing classification of organisms. While Deuteromycota was once a classification taxon, recent molecular analysis has shown that some of the members classified in this group belong to the Ascomycota (Figure 24.18) or the Basidiomycota. Because some members of this group have not yet been appropriately classified, they are less well described in comparison to members of other fungal taxa. Most imperfect fungi live on land, with a few aquatic exceptions. They form visible mycelia with a fuzzy appearance and are commonly known as **mold**.

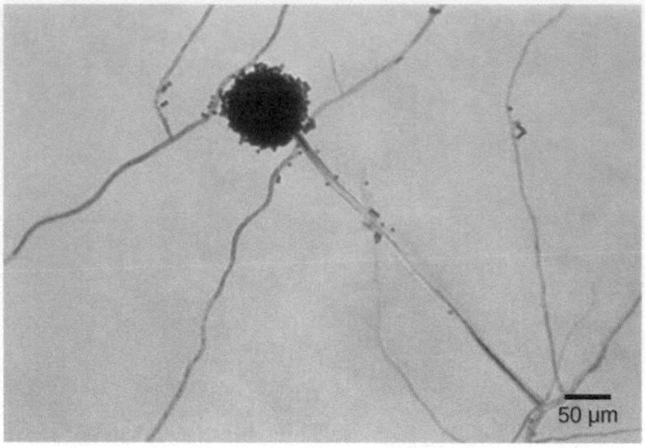

Figure 24.18 Aspergillus. *Aspergillus niger* is an asexually reproducing fungus (phylum Ascomycota) commonly found as a food contaminant. The spherical structure in this light micrograph is an asexual conidiophore. Molecular studies have placed *Aspergillus* with the ascomycetes and sexual cycles have been identified in some species. (credit: modification of work by Dr. Lucille Georg, CDC; scale-bar data from Matt Russell)

The fungi in this group have a large impact on everyday human life. The food industry relies on them for ripening some cheeses. The blue veins in Roquefort cheese and the white crust on Camembert are the result of fungal growth. The antibiotic penicillin was originally discovered on an overgrown Petri plate, on which a colony of *Penicillium* fungi had killed the bacterial growth surrounding it. Other fungi in this group cause serious diseases, either directly as parasites (which infect both plants and humans), or as producers of potent toxic compounds, as seen in the aflatoxins released by fungi of the genus *Aspergillus*.

Glomeromycota

The **Glomeromycota** is a newly established phylum that comprises about 230 species, all of which are involved in close associations with the roots of trees. Fossil records indicate that trees and their root symbionts share a long evolutionary history. It appears that nearly all members of this family form **arbuscular mycorrhizae**: the hyphae interact with the root cells forming a mutually beneficial association in which the plants supply the carbon source and energy in the form of carbohydrates to the fungus, and the fungus supplies essential minerals from the soil to the plant. The exception is *Geosiphon pyriformis*, which hosts the cyanobacterium *Nostoc* as an endosymbiont.

The glomeromycetes do not reproduce sexually and do not survive without the presence of plant roots. Although they have coenocytic hyphae like the zygomycetes, they do not form zygospores. DNA analysis shows that all glomeromycetes probably descended from a common ancestor, making them a monophyletic lineage.

24.3 Ecology of Fungi

By the end of this section, you will be able to do the following:

- Describe the role of fungi in various ecosystems
- Describe mutualistic relationships of fungi with plant roots and photosynthetic organisms
- Describe the beneficial relationship between some fungi and insects

Fungi play a crucial role in the constantly changing "balance" of ecosystems. They colonize most habitats on Earth, preferring dark, moist conditions. They can thrive in seemingly hostile environments, such as the tundra, thanks to a most successful symbiosis with photosynthetic organisms like algae to produce lichens. Within their communities, fungi are not as obvious as are large animals or tall treas. Like bacteria, they act behind the scene as major decomposers. With their versatile metabolism, fungi break down organic matter, which would otherwise not be recycled.

Habitats

Although fungi are primarily associated with humid and cool environments that provide a supply of organic matter, they colonize a surprising diversity of habitats, from seawater to human skin and mucous membranes. Chytrids are found primarily in aquatic environments. Other fungi, such as *Coccidioides immitis*, which causes pneumonia when its spores are inhaled, thrive in the dry and sandy soil of the southwestern United States. Fungi that parasitize coral reefs live in the ocean. However, most members of the Kingdom Fungi grow on the forest floor, where the dark and damp environment is rich in decaying debris from plants and animals. In these environments, fungi play a major role as decomposers and recyclers, making it possible for members of the other kingdoms to be supplied with nutrients and live.

Decomposers and Recyclers

The food web would be incomplete without organisms that decompose organic matter (Figure 24.19). Some elements—such as nitrogen and phosphorus—are required in large quantities by biological systems, and yet are not abundant in the environment. The action of fungi releases these elements from decaying matter, making them available to other living organisms. Trace elements present in low amounts in many habitats are essential for growth, and would remain tied up in rotting organic matter if fungi and bacteria did not return them to the environment via their metabolic activity.

Figure 24.19 Bracket fungi. Fungi are an important part of ecosystem nutrient cycles. These bracket fungi growing on the side of a tree are the fruiting structures of a basidiomycete. They receive their nutrients through their hyphae, which invade and decay the tree trunk. (credit: Cory Zanker)

The ability of fungi to degrade many large and insoluble molecules is due to their mode of nutrition. As seen earlier, digestion precedes ingestion. Fungi produce a variety of exoenzymes to digest nutrients. The enzymes are either released into the substrate or remain bound to the outside of the fungal cell wall. Large molecules are broken down into small molecules, which are transported into the cell by a system of protein carriers embedded in the cell membrane. Because the movement of small molecules and enzymes is dependent on the presence of water, active growth depends on a relatively high percentage of moisture in the environment.

As saprobes, fungi help maintain a sustainable ecosystem for the animals and plants that share the same habitat. In addition to replenishing the environment with nutrients, fungi interact directly with other organisms in beneficial, and sometimes damaging, ways (Figure 24.20).

Figure 24.20 Shelf fungi. Shelf fungi, so called because they grow on trees in a stack, attack and digest the trunk or branches of a tree. While some shelf fungi are found only on dead trees, others can parasitize living trees and cause eventual death, so they are considered serious tree pathogens. (credit: Cory Zanker)

Mutualistic Relationships

Symbiosis is the ecological interaction between two organisms that live together. This definition does not describe the type or quality of the interaction. When both members of the association benefit, the symbiotic relationship is called mutualistic. Fungi form mutualistic associations with many types of organisms, including cyanobacteria, algae, plants, and animals.

Fungus/Plant Mutualism

One of the most remarkable associations between fungi and plants is the establishment of *mycorrhizae*. **Mycorrhiza**, which is derived from the Greek words *myco* meaning fungus and *rhizo* meaning root, refers to the fungal partner of a mutualistic association between vascular plant roots and their symbiotic fungi. Nearly 90 percent of all vascular plant species have mycorrhizal partners. In a mycorrhizal association, the fungal mycelia use their extensive network of hyphae and large surface area in contact with the soil to channel water and minerals from the soil into the plant. In exchange, the plant supplies the products of photosynthesis to fuel the metabolism of the fungus.

There are several basic types of mycorrhizae. **Ectomycorrhizae** ("outside" mycorrhizae) depend on fungi enveloping the roots in a sheath (called a mantle). Hyphae grow from the mantle into the root and envelope the outer layers of the root cells in a network of hyphae called a *Hartig net* (Figure 24.21). The fungal partner can belong to the Ascomycota, Basidiomycota or Zygomycota. **Endomycorrhizae** ("inside" mycorrhizae), also called *arbuscular mycorrhizae*, are produced when the fungi grow inside the root in a branched structure called an *arbuscule* (from the Latin for "little trees"). The fungal partners of endomycorrhizal associates all belong to the Glomeromycota. The fungal arbuscules penetrate root cells between the cell wall and the plasma membrane and are the site of the metabolic exchanges between the fungus and the host plant (Figure 24.21b and Figure 24.22b). Orchids rely on a third type of mycorrhiza. Orchids are epiphytes that typically produce very small airborne seeds without much storage to sustain germination and growth. Their seeds will not germinate without a mycorrhizal partner (usually a Basidiomycete). After nutrients in the seed are depleted, fungal symbionts support the growth of the orchid by providing necessary carbohydrates and minerals. Some orchids continue to be mycorrhizal throughout their life cycle.

VISUAL CONNECTION

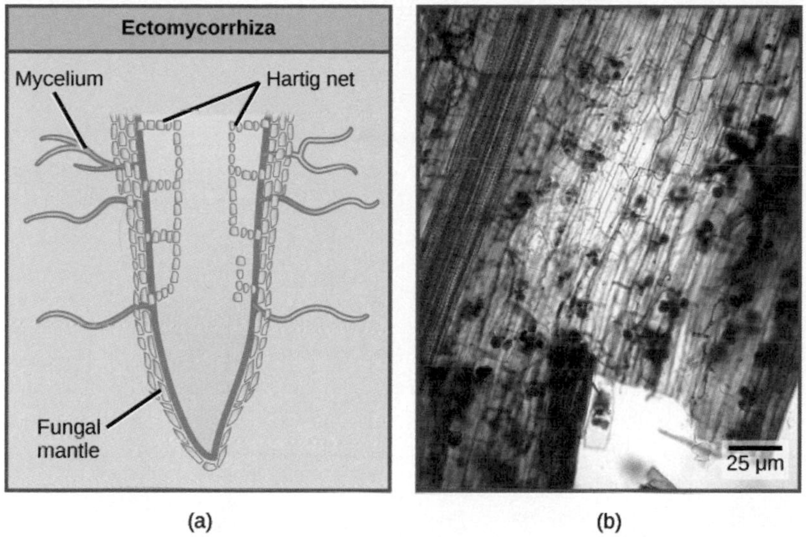

Figure 24.21 Two types of mycorrhizae. (a) Ectomycorrhizae and (b) arbuscular or endomycorrhizae have different mechanisms for interacting with the roots of plants. (credit b: MS Turmel, University of Manitoba, Plant Science Department)

If symbiotic fungi were absent from the soil, what impact do you think this would have on plant growth?

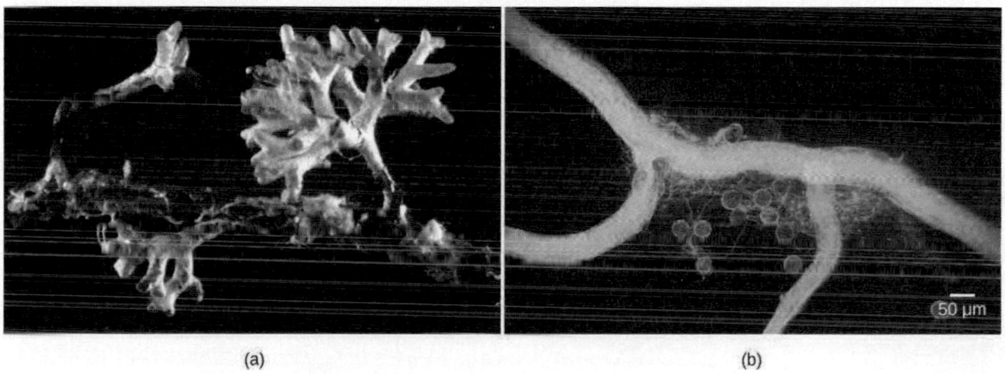

Figure 24.22 Mycorrhizae. The (a) infection of *Pinus radiata* (Monterey pine) roots by the hyphae of *Amanita muscaria* (fly amanita) causes the pine tree to produce many small, branched rootlets. The *Amanita* hyphae cover these small roots with a white mantle. (b) Spores (the round bodies) and hyphae (thread-like structures) are evident in this light micrograph of an arbuscular mycorrhiza by a fungus on the root of a corn plant. (credit a: modification of work by Randy Molina, USDA; credit b: modification of work by Sara Wright, USDA-ARS; scale-bar data from Matt Russell)

Other examples of fungus–plant mutualism include the endophytes: fungi that live inside tissue without damaging the host plant. Endophytes release toxins that repel herbivores, or confer resistance to environmental stress factors, such as infection by microorganisms, drought, or heavy metals in soil.

EVOLUTION CONNECTION

Coevolution of Land Plants and Mycorrhizae

As we have seen, mycorrhizae are the fungal partners of a mutually beneficial symbiotic association that coevolved between roots of vascular plants and fungi. A well-supported theory proposes that fungi were instrumental in the evolution of the root system in plants and contributed to the success of Angiosperms. The bryophytes (mosses and liverworts), which are considered the most ancestral plants and the first to survive and adapt on land, have simple underground rhizoids, rather than a true root system, and therefore cannot survive in dry areas. However, some bryophytes have arbuscular mycorrhizae and some do not.

True roots first appeared in the ancestral vascular plants: Vascular plants that developed a system of thin extensions from their roots would have had a selective advantage over nonvascular plants because they had a greater surface area of contact with the fungal partners than did the rhizoids of mosses and liverworts. The first true roots would have allowed vascular plants to obtain more water and nutrients in the ground.

Fossil records indicate that fungi actually preceded the invasion of ancestral freshwater plants onto dry land. The first association between fungi and photosynthetic organisms on land involved moss-like plants and endophytes. These early associations developed before roots appeared in plants. Slowly, the benefits of the endophyte and rhizoid interactions for both partners led to present-day mycorrhizae: About 90 percent of today's vascular plants have associations with fungi in their rhizosphere.

The fungi involved in mycorrhizae display many characteristics of ancestral fungi; they produce simple spores, show little diversification, do not have a sexual reproductive cycle, and cannot live outside of a mycorrhizal association. The plants benefited from the association because mycorrhizae allowed them to move into new habitats and allowed the increased uptake of nutrients, which gave them an enormous selective advantage over plants that did not establish symbiotic relationships.

Lichens

Lichens display a range of colors and textures (Figure 24.23) and can survive in the most unusual and hostile habitats. They cover rocks, gravestones, tree bark, and the ground in the tundra where plant roots cannot penetrate. Lichens can survive extended periods of drought, when they become completely desiccated, and then rapidly become active once water is available again.

LINK TO LEARNING

Explore the world of lichens using this site (http://openstax.org/l/lichenland) from Oregon State University.

Figure 24.23 Lichens. Lichens have many forms. They may be (a) crust-like, (b) hair-like, or (c) leaf-like. (credit a: modification of work by Jo Naylor; credit b: modification of work by "djpmapleferryman"/Flickr; credit c: modification of work by Cory Zanker)

It is important to note that lichens are *not* a single organism, but rather another wonderful example of a mutualism, in which a fungus (usually a member of the Ascomycota or Basidiomycota) lives in a physical and physiological relationship with a photosynthetic organism (a eukaryotic alga or a prokaryotic cyanobacterium) (Figure 24.24). Generally, neither the fungus nor the photosynthetic organism can survive alone outside of the symbiotic relationship. The body of a lichen, referred to as a thallus, is formed of hyphae wrapped around the photosynthetic partner. The photosynthetic organism provides carbon and energy in the form of carbohydrates. Some cyanobacteria additionally fix nitrogen from the atmosphere, contributing nitrogenous compounds to the association. In return, the fungus supplies minerals and protection from dryness and excessive light by encasing the algae in its mycelium. The fungus also attaches the lichen to its substrate.

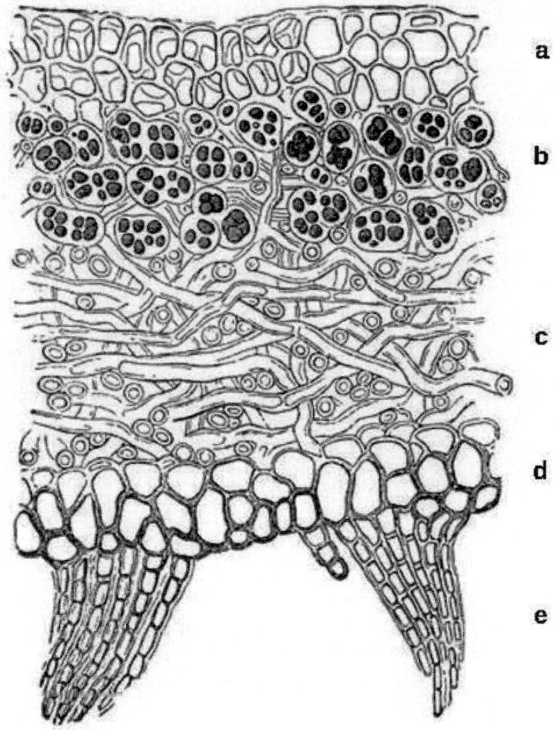

Figure 24.24 Structure of a lichen. This cross-section of a lichen thallus shows the (a) upper cortex of fungal hyphae, which provides protection; the (b) algal zone where photosynthesis occurs, the (c) medulla of fungal hyphae, and the (d) lower cortex, which also provides protection and may have (e) *rhizines* to anchor the thallus to the substrate.

The thallus of lichens grows very slowly, expanding its diameter a few millimeters per year. Both the fungus and the alga participate in the formation of dispersal units, called soredia—clusters of algal cells surrounded by mycelia. Soredia are dispersed by wind and water and form new lichens.

Lichens are extremely sensitive to air pollution, especially to abnormal levels of nitrogenous and sulfurous compounds. The U.S. Forest Service and National Park Service can monitor air quality by measuring the relative abundance and health of the lichen population in an area. Lichens fulfill many ecological roles. Caribou and reindeer eat lichens, and they provide cover for small invertebrates that hide in the mycelium. In the production of textiles, weavers used lichens to dye wool for many centuries until the advent of synthetic dyes. The pigments used in litmus paper are also extracted from lichens.

LINK TO LEARNING

Lichens are used to monitor the quality of air. Read more on this site (http://openstax.org/l/lichen_monitrng) from the United States Forest Service.

Fungus/Animal Mutualism

Fungi have evolved mutualisms with numerous insects in Phylum Arthropoda: joint-legged invertebrates with a chitinous exoskeleton. Arthropods depend on the fungus for protection from predators and pathogens, while the fungus obtains nutrients and a way to disseminate spores into new environments. The association between species of Basidiomycota and scale insects is one example. The fungal mycelium covers and protects the insect colonies. The scale insects foster a flow of nutrients from the parasitized plant to the fungus.

In a second example, leaf-cutter ants of Central and South America literally farm fungi. They cut disks of leaves from plants and pile them up in subterranean gardens (Figure 24.25). Fungi are cultivated in these disk gardens, digesting the cellulose in the leaves that the ants cannot break down. Once smaller sugar molecules are produced and consumed by the fungi, the fungi in turn become a meal for the ants. The insects also patrol their garden, preying on competing fungi. Both ants and fungi benefit from this mutualistic association. The fungus receives a steady supply of leaves and freedom from competition, while the ants feed on the fungi they cultivate.

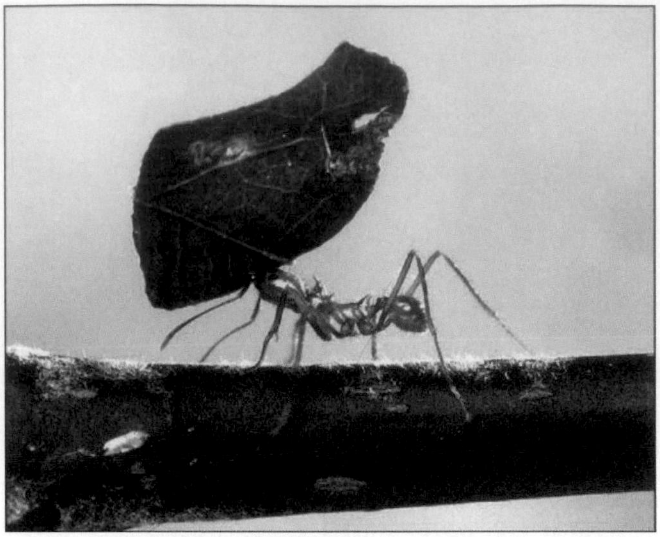

Figure 24.25 Leaf-cutter ant. A leaf-cutter ant transports a leaf that will feed a farmed fungus. (credit: Scott Bauer, USDA-ARS)

Fungivores

Animal dispersal is important for some fungi because an animal may carry fungal spores considerable distances from the source. Fungal spores are rarely completely degraded in the gastrointestinal tract of an animal, and many are able to germinate when they are passed in the feces. Some "dung fungi" actually require passage through the digestive system of herbivores to complete their lifecycle. The black truffle—a prized gourmet delicacy—is the fruiting body of an underground ascomycete. Almost all truffles are ectomycorrhizal, and are usually found in close association with trees. Animals eat truffles and disperse the spores. In Italy and France, truffle hunters use female pigs to sniff out truffles (female pigs are attracted to truffles because the fungus releases a volatile compound closely related to a pheromone produced by male pigs.)

24.4 Fungal Parasites and Pathogens

By the end of this section, you will be able to do the following:

- Describe some fungal parasites and pathogens of plants
- Describe the different types of fungal infections in humans
- Explain why antifungal therapy is hampered by the similarity between fungal and animal cells

Parasitism describes a symbiotic relationship in which one member of the association benefits at the expense of the other. Both parasites and pathogens harm the host; however, pathogens cause disease, damage to host tissues or physiology, whereas parasites usually do not, but can cause serious damage and death by competition for nutrients or other resources. **Commensalism** occurs when one member benefits without affecting the other.

Plant Parasites and Pathogens

The production of sufficient high-quality crops is essential to human existence. Unfortunately, plant diseases have ruined many crops throughout human agricultural history, sometimes creating widespread famine. Many plant pathogens are fungi that cause tissue decay and the eventual death of the host (Figure 24.26). In addition to destroying plant tissue directly, some plant pathogens spoil crops by producing potent toxins that can further damage and kill the host plant. Fungi are also responsible for food spoilage and the rotting of stored crops. For example, the fungus *Claviceps purpurea* causes ergot, a disease of cereal crops (especially of rye). Although the fungus reduces the yield of cereals, the effects of the ergot's alkaloid toxins on humans and animals are of much greater significance. In animals, the disease is referred to as *ergotism*. The most common signs and symptoms are convulsions, hallucination, gangrene, and loss of milk in cattle. The active ingredient of ergot is *lysergic acid,* which is a precursor of the drug LSD. Smuts, rusts, and powdery or downy mildew are other examples of common fungal pathogens that affect crops.

Figure 24.26 Fungal pathogens. Some fungal pathogens include (a) green mold on grapefruit, (b) powdery mildew on a zinnia, (c) stem rust on a sheaf of barley, and (d) grey rot on grapes. In wet conditions *Botrytis cinerea,* the fungus that causes grey rot, can destroy a grape crop. However, controlled infection of grapes by *Botrytis* results in noble rot, a condition that produces strong and much-prized dessert wines. (credit a: modification of work by Scott Bauer, USDA-ARS; credit b: modification of work by Stephen Ausmus, USDA-ARS; credit c: modification of work by David Marshall, USDA-ARS; credit d: modification of work by Joseph Smilanick, USDA-ARS)

Aflatoxins are toxic, carcinogenic compounds released by fungi of the genus *Aspergillus.* Periodically, harvests of nuts and grains are tainted by aflatoxins, leading to massive recall of produce. This sometimes ruins producers and causes food shortages in developing countries.

Animal and Human Parasites and Pathogens

Fungi can affect animals, including humans, in several ways. A **mycosis** is a fungal disease that results from infection and direct damage due to the growth and infiltration of the fungus. Fungi attack animals directly by colonizing and destroying tissues. **Mycotoxicosis** is the poisoning of humans (and other animals) by foods contaminated by fungal toxins (mycotoxins). **Mycetismus** specifically describes the ingestion of preformed toxins in poisonous mushrooms. In addition, individuals who display hypersensitivity to molds and spores may develop strong and dangerous allergic reactions. Fungal infections are generally very difficult to treat because, unlike bacteria, fungi are eukaryotes. Antibiotics only target prokaryotic cells, whereas compounds that kill fungi also harm the eukaryotic animal host.

Many fungal infections are *superficial*; that is, they occur on the animal's skin. Termed *cutaneous* ("skin") *mycoses*, they can have devastating effects. For example, the decline of the world's frog population in recent years is caused (in part) by the chytrid fungus *Batrachochytrium dendrobatidis.* This deadly fungus infects the skin of frogs and presumably interferes with cutaneous gaseous exchange, which is essential for amphibian survival. Similarly, more than a million bats in the United States have been killed by white-nose syndrome, which appears as a white ring around the mouth of the bat. It is caused by the cold-loving fungus *Pseudogymnoascus destructans*, which disseminates its deadly spores in caves where bats hibernate. Mycologists are researching the transmission, mechanism, and control of *P. destructans* to stop its spread.

Fungi that cause the superficial mycoses of the epidermis, hair, and nails rarely spread to the underlying tissue (Figure 24.27). These fungi are often misnamed "dermatophytes", from the Greek words *dermis* meaning skin and *phyte* meaning plant, although they are not plants. Dermatophytes are also called *"ringworms"* because of the red ring they cause on skin. They secrete extracellular enzymes that break down *keratin* (a protein found in hair, skin, and nails), causing conditions such as athlete's foot and jock itch. These conditions are usually treated with over-the-counter topical creams and powders, and are easily cleared. More persistent superficial mycoses may require prescription oral medications.

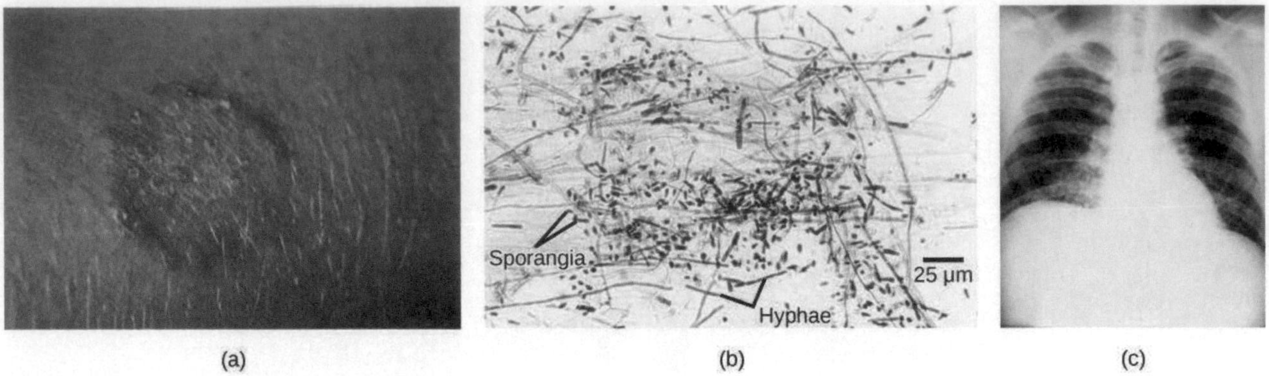

Figure 24.27 Fungal diseases of humans. (a) Ringworm presents as a red ring on skin; (b) *Trichophyton violaceum*, shown in this bright field light micrograph, causes superficial mycoses on the scalp; (c) *Histoplasma capsulatum* is an ascomycete that infects airways and causes symptoms similar to influenza. (credit a: modification of work by Dr. Lucille K. Georg, CDC; credit b: modification of work by Dr. Lucille K. Georg, CDC; credit c: modification of work by M. Renz, CDC; scale-bar data from Matt Russell)

Systemic mycoses spread to internal organs, most commonly entering the body through the respiratory system. For example, *coccidioidomycosis* (often called valley fever) is commonly found in the southwestern United States, but as far north as Washington, where the fungus resides in the dust. Once inhaled, the spores develop in the lungs and cause symptoms similar to those of tuberculosis. *Histoplasmosis* is caused by the dimorphic fungus *Histoplasma capsulatum.* In its human host, *Histoplasma* grows as a yeast, causing pulmonary infections, and in rarer cases, swelling of the membranes of the brain and spinal cord. Treatment of these and many other fungal diseases requires the use of antifungal medications that have serious side effects.

Opportunistic mycoses are fungal infections that are either common in all environments, or part of the normal biota. They mainly affect individuals who have a compromised immune system. Patients in the late stages of AIDS suffer from opportunistic mycoses that can be life threatening. The yeast *Candida* sp., a common member of the natural biota, can grow unchecked and infect the vagina or mouth (oral thrush) if the pH of the surrounding environment, the person's immune defenses, or the normal population of bacteria are altered.

Mycetismus can occur when poisonous mushrooms are eaten. It causes a number of human fatalities during mushroom-picking season. Many edible fruiting bodies of fungi resemble highly poisonous relatives, and amateur mushroom hunters are cautioned to carefully inspect their harvest and avoid eating mushrooms of doubtful origin. The adage "there are bold mushroom pickers and old mushroom pickers, but are there no old, bold mushroom pickers" is unfortunately true.

SCIENTIFIC METHOD CONNECTION

Dutch Elm Disease

Question: Do trees resistant to Dutch elm disease secrete antifungal compounds?

Hypothesis: Construct a hypothesis that addresses this question.

Background: Dutch elm disease is a fungal infestation that affects many species of elm (*Ulmus*) in North America. The fungus infects the vascular system of the tree, which blocks water flow within the plant and mimics drought stress. Accidently introduced to the United States in the early 1930s, it decimated American elm shade trees across the continent. It is caused by the fungus *Ophiostoma ulmi*. The elm bark beetle acts as a vector and transmits the disease from tree to tree. Many European and Asiatic elms are less susceptible to the disease than are American elms.

Test the hypothesis: A researcher testing this hypothesis might do the following. Inoculate several Petri plates containing a medium that supports the growth of fungi with fragments of *Ophiostoma* mycelium. Cut (with a metal punch) several disks from the vascular tissue of susceptible varieties of American elms and resistant European and Asiatic elms. Include control Petri plates inoculated with mycelia without plant tissue to verify that the medium and incubation conditions do not interfere with fungal growth. As a positive control, add paper disks impregnated with a known fungicide to Petri plates inoculated with the mycelium.

Incubate the plates for a set number of days to allow fungal growth and spreading of the mycelium over the surface of the plate. Record the diameter of the zone of clearing, if any, around the tissue samples and the fungicide control disk.

Record your observations in the following table.

Results of Antifungal Testing of Vascular Tissue from Different Species of Elm

Disk	Zone of Inhibition (mm)
Distilled Water	
Fungicide	
Tissue from Susceptible Elm #1	
Tissue from Susceptible Elm #2	
Tissue from Resistant Elm #1	
Tissue from Resistant Elm #2	

Table 24.1

Analyze the data and report the results. Compare the effect of distilled water to the fungicide. These are negative and positive controls that validate the experimental setup. The fungicide should be surrounded by a clear zone where the fungus growth was inhibited. Is there a difference among different species of elm?

Draw a conclusion: Was there antifungal activity as expected from the fungicide? Did the results support the hypothesis? If not, how can this be explained? There are several possible explanations.

24.5 Importance of Fungi in Human Life

By the end of this section, you will be able to do the following:

- Describe the importance of fungi to the balance of the environment
- Summarize the role of fungi in agriculture and food and beverage preparation
- Describe the importance of fungi in the chemical and pharmaceutical industries
- Discuss the role of fungi as model organisms

Although we often think of fungi as organisms that cause disease and rot food, they are vitally important to human life on many levels. As we have seen, fungi influence the well-being of human populations on a large scale because they are part of the nutrient cycle in ecosystems. They have other ecosystem roles as well. As animal pathogens, fungi help to control the population of damaging pests. These fungi are very specific to the insects they attack, and do not infect other animals or plants. Fungi are currently under investigation as potential microbial insecticides, with several already on the market. For example, the fungus *Beauveria bassiana* is being tested as a possible biological control agent for the recent spread of emerald ash borer a beetle that feeds on ash trees. It has been released in Michigan, Illinois, Indiana, Ohio, West Virginia, and Maryland (Figure 24.28).

Figure 24.28 Fungal insect control. The emerald ash borer (*Agrilus planipennis*) is an insect that attacks ash trees. It is in turn parasitized by a pathogenic fungus (*Beauveria bassiana*) that holds promise as a biological insecticide. The parasitic fungus appears as white fuzz on the body of the insect. (credit: Houping Liu, USDA Agricultural Research Service)

The mycorrhizal relationship between fungi and plant roots is essential for the productivity of farm land. Without the fungal partner in root systems, 80–90 percent of trees and grasses would not survive. Mycorrhizal fungal inoculants are available as soil amendments from gardening supply stores and are promoted by supporters of organic agriculture.

We also eat some types of fungi. Mushrooms figure prominently in the human diet. Morels, shiitake mushrooms, chanterelles, and truffles are considered delicacies (Figure 24.29). The humble meadow mushroom, *Agaricus campestris*, appears in many dishes. Molds of the genus *Penicillium* ripen many cheeses. They originate in the natural environment such as the caves of Roquefort, France, where wheels of sheep milk cheese are stacked in order to capture the molds responsible for the blue veins and pungent taste of the cheese.

Figure 24.29 Edible fungi. The morel mushroom (a) is an ascomycete greatly appreciated for its delicate taste. (credit: Jason Hollinger). Basidiocarps of *Agaricus* ready for an omelet (credit: Mary Anne Clark)

Fermentation—of grains to produce beer, and of fruits to produce wine—is an ancient art that humans in most cultures have practiced for millennia. Wild yeasts are acquired from the environment and used to ferment sugars into CO_2 and ethyl alcohol under anaerobic conditions. It is now possible to purchase isolated strains of wild yeasts from different wine-making regions.

Louis Pasteur was instrumental in developing a reliable strain of brewer's yeast, *Saccharomyces cerevisiae*, for the French brewing industry in the late 1850s. This was one of the first examples of biotechnology patenting.

Many secondary metabolites of fungi are of great commercial importance. Antibiotics are naturally produced by fungi to kill or inhibit the growth of bacteria, limiting their competition in the natural environment. Important antibiotics, such as penicillin and the cephalosporins, are isolated from fungi. Valuable drugs isolated from fungi include the immunosuppressant drug *cyclosporine* (which reduces the risk of rejection after organ transplant), the precursors of steroid hormones, and ergot alkaloids used to stop bleeding. Psilocybin is a compound found in fungi such as *Psilocybe semilanceata* and *Gymnopilus junonius,* which have been used for their hallucinogenic properties by various cultures for thousands of years.

As simple eukaryotic organisms, fungi are important model research organisms. Many advances in modern genetics were achieved by the use of the red bread mold *Neurospora crassa*. Additionally, many important genes originally discovered in *S. cerevisiae* served as a starting point in discovering analogous human genes. As a eukaryotic organism, the yeast cell produces and modifies proteins in a manner similar to human cells, as opposed to the bacterium *Eschcrichia coli,* which lacks the internal membrane structures and enzymes to tag proteins for export. This makes yeast a much better organism for use in recombinant DNA technology experiments. Like bacteria, yeasts grow easily in culture, have a short generation time, and are amenable to genetic modification.

KEY TERMS

arbuscular mycorrhiza mycorrhizal association in which the fungal hyphae enter the root cells and form extensive networks

Arbuscular mycorrhizae mycorrhizae commonly involving Glomeromycetes in which the fungal hyphae penetrate the cell walls of the plant root cells (but not the cell membranes)

ascocarp fruiting body of ascomycetes

Ascomycota (also, sac fungi) phylum of fungi that store spores in a sac called ascus

basidiocarp fruiting body that protrudes from the ground and bears the basidia

Basidiomycota (also, club fungi) phylum of fungi that produce club-shaped structures (basidia) that contain spores

basidium club-shaped fruiting body of basidiomycetes

Chytridiomycota (also, chytrids) primitive phylum of fungi that live in water and produce gametes with flagella

coenocytic hypha single hypha that lacks septa and contains many nuclei

commensalism symbiotic relationship in which one member benefits while the other member is not affected

Deuteromycota former form phylum of fungi that do not have a known sexual reproductive cycle (presently members of two phyla: Ascomycota and Basidiomycota)

ectomycorrhiza mycorrhizal fungi that surround the roots with a mantle and have a Hartig net that extends into the roots between cells

Ectomycorrhizae mycorrhizae in which the fungal hyphae do not penetrate the root cells of the plant

facultative anaerobes organisms that can perform both aerobic and anaerobic respiration and can survive in oxygen-rich and oxygen-poor environment

Glomeromycota phylum of fungi that form symbiotic relationships with the roots of trees

haustoria modified hyphae on many parasitic fungi that penetrate the tissues of their hosts, release digestive enzymes, and/or absorb nutrients from the host

heterothallic describes when only one mating type is present in an individual mycelium

homothallic describes when both mating types are present in mycelium

hypha fungal filament composed of one or more cells

karyogamy fusion of nuclei

lichen close association of a fungus with a photosynthetic alga or bacterium that benefits both partners

mold tangle of visible mycelia with a fuzzy appearance

mycelium mass of fungal hyphae

mycetismus ingestion of toxins in poisonous mushrooms

mycology scientific study of fungi

mycorrhiza mutualistic association between fungi and vascular plant roots

mycorrhizae a mutualistic relationship between a plant and a fungus. Mycorrhizae are connections between fungal hyphae, which provide soil minerals to the plant, and plant roots, which provide carbohydrates to the fungus

mycosis fungal infection

mycotoxicosis poisoning by a fungal toxin released in food

obligate aerobes organisms, such as humans, that must perform aerobic respiration to survive

obligate anaerobes organisms that only perform anaerobic respiration and often cannot survive in the presence of oxygen

parasitism symbiotic relationship in which one member of the association benefits at the expense of the other

plasmogamy fusion of cytoplasm

saprobe organism that derives nutrients from decaying organic matter; also saprophyte

septa cell wall division between hyphae

soredia clusters of algal cells and mycelia that allow lichens to propagate

sporangium reproductive sac that contains spores

spore a haploid cell that can undergo mitosis to form a multicellular, haploid individual

thallus vegetative body of a fungus

yeast general term used to describe unicellular fungi

Zygomycota (also, conjugated fungi) phylum of fungi that form a zygote contained in a zygospore

zygospore structure with thick cell wall that contains the zygote in zygomycetes

CHAPTER SUMMARY

24.1 Characteristics of Fungi

Fungi are eukaryotic organisms that appeared on land more than 450 million years ago, but clearly have an evolutionary history far greater. They are heterotrophs and contain neither photosynthetic pigments such as chlorophyll, nor organelles such as chloroplasts. Because fungi feed on decaying and dead matter, they are termed saprobes. Fungi are important decomposers that release essential elements into the environment. External enzymes called exoenzymes digest nutrients that are absorbed by the body of the fungus, which is called a thallus. A thick cell wall made of chitin surrounds the cell. Fungi can be unicellular as yeasts, or develop a network of filaments called a mycelium, which is often described as mold. Most species multiply by asexual and sexual reproductive cycles and display an alternation of generations. In one group of fungi, no sexual cycle has been identified. Sexual reproduction involves plasmogamy (the fusion of the cytoplasm), followed by karyogamy (the fusion of nuclei). Following these processes, meiosis generates haploid spores.

24.2 Classifications of Fungi

Chytridiomycota (chytrids) are considered the most ancestral group of fungi. They are mostly aquatic, and their gametes are the only fungal cells known to have flagella. They reproduce both sexually and asexually; the asexual spores are called zoospores. Zygomycota (conjugated fungi) produce non-septate hyphae with many nuclei. Their hyphae fuse during sexual reproduction to produce a zygospore in a zygosporangium. Ascomycota (sac fungi) form spores in sacs called asci during sexual reproduction. Asexual reproduction is their most common form of reproduction. In the Basidiomycota (club fungi), the sexual phase predominates, producing showy fruiting bodies that contain club-shaped basidia, within which spores form. Most familiar mushrooms belong to this division. Fungi that have no known sexual cycle were originally classified in the "form phylum" Deuteromycota, but many have been classified by comparative molecular analysis with the Ascomycota and Basidiomycota. Glomeromycota form tight associations (called mycorrhizae) with the roots of plants.

24.3 Ecology of Fungi

Fungi have colonized nearly all environments on Earth, but are frequently found in cool, dark, moist places with a supply of decaying material. Fungi are saprobes that decompose organic matter. Many successful mutualistic relationships involve a fungus and another organism. Many fungi establish complex mycorrhizal associations with the roots of plants. Some ants farm fungi as a supply of food. Lichens are a symbiotic relationship between a fungus and a photosynthetic organism, usually an alga or cyanobacterium. The photosynthetic organism provides energy from stored carbohydrates, while the fungus supplies minerals and protection. Some animals that consume fungi help disseminate spores over long distances.

24.4 Fungal Parasites and Pathogens

Fungi establish parasitic relationships with plants and animals. Fungal diseases can decimate crops and spoil food during storage. Compounds produced by fungi can be toxic to humans and other animals. Mycoses are infections caused by fungi. Superficial mycoses affect the skin, whereas systemic mycoses spread through the body. Fungal infections are difficult to cure, since fungi, like their hosts, are eukaryotic, and cladistically related closely to Kingdom Animalia.

24.5 Importance of Fungi in Human Life

Fungi are important to everyday human life. Fungi are important decomposers in most ecosystems. Mycorrhizal fungi are essential for the growth of most plants. Fungi, as food, play a role in human nutrition in the form of mushrooms, and also as agents of fermentation in the production of bread, cheeses, alcoholic beverages, and numerous other food preparations. Secondary metabolites of fungi are used as medicines, such as antibiotics and anticoagulants. Fungi are model organisms for the study of eukaryotic genetics and metabolism

VISUAL CONNECTION QUESTIONS

1. Figure 24.14 Which of the following statements is true?
   a. A dikaryotic ascus that forms in the ascocarp undergoes karyogamy, meiosis, and mitosis to form eight ascospores.
   b. A diploid ascus that forms in the ascocarp undergoes karyogamy, meiosis, and mitosis to form eight ascospores.
   c. A haploid zygote that forms in the ascocarp undergoes karyogamy, meiosis, and mitosis to form eight ascospores.
   d. A dikaryotic ascus that forms in the ascocarp undergoes plasmogamy, meiosis, and mitosis to form eight ascospores.

2. Figure 24.17 Which of the following statements is true?
   a. A basidium is the fruiting body of a mushroom-producing fungus, and it forms four basidiocarps.
   b. The result of the plasmogamy step is four basidiospores.
   c. Karyogamy results directly in the formation of mycelia.
   d. A basidiocarp is the fruiting body of a mushroom-producing fungus.

3. Figure 24.21 If symbiotic fungi are absent from the soil, what impact do you think this would have on plant growth?

REVIEW QUESTIONS

4. Which polysaccharide is usually found in the cell wall of fungi?
   a. starch
   b. glycogen
   c. chitin
   d. cellulose
5. Which of these organelles is not found in a fungal cell?
   a. chloroplast
   b. nucleus
   c. mitochondrion
   d. Golgi apparatus
6. The wall dividing individual cells in a fungal filament is called a
   a. thallus
   b. hypha
   c. mycelium
   d. septum
7. During sexual reproduction, a homothallic mycelium contains
   a. all septated hyphae
   b. all haploid nuclei
   c. both mating types
   d. none of the above
8. The life cycles of perfect fungi are most similar to which other organism?
   a. Hydra that undergo asexual budding
   b. Diploid-dominant pea plants
   c. Haploid-dominant green algae
   d. Bacteria undergoing binary fission
9. The most primitive phylum of fungi is the _______.
   a. Chytridiomycota
   b. Zygomycota
   c. Glomeromycota
   d. Ascomycota
10. Members of which phylum produce a club-shaped structure that contains spores?
    a. Chytridiomycota
    b. Basidiomycota
    c. Glomeromycota
    d. Ascomycota
11. Members of which phylum establish a successful symbiotic relationship with the roots of trees?
    a. Ascomycota
    b. Deuteromycota
    c. Basidiomycota
    d. Glomeromycota
12. The fungi that do not reproduce sexually used to be classified as _______.
    a. Ascomycota
    b. Deuteromycota
    c. Basidiomycota
    d. Glomeromycota
13. A scientist discovers a new species of fungus that introduces genetic diversity during reproduction by creating a diploid zygote. This new species cannot belong to which modern phylum of fungi?
    a. Zygomycota
    b. Glomeromycota
    c. Chytridiomycota
    d. Deuteromycota
14. What term describes the close association of a fungus with the root of a tree?
    a. a rhizoid
    b. a lichen
    c. a mycorrhiza
    d. an endophyte
15. Why are fungi important decomposers?
    a. They produce many spores.
    b. They can grow in many different environments.
    c. They produce mycelia.
    d. They recycle carbon and inorganic minerals by the process of decomposition.
16. Consider an ecosystem where all the fungi not involved in mycorrhizae are eliminated. How would this affect nitrogen intake by plants?
    a. Nitrogen intake would increase.
    b. Nitrogen intake would not change.
    c. Nitrogen intake would decrease.
    d. Nitrogen intake would stop.
17. A fungus that climbs up a tree reaching higher elevation to release its spores in the wind and does not receive any nutrients from the tree or contribute to the tree's welfare is described as a _______.
    a. commensal
    b. mutualist
    c. parasite
    d. pathogen
18. A fungal infection that affects nails and skin is classified as _______.
    a. systemic mycosis
    b. mycetismus
    c. superficial mycosis
    d. mycotoxicosis

19. The targets for anti-fungal drugs are much more limited than antibiotics or anti-viral medications. Why?
    a. There are more bacteria and viruses than fungi.
    b. Fungi can only be targeted during sexual reproduction, while bacteria and viruses can be targeted at any point in their lifespan.
    c. Fungi cause topical infections, while viruses and bacteria cause systemic infections.
    d. Human cells are much more similar to fungi cells than bacteria or viruses.

20. Yeast is a facultative anaerobe. This means that alcohol fermentation takes place only if:
    a. the temperature is close to 37°C
    b. the atmosphere does not contain oxygen
    c. sugar is provided to the cells
    d. light is provided to the cells

21. The advantage of yeast cells over bacterial cells to express human proteins is that:
    a. yeast cells grow faster
    b. yeast cells are easier to manipulate genetically
    c. yeast cells are eukaryotic and modify proteins similarly to human cells
    d. yeast cells are easily lysed to purify the proteins

22. Why are fungal insecticides an attractive alternative to chemical pesticides for growing food crops?
    a. Human consumption of fungal insecticides would not make a person sick, but ingestion of chemical pesticides can be harmful to humans.
    b. A single fungal insecticide would kill a wider variety of insects than a chemical pesticide.
    c. Fungal insecticides can eliminate both harmful insects and plant pathogens, while chemical pesticides only kill insects.
    d. Fungal insecticides will decompose dying plants, enhancing the nitrogen content of the soil, while chemical pesticides are not decomposers.

CRITICAL THINKING QUESTIONS

23. What are the evolutionary advantages for an organism to reproduce both asexually and sexually?
24. Compare plants, animals, and fungi, considering these components: cell wall, chloroplasts, plasma membrane, food source, and polysaccharide storage. Be sure to indicate fungi's similarities and differences to plants and animals.
25. Why is the large surface area of the mycelium essential for nutrient acquisition by fungi?
26. What is the advantage for a basidiomycete to produce a showy and fleshy fruiting body?
27. For each of the four groups of perfect fungi (Chytridiomycota, Zygomycota, Ascomycota, and Basidiomycota), compare the body structure and features, and provide an example.
28. Why does protection from light actually benefit the photosynthetic partner in lichens?
29. Ambrosia bark beetles carry *Ambrosiella* fungal spores to trees, then bore holes and lay their eggs with the fungus. When the new larvae hatch, they eat the fungus that has germinated in the holes. Describe how this relationship can be classified as mutualistic.
30. Ecologists often attempt to introduce new plants to restore degraded land. In an arid climate, scientists recommend introducing plants with arbuscular mycorrhizae. How would the mycorrhizae increase the plants' survival compared to plants without mycorrhizae?
31. Why can superficial mycoses in humans lead to bacterial infections?
32. Explain how the Red Queen Hypothesis describes the continuously evolving relationship between red grapes and *Botrytis cinerea*.
33. Historically, artisanal breads were produced by capturing wild yeasts from the air. Prior to the development of modern yeast strains, the production of artisanal breads was long and laborious because many batches of dough ended up being discarded. Can you explain this fact?
34. How would treating an area of a forest with a broad-spectrum fungicide alter the carbon and nitrogen cycles in the area?

CHAPTER 25 Seedless Plants

Figure 25.1 Seedless plants, like these horsetails (*Equisetum* sp.), thrive in damp, shaded environments under a tree canopy where dryness is rare. (credit: modification of work by Jerry Kirkhart)

INTRODUCTION An incredible variety of seedless plants populates the terrestrial landscape. Mosses may grow on a tree trunk, and horsetails may display their jointed stems and spindly leaves across the forest floor. Today, seedless plants represent only a small fraction of the plants in our environment; yet, 300 million years ago, seedless plants dominated the landscape and grew in the enormous swampy forests of the Carboniferous period. Their decomposition created large deposits of coal that we mine today.

Current evolutionary thought holds that all plants—some green algae as well as land plants—are monophyletic; that is, they are descendants of a single common ancestor. The evolutionary transition from water to land imposed severe constraints on plants. They had to develop strategies to avoid drying out, to disperse reproductive cells in air, for structural support, and for capturing and filtering sunlight. While seed plants have developed adaptations that allow them to populate even the most arid habitats on Earth, full independence from water did not happen in all plants. Most seedless plants still require a moist environment for reproduction.

25.1 Early Plant Life

By the end of this section, you will be able to do the following:

- Discuss the challenges to plant life on land
- Describe the adaptations that allowed plants to colonize the land
- Describe the timeline of plant evolution and the impact of land plants on other living things

The kingdom Plantae constitutes large and varied groups of organisms. There are more than 300,000 species of catalogued plants. Of these, more than 260,000 are seed plants. Mosses, ferns, conifers, and flowering plants are all members of the plant kingdom. Land plants arose within the Archaeplastida, which includes the red algae (Rhodophyta) and two groups of green algae, Chlorophyta and Charaphyta. Most biologists also consider at least some green algae to be plants, although others exclude all algae from the plant kingdom. The reason for this disagreement stems from the fact that only green algae, the **Chlorophytes** and **Charophytes**, share common characteristics with land plants (such as using chlorophyll *a* and *b* plus carotene in the same proportion as plants). These characteristics are absent from other types of algae.

EVOLUTION CONNECTION

Algae and Evolutionary Paths to Photosynthesis

Some scientists consider all algae to be plants, while others assert that only the green algae belong in the kingdom Plantae. Still others include only the Charophytes among the plants. These divergent opinions are related to the different evolutionary paths to photosynthesis selected for in different types of algae. While all algae are photosynthetic—that is, they contain some form of a chloroplast—they didn't all become photosynthetic via the same path.

The ancestors to the Archaeplastida became photosynthetic by forming an endosymbiotic relationship with a green, photosynthetic bacterium about 1.65 billion years ago. That algal line evolved into the red and green algae, and eventually into the modern mosses, ferns, gymnosperms, and angiosperms. Their evolutionary trajectory was relatively straight and monophyletic. In contrast, algae outside of the Archaeplastida, e.g., the brown and golden algae of the stramenopiles, and so on—all became photosynthetic by secondary, or even tertiary, endosymbiotic events; that is, they engulfed cells that already contained an endosymbiotic cyanobacterium. These latecomers to photosynthesis are parallels to the Archaeplastida in terms of autotrophy, but they did not expand to the same extent as the Archaeplastida, nor did they colonize the land.

Scientists who solely track evolutionary straight lines (that is, monophyly), consider only the Charophytes as plants. The common ancestor of Charophytes and land plants excludes the other members of the Archaeplastida. Charophytes also share other features with the land plants. These will be discussed in more detail in another section.

LINK TO LEARNING

Go to this article (http://openstax.org/l/charophytes) to get a more in-depth view of the Charophytes.

Plant Adaptations to Life on Land

As organisms adapted to life on land, they had to contend with several challenges in the terrestrial environment. Water has been described as "the stuff of life." The cell's interior is a thick soup: in this medium, most small molecules dissolve and diffuse, and the majority of the chemical reactions of metabolism take place. Desiccation, or drying out, is a constant danger for an organism exposed to air. Even when parts of a plant are close to a source of water, the aerial structures are likely to dry out. Water also provides buoyancy to organisms. On land, plants need to develop structural support in a medium that does not give the same lift as water. The organism is also subject to bombardment by mutagenic radiation, because air does not filter out ultraviolet rays of sunlight. Additionally, the male gametes must reach the female gametes using new strategies, because swimming is no longer possible. Therefore, both gametes and zygotes must be protected from desiccation. The successful land plants developed strategies to deal with all of these challenges. Not all adaptations appeared at once. Some species never moved very far from the aquatic environment, whereas others went on to conquer the driest environments on Earth.

To balance these survival challenges, life on land offers several advantages. First, sunlight is abundant. Water acts as a filter, altering the spectral quality of light absorbed by the photosynthetic pigment chlorophyll. Second, carbon dioxide is more readily available in air than in water, since it diffuses faster in air. Third, land plants evolved before land animals; therefore, until dry land was colonized by animals, no predators threatened plant life. This situation changed as animals emerged from the water and fed on the abundant sources of nutrients in the established flora. In turn, plants developed strategies to deter predation: from spines and thorns to toxic chemicals.

Early land plants, like the early land animals, did not live very far from an abundant source of water and developed survival strategies to combat dryness. One of these strategies is called tolerance. Many mosses, for example, can dry out to a brown and brittle mat, but as soon as rain or a flood makes water available, mosses will absorb it and are restored to their healthy green appearance. Another strategy is to colonize environments with high humidity, where droughts are uncommon. Ferns, which are considered an early lineage of plants, thrive in damp and cool places such as the understory of temperate forests. Later, plants moved away from moist or aquatic environments using resistance to desiccation, rather than tolerance. These plants, like cacti, minimize the loss of water to such an extent they can survive in extremely dry environments.

The most successful adaptation solution was the development of new structures that gave plants the advantage when colonizing new and dry environments. Four major adaptations contribute to the success of terrestrial plants. The first adaptation is that the life cycle in all land plants exhibits the alternation of generations, a sporophyte in which the spores are formed and a gametophyte that produces gametes. Second is an apical meristem tissue in roots and shoots. Third is the evolution of a waxy cuticle to resist desiccation (absent from some mosses). Finally cell walls with lignin to support structures off the ground. These adaptations all contribute to the success of the land plants, but are noticeably lacking in the closely related green algae—another reason for the debate over their placement in the plant kingdom. They are also not all found in the mosses, which can be regarded as representing an intermediate stage in adaptation to land.

Alternation of Generations

All sexually reproducing organisms have both haploid and diploid cells in their life cycles. In organisms with **haplontic** life cycles, the haploid stage is dominant, while in organisms with a **diplontic** life cycle, the diploid stage is the dominant life stage. *Dominant* in this context means both the stage in which the organism spends most of its time, and the stage in which most mitotic cell reproduction occurs—the multicellular stage. In haplontic life cycles, the only diploid cell is the zygote, which undergoes immediate meiosis to restore the haploid state. In diplontic life cycles, the only haploid cells are the gametes, which combine to restore the diploid state at their earliest convenience. Humans, for example, are diplontic.

Alternation of generations describes a life cycle in which an organism has both haploid and diploid multicellular stages (Figure 25.2). This type of life cycle, which is found in all plants, is described as **haplodiplontic**.

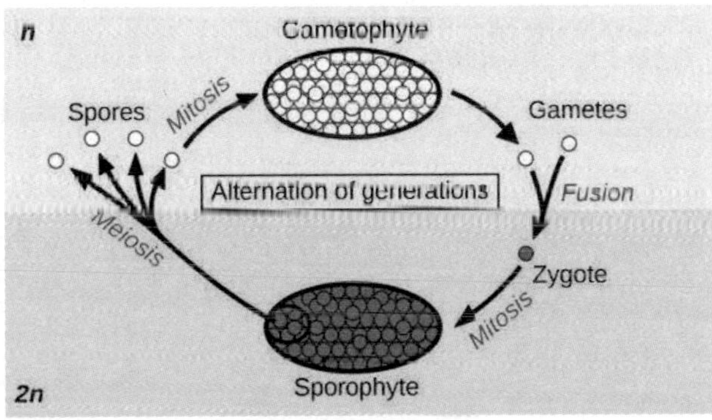

Figure 25.2 Alternation of generations between the 1*n* gametophyte and 2*n* sporophyte is shown. Mitosis occurs in both gametophyte and sporophyte generations. Diploid sporophytes produce haploid spores by meiosis, while haploid gametophytes produce gametes by mitosis. (credit: Peter Coxhead)

In alternation of generations, the multicellular haploid form, known as a gametophyte, is followed in the developmental sequence by a multicellular diploid form, the sporophyte. The gametophyte gives rise to the gametes (reproductive cells) by mitosis. This can be the most obvious phase of the life cycle of the plant, as in the mosses, or it can occur in a microscopic structure, such as a pollen grain, in the seed plants. The evolution of the land plants is marked by increasing prominence of the sporophyte generation. The sporophyte stage is barely noticeable in non-vascular plants (the collective term for the plants that include the liverworts and mosses). In the seed plants, the sporophyte phase can be a towering tree, as in sequoias and pines.

Protection of the embryo is a major requirement for land plants. The vulnerable embryo must be sheltered from desiccation and other environmental hazards. In both seedless and seed plants, the female gametophyte provides protection and nutrients to the embryo as it develops into the new sporophyte. This distinguishing feature of land plants gave the group its alternate name of **embryophytes**.

Sporangia in Seedless Plants

The sporophyte of seedless plants is diploid and results from syngamy (fusion) of two gametes. The sporophyte bears the sporangia (singular, sporangium). The term "sporangia" literally means "a vessel for spores," as it is a reproductive sac in which spores are formed (Figure 25.3). Inside the multicellular sporangia, the diploid **sporocytes**, or mother cells, produce haploid spores by meiosis, during which the $2n$ chromosome number is reduced to $1n$ (note that in many plants, chromosome number is complicated by polyploidy: for example, durum wheat is tetraploid, bread wheat is hexaploid, and some ferns are 1000-ploid). The spores are later released by the sporangia and disperse in the environment. When the haploid spore germinates in a hospitable environment, it generates a multicellular gametophyte by mitosis. The gametophyte supports the zygote formed from the fusion of gametes and the resulting young sporophyte (vegetative form). The cycle then begins anew.

Figure 25.3 Sporangia. Spore-producing sacs called sporangia grow at the ends of long, thin stalks in this photo of the moss *Esporangios bryum*. (credit: Javier Martin)

Plants that produce only one type of spore are called homosporous and the resultant gametophyte produces both male and female gametes, usually on the same individual. Non-vascular plants are homosporous, and the gametophyte is the dominant generation in the life cycle. Plants that produce two types of spores are called heterosporous. The male spores are called microspores, because of their smaller size, and develop into the male gametophyte; the comparatively larger megaspores develop into the female gametophyte. A few seedless vascular plants and all seed plants are heterosporous, and the sporophyte is the dominant generation.

The spores of seedless plants are surrounded by thick cell walls containing a tough polymer known as **sporopollenin**. As the name suggests, it is also found in the walls of pollen grains. This complex substance is characterized by long chains of organic molecules related to fatty acids and carotenoids: hence the yellow color of most pollen. Sporopollenin is unusually resistant to chemical and biological degradation. In seed plants, in which pollen is the male gametophyte, the toughness of sporopollenin explains the existence of well-preserved pollen fossils. Sporopollenin was once thought to be an innovation of land plants; however, the charophyte *Coleochaetes* also forms spores that contain sporopollenin.

Gametangia in Seedless Plants

Gametangia (singular, gametangium) are structures observed on multicellular haploid gametophytes. In the gametangia, precursor cells give rise to gametes by mitosis. The male gametangium (**antheridium**) releases sperm. Seedless plants produce sperm equipped with flagella that enable them to swim in a moist environment to the **archegonium**: the female gametangium. The embryo develops inside the archegonium as the sporophyte. Gametangia are prominent in seedless plants, but are absent or rudimentary in seed plants.

Apical Meristems

Shoots and roots of plants increase in length through rapid cell division in a tissue called the apical meristem, which is a small mitotically active zone of cells found at the shoot tip or root tip (Figure 25.4). The apical meristem is made of undifferentiated cells that continue to proliferate throughout the life of the plant. Meristematic cells give rise to all the specialized tissues of the organism. Elongation of the shoots and roots allows a plant to access additional space and resources: light in the case of the shoot, and water and minerals in the case of roots. A separate meristem, called the lateral meristem, produces cells that increase the diameter of tree trunks.

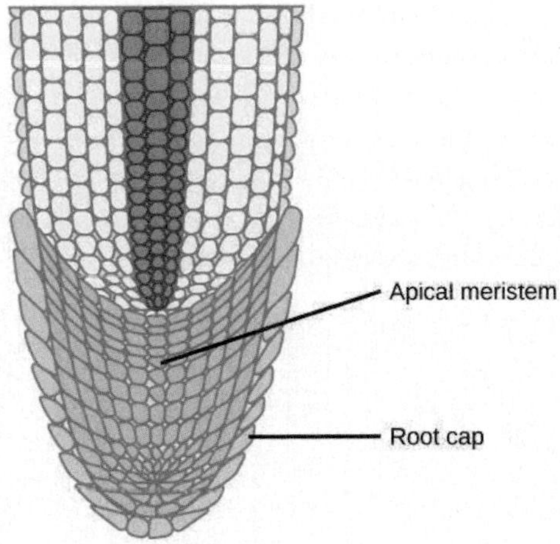

Figure 25.4 Apical meristem at a root tip. Addition of new cells in a root occurs at the apical meristem. Subsequent enlargement of these cells causes the organ to grow and elongate. The root cap protects the fragile apical meristem as the root tip is pushed through the soil by cell elongation.

Additional Land Plant Adaptations

As plants adapted to dry land and became independent from the constant presence of water in damp habitats, new organs and structures made their appearance. Early land plants did not grow more than a few inches off the ground, competing for light on these low mats. By developing a shoot and growing taller, individual plants captured more light. Because air offers substantially less support than water, land plants incorporated more rigid molecules in their stems (and later, tree trunks). In small plants such as single-celled algae, simple diffusion suffices to distribute water and nutrients throughout the organism. However, for plants to evolve larger forms, the evolution of a conductive tissue for the distribution of water and solutes was a prerequisite. The evolution of vascular tissue in plants met both of these needs. The vascular system contains two types of conductive tissue: xylem and phloem. Xylem conducts water and minerals absorbed from the soil up to the shoot, while phloem transports food derived from photosynthesis throughout the entire plant. In xylem, the cells walls are reinforced with lignin, whose tough hydrophobic polymers help prevent the seepage of water across the xylem cell walls. Lignin also adds to the strength of these tissues in supporting the plant. The vascular tissues extend into the root of land plants. The root system evolved to take up water and minerals from the soil, and to anchor the increasingly taller shoot in the soil.

In land plants, a waxy, waterproof cover called a cuticle protects the leaves and stems from desiccation. However, the cuticle also prevents intake of carbon dioxide needed for the synthesis of carbohydrates through photosynthesis. To overcome this, stomata or pores that open and close to regulate traffic of gases and water vapor appeared in plants as they moved away from moist environments into drier habitats.

Water filters ultraviolet-B (UVB) light, which is harmful to all organisms, especially those that must absorb light to survive. This filtering does not occur for land plants. Exposure to damaging radiation presented an additional challenge to land colonization, which was met by the evolution of biosynthetic pathways for the synthesis of protective flavonoids and other pigments that absorb UV wavelengths of light and protect the aerial parts of plants from photodynamic damage.

Plants cannot avoid being eaten by animals. Instead, they synthesize a large range of poisonous secondary metabolites: complex organic molecules such as alkaloids, whose noxious smells and unpleasant taste deter animals. These toxic compounds can also cause severe diseases and even death, thus discouraging predation. Humans have used many of these compounds for centuries as drugs, medications, or spices. In contrast, as plants co-evolved with animals, the development of sweet and nutritious metabolites lured animals into providing valuable assistance in dispersing pollen grains, fruit, or seeds. Plants have been enlisting animals to be their helpers in this way for hundreds of millions of years.

Evolution of Land Plants

No discussion of the evolution of plants on land can be undertaken without a brief review of the timeline of the geological eras. The early era, known as the Paleozoic, is divided into six periods. It starts with the Cambrian period, followed by the Ordovician, Silurian, Devonian, Carboniferous, and Permian. The major event to mark the Ordovician, more than 500 million years ago, was the colonization of land by the ancestors of modern land plants. Fossilized cells, cuticles, and spores of early land plants have been dated as far back as the Ordovician period in the early Paleozoic era. The oldest-known vascular plants have been identified in deposits from the Devonian. One of the richest sources of information is the Rhynie chert, a sedimentary rock deposit found in Rhynie, Scotland (Figure 25.5), where embedded fossils of some of the earliest vascular plants have been identified.

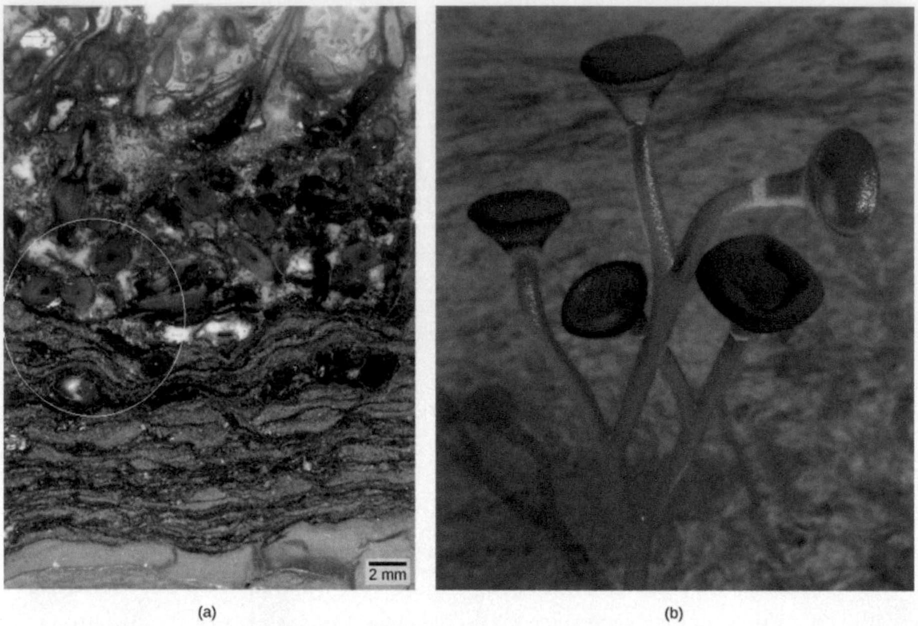

Figure 25.5 Early vascular plant fossils. This Rhynie chert (a) contains fossilized material from vascular plants. Reconstruction of *Cooksonia* (b), the plant forms inside the circle. (credit b: modification of work by Peter Coxhead based on original image by "Smith609"/Wikimedia Commons; scale-bar data from Matt Russell)

Paleobotanists distinguish between **extinct** species, as fossils, and **extant** species, which are still living. The extinct vascular plants most probably lacked true leaves and roots and formed low vegetation mats similar in size to modern-day mosses, although some could reach one meter in height. The later genus *Cooksonia*, which flourished during the Silurian, has been extensively studied from well-preserved examples. Imprints of *Cooksonia* show slender branching stems ending in what appear to be sporangia. From the recovered specimens, it is not possible to establish for certain whether *Cooksonia* possessed vascular tissues. Fossils indicate that by the end of the Devonian period, ferns, horsetails, and seed plants populated the landscape, giving rising to trees and forests. This luxuriant vegetation helped enrich the atmosphere with oxygen, making it easier for air-breathing animals to colonize dry land. Plants also established early symbiotic relationships with fungi, creating mycorrhizae: a relationship in which the fungal network of filaments increases the efficiency of the plant root system, and the plants provide the fungi with byproducts of photosynthesis.

CAREER CONNECTION

Paleobotanist

How organisms acquired traits that allow them to colonize new environments—and how the contemporary ecosystem is shaped—are fundamental questions of evolution. Paleobotany (the study of extinct plants) addresses these questions through the analysis of fossilized specimens retrieved from field studies, reconstituting the morphology of organisms that disappeared long ago. Paleobotanists trace the evolution of plants by following the modifications in plant morphology: shedding light on the connection between existing plants by identifying common ancestors that display the same traits. This field seeks to find transitional species that bridge gaps in the path to the development of modern organisms. Fossils are formed when organisms are trapped in sediments or environments where their shapes are preserved. Paleobotanists collect fossil specimens in the field and place them in the context of the geological sediments and other fossilized organisms surrounding them. The activity requires great care to preserve the integrity of the delicate fossils and the layers of rock in which they are found.

One of the most exciting recent developments in paleobotany is the use of analytical chemistry and molecular biology to study fossils. Preservation of molecular structures requires an environment free of oxygen, since oxidation and degradation of material through the activity of microorganisms depend on its presence. One example of the use of analytical chemistry and molecular biology is the identification of oleanane, a compound that deters pests. Up to this point, oleanane appeared to be unique to flowering plants; however, it has now been recovered from sediments dating from the Permian, much earlier than the current dates given for the appearance of the first flowering plants. Paleobotanists can also study fossil DNA, which can yield a large amount of information, by analyzing and comparing the DNA sequences of extinct plants with those of living and related organisms. Through this analysis, evolutionary relationships can be built for plant lineages.

Some paleobotanists are skeptical of the conclusions drawn from the analysis of molecular fossils. For example, the chemical materials of interest degrade rapidly when exposed to air during their initial isolation, as well as in further manipulations. There is always a high risk of contaminating the specimens with extraneous material, mostly from microorganisms. Nevertheless, as technology is refined, the analysis of DNA from fossilized plants will provide invaluable information on the evolution of plants and their adaptation to an ever-changing environment.

The Major Divisions of Land Plants

The green algae and land plants are grouped together into a subphylum called the Streptophyta, and thus are called Streptophytes. In a further division, land plants are classified into two major groups according to the absence or presence of vascular tissue, as detailed in Figure 25.6. Plants that lack vascular tissue, which is formed of specialized cells for the transport of water and nutrients, are referred to as **non-vascular plants**. Liverworts, mosses, and hornworts are seedless, non-vascular plants that likely appeared early in land plant evolution. Vascular plants developed a network of cells that conduct water and solutes. The first vascular plants appeared in the late Ordovician (500 to 435 MYA) and were probably similar to lycophytes, which include club mosses (not to be confused with the mosses) and the pterophytes (ferns, horsetails, and whisk ferns). Lycophytes and pterophytes are referred to as seedless vascular plants, because they do not produce seeds. The seed plants, or spermatophytes, form the largest group of all existing plants, and hence dominate the landscape. Seed plants include gymnosperms, most notably conifers, which produce "naked seeds," and the most successful of all plants, the flowering plants (angiosperms). Angiosperms protect their seeds inside chambers at the center of a flower; the walls of the chamber later develop into a fruit.

VISUAL CONNECTION

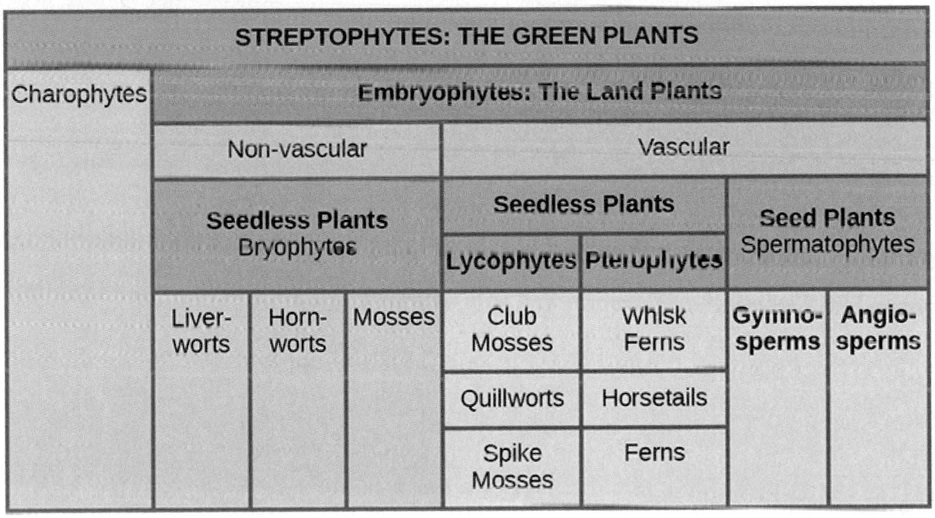

Figure 25.6 Streptophytes. This table shows the major divisions of green plants.

Which of the following statements about plant divisions is false?

a. Lycophytes and pterophytes are seedless vascular plants.
b. All vascular plants produce seeds.
c. All non-vascular embryophytes are bryophytes.
d. Seed plants include angiosperms and gymnosperms.

25.2 Green Algae: Precursors of Land Plants

By the end of this section, you will be able to do the following:

- Describe the traits shared by green algae and land plants
- Explain why charophytes are considered the closest algal relative to land plants
- Explain how current phylogenetic relationships are reshaped by comparative analysis of DNA sequences

Streptophytes

Until recently, all photosynthetic eukaryotes were classified as members of the kingdom Plantae. The brown and golden algae, however, are now reassigned to the protist supergroup Chromalveolata. This is because apart from their ability to capture light energy and fix CO_2, they lack many structural and biochemical traits that are characteristic of plants. The plants are now classified, along with the red and green algae, in the protist supergroup Archaeplastida. Green algae contain the same carotenoids and chlorophyll a and b as land plants, whereas other algae have different accessory pigments and types of chlorophyll molecules in addition to chlorophyll a. Both green algae and land plants also store carbohydrates as starch. Their cells contain chloroplasts that display a dizzying variety of shapes, and their cell walls contain cellulose, as do land plants. Which of the green algae to include among the plants has not been phylogenetically resolved.

Green algae fall into two major groups, the chlorophytes and the charophytes. The chlorophytes include the genera *Chlorella*, *Chlamydomonas*, the "sea lettuce" *Ulva*, and the colonial alga *Volvox*. The charophytes include desmids, as well as the genera *Spirogyra*, *Coleochaete*, and *Chara*. There are familiar green algae in both groups. Some green algae are single cells, such as *Chlamydomonas* and desmids, which adds to the ambiguity of green algae classification, because plants are multicellular. Other green algae, like *Volvox*, form colonies, and some, like *Ulva* are multicellular (Figure 25.7). *Spirogyra* is a long filament of colonial cells. Most members of this genus live in fresh water, brackish water, seawater, or even in snow patches. A few green algae can survive on soil, provided it is covered by a thin film of moisture within which they can live. Periodic dry spells provide a selective advantage to algae that can survive water stress.

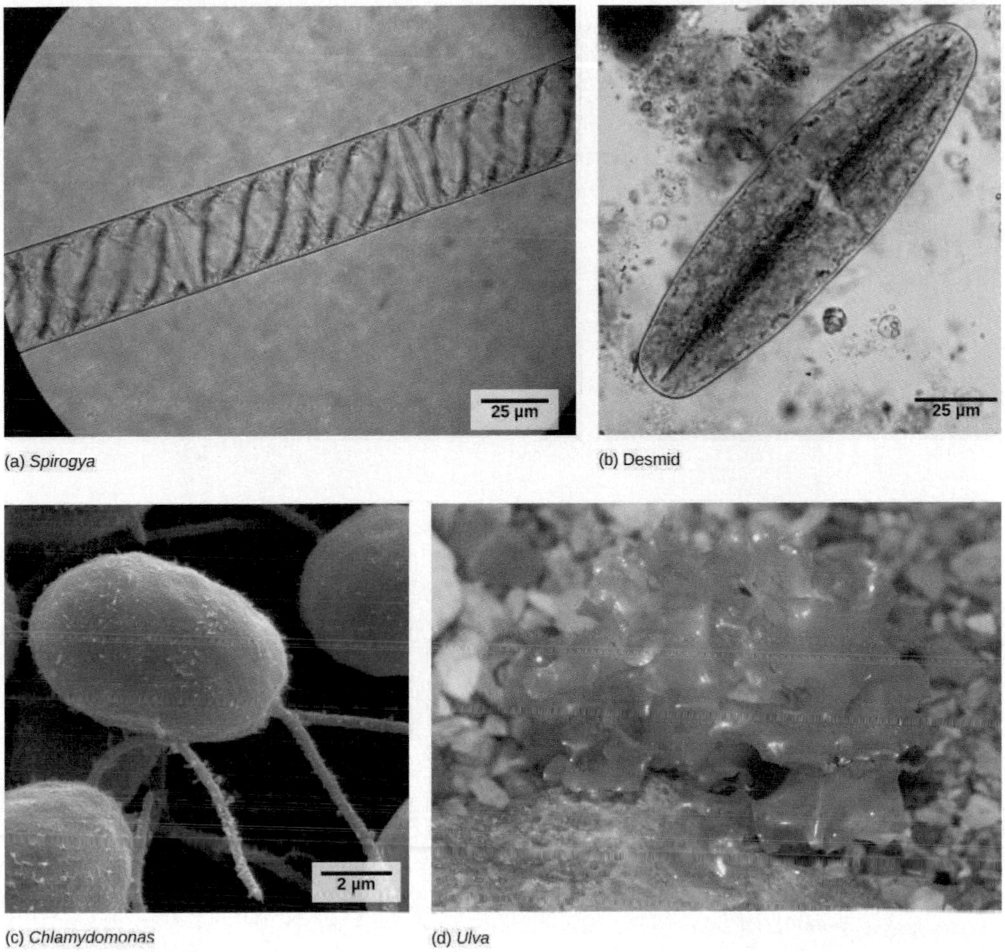

Figure 25.7 Green algae. Charophyta include (a) *Spirogyra* and (b) desmids. Chlorophyta include (c) *Chlamydomonas*, and (d) *Ulva*. Desmids and *Chlamydomonas* are single-celled organisms, *Spirogyra* forms chains of cells, and *Ulva* forms multicellular structures resembling leaves, although the cells are not differentiated as they are in higher plants (credit b: modification of work by Derek Keats; credit c: modification of work by Dartmouth Electron Microscope Facility, Dartmouth College; credit d: modification of work by Holger Krisp; scale-bar data from Matt Russell)

The chlorophytes and the charophytes differ in a few respects that, in addition to molecular analysis, place the land plants as a sister group of the charophytes. First, cells in charophytes and the land plants divide along cell plates called phragmoplasts, in which microtubules parallel to the spindle serve as guides for the vesicles of the forming cell plate. In the chlorophytes, the cell plate is organized by a phycoplast, in which the microtubules are perpendicular to the spindle. Second, only the charophytes and the land plants have plasmodesmata, or intercellular channels that allow the transfer of materials from cell to cell. In the chlorophytes, intercellular connections do not persist in mature multicellular forms. Finally, both charophytes and the land plants show *apical growth*—growth from the tips of the plant rather than throughout the plant body. Consequently, land plants and the charophytes are now part of a new monophyletic group called Streptophyta.

Reproduction of Green Algae

Green algae reproduce both asexually, by fragmentation or dispersal of spores, or sexually, by producing gametes that fuse during fertilization. In a single-celled organism such as *Chlamydomonas*, there is no mitosis after fertilization. In the multicellular *Ulva*, a sporophyte grows by mitosis after fertilization (and thus exhibits alternation of generations). Both *Chlamydomonas* and *Ulva* produce flagellated gametes.

Charophytes

The charophytes include several different algal orders that have each been suggested to be the closest relatives of the land plants: the Charales, the Zygnematales, and the Coleochaetales. The Charales can be traced back 420 million years. They live in a range of freshwater habitats and vary in size from a few millimeters to a meter in length. The representative genus is *Chara* (Figure

25.8), often called muskgrass or skunkweed because of its unpleasant smell. Large cells form the *thallus*: the main stem of the alga. Branches arising from the nodes are made of smaller cells. Male and female reproductive structures are found on the nodes, and the sperm have flagella. Although *Chara* looks superficially like some land plants, a major difference is that the stem has no supportive tissue. However, the Charales exhibit a number of traits that are significant for adaptation to land life. They produce the compounds *lignin* and *sporopollenin*, and form plasmodesmata that connect the cytoplasm of adjacent cells. Although the life cycle of the Charales is haplontic (the main form is haploid, and diploid zygotes are formed but have a brief existence), the egg, and later, the zygote, form in a protected chamber on the haploid parent plant.

Figure 25.8 *Chara*. The representative alga, *Chara*, is a noxious weed in Florida, where it clogs waterways. (credit: South Florida Information Access, U.S. Geological Survey)

The Coleochaetes are branched or disclike multicellular forms. They can produce both sexually and asexually, but the life cycle is basically haplontic. Recent extensive DNA sequence analysis of charophytes indicates that the Zygnematales are more closely related to the embryophytes than the Charales or the Coleochaetales. The Zygnematales include the familiar genus *Spirogyra*, as well as the desmids. As techniques in DNA analysis improve and new information on comparative genomics arises, the phylogenetic connections between the charophytes and the land plants will continued to be examined to produce a satisfactory solution to the mystery of the origin of land plants.

25.3 Bryophytes

By the end of this section, you will be able to do the following:

- Identify the main characteristics of bryophytes
- Describe the distinguishing traits of liverworts, hornworts, and mosses
- Chart the development of land adaptations in the bryophytes
- Describe the events in the bryophyte lifecycle

Bryophytes are the closest extant relatives of early terrestrial plants. The first bryophytes (liverworts) most likely appeared in the Ordovician period, about 450 million years ago. Because they lack lignin and other resistant structures, the likelihood of bryophytes forming fossils is rather small. Some spores protected by *sporopollenin* have survived and are attributed to early bryophytes. By the Silurian period (435 MYA), however, vascular plants had spread through the continents. This compelling fact is used as evidence that non-vascular plants must have preceded the Silurian period.

More than 25,000 species of bryophytes thrive in mostly damp habitats, although some live in deserts. They constitute the major flora of inhospitable environments like the tundra, where their small size and tolerance to desiccation offer distinct advantages. They generally lack lignin and do not have actual tracheids (xylem cells specialized for water conduction). Rather, water and nutrients circulate inside specialized conducting cells. Although the term *non-tracheophyte* is more accurate, bryophytes are commonly called *non-vascular plants*.

In a bryophyte, all the conspicuous vegetative organs—including the photosynthetic leaf-like structures, the thallus ("plant body"), stem, and the rhizoid that anchors the plant to its substrate—belong to the haploid organism or gametophyte. The male gametes formed by bryophytes swim with a flagellum, so fertilization is dependent on the presence of water. The bryophyte embryo also remains attached to the parent plant, which protects and nourishes it. The **sporophyte** that develops from the embryo is barely noticeable. The **sporangium**—the multicellular sexual reproductive structure in which meiosis produces haploid spores—is present in bryophytes and absent in the majority of algae. This is also a characteristic of land plants.

The bryophytes are divided into three phyla: the liverworts or Marchantiophyta, the hornworts or Anthocerotophyta, and the mosses or true Bryophyta.

Liverworts

Liverworts (Marchantiophyta) are currently classified as the plants most closely related to the ancestor of vascular plants that adapted to terrestrial environments. In fact, liverworts have colonized every terrestrial habitat on Earth and diversified to more than 7000 existing species (Figure 25.9). Lobate liverworts form a flat thallus, with lobes that have a vague resemblance to the lobes of the liver (Figure 25.10), which accounts for the name given to the phylum. Leafy liverworts have tiny leaflike structures attached to a stalk. Several leafy liverworts are shown in Figure 25.9.

Figure 25.9 Liverworts. This 1904 drawing shows the variety of forms of Marchantiophyta.

Figure 25.10 Liverwort gametophyte. A liverwort, *Lunularia cruciata*, displays its lobate, flat thallus. The organism in the photograph is in the gametophyte stage, but has not yet produced gametangia. *Lunularia* gametophytes produce crescent-shaped **gemmae** (circled), which contain asexual spores. The tiny white dots on the surface of the thallus are air pores.

Openings in the thallus that allow the movement of gases may be observed in liverworts (Figure 25.10). However, these are *not* stomata, because they do not actively open and close by the action of guard cells. Instead, the thallus takes up water over its entire surface and has no cuticle to prevent desiccation, which explains their preferred wet habitats. Figure 25.11 represents the lifecycle of a lobate liverwort. Haploid spores germinate into flattened thalli attached to the substrate by thin, single-celled filaments. Stalk-like structures (*gametophores*) grow from the thallus and carry male and female gametangia, which may develop on separate, individual plants, or on the same plant, depending on the species. Flagellated male gametes develop within *antheridia* (male gametangia). The female gametes develop within *archegonia* (female gametangia). Once released, the male gametes swim with the aid of their flagella to an archegonium, and fertilization ensues. The zygote grows into a small sporophyte still contained in the archegonium. The diploid zygote will give rise, by meiosis, to the next generation of haploid spores, which can be disseminated by wind or water. In many liverworts, spore dispersal is facilitated by *elaters*—long single cells that suddenly change shape as they dry out and throw adjacent spores out of the spore capsule. Liverwort plants can also reproduce asexually, by the breaking of "branches" or the spreading of leaf fragments called gemmae. In this latter type of reproduction, the **gemmae**—small, intact, complete pieces of plant that are produced in a cup on the surface of the thallus (shown in Figure 25.11 and Figure 25.12)—are splashed out of the cup by raindrops. The gemmae then land nearby and develop into gametophytes.

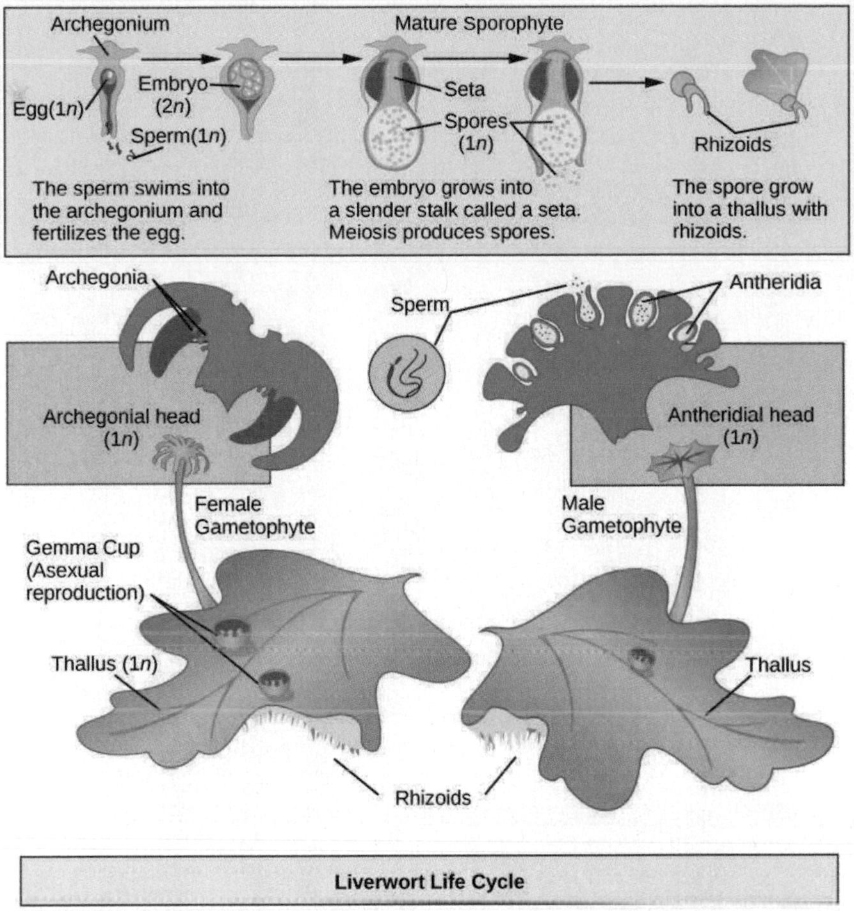

Figure 25.11 Reproductive cycle of liverworts. The life cycle of a typical lobate liverwort is shown. This image shows a liverwort in which antheridia and archegonia are produced on separate gametophytes. (credit: modification of work by Mariana Ruiz Villareal)

Hornworts

The defining characteristic of the hornworts (*Anthocerotophyta*) is the narrow, pipe-like sporophyte. Hornworts have colonized a variety of habitats on land, although they are never far from a source of moisture. The short, blue-green gametophyte is the dominant phase of the life cycle of a hornwort. The sporophytes emerge from the parent gametophyte and continue to grow throughout the life of the plant (Figure 25.12).

Figure 25.12 Hornwort sporophytes. Hornworts grow a tall and slender sporophyte. (credit: modification of work by Jason Hollinger)

Stomata (air pores that can be opened and closed) appear in the hornworts and are abundant on the sporophyte. Photosynthetic cells in the thallus each contain a single chloroplast. Meristem cells at the base of the plant keep dividing and adding to the height of the sporophyte. This growth pattern is unique to the hornworts. Many hornworts establish symbiotic relationships with cyanobacteria that fix nitrogen from the environment.

The lifecycle of hornworts (Figure 25.13) follows the general pattern of *alternation of generations*. The gametophytes grow as flat thalli on the soil with embedded male and female gametangia. Flagellated sperm swim to the archegonia and fertilize eggs. The zygote develops into a long and slender sporophyte that eventually splits open down the side, releasing spores. Thin branched cells called pseudoelaters surround the spores and help propel them farther in the environment. The haploid spores germinate and give rise to the next generation of gametophytes.

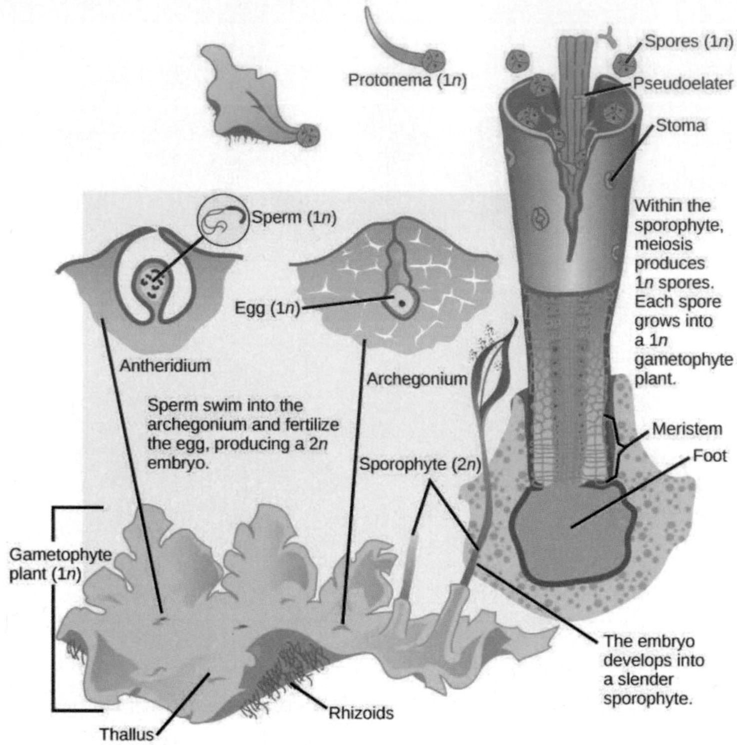

Figure 25.13 Reproductive cycle of hornworts. The alternation of generation in hornworts is shown. (credit: modification of work by "Smith609"/Wikimedia Commons based on original work by Mariana Ruiz Villareal)

Mosses

The mosses are the most numerous of the non-vascular plants. More than 10,000 species of **mosses** have been catalogued. Their habitats vary from the tundra, where they are the main vegetation, to the understory of tropical forests. In the tundra, the mosses' shallow rhizoids allow them to fasten to a substrate without penetrating the frozen soil. Mosses slow down erosion, store moisture and soil nutrients, and provide shelter for small animals as well as food for larger herbivores, such as the musk ox. Mosses are very sensitive to air pollution and are used to monitor air quality. They are also sensitive to copper salts, so these salts are a common ingredient of compounds marketed to eliminate mosses from lawns.

Mosses form diminutive gametophytes, which are the dominant phase of the lifecycle. Green, flat structures with a simple midrib—resembling true leaves, but lacking stomata and vascular tissue—are attached in a spiral to a central stalk. Mosses have stomata only on the sporophyte. Water and nutrients are absorbed directly through the leaflike structures of the gametophyte. Some mosses have small branches. A primitive conductive system that carries water and nutrients runs up the gametophyte's stalk, but does not extend into the leaves. Additionally, mosses are anchored to the substrate—whether it is soil, rock, or roof tiles—by multicellular **rhizoids**, precursors of roots. They originate from the base of the gametophyte, but are not the major route for the absorption of water and minerals. The lack of a true root system explains why it is so easy to rip moss mats from a tree trunk. The mosses therefore occupy a threshold position between other bryophytes and the vascular plants.

The moss lifecycle follows the pattern of alternation of generations as shown in Figure 25.14. The most familiar structure is the haploid gametophyte, which germinates from a haploid spore and forms first a **protonema**—usually, a tangle of single-celled filaments that hug the ground. Cells akin to an apical meristem actively divide and give rise to a gametophore, consisting of a photosynthetic stem and foliage-like structures. Male and female gametangia develop at the tip of separate gametophores. The antheridia (male organs) produce many sperm, whereas the archegonia (the female organs) each form a single egg at the base (venter) of a flask-shaped structure. The archegonium produces attractant substances and at fertilization, the sperm swims down the neck to the venter and unites with the egg inside the archegonium. The zygote, protected by the archegonium, divides and grows into a sporophyte, still attached by its foot to the gametophyte.

VISUAL CONNECTION

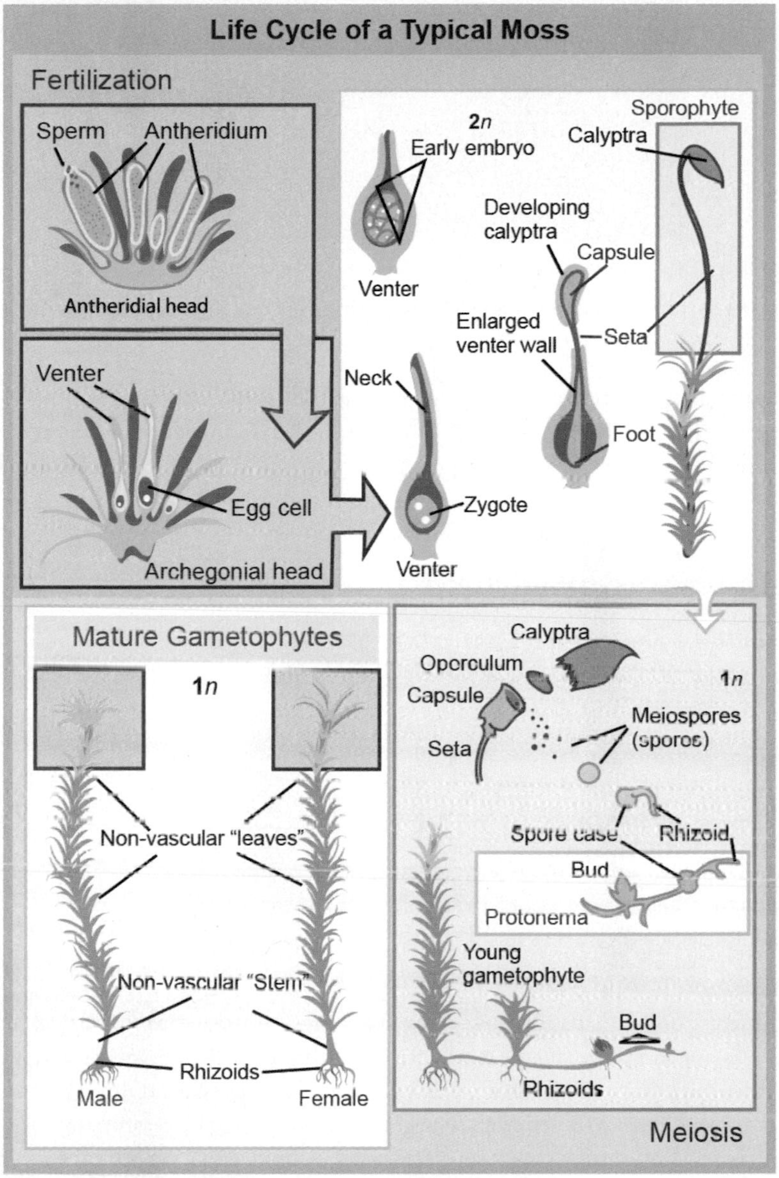

Figure 25.14 Reproductive cycle of mosses. This illustration shows the life cycle of mosses. (credit: modification of work by Mariana Ruiz Villareal)

Which of the following statements about the moss life cycle is false?

a. The mature gametophyte is haploid.
b. The sporophyte produces haploid spores.
c. The calyptra buds to form a mature gametophyte.
d. The zygote is housed in the venter.

The moss sporophyte is dependent on the gametophyte for nutrients. The slender **seta** (plural, setae), as seen in Figure 25.15, contains tubular cells that transfer nutrients from the base of the sporophyte (the foot) to the sporangium or **capsule**.

Figure 25.15 Moss sporophyte. This photograph shows the long slender stems, called setae, connected to capsules of the moss *Thamnobryum alopecurum*. The operculum and remnants of the calyptra are visible in some capsules. (credit: modification of work by Hermann Schachner)

Spore mother cells in the sporangium undergo meiosis to produce haploid spores. The sporophyte has several features that protect the developing spores and aid in their dispersal. The calyptra, derived from the walls of the archegonium, covers the sporangium. A structure called the operculum is at the tip of the spore capsule. The calyptra and operculum fall off when the spores are ready for dispersal. The peristome, tissue around the mouth of the capsule, is made of triangular, close-fitting units like little "teeth." The peristome opens and closes, depending on moisture levels, and periodically releases spores.

25.4 Seedless Vascular Plants

By the end of this section, you will be able to do the following:

- Identify the new traits that first appear in seedless tracheophytes
- Discuss how each trait is important for adaptation to life on land
- Identify the classes of seedless tracheophytes
- Describe the life cycle of a fern
- Explain the role of seedless plants in the ecosystem

The vascular plants, or **tracheophytes**, are the dominant and most conspicuous group of land plants. More than 260,000 species of tracheophytes represent more than 90 percent of Earth's vegetation. Several evolutionary innovations explain their success and their ability to spread to all habitats.

Bryophytes may have been successful at the transition from an aquatic habitat to land, but they are still dependent on water for reproduction, and must absorb moisture and nutrients through the gametophyte surface. The lack of roots for absorbing water and minerals from the soil, as well as a lack of lignin-reinforced conducting cells, limit bryophytes to small sizes. Although they may survive in reasonably dry conditions, they cannot reproduce and expand their habitat range in the absence of water. Vascular plants, on the other hand, can achieve enormous heights, thus competing successfully for light. Photosynthetic organs become leaves, and pipe-like cells or vascular tissues transport water, minerals, and fixed carbon organic compounds throughout the organism.

Throughout plant evolution, there is a progressive increase in the dominance of the sporophyte generation. In seedless vascular plants, the diploid sporophyte is the dominant phase of the life cycle. The gametophyte is now less conspicuous, but still independent of the sporophyte. Seedless vascular plants still depend on water during fertilization, as the flagellated sperm must swim on a layer of moisture to reach the egg. This step in reproduction explains why ferns and their relatives are more abundant in damp environments.

Vascular Tissue: Xylem and Phloem

The first plant fossils that show the presence of vascular tissue date to the Silurian period, about 430 million years ago. The simplest arrangement of conductive cells shows a pattern of xylem at the center surrounded by phloem. **Xylem** is the tissue responsible for the storage and long-distance transport of water and nutrients, as well as the transfer of water-soluble growth factors from the organs of synthesis to the target organs. The tissue consists of conducting cells, known as tracheids, and supportive filler tissue, called parenchyma. Xylem conductive cells incorporate the compound **lignin** into their walls, and are thus described as lignified. Lignin itself is a complex polymer: It is impermeable to water and confers mechanical strength on vascular tissue. With their rigid cell walls, the xylem cells provide support to the plant and allow it to achieve impressive heights. Tall plants have a selective advantage by being able to reach unfiltered sunlight and disperse their spores or seeds away from the parent plant, thus expanding the species' range. By growing higher than other plants, tall trees cast their shadows on shorter plants and thereby outcompete them for water and precious nutrients in the soil.

Phloem is the second type of vascular tissue; it transports sugars, proteins, and other solutes throughout the plant. Phloem cells are divided into sieve elements (conducting cells) and cells that support the *sieve elements*. Together, xylem and phloem tissues form the vascular system of plants (Figure 25.16).

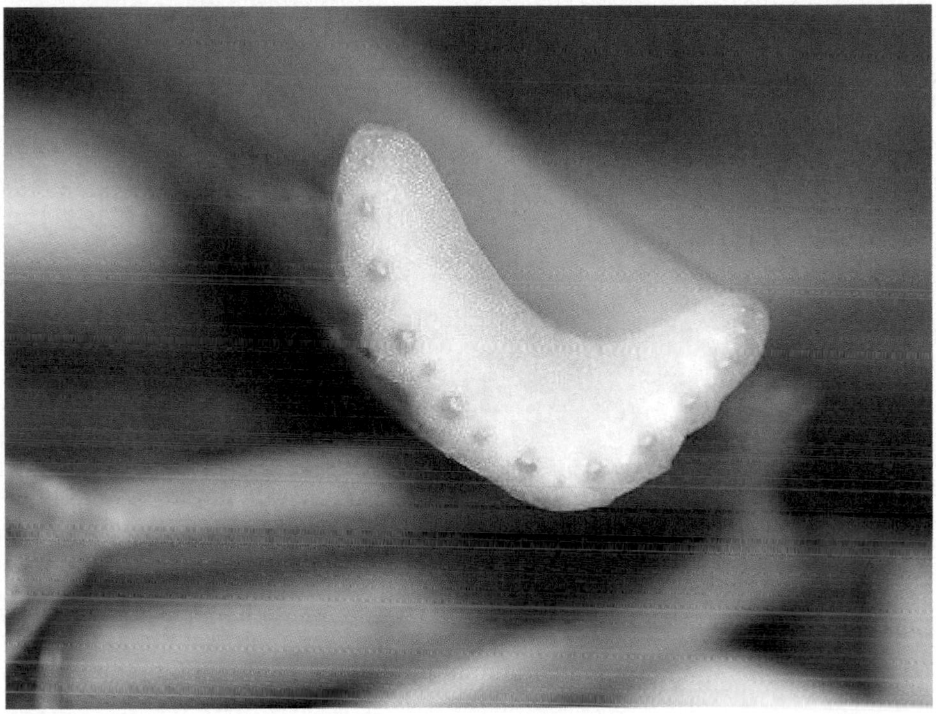

Figure 25.16 Vascular bundles in celery. This cross section of a celery stalk shows a number of vascular bundles. The xylem is on the inner part of each bundle. (credit: fir0002 | flagstaffotos.com.au [GFDL 1.2 (http://www.gnu.org/licenses/old-licenses/fdl-1.2.html)], via Wikimedia Commons. Image modified from source.)

Roots: Support for the Plant

Roots are not well-preserved in the fossil record. Nevertheless, it seems that roots appeared later in evolution than vascular tissue. The development of an extensive network of roots represented a significant new feature of vascular plants. Thin rhizoids attached bryophytes to the substrate, but these rather flimsy filaments did not provide a strong anchor for the plant; nor did they absorb substantial amounts of water and nutrients. In contrast, roots, with their prominent vascular tissue system, transfer water and minerals from the soil to the rest of the plant. The extensive network of roots that penetrates deep into the soil to reach sources of water also stabilizes plants by acting as a ballast or anchor. The majority of roots establish a symbiotic relationship with fungi, forming mutualistic mycorrhizae, which benefit the plant by greatly increasing the surface area for absorption of water, soil minerals, and nutrients.

Leaves, Sporophylls, and Strobili

A third innovation marks the seedless vascular plants. Accompanying the prominence of the sporophyte and the development of vascular tissue, the appearance of true leaves improved their photosynthetic efficiency. Leaves capture more sunlight with their increased surface area by employing more chloroplasts to trap light energy and convert it to chemical energy, which is then used to fix atmospheric carbon dioxide into carbohydrates. The carbohydrates are exported to the rest of the plant by the conductive cells of phloem tissue.

The existence of two types of leaf morphology—*microphylls* and *megaphylls*—suggests that leaves evolved independently in several groups of plants. Microphylls ("little leaves") are small and have a simple vascular system. The first **microphylls** in the fossil record can be dated to 350 million years ago in the late Silurian. A single unbranched **vein**—a bundle of vascular tissue made of xylem and phloem—runs through the center of the leaf. Microphylls may have originated from the flattening of lateral branches, or from sporangia that lost their reproductive capabilities. Microphylls are seen in club mosses. Microphylls probably preceded the development of **megaphylls** ("big leaves"), which are larger leaves with a pattern of multiple veins. Megaphylls most likely appeared independently several times during the course of evolution. Their complex networks of veins suggest that several branches may have combined into a flattened organ, with the gaps between the branches being filled with photosynthetic tissue. Megaphylls are seen in ferns and more derived vascular plants.

In addition to photosynthesis, leaves play another role in the life of the plants. Pine cones, mature fronds of ferns, and flowers are all **sporophylls**—leaves that were modified structurally to bear sporangia. In conifers, the commonly named pine cones, **strobili** are cone-like structures that contain sporangia.

Ferns and Other Seedless Vascular Plants

By the late Devonian period, plants had evolved vascular tissue, well-defined leaves, and root systems. With these advantages, plants increased in height and size. During the Carboniferous period (360 to 300 MYA), swamp forests of club mosses and horsetails—some specimens reaching heights of more than 30 m (100 ft)—covered most of the land. These forests gave rise to the extensive coal deposits that gave the Carboniferous its name. In seedless vascular plants, the sporophyte became the dominant phase of the life cycle.

Water is still required as a medium of sperm transport during the fertilization of seedless vascular plants, and most favor a moist environment. Modern-day seedless tracheophytes include club mosses, horsetails, ferns, and whisk ferns.

Phylum Lycopodiophyta: Club Mosses

The **club mosses**, or phylum **Lycopodiophyta**, are the earliest group of seedless vascular plants. They dominated the landscape of the Carboniferous, growing into tall trees and forming large swamp forests. Today's club mosses are diminutive, evergreen plants consisting of a stem (which may be branched) and microphylls (Figure 25.17). The phylum Lycopodiophyta consists of close to 1,200 species, including the quillworts (*Isoetales*), the club mosses (*Lycopodiales*), and spike mosses (*Selaginellales*), none of which are true mosses or bryophytes.

Lycophytes follow the pattern of alternation of generations seen in the bryophytes, except that the sporophyte is the major stage of the life cycle. Some lycophytes, like the club moss *Lycopodium*, produce gametophytes that are independent of the sporophyte, developing underground or in other locations where they can form mycorrhizal associations with fungi. In many club mosses, the sporophyte gives rise to sporophylls arranged in strobili, cone-like structures that give the class its name. Sporangia develop within the chamber formed by each sporophyll.

Lycophytes can be *homosporous* (spores of the same size) or *heterosporous* (spores of different sizes). The spike moss *Selaginella* is a heterosporous lycophyte. The same strobilus will contain microsporangia, which produce spores that will develop into the male gametophyte, and megasporangia, which produce spores that will develop into the female gametophyte. Both gametophytes develop within the protective strobilus.

Figure 25.17 *Lycopodium*. In the club mosses such as *Lycopodium clavatum*, sporangia are arranged in clusters called strobili. The generic name means "wolf-foot" from the resemblance of the branched sporophyte to a paw. The specific epithet *clavatum* refers to the club-shaped strobilus, and reflects the common name of the phylum. (credit: Cory Zanker)

Phylum Monilophyta: Class Equisetopsida (Horsetails)

Horsetails, whisk ferns, and ferns belong to the phylum Monilophyta, with **horsetails** placed in the class Equisetopsida. The single genus *Equisetum* is the survivor of a large group of plants, known as Arthrophyta, which produced large trees and entire swamp forests in the Carboniferous. The plants are usually found in damp environments and marshes (Figure 25.18).

Figure 25.18 Horsetails. Horsetails, named for the brushy appearance of the sporophyte, thrive in a marsh. (credit: Myriam Feldman)

The stem of a horsetail is characterized by the presence of joints or nodes, hence the name Arthrophyta (arthro- = "joint"; -phyta = "plant"). Leaves and branches come out as whorls from the evenly spaced joints. The needle-shaped leaves do not contribute greatly to photosynthesis, the majority of which takes place in the green stem (Figure 25.19).

Figure 25.19 The jointed stem of a horsetail. Thin leaves originating at the joints are noticeable on the horsetail plant. Because silica deposited in the cell walls made these plants abrasive, horsetails were once used as scrubbing brushes and were nicknamed scouring rushes. (credit: Myriam Feldman)

Silica collected by in the epidermal cells contributes to the stiffness of horsetail plants, but underground stems known as rhizomes anchor the plants to the ground. Modern-day horsetails are homosporous. The spores are attached to elaters—as we have seen, these are coiled threads that spring open in dry weather and casts the spores to a location distant from the parent plants. The spores then germinate to produce small bisexual gametophytes.

Phylum Monilophyta: Class Psilotopsida (Whisk Ferns)

While most ferns form large leaves and branching roots, the **whisk ferns**, class Psilotopsida, lack both roots and leaves, probably lost by reduction. Photosynthesis takes place in their green stems, which branch dichotomously. Small yellow knobs form at the tip of a branch or at branch nodes and contain the sporangia (Figure 25.20). Spores develop into gametophytes that are only a few millimeters across, but which produce both male and female gametangia. Whisk ferns were considered early pterophytes. However, recent comparative DNA analysis suggests that this group may have lost both vascular tissue and roots through evolution, and is more closely related to ferns.

Figure 25.20 *Psilotum*. The whisk fern *Psilotum nudum* has conspicuous green stems with knob-shaped sporangia. (credit: Forest & Kim Starr)

Phylum Monilophyta: Class Polypodiopsida (True Ferns)

With their large fronds, the true **ferns** are perhaps the most readily recognizable seedless vascular plants. They are also considered to be the most advanced seedless vascular plants and display characteristics commonly observed in seed plants. More than 20,000 species of ferns live in environments ranging from the tropics to temperate forests. Although some species survive in dry environments, most ferns are restricted to moist, shaded places. Ferns made their appearance in the fossil record during the Devonian period (420 MYA) and expanded during the Carboniferous (360 to 300 MYA).

The dominant stage of the life cycle of a fern is the sporophyte, which consists of large compound leaves called fronds. Fronds may be either finely divided or broadly lobed. Fronds fulfill a double role; they are photosynthetic organs that also carry reproductive organs. The stem may be buried underground as a rhizome, from which adventitious roots grow to absorb water and nutrients from the soil; or, they may grow above ground as a trunk in tree ferns (Figure 25.21). **Adventitious** organs are those that grow in unusual places, such as roots growing from the side of a stem.

Figure 25.21 A tree fern. Some specimens of this short tree-fern species can grow very tall. (credit: Adrian Pingstone)

The tip of a developing fern frond is rolled into a crozier, or fiddlehead (Figure 25.22). Fiddleheads unroll as the frond develops.

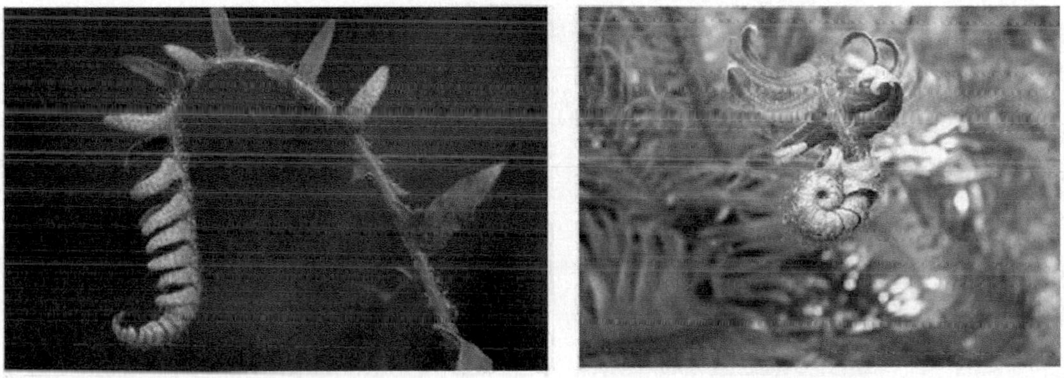

(a) (b)

Figure 25.22 Fern fiddleheads. Croziers, or fiddleheads, are the tips of fern fronds. (credit a: modification of work by Cory Zanker; credit b: modification of work by Myriam Feldman)

On the underside of each mature fern frond are groups of sporangia called sori (Figure 25.23a). Most ferns are homosporous. Spores are produced by meiosis and are released into the air from the sporangium. Those that land on a suitable substrate germinate and form a heart-shaped gametophyte, or prothallus, which is attached to the ground by thin filamentous rhizoids (Figure 25.23b). Gametophytes produce both antheridia and archegonia. Like the sperm cells of other pterophytes, fern sperm have multiple flagella and must swim to the archegonium, which releases a chemoattractant to guide them. The zygote develops into a fern sporophyte, which emerges from the archegonium of the gametophyte. Maturation of antheridia and archegonia at different times encourages cross-fertilization. The full life cycle of a fern is depicted in Figure 25.24.

Figure 25.23 Fern reproductive stages. Sori (a) appear as small bumps on the underside of a fern frond. (credit: Myriam Feldman). (b) Fern gametophyte and young sporophyte. The sporophyte and gametophyte are labeled. (credit: modification of work by "Vlmastra"/Wikimedia Commons)

VISUAL CONNECTION

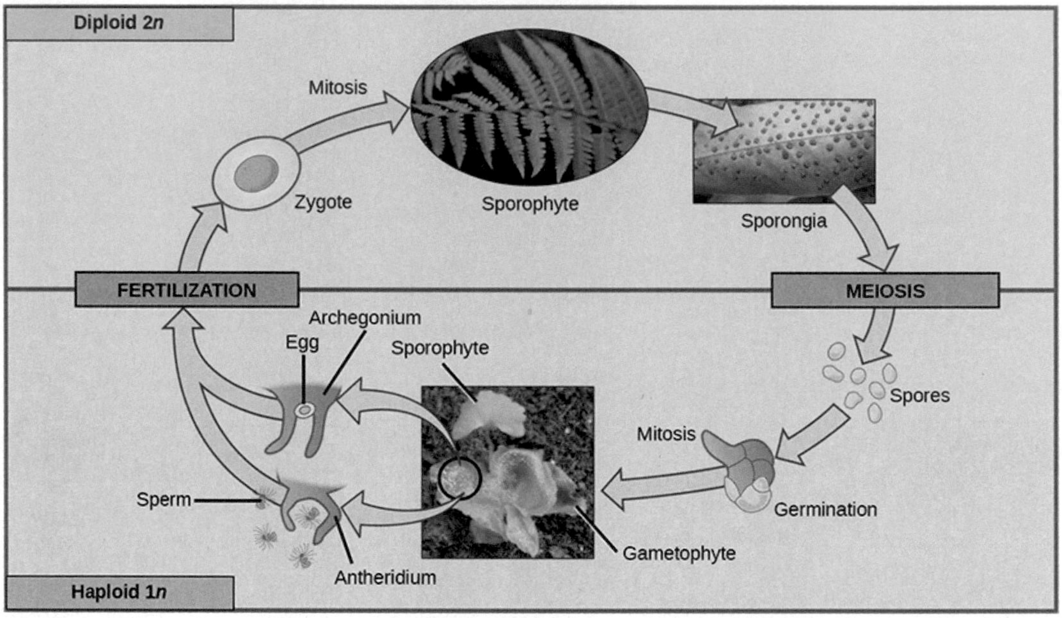

Figure 25.24 Reproductive cycle of a fern. This life cycle of a fern shows alternation of generations with a dominant sporophyte stage. (credit "fern": modification of work by Cory Zanker; credit "gametophyte": modification of work by "Vlmastra"/Wikimedia Commons)

Which of the following statements about the fern life cycle is false?

a. Sporangia produce haploid spores.
b. The sporophyte grows from a gametophyte.
c. The sporophyte is diploid and the gametophyte is haploid.
d. Sporangia form on the underside of the gametophyte.

LINK TO LEARNING

To see an animation of the life cycle of a fern and to test your knowledge, go to the website (http://openstax.org/l/fern_life_cycle)
.

CAREER CONNECTION

Landscape Designer

Looking at the ornamental arrangement of flower beds and fountains typical of the grounds of royal castles and historic houses of Europe, it's clear that the gardens' creators knew about more than art and design. They were also familiar with the biology of the plants they chose. Landscape design also has strong roots in the United States' tradition. A prime example of early American classical design is Monticello, Thomas Jefferson's private estate. Among his many interests, Jefferson maintained a strong passion for botany. Landscape layout can encompass a small private space like a backyard garden, public gathering places such as Central Park in New York City, or an entire city plan like Pierre L'Enfant's design for Washington, DC.

A landscape designer will plan traditional public spaces—such as botanical gardens, parks, college campuses, gardens, and larger developments—as well as natural areas and private gardens. The restoration of natural places encroached on by human intervention, such as wetlands, also requires the expertise of a landscape designer.

With such an array of necessary skills, a landscape designer's education should include a solid background in botany, soil science, plant pathology, entomology, and horticulture. Coursework in architecture and design software is also required for the completion of the degree. The successful design of a landscape rests on an extensive knowledge of plant growth requirements such as light and shade, moisture levels, compatibility of different species, and susceptibility to pathogens and pests. Mosses and ferns will thrive in a shaded area, where fountains provide moisture; cacti, on the other hand, would not fare well in that environment. The future growth of individual plants must be taken into account, to avoid crowding and competition for light and nutrients. The appearance of the space over time is also of concern. Shapes, colors, and biology must be balanced for a well-maintained and sustainable green space. Art, architecture, and biology blend in a beautifully designed and implemented landscape (Figure 25.25).

Figure 25.25 This landscaped border at a college campus was designed by students in the horticulture and landscaping department of the college. (credit: Myriam Feldman)

The Importance of Seedless Plants

Mosses and liverworts are often the first macroscopic organisms to colonize an area, both in a primary succession—where bare land is settled for the first time by living organisms, or in a secondary succession—where soil remains intact after a catastrophic event wipes out many existing species. Their spores are carried by the wind, birds, or insects. Once mosses and liverworts are established, they provide food and shelter for other plant species. In a hostile environment, like the tundra where the soil is frozen, bryophytes grow well because they do not have roots and can dry and rehydrate quickly once water is again available. Mosses are at the base of the food chain in the tundra biome. Many species—from small herbivorous insects to musk oxen and reindeer—depend on mosses for food. In turn, predators feed on the herbivores, which are the primary consumers. Some reports indicate that bryophytes make the soil more amenable to colonization by other plants. Because they establish symbiotic relationships with nitrogen-fixing cyanobacteria, mosses replenish the soil with nitrogen.

By the end of the nineteenth century, scientists had observed that lichens and mosses were becoming increasingly rare in urban and suburban areas. Because bryophytes have neither a root system for absorption of water and nutrients, nor a cuticular layer that protects them from desiccation, pollutants in rainwater readily penetrate their tissues as they absorb moisture and nutrients through their entire exposed surfaces. Therefore, pollutants dissolved in rainwater penetrate plant tissues readily and have a larger impact on mosses than on other plants. The disappearance of mosses can be considered a biological indicator for the level of pollution in the environment.

Ferns contribute to the environment by promoting the weathering of rock, accelerating the formation of topsoil, and slowing down erosion as rhizomes spread throughout the soil. The water ferns of the genus *Azolla* harbor nitrogen-fixing cyanobacteria and restore this important nutrient to aquatic habitats.

Seedless plants have historically played a role in human life with uses as tools, fuel, and medicine. For example, dried **peat moss**, *Sphagnum*, is commonly used as fuel in some parts of Europe and is considered a renewable resource. *Sphagnum* bogs (Figure 25.26) are cultivated with cranberry and blueberry bushes. In addition, the ability of *Sphagnum* to hold moisture makes the moss a common soil conditioner. Even florists use blocks of *Sphagnum* to maintain moisture for floral arrangements!

Figure 25.26 Sphagnum moss. *Sphagnum acutifolium* is dried peat moss and can be used as fuel. (credit: Ken Goulding)

The attractive fronds of ferns make them a favorite ornamental plant. Because they thrive in low light, they are well suited as house plants. More importantly, fiddleheads of bracken fern (*Pteridium aquilinum*) are a traditional spring food of Native Americans, and are popular as a side dish in French cuisine. The licorice fern, *Polypodium glycyrrhiza,* is part of the diet of the Pacific Northwest coastal tribes, owing in part to the sweetness of its rhizomes. It has a faint licorice taste and serves as a sweetener. The rhizome also figures in the pharmacopeia of Native Americans for its medicinal properties and is used as a remedy for sore throat.

LINK TO LEARNING

Go to this website (http://openstax.org/l/fiddleheads) to learn how to identify fern species.

By far the greatest impact of seedless vascular plants on human life, however, comes from their extinct progenitors. The tall club mosses, horsetails, and tree-like ferns that flourished in the swampy forests of the Carboniferous period gave rise to large deposits of coal throughout the world. Coal provided an abundant source of energy during the Industrial Revolution, which had tremendous consequences on human societies, including rapid technological progress and growth of large cities, as well as the degradation of the environment. Coal is still a prime source of energy and also a major contributor to global warming.

KEY TERMS

adventitious describes an organ that grows in an unusual place, such as a roots growing from the side of a stem
antheridium male gametangium
archegonium female gametangium
capsule case of the sporangium in mosses
charophyte other term for green algae; considered the closest relative of land plants
club mosses earliest group of seedless vascular plants
diplontic diploid stage is the dominant stage
embryophyte other name for land plant; embryo is protected and nourished by the sporophyte
extant still-living species
extinct no-longer-existing species
fern seedless vascular plant that produces large fronds; the most advanced group of seedless vascular plants
gametangium structure on the gametophyte in which gametes are produced
gemma (plural, gemmae) leaf fragment that spreads for asexual reproduction
haplodiplodontic haploid and diploid stages alternate
haplontic haploid stage is the dominant stage
heterosporous produces two types of spores
homosporous produces one type of spore
hornworts group of non-vascular plants in which stomata appear
horsetail seedless vascular plant characterized by joints
lignin complex polymer impermeable to water
liverworts most primitive group of the non-vascular plants
lycophyte club moss
megaphyll larger leaves with a pattern of branching veins
megaspore female spore
microphyll small size and simple vascular system with a single unbranched vein
microspore male spore
mosses group of bryophytes in which a primitive conductive system appears
non-vascular plant plant that lacks vascular tissue, which is formed of specialized cells for the transport of water and nutrients
peat moss Sphagnum
peristome tissue that surrounds the opening of the capsule and allows periodic release of spores
phloem tissue responsible for transport of sugars, proteins, and other solutes
protonema tangle of single-celled filaments that forms from the haploid spore
rhizoids thin filaments that anchor the plant to the substrate
seedless vascular plant plant that does not produce seeds
seta stalk that supports the capsule in mosses
sporocyte diploid cell that produces spores by meiosis
sporophyll leaf modified structurally to bear sporangia
sporopollenin tough polymer surrounding the spore
streptophytes group that includes green algae and land plants
strobili cone-like structures that contain the sporangia
tracheophyte vascular plant
vascular plant plant containing a network of cells that conducts water and solutes through the organism
vein bundle of vascular tissue made of xylem and phloem
whisk fern seedless vascular plant that lost roots and leaves by reduction
xylem tissue responsible for long-distance transport of water and nutrients

CHAPTER SUMMARY

25.1 Early Plant Life

Land plants acquired traits that made it possible to colonize land and survive out of the water. All land plants share the following characteristics: alternation of generations, with the haploid plant called a gametophyte, and the diploid plant called a sporophyte; formation of haploid spores in a sporangium; formation of gametes in a gametangium; protection of the embryo; and an apical meristem. Vascular tissues, roots, leaves, cuticle cover, and a tough outer layer that protects the spores contributed to the adaptation of plants to dry land. Land plants appeared about 500 million years ago in the Ordovician period.

25.2 Green Algae: Precursors of Land Plants

Charophytes share more traits with land plants than do other algae, according to structural features and DNA analysis. Within the charophytes, the Charales, the Coleochaetales, and the Zygnematales have been each considered as sharing the closest common ancestry with the land plants. Charophytes form sporopollenin and precursors of lignin, phragmoplasts, and have flagellated sperm. They do not exhibit alternation of generations.

25.3 Bryophytes

Seedless non-vascular plants are small, having the gametophyte as the dominant stage of the lifecycle. Without a vascular system and roots, they absorb water and nutrients on all their exposed surfaces. Collectively known as bryophytes, the three main groups include the liverworts, the hornworts, and the mosses. Liverworts are the most primitive plants and are closely related to the first land plants. Hornworts developed stomata and possess a single chloroplast per cell. Mosses have simple conductive cells and are attached to the substrate by rhizoids. They colonize harsh habitats and can regain moisture after drying out. The moss sporangium is a complex structure that allows release of spores away from the parent plant.

25.4 Seedless Vascular Plants

The seedless vascular plants show several features important to living on land: vascular tissue, roots, and leaves. Vascular systems consist of xylem tissue, which transports water and minerals, and phloem tissue, which transports sugars and proteins. With the development of the vascular system, leaves appeared to act as large photosynthetic organs, and roots to access water from the ground. Small uncomplicated leaves are termed microphylls. Large leaves with vein patterns are termed megaphylls. Modified leaves that bear sporangia are called sporophylls. Some sporophylls are arranged in cone structures called strobili.

The support and conductive properties of vascular tissues have allowed the sporophyte generation of vascular plants to become increasingly dominant. The seedless vascular plants include club mosses, which are the most primitive; whisk ferns, which lost leaves and roots by reductive evolution; and horsetails and ferns. Ferns are the most advanced group of seedless vascular plants. They are distinguished by large leaves called fronds and small sporangia-containing structures called sori, which are found on the underside of the fronds.

Both mosses and ferns play an essential role in the balance of the ecosystems. Mosses are pioneering species that colonize bare or devastated environments and make it possible for succession to occur. They contribute to the enrichment of the soil and provide shelter and nutrients for animals in hostile environments. Mosses are important biological indicators of environmental pollution. Ferns are important for providing natural habitats, as soil stabilizers, and as decorative plants. Both mosses and ferns are part of traditional medical practice. In addition to culinary, medical, and decorative purposes, mosses and ferns can be used as fuels, and ancient seedless plants were important contributors to the fossil fuel deposits that we now use as an energy resource.

VISUAL CONNECTION QUESTIONS

1. Figure 25.6 Which of the following statements about plant divisions is false?
   a. Lycophytes and pterophytes are seedless vascular plants.
   b. All vascular plants produce seeds.
   c. All non-vascular embryophytes are bryophytes.
   d. Seed plants include angiosperms and gymnosperms.

2. Figure 25.14 Which of the following statements about the moss life cycle is false?
   a. The mature gametophyte is haploid.
   b. The sporophyte produces haploid spores.
   c. The calyptra buds to form a mature gametophyte.
   d. The zygote is housed in the venter.

3. Figure 25.24 Which of the following statements about the fern life cycle is false?
   a. Sporangia produce haploid spores.
   b. The sporophyte grows from a gametophyte.
   c. The sporophyte is diploid and the gametophyte is haploid.
   d. Sporangia form on the underside of the gametophyte.

REVIEW QUESTIONS

4. The land plants are probably descendants of which of these groups?
   a. green algae
   b. red algae
   c. brown algae
   d. angiosperms

5. Alternation of generations means that plants produce:
   a. only haploid multicellular organisms
   b. only diploid multicellular organisms
   c. only diploid multicellular organisms with single-celled haploid gametes
   d. both haploid and diploid multicellular organisms

6. Which of the following traits of land plants allows them to grow in height?
   a. alternation of generations
   b. waxy cuticle
   c. tracheids
   d. sporopollenin

7. How does a haplontic plant population maintain genetic diversity?
   a. Zygotes are produced by random fusion.
   b. Gametes are created through meiosis.
   c. Diploid spores undergo independent assortment during mitosis.
   d. The zygote undergoes meiosis to generate a haploid sporophyte.

8. What characteristic of Charales would enable them to survive a dry spell?
   a. sperm with flagella
   b. phragmoplasts
   c. sporopollenin
   d. chlorophyll *a*

9. Which one of these characteristics is present in land plants and not in Charales?
   a. alternation of generations
   b. flagellated sperm
   c. phragmoplasts
   d. plasmodesmata

10. A scientist sequences the genome of *Chara*, red algae, and a tomato plant. What result would support the conclusion that Charophytes should be included in the *Plantae* kingdom?
    a. The *Chara* genome is more similar to the red algae than the tomato plant.
    b. All three genomes are distinctly different.
    c. The *Chara* genome is more similar to the tomato plant genome than the red algae genome.
    d. The tomato plant genome is distinct from the red algae genome.

11. Which of the following features does not support the inclusion of Charophytes in the *Plantae* kingdom?
    a. Charophyte chloroplasts contain chlorophyll a and b.
    b. Charophyte plant cell walls contain plasmodesmata to allow transfer between cells within multicellular organisms.
    c. Charophytes do not exhibit growth throughout the entire plant body.
    d. Charophytes are multicellular organisms that lack vascular tissue.

12. Which of the following structures is not found in bryophytes?
    a. a cellulose cell wall
    b. chloroplast
    c. sporangium
    d. root

13. Stomata appear in which group of plants?
    a. Charales
    b. liverworts
    c. hornworts
    d. mosses

14. The chromosome complement in a moss protonema is:
    a. $1n$
    b. $2n$
    c. $3n$
    d. varies with the size of the protonema

15. Why do mosses grow well in the Arctic tundra?
    a. They grow better at cold temperatures.
    b. They do not require moisture.
    c. They do not have true roots and can grow on hard surfaces.
    d. There are no herbivores in the tundra.

16. A botanist travels to an area that has experienced a long, severe drought. While examining the bryophytes in the area, he notices that many are in the same life-cycle stage. Which life-cycle stage *should* be the most common?
    a. zygote
    b. gametophyte
    c. sporophyte
    d. archegonium

17. Microphylls are characteristic of which types of plants?
    a. mosses
    b. liverworts
    c. club mosses
    d. ferns

18. A plant in the understory of a forest displays a segmented stem and slender leaves arranged in a whorl. It is probably a ________.
    a. club moss
    b. whisk fern
    c. fern
    d. horsetail

19. The following structures are found on the underside of fern leaves and contain sporangia:
    a. sori
    b. rhizomes
    c. megaphylls
    d. microphylls

20. The dominant organism in fern is the _______.
    a. sperm
    b. spore
    c. gamete
    d. sporophyte

21. What seedless plant is a renewable source of energy?
    a. club moss
    b. horsetail
    c. sphagnum moss
    d. fern

22. How do mosses contribute to returning nitrogen to the soil?
    a. Mosses fix nitrogen from the air.
    b. Mosses harbor cyanobacteria that fix nitrogen.
    c. Mosses die and return nitrogen to the soil.
    d. Mosses decompose rocks and release nitrogen.

23. The production of megaphylls by many different species of plants is an example of _____.
    a. parallel evolution
    b. analogy
    c. divergent evolution
    d. homology

CRITICAL THINKING QUESTIONS

24. Why did land plants lose some of the accessory pigments present in brown and red algae?
25. What is the difference between extant and extinct?
26. Describe at least two challenges that cactuses had to overcome that cattails did not.
27. Describe a minimum of two ways that plants changed the land environment to support the emergence of land animals.
28. To an alga, what is the main advantage of producing drought-resistant structures?
29. In areas where it rains often, mosses grow on roofs. How do mosses survive on roofs without soil?
30. What are the three classes of bryophytes?
31. Describe two adaptations that are present in mosses, but not hornworts or liverworts, which reflect steps of evolution toward land plants.
32. Bryophytes form a monophyletic group that transitions between green algae and vascular plants. Describe at least one similarity and one difference between bryophyte reproduction and green algae reproduction.
33. How did the development of a vascular system contribute to the increase in size of plants?
34. Which plant is considered the most advanced seedless vascular plant and why?
35. Ferns are simultaneously involved in promoting rock weathering, while preventing soil erosion. Explain how a single plant can perform both these functions, and how these functions are beneficial to its ecosystem.

CHAPTER 26
Seed Plants

Figure 26.1 Seed plants dominate the landscape and play an enormous and integral role in the success of all human societies. Here are a few examples: (a) Palm trees grow along the shoreline, serving numerous purposes for food, shelter, and even transportation; (b) wheat is an important crop grown throughout most of the world; (c) the fruit of the cotton plant produces fibers that are woven into fabric; (d) the potent alkaloids of the beautiful opium poppy have long influenced human life both as a medicinal remedy and as a dangerously addictive drug. (credit a: modification of work by Ryan Kozie; credit b: modification of work by Stephen Ausmus; credit c: modification of work by David Nance; credit d: modification of work by Jolly Janner)

INTRODUCTION The lush palms on tropical shorelines do not depend on water for the dispersal of their pollen, fertilization, or the survival of the zygote—unlike mosses, liverworts, and ferns living within the same terrain. These palms are seed plants, which have broken free from the need to rely on water for their reproductive needs. The seed plants play an integral role in all aspects of life on the planet, shaping the physical terrain, influencing the climate, and maintaining life as we know it. For millennia, human societies have depended on seed plants for nutrition and medicinal compounds. Somewhat more recently, seed plants have served as a source of manufactured products such as timber and paper, dyes, and textiles. As an example, multiple uses have been found for each of the plants shown above. Palms provide materials including rattans, oils, and dates. Grains like wheat are grown to feed both human and animal populations or fermented to produce alcoholic beverages. The fruit of the cotton flower is harvested as a boll, with its fibers transformed into clothing or pulp for paper. The showy opium poppy is valued both as an ornamental flower and as a source of potent opiate compounds.

Chapter Outline

26.1 Evolution of Seed Plants

By the end of this section, you will be able to do the following:

- Describe the two major innovations that allowed seed plants to reproduce in the absence of water
- Explain when seed plants first appeared and when gymnosperms became the dominant plant group
- Discuss the purpose of pollen grains and seeds
- Describe the significance of angiosperms bearing both flowers and fruit

The first plants to colonize land were most likely related to the ancestors of modern day mosses (bryophytes), which are thought to have appeared about 500 million years ago. They were followed by liverworts (also bryophytes) and primitive vascular plants—the pterophytes—from which modern ferns are descended. The life cycle of bryophytes and pterophytes is characterized by the *alternation of generations*, which is also exhibited in the gymnosperms and angiosperms. However, what sets bryophytes and pterophytes apart from gymnosperms and angiosperms is their reproductive requirement for water. The completion of the bryophyte and pterophyte life cycle requires water because the male gametophyte releases *flagellated sperm*, which must swim to reach and fertilize the female gamete or egg. After fertilization, the zygote undergoes cellular division and grows into a *diploid sporophyte*, which in turn will form sporangia or "spore vessels." In the sporangia, mother cells undergo meiosis and produce the haploid spores. Release of spores in a suitable environment will lead to germination and a new generation of gametophytes.

In seed plants, the evolutionary trend led to a dominant sporophyte generation accompanied by a corresponding reduction in the size of the gametophyte from a conspicuous structure to a microscopic cluster of cells enclosed in the tissues of the sporophyte. Whereas lower vascular plants, such as club mosses and ferns, are mostly *homosporous* (producing only one type of spore), all seed plants, or spermatophytes, are *heterosporous*, producing two types of spores: megaspores (female) and microspores (male). Megaspores develop into female gametophytes that produce eggs, and microspores mature into male gametophytes that generate sperm. Because the gametophytes mature within the spores, they are not free-living, as are the gametophytes of other seedless vascular plants.

Ancestral heterosporous seedless plants, represented by modern-day plants such as the spike moss *Selaginella*, are seen as the evolutionary forerunners of seed plants. In the life cycle of *Selaginella*, both male and female sporangia develop within the same stem-like **strobilus**. In each male sporangium, multiple microspores are produced by meiosis. Each microspore produces a small *antheridium* contained within a spore case. As it develops it is released from the strobilus, and a number of flagellated sperm are produced that then leave the spore case. In the female sporangium, a single megaspore mother cell undergoes meiosis to produce four megaspores. Gametophytes develop within each megaspore, consisting of a mass of tissue that will later nourish the embryo and a few archegonia. The female gametophyte may remain within remnants of the spore wall in the megasporangium until after fertilization has occurred and the embryo begins to develop. This combination of an embryo and nutritional cells is a little different from the organization of a seed, since the nutritive *endosperm* in a seed is formed from a single cell rather than multiple cells.

Both seeds and pollen distinguish seed plants from seedless vascular plants. These innovative structures allowed seed plants to reduce or eliminate their dependence on water for gamete fertilization and development of the embryo, and to conquer dry land. Pollen grains are male gametophytes, which contain the sperm (gametes) of the plant. The small haploid ($1n$) cells are encased in a protective coat that prevents desiccation (drying out) and mechanical damage. Pollen grains can travel far from their original sporophyte, spreading the plant's genes. Seeds offer the embryo protection, nourishment, and a mechanism to maintain dormancy for tens or even thousands of years, ensuring that germination can occur when growth conditions are optimal. Seeds therefore allow plants to disperse the next generation through both space and time. With such evolutionary advantages, seed plants have become the most successful and familiar group of plants.

Both adaptations expanded the colonization of land begun by the bryophytes and their ancestors. Fossils place the earliest distinct seed plants at about 350 million years ago. The first reliable record of gymnosperms dates their appearance to the Pennsylvanian period, about 319 million years ago (Figure 26.2). Gymnosperms were preceded by progymnosperms, the first naked seed plants, which arose about 380 million years ago. *Progymnosperms* were a transitional group of plants that superficially resembled conifers (cone bearers) because they produced wood from the secondary growth of the vascular tissues; however, they still reproduced like ferns, releasing spores into the environment. At least some species were heterosporous. Progymnosperms, like the extinct *Archaeopteris* (not to be confused with the ancient bird *Archaeopteryx*), dominated the forests of the late Devonian period. However, by the early (Triassic, c. 240 MYA) and middle (Jurassic, c. 205 MYA) Mesozoic era, the landscape was dominated by the true gymnosperms. Angiosperms surpassed gymnosperms by the middle of the Cretaceous (c. 100 MYA) in the late Mesozoic era, and today are the most abundant and biologically diverse plant group in most terrestrial biomes.

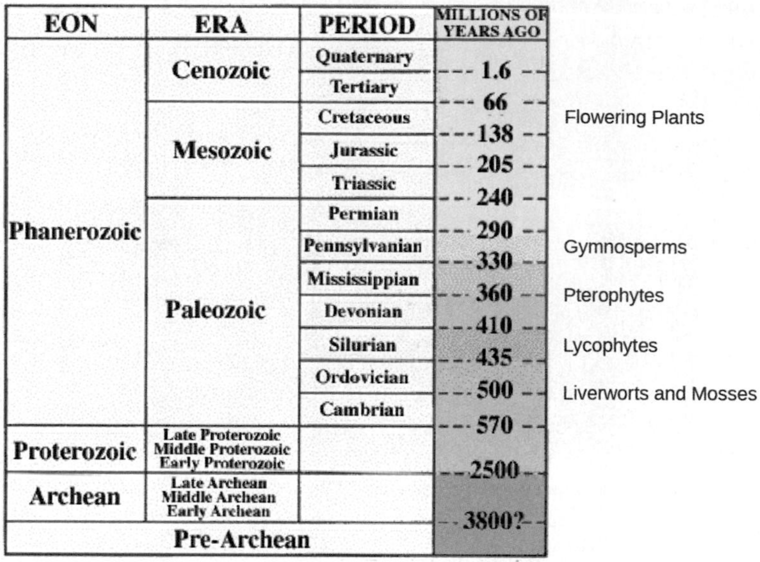

Figure 26.2 Plant timeline. Various plant species evolved in different eras. (credit: United States Geological Survey) Figure modified from source.

Evolution of Gymnosperms

The fossil plant *Elkinsia polymorpha*, a "seed fern" from the Devonian period—about 400 million years ago—is considered the earliest seed plant known to date. Seed ferns (Figure 26.3) produced their seeds along their branches, in structures called cupules that enclosed and protected the ovule—the female gametophyte and associated tissues—which develops into a seed upon fertilization. Seed plants resembling modern tree ferns became more numerous and diverse in the coal swamps of the Carboniferous period.

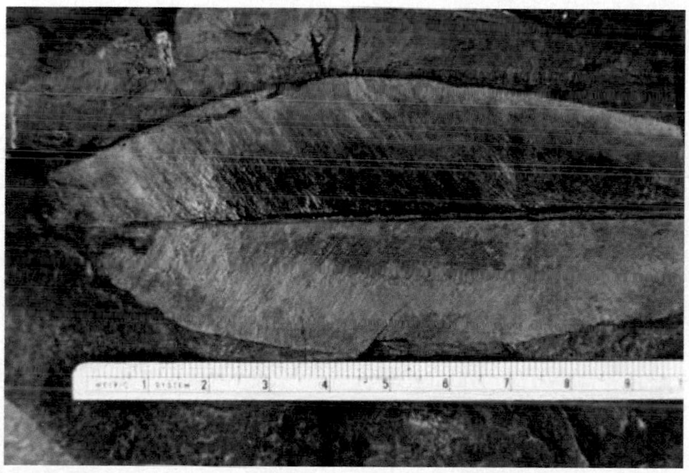

Figure 26.3 Seed fern leaf. This fossilized leaf is from *Glossopteris*, a seed fern that thrived during the Permian age (290–240 million years ago). (credit: D.L. Schmidt, USGS)

Fossil records indicate the first gymnosperms (progymnosperms) most likely originated in the Paleozoic era, during the middle Devonian period: about 390 million years ago. The previous Mississippian and Pennsylvanian periods, were wet and dominated by giant fern trees. But the following Permian period was dry, which gave a reproductive edge to seed plants, which are better adapted to survive dry spells. The Ginkgoales, a group of gymnosperms with only one surviving species—the *Ginkgo biloba*—were the first gymnosperms to appear during the lower Jurassic. Gymnosperms expanded in the Mesozoic era (about 240 million years ago), supplanting ferns in the landscape, and reaching their greatest diversity during this time. The Jurassic period was as much the age of the cycads (palm-tree-like gymnosperms) as the age of the dinosaurs. Ginkgoales and the more familiar conifers also dotted the landscape. Although angiosperms (flowering plants) are the major form of plant life in most biomes, gymnosperms still dominate some ecosystems, such as the taiga (boreal forests) and the alpine forests at higher mountain elevations (Figure 26.4) because of their adaptation to cold and dry growth conditions.

Figure 26.4 Conifers. This boreal forest (taiga) has low-lying plants and conifer trees. (credit: L.B. Brubaker, NOAA)

Seeds and Pollen as an Evolutionary Adaptation to Dry Land

Bryophyte and fern spores are haploid cells dependent on moisture for rapid development of multicellular gametophytes. In the seed plants, the female gametophyte consists of just a few cells: the egg and some supportive cells, including the endosperm-producing cell that will support the growth of the embryo. After fertilization of the egg, the diploid zygote produces an embryo that will grow into the sporophyte when the seed germinates. Storage tissue to sustain growth of the embryo and a protective coat give seeds their superior evolutionary advantage. Several layers of hardened tissue prevent desiccation, and free the embryo from the need for a constant supply of water. Furthermore, seeds remain in a state of dormancy—induced by desiccation and the hormone *abscisic acid*—until conditions for growth become favorable. Whether blown by the wind, floating on water, or carried away by animals, seeds are scattered in an expanding geographic range, thus avoiding competition with the parent plant.

Pollen grains (Figure 26.5) are male gametophytes containing just a few cells and are distributed by wind, water, or an animal pollinator. The whole structure is protected from desiccation and can reach the female organs without depending on water. After reaching a female gametophyte, the pollen grain grows a tube that will deliver a male nucleus to the egg cell. The sperm of modern gymnosperms and all angiosperms lack flagella, but in cycads, *Ginkgo*, and other primitive gymnosperms, the sperm are still motile, and use flagella to swim to the female gamete; however, they are delivered to the female gametophyte enclosed in a pollen grain. The pollen grows or is taken into a fertilization chamber, where the motile sperm are released and swim a short distance to an egg.

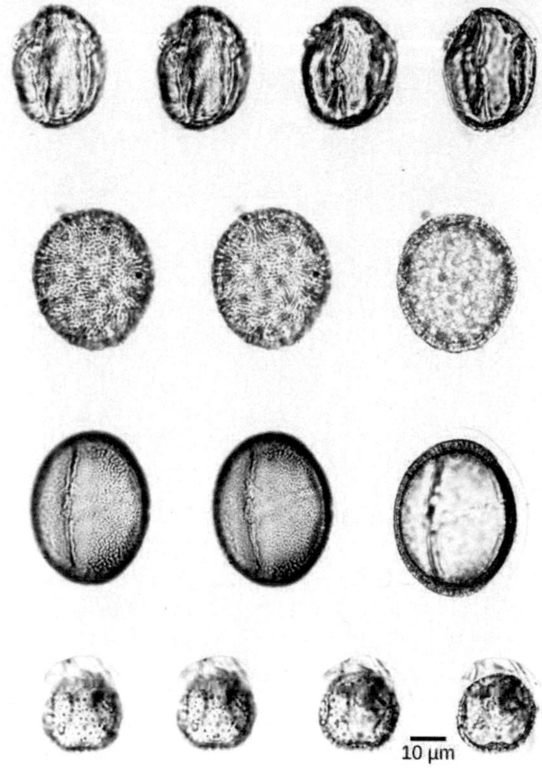

Figure 26.5 Pollen fossils. This fossilized pollen is from a Buckbean fen core found in Yellowstone National Park, Wyoming. The pollen is magnified 1,054 times. (credit: R.G. Baker, USGS; scale-bar data from Matt Russell)

Evolution of Angiosperms

The roughly 200 million years between the appearance of the gymnosperms and the flowering plants gives us some appreciation for the evolutionary experimentation that ultimately produced flowers and fruit. Angiosperms ("seed in a vessel") produce a flower containing male and/or female reproductive structures. Fossil evidence (Figure 26.6) indicates that flowering plants first appeared about 125 million years ago in the Lower Cretaceous (late in the Mesozoic era), and were rapidly diversifying by about 100 million years ago in the Middle Cretaceous. Earlier traces of angiosperms are scarce. Fossilized pollen recovered from Jurassic geological material has been attributed to angiosperms. A few early Cretaceous rocks show clear imprints of leaves resembling angiosperm leaves. By the mid-Cretaceous, a staggering number of diverse flowering plants crowd the fossil record. The same geological period is also marked by the appearance of many modern groups of insects, suggesting that pollinating insects played a key role in the evolution of flowering plants.

New data in comparative genomics and paleobotany (the study of ancient plants) have shed some light on the evolution of angiosperms. Although the angiosperms appeared after the gymnosperms, they are probably not derived from gymnosperm ancestors. Instead, the angiosperms form a sister clade (a species and its descendents) that developed in parallel with the gymnosperms. The two innovative structures of flowers and fruit represent an improved reproductive strategy that served to protect the embryo, while increasing genetic variability and range. There is no current consensus on the origin of the angiosperms. Paleobotanists debate whether angiosperms evolved from small woody bushes, or were related to the ancestors of tropical grasses. Both views draw support from cladistics, and the so-called woody *magnoliid hypothesis*—which proposes that the early ancestors of angiosperms were shrubs like modern magnolia—also offers molecular biological evidence.

The most primitive living angiosperm is considered to be *Amborella trichopoda*, a small plant native to the rainforest of New Caledonia, an island in the South Pacific. Analysis of the genome of *A. trichopoda* has shown that it is related to all existing flowering plants and belongs to the oldest confirmed branch of the angiosperm family tree. The nuclear genome shows evidence of an ancient whole-genome duplication. The mitochondrial genome is large and multichromosomal, containing elements from the mitochondrial genomes of several other species, including algae and a moss. A few other angiosperm groups, called basal angiosperms, are viewed as having ancestral traits because they branched off early from the phylogenetic tree. Most modern angiosperms are classified as either monocots or eudicots, based on the structure of their leaves and embryos. Basal angiosperms, such as water lilies, are considered more ancestral in nature because they share morphological traits with both monocots and eudicots.

Figure 26.6 *Ficus* imprint. This leaf imprint shows a *Ficus speciosissima*, an angiosperm that flourished during the Cretaceous period. (credit: W. T. Lee, USGS)

Flowers and Fruits as an Evolutionary Adaptation

Angiosperms produce their gametes in separate organs, which are usually housed in a flower. Both fertilization and embryo development take place inside an anatomical structure that provides a stable system of sexual reproduction largely sheltered from environmental fluctuations. With about 300,000 species, flowering plants are the most diverse phylum on Earth after insects, which number about 1,200,000 species. Flowers come in a bewildering array of sizes, shapes, colors, smells, and arrangements. Most flowers have a mutualistic pollinator, with the distinctive features of flowers reflecting the nature of the pollination agent. The relationship between pollinator and flower characteristics is one of the great examples of coevolution.

Following fertilization of the egg, the ovule grows into a seed. The surrounding tissues of the ovary thicken, developing into a fruit that will protect the seed and often ensure its dispersal over a wide geographic range. Not all fruits develop completely from an ovary; such "false fruits" or *pseudocarps,* develop from tissues adjacent to the ovary. Like flowers, fruit can vary tremendously in appearance, size, smell, and taste. Tomatoes, green peppers, corn, and avocados are all examples of fruits. Along with pollen and seeds, fruits also act as agents of dispersal. Some may be carried away by the wind. Many attract animals that will eat the fruit and pass the seeds through their digestive systems, then deposit the seeds in another location. Cockleburs are covered with stiff, hooked spines that can hook into fur (or clothing) and hitch a ride on an animal for long distances. The cockleburs that clung to the velvet trousers of an enterprising Swiss hiker, George de Mestral, inspired his invention of the loop and hook fastener he named Velcro.

EVOLUTION CONNECTION

Building Phylogenetic Trees with Analysis of DNA Sequence Alignments

All living organisms display patterns of relationships derived from their evolutionary history. Phylogeny is the science that describes the relative connections between organisms, in terms of ancestral and descendant species. Phylogenetic trees, such as the plant evolutionary history shown in Figure 26.7, are tree-like branching diagrams that depict these relationships. Species are found at the tips of the branches. Each branching point, called a node, is the point at which a single taxonomic group (taxon), such as a species, separates into two or more species.

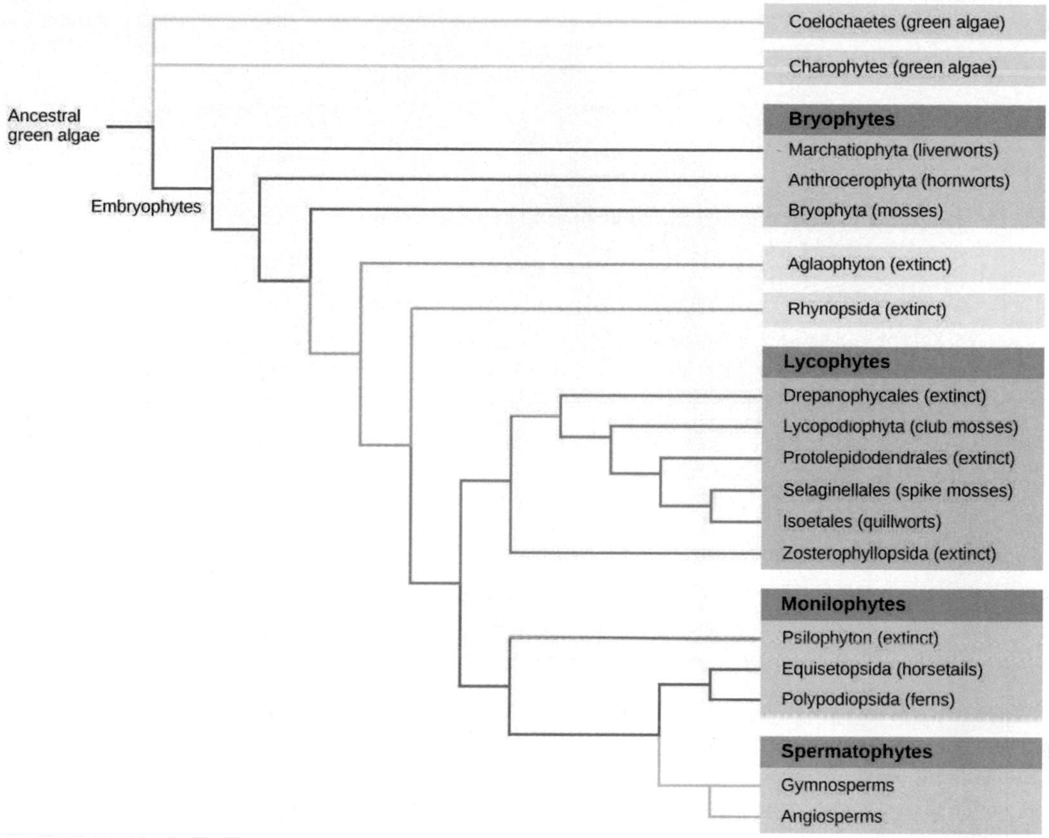

Figure 26.7 Plant phylogeny. This phylogenetic tree shows the evolutionary relationships of plants.

Phylogenetic trees have been built to describe the relationships between species since the first sketch of a tree that appeared in Darwin's *Origin of Species*. Traditional methods involve comparison of homologous anatomical structures and embryonic development, assuming that closely related organisms share anatomical features that emerge during embryo development. Some traits that disappear in the adult are present in the embryo; for example, an early human embryo has a postanal tail, as do all members of the Phylum Chordata. The study of fossil records shows the intermediate stages that link an ancestral form to its descendants. However, many of the approaches to classification based on the fossil record alone are imprecise and lend themselves to multiple interpretations. As the tools of molecular biology and computational analysis have been developed and perfected in recent years, a new generation of tree-building methods has taken shape. The key assumption is that genes for essential proteins or RNA structures, such as the ribosomal RNAs, are inherently conserved because mutations (changes in the DNA sequence) could possibly compromise the survival of the organism. DNA from minute samples of living organisms or fossils can be amplified by *polymerase chain reaction* (PCR) and sequenced, targeting the regions of the genome that are most likely to be conserved between species. The genes encoding the 18S ribosomal RNA from the small subunit and plastid genes are frequently chosen for DNA alignment analysis.

Once the sequences of interest are obtained, they are compared with existing sequences in databases such as GenBank, which is maintained by The National Center for Biotechnology Information. A number of computational tools are available to align and analyze sequences. Sophisticated computer analysis programs determine the percentage of sequence identity or homology. Sequence homology can be used to estimate the evolutionary distance between two DNA sequences and reflect the time elapsed since the genes separated from a common ancestor. Molecular analysis has revolutionized phylogenetic trees. In some cases, prior results from morphological studies have been confirmed: for example, confirming *Amborella trichopoda* as the most primitive angiosperm known. However, some groups and relationships have been rearranged as a result of DNA analysis.

26.2 Gymnosperms

By the end of this section, you will be able to do the following:

- Discuss the type of seeds produced by gymnosperms, as well as other characteristics of gymnosperms
- Identify the geological era dominated by the gymnosperms and describe the conditions to which they were adapted
- List the four groups of modern-day gymnosperms and provide examples of each
- Describe the life cycle of a typical gymnosperm

Gymnosperms, meaning "naked seeds," are a diverse group of seed plants. According to the "anthophyte" hypothesis, the angiosperms are a sister group of one group of gymnosperms (the Gnetales), which makes the gymnosperms a paraphyletic group. Paraphyletic groups are those in which not all descendants of a single common ancestor are included in the group. However , the "netifer" hypothesis suggests that the gnetophytes are sister to the conifers, making the gymnosperms monophyletic and sister to the angiosperms. Further molecular and anatomical studies may clarify these relationships. Characteristics of the gymnosperms include naked seeds, separate female and male gamtophytes, pollen cones and ovulate cones, pollination by wind and insects, and tracheids (which transport water and solutes in the vascular system).

Gymnosperm seeds are not enclosed in an ovary; rather, they are only partially sheltered by modified leaves called **sporophylls**. You may recall the term **strobilus** (plural = strobili) describes a tight arrangement of sporophylls around a central stalk, as seen in pine cones. Some seeds are enveloped by sporophyte tissues upon maturation. The layer of sporophyte tissue that surrounds the megasporangium, and later, the embryo, is called the **integument**.

Gymnosperms were the dominant phylum in the Mesozoic era. They are adapted to live where fresh water is scarce during part of the year, or in the nitrogen-poor soil of a bog. Therefore, they are still the prominent phylum in the coniferous biome or *taiga*, where the evergreen conifers have a selective advantage in cold and dry weather. Evergreen conifers continue low levels of photosynthesis during the cold months, and are ready to take advantage of the first sunny days of spring. One disadvantage is that conifers are more susceptible than deciduous trees to leaf infestations because most conifers do not lose their leaves all at once. They cannot, therefore, shed parasites and restart with a fresh supply of leaves in spring.

The life cycle of a gymnosperm involves alternation of generations, with a dominant sporophyte in which reduced male and female gametophytes reside. All gymnosperms are heterosporous. The male and female reproductive organs can form in cones or strobili. Male and female sporangia are produced either on the same plant, described as **monoecious** ("one home" or bisexual), or on separate plants, referred to as **dioecious** ("two homes" or unisexual) plants. The life cycle of a conifer will serve as our example of reproduction in gymnosperms.

Life Cycle of a Conifer

Pine trees are conifers (coniferous = cone bearing) and carry both male and female sporophylls on the same mature sporophyte. Therefore, they are monoecious plants. Like all gymnosperms, pines are heterosporous and generate two different types of spores (male microspores and female megaspores). Male and female spores develop in different strobili, with small male cones and larger female cones. In the male cones, or staminate cones, the **microsporocytes** undergo meiosis and the resultant haploid microspores give rise to male gametophytes or "pollen grains" by mitosis. Each pollen grain consists of just a few haploid cells enclosed in a tough wall reinforced with sporopollenin. In the spring, large amounts of yellow pollen are released and carried by the wind. Some gametophytes will land on a female cone. Pollination is defined as the initiation of pollen tube growth. The pollen tube develops slowly, and the generative cell in the pollen grain produces two haploid sperm or generative nuclei by mitosis. At fertilization, one of the haploid sperm nuclei will unite with the haploid nucleus of an egg cell.

Female cones, or **ovulate cones**, contain two ovules per scale. Each ovule has a narrow passage that opens near the base of the sporophyll. This passage is the micropyle, through which a pollen tube will later grow. One megaspore mother cell, or **megasporocyte**, undergoes meiosis in each ovule. Three of the four cells break down; only a single surviving cell will develop into a female multicellular gametophyte, which encloses archegonia (an archegonium is a reproductive organ that contains a single large egg). As the female gametophyte begins to develop, a sticky pollination drop traps windblown pollen grains near the opening of the micropyle. A pollen tube is formed and grows toward the developing gametophyte. One of the generative or sperm nuclei from the pollen tube will enter the egg and fuse with the egg nucleus as the egg matures. Upon fertilization, the diploid egg will give rise to the embryo, which is enclosed in a seed coat of tissue from the parent plant. Although several eggs may be formed and even fertilized, there is usually a single surviving embryo in each ovule. Fertilization and seed development is a long process in pine trees: it may take up to two years after pollination. The seed that is formed contains three generations of tissues: the seed coat that originates from the sporophyte tissue, the gametophyte tissue that will provide nutrients, and the embryo itself.

Figure 26.8 illustrates the life cycle of a conifer. The sporophyte (2*n*) phase is the longest phase in the life of a gymnosperm. The gametophytes (1*n*)—produced by microspores and megaspores—are reduced in size. It may take more than a year between pollination and fertilization while the pollen tube grows towards the growing female gametophyte (1*n*), which develops from a single megaspore. The slow growth of the pollen tube allows the female gametophyte time to produce eggs (1*n*).

VISUAL CONNECTION

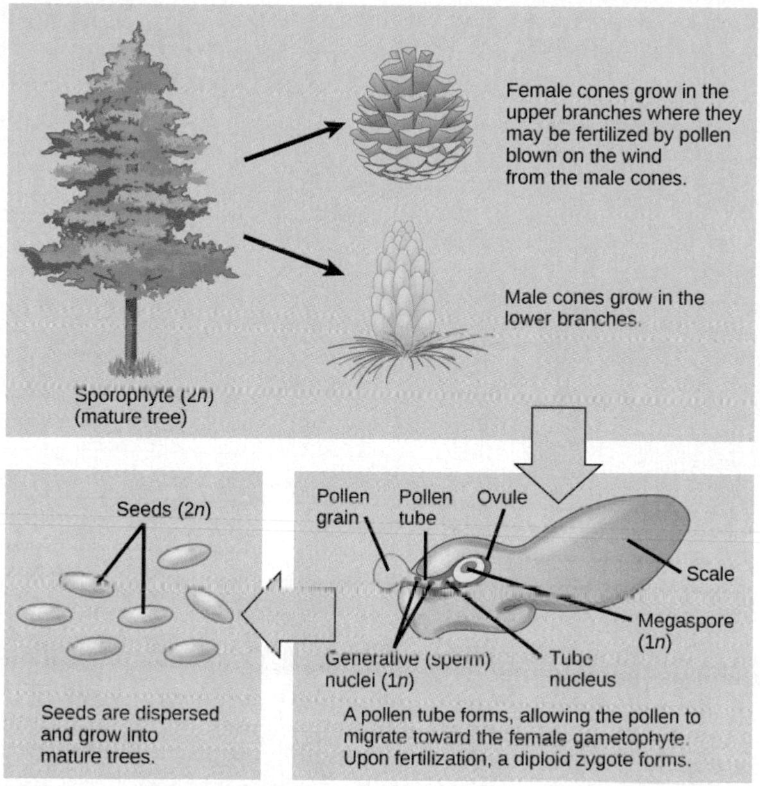

Figure 26.8 Conifer life cycle. This image shows the life cycle of a conifer. Pollen from male cones blows up into upper branches, where it fertilizes female cones. The megaspore shown in the image develops into the female gametophyte as the pollen tube slowly grows toward it, eventually fusing with the egg and delivering a male nucleus, which combines with the female nucleus of the mature egg.

At what stage does the diploid zygote form?

a. when the female cone begins to bud from the tree
b. at fertilization
c. when the seeds drop from the tree
d. when the pollen tube begins to grow

LINK TO LEARNING

Watch this video to see the process of seed production in gymnosperms.

Click to view content (https://www.openstax.org/l/gymnosperm2)

Diversity of Gymnosperms

Modern gymnosperms are classified into four phyla. Coniferophyta, Cycadophyta, and Ginkgophyta are similar in their pattern of seed development and also in their production of *secondary cambium* (cells that generate the vascular system of the trunk or stem and are partially specialized for water transportation). However, the three phyla are not closely related phylogenetically to each other. Gnetophyta are considered the closest group to angiosperms because they produce true xylem tissue, with vessels as well as the tracheids found in the rest of the gymnosperms. It is possible that vessel elements arose independently in the two groups

Conifers (Coniferophyta)

Conifers are the dominant phylum of gymnosperms, with the greatest variety of species (Figure 26.9). Typical conifers are tall trees that bear scale-like or needle-like leaves. Water evaporation from leaves is reduced by their narrow shape and a thick cuticle. Snow easily slides off needle-shaped leaves, keeping the snow load light, thus reducing broken branches. Such adaptations to cold and dry weather explain the predominance of conifers at high altitudes and in cold climates. Conifers include familiar evergreen trees such as pines, spruces, firs, cedars, sequoias, and yews. A few species are deciduous and lose their leaves in fall. The bald cypress, dawn redwood, European larch and the tamarack (Figure 26.9c) are examples of deciduous conifers. Many coniferous trees are harvested for paper pulp and timber. The wood of conifers is more primitive than the wood of angiosperms; it contains *tracheids*, but no vessel elements, and is therefore referred to as "soft wood."

Figure 26.9 Conifers. Conifers are the dominant form of vegetation in cold or arid environments and at high altitudes. Shown here are the (a) evergreen spruce *Picea* sp., (b) juniper *Juniperus* sp., (c) coastal redwood or sequoia *Sequoia sempervirens*, and (d) the tamarack *Larix larcinia*. Notice the deciduous yellow leaves of the tamarack. (credit a: modification of work by Rosendahl; credit b: modification of work by Alan Levine; credit c: modification of work by Wendy McCormic; credit d: modification of work by Micky Zlimen)

Cycads

Cycads thrive in mild climates, and are often mistaken for palms because of the shape of their large, compound leaves. Cycads bear large strobili or cones (Figure 26.10), and may be pollinated by beetles rather than wind, which is unusual for a gymnosperm. Large cycads dominated the landscape during the age of dinosaurs in the Mesozoic, but only a hundred or so smaller species persisted to modern times. They face possible extinction, and several species are protected through international conventions. Because of their attractive shape, they are often used as ornamental plants in gardens in the tropics and subtropics.

Figure 26.10 Cycad. This cycad, *Encephalartos ferox*, has large cones and broad, fern-like leaves. (credit: Wendy Cutler)

Ginkgophytes

The single surviving species of the **ginkgophytes** group is *Ginkgo biloba* (Figure 26.11). Its fan-shaped leaves—unique among seed plants because they feature a dichotomous venation pattern—turn yellow in autumn and fall from the tree. For centuries, *G. biloba* was cultivated by Chinese Buddhist monks in monasteries, which ensured its preservation. It is planted in public spaces because it is unusually resistant to pollution. Male and female organs are produced on separate plants. Typically, gardeners plant only male trees because the seeds produced by the female plant have an off-putting smell of rancid butter.

Figure 26.11 *Ginkgo*. This plate from the 1870 book *Flora Japonica, Sectio Prima (Tafelband)* depicts the leaves and fruit of *Ginkgo biloba*, as drawn by Philipp Franz von Siebold and Joseph Gerhard Zuccarini.

Gnetophytes

The phylogenetic position of the **gnetophytes** is not currently resolved. Their possession of vessel elements suggests they are the closest relative to modern angiosperms. However, molecular analysis places them closer to the conifers. The three living genera are quite dissimilar: *Ephedra*, *Gnetum*, and *Welwitschia* (Figure 26.12), which may indicate that the group is not monophyletic. Like angiosperms, they have broad leaves. *Ephedra* (Figure 26.12a) occurs in dry areas of the West Coast of the United States and

Mexico. *Ephedra*'s small, scale-like leaves are the source of the compound *ephedrine*, which is used in medicine as a potent decongestant. Because ephedrine is similar to amphetamines, both in chemical structure and neurological effects, its use is restricted to prescription drugs. *Gnetum* species (Figure 26.12b) are found in some parts of Africa, South America, and Southeast Asia, and include trees, shrubs and vines. *Welwitschia* (Figure 26.12c) is found in the Namib desert, and is possibly the oddest member of the group. It produces only two leaves, which grow continuously throughout the life of the plant (some plants are hundreds of years old). Like the ginkgos, *Welwitschia* produces male and female gametes on separate plants.

Figure 26.12 (a) *Ephedra viridis*, known by the common name *Mormon tea*, grows on the West Coast of the United States and Mexico. (b) *Gnetum gnemon* grows in Malaysia. (c) The large *Welwitschia mirabilis* can be found in the Namibian desert. (credit a: modification of work by USDA; credit b: modification of work by Malcolm Manners; credit c: modification of work by Derek Keats)

LINK TO LEARNING

Watch this BBC video describing the amazing strangeness of Welwitschia.

Click to view content (https://www.openstax.org/l/welwitschia2)

26.3 Angiosperms

By the end of this section, you will be able to do the following:

- Explain why angiosperms are the dominant form of plant life in most terrestrial ecosystems
- Describe the main parts of a flower and their functions
- Detail the life cycle of a typical gymnosperm and angiosperm
- Discuss the similarities and differences between the two main groups of flowering plants

From their humble and still obscure beginning during the early Jurassic period, the angiosperms—or flowering plants—have evolved to dominate most terrestrial ecosystems (Figure 26.13). With more than 300,000 species, the angiosperm phylum (Anthophyta) is second only to insects in terms of diversification.

Figure 26.13 Flowers. These flowers grow in a botanical garden border in Bellevue, WA. Flowering plants dominate terrestrial landscapes. The vivid colors of flowers and enticing fragrance of flowers are adaptations to pollination by animals like insects, birds, and bats. (credit:

Myriam Feldman)

The success of angiosperms is due to two novel reproductive structures: flowers and fruits. The function of the flower is to ensure pollination, often by arthropods, as well as to protect a developing embryo. The colors and patterns on flowers offer specific signals to many pollinating insects or birds and bats that have coevolved with them. For example, some patterns are visible only in the ultraviolet range of light, which can be seen by arthropod pollinators. For some pollinators, flowers advertise themselves as a reliable source of nectar. Flower scent also helps to select its pollinators. Sweet scents tend to attract bees and butterflies and moths, but some flies and beetles might prefer scents that signal fermentation or putrefaction. Flowers also provide protection for the ovule and developing embryo inside a receptacle. The function of the fruit is seed protection and dispersal. Different fruit structures or tissues on fruit—such as sweet flesh, wings, parachutes, or spines that grab—reflect the dispersal strategies that help spread seeds.

Flowers

Flowers are modified leaves, or sporophylls, organized around a central receptacle. Although they vary greatly in appearance, virtually all flowers contain the same structures: sepals, petals, carpels, and stamens. The peduncle typically attaches the flower to the plant proper. A whorl of **sepals** (collectively called the **calyx**) is located at the base of the peduncle and encloses the unopened floral bud. Sepals are usually photosynthetic organs, although there are some exceptions. For example, the corolla in lilies and tulips consists of three sepals and three petals that look virtually identical. **Petals**, collectively the **corolla**, are located inside the whorl of sepals and may display vivid colors to attract pollinators. Sepals and petals together form the **perianth**. The sexual organs, the *female gynoecium* and *male androecium* are located at the center of the flower. Typically, the sepals, petals, and stamens are attached to the receptacle at the base of the gynoecium, but the gynoecium may also be located deeper in the receptacle, with the other floral structures attached above it.

As illustrated in Figure 26.14, the innermost part of a perfect flower is the **gynoecium**, the location in the flower where the eggs will form. The female reproductive unit consists of one or more carpels, each of which has a stigma, style, and ovary. The **stigma** is the location where the pollen is deposited either by wind or a pollinating arthropod. The sticky surface of the stigma traps pollen grains, and the **style** is a connecting structure through which the pollen tube will grow to reach the ovary. The **ovary** houses one or more ovules, each of which will ultimately develop into a **seed**. Flower structure is very diverse, and carpels may be singular, multiple, or fused. (Multiple fused carpels comprise a **pistil**.) The **androecium**, or male reproductive region is composed of multiple stamens surrounding the central carpel. **Stamens** are composed of a thin stalk called a **filament** and a sac-like structure called the **anther**. The filament supports the anther, where the microspores are produced by meiosis and develop into haploid pollen grains, or male gametophytes.

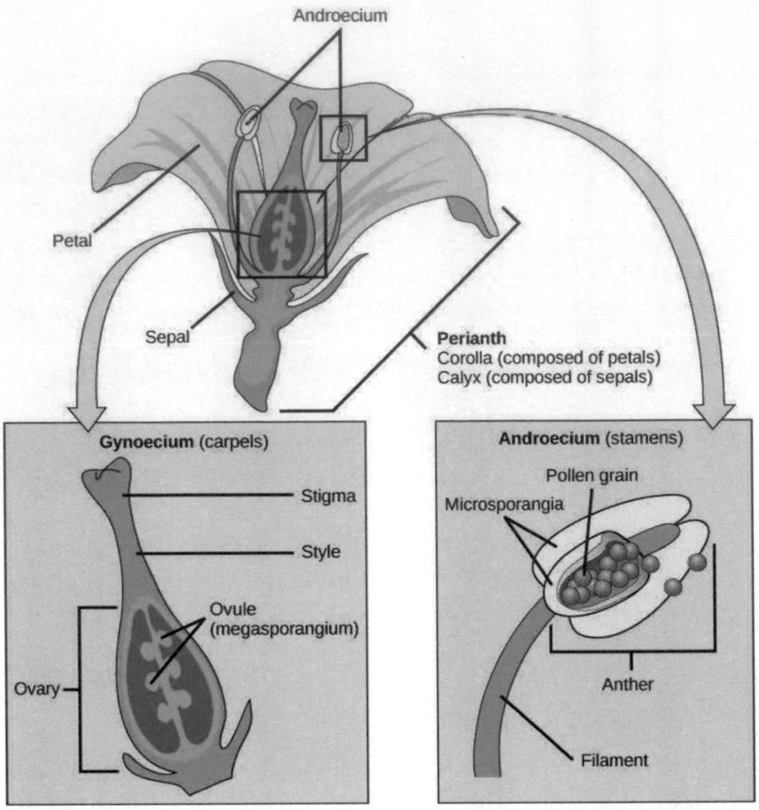

Figure 26.14 Flower structure. This image depicts the structure of a perfect flower. Perfect flowers produce both male and female floral organs. The flower shown has only one carpel, but some flowers have a cluster of carpels. Together, all the carpels make up the gynoecium. (credit: modification of work by Mariana Ruiz Villareal)

The Life Cycle of an Angiosperm

The adult or sporophyte phase is the main phase of an angiosperm's life cycle (Figure 26.15). Like gymnosperms, angiosperms are heterosporous. Therefore, they produce microspores, which will generate pollen grains as the male gametophytes, and *megaspores*, which will form an ovule that contains female gametophytes. Inside the anther's microsporangia, male sporocytes divide by meiosis to generate haploid microspores, which, in turn, undergo mitosis and give rise to pollen grains. Each pollen grain contains two cells: one generative cell that will divide into two sperm and a second cell that will become the pollen tube cell.

VISUAL CONNECTION

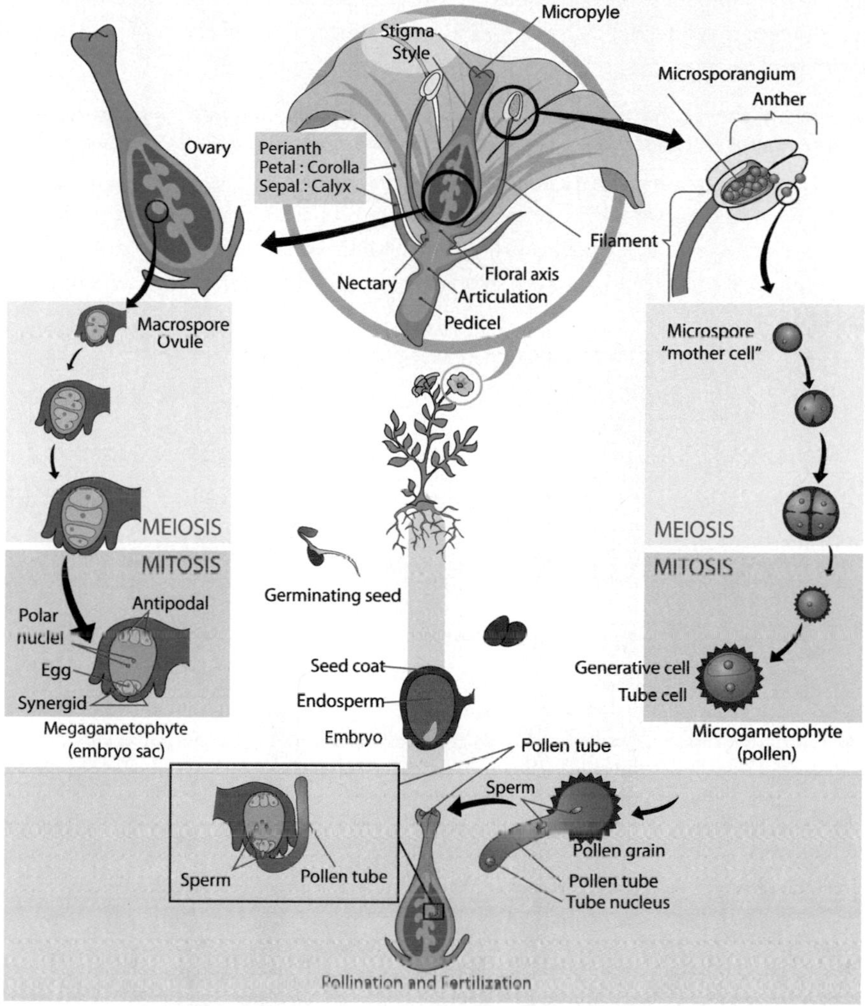

Figure 26.15 Angiosperm life cycle. The life cycle of an angiosperm is shown. Anthers and carpels are structures that shelter the actual gametophytes: the pollen grain and embryo sac. Double fertilization is a process unique to angiosperms. (credit: modification of work by Mariana Ruiz Villareal)

Question: If a flower lacked a megasporangium, what type of gamete would not form? If the flower lacked a microsporangium, what type of gamete would not form?

The ovule, sheltered within the ovary of the carpel, contains the megasporangium protected by two layers of integuments and the ovary wall. Within each megasporangium, a diploid megasporocyte undergoes meiosis, generating four haploid megaspores—three small and one large. Only the large megaspore survives; it divides mitotically three times to produce eight nuclei distributed among the seven cells of the female gametophyte or embryo sac. Three of these cells are located at each pole of the embryo sac. The three cells at one pole become the egg and two *synergids*. The three cells at the opposite pole become antipodal cells. The center cell contains the remaining two nuclei (polar nuclei). This cell will eventually produce the endosperm of the seed. The mature embryo sac then contains one egg cell, two synergids or "helper" cells, three antipodal cells (which eventually degenerate), and a central cell with two polar nuclei. When a pollen grain reaches the stigma, a pollen tube extends from the grain, grows down the style, and enters through the micropyle: an opening in the integuments of the ovule. The two sperm are deposited in the embryo sac.

A double fertilization event then occurs. One sperm and the egg combine, forming a diploid zygote—*the future embryo*. The other sperm fuses with the polar nuclei, forming a triploid cell that will develop into the **endosperm**—the tissue that serves as a food reserve for the developing embryo. The zygote develops into an embryo with a *radicle*, or small root, and one (monocot) or two (dicot) leaf-like organs called *cotyledons*. This difference in the number of embryonic leaves is the basis for the two major groups of angiosperms: the *monocots* and the *eudicots*. Seed food reserves are stored outside the embryo, in the form of complex carbohydrates, lipids, or proteins. The **cotyledons** serve as conduits to transmit the broken-down food reserves from their storage site inside the seed to the developing embryo. The seed consists of a toughened layer of integuments forming the coat, the endosperm with food reserves, and at the center, the well-protected embryo.

Most angiosperms have **perfect flowers**, which means that each flower carries both stamens and carpels (Figure 26.15). In monoecious plants, male (staminate) and female (pistillate) flowers are separate, but carried on the same plant. Sweetgums (*Liquidambar* spp.) and beeches (*Betula* spp. are monoecious (Figure 26.16). In dioecious plants, male and female flowers are found on separate plants. Willows (*Salix* spp.) and poplars (*Populus* spp.) are dioecious. In spite of the predominance of perfect flowers, only a few species of angiosperms self-pollinate. Both anatomical and environmental barriers promote cross-pollination mediated by a physical agent (wind or water), or an animal, such as an insect or bird. Cross-pollination increases genetic diversity in a species.

Figure 26.16 Beech inflorescences. The female inflorescence is at the upper left. The male inflorescence is at the lower right. (credit: Stephen J. Baskauf, 2002. http://bioimages.vanderbilt.edu/baskauf/10593 (http://openstax.org/l/betula) . Morphbank :: Biological Imaging (http://www.morphbank.net/ (http://openstax.org/l/morphbank) , 29 June 2017). Florida State University, Department of Scientific Computing, Tallahassee, FL 32306-4026 USA)

Fruit

As the seed develops, the walls of the ovary thicken and form the fruit. The seed forms in an ovary, which also enlarges as the seeds grow. Many foods commonly called vegetables are actually fruits. Eggplants, zucchini, string beans, tomatoes, and bell peppers are all technically fruits because they contain seeds and are derived from the thick ovary tissue. Acorns are true nuts, and winged maple "helicopter seeds" or whirligigs (whose botanical name is *samara*) are also fruits. Botanists classify fruit into more than two dozen different categories, only a few of which are actually fleshy and sweet.

Mature fruit can be fleshy or dry. *Fleshy fruit* include the familiar berries, peaches, apples, grapes, and tomatoes. Rice, wheat, and nuts are examples of *dry fruit*. Another subtle distinction is that not all fruits are derived from just the ovary. For instance, strawberries are derived from the ovary as well as the receptacle, and apples are formed from the ovary and the pericarp, or *hypanthium*. Some fruits are derived from separate ovaries in a single flower, such as the raspberry. Other fruits, such as the pineapple, form from clusters of flowers. Additionally, some fruits, like watermelon and orange, have rinds. Regardless of how they are formed, fruits are an agent of seed dispersal. The variety of shapes and characteristics reflect the mode of dispersal. Wind carries the light dry fruits of trees and dandelions. Water transports floating coconuts. Some fruits attract herbivores with their color or scent, or as food. Once eaten, tough, undigested seeds are dispersed through the herbivore's feces (*endozoochory*). Other fruits have burrs and hooks to cling to fur and hitch rides on animals (*epizoochory*).

Diversity of Angiosperms

Angiosperms are classified in a single phylum: the **Anthophyta**. Modern angiosperms appear to be a monophyletic group, which as you may recall means that they originated from a single ancestor. Within the angiosperms are three major groups: basal angiosperms, monocots, and dicots. Basal angiosperms are a group of plants that are believed to have branched off before the separation of the monocots and eudicots, because they exhibit traits from both groups. They are categorized separately in most classification schemes. The basal angiosperms include *Amborella*, water lilies, the Magnoliids (magnolia trees, laurels, and spice peppers), and a group called the Austrobaileyales, which includes the star anise. The monocots and dicots are differentiated on the basis of the structure of the cotyledons, pollen grains, and other structures. Monocots include grasses and lilies, and the dicots form a multi-branched group that includes (among many others) roses, cabbages, sunflowers, and mints.

Basal Angiosperms

The Magnoliidae are represented by the magnolias, laurels, and peppers. Magnolias are tall trees bearing dark, shiny leaves, and large, fragrant flowers with many parts, and are considered archaic (Figure 26.17). In the outer whorl of the magnolia flower the sepals and petals are undifferentiated and are collectively called tepals. The reproductive parts are arranged in a spiral around a cone-shaped receptacle, with the carpels located above the stamens (Figure 26.17). The aggregate fruit, with one seed formed from each carpel, is seen in Figure 26.18d. Laurel trees produce fragrant leaves and small, inconspicuous flowers. The *Laurales* grow mostly in warmer climates and are small trees and shrubs. Familiar plants in this group include the bay laurel, cinnamon, spice bush (Figure 26.18a), and avocado tree.

Figure 26.17 *Magnolia grandiflora*. A cluster of carpels can be seen above the stamens, which have shed their pollen and begun to drop from the inflorescence. In the flower, the sepals and petals are undifferentiated and are collectively called tepals. (credit: Ianaré Sévi. http://bioimages.vanderbilt.edu/baskauf/10949 (http://openstax.org/l/grandiflora))

Figure 26.18 Basal angiosperms. The (a) common spicebush belongs to the *Laurales*, the same family as cinnamon and bay laurel. The fruit of (b) the *Piper nigrum* plant is black pepper, the main product that was traded along spice routes. Notice the small, unobtrusive, clustered flowers. The leaf venation resembles that of both the monocots (parallel) and the dicots (branched). (c) Water lilies, *Nymphaea lotus*. Although the leaves of the plant float on the surface of the water, their roots are in the underlying soil at the bottom of the lake. The aggregate fruit of a magnolia (d). The fruit is in its final stage, with its red seeds just starting to appear. (credit a: modification of work by Cory Zanker; credit b: modification of work by Franz Eugen Köhler; credit c: modification of work by Rl/Wikimedia Commons. d: modification of work by "Coastside2"/Wikimedia Commons).

Monocots

Plants in the **monocot** group are primarily identified by the presence of a single cotyledon in the seedling. Other anatomical features shared by monocots include veins that run parallel to and along the length of the leaves, and flower parts that are arranged in a three- or six-fold symmetry. *True woody tissue* is rarely found in monocots. In palm trees, vascular and parenchyma tissues produced by the primary and secondary thickening meristems form the trunk. The pollen from the first angiosperms was likely **monosulcate**, containing a single furrow or pore through the outer layer. This feature is still seen in the modern monocots. Vascular tissue of the stem is scattered, not arranged in any particular pattern, but is organized in a ring in the roots. The root system consists of multiple fibrous roots, with no major tap root. *Adventitious roots* often emerge from the stem or leaves. The monocots include familiar plants such as the true lilies (Liliopsida), orchids, yucca, asparagus, grasses, and palms. Many important crops are monocots, such as rice and other cereals, corn, sugar cane, and tropical fruits like bananas and pineapples (Figure 26.19a,**b**,**c**).

Figure 26.19 Monocot and dicot crop plants. The world's major crops are flowering plants. (a) Rice, (b) wheat, and (c) bananas are monocots, while (d) cabbage, (e) beans, and (f) peaches are dicots. (credit a: modification of work by David Nance, USDA ARS; credit b, c: modification of work by Rosendahl; credit d: modification of work by Bill Tarpenning, USDA; credit e: modification of work by Scott Bauer, USDA ARS; credit f: modification of work by Keith Weller, USDA)

Eudicots

Eudicots, or true **dicots**, are characterized by the presence of two cotyledons in the developing shoot. Veins form a network in leaves, and flower parts come in four, five, or many whorls. Vascular tissue forms a ring in the stem; in monocots, vascular tissue is scattered in the stem. Eudicots can be **herbaceous** (not woody), or produce woody tissues. Most eudicots produce pollen that is trisulcate or triporate, with three furrows or pores. The root system is usually anchored by one main root developed from the embryonic radicle. Eudicots comprise two-thirds of all flowering plants. The major differences between monocots and eudicots are summarized in Table 26.1. However, some species may exhibit characteristics usually associated with the other group, so identification of a plant as a monocot or a eudicot is not always straightforward.

Comparison of Structural Characteristics of Monocots and Eudicots

Characteristic	Monocot	Eudicot
Cotyledon	One	Two
Veins in Leaves	Parallel	Network (branched)
Stem Vascular Tissue	Scattered	Arranged in ring pattern
Roots	Network of fibrous roots	Tap root with many lateral roots
Pollen	Monosulcate	Trisulcate
Flower Parts	Three or multiple of three	Four, five, multiple of four or five and whorls

Table 26.1

26.4 The Role of Seed Plants

By the end of this section, you will be able to do the following:

- Explain how angiosperm diversity is due, in part, to multiple complex interactions with animals
- Describe ways in which pollination occurs
- Discuss the roles that plants play in ecosystems and how deforestation threatens plant biodiversity

Without seed plants, life as we know it would not be possible. Plants play a key role in the maintenance of terrestrial ecosystems through the stabilization of soils, cycling of carbon, and climate moderation. Large tropical forests release oxygen and act as carbon dioxide "sinks." Seed plants provide shelter to many life forms, as well as food for herbivores, thereby indirectly feeding carnivores. Plant secondary metabolites are used for medicinal purposes and industrial production. Virtually all animal life is dependent on plants for survival.

Animals and Plants: Herbivory

Coevolution of flowering plants and insects is a hypothesis that has received much attention and support, especially because both angiosperms and insects diversified at about the same time in the middle Mesozoic. Many authors have attributed the diversity of plants and insects to both pollination and **herbivory**, or the consumption of plants by insects and other animals. Herbivory is believed to have been as much a driving force as pollination. Coevolution of herbivores and plant defenses is easily and commonly observed in nature. Unlike animals, most plants cannot outrun predators or use mimicry to hide from hungry animals (although mimicry has been used to entice pollinators). A sort of arms race exists between plants and herbivores. To "combat" herbivores, some plant seeds—such as acorn and unripened persimmon—are high in alkaloids and therefore unsavory to some animals. Other plants are protected by bark, although some animals developed specialized mouth pieces to tear and chew vegetal material. Spines and thorns (Figure 26.20) deter most animals, except for mammals with thick fur, and some birds have specialized beaks to get past such defenses.

Figure 26.20 Plant defenses. (a) Spines and (b) thorns are examples of plant defenses. (credit a: modification of work by Jon Sullivan; credit b: modification of work by I. Sáček, Sr.)

Herbivory has been exploited by seed plants for their own benefit. The dispersal of fruits by herbivorous animals is a striking example of mutualistic relationships. The plant offers to the herbivore a nutritious source of food in return for spreading the plant's genetic material to a wider area.

An extreme example of coevolution (discovered by Daniel Janzen) between an animal and a plant is exemplified by Mexican acacia trees and their attendant acacia ants *Pseudomyrmex* spp. (this is termed *myrmecophytism*). The trees support the ants with shelter and food: The ants nest in the hollows of large thorns produced by the tree and feed on sugary secretions produced at the ends of the leaves. The sugar pellets also help to keep the ants from interfering with insect pollinators. In return, ants discourage herbivores, both invertebrates and vertebrates, by stinging and attacking leaf-eaters and insects ovipositing on the plants. The ants also help to remove potential plant pathogens, such as fungal growths. Another case of insect-plant coevolution is found in bracken fern (*Pteridium aquinilum*), whose subspecies are found throughout the world. Bracken ferns produce a number of "secondary plant compounds" in their adult fronds that serve as defensive compounds against nonadapted insect attack (these compounds include cyanogenic glucosides, tannins, and phenolics). However, during the "fiddlehead" or crozier stage, bracken secretes nutritious sugary and proteinaceous compounds from special "nectaries" that attract ants and even species of jumping spiders, all of which defend the plant's croziers until they are fully unfolded. These opportunistic groups of protective arthropods greatly reduce the damage that otherwise would occur during the early stages of growth.

Animals and Plants: Pollination

Flowers pollinated by wind are usually small, feathery, and visually inconspicuous. *Grasses* are a successful group of flowering plants that are wind pollinated. They produce large amounts of powdery pollen carried over large distances by the wind. Some large trees such as oaks, maples, and birches are also wind pollinated.

LINK TO LEARNING

Explore this website (http://openstax.org/l/pollinators2) for additional information on pollinators.

More than 80 percent of angiosperms depend on animals for **pollination** (technically the transfer of pollen from the anther to the stigma). Consequently, plants have developed many adaptations to attract pollinators. With over 200,000 different plants dependent on animal pollination, the plant needs to advertise to its pollinators with some specificity. The specificity of specialized plant structures that target animals can be very surprising. It is possible, for example, to determine the general type of pollinators favored by a plant by observing the flower's physical characteristics. Many bird or insect-pollinated flowers secrete nectar, which is a sugary liquid. They also produce both fertile pollen, for reproduction, and sterile pollen rich in nutrients for birds and insects. Many butterflies and bees can detect ultraviolet light, and flowers that attract these pollinators usually display a pattern of ultraviolet reflectance that helps them quickly locate the flower's center. In this manner, pollinating insects collect nectar while at the same time are dusted with pollen (Figure 26.21). Large, red flowers with little smell and a long funnel shape are preferred by hummingbirds, who have good color perception, a poor sense of smell, and need a strong perch. White flowers that open at night attract moths. Other animals—such as bats, lemurs, and lizards—can also act as pollinating agents. Any disruption to these interactions, such as the disappearance of bees, for example as a consequence of colony collapse disorders, can lead to disaster for agricultural industries that depend heavily on pollinated crops.

Figure 26.21 Pollination. As a bee collects nectar from a flower, it is dusted by pollen, which it then disperses to other flowers. (credit: John Severns)

SCIENTIFIC METHOD CONNECTION

Testing Attraction of Flies by Rotting Flesh Smell

Question: Will flowers that offer cues to bees attract carrion flies if sprayed with compounds that smell like rotten flesh?

Background: Visitation of flowers by pollinating flies is a function mostly of smell. Flies are attracted by rotting flesh and carrions. The putrid odor seems to be the major attractant. The polyamines putrescine and cadaverine, which are the products of protein breakdown after animal death, are the source of the pungent smell of decaying meat. Some plants strategically attract flies by synthesizing polyamines similar to those generated by decaying flesh and thereby attract carrion flies.

Flies seek out dead animals because they normally lay their eggs on them and their maggots feed on the decaying flesh. Interestingly, time of death can be determined by a forensic entomologist based on the stages and type of maggots recovered from cadavers.

Hypothesis: Because flies are drawn to other organisms based on smell and not sight, a flower that is normally attractive to bees because of its colors will attract flies if it is sprayed with polyamines similar to those generated by decaying flesh.

Test the hypothesis:

1. Select flowers usually pollinated by bees. White petunia may be a good choice.
2. Divide the flowers into two groups, and while wearing eye protection and gloves, spray one group with a solution of either putrescine or cadaverine. (Putrescine dihydrochloride is typically available in 98 percent concentration; this can be diluted to approximately 50 percent for this experiment.)
3. Place the flowers in a location where flies are present, keeping the sprayed and unsprayed flowers separated.
4. Observe the movement of the flies for one hour. Record the number of visits to the flowers using a table similar to Table 26.2. Given the rapid movement of flies, it may be beneficial to use a video camera to record the fly–flower interaction. Replay the video in slow motion to obtain an accurate record of the number of fly visits to the flowers.
5. Repeat the experiment four more times with the same species of flower, but using different specimens.
6. Repeat the entire experiment with a different type of flower that is normally pollinated by bees.

Results of Number of Visits by Flies to Sprayed and Control/Unsprayed Flowers

Trial #	Sprayed Flowers	Unsprayed Flowers
1		
2		
3		
4		
5		

Table 26.2

Analyze your data: Review the data you have recorded. Average the number of visits that flies made to sprayed flowers over the course of the five trials (on the first flower type) and compare and contrast them to the average number of visits that flies made to the unsprayed/control flowers. Can you draw any conclusions regarding the attraction of the flies to the sprayed flowers?

For the second flower type used, average the number of visits that flies made to sprayed flowers over the course of the five trials and compare and contrast them to the average number of visits that flies made to the unsprayed/control flowers. Can you draw any conclusions regarding the attraction of the flies to the sprayed flowers?

Compare and contrast the average number of visits that flies made to the two flower types. Can you draw any conclusions about whether the appearance of the flower had any impact on the attraction of flies? Did smell override any appearance differences, or were the flies attracted to one flower type more than another?

Form a conclusion: Do the results support the hypothesis? If not, how can your observations be explained?

The Importance of Seed Plants in Human Life

Seed plants are the foundation of human diets across the world (Figure 26.22). Many societies eat almost exclusively vegetarian fare and depend solely on seed plants for their nutritional needs. A few crops (rice, wheat, and potatoes) dominate the agricultural landscape. Many crops were developed during the agricultural revolution, when human societies made the transition from nomadic hunter–gatherers to horticulture and agriculture. Cereals, rich in carbohydrates, provide the staple of many human diets. Beans and nuts supply proteins. Fats are derived from crushed seeds, as is the case for peanut and rapeseed (canola) oils, or fruits such as olives. Animal husbandry also consumes large quantities of crop plants.

Staple crops are not the only food derived from seed plants. Various fruits and vegetables provide nutrient macromolecules, vitamins, minerals, and fiber. Sugar, to sweeten dishes, is produced from the monocot sugarcane and the eudicot sugar beet. Drinks are made from infusions of tea leaves, chamomile flowers, crushed coffee beans, or powdered cocoa beans. Spices come from many different plant parts: saffron and cloves are stamens and buds, black pepper and vanilla are seeds, the bark of a bush in the *Laurales* family supplies cinnamon, and the herbs that flavor many dishes come from dried leaves and fruit, such as the pungent red chili pepper. The volatile oils of a number of flowers and bark provide the scent of perfumes.

Additionally, no discussion of seed plant contribution to human diet would be complete without the mention of alcohol. Fermentation of plant-derived sugars and starches is used to produce alcoholic beverages in all societies. In some cases, the beverages are derived from the fermentation of sugars from fruit, as with wines and, in other cases, from the fermentation of carbohydrates derived from seeds, as with beers. The sharing of foods and beverages also contributes to human social ritual.

Seed plants have many other uses, including providing wood as a source of timber for construction, fuel, and material to build furniture. Most paper is derived from the pulp of coniferous trees. Fibers of seed plants such as cotton, flax, and hemp are woven into cloth. Textile dyes, such as indigo, were mostly of plant origin until the advent of synthetic chemical dyes.

Lastly, it is more difficult to quantify the benefits of ornamental seed plants. These grace private and public spaces, adding beauty and serenity to human lives and inspiring painters and poets alike.

Figure 26.22 Human uses of plants. Humans rely on plants for a variety of reasons. (a) Cacao beans were introduced to Europe from the New World, where they were used by Mesoamerican civilizations. Combined with sugar, another plant product, chocolate is a popular food. (b) Flowers like the tulip are cultivated for their beauty. (c) Quinine, extracted from cinchona trees, is used to treat malaria, to reduce fever, and to alleviate pain. (d) This violin is made of wood. (credit a: modification of work by "Everjean"/Flickr; credit b: modification of work by Rosendahl; credit c: modification of work by Franz Eugen Köhler)

The medicinal properties of plants have been known to human societies since ancient times. There are references to the use of plants' curative properties in Egyptian, Babylonian, and Chinese writings from 5,000 years ago. Many modern synthetic therapeutic drugs are derived or synthesized *de novo* from plant secondary metabolites. It is important to note that the same plant extract can be a therapeutic remedy at low concentrations, become an addictive drug at higher doses, and can potentially kill at high concentrations. Table 26.3 presents a few drugs, their plants of origin, and their medicinal applications.

Plant Origin of Medicinal Compounds and Medical Applications

Plant	Compound	Application
Deadly nightshade (*Atropa belladonna*)	Atropine	Dilate eye pupils for eye exams
Foxglove (*Digitalis purpurea*)	Digitalis	Heart disease, stimulates heart beat
Yam (*Dioscorea* spp.)	Steroids	Steroid hormones: contraceptive pill and cortisone
Ephedra (*Ephedra* spp.)	Ephedrine	Decongestant and bronchiole dilator
Pacific yew (*Taxus brevifolia*)	Taxol	Cancer chemotherapy; inhibits mitosis
Opium poppy (*Papaver somniferum*)	Opioids	Analgesic (reduces pain without loss of consciousness) and narcotic (reduces pain with drowsiness and loss of consciousness) in higher doses
Quinine tree (*Cinchona* spp.)	Quinine	Antipyretic (lowers body temperature) and antimalarial
Willow (*Salix* spp.)	Salicylic acid (aspirin)	Analgesic and antipyretic

Table 26.3

CAREER CONNECTION

Ethnobotanist

The relatively new field of ethnobotany studies the interaction between a particular culture and the plants native to the region. Seed plants have a large influence on day-to-day human life. Not only are plants the major source of food and medicine, they also influence many other aspects of society, from clothing to industry. The medicinal properties of plants were recognized early on in human cultures. From the mid-1900s, synthetic chemicals began to supplant plant-based remedies.

Pharmacognosy is the branch of pharmacology that focuses on medicines derived from natural sources. With massive globalization and industrialization, it is possible that much human knowledge of plants and their medicinal purposes will disappear with the cultures that fostered them. This is where ethnobotanists come in. To learn about and understand the use of plants in a particular culture, an ethnobotanist must bring in knowledge of plant life and an understanding and appreciation of diverse cultures and traditions. The Amazon forest is home to an incredible diversity of vegetation and is considered an untapped resource of medicinal plants; yet, both the ecosystem and its indigenous cultures are threatened with extinction.

To become an ethnobotanist, a person must acquire a broad knowledge of plant biology, ecology, and sociology. Not only are the plant specimens studied and collected, but also the stories, recipes, and traditions that are linked to them. For ethnobotanists, plants are not viewed solely as biological organisms to be studied in a laboratory, but as an integral part of human culture. The convergence of molecular biology, anthropology, and ecology make the field of ethnobotany a truly multidisciplinary science.

Biodiversity of Plants

Biodiversity ensures a resource for new food crops and medicines. Plant life balances ecosystems, protects watersheds, mitigates erosion, moderates our climate, and provides shelter for many animal species. Threats to plant diversity, however, come from many sources. The explosion of the human population, especially in tropical countries where birth rates are highest and economic development is in full swing, is leading to devastating human encroachment into forested areas. To feed the growing population, humans need to obtain arable land, so there has been and continues to be massive clearing of trees. The need for more energy to power larger cities and economic growth therein leads to the construction of dams, the consequent flooding of ecosystems, and increased emissions of pollutants. Other threats to tropical forests come from poachers, who log trees for their precious wood. Ebony and Brazilian rosewood, both on the endangered list, are examples of tree species driven almost to extinction by indiscriminate logging. This unfortunate practice continues unabated today largely due to lack of population control and political willpower.

The number of plant species becoming extinct is increasing at an alarming rate. Because ecosystems are in a delicate balance, and seed plants maintain close symbiotic relationships with animals—whether predators or pollinators—the disappearance of a single plant can lead to the extinction of connected animal species. A real and pressing issue is that many plant species have not yet been catalogued, and so their place in the ecosystem is unknown. These unknown species are threatened by logging, habitat destruction, and loss of pollinators. They may become extinct before we have the chance to begin to understand the possible impacts from their disappearance. Efforts to preserve biodiversity take several lines of action, from preserving heirloom seeds to barcoding species. **Heirloom seeds** come from plants that were traditionally grown in human populations, as opposed to the seeds used for large-scale agricultural production. **Barcoding** is a technique in which one or more short gene sequences, taken from a well-characterized portion of DNA found in most genomes, are used to identify a species through DNA analysis.

KEY TERMS

anther sac-like structure at the tip of the stamen in which pollen grains are produced
Anthophyta phylum to which angiosperms belong
barcoding molecular biology technique in which one or more short gene sequences taken from a well-characterized portion of the genome is used to identify a species
basal angiosperms a group of plants that probably branched off before the separation of monocots and eudicots
calyx whorl of sepals
carpel single unit of the pistil
conifer dominant phylum of gymnosperms with the greatest variety of trees
corolla collection of petals
cotyledon primitive leaf that develops in the zygote; monocots have one cotyledon, and dicots have two cotyledons
crop cultivated plant
cycad gymnosperm that grows in tropical climates and resembles a palm tree; member of the phylum Cycadophyta
dicot (also, eudicot) related group of angiosperms whose embryos possess two cotyledons
dioecious describes a species in which the male and female reproductive organs are carried on separate specimens
filament thin stalk that links the anther to the base of the flower
flower branches specialized for reproduction found in some seed-bearing plants, containing either specialized male or female organs or both male and female organs
fruit thickened tissue derived from ovary wall that protects the embryo after fertilization and facilitates seed dispersal
ginkgophyte gymnosperm with one extant species, the *Ginkgo biloba*: a tree with fan-shaped leaves
gnetophyte gymnosperm shrub with varied morphological features that produces vessel elements in its woody tissues; the phylum includes the genera *Ephedra, Gnetum*, and *Welwitschia*
gymnosperm seed plant with naked seeds (seeds exposed on modified leaves or in cones)
gynoecium (also, carpel) structure that constitutes the female reproductive organ
heirloom seed seed from a plant that was grown historically, but has not been used in modern agriculture on a large scale
herbaceous grass-like plant noticeable by the absence of woody tissue
herbivory consumption of plants by insects and other animals
integument layer of sporophyte tissue that surrounds the megasporangium, and later, the embryo
megasporocyte megaspore mother cell; larger spore that germinates into a female gametophyte in a heterosporous plant
microsporocyte smaller spore that produces a male gametophyte in a heterosporous plant
monocot related group of angiosperms that produce embryos with one cotyledon and pollen with a single ridge
monoecious describes a species in which the male and female reproductive organs are on the same plant
nectar liquid rich in sugars produced by flowers to attract animal pollinators
ovary chamber that contains and protects the ovule or female megasporangium
ovulate cone cone containing two ovules per scale
ovule female gametophyte
perianth part of the plant consisting of the calyx (sepals) and corolla (petals)
petal modified leaf interior to the sepals; colorful petals attract animal pollinators
pistil fused group of carpels
pollen grain structure containing the male gametophyte of the plant
pollen tube extension from the pollen grain that delivers sperm to the egg cell
pollination transfer of pollen from the anther to the stigma
progymnosperm transitional group of plants that resembled conifers because they produced wood, yet still reproduced like ferns
seed structure containing the embryo, storage tissue, and protective coat
sepal modified leaf that encloses the bud; outermost structure of a flower
spermatophyte seed plant; from the Greek *sperm* (seed) and *phyte* (plant)
stamen structure that contains the male reproductive organs
stigma uppermost structure of the carpel where pollen is deposited
strobilus plant structure with a tight arrangement of sporophylls around a central stalk, as seen in cones or flowers; the male strobilus produces pollen, and the female strobilus produces eggs
style long, thin structure that links the stigma to the ovary

CHAPTER SUMMARY

26.1 Evolution of Seed Plants

Seed plants appeared about 350 million years ago, during the Carboniferous period. Two major innovations were seeds and pollen. Seeds protect the embryo from desiccation and provide it with a store of nutrients to support the early growth of the sporophyte. Seeds are also equipped to delay germination until growth conditions are optimal. Pollen allows seed plants to reproduce in the absence of water. The gametophytes of seed plants shrank, while the sporophytes became prominent structures and the diploid stage became the longest phase of the life cycle.

In the gymnosperms, which appeared during the drier Permian period and became the dominant group during the Triassic, pollen was dispersed by wind, and their naked seeds developed in the sporophylls of a strobilus. Angiosperms bear both flowers and fruit. Flowers expand the possibilities for pollination, especially by insects, who have coevolved with the flowering plants. Fruits offer additional protection to the embryo during its development, and also assist with seed dispersal. Angiosperms appeared during the Mesozoic era and have become the dominant plant life in terrestrial habitats.

26.2 Gymnosperms

Gymnosperms are heterosporous seed plants that produce naked seeds. They appeared in the Paleozoic period and were the dominant plant life during the Mesozoic. Modern-day gymnosperms belong to four phyla. The largest phylum, Coniferophyta, is represented by conifers, the predominant plants at high altitude and latitude. Cycads (phylum Cycadophyta) resemble palm trees and grow in tropical climates. Ginkgophyta is represented today by a single species, *Ginkgo biloba*. The last phylum, Gnetophyta, is a diverse group of plants that produce vessel elements in their wood.

26.3 Angiosperms

Angiosperms are the dominant form of plant life in most terrestrial ecosystems, comprising about 90 percent of all plant species. Most crops and ornamental plants are angiosperms. Their success comes from two innovative structures that protect reproduction from variability in the environment: the flower and the fruit. Flowers were derived from modified leaves; their color and fragrance encourages species-specific pollination. The main parts of a flower are the sepals and petals, which protect the reproductive parts: the stamens and the carpels. The stamens produce the male gametes in pollen grains. The carpels contain the female gametes (the eggs inside the ovules), which are within the ovary of a carpel. The walls of the ovary thicken after fertilization, ripening into fruit that ensures dispersal by wind, water, or animals.

The angiosperm life cycle is dominated by the sporophyte stage. Double fertilization is an event unique to angiosperms. One sperm in the pollen fertilizes the egg, forming a diploid zygote, while the other combines with the two polar nuclei, forming a triploid cell that develops into a food storage tissue called the endosperm. Flowering plants are divided into two main groups, the monocots and eudicots, according to the number of cotyledons in the seedlings. Basal angiosperms belong to an older lineage than monocots and eudicots.

26.4 The Role of Seed Plants

Angiosperm diversity is due in part to multiple interactions with animals. Herbivory has favored the development of defense mechanisms in plants, and avoidance of those defense mechanisms in animals. Conversely, seed dispersal can be aided by animals that eat plant fruits. Pollination (the transfer of pollen to a carpel) is mainly carried out by wind and animals, and angiosperm fruits and seeds have evolved numerous adaptations to capture the wind or attract specific classes of animals.

Plants play a key role in ecosystems. They are a source of food and medicinal compounds, and provide raw materials for many industries. Rapid deforestation and industrialization, however, threaten plant biodiversity. In turn, this threatens the ecosystem.

VISUAL CONNECTION QUESTIONS

1. Figure 26.8 At what stage does the diploid zygote form?
   a. when the female cone begins to bud from the tree
   b. at fertilization
   c. when the seeds drop from the tree
   d. when the pollen tube begins to grow

2. Figure 26.15 If a flower lacked a megasporangium, what type of gamete would not form? If the flower lacked a microsporangium, what type of gamete would not form?

REVIEW QUESTIONS

3. Seed plants are _______.
   a. all homosporous
   b. mostly homosporous with some heterosporous
   c. mostly heterosporous with some homosporous
   d. all heterosporous

4. Besides the seed, what other major structure diminishes a plant's reliance on water for reproduction?
   a. flower
   b. fruit
   c. pollen
   d. spore

5. In which of the following geological periods would gymnosperms dominate the landscape?
   a. Carboniferous
   b. Permian
   c. Triassic
   d. Eocene (present)

6. Which of the following structures widens the geographic range of a species and is an agent of dispersal?
   a. seed
   b. flower
   c. leaf
   d. root

7. Which of the following traits characterizes gymnosperms?
   a. The plants carry exposed seeds on modified leaves.
   b. Reproductive structures are located in a flower.
   c. After fertilization, the ovary thickens and forms a fruit.
   d. The gametophyte is the longest phase of the life cycle.

8. Megasporocytes will eventually produce which of the following?
   a. pollen grain
   b. sporophytes
   c. male gametophytes
   d. female gametophytes

9. What is the ploidy of the following structures: gametophyte, seed, spore, sporophyte?
   a. $1n, 1n, 2n, 2n$
   b. $1n, 2n, 1n, 2n$
   c. $2n, 1n, 2n, 1n$
   d. $2n, 2n, 1n, 1n$

10. In the northern forests of Siberia, a tall tree is most likely a:
    a. conifer
    b. cycad
    c. *Ginkgo biloba*
    d. gnetophyte

11. Which of the following structures in a flower is not directly involved in reproduction?
    a. the style
    b. the stamen
    c. the sepal
    d. the anther

12. Pollen grains develop in which structure?
    a. the anther
    b. the stigma
    c. the filament
    d. the carpel

13. In the course of double fertilization, one sperm cell fuses with the egg and the second one fuses with _______.
    a. the synergids
    b. the polar nuclei of the center cell
    c. the egg as well
    d. the antipodal cells

14. Corn develops from a seedling with a single cotyledon, displays parallel veins on its leaves, and produces monosulcate pollen. It is most likely:
    a. a gymnosperm
    b. a monocot
    c. a eudicot
    d. a basal angiosperm

15. Which of the following plant structures is not a defense against herbivory?
    a. thorns
    b. spines
    c. nectar
    d. alkaloids

16. White and sweet-smelling flowers with abundant nectar are probably pollinated by
    a. bees and butterflies
    b. flies
    c. birds
    d. wind

17. Abundant and powdery pollen produced by small, indistinct flowers is probably transported by:
    a. bees and butterflies
    b. flies
    c. birds
    d. wind

18. Plants are a source of ________.
    a. food
    b. fuel
    c. medicine
    d. all of the above

CRITICAL THINKING QUESTIONS

19. The Cretaceous Period was marked by the increase in number and variety of angiosperms. Insects also diversified enormously during the same period. Can you propose the reason or reasons that could foster coevolution?
20. What role did the adaptations of seed and pollen play in the development and expansion of seed plants?
21. The Mediterranean landscape along the sea shore is dotted with pines and cypresses. The weather is not cold, and the trees grow at sea level. What evolutionary adaptation of conifers makes them suitable to the Mediterranean climate?
22. What are the four modern-day phyla of gymnosperms?
23. Some cycads are considered endangered species and their trade is severely restricted. Customs officials stop suspected smugglers who claim that the plants in their possession are palm trees, not cycads. How would a botanist distinguish between the two types of plants?
24. What are the two structures that allow angiosperms to be the dominant form of plant life in most terrestrial ecosystems?
25. Biosynthesis of nectar and nutrient-rich pollen is energetically very expensive for a plant. Yet, plants funnel large amounts of energy into animal pollination. What are the evolutionary advantages that offset the cost of attracting animal pollinators?
26. What is biodiversity and why is it important to an ecosystem?

CHAPTER 27
Introduction to Animal Diversity

Figure 27.1 The leaf chameleon (*Brookesia micra*) was discovered in northern Madagascar in 2012. At just over one inch long, it is the smallest known chameleon. (credit: modification of work by Frank Glaw, et al., PLOS)

INTRODUCTION Animal evolution began in the ocean over 600 million years ago with tiny creatures that probably do not resemble any living organism today. Since then, animals have evolved into a highly diverse kingdom. Although over one million extant (currently living) species of animals have been identified, scientists are continually discovering more species as they explore ecosystems around the world. The number of extant species is estimated to be between 3 and 30 million.

But what is an animal? While we can easily identify dogs, birds, fish, spiders, and worms as animals, other organisms, such as corals and sponges, are not as easy to classify. Animals vary in complexity—from sea sponges to crickets to chimpanzees—and scientists are faced with the difficult task of classifying them within a unified system. They must identify traits that are common to all animals as well as traits that can be used to distinguish among related groups of animals. The animal classification system characterizes animals based on their anatomy, morphology, evolutionary history, features of embryological development, and genetic makeup. This classification scheme is constantly developing as new information about species arises. Understanding and classifying the great variety of living species help us better understand how to conserve the diversity of life on earth.

Chapter Outline

27.1 Features of the Animal Kingdom

By the end of this section, you will be able to do the following:

- List the features that distinguish the kingdom Animalia from other kingdoms
- Explain the processes of animal reproduction and embryonic development
- Describe the roles that Hox genes play in development

Two different groups within the Domain Eukaryota have produced complex multicellular organisms: The plants arose within the Archaeplastida, whereas the animals (and their close relatives, the fungi) arose within the Opisthokonta. However, plants and animals not only have different life styles, they also have different cellular histories as eukaryotes. The opisthokonts share the possession of a single posterior flagellum in flagellated cells, e.g., sperm cells.

Most animals also share other features that distinguish them from organisms in other kingdoms. All animals require a source of food and are therefore *heterotrophic*, ingesting other living or dead organisms. This feature distinguishes them from *autotrophic* organisms, such as most plants, which synthesize their own nutrients through photosynthesis. As heterotrophs, animals may be carnivores, herbivores, omnivores, or parasites (Figure 27.2a,b). As with plants, almost all animals have a complex tissue structure with differentiated and specialized tissues. The necessity to collect food has made most animals motile, at least during certain life stages. The typical life cycle in animals is diplontic (like you, the diploid state is multicellular, whereas the haploid state is gametic, such as sperm or egg). We should note that the alternation of generations characteristic of the land plants is typically not found in animals. In animals whose life histories include several to multiple body forms (e.g., insect larvae or the medusae of some Cnidarians), all body forms are diploid. Animal embryos pass through a series of developmental stages that establish a determined and fixed body plan. The **body plan** refers to the morphology of an animal, determined by developmental cues.

(a) (b)

Figure 27.2 Heterotrophy. All animals are heterotrophs and thus derive energy from a variety of food sources. The (a) black bear is an omnivore, eating both plants and animals. The (b) heartworm *Dirofilaria immitis* is a parasite that derives energy from its hosts. It spends its larval stage in mosquitoes and its adult stage infesting the heart of dogs and other mammals, as shown here. (credit a: modification of work by USDA Forest Service; credit b: modification of work by Clyde Robinson)

Complex Tissue Structure

Many of the specialized tissues of animals are associated with the requirements and hazards of seeking and processing food. This explains why animals typically have evolved special structures associated with specific methods of food capture and complex digestive systems supported by accessory organs. Sensory structures help animals navigate their environment, detect food sources (and avoid becoming a food source for other animals!). Movement is driven by muscle tissue attached to supportive structures like bone or chitin, and is coordinated by neural communication. Animal cells may also have unique structures for intercellular communication (such as gap junctions). The evolution of nerve tissues and muscle tissues has resulted in animals' unique ability to rapidly sense and respond to changes in their environment. This allows animals to survive in environments where they must compete with other species to meet their nutritional demands.

The tissues of animals differ from those of the other major multicellular eukaryotes, plants and fungi, because their cells don't have cell walls. However, cells of animal tissues may be embedded in an extracellular matrix (e.g., mature bone cells reside within a mineralized organic matrix secreted by the cells). In vertebrates, bone tissue is a type of connective tissue that supports the entire body structure. The complex bodies and activities of vertebrates demand such supportive tissues. Epithelial tissues cover and protect both external and internal body surfaces, and may also have secretory functions. Epithelial tissues include the epidermis of the integument, the lining of the digestive tract and trachea, as well as the layers of cells that make up the ducts of the liver and glands of advanced animals, for example. The different types of tissues in true animals are responsible for carrying out specific functions for the organism. This differentiation and specialization of tissues is part of what allows for such incredible animal diversity.

Just as there are multiple ways to be a eukaryote, there are multiple ways to be a multicellular animal. The animal kingdom is currently divided into five monophyletic clades: Parazoa or Porifera (sponges), Placozoa (tiny parasitic creatures that resemble multicellular amoebae), Cnidaria (jellyfish and their relatives), Ctenophora (the comb jellies), and Bilateria (all other animals). The Placozoa ("flat animal") and Parazoa ("beside animal") do not have specialized tissues derived from germ layers of the embryo; although they do possess specialized cells that act *functionally* like tissues. The Placozoa have only four cell types, while the sponges have nearly two dozen. The three other clades do include animals with specialized tissues derived from the germ layers of the embryo. In spite of their superficial similarity to Cnidarian medusae, recent molecular studies indicate that the Ctenophores are only distantly related to the Cnidarians, which together with the Bilateria constitute the Eumetazoa ("true animals"). When we think of animals, we usually think of Eumetazoa, since most animals fall into this category.

LINK TO LEARNING

Watch a presentation (http://openstax.org/l/saving_life) by biologist E.O. Wilson on the importance of diversity.

Animal Reproduction and Development

Most animals are diploid organisms, meaning that their body (somatic) cells are diploid and haploid reproductive (gamete) cells are produced through meiosis. Some exceptions exist: for example, in bees, wasps, and ants, the male is haploid because it develops from unfertilized eggs. Most animals undergo sexual reproduction. However, a few groups, such as cnidarians, flatworms, and roundworms, may also undergo asexual reproduction, in which offspring originate from part of the parental body.

Processes of Animal Reproduction and Embryonic Development

During sexual reproduction, the haploid gametes of the male and female individuals of a species combine in a process called fertilization. Typically, both male and female gametes are required: the small, motile male sperm fertilizes the typically much larger, sessile female egg. This process produces a diploid fertilized egg called a zygote.

Some animal species—including sea stars and sea anemones—are capable of asexual reproduction. The most common forms of asexual reproduction for stationary aquatic animals include *budding* and *fragmentation*, where part of a parent individual can separate and grow into a new individual. This type of asexual reproduction produces genetically identical offspring, which would appear to be disadvantageous from the perspective of evolutionary adaptability, simply because of the potential buildup of deleterious mutations.

In contrast, a form of uniparental reproduction found in some insects and a few vertebrates is called parthenogenesis (or "virgin beginning"). In this case, progeny develop from a gamete, but without fertilization. Because of the nutrients stored in eggs, only females produce parthenogenetic offspring. In some insects, unfertilized eggs develop into new male offspring. This type of sex determination is called haplodiploidy, since females are diploid (with both maternal and paternal chromosomes) and males are haploid (with only maternal chromosomes). A few vertebrates, e.g., some fish, turkeys, rattlesnakes, and whiptail lizards, are also capable of parthenogenesis. In the case of turkeys and rattlesnakes, parthenogenetically reproducing females also produce only male offspring, but not because the males are haploid. In birds and rattlesnakes, the female is the heterogametic (ZW) sex, so the only surviving progeny of post-meiotic parthenogenesis would be ZZ males. In the whiptail lizards, on the other hand, only female progeny are produced by parthenogenesis. These animals may not be identical to their parent, although they have only maternal chromosomes. However, for animals that are limited in their access to mates, uniparental reproduction can ensure genetic propagation.

In animals, the zygote progresses through a series of developmental stages, during which primary **germ layers** (ectoderm, endoderm, and mesoderm) are established and reorganize to form an embryo. During this process, animal tissues begin to specialize and organize into organs and organ systems, determining their future morphology and physiology.

Animal development begins with **cleavage**, a series of mitotic cell divisions, of the zygote (Figure 27.3). Cleavage differs from somatic cell division in that the egg is subdivided by successive cleavages into smaller and smaller cells, with *no* actual cell growth. The cells resulting from subdivision of the material of the egg in this way are called **blastomeres**. Three cell divisions transform the single-celled zygote into an eight-celled structure. After further cell division and rearrangement of existing cells, a solid morula is formed, followed by a hollow structure called a **blastula**. The blastula is hollow only in invertebrates whose eggs have relatively small amounts of yolk. In very yolky eggs of vertebrates, the yolk remains undivided, with most cells forming an embryonic layer on the surface of the yolk (imagine a chicken embryo growing over the egg's yolk), which serve as food for the developing embryo.

Further cell division and cellular rearrangement leads to a process called gastrulation. **Gastrulation** results in two important events: the formation of the primitive gut (archenteron) or digestive cavity, and the formation of the embryonic germ layers, as we have discussed above. These germ layers are programmed to develop into certain tissue types, organs, and organ systems during a process called **organogenesis**. *Diploblastic* organisms have two germ layers, endoderm and ectoderm. Endoderm forms the wall of the digestive tract, and ectoderm covers the surface of the animal. In *triploblastic* animals, a third layer forms: mesoderm, which differentiates into various structures between the ectoderm and endoderm, including the lining of the body cavity.

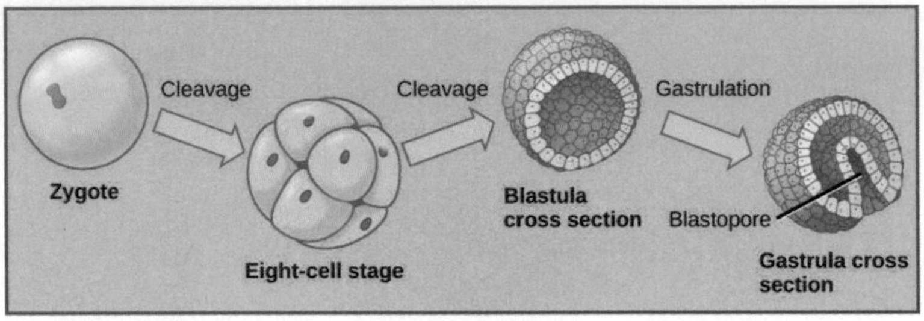

Figure 27.3 Development of a simple embryo. During embryonic development, the zygote undergoes a series of mitotic cell divisions, or cleavages, that subdivide the egg into smaller and smaller blastomeres. Note that the 8-cell stage and the blastula are about the same size as the original zygote. In many invertebrates, the blastula consists of a single layer of cells around a hollow space. During a process called gastrulation, the cells from the blastula move inward on one side to form an inner cavity. This inner cavity becomes the primitive gut (archenteron) of the gastrula ("little gut") stage. The opening into this cavity is called the *blastopore*, and in some invertebrates it is destined to form the mouth.

Some animals produce larval forms that are different from the adult. In insects with *incomplete metamorphosis*, such as grasshoppers, the young resemble wingless adults, but gradually produce larger and larger wing buds during successive molts, until finally producing functional wings and sex organs during the last molt. Other animals, such as some insects and echinoderms, undergo complete metamorphosis in which the embryo develops into one or more feeding larval stages that may differ greatly in structure and function from the adult (Figure 27.4). The adult body then develops from one or more regions of larval tissue. For animals with complete metamorphosis, the larva and the adult may have different diets, limiting competition for food between them. Regardless of whether a species undergoes complete or incomplete metamorphosis, the series of developmental stages of the embryo remains largely the same for most members of the animal kingdom.

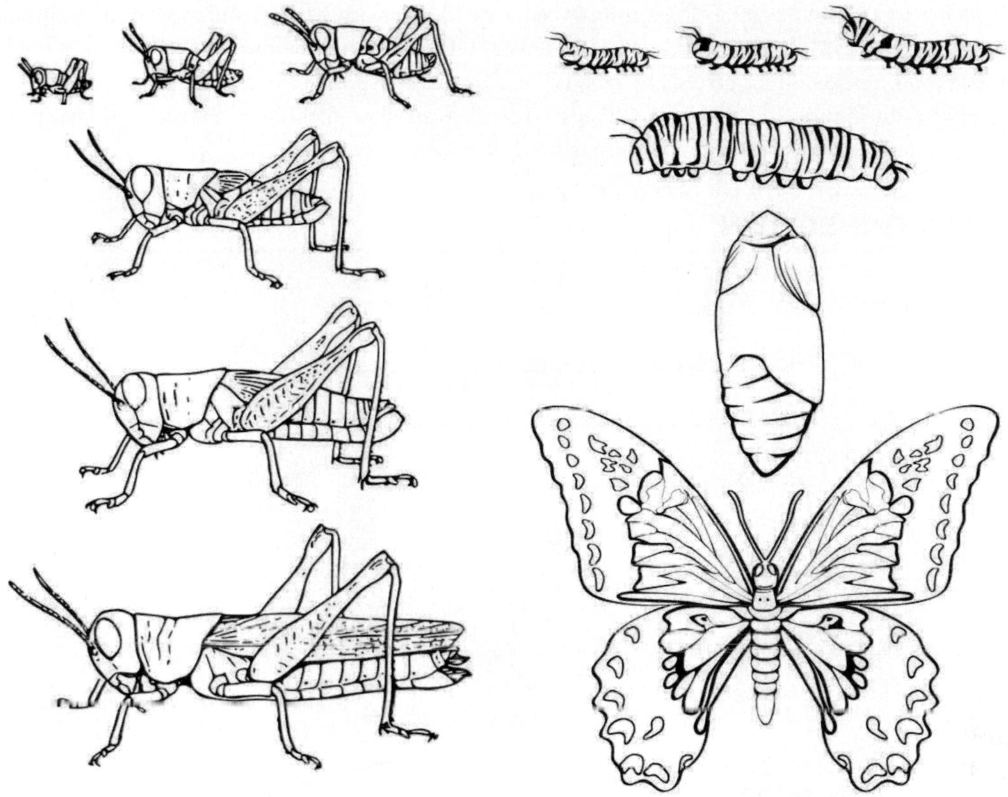

Figure 27.4 Insect metamorphosis. (a) The grasshopper undergoes incomplete metamorphosis. (b) The butterfly undergoes complete metamorphosis. (credit: S.E. Snodgrass, USDA)

LINK TO LEARNING

Watch the following video (http://openstax.org/l/embryo_evol) to see how human embryonic development (after the blastula and gastrula stages of development) reflects evolution.

The Role of Homeobox (*Hox*) Genes in Animal Development

Since the early nineteenth century, scientists have observed that many animals, from the very simple to the complex, shared similar embryonic morphology and development. Surprisingly, a human embryo and a frog embryo, at a certain stage of embryonic development, look remarkably alike! For a long time, scientists did not understand why so many animal species looked similar during embryonic development but were very different as adults. They wondered what dictated the developmental direction that a fly, mouse, frog, or human embryo would take. Near the end of the twentieth century, a particular class of genes was discovered that had this very job. These genes that determine animal structure are called "homeotic genes," and they contain DNA sequences called homeoboxes. Genes with homeoboxes encode protein transcription factors. One group of animal genes containing homeobox sequences is specifically referred to as ***Hox* genes**. This cluster of genes is responsible for determining the general body plan, such as the number of body segments of an animal, the number and placement of appendages, and animal head-tail directionality. The first *Hox* genes to be sequenced were those from the fruit fly (*Drosophila melanogaster*). A single *Hox* mutation in the fruit fly can result in an extra pair of wings or even legs growing from the head in place of antennae (this is because antennae and legs are embryologic homologous structures and their appearance as antennae or legs is dictated by their origination within specific body segments of the head and thorax during development). Now, *Hox* genes are known from virtually all other animals as well.

While there are a great many genes that play roles in the morphological development of an animal, including other homeobox-containing genes, what makes *Hox* genes so powerful is that they serve as "master control genes" that can turn on or off large numbers of other genes. *Hox* genes do this by encoding transcription factors that control the expression of numerous other genes. *Hox* genes are homologous across the animal kingdom, that is, the genetic sequences of *Hox* genes and their positions on chromosomes are remarkably similar across most animals because of their presence in a common ancestor, from worms to flies, mice, and humans (Figure 27.5). In addition, the order of the genes reflects the anterior-posterior axis of the animal's body. One of the contributions to increased animal body complexity is that *Hox* genes have undergone at least two and perhaps as many as four duplication events during animal evolution, with the additional genes allowing for more complex body types to evolve. All vertebrates have four (or more) sets of *Hox* genes, while invertebrates have only one set.

VISUAL CONNECTION

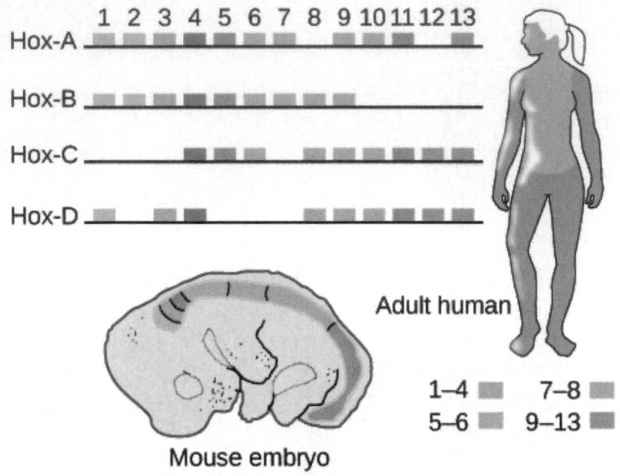

Figure 27.5 *Hox* genes. *Hox* genes are *highly conserved genes* encoding transcription factors that determine the course of embryonic development in animals. In vertebrates, the genes have been duplicated into four clusters on *different chromosomes*: *Hox-A*, *Hox-B*, *Hox-C*, and *Hox-D*. Genes within these clusters are expressed in certain body segments at certain stages of development. Shown here is the homology between *Hox* genes in mice and humans. Note how *Hox* gene expression, as indicated with orange, pink, blue, and green shading, occurs in the same body segments in both the mouse and the human. While at least one copy of each Hox gene is present in humans and other vertebrates, some *Hox* genes are missing in some chromosomal sets.

If a *Hox 13* gene in a mouse was replaced with a *Hox 1* gene, how might this alter animal development?

Two of the five clades within the animal kingdom do *not* have *Hox* genes: the Ctenophora and the Porifera. In spite of the superficial similarities between the Cnidaria and the Ctenophora, the Cnidaria have a number of *Hox* genes, but the Ctenophora have none. The absence of *Hox* genes from the ctenophores has led to the suggestion that they might be "basal" animals, in spite of their tissue differentiation. Ironically, the Placozoa, which have only a few cell types, do have at least one *Hox* gene. The presence of a *Hox* gene in the Placozoa, in addition to similarities in the genomic organization of the Placozoa, Cnidaria and Bilateria, has led to the inclusion of the three groups in a "Parahoxozoa" clade. However, we should note that at this time the reclassification of the Animal Kingdom is still tentative and requires much more study.

27.2 Features Used to Classify Animals

By the end of this section, you will be able to do the following:

- Explain the differences in animal body plans that support basic animal classification
- Compare and contrast the embryonic development of protostomes and deuterostomes

Scientists have developed a classification scheme that categorizes all members of the animal kingdom, although there are exceptions to most "rules" governing animal classification (Figure 27.6). Animals have been traditionally classified according to two characteristics: body plan and developmental pathway. The major feature of the body plan is its symmetry: how the body parts are distributed along the major body axis. Symmetrical animals can be divided into roughly equivalent halves along at least one axis. Developmental characteristics include the number of germ tissue layers formed during development, the origin of the mouth and anus, the presence or absence of an internal body cavity, and other features of embryological development, such as larval types or whether or not periods of growth are interspersed with molting.

VISUAL CONNECTION

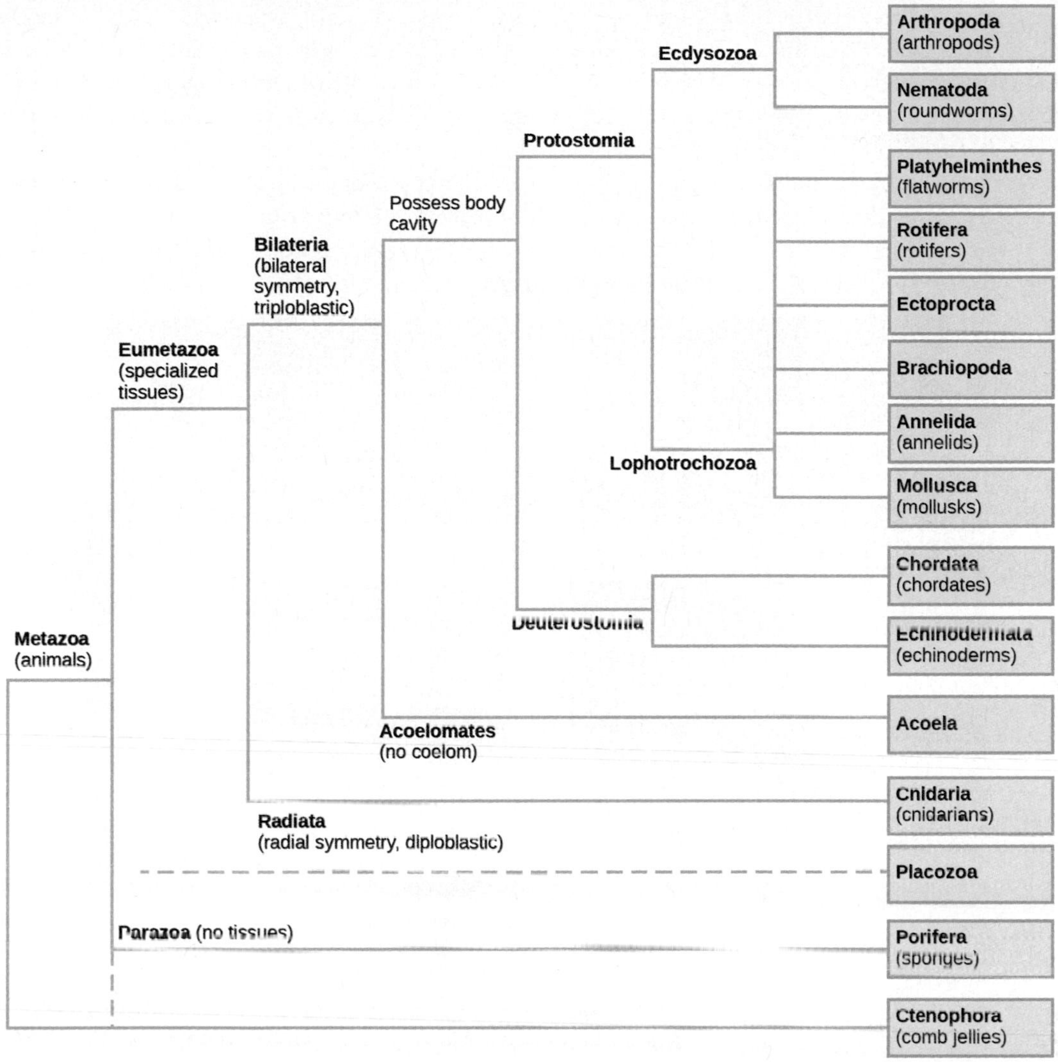

Figure 27.6 Animal phylogeny. The phylogenetic tree of animals is based on morphological, fossil, and genetic evidence. The Ctenophora and Porifera are both considered to be basal because of the absence of Hox genes in this group, but how they are related to the "Parahoxozoa" (Placozoa + Eumetazoa) or to each other, continues to be a matter of debate.

Which of the following statements is false?

a. Eumetazoans have specialized tissues and parazoans don't.
b. Lophotrochozoa and Ecdysozoa are both Bilataria.
c. Acoela and Cnidaria both possess radial symmetry.
d. Arthropods are more closely related to nematodes than they are to annelids.

Animal Characterization Based on Body Symmetry

At a very basic level of classification, true animals can be largely divided into three groups based on the type of symmetry of their body plan: radially symmetrical, bilaterally symmetrical, and asymmetrical. Asymmetry is seen in two modern clades, the Parazoa (Figure 27.7a) and Placozoa. (Although we should note that the ancestral fossils of the Parazoa apparently exhibited bilateral symmetry.) One clade, the Cnidaria (Figure 27.7b,c), exhibits radial or biradial symmetry: Ctenophores have rotational symmetry (Figure 27.7e). Bilateral symmetry is seen in the largest of the clades, the Bilateria (Figure 27.7d); however the Echinodermata are bilateral as larvae and metamorphose secondarily into radial adults. All types of symmetry are well suited to meet the unique demands of a particular animal's lifestyle.

Radial symmetry is the arrangement of body parts around a central axis, as is seen in a bicycle wheel or pie. It results in animals having top and bottom surfaces but no left and right sides, nor front or back. If a radially symmetrical animal is divided in any direction along the oral/aboral axis (the side with a mouth is "oral side," and the side without a mouth is the "aboral side"), the two halves will be mirror images. This form of symmetry marks the body plans of many animals in the phyla Cnidaria, including jellyfish and adult sea anemones (Figure 27.7b, c). Radial symmetry equips these sea creatures (which may be sedentary or only capable of slow movement or floating) to experience the environment equally from all directions. Bilaterally symmetrical animals, like butterflies (Figure 27.7d) have only a single plane along which the body can be divided into equivalent halves. The Ctenophora (Figure 27.7e), although they look similar to jellyfish, are considered to have rotational symmetry rather than radial or biradial symmetry because division of the body into two halves along the oral/aboral axis divides them into two copies of the same half, with one copy rotated 180°, rather than two mirror images.

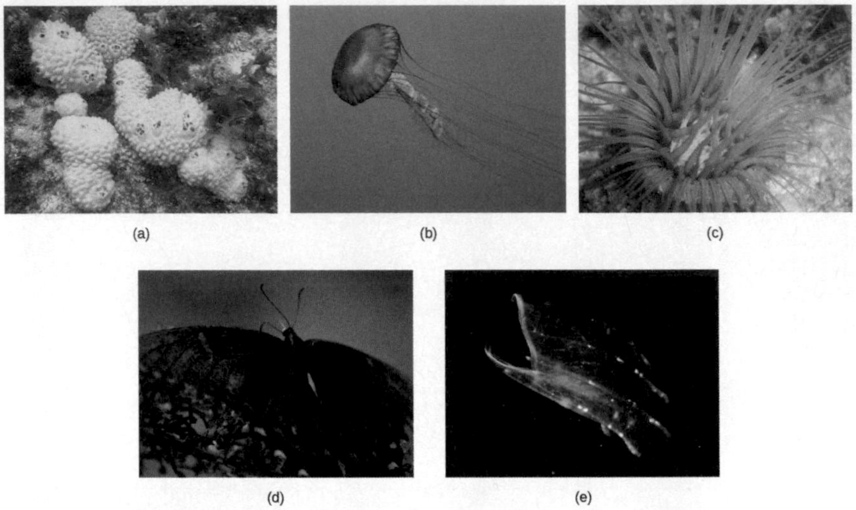

Figure 27.7 Symmetry in animals. The (a) sponge is asymmetrical. The (b) jellyfish and (c) anemone are radially symmetrical, the (d) butterfly is bilaterally symmetrical. Rotational symmetry (e) is seen in the ctenophore *Beroe,* shown swimming open-mouthed. (credit a: modification of work by Andrew Turner; credit b: modification of work by Robert Freiburger; credit c: modification of work by Samuel Chow; credit d: modification of work by Cory Zanker; credit e: modification of work by NOAA)

Bilateral symmetry involves the division of the animal through a midsagittal plane, resulting in two superficially mirror images, right and left halves, such as those of a butterfly (Figure 27.7d), crab, or human body. Animals with bilateral symmetry have a "head" and "tail" (anterior vs. posterior), front and back (dorsal vs. ventral), and right and left sides (Figure 27.8). All Eumetazoa except those with secondary radial symmetry are bilaterally symmetrical. The evolution of bilateral symmetry that allowed for the formation of anterior and posterior (head and tail) ends promoted a phenomenon called cephalization, which refers to the collection of an organized nervous system at the animal's anterior end. In contrast to radial symmetry, which is best suited for stationary or limited-motion lifestyles, bilateral symmetry allows for streamlined and directional motion. In evolutionary terms, this simple form of symmetry promoted active and controlled directional mobility and increased sophistication of resource-seeking and predator-prey relationships.

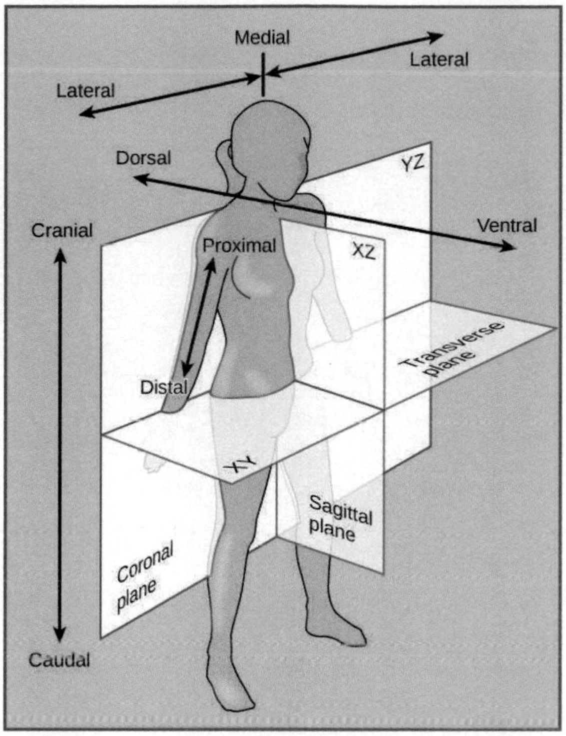

Figure 27.8 Bilateral symmetry. The bilaterally symmetrical human body can be divided by several planes.

Animals in the phylum Echinodermata (such as sea stars, sand dollars, and sea urchins) display modified radial symmetry as adults, but as we have noted, their larval stages (such as the bipinnaria) initially exhibit bilateral symmetry until they metamorphose in animals with radial symmetry (this is termed secondary radial symmetry). Echinoderms evolved from bilaterally symmetrical animals; thus, they are classified as bilaterally symmetrical.

LINK TO LEARNING

Watch this video to see a quick sketch of the different types of body symmetry.

Click to view content (https://www.openstax.org/l/symmetry)

Animal Characterization Based on Features of Embryological Development

Most animal species undergo a separation of tissues into germ layers during embryonic development. Recall that these germ layers are formed during *gastrulation*, and that each germ layer typically gives rise to specific types of embryonic tissues and organs. Animals develop either two or three embryonic germ layers (Figure 27.9). The animals that display radial, biradial, or rotational symmetry develop two germ layers, an inner layer (*endoderm* or *mesendoderm*) and an outer layer (*ectoderm*). These animals are called **diploblasts**, and have a nonliving middle layer between the endoderm and ectoderm (although individual cells may be distributed through this middle layer, there is no *coherent* third layer of tissue). The four clades considered to be diploblastic have different levels of complexity and different developmental pathways, although there is little information about development in Placozoa. More complex animals (usually those with bilateral symmetry) develop three tissue layers: an inner layer (endoderm), an outer layer (ectoderm), and a middle layer (mesoderm). Animals with three tissue layers are called **triploblasts**.

VISUAL CONNECTION

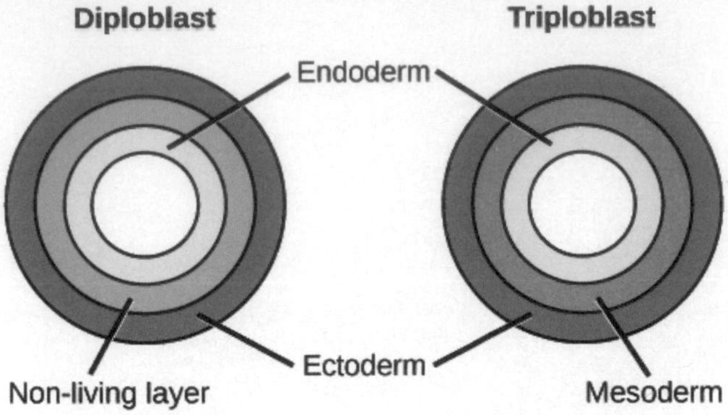

Figure 27.9 Diploblastic and triploblastic embryos. During embryogenesis, diploblasts develop two embryonic germ layers: an ectoderm and an endoderm or mesendoderm. Triploblasts develop a third layer—the mesoderm—which arises from mesendoderm and resides between the endoderm and ectoderm.

Which of the following statements about diploblasts and triploblasts is false?

a. Animals that display only radial symmetry during their lifespans are diploblasts.
b. Animals that display bilateral symmetry are triploblasts.
c. The endoderm gives rise to the lining of the digestive tract and the respiratory tract.
d. The mesoderm gives rise to the central nervous system.

Each of the three germ layers is programmed to give rise to specific body tissues and organs, although there are variations on these themes. Generally speaking, the endoderm gives rise to the lining of the digestive tract (including the stomach, intestines, liver, and pancreas), as well as to the lining of the trachea, bronchi, and lungs of the respiratory tract, along with a few other structures. The ectoderm develops into the outer epithelial covering of the body surface, the central nervous system, and a few other structures. The mesoderm is the third germ layer; it forms between the endoderm and ectoderm in triploblasts. This germ layer gives rise to all specialized muscle tissues (including the cardiac tissues and muscles of the intestines), connective tissues such as the skeleton and blood cells, and most other visceral organs such as the kidneys and the spleen. Diploblastic animals may have cell types that serve multiple functions, such as epitheliomuscular cells, which serve as a covering as well as contractile cells.

Presence or Absence of a Coelom

Further subdivision of animals with three germ layers (triploblasts) results in the separation of animals that may develop an internal *body cavity* derived from mesoderm, called a **coelom**, and those that do not. This epithelial cell-lined *coelomic cavity*, usually filled with fluid, lies between the visceral organs and the body wall. It houses many organs such as the digestive, urinary, and reproductive systems, the heart and lungs, and also contains the major arteries and veins of the circulatory system. In mammals, the body cavity is divided into the thoracic cavity, which houses the heart and lungs, and the abdominal cavity, which houses the digestive organs. In the thoracic cavity further subdivision produces the pleural cavity, which provides space for the lungs to expand during breathing, and the pericardial cavity, which provides room for movements of the heart. The evolution of the coelom is associated with many functional advantages. For example, the coelom provides cushioning and shock absorption for the major organ systems that it encloses. In addition, organs housed within the coelom can grow and move freely, which promotes optimal organ development and placement. The coelom also provides space for the diffusion of gases and nutrients, as well as body flexibility, promoting improved animal motility.

Triploblasts that do not develop a coelom are called **acoelomates**, and their mesoderm region is completely filled with tissue, although they do still have a gut cavity. Examples of acoelomates include animals in the phylum Platyhelminthes, also known as flatworms. Animals with a true coelom are called **eucoelomates** (or coelomates) (Figure 27.10). In such cases, a true coelom arises entirely within the mesoderm germ layer and is lined by an epithelial membrane. This membrane also lines the organs within the coelom, connecting and holding them in position while allowing them some freedom of movement. Annelids, mollusks, arthropods, echinoderms, and chordates are all eucoelomates. A third group of triploblasts has a slightly different coelom lined partly by mesoderm and partly by endoderm. Although still functionally a coelom, these are considered "false" coeloms, and so we call these animals **pseudocoelomates**. The phylum Nematoda (roundworms) is an example of a pseudocoelomate. True coelomates can be further characterized based on other features of their early embryological development.

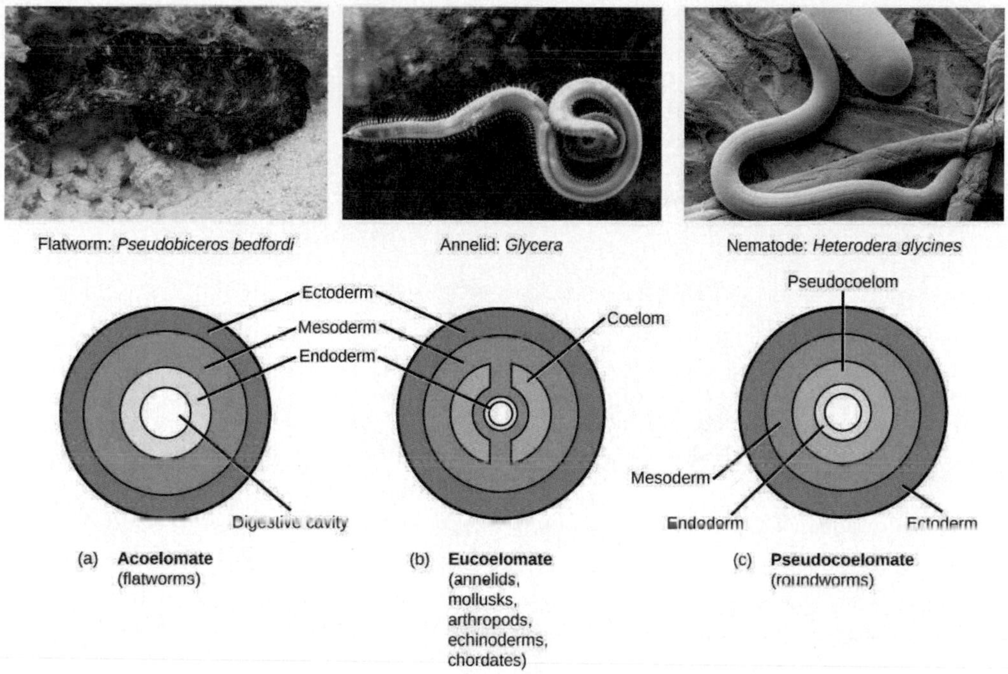

Figure 27.10 Body cavities. Triploblasts may be (a) acoelomates, (b) eucoelomates, or (c) pseudocoelomates. Acoelomates have no body cavity. Eucoelomates have a body cavity within the mesoderm, called a coelom, in which both the gut and the body wall are lined with mesoderm. Pseudocoelomates also have a body cavity, but only the body wall is lined with mesoderm. (credit a: modification of work by Jan Derk; credit b: modification of work by NOAA; credit c: modification of work by USDA, ARS)

Embryonic Development of the Mouth

Bilaterally symmetrical, tribloblastic eucoelomates can be further divided into two groups based on differences in the origin of the mouth. When the primitive gut forms, the opening that first connects the gut cavity to the outside of the embryo is called the blastopore. Most animals have openings at both ends of the gut: mouth at one end and anus at the other. One of these openings will develop at or near the site of the **blastopore**. In **Protostomes** ("mouth first"), the mouth develops at the blastopore (Figure 27.11). In **Deuterostomes** ("mouth second"), the mouth develops at the other end of the gut (Figure 27.11) and the anus develops at the site of the blastopore. Protostomes include arthropods, mollusks, and annelids. Deuterostomes include more complex animals such as chordates but also some "simple" animals such as echinoderms. Recent evidence has challenged this simple view of the relationship between the location of the blastopore and the formation of the mouth, however, and the theory remains under debate. Nevertheless, these details of mouth and anus formation reflect *general* differences in the organization of protostome and deuterostome embryos, which are also expressed in other developmental features.

One of these differences between protostomes and deuterostomes is the method of coelom formation, beginning from the gastrula stage. Since body cavity formation tends to accompany the formation of the mesoderm, the mesoderm of protostomes and deuterostomes forms differently. The coelom of most protostomes is formed through a process called **schizocoely**. The mesoderm in these organisms is usually the product of specific *blastomeres*, which migrate into the interior of the embryo and form two clumps of mesodermal tissue. Within each clump, cavities develop and merge to form the hollow opening of the coelom. Deuterostomes differ in that their coelom forms through a process called **enterocoely**. Here, the mesoderm develops as pouches that are pinched off from the endoderm tissue. These pouches eventually fuse and expand to fill the space between the gut and the body wall, giving rise to the coelom.

Another difference in organization of protostome and deuterostome embryos is expressed during cleavage. Protostomes undergo **spiral cleavage**, meaning that the cells of one pole of the embryo are rotated, and thus misaligned, with respect to the cells of the opposite pole. This is due to the oblique angle of cleavage relative to the two poles of the embryo. Deuterostomes undergo **radial cleavage**, where the cleavage axes are either parallel or perpendicular to the polar axis, resulting in the parallel (up-and-down) alignment of the cells between the two poles.

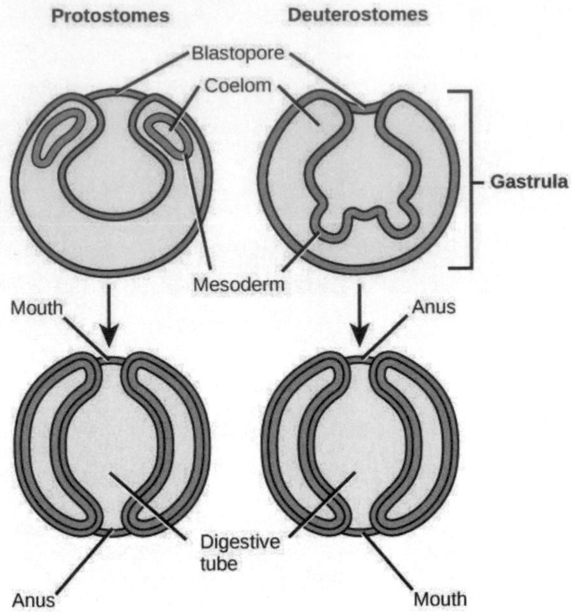

Figure 27.11 Protostomes and deuterostomes. Eucoelomates can be divided into two groups based on their early embryonic development. In protostomes, the mouth forms at or near the site of the blastopore and the body cavity forms by splitting the mesodermal mass during the process of schizocoely. In deuterostomes, the mouth forms at a site opposite the blastopore end of the embryo and the mesoderm pinches off to form the coelom during the process of enterocoely.

A second distinction between the types of cleavage in protostomes and deuterostomes relates to the fate of the resultant *blastomeres* (cells produced by cleavage). In addition to spiral cleavage, protostomes also undergo **determinate cleavage**. This means that even at this early stage, the developmental fate of each embryonic cell is already determined. A given cell does not have the ability to develop into any cell type other than its original destination. Removal of a blastomere from an embryo with determinate cleavage can result in missing structures, and embryos that fail to develop. In contrast, deuterostomes undergo **indeterminate cleavage**, in which cells are not yet fully committed at this early stage to develop into specific cell types. Removal of individual blastomeres from these embryos does not result in the loss of embryonic structures. In fact, twins (clones) can be produced as a result from blastomeres that have been separated from the original mass of blastomere cells. Unlike protostomes, however, if some blastomeres are damaged during embryogenesis, adjacent cells are able to compensate for the missing cells, and the embryo is not damaged. These cells are referred to as undetermined cells. This characteristic of deuterostomes is reflected in the existence of familiar *embryonic stem cells*, which have the ability to develop into any cell type until their fate is programmed at a later developmental stage.

EVOLUTION CONNECTION

The Evolution of the Coelom

One of the first steps in the classification of animals is to examine the animal's body. One structure that is used in classification of animals is the body cavity or coelom. The body cavity develops *within* the mesoderm, so only triploblastic animals can have body cavities. Therefore body cavities are found only within the Bilateria. In other animal clades, the gut is either close to the body wall or separated from it by a jelly-like material. The body cavity is important for two reasons. Fluid within the body cavity protects the organs from shock and compression. In addition, since in triploblastic embryos, most muscle, connective tissue, and blood vessels develop from mesoderm, these tissues developing within the lining of the body cavity can reinforce the gut and body wall, aid in motility, and efficiently circulate nutrients.

To recap what we have discussed above, animals that do not have a coelom are called *acoelomates*. The major acoelomate group in the Bilateria is the flatworms, including both free-living and parasitic forms such as tapeworms. In these animals, mesenchyme fills the space between the gut and the body wall. Although two layers of muscle are found just under the epidermis, there is no muscle or other mesodermal tissue around the gut. Flatworms rely on passive diffusion for nutrient transport across their body.

In *pseudocoelomates*, there is a body cavity between the gut and the body wall, but only the body wall has mesodermal tissue. In these animals, the mesoderm forms, but does not develop cavities within it. Major pseudocoelomate phyla are the rotifers and nematodes. Animals that have a true coelom are called *eucoelomates*; all vertebrates, as well as molluscs, annelids, arthropods, and echinoderms, are eucoelomates. The coelom develops within the mesoderm during embryogenesis. Of the major bilaterian phyla, the molluscs, annelids, and arthropods are *schizocoels*, in which the mesoderm splits to form the body cavity, while the echinoderms and chordates are *enterocoels*, in which the mesoderm forms as two or more buds off of the gut. These buds separate from the gut and coalesce to form the body cavity. In the vertebrates, mammals have a subdivided body cavity, with the thoracic cavity separated from the abdominal cavity. The pseudocoelomates may have had eucoelomate ancestors and may have lost their ability to form a complete coelom through genetic mutations. Thus, this step in early embryogenesis—the formation of the coelom—has had a large evolutionary impact on the various species of the animal kingdom.

27.3 Animal Phylogeny

By the end of this section, you will be able to do the following:

- Interpret the metazoan phylogenetic tree
- Describe the types of data that scientists use to construct and revise animal phylogeny
- List some of the relationships within the modern phylogenetic tree that have been discovered as a result of modern molecular data

Biologists strive to understand the evolutionary history and relationships of members of the animal kingdom, and all of life, for that matter. The study of *phylogeny* (the branching sequence of evolution) aims to determine the evolutionary relationships between phyla. Currently, most biologists divide the animal kingdom into 35 to 40 phyla. Scientists develop phylogenetic trees, which serve as *hypotheses* about which species have evolved from which ancestors.

Recall that until recently, only morphological characteristics and the fossil record were used to determine phylogenetic relationships among animals. Scientific understanding of the distinctions and hierarchies between anatomical characteristics provided much of this knowledge. Used alone, however, this information can be misleading. Morphological characteristics (such as skin color, body shape, etc.) may evolve multiple times, and independently, through evolutionary history. Analogous characteristics may appear similar between animals, but their underlying evolution may be very different. With the advancement of *molecular technologies*, modern phylogenetics is now informed by genetic and molecular analyses, in addition to traditional morphological and fossil data. With a growing understanding of genetics, the animal evolutionary tree has changed substantially and continues to change as new DNA and RNA analyses are performed on additional animal species.

Constructing an Animal Phylogenetic Tree

The current understanding of evolutionary relationships among animal, or **Metazoa**, phyla begins with the distinction between animals with *true differentiated tissues*, called **Eumetazoa**, and animal phyla that do not have true differentiated tissues, such as the sponges (**Porifera**) and the Placozoa. Similarities between the feeding cells of sponges (choanocytes) and choanoflagellate protists (Figure 27.12) have been used to suggest that Metazoa evolved from a common ancestral organism that resembled the moderncolonial choanoflagellates.

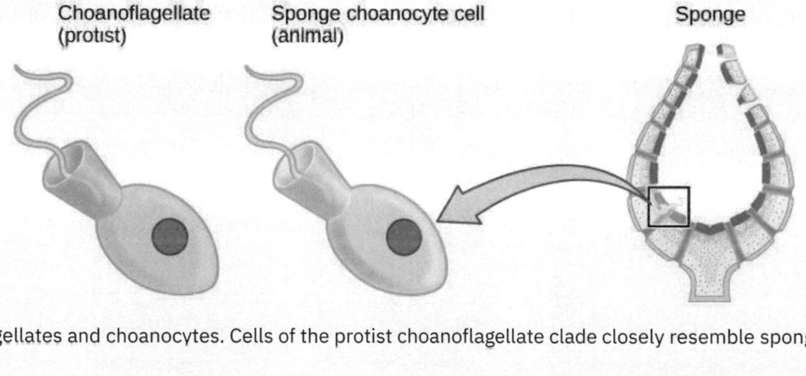

Figure 27.12 Choanoflagellates and choanocytes. Cells of the protist choanoflagellate clade closely resemble sponge choanocyte cells.

Beating of choanocyte flagella draws water through the sponge so that nutrients can be extracted and waste removed.

Eumetazoa are subdivided into radially symmetrical animals and bilaterally symmetrical animals, and are thus classified into the clades Bilateria and Radiata, respectively. As mentioned earlier, the cnidarians and ctenophores are animal phyla with true radial, biradial, or rotational symmetry. All other Eumetazoa are members of the Bilateria clade. The bilaterally symmetrical animals are further divided into deuterostomes (including chordates and echinoderms) and two distinct clades of protostomes (including ecdysozoans and lophotrochozoans) (Figure 27.13a,b). **Ecdysozoa** includes nematodes and arthropods; they are so named for a commonly found characteristic among the group: the physiological process of exoskeletal *molting* followed by the "stripping" of the outer cuticular layer, called *ecdysis*. **Lophotrochozoa** is named for two structural features, each common to certain phyla within the clade. Some lophotrochozoan phyla are characterized by a larval stage called *trochophore larvae*, and other phyla are characterized by the presence of a feeding structure called a *lophophore* (thus, the shorter term, "lopho-trocho-zoa").

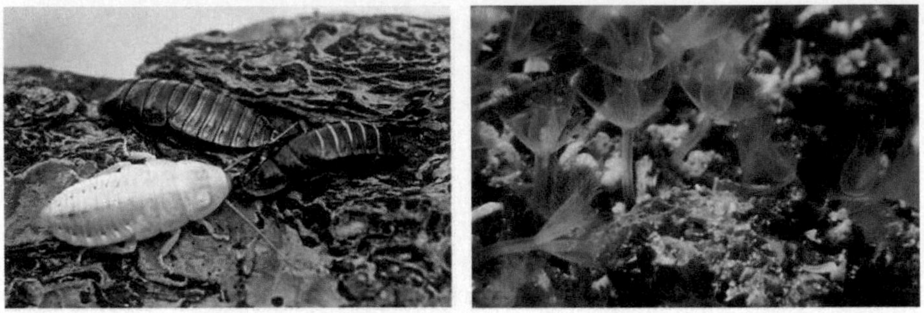

(a) (b)

Figure 27.13 Ecdysozoa. Animals that molt their exoskeletons, such as these (a) Madagascar hissing cockroaches, are in the clade Ecdysozoa. (b) Phoronids are in the clade Lophotrochozoa. The tentacles are part of a feeding structure called a lophophore. (credit a: modification of work by Whitney Cranshaw, Colorado State University, Bugwood.org; credit b: modification of work by NOAA)

LINK TO LEARNING

Explore an interactive tree (http://openstax.org/l/tree_of_life2) of life here. Zoom in and out and click to learn more about the organisms and their evolutionary relationships.

Modern Advances in Phylogenetic Understanding Come from Molecular Analyses

The phylogenetic groupings are continually being debated and refined by evolutionary biologists. Each year, new evidence emerges that further alters the relationships described by a phylogenetic tree diagram.

LINK TO LEARNING

Watch the following video (http://openstax.org/l/build_phylogeny) to learn how biologists use genetic data to determine relationships among organisms.

Nucleic acid and protein analyses have greatly modified and refined the modern phylogenetic animal tree. These data come from a variety of molecular sources, such as mitochondrial DNA, nuclear DNA, ribosomal RNA (rRNA), and certain cellular proteins. Many evolutionary relationships in the modern tree have only recently been determined from the molecular evidence. For example, a previously classified group of animals called lophophorates, which included brachiopods and bryozoans, were long-thought to be primitive deuterostomes. Extensive molecular analysis using rRNA data found these animals are actually protostomes, more closely related to annelids and mollusks. This discovery allowed for the distinction of the protostome clade Lophotrochozoa. Molecular data have also shed light on some differences within the lophotrochozoan group, and the placement of the Platyhelminthes is particularly problematic. Some scientists believe that the phyla Platyhelminthes and Rotifera should actually belong to their own clade of protostomes termed Platyzoa.

Molecular research similar to the discoveries that brought about the distinction of the lophotrochozoan clade has also revealed a dramatic rearrangement of the relationships between mollusks, annelids, arthropods, and nematodes, and as a result, a new ecdysozoan clade was formed. Due to morphological similarities in their segmented body types, annelids and arthropods were once thought to be closely related. However, molecular evidence has revealed that arthropods are actually more closely related to nematodes, now comprising the ecdysozoan clade, and annelids are more closely related to mollusks, brachiopods, and other phyla in the lophotrochozoan clade. These two clades now make up the protostomes.

Another change to former phylogenetic groupings because of modern molecular analyses includes the emergence of an entirely new phylum of worm called Acoelomorpha. These acoel flatworms were long thought to belong to the phylum Platyhelminthes because of their similar "flatworm" morphology. However, molecular analyses revealed this to be a false relationship and originally suggested that acoels represented living species of some of the earliest divergent bilaterians. More recent research into the acoelomorphs has called this hypothesis into question and suggested that the acoels are more closely related to deuterostomes. The placement of this new phylum remains disputed, but scientists agree that with sufficient molecular data, their true phylogeny will be determined.

Another example of phylogenetic reorganization involves the identification of the Ctenophora as the basal clade of the animal kingdom. Ctenophora, or comb jellies, were once considered to be a sister group of the Cnidaria, and the sponges (Porifera) were placed as the basal animal group, sister to other animals. The presence of nerve and muscle cells in both the Ctenophores and the Cnidaria and their absence in the Porifera strengthened this view of the relationships among simple animal forms. However, recent molecular analysis has shown that many of the genes that support neural development in other animals are absent from the Ctenophore genome. The muscle cells are restricted to the mouth and tentacles and are derived from cells in the mesoglea. The mitochondrial genome of the Ctenophores is small and lacks many genes found in other animal mitochondrial genomes. These features plus the absence of *Hox* genes from the Ctenophores have been used to argue that the Ctenophores should be considered basal or as a sister group of the Porifera, and that the evolution of specialized nerve and muscle tissue may have occurred more than once in the history of animal life. Although Ctenophores have been shown as basal to other animals in the phylogeny presented in Chapter 27.2, debate on this issue is likely to continue as Ctenophores are more closely studied.

Changes to the phylogenetic tree can be difficult to track and understand, and are evidence of the process of science. Data and analytical methods play a significant role in the development of phylogenies. For this reason – because molecular analysis and reanalysis are not complete -- we cannot necessarily dismiss a former phylogenetic tree as inaccurate. A recent reanalysis of molecular evidence by an international group of evolutionary biologists refuted the proposition that comb jellies are the phylogenetically oldest extant metazoan group. The study, which relied on more sophisticated methods of analyzing the original genetic data, reaffirms the traditional view that the sponges were indeed the first phylum to diverge from the common ancestor of metazoans. The ongoing discussion concerning the location of sponges and comb jellies on the animal "family tree" is an example of what drives science forward.

27.4 The Evolutionary History of the Animal Kingdom

By the end of this section, you will be able to do the following:

- Describe the features that characterized the earliest animals and approximately when they appeared on earth
- Explain the significance of the Cambrian period for animal evolution and the changes in animal diversity that took place during that time
- Describe some of the unresolved questions surrounding the Cambrian explosion
- Discuss the implications of mass animal extinctions that have occurred in evolutionary history

Many questions regarding the origins and evolutionary history of the animal kingdom continue to be researched and debated, as new fossil and molecular evidence change prevailing theories. Some of these questions include the following: How long have animals existed on Earth? What were the earliest members of the animal kingdom, and what organism was their common ancestor? While animal diversity increased during the Cambrian period of the Paleozoic era, 530 million years ago, modern fossil evidence suggests that primitive animal species existed much earlier.

Pre-Cambrian Animal Life

The time before the Cambrian period is known as the **Ediacaran Period** (from about 635 million years ago to 543 million years ago), the final period of the late Proterozoic Neoproterozoic Era (Figure 27.14). Ediacaran fossils were first found in the Ediacaran hills of Southern Australia. There are no living representatives of these species, which have left impressions that look like those of feathers or coins (Figure 27.15). It is believed that early animal life, termed *Ediacaran biota*, evolved from protists at this time.

EON	ERA	PERIOD	MILLIONS OF YEARS AGO
Phanerozoic	Cenozoic	Quaternary	1.6
		Tertiary	66
	Mesozoic	Cretaceous	138
		Jurassic	205
		Triassic	240
	Paleozoic	Permian	290
		Pennsylvanian	330
		Mississippian	360
		Devonian	410
		Silurian	435
		Ordovician	500
		Cambrian	540
Proterozoic	Late Proterozoic Middle Proterozoic Early Proterozoic	Ediacaran 635-543 MYA	2500
Archean	Late Archean Middle Archean Early Archean		3800?
Pre-Archean			

(a)

(b)

Figure 27.14 An evolutionary timeline. (a) Earth's history is divided into eons, eras, and periods. Note that the Ediacaran period starts in the Proterozoic eon and ends in the Cambrian period of the Phanerozoic eon. (b) Stages on the geological time scale are represented as a spiral. (credit: modification of work by USGS)

Most Ediacaran biota were just a few mm or cm long, but some of the feather-like forms could reach lengths of over a meter. Recently there has been increasing scientific evidence suggesting that more varied and complex animal species lived during this time, and likely even before the Ediacaran period.

Fossils believed to represent the oldest animals with hard body parts were recently discovered in South Australia. These sponge-like fossils, named *Coronacollina acula*, date back as far as 560 million years, and are believed to show the existence of hard body parts and spicules that extended 20–40 cm from the thimble-shaped body (estimated about 5 cm long). Other fossils from the Ediacaran period are shown in Figure 27.15a, b, c.

(a) (b) (c)

Figure 27.15 Ediacaran fauna. Fossils of (a) *Cyclomedusa (up to 20 cm)*, (b) *Dickinsonia (up to 1.4 m)*, (and (c) *Spriggina* (up to 5 cm) date to the Ediacaran period (543-635 MYA). (credit: modification of work by "Smith609"/Wikimedia Commons)

Another recent fossil discovery may represent the earliest animal species ever found. While the validity of this claim is still under investigation, these primitive fossils appear to be small, one-centimeter long, sponge-like creatures, irregularly shaped and with internal tubes or canals. These ancient fossils from South Australia date back 650 million years, actually placing the putative animal before the great ice age extinction event that marked the transition between the **Cryogenian period** and the Ediacaran period. Until this discovery, most scientists believed that there was no animal life prior to the Ediacaran period. Many scientists now believe that animals may in fact have evolved during the Cryogenian period.

The Cambrian Explosion of Animal Life

If the fossils of the Ediacaran and Cryogenian periods are enigmatic, those of the following Cambrian period are far less so, and include body forms similar to those living today. The Cambrian period, occurring between approximately 542–488 million years ago, marks the most rapid evolution of new animal phyla and animal diversity in Earth's history. The rapid diversification of animals that appeared during this period, including most of the animal phyla in existence today, is often referred to as the **Cambrian explosion** (Figure 27.16). Animals resembling echinoderms, mollusks, worms, arthropods, and chordates arose during this period. What may have been a top predator of this period was an arthropod-like creature named *Anomalocaris*, over a meter long, with compound eyes and spiky tentacles. Obviously, all these Cambrian animals already exhibited complex structures, so their ancestors must have existed much earlier.

Figure 27.16 Fauna of the Burgess Shale. An artist's rendition depicts some organisms from the Cambrian period. *Anomalocaris* is seen in the upper left quadrant of the picture.

One of the most dominant species during the Cambrian period was the trilobite, an arthropod that was among the first animals to exhibit a sense of vision (Figure 27.17a,b,c,d). Trilobites were somewhat similar to modern horseshoe crabs. Thousands of different species have been identified in fossil sediments of the Cambrian period; not a single species survives today.

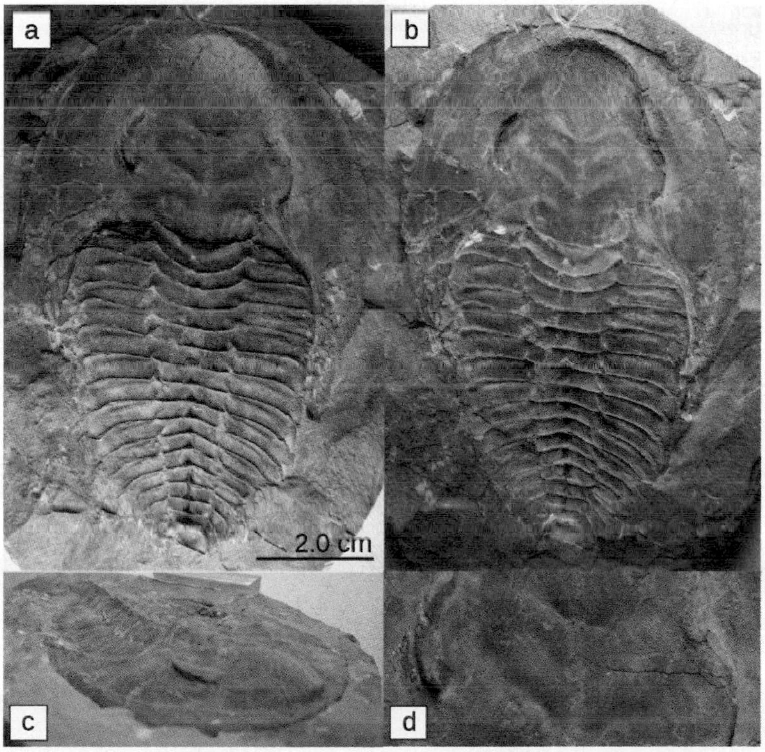

Figure 27.17 Trilobites. These fossils (a–d) belong to trilobites, extinct arthropods that appeared in the early Cambrian period, 525 million years ago, and disappeared from the fossil record during a mass extinction at the end of the Permian period, about 250 million years ago.

The cause of the Cambrian explosion is still debated, and in fact, it may be that a number of interacting causes ushered in this incredible explosion of animal diversity. For this reason, there are a number of hypotheses that attempt to answer this question. Environmental changes may have created a more suitable environment for animal life. Examples of these changes include rising atmospheric oxygen levels (Figure 27.18) and large increases in oceanic calcium concentrations that preceded the Cambrian period. Some scientists believe that an expansive, continental shelf with numerous shallow lagoons or pools provided the necessary living space for larger numbers of different types of animals to coexist. There is also support for hypotheses that argue that ecological relationships between species, such as changes in the food web, competition for food and space, and predator-prey relationships, were primed to promote a sudden massive coevolution of species. Yet other hypotheses claim genetic and developmental reasons for the Cambrian explosion. The morphological flexibility and complexity of animal development afforded by the evolution of *Hox* control genes may have provided the necessary opportunities for increases in possible animal morphologies at the time of the Cambrian period. Hypotheses that attempt to explain why the Cambrian explosion happened must be able to provide valid reasons for the massive animal diversification, as well as explain why it happened *when* it did. There is evidence that both supports and refutes each of the hypotheses described above, and the answer may very well be a combination of these and other theories.

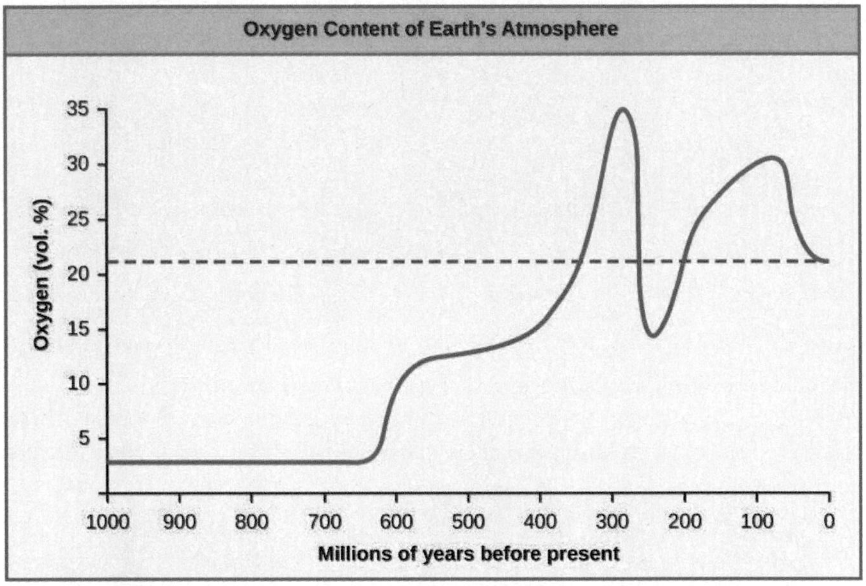

Figure 27.18 Atmospheric oxygen over time. The oxygen concentration in Earth's atmosphere rose sharply around 300 million years ago.

However, unresolved questions about the animal diversification that took place during the Cambrian period remain. For example, we do not understand how the evolution of so many species occurred in such a short period of time. Was there really an "explosion" of life at this particular time? Some scientists question the validity of this idea, because there is increasing evidence to suggest that more animal life existed prior to the Cambrian period and that other similar species' so-called explosions (or radiations) occurred later in history as well. Furthermore, the vast diversification of animal species that appears to have begun during the Cambrian period continued well into the following Ordovician period. Despite some of these arguments, most scientists agree that the Cambrian period marked a time of impressively rapid animal evolution and diversification of body forms that is unmatched for any other time period.

LINK TO LEARNING

View an animation of what ocean life may have been like during the Cambrian explosion.

Click to view content (https://www.openstax.org/l/ocean_life)

Post-Cambrian Evolution and Mass Extinctions

The periods that followed the Cambrian during the Paleozoic Era are marked by further animal evolution and the emergence of many new orders, families, and species. As animal phyla continued to diversify, new species adapted to new ecological niches. During the Ordovician period, which followed the Cambrian period, plant life first appeared on land. This change allowed formerly aquatic animal species to invade land, feeding directly on plants or decaying vegetation. Continual changes in temperature and moisture throughout the remainder of the Paleozoic Era due to continental plate movements encouraged the development of new adaptations to terrestrial existence in animals, such as limbed appendages in amphibians and epidermal scales in reptiles.

Changes in the environment often create new niches (diversified living spaces) that invite rapid speciation and increased diversity. On the other hand, cataclysmic events, such as volcanic eruptions and meteor strikes that obliterate life, can result in devastating losses of diversity to some clades, yet provide new opportunities for others to "fill in the gaps" and speciate. Such periods of **mass extinction** (Figure 27.19) have occurred repeatedly in the evolutionary record of life, erasing some genetic lines while creating room for others to evolve into the empty niches left behind. The end of the Permian period (and the Paleozoic Era) was marked by the largest mass extinction event in Earth's history, a loss of an estimated 95 percent of the extant species at that time. Some of the dominant phyla in the world's oceans, such as the trilobites, disappeared completely. On land, the disappearance of some dominant species of Permian reptiles made it possible for a new line of reptiles to emerge, the dinosaurs. The warm and stable climatic conditions of the ensuing Mesozoic Era promoted an explosive diversification of dinosaurs into every conceivable niche in land, air, and water. Plants, too, radiated into new landscapes and empty niches, creating complex communities of producers and consumers, some of which became very large on the abundant food available.

Another mass extinction event occurred at the end of the Cretaceous period, bringing the Mesozoic Era to an end. Skies darkened and temperatures fell after a large meteor impact and tons of volcanic ash ejected into the atmosphere blocked incoming sunlight. Plants died, herbivores and carnivores starved, and the dinosaurs ceded their dominance of the landscape to the more warm-blooded mammals. In the following Cenozoic Era, mammals radiated into terrestrial and aquatic niches once occupied by dinosaurs, and birds—the warm-blooded direct descendants of one line of the ruling reptiles—became aerial specialists. The appearance and dominance of flowering plants in the Cenozoic Era created new niches for pollinating insects, as well as for birds and mammals. Changes in animal species diversity during the late Cretaceous and early Cenozoic were also promoted by a dramatic shift in Earth's geography, as continental plates slid over the crust into their current positions, leaving some animal groups isolated on islands and continents, or separated by mountain ranges or inland seas from other competitors. Early in the Cenozoic, new ecosystems appeared, with the evolution of grasses and coral reefs. Late in the Cenozoic, further extinctions followed by speciation occurred during ice ages that covered high latitudes with ice and then retreated, leaving new open spaces for colonization.

LINK TO LEARNING

Watch the following video (http://openstax.org/l/mass_extinction) to learn more about the mass extinctions.

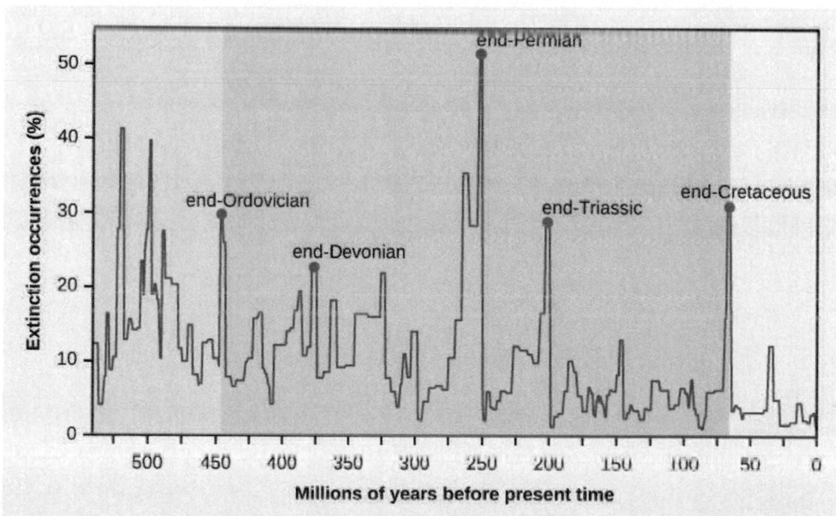

Figure 27.19 Extinctions. Mass extinctions have occurred repeatedly over geological time.

CAREER CONNECTION

Paleontologist

Natural history museums contain the fossils of extinct animals as well as information about how these animals evolved, lived, and died. Paleontologists are scientists who study prehistoric life. They use fossils to observe and explain how life evolved on Earth and how species interacted with each other and with the environment. A paleontologist needs to be knowledgeable in mathematics, biology, ecology, chemistry, geology, and many other scientific disciplines. A paleontologist's work may involve field studies: searching for and studying fossils. In addition to digging for and finding fossils, paleontologists also prepare fossils for further study and analysis. Although dinosaurs are probably the first animals that come to mind when thinking about ancient life, paleontologists study a variety of life forms, from plants, fungi and invertebrates to the vertebrate fishes, amphibians, reptiles, birds and mammals.

An undergraduate degree in earth science or biology is a good place to start toward the career path of becoming a paleontologist. Most often, a graduate degree is necessary. Additionally, work experience in a museum or in a paleontology lab is useful.

KEY TERMS

acoelomate animal without a body cavity
bilateral symmetry type of symmetry in which there is only one plane of symmetry, so the left and right halves of an animal are mirror images
blastopore opening into the archenteron that forms during gastrulation
blastula 16–32 cell stage of development of an animal embryo
body plan morphology or defining shape of an organism
Cambrian explosion time during the Cambrian period (542–488 million years ago) when most of the animal phyla in existence today evolved
cleavage cell divisions subdividing a fertilized egg (zygote) to form a multicellular embryo
coelom lined body cavity
Cryogenian period geologic period (850–630 million years ago) characterized by a very cold global climate
determinate cleavage cleavage pattern in which developmental fate of each blastomere is tightly defined
deuterostome blastopore develops into the anus, with the second opening developing into the mouth
diploblast animal that develops from two germ layers
Ecdysozoa clade of protostomes that exhibit exoskeletal molting (ecdysis)
Ediacaran period geological period (630–542 million years ago) when the oldest definite multicellular organisms with tissues evolved
enterocoely mesoderm of deuterostomes develops as pouches that are pinched off from endodermal tissue, cavity contained within the pouches becomes coelom
eucoelomate animal with a body cavity completely lined with mesodermal tissue
Eumetazoa group of animals with true differentiated tissues
gastrula stage of animal development characterized by the formation of the digestive cavity
germ layer collection of cells formed during embryogenesis that will give rise to future body tissues, more pronounced in vertebrate embryogenesis
Hox gene (also, homeobox gene) master control gene that can turn on or off large numbers of other genes during embryogenesis
indeterminate cleavage cleavage pattern in which individual blastomeres have the character of "stem cells," and are not yet predetermined to develop into specific cell types
Lophotrochozoa clade of protostomes that exhibit a trochophore larvae stage or a lophophore feeding structure
mass extinction event or environmental condition that wipes out the majority of species within a relatively short geological time period
Metazoa group containing all animals
organogenesis formation of organs in animal embryogenesis
Parazoa group of animals without true differentiated tissues
protostome blastopore develops into the mouth of protostomes, with the second opening developing into the anus
pseudocoelomate animal with a body cavity located between the mesoderm and endoderm
radial cleavage cleavage axes are parallel or perpendicular to the polar axis, resulting in the alignment of cells between the two poles
radial symmetry type of symmetry with multiple planes of symmetry, with body parts (rays) arranged around a central disk
schizocoely during development of protostomes, a solid mass of mesoderm splits apart and forms the hollow opening of the coelom
spiral cleavage cells of one pole of the embryo are rotated or misaligned with respect to the cells of the opposite pole
triploblast animal that develops from three germ layers

CHAPTER SUMMARY

27.1 Features of the Animal Kingdom

Animals constitute an incredibly diverse kingdom of organisms. Although animals range in complexity from simple sea sponges to human beings, most members of the animal kingdom share certain features. Animals are eukaryotic, multicellular, heterotrophic organisms that ingest their food and usually develop into motile creatures with a fixed body plan. A major characteristic unique to the animal kingdom is the presence of differentiated tissues, such as nerve, muscle, and connective tissues, which are specialized to perform specific functions. Most animals undergo sexual reproduction, leading to a series of developmental embryonic stages that are relatively similar across the animal kingdom. A class of transcriptional control genes called *Hox* genes directs the organization of the major animal body plans, and these genes are strongly homologous across the animal kingdom.

27.2 Features Used to Classify Animals

Organisms in the animal kingdom are classified based on their body morphology, their developmental pathways, and their genetic affinities. The relationships between the Eumetazoa and more basal clades (Ctenophora, Porifera, and Placozoa) are still being debated. The Eumetazoa ("true animals") are divided into those with radial versus bilateral symmetry. Generally, the simpler and often nonmotile animals display radial symmetry, which allows them to explore their environment in all directions. Animals with radial symmetry are also generally characterized by the development of two embryological germ layers, the endoderm and ectoderm, whereas animals with bilateral symmetry are generally characterized by the development of a third embryologic germ layer, the mesoderm. Animals with three germ layers, called triploblasts, are further characterized by the presence or absence of an internal body cavity called a coelom. The presence of a coelom affords many advantages, and animals with a coelom may be termed true coelomates or pseudocoelomates, depending the extent to which mesoderm lines the body cavity. Coelomates are further divided into one of two groups called protostomes and deuterostomes, based on a number of developmental characteristics, including differences in zygote cleavage, the method of coelom formation, and the rigidity of the developmental fate of blastomeres.

27.3 Animal Phylogeny

Scientists are interested in the evolutionary history of animals and the evolutionary relationships among them. There are three main sources of data that scientists use to create phylogenetic evolutionary tree diagrams that illustrate such relationships: morphological information (which includes developmental morphologies), fossil record data, and, most recently, molecular data. The details of the modern phylogenetic tree change frequently as new data are gathered, and molecular data has recently contributed to many substantial modifications of the understanding of relationships between animal phyla.

27.4 The Evolutionary History of the Animal Kingdom

The most rapid documented diversification and evolution of animal species in all of history occurred during the Cambrian period of the Paleozoic Era, a phenomenon known as the Cambrian explosion. Until recently, scientists believed that there were only very few tiny and simplistic animal species in existence before this period. However, recent fossil discoveries have revealed that additional, larger, and more complex animals existed during the Ediacaran period, and even possibly earlier, during the Cryogenian period. Still, the Cambrian period undoubtedly witnessed the emergence of the majority of animal phyla that we know today, although many questions remain unresolved about this historical phenomenon.

The remainder of the Paleozoic Era is marked by the growing appearance of new classes, families, and species, and the early colonization of land by certain marine animals and semiaquatic arthropods, both freshwater and marine. The evolutionary history of animals is also marked by numerous major extinction events, each of which wiped out a majority of extant species. Some species of most animal phyla survived these extinctions, allowing the phyla to persist and continue to evolve into species that we see today.

VISUAL CONNECTION QUESTIONS

1. Figure 27.5 If a *Hox 13* gene in a mouse was replaced with a *Hox 1* gene, how might this alter animal development?

2. Figure 27.6 Which of the following statements is false?
   a. Eumetazoans have specialized tissues and parazoans don't.
   b. Lophotrochozoa and Ecdysozoa are both Bilataria.
   c. Acoela and Cnidaria both possess radial symmetry.
   d. Arthropods are more closely related to nematodes than they are to annelids.

3. Figure 27.9 Which of the following statements about diploblasts and triploblasts is false?
   a. Animals that display radial symmetry are diploblasts.
   b. Animals that display bilateral symmetry are triploblasts.
   c. The endoderm gives rise to the lining of the digestive tract and the respiratory tract.
   d. The mesoderm gives rise to the central nervous system.

REVIEW QUESTIONS

4. Which of the following is not a feature common to *most* animals?
   a. development into a fixed body plan
   b. asexual reproduction
   c. specialized tissues
   d. heterotrophic nutrient sourcing

5. During embryonic development, unique cell layers develop into specific groups of tissues or organs during a stage called ________.
   a. the blastula stage
   b. the germ layer stage
   c. the gastrula stage
   d. the organogenesis stage

6. Which of the following phenotypes would most likely be the result of a *Hox* gene mutation?
   a. abnormal body length or height
   b. two different eye colors
   c. the contraction of a genetic illness
   d. two fewer appendages than normal

7. Which of the following organisms is most likely to be a diploblast?
   a. sea star
   b. shrimp
   c. jellyfish
   d. insect

8. Which of the following is not possible?
   a. radially symmetrical diploblast
   b. diploblastic eucoelomate
   c. protostomic coelomate
   d. bilaterally symmetrical deuterostome

9. An animal whose development is marked by radial cleavage and enterocoely is ________.
   a. a deuterostome
   b. an annelid or mollusk
   c. either an acoelomate or eucoelomate
   d. none of the above

10. Consulting the modern phylogenetic tree of animals, which of the following would not constitute a clade?
   a. deuterostomes
   b. lophotrochozoans
   c. Parazoa
   d. Bilateria

11. Which of the following is thought to be the most closely related to the common animal ancestor?
   a. fungal cells
   b. protist cells
   c. plant cells
   d. bacterial cells

12. As with the emergence of the Acoelomorpha phylum, it is common for ____ data to misplace animals in close relation to other species, whereas ____ data often reveals a different and more accurate evolutionary relationship.
   a. molecular : morphological
   b. molecular : fossil record
   c. fossil record : morphological
   d. morphological : molecular

13. Which of the following periods is the earliest during which animals may have appeared?
   a. Ordovician period
   b. Cambrian period
   c. Ediacaran period
   d. Cryogenian period

14. What type of data is primarily used to determine the existence and appearance of early animal species?
   a. molecular data
   b. fossil data
   c. morphological data
   d. embryological development data

15. The time between 542–488 million years ago marks which period?
   a. Cambrian period
   b. Silurian period
   c. Ediacaran period
   d. Devonian period

16. Until recent discoveries suggested otherwise, animals existing before the Cambrian period were believed to be:
   a. small and ocean dwelling
   b. small and nonmotile
   c. small and soft-bodied
   d. small and radially symmetrical or asymmetrical

17. Plant life first appeared on land during which of the following periods?
   a. Cambrian period
   b. Ordovician period
   c. Silurian period
   d. Devonian period

18. Approximately how many mass extinction events occurred throughout the evolutionary history of animals?
   a. 3
   b. 4
   c. 5
   d. more than 5

CRITICAL THINKING QUESTIONS

19. Why might the evolution of specialized tissues be important for animal function and complexity?

20. Describe and give examples of how humans display all of the features common to the animal kingdom.

21. How have *Hox* genes contributed to the diversity of animal body plans?
22. Using the following terms, explain what classifications and groups humans fall into, from the most general to the most specific: symmetry, germ layers, coelom, cleavage, embryological development.
23. Explain some of the advantages brought about through the evolution of bilateral symmetry and coelom formation.
24. Describe at least two major changes to the animal phylogenetic tree that have come about due to molecular or genetic findings.
25. How is it that morphological data alone might lead scientists to group animals into erroneous evolutionary relationships?
26. Briefly describe at least two theories that attempt to explain the cause of the Cambrian explosion.
27. How is it that most, if not all, of the extant animal phyla today evolved during the Cambrian period if so many massive extinction events have taken place since then?

CHAPTER 28
Invertebrates

Figure 28.1 Nearly 97 percent of animal species are invertebrates, including this sea star (*Novodinia antillensis*) (credit: NOAA's National Ocean Service)

INTRODUCTION A brief look at any magazine pertaining to our natural world, such as *National Geographic*, would show a rich variety of vertebrates, especially mammals and birds. To most people, these are the animals that attract our attention. Concentrating on vertebrates, however, gives us a rather biased and limited view of animal diversity, because it ignores nearly 97 percent of the animal kingdom—the invertebrates—animals that lack a cranium and a defined vertebral column or spine.

The invertebrate animal phyla exhibit an enormous variety of cells and tissues adapted for specific purposes, and frequently these tissues are unique to their phyla. These specializations show the range of cellular differentiation possible within the clade Opisthokonta, which has both unicellular and multicellular members. Cellular and structural specializations include cuticles for protection, spines and tiny harpoons for defense, toothy structures for feeding, and wings for flight. An exoskeleton may be adapted for movement or for the attachment of muscles as in the clams and insects. Secretory cells can produce venom, mucus, or digestive enzymes. The body plans of some phyla, such as those of the molluscs, annelids, arthropods, and echinoderms, have been modified and adapted throughout evolution to produce thousands of different forms. Perhaps you will find it amazing that an enormous number of both aquatic and terrestrial invertebrates—perhaps millions of species—have not yet been scientifically classified. As a result, the phylogenetic relationships among the invertebrates are constantly being updated as new information is collected about the organisms of each phylum.

Chapter Outline

28.1 Phylum Porifera

By the end of this section, you will be able to do the following:

- Describe the organizational features of the simplest multicellular organisms
- Explain the various body forms and bodily functions of sponges

As we have seen, the vast majority of invertebrate animals do *not* possess a defined bony vertebral endoskeleton, or a bony cranium. However, one of the most ancestral groups of deuterostome invertebrates, the Echinodermata, do produce tiny skeletal "bones" called *ossicles* that make up a true **endoskeleton**, or internal skeleton, covered by an epidermis.

We will start our investigation with the simplest of all the invertebrates—animals sometimes classified within the clade Parazoa ("beside the animals"). This clade currently includes only the phylum Placozoa (containing a single species, *Trichoplax adhaerens*), and the phylum **Porifera**, containing the more familiar sponges (Figure 28.2). The split between the Parazoa and the Eumetazoa (all animal clades above Parazoa) likely took place over a billion years ago.

We should reiterate here that the Porifera do not possess "true" tissues that are embryologically homologous to those of all other derived animal groups such as the insects and mammals. This is because they do not create a true gastrula during embryogenesis, and as a result do not produce a true endoderm or ectoderm. But even though they are not considered to have true tissues, they do have specialized cells that perform specific functions *like tissues* (for example, the external "pinacoderm" of a sponge acts like our epidermis). Thus, *functionally*, the poriferans can be said to have tissues; however, these tissues are likely not *embryologically* homologous to our own.

Sponge larvae (e.g, parenchymula and amphiblastula) are flagellated and able to swim; however, adults are non-motile and spend their life attached to a substratum. Since water is vital to sponges for feeding, excretion, and gas exchange, their body structure facilitates the movement of water through the sponge. Various canals, chambers, and cavities enable water to move through the sponge to allow the exchange of food and waste as well as the exchange of gases to nearly all body cells.

Figure 28.2 Sponges. Sponges are members of the phylum Porifera, which contains the simplest invertebrates. (credit: Andrew Turner)

Morphology of Sponges

There are at least 5,000 named species of sponges, likely with thousands more yet to be classified. The morphology of the simplest sponges takes the shape of an irregular cylinder with a large central cavity, the **spongocoel**, occupying the inside of the cylinder (Figure 28.3). Water enters into the spongocoel through numerous pores, or *ostia*, that create openings in the body wall. Water entering the spongocoel is expelled via a large common opening called the **osculum**. However, we should note that sponges exhibit a range of diversity in body forms, including variations in the size and shape of the spongocoel, as well as the number and arrangement of feeding chambers within the body wall. In some sponges, multiple feeding chambers open off of a central spongocoel and in others, several feeding chambers connecting to one another may lie between the entry pores and the spongocoel.

While sponges do not exhibit true tissue-layer organization, they do have a number of functional "tissues" composed of different cell types specialized for distinct functions. For example, epithelial-like cells called *pinacocytes* form the outermost body, called a *pinacoderm*, that serves a protective function similar that of our epidermis. Scattered among the pinacoderm are the ostia that allow entry of water into the body of the sponge. These pores have given the sponges their phylum name Porifera—pore-bearers. In some sponges, ostia are formed by *porocytes*, single tube-shaped cells that act as valves to regulate the flow of water into the spongocoel. In other sponges, ostia are formed by folds in the body wall of the sponge. Between the outer layer and the feeding chambers of the sponge is a jelly-like substance called the **mesohyl**, which contains collagenous fibers. Various cell types reside within the mesohyl, including **amoebocytes**, the "stem cells" of sponges, and *sclerocytes*, which produce skeletal materials. The gel-like consistency of mesohyl acts like an endoskeleton and maintains the tubular morphology of sponges.

The feeding chambers inside the sponge are lined by choanocytes ("collar cells"). The structure of a choanocyte is critical to its function, which is to generate a *directed* water current through the sponge and to trap and ingest microscopic food particles by phagocytosis. These feeding cells are similar in appearance to unicellular *choanoflagellates* (Protista). This similarity suggests that sponges and choanoflagellates are closely related and likely share common ancestry. The body of the choanocyte is embedded in mesohyl and contains all the organelles required for normal cell function. Protruding into the "open space" inside the feeding chamber is a mesh-like collar composed of microvilli with a single flagellum in the center of the column. The beating of the flagella from all choanocytes draws water into the sponge through the numerous ostia, into the spaces lined by choanocytes, and eventually out through the osculum (or osculi, if the sponge consists of a colony of attached sponges). Food particles, including waterborne bacteria and unicellular organisms such as algae and various animal-like protists, are trapped by the sieve-like collar of the choanocytes, slide down toward the body of the cell, and are ingested by phagocytosis. Choanocytes also serve another surprising function: They can differentiate into sperm for sexual reproduction, at which time they become dislodged from the mesohyl and leave the sponge with expelled water through the osculum.

LINK TO LEARNING

Watch this video to see the movement of water through the sponge body.

Click to view content (https://www.openstax.org/l/filter_sponges)

The **amoebocytes** (derived from stem-cell-like archaeocytes), are so named because they move throughout the mesohyl in an amoeba-like fashion. They have a variety of functions: In addition to delivering nutrients from choanocytes to other cells within the sponge, they also give rise to eggs for sexual reproduction. (The eggs remain in the mesohyl, whereas the sperm cells are released into the water.) The amoebocytes can differentiate into other cell types of the sponge, such as collenocytes and lophocytes, which produce the collagen-like protein that support the mesohyl. Amoebocytes can also give rise to sclerocytes, which produce *spicules* (skeletal spikes of silica or calcium carbonate) in some sponges, and spongocytes, which produce the protein spongin in the majority of sponges. These different cell types in sponges are shown in Figure 28.3.

VISUAL CONNECTION

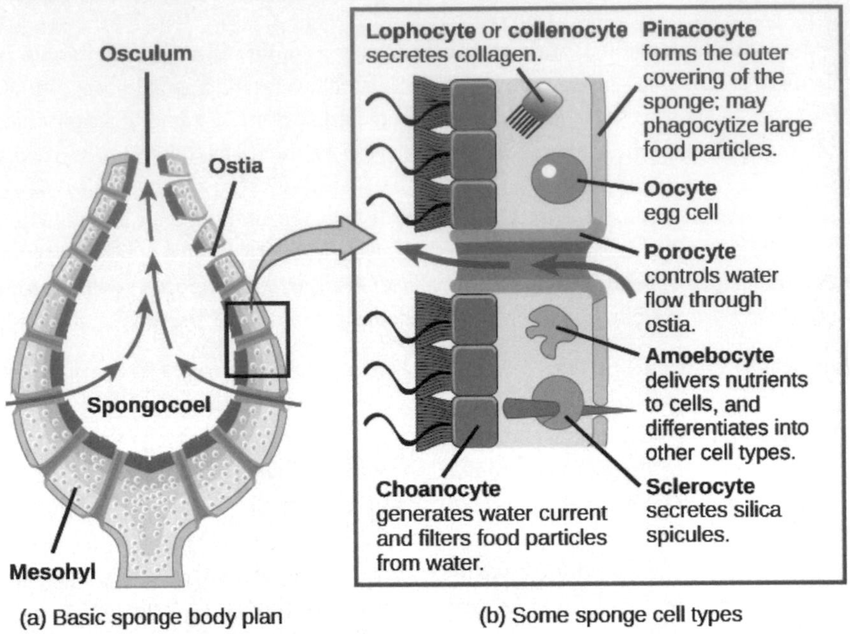

Figure 28.3 Simple sponge body plan and cell types. The sponge's (a) basic body plan and (b) some of the specialized cell types found in sponges are shown.

Which of the following statements is false?

a. Choanocytes have flagella that propel water through the body.
b. Pinacocytes can transform into any cell type.
c. Lophocytes secrete collagen.
d. Porocytes control the flow of water through pores in the sponge body.

LINK TO LEARNING

Take an up-close tour (http://openstax.org/l/sponge_ride) through the sponge and its cells.

As we've seen, most sponges are supported by small bone-like *spicules* (usually tiny pointed structures made of calcium carbonate or silica) in the mesohyl. Spicules provide support for the body of the sponge, and may also deter predation. The presence and composition of spicules form the basis for differentiating three of the four classes of sponges (Figure 28.4). Sponges in class Calcarea produce calcium carbonate spicules and no spongin; those in class Hexactinellida produce six-rayed siliceous (glassy) spicules and no spongin; and those in class Demospongia contain spongin and may or may not have spicules; if present, those spicules are siliceous. Sponges in this last class have been used as bath sponges. Spicules are most conspicuously present in the glass sponges, class Hexactinellida. Some of the spicules may attain gigantic proportions. For example, relative to typical glass sponge spicules, whose size generally ranges from 3 to 10 mm, some of the *basal spicules* of the hexactinellid *Monorhaphis chuni* are enormous and grow up to 3 meters long! The glass sponges are also unusual in that most of their body cells are fused together to form a *multinucleate syncytium*. Because their cells are interconnected in this way, the hexactinellid sponges have no mesohyl. A fourth class of sponges, the Sclerospongiae, was described from species discovered in underwater tunnels. These are also called coralline sponges after their multilayered calcium carbonate skeletons. Dating based on the rate of deposition of the skeletal layers suggests that some of these sponges are hundreds of years old.

Figure 28.4 Several classes of sponges. (a) *Clathrina clathrus* belongs to class Calcarea, (b) *Staurocalyptus* spp. (common name: yellow Picasso sponge) belongs to class Hexactinellida, and (c) *Acarnus erithacus* belongs to class Demospongia. (credit a: modification of work by Parent Géry; credit b: modification of work by Monterey Bay Aquarium Research Institute, NOAA; credit c: modification of work by Sanctuary Integrated Monitoring Network, Monterey Bay National Marine Sanctuary, NOAA)

LINK TO LEARNING

Use the Interactive Sponge Guide (http://openstax.org/l/id_sponges) to identify species of sponges based on their external form, mineral skeleton, fiber, and skeletal architecture.

Physiological Processes in Sponges

Sponges, despite being simple organisms, regulate their different physiological processes through a variety of mechanisms. These processes regulate their metabolism, reproduction, and locomotion.

Digestion

Sponges lack complex digestive, respiratory, circulatory, and nervous systems. Their food is trapped as water passes through the ostia and out through the osculum. Bacteria smaller than 0.5 microns in size are trapped by choanocytes, which are the principal cells engaged in feeding, and are ingested by phagocytosis. However, particles that are larger than the ostia may be phagocytized at the sponge's surface by pinacocytes. In some sponges, amoebocytes transport food from cells that have ingested food particles to those that do not. In sponges, in spite of what looks like a large digestive cavity, all digestion is *intracellular*. The limit of this type of digestion is that food particles must be smaller than individual sponge cells.

All other major body functions in the sponge (gas exchange, circulation, excretion) are performed by diffusion between the cells that line the openings within the sponge and the water that is passing through those openings. All cell types within the sponge obtain oxygen from water through diffusion. Likewise, carbon dioxide is released into seawater by diffusion. In addition, nitrogenous waste produced as a byproduct of protein metabolism is excreted via diffusion by individual cells into the water as it passes through the sponge.

Some sponges host green algae or cyanobacteria as endosymbionts within archeocytes and other cells. It may be a surprise to learn that there are nearly 150 species of carnivorous sponges, which feed primarily on tiny crustaceans, snaring them through sticky threads or hooked spicules!

Although there is no specialized nervous system in sponges, there is intercellular communication that can regulate events like contraction of the sponge's body or the activity of the choanocytes.

Reproduction

Sponges reproduce by sexual as well as asexual methods. The typical means of asexual reproduction is either fragmentation (during this process, a piece of the sponge breaks off, settles on a new substrate, and develops into a new individual), or budding (a genetically identical outgrowth grows from the parent and eventually detaches or remains attached to form a colony). An atypical type of asexual reproduction is found only in freshwater sponges and occurs through the formation of *gemmules*. **Gemmules** are environmentally resistant structures produced by adult sponges (e.g., in the freshwater sponge *Spongilla*). In gemmules, an inner layer of archeocytes (amoebocytes) is surrounded by a pneumatic cellular layer that may be reinforced with spicules. In freshwater sponges, gemmules may survive hostile environmental conditions like changes in temperature, and then serve to recolonize the habitat once environmental conditions improve and stabilize. Gemmules are capable of attaching to a substratum and generating a new sponge. Since gemmules can withstand harsh environments, are resistant to desiccation, and remain dormant for long periods, they are an excellent means of colonization for a sessile organism.

Sexual reproduction in sponges occurs when gametes are generated. Oocytes arise by the differentiation of amoebocytes and are retained within the spongocoel, whereas spermatozoa result from the differentiation of choanocytes and are ejected via the osculum. Sponges are **monoecious** (hermaphroditic), which means that one individual can produce both gametes (eggs and sperm) simultaneously. In some sponges, production of gametes may occur throughout the year, whereas other sponges may show sexual cycles depending upon water temperature. Sponges may also become *sequentially hermaphroditic*, producing oocytes first and spermatozoa later. This temporal separation of gametes produced by the same sponge helps to encourage cross-fertilization and genetic diversity. Spermatozoa carried along by water currents can fertilize the oocytes borne in the mesohyl of other sponges. Early larval development occurs within the sponge, and free-swimming larvae (such as flagellated *parenchymula*) are then released via the osculum.

Locomotion

Sponges are generally sessile as adults and spend their lives attached to a fixed substratum. They do not show movement over large distances like other free-swimming marine invertebrates. However, sponge cells are capable of creeping along substrata via *organizational plasticity*, i.e., rearranging their cells. Under experimental conditions, researchers have shown that sponge cells spread on a physical support demonstrate a leading edge for directed movement. It has been speculated that this localized creeping movement may help sponges adjust to microenvironments near the point of attachment. It must be noted, however, that this pattern of movement has been documented in laboratories, it remains to be observed in natural sponge habitats.

LINK TO LEARNING

Watch this BBC video (http://openstax.org/l/sea_sponges) showing the array of sponges seen along the Cayman Wall during a submersible dive.

28.2 Phylum Cnidaria

By the end of this section, you will be able to do the following:

- Compare structural and organization characteristics of Porifera and Cnidaria
- Describe the progressive development of tissues and their relevance to animal complexity
- Identify the two general body forms found in the Cnidaria
- Describe the identifying features of the major cnidarian classes

Phylum **Cnidaria** includes animals that exhibit radial or biradial symmetry and are diploblastic, meaning that they develop from two embryonic layers, ectoderm and endoderm. Nearly all (about 99 percent) cnidarians are marine species.

Whereas the defining cell type for the sponges is the choanocyte, the defining cell type for the cnidarians is the **cnidocyte**, or stinging cell. These cells are located around the mouth and on the tentacles, and serve to capture prey or repel predators. Cnidocytes have large stinging organelles called **nematocysts**, which usually contain barbs at the base of a long coiled thread. The outer wall of the cell has a hairlike projection called a *cnidocil*, which is sensitive to tactile stimulation. If the cnidocils are touched, the hollow threads evert with enormous acceleration, approaching 40,000 times that of gravity. The microscopic threads then either entangle the prey or instantly penetrate the flesh of the prey or predator, releasing toxins (including neurotoxins and pore-forming toxins that can lead to cell lysis) into the target, thereby immobilizing it or paralyzing it (see Figure 28.5).

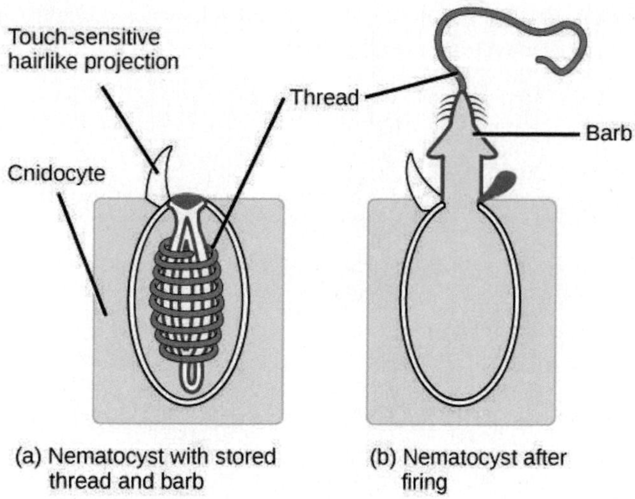

Figure 28.5 Cnidocytes. Animals from the phylum Cnidaria have stinging cells called cnidocytes. Cnidocytes contain large organelles called (a) nematocysts that store a coiled thread and barb, the nematocyst. When the hairlike cnidocil on the cell surface is touched, even lightly, (b) the thread, barb, and a toxin are fired from the organelle.

LINK TO LEARNING

View this video (https://www.openstax.org/l/nematocyst) animation showing two anemones engaged in a battle.

Two distinct body plans are found in Cnidarians: the polyp or tuliplike "stalk" form and the medusa or "bell" form. (Figure 28.6). An example of the polyp form is found in the genus *Hydra*, whereas the most typical form of medusa is found in the group called the "sea jellies" (jellyfish). Polyp forms are sessile as adults, with a single opening (the mouth/anus) to the digestive cavity facing up with tentacles surrounding it. Medusa forms are motile, with the mouth and tentacles hanging down from an umbrella-shaped bell.

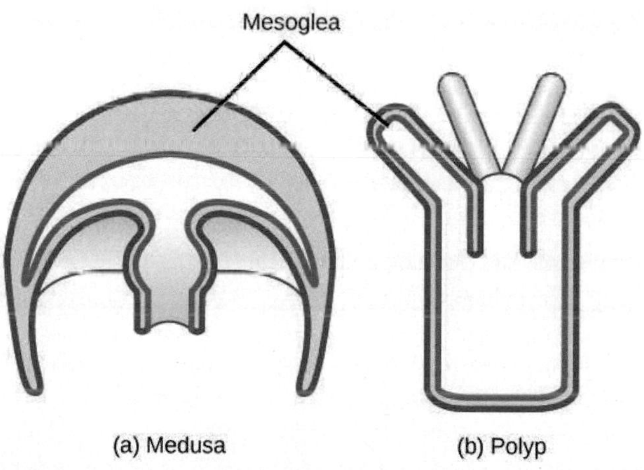

Figure 28.6 Cnidarian body forms. Cnidarians have two distinct body plans, the medusa (a) and the polyp (b). All cnidarians have two membrane layers, with a jelly-like mesoglea between them.

Some cnidarians are **dimorphic**, that is, they exhibit both body plans during their life cycle. In these species, the polyp serves as the asexual phase, while the medusa serves as the sexual stage and produces gametes. However, both body forms are diploid.

An example of cnidarian dimorphism can be seen in the colonial hydroid *Obelia*. The sessile asexual colony has two types of polyps, shown in Figure 28.7. The first is the *gastrozooid*, which is adapted for capturing prey and feeding. In *Obelia*, all polyps are connected through a common digestive cavity called a *coenosarc*. The other type of polyp is the *gonozooid*, adapted for the asexual budding and the production of sexual medusae. The reproductive buds from the gonozooid break off and mature into free-swimming medusae, which are either male or female (dioecious). Each medusa has either several testes or several ovaries in which meiosis occurs to produce sperm or egg cells. Interestingly, the gamete-producing cells do not arise within the gonad itself, but migrate into it from the tissues in the gonozooid. This separate origin of gonad and gametes is common throughout the eumetazoa. The gametes are released into the surrounding water, and after fertilization, the zygote develops into a blastula, which soon develops into a ciliated, bilaterally symmetrical planula larva. The planula swims freely for a while, but eventually attaches to a substrate and becomes a single polyp, from which a new colony of polyps is formed by budding.

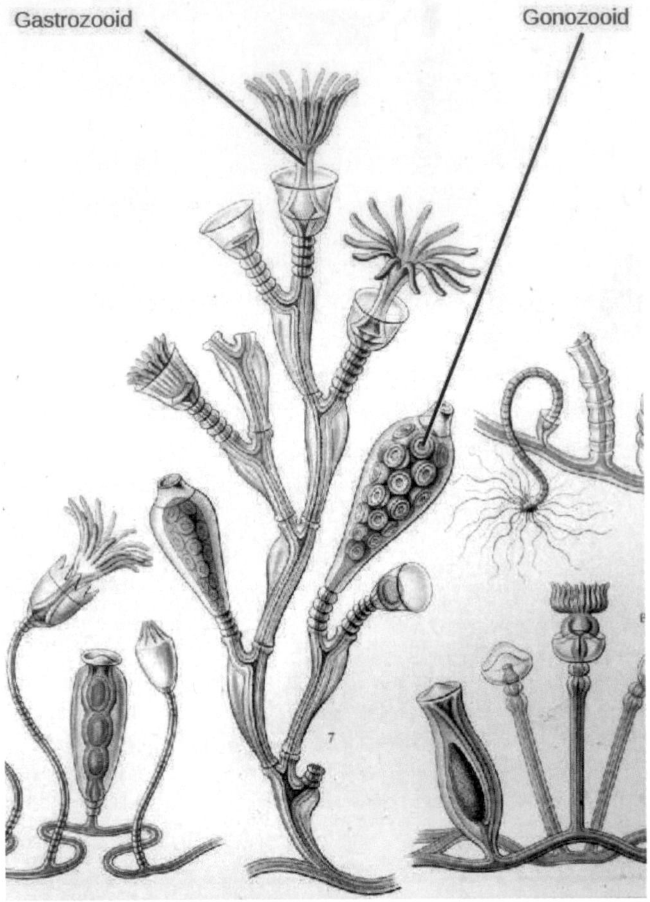

Figure 28.7 *Obelia.* The colonial sessile form of *Obelia geniculata* has two types of polyps: gastrozooids, which are adapted for capturing prey, and gonozooids, which asexually bud to produce medusae.

LINK TO LEARNING

Click here to follow an *Obelia* life cycle (http://openstax.org/l/obelia) animation.

All cnidarians are diploblastic and thus have two "epithelial" layers in the body that are derived from the endoderm and ectoderm of the embryo. The outer layer (from ectoderm) is called the *epidermis* and lines the outside of the animal, whereas the inner layer (from endoderm) is called the *gastrodermis* and lines the digestive cavity. In the planula larva, a layer of ectoderm surrounds a solid mass of endoderm, but as the polyp develops, the digestive or gastrovascular cavity opens within the endoderm. A non-living, jelly-like mesoglea lies between these two epithelial layers. In terms of cellular complexity, cnidarians show the presence of differentiated cell types in each tissue layer, such as nerve cells, contractile epithelial cells, enzyme-secreting cells, and nutrient-absorbing cells, as well as the presence of intercellular connections. However, with a few notable exceptions such as *statocysts* and *rhopalia* (see below), the development of organs or organ systems is not advanced in this phylum.

The nervous system is rudimentary, with nerve cells organized in a network scattered across the body. This **nerve net** may show the presence of groups of cells that form nerve plexi (singular: plexus) or nerve cords. Organization of the nervous system in the motile medusa is more complex than that of the sessile polyp, with a nerve ring around the edge of the medusa bell that controls the action of the tentacles. Cnidarian nerve cells show mixed characteristics of motor and sensory neurons. The predominant signaling molecules in these primitive nervous systems are peptides, which perform both excitatory and inhibitory functions.

Despite the simplicity of the nervous system, it is remarkable that it coordinates the complicated movement of the tentacles, the drawing of captured prey to the mouth, the digestion of food, and the expulsion of waste.

The gastrovascular cavity has only one opening that serves as both a mouth and an anus; this arrangement is called an incomplete digestive system. In the gastrovascular cavity, extracellular digestion occurs as food is taken into the gastrovascular cavity, enzymes are secreted into the cavity, and the cells lining the cavity absorb nutrients. However, some intracellular digestion also occurs. The gastrovascular cavity distributes nutrients throughout the body of the animal, with nutrients passing from the digestive cavity across the mesoglea to the epidermal cells. Thus, this cavity serves both digestive and circulatory functions.

Cnidarian cells exchange oxygen and carbon dioxide by diffusion between cells in the epidermis and water in the environment, and between cells in the gastrodermis and water in the gastrovascular cavity. The lack of a circulatory system to move dissolved gases limits the thickness of the body wall and necessitates a non-living mesoglea between the layers. In the cnidarians with a thicker mesoglea, a number of canals help to distribute both nutrients and gases. There is neither an excretory system nor organs, and nitrogenous wastes simply diffuse from the cells into the water outside the animal or into the gastrovascular cavity.

The phylum Cnidaria contains about 10,000 described species divided into two monophyletic clades: the Anthozoa and the Medusozoa. The Anthozoa include the corals, sea fans, sea whips, and the sea anemones. The Medusozoa include several classes of Cnidaria in two clades: The Hydrozoa include sessile forms, some medusoid forms, and swimming colonial forms like the Portuguese man-of-war. The other clade contains various types of jellies including both Scyphozoa and Cubozoa. The Anthozoa contain only sessile polyp forms, while the Medusozoa include species with both polyp and medusa forms in their life cycle.

Class Anthozoa

The class Anthozoa ("flower animals") includes sea anemones (Figure 28.8), sea pens, and corals, with an estimated number of 6,100 described species. Sea anemones are usually brightly colored and can attain a size of 1.8 to 10 cm in diameter. Individual animals are cylindrical in shape and are attached directly to a substrate.

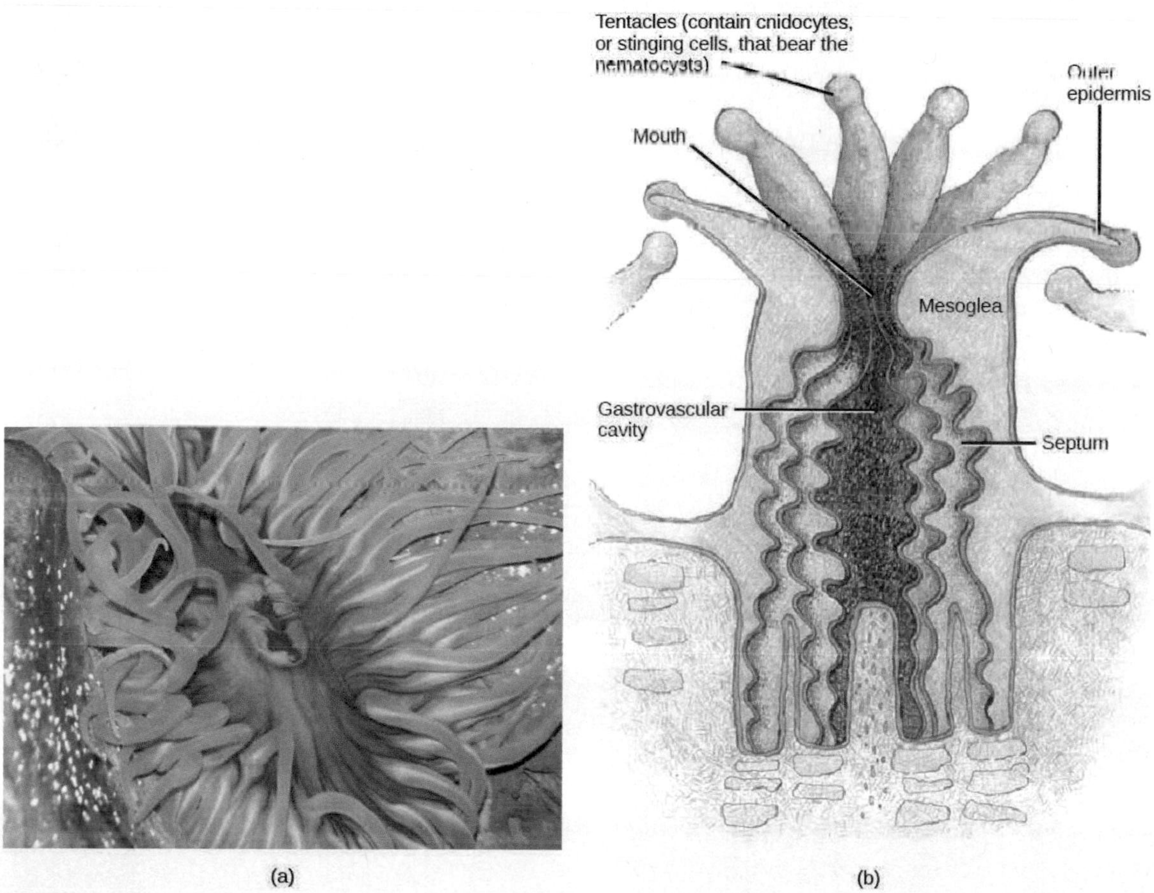

Figure 28.8 Sea anemone. The sea anemone is shown (a) photographed and (b) in a diagram illustrating its morphology. (credit a: modification of work by "Dancing With Ghosts"/Flickr; credit b: modification of work by NOAA)

The mouth of a sea anemone is surrounded by tentacles that bear cnidocytes. The slit-like mouth opening and flattened pharynx are lined with ectoderm. This structure of the pharynx makes anemones bilaterally symmetrical. A ciliated groove called a siphonoglyph is found on two opposite sides of the pharynx and directs water into it. The pharynx is the muscular part of the digestive system that serves to ingest as well as egest food, and may extend for up to two-thirds the length of the body before opening into the gastrovascular cavity. This cavity is divided into several chambers by longitudinal **septa** called *mesenteries*. Each mesentery consists of a fold of gastrodermal tissue with a layer of mesoglea between the sheets of gastrodermis. Mesenteries do not divide the gastrovascular cavity completely, and the smaller cavities coalesce at the pharyngeal opening. The adaptive benefit of the mesenteries appears to be an increase in surface area for absorption of nutrients and gas exchange, as well as additional mechanical support for the body of the anemone.

Sea anemones feed on small fish and shrimp, usually by immobilizing their prey with nematocysts. Some sea anemones establish a mutualistic relationship with hermit crabs when the crab seizes and attaches them to their shell. In this relationship, the anemone gets food particles from prey caught by the crab, and the crab is protected from the predators by the stinging cells of the anemone. Some species of anemone fish, or clownfish, are also able to live with sea anemones because they build up an acquired immunity to the toxins contained within the nematocysts and also secrete a protective mucus that prevents them from being stung.

The structure of coral polyps is similar to that of anemones, although the individual polyps are usually smaller and part of a colony, some of which are massive and the size of small buildings. Coral polyps feed on smaller planktonic organisms, including algae, bacteria, and invertebrate larvae. Some anthozoans have symbiotic associations with dinoflagellate algae called zooxanthellae. The mutually beneficial relationship between zooxanthellae and modern corals—which provides the algae with shelter—gives coral reefs their colors and supplies both organisms with nutrients. This complex mutualistic association began more than 210 million years ago, according to a new study by an international team of scientists. That this symbiotic relationship arose during a time of massive worldwide coral-reef expansion suggests that the interconnection of algae and coral is crucial for the health of coral reefs, which provide habitat for roughly one-fourth of all marine life. Reefs are threatened by a trend in ocean warming that has caused corals to expel their zooxanthellae algae and turn white, a process called coral bleaching.

Anthozoans remain *polypoid* (note that this term is easily confused with "polyploid") throughout their lives and can reproduce asexually by budding or fragmentation, or sexually by producing gametes. Male or female gametes produced by a polyp fuse to give rise to a free-swimming planula larva. The larva settles on a suitable substratum and develops into a sessile polyp.

Class Scyphozoa

Class Scyphozoa ("cup animals") includes all (and only) the marine jellies, with about 200 known species. The medusa is the prominent stage in the life cycle, although there is a polyp stage in the life cycle of most species. Most jellies range from 2 to 40 cm in length but the largest scyphozoan species, *Cyanea capillata*, can reach a size of two meters in diameter. Scyphozoans display a characteristic bell-like morphology (Figure 28.9).

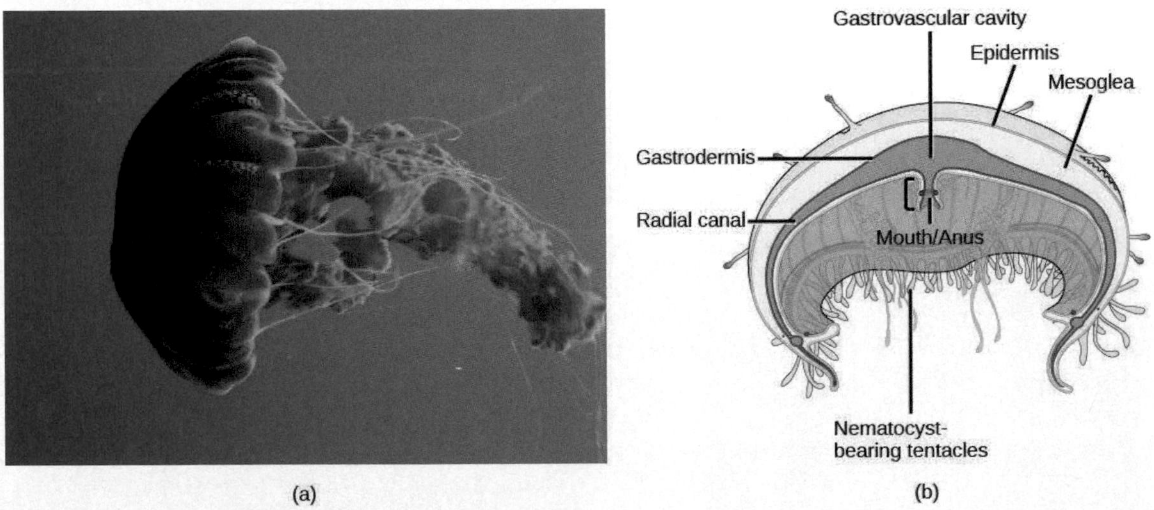

Figure 28.9 A sea jelly. A jelly is shown (a) photographed and (b) in a diagram illustrating its morphology. (credit a: modification of work by "Jimg944"/Flickr; credit b: modification of work by Mariana Ruiz Villareal)

In the sea jelly, a mouth opening is present on the underside of the animal, surrounded by hollow tentacles bearing nematocysts. Scyphozoans live most of their life cycle as free-swimming, solitary carnivores. The mouth leads to the gastrovascular cavity, which may be sectioned into four interconnected sacs, called *diverticuli*. In some species, the digestive system may branch further into *radial canals*. Like the septa in anthozoans, the branched gastrovascular cells serve two functions: to increase the surface area for nutrient absorption and diffusion, and to support the body of the animal.

In scyphozoans, nerve cells are organized in a nerve net that extends over the entire body, with a nerve ring around the edge of the bell. Clusters of sensory organs called rhopalia may be present in pockets in the edge of the bell. Jellies have a ring of muscles lining the dome of the body, which provides the contractile force required to swim through water, as well as to draw in food from the water as they swim. Scyphozoans have separate sexes. The gonads are formed from the gastrodermis and gametes are expelled through the mouth. Planula larvae are formed by external fertilization; they settle on a substratum in a polypoid form. These polyps may bud to form additional polyps or begin immediately to produce medusa buds. In a few species, the planula larva may develop directly into the medusa. The life cycle (Figure 28.10) of most scyphozoans includes both sexual medusoid and asexual polypoid body forms.

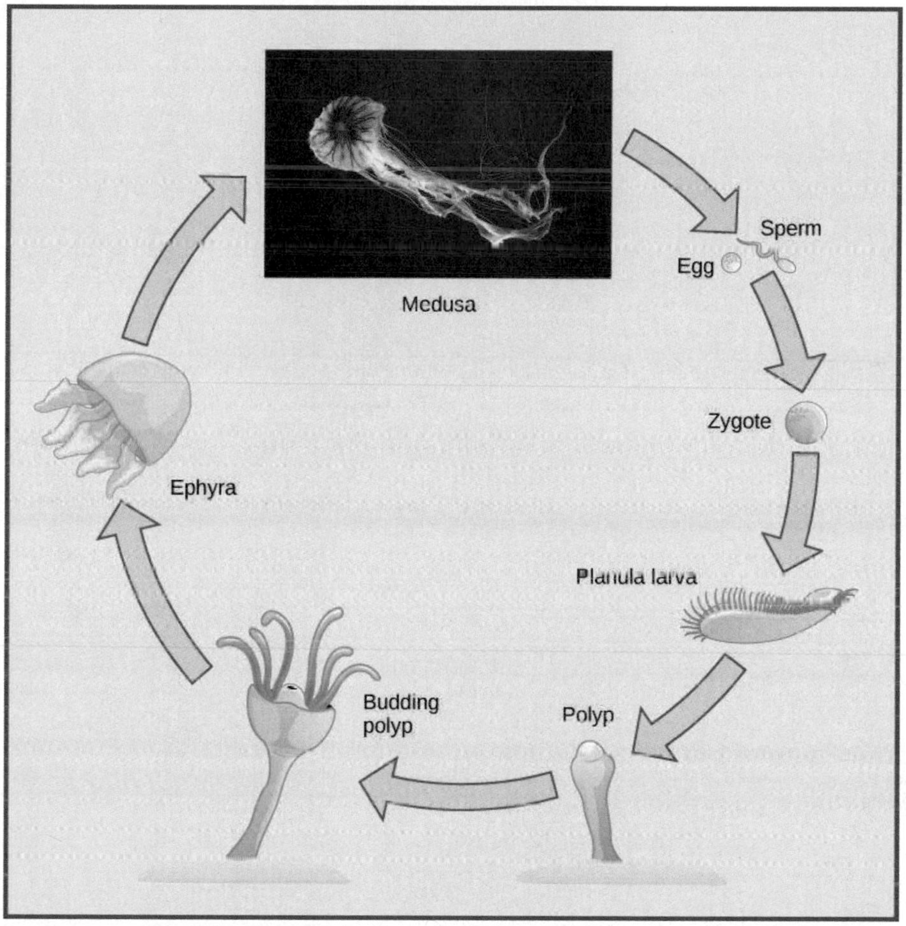

Figure 28.10 Scyphozoan life cycle. The lifecycle of most jellyfish includes two stages: the medusa stage and the polyp stage. The polyp reproduces asexually by budding, and the medusa reproduces sexually. (credit "medusa": modification of work by Francesco Crippa)

Class Cubozoa

This class includes jellies that have a box-shaped medusa, or a bell that is square in cross-section, and are colloquially known as "box jellyfish." These species may achieve sizes of 15 to 25 cm, but typically members of the Cubozoa are not as large as those of the Scyphozoa. However, cubozoans display overall morphological and anatomical characteristics that are similar to those of the scyphozoans. A prominent difference between the two classes is the arrangement of tentacles. The cubozoans contain muscular pads called **pedalia** at the corners of the square bell canopy, with one or more tentacles attached to each pedalium. In some cases, the digestive system may extend into the pedalia. Nematocysts may be arranged in a spiral configuration along the tentacles; this arrangement helps to effectively subdue and capture prey. Cubozoans include the most venomous of all the cnidarians (Figure 28.11).

These animals are unusual in having image-forming eyes, including a cornea, lens, and retina. Because these structures are made from a number of interactive tissues, they can be called *true organs*. Eyes are located in four clusters between each pair of pedalia. Each cluster consists of four simple eye spots plus two image-forming eyes oriented in different directions. How images formed by these very complex eyes are processed remains a mystery, since cubozoans have extensive nerve nets but no distinct brain. Nontheless, the presence of eyes helps the cubozoans to be active and effective hunters of small marine animals like worms, arthropods, and fish.

Cubozoans have separate sexes and fertilization occurs inside the female. Planula larvae may develop inside the female or be released, depending on species. Each planula develops into a polyp. These polyps may bud to form more polyps to create a colony; each polyp then transforms into a single medusa.

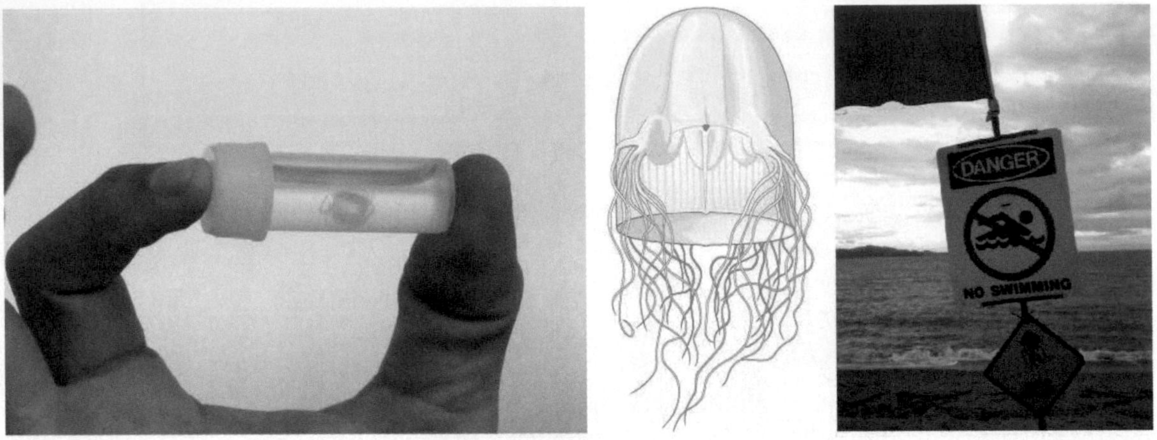

Figure 28.11 A cubozoan. The (a) tiny cubozoan jelly *Malo kingi* is thimble-shaped and, like all cubozoan jellies, (b) has four muscular pedalia to which the tentacles attach. *M. kingi* is one of two species of jellies known to cause Irukandji syndrome, a condition characterized by excruciating muscle pain, vomiting, increased heart rate, and psychological symptoms. Two people in Australia, where Irukandji jellies are most commonly found, are believed to have died from Irukandji stings. (c) A sign on a beach in northern Australia warns swimmers of the danger. (credit c: modification of work by Peter Shanks)

Class Hydrozoa

Hydrozoa is a diverse group that includes nearly 3,200 species; most are marine, although some freshwater species are known (Figure 28.12). Most species exhibit both polypoid and medusoid forms in their lifecycles, although the familiar *Hydra* has only the polyp form. The medusoid form has a muscular veil or **velum** below the margin of the bell and for this reason is called a *hydromedusa*. In contrast, the medusoid form of Scyphozoa lacks a velum and is termed a *scyphomedusa*.

The polyp form in these animals often shows a cylindrical morphology with a central gastrovascular cavity lined by the gastrodermis. The gastrodermis and epidermis have a simple layer of mesoglea sandwiched between them. A mouth opening, surrounded by tentacles, is present at the oral end of the animal. Many hydrozoans form sessile, branched colonies of specialized polyps that share a common, branching gastrovascular cavity (coenosarc), such as is found in the colonial hydroid *Obelia*.

Free-floating colonial species called **siphonophores** contain both medusoid and polypoid individuals that are specialized for feeding, defense, or reproduction. The distinctive rainbow-hued float of the Portuguese man o' war (*Physalia physalis*) creates a pneumatophore with which it regulates buoyancy by filling and expelling carbon monoxide gas. At first glance, these complex superorganisms appear to be a single organism; but the reality is that even the tentacles are actually composed of zooids laden with nematocysts. Thus, although it superficially resembles a typical medusozoan jellyfish, *P. physalis* is a free-floating hydrozoan *colony*; each specimen is made up of many hundreds of organisms, each specialized for a certain function, including motility and buoyancy, feeding, reproduction and defense. Although they are carnivorous and feed on many soft bodied marine animals, *P. physalis* lack stomachs and instead have specialized polyps called gastrozooids that they use to digest their prey in the open water.

Physalia has male and female colonies, which release their gametes into the water. The zygote develops into a single individual, which then buds asexually to form a new colony. Siphonophores include the largest known floating cnidarian colonies such as

Praya dubia, whose chain of zoids can get up to 50 meters (165 feet) long. Other hydrozoan species are solitary polyps (*Hydra*) or solitary hydromedusae (*Gonionemus*). One defining characteristic shared by the hydrozoans is that their gonads are derived from epidermal tissue, whereas in all other cnidarians they are derived from gastrodermal tissue.

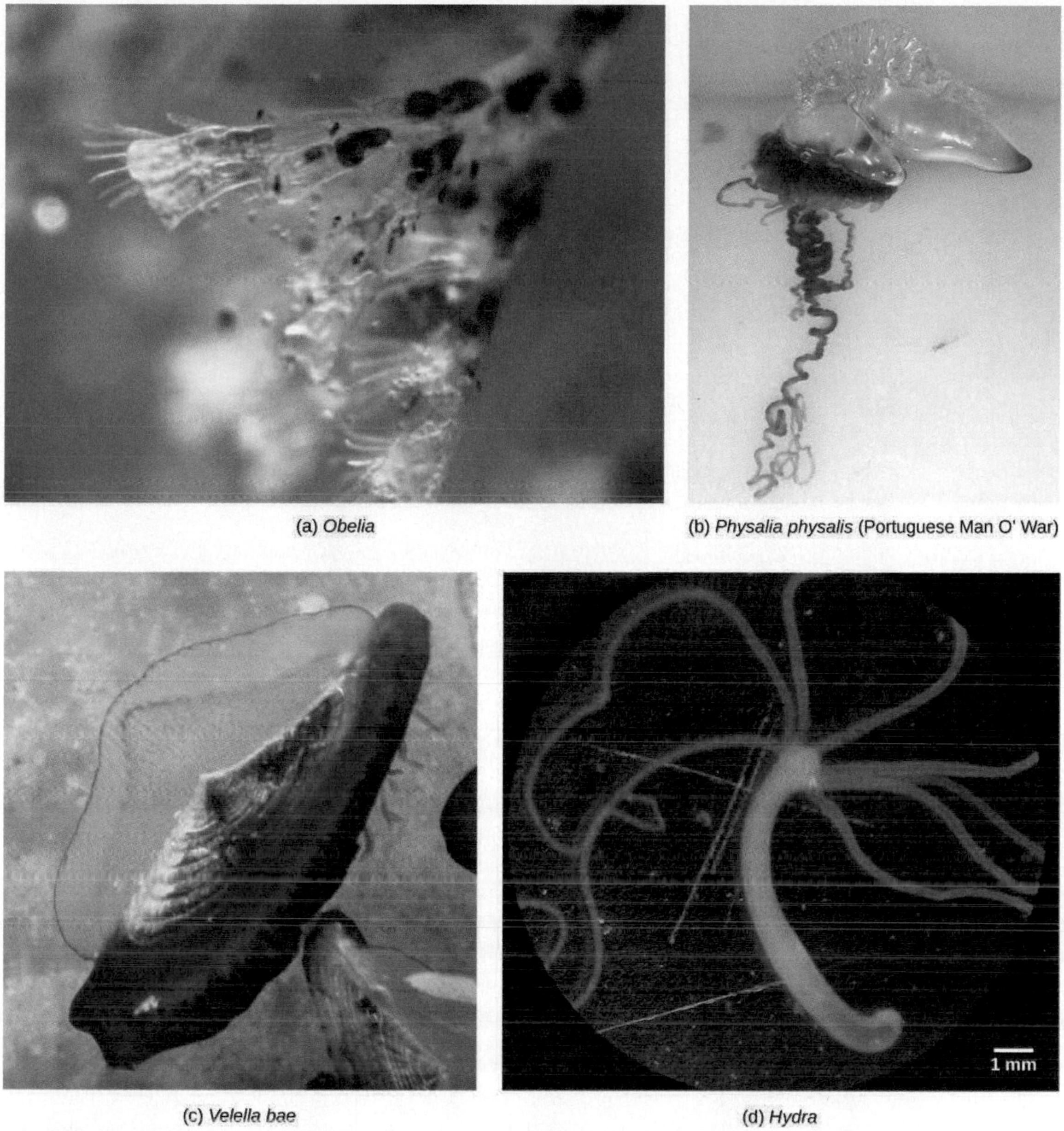

Figure 28.12 Hydrozoans. The polyp colony *Obelia* (a), siphonophore colonies *Physalia* (b) *physalis,* known as the Portuguese man o' war and *Velella bae* (c), and the solitary polyp *Hydra* (d) have different body shapes but all belong to the family Hydrozoa. (credit b: modification of work by NOAA; scale-bar data from Matt Russell)

28.3 Superphylum Lophotrochozoa: Flatworms, Rotifers, and Nemerteans

By the end of this section, you will be able to do the following:

- Describe the unique anatomical and morphological features of flatworms, rotifers, and Nemertea
- Identify an important extracoelomic cavity found in Nemertea
- Explain the key features of Platyhelminthes and their importance as parasites

Animals belonging to superphylum Lophotrochozoa are triploblastic (have three germ layers) and unlike the cnidarians, they possess an embryonic mesoderm sandwiched between the ectoderm and endoderm. These phyla are also bilaterally symmetrical, meaning that a longitudinal section will divide them into right and left sides that are superficially symmetrical. In these phyla, we also see the beginning of cephalization, the evolution of a concentration of nervous tissues and sensory organs in the head of the organism—exactly where a mobile bilaterally symmetrical organism first encounters its environment.

Lophotrochozoa are also protostomes, in which the blastopore, or the point of invagination of the ectoderm (outer germ layer), becomes the mouth opening into the alimentary canal. This developmental pattern is called protostomy or “first mouth.” Protostomes include acoelomate, pseudocoelomate, and eucoelomate phyla. The **coelom** is a cavity that separates the ectoderm from the endoderm. In acoelomates, a *solid mass* of mesoderm is sandwiched between the ectoderm and endoderm and does not form a cavity or “coelom,” leaving little room for organ development; in pseudocoelomates, there is a cavity or *pseudocoelom* that replaces the blastocoel (the cavity within the blastula), but it is only lined by mesoderm on the outside of the cavity, leaving the gut tube and organs unlined; in eucoelomates, the cavity that obliterates the blastocoel as the coelom develops is lined both on the outside of the cavity (parietal layer) *and* also on the inside of the cavity, around the gut tube and the internal organs (visceral layer).

Eucoelmate protostomes are schizocoels, in which mesoderm-producing cells typically migrate into the blastocoel during gastrulation and multiply to form a solid mass of cells. Cavities then develop within the cell mass to form the coelom. Since the forming body cavity splits the mesoderm, this protostomic coelom is termed a **schizocoelom**. As we will see later in this chapter, chordates, the phylum to which we belong, generally develop a coelom by enterocoely: pouches of mesoderm pinch off the invaginating primitive gut, or *archenteron*, and then fuse to form a complete coelom. We should note here that a eucoelomate can form its “true coelom” by either schizocoely or enterocoely. The process that produces the coelom is different and of taxonomic importance, but the result is the same: a complete, mesodermally lined coelom.

Most organisms placed in the superphylum Lophotrochozoa possess either a *lophophore* feeding apparatus or a *trochophore* larvae (thus the contracted name, “lopho-trocho-zoa”). The lophophore is a feeding structure composed of a set of ciliated tentacles surrounding the mouth. A trochophore is a free-swimming larva characterized by two bands of cilia surrounding a top-like body. Some of the phyla classified as Lophotrochozoa may be missing one or both of these defining structures. Nevertheless their placement with the Lophotrochozoa is upheld when ribosomal RNA and other gene sequences are compared. The systematics of this complex group is still unclear and much more work remains to resolve the cladistic relationships among them.

Phylum Platyhelminthes

The flatworms are acoelomate organisms that include many free-living and parasitic forms. The flatworms possess neither a lophophore nor trochophore larvae, although the larvae of one group of flatworms, the Polycladida (named after its many-branched digestive tract), are considered to be homologous to trochophore larvae. Spiral cleavage is also seen in the polycladids and other basal flatworm groups. The developmental pattern of some of the free-living forms is obscured by a phenomenon called "*blastomere anarchy*," in which a sort of temporary feeding larva forms, followed by a regrouping of cells within the embryo that gives rise to a second-stage embryo. However, both the monophyly of the flatworms and their placement in the Lophotrochozoa has been supported by molecular analyses.

The Platyhelminthes consist of two monophyletic lineages: the Catenulida and the Rhabditophora. The Catenulida, or "chain worms," is a small clade of just over 100 species. These worms typically reproduce asexually by budding. However, the offspring do not fully detach from the parents and the formation resembles a chain in appearance. All of the flatworms discussed here are part of the Rhabditophora ("rhabdite bearers"). **Rhabdites** are rodlike structures discharged in the mucus produced by some free-living flatworms; Eucoelmate protostomes are schizocoels, in which mesoderm-producing cells typically migrate into the blastocoel during gastrulation likely serve in both defense and to provide traction for ciliary gliding along the substrate. Unlike free-living flatworms, many species of trematodes and cestodes are parasitic, including important parasites of humans.

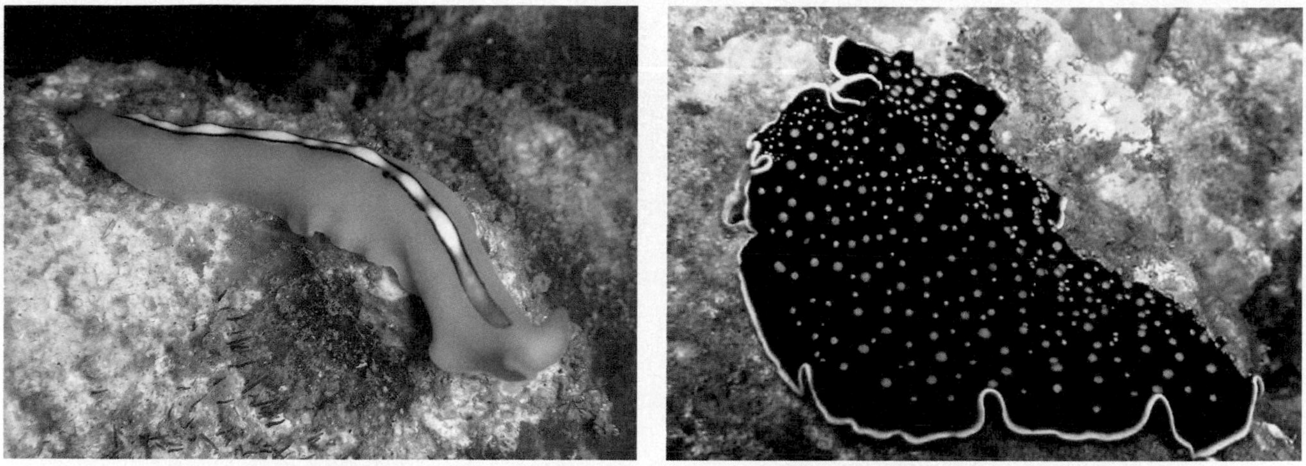

Figure 28.13 Flatworms exhibit significant diversity. (a) A blue Pseudoceros flatworm *(Pseudoceros bifurcus)*; (b) gold speckled flatworm *(Thysanozoon nigropapillosum)*. (credit a: modification of work by Stephen Childs; b: modification of work by Pril Fish.)

Flatworms have three embryonic tissue layers that give rise to epidermal tissues (from ectoderm), the lining of the digestive system (from endoderm), and other internal tissues (from mesoderm). The epidermal tissue is a single layer of cells or a layer of fused cells (**syncytium**) that covers two layers of muscle, one circular and the other longitudinal. The mesodermal tissues include mesenchymal cells that contain collagen and support secretory cells that produce mucus and other materials at the surface. Because flatworms are acoelomates, the mesodermal layer forms a solid mass between the outer epidermal surface and the cavity of the digestive system.

Physiological Processes of Flatworms

The free-living species of flatworms are predators or scavengers. Parasitic forms feed by absorbing nutrients provided by their hosts. Most flatworms, such as the planarian shown in Figure 28.14, have a branching gastrovascular cavity rather than a complete digestive system. In such animals, the "mouth" is also used to expel waste materials from the digestive system, and thus also serves as an anus. (A few species may have a second anal pore or opening.) The gut may be a simple sac or highly branched. Digestion is primarily extracellular, with digested materials taken into the cells of the gut lining by phagocytosis. One parasitic group, the tapeworms (cestodes), lacks a digestive system altogether, and absorb digested food from the host.

Flatworms have an excretory system with a network of tubules attached to **flame cells**, whose cilia beat to direct waste fluids concentrated in the tubules out of the body through excretory pores. The system is responsible for the regulation of dissolved salts and the excretion of nitrogenous wastes. The nervous system consists of a pair of lateral nerve cords running the length of the body with transverse connections between them. Two large cerebral ganglia—concentrations of nerve cell bodies at the anterior end of the worm—are associated with photosensory and chemosensory cells. There is neither a circulatory nor a respiratory system, with gas and nutrient exchange dependent on diffusion and cell-to-cell junctions. This necessarily limits the thickness of the body in these organisms, constraining them to be "flat" worms. Most flatworm species are *monoecious* (both male and female reproductive organs are found in the same individual), and fertilization is typically internal. Asexual reproduction by fission is common in some groups.

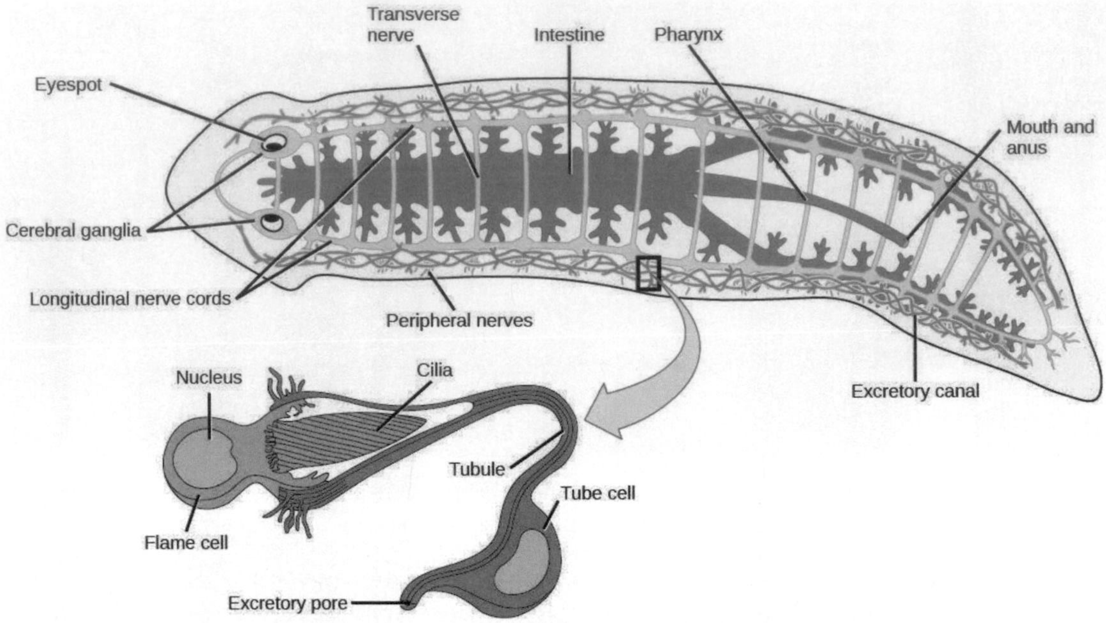

Figure 28.14 *Planaria*, a free-living flatworm. The planarian is a flatworm that has a gastrovascular cavity with one opening that serves as both mouth and anus. The excretory system is made up of flame cells and tubules connected to excretory pores on both sides of the body. The nervous system is composed of two interconnected nerve cords running the length of the body, with cerebral ganglia and eyespots at the anterior end.

Diversity of Flatworms

The flatworms have been traditionally divided into four classes: Turbellaria, Monogenea, Trematoda, and Cestoda (Figure 28.15). However, the relationships among members of these classes has recently been reassessed, with the turbellarians in particular now viewed as paraphyletic, since its descendants may also include members of the other three classes. Members of the clade or class Rhabditophora are now dispersed among multiple orders of Platyhelminthes, the most familiar of these being the Polycladida, which contains the large marine flatworms; the Tricladida (which includes *Dugesia* ["planaria"] and *Planaria* and its relatives); and the major parasitic orders: Monogenea (fish ectoparasites), Trematoda (flukes), and Cestoda (tapeworms), which together form a monophyletic clade.

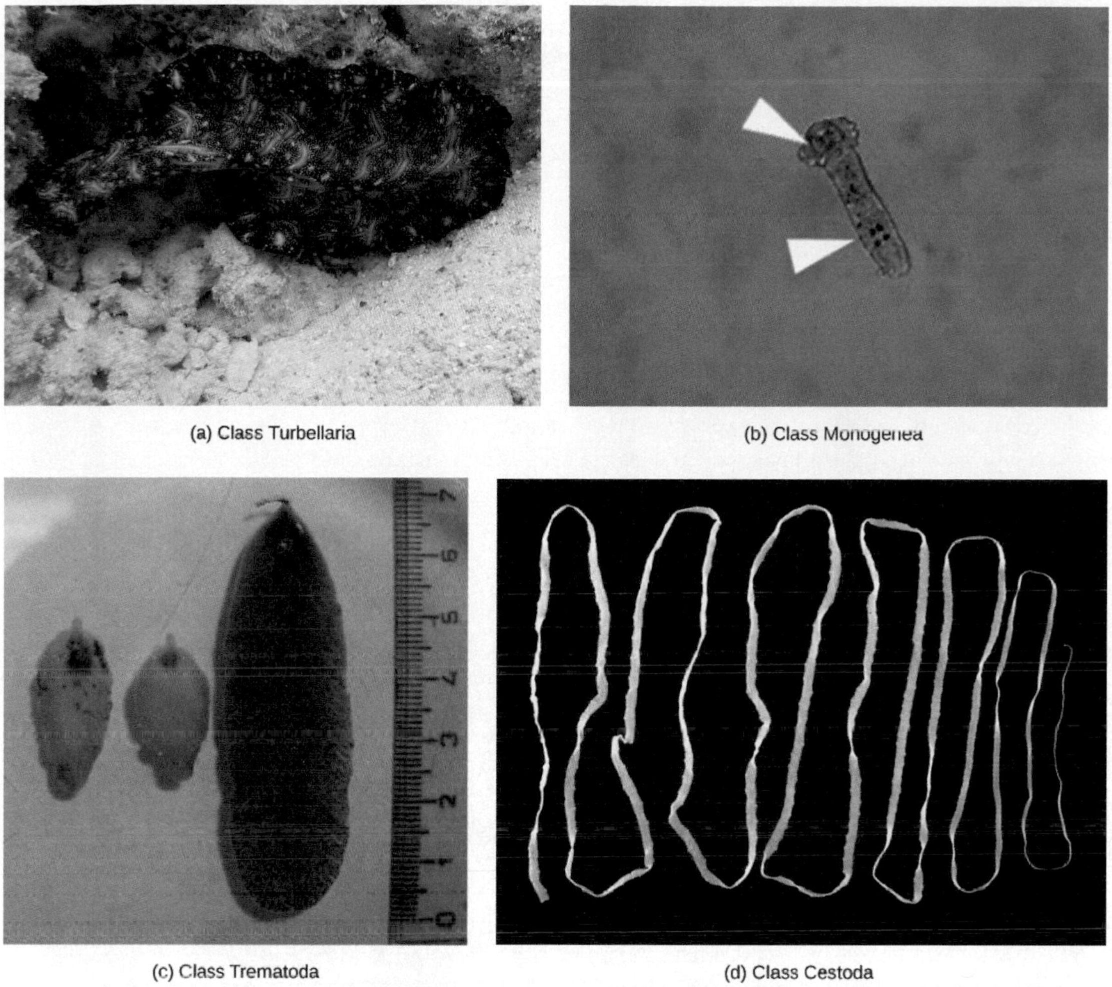

Figure 28.15 Traditional flatworm classes. Phylum Platyhelminthes was previously divided into four classes. (a) Class Turbellaria includes the tree-living polycladid Bedford's flatworm (*Pseudobiceros bedfordi*), which is about 8 to 10 cm in length. (b) The parasitic class Monogenea includes *Dactylogyrus* spp, commonly called gill flukes, which are about 0.2 mm in length and have two anchors, indicated by arrows used to attach the parasite on to the gills of host fish. (c) The class Trematoda includes *Fascioloides magna* (right) and *Fasciola hepatica* (two specimens on left, also known as the common liver fluke). (d) Class Cestoda includes tapeworms such as this *Taenia saginata*, infects both cattle and humans, and can reach 4 to 10 meters in length; the specimen shown here is about four meters long. (credit a: modification of work by Jan Derk; credit d: modification of work by CDC)

Most free-living flatworms are marine polycladids, although tricladid species live in freshwater or moist terrestrial environments, and there are a number of members from other orders in both environments. The ventral epidermis of free-living flatworms is ciliated, which facilitates their locomotion. Some free-living flatworms are capable of remarkable feats of regeneration in which an individual may regrow its head or tail after being severed, or even several heads if the planaria is cut lengthwise.

The monogeneans are *ectoparasites*, mostly of fish, with simple life cycles that consist of a free-swimming larva that attaches to a fish, prior to its transformation to the ectoparasitic adult form. The parasite has only one host and that host is usually very specific. The worms may produce enzymes that digest the host tissues, or they may simply graze on surface mucus and skin particles. Most monogeneans are hermaphroditic, but the male gametes develop first and so cross-fertilization is quite common.

The trematodes, or flukes, are internal parasites of mollusks and many other groups, including humans. Trematodes have complex life cycles that involve a primary host in which sexual reproduction occurs, and one or more secondary hosts in which asexual reproduction occurs. The primary host is usually a vertebrate and the secondary host is almost always a mollusk, in which multiple larvae are produced asexually. Trematodes, which attached internally to the host via an oral and medial sucker, are responsible for serious human diseases including schistosomiasis, caused by several species of the blood fluke, *Schistosoma* spp. Various forms of schistosomiasis infect an estimated 200 million people in the tropics, leading to organ damage, secondary infection by bacteria, and chronic symptoms like fatigue. Infection occurs when the human enters the water and metacercaria larvae, released from the snail host, locate and penetrate the skin. The parasite infects various organs in the body and feeds on red blood cells before reproducing.

Many of the eggs are released in feces and find their way into a waterway, where they are able to reinfect the snail host. The eggs, which have a barb on them, can damage the vascular system of the human host, causing ulceration, abscesses, and bloody diarrhea, wherever they reside, thereby allowing other pathogens to cause secondary infections. In fact, it is the parasite's eggs that produce most of the main ill effects of schistosomiasis. Many eggs do not make the transit through the veins of the host for elimination, and are swept by blood flow back to the liver and other locations, where they can cause severe inflammation. In the liver, the errant eggs may impede circulation and cause cirrhosis. Control is difficult in impoverished areas in unsanitary, crowded conditions, and prognosis is poor in people with heavy infections of *Schistosoma japonicum*, without early treatment.

The cestodes, or tapeworms, are also internal parasites, mainly of vertebrates (Figure 28.16). Tapeworms, such as those of *Taenia* spp, live in the intestinal tract of the primary host and remain fixed using a sucker or hooks on the anterior end, or scolex, of the tapeworm body, which is essentially a colony of similar subunits called proglottids. Each proglottid may contain an excretory system with flame cells, along with reproductive structures, both male and female. Because they are so long and flat, tapeworms do not need a digestive system; instead, they absorb nutrients from the food matter surrounding them in the host's intestine by diffusion.

Proglottids are produced at the scolex and gradually migrate to the end of the tapeworm; at this point, they are "mature" and all structures except fertilized eggs have degenerated. Most reproduction occurs by cross-fertilization between different worms in the same host, but may also occur between proglottids. The mature proglottids detach from the body of the worm and are released into the feces of the organism. The eggs are eaten by an intermediate host, typically another vertebrate. The juvenile worm infects the intermediate host and takes up residence, usually in muscle tissue. When the muscle tissue is consumed by the primary host, the cycle is completed. There are several tapeworm parasites of humans that are transmitted by eating uncooked or poorly cooked pork, beef, or fish.

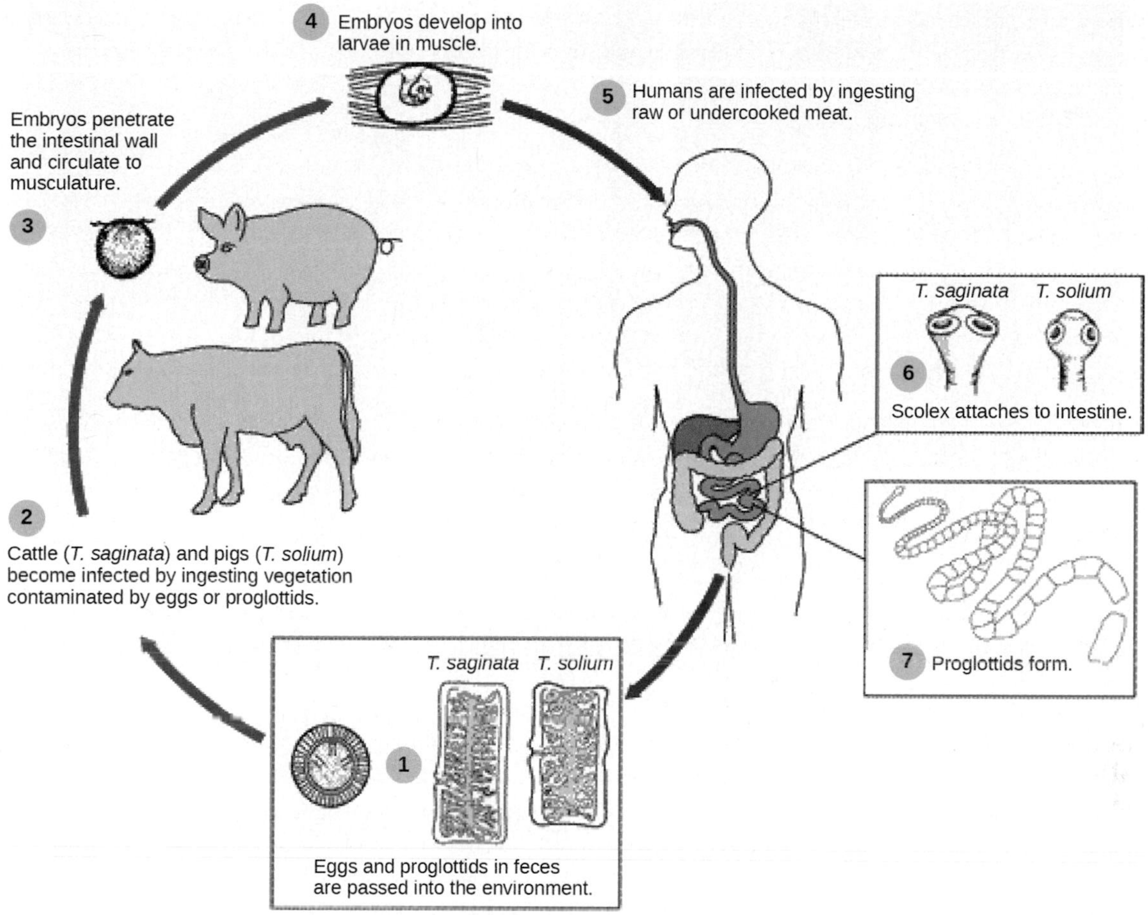

Figure 28.16 Tapeworm life cycle. Tapeworm (*Taenia* spp.) infections occur when humans consume raw or undercooked infected meat. (credit: modification of work by CDC)

Phylum Rotifera

The rotifers ("wheel-bearer") belong to a group of microscopic (about 100 μm to 2 mm) mostly aquatic animals that get their name from the **corona**—a pair of ciliated feeding structures that appear to rotate when viewed under the light microscope (Figure 28.17). Although their taxonomic status is currently in flux, one treatment places the rotifers in three classes: Bdelloidea, Monogononta, and Seisonidea. In addition, the parasitic "spiny headed worms" currently in phylum Acanthocephala, appear to be modified rotifers and will probably be placed into the group in the near future. Undoubtedly the rotifers will continue to be revised as more phylogenetic evidence becomes available.

The pseudocoelomate body of a rotifer is remarkably complex for such a small animal (roughly the size of a *Paramecium*) and is divided into three sections: a *head* (which contains the corona), a *trunk* (which contains most of the internal organs), and the *foot*. A cuticle, rigid in some species and flexible in others, covers the body surface. They have both skeletal muscle associated with locomotion and visceral muscles associated with the gut, both composed of single cells. Rotifers are typically free-swimming or planktonic (drifting) organisms, but the toes or extensions of the foot can secrete a sticky material to help them adhere to surfaces. The head contains a number of eyespots and a bilobed "brain," with nerves extending into the body.

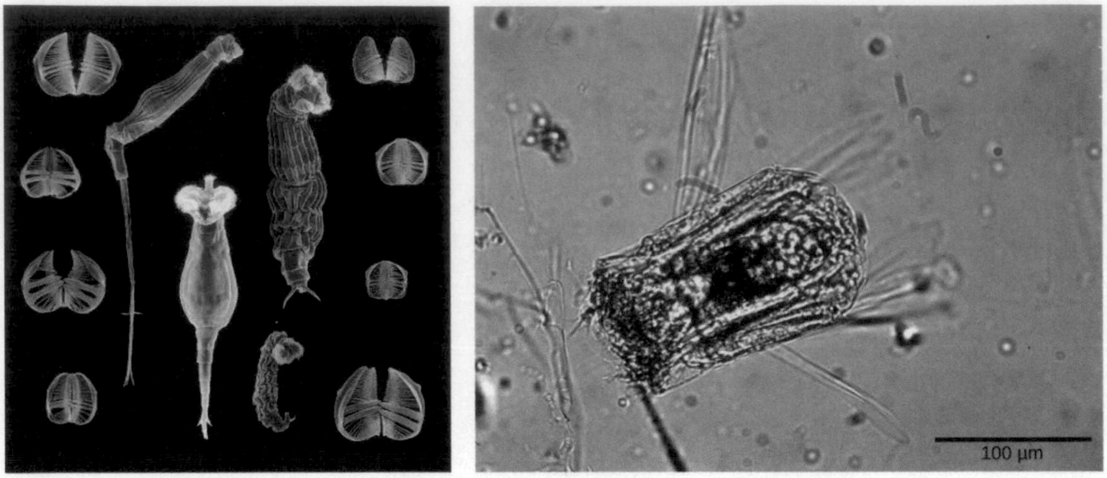

(a) Bdelloidea (a) Monogonota

Figure 28.17 Rotifers. Shown are examples from two of the three classes of rotifer. (a) Species from the class Bdelloidea are characterized by a large corona. The whole animals in the center of this scanning electron micrograph are shown surrounded by several sets of jaws from the mastax of rotifers. (b) *Polyarthra*, from the largest rotifer class Monogononta, has a smaller corona than bdelloid rotifers, and a single gonad, which give the class its name. (credit a: modification of work by Diego Fontaneto; credit b: modification of work by U.S. EPA; scale-bar data from Cory Zanker)

Rotifers are commonly found in freshwater and some saltwater environments throughout the world. As filter feeders, they will eat dead material, algae, and other microscopic living organisms, and are therefore very important components of aquatic food webs. A rotifer's food is directed toward the mouth by the current created from the movement of the coronal cilia. The food particles enter the mouth and travel first to the **mastax**—a muscular pharynx with toothy jaw-like structures. Examples of the jaws of various rotifers are seen in Figure 28.17a. Masticated food passes near digestive and salivary glands, into the stomach, and then to the intestines. Digestive and excretory wastes are collected in a cloacal bladder before being released out the anus.

LINK TO LEARNING

Watch this video (http://openstax.org/l/rotifers) to see rotifers feeding.

About 2,200 species of rotifers have been identified. Figure 28.18 shows the anatomy of a rotifer belonging to class Bdelloidea. Some rotifers are dioecious organisms and exhibit sexual dimorphism (males and females have different forms). In many dioecious species, males are short-lived and smaller with no digestive system and a single testis. Many rotifer species exhibit haplodiploidy, a method of sex determination in which a fertilized egg develops into a female and an unfertilized egg develops into a male. However, reproduction in the bdelloid rotifers is exclusively parthenogenetic and appears to have been so for millions of years: Thus, all bdelloid rotifers and their progeny are female! The bdelloids may compensate for this genetic insularity by borrowing genes from the DNA of other species. Up to 10% of a bdelloid genome comprises genes imported from related species. Some rotifer eggs are capable of extended dormancy for protection during harsh environmental conditions.

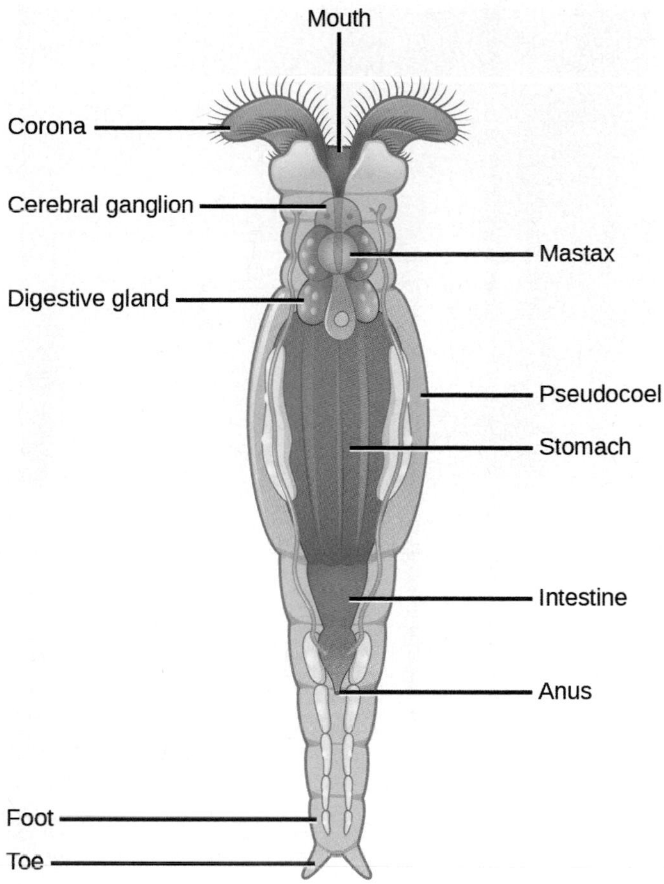

Figure 28.18 A bdelloid rotifer. This illustration shows the anatomy of a bdelloid rotifer.

Phylum Nemertea

The Nemertea are colloquially known as *ribbon worms* or proboscis worms. Most species of phylum Nemertea are marine and predominantly benthic (bottom dwellers), with an estimated 900 known species. However, nemerteans have been recorded in freshwater and very damp terrestrial habitats as well. Most nemerteans are carnivores, feeding on worms, clams, and crustaceans. Some species are scavengers, and some, like *Malacobdella grossa,* have also evolved commensal relationships with mollusks. Economically important species have at times devastated commercial fishing of clams and crabs. Nemerteans have almost no predators and two species are sold as fish bait.

Morphology

Nemerteans vary in size from 1 cm to several meters. They show bilateral symmetry and remarkable contractile properties. Because of their contractility, they can change their morphological presentation in response to environmental cues. Animals in phylum Nemertea are soft and unsegmented animals, with a morphology like a flattened tube. (Figure 28.19).

Figure 28.19 A proboscis worm. The proboscis worm (*Parborlasia corrugatus*) is a scavenger that combs the sea floor for food. The species is a member of the phylum Nemertea. The specimen shown here was photographed in the Ross Sea, Antarctica. (credit: Henry Kaiser, National Science Foundation)

A unique characteristic of this phylum is the presence of an eversible proboscis enclosed in a pocket called a **rhynchocoel** (not part of the animal's actual coelom). The proboscis is located dorsal to the gut and serves as a harpoon or tentacle for food capture. In some species it is ornamented with barbs. The rhynchocoel is a fluid-filled cavity that extends from the head to nearly two-thirds of the length of the gut in these animals (Figure 28.20). The proboscis may be extended by hydrostatic pressure generated by contraction of muscle of the rhynchocoel and retracted by a retractor muscle attached to the rear wall of the rhynchocoel.

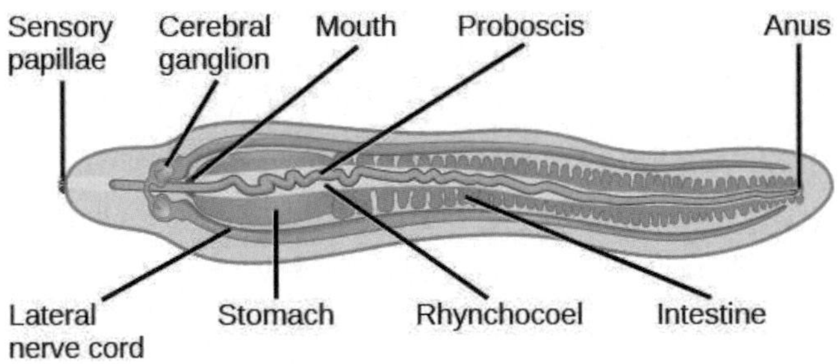

Figure 28.20 The anatomy of a Nemertean is shown.

LINK TO LEARNING

Watch this video (https://www.openstax.org/l/nemertean) to see a nemertean attack a polychaete with its proboscis.

Digestive System

The nemerteans, which are primarily predators of annelids and crustaceans, have a well-developed digestive system. A mouth opening that is ventral to the rhynchocoel leads into the foregut, followed by the intestine. The intestine is present in the form of diverticular pouches and ends in a rectum that opens via an anus. Gonads are interspersed with the intestinal diverticular pouches and open outward via genital pores.

Nemerteans are sometimes classified as acoels, but because their closed circulatory system is derived from the coelomic cavity of the embryo, they may be regarded as coelomic. Their circulatory system consists of a closed loop formed by a connected pair of lateral blood vessels. Some species may also have a dorsal vessel or cross-connecting vessels in addition to lateral ones. Although the circulatory fluid contains cells, it is often colorless. However, the blood cells of some species bear hemoglobin as well as other yellow or green pigments. The blood vessels are contractile, although there is usually no regular circulatory pathway, and movement of blood is also facilitated by the contraction of muscles in the body wall. The circulation of fluids in the rhychocoel is more or less independent of the blood circulation, although blind branches from the blood vessels into the rhyncocoel wall can mediate exchange of materials between them. A pair of **protonephridia**, or excretory tubules, is present in these animals to facilitate osmoregulation. Gaseous exchange occurs through the skin.

Nervous System

Nemerteans have a "brain" composed of four ganglia situated at the anterior end, around the rhynchocoel. Paired longitudinal nerve cords emerge from the brain ganglia and extend to the posterior end. Additional nerve cords are found in some species. Interestingly, the brain can contain hemoglobin, which acts as an oxygen reserve. *Ocelli* or eyespots are present in pairs, in multiples of two in the anterior portion of the body. It is speculated that the eyespots originate from neural tissue and not from the epidermis.

Reproduction

Nemerteans, like flatworms, have excellent powers of regeneration, and asexual reproduction by fragmentation is seen in some species. Most animals in phylum Nemertea are dioecious, although freshwater species may be hermaphroditic. Stem cells that become gametes aggregate within gonads placed along the digestive tract. Eggs and sperm are released into the water, and fertilization occurs externally. Like most lophotrochozoan protostomes, cleavage is spiral, and development is usually direct, although some species have a trochophore-like larva, in which a young worm is constructed from a series of imaginal discs that begin as invaginations from the body surface of the larva.

28.4 Superphylum Lophotrochozoa: Molluscs and Annelids

By the end of this section, you will be able to do the following:

- Describe the unique anatomical and morphological features of molluscs and annelids
- Describe the formation of the coelom
- Identify an important extracoelomic cavity in molluscs
- Describe the major body regions of Mollusca and how they vary in different molluscan classes
- Discuss the advantages of true body segmentation
- Describe the features of animals classified in phylum Annelida

The annelids and the mollusks are the most familiar of the lophotrochozoan protostomes. They are also more "typical" lophotrochozoans, since both groups include aquatic species with trochophore larvae, which unite both taxa in common ancestry. These phyla show how a flexible body plan can lead to biological success in terms of abundance and species diversity. The phylum Mollusca has the second greatest number of species of all animal phyla with nearly 100,000 described extant species, and about 80,000 described extinct species. In fact, it is estimated that about 25 percent of all known marine species are mollusks! The annelids and mollusca are both bilaterally symmetrical, cephalized, triploblastic, schizocoelous eucoeolomates They include animals you are likely to see in your backyard or on your dinner plate!

Phylum Mollusca

The name "Mollusca" means "soft" body, since the earliest descriptions of molluscs came from observations of "squishy," unshelled cuttlefish. Molluscs are predominantly a marine group of animals; however, they are also known to inhabit freshwater as well as terrestrial habitats. This enormous phylum includes chitons, tusk shells, snails, slugs, nudibranchs, sea butterflies, clams, mussels, oysters, squids, octopuses, and nautiluses. Molluscs display a wide range of morphologies in each class and subclass, but share a few key characteristics (Figure 28.21). The chief locomotor structure is usually a muscular **foot**. Most internal organs are contained in a region called the **visceral mass**. Overlying the visceral mass is a fold of tissue called the **mantle**; within the cavity formed by the mantle are respiratory structures called **gills**, that typically fold over the visceral mass. The mouths of most mollusks, except bivalves (e.g., clams) contain a specialized feeding organ called a **radula**, an abrasive tonguelike structure. Finally, the mantle secretes a calcium-carbonate-hardened shell in most molluscs, although this is greatly reduced in the class Cephalopoda, which contains the octopuses and squids.

VISUAL CONNECTION

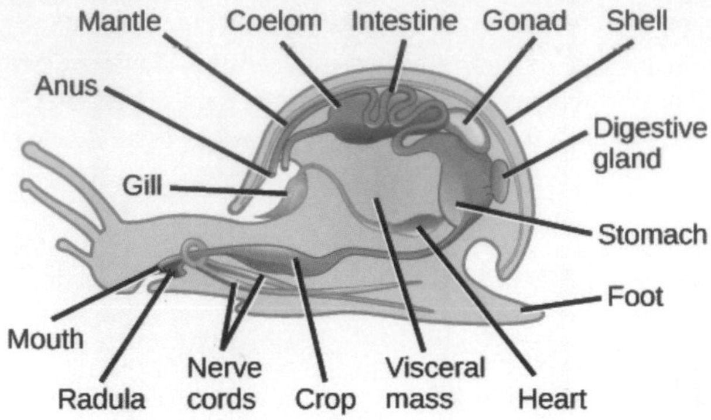

Figure 28.21 Molluscan body regions. There are many species and variations of molluscs; this illustration shows the anatomy of an aquatic gastropod. In a terrestrial gastropod, the mantle cavity itself would serve as a respiratory organ.

Which of the following statements about the anatomy of a mollusc is false?

1. Most molluscs have a radula for grinding food.
2. A digestive gland is connected to the stomach.
3. The tissue beneath the shell is called the mantle.
4. The digestive system includes a gizzard, a stomach, a digestive gland, and the intestine.

The muscular **foot** is the ventral-most organ, whereas the **mantle** is the limiting dorsal organ that folds over the **visceral mass**. The foot, which is used for locomotion and anchorage, varies in shape and function, depending on the type of mollusk under study. In shelled mollusks, the foot is usually the same size as the opening of the shell. The foot is both retractable and extendable. In the class Cephalopoda ("head-foot"), the foot takes the form of a funnel for expelling water at high velocity from the mantle cavity; and the anterior margin of the foot has been modified into a circle of *arms* and *tentacles*.

The visceral mass is present above the foot, in the visceral hump. This mass contains digestive, nervous, excretory, reproductive, and respiratory systems. Molluscan species that are exclusively aquatic have gills that extend into the mantle cavity, whereas some terrestrial species have "lungs" formed from the lining of the mantle cavity. Mollusks are schizocoelous eucoelomates, but the coelomic cavity in adult animals has been largely reduced to a cavity around the heart. However, a reduced coelom sometimes surrounds the gonads, part of the kidneys, and intestine as well. This overall coelomic reduction makes the mantle cavity the major internal body chamber.

Most mollusks have a special rasp-like organ, the **radula**, which bears chitinous filelike teeth. The radula is present in all groups except the bivalves, and serves to shred or scrape food before it enters the digestive tract. The **mantle** (also known as the *pallium*) is the dorsal epidermis in mollusks; all mollusks except some cephalopods are specialized to secrete a calcareous shell that protects the animal's soft body.

Most mollusks are dioecious animals and fertilization occurs externally, although this is not the case in terrestrial mollusks, such as snails and slugs, or in cephalopods. In most aquatic mollusks, the zygote hatches and produces a **trochophore larva**, with several bands of cilia around a toplike body, and an additional apical tuft of cilia. In some species, the trochophore may be followed by additional larval stages, such as a **veliger** larvae, before the final metamorphosis to the adult form. Most cephalopods develop directly into small versions of their adult form.

Classification of Phylum Mollusca

Phylum Mollusca comprises a very diverse group of organisms that exhibits a dramatic variety of forms, ranging from chitons to snails to squids, the latter of which typically show a high degree of intelligence. This variability is a consequence of modification of the basic body regions, especially the foot and mantle. The phylum is organized into eight classes: Caudofoveata, Solenogastres, Monoplacophora, Polyplacophora, Gastropoda, Cephalopoda, Bivalvia, and Scaphopoda. Although each molluscan class appears to be monophyletic, their relationship to one another is unclear and still being reviewed.

Both the Caudofoveata and the Solenogastres include shell-less, worm-like animals primarily found in benthic marine habitats.

Although these animals lack a calcareous shell, they get some protection from calcareous spicules embedded in a cuticle that covers their epidermis. The mantle cavity is reduced, and both groups lack eyes, tentacles, and nephridia (excretory organs). The Caudofoveata possess a radula, but the Solenogastres do not have a radula or gills. The foot is also reduced in the Solenogastres and absent from the Caudofoveata.

Long thought to be extinct, the first living specimens of Monoplacophora, *Neopilina galatheae*, were discovered in 1952 on the ocean bottom near the west coast of Costa Rica. Today there are over 25 described species. Members of class Monoplacophora ("bearing one plate") possess a single, cap-like shell that covers the dorsal body. The morphology of the shell and the underlying animal can vary from circular to ovate. They have a simple radula, a looped digestive system, multiple pairs of excretory organs, and a pair of gonads. Multiple gills are located between the foot and the edge of the mantle.

Animals in class Polyplacophora ("bearing many plates") are commonly known as "chitons" and bear eight limy plates that make up the dorsal shell (Figure 28.22). These animals have a broad, ventral foot that is adapted for suction onto rocks and other substrates, and a mantle that extends beyond the edge of the shell. Calcareous spines on the exposed mantle edge provide protection from predators. Respiration is facilitated by multiple pairs of gills in the mantle cavity. Blood from the gills is collected in a posterior heart, and then sent to the rest of the body in a **hemocoel**—an *open circulation system* in which the blood is contained in connected chambers surrounding various organs rather than within individual blood vessels. The radula, which has teeth composed of an ultra-hard magnetite, is used to scrape food organisms off rocky surfaces. Chiton teeth have been shown to exhibit the greatest hardness and stiffness of any biomineral material reported to date, being as much as three-times harder than human enamel and the calcium carbonate-based shells of mollusks.

The nervous system is rudimentary with only buccal or "cheek" ganglia present at the anterior end. Multiple tiny sensory structures, including photosensors, extend from the mantle into channels in the upper layer of the shell. These structures are called **esthetes** and are unique to the chitons. Another sensory structure under the radula is used to sample the feeding environment. A single pair of nephridia is used for the excretion of nitrogenous wastes.

Figure 28.22 A chiton. This chiton from the class Polyplacaphora has the eight-plated shell for which its class is named. (credit: Jerry Kirkhart)

Class Bivalvia ("two-valves") includes clams, oysters, mussels, scallops, geoducks, and shipworms. Some bivalves are almost microscopic, while others, in the genus *Tridacna,* may be one meter in length and weigh 225 kilograms. Members of this class are found in marine as well as freshwater habitats. As the name suggests, bivalves are enclosed in two-part valves or shells (Figure 28.23a) fused on the dorsal side by hinge ligaments as well as shell teeth on the ventral side that keep the two halves aligned. The two shells, which consist of an outer organic layer, a middle prismatic layer, and a very smooth nacreous layer, are joined at the oldest part of the shell called the umbo. Anterior and posterior adductor and abductor muscles close and open the shell respectively.

The overall body of the bivalve is laterally flattened; the foot is wedge-shaped; and the head region is poorly developed (with no obvious mouth). Bivalves are filter-feeders, and a radula is absent in this class of mollusks. The mantle cavity is fused along the edges except for openings for the foot and for the intake and expulsion of water, which is circulated through the mantle cavity by the actions of the incurrent and excurrent siphons. During water intake by the incurrent siphon, food particles are captured by the paired posterior gills (ctenidia) and then carried by the movement of cilia forward to the mouth. Excretion and osmoregulation are performed by a pair of nephridia. Eyespots and other sensory structures are located along the edge of the mantle in some species. The "eyes" are especially conspicuous in scallops (Figure 28.23b). Three pairs of connected ganglia regulate activity of different body structures.

Figure 28.23 Bivalves. These mussels (a), found in the intertidal zone in Cornwall, England, show the bivalve shell. The scallop *Argopecten irradians* (b) has a fluted shell and conspicuous eyespots. (credit (a): Mark A. Wilson. credit (b) Rachael Norris and Marina Freudzon. https://commons.wikimedia.org/w/index.php?curid=17251065 (http://openstax.org/l/scallop_eyes))

One of the functions of the mantle is to secrete the shell. Some bivalves, like oysters and mussels, possess the unique ability to secrete and deposit a calcareous nacre or "mother of pearl" around foreign particles that may enter the mantle cavity. This property has been commercially exploited to produce pearls.

LINK TO LEARNING

Watch the animations of bivalves feeding: View the process in clams (http://openstax.org/l/clams) and mussels (http://openstax.org/l/mussels) at these sites.

More than half of molluscan species are in the class Gastropoda ("stomach foot"), which includes well-known mollusks like snails, slugs, conchs, cowries, limpets, and whelks. Aquatic gastropods include both marine and freshwater species, and all terrestrial mollusks are gastropods. Gastropoda includes shell-bearing species as well as species without shells. Gastropod bodies are asymmetrical and usually present a coiled shell (Figure 28.24a). Shells may be *planospiral* (like a garden hose wound up), commonly seen in garden snails, or *conispiral*, (like a spiral staircase), commonly seen in marine conches. Cowrie shells have a polished surface because the mantle extends up over the top of the shell as it is secreted.

Figure 28.24 Gastropods. Snails(a) and slugs(b) are both gastropods, but slugs lack a shell. (credit a: modification of work by Murray Stevenson; credit b: modification of work by Rosendahl)

A key characteristic of some gastropods is the embryonic development of **torsion**. During this process, the mantle and visceral mass are rotated around the perpendicular axis over the center of the foot to bring the anal opening forward just behind the head (Figure 28.25), creating a very peculiar situation. The left gill, kidney, and heart atrium are now on the right side, whereas the original right gill, kidney, and heart atrium are on the left side. Even stranger, the nerve cords have been twisted and contorted into a figure-eight pattern. Because of the space made available by torsion in the mantle cavity, the animal's sensitive head end can now be withdrawn into the protection of the shell, and the tougher foot (and sometimes the protective covering or **operculum**) forms a barrier to the outside. The strange arrangement that results from torsion poses a serious sanitation problem by creating the possibility of wastes being washed back over the gills, causing **fouling.** There is actually no really perfect explanation for the embryonic development of torsion, and some groups that formerly exhibited torsion in their ancestral groups are now known to have reversed the process.

Gastropods also have a foot that is modified for crawling. Most gastropods have a well-defined head with tentacles and eyes. A complex radula is used to scrape up food particles. In aquatic gastropods, the mantle cavity encloses the gills (ctenidia), but in land gastropods, the mantle itself is the major respiratory structure, acting as a kind of lung. **Nephridia** ("kidneys") are also found in the mantle cavity.

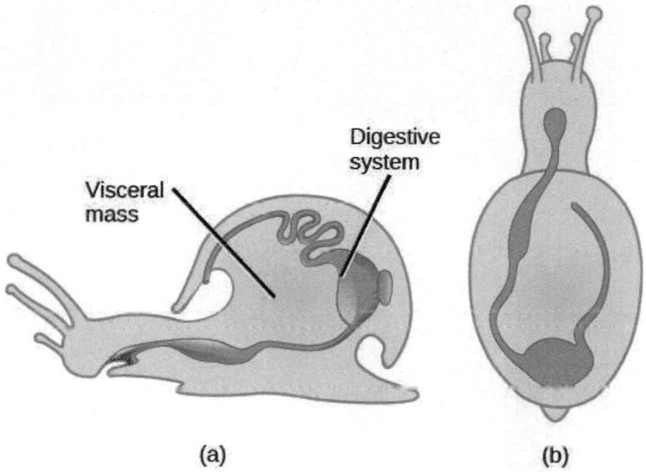

Figure 28.25 Torsion in gastropods. During embryonic development of some gastropods, the visceral mass undergoes torsion, or counterclockwise rotation of the visceral anatomical features. As a result, the anus of the adult animal is located over the head. Although torsion is always counterclockwise, the shell may coil in either direction; thus coiling of a shell is *not* the same as torsion of the visceral mass.

Everyday Connection

Can Snail Venom Be Used as a Pharmacological Painkiller?

Marine snails of the genus *Conus* (Figure 28.26) attack prey with a venomous stinger, modified from the radula. The toxin released, known as conotoxin, is a peptide with internal disulfide linkages. Conotoxins can bring about paralysis in humans, indicating that this toxin attacks neurological targets. Some conotoxins have been shown to block neuronal ion channels. These findings have led researchers to study conotoxins for possible medical applications.

Conotoxins are an exciting area of potential pharmacological development, since these peptides may be possibly modified and used in specific medical conditions to inhibit the activity of specific neurons. For example, conotoxins or modifications of them may be used to induce paralysis in muscles in specific health applications, similar to the use of botulinum toxin Since the entire spectrum of conotoxins, as well as their mechanisms of action, is not completely known, the study of their potential applications is still in its infancy. Most research to date has focused on their use to treat neurological diseases. They have also shown some efficacy in relieving chronic pain, and the pain associated with conditions like sciatica and shingles. The study and use of biotoxins—toxins derived from living organisms—are an excellent example of the application of biological science to modern medicine.

Figure 28.26 *Conus.* Members of the genus *Conus* produce neurotoxins that may one day have medical uses. The tube above the eyes is a siphon used both to circulate water over the gills and to sample the water for chemical evidence of prey nearby. Note the eyes below the siphon. The proboscis, through which the venomous harpoon is projected, is located between the eyes. (credit: David Burdick, NOAA)

Class Cephalopoda ("head foot" animals), includes octopuses, squids, cuttlefish, and nautiluses. Cephalopods include both animals with shells as well as animals in which the shell is reduced or absent. In the shell-bearing *Nautilus,* the spiral shell is multi-chambered. These chambers are filled with gas or water to regulate buoyancy. A siphuncle runs through the chambers, and it is this tube that regulates the amount of water and gases (nitrogen, carbon dioxide, and oxygen mixture) present in the chambers. Ammonites and other nautiloid shells are commonly seen in the fossil record. The shell structure in squids and cuttlefish is reduced and is present internally in the form of a squid pen and cuttlefish bone, respectively. Cuttle bone is sold in pet stores to help smooth the beaks of birds and also to provide birds such as egg-laying chickens and quail with an inexpensive natural source of calcium carbonate. Examples of cephalopods are shown in Figure 28.27.

Cephalopods can display vivid and rapidly changing coloration, almost like flashing neon signs. Typically these flashing displays are seen in squids and octopuses, where they may be used for camouflage and possibly as signals for mating displays. We should note, however, that researchers are not entirely sure if squid can actually see color, or see color in the same way as we do. We know that pigments in the skin are contained in special pigment cells (**chromatophores**), which can expand or contract to produce different color patterns. But chromatophores can only make yellow, red, brown, and black pigmentation; however, underneath them is a whole different set of elements called **iridophores** and **leucophores** that reflect light and can make blue, green, and white. It is possible that squid skin might actually be able to detect some light on its own, without even needing its eyes!

All animals in this class are carnivorous predators and have beak-like jaws in addition to the radula. Cephalopods include the most intelligent of the mollusks, and have a well-developed nervous system along with image-forming eyes. Unlike other mollusks, they have a closed circulatory system, in which the blood is entirely contained in vessels rather than in a hemocoel.

The foot is lobed and subdivided into arms and tentacles. Suckers with chitinized rings are present on the arms and tentacles of octopuses and squid. Siphons are well developed and the expulsion of water is used as their primary mode of locomotion, which resembles jet propulsion. Gills (ctenidia) are attached to the wall of the mantle cavity and are serviced by large blood vessels, each with its own heart. A pair of nephridia is present within the mantle cavity for water balance and excretion of nitrogenous wastes. Cephalopods such as squids and octopuses also produce sepia or a dark ink, which contains melanin. The ink gland is located between the gills and can be released into the excurrent water stream. Ink clouds can be used either as a "smoke screen" to hide the animal from predators during a quick attempt at escape, or to create a fake image to distract predators.

Cephalopods are dioecious. Members of a species mate, and the female then lays the eggs in a secluded and protected niche. Females of some species care for the eggs for an extended period of time and may end up dying during that time period. While most other aquatic mollusks produce trochophore larvae, cephalopod eggs develop directly into a juvenile without an intervening larval stage.

Figure 28.27 Cephalopods. The (a) nautilus, (b) giant cuttlefish, (c) reef squid, and (d) blue-ring octopus are all members of the class Cephalopoda. (credit a: modification of work by J. Baecker; credit b: modification of work by Adrian Mohedano; credit c: modification of work by Silke Baron; credit d: modification of work by Angell Williams)

Members of class Scaphopoda ("boat feet") are known colloquially as "tusk shells" or "tooth shells," as evident when examining *Dentalium*, one of the few remaining scaphopod genera (Figure 28.28). Scaphopods are usually buried in sand with the anterior opening exposed to water. These animals have a single conical shell, which is open on both ends. The head is not well developed, but the mouth, containing a radula, opens among a group of tentacles that terminate in ciliated bulbs used to catch and manipulate prey. Scaphopods also have a foot similar to that seen in bivalves. Ctenidia are absent in these animals; the mantle cavity forms a tube open at both ends and serves as the respiratory structure in these animals.

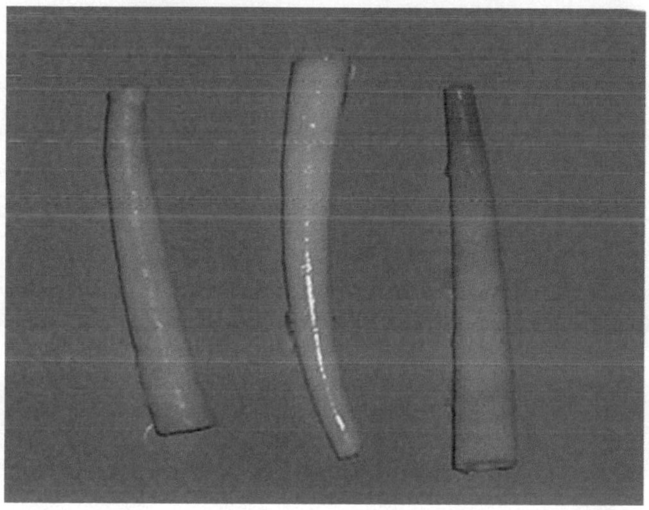

Figure 28.28 Tooth shells. *Antalis vulgaris* shows the classic Dentaliidae shape that gives these animals their common name of "tusk shell." (credit: Georges Jansoone)

Phylum Annelida

Phylum Annelida comprises the true, segmented worms. These animals are found in marine, terrestrial, and freshwater habitats, but the presence of water or humidity is a critical factor for their survival in terrestrial habitats. The annelids are often called "segmented worms" due to their key characteristic of **metamerism**, or true segmentation. Approximately 22,000 species have been described in phylum Annelida, which includes polychaete worms (marine annelids with multiple appendages), and oligochaetes (earthworms and leeches). Some animals in this phylum show parasitic and commensal symbioses with other species in their habitat.

Morphology

Annelids display bilateral symmetry and are worm-like in overall morphology. The name of the phylum is derived from the Latin word *annullus*, which means a small ring, an apt description of the ring-like segmentation of the body. Annelids have a body plan with metameric segmentation, in which several internal and external morphological features are repeated in each body segment. Metamerism allows animals to become bigger by adding "compartments," while making their movement more efficient. The overall body can be divided into head, body, and pygidium (or tail). During development, the segments behind the head arise sequentially from a growth region anterior to the pygidium, a pattern called **teloblastic growth**. In the Oligochaetes, the **clitellum** is a reproductive structure that generates mucus to aid sperm transfer and also produces a "cocoon," within which fertilization occurs; it appears as a permanent, fused band located on the anterior third of the animal (Figure 28.29).

Figure 28.29 The clitellum of an earthworm. The clitellum, seen here as a protruding segment with different coloration than the rest of the body, is a structure that aids in oligochaete reproduction. (credit: Rob Hille)

Anatomy

The epidermis is protected by a collagenous, external cuticle, which is much thinner than the cuticle found in the ecdysozoans and does not require periodic shedding for growth. Circular as well as longitudinal muscles are located interior to the epidermis. Chitinous bristles called **setae** (or chaetae) are anchored in the epidermis, each with its own muscle. In the polychaetes, the setae are borne on paired appendages called **parapodia**.

Most annelids have a well-developed and complete digestive system. Feeding mechanisms vary widely across the phylum. Some polychaetes are filter-feeders that use feather-like appendages to collect small organisms. Others have tentacles, jaws, or an eversible pharynx to capture prey. Earthworms collect small organisms from soil as they burrow through it, and most leeches are blood-feeders armed with teeth or a muscular proboscis. In earthworms, the digestive tract includes a mouth, muscular pharynx, esophagus, crop, and muscular gizzard. The gizzard leads to the intestine, which ends in an anal opening in the terminal segment. A cross-sectional view of a body segment of an earthworm is shown in Figure 28.30; each segment is limited by a membranous septum that divides the coelomic cavity into a series of compartments.

Most annelids possess a closed circulatory system of dorsal and ventral blood vessels that run parallel to the alimentary canal as well as capillaries that service individual tissues. In addition, the dorsal and ventral vessels are connected by transverse loops in every segment. Some polychaetes and leeches have an open system in which the major blood vessels open into a hemocoel. In many species, the blood contains hemoglobin, but not contained in cells. Annelids lack a well-developed respiratory system, and gas exchange occurs across the moist body surface. In the polychaetes, the parapodia are highly vascular and serve as respiratory structures. Excretion is facilitated by a pair of metanephridia (a type of primitive "kidney" that consists of a convoluted tubule and an open, ciliated funnel) that is present in every segment toward the ventral side. Annelids show well-developed nervous systems with a ring of fused ganglia present around the pharynx. The nerve cord is ventral in position and bears enlarged nodes or ganglia in each segment.

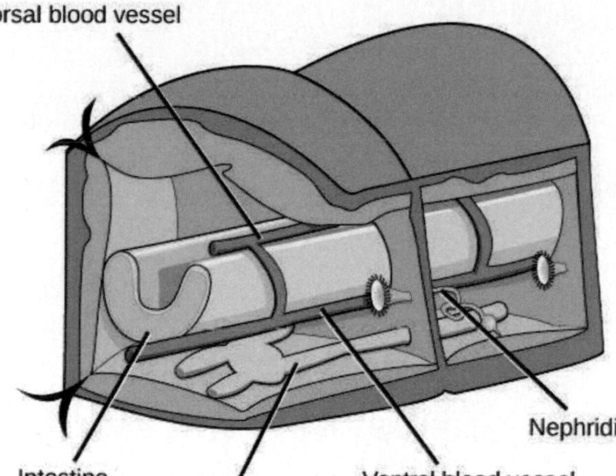

Figure 28.30 Segmental anatomy of an earthworm. This schematic drawing shows the basic anatomy of annelids in a cross-sectional view.

Annelids may be either monoecious with permanent gonads (as in earthworms and leeches) or dioecious with temporary or seasonal gonads (as in polychaetes). However, cross-fertilization is preferred even in hermaphroditic animals. Earthworms may show simultaneous mutual fertilization when they are aligned for copulation. Some leeches change their sex over their reproductive lifetimes. In most polychaetes, fertilization is external and development includes a trochophore larva, which then metamorphoizes to the adult form. In oligochaetes, fertilization is typically internal and the fertilized eggs develop in a cocoon produced by the clitellum; development is direct. Polychaetes are excellent regenerators and some even reproduce asexually by budding or fragmentation.

LINK TO LEARNING

This combination video and animation (http://shapeoflife.org/video/animation/annelid-animation-body-plan) provides a close-up look at annelid anatomy.

Classification of Phylum Annelida

Phylum Annelida contains the class Polychaeta (the polychaetes) and the class Oligochaeta (the earthworms, leeches, and their relatives). The earthworms and the leeches form a monophyletic clade within the polychaetes, which are therefore paraphyletic as a group.

There are more than 22,000 different species of annelids, and more than half of these are marine polychaetes ("many bristles"). In the polychaetes, bristles are arranged in clusters on their parapodia—fleshy, flat, paired appendages that protrude from each segment. Many polychaetes use their parapodia to crawl along the sea floor, but others are adapted for swimming or floating. Some are sessile, living in tubes. Some polychaetes live near hydrothermal vents. These deepwater tubeworms have no digestive tract, but have a symbiotic relationship with bacteria living in their bodies.

Earthworms are the most abundant members of the class Oligochaeta ("few bristles"), distinguished by the presence of a permanent clitellum as well as the small number of reduced chaetae on each segment. (Recall that oligochaetes do not have parapodia.) The oligochaete subclass Hirudinea, includes leeches such as the medicinal leech, *Hirudo medicinalis*, which is effective at increasing blood circulation and breaking up blood clots, and thus can be used to treat some circulatory disorders and cardiovascular diseases. Their use goes back thousands of years. These animals produce a seasonal clitellum, unlike the permanent clitellum of other oligochaetes. A significant difference between leeches and other annelids is the lack of setae and the development of suckers at the anterior and posterior ends, which are used to attach to the host animal. Additionally, in leeches, the segmentation of the body wall may not correspond to the internal segmentation of the coelomic cavity. This adaptation possibly helps the leeches to elongate when they ingest copious quantities of blood from host vertebrates, a condition in which they are said to be "engorged." The subclass Brachiobdella includes tiny leechlike worms that attach themselves to the gills or body surface of crayfish.

(a) (b) (c)

Figure 28.31 Annelid groups. The (a) earthworm, (b) leech, and (c) featherduster are all annelids. The earthworm and leech are oligochaetes, while the featherduster worm is a tube-dwelling filter-feeding polychaete. (credit a: modification of work by S. Shepherd; credit b: modification of work by "Sarah G..."/Flickr; credit c: modification of work by Chris Gotschalk, NOAA)

28.5 Superphylum Ecdysozoa: Nematodes and Tardigrades

By the end of this section, you will be able to do the following:

- Describe the structural organization of nematodes
- Describe the importance of Caenorhabditis elegans in research
- Describe the features of Tardigrades

Superphylum Ecdysozoa

The superphylum Ecdysozoa contains an incredibly large number of species. This is because it contains two of the most diverse animal groups: phylum Nematoda (the roundworms) and phylum Arthropoda (the arthropods). The most prominent distinguishing feature of ecdysozoans is the cuticle—a tough, but flexible exoskeleton that protects these animals from water loss, predators, and other dangers of the external environment. One small phylum within the Ecdysozoa, with exceptional resistance to desiccation and other environmental hazards, is the Tardigrada. The nematodes, tardigrades, and arthropods all belong to the superphylum Ecdysozoa, which is believed to be **monophyletic**—a clade consisting of all evolutionary descendants from one common ancestor. All members of this superphylum periodically go through a molting process that culminates in ecdysis—the actual shedding of the old exoskeleton. (The term "ecdysis" translates roughly as "take off" or "strip.") During the molting process, old cuticle is replaced by a new cuticle, which is secreted beneath it, and which will last until the next growth period.

Phylum Nematoda

The Nematoda, like other members of the superphylum Ecdysozoa, are triploblastic and possess an embryonic mesoderm that is sandwiched between the ectoderm and endoderm. They are also bilaterally symmetrical, meaning that a longitudinal section will divide them into right and left sides that are superficially symmetrical. In contrast with flatworms, nematodes are pseudocoelomates and show a tubular morphology and circular cross-section. Nematodes include both free-living and parasitic forms.

In 1914, N.A. Cobb said, "In short, if all the matter in the universe except the nematodes were swept away, our world would still be dimly recognizable, and if, as disembodied spirits, we could then investigate it, we should find its mountains, hills, vales, rivers, lakes and oceans represented by a thin film of nematodes..." To paraphrase Cobb, nematodes are so abundant that if all the non-nematode matter of the biosphere were removed, there would still remain a shadow of the former world outlined by nematodes![1] The phylum Nematoda includes more than 28,000 species with an estimated 16,000 being parasitic in nature. However, nematologists believe there may be over one million unclassified species.

The name Nematoda is derived from the Greek word "Nemos," which means "thread," and includes all true roundworms. Nematodes are present in all habitats, typically with each species occurring in great abundance. The free-living nematode, *Caenorhabditis elegans*, has been extensively used as a model system for many different avenues of biological inquiry in laboratories all over the world.

1Stoll, N. R., "This wormy world. 1947," *Journal of Parasitology* 85(3) (1999): 392-396.

Morphology

The cylindrical body form of the nematodes is seen in Figure 28.32. These animals have a complete digestive system with a distinct mouth and anus, whereas only one opening is present in the digestive tract of flatworms. The mouth opens into a muscular pharynx and intestine, which leads to a rectum and anal opening at the posterior end. The epidermis can be either a single layer of cells or a **syncytium**—a multinucleated tissue that in this case is formed by the fusion of many single cells. The cuticle of nematodes is rich in collagen and a polymer called **chitin**, which forms a protective armor outside the epidermis. The cuticle extends into both ends of the digestive tract, the pharynx, and rectum. In the head, an anterior mouth opening is composed of three (or six) "lips" as well as teeth derived from the cuticle (in some species). Some nematodes may present other modifications of the cuticle such as rings, head shields, or warts. These external rings, however, do not reflect *true* internal body segmentation, which as we have seen is a hallmark of phylum Annelida. The attachment of the muscles of nematodes differs from that of most animals: they have a longitudinal layer only, and their direct attachment to the dorsal and ventral nerve cords creates a strong muscular contraction that results in a whiplike, almost spastic, body movement.

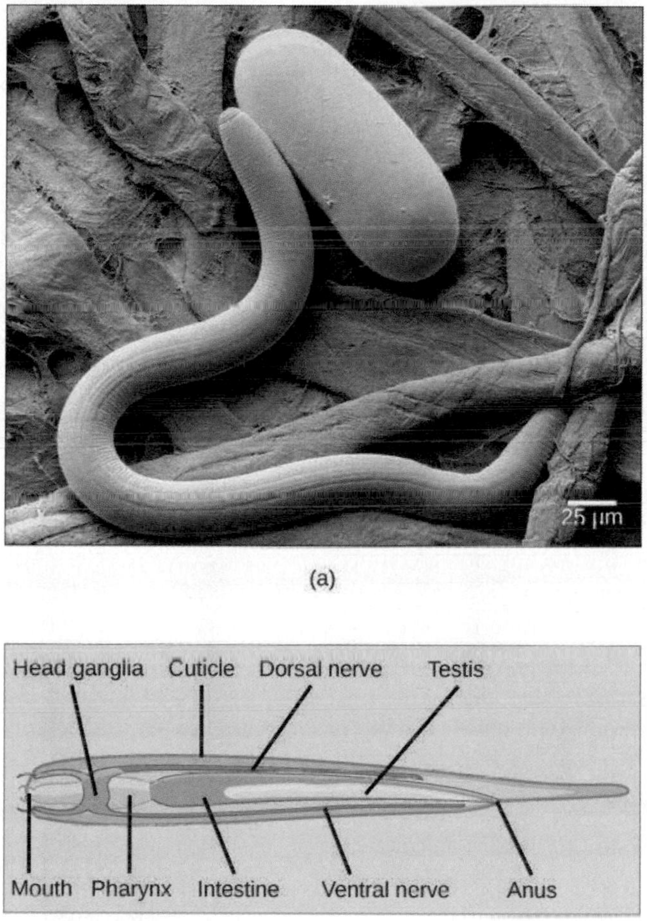

Figure 28.32 Nematode morphology. Scanning electron micrograph shows (a) the soybean cyst nematode (*Heterodera glycines*) and a nematode egg. (b) A schematic representation shows the anatomy of a typical nematode. (credit a: modification of work by USDA ARS; scale-bar data from Matt Russell)

Excretory System

In nematodes, specialized excretory systems are not well developed. Nitrogenous wastes, largely in the form of *ammonia*, are released directly across the body wall. In some nematodes, osmoregulation and salt balance are performed by simple excretory cells or glands that may be connected to paired canals that release wastes through an anterior pore. In marine nematodes, the excretory cells are called **renette cells**, which are unique to nematodes.

Nervous system

Most nematodes have four longitudinal nerve cords that run along the length of the body in dorsal, ventral, and lateral positions. The ventral nerve cord is better developed than the dorsal and lateral cords. Nonetheless, all nerve cords fuse at the anterior end, to form a **pharyngeal nerve ring** around the pharynx, which acts as the head ganglion or the "brain" of the roundworm. A similar fusion forms a posterior ganglion at the tail. In *C. elegans*, the nervous system accounts for nearly one-third of the total number of cells in the animal!

Reproduction

Nematodes employ a variety of reproductive strategies ranging from monoecious to dioecious to parthenogenetic, depending upon the species. *C. elegans* is a mostly monoecious species with both self-fertilizing hermaphrodites and some males. In the hermaphrodites, ova and sperm develop at different times in the same gonad. Ova are contained in a uterus and amoeboid sperm are contained in a spermatheca ("sperm receptacle"). The uterus has an external opening known as the vulva. The female genital pore is near the middle of the body, whereas the male genital pore is nearer to the tip. In anatomical males, specialized structures called *copulatory spicules* at the tail of the male keep him in place and open the vulva of the female into which the amoeboid sperm travel into the spermatheca.

Fertilization is internal, and embryonic development starts very soon after fertilization. The embryo is released from the vulva during the gastrulation stage. The embryonic development stage lasts for 14 hours; development then continues through four successive larval stages with molting and ecdysis taking place between each stage—L1, L2, L3, and L4—ultimately leading to the development of a young adult worm. Adverse environmental conditions such as overcrowding or lack of food can result in the formation of an intermediate larval stage known as the *dauer larva*. An unusual feature of some nematodes is eutely: the body of a given species contains a specific number of cells as the consequence of a rigid developmental pathway.

Everyday Connection

C. elegans: The Model System for Linking Developmental Studies with Genetics

If biologists wanted to research how nicotine dependence develops in the body, how lipids are regulated, or observe the attractant or repellant properties of certain odors, they would clearly need to design three very different experiments. However, they might only need one subject of study: *Caenorhabditis elegans*. The nematode *C. elegans* was brought into the focus of mainstream biological research by Dr. Sydney Brenner. Since 1963, Dr. Brenner and scientists worldwide have used this animal as a model system to study many different physiological and developmental mechanisms.

C. elegans is a free-living nematode found in soil. Only about a millimeter long, it can be cultured on agar plates (10,000 worms/plate!), feeding on the common intestinal bacterium *Escherichia coli* (another long-term resident of biological laboratories worldwide), and therefore can be readily grown and maintained in a laboratory. The biggest asset of this nematode is its transparency, which helps researchers to observe and monitor changes within the animal with ease. It is also a simple organism with about 1,000 cells and a genome of only 20,000 genes. Its chromosomes are organized into five pairs of autosomes plus a pair of sex chromosomes, making it an ideal candidate with which to study genetics. Since every cell can be visualized and identified, this organism is useful for studying cellular phenomena like cell-to-cell interactions, cell-fate determinations, cell division, apoptosis (cell death), and intracellular transport.

Another tremendous asset is the short life cycle of this worm (Figure 28.33). It takes only three days to achieve the "egg to adult to daughter egg"; therefore, the developmental consequences of genetic changes can be quickly identified. The total life span of *C. elegans* is two to three weeks; hence, age-related phenomena are also easy to observe. There are two sexes in this species: hermaphrodites (XX) and males (XO). However, anatomical males are relatively infrequently obtained from matings between hermaphrodites, since their XO chromosome composition requires meiotic nondisjunction when both parents are XX. Another feature that makes *C. elegans* an excellent experimental model is that the position and number of the 959 cells present in adult hermaphrodites of this organism is *constant*. This feature is extremely significant when studying cell differentiation, cell-to-cell communication, and apoptosis. Lastly, *C. elegans* is also amenable to genetic manipulations using molecular methods, rounding off its usefulness as a model system.

Biologists worldwide have created information banks and groups dedicated to research using *C. elegans*. Their findings have led, for example, to better understandings of cell communication during development, neuronal signaling, and insight into lipid regulation (which is important in addressing health issues like the development of obesity and diabetes). In recent years, studies have enlightened the medical community with a better understanding of polycystic kidney disease. This simple organism has led biologists to complex and significant findings, growing the field of science in ways that touch the everyday world.

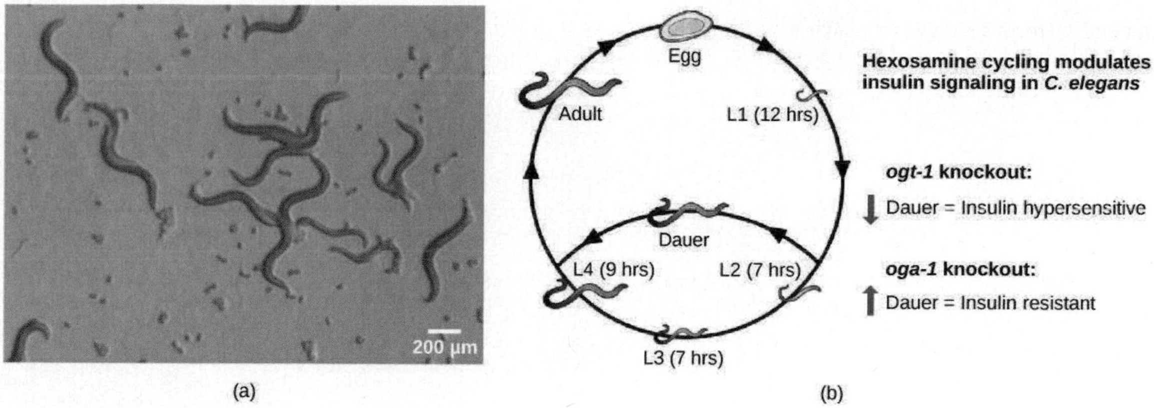

Figure 28.33 *Caenorhabditis elegans.* (a) This light micrograph shows the bodies of a group of roundworms. These hermaphrodites consist of exactly 959 cells. (b) The life cycle of *C. elegans* has four juvenile stages (L1 through L4) and an adult stage. Under ideal conditions, the nematode spends a set amount of time at each juvenile stage, but under stressful conditions, it may enter a *dauer state* that does not age significantly and is somewhat analogous to the diapausing state of some insects. (credit a: modification of work by "snickclunk"/Flickr: credit b: modification of work by NIDDK, NIH; scale-bar data from Matt Russell)

Parasitic Nematodes

A number of common parasitic nematodes serve as prime examples of parasitism (endoparasitism). These economically and medically important animals exhibit complex life cycles that often involve multiple hosts, and they can have significant medical and veterinary impacts. Here is a partial list of nasty nematodes: Humans may become infected by *Dracunculus medinensis*, known as guinea worms, when they drink unfiltered water containing copepods (Figure 28.34), an intermediate crustacean host. Hookworms, such as *Ancylostoma* and *Necator*, infest the intestines and feed on the blood of mammals, especially of dogs, cats, and humans. Trichina worms (*Trichinella*) are the causal organism of trichinosis in humans, often resulting from the consumption of undercooked pork; *Trichinella* can infect other mammalian hosts as well. *Ascaris*, a large intestinal roundworm, steals nutrition from its human host and may create physical blockage of the intestines. The filarial worms, such as *Dirofilaria* and *Wuchereria*, are commonly vectored by mosquitoes, which pass the infective agents among mammals through their blood-sucking activity. One species, *Wuchereria bancrofti*, infects the lymph nodes of over 120 million people worldwide, usually producing a non-lethal but deforming condition called *elephantiasis*. In this disease, parts of the body often swell to gigantic proportions due to obstruction of lymphatic drainage, inflammation of lymphatic tissues, and resulting edema. *Dirofilaria immitis*, a blood-infective parasite, is the notorious dog heartworm species.

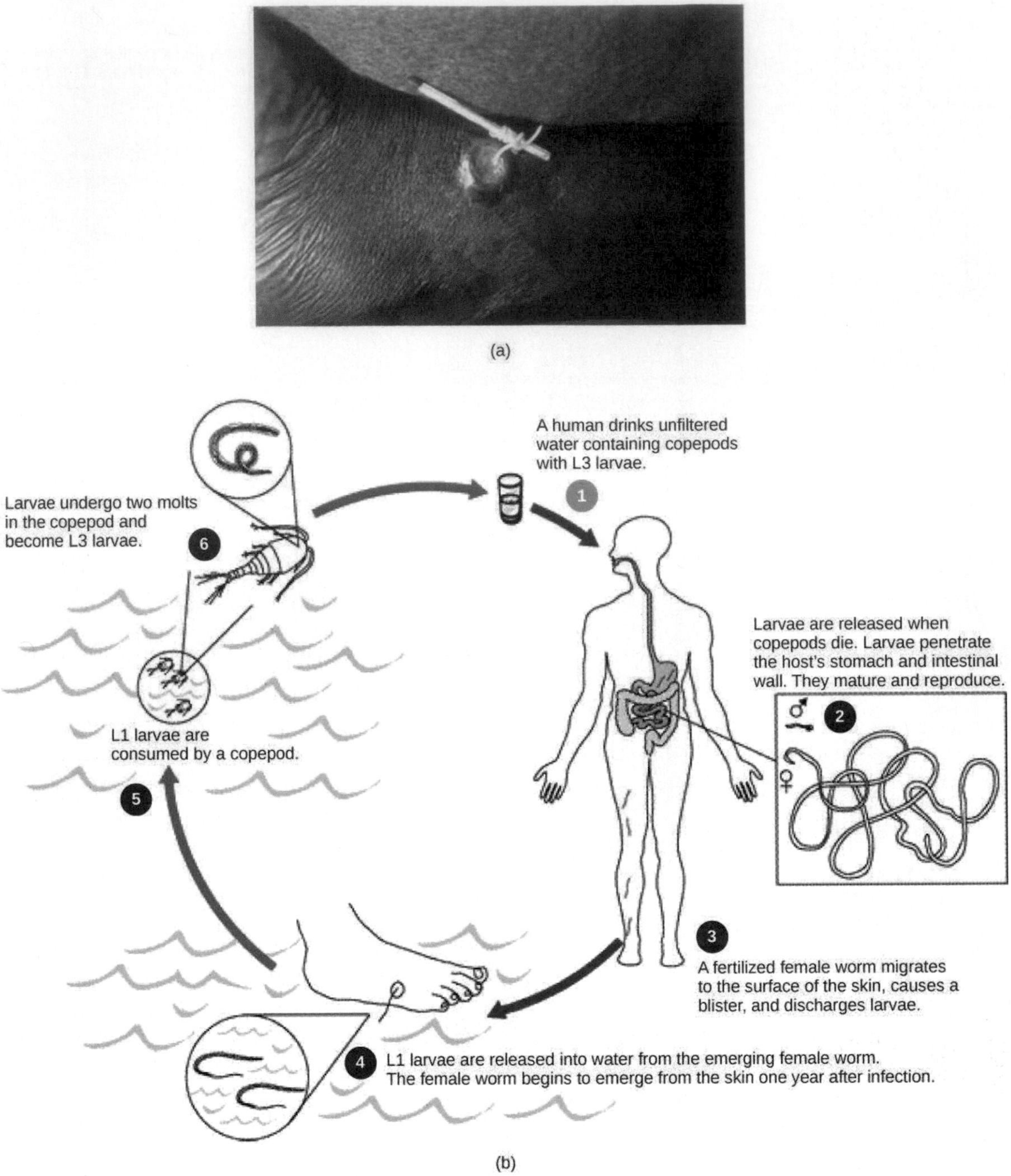

Figure 28.34 Life cycle of the guinea worm. The guinea worm *Dracunculus medinensis* infects about 3.5 million people annually, mostly in Africa. (a) Here, the worm is wrapped around a stick so it can be slowly extracted. (b) Infection occurs when people consume water contaminated by infected copepods, but this can easily be prevented by simple filtration systems. (credit: modification of work by CDC)

Phylum Tardigrada

The tardigrades ("slow-steppers") comprise a phylum of inconspicuous little animals living in marine, freshwater, or damp terrestrial environments throughout the world. They are commonly called "water bears" because of their plump bodies and the large claws on their stubby legs. There are over 1,000 species, most of which are less than 1 mm in length. A chitinous cuticle covers the body surface and may be divided into plates (Figure 28.35). Tardigrades are known for their ability to enter a state called **cryptobiosis**, which provides them with are resistance to multiple environmental challenges, including desiccation, very low temperatures, vacuum, high pressure, and radiation. They can suspend their metabolic activity for years, and survive the loss of up to 99% of their water content. Their remarkable resistance has recently been attributed to unique proteins that replace water in their cells and protect their internal cell structure and their DNA from damage.

Figure 28.35 Scanning electronmicrograph of *Milnesium tardigradum*. (credit: Schokraie E, Warnken U, Hotz-Wagenblatt A, Grohme MA, Hengherr S, et al. (2012) -https://commons.wikimedia.org/w/index.php?curid=22716809 (http://openstax.org/l/waterbear))

Morphology and Physiology

Tardigrades have cylindrical bodies, with four pairs of legs terminating in a number of claws. The cuticle is periodically shed, including the cuticular covering of the claws. The first three pairs of legs are used for walking, and the posterior pair for clinging to the substrate. A circular mouth leads to a muscular pharynx and salivary glands. Tardigrades feed on plants, algae, or small animals. Plant cells are pierced with a chitinous stylet and the cellular contents are then sucked into the gut by the muscular pharynx. Bands of single muscle cells are attached to the various points of the epidermis and extend into the legs to provide ambulatory movement. The major body cavity is a *hemocoel*, but there are no specialized circulatory structures for moving the blood, nor are there specialized respiratory structures. Malpighian tubules in the hemocoel remove metabolic wastes and transport them to the gut. A dorsal brain is connected to a ventral nerve cord with segmental ganglia associated with the appendages. Sensory structures are greatly reduced, but there is a pair of simple eyespots on the head, and sensory cilia or bristles concentrated toward the head end of the animal.

Reproduction

Most tardigrades are dioecious, and males and females each have a single gonad. Mating usually occurs at the time of a molt and fertilization is external. Eggs may be deposited in a molted cuticle or attached to other objects. Development is direct, and the animal may molt a dozen times during its lifetime. In tardigrades, like nematodes, development produces a fixed number of cells, with the actual number of cells being dependent on the species. Further growth occurs by enlarging the cells, not by multiplying them.

28.6 Superphylum Ecdysozoa: Arthropods

By the end of this section, you will be able to do the following:

- Compare the internal systems and appendage specializations of phylum Arthropoda
- Discuss the environmental importance of arthropods
- Discuss the reasons for arthropod success and abundance

The superphylum Ecdysozoa also includes the phylum Arthropoda, one of the most successful clades of animals on the planet. Arthropods are coelomate organisms characterized by a sturdy chitinous exoskeleton and jointed appendages. There are well over a million arthropod species described, and systematists believe that there are millions of species awaiting proper classification. Like other Ecdysozoa, all arthropods periodically go through the physiological process of **molting**, followed by **ecdysis** (the actual shedding of the exoskeleton), as they grow. Arthropods are eucoelomate, protostomic organisms, often with incredibly complicated life cycles.

Phylum Arthropoda

The name "arthropoda" means "jointed feet." The name aptly describes the invertebrates included in this phylum. Arthropods have probably always dominated the animal kingdom in terms of number of species and likely will continue to do so: An estimated 85 percent of all known species are included in this phylum! In effect, life on Earth could conceivably be called the Age of Arthropods beginning nearly 500 million years ago.

The principal characteristics of all the animals in this phylum are the structural and functional segmentation of the body and the presence of jointed appendages. Arthropods have an exoskeleton made principally of chitin—a waterproof, tough polysaccharide composed of N-acetylglucosamine. Phylum Arthropoda is the most speciose clade in the animal world (Table 28.1), and insects form the single largest class within this phylum. For comparison, refer to the approximate numbers of species in the phyla listed below.

Phylum	# species
Ctenophora	100
Porifera	5,000
Cnidaria	11,000
Platyhelminthes	25,000
Rotifera	2,000
Nemertea	1,200
Annelida	22,000
Mollusca	112,000
Nematoda	28,000+
Tardigrades	>1,000
Arthropoda	1,134,000
Echinodermata	7,000
Chordata	100,000

Table 28.1

Phylum Arthropoda includes animals that have been successful in colonizing terrestrial, aquatic, and aerial habitats. This phylum is further classified into five subphyla: Trilobita (trilobites, all extinct), Chelicerata (horseshoe crabs, spiders, scorpions, ticks, mites, and daddy longlegs or harvestmen), Myriapoda (millipedes, centipedes, and their relatives), Crustacea (crabs, lobsters, crayfish, isopods, barnacles, and some zooplankton), and Hexapoda (insects and their six-legged relatives). Trilobites, an extinct group of arthropods found chiefly in the pre-Cambrian Era (about 500 million years ago), are probably most closely related to the Chelicerata. These are identified based on their fossils; they were quite diverse and radiated significantly into thousands of species before their complete extinction at the end of the Permian about 240 million years ago (Figure 28.36).

Figure 28.36 A trilobite. Trilobites, like the one in this fossil, are an extinct group of arthropods. Their name "trilobite" refers to the *three longitudinal lobes* making up the body: right pleural lobe, axial lobe, and left pleural lobe (credit: Kevin Walsh).

Morphology

Characteristic features of the arthropods include the presence of jointed appendages, body segmentation, and chitinized exoskeleton. Fusion of adjacent groups of segments gave rise to functional body regions called tagmata (singular = tagma). Tagmata may be in the form of a head, thorax, and abdomen, or a cephalothorax and abdomen, or a head and trunk, depending on the taxon. Commonly described tagmata may be composed of different numbers of segments; for example, the head of most insects results from the fusion of six ancestral segments, whereas the "head" of another arthropod may be made of fewer ancestral segments, due to independent evolutionary events. Jointed arthropod appendages, often in segmental pairs, have been specialized for various functions: sensing their environment (antennae), capturing and manipulating food (mandibles and maxillae), as well as for walking, jumping, digging, and swimming.

In the arthropod body, a central cavity, called the **hemocoel** (or blood cavity), is present, and the hemocoel fluids are moved by contraction of regions of the tubular dorsal blood vessel called "hearts." Groups of arthropods also differ in the organs used for nitrogenous waste excretion, with crustaceans possessing *green glands* and insects using *Malpighian tubules*, which work in conjunction with the hindgut to reabsorb water while ridding the body of nitrogenous waste. The nervous system tends to be distributed among the segments, with larger ganglia in segments with sensory structures or appendages. The ganglia are connected by a ventral nerve cord.

Respiratory systems vary depending on the group of arthropod. Insects and myriapods use a series of tubes (tracheae) that branch through the body, ending in minute tracheoles. These fine respiratory tubes perform gas exchange *directly* between the air and cells within the organism. The major tracheae open to the surface of the cuticle via apertures called **spiracles.** We should note that these tracheal systems of ventilation have evolved independently in hexapods, myriapods, and arachnids. Although the tracheal system works extremely well in terrestrial environments, it also works well in freshwater aquatic environments: In fact, numerous species of aquatic insects in both immature and adult stages possess tracheal systems. However, although there are insects that live on the surface of marine environments, none is strictly marine—meaning that they complete their entire metamorphosis *in* salt water.

In contrast, aquatic crustaceans utilize gills, terrestrial chelicerates employ book lungs, and aquatic chelicerates use book gills (Figure 28.37). The book lungs of arachnids (scorpions, spiders, ticks, and mites) contain a vertical stack of hemocoel wall tissue that somewhat resembles the pages of a book. Between each of the "pages" of tissue is an air space. This allows both sides of the tissue to be in contact with the air at all times, greatly increasing the efficiency of gas exchange. The gills of crustaceans are filamentous structures that exchange gases with the surrounding water.

The **cuticle** is the hard "covering" of an arthropod. It is made up of two layers: the *epicuticle*, which is a thin, waxy, water-resistant outer layer containing no chitin, and the layer beneath it, the *chitinous procuticle*, which itself is composed of an *exocuticle* and a lower *endocuticle*, all supported ultimately by a *basement membrane*. The exoskeleton is very protective (it is sometimes difficult to squish a big beetle!), but does not sacrifice flexibility or mobility. Both the exocuticle (which is secreted *before* a molt), and an endocuticle, (which is secreted *after* a molt), are composed of chitin bound with a protein; chitin is insoluble in water, alkalis, and weak acids. The procuticle is not only flexible and lightweight, but also provides protection against dehydration and other biological and physical stresses. Some hexapods, such as the crustaceans, add calcium salts to their exoskeleton, which increases the strength of the cuticle, but does reduce its flexibility. In some cases, such as lobsters, the amount of calcium salt deposited within the chitin is extreme. In order to grow, the arthropod must "shed" the exoskeleton during the physiological process called **molting**, following by the actual stripping of the outer cuticle, called **ecdysis** ("to strip off"). At first, this seems to be a dangerous method of growth, because while the new exoskeleton is hardening, the animal is vulnerable to predation; however, molting and ecdysis also allow for growth and change in morphology, as well as for great diversification in size, simply because the numbers of molts can be modified through evolution.

The characteristic morphology of representative animals from each subphylum is described below.

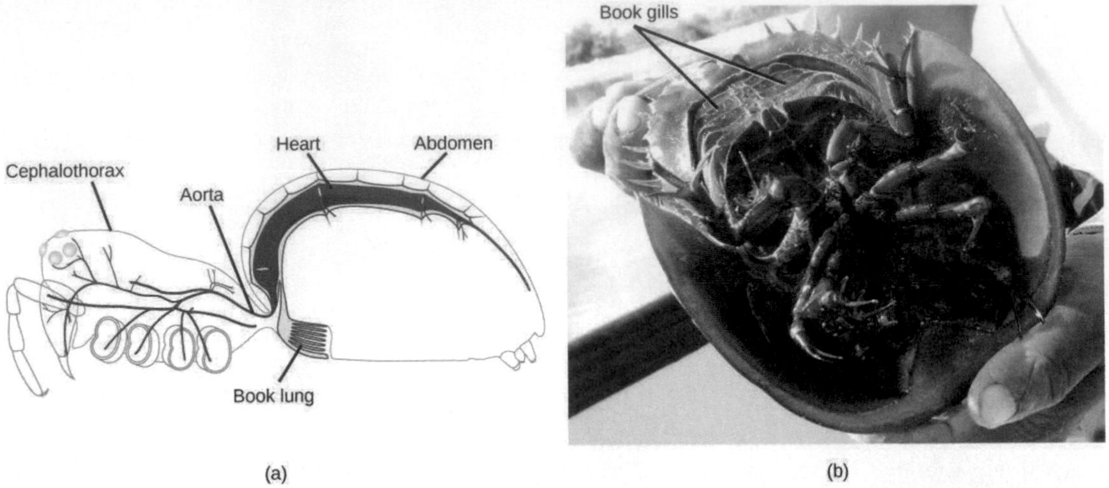

Figure 28.37 Arthropod respiratory structures. The book lungs of (a) arachnids are made up of alternating air pockets and hemocoel tissue shaped like a stack of books (hence the name, "book lung"). The book gills of (b) horseshoe crabs are similar to book lungs but are external so that gas exchange can occur with the surrounding water. (credit a: modification of work by Ryan Wilson based on original work by John Henry Comstock; credit b: modification of work by Angel Schatz)

Subphylum Chelicerata

This subphylum includes animals such as horseshoe crabs, sea spiders, spiders, mites, ticks, scorpions, whip scorpions, and harvestmen. Chelicerates are predominantly terrestrial, although some freshwater and marine species also exist. An estimated 77,000 species of chelicerates can be found in almost all terrestrial habitats.

The body of chelicerates is divided into two tagmata: prosoma and opisthosoma, which are basically the equivalents of a *cephalothorax* (usually smaller) and an *abdomen* (usually larger). A distinct "head" tagma is not usually discernible. The phylum derives its name from the first pair of appendages: the **chelicerae** (Figure 28.38), which serve as specialized clawlike or fanglike mouthparts. We should note here that chelicerae are actually *modified legs*, but they are not the exact serial equivalent of mandibles, which are the modified leglike chewing mouthparts of insects and crustaceans: The chelicerae are borne on the *first segment* making up the prosoma, whereas the mandibles are embryonically on the *fourth segment* of the mandibulate head. The chelicerates have secondarily lost their antennae and hence do not have them. Some of the functions of the antennae (such as touch) are now performed by the second pair of appendages— the **pedipalps**, which may also be used for general sensing the environment as well as the manipulation of food. In some species, such as sea spiders, an additional pair of derived leg appendages, called **ovigers,** is present between the chelicerae and pedipalps. Ovigers are used for grooming and by males to carry eggs. In spiders, the chelicerae are often modified and terminate in fangs that inject venom into their prey before feeding (Figure 28.39).

Figure 28.38 Chelicerae. The chelicerae (first set of appendages) are well developed in the scorpion. (credit: Kevin Walsh)

Most chelicerates ingest food using a preoral cavity formed by the chelicerae and pedipalps. Some chelicerates may secrete digestive enzymes to pre-digest food before ingesting it. Parasitic chelicerates like ticks and mites have evolved blood-sucking apparatuses. Members of this subphylum have an open circulatory system with a heart that pumps blood into the hemocoel. Aquatic species, like horseshoe crabs, have gills, whereas terrestrial species have either tracheae or book lungs for gaseous exchange. Chelicerate hemolymph contains hemocyanin a copper-containing oxygen transport protein.

Figure 28.39 Spider. The trapdoor spider, like all spiders, is a member of the subphylum Chelicerata. (credit: Marshal Hedin)

The nervous system in chelicerates consists of a brain and two ventral nerve cords. Chelicerates are dioecious, meaning that the sexes are separate. These animals use external fertilization as well as internal fertilization strategies for reproduction, depending upon the species and its habitat. Parental care for the young ranges from absolutely none to relatively prolonged care.

LINK TO LEARNING

Visit this site (http://openstax.org/l/arthropodstory) to click through a lesson on arthropods, including interactive habitat maps, and more.

Subphylum Myriapoda

Subphylum Myriapoda comprises arthropods with numerous legs. Although the name is misleading, suggesting that thousands of legs are present in these invertebrates, the number of legs typically varies from 10 to 750. This subphylum includes 16,000 species; the most commonly found examples are millipedes and centipedes. Virtually all myriapods are terrestrial animals and prefer a humid environment. Ancient myriapods (or myriapod-like arthropods) from the Silurian to the Devonian grew up to 10 feet in length (three meters). Unfortunately, they are all extinct!

Myriapods are typically found in moist soils, decaying biological material, and leaf litter. Subphylum Myriapoda is divided into four classes: Chilopoda, Symphyla, Diplopoda, and Pauropoda. Centipedes like *Scutigera coleoptrata* (Figure 28.40) are classified as chilopods. These animals bear one pair of legs per segment, mandibles as mouthparts, and are somewhat dorsoventrally flattened. The legs in the first segment are modified to form forcipules (poison claws) that deliver poison to prey like spiders and cockroaches, as these animals are all predatory. Symphyla are similar to centipedes, but lack the poison claws and are vegetarian. Millipedes bear two pairs of legs per diplosegment—a feature that results from the embryonic fusion of adjacent pairs of body segments. These arthropods are usually rounder in cross-section than centipedes, and are herbivores or detritivores. Millipedes have visibly more numbers of legs as compared to centipedes, although they do *not* have a thousand legs (Figure 28.40b). The Pauropods are similar to millipedes, but have fewer segments.

(a) (b)

Figure 28.40 Myriapods. The centipede *Scutigera coleoptrata* (a) has up to 15 pairs of legs. The North American millipede *Narceus americanus* (b) bears many legs, although not a thousand, as its name might suggest. (credit a: modification of work by Bruce Marlin; credit b: modification of work by Cory Zanker)

Subphylum Crustacea

Crustaceans are the most dominant aquatic (both freshwater and marine) arthropods, with the total number of marine crustaceans standing at about 70,000 species. Krill, shrimp, lobsters, crabs, and crayfish are examples of crustaceans (Figure 28.41). However, there are also a number of terrestrial crustacean species as well: Terrestrial species like the wood lice (*Armadillidium* spp), also called pill bugs, roly-polies, potato bugs, or isopods, are also crustaceans. Nonetheless, the number of terrestrial species in this subphylum is relatively low.

(a) (b) (c)

Figure 28.41 Crustaceans. The (a) crab and (b) shrimp krill are both aquatic crustaceans. The pill bug *Armadillidium* is a terrestrial crustacean. (credit a: modification of work by William Warby; credit b: modification of work by Jon Sullivan credit c: modification of work by Franco Folini. https://commons.wikimedia.org/w/index.php?curid=789616 (http://openstax.org/l/pillbug))

Crustaceans typically possess two pairs of antennae, mandibles as mouthparts, and *biramous* ("two branched") appendages, which means that their legs are formed in two parts called endopods and exopods, which appear superficially distinct from the *uniramous* ("one branched") legs of myriapods and hexapods (Figure 28.42). Since biramous appendages are also seen in the trilobites, biramous appendages represent the ancestral condition in the arthropods. Currently, we describe various arthropods as having uniramous or biramous appendages, but these are descriptive only, and do not necessarily reflect evolutionary relationships other than that all jointed legs of arthropods share common ancestry.

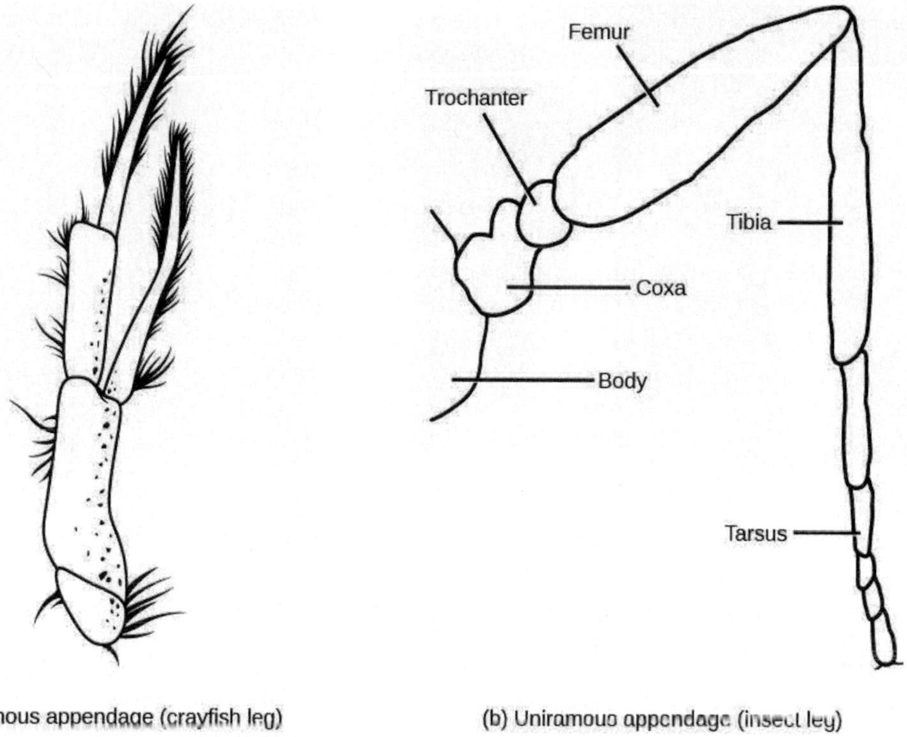

Figure 28.42 Arthropod appendages. Arthropods may have (a) biramous (two-branched) appendages or (b) uniramous (one-branched) appendages. (credit b: modification of work by Nicholas W. Beeson)

In most crustaceans, the head and thorax is fused to form a **cephalothorax** (Figure 28.43), which is covered by a plate called the **carapace**, thus producing a body plan comprising two tagmata: **cephalophorax** and **abdomen**. Crustaceans have a chitinous exoskeleton that is shed by molting and ecdysis whenever the animal requires an increase in size or the next stage of development. The exoskeletons of many aquatic species are also infused with calcium carbonate, which makes them even stronger than those of other arthropods. Crustaceans have an open circulatory system where blood is pumped into the hemocoel by the dorsally located heart. Hemocyanin is the major respiratory pigment present in crustaceans, but hemoglobin is found in a few species and both are dissolved in the hemolymph rather than carried in cells.

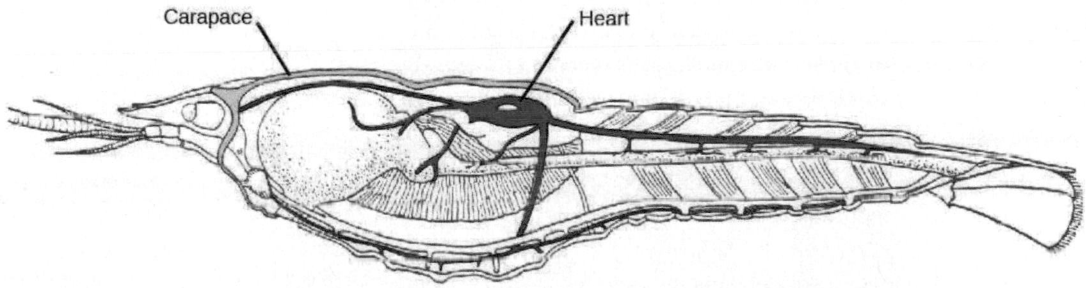

Figure 28.43 Crustacean anatomy. The crayfish is an example of a crustacean. It has a carapace around the cephalothorax and the heart in the dorsal thorax area. (credit: Jane Whitney)

As in the chelicerates, most crustaceans are dioecious. However, some species like barnacles may be hermaphrodites. **Serial hermaphroditism**, where the gonad can switch from producing sperm to ova, is also exhibited in some species. Fertilized eggs may be held within the female of the species or may be released in the water. Terrestrial crustaceans seek out damp spaces in their habitats to lay eggs.

Larval stages—nauplius or zoea—are seen in the early development of aquatic crustaceans. A cypris larva is also seen in the early development of barnacles (Figure 28.44).

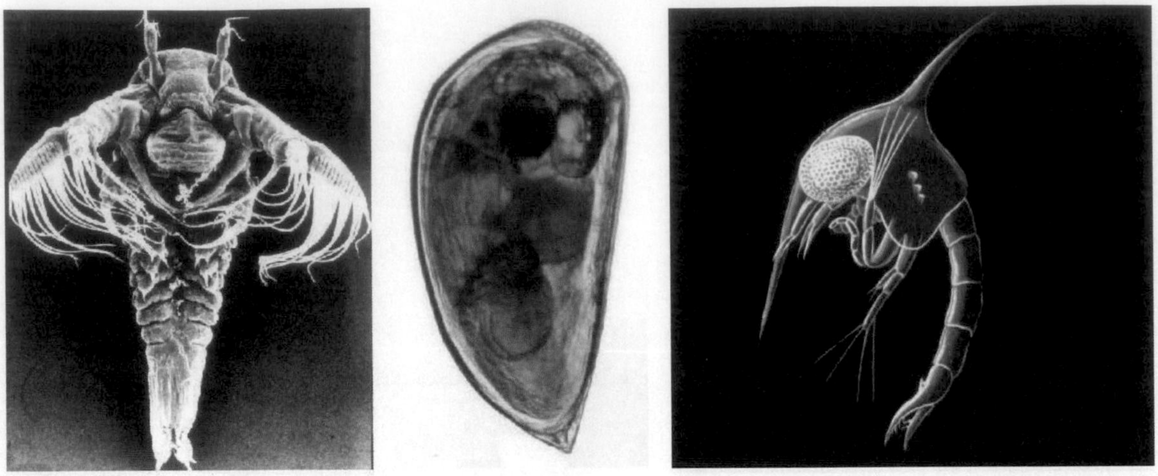

Figure 28.44 Crustacean larvae. All crustaceans go through different larval stages. Shown are (a) the nauplius larval stage of a tadpole shrimp, (b) the cypris larval stage of a barnacle, and (c) the zoea larval stage of a green crab. (credit a: modification of work by USGS; credit b: modification of work by Mª. C. Mingorance Rodríguez; credit c: modification of work by B. Kimmel based on original work by Ernst Haeckel)

Crustaceans possess a brain formed by the fusion of the first three segmental ganglia, as well as two compound eyes. A ventral nerve cord connects additional segmental ganglia. Most crustaceans are carnivorous, but herbivorous and detritivorous species, and even endoparasitic species are known. A highly evolved endoparasitic species, such as *Sacculina* spp, parasitizes its crab host and ultimately destroys it after it forces the host to incubate the parasite's eggs! Crustaceans may also be cannibalistic when extremely high populations of these organisms are present.

Subphylum Hexapoda

The insects comprise the largest class of arthropods in terms of species diversity as well as in terms of biomass—at least in terrestrial habitats.

The name Hexapoda describes the presence of six legs (three pairs) in these animals, which differentiates them from other groups of arthropods that have different numbers of legs. In some cases, however, the number of legs has been evolutionarily reduced, or the legs have been highly modified to accommodate specific conditions, such as endoparasitism. Hexapod bodies are organized into three tagmata: head, thorax, and abdomen. Individual segments of the head have mouthparts derived from jointed legs, and the thorax has three pairs of jointed appendages, and also wings, in most derived groups. For example, in the **pterygotes** (winged insects), in addition to a pair of jointed legs on all three segments comprising the thorax: prothorax, mesothorax, and metathorax.

Appendages found on other body segments are also evolutionarily derived from modified legs. Typically, the head bears an upper "lip" or labrum and mandibles (or derivation of mandibles) that serve as mouthparts; maxillae, and a lower "lip" called a labium: both of which manipulate food. The head also has one pair of sensory antennae, as well as sensory organs such as a pair of compound eyes, ocelli (simple eyes), and numerous sensory hairs. The abdomen usually has 11 segments and bears external reproductive apertures. The subphylum Hexapoda includes some insects that are winged (such as fruit flies) and others that are secondarily wingless (such as fleas). The only order of "primitively wingless" insects is the Thysanura, the bristletails. All other orders are winged or are descendants of formally winged insects.

The evolution of wings is a major, unsolved mystery. Unlike vertebrates, whose "wings" are simply preadaptations of "arms" that served as the structural foundations for the evolution of functional wings (this has occurred independently in pterosaurs, dinosaurs [birds], and bats), the evolution of wings in insects is a what we call a *de novo* (new) development that has given the pteryogotes domination over the Earth. Winged insects existed over 425 million years ago, and by the Carboniferous, several orders of winged insects (**Paleoptera**), most of which are now extinct, had evolved. There is good physical evidence that Paleozoic nymphs with thoracic winglets (perhaps hinged, former gill covers of semi-aquatic species) used these devices on land to elevate the thoracic temperature (the thorax is where the legs are located) to levels that would enable them to escape predators faster, find more food resources and mates, and disperse more easily. The thoracic winglets (which can be found on fossilized insects preceding the advent of truly winged insects) could have easily been selected for thermoregulatory purposes prior to reaching a size that would have allowed them the capacity for gliding or actual flapping flight. Even modern insects with broadly attached wings, such as butterflies, use the basal one-third of their wings (the area next to the thorax) for thermoregulation, and the outer two-thirds for flight, camouflage, and mate selection.

Many of the common insects we encounter on a daily basis—including ants, beetles, cockroaches, butterflies, crickets and flies—are examples of Hexapoda. Among these, adult ants, beetles, flies, and butterflies develop by complete metamorphosis from grub-like or caterpillar-like larvae, whereas adult cockroaches and crickets develop through a gradual or incomplete metamorphosis from wingless immatures. All growth occurs during the juvenile stages. Adults do not grow further (but may become larger) after their final molt. Variations in wing, leg, and mouthpart morphology all contribute to the enormous variety seen in the insects. Insect variability was also encouraged by their activity as pollinators and their coevolution with flowering plants. Some insects, especially termites, ants, bees, and wasps, are eusocial, meaning that they live in large groups with individuals assigned to specific roles or castes, like queen, drone, and worker. Social insects use **pheromones**—external chemical signals—to communicate and maintain group structure as well as a cohesive colony.

VISUAL CONNECTION

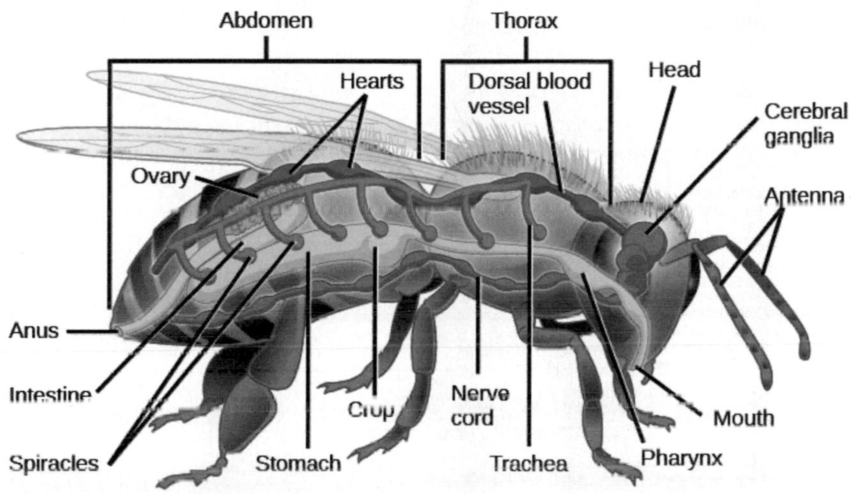

Figure 28.45 Insect anatomy. In this basic anatomy of a hexapod insect, note that insects have a well-developed digestive system (yellow), a respiratory system (blue), a circulatory system (red), and a nervous system (purple). Note the multiple "hearts" and the segmental ganglia.

Which of the following statements about insects is false?

a. Insects have both dorsal and ventral blood vessels.
b. Insects have spiracles, openings that allow air to enter into the tracheal system.
c. The trachea is part of the digestive system.
d. Most insects have a well-developed digestive system with a mouth, crop, and intestine.

28.7 Superphylum Deuterostomia

By the end of this section, you will be able to do the following:

- Describe the distinguishing characteristics of echinoderms
- Describe the distinguishing characteristics of chordates

The phyla Echinodermata and Chordata (the phylum that includes humans) both belong to the superphylum Deuterostomia. Recall that protostomes and deuterostomes differ in certain aspects of their embryonic development, and they are named based on which opening of the **archenteron** (primitive gut tube) develops first. The word deuterostome comes from the Greek word meaning "mouth second," indicating that the mouth develops as a secondary structure opposite the location of the blastopore, which becomes the anus. In protostomes ("mouth first"), the first embryonic opening becomes the mouth, and the second opening becomes the anus.

There are a series of other developmental characteristics that differ between protostomes and deuterostomes, including the type of early cleavage (embryonic cell division) and the mode of formation of the coelom of the embryo: Protosomes typically exhibit spiral mosaic cleavage whereas deuterostomes exhibit radial regulative cleavage. In deuterostomes, the endodermal lining of the archenteron usually forms buds called **coelomic pouches** that expand and ultimately obliterate the embryonic blastocoel (the cavity within the blastula and early gastrula) to become the **embryonic mesoderm**, the third germ layer. This happens when the mesodermal pouches become separated from the invaginating endodermal layer forming the archenteron, then expand and fuse to form the coelomic cavity. The resulting coelom is termed an **enterocoelom**. The archenteron develops into the alimentary canal, and a mouth opening is formed by invagination of ectoderm at the pole opposite the blastopore of the gastrula. The blastopore forms the anus of the alimentary system in the juvenile and adult forms. Cleavage in most deuterostomes is also *indeterminant*, meaning that the developmental fates of early embryonic cells are not decided at that point of embryonic development (this is why we could potentially clone most deuterostomes, including ourselves).

The deuterostomes consist of two major clades—the Chordata and the Ambulacraria. The Chordata include the vertebrates and two invertebrate subphyla, the urochordates and the cephalochordates. The Ambulacraria include the echinoderms and the hemichordates, which were once considered to be a chordate subphylum (Figure 28.46). The two clades, in addition to being deuterostomes, have some other interesting features in common. As we have seen, the vast majority of invertebrate animals do *not* possess a defined bony vertebral endoskeleton, or a bony cranium. However, one of the most ancestral groups of deuterostome invertebrates, the Echinodermata, do produce tiny skeletal "bones" called *ossicles* that make up a true **endoskeleton**, or internal skeleton, covered by an epidermis. The Hemichordata (acorn worms and pterobranchs) will not be covered here, but share with the echinoderms a three-part (tripartite) coelom, similar larval forms, and a derived metanephridium that rids the animals of nitrogenous wastes. They also share pharyngeal slits with the chordates (Figure 28.46). In addition, hemichordates have a dorsal nerve cord in the midline of the epidermis, but lack a neural tube, a true notochord and the endostyle and post-anal tail characteristic of chordates.

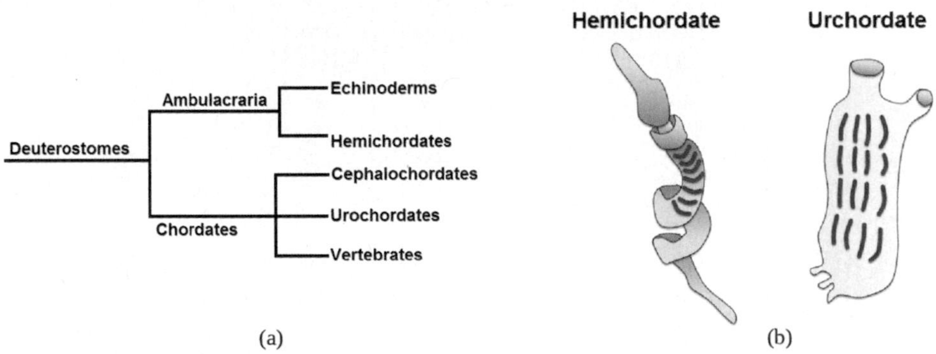

Figure 28.46 Ambulacraria and Chordata. (a) The major deuterostome taxa. (b) pharyngeal slits in hemichordates and urochordates. (credit a MAC; credit b modification of Gill Slits By Own work by Zebra.element [Public domain], via Wikimedia Commons)

Phylum Echinodermata

Echinodermata are named after their "prickly skin" (from the Greek "echinos" meaning "prickly" and "dermos" meaning "skin"). This phylum is a collection of about 7,000 described living species of exclusively marine, bottom-dwelling organisms. Sea stars (Figure 28.47), sea cucumbers, sea urchins, sand dollars, and brittle stars are all examples of echinoderms.

Morphology and Anatomy

Despite the adaptive value of bilaterality for most free-living cephalized animals, adult echinoderms exhibit pentaradial symmetry (with "arms" typically arrayed in multiples of five around a central axis). Echinoderms have an endoskeleton made of calcareous ossicles (small bony plates), covered by the epidermis. For this reason, it is an endoskeleton like our own, not an exoskeleton like that of arthropods. The ossicles may be fused together, embedded separately in the connective tissue of the dermis, or be reduced to minute spicules of bone as in sea cucumbers. The spines for which the echinoderms are named are connected to some of the plates. The spines may be moved by small muscles, but they can also be locked into place for defense. In some species, the spines are surrounded by tiny stalked claws called **pedicellaria**, which help keep the animal's surface clean of debris, protect *papulae* used in respiration, and sometimes aid in food capture.

The endoskeleton is produced by dermal cells, which also produce several kinds of pigments, imparting vivid colors to these animals. In sea stars, fingerlike projections (papillae) of dermal tissue extend through the endoskeleton and function as gills.

Some cells are glandular, and may produce toxins. Each arm or section of the animal contains several different structures: for example, digestive glands, gonads, and the tube feet that are unique to the echinoderms. In echinoderms like sea stars, every arm bears two rows of *tube feet* on the oral side, running along an external ambulacral groove. These tube feet assist in locomotion, feeding, and chemical sensations, as well as serve to attach some species to the substratum.

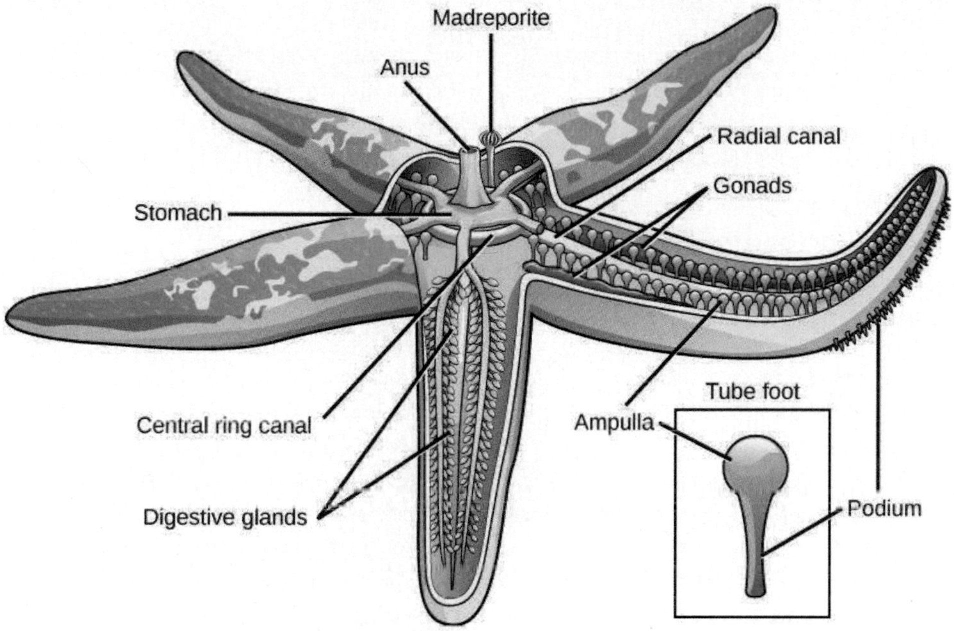

Figure 28.47 Anatomy of a sea star. This diagram of a sea star shows the pentaradial pattern typical of adult echinoderms, and the water vascular system that is their defining characteristic.

Water Vascular and Hemal Systems

Echinoderms have a unique **ambulacral (water vascular) system**, derived from part of the **coelom**, or "body cavity." The water vascular system consists of a central ring canal and radial canals that extend along each arm. Each radial canal is connected to a double row of **tube feet**, which project through holes in the endoskeleton, and function as tactile and ambulatory structures. These tube feet can extend or retract based on the volume of water present in the system of that arm, allowing the animal to move and also allowing it to capture prey with their suckerlike action. Individual tube feet are controlled by bulblike **ampullae**. Seawater enters the system through an **aboral madreporite** (opposite the oral area where the mouth is located) and passes to the ring canal through a short **stone canal**. Water circulating through these structures facilitates gaseous exchange and provides a hydrostatic source for locomotion and prey manipulation. A **hemal system**, consisting of oral, gastric, and aboral rings, as well as other vessels running roughly parallel to the water vascular system, circulates nutrients. Transport of nutrients and gases is shared by the water vascular and hemal systems in addition to the visceral body cavity that surrounds the major organs.

Nervous System

The nervous system in these animals is a relatively simple, comprising a circumoral nerve ring at the center and five radial nerves extending outward along the arms. In addition, several networks of nerves are located in different parts of the body. However, structures analogous to a brain or large ganglia are not present in these animals. Depending on the group, echinoderms may have well-developed sensory organs for touch and chemoreception (e.g., within the tube feet and on tentacles at the tips of the arms), as well as photoreceptors and statocysts.

Digestive and Excretory Systems

A mouth, located on the oral (ventral) side, opens through a short esophagus to a large, baglike stomach. The so-called "cardiac" stomach can be everted through the mouth during feeding (for example, when a starfish everts its stomach into a bivalve prey item to digest the animal—*alive*—within its own shell!) There are masses of digestive glands (**pyloric caeca**) in each arm, running dorsally along the arms and overlying the reproductive glands below them. After passing through the pyloric caeca in each arm, the digested food is channeled to a small anus, if one exists.

Podocytes—cells specialized for ultrafiltration of bodily fluids—are present near the center of the echinoderm disc, at the junction of the water vascular and hemal systems. These podocytes are connected by an internal system of canals to the

madreporite, where water enters the stone canal. The adult echinoderm typically has a spacious and fluid-filled coelom. Cilia aid in circulating the fluid within the body cavity, and lead to the fluid-filled **papulae**, where the exchange of oxygen and carbon dioxide takes place, as well as the secretion of nitrogenous waste such as ammonia, by diffusion.

Reproduction

Echinoderms are dioecious, but males and females are indistinguishable apart from their gametes. Males and females release their gametes into water at the same time and fertilization is external. The early larval stages of all echinoderms (e.g., the *bipinnaria* of asteroid echinoderms such as sea stars) have bilateral symmetry, although each class of echinoderms has its own larval form. The radially symmetrical adult forms from a cluster of cells in the larva. Sea stars, brittle stars, and sea cucumbers may also reproduce asexually by **fragmentation**, as well as regenerate body parts lost in trauma, even when over 75 percent of their body mass is lost!

Classes of Echinoderms

This phylum is divided into five extant classes: Asteroidea (sea stars), Ophiuroidea (brittle stars), Echinoidea (sea urchins and sand dollars), Crinoidea (sea lilies or feather stars), and Holothuroidea (sea cucumbers) (Figure 28.48).

The most well-known echinoderms are members of class Asteroidea, or sea stars. They come in a large variety of shapes, colors, and sizes, with more than 1,800 species known so far. The key characteristic of sea stars that distinguishes them from other echinoderm classes includes thick arms that extend from a central disk from which various body organs branch into the arms. At the end of each arm are simple eye spots and tentacles that serve as touch receptors. Sea stars use their rows of tube feet not only for gripping surfaces but also for grasping prey. Most sea stars are carnivores and their major prey are in the phylum Mollusca. By manipulating its tube feet, a sea star can open molluscan shells. Sea stars have two stomachs, one of which can protrude through their mouths and secrete digestive juices into or onto prey, even before ingestion. A sea star eating a clam can partially open the shell, and then evert its stomach into the shell, introducing digestive enzymes into the interior of the mollusk. This process can both weaken the strong adductor (closing) muscles of a bivalve and begin the process of digestion.

LINK TO LEARNING

Explore the sea star's body plan (http://openstax.org/l/sea_star) up close, watch one move across the sea floor, and see it devour a mussel.

Brittle stars belong to the class Ophiuroidea ("snake-tails"). Unlike sea stars, which have plump arms, brittle stars have long, thin, flexible arms that are sharply demarcated from the central disk. Brittle stars move by lashing out their arms or wrapping them around objects and pulling themselves forward. Their arms are also used for grasping prey. The water vascular system in ophiuroids is not used for locomotion.

Sea urchins and sand dollars are examples of Echinoidea ("prickly"). These echinoderms do not have arms, but are hemispherical or flattened with five rows of tube feet that extend through five rows of pores in a continuous internal shell called a *test*. Their tube feet are used to keep the body surface clean. Skeletal plates around the mouth are organized into a complex multipart feeding structure called *"Aristotle's lantern."* Most echinoids graze on algae, but some are suspension feeders, and others may feed on small animals or organic *detritus*—the fragmentary remains of plants or animals.

Sea lilies and feather stars are examples of Crinoidea. Sea lilies are *sessile*, with the body attached to a stalk, but the feather stars can actively move about using leglike *cirri* that emerge from the aboral surface. Both types of crinoid are suspension feeders, collecting small food organisms along the ambulacral grooves of their feather-like arms. The "feathers" consisted of branched arms lined with tube feet. The tube feet are used to move captured food toward the mouth. There are only about 600 extant species of crinoids, but they were far more numerous and abundant in ancient oceans. Many crinoids are deep-water species, but feather stars typically inhabit shallow areas, especially in substropical and tropical waters.

Sea cucumbers of class Holothuroidea exhibit an extended oral-aboral axis. These are the only echinoderms that demonstrate “functional” bilateral symmetry as adults, because the extended oral-aboral axis compels the animal to lie horizontally rather than stand vertically. The tube feet are reduced or absent, except on the side on which the animal lies. They have a single gonad and the digestive tract is more typical of a bilaterally symmetrical animal. A pair of gill-like structures called **respiratory trees** branch from the posterior gut; muscles around the cloaca pump water in and out of these trees. There are clusters of tentacles around the mouth. Some sea cucumbers feed on detritus, while others are suspension feeders, sifting out small organisms with their oral tentacles. Some species of sea cucumbers are unique among the echinoderms in that cells containing *hemoglobin* circulate in the coelomic fluid, the water vascular system and/or the hemal system.

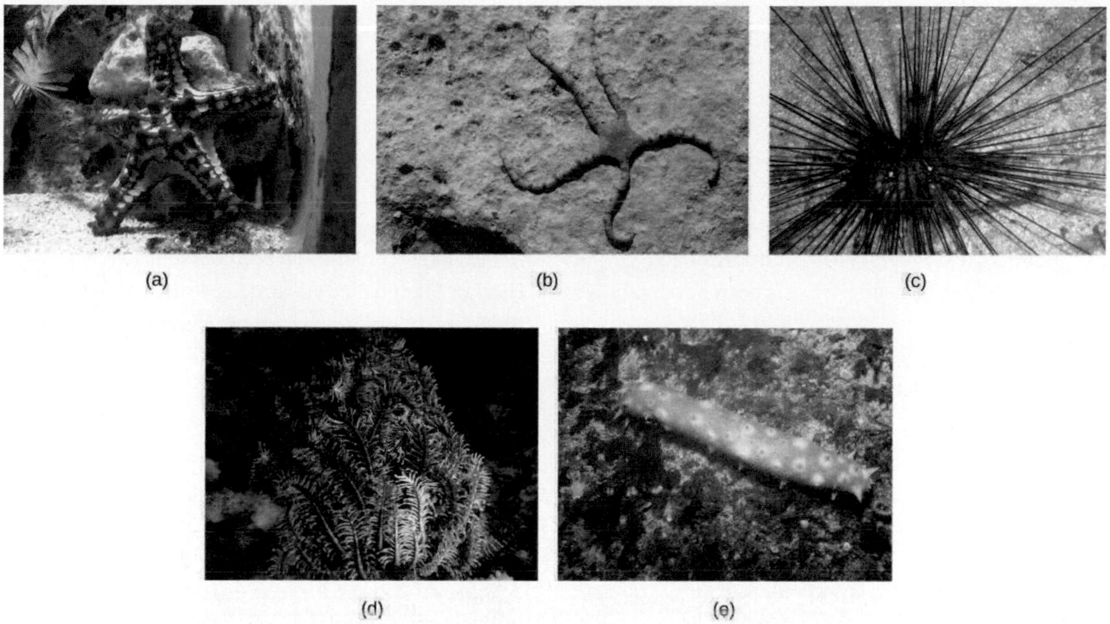

Figure 28.48 Classes of echinoderms. Different members of Echinodermata include the (a) sea star of class Asteroidea, (b) the brittle star of class Ophiuroidea, (c) the sea urchins of class Echinoidea, (d) the sea lilies belonging to class Crinoidea, and (e) sea cucumbers, representing class Holothuroidea. (credit a: modification of work by Adrian Pingstone; credit b: modification of work by Joshua Ganderson; credit c: modification of work by Samuel Chow; credit d: modification of work by Sarah Depper; credit e: modification of work by Ed Bierman)

Phylum Chordata

Animals in the phylum Chordata share five key features that appear at some stage of their development: a notochord, a dorsal hollow nerve cord, pharyngeal slits, a post-anal tail, and an endostyle/thyroid gland that secretes iodinated hormones. In some groups, some of these traits are present only during embryonic development. In addition to containing vertebrate classes, the phylum Chordata contains two clades of "invertebrates": Urochordata (tunicates, salps, and larvaceans) and Cephalochordata (lancelets). Most tunicates live on the ocean floor and are suspension feeders. Lancelets are suspension feeders that feed on phytoplankton and other microorganisms. The invertebrate chordates will be discussed more extensively in the following chapter.

KEY TERMS

amoebocyte sponge cell with multiple functions, including nutrient delivery, egg formation, sperm delivery, and cell differentiation

Annelida phylum of vermiform animals with metamerism

archenteron primitive gut cavity within the gastrula that opens outward via the blastopore

Arthropoda phylum of animals with jointed appendages

biramous referring to two branches per appendage

captacula tentacle-like projection that is present in tusks shells to catch prey

cephalothorax fused head and thorax in some species

chelicera modified first pair of appendages in subphylum Chelicerata

choanocyte (also, collar cell) sponge cell that functions to generate a water current and to trap and ingest food particles via phagocytosis

Chordata phylum of animals distinguished by their possession of a notochord, a dorsal, hollow nerve cord, an endostyle, pharyngeal slits, and a post-anal tail at some point in their development

clitellum specialized band of fused segments, which aids in reproduction

Cnidaria phylum of animals that are diploblastic and have radial symmetry

cnidocyte specialized stinging cell found in Cnidaria

conispiral shell shape coiled around a horizontal axis

corona wheel-like structure on the anterior portion of the rotifer that contains cilia and moves food and water toward the mouth

ctenidium specialized gill structure in mollusks

cuticle (animal) the tough, external layer possessed by members of the invertebrate class Ecdysozoa that is periodically molted and replaced

cypris larval stage in the early development of crustaceans

Echinodermata phylum of deuterostomes with spiny skin; exclusively marine organisms

enterocoelom coelom formed by fusion of coelomic pouches budded from the endodermal lining of the archenteron

epidermis outer layer (from ectoderm) that lines the outside of the animal

extracellular digestion food is taken into the gastrovascular cavity, enzymes are secreted into the cavity, and the cells lining the cavity absorb nutrients

gastrodermis inner layer (from endoderm) that lines the digestive cavity

gastrovascular cavity opening that serves as both a mouth and an anus, which is termed an incomplete digestive system

gemmule structure produced by asexual reproduction in freshwater sponges where the morphology is inverted

hemocoel internal body cavity seen in arthropods

hermaphrodite referring to an animal where both male and female gonads are present in the same individual

invertebrata (also, invertebrates) category of animals that do not possess a cranium or vertebral column

madreporite pore for regulating entry and exit of water into the water vascular system

mantle (also, pallium) specialized epidermis that encloses all visceral organs and secretes shells

mastax jawed pharynx unique to the rotifers

medusa free-floating cnidarian body plan with mouth on underside and tentacles hanging down from a bell

mesoglea non-living, gel-like matrix present between ectoderm and endoderm in cnidarians

mesohyl collagen-like gel containing suspended cells that perform various functions in the sponge

metamerism series of body structures that are similar internally and externally, such as segments

Mollusca phylum of protostomes with soft bodies and no segmentation

nacre calcareous secretion produced by bivalves to line the inner side of shells as well as to coat intruding particulate matter

nauplius larval stage in the early development of crustaceans

nematocyst harpoon-like organelle within cnidocyte with pointed projectile and poison to stun and entangle prey

Nematoda phylum of worm-like animals that are triploblastic, pseudocoelomates that can be free-living or parasitic

Nemertea phylum of dorsoventrally flattened protostomes known as ribbon worms

osculum large opening in the sponge's body through which water leaves

ostium pore present on the sponge's body through which water enters

oviger additional pair of appendages present on some arthropods between the chelicerae and pedipalps

parapodium fleshy, flat, appendage that protrudes in pairs from each segment of polychaetes

pedipalp second pair of appendages in Chelicerata

pilidium larval form found in some nemertine species

pinacocyte epithelial-like cell that forms the outermost layer of sponges and encloses a jelly-like substance called mesohyl

planospiral shell shape coiled around a vertical axis

planuliform larval form found in phylum Nemertea

polymorphic possessing multiple body plans within the lifecycle of a group of organisms

polyp stalk-like sessile life form of a cnidarians with mouth and tentacles facing upward, usually sessile but may be able to glide along surface

Porifera phylum of animals with no true tissues, but a

porous body with rudimentary endoskeleton

radula tongue-like organ with chitinous ornamentation

rhynchocoel cavity present above the mouth that houses the proboscis

schizocoelom coelom formed by groups of cells that split from the endodermal layer

sclerocyte cell that secretes silica spicules into the mesohyl

seta/chaeta chitinous projection from the cuticle

siphon tubular structure that serves as an inlet for water into the mantle cavity

spicule structure made of silica or calcium carbonate that provides structural support for sponges

spongocoel central cavity within the body of some sponges

trochophore first of the two larval stages in mollusks

uniramous referring to one branch per appendage

veliger second of the two larval stages in mollusks

water vascular system system in echinoderms where water is the circulatory fluid

zoea larval stage in the early development of crustaceans

CHAPTER SUMMARY

28.1 Phylum Porifera

Animals included in phylum Porifera are parazoans because they do not show the formation of true embryonically derived tissues, although they have a number of specific cell types and "functional" tissues such as pinacoderm. These organisms show very simple organization, with a rudimentary endoskeleton of spicules and spongin fibers. Glass sponge cells are connected together in a multinucleated syncytium. Although sponges are very simple in organization, they perform most of the physiological functions typical of more complex animals.

28.2 Phylum Cnidaria

Cnidarians represent a more complex level of organization than Porifera. They possess outer and inner tissue layers that sandwich a noncellular mesoglea between them. Cnidarians possess a well-formed digestive system and carry out extracellular digestion in a digestive cavity that extends through much of the animal. The mouth is surrounded by tentacles that contain large numbers of cnidocytes—specialized cells bearing nematocysts used for stinging and capturing prey as well as discouraging predators. Cnidarians have separate sexes and many have a lifecycle that involves two distinct morphological forms—medusoid and polypoid—at various stages in their life cycles. In species with both forms, the medusa is the sexual, gamete-producing stage and the polyp is the asexual stage. Cnidarian species include individual or colonial polypoid forms, floating colonies, or large individual medusa forms (sea jellies).

28.3 Superphylum Lophotrochozoa: Flatworms, Rotifers, and Nemerteans

This section describes three phyla of relatively simple invertebrates: one acoelomate, one pseudocoelomate, and one eucoelomate. Flatworms are acoelomate, triploblastic animals. They lack circulatory and respiratory systems, and have a rudimentary excretory system. This digestive system is incomplete in most species, and absent in tapeworms. There are four traditional groups of flatworms, the largely free-living turbellarians, which include polycladid marine worms and tricladid freshwater species, the ectoparasitic monogeneans, and the endoparasitic trematodes and cestodes. Trematodes have complex life cycles involving a molluscan secondary host and a primary host in which sexual reproduction takes place. Cestodes, or tapeworms, infect the digestive systems of their primary vertebrate hosts.

Rotifers are microscopic, multicellular, mostly aquatic organisms that are currently under taxonomic revision. The group is characterized by the ciliated, wheel-like corona, located on their head. Food collected by the corona is passed to another structure unique to this group of organisms—the mastax or jawed pharynx.

The nemerteans are probably simple eucoelomates. These ribbon-shaped animals also bear a specialized proboscis enclosed within a rhynchocoel. The development of a closed circulatory system derived from the coelom is a significant difference seen in this species compared to other phyla described here. Alimentary, nervous, and excretory systems are more developed in the nemerteans than in the flatworms or rotifers. Embryonic development of nemertean worms proceeds via a planuliform or trochophore-like larval stage.

28.4 Superphylum Lophotrochozoa: Molluscs and Annelids

Phylum Mollusca is a large, group of protostome schizocoelous invertebrates that occupy marine, freshwater, and terrestrial habitats. Mollusks can be divided into seven classes, each of which exhibits variations on the basic molluscan body plan. Two defining features are the mantle, which secretes a protective calcareous shell in many species, and the radula, a rasping feeding organ found in most classes. Some mollusks have evolved a reduced shell, and others have no radula. The mantle also covers the body and forms a mantle cavity, which is quite distinct from the coelomic cavity—typically reduced to the area surrounding the heart, kidneys, and intestine. In aquatic mollusks, respiration is facilitated by gills (ctenidia) in the mantle cavity. In terrestrial mollusks, the mantle cavity itself serves as an organ of gas exchange. Mollusks also have a muscular foot, which is modified in various ways for locomotion or food capture. Most mollusks have separate sexes. Early development in aquatic species occurs via one or more larval stages, including a trochophore larva, that precedes a veliger larva in some groups.

Phylum Annelida includes vermiform, segmented animals. Segmentation is metameric (i.e., each segment is partitioned internally as well as externally, with various structures repeated in each segment). These animals have well-developed neuronal, circulatory, and digestive systems. The two major groups of annelids are the polychaetes, which have parapodia with multiple bristles, and oligochaetes, which have no parapodia and fewer bristles or no bristles. Oligochaetes, which include earthworms and leeches, have a specialized band of segments known as a clitellum, which secretes a cocoon and protects gametes during reproduction. The leeches do not have full internal segmentation. Reproductive strategies include separate sexes, hermaphroditism, and serial hermaphroditism. Polychaetes typically have trochophore larvae, while the oligochaetes develop more directly.

28.5 Superphylum Ecdysozoa: Nematodes and Tardigrades

The defining feature of the Ecdysozoa is a collagenous/chitinous cuticle that covers the body, and the necessity to molt the cuticle periodically during growth. Nematodes are roundworms, with a pseudocoel body cavity. They have a complete digestive system, a differentiated nervous system, and a rudimentory excretory system. The phylum includes free-living species like *Caenorhabditis elegans* as well as many species of endoparasitic organisms such as *Ascaris* spp. They include dioeceous as well as hermaphroditic species. Embryonic development proceeds via several larval stages, and most adults have a fixed number of cells.

The tardigrades, sometimes called "water bears," are a widespread group of tiny animals with a segmented cuticle covering the epidermis and four pairs of clawed legs. Like the nematodes, they are pseudocoelomates and have a fixed number of cells as adults. Specialized proteins enable them to enter cryptobiosis, a kind of suspended animation during which they can resist a number of adverse environmental conditions.

28.6 Superphylum Ecdysozoa: Arthropods

Arthropods represent the most successful animal phylum on Earth, both in terms of the number of species and the number of individuals. As members of the Ecdysozoa, all arthropods have a protective chitinous cuticle that must be periodically molted and shed during development or growth. Arthropods are characterized by a segmented body as well as the presence of jointed appendages. In the basic body plan, a pair of appendages is present per body segment. Within the phylum, traditional classification is based on mouthparts, body subdivisions, number of appendages, and modifications of appendages present. In aquatic arthropods, the chitinous exoskeleton may be calcified. Gills, tracheae, and book lungs facilitate respiration. Unique larval stages are commonly seen in both aquatic and terrestrial groups of arthropods.

28.7 Superphylum Deuterostomia

Echinoderms are deuterostome marine organisms, whose adults show five-fold symmetry. This phylum of animals has a calcareous endoskeleton composed of ossicles, or body plates. Epidermal spines are attached to some ossicles and serve in a protective capacity. Echinoderms possess a water-vascular system that serves both for respiration and for locomotion, although other respiratory structures such as papulae and respiratory trees are found in some species. A large aboral madreporite is the point of entry and exit for sea water pumped into the water vascular system. Echinoderms have a variety of feeding techniques ranging from predation to suspension feeding. Osmoregulation is carried out by specialized cells known as podocytes associated with the hemal system.

The characteristic features of the Chordata are a notochord, a dorsal hollow nerve cord, pharyngeal slits, a post-anal tail, and an endostyle/thyroid that secretes iodinated hormones. The phylum Chordata contains two clades of invertebrates: Urochordata (tunicates, salps, and larvaceans) and Cephalochordata (lancelets), together with the vertebrates in the Vertebrata. Most tunicates live on the ocean floor and are suspension feeders. Lancelets are suspension feeders that feed on phytoplankton and other microorganisms. The sister taxon of the Chordates is the Ambulacraria, which includes both the Echinoderms and the hemichordates, which share pharyngeal slits with the chordates.

VISUAL CONNECTION QUESTIONS

1. Figure 28.3 Which of the following statements is false?
   a. Choanocytes have flagella that propel water through the body.
   b. Pinacocytes can transform into any cell type.
   c. Lophocytes secrete collagen.
   d. Porocytes control the flow of water through pores in the sponge body.

2. Figure 28.21 Which of the following statements about the anatomy of a mollusk is false?
   a. Mollusks have a radula for grinding food.
   b. A digestive gland is connected to the stomach.
   c. The tissue beneath the shell is called the mantle.
   d. The digestive system includes a gizzard, a stomach, a digestive gland, and the intestine.

3. Figure 28.45 Which of the following statements about insects is false?
   a. Insects have both dorsal and ventral blood vessels.
   b. Insects have spiracles, openings that allow air to enter into the tracheal system.
   c. The trachea is part of the digestive system.
   d. Most insects have a well-developed digestive system with a mouth, crop, and intestine.

REVIEW QUESTIONS

4. Mesohyl contains:
   a. a polysaccharide gel and dead cells
   b. a collagen-like gel and suspended cells for various functions.
   c. spicules composed of silica or calcium carbonate.
   d. multiple pores.

5. The large central opening in the parazoan body is called the:
   a. gemmule.
   b. spicule.
   c. ostia.
   d. osculum.

6. Most sponge body plans are slight variations on a simple tube-within-a-tube design. Which of the following is a key limitation of sponge body plans?
   a. Sponges lack the specialized cell types needed to produce more complex body plans.
   b. The reliance on osmosis/diffusion requires a design that maximizes the surface area to volume ratio of the sponge.
   c. Choanocytes must be protected from the hostile exterior environment.
   d. Spongin cannot support heavy bodies.

7. Cnidocytes are found in _____.
   a. phylum Porifera
   b. phylum Nemertea
   c. phylum Nematoda
   d. phylum Cnidaria

8. Cubozoans are ________.
   a. polyps
   b. medusoids
   c. polymorphs
   d. sponges

9. While collecting specimens, a marine biologist finds a sessile Cnidarian. The medusas that bud from it swim by contracting a ring of muscle in their bells. To which class does this specimen belong?
   a. Class Hydrozoa
   b. Class Cubozoa
   c. Class Scyphozoa
   d. Class Anthozoa

10. Which group of flatworms are primarily ectoparasites of fish?
    a. monogeneans
    b. trematodes
    c. cestodes
    d. turbellarians

11. The rhynchocoel is a ________.
    a. circulatory system
    b. fluid-filled cavity
    c. primitive excretory system
    d. proboscis

12. Annelids have (a):
    a. pseudocoelom.
    b. true coelom.
    c. no coelom.
    d. none of the above

13. A mantle and mantle cavity are present in:
    a. phylum Echinodermata.
    b. phylum Adversoidea.
    c. phylum Mollusca.
    d. phylum Nemertea.

14. How does segmentation enhance annelid locomotion?
    a. Segmentation creates repeating body structures so the entire organism functions in synchrony.
    b. Segmentation allows specialization of different body regions.
    c. Neural segmentation allows annelids to localize sensations.
    d. Muscle contractions can be localized to specific regions of the body to coordinate movement.

15. The embryonic development in nematodes can have up to _________ larval stages.
    a. one
    b. two
    c. three
    d. four

16. The nematode cuticle contains _____.
    a. glucose
    b. skin cells
    c. chitin
    d. nerve cells

17. Crustaceans are _____.
    a. ecdysozoans
    b. nematodes
    c. arachnids
    d. parazoans

18. Flies are_______.
    a. chelicerates
    b. hexapods
    c. arachnids
    d. crustaceans

19. Which of the following is **not** a key advantage provided by the exoskeleton of terrestrial arthropods?
    a. Prevents dessication
    b. Protects internal tissue
    c. Provides mechanical support
    d. Grows with the arthropod throughout its life

20. Echinoderms have _____.
    a. triangular symmetry
    b. radial symmetry
    c. hexagonal symmetry
    d. pentaradial symmetry

21. The circulatory fluid in echinoderms is _____.
    a. blood
    b. mesohyl
    c. water
    d. saline

22. Which of the following features does not distinguish humans as a member of phylum Chordata?
    a. Human embryos undergo indeterminate cleavage.
    b. A spinal cord runs along an adult human's dorsal side.
    c. Human embryos exhibit pharyngeal arches and gill slits.
    d. The human coccyx forms from an embryonic tail.

23. The sister taxon of the Chordata is the _____.
    a. Mollusca
    b. Arthropoda
    c. Ambulacraria
    d. Rotifera

CRITICAL THINKING QUESTIONS

24. Describe the different cell types and their functions in sponges.
25. Describe the feeding mechanism of sponges and identify how it is different from other animals.
26. Explain the function of nematocysts in cnidarians.
27. Compare the structural differences between Porifera and Cnidaria.
28. Compare the differences in sexual reproduction between Porifera and Cubozoans. How does the difference in fertilization provide an evolutionary advantage to the Cubozoans?
29. How does the tapeworm body plan support widespread dissemination of the parasite?
30. Describe the morphology and anatomy of mollusks.
31. What are the anatomical differences between nemertines and mollusks?
32. How does a change in the circulatory system organization support the body designs in cephalopods compared to other mollusks?
33. Enumerate features of *Caenorhabditis elegans* that make it a valuable model system for biologists.
34. What are the different ways in which nematodes can reproduce?

35. Why are tardigrades essential to recolonizing habits following destruction or mass extinction?

36. Describe the various superclasses that phylum Arthropoda can be divided into.

37. Compare and contrast the segmentation seen in phylum Annelida with that seen in phylum Arthropoda.

38. How do terrestrial arthropods of the subphylum Hexapoda impact the world's food supply? Provide at least two positive and two negative effects.

39. Describe the different classes of echinoderms using examples.

CHAPTER 29
Vertebrates

Figure 29.1 Examples of critically endangered vertebrate species include (a) the Siberian tiger (*Panthera tigris*), (b) the mountain gorilla (*Gorilla beringei*), and (c) the harpy eagle (*Harpia harpyja*). (The harpy eagle is considered "near threatened" globally, but critically endangered in much of its former range in Mexico and Central America.) (credit a: modification of work by Dave Pape; credit b: modification of work by Dave Proffer; credit c: modification of work by Haui Ared)

INTRODUCTION Vertebrates are among the most recognizable organisms of the Animal Kingdom, and more than 62,000 vertebrate species have been identified. The vertebrate species now living represent only a small portion of the vertebrates that have existed in the past. The best-known extinct vertebrates are the dinosaurs, a unique group of reptiles, some of which reached sizes not seen before or after in *terrestrial* animals. In fact, they were the dominant terrestrial animals for 150 million years, until most of them died out in a mass extinction near the end of the Cretaceous period (except for the feathered theropod ancestors of modern birds, whose direct descendents now number nearly 10,000 species). Although it is not known with certainty what caused this mass extinction (not only of dinosaurs, but of many other groups of organisms), a great deal is known about the anatomy of the dinosaurs and early birds, given the preservation of numerous skeletal elements, nests, eggs, and embryos in the fossil record.

Chapter Outline

29.1 Chordates

By the end of this section, you will be able to do the following:

- Describe the distinguishing characteristics of chordates
- Identify the derived characters of craniates that sets them apart from other chordates
- Describe the developmental fate of the notochord in vertebrates

The vertebrates exhibit two major innovations in their evolution from the invertebrate chordates. These innovations may be associated with the whole genome duplications that resulted in a quadruplication of the basic chordate genome, including the *Hox* gene loci that regulate the placement of structures along the three axes of the body. One of the first major steps was the emergence of the quadrupeds in the form of the amphibians. A second step was the evolution of the amniotic egg, which, similar to the evolution of pollen and seeds in plants, freed terrestrial animals from their dependence on water for fertilization and embryonic development. Within the amniotes, modifications of keratinous epidermal structures have given rise to scales, claws, hair, and feathers. The scales of reptiles sealed their skins against water loss, while hair and feathers provided insulation to support the evolution of endothermy, as well as served other functions such as camouflage and mate attraction in the vertebrate lineages that led to birds and mammals.

Currently, a number of vertebrate species face extinction primarily due to habitat loss and pollution. According to the International Union for the Conservation of Nature, more than 6,000 vertebrate species are classified as threatened. Amphibians and mammals are the classes with the greatest percentage of threatened species, with 29 percent of all amphibians and 21 percent of all mammals classified as threatened. Attempts are being made around the world to prevent the extinction of threatened species. For example, the Biodiversity Action Plan is an international program, ratified by 188 countries, which is designed to protect species and habitats.

Vertebrates are members of the kingdom Animalia and the phylum Chordata (Figure 29.2). Recall that animals that possess bilateral symmetry can be divided into two groups—protostomes and deuterostomes—based on their patterns of embryonic development. The deuterostomes, whose name translates as "second mouth," consist of two major phyla: Echinodermata and Chordata. Echinoderms are invertebrate marine animals that have pentaradial symmetry and a spiny body covering, a group that includes sea stars, sea urchins, and sea cucumbers. The most conspicuous and familiar members of Chordata are vertebrates, but this phylum also includes two groups of invertebrate chordates.

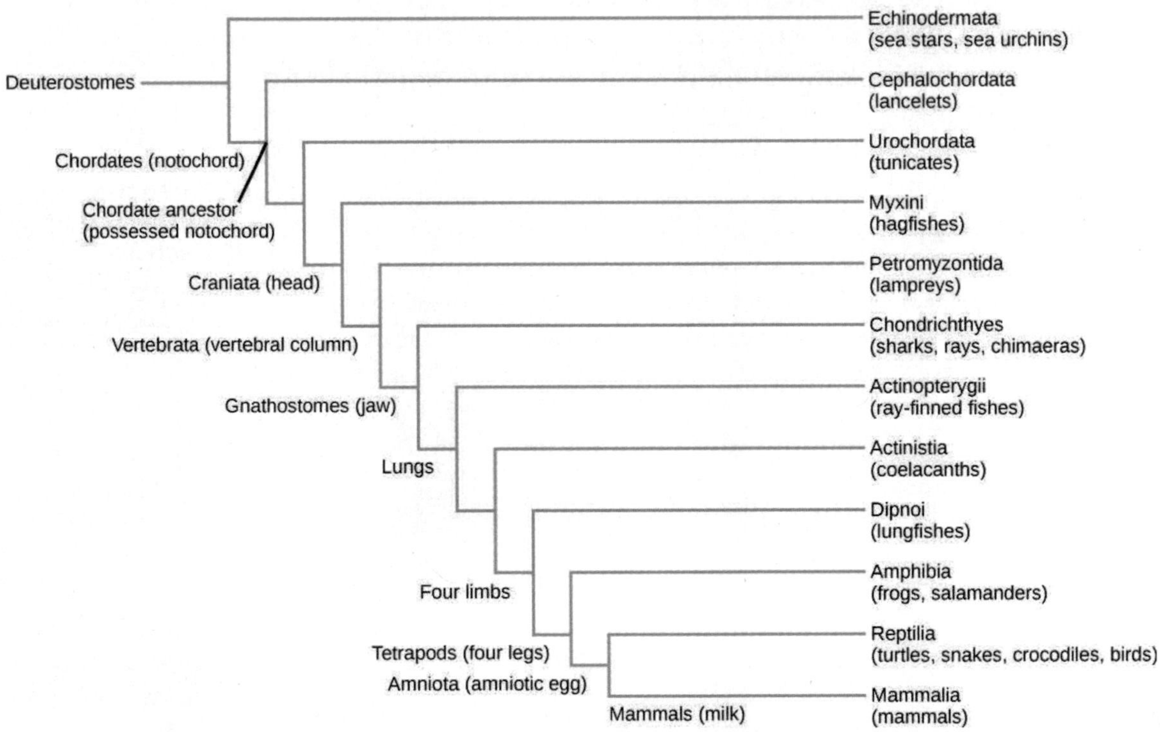

Figure 29.2 Deuterostome phylogeny. All chordates are deuterostomes possessing a notochord at some stage of their life cycle.

Characteristics of Chordata

Animals in the phylum **Chordata** share five key chacteristics that appear at some stage during their development: a notochord, a dorsal hollow (tubular) nerve cord, pharyngeal gill arches or slits, a post-anal tail, and an endostyle/thyroid gland (Figure 29.3). In some groups, some of these key chacteristics are present only during embryonic development.

The chordates are named for the **notochord**, which is a flexible, rod-shaped mesodermal structure that is found in the embryonic stage of all chordates and in the adult stage of some chordate species. It is

strengthened with glycoproteins similar to cartilage and covered with a collagenous sheath. The notocord is located between the digestive tube and the nerve cord, and provides rigid skeletal support as well as a flexible location for attachment of axial muscles. In some chordates, the notochord acts as the primary axial support of the body throughout the animal's lifetime. However, in vertebrates (craniates), the notochord is present only during embryonic development, at which time it induces the development of the neural tube and serves as a support for the developing embryonic body. The notochord, however, is not found in the postembryonic stages of vertebrates; at this point, it has been replaced by the vertebral column (that is, the spine).

VISUAL CONNECTION

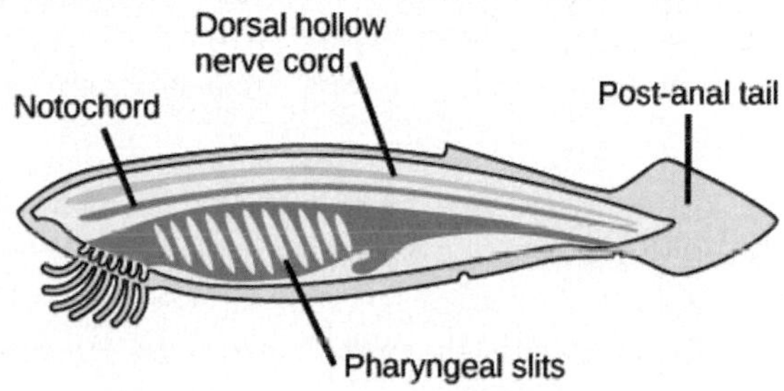

Figure 29.3 Chordate features. In chordates, four common features appear at some point during development: a notochord, a dorsal hollow nerve cord, pharyngeal slits, and a post-anal tail. The endostyle is embedded in the floor of the pharynx.

Which of the following statements about common features of chordates is true?

1. The dorsal hollow nerve cord is part of the chordate central nervous system.
2. In vertebrate fishes, the pharyngeal slits become the gills.
3. Humans are not chordates because humans do not have a tail.
4. Vertebrates do not have a notochord at any point in their development; instead, they have a vertebral column.
5. The endostyle secretes steroid hormones.

The **dorsal hollow nerve cord** is derived from ectoderm that rolls into a hollow tube during development. In chordates, it is located dorsally to the notochord. In contrast, the nervous system in protostome animal phyla is characterized by solid nerve cords that are located either ventrally and/or laterally to the gut. In vertebrates, the neural tube develops into the brain and spinal cord, which together comprise the central nervous system (CNS). The peripheral nervous system (PNS) refers to the peripheral nerves (including the cranial nerves) lying outside of the brain and spinal cord.

Pharyngeal slits are openings in the pharynx (the region just posterior to the mouth) that extend to the outside environment. In organisms that live in aquatic environments, pharyngeal slits allow for the exit of water that enters the mouth during feeding. Some invertebrate chordates use the pharyngeal slits to filter food out of the water that enters the mouth. The endostyle is a strip of ciliated mucus-producing tissue in the floor of the pharynx. Food particles trapped in the mucus are moved along the endostyle toward the gut. The endostyle also produces substances similar to thyroid hormones and is homologous with the thyroid gland in vertebrates. In vertebrate fishes, the pharyngeal slits are modified into gill supports, and in jawed fishes, into jaw supports. In tetrapods (land vertebrates), the slits are highly modified into components of the ear, and tonsils and thymus glands. In other vertebrates, pharyngeal arches, derived from all three germ layers, give rise to the oral jaw from the first pharyngeal arch, with the second arch becoming the hyoid and jaw support.

The **post-anal tail** is a posterior elongation of the body, extending beyond the anus. The tail contains skeletal elements and muscles, which provide a source of locomotion in aquatic species, such as fishes. In some terrestrial vertebrates, the tail also helps with balance, courting, and signaling when danger is near. In humans and other great apes, the post-anal tail is reduced to a vestigial coccyx ("tail bone") that aids in balance during sitting.

LINK TO LEARNING

Click for a video discussing the evolution of chordates and five characteristics that they share.

Click to view content (https://www.openstax.org/l/chordate_evol)

Chordates and the Evolution of Vertebrates

Two clades of chordates are invertebrates: Cephalochordata and Urochordata. Members of these groups also possess the five distinctive features of chordates at some point during their development.

Cephalochordata

Members of **Cephalochordata** possess a notochord, dorsal hollow tubular nerve cord, pharyngeal slits, endostyle/thyroid gland, and a post-anal tail in the adult stage (Figure 29.4). The notochord extends into the head, which gives the subphylum its name. Although the neural tube also extends into the head region, there is no well-defined brain, and the nervous system is centered around a hollow nerve cord lying above the notochord. Extinct members of this subphylum include *Pikaia*, which is the oldest known cephalochordate. Excellently preserved *Pikaia* fossils were recovered from the Burgess shales of Canada and date to the middle of the Cambrian age, making them more than 500 million years old. Its anatomy of *Pikaia* closely resembles that of the extant lancelet in the genus *Branchiostoma*.

The lancelets are named for their bladelike shape. Lancelets are only a few centimeters long and are usually found buried in sand at the bottom of warm temperate and tropical seas. Cephalochordates are suspension feeders. A water current is created by cilia in the mouth, and is filtered through oral tentacles. Water from the mouth then enters the pharyngeal slits, which filter out food particles. The filtered water collects in a gill chamber called the **atrium** and exits through the **atriopore**. Trapped food particles are caught in a stream of mucus produced by the endostyle in a ventral ciliated fold (or groove) of the pharynx and carried to the gut. Most gas exchange occurs across the body surface. Sexes are separate and gametes are released into the water through the atriopore for external fertilization.

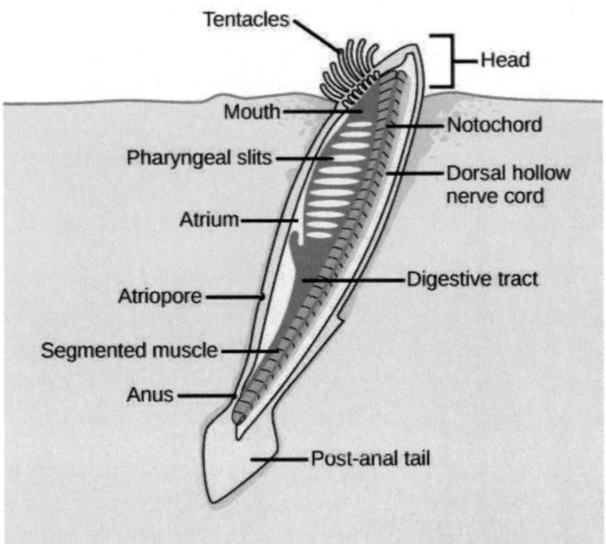

Figure 29.4 Cephalochordate anatomy. In the lancelet and other cephalochordates, the notochord extends into the head region. Adult lancelets retain all five key characteristics of chordates: a notochord, a dorsal hollow nerve cord, pharyngeal slits, an endostyle, and a post-anal tail.

Urochordata

The 1,600 species of **Urochordata** are also known as **tunicates** (Figure 29.5). The name tunicate derives from the cellulose-like carbohydrate material, called the **tunic**, which covers the outer body of tunicates. Although tunicates are classified as chordates, the adults do *not* have a notochord, a dorsal hollow nerve cord, or a post-anal tail, although they do have pharyngeal slits and an endostyle. The "tadpole" larval form, however, possesses all five structures. Most tunicates are hermaphrodites; their larvae hatch from eggs inside the adult tunicate's body. After hatching, a tunicate larva (possessing all five chordate features) swims for a few days until it finds a suitable surface on which it can attach, usually in a dark or shaded location. It then attaches via the head to the surface and undergoes metamorphosis into the adult form, at which point the notochord, nerve cord, and tail disappear, leaving the pharyngeal gill slits and the endostyle as the two remaining features of its chordate morphology.

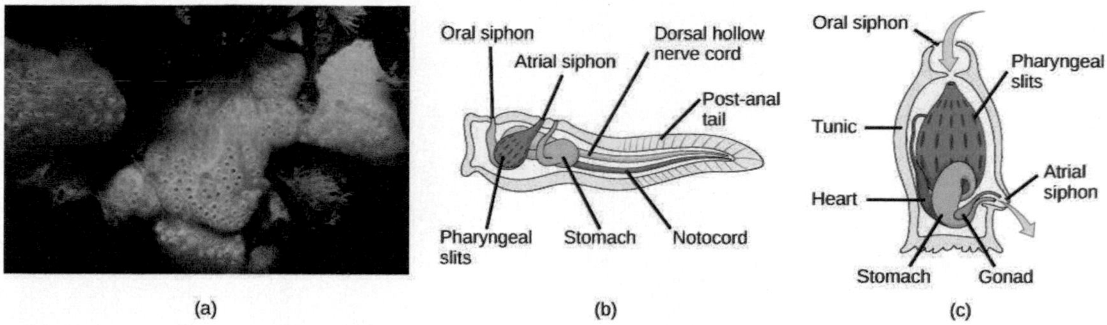

Figure 29.5 Urochordate anatomy. (a) This photograph shows a colony of the tunicate *Botrylloides violaceus*. (b) The larval stage of the tunicate possesses all of the features characteristic of chordates: a notochord, a dorsal hollow nerve cord, pharyngeal slits, an endostyle, and a post-anal tail. (c) In the adult stage, the notochord, nerve cord, and tail disappear, leaving just the pharyngeal slits and endostyle. (credit: modification of work by Dann Blackwood, USGS)

Adult tunicates may be either solitary or colonial forms, and some species may reproduce by budding. Most tunicates live a sessile existence on the ocean floor and are *suspension feeders*. However, chains of thaliacean tunicates called *salps* (Figure 29.6) can swim actively while feeding, propelling themselves as they move water through the pharyngeal slits. The primary foods of tunicates are plankton and detritus. Seawater enters the tunicate's body through its incurrent siphon. Suspended material is filtered out of this water by a mucous net produced by the endostyle and is passed into the intestine via the action of cilia. The anus empties into the excurrent siphon, which expels wastes and water. Tunicates are found in shallow ocean waters around the world.

Figure 29.6 Salps. These colonial tunicates feed on phytoplankton. Salps are sequential hermaphrodites, with younger female colonies fertilized by older male colonies. (credit: Oregon Department of Fish & Wildlife via Wikimedia Commons)

Subphylum Vertebrata (Craniata)

A **cranium** is a bony, cartilaginous, or fibrous structure surrounding the brain, jaw, and facial bones (Figure 29.7). Most bilaterally symmetrical animals have a head; of these, those that have a cranium comprise the clade Craniata/Vertebrata, which includes the primitively jawless Myxini (hagfishes), Petromyzontida (lampreys), and all of the organisms called "vertebrates." (We should note that the Myxini have a cranium but lack a backbone.)

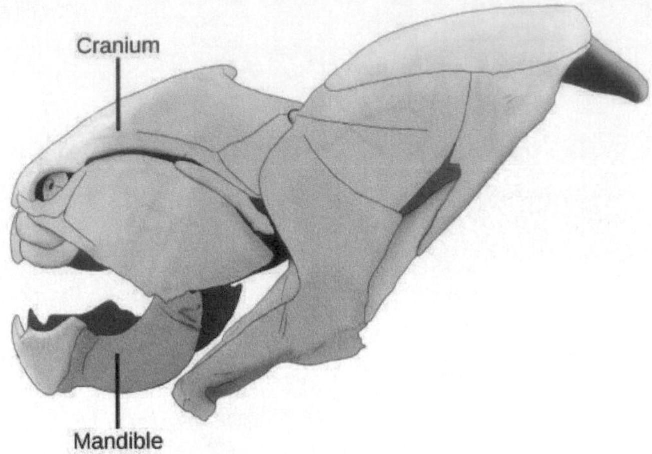

Figure 29.7 A craniate skull. The subphylum **Craniata (or Vertebrata)**, including this placoderm fish (*Dunkleosteus* sp.), are characterized by the presence of a cranium, mandible, and other facial bones. (credit: "Steveoc 86"/Wikimedia Commons)

Members of the phylum Craniata/Vertebrata display the five characteristic features of the chordates; however, members of this group also share derived characteristics that distinguish them from invertebrate chordates. Vertebrates are named for the vertebral column, composed of **vertebrae**—a series of separate, irregularly shaped bones joined together to form a backbone (Figure 29.8). Initially, the vertebrae form in segments around the embryonic notochord, but eventually replace it in adults. In most derived vertebrates, the notochord becomes the *nucleus pulposus* of the intervertebral discs that cushion and support adjacent vertebrae.

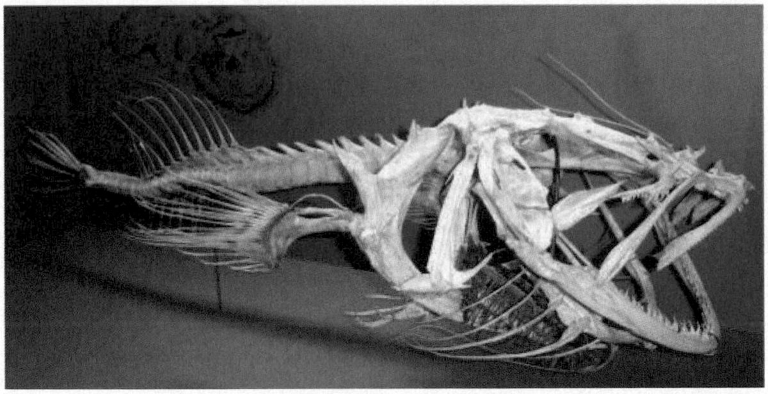

Figure 29.8 A vertebrate skeleton. Vertebrata are characterized by the presence of a backbone, such as the one that runs through the middle of this fish. All vertebrates are in the Craniata clade and have a cranium. (credit: Ernest V. More; taken at Smithsonian Museum of Natural History, Washington, D.C.)

The relationship of the vertebrates to the invertebrate chordates has been a matter of contention, but although these cladistic relationships are still being examined, it appears that the Craniata/Vertebrata are a monophyletic group that shares the five basic chordate characteristics with the other two subphyla, Urochordata and Cephalochordata. Traditional phylogenies place the cephalochordates as a sister clade to the chordates, a view that has been supported by most current molecular analyses. This hypothesis is further supported by the discovery of a fossil in China from the genus *Haikouella*. This organism seems to be an intermediate form between cephalochordates and vertebrates. The *Haikouella* fossils are about 530 million years old and appear similar to modern lancelets. These organisms had a brain and eyes, as do vertebrates, but lack the skull found in craniates.[1] This evidence suggests that vertebrates arose during the Cambrian explosion.

Vertebrates are the largest group of chordates, with more than 62,000 living species, which are grouped based on anatomical and physiological traits. More than one classification and naming scheme is used for these animals. Here we will consider the *traditional groups* Agnatha, Chondrichthyes, Osteichthyes, Amphibia, Reptilia, Aves, and Mammalia, which constitute classes in the subphylum Vertebrata/Craniata. Virtually all modern cladists classify birds within Reptilia, which correctly reflects their

1Chen, J. Y., Huang, D. Y., and Li, C. W., "An early Cambrian craniate-like chordate," *Nature* 402 (1999): 518–522, doi:10.1038/990080.

evolutionary heritage. Thus, we now have the nonavian reptiles and the avian reptiles in our reptilian classification. We consider them separately only for convenience. Further, we will consider hagfishes and lampreys together as jawless fishes, the **Agnatha**, although emerging classification schemes separate them into chordate jawless fishes (the hagfishes) and vertebrate jawless fishes (the lampreys).

Animals that possess jaws are known as **gnathostomes**, which means "jawed mouth." Gnathostomes include fishes and tetrapods. **Tetrapod** literally means "four-footed," which refers to the phylogenetic history of various land vertebrates, even though in some of the tetrapods, the limbs may have been modified for purposes other than walking. Tetrapods include amphibians, reptiles, birds, and mammals, and technically could also refer to the extinct fishlike groups that gave rise to the tetrapods. Tetrapods can be further divided into two groups: amphibians and amniotes. Amniotes are animals whose eggs contain four extraembryonic membranes (yolk sac, amnion, chorion, and allantois) that provide nutrition and a water-retaining environment for their embryos. Amniotes are adapted for terrestrial living, and include mammals, reptiles, and birds.

29.2 Fishes

By the end of this section, you will be able to do the following:

- Describe the difference between jawless and jawed fishes
- Discuss the distinguishing features of sharks and rays compared to other modern fishes

Modern fishes include an estimated 31,000 species, by far the most of all clades within the Vertebrata. Fishes were the earliest vertebrates, with jawless species being the earliest forms and jawed species evolving later. They are active feeders, rather than sessile, suspension feeders. The Agnatha (jawless fishes)—the hagfishes and lampreys—have a distinct cranium and complex sense organs including eyes, that distinguish them from the invertebrate chordates, the urochordates and cephalochordates.

Jawless Fishes: Superclass Agnatha

Jawless fishes (Agnatha) are craniates representing an ancient vertebrate lineage that arose over 550 million years ago. In the past, hagfishes and lampreys were sometimes recognized as separate clades within the Agnatha, primarily because lampreys were regarded as true vertebrates, whereas hagfishes were not. However, recent molecular data, both from rRNA and mtDNA, as well as embryological data, provide strong support for the hypothesis that living agnathans—previously called *cyclostomes*—are monophyletic, and thus share recent common ancestry. The discussion below, for convenience, separates the modern "cyclostomes" into the class Myxini and class Petromyzontida. The defining features of the living jawless fishes are the lack of jaws and lack of paired lateral appendages (fins). They also lack internal ossification and scales, although these are not defining features of the clade.

Some of the earliest jawless fishes were the armored ostracoderms (which translates to "shell-skin"): vertebrate fishes encased in bony armor—unlike present-day jawless fishes, which lack bone in their scales. Some ostracoderms, also unlike living jawless fishes, may have had paired fins. We should note, however, that the "ostracoderms" represent an assemblage of heavily armored extinct jawless fishes that may not form a natural evolutionary group. Fossils of the genus *Haikouichthys* from China, with an age of about 530 million years, show many typical vertebrate characteristics including paired eyes, auditory capsules, and rudimentary vertebrae.

Class Myxini: Hagfishes

The class **Myxini** includes at least 70 species of hagfishes—eel-like scavengers that live on the ocean floor and feed on living or dead invertebrates, fishes, and marine mammals (Figure 29.9). Although they are almost completely blind, sensory barbels around the mouth help them locate food by smell and touch. They feed using keratinized teeth on a movable cartilaginous plate in the mouth, which rasp pieces of flesh from their prey. These feeding structures allow the gills to be used exclusively for respiration, *not* for filter feeding as in the urochordates and cephalochordates. Hagfishes are entirely marine and are found in oceans around the world, except for the polar regions. Unique slime glands beneath the skin release a milky mucus (through surface pores) that upon contact with water becomes incredibly slippery, making the animal almost impossible to hold. This slippery mucus thus allows the hagfish to escape from the grip of predators. Hagfish can also twist their bodies into a knot, which provides additional leverage to feed. Sometimes hagfish enter the bodies of dead animals and eat carcasses from the inside out! Interestingly, they do not have a stomach!

Figure 29.9 Hagfish. Pacific hagfish are scavengers that live on the ocean floor. (credit: Linda Snook, NOAA/CBNMS)

Hagfishes have a cartilaginous skull, as well as a fibrous and cartilaginous skeleton, but the major supportive structure is the notochord that runs the length of the body. In hagfishes, the notochord is *not* replaced by the vertebral column, as it is in true vertebrates, and thus they may (morphologically) represent a sister group to the true vertebrates, making them the most basal clade among the skull-bearing chordates.

Class Petromyzontida: Lampreys

The class **Petromyzontida** includes approximately 40 species of lampreys, which are superficially similar to hagfishes in size and shape. However, lampreys possess extrinsic eye muscles, at least two semicircular canals, and a true cerebellum, as well as simple vertebral elements, called *arcualia*—cartilaginous structures arranged above the notochord. These features are also shared with the *gnathostomes*—vertebrates with jawed mouths and paired appendages (see below). Lampreys also have a dorsal tubular nerve cord with a well-differentiated brain, a small cerebellum, and 10 pairs of nerves. The classification of lampreys is still debated, but they clearly represent one of the oldest divergences of the vertebrate lineage. Lampreys lack paired appendages, as do the hagfishes, although they have one or two fleshy dorsal fins. As adults, lampreys are characterized by a rasping tongue within a toothed, funnel-like sucking mouth. Many species have a parasitic stage of their life cycle during which they are fish ectoparasites (some call them predators because they attack and eventually fall off) (Figure 29.10).

Figure 29.10 Lamprey. These parasitic sea lampreys, *Petromyzon marinus,* attach by suction to their lake trout host, and use their rough tongues to rasp away flesh in order to feed on the trout's blood. (credit: USGS)

Lampreys live primarily in coastal and freshwater environments, and have a worldwide distribution, except for the tropics and polar regions. Some species are marine, but all species spawn in fresh water. Interestingly, northern lampreys in the family Petromyzontidae, have the highest number of chromosomes (164 to 174) among the vertebrates. Eggs are fertilized externally, and the larvae (called *ammocoetes*) differ greatly from the adult form, closely resembling the adult cephalocordate *amphioxus*. After spending three to 15 years as suspension feeders in rivers and streams, they attain sexual maturity. Shortly afterward, the adults swim upstream, reproduce, and die within days.

Gnathostomes: Jawed Fishes

Gnathostomes, or "jaw-mouths," are vertebrates that possess true jaws—a milestone in the evolution of the vertebrates. In fact, one of the most significant developments in early vertebrate evolution was the development of the jaw: a hinged structure attached to the cranium that allows an animal to grasp and tear its food. Jaws were probably derived from the first pair of gill arches supporting the gills of jawless fishes.

Early gnathostomes also possessed two sets of paired fins, allowing the fishes to maneuver accurately. Pectoral fins are typically located on the anterior body, and pelvic fins on the posterior. Evolution of the jaw and paired fins permitted gnathostomes to expand their food options from the scavenging and suspension feeding of jawless fishes to active predation. The ability of gnathostomes to exploit new nutrient sources probably contributed to their replacing most jawless fishes during the Devonian period. Two early groups of gnathostomes were the *acanthodians* and *placoderms* (Figure 29.11), which arose in the late Silurian period and are now extinct. Most modern fishes are gnathostomes that belong to the clades Chondrichthyes and Osteichthyes (which include the class Actinoptertygii and class Sarcopterygii).

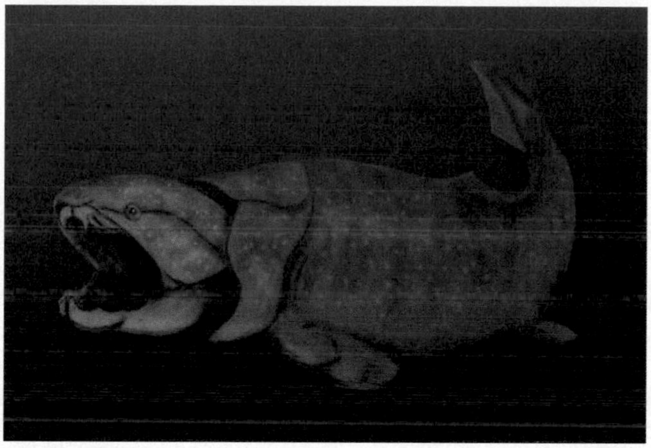

Figure 29.11 A placoderm. *Dunkleosteus* was an enormous placoderm from the Devonian period, 380 to 360 million years ago. It measured up to 10 meters in length and weighed up to 3.6 tons. Its head and neck were armored with heavy bony plates. Although *Dunkleosteus* had no true teeth, the edge of the jaw was armed with sharp bony blades. (credit: Nobu Tamura)

Class Chondrichthyes: Cartilaginous Fishes

The class Chondrichthyes (about 1,000 species) is a morphologically diverse clade, consisting of subclass Elasmobranchii (sharks [Figure 29.12], rays, and skates, together with the obscure and critically endangered sawfishes), and a few dozen species of fishes called *chimaeras*, or "ghost sharks" in the subclass Holocephali. Chondrichthyes are jawed fishes that possess paired fins and a skeleton made of cartilage. This clade arose approximately 370 million years ago in the early or middle Devonian. They are thought to be descended from the placoderms, which had endoskeletons made of bone; thus, the lighter cartilaginous skeleton of Chondrichthyes is a secondarily derived evolutionary development. Parts of shark skeleton are strengthened by granules of calcium carbonate, but this is not the same as bone.

Most cartilaginous fishes live in marine habitats, with a few species living in fresh water for a part or all of their lives. Most sharks are carnivores that feed on live prey, either swallowing it whole or using their jaws and teeth to tear it into smaller pieces. Sharks have abrasive skin covered with tooth-like scales called placoid scales. Shark teeth probably evolved from rows of these scales lining the mouth. A few species of sharks and rays, like the enormous whale shark (Figure 29.13), are suspension feeders that feed on plankton. The sawfishes have an extended rostrum that looks like a double-edged saw. The rostrum is covered with electrosensitive pores that allow the sawfish to detect slight movements of prey hiding in the muddy sea floor. The teeth in the rostrum are actually modified tooth-like structures called denticles, similar to scales.

Figure 29.12 Shark. Hammerhead sharks tend to school during the day and hunt prey at night. (credit: Masashi Sugawara)

Sharks have well-developed sense organs that aid them in locating prey, including a keen sense of smell and the ability to detect electromagnetic fields. Electroreceptors called **ampullae of Lorenzini** allow sharks to detect the electromagnetic fields that are produced by all living things, including their prey. (Electroreception has only been observed in aquatic or amphibious animals and sharks have perhaps the most sensitive electroreceptors of any animal.) Sharks, together with most fishes and aquatic and larval amphibians, also have a row of sensory structures called the **lateral line**, which is used to detect movement and vibration in the surrounding water, and is often considered to be functionally similar to the sense of "hearing" in terrestrial vertebrates. The lateral line is visible as a darker stripe that runs along the length of a fish's body. Sharks have no mechanism for maintaining neutral buoyancy and must swim continuously to stay suspended in the water. Some must also swim in order to ventilate their gills but others have muscular pumps in their mouths to keep water flowing over the gills.

Figure 29.13 Whale shark in the Georgia Aquarium. Whale sharks are filter-feeders and can grow to be over 10 meters long. Whale sharks, like most other sharks, are ovoviviparous. (credit: modified from Zac Wolf [Own work] [CC BY-SA 2.5 (http://creativecommons.org/licenses/by-sa/2.5 (http://openstax.org/l/CCSA))], via Wikimedia Commons)

Sharks reproduce sexually, and eggs are fertilized internally. Most species are *ovoviviparous*: The fertilized egg is retained in the oviduct of the mother's body and the embryo is nourished by the egg yolk. The eggs hatch in the uterus, and young are born alive and fully functional. Some species of sharks are oviparous: They lay eggs that hatch outside of the mother's body. Embryos are protected by a shark egg case or "mermaid's purse" (Figure 29.14) that has the consistency of leather. The shark egg case has tentacles that snag in seaweed and give the newborn shark cover. A few species of sharks, e.g., tiger sharks and hammerheads, are viviparous: the yolk sac that initially contains the egg yolk and transfers its nutrients to the growing embryo becomes attached to the oviduct of the female, and nutrients are transferred directly from the mother to the growing embryo. In both viviparous and ovoviviparous sharks, gas exchange uses this yolk sac transport.

Figure 29.14 Shark egg cases. Shark embryos are clearly visible through these transparent egg cases. The round structure is the yolk that nourishes the growing embryo. (credit: Jek Bacarisas)

In general, the Chondrichthyes have a fusiform or dorsoventrally flattened body, a *heterocercal* caudal fin or tail (unequally sized fin lobes, with the tail vertebrae extending into the larger upper lobe) paired pectoral and pelvic fins (in males these may be modified as claspers), exposed gill slits (*elasmobranch*), and an intestine with a spiral valve that condenses the length of the intestine. They also have three pairs of semicircular canals, and excellent senses of smell, vibration, vision, and electroreception. A very large lobed liver produces *squalene oil* (a lightweight biochemical precursor to steroids) that serves to aid in buoyancy (because with a specific gravity of 0.855, it is lighter than that of water).

Rays and skates comprise more than 500 species. They are closely related to sharks but can be distinguished from sharks by their flattened bodies, pectoral fins that are enlarged and fused to the head, and gill slits on their ventral surface (Figure 29.15). Like sharks, rays and skates have a cartilaginous skeleton. Most species are marine and live on the sea floor, with nearly a worldwide distribution.

Unlike the stereotypical sharks and rays, the Holocephali (chimaeras or ratfish) have a *diphycercal* tail (equally sized fin lobes, with the tail vertebrae located between them), lack scales (lost secondarily in evolution), and have teeth modified as grinding plates that are used to feed on mollusks and other invertebrates (Figure 29.15b). Unlike sharks with elasmobranch or naked gills, chimaeras have four pairs of gills covered by an operculum. Many species have a pearly iridescence and are extremely pretty.

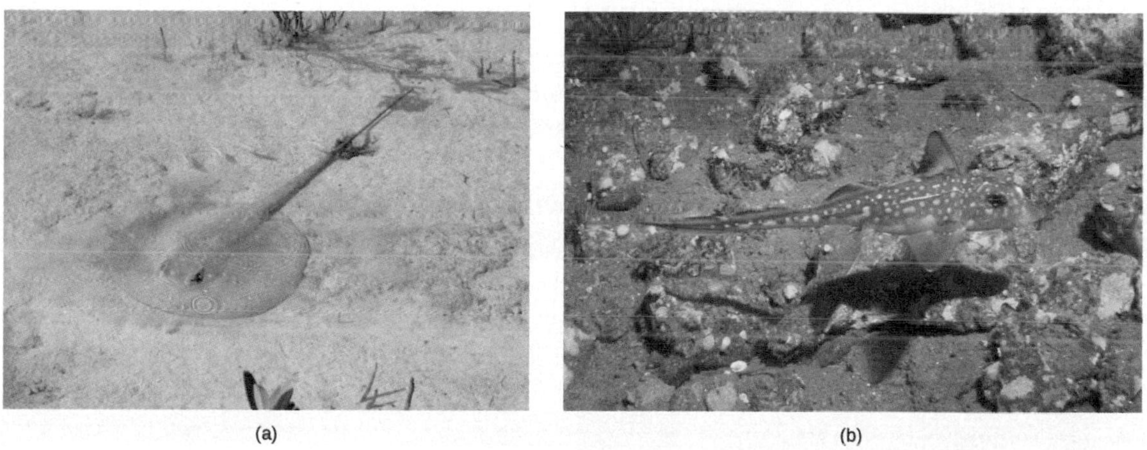

Figure 29.15 Cartilaginous fish. (a) Stingray. This stingray blends into the sandy bottom of the ocean floor. A spotted ratfish (b) Hydrolagus colliei credit a "Sailn1"/Flickr; (credit: a "Sailn1"/Flickr b: Linda Snook / MBNMS [Public domain], via Wikimedia Commons.)

Osteichthyes: Bony Fishes

Members of the clade **Osteichthyes**, also called bony fishes, are characterized by a bony skeleton. The vast majority of present-day fishes belong to this group, which consists of approximately 30,000 species, making it the largest class of vertebrates in existence today.

Nearly all bony fishes have an *ossified skeleton* with specialized bone cells (osteocytes) that produce and maintain a calcium phosphate matrix. This characteristic has been reversed only in a few groups of Osteichthyes, such as sturgeons and paddlefish, which have primarily cartilaginous skeletons. The skin of bony fishes is often covered by overlapping scales, and glands in the skin secrete mucus that reduces drag when swimming and aids the fish in osmoregulation. Like sharks, bony fishes have a lateral line system that detects vibrations in water.

All bony fishes use gills to breathe. Water is drawn over gills that are located in chambers covered and ventilated by a protective, muscular flap called the operculum. Many bony fishes also have a **swim bladder**, a gas-filled organ derived as a pouch from the gut. The swim bladder helps to control the buoyancy of the fish. In most bony fish, the gases of the swim bladder are exchanged directly with the blood. The swim bladder is believed to be homologous to the lungs of lungfish and the lungs of land vertebrates.

Bony fishes are further divided into two extant clades: Class **Actinopterygii** (ray-finned fishes) and Class **Sarcopterygii** (lobe-finned fishes).

Actinopterygii (Figure 29.16a), the ray-finned fishes, include many familiar fishes—tuna, bass, trout, and salmon among others—and represent about half of all vertebrate species. Ray-finned fishes are named for the fan of slender bones that supports their fins.

In contrast, the fins of Sarcopterygii (Figure 29.16b) are fleshy and lobed, supported by bones that are similar in type and arrangement to the bones in the limbs of early tetrapods. The few extant members of this clade include several species of lungfishes and the less familiar coelacanths, which were thought to be extinct until living specimens were discovered between Africa and Madagascar. Currently, two species of coelocanths have been described.

Figure 29.16 Osteichthyes. The (a) sockeye salmon and (b) coelacanth are both bony fishes of the Osteichthyes clade. The coelacanth, sometimes called a lobe-finned fish, was thought to have gone extinct in the Late Cretaceous period, 100 million years ago, until one was discovered in 1938 near the Comoros Islands between Africa and Madagascar. (credit a: modification of work by Timothy Knepp, USFWS; credit b: modification of work by Robbie Cada)

29.3 Amphibians

By the end of this section, you will be able to do the following:

- Describe the important difference between the life cycle of amphibians and the life cycles of other vertebrates
- Distinguish between the characteristics of Urodela, Anura, and Apoda
- Describe the evolutionary history of amphibians

Amphibians are vertebrate tetrapods ("four limbs"), and include frogs, salamanders, and caecilians. The term "amphibian" loosely translates from the Greek as "dual life," which is a reference to the metamorphosis that many frogs and salamanders undergo and the unique mix of aquatic and terrestrial phases that are required in their life cycle. In fact, they cannot stray far from water because their reproduction is intimately tied to aqueous environments. Amphibians evolved during the Devonian period and were the earliest terrestrial tetrapods. They represent an evolutionary transition from water to land that occurred over many millions of years. Thus, the Amphibia are the only living true vertebrates that have made a transition from water to land in both their ontogeny (life development) and phylogeny (evolution). They have not changed much in morphology over the past 350 million years!

LINK TO LEARNING

Watch this series of five Animal Planet videos on tetrapod evolution:

- 1: The evolution from fish to earliest tetrapod
  Click to view content (https://www.openstax.org/l/tetrapod_evol1)
- 2: Fish to Earliest Tetrapod
  Click to view content (https://www.openstax.org/l/tetrapod_evol2)
- 3: The discovery of coelacanth and *Acanthostega* fossils

Click to view content (https://www.openstax.org/l/tetrapod_evol3)

- 4: The number of fingers on "legs"
  Click to view content (https://www.openstax.org/l/tetrapod_evol4)
- 5: Reconstructing the environment of early tetrapods
  Click to view content (https://www.openstax.org/l/tetrapod_evol5)

Characteristics of Amphibians

As tetrapods, most amphibians are characterized by four well-developed limbs. In some species of salamanders, hindlimbs are reduced or absent, but all caecilians are (secondarily) limbless. An important characteristic of extant amphibians is a moist, permeable skin that is achieved via mucus glands. Most water is taken in across the skin rather than by drinking. The skin is also one of three respiratory surfaces used by amphibians. The other two are the lungs and the buccal (mouth) cavity. Air is taken first into the mouth through the nostrils, and then pushed by positive pressure into the lungs by elevating the throat and closing the nostrils.

All extant adult amphibians are carnivorous, and some terrestrial amphibians have a sticky tongue used to capture prey. Amphibians also have multiple small teeth at the edge of the jaws. In salamanders and caecilians, teeth are present in both jaws, sometimes in multiple rows. In frogs and toads, teeth are seen only in the upper jaw. Additional teeth, called **vomerine teeth**, may be found in the roof of the mouth. Amphibian teeth are *pedicellate*, which means that the root and crown are calcified, separated by a zone of noncalcified tissue.

Amphibians have image-forming eyes and color vision. Ears are best developed in frogs and toads, which vocalize to communicate. Frogs use separate regions of the inner ear for detecting higher and lower sounds: the *papilla amphibiorum*, which is sensitive to frequencies below 10,000 hertz and unique to amphibians, and the *papilla basilaris*, which is sensitive to higher frequencies, including mating calls, transmitted from the eardrum through the stapes bone. Amphibians also have an extra bone in the ear, the operculum, which transmits low-frequency vibrations from the forelimbs and shoulders to the inner ear, and may be used for the detection of seismic signals.

Evolution of Amphibians

The fossil record provides evidence of the first tetrapods: now-extinct amphibian species dating to nearly 400 million years ago. Evolution of tetrapods from lobe-finned freshwater fishes (similar to coelacanths and lungfish) represented a significant change in body plan from one suited to organisms that respired and swam in water, to organisms that breathed air and moved onto land; these changes occurred over a span of 50 million years during the Devonian period.

Aquatic tetrapods of the Devonian period include *Ichthyostega* and *Acanthostega*. Both were aquatic, and may have had both gills and lungs. They also had four limbs, with the skeletal structure of limbs found in present-day tetrapods, including amphibians. However, the limbs could not be pulled in under the body and would not have supported their bodies well out of water. They probably lived in shallow freshwater environments, and may have taken brief terrestrial excursions, much like "walking" catfish do today in Florida. In *Ichthyostega*, the forelimbs were more developed than the hind limbs, so it might have dragged itself along when it ventured onto land. What preceded *Acanthostega* and *Ichthyostega*?

In 2006, researchers published news of their discovery of a fossil of a "tetrapod-like fish," *Tiktaalik roseae*, which seems to be a morphologically "intermediate form" between sarcopterygian fishes having feet-like fins and early tetrapods having true limbs (Figure 29.17). *Tiktaalik* likely lived in a shallow water environment about 375 million years ago.[2] Tiktaalik also had gills and lungs, but the loss of some gill elements gave it a neck, which would have allowed its head to move sideways for feeding. The eyes were on top of the head. It had fins, but the attachment of the fin bones to the shoulder suggested they might be weight-bearing. Tiktaalik preceded *Acanthostega* and *Ichthyostega*, with their four limbs, by about 10 million years and is considered to be a true intermediate clade between fish and amphibians.

2Daeschler, E. B., Shubin, N. H., and Jenkins, F. J. "A Devonian tetrapod-like fish and the evolution of the tetrapod body plan," *Nature* 440 (2006): 757–763, doi:10.1038/nature04639, http://www.nature.com/nature/journal/v440/n7085/abs/nature04639.html (http://openstax.org/l/tetrapod) .

Figure 29.17 Tiktaalik. The recent fossil discovery of *Tiktaalik roseae* suggests evidence for an animal intermediate to finned fish and legged tetrapods, sometimes called a "fishapod." (credit: Zina Deretsky, National Science Foundation)

The early tetrapods that moved onto land had access to new nutrient sources and relatively few predators. This led to the widespread distribution of tetrapods during the early Carboniferous period, a period sometimes called the "age of the amphibians."

Modern Amphibians

Amphibia comprises an estimated 6,770 extant species that inhabit tropical and temperate regions around the world. All living species are classified in the subclass Lissamphibia ("smooth-amphibian"), which is divided into three clades: **Urodela** ("tailed"), the salamanders; **Anura** ("tail-less"), the frogs; and **Apoda** ("legless ones"), the caecilians.

Urodela: Salamanders

Salamanders are amphibians that belong to the order Urodela (or Caudata). These animals are probably the most similar to ancestral amphibians. Living salamanders (Figure 29.18) include approximately 620 species, some of which are aquatic, others terrestrial, and some that live on land only as adults. Most adult salamanders have a generalized tetrapod body plan with four limbs and a tail. The placement of their legs makes it difficult to lift their bodies off the ground and they move by bending their bodies from side to side, called *lateral undulation*, in a fish-like manner while "walking" their arms and legs fore-and-aft. It is thought that their gait is similar to that used by early tetrapods. The majority of salamanders are lungless, and respiration occurs through the skin or through external gills in aquatic species. Some terrestrial salamanders have primitive lungs; a few species have both gills and lungs. The giant Japanese salamander, the largest living amphibian, has additional folds in its skin that enlarge its respiratory surface.

Most salamanders reproduce using an unusual process of internal fertilization of the eggs. Mating between salamanders typically involves an elaborate and often prolonged courtship. Such a courtship ends in the deposition of sperm by the males in a packet called a **spermatophore**, which is subsequently picked up by the female, thus ultimately fertilization is internal. All salamanders except one, the fire salamander, are oviparous. Aquatic salamanders lay their eggs in water, where they develop into legless larvae called *efts*. Terrestrial salamanders lay their eggs in damp nests, where the eggs are guarded by their mothers. These embryos go through the larval stage and complete metamorphosis before hatching into tiny adult forms. One aquatic salamander, the Mexican *axolotl*, never leaves the larval stage, becoming sexually mature without metamorphosis.

Figure 29.18 Salamander. Most salamanders have legs and a tail, but respiration varies among species. (credit: Valentina Storti)

LINK TO LEARNING

View River Monsters: Fish With Arms and Hands? (http://openstax.org/l/river_monster) to see a video about an unusually large salamander species.

Anura: Frogs

Frogs (Figure 29.19) are amphibians that belong to the order Anura or Salientia ("jumpers"). Anurans are among the most diverse groups of vertebrates, with approximately 5,965 species that occur on all of the continents except Antarctica. Anurans, ranging from the minute New Guinea frog at 7 mm to the huge goliath frog at 32 cm from tropical Africa, have a body plan that is more specialized for movement. Adult frogs use their hind limbs and their arrow-like endoskeleton to jump accurately to capture prey on land. Tree frogs have hands adapted for grasping branches as they climb. In tropical areas, "flying frogs" can glide from perch to perch on the extended webs of their feet. Frogs have a number of modifications that allow them to avoid predators, including skin that acts as camouflage. Many species of frogs and salamanders also release defensive chemicals that are poisonous to predators from glands in the skin. Frogs with more toxic skins have bright warning (*aposematic*) coloration.

Figure 29.19 Tree frog. The Australian green tree frog is a nocturnal predator that lives in the canopies of trees near a water source.

Frog eggs are fertilized externally, and like other amphibians, frogs generally lay their eggs in moist environments. Although amphibian eggs are protected by a thick jelly layer, they would still dehydrate quickly in a dry environment. Frogs demonstrate a great diversity of parental behaviors, with some species laying many eggs and exhibiting little parental care, to species that carry eggs and tadpoles on their hind legs or embedded in their backs. The males of Darwin's frog carry tadpoles in their vocal sac. Many tree frogs lay their eggs off the ground in a folded leaf located over water so that the tadpoles can drop into the water as they hatch.

The life cycle of most frogs, as other amphibians, consists of two distinct stages: the larval stage followed by metamorphosis to an adult stage. However, the eggs of frogs in the genus *Eleutherodactylus* develop directly into little froglets, guarded by a parent. The larval stage of a frog, the tadpole, is often a filter-feeding herbivore. Tadpoles usually have gills, a lateral line system, longfinned tails, and lack limbs. At the end of the tadpole stage, frogs undergo metamorphosis into the adult form (Figure 29.20). During this stage, the gills, tail, and lateral line system disappear, and four limbs develop. The jaws become larger and are suited for carnivorous feeding, and the digestive system transforms into the typical short gut of a predator. An eardrum and air-breathing lungs also develop. These changes during metamorphosis allow the larvae to move onto land in the adult stage.

Figure 29.20 Amphibian metapmorphosis. A juvenile frog metamorphoses into a frog. Here, the frog has started to develop limbs, but its tadpole tail is still evident.

Apoda: Caecilians

An estimated 185 species comprise the **caecilians**, a group of amphibians that belong to the order Apoda. They have no limbs, although they evolved from a legged vertebrate ancestor. The complete lack of limbs makes them resemble earthworms. This resemblance is enhanced by folds of skin that look like the segments of an earthworm. However, unlike earthworms, they have teeth in both jaws, and feed on a variety of small organisms found in soil, including earthworms! Caecilians are adapted for a burrowing or aquatic lifestyle, and they are nearly blind, with their tiny eyes sometimes covered by skin. Although they have a single lung, they also depend on cutaneous respiration. These animals are found in the tropics of South America, Africa, and Southern Asia. In the caecelians, the only amphibians in which the males have copulatory structures, fertilization is internal. Some caecilians are oviparous, but most bear live young. In these cases, the females help nourish their young with tissue from their oviduct before birth and from their skin after birth.

EVOLUTION CONNECTION

The Paleozoic Era and the Evolution of Vertebrates

When the vertebrates arose during the Paleozoic Era (542 to 251 MYA), the climate and geography of Earth was vastly different. The distribution of landmasses on Earth were also very different from that of today. Near the equator were two large supercontinents, **Laurentia** and **Gondwana**, which included most of today's continents, but in a radically different configuration (Figure 29.21). At this time, sea levels were very high, probably at a level that hasn't been reached since. As the Paleozoic progressed, glaciations created a cool global climate, but conditions warmed near the end of the first half of the Paleozoic. During the latter half of the Paleozoic, the landmasses began moving together, with the initial formation of a large northern block called **Laurasia**, which contained parts of what is now North America, along with Greenland, parts of Europe, and Siberia. Eventually, a single supercontinent, called **Pangaea**, was formed, starting in the latter third of the Paleozoic. Glaciations then began to affect Pangaea's climate and the distribution of vertebrate life.

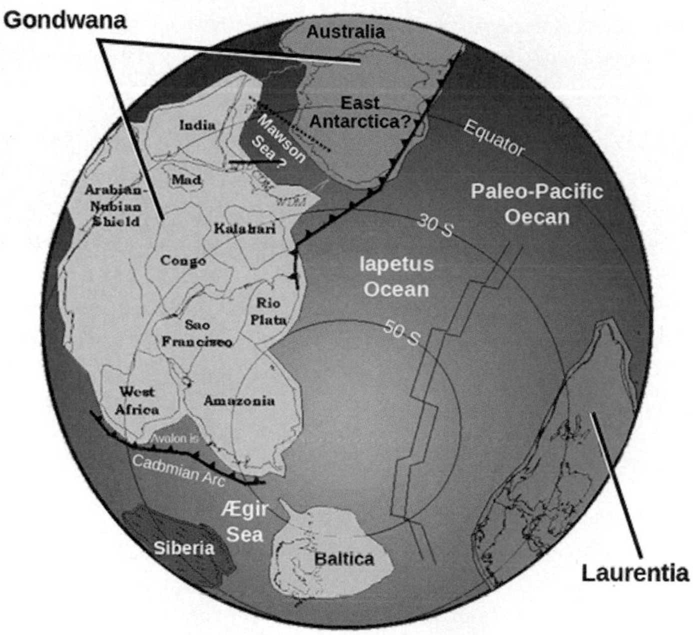

Figure 29.21 Paleozoic continents. During the Paleozoic Era, around 550 million years ago, the continent Gondwana formed. Both Gondwana and the continent Laurentia were located near the equator.

During the early Paleozoic, the amount of carbon dioxide in the atmosphere was much greater than it is today. This may have begun to change later, as land plants became more common. As the roots of land plants began to infiltrate rock and soil began to form, carbon dioxide was drawn out of the atmosphere and became trapped in the rock. This reduced the levels of carbon dioxide and increased the levels of oxygen in the atmosphere, so that by the end of the Paleozoic, atmospheric conditions were similar to those of today.

As plants became more common through the latter half of the Paleozoic, microclimates began to emerge and ecosystems began to change. As plants and ecosystems continued to grow and become more complex, vertebrates moved from the water to land. The presence of shoreline vegetation may have contributed to the movement of vertebrates onto land. One hypothesis suggests that the fins of aquatic vertebrates were used to maneuver through this vegetation, providing a precursor to the movement of fins on land and the further development of limbs. The late Paleozoic was a time of diversification of vertebrates, as amniotes emerged and became two different lines that gave rise, on one hand, to synapsids and mammals, and, on the other hand, to the codonts, reptiles, dinosaurs, and birds. Many marine vertebrates became extinct near the end of the Devonian period, which ended about 360 million years ago, and both marine and terrestrial vertebrates were decimated by a mass extinction in the early Permian period about 250 million years ago.

LINK TO LEARNING

View Earth's Paleogeography: Continental Movements Through Time (http://openstax.org/l/paleogeography) to see changes in Earth as life evolved.

29.4 Reptiles

By the end of this section, you will be able to do the following:

- Describe the main characteristics of amniotes
- Explain the difference between anapsids, synapsids, and diapsids, and give an example of each
- Identify the characteristics of reptiles
- Discuss the evolution of reptiles

The reptiles (including dinosaurs and birds) are distinguished from amphibians by their terrestrially adapted egg, which is supported by four *extraembryonic membranes*: the yolk sac, the amnion, the chorion, and the allantois (Figure 29.22). The chorion and amnion develop from folds in the body wall, and the yolk sac and allantois are extensions of the midgut and hindgut respectively. The amnion forms a fluid-filled cavity that provides the embryo with its own internal aquatic environment. The evolution of the extraembryonic membranes led to less dependence on water for development and thus allowed the amniotes to branch out into drier environments.

In addition to these membranes, the eggs of birds, reptiles, and a few mammals have shells. An **amniote embryo** was then enclosed in the amnion, which was in turn encased in an extra-embryonic coelom contained within the chorion. Between the shell and the chorion was the albumin of the egg, which provided additional fluid and cushioning. This was a significant development that further distinguishes the amniotes from amphibians, which were and continue to be restricted to moist environments due their shell-less eggs. Although the shells of various reptilian amniotic species vary significantly, they all permit the retention of water and nutrients for the developing embryo. The egg shells of bird (avian reptiles) are hardened with calcium carbonate, making them rigid, but fragile. The shells of most nonavian reptile eggs, such as turtles, are leathery and require a moist environment. Most mammals do not lay eggs (except for monotremes such as the echindnas and platypuses). Instead, the embryo grows within the mother's body, with the placenta derived from two of the extraembryonic membranes.

Characteristics of Amniotes

The **amniotic egg** is the key characteristic of amniotes. In amniotes that lay eggs, the shell of the egg provides protection for the developing embryo while being permeable enough to allow for the exchange of carbon dioxide and oxygen. The *albumin*, or egg white, outside of the chorion provides the embryo with water and protein, whereas the fattier egg yolk contained in the yolk sac provides nutrients for the embryo, as is the case with the eggs of many other animals, such as amphibians. Here are the functions of the extraembryonic membranes:

1. Blood vessels in the **yolk sac** transport yolk nutrients to the circulatory system of the embryo.
2. The **chorion** facilitates exchange of oxygen and carbon dioxide between the embryo and the egg's external environment.
3. The **allantois** stores nitrogenous wastes produced by the embryo and also facilitates respiration.
4. The **amnion** protects the embryo from mechanical shock and supports hydration.

In mammals, the yolk sac is very reduced, but the embryo is still cushioned and enclosed within the amnion. The *placenta,* which transports nutrients and functions in gas exchange and waste management, is derived from the chorion and allantois.

VISUAL CONNECTION

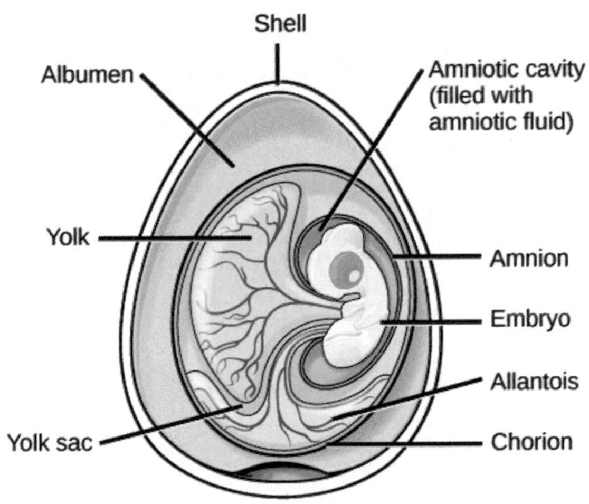

Figure 29.22 An amniotic egg. The key features of an amniotic egg are shown.

Which of the following statements about the parts of an egg are false?

1. The allantois stores nitrogenous waste and facilitates respiration.
2. The chorion facilitates gas exchange.
3. The yolk provides food for the growing embryo.
4. The amniotic cavity is filled with albumen.

Additional derived characteristics of amniotes include a waterproof skin, accessory keratinized structures, and costal (rib) ventilation of the lungs.

Evolution of Amniotes

The first amniotes evolved from tetrapod ancestors approximately 340 million years ago during the Carboniferous period. The early amniotes quickly diverged into two main lines: *synapsids* and *sauropsids*. **Synapsids** included the *therapsids*, a clade from which mammals evolved. **Sauropsids** were further divided into *anapsids* and *diapsids*. Diapsids gave rise to the reptiles, including the dinosaurs and birds. The key differences between the synapsids, anapsids, and diapsids are the structures of the skull and the number of *temporal fenestrae* ("windows") behind each eye (Figure 29.23). **Temporal fenestrae** are post-orbital openings in the skull that allow muscles to expand and lengthen. Anapsids have no temporal fenestrae, synapsids have one (fused ancestrally from two fenestrae), and diapsids have two (although many diapsids such as birds have highly modified diapsid skulls). Anapsids include extinct organisms and traditionally included turtles. However, more recent molecular and fossil evidence clearly shows that turtles arose within the diapsid line and secondarily lost the temporal fenestrae; thus they appear to be anapsids because modern turtles do not have fenestrae in the temporal bones of the skull. The canonical diapsids include dinosaurs, birds, and all other extinct and living reptiles.

Figure 29.23 Amniote skulls. Compare the skulls and temporal fenestrae of anapsids, synapsids, and diapsids. Anapsids have no openings, synapsids have one opening, and diapsids have two openings.

The diapsids in turn diverged into two groups, the *Archosauromorpha* ("ancient lizard form") and the *Lepidosauromorpha* ("scaly lizard form") during the Mesozoic period (Figure 29.24). The **lepidosaurs** include modern lizards, snakes, and tuataras. The **archosaurs** include modern crocodiles and alligators, and the extinct ichthyosaurs ("fish lizards" superficially resembling dolphins), pterosaurs ("winged lizard"), dinosaurs ("terrible lizard"), and birds. (We should note that clade Dinosauria includes birds, which evolved from a branch of maniraptoran theropod dinosaurs in the Mesozoic.)

The evolutionarily derived characteristics of amniotes include the amniotic egg and its four extraembryonic membranes, a thicker and more waterproof skin, and rib ventilation of the lungs (ventilation is performed by drawing air into and out of the lungs by muscles such as the costal rib muscles and the diaphragm).

VISUAL CONNECTION

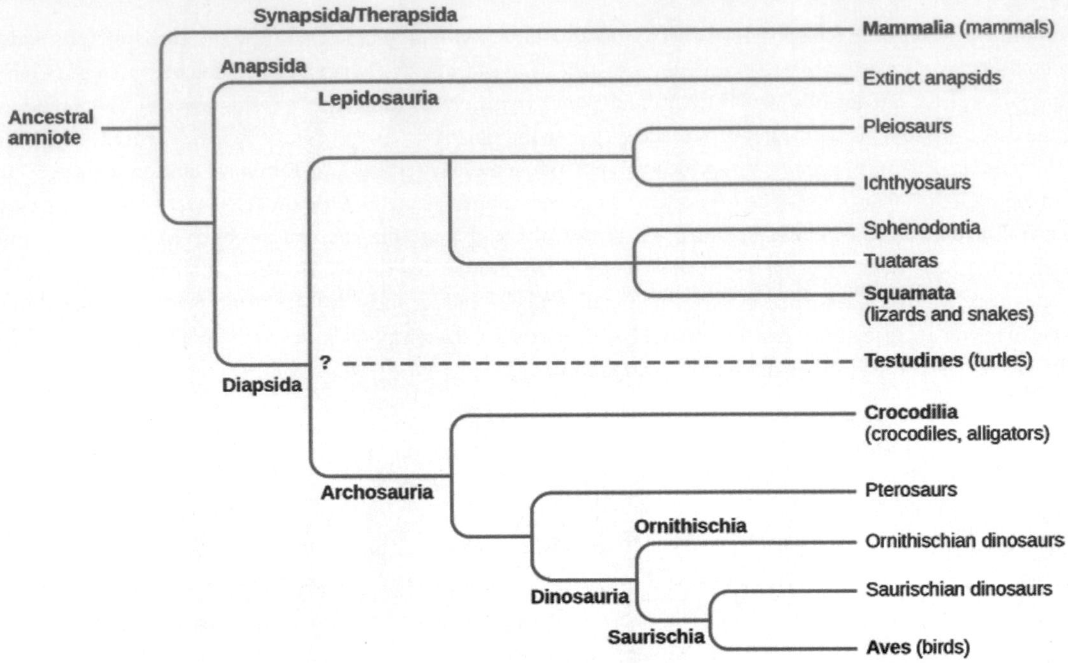

Figure 29.24 Amniote phylogeny. This chart shows the evolution of amniotes. The placement of Testudines (turtles) is currently still debated.

Question: Members of the order Testudines have an anapsid-like skull without obvious temporal fenestrae. However, molecular studies clearly indicate that turtles descended from a diapsid ancestor. Why might this be the case?

In the past, the most common division of amniotes has been into the classes Mammalia, Reptilia, and Aves. However, both birds and mammals are descended from different amniote branches: the synapsids giving rise to the therapsids and mammals, and the diapsids giving rise to the lepidosaurs and archosaurs. We will consider both the birds and the mammals as groups distinct from reptiles for the purpose of this discussion with the understanding that this does not accurately reflect phylogenetic history and relationships.

Characteristics of Reptiles

Reptiles are tetrapods. Limbless reptiles—snakes and legless lizards—are classified as tetrapods because they are descended from four-limbed ancestors. Reptiles lay calcareous or leathery eggs enclosed in shells on land. Even aquatic reptiles return to the land to lay eggs. They usually reproduce sexually with internal fertilization. Some species display ovoviviparity, with the eggs remaining in the mother's body until they are ready to hatch. In ovoviviparous reptiles, most nutrients are supplied by the egg yolk, while the chorioallantois assists with respiration. Other species are viviparous, with the offspring born alive, with their development supported by a yolk sac-placenta, a chorioallantoic-placenta, or both.

One of the key adaptations that permitted reptiles to live on land was the development of their *scaly skin*, containing the protein keratin and waxy lipids, which reduced water loss from the skin. A number of keratinous epidermal structures have emerged in the descendants of various reptilian lineages and some have become defining characters for these lineages: scales, claws, nails, horns, feathers, and hair. Their occlusive skin means that reptiles cannot use their skin for respiration, like amphibians, and thus all amniotes breathe with lungs. All reptiles grow throughout their lives and regularly shed their skin, both to accommodate their growth and to rid themselves of ectoparasites. Snakes tend to shed the entire skin at one time, but other reptiles shed their skins in patches.

Reptiles ventilate their lungs using various muscular mechanisms to produce *negative pressure* (low pressure) within the lungs that allows them to expand and draw in air. In snakes and lizards, the muscles of the spine and ribs are used to expand or contract the rib cage. Since walking or running interferes with this activity, the squamates cannot breathe effectively while running. Some squamates can supplement rib movement with buccal pumping through the nose, with the mouth closed. In crocodilians, the lung chamber is expanded and contracted by moving the liver, which is attached to the pelvis. Turtles have a special problem with breathing, because their rib cage cannot expand. However, they can change the pressure around the lungs by pulling their limbs in and out of the shell, and by moving their internal organs. Some turtles also have a posterior respiratory sac that opens off the hindgut that aids in the diffusion of gases.

Most reptiles are **ectotherms**, animals whose main source of body heat comes from the environment; however, some crocodilians maintain elevated *thoracic* temperatures and thus appear to be at least regional **endotherms**. This is in contrast to true endotherms, which use heat produced by metabolism and muscle contraction to regulate body temperature over a very narrow temperature range, and thus are properly referred to as **homeotherms**. Reptiles have behavioral adaptations to help regulate body temperature, such as basking in sunny places to warm up through the absorption of solar radiation, or finding shady spots or going underground to minimize the absorption of solar radiation, which allows them to cool down and prevent overheating. The advantage of ectothermy is that metabolic energy from food is not required to heat the body; therefore, reptiles can survive on about 10 percent of the calories required by a similarly sized endotherm. In cold weather, some reptiles such as the garter snake *brumate*. **Brumation** is similar to hibernation in that the animal becomes less active and can go for long periods without eating, but differs from hibernation in that brumating reptiles are not asleep or living off fat reserves. Rather, their metabolism is slowed in response to cold temperatures, and the animal is very sluggish.

Evolution of Reptiles

Reptiles originated approximately 300 million years ago during the Carboniferous period. One of the oldest known amniotes is *Casineria*, which had both amphibian and reptilian characteristics. One of the earliest undisputed reptile fossils was *Hylonomus*, a lizardlike animal about 20 cm long. Soon after the first amniotes appeared, they diverged into three groups—synapsids, anapsids, and diapsids—during the Permian period. The Permian period also saw a second major divergence of diapsid reptiles into stem archosaurs (predecessors of thecodonts, crocodilians, dinosaurs, and birds) and lepidosaurs (predecessors of snakes and lizards). These groups remained inconspicuous until the Triassic period, when the archosaurs became the dominant terrestrial group possibly due to the extinction of large-bodied anapsids and synapsids during the Permian-Triassic extinction. About 250 million years ago, archosaurs radiated into the pterosaurs and both saurischian "lizard hip" and ornithischian "bird-hip" dinosaurs (see below).

Although they are sometimes mistakenly called dinosaurs, the **pterosaurs** were distinct from true dinosaurs (Figure 29.25). Pterosaurs had a number of adaptations that allowed for flight, including hollow bones (birds also exhibit hollow bones, a case of convergent evolution). Their wings were formed by membranes of skin that attached to the long, fourth finger of each arm and extended along the body to the legs.

Figure 29.25 Pterosaurs. Pterosaurs, such as this *Quetzalcoatlus*, which existed from the late Triassic to the Cretaceous period (230 to 65.5 million years ago), possessed wings but are not believed to have been capable of powered flight. Instead, they may have been able to soar after launching from cliffs. (credit: Mark Witton, Darren Naish)

Archosaurs: Dinosaurs

Dinosaurs ("fearfully-great lizard") include the Saurischia ("lizard-hipped") with a simple, three-pronged pelvis, and

Ornithischia ("bird-hipped") dinosaurs with a more complex pelvis, superficially similar to that of birds. However, it is a fact that birds evolved from the saurischian "lizard hipped" lineage, *not* the ornithischian "bird hip" lineage. Dinosaurs and their theropod descendants, the birds, are remnants of what was formerly a hugely diverse group of reptiles, some of which like *Argentinosaurus* were nearly 40 meters (130 feet) in length and weighed at least 80,000 kg (88 tons). They were the largest land animals to have lived, challenging and perhaps exceeding the great blue whale in size, but probably not weight—which could be greater than 200 tons.

Herrerasaurus, a bipedal dinosaur from Argentina, was one of the earliest dinosaurs that walked upright with the legs positioned directly below the pelvis, rather than splayed outward to the sides as in the crocodilians. The Ornithischia were all herbivores, and sometimes evolved into crazy shapes, such as ankylosaur "armored tanks" and horned dinosaurs such as *Triceratops*. Some, such as *Parasaurolophus*, lived in great herds and may have amplified their species-specific calls through elaborate crests on their heads.

Both the ornithischian and saurischian dinosaurs provided parental care for their broods, just as crocodilians and birds do today. The end of the age of dinosaurs came about 65 million years ago, during the Mesozoic, coinciding with the impact of a large asteroid (that produced the Chicxulub crater) in what is now the Yucatan Peninsula of Mexico. Besides the immediate environmental disasters associated with this asteroid impacting the Earth at about 45,000 miles per hour, the impact may also have helped generate an enormous series of volcanic eruptions that changed the distribution and abundance of plant life worldwide, as well as its climate. At the end of the Triassic, massive volcanic activity across North America, South America, Africa, and southwest Europe ultimately would lead to the break-up of Pangea and the opening of the Atlantic Ocean. The formerly incredibly diverse dinosaurs (save for the evolving birds) met their extinction during this time period.

Archosaurs: Pterosaurs

More than 200 species of pterosaurs have been described, and in their day, beginning about 230 million years ago, they were the undisputed rulers of the Mesozoic skies for over 170 million years. Recent fossils suggest that hundreds of pterosaur species may have lived during any given period, dividing up the environment much like birds do today. Pterosaurs came in amazing sizes and shapes, ranging in size from that of a small song bird to that of the enormous *Quetzalcoatlus northropi*, which stood nearly 6 meters (19 feet) high and had a wingspan of nearly 14 meters (40 feet). This monstrous pterosaur, named after the Aztec god Quetzalcoatl, the feathered flying serpent that contributed largely to the creation of humankind, may have been the largest flying animal that ever evolved!

Some male pterosaurs apparently had brightly colored crests that may have served in sexual displays; some of these crests were much higher than the actual head! Pterosaurs had ultralight skeletons, with a *pteroid bone*, unique to pterosaurs, that strengthened the forewing membrane. Much of their wing span was exaggerated by a greatly elongated fourth finger that supported perhaps half of the wing. It is tempting to relate to them in terms of bird characteristics, but in reality, their proportions were decidedly not like birds at all. For example, it is common to find specimens, such as *Quetzalcoatlus*, with a head and neck region that together was three to four *times* as large as the torso. In addition, unlike the feathered bird wing, the reptilian wing had a layer of muscles, connective tissue, and blood vessels, all reinforced with a webbing of fibrous cords.

In contrast to the aerial pterosaurs, the dinosaurs were a diverse group of terrestrial reptiles with more than 1,000 species classified to date. Paleontologists continue to discover new species of dinosaurs. Some dinosaurs were quadrupeds (Figure 29.26); others were bipeds. Some were carnivorous, whereas others were herbivorous. Dinosaurs laid eggs, and a number of nests containing fossilized eggs, with intact embryos, have been found. It is not known with certainty whether dinosaurs were homeotherms or facultative endotherms. However, given that modern birds are endothermic, the dinosaurs that were the immediate ancestors to birds likely were endothermic as well. Some fossil evidence exists for dinosaurian parental care, and comparative biology supports this hypothesis since the archosaur birds and crocodilians both display extensive parental care.

(a)

(b)

Figure 29.26 Ornithischian and saurischian Dinosaurs. *Edmontonia* was an armored dinosaur that lived in the Late Cretaceous period, 145.5 to 65.6 million years ago. Herrerrasaurus and Eoraptor (b) were late Triassic saurischian dinosaurs dating to about 230 million years ago. (credit: a Mariana Ruiz Villareal b Zach Tirrell from Plymouth, USA, Dino Origins)

Dinosaurs dominated the Mesozoic era, which was known as the "Age of Reptiles." The dominance of dinosaurs lasted until the end of the Cretaceous, the last period of the Mesozoic era. The Cretaceous-Tertiary extinction resulted in the loss of most of the large-bodied animals of the Mesozoic era. Birds are the only living descendants of one of the major clades of theropod dinosaurs.

LINK TO LEARNING

Visit this site to see a video (http://openstax.org/l/K-T_extinction) discussing the hypothesis that an asteroid caused the Cretaceous-Triassic (KT) extinction.

Modern Reptiles

Class Reptilia includes many diverse species that are classified into four living clades. There are the 25 species of Crocodilia, two species of Sphenodontia, approximately 9,200 Squamata species, and about 325 species of Testudines.

Crocodilia

Crocodilia ("small lizard") arose as a distinct lineage by the middle Triassic; extant species include alligators, crocodiles, gharials, and caimans. Crocodilians (Figure 29.27) live throughout the tropics and subtropics of Africa, South America, Southern Florida, Asia, and Australia. They are found in freshwater, saltwater, and brackish habitats, such as rivers and lakes, and spend most of their time in water. Crocodiles are descended from terrestrial reptiles and can still walk and run well on land. They often move on their bellies in a swimming motion, propelled by alternate movements of their legs. However, some species can lift their bodies off the ground, pulling their legs in under the body with their feet rotated to face forward. This mode of locomotion takes a lot of energy, and seems to be used primarily to clear ground obstacles. Amazingly, some crocodiles can also gallop, pushing off with their hind legs and moving their hind and forelegs alternately in pairs. Galloping crocodiles have been clocked at speeds over 17 kph and, over short distances, in an ambush situation, they can easily chase down most humans if they are taken by surprise. However, they are short distance runners, not interested in a long chase, and most fit humans can probably outrun them in a sprint (assuming they respond quickly to the ambush!).

Figure 29.27 A crocodilian. Crocodilians, such as this Siamese crocodile (*Crocodylus siamensis*), provide parental care for their offspring. (credit: Keshav Mukund Kandhadai)

Sphenodontia

Sphenodontia ("wedge tooth") arose in the early Mesozoic era, when they had a moderate radiation, but now are represented by only two living species: *Sphenodon punctatus* and *Sphenodon guntheri,* both found on offshore islands in New Zealand (Figure 29.28). The common name "tuatara" comes from a Maori word describing the crest along its back. Tuataras have a primitive diapsid skull with biconcave vertebrae. They measure up to 80 centimeters and weigh about 1 kilogram. Although superficially similar to an iguanid lizard, several unique features of the skull and jaws clearly define them and distinguish this group from the Squamata. They have no external ears. Tuataras briefly have a third (parietal) eye—with a lens, retina, and cornea—in the middle of the forehead. The eye is visible only in very young animals; it is soon covered with skin. Parietal eyes can sense light, but have limited color discrimination. Similar light-sensing structures are also seen in some other lizards. In their jaws, tuataras have two rows of teeth in the upper jaw that bracket a single row of teeth in the lower jaw. These teeth are actually projections from the jawbones, and are not replaced as they wear down.

Figure 29.28 A tuatara. This tuatara from New Zealand may resemble a lizard but belongs to a distinct lineage, the Sphenodontidae family. (credit: Sid Mosdell)

Squamata

The **Squamata** ("scaly or having scales") arose in the late Permian, and extant species include lizards and snakes. Both are found on all continents except Antarctica. Lizards and snakes are most closely related to tuataras, both groups having evolved from a lepidosaurian ancestor. Squamata is the largest extant clade of reptiles (Figure 29.29).

Figure 29.29 A chameleon. This Jackson's chameleon (*Trioceros jacksonii*) blends in with its surroundings.

Most lizards differ from snakes by having four limbs, although these have often been lost or significantly reduced in at least 60 lineages. Snakes lack eyelids and external ears, which are both present in lizards. There are about 6,000 species of lizards, ranging in size from tiny chameleons and geckos, some of which are only a few centimeters in length, to the Komodo dragon, which is about 3 meters in length.

Some lizards are extravagantly decorated with spines, crests, and frills, and many are brightly colored. Some lizards, like chameleons (Figure 29.29), can change their skin color by redistributing pigment within *chromatophores* in their skins. Chameleons change color both for camouflage and for social signaling. Lizards have multiple-colored oil droplets in their retinal cells that give them a good range of color vision. Lizards, unlike snakes, can focus their eyes by changing the shape of the lens. The eyes of chameleons can move independently. Several species of lizards have a "hidden" *parietal eye*, similar to that in the tuatara. Both lizards and snakes use their tongues to sample the environment and a pit in the roof of the mouth, Jacobson's organ, is used to evaluate the collected sample.

Most lizards are carnivorous, but some large species, such as iguanas, are herbivores. Some predatory lizards are ambush predators, waiting quietly until their prey is close enough for a quick grab. Others are patient foragers, moving slowly through their environment to detect possible prey. Lizard tongues are long and sticky and can be extended at high speed for capturing insects or other small prey. Traditionally, the only venomous lizards are the Gila monster and the beaded lizard. However, venom glands have also been identified in several species of monitors and iguanids, but the venom is not injected directly and should probably be regarded as a toxin delivered with the bite.

Specialized features of the jaw are related to adaptations for feeding that have evolved to feed on relatively large prey (even though some current species have reversed this trend). Snakes are thought to have descended from either burrowing or aquatic lizards over 100 million years ago (Figure 29.30). They include about 3,600 species, ranging in size from 10 centimeter-long thread snakes to 10 meter-long pythons and anacondas. All snakes are legless, except for boids (e.g., boa constrictors), which have vestigial hindlimbs in the form of *pelvic spurs*. Like caecilian amphibians, the narrow bodies of most snakes have only a single functional lung. All snakes are carnivorous and eat small animals, birds, eggs, fish, and insects.

Figure 29.30 A nonvenomous snake. The garter snake belongs to the genus *Thamnophis*, the most widely distributed reptile genus in North America. (credit: Steve Jurvetson)

Most snakes have a skull that is very flexible, involving eight rotational joints. They also differ from other squamates by having mandibles (lower jaws) without either bony or ligamentous attachment anteriorly. Having this connection via skin and muscle allows for great dynamic expansion of the gape and independent motion of the two sides—both advantages in swallowing big prey. Most snakes are nonvenomous and simply swallow their prey alive, or subdue it by constriction before swallowing it. Venomous snakes use their venom both to kill or immobilize their prey, and to help digest it.

Although snakes have no eyelids, their eyes are protected with a transparent scale. Their retinas have both rods and cones, and like many animals, they do not have receptor pigments for red light. Some species, however, can see in the ultraviolet, which allows them to track ultraviolet signals in rodent trails. Snakes adjust focus by moving their heads. They have lost both external and middle ears, although their inner ears are sensitive to ground vibrations. Snakes have a number of sensory structures that assist in tracking prey. In pit vipers, like rattlesnakes, a sensory pit between the eye and nostrils is sensitive to infrared ("heat") emissions from warm-blooded prey. A row of similar pits is located on the upper lip of boids. As noted above, snakes also use **Jacobson's organ** for detecting olfactory signals.

Testudines

The turtles, terrapins, and tortoises are members of the clade **Testudines** ("having a shell") (Figure 29.31), and are characterized by a bony or cartilaginous shell. The shell in turtles is not just an epidermal covering, but is incorporated *into* the skeletal system. The dorsal shell is called the **carapace** and includes the backbone and ribs; the ventral shell is called the **plastron**. Both shells are covered with keratinous plates or **scutes**, and the two shells are held together by a bridge. In some turtles, the plastron is hinged to allow the head and legs to be withdrawn under the shell.

The two living groups of turtles, Pleurodira and Cryptodira, have significant anatomical differences and are most easily recognized by how they retract their necks. The more common Cryptodira retract their neck in a vertical S-curve; they appear to simply pull their head backward when retracting. Less common Pleurodira ("side-neck") retract their neck with a horizontal curve, basically folding their neck to the side.

The Testudines arose approximately 200 million years ago, predating crocodiles, lizards, and snakes. There are about 325 living species of turtles and tortoises. Like other reptiles, turtles are ectotherms. All turtles are oviparous, laying their eggs on land, although many species live in or near water. None exhibit parental care. Turtles range in size from the speckled padloper tortoise at 8 centimeters (3.1 inches) to the leatherback sea turtle at 200 centimeters (over 6 feet). The term "turtle" is sometimes used to describe only those species of Testudines that live in the sea, with the terms "tortoise" and "terrapin" used to refer to species that live on land and in fresh water, respectively.

Figure 29.31 A tortoise. The African spurred tortoise (*Geochelone sulcata*) lives at the southern edge of the Sahara Desert. It is the third largest tortoise in the world. (credit: Jim Bowen)

29.5 Birds

By the end of this section, you will be able to do the following:

- Describe the evolutionary history of birds
- Describe the derived characteristics in birds that facilitate flight

With over 10,000 identified species, the birds are the most speciose of the land vertebrate classes. Abundant research has shown that birds are really an extant clade that evolved from maniraptoran theropod dinosaurs about 150 million years ago. Thus, even though the most obvious characteristic that *seems* to set birds apart from other extant vertebrates is the presence of feathers, we now know that feathers probably appeared in the common ancestor of both ornithischian and saurischian lineages of dinosaurs. Feathers in these clades are also homologous to reptilian scales and mammalian hair, according to the most recent research. While the wings of vertebrates like bats function without feathers, birds rely on feathers, and wings, along with other modifications of body structure and physiology, for flight, as we shall see.

Characteristics of Birds

Birds are endothermic, and more specifically, **homeothermic**—meaning that they usually maintain an elevated and constant body temperature, which is significantly above the average body temperature of most mammals. This is, in part, due to the fact that active flight—especially the hovering skills of birds such as hummingbirds—requires enormous amounts of energy, which in turn necessitates a high metabolic rate. Like mammals (which are also endothermic and homeothermic and covered with an insulating pelage), birds have several different types of feathers that together keep "heat" (infrared energy) within the core of the body, away from the surface where it can be lost by radiation and convection to the environment.

Modern birds produce two main types of feathers: *contour feathers* and *down feathers*. **Contour feathers** have a number of parallel *barbs* that branch from a *central shaft*. The barbs in turn have microscopic branches called *barbules* that are linked together by minute hooks, making the vane of a feather a strong, flexible, and uninterrupted surface. In contrast, the barbules of **down feathers** do not interlock, making these feathers especially good for insulation, trapping air in spaces between the loose, interlocking barbules of adjacent feathers to decrease the rate of heat loss by convection and radiation. Certain parts of a bird's body are covered in down feathers, and the base of other feathers has a downy portion, whereas newly hatched birds are covered almost entirely in down, which serves as an excellent coat of insulation, increasing the *thermal boundary layer* between the skin and the outside environment.

Feathers not only provide insulation, but also allow for flight, producing the lift and thrust necessary for flying birds to become and stay airborne. The feathers on a wing are *flexible*, so the feathers at the end of the wing separate as air moves over them, reducing the drag on the wing. Flight feathers are also asymmetrical and curved, so that air flowing over them generates lift. Two types of flight feathers are found on the wings, *primary feathers* and *secondary feathers* (Figure 29.32). **Primary feathers** are located at the tip of the wing and provide thrust as the bird moves its wings downward, using the pectoralis major muscles.

Secondary feathers are located closer to the body, in the forearm portion of the wing, and provide lift. In contrast to primary and secondary feathers, contour feathers are found on the body, where they help reduce form drag produced by wind resistance against the body during flight. They create a smooth, aerodynamic surface so that air moves swiftly over the bird's body, preventing turbulence and creating ideal aerodynamic conditions for efficient flight.

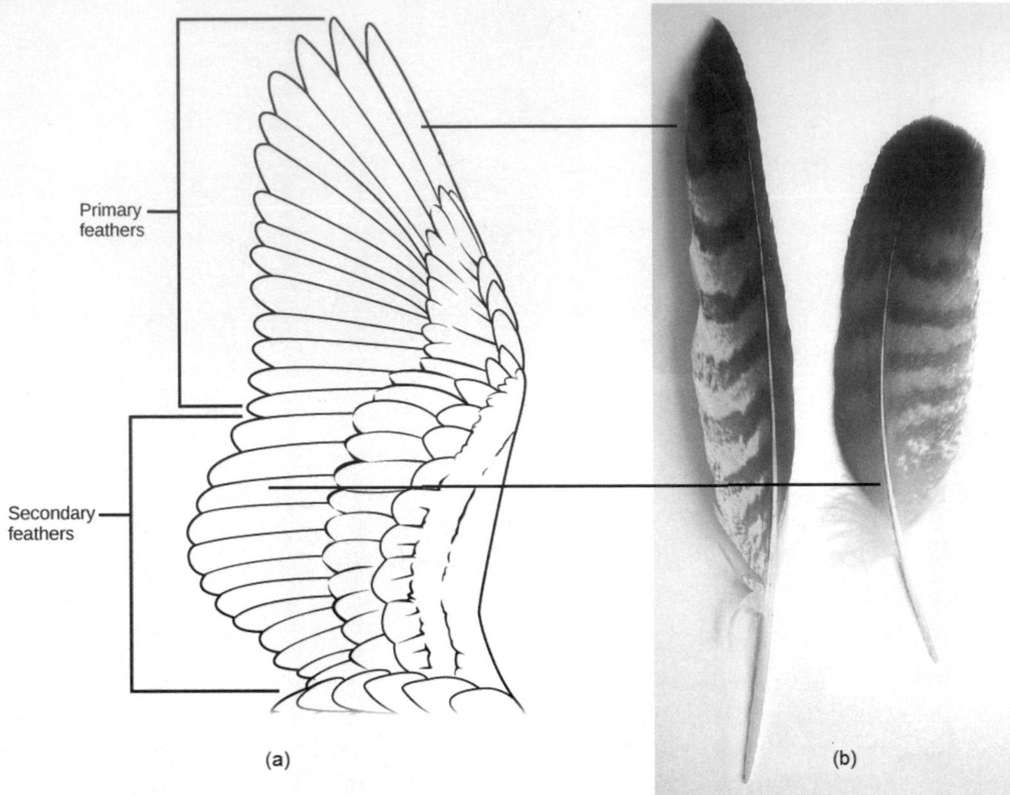

Figure 29.32 Flight feathers. (a) Primary feathers are located at the wing tip and provide thrust; secondary feathers are located close to the body and provide lift. (b) Primary and secondary feathers from a common buzzard (*Buteo buteo*). (Credit b: Mod. from S. Seyfert https://commons.wikimedia.org/w/index.php?curid=613813 (http://openstax.org/l/buzzard_feathers))

Flapping of the entire wing occurs primarily through the actions of the chest muscles: Specifically, the contraction of the pectoralis major muscles moves the wings downward (downstroke), whereas contraction of the supracoracoideus muscles moves the wings upward (upstroke) via a tough tendon that passes over the coracoid bone and the top of the humerus. Both muscles are attached to the keel of the sternum, and these are the muscles that humans eat on holidays (this is why the back of the bird offers little meat!). These muscles are highly developed in birds and account for a higher percentage of body mass than in most mammals. The flight muscles attach to a blade-shaped keel projecting ventrally from the sternum, like the keel of a boat. The sternum of birds is deeper than that of other vertebrates, which accommodates the large flight muscles. The flight muscles of birds who are active flyers are rich with oxygen-storing *myoglobin*. Another skeletal modification found in most birds is the fusion of the two clavicles (collarbones), forming the **furcula** or wishbone. The furcula is flexible enough to bend and provide support to the shoulder girdle during flapping.

An important requirement for flight is a low body weight. As body weight increases, the muscle output required for flying increases. The largest living bird is the ostrich, and while it is much smaller than the largest mammals, it is secondarily flightless. For birds that do fly, reduction in body weight makes flight easier. Several modifications are found in birds to reduce body weight, including *pneumatization of bones*. **Pneumatic bones** (Figure 29.33) are bones that are hollow, rather than filled with tissue; *cross struts* of bone called *trabeculae* provide structural reinforcement. Pneumatic bones are not found in all birds, and they are more extensive in large birds than in small birds. Not all bones of the skeleton are pneumatic, although the skulls of almost all birds are. The jaw is also lightened by the replacement of heavy jawbones and teeth with a beak made of keratin (just as hair, scales, and feathers are).

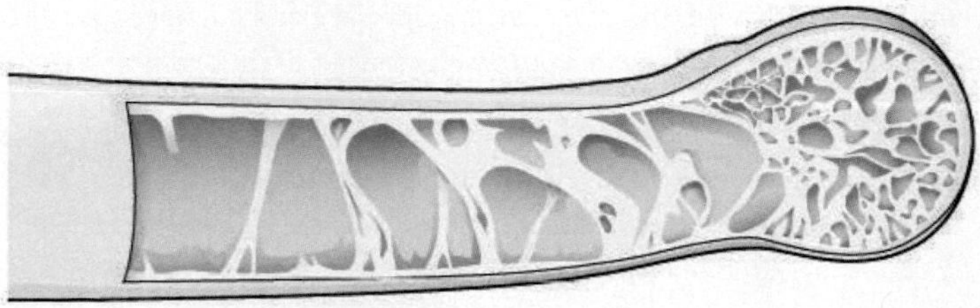

Figure 29.33 Pneumatic bone. Many birds have hollow, pneumatic bones, which make flight easier.

Other modifications that reduce weight include the lack of a urinary bladder. Birds possess a **cloaca**, an external body cavity into which the intestinal, urinary, and genital orifices empty in reptiles, birds, and the monotreme mammals. The cloaca allows water to be reabsorbed from waste back into the bloodstream. Thus, uric acid is *not* eliminated as a liquid but is concentrated into **urate salts**, which are expelled along with fecal matter. In this way, water is not held in a urinary bladder, which would increase body weight. In addition, the females of most bird species only possess one functional (left) ovary rather than two, further reducing body mass.

The respiratory system of birds is dramatically different from that of reptiles and mammals, and is well adapted for the high metabolic rate required for flight. To begin, the air spaces of pneumatic bone are sometimes connected to **air sacs** in the body cavity, which replace coelomic fluid and also lighten the body. These air sacs are also connected to the path of airflow through the bird's body, and function in respiration. Unlike mammalian lungs in which air flows in two directions, as it is breathed in and out, diluting the concentration of oxygen, airflow through bird lungs is unidirectional (Figure 29.34). Gas exchange occurs in "air capillaries" or microscopic air passages within the lungs. The arrangement of air capillaries in the lungs creates a counter-current exchange system with the pulmonary blood. In a counter-current system, the air flows in one direction and the blood flows in the opposite direction, producing a favorable diffusion gradient and creating an efficient means of gas exchange. This very effective oxygen-delivery system of birds supports their higher metabolic activity. In effect, ventilation is provided by the **parabronchi** (minimally expandible lungs) with thin air sacs located among the visceral organs and the skeleton. A **syrinx** (voice box) resides near the junction of the trachea and bronchi. The syrinx, however, is *not* homologous to the mammalian larynx, which resides within the upper part of the trachea.

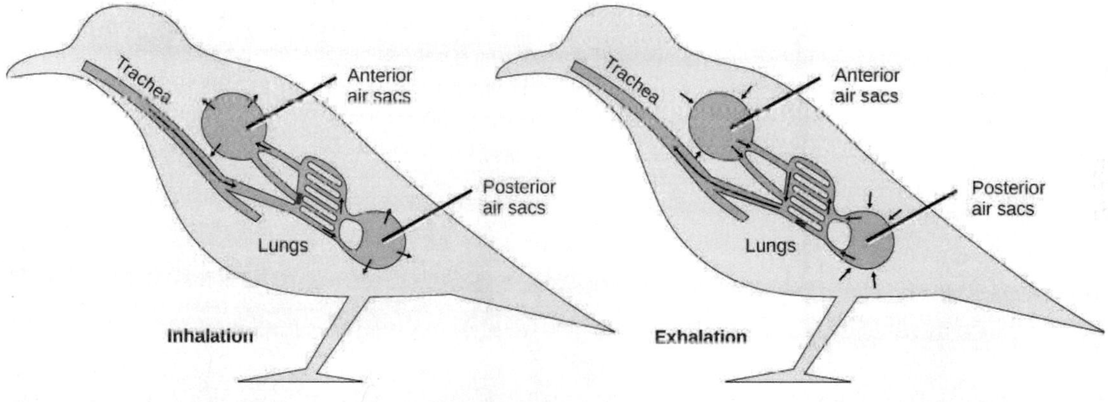

Figure 29.34 Air flow in bird lungs. Avian respiration is an efficient system of gas exchange with air flowing *unidirectionally.* A full ventilation cycle takes two breathing cycles. During the first inhalation, air passes from the trachea into posterior air sacs, then during the first exhalation into the lungs. The second inhalation moves the air in the lungs to the anterior air sacs, and the second exhalation moves the air in the anterior air sacs out of the body. Overall, each inhalation moves air into the air sacs, while each exhalation moves fresh air through the lungs and "used" air out of the body. The air sacs are connected to the hollow interior of bones. (credit: modification of work by L. Shyamal)

Beyond the unique characteristics discussed above, birds are also unusual vertebrates because of a number of other features. First, they typically have an elongate (very "dinosaurian") S-shaped neck, but a short tail or *pygostyle,* produced from the fusion of the caudal vertebrae. Unlike mammals, birds have only one occipital condyle, allowing them extensive movement of the head and neck. They also have a very thin epidermis without sweat glands, and a specialized **uropygial gland** or sebaceous "preening gland" found at the dorsal base of the tail. This gland is an essential to **preening** (a virtually continuous activity) in most birds because it produces an oily substance that birds use to help waterproof their feathers as well as keep them flexible for flight. A number of birds, such as pigeons, parrots, hawks, and owls, lack a uropygial gland but have specialized feathers that "disintegrate" into a powdery down, which serves the same purpose as the oils of the uropygial gland.

Like mammals, birds have 12 pairs of cranial nerves, and a very large cerebellum and optic lobes, but only a single bone in the middle ear called the **columella** (the stapes in mammals). They have a closed circulatory system with two atria and two ventricles, but rather than a "left-bending" aortic arch like that of mammals, they have a "right-bending" aortic arch, and nucleated red blood cells (unlike the enucleated red blood cells of mammals).

All these unique and highly derived characteristics make birds one of the most conspicuous and successful groups of vertebrate animals, filling a range of ecological niches, and ranging in size from the tiny bee hummingbird of Cuba (about 2 grams) to the ostrich (about 140,000 grams). Their large brains, keen senses, and the abilities of many species to imitate vocalization and use tools make them some of the most intelligent vertebrates on Earth.

Evolution of Birds

Thanks to amazing new fossil discoveries in China, the evolutionary history of birds has become clearer, even though bird bones do not fossilize as well as those of other vertebrates. As we've seen earlier, birds are highly modified diapsids, but rather than having two fenestrations or openings in their skulls behind the eye, the skulls of modern birds are so specialized that it is difficult to see any trace of the original diapsid condition.

Birds belong to a group of diapsids called the **archosaurs**, which includes three other groups: living crocodilians, pterosaurs, and dinosaurs. Overwhelming evidence shows that birds evolved within the clade Dinosauria, which is further subdivided into two groups, the Saurischia ("lizard hips") and the Ornithischia ("bird hips"). Despite the names of these groups, it was not the bird-hipped dinosaurs that gave rise to modern birds. Rather, Saurischia diverged into two groups: One included the long-necked herbivorous dinosaurs, such as *Apatosaurus*. The second group, bipedal predators called theropods, gave rise to birds. This course of evolution is highlighted by numerous similarities between late (maniraptoran) theropod fossils and birds, specifically in the structure of the hip and wrist bones, as well as the presence of the wishbone, formed by the fusion of the clavicles.

The clade **Neornithes** includes the avian crown group, which comprises all living birds and the descendants from their most recent common maniraptoran ancestor. One well-known and important fossil of an animal that appears "intermediate" between dinosaurs and birds is *Archaeopteryx* (Figure 29.35), which is from the Jurassic period (200 to 145 MYA). *Archaeopteryx* has characteristics of both maniraptoran dinosaurs and modern birds. Some scientists propose classifying it as a bird, but others prefer to classify it as a dinosaur. Traits in skeletons of *Archaeopteryx* like those of a dinosaur included a jaw with teeth and a long bony tail. Like birds, it had feathers modified for flight, both on the forelimbs and on the tail, a trait associated only with birds among modern animals. Fossils of older feathered dinosaurs exist, but the feathers may not have had the characteristics of modern flight feathers.

Figure 29.35 *Archaeopteryx*. (a) *Archaeopteryx* lived in the late Jurassic period around 150 million years ago. It had cuplike thecodont teeth like a dinosaur, but had (b) flight feathers like modern birds, which can be seen in this fossil. Note the claws on the wings, which are still found in a number of birds, such as the newborn chicks of the South American Hoatzin.

The Evolution of Flight in Birds

There are two basic hypotheses that explain how flight may have evolved in birds: the *arboreal ("tree") hypothesis* and the *terrestrial ("land") hypothesis*. The **arboreal hypothesis** posits that tree-dwelling precursors to modern birds jumped from branch to branch using their feathers for gliding before becoming fully capable of flapping flight. In contrast to this, the **terrestrial hypothesis** holds that running (perhaps pursuing active prey such as small cursorial animals) was the stimulus for flight. In this scenario, wings could be used to capture prey and were preadapted for balance and flapping flight. Ostriches, which are large flightless birds, hold their wings out when they run, possibly for balance. However, this condition may represent a behavioral relict of the clade of flying birds that were their ancestors. It seems more likely that small feathered arboreal dinosaurs, were capable of gliding (and flapping) from tree to tree and branch to branch, improving the chances of escaping enemies, finding mates, and obtaining prey such as flying insects. This early flight behavior would have also greatly increased the opportunity for species dispersal.

Although we have a good understanding of how feathers and flight may have evolved, the question of how endothermy evolved in birds (and other lineages) remains unanswered. Feathers provide insulation, but this is only beneficial for *thermoregulatory purposes* if body heat is being produced internally. Similarly, internal heat production is only viable for the evolution of endothermy if *insulation is present* to retain that infrared energy. It has been suggested that one or the other—feathers or endothermy—evolved first in response to some other selective pressure (e.g., the ability to be active at night, provide camouflage, repel water, or serve as signals for mate selection). It seems probable that feathers and endothermy coevolved together, the improvement and evolutionary advancement of feathers reinforcing the evolutionary advancement of endothermy, and so on.

During the Cretaceous period (145 to 66 MYA), a group known as the Enantiornithes was the dominant bird type (Figure 29.36). Enantiornithes means "opposite birds," which refers to the fact that certain bones of the shoulder are joined differently than the way the bones are joined in modern birds. Like *Archaeopteryx*, these birds retained teeth in their jaws, but did have a shortened tail, and at least some fossils have preserved "fans" of tail feathers. These birds formed an evolutionary lineage separate from that of modern birds, and they did not survive past the Cretaceous. Along with the Enantiornithes, however, another group of birds—the Ornithurae ("bird tails"), with a short, fused tail or **pygostyle**—emerged from the evolutionary line that includes modern birds. This clade was also present in the Cretaceous.

After the extinction of Enantiornithes, the **Ornithurae** became the dominant birds, with a large and rapid radiation occurring after the extinction of the dinosaurs during the Cenozoic era (66 MYA to the present). Molecular analysis based on very large data sets has produced our current understanding of the relationships among living birds. There are three major clades: the Paleognathae, the Galloanserae, and the Neoaves. The Paleognathae ("old jaw") or ratites (polyphyletic) are a group of flightless birds including ostriches, emus, rheas, and kiwis. The Galloanserae include pheasants, ducks, geese and swans. The Neoaves ("new birds") includes all other birds. The Neoaves themselves have been distributed among five clades:[3] Strisores (nightjars, swifts, and hummingbirds), Columbaves (turacos, bustards, cuckoos, pigeons, and doves), Gruiformes (cranes), Aequorlitornithes (diving birds, wading birds, and shorebirds), and Inopinaves (a very large clade of land birds including hawks, owls, woodpeckers, parrots, falcons, crows, and songbirds). Despite the current classification scheme, it is important to understand that phylogenetic revisions, even for the extant birds, are still taking place.

3Prum, RO et al. 2015. A comprehensive phylogeny of birds (Aves) using targeted next-generation DNA sequencing. Nature 526: 569 - 573. http://dx.doi.org/10.1038/nature15697 (http://openstax.org/l/bird_phylogeny)

Figure 29.36 Enantiornithean bird. *Shanweiniao cooperorum* was a species of Enantiornithes that did not survive past the Cretaceous period. (credit: Nobu Tamura)

CAREER CONNECTION

Veterinarian

Veterinarians are concerned with diseases, disorders, and injuries in animals, primarily vertebrates. They treat pets, livestock, and animals in zoos and laboratories. Veterinarians often treat dogs and cats, but also take care of birds, reptiles, rabbits, and other animals that are kept as pets. Veterinarians that work with farms and ranches care for pigs, goats, cows, sheep, and horses.

Veterinarians are required to complete a degree in veterinary medicine, which includes taking courses in comparative zoology, animal anatomy and physiology, microbiology, and pathology, among many other courses in chemistry, physics, and mathematics.

Veterinarians are also trained to perform surgery on many different vertebrate species, which requires an understanding of the vastly different anatomies of various species. For example, the stomach of ruminants like cows has four "compartments" versus one compartment for non-ruminants. As we have seen, birds also have unique anatomical adaptations that allow for flight, which requires additional training and care.

Some veterinarians conduct research in academic settings, broadening our knowledge of animals and medical science. One area of research involves understanding the transmission of animal diseases to humans, called **zoonotic diseases**. For example, one area of great concern is the transmission of the avian flu virus to humans. One type of avian flu virus, H5N1, is a highly pathogenic strain that has been spreading in birds in Asia, Europe, Africa, and the Middle East. Although the virus does not cross over easily to humans, there have been cases of bird-to-human transmission. More research is needed to understand how this virus can cross the species barrier and how its spread can be prevented.

29.6 Mammals

By the end of this section, you will be able to do the following:

- Name and describe the distinguishing features of the three main groups of mammals
- Describe the likely line of evolutionary descent that produced mammals
- List some derived features that may have arisen in response to mammals' need for constant, high-level metabolism
- Identify the major clades of eutherian mammals

Mammals, comprising about 5,200 species, are vertebrates that possess hair and mammary glands. Several other characteristics are distinctive to mammals, including certain features of the jaw, skeleton, integument, and internal anatomy. Modern mammals belong to three clades: monotremes, marsupials, and eutherians (or placental mammals).

Characteristics of Mammals

The presence of **hair**, composed of the protein **keratin**, is one of the most obvious characteristics of mammals. Although it is not very extensive or obvious on some species (such as whales), hair has many important functions for most mammals. Mammals are endothermic, and hair traps a boundary layer of air close to the body, retaining heat generated by metabolic activity. Along with insulation, hair can serve as a sensory mechanism via specialized hairs called *vibrissae*, better known as whiskers. Vibrissae attach to nerves that transmit information about tactile vibration produced by sound sensation, which is particularly useful to nocturnal or burrowing mammals. Hair can also provide protective coloration or be part of social signaling, such as when an animal's hair stands "on end" to warn enemies, or possibly to make the mammal "look bigger" to predators.

Unlike the skin of birds, the integument (skin) of mammals, includes a number of different types of secretory glands. **Sebaceous glands** produce a lipid mixture called *sebum* that is secreted onto the hair and skin, providing water resistance and lubrication for hair. Sebaceous glands are located over most of the body. **Eccrine glands** produce sweat, or perspiration, which is mainly composed of water, but also contains metabolic waste products, and sometimes compounds with antibiotic activity. In most mammals, eccrine glands are limited to certain areas of the body, and some mammals do not possess them at all. However, in primates, especially humans, sweat glands are located over most of the body surface and figure prominently in regulating the body temperature through evaporative cooling. **Apocrine glands**, or *scent glands*, secrete substances that are used for chemical communication, such as in skunks. **Mammary glands** produce milk that is used to feed newborns. In both monotremes and eutherians, both males and females possess mammary glands, while in marsupials, mammary glands have been found only in some opossums. Mammary glands likely are modified sebaceous or eccrine glands, but their evolutionary origin is not entirely clear.

The skeletal system of mammals possesses many unique features. The lower jaw of mammals consists of only one bone, the **dentary**, and the jaw hinge connects the dentary to the squamosal (flat) part of the temporal bone in the skull. The jaws of other vertebrates are composed of several bones, including the quadrate bone at the back of the skull and the articular bone at the back of the jaw, with the jaw connected between the quadrate and articular bones. In the ear of other vertebrates, vibrations are transmitted to the inner ear by a single bone, the *stapes*. In mammals, the quadrate and articular bones have moved into the middle ear (Figure 29.37). The malleus is derived from the articular bone, whereas the incus originated from the quadrate bone. This arrangement of jaw and ear bones aids in distinguishing fossil mammals from fossils of other synapsids.

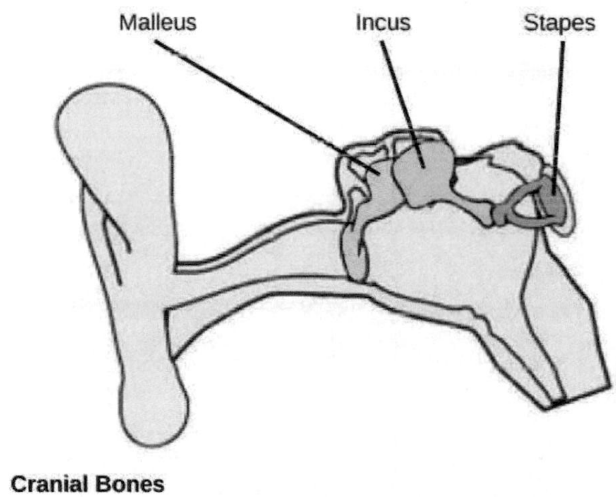

Figure 29.37 Mammalian ear bones. Bones of the mammalian middle ear are modified from bones of the jaw and skull in reptiles. The stapes is found in other vertebrates (e.g., the columella of birds) whereas in mammals, the malleus and incus are derived from the articular and quadrate bones, respectively. (credit: NCI)

The adductor muscles that close the jaw comprise two major muscles in mammals: the *temporalis* and the *masseter*. Working together, these muscles permit up-and-down and side-to-side movements of the jaw, making chewing possible—which is unique to mammals. Most mammals have *heterodont teeth*, meaning that they have different types and shapes of teeth (incisors, canines, premolars, and molars) rather than just one type and shape of tooth. Most mammals are also **diphyodonts**, meaning that they have two sets of teeth in their lifetime: deciduous or "baby" teeth, and permanent teeth. Most other vertebrates with teeth are *polyphyodonts*, that is, their teeth are replaced throughout their entire life.

Mammals, like birds, possess a four-chambered heart; however, the hearts of birds and mammals are an example of convergent evolution, since mammals clearly arose independently from different groups of tetrapod ancestors. Mammals also have a specialized group of cardiac cells (fibers) located in the walls of their right atrium called the sinoatrial node, or pacemaker, which determines the rate at which the heart beats. Mammalian erythrocytes (red blood cells) do *not* have nuclei, whereas the erythrocytes of other vertebrates are nucleated.

The kidneys of mammals have a portion of the nephron called the loop of Henle or nephritic loop, which allows mammals to produce urine with a high concentration of solutes—higher than that of the blood. Mammals lack a renal portal system, which is a system of veins that moves blood from the hind or lower limbs and region of the tail to the kidneys. Renal portal systems are present in all other vertebrates except jawless fishes. A urinary bladder is present in all mammals.

Unlike birds, the skulls of mammals have two occipital condyles, bones at the base of the skull that articulate with the first vertebra, as well as a secondary palate at the rear of the pharynx that helps to separate the pathway of swallowing from that of breathing. Turbinate bones (chonchae in humans) are located along the sides of the nasal cavity, and help warm and moisten air as it is inhaled. The pelvic bones are fused in mammals, and there are typically seven cervical vertebrae (except for some edentates and manatees). Mammals have movable eyelids and fleshy external ears (pinnae), quite unlike the naked external auditory openings of birds. Mammals also have a muscular diaphragm that is lacking in birds.

Mammalian brains also have certain characteristics that differ from the brains of other vertebrates. In some, but not all mammals, the *cerebral cortex*, the outermost part of the cerebrum, is highly convoluted and folded, allowing for a greater surface area than is possible with a smooth cortex. The optic lobes, located in the midbrain, are divided into two parts in mammals, while other vertebrates possess a single, undivided lobe. Eutherian mammals also possess a specialized structure, the corpus callosum, which links the two cerebral hemispheres together. The corpus callosum functions to integrate motor, sensory, and cognitive functions between the left and right cerebral cortexes.

Evolution of Mammals

Mammals are synapsids, meaning they have a single, ancestrally fused, postorbital opening in the skull. They are the only living synapsids, as earlier forms became extinct by the Jurassic period. The early non-mammalian synapsids can be divided into two groups, the pelycosaurs and the therapsids. Within the therapsids, a group called the cynodonts are thought to have been the ancestors of mammals (Figure 29.38).

Figure 29.38 Cynodont. Cynodonts ("dog teeth"), which first appeared in the Late Permian period 260 million years ago, are thought to be the ancestors of modern mammals. Holes in the upper jaws of cynodonts suggest that they had whiskers, which might also indicate the presence of hair. (credit: Nobu Tamura)

As with birds, a key characteristic of synapsids is endothermy, rather than the ectothermy seen in many other vertebrates (such as fish, amphibians, and most reptiles). The increased metabolic rate required to internally modify body temperature likely went hand-in-hand with changes to certain skeletal structures that improved food processing and ambulation. The later synapsids, which had more evolved characteristics unique to mammals, possess cheeks for holding food and heterodont teeth, which are specialized for chewing, mechanically breaking down food to speed digestion, and releasing the energy needed to produce heat. Chewing also requires the ability to breathe at the same time, which is facilitated by the presence of a *secondary palate* (comprising the bony palate and the posterior continuation of the soft palate). The secondary palate separates the area of the mouth where chewing occurs from the area above where respiration occurs, allowing breathing to proceed uninterrupted while the animal is chewing. A secondary palate is not found in pelycosaurs but *is* present in cynodonts and mammals. The jawbone also shows changes from early synapsids to later ones. The zygomatic arch, or cheekbone, is present in mammals and advanced therapsids such as cynodonts, but is not present in pelycosaurs. The presence of the zygomatic arch suggests the presence of masseter muscles, which close the jaw and function in chewing.

In the appendicular skeleton, the shoulder girdle of therian mammals is modified from that of other vertebrates in that it does not possess a procoracoid bone or an interclavicle, and the scapula is the dominant bone.

Mammals evolved from therapsids in the late Triassic period, as the earliest known mammal fossils are from the early Jurassic period, some 205 million years ago. One group of transitional mammals was the **morganucodonts**, small nocturnal insectivores. The jaws of morganucodonts were "transitional," with features of both reptilian and mammalian jaws (Figure 29.39). Like modern mammals, the morganucodonts had differentiated teeth and were diphyodonts. Mammals first began to diversify in the Mesozoic era, from the Jurassic to the Cretaceous periods. Even some small gliding mammals appear in the fossil record during this time period. However, most of the Jurassic mammals were extinct by the end of the Mesozoic. During the Cretaceous period, another radiation of mammals began and continued through the Cenozoic era, about 65 million years ago.

Figure 29.39 A morganucodont. This morganucodont *Megazostrodon*, an extinct basal mammal, may have been nocturnal and insectivorous. Inset: Jaw of a morganucodont, showing a double hinge, one between the dentary and squamosal and one between the articular (yellow) and quadrate (blue) bones. In living mammals, the articular and quadrate bones have been incorporated into the middle ear. (Credit: By Nordelch [Megazostrodon Natural History Museum] Wikimedia Commons. Credit inset: Mod from Philcha. https://commons.wikimedia.org/w/index.php? curid=3631949 (http://openstax.org/l/jaw_joint))

Living Mammals

There are three major groups of living mammals: *monotremes* (*prototheria*), *marsupials* (*metatheria*), and *placental* (*eutheria*) mammals. The eutherians and the marsupials together comprise a clade of therian mammals, with the monotremes forming a sister clade to both metatherians and eutherians.

There are very few living species of **monotremes**: the platypus and four species of echidnas, or spiny anteaters. The leathery-beaked platypus belongs to the family **Ornithorhynchidae** ("bird beak"), whereas echidnas belong to the family **Tachyglossidae** ("sticky tongue") (Figure 29.40). The platypus and one species of echidna are found in Australia, and the other species of echidna are found in New Guinea. Monotremes are unique among mammals because they lay eggs, rather than giving birth to live young. The shells of their eggs are not like the hard shells of birds, but have a leathery shell, similar to the shells of reptile eggs. Monotremes retain their eggs through about two-thirds of the developmental period, and then lay them in nests. A yolk-sac placenta helps support development. The babies hatch in a fetal state and complete their development in the nest, nourished by milk secreted by mammary glands opening directly to the skin. Monotremes, except for young platypuses, do not have teeth. Body temperature in the three monotreme species is maintained at about 30°C, considerably lower than the average body temperature of marsupial and placental mammals, which are typically between 35 and 38°C.

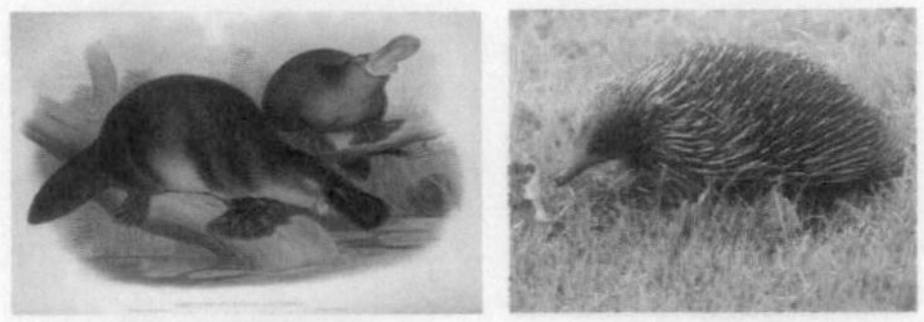

Figure 29.40 Egg-laying mammals. (a) The platypus, a monotreme, possesses a leathery beak and lays eggs rather than giving birth to live young. (b) The echidna is another monotreme, with long hairs modified into spines. (credit b: modification of work by Barry Thomas)

Over 2/3 of the approximately 330 living species of marsupials are found in Australia, with the rest, nearly all various types of opossum, found in the Americas, especially South America. Australian marsupials include the kangaroo, koala, bandicoot, Tasmanian devil (Figure 29.41), and several other species. Like monotremes, the embryos of marsupials are nourished during a short gestational period (about a month in kangaroos) by a yolk-sac placenta, but with no intervening egg shell. Some marsupial embryos can enter an embryonic diapause, and delay implantation, suspending development until implantation is completed. Marsupial young are also effectively fetal at birth. Most, but not all, species of marsupials possess a pouch in which the very premature young reside, receiving milk and continuing their development. In kangaroos, the young joeys continue to nurse for about a year and a half.

Figure 29.41 A marsupial mammal. The Tasmanian devil is one of several marsupials native to Australia. (credit: Wayne McLean)

Eutherians (placentals) are the most widespread and numerous of the mammals, occurring throughout the world. Eutherian mammals are sometimes called "placental mammals" because all species possess a complex **chorioallantoic placenta** that connects a fetus to the mother, allowing for gas, fluid, and nutrient exchange. There are about 4,000 species of placental mammals in 18 to 20 orders with various adaptations for burrowing, flying, swimming, hunting, running, and climbing. In the evolutionary sense, they have been incredibly successful in form, diversity, and abundance. The eutherian mammals are classified in two major clades, the Atlantogenata and the Boreoeutheria. The Atlantogeneta include the Afrotheria (e.g., elephants, hyraxes, and manatees) and the Xenarthra (anteaters, armadillos, and sloths). The Boreoeutheria contain two large groups, the Euarchontoglires and the Laurasiatheria. Familiar orders in the Euarchontoglires are the Scandentia (tree shrews), Rodentia (rats, mice, squirrels, porcupines), Lagomorpha (rabbits and hares), and the Primates (including humans). Major Laurasiatherian orders include the Perissodactyla (e.g., horses and rhinos), the Cetartiodactyla (e.g., cows, giraffes, pigs, hippos, and whales), the Carnivora (e.g., cats, dogs, and bears), and the Chiroptera (bats and flying foxes). The two largest orders are the rodents (2,000 species) and bats (about 1,000 species), which together constitute approximately 60 percent of all eutherian species.

29.7 The Evolution of Primates

By the end of this section, you will be able to do the following:

- Describe the derived features that distinguish primates from other animals
- Describe the defining features of the major groups of primates
- Identify the major hominin precursors to modern humans
- Explain why scientists are having difficulty determining the true lines of descent in hominids

Order Primates of class Mammalia includes lemurs, tarsiers, monkeys, apes, and humans. Non-human primates live primarily in the tropical or subtropical regions of South America, Africa, and Asia. They range in size from the mouse lemur at 30 grams (1 ounce) to the mountain gorilla at 200 kilograms (441 pounds). The characteristics and evolution of primates are of particular interest to us as they allow us to understand the evolution of our own species.

Characteristics of Primates

All primate species possess adaptations for climbing trees, as they all descended from tree-dwellers. This arboreal heritage of primates has resulted in hands and feet that are adapted for climbing, or brachiation (swinging through trees using the arms). These adaptations include, but are not limited to: 1) a rotating shoulder joint, 2) a big toe that is widely separated from the other toes (except humans) and thumbs sufficiently separated from fingers to allow for gripping branches, and 3) **stereoscopic vision**, two overlapping fields of vision from the eyes, which allows for the perception of depth and gauging distance. Other characteristics of primates are brains that are larger than those of most other mammals, claws that have been modified into flattened nails, typically only one offspring per pregnancy, and a trend toward holding the body upright.

Order Primates is divided into two groups: Strepsirrhini ("turned-nosed") and Haplorhini ("simple-nosed") primates. Strepsirrhines, also called the wet-nosed primates, include prosimians like the bush babies and pottos of Africa, the lemurs of Madagascar, and the lorises of Southeast Asia. Haplorhines, or dry-nosed primates, include tarsiers (Figure 29.42) and simians (New World monkeys, Old World monkeys, apes, and humans). In general, strepsirrhines tend to be nocturnal, have larger olfactory centers in the brain, and exhibit a smaller size and smaller brain than anthropoids. Haplorhines, with a few exceptions, are diurnal, and depend more on their vision. Another interesting difference between the strepsirrhines and haplorhines is that strepsirrhines have the enzymes for making vitamin C, while haplorhines have to get it from their food.

Figure 29.42 A Philippine tarsier. This tarsier, *Carlito syrichta*, is one of the smallest primates—about 5 inches long, from nose to the base of the tail. The tail is not shown, but is about twice the length of the body. Note the large eyes, each of which is about the same size as the animal's brain, and the long hind legs. (credit: mtoz (http://creativecommons.org/licenses/by-sa/2.0 (http://openstax.org/l/CCSA_2)), via Wikimedia Commons)

Evolution of Primates

The first primate-like mammals are referred to as proto-primates. They were roughly similar to squirrels and tree shrews in size and appearance. The existing fossil evidence (mostly from North Africa) is very fragmented. These proto-primates remain largely mysterious creatures until more fossil evidence becomes available. Although genetic evidence suggests that primates diverged from other mammals about 85 MYA, the oldest known primate-like mammals with a relatively robust fossil record date to about 65 MYA. Fossils like the proto-primate *Plesiadapis* (although some researchers do not agree that *Plesiadapis* was a proto-primate) had some features of the teeth and skeleton in common with true primates. They were found in North America and Europe in the Cenozoic and went extinct by the end of the Eocene.

The first true primates date to about 55 MYA in the Eocene epoch. They were found in North America, Europe, Asia, and Africa. These early primates resembled present-day prosimians such as lemurs. Evolutionary changes continued in these early primates, with larger brains and eyes, and smaller muzzles being the trend. By the end of the Eocene epoch, many of the early prosimian species went extinct due either to cooler temperatures or competition from the first monkeys.

Anthropoid monkeys evolved from prosimians during the Oligocene epoch. By 40 million years ago, evidence indicates that monkeys were present in the New World (South America) and the Old World (Africa and Asia). New World monkeys are also called Platyrrhini—a reference to their broad noses (Figure 29.43). Old World monkeys are called Catarrhini—a reference to their narrow, downward-pointed noses. There is still quite a bit of uncertainty about the origins of the New World monkeys. At the time the platyrrhines arose, the continents of South American and Africa had drifted apart. Therefore, it is thought that monkeys arose in the Old World and reached the New World either by drifting on log rafts or by crossing land bridges. Due to this reproductive isolation, New World monkeys and Old World monkeys underwent separate adaptive radiations over millions of years. The New World monkeys are all arboreal, whereas Old World monkeys include both arboreal and ground-dwelling species. The arboreal habits of the New World monkeys are reflected in the possession of prehensile or grasping tails by most species. The tails of Old World monkeys are never prehensile and are often reduced, and some species have ischial callosities—thickened patches of skin on their seats.

Figure 29.43 A New World monkey. The howler monkey is native to Central and South America. It makes a call that sounds like a lion roaring. (credit: Xavi Talleda)

Apes evolved from the catarrhines in Africa midway through the Cenozoic, approximately 25 million years ago. Apes are generally larger than monkeys and they do not possess a tail. All apes are capable of moving through trees, although many species spend most their time on the ground. When walking quadrupedally, monkeys walk on their palms, while apes support the upper body on their knuckles. Apes are more intelligent than monkeys, and they have larger brains relative to body size. The apes are divided into two groups. The lesser apes comprise the family **Hylobatidae**, including gibbons and siamangs. The great apes include the genera ***Pan*** (chimpanzees and bonobos) ***Gorilla*** (gorillas), ***Pongo*** (orangutans), and ***Homo*** (humans) (Figure 29.44).

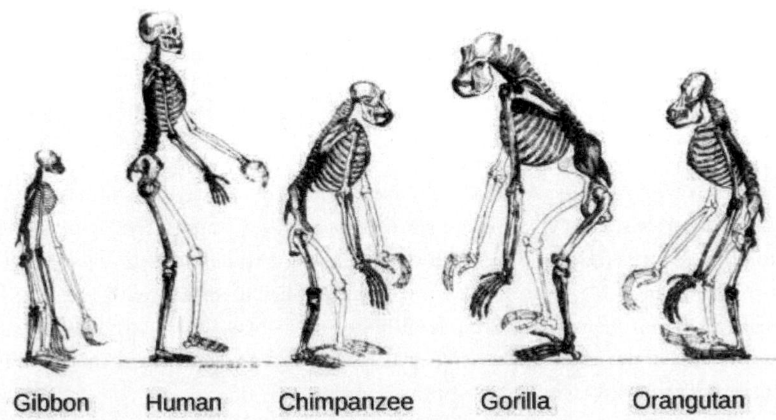

Figure 29.44 Primate skeletons. All great apes have a similar skeletal structure. (credit: modification of work by Tim Vickers)

The very arboreal gibbons are smaller than the great apes; they have low sexual dimorphism (that is, the sexes are not markedly different in size), although in some species, the sexes differ in color; and they have relatively longer arms used for swinging through trees (Figure 29.45a). Two species of orangutan are native to different islands in Indonesia: Borneo (*P. pygmaeus*) and Sumatra (*P. abelii*). A third orangutan species, *Pongo tapanuliensis*, was reported in 2017 from the Batang Toru forest in Sumatra. Orangutans are arboreal and solitary. Males are much larger than females and have cheek and throat pouches when mature. Gorillas all live in Central Africa. The eastern and western populations are recognized as separate species, *G. berengei* and *G. gorilla*. Gorillas are strongly sexually dimorphic, with males about twice the size of females. In older males, called silverbacks, the hair on the back turns white or gray. Chimpanzees (Figure 29.45b) are the species considered to be most closely related to humans. However, the species most closely related to the chimpanzee is the bonobo. Genetic evidence suggests that chimpanzee and human lineages separated 5 to 7 MYA, while chimpanzee (*Pan troglodytes*) and bonobo (*Pan paniscus*) lineages separated about 2 MYA. Chimpanzees and bonobos both live in Central Africa, but the two species are separated by the Congo River, a significant geographic barrier. Bonobos are slighter than chimpanzees, but have longer legs and more hair on their heads. In chimpanzees, white tail tufts identify juveniles, while bonobos keep their white tail tufts for life. Bonobos also have higher-pitched voices than chimpanzees. Chimpanzees are more aggressive and sometimes kill animals from other groups, while bonobos are not known to do so. Both chimpanzees and bonobos are omnivorous. Orangutan and gorilla diets also include foods from multiple sources, although the predominant food items are fruits for orangutans and foliage for gorillas.

Figure 29.45 Lesser and great apes. This white-cheeked gibbon (a) is a lesser ape. In gibbons of this species, females and infants are buff and males are black. This young chimpanzee (b) is one of the great apes. It possesses a relatively large brain and has no tail. (credit a: MAC. credit b: modification of work by Aaron Logan)

Human Evolution

The family Hominidae of order Primates includes the hominoids: the great apes and humans (Figure 29.46). Evidence from the fossil record and from a comparison of human and chimpanzee DNA suggests that humans and chimpanzees diverged from a common hominoid ancestor approximately six million years ago. Several species evolved from the evolutionary branch that includes humans, although our species is the only surviving member. The term hominin is used to refer to those species that evolved after this split of the primate line, thereby designating species that are more closely related to humans than to chimpanzees. A number of marker features differentiate humans from the other hominoids, including bipedalism or upright posture, increase in the size of the brain, and a fully opposable thumb that can touch the little finger. Bipedal hominins include several groups that were probably part of the modern human lineage—*Australopithecus*, *Homo habilis*, and *Homo erectus*—and several non-ancestral groups that can be considered "cousins" of modern humans, such as Neanderthals and Denisovans.

Determining the true lines of descent in hominins is difficult. In years past, when relatively few hominin fossils had been recovered, some scientists believed that considering them in order, from oldest to youngest, would demonstrate the course of evolution from early hominins to modern humans. In the past several years, however, many new fossils have been found, and it is clear that there was often more than one species alive at any one time and that many of the fossils found (and species named) represent hominin species that died out and are not ancestral to modern humans.

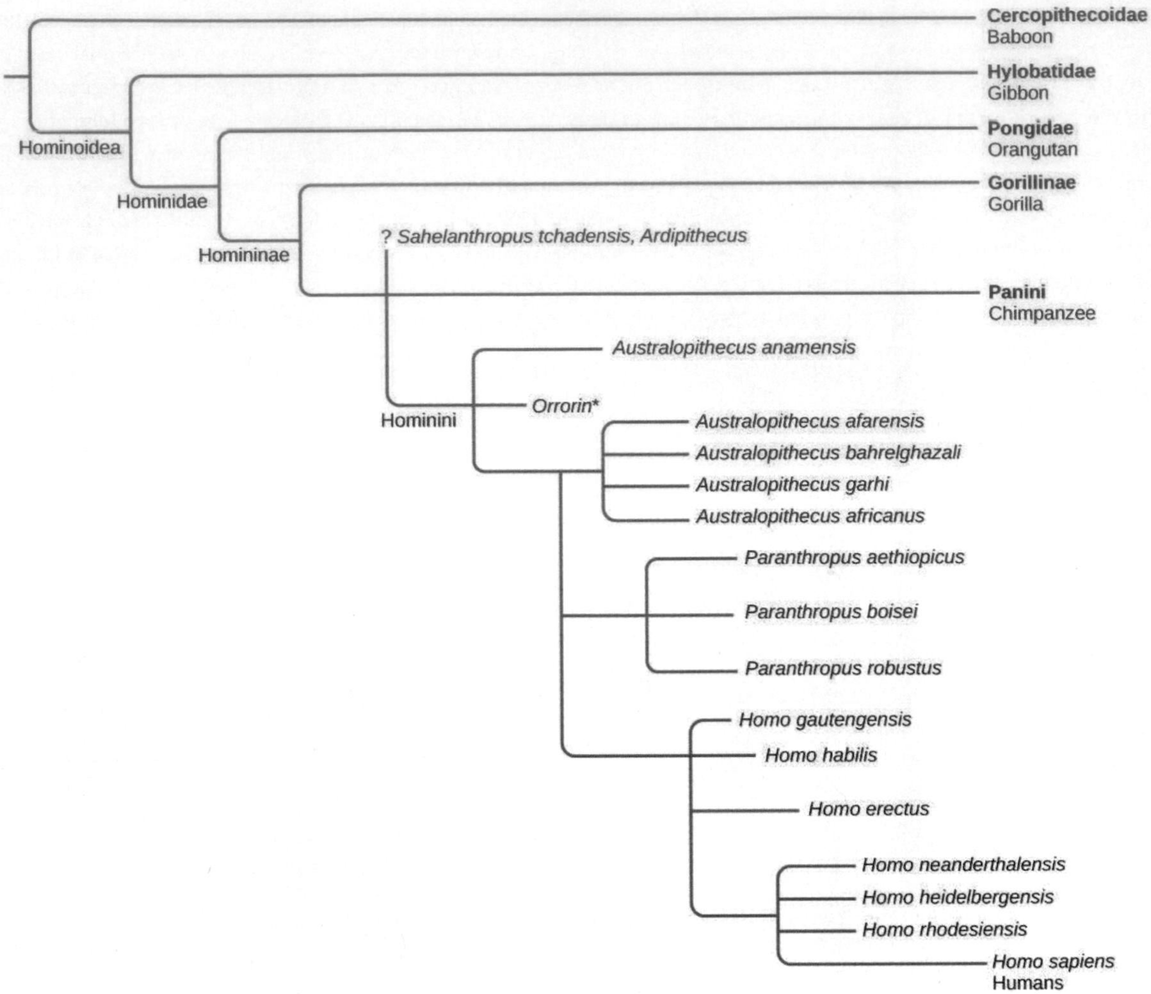

Figure 29.46 Hominin phylogeny. This chart shows evolutionary relationship among Hominins and hypothesized relation to modern humans. (*still debated phylogeny position).

Very Early Hominins

Three species of very early hominids have made news in the late 20th and early 21st centuries: *Ardipithecus*, *Sahelanthropus*, and *Orrorin*. The youngest of the three species, *Ardipithecus*, was discovered in the 1990s, and dates to about 4.4 MYA. Although the bipedality of the early specimens was uncertain, several more specimens of *Ardipithecus* were discovered in the intervening years and demonstrated that the organism was bipedal. Two different species of *Ardipithecus* have been identified, *A. ramidus* and *A. kadabba*, whose specimens are older, dating to 5.6 MYA. However, the status of this genus as a human ancestor is uncertain.

The oldest of the three, *Sahelanthropus tchadensis*, was discovered in 2001-2002 and has been dated to nearly seven million years ago. There is a single specimen of this genus, a skull that was a surface find in Chad. The fossil, informally called "Toumai," is a mosaic of primitive and evolved characteristics, and it is unclear how this fossil fits with the picture given by molecular data, namely that the line leading to modern humans and modern chimpanzees apparently bifurcated about six million years ago. It is not thought at this time that this species was an ancestor of modern humans.

A younger (c. 6 MYA) species, *Orrorin tugenensis*, is also a relatively recent discovery, found in 2000. There are several specimens of *Orrorin*. Some features of *Orrorin* are more similar to those of modern humans than are the australopithicenes, although *Orrorin* is much older. If *Orrorin* is a human ancestor, then the australopithicenes may not be in the direct human lineage. Additional specimens of these species may help to clarify their role.

Early Hominins: Genus *Australopithecus*

Australopithecus ("southern ape") is a genus of hominin that evolved in eastern Africa approximately four million years ago and went extinct about two million years ago. This genus is of particular interest to us as it is thought that our genus, genus *Homo*, evolved from a common ancestor shared with *Australopithecus* about two million years ago (after likely passing through some transitional states). *Australopithecus* had a number of characteristics that were more similar to the great apes than to modern humans. For example, sexual dimorphism was more exaggerated than in modern humans. Males were up to 50 percent larger than females, a ratio that is similar to that seen in modern gorillas and orangutans. In contrast, modern human males are approximately 15 to 20 percent larger than females. The brain size of *Australopithecus* relative to its body mass was also smaller than in modern humans and more similar to that seen in the great apes. A key feature that *Australopithecus* had in common with modern humans was bipedalism, although it is likely that *Australopithecus* also spent time in trees. Hominin footprints, similar to those of modern humans, were found in Lactoli, Tanzania and dated to 3.6 million years ago. They showed that hominins at the time of *Australopithecus* were walking upright.

There were a number of *Australopithecus* species, which are often referred to as *australopiths*. *Australopithecus anamensis* lived about 4.2 million years ago. More is known about another early species, *Australopithecus afarensis*, which lived between 3.9 and 2.9 million years ago. This species demonstrates a trend in human evolution: the reduction of the dentition and jaw in size. *A. afarensis* (Figure 29.47a) had smaller canines and molars compared to apes, but these were larger than those of modern humans. Its brain size was 380 to 450 cubic centimeters, approximately the size of a modern chimpanzee brain. It also had prognathic jaws, which is a relatively longer jaw than that of modern humans. In the mid-1970s, the fossil of an adult female *A. afarensis* was found in the Afar region of Ethiopia and dated to 3.24 million years ago (Figure 29.48). The fossil, which is informally called "Lucy," is significant because it was the most complete australopith fossil found, with 40 percent of the skeleton recovered.

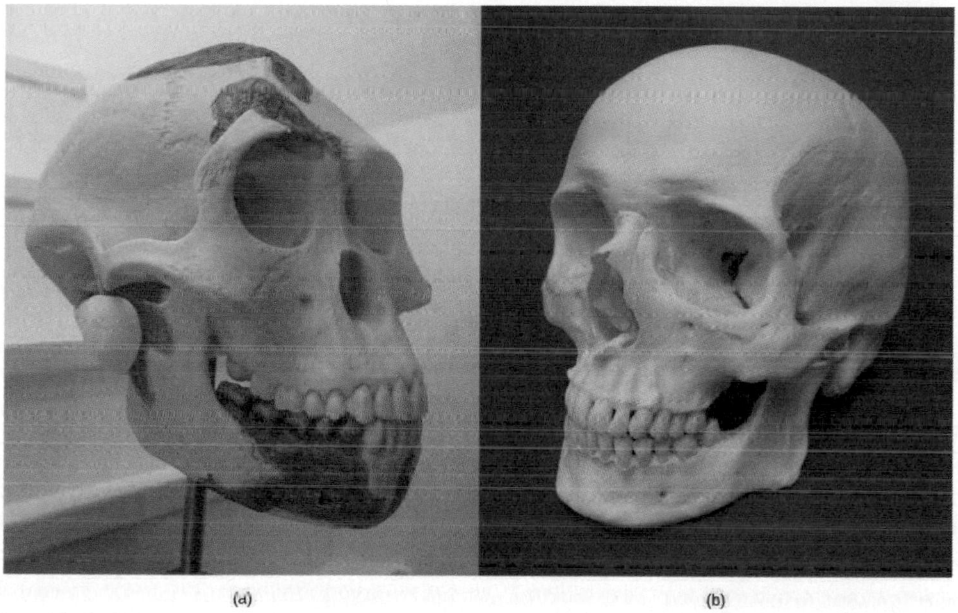

Figure 29.47 Australopithicene and modern human skulls. The skull of (a) *Australopithecus afarensis*, an early hominid that lived between two and three million years ago, resembled that of (b) modern humans but was smaller with a sloped forehead, larger teeth, and a prominent jaw.

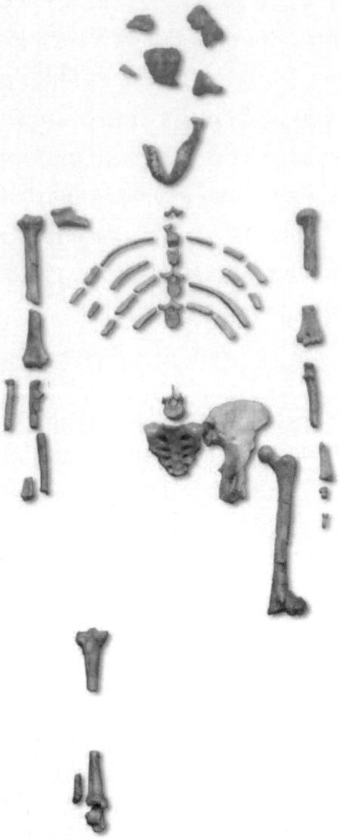

Figure 29.48 Lucy. This adult female *Australopithecus afarensis* skeleton, nicknamed Lucy, was discovered in the mid-1970s. (credit: "120"/Wikimedia Commons)

Australopithecus africanus lived between two and three million years ago. It had a slender build and was bipedal, but had robust arm bones and, like other early hominids, may have spent significant time in trees. Its brain was larger than that of *A. afarensis* at 500 cubic centimeters, which is slightly less than one-third the size of modern human brains. Two other species, *Australopithecus bahrelghazali* and *Australopithecus garhi*, have been added to the roster of australopiths in recent years. A. *bahrelghazali* is unusual in being the only australopith found in Central Africa.

A Dead End: Genus *Paranthropus*

The australopiths had a relatively slender build and teeth that were suited for soft food. In the past several years, fossils of hominids of a different body type have been found and dated to approximately 2.5 million years ago. These hominids, of the genus *Paranthropus*, were muscular, stood 1.3 to 1.4 meters tall, and had large grinding teeth. Their molars showed heavy wear, suggesting that they had a coarse and fibrous vegetarian diet as opposed to the partially carnivorous diet of the australopiths. *Paranthropus* includes *Paranthropus robustus* of South Africa, and *Paranthropus aethiopicus* and *Paranthropus boisei* of East Africa. The hominids in this genus went extinct more than one million years ago and are not thought to be ancestral to modern humans, but rather members of an evolutionary branch on the hominin tree that left no descendants.

Early Hominins: Genus *Homo*

The human genus, *Homo*, first appeared between 2.5 and three million years ago. For many years, fossils of a species called *H. habilis* were the oldest examples in the genus *Homo*, but in 2010, a new species called *Homo gautengensis* was discovered and may be older. Compared to *A. africanus*, *H. habilis* had a number of features more similar to modern humans. *H. habilis* had a jaw that was less prognathic than the australopiths and a larger brain, at 600 to 750 cubic centimeters. However, *H. habilis* retained some features of older hominin species, such as long arms. The name *H. habilis* means "handy man," which is a reference to the stone tools that have been found with its remains.

LINK TO LEARNING

Watch this video about Smithsonian paleontologist Briana Pobiner explaining the link between hominin eating of meat and evolutionary trends.

Click to view content (https://www.openstax.org/l/diet_detective)

H. erectus appeared approximately 1.8 million years ago (Figure 29.49). It is believed to have originated in East Africa and was the first hominin species to migrate out of Africa. Fossils of *H. erectus* have been found in India, China, Java, and Europe, and were known in the past as "Java Man" or "Peking Man." *H. erectus* had a number of features that were more similar to modern humans than those of *H. habilis*. *H. erectus* was larger in size than earlier hominins, reaching heights up to 1.85 meters and weighing up to 65 kilograms, which are sizes similar to those of modern humans. Its degree of sexual dimorphism was less than in earlier species, with males being 20 to 30 percent larger than females, which is close to the size difference seen in our own species. *H. erectus* had a larger brain than earlier species at 775 to 1,100 cubic centimeters, which compares to the 1,130 to 1,260 cubic centimeters seen in modern human brains. *H. erectus* also had a nose with downward-facing nostrils similar to modern humans, rather than the forward-facing nostrils found in other primates. Longer, downward-facing nostrils allow for the warming of cold air before it enters the lungs and may have been an adaptation to colder climates. Artifacts found with fossils of *H. erectus* suggest that it was the first hominin to use fire, hunt, and have a home base. *H. erectus* is generally thought to have lived until about 50,000 years ago.

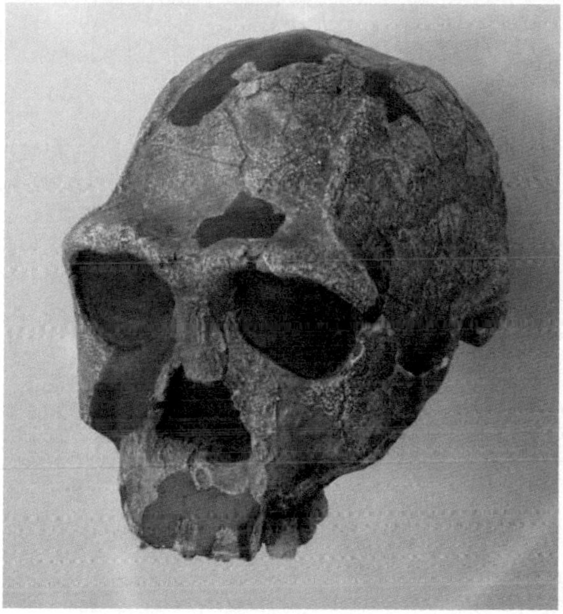

Figure 29.49 *Homo erectus. Homo erectus* had a prominent brow and a nose that pointed downward rather than forward.

Humans: *Homo sapiens*

A number of species, sometimes called archaic *Homo sapiens*, apparently evolved from *H. erectus* starting about 500,000 years ago. These species include *Homo heidelbergensis*, *Homo rhodesiensis*, and *Homo neanderthalensis*. These archaic *H. sapiens* had a brain size similar to that of modern humans, averaging 1,200 to 1,400 cubic centimeters. They differed from modern humans by having a thick skull, a prominent brow ridge, and a receding chin. Some of these species survived until 30,000 to 10,000 years ago, overlapping with modern humans (Figure 29.50).

Figure 29.50 Neanderthal. The *Homo neanderthalensis* used tools and may have worn clothing.

There is considerable debate about the origins of anatomically modern humans or ***Homo sapiens sapiens***. As discussed earlier, *H. erectus* migrated out of Africa and into Asia and Europe in the first major wave of migration about 1.5 million years ago. It is thought that modern humans arose in Africa from *H. erectus* and migrated out of Africa about 100,000 years ago in a second major migration wave. Then, modern humans replaced *H. erectus* species that had migrated into Asia and Europe in the first wave.

This evolutionary timeline is supported by molecular evidence. One approach to studying the origins of modern humans is to examine mitochondrial DNA (mtDNA) from populations around the world. Because a fetus develops from an egg containing its mother's mitochondria (which have their own, non-nuclear DNA), mtDNA is passed entirely through the maternal line. Mutations in mtDNA can now be used to estimate the timeline of genetic divergence. The resulting evidence suggests that all modern humans have mtDNA inherited from a common ancestor that lived in Africa about 160,000 years ago. Another approach to the molecular understanding of human evolution is to examine the Y chromosome, which is passed from father to son. This evidence suggests that all men today inherited a Y chromosome from a male that lived in Africa about 140,000 years ago.

The study of mitochondrial DNA led to the identification of another human species or subspecies, the Denisovans. DNA from teeth and finger bones suggested two things. First, the mitochondrial DNA was different from that of both modern humans and Neanderthals. Second, the genomic DNA suggested that the Denisovans shared a common ancestor with the Neanderthals. Genes from both Neanderthals and Denisovans have been identified in modern human populations, indicating that interbreeding among the three groups occurred over part of their range.

KEY TERMS

Acanthostega one of the earliest known tetrapods

Actinopterygii ray-finned fishes

allantois membrane of the egg that stores nitrogenous wastes produced by the embryo; also facilitates respiration

amnion membrane of the egg that protects the embryo from mechanical shock and prevents dehydration

amniote animal that produces a terrestrially adapted egg protected by amniotic membranes

Amphibia frogs, salamanders, and caecilians

ampulla of Lorenzini sensory organ that allows sharks to detect electromagnetic fields produced by living things

anapsid animal having no temporal fenestrae in the cranium

anthropoid monkeys, apes, and humans

Anura frogs

apocrine gland scent gland that secretes substances that are used for chemical communication

Apoda caecilians

Archaeopteryx transition species from dinosaur to bird from the Jurassic period

archosaur modern crocodilian or bird, or an extinct pterosaur or dinosaur

Australopithecus genus of hominins that evolved in eastern Africa approximately four million years ago

brachiation movement through trees branches via suspension from the arms

brumation period of much reduced metabolism and torpor that occurs in any ectotherm in cold weather

caecilian legless amphibian that belongs to the clade Apoda

Casineria one of the oldest known amniotes; had both amphibian and reptilian characteristics

Catarrhini clade of Old World monkeys

Cephalochordata chordate clade whose members possess a notochord, dorsal hollow nerve cord, pharyngeal slits, and a post-anal tail in the adult stage

Chondrichthyes jawed fish with paired fins and a skeleton made of cartilage

Chordata phylum of animals distinguished by their possession of a notochord, a dorsal hollow nerve cord, pharyngeal slits, and a post-anal tail at some point during their development

chorion membrane of the egg that surrounds the embryo and yolk sac

contour feather feather that creates an aerodynamic surface for efficient flight

Craniata clade composed of chordates that possess a cranium; includes Vertebrata together with hagfishes

cranium bony, cartilaginous, or fibrous structure surrounding the brain, jaw, and facial bones

Crocodilia crocodiles and alligators

cutaneous respiration gas exchange through the skin

dentary single bone that comprises the lower jaw of mammals

diapsid animal having two temporal fenestrae in the cranium

diphyodont refers to the possession of two sets of teeth in a lifetime

dorsal hollow nerve cord hollow, tubular structure derived from ectoderm, which is located dorsal to the notochord in chordates

down feather feather specialized for insulation

eccrine gland sweat gland

Enantiornithes dominant bird group during the Cretaceous period

eutherian mammal mammal that possesses a complex placenta, which connects a fetus to the mother; sometimes called placental mammals

flight feather feather specialized for flight

frog tail-less amphibian that belongs to the clade Anura

furcula wishbone formed by the fusing of the clavicles

gnathostome jawed fish

Gorilla genus of gorillas

hagfish eel-like jawless fish that live on the ocean floor and are scavengers

heterodont tooth different types of teeth that are modified for different purposes

hominin species that are more closely related to humans than chimpanzees

hominoid pertaining to great apes and humans

Homo genus of humans

Homo sapiens sapiens anatomically modern humans

Hylobatidae family of gibbons

Hylonomus one of the earliest reptiles

lamprey jawless fish characterized by a toothed, funnel-like, sucking mouth

lancelet member of Cephalochordata; named for its blade-like shape

lateral line sense organ that runs the length of a fish's body; used to detect vibration in the water

lepidosaur modern lizards, snakes, and tuataras

mammal one of the groups of endothermic vertebrates that possesses hair and mammary glands

mammary gland in female mammals, a gland that produces milk for newborns

marsupial one of the groups of mammals that includes the kangaroo, koala, bandicoot, Tasmanian devil, and several other species; young develop within a pouch

monotreme egg-laying mammal

Myxini hagfishes

Neognathae birds other than the Paleognathae

Neornithes modern birds

notochord flexible, rod-shaped support structure that is found in the embryonic stage of all chordates and in the adult stage of some chordates

Ornithorhynchidae clade that includes the duck-billed platypus

Osteichthyes bony fish

ostracoderm one of the earliest jawless fish covered in bone

Paleognathae ratites; flightless birds, including ostriches and emus

Pan genus of chimpanzees and bonobos

Petromyzontidae clade of lampreys

pharyngeal slit opening in the pharynx

Platyrrhini clade of New World monkeys

Plesiadapis oldest known primate-like mammal

pneumatic bone air-filled bone

Pongo genus of orangutans

post-anal tail muscular, posterior elongation of the body extending beyond the anus in chordates

primary feather feather located at the tip of the wing that provides thrust

Primates order of lemurs, tarsiers, monkeys, apes, and humans

prognathic jaw long jaw

prosimian division of primates that includes bush babies and pottos of Africa, lemurs of Madagascar, and lorises of Southeast Asia

salamander tailed amphibian that belongs to the clade Urodela

Sarcopterygii lobe-finned fish

sauropsid reptile or bird

sebaceous gland in mammals, a skin gland that produce a lipid mixture called *sebum*

secondary feather feather located at the base of the wing that provides lift

Sphenodontia clade of tuataras

Squamata clade of lizards and snakes

stereoscopic vision two overlapping fields of vision from the eyes that produces depth perception

swim bladder in fishes, a gas filled organ that helps to control the buoyancy of the fish

synapsid mammal having one temporal fenestra

Tachyglossidae clade that includes the echidna or spiny anteater

tadpole larval stage of a frog

temporal fenestra non-orbital opening in the skull that may allow muscles to expand and lengthen

Testudines order of turtles

tetrapod phylogenetic reference to an organism with a four-footed evolutionary history; includes amphibians, reptiles, birds, and mammals

theropod dinosaur group ancestral to birds

tunicate sessile chordate that is a member of Urochordata

Urochordata clade composed of tunicates

Urodela salamanders

vertebral column series of separate bones joined together as a backbone

Vertebrata members of the phylum Chordata that possess a backbone

CHAPTER SUMMARY

29.1 Chordates

The five characteristic features of chordates present during some time of their life cycles are a notochord, a dorsal hollow tubular nerve cord, pharyngeal slits, endostyle/thyroid gland, and a post-anal tail. Chordata contains two clades of invertebrates: Urochordata (tunicates) and Cephalochordata (lancelets), together with the vertebrates in the Vertebrata/Craniata. Lancelets are suspension feeders that feed on phytoplankton and other microorganisms. Most tunicates live on the ocean floor and are suspension feeders. Which of the two invertebrate chordate clades is more closely related to the vertebrates continues to be debated. Vertebrata is named for the vertebral column, which is a feature of almost all members of this clade. The name Craniata (organisms with a cranium) is considered to be synonymous with Vertebrata.

29.2 Fishes

The earliest vertebrates that diverged from the invertebrate chordates were the agnathan jawless fishes, whose extant members include the hagfishes and lampreys. Hagfishes are eel-like scavengers that feed on dead invertebrates and other fishes. Lampreys are characterized by a toothed, funnel-like sucking mouth, and most species are parasitic or predaceous on other fishes. Fishes with jaws (gnathostomes) evolved later. Jaws allowed early gnathostomes to exploit new food sources.

Gnathostomes include the cartilaginous fishes and the bony fishes, as well as all other tetrapods (amphibians, reptiles, mammals). Cartilaginous fishes include sharks, rays, skates, and ghost sharks. Most cartilaginous fishes live in marine habitats, with a few species living in fresh water for part or all of their lives. The vast majority of present-day fishes belong to the clade Osteichthyes, which consists of approximately 30,000 species. Bony fishes (Osteichthyes) can be divided into two clades: Actinopterygii (ray-finned fishes, virtually all extant species) and Sarcopterygii (lobe-finned fishes, comprising fewer than 10 extant species, but form the sister group of the tetrapods).

29.3 Amphibians

As tetrapods, most amphibians are characterized by four well-developed limbs, although some species of salamanders and all caecilians are limbless. The most important characteristic of extant amphibians is a moist, permeable skin used for cutaneous respiration, although lungs are found in the adults of many species.

All amphibians are carnivores and possess many small teeth. The fossil record provides evidence of amphibian species, now extinct, that arose over 400 million years ago as the first tetrapods. Living Amphibia can be divided into three classes: salamanders (Urodela), frogs (Anura), and caecilians (Apoda). In the majority of amphibians, development occurs in two distinct stages: a gilled aquatic larval stage that metamorphoses into an adult stage, acquiring lungs and legs, and losing the tail in Anurans. A few species in all three clades bypass a free-living larval stage. Various levels of parental care are seen in the amphibians.

29.4 Reptiles

The amniotes are distinguished from amphibians by the presence of a terrestrially adapted egg protected by four extra-embryonic membranes. The amniotes include reptiles, birds, and mammals. The early amniotes diverged into two main lines soon after the first amniotes arose. The initial split was into synapsids (mammals) and sauropsids. Sauropsids can be further divided into anapsids and diapsids (crocodiles, dinosaurs, birds, and modern reptiles).

Reptiles are tetrapods that ancestrally had four limbs; however, a number of extant species have secondarily lost them or greatly reduced them over evolutionary time. For example, limbless reptiles (e.g., snakes) are classified as tetrapods, because they descended from ancestors with four limbs. One of the key adaptations that permitted reptiles to live on land was the development of scaly skin containing the protein keratin, which prevented water loss from the skin. Reptilia includes four living clades of nonavian organisms: Crocodilia (crocodiles and alligators), Sphenodontia (tuataras), Squamata (lizards and snakes), and Testudines (turtles). Currently, this classification is paraphyletic, leaving out the birds, which are now classified as avian reptiles in the class Reptilia.

29.5 Birds

Birds are the most speciose group of land vertebrates and display a number of adaptations related to their ability to fly, which were first present in their theropod (maniraptoran) ancestors. Birds are endothermic (and homeothermic), meaning they have a very high metabolism that produces a considerable amount of heat, as well as structures such as feathers that allow them to retain their own body heat. These adaptations are used to regulate their internal temperature, making it largely independent of ambient thermal conditions.

Birds have feathers, which allow for insulation and flight, as well as for mating and warning signals. Flight feathers have a broad and continuously curved vane that produces lift. Some birds have pneumatic bones containing air spaces that are sometimes connected to air sacs in the body cavity. Airflow through bird lungs travels in one direction, creating a counter-current gas exchange with the blood.

Birds are highly modified diapsids and belong to a group called the archosaurs. Within the archosaurs, birds are most likely evolved from theropod (maniraptoran) dinosaurs. One of the oldest known fossils (and best known) of a "dinosaur-bird" is that of *Archaeopteryx*, which is dated from the Jurassic period. Modern birds are now classified into three groups: Paleognathae, Galloanserae, and Neoaves.

29.6 Mammals

Mammals are vertebrates that possess hair and mammary glands. The mammalian integument includes various secretory glands, including sebaceous glands, eccrine glands, apocrine glands, and mammary glands.

Mammals are synapsids, meaning that they have a single opening in the skull behind the eye. Mammals probably evolved from therapsids in the late Triassic period, as the earliest known mammal fossils are from the early Jurassic period. A key characteristic of synapsids is endothermy, and most mammals are homeothermic.

There are three groups of mammals living today: monotremes, marsupials, and eutherians. Monotremes are unique among mammals as they lay eggs, rather than giving birth to young. Marsupials give birth to very immature young, which typically complete their development in a pouch. Eutherian mammals are sometimes called placental mammals, because all species possess a complex placenta that connects a fetus to the mother, allowing for gas, fluid, and nutrient exchange. All mammals nourish their young with milk, which is derived from modified sweat or sebaceous glands.

29.7 The Evolution of Primates

All primate species possess adaptations for climbing trees and probably descended from arboreal ancestors, although not all living species are arboreal. Other characteristics of primates are brains that are larger, relative to body size, than those of other mammals, claws that have been modified into flattened nails, typically only one young per pregnancy, stereoscopic vision, and a trend toward holding the body upright. Primates are divided into two groups: strepsirrhines, which include most prosimians, and haplorhines, which include simians. Monkeys evolved from prosimians during the Oligocene epoch. The simian line includes both platyrrhine and catarrhine branches. Apes evolved from catarrhines in Africa during the Miocene epoch. Apes are divided into the lesser apes and the greater apes. Hominins include those groups that gave rise to our own species, such as *Australopithecus* and *H. erectus*, and those groups that can be considered "cousins" of humans, such as Neanderthals and Denisovans. Fossil evidence shows that hominins at the time of *Australopithecus* were walking upright, the first evidence of bipedal hominins. A number of species, sometimes called archaic *H. sapiens*, evolved from *H. erectus* approximately 500,000 years ago. There is considerable debate about the origins of anatomically modern humans or *H. sapiens sapiens*, and the discussion will continue, as new evidence from fossil finds and genetic analysis emerges.

VISUAL CONNECTION QUESTIONS

1. Figure 29.3 Which of the following statements about common features of chordates is true?
   a. The dorsal hollow nerve cord is part of the chordate central nervous system.
   b. In vertebrate fishes, the pharyngeal slits become the gills.
   c. Humans are not chordates because humans do not have a tail.
   d. Vertebrates do not have a notochord at any point in their development; instead, they have a vertebral column.

2. Figure 29.22 Which of the following statements about the parts of an amniotic egg are false?
   a. The allantois stores nitrogenous waste and facilitates respiration.
   b. The chorion facilitates gas exchange.
   c. The yolk provides food for the growing embryo.
   d. The amniotic cavity is filled with albumen.

3. Figure 29.24 Members of the order Testudines have an anapsid-like skull without obvious temporal fenestrae. However, molecular studies indicate that turtles descended from a diapsid ancestor. Why might this be the case?

REVIEW QUESTIONS

4. Which of the following is *not* contained in phylum Chordata?
   a. Cephalochordata
   b. Echinodermata
   c. Urochordata
   d. Vertebrata

5. Which group of invertebrates is most closely related to vertebrates?
   a. cephalochordates
   b. echinoderms
   c. arthropods
   d. urochordates

6. Hagfish, lampreys, sharks, and tuna are all chordates that can also be classified into which group?
   a. Craniates
   b. Vertebrates
   c. Cartilaginous fish
   d. Cephalocordata

7. Members of Chondrichthyes differ from members of Osteichthyes by having (a) _______.
   a. jaw
   b. bony skeleton
   c. cartilaginous skeleton
   d. two sets of paired fins

8. Members of Chondrichthyes are thought to be descended from fishes that had _______.
   a. a cartilaginous skeleton
   b. a bony skeleton
   c. mucus glands
   d. slime glands

9. A marine biologist catches a species of fish she has never seen before. Upon examination, she determines that the species has a predominantly cartilaginous skeleton and a swim bladder. If its pectoral fins are not fused with its head, to which category of fish does the specimen belong?
   a. Rays
   b. Osteichthyes
   c. Sharks
   d. Hagfish

10. Which of the following is *not* true of *Acanthostega*?
    a. It was aquatic.
    b. It had gills.
    c. It had four limbs.
    d. It laid shelled eggs.

11. Frogs belong to which order?
    a. Anura
    b. Urodela
    c. Caudata
    d. Apoda

12. During the Mesozoic period, diapsids diverged into_______.
    a. pterosaurs and dinosaurs
    b. mammals and reptiles
    c. lepidosaurs and archosaurs
    d. Testudines and Sphenodontia

13. Squamata includes_______.
    a. crocodiles and alligators
    b. turtles
    c. tuataras
    d. lizards and snakes

14. Which of the following reptile groups gave rise to modern birds?
    a. Lepidosaurs
    b. Pterosaurs
    c. Anapsids
    d. Archosaurs

15. A bird or feathered dinosaur is ________.
    a. Neornithes
    b. *Archaeopteryx*
    c. Enantiornithes
    d. Paleognathae

16. Which of the following feather types helps to reduce drag produced by wind resistance during flight?
    a. Flight feathers
    b. Primary feathers
    c. Secondary feathers
    d. Contour feathers

17. Eccrine glands produce ________.
    a. sweat
    b. lipids
    c. scents
    d. milk

18. Monotremes include:
    a. kangaroos.
    b. koalas.
    c. bandicoots.
    d. platypuses.

19. The evolution of which of the following features of mammals is hardest to trace through the fossil record?
    a. Jaw structure
    b. Mammary glands
    c. Middle ear structure
    d. Development of hair

20. Which of the following is *not* an anthropoid?
    a. Lemurs
    b. Monkeys
    c. Apes
    d. Humans

21. Which of the following is part of a clade believed to have died out, leaving no descendants?
    a. *Paranthropus robustus*
    b. *Australopithecus africanus*
    c. *Homo erectus*
    d. *Homo sapiens sapiens*

22. Which of the following human traits is not a shared characteristic of primates?
    a. Hip structure supporting bipedalism
    b. Detection and processing of three-color vision
    c. Nails at the end of each digit
    d. Enlarged brain area associated with vision, and reduced area associated with smell

CRITICAL THINKING QUESTIONS

23. What are the characteristic features of the chordates?
24. What is the structural advantage of the notochord in the human embryo? Be sure to compare the notochord with the corresponding structure in adults.
25. What can be inferred about the evolution of the cranium and vertebral column from examining hagfishes and lampreys?
26. Why did gnathostomes replace most agnathans?
27. Explain why frogs are restricted to a moist environment.
28. Describe the differences between the larval and adult stages of frogs.
29. Describe how metamorphosis changes the structures involved in gas exchange over the life cycle of animals in the clade Anura, and what evolutionary advantage this change provides.
30. Describe the functions of the three extra-embryonic membranes present in amniotic eggs.
31. What characteristics differentiate lizards and snakes?
32. Based on how reptiles thermoregulate, which climates would you predict to have the highest reptile population density, and why?
33. Explain why birds are thought to have evolved from theropod dinosaurs.

34. Describe three skeletal adaptations that allow for flight in birds.
35. How would the chest structure differ between ostriches, penguins, and terns?
36. Describe three unique features of the mammalian skeletal system.
37. Describe three characteristics of the mammalian brain that differ from other vertebrates.
38. How did the evolution of jaw musculature allow mammals to spread?
39. How did archaic *Homo sapiens* differ from anatomically modern humans?
40. Why is it so difficult to determine the sequence of hominin ancestors that have led to modern *Homo sapiens*?

CHAPTER 30
Plant Form and Physiology

Figure 30.1 A locust leaf consists of leaflets arrayed along a central midrib. Each leaflet is a complex photosynthetic machine, exquisitely adapted to capture sunlight and carbon dioxide. An intricate vascular system supplies the leaf with water and minerals, and exports the products of photosynthesis. (credit: modification of work by Todd Petit)

INTRODUCTION Plants are as essential to human existence as land, water, and air. Without plants, our day-to-day lives would be impossible because without oxygen from photosynthesis, aerobic life cannot be sustained. From providing food and shelter to serving as a source of medicines, oils, perfumes, and industrial products, plants provide humans with numerous valuable resources.

When you think of plants, most of the organisms that come to mind are vascular plants. These plants have tissues that conduct food and water, and most of them have seeds. Seed plants are divided into gymnosperms and angiosperms. Gymnosperms include the needle-leaved conifers—spruce, fir, and pine—as well as less familiar plants, such as ginkgos and cycads. Their seeds are not enclosed by a fleshy fruit. Angiosperms, also called flowering plants, constitute the majority of seed plants. They include broadleaved trees (such as maple, oak, and elm), vegetables (such as potatoes, lettuce, and carrots), grasses, and plants known for the beauty of their flowers (roses, irises, and daffodils, for example).

While individual plant species are unique, all share a common structure: a plant body consisting of stems, roots, and leaves. They all transport water, minerals, and sugars produced through photosynthesis through the plant body in a similar manner. All plant species also respond to environmental factors, such as light, gravity, competition, temperature, and predation.

30.1 The Plant Body

By the end of this section, you will be able to do the following:

- Describe the shoot organ system and the root organ system
- Distinguish between meristematic tissue and permanent tissue
- Identify and describe the three regions where plant growth occurs
- Summarize the roles of dermal tissue, vascular tissue, and ground tissue
- Compare simple plant tissue with complex plant tissue

Like animals, plants contain cells with organelles in which specific metabolic activities take place. Unlike animals, however, plants use energy from sunlight to form sugars during photosynthesis. In addition, plant cells have cell walls, plastids, and a large central vacuole: structures that are not found in animal cells. Each of these cellular structures plays a specific role in plant structure and function.

LINK TO LEARNING

Watch *Botany Without Borders* (http://openstax.org/l/botany_wo_bord) , a video produced by the Botanical Society of America about the importance of plants.

Plant Organ Systems

In plants, just as in animals, similar cells working together form a tissue. When different types of tissues work together to perform a unique function, they form an organ; organs working together form organ systems. Vascular plants have two distinct organ systems: a shoot system, and a root system. The **shoot system** consists of two portions: the vegetative (non-reproductive) parts of the plant, such as the leaves and the stems, and the reproductive parts of the plant, which include flowers and fruits. The shoot system generally grows above ground, where it absorbs the light needed for photosynthesis. The **root system**, which supports the plants and absorbs water and minerals, is usually underground. Figure 30.2 shows the organ systems of a typical plant.

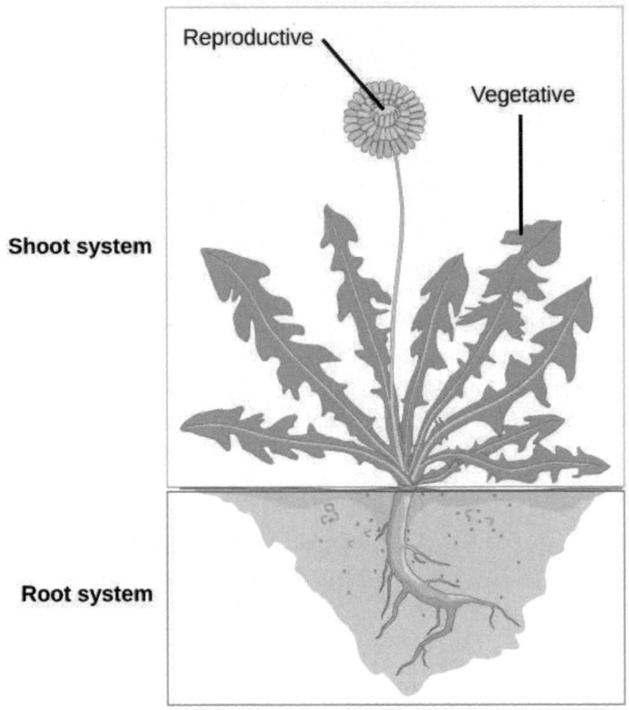

Figure 30.2 The shoot system of a plant consists of leaves, stems, flowers, and fruits. The root system anchors the plant while absorbing water and minerals from the soil.

Plant Tissues

Plants are multicellular eukaryotes with tissue systems made of various cell types that carry out specific functions. Plant tissue systems fall into one of two general types: meristematic tissue, and permanent (or non-meristematic) tissue. Cells of the meristematic tissue are found in **meristems**, which are plant regions of continuous cell division and growth. **Meristematic tissue** cells are either undifferentiated or incompletely differentiated, and they continue to divide and contribute to the growth of the plant. In contrast, **permanent tissue** consists of plant cells that are no longer actively dividing.

Meristematic tissues consist of three types, based on their location in the plant. **Apical meristems** contain meristematic tissue located at the tips of stems and roots, which enable a plant to extend in length. **Lateral**

meristems facilitate growth in thickness or girth in a maturing plant. **Intercalary meristems** occur only in monocots, at the bases of leaf blades and at nodes (the areas where leaves attach to a stem). This tissue enables the monocot leaf blade to increase in length from the leaf base; for example, it allows lawn grass leaves to elongate even after repeated mowing.

Meristems produce cells that quickly differentiate, or specialize, and become permanent tissue. Such cells take on specific roles and lose their ability to divide further. They differentiate into three main types: dermal, vascular, and ground tissue. **Dermal tissue** covers and protects the plant, and **vascular tissue** transports water, minerals, and sugars to different parts of the plant. **Ground tissue** serves as a site for photosynthesis, provides a supporting matrix for the vascular tissue, and helps to store water and sugars.

Secondary tissues are either simple (composed of similar cell types) or complex (composed of different cell types). Dermal tissue, for example, is a simple tissue that covers the outer surface of the plant and controls gas exchange. Vascular tissue is an example of a complex tissue, and is made of two specialized conducting tissues: xylem and phloem. Xylem tissue transports water and nutrients from the roots to different parts of the plant, and includes three different cell types: vessel elements and tracheids (both of which conduct water), and xylem parenchyma. Phloem tissue, which transports organic compounds from the site of photosynthesis to other parts of the plant, consists of four different cell types: sieve cells (which conduct photosynthates), companion cells, phloem parenchyma, and phloem fibers. Unlike xylem conducting cells, phloem conducting cells are alive at maturity. The xylem and phloem always lie adjacent to each other (Figure 30.3). In stems, the xylem and the phloem form a structure called a **vascular bundle**; in roots, this is termed the **vascular stele** or **vascular cylinder**.

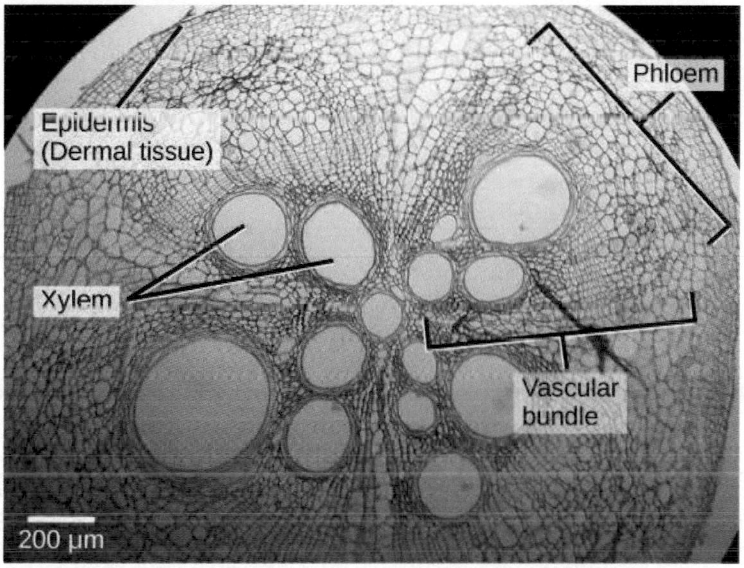

Figure 30.3 This light micrograph shows a cross section of a squash (*Curcurbita maxima*) stem. Each teardrop-shaped vascular bundle consists of large xylem vessels toward the inside and smaller phloem cells toward the outside. Xylem cells, which transport water and nutrients from the roots to the rest of the plant, are dead at functional maturity. Phloem cells, which transport sugars and other organic compounds from photosynthetic tissue to the rest of the plant, are living. The vascular bundles are encased in ground tissue and surrounded by dermal tissue. (credit: modification of work by "(biophotos)"/Flickr; scale-bar data from Matt Russell)

30.2 Stems

By the end of this section, you will be able to do the following:

- Describe the main function and basic structure of stems
- Compare and contrast the roles of dermal tissue, vascular tissue, and ground tissue
- Distinguish between primary growth and secondary growth in stems
- Summarize the origin of annual rings
- List and describe examples of modified stems

Stems are a part of the shoot system of a plant. They may range in length from a few millimeters to hundreds of meters, and also vary in diameter, depending on the plant type. Stems are usually above ground, although the stems of some plants, such as the potato, also grow underground. Stems may be herbaceous (soft) or woody in nature. Their main function is to provide support to the plant, holding leaves, flowers and buds; in some cases, stems also store food for the plant. A stem may be unbranched, like that of a palm tree, or it may be highly branched, like that of a magnolia tree. The stem of the plant connects the roots to the leaves, helping to transport absorbed water and minerals to different parts of the plant. It also helps to transport the products of photosynthesis, namely sugars, from the leaves to the rest of the plant.

Plant stems, whether above or below ground, are characterized by the presence of nodes and internodes (Figure 30.4). **Nodes** are points of attachment for leaves, aerial roots, and flowers. The stem region between two nodes is called an **internode**. The stalk that extends from the stem to the base of the leaf is the petiole. An **axillary bud** is usually found in the axil—the area between the base of a leaf and the stem—where it can give rise to a branch or a flower. The apex (tip) of the shoot contains the apical meristem within the **apical bud**.

Figure 30.4 Leaves are attached to the plant stem at areas called nodes. An internode is the stem region between two nodes. The petiole is the stalk connecting the leaf to the stem. The leaves just above the nodes arose from axillary buds.

Stem Anatomy

The stem and other plant organs arise from the ground tissue, and are primarily made up of simple tissues formed from three types of cells: parenchyma, collenchyma, and sclerenchyma cells.

Parenchyma cells are the most common plant cells (Figure 30.5). They are found in the stem, the root, the inside of the leaf, and the pulp of the fruit. Parenchyma cells are responsible for metabolic functions, such as photosynthesis, and they help repair and heal wounds. Some parenchyma cells also store starch.

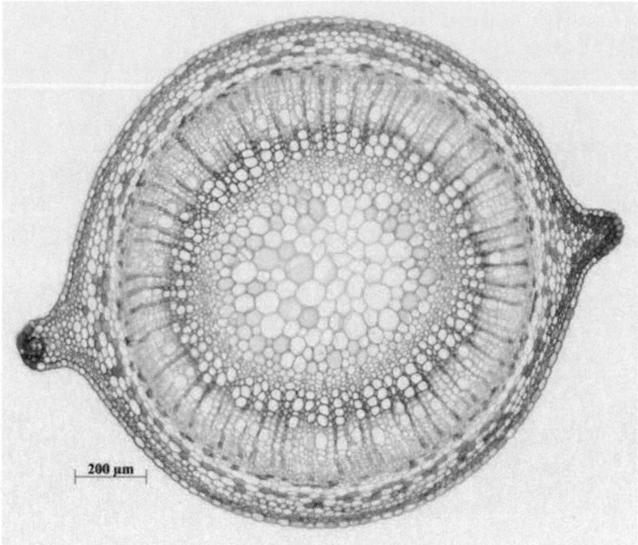

Figure 30.5 The stem of common St John's Wort (*Hypericum perforatum*) is shown in cross section in this light micrograph. The central pith (greenish-blue, in the center) and peripheral cortex (narrow zone 3–5 cells thick just inside the epidermis) are composed of parenchyma cells. Vascular tissue composed of xylem (red) and phloem tissue (green, between the xylem and cortex) surrounds the pith. (credit: Rolf-Dieter Mueller)

Collenchyma cells are elongated cells with unevenly thickened walls (Figure 30.6). They provide structural support, mainly to the stem and leaves. These cells are alive at maturity and are usually found below the epidermis. The "strings" of a celery stalk are an example of collenchyma cells.

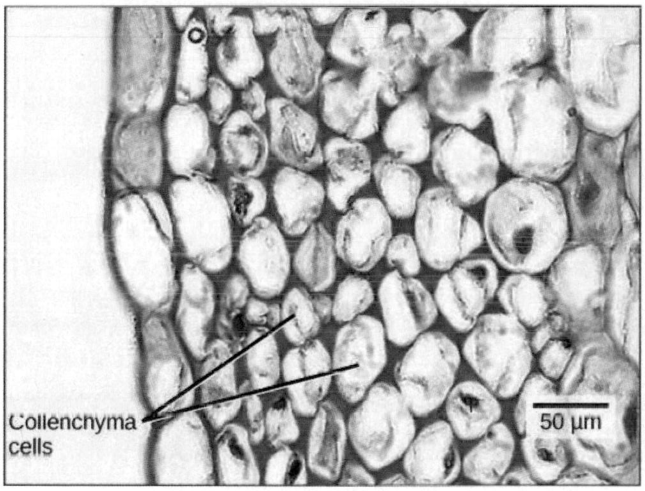

Figure 30.6 Collenchyma cell walls are uneven in thickness, as seen in this light micrograph. They provide support to plant structures. (credit: modification of work by Carl Szczerski; scale-bar data from Matt Russell)

Sclerenchyma cells also provide support to the plant, but unlike collenchyma cells, many of them are dead at maturity. There are two types of sclerenchyma cells: fibers and sclereids. Both types have secondary cell walls that are thickened with deposits of lignin, an organic compound that is a key component of wood. Fibers are long, slender cells; sclereids are smaller-sized. Sclereids give pears their gritty texture. Humans use sclerenchyma fibers to make linen and rope (Figure 30.7).

VISUAL CONNECTION

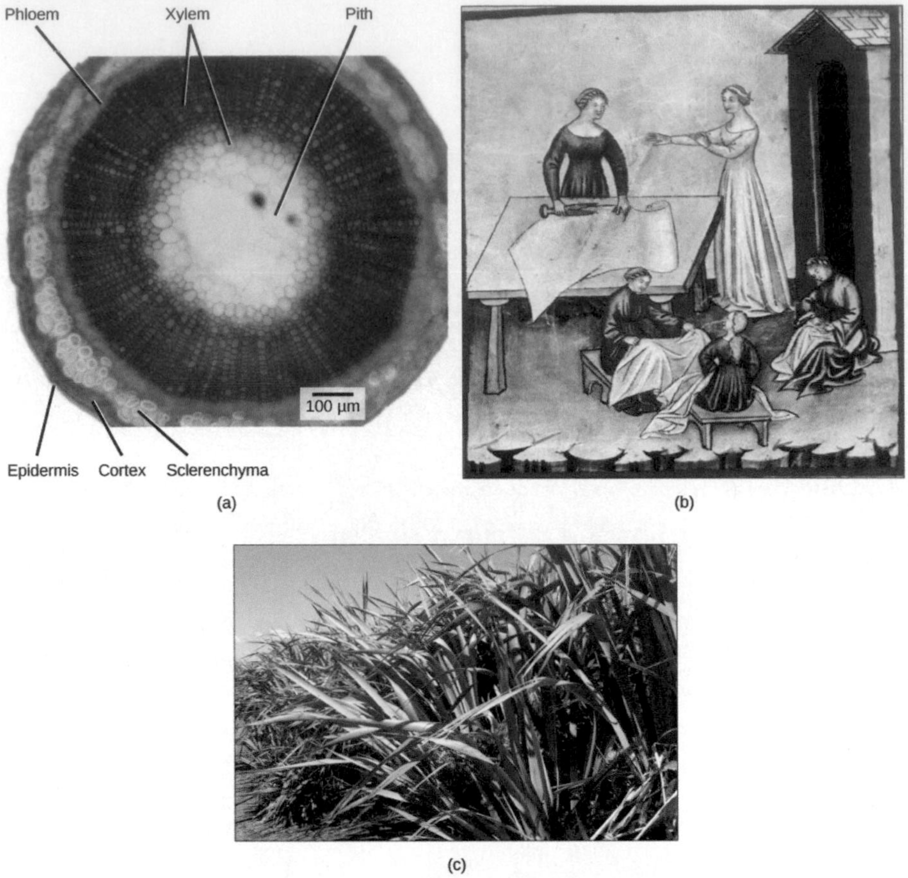

Figure 30.7 The central pith and outer cortex of the (a) flax stem are made up of parenchyma cells. Inside the cortex is a layer of sclerenchyma cells, which make up the fibers in flax rope and clothing. Humans have grown and harvested flax for thousands of years. In (b) this drawing, fourteenth-century women prepare linen. The (c) flax plant is grown and harvested for its fibers, which are used to weave linen, and for its seeds, which are the source of linseed oil. (credit a: modification of work by Emmanuel Boutet based on original work by Ryan R. MacKenzie; credit c: modification of work by Brian Dearth; scale-bar data from Matt Russell)

Which layers of the stem are made of parenchyma cells?

a. cortex and pith
b. phloem
c. sclerenchyma
d. xylem

Like the rest of the plant, the stem has three tissue systems: dermal, vascular, and ground tissue. Each is distinguished by characteristic cell types that perform specific tasks necessary for the plant's growth and survival.

Dermal Tissue

The dermal tissue of the stem consists primarily of **epidermis**, a single layer of cells covering and protecting the underlying tissue. Woody plants have a tough, waterproof outer layer of cork cells commonly known as **bark**, which further protects the plant from damage. Epidermal cells are the most numerous and least differentiated of the cells in the epidermis. The epidermis of a leaf also contains openings known as stomata, through which the exchange of gases takes place (Figure 30.8). Two cells, known as **guard cells**, surround each leaf stoma, controlling its opening and closing and thus regulating the uptake of carbon dioxide and the release of oxygen and water vapor. **Trichomes** are hair-like structures on the epidermal surface. They help to reduce **transpiration** (the loss of water by aboveground plant parts), increase solar reflectance, and store compounds that defend the leaves against predation by herbivores.

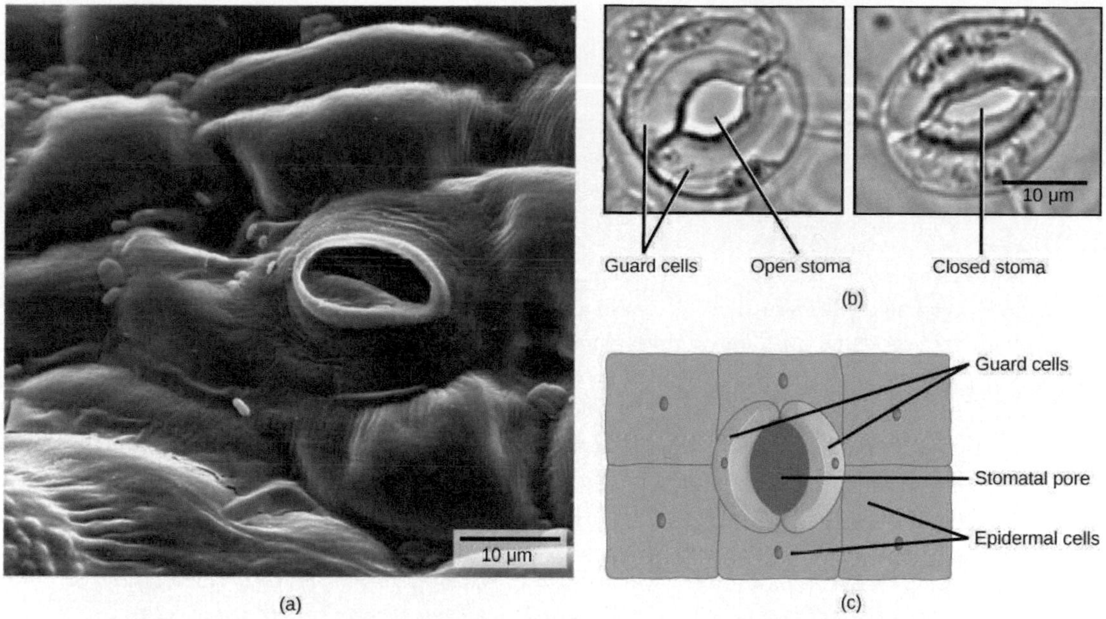

Figure 30.8 Openings called stomata (singular: stoma) allow a plant to take up carbon dioxide and release oxygen and water vapor. The (a) colorized scanning-electron micrograph shows a closed stoma of a dicot. Each stoma is flanked by two guard cells that regulate its (b) opening and closing. The (c) guard cells sit within the layer of epidermal cells. (credit a: modification of work by Louisa Howard, Rippel Electron Microscope Facility, Dartmouth College; credit b: modification of work by June Kwak, University of Maryland; scale-bar data from Matt Russell)

Vascular Tissue

The xylem and phloem that make up the vascular tissue of the stem are arranged in distinct strands called vascular bundles, which run up and down the length of the stem. When the stem is viewed in cross section, the vascular bundles of dicot stems are arranged in a ring. In plants with stems that live for more than one year, the individual bundles grow together and produce the characteristic growth rings. In monocot stems, the vascular bundles are randomly scattered throughout the ground tissue (Figure 30.9).

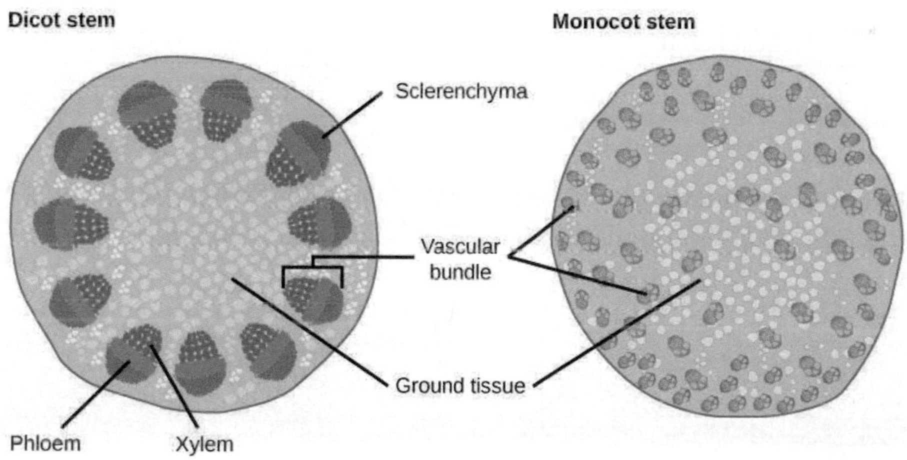

Figure 30.9 In (a) dicot stems, vascular bundles are arranged around the periphery of the ground tissue. The xylem tissue is located toward the interior of the vascular bundle, and phloem is located toward the exterior. Sclerenchyma fibers cap the vascular bundles. In (b) monocot stems, vascular bundles composed of xylem and phloem tissues are scattered throughout the ground tissue.

Xylem tissue has three types of cells: xylem parenchyma, tracheids, and vessel elements. The latter two types conduct water and are dead at maturity. **Tracheids** are xylem cells with thick secondary cell walls that are lignified. Water moves from one tracheid to another through regions on the side walls known as pits, where secondary walls are absent. **Vessel elements** are xylem cells with thinner walls; they are shorter than tracheids. Each vessel element is connected to the next by means of a perforation plate at the end walls of the element. Water moves through the perforation plates to travel up the plant.

Phloem tissue is composed of sieve-tube cells, companion cells, phloem parenchyma, and phloem fibers. A series of **sieve-tube cells** (also called sieve-tube elements) are arranged end to end to make up a long sieve tube, which transports organic substances such as sugars and amino acids. The sugars flow from one sieve-tube cell to the next through perforated sieve plates, which are found at the end junctions between two cells. Although still alive at maturity, the nucleus and other cell components of the sieve-tube cells have disintegrated. **Companion cells** are found alongside the sieve-tube cells, providing them with metabolic support. The companion cells contain more ribosomes and mitochondria than the sieve-tube cells, which lack some cellular organelles.

Ground Tissue

Ground tissue is mostly made up of parenchyma cells, but may also contain collenchyma and sclerenchyma cells that help support the stem. The ground tissue towards the interior of the vascular tissue in a stem or root is known as **pith**, while the layer of tissue between the vascular tissue and the epidermis is known as the **cortex**.

Growth in Stems

Growth in plants occurs as the stems and roots lengthen. Some plants, especially those that are woody, also increase in thickness during their life span. The increase in length of the shoot and the root is referred to as **primary growth**, and is the result of cell division in the shoot apical meristem. **Secondary growth** is characterized by an increase in thickness or girth of the plant, and is caused by cell division in the lateral meristem. Figure 30.10 shows the areas of primary and secondary growth in a plant. Herbaceous plants mostly undergo primary growth, with hardly any secondary growth or increase in thickness. Secondary growth or "wood" is noticeable in woody plants; it occurs in some dicots, but occurs very rarely in monocots.

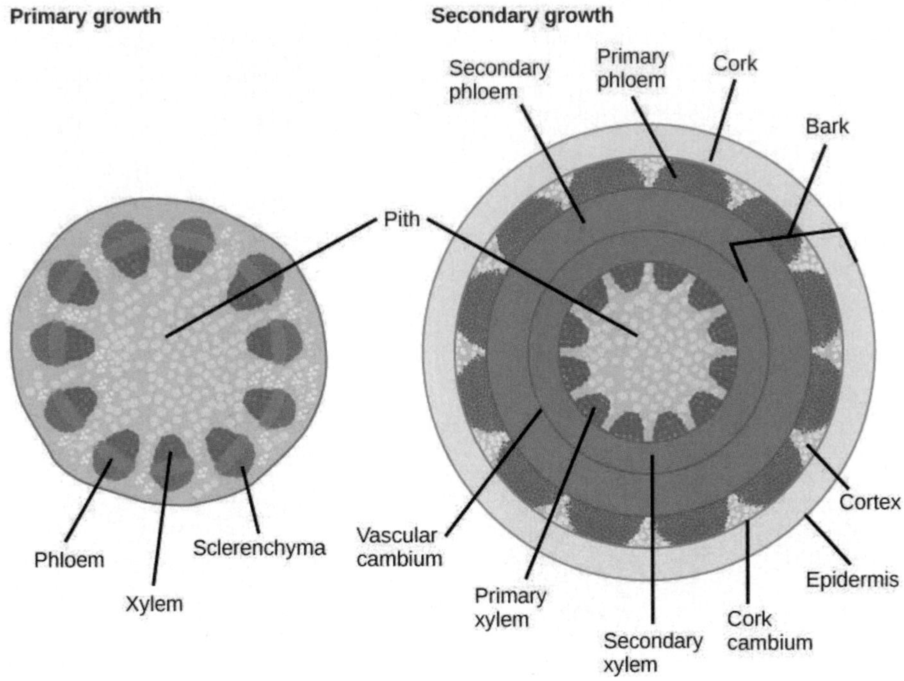

Figure 30.10 In woody plants, primary growth is followed by secondary growth, which allows the plant stem to increase in thickness or girth. Secondary vascular tissue is added as the plant grows, as well as a cork layer. The bark of a tree extends from the vascular cambium to the epidermis.

Some plant parts, such as stems and roots, continue to grow throughout a plant's life: a phenomenon called indeterminate growth. Other plant parts, such as leaves and flowers, exhibit determinate growth, which ceases when a plant part reaches a particular size.

Primary Growth

Most primary growth occurs at the apices, or tips, of stems and roots. Primary growth is a result of rapidly dividing cells in the apical meristems at the shoot tip and root tip. Subsequent cell elongation also contributes to primary growth. The growth of shoots and roots during primary growth enables plants to continuously seek water (roots) or sunlight (shoots).

The influence of the apical bud on overall plant growth is known as apical dominance, which diminishes the growth of axillary buds that form along the sides of branches and stems. Most coniferous trees exhibit strong apical dominance, thus producing the typical conical Christmas tree shape. If the apical bud is removed, then the axillary buds will start forming lateral branches. Gardeners make use of this fact when they prune plants by cutting off the tops of branches, thus encouraging the axillary buds to grow out, giving the plant a bushy shape.

LINK TO LEARNING

Watch this BBC Nature video (http://openstax.org/l/motion_plants) showing how time-lapse photography captures plant growth at high speed.

Secondary Growth

The increase in stem thickness that results from secondary growth is due to the activity of the lateral meristems, which are lacking in herbaceous plants. Lateral meristems include the vascular cambium and, in woody plants, the cork cambium (see Figure 30.10). The vascular cambium is located just outside the primary xylem and to the interior of the primary phloem. The cells of the vascular cambium divide and form secondary xylem (tracheids and vessel elements) to the inside, and secondary phloem (sieve elements and companion cells) to the outside. The thickening of the stem that occurs in secondary growth is due to the formation of secondary phloem and secondary xylem by the vascular cambium, plus the action of cork cambium, which forms the tough outermost layer of the stem. The cells of the secondary xylem contain lignin, which provides hardiness and strength.

In woody plants, cork cambium is the outermost lateral meristem. It produces cork cells (bark) containing a waxy substance known as suberin that can repel water. The bark protects the plant against physical damage and helps reduce water loss. The cork cambium also produces a layer of cells known as phelloderm, which grows inward from the cambium. The cork cambium, cork cells, and phelloderm are collectively termed the **periderm**. The periderm substitutes for the epidermis in mature plants. In some plants, the periderm has many openings, known as **lenticels**, which allow the interior cells to exchange gases with the outside atmosphere (Figure 30.11). This supplies oxygen to the living and metabolically active cells of the cortex, xylem, and phloem.

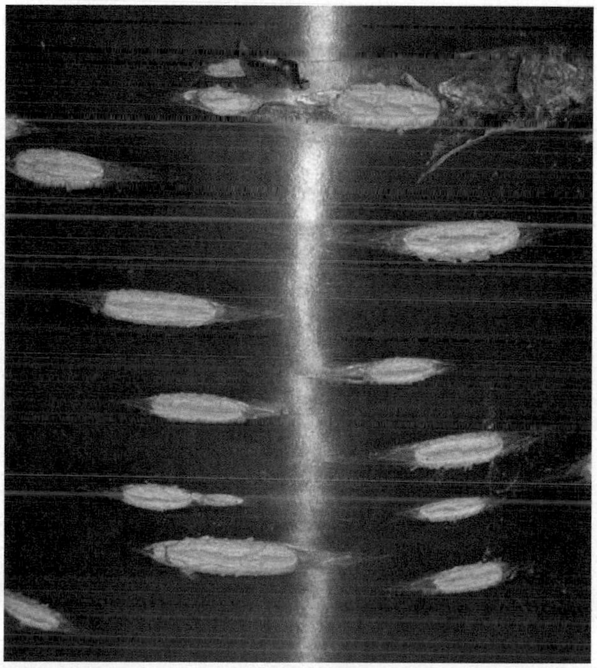

Figure 30.11 Lenticels on the bark of this cherry tree enable the woody stem to exchange gases with the surrounding atmosphere. (credit: Roger Griffith)

Annual Rings

The activity of the vascular cambium gives rise to annual growth rings. During the spring growing season, cells of the secondary xylem have a large internal diameter and their primary cell walls are not extensively thickened. This is known as early wood, or spring wood. During the fall season, the secondary xylem develops thickened cell walls, forming late wood, or autumn wood, which is denser than early wood. This alternation of early and late wood is due largely to a seasonal decrease in the number of vessel elements and a seasonal increase in the number of tracheids. It results in the formation of an annual ring, which can be seen as a circular ring in the cross section of the stem (Figure 30.12). An examination of the number of annual rings and their nature (such as their size and cell wall thickness) can reveal the age of the tree and the prevailing climatic conditions during each season.

Figure 30.12 The rate of wood growth increases in summer and decreases in winter, producing a characteristic ring for each year of growth. Seasonal changes in weather patterns can also affect the growth rate—note how the rings vary in thickness. (credit: Adrian Pingstone)

Stem Modifications

Some plant species have modified stems that are especially suited to a particular habitat and environment (Figure 30.13). A **rhizome** is a modified stem that grows horizontally underground and has nodes and internodes. Vertical shoots may arise from the buds on the rhizome of some plants, such as ginger and ferns. **Corms** are similar to rhizomes, except they are more rounded and fleshy (such as in gladiolus). Corms contain stored food that enables some plants to survive the winter. **Stolons** are stems that run almost parallel to the ground, or just below the surface, and can give rise to new plants at the nodes. **Runners** are a type of stolon that runs above the ground and produces new clone plants at nodes at varying intervals: strawberries are an example. **Tubers** are modified stems that may store starch, as seen in the potato (*Solanum* sp.). Tubers arise as swollen ends of stolons, and contain many adventitious or unusual buds (familiar to us as the "eyes" on potatoes). A **bulb**, which functions as an underground storage unit, is a modification of a stem that has the appearance of enlarged fleshy leaves emerging from the stem or surrounding the base of the stem, as seen in the iris.

Figure 30.13 Stem modifications enable plants to thrive in a variety of environments. Shown are (a) ginger (*Zingiber officinale*) rhizomes, (b) a carrion flower (*Amorphophallus titanum*) corm, (c) Rhodes grass (*Chloris gayana*) stolons, (d) strawberry (*Fragaria ananassa*) runners, (e) potato (*Solanum tuberosum*) tubers, and (f) red onion (*Allium*) bulbs. (credit a: modification of work by Maja Dumat; credit c: modification of work by Harry Rose; credit d: modification of work by Rebecca Siegel; credit e: modification of work by Scott Bauer, USDA ARS; credit f:

modification of work by Stephen Ausmus, USDA ARS)

LINK TO LEARNING

Watch botanist Wendy Hodgson, of Desert Botanical Garden in Phoenix, Arizona, explain how agave plants were cultivated for food hundreds of years ago in the Arizona desert in this video: (http://openstax.org/l/ancient_crop) *Finding the Roots of an Ancient Crop.*

Some aerial modifications of stems are tendrils and thorns (Figure 30.14). **Tendrils** are slender, twining strands that enable a plant (like a vine or pumpkin) to seek support by climbing on other surfaces. **Thorns** are modified branches appearing as sharp outgrowths that protect the plant; common examples include roses, Osage orange, and devil's walking stick.

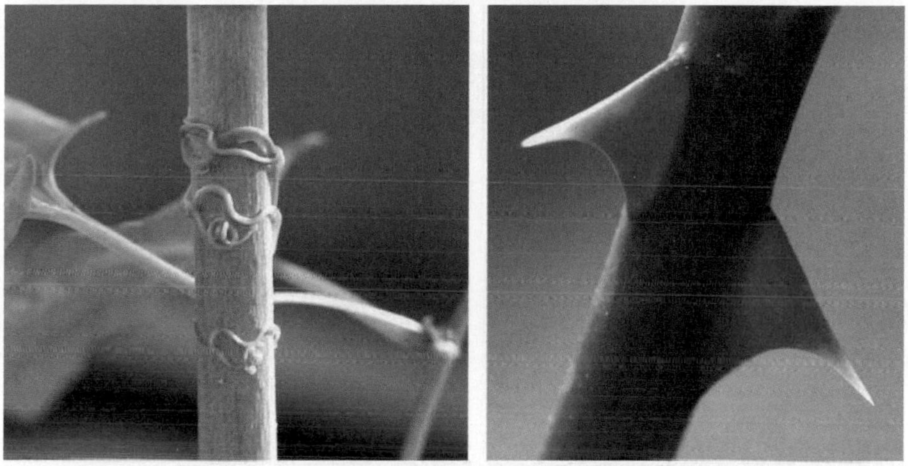

(a) (b)

Figure 30.14 Found in southeastern United States, (a) buckwheat vine (*Brunnichia ovata*) is a weedy plant that climbs with the aid of tendrils. This one is shown climbing up a wooden stake. (b) Thorns are modified branches. (credit a: modification of work by Christopher Meloche, USDA ARS; credit b: modification of work by "macrophile"/Flickr)

30.3 Roots

By the end of this section, you will be able to do the following:

- Identify the two types of root systems
- Describe the three zones of the root tip and summarize the role of each zone in root growth
- Describe the structure of the root
- List and describe examples of modified roots

The roots of seed plants have three major functions: anchoring the plant to the soil, absorbing water and minerals and transporting them upwards, and storing the products of photosynthesis. Some roots are modified to absorb moisture and exchange gases. Most roots are underground. Some plants, however, also have **adventitious roots**, which emerge above the ground from the shoot.

Types of Root Systems

Root systems are mainly of two types (Figure 30.15). Dicots have a tap root system, while monocots have a fibrous root system. A **tap root system** has a main root that grows down vertically, and from which many smaller lateral roots arise. Dandelions are a good example; their tap roots usually break off when trying to pull these weeds, and they can regrow another shoot from the remaining root. A tap root system penetrates deep into the soil. In contrast, a **fibrous root system** is located closer to the soil surface, and forms a dense network of roots that also helps prevent soil erosion (lawn grasses are a good example, as are wheat, rice, and corn). Some plants have a combination of tap roots and fibrous roots. Plants that grow in dry areas often have deep root systems, whereas plants growing in areas with abundant water are likely to have shallower root systems.

Figure 30.15 (a) Tap root systems have a main root that grows down, while (b) fibrous root systems consist of many small roots. (credit b: modification of work by "Austen Squarepants"/Flickr)

Root Growth and Anatomy

Root growth begins with seed germination. When the plant embryo emerges from the seed, the radicle of the embryo forms the root system. The tip of the root is protected by the **root cap**, a structure exclusive to roots and unlike any other plant structure. The root cap is continuously replaced because it gets damaged easily as the root pushes through soil. The root tip can be divided into three zones: a zone of cell division, a zone of elongation, and a zone of maturation and differentiation (Figure 30.16). The zone of cell division is closest to the root tip; it is made up of the actively dividing cells of the root meristem. The zone of elongation is where the newly formed cells increase in length, thereby lengthening the root. Beginning at the first root hair is the zone of cell maturation where the root cells begin to differentiate into special cell types. All three zones are in the first centimeter or so of the root tip.

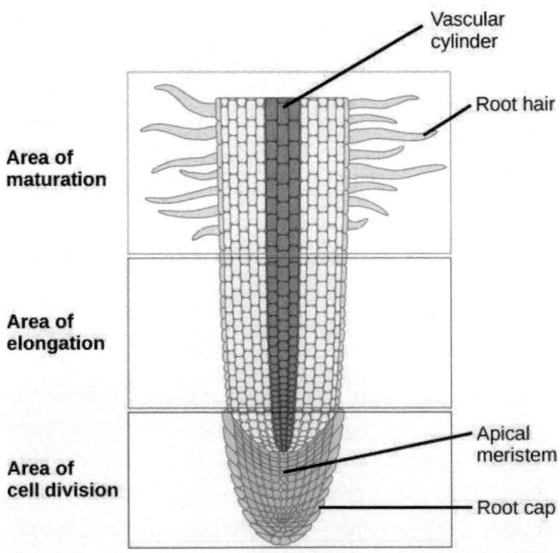

Figure 30.16 A longitudinal view of the root reveals the zones of cell division, elongation, and maturation. Cell division occurs in the apical meristem.

The root has an outer layer of cells called the epidermis, which surrounds areas of ground tissue and vascular tissue. The epidermis provides protection and helps in absorption. **Root hairs**, which are extensions of root epidermal cells, increase the surface area of the root, greatly contributing to the absorption of water and minerals.

Inside the root, the ground tissue forms two regions: the cortex and the pith (Figure 30.17). Compared to stems, roots have lots of cortex and little pith. Both regions include cells that store photosynthetic products. The cortex is between the epidermis and the vascular tissue, whereas the pith lies between the vascular tissue and the center of the root.

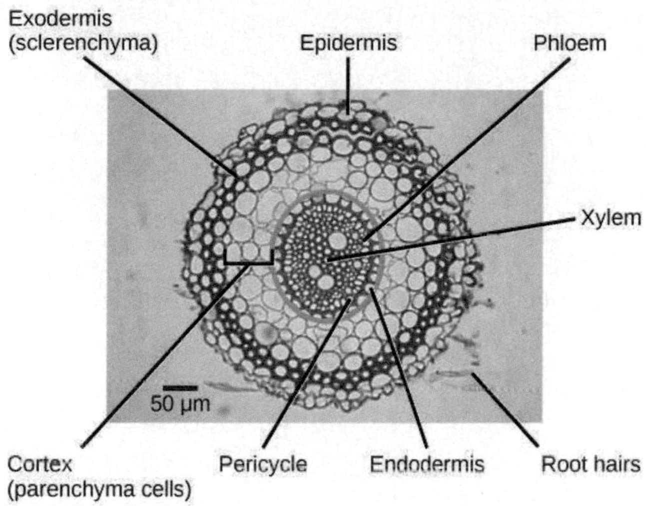

Figure 30.17 Staining reveals different cell types in this light micrograph of a wheat (*Triticum*) root cross section. Sclerenchyma cells of the exodermis and xylem cells stain red, and phloem cells stain blue. Other cell types stain black. The stele, or vascular tissue, is the area inside endodermis (indicated by a green ring). Root hairs are visible outside the epidermis. (credit: scale-bar data from Matt Russell)

The vascular tissue in the root is arranged in the inner portion of the root, which is called the **stele** (Figure 30.18). A layer of cells known as the **endodermis** separates the stele from the ground tissue in the outer portion of the root. The endodermis is exclusive to roots, and serves as a checkpoint for materials entering the root's vascular system. A waxy substance called suberin is present on the walls of the endodermal cells. This waxy region, known as the **Casparian strip**, forces water and solutes to cross the plasma membranes of endodermal cells instead of slipping between the cells. This ensures that only materials required by the root pass through the endodermis, while toxic substances and pathogens are generally excluded. The outermost cell layer of the root's vascular tissue is the **pericycle**, an area that can give rise to lateral roots. In dicot roots, the xylem and phloem of the stele are arranged alternately in an X shape, whereas in monocot roots, the vascular tissue is arranged in a ring around the pith.

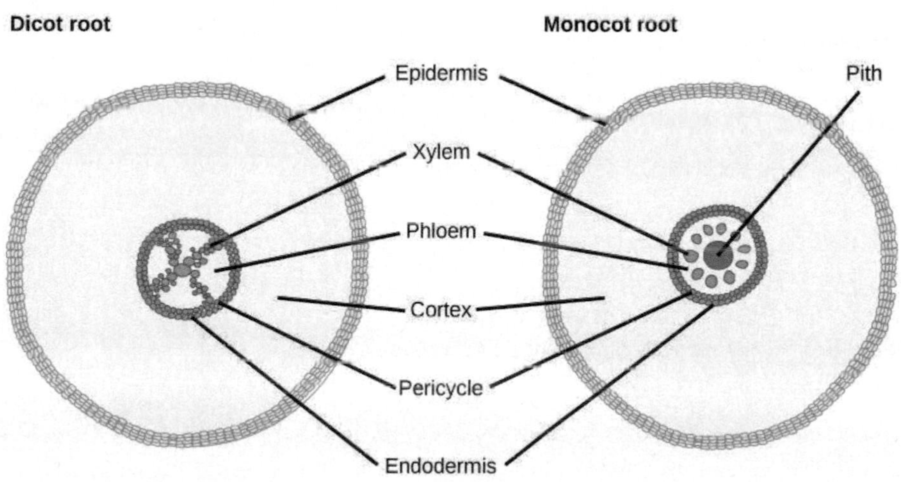

Figure 30.18 In (left) typical dicots, the vascular tissue forms an X shape in the center of the root. In (right) typical monocots, the phloem cells and the larger xylem cells form a characteristic ring around the central pith.

Root Modifications

Root structures may be modified for specific purposes. For example, some roots are bulbous and store starch. Aerial roots and prop roots are two forms of aboveground roots that provide additional support to anchor the plant. Tap roots, such as carrots, turnips, and beets, are examples of roots that are modified for food storage (Figure 30.19).

Figure 30.19 Many vegetables are modified roots.

Epiphytic roots enable a plant to grow on another plant. For example, the epiphytic roots of orchids develop a spongy tissue to absorb moisture. The banyan tree (*Ficus* sp.) begins as an epiphyte, germinating in the branches of a host tree; aerial roots develop from the branches and eventually reach the ground, providing additional support (Figure 30.20). In screwpine (*Pandanus* sp.), a palm-like tree that grows in sandy tropical soils, aboveground prop roots develop from the nodes to provide additional support.

(a) (b)

Figure 30.20 The (a) banyan tree, also known as the strangler fig, begins life as an epiphyte in a host tree. Aerial roots extend to the ground and support the growing plant, which eventually strangles the host tree. The (b) screwpine develops aboveground roots that help support the plant in sandy soils. (credit a: modification of work by "psyberartist"/Flickr; credit b: modification of work by David Eikhoff)

30.4 Leaves

By the end of this section, you will be able to do the following:

- Identify the parts of a typical leaf
- Describe the internal structure and function of a leaf
- Compare and contrast simple leaves and compound leaves
- List and describe examples of modified leaves

Leaves are the main sites for photosynthesis: the process by which plants synthesize food. Most leaves are usually green, due to the presence of chlorophyll in the leaf cells. However, some leaves may have different colors, caused by other plant pigments that mask the green chlorophyll.

The thickness, shape, and size of leaves are adapted to the environment. Each variation helps a plant species maximize its chances of survival in a particular habitat. Usually, the leaves of plants growing in tropical rainforests have larger surface areas than those of plants growing in deserts or very cold conditions, which are likely to have a smaller surface area to minimize water loss.

Structure of a Typical Leaf

Each leaf typically has a leaf blade called the **lamina**, which is also the widest part of the leaf. Some leaves are attached to the plant stem by a **petiole**. Leaves that do not have a petiole and are directly attached to the plant stem are called **sessile** leaves. Small green appendages usually found at the base of the petiole are known as **stipules**. Most leaves have a midrib, which travels the length of the leaf and branches to each side to produce veins of vascular tissue. The edge of the leaf is called the margin. Figure 30.21 shows the structure of a typical eudicot leaf.

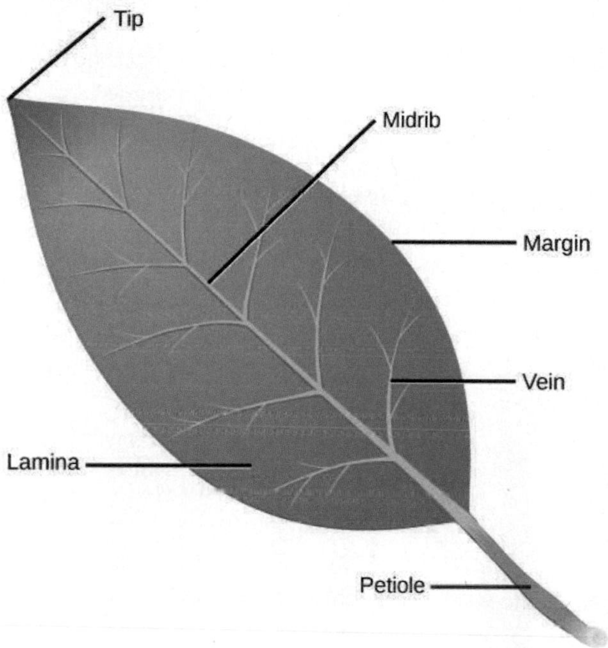

Figure 30.21 Deceptively simple in appearance, a leaf is a highly efficient structure.

Within each leaf, the vascular tissue forms veins. The arrangement of veins in a leaf is called the **venation** pattern. Monocots and dicots differ in their patterns of venation (Figure 30.22). Monocots have parallel venation; the veins run in straight lines across the length of the leaf without converging at a point. In dicots, however, the veins of the leaf have a net-like appearance, forming a pattern known as reticulate venation. One extant plant, the *Ginkgo biloba*, has dichotomous venation where the veins fork.

Figure 30.22 (a) Tulip (*Tulipa*), a monocot, has leaves with parallel venation. The netlike venation in this (b) linden (*Tilia cordata*) leaf distinguishes it as a dicot. The (c) *Ginkgo biloba* tree has dichotomous venation. (credit a photo: modification of work by

"Drewboy64"/Wikimedia Commons; credit b photo: modification of work by Roger Griffith; credit c photo: modification of work by "geishaboy500"/Flickr; credit abc illustrations: modification of work by Agnieszka Kwiecień)

Leaf Arrangement

The arrangement of leaves on a stem is known as **phyllotaxy**. The number and placement of a plant's leaves will vary depending on the species, with each species exhibiting a characteristic leaf arrangement. Leaves are classified as either alternate, spiral, or opposite. Plants that have only one leaf per node have leaves that are said to be either alternate—meaning the leaves alternate on each side of the stem in a flat plane—or spiral, meaning the leaves are arrayed in a spiral along the stem. In an opposite leaf arrangement, two leaves arise at the same point, with the leaves connecting opposite each other along the branch. If there are three or more leaves connected at a node, the leaf arrangement is classified as **whorled**.

Leaf Form

Leaves may be simple or compound (Figure 30.23). In a **simple leaf**, the blade is either completely undivided—as in the banana leaf—or it has lobes, but the separation does not reach the midrib, as in the maple leaf. In a **compound leaf**, the leaf blade is completely divided, forming leaflets, as in the locust tree. Each leaflet may have its own stalk, but is attached to the rachis. A **palmately compound leaf** resembles the palm of a hand, with leaflets radiating outwards from one point. Examples include the leaves of poison ivy, the buckeye tree, or the familiar houseplant *Schefflera* sp. (common name "umbrella plant"). **Pinnately compound leaves** take their name from their feather-like appearance; the leaflets are arranged along the midrib, as in rose leaves (*Rosa* sp.), or the leaves of hickory, pecan, ash, or walnut trees.

Figure 30.23 Leaves may be simple or compound. In simple leaves, the lamina is continuous. The (a) banana plant (*Musa* sp.) has simple leaves. In compound leaves, the lamina is separated into leaflets. Compound leaves may be palmate or pinnate. In (b) palmately compound leaves, such as those of the horse chestnut (*Aesculus hippocastanum*), the leaflets branch from the petiole. In (c) pinnately compound leaves, the leaflets branch from the midrib, as on a scrub hickory (*Carya floridana*). The (d) honey locust has double compound leaves, in which leaflets branch from the veins. (credit a: modification of work by "BazzaDaRambler"/Flickr; credit b: modification of work by Roberto Verzo; credit c: modification of work by Eric Dion; credit d: modification of work by Valerie Lykes)

Leaf Structure and Function

The outermost layer of the leaf is the epidermis; it is present on both sides of the leaf and is called the upper and lower epidermis, respectively. Botanists call the upper side the adaxial surface (or adaxis) and the lower side the abaxial surface (or abaxis). The epidermis helps in the regulation of gas exchange. It contains stomata (Figure 30.24): openings through which the exchange of gases takes place. Two guard cells surround each stoma, regulating its opening and closing.

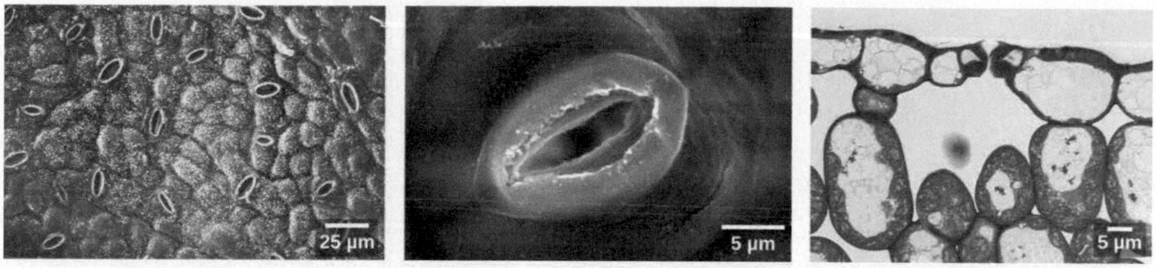

Figure 30.24 Visualized at 500x with a scanning electron microscope, several stomata are clearly visible on (a) the surface of this sumac (*Rhus glabra*) leaf. At 5,000x magnification, the guard cells of (b) a single stoma from lyre-leaved sand cress (*Arabidopsis lyrata)* have the appearance of lips that surround the opening. In this (c) light micrograph cross-section of an *A. lyrata* leaf, the guard cell pair is visible along with the large, sub-stomatal air space in the leaf. (credit: modification of work by Robert R. Wise; part c scale-bar data from Matt Russell)

The epidermis is usually one cell layer thick; however, in plants that grow in very hot or very cold conditions, the epidermis may be several layers thick to protect against excessive water loss from transpiration. A waxy layer known as the **cuticle** covers the leaves of all plant species. The cuticle reduces the rate of water loss from the leaf surface. Other leaves may have small hairs (trichomes) on the leaf surface. Trichomes help to deter herbivory by restricting insect movements, or by storing toxic or bad-tasting compounds; they can also reduce the rate of transpiration by blocking air flow across the leaf surface (Figure 30.25).

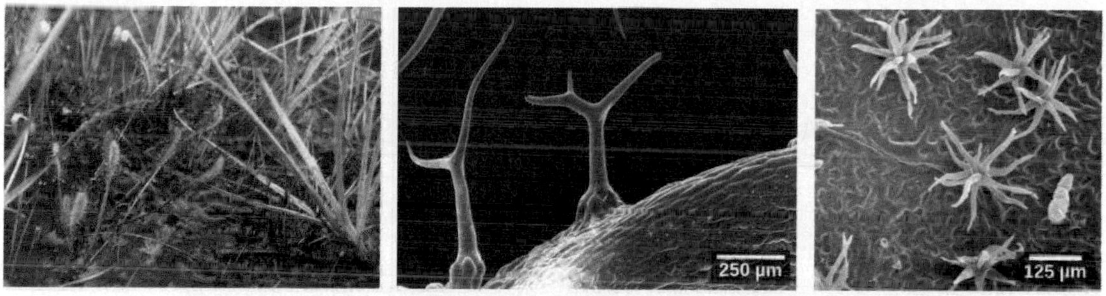

Figure 30.25 Trichomes give leaves a fuzzy appearance as in this (a) sundew (*Drosera* sp.). Leaf trichomes include (b) branched trichomes on the leaf of *Arabidopsis lyrata* and (c) multibranched trichomes on a mature *Quercus marilandica* leaf. (credit a: John Freeland; credit b, c: modification of work by Robert R. Wise; scale-bar data from Matt Russell)

Below the epidermis of dicot leaves are layers of cells known as the mesophyll, or "middle leaf." The mesophyll of most leaves typically contains two arrangements of parenchyma cells: the palisade parenchyma and spongy parenchyma (Figure 30.26). The palisade parenchyma (also called the palisade mesophyll) has column-shaped, tightly packed cells, and may be present in one, two, or three layers. Below the palisade parenchyma are loosely arranged cells of an irregular shape. These are the cells of the spongy parenchyma (or spongy mesophyll). The air space found between the spongy parenchyma cells allows gaseous exchange between the leaf and the outside atmosphere through the stomata. In aquatic plants, the intercellular spaces in the spongy parenchyma help the leaf float. Both layers of the mesophyll contain many chloroplasts. Guard cells are the only epidermal cells to contain chloroplasts.

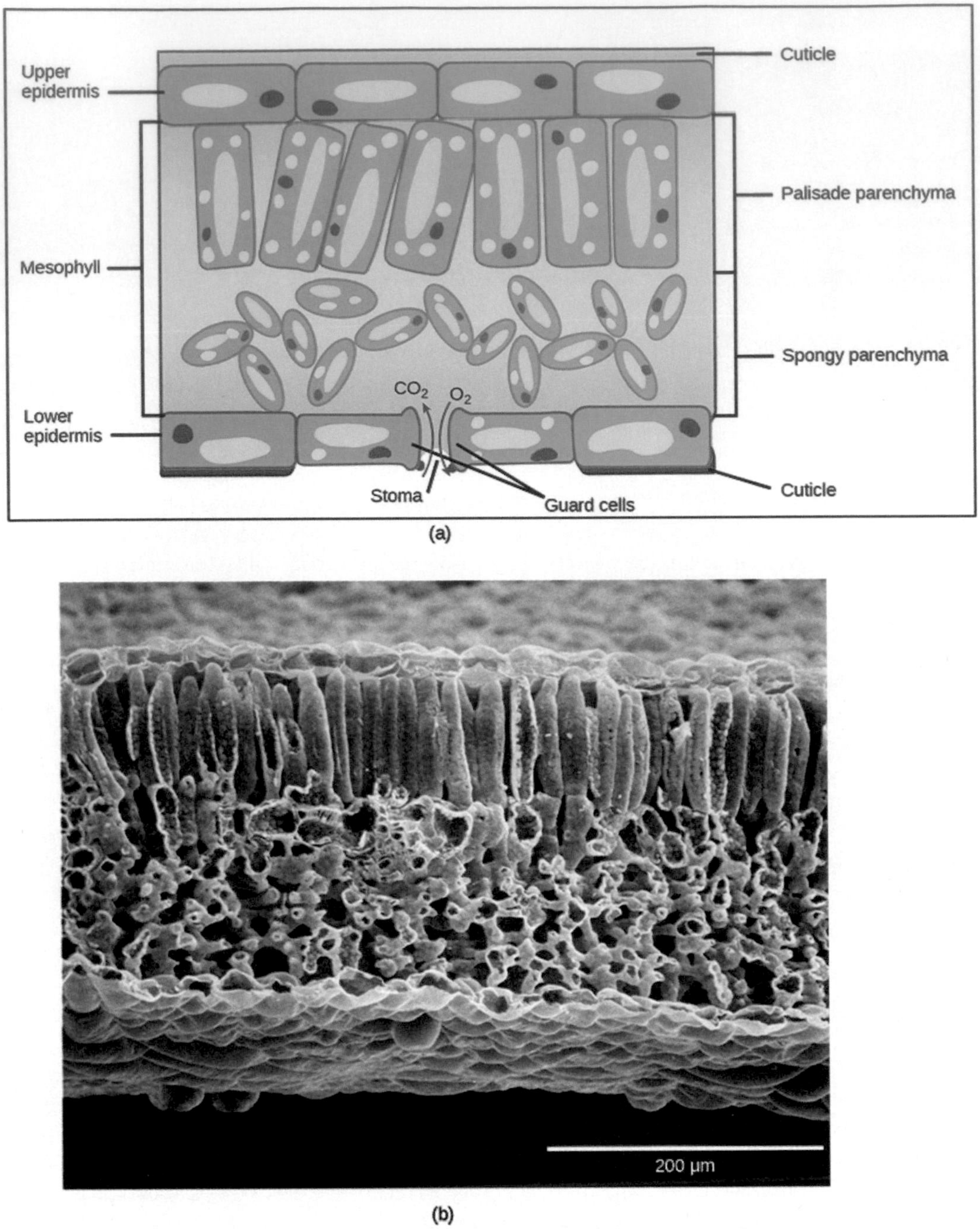

Figure 30.26 In the (a) leaf drawing, the central mesophyll is sandwiched between an upper and lower epidermis. The mesophyll has two layers: an upper palisade layer comprised of tightly packed, columnar cells, and a lower spongy layer, comprised of loosely packed, irregularly shaped cells. Stomata on the leaf underside allow gas exchange. A waxy cuticle covers all aerial surfaces of land plants to minimize water loss. These leaf layers are clearly visible in the (b) scanning electron micrograph. The numerous small bumps in the palisade parenchyma cells are chloroplasts. Chloroplasts are also present in the spongy parenchyma, but are not as obvious. The bumps protruding from the lower surface of the leave are glandular trichomes, which differ in structure from the stalked trichomes in Figure 30.25. (credit b: modification of work by Robert R. Wise)

Like the stem, the leaf contains vascular bundles composed of xylem and phloem (Figure 30.27). The xylem consists of tracheids and vessels, which transport water and minerals to the leaves. The phloem transports the photosynthetic products from the leaf to the other parts of the plant. A single vascular bundle, no matter how large or small, always contains both xylem and phloem tissues.

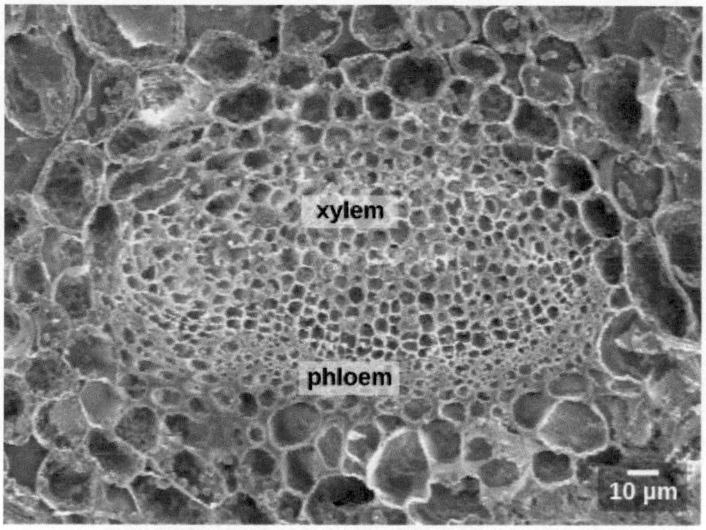

Figure 30.27 This scanning electron micrograph shows xylem and phloem in the leaf vascular bundle from the lyre-leaved sand cress (*Arabidopsis lyrata*). (credit: modification of work by Robert R. Wise; scale-bar data from Matt Russell)

Leaf Adaptations

Coniferous plant species that thrive in cold environments, like spruce, fir, and pine, have leaves that are reduced in size and needle-like in appearance. These needle-like leaves have sunken stomata and a smaller surface area: two attributes that aid in reducing water loss. In hot climates, plants such as cacti have leaves that are reduced to spines, which in combination with their succulent stems, help to conserve water. Many aquatic plants have leaves with wide lamina that can float on the surface of the water, and a thick waxy cuticle on the leaf surface that repels water.

LINK TO LEARNING

Watch "The Pale Pitcher Plant" episode of the video (http://openstax.org/l/plants_cool_too) series *Plants Are Cool, Too,* a Botanical Society of America video about a carnivorous plant species found in Louisiana.

EVOLUTION CONNECTION

Plant Adaptations in Resource-Deficient Environments

Roots, stems, and leaves are structured to ensure that a plant can obtain the required sunlight, water, soil nutrients, and oxygen resources. Some remarkable adaptations have evolved to enable plant species to thrive in less than ideal habitats, where one or more of these resources is in short supply.

In tropical rainforests, light is often scarce, since many trees and plants grow close together and block much of the sunlight from reaching the forest floor. Many tropical plant species have exceptionally broad leaves to maximize the capture of sunlight. Other species are epiphytes: plants that grow on other plants that serve as a physical support. Such plants are able to grow high up in the canopy atop the branches of other trees, where sunlight is more plentiful. Epiphytes live on rain and minerals collected in the branches and leaves of the supporting plant. Bromeliads (members of the pineapple family), ferns, and orchids are examples of tropical epiphytes (Figure 30.28). Many epiphytes have specialized tissues that enable them to efficiently capture and store water.

Figure 30.28 One of the most well known bromeliads is Spanish moss (*Tillandsia usneoides*), seen here in an oak tree. (credit: Kristine Paulus)

Some plants have special adaptations that help them to survive in nutrient-poor environments. Carnivorous plants, such as the Venus flytrap and the pitcher plant (Figure 30.29), grow in bogs where the soil is low in nitrogen. In these plants, leaves are modified to capture insects. The insect-capturing leaves may have evolved to provide these plants with a supplementary source of much-needed nitrogen.

(a) (b)

Figure 30.29 The (a) Venus flytrap has modified leaves that can capture insects. When an unlucky insect touches the trigger hairs inside the leaf, the trap suddenly closes. The opening of the (b) pitcher plant is lined with a slippery wax. Insects crawling on the lip slip and fall into a pool of water in the bottom of the pitcher, where they are digested by bacteria. The plant then absorbs the smaller molecules. (credit a: modification of work by Peter Shanks; credit b: modification of work by Tim Mansfield)

Many swamp plants have adaptations that enable them to thrive in wet areas, where their roots grow submerged underwater. In these aquatic areas, the soil is unstable and little oxygen is available to reach the roots. Trees such as mangroves (*Rhizophora* sp.) growing in coastal waters produce aboveground roots that help support the tree (Figure 30.30). Some species of mangroves, as well as cypress trees, have pneumatophores: upward-growing roots containing pores and pockets of tissue specialized for gas exchange. Wild rice is an aquatic plant with large air spaces in the root cortex. The air-filled tissue—called aerenchyma—provides a path for oxygen to diffuse down to the root tips, which are embedded in oxygen-poor bottom sediments.

(a) (b) (c)

Figure 30.30 The branches of (a) mangrove trees develop aerial roots, which descend to the ground and help to anchor the trees. (b) Cypress trees and some mangrove species have upward-growing roots called pneumatophores that are involved in gas exchange. Aquatic plants such as (c) wild rice have large spaces in the root cortex called aerenchyma, visualized here using scanning electron microscopy. (credit a: modification of work by Roberto Verzo; credit b: modification of work by Duane Burdick; credit c: modification of work by Robert R. Wise)

LINK TO LEARNING

Watch *Venus Flytraps: Jaws of Death*, an extraordinary BBC close-up of the Venus flytrap in action.

Click to view content (https://www.openstax.org/l/venus_flytrap)

30.5 Transport of Water and Solutes in Plants

By the end of this section, you will be able to do the following:

- Define water potential and explain how it is influenced by solutes, pressure, gravity, and the matric potential
- Describe how water potential, evapotranspiration, and stomatal regulation influence how water is transported in plants
- Explain how photosynthates are transported in plants

The structure of plant roots, stems, and leaves facilitates the transport of water, nutrients, and photosynthates throughout the plant. The phloem and xylem are the main tissues responsible for this movement. Water potential, evapotranspiration, and stomatal regulation influence how water and nutrients are transported in plants. To understand how these processes work, we must first understand the energetics of water potential.

Water Potential

Plants are phenomenal hydraulic engineers. Using only the basic laws of physics and the simple manipulation of potential energy, plants can move water to the top of a 116-meter-tall tree (Figure 30.31a). Plants can also use hydraulics to generate enough force to split rocks and buckle sidewalks (Figure 30.31b). Plants achieve this because of water potential.

(a) (b)

Figure 30.31 With heights nearing 116 meters, (a) coastal redwoods (*Sequoia sempervirens*) are the tallest trees in the world. Plant roots can easily generate enough force to (b) buckle and break concrete sidewalks, much to the dismay of homeowners and city maintenance departments. (credit a: modification of work by Bernt Rostad; credit b: modification of work by Pedestrians Educating Drivers on Safety, Inc.)

Water potential is a measure of the potential energy in water. Plant physiologists are not interested in the energy in any one particular aqueous system, but are very interested in water movement between two systems. In practical terms, therefore, water potential is the difference in potential energy between a given water sample and pure water (at atmospheric pressure and ambient temperature). Water potential is denoted by the Greek letter ψ (*psi*) and is expressed in units of pressure (pressure is a form of energy) called **megapascals** (MPa). The potential of pure water ($\Psi_w^{\text{pure H2O}}$) is, by convenience of definition, designated a value of zero (even though pure water contains plenty of potential energy, that energy is ignored). Water potential values for the water in a plant root, stem, or leaf are therefore expressed relative to $\Psi_w^{\text{pure H2O}}$.

The water potential in plant solutions is influenced by solute concentration, pressure, gravity, and factors called matrix effects. Water potential can be broken down into its individual components using the following equation:

$$\Psi_{\text{system}} = \Psi_{\text{total}} = \Psi_s + \Psi_p + \Psi_g + \Psi_m$$

where Ψ_s, Ψ_p, Ψ_g, and Ψ_m refer to the solute, pressure, gravity, and matric potentials, respectively. "System" can refer to the water potential of the soil water (Ψ^{soil}), root water (Ψ^{root}), stem water (Ψ^{stem}), leaf water (Ψ^{leaf}) or the water in the atmosphere ($\Psi^{\text{atmosphere}}$): whichever aqueous system is under consideration. As the individual components change, they raise or lower the total water potential of a system. When this happens, water moves to equilibrate, moving from the system or compartment with a higher water potential to the system or compartment with a lower water potential. This brings the difference in water potential between the two systems ($\Delta\Psi$) back to zero ($\Delta\Psi = 0$). Therefore, for water to move through the plant from the soil to the air (a process called transpiration), Ψ^{soil} must be > $\Psi^{\text{root}} > \Psi^{\text{stem}} > \Psi^{\text{leaf}} > \Psi^{\text{atmosphere}}$.

Water only moves in response to $\Delta\Psi$, not in response to the individual components. However, because the individual components influence the total Ψ_{system}, by manipulating the individual components (especially Ψ_s), a plant can control water movement.

Solute Potential

Solute potential (Ψ_s), also called osmotic potential, is related to the solute concentration (in molarity). That relationship is given by the van 't Hoff equation: $\Psi_s = -Mi\,RT$; where M is the molar concentration of the solute, *i* is the van 't Hoff factor (the ratio of the amount of particles in the solution to amount of formula units dissolved), R is the ideal gas constant, and T is temperature in Kelvin degrees. The solute potential is negative in a plant cell and zero in distilled water. Typical values for cell cytoplasm are −0.5 to −1.0 MPa. Solutes reduce water potential (resulting in a negative Ψ_w) by consuming some of the potential energy available in the water. Solute molecules can dissolve in water because water molecules can bind to them via hydrogen bonds; a hydrophobic molecule like oil, which cannot bind to water, cannot go into solution. The energy in the hydrogen bonds between solute molecules and water is no longer available to do work in the system because it is tied up in the bond. In other words, the amount of available potential energy is reduced when solutes are added to an aqueous system. Thus, Ψ_s decreases with increasing solute concentration. Because Ψ_s is one of the four components of Ψ_{system} or Ψ_{total}, a decrease in Ψ_s will cause a decrease in Ψ_{total}. The internal water potential of a plant cell is more negative than pure water because of the cytoplasm's high solute content (Figure 30.32). Because of this difference in water potential water will move from the soil into a plant's root cells via the process of osmosis. This is why solute potential is sometimes called osmotic potential.

Plant cells can metabolically manipulate Ψ_s (and by extension, Ψ_{total}) by adding or removing solute molecules. Therefore, plants have control over Ψ_{total} via their ability to exert metabolic control over Ψ_s.

VISUAL CONNECTION

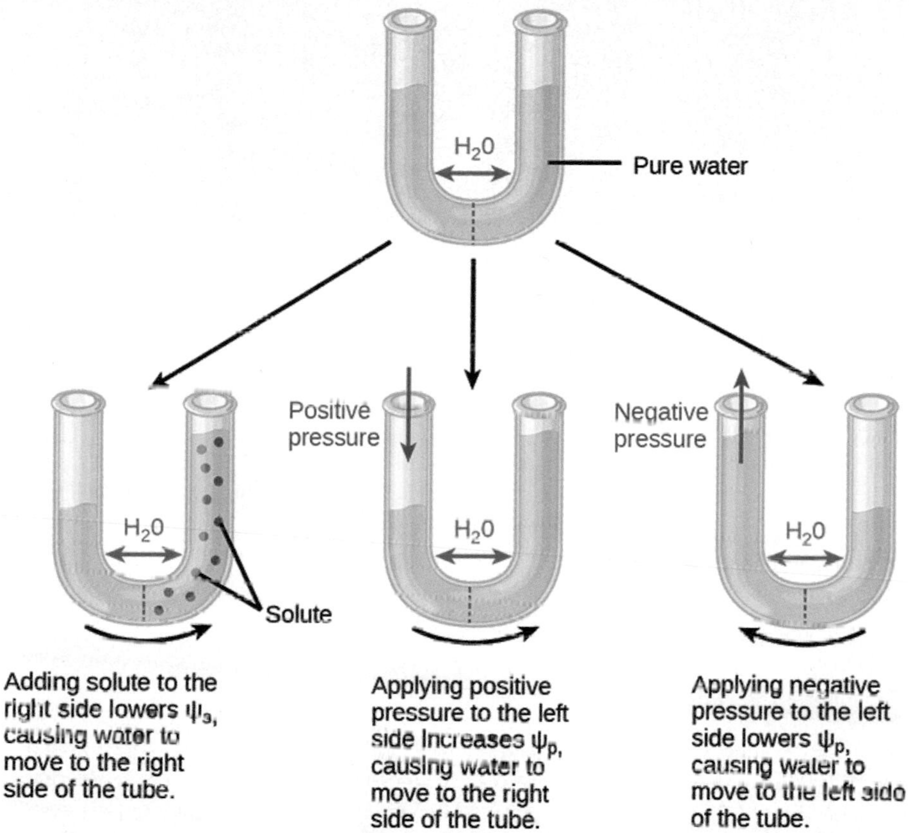

Figure 30.32 In this example with a semipermeable membrane between two aqueous systems, water will move from a region of higher to lower water potential until equilibrium is reached. Solutes (Ψ_s), pressure (Ψ_p), and gravity (Ψ_g) influence total water potential for each side of the tube (Ψ_{total} $^{right\ or\ left}$), and therefore, the difference between Ψ_{total} on each side ($\Delta\Psi$). (Ψ_m, the potential due to interaction of water with solid substrates, is ignored in this example because glass is not especially hydrophilic). Water moves in response to the difference in water potential between two systems (the left and right sides of the tube).

Positive water potential is placed on the left side of the tube by increasing Ψ_p such that the water level rises on the right side. Could you equalize the water level on each side of the tube by adding solute, and if so, how?

Pressure Potential

Pressure potential (Ψ_p), also called turgor potential, may be positive or negative (Figure 30.32). Because pressure is an expression of energy, the higher the pressure, the more potential energy in a system, and vice versa. Therefore, a positive Ψp (compression) increases Ψ_{total}, and a negative Ψ_p (tension) decreases Ψ_{total}. Positive pressure inside cells is contained by the cell wall, producing turgor pressure. Pressure potentials are typically around 0.6–0.8 MPa, but can reach as high as 1.5 MPa in a well-watered plant. A Ψ_p of 1.5 MPa equates to 210 pounds per square inch (1.5 MPa x 140 lb/in^{-2} MPa^{-1} = 210 lb/in^{-2}). As a comparison, most automobile tires are kept at a pressure of 30–34 psi. An example of the effect of turgor pressure is the wilting of leaves and their restoration after the plant has been watered (Figure 30.33). Water is lost from the leaves via transpiration (approaching Ψ_p = 0 MPa at the wilting point) and restored by uptake via the roots.

A plant can manipulate Ψ_p via its ability to manipulate Ψ_s and by the process of osmosis. If a plant cell increases the cytoplasmic solute concentration, Ψ_s will decline, Ψ_{total} will decline, the $\Delta\Psi$ between the cell and the surrounding tissue will decline, water will move into the cell by osmosis, and Ψ_p will increase. Ψ_p is also under indirect plant control via the opening and closing of stomata. Stomatal openings allow water to evaporate from the leaf, reducing Ψ_p and Ψ_{total} of the leaf and increasing Ψ between the water in the leaf and the petiole, thereby allowing water to flow from the petiole into the leaf.

Figure 30.33 When (a) total water potential (Ψ_{total}) is lower outside the cells than inside, water moves out of the cells and the plant wilts. When (b) the total water potential is higher outside the plant cells than inside, water moves into the cells, resulting in turgor pressure (Ψ_p) and keeping the plant erect. (credit: modification of work by Victor M. Vicente Selvas)

Gravity Potential

Gravity potential (Ψ_g) is always negative to zero in a plant with no height. It always removes or consumes potential energy from the system. The force of gravity pulls water downwards to the soil, reducing the total amount of potential energy in the water in the plant (Ψ_{total}). The taller the plant, the taller the water column, and the more influential Ψ_g becomes. On a cellular scale and in short plants, this effect is negligible and easily ignored. However, over the height of a tall tree like a giant coastal redwood, the gravitational pull of −0.1 MPa m^{-1} is equivalent to an extra 1 MPa of resistance that must be overcome for water to reach the leaves of the tallest trees. Plants are unable to manipulate Ψ_g.

Matric Potential

Matric potential (Ψ_m) is always negative to zero. In a dry system, it can be as low as −2 MPa in a dry seed, and it is zero in a water-saturated system. The binding of water to a matrix always removes or consumes potential energy from the system. Ψ_m is similar to solute potential because it involves tying up the energy in an aqueous system by forming hydrogen bonds between the water and some other component. However, in solute potential, the other components are soluble, hydrophilic solute molecules, whereas in Ψ_m, the other components are insoluble, hydrophilic molecules of the plant cell wall. Every plant cell has a cellulosic cell wall and the cellulose in the cell walls is hydrophilic, producing a matrix for adhesion of water: hence the name matric potential. Ψ_m is very large (negative) in dry tissues such as seeds or drought-affected soils. However, it quickly goes to zero as the seed takes up water or the soil hydrates. Ψ_m cannot be manipulated by the plant and is typically ignored in well-watered roots, stems, and leaves.

Movement of Water and Minerals in the Xylem

Solutes, pressure, gravity, and matric potential are all important for the transport of water in plants. Water moves from an area of higher total water potential (higher Gibbs free energy) to an area of lower total water potential. Gibbs free energy is the energy associated with a chemical reaction that can be used to do work. This is expressed as $\Delta\Psi$.

Transpiration is the loss of water from the plant through evaporation at the leaf surface. It is the main driver of water movement in the xylem. Transpiration is caused by the evaporation of water at the leaf–atmosphere interface; it creates negative pressure (tension) equivalent to −2 MPa at the leaf surface. This value varies greatly depending on the vapor pressure deficit, which can be negligible at high relative humidity (RH) and substantial at low RH. Water from the roots is pulled up by this tension. At night, when stomata shut and transpiration stops, the water is held in the stem and leaf by the adhesion of water to the cell walls of the xylem vessels and tracheids, and the cohesion of water molecules to each other. This is called the cohesion–tension theory of sap ascent.

Inside the leaf at the cellular level, water on the surface of mesophyll cells saturates the cellulose microfibrils of the primary cell wall. The leaf contains many large intercellular air spaces for the exchange of oxygen for carbon dioxide, which is required for photosynthesis. The wet cell wall is exposed to this leaf internal air space, and the water on the surface of the cells evaporates into the air spaces, decreasing the thin film on the surface of the mesophyll cells. This decrease creates a greater tension on the water in the mesophyll cells (Figure 30.34), thereby increasing the pull on the water in the xylem vessels. The xylem vessels and tracheids are structurally adapted to cope with large changes in pressure. Rings in the vessels maintain their tubular shape, much like the rings on a vacuum cleaner hose keep the hose open while it is under pressure. Small perforations between vessel elements reduce the number and size of gas bubbles that can form via a process called cavitation. The formation of gas bubbles in xylem interrupts the continuous stream of water from the base to the top of the plant, causing a break termed an embolism in the flow of xylem sap. The taller the tree, the greater the tension forces needed to pull water, and the more cavitation events. In larger trees, the resulting embolisms can plug xylem vessels, making them nonfunctional.

VISUAL CONNECTION

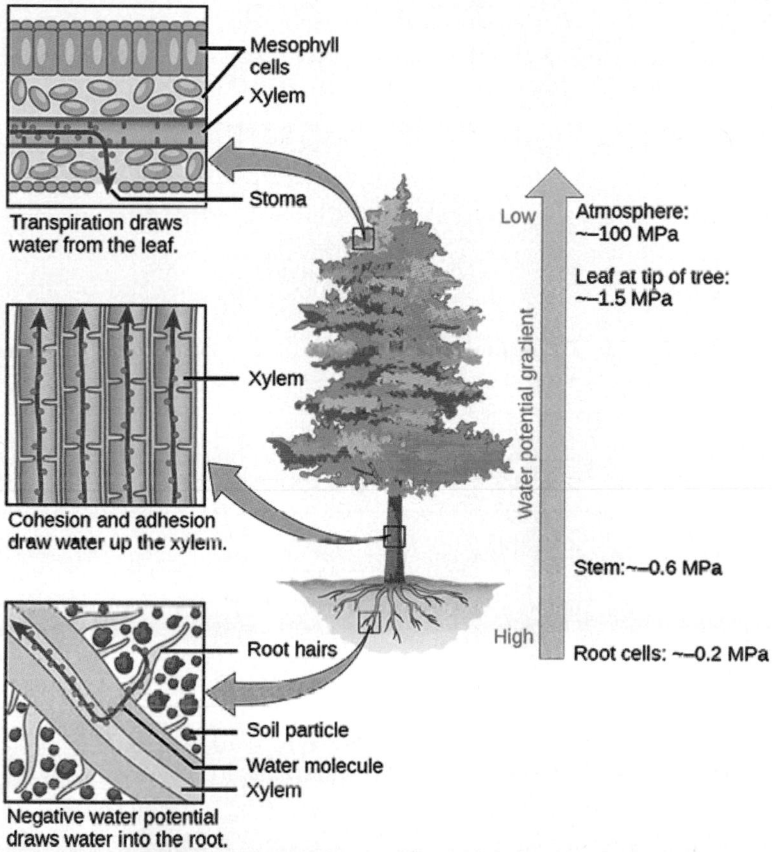

Figure 30.34 The cohesion–tension theory of sap ascent is shown. Evaporation from the mesophyll cells produces a negative water potential gradient that causes water to move upwards from the roots through the xylem.

Which of the following statements is false?

a. Negative water potential draws water into the root hairs. Cohesion and adhesion draw water up the xylem. Transpiration draws water from the leaf.
b. Negative water potential draws water into the root hairs. Cohesion and adhesion draw water up the phloem. Transpiration draws water from the leaf.
c. Water potential decreases from the roots to the top of the plant.
d. Water enters the plants through root hairs and exits through stoma.

Transpiration—the loss of water vapor to the atmosphere through stomata—is a passive process, meaning that metabolic energy in the form of ATP is not required for water movement. The energy driving transpiration is the difference in energy between the water in the soil and the water in the atmosphere. However, transpiration is tightly controlled.

Control of Transpiration

The atmosphere to which the leaf is exposed drives transpiration, but also causes massive water loss from the plant. Up to 90 percent of the water taken up by roots may be lost through transpiration.

Leaves are covered by a waxy **cuticle** on the outer surface that prevents the loss of water. Regulation of transpiration, therefore, is achieved primarily through the opening and closing of stomata on the leaf surface. Stomata are surrounded by two specialized cells called guard cells, which open and close in response to environmental cues such as light intensity and quality, leaf water status, and carbon dioxide concentrations. Stomata must open to allow air containing carbon dioxide and oxygen to diffuse into the leaf for photosynthesis and respiration. When stomata are open, however, water vapor is lost to the external environment, increasing the rate of transpiration. Therefore, plants must maintain a balance between efficient photosynthesis and water loss.

Plants have evolved over time to adapt to their local environment and reduce transpiration (Figure 30.35). Desert plant (xerophytes) and plants that grow on other plants (epiphytes) have limited access to water. Such plants usually have a much thicker waxy cuticle than those growing in more moderate, well-watered environments (mesophytes). Aquatic plants (hydrophytes) also have their own set of anatomical and morphological leaf adaptations.

Figure 30.35 Plants are suited to their local environment. (a) Xerophytes, like this prickly pear cactus (*Opuntia sp.*) and (b) epiphytes such as this tropical *Aeschynanthus perrottetii* have adapted to very limited water resources. The leaves of a prickly pear are modified into spines, which lowers the surface-to-volume ratio and reduces water loss. Photosynthesis takes place in the stem, which also stores water. (b) *A. perottetii* leaves have a waxy cuticle that prevents water loss. (c) Goldenrod (*Solidago sp.*) is a mesophyte, well suited for moderate environments. (d) Hydrophytes, like this fragrant water lily (*Nymphaea odorata*), are adapted to thrive in aquatic environments. (credit a: modification of work by Jon Sullivan; credit b: modification of work by L. Shyamal/Wikimedia Commons; credit c: modification of work by

Huw Williams; credit d: modification of work by Jason Hollinger)

Xerophytes and epiphytes often have a thick covering of trichomes or of stomata that are sunken below the leaf's surface. Trichomes are specialized hair-like epidermal cells that secrete oils and substances. These adaptations impede air flow across the stomatal pore and reduce transpiration. Multiple epidermal layers are also commonly found in these types of plants.

Transportation of Photosynthates in the Phloem

Plants need an energy source to grow. In seeds and bulbs, food is stored in polymers (such as starch) that are converted by metabolic processes into sucrose for newly developing plants. Once green shoots and leaves are growing, plants are able to produce their own food by photosynthesizing. The products of photosynthesis are called photosynthates, which are usually in the form of simple sugars such as sucrose.

Structures that produce photosynthates for the growing plant are referred to as **sources**. Sugars produced in sources, such as leaves, need to be delivered to growing parts of the plant via the phloem in a process called **translocation**. The points of sugar delivery, such as roots, young shoots, and developing seeds, are called **sinks**. Seeds, tubers, and bulbs can be either a source or a sink, depending on the plant's stage of development and the season.

The products from the source are usually translocated to the nearest sink through the phloem. For example, the highest leaves will send photosynthates upward to the growing shoot tip, whereas lower leaves will direct photosynthates downward to the roots. Intermediate leaves will send products in both directions, unlike the flow in the xylem, which is always unidirectional (soil to leaf to atmosphere). The pattern of photosynthate flow changes as the plant grows and develops. Photosynthates are directed primarily to the roots early on, to shoots and leaves during vegetative growth, and to seeds and fruits during reproductive development. They are also directed to tubers for storage.

Translocation: Transport from Source to Sink

Photosynthates, such as sucrose, are produced in the mesophyll cells of photosynthesizing leaves. From there they are translocated through the phloem to where they are used or stored. Mesophyll cells are connected by cytoplasmic channels called plasmodesmata. Photosynthates move through these channels to reach phloem sieve-tube elements (STEs) in the vascular bundles. From the mesophyll cells, the photosynthates are loaded into the phloem STEs. The sucrose is actively transported against its concentration gradient (a process requiring ATP) into the phloem cells using the electrochemical potential of the proton gradient. This is coupled to the uptake of sucrose with a carrier protein called the sucrose-H^+ symporter.

Phloem STEs have reduced cytoplasmic contents, and are connected by a sieve plate with pores that allow for pressure-driven bulk flow, or translocation, of phloem sap. Companion cells are associated with STEs. They assist with metabolic activities and produce energy for the STEs (Figure 30.36).

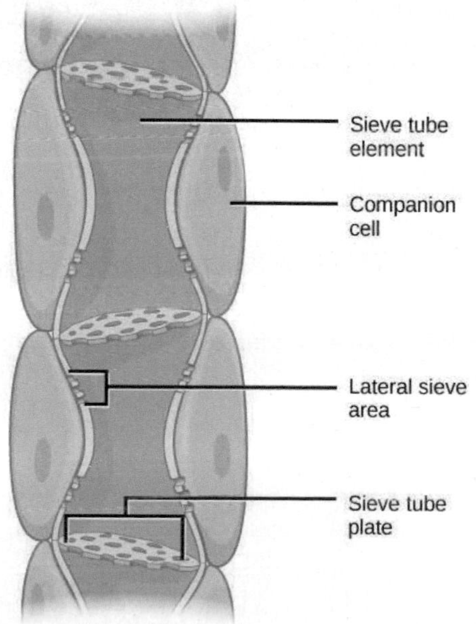

Figure 30.36 Phloem is comprised of cells called sieve-tube elements. Phloem sap travels through perforations called sieve tube plates.

Neighboring companion cells carry out metabolic functions for the sieve-tube elements and provide them with energy. Lateral sieve areas connect the sieve-tube elements to the companion cells.

Once in the phloem, the photosynthates are translocated to the closest sink. Phloem sap is an aqueous solution that contains up to 30 percent sugar, minerals, amino acids, and plant growth regulators. The high percentage of sugar decreases Ψ_s, which decreases the total water potential and causes water to move by osmosis from the adjacent xylem into the phloem tubes, thereby increasing pressure. This increase in total water potential causes the bulk flow of phloem from source to sink (Figure 30.37). Sucrose concentration in the sink cells is lower than in the phloem STEs because the sink sucrose has been metabolized for growth, or converted to starch for storage or other polymers, such as cellulose, for structural integrity. Unloading at the sink end of the phloem tube occurs by either diffusion or active transport of sucrose molecules from an area of high concentration to one of low concentration. Water diffuses from the phloem by osmosis and is then transpired or recycled via the xylem back into the phloem sap.

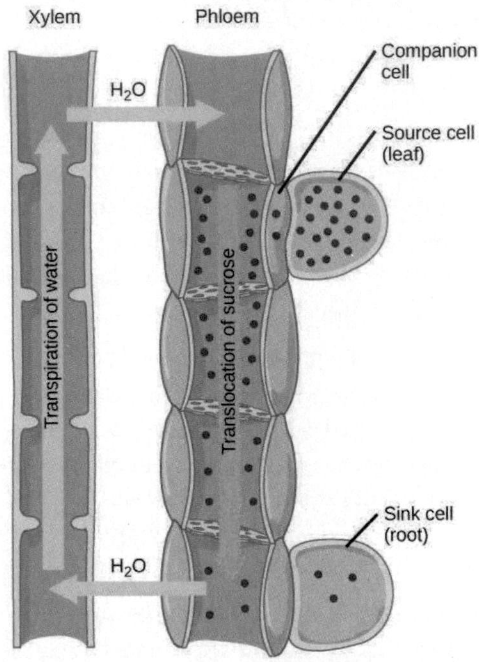

Figure 30.37 Sucrose is actively transported from source cells into companion cells and then into the sieve-tube elements. This reduces the water potential, which causes water to enter the phloem from the xylem. The resulting positive pressure forces the sucrose-water mixture down toward the roots, where sucrose is unloaded. Transpiration causes water to return to the leaves through the xylem vessels.

30.6 Plant Sensory Systems and Responses

By the end of this section, you will be able to do the following:

- Describe how red and blue light affect plant growth and metabolic activities
- Discuss gravitropism
- Understand how hormones affect plant growth and development
- Describe thigmotropism, thigmonastism, and thigmogenesis
- Explain how plants defend themselves from predators and respond to wounds

Animals can respond to environmental factors by moving to a new location. Plants, however, are rooted in place and must respond to the surrounding environmental factors. Plants have sophisticated systems to detect and respond to light, gravity, temperature, and physical touch. Receptors sense environmental factors and relay the information to effector systems—often through intermediate chemical messengers—to bring about plant responses.

Plant Responses to Light

Plants have a number of sophisticated uses for light that go far beyond their ability to photosynthesize low-molecular-weight sugars using only carbon dioxide, light, and water. **Photomorphogenesis** is the growth and development of plants in response to light. It allows plants to optimize their use of light and space. **Photoperiodism** is the ability to use light to track time. Plants can tell the time of day and time of year by sensing and using various wavelengths of sunlight. **Phototropism** is a directional response that allows plants to grow towards, or even away from, light.

The sensing of light in the environment is important to plants; it can be crucial for competition and survival. The response of plants to light is mediated by different photoreceptors, which are comprised of a protein covalently bonded to a light-absorbing pigment called a **chromophore**. Together, the two are called a chromoprotein.

The red/far-red and violet-blue regions of the visible light spectrum trigger structural development in plants. Sensory photoreceptors absorb light in these particular regions of the visible light spectrum because of the quality of light available in the daylight spectrum. In terrestrial habitats, light absorption by chlorophylls peaks in the blue and red regions of the spectrum. As light filters through the canopy and the blue and red wavelengths are absorbed, the spectrum shifts to the far-red end, shifting the plant community to those plants better adapted to respond to far-red light. Blue-light receptors allow plants to gauge the direction and abundance of sunlight, which is rich in blue–green emissions. Water absorbs red light, which makes the detection of blue light essential for algae and aquatic plants.

The Phytochrome System and the Red/Far-Red Response

The **phytochromes** are a family of chromoproteins with a linear tetrapyrrole chromophore, similar to the ringed tetrapyrrole light-absorbing head group of chlorophyll. Phytochromes have two photo-interconvertible forms: Pr and Pfr. Pr absorbs red light (~667 nm) and is immediately converted to Pfr. Pfr absorbs far-red light (~730 nm) and is quickly converted back to Pr. Absorption of red or far-red light causes a massive change to the shape of the chromophore, altering the conformation and activity of the phytochrome protein to which it is bound. Pfr is the physiologically active form of the protein; therefore, exposure to red light yields physiological activity. Exposure to far-red light inhibits phytochrome activity. Together, the two forms represent the phytochrome system (Figure 30.38).

The phytochrome system acts as a biological light switch. It monitors the level, intensity, duration, and color of environmental light. The effect of red light is reversible by immediately shining far-red light on the sample, which converts the chromoprotein to the inactive Pr form. Additionally, Pfr can slowly revert to Pr in the dark, or break down over time. In all instances, the physiological response induced by red light is reversed. The active form of phytochrome (Pfr) can directly activate other molecules in the cytoplasm, or it can be trafficked to the nucleus, where it directly activates or represses specific gene expression.

Once the phytochrome system evolved, plants adapted it to serve a variety of needs. Unfiltered, full sunlight contains much more red light than far-red light. Because chlorophyll absorbs strongly in the red region of the visible spectrum, but not in the far-red region, any plant in the shade of another plant on the forest floor will be exposed to red-depleted, far-red-enriched light. The preponderance of far-red light converts phytochrome in the shaded leaves to the Pr (inactive) form, slowing growth. The nearest non-shaded (or even less-shaded) areas on the forest floor have more red light; leaves exposed to these areas sense the red light, which activates the Pfr form and induces growth. In short, plant shoots use the phytochrome system to grow away from shade and towards light. Because competition for light is so fierce in a dense plant community, the evolutionary advantages of the phytochrome system are obvious.

In seeds, the phytochrome system is not used to determine direction and quality of light (shaded versus unshaded). Instead, is it used merely to determine if there is any light at all. This is especially important in species with very small seeds, such as lettuce. Because of their size, lettuce seeds have few food reserves. Their seedlings cannot grow for long before they run out of fuel. If they germinated even a centimeter under the soil surface, the seedling would never make it into the sunlight and would die. In the dark, phytochrome is in the Pr (inactive form) and the seed will not germinate; it will only germinate if exposed to light at the surface of the soil. Upon exposure to light, Pr is converted to Pfr and germination proceeds.

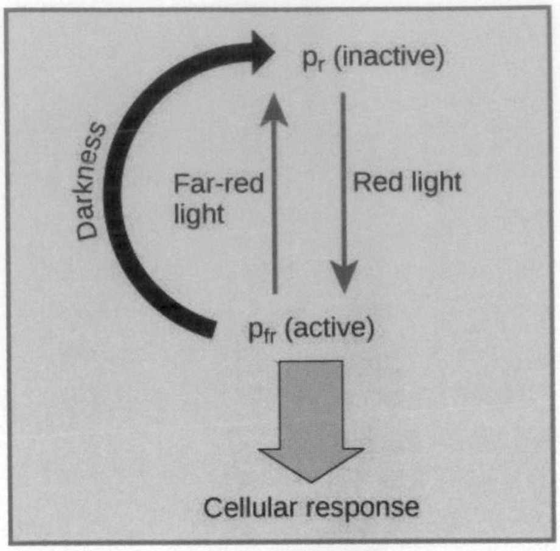

Figure 30.38 The biologically inactive form of phytochrome (Pr) is converted to the biologically active form Pfr under illumination with red light. Far-red light and darkness convert the molecule back to the inactive form.

Plants also use the phytochrome system to sense the change of season. Photoperiodism is a biological response to the timing and duration of day and night. It controls flowering, setting of winter buds, and vegetative growth. Detection of seasonal changes is crucial to plant survival. Although temperature and light intensity influence plant growth, they are not reliable indicators of season because they may vary from one year to the next. Day length is a better indicator of the time of year.

As stated above, unfiltered sunlight is rich in red light but deficient in far-red light. Therefore, at dawn, all the phytochrome molecules in a leaf quickly convert to the active Pfr form, and remain in that form until sunset. In the dark, the Pfr form takes hours to slowly revert back to the Pr form. If the night is long (as in winter), all of the Pfr form reverts. If the night is short (as in summer), a considerable amount of Pfr may remain at sunrise. By sensing the Pr/Pfr ratio at dawn, a plant can determine the length of the day/night cycle. In addition, leaves retain that information for several days, allowing a comparison between the length of the previous night and the preceding several nights. Shorter nights indicate springtime to the plant; when the nights become longer, autumn is approaching. This information, along with sensing temperature and water availability, allows plants to determine the time of the year and adjust their physiology accordingly. Short-day (long-night) plants use this information to flower in the late summer and early fall, when nights exceed a critical length (often eight or fewer hours). Long-day (short-night) plants flower during the spring, when darkness is less than a critical length (often eight to 15 hours). Not all plants use the phytochrome system in this way. Flowering in day-neutral plants is not regulated by daylength.

CAREER CONNECTION

Horticulturalist

The word "horticulturist" comes from the Latin words for garden (*hortus*) and culture (*cultura*). This career has been revolutionized by progress made in the understanding of plant responses to environmental stimuli. Growers of crops, fruit, vegetables, and flowers were previously constrained by having to time their sowing and harvesting according to the season. Now, horticulturists can manipulate plants to increase leaf, flower, or fruit production by understanding how environmental factors affect plant growth and development.

Greenhouse management is an essential component of a horticulturist's education. To lengthen the night, plants are covered with a blackout shade cloth. Long-day plants are irradiated with red light in winter to promote early flowering. For example, fluorescent (cool white) light high in blue wavelengths encourages leafy growth and is excellent for starting seedlings. Incandescent lamps (standard light bulbs) are rich in red light, and promote flowering in some plants. The timing of fruit ripening can be increased or delayed by applying plant hormones. Recently, considerable progress has been made in the development of plant breeds that are suited to different climates and resistant to pests and transportation damage. Both crop yield and quality have increased as a result of practical applications of the knowledge of plant responses to external stimuli and hormones.

Horticulturists find employment in private and governmental laboratories, greenhouses, botanical gardens, and in the production or research fields. They improve crops by applying their knowledge of genetics and plant physiology. To prepare for a horticulture career, students take classes in botany, plant physiology, plant pathology, landscape design, and plant breeding. To complement these traditional courses, horticulture majors add studies in economics, business, computer science, and communications.

The Blue Light Responses

Phototropism—the directional bending of a plant toward or away from a light source—is a response to blue wavelengths of light. Positive phototropism is growth towards a light source (Figure 30.39), while negative phototropism (also called skototropism) is growth away from light.

The aptly-named **phototropins** are protein-based receptors responsible for mediating the phototropic response. Like all plant photoreceptors, phototropins consist of a protein portion and a light-absorbing portion, called the chromophore. In phototropins, the chromophore is a covalently-bound molecule of flavin; hence, phototropins belong to a class of proteins called flavoproteins.

Other responses under the control of phototropins are leaf opening and closing, chloroplast movement, and the opening of stomata. However, of all responses controlled by phototropins, phototropism has been studied the longest and is the best understood.

In their 1880 treatise *The Power of Movements in Plants*, Charles Darwin and his son Francis first described phototropism as the bending of seedlings toward light. Darwin observed that light was perceived by the tip of the plant (the apical meristem), but that the response (bending) took place in a different part of the plant. They concluded that the signal had to travel from the apical meristem to the base of the plant.

Figure 30.39 Azure bluets (*Houstonia caerulea*) display a phototropic response by bending toward the light. (credit: Cory Zanker)

In 1913, Peter Boysen-Jensen demonstrated that a chemical signal produced in the plant tip was responsible for the bending at the base. He cut off the tip of a seedling, covered the cut section with a layer of gelatin, and then replaced the tip. The seedling bent toward the light when illuminated. However, when impermeable mica flakes were inserted between the tip and the cut base, the seedling did not bend. A refinement of the experiment showed that the signal traveled on the shaded side of the seedling. When the mica plate was inserted on the illuminated side, the plant did bend towards the light. Therefore, the chemical signal was a growth stimulant because the phototropic response involved faster cell elongation on the shaded side than on the illuminated side. We now know that as light passes through a plant stem, it is diffracted and generates phototropin activation across the stem. Most activation occurs on the lit side, causing the plant hormone indole acetic acid (IAA) to accumulate on the shaded side. Stem cells elongate under influence of IAA.

Cryptochromes are another class of blue-light absorbing photoreceptors that also contain a flavin-based chromophore. Cryptochromes set the plants' 24-hour activity cycle, also know as its circadian rhythem, using blue light cues. There is some evidence that cryptochromes work together with phototropins to mediate the phototropic response.

LINK TO LEARNING

Use the navigation menu in the left panel of this website (http://openstax.org/l/plnts_n_motion) to view images of plants in motion.

Plant Responses to Gravity

Whether or not they germinate in the light or in total darkness, shoots usually sprout up from the ground, and roots grow downward into the ground. A plant laid on its side in the dark will send shoots upward when given enough time. Gravitropism ensures that roots grow into the soil and that shoots grow toward sunlight. Growth of the shoot apical tip upward is called **negative gravitropism**, whereas growth of the roots downward is called **positive gravitropism**.

Amyloplasts (also known as **statoliths**) are specialized plastids that contain starch granules and settle downward in response to gravity. Amyloplasts are found in shoots and in specialized cells of the root cap. When a plant is tilted, the statoliths drop to the new bottom cell wall. A few hours later, the shoot or root will show growth in the new vertical direction.

The mechanism that mediates gravitropism is reasonably well understood. When amyloplasts settle to the bottom of the gravity-sensing cells in the root or shoot, they physically contact the endoplasmic reticulum (ER), causing the release of calcium ions from inside the ER. This calcium signaling in the cells causes polar transport of the plant hormone IAA to the bottom of the cell. In roots, a high concentration of IAA inhibits cell elongation. The effect slows growth on the lower side of the root, while cells develop normally on the upper side. IAA has the opposite effect in shoots, where a higher concentration at the lower side of the shoot stimulates cell expansion, causing the shoot to grow up. After the shoot or root begin to grow vertically, the amyloplasts return to their normal position. Other hypotheses—involving the entire cell in the gravitropism effect—have been proposed to explain why some mutants that lack amyloplasts may still exhibit a weak gravitropic response.

Growth Responses

A plant's sensory response to external stimuli relies on chemical messengers (hormones). Plant hormones affect all aspects of plant life, from flowering to fruit setting and maturation, and from phototropism to leaf fall. Potentially every cell in a plant can produce plant hormones. They can act in their cell of origin or be transported to other portions of the plant body, with many plant responses involving the synergistic or antagonistic interaction of two or more hormones. In contrast, animal hormones are produced in specific glands and transported to a distant site for action, and they act alone.

Plant hormones are a group of unrelated chemical substances that affect plant morphogenesis. Five major plant hormones are traditionally described: auxins (particularly IAA), cytokinins, gibberellins, ethylene, and abscisic acid. In addition, other nutrients and environmental conditions can be characterized as growth factors.

Auxins

The term auxin is derived from the Greek word *auxein*, which means "to grow." **Auxins** are the main hormones responsible for cell elongation in phototropism and gravitropism. They also control the differentiation of meristem into vascular tissue, and promote leaf development and arrangement. While many synthetic auxins are used as herbicides, IAA is the only naturally occurring auxin that shows physiological activity. Apical dominance—the inhibition of lateral bud formation—is triggered by auxins produced in the apical meristem. Flowering, fruit setting and ripening, and inhibition of **abscission** (leaf falling) are other plant responses under the direct or indirect control of auxins. Auxins also act as a relay for the effects of the blue light and red/far-red responses.

Commercial use of auxins is widespread in plant nurseries and for crop production. IAA is used as a rooting hormone to promote growth of adventitious roots on cuttings and detached leaves. Applying synthetic auxins to tomato plants in greenhouses promotes normal fruit development. Outdoor application of auxin promotes synchronization of fruit setting and dropping to coordinate the harvesting season. Fruits such as seedless cucumbers can be induced to set fruit by treating unfertilized plant flowers with auxins.

Cytokinins

The effect of cytokinins was first reported when it was found that adding the liquid endosperm of coconuts to developing plant embryos in culture stimulated their growth. The stimulating growth factor was found to be **cytokinin**, a hormone that promotes cytokinesis (cell division). Almost 200 naturally occurring or synthetic cytokinins are known to date. Cytokinins are most abundant in growing tissues, such as roots, embryos, and fruits, where cell division is occurring. Cytokinins are known to delay senescence in leaf tissues, promote mitosis, and stimulate differentiation of the meristem in shoots and roots. Many effects on plant development are under the influence of cytokinins, either in conjunction with auxin or another hormone. For example, apical dominance seems to result from a balance between auxins that inhibit lateral buds, and cytokinins that promote bushier growth.

Gibberellins

Gibberellins (GAs) are a group of about 125 closely related plant hormones that stimulate shoot elongation, seed germination, and fruit and flower maturation. GAs are synthesized in the root and stem apical meristems, young leaves, and seed embryos. In urban areas, GA antagonists are sometimes applied to trees under power lines to control growth and reduce the frequency of pruning.

GAs break dormancy (a state of inhibited growth and development) in the seeds of plants that require exposure to cold or light to germinate. Abscisic acid is a strong antagonist of GA action. Other effects of GAs include gender expression, seedless fruit development, and the delay of senescence in leaves and fruit. Seedless grapes are obtained through standard breeding methods and contain inconspicuous seeds that fail to develop. Because GAs are produced by the seeds, and because fruit development and stem elongation are under GA control, these varieties of grapes would normally produce small fruit in compact clusters. Maturing grapes are routinely treated with GA to promote larger fruit size, as well as looser bunches (longer stems), which reduces the instance of mildew infection (Figure 30.40).

Figure 30.40 In grapes, application of gibberellic acid increases the size of fruit and loosens clustering. (credit: Bob Nichols, USDA)

Abscisic Acid

The plant hormone **abscisic acid** (ABA) was first discovered as the agent that causes the abscission or dropping of cotton bolls. However, more recent studies indicate that ABA plays only a minor role in the abscission process. ABA accumulates as a response to stressful environmental conditions, such as dehydration, cold temperatures, or shortened day lengths. Its activity counters many of the growth-promoting effects of GAs and auxins. ABA inhibits stem elongation and induces dormancy in lateral buds.

ABA induces dormancy in seeds by blocking germination and promoting the synthesis of storage proteins. Plants adapted to temperate climates require a long period of cold temperature before seeds germinate. This mechanism protects young plants from sprouting too early during unseasonably warm weather in winter. As the hormone gradually breaks down over winter, the seed is released from dormancy and germinates when conditions are favorable in spring. Another effect of ABA is to promote the development of winter buds; it mediates the conversion of the apical meristem into a dormant bud. Low soil moisture causes an increase in ABA, which causes stomata to close, reducing water loss in winter buds.

Ethylene

Ethylene is associated with fruit ripening, flower wilting, and leaf fall. Ethylene is unusual because it is a volatile gas (C_2H_4). Hundreds of years ago, when gas street lamps were installed in city streets, trees that grew close to lamp posts developed twisted, thickened trunks and shed their leaves earlier than expected. These effects were caused by ethylene volatilizing from the lamps.

Aging tissues (especially senescing leaves) and nodes of stems produce ethylene. The best-known effect of the hormone, however, is the promotion of fruit ripening. Ethylene stimulates the conversion of starch and acids to sugars. Some people store unripe fruit, such as avocadoes, in a sealed paper bag to accelerate ripening; the gas released by the first fruit to mature will speed up the maturation of the remaining fruit. Ethylene also triggers leaf and fruit abscission, flower fading and dropping, and promotes germination in some cereals and sprouting of bulbs and potatoes.

Ethylene is widely used in agriculture. Commercial fruit growers control the timing of fruit ripening with application of the gas. Horticulturalists inhibit leaf dropping in ornamental plants by removing ethylene from greenhouses using fans and ventilation.

Nontraditional Hormones

Recent research has discovered a number of compounds that also influence plant development. Their roles are less understood than the effects of the major hormones described so far.

Jasmonates play a major role in defense responses to herbivory. Their levels increase when a plant is wounded by a predator, resulting in an increase in toxic secondary metabolites. They contribute to the production of volatile compounds that attract natural enemies of predators. For example, chewing of tomato plants by caterpillars leads to an increase in jasmonic acid levels, which in turn triggers the release of volatile compounds that attract predators of the pest.

Oligosaccharins also play a role in plant defense against bacterial and fungal infections. They act locally at the site of injury, and can also be transported to other tissues. **Strigolactones** promote seed germination in some species and inhibit lateral apical development in the absence of auxins. Strigolactones also play a role in the establishment of mycorrhizae, a mutualistic association of plant roots and fungi. Brassinosteroids are important to many developmental and physiological processes. Signals between these compounds and other hormones, notably auxin and GAs, amplifies their physiological effect. Apical dominance, seed germination, gravitropism, and resistance to freezing are all positively influenced by hormones. Root growth and fruit dropping are inhibited by steroids.

Plant Responses to Wind and Touch

The shoot of a pea plant winds around a trellis, while a tree grows on an angle in response to strong prevailing winds. These are examples of how plants respond to touch or wind.

The movement of a plant subjected to constant directional pressure is called **thigmotropism**, from the Greek words *thigma* meaning "touch," and *tropism* implying "direction." Tendrils are one example of this. The meristematic region of tendrils is very touch sensitive; light touch will evoke a quick coiling response. Cells in contact with a support surface contract, whereas cells on the opposite side of the support expand (Figure 30.14). Application of jasmonic acid is sufficient to trigger tendril coiling without a mechanical stimulus.

A **thigmonastic** response is a touch response independent of the direction of stimulus Figure 30.24. In the Venus flytrap, two modified leaves are joined at a hinge and lined with thin fork-like tines along the outer edges. Tiny hairs are located inside the trap. When an insect brushes against these trigger hairs, touching two or more of them in succession, the leaves close quickly, trapping the prey. Glands on the leaf surface secrete enzymes that slowly digest the insect. The released nutrients are absorbed by the leaves, which reopen for the next meal.

Thigmomorphogenesis is a slow developmental change in the shape of a plant subjected to continuous mechanical stress. When trees bend in the wind, for example, growth is usually stunted and the trunk thickens. Strengthening tissue, especially xylem, is produced to add stiffness to resist the wind's force. Researchers hypothesize that mechanical strain induces growth and differentiation to strengthen the tissues. Ethylene and jasmonate are likely involved in thigmomorphogenesis.

LINK TO LEARNING

Use the menu at the left to navigate to three short movies: (http://openstax.org/l/nastic_mvmt) a Venus fly trap capturing prey, the progressive closing of sensitive plant leaflets, and the twining of tendrils.

Defense Responses against Herbivores and Pathogens

Plants face two types of enemies: herbivores and pathogens. Herbivores both large and small use plants as food, and actively chew them. Pathogens are agents of disease. These infectious microorganisms, such as fungi, bacteria, and nematodes, live off of the plant and damage its tissues. Plants have developed a variety of strategies to discourage or kill attackers.

The first line of defense in plants is an intact and impenetrable barrier. Bark and the waxy cuticle can protect against predators. Other adaptations against herbivory include thorns, which are modified branches, and spines, which are modified leaves. They discourage animals by causing physical damage and inducing rashes and allergic reactions. A plant's exterior protection can be compromised by mechanical damage, which may provide an entry point for pathogens. If the first line of defense is breached, the plant must resort to a different set of defense mechanisms, such as toxins and enzymes.

Secondary metabolites are compounds that are not directly derived from photosynthesis and are not necessary for respiration or plant growth and development. Many metabolites are toxic, and can even be lethal to animals that ingest them. Some metabolites are alkaloids, which discourage predators with noxious odors (such as the volatile oils of mint and sage) or repellent tastes (like the bitterness of quinine). Other alkaloids affect herbivores by causing either excessive stimulation (caffeine is one example) or the lethargy associated with opioids. Some compounds become toxic after ingestion. For instance, glycol cyanide in the cassava root releases cyanide only upon ingestion; the nearly 500 million humans who rely on cassava for nutrition must be certain to process the root properly before eating.

Mechanical wounding and predator attacks activate defense and protection mechanisms both in the damaged tissue and at sites farther from the injury location. Some defense reactions occur within minutes: others over several hours. The infected and surrounding cells may die, thereby stopping the spread of infection.

Long-distance signaling elicits a systemic response aimed at deterring the predator. As tissue is damaged, jasmonates may promote the synthesis of compounds that are toxic to predators. Jasmonates also elicit the synthesis of volatile compounds that attract parasitoids, which are insects that spend their developing stages in or on another insect, and eventually kill their host. The plant may activate abscission of injured tissue if it is damaged beyond repair.

KEY TERMS

abscisic acid (ABA) plant hormone that induces dormancy in seeds and other organs

abscission physiological process that leads to the fall of a plant organ (such as leaf or petal drop)

adventitious root aboveground root that arises from a plant part other than the radicle of the plant embryo

apical bud bud formed at the tip of the shoot

apical meristem meristematic tissue located at the tips of stems and roots; enables a plant to extend in length

auxin plant hormone that influences cell elongation (in phototropism), gravitropism, apical dominance, and root growth

axillary bud bud located in the axil: the stem area where the petiole connects to the stem

bark tough, waterproof, outer epidermal layer of cork cells

bulb modified underground stem that consists of a large bud surrounded by numerous leaf scales

Casparian strip waxy coating that forces water to cross endodermal plasma membranes before entering the vascular cylinder, instead of moving between endodermal cells

chromophore molecule that absorbs light

collenchyma cell elongated plant cell with unevenly thickened walls; provides structural support to the stem and leaves

companion cell phloem cell that is connected to sieve-tube cells; has large amounts of ribosomes and mitochondria

compound leaf leaf in which the leaf blade is subdivided to form leaflets, all attached to the midrib

corm rounded, fleshy underground stem that contains stored food

cortex ground tissue found between the vascular tissue and the epidermis in a stem or root

cryptochrome protein that absorbs light in the blue and ultraviolet regions of the light spectrum

cuticle waxy protective layer on the leaf surface

cuticle waxy covering on the outside of the leaf and stem that prevents the loss of water

cytokinin plant hormone that promotes cell division

dermal tissue protective plant tissue covering the outermost part of the plant; controls gas exchange

endodermis layer of cells in the root that forms a selective barrier between the ground tissue and the vascular tissue, allowing water and minerals to enter the root while excluding toxins and pathogens

epidermis single layer of cells found in plant dermal tissue; covers and protects underlying tissue

ethylene volatile plant hormone that is associated with fruit ripening, flower wilting, and leaf fall

fibrous root system type of root system in which the roots arise from the base of the stem in a cluster, forming a dense network of roots; found in monocots

gibberellin (GA) plant hormone that stimulates shoot elongation, seed germination, and the maturation and dropping of fruit and flowers

ground tissue plant tissue involved in photosynthesis; provides support, and stores water and sugars

guard cells paired cells on either side of a stoma that control stomatal opening and thereby regulate the movement of gases and water vapor

intercalary meristem meristematic tissue located at nodes and the bases of leaf blades; found only in monocots

internode region between nodes on the stem

jasmonates small family of compounds derived from the fatty acid linoleic acid

lamina leaf blade

lateral meristem meristematic tissue that enables a plant to increase in thickness or girth

lenticel opening on the surface of mature woody stems that facilitates gas exchange

megapascal (MPa) pressure units that measure water potential

meristem plant region of continuous growth

meristematic tissue tissue containing cells that constantly divide; contributes to plant growth

negative gravitropism growth away from Earth's gravity

node point along the stem at which leaves, flowers, or aerial roots originate

oligosaccharin hormone important in plant defenses against bacterial and fungal infections

palmately compound leaf leaf type with leaflets that emerge from a point, resembling the palm of a hand

parenchyma cell most common type of plant cell; found in the stem, root, leaf, and in fruit pulp; site of photosynthesis and starch storage

pericycle outer boundary of the stele from which lateral roots can arise

periderm outermost covering of woody stems; consists of the cork cambium, cork cells, and the phelloderm

permanent tissue plant tissue composed of cells that are no longer actively dividing

petiole stalk of the leaf

photomorphogenesis growth and development of plants in response to light

photoperiodism occurrence of plant processes, such as germination and flowering, according to the time of year

phototropin blue-light receptor that promotes phototropism, stomatal opening and closing, and other responses that promote photosynthesis

phototropism directional bending of a plant toward a light source

phyllotaxy arrangement of leaves on a stem

phytochrome plant pigment protein that exists in two reversible forms (Pr and Pfr) and mediates morphologic changes in response to red light

pinnately compound leaf leaf type with a divided leaf blade consisting of leaflets arranged on both sides of the midrib

pith ground tissue found towards the interior of the vascular tissue in a stem or root

positive gravitropism growth toward Earth's gravitational center

primary growth growth resulting in an increase in length of the stem and the root; caused by cell division in the shoot or root apical meristem

rhizome modified underground stem that grows horizontally to the soil surface and has nodes and internodes

root cap protective cells covering the tip of the growing root

root hair hair-like structure that is an extension of epidermal cells; increases the root surface area and aids in absorption of water and minerals

root system belowground portion of the plant that supports the plant and absorbs water and minerals

runner stolon that runs above the ground and produces new clone plants at nodes

sclerenchyma cell plant cell that has thick secondary walls and provides structural support; usually dead at maturity

secondary growth growth resulting in an increase in thickness or girth; caused by the lateral meristem and cork cambium

sessile leaf without a petiole that is attached directly to the plant stem

shoot system aboveground portion of the plant; consists of nonreproductive plant parts, such as leaves and stems, and reproductive parts, such as flowers and fruits

sieve-tube cell phloem cell arranged end to end to form a sieve tube that transports organic substances such as sugars and amino acids

simple leaf leaf type in which the lamina is completely undivided or merely lobed

sink growing parts of a plant, such as roots and young leaves, which require photosynthate

source organ that produces photosynthate for a plant

statolith (also, **amyloplast**) plant organelle that contains heavy starch granules

stele inner portion of the root containing the vascular tissue; surrounded by the endodermis

stipule small green structure found on either side of the leaf stalk or petiole

stolon modified stem that runs parallel to the ground and can give rise to new plants at the nodes

strigolactone hormone that promotes seed germination in some species and inhibits lateral apical development in the absence of auxins

tap root system type of root system with a main root that grows vertically with few lateral roots; found in dicots

tendril modified stem consisting of slender, twining strands used for support or climbing

thigmomorphogenesis developmental response to touch

thigmonastic directional growth of a plant independent of the direction in which contact is applied

thigmotropism directional growth of a plant in response to constant contact

thorn modified stem branch appearing as a sharp outgrowth that protects the plant

tracheid xylem cell with thick secondary walls that helps transport water

translocation mass transport of photosynthates from source to sink in vascular plants

transpiration loss of water vapor to the atmosphere through stomata

trichome hair-like structure on the epidermal surface

tuber modified underground stem adapted for starch storage; has many adventitious buds

vascular bundle strands of stem tissue made up of xylem and phloem

vascular stele strands of root tissue made up of xylem and phloem

vascular tissue tissue made up of xylem and phloem that transports food and water throughout the plant

venation pattern of veins in a leaf; may be parallel (as in monocots), reticulate (as in dicots), or dichotomous (as in *Gingko biloba*)

vessel element xylem cell that is shorter than a tracheid and has thinner walls

water potential (Ψ_w) the potential energy of a water solution per unit volume in relation to pure water at atmospheric pressure and ambient temperature

whorled pattern of leaf arrangement in which three or more leaves are connected at a node

CHAPTER SUMMARY

30.1 The Plant Body

A vascular plant consists of two organ systems: the shoot system and the root system. The shoot system includes the aboveground vegetative portions (stems and leaves) and reproductive parts (flowers and fruits). The root system supports the plant and is usually underground. A plant is composed of two main types of tissue: meristematic tissue and permanent tissue. Meristematic tissue consists of actively dividing cells found in root and shoot tips. As growth occurs, meristematic tissue differentiates into permanent tissue, which is categorized as either simple or complex. Simple tissues are made up of similar cell types; examples include dermal tissue and ground tissue. Dermal tissue provides the outer covering of the plant. Ground tissue is responsible for photosynthesis; it also supports vascular tissue and may store water and sugars. Complex tissues are made up of different cell types. Vascular tissue, for example, is made up of xylem and phloem cells.

30.2 Stems

The stem of a plant bears the leaves, flowers, and fruits. Stems are characterized by the presence of nodes (the points of attachment for leaves or branches) and internodes (regions between nodes).

Plant organs are made up of simple and complex tissues. The stem has three tissue systems: dermal, vascular, and ground tissue. Dermal tissue is the outer covering of the plant. It contains epidermal cells, stomata, guard cells, and trichomes. Vascular tissue is made up of xylem and phloem tissues and conducts water, minerals, and photosynthetic products. Ground tissue is responsible for photosynthesis and support and is composed of parenchyma, collenchyma, and sclerenchyma cells.

Primary growth occurs at the tips of roots and shoots, causing an increase in length. Woody plants may also exhibit secondary growth, or increase in thickness. In woody plants, especially trees, annual rings may form as growth slows at the end of each season. Some plant species have modified stems that help to store food, propagate new plants, or discourage predators. Rhizomes, corms, stolons, runners, tubers, bulbs, tendrils, and thorns are examples of modified stems.

30.3 Roots

Roots help to anchor a plant, absorb water and minerals, and serve as storage sites for food. Taproots and fibrous roots are the two main types of root systems. In a taproot system, a main root grows vertically downward with a few lateral roots. Fibrous root systems arise at the base of the stem, where a cluster of roots forms a dense network that is shallower than a taproot. The growing root tip is protected by a root cap. The root tip has three main zones: a zone of cell division (cells are actively dividing), a zone of elongation (cells increase in length), and a zone of maturation (cells differentiate to form different kinds of cells). Root vascular tissue conducts water, minerals, and sugars. In some habitats, the roots of certain plants may be modified to form aerial roots or epiphytic roots.

30.4 Leaves

Leaves are the main site of photosynthesis. A typical leaf consists of a lamina (the broad part of the leaf, also called the blade) and a petiole (the stalk that attaches the leaf to a stem). The arrangement of leaves on a stem, known as phyllotaxy, enables maximum exposure to sunlight. Each plant species has a characteristic leaf arrangement and form. The pattern of leaf arrangement may be alternate, opposite, or spiral, while leaf form may be simple or compound. Leaf tissue consists of the epidermis, which forms the outermost cell layer, and mesophyll and vascular tissue, which make up the inner portion of the leaf. In some plant species, leaf form is modified to form structures such as tendrils, spines, bud scales, and needles.

30.5 Transport of Water and Solutes in Plants

Water potential (Ψ) is a measure of the difference in potential energy between a water sample and pure water. The water potential in plant solutions is influenced by solute concentration, pressure, gravity, and matric potential. Water potential and transpiration influence how water is transported through the xylem in plants. These processes are regulated by stomatal opening and closing. Photosynthates (mainly sucrose) move from sources to sinks through the plant's phloem. Sucrose is actively loaded into the sieve-tube elements of the phloem. The increased solute concentration causes water to move by osmosis from the xylem into the phloem. The positive pressure that is produced pushes water and solutes down the pressure gradient. The sucrose is unloaded into the sink, and the water returns to the xylem vessels.

30.6 Plant Sensory Systems and Responses

Plants respond to light by changes in morphology and activity. Irradiation by red light converts the photoreceptor phytochrome to its far-red light-absorbing form—Pfr. This form controls germination and flowering in response to length of day, as well as triggers photosynthesis in dormant plants or those that just emerged from the soil. Blue-light receptors, cryptochromes, and phototropins are responsible for phototropism. Amyloplasts, which contain heavy starch granules, sense gravity. Shoots exhibit negative gravitropism, whereas roots exhibit positive gravitropism. Plant hormones—naturally occurring compounds synthesized in small amounts—can act both in the cells that produce them and in distant tissues and organs. Auxins are responsible for apical dominance, root growth, directional growth toward light, and many other growth responses. Cytokinins stimulate cell division and counter apical dominance in shoots. Gibberellins inhibit dormancy of seeds and promote stem growth. Abscisic acid induces dormancy in seeds and buds, and protects plants from excessive water loss by promoting stomatal closure. Ethylene gas speeds up fruit ripening and dropping of leaves. Plants respond to touch by rapid movements (thigmotropy and thigmonasty) and slow differential growth (thigmomorphogenesis). Plants have evolved defense mechanisms against predators and pathogens. Physical barriers like bark and spines protect tender tissues. Plants also have chemical defenses, including toxic secondary metabolites and hormones, which elicit additional defense mechanisms.

VISUAL CONNECTION QUESTIONS

1. Figure 30.7 Which layers of the stem are made of parenchyma cells?
   A. cortex and pith
   B. phloem
   C. sclerenchyma
   D. xylem

2. Figure 30.32 Positive water potential is placed on the left side of the tube by increasing Ψ_p such that the water level rises on the right side. Could you equalize the water level on each side of the tube by adding solute, and if so, how?

3. Figure 30.34 Which of the following statements is false?
   a. Negative water potential draws water into the root hairs. Cohesion and adhesion draw water up the xylem. Transpiration draws water from the leaf.
   b. Negative water potential draws water into the root hairs. Cohesion and adhesion draw water up the phloem. Transpiration draws water from the leaf.
   c. Water potential decreases from the roots to the top of the plant.
   d. Water enters the plants through root hairs and exits through stoma.

REVIEW QUESTIONS

4. Plant regions of continuous growth are made up of ______.
   a. dermal tissue
   b. vascular tissue
   c. meristematic tissue
   d. permanent tissue

5. Which of the following is the major site of photosynthesis?
   a. apical meristem
   b. ground tissue
   c. xylem cells
   d. phloem cells

6. Stem regions at which leaves are attached are called ______.
   a. trichomes
   b. lenticels
   c. nodes
   d. internodes

7. Which of the following cell types forms most of the inside of a plant?
   a. meristem cells
   b. collenchyma cells
   c. sclerenchyma cells
   d. parenchyma cells

8. Tracheids, vessel elements, sieve-tube cells, and companion cells are components of ______.
   a. vascular tissue
   b. meristematic tissue
   c. ground tissue
   d. dermal tissue

9. The primary growth of a plant is due to the action of the ______.
   a. lateral meristem
   b. vascular cambium
   c. apical meristem
   d. cork cambium

10. Which of the following is an example of secondary growth?
    a. increase in length
    b. increase in thickness or girth
    c. increase in root hairs
    d. increase in leaf number

11. Secondary growth in stems is usually seen in ______.
    a. monocots
    b. dicots
    c. both monocots and dicots
    d. neither monocots nor dicots

12. Roots that enable a plant to grow on another plant are called ______.
    a. epiphytic roots
    b. prop roots
    c. adventitious roots
    d. aerial roots

13. The ______ forces selective uptake of minerals in the root.
    a. pericycle
    b. epidermis
    c. endodermis
    d. root cap

14. Newly-formed root cells begin to form different cell types in the _______.
    a. zone of elongation
    b. zone of maturation
    c. root meristem
    d. zone of cell division

15. The stalk of a leaf is known as the _______.
    a. petiole
    b. lamina
    c. stipule
    d. rachis

16. Leaflets are a characteristic of _______ leaves.
    a. alternate
    b. whorled
    c. compound
    d. opposite

17. Cells of the _______ contain chloroplasts.
    a. epidermis
    b. vascular tissue
    c. stomata
    d. mesophyll

18. Which of the following is most likely to be found in a desert environment?
    a. broad leaves to capture sunlight
    b. spines instead of leaves
    c. needle-like leaves
    d. wide, flat leaves that can float

19. When stomata open, what occurs?
    a. Water vapor is lost to the external environment, increasing the rate of transpiration.
    b. Water vapor is lost to the external environment, decreasing the rate of transpiration.
    c. Water vapor enters the spaces in the mesophyll, increasing the rate of transpiration.
    d. Water vapor enters the spaces in the mesophyll, decreasing the rate of transpiration.

20. Which cells are responsible for the movement of photosynthates through a plant?
    a. tracheids, vessel elements
    b. tracheids, companion cells
    c. vessel elements, companion cells
    d. sieve-tube elements, companion cells

21. The main photoreceptor that triggers phototropism is a _______.
    a. phytochrome
    b. cryptochrome
    c. phototropin
    d. carotenoid

22. Phytochrome is a plant pigment protein that:
    a. mediates plant infection
    b. promotes plant growth
    c. mediates morphological changes in response to red and far-red light
    d. inhibits plant growth

23. A mutant plant has roots that grow in all directions. Which of the following organelles would you expect to be missing in the cell?
    a. mitochondria
    b. amyloplast
    c. chloroplast
    d. nucleus

24. After buying green bananas or unripe avocadoes, they can be kept in a brown bag to ripen. The hormone released by the fruit and trapped in the bag is probably:
    a. abscisic acid
    b. cytokinin
    c. ethylene
    d. gibberellic acid

25. A decrease in the level of which hormone releases seeds from dormancy?
    a. abscisic acid
    b. cytokinin
    c. ethylene
    d. gibberellic acid

26. A seedling germinating under a stone grows at an angle away from the stone and upward. This response to touch is called _______.
    a. gravitropism
    b. thigmonasty
    c. thigmotropism
    d. skototropism

CRITICAL THINKING QUESTIONS

27. What type of meristem is found only in monocots, such as lawn grasses? Explain how this type of meristematic tissue is beneficial in lawn grasses that are mowed each week.

28. Which plant part is responsible for transporting water, minerals, and sugars to different parts of the plant? Name the two types of tissue that make up this overall tissue, and explain the role of each.

29. Describe the roles played by stomata and guard cells. What would happen to a plant if these cells did not function correctly?
30. Compare the structure and function of xylem to that of phloem.
31. Explain the role of the cork cambium in woody plants.
32. What is the function of lenticels?
33. Besides the age of a tree, what additional information can annual rings reveal?
34. Give two examples of modified stems and explain how each example benefits the plant.
35. Compare a tap root system with a fibrous root system. For each type, name a plant that provides a food in the human diet. Which type of root system is found in monocots? Which type of root system is found in dicots?
36. What might happen to a root if the pericycle disappeared?
37. How do dicots differ from monocots in terms of leaf structure?
38. Describe an example of a plant with leaves that are adapted to cold temperatures.
39. The process of bulk flow transports fluids in a plant. Describe the two main bulk flow processes.
40. Owners and managers of plant nurseries have to plan lighting schedules for a long-day plant that will flower in February. What lighting periods will be most effective? What color of light should be chosen?
41. What are the major benefits of gravitropism for a germinating seedling?
42. Fruit and vegetable storage facilities are usually refrigerated and well ventilated. Why are these conditions advantageous?
43. Stomata close in response to bacterial infection. Why is this response a mechanism of defense for the plant? Which hormone is most likely to mediate this response?

CHAPTER 31
Soil and Plant Nutrition

Figure 31.1 For this (a) squash seedling (*Cucurbita maxima*) to develop into a mature plant bearing its (b) fruit, numerous nutritional requirements must be met. (credit a: modification of work by Julian Colton; credit b: modification of work by "Wildfeuer"/Wikimedia Commons)

INTRODUCTION Cucurbitaceae is a family of plants first cultivated in Mesoamerica, although several species are native to North America. The family includes many edible species, such as squash and pumpkin, as well as inedible gourds. In order to grow and develop into mature, fruit-bearing plants, many requirements must be met and events must be coordinated. Seeds must germinate under the right conditions in the soil; therefore, temperature, moisture, and soil quality are important factors that play a role in germination and seedling development. Soil quality and climate are significant to plant distribution and growth. The young seedling will eventually grow into a mature plant, and the roots will absorb nutrients and water from the soil. At the same time, the aboveground parts of the plant will absorb carbon dioxide from the atmosphere and use energy from sunlight to produce organic compounds through photosynthesis. This chapter will explore the complex dynamics between plants and soils, and the adaptations that plants have evolved to make better use of nutritional resources.

Chapter Outline

31.1 Nutritional Requirements of Plants

By the end of this section, you will be able to do the following:

- Describe how plants obtain nutrients
- List the elements and compounds required for proper plant nutrition
- Describe an essential nutrient

Plants are unique organisms that can absorb nutrients and water through their root system, as well as carbon dioxide from the atmosphere. Soil quality and climate are the major determinants of plant distribution and growth. The combination of soil nutrients, water, and carbon dioxide, along with sunlight, allows plants to grow.

The Chemical Composition of Plants

Since plants require nutrients in the form of elements such as carbon and potassium, it is important to understand the chemical composition of plants. The majority of volume in a plant cell is water; it typically comprises 80 to 90 percent of the plant's total weight. Soil is the water source for land plants, and can be an abundant source of water, even if it appears dry. Plant roots absorb water from the soil through root hairs and transport it up to the leaves through the xylem. As water vapor is lost from the leaves, the process of transpiration and the polarity of water molecules (which enables them to form hydrogen bonds) draws more water from the roots up through the plant to the leaves (Figure 31.2). Plants need water to support cell structure, for metabolic functions, to carry nutrients, and for photosynthesis.

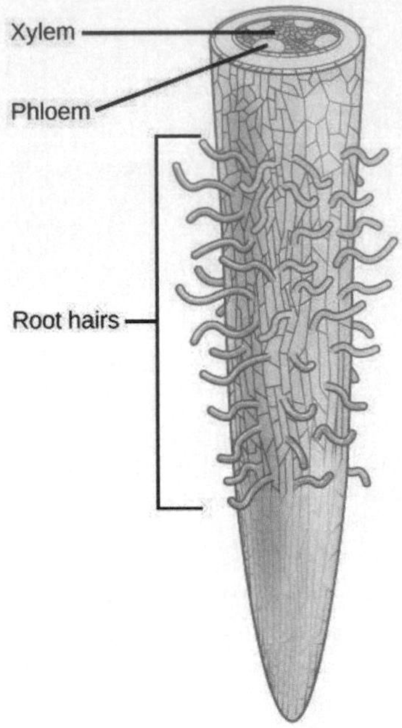

Figure 31.2 Water is absorbed through the root hairs and moves up the xylem to the leaves.

Plant cells need essential substances, collectively called nutrients, to sustain life. Plant nutrients may be composed of either organic or inorganic compounds. An **organic compound** is a chemical compound that contains carbon, such as carbon dioxide obtained from the atmosphere. Carbon that was obtained from atmospheric CO_2 composes the majority of the dry mass within most plants. An **inorganic compound** does not contain carbon and is not part of, or produced by, a living organism. Inorganic substances, which form the majority of the soil solution, are commonly called minerals: those required by plants include nitrogen (N) and potassium (K) for structure and regulation.

Essential Nutrients

Plants require only light, water, and about 20 elements to support all their biochemical needs: these 20 elements are called essential nutrients (Table 31.1). For an element to be regarded as **essential**, three criteria are required: 1) a plant cannot complete its life cycle without the element; 2) no other element can perform the function of the element; and 3) the element is directly involved in plant nutrition.

Essential Elements for Plant Growth

Macronutrients	Micronutrients
Carbon (C)	Iron (Fe)
Hydrogen (H)	Manganese (Mn)
Oxygen (O)	Boron (B)

Table 31.1

Macronutrients	Micronutrients
Nitrogen (N)	Molybdenum (Mo)
Phosphorus (P)	Copper (Cu)
Potassium (K)	Zinc (Zn)
Calcium (Ca)	Chlorine (Cl)
Magnesium (Mg)	Nickel (Ni)
Sulfur (S)	Cobalt (Co)
	Sodium (Na)
	Silicon (Si)

Table 31.1

Macronutrients and Micronutrients

The essential elements can be divided into two groups: macronutrients and micronutrients. Nutrients that plants require in larger amounts are called **macronutrients**. About half of the essential elements are considered macronutrients: carbon, hydrogen, oxygen, nitrogen, phosphorus, potassium, calcium, magnesium and sulfur. The first of these macronutrients, carbon (C), is required to form carbohydrates, proteins, nucleic acids, and many other compounds; it is therefore present in all macromolecules. On average, the dry weight (excluding water) of a cell is 50 percent carbon. As shown in Figure 31.3, carbon is a key part of plant biomolecules.

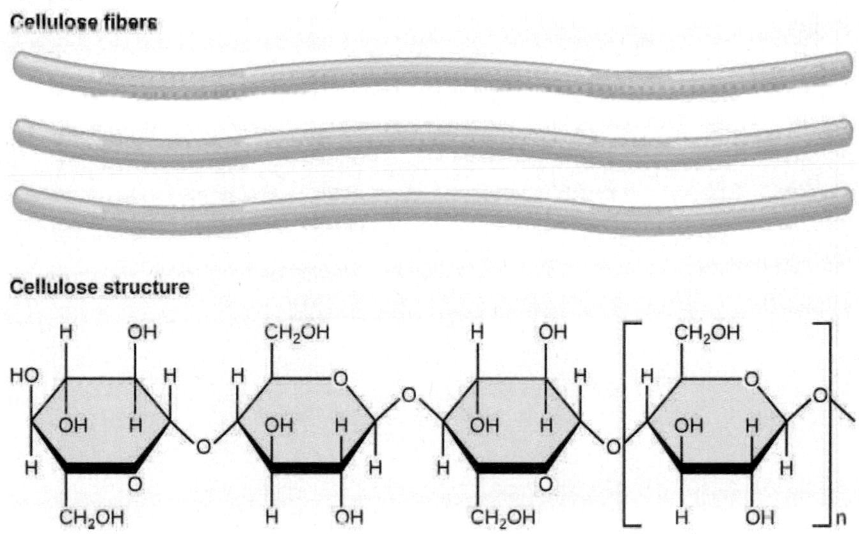

Figure 31.3 Cellulose, the main structural component of the plant cell wall, makes up over thirty percent of plant matter. It is the most abundant organic compound on earth.

The next most abundant element in plant cells is nitrogen (N); it is part of proteins and nucleic acids. Nitrogen is also used in the synthesis of some vitamins. Hydrogen and oxygen are macronutrients that are part of many organic compounds, and also form water. Oxygen is necessary for cellular respiration; plants use oxygen to store energy in the form of ATP. Phosphorus (P), another macromolecule, is necessary to synthesize nucleic acids and phospholipids. As part of ATP, phosphorus enables food energy to be converted into chemical energy through oxidative phosphorylation. Likewise, light energy is converted into chemical energy during photophosphorylation in photosynthesis, and into chemical energy to be extracted during respiration. Sulfur is part of certain amino acids, such as cysteine and methionine, and is present in several coenzymes. Sulfur also plays a role in photosynthesis as part of the electron transport chain, where hydrogen gradients play a key role in the conversion of light energy into ATP. Potassium (K) is important because of its role in regulating stomatal opening and closing. As the openings for gas exchange, stomata help maintain a healthy water balance; a potassium ion pump supports this process.

Magnesium (Mg) and calcium (Ca) are also important macronutrients. The role of calcium is twofold: to regulate nutrient transport, and to support many enzyme functions. Magnesium is important to the photosynthetic process. These minerals, along with the micronutrients, which are described below, also contribute to the plant's ionic balance.

In addition to macronutrients, organisms require various elements in small amounts. These **micronutrients**, or trace elements, are present in very small quantities. They include boron (B), chlorine (Cl), manganese (Mn), iron (Fe), zinc (Zn), copper (Cu), molybdenum (Mo), nickel (Ni), silicon (Si), and sodium (Na).

Deficiencies in any of these nutrients—particularly the macronutrients—can adversely affect plant growth (Figure 31.4). Depending on the specific nutrient, a lack can cause stunted growth, slow growth, or chlorosis (yellowing of the leaves). Extreme deficiencies may result in leaves showing signs of cell death.

LINK TO LEARNING

Visit this website (http://openstax.org/l/plant_mineral) to participate in an interactive experiment on plant nutrient deficiencies. You can adjust the amounts of N, P, K, Ca, Mg, and Fe that plants receive . . . and see what happens.

Figure 31.4 Nutrient deficiency is evident in the symptoms these plants show. This (a) grape tomato suffers from blossom end rot caused by calcium deficiency. The yellowing in this (b) *Frangula alnus* results from magnesium deficiency. Inadequate magnesium also leads to (c) intervenal chlorosis, seen here in a sweetgum leaf. This (d) palm is affected by potassium deficiency. (credit c: modification of work by Jim Conrad; credit d: modification of work by Malcolm Manners)

Everyday Connection

Hydroponics

Figure 31.5 Plant physiologist Ray Wheeler checks onions being grown using hydroponic techniques. The other plants are Bibb lettuce (left) and radishes (right). Credit: NASA

Hydroponics is a method of growing plants in a water-nutrient solution instead of soil. Since its advent, hydroponics has developed into a growing process that researchers often use. Scientists who are interested in studying plant nutrient deficiencies can use hydroponics to study the effects of different nutrient combinations under strictly controlled conditions. Hydroponics has also developed as a way to grow flowers, vegetables, and other crops in greenhouse environments. You might find hydroponically grown produce at your local grocery store. Today, many lettuces and tomatoes in your market have been hydroponically grown.

31.2 The Soil

By the end of this section, you will be able to do the following:

- Describe how soils are formed
- Explain soil composition
- Describe a soil profile

Plants obtain inorganic elements from the soil, which serves as a natural medium for land plants. **Soil** is the outer loose layer that covers the surface of Earth. Soil quality is a major determinant, along with climate, of plant distribution and growth. Soil quality depends not only on the chemical composition of the soil, but also the topography (regional surface features) and the presence of living organisms. In agriculture, the history of the soil, such as the cultivating practices and previous crops, modify the characteristics and fertility of that soil.

Soil develops very slowly over long periods of time, and its formation results from natural and environmental forces acting on mineral, rock, and organic compounds. Soils can be divided into two groups: **organic soils** are those that are formed from sedimentation and primarily composed of organic matter, while those that are formed from the weathering of rocks and are primarily composed of inorganic material are called **mineral soils**. Mineral soils are predominant in terrestrial ecosystems, where soils may be covered by water for part of the year or exposed to the atmosphere.

Soil Composition

Soil consists of these major components (Figure 31.6):

- inorganic mineral matter, about 40 to 45 percent of the soil volume
- organic matter, about 5 percent of the soil volume
- water and air, about 50 percent of the soil volume

The amount of each of the four major components of soil depends on the amount of vegetation, soil compaction, and water present in the soil. A good healthy soil has sufficient air, water, minerals, and organic material to promote and sustain plant life.

VISUAL CONNECTION

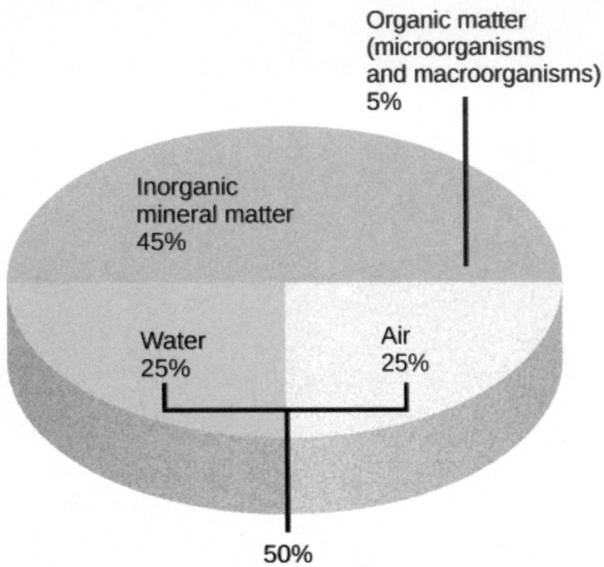

Figure 31.6 The four major components of soil are shown: inorganic minerals, organic matter, water, and air.

Soil compaction can result when soil is compressed by heavy machinery or even foot traffic. How might this compaction change the soil composition?

The organic material of soil, called **humus**, is made up of microorganisms (dead and alive), and dead animals and plants in varying stages of decay. Humus improves soil structure and provides plants with water and minerals. The inorganic material of soil consists of rock, slowly broken down into smaller particles that vary in size. Soil particles that are 0.1 to 2 mm in diameter are **sand**. Soil particles between 0.002 and 0.1 mm are called **silt**, and even smaller particles, less than 0.002 mm in diameter, are called **clay**. Some soils have no dominant particle size and contain a mixture of sand, silt, and humus; these soils are called **loams**.

LINK TO LEARNING

Explore this interactive map (http://openstax.org/l/soil_survey_map) from the USDA's National Cooperative Soil Survey to access soil data for almost any region in the United States.

Soil Formation

Soil formation is the consequence of a combination of biological, physical, and chemical processes. Soil should ideally contain 50 percent solid material and 50 percent pore space. About one-half of the pore space should contain water, and the other half should contain air. The organic component of soil serves as a cementing agent, returns nutrients to the plant, allows soil to store moisture, makes soil tillable for farming, and provides energy for soil microorganisms. Most soil microorganisms—bacteria, algae, or fungi—are dormant in dry soil, but become active once moisture is available.

Soil distribution is not homogenous because its formation results in the production of layers; together, the vertical section of a soil is called the **soil profile**. Within the soil profile, soil scientists define zones called horizons. A **horizon** is a soil layer with distinct physical and chemical properties that differ from those of other layers. Five factors account for soil formation: parent material, climate, topography, biological factors, and time.

Parent Material

The organic and inorganic material in which soils form is the **parent material**. Mineral soils form directly from the weathering of **bedrock**, the solid rock that lies beneath the soil, and therefore, they have a similar composition to the original rock. Other soils form in materials that came from elsewhere, such as sand and glacial drift. Materials located in the depth of the soil are relatively unchanged compared with the deposited material. Sediments in rivers may have different characteristics, depending on whether the stream moves quickly or slowly. A fast-moving river could have sediments of rocks and sand, whereas a slow-moving river could have fine-textured material, such as clay.

Climate

Temperature, moisture, and wind cause different patterns of weathering and therefore affect soil characteristics. The presence of moisture and nutrients from weathering will also promote biological activity: a key component of a quality soil.

Topography

Regional surface features (familiarly called "the lay of the land") can have a major influence on the characteristics and fertility of a soil. Topography affects water runoff, which strips away parent material and affects plant growth. Steeps soils are more prone to erosion and may be thinner than soils that are relatively flat or level.

Biological factors

The presence of living organisms greatly affects soil formation and structure. Animals and microorganisms can produce pores and crevices, and plant roots can penetrate into crevices to produce more fragmentation. Plant secretions promote the development of microorganisms around the root, in an area known as the **rhizosphere**. Additionally, leaves and other material that fall from plants decompose and contribute to soil composition.

Time

Time is an important factor in soil formation because soils develop over long periods. Soil formation is a dynamic process. Materials are deposited over time, decompose, and transform into other materials that can be used by living organisms or deposited onto the surface of the soil.

Physical Properties of the Soil

Soils are named and classified based on their horizons. The soil profile has four distinct layers: 1) O horizon; 2) A horizon; 3) B horizon, or subsoil; and 4) C horizon, or soil base (Figure 31.7). The **O horizon** has freshly decomposing organic matter—humus—at its surface, with decomposed vegetation at its base. Humus enriches the soil with nutrients and enhances soil moisture retention. Topsoil—the top layer of soil—is usually two to three inches deep, but this depth can vary considerably. For instance, river deltas like the Mississippi River delta have deep layers of topsoil. Topsoil is rich in organic material; microbial processes occur there, and it is the "workhorse" of plant production. The **A horizon** consists of a mixture of organic material with inorganic products of weathering, and it is therefore the beginning of true mineral soil. This horizon is typically darkly colored because of the presence of organic matter. In this area, rainwater percolates through the soil and carries materials from the surface. The **B horizon** is an accumulation of mostly fine material that has moved downward, resulting in a dense layer in the soil. In some soils, the B horizon contains nodules or a layer of calcium carbonate. The **C horizon**, or soil base, includes the parent material, plus the organic and inorganic material that is broken down to form soil. The parent material may be either created in its natural place, or transported from elsewhere to its present location. Beneath the C horizon lies bedrock.

VISUAL CONNECTION

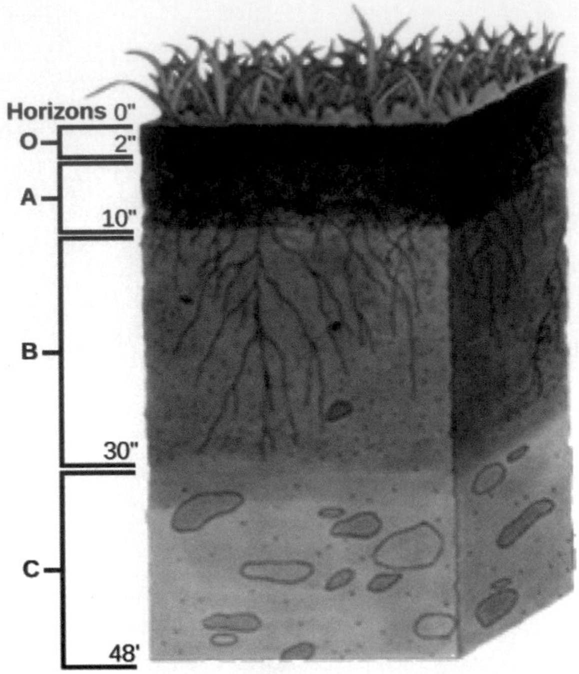

Figure 31.7 This soil profile shows the different soil layers (O horizon, A horizon, B horizon, and C horizon) found in typical soils. (credit: modification of work by USDA)

Which horizon is considered the topsoil, and which is considered the subsoil?

Some soils may have additional layers, or lack one of these layers. The thickness of the layers is also variable, and depends on the factors that influence soil formation. In general, immature soils may have O, A, and C horizons, whereas mature soils may display all of these, plus additional layers (Figure 31.8).

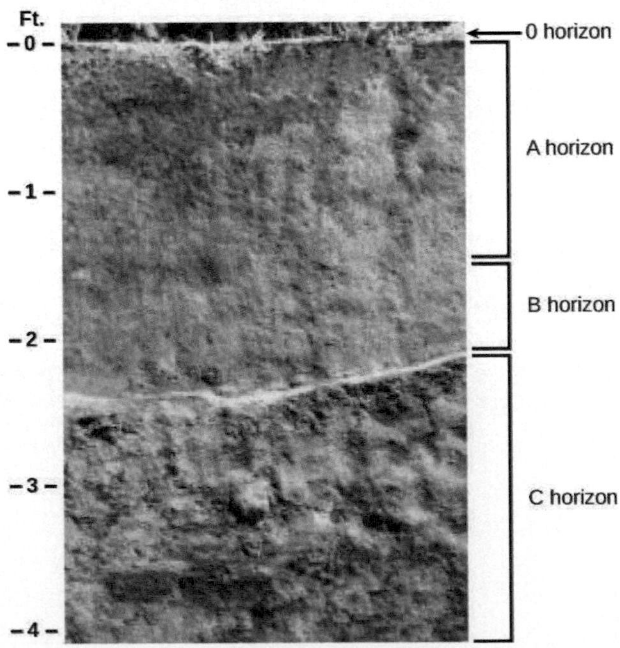

Figure 31.8 The San Joaquin soil profile has an O horizon, A horizon, B horizon, and C horizon. (credit: modification of work by USDA)

CAREER CONNECTION

Soil Scientist

A soil scientist studies the biological components, physical and chemical properties, distribution, formation, and morphology of soils. Soil scientists need to have a strong background in physical and life sciences, plus a foundation in mathematics. They may work for federal or state agencies, academia, or the private sector. Their work may involve collecting data, carrying out research, interpreting results, inspecting soils, conducting soil surveys, and recommending soil management programs.

Figure 31.9 This soil scientist is studying the horizons and composition of soil at a research site. (credit: USDA)

Many soil scientists work both in an office and in the field. According to the United States Department of Agriculture (USDA): "a soil scientist needs good observation skills to analyze and determine the characteristics of different types of soils. Soil types are complex and the geographical areas a soil scientist may survey are varied. Aerial photos or various satellite images are often used to research the areas. Computer skills and geographic information systems (GIS) help the scientist to analyze the multiple facets of geomorphology, topography, vegetation, and climate to discover the patterns left on the landscape."[1] Soil scientists play a key role in understanding the soil's past, analyzing present conditions, and making recommendations for future soil-related practices.

31.3 Nutritional Adaptations of Plants

By the end of this section, you will be able to do the following:

- Understand the nutritional adaptations of plants
- Describe mycorrhizae
- Explain nitrogen fixation

Plants obtain food in two different ways. Autotrophic plants can make their own food from inorganic raw materials, such as carbon dioxide and water, through photosynthesis in the presence of sunlight. Green plants are included in this group. Some plants, however, are heterotrophic: they are totally parasitic and lacking in chlorophyll. These plants, referred to as holo-parasitic plants, are unable to synthesize organic carbon and draw all of their nutrients from the host plant.

1National Resources Conservation Service / United States Department of Agriculture. "Careers in Soil Science." http://openstax.org/l/NRCS (http://openstax.org/l/NRCS)

Plants may also enlist the help of microbial partners in nutrient acquisition. Particular species of bacteria and fungi have evolved along with certain plants to create a mutualistic symbiotic relationship with roots. This improves the nutrition of both the plant and the microbe. The formation of nodules in legume plants and mycorrhization can be considered among the nutritional adaptations of plants. However, these are not the only type of adaptations that we may find; many plants have other adaptations that allow them to thrive under specific conditions.

LINK TO LEARNING

This video (http://openstax.org/l/basic_photosyn) reviews basic concepts about photosynthesis. In the left panel, click each tab to select a topic for review.

Nitrogen Fixation: Root and Bacteria Interactions

Nitrogen is an important macronutrient because it is part of nucleic acids and proteins. Atmospheric nitrogen, which is the diatomic molecule N_2, or dinitrogen, is the largest pool of nitrogen in terrestrial ecosystems. However, plants cannot take advantage of this nitrogen because they do not have the necessary enzymes to convert it into biologically useful forms. However, nitrogen can be "fixed," which means that it can be converted to ammonia (NH_3) through biological, physical, or chemical processes. As you have learned, biological nitrogen fixation (BNF) is the conversion of atmospheric nitrogen (N_2) into ammonia (NH_3), exclusively carried out by prokaryotes such as soil bacteria or cyanobacteria. Biological processes contribute 65 percent of the nitrogen used in agriculture. The following equation represents the process:

$$N_2 + 16\,ATP + 8\,e^- + 8\,H^+ \rightarrow 2NH_3 + 16\,ADP + 16\,Pi + H_2$$

The most important source of BNF is the symbiotic interaction between soil bacteria and legume plants, including many crops important to humans (Figure 31.10). The NH_3 resulting from fixation can be transported into plant tissue and incorporated into amino acids, which are then made into plant proteins. Some legume seeds, such as soybeans and peanuts, contain high levels of protein, and serve among the most important agricultural sources of protein in the world.

VISUAL CONNECTION

(a) (b) (c)

Figure 31.10 Some common edible legumes—like (a) peanuts, (b) beans, and (c) chickpeas—are able to interact symbiotically with soil bacteria that fix nitrogen. (credit a: modification of work by Jules Clancy; credit b: modification of work by USDA)

Farmers often rotate corn (a cereal crop) and soy beans (a legume), planting a field with each crop in alternate seasons. What advantage might this crop rotation confer?

Soil bacteria, collectively called **rhizobia**, symbiotically interact with legume roots to form specialized structures called **nodules**, in which nitrogen fixation takes place. This process entails the reduction of atmospheric nitrogen to ammonia, by means of the enzyme **nitrogenase**. Therefore, using rhizobia is a natural and environmentally friendly way to fertilize plants, as opposed to chemical fertilization that uses a nonrenewable resource, such as natural gas. Through symbiotic nitrogen fixation, the plant benefits from using an endless source of nitrogen from the atmosphere. The process simultaneously contributes to soil fertility because the plant root system leaves behind some of the biologically available nitrogen. As in any symbiosis, both organisms benefit from the interaction: the plant obtains ammonia, and bacteria obtain carbon compounds generated through photosynthesis, as well as a protected niche in which to grow (Figure 31.11).

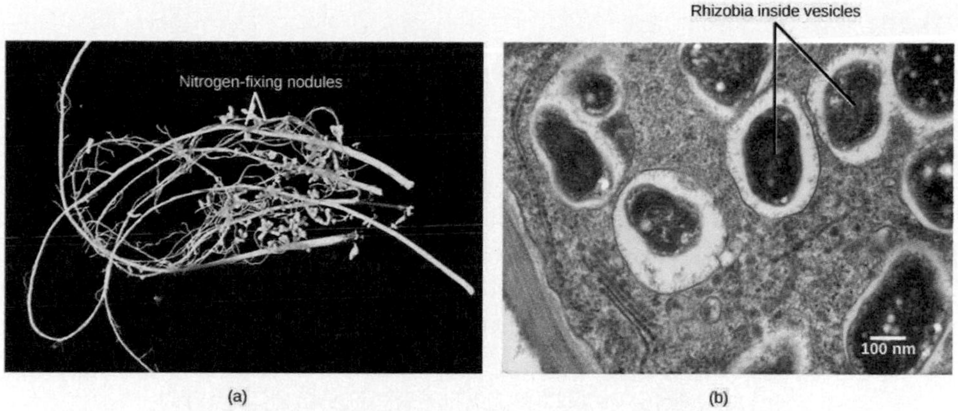

Figure 31.11 Soybean roots contain (a) nitrogen-fixing nodules. Cells within the nodules are infected with *Bradyrhyzobium japonicum*, a rhizobia or "root-loving" bacterium. The bacteria are encased in (b) vesicles inside the cell, as can be seen in this transmission electron micrograph. (credit a: modification of work by USDA; credit b: modification of work by Louisa Howard, Dartmouth Electron Microscope Facility; scale-bar data from Matt Russell)

Mycorrhizae: The Symbiotic Relationship between Fungi and Roots

A nutrient depletion zone can develop when there is rapid soil solution uptake, low nutrient concentration, low diffusion rate, or low soil moisture. These conditions are very common; therefore, most plants rely on fungi to facilitate the uptake of minerals from the soil. Fungi form symbiotic associations called mycorrhizae with plant roots, in which the fungi actually are integrated into the physical structure of the root. The fungi colonize the living root tissue during active plant growth.

Through mycorrhization, the plant obtains mainly phosphate and other minerals, such as zinc and copper, from the soil. The fungus obtains nutrients, such as sugars, from the plant root (Figure 31.12). Mycorrhizae help increase the surface area of the plant root system because hyphae, which are narrow, can spread beyond the nutrient depletion zone. Hyphae can grow into small soil pores that allow access to phosphorus that would otherwise be unavailable to the plant. The beneficial effect on the plant is best observed in poor soils. The benefit to fungi is that they can obtain up to 20 percent of the total carbon accessed by plants. Mycorrhizae functions as a physical barrier to pathogens. It also provides an induction of generalized host defense mechanisms, and sometimes involves production of antibiotic compounds by the fungi.

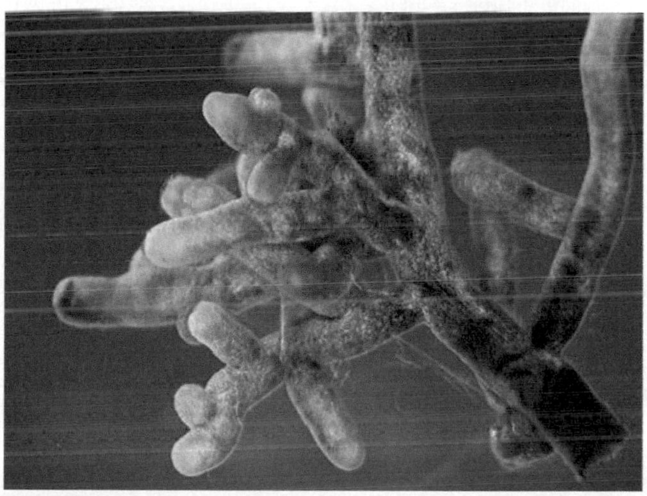

Figure 31.12 Root tips proliferate in the presence of mycorrhizal infection, which appears as off-white fuzz in this image. (credit: modification of work by Nilsson et al., BMC Bioinformatics 2005)

There are two types of mycorrhizae: ectomycorrhizae and endomycorrhizae. Ectomycorrhizae form an extensive dense sheath around the roots, called a mantle. Hyphae from the fungi extend from the mantle into the soil, which increases the surface area for water and mineral absorption. This type of mycorrhizae is found in forest trees, especially conifers, birches, and oaks. Endomycorrhizae, also called arbuscular mycorrhizae, do not form a dense sheath over the root. Instead, the fungal mycelium is embedded within the root tissue. Endomycorrhizae are found in the roots of more than 80 percent of terrestrial plants.

Nutrients from Other Sources

Some plants cannot produce their own food and must obtain their nutrition from outside sources. This may occur with plants that are parasitic or saprophytic. Some plants are mutualistic symbionts, epiphytes, or insectivorous.

Plant Parasites

A **parasitic plant** depends on its host for survival. Some parasitic plants have no leaves. An example of this is the dodder (Figure 31.13), which has a weak, cylindrical stem that coils around the host and forms suckers. From these suckers, cells invade the host stem and grow to connect with the vascular bundles of the host. The parasitic plant obtains water and nutrients through these connections. The plant is a total parasite (a holoparasite) because it is completely dependent on its host. Other parasitic plants (hemiparasites) are fully photosynthetic and only use the host for water and minerals. There are about 4,100 species of parasitic plants.

Figure 31.13 The dodder is a holoparasite that penetrates the host's vascular tissue and diverts nutrients for its own growth. Note that the vines of the dodder, which has white flowers, are beige. The dodder has no chlorophyll and cannot produce its own food. (credit: "Lalithamba"/Flickr)

Saprophytes

A **saprophyte** is a plant that does not have chlorophyll and gets its food from dead matter, similar to bacteria and fungi (note that fungi are often called saprophytes, which is incorrect, because fungi are not plants). Plants like these use enzymes to convert organic food materials into simpler forms from which they can absorb nutrients (Figure 31.14). Most saprophytes do not directly digest dead matter: instead, they parasitize fungi that digest dead matter, or are mycorrhizal, ultimately obtaining photosynthate from a fungus that derived photosynthate from its host. Saprophytic plants are uncommon; only a few species are described.

Figure 31.14 Saprophytes, like this Dutchmen's pipe (*Monotropa hypopitys*), obtain their food from dead matter and do not have chlorophyll. (credit: modification of work by Iwona Erskine-Kellie)

Symbionts

A **symbiont** is a plant in a symbiotic relationship, with special adaptations such as mycorrhizae or nodule formation. Fungi also form symbiotic associations with cyanobacteria and green algae (called lichens). Lichens can sometimes be seen as colorful growths on the surface of rocks and trees (Figure 31.15). The algal partner (phycobiont) makes food autotrophically, some of which it shares with the fungus; the fungal partner (mycobiont) absorbs water and minerals from the environment, which are made available to the green alga. If one partner was separated from the other, they would both die.

Figure 31.15 Lichens, which often have symbiotic relationships with other plants, can sometimes be found growing on trees. (credit: "benketaro"/Flickr)

Epiphytes

An **epiphyte** is a plant that grows on other plants, but is not dependent upon the other plant for nutrition (Figure 31.16). Epiphytes have two types of roots: clinging aerial roots, which absorb nutrients from humus that accumulates in the crevices of trees; and aerial roots, which absorb moisture from the atmosphere.

Figure 31.16 These epiphyte plants grow in the main greenhouse of the *Jardin des Plantes* in Paris.

Insectivorous Plants

An **insectivorous** plant has specialized leaves to attract and digest insects. The Venus flytrap is popularly known for its insectivorous mode of nutrition, and has leaves that work as traps (Figure 31.17). The minerals it obtains from prey compensate for those lacking in the boggy (low pH) soil of its native North Carolina coastal plains. There are three sensitive hairs in the center of each half of each leaf. The edges of each leaf are covered with long spines. Nectar secreted by the plant attracts flies to the leaf. When a fly touches the sensory hairs, the leaf immediately closes. Next, fluids and enzymes break down the prey and minerals are absorbed by the leaf. Since this plant is popular in the horticultural trade, it is threatened in its original habitat.

Figure 31.17 A Venus flytrap has specialized leaves to trap insects. (credit: "Selena N. B. H."/Flickr)

KEY TERMS

A horizon consists of a mixture of organic material with inorganic products of weathering
B horizon soil layer that is an accumulation of mostly fine material that has moved downward
bedrock solid rock that lies beneath the soil
C horizon layer of soil that contains the parent material, and the organic and inorganic material that is broken down to form soil; also known as the soil base
clay soil particles that are less than 0.002 mm in diameter
epiphyte plant that grows on other plants but is not dependent upon other plants for nutrition
horizon soil layer with distinct physical and chemical properties, which differs from other layers depending on how and when it was formed
humus organic material of soil; made up of microorganisms, dead animals, and plants in varying stages of decay
inorganic compound chemical compound that does not contain carbon; it is not part of or produced by a living organism
insectivorous plant plant that has specialized leaves to attract and digest insects
loam soil that has no dominant particle size
macronutrient nutrient that is required in large amounts for plant growth; carbon, hydrogen, oxygen, nitrogen, phosphorus, potassium, calcium, magnesium, and sulfur
micronutrient nutrient required in small amounts; also called trace element
mineral soil type of soil that is formed from the weathering of rocks and inorganic material; composed primarily of sand, silt, and clay
nitrogenase enzyme that is responsible for the reduction of atmospheric nitrogen to ammonia
nodules specialized structures that contain *Rhizobia* bacteria where nitrogen fixation takes place
O horizon layer of soil with humus at the surface and decomposed vegetation at the base
organic compound chemical compound that contains carbon
organic soil type of soil that is formed from sedimentation; composed primarily of organic material
parasitic plant plant that is dependent on its host for survival
parent material organic and inorganic material in which soils form
rhizobia soil bacteria that symbiotically interact with legume roots to form nodules and fix nitrogen
rhizosphere area of soil affected by root secretions and microorganisms
sand soil particles between 0.1–2 mm in diameter
saprophyte plant that does not have chlorophyll and gets its food from dead matter
silt soil particles between 0.002 and 0.1 mm in diameter
soil outer loose layer that covers the surface of Earth
soil profile vertical section of a soil
symbiont plant in a symbiotic relationship with bacteria or fungi

CHAPTER SUMMARY

31.1 Nutritional Requirements of Plants

Plants can absorb inorganic nutrients and water through their root system, and carbon dioxide from the environment. The combination of organic compounds, along with water, carbon dioxide, and sunlight, produce the energy that allows plants to grow. Inorganic compounds form the majority of the soil solution. Plants access water though the soil. Water is absorbed by the plant root, transports nutrients throughout the plant, and maintains the structure of the plant. Essential elements are indispensable elements for plant growth. They are divided into macronutrients and micronutrients. The macronutrients plants require are carbon, nitrogen, hydrogen, oxygen, phosphorus, potassium, calcium, magnesium, and sulfur. Important micronutrients include iron, manganese, boron, molybdenum, copper, zinc, chlorine, nickel, cobalt, silicon, and sodium.

31.2 The Soil

Plants obtain mineral nutrients from the soil. Soil is the outer loose layer that covers the surface of Earth. Soil quality depends on the chemical composition of the soil, the topography, the presence of living organisms, the climate, and time. Agricultural practice and history may also modify the characteristics and fertility of soil. Soil consists of four major components: 1) inorganic mineral matter, 2) organic matter, 3) water and air, and 4) living matter. The organic material of soil is made of humus, which improves soil structure and provides water and minerals. Soil inorganic material consists of rock slowly broken down into smaller particles that vary in size, such as sand, silt, and loam.

Soil formation results from a combination of biological, physical, and chemical processes. Soil is not homogenous because its formation results in the production of layers called a soil profile. Factors that affect soil formation include: parent material, climate, topography, biological factors, and time. Soils are classified based on their horizons, soil particle size, and proportions. Most soils have four distinct horizons: O, A, B, and C.

31.3 Nutritional Adaptations of Plants

Atmospheric nitrogen is the largest pool of available nitrogen in terrestrial ecosystems. However, plants cannot use this nitrogen because they do not have the necessary enzymes. Biological nitrogen fixation (BNF) is the conversion of atmospheric nitrogen to ammonia. The most important source of BNF is the symbiotic interaction between soil bacteria and legumes. The bacteria form nodules on the legume's roots in which nitrogen fixation takes place. Fungi form symbiotic associations (mycorrhizae) with plants, becoming integrated into the physical structure of the root. Through mycorrhization, the plant obtains minerals from the soil and the fungus obtains photosynthate from the plant root. Ectomycorrhizae form an extensive dense sheath around the root, while endomycorrhizae are embedded within the root tissue. Some plants—parasites, saprophytes, symbionts, epiphytes, and insectivores—have evolved adaptations to obtain their organic or mineral nutrition from various sources.

VISUAL CONNECTION QUESTIONS

1. Figure 31.6 Soil compaction can result when soil is compressed by heavy machinery or even foot traffic. How might this compaction change the soil composition?
2. Figure 31.7 Which horizon is considered the topsoil, and which is considered the subsoil?
3. Figure 31.10 Farmers often rotate corn (a cereal crop) and soy beans (a legume) planting a field with each crop in alternate seasons. What advantage might this crop rotation confer?

REVIEW QUESTIONS

4. For an element to be regarded as essential, all of the following criteria must be met, except:
   a. No other element can perform the function.
   b. The element is directly involved in plant nutrition.
   c. The element is inorganic.
   d. The plant cannot complete its lifecycle without the element.
5. The nutrient that is part of carbohydrates, proteins, and nucleic acids, and that forms biomolecules, is _______.
   a. nitrogen
   b. carbon
   c. magnesium
   d. iron
6. Most _______ are necessary for enzyme function.
   a. micronutrients
   b. macronutrients
   c. biomolecules
   d. essential nutrients
7. What is the main water source for land plants?
   a. rain
   b. soil
   c. biomolecules
   d. essential nutrients
8. Which factors affect soil quality?
   a. chemical composition
   b. history of the soil
   c. presence of living organisms and topography
   d. all of the above
9. Soil particles that are 0.1 to 2 mm in diameter are called _______.
   a. sand
   b. silt
   c. clay
   d. loam
10. A soil consists of layers called _______ that taken together are called a _______.
   a. soil profiles : horizon
   b. horizons : soil profile
   c. horizons : humus
   d. humus : soil profile
11. What is the term used to describe the solid rock that lies beneath the soil?
   a. sand
   b. bedrock
   c. clay
   d. loam
12. Which process produces an inorganic compound that plants can easily use?
   a. photosynthesis
   b. nitrogen fixation
   c. mycorrhization
   d. Calvin cycle

13. Through mycorrhization, a plant obtains important nutrients such as ________.
    a. phosphorus, zinc, and copper
    b. phosphorus, zinc, and calcium
    c. nickel, calcium, and zinc
    d. all of the above
14. What term describes a plant that requires nutrition from a living host plant?
    a. parasite
    b. saprophyte
    c. epiphyte
    d. insectivorous
15. What is the term for the symbiotic association between fungi and cyanobacteria?
    a. lichen
    b. mycorrhizae
    c. epiphyte
    d. nitrogen-fixing nodule

CRITICAL THINKING QUESTIONS

16. What type of plant problems result from nitrogen and calcium deficiencies?
17. Research the life of Jan Babtista van Helmont. What did the van Helmont experiment show?
18. List two essential macronutrients and two essential micro nutrients.
19. Describe the main differences between a mineral soil and an organic soil.
20. Name and briefly explain the factors that affect soil formation.
21. Describe how topography influences the characteristics and fertility of a soil.
22. Why is biological nitrogen fixation an environmentally friendly way of fertilizing plants?
23. What is the main difference, from an energy point of view, between photosynthesis and biological nitrogen fixation?
24. Why is a root nodule a nutritional adaptation of a plant?

CHAPTER 32
Plant Reproduction

Figure 32.1 Plants that reproduce sexually often achieve fertilization with the help of pollinators such as (a) bees, (b) birds, and (c) butterflies. (credit a: modification of work by John Severns; credit b: modification of work by Charles J. Sharp; credit c: modification of work by "Galawebdesign"/Flickr)

INTRODUCTION Plants have evolved different reproductive strategies for the continuation of their species. Some plants reproduce sexually, and others asexually, in contrast to animal species, which rely almost exclusively on sexual reproduction. Plant sexual reproduction usually depends on pollinating agents, while asexual reproduction is independent of these agents. Flowers are often the showiest or most strongly scented part of plants. With their bright colors, fragrances, and interesting shapes and sizes, flowers attract insects, birds, and animals to serve their pollination needs. Other plants pollinate via wind or water; still others self-pollinate.

32.1 Reproductive Development and Structure

By the end of this section, you will be able to do the following:

- Describe the two stages of a plant's lifecycle
- Compare and contrast male and female gametophytes and explain how they form in angiosperms
- Describe the reproductive structures of a plant
- Describe the components of a complete flower
- Describe the development of microsporangium and megasporangium in gymnosperms

Sexual reproduction takes place with slight variations in different groups of plants. Plants have two distinct stages in their lifecycle: the gametophyte stage and the sporophyte stage. The haploid **gametophyte** produces the male and female gametes by mitosis in distinct multicellular structures. Fusion of the male and females gametes forms the diploid zygote, which develops into the **sporophyte**. After reaching maturity, the diploid sporophyte produces spores by meiosis, which in turn divide by mitosis to produce the haploid gametophyte. The new gametophyte produces gametes, and the cycle continues. This is the alternation of generations, and is typical of plant reproduction (Figure 32.2).

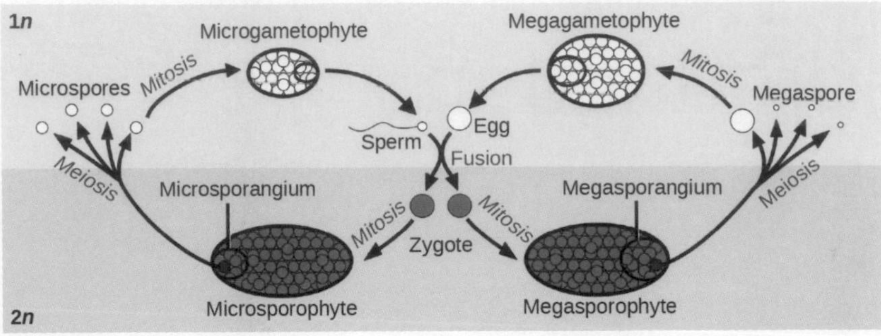

Figure 32.2 The alternation of generations in angiosperms is depicted in this diagram. (credit: modification of work by Peter Coxhead)

The life cycle of higher plants is dominated by the sporophyte stage, with the gametophyte borne on the sporophyte. In ferns, the gametophyte is free-living and very distinct in structure from the diploid sporophyte. In bryophytes, such as mosses, the haploid gametophyte is more developed than the sporophyte.

During the vegetative phase of growth, plants increase in size and produce a shoot system and a root system. As they enter the reproductive phase, some of the branches start to bear flowers. Many flowers are borne singly, whereas some are borne in clusters. The flower is borne on a stalk known as a receptacle. Flower shape, color, and size are unique to each species, and are often used by taxonomists to classify plants.

Sexual Reproduction in Angiosperms

The lifecycle of angiosperms follows the alternation of generations explained previously. The haploid gametophyte alternates with the diploid sporophyte during the sexual reproduction process of angiosperms. Flowers contain the plant's reproductive structures.

Flower Structure

A typical flower has four main parts—or whorls—known as the calyx, corolla, androecium, and gynoecium (Figure 32.3). The outermost whorl of the flower has green, leafy structures known as sepals. The sepals, collectively called the calyx, help to protect the unopened bud. The second whorl is comprised of petals—usually, brightly colored—collectively called the corolla. The number of sepals and petals varies depending on whether the plant is a monocot or dicot. In monocots, petals usually number three or multiples of three; in dicots, the number of petals is four or five, or multiples of four and five. Together, the calyx and corolla are known as the **perianth**. The third whorl contains the male reproductive structures and is known as the androecium. The **androecium** has stamens with anthers that contain the microsporangia. The innermost group of structures in the flower is the **gynoecium**, or the female reproductive component(s). The carpel is the individual unit of the gynoecium and has a stigma, style, and ovary. A flower may have one or multiple carpels.

VISUAL CONNECTION

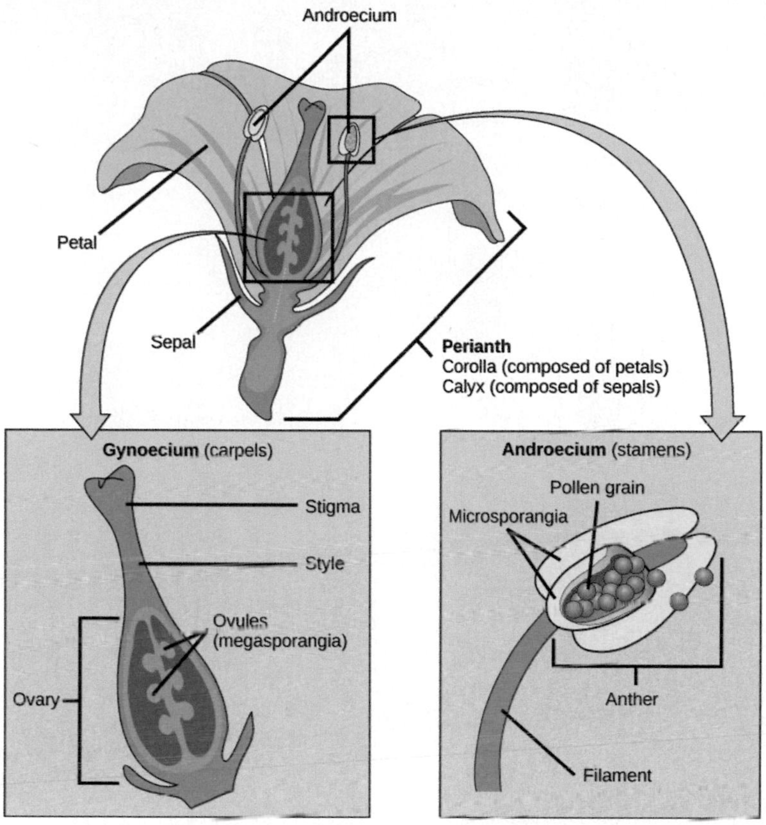

Figure 32.3 The four main parts of the flower are the calyx, corolla, androecium, and gynoecium. The androecium is the sum of all the male reproductive organs, and the gynoecium is the sum of the female reproductive organs. (credit: modification of work by Mariana Ruiz Villareal)

If the anther is missing, what type of reproductive structure will the flower be unable to produce? What term is used to describe an incomplete flower lacking the androecium? What term describes an incomplete flower lacking a gynoecium?

If all four whorls (the calyx, corolla, androecium, and gynoecium) are present, the flower is described as complete. If any of the four parts is missing, the flower is known as incomplete. Flowers that contain both an androecium and a gynoecium are called perfect, androgynous or hermaphrodites. There are two types of incomplete flowers: staminate flowers contain only an androecium, and carpellate flowers have only a gynoecium (Figure 32.4).

Figure 32.4 The corn plant has both staminate (male) and carpellate (female) flowers. Staminate flowers, which are clustered in the tassel at the tip of the stem, produce pollen grains. Carpellate flowers are clustered in the immature ears. Each strand of silk is a stigma. The corn kernels are seeds that develop on the ear after fertilization. Also shown is the lower stem and root.

If both male and female flowers are borne on the same plant, the species is called monoecious (meaning "one home"): examples are corn and pea. Species with male and female flowers borne on separate plants are termed dioecious, or "two homes," examples of which are *C. papaya* and *Cannabis*. The ovary, which may contain one or multiple ovules, may be placed above other flower parts, which is referred to as superior; or, it may be placed below the other flower parts, referred to as inferior (Figure 32.5).

Figure 32.5 The (a) lily is a superior flower, which has the ovary above the other flower parts. (b) Fuchsia is an inferior flower, which has the ovary beneath other flower parts. (credit a photo: modification of work by Benjamin Zwittnig; credit b photo: modification of work by "Koshy Koshy"/Flickr)

Male Gametophyte (The Pollen Grain)

The male gametophyte develops and reaches maturity in an immature anther. In a plant's male reproductive organs, development of pollen takes place in a structure known as the **microsporangium** (Figure 32.6). The microsporangia, which are usually bilobed, are pollen sacs in which the microspores develop into pollen grains. These are found in the anther, which is at the end of the stamen—the long filament that supports the anther.

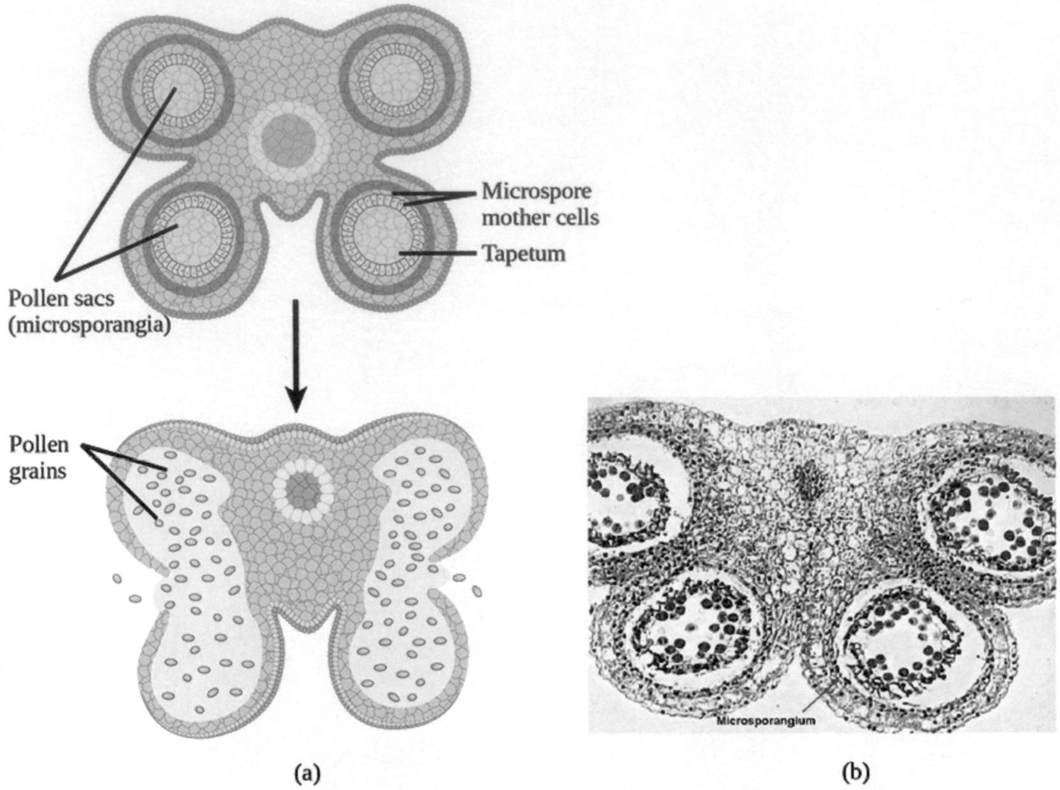

Figure 32.6 Shown is (a) a cross section of an anther at two developmental stages. The immature anther (top) contains four microsporangia, or pollen sacs. Each microsporangium contains hundreds of microspore mother cells that will each give rise to four pollen grains. The tapetum supports the development and maturation of the pollen grains. Upon maturation of the pollen (bottom), the pollen sac walls split open and the pollen grains (male gametophytes) are released, as shown in the (b) micrograph of an immature lily anther. In these scanning electron micrographs, pollen sacs are ready to burst, releasing their grains. (credit a: modification of work by LibreTexts; b: modification of work by Robert R. Wise; scale-bar data from Matt Russell)

Within the microsporangium, each of the microspore mother cells divides by meiosis to give rise to four microspores, each of which will ultimately form a pollen grain (Figure 32.7). An inner layer of cells, known as the tapetum, provides nutrition to the developing microspores and contributes key components to the pollen wall. Mature pollen grains contain two cells: a generative cell and a pollen tube cell. The generative cell is contained within the larger pollen tube cell. Upon germination, the tube cell forms the pollen tube through which the generative cell migrates to enter the ovary. During its transit inside the pollen tube, the generative cell divides to form two male gametes (sperm cells). Upon maturity, the microsporangia burst, releasing the pollen grains from the anther.

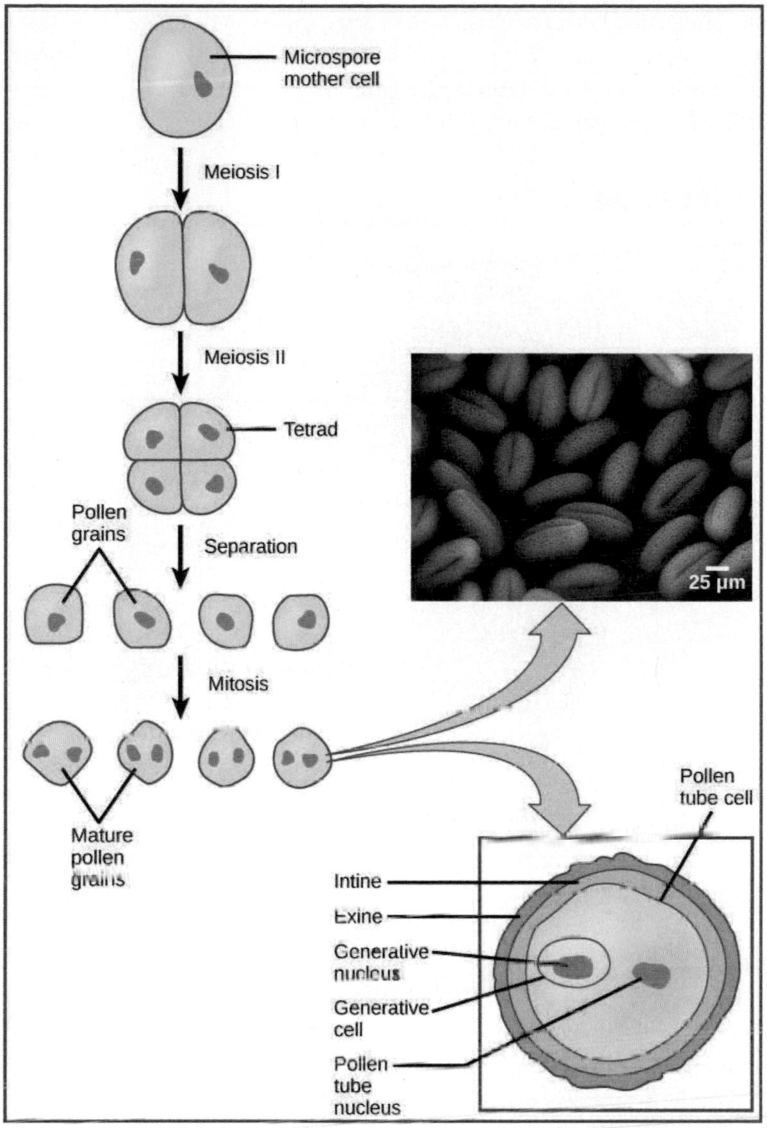

Figure 32.7 Pollen develops from the microspore mother cells. The mature pollen grain is composed of two cells: the pollen tube cell and the generative cell, which is inside the tube cell. The pollen grain has two coverings: an inner layer (intine) and an outer layer (exine). The inset scanning electron micrograph shows *Arabidopsis lyrata* pollen grains. (credit "pollen micrograph": modification of work by Robert R. Wise; scale-bar data from Matt Russell)

Each pollen grain has two coverings: the **exine** (thicker, outer layer) and the **intine** (Figure 32.7). The exine contains sporopollenin, a complex waterproofing substance supplied by the tapetal cells. Sporopollenin allows the pollen to survive under unfavorable conditions and to be carried by wind, water, or biological agents without undergoing damage.

Female Gametophyte (The Embryo Sac)

While the details may vary between species, the overall development of the female gametophyte has two distinct phases. First, in the process of **megasporogenesis**, a single cell in the diploid **megasporangium**—an area of tissue in the ovules—undergoes meiosis to produce four megaspores, only one of which survives. During the second phase, **megagametogenesis**, the surviving haploid megaspore undergoes mitosis to produce an eight-nucleate, seven-cell female gametophyte, also known as the megagametophyte or embryo sac. Two of the nuclei—the **polar nuclei**—move to the equator and fuse, forming a single, diploid central cell. This central cell later fuses with a sperm to form the triploid endosperm. Three nuclei position themselves on the end of the embryo sac opposite the micropyle and develop into the **antipodal** cells, which later degenerate. The nucleus closest to the micropyle becomes the female gamete, or egg cell, and the two adjacent nuclei develop into **synergid** cells (Figure 32.8). The synergids help guide the pollen tube for successful fertilization, after which they disintegrate. Once fertilization is complete, the resulting diploid zygote develops into the embryo, and the fertilized ovule forms the other tissues of the seed.

A double-layered integument protects the megasporangium and, later, the embryo sac. The integument will develop into the seed coat after fertilization and protect the entire seed. The ovule wall will become part of the fruit. The integuments, while protecting the megasporangium, do not enclose it completely, but leave an opening called the **micropyle**. The micropyle allows the pollen tube to enter the female gametophyte for fertilization.

VISUAL CONNECTION

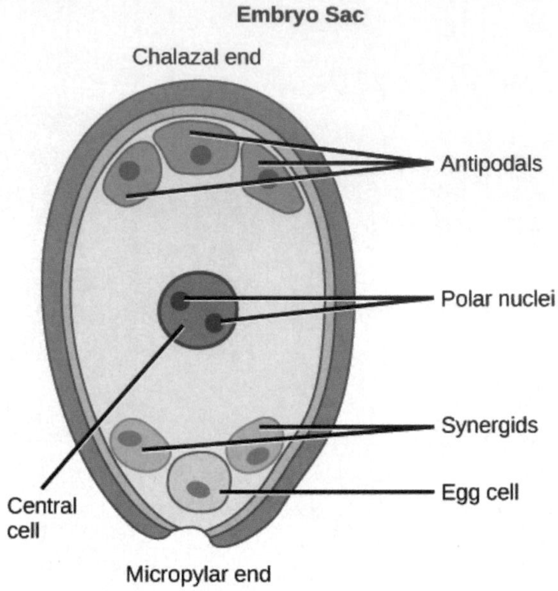

Figure 32.8 As shown in this diagram of the embryo sac in angiosperms, the ovule is covered by integuments and has an opening called a micropyle. Inside the embryo sac are three antipodal cells, two synergids, a central cell, and the egg cell.

An embryo sac is missing the synergids. What specific impact would you expect this to have on fertilization?

a. The pollen tube will be unable to form.
b. The pollen tube will form but will not be guided toward the egg.
c. Fertilization will not occur because the synergid is the egg.
d. Fertilization will occur but the embryo will not be able to grow.

Sexual Reproduction in Gymnosperms

As with angiosperms, the lifecycle of a gymnosperm is also characterized by alternation of generations. In conifers such as pines, the green leafy part of the plant is the sporophyte, and the cones contain the male and female gametophytes (Figure 32.9). The female cones are larger than the male cones and are positioned towards the top of the tree; the small, male cones are located in the lower region of the tree. Because the pollen is shed and blown by the wind, this arrangement makes it difficult for a gymnosperm to self-pollinate.

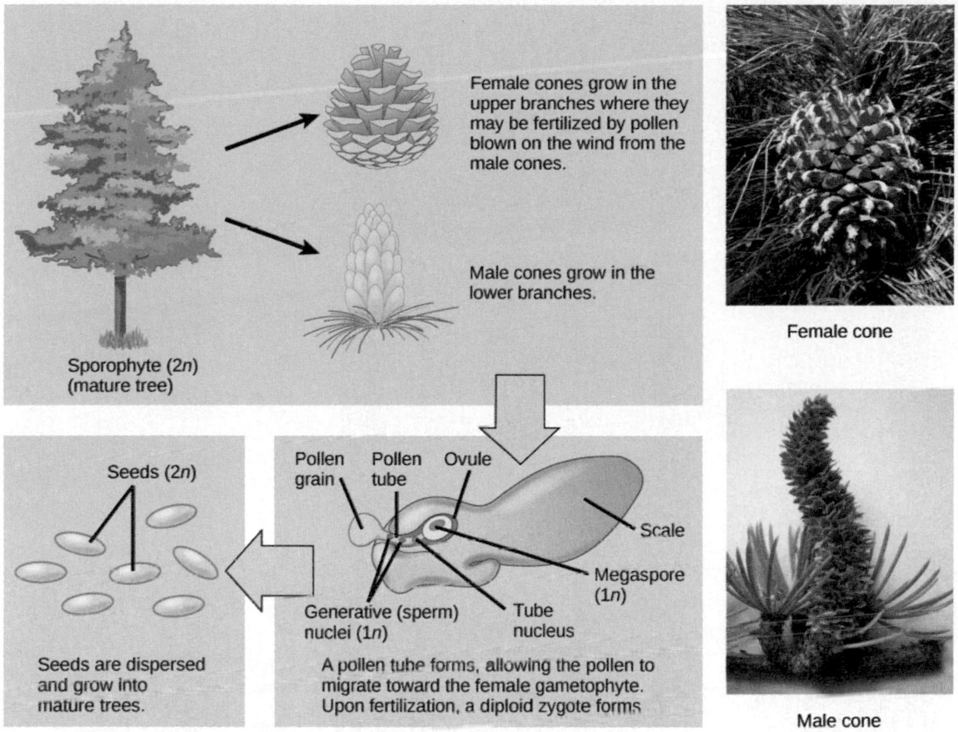

Figure 32.9 This image shows the lifecycle of a conifer. Pollen from male cones blows up into upper branches, where it fertilizes female cones. Examples are shown of female and male cones. (credit "female": modification of work by "Geographer"/Wikimedia Commons; credit "male": modification of work by Roger Griffith)

Male Gametophyte

A male cone has a central axis on which bracts, a type of modified leaf, are attached. The bracts are known as **microsporophylls** (Figure 32.10) and are the sites where microspores will develop. The microspores develop inside the microsporangium. Within the microsporangium, cells known as microsporocytes divide by meiosis to produce four haploid microspores. Further mitosis of the microspore produces two nuclei: the generative nucleus, and the tube nucleus. Upon maturity, the male gametophyte (pollen) is released from the male cones and is carried by the wind to land on the female cone.

LINK TO LEARNING

Watch this video to see a cedar releasing its pollen in the wind.

Click to view content (https://www.openstax.org/l/pollen_release)

Female Gametophyte

The female cone also has a central axis on which bracts known as **megasporophylls** (Figure 32.10) are present. In the female cone, megaspore mother cells are present in the megasporangium. The megaspore mother cell divides by meiosis to produce four haploid megaspores. One of the megaspores divides to form the multicellular female gametophyte, while the others divide to form the rest of the structure. The female gametophyte is contained within a structure called the archegonium.

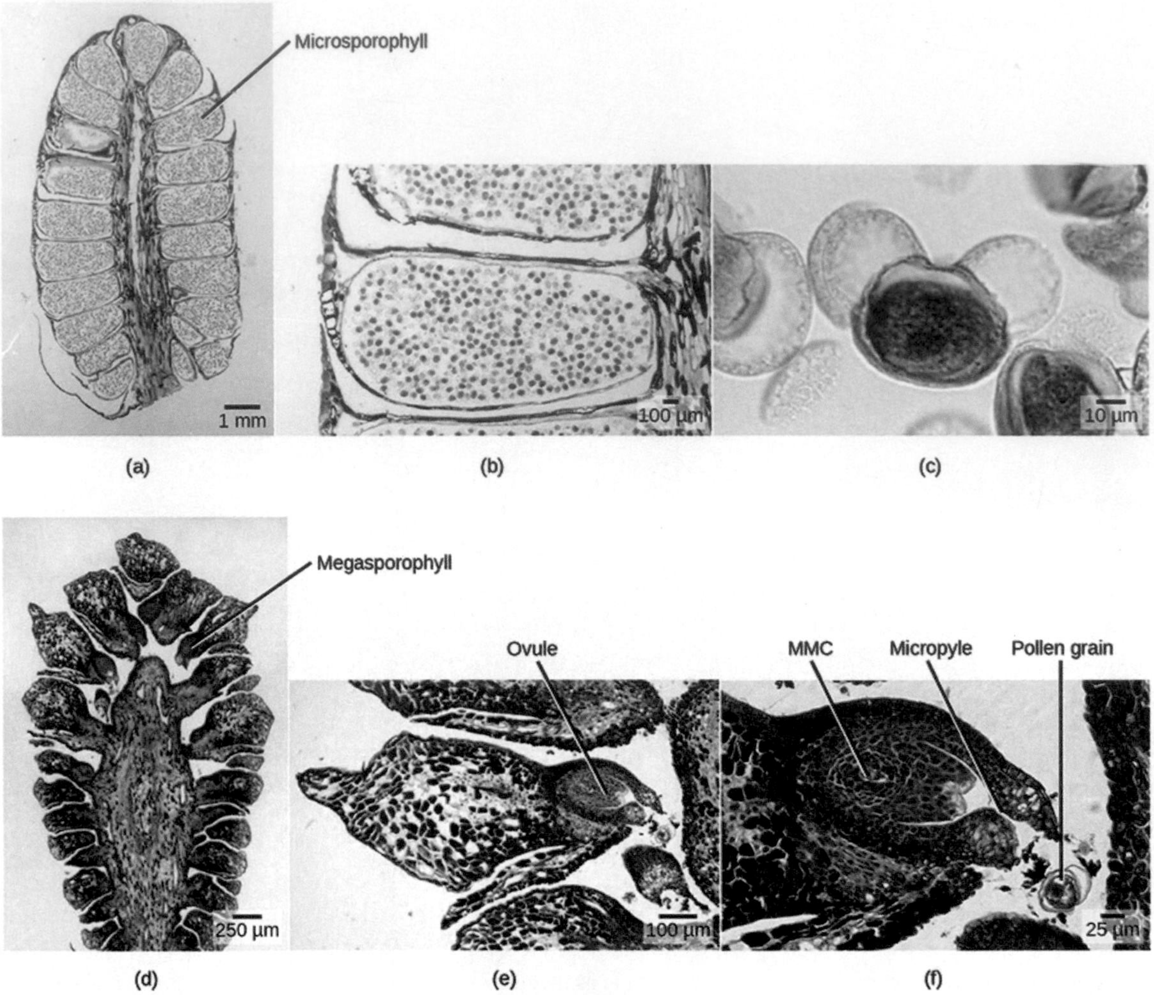

Figure 32.10 This series of micrographs shows male and female gymnosperm gametophytes. (a) This male cone, shown in cross section, has approximately 20 microsporophylls, each of which produces hundreds of male gametophytes (pollen grains). (b) Pollen grains are visible in this single microsporophyll. (c) This micrograph shows an individual pollen grain. (d) This cross section of a female cone shows portions of about 15 megasporophylls. (e) The ovule can be seen in this single megasporophyll. (f) Within this single ovule are the megaspore mother cell (MMC), micropyle, and a pollen grain. (credit: modification of work by Robert R. Wise; scale-bar data from Matt Russell)

Reproductive Process

Upon landing on the female cone, the tube cell of the pollen forms the pollen tube, through which the generative cell migrates towards the female gametophyte through the micropyle. It takes approximately one year for the pollen tube to grow and migrate towards the female gametophyte. The male gametophyte containing the generative cell splits into two sperm nuclei, one of which fuses with the egg, while the other degenerates. After fertilization of the egg, the diploid zygote is formed, which divides by mitosis to form the embryo. The scales of the cones are closed during development of the seed. The seed is covered by a seed coat, which is derived from the female sporophyte. Seed development takes another one to two years. Once the seed is ready to be dispersed, the bracts of the female cones open to allow the dispersal of seed; no fruit formation takes place because gymnosperm seeds have no covering.

Angiosperms versus Gymnosperms

Gymnosperm reproduction differs from that of angiosperms in several ways (Figure 32.11). In angiosperms, the female gametophyte exists in an enclosed structure—the ovule—which is within the ovary; in gymnosperms, the female gametophyte is present on exposed bracts of the female cone. Double fertilization is a key event in the lifecycle of angiosperms, but is completely absent in gymnosperms. The male and female gametophyte structures are present on separate male and female cones in gymnosperms, whereas in angiosperms, they are a part of the flower. Lastly, wind plays an important role in pollination in gymnosperms because pollen is blown by the wind to land on the female cones. Although many angiosperms are also wind-pollinated, animal pollination is more common.

Figure 32.11 (a) Angiosperms are flowering plants, and include grasses, herbs, shrubs and most deciduous trees, while (b) gymnosperms are conifers. Both produce seeds but have different reproductive strategies. (credit a: modification of work by Wendy Cutler; credit b: modification of work by Lews Castle UHI)

LINK TO LEARNING

View an animation of the double fertilization process of angiosperms.

Click to view content (https://www.openstax.org/l/angiosperms)

32.2 Pollination and Fertilization

By the end of this section, you will be able to do the following:

- Describe what must occur for plant fertilization
- Explain cross-pollination and the ways in which it takes place
- Describe the process that leads to the development of a seed
- Define double fertilization

In angiosperms, **pollination** is defined as the placement or transfer of pollen from the anther to the stigma of the same flower or another flower. In gymnosperms, pollination involves pollen transfer from the male cone to the female cone. Upon transfer, the pollen germinates to form the pollen tube and the sperm for fertilizing the egg. Pollination has been well studied since the time of Gregor Mendel. Mendel successfully carried out self- as well as cross-pollination in garden peas while studying how characteristics were passed on from one generation to the next. Today's crops are a result of plant breeding, which employs artificial selection to produce the present-day cultivars. A case in point is today's corn, which is a result of years of breeding that started with its ancestor, teosinte. The teosinte that the ancient Mayans originally began cultivating had tiny seeds—vastly different from today's relatively giant ears of corn. Interestingly, though these two plants appear to be entirely different, the genetic difference between them is miniscule.

Pollination takes two forms: self-pollination and cross-pollination. **Self-pollination** occurs when the pollen from the anther is deposited on the stigma of the same flower, or another flower on the same plant. **Cross-pollination** is the transfer of pollen from the anther of one flower to the stigma of another flower on a different individual of the same species. Self-pollination occurs in flowers where the stamen and carpel mature at the same time, and are positioned so that the pollen can land on the flower's stigma. This method of pollination does not require an investment from the plant to provide nectar and pollen as food for pollinators.

LINK TO LEARNING

Explore this interactive website (http://openstax.org/l/pollination) to review self-pollination and cross-pollination.

Living species are designed to ensure survival of their progeny; those that fail become extinct. Genetic diversity is therefore required so that in changing environmental or stress conditions, some of the progeny can survive. Self-pollination leads to the production of plants with less genetic diversity, since genetic material from the same plant is used to form gametes, and eventually, the zygote. In contrast, cross-pollination—or out-crossing—leads to greater genetic diversity because the microgametophyte and megagametophyte are derived from different plants.

Because cross-pollination allows for more genetic diversity, plants have developed many ways to avoid self-pollination. In some species, the pollen and the ovary mature at different times. These flowers make self-pollination nearly impossible. By the time pollen matures and has been shed, the stigma of this flower is mature and can only be pollinated by pollen from another flower. Some flowers have developed physical features that prevent self-pollination. The primrose is one such flower. Primroses have evolved two flower types with differences in anther and stigma length: the pin-eyed flower has anthers positioned at the pollen tube's halfway point, and the thrum-eyed flower's stigma is likewise located at the halfway point. Insects easily cross-pollinate while seeking the nectar at the bottom of the pollen tube. This phenomenon is also known as heterostyly. Many plants, such as cucumber, have male and female flowers located on different parts of the plant, thus making self-pollination difficult. In yet other species, the male and female flowers are borne on different plants (dioecious). All of these are barriers to self-pollination; therefore, the plants depend on pollinators to transfer pollen. The majority of pollinators are biotic agents such as insects (like bees, flies, and butterflies), bats, birds, and other animals. Other plant species are pollinated by abiotic agents, such as wind and water.

Everyday Connection

Incompatibility Genes in Flowers

In recent decades, incompatibility genes—which prevent pollen from germinating or growing into the stigma of a flower—have been discovered in many angiosperm species. If plants do not have compatible genes, the pollen tube stops growing. Self-incompatibility is controlled by the S (sterility) locus. Pollen tubes have to grow through the tissue of the stigma and style before they can enter the ovule. The carpel is selective in the type of pollen it allows to grow inside. The interaction is primarily between the pollen and the stigma epidermal cells. In some plants, like cabbage, the pollen is rejected at the surface of the stigma, and the unwanted pollen does not germinate. In other plants, pollen tube germination is arrested after growing one-third the length of the style, leading to pollen tube death. Pollen tube death is due either to apoptosis (programmed cell death) or to degradation of pollen tube RNA. The degradation results from the activity of a ribonuclease encoded by the S locus. The ribonuclease is secreted from the cells of the style in the extracellular matrix, which lies alongside the growing pollen tube.

In summary, self-incompatibility is a mechanism that prevents self-fertilization in many flowering plant species. The working of this self-incompatibility mechanism has important consequences for plant breeders because it inhibits the production of inbred and hybrid plants.

Pollination by Insects

Bees are perhaps the most important pollinator of many garden plants and most commercial fruit trees (Figure 32.12). The most common species of bees are bumblebees and honeybees. Since bees cannot see the color red, bee-pollinated flowers usually have shades of blue, yellow, or other colors. Bees collect energy-rich pollen or nectar for their survival and energy needs. They visit flowers that are open during the day, are brightly colored, have a strong aroma or scent, and have a tubular shape, typically with the presence of a nectar guide. A **nectar guide** includes regions on the flower petals that are visible only to bees, and not to humans; it helps to guide bees to the center of the flower, thus making the pollination process more efficient. The pollen sticks to the bees' fuzzy hair, and when the bee visits another flower, some of the pollen is transferred to the second flower. Recently, there have been many reports about the declining population of honeybees. Many flowers will remain unpollinated and not bear seed if honeybees disappear. The impact on commercial fruit growers could be devastating.

Figure 32.12 Insects, such as bees, are important agents of pollination. (credit: modification of work by Jon Sullivan)

Many flies are attracted to flowers that have a decaying smell or an odor of rotting flesh. These flowers, which produce nectar, usually have dull colors, such as brown or purple. They are found on the corpse flower or voodoo lily (*Amorphophallus*), dragon arum (*Dracunculus*), and carrion flower (*Stapleia, Rafflesia*). The nectar provides energy, whereas the pollen provides protein. Wasps are also important insect pollinators, and pollinate many species of figs.

Butterflies, such as the monarch, pollinate many garden flowers and wildflowers, which usually occur in clusters. These flowers are brightly colored, have a strong fragrance, are open during the day, and have nectar guides to make access to nectar easier. The pollen is picked up and carried on the butterfly's limbs. Moths, on the other hand, pollinate flowers during the late afternoon and night. The flowers pollinated by moths are pale or white and are flat, enabling the moths to land. One well-studied example of a moth-pollinated plant is the yucca plant, which is pollinated by the yucca moth. The shape of the flower and moth have adapted in such a way as to allow successful pollination. The moth deposits pollen on the sticky stigma for fertilization to occur later. The female moth also deposits eggs into the ovary. As the eggs develop into larvae, they obtain food from the flower and developing seeds. Thus, both the insect and flower benefit from each other in this symbiotic relationship. The corn earworm moth and Gaura plant have a similar relationship (Figure 32.13).

Figure 32.13 A corn earworm sips nectar from a night-blooming Gaura plant. (credit: Juan Lopez, USDA ARS)

Pollination by Bats

In the tropics and deserts, bats are often the pollinators of nocturnal flowers such as agave, guava, and morning glory. The flowers are usually large and white or pale-colored; thus, they can be distinguished from the dark surroundings at night. The flowers have a strong, fruity, or musky fragrance and produce large amounts of nectar. They are naturally large and wide-mouthed to accommodate the head of the bat. As the bats seek the nectar, their faces and heads become covered with pollen, which is then transferred to the next flower.

Pollination by Birds

Many species of small birds, such as the hummingbird (Figure 32.14) and sun birds, are pollinators for plants such as orchids and other wildflowers. Flowers visited by birds are usually sturdy and are oriented in such a way as to allow the birds to stay near the flower without getting their wings entangled in the nearby flowers. The flower typically has a curved, tubular shape, which allows access for the bird's beak. Brightly colored, odorless flowers that are open during the day are pollinated by birds. As a bird seeks energy-rich nectar, pollen is deposited on the bird's head and neck and is then transferred to the next flower it visits. Botanists have been known to determine the range of extinct plants by collecting and identifying pollen from 200-year-old bird specimens from the same site.

Figure 32.14 Hummingbirds have adaptations that allow them to reach the nectar of certain tubular flowers. (credit: Lori Branham)

Pollination by Wind

Most species of conifers, and many angiosperms, such as grasses, maples and oaks, are pollinated by wind. Pine cones are brown and unscented, while the flowers of wind-pollinated angiosperm species are usually green, small, may have small or no petals, and produce large amounts of pollen. Unlike the typical insect-pollinated flowers, flowers adapted to pollination by wind do not produce nectar or scent. In wind-pollinated species, the microsporangia hang out of the flower, and, as the wind blows, the lightweight pollen is carried with it (Figure 32.15). The flowers usually emerge early in the spring, before the leaves, so that the leaves do not block the movement of the wind. The pollen is deposited on the exposed feathery stigma of the flower (Figure 32.16).

Figure 32.15 A person knocks pollen from a pine tree.

Figure 32.16 These male (a) and female (b) catkins are from the goat willow tree (*Salix caprea*). Note how both structures are light and feathery to better disperse and catch the wind-blown pollen.

Pollination by Water

Some weeds, such as Australian sea grass and pond weeds, are pollinated by water. The pollen floats on water, and when it comes into contact with the flower, it is deposited inside the flower.

EVOLUTION CONNECTION

Pollination by Deception

Orchids are highly valued flowers, with many rare varieties (Figure 32.17). They grow in a range of specific habitats, mainly in the tropics of Asia, South America, and Central America. At least 25,000 species of orchids have been identified.

Figure 32.17 Certain orchids use food deception or sexual deception to attract pollinators. Shown here is a bee orchid (*Ophrys apifera*). (credit: David Evans)

Flowers often attract pollinators with food rewards, in the form of nectar. However, some species of orchid are an exception to this standard: they have evolved different ways to attract the desired pollinators. They use a method known as food deception, in which bright colors and perfumes are offered, but no food. *Anacamptis morio*, commonly known as the green-winged orchid, bears bright purple flowers and emits a strong scent. The bumblebee, its main pollinator, is attracted to the flower because of the strong scent—which usually indicates food for a bee—and in the process, picks up the pollen to be transported to another flower.

Other orchids use sexual deception. *Chiloglottis trapeziformis* emits a compound that smells the same as the pheromone emitted by a female wasp to attract male wasps. The male wasp is attracted to the scent, lands on the orchid flower, and in the process, transfers pollen. Some orchids, like the Australian hammer orchid, use scent as well as visual trickery in yet another sexual deception strategy to attract wasps. The flower of this orchid mimics the appearance of a female wasp and emits a pheromone. The male wasp tries to mate with what appears to be a female wasp, and in the process, picks up pollen, which it then transfers to the next counterfeit mate.

Double Fertilization

After pollen is deposited on the stigma, it must germinate and grow through the style to reach the ovule. The microspores, or the pollen, contain two cells: the pollen tube cell and the generative cell. The pollen tube cell grows into a pollen tube through which the generative cell travels. The germination of the pollen tube requires water, oxygen, and certain chemical signals. As it travels through the style to reach the embryo sac, the pollen tube's growth is supported by the tissues of the style. In the meantime, if the generative cell has not already split into two cells, it now divides to form two sperm cells. The pollen tube is guided by the chemicals secreted by the synergids present in the embryo sac, and it enters the ovule sac through the micropyle. Of the two sperm cells, one sperm fertilizes the egg cell, forming a diploid zygote; the other sperm fuses with the two polar nuclei, forming a triploid cell that develops into the **endosperm**. Together, these two fertilization events in angiosperms are known as **double fertilization** (Figure 32.18). After fertilization is complete, no other sperm can enter. The fertilized ovule forms the seed, whereas the tissues of the ovary become the fruit, usually enveloping the seed.

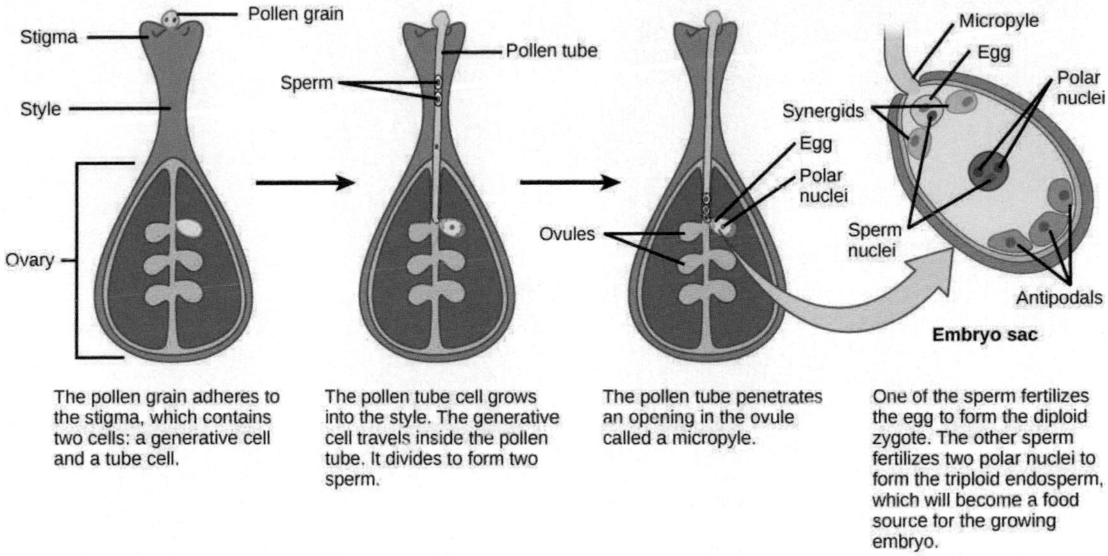

Figure 32.18 In angiosperms, one sperm fertilizes the egg to form the 2*n* zygote, and the other sperm fertilizes the central cell to form the 3*n* endosperm. This is called a double fertilization.

After fertilization, the zygote divides to form two cells: the upper cell, or terminal cell, and the lower, or basal, cell. The division of the basal cell gives rise to the **suspensor**, which eventually makes connection with the maternal tissue. The suspensor provides a route for nutrition to be transported from the mother plant to the growing embryo. The terminal cell also divides, giving rise to a globular-shaped proembryo (Figure 32.19a). In dicots (eudicots), the developing embryo has a heart shape, due to the presence of the two rudimentary **cotyledons** (Figure 32.19b). In non-endospermic dicots, such as *Capsella bursa*, the endosperm develops initially, but is then digested, and the food reserves are moved into the two cotyledons. As the embryo and cotyledons enlarge, they run out of room inside the developing seed, and are forced to bend (Figure 32.19c). Ultimately, the embryo and cotyledons fill the seed (Figure 32.19d), and the seed is ready for dispersal. Embryonic development is suspended after some time, and growth is resumed only when the seed germinates. The developing seedling will rely on the food reserves stored in the cotyledons until the first set of leaves begin photosynthesis.

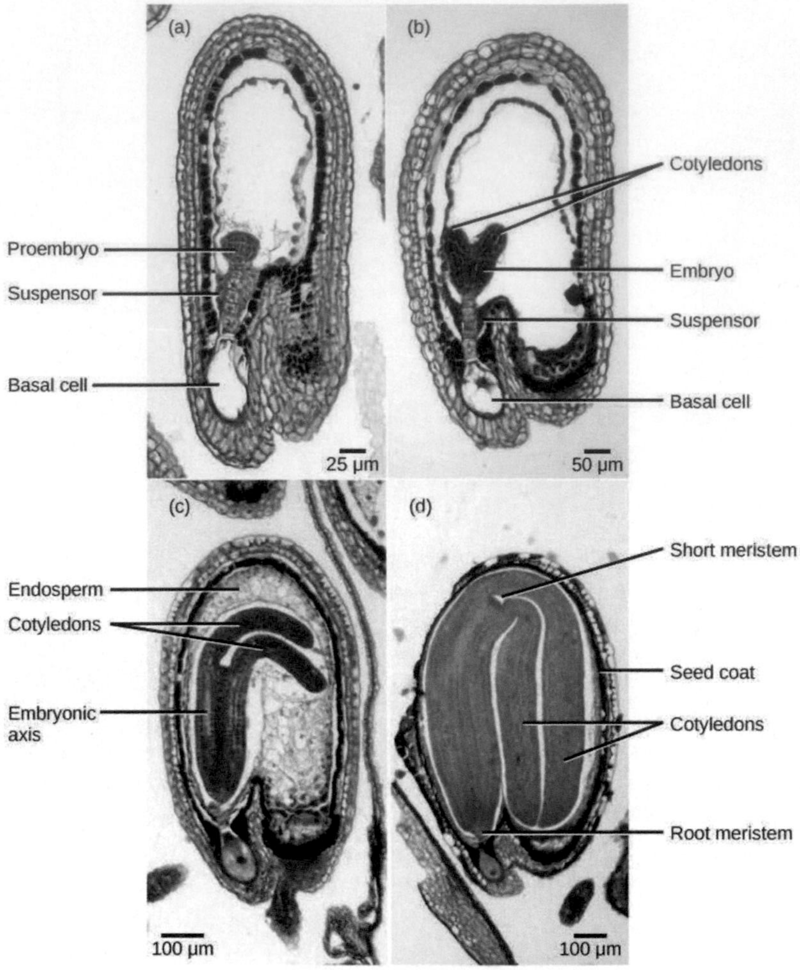

Figure 32.19 Shown are the stages of embryo development in the ovule of a shepherd's purse (*Capsella bursa*). After fertilization, the zygote divides to form an upper terminal cell and a lower basal cell. (a) In the first stage of development, the terminal cell divides, forming a globular pro-embryo. The basal cell also divides, giving rise to the suspensor. (b) In the second stage, the developing embryo has a heart shape due to the presence of cotyledons. (c) In the third stage, the growing embryo runs out of room and starts to bend. (d) Eventually, it completely fills the seed. (credit: modification of work by Robert R. Wise; scale-bar data from Matt Russell)

Development of the Seed

The mature ovule develops into the seed. A typical seed contains a seed coat, cotyledons, endosperm, and a single embryo (Figure 32.20).

VISUAL CONNECTION

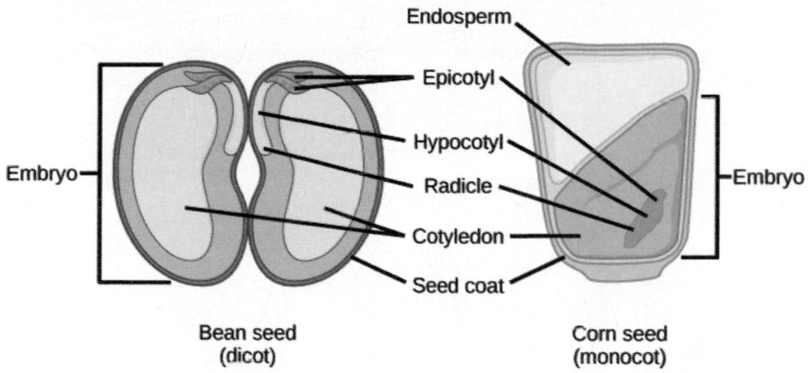

Figure 32.20 The structures of dicot and monocot seeds are shown. Dicots (left) have two cotyledons. Monocots, such as corn (right), have one cotyledon, called the scutellum; it channels nutrition to the growing embryo. Both monocot and dicot embryos have a plumule that forms the leaves, a hypocotyl that forms the stem, and a radicle that forms the root. The embryonic axis comprises everything between the plumule and the radicle, not including the cotyledon(s).

What of the following statements is true?

a. Both monocots and dicots have an endosperm.
b. The radicle develops into the root.
c. The plumule is part of the epicotyl.
d. The endosperm is part of the embryo.

The storage of food reserves in angiosperm seeds differs between monocots and dicots. In monocots, such as corn and wheat, the single cotyledon is called a **scutellum**; the scutellum is connected directly to the embryo via vascular tissue (xylem and phloem). Food reserves are stored in the large endosperm. Upon germination, enzymes are secreted by the **aleurone**, a single layer of cells just inside the seed coat that surrounds the endosperm and embryo. The enzymes degrade the stored carbohydrates, proteins and lipids, the products of which are absorbed by the scutellum and transported via a vasculature strand to the developing embryo. Therefore, the scutellum can be seen to be an absorptive organ, not a storage organ.

The two cotyledons in the dicot seed also have vascular connections to the embryo. In **endospermic dicots**, the food reserves are stored in the endosperm. During germination, the two cotyledons therefore act as absorptive organs to take up the enzymatically released food reserves, much like in monocots (monocots, by definition, also have endospermic seeds). Tobacco (*Nicotiana tabaccum*), tomato (*Solanum lycopersicum*), and pepper (*Capsicum annuum*) are examples of endospermic dicots. In **non-endospermic dicots**, the triploid endosperm develops normally following double fertilization, but the endosperm food reserves are quickly remobilized and moved into the developing cotyledon for storage. The two halves of a peanut seed (*Arachis hypogaea*) and the split peas (*Pisum sativum*) of split pea soup are individual cotyledons loaded with food reserves.

The seed, along with the ovule, is protected by a seed coat that is formed from the integuments of the ovule sac. In dicots, the seed coat is further divided into an outer coat known as the **testa** and inner coat known as the **tegmen**.

The embryonic axis consists of three parts: the plumule, the radicle, and the hypocotyl. The portion of the embryo between the cotyledon attachment point and the radicle is known as the **hypocotyl** (hypocotyl means "below the cotyledons"). The embryonic axis terminates in a **radicle** (the embryonic root), which is the region from which the root will develop. In dicots, the hypocotyls extend above ground, giving rise to the stem of the plant. In monocots, the hypocotyl does not show above ground because monocots do not exhibit stem elongation. The part of the embryonic axis that projects above the cotyledons is known as the **epicotyl**. The **plumule** is composed of the epicotyl, young leaves, and the shoot apical meristem.

Upon germination in dicot seeds, the epicotyl is shaped like a hook with the plumule pointing downwards. This shape is called the plumule hook, and it persists as long as germination proceeds in the dark. Therefore, as the epicotyl pushes through the tough and abrasive soil, the plumule is protected from damage. Upon exposure to light, the hypocotyl hook straightens out, the young foliage leaves face the sun and expand, and the epicotyl continues to elongate. During this time, the radicle is also growing and producing the primary root. As it grows downward to form the tap root, lateral roots branch off to all sides, producing the typical dicot tap root system.

In monocot seeds (Figure 32.21), the testa and tegmen of the seed coat are fused. As the seed germinates, the primary root emerges, protected by the root-tip covering: the **coleorhiza**. Next, the primary shoot emerges, protected by the **coleoptile**: the covering of the shoot tip. Upon exposure to light (i.e., when the plumule has exited the soil and the protective coleoptile is no longer needed), elongation of the coleoptile ceases and the leaves expand and unfold. At the other end of the embryonic axis, the primary root soon dies, while other, adventitious roots (roots that do not arise from the usual place – i.e., the root) emerge from the base of the stem. This gives the monocot a fibrous root system.

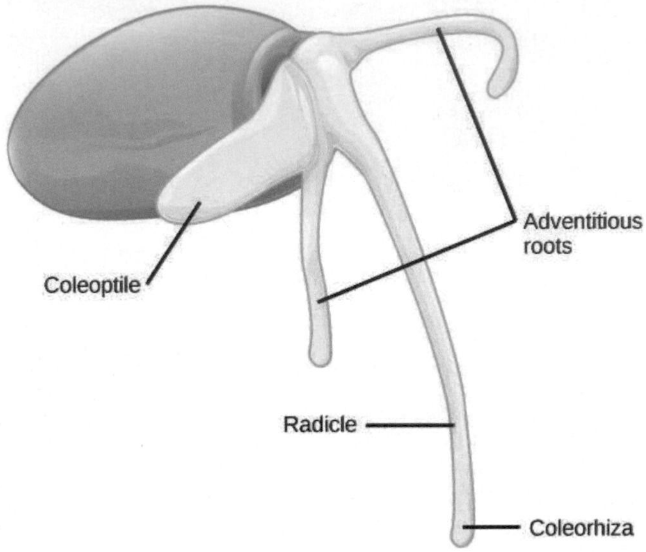

Figure 32.21 As this monocot grass seed germinates, the primary root, or radicle, emerges first, followed by the primary shoot, or coleoptile, and the adventitious roots.

Seed Germination

Many mature seeds enter a period of inactivity, or extremely low metabolic activity: a process known as **dormancy**, which may last for months, years, or even centuries. Dormancy helps keep seeds viable during unfavorable conditions. Upon a return to favorable conditions, seed germination takes place. Favorable conditions could be as diverse as moisture, light, cold, fire, or chemical treatments. After heavy rains, many new seedlings emerge. Forest fires also lead to the emergence of new seedlings. Some seeds require **vernalization** (cold treatment) before they can germinate. This guarantees that seeds produced by plants in temperate climates will not germinate until the spring. Plants growing in hot climates may have seeds that need a heat treatment in order to germinate, to avoid germination in the hot, dry summers. In many seeds, the presence of a thick seed coat retards the ability to germinate. **Scarification**, which includes mechanical or chemical processes to soften the seed coat, is often employed before germination. Presoaking in hot water, or passing through an acid environment, such as an animal's digestive tract, may also be employed.

Depending on seed size, the time taken for a seedling to emerge may vary. Species with large seeds have enough food reserves to germinate deep below ground, and still extend their epicotyl all the way to the soil surface. Seeds of small-seeded species usually require light as a germination cue. This ensures the seeds only germinate at or near the soil surface (where the light is greatest). If they were to germinate too far underneath the surface, the developing seedling would not have enough food reserves to reach the sunlight.

Development of Fruit and Fruit Types

After fertilization, the ovary of the flower usually develops into the fruit. Fruits are usually associated with having a sweet taste; however, not all fruits are sweet. Botanically, the term "fruit" is used for a ripened ovary. In most cases, flowers in which fertilization has taken place will develop into fruits, and flowers in which fertilization has not taken place will not. Some fruits develop from the ovary and are known as true fruits, whereas others develop from other parts of the female gametophyte and are known as accessory fruits. The fruit encloses the seeds and the developing embryo, thereby providing it with protection. Fruits are of many types, depending on their origin and texture. The sweet tissue of the blackberry, the red flesh of the tomato, the shell of the peanut, and the hull of corn (the tough, thin part that gets stuck in your teeth when you eat popcorn) are all fruits. As the fruit matures, the seeds also mature.

Fruits may be classified as simple, aggregate, multiple, or accessory, depending on their origin (Figure 32.22). If the fruit develops from a single carpel or fused carpels of a single ovary, it is known as a **simple fruit**, as seen in nuts and beans. An **aggregate fruit** is one that develops from more than one carpel, but all are in the same flower: the mature carpels fuse together to form the entire fruit, as seen in the raspberry. **Multiple fruit** develops from an inflorescence or a cluster of flowers. An example is the pineapple, where the flowers fuse together to form the fruit. **Accessory fruits** (sometimes called false fruits) are not derived from the ovary, but from another part of the flower, such as the receptacle (strawberry) or the hypanthium (apples and pears).

Figure 32.22 There are four main types of fruits. Simple fruits, such as these nuts, are derived from a single ovary. Aggregate fruits, like raspberries, form from many carpels that fuse together. Multiple fruits, such as pineapple, form from a cluster of flowers called an inflorescence. Accessory fruit, like the apple, are formed from a part of the plant other than the ovary. (credit "nuts": modification of work by Petr Kratochvil; credit "raspberries": modification of work by Cory Zanker; credit "pineapple": modification of work by Howie Le; credit "apple": modification of work by Paolo Neo)

Fruits generally have three parts: the **exocarp** (the outermost skin or covering), the **mesocarp** (middle part of the fruit), and the **endocarp** (the inner part of the fruit). Together, all three are known as the **pericarp**. The mesocarp is usually the fleshy, edible part of the fruit; however, in some fruits, such as the almond, the endocarp is the edible part. In many fruits, two or all three of the layers are fused, and are indistinguishable at maturity. Fruits can be dry or fleshy. Furthermore, fruits can be divided into dehiscent or indehiscent types. Dehiscent fruits, such as peas, readily release their seeds, while indehiscent fruits, like peaches, rely on decay to release their seeds.

Fruit and Seed Dispersal

The fruit has a single purpose: seed dispersal. Seeds contained within fruits need to be dispersed far from the mother plant, so they may find favorable and less competitive conditions in which to germinate and grow.

Some fruit have built-in mechanisms so they can disperse by themselves, whereas others require the help of agents like wind, water, and animals (Figure 32.23). Modifications in seed structure, composition, and size help in dispersal. Wind-dispersed fruit are lightweight and may have wing-like appendages that allow them to be carried by the wind. Some have a parachute-like structure to keep them afloat. Some fruits—for example, the dandelion—have hairy, weightless structures that are suited to dispersal by wind.

Seeds dispersed by water are contained in light and buoyant fruit, giving them the ability to float. Coconuts are well known for their ability to float on water to reach land where they can germinate. Similarly, willow and silver birches produce lightweight fruit that can float on water.

Animals and birds eat fruits, and the seeds that are not digested are excreted in their droppings some distance away. Some animals, like squirrels, bury seed-containing fruits for later use; if the squirrel does not find its stash of fruit, and if conditions are favorable, the seeds germinate. Some fruits, like the cocklebur, have hooks or sticky structures that stick to an animal's coat and are then transported to another place. Humans also play a big role in dispersing seeds when they carry fruits to new places and throw away the inedible part that contains the seeds.

All of the above mechanisms allow for seeds to be dispersed through space, much like an animal's offspring can move to a new location. Seed dormancy, which was described earlier, allows plants to disperse their progeny through time: something animals cannot do. Dormant seeds can wait months, years, or even decades for the proper conditions for germination and propagation of the species.

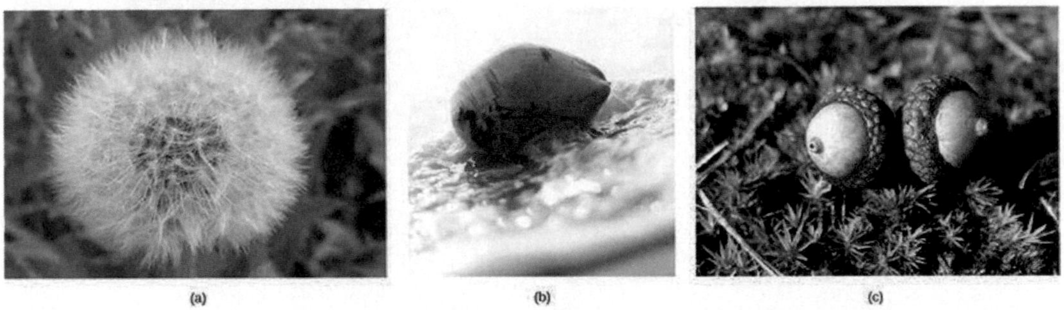

Figure 32.23 Fruits and seeds are dispersed by various means. (a) Dandelion seeds are dispersed by wind, the (b) coconut seed is dispersed by water, and the (c) acorn is dispersed by animals that cache and then forget it. (credit a: modification of work by "Rosendahl"/Flickr; credit b: modification of work by Shine Oa; credit c: modification of work by Paolo Neo)

32.3 Asexual Reproduction

By the end of this section, you will be able to do the following:

- Compare the mechanisms and methods of natural and artificial asexual reproduction
- Describe the advantages and disadvantages of natural and artificial asexual reproduction
- Discuss plant life spans

Many plants are able to propagate themselves using asexual reproduction. This method does not require the investment required to produce a flower, attract pollinators, or find a means of seed dispersal. Asexual reproduction produces plants that are genetically identical to the parent plant because no mixing of male and female gametes takes place. Traditionally, these plants survive well under stable environmental conditions when compared with plants produced from sexual reproduction because they carry genes identical to those of their parents.

Many different types of roots exhibit asexual reproduction (Figure 32.24). The corm is used by gladiolus and garlic. Bulbs, such as a scaly bulb in lilies and a tunicate bulb in daffodils, are other common examples. A potato is a stem tuber, while parsnip propagates from a taproot. Ginger and iris produce rhizomes, while ivy uses an adventitious root (a root arising from a plant part other than the main or primary root), and the strawberry plant has a stolon, which is also called a runner.

Figure 32.24 Different types of stems allow for asexual reproduction. (a) The corm of a garlic plant looks similar to (b) a tulip bulb, but the corm is solid tissue, while the bulb consists of layers of modified leaves that surround an underground stem. Both corms and bulbs can self-propagate, giving rise to new plants. (c) Ginger forms masses of stems called rhizomes that can give rise to multiple plants. (d) Potato plants form fleshy stem tubers. Each eye in the stem tuber can give rise to a new plant. (e) Strawberry plants form stolons: stems that grow at the soil surface or just below ground and can give rise to new plants. (credit a: modification of work by Dwight Sipler; credit c: modification of work by Albert Cahalan, USDA ARS; credit d: modification of work by Richard North; credit e: modification of work by Julie Magro)

Some plants can produce seeds without fertilization. Either the ovule or part of the ovary, which is diploid in nature, gives rise to a new seed. This method of reproduction is known as **apomixis**.

An advantage of asexual reproduction is that the resulting plant will reach maturity faster. Since the new plant is arising from an adult plant or plant parts, it will also be sturdier than a seedling. Asexual reproduction can take place by natural or artificial (assisted by humans) means.

Natural Methods of Asexual Reproduction

Natural methods of asexual reproduction include strategies that plants have developed to self-propagate. Many plants—like ginger, onion, gladioli, and dahlia—continue to grow from buds that are present on the surface of the stem. In some plants, such as the sweet potato, adventitious roots or runners can give rise to new plants (Figure 32.25). In *Bryophyllum* and kalanchoe, the leaves have small buds on their margins. When these are detached from the plant, they grow into independent plants; or, they may start growing into independent plants if the leaf touches the soil. Some plants can be propagated through cuttings alone.

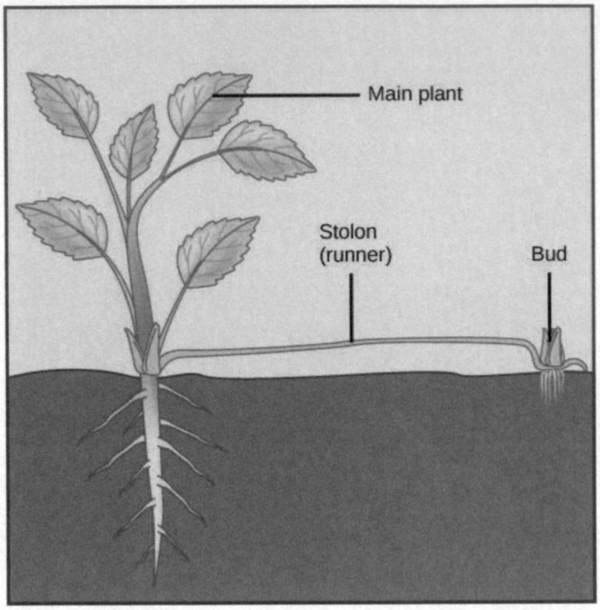

Figure 32.25 A stolon, or runner, is a stem that runs along the ground. At the nodes, it forms adventitious roots and buds that grow into a new plant.

Artificial Methods of Asexual Reproduction

These methods are frequently employed to give rise to new, and sometimes novel, plants. They include grafting, cutting, layering, and micropropagation.

Grafting

Grafting has long been used to produce novel varieties of roses, citrus species, and other plants. In **grafting**, two plant species are used; part of the stem of the desirable plant is grafted onto a rooted plant called the stock. The part that is grafted or attached is called the **scion**. Both are cut at an oblique angle (any angle other than a right angle), placed in close contact with each other, and are then held together (Figure 32.26). Matching up these two surfaces as closely as possible is extremely important because these will be holding the plant together. The vascular systems of the two plants grow and fuse, forming a graft. After a period of time, the scion starts producing shoots, and eventually starts bearing flowers and fruits. Grafting is widely used in viticulture (grape growing) and the citrus industry. Scions capable of producing a particular fruit variety are grafted onto root stock with specific resistance to disease.

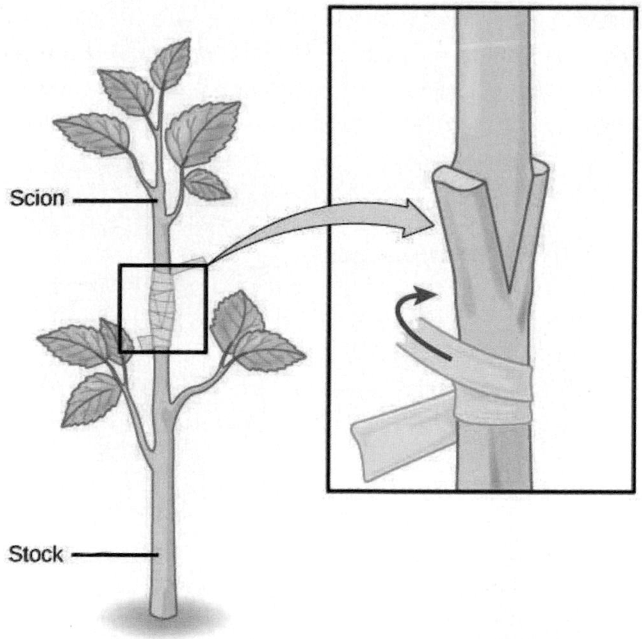

Figure 32.26 Grafting is an artificial method of asexual reproduction used to produce plants combining favorable stem characteristics with favorable root characteristics. The stem of the plant to be grafted is known as the scion, and the root is called the stock.

Cutting

Plants such as coleus and money plant are propagated through stem **cuttings**, where a portion of the stem containing nodes and internodes is placed in moist soil and allowed to root. In some species, stems can start producing a root even when placed only in water. For example, leaves of the African violet will root if kept in water undisturbed for several weeks.

Layering

Layering is a method in which a stem attached to the plant is bent and covered with soil. Young stems that can be bent easily without any injury are preferred. Jasmine and bougainvillea (paper flower) can be propagated this way (Figure 32.27). In some plants, a modified form of layering known as air layering is employed. A portion of the bark or outermost covering of the stem is removed and covered with moss, which is then taped. Some gardeners also apply rooting hormone. After some time, roots will appear, and this portion of the plant can be removed and transplanted into a separate pot.

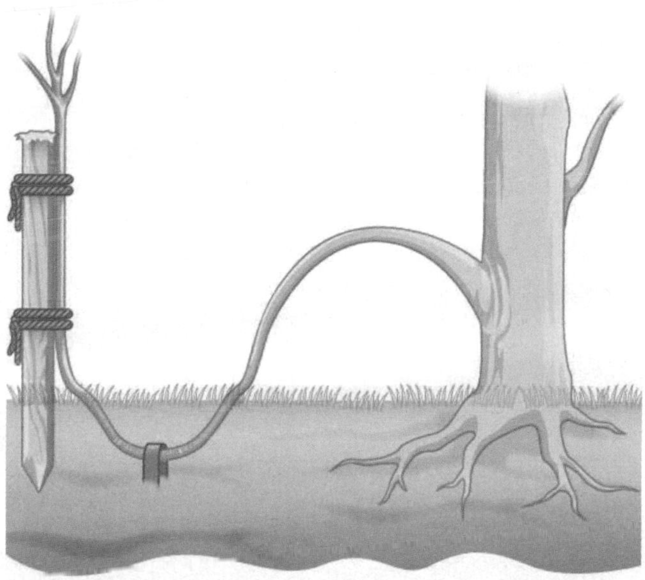

Figure 32.27 In layering, a part of the stem is buried so that it forms a new plant. (credit: modification of work by Pearson Scott Foresman, donated to the Wikimedia Foundation)

Micropropagation

Micropropagation (also called plant tissue culture) is a method of propagating a large number of plants from a single plant in a short time under laboratory conditions (Figure 32.28). This method allows propagation of rare, endangered species that may be difficult to grow under natural conditions, are economically important, or are in demand as disease-free plants.

Figure 32.28 Micropropagation is used to propagate plants in sterile conditions. (credit: Nikhilesh Sanyal)

To start plant tissue culture, a part of the plant such as a stem, leaf, embryo, anther, or seed can be used. The plant material is thoroughly sterilized using a combination of chemical treatments standardized for that species. Under sterile conditions, the plant material is placed on a plant tissue culture medium that contains all the minerals, vitamins, and hormones required by the plant. The plant part often gives rise to an undifferentiated mass known as callus, from which individual plantlets begin to grow after a period of time. These can be separated and are first grown under greenhouse conditions before they are moved to field conditions.

Plant Life Spans

The length of time from the beginning of development to the death of a plant is called its life span. The life cycle, on the other hand, is the sequence of stages a plant goes through from seed germination to seed production of the mature plant. Some plants, such as annuals, only need a few weeks to grow, produce seeds and die. Other plants, such as the bristlecone pine, live for thousands of years. Some bristlecone pines have a documented age of 4,500 years (Figure 32.29). Even as some parts of a plant, such as regions containing meristematic tissue—the area of active plant growth consisting of undifferentiated cells capable of cell division—continue to grow, some parts undergo programmed cell death (apoptosis). The cork found on stems, and the water-conducting tissue of the xylem, for example, are composed of dead cells.

Figure 32.29 The bristlecone pine, shown here in the Ancient Bristlecone Pine Forest in the White Mountains of eastern California, has been known to live for 4,500 years. (credit: Rick Goldwaser)

Plant species that complete their lifecycle in one season are known as annuals, an example of which is *Arabidopsis*, or mouse-ear cress. Biennials such as carrots complete their lifecycle in two seasons. In a biennial's first season, the plant has a vegetative phase, whereas in the next season, it completes its reproductive phase. Commercial growers harvest the carrot roots after the first year of growth, and do not allow the plants to flower. Perennials, such as the magnolia, complete their lifecycle in two years or more.

In another classification based on flowering frequency, **monocarpic** plants flower only once in their lifetime; examples include bamboo and yucca. During the vegetative period of their life cycle (which may be as long as 120 years in some bamboo species), these plants may reproduce asexually and accumulate a great deal of food material that will be required during their once-in-a-lifetime flowering and setting of seed after fertilization. Soon after flowering, these plants die. **Polycarpic** plants form flowers many times during their lifetime. Fruit trees, such as apple and orange trees, are polycarpic; they flower every year. Other polycarpic species, such as perennials, flower several times during their life span, but not each year. By this means, the plant does not require all its nutrients to be channelled towards flowering each year.

As is the case with all living organisms, genetics and environmental conditions have a role to play in determining how long a plant will live. Susceptibility to disease, changing environmental conditions, drought, cold, and competition for nutrients are some of the factors that determine the survival of a plant. Plants continue to grow, despite the presence of dead tissue such as cork. Individual parts of plants, such as flowers and leaves, have different rates of survival. In many trees, the older leaves turn yellow and eventually fall from the tree. Leaf fall is triggered by factors such as a decrease in photosynthetic efficiency, due to shading by upper leaves, or oxidative damage incurred as a result of photosynthetic reactions. The components of the part to be shed are recycled by the plant for use in other processes, such as development of seed and storage. This process is known as nutrient recycling.

The aging of a plant and all the associated processes is known as **senescence**, which is marked by several complex biochemical changes. One of the characteristics of senescence is the breakdown of chloroplasts, which is characterized by the yellowing of leaves. The chloroplasts contain components of photosynthetic machinery such as membranes and proteins. Chloroplasts also contain DNA. The proteins, lipids, and nucleic acids are broken down by specific enzymes into smaller molecules and salvaged by the plant to support the growth of other plant tissues.

The complex pathways of nutrient recycling within a plant are not well understood. Hormones are known to play a role in senescence. Applications of cytokinins and ethylene delay or prevent senescence; in contrast, abscissic acid causes premature onset of senescence.

KEY TERMS

accessory fruit fruit derived from tissues other than the ovary

aggregate fruit fruit that develops from multiple carpels in the same flower

aleurone single layer of cells just inside the seed coat that secretes enzymes upon germination

androecium sum of all the stamens in a flower

antipodals the three cells away from the micropyle

apomixis process by which seeds are produced without fertilization of sperm and egg

coleoptile covering of the shoot tip, found in germinating monocot seeds

coleorhiza covering of the root tip, found in germinating monocot seeds

cotyledon fleshy part of seed that provides nutrition to the seed

cross-pollination transfer of pollen from the anther of one flower to the stigma of a different flower

cutting method of asexual reproduction where a portion of the stem contains nodes and internodes is placed in moist soil and allowed to root

dormancy period of no growth and very slow metabolic processes

double fertilization two fertilization events in angiosperms; one sperm fuses with the egg, forming the zygote, whereas the other sperm fuses with the polar nuclei, forming endosperm

endocarp innermost part of fruit

endosperm triploid structure resulting from fusion of a sperm with polar nuclei, which serves as a nutritive tissue for embryo

endospermic dicot dicot that stores food reserves in the endosperm

epicotyl embryonic shoot above the cotyledons

exine outermost covering of pollen

exocarp outermost covering of a fruit

gametophyte multicellular stage of the plant that gives rise to haploid gametes or spores

grafting method of asexual reproduction where the stem from one plant species is spliced to a different plant

gravitropism response of a plant growth in the same direction as gravity

gynoecium the sum of all the carpels in a flower

hypocotyl embryonic axis above the cotyledons

intine inner lining of the pollen

layering method of propagating plants by bending a stem under the soil

megagametogenesis second phase of female gametophyte development, during which the surviving haploid megaspore undergoes mitosis to produce an eight-nucleate, seven-cell female gametophyte, also known as the megagametophyte or embryo sac.

megasporangium tissue found in the ovary that gives rise to the female gamete or egg

megasporogenesis first phase of female gametophyte development, during which a single cell in the diploid megasporangium undergoes meiosis to produce four megaspores, only one of which survives

megasporophyll bract (a type of modified leaf) on the central axis of a female gametophyte

mesocarp middle part of a fruit

micropropagation propagation of desirable plants from a plant part; carried out in a laboratory

micropyle opening on the ovule sac through which the pollen tube can gain entry

microsporangium tissue that gives rise to the microspores or the pollen grain

microsporophyll central axis of a male cone on which bracts (a type of modified leaf) are attached

monocarpic plants that flower once in their lifetime

multiple fruit fruit that develops from multiple flowers on an inflorescence

nectar guide pigment pattern on a flower that guides an insect to the nectaries

non-endospermic dicot dicot that stores food reserves in the developing cotyledon

perianth (also, petal or sepal) part of the flower consisting of the calyx and/or corolla; forms the outer envelope of the flower

pericarp collective term describing the exocarp, mesocarp, and endocarp; the structure that encloses the seed and is a part of the fruit

plumule shoot that develops from the germinating seed

polar nuclei found in the ovule sac; fusion with one sperm cell forms the endosperm

pollination transfer of pollen to the stigma

polycarpic plants that flower several times in their lifetime

radicle original root that develops from the germinating seed

scarification mechanical or chemical processes to soften the seed coat

scion the part of a plant that is grafted onto the root stock of another plant

scutellum type of cotyledon found in monocots, as in grass seeds

self-pollination transfer of pollen from the anther to the stigma of the same flower

senescence process that describes aging in plant tissues

simple fruit fruit that develops from a single carpel or fused carpels

sporophyte multicellular diploid stage in plants that is formed after the fusion of male and female gametes

suspensor part of the growing embryo that makes connection with the maternal tissues

synergid type of cell found in the ovule sac that secretes chemicals to guide the pollen tube towards the egg

tegmen inner layer of the seed coat

testa outer layer of the seed coat

vernalization exposure to cold required by some seeds before they can germinate

CHAPTER SUMMARY

32.1 Reproductive Development and Structure

The flower contains the reproductive structures of a plant. All complete flowers contain four whorls: the calyx, corolla, androecium, and gynoecium. The stamens are made up of anthers, in which pollen grains are produced, and a supportive strand called the filament. The pollen contains two cells— a generative cell and a tube cell—and is covered by two layers called the intine and the exine. The carpels, which are the female reproductive structures, consist of the stigma, style, and ovary. The female gametophyte is formed from mitotic divisions of the megaspore, forming an eight-nuclei ovule sac. This is covered by a layer known as the integument. The integument contains an opening called the micropyle, through which the pollen tube enters the embryo sac.

The diploid sporophyte of angiosperms and gymnosperms is the conspicuous and long-lived stage of the life cycle. The sporophytes differentiate specialized reproductive structures called sporangia, which are dedicated to the production of spores. The microsporangium contains microspore mother cells, which divide by meiosis to produce haploid microspores. The microspores develop into male gametophytes that are released as pollen. The megasporangium contains megaspore mother cells, which divide by meiosis to produce haploid megaspores. A megaspore develops into a female gametophyte containing a haploid egg. A new diploid sporophyte is formed when a male gamete from a pollen grain enters the ovule sac and fertilizes this egg.

32.2 Pollination and Fertilization

For fertilization to occur in angiosperms, pollen has to be transferred to the stigma of a flower: a process known as pollination. Gymnosperm pollination involves the transfer of pollen from a male cone to a female cone. When the pollen of the flower is transferred to the stigma of the same or another flower on the same plant, it is called self-pollination. Cross-pollination occurs when pollen is transferred from one flower to another flower of another plant. Cross-pollination requires pollinating agents such as water, wind, or animals, and increases genetic diversity. After the pollen lands on the stigma, the tube cell gives rise to the pollen tube, through which the generative nucleus migrates. The pollen tube gains entry through the micropyle on the ovule sac. The generative cell divides to form two sperm cells: one fuses with the egg to form the diploid zygote, and the other fuses with the polar nuclei to form the endosperm, which is triploid in nature. This is known as double fertilization. After fertilization, the zygote divides to form the embryo and the fertilized ovule forms the seed. The walls of the ovary form the fruit in which the seeds develop. The seed, when mature, will germinate under favorable conditions and give rise to the diploid sporophyte.

32.3 Asexual Reproduction

Many plants reproduce asexually as well as sexually. In asexual reproduction, part of the parent plant is used to generate a new plant. Grafting, layering, and micropropagation are some methods used for artificial asexual reproduction. The new plant is genetically identical to the parent plant from which the stock has been taken. Asexually reproducing plants thrive well in stable environments.

Plants have different life spans, dependent on species, genotype, and environmental conditions. Parts of the plant, such as regions containing meristematic tissue, continue to grow, while other parts experience programmed cell death. Leaves that are no longer photosynthetically active are shed from the plant as part of senescence, and the nutrients from these leaves are recycled by the plant. Other factors, including the presence of hormones, are known to play a role in delaying senescence.

VISUAL CONNECTION QUESTIONS

1. Figure 32.3 If the anther is missing, what type of reproductive structure will the flower be unable to produce? What term is used to describe an incomplete flower lacking the androecium? What term describes an incomplete flower lacking a gynoecium?

2. Figure 32.8 An embryo sac is missing the synergids. What specific impact would you expect this to have on fertilization?
   a. The pollen tube will be unable to form.
   b. The pollen tube will form but will not be guided toward the egg.
   c. Fertilization will not occur because the synergid is the egg.
   d. Fertilization will occur but the embryo will not be able to grow.

3. Figure 32.20 What is the function of the cotyledon?
   a. It develops into the root.
   b. It provides nutrition for the embryo.
   c. It forms the embryo.
   d. It protects the embryo.

REVIEW QUESTIONS

4. In a plant's male reproductive organs, development of pollen takes place in a structure known as the _______.
   a. stamen
   b. microsporangium
   c. anther
   d. tapetum

5. The stamen consists of a long stalk called the filament that supports the _______.
   a. stigma
   b. sepal
   c. style
   d. anther

6. The _______ are collectively called the calyx.
   a. sepals
   b. petals
   c. tepals
   d. stamens

7. The pollen lands on which part of the flower?
   a. stigma
   b. style
   c. ovule
   d. integument

8. After double fertilization, a zygote and _______ form.
   a. an ovule
   b. endosperm
   c. a cotyledon
   d. a suspensor

9. The fertilized ovule gives rise to the _______.
   a. fruit
   b. seed
   c. endosperm
   d. embryo

10. What is the term for a fruit that develops from tissues other than the ovary?
    a. simple fruit
    b. aggregate fruit
    c. multiple fruit
    d. accessory fruit

11. The _______ is the outermost covering of a fruit.
    a. endocarp
    b. pericarp
    c. exocarp
    d. mesocarp

12. _______ is a useful method of asexual reproduction for propagating hard-to-root plants.
    a. grafting
    b. layering
    c. cuttings
    d. budding

13. Which of the following is an advantage of asexual reproduction?
    a. Cuttings taken from an adult plant show increased resistance to diseases.
    b. Grafted plants can more successfully endure drought.
    c. When cuttings or buds are taken from an adult plant or plant parts, the resulting plant will grow into an adult faster than a seedling.
    d. Asexual reproduction takes advantage of a more diverse gene pool.

14. Plants that flower once in their lifetime are known as _______.
    a. monoecious
    b. dioecious
    c. polycarpic
    d. monocarpic

15. Plant species that complete their lifecycle in one season are known as ________.
    a. biennials
    b. perennials
    c. annuals
    d. polycarpic

CRITICAL THINKING QUESTIONS

16. Describe the reproductive organs inside a flower.
17. Describe the two-stage lifecycle of plants: the gametophyte stage and the sporophyte stage.
18. Describe the four main parts, or whorls, of a flower.
19. Discuss the differences between a complete flower and an incomplete flower.
20. Why do some seeds undergo a period of dormancy, and how do they break dormancy?
21. Discuss some ways in which fruit seeds are dispersed.
22. What are some advantages of asexual reproduction in plants?
23. Describe natural and artificial methods of asexual reproduction in plants.
24. Discuss the life cycles of various plants.
25. How are plants classified on the basis of flowering frequency?

CHAPTER 33
The Animal Body: Basic Form and Function

Figure 33.1 An arctic fox is a complex animal, well adapted to its environment. It changes coat color with the seasons, and has longer fur in winter to trap heat. (credit: modification of work by Keith Morehouse, USFWS)

INTRODUCTION The arctic fox is an example of a complex animal that has adapted to its environment and illustrates the relationships between an animal's form and function. The structures of animals consist of primary tissues that make up more complex organs and organ systems. Homeostasis allows an animal to maintain a balance between its internal and external environments.

Chapter Outline

33.1 Animal Form and Function

By the end of this section, you will be able to do the following:

- Describe the various types of body plans that occur in animals
- Describe limits on animal size and shape
- Relate bioenergetics to body size, levels of activity, and the environment

Animals vary in form and function. From a sponge to a worm to a goat, an organism has a distinct body plan that limits its size and shape. Animals' bodies are also designed to interact with their environments, whether in the deep sea, a rainforest canopy, or the desert. Therefore, a large amount of information about the structure of an organism's body (anatomy) and the function of its cells, tissues and organs (physiology) can be learned by studying that organism's environment.

Body Plans

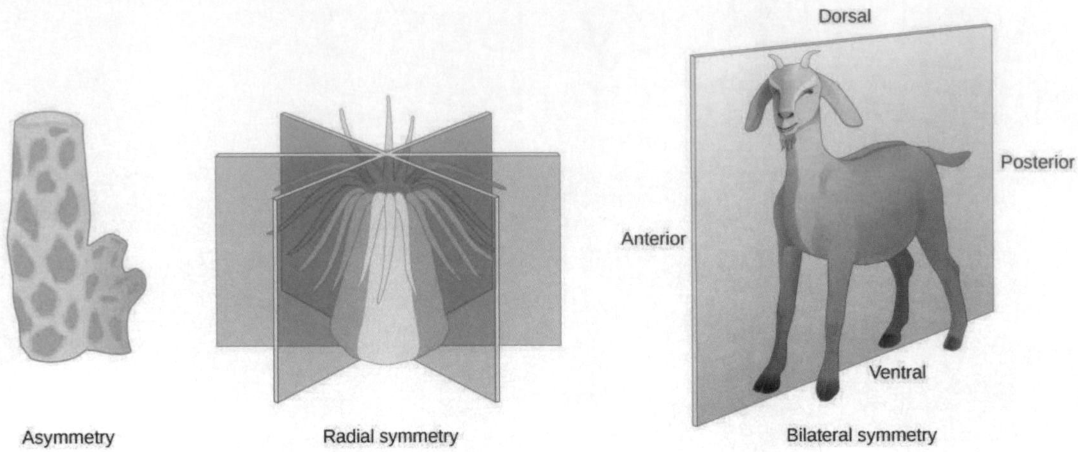

Figure 33.2 Animals exhibit different types of body symmetry. The sponge is asymmetrical, the sea anemone has radial symmetry, and the goat has bilateral symmetry.

Animal body plans follow set patterns related to symmetry. They are asymmetrical, radial, or bilateral in form as illustrated in Figure 33.2. **Asymmetrical** animals are animals with no pattern or symmetry; an example of an asymmetrical animal is a sponge. Radial symmetry, as illustrated in Figure 33.2, describes when an animal has an up-and-down orientation: any plane cut along its longitudinal axis through the organism produces equal halves, but not a definite right or left side. This plan is found mostly in aquatic animals, especially organisms that attach themselves to a base, like a rock or a boat, and extract their food from the surrounding water as it flows around the organism. Bilateral symmetry is illustrated in the same figure by a goat. The goat also has an upper and lower component to it, but a plane cut from front to back separates the animal into definite right and left sides. Additional terms used when describing positions in the body are anterior (front), posterior (rear), dorsal (toward the back), and ventral (toward the stomach). Bilateral symmetry is found in both land-based and aquatic animals; it enables a high level of mobility.

Limits on Animal Size and Shape

Animals with bilateral symmetry that live in water tend to have a **fusiform** shape: this is a tubular shaped body that is tapered at both ends. This shape decreases the drag on the body as it moves through water and allows the animal to swim at high speeds. Table 33.1 lists the maximum speed of various animals. Certain types of sharks can swim at fifty kilometers per hour and some dolphins at 32 to 40 kilometers per hour. Land animals frequently travel faster, although the tortoise and snail are significantly slower than cheetahs. Another difference in the adaptations of aquatic and land-dwelling organisms is that aquatic organisms are constrained in shape by the forces of drag in the water since water has higher viscosity than air. On the other hand, land-dwelling organisms are constrained mainly by gravity, and drag is relatively unimportant. For example, most adaptations in birds are for gravity not for drag.

Maximum Speed of Assorted Land & Marine Animals

Animal	Speed (kmh)	Speed (mph)
Cheetah	113	70
Quarter horse	77	48
Fox	68	42

Table 33.1

Animal	Speed (kmh)	Speed (mph)
Shortfin mako shark	50	31
Domestic house cat	48	30
Human	45	28
Dolphin	32–40	20–25
Mouse	13	8
Snail	0.05	0.03

Table 33.1

Most animals have an exoskeleton, including insects, spiders, scorpions, horseshoe crabs, centipedes, and crustaceans. Scientists estimate that, of insects alone, there are over 30 million species on our planet. The exoskeleton is a hard covering or shell that provides benefits to the animal, such as protection against damage from predators and from water loss (for land animals); it also provides for the attachments of muscles.

As the tough and resistant outer cover of an arthropod, the exoskeleton may be constructed of a tough polymer such as chitin and is often biomineralized with materials such as calcium carbonate. This is fused to the animal's epidermis. Ingrowths of the exoskeleton, called **apodemes**, function as attachment sites for muscles, similar to tendons in more advanced animals (Figure 33.3). In order to grow, the animal must first synthesize a new exoskeleton underneath the old one and then shed or molt the original covering. This limits the animal's ability to grow continually, and may limit the individual's ability to mature if molting does not occur at the proper time. The thickness of the exoskeleton must be increased significantly to accommodate any increase in weight. It is estimated that a doubling of body size increases body weight by a factor of eight. The increasing thickness of the chitin necessary to support this weight limits most animals with an exoskeleton to a relatively small size. The same principles apply to endoskeletons, but they are more efficient because muscles are attached on the outside, making it easier to compensate for increased mass.

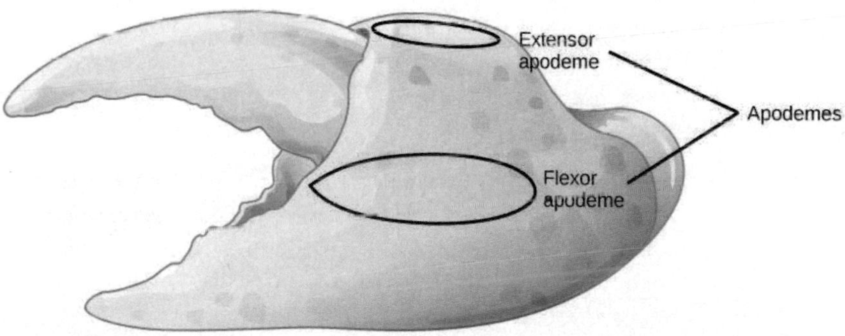

Figure 33.3 Apodemes are ingrowths on arthropod exoskeletons to which muscles attach. The apodemes on this crab leg are located above and below the fulcrum of the claw. Contraction of muscles attached to the apodemes pulls the claw closed.

An animal with an endoskeleton has its size determined by the amount of skeletal system it needs in order to support the other tissues and the amount of muscle it needs for movement. As the body size increases, both bone and muscle mass increase. The speed achievable by the animal is a balance between its overall size and the bone and muscle that provide support and movement.

Limiting Effects of Diffusion on Size and Development

The exchange of nutrients and wastes between a cell and its watery environment occurs through the process of diffusion. All living cells are bathed in liquid, whether they are in a single-celled organism or a multicellular one. Diffusion is effective over a specific distance and limits the size that an individual cell can attain. If a cell is a single-celled microorganism, such as an amoeba, it can satisfy all of its nutrient and waste needs through diffusion. If the cell is too large, then diffusion is ineffective and the center of the cell does not receive adequate nutrients nor is it able to effectively dispel its waste.

An important concept in understanding how efficient diffusion is as a means of transport is the surface to volume ratio. Recall that any three-dimensional object has a surface area and volume; the ratio of these two quantities is the surface-to-volume ratio. Consider a cell shaped like a perfect sphere: it has a surface area of $4\pi r^2$, and a volume of $(4/3)\pi r^3$. The surface-to-volume ratio of a sphere is 3/r; as the cell gets bigger, its surface to volume ratio decreases, making diffusion less efficient. The larger the size of the sphere, or animal, the less surface area for diffusion it possesses.

The solution to producing larger organisms is for them to become multicellular. Specialization occurs in complex organisms, allowing cells to become more efficient at doing fewer tasks. For example, circulatory systems bring nutrients and remove waste, while respiratory systems provide oxygen for the cells and remove carbon dioxide from them. Other organ systems have developed further specialization of cells and tissues and efficiently control body functions. Moreover, surface-to-volume ratio applies to other areas of animal development, such as the relationship between muscle mass and cross-sectional surface area in supporting skeletons, and in the relationship between muscle mass and the generation and dissipation of heat.

LINK TO LEARNING

Visit this interactive site (http://openstax.org/l/nanoscopy) to see an entire animal (a zebrafish embryo) at the cellular and sub-cellular level. Use the zoom and navigation functions for a virtual nanoscopy exploration.

Animal Bioenergetics

All animals must obtain their energy from food they ingest or absorb. These nutrients are converted to adenosine triphosphate (ATP) for short-term storage and use by all cells. Some animals store energy for slightly longer times as glycogen, and others store energy for much longer times in the form of triglycerides housed in specialized adipose tissues. No energy system is one hundred percent efficient, and an animal's metabolism produces waste energy in the form of heat. If an animal can conserve that heat and maintain a relatively constant body temperature, it is classified as a warm-blooded animal and called an **endotherm**. The insulation used to conserve the body heat comes in the forms of fur, fat, or feathers. The absence of insulation in **ectothermic** animals increases their dependence on the environment for body heat.

The amount of energy expended by an animal over a specific time is called its metabolic rate. The rate is measured variously in joules, calories, or kilocalories (1000 calories). Carbohydrates and proteins contain about 4.5 to 5 kcal/g, and fat contains about 9 kcal/g. Metabolic rate is estimated as the **basal metabolic rate (BMR)** in endothermic animals at rest and as the **standard metabolic rate (SMR)** in ectotherms. Human males have a BMR of 1600 to 1800 kcal/day, and human females have a BMR of 1300 to 1500 kcal/day. Even with insulation, endothermal animals require extensive amounts of energy to maintain a constant body temperature. An ectotherm such as an alligator has an SMR of 60 kcal/day.

Energy Requirements Related to Body Size

Smaller endothermic animals have a greater surface area for their mass than larger ones (Figure 33.4). Therefore, smaller animals lose heat at a faster rate than larger animals and require more energy to maintain a constant internal temperature. This results in a smaller endothermic animal having a higher BMR, per body weight, than a larger endothermic animal.

Species		
Mass	35 g	4,500,000 g
Metabolic rate	890 mm^3 0_2/g body mass/hr	75 mm^3 0_2/g body mass/hr

Figure 33.4 The mouse has a much higher metabolic rate than the elephant. (credit "mouse": modification of work by Magnus Kjaergaard;

credit "elephant": modification of work by "TheLizardQueen"/Flickr)

Energy Requirements Related to Levels of Activity

The more active an animal is, the more energy is needed to maintain that activity, and the higher its BMR or SMR. The average daily rate of energy consumption is about two to four times an animal's BMR or SMR. Humans are more sedentary than most animals and have an average daily rate of only 1.5 times the BMR. The diet of an endothermic animal is determined by its BMR. For example: the type of grasses, leaves, or shrubs that an herbivore eats affects the number of calories that it takes in. The relative caloric content of herbivore foods, in descending order, is tall grasses > legumes > short grasses > forbs (any broad-leaved plant, not a grass) > subshrubs > annuals/biennials.

Energy Requirements Related to Environment

Animals adapt to extremes of temperature or food availability through torpor. **Torpor** is a process that leads to a decrease in activity and metabolism and allows animals to survive adverse conditions. Torpor can be used by animals for long periods, such as entering a state of **hibernation** during the winter months, in which case it enables them to maintain a reduced body temperature. During hibernation, ground squirrels can achieve an abdominal temperature of 0° C (32° F), while a bear's internal temperature is maintained higher at about 37° C (99° F).

If torpor occurs during the summer months with high temperatures and little water, it is called **estivation**. Some desert animals use this to survive the harshest months of the year. Torpor can occur on a daily basis; this is seen in bats and hummingbirds. While endothermy is limited in smaller animals by surface to volume ratio, some organisms can be smaller and still be endotherms because they employ daily torpor during the part of the day that is coldest. This allows them to conserve energy during the colder parts of the day, when they consume more energy to maintain their body temperature.

Animal Body Planes and Cavities

A standing vertebrate animal can be divided by several planes. A **sagittal plane** divides the body into right and left portions. A **midsagittal plane** divides the body exactly in the middle, making two equal right and left halves. A **frontal plane** (also called a coronal plane) separates the front from the back. A **transverse plane** (or, horizontal plane) divides the animal into upper and lower portions. This is sometimes called a cross section, and, if the transverse cut is at an angle, it is called an oblique plane. Figure 33.5 illustrates these planes on a goat (a four-legged animal) and a human being.

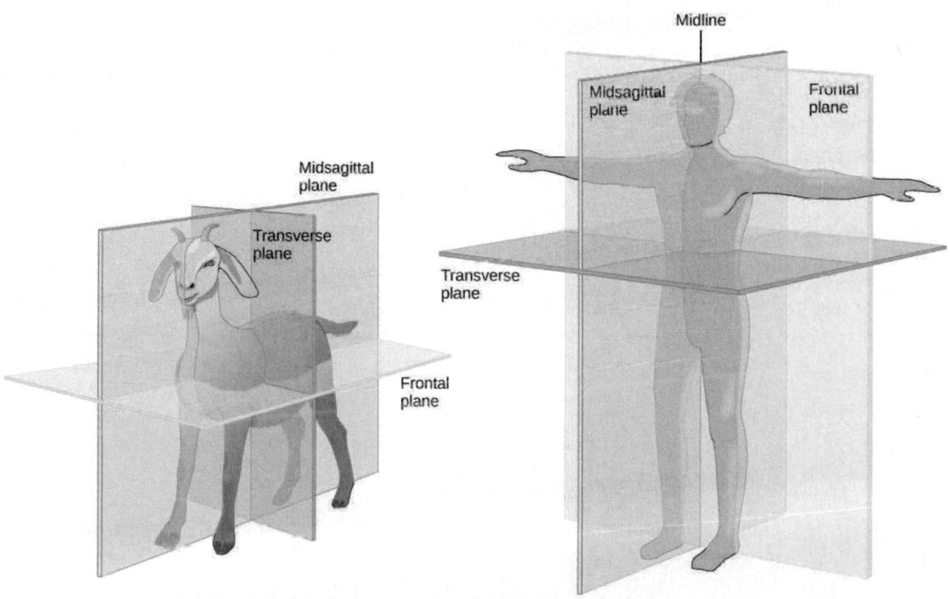

Figure 33.5 Shown are the planes of a quadrupedal goat and a bipedal human. The midsagittal plane divides the body exactly in half, into right and left portions. The frontal plane divides the front and back, and the transverse plane divides the body into upper and lower portions.

Vertebrate animals have a number of defined body cavities, as illustrated in Figure 33.6. Two of these are major cavities that contain smaller cavities within them. The **dorsal cavity** contains the cranial and the vertebral (or spinal) cavities. The **ventral cavity** contains the thoracic cavity, which in turn contains the pleural cavity around the lungs and the pericardial cavity, which surrounds the heart. The ventral cavity also contains the abdominopelvic cavity, which can be separated into the abdominal and the pelvic cavities.

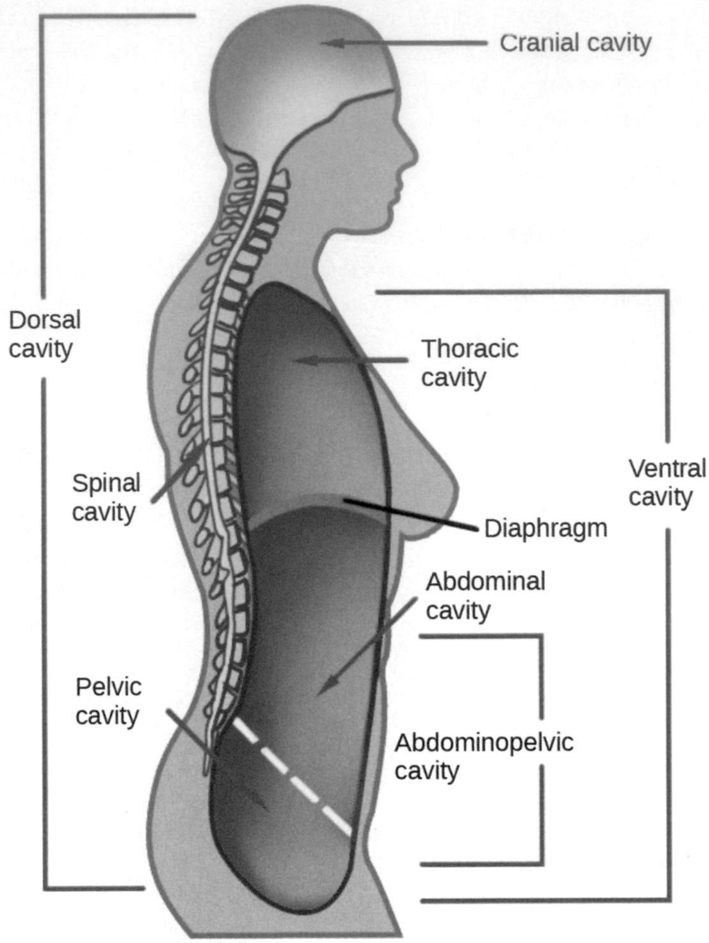

Figure 33.6 Vertebrate animals have two major body cavities. The dorsal cavity contains the cranial and the spinal cavity. The ventral cavity contains the thoracic cavity and the abdominopelvic cavity. The thoracic cavity is separated from the abdominopelvic cavity by the diaphragm. The abdominopelvic cavity is separated into the abdominal cavity and the pelvic cavity by an imaginary line parallel to the pelvis bones. (credit: modification of work by NCI)

CAREER CONNECTION

Physical Anthropologist

Physical anthropologists study the adaption, variability, and evolution of human beings, plus their living and fossil relatives. They can work in a variety of settings, although most will have an academic appointment at a university, usually in an anthropology department or a biology, genetics, or zoology department.

Nonacademic positions are available in the automotive and aerospace industries where the focus is on human size, shape, and anatomy. Research by these professionals might range from studies of how the human body reacts to car crashes to exploring how to make seats more comfortable. Other nonacademic positions can be obtained in museums of natural history, anthropology, archaeology, or science and technology. These positions involve educating students from grade school through graduate school. Physical anthropologists serve as education coordinators, collection managers, writers for museum publications, and as administrators. Zoos employ these professionals, especially if they have an expertise in primate biology; they work in collection management and captive breeding programs for endangered species. Forensic science utilizes physical anthropology expertise in identifying human and animal remains, assisting in determining the cause of death, and for expert testimony in trials.

33.2 Animal Primary Tissues

By the end of this section, you will be able to do the following:

- Describe epithelial tissues
- Discuss the different types of connective tissues in animals
- Describe three types of muscle tissues
- Describe nervous tissue

The tissues of multicellular, complex animals are four primary types: epithelial, connective, muscle, and nervous. Recall that tissues are groups of similar cells (cells carrying out related functions). These tissues combine to form organs—like the skin or kidney—that have specific, specialized functions within the body. Organs are organized into organ systems to perform functions; examples include the circulatory system, which consists of the heart and blood vessels, and the digestive system, consisting of several organs, including the stomach, intestines, liver, and pancreas. Organ systems come together to create an entire organism.

Epithelial Tissues

Epithelial tissues cover the outside of organs and structures in the body and line the lumens of organs in a single layer or multiple layers of cells. The types of epithelia are classified by the shapes of cells present and the number of layers of cells. Epithelia composed of a single layer of cells is called **simple epithelia**; epithelial tissue composed of multiple layers is called **stratified epithelia**. Table 33.2 summarizes the different types of epithelial tissues.

Different Types of Epithelial Tissues

Cell shape	Description	Location
squamous	flat, irregular round shape	simple: lung alveoli, capillaries; stratified: skin, mouth, vagina
cuboidal	cube shaped, central nucleus	glands, renal tubules
columnar	tall, narrow, nucleus toward base; tall, narrow, nucleus along cell	simple: digestive tract; pseudostratified: respiratory tract
transitional	round, simple but appear stratified	urinary bladder

Table 33.2

Squamous Epithelia

Squamous epithelial cells are generally round, flat, and have a small, centrally located nucleus. The cell outline is slightly irregular, and cells fit together to form a covering or lining. When the cells are arranged in a single layer (simple epithelia), they facilitate diffusion in tissues, such as the areas of gas exchange in the lungs and the exchange of nutrients and waste at blood capillaries.

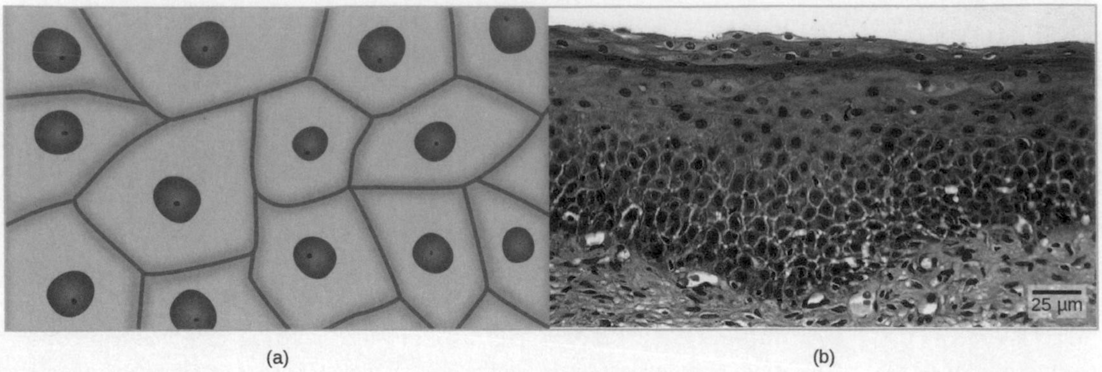

Figure 33.7 Squamous epithelia cells (a) have a slightly irregular shape, and a small, centrally located nucleus. These cells can be stratified into layers, as in (b) this human cervix specimen. (credit b: modification of work by Ed Uthman; scale-bar data from Matt Russell)

Figure 33.7a illustrates a layer of squamous cells with their membranes joined together to form an epithelium. Image Figure 33.7b illustrates squamous epithelial cells arranged in stratified layers, where protection is needed on the body from outside abrasion and damage. This is called a stratified squamous epithelium and occurs in the skin and in tissues lining the mouth and vagina.

Cuboidal Epithelia

Cuboidal epithelial cells, shown in Figure 33.8, are cube-shaped with a single, central nucleus. They are most commonly found in a single layer representing a simple epithelia in glandular tissues throughout the body where they prepare and secrete glandular material. They are also found in the walls of tubules and in the ducts of the kidney and liver.

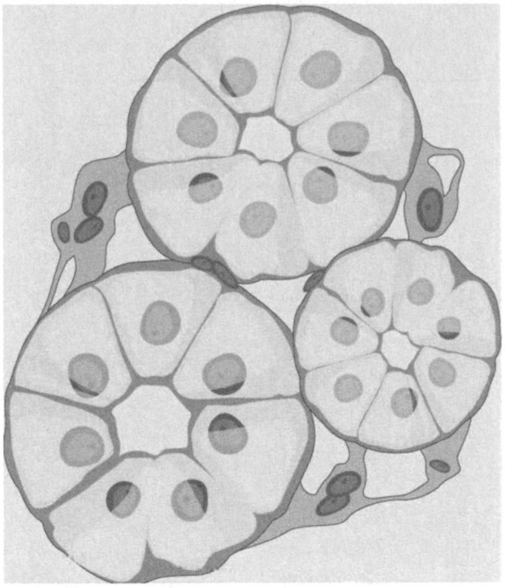

Figure 33.8 Simple cuboidal epithelial cells line tubules in the mammalian kidney, where they are involved in filtering the blood.

Columnar Epithelia

Columnar epithelial cells are taller than they are wide: they resemble a stack of columns in an epithelial layer, and are most commonly found in a single-layer arrangement. The nuclei of columnar epithelial cells in the digestive tract appear to be lined up at the base of the cells, as illustrated in Figure 33.9. These cells absorb material from the lumen of the digestive tract and prepare it for entry into the body through the circulatory and lymphatic systems.

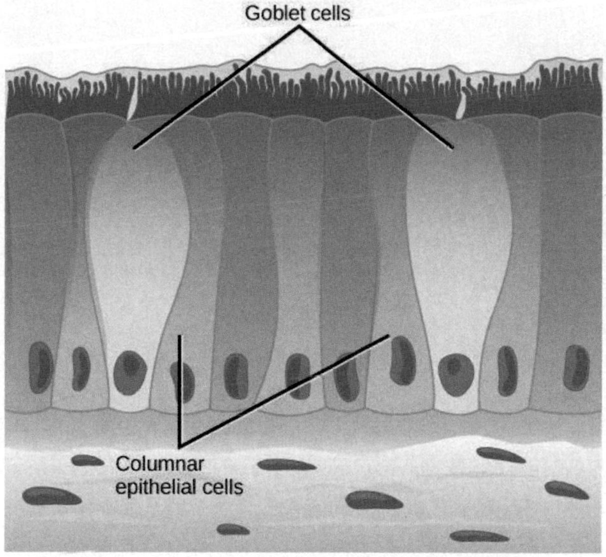

Figure 33.9 Simple columnar epithelial cells absorb material from the digestive tract. Goblet cells secrete mucus into the digestive tract lumen.

Columnar epithelial cells lining the respiratory tract appear to be stratified. However, each cell is attached to the base membrane of the tissue and, therefore, they are simple tissues. The nuclei are arranged at different levels in the layer of cells, making it appear as though there is more than one layer, as seen in Figure 33.10. This is called **pseudostratified**, columnar epithelia. This cellular covering has cilia at the apical, or free, surface of the cells. The cilia enhance the movement of mucus and trapped particles out of the respiratory tract, helping to protect the system from invasive microorganisms and harmful material that has been breathed into the body. Goblet cells are interspersed in some tissues (such as the lining of the trachea). The goblet cells contain mucus that traps irritants, which in the case of the trachea keep these irritants from getting into the lungs.

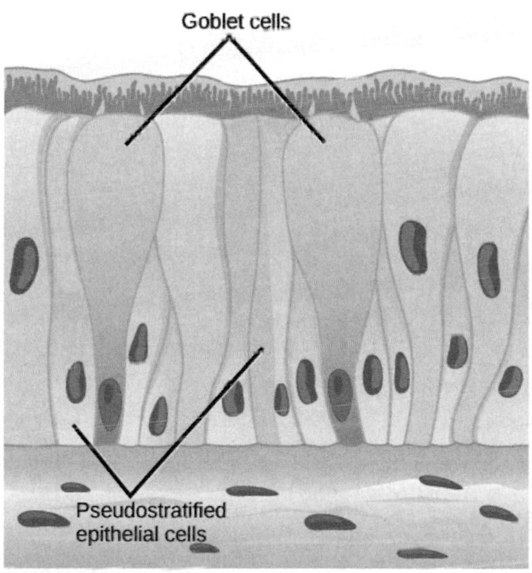

Figure 33.10 Pseudostratified columnar epithelia line the respiratory tract. They exist in one layer, but the arrangement of nuclei at different levels makes it appear that there is more than one layer. Goblet cells interspersed between the columnar epithelial cells secrete mucus into the respiratory tract.

Transitional Epithelia

Transitional or uroepithelial cells appear only in the urinary system, primarily in the bladder and ureter. These cells are arranged in a stratified layer, but they have the capability of appearing to pile up on top of each other in a relaxed, empty bladder, as illustrated in Figure 33.11. As the urinary bladder fills, the epithelial layer unfolds and expands to hold the volume of urine introduced into it. As the bladder fills, it expands and the lining becomes thinner. In other words, the tissue transitions from thick to thin.

VISUAL CONNECTION

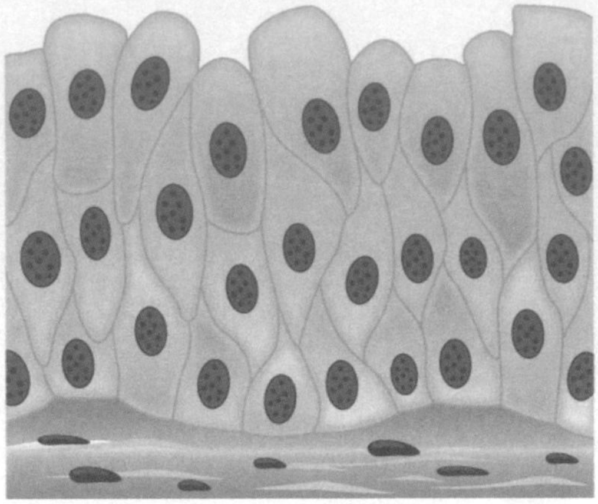

Figure 33.11 Transitional epithelia of the urinary bladder undergo changes in thickness depending on how full the bladder is.

Which of the following statements about types of epithelial cells is false?

a. Simple columnar epithelial cells line the tissue of the lung.
b. Simple cuboidal epithelial cells are involved in the filtering of blood in the kidney.
c. Pseudostratisfied columnar epithilia occur in a single layer, but the arrangement of nuclei makes it appear that more than one layer is present.
d. Transitional epithelia change in thickness depending on how full the bladder is.

Connective Tissues

Connective tissues are made up of a matrix consisting of living cells and a nonliving substance, called the ground substance. The ground substance is made of an organic substance (usually a protein) and an inorganic substance (usually a mineral or water). The principal cell of connective tissues is the fibroblast. This cell makes the fibers found in nearly all of the connective tissues. Fibroblasts are motile, able to carry out mitosis, and can synthesize whichever connective tissue is needed. Macrophages, lymphocytes, and, occasionally, leukocytes can be found in some of the tissues. Some tissues have specialized cells that are not found in the others. The **matrix** in connective tissues gives the tissue its density. When a connective tissue has a high concentration of cells or fibers, it has proportionally a less dense matrix.

The organic portion or protein fibers found in connective tissues are either collagen, elastic, or reticular fibers. Collagen fibers provide strength to the tissue, preventing it from being torn or separated from the surrounding tissues. Elastic fibers are made of the protein elastin; this fiber can stretch to one and one half of its length and return to its original size and shape. Elastic fibers provide flexibility to the tissues. Reticular fibers are the third type of protein fiber found in connective tissues. This fiber consists of thin strands of collagen that form a network of fibers to support the tissue and other organs to which it is connected. The various types of connective tissues, the types of cells and fibers they are made of, and sample locations of the tissues is summarized in Table 33.3.

Connective Tissues

Tissue	Cells	Fibers	Location
loose/areolar	fibroblasts, macrophages, some lymphocytes, some neutrophils	few: collagen, elastic, reticular	around blood vessels; anchors epithelia

Table 33.3

Tissue	Cells	Fibers	Location
dense, fibrous connective tissue	fibroblasts, macrophages	mostly collagen	irregular: skin; regular: tendons, ligaments
cartilage	chondrocytes, chondroblasts	hyaline: few: collagen fibrocartilage: large amount of collagen	shark skeleton, fetal bones, human ears, intervertebral discs
bone	osteoblasts, osteocytes, osteoclasts	some: collagen, elastic	vertebrate skeletons
adipose	adipocytes	few	adipose (fat)
blood	red blood cells, white blood cells	none	blood

Table 33.3

Loose/Areolar Connective Tissue

Loose connective tissue, also called areolar connective tissue, has a sampling of all of the components of a connective tissue. As illustrated in Figure 33.12, loose connective tissue has some fibroblasts; macrophages are present as well. Collagen fibers are relatively wide and stain a light pink, while elastic fibers are thin and stain dark blue to black. The space between the formed elements of the tissue is filled with the matrix. The material in the connective tissue gives it a loose consistency similar to a cotton ball that has been pulled apart. Loose connective tissue is found around every blood vessel and helps to keep the vessel in place. The tissue is also found around and between most body organs. In summary, areolar tissue is tough, yet flexible, and comprises membranes.

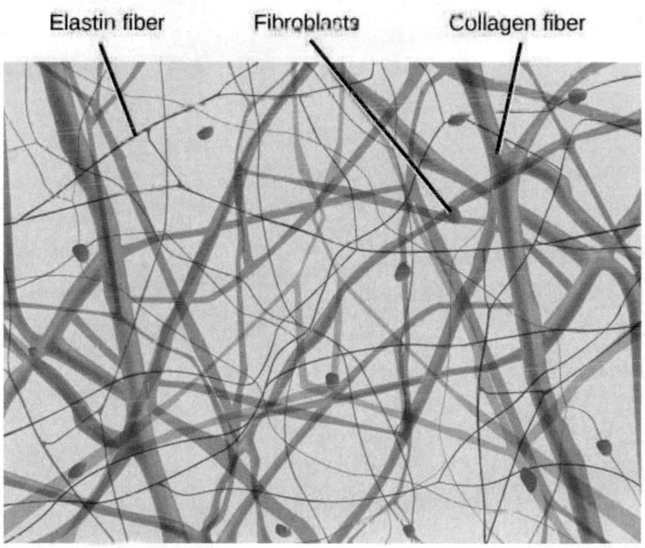

Figure 33.12 Loose connective tissue is composed of loosely woven collagen and elastic fibers. The fibers and other components of the connective tissue matrix are secreted by fibroblasts.

Fibrous Connective Tissue

Fibrous connective tissues contain large amounts of collagen fibers and few cells or matrix material. The fibers can be arranged irregularly or regularly with the strands lined up in parallel. Irregularly arranged fibrous connective tissues are found in areas of the body where stress occurs from all directions, such as the dermis of the skin. Regular fibrous connective tissue, shown in Figure 33.13, is found in tendons (which connect muscles to bones) and ligaments (which connect bones to bones).

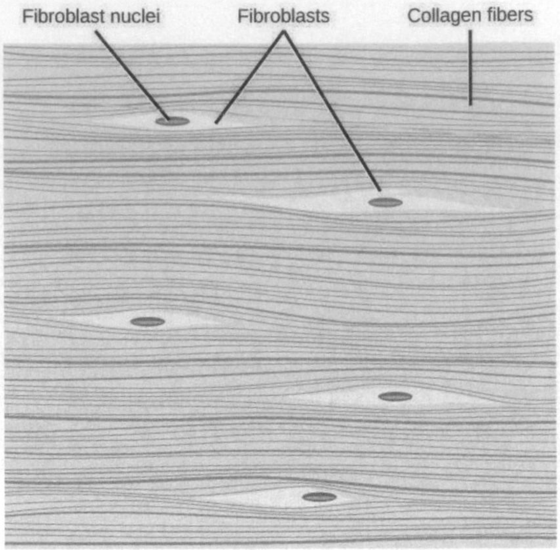

Figure 33.13 Fibrous connective tissue from the tendon has strands of collagen fibers lined up in parallel.

Cartilage

Cartilage is a connective tissue with a large amount of the matrix and variable amounts of fibers. The cells, called **chondrocytes**, make the matrix and fibers of the tissue. Chondrocytes are found in spaces within the tissue called **lacunae**.

A cartilage with few collagen and elastic fibers is hyaline cartilage, illustrated in Figure 33.14. The lacunae are randomly scattered throughout the tissue and the matrix takes on a milky or scrubbed appearance with routine histological stains. Sharks have cartilaginous skeletons, as does nearly the entire human skeleton during a specific pre-birth developmental stage. A remnant of this cartilage persists in the outer portion of the human nose. Hyaline cartilage is also found at the ends of long bones, reducing friction and cushioning the articulations of these bones.

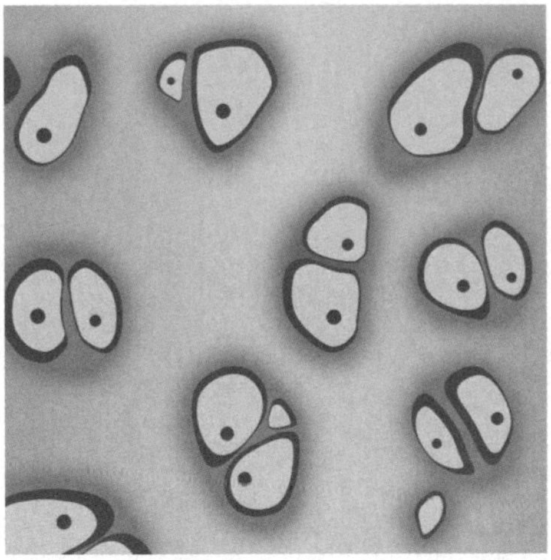

Figure 33.14 Hyaline cartilage consists of a matrix with cells called chondrocytes embedded in it. The chondrocytes exist in cavities in the matrix called lacunae.

Elastic cartilage has a large amount of elastic fibers, giving it tremendous flexibility. The ears of most vertebrate animals contain this cartilage as do portions of the larynx, or voice box. Fibrocartilage contains a large amount of collagen fibers, giving the tissue tremendous strength. Fibrocartilage comprises the intervertebral discs in vertebrate animals. Hyaline cartilage found in movable joints such as the knee and shoulder becomes damaged as a result of age or trauma. Damaged hyaline cartilage is replaced by fibrocartilage and results in the joints becoming “stiff.”

Bone

Bone, or osseous tissue, is a connective tissue that has a large amount of two different types of matrix material. The organic matrix is similar to the matrix material found in other connective tissues, including some amount of collagen and elastic fibers. This gives strength and flexibility to the tissue. The inorganic matrix consists of mineral salts—mostly calcium salts—that give the tissue hardness. Without adequate organic material in the matrix, the tissue breaks; without adequate inorganic material in the matrix, the tissue bends.

There are three types of cells in bone: osteoblasts, osteocytes, and osteoclasts. Osteoblasts are active in making bone for growth and remodeling. Osteoblasts deposit bone material into the matrix and, after the matrix surrounds them, they continue to live, but in a reduced metabolic state as osteocytes. Osteocytes are found in lacunae of the bone. Osteoclasts are active in breaking down bone for bone remodeling, and they provide access to calcium stored in tissues. Osteoclasts are usually found on the surface of the tissue.

Bone can be divided into two types: compact and spongy. Compact bone is found in the shaft (or diaphysis) of a long bone and the surface of the flat bones, while spongy bone is found in the end (or epiphysis) of a long bone. Compact bone is organized into subunits called **osteons**, as illustrated in Figure 33.15. A blood vessel and a nerve are found in the center of the structure within the Haversian canal, with radiating circles of lacunae around it known as lamellae. The wavy lines seen between the lacunae are microchannels called **canaliculi**; they connect the lacunae to aid diffusion between the cells. Spongy bone is made of tiny plates called **trabeculae**; these plates serve as struts to give the spongy bone strength. Over time, these plates can break causing the bone to become less resilient. Bone tissue forms the internal skeleton of vertebrate animals, providing structure to the animal and points of attachment for tendons.

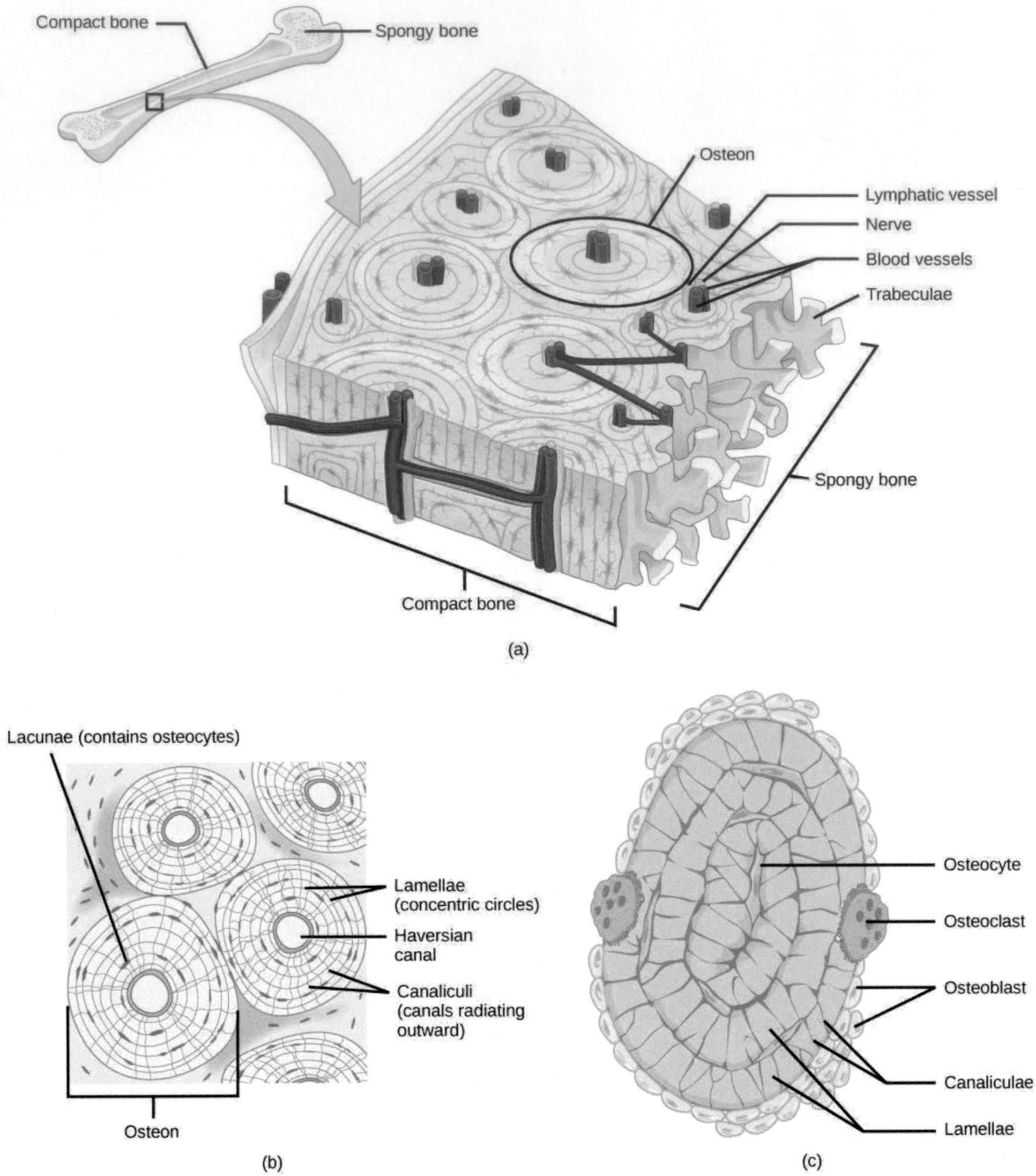

Figure 33.15 (a) Compact bone is a dense matrix on the outer surface of bone. Spongy bone, inside the compact bone, is porous with web-like trabeculae. (b) Compact bone is organized into rings called osteons. Blood vessels, nerves, and lymphatic vessels are found in the central Haversian canal. Rings of lamellae surround the Haversian canal. Between the lamellae are cavities called lacunae. Canaliculi are microchannels connecting the lacunae together. (c) Osteoblasts surround the exterior of the bone. Osteoclasts bore tunnels into the bone and osteocytes are found in the lacunae.

Adipose Tissue

Adipose tissue, or fat tissue, is considered a connective tissue even though it does not have fibroblasts or a real matrix and only has a few fibers. Adipose tissue is made up of cells called adipocytes that collect and store fat in the form of triglycerides, for energy metabolism. Adipose tissues additionally serve as insulation to help maintain body temperatures, allowing animals to be endothermic, and they function as cushioning against damage to body organs. Under a microscope, adipose tissue cells appear empty due to the extraction of fat during the processing of the material for viewing, as seen in Figure 33.16. The thin lines in the image are the cell membranes, and the nuclei are the small, black dots at the edges of the cells.

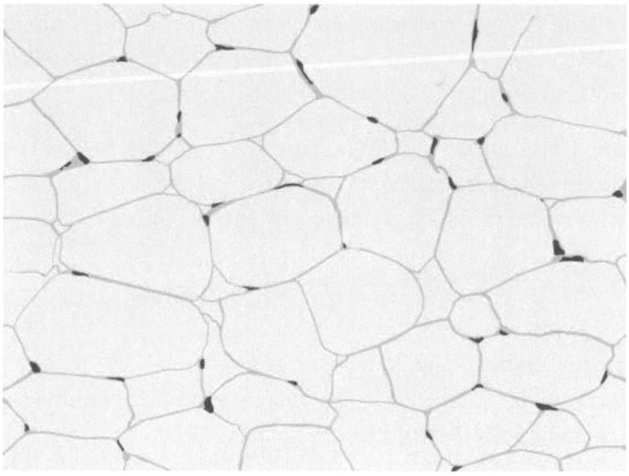

Figure 33.16 Adipose is a connective tissue is made up of cells called adipocytes. Adipocytes have small nuclei localized at the cell edge.

Blood

Blood is considered a connective tissue because it has a matrix, as shown in Figure 33.17. The living cell types are red blood cells (RBC), also called erythrocytes, and white blood cells (WBC), also called leukocytes. The fluid portion of whole blood, its matrix, is commonly called plasma.

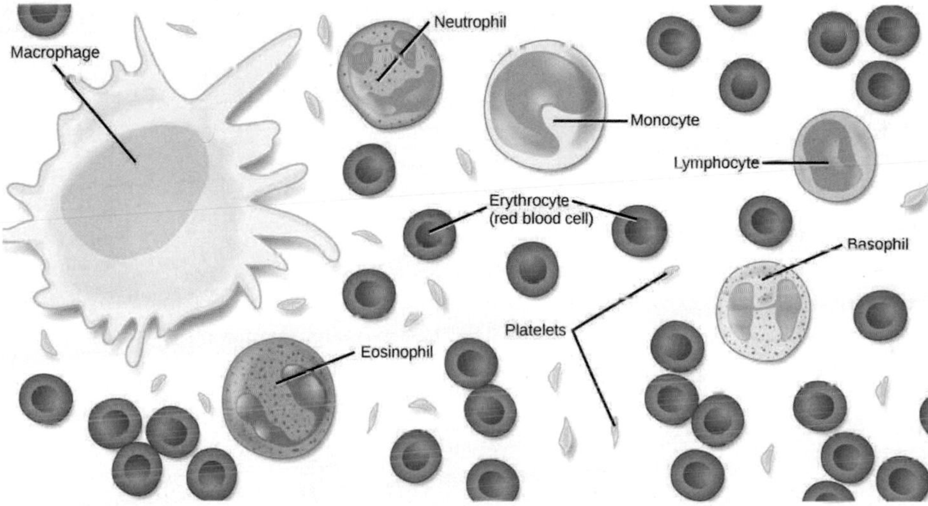

Figure 33.17 Blood is a connective tissue that has a fluid matrix, called plasma, and no fibers. Erythrocytes (red blood cells), the predominant cell type, are involved in the transport of oxygen and carbon dioxide. Also present are various leukocytes (white blood cells) involved in immune response.

The cell found in greatest abundance in blood is the erythrocyte. Erythrocytes are counted in millions in a blood sample: the average number of red blood cells in primates is 4.7 to 5.5 million cells per microliter. Erythrocytes are consistently the same size in a species, but vary in size between species. For example, the average diameter of a primate red blood cell is 7.5 µl, a dog is close at 7.0 µl, but a cat's RBC diameter is 5.9 µl. Sheep erythrocytes are even smaller at 4.6 µl. Mammalian erythrocytes lose their nuclei and mitochondria when they are released from the bone marrow where they are made. Fish, amphibian, and avian red blood cells maintain their nuclei and mitochondria throughout the cell's life. The principal job of an erythrocyte is to carry and deliver oxygen to the tissues.

Leukocytes are the predominant white blood cells found in the peripheral blood. Leukocytes are counted in the thousands in the blood with measurements expressed as ranges: primate counts range from 4,800 to 10,800 cells per µl, dogs from 5,600 to 19,200 cells per µl, cats from 8,000 to 25,000 cells per µl, cattle from 4,000 to 12,000 cells per µl, and pigs from 11,000 to 22,000 cells per µl.

Lymphocytes function primarily in the immune response to foreign antigens or material. Different types of lymphocytes make antibodies tailored to the foreign antigens and control the production of those antibodies. Neutrophils are phagocytic cells and they participate in one of the early lines of defense against microbial invaders, aiding in the removal of bacteria that has entered the body. Another leukocyte that is found in the peripheral blood is the monocyte. Monocytes give rise to phagocytic macrophages that clean up dead and damaged cells in the body, whether they are foreign or from the host animal. Two additional leukocytes in the blood are eosinophils and basophils—both help to facilitate the inflammatory response.

The slightly granular material among the cells is a cytoplasmic fragment of a cell in the bone marrow. This is called a platelet or thrombocyte. Platelets participate in the stages leading up to coagulation of the blood to stop bleeding through damaged blood vessels. Blood has a number of functions, but primarily it transports material through the body to bring nutrients to cells and remove waste material from them.

Muscle Tissues

There are three types of muscle in animal bodies: smooth, skeletal, and cardiac. They differ by the presence or absence of striations or bands, the number and location of nuclei, whether they are voluntarily or involuntarily controlled, and their location within the body. Table 33.4 summarizes these differences.

Types of Muscles

Type of Muscle	Striations	Nuclei	Control	Location
smooth	no	single, in center	involuntary	visceral organs
skeletal	yes	many, at periphery	voluntary	skeletal muscles
cardiac	yes	single, in center	involuntary	heart

Table 33.4

Smooth Muscle

Smooth muscle does not have striations in its cells. It has a single, centrally located nucleus, as shown in Figure 33.18. Constriction of smooth muscle occurs under involuntary, autonomic nervous control and in response to local conditions in the tissues. Smooth muscle tissue is also called non-striated as it lacks the banded appearance of skeletal and cardiac muscle. The walls of blood vessels, the tubes of the digestive system, and the tubes of the reproductive systems are composed of mostly smooth muscle.

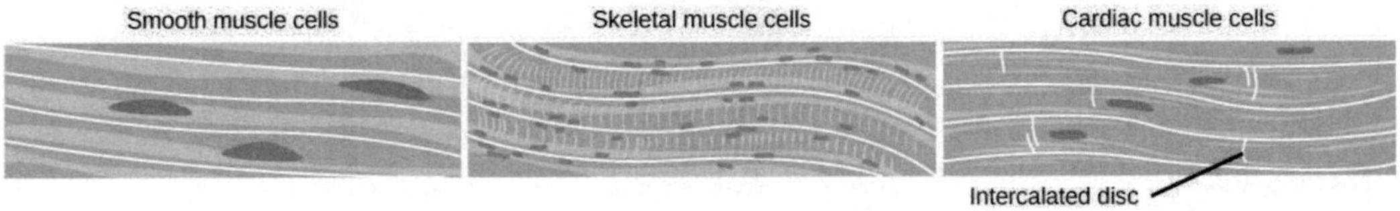

Figure 33.18 Smooth muscle cells do not have striations, while skeletal muscle cells do. Cardiac muscle cells have striations, but, unlike the multinucleate skeletal cells, they have only one nucleus. Cardiac muscle tissue also has intercalated discs, specialized regions running along the plasma membrane that join adjacent cardiac muscle cells and assist in passing an electrical impulse from cell to cell.

Skeletal Muscle

Skeletal muscle has striations across its cells caused by the arrangement of the contractile proteins actin and myosin. These muscle cells are relatively long and have multiple nuclei along the edge of the cell. Skeletal muscle is under voluntary, somatic nervous system control and is found in the muscles that move bones. Figure 33.18 illustrates the histology of skeletal muscle.

Cardiac Muscle

Cardiac muscle, shown in Figure 33.18, is found only in the heart. Like skeletal muscle, it has cross striations in its cells, but cardiac muscle has a single, centrally located nucleus. Cardiac muscle is not under voluntary control but can be influenced by the autonomic nervous system to speed up or slow down. An added feature to cardiac muscle cells is a line than extends along the end of the cell as it abuts the next cardiac cell in the row. This line is called an intercalated disc: it assists in passing electrical impulse efficiently from one cell to the next and maintains the strong connection between neighboring cardiac cells.

Nervous Tissues

Nervous tissues are made of cells specialized to receive and transmit electrical impulses from specific areas of the body and to send them to specific locations in the body. The main cell of the nervous system is the neuron, illustrated in Figure 33.19. The large structure with a central nucleus is the cell body of the neuron. Projections from the cell body are either dendrites specialized in receiving input or a single axon specialized in transmitting impulses. Some glial cells are also shown. Astrocytes regulate the chemical environment of the nerve cell, and oligodendrocytes insulate the axon so the electrical nerve impulse is transferred more efficiently. Other glial cells that are not shown support the nutritional and waste requirements of the neuron. Some of the glial cells are phagocytic and remove debris or damaged cells from the tissue. A nerve consists of neurons and glial cells.

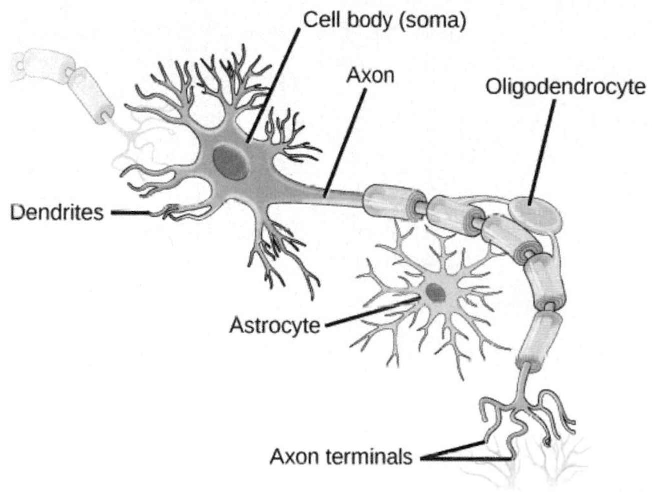

Figure 33.19 The neuron has projections called dendrites that receive signals and projections called axons that send signals. Also shown are two types of glial cells: astrocytes regulate the chemical environment of the nerve cell, and oligodendrocytes insulate the axon so the electrical nerve impulse is transferred more efficiently.

LINK TO LEARNING

Click through the interactive review (http://openstax.org/l/tissues) to learn more about epithelial tissues.

CAREER CONNECTION

Pathologist

A pathologist is a medical doctor or veterinarian who has specialized in the laboratory detection of disease in animals, including humans. These professionals complete medical school education and follow it with an extensive post-graduate residency at a medical center. A pathologist may oversee clinical laboratories for the evaluation of body tissue and blood samples for the detection of disease or infection. They examine tissue specimens through a microscope to identify cancers and other diseases. Some pathologists perform autopsies to determine the cause of death and the progression of disease.

33.3 Homeostasis

By the end of this section, you will be able to do the following:

- Define homeostasis
- Describe the factors affecting homeostasis
- Discuss positive and negative feedback mechanisms used in homeostasis
- Describe thermoregulation of endothermic and ectothermic animals

Animal organs and organ systems constantly adjust to internal and external changes through a process called homeostasis ("steady state"). These changes might be in the level of glucose or calcium in blood or in external temperatures. **Homeostasis** means to maintain dynamic equilibrium in the body. It is dynamic because it is constantly adjusting to the changes that the body's systems encounter. It is equilibrium because body functions are kept within specific ranges. Even an animal that is apparently inactive is maintaining this homeostatic equilibrium.

Homeostatic Process

The goal of homeostasis is the maintenance of equilibrium around a point or value called a **set point**. While there are normal fluctuations from the set point, the body's systems will usually attempt to go back to this point. A change in the internal or external environment is called a stimulus and is detected by a receptor; the response of the system is to adjust the deviation parameter toward the set point. For instance, if the body becomes too warm, adjustments are made to cool the animal. If the blood's glucose rises after a meal, adjustments are made to lower the blood glucose level by getting the nutrient into tissues that need it or to store it for later use.

Control of Homeostasis

When a change occurs in an animal's environment, an adjustment must be made. The receptor senses the change in the environment, then sends a signal to the control center (in most cases, the brain) which in turn generates a response that is signaled to an effector. The effector is a muscle (that contracts or relaxes) or a gland that secretes. Homeostatsis is maintained by negative feedback loops. Positive feedback loops actually push the organism further out of homeostasis, but may be necessary for life to occur. Homeostasis is controlled by the nervous and endocrine system of mammals.

Negative Feedback Mechanisms

Any homeostatic process that changes the direction of the stimulus is a **negative feedback loop**. It may either increase or decrease the stimulus, but the stimulus is not allowed to continue as it did before the receptor sensed it. In other words, if a level is too high, the body does something to bring it down, and conversely, if a level is too low, the body does something to make it go up. Hence the term negative feedback. An example is animal maintenance of blood glucose levels. When an animal has eaten, blood glucose levels rise. This is sensed by the nervous system. Specialized cells in the pancreas sense this, and the hormone insulin is released by the endocrine system. Insulin causes blood glucose levels to decrease, as would be expected in a negative feedback system, as illustrated in Figure 33.20. However, if an animal has not eaten and blood glucose levels decrease, this is sensed in another group of cells in the pancreas, and the hormone glucagon is released causing glucose levels to increase. This is still a negative feedback loop, but not in the direction expected by the use of the term "negative." Another example of an increase as a result of the feedback loop is the control of blood calcium. If calcium levels decrease, specialized cells in the parathyroid gland sense this and release parathyroid hormone (PTH), causing an increased absorption of calcium through the intestines and kidneys and, possibly, the breakdown of bone in order to liberate calcium. The effects of PTH are to raise blood levels of the element. Negative feedback loops are the predominant mechanism used in homeostasis.

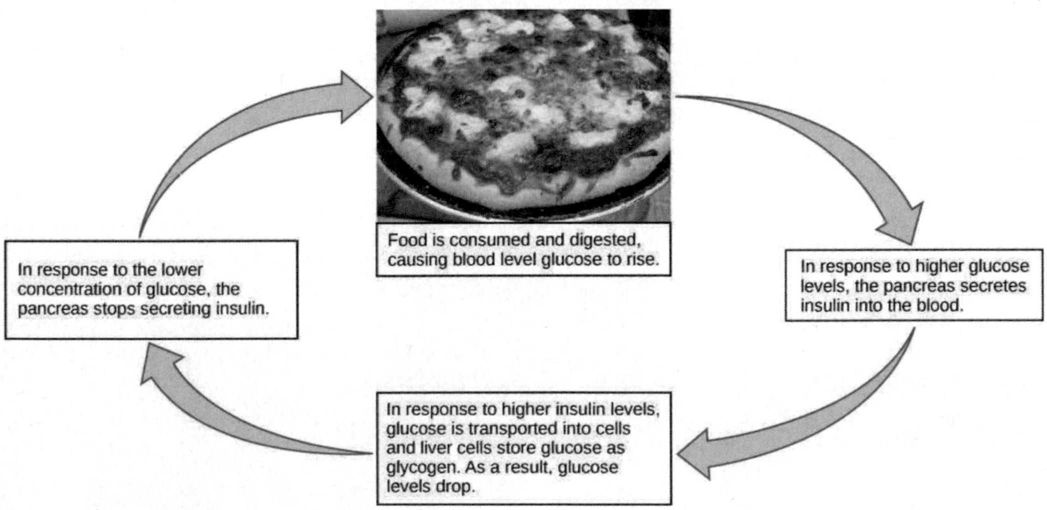

Figure 33.20 Blood sugar levels are controlled by a negative feedback loop. (credit: modification of work by Jon Sullivan)

Positive Feedback Loop

A **positive feedback loop** maintains the direction of the stimulus, possibly accelerating it. Few examples of positive feedback loops exist in animal bodies, but one is found in the cascade of chemical reactions that result in blood clotting, or coagulation.

As one clotting factor is activated, it activates the next factor in sequence until a fibrin clot is achieved. The direction is maintained, not changed, so this is positive feedback. Another example of positive feedback is uterine contractions during childbirth, as illustrated in Figure 33.21. The hormone oxytocin, made by the endocrine system, stimulates the contraction of the uterus. This produces pain sensed by the nervous system. Instead of lowering the oxytocin and causing the pain to subside, more oxytocin is produced until the contractions are powerful enough to produce childbirth.

VISUAL CONNECTION

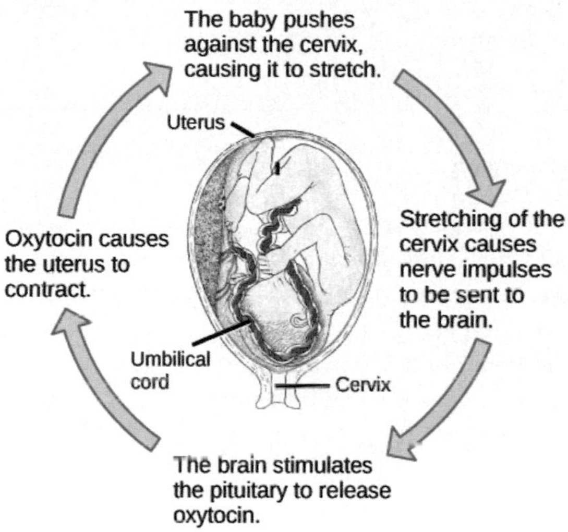

Figure 33.21 The birth of a human infant is the result of positive feedback.

State whether each of the following processes is regulated by a positive feedback loop or a negative feedback loop.

a. A person feels satiated after eating a large meal.
b. The blood has plenty of red blood cells. As a result, erythropoietin, a hormone that stimulates the production of new red blood cells, is no longer released from the kidney.

Set Point

It is possible to adjust a system's set point. When this happens, the feedback loop works to maintain the new setting. An example of this is blood pressure: over time, the normal or set point for blood pressure can increase as a result of continued increases in blood pressure. The body no longer recognizes the elevation as abnormal and no attempt is made to return to the lower set point. The result is the maintenance of an elevated blood pressure that can have harmful effects on the body. Medication can lower blood pressure and lower the set point in the system to a more healthy level. This is called a process of **alteration** of the set point in a feedback loop.

Changes can be made in a group of body organ systems in order to maintain a set point in another system. This is called **acclimatization**. This occurs, for instance, when an animal migrates to a higher altitude than that to which it is accustomed. In order to adjust to the lower oxygen levels at the new altitude, the body increases the number of red blood cells circulating in the blood to ensure adequate oxygen delivery to the tissues. Another example of acclimatization is animals that have seasonal changes in their coats: a heavier coat in the winter ensures adequate heat retention, and a light coat in summer assists in keeping body temperature from rising to harmful levels.

LINK TO LEARNING

Feedback mechanisms can be understood in terms of driving a race car along a track: watch a short video lesson on positive and negative feedback loops.

Click to view content (https://www.openstax.org/l/feedback_loops)

Homeostasis: Thermoregulation

Body temperature affects body activities. Generally, as body temperature rises, enzyme activity rises as well. For every ten degree centigrade rise in temperature, enzyme activity doubles, up to a point. Body proteins, including enzymes, begin to denature and lose their function with high heat (around 50°C for mammals). Enzyme activity will decrease by half for every ten degree centigrade drop in temperature, to the point of freezing, with a few exceptions. Some fish can withstand freezing solid and return to normal with thawing.

LINK TO LEARNING

Watch this Discovery Channel video on thermoregulation to see illustrations of this process in a variety of animals.

Click to view content (https://www.openstax.org/l/thermoregulate)

Endotherms and Ectotherms

Animals can be divided into two groups: some maintain a constant body temperature in the face of differing environmental temperatures, while others have a body temperature that is the same as their environment and thus varies with the environment. Animals that rely on external temperatures to set their body temperature are ectotherms. This group has been called cold-blooded, but the term may not apply to an animal in the desert with a very warm body temperature. In contrast to ectotherms, poikilotherms are animals with constantly varying internal temperatures. An animal that maintains a constant body temperature in the face of environmental changes is called a homeotherm. Endotherms are animals that rely on internal sources for maintenance of relatively constant body temperature in varying environmental temperatures. These animals are able to maintain a level of metabolic activity at cooler temperature, which an ectotherm cannot due to differing enzyme levels of activity. It is worth mentioning that some ectotherms and poikilotherms have relatively constant body temperatures due to the constant environmental temperatures in their habitats. These animals are so-called ectothermic homeotherms, like some deep sea fish species.

Heat can be exchanged between an animal and its environment through four mechanisms: radiation, evaporation, convection, and conduction (Figure 33.22). Radiation is the emission of electromagnetic "heat" waves. Heat comes from the sun in this manner and radiates from dry skin the same way. Heat can be removed with liquid from a surface during evaporation. This occurs when a mammal sweats. Convection currents of air remove heat from the surface of dry skin as the air passes over it. Heat will be conducted from one surface to another during direct contact with the surfaces, such as an animal resting on a warm rock.

Figure 33.22 Heat can be exchanged by four mechanisms: (a) radiation, (b) evaporation, (c) convection, or (d) conduction. (credit b: modification of work by "Kullez"/Flickr; credit c: modification of work by Chad Rosenthal; credit d: modification of work by "stacey.d"/Flickr)

Heat Conservation and Dissipation

Animals conserve or dissipate heat in a variety of ways. In certain climates, endothermic animals have some form of insulation, such as fur, fat, feathers, or some combination thereof. Animals with thick fur or feathers create an insulating layer of air between their skin and internal organs. Polar bears and seals live and swim in a subfreezing environment and yet maintain a constant, warm, body temperature. The arctic fox, for example, uses its fluffy tail as extra insulation when it curls up to sleep in cold weather. Mammals have a residual effect from shivering and increased muscle activity: arrector pili muscles cause "goose bumps," causing small hairs to stand up when the individual is cold; this has the intended effect of increasing body temperature. Mammals use layers of fat to achieve the same end. Loss of significant amounts of body fat will compromise an individual's ability to conserve heat.

Endotherms use their circulatory systems to help maintain body temperature. Vasodilation brings more blood and heat to the body surface, facilitating radiation and evaporative heat loss, which helps to cool the body. Vasoconstriction reduces blood flow in peripheral blood vessels, forcing blood toward the core and the vital organs found there, and conserving heat. Some animals have adaptions to their circulatory system that enable them to transfer heat from arteries to veins, warming blood returning to the heart. This is called a countercurrent heat exchange; it prevents the cold venous blood from cooling the heart and other internal organs. This adaption can be shut down in some animals to prevent overheating the internal organs. The countercurrent adaption is found in many animals, including dolphins, sharks, bony fish, bees, and hummingbirds. In contrast, similar adaptations can help cool endotherms when needed, such as dolphin flukes and elephant ears.

Some ectothermic animals use changes in their behavior to help regulate body temperature. For example, a desert ectothermic animal may simply seek cooler areas during the hottest part of the day in the desert to keep from getting too warm. The same animals may climb onto rocks to capture heat during a cold desert night. Some animals seek water to aid evaporation in cooling them, as seen with reptiles. Other ectotherms use group activity such as the activity of bees to warm a hive to survive winter.

Many animals, especially mammals, use metabolic waste heat as a heat source. When muscles are contracted, most of the energy from the ATP used in muscle actions is wasted energy that translates into heat. Severe cold elicits a shivering reflex that generates heat for the body. Many species also have a type of adipose tissue called brown fat that specializes in generating heat.

Neural Control of Thermoregulation

The nervous system is important to **thermoregulation**, as illustrated in Figure 33.22. The processes of homeostasis and temperature control are centered in the hypothalamus of the advanced animal brain.

VISUAL CONNECTION

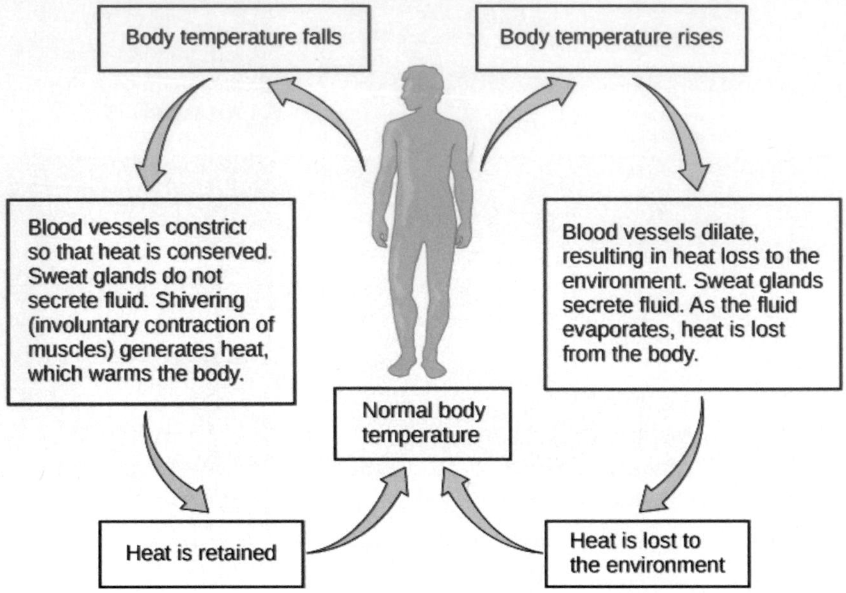

Figure 33.23 The body is able to regulate temperature in response to signals from the nervous system.

When bacteria are destroyed by leuckocytes, pyrogens are released into the blood. Pyrogens reset the body's thermostat to a higher temperature, resulting in fever. How might pyrogens cause the body temperature to rise?

The hypothalamus maintains the set point for body temperature through reflexes that cause vasodilation and sweating when the body is too warm, or vasoconstriction and shivering when the body is too cold. It responds to chemicals from the body. When a bacterium is destroyed by phagocytic leukocytes, chemicals called endogenous pyrogens are released into the blood. These pyrogens circulate to the hypothalamus and reset the thermostat. This allows the body's temperature to increase in what is commonly called a fever. An increase in body temperature causes iron to be conserved, which reduces a nutrient needed by bacteria. An increase in body heat also increases the activity of the animal's enzymes and protective cells while inhibiting the enzymes and activity of the invading microorganisms. Finally, heat itself may also kill the pathogen. A fever that was once thought to be a complication of an infection is now understood to be a normal defense mechanism.

KEY TERMS

acclimatization alteration in a body system in response to environmental change

alteration change of the set point in a homeostatic system

apodeme ingrowth of an animal's exoskeleton that functions as an attachment site for muscles

asymmetrical describes animals with no axis of symmetry in their body pattern

basal metabolic rate (BMR) metabolic rate at rest in endothermic animals

canaliculus microchannel that connects the lacunae and aids diffusion between cells

cartilage type of connective tissue with a large amount of ground substance matrix, cells called chondrocytes, and some amount of fibers

chondrocyte cell found in cartilage

columnar epithelia epithelia made of cells taller than they are wide, specialized in absorption

connective tissue type of tissue made of cells, ground substance matrix, and fibers

cuboidal epithelia epithelia made of cube-shaped cells, specialized in glandular functions

dorsal cavity body cavity on the posterior or back portion of an animal; includes the cranial and vertebral cavities

ectotherm animal incapable of maintaining a relatively constant internal body temperature

endotherm animal capable of maintaining a relatively constant internal body temperature

epithelial tissue tissue that either lines or covers organs or other tissues

estivation torpor in response to extremely high temperatures and low water availability

fibrous connective tissue type of connective tissue with a high concentration of fibers

frontal (coronal) plane plane cutting through an animal separating the individual into front and back portions

fusiform animal body shape that is tubular and tapered at both ends

hibernation torpor over a long period of time, such as a winter

homeostasis dynamic equilibrium maintaining appropriate body functions

lacuna space in cartilage and bone that contains living cells

loose (areolar) connective tissue type of connective tissue with small amounts of cells, matrix, and fibers; found around blood vessels

matrix component of connective tissue made of both living and nonliving (ground substances) cells

midsagittal plane plane cutting through an animal separating the individual into even right and left sides

negative feedback loop feedback to a control mechanism that increases or decreases a stimulus instead of maintaining it

osteon subunit of compact bone

positive feedback loop feedback to a control mechanism that continues the direction of a stimulus

pseudostratified layer of epithelia that appears multilayered, but is a simple covering

sagittal plane plane cutting through an animal separating the individual into right and left sides

set point midpoint or target point in homeostasis

simple epithelia single layer of epithelial cells

squamous epithelia type of epithelia made of flat cells, specialized in aiding diffusion or preventing abrasion

standard metabolic rate (SMR) metabolic rate at rest in ectothermic animals

stratified epithelia multiple layers of epithelial cells

thermoregulation regulation of body temperature

torpor decrease in activity and metabolism that allows an animal to survive adverse conditions

trabecula tiny plate that makes up spongy bone and gives it strength

transitional epithelia epithelia that can transition for appearing multilayered to simple; also called uroepithelial

transverse (horizontal) plane plane cutting through an animal separating the individual into upper and lower portions

ventral cavity body cavity on the anterior or front portion of an animal that includes the thoracic cavities and the abdominopelvic cavities

CHAPTER SUMMARY

33.1 Animal Form and Function

Animal bodies come in a variety of sizes and shapes. Limits on animal size and shape include impacts to their movement. Diffusion affects their size and development. Bioenergetics describes how animals use and obtain energy in relation to their body size, activity level, and environment.

33.2 Animal Primary Tissues

The basic building blocks of complex animals are four primary tissues. These are combined to form organs, which have a specific, specialized function within the body, such as the skin or kidney. Organs are organized together to perform common functions in the form of systems. The four primary tissues are epithelia, connective tissues, muscle tissues, and nervous tissues.

33.3 Homeostasis

Homeostasis is a dynamic equilibrium that is maintained in body tissues and organs. It is dynamic because it is constantly adjusting to the changes that the systems encounter. It is in equilibrium because body functions are kept within a normal range, with some fluctuations around a set point for the processes.

VISUAL CONNECTION QUESTIONS

1. Figure 33.11 Which of the following statements about types of epithelial cells is false?
   a. Simple columnar epithelial cells line the tissue of the lung.
   b. Simple cuboidal epithelial cells are involved in the filtering of blood in the kidney.
   c. Pseudostratisfied columnar epithilia occur in a single layer, but the arrangement of nuclei makes it appear that more than one layer is present.
   d. Transitional epithelia change in thickness depending on how full the bladder is.

2. Figure 33.21 State whether each of the following processes are regulated by a positive feedback loop or a negative feedback loop.
   a. A person feels satiated after eating a large meal.
   b. The blood has plenty of red blood cells. As a result, erythropoietin, a hormone that stimulates the production of new red blood cells, is no longer released from the kidney.

3. Figure 33.23 When bacteria are destroyed by leuckocytes, pyrogens are released into the blood. Pyrogens reset the body's thermostat to a higher temperature, resulting in fever. How might pyrogens cause the body temperature to rise?

REVIEW QUESTIONS

4. Which type of animal maintains a constant internal body temperature?
   a. endotherm
   b. ectotherm
   c. coelomate
   d. mesoderm

5. The symmetry found in animals that move swiftly is _______.
   a. radial
   b. bilateral
   c. sequential
   d. interrupted

6. What term describes the condition of a desert mouse that lowers its metabolic rate and "sleeps" during the hot day?
   a. turgid
   b. hibernation
   c. estivation
   d. normal sleep pattern

7. A plane that divides an animal into equal right and left portions is _______.
   a. diagonal
   b. midsagittal
   c. coronal
   d. transverse

8. A plane that divides an animal into dorsal and ventral portions is _______.
   a. sagittal
   b. midsagittal
   c. coronal
   d. transverse

9. The pleural cavity is a part of which cavity?
   a. dorsal cavity
   b. thoracic cavity
   c. abdominal cavity
   d. pericardial cavity

10. How could the increasing global temperature associated with climate change impact ectotherms?
    a. Ectotherm diversity will decrease in cool regions.
    b. Ectotherms will be able to be active all day in the tropics.
    c. Ectotherms will have to expend more energy to cool their body temperatures.
    d. Ectotherms will be able to expand into new habitats.

11. Although most animals are bilaterally symmetrical, a few exhibit radial symmetry. What is an advantage of radial symmetry?
    a. It confuses predators.
    b. It allows the animal to gather food from all sides.
    c. It allows the animal to undergo rapid, purposeful movement in any direction.
    d. It lets an animal use its dorsal surface to sense its environment.

12. Which type of epithelial cell is best adapted to aid diffusion?
    a. squamous
    b. cuboidal
    c. columnar
    d. transitional

13. Which type of epithelial cell is found in glands?
    a. squamous
    b. cuboidal
    c. columnar
    d. transitional

14. Which type of epithelial cell is found in the urinary bladder?
    a. squamous
    b. cuboidal
    c. columnar
    d. transitional

15. Which type of connective tissue has the most fibers?
    a. loose connective tissue
    b. fibrous connective tissue
    c. cartilage
    d. bone

16. Which type of connective tissue has a mineralized different matrix?
    a. loose connective tissue
    b. fibrous connective tissue
    c. cartilage
    d. bone

17. The cell found in bone that breaks it down is called an ________.
    a. osteoblast
    b. osteocyte
    c. osteoclast
    d. osteon

18. The cell found in bone that makes the bone is called an ________.
    a. osteoblast
    b. osteocyte
    c. osteoclast
    d. osteon

19. Plasma is the ________.
    a. fibers in blood
    b. matrix of blood
    c. cell that phagocytizes bacteria
    d. cell fragment found in the tissue

20. The type of muscle cell under voluntary control is the ________.
    a. smooth muscle
    b. skeletal muscle
    c. cardiac muscle
    d. visceral muscle

21. The part of a neuron that contains the nucleus is the
    a. cell body
    b. dendrite
    c. axon
    d. glial

22. Why are intercalated discs essential to the function of cardiac muscle?
    a. The discs maintain the barriers between the cells.
    b. The discs pass nutrients between cells.
    c. The discs ensure that all the cardiac muscle cells beat as a single unit.
    d. The discs control the heart rate.

23. When faced with a sudden drop in environmental temperature, an endothermic animal will:
    a. experience a drop in its body temperature
    b. wait to see if it goes lower
    c. increase muscle activity to generate heat
    d. add fur or fat to increase insulation

24. Which is an example of negative feedback?
    a. lowering of blood glucose after a meal
    b. blood clotting after an injury
    c. lactation during nursing
    d. uterine contractions during labor

25. Which method of heat exchange occurs during direct contact between the source and animal?
    a. radiation
    b. evaporation
    c. convection
    d. conduction

26. The body's thermostat is located in the ________.
    a. homeostatic receptor
    b. hypothalamus
    c. medulla
    d. vasodilation center

27. Which of the following is **not** true about acclimatization?
    a. Acclimatization allows animals to compensate for changes in their environment.
    b. Acclimatization improves function in a new environment.
    c. Acclimatization occurs when an animal tries to reestablish a homeostatic set point.
    d. Acclimatization is passed on to offspring of acclimated individuals.

28. Which of the following is **not** a way that ectotherms can change their body temperatures?
    a. Sweating for evaporative cooling.
    b. Adjusting the timing of their daily activities.
    c. Seek out or avoid direct sunlight.
    d. Huddle in a group.

CRITICAL THINKING QUESTIONS

29. How does diffusion limit the size of an organism? How is this counteracted?
30. What is the relationship between BMR and body size? Why?
31. Explain how using an open circulatory system constrains the size of animals.
32. Describe one key environmental constraint for ectotherms and one for endotherms. Why are they limited by different factors?
33. How can squamous epithelia both facilitate diffusion and prevent damage from abrasion?
34. What are the similarities between cartilage and bone?
35. Multiple sclerosis is a debilitating autoimmune disease that results in the loss of the insulation around neuron axons. What cell type is the immune system attacking, and how does this disrupt the transfer of messages by the nervous system?
36. When a person leads a sedentary life his skeletal muscles atrophy, but his smooth muscles do not. Why?
37. Why are negative feedback loops used to control body homeostasis?
38. Why is a fever a "good thing" during a bacterial infection?
39. How is a condition such as diabetes a good example of the failure of a set point in humans?
40. On a molecular level, how can endotherms produce their own heat by adjusting processes associated with cellular respiration? If needed, review Ch. 7 for details on respiration.

CHAPTER 34
Animal Nutrition and the Digestive System

Figure 34.1 For humans, fruits and vegetables are important in maintaining a balanced diet. (credit: modification of work by Julie Ryharczyk)

Chapter Outline

INTRODUCTION All living organisms need nutrients to survive. While plants can obtain the molecules required for cellular function through the process of photosynthesis, most animals obtain their nutrients by the consumption of other organisms. At the cellular level, the biological molecules necessary for animal function are amino acids, lipid molecules, nucleotides, and simple sugars. However, the food consumed consists of protein, fat, and complex carbohydrates. Animals must convert these macromolecules into the simple molecules required for maintaining cellular functions, such as assembling new molecules, cells, and tissues. The conversion of the food consumed to the nutrients required is a multistep process involving digestion and absorption. During digestion, food particles are broken down to smaller components, and later, they are absorbed by the body.

One of the challenges in human nutrition is maintaining a balance between food intake, storage, and energy expenditure. Imbalances can have serious health consequences. For example, eating too much food while not expending much energy leads to obesity, which in turn will increase the risk of developing illnesses such as type-2 diabetes and cardiovascular disease. The recent rise in obesity and related diseases makes understanding the role of diet and nutrition in maintaining good health all the more important.

34.1 Digestive Systems

By the end of this section, you will be able to do the following:

- Explain the processes of digestion and absorption
- Compare and contrast different types of digestive systems
- Explain the specialized functions of the organs involved in processing food in the body
- Describe the ways in which organs work together to digest food and absorb nutrients

Animals obtain their nutrition from the consumption of other organisms. Depending on their diet, animals can be classified into the following categories: plant eaters (herbivores), meat eaters (carnivores), and those that eat both plants and animals (omnivores). The nutrients and macromolecules present in food are not immediately accessible to the cells. There are a number of processes that modify food within the animal body in order to make the nutrients and organic molecules accessible for cellular function. As animals evolved in complexity of form and function, their digestive systems have also evolved to accommodate their various dietary needs.

Herbivores, Omnivores, and Carnivores

Herbivores are animals whose primary food source is plant-based. Examples of herbivores, as shown in Figure 34.2 include vertebrates like deer, koalas, and some bird species, as well as invertebrates such as crickets and caterpillars. These animals have evolved digestive systems capable of handling large amounts of plant material. Herbivores can be further classified into frugivores (fruit-eaters), granivores (seed eaters), nectivores (nectar feeders), and folivores (leaf eaters).

Figure 34.2 Herbivores, like this (a) mule deer and (b) monarch caterpillar, eat primarily plant material. (credit a: modification of work by Bill Ebbesen; credit b: modification of work by Doug Bowman)

Carnivores are animals that eat other animals. The word carnivore is derived from Latin and literally means "meat eater." Wild cats such as lions, shown in Figure 34.3a and tigers are examples of vertebrate carnivores, as are snakes and sharks, while invertebrate carnivores include sea stars, spiders, and ladybugs, shown in Figure 34.3b. Obligate carnivores are those that rely entirely on animal flesh to obtain their nutrients; examples of obligate carnivores are members of the cat family, such as lions and cheetahs. Facultative carnivores are those that also eat non-animal food in addition to animal food. Note that there is no clear line that differentiates facultative carnivores from omnivores; dogs would be considered facultative carnivores.

Figure 34.3 Carnivores like the (a) lion eat primarily meat. The (b) ladybug is also a carnivore that consumes small insects called aphids. (credit a: modification of work by Kevin Pluck; credit b: modification of work by Jon Sullivan)

Omnivores are animals that eat both plant- and animal-derived food. In Latin, omnivore means to eat everything. Humans, bears (shown in Figure 34.4a), and chickens are example of vertebrate omnivores; invertebrate omnivores include cockroaches and crayfish (shown in Figure 34.4b).

Figure 34.4 Omnivores like the (a) bear and (b) crayfish eat both plant and animal based food. (credit a: modification of work by Dave Menke; credit b: modification of work by Jon Sullivan)

Invertebrate Digestive Systems

Animals have evolved different types of digestive systems to aid in the digestion of the different foods they consume. The simplest example is that of a **gastrovascular cavity** and is found in organisms with only one opening for digestion. Platyhelminthes (flatworms), Ctenophora (comb jellies), and Cnidaria (coral, jelly fish, and sea anemones) use this type of digestion. Gastrovascular cavities, as shown in Figure 34.5a, are typically a blind tube or cavity with only one opening, the "mouth", which also serves as an "anus". Ingested material enters the mouth and passes through a hollow, tubular cavity. Cells within the cavity secrete digestive enzymes that breakdown the food. The food particles are engulfed by the cells lining the gastrovascular cavity.

The **alimentary canal**, shown in Figure 34.5b, is a more advanced system: it consists of one tube with a mouth at one end and an anus at the other. Earthworms are an example of an animal with an alimentary canal. Once the food is ingested through the mouth, it passes through the esophagus and is stored in an organ called the crop; then it passes into the gizzard where it is churned and digested. From the gizzard, the food passes through the intestine, the nutrients are absorbed, and the waste is eliminated as feces, called castings, through the anus.

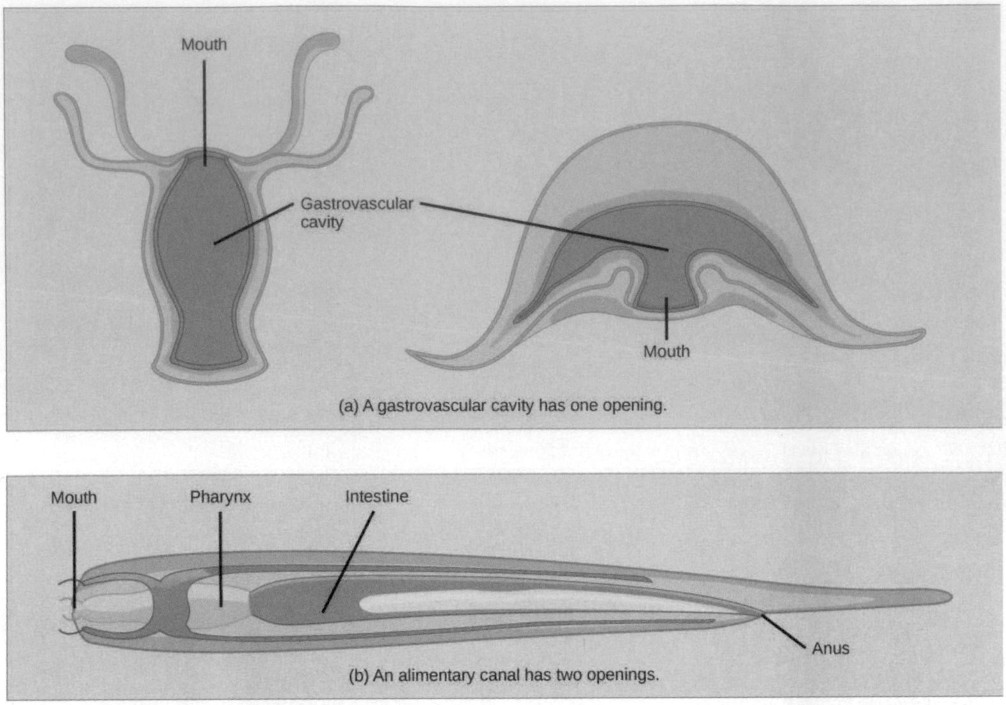

Figure 34.5 (a) A gastrovascular cavity has a single opening through which food is ingested and waste is excreted, as shown in this hydra and in this jellyfish medusa. (b) An alimentary canal has two openings: a mouth for ingesting food, and an anus for eliminating waste, as shown in this nematode.

Vertebrate Digestive Systems

Vertebrates have evolved more complex digestive systems to adapt to their dietary needs. Some animals have a single stomach, while others have multi-chambered stomachs. Birds have developed a digestive system adapted to eating unmasticated food.

Monogastric: Single-chambered Stomach

As the word **monogastric** suggests, this type of digestive system consists of one ("mono") stomach chamber ("gastric"). Humans and many animals have a monogastric digestive system as illustrated in Figure 34.6ab. The process of digestion begins with the mouth and the intake of food. The teeth play an important role in masticating (chewing) or physically breaking down food into smaller particles. The enzymes present in saliva also begin to chemically breakdown food. The esophagus is a long tube that connects the mouth to the stomach. Using peristalsis, or wave-like smooth muscle contractions, the muscles of the esophagus push the food towards the stomach. In order to speed up the actions of enzymes in the stomach, the stomach is an extremely acidic environment, with a pH between 1.5 and 2.5. The gastric juices, which include enzymes in the stomach, act on the food particles and continue the process of digestion. Further breakdown of food takes place in the small intestine where enzymes produced by the liver, the small intestine, and the pancreas continue the process of digestion. The nutrients are absorbed into the bloodstream across the epithelial cells lining the walls of the small intestines. The waste material travels on to the large intestine where water is absorbed and the drier waste material is compacted into feces; it is stored until it is excreted through the rectum.

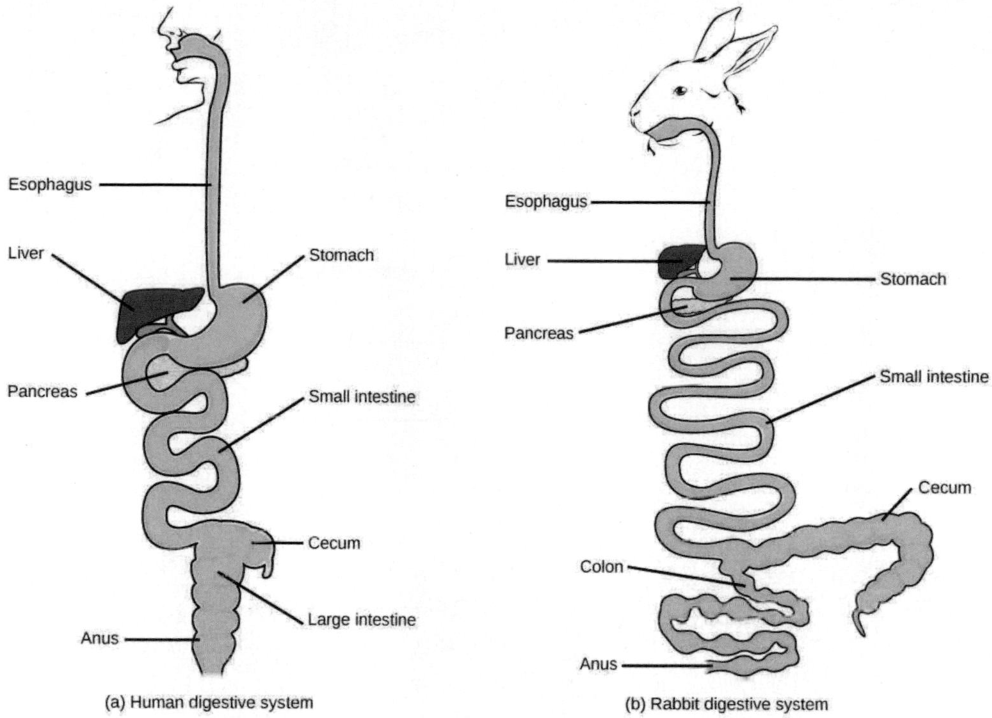

Figure 34.6 (a) Humans and herbivores, such as the (b) rabbit, have a monogastric digestive system. However, in the rabbit the small intestine and cecum are enlarged to allow more time to digest plant material. The enlarged organ provides more surface area for absorption of nutrients. Rabbits digest their food twice: the first time food passes through the digestive system, it collects in the cecum, and then it passes as soft feces called cecotrophes. The rabbit re-ingests these cecotrophes to further digest them.

Avian

Birds face special challenges when it comes to obtaining nutrition from food. They do not have teeth and so their digestive system, shown in Figure 34.7, must be able to process un-masticated food. Birds have evolved a variety of beak types that reflect the vast variety in their diet, ranging from seeds and insects to fruits and nuts. Because most birds fly, their metabolic rates are high in order to efficiently process food and keep their body weight low. The stomach of birds has two chambers: the **proventriculus**, where gastric juices are produced to digest the food before it enters the stomach, and the **gizzard**, where the food is stored, soaked, and mechanically ground. The undigested material forms food pellets that are sometimes regurgitated. Most of the chemical digestion and absorption happens in the intestine and the waste is excreted through the cloaca.

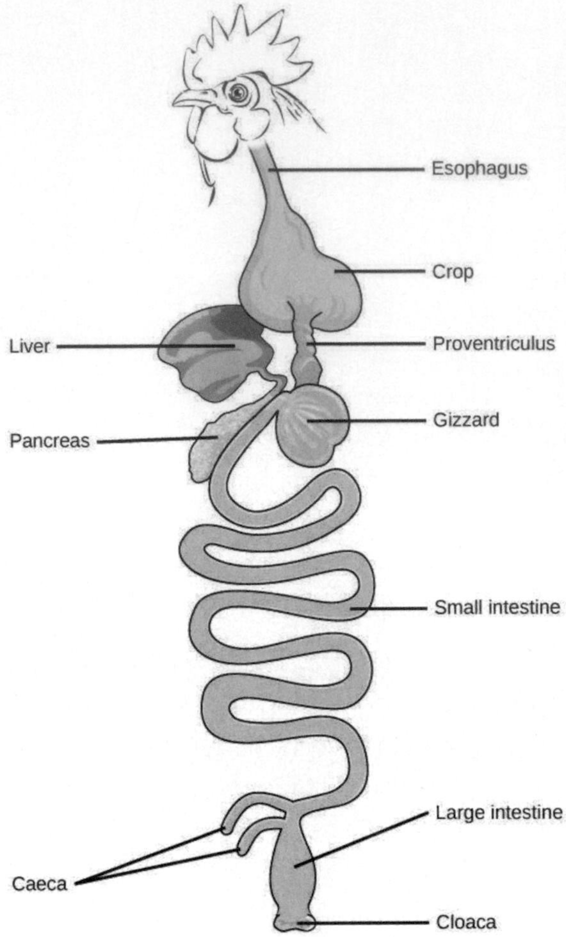

Figure 34.7 The avian esophagus has a pouch, called a crop, which stores food. Food passes from the crop to the first of two stomachs, called the proventriculus, which contains digestive juices that breakdown food. From the proventriculus, the food enters the second stomach, called the gizzard, which grinds food. Some birds swallow stones or grit, which are stored in the gizzard, to aid the grinding process. Birds do not have separate openings to excrete urine and feces. Instead, uric acid from the kidneys is secreted into the large intestine and combined with waste from the digestive process. This waste is excreted through an opening called the cloaca.

EVOLUTION CONNECTION

Avian Adaptations

Birds have a highly efficient, simplified digestive system. Recent fossil evidence has shown that the evolutionary divergence of birds from other land animals was characterized by streamlining and simplifying the digestive system. Unlike many other animals, birds do not have teeth to chew their food. In place of lips, they have sharp pointy beaks. The horny beak, lack of jaws, and the smaller tongue of the birds can be traced back to their dinosaur ancestors. The emergence of these changes seems to coincide with the inclusion of seeds in the bird diet. Seed-eating birds have beaks that are shaped for grabbing seeds and the two-compartment stomach allows for delegation of tasks. Since birds need to remain light in order to fly, their metabolic rates are very high, which means they digest their food very quickly and need to eat often. Contrast this with the ruminants, where the digestion of plant matter takes a very long time.

Ruminants

Ruminants are mainly herbivores like cows, sheep, and goats, whose entire diet consists of eating large amounts of **roughage** or fiber. They have evolved digestive systems that help them digest vast amounts of cellulose. An interesting feature of the ruminants' mouth is that they do not have upper incisor teeth. They use their lower teeth, tongue and lips to tear and chew their food. From the mouth, the food travels to the esophagus and on to the stomach.

To help digest the large amount of plant material, the stomach of the ruminants is a multi-chambered organ, as illustrated in Figure 34.8. The four compartments of the stomach are called the rumen, reticulum, omasum, and abomasum. These chambers contain many microbes that breakdown cellulose and ferment ingested food. The abomasum is the "true" stomach and is the equivalent of the monogastric stomach chamber where gastric juices are secreted. The four-compartment gastric chamber provides larger space and the microbial support necessary to digest plant material in ruminants. The fermentation process produces large amounts of gas in the stomach chamber, which must be eliminated. As in other animals, the small intestine plays an important role in nutrient absorption, and the large intestine helps in the elimination of waste.

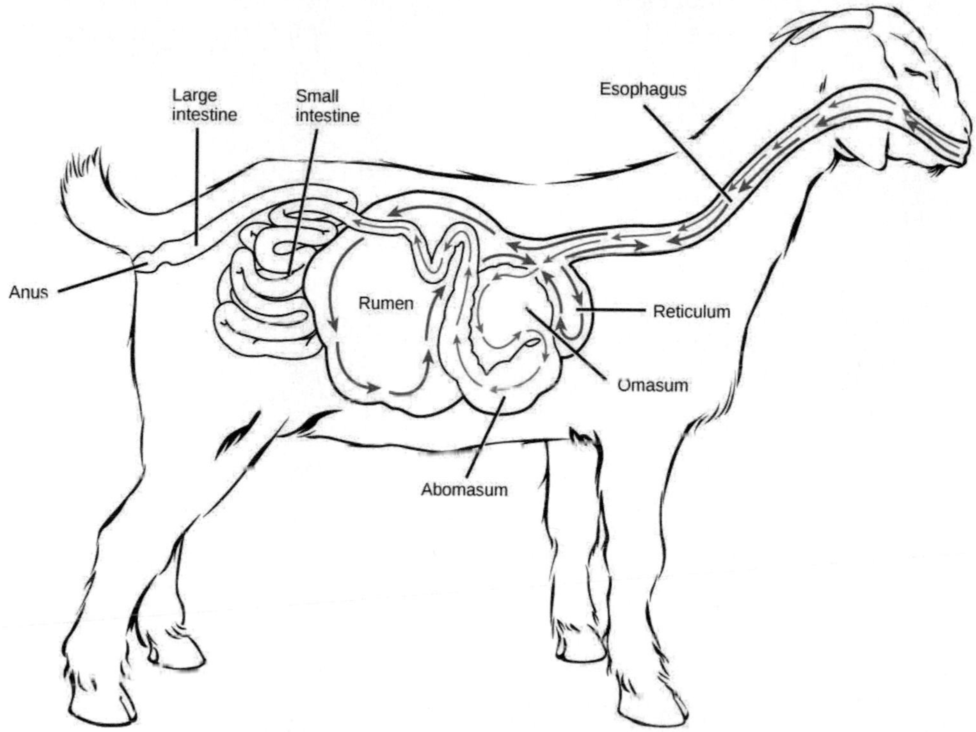

Figure 34.8 Ruminant animals, such as goats and cows, have four stomachs. The first two stomachs, the rumen and the reticulum, contain prokaryotes and protists that are able to digest cellulose fiber. The ruminant regurgitates cud from the reticulum, chews it, and swallows it into a third stomach, the omasum, which removes water. The cud then passes onto the fourth stomach, the abomasum, where it is digested by enzymes produced by the ruminant.

Pseudo-ruminants

Some animals, such as camels and alpacas, are pseudo-ruminants. They eat a lot of plant material and roughage. Digesting plant material is not easy because plant cell walls contain the polymeric sugar molecule cellulose. The digestive enzymes of these animals cannot breakdown cellulose, but microorganisms present in the digestive system can. Therefore, the digestive system must be able to handle large amounts of roughage and breakdown the cellulose. Pseudo-ruminants have a three-chamber stomach in the digestive system. However, their cecum—a pouched organ at the beginning of the large intestine containing many microorganisms that are necessary for the digestion of plant materials—is large and is the site where the roughage is fermented and digested. These animals do not have a rumen but have an omasum, abomasum, and reticulum.

Parts of the Digestive System

The vertebrate digestive system is designed to facilitate the transformation of food matter into the nutrient components that sustain organisms.

Oral Cavity

The oral cavity, or mouth, is the point of entry of food into the digestive system, illustrated in Figure 34.9. The food consumed is broken into smaller particles by mastication, the chewing action of the teeth. All mammals have teeth and can chew their food.

The extensive chemical process of digestion begins in the mouth. As food is being chewed, saliva, produced by the salivary glands, mixes with the food. Saliva is a watery substance produced in the mouths of many animals. There are three major glands that secrete saliva—the parotid, the submandibular, and the sublingual. Saliva contains mucus that moistens food and buffers the pH of the food. Saliva also contains immunoglobulins and lysozymes, which have antibacterial action to reduce tooth decay by inhibiting growth of some bacteria. Saliva also contains an enzyme called **salivary amylase** that begins the process of converting starches in the food into a disaccharide called maltose. Another enzyme called **lipase** is produced by the cells in the tongue. Lipases are a class of enzymes that can breakdown triglycerides. The lingual lipase begins the breakdown of fat components in the food. The chewing and wetting action provided by the teeth and saliva prepare the food into a mass called the **bolus** for swallowing. The tongue helps in swallowing—moving the bolus from the mouth into the pharynx. The pharynx opens to two passageways: the trachea, which leads to the lungs, and the esophagus, which leads to the stomach. The trachea has an opening called the glottis, which is covered by a cartilaginous flap called the epiglottis. When swallowing, the epiglottis closes the glottis and food passes into the esophagus and not the trachea. This arrangement allows food to be kept out of the trachea.

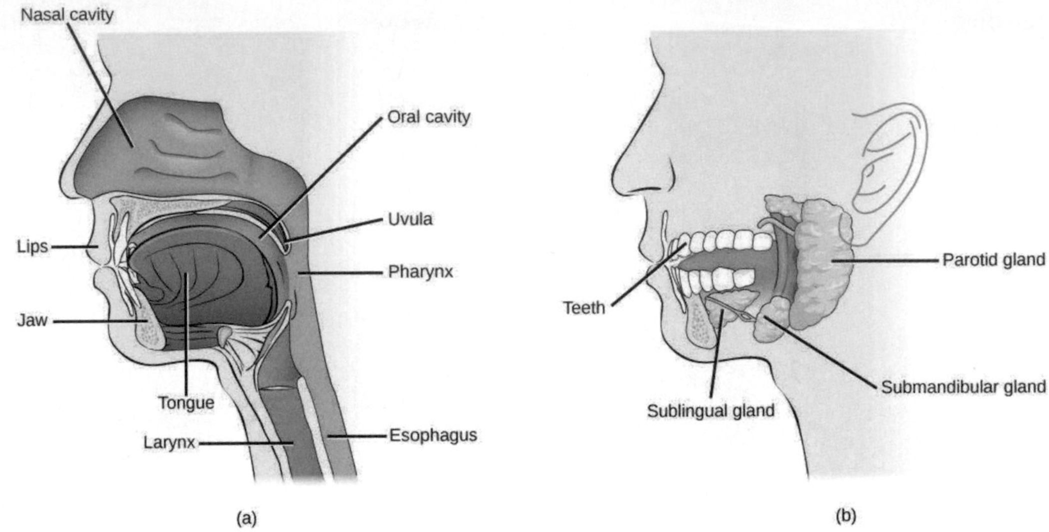

Figure 34.9 Digestion of food begins in the (a) oral cavity. Food is masticated by teeth and moistened by saliva secreted from the (b) salivary glands. Enzymes in the saliva begin to digest starches and fats. With the help of the tongue, the resulting bolus is moved into the esophagus by swallowing. (credit: modification of work by the National Cancer Institute)

Esophagus

The **esophagus** is a tubular organ that connects the mouth to the stomach. The chewed and softened food passes through the esophagus after being swallowed. The smooth muscles of the esophagus undergo a series of wave like movements called **peristalsis** that push the food toward the stomach, as illustrated in Figure 34.10. The peristalsis wave is unidirectional—it moves food from the mouth to the stomach, and reverse movement is not possible. The peristaltic movement of the esophagus is an involuntary reflex; it takes place in response to the act of swallowing.

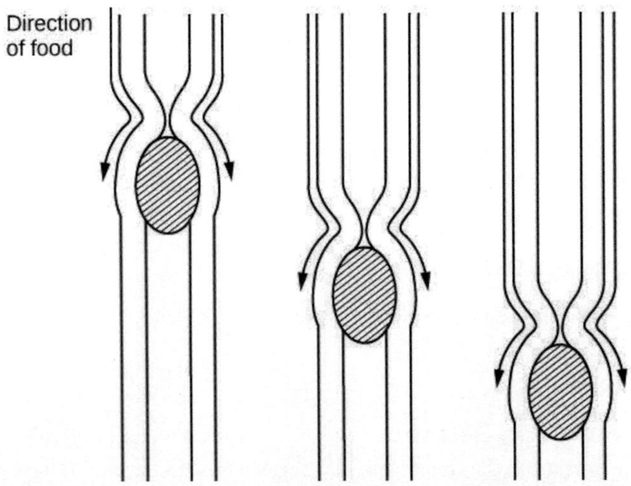

Figure 34.10 The esophagus transfers food from the mouth to the stomach through peristaltic movements.

A ring-like muscle called a **sphincter** forms valves in the digestive system. The gastro-esophageal sphincter is located at the stomach end of the esophagus. In response to swallowing and the pressure exerted by the bolus of food, this sphincter opens, and the bolus enters the stomach. When there is no swallowing action, this sphincter is shut and prevents the contents of the stomach from traveling up the esophagus. Many animals have a true sphincter; however, in humans, there is no true sphincter, but the esophagus remains closed when there is no swallowing action. Acid reflux or "heartburn" occurs when the acidic digestive juices escape into the esophagus.

Stomach

A large part of digestion occurs in the stomach, shown in Figure 34.11. The **stomach** is a saclike organ that secretes gastric digestive juices. The pH in the stomach is between 1.5 and 2.5. This highly acidic environment is required for the chemical breakdown of food and the extraction of nutrients. When empty, the stomach is a rather small organ; however, it can expand to up to 20 times its resting size when filled with food. This characteristic is particularly useful for animals that need to eat when food is available.

VISUAL CONNECTION

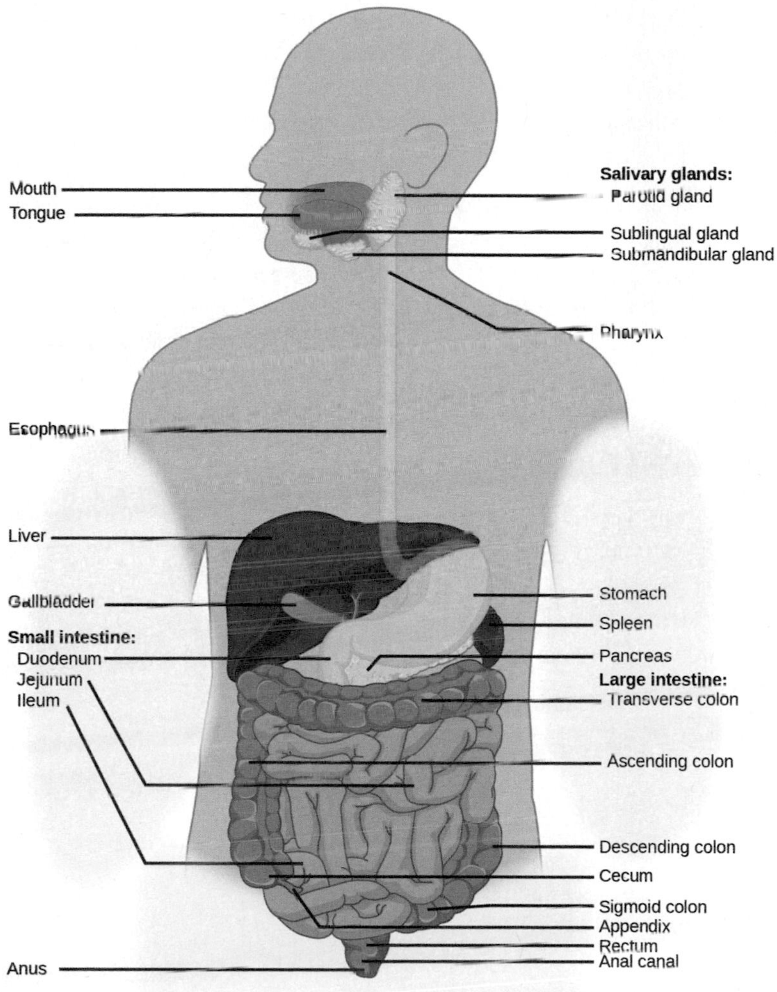

Figure 34.11 The human stomach has an extremely acidic environment where most of the protein gets digested. (credit: modification of work by Mariana Ruiz Villareal)

Which of the following statements about the digestive system is false?

a. Chyme is a mixture of food and digestive juices that is produced in the stomach.
b. Food enters the large intestine before the small intestine.
c. In the small intestine, chyme mixes with bile, which emulsifies fats.
d. The stomach is separated from the small intestine by the pyloric sphincter.

The stomach is also the major site for protein digestion in animals other than ruminants. Protein digestion is mediated by an enzyme called pepsin in the stomach chamber. **Pepsin** is secreted by the chief cells in the stomach in an inactive form called **pepsinogen**. Pepsin breaks peptide bonds and cleaves proteins into smaller polypeptides; it also helps activate more pepsinogen, starting a positive feedback mechanism that generates more pepsin. Another cell type—parietal cells—secrete hydrogen and chloride ions, which combine in the lumen to form hydrochloric acid, the primary acidic component of the stomach juices. Hydrochloric acid helps to convert the inactive pepsinogen to pepsin. The highly acidic environment also kills many microorganisms in the food and, combined with the action of the enzyme pepsin, results in the hydrolysis of protein in the food. Chemical digestion is facilitated by the churning action of the stomach. Contraction and relaxation of smooth muscles mixes the stomach contents about every 20 minutes. The partially digested food and gastric juice mixture is called **chyme**. Chyme passes from the stomach to the small intestine. Further protein digestion takes place in the small intestine. Gastric emptying occurs within two to six hours after a meal. Only a small amount of chyme is released into the small intestine at a time. The movement of chyme from the stomach into the small intestine is regulated by the pyloric sphincter.

When digesting protein and some fats, the stomach lining must be protected from getting digested by pepsin. There are two points to consider when describing how the stomach lining is protected. First, as previously mentioned, the enzyme pepsin is synthesized in the inactive form. This protects the chief cells, because pepsinogen does not have the same enzyme functionality of pepsin. Second, the stomach has a thick mucus lining that protects the underlying tissue from the action of the digestive juices. When this mucus lining is ruptured, ulcers can form in the stomach. Ulcers are open wounds in or on an organ caused by bacteria (*Helicobacter pylori*) when the mucus lining is ruptured and fails to reform.

Small Intestine

Chyme moves from the stomach to the small intestine. The **small intestine** is the organ where the digestion of protein, fats, and carbohydrates is completed. The small intestine is a long tube-like organ with a highly folded surface containing finger-like projections called the **villi**. The apical surface of each villus has many microscopic projections called microvilli. These structures, illustrated in Figure 34.12, are lined with epithelial cells on the luminal side and allow for the nutrients to be absorbed from the digested food and absorbed into the bloodstream on the other side. The villi and microvilli, with their many folds, increase the surface area of the intestine and increase absorption efficiency of the nutrients. Absorbed nutrients in the blood are carried into the hepatic portal vein, which leads to the liver. There, the liver regulates the distribution of nutrients to the rest of the body and removes toxic substances, including drugs, alcohol, and some pathogens.

VISUAL CONNECTION

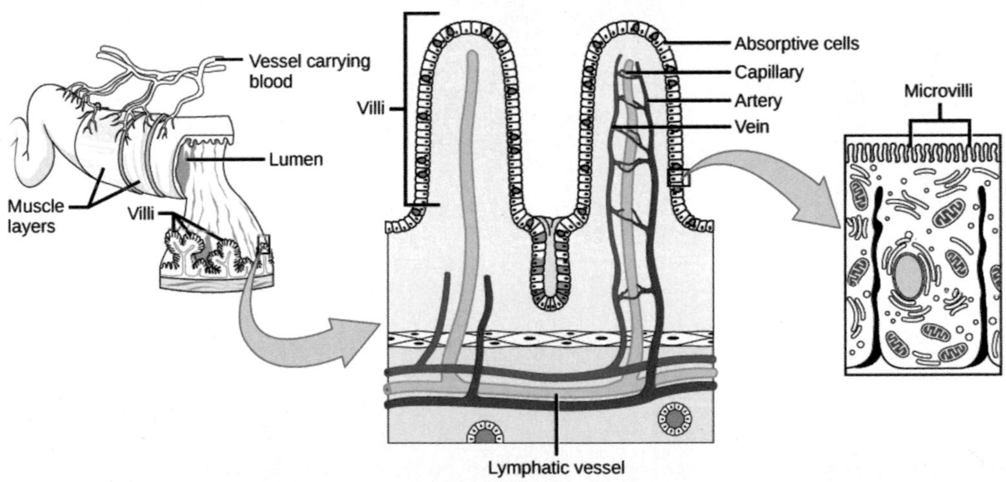

Figure 34.12 Villi are folds on the small intestine lining that increase the surface area to facilitate the absorption of nutrients.

Which of the following statements about the small intestine is false?

a. Absorptive cells that line the small intestine have microvilli, small projections that increase surface area and aid in the absorption of food.
b. The inside of the small intestine has many folds, called villi.
c. Microvilli are lined with blood vessels as well as lymphatic vessels.

d. The inside of the small intestine is called the lumen.

The human small intestine is over 6m long and is divided into three parts: the duodenum, the jejunum, and the ileum. The "C-shaped," fixed part of the small intestine is called the **duodenum** and is shown in Figure 34.11. The duodenum is separated from the stomach by the pyloric sphincter which opens to allow chyme to move from the stomach to the duodenum. In the duodenum, chyme is mixed with pancreatic juices in an alkaline solution rich in bicarbonate that neutralizes the acidity of chyme and acts as a buffer. Pancreatic juices also contain several digestive enzymes. Digestive juices from the pancreas, liver, and gallbladder, as well as from gland cells of the intestinal wall itself, enter the duodenum. **Bile** is produced in the liver and stored and concentrated in the gallbladder. Bile contains bile salts which emulsify lipids while the pancreas produces enzymes that catabolize starches, disaccharides, proteins, and fats. These digestive juices breakdown the food particles in the chyme into glucose, triglycerides, and amino acids. Some chemical digestion of food takes place in the duodenum. Absorption of fatty acids also takes place in the duodenum.

The second part of the small intestine is called the **jejunum**, shown in Figure 34.11. Here, hydrolysis of nutrients is continued while most of the carbohydrates and amino acids are absorbed through the intestinal lining. The bulk of chemical digestion and nutrient absorption occurs in the jejunum.

The **ileum**, also illustrated in Figure 34.11 is the last part of the small intestine and here the bile salts and vitamins are absorbed into the bloodstream. The undigested food is sent to the colon from the ileum via peristaltic movements of the muscle. The ileum ends and the large intestine begins at the ileocecal valve. The vermiform, "worm-like," appendix is located at the ileocecal valve. The appendix of humans secretes no enzymes and has an insignificant role in immunity.

Large Intestine

The **large intestine**, illustrated in Figure 34.13, reabsorbs the water from the undigested food material and processes the waste material. The human large intestine is much smaller in length compared to the small intestine but larger in diameter. It has three parts: the cecum, the colon, and the rectum. The cecum joins the ileum to the colon and is the receiving pouch for the waste matter. The colon is home to many bacteria or "intestinal flora" that aid in the digestive processes. The colon can be divided into four regions, the ascending colon, the transverse colon, the descending colon, and the sigmoid colon. The main functions of the colon are to extract the water and mineral salts from undigested food, and to store waste material. Carnivorous mammals have a shorter large intestine compared to herbivorous mammals due to their diet.

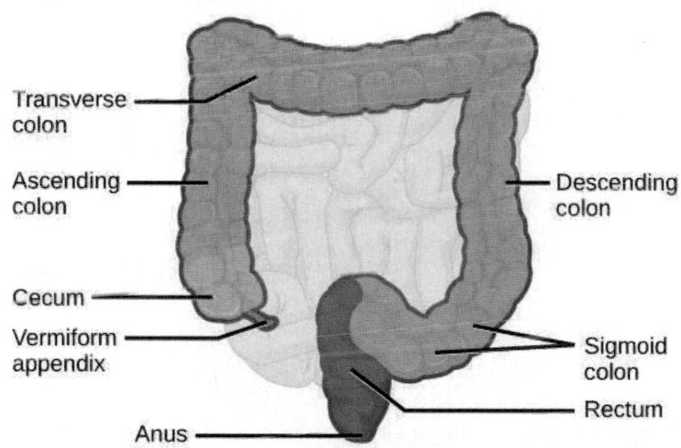

Figure 34.13 The large intestine reabsorbs water from undigested food and stores waste material until it is eliminated.

Rectum and Anus

The **rectum** is the terminal end of the large intestine, as shown in Figure 34.13. The primary role of the rectum is to store the feces until defecation. The feces are propelled using peristaltic movements during elimination. The **anus** is an opening at the far-end of the digestive tract and is the exit point for the waste material. Two sphincters between the rectum and anus control elimination: the inner sphincter is involuntary and the outer sphincter is voluntary.

Accessory Organs

The organs discussed above are the organs of the digestive tract through which food passes. Accessory organs are organs that add secretions (enzymes) that catabolize food into nutrients. Accessory organs include salivary glands, the liver, the pancreas, and the gallbladder. The liver, pancreas, and gallbladder are regulated by hormones in response to the food consumed.

The **liver** is the largest internal organ in humans and it plays a very important role in digestion of fats and detoxifying blood. The liver produces bile, a digestive juice that is required for the breakdown of fatty components of the food in the duodenum. The liver also processes the vitamins and fats and synthesizes many plasma proteins.

The **pancreas** is another important gland that secretes digestive juices. The chyme produced from the stomach is highly acidic in nature; the pancreatic juices contain high levels of bicarbonate, an alkali that neutralizes the acidic chyme. Additionally, the pancreatic juices contain a large variety of enzymes that are required for the digestion of protein and carbohydrates.

The **gallbladder** is a small organ that aids the liver by storing bile and concentrating bile salts. When chyme containing fatty acids enters the duodenum, the bile is secreted from the gallbladder into the duodenum.

34.2 Nutrition and Energy Production

By the end of this section, you will be able to do the following:

- Explain why an animal's diet should be balanced and meet the needs of the body
- Define the primary components of food
- Describe the essential nutrients required for cellular function that cannot be synthesized by the animal body
- Explain how energy is produced through diet and digestion
- Describe how excess carbohydrates and energy are stored in the body

Given the diversity of animal life on our planet, it is not surprising that the animal diet would also vary substantially. The animal diet is the source of materials needed for building DNA and other complex molecules needed for growth, maintenance, and reproduction; collectively these processes are called biosynthesis. The diet is also the source of materials for ATP production in the cells. The diet must be balanced to provide the minerals and vitamins that are required for cellular function.

Food Requirements

What are the fundamental requirements of the animal diet? The animal diet should be well balanced and provide nutrients required for bodily function and the minerals and vitamins required for maintaining structure and regulation necessary for good health and reproductive capability. These requirements for a human are illustrated graphically in Figure 34.14

Figure 34.14 For humans, a balanced diet includes fruits, vegetables, grains, and protein. (credit: USDA)

LINK TO LEARNING

The first step in ensuring that you are meeting the food requirements of your body is an awareness of the food groups and the nutrients they provide. To learn more about each food group and the recommended daily amounts, explore this interactive site (http://openstax.org/l/food_groups) by the United States Department of Agriculture.

Everyday Connection

Let's Move! Campaign

Obesity is a growing epidemic and the rate of obesity among children is rapidly rising in the United States. To combat childhood obesity and ensure that children get a healthy start in life, first lady Michelle Obama has launched the Let's Move! campaign. The goal of this campaign is to educate parents and caregivers on providing healthy nutrition and encouraging active lifestyles to future generations. This program aims to involve the entire community, including parents, teachers, and healthcare providers to ensure that children have access to healthy foods—more fruits, vegetables, and whole grains—and consume fewer calories from processed foods. Another goal is to ensure that children get physical activity. With the increase in television viewing and stationary pursuits such as video games, sedentary lifestyles have become the norm. Learn more at https://letsmove.obamawhitehouse.archives.gov (http://openstax.org/l/Letsmove) .

Organic Precursors

The organic molecules required for building cellular material and tissues must come from food. Carbohydrates or sugars are the primary source of organic carbons in the animal body. During digestion, digestible carbohydrates are ultimately broken down into glucose and used to provide energy through metabolic pathways. Complex carbohydrates, including polysaccharides, can be broken down into glucose through biochemical modification; however, humans do not produce the enzyme cellulase and lack the ability to derive glucose from the polysaccharide cellulose. In humans, these molecules provide the fiber required for moving waste through the large intestine and a healthy colon. The intestinal flora in the human gut are able to extract some nutrition from these plant fibers. The excess sugars in the body are converted into glycogen and stored in the liver and muscles for later use. Glycogen stores are used to fuel prolonged exertions, such as long-distance running, and to provide energy during food shortage. Excess glycogen can be converted to fats, which are stored in the lower layer of the skin of mammals for insulation and energy storage. Excess digestible carbohydrates are stored by mammals in order to survive famine and aid in mobility.

Another important requirement is that of nitrogen. Protein catabolism provides a source of organic nitrogen. Amino acids are the building blocks of proteins and protein breakdown provides amino acids that are used for cellular function. The carbon and nitrogen derived from these become the building block for nucleotides, nucleic acids, proteins, cells, and tissues. Excess nitrogen must be excreted as it is toxic. Fats add flavor to food and promote a sense of satiety or fullness. Fatty foods are also significant sources of energy because one gram of fat contains nine calories. Fats are required in the diet to aid the absorption of fat-soluble vitamins and the production of fat-soluble hormones.

Essential Nutrients

While the animal body can synthesize many of the molecules required for function from the organic precursors, there are some nutrients that need to be consumed from food. These nutrients are termed **essential nutrients**, meaning they must be eaten, and the body cannot produce them.

The omega-3 alpha-linolenic acid and the omega-6 linoleic acid are essential fatty acids needed to make some membrane phospholipids. **Vitamins** are another class of essential organic molecules that are required in small quantities for many enzymes to function and, for this reason, are considered to be coenzymes. Absence or low levels of vitamins can have a dramatic effect on health, as outlined in Table 34.1 and Table 34.2. Both fat-soluble and water-soluble vitamins must be obtained from food. **Minerals**, listed in Table 34.3, are inorganic essential nutrients that must be obtained from food. Among their many functions, minerals help in structure and regulation and are considered cofactors. Certain amino acids also must be procured from food and cannot be synthesized by the body. These amino acids are the "essential" amino acids. The human body can synthesize only 11 of the 20 required amino acids; the rest must be obtained from food. The essential amino acids are listed in Table 34.4.

Water-soluble Essential Vitamins

Vitamin	Function	Deficiencies Can Lead To	Sources
Vitamin B_1 (Thiamine)	Needed by the body to process lipids, proteins, and carbohydrates; coenzyme removes CO_2 from organic compounds	Muscle weakness, Beriberi: reduced heart function, CNS problems	Milk, meat, dried beans, whole grains
Vitamin B_2 (Riboflavin)	Takes an active role in metabolism, aiding in the conversion of food to energy (FAD and FMN)	Cracks or sores on the outer surface of the lips (cheliosis); inflammation and redness of the tongue; moist, scaly skin inflammation (seborrheic dermatitis)	Meat, eggs, enriched grains, vegetables
Vitamin B_3 (Niacin)	Used by the body to release energy from carbohydrates and to process alcohol; required for the synthesis of sex hormones; component of coenzyme NAD^+ and $NADP^+$	Pellagra, which can result in dermatitis, diarrhea, dementia, and death	Meat, eggs, grains, nuts, potatoes
Vitamin B_5 (Pantothenic acid)	Assists in producing energy from foods (lipids, in particular); component of coenzyme A	Fatigue, poor coordination, retarded growth, numbness, tingling of hands and feet	Meat, whole grains, milk, fruits, vegetables
Vitamin B_6 (Pyridoxine)	The principal vitamin for processing amino acids and lipids; also helps convert nutrients into energy	Irritability, depression, confusion, mouth sores or ulcers, anemia, muscular twitching	Meat, dairy products, whole grains, orange juice
Vitamin B_7 (Biotin)	Used in energy and amino acid metabolism, fat synthesis, and fat breakdown; helps the body use blood sugar	Hair loss, dermatitis, depression, numbness and tingling in the extremities; neuromuscular disorders	Meat, eggs, legumes and other vegetables
Vitamin B_9 (Folic acid)	Assists the normal development of cells, especially during fetal development; helps metabolize nucleic and amino acids	Deficiency during pregnancy is associated with birth defects, such as neural tube defects and anemia	Leafy green vegetables, whole wheat, fruits, nuts, legumes
Vitamin B_{12} (Cobalamin)	Maintains healthy nervous system and assists with blood cell formation; coenzyme in nucleic acid metabolism	Anemia, neurological disorders, numbness, loss of balance	Meat, eggs, animal products
Vitamin C (Ascorbic acid)	Helps maintain connective tissue: bone, cartilage, and dentin; boosts the immune system	Scurvy, which results in bleeding, hair and tooth loss; joint pain and swelling; delayed wound healing	Citrus fruits, broccoli, tomatoes, red sweet bell peppers

Table 34.1

Fat-soluble Essential Vitamins

Vitamin	Function	Deficiencies Can Lead To	Sources
Vitamin A (Retinol)	Critical to the development of bones, teeth, and skin; helps maintain eyesight, enhances the immune system, fetal development, gene expression	Night-blindness, skin disorders, impaired immunity	Dark green leafy vegetables, yellow-orange vegetables, fruits, milk, butter
Vitamin D	Critical for calcium absorption for bone development and strength; maintains a stable nervous system; maintains a normal and strong heartbeat; helps in blood clotting	Rickets, osteomalacia, immunity	Cod liver oil, milk, egg yolk
Vitamin E (Tocopherol)	Lessens oxidative damage of cells and prevents lung damage from pollutants; vital to the immune system	Deficiency is rare; anemia, nervous system degeneration	Wheat germ oil, unrefined vegetable oils, nuts, seeds, grains
Vitamin K (Phylloquinone)	Essential to blood clotting	Bleeding and easy bruising	Leafy green vegetables, tea

Table 34.2

Figure 34.15 A healthy diet should include a variety of foods to ensure that needs for essential nutrients are met. (credit: Keith Weller, USDA ARS)

Minerals and Their Function in the Human Body

Mineral	Function	Deficiencies Can Lead To	Sources
*Calcium	Needed for muscle and neuron function; heart health; builds bone and supports synthesis and function of blood cells; nerve function	Osteoporosis, rickets, muscle spasms, impaired growth	Milk, yogurt, fish, green leafy vegetables, legumes
*Chlorine	Needed for production of hydrochloric acid (HCl) in the stomach and nerve function; osmotic balance	Muscle cramps, mood disturbances, reduced appetite	Table salt
Copper (trace amounts)	Required component of many redox enzymes, including cytochrome c oxidase; cofactor for hemoglobin synthesis	Copper deficiency is rare	Liver, oysters, cocoa, chocolate, sesame, nuts
Iodine	Required for the synthesis of thyroid hormones	Goiter	Seafood, iodized salt, dairy products
Iron	Required for many proteins and enzymes, notably hemoglobin, to prevent anemia	Anemia, which causes poor concentration, fatigue, and poor immune function	Red meat, leafy green vegetables, fish (tuna, salmon), eggs, dried fruits, beans, whole grains
*Magnesium	Required cofactor for ATP formation; bone formation; normal membrane functions; muscle function	Mood disturbances, muscle spasms	Whole grains, leafy green vegetables
Manganese (trace amounts)	A cofactor in enzyme functions; trace amounts are required	Manganese deficiency is rare	Common in most foods
Molybdenum (trace amounts)	Acts as a cofactor for three essential enzymes in humans: sulfite oxidase, xanthine oxidase, and aldehyde oxidase	Molybdenum deficiency is rare	
*Phosphorus	A component of bones and teeth; helps regulate acid-base balance; nucleotide synthesis	Weakness, bone abnormalities, calcium loss	Milk, hard cheese, whole grains, meats
*Potassium	Vital for muscles, heart, and nerve function	Cardiac rhythm disturbance, muscle weakness	Legumes, potato skin, tomatoes, bananas
Selenium (trace amounts)	A cofactor essential to activity of antioxidant enzymes like glutathione peroxidase; trace amounts are required	Selenium deficiency is rare	Common in most foods

Table 34.3

Mineral	Function	Deficiencies Can Lead To	Sources
*Sodium	Systemic electrolyte required for many functions; acid-base balance; water balance; nerve function	Muscle cramps, fatigue, reduced appetite	Table salt
Zinc (trace amounts)	Required for several enzymes such as carboxypeptidase, liver alcohol dehydrogenase, and carbonic anhydrase	Anemia, poor wound healing, can lead to short stature	Common in most foods

*Greater than 200mg/day required

Table 34.3

Essential Amino Acids

Amino acids that must be consumed	Amino acids anabolized by the body
isoleucine	alanine
leucine	selenocysteine
lysine	aspartate
methionine	cysteine
phenylalanine	glutamate
tryptophan	glycine
valine	proline
histidine*	serine
threonine	tyrosine
arginine*	asparagine

*The human body can synthesize histidine and arginine, but not in the quantities required, especially for growing children.

Table 34.4

Food Energy and ATP

Animals need food to obtain energy and maintain homeostasis. Homeostasis is the ability of a system to maintain a stable internal environment even in the face of external changes to the environment. For example, the normal body temperature of humans is 37°C (98.6°F). Humans maintain this temperature even when the external temperature is hot or cold. It takes energy to maintain this body temperature, and animals obtain this energy from food.

The primary source of energy for animals is carbohydrates, mainly glucose. Glucose is called the body's fuel. The digestible carbohydrates in an animal's diet are converted to glucose molecules through a series of catabolic chemical reactions.

Adenosine triphosphate, or ATP, is the primary energy currency in cells; ATP stores energy in phosphate ester bonds. ATP releases energy when the phosphodiester bonds are broken and ATP is converted to ADP and a phosphate group. ATP is produced by the oxidative reactions in the cytoplasm and mitochondrion of the cell, where carbohydrates, proteins, and fats undergo a series of metabolic reactions collectively called cellular respiration. For example, glycolysis is a series of reactions in which glucose is converted to pyruvic acid and some of its chemical potential energy is transferred to NADH and ATP.

ATP is required for all cellular functions. It is used to build the organic molecules that are required for cells and tissues; it provides energy for muscle contraction and for the transmission of electrical signals in the nervous system. When the amount of ATP is available in excess of the body's requirements, the liver uses the excess ATP and excess glucose to produce molecules called glycogen. Glycogen is a polymeric form of glucose and is stored in the liver and skeletal muscle cells. When blood sugar drops, the liver releases glucose from stores of glycogen. Skeletal muscle converts glycogen to glucose during intense exercise. The process of converting glucose and excess ATP to glycogen and the storage of excess energy is an evolutionarily important step in helping animals deal with mobility, food shortages, and famine.

Everyday Connection

Obesity

Obesity is a major health concern in the United States, and there is a growing focus on reducing obesity and the diseases it may lead to, such as type-2 diabetes, cancers of the colon and breast, and cardiovascular disease. How does the food consumed contribute to obesity?

Fatty foods are calorie-dense, meaning that they have more calories per unit mass than carbohydrates or proteins. One gram of carbohydrates has four calories, one gram of protein has four calories, and one gram of fat has nine calories. Animals tend to seek lipid-rich food for their higher energy content.

The signals of hunger ("time to eat") and satiety ("time to stop eating") are controlled in the hypothalamus region of the brain. Foods that are rich in fatty acids tend to promote satiety more than foods that are rich only in carbohydrates.

Excess carbohydrate and ATP are used by the liver to synthesize glycogen. The pyruvate produced during glycolysis is used to synthesize fatty acids. When there is more glucose in the body than required, the resulting excess pyruvate is converted into molecules that eventually result in the synthesis of fatty acids within the body. These fatty acids are stored in adipose cells—the fat cells in the mammalian body whose primary role is to store fat for later use.

It is important to note that some animals benefit from obesity. Polar bears and seals need body fat for insulation and to keep them from losing body heat during Arctic winters. When food is scarce, stored body fat provides energy for maintaining homeostasis. Fats prevent famine in mammals, allowing them to access energy when food is not available on a daily basis; fats are stored when a large kill is made or lots of food is available.

34.3 Digestive System Processes

By the end of this section, you will be able to do the following:

- Describe the process of digestion
- Detail the steps involved in digestion and absorption
- Define elimination
- Explain the role of both the small and large intestines in absorption

Obtaining nutrition and energy from food is a multistep process. For true animals, the first step is ingestion, the act of taking in food. This is followed by digestion, absorption, and elimination. In the following sections, each of these steps will be discussed in detail.

Ingestion

The large molecules found in intact food cannot pass through the cell membranes. Food needs to be broken into smaller particles so that animals can harness the nutrients and organic molecules. The first step in this process is **ingestion**. Ingestion is the process of taking in food through the mouth. In vertebrates, the teeth, saliva, and tongue play important roles in mastication (preparing the food into bolus). While the food is being mechanically broken down, the enzymes in saliva begin to chemically process the food as well. The combined action of these processes modifies the food from large particles to a soft mass that can be swallowed and can travel the length of the esophagus.

Digestion and Absorption

Digestion is the mechanical and chemical breakdown of food into small organic fragments. It is important to breakdown macromolecules into smaller fragments that are of suitable size for absorption across the digestive epithelium. Large, complex molecules of proteins, polysaccharides, and lipids must be reduced to simpler particles such as simple sugar before they can be absorbed by the digestive epithelial cells. Different organs play specific roles in the digestive process. The animal diet needs carbohydrates, protein, and fat, as well as vitamins and inorganic components for nutritional balance. How each of these components is digested is discussed in the following sections.

Carbohydrates

The digestion of carbohydrates begins in the mouth. The salivary enzyme amylase begins the breakdown of food starches into maltose, a disaccharide. As the bolus of food travels through the esophagus to the stomach, no significant digestion of carbohydrates takes place. The esophagus produces no digestive enzymes but does produce mucous for lubrication. The acidic environment in the stomach stops the action of the amylase enzyme.

The next step of carbohydrate digestion takes place in the duodenum. Recall that the chyme from the stomach enters the duodenum and mixes with the digestive secretion from the pancreas, liver, and gallbladder. Pancreatic juices also contain amylase, which continues the breakdown of starch and glycogen into maltose, a disaccharide. The disaccharides are broken down into monosaccharides by enzymes called **maltases**, **sucrases**, and **lactases**, which are also present in the brush border of the small intestinal wall. Maltase breaks down maltose into glucose. Other disaccharides, such as sucrose and lactose are broken down by sucrase and lactase, respectively. Sucrase breaks down sucrose (or "table sugar") into glucose and fructose, and lactase breaks down lactose (or "milk sugar") into glucose and galactose. The monosaccharides (glucose) thus produced are absorbed and then can be used in metabolic pathways to harness energy. The monosaccharides are transported across the intestinal epithelium into the bloodstream to be transported to the different cells in the body. The steps in carbohydrate digestion are summarized in Figure 34.16 and Table 34.5.

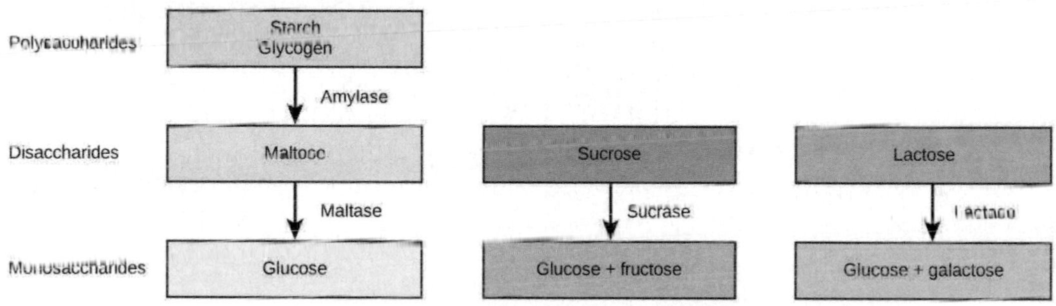

Figure 34.16 Digestion of carbohydrates is performed by several enzymes. Starch and glycogen are broken down into glucose by amylase and maltase. Sucrose (table sugar) and lactose (milk sugar) are broken down by sucrase and lactase, respectively.

Digestion of Carbohydrates

Enzyme	Produced By	Site of Action	Substrate Acting On	End Products
Salivary amylase	Salivary glands	Mouth	Polysaccharides (Starch)	Disaccharides (maltose), oligosaccharides
Pancreatic amylase	Pancreas	Small intestine	Polysaccharides (starch)	Disaccharides (maltose), monosaccharides
Oligosaccharidases	Lining of the intestine; brush border membrane	Small intestine	Disaccharides	Monosaccharides (e.g., glucose, fructose, galactose)

Table 34.5

Protein

A large part of protein digestion takes place in the stomach. The enzyme pepsin plays an important role in the digestion of proteins by breaking down the intact protein to peptides, which are short chains of four to nine amino acids. In the duodenum, other enzymes—**trypsin**, **elastase**, and **chymotrypsin**—act on the peptides reducing them to smaller peptides. Trypsin elastase, carboxypeptidase, and chymotrypsin are produced by the pancreas and released into the duodenum where they act on the chyme. Further breakdown of peptides to single amino acids is aided by enzymes called peptidases (those that breakdown peptides). Specifically, **carboxypeptidase**, **dipeptidase**, and **aminopeptidase** play important roles in reducing the peptides to free amino acids. The amino acids are absorbed into the bloodstream through the small intestines. The steps in protein digestion are summarized in Figure 34.17 and Table 34.6.

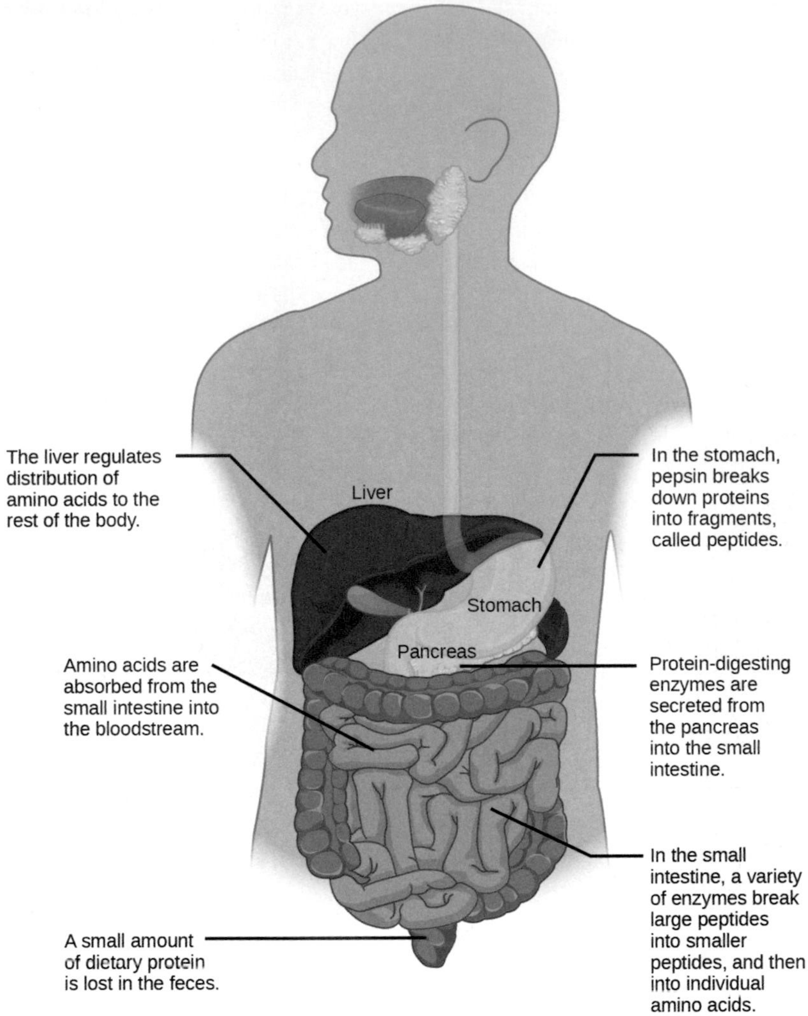

Figure 34.17 Protein digestion is a multistep process that begins in the stomach and continues through the intestines.

Digestion of Protein

Enzyme	Produced By	Site of Action	Substrate Acting On	End Products
Pepsin	Stomach chief cells	Stomach	Proteins	Peptides
Trypsin Elastase Chymotrypsin	Pancreas	Small intestine	Proteins	Peptides

Table 34.6

Enzyme	Produced By	Site of Action	Substrate Acting On	End Products
Carboxypeptidase	Pancreas	Small intestine	Peptides	Amino acids and peptides
Aminopeptidase Dipeptidase	Lining of intestine	Small intestine	Peptides	Amino acids

Table 34.6

Lipids

Lipid digestion begins in the stomach with the aid of lingual lipase and gastric lipase. However, the bulk of lipid digestion occurs in the small intestine due to pancreatic lipase. When chyme enters the duodenum, the hormonal responses trigger the release of bile, which is produced in the liver and stored in the gallbladder. Bile aids in the digestion of lipids, primarily triglycerides by emulsification. Emulsification is a process in which large lipid globules are broken down into several small lipid globules. These small globules are more widely distributed in the chyme rather than forming large aggregates. Lipids are hydrophobic substances: in the presence of water, they will aggregate to form globules to minimize exposure to water. Bile contains bile salts, which are amphipathic, meaning they contain hydrophobic and hydrophilic parts. Thus, the bile salts hydrophilic side can interface with water on one side and the hydrophobic side interfaces with lipids on the other. By doing so, bile salts emulsify large lipid globules into small lipid globules.

Why is emulsification important for digestion of lipids? Pancreatic juices contain enzymes called lipases (enzymes that breakdown lipids). If the lipid in the chyme aggregates into large globules, very little surface area of the lipids is available for the lipases to act on, leaving lipid digestion incomplete. By forming an emulsion, bile salts increase the available surface area of the lipids many fold. The pancreatic lipases can then act on the lipids more efficiently and digest them, as detailed in Figure 34.18. Lipases breakdown the lipids into fatty acids and glycerides. These molecules can pass through the plasma membrane of the cell and enter the epithelial cells of the intestinal lining. The bile salts surround long-chain fatty acids and monoglycerides forming tiny spheres called micelles. The micelles move into the brush border of the small intestine absorptive cells where the long-chain fatty acids and monoglycerides diffuse out of the micelles into the absorptive cells leaving the micelles behind in the chyme. The long-chain fatty acids and monoglycerides recombine in the absorptive cells to form triglycerides, which aggregate into globules and become coated with proteins. These large spheres are called **chylomicrons**. Chylomicrons contain triglycerides, cholesterol, and other lipids and have proteins on their surface. The surface is also composed of the hydrophilic phosphate "heads" of phospholipids. Together, they enable the chylomicron to move in an aqueous environment without exposing the lipids to water. Chylomicrons leave the absorptive cells via exocytosis. Chylomicrons enter the lymphatic vessels, and then enter the blood in the subclavian vein.

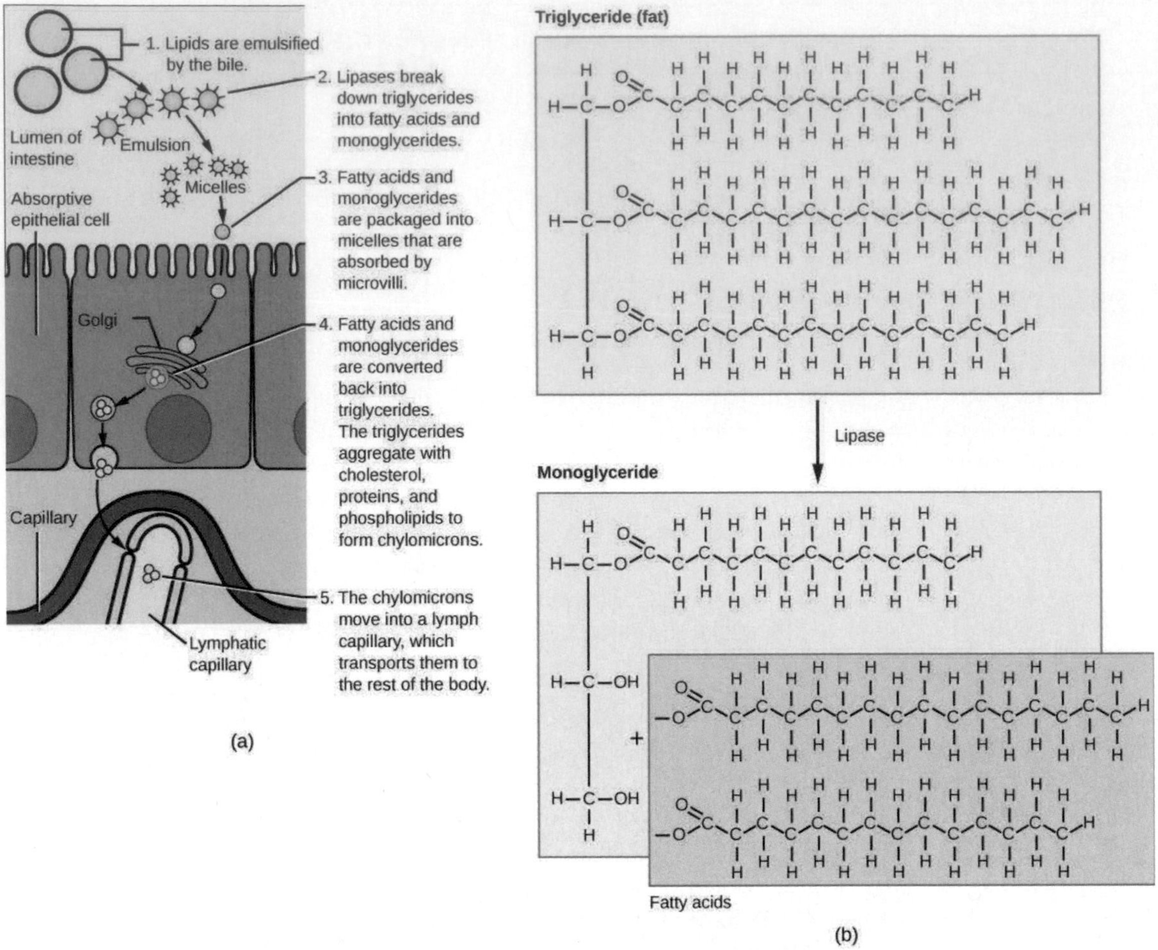

Figure 34.18 Lipids are digested and absorbed in the small intestine.

Vitamins

Vitamins can be either water-soluble or lipid-soluble. Fat soluble vitamins are absorbed in the same manner as lipids. It is important to consume some amount of dietary lipid to aid the absorption of lipid-soluble vitamins. Water-soluble vitamins can be directly absorbed into the bloodstream from the intestine.

LINK TO LEARNING

This website (http://openstax.org/l/digest_enzymes) has an overview of the digestion of protein, fat, and carbohydrates.

VISUAL CONNECTION

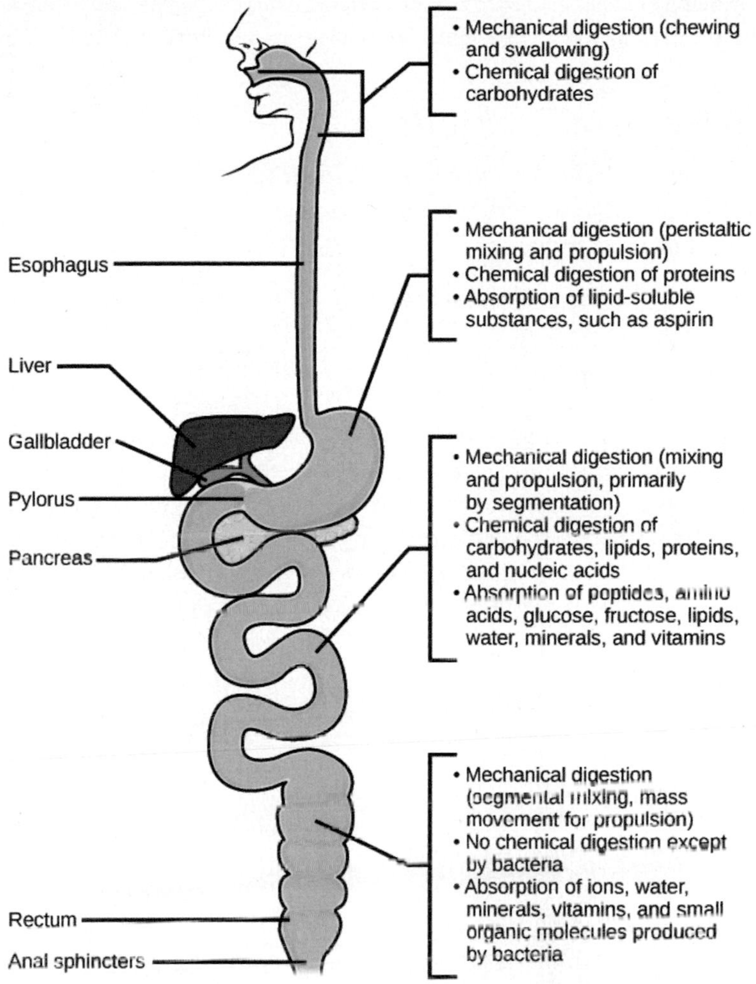

Figure 34.19 Mechanical and chemical digestion of food takes place in many steps, beginning in the mouth and ending in the rectum.

Which of the following statements about digestive processes is true?

a. Amylase, maltase, and lactase in the mouth digest carbohydrates.
b. Trypsin and lipase in the stomach digest protein.
c. Bile emulsifies lipids in the small intestine.
d. No food is absorbed until the small intestine.

Elimination

The final step in digestion is the elimination of undigested food content and waste products. The undigested food material enters the colon, where most of the water is reabsorbed. Recall that the colon is also home to the microflora called "intestinal flora" that aid in the digestion process. The semi-solid waste is moved through the colon by peristaltic movements of the muscle and is stored in the rectum. As the rectum expands in response to storage of fecal matter, it triggers the neural signals required to set up the urge to eliminate. The solid waste is eliminated through the anus using peristaltic movements of the rectum.

Common Problems with Elimination

Diarrhea and constipation are some of the most common health concerns that affect digestion. Constipation is a condition where the feces are hardened because of excess water removal in the colon. In contrast, if enough water is not removed from the feces, it results in diarrhea. Many bacteria, including the ones that cause cholera, affect the proteins involved in water reabsorption in the colon and result in excessive diarrhea.

Emesis

Emesis, or vomiting, is elimination of food by forceful expulsion through the mouth. It is often in response to an irritant that affects the digestive tract, including but not limited to viruses, bacteria, emotions, sights, and food poisoning. This forceful expulsion of the food is due to the strong contractions produced by the stomach muscles. The process of emesis is regulated by the medulla.

34.4 Digestive System Regulation

By the end of this section, you will be able to do the following:

- Discuss the role of neural regulation in digestive processes
- Explain how hormones regulate digestion

The brain is the control center for the sensation of hunger and satiety. The functions of the digestive system are regulated through neural and hormonal responses.

Neural Responses to Food

In reaction to the smell, sight, or thought of food, like that shown in Figure 34.20, the first response is that of salivation. The salivary glands secrete more saliva in response to stimulation by the autonomic nervous system triggered by food in preparation for digestion. Simultaneously, the stomach begins to produce hydrochloric acid to digest the food. Recall that the peristaltic movements of the esophagus and other organs of the digestive tract are under the control of the brain. The brain prepares these muscles for movement as well. When the stomach is full, the part of the brain that detects satiety signals fullness. There are three overlapping phases of gastric control—the cephalic phase, the gastric phase, and the intestinal phase—each requires many enzymes and is under neural control as well.

Figure 34.20 Seeing a plate of food triggers the secretion of saliva in the mouth and the production of HCL in the stomach. (credit: Kelly Bailey)

Digestive Phases

The response to food begins even before food enters the mouth. The first phase of ingestion, called the **cephalic phase**, is controlled by the neural response to the stimulus provided by food. All aspects—such as sight, sense, and smell—trigger the neural responses resulting in salivation and secretion of gastric juices. The gastric and salivary secretion in the cephalic phase can also take place due to the thought of food. Right now, if you think about a piece of chocolate or a crispy potato chip, the increase in salivation is a cephalic phase response to the thought. The central nervous system prepares the stomach to receive food.

The **gastric phase** begins once the food arrives in the stomach. It builds on the stimulation provided during the cephalic phase. Gastric acids and enzymes process the ingested materials. The gastric phase is stimulated by (1) distension of the stomach, (2) a decrease in the pH of the gastric contents, and (3) the presence of undigested material. This phase consists of local, hormonal, and neural responses. These responses stimulate secretions and powerful contractions.

The **intestinal phase** begins when chyme enters the small intestine triggering digestive secretions. This phase controls the rate of gastric emptying. In addition to gastrin emptying, when chyme enters the small intestine, it triggers other hormonal and neural events that coordinate the activities of the intestinal tract, pancreas, liver, and gallbladder.

Hormonal Responses to Food

The **endocrine system** controls the response of the various glands in the body and the release of hormones at the appropriate times.

One of the important factors under hormonal control is the stomach acid environment. During the gastric phase, the hormone **gastrin** is secreted by G cells in the stomach in response to the presence of proteins. Gastrin stimulates the release of stomach acid, or hydrochloric acid (HCl) which aids in the digestion of the proteins. However, when the stomach is emptied, the acidic environment need not be maintained and a hormone called **somatostatin** stops the release of hydrochloric acid. This is controlled by a negative feedback mechanism.

In the duodenum, digestive secretions from the liver, pancreas, and gallbladder play an important role in digesting chyme during the intestinal phase. In order to neutralize the acidic chyme, a hormone called **secretin** stimulates the pancreas to produce alkaline bicarbonate solution and deliver it to the duodenum. Secretin acts in tandem with another hormone called **cholecystokinin** (CCK). Not only does CCK stimulate the pancreas to produce the requisite pancreatic juices, it also stimulates the gallbladder to release bile into the duodenum.

LINK TO LEARNING

Visit this website (http://openstax.org/l/enteric_endo) to learn more about the endocrine system. Review the text and watch the animation of how control is implemented in the endocrine system.

Another level of hormonal control occurs in response to the composition of food. Foods high in lipids take a long time to digest. A hormone called **gastric inhibitory peptide** is secreted by the small intestine to slow down the peristaltic movements of the intestine to allow fatty foods more time to be digested and absorbed.

Understanding the hormonal control of the digestive system is an important area of ongoing research. Scientists are exploring the role of each hormone in the digestive process and developing ways to target these hormones. Advances could lead to knowledge that may help to battle the obesity epidemic.

KEY TERMS

alimentary canal tubular digestive system with a mouth and anus

aminopeptidase protease that breaks down peptides to single amino acids; secreted by the brush border of small intestine

anus exit point for waste material

bile digestive juice produced by the liver; important for digestion of lipids

bolus mass of food resulting from chewing action and wetting by saliva

carboxypeptidase protease that breaks down peptides to single amino acids; secreted by the brush border of the small intestine

carnivore animal that consumes animal flesh

cephalic phase first phase of digestion, controlled by the neural response to the stimulus provided by food

cholecystokinin hormone that stimulates the contraction of the gallbladder to release bile

chylomicron small lipid globule

chyme mixture of partially digested food and stomach juices

chymotrypsin pancreatic protease

digestion mechanical and chemical breakdown of food into small organic fragments

dipeptidase protease that breaks down peptides to single amino acids; secreted by the brush border of small intestine

duodenum first part of the small intestine where a large part of digestion of carbohydrates and fats occurs

elastase pancreatic protease

endocrine system system that controls the response of the various glands in the body and the release of hormones at the appropriate times

esophagus tubular organ that connects the mouth to the stomach

essential nutrient nutrient that cannot be synthesized by the body; it must be obtained from food

gallbladder organ that stores and concentrates bile

gastric inhibitory peptide hormone secreted by the small intestine in the presence of fatty acids and sugars; it also inhibits acid production and peristalsis in order to slow down the rate at which food enters the small intestine

gastric phase digestive phase beginning once food enters the stomach; gastric acids and enzymes process the ingested materials

gastrin hormone which stimulates hydrochloric acid secretion in the stomach

gastrovascular cavity digestive system consisting of a single opening

gizzard muscular organ that grinds food

herbivore animal that consumes a strictly plant diet

ileum last part of the small intestine; connects the small intestine to the large intestine; important for absorption of B-12

ingestion act of taking in food

intestinal phase third digestive phase; begins when chyme enters the small intestine triggering digestive secretions and controlling the rate of gastric emptying

jejunum second part of the small intestine

lactase enzyme that breaks down lactose into glucose and galactose

large intestine digestive system organ that reabsorbs water from undigested material and processes waste matter

lipase enzyme that chemically breaks down lipids

liver organ that produces bile for digestion and processes vitamins and lipids

maltase enzyme that breaks down maltose into glucose

mineral inorganic, elemental molecule that carries out important roles in the body

monogastric digestive system that consists of a single-chambered stomach

omnivore animal that consumes both plants and animals

pancreas gland that secretes digestive juices

pepsin enzyme found in the stomach whose main role is protein digestion

pepsinogen inactive form of pepsin

peristalsis wave-like movements of muscle tissue

proventriculus glandular part of a bird's stomach

rectum area of the body where feces is stored until elimination

roughage component of food that is low in energy and high in fiber

ruminant animal with a stomach divided into four compartments

salivary amylase enzyme found in saliva, which converts carbohydrates to maltose

secretin hormone which stimulates sodium bicarbonate secretion in the small intestine

small intestine organ where digestion of protein, fats, and carbohydrates is completed

somatostatin hormone released to stop acid secretion when the stomach is empty

sphincter band of muscle that controls movement of materials throughout the digestive tract

stomach saclike organ containing acidic digestive juices

sucrase enzyme that breaks down sucrose into glucose and fructose

trypsin pancreatic protease that breaks down protein

villi folds on the inner surface of the small intestine whose role is to increase absorption area

vitamin organic substance necessary in small amounts to sustain life

CHAPTER SUMMARY

34.1 Digestive Systems

Different animals have evolved different types of digestive systems specialized to meet their dietary needs. Humans and many other animals have monogastric digestive systems with a single-chambered stomach. Birds have evolved a digestive system that includes a gizzard where the food is crushed into smaller pieces. This compensates for their inability to masticate. Ruminants that consume large amounts of plant material have a multi-chambered stomach that digests roughage. Pseudo-ruminants have similar digestive processes as ruminants but do not have the four-compartment stomach. Processing food involves ingestion (eating), digestion (mechanical and enzymatic breakdown of large molecules), absorption (cellular uptake of nutrients), and elimination (removal of undigested waste as feces).

Many organs work together to digest food and absorb nutrients. The mouth is the point of ingestion and the location where both mechanical and chemical breakdown of food begins. Saliva contains an enzyme called amylase that breaks down carbohydrates. The food bolus travels through the esophagus by peristaltic movements to the stomach. The stomach has an extremely acidic environment. An enzyme called pepsin digests protein in the stomach. Further digestion and absorption take place in the small intestine. The large intestine reabsorbs water from the undigested food and stores waste until elimination.

34.2 Nutrition and Energy Production

Animal diet should be balanced and meet the needs of the body. Carbohydrates, proteins, and fats are the primary components of food. Some essential nutrients are required for cellular function but cannot be produced by the animal body. These include vitamins, minerals, some fatty acids, and some amino acids. Food intake in more than necessary amounts is stored as glycogen in the liver and muscle cells, and in fat cells. Excess adipose storage can lead to obesity and serious health problems. ATP is the energy currency of the cell and is obtained from the metabolic pathways. Excess carbohydrates and energy are stored as glycogen in the body.

34.3 Digestive System Processes

Digestion begins with ingestion, where the food is taken in the mouth. Digestion and absorption take place in a series of steps with special enzymes playing important roles in digesting carbohydrates, proteins, and lipids. Elimination describes removal of undigested food contents and waste products from the body. While most absorption occurs in the small intestines, the large intestine is responsible for the final removal of water that remains after the absorptive process of the small intestines. The cells that line the large intestine absorb some vitamins as well as any leftover salts and water. The large intestine (colon) is also where feces is formed.

34.4 Digestive System Regulation

The brain and the endocrine system control digestive processes. The brain controls the responses of hunger and satiety. The endocrine system controls the release of hormones and enzymes required for digestion of food in the digestive tract.

VISUAL CONNECTION QUESTIONS

1. Figure 34.11 Which of the following statements about the digestive system is false?
   a. Chyme is a mixture of food and digestive juices that is produced in the stomach.
   b. Food enters the large intestine before the small intestine.
   c. In the small intestine, chyme mixes with bile, which emulsifies fats.
   d. The stomach is separated from the small intestine by the pyloric sphincter.

2. Figure 34.12 Which of the following statements about the small intestine is false?
   a. Absorptive cells that line the small intestine have microvilli, small projections that increase surface area and aid in the absorption of food.
   b. The inside of the small intestine has many folds, called villi.
   c. Microvilli are lined with blood vessels as well as lymphatic vessels.
   d. The inside of the small intestine is called the lumen.

3. Figure 34.19 Which of the following statements about digestive processes is true?
   a. Amylase, maltase, and lactase in the mouth digest carbohydrates.
   b. Trypsin and lipase in the stomach digest protein.
   c. Bile emulsifies lipids in the small intestine.
   d. No food is absorbed until the small intestine.

REVIEW QUESTIONS

4. Which of the following is a pseudo-ruminant?
   a. cow
   b. pig
   c. crow
   d. horse

5. Which of the following statements is untrue?
   a. Roughage takes a long time to digest.
   b. Birds eat large quantities at one time so that they can fly long distances.
   c. Cows do not have upper teeth.
   d. In pseudo-ruminants, roughage is digested in the cecum.

6. The acidic nature of chyme is neutralized by ________.
   a. potassium hydroxide
   b. sodium hydroxide
   c. bicarbonates
   d. vinegar

7. The digestive juices from the liver are delivered to the ________.
   a. stomach
   b. liver
   c. duodenum
   d. colon

8. A scientist dissects a new species of animal. If the animal's digestive system has a single stomach with an extended small intestine, to which animal could the dissected specimen be closely related?
   a. lion
   b. snowshoe hare
   c. earthworm
   d. eagle

9. Which of the following statements is not true?
   a. Essential nutrients can be synthesized by the body.
   b. Vitamins are required in small quantities for bodily function.
   c. Some amino acids can be synthesized by the body, while others need to be obtained from diet.
   d. Vitamins come in two categories: fat-soluble and water-soluble.

10. Which of the following is a water-soluble vitamin?
    a. vitamin A
    b. vitamin E
    c. vitamin K
    d. vitamin C

11. What is the primary fuel for the body?
    a. carbohydrates
    b. lipids
    c. protein
    d. glycogen

12. Excess glucose is stored as ________.
    a. fat
    b. glucagon
    c. glycogen
    d. it is not stored in the body

13. Many distance runners "carb load" the day before a big race. How does this eating strategy provide an advantage to the runner?
    a. The carbohydrates cause the release of insulin.
    b. The excess carbohydrates are converted to fats, which have a higher calorie density.
    c. The glucose from the carbohydrates lets the muscles make excess ATP overnight.
    d. The excess carbohydrates can be stored in the muscles as glycogen.

14. Where does the majority of protein digestion take place?
    a. stomach
    b. duodenum
    c. mouth
    d. jejunum

15. Lipases are enzymes that breakdown ________.
    a. disaccharides
    b. lipids
    c. proteins
    d. cellulose

16. Which of the following conditions is most likely to cause constipation?
    a. bacterial infection
    b. dehydration
    c. ulcer
    d. excessive cellulose consumption

17. Which hormone controls the release of bile from the gallbladder
    a. pepsin
    b. amylase
    c. CCK
    d. gastrin

18. Which hormone stops acid secretion in the stomach?
    a. gastrin
    b. somatostatin
    c. gastric inhibitory peptide
    d. CCK

19. In the famous conditioning experiment, Pavlov demonstrated that his dogs started drooling in response to a bell sounding. What part of the digestive process did he stimulate?
    a. cephalic phase
    b. gastric phase
    c. intestinal phase
    d. elimination phase

CRITICAL THINKING QUESTIONS

20. How does the polygastric digestive system aid in digesting roughage?
21. How do birds digest their food in the absence of teeth?
22. What is the role of the accessory organs in digestion?
23. Explain how the villi and microvilli aid in absorption.
24. Name two components of the digestive system that perform mechanical digestion. Describe how mechanical digestion contributes to acquiring nutrients from food.
25. What are essential nutrients?
26. What is the role of minerals in maintaining good health?
27. Discuss why obesity is a growing epidemic.
28. There are several nations where malnourishment is a common occurrence. What may be some of the health challenges posed by malnutrition?
29. Generally describe how a piece of bread can power your legs as you walk up a flight of stairs.
30. In the 1990s fat-free foods became popular among people trying to lose weight. However, many dieticians now conclude that the fat-free trend made people less healthy and heavier. Describe how this could occur.
31. Explain why some dietary lipid is a necessary part of a balanced diet.
32. The gut microbiome (the bacterial colonies in the intestines) have become a popular area of study in biomedical research. How could varying gut microbiomes impact a person's nutrition?
33. Many mammals become ill if they drink milk as adults even though they could consume it as babies. What causes this digestive issue?
34. Describe how hormones regulate digestion.
35. Describe one or more scenarios where loss of hormonal regulation of digestion can lead to diseases.
36. A scientist is studying a model that has a mutation in the receptor for somatostatin that prevents hormone binding. How would this mutation affect the structure and function of the digestive system?

CHAPTER 35
The Nervous System

Figure 35.1 An athlete's nervous system is hard at work during the planning and execution of a movement as precise as a high jump. Parts of the nervous system are involved in determining how hard to push off and when to turn, as well as controlling the muscles throughout the body that make this complicated movement possible without knocking the bar down—all in just a few seconds. (credit: Scott Ray)

INTRODUCTION When you're reading this book, your nervous system is performing several functions simultaneously. The visual system is processing what is seen on the page; the motor system controls the turn of the pages (or click of the mouse); the prefrontal cortex maintains attention. Even fundamental functions, like breathing and regulation of body temperature, are controlled by the nervous system. A nervous system is an organism's control center: it processes sensory information from outside (and inside) the body and controls all behaviors—from eating to sleeping to finding a mate.

Chapter Outline

35.1 Neurons and Glial Cells

By the end of this section, you will be able to do the following:

- List and describe the functions of the structural components of a neuron
- List and describe the four main types of neurons
- Compare the functions of different types of glial cells

Nervous systems throughout the animal kingdom vary in structure and complexity, as illustrated by the variety of animals shown in Figure 35.2. Some organisms, like sea sponges, lack a true nervous system. Others, like jellyfish, lack a true brain and instead have a system of separate but connected nerve cells (neurons) called a "nerve net." Echinoderms such as sea stars have nerve cells that are bundled into fibers called nerves. Flatworms of the phylum Platyhelminthes have both a central nervous system (CNS), made up of a small "brain" and two nerve cords, and a peripheral nervous system (PNS) containing a system of nerves that extend throughout the body. The insect nervous system is more complex but also fairly decentralized. It contains a brain, ventral nerve cord, and ganglia (clusters of connected neurons). These ganglia can control movements and behaviors without input from the brain. Octopi may have the most complicated of invertebrate nervous systems—they have neurons that are organized in specialized lobes and eyes that are structurally similar to vertebrate species.

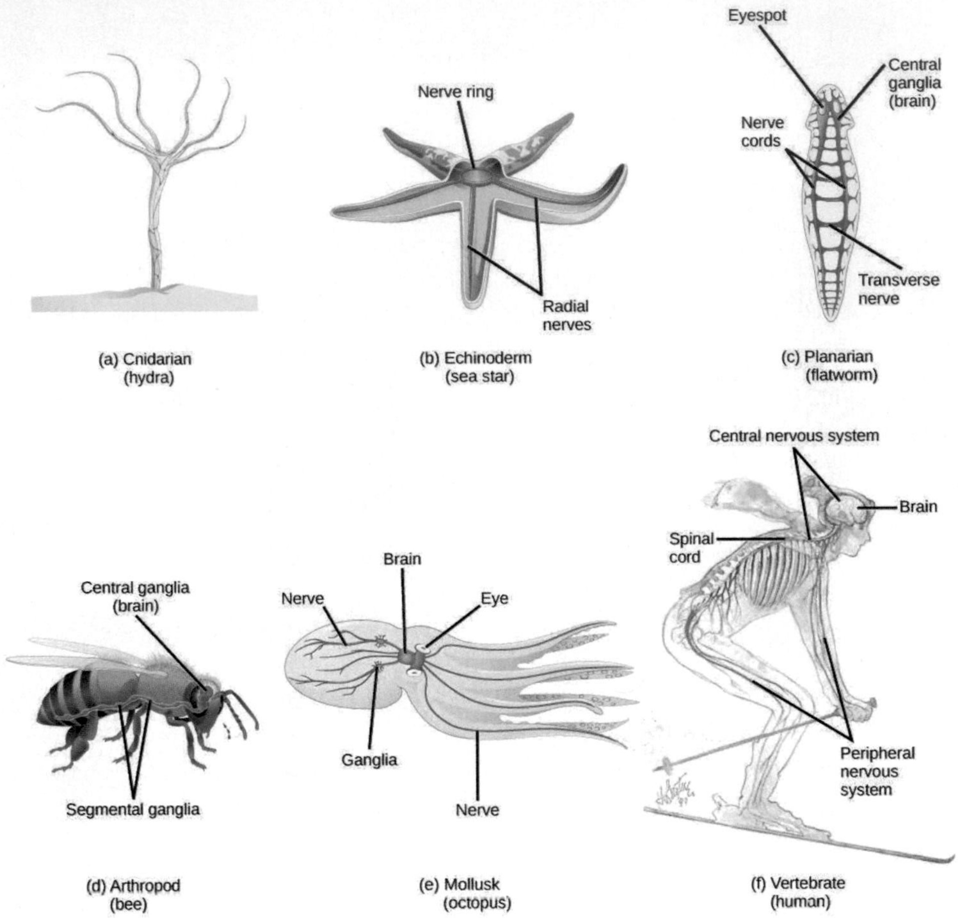

Figure 35.2 Nervous systems vary in structure and complexity. In (a) cnidarians, nerve cells form a decentralized nerve net. In (b) echinoderms, nerve cells are bundled into fibers called nerves. In animals exhibiting bilateral symmetry such as (c) planarians, neurons cluster into an anterior brain that processes information. In addition to a brain, (d) arthropods have clusters of nerve cell bodies, called peripheral ganglia, located along the ventral nerve cord. Mollusks such as squid and (e) octopi, which must hunt to survive, have complex brains containing millions of neurons. In (f) vertebrates, the brain and spinal cord comprise the central nervous system, while neurons extending into the rest of the body comprise the peripheral nervous system. (credit e: modification of work by Michael Vecchione, Clyde F.E. Roper, and Michael J. Sweeney, NOAA; credit f: modification of work by NIH)

Compared to invertebrates, vertebrate nervous systems are more complex, centralized, and specialized. While there is great diversity among different vertebrate nervous systems, they all share a basic structure: a CNS that contains a brain and spinal cord and a PNS made up of peripheral sensory and motor nerves. One interesting difference between the nervous systems of invertebrates and vertebrates is that the nerve cords of many invertebrates are located ventrally whereas the vertebrate spinal cords are located dorsally. There is debate among evolutionary biologists as to whether these different nervous system plans evolved separately or whether the invertebrate body plan arrangement somehow "flipped" during the evolution of vertebrates.

LINK TO LEARNING

Watch this video of biologist Mark Kirschner discussing the "flipping" phenomenon of vertebrate evolution.

Click to view content (https://www.openstax.org/l/vertebrate_evol)

The nervous system is made up of **neurons**, specialized cells that can receive and transmit chemical or electrical signals, and **glia**, cells that provide support functions for the neurons by playing an information processing role that is complementary to neurons. A neuron can be compared to an electrical wire—it transmits a signal from one place to another. Glia can be compared to the workers at the electric company who make sure wires go to the right places, maintain the wires, and take down wires that are broken. Although glia have been compared to workers, recent evidence suggests that they also usurp some of the signaling functions of neurons.

There is great diversity in the types of neurons and glia that are present in different parts of the nervous system. There are four major types of neurons, and they share several important cellular components.

Neurons

The nervous system of the common laboratory fly, *Drosophila melanogaster*, contains around 100,000 neurons, the same number as a lobster. This number compares to 75 million in the mouse and 300 million in the octopus. A human brain contains around 86 billion neurons. Despite these very different numbers, the nervous systems of these animals control many of the same behaviors—from basic reflexes to more complicated behaviors like finding food and courting mates. The ability of neurons to communicate with each other as well as with other types of cells underlies all of these behaviors.

Most neurons share the same cellular components. But neurons are also highly specialized—different types of neurons have different sizes and shapes that relate to their functional roles.

Parts of a Neuron

Like other cells, each neuron has a cell body (or soma) that contains a nucleus, smooth and rough endoplasmic reticulum, Golgi apparatus, mitochondria, and other cellular components. Neurons also contain unique structures, illustrated in Figure 35.3 for receiving and sending the electrical signals that make neuronal communication possible. **Dendrites** are tree-like structures that extend away from the cell body to receive messages from other neurons at specialized junctions called **synapses**. Although some neurons do not have any dendrites, some types of neurons have multiple dendrites. Dendrites can have small protrusions called dendritic spines, which further increase surface area for possible synaptic connections.

Once a signal is received by the dendrite, it then travels passively to the cell body. The cell body contains a specialized structure, the **axon hillock** that integrates signals from multiple synapses and serves as a junction between the cell body and an **axon**. An axon is a tube-like structure that propagates the integrated signal to specialized endings called **axon terminals**. These terminals in turn synapse on other neurons, muscle, or target organs. Chemicals released at axon terminals allow signals to be communicated to these other cells. Neurons usually have one or two axons, but some neurons, like amacrine cells in the retina, do not contain any axons. Some axons are covered with **myelin**, which acts as an insulator to minimize dissipation of the electrical signal as it travels down the axon, greatly increasing the speed of conduction. This insulation is important as the axon from a human motor neuron can be as long as a meter—from the base of the spine to the toes. The myelin sheath is not actually part of the neuron. Myelin is produced by glial cells. Along the axon there are periodic gaps in the myelin sheath. These gaps are called **nodes of Ranvier** and are sites where the signal is "recharged" as it travels along the axon.

It is important to note that a single neuron does not act alone—neuronal communication depends on the connections that neurons make with one another (as well as with other cells, like muscle cells). Dendrites from a single neuron may receive synaptic contact from many other neurons. For example, dendrites from a Purkinje cell in the cerebellum are thought to receive contact from as many as 200,000 other neurons.

VISUAL CONNECTION

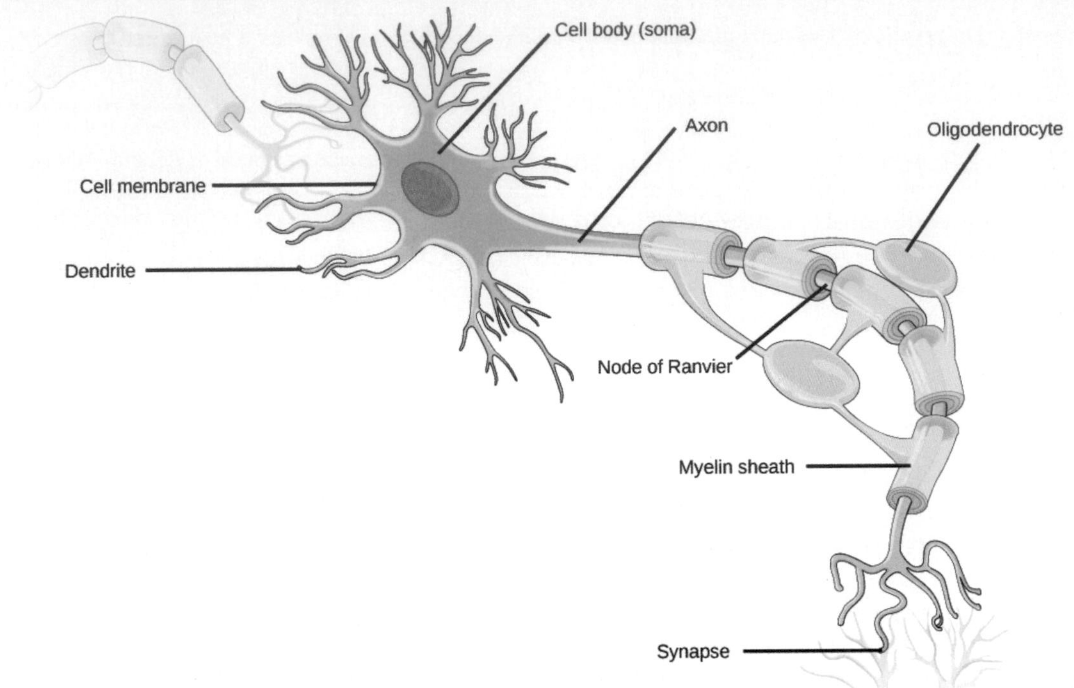

Figure 35.3 Neurons contain organelles common to many other cells, such as a nucleus and mitochondria. They also have more specialized structures, including dendrites and axons.

Which of the following statements is false?

a. The soma is the cell body of a nerve cell.
b. Myelin sheath provides an insulating layer to the dendrites.
c. Axons carry the signal from the soma to the target.
d. Dendrites carry the signal to the soma.

Types of Neurons

There are different types of neurons, and the functional role of a given neuron is intimately dependent on its structure. There is an amazing diversity of neuron shapes and sizes found in different parts of the nervous system (and across species), as illustrated by the neurons shown in Figure 35.4.

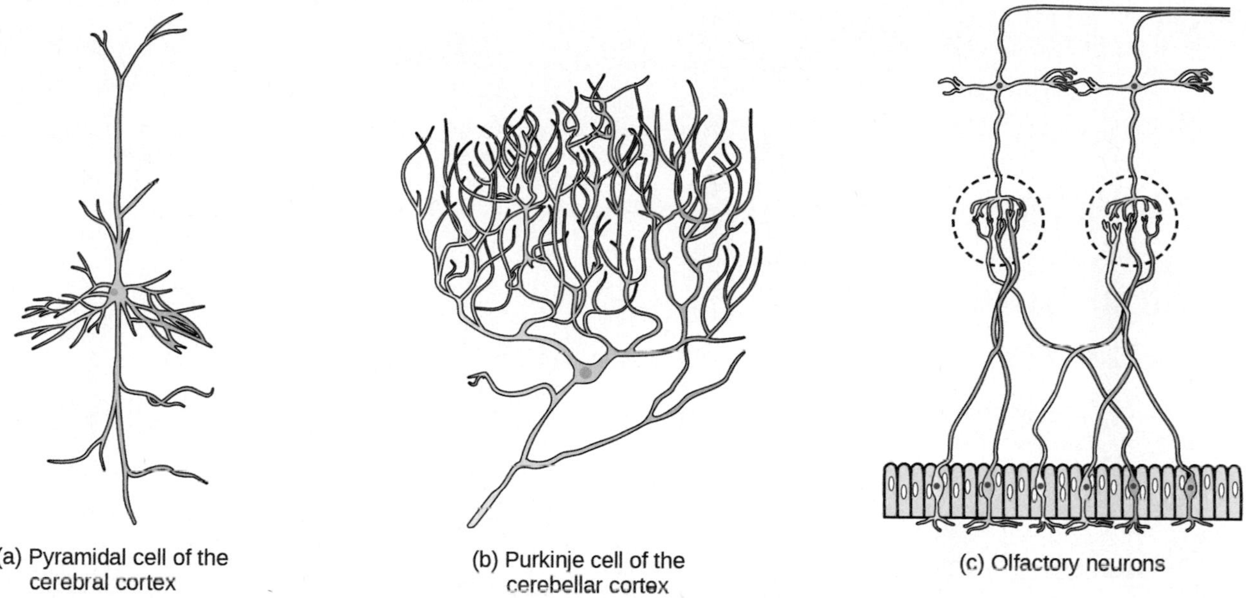

Figure 35.4 There is great diversity in the size and shape of neurons throughout the nervous system. Examples include (a) a pyramidal cell from the cerebral cortex, (b) a Purkinje cell from the cerebellar cortex, and (c) olfactory cells from the olfactory epithelium and olfactory bulb.

While there are many defined neuron cell subtypes, neurons are broadly divided into four basic types: unipolar, bipolar, multipolar, and pseudounipolar. Figure 35.5 illustrates these four basic neuron types. Unipolar neurons have only one structure that extends away from the soma. These neurons are not found in vertebrates but are found in insects where they stimulate muscles or glands. A bipolar neuron has one axon and one dendrite extending from the soma. An example of a bipolar neuron is a retinal bipolar cell, which receives signals from photoreceptor cells that are sensitive to light and transmits these signals to ganglion cells that carry the signal to the brain. Multipolar neurons are the most common type of neuron. Each multipolar neuron contains one axon and multiple dendrites. Multipolar neurons can be found in the central nervous system (brain and spinal cord). An example of a multipolar neuron is a Purkinje cell in the cerebellum, which has many branching dendrites but only one axon. Pseudounipolar cells share characteristics with both unipolar and bipolar cells. A pseudounipolar cell has a single process that extends from the soma, like a unipolar cell, but this process later branches into two distinct structures, like a bipolar cell. Most sensory neurons are pseudounipolar and have an axon that branches into two extensions: one connected to dendrites that receive sensory information and another that transmits this information to the spinal cord.

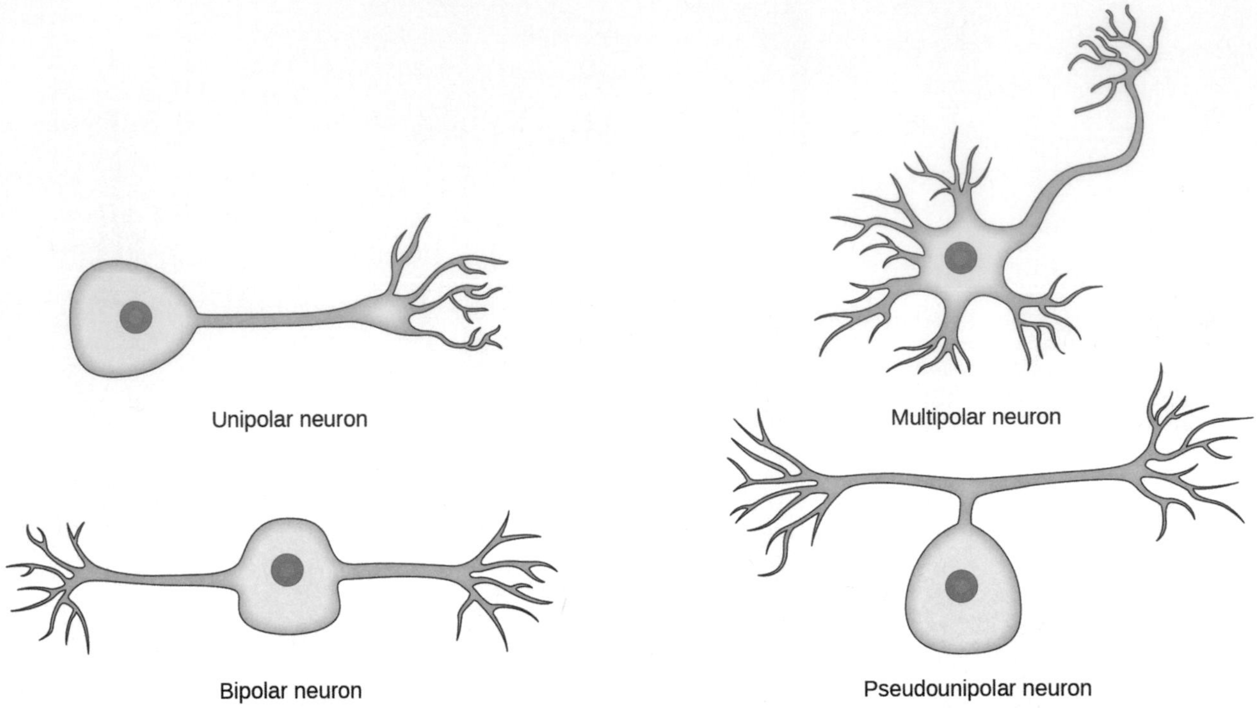

Figure 35.5 Neurons are broadly divided into four main types based on the number and placement of axons: (1) unipolar, (2) bipolar, (3) multipolar, and (4) pseudounipolar.

Everyday Connection

Neurogenesis

At one time, scientists believed that people were born with all the neurons they would ever have. Research performed during the last few decades indicates that neurogenesis, the birth of new neurons, continues into adulthood. Neurogenesis was first discovered in songbirds that produce new neurons while learning songs. For mammals, new neurons also play an important role in learning: about 1000 new neurons develop in the hippocampus (a brain structure involved in learning and memory) each day. While most of the new neurons will die, researchers found that an increase in the number of surviving new neurons in the hippocampus correlated with how well rats learned a new task. Interestingly, both exercise and some antidepressant medications also promote neurogenesis in the hippocampus. Stress has the opposite effect. While neurogenesis is quite limited compared to regeneration in other tissues, research in this area may lead to new treatments for disorders such as Alzheimer's, stroke, and epilepsy.

How do scientists identify new neurons? A researcher can inject a compound called bromodeoxyuridine (BrdU) into the brain of an animal. While all cells will be exposed to BrdU, BrdU will only be incorporated into the DNA of newly generated cells that are in S phase. A technique called immunohistochemistry can be used to attach a fluorescent label to the incorporated BrdU, and a researcher can use fluorescent microscopy to visualize the presence of BrdU, and thus new neurons, in brain tissue. Figure 35.6 is a micrograph which shows fluorescently labeled neurons in the hippocampus of a rat.

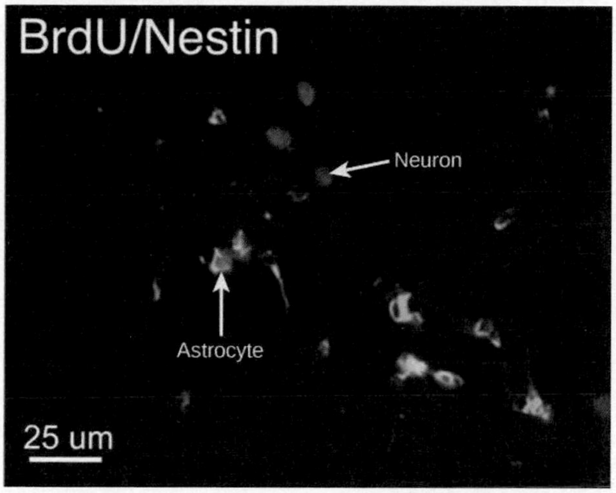

Figure 35.6 This micrograph shows fluorescently labeled new neurons in a rat hippocampus. Cells that are actively dividing have bromodoxyuridine (BrdU) incorporated into their DNA and are labeled in red. Cells that express glial fibrillary acidic protein (GFAP) are labeled in green. Astrocytes, but not neurons, express GFAP. Thus, cells that are labeled both red and green are actively dividing astrocytes, whereas cells labeled red only are actively dividing neurons. (credit: modification of work by Dr. Maryam Faiz, et. al., University of Barcelona; scale bar data from Matt Russell)

LINK TO LEARNING

This site (http://openstax.org/l/neurogenesis) contains more information about neurogenesis, including an interactive laboratory simulation and a video that explains how BrdU labels new cells.

Glia

While glia are often thought of as the supporting cast of the nervous system, the number of glial cells in the brain actually outnumbers the number of neurons by a factor of ten. Neurons would be unable to function without the vital roles that are fulfilled by these glial cells. Glia guide developing neurons to their destinations, buffer ions and chemicals that would otherwise harm neurons, and provide myelin sheaths around axons. Scientists have recently discovered that they also play a role in responding to nerve activity and modulating communication between nerve cells. When glia do not function properly, the result can be disastrous—most brain tumors are caused by mutations in glia.

Types of Glia

There are several different types of glia with different functions, two of which are shown in Figure 35.7. **Astrocytes**, shown in Figure 35.8a make contact with both capillaries and neurons in the CNS. They provide nutrients and other substances to neurons, regulate the concentrations of ions and chemicals in the extracellular fluid, and provide structural support for synapses. Astrocytes also form the blood-brain barrier—a structure that blocks entrance of toxic substances into the brain. Astrocytes, in particular, have been shown through calcium imaging experiments to become active in response to nerve activity, transmit calcium waves between astrocytes, and modulate the activity of surrounding synapses. **Satellite glia** provide nutrients and structural support for neurons in the PNS. **Microglia** scavenge and degrade dead cells and protect the brain from invading microorganisms. **Oligodendrocytes**, shown in Figure 35.8b form myelin sheaths around axons in the CNS. One axon can be myelinated by several oligodendrocytes, and one oligodendrocyte can provide myelin for multiple neurons. This is distinctive from the PNS where a single **Schwann cell** provides myelin for only one axon as the entire Schwann cell surrounds the axon. **Radial glia** serve as scaffolds for developing neurons as they migrate to their end destinations. **Ependymal** cells line fluid-filled ventricles of the brain and the central canal of the spinal cord. They are involved in the production of cerebrospinal fluid, which serves as a cushion for the brain, moves the fluid between the spinal cord and the brain, and is a component for the choroid plexus.

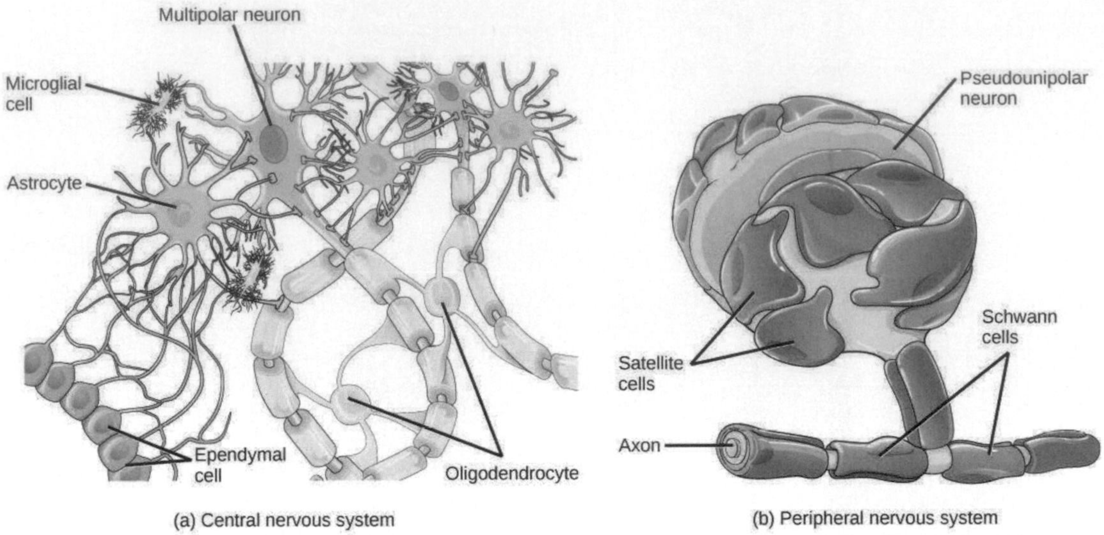

Figure 35.7 Glial cells support neurons and maintain their environment. Glial cells of the (a) central nervous system include oligodendrocytes, astrocytes, ependymal cells, and microglial cells. Oligodendrocytes form the myelin sheath around axons. Astrocytes provide nutrients to neurons, maintain their extracellular environment, and provide structural support. Microglia scavenge pathogens and dead cells. Ependymal cells produce cerebrospinal fluid that cushions the neurons. Glial cells of the (b) peripheral nervous system include Schwann cells, which form the myelin sheath, and satellite cells, which provide nutrients and structural support to neurons.

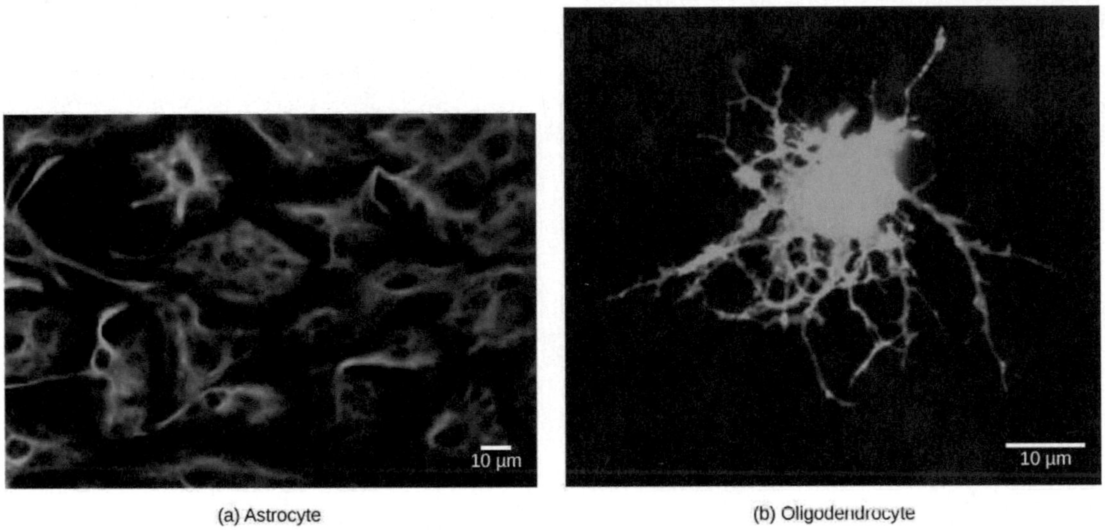

Figure 35.8 (a) Astrocytes and (b) oligodendrocytes are glial cells of the central nervous system. (credit a: modification of work by Uniformed Services University; credit b: modification of work by Jurjen Broeke; scale-bar data from Matt Russell)

35.2 How Neurons Communicate

By the end of this section, you will be able to do the following:

- Describe the basis of the resting membrane potential
- Explain the stages of an action potential and how action potentials are propagated
- Explain the similarities and differences between chemical and electrical synapses
- Describe long-term potentiation and long-term depression

All functions performed by the nervous system—from a simple motor reflex to more advanced functions like making a memory or a decision—require neurons to communicate with one another. While humans use words and body language to communicate, neurons use electrical and chemical signals. Just like a person in a committee, one neuron usually receives and synthesizes messages from multiple other neurons before "making the decision" to send the message on to other neurons.

Nerve Impulse Transmission within a Neuron

For the nervous system to function, neurons must be able to send and receive signals. These signals are possible because each neuron has a charged cellular membrane (a voltage difference between the inside and the outside), and the charge of this membrane can change in response to neurotransmitter molecules released from other neurons and environmental stimuli. To understand how neurons communicate, one must first understand the basis of the baseline or 'resting' membrane charge.

Neuronal Charged Membranes

The lipid bilayer membrane that surrounds a neuron is impermeable to charged molecules or ions. To enter or exit the neuron, ions must pass through special proteins called ion channels that span the membrane. Ion channels have different configurations: open, closed, and inactive, as illustrated in Figure 35.9. Some ion channels need to be activated in order to open and allow ions to pass into or out of the cell. These ion channels are sensitive to the environment and can change their shape accordingly. Ion channels that change their structure in response to voltage changes are called voltage-gated ion channels. Voltage-gated ion channels regulate the relative concentrations of different ions inside and outside the cell. The difference in total charge between the inside and outside of the cell is called the **membrane potential**.

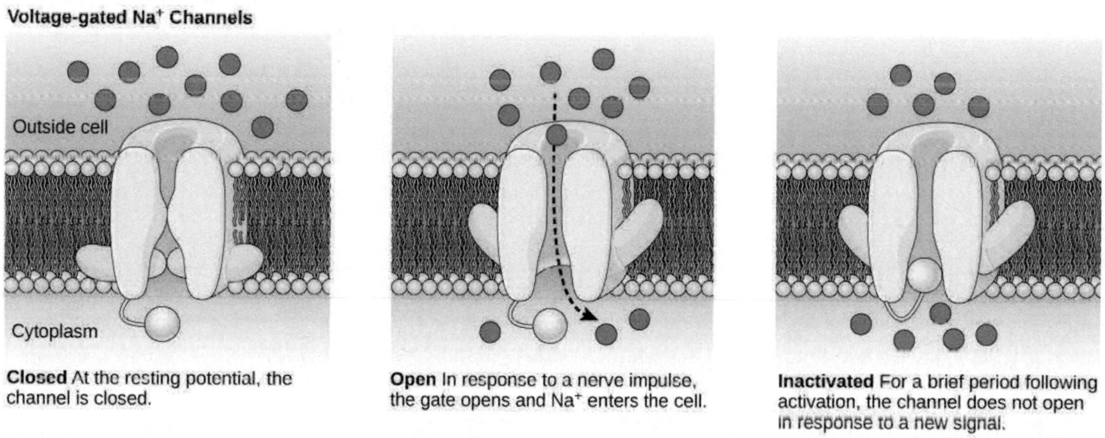

Figure 35.9 Voltage-gated ion channels open in response to changes in membrane voltage. After activation, they become inactivated for a brief period and will no longer open in response to a signal.

LINK TO LEARNING

This video discusses the basis of the resting membrane potential.

Click to view content (https://www.openstax.org/l/resting_neuron)

Resting Membrane Potential

A neuron at rest is negatively charged: the inside of a cell is approximately 70 millivolts more negative than the outside (–70 mV, note that this number varies by neuron type and by species). This voltage is called the resting membrane potential; it is caused by differences in the concentrations of ions inside and outside the cell. If the membrane were equally permeable to all ions, each type of ion would flow across the membrane and the system would reach equilibrium. Because ions cannot simply cross the membrane at will, there are different concentrations of several ions inside and outside the cell, as shown in Table 35.1. The difference in the number of positively charged potassium ions (K^+) inside and outside the cell dominates the resting membrane potential (Figure 35.10). When the membrane is at rest, K^+ ions accumulate inside the cell due to a net movement with the concentration gradient. The negative resting membrane potential is created and maintained by increasing the concentration of cations outside the cell (in the extracellular fluid) relative to inside the cell (in the cytoplasm). The negative charge within the cell is created by the cell membrane being more permeable to potassium ion movement than sodium ion movement. In neurons, potassium ions are maintained at high concentrations within the cell while sodium ions are maintained at high concentrations outside of the cell. The cell possesses potassium and sodium leakage channels that allow the two cations to diffuse down their concentration gradient. However, the neurons have far more potassium leakage channels than sodium leakage channels. Therefore, potassium diffuses out of the cell at a much faster rate than sodium leaks in. Because more cations are leaving the cell than are entering, this causes the interior of the cell to be negatively charged relative to the outside of the cell. The actions of the sodium potassium pump help to maintain the resting potential, once established. Recall that sodium potassium pumps brings two K^+ ions into the cell while removing three Na^+ ions per ATP consumed. As more cations are expelled from the cell than taken in, the inside of the cell remains negatively charged relative to the extracellular fluid. It should be noted that chloride ions (Cl^-) tend to accumulate outside of the cell because they are repelled by negatively-charged proteins within the cytoplasm.

Ion Concentration Inside and Outside Neurons

Ion	Extracellular concentration (mM)	Intracellular concentration (mM)	Ratio outside/ inside
Na^+	145	12	12
K+	4	155	0.026
Cl^-	120	4	30
Organic anions (A–)	—	100	

Table 35.1 The resting membrane potential is a result of different concentrations inside and outside the cell.

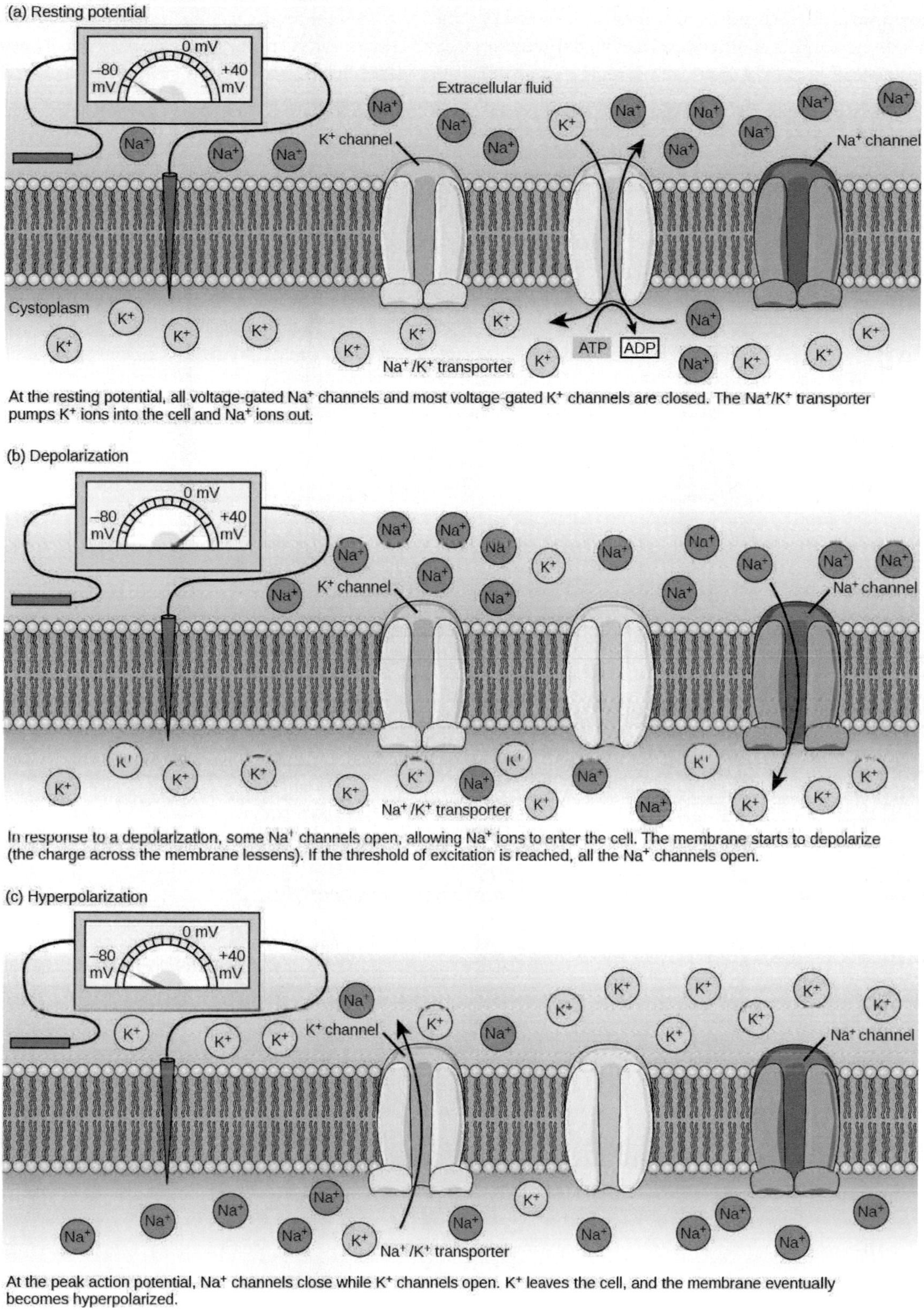

Figure 35.10 The (a) resting membrane potential is a result of different concentrations of Na^+ and K^+ ions inside and outside the cell. A nerve impulse causes Na^+ to enter the cell, resulting in (b) depolarization. At the peak action potential, K^+ channels open and the cell becomes (c) hyperpolarized.

Action Potential

A neuron can receive input from other neurons and, if this input is strong enough, send the signal to downstream neurons. Transmission of a signal between neurons is generally carried by a chemical called a neurotransmitter. Transmission of a signal within a neuron (from dendrite to axon terminal) is carried by a brief reversal of the resting membrane potential called an **action potential**. When neurotransmitter molecules bind to receptors located on a neuron's dendrites, ion channels open. At excitatory synapses, this opening allows positive ions to enter the neuron and results in **depolarization** of the membrane—a decrease in the difference in voltage between the inside and outside of the neuron. A stimulus from a sensory cell or another neuron depolarizes the target neuron to its threshold potential (-55 mV). Na^+ channels in the axon hillock open, allowing positive ions to enter the cell (Figure 35.10 and Figure 35.11). Once the sodium channels open, the neuron completely depolarizes to a membrane potential of about +40 mV. Action potentials are considered an "all-or nothing" event, in that, once the threshold potential is reached, the neuron always completely depolarizes. Once depolarization is complete, the cell must now "reset" its membrane voltage back to the resting potential. To accomplish this, the Na^+ channels close and cannot be opened. This begins the neuron's **refractory period**, in which it cannot produce another action potential because its sodium channels will not open. At the same time, voltage-gated K^+ channels open, allowing K^+ to leave the cell. As K^+ ions leave the cell, the membrane potential once again becomes negative. The diffusion of K^+ out of the cell actually **hyperpolarizes** the cell, in that the membrane potential becomes more negative than the cell's normal resting potential. At this point, the sodium channels will return to their resting state, meaning they are ready to open again if the membrane potential again exceeds the threshold potential. Eventually the extra K^+ ions diffuse out of the cell through the potassium leakage channels, bringing the cell from its hyperpolarized state, back to its resting membrane potential.

VISUAL CONNECTION

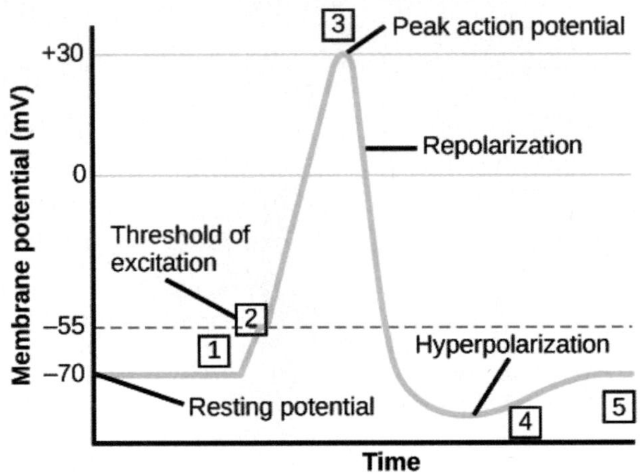

Figure 35.11 The formation of an action potential can be divided into five steps: (1) A stimulus from a sensory cell or another neuron causes the target cell to depolarize toward the threshold potential. (2) If the threshold of excitation is reached, all Na^+ channels open and the membrane depolarizes. (3) At the peak action potential, K^+ channels open and K^+ begins to leave the cell. At the same time, Na^+ channels close. (4) The membrane becomes hyperpolarized as K^+ ions continue to leave the cell. The hyperpolarized membrane is in a refractory period and cannot fire. (5) The K^+ channels close and the Na^+/K^+ transporter restores the resting potential.

Potassium channel blockers, such as amiodarone and procainamide, which are used to treat abnormal electrical activity in the heart, called cardiac dysrhythmia, impede the movement of K^+ through voltage-gated K^+ channels. Which part of the action potential would you expect potassium channels to affect?

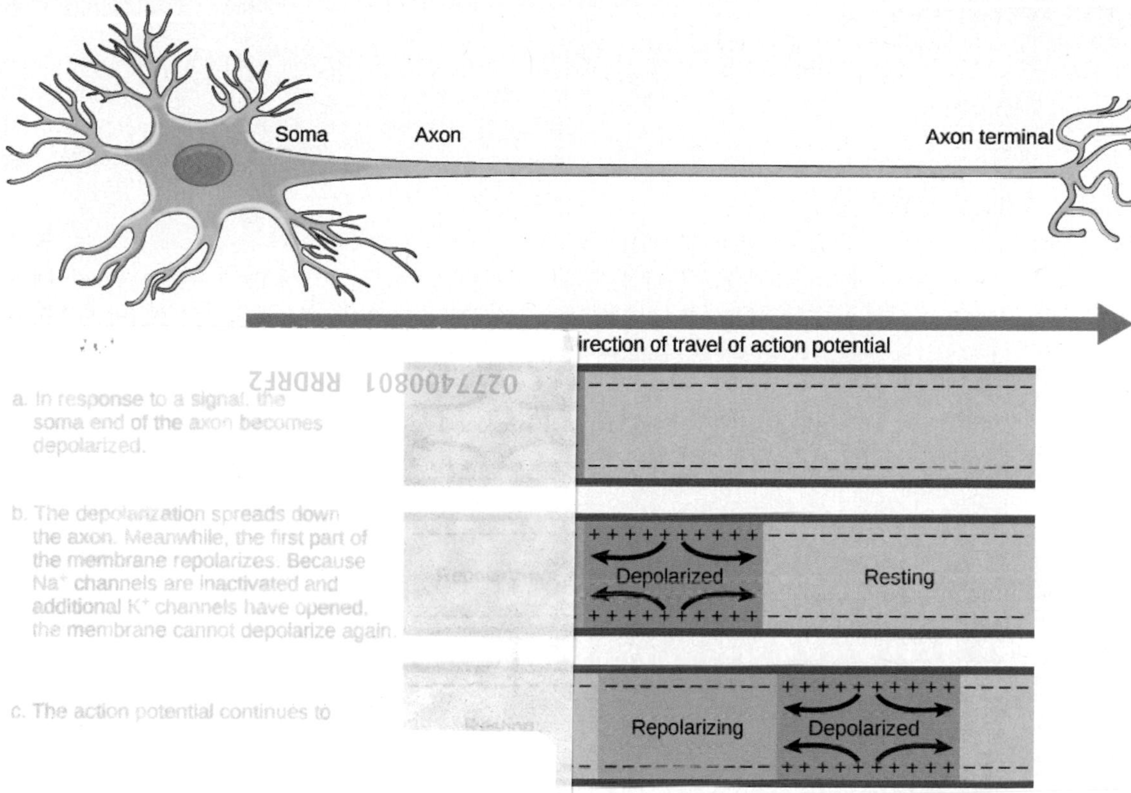

axon as the axon membrane depolarizes, then repolarizes.

rerview of action potential.

ntial

neuron, it must travel along the axon and reach the axon
eed of conduction of an action potential along an axon is
tance to current leak. Myelin acts as an insulator that prevents
potential conduction. In demyelinating diseases like multiple
ks from previously insulated axon areas. The nodes of Ranvier,
e axon. These unmyelinated spaces are about one micrometer
ns through these channels, particularly the Na^+ channels,
xon. This 'jumping' of the action potential from one node to the
t present along an axon, the action potential would propagate
ly regenerate action potentials at every point along the axon
or the neuron since the channels only need to be present at the

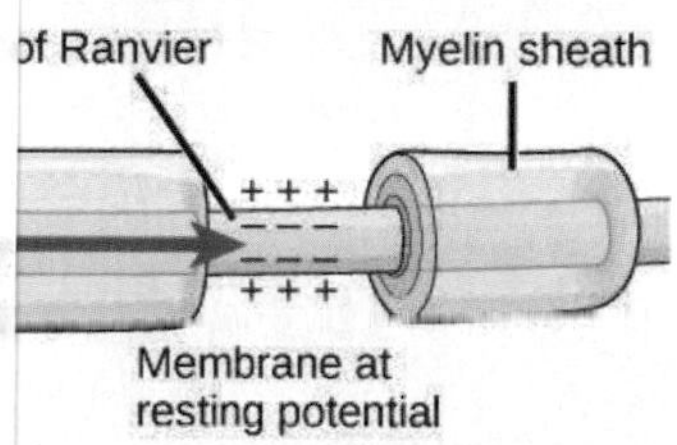

ns. Nodes contain voltage-gated K^+ and Na^+ channels. Action
t.

Synaptic Transmission

The synapse or "gap" is the place where information is transmitted from one neuron to another. Synapses usually form between axon terminals and dendritic spines, but this is not universally true. There are also axon-to-axon, dendrite-to-dendrite, and axon-to-cell body synapses. The neuron transmitting the signal is called the presynaptic neuron, and the neuron receiving the signal is called the postsynaptic neuron. Note that these designations are relative to a particular synapse—most neurons are both presynaptic and postsynaptic. There are two types of synapses: chemical and electrical.

Chemical Synapse

When an action potential reaches the axon terminal it depolarizes the membrane and opens voltage-gated Na^+ channels. Na^+ ions enter the cell, further depolarizing the presynaptic membrane. This depolarization causes voltage-gated Ca^{2+} channels to open. Calcium ions entering the cell initiate a signaling cascade that causes small membrane-bound vesicles, called **synaptic vesicles**, containing neurotransmitter molecules to fuse with the presynaptic membrane. Synaptic vesicles are shown in Figure 35.14, which is an image from a scanning electron microscope.

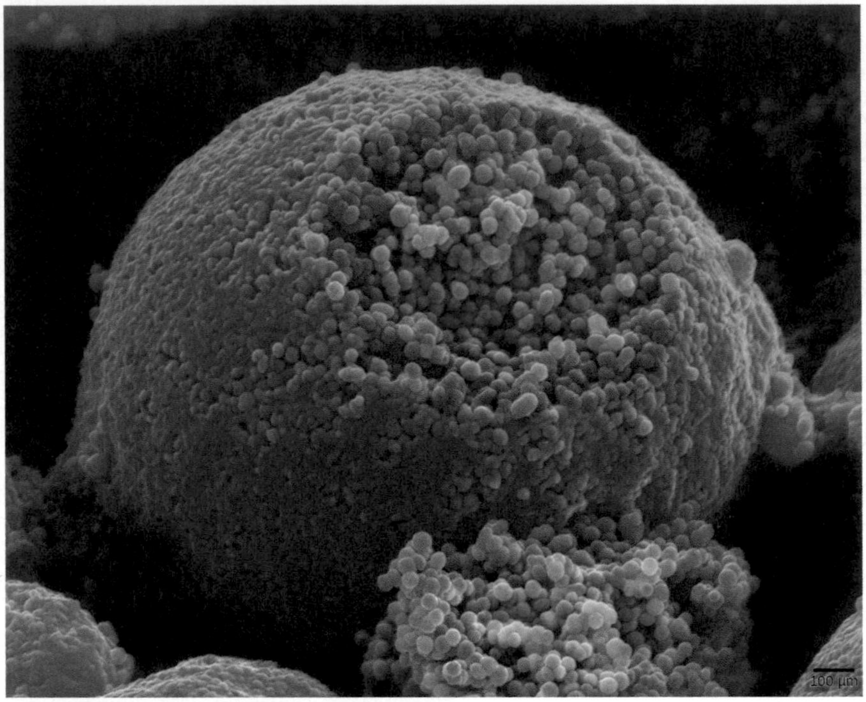

Figure 35.14 This pseudocolored image taken with a scanning electron microscope shows an axon terminal that was broken open to reveal synaptic vesicles (blue and orange) inside the neuron. (credit: modification of work by Tina Carvalho, NIH-NIGMS; scale-bar data from Matt Russell)

Fusion of a vesicle with the presynaptic membrane causes neurotransmitter to be released into the **synaptic cleft**, the extracellular space between the presynaptic and postsynaptic membranes, as illustrated in Figure 35.15. The neurotransmitter diffuses across the synaptic cleft and binds to receptor proteins on the postsynaptic membrane.

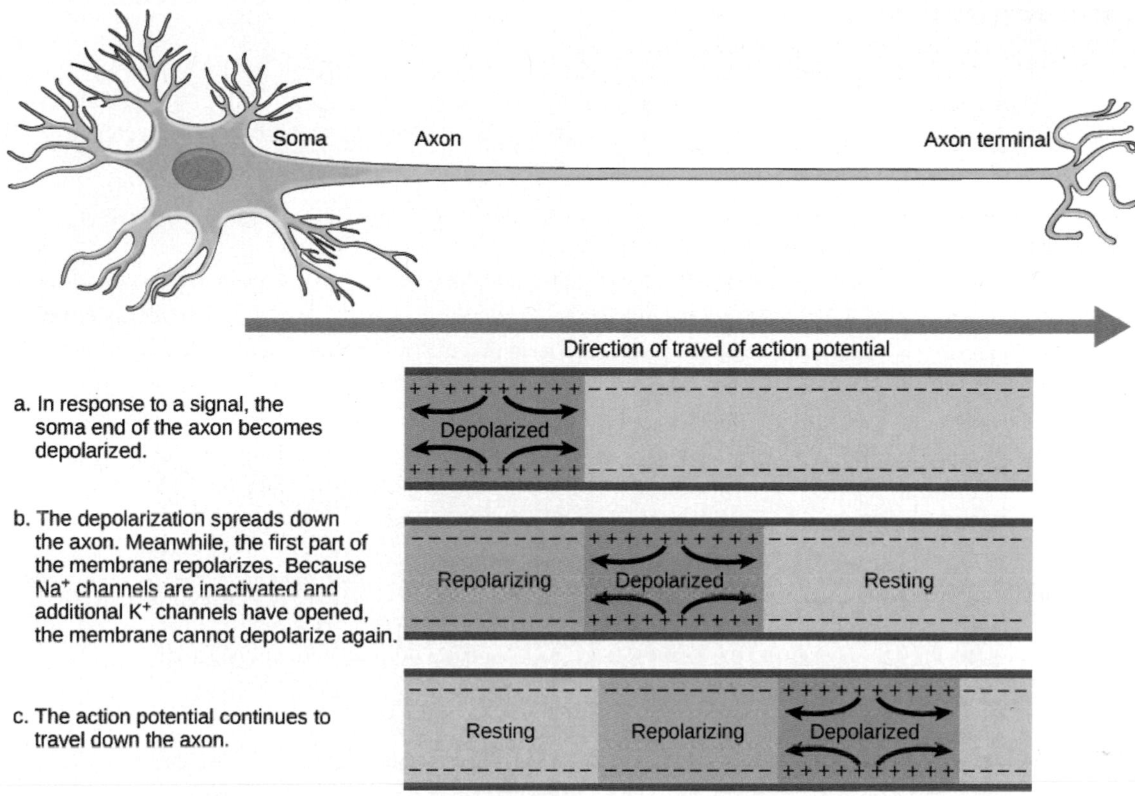

Figure 35.12 The action potential is conducted down the axon as the axon membrane depolarizes, then repolarizes.

LINK TO LEARNING

This video (http://openstax.org/l/actionpotential) presents an overview of action potential.

Myelin and the Propagation of the Action Potential

For an action potential to communicate information to another neuron, it must travel along the axon and reach the axon terminals where it can initiate neurotransmitter release. The speed of conduction of an action potential along an axon is influenced by both the diameter of the axon and the axon's resistance to current leak. Myelin acts as an insulator that prevents current from leaving the axon; this increases the speed of action potential conduction. In demyelinating diseases like multiple sclerosis, action potential conduction slows because current leaks from previously insulated axon areas. The nodes of Ranvier, illustrated in Figure 35.13 are gaps in the myelin sheath along the axon. These unmyelinated spaces are about one micrometer long and contain voltage-gated Na^+ and K^+ channels. Flow of ions through these channels, particularly the Na^+ channels, regenerates the action potential over and over again along the axon. This 'jumping' of the action potential from one node to the next is called **saltatory conduction**. If nodes of Ranvier were not present along an axon, the action potential would propagate very slowly since Na^+ and K^+ channels would have to continuously regenerate action potentials at every point along the axon instead of at specific points. Nodes of Ranvier also save energy for the neuron since the channels only need to be present at the nodes and not along the entire axon.

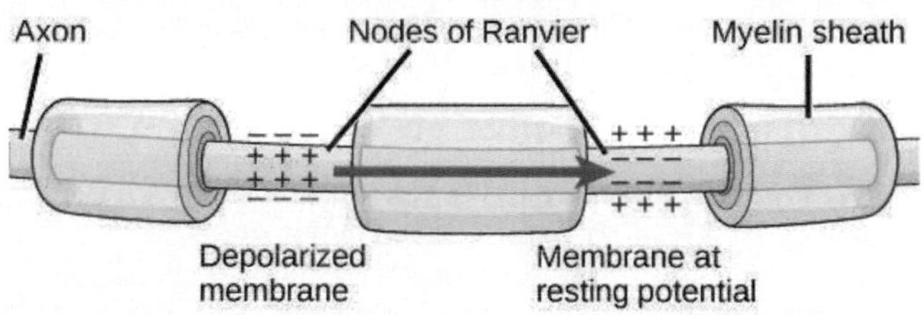

Figure 35.13 Nodes of Ranvier are gaps in myelin coverage along axons. Nodes contain voltage-gated K^+ and Na^+ channels. Action potentials travel down the axon by jumping from one node to the next.

Synaptic Transmission

The synapse or "gap" is the place where information is transmitted from one neuron to another. Synapses usually form between axon terminals and dendritic spines, but this is not universally true. There are also axon-to-axon, dendrite-to-dendrite, and axon-to-cell body synapses. The neuron transmitting the signal is called the presynaptic neuron, and the neuron receiving the signal is called the postsynaptic neuron. Note that these designations are relative to a particular synapse—most neurons are both presynaptic and postsynaptic. There are two types of synapses: chemical and electrical.

Chemical Synapse

When an action potential reaches the axon terminal it depolarizes the membrane and opens voltage-gated Na^+ channels. Na^+ ions enter the cell, further depolarizing the presynaptic membrane. This depolarization causes voltage-gated Ca^{2+} channels to open. Calcium ions entering the cell initiate a signaling cascade that causes small membrane-bound vesicles, called **synaptic vesicles**, containing neurotransmitter molecules to fuse with the presynaptic membrane. Synaptic vesicles are shown in Figure 35.14, which is an image from a scanning electron microscope.

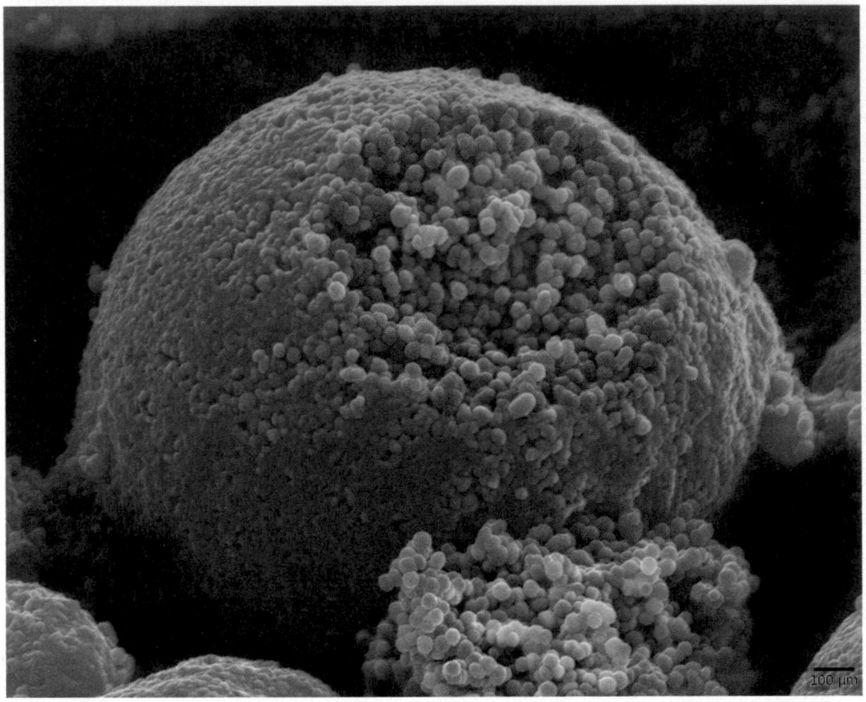

Figure 35.14 This pseudocolored image taken with a scanning electron microscope shows an axon terminal that was broken open to reveal synaptic vesicles (blue and orange) inside the neuron. (credit: modification of work by Tina Carvalho, NIH-NIGMS; scale-bar data from Matt Russell)

Fusion of a vesicle with the presynaptic membrane causes neurotransmitter to be released into the **synaptic cleft**, the extracellular space between the presynaptic and postsynaptic membranes, as illustrated in Figure 35.15. The neurotransmitter diffuses across the synaptic cleft and binds to receptor proteins on the postsynaptic membrane.

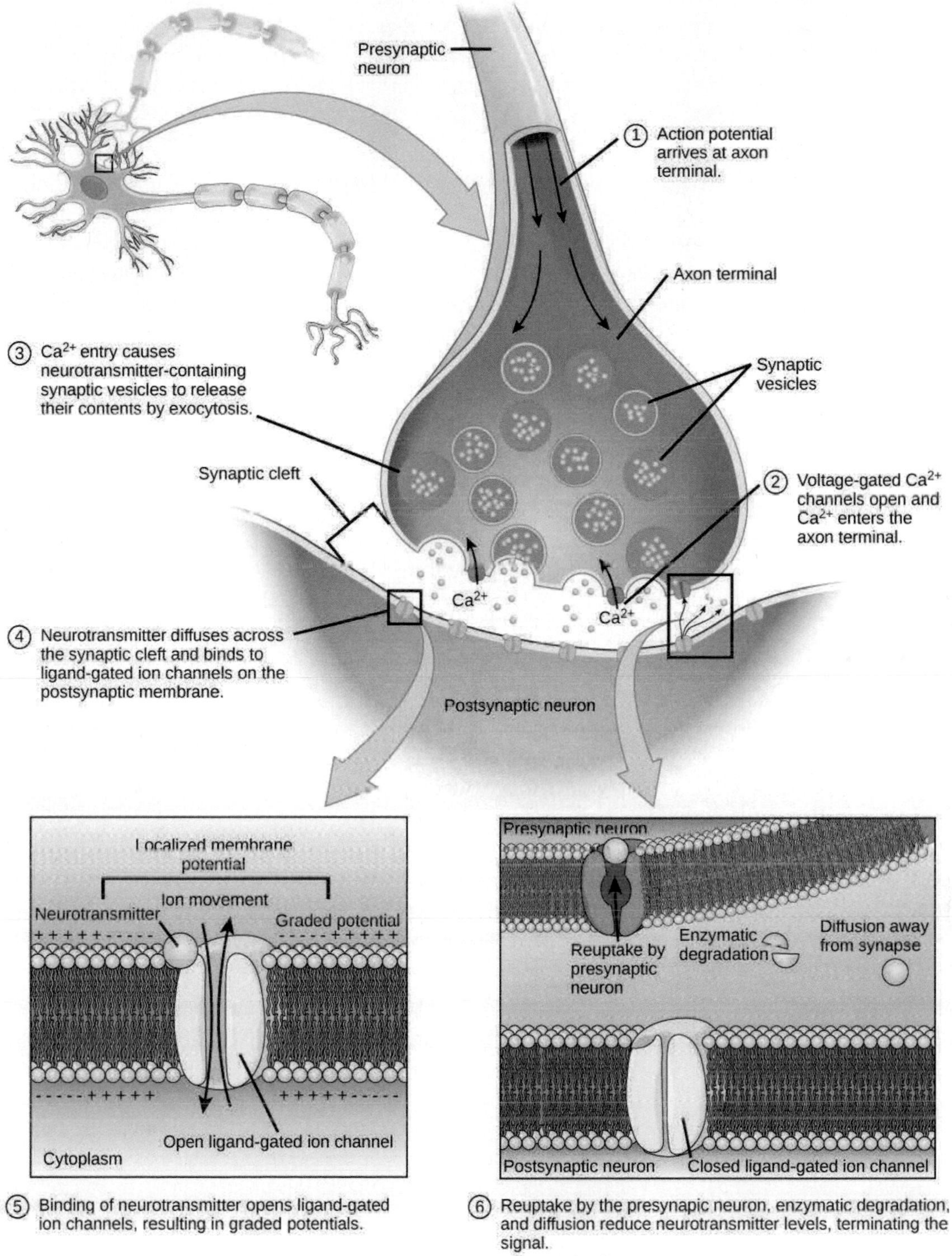

Figure 35.15 Communication at chemical synapses requires release of neurotransmitters. When the presynaptic membrane is depolarized, voltage-gated Ca^{2+} channels open and allow Ca^{2+} to enter the cell. The calcium entry causes synaptic vesicles to fuse with the membrane and release neurotransmitter molecules into the synaptic cleft. The neurotransmitter diffuses across the synaptic cleft and binds to ligand-gated ion channels in the postsynaptic membrane, resulting in a localized depolarization or hyperpolarization of the postsynaptic neuron.

The binding of a specific neurotransmitter causes particular ion channels, in this case ligand-gated channels, on the postsynaptic membrane to open. Neurotransmitters can either have excitatory or inhibitory effects on the postsynaptic membrane, as detailed in Table 35.1. For example, when acetylcholine is released at the synapse between a nerve and muscle (called the neuromuscular junction) by a presynaptic neuron, it causes postsynaptic Na^+ channels to open. Na^+ enters the postsynaptic cell and causes the postsynaptic membrane to depolarize. This depolarization is called an **excitatory postsynaptic**

potential (EPSP) and makes the postsynaptic neuron more likely to fire an action potential. Release of neurotransmitter at inhibitory synapses causes **inhibitory postsynaptic potentials (IPSPs)**, a hyperpolarization of the presynaptic membrane. For example, when the neurotransmitter GABA (gamma-aminobutyric acid) is released from a presynaptic neuron, it binds to and opens Cl^- channels. Cl^- ions enter the cell and hyperpolarizes the membrane, making the neuron less likely to fire an action potential.

Once neurotransmission has occurred, the neurotransmitter must be removed from the synaptic cleft so the postsynaptic membrane can "reset" and be ready to receive another signal. This can be accomplished in three ways: the neurotransmitter can diffuse away from the synaptic cleft, it can be degraded by enzymes in the synaptic cleft, or it can be recycled (sometimes called reuptake) by the presynaptic neuron. Several drugs act at this step of neurotransmission. For example, some drugs that are given to Alzheimer's patients work by inhibiting acetylcholinesterase, the enzyme that degrades acetylcholine. This inhibition of the enzyme essentially increases neurotransmission at synapses that release acetylcholine. Once released, the acetylcholine stays in the cleft and can continually bind and unbind to postsynaptic receptors.

Neurotransmitter Function and Location

Neurotransmitter	Example	Location
Acetylcholine	—	CNS and/or PNS
Biogenic amine	Dopamine, serotonin, norepinephrine	CNS and/or PNS
Amino acid	Glycine, glutamate, aspartate, gamma aminobutyric acid	CNS
Neuropeptide	Substance P, endorphins	CNS and/or PNS

Table 35.2

Electrical Synapse

While electrical synapses are fewer in number than chemical synapses, they are found in all nervous systems and play important and unique roles. The mode of neurotransmission in electrical synapses is quite different from that in chemical synapses. In an electrical synapse, the presynaptic and postsynaptic membranes are very close together and are actually physically connected by channel proteins forming gap junctions. Gap junctions allow current to pass directly from one cell to the next. In addition to the ions that carry this current, other molecules, such as ATP, can diffuse through the large gap junction pores.

There are key differences between chemical and electrical synapses. Because chemical synapses depend on the release of neurotransmitter molecules from synaptic vesicles to pass on their signal, there is an approximately one millisecond delay between when the axon potential reaches the presynaptic terminal and when the neurotransmitter leads to opening of postsynaptic ion channels. Additionally, this signaling is unidirectional. Signaling in electrical synapses, in contrast, is virtually instantaneous (which is important for synapses involved in key reflexes), and some electrical synapses are bidirectional. Electrical synapses are also more reliable as they are less likely to be blocked, and they are important for synchronizing the electrical activity of a group of neurons. For example, electrical synapses in the thalamus are thought to regulate slow-wave sleep, and disruption of these synapses can cause seizures.

Signal Summation

Sometimes a single EPSP is strong enough to induce an action potential in the postsynaptic neuron, but often multiple presynaptic inputs must create EPSPs around the same time for the postsynaptic neuron to be sufficiently depolarized to fire an action potential. This process is called **summation** and occurs at the axon hillock, as illustrated in Figure 35.16. Additionally, one neuron often has inputs from many presynaptic neurons—some excitatory and some inhibitory—so IPSPs can cancel out EPSPs and vice versa. It is the net change in postsynaptic membrane voltage that determines whether the postsynaptic cell has reached its threshold of excitation needed to fire an action potential. Together, synaptic summation and the threshold for excitation act as a filter so that random "noise" in the system is not transmitted as important information.

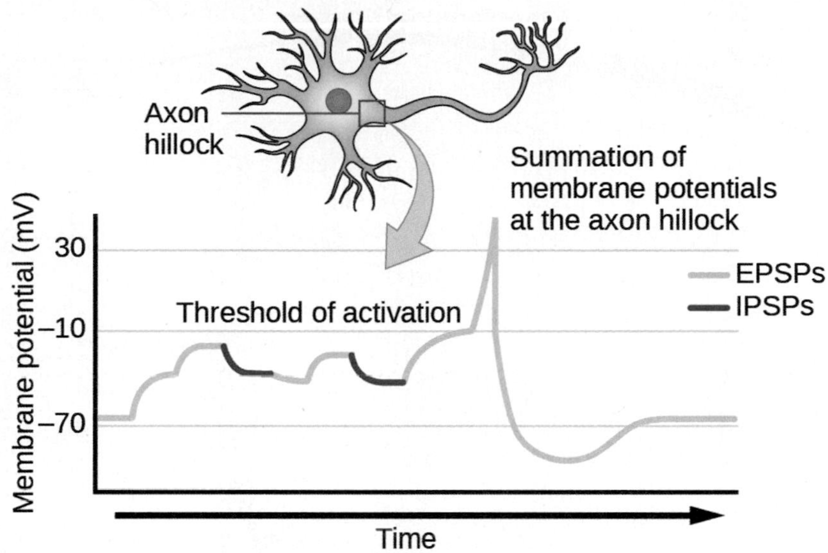

Figure 35.16 A single neuron can receive both excitatory and inhibitory inputs from multiple neurons, resulting in local membrane depolarization (EPSP input) and hyperpolarization (IPSP input). All these inputs are added together at the axon hillock. If the EPSPs are strong enough to overcome the IPSPs and reach the threshold of excitation, the neuron will fire.

Everyday Connection

Brain-computer interface

Amyotrophic lateral sclerosis (ALS, also called Lou Gehrig's Disease) is a neurological disease characterized by the degeneration of the motor neurons that control voluntary movements. The disease begins with muscle weakening and lack of coordination and eventually destroys the neurons that control speech, breathing, and swallowing; in the end, the disease can lead to paralysis. At that point, patients require assistance from machines to be able to breathe and to communicate. Several special technologies have been developed to allow "locked-in" patients to communicate with the rest of the world. One technology, for example, allows patients to type out sentences by twitching their cheek. These sentences can then be read aloud by a computer.

A relatively new line of research for helping paralyzed patients, including those with ALS, to communicate and retain a degree of self-sufficiency is called brain-computer interface (BCI) technology and is illustrated in Figure 35.17. This technology sounds like something out of science fiction: it allows paralyzed patients to control a computer using only their thoughts. There are several forms of BCI. Some forms use EEG recordings from electrodes taped onto the skull. These recordings contain information from large populations of neurons that can be decoded by a computer. Other forms of BCI require the implantation of an array of electrodes smaller than a postage stamp in the arm and hand area of the motor cortex. This form of BCI, while more invasive, is very powerful as each electrode can record actual action potentials from one or more neurons. These signals are then sent to a computer, which has been trained to decode the signal and feed it to a tool—such as a cursor on a computer screen. This means that a patient with ALS can use e-mail, read the Internet, and communicate with others by thinking of moving his or her hand or arm (even though the paralyzed patient cannot make that bodily movement). Recent advances have allowed a paralyzed locked-in patient who suffered a stroke 15 years ago to control a robotic arm and even to feed herself coffee using BCI technology.

Despite the amazing advancements in BCI technology, it also has limitations. The technology can require many hours of training and long periods of intense concentration for the patient; it can also require brain surgery to implant the devices.

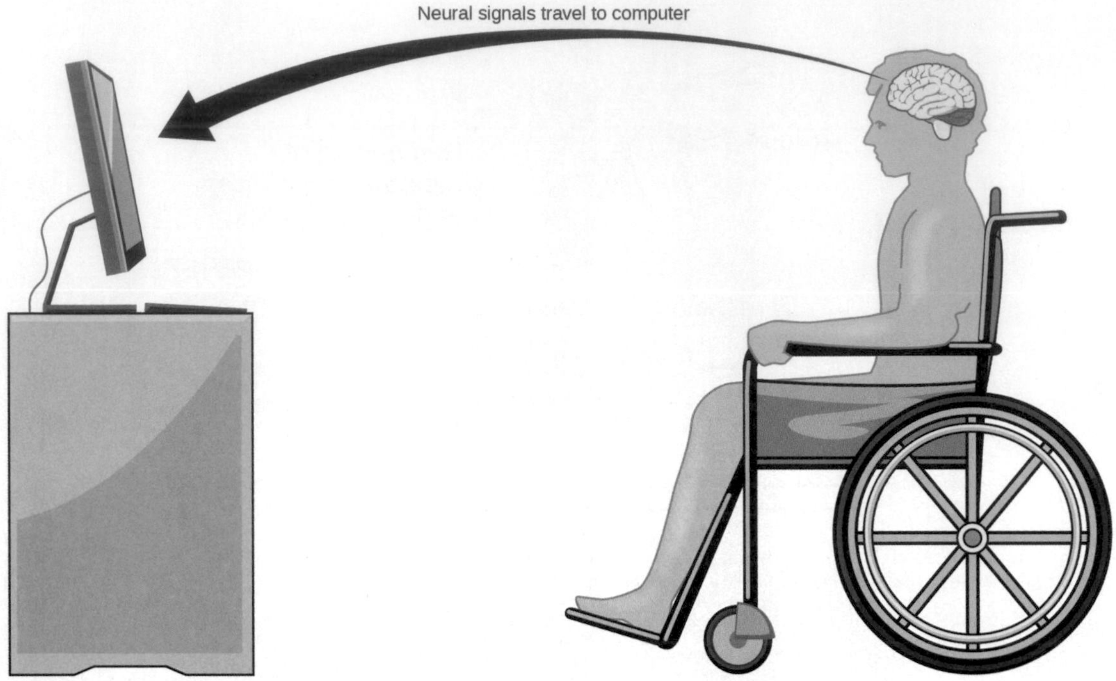

Figure 35.17 With brain-computer interface technology, neural signals from a paralyzed patient are collected, decoded, and then fed to a tool, such as a computer, a wheelchair, or a robotic arm.

LINK TO LEARNING

Watch this video (http://openstax.org/l/paralyzation) in which a paralyzed woman uses a brain-controlled robotic arm to bring a drink to her mouth, among other images of brain-computer interface technology in action.

Click to view content (https://www.openstax.org/l/paralyzation)

Synaptic Plasticity

Synapses are not static structures. They can be weakened or strengthened. They can be broken, and new synapses can be made. Synaptic plasticity allows for these changes, which are all needed for a functioning nervous system. In fact, synaptic plasticity is the basis of learning and memory. Two processes in particular, long-term potentiation (LTP) and long-term depression (LTD) are important forms of synaptic plasticity that occur in synapses in the hippocampus, a brain region that is involved in storing memories.

Long-term Potentiation (LTP)

Long-term potentiation (LTP) is a persistent strengthening of a synaptic connection. LTP is based on the Hebbian principle: cells that fire together wire together. There are various mechanisms, none fully understood, behind the synaptic strengthening seen with LTP. One known mechanism involves a type of postsynaptic glutamate receptor, called NMDA (N-Methyl-D-aspartate) receptors, shown in Figure 35.18. These receptors are normally blocked by magnesium ions; however, when the postsynaptic neuron is depolarized by multiple presynaptic inputs in quick succession (either from one neuron or multiple neurons), the magnesium ions are forced out allowing Ca ions to pass into the postsynaptic cell. Next, Ca^{2+} ions entering the cell initiate a signaling cascade that causes a different type of glutamate receptor, called AMPA (α-amino-3-hydroxy-5-methyl-4-isoxazolepropionic acid) receptors, to be inserted into the postsynaptic membrane, since activated AMPA receptors allow positive ions to enter the cell. So, the next time glutamate is released from the presynaptic membrane, it will have a larger excitatory effect (EPSP) on the postsynaptic cell because the binding of glutamate to these AMPA receptors will allow more positive ions into the cell. The insertion of additional AMPA receptors strengthens the synapse and means that the postsynaptic neuron is more likely to fire in response to presynaptic neurotransmitter release. Some drugs of abuse co-opt the LTP pathway, and this synaptic strengthening can lead to addiction.

Long-term Depression (LTD)

Long-term depression (LTD) is essentially the reverse of LTP: it is a long-term weakening of a synaptic connection. One mechanism known to cause LTD also involves AMPA receptors. In this situation, calcium that enters through NMDA receptors initiates a different signaling cascade, which results in the removal of AMPA receptors from the postsynaptic membrane, as illustrated in Figure 35.18. The decrease in AMPA receptors in the membrane makes the postsynaptic neuron less responsive to glutamate released from the presynaptic neuron. While it may seem counterintuitive, LTD may be just as important for learning and memory as LTP. The weakening and pruning of unused synapses allows for unimportant connections to be lost and makes the synapses that have undergone LTP that much stronger by comparison.

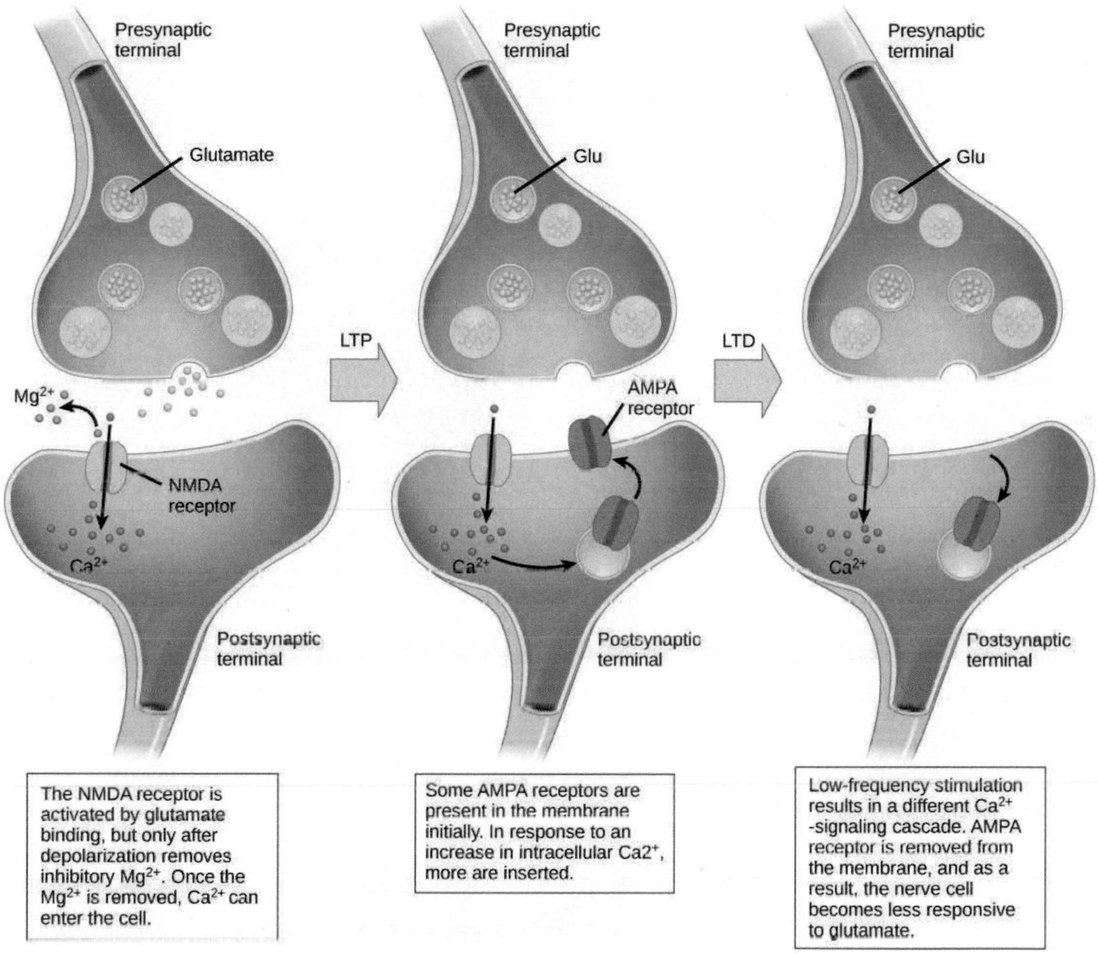

Figure 35.18 Calcium entry through postsynaptic NMDA receptors can initiate two different forms of synaptic plasticity: long-term potentiation (LTP) and long-term depression (LTD). LTP arises when a single synapse is repeatedly stimulated. This stimulation causes a calcium- and CaMKII-dependent cellular cascade, which results in the insertion of more AMPA receptors into the postsynaptic membrane. The next time glutamate is released from the presynaptic cell, it will bind to both NMDA and the newly inserted AMPA receptors, thus depolarizing the membrane more efficiently. LTD occurs when few glutamate molecules bind to NMDA receptors at a synapse (due to a low firing rate of the presynaptic neuron). The calcium that does flow through NMDA receptors initiates a different calcineurin and protein phosphatase 1-dependent cascade, which results in the endocytosis of AMPA receptors. This makes the postsynaptic neuron less responsive to glutamate released from the presynaptic neuron.

35.3 The Central Nervous System

By the end of this section, you will be able to do the following:

- Identify the spinal cord, cerebral lobes, and other brain areas on a diagram of the brain
- Describe the basic functions of the spinal cord, cerebral lobes, and other brain areas

The central nervous system (CNS) is made up of the brain, a part of which is shown in Figure 35.19 and spinal cord and is covered with three layers of protective coverings called **meninges** (from the Greek word for membrane). The outermost layer is the **dura**

mater (Latin for "hard mother"). As the Latin suggests, the primary function for this thick layer is to protect the brain and spinal cord. The dura mater also contains vein-like structures that carry blood from the brain back to the heart. The middle layer is the web-like **arachnoid mater**. The last layer is the **pia mater** (Latin for "soft mother"), which directly contacts and covers the brain and spinal cord like plastic wrap. The space between the arachnoid and pia maters is filled with **cerebrospinal fluid (CSF)**. CSF is produced by a tissue called **choroid plexus** in fluid-filled compartments in the CNS called **ventricles**. The brain floats in CSF, which acts as a cushion and shock absorber and makes the brain neutrally buoyant. CSF also functions to circulate chemical substances throughout the brain and into the spinal cord.

The entire brain contains only about 8.5 tablespoons of CSF, but CSF is constantly produced in the ventricles. This creates a problem when a ventricle is blocked—the CSF builds up and creates swelling and the brain is pushed against the skull. This swelling condition is called hydrocephalus ("water head") and can cause seizures, cognitive problems, and even death if a shunt is not inserted to remove the fluid and pressure.

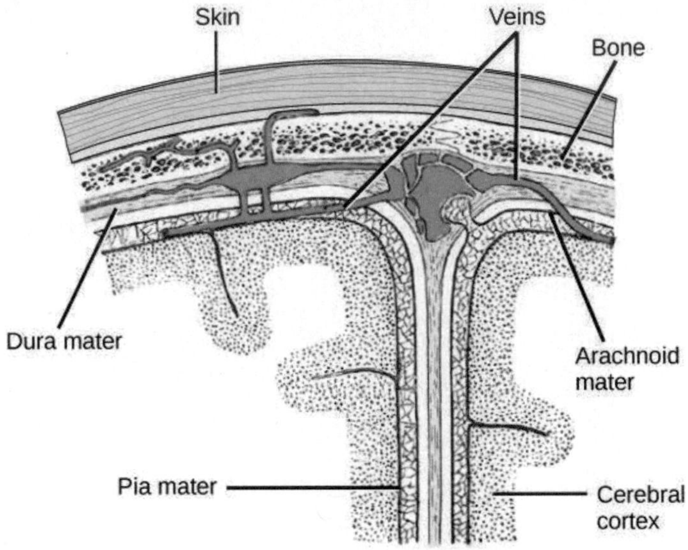

Figure 35.19 The cerebral cortex is covered by three layers of meninges: the dura, arachnoid, and pia maters. (credit: modification of work by Gray's Anatomy)

Brain

The brain is the part of the central nervous system that is contained in the cranial cavity of the skull. It includes the cerebral cortex, limbic system, basal ganglia, thalamus, hypothalamus, and cerebellum. There are three different ways that a brain can be sectioned in order to view internal structures: a sagittal section cuts the brain left to right, as shown in Figure 35.21b, a coronal section cuts the brain front to back, as shown in Figure 35.20a, and a horizontal section cuts the brain top to bottom.

Cerebral Cortex

The outermost part of the brain is a thick piece of nervous system tissue called the **cerebral cortex**, which is folded into hills called **gyri** (singular: gyrus) and valleys called **sulci** (singular: sulcus). The cortex is made up of two hemispheres—right and left—which are separated by a large sulcus. A thick fiber bundle called the **corpus callosum** (Latin: "tough body") connects the two hemispheres and allows information to be passed from one side to the other. Although there are some brain functions that are localized more to one hemisphere than the other, the functions of the two hemispheres are largely redundant. In fact, sometimes (very rarely) an entire hemisphere is removed to treat severe epilepsy. While patients do suffer some deficits following the surgery, they can have surprisingly few problems, especially when the surgery is performed on children who have very immature nervous systems.

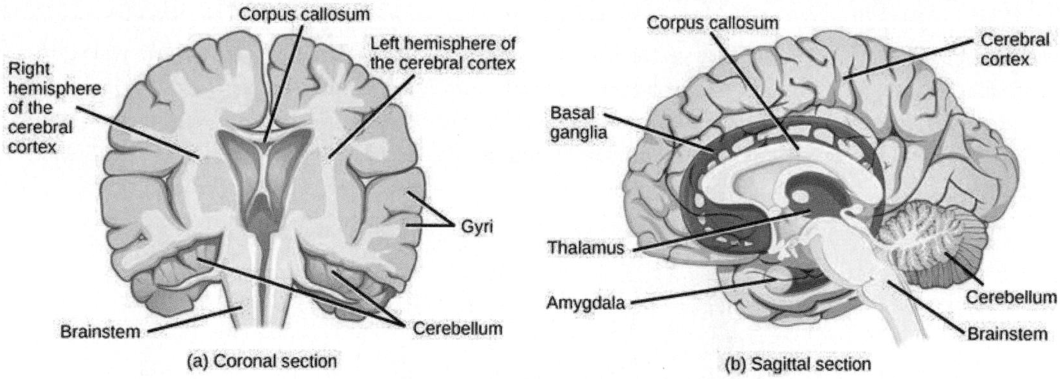

Figure 35.20 These illustrations show the (a) coronal and (b) sagittal sections of the human brain.

In other surgeries to treat severe epilepsy, the corpus callosum is cut instead of removing an entire hemisphere. This causes a condition called split-brain, which gives insights into unique functions of the two hemispheres. For example, when an object is presented to patients' left visual field, they may be unable to verbally name the object (and may claim to not have seen an object at all). This is because the visual input from the left visual field crosses and enters the right hemisphere and cannot then signal to the speech center, which generally is found in the left side of the brain. Remarkably, if a split-brain patient is asked to pick up a specific object out of a group of objects with the left hand, the patient will be able to do so but will still be unable to vocally identify it.

LINK TO LEARNING

See this website (http://openstax.org/l/split-brain) to learn more about split-brain patients and to play a game where you can model the split-brain experiments yourself.

Each cortical hemisphere contains regions called lobes that are involved in different functions. Scientists use various techniques to determine what brain areas are involved in different functions: they examine patients who have had injuries or diseases that affect specific areas and see how those areas are related to functional deficits. They also conduct animal studies where they stimulate brain areas and see if there are any behavioral changes. They use a technique called transcranial magnetic stimulation (TMS) to temporarily deactivate specific parts of the cortex using strong magnets placed outside the head; and they use functional magnetic resonance imaging (fMRI) to look at changes in oxygenated blood flow in particular brain regions that correlate with specific behavioral tasks. These techniques, and others, have given great insight into the functions of different brain regions but have also showed that any given brain area can be involved in more than one behavior or process, and any given behavior or process generally involves neurons in multiple brain areas. That being said, each hemisphere of the mammalian cerebral cortex can be broken down into four functionally and spatially defined lobes: frontal, parietal, temporal, and occipital. Figure 35.21 illustrates these four lobes of the human cerebral cortex.

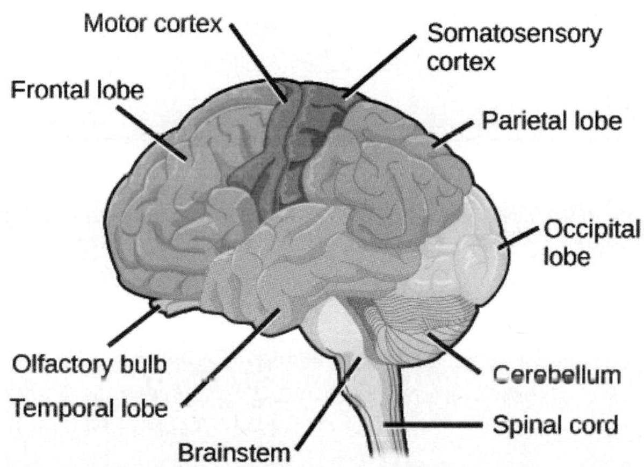

Figure 35.21 The human cerebral cortex includes the frontal, parietal, temporal, and occipital lobes.

The **frontal lobe** is located at the front of the brain, over the eyes. This lobe contains the olfactory bulb, which processes smells. The frontal lobe also contains the motor cortex, which is important for planning and implementing movement. Areas within the motor cortex map to different muscle groups, and there is some organization to this map, as shown in Figure 35.22. For example, the neurons that control movement of the fingers are next to the neurons that control movement of the hand. Neurons in the frontal lobe also control cognitive functions like maintaining attention, speech, and decision-making. Studies of humans who have damaged their frontal lobes show that parts of this area are involved in personality, socialization, and assessing risk.

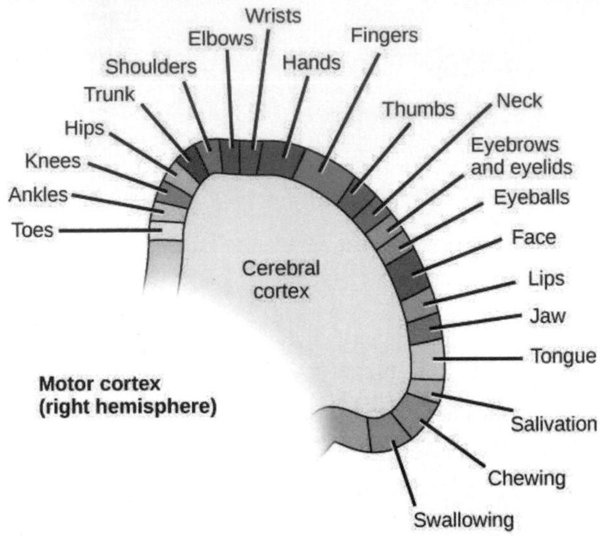

Figure 35.22 Different parts of the motor cortex control different muscle groups. Muscle groups that are neighbors in the body are generally controlled by neighboring regions of the motor cortex as well. For example, the neurons that control finger movement are near the neurons that control hand movement.

The **parietal lobe** is located at the top of the brain. Neurons in the parietal lobe are involved in speech and also reading. Two of the parietal lobe's main functions are processing **somatosensation**—touch sensations like pressure, pain, heat, cold—and processing **proprioception**—the sense of how parts of the body are oriented in space. The parietal lobe contains a somatosensory map of the body similar to the motor cortex.

The **occipital lobe** is located at the back of the brain. It is primarily involved in vision—seeing, recognizing, and identifying the visual world.

The **temporal lobe** is located at the base of the brain by your ears and is primarily involved in processing and interpreting sounds. It also contains the **hippocampus** (Greek for "seahorse")—a structure that processes memory formation. The hippocampus is illustrated in Figure 35.24. The role of the hippocampus in memory was partially determined by studying one famous epileptic patient, HM, who had both sides of his hippocampus removed in an attempt to cure his epilepsy. His seizures went away, but he could no longer form new memories (although he could remember some facts from before his surgery and could learn new motor tasks).

EVOLUTION CONNECTION

Cerebral Cortex

Compared to other vertebrates, mammals have exceptionally large brains for their body size. An entire alligator's brain, for example, would fill about one and a half teaspoons. This increase in brain to body size ratio is especially pronounced in apes, whales, and dolphins. While this increase in overall brain size doubtlessly played a role in the evolution of complex behaviors unique to mammals, it does not tell the whole story. Scientists have found a relationship between the relatively high surface area of the cortex and the intelligence and complex social behaviors exhibited by some mammals. This increased surface area is due, in part, to increased folding of the cortical sheet (more sulci and gyri). For example, a rat cortex is very smooth with very few sulci and gyri. Cat and sheep cortices have more sulci and gyri. Chimps, humans, and dolphins have even more.

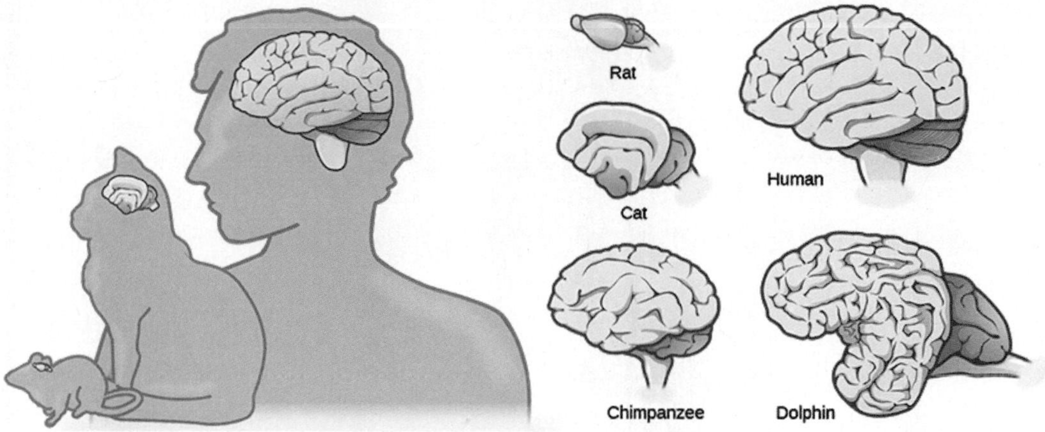

Figure 35.23 Mammals have larger brain-to-body ratios than other vertebrates. Within mammals, increased cortical folding and surface area is correlated with complex behavior.

Basal Ganglia

Interconnected brain areas called the **basal ganglia** (or **basal nuclei**), shown in Figure 35.20b, play important roles in movement control and posture. Damage to the basal ganglia, as in Parkinson's disease, leads to motor impairments like a shuffling gait when walking. The basal ganglia also regulate motivation. For example, when a wasp sting led to bilateral basal ganglia damage in a 25-year-old businessman, he began to spend all his days in bed and showed no interest in anything or anybody. But when he was externally stimulated—as when someone asked to play a card game with him he was able to function normally. Interestingly, he and other similar patients do not report feeling bored or frustrated by their state.

Thalamus

The **thalamus** (Greek for "inner chamber"), illustrated in Figure 35.24, acts as a gateway to and from the cortex. It receives sensory and motor inputs from the body and also receives feedback from the cortex. This feedback mechanism can modulate conscious awareness of sensory and motor inputs depending on the attention and arousal state of the animal. The thalamus helps regulate consciousness, arousal, and sleep states. A rare genetic disorder called fatal familial insomnia causes the degeneration of thalamic neurons and glia. This disorder prevents affected patients from being able to sleep, among other symptoms, and is eventually fatal.

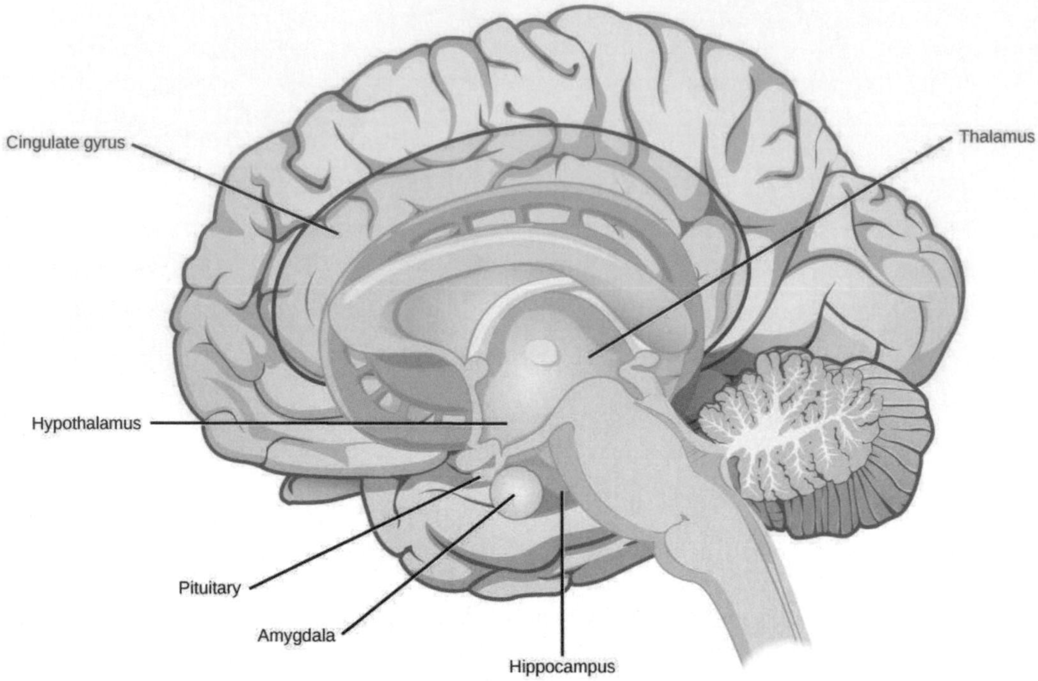

Figure 35.24 The limbic system regulates emotion and other behaviors. It includes parts of the cerebral cortex located near the center of the brain, including the cingulate gyrus and the hippocampus as well as the thalamus, hypothalamus, and amygdala.

Hypothalamus

Below the thalamus is the **hypothalamus**, shown in Figure 35.24. The hypothalamus controls the endocrine system by sending signals to the pituitary gland, a pea-sized endocrine gland that releases several different hormones that affect other glands as well as other cells. This relationship means that the hypothalamus regulates important behaviors that are controlled by these hormones. The hypothalamus is the body's thermostat—it makes sure key functions like food and water intake, energy expenditure, and body temperature are kept at appropriate levels. Neurons within the hypothalamus also regulate circadian rhythms, sometimes called sleep cycles.

Limbic System

The **limbic system** is a connected set of structures that regulates emotion, as well as behaviors related to fear and motivation. It plays a role in memory formation and includes parts of the thalamus and hypothalamus as well as the hippocampus. One important structure within the limbic system is a temporal lobe structure called the **amygdala** (Greek for "almond"), illustrated in Figure 35.24. The two amygdala are important both for the sensation of fear and for recognizing fearful faces. The **cingulate gyrus** helps regulate emotions and pain.

Cerebellum

The **cerebellum** (Latin for "little brain"), shown in Figure 35.21, sits at the base of the brain on top of the brainstem. The cerebellum controls balance and aids in coordinating movement and learning new motor tasks.

Brainstem

The **brainstem**, illustrated in Figure 35.21, connects the rest of the brain with the spinal cord. It consists of the midbrain, medulla oblongata, and the pons. Motor and sensory neurons extend through the brainstem allowing for the relay of signals between the brain and spinal cord. Ascending neural pathways cross in this section of the brain allowing the left hemisphere of the cerebrum to control the right side of the body and vice versa. The brainstem coordinates motor control signals sent from the brain to the body. The brainstem controls several important functions of the body including alertness, arousal, breathing, blood pressure, digestion, heart rate, swallowing, walking, and sensory and motor information integration.

Spinal Cord

Connecting to the brainstem and extending down the body through the spinal column is the **spinal cord**, shown in Figure 35.21. The spinal cord is a thick bundle of nerve tissue that carries information about the body to the brain and from the brain to the body. The spinal cord is contained within the bones of the vertebrate column but is able to communicate signals to and from the body through its connections with spinal nerves (part of the peripheral nervous system). A cross-section of the spinal cord looks like a white oval containing a gray butterfly-shape, as illustrated in Figure 35.25. Myelinated axons make up the "white matter" and neuron and glial cell bodies make up the "gray matter." Gray matter is also composed of interneurons, which connect two neurons each located in different parts of the body. Axons and cell bodies in the dorsal (facing the back of the animal) spinal cord convey mostly sensory information from the body to the brain. Axons and cell bodies in the ventral (facing the front of the animal) spinal cord primarily transmit signals controlling movement from the brain to the body.

The spinal cord also controls motor reflexes. These reflexes are quick, unconscious movements—like automatically removing a hand from a hot object. Reflexes are so fast because they involve local synaptic connections. For example, the knee reflex that a doctor tests during a routine physical is controlled by a single synapse between a sensory neuron and a motor neuron. While a reflex may only require the involvement of one or two synapses, synapses with interneurons in the spinal column transmit information to the brain to convey what happened (the knee jerked, or the hand was hot).

In the United States, there around 10,000 spinal cord injuries each year. Because the spinal cord is the information superhighway connecting the brain with the body, damage to the spinal cord can lead to paralysis. The extent of the paralysis depends on the location of the injury along the spinal cord and whether the spinal cord was completely severed. For example, if the spinal cord is damaged at the level of the neck, it can cause paralysis from the neck down, whereas damage to the spinal column further down may limit paralysis to the legs. Spinal cord injuries are notoriously difficult to treat because spinal nerves do not regenerate, although ongoing research suggests that stem cell transplants may be able to act as a bridge to reconnect severed nerves. Researchers are also looking at ways to prevent the inflammation that worsens nerve damage after injury. One such treatment is to pump the body with cold saline to induce hypothermia. This cooling can prevent swelling and other processes that are thought to worsen spinal cord injuries.

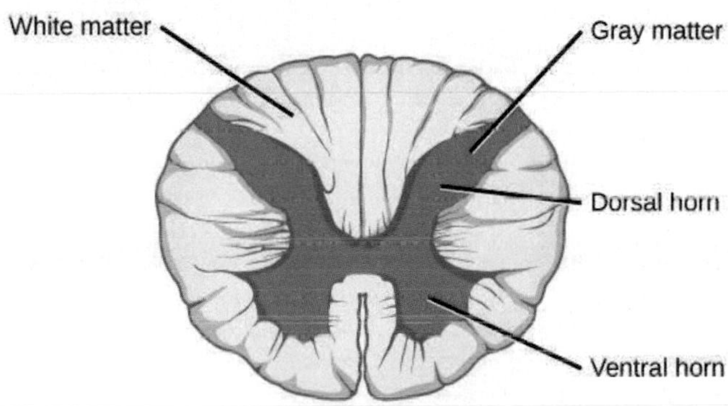

Figure 35.25 A cross-section of the spinal cord shows gray matter (containing cell bodies and interneurons) and white matter (containing axons).

35.4 The Peripheral Nervous System

By the end of this section, you will be able to do the following:

- Describe the organization and functions of the sympathetic and parasympathetic nervous systems
- Describe the organization and function of the sensory-somatic nervous system

The peripheral nervous system (PNS) is the connection between the central nervous system and the rest of the body. The CNS is like the power plant of the nervous system. It creates the signals that control the functions of the body. The PNS is like the wires that go to individual houses. Without those "wires," the signals produced by the CNS could not control the body (and the CNS would not be able to receive sensory information from the body either).

The PNS can be broken down into the **autonomic nervous system**, which controls bodily functions without conscious control, and the **sensory-somatic nervous system**, which transmits sensory information from the skin, muscles, and sensory organs to the CNS and sends motor commands from the CNS to the muscles.

Autonomic Nervous System

VISUAL CONNECTION

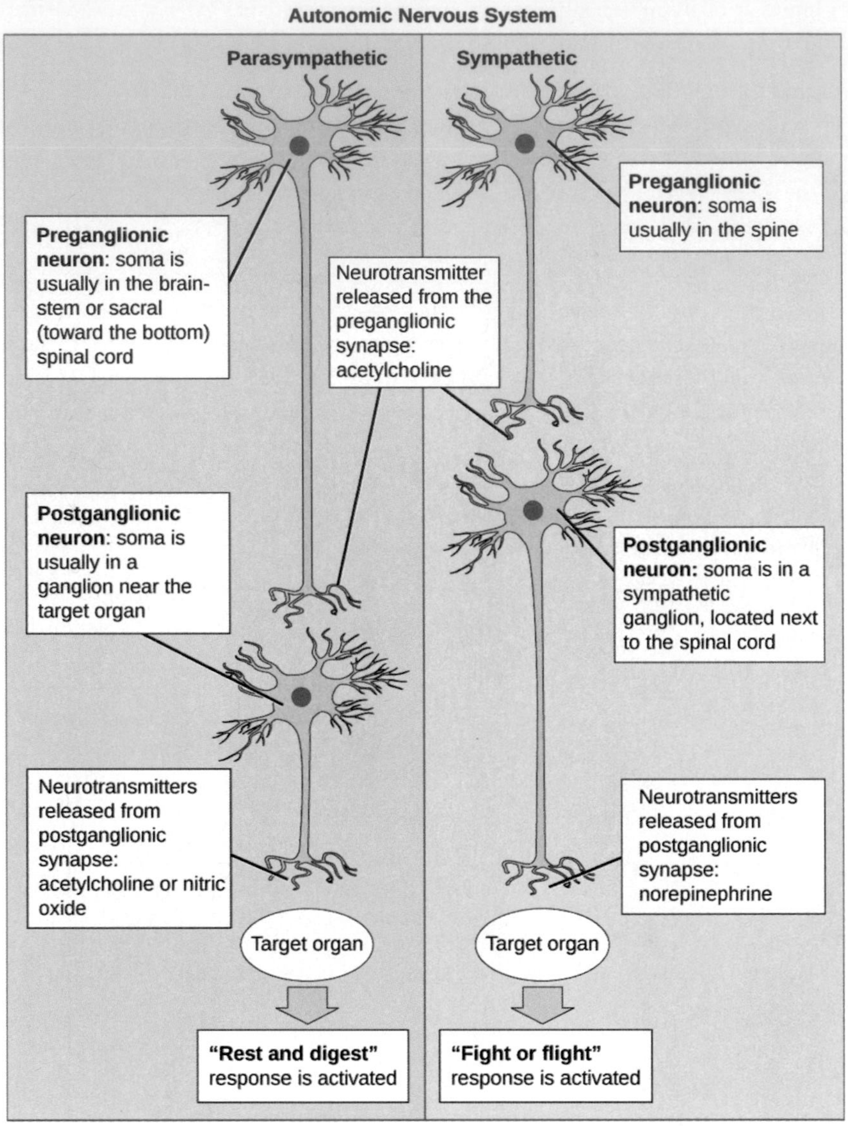

Figure 35.26 In the autonomic nervous system, a preganglionic neuron of the CNS synapses with a postganglionic neuron of the PNS. The postganglionic neuron, in turn, acts on a target organ. Autonomic responses are mediated by the sympathetic and the parasympathetic systems, which are antagonistic to one another. The sympathetic system activates the "fight or flight" response, while the parasympathetic system activates the "rest and digest" response.

Which of the following statements is false?

a. The parasympathetic pathway is responsible for resting the body, while the sympathetic pathway is responsible for preparing for an emergency.
b. Most preganglionic neurons in the sympathetic pathway originate in the spinal cord.
c. Slowing of the heartbeat is a parasympathetic response.
d. Parasympathetic neurons are responsible for releasing norepinephrine on the target organ, while sympathetic neurons are responsible for releasing acetylcholine.

The autonomic nervous system serves as the relay between the CNS and the internal organs. It controls the lungs, the heart, smooth muscle, and exocrine and endocrine glands. The autonomic nervous system controls these organs largely without conscious control; it can continuously monitor the conditions of these different systems and implement changes as needed. Signaling to the target tissue usually involves two synapses: a preganglionic neuron (originating in the CNS) synapses to a neuron in a ganglion that, in turn, synapses on the target organ, as illustrated in Figure 35.26. There are two divisions of the autonomic nervous system that often have opposing effects: the sympathetic nervous system and the parasympathetic nervous system.

Sympathetic Nervous System

The **sympathetic nervous system** is responsible for the "fight or flight" response that occurs when an animal encounters a dangerous situation. One way to remember this is to think of the surprise a person feels when encountering a snake ("snake" and "sympathetic" both begin with "s"). Examples of functions controlled by the sympathetic nervous system include an accelerated heart rate and inhibited digestion. These functions help prepare an organism's body for the physical strain required to escape a potentially dangerous situation or to fend off a predator.

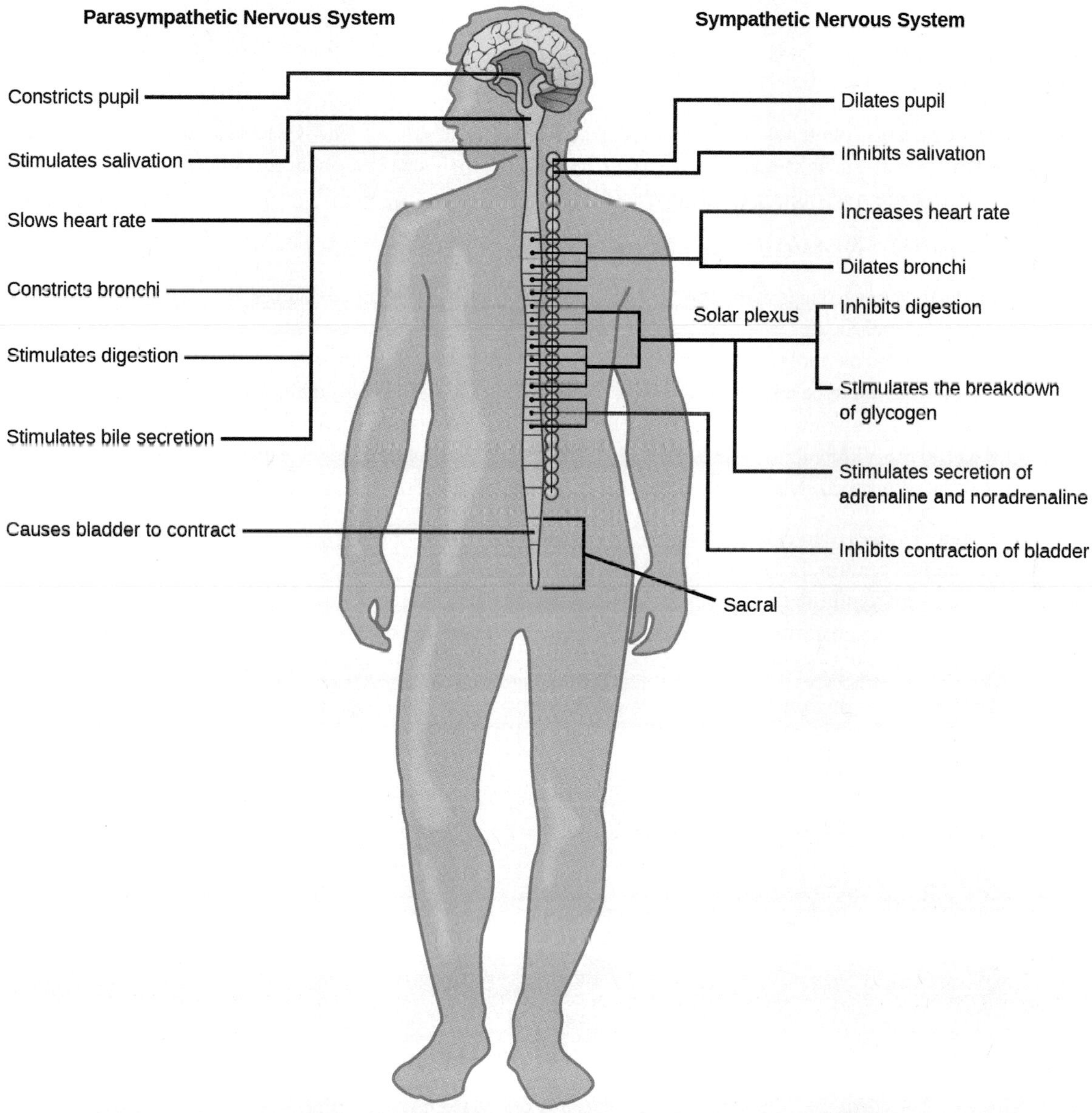

Figure 35.27 The sympathetic and parasympathetic nervous systems often have opposing effects on target organs.

Most preganglionic neurons in the sympathetic nervous system originate in the spinal cord, as illustrated in Figure 35.27. The axons of these neurons release **acetylcholine** on postganglionic neurons within sympathetic ganglia (the sympathetic ganglia form a chain that extends alongside the spinal cord). The acetylcholine activates the postganglionic neurons. Postganglionic neurons then release **norepinephrine** onto target organs. As anyone who has ever felt a rush before a big test, speech, or athletic event can attest, the effects of the sympathetic nervous system are quite pervasive. This is both because one preganglionic neuron synapses on multiple postganglionic neurons, amplifying the effect of the original synapse, and because the adrenal gland also releases norepinephrine (and the closely related hormone epinephrine) into the bloodstream. The physiological effects of this norepinephrine release include dilating the trachea and bronchi (making it easier for the animal to breathe), increasing heart rate, and moving blood from the skin to the heart, muscles, and brain (so the animal can think and run). The strength and speed of the sympathetic response helps an organism avoid danger, and scientists have found evidence that it may also increase LTP—allowing the animal to remember the dangerous situation and avoid it in the future.

Parasympathetic Nervous System

While the sympathetic nervous system is activated in stressful situations, the **parasympathetic nervous system** allows an animal to "rest and digest." One way to remember this is to think that during a restful situation like a picnic, the parasympathetic nervous system is in control ("picnic" and "parasympathetic" both start with "p"). Parasympathetic preganglionic neurons have cell bodies located in the brainstem and in the sacral (toward the bottom) spinal cord, as shown in Figure 35.27. The axons of the preganglionic neurons release acetylcholine on the postganglionic neurons, which are generally located very near the target organs. Most postganglionic neurons release acetylcholine onto target organs, although some release nitric oxide.

The parasympathetic nervous system resets organ function after the sympathetic nervous system is activated (the common adrenaline dump you feel after a 'fight-or-flight' event). Effects of acetylcholine release on target organs include slowing of heart rate, lowered blood pressure, and stimulation of digestion.

Sensory-Somatic Nervous System

The sensory-somatic nervous system is made up of cranial and spinal nerves and contains both sensory and motor neurons. Sensory neurons transmit sensory information from the skin, skeletal muscle, and sensory organs to the CNS. Motor neurons transmit messages about desired movement from the CNS to the muscles to make them contract. Without its sensory-somatic nervous system, an animal would be unable to process any information about its environment (what it sees, feels, hears, and so on) and could not control motor movements. Unlike the autonomic nervous system, which has two synapses between the CNS and the target organ, sensory and motor neurons have only one synapse—one ending of the neuron is at the organ and the other directly contacts a CNS neuron. Acetylcholine is the main neurotransmitter released at these synapses.

Humans have 12 **cranial nerves**, nerves that emerge from or enter the skull (cranium), as opposed to the spinal nerves, which emerge from the vertebral column. Each cranial nerve is accorded a name, which are detailed in Figure 35.28. Some cranial nerves transmit only sensory information. For example, the olfactory nerve transmits information about smells from the nose to the brainstem. Other cranial nerves transmit almost solely motor information. For example, the oculomotor nerve controls the opening and closing of the eyelid and some eye movements. Other cranial nerves contain a mix of sensory and motor fibers. For example, the glossopharyngeal nerve has a role in both taste (sensory) and swallowing (motor).

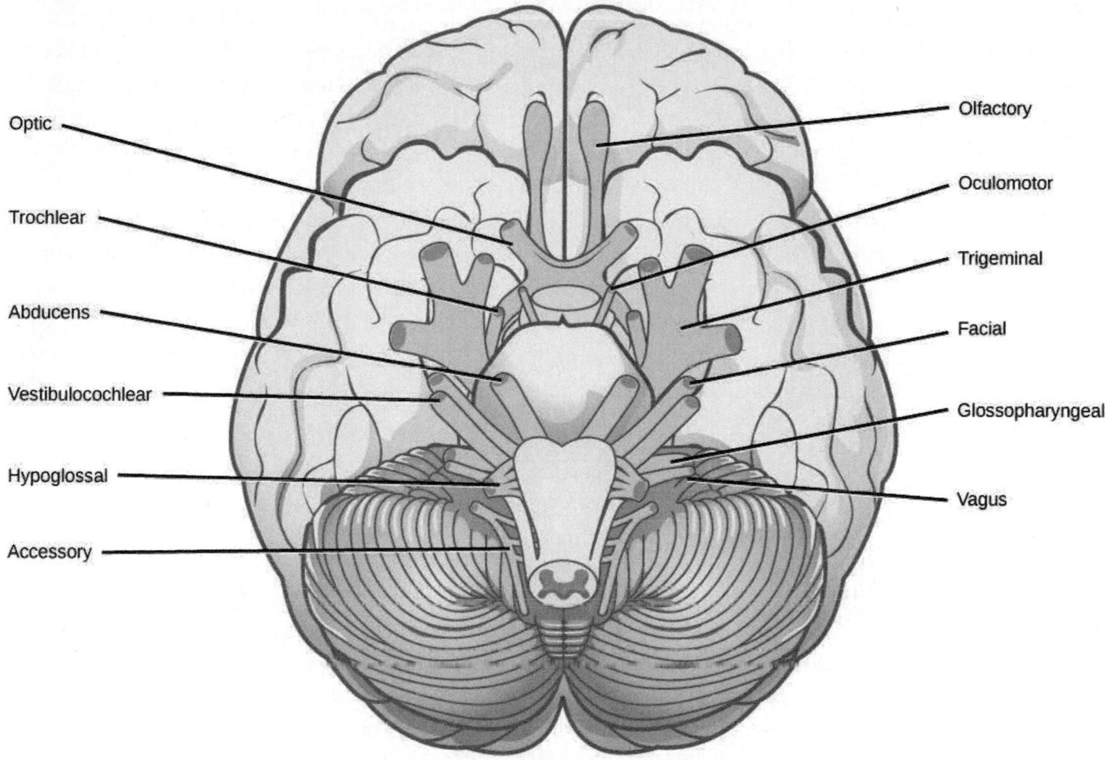

Figure 35.28 The human brain contains 12 cranial nerves that receive sensory input and control motor output for the head and neck.

Spinal nerves transmit sensory and motor information between the spinal cord and the rest of the body. Each of the 31 spinal nerves (in humans) contains both sensory and motor axons. The sensory neuron cell bodies are grouped in structures called dorsal root ganglia and are shown in Figure 35.29. Each sensory neuron has one projection—with a sensory receptor ending in skin, muscle, or sensory organs—and another that synapses with a neuron in the dorsal spinal cord. Motor neurons have cell bodies in the ventral gray matter of the spinal cord that project to muscle through the ventral root. These neurons are usually stimulated by interneurons within the spinal cord but are sometimes directly stimulated by sensory neurons.

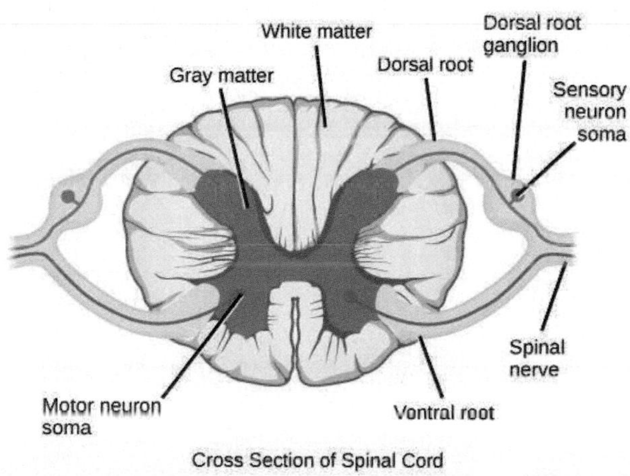

Figure 35.29 Spinal nerves contain both sensory and motor axons. The somas of sensory neurons are located in dorsal root ganglia. The somas of motor neurons are found in the ventral portion of the gray matter of the spinal cord.

35.5 Nervous System Disorders

By the end of this section, you will be able to do the following:

- Describe the symptoms, potential causes, and treatment of several examples of nervous system disorders

A nervous system that functions correctly is a fantastically complex, well-oiled machine—synapses fire appropriately, muscles move when needed, memories are formed and stored, and emotions are well regulated. Unfortunately, each year millions of people in the United States deal with some sort of nervous system disorder. While scientists have discovered potential causes of many of these diseases, and viable treatments for some, ongoing research seeks to find ways to better prevent and treat all of these disorders.

Neurodegenerative Disorders

Neurodegenerative disorders are illnesses characterized by a loss of nervous system functioning that are usually caused by neuronal death. These diseases generally worsen over time as more and more neurons die. The symptoms of a particular neurodegenerative disease are related to where in the nervous system the death of neurons occurs. Spinocerebellar ataxia, for example, leads to neuronal death in the cerebellum. The death of these neurons causes problems in balance and walking. Neurodegenerative disorders include Huntington's disease, amyotrophic lateral sclerosis, Alzheimer's disease and other types of dementia disorders, and Parkinson's disease. Here, Alzheimer's and Parkinson's disease will be discussed in more depth.

Alzheimer's Disease

Alzheimer's disease is the most common cause of dementia in the elderly. In 2012, an estimated 5.4 million Americans suffered from Alzheimer's disease, and payments for their care are estimated at $200 billion. Roughly one in every eight people age 65 or older has the disease. Due to the aging of the baby-boomer generation, there are projected to be as many as 13 million Alzheimer's patients in the United States in the year 2050.

Symptoms of Alzheimer's disease include disruptive memory loss, confusion about time or place, difficulty planning or executing tasks, poor judgment, and personality changes. Problems smelling certain scents can also be indicative of Alzheimer's disease and may serve as an early warning sign. Many of these symptoms are also common in people who are aging normally, so it is the severity and longevity of the symptoms that determine whether a person is suffering from Alzheimer's.

Alzheimer's disease was named for Alois Alzheimer, a German psychiatrist who published a report in 1911 about a woman who showed severe dementia symptoms. Along with his colleagues, he examined the woman's brain following her death and reported the presence of abnormal clumps, which are now called amyloid plaques, along with tangled brain fibers called neurofibrillary tangles. Amyloid plaques, neurofibrillary tangles, and an overall shrinking of brain volume are commonly seen in the brains of Alzheimer's patients. Loss of neurons in the hippocampus is especially severe in advanced Alzheimer's patients. Figure 35.30 compares a normal brain to the brain of an Alzheimer's patient. Many research groups are examining the causes of these hallmarks of the disease.

One form of the disease is usually caused by mutations in one of three known genes. This rare form of early onset Alzheimer's disease affects fewer than five percent of patients with the disease and causes dementia beginning between the ages of 30 and 60. The more prevalent, late-onset form of the disease likely also has a genetic component. One particular gene, apolipoprotein E (APOE) has a variant (E4) that increases a carrier's likelihood of getting the disease. Many other genes have been identified that might be involved in the pathology.

LINK TO LEARNING

Visit this website (http://openstax.org/l/alzheimers) for video links discussing genetics and Alzheimer's disease.

Unfortunately, there is no cure for Alzheimer's disease. Current treatments focus on managing the symptoms of the disease. Because decrease in the activity of cholinergic neurons (neurons that use the neurotransmitter acetylcholine) is common in Alzheimer's disease, several drugs used to treat the disease work by increasing acetylcholine neurotransmission, often by inhibiting the enzyme that breaks down acetylcholine in the synaptic cleft. Other clinical interventions focus on behavioral therapies like psychotherapy, sensory therapy, and cognitive exercises. Since Alzheimer's disease appears to hijack the normal aging process, research into prevention is prevalent. Smoking, obesity, and cardiovascular problems may be risk factors for the disease, so treatments for those may also help to prevent Alzheimer's disease. Some studies have shown that people who remain intellectually active by playing games, reading, playing musical instruments, and being socially active in later life have a reduced risk of developing the disease.

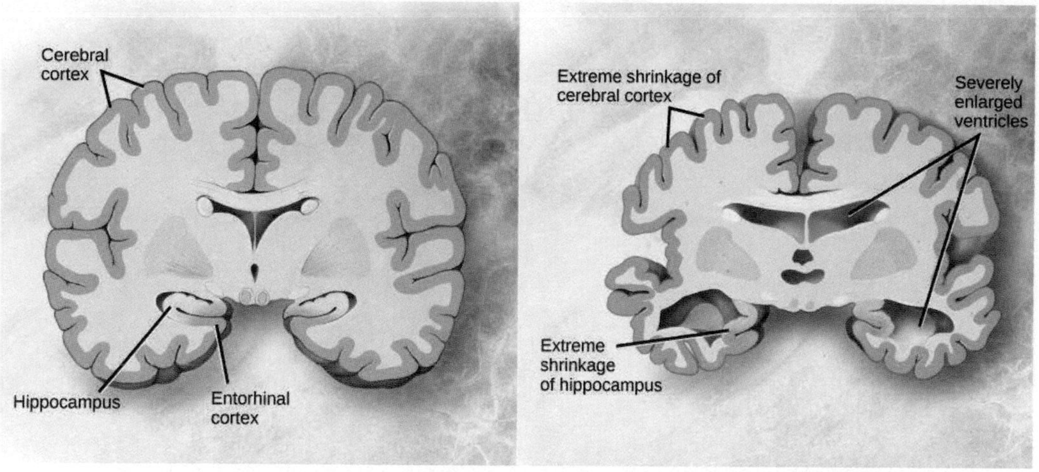

Figure 35.30 Compared to a normal brain (left), the brain from a patient with Alzheimer's disease (right) shows a dramatic neurodegeneration, particularly within the ventricles and hippocampus. (credit: modification of work by "Garrando"/Wikimedia Commons based on original images by ADEAR: "Alzheimer's Disease Education and Referral Center, a service of the National Institute on Aging")

Parkinson's Disease

Like Alzheimer's disease, **Parkinson's disease** is a neurodegenerative disease. It was first characterized by James Parkinson in 1817. Each year, 50,000-60,000 people in the United States are diagnosed with the disease. Parkinson's disease causes the loss of dopamine neurons in the substantia nigra, a midbrain structure that regulates movement. Loss of these neurons causes many symptoms including tremor (shaking of fingers or a limb), slowed movement, speech changes, balance and posture problems, and rigid muscles. The combination of these symptoms often causes a characteristic slow hunched shuffling walk, illustrated in Figure 35.31. Patients with Parkinson's disease can also exhibit psychological symptoms, such as dementia or emotional problems.

Although some patients have a form of the disease known to be caused by a single mutation, for most patients the exact causes of Parkinson's disease remain unknown: the disease likely results from a combination of genetic and environmental factors (similar to Alzheimer's disease). Post-mortem analysis of brains from Parkinson's patients shows the presence of Lewy bodies—abnormal protein clumps—in dopaminergic neurons. The prevalence of these Lewy bodies often correlates with the severity of the disease.

There is no cure for Parkinson's disease, and treatment is focused on easing symptoms. One of the most commonly prescribed drugs for Parkinson's is L-DOPA, which is a chemical that is converted into dopamine by neurons in the brain. This conversion increases the overall level of dopamine neurotransmission and can help compensate for the loss of dopaminergic neurons in the substantia nigra. Other drugs work by inhibiting the enzyme that breaks down dopamine.

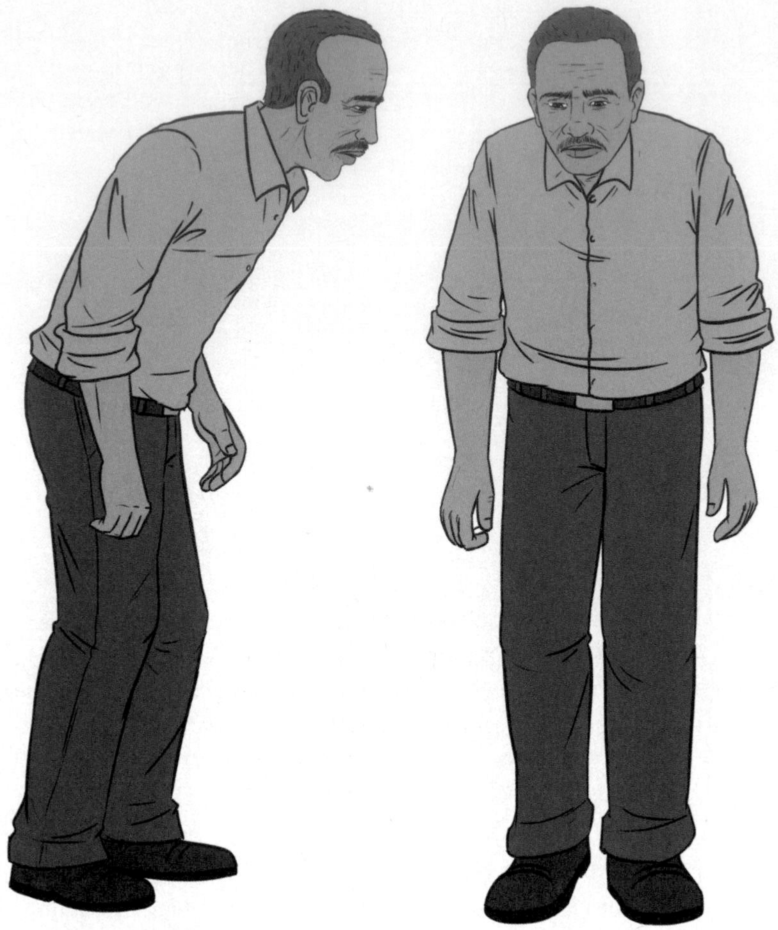

Figure 35.31 Parkinson's patients often have a characteristic hunched walk.

Neurodevelopmental Disorders

Neurodevelopmental disorders occur when the development of the nervous system is disturbed. There are several different classes of neurodevelopmental disorders. Some, like Down Syndrome, cause intellectual deficits. Others specifically affect communication, learning, or the motor system. Some disorders like autism spectrum disorder and attention deficit/hyperactivity disorder have complex symptoms.

Autism

Autism spectrum disorder (ASD) is a neurodevelopmental disorder. Its severity differs from person to person. Estimates for the prevalence of the disorder have changed rapidly in the past few decades. Current estimates suggest that one in 88 children will develop the disorder. ASD is four times more prevalent in males than females.

LINK TO LEARNING

This video (http://openstax.org/l/autism) discusses possible reasons why there has been a recent increase in the number of people diagnosed with autism.

A characteristic symptom of ASD is impaired social skills. Children with autism may have difficulty making and maintaining eye contact and reading social cues. They also may have problems feeling empathy for others. Other symptoms of ASD include repetitive motor behaviors (such as rocking back and forth), preoccupation with specific subjects, strict adherence to certain rituals, and unusual language use. Up to 30 percent of patients with ASD develop epilepsy, and patients with some forms of the disorder (like Fragile X) also have intellectual disability. Because it is a spectrum disorder, other ASD patients are very functional and have good-to-excellent language skills. Many of these patients do not feel that they suffer from a disorder and instead think that their brains just process information differently.

Except for some well-characterized, clearly genetic forms of autism (like Fragile X and Rett's Syndrome), the causes of ASD are largely unknown. Variants of several genes correlate with the presence of ASD, but for any given patient, many different mutations in different genes may be required for the disease to develop. At a general level, ASD is thought to be a disease of "incorrect" wiring. Accordingly, brains of some ASD patients lack the same level of synaptic pruning that occurs in non-affected people. In the 1990s, a research paper linked autism to a common vaccine given to children. This paper was retracted when it was discovered that the author falsified data, and follow-up studies showed no connection between vaccines and autism.

Treatment for autism usually combines behavioral therapies and interventions, along with medications to treat other disorders common to people with autism (depression, anxiety, obsessive compulsive disorder). Although early interventions can help mitigate the effects of the disease, there is currently no cure for ASD.

Attention Deficit Hyperactivity Disorder (ADHD)

Approximately three to five percent of children and adults are affected by **attention deficit/hyperactivity disorder (ADHD)**. Like ASD, ADHD is more prevalent in males than females. Symptoms of the disorder include inattention (lack of focus), executive functioning difficulties, impulsivity, and hyperactivity beyond what is characteristic of the normal developmental stage. Some patients do not have the hyperactive component of symptoms and are diagnosed with a subtype of ADHD: attention deficit disorder (ADD). Many people with ADHD also show comorbitity, in that they develop secondary disorders in addition to ADHD. Examples include depression or obsessive compulsive disorder (OCD). Figure 35.32 provides some statistics concerning comorbidity with ADHD.

The cause of ADHD is unknown, although research points to a delay and dysfunction in the development of the prefrontal cortex and disturbances in neurotransmission. According to studies of twins, the disorder has a strong genetic component. There are several candidate genes that may contribute to the disorder, but no definitive links have been discovered. Environmental factors, including exposure to certain pesticides, may also contribute to the development of ADHD in some patients. Treatment for ADHD often involves behavioral therapies and the prescription of stimulant medications, which paradoxically cause a calming effect in these patients.

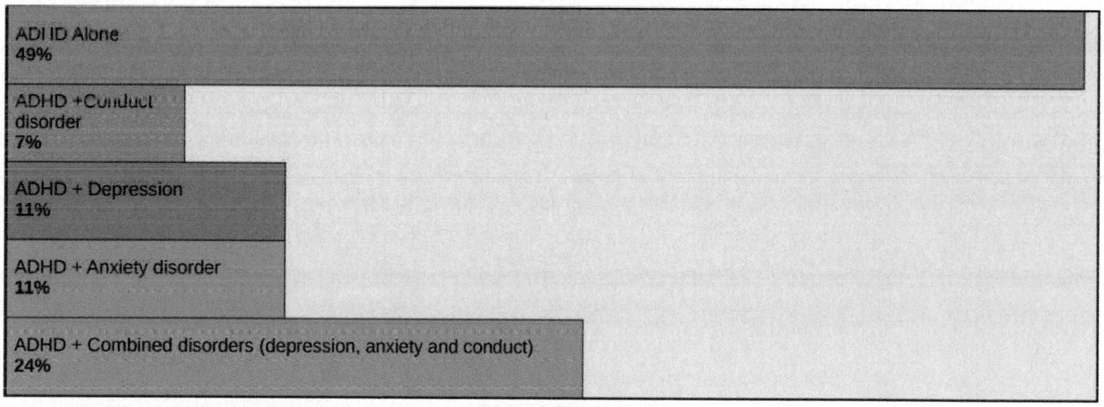

Figure 35.32 Many people with ADHD have one or more other neurological disorders. (credit "chart design and illustration": modification of work by Leigh Coriale; credit "data": Drs. Biederman and Faraone, Massachusetts General Hospital).

CAREER CONNECTION

Neurologist

Neurologists are physicians who specialize in disorders of the nervous system. They diagnose and treat disorders such as epilepsy, stroke, dementia, nervous system injuries, Parkinson's disease, sleep disorders, and multiple sclerosis. Neurologists are medical doctors who have attended college, medical school, and completed three to four years of neurology residency.

When examining a new patient, a neurologist takes a full medical history and performs a complete physical exam. The physical exam contains specific tasks that are used to determine what areas of the brain, spinal cord, or peripheral nervous system may be damaged. For example, to check whether the hypoglossal nerve is functioning correctly, the neurologist will ask the patient to move his or her tongue in different ways. If the patient does not have full control over tongue movements, then the hypoglossal nerve may be damaged or there may be a lesion in the brainstem where the cell bodies of these neurons reside (or there could be damage to the tongue muscle itself).

Neurologists have other tools besides a physical exam they can use to diagnose particular problems in the nervous system. If the patient has had a seizure, for example, the neurologist can use electroencephalography (EEG), which involves taping electrodes to the scalp to record brain activity, to try to determine which brain regions are involved in the seizure. In suspected stroke patients, a neurologist can use a computerized tomography (CT) scan, which is a type of X-ray, to look for bleeding in the brain or a possible brain tumor. To treat patients with neurological problems, neurologists can prescribe medications or refer the patient to a neurosurgeon for surgery.

LINK TO LEARNING

This website (http://openstax.org/l/neurologic_exam) allows you to see the different tests a neurologist might use to see what regions of the nervous system may be damaged in a patient.

Mental Illnesses

Mental illnesses are nervous system disorders that result in problems with thinking, mood, or relating with other people. These disorders are severe enough to affect a person's quality of life and often make it difficult for people to perform the routine tasks of daily living. Debilitating mental disorders plague approximately 12.5 million Americans (about 1 in 17 people) at an annual cost of more than $300 billion. There are several types of mental disorders including schizophrenia, major depression, bipolar disorder, anxiety disorders and phobias, post-traumatic stress disorders, and obsessive-compulsive disorder (OCD), among others. The American Psychiatric Association publishes the Diagnostic and Statistical Manual of Mental Disorders (or DSM), which describes the symptoms required for a patient to be diagnosed with a particular mental disorder. Each newly released version of the DSM contains different symptoms and classifications as scientists learn more about these disorders, their causes, and how they relate to each other. A more detailed discussion of two mental illnesses—schizophrenia and major depression—is given below.

Schizophrenia

Schizophrenia is a serious and often debilitating mental illness affecting one percent of people in the United States. Symptoms of the disease include the inability to differentiate between reality and imagination, inappropriate and unregulated emotional responses, difficulty thinking, and problems with social situations. People with schizophrenia can suffer from hallucinations and hear voices; they may also suffer from delusions. Patients also have so-called "negative" symptoms like a flattened emotional state, loss of pleasure, and loss of basic drives. Many schizophrenic patients are diagnosed in their late adolescence or early 20s. The development of schizophrenia is thought to involve malfunctioning dopaminergic neurons and may also involve problems with glutamate signaling. Treatment for the disease usually requires antipsychotic medications that work by blocking dopamine receptors and decreasing dopamine neurotransmission in the brain. This decrease in dopamine can cause Parkinson's disease-like symptoms in some patients. While some classes of antipsychotics can be quite effective at treating the disease, they are not a cure, and most patients must remain medicated for the rest of their lives.

Depression

Major depression affects approximately 6.7 percent of the adults in the United States each year and is one of the most common mental disorders. To be diagnosed with major depressive disorder, a person must have experienced a severely depressed mood lasting longer than two weeks along with other symptoms including a loss of enjoyment in activities that were previously enjoyed, changes in appetite and sleep schedules, difficulty concentrating, feelings of worthlessness, and suicidal thoughts. The exact causes of major depression are unknown and likely include both genetic and environmental risk factors. Some research supports the "classic monoamine hypothesis," which suggests that depression is caused by a decrease in norepinephrine and serotonin neurotransmission. One argument against this hypothesis is the fact that some antidepressant medications cause an increase in norepinephrine and serotonin release within a few hours of beginning treatment—but clinical results of these medications are not seen until weeks later. This has led to alternative hypotheses: for example, dopamine may also be decreased in depressed patients, or it may actually be an increase in norepinephrine and serotonin that causes the disease, and antidepressants force a feedback loop that decreases this release. Treatments for depression include psychotherapy, electroconvulsive therapy, deep-brain stimulation, and prescription medications. There are several classes of antidepressant medications that work through different mechanisms. For example, monoamine oxidase inhibitors (MAO inhibitors) block the enzyme that degrades many neurotransmitters (including dopamine, serotonin, norepinephrine), resulting in increased neurotransmitter in the synaptic cleft. Selective serotonin reuptake inhibitors (SSRIs) block the reuptake of serotonin into the presynaptic neuron. This blockage results in an increase in serotonin in the synaptic cleft. Other types of drugs such as norepinephrine-dopamine reuptake inhibitors and norepinephrine-serotonin reuptake inhibitors are also used to treat depression.

Other Neurological Disorders

There are several other neurological disorders that cannot be easily placed in the above categories. These include chronic pain conditions, cancers of the nervous system, epilepsy disorders, and stroke. Epilepsy and stroke are discussed below.

Epilepsy

Estimates suggest that up to three percent of people in the United States will be diagnosed with **epilepsy** in their lifetime. While there are several different types of epilepsy, all are characterized by recurrent seizures. Epilepsy itself can be a symptom of a brain injury, disease, or other illness. For example, people who have intellectual disability or ASD can experience seizures, presumably because the developmental wiring malfunctions that caused their disorders also put them at risk for epilepsy. For many patients, however, the cause of their epilepsy is never identified and is likely to be a combination of genetic and environmental factors. Often, seizures can be controlled with anticonvulsant medications. However, for very severe cases, patients may undergo brain surgery to remove the brain area where seizures originate.

Stroke

A stroke results when blood fails to reach a portion of the brain for a long enough time to cause damage. Without the oxygen supplied by blood flow, neurons in this brain region die. This neuronal death can cause many different symptoms—depending on the brain area affected— including headache, muscle weakness or paralysis, speech disturbances, sensory problems, memory loss, and confusion. Stroke is often caused by blood clots and can also be caused by the bursting of a weak blood vessel. Strokes are extremely common and are the third most common cause of death in the United States. On average one person experiences a stroke every 40 seconds in the United States. Approximately 75 percent of strokes occur in people older than 65. Risk factors for stroke include high blood pressure, diabetes, high cholesterol, and a family history of stroke. Smoking doubles the risk of stroke. Because a stroke is a medical emergency, patients with symptoms of a stroke should immediately go to the emergency room, where they can receive drugs that will dissolve any clot that may have formed. These drugs will not work if the stroke was caused by a burst blood vessel or if the stroke occurred more than three hours before arriving at the hospital. Treatment following a stroke can include blood pressure medication (to prevent future strokes) and (sometimes intense) physical therapy.

KEY TERMS

acetylcholine neurotransmitter released by neurons in the central nervous system and peripheral nervous system

action potential self-propagating momentary change in the electrical potential of a neuron (or muscle) membrane

Alzheimer's disease neurodegenerative disorder characterized by problems with memory and thinking

amygdala structure within the limbic system that processes fear

arachnoid mater spiderweb-like middle layer of the meninges that cover the central nervous system

astrocyte glial cell in the central nervous system that provide nutrients, extracellular buffering, and structural support for neurons; also makes up the blood-brain barrier

attention deficit hyperactivity disorder (ADHD) neurodevelopmental disorder characterized by difficulty maintaining attention and controlling impulses

autism spectrum disorder (ASD) neurodevelopmental disorder characterized by impaired social interaction and communication abilities

autonomic nervous system part of the peripheral nervous system that controls bodily functions

axon tube-like structure that propagates a signal from a neuron's cell body to axon terminals

axon hillock electrically sensitive structure on the cell body of a neuron that integrates signals from multiple neuronal connections

axon terminal structure on the end of an axon that can form a synapse with another neuron

basal ganglia interconnected collections of cells in the brain that are involved in movement and motivation; also known as basal nuclei

basal nuclei see basal ganglia

brainstem portion of the brain that connects with the spinal cord; controls basic nervous system functions like breathing, heart rate, and swallowing

cerebellum brain structure involved in posture, motor coordination, and learning new motor actions

cerebral cortex outermost sheet of brain tissue; involved in many higher-order functions

cerebrospinal fluid (CSF) clear liquid that surrounds the brain and spinal cord and fills the ventricles and central canal; acts as a shock absorber and circulates material throughout the brain and spinal cord

choroid plexus spongy tissue within ventricles that produces cerebrospinal fluid

cingulate gyrus helps regulate emotions and pain; thought to directly drive the body's conscious response to unpleasant experiences

corpus callosum thick fiber bundle that connects the cerebral hemispheres

cranial nerve sensory and/or motor nerve that emanates from the brain

dendrite structure that extends away from the cell body to receive messages from other neurons

depolarization change in the membrane potential to a less negative value

dura mater tough outermost layer that covers the central nervous system

ependymal cell that lines fluid-filled ventricles of the brain and the central canal of the spinal cord; involved in production of cerebrospinal fluid

epilepsy neurological disorder characterized by recurrent seizures

excitatory postsynaptic potential (EPSP) depolarization of a postsynaptic membrane caused by neurotransmitter molecules released from a presynaptic cell

frontal lobe part of the cerebral cortex that contains the motor cortex and areas involved in planning, attention, and language

glia (also, glial cells) cells that provide support functions for neurons

gyrus (plural: gyri) ridged protrusions in the cortex

hippocampus brain structure in the temporal lobe involved in processing memories

hyperpolarization change in the membrane potential to a more negative value

hypothalamus brain structure that controls hormone release and body homeostasis

inhibitory postsynaptic potential (IPSP) hyperpolarization of a postsynaptic membrane caused by neurotransmitter molecules released from a presynaptic cell

limbic system connected brain areas that process emotion and motivation

long-term depression (LTD) prolonged decrease in synaptic coupling between a pre- and postsynaptic cell

long-term potentiation (LTP) prolonged increase in synaptic coupling between a pre-and postsynaptic cell

major depression mental illness characterized by prolonged periods of sadness

membrane potential difference in electrical potential between the inside and outside of a cell

meninge membrane that covers and protects the central nervous system

microglia glia that scavenge and degrade dead cells and protect the brain from invading microorganisms

myelin fatty substance produced by glia that insulates axons

neurodegenerative disorder nervous system disorder characterized by the progressive loss of neurological functioning, usually caused by neuron death

neuron specialized cell that can receive and transmit electrical and chemical signals

nodes of Ranvier gaps in the myelin sheath where the

signal is recharged

norepinephrine neurotransmitter and hormone released by activation of the sympathetic nervous system

occipital lobe part of the cerebral cortex that contains visual cortex and processes visual stimuli

oligodendrocyte glial cell that myelinates central nervous system neuron axons

parasympathetic nervous system division of autonomic nervous system that regulates visceral functions during rest and digestion

parietal lobe part of the cerebral cortex involved in processing touch and the sense of the body in space

Parkinson's disease neurodegenerative disorder that affects the control of movement

pia mater thin membrane layer directly covering the brain and spinal cord

proprioception sense about how parts of the body are oriented in space

radial glia glia that serve as scaffolds for developing neurons as they migrate to their final destinations

refractory period period after an action potential when it is more difficult or impossible for an action potential to be fired; caused by inactivation of sodium channels and activation of additional potassium channels of the membrane

saltatory conduction "jumping" of an action potential along an axon from one node of Ranvier to the next

satellite glia glial cell that provides nutrients and structural support for neurons in the peripheral nervous system

schizophrenia mental disorder characterized by the inability to accurately perceive reality; patients often have difficulty thinking clearly and can suffer from delusions

Schwann cell glial cell that creates myelin sheath around a peripheral nervous system neuron axon

sensory-somatic nervous system system of sensory and motor nerves

somatosensation sense of touch

spinal cord thick fiber bundle that connects the brain with peripheral nerves; transmits sensory and motor information; contains neurons that control motor reflexes

spinal nerve nerve projecting between skin or muscle and spinal cord

sulcus (plural: sulci) indents or "valleys" in the cortex

summation process of multiple presynaptic inputs creating EPSPs around the same time for the postsynaptic neuron to be sufficiently depolarized to fire an action potential

sympathetic nervous system division of autonomic nervous system activated during stressful "fight or flight" situations

synapse junction between two neurons where neuronal signals are communicated

synaptic cleft space between the presynaptic and postsynaptic membranes

synaptic vesicle spherical structure that contains a neurotransmitter

temporal lobe part of the cerebral cortex that processes auditory input; parts of the temporal lobe are involved in speech, memory, and emotion processing

thalamus brain area that relays sensory information to the cortex

threshold of excitation level of depolarization needed for an action potential to fire

ventricle cavity within brain that contains cerebrospinal fluid

CHAPTER SUMMARY

35.1 Neurons and Glial Cells

The nervous system is made up of neurons and glia. Neurons are specialized cells that are capable of sending electrical as well as chemical signals. Most neurons contain dendrites, which receive these signals, and axons that send signals to other neurons or tissues. There are four main types of neurons: unipolar, bipolar, multipolar, and pseudounipolar neurons. Glia are non-neuronal cells in the nervous system that support neuronal development and signaling. There are several types of glia that serve different functions.

35.2 How Neurons Communicate

Neurons have charged membranes because there are different concentrations of ions inside and outside of the cell. Voltage-gated ion channels control the movement of ions into and out of a neuron. When a neuronal membrane is depolarized to at least the threshold of excitation, an action potential is fired. The action potential is then propagated along a myelinated axon to the axon terminals. In a chemical synapse, the action potential causes release of neurotransmitter molecules into the synaptic cleft. Through binding to postsynaptic receptors, the neurotransmitter can cause excitatory or inhibitory postsynaptic potentials by depolarizing or hyperpolarizing, respectively, the postsynaptic membrane. In electrical synapses, the action potential is directly communicated to the postsynaptic cell through gap junctions—large channel proteins that connect the pre-and postsynaptic membranes. Synapses are not static structures and can be strengthened and weakened. Two mechanisms of synaptic plasticity are long-term potentiation and long-term depression.

35.3 The Central Nervous System

The vertebrate central nervous system contains the brain and the spinal cord, which are covered and protected by three meninges. The brain contains structurally and functionally defined regions. In mammals, these include the cortex (which can be broken down into four primary functional lobes: frontal, temporal, occipital, and parietal), basal ganglia, thalamus, hypothalamus, limbic system, cerebellum, and brainstem—although structures in some of these designations overlap. While functions may be primarily localized to one structure in the brain, most complex functions, like language and sleep, involve neurons in multiple brain regions. The spinal cord is the information superhighway that connects the brain with the rest of the body through its connections with peripheral nerves. It transmits sensory and motor input and also controls motor reflexes.

35.4 The Peripheral Nervous System

The peripheral nervous system contains both the autonomic and sensory-somatic nervous systems. The autonomic nervous system provides unconscious control over visceral functions and has two divisions: the sympathetic and parasympathetic nervous systems. The sympathetic nervous system is activated in stressful situations to prepare the animal for a "fight or flight" response. The parasympathetic nervous system is active during restful periods. The sensory-somatic nervous system is made of cranial and spinal nerves that transmit sensory information from skin and muscle to the CNS and motor commands from the CNS to the muscles.

35.5 Nervous System Disorders

Some general themes emerge from the sampling of nervous system disorders presented above. The causes for most disorders are not fully understood—at least not for all patients—and likely involve a combination of nature (genetic mutations that become risk factors) and nurture (emotional trauma, stress, hazardous chemical exposure). Because the causes have yet to be fully determined, treatment options are often lacking and only address symptoms.

VISUAL CONNECTION QUESTIONS

1. Figure 35.3 Which of the following statements is false?
   a. The soma is the cell body of a nerve cell.
   b. Myelin sheath provides an insulating layer to the dendrites.
   c. Axons carry the signal from the soma to the target.
   d. Dendrites carry the signal to the soma.

2. Figure 35.11 Potassium channel blockers, such as amiodarone and procainamide, which are used to treat abnormal electrical activity in the heart, called cardiac dysrhythmia, impede the movement of K+ through voltage-gated K+ channels. Which part of the action potential would you expect potassium channels to affect?

3. Figure 35.26 Which of the following statements is false?
   a. The parasympathetic pathway is responsible for relaxing the body, while the sympathetic pathway is responsible for preparing for an emergency.
   b. Most preganglionic neurons in the sympathetic pathway originate in the spinal cord.
   c. Slowing of the heartbeat is a parasympathetic response.
   d. Parasympathetic neurons are responsible for releasing norepinephrine on the target organ, while sympathetic neurons are responsible for releasing acetylcholine.

REVIEW QUESTIONS

4. Neurons contain ________, which can receive signals from other neurons.
   a. axons
   b. mitochondria
   c. dendrites
   d. Golgi bodies

5. A(n) ________ neuron has one axon and one dendrite extending directly from the cell body.
   a. unipolar
   b. bipolar
   c. multipolar
   d. pseudounipolar

6. Glia that provide myelin for neurons in the brain are called ________.
   a. Schwann cells
   b. oligodendrocytes
   c. microglia
   d. astrocytes

7. Meningitis is a viral or bacterial infection of the brain. Which cell type is the first to have its function disrupted during meningitis?
   a. astrocytes
   b. microglia
   c. neurons
   d. satellite glia

8. For a neuron to fire an action potential, its membrane must reach _______.
   a. hyperpolarization
   b. the threshold of excitation
   c. the refractory period
   d. inhibitory postsynaptic potential

9. After an action potential, the opening of additional voltage-gated _______ channels and the inactivation of sodium channels, cause the membrane to return to its resting membrane potential.
   a. sodium
   b. potassium
   c. calcium
   d. chloride

10. What is the term for protein channels that connect two neurons at an electrical synapse?
   a. synaptic vesicles
   b. voltage-gated ion channels
   c. gap junction protein
   d. sodium-potassium exchange pumps

11. Which of the following molecules is **not** involved in the maintenance of the resting membrane potential?
   a. potassium cations
   b. ATP
   c. voltage-gated ion channels
   d. calcium cations

12. The _______ lobe contains the visual cortex.
   a. frontal
   b. parietal
   c. temporal
   d. occipital

13. The _______ connects the two cerebral hemispheres.
   a. limbic system
   b. corpus callosum
   c. cerebellum
   d. pituitary

14. Neurons in the _______ control motor reflexes.
   a. thalamus
   b. spinal cord
   c. parietal lobe
   d. hippocampus

15. Phineas Gage was a 19th century railroad worker who survived an accident that drove a large iron rod through his head. If the injury resulted in him becoming temperamental and capricious what part of his brain was damaged?
   a. frontal lobe
   b. hippocampus
   c. parietal lobe
   d. temporal lobe

16. Activation of the sympathetic nervous system causes:
   a. increased blood flow into the skin
   b. a decreased heart rate
   c. an increased heart rate
   d. increased digestion

17. Where are parasympathetic preganglionic cell bodies located?
   a. cerebellum
   b. brainstem
   c. dorsal root ganglia
   d. skin

18. _______ is released by motor nerve endings onto muscle.
   a. Acetylcholine
   b. Norepinephrine
   c. Dopamine
   d. Serotonin

19. Parkinson's disease is a caused by the degeneration of neurons that release _______.
   a. serotonin
   b. dopamine
   c. glutamate
   d. norepinephrine

20. _______ medications are often used to treat patients with ADHD.
   a. Tranquilizer
   b. Antibiotic
   c. Stimulant
   d. Anti-seizure

21. Strokes are often caused by _______.
   a. neurodegeneration
   b. blood clots or burst blood vessels
   c. seizures
   d. viruses

22. Why is it difficult to identify the cause of many nervous system disorders?
    a. The genes associated with the diseases are not known.
    b. There are no obvious defects in brain structure.
    c. The onset and display of symptoms varies between patients.
    d. all of the above

23. Why do many patients with neurodevelopmental disorders develop secondary disorders?
    a. Their genes predispose them to schizophrenia.
    b. Stimulant medications cause new behavioral disorders.
    c. Behavioral therapies only improve neurodevelopmental disorders.
    d. Dysfunction in the brain can affect many aspects of the body.

CRITICAL THINKING QUESTIONS

24. How are neurons similar to other cells? How are they unique?
25. Multiple sclerosis causes demyelination of axons in the brain and spinal cord. Why is this problematic?
26. Many neurons have only a single axon, but many terminals at the end of the axon. How does this end structure of the axon support its function?
27. How does myelin aid propagation of an action potential along an axon? How do the nodes of Ranvier help this process?
28. What are the main steps in chemical neurotransmission?
29. Describe how long-term potentiation can lead to a nicotine addiction.
30. What methods can be used to determine the function of a particular brain region?
31. What are the main functions of the spinal cord?
32. Alzheimer's disease involves three of the four lobes of the brain. Identify one of the involved lobes and describe the lobe's symptoms associated with the disease.
33. What are the main differences between the sympathetic and parasympathetic branches of the autonomic nervous system?
34. What are the main functions of the sensory-somatic nervous system?
35. Describe how the sensory-somatic nervous system reacts by reflex to a person touching something hot. How does this allow for rapid responses in potentially dangerous situations?
36. Scientists have suggested that the autonomic nervous system is not well-adapted to modern human life. How is the sympathetic nervous system an ineffective response to the everyday challenges faced by modern humans?
37. What are the main symptoms of Alzheimer's disease?
38. What are possible treatments for patients with major depression?

CHAPTER 36
Sensory Systems

Figure 36.1 This shark uses its senses of sight, vibration (lateral-line system), and smell to hunt, but it also relies on its ability to sense the electric fields of prey, a sense not present in most land animals. (credit: modification of work by Hermanus Backpackers Hostel, South Africa)

INTRODUCTION In more advanced animals, the senses are constantly at work, making the animal aware of stimuli—such as light, or sound, or the presence of a chemical substance in the external environment—and monitoring information about the organism's internal environment. All bilaterally symmetric animals have a sensory system, and the development of any species' sensory system has been driven by natural selection; thus, sensory systems differ among species according to the demands of their environments. The shark, unlike most fish predators, is electrosensitive—that is, sensitive to electrical fields produced by other animals in its environment. While it is helpful to this underwater predator, electrosensitivity is a sense not found in most land animals.

Chapter Outline

36.1 Sensory Processes

By the end of this section, you will be able to do the following:

- Identify the general and special senses in humans
- Describe three important steps in sensory perception
- Explain the concept of just-noticeable difference in sensory perception

Senses provide information about the body and its environment. Humans have five special senses: olfaction (smell), gustation (taste), equilibrium (balance and body position), vision, and hearing. Additionally, we possess general senses, also called somatosensation, which respond to stimuli like temperature, pain, pressure, and vibration. **Vestibular sensation**, which is an organism's sense of spatial orientation and balance, **proprioception** (position of bones, joints, and muscles), and the sense of limb position that is used to track **kinesthesia** (limb movement) are part of somatosensation. Although the sensory systems associated with these senses are very different, all share a common function: to convert a stimulus (such as light, or sound, or the position of the body) into an electrical signal in the nervous system. This process is called **sensory transduction**.

There are two broad types of cellular systems that perform sensory transduction. In one, a neuron works with a **sensory receptor**, a cell, or cell process that is specialized to engage with and detect a specific stimulus. Stimulation of the sensory receptor activates the associated afferent neuron, which carries information about the stimulus to the central nervous system. In the second type of sensory transduction, a sensory nerve ending responds to a stimulus in the internal or external environment: this neuron constitutes the sensory receptor. Free nerve endings can be stimulated by several different stimuli, thus showing little receptor specificity. For example, pain receptors in your gums and teeth may be stimulated by temperature changes, chemical stimulation, or pressure.

Reception

The first step in sensation is **reception**, which is the activation of sensory receptors by stimuli such as mechanical stimuli (being bent or squished, for example), chemicals, or temperature. The receptor can then respond to the stimuli. The region in space in which a given sensory receptor can respond to a stimulus, be it far away or in contact with the body, is that receptor's **receptive field**. Think for a moment about the differences in receptive fields for the different senses. For the sense of touch, a stimulus must come into contact with the body. For the sense of hearing, a stimulus can be a moderate distance away (some baleen whale sounds can propagate for many kilometers). For vision, a stimulus can be very far away; for example, the visual system perceives light from stars at enormous distances.

Transduction

The most fundamental function of a sensory system is the translation of a sensory signal to an electrical signal in the nervous system. This takes place at the sensory receptor, and the change in electrical potential that is produced is called the **receptor potential**. How is sensory input, such as pressure on the skin, changed to a receptor potential? In this example, a type of receptor called a **mechanoreceptor** (as shown in Figure 36.2) possesses specialized membranes that respond to pressure. Disturbance of these dendrites by compressing them or bending them opens gated ion channels in the plasma membrane of the sensory neuron, changing its electrical potential. Recall that in the nervous system, a positive change of a neuron's electrical potential (also called the membrane potential), depolarizes the neuron. Receptor potentials are graded potentials: the magnitude of these graded (receptor) potentials varies with the strength of the stimulus. If the magnitude of depolarization is sufficient (that is, if membrane potential reaches a threshold), the neuron will fire an action potential. In most cases, the correct stimulus impinging on a sensory receptor will drive membrane potential in a positive direction, although for some receptors, such as those in the visual system, this is not always the case.

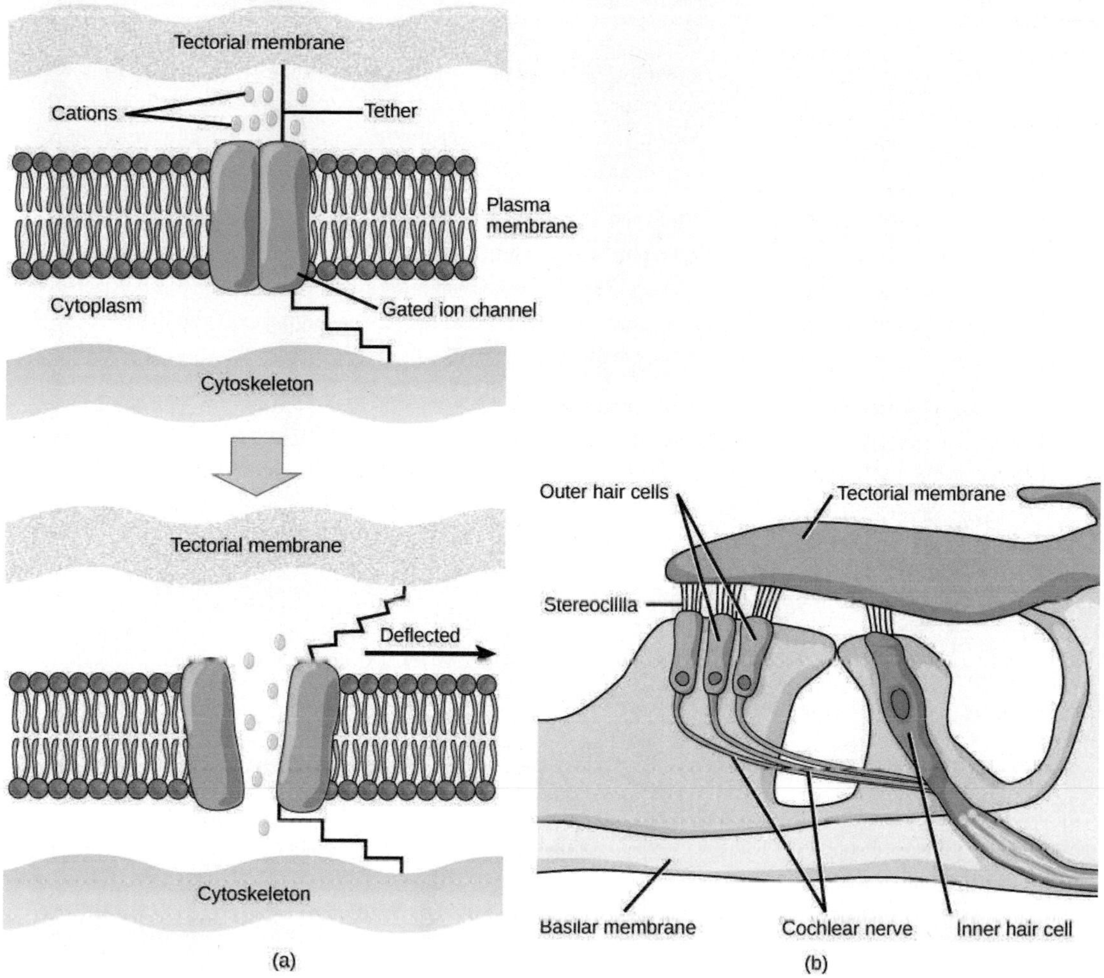

Figure 36.2 (a) Mechanosensitive ion channels are gated ion channels that respond to mechanical deformation of the plasma membrane. A mechanosensitive channel is connected to the plasma membrane and the cytoskeleton by hair-like tethers. When pressure causes the extracellular matrix to move, the channel opens, allowing ions to enter or exit the cell. (b) Stereocilia in the human ear are connected to mechanosensitive ion channels. When a sound causes the stereocilia to move, mechanosensitive ion channels transduce the signal to the cochlear nerve.

Sensory receptors for different senses are very different from each other, and they are specialized according to the type of stimulus they sense: they have receptor specificity. For example, touch receptors, light receptors, and sound receptors are each activated by different stimuli. Touch receptors are not sensitive to light or sound; they are sensitive only to touch or pressure. However, stimuli may be combined at higher levels in the brain, as happens with olfaction, contributing to our sense of taste.

Encoding and Transmission of Sensory Information

Four aspects of sensory information are encoded by sensory systems: the type of stimulus, the location of the stimulus in the receptive field, the duration of the stimulus, and the relative intensity of the stimulus. Thus, action potentials transmitted over a sensory receptor's afferent axons encode one type of stimulus, and this segregation of the senses is preserved in other sensory circuits. For example, auditory receptors transmit signals over their own dedicated system, and electrical activity in the axons of the auditory receptors will be interpreted by the brain as an auditory stimulus—a sound.

The intensity of a stimulus is often encoded in the rate of action potentials produced by the sensory receptor. Thus, an intense stimulus will produce a more rapid train of action potentials, and reducing the stimulus will likewise slow the rate of production of action potentials. A second way in which intensity is encoded is by the number of receptors activated. An intense stimulus might initiate action potentials in a large number of adjacent receptors, while a less intense stimulus might stimulate fewer receptors. Integration of sensory information begins as soon as the information is received in the CNS, and the brain will further process incoming signals.

Perception

Perception is an individual's interpretation of a sensation. Although perception relies on the activation of sensory receptors, perception happens not at the level of the sensory receptor, but at higher levels in the nervous system, in the brain. The brain distinguishes sensory stimuli through a sensory pathway: action potentials from sensory receptors travel along neurons that are dedicated to a particular stimulus. These neurons are dedicated to that particular stimulus and synapse with particular neurons in the brain or spinal cord.

All sensory signals, except those from the olfactory system, are transmitted though the central nervous system and are routed to the thalamus and to the appropriate region of the cortex. Recall that the thalamus is a structure in the forebrain that serves as a clearinghouse and relay station for sensory (as well as motor) signals. When the sensory signal exits the thalamus, it is conducted to the specific area of the cortex (Figure 36.3) dedicated to processing that particular sense.

How are neural signals interpreted? Interpretation of sensory signals between individuals of the same species is largely similar, owing to the inherited similarity of their nervous systems; however, there are some individual differences. A good example of this is individual tolerances to a painful stimulus, such as dental pain, which certainly differ.

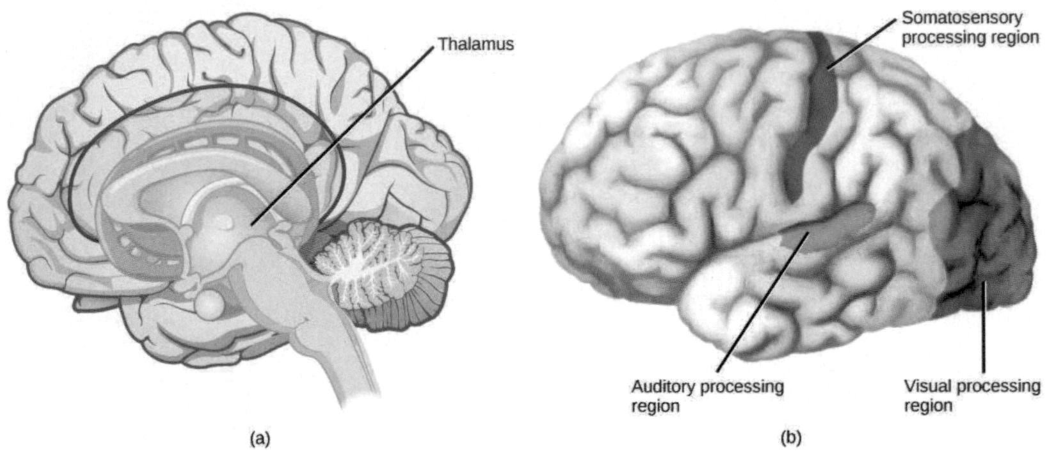

Figure 36.3 In humans, with the exception of olfaction, all sensory signals are routed from the (a) thalamus to (b) final processing regions in the cortex of the brain. (credit b: modification of work by Polina Tishina)

SCIENTIFIC METHOD CONNECTION

Just-Noticeable Difference

It is easy to differentiate between a one-pound bag of rice and a two-pound bag of rice. There is a one-pound difference, and one bag is twice as heavy as the other. However, would it be as easy to differentiate between a 20- and a 21-pound bag?

Question: What is the smallest detectible weight difference between a one-pound bag of rice and a larger bag? What is the smallest detectible difference between a 20-pound bag and a larger bag? In both cases, at what weights are the differences detected? This smallest detectible difference in stimuli is known as the just-noticeable difference (JND).

Background: Research background literature on JND and on Weber's Law, a description of a proposed mathematical relationship between the overall magnitude of the stimulus and the JND. You will be testing JND of different weights of rice in bags. Choose a convenient increment that is to be stepped through while testing. For example, you could choose 10 percent increments between one and two pounds (1.1, 1.2, 1.3, 1.4, and so on) or 20 percent increments (1.2, 1.4, 1.6, and 1.8).

Hypothesis: Develop a hypothesis about JND in terms of percentage of the whole weight being tested (such as "the JND between the two small bags and between the two large bags is proportionally the same," or ". . . is not proportionally the same.") So, for the first hypothesis, if the JND between the one-pound bag and a larger bag is 0.2 pounds (that is, 20 percent; 1.0 pound feels the same as 1.1 pounds, but 1.0 pound feels less than 1.2 pounds), then the JND between the 20-pound bag and a larger bag will also be 20 percent. (So, 20 pounds feels the same as 22 pounds or 23 pounds, but 20 pounds feels less than 24 pounds.)

Test the hypothesis: Enlist 24 participants, and split them into two groups of 12. To set up the demonstration, assuming a 10 percent increment was selected, have the first group be the one-pound group. As a counter-balancing measure against a systematic error, however, six of the first group will compare one pound to two pounds, and step down in weight (1.0 to 2.0, 1.0 to 1.9, and so on), while the other six will step up (1.0 to 1.1, 1.0 to 1.2, and so on). Apply the same principle to the 20-pound group (20 to 40, 20 to 38, and so on, and 20 to 22, 20 to 24, and so on). Given the large difference between 20 and 40 pounds, you may wish to use 30 pounds as your larger weight. In any case, use two weights that are easily detectable as different.

Record the observations: Record the data in a table similar to the table below. For the one-pound and 20-pound groups (base weights) record a plus sign (+) for each participant that detects a difference between the base weight and the step weight. Record a minus sign (-) for each participant that finds no difference. If one-tenth steps were not used, then replace the steps in the "Step Weight" columns with the step you are using.

Results of JND Testing (+ = difference; – = no difference)

Step Weight	One pound	20 pounds	Step Weight
1.1			22
1.2			24
1.3			26
1.4			28
1.5			30
1.6			32
1.7			34
1.8			36
1.9			38
2.0			40

Table 36.1

Analyze the data/report the results: What step weight did all participants find to be equal with one-pound base weight? What about the 20-pound group?

Draw a conclusion: Did the data support the hypothesis? Are the final weights proportionally the same? If not, why not? Do the findings adhere to Weber's Law? Weber's Law states that the concept that a just-noticeable difference in a stimulus is proportional to the magnitude of the original stimulus.

36.2 Somatosensation

By the end of this section, you will be able to do the following:

- Describe four important mechanoreceptors in human skin
- Describe the topographical distribution of somatosensory receptors between glabrous and hairy skin
- Explain why the perception of pain is subjective

Somatosensation is a mixed sensory category and includes all sensation received from the skin and mucous membranes, as well from as the limbs and joints. Somatosensation is also known as tactile sense, or more familiarly, as the sense of touch. Somatosensation occurs all over the exterior of the body and at some interior locations as well. A variety of receptor types—embedded in the skin, mucous membranes, muscles, joints, internal organs, and cardiovascular system—play a role.

Recall that the epidermis is the outermost layer of skin in mammals. It is relatively thin, is composed of keratin-filled cells, and has no blood supply. The epidermis serves as a barrier to water and to invasion by pathogens. Below this, the much thicker dermis contains blood vessels, sweat glands, hair follicles, lymph vessels, and lipid-secreting sebaceous glands (Figure 36.4). Below the epidermis and dermis is the subcutaneous tissue, or hypodermis, the fatty layer that contains blood vessels, connective tissue, and the axons of sensory neurons. The hypodermis, which holds about 50 percent of the body's fat, attaches the dermis to the bone and muscle, and supplies nerves and blood vessels to the dermis.

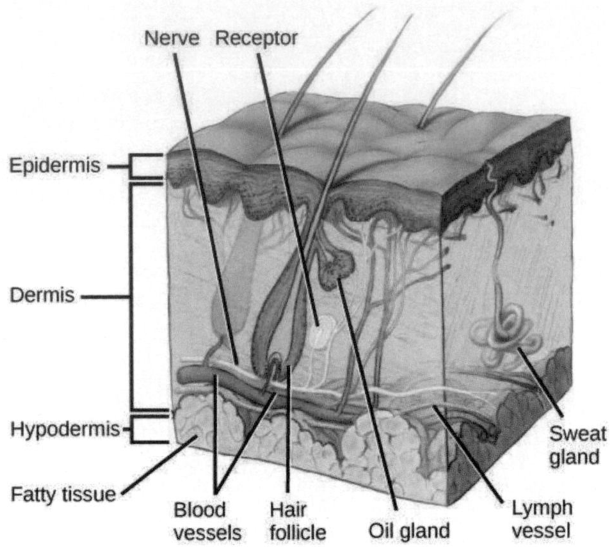

Figure 36.4 Mammalian skin has three layers: an epidermis, a dermis, and a hypodermis. (credit: modification of work by Don Bliss, National Cancer Institute)

Somatosensory Receptors

Sensory receptors are classified into five categories: mechanoreceptors, thermoreceptors, proprioceptors, pain receptors, and chemoreceptors. These categories are based on the nature of stimuli each receptor class transduces. What is commonly referred to as "touch" involves more than one kind of stimulus and more than one kind of receptor. Mechanoreceptors in the skin are described as encapsulated (that is, surrounded by a capsule) or unencapsulated (a group that includes free nerve endings). A **free nerve ending**, as its name implies, is an unencapsulated dendrite of a sensory neuron. Free nerve endings are the most common nerve endings in skin, and they extend into the middle of the epidermis. Free nerve endings are sensitive to painful stimuli, to hot and cold, and to light touch. They are slow to adjust to a stimulus and so are less sensitive to abrupt changes in stimulation.

There are three classes of mechanoreceptors: tactile, proprioceptors, and baroreceptors. Mechanoreceptors sense stimuli due to physical deformation of their plasma membranes. They contain mechanically gated ion channels whose gates open or close in response to pressure, touch, stretching, and sound." There are four primary tactile mechanoreceptors in human skin: Merkel's disks, Meissner's corpuscles, Ruffini endings, and Pacinian corpuscles; two are located toward the surface of the skin and two are located deeper. A fifth type of mechanoreceptor, Krause end bulbs, are found only in specialized regions. **Merkel's disks** (shown in Figure 36.5) are found in the upper layers of skin near the base of the epidermis, both in skin that has hair and on **glabrous** skin, that is, the hairless skin found on the palms and fingers, the soles of the feet, and the lips of humans and other primates. Merkel's disks are densely distributed in the fingertips and lips. They are slow-adapting, encapsulated nerve endings, and they respond to light touch. Light touch, also known as discriminative touch, is a light pressure that allows the location of a stimulus to be pinpointed. The receptive fields of Merkel's disks are small with well-defined borders. That makes them finely sensitive to edges and they come into use in tasks such as typing on a keyboard.

VISUAL CONNECTION

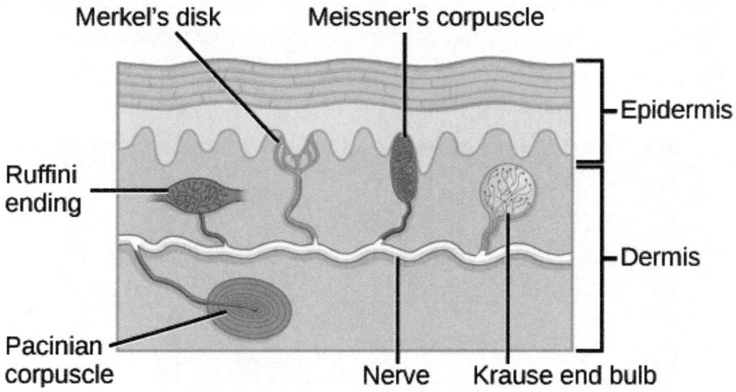

Figure 36.5 Four of the primary mechanoreceptors in human skin are shown. Merkel's disks, which are unencapsulated, respond to light touch. Meissner's corpuscles, Ruffini endings, Pacinian corpuscles, and Krause end bulbs are all encapsulated. Meissner's corpuscles respond to touch and low-frequency vibration. Ruffini endings detect stretch, deformation within joints, and warmth. Pacinian corpuscles detect transient pressure and high-frequency vibration. Krause end bulbs detect cold.

Which of the following statements about mechanoreceptors is false?

a. Pacinian corpuscles are found in both glabrous and hairy skin.
b. Merkel's disks are abundant on the fingertips and lips.
c. Ruffini endings are encapsulated mechanoreceptors.
d. Meissner's corpuscles extend into the lower dermis.

Meissner's corpuscles, (shown in Figure 36.6) also known as tactile corpuscles, are found in the upper dermis, but they project into the epidermis. They, too, are found primarily in the glabrous skin on the fingertips and eyelids. They respond to fine touch and pressure, but they also respond to low-frequency vibration or flutter. They are rapidly adapting, fluid-filled, encapsulated neurons with small, well-defined borders and are responsive to fine details. Like Merkel's disks, Meissner's corpuscles are not as plentiful in the palms as they are in the fingertips.

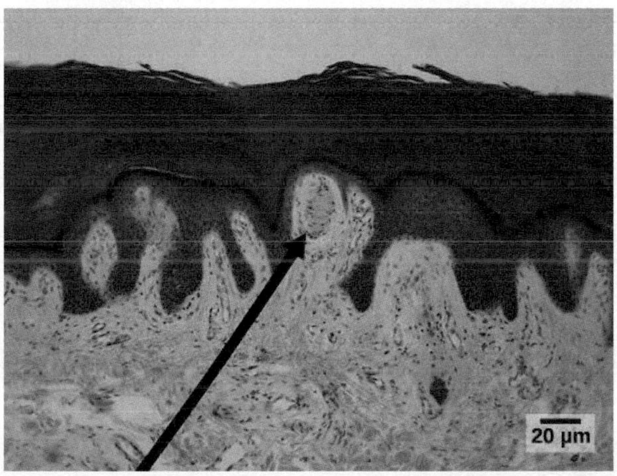

Figure 36.6 Meissner corpuscles in the fingertips, such as the one viewed here using bright field light microscopy, allow for touch discrimination of fine detail. (credit: modification of work by "Wbensmith"/Wikimedia Commons; scale-bar data from Matt Russell)

Deeper in the epidermis, near the base, are **Ruffini endings**, which are also known as bulbous corpuscles. They are found in both glabrous and hairy skin. These are slow-adapting, encapsulated mechanoreceptors that detect skin stretch and deformations within joints, so they provide valuable feedback for gripping objects and controlling finger position and movement. Thus, they also contribute to proprioception and kinesthesia. Ruffini endings also detect warmth. Note that these warmth detectors are situated deeper in the skin than are the cold detectors. It is not surprising, then, that humans detect cold stimuli before they detect warm stimuli.

Pacinian corpuscles (seen in Figure 36.7) are located deep in the dermis of both glabrous and hairy skin and are structurally similar to Meissner's corpuscles; they are found in the bone periosteum, joint capsules, pancreas and other viscera, breast, and genitals. They are rapidly adapting mechanoreceptors that sense deep transient (but not prolonged) pressure and high-frequency vibration. Pacinian receptors detect pressure and vibration by being compressed, stimulating their internal dendrites. There are fewer Pacinian corpuscles and Ruffini endings in skin than there are Merkel's disks and Meissner's corpuscles.

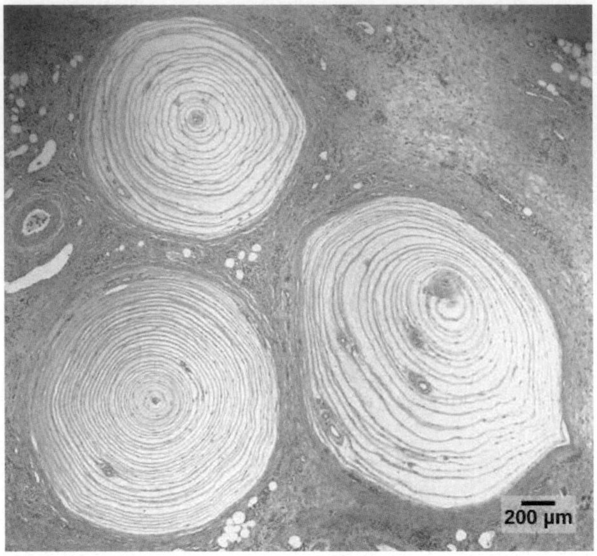

Figure 36.7 Pacinian corpuscles, such as these visualized using bright field light microscopy, detect pressure (touch) and high-frequency vibration. (credit: modification of work by Ed Uthman; scale-bar data from Matt Russell)

In proprioception, proprioceptive and kinesthetic signals travel through myelinated afferent neurons running from the spinal cord to the medulla. Neurons are not physically connected, but communicate via neurotransmitters secreted into synapses or "gaps" between communicating neurons. Once in the medulla, the neurons continue carrying the signals to the thalamus.

Muscle spindles are stretch receptors that detect the amount of stretch, or lengthening of muscles. Related to these are **Golgi tendon organs**, which are tension receptors that detect the force of muscle contraction. Proprioceptive and kinesthetic signals come from limbs. Unconscious proprioceptive signals run from the spinal cord to the cerebellum, the brain region that coordinates muscle contraction, rather than to the thalamus, like most other sensory information.

Baroreceptors detect pressure changes in an organ. They are found in the walls of the carotid artery and the aorta where they monitor blood pressure, and in the lungs where they detect the degree of lung expansion. Stretch receptors are found at various sites in the digestive and urinary systems.

In addition to these two types of deeper receptors, there are also rapidly adapting hair receptors, which are found on nerve endings that wrap around the base of hair follicles. There are a few types of hair receptors that detect slow and rapid hair movement, and they differ in their sensitivity to movement. Some hair receptors also detect skin deflection, and certain rapidly adapting hair receptors allow detection of stimuli that have not yet touched the skin.

Integration of Signals from Mechanoreceptors

The configuration of the different types of receptors working in concert in human skin results in a very refined sense of touch. The nociceptive receptors—those that detect pain—are located near the surface. Small, finely calibrated mechanoreceptors—Merkel's disks and Meissner's corpuscles—are located in the upper layers and can precisely localize even gentle touch. The large mechanoreceptors—Pacinian corpuscles and Ruffini endings—are located in the lower layers and respond to deeper touch. (Consider that the deep pressure that reaches those deeper receptors would not need to be finely localized.) Both the upper and lower layers of the skin hold rapidly and slowly adapting receptors. Both primary somatosensory cortex and secondary cortical areas are responsible for processing the complex picture of stimuli transmitted from the interplay of mechanoreceptors.

Density of Mechanoreceptors

The distribution of touch receptors in human skin is not consistent over the body. In humans, touch receptors are less dense in skin covered with any type of hair, such as the arms, legs, torso, and face. Touch receptors are denser in glabrous skin (the type found on human fingertips and lips, for example), which is typically more sensitive and is thicker than hairy skin (4 to 5 mm versus 2 to 3 mm).

How is receptor density estimated in a human subject? The relative density of pressure receptors in different locations on the body can be demonstrated experimentally using a two-point discrimination test. In this demonstration, two sharp points, such as two thumbtacks, are brought into contact with the subject's skin (though not hard enough to cause pain or break the skin). The subject reports if he or she feels one point or two points. If the two points are felt as one point, it can be inferred that the two points are both in the receptive field of a single sensory receptor. If two points are felt as two separate points, each is in the receptive field of two separate sensory receptors. The points could then be moved closer and retested until the subject reports feeling only one point, and the size of the receptive field of a single receptor could be estimated from that distance.

Thermoreception

In addition to Krause end bulbs that detect cold and Ruffini endings that detect warmth, there are different types of cold receptors on some free nerve endings: thermoreceptors, located in the dermis, skeletal muscles, liver, and hypothalamus, that are activated by different temperatures. Their pathways into the brain run from the spinal cord through the thalamus to the primary somatosensory cortex. Warmth and cold information from the face travels through one of the cranial nerves to the brain. You know from experience that a tolerably cold or hot stimulus can quickly progress to a much more intense stimulus that is no longer tolerable. Any stimulus that is too intense can be perceived as pain because temperature sensations are conducted along the same pathways that carry pain sensations.

Pain

Pain is the name given to **nociception**, which is the neural processing of injurious stimuli in response to tissue damage. Pain is caused by true sources of injury, such as contact with a heat source that causes a thermal burn or contact with a corrosive chemical. But pain also can be caused by harmless stimuli that mimic the action of damaging stimuli, such as contact with capsaicins, the compounds that cause peppers to taste hot and which are used in self-defense pepper sprays and certain topical medications. Peppers taste "hot" because the protein receptors that bind capsaicin open the same calcium channels that are activated by warm receptors.

Nociception starts at the sensory receptors, but pain, inasmuch as it is the perception of nociception, does not start until it is communicated to the brain. There are several nociceptive pathways to and through the brain. Most axons carrying nociceptive information into the brain from the spinal cord project to the thalamus (as do other sensory neurons) and the neural signal undergoes final processing in the primary somatosensory cortex. Interestingly, one nociceptive pathway projects not to the thalamus but directly to the hypothalamus in the forebrain, which modulates the cardiovascular and neuroendocrine functions of the autonomic nervous system. Recall that threatening—or painful—stimuli stimulate the sympathetic branch of the visceral sensory system, readying a fight-or-flight response.

LINK TO LEARNING

View this video (http://openstax.org/l/nociceptive) that animates the five phases of nociceptive pain.

Click to view content (https://www.openstax.org/l/nociceptive)

36.3 Taste and Smell

By the end of this section, you will be able to do the following:

- Explain in what way smell and taste stimuli differ from other sensory stimuli
- Identify the five primary tastes that can be distinguished by humans
- Explain in anatomical terms why a dog's sense of smell is more acute than a human's

Taste, also called **gustation**, and smell, also called **olfaction**, are the most interconnected senses in that both involve molecules of the stimulus entering the body and bonding to receptors. Smell lets an animal sense the presence of food or other animals—whether potential mates, predators, or prey—or other chemicals in the environment that can impact their survival. Similarly, the sense of taste allows animals to discriminate between types of foods. While the value of a sense of smell is obvious, what is the value of a sense of taste? Different tasting foods have different attributes, both helpful and harmful. For example, sweet-tasting substances tend to be highly caloric, which could be necessary for survival in lean times. Bitterness is associated with toxicity, and sourness is associated with spoiled food. Salty foods are valuable in maintaining homeostasis by helping the body retain water and by providing ions necessary for cells to function.

Tastes and Odors

Both taste and odor stimuli are molecules taken in from the environment. The primary tastes detected by humans are sweet, sour, bitter, salty, and umami. The first four tastes need little explanation. The identification of **umami** as a fundamental taste occurred fairly recently—it was identified in 1908 by Japanese scientist Kikunae Ikeda while he worked with seaweed broth, but it was not widely accepted as a taste that could be physiologically distinguished until many years later. The taste of umami, also known as savoriness, is attributable to the taste of the amino acid L-glutamate. In fact, monosodium glutamate, or MSG, is often used in cooking to enhance the savory taste of certain foods. What is the adaptive value of being able to distinguish umami? Savory substances tend to be high in protein.

All odors that we perceive are molecules in the air we breathe. If a substance does not release molecules into the air from its surface, it has no smell. And if a human or other animal does not have a receptor that recognizes a specific molecule, then that molecule has no smell. Humans have about 350 olfactory receptor subtypes that work in various combinations to allow us to sense about 10,000 different odors. Compare that to mice, for example, which have about 1,300 olfactory receptor types, and therefore probably sense more odors. Both odors and tastes involve molecules that stimulate specific chemoreceptors. Although humans commonly distinguish taste as one sense and smell as another, they work together to create the perception of flavor. A person's perception of flavor is reduced if he or she has congested nasal passages.

Reception and Transduction

Odorants (odor molecules) enter the nose and dissolve in the olfactory epithelium, the mucosa at the back of the nasal cavity (as illustrated in Figure 36.8). The **olfactory epithelium** is a collection of specialized olfactory receptors in the back of the nasal cavity that spans an area about 5 cm^2 in humans. Recall that sensory cells are neurons. An **olfactory receptor**, which is a dendrite of a specialized neuron, responds when it binds certain molecules inhaled from the environment by sending impulses directly to the olfactory bulb of the brain. Humans have about 12 million olfactory receptors, distributed among hundreds of different receptor types that respond to different odors. Twelve million seems like a large number of receptors, but compare that to other animals: rabbits have about 100 million, most dogs have about 1 billion, and bloodhounds—dogs selectively bred for their sense of smell—have about 4 billion. The overall size of the olfactory epithelium also differs between species, with that of bloodhounds, for example, being many times larger than that of humans.

Olfactory neurons are **bipolar neurons** (neurons with two processes from the cell body). Each neuron has a single dendrite buried in the olfactory epithelium, and extending from this dendrite are 5 to 20 receptor-laden, hair-like cilia that trap odorant molecules. The sensory receptors on the cilia are proteins, and it is the variations in their amino acid chains that make the receptors sensitive to different odorants. Each olfactory sensory neuron has only one type of receptor on its cilia, and the receptors are specialized to detect specific odorants, so the bipolar neurons themselves are specialized. When an odorant binds with a receptor that recognizes it, the sensory neuron associated with the receptor is stimulated. Olfactory stimulation is the only sensory information that directly reaches the cerebral cortex, whereas other sensations are relayed through the thalamus.

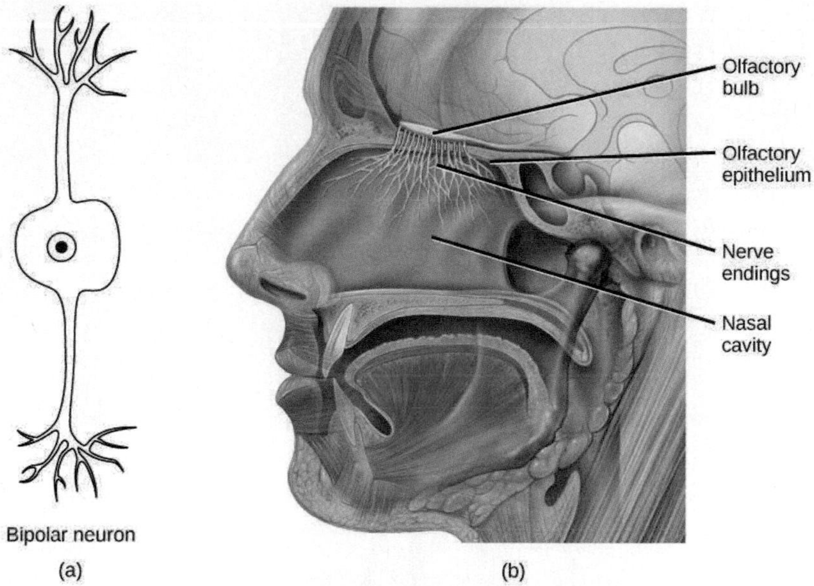

Figure 36.8 In the human olfactory system, (a) bipolar olfactory neurons extend from (b) the olfactory epithelium, where olfactory receptors are located, to the olfactory bulb. (credit: modification of work by Patrick J. Lynch, medical illustrator; C. Carl Jaffe, MD, cardiologist)

EVOLUTION CONNECTION

Pheromones

A **pheromone** is a chemical released by an animal that affects the behavior or physiology of animals of the same species. Pheromonal signals can have profound effects on animals that inhale them, but pheromones apparently are not consciously perceived in the same way as other odors. There are several different types of pheromones, which are released in urine or as glandular secretions. Certain pheromones are attractants to potential mates, others are repellants to potential competitors of the same sex, and still others play roles in mother-infant attachment. Some pheromones can also influence the timing of puberty, modify reproductive cycles, and even prevent embryonic implantation. While the roles of pheromones in many nonhuman species are important, pheromones have become less important in human behavior over evolutionary time compared to their importance to organisms with more limited behavioral repertoires.

The vomeronasal organ (VNO, or Jacobson's organ) is a tubular, fluid-filled, olfactory organ present in many vertebrate animals that sits adjacent to the nasal cavity. It is very sensitive to pheromones and is connected to the nasal cavity by a duct. When molecules dissolve in the mucosa of the nasal cavity, they then enter the VNO where the pheromone molecules among them bind with specialized pheromone receptors. Upon exposure to pheromones from their own species or others, many animals, including cats, may display the flehmen response (shown in Figure 36.9), a curling of the upper lip that helps pheromone molecules enter the VNO.

Pheromonal signals are sent, not to the main olfactory bulb, but to a different neural structure that projects directly to the amygdala (recall that the amygdala is a brain center important in emotional reactions, such as fear). The pheromonal signal then continues to areas of the hypothalamus that are key to reproductive physiology and behavior. While some scientists assert that the VNO is apparently functionally vestigial in humans, even though there is a similar structure located near human nasal cavities, others are researching it as a possible functional system that may, for example, contribute to synchronization of menstrual cycles in women living in close proximity.

Figure 36.9 The flehmen response in this tiger results in the curling of the upper lip and helps airborne pheromone molecules enter the vomeronasal organ. (credit: modification of work by "chadh"/Flickr)

Taste

Detecting a taste (gustation) is fairly similar to detecting an odor (olfaction), given that both taste and smell rely on chemical receptors being stimulated by certain molecules. The primary organ of taste is the taste bud. A **taste bud** is a cluster of gustatory receptors (taste cells) that are located within the bumps on the tongue called **papillae** (singular: papilla) (illustrated in Figure 36.11). There are several structurally distinct papillae. Filiform papillae, which are located across the tongue, are tactile, providing friction that helps the tongue move substances, and contain no taste cells. In contrast, fungiform papillae, which are located mainly on the anterior two-thirds of the tongue, each contain one to eight taste buds and also have receptors for pressure and temperature. The large circumvallate papillae contain up to 250 taste buds and form a V near the posterior margin of the tongue.

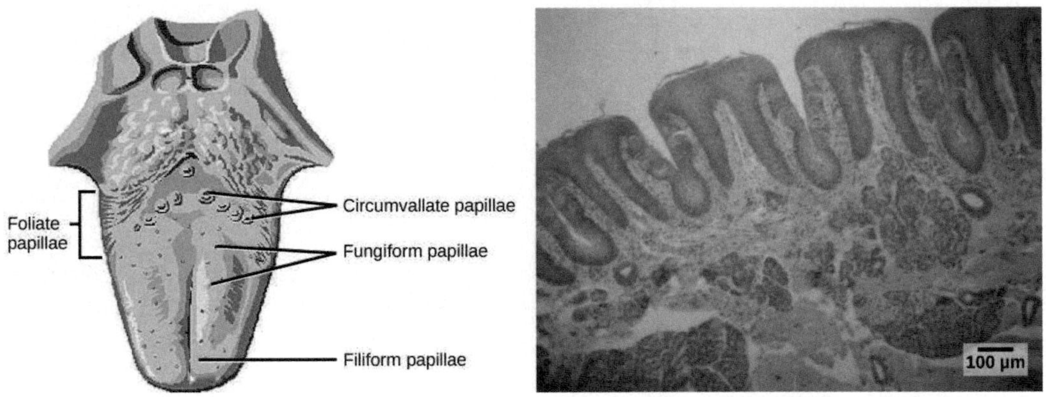

Figure 36.10 (a) Foliate, circumvallate, and fungiform papillae are located on different regions of the tongue. (b) Foliate papillae are prominent protrusions on this light micrograph. (credit a: modification of work by NCI; scale-bar data from Matt Russell)

In addition to those two types of chemically and mechanically sensitive papillae are foliate papillae—leaf-like papillae located in parallel folds along the edges and toward the back of the tongue, as seen in the Figure 36.10 micrograph. Foliate papillae contain about 1,300 taste buds within their folds. Finally, there are circumvallate papillae, which are wall-like papillae in the shape of an inverted "V" at the back of the tongue. Each of these papillae is surrounded by a groove and contains about 250 taste buds.

Each taste bud's taste cells are replaced every 10 to 14 days. These are elongated cells with hair-like processes called microvilli at the tips that extend into the taste bud pore (illustrated in Figure 36.11). Food molecules (**tastants**) are dissolved in saliva, and they bind with and stimulate the receptors on the microvilli. The receptors for tastants are located across the outer portion and front of the tongue, outside of the middle area where the filiform papillae are most prominent.

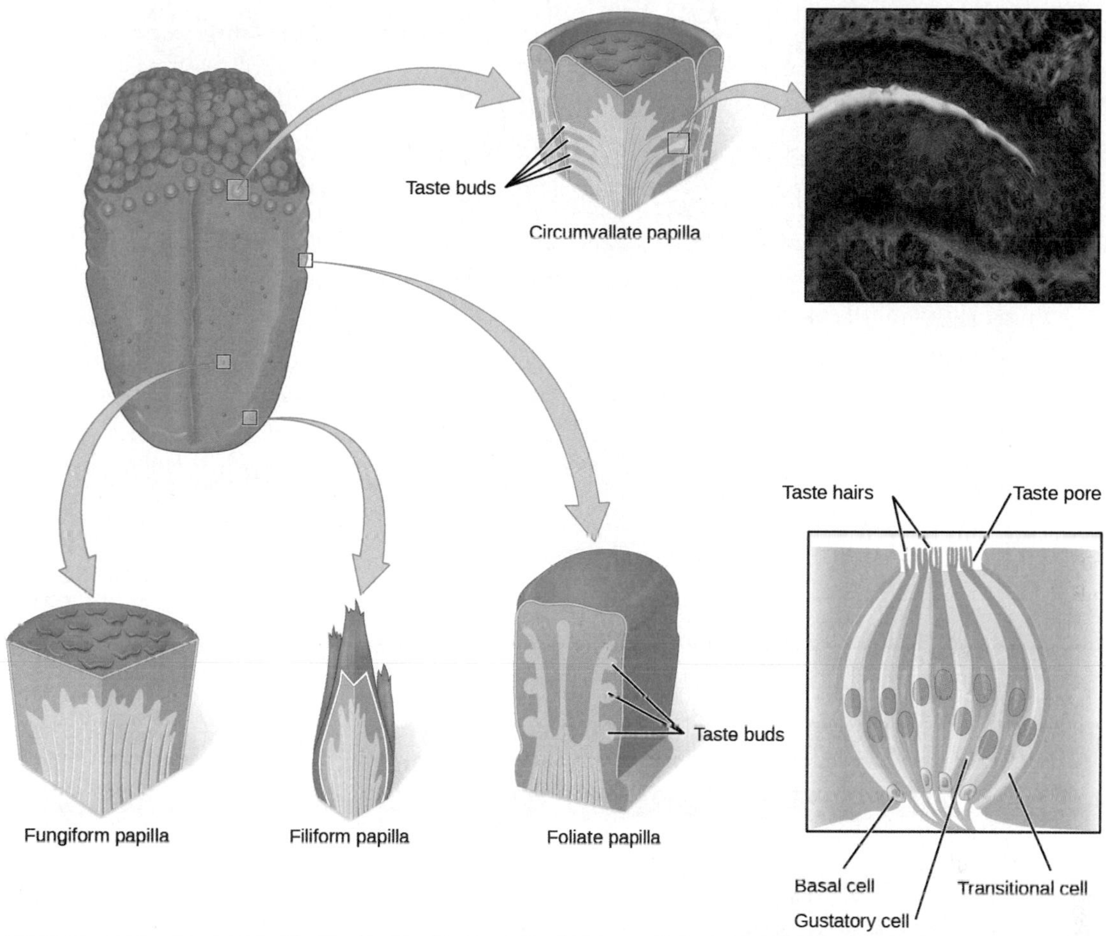

Figure 36.11 Pores in the tongue allow tastants to enter taste pores in the tongue. (credit: modification of work by Vincenzo Rizzo)

In humans, there are five primary tastes, and each taste has only one corresponding type of receptor. Thus, like olfaction, each receptor is specific to its stimulus (tastant). Transduction of the five tastes happens through different mechanisms that reflect the molecular composition of the tastant. A salty tastant (containing NaCl) provides the sodium ions (Na^+) that enter the taste neurons and excite them directly. Sour tastants are acids and belong to the thermoreceptor protein family. Binding of an acid or other sour-tasting molecule triggers a change in the ion channel and these increase hydrogen ion (H^+) concentrations in the taste neurons, thus depolarizing them. Sweet, bitter, and umami tastants require a G-protein coupled receptor. These tastants bind to their respective receptors, thereby exciting the specialized neurons associated with them.

Both tasting abilities and sense of smell change with age. In humans, the senses decline dramatically by age 50 and continue to decline. A child may find a food to be too spicy, whereas an elderly person may find the same food to be bland and unappetizing.

LINK TO LEARNING

View this animation (http://openstax.org/l/taste) that shows how the sense of taste works.

Smell and Taste in the Brain

Olfactory neurons project from the olfactory epithelium to the olfactory bulb as thin, unmyelinated axons. The **olfactory bulb** is composed of neural clusters called **glomeruli**, and each glomerulus receives signals from one type of olfactory receptor, so each glomerulus is specific to one odorant. From glomeruli, olfactory signals travel directly to the olfactory cortex and then to the frontal cortex and the thalamus. Recall that this is a different path from most other sensory information, which is sent directly to the thalamus before ending up in the cortex. Olfactory signals also travel directly to the amygdala, thereafter reaching the hypothalamus, thalamus, and frontal cortex. The last structure that olfactory signals directly travel to is a cortical center in the temporal lobe structure important in spatial, autobiographical, declarative, and episodic memories. Olfaction is finally processed by areas of the brain that deal with memory, emotions, reproduction, and thought.

Taste neurons project from taste cells in the tongue, esophagus, and palate to the medulla, in the brainstem. From the medulla, taste signals travel to the thalamus and then to the primary gustatory cortex. Information from different regions of the tongue is segregated in the medulla, thalamus, and cortex.

36.4 Hearing and Vestibular Sensation

By the end of this section, you will be able to do the following:

- Describe the relationship of amplitude and frequency of a sound wave to attributes of sound
- Trace the path of sound through the auditory system to the site of transduction of sound
- Identify the structures of the vestibular system that respond to gravity

Audition, or hearing, is important to humans and to other animals for many different interactions. It enables an organism to detect and receive information about danger, such as an approaching predator, and to participate in communal exchanges like those concerning territories or mating. On the other hand, although it is physically linked to the auditory system, the vestibular system is not involved in hearing. Instead, an animal's vestibular system detects its own movement, both linear and angular acceleration and deceleration, and balance.

Sound

Auditory stimuli are sound waves, which are mechanical, pressure waves that move through a medium, such as air or water. There are no sound waves in a vacuum since there are no air molecules to move in waves. The speed of sound waves differs, based on altitude, temperature, and medium, but at sea level and a temperature of 20° C (68° F), sound waves travel in the air at about 343 meters per second.

As is true for all waves, there are four main characteristics of a sound wave: frequency, wavelength, period, and amplitude. Frequency is the number of waves per unit of time, and in sound is heard as pitch. High-frequency (≥15.000Hz) sounds are higher-pitched (short wavelength) than low-frequency (long wavelengths; ≤100Hz) sounds. Frequency is measured in cycles per second, and for sound, the most commonly used unit is hertz (Hz), or cycles per second. Most humans can perceive sounds with frequencies between 30 and 20,000 Hz. Women are typically better at hearing high frequencies, but everyone's ability to hear high frequencies decreases with age. Dogs detect up to about 40,000 Hz; cats, 60,000 Hz; bats, 100,000 Hz; and dolphins 150,000 Hz, and American shad (*Alosa sapidissima*), a fish, can hear 180,000 Hz. Those frequencies above the human range are called **ultrasound**.

Amplitude, or the dimension of a wave from peak to trough, in sound is heard as volume and is illustrated in Figure 36.12. The sound waves of louder sounds have greater amplitude than those of softer sounds. For sound, volume is measured in decibels (dB). The softest sound that a human can hear is the zero point. Humans speak normally at 60 decibels.

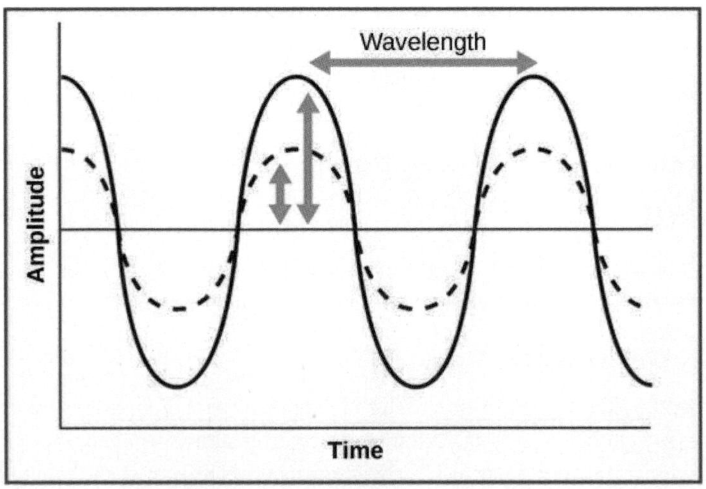

Figure 36.12 For sound waves, wavelength corresponds to pitch. Amplitude of the wave corresponds to volume. The sound wave shown

with a dashed line is softer in volume than the sound wave shown with a solid line. (credit: NIH)

Reception of Sound

In mammals, sound waves are collected by the external, cartilaginous part of the ear called the **pinna**, then travel through the auditory canal and cause vibration of the thin diaphragm called the **tympanum** or ear drum, the innermost part of the **outer ear** (illustrated in Figure 36.13). Interior to the tympanum is the **middle ear**. The middle ear holds three small bones called the **ossicles**, which transfer energy from the moving tympanum to the inner ear. The three ossicles are the **malleus** (also known as the hammer), the **incus** (the anvil), and **stapes** (the stirrup). The aptly named stapes looks very much like a stirrup. The three ossicles are unique to mammals, and each plays a role in hearing. The malleus attaches at three points to the interior surface of the tympanic membrane. The incus attaches the malleus to the stapes. In humans, the stapes is not long enough to reach the tympanum. If we did not have the malleus and the incus, then the vibrations of the tympanum would never reach the inner ear. These bones also function to collect force and amplify sounds. The ear ossicles are homologous to bones in a fish mouth: the bones that support gills in fish are thought to be adapted for use in the vertebrate ear over evolutionary time. Many animals (frogs, reptiles, and birds, for example) use the stapes of the middle ear to transmit vibrations to the middle ear.

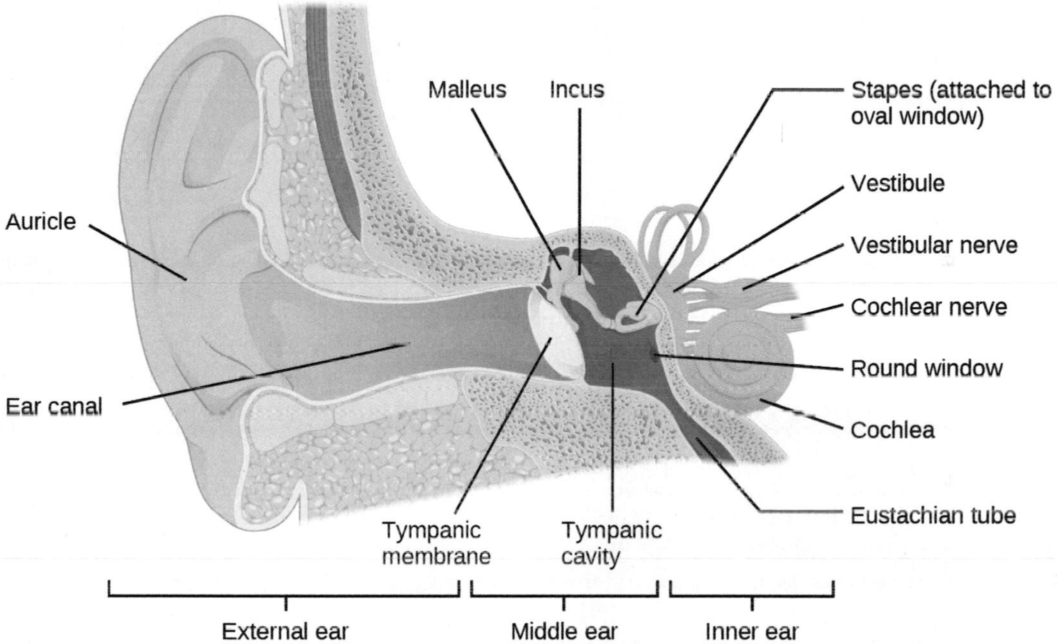

Figure 36.13 Sound travels through the outer ear to the middle ear, which is bounded on its exterior by the tympanic membrane. The middle ear contains three bones called ossicles that transfer the sound wave to the oval window, the exterior boundary of the inner ear. The organ of Corti, which is the organ of sound transduction, lies inside the cochlea.

Transduction of Sound

Vibrating objects, such as vocal cords, create sound waves or pressure waves in the air. When these pressure waves reach the ear, the ear transduces this mechanical stimulus (pressure wave) into a nerve impulse (electrical signal) that the brain perceives as sound. The pressure waves strike the tympanum, causing it to vibrate. The mechanical energy from the moving tympanum transmits the vibrations to the three bones of the middle ear. The stapes transmits the vibrations to a thin diaphragm called the **oval window**, which is the outermost structure of the **inner ear**. The structures of the inner ear are found in the **labyrinth**, a bony, hollow structure that is the most interior portion of the ear. Here, the energy from the sound wave is transferred from the stapes through the flexible oval window and to the fluid of the cochlea. The vibrations of the oval window create pressure waves in the fluid (perilymph) inside the cochlea. The **cochlea** is a whorled structure, like the shell of a snail, and it contains receptors for transduction of the mechanical wave into an electrical signal (as illustrated in Figure 36.14). Inside the cochlea, the **basilar membrane** is a mechanical analyzer that runs the length of the cochlea, curling toward the cochlea's center.

The mechanical properties of the basilar membrane change along its length, such that it is thicker, tauter, and narrower at the outside of the whorl (where the cochlea is largest), and thinner, floppier, and broader toward the apex, or center, of the whorl (where the cochlea is smallest). Different regions of the basilar membrane vibrate according to the frequency of the sound wave conducted through the fluid in the cochlea. For these reasons, the fluid-filled cochlea detects different wave frequencies (pitches) at different regions of the membrane. When the sound waves in the cochlear fluid contact the basilar membrane, it flexes back and forth in a wave-like fashion. Above the basilar membrane is the **tectorial membrane**.

VISUAL CONNECTION

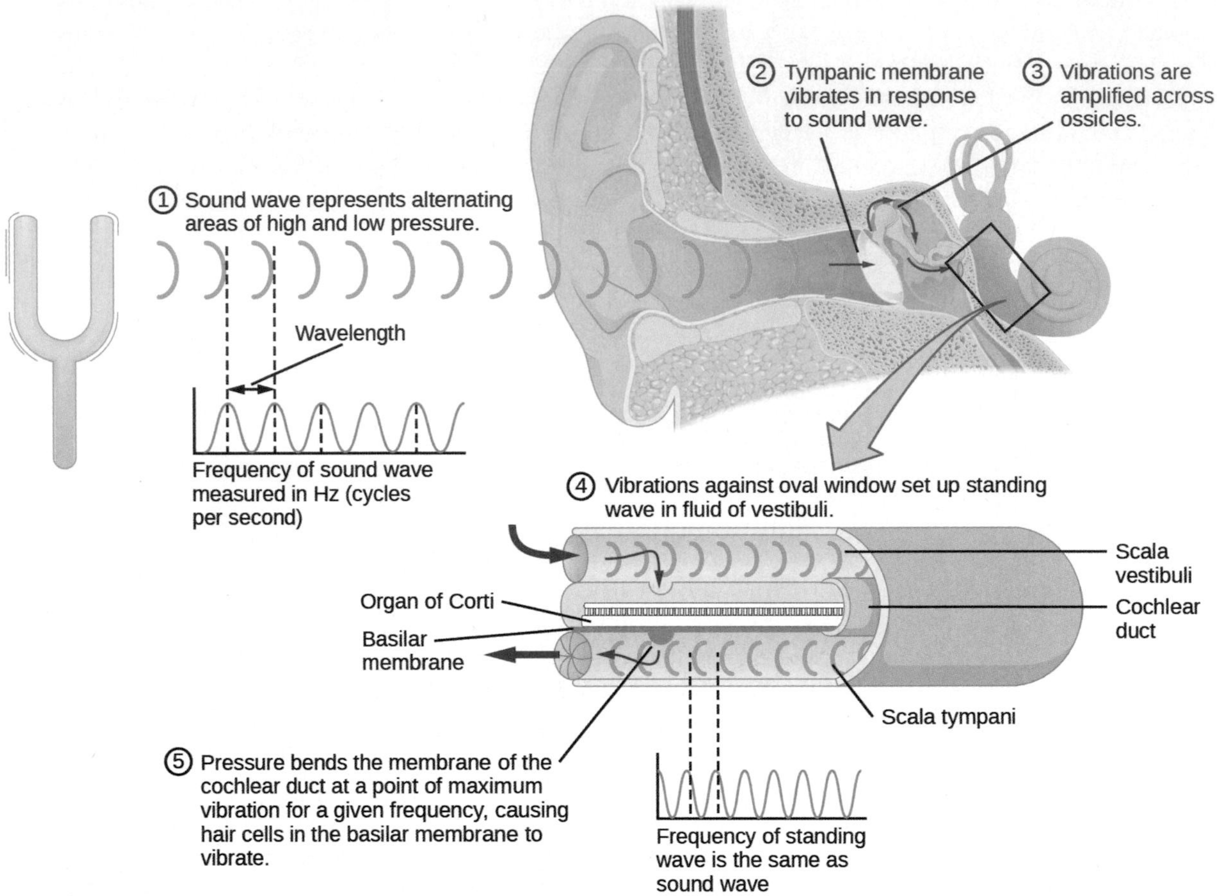

Figure 36.14 A sound wave causes the tympanic membrane to vibrate. This vibration is amplified as it moves across the malleus, incus, and stapes. The amplified vibration is picked up by the oval window causing pressure waves in the fluid of the scala vestibuli and scala tympani. The complexity of the pressure waves is determined by the changes in amplitude and frequency of the sound waves entering the ear.

Cochlear implants can restore hearing in people who have a nonfunctional cochlea. The implant consists of a microphone that picks up sound. A speech processor selects sounds in the range of human speech, and a transmitter converts these sounds to electrical impulses, which are then sent to the auditory nerve. Which of the following types of hearing loss would not be restored by a cochlear implant?

a. Hearing loss resulting from absence or loss of hair cells in the organ of Corti.
b. Hearing loss resulting from an abnormal auditory nerve.
c. Hearing loss resulting from fracture of the cochlea.
d. Hearing loss resulting from damage to bones of the middle ear.

The site of transduction is in the **organ of Corti** (spiral organ). It is composed of hair cells held in place above the basilar membrane like flowers projecting up from soil, with their exposed short, hair-like **stereocilia** contacting or embedded in the tectorial membrane above them. The inner hair cells are the primary auditory receptors and exist in a single row, numbering approximately 3,500. The stereocilia from inner hair cells extend into small dimples on the tectorial membrane's lower surface. The outer hair cells are arranged in three or four rows. They number approximately 12,000, and they function to fine tune incoming sound waves. The longer stereocilia that project from the outer hair cells actually attach to the tectorial membrane. All of the stereocilia are mechanoreceptors, and when bent by vibrations they respond by opening a gated ion channel (refer to Figure 36.15). As a result, the hair cell membrane is depolarized, and a signal is transmitted to the chochlear nerve. Intensity (volume) of sound is determined by how many hair cells at a particular location are stimulated.

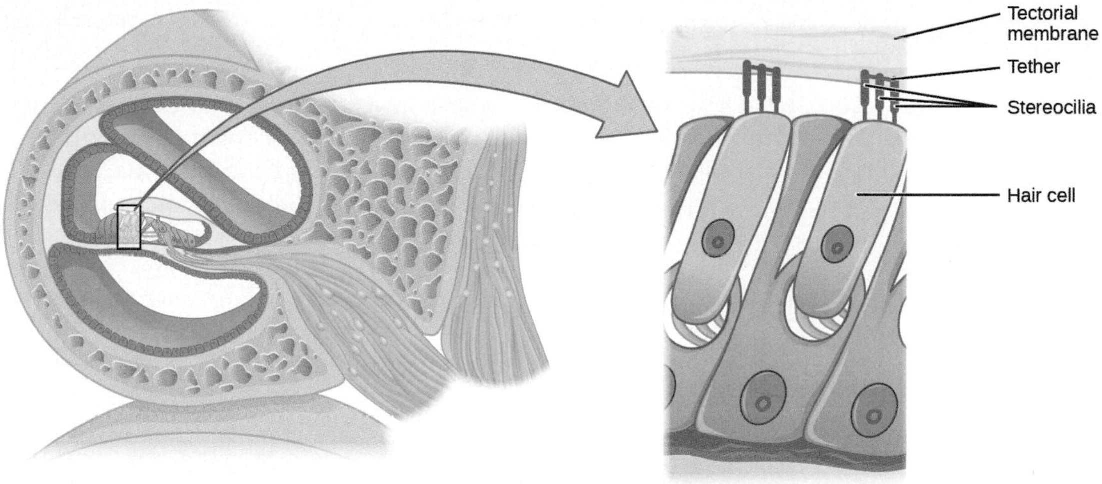

Figure 36.15 The hair cell is a mechanoreceptor with an array of stereocilia emerging from its apical surface. The stereocilia are tethered together by proteins that open ion channels when the array is bent toward the tallest member of their array, and closed when the array is bent toward the shortest member of their array.

The hair cells are arranged on the basilar membrane in an orderly way. The basilar membrane vibrates in different regions, according to the frequency of the sound waves impinging on it. Likewise, the hair cells that lay above it are most sensitive to a specific frequency of sound waves. Hair cells can respond to a small range of similar frequencies, but they require stimulation of greater intensity to fire at frequencies outside of their optimal range. The difference in response frequency between adjacent inner hair cells is about 0.2 percent. Compare that to adjacent piano strings, which are about six percent different. Place theory, which is the model for how biologists think pitch detection works in the human ear, states that high frequency sounds selectively vibrate the basilar membrane of the inner ear near the entrance port (the oval window). Lower frequencies travel farther along the membrane before causing appreciable excitation of the membrane. The basic pitch-determining mechanism is based on the location along the membrane where the hair cells are stimulated. The place theory is the first step toward an understanding of pitch perception. Considering the extreme pitch sensitivity of the human ear, it is thought that there must be some auditory "sharpening" mechanism to enhance the pitch resolution.

When sound waves produce fluid waves inside the cochlea, the basilar membrane flexes, bending the stereocilia that attach to the tectorial membrane. Their bending results in action potentials in the hair cells, and auditory information travels along the neural endings of the bipolar neurons of the hair cells (collectively, the auditory nerve) to the brain. When the hairs bend, they release an excitatory neurotransmitter at a synapse with a sensory neuron, which then conducts action potentials to the central nervous system. The cochlear branch of the vestibulocochlear cranial nerve sends information on hearing. The auditory system is very refined, and there is some modulation or "sharpening" built in. The brain can send signals back to the cochlea, resulting in a change of length in the outer hair cells, sharpening or dampening the hair cells' response to certain frequencies.

LINK TO LEARNING

Watch an animation (http://openstax.org/l/hearing) of sound entering the outer ear, moving through the ear structure, stimulating cochlear nerve impulses, and eventually sending signals to the temporal lobe.

Higher Processing

The inner hair cells are most important for conveying auditory information to the brain. About 90 percent of the afferent neurons carry information from inner hair cells, with each hair cell synapsing with 10 or so neurons. Outer hair cells connect to only 10 percent of the afferent neurons, and each afferent neuron innervates many hair cells. The afferent, bipolar neurons that convey auditory information travel from the cochlea to the medulla, through the pons and midbrain in the brainstem, finally reaching the primary auditory cortex in the temporal lobe.

Vestibular Information

The stimuli associated with the vestibular system are linear acceleration (gravity) and angular acceleration and deceleration. Gravity, acceleration, and deceleration are detected by evaluating the inertia on receptive cells in the vestibular system. Gravity is detected through head position. Angular acceleration and deceleration are expressed through turning or tilting of the head.

The vestibular system has some similarities with the auditory system. It utilizes hair cells just like the auditory system, but it excites them in different ways. There are five vestibular receptor organs in the inner ear: the utricle, the saccule, and three semicircular canals. Together, they make up what's known as the vestibular labyrinth that is shown in Figure 36.16. The utricle and saccule respond to acceleration in a straight line, such as gravity. The roughly 30,000 hair cells in the utricle and 16,000 hair cells in the saccule lie below a gelatinous layer, with their stereocilia projecting into the gelatin. Embedded in this gelatin are calcium carbonate crystals—like tiny rocks. When the head is tilted, the crystals continue to be pulled straight down by gravity, but the new angle of the head causes the gelatin to shift, thereby bending the stereocilia. The bending of the stereocilia stimulates the neurons, and they signal to the brain that the head is tilted, allowing the maintenance of balance. It is the vestibular branch of the vestibulocochlear cranial nerve that deals with balance.

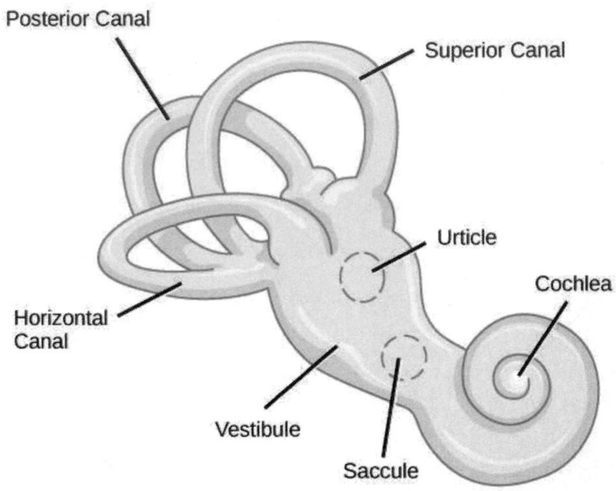

Figure 36.16 The structure of the vestibular labyrinth is shown. (credit: modification of work by NIH)

The fluid-filled **semicircular canals** are tubular loops set at oblique angles. They are arranged in three spatial planes. The base of each canal has a swelling that contains a cluster of hair cells. The hairs project into a gelatinous cap called the cupula and monitor angular acceleration and deceleration from rotation. They would be stimulated by driving your car around a corner, turning your head, or falling forward. One canal lies horizontally, while the other two lie at about 45 degree angles to the horizontal axis, as illustrated in Figure 36.16. When the brain processes input from all three canals together, it can detect angular acceleration or deceleration in three dimensions. When the head turns, the fluid in the canals shifts, thereby bending stereocilia and sending signals to the brain. Upon cessation accelerating or decelerating—or just moving—the movement of the fluid within the canals slows or stops. For example, imagine holding a glass of water. When moving forward, water may splash backwards onto the hand, and when motion has stopped, water may splash forward onto the fingers. While in motion, the water settles in the glass and does not splash. Note that the canals are not sensitive to velocity itself, but to changes in velocity, so moving forward at 60mph with your eyes closed would not give the sensation of movement, but suddenly accelerating or braking would stimulate the receptors.

Higher Processing

Hair cells from the utricle, saccule, and semicircular canals also communicate through bipolar neurons to the cochlear nucleus in the medulla. Cochlear neurons send descending projections to the spinal cord and ascending projections to the pons, thalamus, and cerebellum. Connections to the cerebellum are important for coordinated movements. There are also projections to the temporal cortex, which account for feelings of dizziness; projections to autonomic nervous system areas in the brainstem, which account for motion sickness; and projections to the primary somatosensory cortex, which monitors subjective measurements of the external world and self-movement. People with lesions in the vestibular area of the somatosensory cortex see vertical objects in the world as being tilted. Finally, the vestibular signals project to certain optic muscles to coordinate eye and head movements.

LINK TO LEARNING

Click through this interactive tutorial (http://openstax.org/l/ear_anatomy) to review the parts of the ear and how they function to process sound.

36.5 Vision

By the end of this section, you will be able to do the following:

- Explain how electromagnetic waves differ from sound waves
- Trace the path of light through the eye to the point of the optic nerve
- Explain tonic activity as it is manifested in photoreceptors in the retina

Vision is the ability to detect light patterns from the outside environment and interpret them into images. Animals are bombarded with sensory information, and the sheer volume of visual information can be problematic. Fortunately, the visual systems of species have evolved to attend to the most-important stimuli. The importance of vision to humans is further substantiated by the fact that about one-third of the human cerebral cortex is dedicated to analyzing and perceiving visual information.

Light

As with auditory stimuli, light travels in waves. The compression waves that compose sound must travel in a medium—a gas, a liquid, or a solid. In contrast, light is composed of electromagnetic waves and needs no medium; light can travel in a vacuum (Figure 36.17). The behavior of light can be discussed in terms of the behavior of waves and also in terms of the behavior of the fundamental unit of light—a packet of electromagnetic radiation called a photon. A glance at the electromagnetic spectrum shows that visible light for humans is just a small slice of the entire spectrum, which includes radiation that we cannot see as light because it is below the frequency of visible red light and above the frequency of visible violet light.

Certain variables are important when discussing perception of light. Wavelength (which varies inversely with frequency) manifests itself as hue. Light at the red end of the visible spectrum has longer wavelengths (and is lower frequency), while light at the violet end has shorter wavelengths (and is higher frequency). The wavelength of light is expressed in nanometers (nm); one nanometer is one billionth of a meter. Humans perceive light that ranges between approximately 380 nm and 740 nm. Some other animals, though, can detect wavelengths outside of the human range. For example, bees see near-ultraviolet light in order to locate nectar guides on flowers, and some non-avian reptiles sense infrared light (heat that prey gives off).

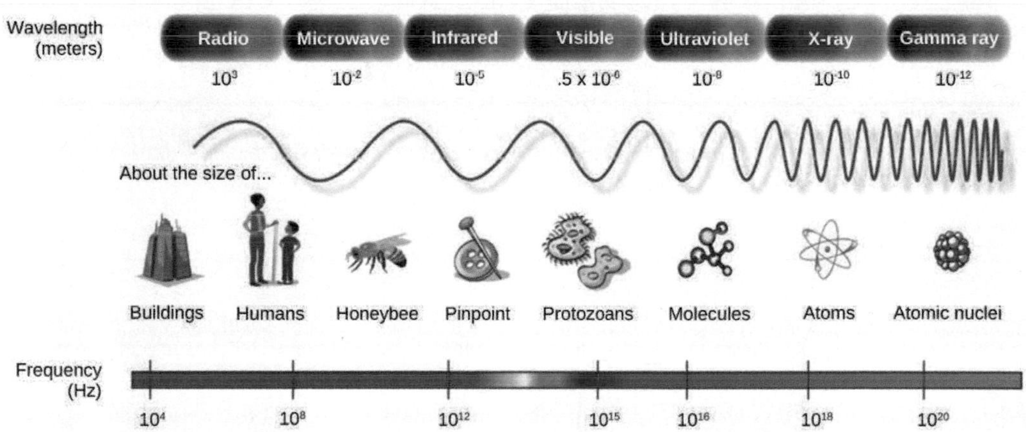

Figure 36.17 In the electromagnetic spectrum, visible light lies between 380 nm and 740 nm. (credit: modification of work by NASA)

Wave amplitude is perceived as luminous intensity, or brightness. The standard unit of intensity of light is the **candela**, which is approximately the luminous intensity of one common candle.

Light waves travel 299,792 km per second in a vacuum, (and somewhat slower in various media such as air and water), and those waves arrive at the eye as long (red), medium (green), and short (blue) waves. What is termed "white light" is light that is perceived as white by the human eye. This effect is produced by light that stimulates equally the color receptors in the human eye. The apparent color of an object is the color (or colors) that the object reflects. Thus a red object reflects the red wavelengths in mixed (white) light and absorbs all other wavelengths of light.

Anatomy of the Eye

The photoreceptive cells of the eye, where transduction of light to nervous impulses occurs, are located in the **retina** (shown in Figure 36.18) on the inner surface of the back of the eye. But light does not impinge on the retina unaltered. It passes through other layers that process it so that it can be interpreted by the retina (Figure 36.18b). The **cornea**, the front transparent layer of the eye, and the crystalline **lens**, a transparent convex structure behind the cornea, both refract (bend) light to focus the image on the retina. The **iris**, which is conspicuous as the colored part of the eye, is a circular muscular ring lying between the lens and cornea that regulates the amount of light entering the eye. In conditions of high ambient light, the iris contracts, reducing the size of the pupil at its center. In conditions of low light, the iris relaxes and the pupil enlarges.

VISUAL CONNECTION

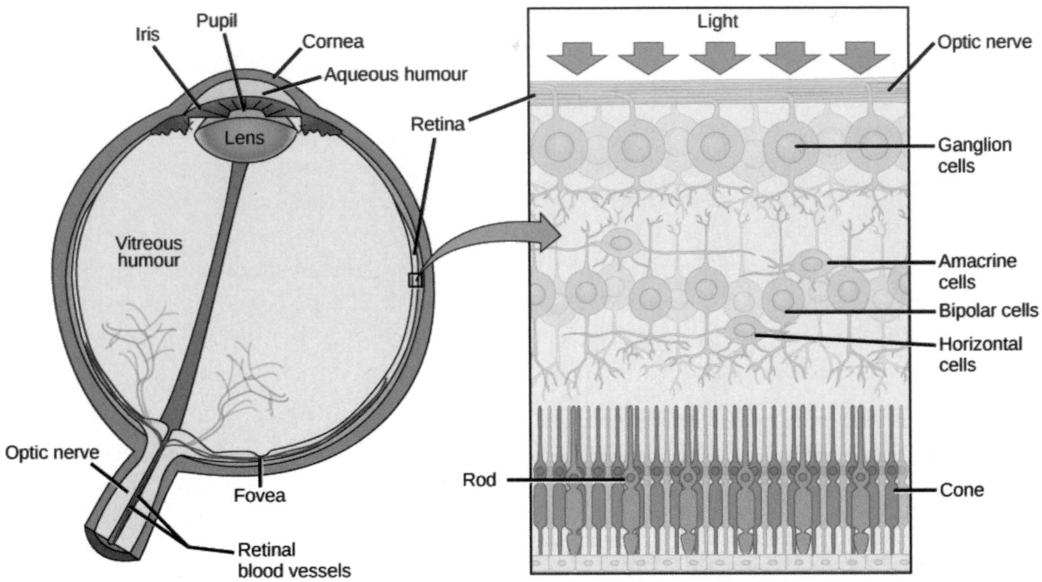

Figure 36.18 (a) The human eye is shown in cross section. (b) A blowup shows the layers of the retina.

Which of the following statements about the human eye is false?

a. Rods detect color, while cones detect only shades of gray.
b. When light enters the retina, it passes the ganglion cells and bipolar cells before reaching photoreceptors at the rear of the eye.
c. The iris adjusts the amount of light coming into the eye.
d. The cornea is a protective layer on the front of the eye.

The main function of the lens is to focus light on the retina and fovea centralis. The lens is dynamic, focusing and re-focusing light as the eye rests on near and far objects in the visual field. The lens is operated by muscles that stretch it flat or allow it to thicken, changing the focal length of light coming through it to focus it sharply on the retina. With age comes the loss of the flexibility of the lens, and a form of farsightedness called **presbyopia** results. Presbyopia occurs because the image focuses behind the retina. Presbyopia is a deficit similar to a different type of farsightedness called **hyperopia** caused by an eyeball that is too short. For both defects, images in the distance are clear but images nearby are blurry. **Myopia** (nearsightedness) occurs when an eyeball is elongated and the image focus falls in front of the retina. In this case, images in the distance are blurry but images nearby are clear.

There are two types of photoreceptors in the retina: **rods** and **cones**, named for their general appearance as illustrated in Figure 36.19. Rods are strongly photosensitive and are located in the outer edges of the retina. They detect dim light and are used primarily for peripheral and nighttime vision. Cones are weakly photosensitive and are located near the center of the retina. They respond to bright light, and their primary role is in daytime, color vision.

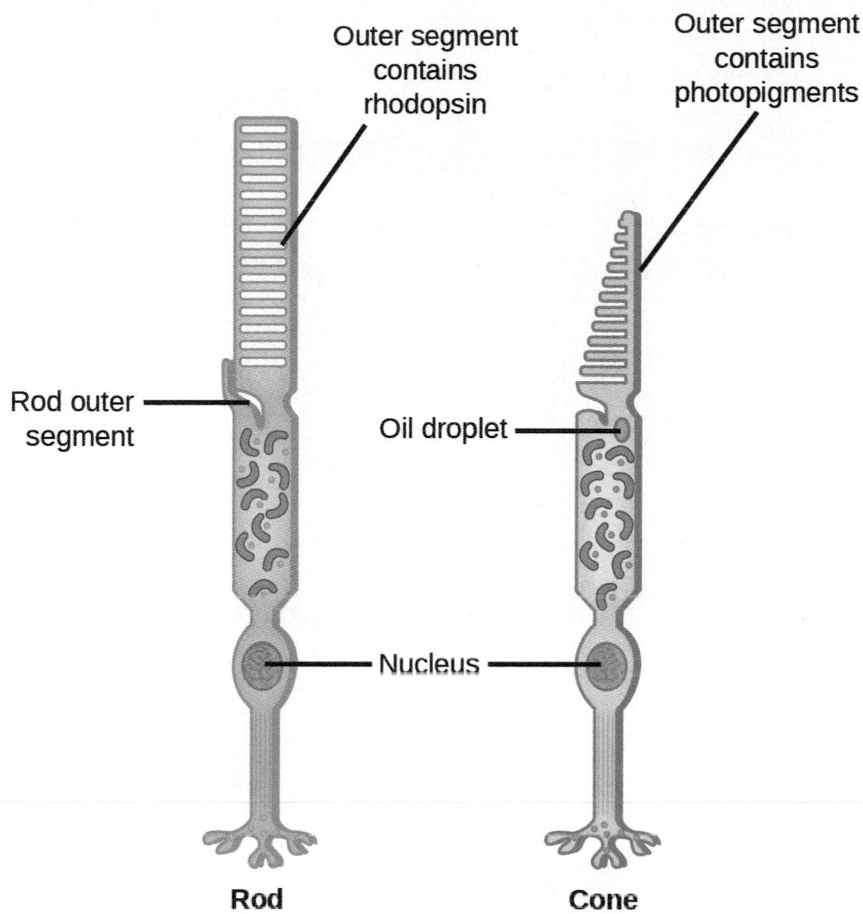

Figure 36.19 Rods and cones are photoreceptors in the retina. Rods respond in low light and can detect only shades of gray. Cones respond in intense light and are responsible for color vision. (credit: modification of work by Piotr Sliwa)

The **fovea** is the region in the center back of the eye that is responsible for acute vision. The fovea has a high density of cones. When you bring your gaze to an object to examine it intently in bright light, the eyes orient so that the object's image falls on the fovea. However, when looking at a star in the night sky or other object in dim light, the object can be better viewed by the peripheral vision because it is the rods at the edges of the retina, rather than the cones at the center, that operate better in low light. In humans, cones far outnumber rods in the fovea.

LINK TO LEARNING

Review the anatomical structure (http://openstax.org/l/eye_diagram) of the eye, clicking on each part to practice identification.

Transduction of Light

The rods and cones are the site of transduction of light to a neural signal. Both rods and cones contain photopigments. In vertebrates, the main photopigment, **rhodopsin**, has two main parts (Figure 36.20): an opsin, which is a membrane protein (in the form of a cluster of α-helices that span the membrane), and retinal—a molecule that absorbs light. When light hits a photoreceptor, it causes a shape change in the retinal, altering its structure from a bent (*cis*) form of the molecule to its linear (*trans*) isomer. This isomerization of retinal activates the rhodopsin, starting a cascade of events that ends with the closing of Na^+ channels in the membrane of the photoreceptor. Thus, unlike most other sensory neurons (which become depolarized by exposure to a stimulus) visual receptors become hyperpolarized and thus driven away from threshold (Figure 36.21).

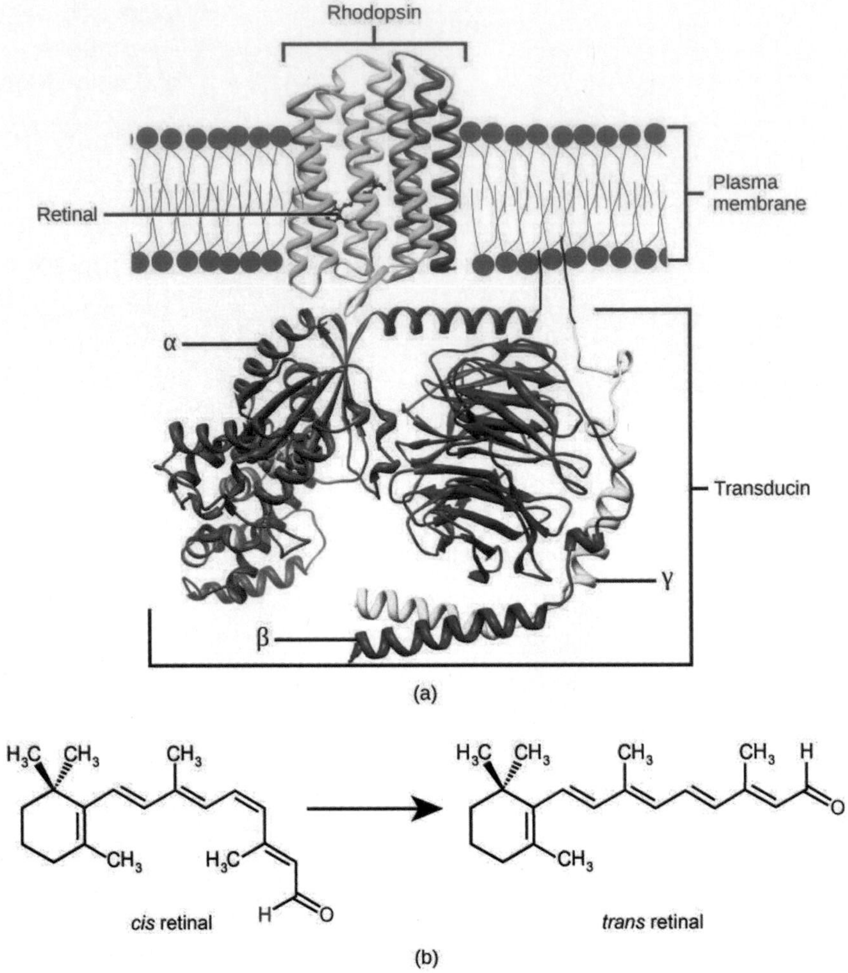

Figure 36.20 (a) Rhodopsin, the photoreceptor in vertebrates, has two parts: the trans-membrane protein opsin, and retinal. When light strikes retinal, it changes shape from (b) a *cis* to a *trans* form. The signal is passed to a G-protein called transducin, triggering a series of downstream events.

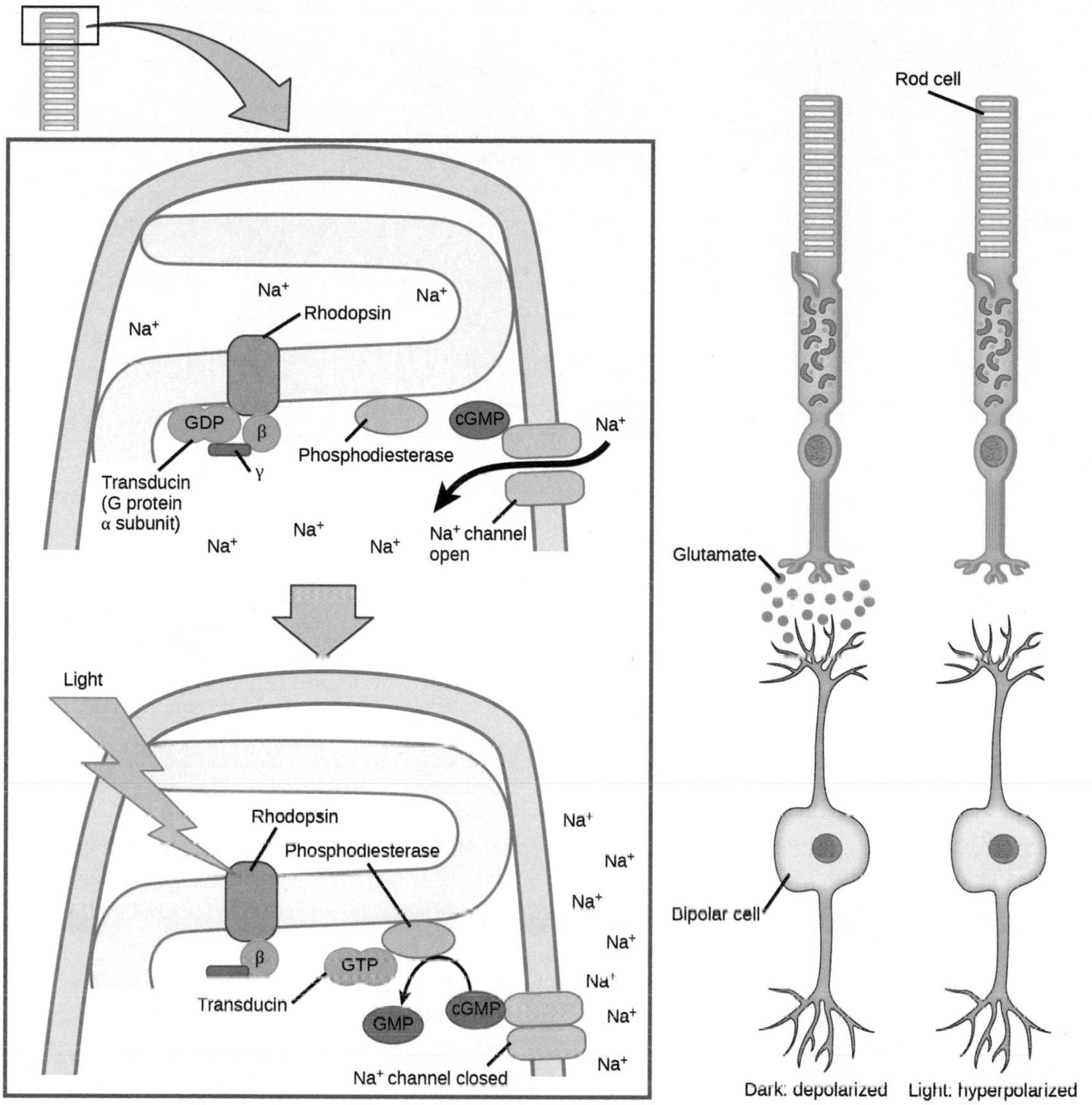

Figure 36.21 When light strikes rhodopsin, the G-protein transducin is activated, which in turn activates phosphodiesterase. Phosphodiesterase converts cGMP to GMP, thereby closing sodium channels. As a result, the membrane becomes hyperpolarized. The hyperpolarized membrane does not release glutamate to the bipolar cell.

Trichromatic Coding

There are three types of cones (with different photopsins), and they differ in the wavelength to which they are most responsive, as shown in Figure 36.22. Some cones are maximally responsive to short light waves of 420 nm, so they are called S cones ("S" for "short"); others respond maximally to waves of 530 nm (M cones, for "medium"); a third group responds maximally to light of longer wavelengths, at 560 nm (L, or "long" cones). With only one type of cone, color vision would not be possible, and a two-cone (dichromatic) system has limitations. Primates use a three-cone (trichromatic) system, resulting in full color vision.

The color we perceive is a result of the ratio of activity of our three types of cones. The colors of the visual spectrum, running from long-wavelength light to short, are red (700 nm), orange (600 nm), yellow (565 nm), green (497 nm), blue (470 nm), indigo (450 nm), and violet (425 nm). Humans have very sensitive perception of color and can distinguish about 500 levels of brightness, 200 different hues, and 20 steps of saturation, or about 2 million distinct colors.

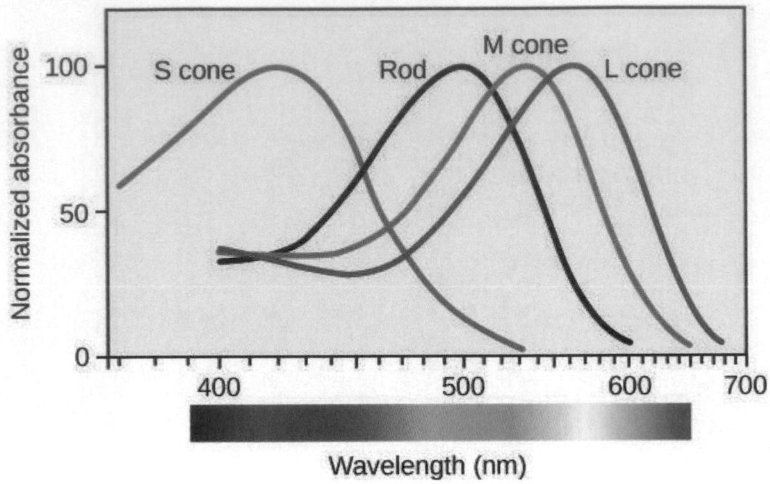

Figure 36.22 Human rod cells and the different types of cone cells each have an optimal wavelength. However, there is considerable overlap in the wavelengths of light detected.

Retinal Processing

Visual signals leave the cones and rods, travel to the bipolar cells, and then to ganglion cells. A large degree of processing of visual information occurs in the retina itself, before visual information is sent to the brain.

Photoreceptors in the retina continuously undergo **tonic activity**. That is, they are always slightly active even when not stimulated by light. In neurons that exhibit tonic activity, the absence of stimuli maintains a firing rate at a baseline; while some stimuli increase firing rate from the baseline, and other stimuli decrease firing rate. In the absence of light, the bipolar neurons that connect rods and cones to ganglion cells are continuously and actively inhibited by the rods and cones. Exposure of the retina to light hyperpolarizes the rods and cones and removes their inhibition of bipolar cells. The now active bipolar cells in turn stimulate the ganglion cells, which send action potentials along their axons (which leave the eye as the optic nerve). Thus, the visual system relies on change in retinal activity, rather than the absence or presence of activity, to encode visual signals for the brain. Sometimes horizontal cells carry signals from one rod or cone to other photoreceptors and to several bipolar cells. When a rod or cone stimulates a horizontal cell, the horizontal cell inhibits more distant photoreceptors and bipolar cells, creating lateral inhibition. This inhibition sharpens edges and enhances contrast in the images by making regions receiving light appear lighter and dark surroundings appear darker. Amacrine cells can distribute information from one bipolar cell to many ganglion cells.

You can demonstrate this using an easy demonstration to "trick" your retina and brain about the colors you are observing in your visual field. Look fixedly at Figure 36.23 for about 45 seconds. Then quickly shift your gaze to a sheet of blank white paper or a white wall. You should see an afterimage of the Norwegian flag in its correct colors. At this point, close your eyes for a moment, then reopen them, looking again at the white paper or wall; the afterimage of the flag should continue to appear as red, white, and blue. What causes this? According to an explanation called opponent process theory, as you gazed fixedly at the green, black, and yellow flag, your retinal ganglion cells that respond positively to green, black, and yellow increased their firing dramatically. When you shifted your gaze to the neutral white ground, these ganglion cells abruptly decreased their activity and the brain interpreted this abrupt downshift as if the ganglion cells were responding now to their "opponent" colors: red, white, and blue, respectively, in the visual field. Once the ganglion cells return to their baseline activity state, the false perception of color will disappear.

Figure 36.23 View this flag to understand how retinal processing works. Stare at the center of the flag (indicated by the white dot) for 45 seconds, and then quickly look at a white background, noticing how colors appear.

Higher Processing

The myelinated axons of ganglion cells make up the optic nerves. Within the nerves, different axons carry different qualities of the visual signal. Some axons constitute the magnocellular (big cell) pathway, which carries information about form, movement, depth, and differences in brightness. Other axons constitute the parvocellular (small cell) pathway, which carries information on color and fine detail. Some visual information projects directly back into the brain, while other information crosses to the opposite side of the brain. This crossing of optical pathways produces the distinctive optic chiasma (Greek, for "crossing") found at the base of the brain and allows us to coordinate information from both eyes.

Once in the brain, visual information is processed in several places, and its routes reflect the complexity and importance of visual information to humans and other animals. One route takes the signals to the thalamus, which serves as the routing station for all incoming sensory impulses except olfaction. In the thalamus, the magnocellular and parvocellular distinctions remain intact, and there are different layers of the thalamus dedicated to each. When visual signals leave the thalamus, they travel to the primary visual cortex at the rear of the brain. From the visual cortex, the visual signals travel in two directions. One stream that projects to the parietal lobe, in the side of the brain, carries magnocellular ("where") information. A second stream projects to the temporal lobe and carries both magnocellular ("where") and parvocellular ("what") information.

Another important visual route is a pathway from the retina to the **superior colliculus** in the midbrain, where eye movements are coordinated and integrated with auditory information. Finally, there is the pathway from the retina to the **suprachiasmatic nucleus** (SCN) of the hypothalamus. The SCN is a cluster of cells that is considered to be the body's internal clock, which controls our **circadian** (day-long) cycle. The SCN sends information to the pineal gland, which is important in sleep/wake patterns and annual cycles.

LINK TO LEARNING

View this interactive presentation (http://openstax.org/l/sense_of_sight) to review what you have learned about how vision functions.

KEY TERMS

audition sense of hearing

basilar membrane stiff structure in the cochlea that indirectly anchors auditory receptors

bipolar neuron neuron with two processes from the cell body, typically in opposite directions

candela (cd) unit of measurement of luminous intensity (brightness)

circadian describes a time cycle about one day in length

cochlea whorled structure that contains receptors for transduction of the mechanical wave into an electrical signal

cone weakly photosensitive, chromatic, cone-shaped neuron in the fovea of the retina that detects bright light and is used in daytime color vision

cornea transparent layer over the front of the eye that helps focus light waves

fovea region in the center of the retina with a high density of photoreceptors and which is responsible for acute vision

free nerve ending ending of an afferent neuron that lacks a specialized structure for detection of sensory stimuli; some respond to touch, pain, or temperature

glabrous describes the non-hairy skin found on palms and fingers, soles of feet, and lips of humans and other primates

glomerulus in the olfactory bulb, one of the two neural clusters that receives signals from one type of olfactory receptor

Golgi tendon organ muscular proprioceptive tension receptor that provides the sensory component of the Golgi tendon reflex

gustation sense of taste

hyperopia (also, farsightedness) visual defect in which the image focus falls behind the retina, thereby making images in the distance clear, but close-up images blurry

incus (also, anvil) second of the three bones of the middle ear

inner ear innermost part of the ear; consists of the cochlea and the vestibular system

iris pigmented, circular muscle at the front of the eye that regulates the amount of light entering the eye

kinesthesia sense of body movement

labyrinth bony, hollow structure that is the most internal part of the ear; contains the sites of transduction of auditory and vestibular information

lens transparent, convex structure behind the cornea that helps focus light waves on the retina

malleus (also, hammer) first of the three bones of the middle ear

mechanoreceptor sensory receptor modified to respond to mechanical disturbance such as being bent, touch, pressure, motion, and sound

Meissner's corpuscle (also, tactile corpuscle) encapsulated, rapidly-adapting mechanoreceptor in the skin that responds to light touch

Merkel's disk unencapsulated, slowly-adapting mechanoreceptor in the skin that responds to touch

middle ear part of the hearing apparatus that functions to transfer energy from the tympanum to the oval window of the inner ear

muscle spindle proprioceptive stretch receptor that lies within a muscle and that shortens the muscle to an optimal length for efficient contraction

myopia (also, nearsightedness) visual defect in which the image focus falls in front of the retina, thereby making images in the distance blurry, but close-up images clear

nociception neural processing of noxious (such as damaging) stimuli

odorant airborne molecule that stimulates an olfactory receptor

olfaction sense of smell

olfactory bulb neural structure in the vertebrate brain that receives signals from olfactory receptors

olfactory epithelium specialized tissue in the nasal cavity where olfactory receptors are located

olfactory receptor dendrite of a specialized neuron

organ of Corti in the basilar membrane, the site of the transduction of sound, a mechanical wave, to a neural signal

ossicle one of the three bones of the middle ear

outer ear part of the ear that consists of the pinna, ear canal, and tympanum and which conducts sound waves into the middle ear

oval window thin diaphragm between the middle and inner ears that receives sound waves from contact with the stapes bone of the middle ear

Pacinian corpuscle encapsulated mechanoreceptor in the skin that responds to deep pressure and vibration

papilla one of the small bump-like projections from the tongue

perception individual interpretation of a sensation; a brain function

pheromone substance released by an animal that can affect the physiology or behavior of other animals

pinna cartilaginous outer ear

presbyopia visual defect in which the image focus falls behind the retina, thereby making images in the distance clear, but close-up images blurry; caused by age-based changes in the lens

proprioception sense of limb position; used to track kinesthesia

pupil small opening though which light enters

reception receipt of a signal (such as light or sound) by sensory receptors

receptive field region in space in which a stimulus can activate a given sensory receptor

receptor potential membrane potential in a sensory receptor in response to detection of a stimulus

retina layer of photoreceptive and supporting cells on the inner surface of the back of the eye

rhodopsin main photopigment in vertebrates

rod strongly photosensitive, achromatic, cylindrical neuron in the outer edges of the retina that detects dim light and is used in peripheral and nighttime vision

Ruffini ending (also, bulbous corpuscle) slowly-adapting mechanoreceptor in the skin that responds to skin stretch and joint position

semicircular canal one of three half-circular, fluid-filled tubes in the vestibular labyrinth that monitors angular acceleration and deceleration

sensory receptor specialized neuron or other cells associated with a neuron that is modified to receive specific sensory input

sensory transduction conversion of a sensory stimulus into electrical energy in the nervous system by a change in the membrane potential

stapes (also, stirrup) third of the three bones of the middle ear

stereocilia in the auditory system, hair-like projections from hair cells that help detect sound waves

superior colliculus paired structure in the top of the midbrain, which manages eye movements and auditory integration

suprachiasmatic nucleus cluster of cells in the hypothalamus that plays a role in the circadian cycle

tastant food molecule that stimulates gustatory receptors

taste bud clusters of taste cells

tectorial membrane cochlear structure that lies above the hair cells and participates in the transduction of sound at the hair cells

tonic activity in a neuron, slight continuous activity while at rest

tympanum (also, tympanic membrane or ear drum) thin diaphragm between the outer and middle ears

ultrasound sound frequencies above the human detectable ceiling of approximately 20,000 Hz

umami one of the five basic tastes, which is described as "savory" and which may be largely the taste of L glutamate

vestibular sense sense of spatial orientation and balance

vision sense of sight

CHAPTER SUMMARY

36.1 Sensory Processes

A sensory activation occurs when a physical or chemical stimulus is processed into a neural signal (sensory transduction) by a sensory receptor. Perception is an individual interpretation of a sensation and is a brain function. Humans have special senses: olfaction, gustation, equilibrium, and hearing, plus the general senses of somatosensation.

Sensory receptors are either specialized cells associated with sensory neurons or the specialized ends of sensory neurons that are a part of the peripheral nervous system, and they are used to receive information about the environment (internal or external). Each sensory receptor is modified for the type of stimulus it detects. For example, neither gustatory receptors nor auditory receptors are sensitive to light. Each sensory receptor is responsive to stimuli within a specific region in space, which is known as that receptor's receptive field. The most fundamental function of a sensory system is the translation of a sensory signal to an electrical signal in the nervous system.

All sensory signals, except those from the olfactory system, enter the central nervous system and are routed to the thalamus. When the sensory signal exits the thalamus, it is conducted to the specific area of the cortex dedicated to processing that particular sense.

36.2 Somatosensation

Somatosensation includes all sensation received from the skin and mucous membranes, as well as from the limbs and joints. Somatosensation occurs all over the exterior of the body and at some interior locations as well, and a variety of receptor types, embedded in the skin and mucous membranes, play a role.

There are several types of specialized sensory receptors. Rapidly adapting free nerve endings detect nociception, hot and cold, and light touch. Slowly adapting, encapsulated Merkel's disks are found in fingertips and lips, and respond to light touch. Meissner's corpuscles, found in glabrous skin, are rapidly adapting, encapsulated receptors that detect touch, low-frequency vibration, and flutter. Ruffini endings are slowly adapting, encapsulated receptors that detect skin stretch, joint activity, and warmth. Hair receptors are rapidly adapting nerve endings wrapped around the base of hair follicles that detect hair movement and skin deflection. Finally, Pacinian corpuscles are encapsulated, rapidly adapting receptors that detect transient pressure and high-frequency vibration.

36.3 Taste and Smell

There are five primary tastes in humans: sweet, sour, bitter, salty, and umami. Each taste has its own receptor type that responds only to that taste. Tastants enter the body and are dissolved in saliva. Taste cells are located within taste buds, which are found on three of the four types of papillae in the mouth.

Regarding olfaction, there are many thousands of odorants, but humans detect only about 10,000. Like taste receptors, olfactory receptors are each responsive to only one odorant. Odorants dissolve in nasal mucosa, where they excite their corresponding olfactory sensory cells. When these cells detect an odorant, they send their signals to the main olfactory bulb and then to other locations in the brain, including the olfactory cortex.

36.4 Hearing and Vestibular Sensation

Audition is important for territory defense, predation, predator defense, and communal exchanges. The vestibular system, which is not auditory, detects linear acceleration and angular acceleration and deceleration. Both the auditory system and vestibular system use hair cells as their receptors.

Auditory stimuli are sound waves. The sound wave energy reaches the outer ear (pinna, canal, tympanum), and vibrations of the tympanum send the energy to the middle ear. The middle ear bones shift and the stapes transfers mechanical energy to the oval window of the fluid-filled inner ear cochlea. Once in the cochlea, the energy causes the basilar membrane to flex, thereby bending the stereocilia on receptor hair cells. This activates the receptors, which send their auditory neural signals to the brain.

The vestibular system has five parts that work together to provide the sense of direction, thus helping to maintain balance. The utricle and saccule measure head orientation: their calcium carbonate crystals shift when the head is tilted, thereby activating hair cells. The semicircular canals work similarly, such that when the head is turned, the fluid in the canals bends stereocilia on hair cells. The vestibular hair cells also send signals to the thalamus and to the somatosensory cortex, but also to the cerebellum, the structure above the brainstem that plays a large role in timing and coordination of movement.

36.5 Vision

Vision is the only photo responsive sense. Visible light travels in waves and is a very small slice of the electromagnetic radiation spectrum. Light waves differ based on their frequency (wavelength = hue) and amplitude (intensity = brightness).

In the vertebrate retina, there are two types of light receptors (photoreceptors): cones and rods. Cones, which are the source of color vision, exist in three forms—L, M, and S—and they are differentially sensitive to different wavelengths. Cones are located in the retina, along with the dim-light, achromatic receptors (rods). Cones are found in the fovea, the central region of the retina, whereas rods are found in the peripheral regions of the retina.

Visual signals travel from the eye over the axons of retinal ganglion cells, which make up the optic nerves. Ganglion cells come in several versions. Some ganglion cell axons carry information on form, movement, depth, and brightness, while other axons carry information on color and fine detail. Visual information is sent to the superior colliculi in the midbrain, where coordination of eye movements and integration of auditory information takes place. Visual information is also sent to the suprachiasmatic nucleus (SCN) of the hypothalamus, which plays a role in the circadian cycle.

VISUAL CONNECTION QUESTIONS

1. Figure 36.5 Which of the following statements about mechanoreceptors is false?
   a. Pacini corpuscles are found in both glabrous and hairy skin.
   b. Merkel's disks are abundant on the fingertips and lips.
   c. Ruffini endings are encapsulated mechanoreceptors.
   d. Meissner's corpuscles extend into the lower dermis.
2. Figure 36.14 Cochlear implants can restore hearing in people who have a nonfunctional cochlea. The implant consists of a microphone that picks up sound. A speech processor selects sounds in the range of human speech, and a transmitter converts these sounds to electrical impulses, which are then sent to the auditory nerve. Which of the following types of hearing loss would not be restored by a cochlear implant?
   a. Hearing loss resulting from absence or loss of hair cells in the organ of Corti.
   b. Hearing loss resulting from an abnormal auditory nerve.
   c. Hearing loss resulting from fracture of the cochlea.
   d. Hearing loss resulting from damage to bones of the middle ear.

3. Figure 36.18 Which of the following statements about the human eye is false?
    a. Rods detect color, while cones detect only shades of gray.
    b. When light enters the retina, it passes the ganglion cells and bipolar cells before reaching photoreceptors at the rear of the eye.
    c. The iris adjusts the amount of light coming into the eye.
    d. The cornea is a protective layer on the front of the eye.

REVIEW QUESTIONS

4. Where does perception occur?
    a. spinal cord
    b. cerebral cortex
    c. receptors
    d. thalamus

5. If a person's cold receptors no longer convert cold stimuli into sensory signals, that person has a problem with the process of ______.
    a. reception
    b. transmission
    c. perception
    d. transduction

6. After somatosensory transduction, the sensory signal travels through the brain as a(n) _____ signal.
    a. electrical
    b. pressure
    c. optical
    d. thermal

7. Many people experience motion sickness while traveling in a car. This sensation results from contradictory inputs arising from which senses?
    a. Proprioception and Kinesthesia
    b. Somatosensation and Equilibrium
    c. Gustation and Vibration
    d. Vision and Vestibular System

8. _____ are found only in _____ skin, and detect skin deflection.
    a. Meissner's corpuscles; hairy
    b. Merkel's disks; glabrous
    c. hair receptors; hairy
    d. Krause end bulbs; hairy

9. If you were to burn your epidermis, what receptor type would you most likely burn?
    a. free nerve endings
    b. Ruffini endings
    c. Pacinian corpuscle
    d. hair receptors

10. Many diabetic patients are warned by their doctors to test their glucose levels by pricking the sides of their fingers rather than the pads. Pricking the sides avoids stimulating which receptor?
    a. Krause end bulbs
    b. Meissner's corpuscles
    c. Ruffini ending
    d. Nociceptors

11. Which of the following has the fewest taste receptors?
    a. fungiform papillae
    b. circumvallate papillae
    c. foliate papillae
    d. filiform papillae

12. How many different taste molecules do taste cells each detect?
    a. one
    b. five
    c. ten
    d. It depends on the spot on the tongue.

13. Salty foods activate the taste cells by _____.
    a. exciting the taste cell directly
    b. causing hydrogen ions to enter the cell
    c. causing sodium channels to close
    d. binding directly to the receptors

14. All sensory signals except _____ travel to the _____ in the brain before the cerebral cortex.
    a. vision; thalamus
    b. olfaction; thalamus
    c. vision; cranial nerves
    d. olfaction; cranial nerves

15. How is the ability to recognize the umami taste an evolutionary advantage?
    a. Umami identifies healthy foods that are low in salt and sugar.
    b. Umami enhances the flavor of bland foods.
    c. Umami identifies foods that might contain essential amino acids.
    d. Umami identifies foods that help maintain electrolyte balance.

16. In sound, pitch is measured in _____, and volume is measured in _____.
    a. nanometers (nm); decibels (dB)
    b. decibels (dB); nanometers (nm)
    c. decibels (dB); hertz (Hz)
    d. hertz (Hz); decibels (dB)

17. Auditory hair cells are indirectly anchored to the _____.
    a. basilar membrane
    b. oval window
    c. tectorial membrane
    d. ossicles

18. Which of the following are found both in the auditory system and the vestibular system?
    a. basilar membrane
    b. hair cells
    c. semicircular canals
    d. ossicles

19. Benign Paroxysmal Positional Vertigo is a disorder where some of the calcium carbonate crystals in the utricle migrate into the semicircular canals. Why does this condition cause periods of dizziness?
    a. The hair cells in the semicircular canals will be constantly activated.
    b. The hair cells in the semicircular canals will now be stimulated by gravity.
    c. The utricle will no longer recognize acceleration.
    d. There will be too much volume in the semicircular canals for them to detect motion.

20. Why do people over 55 often need reading glasses?
    a. Their cornea no longer focuses correctly.
    b. Their lens no longer focuses correctly.
    c. Their eyeball has elongated with age, causing images to focus in front of their retina.
    d. Their retina has thinned with age, making vision more difficult.

21. Why is it easier to see images at night using peripheral, rather than the central, vision?
    a. Cones are denser in the periphery of the retina.
    b. Bipolar cells are denser in the periphery of the retina.
    c. Rods are denser in the periphery of the retina.
    d. The optic nerve exits at the periphery of the retina.

22. A person catching a ball must coordinate her head and eyes. What part of the brain is helping to do this?
    a. hypothalamus
    b. pineal gland
    c. thalamus
    d. superior colliculus

23. A satellite is launched into space, but explodes after exiting the Earth's atmosphere. Which statement accurately reflects the observations made by an astronaut on a space walk outside the International Space Station during the explosion?
    a. The astronaut would see the explosion, but would not hear a boom.
    b. The astronaut will not sense the explosion.
    c. The astronaut will see the explosion, and then hear the boom.
    d. The astronaut will feel the concussive force of the explosion, but will not see it.

CRITICAL THINKING QUESTIONS

24. If a person sustains damage to axons leading from sensory receptors to the central nervous system, which step or steps of sensory perception will be affected?
25. In what way does the overall magnitude of a stimulus affect the just-noticeable difference in the perception of that stimulus?
26. Describe the difference in the localization of the sensory receptors for general and special senses in humans.
27. What can be inferred about the relative sizes of the areas of cortex that process signals from skin not densely innervated with sensory receptors and skin that is densely innervated with sensory receptors?
28. Many studies have demonstrated that women are able to tolerate the same painful stimuli for longer than men. Why don't all people experience pain the same way?
29. From the perspective of the recipient of the signal, in what ways do pheromones differ from other odorants?

30. What might be the effect on an animal of not being able to perceive taste?
31. A few recent cancer detection studies have used trained dogs to detect lung cancer in urine samples. What is the hypothesis behind this study? Why are dogs a better choice of detectors in this study than humans?
32. How would a rise in altitude likely affect the speed of a sound transmitted through air? Why?
33. How might being in a place with less gravity than Earth has (such as Earth's moon) affect vestibular sensation, and why?
34. How does the structure of the ear allow a person to determine where a sound originates?
35. How could the pineal gland, the brain structure that plays a role in annual cycles, use visual information from the suprachiasmatic nucleus of the hypothalamus?
36. How is the relationship between photoreceptors and bipolar cells different from other sensory receptors and adjacent cells?
37. Cataracts, the medical condition where the lens of the eye becomes cloudy, are a leading cause of blindness. Describe how developing a cataract would change the path of light through the eye.

CHAPTER 37
The Endocrine System

Figure 37.1 The process of amphibian metamorphosis, as seen in the tadpole-to-frog stages shown here, is driven by hormones. (credit "tadpole": modification of work by Brian Gratwicke)

INTRODUCTION An animal's endocrine system controls body processes through the production, secretion, and regulation of hormones, which serve as chemical "messengers" functioning in cellular and organ activity and, ultimately, maintaining the body's homeostasis. The endocrine system plays a role in growth, metabolism, and sexual development. In humans, common endocrine system diseases include thyroid disease and diabetes mellitus. In organisms that undergo metamorphosis, the process is controlled by the endocrine system. The transformation from tadpole to frog, for example, is complex and nuanced to adapt to specific environments and ecological circumstances.

Chapter Outline

37.1 Types of Hormones

By the end of this section, you will be able to do the following:

- List the different types of hormones
- Explain their role in maintaining homeostasis

Maintaining homeostasis within the body requires the coordination of many different systems and organs. Communication between neighboring cells, and between cells and tissues in distant parts of the body, occurs through the release of chemicals called hormones. Hormones are released into body fluids (usually blood) that carry these chemicals to their target cells. At the target cells, which are cells that have a receptor for a signal or ligand from a signal cell, the hormones elicit a response. The cells, tissues, and organs that secrete hormones make up the endocrine system. Examples of glands of the endocrine system include the adrenal glands, which produce hormones such as epinephrine and norepinephrine that regulate responses to stress, and the thyroid gland, which produces thyroid hormones that regulate metabolic rates.

Although there are many different hormones in the human body, they can be divided into three classes based on their chemical structure: lipid-derived, amino acid-derived, and peptide (peptide and proteins) hormones. One of the key distinguishing features of lipid-derived hormones is that they can diffuse across plasma membranes whereas the amino acid-derived and peptide hormones cannot.

Lipid-Derived Hormones (or Lipid-soluble Hormones)

Most **lipid hormones** are derived from cholesterol and thus are structurally similar to it, as illustrated in Figure 37.2. The primary class of lipid hormones in humans is the steroid hormones. Chemically, these hormones are usually ketones or alcohols; their chemical names will end in "-ol" for alcohols or "-one" for ketones. Examples of steroid hormones include estradiol, which is an **estrogen**, or female sex hormone, and testosterone, which is an androgen, or male sex hormone. These two hormones are released by the female and male reproductive organs, respectively. Other steroid hormones include aldosterone and cortisol, which are released by the adrenal glands along with some other types of androgens. Steroid hormones are insoluble in water, and they are transported by transport proteins in blood. As a result, they remain in circulation longer than peptide hormones. For example, cortisol has a half-life of 60 to 90 minutes, while epinephrine, an amino acid derived-hormone, has a half-life of approximately one minute.

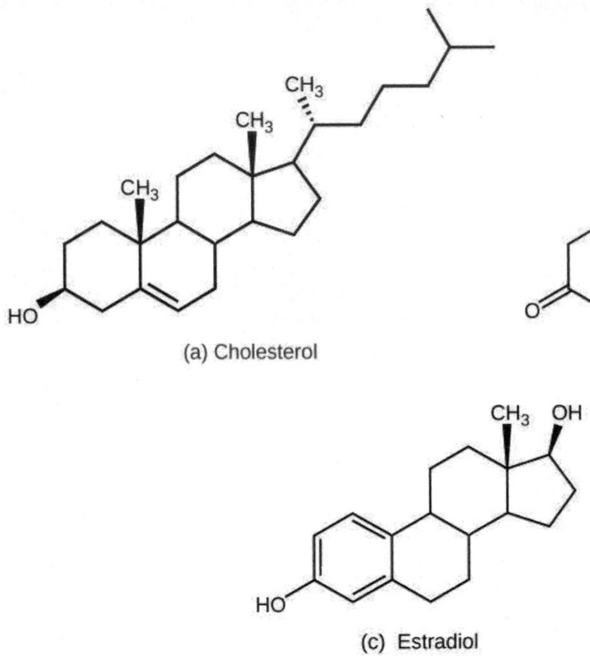

Figure 37.2 The structures shown here represent (a) cholesterol, plus the steroid hormones (b) testosterone and (c) estradiol.

Amino Acid-Derived Hormones

The **amino acid-derived hormones** are relatively small molecules that are derived from the amino acids tyrosine and tryptophan, shown in Figure 37.3. If a hormone is amino acid-derived, its chemical name will end in "-ine". Examples of amino acid-derived hormones include epinephrine and norepinephrine, which are synthesized in the medulla of the adrenal glands, and thyroxine, which is produced by the thyroid gland. The pineal gland in the brain makes and secretes melatonin which regulates sleep cycles.

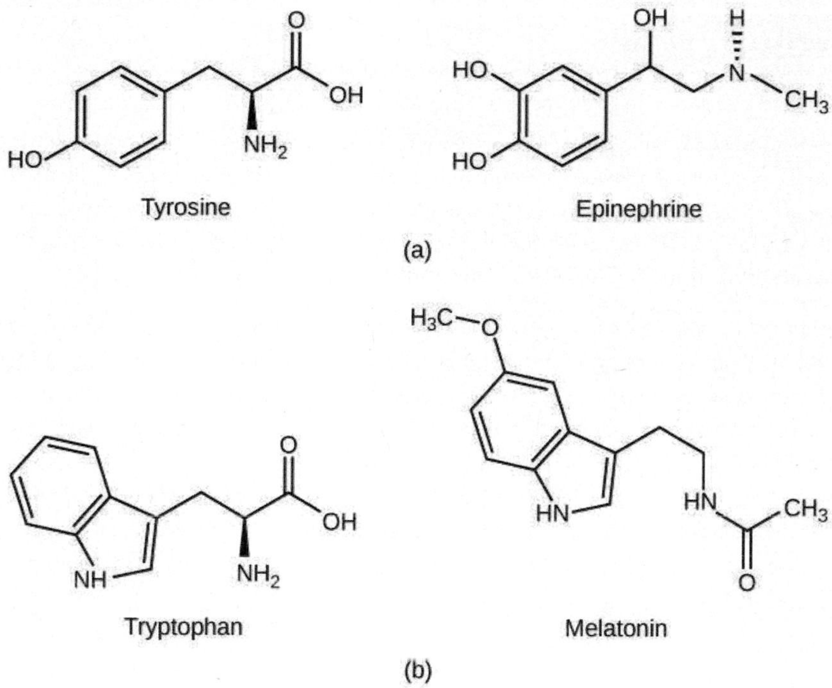

Figure 37.3 (a) The hormone epinephrine, which triggers the fight-or-flight response, is derived from the amino acid tyrosine. (b) The hormone melatonin, which regulates circadian rhythms, is derived from the amino acid tryptophan.

Peptide Hormones

The structure of **peptide hormones** is that of a polypeptide chain (chain of amino acids). The peptide hormones include molecules that are short polypeptide chains, such as antidiuretic hormone and oxytocin produced in the brain and released into the blood in the posterior pituitary gland. This class also includes small proteins, like growth hormones produced by the pituitary, and large glycoproteins such as follicle-stimulating hormone produced by the pituitary. Figure 37.4 illustrates these peptide hormones.

Secreted peptides like insulin are stored within vesicles in the cells that synthesize them. They are then released in response to stimuli such as high blood glucose levels in the case of insulin. Amino acid-derived and polypeptide hormones are water-soluble and insoluble in lipids. These hormones cannot pass through plasma membranes of cells; therefore, their receptors are found on the surface of the target cells.

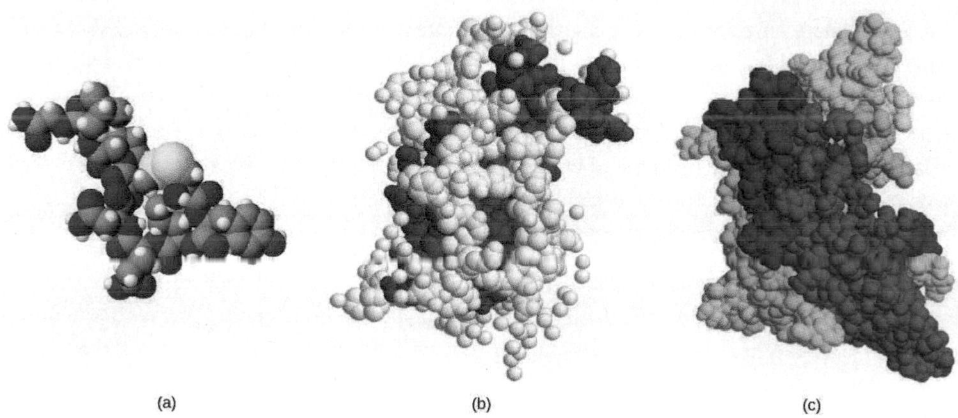

Figure 37.4 The structures of peptide hormones (a) oxytocin, (b) growth hormone, and (c) follicle-stimulating hormone are shown. These peptide hormones are much larger than those derived from cholesterol or amino acids.

CAREER CONNECTION

Endocrinologist

An endocrinologist is a medical doctor who specializes in treating disorders of the endocrine glands, hormone systems, and glucose and lipid metabolic pathways. An endocrine surgeon specializes in the surgical treatment of endocrine diseases and glands. Some of the diseases that are managed by endocrinologists: disorders of the pancreas (diabetes mellitus), disorders of the pituitary (gigantism, acromegaly, and pituitary dwarfism), disorders of the thyroid gland (goiter and Graves' disease), and disorders of the adrenal glands (Cushing's disease and Addison's disease).

Endocrinologists are required to assess patients and diagnose endocrine disorders through extensive use of laboratory tests. Many endocrine diseases are diagnosed using tests that stimulate or suppress endocrine organ functioning. Blood samples are then drawn to determine the effect of stimulating or suppressing an endocrine organ on the production of hormones. For example, to diagnose diabetes mellitus, patients are required to fast for 12 to 24 hours. They are then given a sugary drink, which stimulates the pancreas to produce insulin to decrease blood glucose levels. A blood sample is taken one to two hours after the sugar drink is consumed. If the pancreas is functioning properly, the blood glucose level will be within a normal range. Another example is the A1C test, which can be performed during blood screening. The A1C test measures average blood glucose levels over the past two to three months by examining how well the blood glucose is being managed over a long time.

Once a disease has been diagnosed, endocrinologists can prescribe lifestyle changes and/or medications to treat the disease. Some cases of diabetes mellitus can be managed by exercise, weight loss, and a healthy diet; in other cases, medications may be required to enhance insulin release. If the disease cannot be controlled by these means, the endocrinologist may prescribe insulin injections.

In addition to clinical practice, endocrinologists may also be involved in primary research and development activities. For example, ongoing islet transplant research is investigating how healthy pancreas islet cells may be transplanted into diabetic patients. Successful islet transplants may allow patients to stop taking insulin injections.

37.2 How Hormones Work

By the end of this section, you will be able to do the following:

- Explain how hormones work
- Discuss the role of different types of hormone receptors

Hormones mediate changes in target cells by binding to specific **hormone receptors**. In this way, even though hormones circulate throughout the body and come into contact with many different cell types, they only affect cells that possess the necessary receptors. Receptors for a specific hormone may be found on many different cells or may be limited to a small number of specialized cells. For example, thyroid hormones act on many different tissue types, stimulating metabolic activity throughout the body. Cells can have many receptors for the same hormone but often also possess receptors for different types of hormones. The number of receptors that respond to a hormone determines the cell's sensitivity to that hormone, and the resulting cellular response. Additionally, the number of receptors that respond to a hormone can change over time, resulting in increased or decreased cell sensitivity. In **up-regulation**, the number of receptors increases in response to rising hormone levels, making the cell more sensitive to the hormone and allowing for more cellular activity. When the number of receptors decreases in response to rising hormone levels, called **down-regulation**, cellular activity is reduced.

Receptor binding alters cellular activity and results in an increase or decrease in normal body processes. Depending on the location of the protein receptor on the target cell and the chemical structure of the hormone, hormones can mediate changes directly by binding to **intracellular hormone receptors** and modulating gene transcription, or indirectly by binding to cell surface receptors and stimulating signaling pathways.

Intracellular Hormone Receptors

Lipid-derived (soluble) hormones such as steroid hormones diffuse across the membranes of the endocrine cell. Once outside the cell, they bind to transport proteins that keep them soluble in the bloodstream. At the target cell, the hormones are released from the carrier protein and diffuse across the lipid bilayer of the plasma membrane of cells. The steroid hormones pass through the plasma membrane of a target cell and adhere to intracellular receptors residing in the cytoplasm or in the nucleus. The cell signaling pathways induced by the steroid hormones regulate specific genes on the cell's DNA. The hormones and receptor complex act as transcription regulators by increasing or decreasing the synthesis of mRNA molecules of specific genes. This, in turn, determines the amount of corresponding protein that is synthesized by altering gene expression. This protein can be used either to change the structure of the cell or to produce enzymes that catalyze chemical reactions. In this way, the steroid hormone regulates specific cell processes as illustrated in Figure 37.5.

VISUAL CONNECTION

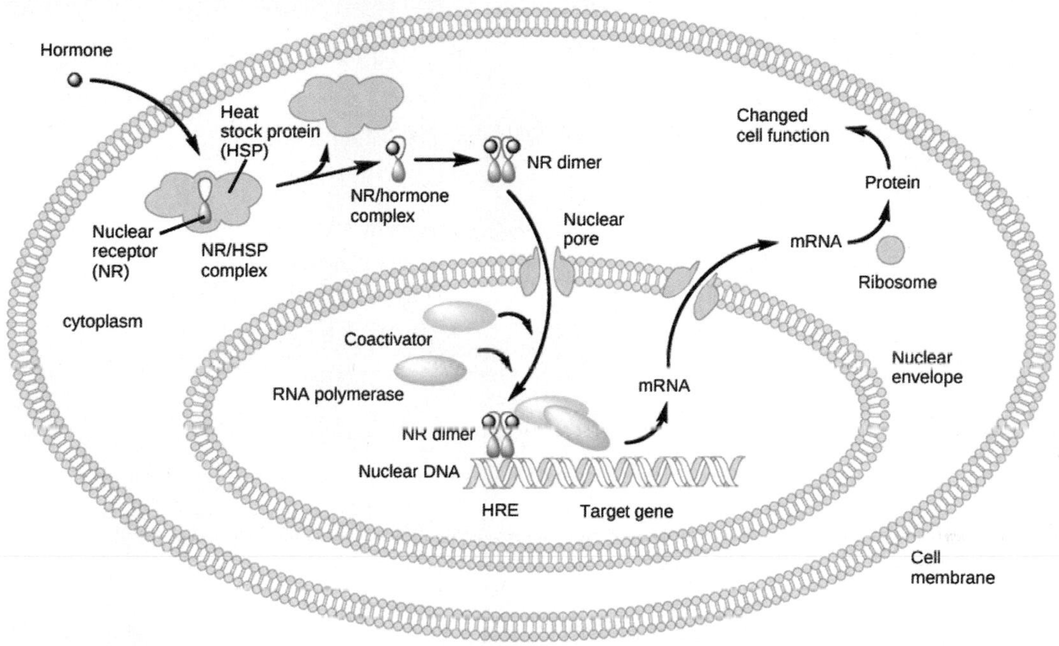

Figure 37.5 An intracellular nuclear receptor (NR) is located in the cytoplasm bound to a heat shock protein (HSP). Upon hormone binding, the receptor dissociates from the heat shock protein and translocates to the nucleus. In the nucleus, the hormone-receptor complex binds to a DNA sequence called a hormone response element (HRE), which triggers gene transcription and translation. The corresponding protein product can then mediate changes in cell function.

Heat shock proteins (HSP) are so named because they help refold misfolded proteins. In response to increased temperature (a "heat shock"), heat shock proteins are activated by release from the NR/HSP complex. At the same time, transcription of HSP genes is activated. Why do you think the cell responds to a heat shock by increasing the activity of proteins that help refold misfolded proteins?

Other lipid-soluble hormones that are not steroid hormones, such as vitamin D and thyroxine, have receptors located in the nucleus. While thyroxine is mostly hydrophobic, its passage across the membrane is dependent on transporter protein. Vitamin D diffuses across both the plasma membrane and the nuclear envelope. Once in the cell, both hormones bind to receptors in the nucleus. The hormone-receptor complex stimulates transcription of specific genes.

Plasma Membrane Hormone Receptors

Amino acid-derived hormones (with the exception of thyroxine) and polypeptide hormones are not lipid-derived (lipid-soluble) and therefore cannot diffuse through the plasma membrane of cells. Lipid insoluble hormones bind to receptors on the outer surface of the plasma membrane, via **plasma membrane hormone receptors**. Unlike steroid hormones, lipid insoluble hormones do not directly affect the target cell because they cannot enter the cell and act directly on DNA. Binding of these hormones to a cell surface receptor results in activation of a signaling pathway; this triggers intracellular activity and carries out the specific effects associated with the hormone. In this way, nothing passes through the cell membrane; the hormone that binds at the surface remains at the surface of the cell while the intracellular product remains inside the cell. The hormone that initiates the signaling pathway is called a **first messenger**, which activates a second messenger in the cytoplasm, as illustrated in Figure 37.6.

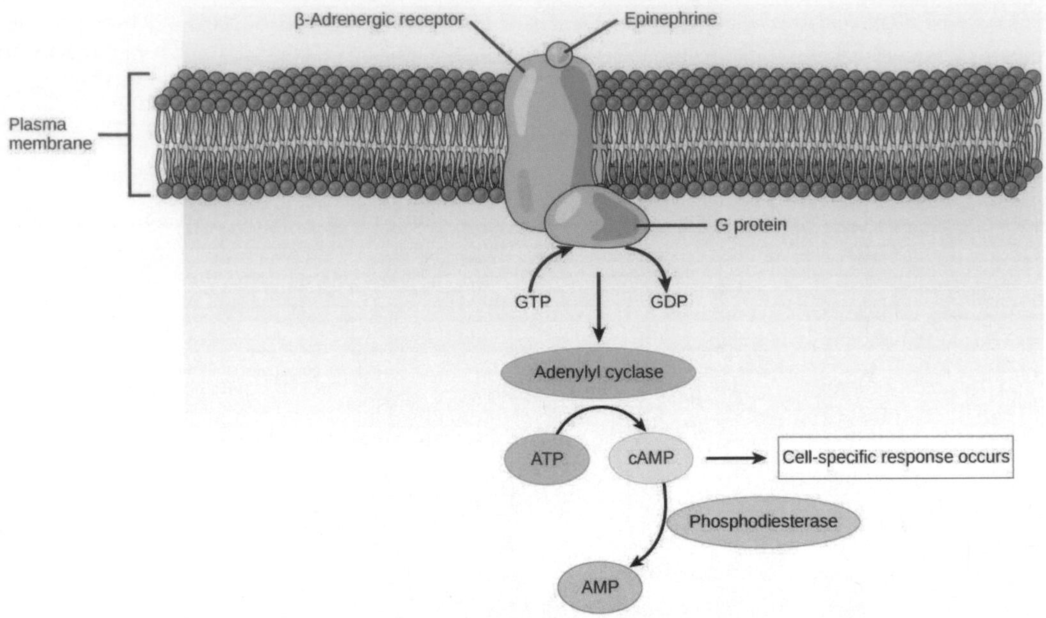

Figure 37.6 The amino acid-derived hormones epinephrine and norepinephrine bind to beta-adrenergic receptors on the plasma membrane of cells. Hormone binding to receptor activates a G-protein, which in turn activates adenylyl cyclase, converting ATP to cAMP. cAMP is a second messenger that mediates a cell-specific response. An enzyme called phosphodiesterase breaks down cAMP, terminating the signal.

One very important second messenger is cyclic AMP (cAMP). When a hormone binds to its membrane receptor, a **G-protein** that is associated with the receptor is activated; G-proteins are proteins separate from receptors that are found in the cell membrane. When a hormone is not bound to the receptor, the G-protein is inactive and is bound to guanosine diphosphate, or GDP. When a hormone binds to the receptor, the G-protein is activated by binding guanosine triphosphate, or GTP, in place of GDP. After binding, GTP is hydrolysed by the G-protein into GDP and becomes inactive.

The activated G-protein in turn activates a membrane-bound enzyme called **adenylyl cyclase**. Adenylyl cyclase catalyzes the conversion of ATP to cAMP. cAMP, in turn, activates a group of proteins called protein kinases, which transfer a phosphate group from ATP to a substrate molecule in a process called phosphorylation. The phosphorylation of a substrate molecule changes its structural orientation, thereby activating it. These activated molecules can then mediate changes in cellular processes.

The effect of a hormone is amplified as the signaling pathway progresses. The binding of a hormone at a single receptor causes the activation of many G-proteins, which activates adenylyl cyclase. Each molecule of adenylyl cyclase then triggers the formation of many molecules of cAMP. Further amplification occurs as protein kinases, once activated by cAMP, can catalyze many reactions. In this way, a small amount of hormone can trigger the formation of a large amount of cellular product. To stop hormone activity, cAMP is deactivated by the cytoplasmic enzyme **phosphodiesterase**, or PDE. PDE is always present in the cell and breaks down cAMP to control hormone activity, preventing overproduction of cellular products.

The specific response of a cell to a lipid insoluble hormone depends on the type of receptors that are present on the cell membrane and the substrate molecules present in the cell cytoplasm. Cellular responses to hormone binding of a receptor include altering membrane permeability and metabolic pathways, stimulating synthesis of proteins and enzymes, and activating hormone release.

37.3 Regulation of Body Processes

By the end of this section, you will be able to do the following:

- Explain how hormones regulate the excretory system
- Discuss the role of hormones in the reproductive system
- Describe how hormones regulate metabolism
- Explain the role of hormones in different diseases

Hormones have a wide range of effects and modulate many different body processes. The key regulatory processes that will be examined here are those affecting the excretory system, the reproductive system, metabolism, blood calcium concentrations, growth, and the stress response.

Hormonal Regulation of the Excretory System

Maintaining a proper water balance in the body is important to avoid dehydration or over-hydration (hyponatremia). The water concentration of the body is monitored by **osmoreceptors** in the hypothalamus, which detect the concentration of electrolytes in the extracellular fluid. The concentration of electrolytes in the blood rises when there is water loss caused by excessive perspiration, inadequate water intake, or low blood volume due to blood loss. An increase in blood electrolyte levels results in a neuronal signal being sent from the osmoreceptors in hypothalamic nuclei. The pituitary gland has two components: anterior and posterior. The anterior pituitary is composed of glandular cells that secrete protein hormones. The posterior pituitary is an extension of the hypothalamus. It is composed largely of neurons that are continuous with the hypothalamus.

The hypothalamus produces a polypeptide hormone known as **antidiuretic hormone (ADH)**, which is transported to and released from the posterior pituitary gland. The principal action of ADH is to regulate the amount of water excreted by the kidneys. As ADH (which is also known as vasopressin) causes direct water reabsorption from the kidney tubules, salts and wastes are concentrated in what will eventually be excreted as urine. The hypothalamus controls the mechanisms of ADH secretion, either by regulating blood volume or the concentration of water in the blood. Dehydration or physiological stress can cause an increase of osmolarity above 300 mOsm/L, which in turn, raises ADH secretion and water will be retained, causing an increase in blood pressure. ADH travels in the bloodstream to the kidneys. Once at the kidneys, ADH changes the kidneys to become more permeable to water by temporarily inserting water channels, aquaporins, into the kidney tubules. Water moves out of the kidney tubules through the aquaporins, reducing urine volume. The water is reabsorbed into the capillaries lowering blood osmolarity back toward normal. As blood osmolarity decreases, a negative feedback mechanism reduces osmoreceptor activity in the hypothalamus, and ADH secretion is reduced. ADH release can be reduced by certain substances, including alcohol, which can cause increased urine production and dehydration.

Chronic underproduction of ADH or a mutation in the ADH receptor results in **diabetes insipidus**. If the posterior pituitary does not release enough ADH, water cannot be retained by the kidneys and is lost as urine. This causes increased thirst, but water taken in is lost again and must be continually consumed. If the condition is not severe, dehydration may not occur, but severe cases can lead to electrolyte imbalances due to dehydration.

Another hormone responsible for maintaining electrolyte concentrations in extracellular fluids is **aldosterone**, a steroid hormone that is produced by the adrenal cortex. In contrast to ADH, which promotes the reabsorption of water to maintain proper water balance, aldosterone maintains proper water balance by enhancing Na^+ reabsorption and K^+ secretion from extracellular fluid of the cells in kidney tubules. Because it is produced in the cortex of the adrenal gland and affects the concentrations of minerals Na^+ and K^+, aldosterone is referred to as a **mineralocorticoid**, a corticosteroid that affects ion and water balance. Aldosterone release is stimulated by a decrease in blood sodium levels, blood volume, or blood pressure, or an increase in blood potassium levels. It also prevents the loss of Na^+ from sweat, saliva, and gastric juice. The reabsorption of Na^+ also results in the osmotic reabsorption of water, which alters blood volume and blood pressure.

Aldosterone production can be stimulated by low blood pressure, which triggers a sequence of chemical release, as illustrated in Figure 37.7. When blood pressure drops, the renin-angiotensin-aldosterone system (RAAS) is activated. Cells in the juxtaglomerular apparatus, which regulates the functions of the nephrons of the kidney, detect this and release **renin**. Renin, an enzyme, circulates in the blood and reacts with a plasma protein produced by the liver called angiotensinogen. When angiotensinogen is cleaved by renin, it produces angiotensin I, which is then converted into angiotensin II in the lungs. Angiotensin II functions as a hormone and then causes the release of the hormone aldosterone by the adrenal cortex, resulting in increased Na^+ reabsorption, water retention, and an increase in blood pressure. Angiotensin II in addition to being a potent vasoconstrictor also causes an increase in ADH and increased thirst, both of which help to raise blood pressure.

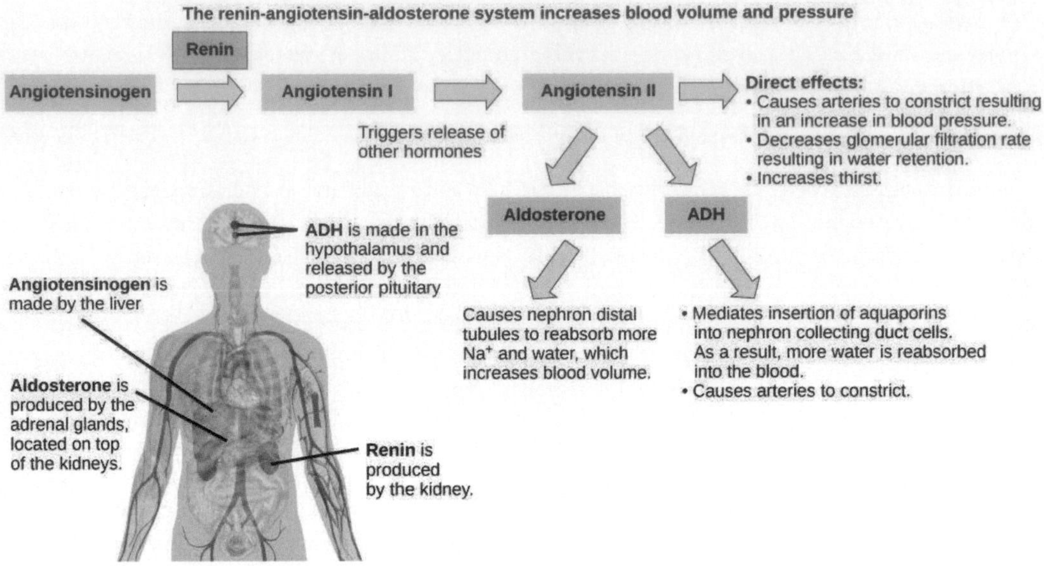

Figure 37.7 ADH and aldosterone increase blood pressure and volume. Angiotensin II stimulates release of these hormones. Angiotensin II, in turn, is formed when renin cleaves angiotensinogen. (credit: modification of work by Mikael Häggström)

Hormonal Regulation of the Reproductive System

Regulation of the reproductive system is a process that requires the action of hormones from the pituitary gland, the adrenal cortex, and the gonads. During puberty in both males and females, the hypothalamus produces gonadotropin-releasing hormone (GnRH), which stimulates the production and release of **follicle-stimulating hormone (FSH)** and luteinizing hormone (LH) from the anterior pituitary gland. These hormones regulate the gonads (testes in males and ovaries in females) and therefore are called **gonadotropins**. In both males and females, FSH stimulates gamete production and LH stimulates production of hormones by the gonads. An increase in gonad hormone levels inhibits GnRH production through a negative feedback loop.

Regulation of the Male Reproductive System

In males, FSH stimulates the maturation of sperm cells. FSH production is inhibited by the hormone inhibin, which is released by the testes. LH stimulates production of the sex hormones (**androgens**) by the interstitial cells of the testes and therefore is also called interstitial cell-stimulating hormone.

The most widely known androgen in males is testosterone. Testosterone promotes the production of sperm and masculine characteristics. The adrenal cortex also produces small amounts of testosterone precursor, although the role of this additional hormone production is not fully understood.

Everyday Connection

The Dangers of Synthetic Hormones

Figure 37.8 Professional baseball player Jason Giambi publically admitted to, and apologized for, his use of anabolic steroids supplied by a trainer. (credit: Bryce Edwards)

Some athletes attempt to boost their performance by using artificial hormones that enhance muscle performance. Anabolic steroids, a form of the male sex hormone testosterone, are one of the most widely known performance-enhancing drugs. Steroids are used to help build muscle mass. Other hormones that are used to enhance athletic performance include erythropoietin, which triggers the production of red blood cells, and human growth hormone, which can help in building muscle mass. Most performance enhancing drugs are illegal for nonmedical purposes. They are also banned by national and international governing bodies including the International Olympic Committee, the U.S. Olympic Committee, the National Collegiate Athletic Association, the Major League Baseball, and the National Football League.

The side effects of synthetic hormones are often significant and nonreversible, and in some cases, fatal. Androgens produce several complications such as liver dysfunctions and liver tumors, prostate gland enlargement, difficulty urinating, premature closure of epiphyseal cartilages, testicular atrophy, infertility, and immune system depression. The physiological strain caused by these substances is often greater than what the body can handle, leading to unpredictable and dangerous effects and linking their use to heart attacks, strokes, and impaired cardiac function.

Regulation of the Female Reproductive System

In females, FSH stimulates development of egg cells, called ova, which develop in structures called follicles. Follicle cells produce the hormone inhibin, which inhibits FSH production. LH also plays a role in the development of ova, induction of ovulation, and stimulation of estradiol and progesterone production by the ovaries, as illustrated in Figure 37.9. Estradiol and progesterone are steroid hormones that prepare the body for pregnancy. Estradiol produces secondary sex characteristics in females, while both estradiol and progesterone regulate the menstrual cycle.

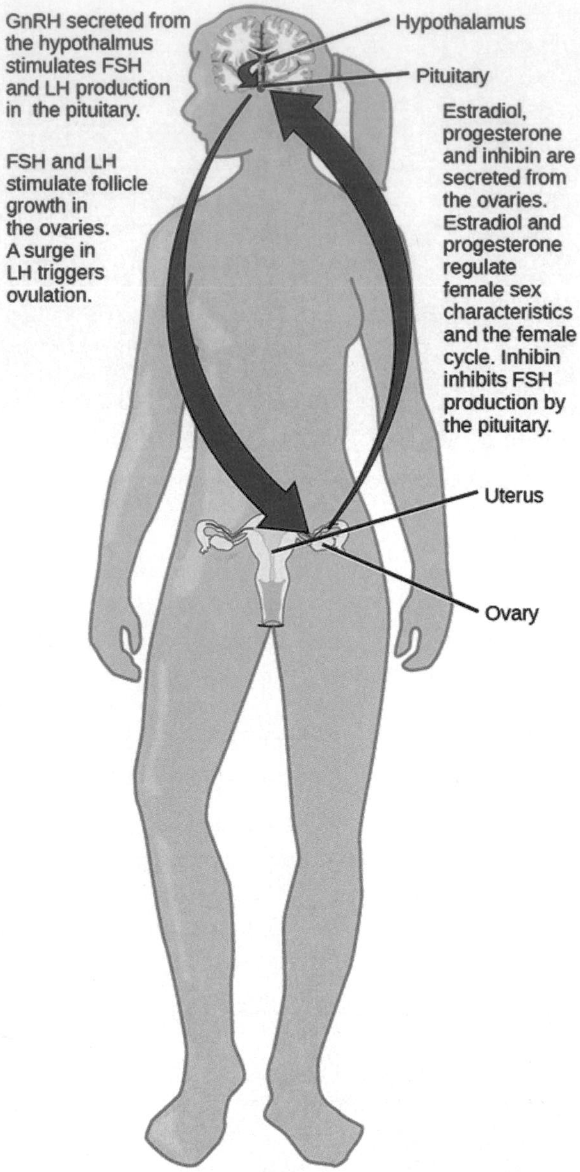

Figure 37.9 Hormonal regulation of the female reproductive system involves hormones from the hypothalamus, pituitary, and ovaries.

In addition to producing FSH and LH, the anterior portion of the pituitary gland also produces the hormone **prolactin (PRL)** in females. Prolactin stimulates the production of milk by the mammary glands following childbirth. Prolactin levels are regulated by the hypothalamic hormones **prolactin-releasing hormone (PRH)** and **prolactin-inhibiting hormone (PIH)**, which is now known to be dopamine. PRH stimulates the release of prolactin and PIH inhibits it.

The posterior pituitary releases the hormone **oxytocin**, which stimulates uterine contractions during childbirth. The uterine smooth muscles are not very sensitive to oxytocin until late in pregnancy when the number of oxytocin receptors in the uterus peaks. Stretching of tissues in the uterus and cervix stimulates oxytocin release during childbirth. Contractions increase in intensity as blood levels of oxytocin rise via a positive feedback mechanism until the birth is complete. Oxytocin also stimulates the contraction of myoepithelial cells around the milk-producing mammary glands. As these cells contract, milk is forced from the secretory alveoli into milk ducts and is ejected from the breasts in milk ejection ("let-down") reflex. Oxytocin release is stimulated by the suckling of an infant, which triggers the synthesis of oxytocin in the hypothalamus and its release into circulation at the posterior pituitary.

Hormonal Regulation of Metabolism

Blood glucose levels vary widely over the course of a day as periods of food consumption alternate with periods of fasting. Insulin and glucagon are the two hormones primarily responsible for maintaining homeostasis of blood glucose levels.

Additional regulation is mediated by the thyroid hormones.

Regulation of Blood Glucose Levels by Insulin and Glucagon

Cells of the body require nutrients in order to function, and these nutrients are obtained through feeding. In order to manage nutrient intake, storing excess intake and utilizing reserves when necessary, the body uses hormones to moderate energy stores. **Insulin** is produced by the beta cells of the pancreas, which are stimulated to release insulin as blood glucose levels rise (for example, after a meal is consumed). Insulin lowers blood glucose levels by enhancing the rate of glucose uptake and utilization by target cells, which use glucose for ATP production. It also stimulates the liver to convert glucose to glycogen, which is then stored by cells for later use. Insulin also increases glucose transport into certain cells, such as muscle cells and the liver. This results from an insulin-mediated increase in the number of glucose transporter proteins in cell membranes, which remove glucose from circulation by facilitated diffusion. As insulin binds to its target cell via insulin receptors and signal transduction, it triggers the cell to incorporate glucose transport proteins into its membrane. This allows glucose to enter the cell, where it can be used as an energy source. However, this does not occur in all cells: some cells, including those in the kidneys and brain, can access glucose without the use of insulin. Insulin also stimulates the conversion of glucose to fat in adipocytes and the synthesis of proteins. These actions mediated by insulin cause blood glucose concentrations to fall, called a hypoglycemic "low sugar" effect, which inhibits further insulin release from beta cells through a negative feedback loop.

LINK TO LEARNING

This animation describes the role of insulin and the pancreas in diabetes.

Click to view content (https://www.openstax.org/l/insulin)

Impaired insulin function can lead to a condition called **diabetes mellitus**, the main symptoms of which are illustrated in Figure 37.10. This can be caused by low levels of insulin production by the beta cells of the pancreas, or by reduced sensitivity of tissue cells to insulin. This prevents glucose from being absorbed by cells, causing high levels of blood glucose, or **hyperglycemia** (high sugar). High blood glucose levels make it difficult for the kidneys to recover all the glucose from nascent urine, resulting in glucose being lost in urine. High glucose levels also result in less water being reabsorbed by the kidneys, causing high amounts of urine to be produced; this may result in dehydration. Over time, high blood glucose levels can cause nerve damage to the eyes and peripheral body tissues, as well as damage to the kidneys and cardiovascular system. Oversecretion of insulin can cause **hypoglycemia**, low blood glucose levels. This causes insufficient glucose availability to cells, often leading to muscle weakness, and can sometimes cause unconsciousness or death if left untreated.

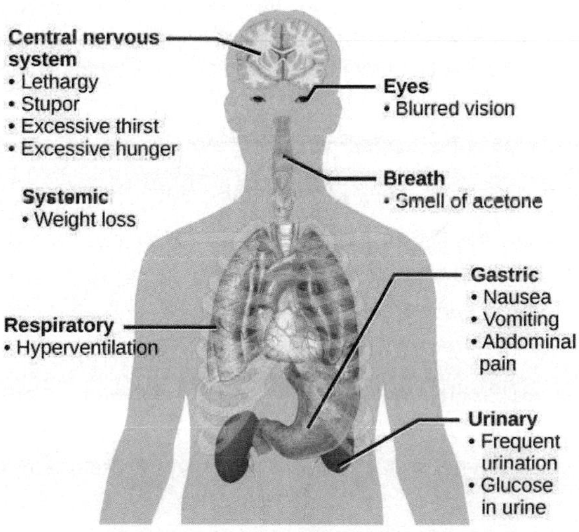

Figure 37.10 The main symptoms of diabetes are shown. (credit: modification of work by Mikael Häggström)

When blood glucose levels decline below normal levels, for example between meals or when glucose is utilized rapidly during exercise, the hormone **glucagon** is released from the alpha cells of the pancreas. Glucagon raises blood glucose levels, eliciting what is called a hyperglycemic effect, by stimulating the breakdown of glycogen to glucose in skeletal muscle cells and liver cells in a process called **glycogenolysis**. Glucose can then be utilized as energy by muscle cells and released into circulation by the liver cells. Glucagon also stimulates absorption of amino acids from the blood by the liver, which then converts them to glucose. This process of glucose synthesis is called **gluconeogenesis**. Glucagon also stimulates adipose cells to release fatty acids into the blood. These actions mediated by glucagon result in an increase in blood glucose levels to normal homeostatic levels. Rising blood glucose levels inhibit further glucagon release by the pancreas via a negative feedback mechanism. In this way, insulin and glucagon work together to maintain homeostatic glucose levels, as shown in Figure 37.11.

VISUAL CONNECTION

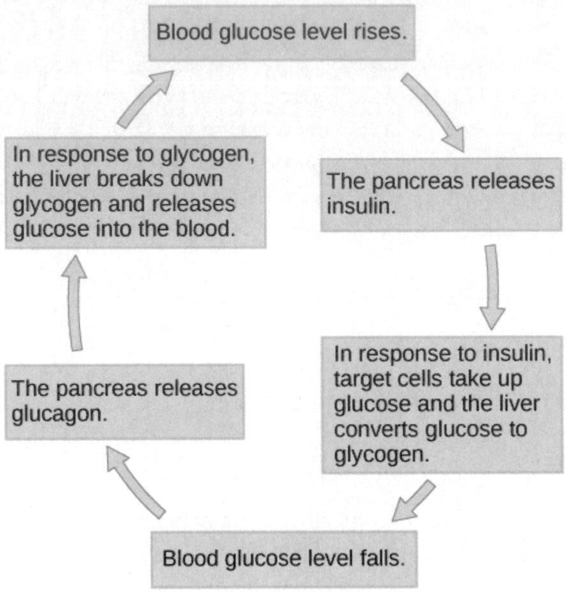

Figure 37.11 Insulin and glucagon regulate blood glucose levels.

Pancreatic tumors may cause excess secretion of glucagon. Type I diabetes results from the failure of the pancreas to produce insulin. Which of the following statement about these two conditions is true?

a. A pancreatic tumor and type I diabetes will have the opposite effects on blood sugar levels.
b. A pancreatic tumor and type I diabetes will both cause hyperglycemia.
c. A pancreatic tumor and type I diabetes will both cause hypoglycemia.
d. Both pancreatic tumors and type I diabetes result in the inability of cells to take up glucose.

Regulation of Blood Glucose Levels by Thyroid Hormones

The basal metabolic rate, which is the amount of calories required by the body at rest, is determined by two hormones produced by the thyroid gland: **thyroxine**, also known as tetraiodothyronine or T_4, and **triiodothyronine**, also known as T_3. These hormones affect nearly every cell in the body except for the adult brain, uterus, testes, blood cells, and spleen. They are transported across the plasma membrane of target cells and bind to receptors on the mitochondria resulting in increased ATP production. In the nucleus, T_3 and T_4 activate genes involved in energy production and glucose oxidation. This results in increased rates of metabolism and body heat production, which is known as the hormone's calorigenic effect.

T_3 and T_4 release from the thyroid gland is stimulated by **thyroid-stimulating hormone (TSH)**, which is produced by the anterior pituitary. TSH binding at the receptors of the follicle of the thyroid triggers the production of T_3 and T_4 from a glycoprotein called **thyroglobulin**. Thyroglobulin is present in the follicles of the thyroid, and is converted into thyroid hormones with the addition of iodine. Iodine is formed from iodide ions that are actively transported into the thyroid follicle from the bloodstream. A peroxidase enzyme then attaches the iodine to the tyrosine amino acid found in thyroglobulin. T_3 has three iodine ions attached, while T_4 has four iodine ions attached. T_3 and T_4 are then released into the bloodstream, with T_4 being released in much greater amounts than T_3. As T_3 is more active than T_4 and is responsible for most of the effects of thyroid hormones, tissues of the body convert T_4 to T_3 by the removal of an iodine ion. Most of the released T_3 and T_4 becomes attached to transport proteins in the bloodstream and is unable to cross the plasma membrane of cells. These protein-bound molecules are only released when blood levels of the unattached hormone begin to decline. In this way, a week's worth of reserve hormone is maintained in the blood. Increased T_3 and T_4 levels in the blood inhibit the release of TSH, which results in lower T_3 and T_4 release from the thyroid.

The follicular cells of the thyroid require iodides (anions of iodine) in order to synthesize T_3 and T_4. Iodides obtained from the diet are actively transported into follicle cells resulting in a concentration that is approximately 30 times higher than in blood. The typical diet in North America provides more iodine than required due to the addition of iodide to table salt. Inadequate iodine intake, which occurs in many developing countries, results in an inability to synthesize T_3 and T_4 hormones. The thyroid gland enlarges in a condition called **goiter**, which is caused by overproduction of TSH without the formation of thyroid hormone. Thyroglobulin is contained in a fluid called colloid, and TSH stimulation results in higher levels of colloid accumulation in the thyroid. In the absence of iodine, this is not converted to thyroid hormone, and colloid begins to accumulate more and more in the thyroid gland, leading to goiter.

Disorders can arise from both the underproduction and overproduction of thyroid hormones. **Hypothyroidism**, underproduction of the thyroid hormones, can cause a low metabolic rate leading to weight gain, sensitivity to cold, and reduced mental activity, among other symptoms. In children, hypothyroidism can cause cretinism, which can lead to mental retardation and growth defects. **Hyperthyroidism**, the overproduction of thyroid hormones, can lead to an increased metabolic rate and its effects: weight loss, excess heat production, sweating, and an increased heart rate. Graves' disease is one example of a hyperthyroid condition.

Hormonal Control of Blood Calcium Levels

Regulation of blood calcium concentrations is important for generation of muscle contractions and nerve impulses, which are electrically stimulated. If calcium levels get too high, membrane permeability to sodium decreases and membranes become less responsive. If calcium levels get too low, membrane permeability to sodium increases and convulsions or muscle spasms can result.

Blood calcium levels are regulated by **parathyroid hormone (PTH)**, which is produced by the parathyroid glands, as illustrated in Figure 37.12. PTH is released in response to low blood Ca^{2+} levels. PTH increases Ca^{2+} levels by targeting the skeleton, the kidneys, and the intestine. In the skeleton, PTH stimulates osteoclasts, which causes bone to be reabsorbed, releasing Ca^{2+} from bone into the blood. PTH also inhibits osteoblasts, reducing Ca^{2+} deposition in bone. In the intestines, PTH increases dietary Ca^{2+} absorption, and in the kidneys, PTH stimulates reabsorption of the CA^{2+}. While PTH acts directly on the kidneys to increase Ca^{2+} reabsorption, its effects on the intestine are indirect. PTH triggers the formation of calcitriol, an active form of vitamin D, which acts on the intestines to increase absorption of dietary calcium. PTH release is inhibited by rising blood calcium levels.

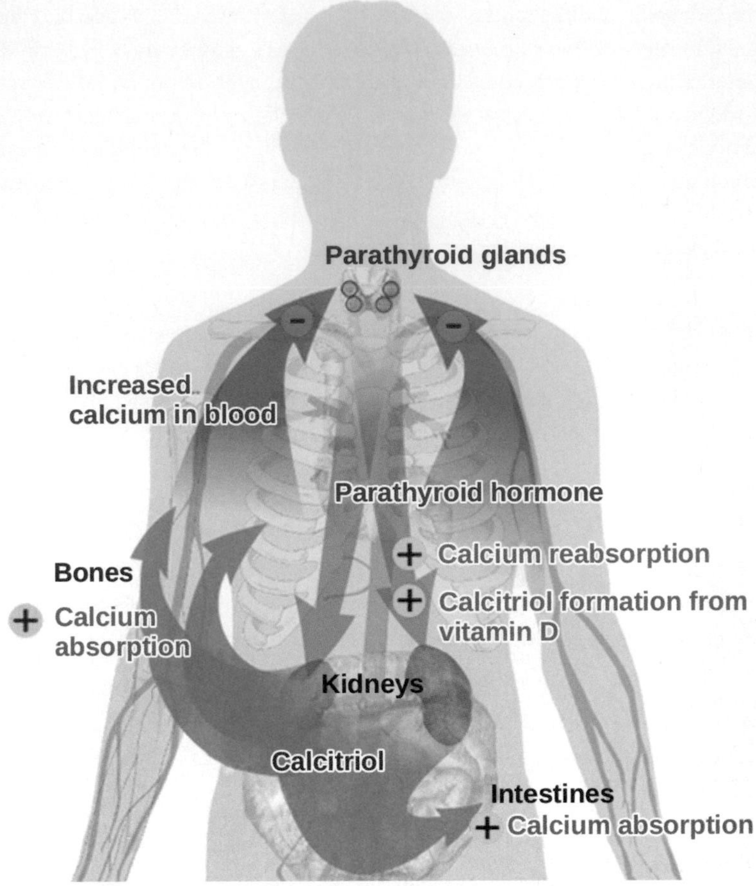

Figure 37.12 Parathyroid hormone (PTH) is released in response to low blood calcium levels. It increases blood calcium levels by targeting the skeleton, the kidneys, and the intestine. (credit: modification of work by Mikael Häggström)

Hyperparathyroidism results from an overproduction of parathyroid hormone. This results in excessive calcium being removed from bones and introduced into blood circulation, producing structural weakness of the bones, which can lead to deformation and fractures, plus nervous system impairment due to high blood calcium levels. Hypoparathyroidism, the underproduction of PTH, results in extremely low levels of blood calcium, which causes impaired muscle function and may result in tetany (severe sustained muscle contraction).

The hormone **calcitonin**, which is produced by the parafollicular or C cells of the thyroid, has the opposite effect on blood calcium levels as does PTH. Calcitonin decreases blood calcium levels by inhibiting osteoclasts, stimulating osteoblasts, and stimulating calcium excretion by the kidneys. This results in calcium being added to the bones to promote structural integrity. Calcitonin is most important in children (when it stimulates bone growth), during pregnancy (when it reduces maternal bone loss), and during prolonged starvation (because it reduces bone mass loss). In healthy nonpregnant, unstarved adults, the role of calcitonin is unclear.

Hormonal Regulation of Growth

Hormonal regulation is required for the growth and replication of most cells in the body. **Growth hormone (GH)**, produced by the anterior portion of the pituitary gland, accelerates the rate of protein synthesis, particularly in skeletal muscle and bones. Growth hormone has direct and indirect mechanisms of action. The first direct action of GH is stimulation of triglyceride breakdown (lipolysis) and release into the blood by adipocytes. This results in a switch by most tissues from utilizing glucose as an energy source to utilizing fatty acids. This process is called a **glucose-sparing effect**. In another direct mechanism, GH stimulates glycogen breakdown in the liver; the glycogen is then released into the blood as glucose. Blood glucose levels increase as most tissues are utilizing fatty acids instead of glucose for their energy needs. The GH mediated increase in blood glucose levels is called a **diabetogenic effect** because it is similar to the high blood glucose levels seen in diabetes mellitus.

The indirect mechanism of GH action is mediated by **insulin-like growth factors (IGFs)** or somatomedins, which are a family of growth-promoting proteins produced by the liver, which stimulates tissue growth. IGFs stimulate the uptake of amino acids from the blood, allowing the formation of new proteins, particularly in skeletal muscle cells, cartilage cells, and other target cells, as shown in Figure 37.13. This is especially important after a meal, when glucose and amino acid concentration levels are high in the blood. GH levels are regulated by two hormones produced by the hypothalamus. GH release is stimulated by **growth hormone-releasing hormone (GHRH)** and is inhibited by **growth hormone-inhibiting hormone (GHIH)**, also called somatostatin.

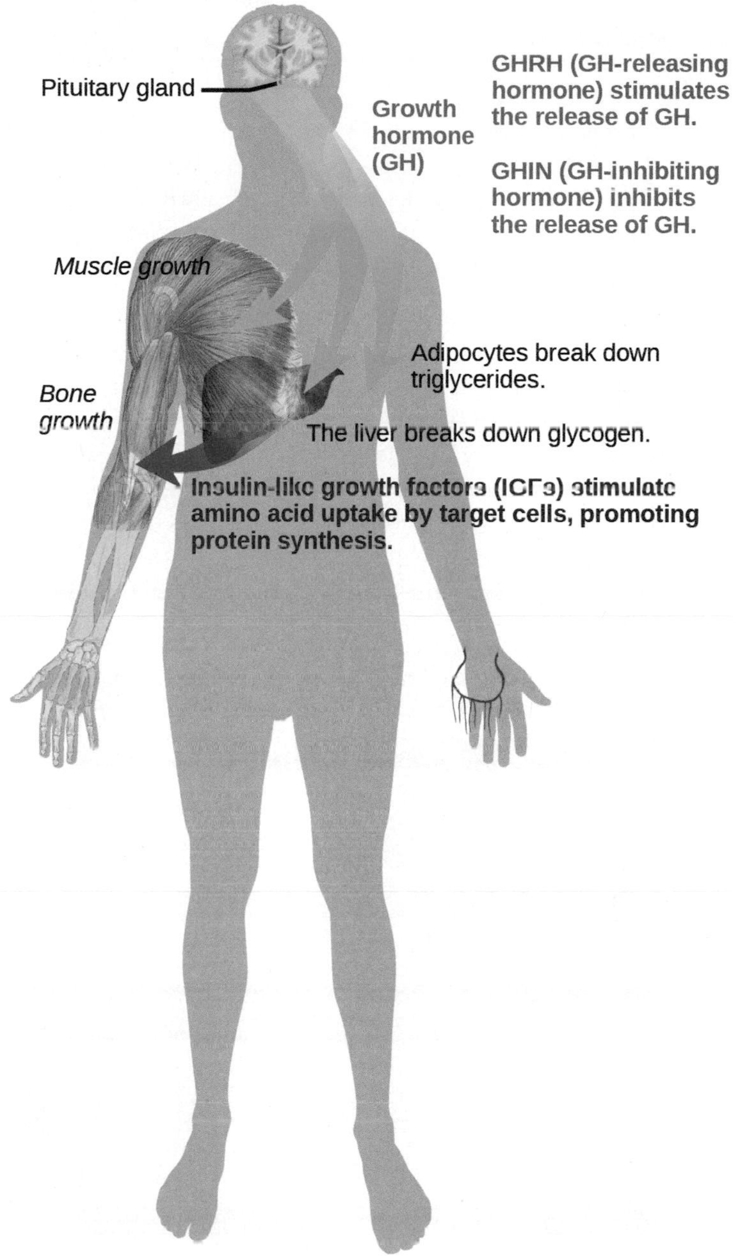

Figure 37.13 Growth hormone directly accelerates the rate of protein synthesis in skeletal muscle and bones. Insulin-like growth factor 1 (IGF-1) is activated by growth hormone and also allows formation of new proteins in muscle cells and bone. (credit: modification of work by Mikael Häggström)

A balanced production of growth hormone is critical for proper development. Underproduction of GH in adults does not appear to cause any abnormalities, but in children it can result in **pituitary dwarfism**, in which growth is reduced. Pituitary dwarfism is characterized by symmetric body formation. In some cases, individuals are under 30 inches in height. Oversecretion of growth hormone can lead to **gigantism** in children, causing excessive growth. In some documented cases, individuals can reach heights of over eight feet. In adults, excessive GH can lead to **acromegaly**, a condition in which there is enlargement of bones in the face, hands, and feet that are still capable of growth.

Hormonal Regulation of Stress

When a threat or danger is perceived, the body responds by releasing hormones that will ready it for the "fight-or-flight" response. The effects of this response are familiar to anyone who has been in a stressful situation: increased heart rate, dry mouth, and hair standing up.

EVOLUTION CONNECTION

Fight-or-Flight Response

Interactions of the endocrine hormones have evolved to ensure the body's internal environment remains stable. Stressors are stimuli that disrupt homeostasis. The sympathetic division of the vertebrate autonomic nervous system has evolved the fight-or-flight response to counter stress-induced disruptions of homeostasis. In the initial alarm phase, the sympathetic nervous system stimulates an increase in energy levels through increased blood glucose levels. This prepares the body for physical activity that may be required to respond to stress: to either fight for survival or to flee from danger.

However, some stresses, such as illness or injury, can last for a long time. Glycogen reserves, which provide energy in the short-term response to stress, are exhausted after several hours and cannot meet long-term energy needs. If glycogen reserves were the only energy source available, neural functioning could not be maintained once the reserves became depleted due to the nervous system's high requirement for glucose. In this situation, the body has evolved a response to counter long-term stress through the actions of the glucocorticoids, which ensure that long-term energy requirements can be met. The glucocorticoids mobilize lipid and protein reserves, stimulate gluconeogenesis, conserve glucose for use by neural tissue, and stimulate the conservation of salts and water. The mechanisms to maintain homeostasis that are described here are those observed in the human body. However, the fight-or-flight response exists in some form in all vertebrates.

The sympathetic nervous system regulates the stress response via the hypothalamus. Stressful stimuli cause the hypothalamus to signal the adrenal medulla (which mediates short-term stress responses) via nerve impulses, and the adrenal cortex, which mediates long-term stress responses, via the hormone **adrenocorticotropic hormone (ACTH)**, which is produced by the anterior pituitary.

Short-term Stress Response

When presented with a stressful situation, the body responds by calling for the release of hormones that provide a burst of energy. The hormones **epinephrine** (also known as adrenaline) and **norepinephrine** (also known as noradrenaline) are released by the adrenal medulla. How do these hormones provide a burst of energy? Epinephrine and norepinephrine increase blood glucose levels by stimulating the liver and skeletal muscles to break down glycogen and by stimulating glucose release by liver cells. Additionally, these hormones increase oxygen availability to cells by increasing the heart rate and dilating the bronchioles. The hormones also prioritize body function by increasing blood supply to essential organs such as the heart, brain, and skeletal muscles, while restricting blood flow to organs not in immediate need, such as the skin, digestive system, and kidneys. Epinephrine and norepinephrine are collectively called catecholamines.

LINK TO LEARNING

Watch this Discovery Channel animation (http://openstax.org/l/adrenaline) describing the flight-or-flight response.

Long-term Stress Response

Long-term stress response differs from short-term stress response. The body cannot sustain the bursts of energy mediated by epinephrine and norepinephrine for long times. Instead, other hormones come into play. In a long-term stress response, the hypothalamus triggers the release of ACTH from the anterior pituitary gland. The adrenal cortex is stimulated by ACTH to release steroid hormones called **corticosteroids**. Corticosteroids turn on transcription of certain genes in the nuclei of target cells. They change enzyme concentrations in the cytoplasm and affect cellular metabolism. There are two main corticosteroids: glucocorticoids such as **cortisol**, and mineralocorticoids such as aldosterone. These hormones target the breakdown of fat into fatty acids in the adipose tissue. The fatty acids are released into the bloodstream for other tissues to use for ATP production. The **glucocorticoids** primarily affect glucose metabolism by stimulating glucose synthesis. Glucocorticoids also have anti-inflammatory properties through inhibition of the immune system. For example, cortisone is used as an anti-inflammatory medication; however, it cannot be used long term as it increases susceptibility to disease due to its immune-suppressing effects.

Mineralocorticoids function to regulate ion and water balance of the body. The hormone aldosterone stimulates the reabsorption of water and sodium ions in the kidney, which results in increased blood pressure and volume.

Hypersecretion of glucocorticoids can cause a condition known as **Cushing's disease**, characterized by a shifting of fat storage areas of the body. This can cause the accumulation of adipose tissue in the face and neck, and excessive glucose in the blood. Hyposecretion of the corticosteroids can cause **Addison's disease**, which may result in bronzing of the skin, hypoglycemia, and low electrolyte levels in the blood.

37.4 Regulation of Hormone Production

By the end of this section, you will be able to do the following:

- Explain how hormone production is regulated
- Discuss the different stimuli that control hormone levels in the body

Hormone production and release are primarily controlled by negative feedback. In negative feedback systems, a stimulus elicits the release of a substance; once the substance reaches a certain level, it sends a signal that stops further release of the substance. In this way, the concentration of hormones in blood is maintained within a narrow range. For example, the anterior pituitary signals the thyroid to release thyroid hormones. Increasing levels of these hormones in the blood then give feedback to the hypothalamus and anterior pituitary to inhibit further signaling to the thyroid gland, as illustrated in Figure 37.14. There are three mechanisms by which endocrine glands are stimulated to synthesize and release hormones: humoral stimuli, hormonal stimuli, and neural stimuli.

VISUAL CONNECTION

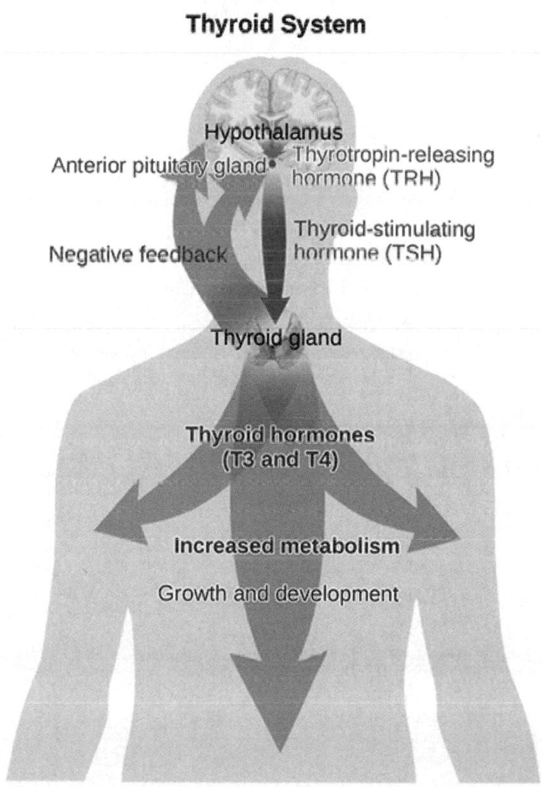

Figure 37.14 The anterior pituitary stimulates the thyroid gland to release thyroid hormones T_3 and T_4. Increasing levels of these hormones in the blood results in feedback to the hypothalamus and anterior pituitary to inhibit further signaling to the thyroid gland. (credit: modification of work by Mikael Häggström)

Hyperthyroidism is a condition in which the thyroid gland is overactive. Hypothyroidism is a condition in which the thyroid gland is underactive. Which of the conditions are the following two patients most likely to have?

Patient A has symptoms including weight gain, cold sensitivity, low heart rate, and fatigue.

Patient B has symptoms including weight loss, profuse sweating, increased heart rate, and difficulty sleeping.

Humoral Stimuli

The term “humoral” is derived from the term “humor,” which refers to bodily fluids such as blood. A **humoral stimulus** refers to the control of hormone release in response to changes in extracellular fluids such as blood or the ion concentration in the blood. For example, a rise in blood glucose levels triggers the pancreatic release of insulin. Insulin causes blood glucose levels to drop, which signals the pancreas to stop producing insulin in a negative feedback loop.

Hormonal Stimuli

Hormonal stimuli refers to the release of a hormone in response to another hormone. A number of endocrine glands release hormones when stimulated by hormones released by other endocrine glands. For example, the hypothalamus produces hormones that stimulate the anterior portion of the pituitary gland. The anterior pituitary in turn releases hormones that regulate hormone production by other endocrine glands. The anterior pituitary releases the thyroid-stimulating hormone, which then stimulates the thyroid gland to produce the hormones T_3 and T_4. As blood concentrations of T_3 and T_4 rise, they inhibit both the pituitary and the hypothalamus in a negative feedback loop.

Neural Stimuli

In some cases, the nervous system directly stimulates endocrine glands to release hormones, which is referred to as **neural stimuli**. Recall that in a short-term stress response, the hormones epinephrine and norepinephrine are important for providing the bursts of energy required for the body to respond. Here, neuronal signaling from the sympathetic nervous system directly stimulates the adrenal medulla to release the hormones epinephrine and norepinephrine in response to stress.

37.5 Endocrine Glands

By the end of this section, you will be able to do the following:

- Describe the role of different glands in the endocrine system
- Explain how the different glands work together to maintain homeostasis

Both the endocrine and nervous systems use chemical signals to communicate and regulate the body's physiology. The endocrine system releases hormones that act on target cells to regulate development, growth, energy metabolism, reproduction, and many behaviors. The nervous system releases neurotransmitters or neurohormones that regulate neurons, muscle cells, and endocrine cells. Because the neurons can regulate the release of hormones, the nervous and endocrine systems work in a coordinated manner to regulate the body's physiology.

Hypothalamic-Pituitary Axis

The **hypothalamus** in vertebrates integrates the endocrine and nervous systems. The hypothalamus is an endocrine organ located in the diencephalon of the brain. It receives input from the body and other brain areas and initiates endocrine responses to environmental changes. The hypothalamus acts as an endocrine organ, synthesizing hormones and transporting them along axons to the posterior pituitary gland. It synthesizes and secretes regulatory hormones that control the endocrine cells in the anterior pituitary gland. The hypothalamus contains autonomic centers that control endocrine cells in the adrenal medulla via neuronal control.

The **pituitary gland**, sometimes called the hypophysis or “master gland” is located at the base of the brain in the sella turcica, a groove of the sphenoid bone of the skull, illustrated in Figure 37.15. It is attached to the hypothalamus via a stalk called the **pituitary stalk** (or infundibulum). The anterior portion of the pituitary gland is regulated by releasing or release-inhibiting hormones produced by the hypothalamus, and the posterior pituitary receives signals via neurosecretory cells to release hormones produced by the hypothalamus. The pituitary has two distinct regions—the anterior pituitary and the posterior pituitary—which between them secrete nine different peptide or protein hormones. The posterior lobe of the pituitary gland contains axons of the hypothalamic neurons.

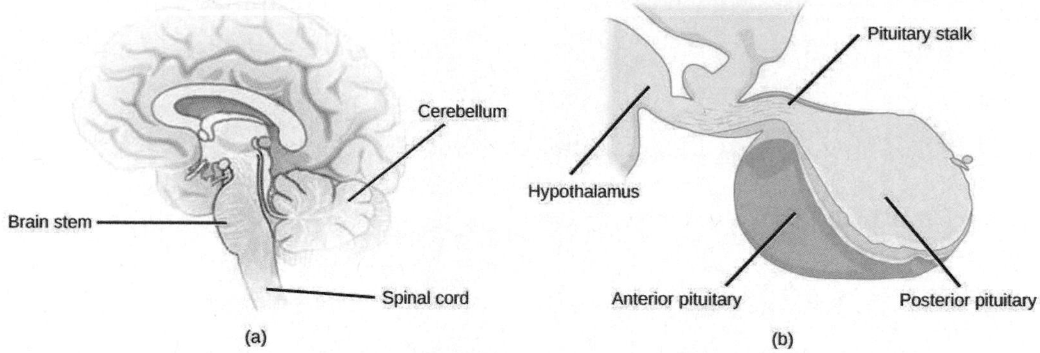

Figure 37.15 The pituitary gland is located at (a) the base of the brain and (b) connected to the hypothalamus by the pituitary stalk. (credit a: modification of work by NCI; credit b: modification of work by Gray's Anatomy)

Anterior Pituitary

The **anterior pituitary** gland, or adenohypophysis, is surrounded by a capillary network that extends from the hypothalamus, down along the infundibulum, and to the anterior pituitary. This capillary network is a part of the **hypophyseal portal system** that carries substances from the hypothalamus to the anterior pituitary and hormones from the anterior pituitary into the circulatory system. A portal system carries blood from one capillary network to another; therefore, the hypophyseal portal system allows hormones produced by the hypothalamus to be carried directly to the anterior pituitary without first entering the circulatory system.

The anterior pituitary produces seven hormones: growth hormone (GH), prolactin (PRL), thyroid-stimulating hormone (TSH), melanin-stimulating hormone (MSH), adrenocorticotropic hormone (ACTH), follicle-stimulating hormone (FSH), and luteinizing hormone (LH). Anterior pituitary hormones are sometimes referred to as tropic hormones, because they control the functioning of other organs. While these hormones are produced by the anterior pituitary, their production is controlled by regulatory hormones produced by the hypothalamus. These regulatory hormones can be releasing hormones or inhibiting hormones, causing more or less of the anterior pituitary hormones to be secreted. These travel from the hypothalamus through the hypophyseal portal system to the anterior pituitary where they exert their effect. Negative feedback then regulates how much of these regulatory hormones are released and how much anterior pituitary hormone is secreted.

Posterior Pituitary

The **posterior pituitary** is significantly different in structure from the anterior pituitary. It is a part of the brain, extending down from the hypothalamus, and contains mostly nerve fibers and neuroglial cells, which support axons that extend from the hypothalamus to the posterior pituitary. The posterior pituitary and the infundibulum together are referred to as the neurohypophysis.

The hormones antidiuretic hormone (ADH), also known as vasopressin, and oxytocin are produced by neurons in the hypothalamus and transported within these axons along the infundibulum to the posterior pituitary. They are released into the circulatory system via neural signaling from the hypothalamus. These hormones are considered to be posterior pituitary hormones, even though they are produced by the hypothalamus, because that is where they are released into the circulatory system. The posterior pituitary itself does not produce hormones, but instead stores hormones produced by the hypothalamus and releases them into the bloodstream.

Thyroid Gland

The **thyroid gland** is located in the neck, just below the larynx and in front of the trachea, as shown in Figure 37.16. It is a butterfly-shaped gland with two lobes that are connected by the **isthmus**. It has a dark red color due to its extensive vascular system. When the thyroid swells due to dysfunction, it can be felt under the skin of the neck.

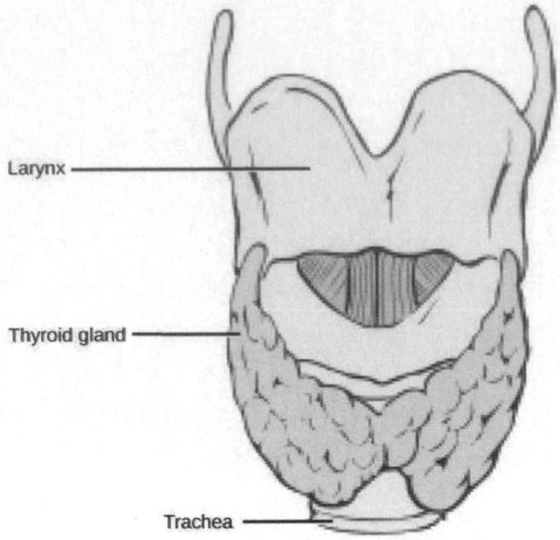

Figure 37.16 This illustration shows the location of the thyroid gland.

The thyroid gland is made up of many spherical thyroid follicles, which are lined with a simple cuboidal epithelium. These follicles contain a viscous fluid, called **colloid**, which stores the glycoprotein thyroglobulin, the precursor to the thyroid hormones. The follicles produce hormones that can be stored in the colloid or released into the surrounding capillary network for transport to the rest of the body via the circulatory system.

Thyroid follicle cells synthesize the hormone thyroxine, which is also known as T_4 because it contains four atoms of iodine, and triiodothyronine, also known as T_3 because it contains three atoms of iodine. Follicle cells are stimulated to release stored T_3 and T_4 by thyroid stimulating hormone (TSH), which is produced by the anterior pituitary. These thyroid hormones increase the rates of mitochondrial ATP production.

A third hormone, calcitonin, is produced by **parafollicular cells** of the thyroid either releasing hormones or inhibiting hormones. Calcitonin release is not controlled by TSH, but instead is released when calcium ion concentrations in the blood rise. Calcitonin functions to help regulate calcium concentrations in body fluids. It acts in the bones to inhibit osteoclast activity and in the kidneys to stimulate excretion of calcium. The combination of these two events lowers body fluid levels of calcium.

Parathyroid Glands

Most people have four **parathyroid glands**; however, the number can vary from two to six. These glands are located on the posterior surface of the thyroid gland, as shown in Figure 37.17. Normally, there is a superior gland and an inferior gland associated with each of the thyroid's two lobes. Each parathyroid gland is covered by connective tissue and contains many secretory cells that are associated with a capillary network.

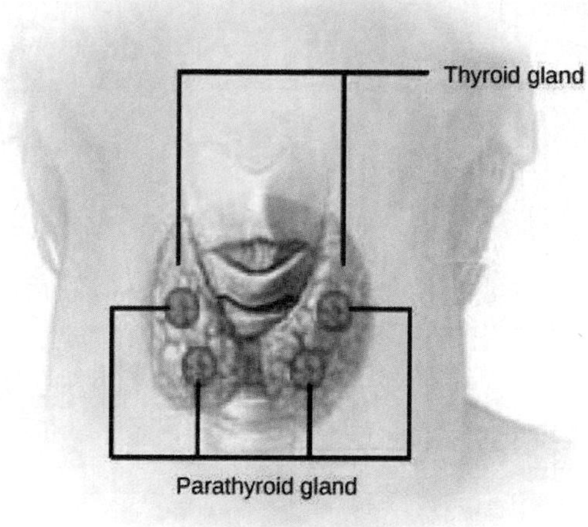

Figure 37.17 The parathyroid glands are located on the posterior of the thyroid gland. (credit: modification of work by NCI)

The parathyroid glands produce parathyroid hormone (PTH). PTH increases blood calcium concentrations when calcium ion levels fall below normal. PTH (1) enhances reabsorption of Ca^{2+} by the kidneys, (2) stimulates osteoclast activity and inhibits osteoblast activity, and (3) it stimulates synthesis and secretion of calcitriol by the kidneys, which enhances Ca^{2+} absorption by the digestive system. PTH is produced by chief cells of the parathyroid. PTH and calcitonin work in opposition to one another to maintain homeostatic Ca^{2+} levels in body fluids. Another type of cells, oxyphil cells, exist in the parathyroid but their function is not known. These hormones encourage bone growth, muscle mass, and blood cell formation in children and women.

Adrenal Glands

The **adrenal glands** are associated with the kidneys; one gland is located on top of each kidney as illustrated in Figure 37.18. The adrenal glands consist of an outer adrenal cortex and an inner adrenal medulla. These regions secrete different hormones.

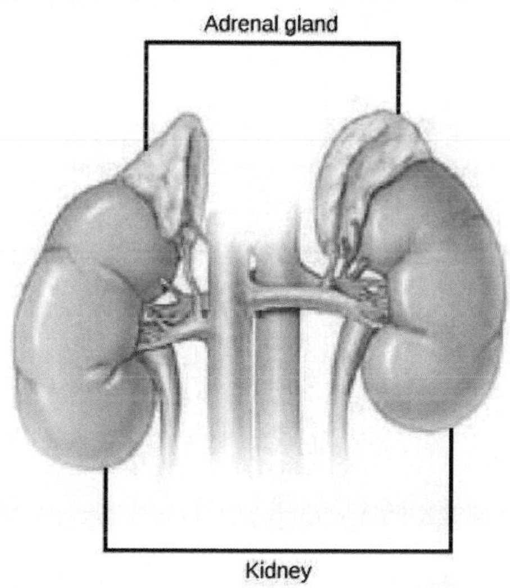

Figure 37.18 The location of the adrenal glands on top of the kidneys is shown. (credit: modification of work by NCI)

Adrenal Cortex

The **adrenal cortex** is made up of layers of epithelial cells and associated capillary networks. These layers form three distinct regions: an outer zona glomerulosa that produces mineralocorticoids, a middle zona fasciculata that produces glucocorticoids, and an inner zona reticularis that produces androgens.

The main mineralocorticoid is aldosterone, which regulates the concentration of Na^+ ions in urine, sweat, pancreas, and saliva. Aldosterone release from the adrenal cortex is stimulated by a decrease in blood concentrations of sodium ions, blood volume, or blood pressure, or by an increase in blood potassium levels.

The three main glucocorticoids are cortisol, corticosterone, and cortisone. The glucocorticoids stimulate the synthesis of glucose and gluconeogenesis (converting a non-carbohydrate to glucose) by liver cells and they promote the release of fatty acids from adipose tissue. These hormones increase blood glucose levels to maintain levels within a normal range between meals. These hormones are secreted in response to ACTH and levels are regulated by negative feedback.

Androgens are sex hormones that promote masculinity. They are produced in small amounts by the adrenal cortex in both males and females. They do not affect sexual characteristics and may supplement sex hormones released from the gonads.

Adrenal Medulla

The **adrenal medulla** contains large, irregularly shaped cells that are closely associated with blood vessels. These cells are innervated by preganglionic autonomic nerve fibers from the central nervous system.

The adrenal medulla contains two types of secretory cells: one that produces epinephrine (adrenaline) and another that produces norepinephrine (noradrenaline). Epinephrine is the primary adrenal medulla hormone accounting for 75 to 80 percent of its secretions. Epinephrine and norepinephrine increase heart rate, breathing rate, cardiac muscle contractions, blood pressure, and blood glucose levels. They also accelerate the breakdown of glucose in skeletal muscles and stored fats in adipose tissue.

The release of epinephrine and norepinephrine is stimulated by neural impulses from the sympathetic nervous system. Secretion of these hormones is stimulated by acetylcholine release from preganglionic sympathetic fibers innervating the adrenal medulla. These neural impulses originate from the hypothalamus in response to stress to prepare the body for the fight-or-flight response.

Pancreas

The **pancreas**, illustrated in Figure 37.19, is an elongated organ that is located between the stomach and the proximal portion of the small intestine. It contains both exocrine cells that excrete digestive enzymes and endocrine cells that release hormones. It is sometimes referred to as a heterocrine gland because it has both endocrine and exocrine functions.

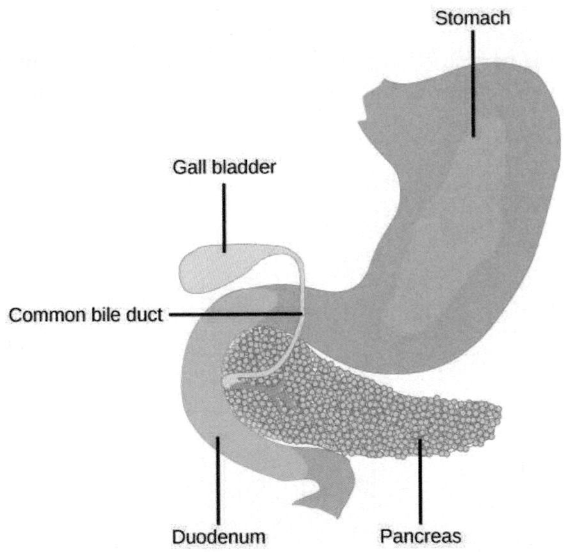

Figure 37.19 The pancreas is found underneath the stomach and points toward the spleen. (credit: modification of work by NCI)

The endocrine cells of the pancreas form clusters called pancreatic islets or the **islets of Langerhans**, as visible in the micrograph shown in Figure 37.20. The pancreatic islets contain two primary cell types: **alpha cells**, which produce the hormone glucagon, and **beta cells**, which produce the hormone insulin. These hormones regulate blood glucose levels. As blood glucose levels decline, alpha cells release glucagon to raise the blood glucose levels by increasing rates of glycogen breakdown and glucose release by the liver. When blood glucose levels rise, such as after a meal, beta cells release insulin to lower blood glucose levels by increasing the rate of glucose uptake in most body cells, and by increasing glycogen synthesis in skeletal muscles and the liver. Together, glucagon and insulin regulate blood glucose levels.

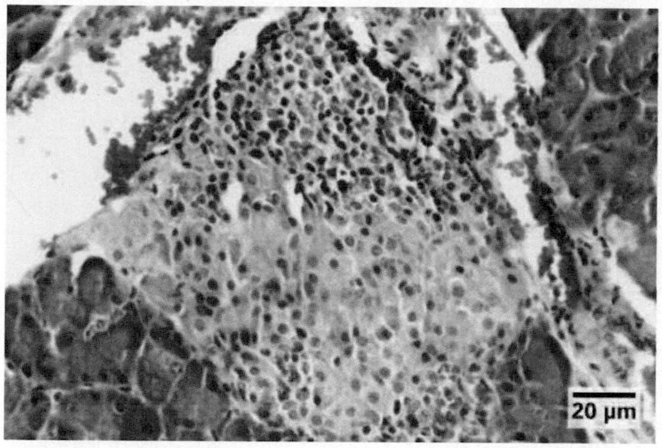

Figure 37.20 The islets of Langerhans are clusters of endocrine cells found in the pancreas; they stain lighter than surrounding cells. (credit: modification of work by Muhammad T. Tabiin, Christopher P. White, Grant Morahan, and Bernard E. Tuch; scale-bar data from Matt Russell)

Pineal Gland

The pineal gland produces melatonin. The rate of melatonin production is affected by the photoperiod. Collaterals from the visual pathways innervate the pineal gland. During the day photoperiod, little melatonin is produced; however, melatonin production increases during the dark photoperiod (night). In some mammals, melatonin has an inhibitory affect on reproductive functions by decreasing production and maturation of sperm, oocytes, and reproductive organs. Melatonin is an effective antioxidant, protecting the CNS from free radicals such as nitric oxide and hydrogen peroxide. Lastly, melatonin is involved in biological rhythms, particularly circadian rhythms such as the sleep-wake cycle and eating habits.

Gonads

The gonads—the male testes and female ovaries—produce steroid hormones. The testes produce androgens, testosterone being the most prominent, which allow for the development of secondary sex characteristics and the production of sperm cells. The ovaries produce estradiol and progesterone, which cause secondary sex characteristics and prepare the body for childbirth.

Endocrine Glands and their Associated Hormones

Endocrine Gland	Associated Hormones	Effect
Hypothalamus	releasing and inhibiting hormones	regulate hormone release from pituitary gland; produce oxytocin; produce uterine contractions and milk secretion in females
	antidiuretic hormone (ADH)	water reabsorption from kidneys; vasoconstriction to increase blood pressure
Pituitary (Anterior)	growth hormone (GH)	promotes growth of body tissues, protein synthesis; metabolic functions
	prolactin (PRL)	promotes milk production
	thyroid stimulating hormone (TSH)	stimulates thyroid hormone release
	adrenocorticotropic hormone (ACTH)	stimulates hormone release by adrenal cortex, glucocorticoids

Table 37.1

Endocrine Gland	Associated Hormones	Effect
	follicle-stimulating hormone (FSH)	stimulates gamete production (both ova and sperm); secretion of estradiol
	luteinizing hormone (LH)	stimulates androgen production by gonads; ovulation, secretion of progesterone
	melanocyte-stimulating hormone (MSH)	stimulates melanocytes of the skin increasing melanin pigment production.
Pituitary (Posterior)	antidiuretic hormone (ADH)	stimulates water reabsorption by kidneys
	oxytocin	stimulates uterine contractions during childbirth; milk ejection; stimulates ductus deferens and prostate gland contraction during emission
Thyroid	thyroxine, triiodothyronine	stimulate and maintain metabolism; growth and development
	calcitonin	reduces blood Ca^{2+} levels
Parathyroid	parathyroid hormone (PTH)	increases blood Ca^{2+} levels
Adrenal (Cortex)	aldosterone	increases blood Na^{+} levels; increases K^{+} secretion
	cortisol, corticosterone, cortisone	increase blood glucose levels; anti-inflammatory effects
Adrenal (Medulla)	epinephrine, norepinephrine	stimulate fight-or-flight response; increase blood gluclose levels; increase metabolic activities
Pancreas	insulin	reduces blood glucose levels
	glucagon	increases blood glucose levels
Pineal gland	melatonin	regulates some biological rhythms and protects CNS from free radicals
Testes	androgens	regulate, promote, increase or maintain sperm production; male secondary sexual characteristics
Ovaries	estrogen	promotes uterine lining growth; female secondary sexual characteristics
	progestins	promote and maintain uterine lining growth

Table 37.1

Organs with Secondary Endocrine Functions

There are several organs whose primary functions are non-endocrine but that also possess endocrine functions. These include the heart, kidneys, intestines, thymus, gonads, and adipose tissue.

The heart possesses endocrine cells in the walls of the atria that are specialized cardiac muscle cells. These cells release the hormone **atrial natriuretic peptide (ANP)** in response to increased blood volume. High blood volume causes the cells to be stretched, resulting in hormone release. ANP acts on the kidneys to reduce the reabsorption of Na^+, causing Na^+ and water to be excreted in the urine. ANP also reduces the amounts of renin released by the kidneys and aldosterone released by the adrenal cortex, further preventing the retention of water. In this way, ANP causes a reduction in blood volume and blood pressure, and reduces the concentration of Na^+ in the blood.

The gastrointestinal tract produces several hormones that aid in digestion. The endocrine cells are located in the mucosa of the GI tract throughout the stomach and small intestine. Some of the hormones produced include gastrin, secretin, and cholecystokinin, which are secreted in the presence of food, and some of which act on other organs such as the pancreas, gallbladder, and liver. They trigger the release of gastric juices, which help to break down and digest food in the GI tract.

While the adrenal glands associated with the kidneys are major **endocrine glands**, the kidneys themselves also possess endocrine function. Renin is released in response to decreased blood volume or pressure and is part of the renin-angiotensin-aldosterone system that leads to the release of aldosterone. Aldosterone then causes the retention of Na^+ and water, raising blood volume. The kidneys also release calcitriol, which aids in the absorption of Ca^{2+} and phosphate ions. **Erythropoietin (EPO)** is a protein hormone that triggers the formation of red blood cells in the bone marrow. EPO is released in response to low oxygen levels. Because red blood cells are oxygen carriers, increased production results in greater oxygen delivery throughout the body. EPO has been used by athletes to improve performance, as greater oxygen delivery to muscle cells allows for greater endurance. Because red blood cells increase the viscosity of blood, artificially high levels of EPO can cause severe health risks.

The **thymus** is found behind the sternum; it is most prominent in infants, becoming smaller in size through adulthood. The thymus produces hormones referred to as thymosins, which contribute to the development of the immune response.

Adipose tissue is a connective tissue found throughout the body. It produces the hormone **leptin** in response to food intake. Leptin increases the activity of anorexigenic neurons and decreases that of orexigenic neurons, producing a feeling of satiety after eating, thus affecting appetite and reducing the urge for further eating. Leptin is also associated with reproduction. It must be present for GnRH and gonadotropin synthesis to occur. Extremely thin females may enter puberty late; however, if adipose levels increase, more leptin will be produced, improving fertility.

KEY TERMS

acromegaly condition caused by overproduction of GH in adults

Addison's disease disorder caused by the hyposecretion of corticosteroids

adenylate cyclase an enzyme that catalyzes the conversion of ATP to cyclic AMP

adrenal cortex outer portion of adrenal glands that produces corticosteroids

adrenal gland endocrine glands associated with the kidneys

adrenal medulla inner portion of adrenal glands that produces epinephrine and norepinephrine

adrenocorticotropic hormone (ACTH) hormone released by the anterior pituitary, which stimulates the adrenal cortex to release corticosteroids during the long-term stress response

aldosterone steroid hormone produced by the adrenal cortex that stimulates the reabsorption of Na^+ from extracellular fluids and secretion of K^+

alpha cell endocrine cell of the pancreatic islets that produces the hormone glucagon

amino acid-derived hormone hormone derived from amino acids

androgen male sex hormone such as testosterone

anterior pituitary portion of the pituitary gland that produces six hormones; also called adenohypophysis

antidiuretic hormone (ADH) hormone produced by the hypothalamus and released by the posterior pituitary that increases water reabsorption by the kidneys

atrial natriuretic peptide (ANP) hormone produced by the heart to reduce blood volume, pressure, and Na^+ concentration

beta cell endocrine cell of the pancreatic islets that produces the hormone insulin

calcitonin hormone produced by the parafollicular cells of the thyroid gland that functions to lower blood Ca^{2+} levels and promote bone growth

colloid fluid inside the thyroid gland that contains the glycoprotein thyroglobulin

corticosteroid hormone released by the adrenal cortex in response to long-term stress

cortisol glucocorticoid produced in response to stress

Cushing's disease disorder caused by the hypersecretion of glucocorticoids

diabetes insipidus disorder caused by underproduction of ADH

diabetes mellitus disorder caused by low levels of insulin activity

diabetogenic effect effect of GH that causes blood glucose levels to rise similar to diabetes mellitus

down-regulation a decrease in the number of hormone receptors in response to increased hormone levels

endocrine gland gland that secretes hormones into the surrounding interstitial fluid, which then diffuse into blood and are carried to various organs and tissues within the body

epinephrine hormone released by the adrenal medulla in response to a short term stress

erythropoietin (EPO) hormone produced by the kidneys to stimulate red blood cell production in the bone marrow

estrogens a group of steroid hormones, including estradiol and several others, that are produced by the ovaries and elicit secondary sex characteristics in females as well as control the maturation of the ova

first messenger the hormone that binds to a plasma membrane hormone receptor to trigger a signal transduction pathway

follicle-stimulating hormone (FSH) hormone produced by the anterior pituitary that stimulates gamete production

G-protein a membrane protein activated by the hormone first messenger to activate formation of cyclic AMP

gigantism condition caused by overproduction of GH in children

glucagon hormone produced by the alpha cells of the pancreas in response to low blood sugar; functions to raise blood sugar levels

glucocorticoid corticosteroid that affects glucose metabolism

gluconeogenesis synthesis of glucose from amino acids

glucose-sparing effect effect of GH that causes tissues to use fatty acids instead of glucose as an energy source

glycogenolysis breakdown of glycogen into glucose

goiter enlargement of the thyroid gland caused by insufficient dietary iodine levels

gonadotropin hormone that regulates the gonads, including FSH and LH

growth hormone (GH) hormone produced by the anterior pituitary that promotes protein synthesis and body growth

growth hormone-inhibiting hormone (GHIH) hormone produced by the hypothalamus that inhibits growth hormone production, also called somatostatin

growth hormone-releasing hormone (GHRH) hormone released by the hypothalamus that triggers the release of GH

hormonal stimuli release of a hormone in response to another hormone

hormone receptor the cellular protein that binds to a hormone

humoral stimuli control of hormone release in response to changes in extracellular fluids such as blood or the ion concentration in the blood

hyperglycemia high blood sugar level

hyperthyroidism overactivity of the thyroid gland

hypoglycemia low blood sugar level

hypophyseal portal system system of blood vessels that carries hormones from the hypothalamus to the anterior pituitary

hypothyroidism underactivity of the thyroid gland

insulin hormone produced by the beta cells of the pancreas in response to high blood glucose levels; functions to lower blood glucose levels

insulin-like growth factor (IGF) growth-promoting protein produced by the liver

intracellular hormone receptor a hormone receptor in the cytoplasm or nucleus of a cell

islets of Langerhans (pancreatic islets) endocrine cells of the pancreas

isthmus tissue mass that connects the two lobes of the thyroid gland

leptin hormone produced by adipose tissue that promotes feelings of satiety and reduces hunger

lipid-derived hormone hormone derived mostly from cholesterol

mineralocorticoid corticosteroid that affects ion and water balance

neural stimuli stimulation of endocrine glands by the nervous system

norepinephrine hormone released by the adrenal medulla in response to a short-term stress hormone production by the gonads

osmoreceptor receptor in the hypothalamus that monitors the concentration of electrolytes in the blood

oxytocin hormone released by the posterior pituitary to stimulate uterine contractions during childbirth and milk let-down in the mammary glands

pancreas organ located between the stomach and the small intestine that contains exocrine and endocrine cells

parafollicular cell thyroid cell that produces the hormone calcitonin

parathyroid gland gland located on the surface of the thyroid that produces parathyroid hormone

parathyroid hormone (PTH) hormone produced by the parathyroid glands in response to low blood Ca^{2+} levels; functions to raise blood Ca^{2+} levels

peptide hormone hormone composed of a polypeptide chain

phosphodiesterase (PDE) enzyme that deactivates cAMP, stopping hormone activity

pituitary dwarfism condition caused by underproduction of GH in children

pituitary gland endocrine gland located at the base of the brain composed of an anterior and posterior region; also called hypophysis

pituitary stalk (also, infundibulum) stalk that connects the pituitary gland to the hypothalamus

plasma membrane hormone receptor a hormone receptor on the surface of the plasma membrane of a cell

posterior pituitary extension of the brain that releases hormones produced by the hypothalamus; along with the infundibulum, it is also referred to as the neurohypophysis

prolactin (PRL) hormone produced by the anterior pituitary that stimulates milk production

prolactin-inhibiting hormone hormone produced by the hypothalamus that inhibits the release of prolactin

prolactin-releasing hormone hormone produced by the hypothalamus that stimulates the release of prolactin

renin enzyme produced by the juxtaglomerular apparatus of the kidneys that reacts with angiotensinogen to cause the release of aldosterone

thymus gland located behind the sternum that produces thymosin hormones that contribute to the development of the immune system

thyroglobulin glycoprotein found in the thyroid that is converted into thyroid hormone

thyroid gland endocrine gland located in the neck that produces thyroid hormones thyroxine and triiodothyronine

thyroid-stimulating hormone (TSH) hormone produced by the anterior pituitary that controls the release of T_3 and T_4 from the thyroid gland

thyroxine (tetraiodothyronine, T_4) thyroid hormone containing 4 iodines that controls the basal metabolic rate

triiodothyronine (T_3) thyroid hormone containing 3 iodines that controls the basal metabolic rate

up-regulation an increase in the number of hormone receptors in response to increased hormone levels

CHAPTER SUMMARY

37.1 Types of Hormones

There are three basic types of hormones: lipid-derived, amino acid-derived, and peptide. Lipid-derived hormones are structurally similar to cholesterol and include steroid hormones such as estradiol and testosterone. Amino acid-derived hormones are relatively small molecules and include the adrenal hormones epinephrine and norepinephrine. Peptide hormones are polypeptide chains or proteins and include the pituitary hormones, antidiuretic hormone (vasopressin), and oxytocin.

37.2 How Hormones Work

Hormones cause cellular changes by binding to receptors on target cells. The number of receptors on a target cell can increase or decrease in response to hormone activity. Hormones can affect cells directly through intracellular hormone receptors or indirectly through plasma membrane hormone receptors.

Lipid-derived (soluble) hormones can enter the cell by diffusing across the plasma membrane and binding to DNA to regulate gene transcription and to change the cell's activities by inducing production of proteins that affect, in general, the long-term structure and function of the cell. Lipid insoluble hormones bind to receptors on the plasma membrane surface and trigger a signaling pathway to change the cell's activities by inducing production of various cell products that affect the cell in the short-term. The hormone is called a first messenger and the cellular component is called a second messenger. G-proteins activate the second messenger (cyclic AMP), triggering the cellular response. Response to hormone binding is amplified as the signaling pathway progresses. Cellular responses to hormones include the production of proteins and enzymes and altered membrane permeability.

37.3 Regulation of Body Processes

Water levels in the body are controlled by antidiuretic hormone (ADH), which is produced in the hypothalamus and triggers the reabsorption of water by the kidneys. Underproduction of ADH can cause diabetes insipidus. Aldosterone, a hormone produced by the adrenal cortex of the kidneys, enhances Na^+ reabsorption from the extracellular fluids and subsequent water reabsorption by diffusion. The renin-angiotensin-aldosterone system is one way that aldosterone release is controlled.

The reproductive system is controlled by the gonadotropins follicle-stimulating hormone (FSH) and luteinizing hormone (LH), which are produced by the pituitary gland. Gonadotropin release is controlled by the hypothalamic hormone gonadotropin-releasing hormone (GnRH). FSH stimulates the maturation of sperm cells in males and is inhibited by the hormone inhibin, while LH stimulates the production of the androgen testosterone. FSH stimulates egg maturation in females, while LH stimulates the production of estrogens and progesterone. Estrogens are a group of steroid hormones produced by the ovaries that trigger the development of secondary sex characteristics in females as well as control the maturation of the ova. In females, the pituitary also produces prolactin, which stimulates milk production after childbirth, and oxytocin, which stimulates uterine contraction during childbirth and milk let-down during suckling.

Insulin is produced by the pancreas in response to rising blood glucose levels and allows cells to utilize blood glucose and store excess glucose for later use. Diabetes mellitus is caused by reduced insulin activity and causes high blood glucose levels, or hyperglycemia. Glucagon is released by the pancreas in response to low blood glucose levels and stimulates the breakdown of glycogen into glucose, which can be used by the body. The body's basal metabolic rate is controlled by the thyroid hormones thyroxine (T_4) and triiodothyronine (T_3). The anterior pituitary produces thyroid stimulating hormone (TSH), which controls the release of T_3 and T_4 from the thyroid gland. Iodine is necessary in the production of thyroid hormone, and the lack of iodine can lead to a condition called goiter.

Parathyroid hormone (PTH) is produced by the parathyroid glands in response to low blood Ca^{2+} levels. The parafollicular cells of the thyroid produce calcitonin, which reduces blood Ca^{2+} levels. Growth hormone (GH) is produced by the anterior pituitary and controls the growth rate of muscle and bone. GH action is indirectly mediated by insulin-like growth factors (IGFs). Short-term stress causes the hypothalamus to trigger the adrenal medulla to release epinephrine and norepinephrine, which trigger the fight or flight response. Long-term stress causes the hypothalamus to trigger the anterior pituitary to release adrenocorticotropic hormone (ACTH), which causes the release of corticosteroids, glucocorticoids, and mineralocorticoids, from the adrenal cortex.

37.4 Regulation of Hormone Production

Hormone levels are primarily controlled through negative feedback, in which rising levels of a hormone inhibit its further release. The three mechanisms of hormonal release are humoral stimuli, hormonal stimuli, and neural stimuli. Humoral stimuli refers to the control of hormonal release in response to changes in extracellular fluid levels or ion levels. Hormonal stimuli refers to the release of hormones in response to hormones released by other endocrine glands. Neural stimuli refers to the release of hormones in response to neural stimulation.

37.5 Endocrine Glands

The pituitary gland is located at the base of the brain and is attached to the hypothalamus by the infundibulum. The anterior pituitary receives products from the hypothalamus by the hypophyseal portal system and produces six hormones. The posterior pituitary is an extension of the brain and releases hormones (antidiuretic hormone and oxytocin) produced by the hypothalamus.

The thyroid gland is located in the neck and is composed of two lobes connected by the isthmus. The thyroid is made up of follicle cells that produce the hormones thyroxine and triiodothyronine. Parafollicular cells of the thyroid produce calcitonin. The parathyroid glands lie on the posterior surface of the thyroid gland and produce parathyroid hormone.

The adrenal glands are located on top of the kidneys and consist of the renal cortex and renal medulla. The adrenal cortex is the outer part of the adrenal gland and produces the corticosteroids, glucocorticoids, and mineralocorticoids. The adrenal medulla is the inner part of the adrenal gland and produces the catecholamines epinephrine and norepinephrine.

The pancreas lies in the abdomen between the stomach and the small intestine. Clusters of endocrine cells in the pancreas form the islets of Langerhans, which are composed of alpha cells that release glucagon and beta cells that release insulin.

Some organs possess endocrine activity as a secondary function but have another primary function. The heart produces the hormone atrial natriuretic peptide, which functions to reduce blood volume, pressure, and Na^+ concentration. The gastrointestinal tract produces various hormones that aid in digestion. The kidneys produce renin, calcitriol, and erythropoietin. Adipose tissue produces leptin, which promotes satiety signals in the brain.

VISUAL CONNECTION QUESTIONS

1. Figure 37.5 Heat shock proteins (HSP) are so named because they help refold misfolded proteins. In response to increased temperature (a "heat shock"), heat shock proteins are activated by release from the NR/HSP complex. At the same time, transcription of HSP genes is activated. Why do you think the cell responds to a heat shock by increasing the activity of proteins that help refold misfolded proteins?

2. Figure 37.11 Pancreatic tumors may cause excess secretion of glucagon. Type I diabetes results from the failure of the pancreas to produce insulin. Which of the following statement about these two conditions is true?
   a. A pancreatic tumor and type I diabetes will have the opposite effects on blood sugar levels.
   b. A pancreatic tumor and type I diabetes will both cause hyperglycemia.
   c. A pancreatic tumor and type I diabetes will both cause hypoglycemia.
   d. Both pancreatic tumors and type I diabetes result in the inability of cells to take up glucose.

3. Figure 37.14 Hyperthyroidism is a condition in which the thyroid gland is overactive. Hypothyroidism is a condition in which the thyroid gland is underactive. Which of the conditions are the following two patients most likely to have?
   Patient A has symptoms including weight gain, cold sensitivity, low heart rate, and fatigue.
   Patient B has symptoms including weight loss, profuse sweating, increased heart rate, and difficulty sleeping.

REVIEW QUESTIONS

4. A newly discovered hormone contains four amino acids linked together. Under which chemical class would this hormone be classified?
   a. lipid-derived hormone
   b. amino acid-derived hormone
   c. peptide hormone
   d. glycoprotein

5. Which class of hormones can diffuse through plasma membranes?
   a. lipid-derived hormones
   b. amino acid-derived hormones
   c. peptide hormones
   d. glycoprotein hormones

6. Why are steroids able to diffuse across the plasma membrane?
   a. Their transport protein moves them through the membrane.
   b. They are amphipathic, allowing them to interact with the entire phospholipid.
   c. Cells express channels that let hormones flow down their concentration gradient into the cells.
   d. They are non-polar molecules.

7. A new antagonist molecule has been discovered that binds to and blocks plasma membrane receptors. What effect will this antagonist have on testosterone, a steroid hormone?
   a. It will block testosterone from binding to its receptor.
   b. It will block testosterone from activating cAMP signaling.
   c. It will increase testosterone-mediated signaling.
   d. It will not affect testosterone-mediated signaling.

8. What effect will a cAMP inhibitor have on a peptide hormone-mediated signaling pathway?
   a. It will prevent the hormone from binding its receptor.
   b. It will prevent activation of a G-protein.
   c. It will prevent activation of adenylate cyclase.
   d. It will prevent activation of protein kinases.

9. When insulin binds to its receptor, the complex is endocytosed into the cell. This is an example of ______ in response to hormone signaling.
   a. cAMP activation
   b. generating an intracellular receptor
   c. activation of a hormone response element
   d. receptor down-regulation

10. Drinking alcoholic beverages causes an increase in urine output. This most likely occurs because alcohol:
    a. inhibits ADH release.
    b. stimulates ADH release.
    c. inhibits TSH release.
    d. stimulates TSH release.

11. FSH and LH release from the anterior pituitary is stimulated by ________.
    a. TSH
    b. GnRH
    c. T_3
    d. PTH

12. What hormone is produced by beta cells of the pancreas?
    a. T_3
    b. glucagon
    c. insulin
    d. T_4

13. When blood calcium levels are low, PTH stimulates:
    a. excretion of calcium from the kidneys.
    b. excretion of calcium from the intestines.
    c. osteoblasts.
    d. osteoclasts.

14. How would mutations that completely ablate the function of the androgen receptor impact the phenotypic development of humans with XY chromosomes?
    a. Patients would appear phenotypically female.
    b. Patients would appear phenotypically male with underdeveloped secondary sex characteristics.
    c. Patients would appear phenotypically male, but cannot produce sperm.
    d. Patients would express both male and female secondary sex characteristics.

15. A rise in blood glucose levels triggers release of insulin from the pancreas. This mechanism of hormone production is stimulated by:
    a. humoral stimuli
    b. hormonal stimuli
    c. neural stimuli
    d. negative stimuli

16. Which mechanism of hormonal stimulation would be affected if signaling and hormone release from the hypothalamus was blocked?
    a. humoral and hormonal stimuli
    b. hormonal and neural stimuli
    c. neural and humoral stimuli
    d. hormonal and negative stimuli

17. A scientist hypothesizes that the pancreas's hormone production is controlled by neural stimuli. Which observation would support this hypothesis?
    a. Insulin is produced in response to sudden stress without a rise in blood glucose.
    b. Insulin is produced in response to a rise in glucagon levels.
    c. Beta cells express epinephrine receptors.
    d. Insulin is produced in response to a rise in blood glucose in the brain.

18. Which endocrine glands are associated with the kidneys?
    a. thyroid glands
    b. pituitary glands
    c. adrenal glands
    d. gonads

19. Which of the following hormones is not produced by the anterior pituitary?
    a. oxytocin
    b. growth hormone
    c. prolactin
    d. thyroid-stimulating hormone

20. Recent studies suggest that blue light exposure can impact human circadian rhythms. This suggests that blue light disrupts the function of the _____ gland(s).
    a. adrenal
    b. pituitary
    c. pineal
    d. thyroid

CRITICAL THINKING QUESTIONS

21. Although there are many different hormones in the human body, they can be divided into three classes based on their chemical structure. What are these classes and what is one factor that distinguishes them?
22. Where is insulin stored, and why would it be released?
23. Glucagon is the peptide hormone that signals for the body to release glucose into the bloodstream. How does glucagon contribute to maintaining homeostasis throughout the body? What other hormones are involved in regulating the blood glucose cycle?
24. Name two important functions of hormone receptors.
25. How can hormones mediate changes?
26. Why is cAMP-mediated signal amplification not required in steroid hormone signaling? Describe how steroid signaling is amplified instead.
27. Name and describe a function of one hormone produced by the anterior pituitary and one hormone produced by the posterior pituitary.
28. Describe one direct action of growth hormone (GH).
29. Researchers have recently demonstrated that stressed people are more susceptible to contracting the common cold than people who are not stressed. What kind of stress must the infected patients be experiencing, and why does it make them more susceptible to the virus?
30. How is hormone production and release primarily controlled?
31. Compare and contrast hormonal and humoral stimuli.
32. Oral contraceptive pills work by delivering synthetic progestins to a woman every day. Describe why this is an effective method of birth control.
33. What does aldosterone regulate, and how is it stimulated?
34. The adrenal medulla contains two types of secretory cells, what are they and what are their functions?
35. How would damage to the posterior pituitary gland affect the production and release of ADH and inhibiting hormones?

CHAPTER 38
The Musculoskeletal System

Figure 38.1 Improvements in the design of prostheses have allowed for a wider range of activities in recipients. (credit: modification of work by Stuart Grout)

Chapter Outline

INTRODUCTION The muscular and skeletal systems provide support to the body and allow for a wide range of movement. The bones of the skeletal system protect the body's internal organs and support the weight of the body. The muscles of the muscular system contract and pull on the bones, allowing for movements as diverse as standing, walking, running, and grasping items.

Injury or disease affecting the musculoskeletal system can be very debilitating. In humans, the most common musculoskeletal diseases worldwide are caused by malnutrition. Ailments that affect the joints are also widespread, such as arthritis, which can make movement difficult and—in advanced cases—completely impair mobility. In severe cases in which the joint has suffered extensive damage, joint replacement surgery may be needed.

Progress in the science of prosthesis design has resulted in the development of artificial joints, with joint replacement surgery in the hips and knees being the most common. Replacement joints for shoulders, elbows, and fingers are also available. Even with this progress, there is still room for improvement in the design of prostheses. The state-of-the-art prostheses have limited durability and therefore wear out quickly, particularly in young or active individuals. Current research is focused on the use of new materials, such as carbon fiber, that may make prostheses more durable.

38.1 Types of Skeletal Systems

By the end of this section, you will be able to do the following:

- Discuss the different types of skeletal systems
- Explain the role of the human skeletal system
- Compare and contrast different skeletal systems

A skeletal system is necessary to support the body, protect internal organs, and allow for the movement of an organism. There are three different skeleton designs that fulfill these functions: hydrostatic skeleton, exoskeleton, and endoskeleton.

Hydrostatic Skeleton

A **hydrostatic skeleton** is a skeleton formed by a fluid-filled compartment within the body, called the coelom. The organs of the coelom are supported by the aqueous fluid, which also resists external compression. This compartment is under hydrostatic pressure because of the fluid and supports the other organs of the organism. This type of skeletal system is found in soft-bodied animals such as sea anemones, earthworms, Cnidaria, and other invertebrates (Figure 38.2).

Figure 38.2 The skeleton of the red-knobbed sea star (*Protoreaster linckii*) is an example of a hydrostatic skeleton. (credit: "Amada44"/Wikimedia Commons)

Movement in a hydrostatic skeleton is provided by muscles that surround the coelom. The muscles in a hydrostatic skeleton contract to change the shape of the coelom; the pressure of the fluid in the coelom produces movement. For example, earthworms move by waves of muscular contractions of the skeletal muscle of the body wall hydrostatic skeleton, called peristalsis, which alternately shorten and lengthen the body. Lengthening the body extends the anterior end of the organism. Most organisms have a mechanism to fix themselves in the substrate. Shortening the muscles then draws the posterior portion of the body forward. Although a hydrostatic skeleton is well-suited to invertebrate organisms such as earthworms and some aquatic organisms, it is not an efficient skeleton for terrestrial animals.

Exoskeleton

An **exoskeleton** is an external skeleton that consists of a hard encasement on the surface of an organism. For example, the shells of crabs and insects are exoskeletons (Figure 38.3). This skeleton type provides defence against predators, supports the body, and allows for movement through the contraction of attached muscles. As with vertebrates, muscles must cross a joint inside the exoskeleton. Shortening of the muscle changes the relationship of the two segments of the exoskeleton. Arthropods such as crabs and lobsters have exoskeletons that consist of 30–50 percent chitin, a polysaccharide derivative of glucose that is a strong but flexible material. Chitin is secreted by the epidermal cells. The exoskeleton is further strengthened by the addition of calcium carbonate in organisms such as the lobster. Because the exoskeleton is acellular, arthropods must periodically shed their exoskeletons because the exoskeleton does not grow as the organism grows.

Figure 38.3 Muscles attached to the exoskeleton of the Halloween crab (*Gecarcinus quadratus*) allow it to move.

Endoskeleton

An **endoskeleton** is a skeleton that consists of hard, mineralized structures located within the soft tissue of organisms. An example of a primitive endoskeletal structure is the spicules of sponges. The bones of vertebrates are composed of tissues, whereas sponges have no true tissues (Figure 38.4). Endoskeletons provide support for the body, protect internal organs, and allow for movement through contraction of muscles attached to the skeleton.

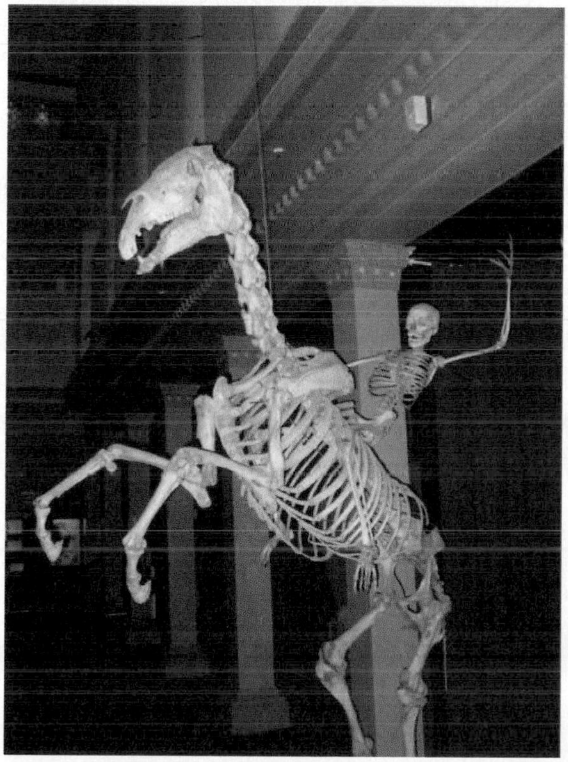

Figure 38.4 The skeletons of humans and horses are examples of endoskeletons. (credit: Ross Murphy)

The human skeleton is an endoskeleton that consists of 206 bones in the adult. It has five main functions: providing support to the body, storing minerals and lipids, producing blood cells, protecting internal organs, and allowing for movement. The skeletal system in vertebrates is divided into the axial skeleton (which consists of the skull, vertebral column, and rib cage), and the appendicular skeleton (which consists of the shoulders, limb bones, the pectoral girdle, and the pelvic girdle).

LINK TO LEARNING

Visit the interactive body (http://openstax.org/l/virt_skeleton) site to build a virtual skeleton: select "skeleton" and click through the activity to place each bone.

Human Axial Skeleton

The **axial skeleton** forms the central axis of the body and includes the bones of the skull, ossicles of the middle ear, hyoid bone of the throat, vertebral column, and the thoracic cage (ribcage) (Figure 38.5). The function of the axial skeleton is to provide support and protection for the brain, the spinal cord, and the organs in the ventral body cavity. It provides a surface for the attachment of muscles that move the head, neck, and trunk, performs respiratory movements, and stabilizes parts of the appendicular skeleton.

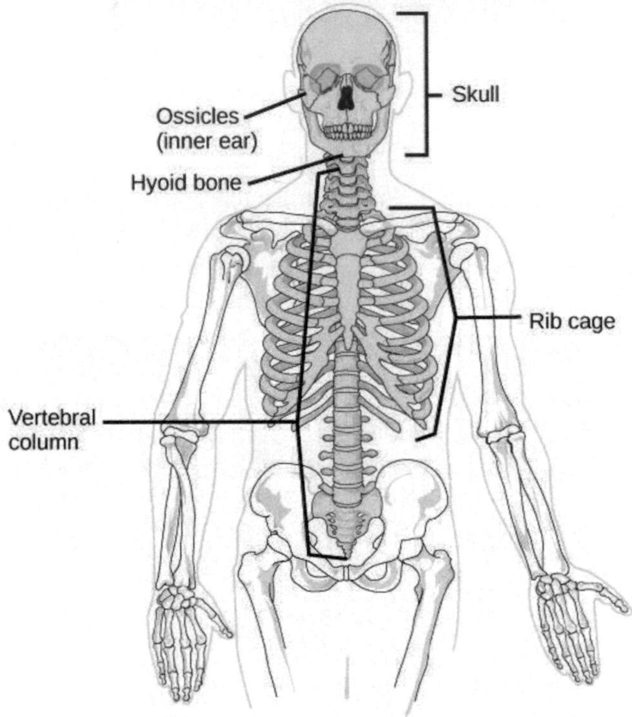

Figure 38.5 The axial skeleton consists of the bones of the skull, ossicles of the middle ear, hyoid bone, vertebral column, and rib cage. (credit: modification of work by Mariana Ruiz Villareal)

The Skull

The bones of the **skull** support the structures of the face and protect the brain. The skull consists of 22 bones, which are divided into two categories: cranial bones and facial bones. The **cranial bones** are eight bones that form the cranial cavity, which encloses the brain and serves as an attachment site for the muscles of the head and neck. The eight cranial bones are the frontal bone, two parietal bones, two temporal bones, occipital bone, sphenoid bone, and the ethmoid bone. Although the bones developed separately in the embryo and fetus, in the adult, they are tightly fused with connective tissue and adjoining bones do not move (Figure 38.6).

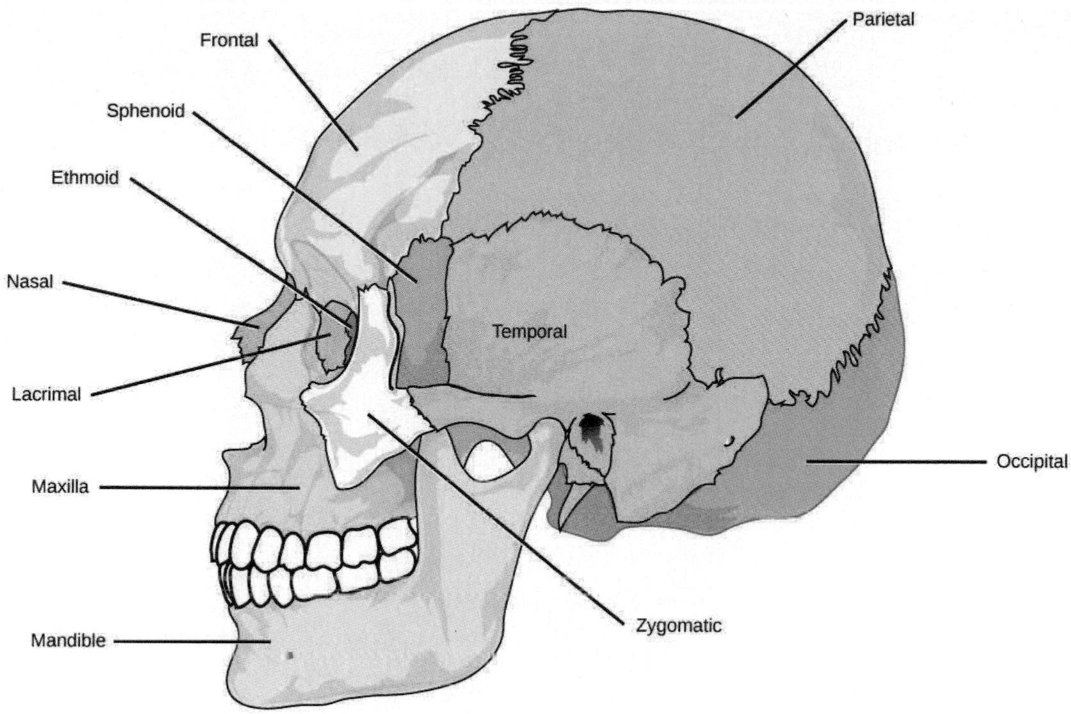

Figure 38.6 The bones of the skull support the structures of the face and protect the brain. (credit: modification of work by Mariana Ruiz Villareal)

The **auditory ossicles** of the middle ear transmit sounds from the air as vibrations to the fluid-filled cochlea. The auditory ossicles consist of three bones each: the malleus, incus, and stapes. These are the smallest bones in the body and are unique to mammals.

Fourteen **facial bones** form the face, provide cavities for the sense organs (eyes, mouth, and nose), protect the entrances to the digestive and respiratory tracts, and serve as attachment points for facial muscles. The 14 facial bones are the nasal bones, the maxillary bones, zygomatic bones, palatine, vomer, lacrimal bones, the inferior nasal conchae, and the mandible. All of these bones occur in pairs except for the mandible and the vomer (Figure 38.7).

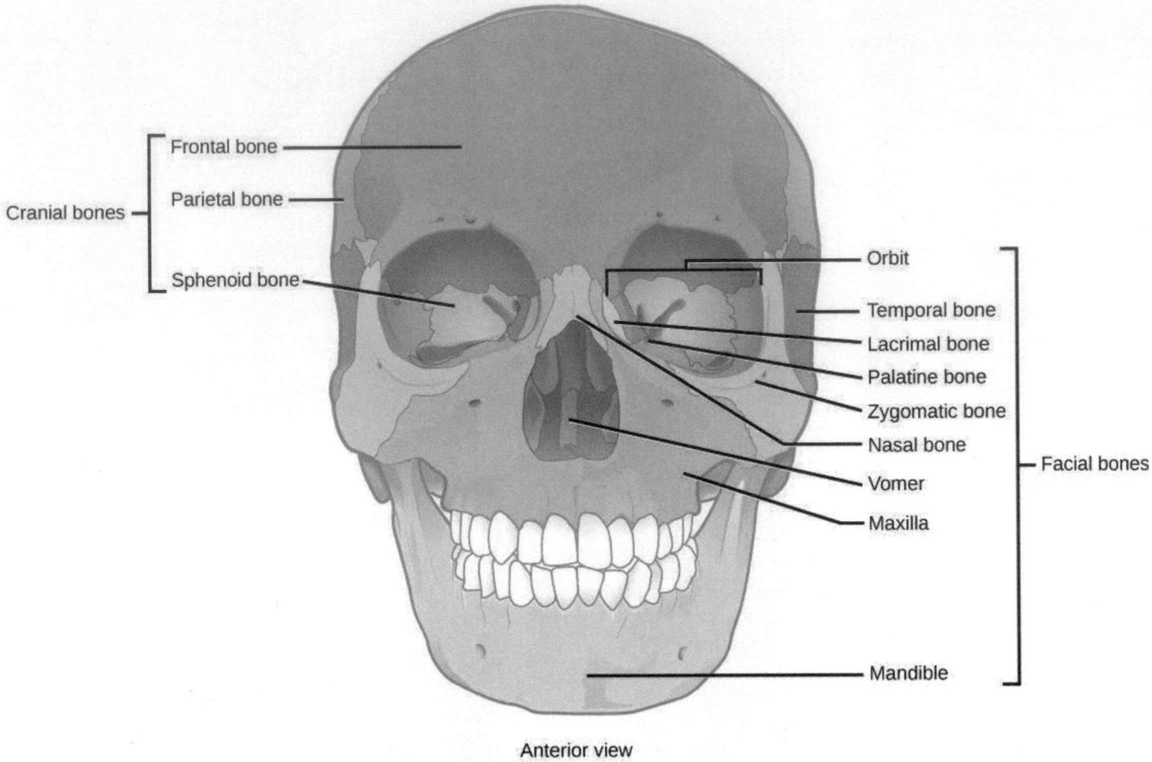

Figure 38.7 The cranial bones, including the frontal, parietal, and sphenoid bones, cover the top of the head. The facial bones of the skull form the face and provide cavities for the eyes, nose, and mouth.

Although it is not found in the skull, the hyoid bone is considered a component of the axial skeleton. The **hyoid bone** lies below the mandible in the front of the neck. It acts as a movable base for the tongue and is connected to muscles of the jaw, larynx, and tongue. The mandible articulates with the base of the skull. The mandible controls the opening to the airway and gut. In animals with teeth, the mandible brings the surfaces of the teeth in contact with the maxillary teeth.

The Vertebral Column

The **vertebral column**, or spinal column, surrounds and protects the spinal cord, supports the head, and acts as an attachment point for the ribs and muscles of the back and neck. The adult vertebral column comprises 26 bones: the 24 vertebrae, the sacrum, and the coccyx bones. In the adult, the sacrum is typically composed of five vertebrae that fuse into one. The coccyx is typically 3–4 vertebrae that fuse into one. Around the age of 70, the sacrum and the coccyx may fuse together. We begin life with approximately 33 vertebrae, but as we grow, several vertebrae fuse together. The adult vertebrae are further divided into the 7 cervical vertebrae, 12 thoracic vertebrae, and 5 lumbar vertebrae (Figure 38.8).

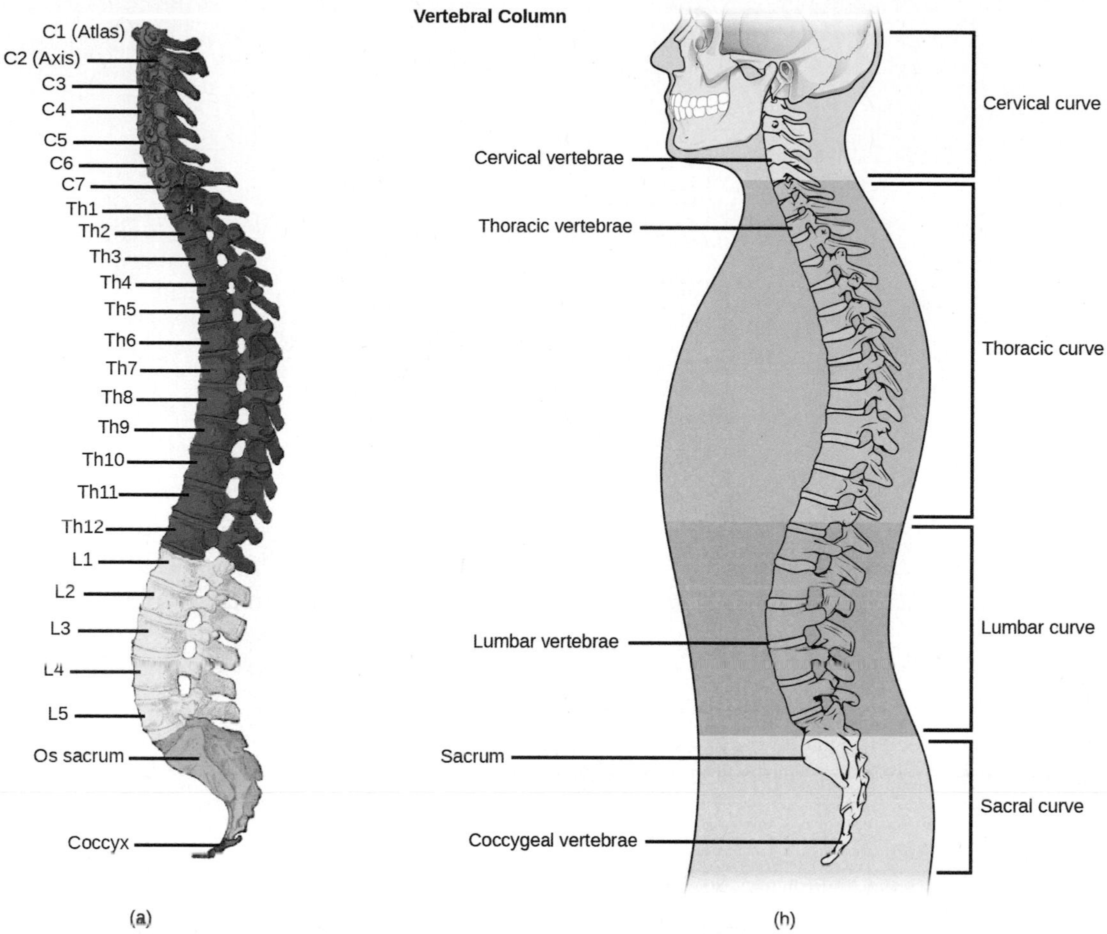

Figure 38.8 (a) The vertebral column consists of seven cervical vertebrae (C1–7) twelve thoracic vertebrae (Th1–12), five lumbar vertebrae (L1–5), the os sacrum, and the coccyx. (b) Spinal curves increase the strength and flexibility of the spine. (credit a: modification of work by Uwe Gille based on original work by Gray's Anatomy; credit b: modification of work by NCI, NIH)

Each vertebral body has a large hole in the center through which the nerves of the spinal cord pass. There is also a notch on each side through which the spinal nerves, which serve the body at that level, can exit from the spinal cord. The vertebral column is approximately 71 cm (28 inches) in adult male humans and is curved, which can be seen from a side view. The names of the spinal curves correspond to the region of the spine in which they occur. The thoracic and sacral curves are concave (curve inwards relative to the front of the body) and the cervical and lumbar curves are convex (curve outwards relative to the front of the body). The arched curvature of the vertebral column increases its strength and flexibility, allowing it to absorb shocks like a spring (Figure 38.8).

Intervertebral discs composed of fibrous cartilage lie between adjacent vertebral bodies from the second cervical vertebra to the sacrum. Each disc is part of a joint that allows for some movement of the spine and acts as a cushion to absorb shocks from movements such as walking and running. Intervertebral discs also act as ligaments to bind vertebrae together. The inner part of discs, the nucleus pulposus, hardens as people age and becomes less elastic. This loss of elasticity diminishes its ability to absorb shocks.

The Thoracic Cage

The **thoracic cage**, also known as the ribcage, is the skeleton of the chest, and consists of the ribs, sternum, thoracic vertebrae, and costal cartilages (Figure 38.9). The thoracic cage encloses and protects the organs of the thoracic cavity, including the heart and lungs. It also provides support for the shoulder girdles and upper limbs, and serves as the attachment point for the diaphragm, muscles of the back, chest, neck, and shoulders. Changes in the volume of the thorax enable breathing.

The **sternum**, or breastbone, is a long, flat bone located at the anterior of the chest. It is formed from three bones that fuse in the adult. The **ribs** are 12 pairs of long, curved bones that attach to the thoracic vertebrae and curve toward the front of the body, forming the ribcage. Costal cartilages connect the anterior ends of the ribs to the sternum, with the exception of rib pairs 11 and

12, which are free-floating ribs.

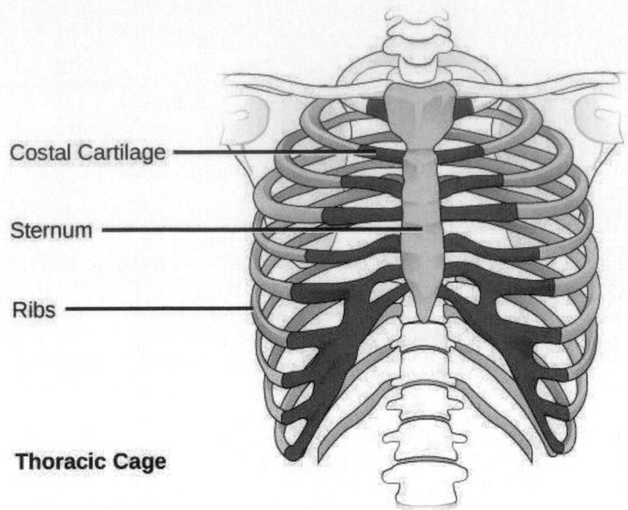

Figure 38.9 The thoracic cage, or rib cage, protects the heart and the lungs. (credit: modification of work by NCI, NIH)

Human Appendicular Skeleton

The **appendicular skeleton** is composed of the bones of the upper limbs (which function to grasp and manipulate objects) and the lower limbs (which permit locomotion). It also includes the pectoral girdle, or shoulder girdle, that attaches the upper limbs to the body, and the pelvic girdle that attaches the lower limbs to the body (Figure 38.10).

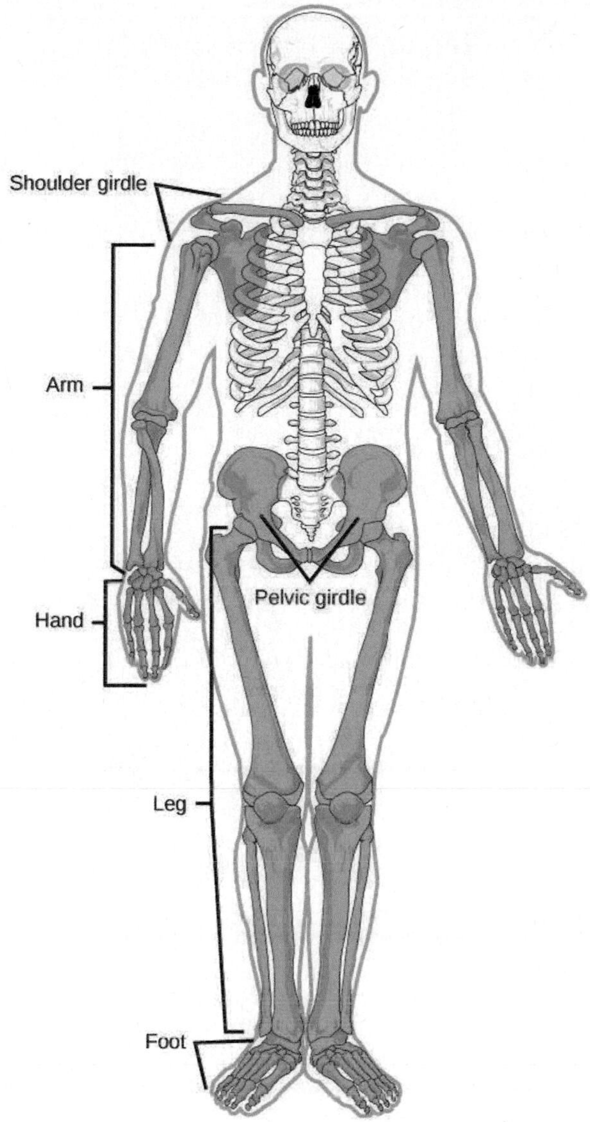

Figure 38.10 The appendicular skeleton is composed of the bones of the pectoral limbs (arm, forearm, hand), the pelvic limbs (thigh, leg, foot), the pectoral girdle, and the pelvic girdle. (credit: modification of work by Mariana Ruiz Villareal)

The Pectoral Girdle

The **pectoral girdle** bones provide the points of attachment of the upper limbs to the axial skeleton. The human pectoral girdle consists of the clavicle (or collarbone) in the anterior, and the scapula (or shoulder blades) in the posterior (Figure 38.11).

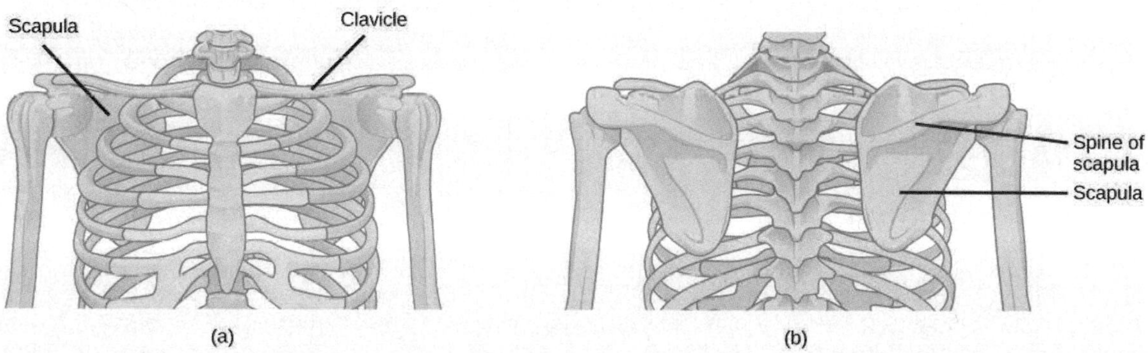

Figure 38.11 (a) The pectoral girdle in primates consists of the clavicles and scapulae. (b) The posterior view reveals the spine of the scapula to which muscle attaches.

The **clavicles** are S-shaped bones that position the arms on the body. The clavicles lie horizontally across the front of the thorax (chest) just above the first rib. These bones are fairly fragile and are susceptible to fractures. For example, a fall with the arms outstretched causes the force to be transmitted to the clavicles, which can break if the force is excessive. The clavicle articulates with the sternum and the scapula.

The **scapulae** are flat, triangular bones that are located at the back of the pectoral girdle. They support the muscles crossing the shoulder joint. A ridge, called the spine, runs across the back of the scapula and can easily be felt through the skin (Figure 38.11). The spine of the scapula is a good example of a bony protrusion that facilitates a broad area of attachment for muscles to bone.

The Upper Limb

The upper limb contains 30 bones in three regions: the arm (shoulder to elbow), the forearm (ulna and radius), and the wrist and hand (Figure 38.12).

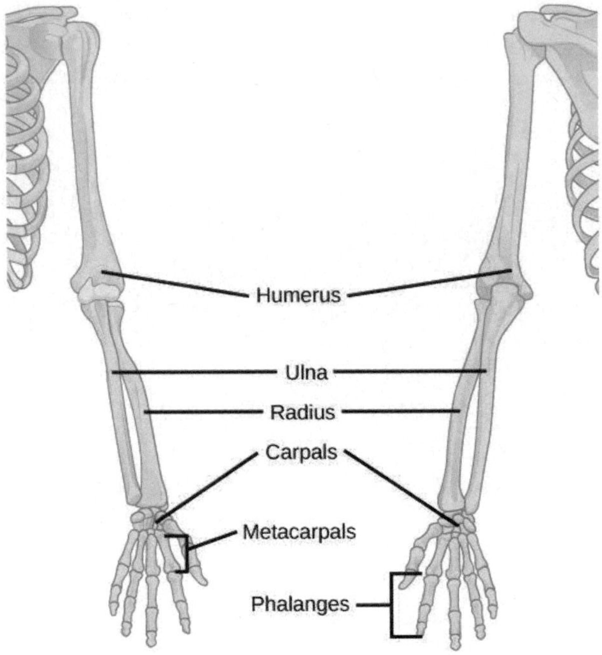

Figure 38.12 The upper limb consists of the humerus of the upper arm, the radius and ulna of the forearm, eight bones of the carpus, five bones of the metacarpus, and 14 bones of the phalanges.

An **articulation** is any place at which two bones are joined. The **humerus** is the largest and longest bone of the upper limb and the only bone of the arm. It articulates with the scapula at the shoulder and with the forearm at the elbow. The **forearm** extends from the elbow to the wrist and consists of two bones: the ulna and the radius. The **radius** is located along the lateral (thumb) side of the forearm and articulates with the humerus at the elbow. The **ulna** is located on the medial aspect (pinky-finger side) of the forearm. It is longer than the radius. The ulna articulates with the humerus at the elbow. The radius and ulna also articulate with the carpal bones and with each other, which in vertebrates enables a variable degree of rotation of the carpus with respect to the long axis of the limb. The hand includes the eight bones of the **carpus** (wrist), the five bones of the **metacarpus** (palm), and the 14 bones of the **phalanges** (digits). Each digit consists of three phalanges, except for the thumb, when present, which has only two.

The Pelvic Girdle

The **pelvic girdle** attaches to the lower limbs of the axial skeleton. Because it is responsible for bearing the weight of the body and for locomotion, the pelvic girdle is securely attached to the axial skeleton by strong ligaments. It also has deep sockets with robust ligaments to securely attach the femur to the body. The pelvic girdle is further strengthened by two large hip bones. In adults, the hip bones, or **coxal bones**, are formed by the fusion of three pairs of bones: the ilium, ischium, and pubis. The pelvis joins together in the anterior of the body at a joint called the pubic symphysis and with the bones of the sacrum at the posterior of the body.

The female pelvis is slightly different from the male pelvis. Over generations of evolution, females with a wider pubic angle and larger diameter pelvic canal reproduced more successfully. Therefore, their offspring also had pelvic anatomy that enabled successful childbirth (Figure 38.13).

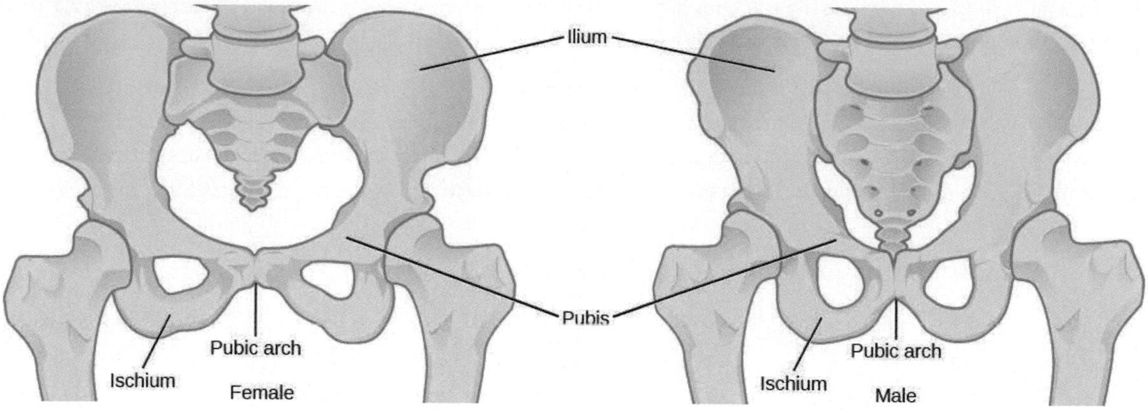

Figure 38.13 To adapt to reproductive fitness, the (a) female pelvis is lighter, wider, shallower, and has a broader angle between the pubic bones than (b) the male pelvis.

The Lower Limb

The **lower limb** consists of the thigh, the leg, and the foot. The bones of the lower limb are the femur (thigh bone), patella (kneecap), tibia and fibula (bones of the leg), tarsals (bones of the ankle), and metatarsals and phalanges (bones of the foot) (Figure 38.14). The bones of the lower limbs are thicker and stronger than the bones of the upper limbs because of the need to support the entire weight of the body and the resulting forces from locomotion. In addition to evolutionary fitness, the bones of an individual will respond to forces exerted upon them.

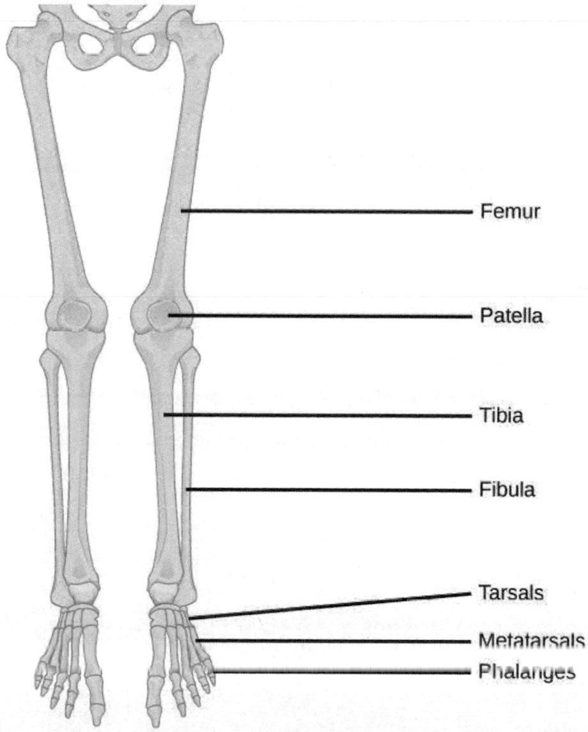

Figure 38.14 The lower limb consists of the thigh (femur), kneecap (patella), leg (tibia and fibula), ankle (tarsals), and foot (metatarsals and phalanges) bones.

The **femur**, or thighbone, is the longest, heaviest, and strongest bone in the body. The femur and pelvis form the hip joint at the proximal end. At the distal end, the femur, tibia, and patella form the knee joint. The **patella**, or kneecap, is a triangular bone that lies anterior to the knee joint. The patella is embedded in the tendon of the femoral extensors (quadriceps). It improves knee extension by reducing friction. The **tibia**, or shinbone, is a large bone of the leg that is located directly below the knee. The tibia articulates with the femur at its proximal end, with the fibula and the tarsal bones at its distal end. It is the second largest bone in the human body and is responsible for transmitting the weight of the body from the femur to the foot. The **fibula**, or calf bone, parallels and articulates with the tibia. It does not articulate with the femur and does not bear weight. The fibula acts as a site for muscle attachment and forms the lateral part of the ankle joint.

The **tarsals** are the seven bones of the ankle. The ankle transmits the weight of the body from the tibia and the fibula to the foot. The **metatarsals** are the five bones of the foot. The phalanges are the 14 bones of the toes. Each toe consists of three phalanges, except for the big toe that has only two (Figure 38.15). Variations exist in other species; for example, the horse's metacarpals and metatarsals are oriented vertically and do not make contact with the substrate.

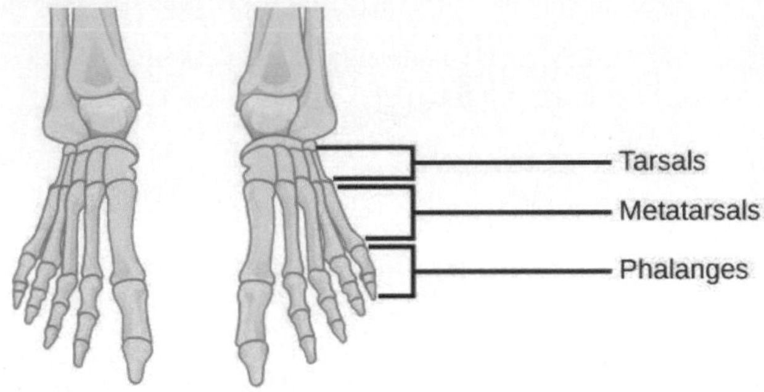

Figure 38.15 This drawing shows the bones of the human foot and ankle, including the metatarsals and the phalanges.

EVOLUTION CONNECTION

Evolution of Body Design for Locomotion on Land

The transition of vertebrates onto land required a number of changes in body design, as movement on land presents a number of challenges for animals that are adapted to movement in water. The buoyancy of water provides a certain amount of lift, and a common form of movement by fish is lateral undulations of the entire body. This back and forth movement pushes the body against the water, creating forward movement. In most fish, the muscles of paired fins attach to girdles within the body, allowing for some control of locomotion. As certain fish began moving onto land, they retained their lateral undulation form of locomotion (anguilliform). However, instead of pushing against water, their fins or flippers became points of contact with the ground, around which they rotated their bodies.

The effect of gravity and the lack of buoyancy on land meant that body weight was suspended on the limbs, leading to increased strengthening and ossification of the limbs. The effect of gravity also required changes to the axial skeleton. Lateral undulations of land animal vertebral columns cause torsional strain. A firmer, more ossified vertebral column became common in terrestrial tetrapods because it reduces strain while providing the strength needed to support the body's weight. In later tetrapods, the vertebrae began allowing for vertical motion rather than lateral flexion. Another change in the axial skeleton was the loss of a direct attachment between the pectoral girdle and the head. This reduced the jarring to the head caused by the impact of the limbs on the ground. The vertebrae of the neck also evolved to allow movement of the head independently of the body.

The appendicular skeleton of land animals is also different from aquatic animals. The shoulders attach to the pectoral girdle through muscles and connective tissue, thus reducing the jarring of the skull. Because of a lateral undulating vertebral column, in early tetrapods, the limbs were splayed out to the side and movement occurred by performing "push-ups." The vertebrae of these animals had to move side-to-side in a similar manner to fish and reptiles. This type of motion requires large muscles to move the limbs toward the midline; it was almost like walking while doing push-ups, and it is not an efficient use of energy. Later tetrapods have their limbs placed under their bodies, so that each stride requires less force to move forward. This resulted in decreased adductor muscle size and an increased range of motion of the scapulae. This also restricts movement primarily to one plane, creating forward motion rather than moving the limbs upward as well as forward. The femur and humerus were also rotated, so that the ends of the limbs and digits were pointed forward, in the direction of motion, rather than out to the side. By placement underneath the body, limbs can swing forward like a pendulum to produce a stride that is more efficient for moving over land.

38.2 Bone

By the end of this section, you will be able to do the following:

- Classify the different types of bones in the skeleton
- Explain the role of the different cell types in bone
- Explain how bone forms during development

Bone, or **osseous tissue**, is a connective tissue that constitutes the endoskeleton. It contains specialized cells and a matrix of mineral salts and collagen fibers.

The mineral salts primarily include hydroxyapatite, a mineral formed from calcium phosphate. **Calcification** is the process of deposition of mineral salts on the collagen fiber matrix that crystallizes and hardens the tissue. The process of calcification only occurs in the presence of collagen fibers.

The bones of the human skeleton are classified by their shape: long bones, short bones, flat bones, sutural bones, sesamoid bones, and irregular bones (Figure 38.16).

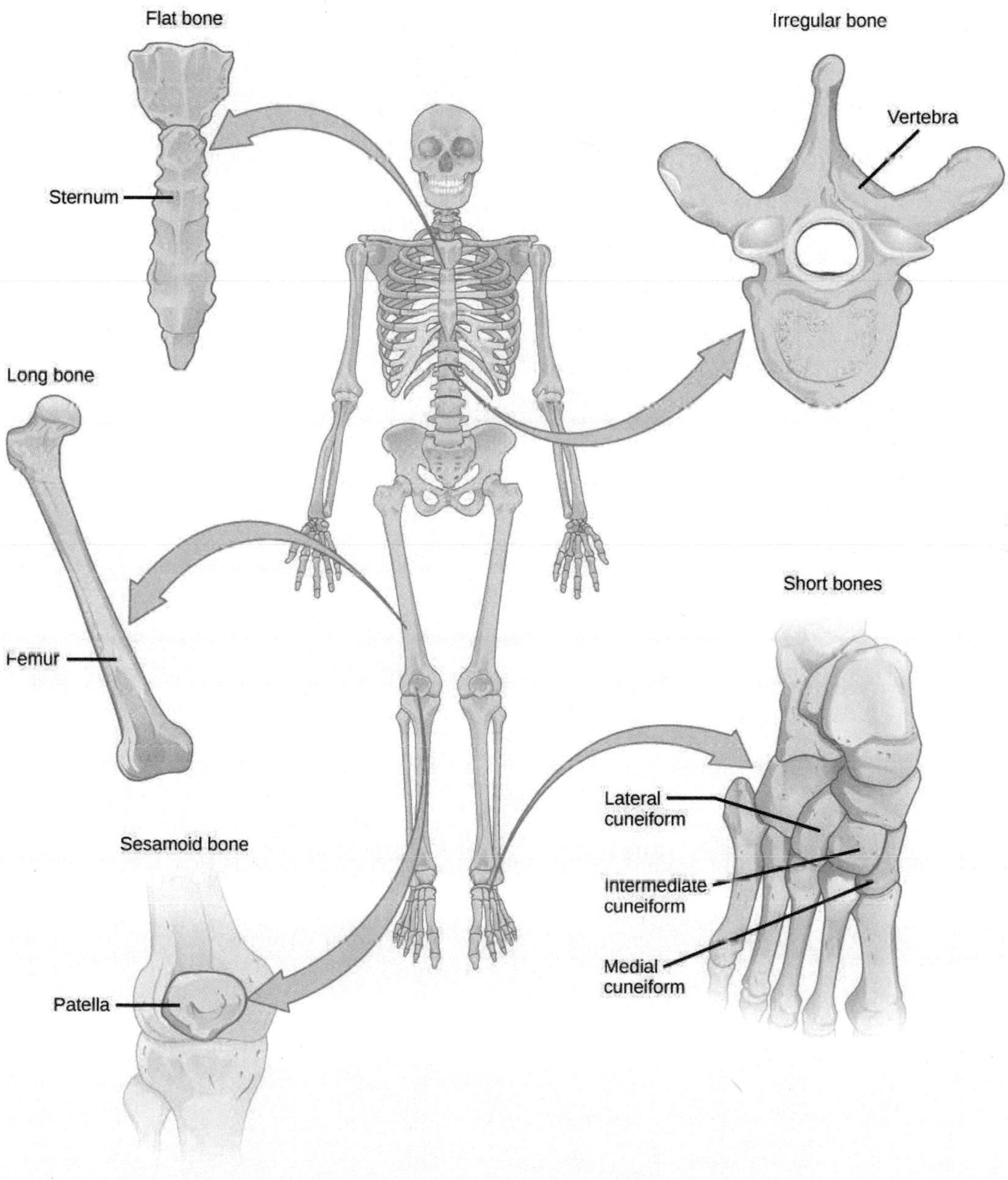

Figure 38.16 Shown are different types of bones: flat, irregular, long, short, and sesamoid.

Long bones are longer than they are wide and have a shaft and two ends. The **diaphysis**, or central shaft, contains bone marrow in a marrow cavity. The rounded ends, the **epiphyses**, are covered with articular cartilage and are filled with red bone marrow, which produces blood cells (Figure 38.17). Most of the limb bones are long bones—for example, the femur, tibia, ulna, and radius. Exceptions to this include the patella and the bones of the wrist and ankle.

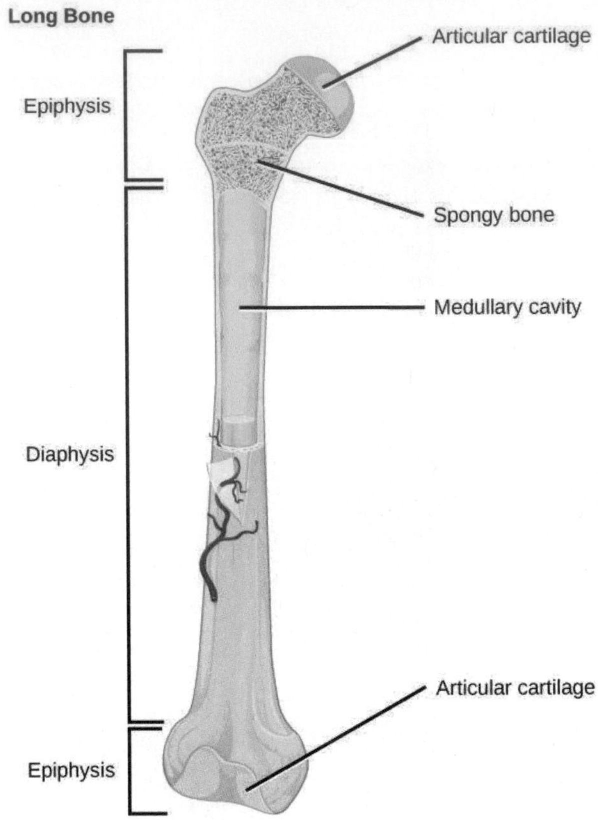

Figure 38.17 The long bone is covered by articular cartilage at either end and contains bone marrow (shown in yellow in this illustration) in the marrow cavity.

Short bones, or cuboidal bones, are bones that are the same width and length, giving them a cube-like shape. For example, the bones of the wrist (carpals) and ankle (tarsals) are short bones (Figure 38.16).

Flat bones are thin and relatively broad bones that are found where extensive protection of organs is required or where broad surfaces of muscle attachment are required. Examples of flat bones are the sternum (breast bone), ribs, scapulae (shoulder blades), and the roof of the skull (Figure 38.16).

Irregular bones are bones with complex shapes. These bones may have short, flat, notched, or ridged surfaces. Examples of irregular bones are the vertebrae, hip bones, and several skull bones.

Sesamoid bones are small, flat bones and are shaped similarly to a sesame seed. The patellae are sesamoid bones (Figure 38.18). Sesamoid bones develop inside tendons and may be found near joints at the knees, hands, and feet.

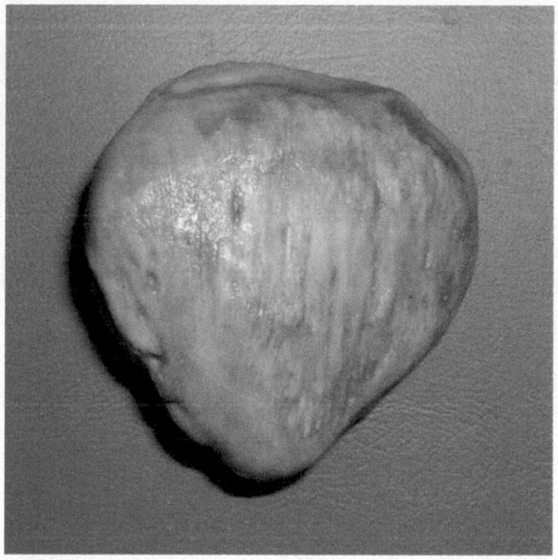

Figure 38.18 The patella of the knee is an example of a sesamoid bone.

Sutural bones are small, flat, irregularly shaped bones. They may be found between the flat bones of the skull. They vary in number, shape, size, and position.

Bone Tissue

Bones are considered organs because they contain various types of tissue, such as blood, connective tissue, nerves, and bone tissue. Osteocytes, the living cells of bone tissue, form the mineral matrix of bones. There are two types of bone tissue: compact and spongy.

Compact Bone Tissue

Compact bone (or cortical bone) forms the hard external layer of all bones and surrounds the medullary cavity, or bone marrow. It provides protection and strength to bones. Compact bone tissue consists of units called osteons or Haversian systems. **Osteons** are cylindrical structures that contain a mineral matrix and living osteocytes connected by canaliculi, which transport blood. They are aligned parallel to the long axis of the bone. Each osteon consists of **lamellae**, which are layers of compact matrix that surround a central canal called the Haversian canal. The **Haversian canal** (osteonic canal) contains the bone's blood vessels and nerve fibers (Figure 38.19). Osteons in compact bone tissue are aligned in the same direction along lines of stress and help the bone resist bending or fracturing. Therefore, compact bone tissue is prominent in areas of bone at which stresses are applied in only a few directions.

VISUAL CONNECTION

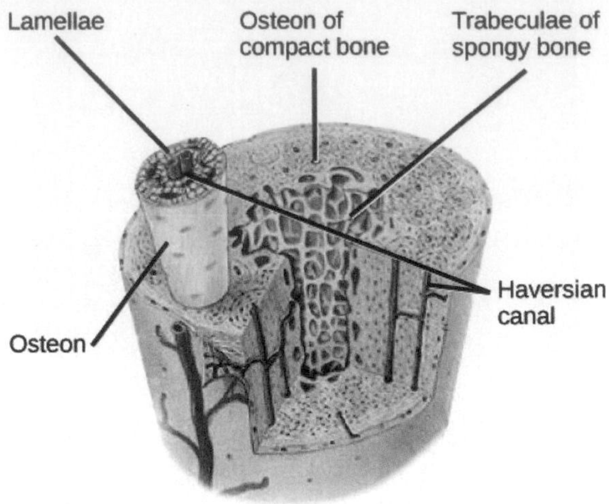

Figure 38.19 Compact bone tissue consists of osteons that are aligned parallel to the long axis of the bone, and the Haversian canal that contains the bone's blood vessels and nerve fibers. The inner layer of bones consists of spongy bone tissue. The small dark ovals in the osteon represent the living osteocytes. (credit: modification of work by NCI, NIH)

Which of the following statements about bone tissue is false?

a. Compact bone tissue is made of cylindrical osteons that are aligned such that they travel the length of the bone.
b. Haversian canals contain blood vessels only.
c. Haversian canals contain blood vessels and nerve fibers.
d. Spongy tissue is found on the interior of the bone, and compact bone tissue is found on the exterior.

Spongy Bone Tissue

Whereas compact bone tissue forms the outer layer of all bones, **spongy bone** or cancellous bone forms the inner layer of all bones. Spongy bone tissue does not contain osteons that constitute compact bone tissue. Instead, it consists of **trabeculae**, which are lamellae that are arranged as rods or plates. Red bone marrow is found between the trabuculae. Blood vessels within this tissue deliver nutrients to osteocytes and remove waste. The red bone marrow of the femur and the interior of other large bones, such as the ilium, forms blood cells.

Spongy bone reduces the density of bone and allows the ends of long bones to compress as the result of stresses applied to the bone. Spongy bone is prominent in areas of bones that are not heavily stressed or where stresses arrive from many directions. The epiphyses of bones, such as the neck of the femur, are subject to stress from many directions. Imagine laying a heavy framed picture flat on the floor. You could hold up one side of the picture with a toothpick if the toothpick was perpendicular to the floor and the picture. Now drill a hole and stick the toothpick into the wall to hang up the picture. In this case, the function of the toothpick is to transmit the downward pressure of the picture to the wall. The force on the picture is straight down to the floor, but the force on the toothpick is both the picture wire pulling down and the bottom of the hole in the wall pushing up. The toothpick will break off right at the wall.

The neck of the femur is horizontal like the toothpick in the wall. The weight of the body pushes it down near the joint, but the vertical diaphysis of the femur pushes it up at the other end. The neck of the femur must be strong enough to transfer the downward force of the body weight horizontally to the vertical shaft of the femur (Figure 38.20).

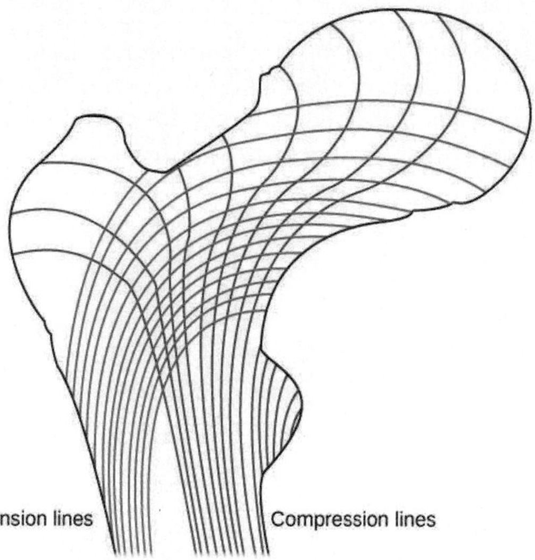

Figure 38.20 Trabeculae in spongy bone are arranged such that one side of the bone bears tension and the other withstands compression.

LINK TO LEARNING

View micrographs (http://openstax.org/l/muscle_tissue) of musculoskeletal tissues as you review the anatomy.

Cell Types in Bones

Bone consists of four types of cells: osteoblasts, osteoclasts, osteocytes, and osteoprogenitor cells. **Osteoblasts** are bone cells that are responsible for bone formation. Osteoblasts synthesize and secrete the organic part and inorganic part of the extracellular matrix of bone tissue, and collagen fibers. Osteoblasts become trapped in these secretions and differentiate into less active osteocytes. **Osteoclasts** are large bone cells with up to 50 nuclei. They remove bone structure by releasing lysosomal enzymes and acids that dissolve the bony matrix. These minerals, released from bones into the blood, help regulate calcium concentrations in body fluids. Bone may also be resorbed for remodeling, if the applied stresses have changed. **Osteocytes** are mature bone cells and are the main cells in bony connective tissue; these cells cannot divide. Osteocytes maintain normal bone structure by recycling the mineral salts in the bony matrix. **Osteoprogenitor cells** are squamous stem cells that divide to produce daughter cells that differentiate into osteoblasts. Osteoprogenitor cells are important in the repair of fractures.

Development of Bone

Ossification, or osteogenesis, is the process of bone formation by osteoblasts. Ossification is distinct from the process of calcification; whereas calcification takes place during the ossification of bones, it can also occur in other tissues. Ossification begins approximately six weeks after fertilization in an embryo. Before this time, the embryonic skeleton consists entirely of fibrous membranes and hyaline cartilage. The development of bone from fibrous membranes is called intramembranous ossification; development from hyaline cartilage is called endochondral ossification. Bone growth continues until approximately age 25. Bones can grow in thickness throughout life, but after age 25, ossification functions primarily in bone remodeling and repair.

Intramembranous Ossification

Intramembranous ossification is the process of bone development from fibrous membranes. It is involved in the formation of the flat bones of the skull, the mandible, and the clavicles. Ossification begins as mesenchymal cells form a template of the future bone. They then differentiate into osteoblasts at the ossification center. Osteoblasts secrete the extracellular matrix and deposit calcium, which hardens the matrix. The non-mineralized portion of the bone or osteoid continues to form around blood vessels, forming spongy bone. Connective tissue in the matrix differentiates into red bone marrow in the fetus. The spongy bone is remodeled into a thin layer of compact bone on the surface of the spongy bone.

Endochondral Ossification

Endochondral ossification is the process of bone development from hyaline cartilage. All of the bones of the body, except for the flat bones of the skull, mandible, and clavicles, are formed through endochondral ossification.

In long bones, chondrocytes form a template of the hyaline cartilage diaphysis. Responding to complex developmental signals, the matrix begins to calcify. This calcification prevents diffusion of nutrients into the matrix, resulting in chondrocytes dying and the opening up of cavities in the diaphysis cartilage. Blood vessels invade the cavities, and osteoblasts and osteoclasts modify the calcified cartilage matrix into spongy bone. Osteoclasts then break down some of the spongy bone to create a marrow, or medullary, cavity in the center of the diaphysis. Dense, irregular connective tissue forms a sheath (periosteum) around the bones. The periosteum assists in attaching the bone to surrounding tissues, tendons, and ligaments. The bone continues to grow and elongate as the cartilage cells at the epiphyses divide.

In the last stage of prenatal bone development, the centers of the epiphyses begin to calcify. Secondary ossification centers form in the epiphyses as blood vessels and osteoblasts enter these areas and convert hyaline cartilage into spongy bone. Until adolescence, hyaline cartilage persists at the **epiphyseal plate** (growth plate), which is the region between the diaphysis and epiphysis that is responsible for the lengthwise growth of long bones (Figure 38.21).

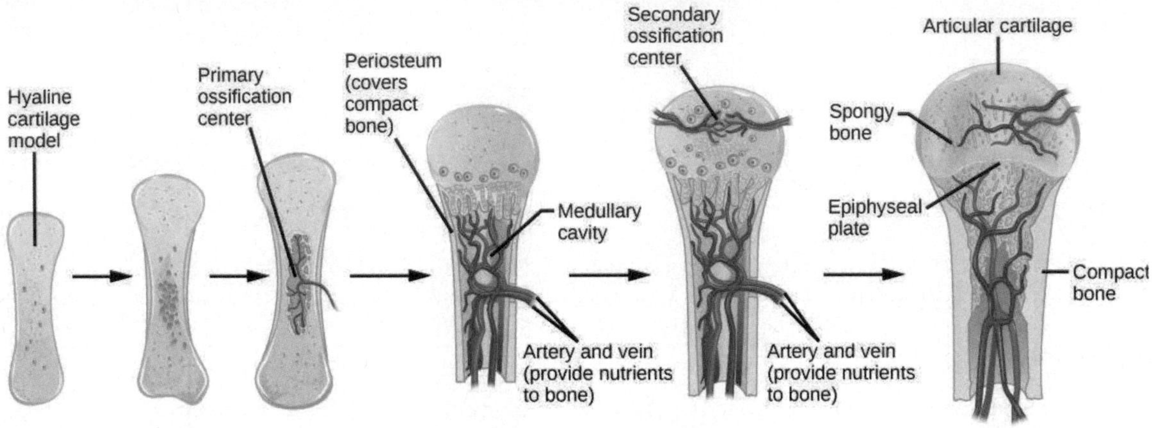

Figure 38.21 Endochondral ossification is the process of bone development from hyaline cartilage. The periosteum is the connective tissue on the outside of bone that acts as the interface between bone, blood vessels, tendons, and ligaments.

Growth of Bone

Long bones continue to lengthen, potentially until adolescence, through the addition of bone tissue at the epiphyseal plate. They also increase in width through appositional growth.

Lengthening of Long Bones

Chondrocytes on the epiphyseal side of the epiphyseal plate divide; one cell remains undifferentiated near the epiphysis, and one cell moves toward the diaphysis. The cells, which are pushed from the epiphysis, mature and are destroyed by calcification. This process replaces cartilage with bone on the diaphyseal side of the plate, resulting in a lengthening of the bone.

Long bones stop growing at around the age of 18 in females and the age of 21 in males in a process called epiphyseal plate closure. During this process, cartilage cells stop dividing and all of the cartilage is replaced by bone. The epiphyseal plate fades, leaving a structure called the epiphyseal line or epiphyseal remnant, and the epiphysis and diaphysis fuse.

Thickening of Long Bones

Appositional growth is the increase in the diameter of bones by the addition of bony tissue at the surface of bones. Osteoblasts at the bone surface secrete bone matrix, and osteoclasts on the inner surface break down bone. The osteoblasts differentiate into osteocytes. A balance between these two processes allows the bone to thicken without becoming too heavy.

Bone Remodeling and Repair

Bone renewal continues after birth into adulthood. **Bone remodeling** is the replacement of old bone tissue by new bone tissue. It involves the processes of bone deposition by osteoblasts and bone resorption by osteoclasts. Normal bone growth requires vitamins D, C, and A, plus minerals such as calcium, phosphorous, and magnesium. Hormones such as parathyroid hormone, growth hormone, and calcitonin are also required for proper bone growth and maintenance.

Bone turnover rates are quite high, with five to seven percent of bone mass being recycled every week. Differences in turnover rate exist in different areas of the skeleton and in different areas of a bone. For example, the bone in the head of the femur may be fully replaced every six months, whereas the bone along the shaft is altered much more slowly.

Bone remodeling allows bones to adapt to stresses by becoming thicker and stronger when subjected to stress. Bones that are not subject to normal stress, for example when a limb is in a cast, will begin to lose mass. A fractured or broken bone undergoes repair through four stages:

1. Blood vessels in the broken bone tear and hemorrhage, resulting in the formation of clotted blood, or a hematoma, at the site of the break. The severed blood vessels at the broken ends of the bone are sealed by the clotting process, and bone cells that are deprived of nutrients begin to die.
2. Within days of the fracture, capillaries grow into the hematoma, and phagocytic cells begin to clear away the dead cells. Though fragments of the blood clot may remain, fibroblasts and osteoblasts enter the area and begin to reform bone. Fibroblasts produce collagen fibers that connect the broken bone ends, and osteoblasts start to form spongy bone. The repair tissue between the broken bone ends is called the fibrocartilaginous callus, as it is composed of both hyaline and fibrocartilage (Figure 38.22). Some bone spicules may also appear at this point.
3. The fibrocartilaginous callus is converted into a bony callus of spongy bone. It takes about two months for the broken bone ends to be firmly joined together after the fracture. This is similar to the endochondral formation of bone, as cartilage becomes ossified; osteoblasts, osteoclasts, and bone matrix are present.
4. The bony callus is then remodelled by osteoclasts and osteoblasts, with excess material on the exterior of the bone and within the medullary cavity being removed. Compact bone is added to create bone tissue that is similar to the original, unbroken bone. This remodeling can take many months, and the bone may remain uneven for years.

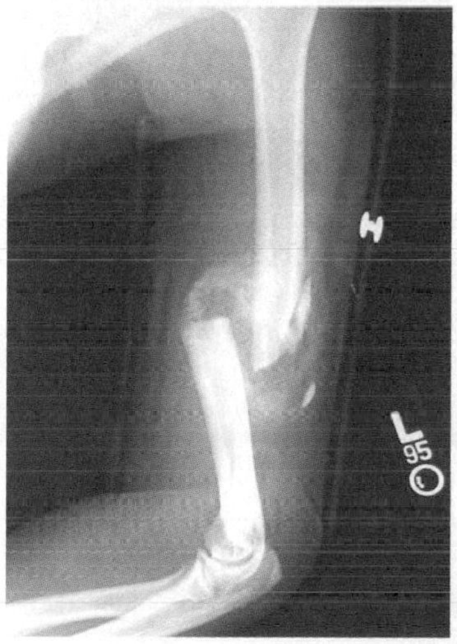

Figure 38.22 After this bone is set, a callus will knit the two ends together. (credit: Bill Rhodes)

SCIENTIFIC METHOD CONNECTION

Decalcification of Bones

Question: What effect does the removal of calcium and collagen have on bone structure?

Background: Conduct a literature search on the role of calcium and collagen in maintaining bone structure. Conduct a literature search on diseases in which bone structure is compromised.

Hypothesis: Develop a hypothesis that states predictions of the flexibility, strength, and mass of bones that have had the calcium and collagen components removed. Develop a hypothesis regarding the attempt to add calcium back to decalcified bones.

Test the hypothesis: Test the prediction by removing calcium from chicken bones by placing them in a jar of vinegar for seven days. Test the hypothesis regarding adding calcium back to decalcified bone by placing the decalcified chicken bones into a jar of water with calcium supplements added. Test the prediction by denaturing the collagen from the bones by baking them at 250°C for three hours.

Analyze the data: Create a table showing the changes in bone flexibility, strength, and mass in the three different environments.

Report the results: Under which conditions was the bone most flexible? Under which conditions was the bone the strongest?

Draw a conclusion: Did the results support or refute the hypothesis? How do the results observed in this experiment correspond to diseases that destroy bone tissue?

38.3 Joints and Skeletal Movement

By the end of this section, you will be able to do the following:

- Classify the different types of joints on the basis of structure
- Explain the role of joints in skeletal movement

The point at which two or more bones meet is called a **joint**, or **articulation**. Joints are responsible for movement, such as the movement of limbs, and stability, such as the stability found in the bones of the skull.

Classification of Joints on the Basis of Structure

There are two ways to classify joints: on the basis of their structure or on the basis of their function. The structural classification divides joints into bony, fibrous, cartilaginous, and synovial joints depending on the material composing the joint and the presence or absence of a cavity in the joint.

Fibrous Joints

The bones of **fibrous joints** are held together by fibrous connective tissue. There is no cavity, or space, present between the bones and so most fibrous joints do not move at all, or are only capable of minor movements. There are three types of fibrous joints: sutures, syndesmoses, and gomphoses. **Sutures** are found only in the skull and possess short fibers of connective tissue that hold the skull bones tightly in place (Figure 38.23).

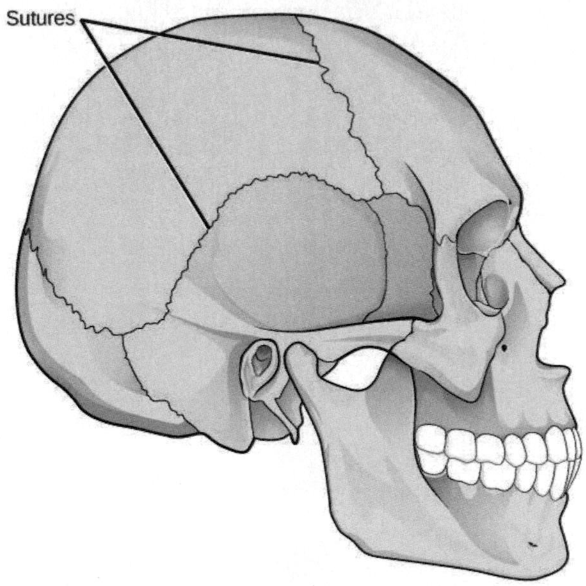

Figure 38.23 Sutures are fibrous joints found only in the skull.

Syndesmoses are joints in which the bones are connected by a band of connective tissue, allowing for more movement than in a suture. An example of a syndesmosis is the joint of the tibia and fibula in the ankle. The amount of movement in these types of joints is determined by the length of the connective tissue fibers. **Gomphoses** occur between teeth and their sockets; the term refers to the way the tooth fits into the socket like a peg (Figure 38.24). The tooth is connected to the socket by a connective tissue referred to as the periodontal ligament.

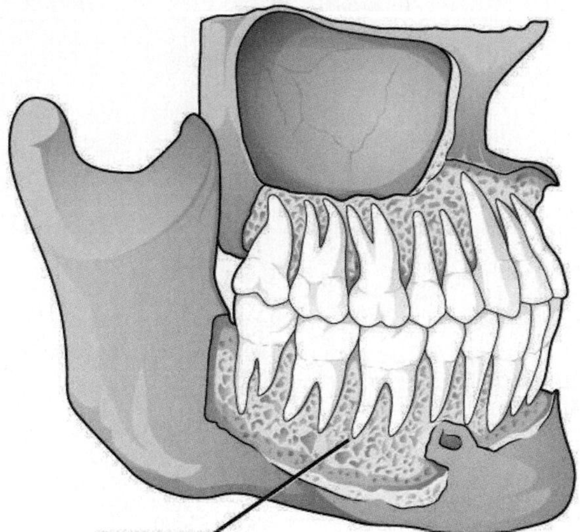

Figure 38.24 Gomphoses are fibrous joints between the teeth and their sockets. (credit: modification of work by Gray's Anatomy)

Cartilaginous Joints

Cartilaginous joints are joints in which the bones are connected by cartilage. There are two types of cartilaginous joints: synchondroses and symphyses. In a **synchondrosis**, the bones are joined by hyaline cartilage. Synchondroses are found in the epiphyseal plates of growing bones in children. In **symphyses**, hyaline cartilage covers the end of the bone but the connection between bones occurs through fibrocartilage. Symphyses are found at the joints between vertebrae. Either type of cartilaginous joint allows for very little movement.

Synovial Joints

Synovial joints are the only joints that have a space between the adjoining bones (Figure 38.25). This space is referred to as the synovial (or joint) cavity and is filled with synovial fluid. Synovial fluid lubricates the joint, reducing friction between the bones and allowing for greater movement. The ends of the bones are covered with articular cartilage, a hyaline cartilage, and the entire joint is surrounded by an articular capsule composed of connective tissue that allows movement of the joint while resisting dislocation. Articular capsules may also possess ligaments that hold the bones together. Synovial joints are capable of the greatest movement of the three structural joint types; however, the more mobile a joint, the weaker the joint. Knees, elbows, and shoulders are examples of synovial joints.

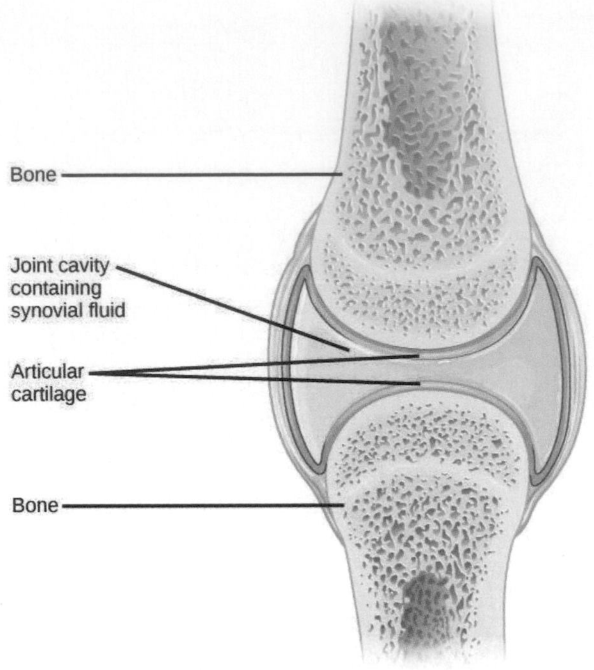

Figure 38.25 Synovial joints are the only joints that have a space or "synovial cavity" in the joint.

Classification of Joints on the Basis of Function

The functional classification divides joints into three categories: synarthroses, amphiarthroses, and diarthroses. A **synarthrosis** is a joint that is immovable. This includes sutures, gomphoses, and synchondroses. **Amphiarthroses** are joints that allow slight movement, including syndesmoses and symphyses. **Diarthroses** are joints that allow for free movement of the joint, as in synovial joints.

Movement at Synovial Joints

The wide range of movement allowed by synovial joints produces different types of movements. The movement of synovial joints can be classified as one of four different types: gliding, angular, rotational, or special movement.

Gliding Movement

Gliding movements occur as relatively flat bone surfaces move past each other. Gliding movements produce very little rotation or angular movement of the bones. The joints of the carpal and tarsal bones are examples of joints that produce gliding movements.

Angular Movement

Angular movements are produced when the angle between the bones of a joint changes. There are several different types of angular movements, including flexion, extension, hyperextension, abduction, adduction, and circumduction. **Flexion**, or bending, occurs when the angle between the bones decreases. Moving the forearm upward at the elbow or moving the wrist to move the hand toward the forearm are examples of flexion. **Extension** is the opposite of flexion in that the angle between the bones of a joint increases. Straightening a limb after flexion is an example of extension. Extension past the regular anatomical position is referred to as **hyperextension**. This includes moving the neck back to look upward, or bending the wrist so that the hand moves away from the forearm.

Abduction occurs when a bone moves away from the midline of the body. Examples of abduction are moving the arms or legs laterally to lift them straight out to the side. **Adduction** is the movement of a bone toward the midline of the body. Movement of the limbs inward after abduction is an example of adduction. **Circumduction** is the movement of a limb in a circular motion, as in moving the arm in a circular motion.

Rotational Movement

Rotational movement is the movement of a bone as it rotates around its longitudinal axis. Rotation can be toward the midline of the body, which is referred to as **medial rotation**, or away from the midline of the body, which is referred to as **lateral rotation**.

Movement of the head from side to side is an example of rotation.

Special Movements

Some movements that cannot be classified as gliding, angular, or rotational are called special movements. **Inversion** involves the soles of the feet moving inward, toward the midline of the body. **Eversion** is the opposite of inversion, movement of the sole of the foot outward, away from the midline of the body. **Protraction** is the anterior movement of a bone in the horizontal plane. **Retraction** occurs as a joint moves back into position after protraction. Protraction and retraction can be seen in the movement of the mandible as the jaw is thrust outwards and then back inwards. **Elevation** is the movement of a bone upward, such as when the shoulders are shrugged, lifting the scapulae. **Depression** is the opposite of elevation—movement downward of a bone, such as after the shoulders are shrugged and the scapulae return to their normal position from an elevated position. **Dorsiflexion** is a bending at the ankle such that the toes are lifted toward the knee. **Plantar flexion** is a bending at the ankle when the heel is lifted, such as when standing on the toes. **Supination** is the movement of the radius and ulna bones of the forearm so that the palm faces forward. **Pronation** is the opposite movement, in which the palm faces backward. **Opposition** is the movement of the thumb toward the fingers of the same hand, making it possible to grasp and hold objects.

Types of Synovial Joints

Synovial joints are further classified into six different categories on the basis of the shape and structure of the joint. The shape of the joint affects the type of movement permitted by the joint (Figure 38.26). These joints can be described as planar, hinge, pivot, condyloid, saddle, or ball-and-socket joints.

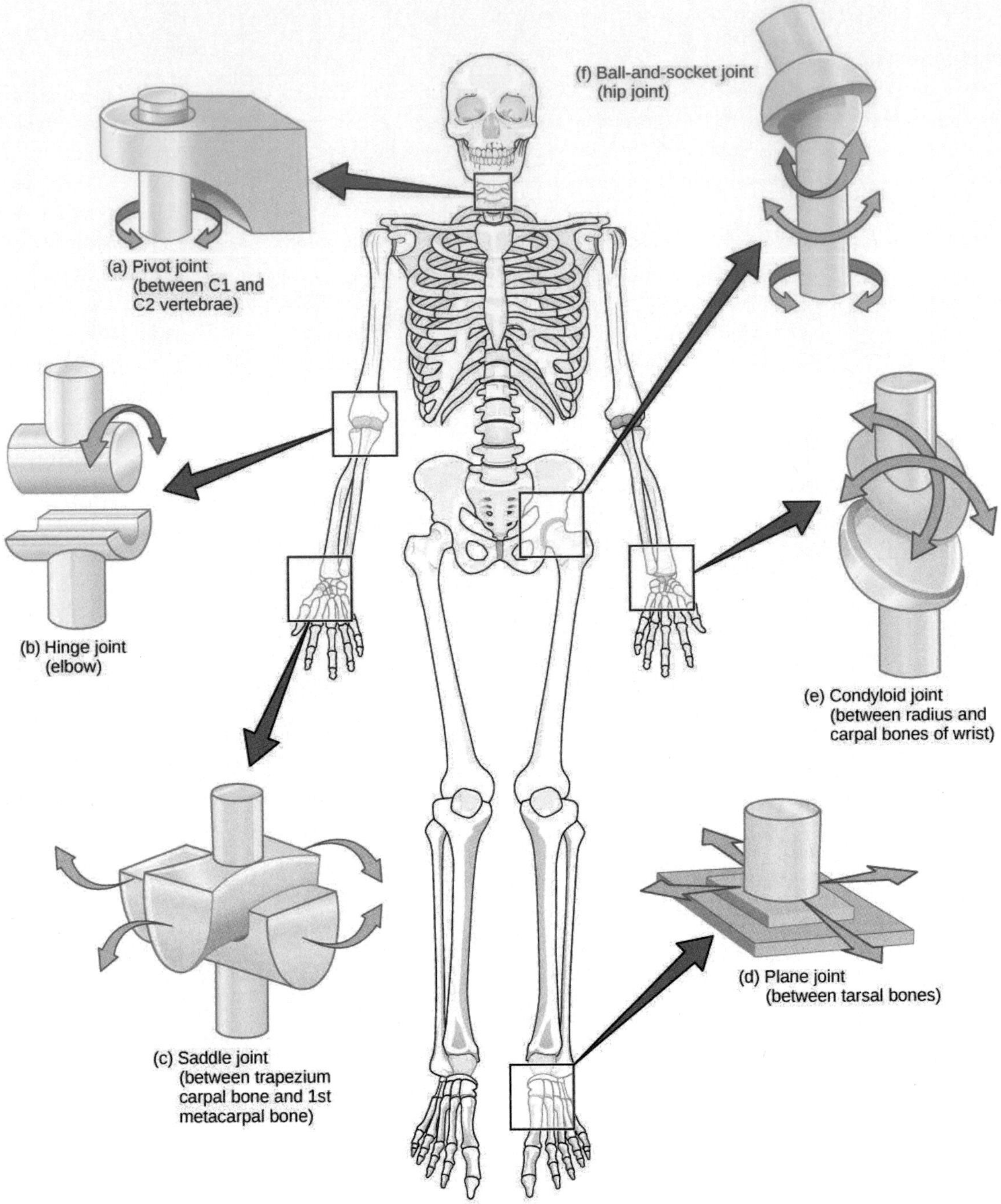

Figure 38.26 Different types of joints allow different types of movement. Planar, hinge, pivot, condyloid, saddle, and ball-and-socket are all types of synovial joints.

Planar Joints

Planar joints have bones with articulating surfaces that are flat or slightly curved faces. These joints allow for gliding movements, and so the joints are sometimes referred to as gliding joints. The range of motion is limited in these joints and does not involve rotation. Planar joints are found in the carpal bones in the hand and the tarsal bones of the foot, as well as between vertebrae (Figure 38.27).

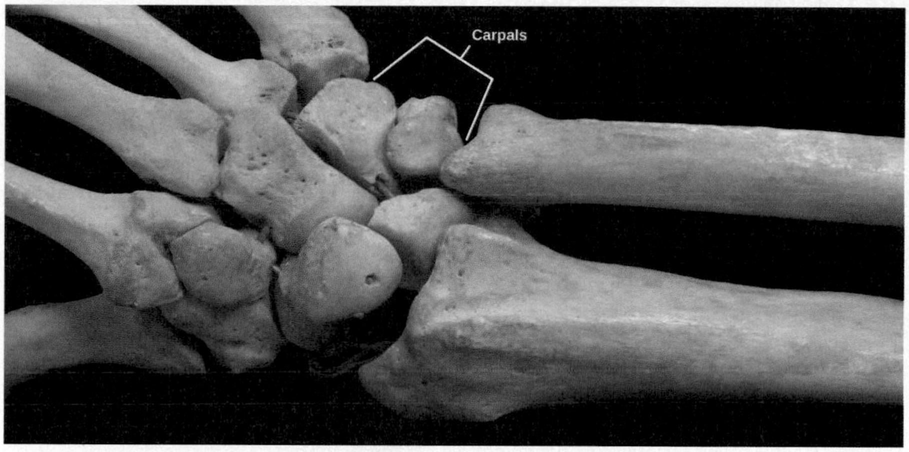

Figure 38.27 The joints of the carpal bones in the wrist are examples of planar joints. (credit: modification of work by Brian C. Goss)

Hinge Joints

In **hinge joints**, the slightly rounded end of one bone fits into the slightly hollow end of the other bone. In this way, one bone moves while the other remains stationary, like the hinge of a door. The elbow is an example of a hinge joint. The knee is sometimes classified as a modified hinge joint (Figure 38.28).

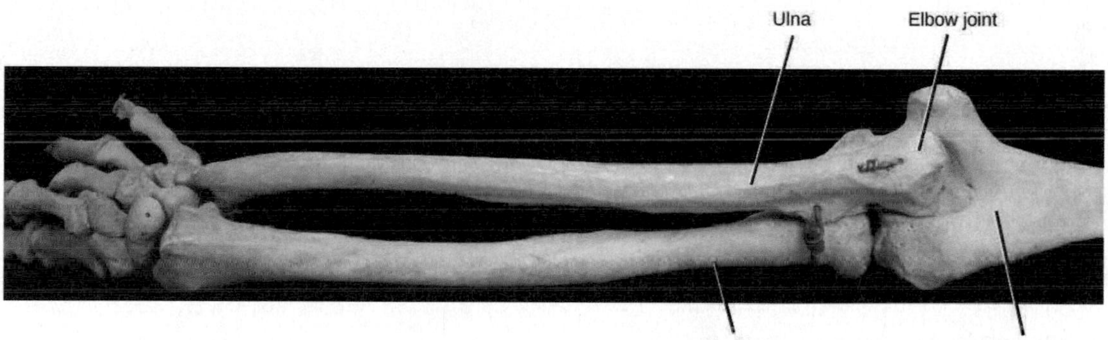

Figure 38.28 The elbow joint, where the radius articulates with the humerus, is an example of a hinge joint. (credit: modification of work by Brian C. Goss)

Pivot Joints

Pivot joints consist of the rounded end of one bone fitting into a ring formed by the other bone. This structure allows rotational movement, as the rounded bone moves around its own axis. An example of a pivot joint is the joint of the first and second vertebrae of the neck that allows the head to move back and forth (Figure 38.29). The joint of the wrist that allows the palm of the hand to be turned up and down is also a pivot joint.

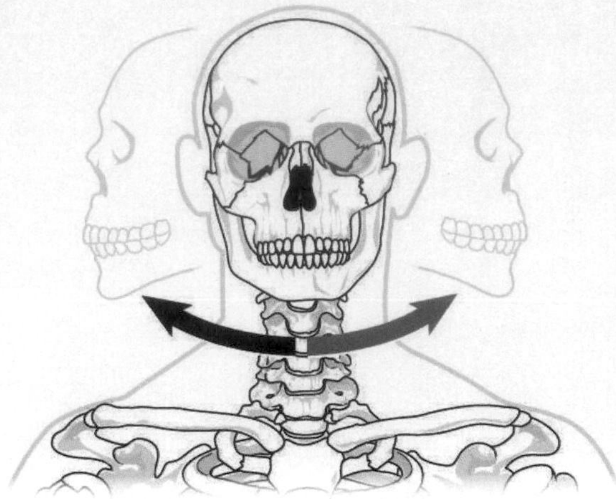

Figure 38.29 The joint in the neck that allows the head to move back and forth is an example of a pivot joint.

Condyloid Joints

Condyloid joints consist of an oval-shaped end of one bone fitting into a similarly oval-shaped hollow of another bone (Figure 38.30). This is also sometimes called an ellipsoidal joint. This type of joint allows angular movement along two axes, as seen in the joints of the wrist and fingers, which can move both side to side and up and down.

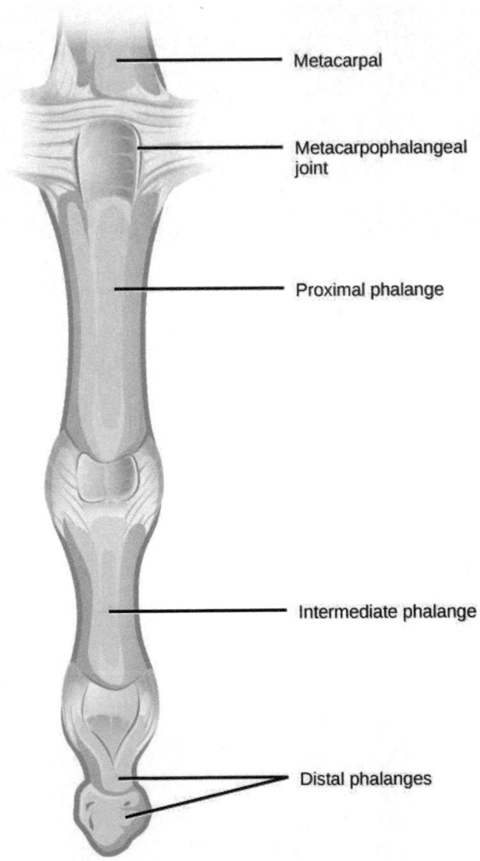

Figure 38.30 The metacarpophalangeal joints in the finger are examples of condyloid joints. (credit: modification of work by Gray's Anatomy)

Saddle Joints

Saddle joints are so named because the ends of each bone resemble a saddle, with concave and convex portions that fit together. Saddle joints allow angular movements similar to condyloid joints but with a greater range of motion. An example of a saddle joint is the thumb joint, which can move back and forth and up and down, but more freely than the wrist or fingers (Figure 38.31).

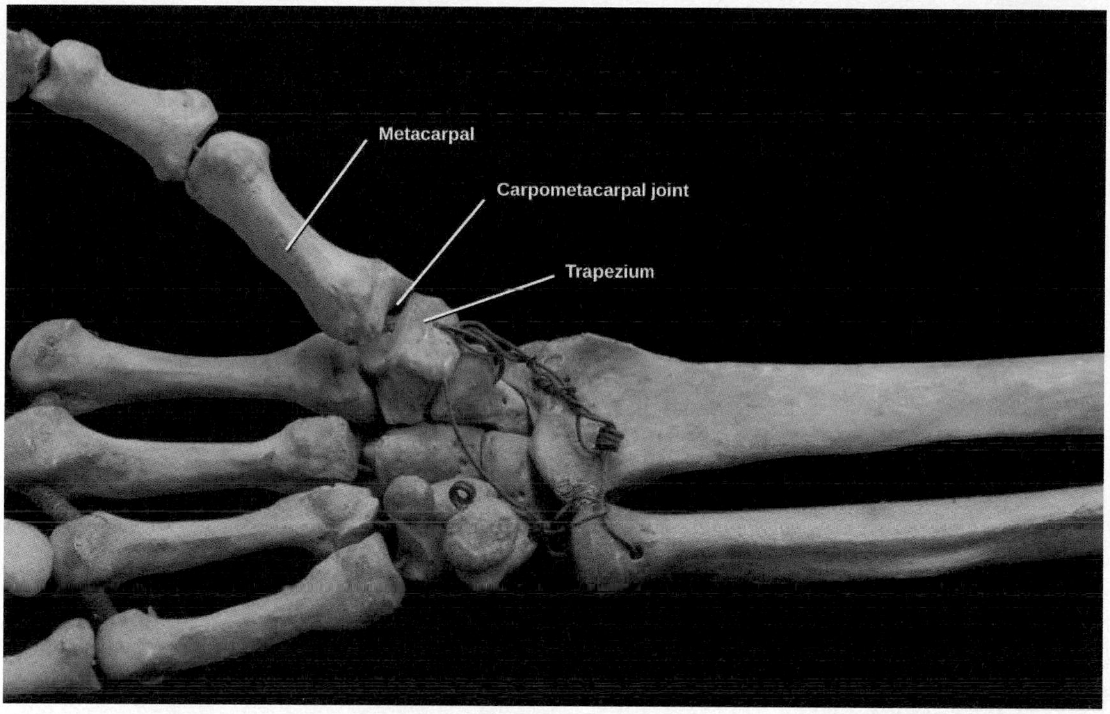

Figure 38.31 The carpometacarpal joints in the thumb are examples of saddle joints. (credit: modification of work by Brian C. Goss)

Ball-and-Socket Joints

Ball-and-socket joints possess a rounded, ball-like end of one bone fitting into a cuplike socket of another bone. This organization allows the greatest range of motion, as all movement types are possible in all directions. Examples of ball-and-socket joints are the shoulder and hip joints (Figure 38.32).

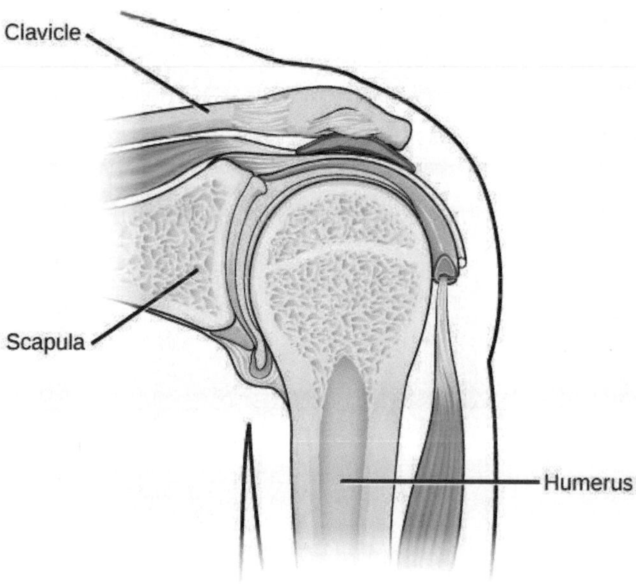

Figure 38.32 The shoulder joint is an example of a ball-and-socket joint.

LINK TO LEARNING

Watch this animation showing the six types of synovial joints.

Click to view content (https://www.openstax.org/l/synovial_joints)

CAREER CONNECTION

Rheumatologist

Rheumatologists are medical doctors who specialize in the diagnosis and treatment of disorders of the joints, muscles, and bones. They diagnose and treat diseases such as arthritis, musculoskeletal disorders, osteoporosis, and autoimmune diseases such as ankylosing spondylitis and rheumatoid arthritis.

Rheumatoid arthritis (RA) is an inflammatory disorder that primarily affects the synovial joints of the hands, feet, and cervical spine. Affected joints become swollen, stiff, and painful. Although it is known that RA is an autoimmune disease in which the body's immune system mistakenly attacks healthy tissue, the cause of RA remains unknown. Immune cells from the blood enter joints and the synovium causing cartilage breakdown, swelling, and inflammation of the joint lining. Breakdown of cartilage causes bones to rub against each other causing pain. RA is more common in women than men and the age of onset is usually 40–50 years of age.

Rheumatologists can diagnose RA on the basis of symptoms such as joint inflammation and pain, X-ray and MRI imaging, and blood tests. Arthrography is a type of medical imaging of joints that uses a contrast agent, such as a dye, that is opaque to X-rays. This allows the soft tissue structures of joints—such as cartilage, tendons, and ligaments—to be visualized. An arthrogram differs from a regular X-ray by showing the surface of soft tissues lining the joint in addition to joint bones. An arthrogram allows early degenerative changes in joint cartilage to be detected before bones become affected.

There is currently no cure for RA; however, rheumatologists have a number of treatment options available. Early stages can be treated with rest of the affected joints by using a cane or by using joint splints that minimize inflammation. When inflammation has decreased, exercise can be used to strengthen the muscles that surround the joint and to maintain joint flexibility. If joint damage is more extensive, medications can be used to relieve pain and decrease inflammation. Anti-inflammatory drugs such as aspirin, topical pain relievers, and corticosteroid injections may be used. Surgery may be required in cases in which joint damage is severe.

38.4 Muscle Contraction and Locomotion

By the end of this section, you will be able to do the following:

- Classify the different types of muscle tissue
- Explain the role of muscles in locomotion

Muscle cells are specialized for contraction. Muscles allow for motions such as walking, and they also facilitate bodily processes such as respiration and digestion. The body contains three types of muscle tissue: skeletal muscle, cardiac muscle, and smooth muscle (Figure 38.33).

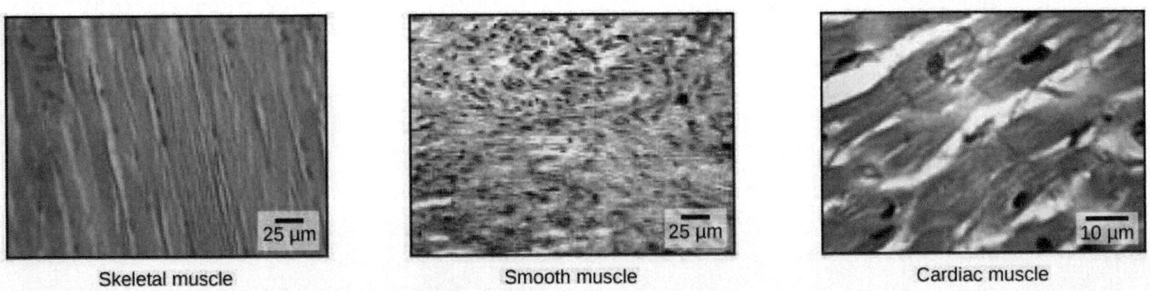

Figure 38.33 The body contains three types of muscle tissue: skeletal muscle, smooth muscle, and cardiac muscle, visualized here using light microscopy. Smooth muscle cells are short, tapered at each end, and have only one plump nucleus in each. Cardiac muscle cells are branched and striated, but short. The cytoplasm may branch, and they have one nucleus in the center of the cell. (credit: modification of work by NCI, NIH; scale-bar data from Matt Russell)

Skeletal muscle tissue forms skeletal muscles, which attach to bones or skin and control locomotion and any movement that can be consciously controlled. Because it can be controlled by thought, skeletal muscle is also called voluntary muscle. Skeletal muscles are long and cylindrical in appearance; when viewed under a microscope, skeletal muscle tissue has a striped or striated appearance. The striations are caused by the regular arrangement of contractile proteins (actin and myosin). **Actin** is a globular contractile protein that interacts with **myosin** for muscle contraction. Skeletal muscle also has multiple nuclei present in a single cell.

Smooth muscle tissue occurs in the walls of hollow organs such as the intestines, stomach, and urinary bladder, and around passages such as the respiratory tract and blood vessels. Smooth muscle has no striations, is not under voluntary control, has only one nucleus per cell, is tapered at both ends, and is called involuntary muscle.

Cardiac muscle tissue is only found in the heart, and cardiac contractions pump blood throughout the body and maintain blood pressure. Like skeletal muscle, cardiac muscle is striated, but unlike skeletal muscle, cardiac muscle cannot be consciously controlled and is called involuntary muscle. It has one nucleus per cell, is branched, and is distinguished by the presence of intercalated disks.

Skeletal Muscle Fiber Structure

Each skeletal muscle fiber is a skeletal muscle cell. These cells are incredibly large, with diameters of up to 100 μm and lengths of up to 30 cm. The plasma membrane of a skeletal muscle fiber is called the **sarcolemma**. The sarcolemma is the site of action potential conduction, which triggers muscle contraction. Within each muscle fiber are **myofibrils**—long cylindrical structures that lie parallel to the muscle fiber. Myofibrils run the entire length of the muscle fiber, and because they are only approximately 1.2 μm in diameter, hundreds to thousands can be found inside one muscle fiber. They attach to the sarcolemma at their ends, so that as myofibrils shorten, the entire muscle cell contracts (Figure 38.34).

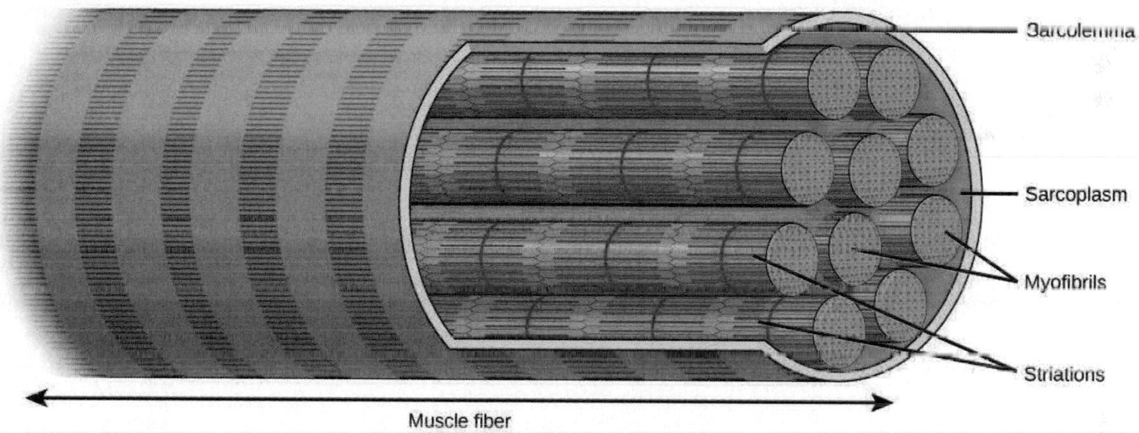

Figure 38.34 A skeletal muscle cell is surrounded by a plasma membrane called the sarcolemma with a cytoplasm called the sarcoplasm. A muscle fiber is composed of many fibrils, packaged into orderly units.

The striated appearance of skeletal muscle tissue is a result of repeating bands of the proteins actin and myosin that are present along the length of myofibrils. Dark A bands and light I bands repeat along myofibrils, and the alignment of myofibrils in the cell causes the entire cell to appear striated or banded.

Each I band has a dense line running vertically through the middle called a Z disc or Z line. The Z discs mark the border of units called **sarcomeres**, which are the functional units of skeletal muscle. One sarcomere is the space between two consecutive Z discs and contains one entire A band and two halves of an I band, one on either side of the A band. A myofibril is composed of many sarcomeres running along its length, and as the sarcomeres individually contract, the myofibrils and muscle cells shorten (Figure 38.35).

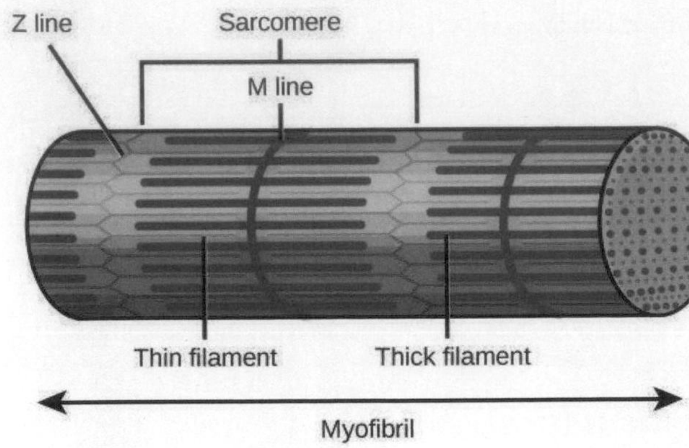

Figure 38.35 A sarcomere is the region from one Z line to the next Z line. Many sarcomeres are present in a myofibril, resulting in the striation pattern characteristic of skeletal muscle.

Myofibrils are composed of smaller structures called **myofilaments**. There are two main types of filaments: thick filaments and thin filaments; each has different compositions and locations. **Thick filaments** occur only in the A band of a myofibril. **Thin filaments** attach to a protein in the Z disc called alpha-actinin and occur across the entire length of the I band and partway into the A band. The region at which thick and thin filaments overlap has a dense appearance, as there is little space between the filaments. Thin filaments do not extend all the way into the A bands, leaving a central region of the A band that only contains thick filaments. This central region of the A band looks slightly lighter than the rest of the A band and is called the H zone. The middle of the H zone has a vertical line called the M line, at which accessory proteins hold together thick filaments. Both the Z disc and the M line hold myofilaments in place to maintain the structural arrangement and layering of the myofibril. Myofibrils are connected to each other by intermediate, or desmin, filaments that attach to the Z disc.

Thick and thin filaments are themselves composed of proteins. Thick filaments are composed of the protein myosin. The tail of a myosin molecule connects with other myosin molecules to form the central region of a thick filament near the M line, whereas the heads align on either side of the thick filament where the thin filaments overlap. The primary component of thin filaments is the actin protein. Two other components of the thin filament are tropomyosin and troponin. Actin has binding sites for myosin attachment. Strands of tropomyosin block the binding sites and prevent actin–myosin interactions when the muscles are at rest. Troponin consists of three globular subunits. One subunit binds to tropomyosin, one subunit binds to actin, and one subunit binds Ca^{2+} ions.

LINK TO LEARNING

View this animation showing the organization of muscle fibers.

Click to view content (https://www.openstax.org/l/skeletal_muscle)

Sliding Filament Model of Contraction

For a muscle cell to contract, the sarcomere must shorten. However, thick and thin filaments—the components of sarcomeres—do not shorten. Instead, they slide by one another, causing the sarcomere to shorten while the filaments remain the same length. The sliding filament theory of muscle contraction was developed to fit the differences observed in the named bands on the sarcomere at different degrees of muscle contraction and relaxation. The mechanism of contraction is the binding of myosin to actin, forming cross-bridges that generate filament movement (Figure 38.36).

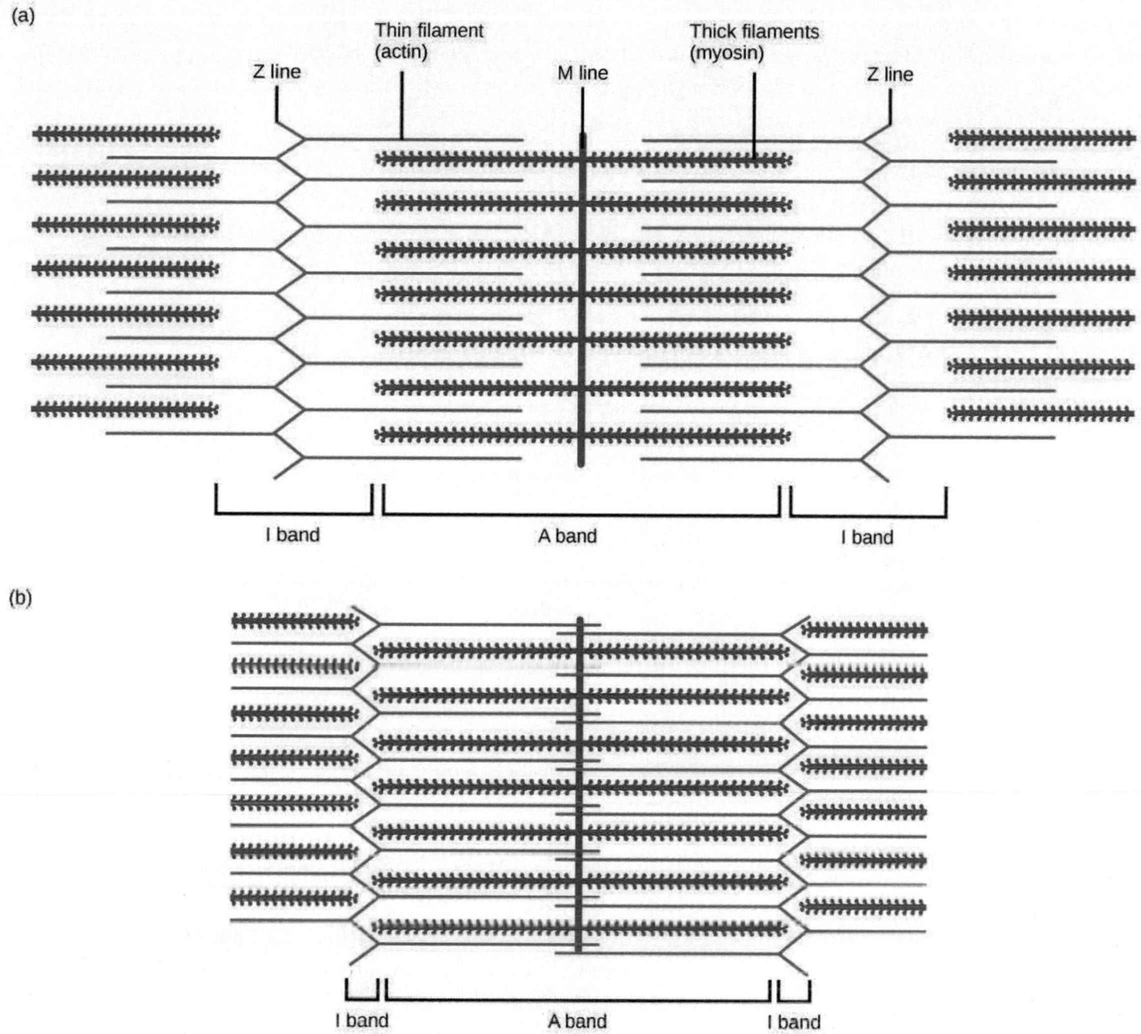

Figure 38.36 When (a) a sarcomere (b) contracts, the Z lines move closer together and the I band gets smaller. The A band stays the same width and, at full contraction, the thin filaments overlap.

When a sarcomere shortens, some regions shorten whereas others stay the same length. A sarcomere is defined as the distance between two consecutive Z discs or Z lines; when a muscle contracts, the distance between the Z discs is reduced. The H zone—the central region of the A zone—contains only thick filaments and is shortened during contraction. The I band contains only thin filaments and also shortens. The A band does not shorten—it remains the same length—but A bands of different sarcomeres move closer together during contraction, eventually disappearing. Thin filaments are pulled by the thick filaments toward the center of the sarcomere until the Z discs approach the thick filaments. The zone of overlap, in which thin filaments and thick filaments occupy the same area, increases as the thin filaments move inward.

ATP and Muscle Contraction

The motion of muscle shortening occurs as myosin heads bind to actin and pull the actin inwards. This action requires energy, which is provided by ATP. Myosin binds to actin at a binding site on the globular actin protein. Myosin has another binding site for ATP at which enzymatic activity hydrolyzes ATP to ADP, releasing an inorganic phosphate molecule and energy.

ATP binding causes myosin to release actin, allowing actin and myosin to detach from each other. After this happens, the newly bound ATP is converted to ADP and inorganic phosphate, P_i. The enzyme at the binding site on myosin is called ATPase. The energy released during ATP hydrolysis changes the angle of the myosin head into a "cocked" position. The myosin head is then in a position for further movement, possessing potential energy, but ADP and P_i are still attached. If actin binding sites are covered and unavailable, the myosin will remain in the high energy configuration with ATP hydrolyzed, but still attached.

If the actin binding sites are uncovered, a cross-bridge will form; that is, the myosin head spans the distance between the actin and myosin molecules. P_i is then released, allowing myosin to expend the stored energy as a conformational change. The myosin head moves toward the M line, pulling the actin along with it. As the actin is pulled, the filaments move approximately 10 nm toward the M line. This movement is called the power stroke, as it is the step at which force is produced. As the actin is pulled toward the M line, the sarcomere shortens and the muscle contracts.

When the myosin head is "cocked," it contains energy and is in a high-energy configuration. This energy is expended as the myosin head moves through the power stroke; at the end of the power stroke, the myosin head is in a low-energy position. After the power stroke, ADP is released; however, the cross-bridge formed is still in place, and actin and myosin are bound together. ATP can then attach to myosin, which allows the cross-bridge cycle to start again and further muscle contraction can occur (Figure 38.37).

LINK TO LEARNING

Watch this video explaining how a muscle contraction is signaled.

Click to view content (https://www.openstax.org/l/contract_muscle)

VISUAL CONNECTION

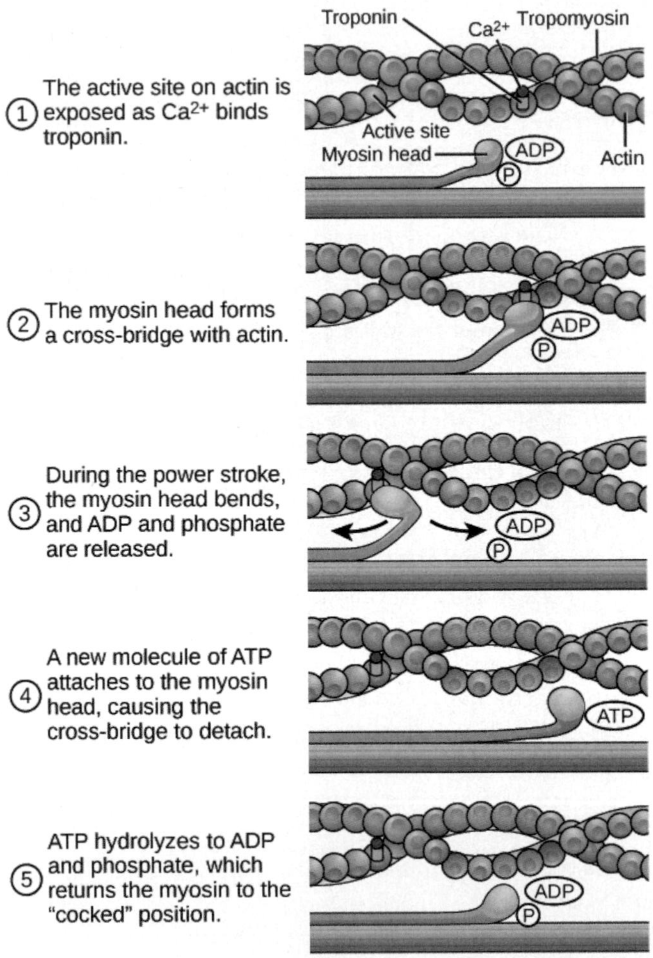

Figure 38.37 The cross-bridge muscle contraction cycle, which is triggered by Ca^{2+} binding to the actin active site, is shown. With each contraction cycle, actin moves relative to myosin.

Which of the following statements about muscle contraction is true?

a. The power stroke occurs when ATP is hydrolyzed to ADP and phosphate.
b. The power stroke occurs when ADP and phosphate dissociate from the myosin head.

c. The power stroke occurs when ADP and phosphate dissociate from the actin active site.
d. The power stroke occurs when Ca^{2+} binds the calcium head.

LINK TO LEARNING

View this animation (http://openstax.org/l/muscle_contract) of the cross-bridge muscle contraction.

Regulatory Proteins

When a muscle is in a resting state, actin and myosin are separated. To keep actin from binding to the active site on myosin, regulatory proteins block the molecular binding sites. **Tropomyosin** blocks myosin binding sites on actin molecules, preventing cross-bridge formation and preventing contraction in a muscle without nervous input. **Troponin** binds to tropomyosin and helps to position it on the actin molecule; it also binds calcium ions.

To enable a muscle contraction, tropomyosin must change conformation, uncovering the myosin-binding site on an actin molecule and allowing cross-bridge formation. This can only happen in the presence of calcium, which is kept at extremely low concentrations in the sarcoplasm. If present, calcium ions bind to troponin, causing conformational changes in troponin that allow tropomyosin to move away from the myosin binding sites on actin. Once the tropomyosin is removed, a cross-bridge can form between actin and myosin, triggering contraction. Cross-bridge cycling continues until Ca^{2+} ions and ATP are no longer available and tropomyosin again covers the binding sites on actin.

Excitation–Contraction Coupling

Excitation–contraction coupling is the link (transduction) between the action potential generated in the sarcolemma and the start of a muscle contraction. The trigger for calcium release from the sarcoplasmic reticulum into the sarcoplasm is a neural signal. Each skeletal muscle fiber is controlled by a motor neuron, which conducts signals from the brain or spinal cord to the muscle. The area of the sarcolemma on the muscle fiber that interacts with the neuron is called the **motor end plate**. The end of the neuron's axon is called the synaptic terminal, and it does not actually contact the motor end plate. A small space called the synaptic cleft separates the synaptic terminal from the motor end plate. Electrical signals travel along the neuron's axon, which branches through the muscle and connects to individual muscle fibers at a neuromuscular junction.

The ability of cells to communicate electrically requires that the cells expend energy to create an electrical gradient across their cell membranes. This charge gradient is carried by ions, which are differentially distributed across the membrane. Each ion exerts an electrical influence and a concentration influence. Just as milk will eventually mix with coffee without the need to stir, ions also distribute themselves evenly, if they are permitted to do so. In this case, they are not permitted to return to an evenly mixed state.

The sodium–potassium ATPase uses cellular energy to move K^+ ions inside the cell and Na^+ ions outside. This alone accumulates a small electrical charge, but a big concentration gradient. There is lots of K^+ in the cell and lots of Na^+ outside the cell. Potassium is able to leave the cell through K^+ channels that are open 90% of the time, and it does. However, Na^+ channels are rarely open, so Na^+ remains outside the cell. When K^+ leaves the cell, obeying its concentration gradient, that effectively leaves a negative charge behind. So at rest, there is a large concentration gradient for Na^+ to enter the cell, and there is an accumulation of negative charges left behind in the cell. This is the resting membrane potential. Potential in this context means a separation of electrical charge that is capable of doing work. It is measured in volts, just like a battery. However, the transmembrane potential is considerably smaller (0.07 V); therefore, the small value is expressed as millivolts (mV) or 70 mV. Because the inside of a cell is negative compared with the outside, a minus sign signifies the excess of negative charges inside the cell, –70 mV.

If an event changes the permeability of the membrane to Na^+ ions, they will enter the cell. That will change the voltage. This is an electrical event, called an action potential, that can be used as a cellular signal. Communication occurs between nerves and muscles through neurotransmitters. Neuron action potentials cause the release of neurotransmitters from the synaptic terminal into the synaptic cleft, where they can then diffuse across the synaptic cleft and bind to a receptor molecule on the motor end plate. The motor end plate possesses junctional folds—folds in the sarcolemma that create a large surface area for the neurotransmitter to bind to receptors. The receptors are actually sodium channels that open to allow the passage of Na^+ into the cell when they receive a neurotransmitter signal.

Acetylcholine (ACh) is a neurotransmitter released by motor neurons that binds to receptors in the motor end plate.

Neurotransmitter release occurs when an action potential travels down the motor neuron's axon, resulting in altered permeability of the synaptic terminal membrane and an influx of calcium. The Ca^{2+} ions allow synaptic vesicles to move to and bind with the presynaptic membrane (on the neuron), and release neurotransmitter from the vesicles into the synaptic cleft. Once released by the synaptic terminal, ACh diffuses across the synaptic cleft to the motor end plate, where it binds with ACh receptors. As a neurotransmitter binds, these ion channels open, and Na^{+} ions cross the membrane into the muscle cell. This reduces the voltage difference between the inside and outside of the cell, which is called depolarization. As ACh binds at the motor end plate, this depolarization is called an end-plate potential. The depolarization then spreads along the sarcolemma, creating an action potential as sodium channels adjacent to the initial depolarization site sense the change in voltage and open. The action potential moves across the entire cell, creating a wave of depolarization.

ACh is broken down by the enzyme **acetylcholinesterase** (AChE) into acetyl and choline. AChE resides in the synaptic cleft, breaking down ACh so that it does not remain bound to ACh receptors, which would cause unwanted extended muscle contraction (Figure 38.38).

VISUAL CONNECTION

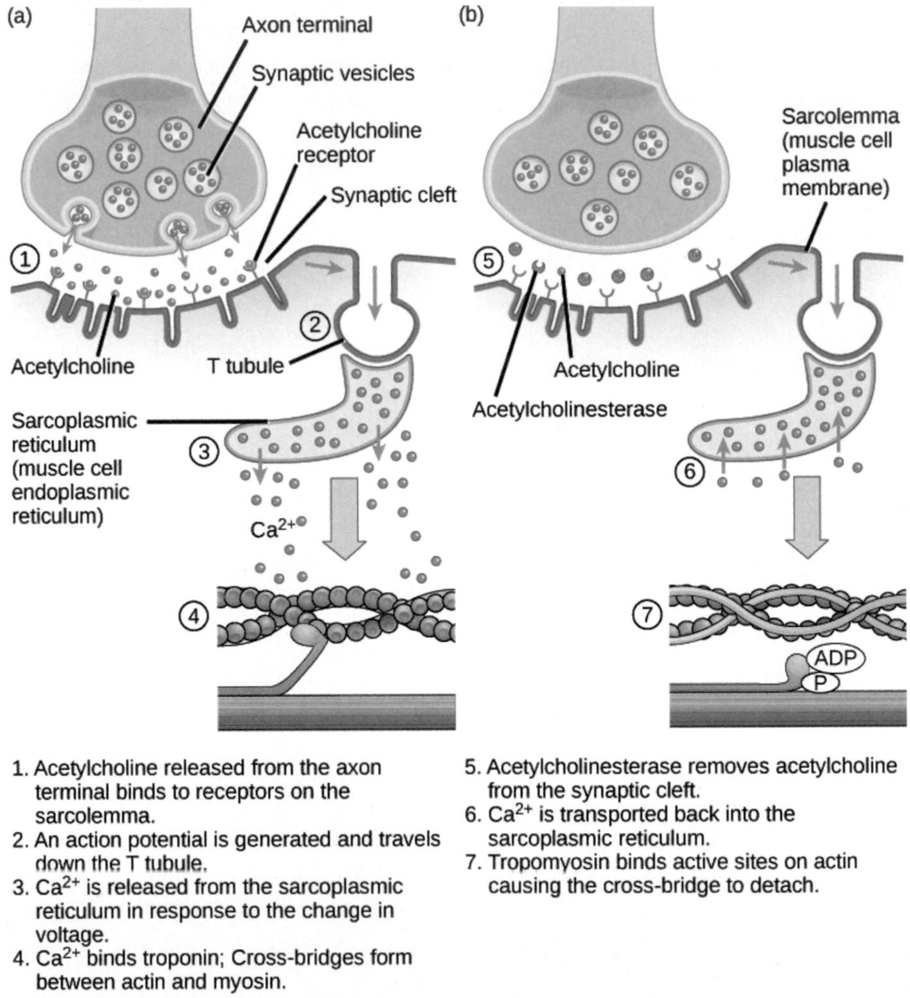

Figure 38.38 This diagram shows excitation-contraction coupling in a skeletal muscle contraction. The sarcoplasmic reticulum is a specialized endoplasmic reticulum found in muscle cells.

The deadly nerve gas Sarin irreversibly inhibits Acetylcholinesterase. What effect would Sarin have on muscle contraction?

After depolarization, the membrane returns to its resting state. This is called repolarization, during which voltage-gated sodium channels close. Potassium channels continue at 90% conductance. Because the plasma membrane sodium–potassium ATPase always transports ions, the resting state (negatively charged inside relative to the outside) is restored. The period immediately following the transmission of an impulse in a nerve or muscle, in which a neuron or muscle cell regains its ability to transmit another impulse, is called the refractory period. During the refractory period, the membrane cannot generate another action potential. The refractory period allows the voltage-sensitive ion channels to return to their resting configurations. The sodium potassium ATPase continually moves Na^+ back out of the cell and K^+ back into the cell, and the K^+ leaks out leaving negative charge behind. Very quickly, the membrane repolarizes, so that it can again be depolarized.

Control of Muscle Tension

Neural control initiates the formation of actin–myosin cross-bridges, leading to the sarcomere shortening involved in muscle contraction. These contractions extend from the muscle fiber through connective tissue to pull on bones, causing skeletal movement. The pull exerted by a muscle is called tension, and the amount of force created by this tension can vary. This enables the same muscles to move very light objects and very heavy objects. In individual muscle fibers, the amount of tension produced depends on the cross-sectional area of the muscle fiber and the frequency of neural stimulation.

The number of cross-bridges formed between actin and myosin determine the amount of tension that a muscle fiber can produce. Cross-bridges can only form where thick and thin filaments overlap, allowing myosin to bind to actin. If more cross-bridges are formed, more myosin will pull on actin, and more tension will be produced.

The ideal length of a sarcomere during production of maximal tension occurs when thick and thin filaments overlap to the greatest degree. If a sarcomere at rest is stretched past an ideal resting length, thick and thin filaments do not overlap to the greatest degree, and fewer cross-bridges can form. This results in fewer myosin heads pulling on actin, and less tension is produced. As a sarcomere is shortened, the zone of overlap is reduced as the thin filaments reach the H zone, which is composed of myosin tails. Because it is myosin heads that form cross-bridges, actin will not bind to myosin in this zone, reducing the tension produced by this myofiber. If the sarcomere is shortened even more, thin filaments begin to overlap with each other—reducing cross-bridge formation even further, and producing even less tension. Conversely, if the sarcomere is stretched to the point at which thick and thin filaments do not overlap at all, no cross-bridges are formed and no tension is produced. This amount of stretching does not usually occur because accessory proteins, internal sensory nerves, and connective tissue oppose extreme stretching.

The primary variable determining force production is the number of myofibers within the muscle that receive an action potential from the neuron that controls that fiber. When using the biceps to pick up a pencil, the motor cortex of the brain only signals a few neurons of the biceps, and only a few myofibers respond. In vertebrates, each myofiber responds fully if stimulated. When picking up a piano, the motor cortex signals all of the neurons in the biceps and every myofiber participates. This is close to the maximum force the muscle can produce. As mentioned above, increasing the frequency of action potentials (the number of signals per second) can increase the force a bit more, because the tropomyosin is flooded with calcium.

KEY TERMS

abduction when a bone moves away from the midline of the body

acetylcholinesterase (AChE) enzyme that breaks down ACh into acetyl and choline

actin globular contractile protein that interacts with myosin for muscle contraction

adduction movement of the limbs inward after abduction

amphiarthrosis joint that allows slight movement; includes syndesmoses and symphyses

angular movement produced when the angle between the bones of a joint changes

appendicular skeleton composed of the bones of the upper limbs, which function to grasp and manipulate objects, and the lower limbs, which permit locomotion

appositional growth increase in the diameter of bones by the addition of bone tissue at the surface of bones

articulation any place where two bones are joined

auditory ossicle (also, middle ear) transduces sounds from the air into vibrations in the fluid-filled cochlea

axial skeleton forms the central axis of the body and includes the bones of the skull, the ossicles of the middle ear, the hyoid bone of the throat, the vertebral column, and the thoracic cage (ribcage)

ball-and-socket joint joint with a rounded, ball-like end of one bone fitting into a cuplike socket of another bone

bone (also, osseous tissue) connective tissue that constitutes the endoskeleton

bone remodeling replacement of old bone tissue by new bone tissue

calcification process of deposition of mineral salts in the collagen fiber matrix that crystallizes and hardens the tissue

cardiac muscle tissue muscle tissue found only in the heart; cardiac contractions pump blood throughout the body and maintain blood pressure

carpus eight bones that comprise the wrist

cartilaginous joint joint in which the bones are connected by cartilage

circumduction movement of a limb in a circular motion

clavicle S-shaped bone that positions the arms laterally

compact bone forms the hard external layer of all bones

condyloid joint oval-shaped end of one bone fitting into a similarly oval-shaped hollow of another bone

coxal bone hip bone

cranial bone one of eight bones that form the cranial cavity that encloses the brain and serves as an attachment site for the muscles of the head and neck

depression movement downward of a bone, such as after the shoulders are shrugged and the scapulae return to their normal position from an elevated position; opposite of elevation

diaphysis central shaft of bone, contains bone marrow in a marrow cavity

diarthrosis joint that allows for free movement of the joint; found in synovial joints

dorsiflexion bending at the ankle such that the toes are lifted toward the knee

elevation movement of a bone upward, such as when the shoulders are shrugged, lifting the scapulae

endochondral ossification process of bone development from hyaline cartilage

endoskeleton skeleton of living cells that produces a hard, mineralized tissue located within the soft tissue of organisms

epiphyseal plate region between the diaphysis and epiphysis that is responsible for the lengthwise growth of long bones

epiphysis rounded end of bone, covered with articular cartilage and filled with red bone marrow, which produces blood cells

eversion movement of the sole of the foot outward, away from the midline of the body; opposite of inversion

exoskeleton a secreted cellular product external skeleton that consists of a hard encasement on the surface of an organism

extension movement in which the angle between the bones of a joint increases; opposite of flexion

facial bone one of the 14 bones that form the face; provides cavities for the sense organs (eyes, mouth, and nose) and attachment points for facial muscles

femur (also, thighbone) longest, heaviest, and strongest bone in the body

fibrous joint joint held together by fibrous connective tissue

fibula (also, calf bone) parallels and articulates with the tibia

flat bone thin and relatively broad bone found where extensive protection of organs is required or where broad surfaces of muscle attachment are required

flexion movement in which the angle between the bones decreases; opposite of extension

forearm extends from the elbow to the wrist and consists of two bones: the ulna and the radius

gliding movement when relatively flat bone surfaces move past each other

gomphosis the joint in which the tooth fits into the socket like a peg

Haversian canal contains the bone's blood vessels and nerve fibers

hinge joint slightly rounded end of one bone fits into the slightly hollow end of the other bone

humerus only bone of the arm

hydrostatic skeleton skeleton that consists of aqueous fluid held under pressure in a closed body compartment

hyoid bone lies below the mandible in the front of the neck

hyperextension extension past the regular anatomical position

intervertebral disc composed of fibrous cartilage; lies between adjacent vertebrae from the second cervical vertebra to the sacrum

intramembranous ossification process of bone development from fibrous membranes

inversion soles of the feet moving inward, toward the midline of the body

irregular bone bone with complex shapes; examples include vertebrae and hip bones

joint point at which two or more bones meet

lamella layer of compact tissue that surrounds a central canal called the Haversian canal

lateral rotation rotation away from the midline of the body

long bone bone that is longer than wide, and has a shaft and two ends

lower limb consists of the thigh, the leg, and the foot

medial rotation rotation toward the midline of the body

metacarpus five bones that comprise the palm

metatarsal one of the five bones of the foot

motor end plate sarcolemma of the muscle fiber that interacts with the neuron

myofibril long cylindrical structures that lie parallel to the muscle fiber

myofilament small structures that make up myofibrils

myosin contractile protein that interacts with actin for muscle contraction

opposition movement of the thumb toward the fingers of the same hand, making it possible to grasp and hold objects

osseous tissue connective tissue that constitutes the endoskeleton

ossification (also, osteogenesis) process of bone formation by osteoblasts

osteoblast bone cell responsible for bone formation

osteoclast large bone cells with up to 50 nuclei, responsible for bone remodeling

osteocyte mature bone cells and the main cell in bone tissue

osteon cylindrical structure aligned parallel to the long axis of the bone

patella (also, kneecap) triangular bone that lies anterior to the knee joint

pectoral girdle bones that transmit the force generated by the upper limbs to the axial skeleton

pelvic girdle bones that transmit the force generated by the lower limbs to the axial skeleton

phalange one of the bones of the fingers or toes

pivot joint joint with the rounded end of one bone fitting into a ring formed by the other bone

planar joint joint with bones whose articulating surfaces are flat

plantar flexion bending at the ankle such that the heel is lifted, such as when standing on the toes

pronation movement in which the palm faces backward

protraction anterior movement of a bone in the horizontal plane

radius bone located along the lateral (thumb) side of the forearm; articulates with the humerus at the elbow

resorption process by which osteoclasts release minerals stored in bones

retraction movement in which a joint moves back into position after protraction

rib one of 12 pairs of long, curved bones that attach to the thoracic vertebrae and curve toward the front of the body to form the ribcage

rotational movement movement of a bone as it rotates around its own longitudinal axis

saddle joint joint with concave and convex portions that fit together; named because the ends of each bone resemble a saddle

sarcolemma plasma membrane of a skeletal muscle fiber

sarcomere functional unit of skeletal muscle

scapula flat, triangular bone located at the posterior pectoral girdle

sesamoid bone small, flat bone shaped like a sesame seed; develops inside tendons

short bone bone that has the same width and length, giving it a cube-like shape

skeletal muscle tissue forms skeletal muscles, which attach to bones and control locomotion and any movement that can be consciously controlled

skull bone that supports the structures of the face and protects the brain

smooth muscle tissue occurs in the walls of hollow organs such as the intestines, stomach, and urinary bladder, and around passages such as the respiratory tract and blood vessels

spongy bone tissue forms the inner layer of all bones

sternum (also, breastbone) long, flat bone located at the front of the chest

supination movement of the radius and ulna bones of the forearm so that the palm faces forward

sutural bone small, flat, irregularly shaped bone that forms between the flat bones of the cranium

suture short fiber of connective tissue that holds the skull bones tightly in place; found only in the skull

symphysis hyaline cartilage covers the end of the bone, but the connection between bones occurs through fibrocartilage; symphyses are found at the joints between vertebrae

synarthrosis joint that is immovable

synchondrosis bones joined by hyaline cartilage; synchondroses are found in the epiphyseal plates of growing bones in children

syndesmosis joint in which the bones are connected by a band of connective tissue, allowing for more movement than in a suture

synovial joint only joint that has a space between the adjoining bones

tarsal one of the seven bones of the ankle

thick filament a group of myosin molecules

thin filament two polymers of actin wound together along with tropomyosin and troponin

thoracic cage (also, ribcage) skeleton of the chest, which consists of the ribs, thoracic vertebrae, sternum, and costal cartilages

tibia (also, shinbone) large bone of the leg that is located directly below the knee

trabeculae lamellae that are arranged as rods or plates

tropomyosin acts to block myosin binding sites on actin molecules, preventing cross-bridge formation and preventing contraction until a muscle receives a neuron signal

troponin binds to tropomyosin and helps to position it on the actin molecule, and also binds calcium ions

ulna bone located on the medial aspect (pinky-finger side) of the forearm

vertebral column (also, spine) surrounds and protects the spinal cord, supports the head, and acts as an attachment point for ribs and muscles of the back and neck

CHAPTER SUMMARY

38.1 Types of Skeletal Systems

The three types of skeleton designs are hydrostatic skeletons, exoskeletons, and endoskeletons. A hydrostatic skeleton is formed by a fluid-filled compartment held under hydrostatic pressure; movement is created by the muscles producing pressure on the fluid. An exoskeleton is a hard external skeleton that protects the outer surface of an organism and enables movement through muscles attached on the inside. An endoskeleton is an internal skeleton composed of hard, mineralized tissue that also enables movement by attachment to muscles. The human skeleton is an endoskeleton that is composed of the axial and appendicular skeleton. The axial skeleton is composed of the bones of the skull, ossicles of the ear, hyoid bone, vertebral column, and ribcage. The skull consists of eight cranial bones and 14 facial bones. Six bones make up the ossicles of the middle ear, while the hyoid bone is located in the neck under the mandible. The vertebral column contains 26 bones, and it surrounds and protects the spinal cord. The thoracic cage consists of the sternum, ribs, thoracic vertebrae, and costal cartilages. The appendicular skeleton is made up of the limbs of the upper and lower limbs. The pectoral girdle is composed of the clavicles and the scapulae. The upper limb contains 30 bones in the arm, the forearm, and the hand. The pelvic girdle attaches the lower limbs to the axial skeleton. The lower limb includes the bones of the thigh, the leg, and the foot.

38.2 Bone

Bone, or osseous tissue, is connective tissue that includes specialized cells, mineral salts, and collagen fibers. The human skeleton can be divided into long bones, short bones, flat bones, and irregular bones. Compact bone tissue is composed of osteons and forms the external layer of all bones. Spongy bone tissue is composed of trabeculae and forms the inner part of all bones. Four types of cells compose bony tissue: osteocytes, osteoclasts, osteoprogenitor cells, and osteoblasts. Ossification is the process of bone formation by osteoblasts. Intramembranous ossification is the process of bone development from fibrous membranes. Endochondral ossification is the process of bone development from hyaline cartilage. Long bones lengthen as chondrocytes divide and secrete hyaline cartilage. Osteoblasts replace cartilage with bone. Appositional growth is the increase in the diameter of bones by the addition of bone tissue at the surface of bones. Bone remodeling involves the processes of bone deposition by osteoblasts and bone resorption by osteoclasts. Bone repair occurs in four stages and can take several months.

38.3 Joints and Skeletal Movement

The structural classification of joints divides them into bony, fibrous, cartilaginous, and synovial joints. The bones of fibrous joints are held together by fibrous connective tissue; the three types of fibrous joints are sutures, syndesomes, and gomphoses. Cartilaginous joints are joints in which the bones are connected by cartilage; the two types of cartilaginous joints are synchondroses and symphyses. Synovial joints are joints that have a space between the adjoining bones. The functional classification divides joints into three categories: synarthroses, amphiarthroses, and diarthroses. The movement of synovial joints can be classified as one of four different types: gliding, angular, rotational, or special movement. Gliding movements occur as relatively flat bone surfaces move past each other. Angular movements are produced when the angle between the bones of a joint changes. Rotational movement is the movement of a bone as it rotates around its own longitudinal axis. Special movements include inversion, eversion, protraction, retraction, elevation, depression, dorsiflexion, plantar flexion, supination, pronation, and opposition. Synovial joints are also classified into six different categories on the basis of the shape and structure of the joint: planar, hinge, pivot, condyloid, saddle, and ball-and-socket.

38.4 Muscle Contraction and Locomotion

The body contains three types of muscle tissue: skeletal muscle, cardiac muscle, and smooth muscle. Skeleton muscle tissue is composed of sarcomeres, the functional units of muscle tissue. Muscle contraction occurs when sarcomeres shorten, as thick and thin filaments slide past each other, which is called the sliding filament model of muscle contraction. ATP provides the energy for cross-bridge formation and filament sliding. Regulatory proteins, such as troponin and tropomyosin, control cross-bridge formation. Excitation–contraction coupling transduces the electrical signal of the neuron, via acetylcholine, to an electrical signal on the muscle membrane, which initiates force production. The number of muscle fibers contracting determines how much force the whole muscle produces.

VISUAL CONNECTION QUESTIONS

1. Figure 38.19 Which of the following statements about bone tissue is false?
    a. Compact bone tissue is made of cylindrical osteons that are aligned such that they travel the length of the bone.
    b. Haversian canals contain blood vessels only.
    c. Haversian canals contain blood vessels and nerve fibers.
    d. Spongy tissue is found on the interior of the bone, and compact bone tissue is found on the exterior.

2. Figure 38.37 Which of the following statements about muscle contraction is true?
    a. The power stroke occurs when ATP is hydrolyzed to ADP and phosphate.
    b. The power stroke occurs when ADP and phosphate dissociate from the myosin head.
    c. The power stroke occurs when ADP and phosphate dissociate from the actin active site.
    d. The power stroke occurs when Ca^{2+} binds the calcium head.

3. Figure 38.38 The deadly nerve gas Sarin irreversibly inhibits Acetylcholinesterase. What effect would Sarin have on muscle contraction?

REVIEW QUESTIONS

4. The forearm consists of the:
    a. radius and ulna
    b. radius and humerus
    c. ulna and humerus
    d. humerus and carpus

5. The pectoral girdle consists of the:
    a. clavicle and sternum
    b. sternum and scapula
    c. clavicle and scapula
    d. clavicle and coccyx

6. All of the following are groups of vertebrae except ________, which is a curvature.
    a. thoracic
    b. cervical
    c. lumbar
    d. pelvic

7. Which of these is a facial bone?
    a. frontal
    b. occipital
    c. lacrimal
    d. temporal

8. Which of the following is **not** a true statement comparing exoskeletons and endoskeletons?
    a. Endoskeletons can support larger organisms.
    b. Only endoskeletons can grow as an organism grows.
    c. Exoskeletons provide greater protection of the internal organs.
    d. Exoskeletons provide less mechanical leverage.

9. The Haversian canal:
    a. is arranged as rods or plates
    b. contains the bone's blood vessels and nerve fibers
    c. is responsible for the lengthwise growth of long bones
    d. synthesizes and secretes matrix

10. The epiphyseal plate:
    a. is arranged as rods or plates
    b. contains the bone's blood vessels and nerve fibers
    c. is responsible for the lengthwise growth of long bones
    d. synthesizes and secretes bone matrix

11. The cells responsible for bone resorption are _______.
    a. osteoclasts
    b. osteoblasts
    c. fibroblasts
    d. osteocytes

12. Compact bone is composed of _______.
    a. trabeculae
    b. compacted collagen
    c. osteons
    d. calcium phosphate only

13. Osteoporosis is a condition where bones become weak and brittle. It is caused by an imbalance in the activity of which cells?
    a. osteoclasts and osteoblasts
    b. osteoclasts and osteocytes
    c. osteoblasts and chondrocytes
    d. osteocytes and chondrocytes

14. While assembling a skeleton of a new species, a scientist points to one of the bones and observes that it looks like the most likely site of leg muscle attachment. What kind of bone did she indicate?
    a. sesamoid bone
    b. long bone
    c. trabecular bone
    d. flat bone

15. Synchondroses and symphyses are:
    a. synovial joints
    b. cartilaginous joints
    c. fibrous joints
    d. condyloid joints

16. The movement of bone away from the midline of the body is called _______.
    a. circumduction
    b. extension
    c. adduction
    d. abduction

17. Which of the following is not a characteristic of the synovial fluid?
    a. lubrication
    b. shock absorption
    c. regulation of water balance in the joint
    d. protection of articular cartilage

18. The elbow is an example of which type of joint?
    a. hinge
    b. pivot
    c. saddle
    d. gliding

19. A high ankle sprain is an injury caused by over-stretching the ligaments connecting the tibia and fibula. What type of joint is involved in this sprain?
    a. ball and socket
    b. gomphosis
    c. syndesmosis
    d. symphysis

20. In relaxed muscle, the myosin-binding site on actin is blocked by _______.
    a. titin
    b. troponin
    c. myoglobin
    d. tropomyosin

21. The cell membrane of a muscle fiber is called a _______.
    a. myofibril
    b. sarcolemma
    c. sarcoplasm
    d. myofilament

22. The muscle relaxes if no new nerve signal arrives. However the neurotransmitter from the previous stimulation is still present in the synapse. The activity of _______ helps to remove this neurotransmitter.
    a. myosin
    b. action potential
    c. tropomyosin
    d. acetylcholinesterase

23. The ability of a muscle to generate tension immediately after stimulation is dependent on:
    a. myosin interaction with the M line
    b. overlap of myosin and actin
    c. actin attachments to the Z line
    d. none of the above

24. Botulinum toxin causes flaccid paralysis of the muscles, and is used for cosmetic purposes under the name Botox. Which of the following is the most likely mechanism of action of Botox?
    a. Botox decreases the production of acetylcholinesterase.
    b. Botox increases calcium release from the sarcoplasmic reticulum.
    c. Botox blocks the ATP binding site in actin.
    d. Botox decreases the release of acetylcholine from motor neurons.

CRITICAL THINKING QUESTIONS

25. What are the major differences between the male pelvis and female pelvis that permit childbirth in females?
26. What are the major differences between the pelvic girdle and the pectoral girdle that allow the pelvic girdle to bear the weight of the body?
27. Both hydrostatic and exoskeletons can protect internal organs from harm. Contrast the ways the skeletons perform these functions.
28. Scoliosis is a medical condition where the spine develops a sideways curvature. How would this change interfere with the normal function of the spine?
29. What are the major differences between spongy bone and compact bone?
30. What are the roles of osteoblasts, osteocytes, and osteoclasts?
31. Thalidomide was a morning sickness drug given to women that caused babies to be born without arm bones. If recent studies have shown that thalidomide prevents the formation of new blood vessels, describe the type of bone development inhibited by the drug and what stage of ossification was affected.
32. What movements occur at the hip joint and knees as you bend down to touch your toes?
33. What movement(s) occur(s) at the scapulae when you shrug your shoulders?
34. Describe the joints and motions involved in taking a step forward if a person is initially standing still. Assume the person holds his foot at the same angle throughout the motion.
35. How would muscle contractions be affected if ATP was completely depleted in a muscle fiber?
36. What factors contribute to the amount of tension produced in an individual muscle fiber?
37. What effect will low blood calcium have on neurons? What effect will low blood calcium have on skeletal muscles?
38. Skeletal muscles can only produce a mechanical force as they are contracted, but a leg flexes and extends while walking. How can muscles perform this task?

CHAPTER 39
The Respiratory System

Figure 39.1 Lungs, which appear as nearly transparent tissue surrounding the heart in this X ray of a dog (left), are the central organs of the respiratory system. The left lung is smaller than the right lung to accommodate space for the heart. A dog's nose (right) has a slit on the side of each nostril. When tracking a scent, the slits open, blocking the front of the nostrils. This allows the dog to exhale though the now-open area on the side of the nostrils without losing the scent that is being followed. (credit a: modification of work by Geoff Stearns; credit b: modification of work by Cory Zanker)

INTRODUCTION Breathing is an involuntary event. How often a breath is taken and how much air is inhaled or exhaled are tightly regulated by the respiratory center in the brain. Humans, when they aren't exerting themselves, breathe approximately 15 times per minute on average. Canines, like the dog in Figure 39.1, have a respiratory rate of about 15–30 breaths per minute. With every inhalation, air fills the lungs, and with every exhalation, air rushes back out. That air is doing more than just inflating and deflating the lungs in the chest cavity. The air contains oxygen that crosses the lung tissue, enters the bloodstream, and travels to organs and tissues. Oxygen (O_2) enters the cells where it is used for metabolic reactions that produce ATP, a high-energy compound. At the same time, these reactions release carbon dioxide (CO_2) as a by-product. CO_2 is toxic and must be eliminated. Carbon dioxide exits the cells, enters the bloodstream, travels back to the lungs, and is expired out of the body during exhalation.

Chapter Outline

39.1 Systems of Gas Exchange

By the end of this section, you will be able to do the following:

- Describe the passage of air from the outside environment to the lungs
- Explain how the lungs are protected from particulate matter

The primary function of the respiratory system is to deliver oxygen to the cells of the body's tissues and remove carbon dioxide, a cell waste product. The main structures of the human respiratory system are the nasal cavity, the trachea, and lungs.

All aerobic organisms require oxygen to carry out their metabolic functions. Along the evolutionary tree, different organisms have devised different means of obtaining oxygen from the surrounding atmosphere. The environment in which the animal lives greatly determines how an animal respires. The complexity of the respiratory system is correlated with the size of the organism. As animal size increases, diffusion distances increase and the ratio of surface area to volume drops. In unicellular organisms, diffusion across the cell membrane is sufficient for supplying oxygen to the cell (Figure 39.2). Diffusion is a slow, passive transport process. In order for diffusion to be a feasible means of providing oxygen to the cell, the rate of oxygen uptake must match the rate of diffusion across the membrane. In other words, if the cell were very large or thick, diffusion would not be able to provide oxygen quickly enough to the inside of the cell. Therefore, dependence on diffusion as a means of obtaining oxygen and removing carbon dioxide remains feasible only for small organisms or those with highly-flattened bodies, such as many flatworms (Platyhelminthes). Larger organisms had to evolve specialized respiratory tissues, such as gills, lungs, and respiratory passages accompanied by complex circulatory systems, to transport oxygen throughout their entire body.

Figure 39.2 The cell of the unicellular alga *Ventricaria ventricosa* is one of the largest known, reaching one to five centimeters in diameter. Like all single-celled organisms, *V. ventricosa* exchanges gases across the cell membrane.

Direct Diffusion

For small multicellular organisms, diffusion across the outer membrane is sufficient to meet their oxygen needs. Gas exchange by direct diffusion across surface membranes is efficient for organisms less than 1 mm in diameter. In simple organisms, such as cnidarians and flatworms, every cell in the body is close to the external environment. Their cells are kept moist and gases diffuse quickly via direct diffusion. Flatworms are small, literally flat worms, which 'breathe' through diffusion across the outer membrane (Figure 39.3). The flat shape of these organisms increases the surface area for diffusion, ensuring that each cell within the body is close to the outer membrane surface and has access to oxygen. If the flatworm had a cylindrical body, then the cells in the center would not be able to get oxygen.

Figure 39.3 This flatworm's process of respiration works by diffusion across the outer membrane. (credit: Stephen Childs)

Skin and Gills

Earthworms and amphibians use their skin (integument) as a respiratory organ. A dense network of capillaries lies just below the skin and facilitates gas exchange between the external environment and the circulatory system. The respiratory surface must be kept moist in order for the gases to dissolve and diffuse across cell membranes.

Organisms that live in water need to obtain oxygen from the water. Oxygen dissolves in water but at a lower concentration than in the atmosphere. The atmosphere has roughly 21 percent oxygen. In water, the oxygen concentration is much lower than that. Fish and many other aquatic organisms have evolved gills to take up the dissolved oxygen from water (Figure 39.4). Gills are thin tissue filaments that are highly branched and folded. When water passes over the gills, the dissolved oxygen in water rapidly diffuses across the gills into the bloodstream. The circulatory system can then carry the oxygenated blood to the other parts of the body. In animals that contain coelomic fluid instead of blood, oxygen diffuses across the gill surfaces into the coelomic fluid. Gills are found in mollusks, annelids, and crustaceans.

Figure 39.4 This common carp, like many other aquatic organisms, has gills that allow it to obtain oxygen from water. (credit: "Guitardude012"/Wikimedia Commons)

The folded surfaces of the gills provide a large surface area to ensure that the fish gets sufficient oxygen. Diffusion is a process in which material travels from regions of high concentration to low concentration until equilibrium is reached. In this case, blood with a low concentration of oxygen molecules circulates through the gills. The concentration of oxygen molecules in water is higher than the concentration of oxygen molecules in gills. As a result, oxygen molecules diffuse from water (high concentration) to blood (low concentration), as shown in Figure 39.5. Similarly, carbon dioxide molecules in the blood diffuse from the blood (high concentration) to water (low concentration).

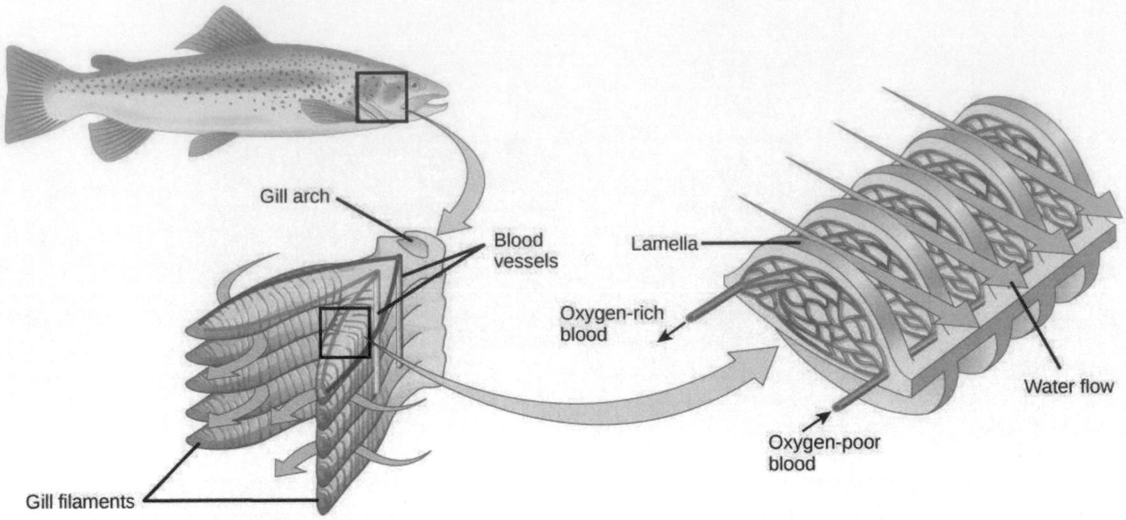

Figure 39.5 As water flows over the gills, oxygen is transferred to blood via the veins. (credit "fish": modification of work by Duane Raver, NOAA)

Tracheal Systems

Insect respiration is independent of its circulatory system; therefore, the blood does not play a direct role in oxygen transport. Insects have a highly specialized type of respiratory system called the tracheal system, which consists of a network of small tubes that carries oxygen to the entire body. The tracheal system is the most direct and efficient respiratory system in active animals. The tubes in the tracheal system are made of a polymeric material called chitin.

Insect bodies have openings, called spiracles, along the thorax and abdomen. These openings connect to the tubular network, allowing oxygen to pass into the body (Figure 39.6) and regulating the diffusion of CO_2 and water vapor. Air enters and leaves the tracheal system through the spiracles. Some insects can ventilate the tracheal system with body movements.

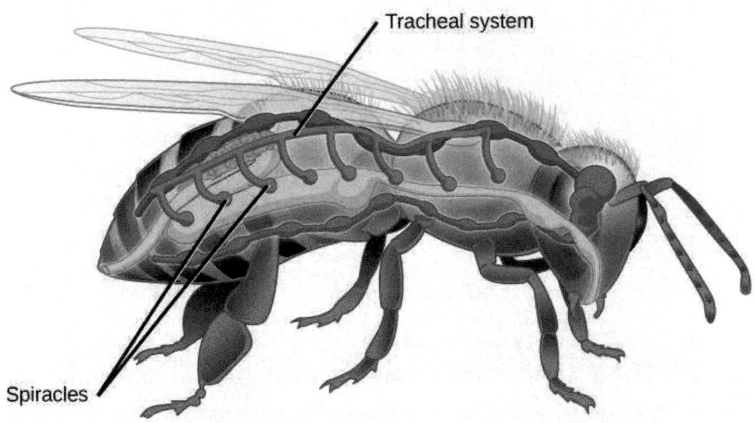

Figure 39.6 Insects perform respiration via a tracheal system.

Mammalian Systems

In mammals, pulmonary ventilation occurs via inhalation (breathing). During inhalation, air enters the body through the **nasal cavity** located just inside the nose (Figure 39.7). As air passes through the nasal cavity, the air is warmed to body temperature and humidified. The respiratory tract is coated with mucus to seal the tissues from direct contact with air. Mucus is high in water. As air crosses these surfaces of the mucous membranes, it picks up water. These processes help equilibrate the air to the body conditions, reducing any damage that cold, dry air can cause. Particulate matter that is floating in the air is removed in the nasal passages via mucus and cilia. The processes of warming, humidifying, and removing particles are important protective mechanisms that prevent damage to the trachea and lungs. Thus, inhalation serves several purposes in addition to bringing oxygen into the respiratory system.

VISUAL CONNECTION

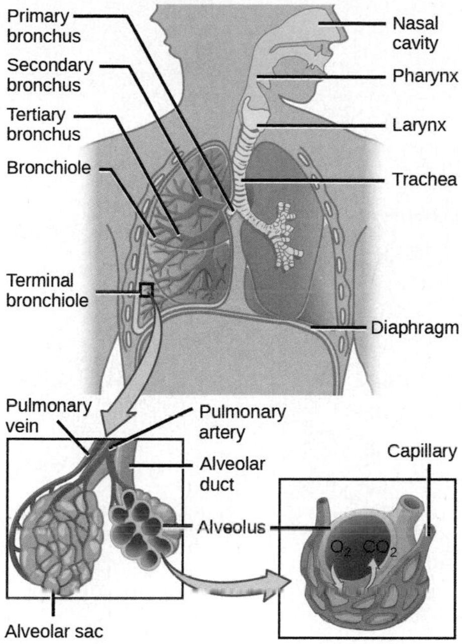

Figure 39.7 Air enters the respiratory system through the nasal cavity and pharynx, and then passes through the trachea and into the bronchi, which bring air into the lungs. (credit: modification of work by NCI)

Which of the following statements about the mammalian respiratory system is false?

a. When we breathe in, air travels from the pharynx to the trachea.
b. The bronchioles branch into bronchi.
c. Alveolar ducts connect to alveolar sacs.
d. Gas exchange between the lung and blood takes place in the alveolus.

From the nasal cavity, air passes through the **pharynx** (throat) and the **larynx** (voice box), as it makes its way to the **trachea** (Figure 39.7). The main function of the trachea is to funnel the inhaled air to the lungs and the exhaled air back out of the body. The human trachea is a cylinder about 10 to 12 cm long and 2 cm in diameter that sits in front of the esophagus and extends from the larynx into the chest cavity where it divides into the two primary bronchi at the midthorax. It is made of incomplete rings of hyaline cartilage and smooth muscle (Figure 39.8). The trachea is lined with mucus-producing goblet cells and ciliated epithelia. The cilia propel foreign particles trapped in the mucus toward the pharynx. The cartilage provides strength and support to the trachea to keep the passage open. The smooth muscle can contract, decreasing the trachea's diameter, which causes expired air to rush upwards from the lungs at a great force. The forced exhalation helps expel mucus when we cough. Smooth muscle can contract or relax, depending on stimuli from the external environment or the body's nervous system.

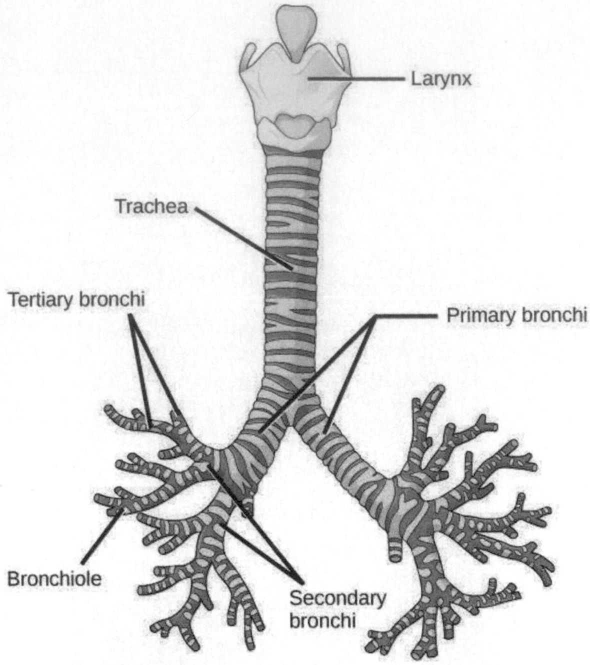

Figure 39.8 The trachea and bronchi are made of incomplete rings of cartilage. (credit: modification of work by Gray's Anatomy)

Lungs: Bronchi and Alveoli

The end of the trachea bifurcates (divides) to the right and left lungs. The lungs are not identical. The right lung is larger and contains three lobes, whereas the smaller left lung contains two lobes (Figure 39.9). The muscular **diaphragm**, which facilitates breathing, is inferior to (below) the lungs and marks the end of the thoracic cavity.

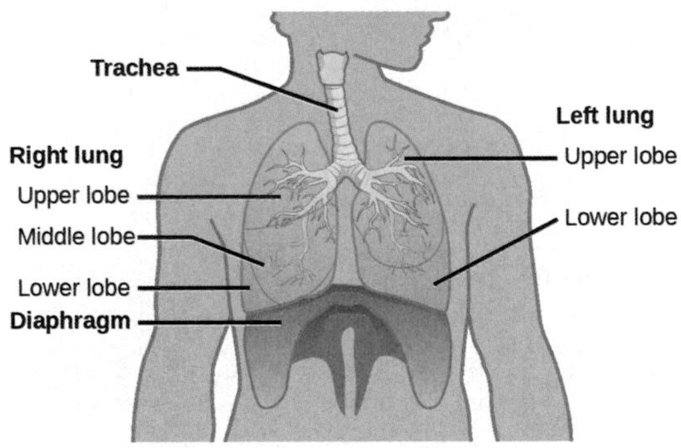

Figure 39.9 The trachea bifurcates into the right and left bronchi in the lungs. The right lung is made of three lobes and is larger. To accommodate the heart, the left lung is smaller and has only two lobes.

In the lungs, air is diverted into smaller and smaller passages, or **bronchi**. Air enters the lungs through the two **primary (main) bronchi** (singular: bronchus). Each bronchus divides into secondary bronchi, then into tertiary bronchi, which in turn divide, creating smaller and smaller diameter **bronchioles** as they split and spread through the lung. Like the trachea, the bronchi are made of cartilage and smooth muscle. At the bronchioles, the cartilage is replaced with elastic fibers. Bronchi are innervated by nerves of both the parasympathetic and sympathetic nervous systems that control muscle contraction (parasympathetic) or relaxation (sympathetic) in the bronchi and bronchioles, depending on the nervous system's cues. In humans, bronchioles with a diameter smaller than 0.5 mm are the **respiratory bronchioles**. They lack cartilage and therefore rely on inhaled air to support their shape. As the passageways decrease in diameter, the relative amount of smooth muscle increases.

The **terminal bronchioles** subdivide into microscopic branches called respiratory bronchioles. The respiratory bronchioles subdivide into several alveolar ducts. Numerous alveoli and alveolar sacs surround the alveolar ducts. The alveolar sacs resemble bunches of grapes tethered to the end of the bronchioles (Figure 39.10). In the acinar region, the **alveolar ducts** are attached to the end of each bronchiole. At the end of each duct are approximately 100 **alveolar sacs**, each containing 20 to 30 **alveoli** that are 200 to 300 microns in diameter. Gas exchange occurs only in alveoli. Alveoli are made of thin-walled parenchymal cells, typically one-cell thick, that look like tiny bubbles within the sacs. Alveoli are in direct contact with capillaries (one-cell thick) of the circulatory system. Such intimate contact ensures that oxygen will diffuse from alveoli into the blood and be distributed to the cells of the body. In addition, the carbon dioxide that was produced by cells as a waste product will diffuse from the blood into alveoli to be exhaled. The anatomical arrangement of capillaries and alveoli emphasizes the structural and functional relationship of the respiratory and circulatory systems. Because there are so many alveoli (~300 million per lung) within each alveolar sac and so many sacs at the end of each alveolar duct, the lungs have a sponge-like consistency. This organization produces a very large surface area that is available for gas exchange. The surface area of alveoli in the lungs is approximately 75 m^2. This large surface area, combined with the thin-walled nature of the alveolar parenchymal cells, allows gases to easily diffuse across the cells.

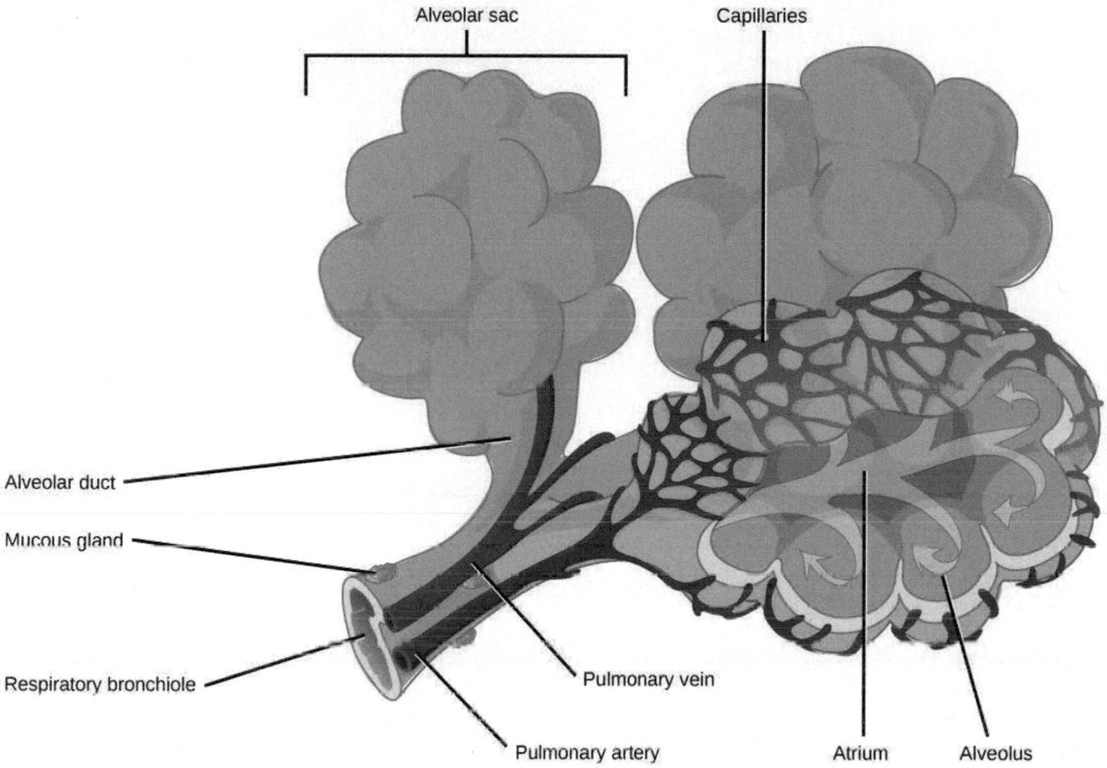

Figure 39.10 Terminal bronchioles are connected by respiratory bronchioles to alveolar ducts and alveolar sacs. Each alveolar sac contains 20 to 30 spherical alveoli and has the appearance of a bunch of grapes. Air flows into the atrium of the alveolar sac, then circulates into alveoli where gas exchange occurs with the capillaries. Mucous glands secrete mucous into the airways, keeping them moist and flexible. (credit: modification of work by Mariana Ruiz Villareal)

LINK TO LEARNING

Watch the following video to review the respiratory system.

Click to view content (https://www.openstax.org/l/lungs_pulmonary)

Protective Mechanisms

The air that organisms breathe contains **particulate matter** such as dust, dirt, viral particles, and bacteria that can damage the lungs or trigger allergic immune responses. The respiratory system contains several protective mechanisms to avoid problems or tissue damage. In the nasal cavity, hairs and mucus trap small particles, viruses, bacteria, dust, and dirt to prevent their entry.

If particulates do make it beyond the nose, or enter through the mouth, the bronchi and bronchioles of the lungs also contain several protective devices. The lungs produce **mucus**—a sticky substance made of **mucin**, a complex glycoprotein, as well as salts and water—that traps particulates. The bronchi and bronchioles contain cilia, small hair-like projections that line the walls of the bronchi and bronchioles (Figure 39.11). These cilia beat in unison and move mucus and particles out of the bronchi and bronchioles back up to the throat where it is swallowed and eliminated via the esophagus.

In humans, for example, tar and other substances in cigarette smoke destroy or paralyze the cilia, making the removal of particles more difficult. In addition, smoking causes the lungs to produce more mucus, which the damaged cilia are not able to move. This causes a persistent cough, as the lungs try to rid themselves of particulate matter, and makes smokers more susceptible to respiratory ailments.

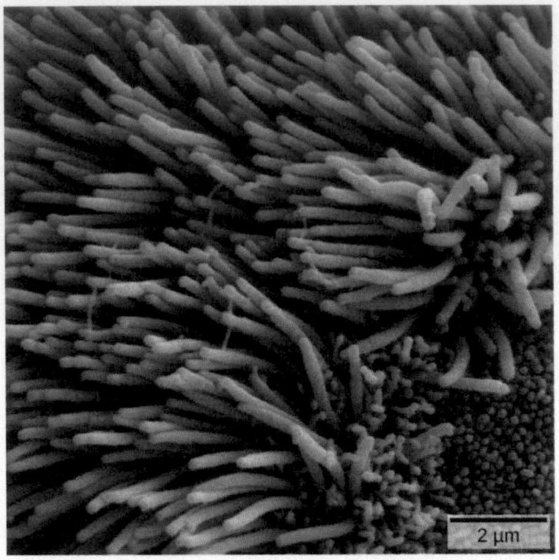

Figure 39.11 The bronchi and bronchioles contain cilia that help move mucus and other particles out of the lungs. (credit: Louisa Howard, modification of work by Dartmouth Electron Microscope Facility)

39.2 Gas Exchange across Respiratory Surfaces

By the end of this section, you will be able to do the following:

- Name and describe lung volumes and capacities
- Understand how gas pressure influences how gases move into and out of the body

The structure of the lung maximizes its surface area to increase gas diffusion. Because of the enormous number of alveoli (approximately 300 million in each human lung), the surface area of the lung is very large (75 m^2). Having such a large surface area increases the amount of gas that can diffuse into and out of the lungs.

Basic Principles of Gas Exchange

Gas exchange during respiration occurs primarily through diffusion. Diffusion is a process in which transport is driven by a concentration gradient. Gas molecules move from a region of high concentration to a region of low concentration. Blood that is low in oxygen concentration and high in carbon dioxide concentration undergoes gas exchange with air in the lungs. The air in the lungs has a higher concentration of oxygen than that of oxygen-depleted blood and a lower concentration of carbon dioxide. This concentration gradient allows for gas exchange during respiration.

Partial pressure is a measure of the concentration of the individual components in a mixture of gases. The total pressure exerted by the mixture is the sum of the partial pressures of the components in the mixture. The rate of diffusion of a gas is proportional to its partial pressure within the total gas mixture. This concept is discussed further in detail below.

Lung Volumes and Capacities

Different animals have different lung capacities based on their activities. Cheetahs have evolved a much higher lung capacity than humans; it helps provide oxygen to all the muscles in the body and allows them to run very fast. Elephants also have a high lung capacity. In this case, it is not because they run fast but because they have a large body and must be able to take up oxygen in accordance with their body size.

Human lung size is determined by genetics, sex, and height. At maximal capacity, an average lung can hold almost six liters of air, but lungs do not usually operate at maximal capacity. Air in the lungs is measured in terms of **lung volumes** and **lung capacities** (Figure 39.12 and Table 39.1). Volume measures the amount of air for one function (such as inhalation or exhalation). Capacity is any two or more volumes (for example, how much can be inhaled from the end of a maximal exhalation).

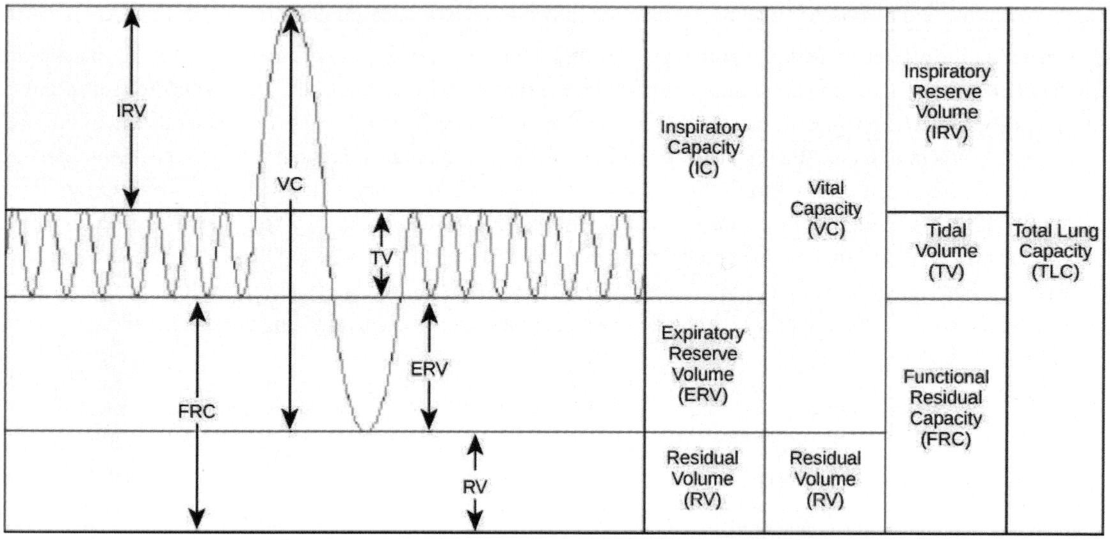

Figure 39.12 Human lung volumes and capacities are shown. The total lung capacity of the adult male is six liters. Tidal volume is the volume of air inhaled in a single, normal breath. Inspiratory capacity is the amount of air taken in during a deep breath, and residual volume is the amount of air left in the lungs after forceful respiration.

Lung Volumes and Capacities (Avg Adult Male)

Volume/Capacity	Definition	Volume (liters)	Equations
Tidal volume (TV)	Amount of air inhaled during a normal breath	0.5	-
Expiratory reserve volume (ERV)	Amount of air that can be exhaled after a normal exhalation	1.2	-
Inspiratory reserve volume (IRV)	Amount of air that can be further inhaled after a normal inhalation	3.1	-
Residual volume (RV)	Air left in the lungs after a forced exhalation	1.2	-
Vital capacity (VC)	Maximum amount of air that can be moved in or out of the lungs in a single respiratory cycle	4.8	ERV+TV+IRV
Inspiratory capacity (IC)	Volume of air that can be inhaled in addition to a normal exhalation	3.6	TV+IRV
Functional residual capacity (FRC)	Volume of air remaining after a normal exhalation	2.4	ERV+RV
Total lung capacity (TLC)	Total volume of air in the lungs after a maximal inspiration	6.0	RV+ERV+TV+IRV
Forced expiratory volume (FEV1)	How much air can be forced out of the lungs over a specific time period, usually one second	~4.1 to 5.5	-

Table 39.1

The volume in the lung can be divided into four units: tidal volume, expiratory reserve volume, inspiratory reserve volume, and residual volume. **Tidal volume (TV)** measures the amount of air that is inspired and expired during a normal breath. On average, this volume is around one-half liter, which is a little less than the capacity of a 20-ounce drink bottle. The **expiratory reserve volume (ERV)** is the additional amount of air that can be exhaled after a normal exhalation. It is the reserve amount that can be exhaled beyond what is normal. Conversely, the **inspiratory reserve volume (IRV)** is the additional amount of air that can be inhaled after a normal inhalation. The **residual volume (RV)** is the amount of air that is left after expiratory reserve volume is exhaled. The lungs are never completely empty: There is always some air left in the lungs after a maximal exhalation. If this residual volume did not exist and the lungs emptied completely, the lung tissues would stick together and the energy necessary to reinflate the lung could be too great to overcome. Therefore, there is always some air remaining in the lungs. Residual volume is also important for preventing large fluctuations in respiratory gases (O_2 and CO_2). The residual volume is the only lung volume that cannot be measured directly because it is impossible to completely empty the lung of air. This volume can only be calculated rather than measured.

Capacities are measurements of two or more volumes. The **vital capacity (VC)** measures the maximum amount of air that can be inhaled or exhaled during a respiratory cycle. It is the sum of the expiratory reserve volume, tidal volume, and inspiratory reserve volume. The **inspiratory capacity (IC)** is the amount of air that can be inhaled after the end of a normal expiration. It is, therefore, the sum of the tidal volume and inspiratory reserve volume. The **functional residual capacity (FRC)** includes the expiratory reserve volume and the residual volume. The FRC measures the amount of additional air that can be exhaled after a normal exhalation. Lastly, the **total lung capacity (TLC)** is a measurement of the total amount of air that the lung can hold. It is the sum of the residual volume, expiratory reserve volume, tidal volume, and inspiratory reserve volume.

Lung volumes are measured by a technique called **spirometry**. An important measurement taken during spirometry is the **forced expiratory volume (FEV)**, which measures how much air can be forced out of the lung over a specific period, usually one second (FEV1). In addition, the forced vital capacity (FVC), which is the total amount of air that can be forcibly exhaled, is measured. The ratio of these values (**FEV1/FVC ratio**) is used to diagnose lung diseases including asthma, emphysema, and fibrosis. If the FEV1/FVC ratio is high, the lungs are not compliant (meaning they are stiff and unable to bend properly), and the patient most likely has lung fibrosis. Patients exhale most of the lung volume very quickly. Conversely, when the FEV1/FVC ratio is low, there is resistance in the lung that is characteristic of asthma. In this instance, it is hard for the patient to get the air out of his or her lungs, and it takes a long time to reach the maximal exhalation volume. In either case, breathing is difficult and complications arise.

CAREER CONNECTION

Respiratory Therapist

Respiratory therapists or respiratory practitioners evaluate and treat patients with lung and cardiovascular diseases. They work as part of a medical team to develop treatment plans for patients. Respiratory therapists may treat premature babies with underdeveloped lungs, patients with chronic conditions such as asthma, or older patients suffering from lung disease such as emphysema and chronic obstructive pulmonary disease (COPD). They may operate advanced equipment such as compressed gas delivery systems, ventilators, blood gas analyzers, and resuscitators. Specialized programs to become a respiratory therapist generally lead to a bachelor's degree with a respiratory therapist specialty. Because of a growing aging population, career opportunities as a respiratory therapist are expected to remain strong.

Gas Pressure and Respiration

The respiratory process can be better understood by examining the properties of gases. Gases move freely, but gas particles are constantly hitting the walls of their vessel, thereby producing gas pressure.

Air is a mixture of gases, primarily nitrogen (N_2; 78.6 percent), oxygen (O_2; 20.9 percent), water vapor (H_2O; 0.5 percent), and carbon dioxide (CO_2; 0.04 percent). Each gas component of that mixture exerts a pressure. The pressure for an individual gas in the mixture is the partial pressure of that gas. Approximately 21 percent of atmospheric gas is oxygen. Carbon dioxide, however, is found in relatively small amounts, 0.04 percent. The partial pressure for oxygen is much greater than that of carbon dioxide. The partial pressure of any gas can be calculated by:

$$P = (P_{atm}) \times \text{(percent content in mixture)}.$$

P_{atm}, the atmospheric pressure, is the sum of all of the partial pressures of the atmospheric gases added together,

$$P_{atm} = P_{N_2} + P_{O_2} + P_{H_2O} + P_{CO_2} = 760 \text{ mm Hg}$$

× (percent content in mixture).

The pressure of the atmosphere at sea level is 760 mm Hg. Therefore, the partial pressure of oxygen is:

$$P_{O_2} = (760 \text{ mm Hg})\ (0.21) = 160 \text{ mm Hg}$$

and for carbon dioxide:

$$P_{CO_2} = (760 \text{ mm Hg})\ (0.0004) = 0.3 \text{ mm Hg}.$$

At high altitudes, P_{atm} decreases but concentration does not change; the partial pressure decrease is due to the reduction in P_{atm}.

When the air mixture reaches the lung, it has been humidified. The pressure of the water vapor in the lung does not change the pressure of the air, but it must be included in the partial pressure equation. For this calculation, the water pressure (47 mm Hg) is subtracted from the atmospheric pressure:

$$760 \text{ mm Hg} - 47 \text{ mm Hg} = 713 \text{ mm Hg}$$

and the partial pressure of oxygen is:

$$(760 \text{ mm Hg} - 47 \text{ mm Hg}) \times 0.21 = 150 \text{ mm Hg}.$$

These pressures determine the gas exchange, or the flow of gas, in the system. Oxygen and carbon dioxide will flow according to their pressure gradient from high to low. Therefore, understanding the partial pressure of each gas will aid in understanding how gases move in the respiratory system.

Gas Exchange across the Alveoli

In the body, oxygen is used by cells of the body's tissues and carbon dioxide is produced as a waste product. The ratio of carbon dioxide production to oxygen consumption is the **respiratory quotient (RQ)**. RQ varies between 0.7 and 1.0. If just glucose were used to fuel the body, the RQ would equal one. One mole of carbon dioxide would be produced for every mole of oxygen consumed. Glucose, however, is not the only fuel for the body. Protein and fat are also used as fuels for the body. Because of this, less carbon dioxide is produced than oxygen is consumed and the RQ is, on average, about 0.7 for fat and about 0.8 for protein.

The RQ is used to calculate the partial pressure of oxygen in the alveolar spaces within the lung, the **alveolar** P_{O_2} . Above, the partial pressure of oxygen in the lungs was calculated to be 150 mm Hg. However, lungs never fully deflate with an exhalation; therefore, the inspired air mixes with this residual air and lowers the partial pressure of oxygen within the alveoli. This means that there is a lower concentration of oxygen in the lungs than is found in the air outside the body. Knowing the RQ, the partial pressure of oxygen in the alveoli can be calculated:

$$\text{alveolar } P_{O_2} = \text{inspired } P_{O_2} - \left(\frac{\text{alveolar } P_{O_2}}{RQ}\right)$$

With an RQ of 0.8 and a P_{CO_2} in the alveoli of 40 mm Hg, the alveolar P_{O_2} is equal to:

$$\text{alveolar } P_{O_2} = 150 \text{ mm Hg} - \left(\frac{40 \text{ mm Hg}}{0.8}\right) = 100 \text{ mm Hg}.$$

Notice that this pressure is less than the external air. Therefore, the oxygen will flow from the inspired air in the lung (P_{O_2} = 150 mm Hg) into the bloodstream (P_{O_2} = 100 mm Hg) (Figure 39.13).

In the lungs, oxygen diffuses out of the alveoli and into the capillaries surrounding the alveoli. Oxygen (about 98 percent) binds reversibly to the respiratory pigment hemoglobin found in red blood cells (RBCs). RBCs carry oxygen to the tissues where oxygen dissociates from the hemoglobin and diffuses into the cells of the tissues. More specifically, alveolar P_{O_2} is higher in the alveoli (P_{ALVO_2} = 100 mm Hg) than blood P_{O_2} (40 mm Hg) in the capillaries. Because this pressure gradient exists, oxygen diffuses down its pressure gradient, moving out of the alveoli and entering the blood of the capillaries where O_2 binds to hemoglobin. At the same time, alveolar P_{CO_2} is lower P_{ALVO_2} = 40 mm Hg than blood P_{CO_2} = (45 mm Hg). CO_2 diffuses down its pressure gradient, moving out of the capillaries and entering the alveoli.

Oxygen and carbon dioxide move independently of each other; they diffuse down their own pressure gradients. As blood leaves the lungs through the pulmonary veins, the **venous** P_{O_2} = 100 mm Hg, whereas the **venous** P_{CO_2} = 40 mm Hg. As blood enters the systemic capillaries, the blood will lose oxygen and gain carbon dioxide because of the pressure difference of the tissues and blood. In systemic capillaries, P_{O_2} = 100 mm Hg, but in the tissue cells, P_{O_2} = 40 mm Hg. This pressure gradient drives the diffusion of oxygen out of the capillaries and into the tissue cells. At the same time, blood P_{CO_2} = 40 mm Hg and systemic tissue P_{CO_2} = 45 mm Hg. The pressure gradient drives CO_2 out of tissue cells and into the capillaries. The blood returning to the lungs through the pulmonary arteries has a venous P_{O_2} = 40 mm Hg and a P_{CO_2} = 45 mm Hg. The blood enters the lung capillaries where the process of exchanging gases between the capillaries and alveoli begins again (Figure 39.13).

VISUAL CONNECTION

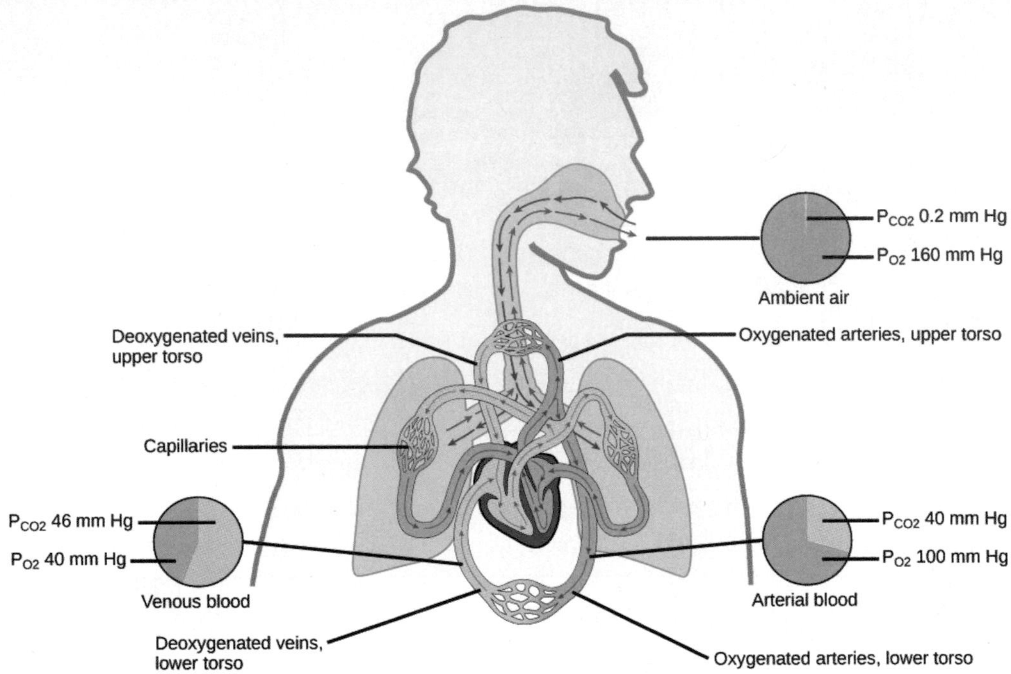

Figure 39.13 The partial pressures of oxygen and carbon dioxide change as blood moves through the body.

Which of the following statements is false?

a. In the tissues, P_{O_2} drops as blood passes from the arteries to the veins, while P_{CO_2} increases.
b. Blood travels from the lungs to the heart to body tissues, then back to the heart, then the lungs.
c. Blood travels from the lungs to the heart to body tissues, then back to the lungs, then the heart.
d. P_{O_2} is higher in air than in the lungs.

In short, the change in partial pressure from the alveoli to the capillaries drives the oxygen into the tissues and the carbon dioxide into the blood from the tissues. The blood is then transported to the lungs where differences in pressure in the alveoli result in the movement of carbon dioxide out of the blood into the lungs, and oxygen into the blood.

LINK TO LEARNING

Watch this video to learn how to carry out spirometry.

Click to view content (https://www.openstax.org/l/spirometry)

39.3 Breathing

By the end of this section, you will be able to do the following:

- Describe how the structures of the lungs and thoracic cavity control the mechanics of breathing
- Explain the importance of compliance and resistance in the lungs
- Discuss problems that may arise due to a V/Q mismatch

Mammalian lungs are located in the thoracic cavity where they are surrounded and protected by the rib cage, intercostal muscles, and bound by the chest wall. The bottom of the lungs is contained by the diaphragm, a skeletal muscle that facilitates breathing. Breathing requires the coordination of the lungs, the chest wall, and most importantly, the diaphragm.

Types of Breathing

Amphibians have evolved multiple ways of breathing. Young amphibians, like tadpoles, use gills to breathe, and they don't leave the water. Some amphibians retain gills for life. As the tadpole grows, the gills disappear and lungs grow. These lungs are primitive and not as evolved as mammalian lungs. Adult amphibians are lacking or have a reduced diaphragm, so breathing via lungs is forced. The other means of breathing for amphibians is diffusion across the skin. To aid this diffusion, amphibian skin must remain moist.

Birds face a unique challenge with respect to breathing: They fly. Flying consumes a great amount of energy; therefore, birds require a lot of oxygen to aid their metabolic processes. Birds have evolved a respiratory system that supplies them with the oxygen needed to enable flying. Similar to mammals, birds have lungs, which are organs specialized for gas exchange. Oxygenated air, taken in during inhalation, diffuses across the surface of the lungs into the bloodstream, and carbon dioxide diffuses from the blood into the lungs and expelled during exhalation. The details of breathing between birds and mammals differ substantially.

In addition to lungs, birds have air sacs inside their body. Air flows in one direction from the posterior air sacs to the lungs and out of the anterior air sacs. The flow of air is in the opposite direction from blood flow, and gas exchange takes place much more efficiently. This type of breathing enables birds to obtain the requisite oxygen, even at higher altitudes where the oxygen concentration is low. This directionality of airflow requires two cycles of air intake and exhalation to completely get the air out of the lungs.

EVOLUTION CONNECTION

Avian Respiration

Birds have evolved a respiratory system that enables them to fly. Flying is a high-energy process and requires a lot of oxygen. Furthermore, many birds fly in high altitudes where the concentration of oxygen is low. How did birds evolve a respiratory system that is so unique?

Decades of research by paleontologists have shown that birds evolved from therapods, meat-eating dinosaurs (Figure 39.14). In fact, fossil evidence shows that meat-eating dinosaurs that lived more than 100 million years ago had a similar flow-through respiratory system with lungs and air sacs. *Archaeopteryx* and *Xiaotingia*, for example, were flying dinosaurs and are believed to be early precursors of birds.

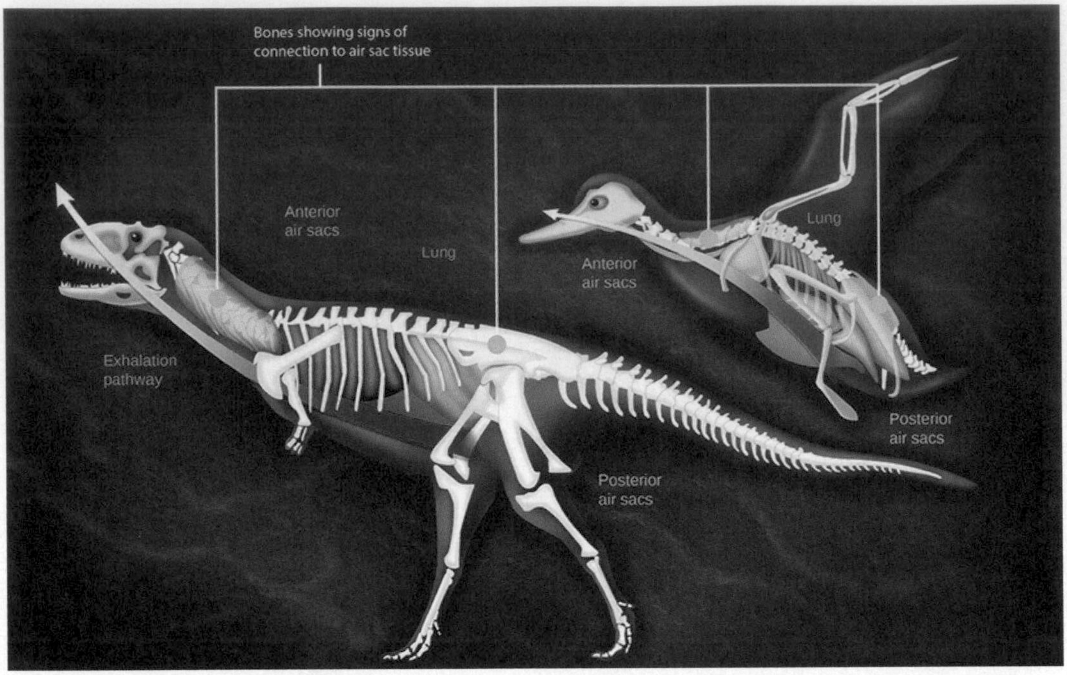

Figure 39.14 Dinosaurs, from which birds descended, have similar hollow bones and are believed to have had a similar respiratory system. (credit b: modification of work by Zina Deretsky, National Science Foundation)

Most of us consider that dinosaurs are extinct. However, modern birds are descendants of avian dinosaurs. The respiratory system of modern birds has been evolving for hundreds of millions of years.

All mammals have lungs that are the main organs for breathing. Lung capacity has evolved to support the animal's activities. During inhalation, the lungs expand with air, and oxygen diffuses across the lung's surface and enters the bloodstream. During exhalation, the lungs expel air and lung volume decreases. In the next few sections, the process of human breathing will be explained.

The Mechanics of Human Breathing

Boyle's Law is the gas law that states that in a closed space, pressure and volume are inversely related. As volume decreases, pressure increases and vice versa (Figure 39.15). The relationship between gas pressure and volume helps to explain the mechanics of breathing.

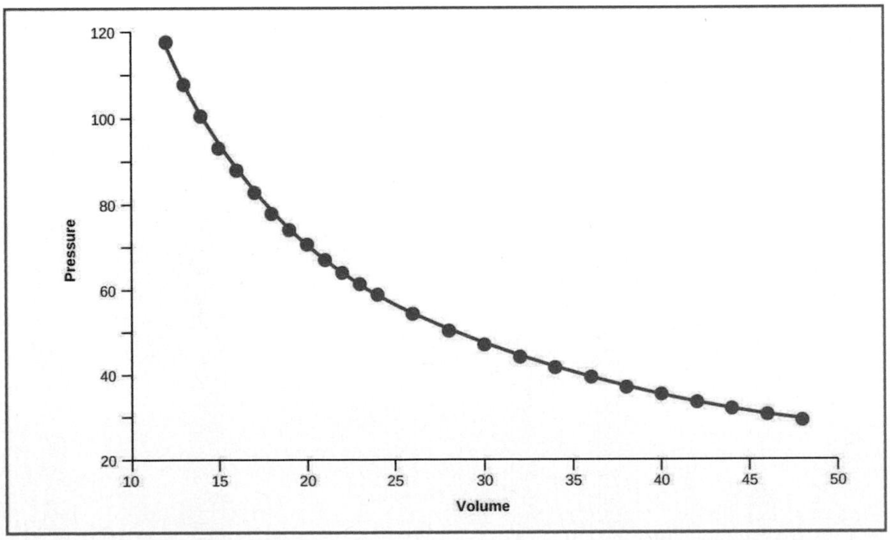

Figure 39.15 This graph shows data from Boyle's original 1662 experiment, which shows that pressure and volume are inversely related.

No units are given as Boyle used arbitrary units in his experiments.

There is always a slightly negative pressure within the thoracic cavity, which aids in keeping the airways of the lungs open. During inhalation, volume increases as a result of contraction of the diaphragm, and pressure decreases (according to Boyle's Law). This decrease of pressure in the thoracic cavity relative to the environment makes the cavity less than the atmosphere (Figure 39.16a). Because of this drop in pressure, air rushes into the respiratory passages. To increase the volume of the lungs, the chest wall expands. This results from the contraction of the **intercostal muscles**, the muscles that are connected to the rib cage. Lung volume expands because the diaphragm contracts and the intercostal muscles contract, thus expanding the thoracic cavity. This increase in the volume of the thoracic cavity lowers pressure compared to the atmosphere, so air rushes into the lungs, thus increasing its volume. The resulting increase in volume is largely attributed to an increase in alveolar space, because the bronchioles and bronchi are stiff structures that do not change in size.

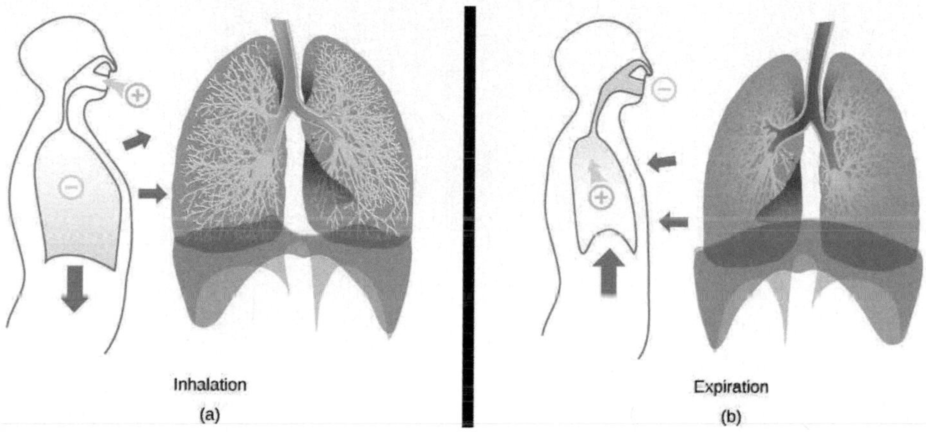

Figure 39.16 The lungs, chest wall, and diaphragm are all involved in respiration, both (a) inhalation and (b) expiration. (credit: modification of work by Mariana Ruiz Villareal)

The chest wall expands out and away from the lungs. The lungs are elastic; therefore, when air fills the lungs, the **elastic recoil** within the tissues of the lung exerts pressure back toward the interior of the lungs. These outward and inward forces compete to inflate and deflate the lung with every breath. Upon exhalation, the lungs recoil to force the air out of the lungs, and the intercostal muscles relax, returning the chest wall back to its original position (Figure 39.16b). The diaphragm also relaxes and moves higher into the thoracic cavity. This increases the pressure within the thoracic cavity relative to the environment, and air rushes out of the lungs. The movement of air out of the lungs is a passive event. No muscles are contracting to expel the air.

Each lung is surrounded by an invaginated sac. The layer of tissue that covers the lung and dips into spaces is called the visceral **pleura**. A second layer of parietal pleura lines the interior of the thorax (Figure 39.17). The space between these layers, the **intrapleural space**, contains a small amount of fluid that protects the tissue and reduces the friction generated from rubbing the tissue layers together as the lungs contract and relax. **Pleurisy** results when these layers of tissue become inflamed; it is painful because the inflammation increases the pressure within the thoracic cavity and reduces the volume of the lung.

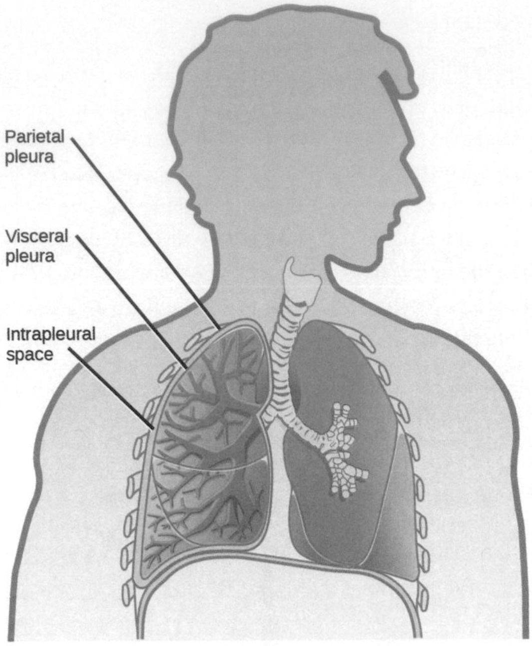

Figure 39.17 A tissue layer called pleura surrounds the lung and interior of the thoracic cavity. (credit: modification of work by NCI)

LINK TO LEARNING

View how Boyle's Law is related to breathing and watch a video (https://www.openstax.org/l/boyles_law) on Boyle's Law.

Click to view content (https://www.openstax.org/l/boyle_breathing)

The Work of Breathing

The number of breaths per minute is the **respiratory rate**. On average, under non-exertion conditions, the human respiratory rate is 12–15 breaths/minute. The respiratory rate contributes to the **alveolar ventilation**, or how much air moves into and out of the alveoli. Alveolar ventilation prevents carbon dioxide buildup in the alveoli. There are two ways to keep the alveolar ventilation constant: increase the respiratory rate while decreasing the tidal volume of air per breath (shallow breathing), or decrease the respiratory rate while increasing the tidal volume per breath. In either case, the ventilation remains the same, but the work done and type of work needed are quite different. Both tidal volume and respiratory rate are closely regulated when oxygen demand increases.

There are two types of work conducted during respiration, flow-resistive and elastic work. **Flow-resistive** refers to the work of the alveoli and tissues in the lung, whereas **elastic work** refers to the work of the intercostal muscles, chest wall, and diaphragm. Increasing the respiration rate increases the flow-resistive work of the airways and decreases the elastic work of the muscles. Decreasing the respiratory rate reverses the type of work required.

Surfactant

The air-tissue/water interface of the alveoli has a high surface tension. This surface tension is similar to the surface tension of water at the liquid-air interface of a water droplet that results in the bonding of the water molecules together. **Surfactant** is a complex mixture of phospholipids and lipoproteins that works to reduce the surface tension that exists between the alveoli tissue and the air found within the alveoli. By lowering the surface tension of the alveolar fluid, it reduces the tendency of alveoli to collapse.

Surfactant works like a detergent to reduce the surface tension and allows for easier inflation of the airways. When a balloon is first inflated, it takes a large amount of effort to stretch the plastic and start to inflate the balloon. If a little bit of detergent was applied to the interior of the balloon, then the amount of effort or work needed to begin to inflate the balloon would decrease, and it would become much easier to start blowing up the balloon. This same principle applies to the airways. A small amount of surfactant to the airway tissues reduces the effort or work needed to inflate those airways. Babies born prematurely sometimes do not produce enough surfactant. As a result, they suffer from **respiratory distress syndrome**, because it requires more effort to inflate their lungs. Surfactant is also important for preventing collapse of small alveoli relative to large alveoli.

Lung Resistance and Compliance

Pulmonary diseases reduce the rate of gas exchange into and out of the lungs. Two main causes of decreased gas exchange are **compliance** (how elastic the lung is) and **resistance** (how much obstruction exists in the airways). A change in either can dramatically alter breathing and the ability to take in oxygen and release carbon dioxide.

Examples of **restrictive diseases** are respiratory distress syndrome and pulmonary fibrosis. In both diseases, the airways are less compliant and they are stiff or fibrotic. There is a decrease in compliance because the lung tissue cannot bend and move. In these types of restrictive diseases, the intrapleural pressure is more positive and the airways collapse upon exhalation, which traps air in the lungs. Forced or **functional vital capacity (FVC)**, which is the amount of air that can be forcibly exhaled after taking the deepest breath possible, is much lower than in normal patients, and the time it takes to exhale most of the air is greatly prolonged (Figure 39.18). A patient suffering from these diseases cannot exhale the normal amount of air.

Obstructive diseases and conditions include emphysema, asthma, and pulmonary edema. In emphysema, which mostly arises from smoking tobacco, the walls of the alveoli are destroyed, decreasing the surface area for gas exchange. The overall compliance of the lungs is increased, because as the alveolar walls are damaged, lung elastic recoil decreases due to a loss of elastic fibers, and more air is trapped in the lungs at the end of exhalation. Asthma is a disease in which inflammation is triggered by environmental factors. Inflammation obstructs the airways. The obstruction may be due to edema (fluid accumulation), smooth muscle spasms in the walls of the bronchioles, increased mucus secretion, damage to the epithelia of the airways, or a combination of these events. Those with asthma or edema experience increased occlusion from increased inflammation of the airways. This tends to block the airways, preventing the proper movement of gases (Figure 39.18). Those with obstructive diseases have large volumes of air trapped after exhalation and breathe at a very high lung volume to compensate for the lack of airway recruitment.

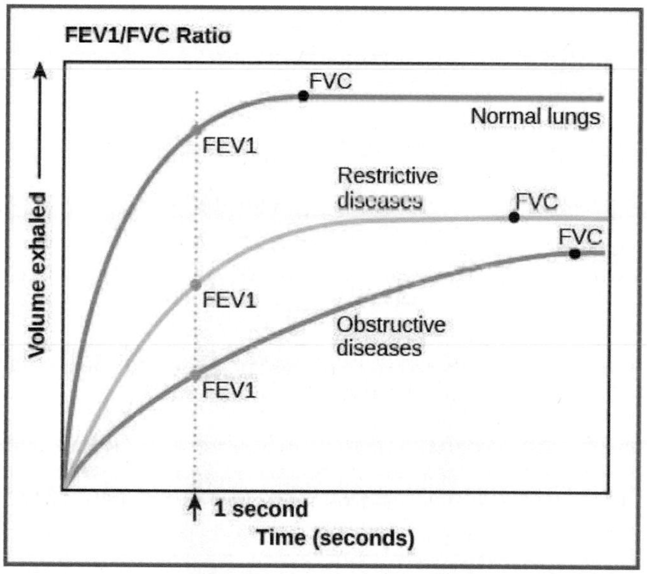

Figure 39.18 The ratio of FEV1 (the amount of air that can be forcibly exhaled in one second after taking a deep breath) to FVC (the total amount of air that can be forcibly exhaled) can be used to diagnose whether a person has restrictive or obstructive lung disease. In restrictive lung disease, FVC is reduced but airways are not obstructed, so the person is able to expel air reasonably fast. In obstructive lung disease, airway obstruction results in slow exhalation as well as reduced FVC. Thus, the FEV1/FVC ratio is lower in persons with obstructive lung disease (less than 69 percent) than in persons with restrictive disease (88 to 90 percent).

Dead Space: V/Q Mismatch

Pulmonary circulation pressure is very low compared to that of the systemic circulation. It is also independent of cardiac output. This is because of a phenomenon called **recruitment**, which is the process of opening airways that normally remain closed when cardiac output increases. As cardiac output increases, the number of capillaries and arteries that are perfused (filled with blood) increases. These capillaries and arteries are not always in use but are ready if needed. At times, however, there is a mismatch between the amount of air (ventilation, V) and the amount of blood (perfusion, Q) in the lungs. This is referred to as **ventilation/perfusion (V/Q) mismatch**.

There are two types of V/Q mismatch. Both produce **dead space**, regions of broken down or blocked lung tissue. Dead spaces can severely impact breathing, because they reduce the surface area available for gas diffusion. As a result, the amount of oxygen in the blood decreases, whereas the carbon dioxide level increases. Dead space is created when no ventilation and/or perfusion takes place. **Anatomical dead space** or anatomical shunt, arises from an anatomical failure, while **physiological dead space** or physiological shunt, arises from a functional impairment of the lung or arteries.

An example of an anatomical shunt is the effect of gravity on the lungs. The lung is particularly susceptible to changes in the magnitude and direction of gravitational forces. When someone is standing or sitting upright, the pleural pressure gradient leads to increased ventilation further down in the lung. As a result, the intrapleural pressure is more negative at the base of the lung than at the top, and more air fills the bottom of the lung than the top. Likewise, it takes less energy to pump blood to the bottom of the lung than to the top when in a prone position. Perfusion of the lung is not uniform while standing or sitting. This is a result of hydrostatic forces combined with the effect of airway pressure. An anatomical shunt develops because the ventilation of the airways does not match the perfusion of the arteries surrounding those airways. As a result, the rate of gas exchange is reduced. Note that this does not occur when lying down, because in this position, gravity does not preferentially pull the bottom of the lung down.

A physiological shunt can develop if there is infection or edema in the lung that obstructs an area. This will decrease ventilation but not affect perfusion; therefore, the V/Q ratio changes and gas exchange is affected.

The lung can compensate for these mismatches in ventilation and perfusion. If ventilation is greater than perfusion, the arterioles dilate and the bronchioles constrict. This increases perfusion and reduces ventilation. Likewise, if ventilation is less than perfusion, the arterioles constrict and the bronchioles dilate to correct the imbalance.

LINK TO LEARNING

View the mechanics of breathing.

Click to view content (https://www.openstax.org/l/breathing)

39.4 Transport of Gases in Human Bodily Fluids

By the end of this section, you will be able to do the following:

- Describe how oxygen is bound to hemoglobin and transported to body tissues
- Explain how carbon dioxide is transported from body tissues to the lungs

Once the oxygen diffuses across the alveoli, it enters the bloodstream and is transported to the tissues where it is unloaded, and carbon dioxide diffuses out of the blood and into the alveoli to be expelled from the body. Although gas exchange is a continuous process, the oxygen and carbon dioxide are transported by different mechanisms.

Transport of Oxygen in the Blood

Although oxygen dissolves in blood, only a small amount of oxygen is transported this way. Only 1.5 percent of oxygen in the blood is dissolved directly into the blood itself. Most oxygen—98.5 percent—is bound to a protein called hemoglobin and carried to the tissues.

Hemoglobin

Hemoglobin, or Hb, is a protein molecule found in red blood cells (erythrocytes) made of four subunits: two alpha subunits and two beta subunits (Figure 39.19). Each subunit surrounds a central **heme group** that contains iron and binds one oxygen molecule, allowing each hemoglobin molecule to bind four oxygen molecules. Molecules with more oxygen bound to the heme groups are brighter red. As a result, oxygenated arterial blood where the Hb is carrying four oxygen molecules is bright red, while venous blood that is deoxygenated is darker red.

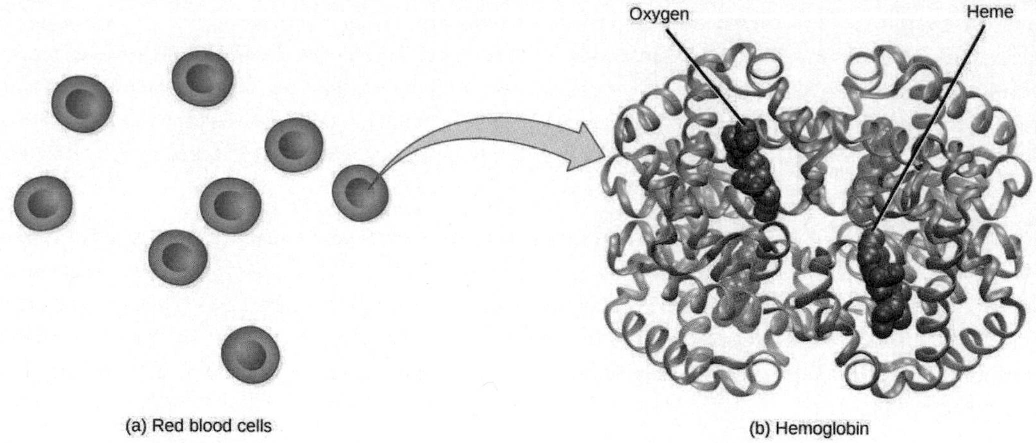

Figure 39.19 The protein inside (a) red blood cells that carries oxygen to cells and carbon dioxide to the lungs is (b) hemoglobin. Hemoglobin is made up of four symmetrical subunits and four heme groups. Iron associated with the heme binds oxygen. It is the iron in hemoglobin that gives blood its red color.

It is easier to bind a second and third oxygen molecule to Hb than the first molecule. This is because the hemoglobin molecule changes its shape, or conformation, as oxygen binds. The fourth oxygen is then more difficult to bind. The binding of oxygen to hemoglobin can be plotted as a function of the partial pressure of oxygen in the blood (x-axis) versus the relative Hb-oxygen saturation (y-axis). The resulting graph—an **oxygen dissociation curve**—is sigmoidal, or S-shaped (Figure 39.20). As the partial pressure of oxygen increases, the hemoglobin becomes increasingly saturated with oxygen.

VISUAL CONNECTION

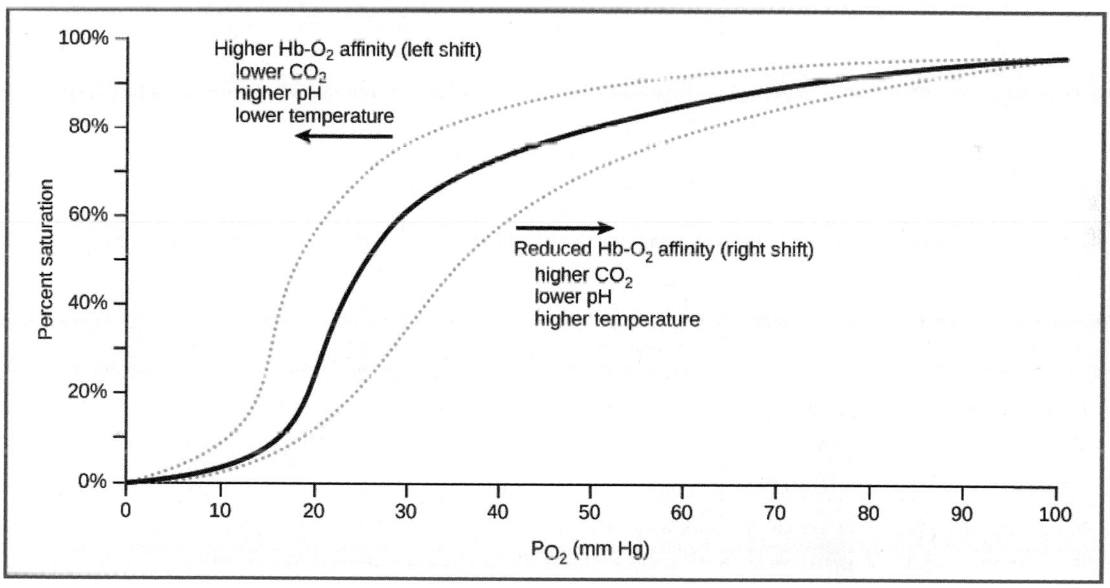

Figure 39.20 The oxygen dissociation curve demonstrates that, as the partial pressure of oxygen increases, more oxygen binds hemoglobin. However, the affinity of hemoglobin for oxygen may shift to the left or the right depending on environmental conditions.

The kidneys are responsible for removing excess H+ ions from the blood. If the kidneys fail, what would happen to blood pH and to hemoglobin affinity for oxygen?

Factors That Affect Oxygen Binding

The **oxygen-carrying capacity** of hemoglobin determines how much oxygen is carried in the blood. In addition to P_{O_2}, other environmental factors and diseases can affect oxygen carrying capacity and delivery.

Carbon dioxide levels, blood pH, and body temperature affect oxygen-carrying capacity (Figure 39.20). When carbon dioxide is in the blood, it reacts with water to form bicarbonate (HCO_3^-) and hydrogen ions (H^+). As the level of carbon dioxide in the blood increases, more H^+ is produced and the pH decreases. This increase in carbon dioxide and subsequent decrease in pH reduce the affinity of hemoglobin for oxygen. The oxygen dissociates from the Hb molecule, shifting the oxygen dissociation curve to the right. Therefore, more oxygen is needed to reach the same hemoglobin saturation level as when the pH was higher. A similar shift in the curve also results from an increase in body temperature. Increased temperature, such as from increased activity of skeletal muscle, causes the affinity of hemoglobin for oxygen to be reduced.

Diseases like sickle cell anemia and thalassemia decrease the blood's ability to deliver oxygen to tissues and its oxygen-carrying capacity. In **sickle cell anemia**, the shape of the red blood cell is crescent-shaped, elongated, and stiffened, reducing its ability to deliver oxygen (Figure 39.21). In this form, red blood cells cannot pass through the capillaries. This is painful when it occurs. **Thalassemia** is a rare genetic disease caused by a defect in either the alpha or the beta subunit of Hb. Patients with thalassemia produce a high number of red blood cells, but these cells have lower-than-normal levels of hemoglobin. Therefore, the oxygen-carrying capacity is diminished.

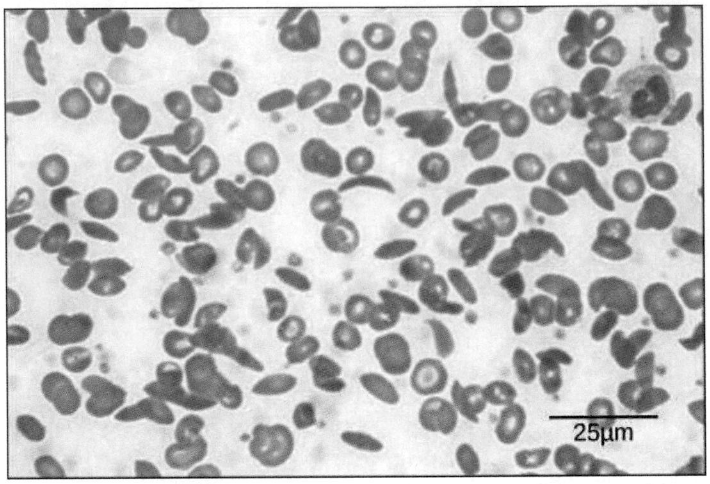

Figure 39.21 Individuals with sickle cell anemia have crescent-shaped red blood cells. (credit: modification of work by Ed Uthman; scale-bar data from Matt Russell)

Transport of Carbon Dioxide in the Blood

Carbon dioxide molecules are transported in the blood from body tissues to the lungs by one of three methods: dissolution directly into the blood, binding to hemoglobin, or carried as a bicarbonate ion. Several properties of carbon dioxide in the blood affect its transport. First, carbon dioxide is more soluble in blood than oxygen. About 5 to 7 percent of all carbon dioxide is dissolved in the plasma. Second, carbon dioxide can bind to plasma proteins or can enter red blood cells and bind to hemoglobin. This form transports about 10 percent of the carbon dioxide. When carbon dioxide binds to hemoglobin, a molecule called **carbaminohemoglobin** is formed. Binding of carbon dioxide to hemoglobin is reversible. Therefore, when it reaches the lungs, the carbon dioxide can freely dissociate from the hemoglobin and be expelled from the body.

Third, the majority of carbon dioxide molecules (85 percent) are carried as part of the **bicarbonate buffer system**. In this system, carbon dioxide diffuses into the red blood cells. **Carbonic anhydrase (CA)** within the red blood cells quickly converts the carbon dioxide into carbonic acid (H_2CO_3). Carbonic acid is an unstable intermediate molecule that immediately dissociates into **bicarbonate ions** (HCO_3^-) and hydrogen (H^+) ions. Since carbon dioxide is quickly converted into bicarbonate ions, this reaction allows for the continued uptake of carbon dioxide into the blood down its concentration gradient. It also results in the production of H^+ ions. If too much H^+ is produced, it can alter blood pH. However, hemoglobin binds to the free H^+ ions and thus limits shifts in pH. The newly synthesized bicarbonate ion is transported out of the red blood cell into the liquid component of the blood in exchange for a chloride ion (Cl^-); this is called the **chloride shift**. When the blood reaches the lungs, the bicarbonate ion is transported back into the red blood cell in exchange for the chloride ion. The H^+ ion dissociates from the hemoglobin and binds to the bicarbonate ion. This produces the carbonic acid intermediate, which is converted back into carbon dioxide through the enzymatic action of CA. The carbon dioxide produced is expelled through the lungs during exhalation.

$$CO_2 + H_2O \leftrightarrow \underset{\text{(carbonic acid)}}{H_2CO_3} \leftrightarrow \underset{\text{(bicarbonate)}}{HCO_3 + H^+}$$

The benefit of the bicarbonate buffer system is that carbon dioxide is "soaked up" into the blood with little change to the pH of the system. This is important because it takes only a small change in the overall pH of the body for severe injury or death to result. The presence of this bicarbonate buffer system also allows for people to travel and live at high altitudes: When the partial pressure of oxygen and carbon dioxide change at high altitudes, the bicarbonate buffer system adjusts to regulate carbon dioxide while maintaining the correct pH in the body.

Carbon Monoxide Poisoning

While carbon dioxide can readily associate and dissociate from hemoglobin, other molecules such as carbon monoxide (CO) cannot. Carbon monoxide has a greater affinity for hemoglobin than oxygen. Therefore, when carbon monoxide is present, it binds to hemoglobin preferentially over oxygen. As a result, oxygen cannot bind to hemoglobin, so very little oxygen is transported through the body (Figure 39.22). Carbon monoxide is a colorless, odorless gas and is therefore difficult to detect. It is produced by gas-powered vehicles and tools. Carbon monoxide can cause headaches, confusion, and nausea; long-term exposure can cause brain damage or death. Administering 100 percent (pure) oxygen is the usual treatment for carbon monoxide poisoning. Administration of pure oxygen speeds up the separation of carbon monoxide from hemoglobin.

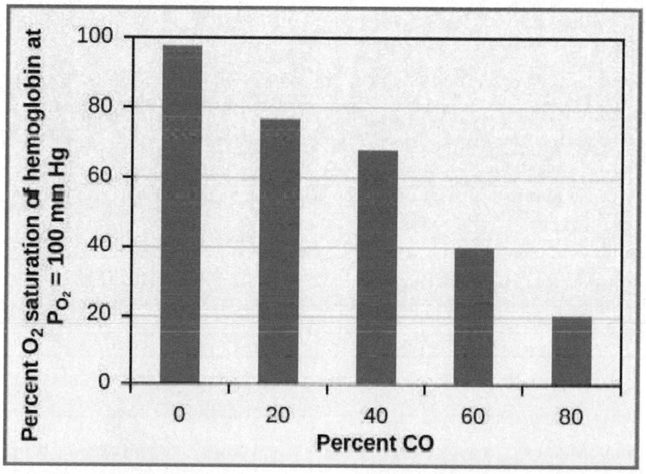

Figure 39.22 As percent CO increases, the oxygen saturation of hemoglobin decreases.

KEY TERMS

alveolar P_{O_2} partial pressure of oxygen in the alveoli (usually around 100 mmHg)
alveolar duct duct that extends from the terminal bronchiole to the alveolar sac
alveolar sac structure consisting of two or more alveoli that share a common opening
alveolar ventilation how much air is in the alveoli
alveolus (plural: alveoli) (also, air sac) terminal region of the lung where gas exchange occurs
anatomical dead space (also, anatomical shunt) region of the lung that lacks proper ventilation/perfusion due to an anatomical block
bicarbonate (HCO_3^-) ion ion created when carbonic acid dissociates into H^+ and (HCO_3^-)
bicarbonate buffer system system in the blood that absorbs carbon dioxide and regulates pH levels
bronchiole airway that extends from the main tertiary bronchi to the alveolar sac
bronchus (plural: bronchi) smaller branch of cartilaginous tissue that stems off of the trachea; air is funneled through the bronchi to the region where gas exchange occurs in alveoli
carbaminohemoglobin molecule that forms when carbon dioxide binds to hemoglobin
carbonic anhydrase (CA) enzyme that catalyzes carbon dioxide and water into carbonic acid
chloride shift exchange of chloride for bicarbonate into or out of the red blood cell
compliance measurement of the elasticity of the lung
dead space area in the lung that lacks proper ventilation or perfusion
diaphragm domed-shaped skeletal muscle located under lungs that separates the thoracic cavity from the abdominal cavity
elastic recoil property of the lung that drives the lung tissue inward
elastic work work conducted by the intercostal muscles, chest wall, and diaphragm
expiratory reserve volume (ERV) amount of additional air that can be exhaled after a normal exhalation
FEV1/FVC ratio ratio of how much air can be forced out of the lung in one second to the total amount that is forced out of the lung; a measurement of lung function that can be used to detect disease states
flow-resistive work of breathing performed by the alveoli and tissues in the lung
forced expiratory volume (FEV) (also, forced vital capacity) measure of how much air can be forced out of the lung from maximal inspiration over a specific amount of time
functional residual capacity (FRC) expiratory reserve volume plus residual volume
functional vital capacity (FVC) amount of air that can be forcibly exhaled after taking the deepest breath possible
heme group centralized iron-containing group that is surrounded by the alpha and beta subunits of hemoglobin
hemoglobin molecule in red blood cells that can bind oxygen, carbon dioxide, and carbon monoxide
inspiratory capacity (IC) tidal volume plus inspiratory reserve volume
inspiratory reserve volume (IRV) amount of additional air that can be inspired after a normal inhalation
intercostal muscle muscle connected to the rib cage that contracts upon inspiration
intrapleural space space between the layers of pleura
larynx voice box, a short passageway connecting the pharynx and the trachea
lung capacity measurement of two or more lung volumes (how much air can be inhaled from the end of an expiration to maximal capacity)
lung volume measurement of air for one lung function (normal inhalation or exhalation)
mucin complex glycoprotein found in mucus
mucus sticky protein-containing fluid secretion in the lung that traps particulate matter to be expelled from the body
nasal cavity opening of the respiratory system to the outside environment
obstructive disease disease (such as emphysema and asthma) that arises from obstruction of the airways; compliance increases in these diseases
oxygen dissociation curve curve depicting the affinity of oxygen for hemoglobin
oxygen-carrying capacity amount of oxygen that can be transported in the blood
partial pressure amount of pressure exerted by one gas within a mixture of gases
particulate matter small particle such as dust, dirt, viral particles, and bacteria that are in the air
pharynx throat; a tube that starts in the internal nares and runs partway down the neck, where it opens into the esophagus and the larynx
physiological dead space (also, physiological shunt) region of the lung that lacks proper ventilation/perfusion due to a physiological change in the lung (like inflammation or edema)
pleura tissue layer that surrounds the lungs and lines the interior of the thoracic cavity
pleurisy painful inflammation of the pleural tissue layers
primary bronchus (also, main bronchus) region of the airway within the lung that attaches to the trachea and bifurcates to each lung where it branches into secondary bronchi
recruitment process of opening airways that normally remain closed when the cardiac output increases
residual volume (RV) amount of air remaining in the lung

after a maximal expiration

resistance measurement of lung obstruction

respiratory bronchiole terminal portion of the bronchiole tree that is attached to the terminal bronchioles and alveoli ducts, alveolar sacs, and alveoli

respiratory distress syndrome disease that arises from a deficient amount of surfactant

respiratory quotient (RQ) ratio of carbon dioxide production to each oxygen molecule consumed

respiratory rate number of breaths per minute

restrictive disease disease that results from a restriction and decreased compliance of the alveoli; respiratory distress syndrome and pulmonary fibrosis are examples

sickle cell anemia genetic disorder that affects the shape of red blood cells, and their ability to transport oxygen and move through capillaries

spirometry method to measure lung volumes and to diagnose lung diseases

surfactant detergent-like liquid in the airways that lowers the surface tension of the alveoli to allow for expansion

terminal bronchiole region of bronchiole that attaches to the respiratory bronchioles

thalassemia rare genetic disorder that results in mutation of the alpha or beta subunits of hemoglobin, creating smaller red blood cells with less hemoglobin

tidal volume (TV) amount of air that is inspired and expired during normal breathing

total lung capacity (TLC) sum of the residual volume, expiratory reserve volume, tidal volume, and inspiratory reserve volume

trachea cartilaginous tube that transports air from the larynx to the primary bronchi

venous P_{CO_2} partial pressure of carbon dioxide in the veins (40 mm Hg in the pulmonary veins)

venous P_{O_2} partial pressure of oxygen in the veins (100 mm Hg in the pulmonary veins)

ventilation/perfusion (V/Q) mismatch region of the lung that lacks proper alveolar ventilation (V) and/or arterial perfusion (Q)

vital capacity (VC) sum of the expiratory reserve volume, tidal volume, and inspiratory reserve volume

CHAPTER SUMMARY

39.1 Systems of Gas Exchange

Animal respiratory systems are designed to facilitate gas exchange. In mammals, air is warmed and humidified in the nasal cavity. Air then travels down the pharynx, through the trachea, and into the lungs. In the lungs, air passes through the branching bronchi, reaching the respiratory bronchioles, which house the first site of gas exchange. The respiratory bronchioles open into the alveolar ducts, alveolar sacs, and alveoli. Because there are so many alveoli and alveolar sacs in the lung, the surface area for gas exchange is very large. Several protective mechanisms are in place to prevent damage or infection. These include the hair and mucus in the nasal cavity that trap dust, dirt, and other particulate matter before they can enter the system. In the lungs, particles are trapped in a mucus layer and transported via cilia up to the esophageal opening at the top of the trachea to be swallowed.

39.2 Gas Exchange across Respiratory Surfaces

The lungs can hold a large volume of air, but they are not usually filled to maximal capacity. Lung volume measurements include tidal volume, expiratory reserve volume, inspiratory reserve volume, and residual volume. The sum of these equals the total lung capacity. Gas movement into or out of the lungs is dependent on the pressure of the gas. Air is a mixture of gases; therefore, the partial pressure of each gas can be calculated to determine how the gas will flow in the lung. The difference between the partial pressure of the gas in the air drives oxygen into the tissues and carbon dioxide out of the body.

39.3 Breathing

The structure of the lungs and thoracic cavity control the mechanics of breathing. Upon inspiration, the diaphragm contracts and lowers. The intercostal muscles contract and expand the chest wall outward. The intrapleural pressure drops, the lungs expand, and air is drawn into the airways. When exhaling, the intercostal muscles and diaphragm relax, returning the intrapleural pressure back to the resting state. The lungs recoil and airways close. The air passively exits the lung. There is high surface tension at the air-airway interface in the lung. Surfactant, a mixture of phospholipids and lipoproteins, acts like a detergent in the airways to reduce surface tension and allow for opening of the alveoli.

Breathing and gas exchange are both altered by changes in the compliance and resistance of the lung. If the compliance of the lung decreases, as occurs in restrictive diseases like fibrosis, the airways stiffen and collapse upon exhalation. Air becomes trapped in the lungs, making breathing more difficult. If resistance increases, as happens with asthma or emphysema, the airways become obstructed, trapping air in the lungs and causing breathing to become difficult. Alterations in the ventilation of the airways or perfusion of the arteries can affect gas exchange. These changes in ventilation and perfusion, called V/Q mismatch, can arise from anatomical or physiological changes.

39.4 Transport of Gases in Human Bodily Fluids

Hemoglobin is a protein found in red blood cells that is comprised of two alpha and two beta subunits that surround an iron-containing heme group. Oxygen readily binds this heme group. The ability of oxygen to bind increases as more oxygen molecules are bound to heme. Disease states and altered conditions in the body can affect the binding ability of oxygen, and increase or decrease its ability to dissociate from hemoglobin.

Carbon dioxide can be transported through the blood via three methods. It is dissolved directly in the blood, bound to plasma proteins or hemoglobin, or converted into bicarbonate. The majority of carbon dioxide is transported as part of the bicarbonate system. Carbon dioxide diffuses into red blood cells. Inside, carbonic anhydrase converts carbon dioxide into carbonic acid (H_2CO_3), which is subsequently hydrolyzed into bicarbonate (HCO_3^-) and H^+. The H^+ ion binds to hemoglobin in red blood cells, and bicarbonate is transported out of the red blood cells in exchange for a chloride ion. This is called the chloride shift. Bicarbonate leaves the red blood cells and enters the blood plasma. In the lungs, bicarbonate is transported back into the red blood cells in exchange for chloride. The H^+ dissociates from hemoglobin and combines with bicarbonate to form carbonic acid with the help of carbonic anhydrase, which further catalyzes the reaction to convert carbonic acid back into carbon dioxide and water. The carbon dioxide is then expelled from the lungs.

VISUAL CONNECTION QUESTIONS

1. Figure 39.7 Which of the following statements about the mammalian respiratory system is false?
   a. When we breathe in, air travels from the pharynx to the trachea.
   b. The bronchioles branch into bronchi.
   c. Alveolar ducts connect to alveolar sacs.
   d. Gas exchange between the lung and blood takes place in the alveolus.

2. Figure 39.13 Which of the following statements is false?
   a. In the tissues, P_{O_2} drops as blood passes from the arteries to the veins, while P_{CO_2} increases.
   b. Blood travels from the lungs to the heart to body tissues, then back to the heart, then the lungs.
   c. Blood travels from the lungs to the heart to body tissues, then back to the lungs, then the heart.
   d. P_{O_2} is higher in air than in the lungs.

3. Figure 39.20 The kidneys are responsible for removing excess H+ ions from the blood. If the kidneys fail, what would happen to blood pH and to hemoglobin affinity for oxygen?

REVIEW QUESTIONS

4. The respiratory system ________.
   a. provides body tissues with oxygen
   b. provides body tissues with oxygen and carbon dioxide
   c. establishes how many breaths are taken per minute
   d. provides the body with carbon dioxide

5. Air is warmed and humidified in the nasal passages. This helps to ________.
   a. ward off infection
   b. decrease sensitivity during breathing
   c. prevent damage to the lungs
   d. all of the above

6. Which is the order of airflow during inhalation?
   a. nasal cavity, trachea, larynx, bronchi, bronchioles, alveoli
   b. nasal cavity, larynx, trachea, bronchi, bronchioles, alveoli
   c. nasal cavity, larynx, trachea, bronchioles, bronchi, alveoli
   d. nasal cavity, trachea, larynx, bronchioles, bronchi, alveoli

7. The inspiratory reserve volume measures the ________.
   a. amount of air remaining in the lung after a maximal exhalation
   b. amount of air that the lung holds
   c. amount of air that can be further exhaled after a normal breath
   d. amount of air that can be further inhaled after a normal breath

8. Of the following, which does not explain why the partial pressure of oxygen is lower in the lung than in the external air?
   a. Air in the lung is humidified; therefore, water vapor pressure alters the pressure.
   b. Carbon dioxide mixes with oxygen.
   c. Oxygen is moved into the blood and is headed to the tissues.
   d. Lungs exert a pressure on the air to reduce the oxygen pressure.

9. The total lung capacity is calculated using which of the following formulas?
   a. residual volume + tidal volume + inspiratory reserve volume
   b. residual volume + expiratory reserve volume + inspiratory reserve volume
   c. expiratory reserve volume + tidal volume + inspiratory reserve volume
   d. residual volume + expiratory reserve volume + tidal volume + inspiratory reserve volume

10. How would paralysis of the diaphragm alter inspiration?
    a. It would prevent contraction of the intercostal muscles.
    b. It would prevent inhalation because the intrapleural pressure would not change.
    c. It would decrease the intrapleural pressure and allow more air to enter the lungs.
    d. It would slow expiration because the lung would not relax.

11. Restrictive airway diseases ________.
    a. increase the compliance of the lung
    b. decrease the compliance of the lung
    c. increase the lung volume
    d. decrease the work of breathing

12. Alveolar ventilation remains constant when ________.
    a. the respiratory rate is increased while the volume of air per breath is decreased
    b. the respiratory rate and the volume of air per breath are increased
    c. the respiratory rate is decreased while increasing the volume per breath
    d. both a and c

13. Which of the following will NOT facilitate the transfer of oxygen to tissues?
    a. decreased body temperature
    b. decreased pH of the blood
    c. increased carbon dioxide
    d. increased exercise

14. The majority of carbon dioxide in the blood is transported by ________.
    a. binding to hemoglobin
    b. dissolution in the blood
    c. conversion to bicarbonate
    d. binding to plasma proteins

15. The majority of oxygen in the blood is transported by ________.
    a. dissolution in the blood
    b. being carried as bicarbonate ions
    c. binding to blood plasma
    d. binding to hemoglobin

CRITICAL THINKING QUESTIONS

16. Describe the function of these terms and describe where they are located: main bronchus, trachea, alveoli, and acinus.
17. How does the structure of alveoli maximize gas exchange?
18. What does FEV1/FVC measure? What factors may affect FEV1/FVC?
19. What is the reason for having residual volume in the lung?
20. How can a decrease in the percent of oxygen in the air affect the movement of oxygen in the body?
21. If a patient has increased resistance in his or her lungs, how can this be detected by a doctor? What does this mean?
22. How would increased airway resistance affect intrapleural pressure during inhalation?
23. Explain how a puncture to the thoracic cavity (from a knife wound, for instance) could alter the ability to inhale.
24. When someone is standing, gravity stretches the bottom of the lung down toward the floor to a greater extent than the top of the lung. What implication could this have on the flow of air in the lungs? Where does gas exchange occur in the lungs?
25. What would happen if no carbonic anhydrase were present in red blood cells?
26. How does the administration of 100 percent oxygen save a patient from carbon monoxide poisoning? Why wouldn't giving carbon dioxide work?

CHAPTER 40
The Circulatory System

Figure 40.1 Just as highway systems transport people and goods through a complex network, the circulatory system transports nutrients, gases, and wastes throughout the animal body. (credit: modification of work by Andrey Belenko)

INTRODUCTION Most animals are complex multicellular organisms that require a mechanism for transporting nutrients throughout their bodies and removing waste products. The circulatory system has evolved over time from simple diffusion through cells in the early evolution of animals to a complex network of blood vessels that reach all parts of the human body. This extensive network supplies the cells, tissues, and organs with oxygen and nutrients, and removes carbon dioxide and waste, which are byproducts of respiration.

At the core of the human circulatory system is the heart. The size of a clenched fist, the human heart is protected beneath the rib cage. Made of specialized and unique cardiac muscle, it pumps blood throughout the body and to the heart itself. Heart contractions are driven by intrinsic electrical impulses that the brain and endocrine hormones help to regulate. Understanding the heart's basic anatomy and function is important to understanding the body's circulatory and respiratory systems.

Gas exchange is one essential function of the circulatory system. A circulatory system is not needed in organisms with no specialized respiratory organs because oxygen and carbon dioxide diffuse directly between their body tissues and the external environment. However, in organisms that possess lungs and gills, oxygen must be transported from these specialized respiratory organs to the body tissues via a circulatory system. Therefore, circulatory systems have had to evolve to accommodate the great diversity of body sizes and body types present among animals.

40.1 Overview of the Circulatory System

By the end of this section, you will be able to do the following:

- Describe an open and closed circulatory system
- Describe interstitial fluid and hemolymph
- Compare and contrast the organization and evolution of the vertebrate circulatory system

In all animals, except a few simple types, the circulatory system is used to transport nutrients and gases through the body. Simple diffusion allows some water, nutrient, waste, and gas exchange into primitive animals that are only a few cell layers thick; however, bulk flow is the only method by which the entire body of larger more complex organisms is accessed.

Circulatory System Architecture

The circulatory system is effectively a network of cylindrical vessels: the arteries, veins, and capillaries that emanate from a pump, the heart. In all vertebrate organisms, as well as some invertebrates, this is a closed-loop system, in which the blood is not free in a cavity. In a **closed circulatory system**, blood is contained inside blood vessels and circulates **unidirectionally** from the heart around the systemic circulatory route, then returns to the heart again, as illustrated in Figure 40.2a. As opposed to a closed system, arthropods—including insects, crustaceans, and most mollusks—have an open circulatory system, as illustrated in Figure 40.2b. In an **open circulatory system**, the blood is not enclosed in the blood vessels but is pumped into a cavity called a **hemocoel** and is called **hemolymph** because the blood mixes with the **interstitial fluid**. As the heart beats and the animal moves, the hemolymph circulates around the organs within the body cavity and then reenters the hearts through openings called **ostia**. This movement allows for gas and nutrient exchange. An open circulatory system does not use as much energy as a closed system to operate or to maintain; however, there is a trade-off with the amount of blood that can be moved to metabolically active organs and tissues that require high levels of oxygen. In fact, one reason that insects with wing spans of up to two feet wide (70 cm) are not around today is probably because they were outcompeted by the arrival of birds 150 million years ago. Birds, having a closed circulatory system, are thought to have moved more agilely, allowing them to get food faster and possibly to prey on the insects.

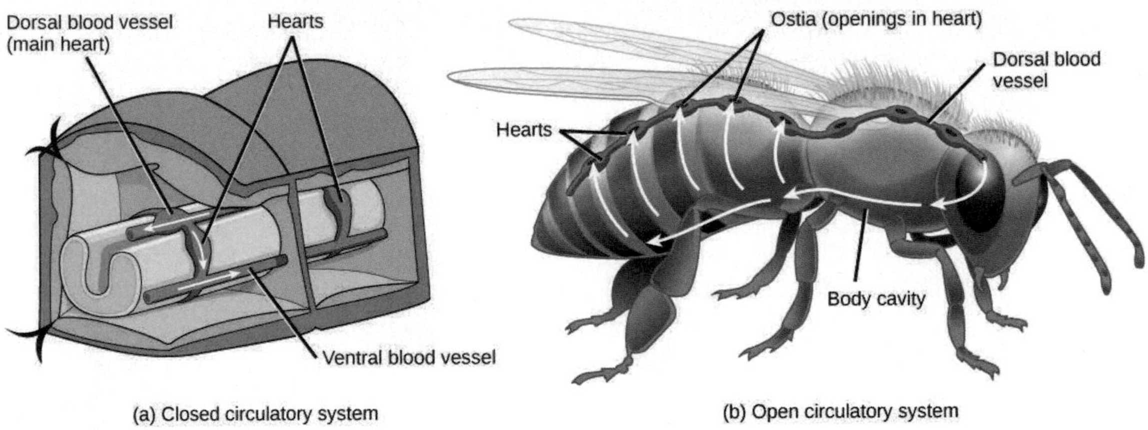

Figure 40.2 In (a) closed circulatory systems, the heart pumps blood through vessels that are separate from the interstitial fluid of the body. Most vertebrates and some invertebrates, like this annelid earthworm, have a closed circulatory system. In (b) open circulatory systems, a fluid called hemolymph is pumped through a blood vessel that empties into the body cavity. Hemolymph returns to the blood vessel through openings called ostia. Arthropods like this bee and most mollusks have open circulatory systems.

Circulatory System Variation in Animals

The circulatory system varies from simple systems in invertebrates to more complex systems in vertebrates. The simplest animals, such as the sponges (Porifera) and rotifers (Rotifera), do not need a circulatory system because diffusion allows adequate exchange of water, nutrients, and waste, as well as dissolved gases, as shown in Figure 40.3a. Organisms that are more complex but still only have two layers of cells in their body plan, such as jellies (Cnidaria) and comb jellies (Ctenophora) also use diffusion through their epidermis and internally through the gastrovascular compartment. Both their internal and external tissues are bathed in an aqueous environment and exchange fluids by diffusion on both sides, as illustrated in Figure 40.3b. Exchange of fluids is assisted by the pulsing of the jellyfish body.

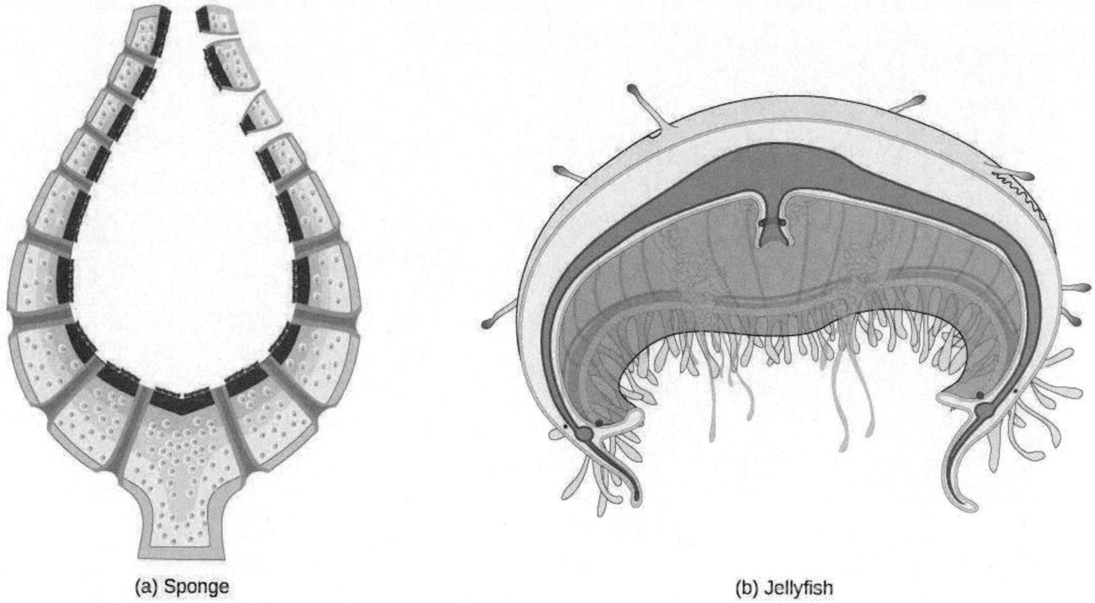

Figure 40.3 Simple animals consisting of a single cell layer such as the (a) sponge or only a few cell layers such as the (b) jellyfish do not have a circulatory system. Instead, gases, nutrients, and wastes are exchanged by diffusion.

For more complex organisms, diffusion is not efficient for cycling gases, nutrients, and waste effectively through the body; therefore, more complex circulatory systems evolved. Most arthropods and many mollusks have open circulatory systems. In an open system, an elongated beating heart pushes the hemolymph through the body and muscle contractions help to move fluids. The larger more complex crustaceans, including lobsters, have developed arterial-like vessels to push blood through their bodies, and the most active mollusks, such as squids, have evolved a closed circulatory system and are able to move rapidly to catch prey. Closed circulatory systems are a characteristic of vertebrates; however, there are significant differences in the structure of the heart and the circulation of blood between the different vertebrate groups due to adaptation during evolution and associated differences in anatomy. Figure 40.4 illustrates the basic circulatory systems of some vertebrates: fish, amphibians, reptiles, and mammals.

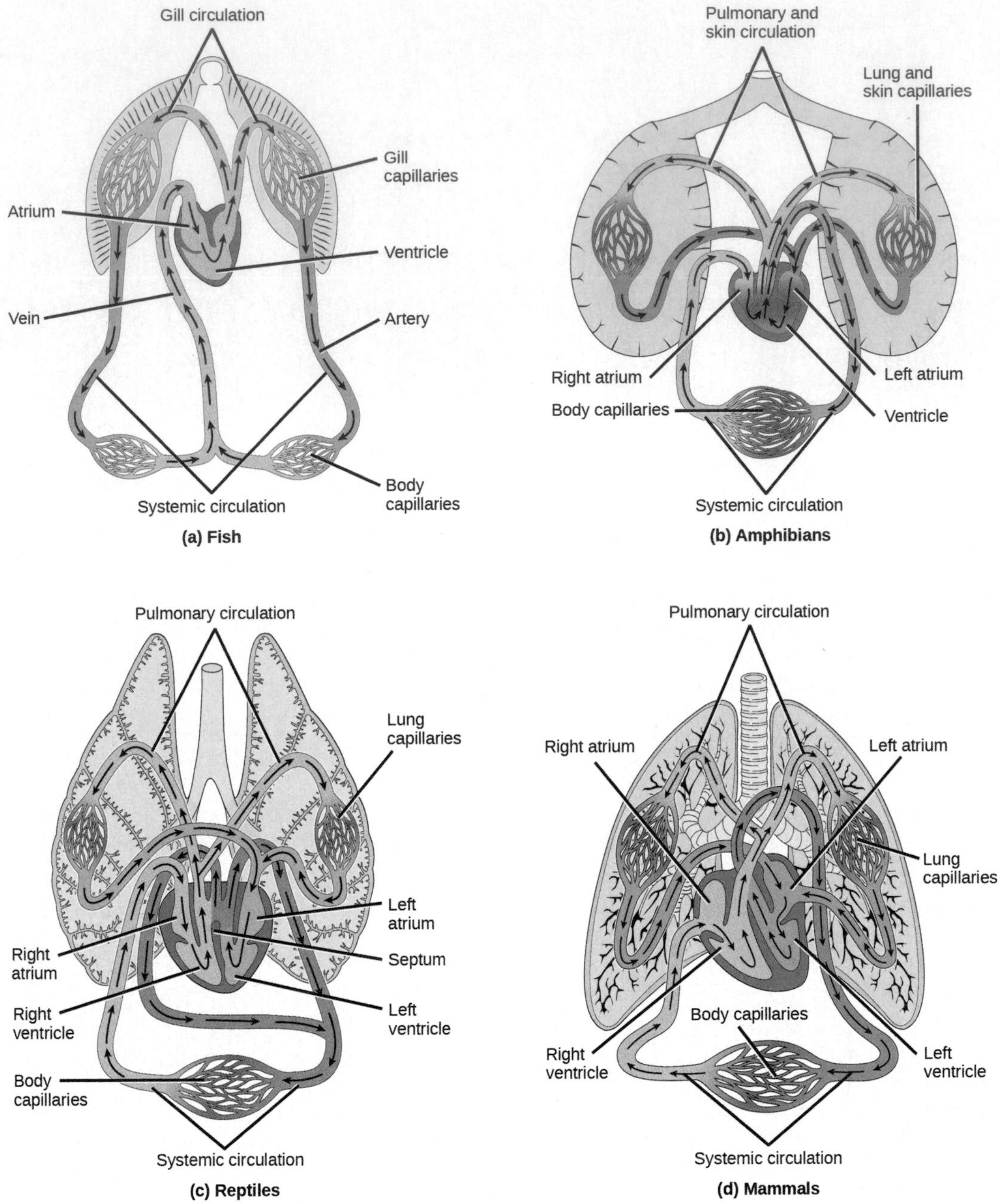

Figure 40.4 (a) Fish have the simplest circulatory systems of the vertebrates: blood flows unidirectionally from the two-chambered heart through the gills and then the rest of the body. (b) Amphibians have two circulatory routes: one for oxygenation of the blood through the lungs and skin, and the other to take oxygen to the rest of the body. The blood is pumped from a three-chambered heart with two atria and a single ventricle. (c) Reptiles also have two circulatory routes; however, blood is only oxygenated through the lungs. The heart is three chambered, but the ventricles are partially separated so some mixing of oxygenated and deoxygenated blood occurs except in crocodilians and birds. (d) Mammals and birds have the most efficient heart with four chambers that completely separate the oxygenated and deoxygenated blood; it pumps only oxygenated blood through the body and deoxygenated blood to the lungs.

As illustrated in Figure 40.4a. Fish have a single circuit for blood flow and a two-chambered heart that has only a single atrium and a single ventricle. The atrium collects blood that has returned from the body and the ventricle pumps the blood to the gills where gas exchange occurs and the blood is re-oxygenated; this is called **gill circulation**. The blood then continues through the rest of the body before arriving back at the atrium; this is called **systemic circulation**. This unidirectional flow of blood produces a gradient of oxygenated to deoxygenated blood around the fish's systemic circuit. The result is a limit in the amount of oxygen that can reach some of the organs and tissues of the body, reducing the overall metabolic capacity of fish.

In amphibians, reptiles, birds, and mammals, blood flow is directed in two circuits: one through the lungs and back to the heart, which is called **pulmonary circulation**, and the other throughout the rest of the body and its organs including the brain (systemic circulation). In amphibians, gas exchange also occurs through the skin during pulmonary circulation and is referred to as **pulmocutaneous circulation**.

As shown in Figure 40.4b, amphibians have a three-chambered heart that has two atria and one ventricle rather than the two-chambered heart of fish. The two **atria** (superior heart chambers) receive blood from the two different circuits (the lungs and the systems), and then there is some mixing of the blood in the heart's **ventricle** (inferior heart chamber), which reduces the efficiency of oxygenation. The advantage to this arrangement is that high pressure in the vessels pushes blood to the lungs and body. The mixing is mitigated by a ridge within the ventricle that diverts oxygen-rich blood through the systemic circulatory system and deoxygenated blood to the pulmocutaneous circuit. For this reason, amphibians are often described as having **double circulation**.

Most reptiles also have a three-chambered heart similar to the amphibian heart that directs blood to the pulmonary and systemic circuits, as shown in Figure 40.4c. The ventricle is divided more effectively by a partial septum, which results in less mixing of oxygenated and deoxygenated blood. Some reptiles (alligators and crocodiles) are the most primitive animals to exhibit a four-chambered heart. Crocodilians have a unique circulatory mechanism where the heart shunts blood from the lungs toward the stomach and other organs during long periods of submergence, for instance, while the animal waits for prey or stays underwater waiting for prey to rot. One adaptation includes two main arteries that leave the same part of the heart: one takes blood to the lungs and the other provides an alternate route to the stomach and other parts of the body. Two other adaptations include a hole in the heart between the two ventricles, called the foramen of Panizza, which allows blood to move from one side of the heart to the other, and specialized connective tissue that slows the blood flow to the lungs. Together these adaptations have made crocodiles and alligators one of the most evolutionarily successful animal groups on earth.

In mammals and birds, the heart is also divided into four chambers: two atria and two ventricles, as illustrated in Figure 40.4d. The oxygenated blood is separated from the deoxygenated blood, which improves the efficiency of double circulation and is probably required for the warm-blooded lifestyle of mammals and birds. The four-chambered heart of birds and mammals evolved independently from a three-chambered heart. The independent evolution of the same or a similar biological trait is referred to as convergent evolution.

40.2 Components of the Blood

By the end of this section, you will be able to do the following:

- List the basic components of the blood
- Compare red and white blood cells
- Describe blood plasma and serum

Hemoglobin is responsible for distributing oxygen, and to a lesser extent, carbon dioxide, throughout the circulatory systems of humans, vertebrates, and many invertebrates. The blood is more than the proteins, though. Blood is actually a term used to describe the liquid that moves through the vessels and includes **plasma** (the liquid portion, which contains water, proteins, salts, lipids, and glucose) and the cells (red and white cells) and cell fragments called **platelets**. Blood plasma is actually the dominant component of blood and contains the water, proteins, electrolytes, lipids, and glucose. The cells are responsible for carrying the gases (red cells) and the immune response (white). The platelets are responsible for blood clotting. Interstitial fluid that surrounds cells is separate from the blood, but in hemolymph, they are combined. In humans, cellular components make up approximately 45 percent of the blood and the liquid plasma 55 percent. Blood is 20 percent of a person's extracellular fluid and eight percent of weight.

The Role of Blood in the Body

Blood, like the human blood illustrated in Figure 40.5 is important for regulation of the body's systems and homeostasis. Blood helps maintain homeostasis by stabilizing pH, temperature, osmotic pressure, and by eliminating excess heat. Blood supports growth by distributing nutrients and hormones, and by removing waste. Blood plays a protective role by transporting clotting factors and platelets to prevent blood loss and transporting the disease-fighting agents or **white blood cells** to sites of infection.

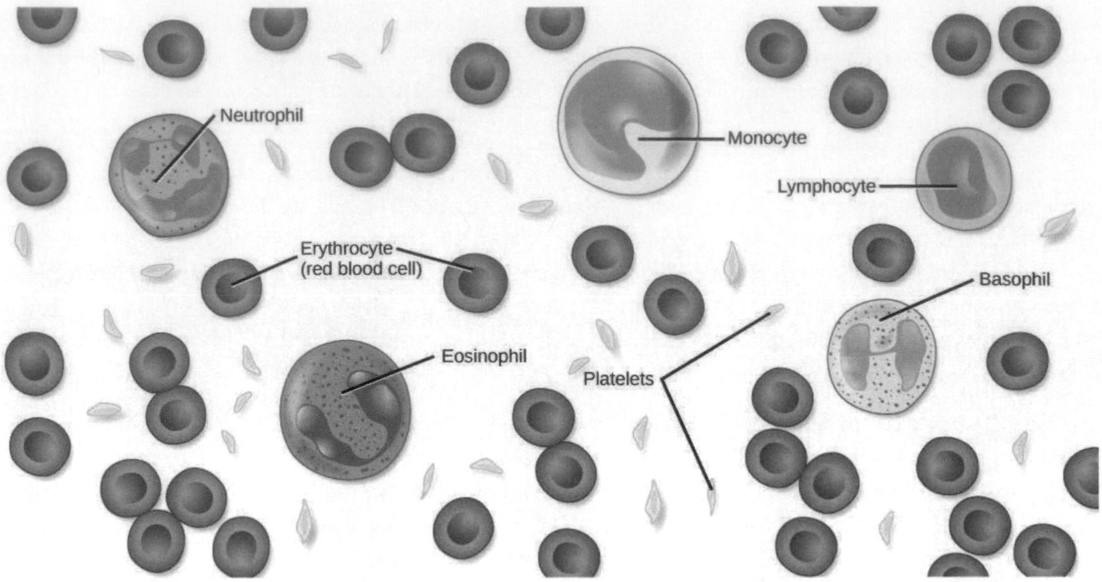

Figure 40.5 The cells and cellular components of human blood are shown. Red blood cells deliver oxygen to the cells and remove carbon dioxide. White blood cells—including neutrophils, monocytes, lymphocytes, eosinophils, and basophils—are involved in the immune response. Platelets form clots that prevent blood loss after injury.

Red Blood Cells

Red blood cells, or erythrocytes (erythro- = "red"; -cyte = "cell"), are specialized cells that circulate through the body delivering oxygen to cells; they are formed from stem cells in the bone marrow. In mammals, red blood cells are small biconcave cells that at maturity do not contain a nucleus or mitochondria and are only 7–8 µm in size. In birds and non-avian reptiles, a nucleus is still maintained in red blood cells.

The red coloring of blood comes from the iron-containing protein hemoglobin, illustrated in Figure 40.6a. The principle job of this protein is to carry oxygen, but it also transports carbon dioxide as well. Hemoglobin is packed into red blood cells at a rate of about 250 million molecules of hemoglobin per cell. Each hemoglobin molecule binds four oxygen molecules so that each red blood cell carries one billion molecules of oxygen. There are approximately 25 trillion red blood cells in the five liters of blood in the human body, which could carry up to 25 sextillion (25×10^{21}) molecules of oxygen in the body at any time. In mammals, the lack of organelles in erythrocytes leaves more room for the hemoglobin molecules, and the lack of mitochondria also prevents use of the oxygen for metabolic respiration. Only mammals have anucleated red blood cells, and some mammals (camels, for instance) even have nucleated red blood cells. The advantage of nucleated red blood cells is that these cells can undergo mitosis. Anucleated red blood cells metabolize anaerobically (without oxygen), making use of a primitive metabolic pathway to produce ATP and increase the efficiency of oxygen transport.

Not all organisms use hemoglobin as the method of oxygen transport. Invertebrates that utilize hemolymph rather than blood use different pigments to bind to the oxygen. These pigments use copper or iron to the oxygen. Invertebrates have a variety of other respiratory pigments. Hemocyanin, a blue-green, copper-containing protein, illustrated in Figure 40.6b is found in mollusks, crustaceans, and some of the arthropods. Chlorocruorin, a green-colored, iron-containing pigment is found in four families of polychaete tubeworms. Hemerythrin, a red, iron-containing protein is found in some polychaete worms and annelids and is illustrated in Figure 40.6c. Despite the name, hemerythrin does not contain a heme group and its oxygen-carrying capacity is poor compared to hemoglobin.

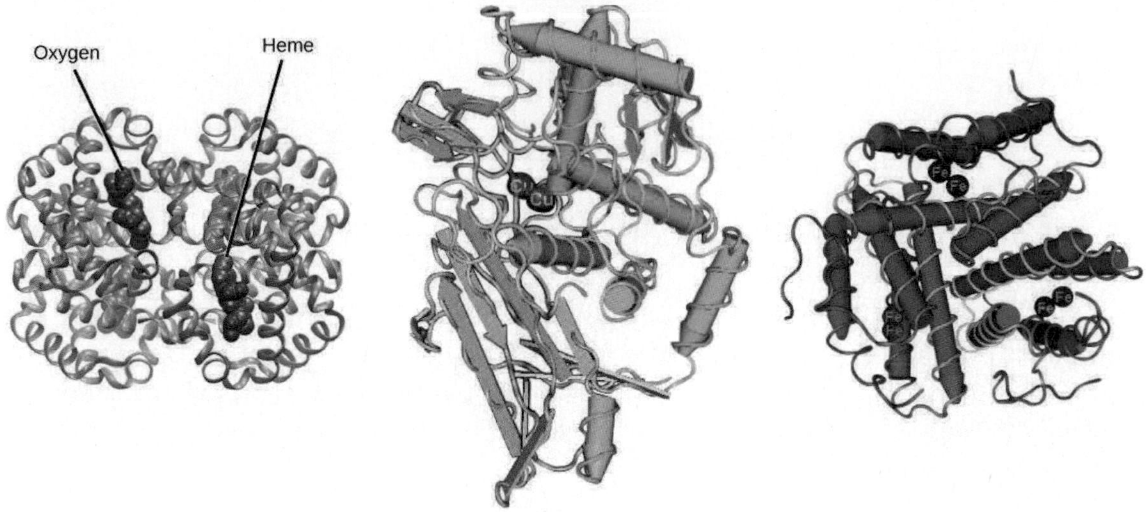

Figure 40.6 In most vertebrates, (a) hemoglobin delivers oxygen to the body and removes some carbon dioxide. Hemoglobin is composed of four protein subunits, two alpha chains and two beta chains, and a heme group that has iron associated with it. The iron reversibly associates with oxygen, and in so doing is oxidized from Fe^{2+} to Fe^{3+}. In most mollusks and some arthropods, (b) hemocyanin delivers oxygen. Unlike hemoglobin, hemolymph is not carried in blood cells, but floats free in the hemolymph. Copper instead of iron binds the oxygen, giving the hemolymph a blue-green color. In annelids, such as the earthworm, and some other invertebrates, (c) hemerythrin carries oxygen. Like hemoglobin, hemerythrin is carried in blood cells and has iron associated with it, but despite its name, hemerythrin does not contain heme.

The small size and large surface area of red blood cells allows for rapid diffusion of oxygen and carbon dioxide across the plasma membrane. In the lungs, carbon dioxide is released and oxygen is taken in by the blood. In the tissues, oxygen is released from the blood and carbon dioxide is bound for transport back to the lungs. Studies have found that hemoglobin also binds nitrous oxide (NO). NO is a vasodilator that relaxes the blood vessels and capillaries and may help with gas exchange and the passage of red blood cells through narrow vessels. Nitroglycerin, a heart medication for angina and heart attacks, is converted to NO to help relax the blood vessels and increase oxygen flow through the body.

A characteristic of red blood cells is their glycolipid and glycoprotein coating; these are lipids and proteins that have carbohydrate molecules attached. In humans, the surface glycoproteins and glycolipids on red blood cells vary between individuals, producing the different blood types, such as A, B, and O. Red blood cells have an average life span of 120 days, at which time they are broken down and recycled in the liver and spleen by phagocytic macrophages, a type of white blood cell.

White Blood Cells

White blood cells, also called leukocytes (leuko = white), make up approximately one percent by volume of the cells in blood. The role of white blood cells is very different than that of red blood cells: they are primarily involved in the immune response to identify and target pathogens, such as invading bacteria, viruses, and other foreign organisms. White blood cells are formed continually; some only live for hours or days, but some live for years.

The morphology of white blood cells differs significantly from red blood cells. They have nuclei and do not contain hemoglobin. The different types of white blood cells are identified by their microscopic appearance after histologic staining, and each has a different specialized function. The two main groups, both illustrated in Figure 40.7 are the granulocytes, which include the neutrophils, eosinophils, and basophils, and the agranulocytes, which include the monocytes and lymphocytes.

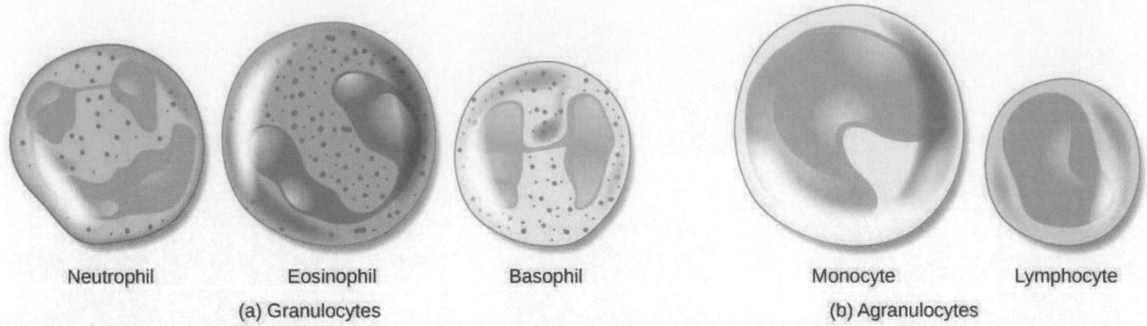

Figure 40.7 (a) Granulocytes—including neutrophils, eosinophils and basophils—are characterized by a lobed nucleus and granular inclusions in the cytoplasm. Granulocytes are typically first-responders during injury or infection. (b) Agranulocytes include lymphocytes and monocytes. Lymphocytes, including B and T cells, are responsible for adaptive immune response. Monocytes differentiate into macrophages and dendritic cells, which in turn respond to infection or injury.

Granulocytes contain granules in their cytoplasm; the agranulocytes are so named because of the lack of granules in their cytoplasm. Some leukocytes become macrophages that either stay at the same site or move through the bloodstream and gather at sites of infection or inflammation where they are attracted by chemical signals from foreign particles and damaged cells. Lymphocytes are the primary cells of the immune system and include B cells, T cells, and natural killer cells. B cells destroy bacteria and inactivate their toxins. They also produce antibodies. T cells attack viruses, fungi, some bacteria, transplanted cells, and cancer cells. T cells attack viruses by releasing toxins that kill the viruses. Natural killer cells attack a variety of infectious microbes and certain tumor cells.

One reason that HIV poses significant management challenges is because the virus directly targets T cells by gaining entry through a receptor. Once inside the cell, HIV then multiplies using the T cell's own genetic machinery. After the HIV virus replicates, it is transmitted directly from the infected T cell to macrophages. The presence of HIV can remain unrecognized for an extensive period of time before full disease symptoms develop.

Platelets and Coagulation Factors

Blood must clot to heal wounds and prevent excess blood loss. Small cell fragments called platelets (thrombocytes) are attracted to the wound site where they adhere by extending many projections and releasing their contents. These contents activate other platelets and also interact with other coagulation factors, which convert fibrinogen, a water-soluble protein present in blood serum into fibrin (a non-water soluble protein), causing the blood to clot. Many of the clotting factors require vitamin K to work, and vitamin K deficiency can lead to problems with blood clotting. Many platelets converge and stick together at the wound site forming a platelet plug (also called a fibrin clot), as illustrated in Figure 40.8b. The plug or clot lasts for a number of days and stops the loss of blood. Platelets are formed from the disintegration of larger cells called megakaryocytes, like that shown in Figure 40.8a. For each megakaryocyte, 2000–3000 platelets are formed with 150,000 to 400,000 platelets present in each cubic millimeter of blood. Each platelet is disc shaped and 2–4 μm in diameter. They contain many small vesicles but do not contain a nucleus.

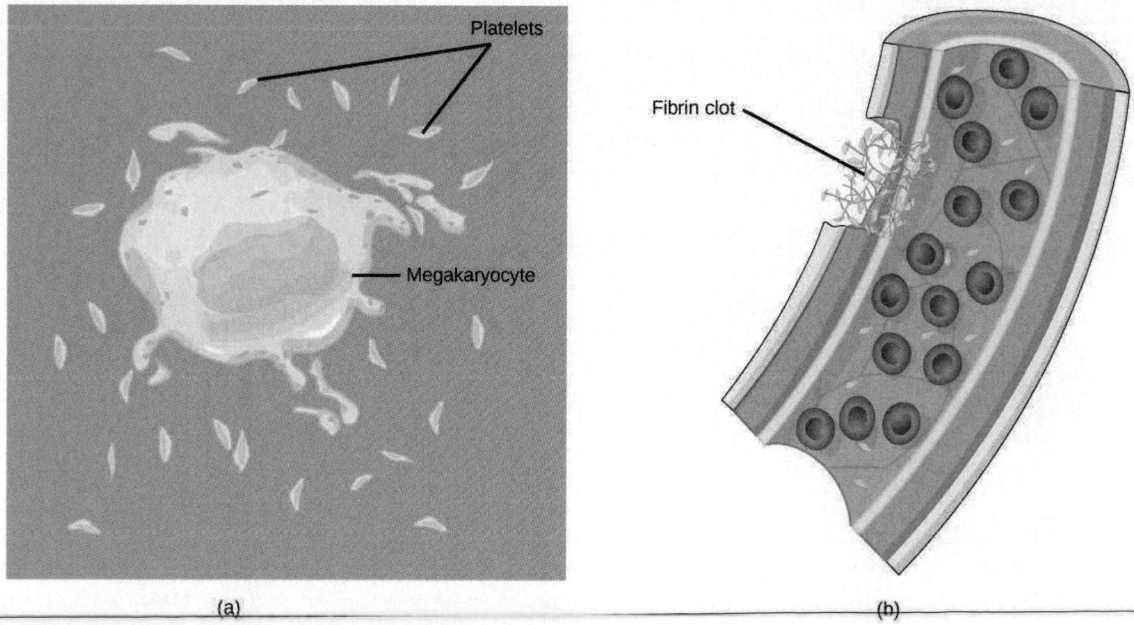

Figure 40.8 (a) Platelets are formed from large cells called megakaryocytes. The megakaryocyte breaks up into thousands of fragments that become platelets. (b) Platelets are required for clotting of the blood. The platelets collect at a wound site in conjunction with other clotting factors, such as fibrinogen, to form a fibrin clot that prevents blood loss and allows the wound to heal.

Plasma and Serum

The liquid component of blood is called plasma, and it is separated by spinning or centrifuging the blood at high rotations (3000 rpm or higher). The blood cells and platelets are separated by centrifugal forces to the bottom of a specimen tube. The upper liquid layer, the plasma, consists of 90 percent water along with various substances required for maintaining the body's pH, osmotic load, and for protecting the body. The plasma also contains the coagulation factors and antibodies.

The plasma component of blood without the coagulation factors is called the **serum**. Serum is similar to interstitial fluid in which the correct composition of key ions acting as electrolytes is essential for normal functioning of muscles and nerves. Other components in the serum include proteins that assist with maintaining pH and osmotic balance while giving viscosity to the blood. The serum also contains antibodies, specialized proteins that are important for defense against viruses and bacteria. Lipids, including cholesterol, are also transported in the serum, along with various other substances including nutrients, hormones, metabolic waste, plus external substances, such as, drugs, viruses, and bacteria.

Human serum albumin is the most abundant protein in human blood plasma and is synthesized in the liver. Albumin, which constitutes about half of the blood serum protein, transports hormones and fatty acids, buffers pH, and maintains osmotic pressures. Immunoglobin is a protein antibody produced in the mucosal lining and plays an important role in antibody mediated immunity.

EVOLUTION CONNECTION

Blood Types Related to Proteins on the Surface of the Red Blood Cells

Red blood cells are coated in antigens made of glycolipids and glycoproteins. The composition of these molecules is determined by genetics, which have evolved over time. In humans, the different surface antigens are grouped into 24 different blood groups with more than 100 different antigens on each red blood cell. The two most well known blood groups are the ABO, shown in Figure 40.9, and Rh systems. The surface antigens in the ABO blood group are glycolipids, called antigen A and antigen B. People with blood type A have antigen A, those with blood type B have antigen B, those with blood type AB have both antigens, and people with blood type O have neither antigen. Antibodies called agglutinougens are found in the blood plasma and react with the A or B antigens, if the two are mixed. When type A and type B blood are combined, agglutination (clumping) of the blood occurs because of antibodies in the plasma that bind with the opposing antigen; this causes clots that coagulate in the kidney causing kidney failure. Type O blood has neither A or B antigens, and therefore, type O blood can be given to all blood types.

Type O negative blood is the universal donor. Type AB positive blood is the universal acceptor because it has both A and B antigen. The ABO blood groups were discovered in 1900 and 1901 by Karl Landsteiner at the University of Vienna.

The Rh blood group was first discovered in Rhesus monkeys. Most people have the Rh antigen (Rh+) and do not have anti-Rh antibodies in their blood. The few people who do not have the Rh antigen and are Rh– can develop anti-Rh antibodies if exposed to Rh+ blood. This can happen after a blood transfusion or after an Rh– woman has an Rh+ baby. The first exposure does not usually cause a reaction; however, at the second exposure, enough antibodies have built up in the blood to produce a reaction that causes agglutination and breakdown of red blood cells. An injection can prevent this reaction.

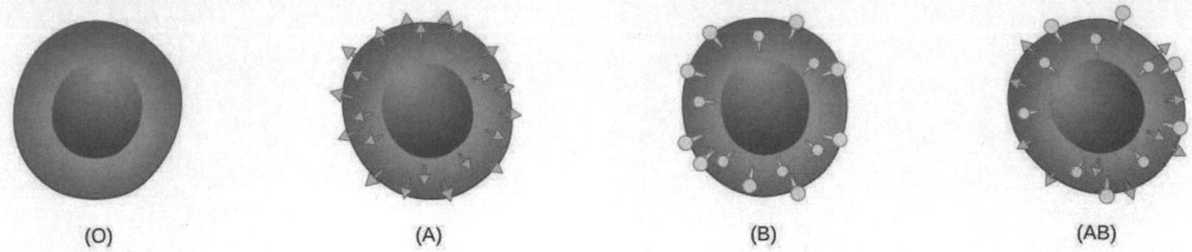

Figure 40.9 Human red blood cells may have either type A or B glycoproteins on their surface, both glycoproteins combined (AB), or neither (O). The glycoproteins serve as antigens and can elicit an immune response in a person who receives a transfusion containing unfamiliar antigens. Type O blood, which has no A or B antigens, does not elicit an immune response when injected into a person of any blood type. Thus, O is considered the universal donor. Persons with type AB blood can accept blood from any blood type, and type AB is considered the universal acceptor.

LINK TO LEARNING

Play a blood typing game on the Nobel Prize website (http://openstax.org/l/blood_typing) to solidify your understanding of blood types.

40.3 Mammalian Heart and Blood Vessels

By the end of this section, you will be able to do the following:

- Describe the structure of the heart and explain how cardiac muscle is different from other muscles
- Describe the cardiac cycle
- Explain the structure of arteries, veins, and capillaries, and how blood flows through the body

The heart is a complex muscle that pumps blood through the three divisions of the circulatory system: the coronary (vessels that serve the heart), pulmonary (heart and lungs), and systemic (systems of the body), as shown in Figure 40.10. Coronary circulation intrinsic to the heart takes blood directly from the main artery (aorta) coming from the heart. For pulmonary and systemic circulation, the heart has to pump blood to the lungs or the rest of the body, respectively. In vertebrates, the lungs are relatively close to the heart in the thoracic cavity. The shorter distance to pump means that the muscle wall on the right side of the heart is not as thick as the left side which must have enough pressure to pump blood all the way to your big toe.

VISUAL CONNECTION

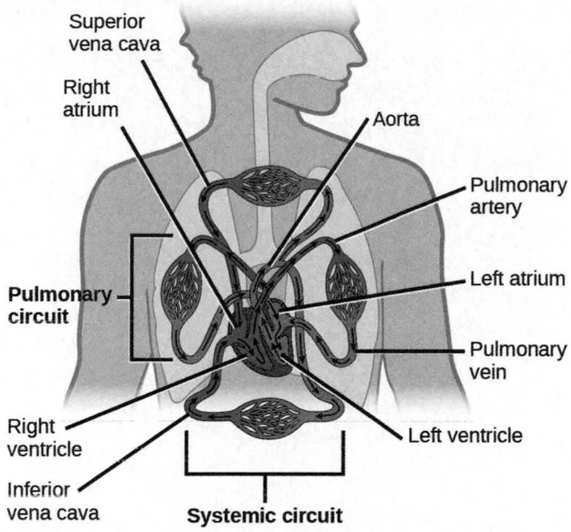

Figure 40.10 The mammalian circulatory system is divided into three circuits: the systemic circuit, the pulmonary circuit, and the coronary circuit. Blood is pumped from veins of the systemic circuit into the right atrium of the heart, then into the right ventricle. Blood then enters the pulmonary circuit, and is oxygenated by the lungs. From the pulmonary circuit, blood reenters the heart through the left atrium. From the left ventricle, blood reenters the systemic circuit through the aorta and is distributed to the rest of the body. The coronary circuit, which provides blood to the heart, is not shown.

Which of the following statements about the circulatory system is false?

a. Blood in the pulmonary vein is deoxygenated.
b. Blood in the inferior vena cava is deoxygenated.
c. Blood in the pulmonary artery is deoxygenated.
d. Blood in the aorta is oxygenated.

Structure of the Heart

The heart muscle is asymmetrical as a result of the distance blood must travel in the pulmonary and systemic circuits. Since the right side of the heart sends blood to the pulmonary circuit it is smaller than the left side which must send blood out to the whole body in the systemic circuit, as shown in Figure 40.11. In humans, the heart is about the size of a clenched fist; it is divided into four chambers: two atria and two ventricles. There is one atrium and one ventricle on the right side and one atrium and one ventricle on the left side. The atria are the chambers that receive blood, and the ventricles are the chambers that pump blood. The right atrium receives deoxygenated blood from the **superior vena cava**, which drains blood from the jugular vein that comes from the brain and from the veins that come from the arms, as well as from the **inferior vena cava** which drains blood from the veins that come from the lower organs and the legs. In addition, the right atrium receives blood from the coronary sinus which drains deoxygenated blood from the heart itself. This deoxygenated blood then passes to the right ventricle through the **atrioventricular valve** or the **tricuspid valve**, a flap of connective tissue that opens in only one direction to prevent the backflow of blood. The valve separating the chambers on the left side of the heart valve is called the biscuspid or mitral valve. After it is filled, the right ventricle pumps the blood through the pulmonary arteries, bypassing the **semilunar valve** (or pulmonic valve) to the lungs for re-oxygenation. After blood passes through the pulmonary arteries, the right semilunar valves close preventing the blood from flowing backwards into the right ventricle. The left atrium then receives the oxygen-rich blood from the lungs via the pulmonary veins. This blood passes through the **bicuspid valve** or mitral valve (the atrioventricular valve on the left side of the heart) to the left ventricle where the blood is pumped out through the **aorta**, the major artery of the body, taking oxygenated blood to the organs and muscles of the body. Once blood is pumped out of the left ventricle and into the aorta, the aortic semilunar valve (or aortic valve) closes preventing blood from flowing backward into the left ventricle. This pattern of pumping is referred to as double circulation and is found in all mammals.

VISUAL CONNECTION

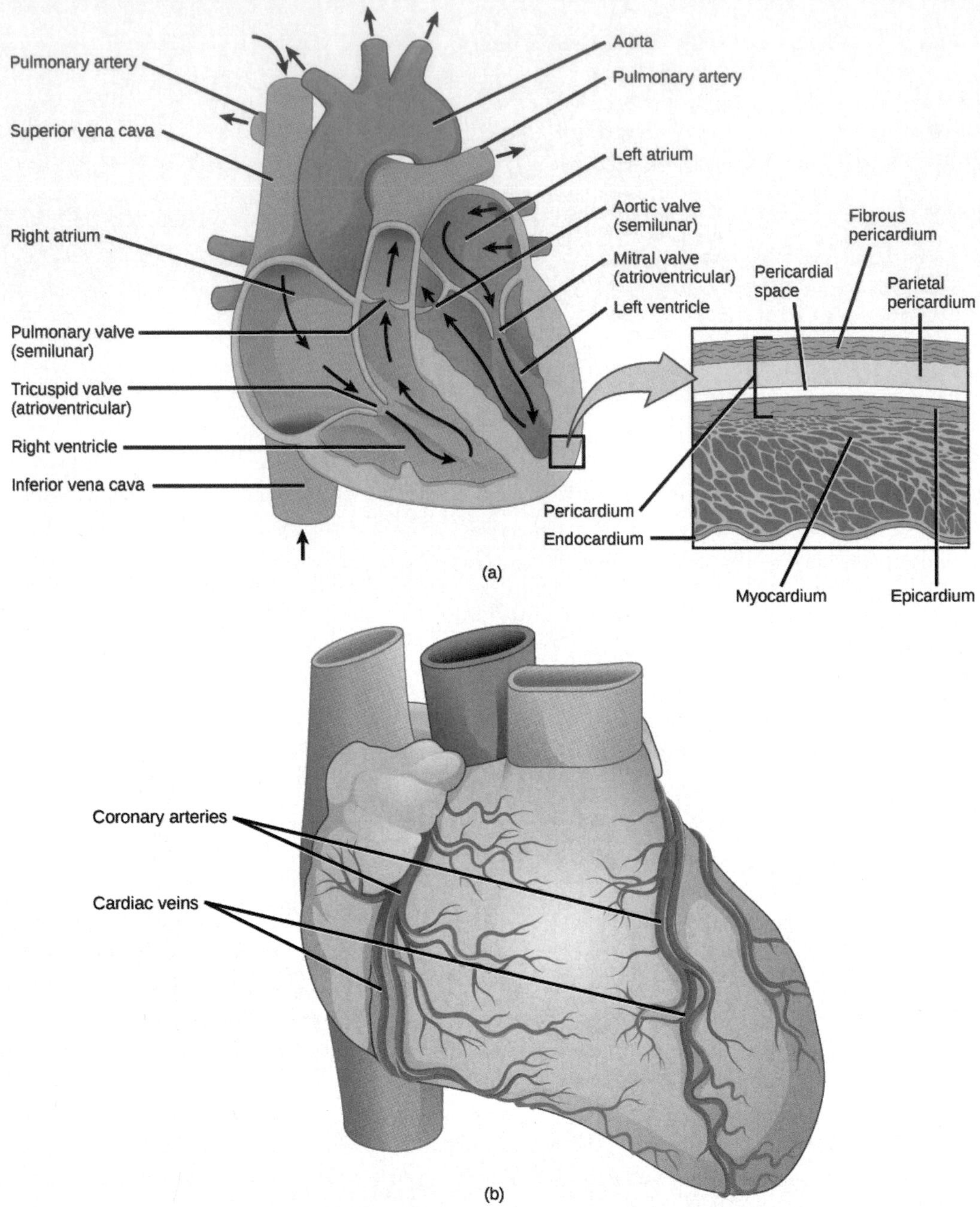

Figure 40.11 (a) The heart is primarily made of a thick muscle layer, called the myocardium, surrounded by membranes. One-way valves separate the four chambers. (b) Blood vessels of the coronary system, including the coronary arteries and veins, keep the heart musculature oxygenated.

Which of the following statements about the heart is false?

a. The mitral valve separates the left ventricle from the left atrium.
b. Blood travels through the bicuspid valve to the left atrium.
c. Both the aortic and the pulmonary valves are semilunar valves.
d. The mitral valve is an atrioventricular valve.

The heart is composed of three layers; the epicardium, the myocardium, and the endocardium, illustrated in Figure 40.11. The inner wall of the heart has a lining called the **endocardium**. The **myocardium** consists of the heart muscle cells that make up the middle layer and the bulk of the heart wall. The outer layer of cells is called the **epicardium**, of which the second layer is a membranous layered structure called the **pericardium** that surrounds and protects the heart; it allows enough room for vigorous pumping but also keeps the heart in place to reduce friction between the heart and other structures.

The heart has its own blood vessels that supply the heart muscle with blood. The **coronary arteries** branch from the aorta and surround the outer surface of the heart like a crown. They diverge into capillaries where the heart muscle is supplied with oxygen before converging again into the **coronary veins** to take the deoxygenated blood back to the right atrium where the blood will be re-oxygenated through the pulmonary circuit. The heart muscle will die without a steady supply of blood. **Atherosclerosis** is the blockage of an artery by the buildup of fatty plaques. Because of the size (narrow) of the coronary arteries and their function in serving the heart itself, atherosclerosis can be deadly in these arteries. The slowdown of blood flow and subsequent oxygen deprivation that results from atherosclerosis causes severe pain, known as **angina**, and complete blockage of the arteries will cause **myocardial infarction**: the death of cardiac muscle tissue, commonly known as a heart attack.

The Cardiac Cycle

The main purpose of the heart is to pump blood through the body; it does so in a repeating sequence called the cardiac cycle. The **cardiac cycle** is the coordination of the filling and emptying of the heart of blood by electrical signals that cause the heart muscles to contract and relax. The human heart beats over 100,000 times per day. In each cardiac cycle, the heart contracts (**systole**), pushing out the blood and pumping it through the body; this is followed by a relaxation phase (**diastole**), where the heart fills with blood, as illustrated in Figure 40.12. The atria contract at the same time, forcing blood through the atrioventricular valves into the ventricles. Closing of the atrioventricular valves produces a monosyllabic "lup" sound. Following a brief delay, the ventricles contract at the same time forcing blood through the semilunar valves into the aorta and the artery transporting blood to the lungs (via the pulmonary artery). Closing of the semilunar valves produces a monosyllabic "dup" sound.

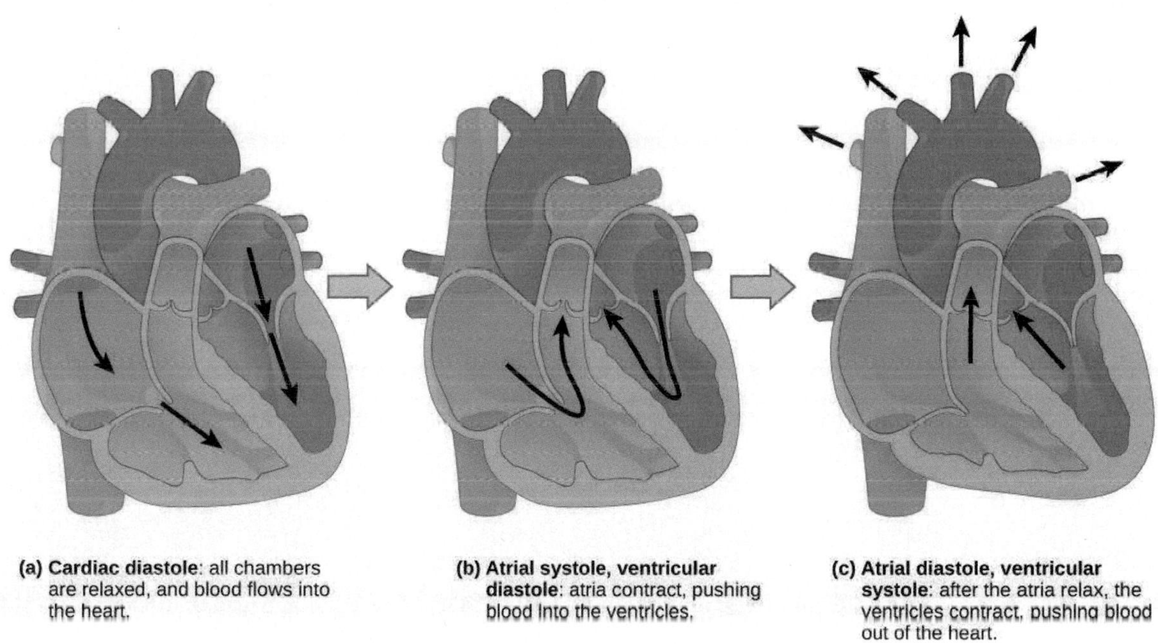

Figure 40.12 During (a) cardiac diastole, the heart muscle is relaxed and blood flows into the heart. During (b) atrial systole, the atria contract, pushing blood into the ventricles. During (c) atrial diastole, the ventricles contract, forcing blood out of the heart.

The pumping of the heart is a function of the cardiac muscle cells, or cardiomyocytes, that make up the heart muscle. **Cardiomyocytes**, shown in Figure 40.13, are distinctive muscle cells that are striated like skeletal muscle but pump rhythmically and involuntarily like smooth muscle; they are connected by intercalated disks exclusive to cardiac muscle. They are self-stimulated for a period of time and isolated cardiomyocytes will beat if given the correct balance of nutrients and electrolytes.

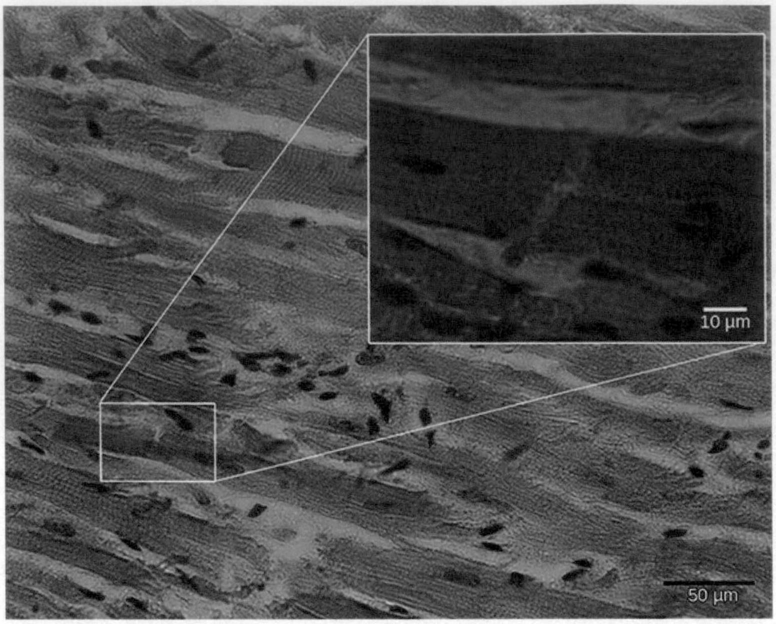

Figure 40.13 Cardiomyocytes are striated muscle cells found in cardiac tissue. (credit: modification of work by Dr. S. Girod, Anton Becker; scale-bar data from Matt Russell)

The autonomous beating of cardiac muscle cells is regulated by the heart's internal pacemaker that uses electrical signals to time the beating of the heart. The electrical signals and mechanical actions, illustrated in Figure 40.14, are intimately intertwined. The internal pacemaker starts at the **sinoatrial (SA) node**, which is located near the wall of the right atrium. Electrical charges spontaneously pulse from the SA node causing the two atria to contract in unison. The pulse reaches a second node, called the atrioventricular (AV) node, between the right atrium and right ventricle where it pauses for approximately 0.1 second before spreading to the walls of the ventricles. From the AV node, the electrical impulse enters the bundle of His, then to the left and right bundle branches extending through the interventricular septum. Finally, the Purkinje fibers conduct the impulse from the apex of the heart up the ventricular myocardium, and then the ventricles contract. This pause allows the atria to empty completely into the ventricles before the ventricles pump out the blood. The electrical impulses in the heart produce electrical currents that flow through the body and can be measured on the skin using electrodes. This information can be observed as an **electrocardiogram (ECG)**—a recording of the electrical impulses of the cardiac muscle.

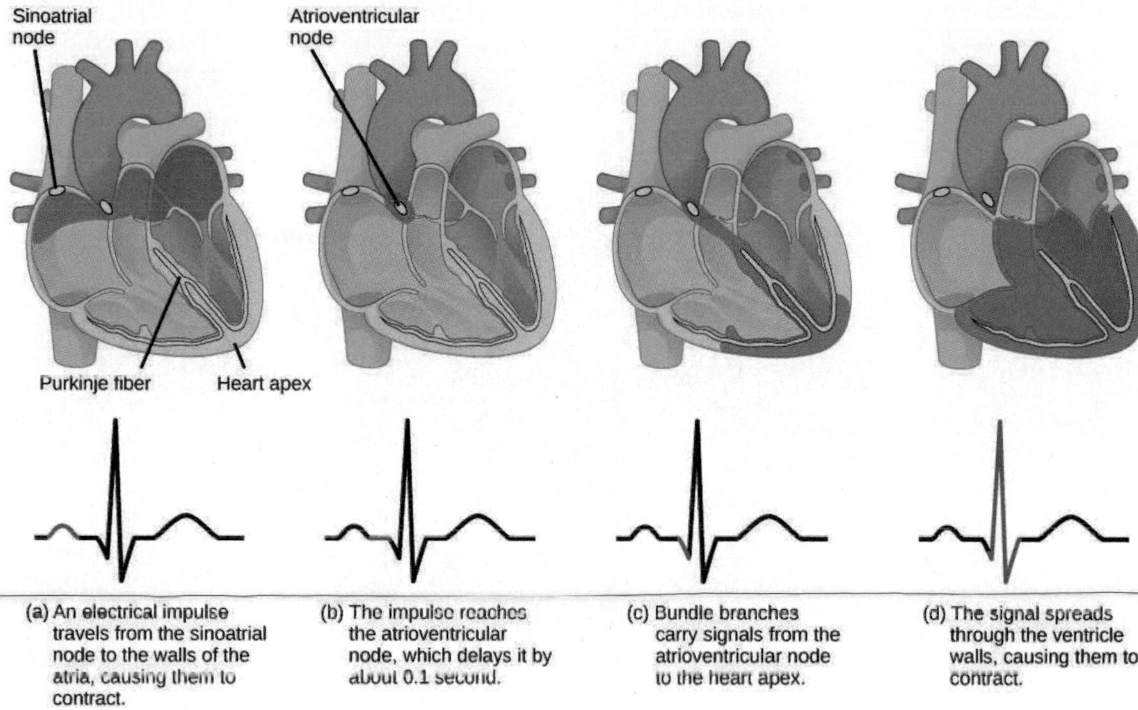

Figure 40.14 The beating of the heart is regulated by an electrical impulse that causes the characteristic reading of an ECG. The signal is initiated at the sinoatrial valve. The signal then (a) spreads to the atria, causing them to contract. The signal is (b) delayed at the atrioventricular node before it is passed on to the (c) heart apex. The delay allows the atria to relax before the (d) ventricles contract. The final part of the ECG cycle prepares the heart for the next beat.

LINK TO LEARNING

Visit this site (http://openstax.org/l/electric_heart) to see the heart's "pacemaker" in action.

Arteries, Veins, and Capillaries

The blood from the heart is carried through the body by a complex network of blood vessels (Figure 40.15). **Arteries** take blood away from the heart. The main artery is the aorta that branches into major arteries that take blood to different limbs and organs. These major arteries include the carotid artery that takes blood to the brain, the brachial arteries that take blood to the arms, and the thoracic artery that takes blood to the thorax and then into the hepatic, renal, and gastric arteries for the liver, kidney, and stomach, respectively. The iliac artery takes blood to the lower limbs. The major arteries diverge into minor arteries, and then smaller vessels called **arterioles**, to reach more deeply into the muscles and organs of the body.

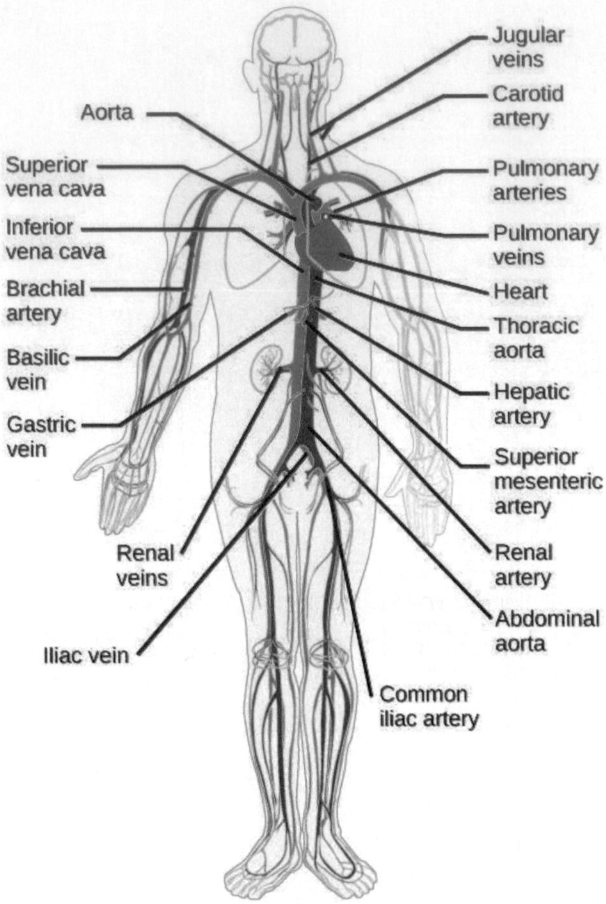

Figure 40.15 The major human arteries and veins are shown. (credit: modification of work by Mariana Ruiz Villareal)

Arterioles diverge into capillary beds. **Capillary beds** contain a large number (10 to 100) of **capillaries** that branch among the cells and tissues of the body. Capillaries are narrow-diameter tubes that can fit red blood cells through in single file and are the sites for the exchange of nutrients, waste, and oxygen with tissues at the cellular level. Fluid also crosses into the interstitial space from the capillaries. The capillaries converge again into **venules** that connect to minor veins that finally connect to major veins that take blood high in carbon dioxide back to the heart. **Veins** are blood vessels that bring blood back to the heart. The major veins drain blood from the same organs and limbs that the major arteries supply. Fluid is also brought back to the heart via the lymphatic system.

The structure of the different types of blood vessels reflects their function or layers. There are three distinct layers, or tunics, that form the walls of blood vessels (Figure 40.16). The first tunic is a smooth, inner lining of endothelial cells that are in contact with the red blood cells. The endothelial tunic is continuous with the endocardium of the heart. In capillaries, this single layer of cells is the location of diffusion of oxygen and carbon dioxide between the endothelial cells and red blood cells, as well as the exchange site via endocytosis and exocytosis. The movement of materials at the site of capillaries is regulated by **vasoconstriction**, narrowing of the blood vessels, and **vasodilation**, widening of the blood vessels; this is important in the overall regulation of blood pressure.

Veins and arteries both have two further tunics that surround the endothelium: the middle tunic is composed of smooth muscle and the outermost layer is connective tissue (collagen and elastic fibers). The elastic connective tissue stretches and supports the blood vessels, and the smooth muscle layer helps regulate blood flow by altering vascular resistance through vasoconstriction and vasodilation. The arteries have thicker smooth muscle and connective tissue than the veins to accommodate the higher pressure and speed of freshly pumped blood. The veins are thinner walled as the pressure and rate of flow are much lower. In addition, veins are structurally different than arteries in that veins have valves to prevent the backflow of blood. Because veins have to work against gravity to get blood back to the heart, contraction of skeletal muscle assists with the flow of blood back to the heart.

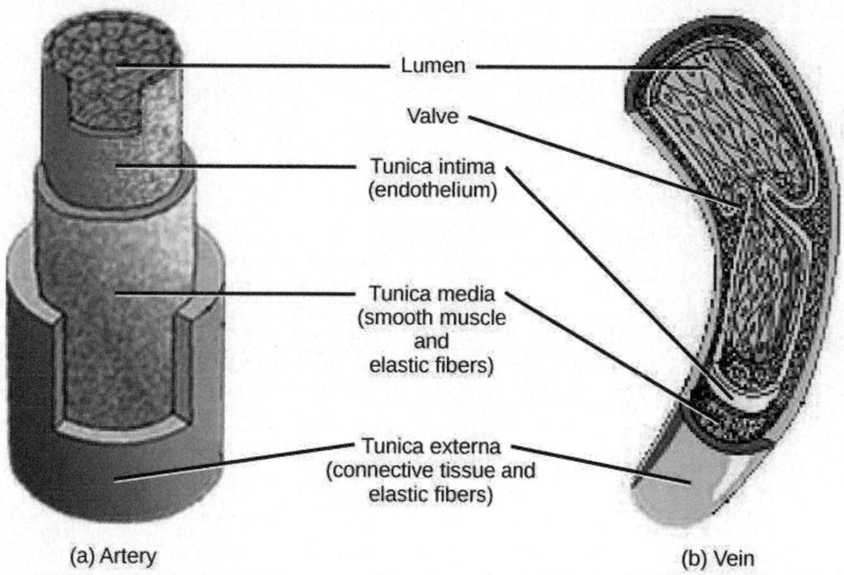

Figure 40.16 Arteries and veins consist of three layers: an outer tunica externa, a middle tunica media, and an inner tunica intima. Capillaries consist of a single layer of epithelial cells, the tunica intima. (credit: modification of work by NCI, NIH)

40.4 Blood Flow and Blood Pressure Regulation

By the end of this section, you will be able to do the following:

- Describe the system of blood flow through the body
- Describe how blood pressure is regulated

Blood pressure (BP) is the pressure exerted by blood on the walls of a blood vessel that helps to push blood through the body. Systolic blood pressure measures the amount of pressure that blood exerts on vessels while the heart is beating. The optimal systolic blood pressure is 120 mmHg. Diastolic blood pressure measures the pressure in the vessels between heartbeats. The optimal diastolic blood pressure is 80 mmHg. Many factors can affect blood pressure, such as hormones, stress, exercise, eating, sitting, and standing. Blood flow through the body is regulated by the size of blood vessels, by the action of smooth muscle, by one-way valves, and by the fluid pressure of the blood itself.

How Blood Flows Through the Body

Blood is pushed through the body by the action of the pumping heart. With each rhythmic pump, blood is pushed under high pressure and velocity away from the heart, initially along the main artery, the aorta. In the aorta, the blood travels at 30 cm/sec. As blood moves into the arteries, arterioles, and ultimately to the capillary beds, the rate of movement slows dramatically to about 0.026 cm/sec, one-thousand times slower than the rate of movement in the aorta. While the diameter of each individual arteriole and capillary is far narrower than the diameter of the aorta, and according to the law of continuity, fluid should travel faster through a narrower diameter tube, the rate is actually slower due to the overall diameter of all the combined capillaries being far greater than the diameter of the individual aorta.

The slow rate of travel through the capillary beds, which reach almost every cell in the body, assists with gas and nutrient exchange and also promotes the diffusion of fluid into the interstitial space. After the blood has passed through the capillary beds to the venules, veins, and finally to the main venae cavae, the rate of flow increases again but is still much slower than the initial rate in the aorta. Blood primarily moves in the veins by the rhythmic movement of smooth muscle in the vessel wall and by the action of the skeletal muscle as the body moves. Because most veins must move blood against the pull of gravity, blood is prevented from flowing backward in the veins by one-way valves. Because skeletal muscle contraction aids in venous blood flow, it is important to get up and move frequently after long periods of sitting so that blood will not pool in the extremities.

Blood flow through the capillary beds is regulated depending on the body's needs and is directed by nerve and hormone signals. For example, after a large meal, most of the blood is diverted to the stomach by vasodilation of vessels of the digestive system and vasoconstriction of other vessels. During exercise, blood is diverted to the skeletal muscles through vasodilation while blood to the digestive system would be lessened through vasoconstriction. The blood entering some capillary beds is controlled by small muscles, called precapillary sphincters, illustrated in Figure 40.17. If the sphincters are open, the blood will flow into the associated branches of the capillary blood. If all of the sphincters are closed, then the blood will flow directly from the arteriole to the venule through the thoroughfare channel (see Figure 40.17). These muscles allow the body to precisely control when capillary beds receive blood flow. At any given moment only about 5–10% of our capillary beds actually have blood flowing through them.

VISUAL CONNECTION

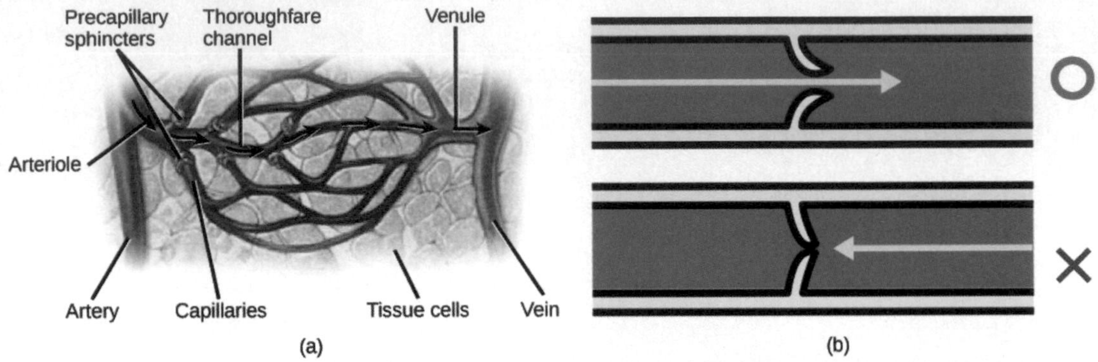

Figure 40.17 (a) Precapillary sphincters are rings of smooth muscle that regulate the flow of blood through capillaries; they help control the location of blood flow to where it is needed. (b) Valves in the veins prevent blood from moving backward. (credit a: modification of work by NCI)

Varicose veins are veins that become enlarged because the valves no longer close properly, allowing blood to flow backward. Varicose veins are often most prominent on the legs. Why do you think this is the case?

LINK TO LEARNING

See the circulatory system's blood flow.

Click to view content (https://www.openstax.org/l/circulation)

Proteins and other large solutes cannot leave the capillaries. The loss of the watery plasma creates a hyperosmotic solution within the capillaries, especially near the venules. This causes about 85% of the plasma that leaves the capillaries to eventually diffuse back into the capillaries near the venules. The remaining 15% of blood plasma drains out from the interstitial fluid into nearby lymphatic vessels (Figure 40.18). The fluid in the lymph is similar in composition to the interstitial fluid. The lymph fluid passes through lymph nodes before it returns to the heart via the vena cava. **Lymph nodes** are specialized organs that filter the lymph by percolation through a maze of connective tissue filled with white blood cells. The white blood cells remove infectious agents, such as bacteria and viruses, to clean the lymph before it returns to the bloodstream. After it is cleaned, the lymph returns to the heart by the action of smooth muscle pumping, skeletal muscle action, and one-way valves joining the returning blood near the junction of the venae cavae entering the right atrium of the heart.

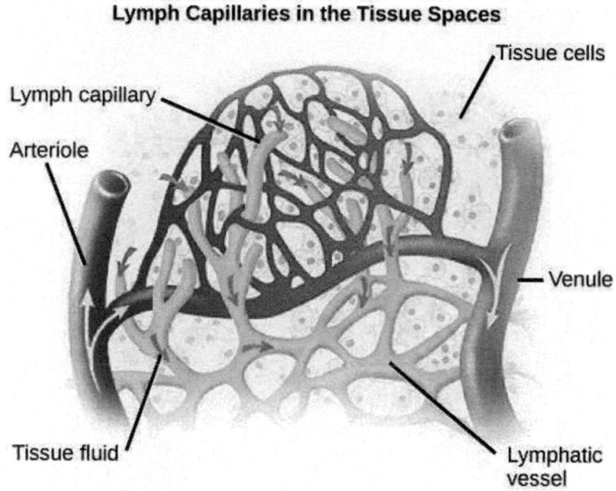

Figure 40.18 Fluid from the capillaries moves into the interstitial space and lymph capillaries by diffusion down a pressure gradient and also by osmosis. Out of 7,200 liters of fluid pumped by the average heart in a day, over 1,500 liters is filtered. (credit: modification of work by NCI, NIH)

EVOLUTION CONNECTION

Vertebrate Diversity in Blood Circulation

Blood circulation has evolved differently in vertebrates and may show variation in different animals for the required amount of pressure, organ and vessel location, and organ size. Animals with longs necks and those that live in cold environments have distinct blood pressure adaptations.

Long necked animals, such as giraffes, need to pump blood upward from the heart against gravity. The blood pressure required from the pumping of the left ventricle would be equivalent to 250 mm Hg (mm Hg = millimeters of mercury, a unit of pressure) to reach the height of a giraffe's head, which is 2.5 meters higher than the heart. However, if checks and balances were not in place, this blood pressure would damage the giraffe's brain, particularly if it was bending down to drink. These checks and balances include valves and feedback mechanisms that reduce the rate of cardiac output. Long-necked dinosaurs such as the sauropods had to pump blood even higher, up to ten meters above the heart. This would have required a blood pressure of more than 600 mm Hg, which could only have been achieved by an enormous heart. Evidence for such an enormous heart does not exist and mechanisms to reduce the blood pressure required include the slowing of metabolism as these animals grew larger. It is likely that they did not routinely feed on tree tops but grazed on the ground.

Living in cold water, whales need to maintain the temperature in their blood. This is achieved by the veins and arteries being close together so that heat exchange can occur. This mechanism is called a countercurrent heat exchanger. The blood vessels and the whole body are also protected by thick layers of blubber to prevent heat loss. In land animals that live in cold environments, thick fur and hibernation are used to retain heat and slow metabolism.

Blood Pressure

The pressure of the blood flow in the body is produced by the hydrostatic pressure of the fluid (blood) against the walls of the blood vessels. Fluid will move from areas of high to low hydrostatic pressures. In the arteries, the hydrostatic pressure near the heart is very high and blood flows to the arterioles where the rate of flow is slowed by the narrow openings of the arterioles. During systole, when new blood is entering the arteries, the artery walls stretch to accommodate the increase of pressure of the extra blood; during diastole, the walls return to normal because of their elastic properties. The blood pressure of the systole phase and the diastole phase, graphed in Figure 40.19, gives the two pressure readings for blood pressure. For example, 120/80 indicates a reading of 120 mm Hg during the systole and 80 mm Hg during diastole. Throughout the cardiac cycle, the blood continues to empty into the arterioles at a relatively even rate. This resistance to blood flow is called **peripheral resistance**.

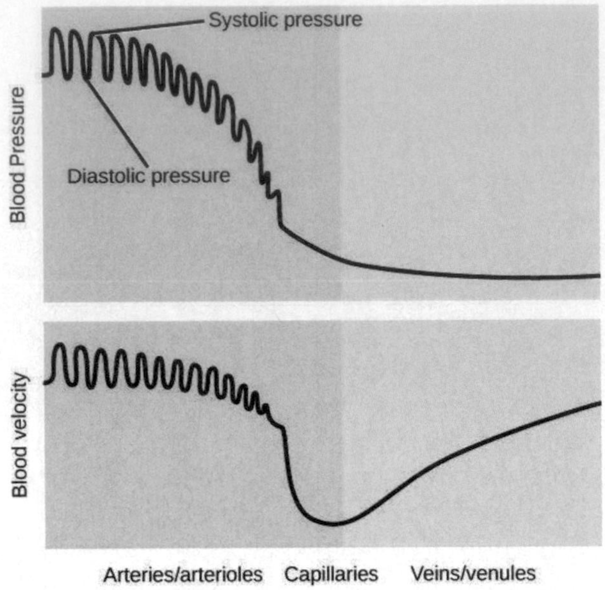

Figure 40.19 Blood pressure is related to the blood velocity in the arteries and arterioles. In the capillaries and veins, the blood pressure continues to decrease but velocity increases.

Blood Pressure Regulation

Cardiac output is the volume of blood pumped by the heart in one minute. It is calculated by multiplying the number of heart contractions that occur per minute (heart rate) times the **stroke volume** (the volume of blood pumped into the aorta per contraction of the left ventricle). Therefore, cardiac output can be increased by increasing heart rate, as when exercising. However, cardiac output can also be increased by increasing stroke volume, such as if the heart contracts with greater strength. Stroke volume can also be increased by speeding blood circulation through the body so that more blood enters the heart between contractions. During heavy exertion, the blood vessels relax and increase in diameter, offsetting the increased heart rate and ensuring adequate oxygenated blood gets to the muscles. Stress triggers a decrease in the diameter of the blood vessels, consequently increasing blood pressure. These changes can also be caused by nerve signals or hormones, and even standing up or lying down can have a great effect on blood pressure.

KEY TERMS

angina pain caused by partial blockage of the coronary arteries by the buildup of plaque and lack of oxygen to the heart muscle

aorta major artery of the body that takes blood away from the heart

arteriole small vessel that connects an artery to a capillary bed

artery blood vessel that takes blood away from the heart

atherosclerosis buildup of fatty plaques in the coronary arteries in the heart

atrioventricular valve one-way membranous flap of connective tissue between the atrium and the ventricle in the right side of the heart; also known as tricuspid valve

atrium (plural: atria) chamber of the heart that receives blood from the veins and sends blood to the ventricles

bicuspid valve (also, mitral valve; left atrioventricular valve) one-way membranous flap between the atrium and the ventricle in the left side of the heart

blood pressure (BP) pressure of blood in the arteries that helps to push blood through the body

capillary smallest blood vessel that allows the passage of individual blood cells and the site of diffusion of oxygen and nutrient exchange

capillary bed large number of capillaries that converge to take blood to a particular organ or tissue

cardiac cycle filling and emptying the heart of blood by electrical signals that cause the heart muscles to contract and relax

cardiac output the volume of blood pumped by the heart in one minute as a product of heart rate multiplied by stroke volume

cardiomyocyte specialized heart muscle cell that is striated but contracts involuntarily like smooth muscle

closed circulatory system system in which the blood is separated from the bodily interstitial fluid and contained in blood vessels

coronary artery vessel that supplies the heart tissue with blood

coronary vein vessel that takes blood away from the heart tissue back to the chambers in the heart

diastole relaxation phase of the cardiac cycle when the heart is relaxed and the ventricles are filling with blood

double circulation flow of blood in two circuits: the pulmonary circuit through the lungs and the systemic circuit through the organs and body

electrocardiogram (ECG) recording of the electrical impulses of the cardiac muscle

endocardium innermost layer of tissue in the heart

epicardium outermost tissue layer of the heart

gill circulation circulatory system that is specific to animals with gills for gas exchange; the blood flows through the gills for oxygenation

hemocoel cavity into which blood is pumped in an open circulatory system

hemolymph mixture of blood and interstitial fluid that is found in insects and other arthropods as well as most mollusks

inferior vena cava drains blood from the veins that come from the lower organs and the legs

interstitial fluid fluid between cells

lymph node specialized organ that contains a large number of macrophages that clean the lymph before the fluid is returned to the heart

myocardial infarction (also, heart attack) complete blockage of the coronary arteries and death of the cardiac muscle tissue

myocardium heart muscle cells that make up the middle layer and the bulk of the heart wall

open circulatory system system in which the blood is mixed with interstitial fluid and directly covers the organs

ostium (plural: ostia) holes between blood vessels that allow the movement of hemolymph through the body of insects, arthropods, and mollusks with open circulatory systems

pericardium membrane layer protecting the heart; also part of the epicardium

peripheral resistance resistance of the artery and blood vessel walls to the pressure placed on them by the force of the heart pumping

plasma liquid component of blood that is left after the cells are removed

platelet (also, thrombocyte) small cellular fragment that collects at wounds, cross-reacts with clotting factors, and forms a plug to prevent blood loss

precapillary sphincter small muscle that controls blood circulation in the capillary beds

pulmocutaneous circulation circulatory system in amphibians; the flow of blood to the lungs and the moist skin for gas exchange

pulmonary circulation flow of blood away from the heart through the lungs where oxygenation occurs and then returns to the heart again

red blood cell small (7–8 µm) biconcave cell without mitochondria (and in mammals without nuclei) that is packed with hemoglobin, giving the cell its red color; transports oxygen through the body

semilunar valve membranous flap of connective tissue between the aorta and a ventricle of the heart (the aortic or pulmonary semilunar valves)

serum plasma without the coagulation factors

sinoatrial (SA) node the heart's internal pacemaker; located near the wall of the right atrium

stroke volume the volume of blood pumped into the aorta

per contraction of the left ventricle

superior vena cava drains blood from the jugular vein that comes from the brain and from the veins that come from the arms

systemic circulation flow of blood away from the heart to the brain, liver, kidneys, stomach, and other organs, the limbs, and the muscles of the body, and then the return of this blood to the heart

systole contraction phase of cardiac cycle when the ventricles are pumping blood into the arteries

tricuspid valve one-way membranous flap of connective tissue between the atrium and the ventricle in the right side of the heart; also known as atrioventricular valve

unidirectional circulation flow of blood in a single circuit; occurs in fish where the blood flows through the gills, then past the organs and the rest of the body, before returning to the heart

vasoconstriction narrowing of a blood vessel

vasodilation widening of a blood vessel

vein blood vessel that brings blood back to the heart

vena cava major vein of the body returning blood from the upper and lower parts of the body; see the superior vena cava and inferior vena cava

ventricle (heart) large inferior chamber of the heart that pumps blood into arteries

venule blood vessel that connects a capillary bed to a vein

white blood cell large (30 µm) cell with nuclei of which there are many types with different roles including the protection of the body from viruses and bacteria, and cleaning up dead cells and other waste

CHAPTER SUMMARY

40.1 Overview of the Circulatory System

In most animals, the circulatory system is used to transport blood through the body. Some primitive animals use diffusion for the exchange of water, nutrients, and gases. However, complex organisms use the circulatory system to carry gases, nutrients, and waste through the body. Circulatory systems may be open (mixed with the interstitial fluid) or closed (separated from the interstitial fluid). Closed circulatory systems are a characteristic of vertebrates; however, there are significant differences in the structure of the heart and the circulation of blood between the different vertebrate groups due to adaptions during evolution and associated differences in anatomy. Fish have a two-chambered heart with unidirectional circulation. Amphibians have a three-chambered heart, which has some mixing of the blood, and they have double circulation. Most non-avian reptiles have a three-chambered heart, but have little mixing of the blood; they have double circulation. Mammals and birds have a four-chambered heart with no mixing of the blood and double circulation.

40.2 Components of the Blood

Specific components of the blood include red blood cells, white blood cells, platelets, and the plasma, which contains coagulation factors and serum. Blood is important for regulation of the body's pH, temperature, osmotic pressure, the circulation of nutrients and removal of waste, the distribution of hormones from endocrine glands, and the elimination of excess heat; it also contains components for blood clotting. Red blood cells are specialized cells that contain hemoglobin and circulate through the body delivering oxygen to cells. White blood cells are involved in the immune response to identify and target invading bacteria, viruses, and other foreign organisms; they also recycle waste components, such as old red blood cells. Platelets and blood clotting factors cause the change of the soluble protein fibrinogen to the insoluble protein fibrin at a wound site forming a plug. Plasma consists of 90 percent water along with various substances, such as coagulation factors and antibodies. The serum is the plasma component of the blood without the coagulation factors.

40.3 Mammalian Heart and Blood Vessels

The heart muscle pumps blood through three divisions of the circulatory system: coronary, pulmonary, and systemic. There is one atrium and one ventricle on the right side and one atrium and one ventricle on the left side. The pumping of the heart is a function of cardiomyocytes, distinctive muscle cells that are striated like skeletal muscle but pump rhythmically and involuntarily like smooth muscle. The internal pacemaker starts at the sinoatrial node, which is located near the wall of the right atrium. Electrical charges pulse from the SA node causing the two atria to contract in unison; then the pulse reaches the atrioventricular node between the right atrium and right ventricle. A pause in the electric signal allows the atria to empty completely into the ventricles before the ventricles pump out the blood. The blood from the heart is carried through the body by a complex network of blood vessels; arteries take blood away from the heart, and veins bring blood back to the heart.

40.4 Blood Flow and Blood Pressure Regulation

Blood primarily moves through the body by the rhythmic movement of smooth muscle in the vessel wall and by the action of the skeletal muscle as the body moves. Blood is prevented from flowing backward in the veins by one-way valves. Blood flow through the capillary beds is controlled by precapillary sphincters to increase and decrease flow depending on the body's needs and is directed by nerve and hormone signals. Lymph vessels take fluid that has leaked out of the blood to the lymph nodes where it is cleaned before returning to the heart. During systole, blood enters the arteries, and the artery walls stretch to accommodate the extra blood. During diastole, the artery walls return to normal. The blood pressure of the systole phase and the diastole phase gives the two pressure readings for blood pressure.

VISUAL CONNECTION QUESTIONS

1. Figure 40.10 Which of the following statements about the circulatory system is false?
   a. Blood in the pulmonary vein is deoxygenated.
   b. Blood in the inferior vena cava is deoxygenated.
   c. Blood in the pulmonary artery is deoxygenated.
   d. Blood in the aorta is oxygenated.

2. Figure 40.11 Which of the following statements about the heart is false?
   a. The mitral valve separates the left ventricle from the left atrium.
   b. Blood travels through the bicuspid valve to the left atrium.
   c. Both the aortic and the pulmonary valves are semilunar valves.
   d. The mitral valve is an atrioventricular valve.

3. Figure 40.17 Varicose veins are veins that become enlarged because the valves no longer close properly, allowing blood to flow backward. Varicose veins are often most prominent on the legs. Why do you think this is the case?

REVIEW QUESTIONS

4. Why are open circulatory systems advantageous to some animals?
   a. They use less metabolic energy.
   b. They help the animal move faster.
   c. They do not need a heart.
   d. They help large insects develop.

5. Some animals use diffusion instead of a circulatory system. Examples include:
   a. birds and jellyfish
   b. flatworms and arthropods
   c. mollusks and jellyfish
   d. none of the above

6. Blood flow that is directed through the lungs and back to the heart is called _______.
   a. unidirectional circulation
   b. gill circulation
   c. pulmonary circulation
   d. pulmocutaneous circulation

7. White blood cells:
   a. can be classified as granulocytes or agranulocytes
   b. defend the body against bacteria and viruses
   c. are also called leucocytes
   d. all of the above

8. Platelet plug formation occurs at which point?
   a. when large megakaryocytes break up into thousands of smaller fragments
   b. when platelets are dispersed through the bloodstream
   c. when platelets are attracted to a site of blood vessel damage
   d. none of the above

9. In humans, the plasma comprises what percentage of the blood?
   a. 45 percent
   b. 55 percent
   c. 25 percent
   d. 90 percent

10. The red blood cells of birds differ from mammalian red blood cells because:
    a. they are white and have nuclei
    b. they do not have nuclei
    c. they have nuclei
    d. they fight disease

11. The heart's internal pacemaker beats by:
    a. an internal implant that sends an electrical impulse through the heart
    b. the excitation of cardiac muscle cells at the sinoatrial node followed by the atrioventricular node
    c. the excitation of cardiac muscle cells at the atrioventricular node followed by the sinoatrial node
    d. the action of the sinus

12. During the systolic phase of the cardiac cycle, the heart is _______.
    a. contracting
    b. relaxing
    c. contracting and relaxing
    d. filling with blood

13. Cardiomyocytes are similar to skeletal muscle because:
    a. they beat involuntarily
    b. they are used for weight lifting
    c. they pulse rhythmically
    d. they are striated

14. How do arteries differ from veins?
    a. Arteries have thicker smooth muscle layers to accommodate the changes in pressure from the heart.
    b. Arteries carry blood.
    c. Arteries have thinner smooth muscle layers and valves and move blood by the action of skeletal muscle.
    d. Arteries are thin walled and are used for gas exchange.

15. High blood pressure would be a result of _______.
    a. a high cardiac output and high peripheral resistance
    b. a high cardiac output and low peripheral resistance
    c. a low cardiac output and high peripheral resistance
    d. a low cardiac output and low peripheral resistance

CRITICAL THINKING QUESTIONS

16. Describe a closed circulatory system.
17. Describe systemic circulation.
18. Describe the cause of different blood type groups.
19. List some of the functions of blood in the body.
20. How does the lymphatic system work with blood flow?
21. Describe the cardiac cycle.
22. What happens in capillaries?
23. How does blood pressure change during heavy exercise?

CHAPTER 41
Osmotic Regulation and Excretion

Figure 41.1 Just as humans recycle what we can and dump the remains into landfills, our bodies use and recycle what they can and excrete the remaining waste products. Our bodies' complex systems have developed ways to treat waste and maintain a balanced internal environment. (credit: modification of work by Redwin Law)

INTRODUCTION The daily intake recommendation for human water consumption is eight to ten glasses of water. In order to achieve a healthy balance, the human body should excrete the eight to ten glasses of water every day. This occurs via the processes of urination, defecation, sweating and, to a small extent, respiration. The organs and tissues of the human body are soaked in fluids that are maintained at constant temperature, pH, and solute concentration, all crucial elements of homeostasis. The solutes in body fluids are mainly mineral salts and sugars, and osmotic regulation is the process by which the mineral salts and water are kept in balance. Osmotic homeostasis is maintained despite the influence of external factors like temperature, diet, and weather conditions.

Chapter Outline

41.1 Osmoregulation and Osmotic Balance

By the end of this section, you will be able to do the following:

- Define osmosis and explain its role within molecules
- Explain why osmoregulation and osmotic balance are important body functions
- Describe active transport mechanisms
- Explain osmolarity and the way in which it is measured
- Describe osmoregulators or osmoconformers and how these tools allow animals to adapt to different environments

Osmosis is the diffusion of water across a membrane in response to **osmotic pressure** caused by an imbalance of molecules on either side of the membrane. **Osmoregulation** is the process of maintenance of salt and water balance (**osmotic balance**) across membranes within the body's fluids, which are composed of water, plus electrolytes and non-electrolytes. An **electrolyte** is a solute that dissociates into ions when dissolved in water. A **non-electrolyte**, in contrast, doesn't dissociate into ions during water dissolution. Both electrolytes and non-electrolytes contribute to the osmotic balance. The body's fluids include blood plasma, the cytosol within cells, and interstitial fluid, the fluid that exists in the spaces between cells and tissues of the body. The membranes of the body (such as the pleural, serous, and cell membranes) are **semi-permeable membranes**. Semi-permeable membranes are permeable (or permissive) to certain types of solutes and water. Solutions on two sides of a semi-permeable membrane tend to equalize in solute concentration by movement of solutes and/or water across the membrane. As seen in Figure 41.2, a cell placed in water tends to swell due to gain of water from the hypotonic or "low salt" environment. A cell placed in a solution with higher salt concentration, on the other hand, tends to make the membrane shrivel up due to loss of water into the hypertonic or "high salt" environment. Isotonic cells have an equal concentration of solutes inside and outside the cell; this equalizes the osmotic pressure on either side of the cell membrane which is a semi-permeable membrane.

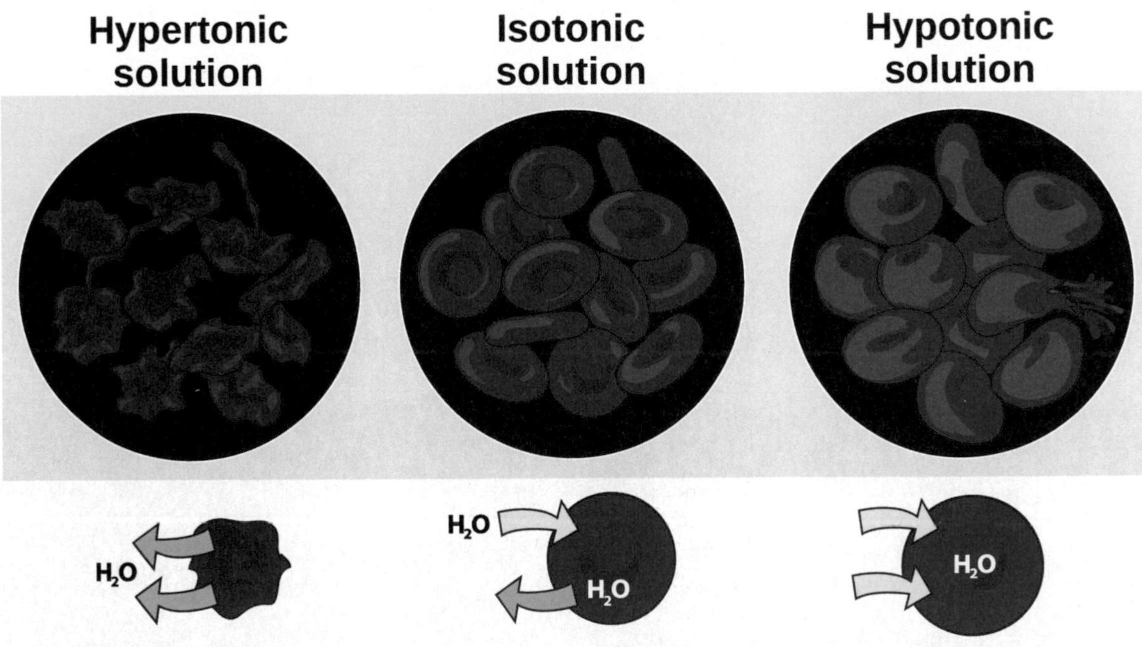

Figure 41.2 Cells placed in a hypertonic environment tend to shrink due to loss of water. In a hypotonic environment, cells tend to swell due to intake of water. The blood maintains an isotonic environment so that cells neither shrink nor swell. (credit: Mariana Ruiz Villareal)

The body does not exist in isolation. There is a constant input of water and electrolytes into the system. While osmoregulation is achieved across membranes within the body, excess electrolytes and wastes are transported to the kidneys and excreted, helping to maintain osmotic balance.

Need for Osmoregulation

Biological systems constantly interact and exchange water and nutrients with the environment by way of consumption of food and water and through excretion in the form of sweat, urine, and feces. Without a mechanism to regulate osmotic pressure, or when a disease damages this mechanism, there is a tendency to accumulate toxic waste and water, which can have dire consequences.

Mammalian systems have evolved to regulate not only the overall osmotic pressure across membranes, but also specific concentrations of important electrolytes in the three major fluid compartments: blood plasma, extracellular fluid, and intracellular fluid. Since osmotic pressure is regulated by the movement of water across membranes, the volume of the fluid compartments can also change temporarily. Because blood plasma is one of the fluid components, osmotic pressures have a direct bearing on blood pressure.

Transport of Electrolytes across Cell Membranes

Electrolytes, such as sodium chloride, ionize in water, meaning that they dissociate into their component ions. In water, sodium chloride (NaCl), dissociates into the sodium ion (Na^+) and the chloride ion (Cl^-). The most important ions, whose concentrations are very closely regulated in body fluids, are the cations sodium (Na^+), potassium (K^+), calcium (Ca^{+2}), magnesium (Mg^{+2}), and the anions chloride (Cl^-), carbonate (CO_3^{-2}), bicarbonate (HCO_3^-), and phosphate(PO_3^-). Electrolytes are lost from the body during urination and perspiration. For this reason, athletes are encouraged to replace electrolytes and fluids during periods of increased activity and perspiration.

Osmotic pressure is influenced by the concentration of solutes in a solution. It is directly proportional to the number of solute atoms or molecules and not dependent on the size of the solute molecules. Because electrolytes dissociate into their component ions, they, in essence, add more solute particles into the solution and have a greater effect on osmotic pressure, per mass than compounds that do not dissociate in water, such as glucose.

Water can pass through membranes by passive diffusion. If electrolyte ions could passively diffuse across membranes, it would be impossible to maintain specific concentrations of ions in each fluid compartment therefore they require special mechanisms to cross the semi-permeable membranes in the body. This movement can be accomplished by facilitated diffusion and active transport. Facilitated diffusion requires protein-based channels for moving the solute. Active transport requires energy in the form of ATP conversion, carrier proteins, or pumps in order to move ions against the concentration gradient.

Concept of Osmolality and Milliequivalent

In order to calculate osmotic pressure, it is necessary to understand how solute concentrations are measured. The unit for measuring solutes is the **mole**. One mole is defined as the gram molecular weight of the solute. For example, the molecular weight of sodium chloride is 58.44. Thus, one mole of sodium chloride weighs 58.44 grams. The **molarity** of a solution is the number of moles of solute per liter of solution. The **molality** of a solution is the number of moles of solute per kilogram of solvent. If the solvent is water, one kilogram of water is equal to one liter of water. While molarity and molality are used to express the concentration of solutions, electrolyte concentrations are usually expressed in terms of milliequivalents per liter (mEq/L): the mEq/L is equal to the ion concentration (in millimoles) multiplied by the number of electrical charges on the ion. The unit of milliequivalent takes into consideration the ions present in the solution (since electrolytes form ions in aqueous solutions) and the charge on the ions.

Thus, for ions that have a charge of one, one milliequivalent is equal to one millimole. For ions that have a charge of two (like calcium), one milliequivalent is equal to 0.5 millimoles. Another unit for the expression of electrolyte concentration is the milliosmole (mOsm), which is the number of milliequivalents of solute per kilogram of solvent. Body fluids are usually maintained within the range of 280 to 300 mOsm.

Osmoregulators and Osmoconformers

Persons lost at sea without any freshwater to drink are at risk of severe dehydration because the human body cannot adapt to drinking seawater, which is hypertonic in comparison to body fluids. Organisms such as goldfish that can tolerate only a relatively narrow range of salinity are referred to as stenohaline. About 90 percent of all bony fish are restricted to either freshwater or seawater. They are incapable of osmotic regulation in the opposite environment. It is possible, however, for a few fishes like salmon to spend part of their life in freshwater and part in seawater. Organisms like the salmon and molly that can tolerate a relatively wide range of salinity are referred to as euryhaline organisms. This is possible because some fish have evolved **osmoregulatory** mechanisms to survive in all kinds of aquatic environments. When they live in freshwater, their bodies tend to take up water because the environment is relatively hypotonic, as illustrated in Figure 41.3a. In such hypotonic environments, these fish do not drink much water. Instead, they pass a lot of very dilute urine, and they achieve electrolyte balance by active transport of salts through the gills. When they move to a hypertonic marine environment, these fish start drinking seawater; they excrete the excess salts through their gills and their urine, as illustrated in Figure 41.3b. Most marine invertebrates, on the other hand, may be isotonic with seawater (**osmoconformers**). Their body fluid concentrations conform to changes in seawater concentration. Cartilaginous fishes' salt composition of the blood is similar to bony fishes; however, the blood of sharks contains the organic compounds urea and trimethylamine oxide (TMAO). This does not mean that their electrolyte composition is similar to that of seawater. They achieve isotonicity with the sea by storing large concentrations of urea. These animals that secrete urea are called ureotelic animals. TMAO stabilizes proteins in the presence of high urea levels, preventing the disruption of peptide bonds that would occur in other animals exposed to similar levels of urea. Sharks are cartilaginous fish with a rectal gland to secrete salt and assist in osmoregulation.

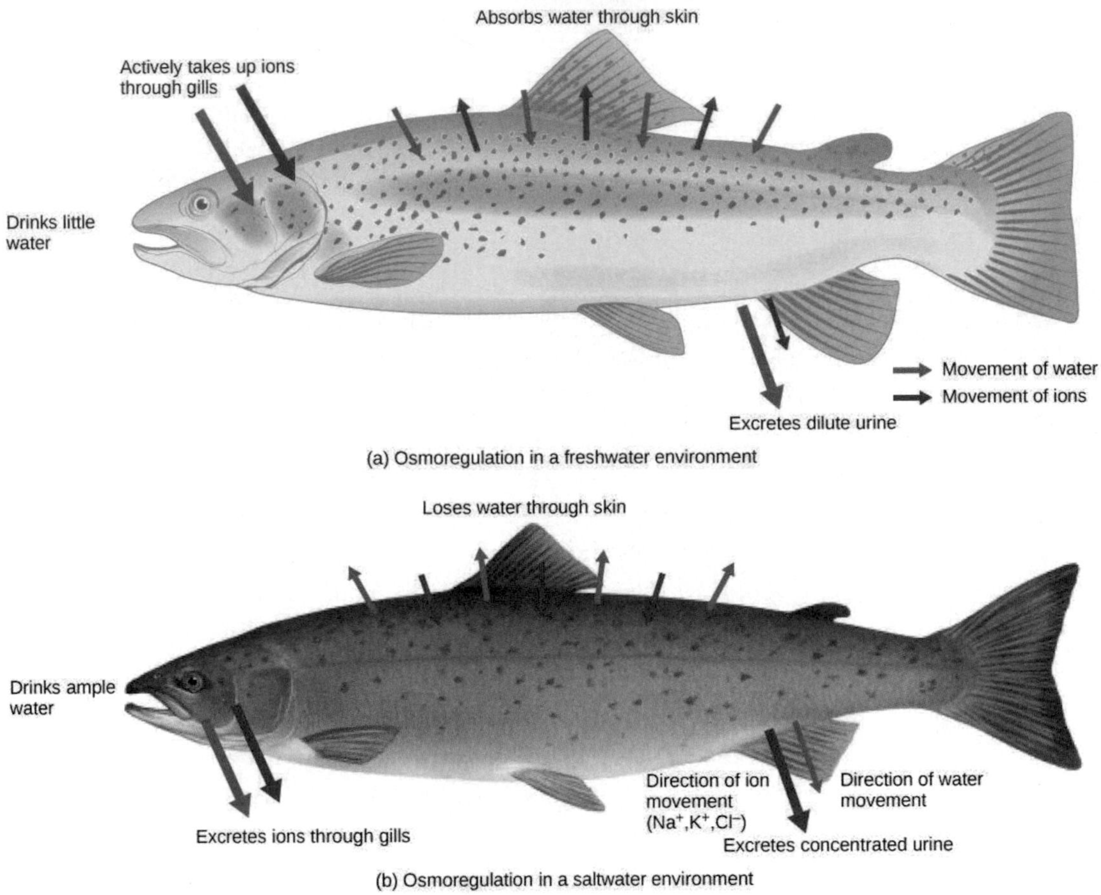

Figure 41.3 Fish are osmoregulators, but must use different mechanisms to survive in (a) freshwater or (b) saltwater environments. (credit: modification of work by Duane Raver, NOAA)

CAREER CONNECTION

Dialysis Technician

Dialysis is a medical process of removing wastes and excess water from the blood by diffusion and ultrafiltration. When kidney function fails, dialysis must be done to artificially rid the body of wastes. This is a vital process to keep patients alive. In some cases, the patients undergo artificial dialysis until they are eligible for a kidney transplant. In others who are not candidates for kidney transplants, dialysis is a life-long necessity.

Dialysis technicians typically work in hospitals and clinics. While some roles in this field include equipment development and maintenance, most dialysis technicians work in direct patient care. Their on-the-job duties, which typically occur under the direct supervision of a registered nurse, focus on providing dialysis treatments. This can include reviewing patient history and current condition, assessing and responding to patient needs before and during treatment, and monitoring the dialysis process. Treatment may include taking and reporting a patient's vital signs and preparing solutions and equipment to ensure accurate and sterile procedures.

41.2 The Kidneys and Osmoregulatory Organs

By the end of this section, you will be able to do the following:

- Explain how the kidneys serve as the main osmoregulatory organs in mammalian systems
- Describe the structure of the kidneys and the functions of the parts of the kidney
- Describe how the nephron is the functional unit of the kidney and explain how it actively filters blood and generates urine
- Detail the three steps in the formation of urine: glomerular filtration, tubular reabsorption, and tubular secretion

Although the kidneys are the major osmoregulatory organ, the skin and lungs also play a role in the process. Water and electrolytes are lost through sweat glands in the skin, which helps moisturize and cool the skin surface, while the lungs expel a small amount of water in the form of mucous secretions and via evaporation of water vapor.

Kidneys: The Main Osmoregulatory Organ

The **kidneys**, illustrated in Figure 41.4, are a pair of bean-shaped structures that are located just below and posterior to the liver in the peritoneal cavity. The adrenal glands sit on top of each kidney and are also called the suprarenal glands. Kidneys filter blood and purify it. All the blood in the human body is filtered many times a day by the kidneys; these organs use up almost 25 percent of the oxygen absorbed through the lungs to perform this function. Oxygen allows the kidney cells to efficiently manufacture chemical energy in the form of ATP through aerobic respiration. The filtrate coming out of the kidneys is called **urine**.

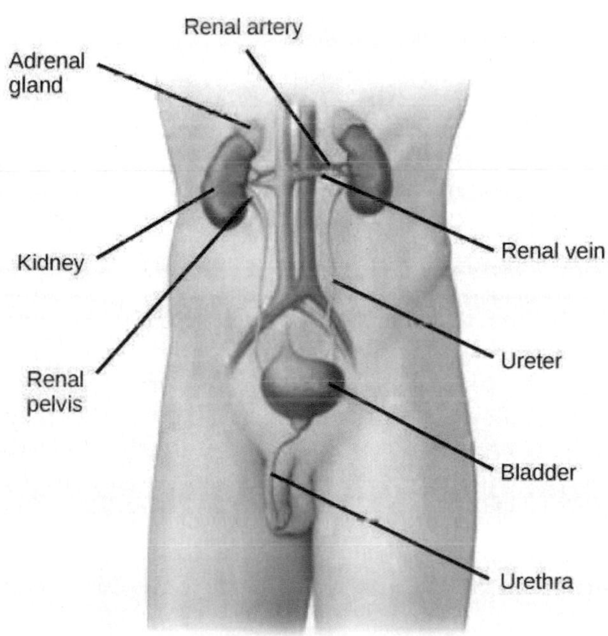

Figure 41.4 Kidneys filter the blood, producing urine that is stored in the bladder prior to elimination through the urethra. (credit: modification of work by NCI)

Kidney Structure

Externally, the kidneys are surrounded by three layers, illustrated in Figure 41.5. The outermost layer is a tough connective tissue layer called the **renal fascia**. The second layer is called the **perirenal fat capsule**, which helps anchor the kidneys in place. The third and innermost layer is the **renal capsule**. Internally, the kidney has three regions—an outer **cortex**, a **medulla** in the middle, and the **renal pelvis** in the region called the **hilum** of the kidney. The hilum is the concave part of the bean-shape where blood vessels and nerves enter and exit the kidney; it is also the point of exit for the ureters. The renal cortex is granular due to the presence of **nephrons**—the functional unit of the kidney. The medulla consists of multiple pyramidal tissue masses, called the **renal pyramids**. In between the pyramids are spaces called **renal columns** through which the blood vessels pass. The tips of the pyramids, called renal papillae, point toward the renal pelvis. There are, on average, eight renal pyramids in each kidney. The renal pyramids along with the adjoining cortical region are called the **lobes of the kidney**. The renal pelvis leads to the **ureter** on the outside of the kidney. On the inside of the kidney, the renal pelvis branches out into two or three extensions called the major **calyces**, which further branch into the minor calyces. The ureters are urine-bearing tubes that exit the kidney and empty into the **urinary bladder**.

VISUAL CONNECTION

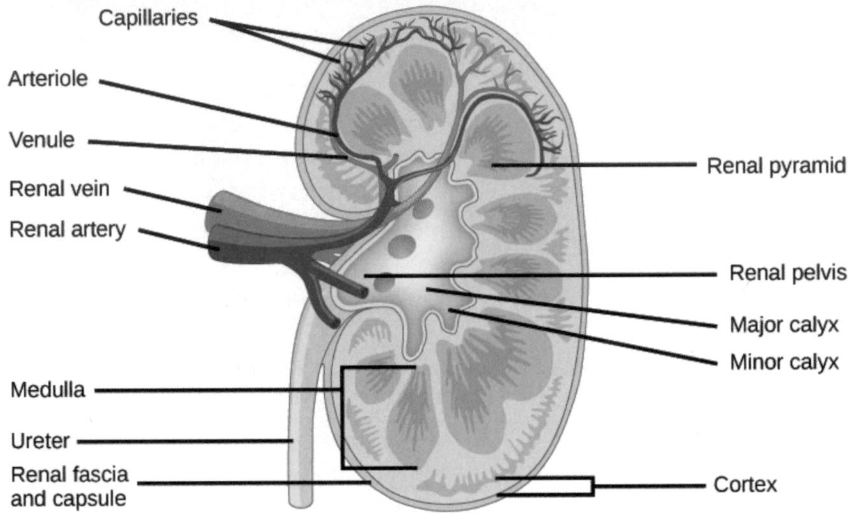

Figure 41.5 The internal structure of the kidney is shown. (credit: modification of work by NCI)

Which of the following statements about the kidney is false?

a. The renal pelvis drains into the ureter.
b. The renal pyramids are in the medulla.
c. The cortex covers the capsule.
d. Nephrons are in the renal cortex.

Because the kidney filters blood, its network of blood vessels is an important component of its structure and function. The arteries, veins, and nerves that supply the kidney enter and exit at the renal hilum. Renal blood supply starts with the branching of the aorta into the **renal arteries** (which are each named based on the region of the kidney they pass through) and ends with the exiting of the **renal veins** to join the **inferior vena cava**. The renal arteries split into several **segmental arteries** upon entering the kidneys. Each segmental artery splits further into several **interlobar arteries** and enters the renal columns, which supply the renal lobes. The interlobar arteries split at the junction of the renal cortex and medulla to form the **arcuate arteries**. The arcuate "bow shaped" arteries form arcs along the base of the medullary pyramids. **Cortical radiate arteries**, as the name suggests, radiate out from the arcuate arteries. The cortical radiate arteries branch into numerous afferent arterioles, and then enter the capillaries supplying the nephrons. Veins trace the path of the arteries and have similar names, except there are no segmental veins.

As mentioned previously, the functional unit of the kidney is the nephron, illustrated in Figure 41.6. Each kidney is made up of over one million nephrons that dot the renal cortex, giving it a granular appearance when sectioned sagittally. There are two types of nephrons—**cortical nephrons** (85 percent), which are deep in the renal cortex, and **juxtamedullary nephrons** (15 percent), which lie in the renal cortex close to the renal medulla. A nephron consists of three parts—a **renal corpuscle**, a **renal tubule**, and the associated capillary network, which originates from the cortical radiate arteries.

VISUAL CONNECTION

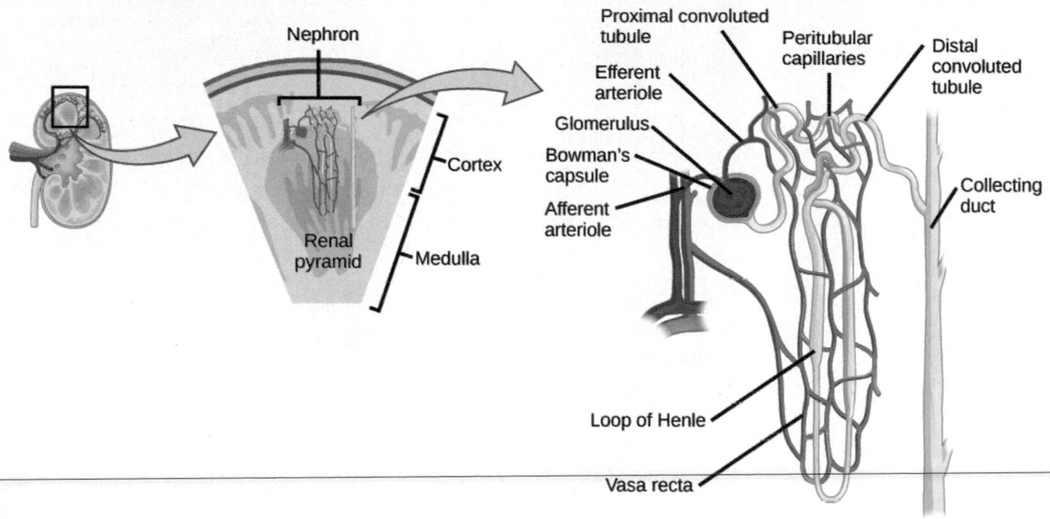

Figure 41.6 The nephron is the functional unit of the kidney. The glomerulus and convoluted tubules are located in the kidney cortex, while collecting ducts are located in the pyramids of the medulla. (credit: modification of work by NIDDK)

Which of the following statements about the nephron is false?

a. The collecting duct empties into the distal convoluted tubule.
b. The Bowman's capsule surrounds the glomerulus.
c. The loop of Henle is between the proximal and distal convoluted tubules.
d. The loop of Henle empties into the distal convoluted tubule.

Renal Corpuscle

The renal corpuscle, located in the renal cortex, is made up of a network of capillaries known as the **glomerulus** and the capsule, a cup-shaped chamber that surrounds it, called the glomerular or **Bowman's capsule**.

Renal Tubule

The renal tubule is a long and convoluted structure that emerges from the glomerulus and can be divided into three parts based on function. The first part is called the **proximal convoluted tubule (PCT)** due to its proximity to the glomerulus; it stays in the renal cortex. The second part is called the **loop of Henle**, or nephritic loop, because it forms a loop (with **descending** and **ascending limbs**) that goes through the renal medulla. The third part of the renal tubule is called the **distal convoluted tubule (DCT)** and this part is also restricted to the renal cortex. The DCT, which is the last part of the nephron, connects and empties its contents into collecting ducts that line the medullary pyramids. The collecting ducts amass contents from multiple nephrons and fuse together as they enter the papillae of the renal medulla.

Capillary Network within the Nephron

The capillary network that originates from the renal arteries supplies the nephron with blood that needs to be filtered. The branch that enters the glomerulus is called the **afferent arteriole**. The branch that exits the glomerulus is called the **efferent arteriole**. Within the glomerulus, the network of capillaries is called the glomerular capillary bed. Once the efferent arteriole exits the glomerulus, it forms the **peritubular capillary network**, which surrounds and interacts with parts of the renal tubule. In cortical nephrons, the peritubular capillary network surrounds the PCT and DCT. In juxtamedullary nephrons, the peritubular capillary network forms a network around the loop of Henle and is called the **vasa recta**.

LINK TO LEARNING

Go to this website (http://openstax.org/l/kidney_section) to see another coronal section of the kidney and to explore an animation of the workings of nephrons.

Kidney Function and Physiology

Kidneys filter blood in a three-step process. First, the nephrons filter blood that runs through the capillary network in the glomerulus. Almost all solutes, except for proteins, are filtered out into the glomerulus by a process called **glomerular filtration**. Second, the filtrate is collected in the renal tubules. Most of the solutes get reabsorbed in the PCT by a process called **tubular reabsorption**. In the loop of Henle, the filtrate continues to exchange solutes and water with the renal medulla and the peritubular capillary network. Water is also reabsorbed during this step. Then, additional solutes and wastes are secreted into the kidney tubules during **tubular secretion**, which is, in essence, the opposite process to tubular reabsorption. The collecting ducts collect filtrate coming from the nephrons and fuse in the medullary papillae. From here, the papillae deliver the filtrate, now called urine, into the minor calyces that eventually connect to the ureters through the renal pelvis. This entire process is illustrated in Figure 41.7.

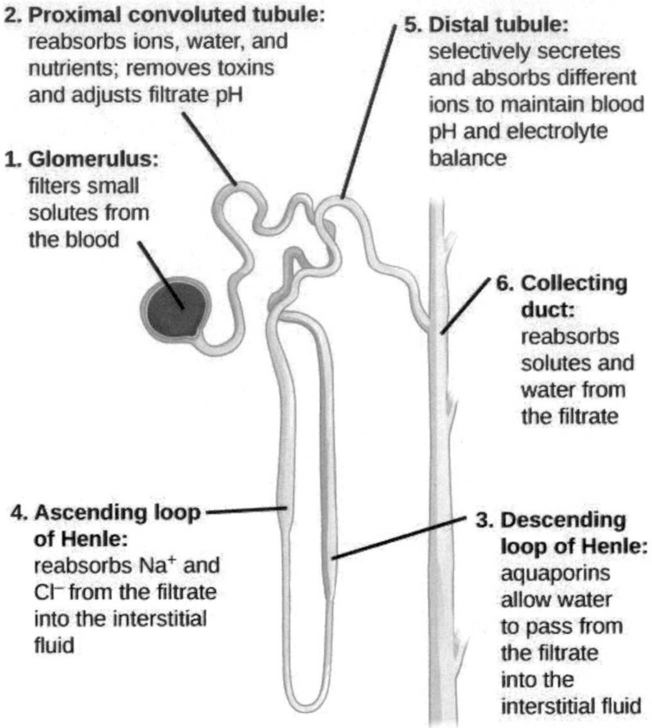

Figure 41.7 Each part of the nephron performs a different function in filtering waste and maintaining homeostatic balance. (1) The glomerulus forces small solutes out of the blood by pressure. (2) The proximal convoluted tubule reabsorbs ions, water, and nutrients from the filtrate into the interstitial fluid, and actively transports toxins and drugs from the interstitial fluid into the filtrate. The proximal convoluted tubule also adjusts blood pH by selectively secreting ammonia (NH_3) into the filtrate, where it reacts with H^+ to form NH_4^+. The more acidic the filtrate, the more ammonia is secreted. (3) The descending loop of Henle is lined with cells containing aquaporins that allow water to pass from the filtrate into the interstitial fluid. (4) In the thin part of the ascending loop of Henle, Na^+ and Cl^- ions diffuse into the interstitial fluid. In the thick part, these same ions are actively transported into the interstitial fluid. Because salt but not water is lost, the filtrate becomes more dilute as it travels up the limb. (5) In the distal convoluted tubule, K^+ and H^+ ions are selectively secreted into the filtrate, while Na^+, Cl^-, and HCO_3^- ions are reabsorbed to maintain pH and electrolyte balance in the blood. (6) The collecting duct reabsorbs solutes and water from the filtrate, forming dilute urine. (credit: modification of work by NIDDK)

Glomerular Filtration

Glomerular filtration filters out most of the solutes due to high blood pressure and specialized membranes in the afferent arteriole. The blood pressure in the glomerulus is maintained independent of factors that affect systemic blood pressure. The "leaky" connections between the endothelial cells of the glomerular capillary network allow solutes to pass through easily. All solutes in the glomerular capillaries, except for macromolecules like proteins, pass through by passive diffusion. There is no energy requirement at this stage of the filtration process. **Glomerular filtration rate (GFR)** is the volume of glomerular filtrate formed per minute by the kidneys. GFR is regulated by multiple mechanisms and is an important indicator of kidney function.

LINK TO LEARNING

To learn more about the vascular system of kidneys, click through this review (http://openstax.org/l/kidneys) and the steps of blood flow.

Tubular Reabsorption and Secretion

Tubular reabsorption occurs in the PCT part of the renal tubule. Almost all nutrients are reabsorbed, and this occurs either by passive or active transport. Reabsorption of water and some key electrolytes are regulated and can be influenced by hormones. Sodium (Na^+) is the most abundant ion and most of it is reabsorbed by active transport and then transported to the peritubular capillaries. Because Na^+ is actively transported out of the tubule, water follows it to even out the osmotic pressure. Water is also independently reabsorbed into the peritubular capillaries due to the presence of aquaporins, or water channels, in the PCT. This occurs due to the low blood pressure and high osmotic pressure in the peritubular capillaries. However, every solute has a **transport maximum** and the excess is not reabsorbed.

In the loop of Henle, the permeability of the membrane changes. The descending limb is permeable to water, not solutes; the opposite is true for the ascending limb. Additionally, the loop of Henle invades the renal medulla, which is naturally high in salt concentration and tends to absorb water from the renal tubule and concentrate the filtrate. The osmotic gradient increases as it moves deeper into the medulla. Because two sides of the loop of Henle perform opposing functions, as illustrated in Figure 41.8, it acts as a **countercurrent multiplier**. The vasa recta around it acts as the **countercurrent exchanger**.

VISUAL CONNECTION

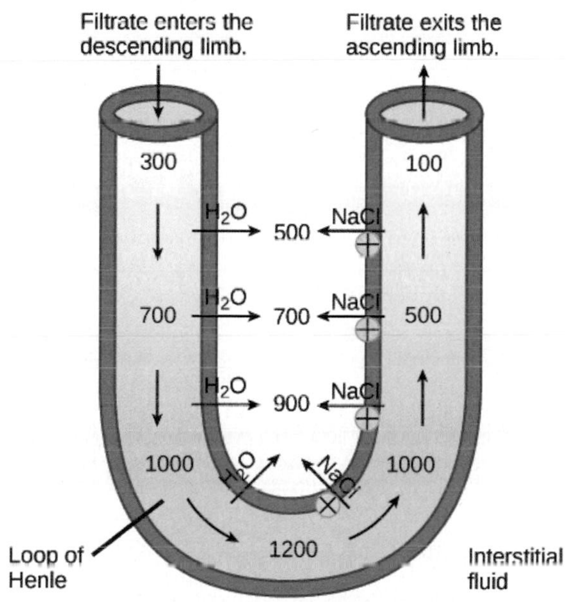

Figure 41.8 The loop of Henle acts as a countercurrent multiplier that uses energy to create concentration gradients. The descending limb is water permeable. Water flows from the filtrate to the interstitial fluid, so osmolality inside the limb increases as it descends into the renal medulla. At the bottom, the osmolality is higher inside the loop than in the interstitial fluid. Thus, as filtrate enters the ascending limb, Na^+ and Cl^- ions exit through ion channels present in the cell membrane. Further up, Na^+ is actively transported out of the filtrate and Cl^- follows. Osmolarity is given in units of milliosmoles per liter (mOsm/L).

Loop diuretics are drugs sometimes used to treat hypertension. These drugs inhibit the reabsorption of Na^+ and Cl^- ions by the ascending limb of the loop of Henle. A side effect is that they increase urination. Why do you think this is the case?

By the time the filtrate reaches the DCT, most of the urine and solutes have been reabsorbed. If the body requires additional water, all of it can be reabsorbed at this point. Further reabsorption is controlled by hormones, which will be discussed in a later section. Excretion of wastes occurs due to lack of reabsorption combined with tubular secretion. Undesirable products like metabolic wastes, urea, uric acid, and certain drugs, are excreted by tubular secretion. Most of the tubular secretion happens in the DCT, but some occurs in the early part of the collecting duct. Kidneys also maintain an acid-base balance by secreting excess H^+ ions.

Although parts of the renal tubules are named proximal and distal, in a cross-section of the kidney, the tubules are placed close together and in contact with each other and the glomerulus. This allows for exchange of chemical messengers between the different cell types. For example, the DCT ascending limb of the loop of Henle has masses of cells called **macula densa**, which are in contact with cells of the afferent arterioles called **juxtaglomerular cells**. Together, the macula densa and juxtaglomerular cells form the juxtaglomerular complex (JGC). The JGC is an endocrine structure that secretes the enzyme renin and the hormone erythropoietin. When hormones trigger the macula densa cells in the DCT due to variations in blood volume, blood pressure, or electrolyte balance, these cells can immediately communicate the problem to the capillaries in the afferent and efferent arterioles, which can constrict or relax to change the glomerular filtration rate of the kidneys.

CAREER CONNECTION

Nephrologist

A nephrologist studies and deals with diseases of the kidneys—both those that cause kidney failure (such as diabetes) and the conditions that are produced by kidney disease (such as hypertension). Blood pressure, blood volume, and changes in electrolyte balance come under the purview of a nephrologist.

Nephrologists usually work with other physicians who refer patients to them or consult with them about specific diagnoses and treatment plans. Patients are usually referred to a nephrologist for symptoms such as blood or protein in the urine, very high blood pressure, kidney stones, or renal failure.

Nephrology is a subspecialty of internal medicine. To become a nephrologist, medical school is followed by additional training to become certified in internal medicine. An additional two or more years is spent specifically studying kidney disorders and their accompanying effects on the body.

41.3 Excretion Systems

By the end of this section, you will be able to do the following:

- Explain how vacuoles, present in microorganisms, work to excrete waste
- Describe the way in which flame cells and nephridia in worms perform excretory functions and maintain osmotic balance
- Explain how insects use Malpighian tubules to excrete wastes and maintain osmotic balance

Microorganisms and invertebrate animals use more primitive and simple mechanisms to get rid of their metabolic wastes than the mammalian system of kidney and urinary function. Three excretory systems evolved in organisms before complex kidneys: vacuoles, flame cells, and Malpighian tubules.

Contractile Vacuoles in Microorganisms

The most fundamental feature of life is the presence of a cell. In other words, a cell is the simplest functional unit of a life. Bacteria are unicellular, prokaryotic organisms that have some of the least complex life processes in place; however, prokaryotes such as bacteria do not contain membrane-bound vacuoles. The cells of microorganisms like bacteria, protozoa, and fungi are bound by cell membranes and use them to interact with the environment. Some cells, including some leucocytes in humans, are able to engulf food by endocytosis—the formation of vesicles by involution of the cell membrane within the cells. The same vesicles are able to interact and exchange metabolites with the intracellular environment. In some unicellular eukaryotic organisms such as the amoeba, shown in Figure 41.9, cellular wastes and excess water are excreted by exocytosis, when the contractile vacuoles merge with the cell membrane and expel wastes into the environment. Contractile vacuoles (CV) should not be confused with vacuoles, which store food or water.

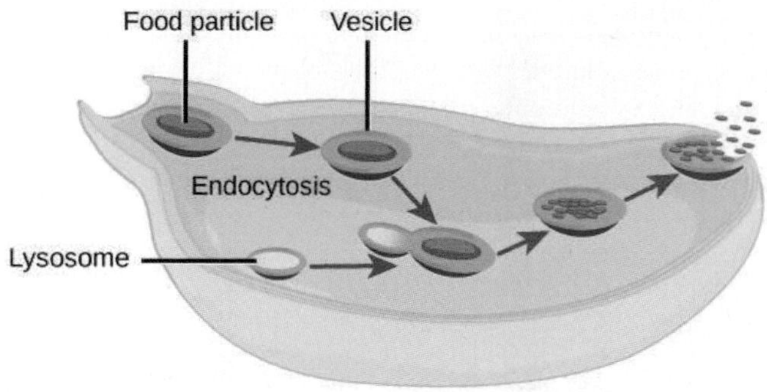

Figure 41.9 Some unicellular organisms, such as the amoeba, ingest food by endocytosis. The food vesicle fuses with a lysosome, which digests the food. Waste is excreted by exocytosis.

Flame Cells of Planaria and Nephridia of Worms

As multicellular systems evolved to have organ systems that divided the metabolic needs of the body, individual organs evolved to perform the excretory function. Planaria are flatworms that live in freshwater. Their excretory system consists of two tubules connected to a highly branched duct system. The cells in the tubules are called **flame cells** (or **protonephridia**) because they have a cluster of cilia that looks like a flickering flame when viewed under the microscope, as illustrated in Figure 41.10a. The cilia propel waste matter down the tubules and out of the body through excretory pores that open on the body surface; cilia also draw water from the interstitial fluid, allowing for filtration. Any valuable metabolites are recovered by reabsorption. Flame cells are found in flatworms, including parasitic tapeworms and free-living planaria. They also maintain the organism's osmotic balance.

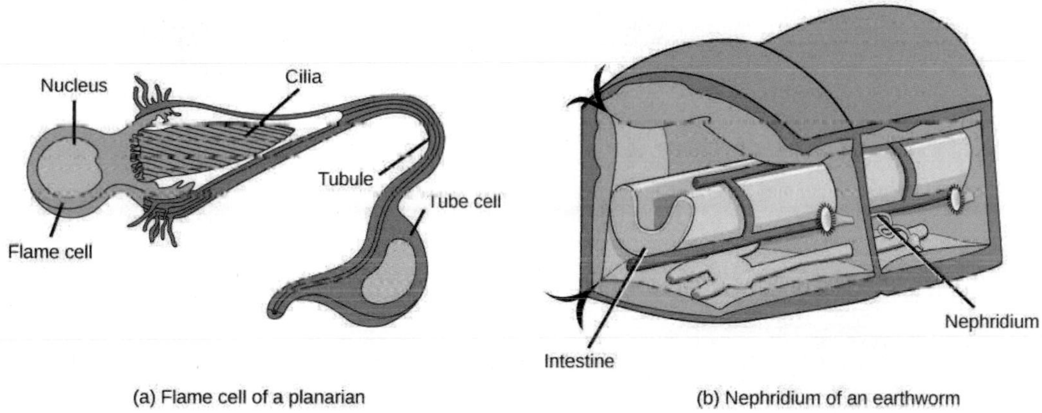

Figure 41.10 In the excretory system of the (a) planaria, cilia of flame cells propel waste through a tubule formed by a tube cell. Tubules are connected into branched structures that lead to pores located all along the sides of the body. The filtrate is secreted through these pores. In (b) annelids such as earthworms, nephridia filter fluid from the coelom, or body cavity. Beating cilia at the opening of the nephridium draw water from the coelom into a tubule. As the filtrate passes down the tubules, nutrients and other solutes are reabsorbed by capillaries. Filtered fluid containing nitrogenous and other wastes is stored in a bladder and then secreted through a pore in the side of the body.

Earthworms (annelids) have slightly more evolved excretory structures called **nephridia**, illustrated in Figure 41.10b. A pair of nephridia is present on each segment of the earthworm. They are similar to flame cells in that they have a tubule with cilia. Excretion occurs through a pore called the **nephridiopore**. They are more evolved than the flame cells in that they have a system for tubular reabsorption by a capillary network before excretion.

Malpighian Tubules of Insects

Malpighian tubules are found lining the gut of some species of arthropods, such as the bee illustrated in Figure 41.11. They are usually found in pairs and the number of tubules varies with the species of insect. Malpighian tubules are convoluted, which increases their surface area, and they are lined with **microvilli** for reabsorption and maintenance of osmotic balance. Malpighian tubules work cooperatively with specialized glands in the wall of the rectum. Body fluids are not filtered as in the case of nephridia; urine is produced by tubular secretion mechanisms by the cells lining the Malpighian tubules that are bathed in hemolymph (a mixture of blood and interstitial fluid that is found in insects and other arthropods as well as most mollusks).

Metabolic wastes like uric acid freely diffuse into the tubules. There are exchange pumps lining the tubules, which actively transport H^+ ions into the cell and K^+ or Na^+ ions out; water passively follows to form urine. The secretion of ions alters the osmotic pressure which draws water, electrolytes, and nitrogenous waste (uric acid) into the tubules. Water and electrolytes are reabsorbed when these organisms are faced with low-water environments, and uric acid is excreted as a thick paste or powder. Not dissolving wastes in water helps these organisms to conserve water; this is especially important for life in dry environments.

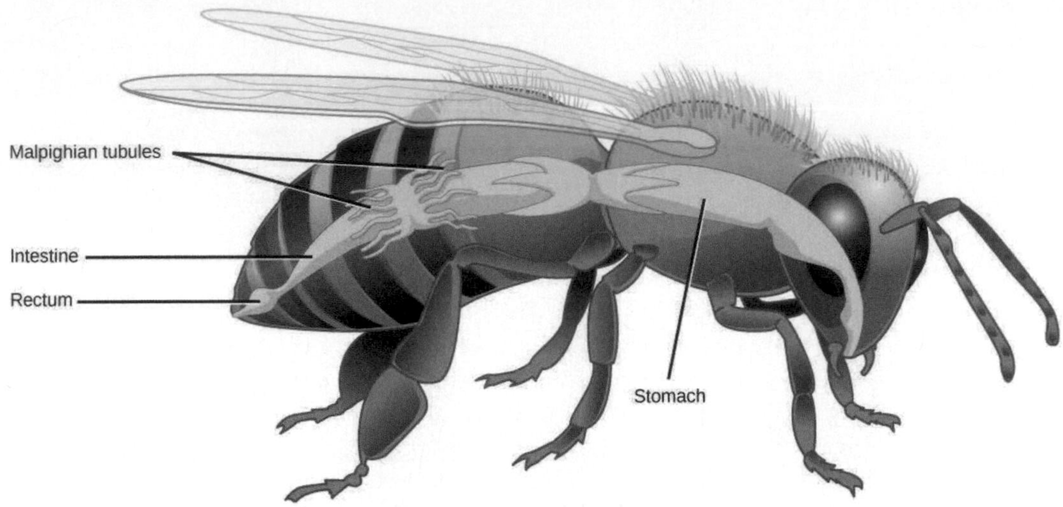

Figure 41.11 Malpighian tubules of insects and other terrestrial arthropods remove nitrogenous wastes and other solutes from the hemolymph. Na^+ and/or K^+ ions are actively transported into the lumen of the tubules. Water then enters the tubules via osmosis, forming urine. The urine passes through the intestine, and into the rectum. There, nutrients diffuse back into the hemolymph. Na^+ and/or K^+ ions are pumped into the hemolymph, and water follows. The concentrated waste is then excreted.

LINK TO LEARNING

See a dissected cockroach, including a close-up look at its Malpighian tubules, in this video (https://openstax.org/l/malpighian)
.

41.4 Nitrogenous Wastes

By the end of this section, you will be able to do the following:

- Compare and contrast the way in which aquatic animals and terrestrial animals can eliminate toxic ammonia from their systems
- Compare the major byproduct of ammonia metabolism in vertebrate animals to that of birds, insects, and reptiles

Of the four major macromolecules in biological systems, both proteins and nucleic acids contain nitrogen. During the catabolism, or breakdown, of nitrogen-containing macromolecules, carbon, hydrogen, and oxygen are extracted and stored in the form of carbohydrates and fats. Excess nitrogen is excreted from the body. Nitrogenous wastes tend to form toxic **ammonia**, which raises the pH of body fluids. The formation of ammonia itself requires energy in the form of ATP and large quantities of water to dilute it out of a biological system. Animals that live in aquatic environments tend to release ammonia into the water. Animals that excrete ammonia are said to be **ammonotelic**. Terrestrial organisms have evolved other mechanisms to excrete nitrogenous wastes. The animals must detoxify ammonia by converting it into a relatively nontoxic form such as urea or uric acid. Mammals, including humans, produce urea, whereas reptiles and many terrestrial invertebrates produce uric acid. Animals that secrete urea as the primary nitrogenous waste material are called **ureotelic** animals.

Nitrogenous Waste in Terrestrial Animals: The Urea Cycle

The **urea cycle** is the primary mechanism by which mammals convert ammonia to urea. Urea is made in the liver and excreted in urine. The overall chemical reaction by which ammonia is converted to urea is $2\ NH_3$ (ammonia) + CO_2 + 3 ATP + $H_2O \rightarrow H_2N\text{-}CO\text{-}NH_2$ (urea) + 2 ADP + 4 P_i + AMP.

The urea cycle utilizes five intermediate steps, catalyzed by five different enzymes, to convert ammonia to urea, as shown in Figure 41.12. The amino acid L-ornithine gets converted into different intermediates before being regenerated at the end of the urea cycle. Hence, the urea cycle is also referred to as the ornithine cycle. The enzyme ornithine transcarbamylase catalyzes a key step in the urea cycle and its deficiency can lead to accumulation of toxic levels of ammonia in the body. The first two reactions occur in the mitochondria and the last three reactions occur in the cytosol. Urea concentration in the blood, called **blood urea nitrogen** or BUN, is used as an indicator of kidney function.

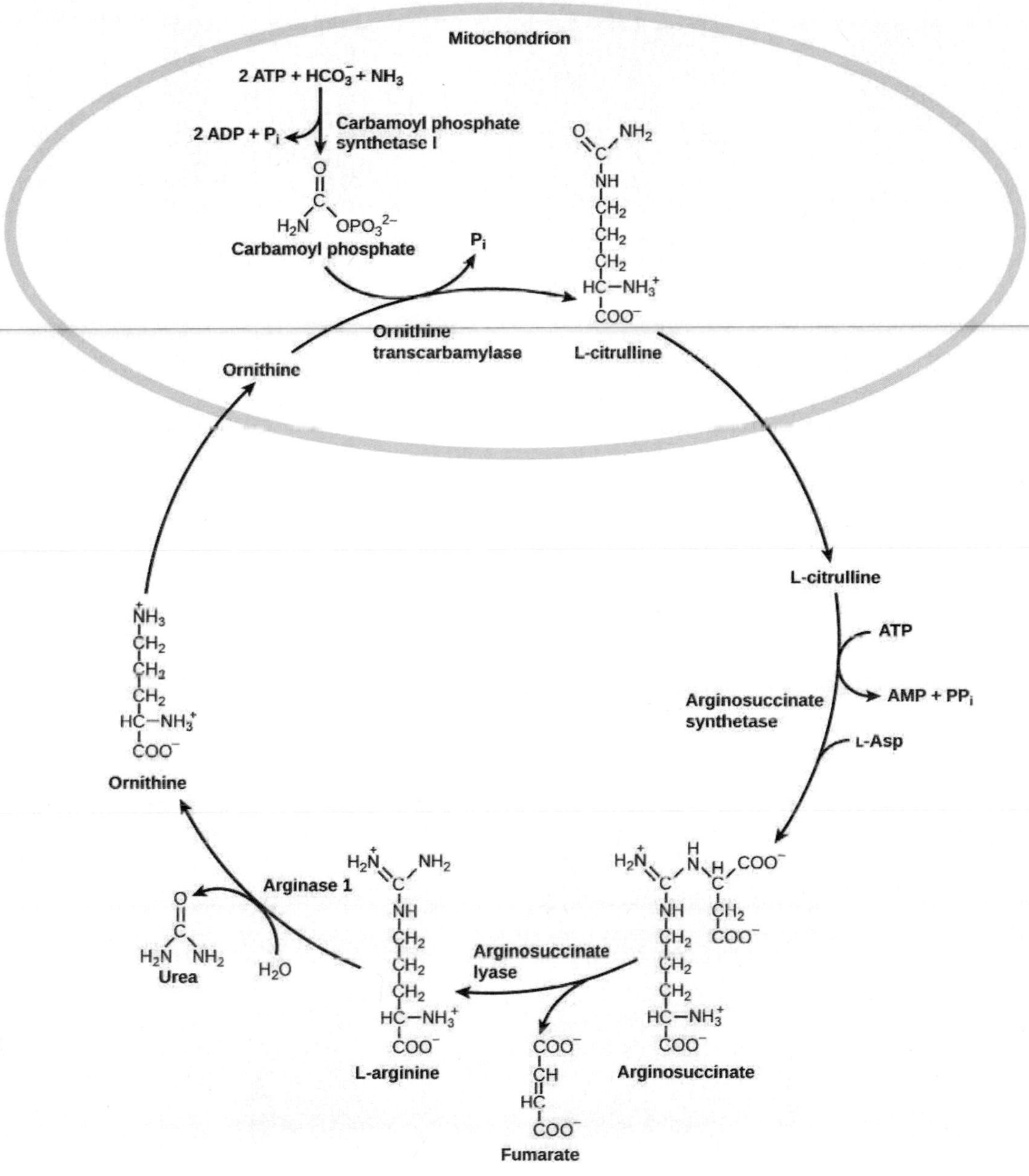

Figure 41.12 The urea cycle converts ammonia to urea.

EVOLUTION CONNECTION

Excretion of Nitrogenous Waste

The theory of evolution proposes that life started in an aquatic environment. It is not surprising to see that biochemical pathways like the urea cycle evolved to adapt to a changing environment when terrestrial life forms evolved. Arid conditions probably led to the evolution of the uric acid pathway as a means of conserving water.

Nitrogenous Waste in Birds and Reptiles: Uric Acid

Birds, reptiles, and most terrestrial arthropods convert toxic ammonia to **uric acid** or the closely related compound guanine (guano) instead of urea. Mammals also form some uric acid during breakdown of nucleic acids. Uric acid is a compound similar to purines found in nucleic acids. It is water insoluble and tends to form a white paste or powder; it is excreted by birds, insects, and reptiles. Conversion of ammonia to uric acid requires more energy and is much more complex than conversion of ammonia to urea Figure 41.13.

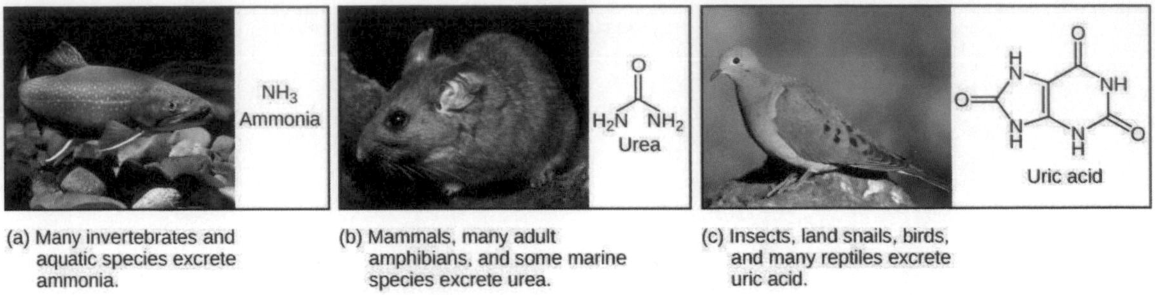

Figure 41.13 Nitrogenous waste is excreted in different forms by different species. These include (a) ammonia, (b) urea, and (c) uric acid. (credit a: modification of work by Eric Engbretson, USFWS; credit b: modification of work by B. "Moose" Peterson, USFWS; credit c: modification of work by Dave Menke, USFWS)

Everyday Connection

Gout

Mammals use uric acid crystals as an **antioxidant** in their cells. However, too much uric acid tends to form kidney stones and may also cause a painful condition called gout, where uric acid crystals accumulate in the joints, as illustrated in Figure 41.14. Food choices that reduce the amount of nitrogenous bases in the diet help reduce the risk of gout. For example, tea, coffee, and chocolate have purine-like compounds, called xanthines, and should be avoided by people with gout and kidney stones.

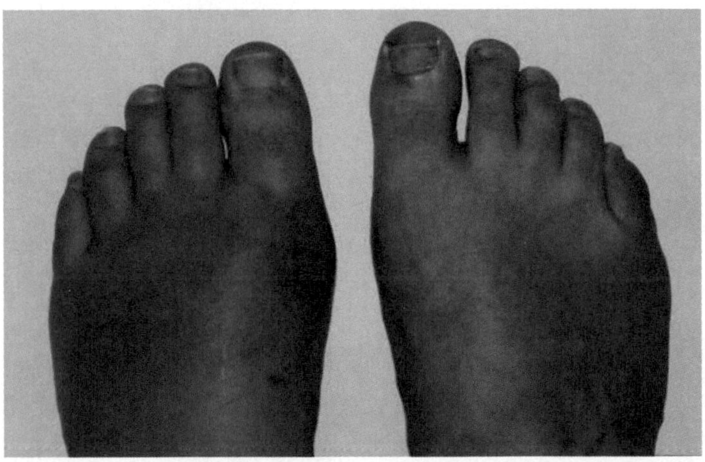

Figure 41.14 Gout causes the inflammation visible in this person's left big toe joint. (credit: "Gonzosft"/Wikimedia Commons)

41.5 Hormonal Control of Osmoregulatory Functions

By the end of this section, you will be able to do the following:

- Explain how hormonal cues help the kidneys synchronize the osmotic needs of the body
- Describe how hormones like epinephrine, norepinephrine, renin-angiotensin, aldosterone, anti-diuretic hormone, and atrial natriuretic peptide help regulate waste elimination, maintain correct osmolarity, and perform other osmoregulatory functions

While the kidneys operate to maintain osmotic balance and blood pressure in the body, they also act in concert with hormones.

Hormones are small molecules that act as messengers within the body. Hormones are typically secreted from one cell and travel in the bloodstream to affect a target cell in another portion of the body. Different regions of the nephron bear specialized cells that have receptors to respond to chemical messengers and hormones. Table 41.1 summarizes the hormones that control the osmoregulatory functions.

Hormones That Affect Osmoregulation

Hormone	Where produced	Function
Epinephrine and Norepinephrine	Adrenal medulla	Can decrease kidney function temporarily by vasoconstriction
Renin	Kidney nephrons	Increases blood pressure by acting on angiotensinogen
Angiotensin	Liver	Angiotensin II affects multiple processes and increases blood pressure
Aldosterone	Adrenal cortex	Prevents loss of sodium and water
Anti-diuretic hormone (vasopressin)	Hypothalamus (stored in the posterior pituitary)	Prevents water loss
Atrial natriuretic peptide	Heart atrium	Decreases blood pressure by acting as a vasodilator and increasing glomerular filtration rate; decreases sodium reabsorption in kidneys

Table 41.1

Epinephrine and Norepinephrine

Epinephrine and norepinephrine are released by the adrenal medulla and nervous system respectively. They are the flight/fight hormones that are released when the body is under extreme stress. During stress, much of the body's energy is used to combat imminent danger. Kidney function is halted temporarily by epinephrine and norepinephrine. These hormones function by acting directly on the smooth muscles of blood vessels to constrict them. Once the afferent arterioles are constricted, blood flow into the nephrons stops. These hormones go one step further and trigger the **renin-angiotensin-aldosterone** system.

Renin-Angiotensin-Aldosterone

The renin-angiotensin-aldosterone system, illustrated in Figure 41.15 proceeds through several steps to produce **angiotensin II**, which acts to stabilize blood pressure and volume. Renin (secreted by a part of the juxtaglomerular complex) is produced by the granular cells of the afferent and efferent arterioles. Thus, the kidneys control blood pressure and volume directly. Renin acts on angiotensinogen, which is made in the liver and converts it to **angiotensin I**. **Angiotensin converting enzyme (ACE)** converts angiotensin I to angiotensin II. Angiotensin II raises blood pressure by constricting blood vessels. It also triggers the release of the mineralocorticoid aldosterone from the adrenal cortex, which in turn stimulates the renal tubules to reabsorb more sodium. Angiotensin II also triggers the release of **anti-diuretic hormone (ADH)** from the hypothalamus, leading to water retention in the kidneys. It acts directly on the nephrons and decreases glomerular filtration rate. Medically, blood pressure can be controlled by drugs that inhibit ACE (called ACE inhibitors).

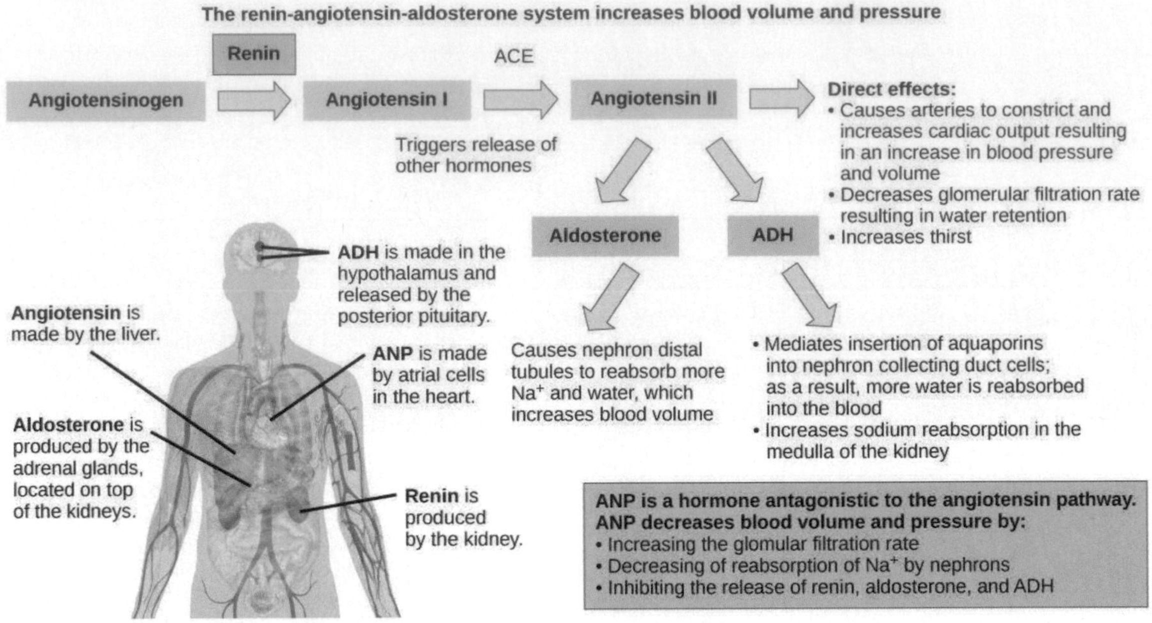

Figure 41.15 The renin-angiotensin-aldosterone system increases blood pressure and volume. The hormone ANP has antagonistic effects. (credit: modification of work by Mikael Häggström)

Mineralocorticoids

Mineralocorticoids are hormones synthesized by the adrenal cortex that affect osmotic balance. Aldosterone is a mineralocorticoid that regulates sodium levels in the blood. Almost all of the sodium in the blood is reclaimed by the renal tubules under the influence of aldosterone. Because sodium is always reabsorbed by active transport and water follows sodium to maintain osmotic balance, aldosterone manages not only sodium levels but also the water levels in body fluids. In contrast, the aldosterone also stimulates potassium secretion concurrently with sodium reabsorption. In contrast, absence of aldosterone means that no sodium gets reabsorbed in the renal tubules and all of it gets excreted in the urine. In addition, the daily dietary potassium load is not secreted and the retention of K^+ can cause a dangerous increase in plasma K^+ concentration. Patients who have Addison's disease have a failing adrenal cortex and cannot produce aldosterone. They lose sodium in their urine constantly, and if the supply is not replenished, the consequences can be fatal.

Antidiuretic Hormone

As previously discussed, antidiuretic hormone or ADH (also called **vasopressin**), as the name suggests, helps the body conserve water when body fluid volume, especially that of blood, is low. It is formed by the hypothalamus and is stored and released from the posterior pituitary. It acts by inserting aquaporins in the collecting ducts and promotes reabsorption of water. ADH also acts as a vasoconstrictor and increases blood pressure during hemorrhaging.

Atrial Natriuretic Peptide Hormone

The atrial natriuretic peptide (ANP) lowers blood pressure by acting as a **vasodilator**. It is released by cells in the atrium of the heart in response to high blood pressure and in patients with sleep apnea. ANP affects salt release, and because water passively follows salt to maintain osmotic balance, it also has a diuretic effect. ANP also prevents sodium reabsorption by the renal tubules, decreasing water reabsorption (thus acting as a diuretic) and lowering blood pressure. Its actions suppress the actions of aldosterone, ADH, and renin.

KEY TERMS

afferent arteriole arteriole that branches from the cortical radiate artery and enters the glomerulus

ammonia compound made of one nitrogen atom and three hydrogen atoms

ammonotelic describes an animal that excretes ammonia as the primary waste material

angiotensin converting enzyme (ACE) enzyme that converts angiotensin I to angiotensin II

angiotensin I product in the renin-angiotensin-aldosterone pathway

angiotensin II molecule that affects different organs to increase blood pressure

anti-diuretic hormone (ADH) hormone that prevents the loss of water

antioxidant agent that prevents cell destruction by reactive oxygen species

arcuate artery artery that branches from the interlobar artery and arches over the base of the renal pyramids

ascending limb part of the loop of Henle that ascends from the renal medulla to the renal cortex

blood urea nitrogen (BUN) estimate of urea in the blood and an indicator of kidney function

Bowman's capsule structure that encloses the glomerulus

calyx structure that connects the renal pelvis to the renal medulla

cortex (animal) outer layer of an organ like the kidney or adrenal gland

cortical nephron nephron that lies in the renal cortex

cortical radiate artery artery that radiates from the arcuate arteries into the renal cortex

countercurrent exchanger peritubular capillary network that allows exchange of solutes and water from the renal tubules

countercurrent multiplier osmotic gradient in the renal medulla that is responsible for concentration of urine

descending limb part of the loop of Henle that descends from the renal cortex into the renal medulla

distal convoluted tubule (DCT) part of the renal tubule that is the most distant from the glomerulus

efferent arteriole arteriole that exits from the glomerulus

electrolyte solute that breaks down into ions when dissolved in water

flame cell (also, protonephridia) excretory cell found in flatworms

glomerular filtration filtration of blood in the glomerular capillary network into the glomerulus

glomerular filtration rate (GFR) amount of filtrate formed by the glomerulus per minute

glomerulus (renal) part of the renal corpuscle that contains the capillary network

hilum region in the renal pelvis where blood vessels, nerves, and ureters bunch before entering or exiting the kidney

inferior vena cava one of the main veins in the human body

interlobar artery artery that branches from the segmental artery and travels in between the renal lobes

juxtaglomerular cell cell in the afferent and efferent arterioles that responds to stimuli from the macula densa

juxtamedullary nephron nephron that lies in the cortex but close to the renal medulla

kidney organ that performs excretory and osmoregulatory functions

lobes of the kidney renal pyramid along with the adjoining cortical region

loop of Henle part of the renal tubule that loops into the renal medulla

macula densa group of cells that senses changes in sodium ion concentration; present in parts of the renal tubule and collecting ducts

Malpighian tubule excretory tubules found in arthropods

medulla middle layer of an organ like the kidney or adrenal gland

microvilli cellular processes that increase the surface area of cells

molality number of moles of solute per kilogram of solvent

molarity number of moles of solute per liter of solution

mole gram equivalent of the molecular weight of a substance

nephridia excretory structures found in annelids

nephridiopore pore found at the end of nephridia

nephron functional unit of the kidney

non-electrolyte solute that does not break down into ions when dissolved in water

osmoconformer organism that changes its tonicity based on its environment

osmoregulation mechanism by which water and solute concentrations are maintained at desired levels

osmoregulator organism that maintains its tonicity irrespective of its environment

osmotic balance balance of the amount of water and salt input and output to and from a biological system without disturbing the desired osmotic pressure and solute concentration in every compartment

osmotic pressure pressure exerted on a membrane to equalize solute concentration on either side

perirenal fat capsule fat layer that suspends the kidneys

peritubular capillary network capillary network that surrounds the renal tubule after the efferent artery exits the glomerulus

proximal convoluted tubule (PCT) part of the renal tubule that lies close to the glomerulus

renal artery branch of the artery that enters the kidney

renal capsule layer that encapsulates the kidneys

renal column area of the kidney through which the

interlobar arteries travel in the process of supplying blood to the renal lobes

renal corpuscle glomerulus and the Bowman's capsule together

renal fascia connective tissue that supports the kidneys

renal pelvis region in the kidney where the calyces join the ureters

renal pyramid conical structure in the renal medulla

renal tubule tubule of the nephron that arises from the glomerulus

renal vein branch of a vein that exits the kidney and joins the inferior vena cava

renin-angiotensin-aldosterone biochemical pathway that activates angiotensin II, which increases blood pressure

segmental artery artery that branches from the renal artery

semi-permeable membrane membrane that allows only certain solutes to pass through

transport maximum maximum amount of solute that can be transported out of the renal tubules during reabsorption

tubular reabsorption reclamation of water and solutes that got filtered out in the glomerulus

tubular secretion process of secretion of wastes that do not get reabsorbed

urea cycle pathway by which ammonia is converted to urea

ureotelic describes animals that secrete urea as the primary nitrogenous waste material

ureter urine-bearing tube coming out of the kidney; carries urine to the bladder

uric acid byproduct of ammonia metabolism in birds, insects, and reptiles

urinary bladder structure that the ureters empty the urine into; stores urine

urine filtrate produced by kidneys that gets excreted out of the body

vasa recta peritubular network that surrounds the loop of Henle of the juxtamedullary nephrons

vasodilator compound that increases the diameter of blood vessels

vasopressin another name for anti-diuretic hormone

CHAPTER SUMMARY

41.1 Osmoregulation and Osmotic Balance

Solute concentrations across semi-permeable membranes influence the movement of water and solutes across the membrane. It is the number of solute molecules and not the molecular size that is important in osmosis. Osmoregulation and osmotic balance are important bodily functions, resulting in water and salt balance. Not all solutes can pass through a semi-permeable membrane. Osmosis is the movement of water across the membrane. Osmosis occurs to equalize the number of solute molecules across a semi-permeable membrane by the movement of water to the side of higher solute concentration. Facilitated diffusion utilizes protein channels to move solute molecules from areas of higher to lower concentration while active transport mechanisms are required to move solutes against concentration gradients. Osmolarity is measured in units of milliequivalents or milliosmoles, both of which take into consideration the number of solute particles and the charge on them. Fish that live in freshwater or saltwater adapt by being osmoregulators or osmoconformers.

41.2 The Kidneys and Osmoregulatory Organs

The kidneys are the main osmoregulatory organs in mammalian systems; they function to filter blood and maintain the osmolarity of body fluids at 300 mOsm. They are surrounded by three layers and are made up internally of three distinct regions—the cortex, medulla, and pelvis.

The blood vessels that transport blood into and out of the kidneys arise from and merge with the aorta and inferior vena cava, respectively. The renal arteries branch out from the aorta and enter the kidney where they further divide into segmental, interlobar, arcuate, and cortical radiate arteries.

The nephron is the functional unit of the kidney, which actively filters blood and generates urine. The nephron is made up of the renal corpuscle and renal tubule. Cortical nephrons are found in the renal cortex, while juxtamedullary nephrons are found in the renal cortex close to the renal medulla. The nephron filters and exchanges water and solutes with two sets of blood vessels and the tissue fluid in the kidneys.

There are three steps in the formation of urine: glomerular filtration, which occurs in the glomerulus; tubular reabsorption, which occurs in the renal tubules; and tubular secretion, which also occurs in the renal tubules.

41.3 Excretion Systems

Many systems have evolved for excreting wastes that are simpler than the kidney and urinary systems of vertebrate animals. The simplest system is that of contractile vacuoles present in microorganisms. Flame cells and nephridia in worms perform excretory functions and maintain osmotic balance. Some insects have evolved Malpighian tubules to excrete wastes and maintain osmotic balance.

41.4 Nitrogenous Wastes

Ammonia is the waste produced by metabolism of nitrogen-containing compounds like proteins and nucleic acids. While aquatic animals can easily excrete ammonia into their watery surroundings, terrestrial animals have evolved special mechanisms to eliminate the toxic ammonia from their systems. Urea is the major byproduct of ammonia metabolism in vertebrate animals. Uric acid is the major byproduct of ammonia metabolism in birds, terrestrial arthropods, and reptiles.

41.5 Hormonal Control of Osmoregulatory Functions

Hormonal cues help the kidneys synchronize the osmotic needs of the body. Hormones like epinephrine, norepinephrine, renin-angiotensin, aldosterone, anti-diuretic hormone, and atrial natriuretic peptide help regulate the needs of the body as well as the communication between the different organ systems.

VISUAL CONNECTION QUESTIONS

1. Figure 41.5 Which of the following statements about the kidney is false?
   a. The renal pelvis drains into the ureter.
   b. The renal pyramids are in the medulla.
   c. The cortex covers the capsule.
   d. Nephrons are in the renal cortex.

2. Figure 41.6 Which of the following statements about the nephron is false?
   a. The collecting duct empties into the distal convoluted tubule.
   b. The Bowman's capsule surrounds the glomerulus.
   c. The loop of Henle is between the proximal and distal convoluted tubules.
   d. The loop of Henle empties into the distal convoluted tubule.

3. Figure 41.8 Loop diuretics are drugs sometimes used to treat hypertension. These drugs inhibit the reabsorption of Na^+ and Cl^- ions by the ascending limb of the loop of Henle. A side effect is that they increase urination. Why do you think this is the case?

REVIEW QUESTIONS

4. When a dehydrated human patient needs to be given fluids intravenously, he or she is given:
   a. water, which is hypotonic with respect to body fluids
   b. saline at a concentration that is isotonic with respect to body fluids
   c. glucose because it is a non-electrolyte
   d. blood

5. The sodium ion is at the highest concentration in:
   a. intracellular fluid
   b. extracellular fluid
   c. blood plasma
   d. none of the above

6. Cells in a hypertonic solution tend to:
   a. shrink due to water loss
   b. swell due to water gain
   c. stay the same size due to water moving into and out of the cell at the same rate
   d. none of the above

7. The macula densa is/are:
   a. present in the renal medulla.
   b. dense tissue present in the outer layer of the kidney.
   c. cells present in the DCT and collecting tubules.
   d. present in blood capillaries.

8. The osmolarity of body fluids is maintained at ________.
   a. 100 mOsm
   b. 300 mOsm
   c. 1000 mOsm
   d. it is not constantly maintained

9. The gland located at the top of the kidney is the ________ gland.
   a. adrenal
   b. pituitary
   c. thyroid
   d. thymus

10. Active transport of K^+ in Malpighian tubules ensures that:
    a. water follows K^+ to make urine
    b. osmotic balance is maintained between waste matter and bodily fluids
    c. both a and b
    d. neither a nor b

11. Contractile vacuoles in microorganisms:
    a. exclusively perform an excretory function
    b. can perform many functions, one of which is excretion of metabolic wastes
    c. originate from the cell membrane
    d. both b and c

12. Flame cells are primitive excretory organs found in ________.
    a. arthropods
    b. annelids
    c. mammals
    d. flatworms

13. BUN is ________.
    a. blood urea nitrogen
    b. blood uric acid nitrogen
    c. an indicator of blood volume
    d. an indicator of blood pressure

14. Human beings accumulate ________ before excreting nitrogenous waste.
    a. nitrogen
    b. ammonia
    c. urea
    d. uric acid

15. Renin is made by ________.
    a. granular cells of the juxtaglomerular apparatus
    b. the kidneys
    c. the nephrons
    d. all of the above

16. Patients with Addison's disease ________.
    a. retain water
    b. retain salts
    c. lose salts and water
    d. have too much aldosterone

17. Which hormone elicits the "fight or flight" response?
    a. epinephrine
    b. mineralcorticoids
    c. anti-diuretic hormone
    d. thyroxine

CRITICAL THINKING QUESTIONS

18. Why is excretion important in order to achieve osmotic balance?
19. Why do electrolyte ions move across membranes by active transport?
20. Why are the loop of Henle and vasa recta important for the formation of concentrated urine?
21. Describe the structure of the kidney.
22. Why might specialized organs have evolved for excretion of wastes?
23. Explain two different excretory systems other than the kidneys.
24. In terms of evolution, why might the urea cycle have evolved in organisms?
25. Compare and contrast the formation of urea and uric acid.
26. Describe how hormones regulate blood pressure, blood volume, and kidney function.
27. How does the renin-angiotensin-aldosterone mechanism function? Why is it controlled by the kidneys?

CHAPTER 42
The Immune System

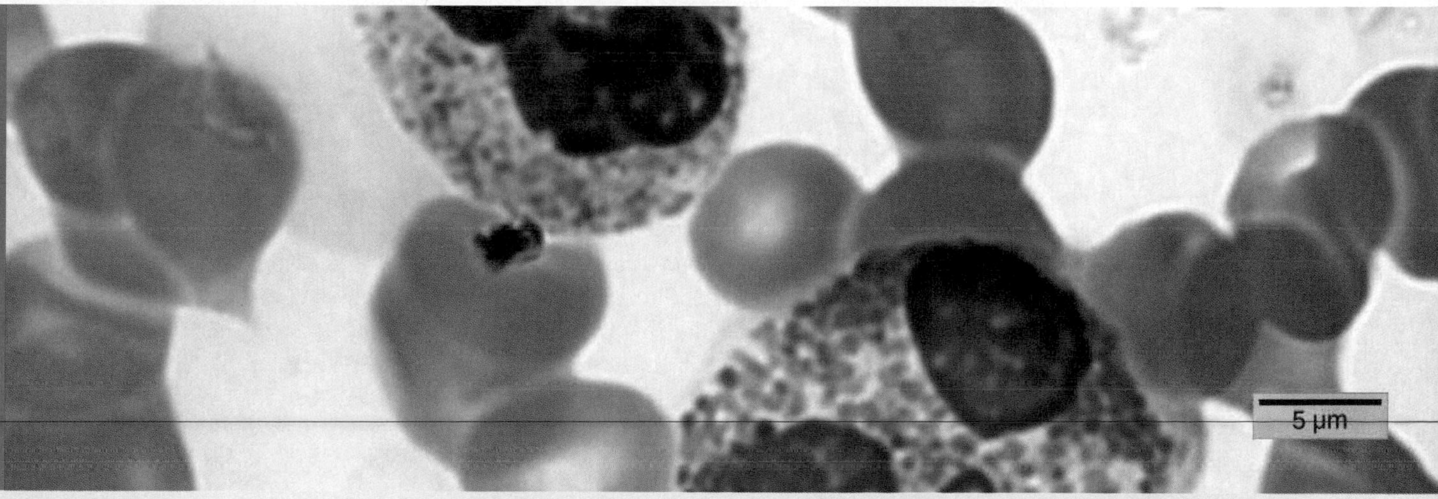

Figure 42.1 In this compound light micrograph purple-stained neutrophil (upper left) and eosinophil (lower right) are white blood cells that float among red blood cells in this blood smear. Neutrophils provide an early, rapid, and nonspecific defense against invading pathogens. Eosinophils play a variety of roles in the immune response. Red blood cells are about 7–8 µm in diameter, and a neutrophil is about 10–12µm. (credit: modification of work by Dr. David Csaba)

INTRODUCTION The environment consists of numerous **pathogens**, which are agents, usually microorganisms, that cause diseases in their hosts. A **host** is the organism that is invaded and often harmed by a pathogen. Pathogens include bacteria, protists, fungi and other infectious organisms. We are constantly exposed to pathogens in food and water, on surfaces, and in the air. Mammalian immune systems evolved for protection from such pathogens; they are composed of an extremely diverse array of specialized cells and soluble molecules that coordinate a rapid and flexible defense system capable of providing protection from a majority of these disease agents.

Components of the immune system constantly search the body for signs of pathogens. When pathogens are found, immune factors are mobilized to the site of an infection. The immune factors identify the nature of the pathogen, strengthen the corresponding cells and molecules to combat it efficiently, and then halt the immune response after the infection is cleared to avoid unnecessary host cell damage. The immune system can remember pathogens to which it has been exposed to create a more efficient response upon reexposure. This memory can last several decades. Features of the immune system, such as pathogen identification, specific response, amplification, retreat, and remembrance are essential for survival against pathogens. The immune response can be classified as either innate or active. The innate immune response is always present and attempts to defend against all pathogens rather than focusing on specific ones. Conversely, the adaptive immune response stores information about past infections and mounts pathogen-specific defenses.

42.1 Innate Immune Response

By the end of this section, you will be able to do the following:

- Describe physical and chemical immune barriers
- Explain immediate and induced innate immune responses
- Discuss natural killer cells
- Describe major histocompatibility class I molecules
- Summarize how the proteins in a complement system function to destroy extracellular pathogens

The immune system comprises both innate and adaptive immune responses. **Innate immunity** occurs naturally because of genetic factors or physiology; it is not induced by infection or vaccination but works to reduce the workload for the adaptive immune response. Both the innate and adaptive levels of the immune response involve secreted proteins, receptor-mediated signaling, and intricate cell-to-cell communication. The innate immune system developed early in animal evolution, roughly a billion years ago, as an essential response to infection. Innate immunity has a limited number of specific targets: any pathogenic threat triggers a consistent sequence of events that can identify the type of pathogen and either clear the infection independently or mobilize a highly specialized adaptive immune response. For example, tears and mucus secretions contain microbicidal factors.

Physical and Chemical Barriers

Before any immune factors are triggered, the skin functions as a continuous, impassable barrier to potentially infectious pathogens. Pathogens are killed or inactivated on the skin by desiccation (drying out) and by the skin's acidity. In addition, beneficial microorganisms that coexist on the skin compete with invading pathogens, preventing infection. Regions of the body that are not protected by skin (such as the eyes and mucus membranes) have alternative methods of defense, such as tears and mucus secretions that trap and rinse away pathogens, and cilia in the nasal passages and respiratory tract that push the mucus with the pathogens out of the body. Throughout the body are other defenses, such as the low pH of the stomach (which inhibits the growth of pathogens), blood proteins that bind and disrupt bacterial cell membranes, and the process of urination (which flushes pathogens from the urinary tract).

Despite these barriers, pathogens may enter the body through skin abrasions or punctures, or by collecting on mucosal surfaces in large numbers that overcome the mucus or cilia. Some pathogens have evolved specific mechanisms that allow them to overcome physical and chemical barriers. When pathogens do enter the body, the innate immune system responds with inflammation, pathogen engulfment, and secretion of immune factors and proteins.

Pathogen Recognition

An infection may be intracellular or extracellular, depending on the pathogen. All viruses infect cells and replicate within those cells (intracellularly), whereas bacteria and other parasites may replicate intracellularly or extracellularly, depending on the species. The innate immune system must respond accordingly: by identifying the extracellular pathogen and/or by identifying host cells that have already been infected. When a pathogen enters the body, cells in the blood and lymph detect the specific **pathogen-associated molecular patterns (PAMPs)** on the pathogen's surface. PAMPs are carbohydrate, polypeptide, and nucleic acid "signatures" that are expressed by viruses, bacteria, and parasites but which differ from molecules on host cells. The immune system has specific cells, described in Figure 42.2 and shown in Figure 42.3, with receptors that recognize these PAMPs. A **macrophage** is a large phagocytic cell that engulfs foreign particles and pathogens. Macrophages recognize PAMPs via complementary **pattern recognition receptors (PRRs)**. PRRs are molecules on macrophages and dendritic cells which are in contact with the external environment. A **monocyte** is a type of white blood cell that circulates in the blood and lymph and differentiates into macrophages after it moves into infected tissue. Dendritic cells bind molecular signatures of pathogens and promote pathogen engulfment and destruction. Toll-like receptors (TLRs) are a type of PRR that recognizes molecules that are shared by pathogens but distinguishable from host molecules. TLRs are present in invertebrates as well as vertebrates, and appear to be one of the most ancient components of the immune system. TLRs have also been identified in the mammalian nervous system.

Cell type	Characteristics	Location	Image
Mast cell	Dilates blood vessels and induces inflammation through release of histamines and heparin. Recruits macrophages and neutrophils. Involved in wound healing and defense against pathogens but can also be responsible for allergic reactions.	Connective tissues, mucous membranes	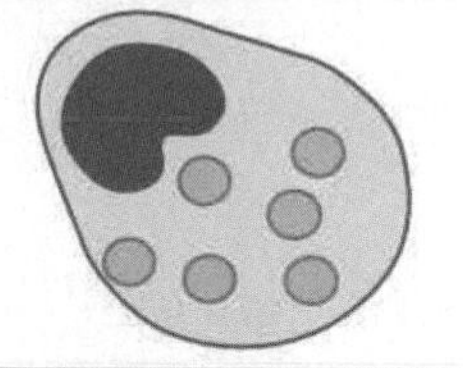
Macrophage	Phagocytic cell that consumes foreign pathogens and cancer cells. Stimulates response of other immune cells.	Migrates from blood vessels into tissues.	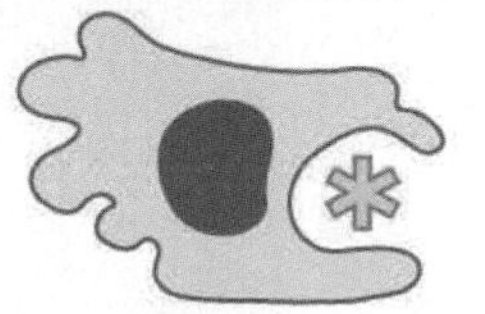
Natural killer cell	Kills tumor cells and virus-infected cells.	Circulates in blood and migrates into tissues.	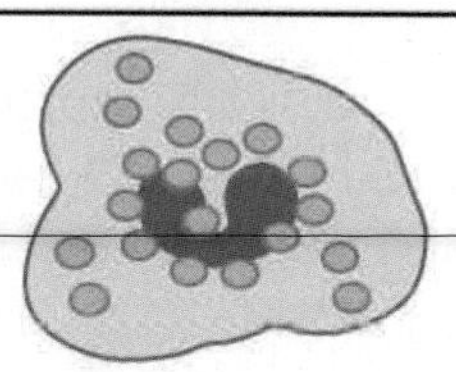
Dendritic cell	Presents antigens on its surface, thereby triggering adaptive immunity.	Present in epithelial tissue, including skin, lung and tissues of the digestive tract. Migrates to lymph nodes upon activation.	
Monocyte	Differentiates into macrophages and dendritic cells in response to inflammation.	Stored in spleen, moves through blood vessels to infected tissues.	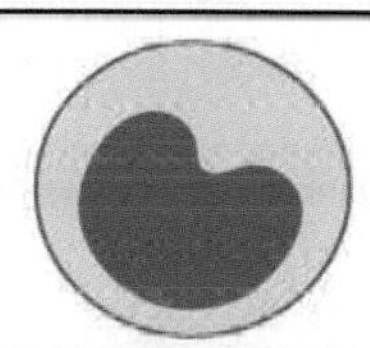
Neutrophil	First responders at the site of infection or trauma, this abundant phagocytic cell represents 50-60 percent of all leukocytes. Releases toxins that kill or inhibit bacteria and fungi and recruits other immune cells to the site of infection.	Migrates from blood vessels into tissues.	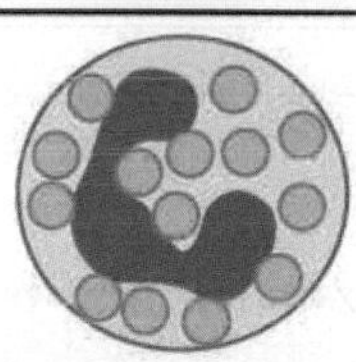
Basophil	Responsible for defense against parasites. Releases histamines that cause inflammation and may be responsible for allergic reactions.	Circulates in blood and migrates to tissues.	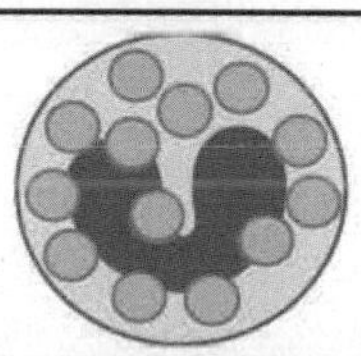
Eosinophil	Releases toxins that kill bacteria and parasites but also causes tissue damage.	Circulates in blood and migrates to tissues.	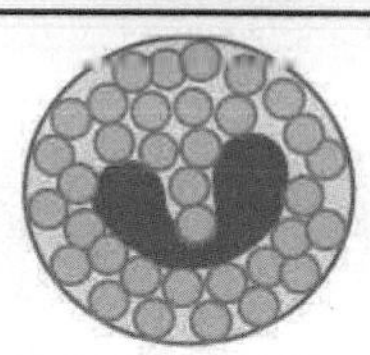

Figure 42.2 The characteristics and location of cells involved in the innate immune system are described. (credit: modification of work by NIH)

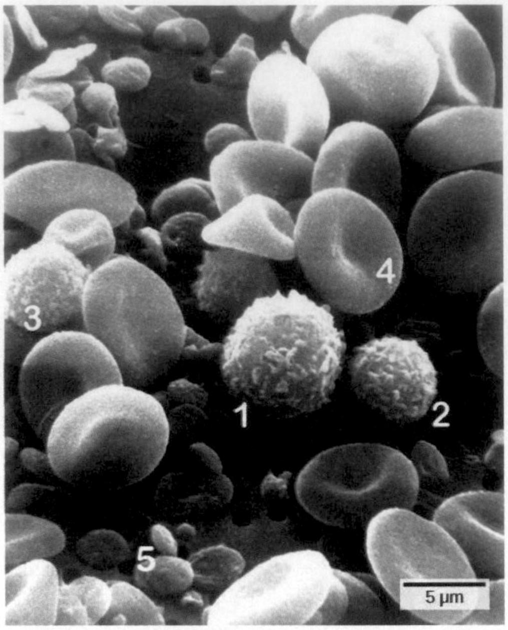

Figure 42.3 Cells of the blood include (1) monocytes, (2) lymphocytes, (3) neutrophils, (4) red blood cells, and (5) platelets. Note the very similar morphologies of the leukocytes (1, 2, 3). (credit: modification of work by Bruce Wetzel, Harry Schaefer, NCI; scale-bar data from Matt Russell)

Cytokine Release Effect

The binding of PRRs with PAMPs triggers the release of cytokines, which signal that a pathogen is present and needs to be destroyed along with any infected cells. A **cytokine** is a chemical messenger that regulates cell differentiation (form and function), proliferation (production), and gene expression to affect immune responses. At least 40 types of cytokines exist in humans that differ in terms of the cell type that produces them, the cell type that responds to them, and the changes they produce. One type of cytokine, interferon, is illustrated in Figure 42.4.

One subclass of cytokines is the interleukin (IL), so named because they mediate interactions between leukocytes (white blood cells). Interleukins are involved in bridging the innate and adaptive immune responses. In addition to being released from cells after PAMP recognition, cytokines are released by the infected cells which bind to nearby uninfected cells and induce those cells to release cytokines, which results in a cytokine burst.

A second class of early-acting cytokines is interferons, which are released by infected cells as a warning to nearby uninfected cells. One of the functions of an **interferon** is to inhibit viral replication. They also have other important functions, such as tumor surveillance. Interferons work by signaling neighboring uninfected cells to destroy RNA and reduce protein synthesis, signaling neighboring infected cells to undergo apoptosis (programmed cell death), and activating immune cells.

In response to interferons, uninfected cells alter their gene expression, which increases the cells' resistance to infection. One effect of interferon-induced gene expression is a sharply reduced cellular protein synthesis. Virally infected cells produce more viruses by synthesizing large quantities of viral proteins. Thus, by reducing protein synthesis, a cell becomes resistant to viral infection.

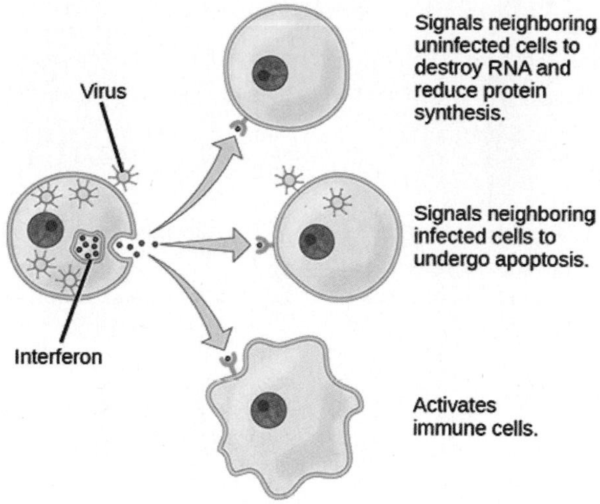

Figure 42.4 Interferons are cytokines that are released by a cell infected with a virus. Response of neighboring cells to interferon helps stem the infection.

Phagocytosis and Inflammation

The first cytokines to be produced are pro-inflammatory; that is, they encourage **inflammation**, the localized redness, swelling, heat, and pain that result from the movement of leukocytes and fluid through increasingly permeable capillaries to a site of infection. The population of leukocytes that arrives at an infection site depends on the nature of the infecting pathogen. Both macrophages and dendritic cells engulf pathogens and cellular debris through phagocytosis. A **neutrophil** is also a phagocytic leukocyte that engulfs and digests pathogens. Neutrophils, shown in Figure 42.3, are the most abundant leukocytes of the immune system. Neutrophils have a nucleus with two to five lobes, and they contain organelles, called lysosomes, that digest engulfed pathogens. An **eosinophil** is a leukocyte that works with other eosinophils to surround a parasite; it is involved in the allergic response and in protection against helminthes (parasitic worms).

Neutrophils and eosinophils are particularly important leukocytes that engulf large pathogens, such as bacteria and fungi. A **mast cell** is a leukocyte that produces inflammatory molecules, such as histamine, in response to large pathogens. A **basophil** is a leukocyte that, like a neutrophil, releases chemicals to stimulate the inflammatory response as illustrated in Figure 42.5. Basophils are also involved in allergy and hypersensitivity responses and induce specific types of inflammatory responses. Eosinophils and basophils produce additional inflammatory mediators to recruit more leukocytes. A hypersensitive immune response to harmless antigens, such as in pollen, often involves the release of histamine by basophils and mast cells.

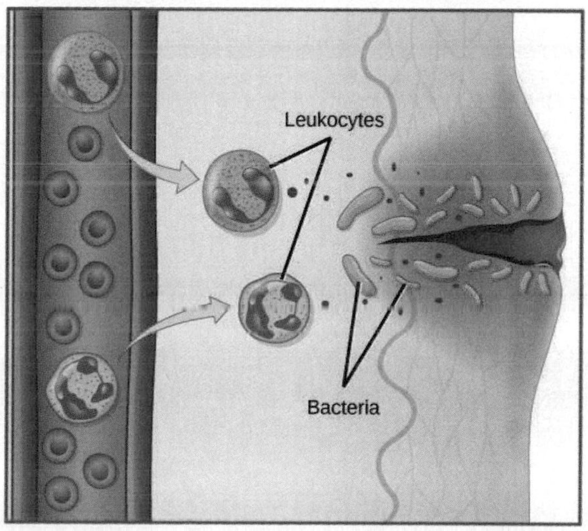

Figure 42.5 In response to a cut, mast cells secrete histamines that cause nearby capillaries to dilate. Neutrophils and monocytes leave the capillaries. Monocytes mature into macrophages. Neutrophils, dendritic cells, and macrophages release chemicals to stimulate the inflammatory response. Neutrophils and macrophages also consume invading bacteria by phagocytosis.

Cytokines also send feedback to cells of the nervous system to bring about the overall symptoms of feeling sick, which include lethargy, muscle pain, and nausea. These effects may have evolved because the symptoms encourage the individual to rest and prevent the spreading of the infection to others. Cytokines also increase the core body temperature, causing a fever, which causes the liver to withhold iron from the blood. Without iron, certain pathogens, such as some bacteria, are unable to replicate; this is called nutritional immunity.

LINK TO LEARNING

Watch this 23-second stop-motion video (http://openstax.org/l/conidia) showing a neutrophil that searches for and engulfs fungus spores during an elapsed time of about 79 minutes.

Natural Killer Cells

Lymphocytes are leukocytes that are histologically identifiable by their large, darkly staining nuclei; they are small cells with very little cytoplasm, as shown in Figure 42.6. Infected cells are identified and destroyed by **natural killer (NK) cells**, lymphocytes that can kill cells infected with viruses or tumor cells (abnormal cells that uncontrollably divide and invade other tissue). T cells and B cells of the adaptive immune system also are classified as lymphocytes. **T cells** are lymphocytes that mature in the thymus gland, and **B cells** are lymphocytes that mature in the bone marrow. NK cells identify intracellular infections, especially from viruses, by the altered expression of **major histocompatibility class (MHC) I molecules** on the surface of infected cells. MHC I molecules are proteins on the surfaces of all nucleated cells, thus they are scarce on red blood cells and platelets which are non-nucleated. The function of MHC I molecules is to display fragments of proteins from the infectious agents within the cell to T cells; healthy cells will be ignored, while "non-self" or foreign proteins will be attacked by the immune system. MHC II molecules are found mainly on cells containing antigens ("non-self proteins") and on lymphocytes. **MHC II molecules** interact with helper T cells to trigger the appropriate immune response, which may include the inflammatory response.

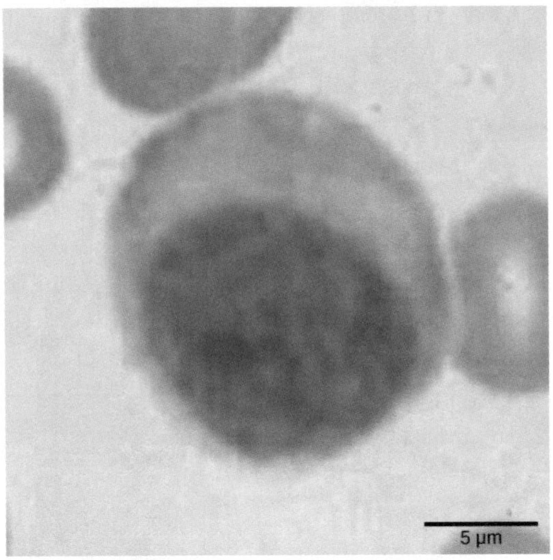

Figure 42.6 Lymphocytes, such as NK cells, are characterized by their large nuclei that actively absorb Wright stain and therefore appear dark colored under a microscope.

An infected cell (or a tumor cell) is usually incapable of synthesizing and displaying MHC I molecules appropriately. The metabolic resources of cells infected by some viruses produce proteins that interfere with MHC I processing and/or trafficking to the cell surface. The reduced MHC I on host cells varies from virus to virus and results from active inhibitors being produced by the viruses. This process can deplete host MHC I molecules on the cell surface, which NK cells detect as "unhealthy" or "abnormal" while searching for cellular MHC I molecules. Similarly, the dramatically altered gene expression of tumor cells leads to expression of extremely deformed or absent MHC I molecules that also signal "unhealthy" or "abnormal."

NK cells are always active; an interaction with normal, intact MHC I molecules on a healthy cell disables the killing sequence, and the NK cell moves on. After the NK cell detects an infected or tumor cell, its cytoplasm secretes granules comprised of **perforin**, a destructive protein that creates a pore in the target cell. Granzymes are released along with the perforin in the immunological synapse. A **granzyme** is a protease that digests cellular proteins and induces the target cell to undergo programmed cell death, or apoptosis. Phagocytic cells then digest the cell debris left behind. NK cells are constantly patrolling the body and are an effective mechanism for controlling potential infections and preventing cancer progression.

Complement

An array of approximately 20 types of soluble proteins, called a **complement system**, functions to destroy extracellular pathogens. Cells of the liver and macrophages synthesize complement proteins continuously; these proteins are abundant in the blood serum and are capable of responding immediately to infecting microorganisms. The complement system is so named because it is complementary to the antibody response of the adaptive immune system. Complement proteins bind to the surfaces of microorganisms and are particularly attracted to pathogens that are already bound by antibodies. Binding of complement proteins occurs in a specific and highly regulated sequence, with each successive protein being activated by cleavage and/or structural changes induced upon binding of the preceding protein(s). After the first few complement proteins bind, a cascade of sequential binding events follows in which the pathogen rapidly becomes coated in complement proteins.

Complement proteins perform several functions. The proteins serve as a marker to indicate the presence of a pathogen to phagocytic cells, such as macrophages and B cells, and enhance engulfment; this process is called **opsonization**. Certain complement proteins can combine to form attack complexes that open pores in microbial cell membranes. These structures destroy pathogens by causing their contents to leak, as illustrated in Figure 42.7.

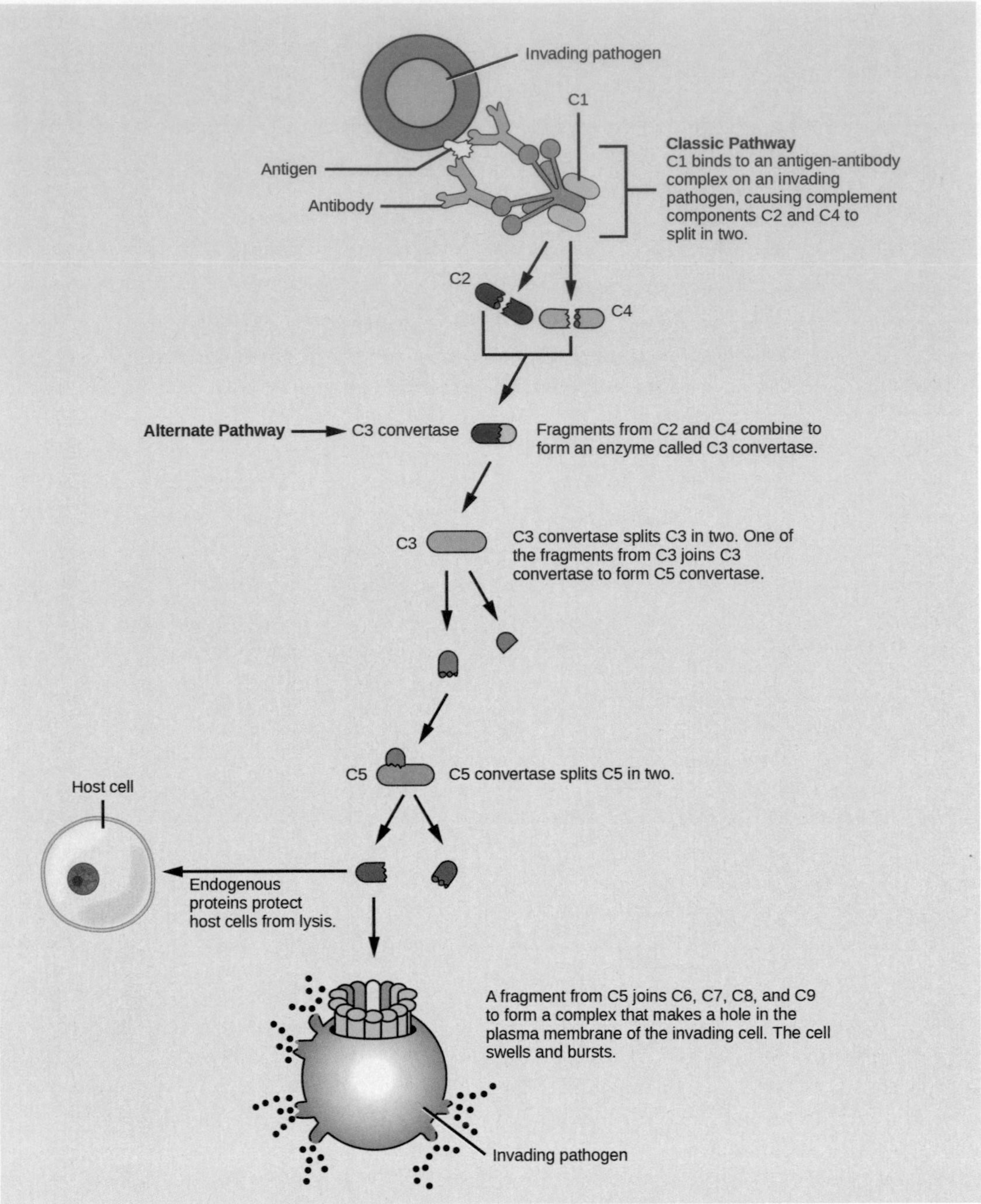

Figure 42.7 The classic pathway for the complement cascade involves the attachment of several initial complement proteins to an antibody-bound pathogen followed by rapid activation and binding of many more complement proteins and the creation of destructive pores in the microbial cell envelope and cell wall. The alternate pathway does not involve antibody activation. Rather, C3 convertase spontaneously breaks down C3. Endogenous regulatory proteins prevent the complement complex from binding to host cells. Pathogens lacking these regulatory proteins are lysed. (credit: modification of work by NIH)

42.2 Adaptive Immune Response

By the end of this section, you will be able to do the following:

- Explain adaptive immunity
- Compare and contrast adaptive and innate immunity
- Describe cell-mediated immune response and humoral immune response
- Describe immune tolerance

The adaptive, or acquired, immune response takes days or even weeks to become established—much longer than the innate response; however, adaptive immunity is more specific to pathogens and has memory. **Adaptive immunity** is an immunity that occurs after exposure to an antigen either from a pathogen or a vaccination. This part of the immune system is activated when the innate immune response is insufficient to control an infection. In fact, without information from the innate immune system, the adaptive response could not be mobilized. There are two types of adaptive responses: the **cell-mediated immune response**, which is carried out by T cells, and the **humoral immune response**, which is controlled by activated B cells and antibodies. Activated T cells and B cells that are specific to molecular structures on the pathogen proliferate and attack the invading pathogen. Their attack can kill pathogens directly or secrete antibodies that enhance the phagocytosis of pathogens and disrupt the infection. Adaptive immunity also involves a memory to provide the host with long-term protection from reinfection with the same type of pathogen; on reexposure, this memory will facilitate an efficient and quick response.

Antigen-presenting Cells

Unlike NK cells of the innate immune system, B cells (B lymphocytes) are a type of white blood cell that gives rise to antibodies, whereas T cells (T lymphocytes) are a type of white blood cell that plays an important role in the immune response. T cells are a key component in the cell-mediated response—the specific immune response that utilizes T cells to neutralize cells that have been infected with viruses and certain bacteria. There are three types of T cells: cytotoxic, helper, and suppressor T cells. Cytotoxic T cells destroy virus-infected cells in the cell-mediated immune response, and helper T cells play a part in activating both the antibody and the cell-mediated immune responses. Suppressor T cells deactivate T cells and B cells when needed, and thus prevent the immune response from becoming too intense.

An **antigen** is a foreign or "non-self" macromolecule that reacts with cells of the immune system. Not all antigens will provoke a response. For instance, individuals produce innumerable "self" antigens and are constantly exposed to harmless foreign antigens, such as food proteins, pollen, or dust components. The suppression of immune responses to harmless macromolecules is highly regulated and typically prevents processes that could be damaging to the host, known as tolerance.

The innate immune system contains cells that detect potentially harmful antigens, and then inform the adaptive immune response about the presence of these antigens. An **antigen-presenting cell (APC)** is an immune cell that detects, engulfs, and informs the adaptive immune response about an infection. When a pathogen is detected, these APCs will phagocytose the pathogen and digest it to form many different fragments of the antigen. Antigen fragments will then be transported to the surface of the APC, where they will serve as an indicator to other immune cells. **Dendritic cells** are immune cells that process antigen material; they are present in the skin (Langerhans cells) and the lining of the nose, lungs, stomach, and intestines. Sometimes a dendritic cell presents on the surface of other cells to induce an immune response, thus functioning as an antigen-presenting cell. Macrophages also function as APCs. Before activation and differentiation, B cells can also function as APCs.

After phagocytosis by APCs, the phagocytic vesicle fuses with an intracellular lysosome forming phagolysosome. Within the phagolysosome, the components are broken down into fragments; the fragments are then loaded onto MHC class I or MHC class II molecules and are transported to the cell surface for antigen presentation, as illustrated in Figure 42.8. Note that T lymphocytes cannot properly respond to the antigen unless it is processed and embedded in an MHC II molecule. APCs express MHC on their surfaces, and when combined with a foreign antigen, these complexes signal a "non-self" invader. Once the fragment of antigen is embedded in the MHC II molecule, the immune cell can respond. Helper T cells are one of the main lymphocytes that respond to antigen-presenting cells. Recall that all other nucleated cells of the body expressed MHC I molecules, which signal "healthy" or "normal."

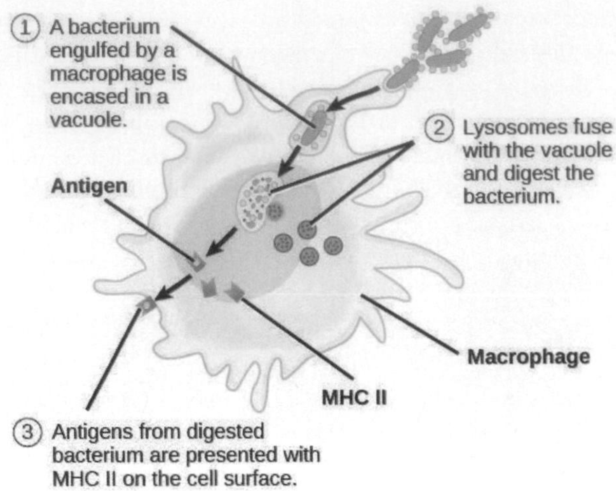

Figure 42.8 An APC, such as a macrophage, engulfs and digests a foreign bacterium. An antigen from the bacterium is presented on the cell surface in conjunction with an MHC II molecule. Lymphocytes of the adaptive immune response interact with antigen-embedded MHC II molecules to mature into functional immune cells.

LINK TO LEARNING

This animation (http://openstax.org/l/immune_system) from Rockefeller University shows how dendritic cells act as sentinels in the body's immune system.

T and B Lymphocytes

Lymphocytes in human circulating blood are approximately 80 to 90 percent T cells, shown in Figure 42.9, and 10 to 20 percent B cells. Recall that the T cells are involved in the cell-mediated immune response, whereas B cells are part of the humoral immune response.

T cells encompass a heterogeneous population of cells with extremely diverse functions. Some T cells respond to APCs of the innate immune system, and indirectly induce immune responses by releasing cytokines. Other T cells stimulate B cells to prepare their own response. Another population of T cells detects APC signals and directly kills the infected cells. Other T cells are involved in suppressing inappropriate immune reactions to harmless or "self" antigens.

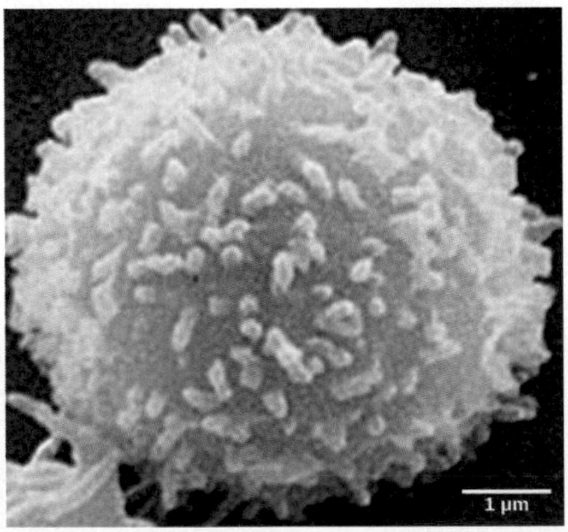

Figure 42.9 This scanning electron micrograph shows a T lymphocyte, which is responsible for the cell-mediated immune response. T cells are able to recognize antigens. (credit: modification of work by NCI; scale-bar data from Matt Russell)

T and B cells exhibit a common theme of recognition/binding of specific antigens via a complementary receptor, followed by activation and self-amplification/maturation to specifically bind to the particular antigen of the infecting pathogen. T and B lymphocytes are also similar in that each cell only expresses one type of antigen receptor. Any individual may possess a population of T and B cells that together express a near limitless variety of antigen receptors that are capable of recognizing virtually any infecting pathogen. T and B cells are activated when they recognize small components of antigens, called **epitopes**, presented by APCs, illustrated in Figure 42.10. Note that recognition occurs at a specific epitope rather than on the entire antigen; for this reason, epitopes are known as "antigenic determinants." In the absence of information from APCs, T and B cells remain inactive, or naïve, and are unable to prepare an immune response. The requirement for information from the APCs of innate immunity to trigger B cell or T cell activation illustrates the essential nature of the innate immune response to the functioning of the entire immune system.

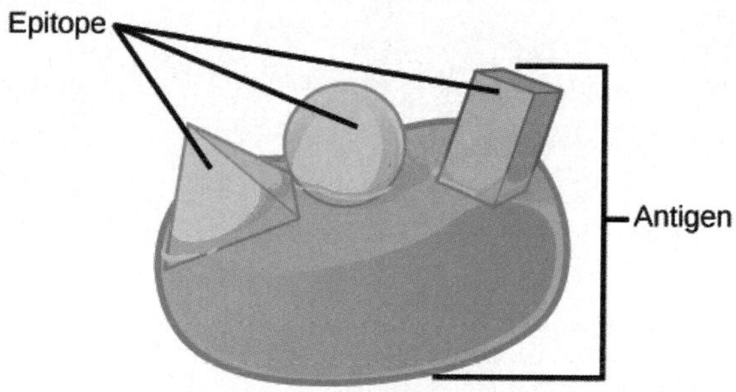

Figure 42.10 An antigen is a macromolecule that reacts with components of the immune system. A given antigen may contain several motifs that are recognized by immune cells. Each motif is an epitope. In this figure, the entire structure is an antigen, and the orange, salmon and green components projecting from it represent potential epitopes.

Naïve T cells can express one of two different molecules, CD4 or CD8, on their surface, as shown in Figure 42.11, and are accordingly classified as $CD4^+$ or $CD8^+$ cells. These molecules are important because they regulate how a T cell will interact with and respond to an APC. Naïve $CD4^+$ cells bind APCs via their antigen-embedded MHC II molecules and are stimulated to become **helper T (T_H) lymphocytes**, cells that go on to stimulate B cells (or cytotoxic T cells) directly or secrete cytokines to inform more and various target cells about the pathogenic threat. In contrast, $CD8^+$ cells engage antigen-embedded MHC I molecules on APCs and are stimulated to become **cytotoxic T lymphocytes (CTLs)**, which directly kill infected cells by apoptosis and emit cytokines to amplify the immune response. The two populations of T cells have different mechanisms of immune protection, but both bind MHC molecules via their antigen receptors called T cell receptors (TCRs). The CD4 or CD8 surface molecules differentiate whether the TCR will engage an MHC II or an MHC I molecule. Because they assist in binding specificity, the CD4 and CD8 molecules are described as coreceptors.

VISUAL CONNECTION

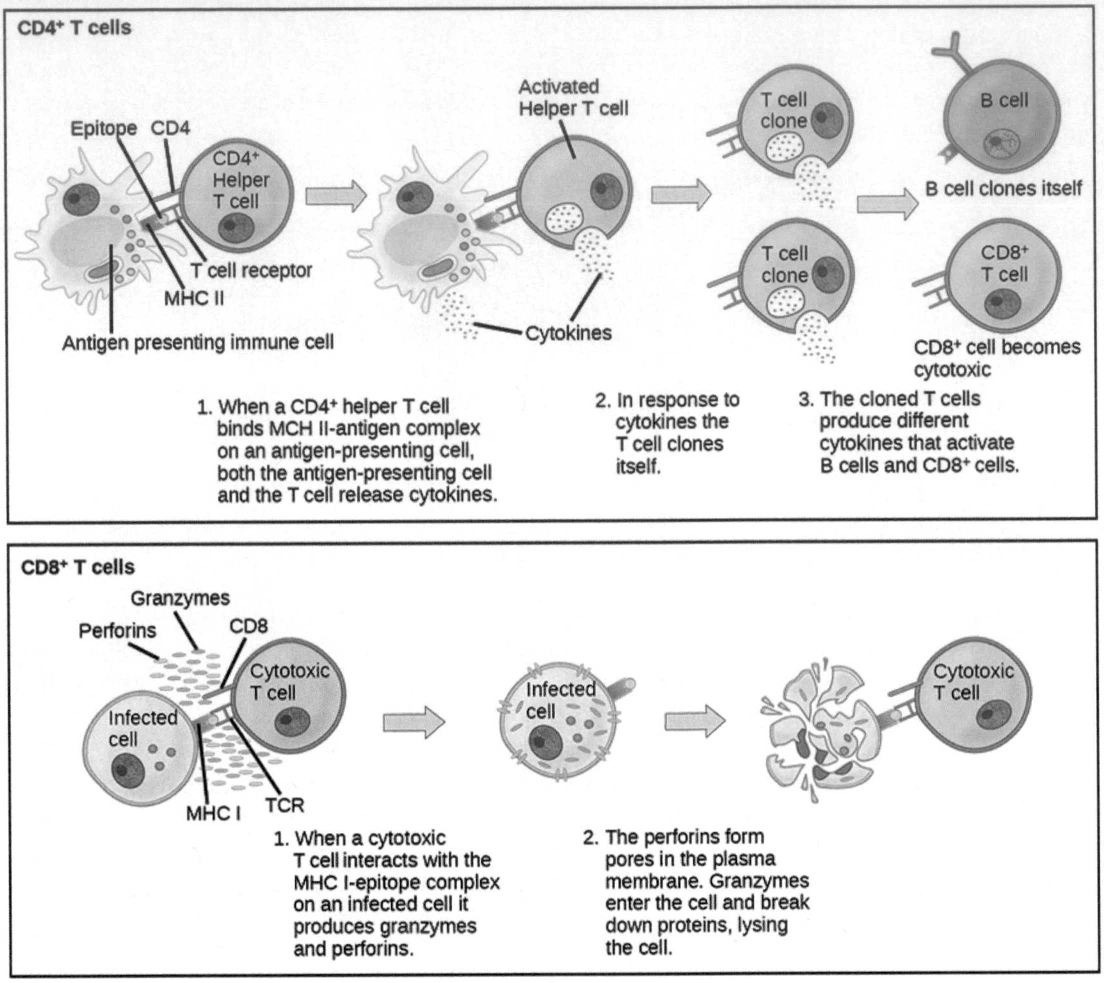

Figure 42.11 Naïve CD4$^+$ T cells engage MHC II molecules on antigen-presenting cells (APCs) and become activated. Clones of the activated helper T cell, in turn, activate B cells and CD8$^+$ T cells, which become cytotoxic T cells. Cytotoxic T cells kill infected cells.

Which of the following statements about T cells is false?

a. Helper T cells release cytokines while cytotoxic T cells kill the infected cell.
b. Helper T cells are CD4$^+$, while cytotoxic T cells are CD8$^+$.
c. MHC II is a receptor found on most body cells, while MHC I is a receptor found on immune cells only.
d. The T cell receptor is found on both CD4$^+$ and CD8$^+$ T cells.

Consider the innumerable possible antigens that an individual will be exposed to during a lifetime. The mammalian adaptive immune system is adept in responding appropriately to each antigen. Mammals have an enormous diversity of T cell populations, resulting from the diversity of TCRs. Each TCR consists of two polypeptide chains that span the T cell membrane, as illustrated in Figure 42.12; the chains are linked by a disulfide bridge. Each polypeptide chain is comprised of a constant domain and a variable domain: a domain, in this sense, is a specific region of a protein that may be regulatory or structural. The intracellular domain is involved in intracellular signaling. A single T cell will express thousands of identical copies of one specific TCR variant on its cell surface. The specificity of the adaptive immune system occurs because it synthesizes millions of different T cell populations, each expressing a TCR that differs in its variable domain. This TCR diversity is achieved by the mutation and recombination of genes that encode these receptors in stem cell precursors of T cells. The binding between an antigen-displaying MHC molecule and a complementary TCR "match" indicates that the adaptive immune system needs to activate and produce that specific T cell because its structure is appropriate to recognize and destroy the invading pathogen.

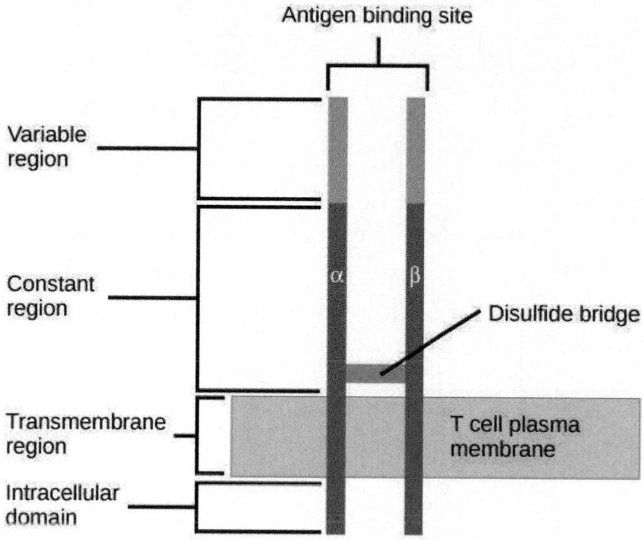

Figure 42.12 A T cell receptor spans the membrane and projects variable binding regions into the extracellular space to bind processed antigens via MHC molecules on APCs.

Helper T Lymphocytes

The T_H lymphocytes function indirectly to identify potential pathogens for other cells of the immune system. These cells are important for extracellular infections, such as those caused by certain bacteria, helminths, and protozoa. T_H lymphocytes recognize specific antigens displayed in the MHC II complexes of APCs. There are two major populations of T_H cells: T_H1 and T_H2. T_H1 cells secrete cytokines to enhance the activities of macrophages and other T cells. T_H1 cells activate the action of cyotoxic T cells, as well as macrophages. T_H2 cells stimulate naïve B cells to destroy foreign invaders via antibody secretion. Whether a T_H1 or a T_H2 immune response develops depends on the specific types of cytokines secreted by cells of the innate immune system, which in turn depends on the nature of the invading pathogen.

The T_H1-mediated response involves macrophages and is associated with inflammation. Recall the frontline defenses of macrophages involved in the innate immune response. Some intracellular bacteria, such as *Mycobacterium tuberculosis*, have evolved to multiply in macrophages after they have been engulfed. These pathogens evade attempts by macrophages to destroy and digest the pathogen. When *M. tuberculosis* infection occurs, macrophages can stimulate naïve T cells to become T_H1 cells. These stimulated T cells secrete specific cytokines that send feedback to the macrophage to stimulate its digestive capabilities and allow it to destroy the colonizing *M. tuberculosis*. In the same manner, T_H1-activated macrophages also become better suited to ingest and kill tumor cells. In summary; T_H1 responses are directed toward intracellular invaders while T_H2 responses are aimed at those that are extracellular.

B Lymphocytes

When stimulated by the T_H2 pathway, naïve B cells differentiate into antibody-secreting plasma cells. A **plasma cell** is an immune cell that secrets antibodies; these cells arise from B cells that were stimulated by antigens. Similar to T cells, naïve B cells initially are coated in thousands of B cell receptors (BCRs), which are membrane-bound forms of Ig (immunoglobulin, or an antibody). The B cell receptor has two heavy chains and two light chains connected by disulfide linkages. Each chain has a constant and a variable region; the latter is involved in antigen binding. Two other membrane proteins, Ig alpha and Ig beta, are involved in signaling. The receptors of any particular B cell, as shown in Figure 42.13 are all the same, but the hundreds of millions of different B cells in an individual have distinct recognition domains that contribute to extensive diversity in the types of molecular structures to which they can bind. In this state, B cells function as APCs. They bind and engulf foreign antigens via their BCRs and then display processed antigens in the context of MHC II molecules to T_H2 cells. When a T_H2 cell detects that a B cell is bound to a relevant antigen, it secretes specific cytokines that induce the B cell to proliferate rapidly, which makes thousands of identical (clonal) copies of it, and then it synthesizes and secretes antibodies with the same antigen recognition pattern as the BCRs. The activation of B cells corresponding to one specific BCR variant and the dramatic proliferation of that variant is known as **clonal selection**. This phenomenon drastically, but briefly, changes the proportions of BCR variants expressed by the immune system, and shifts the balance toward BCRs specific to the infecting pathogen.

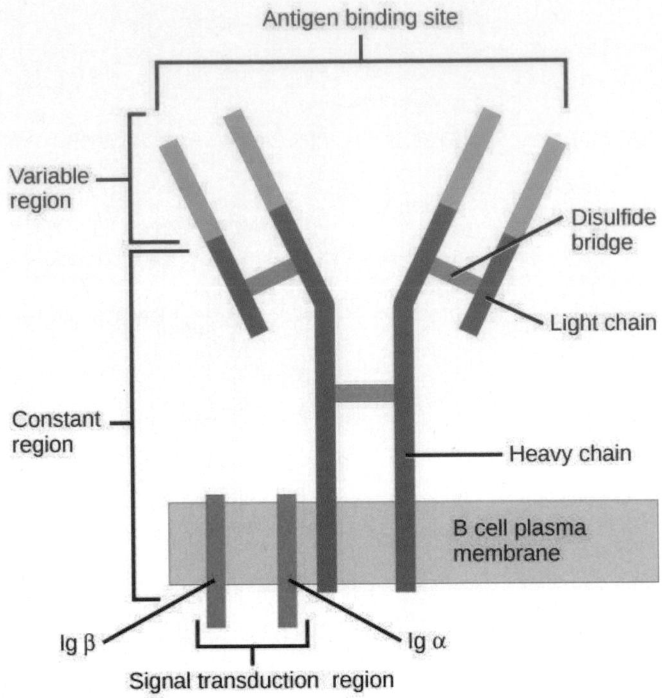

Figure 42.13 B cell receptors are embedded in the membranes of B cells and bind a variety of antigens through their variable regions. The signal transduction region transfers the signal into the cell.

T and B cells differ in one fundamental way: whereas T cells bind antigens that have been digested and embedded in MHC molecules by APCs, B cells function as APCs that bind intact antigens that have not been processed. Although T and B cells both react with molecules that are termed "antigens," these lymphocytes actually respond to very different types of molecules. B cells must be able to bind intact antigens because they secrete antibodies that must recognize the pathogen directly, rather than digested remnants of the pathogen. Bacterial carbohydrate and lipid molecules can activate B cells independently from the T cells.

Cytotoxic T Lymphocytes

CTLs, a subclass of T cells, function to clear infections directly. The cell-mediated part of the adaptive immune system consists of CTLs that attack and destroy infected cells. CTLs are particularly important in protecting against viral infections; this is because viruses replicate within cells where they are shielded from extracellular contact with circulating antibodies. When APCs phagocytize pathogens and present MHC I-embedded antigens to naïve $CD8^+$ T cells that express complementary TCRs, the $CD8^+$ T cells become activated to proliferate according to clonal selection. These resulting CTLs then identify non-APCs displaying the same MHC I-embedded antigens (for example, viral proteins)—for example, the CTLs identify infected host cells.

Intracellularly, infected cells typically die after the infecting pathogen replicates to a sufficient concentration and lyses the cell, as many viruses do. CTLs attempt to identify and destroy infected cells before the pathogen can replicate and escape, thereby halting the progression of intracellular infections. CTLs also support NK lymphocytes to destroy early cancers. Cytokines secreted by the T_H1 response that stimulates macrophages also stimulate CTLs and enhance their ability to identify and destroy infected cells and tumors.

CTLs sense MHC I-embedded antigens by directly interacting with infected cells via their TCRs. Binding of TCRs with antigens activates CTLs to release perforin and granzyme, degradative enzymes that will induce apoptosis of the infected cell. Recall that this is a similar destruction mechanism to that used by NK cells. In this process, the CTL does not become infected and is not harmed by the secretion of perforin and granzymes. In fact, the functions of NK cells and CTLs are complementary and maximize the removal of infected cells, as illustrated in Figure 42.14. If the NK cell cannot identify the "missing self" pattern of down-regulated MHC I molecules, then the CTL can identify it by the complex of MHC I with foreign antigens, which signals "altered self." Similarly, if the CTL cannot detect antigen-embedded MHC I because the receptors are depleted from the cell surface, NK cells will destroy the cell instead. CTLs also emit cytokines, such as interferons, that alter surface protein expression in other infected cells, such that the infected cells can be easily identified and destroyed. Moreover, these interferons can also prevent virally infected cells from releasing virus particles.

VISUAL CONNECTION

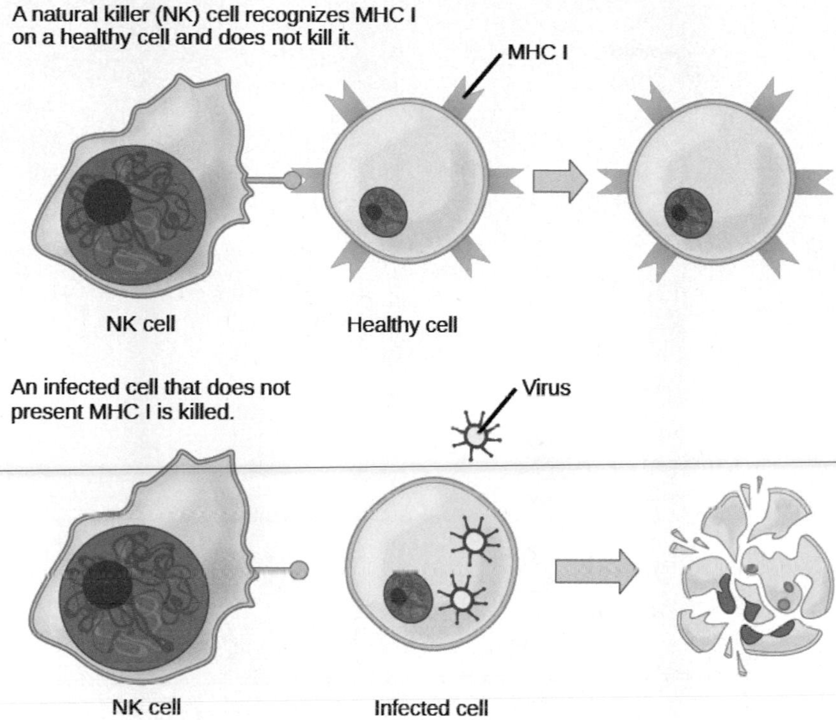

Figure 42.14 Natural killer (NK) cells recognize the MHC I receptor on healthy cells. If MHC I is absent, the cell is lysed.

Based on what you know about MHC receptors, why do you think an organ transplanted from an incompatible donor to a recipient will be rejected?

Plasma cells and CTLs are collectively called **effector cells**: they represent differentiated versions of their naïve counterparts, and they are involved in bringing about the immune defense of killing pathogens and infected host cells.

Mucosal Surfaces and Immune Tolerance

The innate and adaptive immune responses discussed thus far comprise the systemic immune system (affecting the whole body), which is distinct from the mucosal immune system. Mucosal immunity is formed by mucosa-associated lymphoid tissue, which functions independently of the systemic immune system, and which has its own innate and adaptive components. **Mucosa-associated lymphoid tissue (MALT)**, illustrated in Figure 42.15, is a collection of lymphatic tissue that combines with epithelial tissue lining the mucosa throughout the body. This tissue functions as the immune barrier and response in areas of the body with direct contact to the external environment. The systemic and mucosal immune systems use many of the same cell types. Foreign particles that make their way to MALT are taken up by absorptive epithelial cells called M cells and delivered to APCs located directly below the mucosal tissue. M cells function in the transport described, and are located in the Peyer's patch, a lymphoid nodule. APCs of the mucosal immune system are primarily dendritic cells, with B cells and macrophages having minor roles. Processed antigens displayed on APCs are detected by T cells in the MALT and at various mucosal induction sites, such as the tonsils, adenoids, appendix, or the mesenteric lymph nodes of the intestine. Activated T cells then migrate through the lymphatic system and into the circulatory system to mucosal sites of infection.

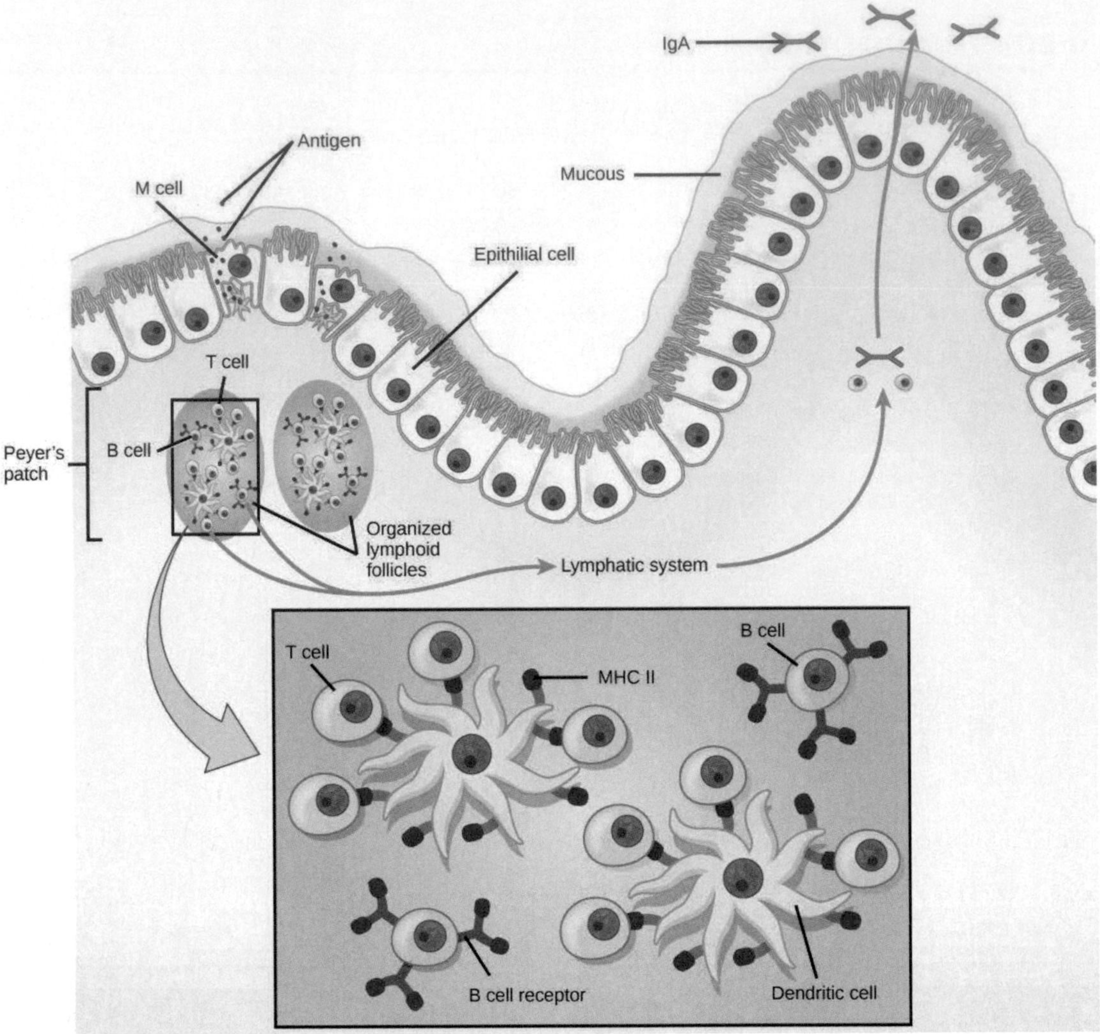

Figure 42.15 The topology and function of intestinal MALT is shown. Pathogens are taken up by M cells in the intestinal epithelium and excreted into a pocket formed by the inner surface of the cell. The pocket contains antigen-presenting cells such as dendritic cells, which engulf the antigens, then present them with MHC II molecules on the cell surface. The dendritic cells migrate to an underlying tissue called a Peyer's patch. Antigen-presenting cells, T cells, and B cells aggregate within the Peyer's patch, forming organized lymphoid follicles. There, some T cells and B cells are activated. Other antigen-loaded dendritic cells migrate through the lymphatic system where they activate B cells, T cells, and plasma cells in the lymph nodes. The activated cells then return to MALT tissue effector sites. IgA and other antibodies are secreted into the intestinal lumen.

MALT is a crucial component of a functional immune system because mucosal surfaces, such as the nasal passages, are the first tissues onto which inhaled or ingested pathogens are deposited. The mucosal tissue includes the mouth, pharynx, and esophagus, and the gastrointestinal, respiratory, and urogenital tracts.

The immune system has to be regulated to prevent wasteful, unnecessary responses to harmless substances, and more importantly so that it does not attack "self." The acquired ability to prevent an unnecessary or harmful immune response to a detected foreign substance known not to cause disease is described as **immune tolerance**. Immune tolerance is crucial for maintaining mucosal homeostasis given the tremendous number of foreign substances (such as food proteins) that APCs of the oral cavity, pharynx, and gastrointestinal mucosa encounter. Immune tolerance is brought about by specialized APCs in the liver, lymph nodes, small intestine, and lung that present harmless antigens to an exceptionally diverse population of **regulatory T (T_{reg}) cells**, specialized lymphocytes that suppress local inflammation and inhibit the secretion of stimulatory immune factors. The combined result of T_{reg} cells is to prevent immunologic activation and inflammation in undesired tissue compartments and to allow the immune system to focus on pathogens instead. In addition to promoting immune tolerance of harmless antigens, other subsets of T_{reg} cells are involved in the prevention of the **autoimmune response**, which is an inappropriate immune response to host cells or self-antigens. Another T_{reg} class suppresses immune responses to harmful pathogens after the infection has cleared to minimize host cell damage induced by inflammation and cell lysis.

Immunological Memory

The adaptive immune system possesses a memory component that allows for an efficient and dramatic response upon reinvasion of the same pathogen. Memory is handled by the adaptive immune system with little reliance on cues from the innate response. During the adaptive immune response to a pathogen that has not been encountered before, called a primary response, plasma cells secreting antibodies and differentiated T cells increase, then plateau over time. As B and T cells mature into effector cells, a subset of the naïve populations differentiates into B and T memory cells with the same antigen specificities, as illustrated in Figure 42.16.

A **memory cell** is an antigen-specific B or T lymphocyte that does not differentiate into effector cells during the primary immune response, but that can immediately become effector cells upon reexposure to the same pathogen. During the primary immune response, memory cells do not respond to antigens and do not contribute to host defenses. As the infection is cleared and pathogenic stimuli subside, the effectors are no longer needed, and they undergo apoptosis. In contrast, the memory cells persist in the circulation.

VISUAL CONNECTION

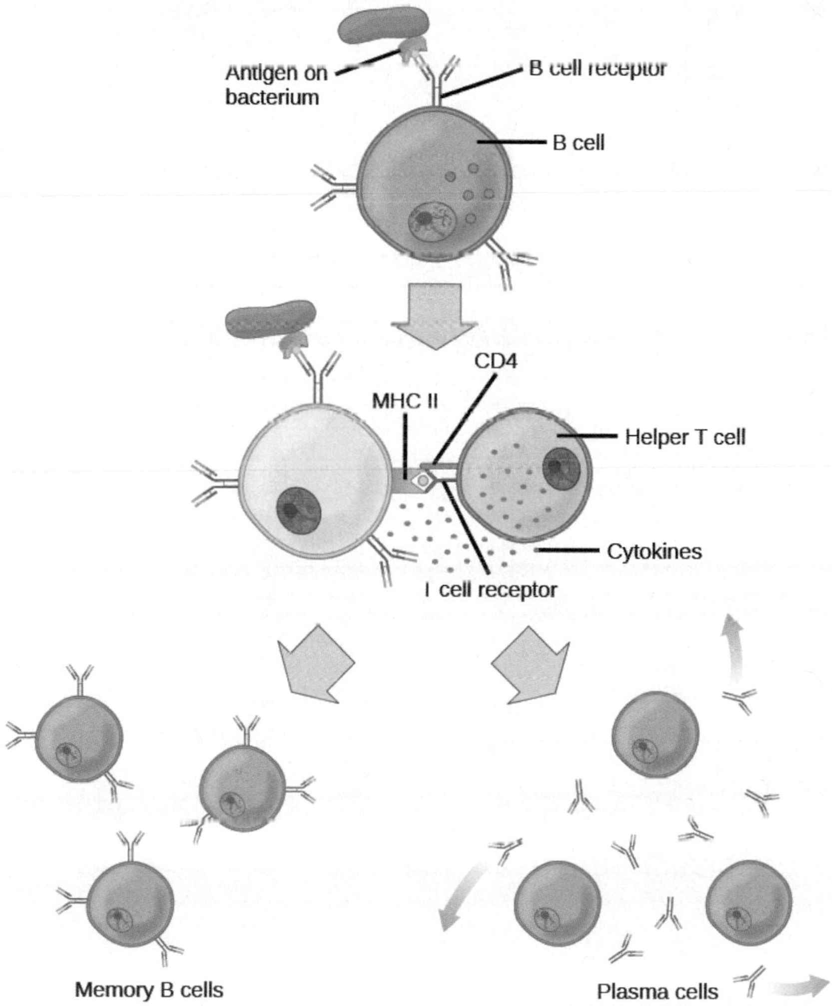

Figure 42.16 After initially binding an antigen to the B cell receptor (BCR), a B cell internalizes the antigen and presents it on MHC II. A helper T cell recognizes the MHC II–antigen complex and activates the B cell. As a result, memory B cells and plasma cells are made.

The Rh antigen is found on Rh-positive red blood cells. An Rh-negative female can usually carry an Rh-positive fetus to term without difficulty. However, if she has a second Rh-positive fetus, her body may launch an immune attack that causes hemolytic disease of the newborn. Why do you think hemolytic disease is only a problem during the second or subsequent pregnancies?

If the pathogen is never encountered again during the individual's lifetime, B and T memory cells will circulate for a few years or even several decades and will gradually die off, having never functioned as effector cells. However, if the host is reexposed to the same pathogen type, circulating memory cells will immediately differentiate into plasma cells and CTLs without input from APCs or T_H cells. One reason the adaptive immune response is delayed is because it takes time for naïve B and T cells with the appropriate antigen specificities to be identified and activated. Upon reinfection, this step is skipped, and the result is a more rapid production of immune defenses. Memory B cells that differentiate into plasma cells output tens to hundreds-fold greater antibody amounts than were secreted during the primary response, as the graph in Figure 42.17 illustrates. This rapid and dramatic antibody response may stop the infection before it can even become established, and the individual may not realize he or she had been exposed.

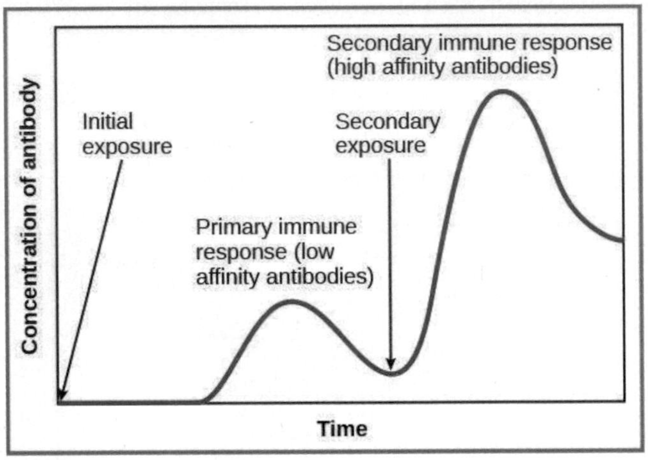

Figure 42.17 In the primary response to infection, antibodies are secreted first from plasma cells. Upon reexposure to the same pathogen, memory cells differentiate into antibody-secreting plasma cells that output a greater amount of antibody for a longer period of time.

Vaccination is based on the knowledge that exposure to noninfectious antigens, derived from known pathogens, generates a mild primary immune response. The immune response to vaccination may not be perceived by the host as illness but still confers immune memory. When exposed to the corresponding pathogen to which an individual was vaccinated, the reaction is similar to a secondary exposure. Because each reinfection generates more memory cells and increased resistance to the pathogen, and because some memory cells die, certain vaccine courses involve one or more booster vaccinations to mimic repeat exposures: for instance, tetanus boosters are necessary every ten years because the memory cells only live that long.

Mucosal Immune Memory

A subset of T and B cells of the mucosal immune system differentiates into memory cells just as in the systemic immune system. Upon reinvasion of the same pathogen type, a pronounced immune response occurs at the mucosal site where the original pathogen deposited, but a collective defense is also organized within interconnected or adjacent mucosal tissue. For instance, the immune memory of an infection in the oral cavity would also elicit a response in the pharynx if the oral cavity was exposed to the same pathogen.

CAREER CONNECTION

Vaccinologist

Vaccination (or immunization) involves the delivery, usually by injection as shown in Figure 42.18, of noninfectious antigen(s) derived from known pathogens. Other components, called adjuvants, are delivered in parallel to help stimulate the immune response. Immunological memory is the reason vaccines work. Ideally, the effect of vaccination is to elicit immunological memory, and thus resistance to specific pathogens without the individual having to experience an infection.

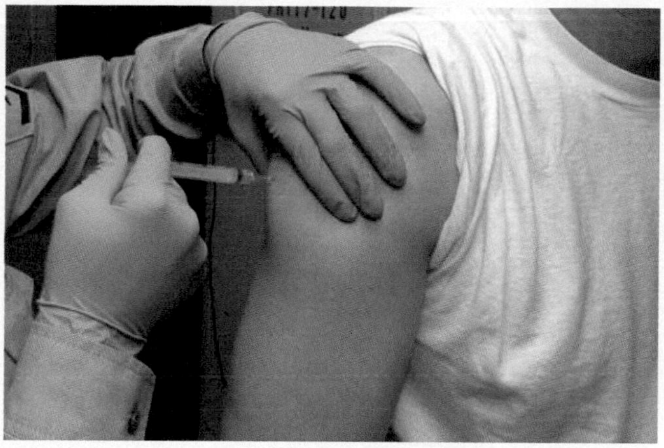

Figure 42.18 Vaccines are often delivered by injection into the arm. (credit: U.S. Navy Photographer's Mate Airman Apprentice Christopher D. Blachly)

Vaccinologists are involved in the process of vaccine development from the initial idea to the availability of the completed vaccine. This process can take decades, can cost millions of dollars, and can involve many obstacles along the way. For instance, injected vaccines stimulate the systemic immune system, eliciting humoral and cell-mediated immunity, but have little effect on the mucosal response, which presents a challenge because many pathogens are deposited and replicate in mucosal compartments, and the injection does not provide the most efficient immune memory for these disease agents. For this reason, vaccinologists are actively involved in developing new vaccines that are applied via intranasal, aerosol, oral, or transcutaneous (absorbed through the skin) delivery methods. Importantly, mucosal-administered vaccines elicit both mucosal and systemic immunity and produce the same level of disease resistance as injected vaccines.

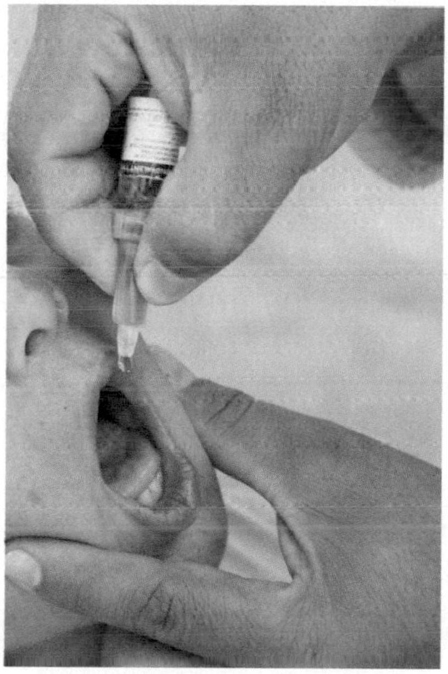

Figure 42.19 The polio vaccine can be administered orally. (credit: modification of work by UNICEF Sverige)

Currently, a version of intranasal influenza vaccine is available, and the polio and typhoid vaccines can be administered orally, as shown in Figure 42.19. Similarly, the measles and rubella vaccines are being adapted to aerosol delivery using inhalation devices. Eventually, transgenic plants may be engineered to produce vaccine antigens that can be eaten to confer disease resistance. Other vaccines may be adapted to rectal or vaginal application to elicit immune responses in rectal, genitourinary, or reproductive mucosa. Finally, vaccine antigens may be adapted to transdermal application in which the skin is lightly scraped and microneedles are used to pierce the outermost layer. In addition to mobilizing the mucosal immune response, this new generation of vaccines may end the anxiety associated with injections and, in turn, improve patient participation.

Primary Centers of the Immune System

Although the immune system is characterized by circulating cells throughout the body, the regulation, maturation, and intercommunication of immune factors occur at specific sites. The blood circulates immune cells, proteins, and other factors through the body. Approximately 0.1 percent of all cells in the blood are leukocytes, which encompass monocytes (the precursor of macrophages) and lymphocytes. The majority of cells in the blood are erythrocytes (red blood cells). **Lymph** is a watery fluid that bathes tissues and organs with protective white blood cells and does not contain erythrocytes. Cells of the immune system can travel between the distinct lymphatic and blood circulatory systems, which are separated by interstitial space, by a process called extravasation (passing through to surrounding tissue).

The cells of the immune system originate from hematopoietic stem cells in the bone marrow. Cytokines stimulate these stem cells to differentiate into immune cells. B cell maturation occurs in the bone marrow, whereas naïve T cells transit from the bone marrow to the thymus for maturation. In the thymus, immature T cells that express TCRs complementary to self-antigens are destroyed. This process helps prevent autoimmune responses.

On maturation, T and B lymphocytes circulate to various destinations. Lymph nodes scattered throughout the body, as illustrated in Figure 42.20, house large populations of T and B cells, dendritic cells, and macrophages. Lymph gathers antigens as it drains from tissues. These antigens then are filtered through lymph nodes before the lymph is returned to circulation. APCs in the lymph nodes capture and process antigens and inform nearby lymphocytes about potential pathogens.

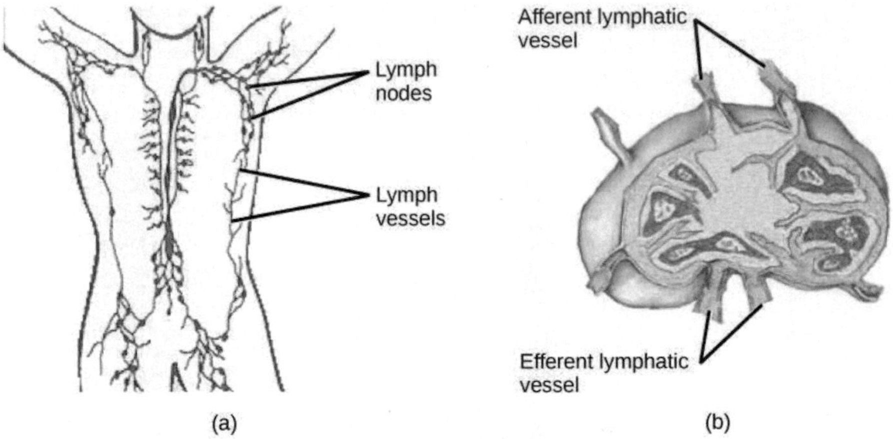

Figure 42.20 (a) Lymphatic vessels carry a clear fluid called lymph throughout the body. The liquid enters (b) lymph nodes through afferent vessels. Lymph nodes are filled with lymphocytes that purge infecting cells. The lymph then exits through efferent vessels. (credit: modification of work by NIH, NCI)

The spleen houses B and T cells, macrophages, dendritic cells, and NK cells. The spleen, shown in Figure 42.21, is the site where APCs that have trapped foreign particles in the blood can communicate with lymphocytes. Antibodies are synthesized and secreted by activated plasma cells in the spleen, and the spleen filters foreign substances and antibody-complexed pathogens from the blood. Functionally, the spleen is to the blood as lymph nodes are to the lymph.

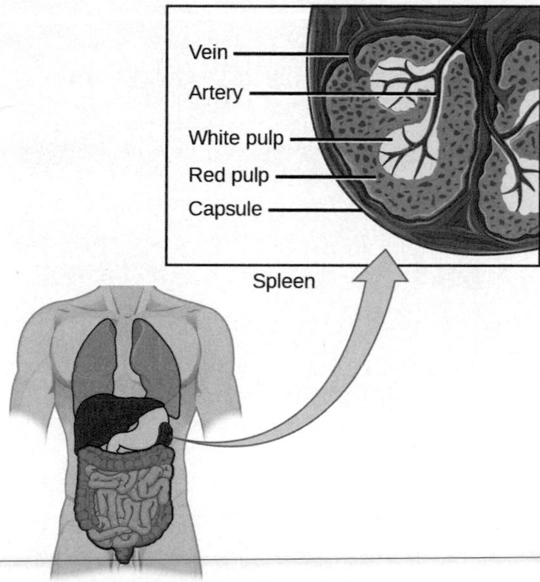

Figure 42.21 The spleen is similar to a lymph node but is much larger and filters blood instead of lymph. Blood enters the spleen through arteries and exits through veins. The spleen contains two types of tissue: red pulp and white pulp. Red pulp consists of cavities that store blood. Within the red pulp, damaged red blood cells are removed and replaced by new ones. White pulp is rich in lymphocytes that remove antigen-coated bacteria from the blood. (credit: modification of work by NCI)

42.3 Antibodies

By the end of this section, you will be able to do the following:

- Explain cross-reactivity
- Describe the structure and function of antibodies
- Discuss antibody production

An **antibody**, also known as an immunoglobulin (Ig), is a protein that is produced by plasma cells after stimulation by an antigen. Antibodies are the functional basis of humoral immunity. Antibodies occur in the blood, in gastric and mucus secretions, and in breast milk. Antibodies in these bodily fluids can bind pathogens and mark them for destruction by phagocytes before they can infect cells.

Antibody Structure

An antibody molecule is comprised of four polypeptides: two identical heavy chains (large peptide units) that are partially bound to each other in a "Y" formation, which are flanked by two identical light chains (small peptide units), as illustrated in Figure 42.22. Bonds between the cysteine amino acids in the antibody molecule attach the polypeptides to each other. The areas where the antigen is recognized on the antibody are variable domains and the antibody base is composed of constant domains.

In germ-line B cells, the variable region of the light chain gene has 40 variable (V) and five joining (J) segments. An enzyme called DNA recombinase randomly excises most of these segments out of the gene, and splices one V segment to one J segment. During RNA processing, all but one V and J segment are spliced out. Recombination and splicing may result in over 10^6 possible VJ combinations. As a result, each differentiated B cell in the human body typically has a unique variable chain. The constant domain, which does not bind antibody, is the same for all antibodies.

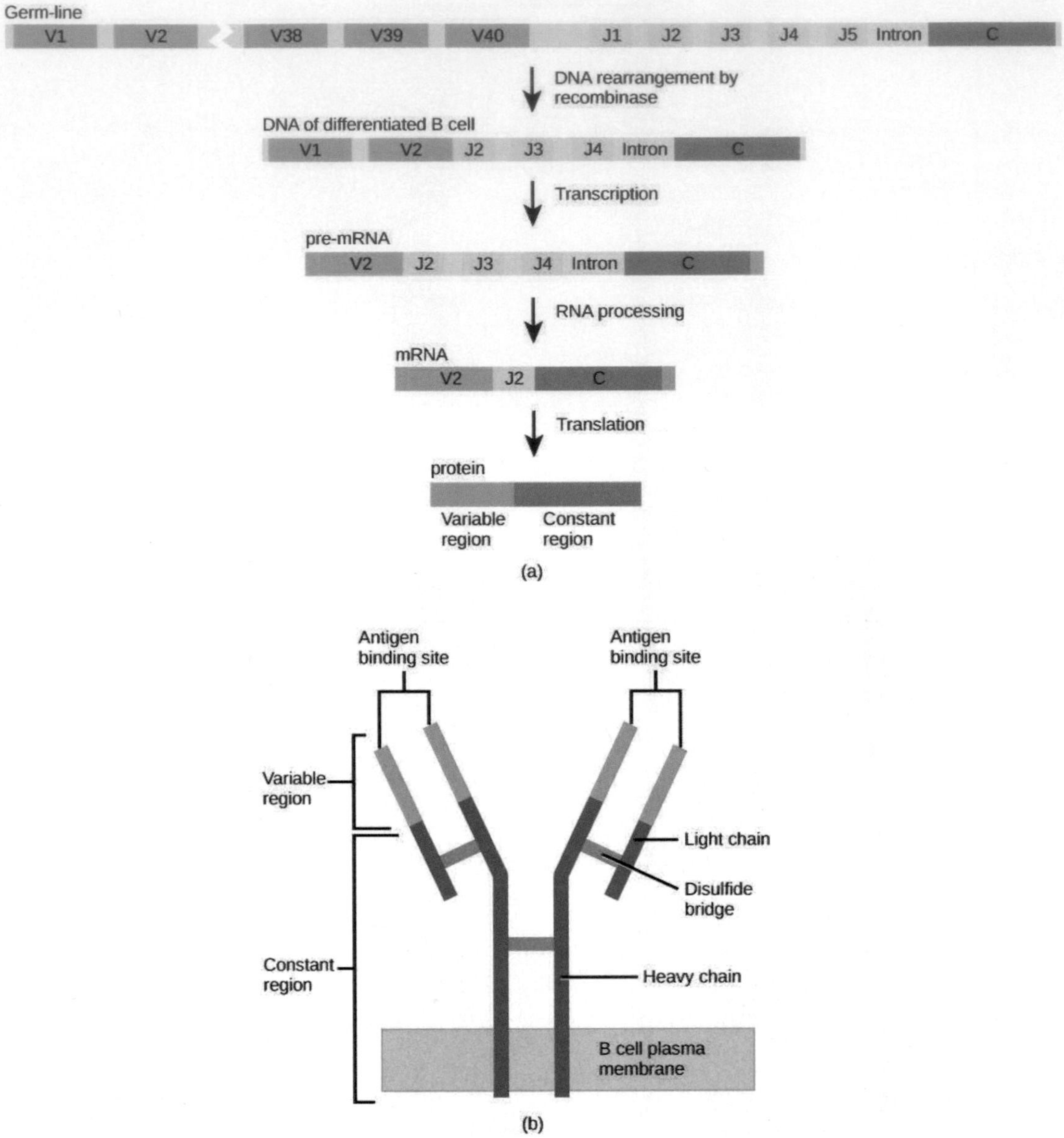

Figure 42.22 (a) As a germ-line B cell matures, an enzyme called DNA recombinase randomly excises V and J segments from the light chain gene. Splicing at the mRNA level results in further gene rearrangement. As a result, (b) each antibody has a unique variable region capable of binding a different antigen.

Similar to TCRs and BCRs, antibody diversity is produced by the mutation and recombination of approximately 300 different gene segments encoding the light and heavy chain variable domains in precursor cells that are destined to become B cells. The variable domains from the heavy and light chains interact to form the binding site through which an antibody can bind a specific epitope on an antigen. The numbers of repeated constant domains in Ig classes are the same for all antibodies corresponding to a specific class. Antibodies are structurally similar to the extracellular component of the BCRs, and B cell maturation to plasma cells can be visualized in simple terms as the cell acquires the ability to secrete the extracellular portion of its BCR in large quantities.

Antibody Classes

Antibodies can be divided into five classes—IgM, IgG, IgA, IgD, IgE—based on their physiochemical, structural, and immunological properties. IgGs, which make up about 80 percent of all antibodies, have heavy chains that consist of one variable domain and three identical constant domains. IgA and IgD also have three constant domains per heavy chain, whereas IgM and IgE each have four constant domains per heavy chain. The variable domain determines binding specificity and the constant domain of the heavy chain determines the immunological mechanism of action of the corresponding antibody class. It is possible for two antibodies to have the same binding specificities but be in different classes and, therefore, to be involved in different functions.

After an adaptive defense is produced against a pathogen, typically plasma cells first secrete IgM into the blood. BCRs on naïve B cells are of the IgM class and occasionally IgD class. IgM molecules make up approximately ten percent of all antibodies. Prior to antibody secretion, plasma cells assemble IgM molecules into pentamers (five individual antibodies) linked by a joining (J) chain, as shown in Figure 42.23. The pentamer arrangement means that these macromolecules can bind ten identical antigens. However, IgM molecules released early in the adaptive immune response do not bind to antigens as stably as IgGs, which are one of the possible types of antibodies secreted in large quantities upon reexposure to the same pathogen. Figure 42.23 summarizes the properties of immunoglobulins and illustrates their basic structures.

Name	Properties	Structure
IgA	Found in mucous, saliva, tears, and breast milk. Protects against pathogens.	
IgD	Part of the B cell receptor. Activates basophils and mast cells.	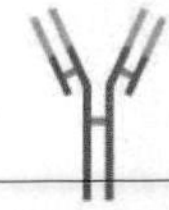
IgE	Protects against parasitic worms. Responsible for allergic reactions.	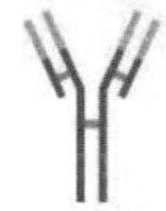
IgG	Secreted by plasma cells in the blood. Able to cross the placenta into the fetus.	
IgM	May be attached to the surface of a B cell or secreted into the blood. Responsible for early stages of immunity.	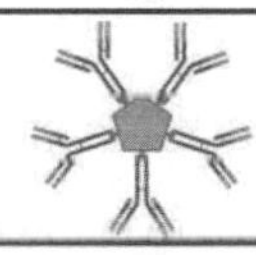

Figure 42.23 Immunoglobulins have different functions, but all are composed of light and heavy chains that form a Y-shaped structure.

IgAs populate the saliva, tears, breast milk, and mucus secretions of the gastrointestinal, respiratory, and genitourinary tracts. Collectively, these bodily fluids coat and protect the extensive mucosa (4000 square feet in humans). The total number of IgA molecules in these bodily secretions is greater than the number of IgG molecules in the blood serum. A small amount of IgA is also secreted into the serum in monomeric form. Conversely, some IgM is secreted into bodily fluids of the mucosa. Similar to IgM, IgA molecules are secreted as polymeric structures linked with a J chain. However, IgAs are secreted mostly as dimeric molecules, not pentamers.

IgE is present in the serum in small quantities and is best characterized in its role as an allergy mediator. IgD is also present in small quantities. Similar to IgM, BCRs of the IgD class are found on the surface of naïve B cells. This class supports antigen recognition and maturation of B cells to plasma cells.

Antibody Functions

Differentiated plasma cells are crucial players in the humoral response, and the antibodies they secrete are particularly significant against extracellular pathogens and toxins. Antibodies circulate freely and act independently of plasma cells. Antibodies can be transferred from one individual to another to temporarily protect against infectious disease. For instance, a person who has recently produced a successful immune response against a particular disease agent can donate blood to a nonimmune recipient and confer temporary immunity through antibodies in the donor's blood serum. This phenomenon is called **passive immunity**; it also occurs naturally during breastfeeding, which makes breastfed infants highly resistant to infections during the first few months of life.

Antibodies coat extracellular pathogens and neutralize them, as illustrated in Figure 42.24, by blocking key sites on the pathogen that enhance their infectivity (such as receptors that "dock" pathogens on host cells). Antibody neutralization can prevent pathogens from entering and infecting host cells, as opposed to the CTL-mediated approach of killing cells that are already infected to prevent progression of an established infection. The neutralized antibody-coated pathogens can then be filtered by the spleen and eliminated in urine or feces.

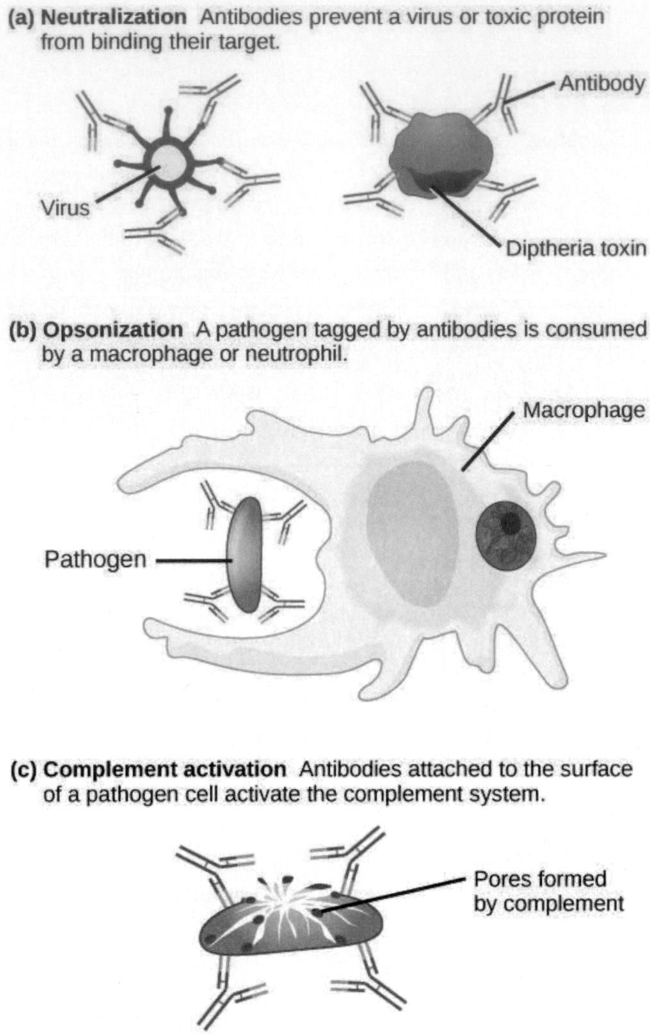

Figure 42.24 Antibodies may inhibit infection by (a) preventing the antigen from binding its target, (b) tagging a pathogen for destruction by macrophages or neutrophils, or (c) activating the complement cascade.

Antibodies also mark pathogens for destruction by phagocytic cells, such as macrophages or neutrophils, because phagocytic cells are highly attracted to macromolecules complexed with antibodies. Phagocytic enhancement by antibodies is called opsonization. In a process called complement fixation, IgM and IgG in serum bind to antigens and provide docking sites onto which sequential complement proteins can bind. The combination of antibodies and complement enhances opsonization even further and promotes rapid clearing of pathogens.

Affinity, Avidity, and Cross Reactivity

Not all antibodies bind with the same strength, specificity, and stability. In fact, antibodies exhibit different **affinities** (attraction) depending on the molecular complementarity between antigen and antibody molecules, as illustrated in Figure 42.25. An antibody with a higher affinity for a particular antigen would bind more strongly and stably, and thus would be expected to present a more challenging defense against the pathogen corresponding to the specific antigen.

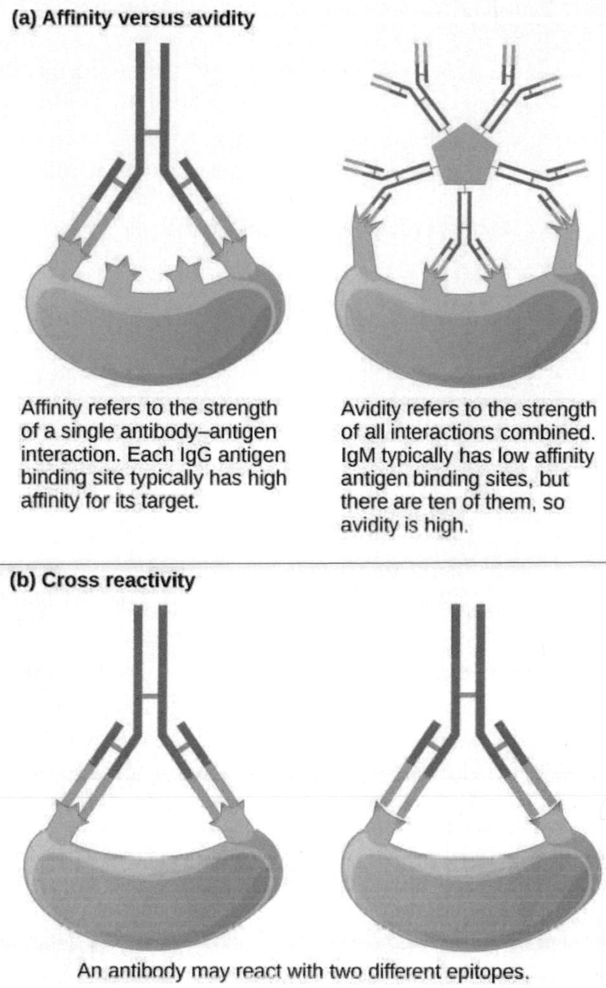

Figure 42.25 (a) Affinity refers to the strength of single interaction between antigen and antibody, while avidity refers to the strength of all interactions combined. (b) An antibody may cross react with different epitopes.

The term **avidity** describes binding by antibody classes that are secreted as joined, multivalent structures (such as IgM and IgA). Although avidity measures the strength of binding, just as affinity does, the avidity is not simply the sum of the affinities of the antibodies in a multimeric structure. The avidity depends on the number of identical binding sites on the antigen being detected, as well as other physical and chemical factors. Typically, multimeric antibodies, such as pentameric IgM, are classified as having lower affinity than monomeric antibodies, but high avidity. Essentially, the fact that multimeric antibodies can bind many antigens simultaneously balances their slightly lower binding strength for each antibody/antigen interaction.

Antibodies secreted after binding to one epitope on an antigen may exhibit cross reactivity for the same or similar epitopes on different antigens. Because an epitope corresponds to such a small region (the surface area of about four to six amino acids), it is possible for different macromolecules to exhibit the same molecular identities and orientations over short regions. **Cross reactivity** describes when an antibody binds not to the antigen that elicited its synthesis and secretion, but to a different antigen.

Cross reactivity can be beneficial if an individual develops immunity to several related pathogens despite having only been exposed to or vaccinated against one of them. For instance, antibody cross reactivity may occur against the similar surface structures of various Gram-negative bacteria. Conversely, antibodies raised against pathogenic molecular components that resemble self molecules may incorrectly mark host cells for destruction and cause autoimmune damage. Patients who develop systemic lupus erythematosus (SLE) commonly exhibit antibodies that react with their own DNA. These antibodies may have been initially raised against the nucleic acid of microorganisms but later cross-reacted with self-antigens. This phenomenon is also called molecular mimicry.

Antibodies of the Mucosal Immune System

Antibodies synthesized by the mucosal immune system include IgA and IgM. Activated B cells differentiate into mucosal plasma cells that synthesize and secrete dimeric IgA, and to a lesser extent, pentameric IgM. Secreted IgA is abundant in tears, saliva, breast milk, and in secretions of the gastrointestinal and respiratory tracts. Antibody secretion results in a local humoral response at epithelial surfaces and prevents infection of the mucosa by binding and neutralizing pathogens.

42.4 Disruptions in the Immune System

By the end of this section, you will be able to do the following:

- Describe hypersensitivity
- Define autoimmunity

A functioning immune system is essential for survival, but even the sophisticated cellular and molecular defenses of the mammalian immune response can be defeated by pathogens at virtually every step. In the competition between immune protection and pathogen evasion, pathogens have the advantage of more rapid evolution because of their shorter generation time and other characteristics. For instance, *Streptococcus pneumoniae* (bacterium that cause pneumonia and meningitis) surrounds itself with a capsule that inhibits phagocytes from engulfing it and displaying antigens to the adaptive immune system. *Staphylococcus aureus* (bacterium that can cause skin infections, abscesses, and meningitis) synthesizes a toxin called leukocidin that kills phagocytes after they engulf the bacterium. Other pathogens can also hinder the adaptive immune system. HIV infects T_H cells via their CD4 surface molecules, gradually depleting the number of T_H cells in the body; this inhibits the adaptive immune system's capacity to generate sufficient responses to infection or tumors. As a result, HIV-infected individuals often suffer from infections that would not cause illness in people with healthy immune systems but which can cause devastating illness to immune-compromised individuals. Maladaptive responses of immune cells and molecules themselves can also disrupt the proper functioning of the entire system, leading to host cell damage that could become fatal.

Immunodeficiency

Failures, insufficiencies, or delays at any level of the immune response can allow pathogens or tumor cells to gain a foothold and replicate or proliferate to high enough levels that the immune system becomes overwhelmed. **Immunodeficiency** is the failure, insufficiency, or delay in the response of the immune system, which may be acquired or inherited. Immunodeficiency can be acquired as a result of infection with certain pathogens (such as HIV), chemical exposure (including certain medical treatments), malnutrition, or possibly by extreme stress. For instance, radiation exposure can destroy populations of lymphocytes and elevate an individual's susceptibility to infections and cancer. Dozens of genetic disorders result in immunodeficiencies, including Severe Combined Immunodeficiency (SCID), Bare lymphocyte syndrome, and MHC II deficiencies. Rarely, primary immunodeficiencies that are present from birth may occur. Neutropenia is one form in which the immune system produces a below-average number of neutrophils, the body's most abundant phagocytes. As a result, bacterial infections may go unrestricted in the blood, causing serious complications.

Hypersensitivities

Maladaptive immune responses toward harmless foreign substances or self antigens that occur after tissue sensitization are termed **hypersensitivities**. The types of hypersensitivities include immediate, delayed, and autoimmunity. A large proportion of the population is affected by one or more types of hypersensitivity.

Allergies

The immune reaction that results from immediate hypersensitivities in which an antibody-mediated immune response occurs within minutes of exposure to a harmless antigen is called an **allergy**. In the United States, 20 percent of the population exhibits symptoms of allergy or asthma, whereas 55 percent test positive against one or more allergens. Upon initial exposure to a potential allergen, an allergic individual synthesizes antibodies of the IgE class via the typical process of APCs presenting processed antigen to T_H cells that stimulate B cells to produce IgE. This class of antibodies also mediates the immune response to parasitic worms. The constant domain of the IgE molecules interact with mast cells embedded in connective tissues. This process primes, or sensitizes, the tissue. Upon subsequent exposure to the same allergen, IgE molecules on mast cells bind the antigen via their variable domains and stimulate the mast cell to release the modified amino acids histamine and serotonin; these chemical mediators then recruit eosinophils which mediate allergic responses. Figure 42.26 shows an example of an allergic response to ragweed pollen. The effects of an allergic reaction range from mild symptoms like sneezing and itchy, watery eyes to more severe or even life-threatening reactions involving intensely itchy welts or hives, airway contraction with severe respiratory distress, and plummeting blood pressure. This extreme reaction is known as anaphylactic shock. If not treated with epinephrine to counter the blood pressure and breathing effects, this condition can be fatal.

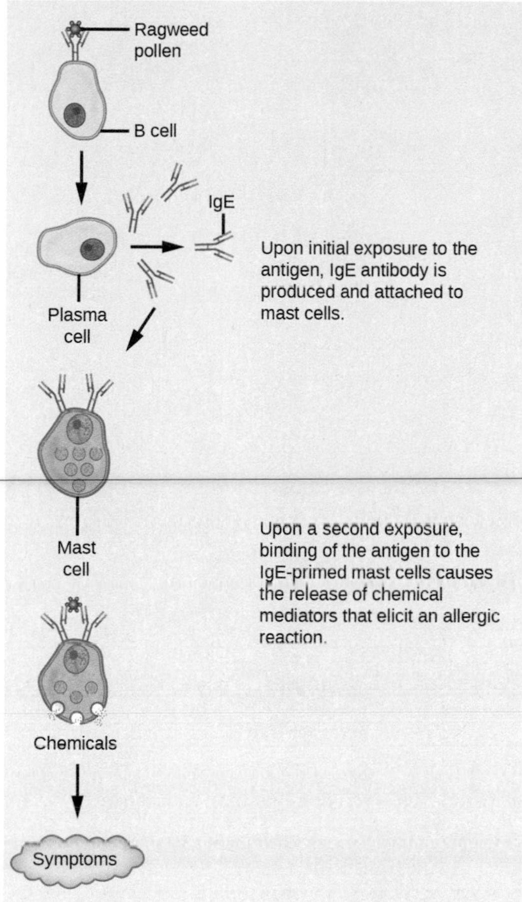

Figure 42.26 On first exposure to an allergen, an IgE antibody is synthesized by plasma cells in response to a harmless antigen. The IgE molecules bind to mast cells, and on secondary exposure, the mast cells release histamines and other modulators that affect the symptoms of allergy. (credit: modification of work by NIH)

Delayed hypersensitivity is a cell-mediated immune response that takes approximately one to two days after secondary exposure for a maximal reaction to be observed. This type of hypersensitivity involves the T_H1 cytokine-mediated inflammatory response and may manifest as local tissue lesions or contact dermatitis (rash or skin irritation). Delayed hypersensitivity occurs in some individuals in response to contact with certain types of jewelry or cosmetics. Delayed hypersensitivity facilitates the immune response to poison ivy and is also the reason why the skin test for tuberculosis results in a small region of inflammation on individuals who were previously exposed to *Mycobacterium tuberculosis*. That is also why cortisone is used to treat such responses: it will inhibit cytokine production.

Autoimmunity

Autoimmunity is a type of hypersensitivity to self antigens that affects approximately five percent of the population. Most types of autoimmunity involve the humoral immune response. Antibodies that inappropriately mark self components as foreign are termed **autoantibodies**. In patients with the autoimmune disease myasthenia gravis, muscle cell receptors that induce contraction in response to acetylcholine are targeted by antibodies. The result is muscle weakness that may include marked difficultly with fine and/or gross motor functions. In systemic lupus erythematosus, a diffuse autoantibody response to the individual's own DNA and proteins results in various systemic diseases. As illustrated in Figure 42.27, systemic lupus erythematosus may affect the heart, joints, lungs, skin, kidneys, central nervous system, or other tissues, causing tissue damage via antibody binding, complement recruitment, lysis, and inflammation.

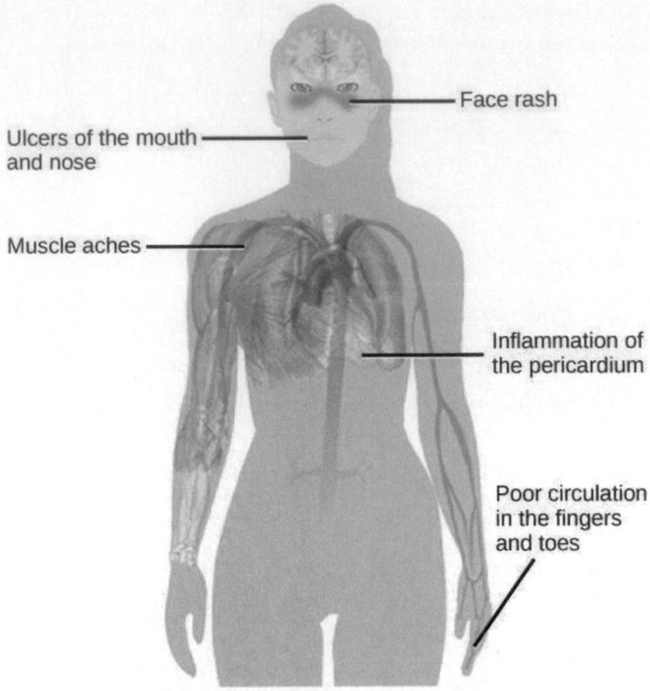

Figure 42.27 Systemic lupus erythematosus is characterized by autoimmunity to the individual's own DNA and/or proteins, which leads to varied dysfunction of the organs. (credit: modification of work by Mikael Häggström)

Autoimmunity can develop with time, and its causes may be rooted in molecular mimicry. Antibodies and TCRs may bind self antigens that are structurally similar to pathogen antigens, which the immune receptors first raised. As an example, infection with *Streptococcus pyogenes* (bacterium that causes strep throat) may generate antibodies or T cells that react with heart muscle, which has a similar structure to the surface of *S. pyogenes*. These antibodies can damage heart muscle with autoimmune attacks, leading to rheumatic fever. Insulin-dependent (Type 1) diabetes mellitus arises from a destructive inflammatory T_H1 response against insulin-producing cells of the pancreas. Patients with this autoimmunity must be injected with insulin that originates from other sources.

KEY TERMS

adaptive immunity immunity that has memory and occurs after exposure to an antigen either from a pathogen or a vaccination

affinity attraction of molecular complementarity between antigen and antibody molecules

allergy immune reaction that results from immediate hypersensitivities in which an antibody-mediated immune response occurs within minutes of exposure to a harmless antigen

antibody protein that is produced by plasma cells after stimulation by an antigen; also known as an immunoglobulin

antigen foreign or "non-self" protein that triggers the immune response

antigen-presenting cell (APC) immune cell that detects, engulfs, and informs the adaptive immune response about an infection by presenting the processed antigen on the cell surface

autoantibody antibody that incorrectly marks "self" components as foreign and stimulates the immune response

autoimmune response inappropriate immune response to host cells or self-antigens

autoimmunity type of hypersensitivity to self antigens

avidity total binding strength of a multivalent antibody with antigen

B cell lymphocyte that matures in the bone marrow and differentiates into antibody-secreting plasma cells

basophil leukocyte that releases chemicals usually involved in the inflammatory response

cell-mediated immune response adaptive immune response that is carried out by T cells

clonal selection activation of B cells corresponding to one specific BCR variant and the dramatic proliferation of that variant

complement system array of approximately 20 soluble proteins of the innate immune system that enhance phagocytosis, bore holes in pathogens, and recruit lymphocytes; enhances the adaptive response when antibodies are produced

cross reactivity binding of an antibody to an epitope corresponding to an antigen that is different from the one the antibody was raised against

cytokine chemical messenger that regulates cell differentiation, proliferation, gene expression, and cell trafficking to effect immune responses

cytotoxic T lymphocyte (CTL) adaptive immune cell that directly kills infected cells via perforin and granzymes, and releases cytokines to enhance the immune response

dendritic cell immune cell that processes antigen material and presents it on the surface of other cells to induce an immune response

effector cell lymphocyte that has differentiated, such as a B cell, plasma cell, or cytotoxic T lymphocyte

eosinophil leukocyte that responds to parasites and is involved in the allergic response

epitope small component of an antigen that is specifically recognized by antibodies, B cells, and T cells; the antigenic determinant

granzyme protease that enters target cells through perforin and induces apoptosis in the target cells; used by NK cells and killer T cells

helper T lymphocyte (T_H) cell of the adaptive immune system that binds APCs via MHC II molecules and stimulates B cells or secretes cytokines to initiate the immune response

host an organism that is invaded by a pathogen or parasite

humoral immune response adaptive immune response that is controlled by activated B cells and antibodies

hypersensitivities spectrum of maladaptive immune responses toward harmless foreign particles or self antigens; occurs after tissue sensitization and includes immediate-type (allergy), delayed-type, and autoimmunity

immune tolerance acquired ability to prevent an unnecessary or harmful immune response to a detected foreign body known not to cause disease or to self-antigens

immunodeficiency failure, insufficiency, or delay at any level of the immune system, which may be acquired or inherited

inflammation localized redness, swelling, heat, and pain that results from the movement of leukocytes and fluid through opened capillaries to a site of infection

innate immunity immunity that occurs naturally because of genetic factors or physiology, and is not induced by infection or vaccination

interferon cytokine that inhibits viral replication and modulates the immune response

lymph watery fluid that bathes tissues and organs with protective white blood cells and does not contain erythrocytes

lymphocyte leukocyte that is histologically identifiable by its large nuclei; it is a small cell with very little cytoplasm

macrophage large phagocytic cell that engulfs foreign particles and pathogens

major histocompatibility class (MHC) I/II molecule protein found on the surface of all nucleated cells (I) or specifically on antigen-presenting cells (II) that signals to immune cells whether the cell is healthy/normal or is infected/cancerous; it provides the appropriate template into which antigens can be loaded for recognition by lymphocytes

mast cell leukocyte that produces inflammatory molecules, such as histamine, in response to large pathogens and allergens

memory cell antigen-specific B or T lymphocyte that does not differentiate into effector cells during the primary immune response but that can immediately become an effector cell upon reexposure to the same pathogen

monocyte type of white blood cell that circulates in the blood and lymph and differentiates into macrophages after it moves into infected tissue

mucosa-associated lymphoid tissue (MALT) collection of lymphatic tissue that combines with epithelial tissue lining the mucosa throughout the body

natural killer (NK) cell lymphocyte that can kill cells infected with viruses or tumor cells

neutrophil phagocytic leukocyte that engulfs and digests pathogens

opsonization process that enhances phagocytosis using proteins to indicate the presence of a pathogen to phagocytic cells

passive immunity transfer of antibodies from one individual to another to provide temporary protection against pathogens

pathogen an agent, usually a microorganism, that causes disease in the organisms that it invades

pathogen-associated molecular pattern (PAMP) carbohydrate, polypeptide, and nucleic acid "signature" that is expressed by viruses, bacteria, and parasites but differs from molecules on host cells

pattern recognition receptor (PRR) molecule on macrophages and dendritic cells that binds molecular signatures of pathogens and promotes pathogen engulfment and destruction

perforin destructive protein that creates a pore in the target cell; used by NK cells and killer T cells

plasma cell immune cell that secrets antibodies; these cells arise from B cells that were stimulated by antigens

regulatory T (T_{reg}) cell specialized lymphocyte that suppresses local inflammation and inhibits the secretion of cytokines, antibodies, and other stimulatory immune factors; involved in immune tolerance

T cell lymphocyte that matures in the thymus gland; one of the main cells involved in the adaptive immune system

CHAPTER SUMMARY

42.1 Innate Immune Response

The innate immune system serves as a first responder to pathogenic threats that bypass natural physical and chemical barriers of the body. Using a combination of cellular and molecular attacks, the innate immune system identifies the nature of a pathogen and responds with inflammation, phagocytosis, cytokine release, destruction by NK cells, and/or a complement system. When innate mechanisms are insufficient to clear an infection, the adaptive immune response is informed and mobilized.

42.2 Adaptive Immune Response

The adaptive immune response is a slower-acting, longer-lasting, and more specific response than the innate response. However, the adaptive response requires information from the innate immune system to function. APCs display antigens via MHC molecules to complementary naïve T cells. In response, the T cells differentiate and proliferate, becoming T_H cells or CTLs. T_H cells stimulate B cells that have engulfed and presented pathogen-derived antigens. B cells differentiate into plasma cells that secrete antibodies, whereas CTLs induce apoptosis in intracellularly infected or cancerous cells. Memory cells persist after a primary exposure to a pathogen. If reexposure occurs, memory cells differentiate into effector cells without input from the innate immune system. The mucosal immune system is largely independent from the systemic immune system but functions in a parallel fashion to protect the extensive mucosal surfaces of the body.

42.3 Antibodies

Antibodies (immunoglobulins) are the molecules secreted from plasma cells that mediate the humoral immune response. There are five antibody classes; an antibody's class determines its mechanism of action and production site but does not control its binding specificity. Antibodies bind antigens via variable domains and can either neutralize pathogens or mark them for phagocytosis or activate the complement cascade.

42.4 Disruptions in the Immune System

Immune disruptions may involve insufficient immune responses or inappropriate immune targets. Immunodeficiency increases an individual's susceptibility to infections and cancers. Hypersensitivities are misdirected responses either to harmless foreign particles, as in the case of allergies, or to host factors, as in the case of autoimmunity. Reactions to self components may be the result of molecular mimicry.

VISUAL CONNECTION QUESTIONS

1. Figure 42.11 Which of the following statements about T cells is false?
   a. Helper T cells release cytokines while cytotoxic T cells kill the infected cell.
   b. Helper T cells are CD4+, while cytotoxic T cells are $CD8^+$.
   c. MHC II is a receptor found on most body cells, while MHC I is a receptor found on immune cells only.
   d. The T cell receptor is found on both $CD4^+$ and $CD8^+$ T cells.
2. Figure 42.14 Based on what you know about MHC receptors, why do you think an organ transplanted from an incompatible donor to a recipient will be rejected?
3. Figure 42.16 The Rh antigen is found on Rh-positive red blood cells. An Rh-negative female can usually carry an Rh-positive fetus to term without difficulty. However, if she has a second Rh-positive fetus, her body may launch an immune attack that causes hemolytic disease of the newborn. Why do you think hemolytic disease is only a problem during the second or subsequent pregnancies?

REVIEW QUESTIONS

4. Which of the following is a barrier against pathogens provided by the skin?
   a. high pH
   b. mucus
   c. tears
   d. desiccation
5. Although interferons have several effects, they are particularly useful against infections with which type of pathogen?
   a. bacteria
   b. viruses
   c. fungi
   d. helminths
6. Which organelle do phagocytes use to digest engulfed particles?
   a. lysosome
   b. nucleus
   c. endoplasmic reticulum
   d. mitochondria
7. Which innate immune system component uses MHC I molecules directly in its defense strategy?
   a. macrophages
   b. neutrophils
   c. NK cells
   d. interferon
8. Which of the following is both a phagocyte and an antigen-presenting cell?
   a. NK cell
   b. eosinophil
   c. neutrophil
   d. macrophage
9. Which immune cells bind MHC molecules on APCs via CD8 coreceptors on their cell surfaces?
   a. T_H cells
   b. CTLs
   c. mast cells
   d. basophils
10. What "self" pattern is identified by NK cells?
    a. altered self
    b. missing self
    c. normal self
    d. non-self
11. The acquired ability to prevent an unnecessary or destructive immune reaction to a harmless foreign particle, such as a food protein, is called ______.
    a. the T_H2 response
    b. allergy
    c. immune tolerance
    d. autoimmunity
12. Upon reexposure to a pathogen, a memory B cell can differentiate to which cell type?
    a. CTL
    b. naïve B cell
    c. memory T cell
    d. plasma cell
13. Foreign particles circulating in the blood are filtered by the ______.
    a. spleen
    b. lymph nodes
    c. MALT
    d. lymph

14. The structure of an antibody is similar to the extracellular component of which receptor?
    a. MHC I
    b. MHC II
    c. BCR
    d. none of the above

15. The first antibody class to appear in the serum in response to a newly encountered pathogen is _______.
    a. IgM
    b. IgA
    c. IgG
    d. IgE

16. What is the most abundant antibody class detected in the serum upon reexposure to a pathogen or in reaction to a vaccine?
    a. IgM
    b. IgA
    c. IgG
    d. IgE

17. Breastfed infants typically are resistant to disease because of _______.
    a. active immunity
    b. passive immunity
    c. immune tolerance
    d. immune memory

18. Allergy to pollen is classified as:
    a. an autoimmune reaction
    b. immunodeficiency
    c. delayed hypersensitivity
    d. immediate hypersensitivity

19. A potential cause of acquired autoimmunity is _______.
    a. tissue hypersensitivity
    b. molecular mimicry
    c. histamine release
    d. radiation exposure

20. Autoantibodies are probably involved in:
    a. reactions to poison ivy
    b. pollen allergies
    c. systemic lupus erythematosus
    d. HIV/AIDS

21. Which of the following diseases is not due to autoimmunity?
    a. rheumatic fever
    b. systemic lupus erythematosus
    c. diabetes mellitus
    d. HIV/AIDS

CRITICAL THINKING QUESTIONS

22. Different MHC I molecules between donor and recipient cells can lead to rejection of a transplanted organ or tissue. Suggest a reason for this.
23. If a series of genetic mutations prevented some, but not all, of the complement proteins from binding antibodies or pathogens, would the entire complement system be compromised?
24. Explain the difference between an epitope and an antigen.
25. What is a naïve B or T cell?
26. How does the T_H1 response differ from the T_H2 response?
27. In mammalian adaptive immune systems, T cell receptors are extraordinarily diverse. What function of the immune system results from this diversity, and how is this diversity achieved?
28. How do B and T cells differ with respect to antigens that they bind?
29. Why is the immune response after reinfection much faster than the adaptive immune response after the initial infection?
30. What are the benefits and costs of antibody cross reactivity?

CHAPTER 43
Animal Reproduction and Development

Figure 43.1 Female seahorses produce eggs for reproduction that are then fertilized by the male. Unlike almost all other animals, the male seahorse then gestates the young until birth. (credit: modification of work by "cliff1066"/Flickr)

INTRODUCTION Animal reproduction is necessary for the survival of a species. In the animal kingdom, there are innumerable ways that species reproduce. Asexual reproduction produces genetically identical organisms (clones), whereas in sexual reproduction, the genetic material of two individuals combines to produce offspring that are genetically different from their parents. During sexual reproduction the male gamete (sperm) may be placed inside the female's body for internal fertilization, or the sperm and eggs may be released into the environment for external fertilization. Seahorses, like the one shown in Figure 43.1, provide an example of the latter. Following a mating dance, the female lays eggs in the male seahorse's abdominal brood pouch where they are fertilized. The eggs hatch and the offspring develop in the pouch for several weeks.

Chapter Outline

43.1 Reproduction Methods

By the end of this section, you will be able to do the following:

- Describe advantages and disadvantages of asexual and sexual reproduction
- Discuss asexual reproduction methods
- Discuss sexual reproduction methods

Animals produce offspring through asexual and/or sexual reproduction. Both methods have advantages and disadvantages. **Asexual reproduction** produces offspring that are genetically identical to the parent because the offspring are all clones of the original parent. A single individual can produce offspring asexually and large numbers of offspring can be produced quickly. In a stable or predictable environment, asexual reproduction is an effective means of reproduction because all the offspring will be adapted to that environment. In an unstable or unpredictable environment asexually-reproducing species may be at a disadvantage because all the offspring are genetically identical and may not have the genetic variation to survive in new or different conditions. On the other hand, the rapid rates of asexual reproduction may allow for a speedy response to environmental changes if individuals have mutations. An additional advantage of asexual reproduction is that colonization of new habitats may be easier when an individual does not need to find a mate to reproduce.

During **sexual reproduction** the genetic material of two individuals is combined to produce genetically diverse offspring that differ from their parents. The genetic diversity of sexually produced offspring is thought to give species a better chance of surviving in an unpredictable or changing environment. Species that reproduce sexually must maintain two different types of individuals, males and females, which can limit the ability to colonize new habitats as both sexes must be present.

Asexual Reproduction

Asexual reproduction occurs in prokaryotic microorganisms (bacteria) and in some eukaryotic single-celled and multi-celled organisms. There are a number of ways that animals reproduce asexually.

Fission

Fission, also called binary fission, occurs in prokaryotic microorganisms and in some invertebrate, multi-celled organisms. After a period of growth, an organism splits into two separate organisms. Some unicellular eukaryotic organisms undergo binary fission by mitosis. In other organisms, part of the individual separates and forms a second individual. This process occurs, for example, in many asteroid echinoderms through splitting of the central disk. Some sea anemones and some coral polyps (Figure 43.2) also reproduce through fission.

Figure 43.2 Coral polyps reproduce asexually by fission. (credit: G. P. Schmahl, NOAA FGBNMS Manager)

Budding

Budding is a form of asexual reproduction that results from the outgrowth of a part of a cell or body region leading to a separation from the original organism into two individuals. Budding occurs commonly in some invertebrate animals such as corals and hydras. In hydras, a bud forms that develops into an adult and breaks away from the main body, as illustrated in Figure 43.3, whereas in coral budding, the bud does not detach and multiplies as part of a new colony.

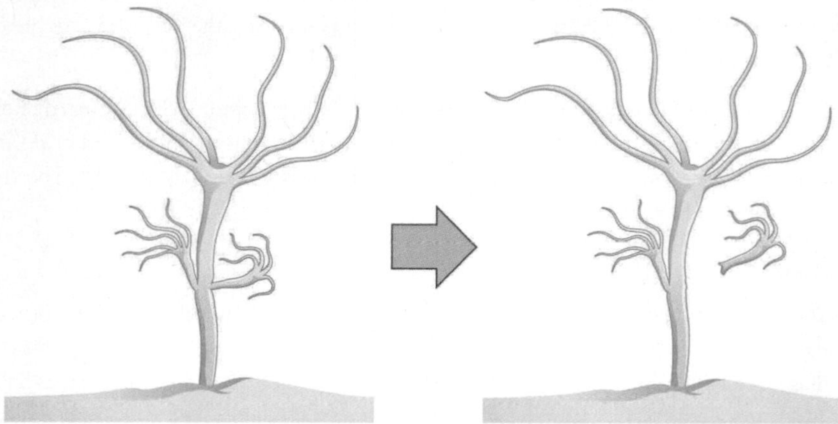

Figure 43.3 Hydra reproduce asexually through budding.

LINK TO LEARNING

Watch a video of a hydra budding.

Click to view content (https://www.openstax.org/l/budding_hydra)

Fragmentation

Fragmentation is the breaking of the body into two parts with subsequent regeneration. If the animal is capable of fragmentation, and the part is big enough, a separate individual will regrow.

For example, in many sea stars, asexual reproduction is accomplished by fragmentation. Figure 43.4 illustrates a sea star for which an arm of the individual is broken off and regenerates a new sea star. Fisheries workers have been known to try to kill the sea stars eating their clam or oyster beds by cutting them in half and throwing them back into the ocean. Unfortunately for the workers, the two parts can each regenerate a new half, resulting in twice as many sea stars to prey upon the oysters and clams. Fragmentation also occurs in annelid worms, turbellarians, and poriferans.

Figure 43.4 Sea stars can reproduce through fragmentation. The large arm, a fragment from another sea star, is developing into a new individual.

Note that in fragmentation, there is generally a noticeable difference in the size of the individuals, whereas in fission, two individuals of approximate size are formed.

Parthenogenesis

Parthenogenesis is a form of asexual reproduction where an egg develops into a complete individual without being fertilized. The resulting offspring can be either haploid or diploid, depending on the process and the species. Parthenogenesis occurs in invertebrates such as water fleas, rotifers, aphids, stick insects, some ants, wasps, and bees. Bees use parthenogenesis to produce haploid males (drones). If eggs are fertilized, diploid females develop, and if the fertilized eggs are fed a special diet (so called royal jelly), a queen is produced.

Some vertebrate animals—such as certain reptiles, amphibians, and fish—also reproduce through parthenogenesis. Although more common in plants, parthenogenesis has been observed in animal species that were segregated by sex in terrestrial or marine zoos. Two female Komodo dragons, a hammerhead shark, and a blacktop shark have produced parthenogenic young when the females have been isolated from males.

Sexual Reproduction

Sexual reproduction is the combination of (usually haploid) reproductive cells from two individuals to form a third (usually diploid) unique offspring. Sexual reproduction produces offspring with novel combinations of genes. This can be an adaptive advantage in unstable or unpredictable environments. As humans, we are used to thinking of animals as having two separate sexes—male and female—determined at conception. However, in the animal kingdom, there are many variations on this theme.

Hermaphroditism

Hermaphroditism occurs in animals where one individual has both male and female reproductive parts. Invertebrates such as earthworms, slugs, tapeworms and snails, shown in Figure 43.5, are often hermaphroditic. Hermaphrodites may self-fertilize or may mate with another of their species, fertilizing each other and both producing offspring. Self fertilization is common in animals that have limited mobility or are not motile, such as barnacles and clams.

Figure 43.5 Many snails are hermaphrodites. When two individuals mate, they can produce up to one hundred eggs each. (credit: Assaf Shtilman)

Sex Determination

Mammalian sex determination is determined genetically by the presence of X and Y chromosomes. Individuals homozygous for X (XX) are female and heterozygous individuals (XY) are male. The presence of a Y chromosome causes the development of male characteristics and its absence results in female characteristics. The XY system is also found in some insects and plants.

Avian sex determination is dependent on the presence of Z and W chromosomes. Homozygous for Z (ZZ) results in a male and heterozygous (ZW) results in a female. The W appears to be essential in determining the sex of the individual, similar to the Y chromosome in mammals. Some fish, crustaceans, insects (such as butterflies and moths), and reptiles use this system.

The sex of some species is not determined by genetics but by some aspect of the environment. Sex determination in some crocodiles and turtles, for example, is often dependent on the temperature during critical periods of egg development. This is referred to as environmental sex determination, or more specifically as temperature-dependent sex determination. In many turtles, cooler temperatures during egg incubation produce males and warm temperatures produce females. In some crocodiles, moderate temperatures produce males and both warm and cool temperatures produce females. In some species, sex is both genetic- and temperature-dependent.

Individuals of some species change their sex during their lives, alternating between male and female. If the individual is female first, it is termed protogyny or "first female," if it is male first, its termed protandry or "first male." Oysters, for example, are born male, grow, and become female and lay eggs; some oyster species change sex multiple times.

43.2 Fertilization

By the end of this section, you will be able to do the following:

- Discuss internal and external methods of fertilization
- Describe the methods used by animals for development of offspring during gestation
- Describe the anatomical adaptions that occurred in animals to facilitate reproduction

Sexual reproduction starts with the combination of a sperm and an egg in a process called fertilization. This can occur either inside (**internal fertilization**) or outside (**external fertilization**) the body of the female. Humans provide an example of the former whereas seahorse reproduction is an example of the latter.

External Fertilization

External fertilization usually occurs in aquatic environments where both eggs and sperm are released into the water. After the sperm reaches the egg, fertilization takes place. Most external fertilization happens during the process of spawning where one or several females release their eggs and the male(s) release sperm in the same area, at the same time. The release of the reproductive material may be triggered by water temperature or the length of daylight. Nearly all fish spawn, as do crustaceans (such as crabs and shrimp), mollusks (such as oysters), squid, and echinoderms (such as sea urchins and sea cucumbers). Figure 43.6 shows salmon spawning in a shallow stream. Frogs, like those shown in Figure 43.7, corals, molluscs, and sea cucumbers also spawn.

Figure 43.6 Salmon reproduce through spawning. (credit: Dan Bennett)

Figure 43.7 During sexual reproduction in toads, the male grasps the female from behind and externally fertilizes the eggs as they are deposited. (credit: "OakleyOriginals"/Flickr)

Pairs of fish that are not broadcast spawners may exhibit courtship behavior. This allows the female to select a particular male. The trigger for egg and sperm release (spawning) causes the egg and sperm to be placed in a small area, enhancing the possibility of fertilization.

External fertilization in an aquatic environment protects the eggs from drying out. Broadcast spawning can result in a greater mixture of the genes within a group, leading to higher genetic diversity and a greater chance of species survival in a hostile environment. For sessile aquatic organisms like sponges, broadcast spawning is the only mechanism for fertilization and colonization of new environments. The presence of the fertilized eggs and developing young in the water provides opportunities for predation resulting in a loss of offspring. Therefore, millions of eggs must be produced by individuals, and the offspring produced through this method must mature rapidly. The survival rate of eggs produced through broadcast spawning is low.

Internal Fertilization

Internal fertilization occurs most often in land-based animals, although some aquatic animals also use this method. There are three ways that offspring are produced following internal fertilization. In **oviparity**, fertilized eggs are laid outside the female's body and develop there, receiving nourishment from the yolk that is a part of the egg. This occurs in most bony fish, many reptiles, some cartilaginous fish, most amphibians, two mammals, and all birds. Reptiles and insects produce leathery eggs, while birds and turtles produce eggs with high concentrations of calcium carbonate in the shell, making them hard. Chicken eggs are an example of this second type.

In **ovoviparity**, fertilized eggs are retained in the female, but the embryo obtains its nourishment from the egg's yolk and the young are fully developed when they are hatched. This occurs in some bony fish (like the guppy *Lebistes reticulatus*), some sharks, some lizards, some snakes (such as the garter snake *Thamnophis sirtalis*), some vipers, and some invertebrate animals (like the Madagascar hissing cockroach *Gromphadorhina portentosa*).

In **viviparity** the young develop within the female, receiving nourishment from the mother's blood through a placenta. The offspring develops in the female and is born alive. This occurs in most mammals, some cartilaginous fish, and a few reptiles.

Internal fertilization has the advantage of protecting the fertilized egg from dehydration on land. The embryo is isolated within the female, which limits predation on the young. Internal fertilization enhances the fertilization of eggs by a specific male. Fewer offspring are produced through this method, but their survival rate is higher than that for external fertilization.

The Evolution of Reproduction

Once multicellular organisms evolved and developed specialized cells, some also developed tissues and organs with specialized functions. An early development in reproduction occurred in the Annelids. These organisms produce sperm and eggs from undifferentiated cells in their coelom and store them in that cavity. When the coelom becomes filled, the cells are released through an excretory opening or by the body splitting open. Reproductive organs evolved with the development of gonads that produce sperm and eggs. These cells went through meiosis, an adaption of mitosis, which reduced the number of chromosomes in each reproductive cell by half, while increasing the number of cells through cell division.

Complete reproductive systems were developed in insects, with separate sexes. Sperm are made in testes and then travel through coiled tubes to the epididymis for storage. Eggs mature in the ovary. When they are released from the ovary, they travel to the uterine tubes for fertilization. Some insects have a specialized sac, called a **spermatheca**, which stores sperm for later use, sometimes up to a year. Fertilization can be timed with environmental or food conditions that are optimal for offspring survival.

Vertebrates have similar structures, with a few differences. Non-mammals, such as birds and reptiles, have a common body opening, called a **cloaca**, for the digestive, excretory and reproductive systems. Coupling between birds usually involves positioning the cloaca openings opposite each other for transfer of sperm. Mammals have separate openings for the systems in the female and a uterus for support of developing offspring. The uterus has two chambers in species that produce large numbers of offspring at a time, while species that produce one offspring, such as primates, have a single uterus.

Sperm transfer from the male to the female during reproduction ranges from releasing the sperm into the watery environment for external fertilization, to the joining of cloaca in birds, to the development of a penis for direct delivery into the female's vagina in mammals.

43.3 Human Reproductive Anatomy and Gametogenesis

By the end of this section, you will be able to do the following:

- Describe human male and female reproductive anatomies
- Discuss the human sexual response
- Describe spermatogenesis and oogenesis and discuss their differences and similarities

As animals became more complex, specific organs and organ systems developed to support specific functions for the organism. The reproductive structures that evolved in land animals allow males and females to mate, fertilize internally, and support the growth and development of offspring.

Human Reproductive Anatomy

The reproductive tissues of male and female humans develop similarly *in utero* until a low level of the hormone testosterone is released from male gonads. Testosterone causes the undeveloped tissues to differentiate into male sexual organs. When testosterone is absent, the tissues develop into female sexual tissues. Primitive gonads become testes or ovaries. Tissues that produce a penis in males produce a clitoris in females. The tissue that will become the scrotum in a male becomes the labia in a female; that is, they are homologous structures.

Male Reproductive Anatomy

In the male reproductive system, the **scrotum** houses the testicles or testes (singular: testis), including providing passage for blood vessels, nerves, and muscles related to testicular function. The **testes** are a pair of male reproductive organs that produce sperm and some reproductive hormones. Each testis is approximately 2.5 by 3.8 cm (1.5 by 1 in.) in size and divided into wedge-shaped lobules by connective tissue called septa. Coiled in each wedge are seminiferous tubules that produce sperm.

Sperm are immobile at body temperature; therefore, the scrotum and penis are external to the body, as illustrated in Figure 43.8 so that a proper temperature is maintained for motility. In land mammals, the pair of testes must be suspended outside the body at about 2° C lower than body temperature to produce viable sperm. Infertility can occur in land mammals when the testes do not descend through the abdominal cavity during fetal development.

VISUAL CONNECTION

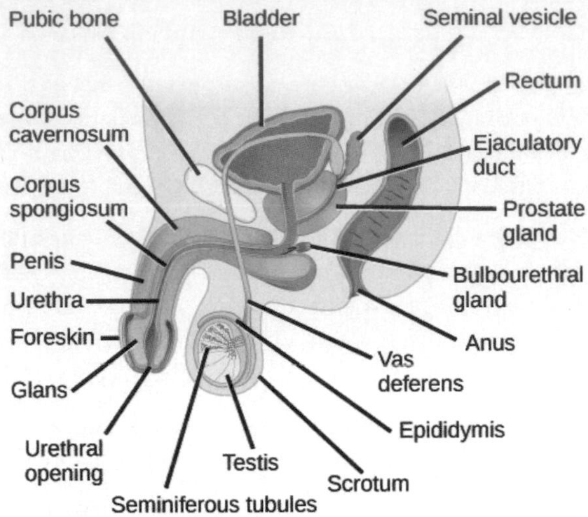

Figure 43.8 The reproductive structures of the human male are shown.

Which of the following statements about the male reproductive system is false?

a. The vas deferens carries sperm from the testes to the penis.
b. Sperm mature in seminiferous tubules in the testes.
c. Both the prostate and the bulbourethral glands produce components of the semen.
d. The prostate gland is located in the testes.

Sperm mature in **seminiferous tubules** that are coiled inside the testes, as illustrated in Figure 43.8. The walls of the seminiferous tubules are made up of the developing sperm cells, with the least developed sperm at the periphery of the tubule and the fully developed sperm in the lumen. The sperm cells are mixed with "nursemaid" cells called Sertoli cells which protect the germ cells and promote their development. Other cells mixed in the wall of the tubules are the interstitial cells of Leydig. These cells produce high levels of testosterone once the male reaches adolescence.

When the sperm have developed flagella and are nearly mature, they leave the testicles and enter the epididymis, shown in Figure 43.8. This structure resembles a comma and lies along the top and posterior portion of the testes; it is the site of sperm maturation. The sperm leave the epididymis and enter the vas deferens (or ductus deferens), which carries the sperm, behind the bladder, and forms the ejaculatory duct with the duct from the seminal vesicles. During a vasectomy, a section of the vas deferens is removed, preventing sperm from being passed out of the body during ejaculation and preventing fertilization.

Semen is a mixture of sperm and spermatic duct secretions (about 10 percent of the total) and fluids from accessory glands that contribute most of the semen's volume. Sperm are haploid cells, consisting of a flagellum as a tail, a neck that contains the cell's energy-producing mitochondria, and a head that contains the genetic material. Figure 43.9 shows a micrograph of human sperm as well as a diagram of the parts of the sperm. An acrosome is found at the top of the head of the sperm. This structure contains lysosomal enzymes that can digest the protective coverings that surround the egg to help the sperm penetrate and fertilize the egg. An ejaculate will contain from two to five milliliters of fluid with from 50–120 million sperm per milliliter.

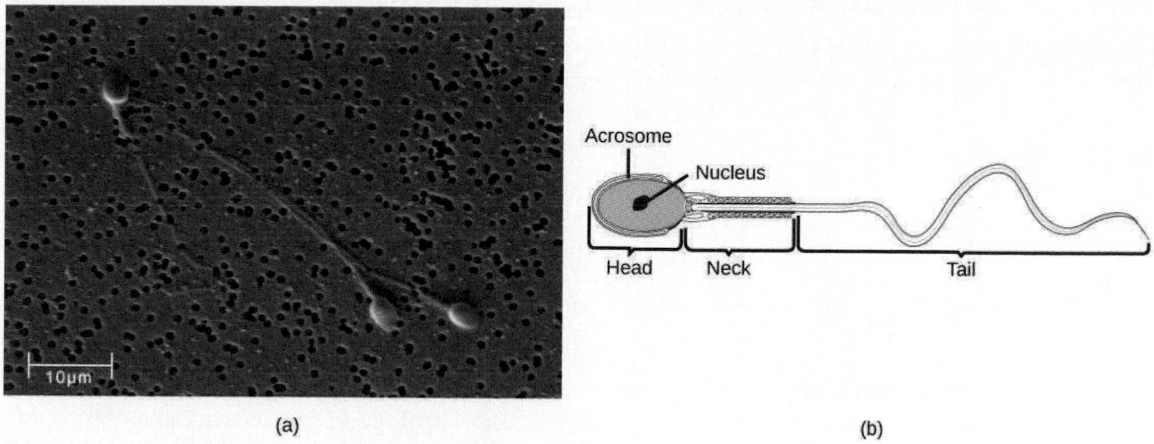

Figure 43.9 Human sperm, visualized using scanning electron microscopy, have a flagellum, neck, and head. (credit b: modification of work by Mariana Ruiz Villareal; scale-bar data from Matt Russell)

The bulk of the semen comes from the accessory glands associated with the male reproductive system. These are the seminal vesicles, the prostate gland, and the bulbourethral gland, all of which are illustrated in Figure 43.8. The **seminal vesicles** are a pair of glands that lie along the posterior border of the urinary bladder. The glands make a solution that is thick, yellowish, and alkaline. As sperm are only motile in an alkaline environment, a basic pH is important to reverse the acidity of the vaginal environment. The solution also contains mucus, fructose (a sperm mitochondrial nutrient), a coagulating enzyme, ascorbic acid, and local-acting hormones called prostaglandins. The seminal vesicle glands account for 60 percent of the bulk of semen.

The **penis**, illustrated in Figure 43.8, is an organ that drains urine from the renal bladder and functions as a copulatory organ during intercourse. The penis contains three tubes of erectile tissue running through the length of the organ. These consist of a pair of tubes on the dorsal side, called the corpus cavernosum, and a single tube of tissue on the ventral side, called the corpus spongiosum. This tissue will become engorged with blood, becoming erect and hard, in preparation for intercourse. The organ is inserted into the vagina culminating with an ejaculation. During intercourse, the smooth muscle sphincters at the opening to the renal bladder close and prevent urine from entering the penis. An orgasm is a two-stage process: first, glands and accessory organs connected to the testes contract, then semen (containing sperm) is expelled through the urethra during ejaculation. After intercourse, the blood drains from the erectile tissue and the penis becomes flaccid.

The walnut-shaped **prostate gland** surrounds the urethra, the connection to the urinary bladder. It has a series of short ducts that directly connect to the urethra. The gland is a mixture of smooth muscle and glandular tissue. The muscle provides much of the force needed for ejaculation to occur. The glandular tissue makes a thin, milky fluid that contains citrate (a nutrient), enzymes, and prostate specific antigen (PSA). PSA is a proteolytic enzyme that helps to liquefy the ejaculate several minutes after release from the male. Prostate gland secretions account for about 30 percent of the bulk of semen.

The **bulbourethral gland**, or Cowper's gland, releases its secretion prior to the release of the bulk of the semen. It neutralizes any acid residue in the urethra left over from urine. This usually accounts for a couple of drops of fluid in the total ejaculate and may contain a few sperm. Withdrawal of the penis from the vagina before ejaculation to prevent pregnancy may not work if sperm are present in the bulbourethral gland secretions. The location and functions of the male reproductive organs are summarized in Table 43.1.

Male Reproductive Anatomy

Organ	Location	Function
Scrotum	External	Carry and support testes
Penis	External	Deliver urine, copulating organ
Testes	Internal	Produce sperm and male hormones

Table 43.1

Organ	Location	Function
Seminal Vesicles	Internal	Contribute to semen production
Prostate Gland	Internal	Contribute to semen production
Bulbourethral Glands	Internal	Clean urethra at ejaculation

Table 43.1

Female Reproductive Anatomy

A number of reproductive structures are exterior to the female's body. These include the breasts and the vulva, which consists of the mons pubis, clitoris, labia majora, labia minora, and the vestibular glands, all illustrated in Figure 43.10. The location and functions of the female reproductive organs are summarized in Table 43.2. The vulva is an area associated with the vestibule which includes the structures found in the inguinal (groin) area of women. The mons pubis is a round, fatty area that overlies the pubic symphysis. The **clitoris** is a structure with erectile tissue that contains a large number of sensory nerves and serves as a source of stimulation during intercourse. The **labia majora** are a pair of elongated folds of tissue that run posterior from the mons pubis and enclose the other components of the vulva. The labia majora derive from the same tissue that produces the scrotum in a male. The **labia minora** are thin folds of tissue centrally located within the labia majora. These labia protect the openings to the vagina and urethra. The mons pubis and the anterior portion of the labia majora become covered with hair during adolescence; the labia minora is hairless. The greater vestibular glands are found at the sides of the vaginal opening and provide lubrication during intercourse.

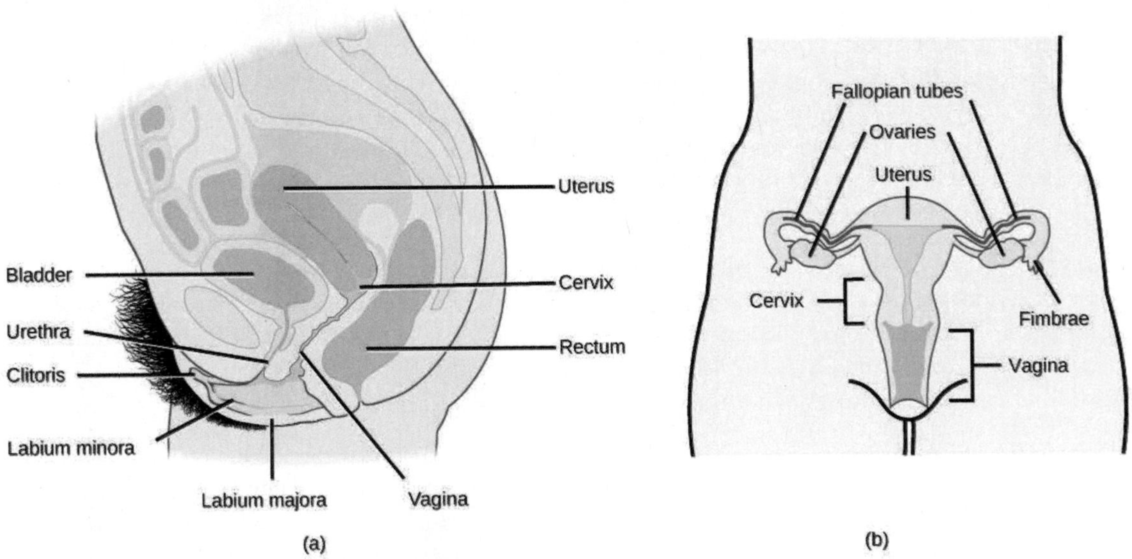

Figure 43.10 The reproductive structures of the human female are shown. (credit a: modification of work by Gray's Anatomy; credit b: modification of work by CDC)

Female Reproductive Anatomy

Organ	Location	Function
Clitoris	External	Sensory organ
Mons pubis	External	Fatty area overlying pubic bone

Table 43.2

Organ	Location	Function
Labia majora	External	Covers labia minora
Labia minora	External	Covers vestibule
Greater vestibular glands	External	Secrete mucus; lubricate vagina
Breasts	External	Produce and deliver milk
Ovaries	Internal	Carry and develop eggs
Oviducts (Fallopian tubes)	Internal	Transport egg to uterus
Uterus	Internal	Support developing embryo
Vagina	Internal	Common tube for intercourse, birth canal, passing menstrual flow

Table 43.2

The breasts consist of mammary glands and fat. The size of the breast is determined by the amount of fat deposited behind the gland. Each gland consists of 15 to 25 lobes that have ducts that empty at the nipple and that supply the nursing child with nutrient- and antibody-rich milk to aid development and protect the child.

Internal female reproductive structures include ovaries, oviducts, the **uterus**, and the vagina, shown in Figure 43.10. The pair of ovaries is held in place in the abdominal cavity by a system of ligaments. Ovaries consist of a medulla and cortex: the medulla contains nerves and blood vessels to supply the cortex with nutrients and remove waste. The outer layers of cells of the cortex are the functional parts of the ovaries. The cortex is made up of follicular cells that surround eggs that develop during fetal development *in utero*. During the menstrual period, a batch of follicular cells develops and prepares the eggs for release. At ovulation, one follicle ruptures and one egg is released, as illustrated in Figure 43.11a.

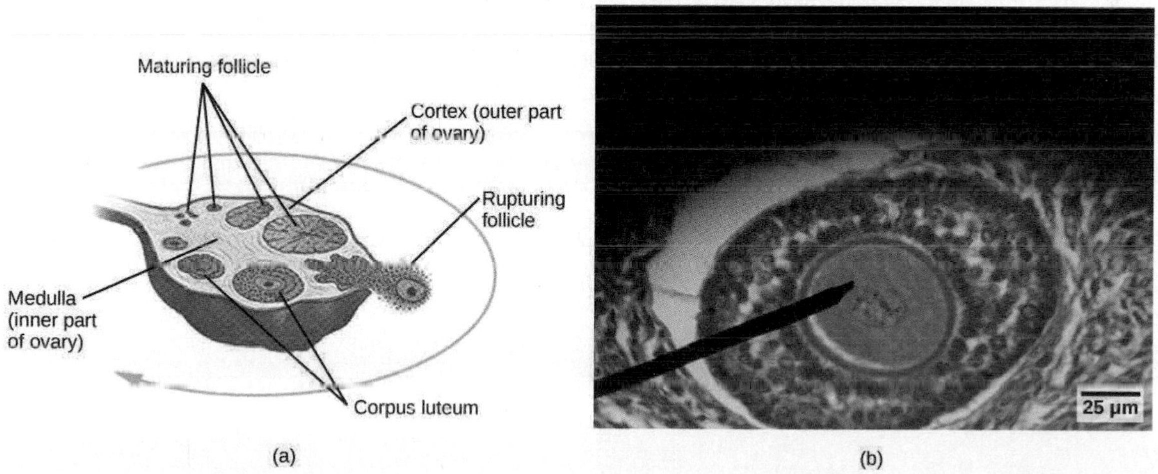

Figure 43.11 Oocytes develop in (a) follicles, located in the ovary. At the beginning of the menstrual cycle, the follicle matures. At ovulation, the follicle ruptures, releasing the egg. The follicle becomes a corpus luteum, which eventually degenerates. The (b) follicle in this light micrograph has an oocyte at its center. (credit a: modification of work by NIH; scale-bar data from Matt Russell)

The **oviducts**, or fallopian tubes, extend from the uterus in the lower abdominal cavity to the ovaries, but they are not in contact with the ovaries. The lateral ends of the oviducts flare out into a trumpet-like structure and have a fringe of finger-like projections called fimbriae, illustrated in Figure 43.10b. When an egg is released at ovulation, the fimbrae help the nonmotile egg enter into the tube and passage to the uterus. The walls of the oviducts are ciliated and are made up mostly of smooth muscle. The cilia beat toward the middle, and the smooth muscle contracts in the same direction, moving the egg toward the uterus. Fertilization usually takes place within the oviducts and the developing embryo is moved toward the uterus for development. It usually takes the egg or embryo a week to travel through the oviduct. Sterilization in women is called a tubal ligation; it is analogous to a vasectomy in males in that the oviducts are severed and sealed.

The uterus is a structure about the size of a woman's fist. This is lined with an endometrium rich in blood vessels and mucus glands. The uterus supports the developing embryo and fetus during gestation. The thickest portion of the wall of the uterus is made of smooth muscle. Contractions of the smooth muscle in the uterus aid in passing the baby through the vagina during labor. A portion of the lining of the uterus sloughs off during each menstrual period, and then builds up again in preparation for an implantation. Part of the uterus, called the cervix, protrudes into the top of the vagina. The cervix functions as the birth canal.

The **vagina** is a muscular tube that serves several purposes. It allows menstrual flow to leave the body. It is the receptacle for the penis during intercourse and the vessel for the delivery of offspring. It is lined by stratified squamous epithelial cells to protect the underlying tissue.

Sexual Response during Intercourse

The sexual response in humans is both psychological and physiological. Both sexes experience sexual arousal through psychological and physical stimulation. There are four phases of the sexual response. During phase one, called excitement, vasodilation leads to vasocongestion in erectile tissues in both men and women. The nipples, clitoris, labia, and penis engorge with blood and become enlarged. Vaginal secretions are released to lubricate the vagina to facilitate intercourse. During the second phase, called the plateau, stimulation continues, the outer third of the vaginal wall enlarges with blood, and breathing and heart rate increase.

During phase three, or orgasm, rhythmic, involuntary contractions of muscles occur in both sexes. In the male, the reproductive accessory glands and tubules constrict placing semen in the urethra, then the urethra contracts expelling the semen through the penis. In women, the uterus and vaginal muscles contract in waves that may last slightly less than a second each. During phase four, or resolution, the processes described in the first three phases reverse themselves and return to their normal state. Men experience a refractory period in which they cannot maintain an erection or ejaculate for a period of time ranging from minutes to hours.

Gametogenesis (Spermatogenesis and Oogenesis)

Gametogenesis, the production of sperm and eggs, takes place through the process of meiosis. During meiosis, two cell divisions separate the paired chromosomes in the nucleus and then separate the chromatids that were made during an earlier stage of the cell's life cycle. Meiosis produces haploid cells with half of each pair of chromosomes normally found in diploid cells. The production of sperm is called **spermatogenesis** and the production of eggs is called **oogenesis**.

Spermatogenesis

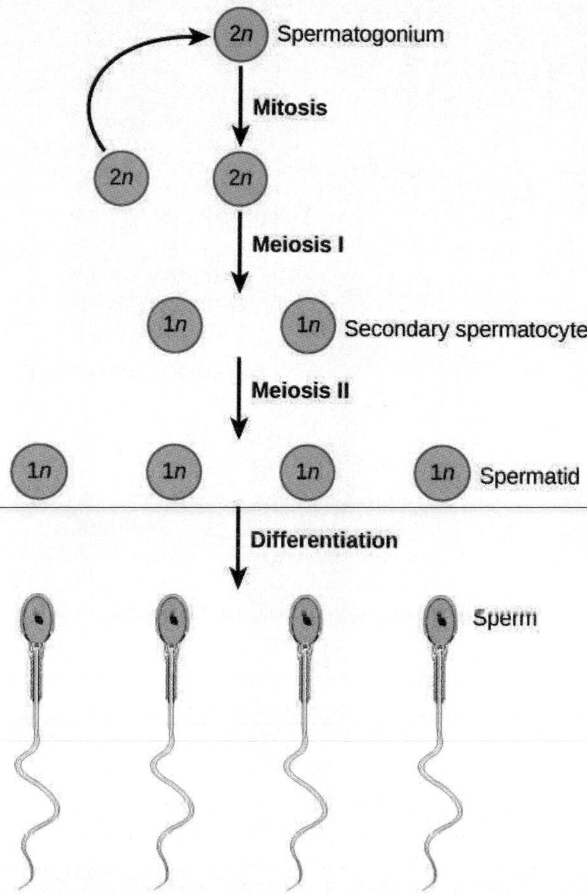

Figure 43.12 During spermatogenesis, four sperm result from each primary spermatocyte.

Spermatogenesis, illustrated in Figure 43.12, occurs in the wall of the seminiferous tubules (Figure 43.8), with stem cells at the periphery of the tube and the spermatozoa at the lumen of the tube. Immediately under the capsule of the tubule are diploid, undifferentiated cells. These stem cells, called spermatogonia (singular: spermatagonium), go through mitosis with one offspring going on to differentiate into a sperm cell and the other giving rise to the next generation of sperm.

Meiosis starts with a cell called a primary spermatocyte. At the end of the first meiotic division, a haploid cell is produced called a secondary spermatocyte. This cell is haploid and must go through another meiotic cell division. The cell produced at the end of meiosis is called a spermatid and when it reaches the lumen of the tubule and grows a flagellum, it is called a sperm cell. Four sperm result from each primary spermatocyte that goes through meiosis.

Stem cells are deposited during gestation and are present at birth through the beginning of adolescence, but in an inactive state. During adolescence, gonadotropic hormones from the anterior pituitary cause the activation of these cells and the production of viable sperm. This continues into old age.

LINK TO LEARNING

Visit this site (http://openstax.org/l/spermatogenesis) to see the process of spermatogenesis.

Oogenesis

Oogenesis, illustrated in Figure 43.13, occurs in the outermost layers of the ovaries. As with sperm production, oogenesis starts with a germ cell, called an oogonium (plural: oogonia), but this cell undergoes mitosis to increase in number, eventually resulting in up to about one to two million cells in the embryo.

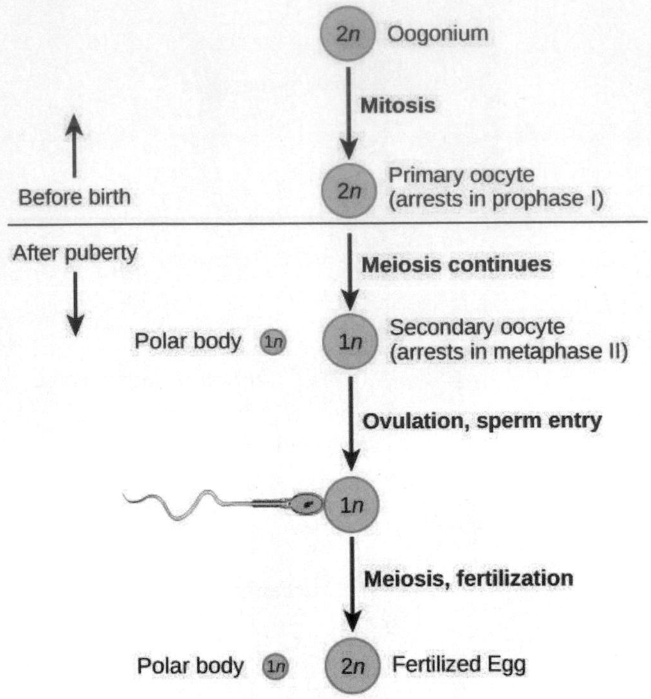

Figure 43.13 The process of oogenesis occurs in the ovary's outermost layer.

The cell starting meiosis is called a primary oocyte, as shown in Figure 43.13. This cell will start the first meiotic division and be arrested in its progress in the first prophase stage. At the time of birth, all future eggs are in the prophase stage. At adolescence, anterior pituitary hormones cause the development of a number of follicles in an ovary. This results in the primary oocyte finishing the first meiotic division. The cell divides unequally, with most of the cellular material and organelles going to one cell, called a secondary oocyte, and only one set of chromosomes and a small amount of cytoplasm going to the other cell. This second cell is called a polar body and usually dies. A secondary meiotic arrest occurs, this time at the metaphase II stage. At ovulation, this secondary oocyte will be released and travel toward the uterus through the oviduct. If the secondary oocyte is fertilized, the cell continues through the meiosis II, producing a second polar body and a fertilized egg containing all 46 chromosomes of a human being, half of them coming from the sperm.

Egg production begins before birth, is arrested during meiosis until puberty, and then individual cells continue through at each menstrual cycle. One egg is produced from each meiotic process, with the extra chromosomes and chromatids going into polar bodies that degenerate and are reabsorbed by the body.

43.4 Hormonal Control of Human Reproduction

By the end of this chapter, you will be able to do the following:

- Describe the roles of male and female reproductive hormones
- Discuss the interplay of the ovarian and menstrual cycles
- Describe the process of menopause

The human male and female reproductive cycles are controlled by the interaction of hormones from the hypothalamus and anterior pituitary with hormones from reproductive tissues and organs. In both sexes, the hypothalamus monitors and causes the release of hormones from the pituitary gland. When the reproductive hormone is required, the hypothalamus sends a **gonadotropin-releasing hormone (GnRH)** to the anterior pituitary. This causes the release of **follicle stimulating hormone (FSH)** and **luteinizing hormone (LH)** from the anterior pituitary into the blood. Note that the body must reach puberty in order for the adrenals to release the hormones that must be present for GnRH to be produced. Although FSH and LH are named after their functions in female reproduction, they are produced in both sexes and play important roles in controlling reproduction. Other hormones have specific functions in the male and female reproductive systems.

Male Hormones

At the onset of puberty, the hypothalamus causes the release of FSH and LH into the male system for the first time. FSH enters the testes and stimulates the **Sertoli cells** to begin facilitating spermatogenesis using negative feedback, as illustrated in Figure 43.14. LH also enters the testes and stimulates the **interstitial cells of Leydig** to make and release testosterone into the testes and the blood.

Testosterone, the hormone responsible for the secondary sexual characteristics that develop in the male during adolescence, stimulates spermatogenesis. These secondary sex characteristics include a deepening of the voice, the growth of facial, axillary, and pubic hair, and the beginnings of the sex drive.

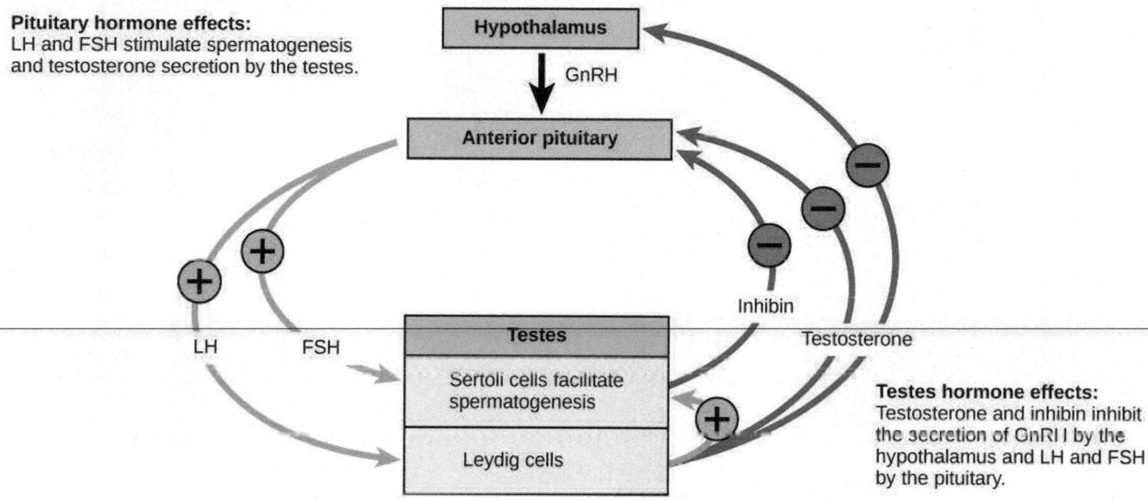

Figure 43.14 Hormones control sperm production in a negative feedback system.

A negative feedback system occurs in the male with rising levels of testosterone acting on the hypothalamus and anterior pituitary to inhibit the release of GnRH, FSH, and LH. The Sertoli cells produce the hormone **inhibin**, which is released into the blood when the sperm count is too high. This inhibits the release of GnRH and FSH, which will cause spermatogenesis to slow down. If the sperm count reaches 20 million/ml, the Sertoli cells cease the release of inhibin, and the sperm count increases.

Female Hormones

The control of reproduction in females is more complex. As with the male, the anterior pituitary hormones cause the release of the hormones FSH and LH. In addition, estrogens and progesterone are released from the developing follicles. **Estrogen** is the reproductive hormone in females that assists in endometrial regrowth, ovulation, and calcium absorption; it is also responsible for the secondary sexual characteristics of females. These include breast development, flaring of the hips, and a shorter period necessary for bone maturation. **Progesterone** assists in endometrial regrowth and inhibition of FSH and LH release.

In females, FSH stimulates development of egg cells, called ova, which develop in structures called follicles. Follicle cells produce the hormone inhibin, which inhibits FSH production. LH also plays a role in the development of ova, induction of ovulation, and stimulation of estradiol and progesterone production by the ovaries. Estradiol and progesterone are steroid hormones that prepare the body for pregnancy. Estradiol produces secondary sex characteristics in females, while both estradiol and progesterone regulate the menstrual cycle.

The Ovarian Cycle and the Menstrual Cycle

The **ovarian cycle** governs the preparation of endocrine tissues and release of eggs, while the **menstrual cycle** governs the preparation and maintenance of the uterine lining. These cycles occur concurrently and are coordinated over a 22–32 day cycle, with an average length of 28 days.

The first half of the ovarian cycle is the follicular phase shown in Figure 43.15. Slowly rising levels of FSH and LH cause the growth of follicles on the surface of the ovary. This process prepares the egg for ovulation. As the follicles grow, they begin releasing estrogens and a low level of progesterone. Progesterone maintains the endometrium to help ensure pregnancy. The trip through the fallopian tube takes about seven days. At this stage of development, called the morula, there are 30-60 cells. If pregnancy implantation does not occur, the lining is sloughed off. After about five days, estrogen levels rise and the menstrual cycle enters the proliferative phase. The endometrium begins to regrow, replacing the blood vessels and glands that deteriorated during the end of the last cycle.

VISUAL CONNECTION

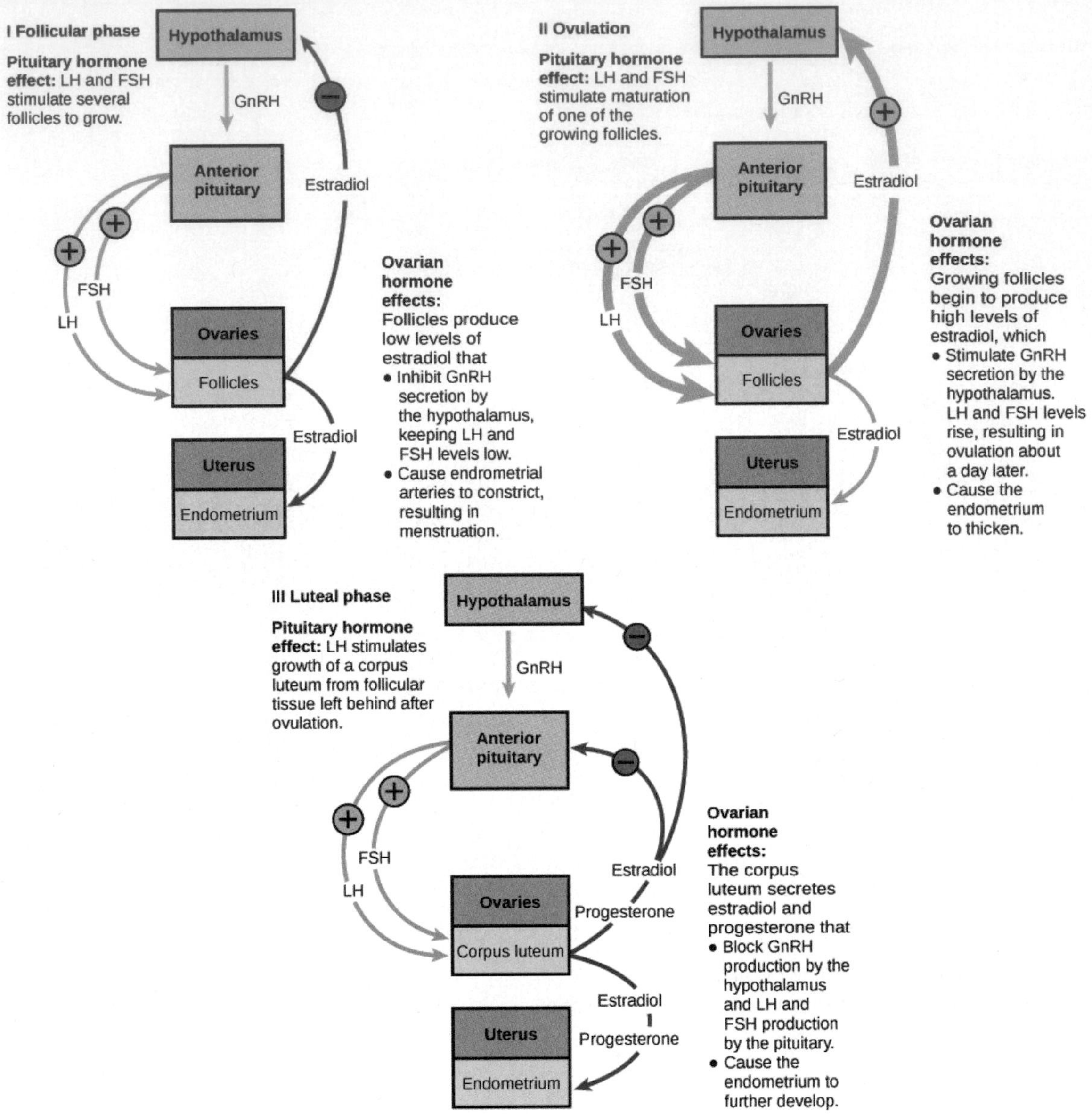

Figure 43.15 The ovarian and menstrual cycles of female reproduction are regulated by hormones produced by the hypothalamus, pituitary, and ovaries.

Which of the following statements about hormone regulation of the female reproductive cycle is false?

a. LH and FSH are produced in the pituitary, and estradiol and progesterone are produced in the ovaries.
b. Estradiol and progesterone secreted from the corpus luteum cause the endometrium to thicken.
c. Both progesterone and estradiol are produced by the follicles.
d. Secretion of GnRH by the hypothalamus is inhibited by low levels of estradiol but stimulated by high levels of estradiol.

Just prior to the middle of the cycle (approximately day 14), the high level of estrogen causes FSH and especially LH to rise rapidly, then fall. The spike in LH causes **ovulation**: the most mature follicle, like that shown in Figure 43.16, ruptures and releases its egg. The follicles that did not rupture degenerate and their eggs are lost. The level of estrogen decreases when the extra follicles degenerate.

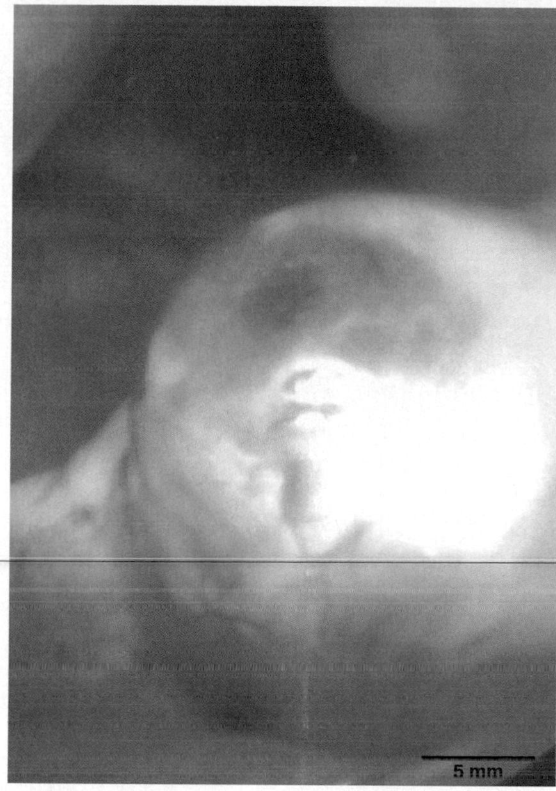

Figure 43.16 This mature egg follicle may rupture and release an egg. (credit: scale-bar data from Matt Russell)

Following ovulation, the ovarian cycle enters its luteal phase, illustrated in Figure 43.15 and the menstrual cycle enters its secretory phase, both of which run from about day 15 to 28. The luteal and secretory phases refer to changes in the ruptured follicle. The cells in the follicle undergo physical changes and produce a structure called a corpus luteum. The corpus luteum produces estrogen and progesterone. The progesterone facilitates the regrowth of the uterine lining and inhibits the release of further FSH and LH. The uterus is being prepared to accept a fertilized egg, should it occur during this cycle. The inhibition of FSH and LH prevents any further eggs and follicles from developing, while the progesterone is elevated. The level of estrogen produced by the corpus luteum increases to a steady level for the next few days.

If no fertilized egg is implanted into the uterus, the corpus luteum degenerates and the levels of estrogen and progesterone decrease. The endometrium begins to degenerate as the progesterone levels drop, initiating the next menstrual cycle. The decrease in progesterone also allows the hypothalamus to send GnRH to the anterior pituitary, releasing FSH and LH and starting the cycles again. Figure 43.17 visually compares the ovarian and uterine cycles as well as the commensurate hormone levels.

VISUAL CONNECTION

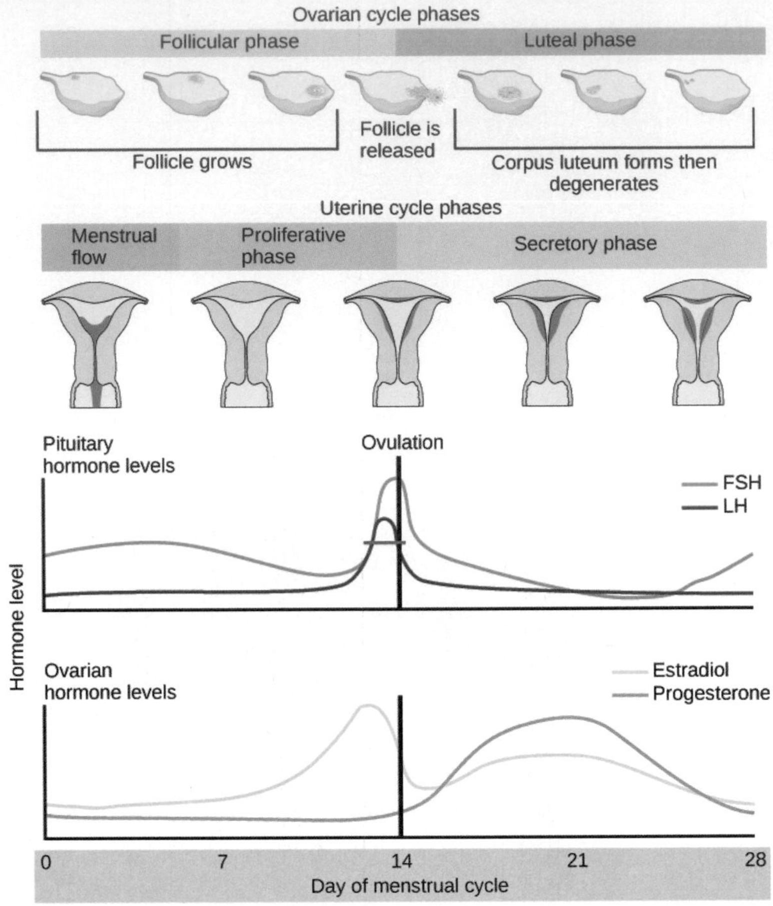

Figure 43.17 Rising and falling hormone levels result in progression of the ovarian and menstrual cycles. (credit: modification of work by Mikael Häggström)

Which of the following statements about the menstrual cycle is false?

a. Progesterone levels rise during the luteal phase of the ovarian cycle and the secretory phase of the uterine cycle.
b. Menstruation occurs just after LH and FSH levels peak.
c. Menstruation occurs after progesterone levels drop.
d. Estrogen levels rise before ovulation, while progesterone levels rise after.

Menopause

As women approach their mid-40s to mid-50s, their ovaries begin to lose their sensitivity to FSH and LH. Menstrual periods become less frequent and finally cease; this is **menopause**. There are still eggs and potential follicles on the ovaries, but without the stimulation of FSH and LH, they will not produce a viable egg to be released. The outcome of this is the inability to have children.

The side effects of menopause include hot flashes, heavy sweating (especially at night), headaches, some hair loss, muscle pain, vaginal dryness, insomnia, depression, weight gain, and mood swings. Estrogen is involved in calcium metabolism and, without it, blood levels of calcium decrease. To replenish the blood, calcium is lost from bone which may decrease the bone density and lead to osteoporosis. Supplementation of estrogen in the form of hormone replacement therapy (HRT) can prevent bone loss, but the therapy can have negative side effects. While HRT is thought to give some protection from colon cancer, osteoporosis, heart disease, macular degeneration, and possibly depression, its negative side effects include increased risk of: stroke or heart attack, blood clots, breast cancer, ovarian cancer, endometrial cancer, gall bladder disease, and possibly dementia.

CAREER CONNECTION

Reproductive Endocrinologist

A reproductive endocrinologist is a physician who treats a variety of hormonal disorders related to reproduction and infertility in both men and women. The disorders include menstrual problems, infertility, pregnancy loss, sexual dysfunction, and menopause. Doctors may use fertility drugs, surgery, or assisted reproductive techniques (ART) in their therapy. ART involves the use of procedures to manipulate the egg or sperm to facilitate reproduction, such as *in vitro* fertilization.

Reproductive endocrinologists undergo extensive medical training, first in a four-year residency in obstetrics and gynecology, then in a three-year fellowship in reproductive endocrinology. To be board certified in this area, the physician must pass written and oral exams in both areas.

43.5 Human Pregnancy and Birth

By the end of this section, you will be able to do the following:

- Explain fetal development during the three trimesters of gestation
- Describe labor and delivery
- Compare the efficacy and duration of various types of contraception
- Discuss causes of infertility and the therapeutic options available

Pregnancy begins with the fertilization of an egg and continues through to the birth of the individual. The length of time of **gestation** varies among animals, but is very similar among the great apes: human gestation is 266 days, while chimpanzee gestation is 237 days, a gorilla's is 257 days, and orangutan gestation is 260 days long. The fox has a 57-day gestation. Dogs and cats have similar gestations averaging 60 days. The longest gestation for a land mammal is an African elephant at 640 days. The longest gestations among marine mammals are the beluga and sperm whales at 460 days.

Human Gestation

Twenty-four hours before fertilization, the egg has finished meiosis and becomes a mature oocyte. When fertilized (at conception) the egg becomes known as a zygote. The zygote travels through the oviduct to the uterus (Figure 43.18). The developing embryo must implant into the wall of the uterus within seven days, or it will deteriorate and die. The outer layers of the zygote (blastocyst) grow into the endometrium by digesting the endometrial cells, and wound healing of the endometrium closes up the blastocyst into the tissue. Another layer of the blastocyst, the chorion, begins releasing a hormone called **human beta chorionic gonadotropin (β-HCG)** which makes its way to the corpus luteum and keeps that structure active. This ensures adequate levels of progesterone that will maintain the endometrium of the uterus for the support of the developing embryo. Pregnancy tests determine the level of β-HCG in urine or serum. If the hormone is present, the test is positive.

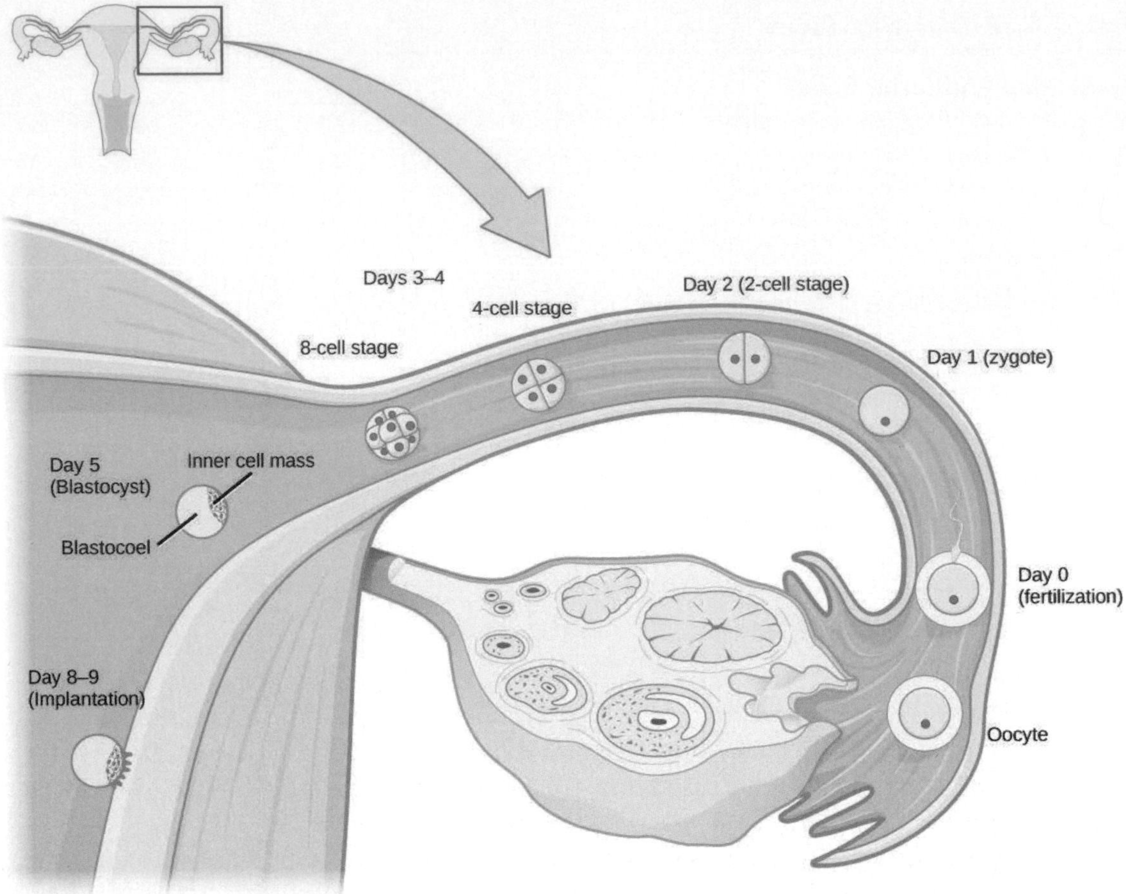

Figure 43.18 In humans, fertilization occurs soon after the oocyte leaves the ovary. Implantation occurs eight or nine days later.(credit: Ed Uthman)

The gestation period is divided into three equal periods or trimesters. During the first two to four weeks of the first trimester, nutrition and waste are handled by the endometrial lining through diffusion. As the trimester progresses, the outer layer of the embryo begins to merge with the endometrium, and the **placenta** forms. This organ takes over the nutrient and waste requirements of the embryo and fetus, with the mother's blood passing nutrients to the placenta and removing waste from it. Chemicals from the fetus, such as bilirubin, are processed by the mother's liver for elimination. Some of the mother's immunoglobulins will pass through the placenta, providing passive immunity against some potential infections.

Internal organs and body structures begin to develop during the first trimester. By five weeks, limb buds, eyes, the heart, and liver have been basically formed. By eight weeks, the term fetus applies, and the body is essentially formed, as shown in Figure 43.19. The individual is about five centimeters (two inches) in length and many of the organs, such as the lungs and liver, are not yet functioning. Exposure to any toxins is especially dangerous during the first trimester, as all of the body's organs and structures are going through initial development. Anything that affects that development can have a severe effect on the fetus' survival.

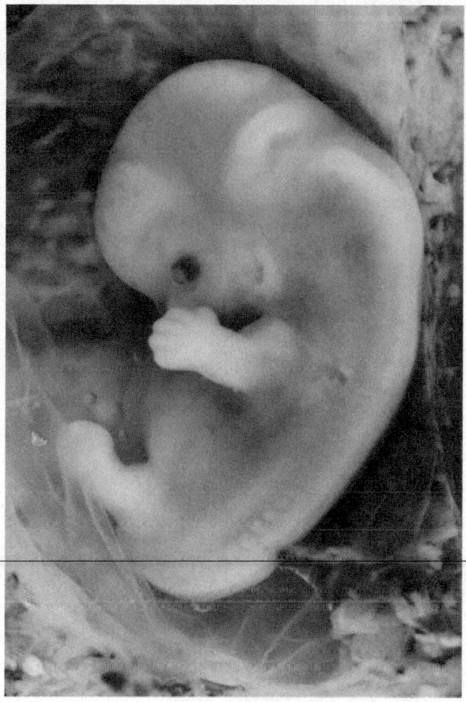

Figure 43.19 Fetal development is shown at nine weeks gestation. (credit: Ed Uthman)

During the second trimester, the fetus grows to about 30 cm (12 inches), as shown in Figure 43.20. It becomes active and the mother usually feels the first movements. All organs and structures continue to develop. The placenta has taken over the functions of nutrition and waste and the production of estrogen and progesterone from the corpus luteum, which has degenerated. The placenta will continue functioning up through the delivery of the baby.

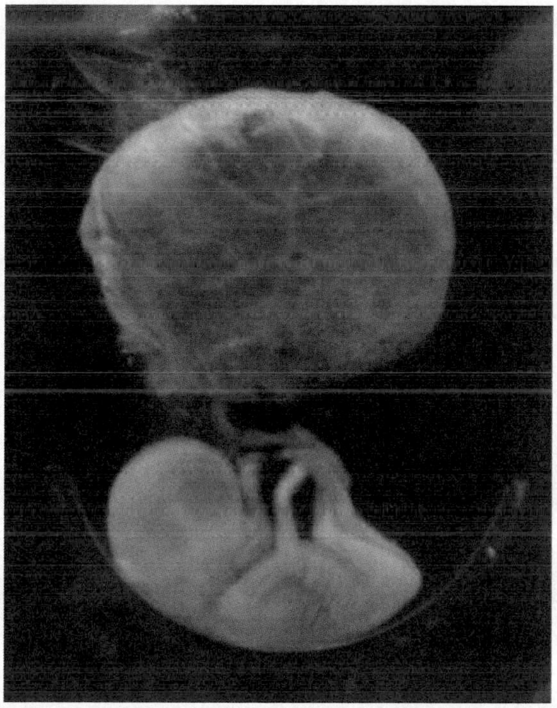

Figure 43.20 This fetus is just entering the second trimester, when the placenta takes over more of the functions performed as the baby develops. (credit: National Museum of Health and Medicine)

During the third trimester, the fetus grows to 3 to 4 kg (6 ½ -8 ½ lbs.) and about 50 cm (19-20 inches) long, as illustrated in Figure 43.21. This is the period of the most rapid growth during the pregnancy. Organ development continues to birth (and some systems, such as the nervous system and liver, continue to develop after birth). The mother will be at her most uncomfortable during this trimester. She may urinate frequently due to pressure on the bladder from the fetus. There may also be intestinal blockage and circulatory problems, especially in her legs. Clots may form in her legs due to pressure from the fetus on returning veins as they enter the abdominal cavity.

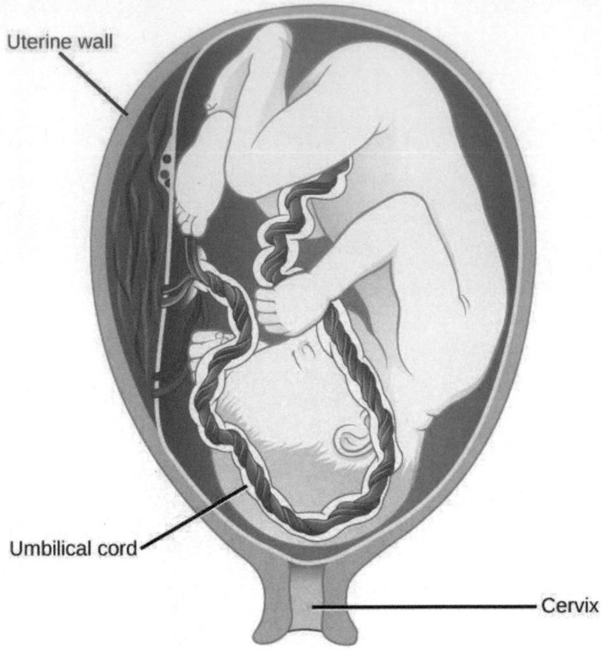

Figure 43.21 There is rapid fetal growth during the third trimester. (credit: modification of work by Gray's Anatomy)

LINK TO LEARNING

Visit this site (http://openstax.org/l/embryo_fetus) to see the stages of human fetal development.

Labor and Birth

Labor is the physical efforts of expulsion of the fetus and the placenta from the uterus during birth (parturition). Toward the end of the third trimester, estrogen causes receptors on the uterine wall to develop and bind the hormone oxytocin. At this time, the baby reorients, facing forward and down with the back or crown of the head engaging the cervix (uterine opening). This causes the cervix to stretch and nerve impulses are sent to the hypothalamus, which signals for the release of oxytocin from the posterior pituitary. The oxytocin causes the smooth muscle in the uterine wall to contract. At the same time, the placenta releases prostaglandins into the uterus, increasing the contractions. A positive feedback relay occurs between the uterus, hypothalamus, and the posterior pituitary to assure an adequate supply of oxytocin. As more smooth muscle cells are recruited, the contractions increase in intensity and force.

There are three stages to labor. During stage one, the cervix thins and dilates. This is necessary for the baby and placenta to be expelled during birth. The cervix will eventually dilate to about 10 cm. During stage two, the baby is expelled from the uterus. The uterus contracts and the mother pushes as she compresses her abdominal muscles to aid the delivery. The last stage is the passage of the placenta after the baby has been born and the organ has completely disengaged from the uterine wall. If labor should stop before stage two is reached, synthetic oxytocin, known as Pitocin, can be administered to restart and maintain labor.

An alternative to labor and delivery is the surgical delivery of the baby through a procedure called a Caesarian section. This is major abdominal surgery and can lead to post-surgical complications for the mother, but in some cases it may be the only way to safely deliver the baby.

The mother's mammary glands go through changes during the third trimester to prepare for lactation and breastfeeding. When the baby begins suckling at the breast, signals are sent to the hypothalamus causing the release of prolactin from the anterior pituitary. Prolactin causes the mammary glands to produce milk. Oxytocin is also released, promoting the release of the milk.

The milk contains nutrients for the baby's development and growth as well as immunoglobulins to protect the child from bacterial and viral infections.

Contraception and Birth Control

The prevention of a pregnancy comes under the terms contraception or birth control. Strictly speaking, **contraception** refers to preventing the sperm and egg from joining. Both terms are, however, frequently used interchangeably.

Contraceptive Methods

Method	Examples	Failure Rate in Typical Use Over 12 Months
Barrier	male condom, female condom, sponge, cervical cap, diaphragm, spermicides	15 to 24%
Hormonal	oral, patch, vaginal ring	8%
	injection	3%
	implant	less than 1%
Other	natural family planning	12 to 25%
	withdrawal	27%
	sterilization	less than 1%

Table 43.3

Table 43.3 lists common methods of contraception. The failure rates listed are not the ideal rates that could be realized, but the typical rates that occur. A failure rate is the number of pregnancies resulting from the method's use over a twelve-month period. Barrier methods, such as condoms, cervical caps, and diaphragms, block sperm from entering the uterus, preventing fertilization. Spermicides are chemicals that are placed in the vagina that kill sperm. Sponges, which are saturated with spermicides, are placed in the vagina at the cervical opening. Combinations of spermicidal chemicals and barrier methods achieve lower failure rates than do the methods when used separately.

Nearly a quarter of the couples using barrier methods, natural family planning, or withdrawal can expect a failure of the method. Natural family planning is based on the monitoring of the menstrual cycle and having intercourse only during times when the egg is not available. A woman's body temperature may rise a degree Celsius at ovulation and the cervical mucus may increase in volume and become more pliable. These changes give a general indication of when intercourse is more or less likely to result in fertilization. Withdrawal involves the removal of the penis from the vagina during intercourse, before ejaculation occurs. This is a risky method with a high failure rate due to the possible presence of sperm in the bulbourethral gland's secretion, which may enter the vagina prior to removing the penis.

Hormonal methods use synthetic progesterone (sometimes in combination with estrogen), to inhibit the hypothalamus from releasing FSH or LH, and thus prevent an egg from being available for fertilization. The method of administering the hormone affects failure rate. The most reliable method, with a failure rate of less than 1 percent, is the implantation of the hormone under the skin. The same rate can be achieved through the sterilization procedures of vasectomy in the man or of tubal ligation in the woman, or by using an intrauterine device (IUD). IUDs are inserted into the uterus and establish an inflammatory condition that prevents fertilized eggs from implanting into the uterine wall.

Compliance with the contraceptive method is a strong contributor to the success or failure rate of any particular method. The only method that is completely effective at preventing conception is abstinence. The choice of contraceptive method depends on the goals of the woman or couple. Tubal ligation and vasectomy are considered permanent prevention, while other methods are reversible and provide short-term contraception.

Termination of an existing pregnancy can be spontaneous or voluntary. Spontaneous termination is a miscarriage and usually occurs very early in the pregnancy, usually within the first few weeks. This occurs when the fetus cannot develop properly and the gestation is naturally terminated. Voluntary termination of a pregnancy is an abortion. Laws regulating abortion vary between states and tend to view fetal viability as the criteria for allowing or preventing the procedure.

Infertility

Infertility is the inability to conceive a child or carry a child to birth. About 75 percent of causes of infertility can be identified; these include diseases, such as sexually transmitted diseases that can cause scarring of the reproductive tubes in either men or women, or developmental problems frequently related to abnormal hormone levels in one of the individuals. Inadequate nutrition, especially starvation, can delay menstruation. Stress can also lead to infertility. Short-term stress can affect hormone levels, while long-term stress can delay puberty and cause less frequent menstrual cycles. Other factors that affect fertility include toxins (such as cadmium), tobacco smoking, marijuana use, gonadal injuries, and aging.

If infertility is identified, several assisted reproductive technologies (ART) are available to aid conception. A common type of ART is *in vitro* fertilization (IVF) where an egg and sperm are combined outside the body and then placed in the uterus. Eggs are obtained from the woman after extensive hormonal treatments that prepare mature eggs for fertilization and prepare the uterus for implantation of the fertilized egg. Sperm are obtained from the man and they are combined with the eggs and supported through several cell divisions to ensure viability of the zygotes. When the embryos have reached the eight-cell stage, one or more is implanted into the woman's uterus. If fertilization is not accomplished by simple IVF, a procedure that injects the sperm into an egg can be used. This is called intracytoplasmic sperm injection (ICSI) and is shown in Figure 43.22. IVF procedures produce a surplus of fertilized eggs and embryos that can be frozen and stored for future use. The procedures can also result in multiple births.

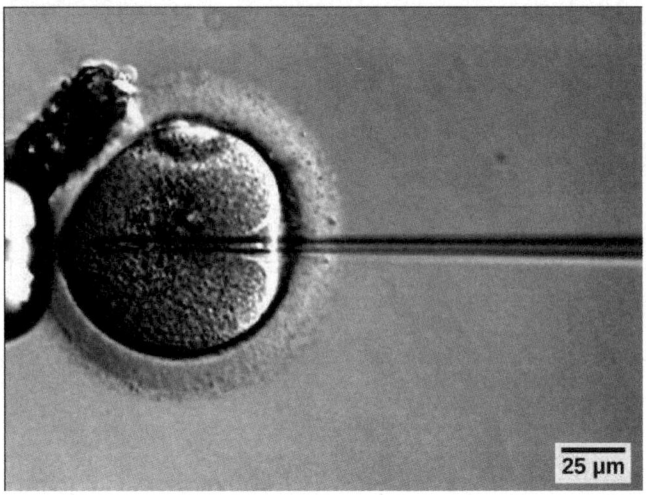

Figure 43.22 A sperm is inserted into an egg for fertilization during intracytoplasmic sperm injection (ICSI). (credit: scale-bar data from Matt Russell)

43.6 Fertilization and Early Embryonic Development

By the end of this section, you will be able to do the following:

- Discuss how fertilization occurs
- Explain how the embryo forms from the zygote
- Discuss the role of cleavage and gastrulation in animal development

The process in which an organism develops from a single-celled zygote to a multi-cellular organism is complex and well-regulated. The early stages of embryonic development are also crucial for ensuring the fitness of the organism.

Fertilization

Fertilization, pictured in Figure 43.23a is the process in which gametes (an egg and sperm) fuse to form a zygote. The egg and sperm each contain one set of chromosomes. To ensure that the offspring has only one complete diploid set of chromosomes, only one sperm must fuse with one egg. In mammals, the egg is protected by a layer of extracellular matrix consisting mainly of glycoproteins called the **zona pellucida**. When a sperm binds to the zona pellucida, a series of biochemical events, called the **acrosomal reactions**, take place. In placental mammals, the acrosome contains digestive enzymes that initiate the degradation of the glycoprotein matrix protecting the egg and allowing the sperm plasma membrane to fuse with the egg plasma membrane, as illustrated in Figure 43.23b. The fusion of these two membranes creates an opening through which the sperm nucleus is transferred into the ovum. The nuclear membranes of the egg and sperm break down and the two haploid genomes condense to form a diploid genome.

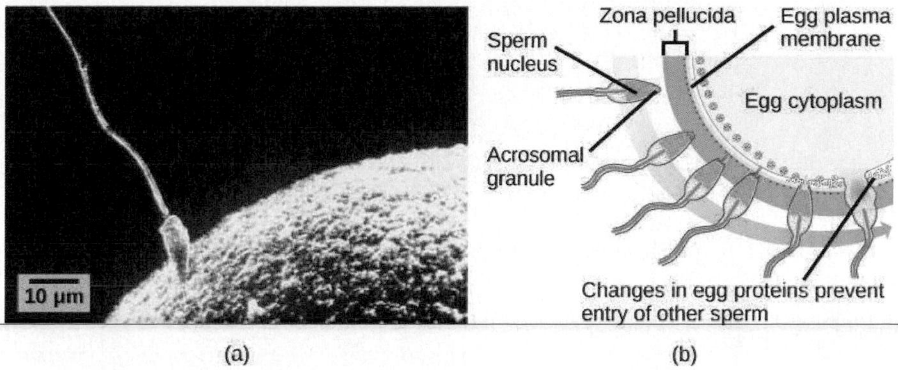

Figure 43.23 (a) Fertilization is the process in which sperm and egg fuse to form a zygote. (b) Acrosomal reactions help the sperm degrade the glycoprotein matrix protecting the egg and allow the sperm to transfer its nucleus. (credit: (b) modification of work by Mariana Ruiz Villareal; scale-bar data from Matt Russell)

To ensure that no more than one sperm fertilizes the egg, once the acrosomal reactions take place at one location of the egg membrane, the egg releases proteins in other locations to prevent other sperm from fusing with the egg. If this mechanism fails, multiple sperm can fuse with the egg, resulting in **polyspermy**. The resulting embryo is not genetically viable and dies within a few days.

Cleavage and Blastula Stage

The development of multi-cellular organisms begins from a single-celled zygote, which undergoes rapid cell division to form the blastula. The rapid, multiple rounds of cell division are termed cleavage. Cleavage is illustrated in (Figure 43.24a). After the cleavage has produced over 100 cells, the embryo is called a blastula. The blastula is usually a spherical layer of cells (the blastoderm) surrounding a fluid-filled or yolk-filled cavity (the blastocoel). Mammals at this stage form a structure called the blastocyst, characterized by an inner cell mass that is distinct from the surrounding blastula, shown in Figure 43.24b. During cleavage, the cells divide without an increase in mass; that is, one large single-celled zygote divides into multiple smaller cells. Each cell within the blastula is called a blastomere.

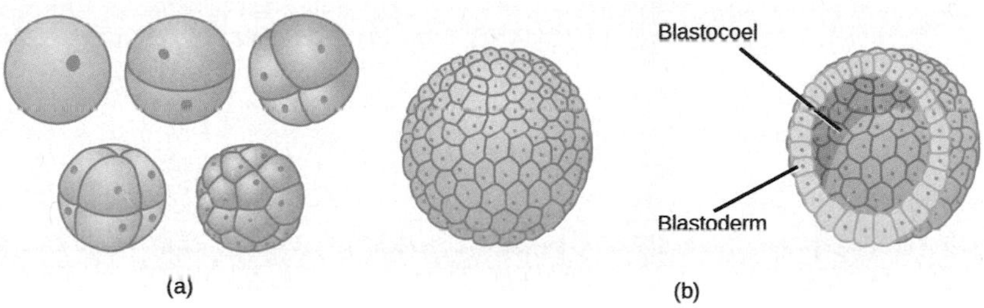

Figure 43.24 (a) During cleavage, the zygote rapidly divides into multiple cells without increasing in size. (b) The cells rearrange themselves to form a hollow ball with a fluid-filled or yolk-filled cavity called the blastula.

Cleavage can take place in two ways: **holoblastic** (total) cleavage or **meroblastic** (partial) cleavage. The type of cleavage depends on the amount of yolk in the eggs. In placental mammals (including humans) where nourishment is provided by the mother's body, the eggs have a very small amount of yolk and undergo holoblastic cleavage. Other species, such as birds, with a lot of yolk in the egg to nourish the embryo during development, undergo meroblastic cleavage.

In mammals, the blastula forms the **blastocyst** in the next stage of development. Here the cells in the blastula arrange themselves in two layers: the **inner cell mass**, and an outer layer called the **trophoblast**. The inner cell mass is also known as the embryoblast and this mass of cells will go on to form the embryo. At this stage of development, illustrated in Figure 43.25 the inner cell mass consists of embryonic stem cells that will differentiate into the different cell types needed by the organism. The trophoblast will contribute to the placenta and nourish the embryo.

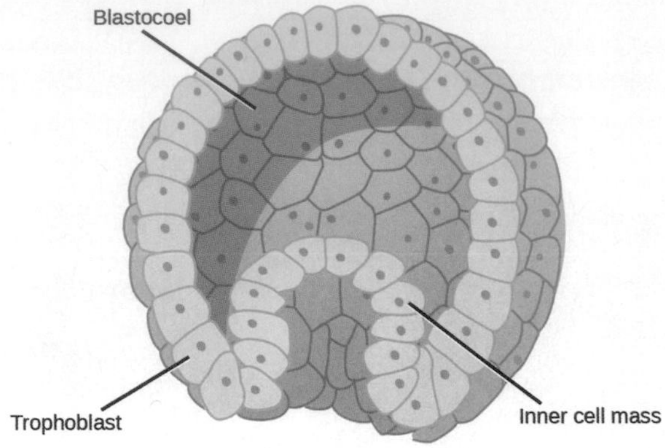

Figure 43.25 The rearrangement of the cells in the mammalian blastula to two layers—the inner cell mass and the trophoblast—results in the formation of the blastocyst.

LINK TO LEARNING

Visit the Virtual Human Embryo project (http://openstax.org/l/human_embryo) at the Endowment for Human Development site to step through an interactive that shows the stages of embryo development, including micrographs and rotating 3-D images.

Gastrulation

The typical blastula is a ball of cells. The next stage in embryonic development is the formation of the body plan. The cells in the blastula rearrange themselves spatially to form three layers of cells. This process is called **gastrulation**. During gastrulation, the blastula folds upon itself to form the three layers of cells. Each of these layers is called a germ layer and each germ layer differentiates into different organ systems.

The three germ layers, shown in Figure 43.26, are the endoderm, the ectoderm, and the mesoderm. The ectoderm gives rise to the nervous system and the epidermis. The mesoderm gives rise to the muscle cells and connective tissue in the body. The endoderm gives rise to columnar cells found in the digestive system and many internal organs.

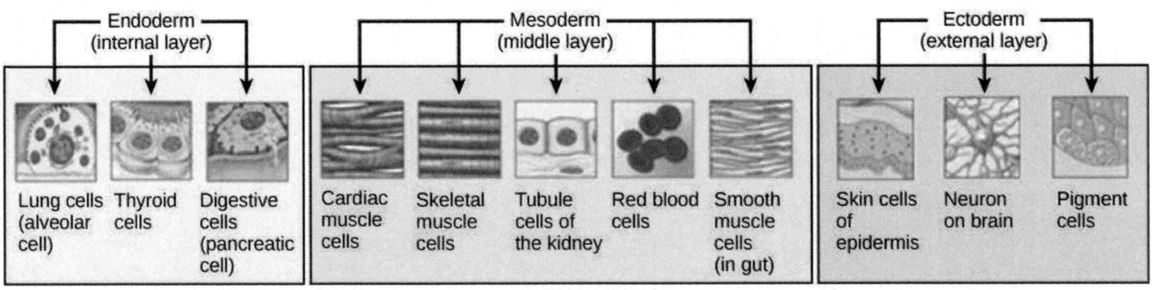

Figure 43.26 The three germ layers give rise to different cell types in the animal body. (credit: modification of work by NIH, NCBI)

Everyday Connection

Are Designer Babies in Our Future?

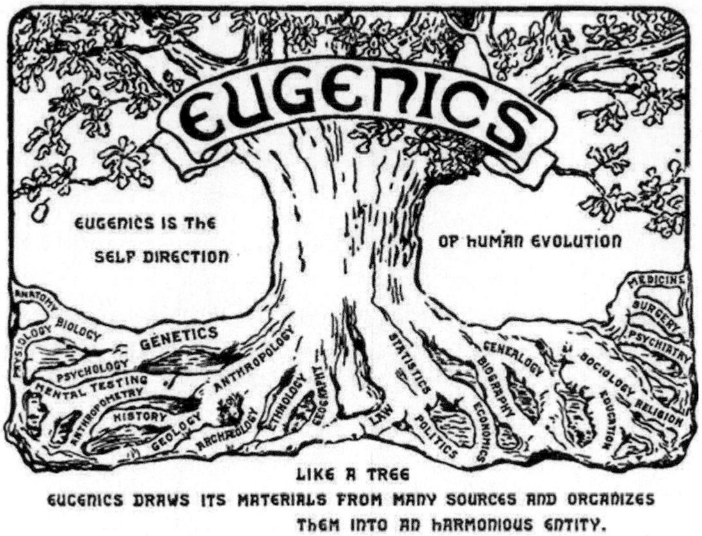

Figure 43.27 This logo from the Second International Eugenics Conference in New York City in September of 1921 shows how eugenics attempted to merge several fields of study with the goal of producing a genetically superior human race.

If you could prevent your child from getting a devastating genetic disease, would you do it? Would you select the sex of your child or select for their attractiveness, strength, or intelligence? How far would you go to maximize the possibility of resistance to disease? The genetic engineering of a human child, the production of "designer babies" with desirable phenotypic characteristics, was once a topic restricted to science fiction. This is the case no longer: science fiction is now overlapping into science fact. Many phenotypic choices for offspring are already available, with many more likely to be possible in the not too distant future. Which traits should be selected and how they should be selected are topics of much debate within the worldwide medical community. The ethical and moral line is not always clear or agreed upon, and some fear that modern reproductive technologies could lead to a new form of eugenics.

Eugenics is the use of information and technology from a variety of sources to improve the genetic makeup of the human race. The goal of creating genetically superior humans was quite prevalent (although controversial) in several countries during the early 20th century, but fell into disrepute when Nazi Germany developed an extensive eugenics program in the 1930s and 40s. As part of their program, the Nazis forcibly sterilized hundreds of thousands of the so-called "unfit" and killed tens of thousands of institutionally disabled people as part of a systematic program to develop a genetically superior race of Germans known as Aryans. Ever since, eugenic ideas have not been as publicly expressed, but there are still those who promote them.

Efforts have been made in the past to control traits in human children using donated sperm from men with desired traits. In fact, eugenicist Robert Klark Graham established a sperm bank in 1980 that included samples exclusively from donors with high IQs. The "genius" sperm bank failed to capture the public's imagination and the operation closed in 1999.

In more recent times, the procedure known as prenatal genetic diagnosis (PGD) has been developed. PGD involves the screening of human embryos as part of the process of *in vitro* fertilization, during which embryos are conceived and grown outside the mother's body for some period of time before they are implanted. The term PGD usually refers to both the diagnosis, selection, and the implantation of the selected embryos.

In the least controversial use of PGD, embryos are tested for the presence of alleles which cause genetic diseases such as sickle cell disease, muscular dystrophy, and hemophilia, in which a single disease-causing allele or pair of alleles has been identified. By excluding embryos containing these alleles from implantation into the mother, the disease is prevented, and the unused embryos are either donated to science or discarded. There are relatively few in the worldwide medical community that question the ethics of this type of procedure, which allows individuals scared to have children because of the alleles they carry to do so successfully. The major limitation to this procedure is its expense. Not usually covered by medical insurance and thus out of reach financially for most couples, only a very small percentage of all live births use such complicated methodologies. Yet, even in cases like these where the ethical issues may seem to be clear-cut, not everyone agrees with the morality of these types of procedures. For example, to those who take the position that human life begins at conception, the discarding of unused embryos, a necessary result of PGD, is unacceptable under any circumstances.

A murkier ethical situation is found in the selection of a child's sex, which is easily performed by PGD. Currently, countries such as Great Britain have banned the selection of a child's sex for reasons other than preventing sex-linked diseases. Other countries allow the procedure for "family balancing", based on the desire of some parents to have at least one child of each sex. Still others, including the United States, have taken a scattershot approach to regulating these practices, essentially leaving it to the individual practicing physician to decide which practices are acceptable and which are not.

Even murkier are rare instances of disabled parents, such as those with deafness or dwarfism, who select embryos via PGD to ensure that they share their disability. These parents usually cite many positive aspects of their disabilities and associated culture as reasons for their choice, which they see as their moral right. To others, to purposely cause a disability in a child violates the basic medical principle of *Primum non nocere,* "first, do no harm." This procedure, although not illegal in most countries, demonstrates the complexity of ethical issues associated with choosing genetic traits in offspring.

Where could this process lead? Will this technology become more affordable and how should it be used? With the ability of technology to progress rapidly and unpredictably, a lack of definitive guidelines for the use of reproductive technologies before they arise might make it difficult for legislators to keep pace once they are in fact realized, assuming the process needs any government regulation at all. Other bioethicists argue that we should only deal with technologies that exist now, and not in some uncertain future. They argue that these types of procedures will always be expensive and rare, so the fears of eugenics and "master" races are unfounded and overstated. The debate continues.

43.7 Organogenesis and Vertebrate Formation

By the end of this section, you will be able to do the following:

- Describe the process of organogenesis
- Identify the anatomical axes formed in vertebrates

Gastrulation leads to the formation of the three germ layers that give rise, during further development, to the different organs in the animal body. This process is called **organogenesis**. Organogenesis is characterized by rapid and precise movements of the cells within the embryo.

Organogenesis

Organs form from the germ layers through the process of differentiation. During differentiation, the embryonic stem cells express specific sets of genes which will determine their ultimate cell type. For example, some cells in the ectoderm will express the genes specific to skin cells. As a result, these cells will differentiate into epidermal cells. The process of differentiation is regulated by cellular signaling cascades.

Scientists study organogenesis extensively in the lab in fruit flies (*Drosophila*) and the nematode *Caenorhabditis elegans. Drosophila* have segments along their bodies, and the patterning associated with the segment formation has allowed scientists to study which genes play important roles in organogenesis along the length of the embryo at different time points. The nematode *C.elegans* has roughly 1000 somatic cells and scientists have studied the fate of each of these cells during their development in the nematode life cycle. There is little variation in patterns of cell lineage between individuals, unlike in mammals where cell development from the embryo is dependent on cellular cues.

In vertebrates, one of the primary steps during organogenesis is the formation of the neural system. The ectoderm forms epithelial cells and tissues, and neuronal tissues. During the formation of the neural system, special signaling molecules called growth factors signal some cells at the edge of the ectoderm to become epidermis cells. The remaining cells in the center form the neural plate. If the signaling by growth factors were disrupted, then the entire ectoderm would differentiate into neural tissue.

The neural plate undergoes a series of cell movements where it rolls up and forms a tube called the **neural tube**, as illustrated in Figure 43.28. In further development, the neural tube will give rise to the brain and the spinal cord.

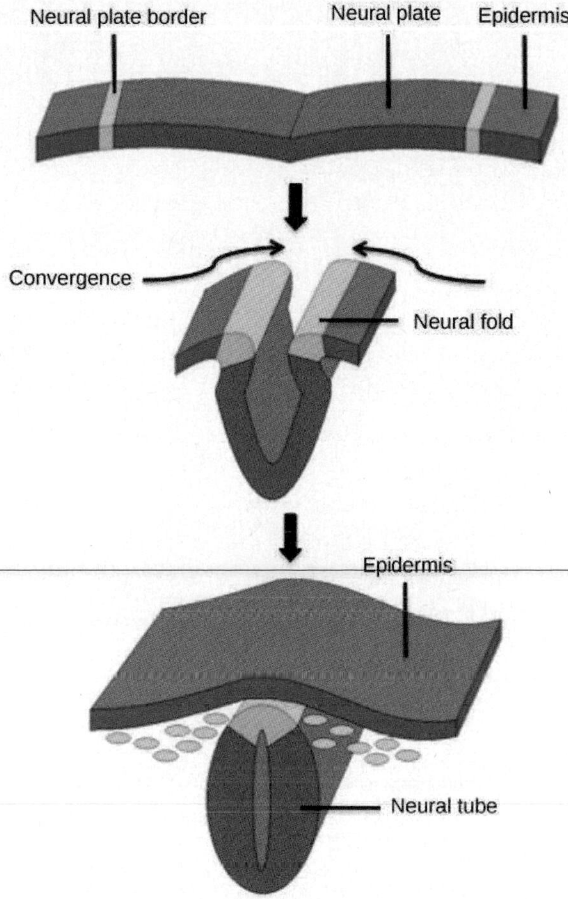

Figure 43.28 The central region of the ectoderm forms the neural tube, which gives rise to the brain and the spinal cord.

The mesoderm that lies on either side of the vertebrate neural tube will develop into the various connective tissues of the animal body. A spatial pattern of gene expression reorganizes the mesoderm into groups of cells called **somites** with spaces between them. The somites illustrated in Figure 43.29 will further develop into the cells that form the vertebrae and ribs, the dermis of the dorsal skin, the skeletal muscles of the back, and the skeletal muscles of the body wall and limbs. The mesoderm also forms a structure called the notochord, which is rod-shaped and forms the central axis of the animal body.

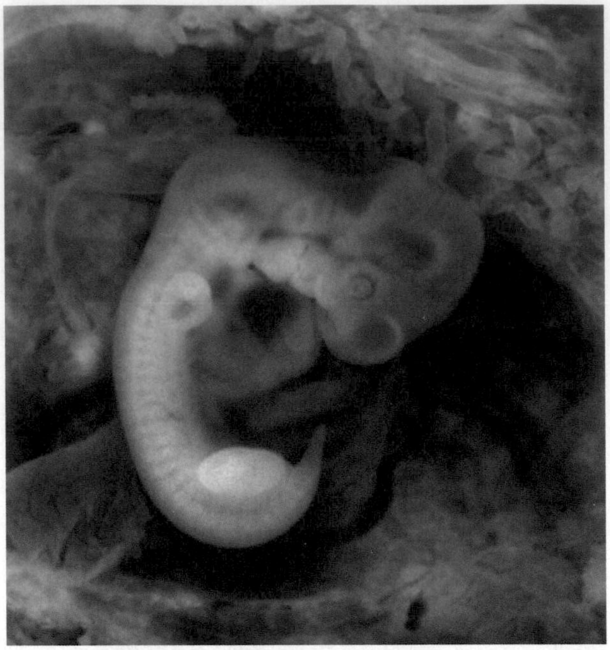

Figure 43.29 In this five-week old human embryo, somites are segments along the length of the body. (credit: modification of work by Ed Uthman)

Vertebrate Axis Formation

Even as the germ layers form, the ball of cells still retains its spherical shape. However, animal bodies have lateral-medial (left-right), dorsal-ventral (back-belly), and anterior-posterior (head-feet) axes, illustrated in Figure 43.30.

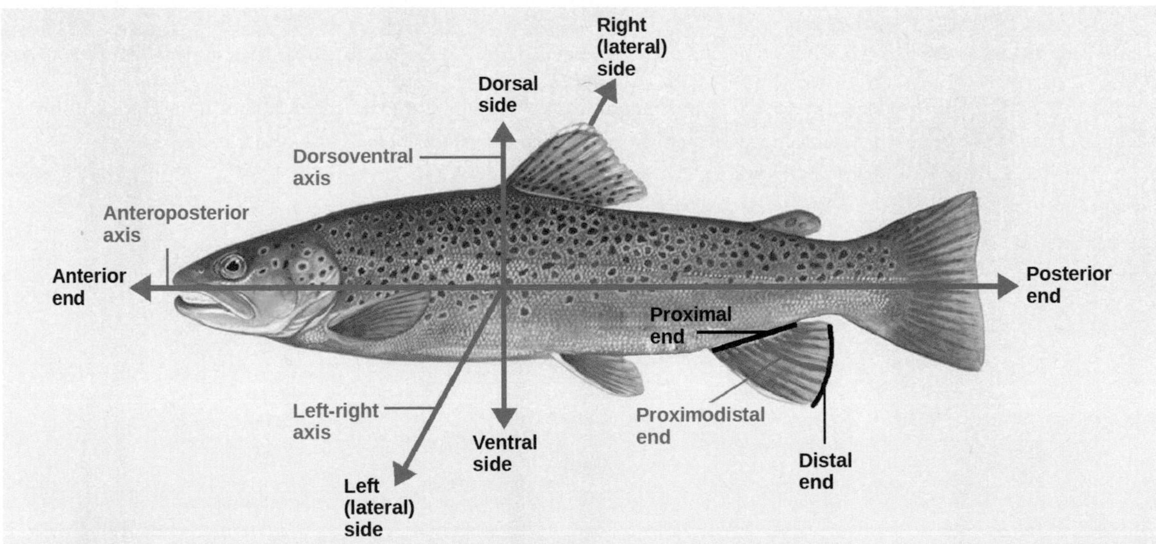

Figure 43.30 Animal bodies have three axes for symmetry. (credit: modification of work by NOAA)

How are these established? In one of the most seminal experiments ever to be carried out in developmental biology, Spemann and Mangold took dorsal cells from one embryo and transplanted them into the belly region of another embryo. They found that the transplanted embryo now had two notochords: one at the dorsal site from the original cells and another at the transplanted site. This suggested that the dorsal cells were genetically programmed to form the notochord and define the axis. Since then, researchers have identified many genes that are responsible for axis formation. Mutations in these genes leads to the loss of symmetry required for organism development.

Animal bodies have externally visible symmetry. However, the internal organs are not symmetric. For example, the heart is on the left side and the liver on the right. The formation of the central left-right axis is an important process during development. This internal asymmetry is established very early during development and involves many genes. Research is still ongoing to fully understand the developmental implications of these genes.

KEY TERMS

acrosomal reaction series of biochemical reactions that the sperm uses to break through the zona pellucida

asexual reproduction form of reproduction that produces offspring that are genetically identical to the parent

blastocyst structure formed when cells in the mammalian blastula separate into an inner and outer layer

budding form of asexual reproduction that results from the outgrowth of a part of a cell leading to a separation from the original animal into two individuals

bulbourethral gland secretion that cleanses the urethra prior to ejaculation

clitoris sensory structure in females; stimulated during sexual arousal

cloaca common body opening for the digestive, excretory, and reproductive systems found in non-mammals, such as birds

contraception (also, birth control) various means used to prevent pregnancy

estrogen reproductive hormone in females that assists in endometrial regrowth, ovulation, and calcium absorption

external fertilization fertilization of egg by sperm outside animal body, often during spawning

fission (also, binary fission) method by which multicellular organisms increase in size or asexual reproduction in which a unicellular organism splits into two separate organisms by mitosis

follicle stimulating hormone (FSH) reproductive hormone that causes sperm production in men and follicle development in women

fragmentation cutting or fragmenting of the original animal into parts and the growth of a separate animal from each part

gastrulation process in which the blastula folds over itself to form the three germ layers

gestation length of time for fetal development to birth

gonadotropin-releasing hormone (GnRH) hormone from the hypothalamus that causes the release of FSH and LH from the anterior pituitary

hermaphroditism state of having both male and female reproductive parts within the same individual

holoblastic complete cleavage; takes place in cells with a small amount of yolk

human beta chorionic gonadotropin (β-HCG) hormone produced by the chorion of the zygote that helps to maintain the corpus luteum and elevated levels of progesterone

infertility inability to conceive, carry, and deliver children

inhibin hormone made by Sertoli cells; provides negative feedback to hypothalamus in control of FSH and GnRH release

inner cell mass inner layer of cells in the blastocyst

internal fertilization fertilization of egg by sperm inside the body of the female

interstitial cell of Leydig cell in seminiferous tubules that makes testosterone

labia majora large folds of tissue covering the inguinal area

labia minora smaller folds of tissue within the labia majora

luteinizing hormone (LH) reproductive hormone in both men and women, causes testosterone production in men and ovulation and lactation in women

menopause loss of reproductive capacity in women due to decreased sensitivity of the ovaries to FSH and LH

menstrual cycle cycle of the degradation and regrowth of the endometrium

meroblastic partial cleavage; takes place in cells with a large amount of yolk

morning sickness condition in the mother during the first trimester; includes feelings of nausea

neural tube tube-like structure that forms from the ectoderm and gives rise to the brain and spinal cord

oogenesis process of producing haploid eggs

organogenesis process of organ formation

ovarian cycle cycle of preparation of egg for ovulation and the conversion of the follicle to the corpus luteum

oviduct (also, fallopian tube) muscular tube connecting the uterus with the ovary area

oviparity process by which fertilized eggs are laid outside the female's body and develop there, receiving nourishment from the yolk that is a part of the egg

ovoviparity process by which fertilized eggs are retained within the female; the embryo obtains its nourishment from the egg's yolk and the young are fully developed when they are hatched

ovulation release of the egg by the most mature follicle

parthenogenesis form of asexual reproduction where an egg develops into a complete individual without being fertilized

penis male reproductive structure for urine elimination and copulation

placenta organ that supports the diffusion of nutrients and waste between the mother's and fetus' blood

polyspermy condition in which one egg is fertilized by multiple sperm

progesterone reproductive hormone in women; assists in endometrial regrowth and inhibition of FSH and LH release

prostate gland structure that is a mixture of smooth muscle and glandular material and that contributes to semen

scrotum sac containing testes; exterior to the body

semen fluid mixture of sperm and supporting materials

seminal vesicle secretory accessory gland in males; contributes to semen

seminiferous tubule site of sperm production in testes

Sertoli cell cell in seminiferous tubules that assists developing sperm and makes inhibin
sexual reproduction mixing of genetic material from two individuals to produce genetically unique offspring
somite group of cells separated by small spaces that form from the mesoderm and give rise to connective tissue
spermatheca specialized sac that stores sperm for later use
spermatogenesis process of producing haploid sperm
testes pair of reproductive organs in males
testosterone reproductive hormone in men that assists in sperm production and promoting secondary sexual characteristics
trophoblast outer layer of cells in the blastocyst
uterus environment for developing embryo and fetus
vagina muscular tube for the passage of menstrual flow, copulation, and birth of offspring
viviparity process in which the young develop within the female, receiving nourishment from the mother's blood through a placenta
zona pellucida protective layer of glycoproteins on the mammalian egg

CHAPTER SUMMARY

43.1 Reproduction Methods

Reproduction may be asexual when one individual produces genetically identical offspring, or sexual when the genetic material from two individuals is combined to produce genetically diverse offspring. Asexual reproduction occurs through fission, budding, and fragmentation. Sexual reproduction may mean the joining of sperm and eggs within animals' bodies or it may mean the release of sperm and eggs into the environment. An individual may be one sex, or both; it may start out as one sex and switch during its life, or it may stay male or female.

43.2 Fertilization

Sexual reproduction starts with the combination of a sperm and an egg in a process called fertilization. This can occur either outside the bodies or inside the female. Both methods have advantages and disadvantages. Once fertilized, the eggs can develop inside the female or outside. If the egg develops outside the body, it usually has a protective covering over it. Animal anatomy evolved various ways to fertilize, hold, or expel the egg. The method of fertilization varies among animals. Some species release the egg and sperm into the environment, some species retain the egg and receive the sperm into the female body and then expel the developing embryo covered with shell, while still other species retain the developing offspring through the gestation period.

43.3 Human Reproductive Anatomy and Gametogenesis

As animals became more complex, specific organs and organ systems developed to support specific functions for the organism. The reproductive structures that evolved in land animals allow males and females to mate, fertilize internally, and support the growth and development of offspring. Processes developed to produce reproductive cells that had exactly half the number of chromosomes of each parent so that new combinations would have the appropriate amount of genetic material. Gametogenesis, the production of sperm (spermatogenesis) and eggs (oogenesis), takes place through the process of meiosis.

43.4 Hormonal Control of Human Reproduction

The male and female reproductive cycles are controlled by hormones released from the hypothalamus and anterior pituitary as well as hormones from reproductive tissues and organs. The hypothalamus monitors the need for the FSH and LH hormones made and released from the anterior pituitary. FSH and LH affect reproductive structures to cause the formation of sperm and the preparation of eggs for release and possible fertilization. In the male, FSH and LH stimulate Sertoli cells and interstitial cells of Leydig in the testes to facilitate sperm production. The Leydig cells produce testosterone, which also is responsible for the secondary sexual characteristics of males. In females, FSH and LH cause estrogen and progesterone to be produced. They regulate the female reproductive system which is divided into the ovarian cycle and the menstrual cycle. Menopause occurs when the ovaries lose their sensitivity to FSH and LH and the female reproductive cycles slow to a stop.

43.5 Human Pregnancy and Birth

Human pregnancy begins with fertilization of an egg and proceeds through the three trimesters of gestation. The labor process has three stages (contractions, delivery of the fetus, expulsion of the placenta), each propelled by hormones. The first trimester lays down the basic structures of the body, including the limb buds, heart, eyes, and the liver. The second trimester continues the development of all of the organs and systems. The third trimester exhibits the greatest growth of the fetus and culminates in labor and delivery. Prevention of a pregnancy can be accomplished through a variety of methods including barriers, hormones, or other means. Assisted reproductive technologies may help individuals who have infertility problems.

43.6 Fertilization and Early Embryonic Development

The early stages of embryonic development begin with fertilization. The process of fertilization is tightly controlled to ensure that only one sperm fuses with one egg. After fertilization, the zygote undergoes cleavage to form the blastula. The blastula, which in some species is a hollow ball of cells, undergoes a process called gastrulation, in which the three germ layers form. The ectoderm gives rise to the nervous system and the epidermal skin cells, the mesoderm gives rise to the muscle cells and connective tissue in the body, and the endoderm gives rise to columnar cells and internal organs.

43.7 Organogenesis and Vertebrate Formation

Organogenesis is the formation of organs from the germ layers. Each germ layer gives rise to specific tissue types. The first stage is the formation of the neural system in the ectoderm. The mesoderm gives rise to somites and the notochord. Formation of vertebrate axis is another important developmental stage.

VISUAL CONNECTION QUESTIONS

1. Figure 43.8 Which of the following statements about the male reproductive system is false?
   a. The vas deferens carries sperm from the testes to the penis.
   b. Sperm mature in seminiferous tubules in the testes.
   c. Both the prostate and the bulbourethral glands produce components of the semen.
   d. The prostate gland is located in the testes.

2. Figure 43.15 Which of the following statements about hormone regulation of the female reproductive cycle is false?
   a. LH and FSH are produced in the pituitary, and estradiol and progesterone are produced in the ovaries.
   b. Estradiol and progesterone secreted from the corpus luteum cause the endometrium to thicken.
   c. Both progesterone and estradiol are produced by the follicles.
   d. Secretion of GnRH by the hypothalamus is inhibited by low levels of estradiol but stimulated by high levels of estradiol.

3. Figure 43.17 Which of the following statements about the menstrual cycle is false?
   a. Progesterone levels rise during the luteal phase of the ovarian cycle and the secretory phase of the uterine cycle.
   b. Menstruation occurs just after LH and FSH levels peak.
   c. Menstruation occurs after progesterone levels drop.
   d. Estrogen levels rise before ovulation, while progesterone levels rise after.

REVIEW QUESTIONS

4. Which form of reproduction is thought to be best in a stable environment?
   a. asexual
   b. sexual
   c. budding
   d. parthenogenesis

5. Which form of reproduction can result from damage to the original animal?
   a. asexual
   b. fragmentation
   c. budding
   d. parthenogenesis

6. Which form of reproduction is useful to an animal with little mobility that reproduces sexually?
   a. fission
   b. budding
   c. parthenogenesis
   d. hermaphroditism

7. Genetically unique individuals are produced through ________.
   a. sexual reproduction
   b. parthenogenesis
   c. budding
   d. fragmentation

8. External fertilization occurs in which type of environment?
   a. aquatic
   b. forested
   c. savanna
   d. steppe

9. Which term applies to egg development within the female with nourishment derived from a yolk?
   a. oviparity
   b. viviparity
   c. ovoviparity
   d. ovovoparity

10. Which term applies to egg development outside the female with nourishment derived from a yolk?
    a. oviparity
    b. viviparity
    c. ovoviparity
    d. ovovoparity

11. Sperm are produced in the ________.
    a. scrotum
    b. seminal vesicles
    c. seminiferous tubules
    d. prostate gland

12. Most of the bulk of semen is made by the ________.
    a. scrotum
    b. seminal vesicles
    c. seminiferous tubules
    d. prostate gland

13. Which of the following cells in spermatogenesis is diploid?
    a. primary spermatocyte
    b. secondary spermatocyte
    c. spermatid
    d. sperm

14. Which female organ has the same embryonic origin as the penis?
    a. clitoris
    b. labia majora
    c. greater vestibular glands
    d. vagina

15. Which female organ has an endometrial lining that will support a developing baby?
    a. labia minora
    b. breast
    c. ovaries
    d. uterus

16. How many eggs are produced as a result of one meiotic series of cell divisions?
    a. one
    b. two
    c. three
    d. four

17. Which hormone causes Leydig cells to make testosterone?
    a. FSH
    b. LH
    c. inhibin
    d. estrogen

18. Which hormone causes FSH and LH to be released?
    a. testosterone
    b. estrogen
    c. GnRH
    d. progesterone

19. Which hormone signals ovulation?
    a. FSH
    b. LH
    c. inhibin
    d. estrogen

20. Which hormone causes the regrowth of the endometrial lining of the uterus?
    a. testosterone
    b. estrogen
    c. GnRH
    d. progesterone

21. Nutrient and waste requirements for the developing fetus are handled during the first few weeks by:
    a. the placenta
    b. diffusion through the endometrium
    c. the chorion
    d. the blastocyst

22. Progesterone is made during the third trimester by the:
    a. placenta
    b. endometrial lining
    c. chorion
    d. corpus luteum

23. Which contraceptive method is 100 percent effective at preventing pregnancy?
    a. condom
    b. oral hormonal methods
    c. sterilization
    d. abstinence

24. Which type of short term contraceptive method is generally more effective than others?
    a. barrier
    b. hormonal
    c. natural family planning
    d. withdrawal
25. Which hormone is primarily responsible for the contractions during labor?
    a. oxytocin
    b. estrogen
    c. β-HCG
    d. progesterone
26. Major organs begin to develop during which part of human gestation?
    a. fertilization
    b. first trimester
    c. second trimester
    d. third trimester
27. Which of the following is false?
    a. The endoderm, mesoderm, ectoderm are germ layers.
    b. The trophoblast is a germ layer.
    c. The inner cell mass is a source of embryonic stem cells.
    d. The blastula is often a hollow ball of cells.
28. During cleavage, the mass of cells:
    a. increases
    b. decreases
    c. doubles with every cell division
    d. does not change significantly
29. Which of the following gives rise to the skin cells?
    a. ectoderm
    b. endoderm
    c. mesoderm
    d. none of the above
30. The ribs form from the _______.
    a. notochord
    b. neural plate
    c. neural tube
    d. somites

CRITICAL THINKING QUESTIONS

31. Why is sexual reproduction useful if only half the animals can produce offspring and two separate cells must be combined to form a third?
32. What determines which sex will result in offspring of birds and mammals?
33. What are the advantages and disadvantages of external and internal forms of fertilization?
34. Why would paired external fertilization be preferable to group spawning?
35. Describe the phases of the human sexual response.
36. Compare spermatogenesis and oogenesis as to timing of the processes and the number and type of cells finally produced.
37. If male reproductive pathways are not cyclical, how are they controlled?
38. Describe the events in the ovarian cycle leading up to ovulation.
39. Describe the major developments during each trimester of human gestation.
40. Describe the stages of labor.
41. What do you think would happen if multiple sperm fused with one egg?
42. Why do mammalian eggs have a small concentration of yolk, while bird and reptile eggs have a large concentration of yolk?
43. Explain how the different germ layers give rise to different tissue types.
44. Explain the role of axis formation in development.

CHAPTER 44
Ecology and the Biosphere

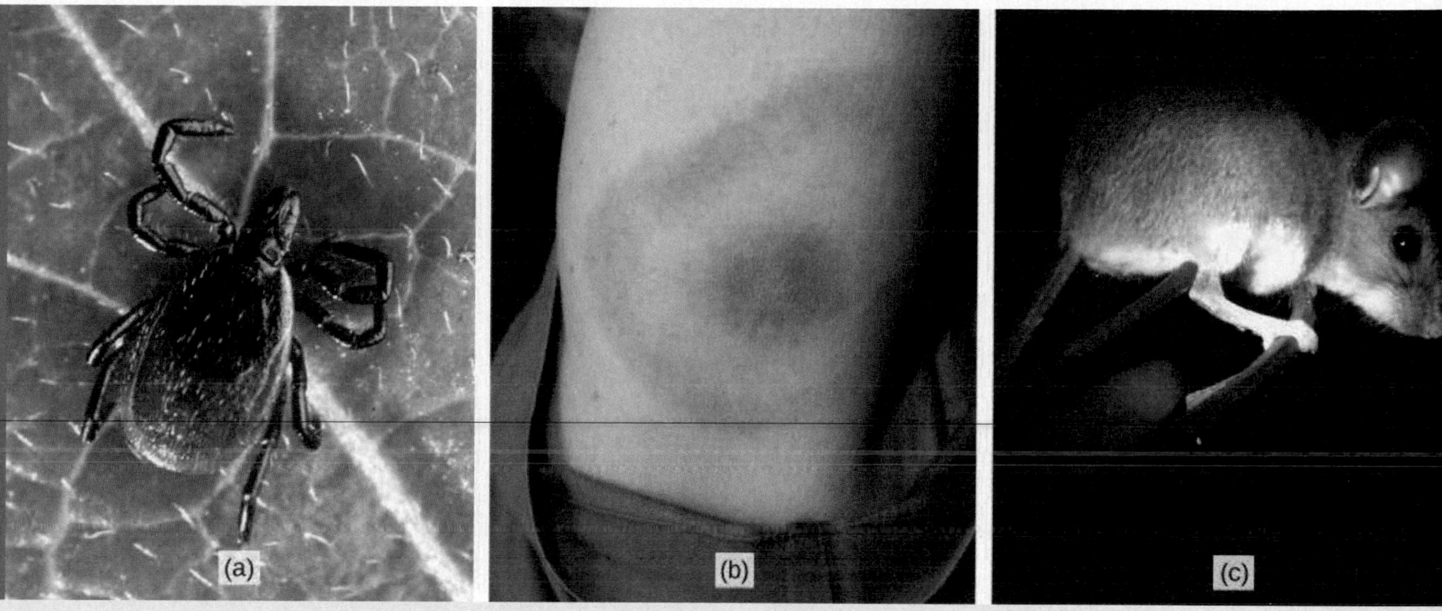

Figure 44.1 The (a) deer tick carries the bacterium that produces Lyme disease in humans, often evident in (b) a symptomatic bull's eye rash. The (c) white-footed mouse is one well-known host to deer ticks carrying the Lyme disease bacterium. (credit a: modification of work by Scott Bauer, USDA ARS; credit b: modification of work by James Gathany, CDC; credit c: modification of work by Rob Ireton)

INTRODUCTION Why study ecology? Perhaps you are interested in learning about the natural world and how living things have adapted to the physical conditions of their environment. Or, perhaps you're a future physician seeking to understand the connection between your patients' health and their environment.

Humans are a part of the ecological landscape, and human health is one important part of human interaction with our physical and living environment. Lyme disease, for instance, serves as one modern-day example of the connection between our health and the natural world (Figure 44.1). More formally known as Lyme borreliosis, Lyme disease is a bacterial infection that can be transmitted to humans when they are bitten by the deer tick (*Ixodes scapularis* in the eastern U.S., and *Ixodes pacificus* along the Pacific coast). Deer ticks are the primary *vectors* (a vector is an organism that transmits a pathogen) for this disease. However, not all ticks carry the pathogen, and not all deer ticks carry the bacteria that will cause Lyme disease in humans. Also, the ticks *I. scapularis* and *pacificus* can have other hosts besides deer. In fact, it turns out that the *probability of infection* depends on the type of host upon which the tick develops: a higher proportion of ticks that live on white-footed mice carry the bacterium than do ticks that live on deer. Knowledge about the environments and population densities in which the host species is abundant would help a physician or an epidemiologist better understand how Lyme disease is transmitted and how its incidence could be reduced.

Chapter Outline

44.1 The Scope of Ecology

By the end of this section, you will be able to do the following:

- Define ecology and the four basic levels of ecological research
- Describe examples of the ways in which ecology requires the integration of different scientific disciplines
- Distinguish between abiotic and biotic components of the environment
- Recognize the relationship between abiotic and biotic components of the environment

Ecology is the study of the interactions of living organisms with their environment. One core goal of ecology is to understand the distribution and abundance of living things in the physical environment. Attainment of this goal requires the integration of scientific disciplines inside and outside of biology, such as mathematics, statistics, biochemistry, molecular biology, physiology, evolution, biodiversity, geology, and climatology.

LINK TO LEARNING

Climate change can alter where organisms live, which can sometimes directly affect human health. Watch the PBS video "Feeling the Effects of Climate Change" (http://openstax.org/l/climate_health) in which researchers discover a pathogenic organism living far outside of its normal range.

Levels of Ecological Study

When a discipline such as biology is studied, it is often helpful to subdivide it into smaller, related areas. For instance, cell biologists interested in cell signaling need to understand the chemistry of the signal molecules (which are usually proteins) as well as the result of cell signaling. Ecologists interested in the factors that influence the survival of an endangered species might use mathematical models to predict how current conservation efforts affect endangered organisms.

To produce a sound set of management options, a *conservation biologist* needs to collect accurate data, including current population size, factors affecting reproduction (like physiology and behavior), habitat requirements (such as plants and soils), and potential human influences on the endangered population and its habitat (which might be derived through studies in sociology and urban ecology). Within the discipline of ecology, researchers work at four general levels, which sometimes overlap. These levels are organism, population, community, and ecosystem (Figure 44.2).

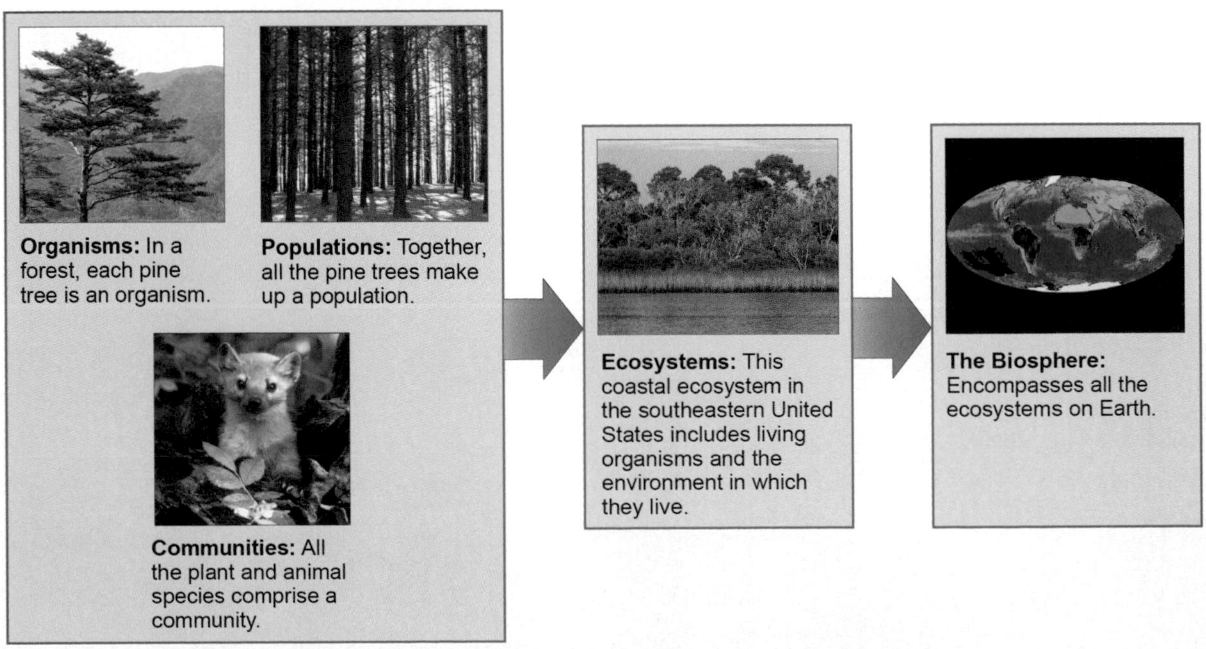

Figure 44.2 Ecologists study within several biological levels of organization. (credit "organisms": modification of work

by yeowatzup"/Flickr; credit "populations": modification of work by "Crystl"/Flickr; credit "communities": modification of work by US Fish and Wildlife Service; credit "ecosystems": modification of work by Tom Carlisle, US Fish and Wildlife Service Headquarters; credit "biosphere": NASA)

Organismal Ecology

Researchers studying ecology at the organismal level are interested in the adaptations that enable individuals to live in specific habitats. These adaptations can be morphological, physiological, and behavioral. For instance, the Karner blue butterfly (*Lycaeides melissa samuelis*) (Figure 44.3) is considered a specialist because the females only *oviposit* (that is, lay eggs) on wild lupine (*Lupinus perennis*). This specific requirement and adaptation means that the Karner blue butterfly is completely dependent on the presence of wild lupine plants for its survival.

Figure 44.3 The Karner blue butterfly (*Lycaeides melissa samuelis*) is a rare butterfly that lives only in open areas with few trees or shrubs, such as pine barrens and oak savannas. It can only lay its eggs on lupine plants. (credit: modification of work by J & K Hollingsworth, USFWS)

After hatching, the (first instar) caterpillars emerge and spend four to six weeks feeding solely on wild lupine (Figure 44.4). The caterpillars pupate as a chrysalis to undergo the final stage of metamorphosis and emerge as butterflies after about four weeks. The adult butterflies feed on the nectar of flowers of wild lupine and other plant species, such as milkweeds. Generally there are two broods of the Karner blue each year.

A researcher interested in studying Karner blue butterflies at the organismal level might, in addition to asking questions about egg laying requirements, ask questions about the butterflies' preferred thoracic flight temperature (a physiological question), or the behavior of the caterpillars when they are at different larval stages (a behavioral question).

Figure 44.4 The wild lupine (*Lupinus perennis*) is the only known host plant for the Karner blue butterfly.

Population Ecology

A **population** is a group of *interbreeding organisms* that are members of the same species living in the same area at the same time. (Organisms that are all members of the same species are called **conspecifics**.) A population is identified, in part, by where it lives, and its area of population may have natural or artificial boundaries. Natural boundaries might be rivers, mountains, or deserts, while artificial boundaries may be mowed grass, manmade structures, or roads. The study of *population ecology* focuses on the number of individuals in an area and how and why population size changes over time.

For example, population ecologists are particularly interested in counting the Karner blue butterfly because it is classified as a federally endangered species. However, the distribution and density of this species is highly influenced by the distribution and abundance of wild lupine, and the biophysical environment around it. Researchers might ask questions about the factors leading to the decline of wild lupine and how these affect Karner blue butterflies. For example, ecologists know that wild lupine thrives in open areas where trees and shrubs are largely absent. In natural settings, intermittent wildfires regularly remove trees and shrubs, helping to maintain the open areas that wild lupine requires. Mathematical models can be used to understand how wildfire suppression by humans has led to the decline of this important plant for the Karner blue butterfly.

Community Ecology

A **biological community** consists of the different species within an area, typically a three-dimensional space, and the interactions within and among these species. Community ecologists are interested in the processes driving these interactions and their consequences. Questions about *conspecific* interactions often focus on competition among members of the same species for a limited resource. Ecologists also study interactions between various species; members of different species are called **heterospecifics**. Examples of heterospecific interactions include predation, parasitism, herbivory, competition, and pollination. These interactions can have regulating effects on population sizes and can impact ecological and evolutionary processes affecting diversity.

For example, Karner blue butterfly larvae form mutualistic relationships with ants (especially *Formica* spp). **Mutualism** is a form of long-term relationship that has coevolved between two species and from which each species benefits. For mutualism to exist between individual organisms, each species must receive *some* benefit from the other as a consequence of the relationship. Researchers have shown that there is an increase in survival when ants protect Karner blue butterfly larvae (caterpillars) from predaceous insects and spiders, an act known as "tending." This might be because the larvae spend less time in each life stage when tended by ants, which provides an advantage for the larvae. Meanwhile, to attract the ants, the Karner blue butterfly larvae secrete ant-like pheromones and a carbohydrate-rich substance that is an important energy source for the ants. Both the Karner blue larvae and the ants benefit from their interaction, although the species of attendant ants may be partially opportunistic and vary over the range of the butterfly.

Ecosystem Ecology

Ecosystem ecology is an extension of organismal, population, and community ecology. The ecosystem is composed of all the **biotic** components (living things) in an area along with the **abiotic** components (nonliving things) of that area. Some of the abiotic components include air, water, and soil. Ecosystem biologists ask questions about how nutrients and energy are stored and how they move among organisms and through the surrounding atmosphere, soil, and water.

The Karner blue butterflies and the wild lupine live in an oak-pine barren habitat. This habitat is characterized by natural disturbance and nutrient-poor soils that are low in nitrogen. The availability of nutrients is an important factor in the distribution of the plants that live in this habitat. Researchers interested in ecosystem ecology could ask questions about the importance of limited resources and the movement of resources, such as nutrients, though the biotic and abiotic portions of the ecosystem.

CAREER CONNECTION

Ecologist

A career in ecology contributes to many facets of human society. Understanding ecological issues can help society meet the basic human needs of food, shelter, and health care. Ecologists can conduct their research in the laboratory and outside in natural environments (Figure 44.5). These natural environments can be as close to home as the stream running through your campus or as far away as the hydrothermal vents at the bottom of the Pacific Ocean. Ecologists manage natural resources such as white-tailed deer populations (*Odocoileus virginianus*) for hunting or aspen (*Populus* spp.) timber stands for paper production. Ecologists also work as educators who teach children and adults at various institutions including universities, high schools, museums, and nature centers. Ecologists may also work in advisory positions assisting local, state, and federal policymakers to develop laws that are ecologically sound, or they may develop those policies and legislation themselves. To become an ecologist requires at least an undergraduate degree, usually in a natural science. The undergraduate degree is often followed by specialized training or an advanced degree, depending on the area of ecology selected. Ecologists should also have a broad background in the physical sciences, as well as a solid foundation in mathematics and statistics.

Figure 44.5 This landscape ecologist is releasing a black-footed ferret into its native habitat as part of a study. (credit: USFWS Mountain Prairie Region, NPS)

LINK TO LEARNING

Visit this site (http://openstax.org/l/ecologist_role) to see Stephen Wing, a marine ecologist from the University of Otago, discuss the role of an ecologist and the types of issues ecologists explore.

44.2 Biogeography

By the end of this section, you will be able to do the following:

- Define biogeography
- List and describe abiotic factors that affect the global distribution of plant and animal species
- Compare the impact of abiotic forces on aquatic and terrestrial environments
- Summarize the effects of abiotic factors on net primary productivity

Many forces influence the communities of living organisms present in different parts of the **biosphere** (all of the parts of Earth inhabited by life). The biosphere extends into the atmosphere (several kilometers above Earth) and into the depths of the oceans. Despite its apparent vastness to an individual human, the biosphere occupies only a minute space when compared to the known universe. Many abiotic forces influence where life can exist and the types of organisms found in different parts of the biosphere. The abiotic factors influence the distribution of **biomes**: large areas of land with similar climate, flora, and fauna.

Biogeography

Biogeography is the study of the geographic distribution of living things and the *abiotic fact*ors that affect their distribution. Abiotic factors such as temperature and rainfall vary based mainly on latitude and elevation. As these abiotic factors change, the composition of plant and animal communities also changes. For example, if you were to begin a journey at the equator and walk north, you would notice gradual changes in plant communities. At the beginning of your journey, you would see tropical wet forests with broad-leaved evergreen trees, which are characteristic of plant communities found near the equator. As you continued to travel north, you would see these broad-leaved evergreen plants eventually give rise to seasonally dry forests with scattered trees. You would also begin to notice changes in temperature and moisture. At about 30 degrees north, these forests would give way to deserts, which are characterized by low precipitation and high insolation (sunlight).

Moving farther north, you would see that deserts are replaced by grasslands or prairies. Eventually, grasslands are replaced by deciduous temperate forests. These deciduous forests give way to the boreal forests and taiga found in the subarctic, the area south of the Arctic Circle. Finally, you would reach the Arctic tundra, which is found at the most northern latitudes. This trek north reveals gradual changes in both climate and the types of organisms that have adapted to environmental factors associated with ecosystems found at different latitudes. However, different ecosystems exist at the same latitude due in part to abiotic factors such as jet streams, the Gulf Stream, and ocean currents. If you were to hike up a mountain, the changes you would see in the vegetation would parallel in many ways those as you move to higher latitudes.

Ecologists who study biogeography examine patterns of *species distribution*. No species exists everywhere; for example, the Venus flytrap (*Dionaea muscipula*) is endemic to a small area in North and South Carolina. An **endemic species** is one which is naturally found only in a specific geographic area that is usually restricted in size. Other species are generalists: species which live in a wide variety of geographic areas; the raccoon (*Procyon* spp) for example, is native to most of North and Central America.

Species distribution patterns are based on biotic and abiotic factors and their influences during the very long periods of time required for species evolution; therefore, early studies of biogeography were closely linked to the emergence of evolutionary thinking in the eighteenth century. Some of the most distinctive assemblages of plants and animals occur in regions that have been physically separated for millions of years by geographic barriers. Biologists estimate that Australia, for example, has between 600,000 and 700,000 species of plants and animals. Approximately 3/4 of living plant and mammal species are endemic species found solely in Australia (Figure 44.6ab).

(a) (b)

Figure 44.6 Australia is home to many endemic species. The (a) wallaby (*Wallabia bicolor*), a medium-sized member of the kangaroo family, is a pouched mammal, or marsupial. The (b) echidna (*Tachyglossus aculeatus*) is an egg-laying mammal. (credit a: modification of work by Derrick Coetzee; credit b: modification of work by Allan Whittome)

Sometimes ecologists discover unique patterns of species distribution by determining where species are *not* found. Despite being tropical, Hawaii, for example, has no native land species of reptiles or amphibians, only a few native species of butterflies, and only one native terrestrial mammal, the hoary bat. Most of New Guinea, as another example, lacks placental mammals.

LINK TO LEARNING

Check out this video (http://openstax.org/l/platypus) to observe a platypus swimming in its natural habitat in New South Wales, Australia.

Like animals, plants can be endemic or generalists: endemic plants are found only on specific regions of the Earth, while generalists are found on many regions. Isolated land masses—such as Australia, Hawaii, and Madagascar—often have large numbers of endemic plant species. Some of these plants are endangered due to human activity. The forest gardenia (*Gardenia brighamii*), for instance, is endemic to Hawaii; only an estimated 15–20 trees are thought to exist (Figure 44.7).

Figure 44.7 Listed as federally endangered, the forest gardenia is a small tree with distinctive flowers. It is found only in five of the Hawaiian Islands in small populations consisting of a few individual specimens. (credit: Forest & Kim Starr)

Energy Sources

Energy from the sun is captured by green plants, algae, cyanobacteria, and photosynthetic protists. These organisms convert solar energy into the chemical energy needed by all living things. Light availability can be an important force directly affecting the evolution of adaptations in photosynthesizers. For instance, plants in the understory of a temperate forest are shaded when the trees above them in the canopy completely leaf out in the late spring. Not surprisingly, understory plants have adaptations to successfully capture available light that passes through the canopy. One such adaptation is the rapid growth of spring ephemeral plants such as the spring beauty (*Claytonia virginica*) (Figure 44.8). These spring flowers achieve much of their growth and finish their life cycle (reproduce) early in the season before the trees in the canopy develop leaves.

Figure 44.8 The spring beauty is an ephemeral spring plant that flowers early in the spring to avoid competing with larger forest trees for sunlight. (credit: John Beetham)

In aquatic ecosystems, the availability of light may be limited because sunlight is absorbed by water, plants, suspended particles, and resident microorganisms. Toward the bottom of a lake, pond, or ocean, there is a zone that light cannot reach (because most wavelengths except for the shortest blues are absorbed by the water column). Photosynthesis cannot take place there and, as a result, a number of adaptations have evolved that enable living things to survive without light. For instance, aquatic plants have photosynthetic tissue near the surface of the water. You can think of the broad, floating leaves of a water lily—water lilies cannot survive without light. In environments such as hydrothermal vents, some bacteria extract energy from inorganic chemicals because there is no light for photosynthesis.

The availability of nutrients in aquatic systems such as oceans is also an important aspect of energy or photosynthesis. Many organisms sink to the bottom of the ocean when they die in the open water; when this occurs, the energy found in that living organism is sequestered for some time unless ocean upwelling occurs. **Ocean upwelling** is the rising of deep ocean waters that occurs when prevailing winds blow along surface waters near a coastline (Figure 44.9). As the wind pushes ocean waters offshore, water from the bottom of the ocean moves up to replace this water. As a result, the nutrients once contained in dead organisms become available for reuse by other living organisms.

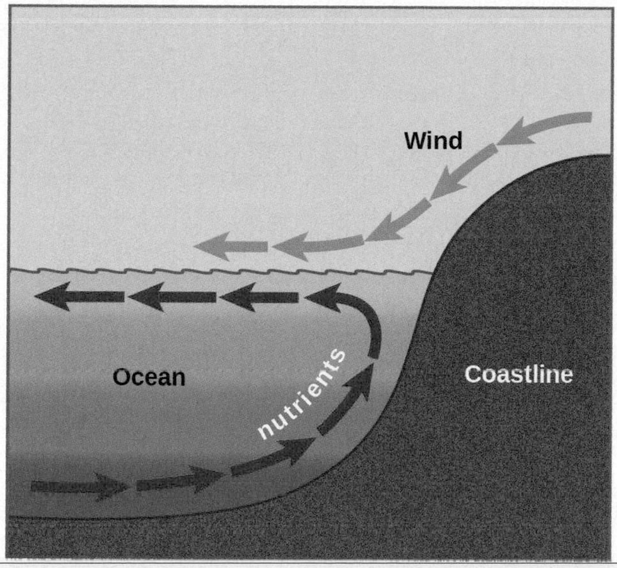

Figure 44.9 Ocean upwelling is an important process that recycles nutrients and energy in the ocean. As wind (green arrows) pushes offshore, it causes water from the ocean bottom (red arrows) to move to the surface, bringing up nutrients from the ocean depths.

In freshwater systems, such as lakes, the recycling of nutrients occurs in response to air temperature and wind changes. The nutrients at the bottom of lakes are recycled twice each year: in the spring and fall turnover. The **spring-and-fall turnover** are seasonal processes that recycle nutrients and oxygen from the bottom of a freshwater lake to the top of the lake (Figure 44.10). These turnovers are caused by the formation of a **thermocline**: layers of water with temperatures that are significantly different from those above and below it.

In wintertime, the surface of lakes found in many northern regions is frozen. However, the water under the ice is slightly warmer, and the water at the bottom of the lake is warmer yet at 4 °C to 5 °C (39.2 °F to 41 °F). Water is densest at about 4 °C; therefore, the deepest water is also the densest. The deepest water is oxygen-poor because the decomposition of organic material at the bottom of the lake uses up available oxygen that cannot be replaced by means of oxygen diffusion into the surface of the water, due to the surface ice layer.

VISUAL CONNECTION

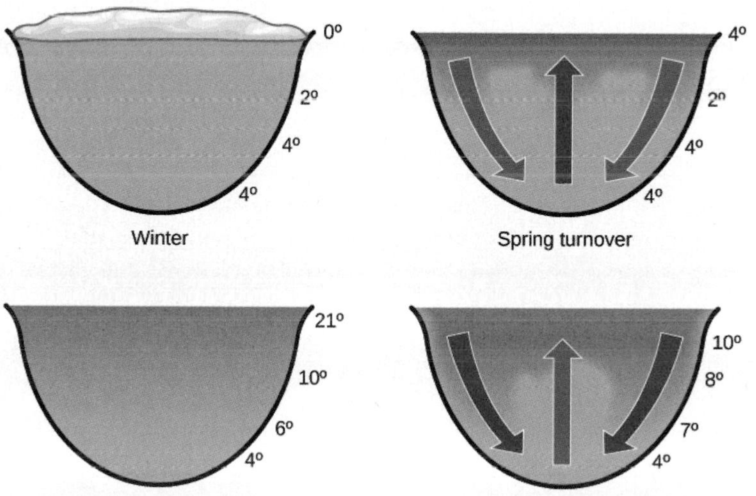

Figure 44.10 The spring and fall turnovers are important processes in freshwater lakes that act to move the nutrients and oxygen at the bottom of deep lakes to the top. Turnover occurs because water has a maximum density at 4 °C. Surface water temperature changes as the seasons progress, and denser water sinks.

How might turnover in tropical lakes differ from turnover in lakes that exist in temperate regions? Think of the variation, or lack of variation, in seasonal temperature change.

In springtime, air temperatures increase and surface ice melts. When the temperature of the surface water begins to approach 4 °C, the water becomes heavier and sinks to the bottom. The water at the bottom of the lake is then displaced by the heavier and denser surface water and, thus, rises to the top. As that water rises to the top, the sediments and nutrients from the lake bottom are brought along with it. This is called the *spring turnover*. During the summer months, the lake water stratifies, or forms layers, with the warmest water at the lake surface.

As air temperatures drop in the fall, the temperature of the lake water cools to 4 °C; therefore, this causes *fall turnover* as the heavy cold water sinks and displaces the water at the bottom. The oxygen-rich water at the surface of the lake then moves to the bottom of the lake, while the nutrients at the bottom of the lake rise to the surface (Figure 44.10). During the winter, the oxygen at the bottom of the lake is used by decomposers and other organisms requiring oxygen, such as fish. It is important to note, however, that the relative transparency of ice also allows the penetration of the shorter wavelengths of visible light so that photosynthesis, especially by algae can continue.

Temperature

Temperature affects the physiology of organisms as well as the density and state of water. Temperature exerts an important influence on living things because few living things can survive at temperatures below 0 °C (32 °F) due to metabolic constraints. It is also rare for living things to survive at temperatures exceeding 45 °C (113 °F); this is a reflection of evolutionary response to typical temperatures near the Earth's surface. Enzymes are most efficient within a narrow and specific range of temperatures; enzyme degradation can occur at higher temperatures. Therefore, organisms either must maintain an internal temperature or they must inhabit an environment that will keep the body within a temperature range that supports metabolism. Some animals have adapted to enable their bodies to survive significant temperature fluctuations, such as seen in hibernation or reptilian torpor. Similarly, some Archaea bacteria have evolved to tolerate extremely hot temperatures such as those found in the geysers within Yellowstone National Park. Such bacteria are examples of *extremophiles*: organisms that thrive in extreme environments.

The temperature (of both water and air) can limit the distribution of living things. Animals faced with temperature fluctuations may respond with adaptations, such as migration, in order to survive. **Migration**, the regular movement from one place to another, is an adaptation found in many animals, including many that inhabit seasonally cold climates. Migration solves problems related to temperature, locating food, and finding a mate. For example, the Arctic Tern (*Sterna paradisaea*) makes a 40,000 km (24,000 mi) round-trip flight each year between its feeding grounds in the southern hemisphere and its breeding grounds in the Arctic Ocean. Monarch butterflies (*Danaus plexippus*) live in the eastern and western United States in the warmer months, where they build up enormous populations, and migrate to areas around Michoacan, Mexico as well as areas along the Pacific Coast, and the southern United States in the wintertime. Some species of mammals also make migratory forays. Reindeer (*Rangifer tarandus*) travel about 5,000 km (3,100 mi) each year to find food. Amphibians and reptiles are more limited in their distribution because they generally lack migratory ability. Not all animals that could migrate do so: migration carries risk and comes at a high-energy cost.

Some animals *hibernate* or *estivate* to survive hostile temperatures. **Hibernation** enables animals to survive cold conditions, and **estivation** allows animals to survive the hostile conditions of a hot, dry climate. Animals that hibernate or estivate enter a state known as *torpor*: a condition in which their metabolic rate is significantly lowered. This enables the animal to wait until its environment better supports its survival. Some amphibians, such as the wood frog (*Rana sylvatica*), have an antifreeze-like chemical in their cells, which retains the cells' integrity and prevents them from freezing and bursting.

Water

Water is required by all living things because it is critical for cellular processes. Since terrestrial organisms lose water to the environment, they have evolved many adaptations to retain water.

- Plants have a number of interesting features on their leaves, such as leaf hairs and a waxy cuticle, that serve to decrease the rate of water loss via transpiration and convection.
- Freshwater organisms are surrounded by water and are constantly in danger of having water rush into their cells because of osmosis. Many adaptations of organisms living in freshwater environments have evolved to ensure that solute concentrations in their bodies remain within appropriate levels. One such adaptation is the excretion of dilute urine.

- Marine organisms are surrounded by water with a higher solute concentration than the organism and, thus, are in danger of losing water to the environment because of osmosis. These organisms have morphological and physiological adaptations to retain water and release solutes into the environment. For example, Marine iguanas (*Amblyrhynchus cristatus*), sneeze out water vapor that is high in salt in order to maintain solute concentrations within an acceptable range while swimming in the ocean and eating marine plants.

Inorganic Nutrients and Soil

Inorganic nutrients, such as nitrogen and phosphorus, are important in determining the distribution and the abundance of living things. Plants obtain these inorganic nutrients from the soil when water moves into the plant through the roots. Therefore, soil structure (particle size of soil components), soil pH, and soil nutrient content together all play an important role in the distribution of plants. Animals obtain inorganic nutrients from the food they consume. Therefore, animal distributions are related to the distribution of what they eat. In some cases, animals will follow their food resource as it moves through the environment.

Other Aquatic Factors

Some abiotic factors, such as oxygen, are important in aquatic ecosystems as well as terrestrial environments. Terrestrial animals obtain oxygen from the air they breathe. Oxygen availability can be an issue for organisms living at very high elevations, however, where there are fewer molecules of oxygen in the air. In aquatic systems, the concentration of dissolved oxygen is related to water temperature and the speed at which the water moves. Cold water has more dissolved oxygen than warmer water. In addition, salinity, currents, and tidal changes can be important abiotic factors in aquatic ecosystems.

Other Terrestrial Factors

Wind can be an important abiotic factor because it influences the rate of evaporation, transpiration, and convective heat loss from the surface of all organisms. The physical force of wind is also important because it can move soil, water, or other abiotic factors, as well as an ecosystem's organisms.

Fire is another terrestrial factor that can be an important agent of disturbance in terrestrial ecosystems. Some organisms are adapted to fire and, thus, require the high heat associated with fire to complete a part of their life cycle. For example, the jack pine (*Pinus banksiana*) requires heat from fire for its seed cones to open (Figure 44.11). Through the burning of pine needles, fire adds nitrogen to the soil and limits competition by destroying undergrowth.

Figure 44.11 The mature cones of the jack pine *(Pinus banksiana)* open only when exposed to high temperatures, such as during a forest fire. A fire is likely to kill most vegetation, so a seedling that germinates after a fire is more likely to receive ample sunlight than one that germinates under normal conditions. (credit: USDA)

Abiotic Factors Influencing Plant Growth

Temperature and moisture are important influences on plant production (primary productivity) and the amount of organic matter available as food (net primary productivity). **Net primary productivity** is an estimation of all of the organic matter available as food; it is calculated as the total amount of carbon fixed per year minus the amount that is oxidized during cellular respiration. In terrestrial environments, net primary productivity is estimated by measuring the **above-ground biomass** per unit area, which is the total mass of living plants, excluding roots (whose mass is very difficult to measure). This means that a large percentage of plant biomass which exists underground is not included in this measurement. Net primary productivity is an important variable when considering differences in biomes. Very productive biomes have a high level of aboveground biomass.

Annual biomass production is directly related to the abiotic components of the environment. Environments with the greatest amount of biomass produce conditions in which photosynthesis, plant growth, and the resulting net primary productivity are optimized. The climate of these areas is warm and wet. Photosynthesis can proceed at a high rate, enzymes can work most efficiently, and stomata can remain open without the risk of excessive transpiration; together, these factors lead to the maximal amount of carbon dioxide (CO_2) moving into the plant, resulting in high biomass production. The above-ground biomass produces several important resources for other living things, including habitat and food. Conversely, dry and cold environments have lower photosynthetic rates and therefore less biomass. The animal communities living there will also be affected by the decrease in available food.

44.3 Terrestrial Biomes

By the end of this section, you will be able to do the following:

- Identify the two major abiotic factors that determine terrestrial biomes
- Recognize distinguishing characteristics of each of the eight major terrestrial biomes

The Earth's biomes are categorized into two major groups: *terrestrial* and *aquatic*. **Terrestrial biomes** are based on land, while **aquatic biomes** include both ocean and freshwater biomes. The eight major terrestrial biomes on Earth are each distinguished by characteristic temperatures and amount of precipitation. Comparing the annual totals of precipitation and fluctuations in precipitation from one biome to another provides clues as to the importance of abiotic factors in the distribution of biomes. Temperature variation on a daily and seasonal basis is also important for predicting the geographic distribution of the biome and the vegetation type in the biome. The distribution of these biomes shows that the same biome can occur in geographically distinct areas with similar climates (Figure 44.12).

VISUAL CONNECTION

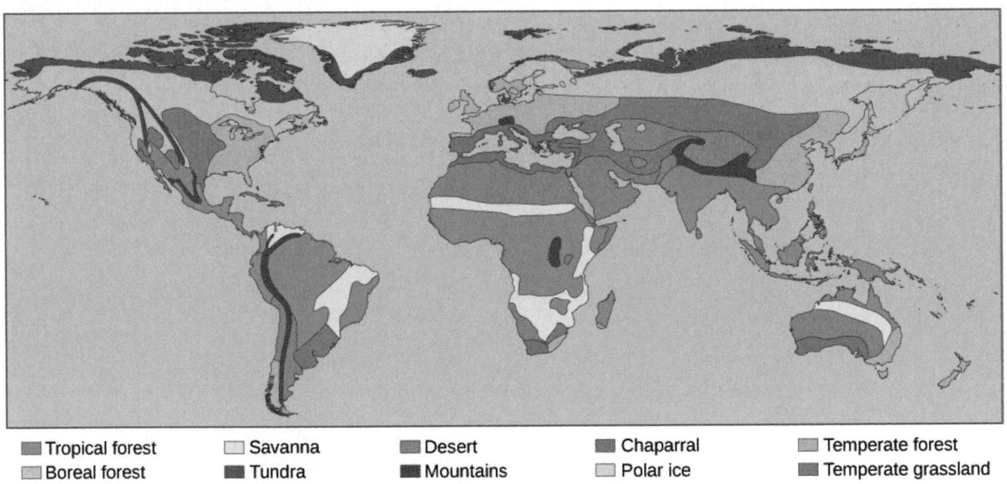

Figure 44.12 Each of the world's major biomes is distinguished by characteristic temperatures and amounts of precipitation. Polar ice and mountains are also shown.

Which of the following statements about biomes is false?

a. Chaparral is dominated by shrubs.
b. Savannas and temperate grasslands are dominated by grasses.
c. Boreal forests are dominated by deciduous trees.
d. Lichens are common in the arctic tundra.

Tropical Wet Forest

Tropical wet forests are also referred to as tropical rainforests. This biome is found in equatorial regions (Figure 44.12). The vegetation is characterized by plants with broad leaves that fall and are replaced throughout the year. Unlike the trees of deciduous forests, the trees in this biome do not have a seasonal loss of leaves associated with variations in temperature and sunlight; these forests are “evergreen” year-round.

The temperature and sunlight profiles of tropical wet forests are very stable in comparison to that of other terrestrial biomes, with the temperatures ranging from 20 °C to 34 °C (68 °F to 93 °F). When one compares the annual temperature variation of tropical wet forests with that of other forest biomes, the lack of seasonal temperature variation in the tropical wet forest becomes apparent. This lack of seasonality leads to year-round plant growth, rather than the seasonal (spring, summer, and fall) growth seen in other more temperate biomes. In contrast to other ecosystems, tropical ecosystems do not have long days and short days during the yearly cycle. Instead, a constant daily amount of sunlight (11–12 hrs per day) provides more solar radiation, thereby, a longer period of time for plant growth.

The annual rainfall in tropical wet forests ranges from 125 cm to 660 cm (50–200 in) with some monthly variation. While sunlight and temperature remain fairly consistent, annual rainfall is highly variable. Tropical wet forests typically have wet months in which there can be more than 30 cm (11–12 in) of precipitation, as well as dry months in which there are fewer than 10 cm (3.5 in) of rainfall. However, the driest month of a tropical wet forest still exceeds the *annual* rainfall of some other biomes, such as deserts.

Tropical wet forests have high net primary productivity because the annual temperatures and precipitation values in these areas are ideal for plant growth. Therefore, the extensive biomass present in the tropical wet forest leads to plant communities with very high species diversities (Figure 44.13). Tropical wet forests have more species of trees than any other biome; on average between 100 and 300 species of trees are present in a single hectare (2.5 acres) of South American Amazonian rain forest. One way to visualize this is to compare the distinctive horizontal layers within the tropical wet forest biome. On the forest floor is a sparse layer of plants and decaying plant matter. Above that is an understory of short shrubby foliage. A layer of trees rises above this understory and is topped by a closed upper **canopy**—the uppermost overhead layer of branches and leaves. Some additional trees emerge through this closed upper canopy. These layers provide diverse and complex habitats for the variety of plants, fungi, animals, and other organisms within the tropical wet forests.

For example, **epiphytes** are plants that grow on other plants, which typically are not harmed. Epiphytes are found throughout tropical wet forest biomes. Many species of animals use the variety of plants and the complex structure of the tropical wet forests for food and shelter. Some organisms live several meters above ground and have adapted to this arboreal lifestyle.

Figure 44.13 Tropical wet forests, such as these forests along the Madre de Dios river, Peru, near the Amazon River, have high species diversity. (credit: Roosevelt Garcia)

Savannas

Savannas are grasslands with scattered trees, and they are located in Africa, South America, and northern Australia (Figure 44.12). Savannas are usually hot, tropical areas with temperatures averaging from 24 °C to 29 °C (75 °F to 84 °F) and an annual rainfall of 10–40 cm (3.9–15.7 in). Savannas have an extensive dry season; for this reason, forest trees do not grow as well as they do in the tropical wet forest (or other forest biomes). As a result, within the grasses and *forbs* (herbaceous flowering plants) that dominate the savanna, there are relatively few trees (Figure 44.14). Since fire is an important source of disturbance in this biome, plants have evolved well-developed root systems that allow them to quickly resprout after a fire.

Figure 44.14 Savannas, like this one in Taita Hills Wildlife Sanctuary in Kenya, are dominated by grasses. (credit: Christopher T. Cooper)

Subtropical Deserts

Subtropical deserts exist between 15° and 30° north and south latitude and are centered on the Tropics of Cancer and Capricorn (Figure 44.12). This biome is very dry; in some years, evaporation exceeds precipitation. Subtropical hot deserts can have daytime soil surface temperatures above 60 °C (140 °F) and nighttime temperatures approaching 0 °C (32 °F). This is largely due to the lack of atmospheric water. In cold deserts, temperatures can be as high as 25 °C and can drop below -30 °C (-22 °F). Subtropical deserts are characterized by low annual precipitation of fewer than 30 cm (12 in) with little monthly variation and lack of predictability in rainfall. In some cases, the annual rainfall can be as low as 2 cm (0.8 in) in subtropical deserts located in central Australia ("the Outback") and northern Africa.

The vegetation and low animal diversity of this biome is closely related to low and unpredictable precipitation. Very dry deserts lack perennial vegetation that lives from one year to the next; instead, many plants are annuals that grow quickly and reproduce when rainfall does occur, and then die. Many other plants in these areas are characterized by having a number of adaptations that conserve water, such as deep roots, reduced foliage, and water-storing stems (Figure 44.15). Seed plants in the desert produce seeds that can be in dormancy for extended periods between rains. Adaptations in desert animals include nocturnal behavior and burrowing.

Figure 44.15 To reduce water loss, many desert plants have tiny leaves or no leaves at all. The leaves of ocotillo (*Fouquieria splendens*), shown here in the Sonora Desert near Gila Bend, Arizona, appear only after rainfall, and then are shed.

Chaparral

The chaparral is also called the scrub forest and is found in California, along the Mediterranean Sea, and along the southern coast of Australia (Figure 44.12). The annual rainfall in this biome ranges from 65 cm to 75 cm (25.6–29.5 in), and the majority of the rain falls in the winter. Summers are very dry and many chaparral plants are dormant during the summertime. The chaparral vegetation, shown in Figure 44.16, is dominated by shrubs adapted to periodic fires, with some plants producing seeds that only germinate after a hot fire. The ashes left behind after a fire are rich in nutrients like nitrogen that fertilize the soil and promote plant regrowth.

Figure 44.16 The chaparral is dominated by shrubs. (credit: Miguel Vieira)

Temperate Grasslands

Temperate grasslands are found throughout central North America, where they are also known as *prairies*; they are also in Eurasia, where they are known as *steppes* (Figure 44.12). Temperate grasslands have pronounced annual fluctuations in temperature with hot summers and cold winters. The annual temperature variation produces specific growing seasons for plants. Plant growth is possible when temperatures are warm enough to sustain plant growth and when ample water is available, which occurs in the spring, summer, and fall. During much of the winter, temperatures are low, and water, which is stored in the form of ice, is not available for plant growth.

Annual precipitation ranges from 25 cm to 75 cm (9.8–29.5 in). Because of relatively lower annual precipitation in temperate grasslands, there are few trees except for those found growing along rivers or streams. The dominant vegetation tends to consist of grasses dense enough to sustain populations of grazing animals Figure 44.17. The vegetation is very dense and the soils are fertile because the subsurface of the soil is packed with the roots and *rhizomes* (underground stems) of these grasses. The roots and rhizomes act to anchor plants into the ground and replenish the organic material (humus) in the soil when they die and decay.

Figure 44.17 The American bison (*Bison bison*), more commonly called the buffalo, is a grazing mammal that once populated American prairies in huge numbers. (credit: Jack Dykinga, USDA Agricultural Research Service)

Fires, mainly caused by lightning, are a *natural disturbance* in temperate grasslands. When fire is suppressed in temperate grasslands, the vegetation eventually converts to scrub and sometimes dense forests with drought-tolerant tree species. Often, the restoration or management of temperate grasslands requires the use of controlled burns to suppress the growth of trees and maintain the grasses.

Temperate Forests

Temperate forests are the most common biome in eastern North America, Western Europe, Eastern Asia, Chile, and New Zealand (Figure 44.12). This biome is found throughout mid-latitude regions. Temperatures range between -30 °C and 30 °C (-22 °F to 86 °F) and drop to below freezing periodically during cold winters. These temperatures mean that temperate forests have defined growing seasons during the spring, summer, and early fall. Precipitation is relatively constant throughout the year and ranges between 75 cm and 150 cm (29.5–59 in).

Because of the moderate annual rainfall and temperatures, deciduous trees are the dominant plant in this biome (Figure 44.18). Deciduous trees lose their leaves each fall and remain leafless in the winter. Thus, no photosynthesis occurs in the deciduous trees during the dormant winter period. Each spring, new leaves appear as the temperature increases. Because of the dormant period, the net primary productivity of temperate forests is less than that of tropical wet forests. In addition, temperate forests show less diversity of tree species than tropical wet forest biomes.

Figure 44.18 Deciduous trees are the dominant plant in the temperate forest. (credit: Oliver Herold)

The trees of the temperate forests leaf out and shade much of the ground; however, this biome is more open than tropical wet forests because most trees in the temperate forests do not grow as tall as the trees in tropical wet forests. The soils of the temperate forests are rich in inorganic and organic nutrients. This is due to the thick layer of leaf litter on forest floors, which does not develop in tropical rainforests. As this leaf litter decays, nutrients are returned to the soil. The leaf litter also protects soil from erosion, insulates the ground, and provides habitats for invertebrates (such as the pill bug or roly-poly, *Armadillidium vulgare*) and their predators, such as the red-backed salamander (*Plethodon cinereus*).

Boreal Forests

The **boreal forest**, also known as **taiga** or coniferous forest, is found south of the Arctic Circle and across most of Canada, Alaska, Russia, and northern Europe (Figure 44.12). This biome has cold, dry winters and short, cool, wet summers. The annual precipitation is from 40 cm to 100 cm (15.7–39 in) and usually takes the form of snow. Little evaporation occurs because of the cold temperatures.

The long and cold winters in the boreal forest have led to the predominance of cold-tolerant cone-bearing (coniferous) plants. These are evergreen coniferous trees like pines, spruce, and fir, which retain their needle-shaped leaves year-round. Evergreen trees can photosynthesize earlier in the spring than deciduous trees because less energy from the sun is required to warm a needle-like leaf than a broad leaf. This benefits evergreen trees, which grow faster than deciduous trees in the boreal forest. In addition, soils in boreal forest regions tend to be acidic with little available nitrogen. Leaves are a nitrogen-rich structure and deciduous trees must produce a new set of these nitrogen-rich structures each year. Therefore, coniferous trees that retain nitrogen-rich needles may have a competitive advantage over the broad-leafed deciduous trees.

The net primary productivity of boreal forests is lower than that of temperate forests and tropical wet forests. The above-ground biomass of boreal forests is high because these slow-growing tree species are long-lived and accumulate a large standing biomass over time. Plant species diversity is less than that seen in temperate forests and tropical wet forests. Boreal forests lack the pronounced elements of the layered forest structure seen in tropical wet forests. The structure of a boreal forest is often only a tree layer and a ground layer (Figure 44.19). When conifer needles are dropped, they decompose more slowly than broad leaves; therefore, fewer nutrients are returned to the soil to fuel plant growth.

Figure 44.19 The boreal forest (taiga) has low lying plants and conifer trees. (credit: L.B. Brubaker)

Arctic Tundra

The Arctic tundra lies north of the subarctic boreal forest and is located throughout the Arctic regions of the northern hemisphere (Figure 44.12). The average winter temperature is –34 °C (–29.2 °F) and the average summer temperature is from 3 °C to 12 °C (37 °F–52 °F). Plants in the arctic tundra have a very short growing season of approximately 10–12 weeks.

However, during this time, there are almost 24 hours of daylight and plant growth is rapid. The annual precipitation of the Arctic tundra is very low with little annual variation in precipitation. And, as in the boreal forests, there is little evaporation due to the cold temperatures.

Plants in the Arctic tundra are generally low to the ground (Figure 44.20). There is little species diversity, low net primary productivity, and low above-ground biomass. The soils of the Arctic tundra may remain in a perennially frozen state referred to as **permafrost**. The permafrost makes it impossible for roots to penetrate deep into the soil and slows the decay of organic matter, which inhibits the release of nutrients from organic matter. During the growing season, the ground of the Arctic tundra can be completely covered with plants or lichens.

Figure 44.20 Low-growing plants such as shrub willow dominate the tundra landscape, shown here in the Arctic National Wildlife Refuge. (credit: USFWS Arctic National Wildlife Refuge)

LINK TO LEARNING

Watch this *Assignment Discovery: Biomes* video (http://openstax.org/l/biomes) for an overview of biomes. To explore further, select one of the biomes on the extended playlist: desert, savanna, temperate forest, temperate grassland, tropic, tundra.

44.4 Aquatic Biomes

By the end of this section, you will be able to do the following:

- Describe the effects of abiotic factors on the composition of plant and animal communities in aquatic biomes
- Compare and contrast the characteristics of the ocean zones
- Summarize the characteristics of standing water and flowing water freshwater biomes

Abiotic Factors Influencing Aquatic Biomes

Like terrestrial biomes, aquatic biomes are influenced by a series of abiotic factors. The aquatic medium—water— has different physical and chemical properties than air, however. Even if the water in a pond or other body of water is perfectly clear (there are no suspended particles), water, on its own, absorbs light. As one descends into a deep body of water, there will eventually be a depth which the sunlight cannot reach. While there are some abiotic and biotic factors in a terrestrial ecosystem that might obscure light (like fog, dust, or insect swarms), usually these are not permanent features of the environment. The importance of light in aquatic biomes is central to the communities of organisms found in both freshwater and marine ecosystems. In freshwater systems, stratification due to differences in density is perhaps the most critical abiotic factor and is related to the energy aspects of light. The thermal properties of water (rates of heating and cooling and the ability to store much larger amounts of energy than the air) are significant to the function of marine systems and have major impacts on global climate and weather patterns. Marine systems are also influenced by large-scale physical water movements, such as currents; these are less important in most freshwater lakes.

The ocean is categorized by several areas or zones (Figure 44.21). All of the ocean's open water is referred to as the **pelagic realm** (or zone). The **benthic realm** (or zone) extends along the ocean bottom from the shoreline to the deepest parts of the ocean floor. Within the pelagic realm is the **photic zone**, which is the portion of the ocean that light can penetrate (approximately 200 m or 650 ft). At depths greater than 200 m, light cannot penetrate; thus, this is referred to as the **aphotic zone**. The majority of the ocean is aphotic and lacks sufficient light for photosynthesis. The deepest part of the ocean, the Challenger Deep (in the Mariana Trench, located in the western Pacific Ocean), is about 11,000 m (about 6.8 mi) deep. To give some perspective on the depth of this trench, the ocean is, on average, 4267 m or 14,000 ft deep. These realms and zones are relevant to freshwater lakes as well.

VISUAL CONNECTION

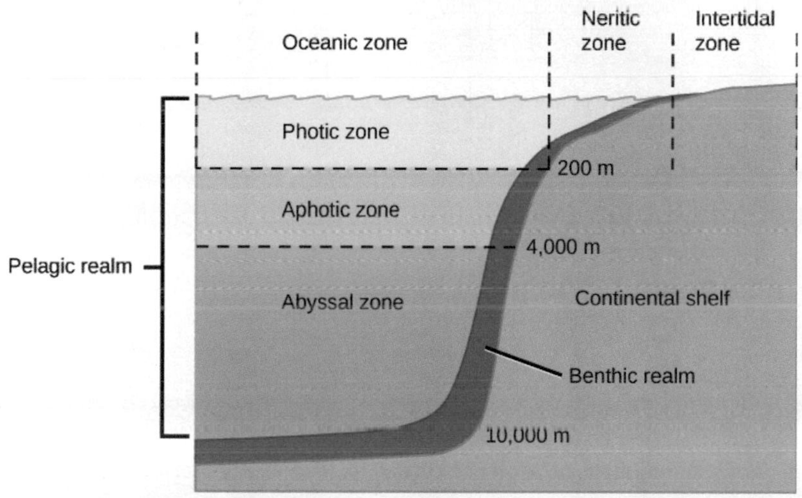

Figure 44.21 The ocean is divided into different zones based on water depth and distance from the shoreline.

In which of the following regions would you expect to find photosynthetic organisms?

a. the aphotic zone, the neritic zone, the oceanic zone, and the benthic realm
b. the photic zone, the intertidal zone, the neritic zone, and the oceanic zone
c. the photic zone, the abyssal zone, the neritic zone, and the oceanic zone
d. the pelagic realm, the aphotic zone, the neritic zone, and the oceanic zone

Marine Biomes

The ocean is the largest marine biome. It is a continuous body of salt water that is relatively uniform in chemical composition; in fact, it is a weak solution of mineral salts and decayed biological matter. Within the ocean, coral reefs are a second kind of marine biome. Estuaries, coastal areas where salt water and fresh water mix, form a third unique marine biome.

Ocean

The physical diversity of the ocean is a significant influence on plants, animals, and other organisms. The ocean is categorized into different zones based on how far light reaches into the water. Each zone has a distinct group of species adapted to the biotic and abiotic conditions particular to that zone.

The **intertidal zone**, which is the zone between high and low tide, is the oceanic region that is closest to land (Figure 44.21). Generally, most people think of this portion of the ocean as a sandy beach. In some cases, the intertidal zone is indeed a sandy beach, but it can also be rocky or muddy. The intertidal zone is an extremely variable environment because of action of tidal ebb and flow. Organisms are exposed to air and sunlight at low tide and are underwater most of the time, especially during high tide. Therefore, living things that thrive in the intertidal zone are adapted to being dry for long periods of time. The shore of the intertidal zone may also be repeatedly struck by waves, and the organisms found there are adapted to withstand damage from their pounding action (Figure 44.22). The exoskeletons of shoreline crustaceans (such as the shore crab, *Carcinus maenas*) are tough and protect them from desiccation (drying out) and wave damage. Another consequence of the pounding waves is that few algae and plants establish themselves in the constantly moving rocks, sand, or mud.

Figure 44.22 Sea urchins, mussel shells, and starfish are often found in the intertidal zone, shown here in Kachemak Bay, Alaska. (credit: NOAA)

The **neritic zone** (Figure 44.21) extends from the intertidal zone to depths of about 200 m (or 650 ft) at the edge of the *continental shelf* (the underwater landmass that extends from a continent). Since light can penetrate this depth, photosynthesis can still occur in the neritic zone. The water here contains silt and is well-oxygenated, low in pressure, and stable in temperature. Phytoplankton and floating ***Sargassum*** (a type of free-floating marine seaweed) provide a habitat for some sea life found in the neritic zone. Zooplankton, protists, small fishes, and shrimp are found in the neritic zone and are the base of the food chain for most of the world's fisheries.

Beyond the neritic zone is the open ocean area known as the **pelagic** or **open oceanic zone** (Figure 44.21). Within the oceanic zone there is *thermal stratification* where warm and cold waters mix because of ocean currents. Abundant plankton serve as the base of the food chain for larger animals such as whales and dolphins. Nutrients are scarce and this is a relatively less productive part of the marine biome. When photosynthetic organisms and the protists and animals that feed on them die, their bodies fall to the bottom of the ocean, where they remain. Unlike freshwater lakes, most of the open ocean lacks a process for bringing the organic nutrients back up to the surface. (Exceptions include major oceanic upwellings within the Humboldt Current along the western coast of South America.) The majority of organisms in the aphotic zone include sea cucumbers (phylum Echinodermata) and other organisms that survive on the nutrients contained in the dead bodies of organisms in the photic zone.

Beneath the pelagic zone is the **benthic realm**, the deep-water region beyond the continental shelf (Figure 44.21). The bottom of the benthic realm is composed of sand, silt, and dead organisms. Temperature decreases, remaining above freezing, as water depth increases. This is a nutrient-rich portion of the ocean because of the dead organisms that fall from the upper layers of the ocean. Because of this high level of nutrients, a diversity of fungi, sponges, sea anemones, marine worms, sea stars, fishes, and bacteria exist.

The deepest part of the ocean is the **abyssal zone**, which is at depths of 4000 m or greater. The abyssal zone (Figure 44.21) is very cold and has very high pressure, high oxygen content, and low nutrient content. There are a variety of invertebrates and fishes found in this zone, but the abyssal zone does not have plants because of the lack of light. Hydrothermal vents are found primarily in the abyssal zone; chemosynthetic bacteria utilize the hydrogen sulfide and other minerals emitted from the vents. These chemosynthetic bacteria use the hydrogen sulfide as an energy source and serve as the base of the food chain found in the abyssal zone.

Coral Reefs

Coral reefs are *ocean ridges* formed by marine invertebrates, comprising mostly cnidarians and molluscs, living in warm shallow waters within the photic zone of the ocean. They are found within 30˚ north and south of the equator. The **Great Barrier Reef** is perhaps the best-known and largest reef system in the world—visible from the International Space Station! This massive and ancient reef is located several miles off the northeastern coast of Australia. Other coral reef systems are fringing islands, which are directly adjacent to land, or atolls, which are circular reef systems surrounding a former landmass that is now underwater. The coral organisms (members of phylum Cnidaria) are colonies of saltwater polyps that secrete a calcium carbonate skeleton. These calcium-rich skeletons slowly accumulate, forming the underwater reef (Figure 44.23). Corals found in shallower waters (at a depth of approximately 60 m or about 200 ft) have a mutualistic relationship with photosynthetic unicellular algae. The relationship provides corals with the majority of the nutrition and the energy they require. The waters in which these corals live are nutritionally poor and, without this mutualism, it would not be possible for large corals to grow. Some corals living in deeper and colder water do not have a mutualistic relationship with algae; these corals attain energy and nutrients using stinging cells called cnidocytes on their tentacles to capture prey.

LINK TO LEARNING

Watch this National Oceanic and Atmospheric Administration (NOAA) video (http://openstax.org/l/marine_biology) to see marine ecologist Dr. Peter Etnoyer discuss his research on coral organisms.

It is estimated that more than 4,000 fish species inhabit coral reefs. These fishes can feed on coral, the **cryptofauna** (invertebrates found within the calcium carbonate substrate of the coral reefs), or the seaweed and algae that are associated with the coral. In addition, some fish species inhabit the boundaries of a coral reef; these species include predators, herbivores, and **planktivores**, which consume planktonic organisms such as bacteria, archaea, algae, and protists floating in the pelagic zone.

Figure 44.23 Coral reefs are formed by the calcium carbonate skeletons of coral organisms, which are marine invertebrates in the phylum

Cnidaria. (credit: Terry Hughes)

EVOLUTION CONNECTION

Global Decline of Coral Reefs

It takes many thousands of years to build a coral reef. The animals that create coral reefs have evolved over millions of years, continuing to slowly deposit the calcium carbonate that forms their characteristic ocean homes. Bathed in warm tropical waters, the coral animals and their symbiotic algal partners evolved to survive at the upper limit of ocean water temperature.

Together, climate change and human activity pose dual threats to the long-term survival of the world's coral reefs. As global warming due to fossil fuel emissions raises ocean temperatures, coral reefs are suffering. The excessive warmth causes the reefs to lose their symbiotic, food-producing algae, resulting in a phenomenon known as **bleaching**. When bleaching occurs, the reefs lose much of their characteristic color as the algae and the coral animals die if loss of the symbiotic zooxanthellae is prolonged.

Rising levels of atmospheric carbon dioxide further threaten the corals in other ways; as CO_2 dissolves in ocean waters, it lowers the pH and increases ocean acidity. As acidity increases, it interferes with the calcification that normally occurs when coral animals build their calcium carbonate shelters.

When a coral reef begins to die, species diversity plummets as animals lose food and shelter. Coral reefs are also economically important tourist destinations, so the decline of coral reefs poses a serious threat to coastal economies.

Human population growth has damaged corals in other ways, too. As human coastal populations increase, the runoff of sediment and agricultural chemicals has increased, as well, causing some of the once-clear tropical waters to become cloudy. At the same time, overfishing of popular fish species has allowed the predator species that eat corals to go unchecked.

Although a rise in global temperatures of 1–2 °C (a conservative scientific projection) in the coming decades may not seem large, it is very significant to this biome. When change occurs rapidly, species can become extinct before evolution can offer new adaptations. Many scientists believe that global warming, with its rapid (in terms of evolutionary time) and inexorable increases in temperature, is tipping the balance beyond the point at which many of the world's coral reefs can recover.

Estuaries: Where the Ocean Meets Fresh Water

Estuaries are biomes that occur where a source of fresh water, such as a river, meets the ocean. Therefore, both fresh water and salt water are found in the same vicinity; mixing results in a diluted (*brackish*) saltwater. Estuaries form protected areas where many of the young offspring of crustaceans, molluscs, and fish begin their lives, which also creates important breeding grounds for other animals. Salinity is a very important factor that influences the organisms and the adaptations of the organisms found in estuaries. The salinity of estuaries varies considerably and is based on the rate of flow of its freshwater sources, which may depend on the seasonal rainfall. Once or twice a day, high tides bring salt water into the estuary. Low tides occurring at the same frequency reverse the current of salt water.

The short-term and rapid variation in salinity due to the mixing of fresh water and salt water is a difficult physiological challenge for the plants and animals that inhabit estuaries. Many estuarine plant species are halophytes: plants that can tolerate salty conditions. Halophytic plants are adapted to deal with the salinity resulting from saltwater on their roots or from sea spray. In some halophytes, filters in the roots remove the salt from the water that the plant absorbs. Other plants are able to pump oxygen into their roots. Animals, such as mussels and clams (phylum Mollusca), have developed behavioral adaptations that expend a lot of energy to function in this rapidly changing environment. When these animals are exposed to low salinity, they stop feeding, close their shells, and switch from aerobic respiration (in which they use gills to remove oxygen from the water) to anaerobic respiration (a process that does not require oxygen and takes place in the cytoplasm of the animal's cells). When high tide returns to the estuary, the salinity and oxygen content of the water increases, and these animals open their shells, begin feeding, and return to aerobic respiration.

Freshwater Biomes

Freshwater biomes include lakes and ponds (standing water) as well as rivers and streams (flowing water). They also include wetlands, which will be discussed later. Humans rely on freshwater biomes to provide ecosystem benefits, which are aquatic resources for drinking water, crop irrigation, sanitation, and industry. Lakes and ponds are connected with abiotic and biotic factors influencing their terrestrial biomes.

Lakes and Ponds

Lakes and ponds can range in area from a few square meters to thousands of square kilometers. Temperature is an important abiotic factor affecting living things found in lakes and ponds. In the summer, as we have seen, thermal stratification of lakes and ponds occurs when the upper layer of water is warmed by the sun and does not mix with deeper, cooler water. Light can penetrate within the photic zone of the lake or pond. **Phytoplankton** (algae and cyanobacteria) are found here and carry out photosynthesis, providing the base of the food web of lakes and ponds. **Zooplankton**, such as rotifers and larvae and adult crustaceans, consume these phytoplankton. At the bottom of lakes and ponds, bacteria in the aphotic zone break down dead organisms that sink to the bottom.

Nitrogen and phosphorus are important limiting nutrients in lakes and ponds. Because of this, they are the determining factors in the amount of phytoplankton growth that takes place in lakes and ponds. When there is a large input of nitrogen and phosphorus (from sewage and runoff from fertilized lawns and farms, for example), the growth of algae skyrockets, resulting in a large accumulation of algae called an **algal bloom**. Algal blooms (Figure 44.24) can become so extensive that they reduce light penetration in water. They may also release toxic byproducts into the water, contaminating any drinking water taken from that source. In addition, the lake or pond becomes aphotic, and photosynthetic plants cannot survive. When the algae die and decompose, severe oxygen depletion of the water occurs. Fishes and other organisms that require oxygen are then more likely to die, resulting in a dead zone. Lake Erie and the Gulf of Mexico represent freshwater and marine habitats where phosphorus control and storm water runoff pose significant environmental challenges.

Figure 44.24 The uncontrolled growth of algae in this lake has resulted in an algal bloom. (credit: Jeremy Nettleton)

Rivers and Streams

Rivers and streams are continuously moving bodies of water that carry large amounts of water from the source, or *headwater*, to a lake or ocean. The largest rivers include the Nile River in Africa, the Amazon River in South America, and the Mississippi River in North America.

Abiotic features of rivers and streams vary along the length of the river or stream. Streams begin at a point of origin referred to as **source water**. The source water is usually cold, low in nutrients, and clear. The **channel** (the width of the river or stream) is narrower than at any other place along the length of the river or stream. Because of this, the current is often faster here than at any other point of the river or stream.

The fast-moving water results in minimal silt accumulation at the bottom of the river or stream; therefore, the water is usually clear and free of debris. Photosynthesis here is mostly attributed to algae that are growing on rocks; the swift current inhibits the growth of phytoplankton. An additional input of energy can come from leaves and other organic material that fall downstream into the river or stream, as well as from trees and other plants that border the water. When the leaves decompose, the organic material and nutrients in the leaves are returned to the water. Plants and animals have adapted to this fast-moving water. For instance, leeches (phylum Annelida) have elongated bodies and suckers on the anterior and ventral areas of the body. These suckers attach to the substrate, keeping the leech anchored in place, and are also used to attach to their prey. Freshwater trout species (phylum Chordata) are an important predator in these fast-moving rivers and streams.

As the river or stream flows away from the source, the width of the channel gradually widens and the current slows. This slow-moving water, caused by the gradient decrease and the volume increase as tributaries unite, has more sedimentation.

Phytoplankton can also be suspended in slow-moving water. Therefore, the water will not be as clear as it is near the source. The water is also warmer. Worms (phylum Annelida) and insects (phylum Arthropoda) can be found burrowing into the mud. The higher order predator vertebrates (phylum Chordata) include waterfowl, frogs, and fishes. These predators must find food in these slow moving, sometimes murky, waters and, unlike the trout in the waters at the source, these vertebrates may not be able to use vision as their primary sense to find food. Instead, they are more likely to use taste or chemical cues to find prey.

Wetlands

Wetlands are environments in which the soil is either permanently or periodically saturated with water. Wetlands are different from lakes because wetlands are shallow bodies of water whereas lakes vary in depth. **Emergent vegetation** consists of wetland plants that are rooted in the soil but have portions of leaves, stems, and flowers extending above the water's surface. There are several types of wetlands including marshes, swamps, bogs, mudflats, and salt marshes (Figure 44.25). The three shared characteristics among these types—what makes them wetlands—are their hydrology, hydrophytic vegetation, and hydric soils.

Figure 44.25 Located in southern Florida, Everglades National Park is vast array of wetland environments, including sawgrass marshes, cypress swamps, and estuarine mangrove forests. Here, a great egret walks among cypress trees. (credit: NPS)

Freshwater marshes and swamps are characterized by slow and steady water flow. Bogs, however, develop in depressions where water flow is low or nonexistent. Bogs usually occur in areas where there is a clay bottom with poor percolation of water. (Percolation is the movement of water through the pores in the soil or rocks.) The water found in a bog is stagnant and oxygen-depleted because the oxygen used during the decomposition of organic matter is not readily replaced. As the oxygen in the water is depleted, decomposition slows. This leads to a buildup of acids and a lower water pH. The lower pH creates challenges for plants because it limits the available nitrogen. As a result, some bog plants (such as sundews, pitcher plants, and Venus flytraps) capture insects in order to extract the nitrogen from their bodies. Bogs have low net primary productivity because the water found in bogs has low levels of nitrogen and oxygen.

44.5 Climate and the Effects of Global Climate Change

By the end of this section, you will be able to do the following:

- Define global climate change
- Summarize the effects of the Industrial Revolution on global atmospheric carbon dioxide concentration
- Describe three natural factors affecting long-term global climate
- List two or more greenhouse gases and describe their role in the greenhouse effect

All biomes are *universally affected* by global conditions, such as climate, that ultimately shape each biome's environment. Scientists who study climate have noted a series of marked changes that have gradually become increasingly evident during the last sixty years. **Global climate change** is the term used to describe altered global weather patterns, especially a worldwide increase in temperature and resulting changes in the climate, due largely to rising levels of atmospheric carbon dioxide.

Climate and Weather

A common misconception about global climate change is that a specific weather event occurring in a particular region (for example, a very cool week in June in central Indiana) provides evidence of global climate change. However, a cold week in June is a weather-related event and not a climate-related one. These misconceptions often arise because of confusion over the terms *climate* and *weather*.

Climate refers to the long-term, *predictable atmospheric conditions* of a specific area. The climate of a biome is characterized by having consistent seasonal temperature and rainfall ranges. Climate does not address the amount of rain that fell on one particular day in a biome or the colder-than-average temperatures that occurred on one day. In contrast, **weather** refers to the conditions of the atmosphere during a short period of time. Weather forecasts are usually made for 48-hour cycles. Long-range weather forecasts are available but can be unreliable.

To better understand the difference between climate and weather, imagine that you are planning an outdoor event in northern Wisconsin. You would be thinking about *climate* when you plan the event in the summer rather than the winter because you have long-term knowledge that any given Saturday in the months of May to August would be a better choice for an outdoor event in Wisconsin than any given Saturday in January. However, you cannot determine the specific day that the event should be held on because it is difficult to accurately predict the weather on a specific day. Climate can be considered "average" weather that takes place over many years.

Global Climate Change

Climate change can be understood by approaching three areas of study:

- evidence of current and past global climate change
- drivers of global climate change
- documented results of climate change

It is helpful to keep these three different aspects of climate change clearly separated when consuming media reports about global climate change. We should note that it is common for reports and discussions about global climate change to confuse the data showing that Earth's climate is changing with the factors that drive this climate change.

Evidence for Global Climate Change

Since scientists cannot go back in time to directly measure climatic variables, such as average temperature and precipitation, they must instead indirectly measure temperature. To do this, scientists rely on *historical evidence of Earth's past climate.*

Antarctic ice cores are a key example of such evidence for climate change. These ice cores are samples of *polar ice* obtained by means of drills that reach thousands of meters into ice sheets or high mountain glaciers. Viewing the ice cores is like traveling backwards through time; the deeper the sample, the earlier the time period. Trapped within the ice are air bubbles and other biological evidence that can reveal temperature and carbon dioxide data. Antarctic ice cores have been collected and analyzed to indirectly estimate the temperature of the Earth over the past 400,000 years (Figure 44.26a). The 0 °C on this graph refers to the long-term average. Temperatures that are greater than 0 °C exceed Earth's long-term average temperature. Conversely, temperatures that are less than 0 °C are less than Earth's average temperature. This figure shows that there have been periodic cycles of increasing and decreasing temperature.

(a) (b)

Figure 44.26 Scientists drill for ice cores in polar regions. The ice contains air bubbles and biological substances that provide important information for researchers. (credit: a: Helle Astrid Kjær; b: National Ice Core Laboratory, USGS)

Before the late 1800s, the Earth has been as much as 9 °C cooler and about 3 °C warmer. Note that the graph in Figure 44.27b shows that the atmospheric concentration of carbon dioxide has also risen and fallen in periodic cycles. Also note the relationship between carbon dioxide concentration and temperature. Figure 44.27b shows that carbon dioxide levels in the atmosphere have historically cycled between 180 and 300 parts per million (ppm) by volume.

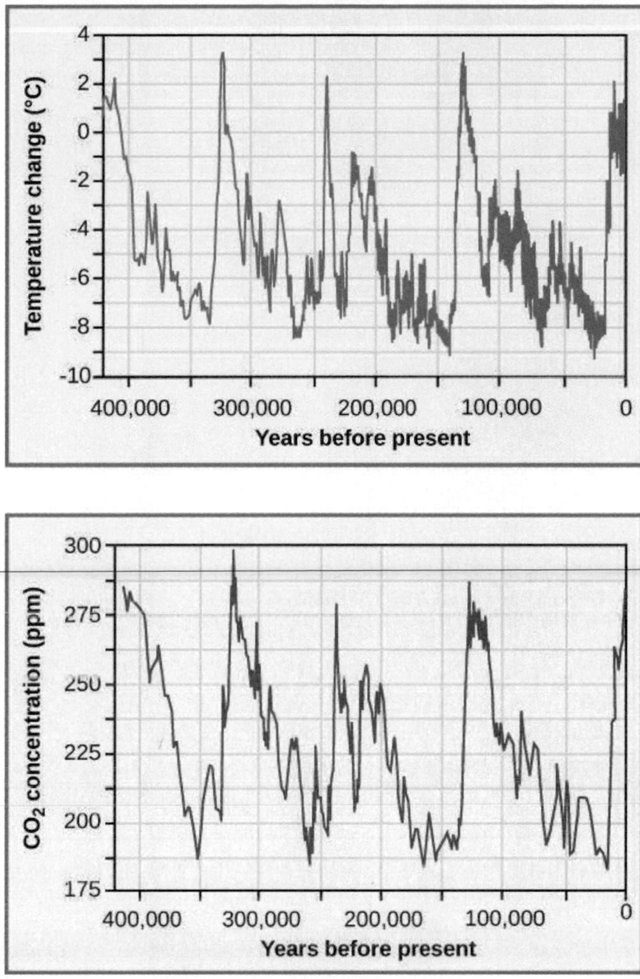

Figure 44.27 Ice at the Russian Vostok station in East Antarctica was laid down over the course of 420,000 years and reached a depth of over 3,000 m. By measuring the amount of CO_2 trapped in the ice, scientists have determined past atmospheric CO_2 concentrations. Temperatures relative to modern day were determined from the amount of *deuterium* (a nonradioactive isotope of hydrogen) present.

Figure 44.27a does not show the last 2,000 years with enough detail to compare the changes of Earth's temperature during the last 400,000 years with the temperature change that has occurred in the more recent past. Two significant temperature anomalies, or *irregularities*, have occurred in the last 2,000 years. These are the *Medieval Climate Anomaly* (or the Medieval Warm Period) and the *Little Ice Age*. A third temperature anomaly aligns with the *Industrial Era*. The Medieval Climate Anomaly occurred between 900 and 1300 AD. During this time period, many climate scientists think that slightly warmer weather conditions prevailed in many parts of the world; the higher-than-average temperature changes varied between 0.10 °C and 0.20 °C above the norm. Although 0.10 °C does not seem large enough to produce any noticeable change, it did free seas of ice. Because of this warming, the Vikings were able to colonize Greenland.

The **Little Ice Age** was a cold period that occurred between 1550 AD and 1850 AD. During this time, a slight cooling of a little less than 1 °C was observed in North America, Europe, and possibly other areas of the Earth. This 1 °C change in global temperature is a seemingly small deviation in temperature (as was observed during the Medieval Climate Anomaly); however, it also resulted in noticeable climatic changes. Historical accounts reveal a time of exceptionally harsh winters with much snow and frost.

The *Industrial Revolution*, which began around 1750, was characterized by changes in much of human society. Advances in agriculture increased the food supply, which improved the standard of living for people in Europe and the United States. New technologies were invented that provided jobs and cheaper goods. These new technologies were powered using fossil fuels, especially coal. The Industrial Revolution starting in the early nineteenth century ushered in the beginning of the Industrial Era. When a fossil fuel is burned, carbon dioxide is released. With the beginning of the Industrial Era, atmospheric carbon dioxide began to rise (Figure 44.28).

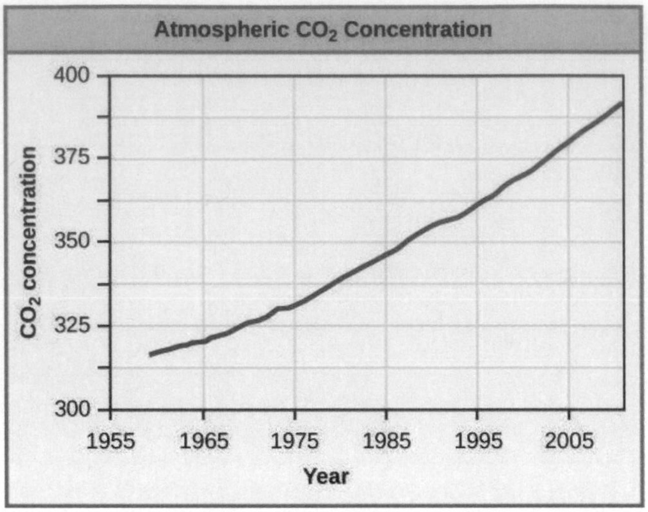

Figure 44.28 The atmospheric concentration of CO_2 has risen steadily since the beginning of industrialization.

Current and Past Drivers of Global Climate Change

Because it is not possible to go back in time to directly observe and measure climate, scientists must use *indirect evidence* to determine the *drivers*, or factors, that may be responsible for climate change. The indirect evidence includes data collected using ice cores, *boreholes* (a narrow shaft bored into the ground), tree rings, glacier lengths, pollen remains, and ocean sediments. The data shows a correlation between the timing of temperature changes and drivers of climate change. Before the Industrial Era (pre-1780), there were three drivers of climate change that were not related to human activity or atmospheric gases. The first of these is the *Milankovitch cycles*. The **Milankovitch cycles** describe the effects of slight changes in the Earth's orbit on Earth's climate. The length of the Milankovitch cycles ranges between 19,000 and 100,000 years. In other words, one could expect to see some predictable changes in the Earth's climate associated with changes in the Earth's orbit at a minimum of every 19,000 years.

The *variation in the sun's intensity* is the second natural factor responsible for climate change. **Solar intensity** is the amount of solar power or energy the sun emits in a given amount of time. There is a direct relationship between solar intensity and temperature. As solar intensity increases (or decreases), the Earth's temperature correspondingly increases (or decreases). Changes in solar intensity have been proposed as one of several possible explanations for the Little Ice Age.

Finally, *volcanic eruptions* are a third natural driver of climate change. Volcanic eruptions can last a few days, but the solids and gases released during an eruption can influence the climate over a period of a few years, causing short-term climate changes. The gases and solids released by volcanic eruptions can include carbon dioxide, water vapor, sulfur dioxide, hydrogen sulfide, hydrogen, and carbon monoxide. Generally, volcanic eruptions cool the climate. This occurred in 1783 when volcanos in Iceland erupted and caused the release of large volumes of sulfuric oxide. This led to **haze-effect cooling**, a global phenomenon that occurs when dust, ash, or other suspended particles block out sunlight and trigger lower global temperatures as a result; haze-effect cooling usually extends for one or more years before dissipating in intensity. In Europe and North America, haze-effect cooling produced some of the lowest average winter temperatures on record in 1783 and 1784.

Greenhouse gases are probably the most significant drivers of the climate. When heat energy from the sun strikes the Earth, gases known as **greenhouse gases** trap the heat in the atmosphere, in a similar manner as do the glass panes of a greenhouse keep heat from escaping. The greenhouse gases that affect Earth include carbon dioxide, methane, water vapor, nitrous oxide, and ozone. Approximately half of the radiation from the sun passes through these gases in the atmosphere and strikes the Earth. This radiation is converted into thermal (infrared) radiation on the Earth's surface, and then a portion of that energy is re-radiated back into the atmosphere. Greenhouse gases, however, reflect much of the thermal energy back to the Earth's surface. The more greenhouse gases there are in the atmosphere, the more thermal energy is reflected back to the Earth's surface, heating it up and the atmosphere immediately above it. Greenhouse gases absorb and emit radiation and are an important factor in the **greenhouse effect**: the warming of Earth due to carbon dioxide and other greenhouse gases in the atmosphere.

Direct evidence supports the relationship between atmospheric concentrations of carbon dioxide and temperature: as carbon dioxide rises, global temperature rises. Since 1950, the concentration of atmospheric carbon dioxide has increased from about 280 ppm to 382 ppm in 2006. In 2011, the atmospheric carbon dioxide concentration was 392 ppm. However, the planet would not be inhabitable by current life forms if water vapor did not produce its drastic greenhouse warming effect.

Scientists look at patterns in data and try to explain differences or deviations from these patterns. The atmospheric carbon dioxide data reveal a historical pattern of carbon dioxide increasing and decreasing, cycling between a low of 180 ppm and a high of 300 ppm. Scientists have concluded that it took around 50,000 years for the atmospheric carbon dioxide level to increase from its low minimum concentration to its higher maximum concentration. However, beginning only a few centuries ago, atmospheric carbon dioxide concentrations have increased beyond the historical maximum of 300 ppm. The current increases in atmospheric carbon dioxide have happened very quickly—in a matter of hundreds of years rather than thousands of years. What is the reason for this difference in the rate of change and the amount of increase in carbon dioxide? A key factor that must be recognized when comparing the historical data and the current data is the presence and industrial activities of modern human society; no other driver of climate change has yielded changes in atmospheric carbon dioxide levels *at this rate or to this magnitude*.

Human activity releases carbon dioxide and methane, two of the most important greenhouse gases, into the atmosphere in several ways. The primary mechanism that releases carbon dioxide is the burning of fossil fuels, such as gasoline, coal, and natural gas (Figure 44.29). Deforestation, cement manufacture, animal agriculture, the clearing of land, and the burning of forests are other human activities that release carbon dioxide. Methane (CH_4) is produced when bacteria break down organic matter under anaerobic conditions. Anaerobic conditions can happen when organic matter is trapped underwater (such as in rice paddies) or in the intestines of herbivores. Methane can also be released from natural gas fields and the decomposition of animal and plant material that occurs in landfills. Another source of methane is the melting of *clathrates*. **Clathrates** are frozen chunks of ice and methane found at the bottom of the ocean. When water warms, these chunks of ice melt and methane is released. As the ocean's water temperature increases, the rate at which clathrates melt is increasing, releasing even more methane. This leads to increased levels of methane in the atmosphere, which further accelerates the rate of global warming. This is an example of the positive feedback loop that is leading to the rapid rate of increase of global temperatures.

Figure 44.29 The burning of fossil fuels in industry and by vehicles releases carbon dioxide and other greenhouse gases into the atmosphere. (credit: "Pöllö"/Wikimedia Commons)

Documented Results of Climate Change: Past and Present

Scientists have geological evidence of the consequences of long-ago climate change. Modern-day phenomena such as retreating glaciers and melting polar ice cause a continual rise in sea level. Meanwhile, changes in climate can negatively affect organisms.

Geological Climate Change

Global warming has been associated with at least one planet-wide extinction event during the geological past. The **Permian extinction** event occurred about 251 million years ago toward the end of the roughly 50-million-year-long geological time span known as the Permian period. This geologic time period was one of the three warmest periods in Earth's geologic history. Scientists estimate that approximately 70 percent of the terrestrial plant and animal species and 84 percent of marine species became extinct, vanishing forever near the end of the Permian period.

Organisms that had adapted to wet and warm climatic conditions, such as annual rainfall of 300–400 cm (118–157 in) and 20 °C–30 °C (68 °F–86 °F) in the tropical wet forest, may not have been able to survive the Permian climate change.

LINK TO LEARNING

Watch this NASA video (http://openstax.org/l/climate_plants) to discover the mixed effects of global warming on plant growth. While scientists found that warmer temperatures in the 1980s and 1990s caused an increase in plant productivity, this advantage has since been counteracted by more frequent droughts.

Present Climate Change

A number of global events have occurred that may be attributed to climate change during our lifetimes. *Glacier National Park* in Montana is undergoing the retreat of many of its glaciers, a phenomenon known as glacier recession. In 1850, the area contained approximately 150 glaciers. By 2010, however, the park contained only about 24 glaciers greater than 25 acres in size. One of these glaciers is the *Grinnell Glacier* (Figure 44.30) at Mount Gould. Between 1966 and 2005, the size of Grinnell Glacier shrank by 40 percent. Similarly, the mass of the ice sheets in Greenland and the Antarctic is decreasing: Greenland lost 150–250 km^3 of ice per year between 2002 and 2006. In addition, the size and thickness of the Arctic sea ice is decreasing.

Figure 44.30 The effect of global warming can be seen in the continuing retreat of Grinnel Glacier. The mean annual temperature in the park has increased 1.33 °C since 1900. The loss of a glacier results in the loss of summer meltwaters, sharply reducing seasonal water supplies and severely affecting local ecosystems. (credit: modification of work by USGS)

This loss of ice is leading to increases in the global sea level. On average, the sea is rising at a rate of 1.8 mm per year. However, between 1993 and 2010 the rate of sea level increase ranged between 2.9 and 3.4 mm per year. A variety of factors affect the volume of water in the ocean, especially the temperature of the water (the density of water is related to its temperature: water volume expands as it warms, thus raising sea levels), as well as the *amount of water* found in rivers, lakes, glaciers, polar ice caps, and sea ice. As glaciers and polar ice caps melt, there is a significant contribution of liquid water that was previously frozen.

In addition to some abiotic conditions changing in response to climate change, many organisms are also being affected by the changes in temperature. Temperature and precipitation play key roles in determining the *geographic distribution* and *phenology* of plants and animals. (**Phenology** is the study of the effects of climatic conditions on the timing of periodic life cycle events, such as flowering in plants or migration in birds.) Researchers have shown that 385 plant species in Great Britain are flowering 4.5 days sooner than was recorded earlier during the previous 40 years. In addition, insect-pollinated species were more likely to flower earlier than wind-pollinated species. The impact of changes in flowering date would be mitigated if the insect pollinators emerged earlier. This mismatched timing of plants and pollinators could result in injurious ecosystem effects because, for continued survival, insect-pollinated plants must flower when their pollinators are present.

KEY TERMS

abiotic nonliving components of the environment

above-ground biomass total mass of aboveground living plants per area

abyssal zone deepest part of the ocean at depths of 4000 m or greater

algal bloom rapid increase of algae in an aquatic system

aphotic zone part of the ocean where no light penetrates

benthic realm (also, benthic zone) part of the ocean that extends along the ocean bottom from the shoreline to the deepest parts of the ocean floor

biogeography study of the geographic distribution of living things and the abiotic factors that affect their distribution

biome ecological community of plants, animals, and other organisms that is adapted to a characteristic set of environmental conditions

biotic living components of the environment

canopy branches and foliage of trees that form a layer of overhead coverage in a forest

channel width of a river or stream from one bank to the other bank

clathrates frozen chunks of ice and methane found at the bottom of the ocean

climate long-term, predictable atmospheric conditions present in a specific area

conspecifics individuals that are members of the same species

coral reef ocean ridges formed by marine invertebrates living in warm, shallow waters within the photic zone

cryptofauna invertebrates found within the calcium carbonate substrate of coral reefs

ecology study of interaction between living things and their environment

ecosystem services human benefits and services provided by natural ecosystems

emergent vegetation wetland plants that are rooted in the soil but have portions of leaves, stems, and flowers extending above the water's surface

endemic species found only in a specific geographic area that is usually restricted in size

estuary biomes where a source of fresh water, such as a river, meets the ocean

fall and spring turnover seasonal process that recycles nutrients and oxygen from the bottom of a freshwater ecosystem to the top

global climate change altered global weather patterns, including a worldwide increase in temperature, due largely to rising levels of atmospheric carbon dioxide

greenhouse effect warming of Earth due to carbon dioxide and other greenhouse gases in the atmosphere

greenhouse gases atmospheric gases such as carbon dioxide and methane that absorb and emit radiation, thus trapping heat in Earth's atmosphere

haze-effect cooling effect of the gases and solids from a volcanic eruption on global climate

heterospecifics individuals that are members of different species

intertidal zone part of the ocean that is closest to land; parts extend above the water at low tide

Milankovitch cycles cyclic changes in the Earth's orbit that may affect climate

neritic zone part of the ocean that extends from low tide to the edge of the continental shelf

net primary productivity measurement of the energy accumulation within an ecosystem, calculated as the total amount of carbon fixed per year minus the amount that is oxidized during cellular respiration

ocean upwelling rising of deep ocean waters that occurs when prevailing winds blow along surface waters near a coastline

oceanic zone part of the ocean that begins offshore where the water measures 200 m deep or deeper

pelagic realm (also, pelagic zone) open ocean waters that are not close to the bottom or near the shore

permafrost perennially frozen portion of the Arctic tundra soil

photic zone portion of the ocean that light can penetrate

planktivore animal species that eats plankton

predator organism that kills and consumes another organism

Sargassum type of free-floating marine seaweed

solar intensity amount of solar power energy the sun emits in a given amount of time

source water point of origin of a river or stream

thermocline layer of water with a temperature that is significantly different from that of the surrounding layers

weather conditions of the atmosphere during a short period of time

CHAPTER SUMMARY

44.1 The Scope of Ecology

Ecology is the study of the interactions of living things with their environment. Ecologists ask questions that comprise four levels of general biological organization—organismal, population, community, and ecosystem. At the organismal level, ecologists study individual organisms and how they interact with their environments. At the population and community levels, ecologists explore, respectively, how a population of organisms changes over time and the ways in which that population interacts with other species in the community. Ecologists studying an ecosystem examine the living species (the biotic components) of the ecosystem as well as the nonliving portions (the abiotic components), such as air, water, and soil, of the environment.

44.2 Biogeography

Biogeography is the study of the geographic distribution of living things as well as the abiotic factors that affect their distribution. Endemic species are species that are naturally found only in a specific geographic area. The distribution of living things is influenced by several environmental factors that are, in part, controlled by the latitude or elevation at which a species is found. Ocean upwellings, and spring and fall turnovers are important processes regulating the distribution of nutrients and other abiotic factors important in aquatic ecosystems. Energy sources, temperature, water, inorganic nutrients, and soil are factors limiting the distribution of living things in terrestrial systems. Net primary productivity is a measure of the amount of biomass produced by a biome.

44.3 Terrestrial Biomes

The Earth has terrestrial biomes and aquatic biomes. Aquatic biomes include both freshwater and marine environments. There are eight major terrestrial biomes: tropical wet forests, savannas, subtropical deserts, chaparral, temperate grasslands, temperate forests, boreal forests, and Arctic tundra. The same biome can occur in different geographic locations with similar climates. Temperature and precipitation, and variations in both, are key abiotic factors that shape the composition of animal and plant communities in terrestrial biomes. Some biomes, such as temperate grasslands and temperate forests, have distinct seasons, with cold weather and hot weather alternating throughout the year. In warm, moist biomes, such as the tropical wet forest, net primary productivity is high, as warm temperatures, abundant water, and a year-round growing season fuel plant growth and supply energy for high diversity throughout the food web. Other biomes, such as deserts and tundras, have low primary productivity due to extreme temperatures and a shortage of available water.

44.4 Aquatic Biomes

Aquatic ecosystems include both saltwater and freshwater biomes. The abiotic factors important for the structuring of aquatic ecosystems can be different than those seen in terrestrial systems. Sunlight is a driving force behind the structure of forests and also is an important factor in bodies of water, especially those that are very deep, because of the role of photosynthesis in sustaining certain organisms.

Density and temperature shape the structure of aquatic systems. Oceans may be thought of as consisting of different zones based on water depth and distance from the shoreline and light penetrance. Different kinds of organisms are adapted to the conditions found in each zone. Coral reefs are unique marine ecosystems that are home to a wide variety of species. Estuaries are found where rivers meet the ocean; their shallow waters provide nourishment and shelter for young crustaceans, mollusks, fishes, and many other species. Freshwater biomes include lakes, ponds, rivers, streams, and wetlands. Bogs are an interesting type of wetland characterized by standing water, lower pH, and a lack of nitrogen.

44.5 Climate and the Effects of Global Climate Change

The Earth has gone through periodic cycles of increases and decreases in temperature. During the past 2,000 years, the Medieval Climate Anomaly was a warmer period, while the Little Ice Age was unusually cool. Both of these irregularities can be explained by natural causes of changes in climate, and, although the temperature changes were small, they had significant effects. Natural drivers of climate change include Milankovitch cycles, changes in solar activity, and volcanic eruptions. None of these factors, however, leads to rapid increases in global temperature or sustained increases in carbon dioxide.

The burning of fossil fuels is an important source of greenhouse gases, which play a major role in the greenhouse effect. Two hundred and fifty million years ago, global warming resulted in the Permian extinction: a large-scale extinction event that is documented in the fossil record. Currently, modern-day climate change is associated with the increased melting of glaciers and polar ice sheets, resulting in a gradual increase in sea level. Plants and animals can also be affected by global climate change when the timing of seasonal events, such as flowering or pollination, is affected by global warming.

VISUAL CONNECTION QUESTIONS

1. Figure 44.10 How might turnover in tropical lakes differ from turnover in lakes that exist in temperate regions? Think of the variation, or lack of variation, in seasonal temperature change.

2. Figure 44.12 Which of the following statements about biomes is false?
   a. Chaparral is dominated by shrubs.
   b. Savannas and temperate grasslands are dominated by grasses.
   c. Boreal forests are dominated by deciduous trees.
   d. Lichens are common in the arctic tundra.

3. Figure 44.21 In which of the following regions would you expect to find photosynthetic organisms?
   a. the aphotic zone, the neritic zone, the oceanic zone, and the benthic realm
   b. the photic zone, the intertidal zone, the neritic zone, and the oceanic zone
   c. the photic zone, the abyssal zone, the neritic zone, and the oceanic zone
   d. the pelagic realm, the aphotic zone, the neritic zone, and the oceanic zone

REVIEW QUESTIONS

4. Which of the following is a biotic factor?
   a. wind
   b. disease-causing microbe
   c. temperature
   d. soil particle size

5. The study of nutrient cycling though the environment is an example of which of the following?
   a. organismal ecology
   b. population ecology
   c. community ecology
   d. ecosystem ecology

6. Understory plants in a temperate forest have adaptations to capture limited ________.
   a. water
   b. nutrients
   c. heat
   d. sunlight

7. An ecologist hiking up a mountain may notice different biomes along the way due to changes in all of the following except:
   a. elevation
   b. rainfall
   c. latitude
   d. temperature

8. Which of the following biomes is characterized by abundant water resources?
   a. deserts
   b. boreal forests
   c. savannas
   d. tropical wet forests

9. Which of the following biomes is characterized by short growing seasons?
   a. deserts
   b. tropical wet forests
   c. Arctic tundras
   d. savannas

10. Where would you expect to find the most photosynthesis in an ocean biome?
    a. aphotic zone
    b. abyssal zone
    c. benthic realm
    d. intertidal zone

11. A key feature of estuaries is:
    a. low light conditions and high productivity
    b. salt water and fresh water
    c. frequent algal blooms
    d. little or no vegetation

12. Which of the following is an example of a weather event?
    a. The hurricane season lasts from June 1 through November 30.
    b. The amount of atmospheric CO_2 has steadily increased during the last century.
    c. A windstorm blew down trees in the Boundary Waters Canoe Area in Minnesota on July 4, 1999.
    d. Deserts are generally dry ecosystems having very little rainfall.

13. Which of the following natural forces is responsible for the release of carbon dioxide and other atmospheric gases?
    a. the Milankovitch cycles
    b. volcanoes
    c. solar intensity
    d. burning of fossil fuels

CRITICAL THINKING QUESTIONS

14. Ecologists often collaborate with other researchers interested in ecological questions. Describe the levels of ecology that would be easier for collaboration because of the similarities of questions asked. What levels of ecology might be more difficult for collaboration?
15. The population is an important unit in ecology as well as other biological sciences. How is a population defined, and what are the strengths and weaknesses of this definition? Are there some species that at certain times or places are not in populations?
16. Compare and contrast ocean upwelling and spring and fall turnovers.
17. Many endemic species are found in areas that are geographically isolated. Suggest a plausible scientific explanation for why this is so.
18. The extremely low precipitation of subtropical desert biomes might lead one to expect fire to be a major disturbance factor; however, fire is more common in the temperate grassland biome than in the subtropical desert biome. Why is this?
19. In what ways are the subtropical desert and the arctic tundra similar?
20. Scientists have discovered the bodies of humans and other living things buried in bogs for hundreds of years, but not yet decomposed. Suggest a possible biological explanation for why such bodies are so well-preserved.
21. Describe the conditions and challenges facing organisms living in the intertidal zone.
22. Compare and contrast how natural- and human-induced processes have influenced global climate change.
23. Predict possible consequences if carbon emissions from fossil fuels continue to rise.

CHAPTER 45
Population and Community Ecology

Figure 45.1 Asian carp jump out of the water in response to electrofishing. The Asian carp in the photograph on the right were harvested from the Little Calumet River in Illinois in May, 2010, using rotenone, a toxin often used as an insecticide, in an effort to learn more about the population of the species. (credit left image: modification of work by USGS; credit right image: modification of work by Lt. David French, USCG)

INTRODUCTION Imagine sailing down a river in a small motorboat on a weekend afternoon; the water is smooth and you are enjoying the warm sunshine and cool breeze when suddenly you are hit in the head by a 20-pound silver carp. This is now a risk on many rivers and canal systems in Illinois and Missouri because of the presence of Asian carp.

This fish—actually a group of species including the silver, black, grass, and big head carp—has been farmed and eaten in China for over 1000 years. It is one of the most important aquaculture food resources worldwide. In the United States, however, Asian carp is considered a dangerous invasive species that disrupts community structure and composition to the point of threatening native species.

Chapter Outline

45.1 Population Demography

By the end of this section, you will be able to do the following:

- Describe how ecologists measure population size and density
- Describe three different patterns of population distribution
- Use life tables to calculate mortality rates
- Describe the three types of survivorship curves and relate them to specific populations

Populations are dynamic entities. Populations consist all of the species living within a specific area, and populations fluctuate based on a number of factors: seasonal and yearly changes in the environment, natural disasters such as forest fires and volcanic eruptions, and competition for resources between and within species. The statistical study of population dynamics, **demography**, uses a series of mathematical tools to investigate how populations respond to changes in their biotic and abiotic environments. Many of these tools were originally designed to study human populations. For example, **life tables**, which detail the life expectancy of individuals within a population, were initially developed by life insurance companies to set insurance rates. In fact, while the term "demographics" is commonly used when discussing humans, all living populations can be studied using this approach.

Population Size and Density

The study of any population usually begins by determining how many individuals of a particular species exist, and how closely associated they are with each other. Within a particular habitat, a population can be characterized by its **population size (*N*)**, the total number of individuals, and its **population density**, the number of individuals within a specific area or volume. Population size and density are the two main characteristics used to describe and understand populations. For example, populations with more individuals may be more stable than smaller populations based on their genetic variability, and thus their potential to adapt to the environment. Alternatively, a member of a population with low population density (more spread out in the habitat), might have more difficulty finding a mate to reproduce compared to a population of higher density. As is shown in Figure 45.2, smaller organisms tend to be more densely distributed than larger organisms.

VISUAL CONNECTION

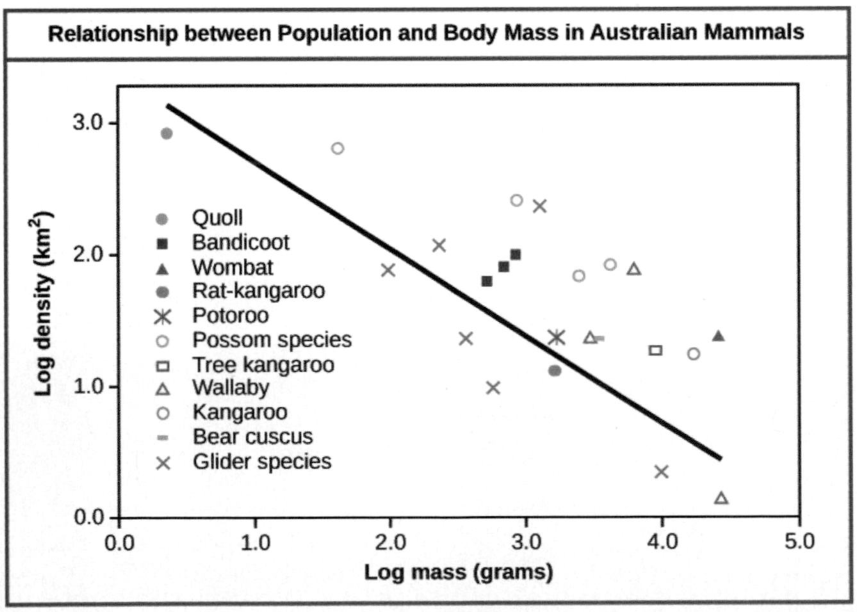

Figure 45.2 Australian mammals show a typical inverse relationship between population density and body size.

As this graph shows, population density typically decreases with increasing body size. Why do you think this is the case?

Population Research Methods

The most accurate way to determine population size is to simply count all of the individuals within the habitat. However, this method is often not logistically or economically feasible, especially when studying large habitats. Thus, scientists usually study populations by sampling a representative portion of each habitat and using this data to make inferences about the habitat as a whole. A variety of methods can be used to sample populations to determine their size and density. For immobile organisms such as plants, or for very small and slow-moving organisms, a **quadrat** may be used (Figure 45.3). A quadrat is a way of marking off square areas within a habitat, either by staking out an area with sticks and string, or by the use of a wood, plastic, or metal square placed on the ground. After setting the quadrats, researchers then count the number of individuals that lie within their boundaries. Multiple quadrat samples are performed throughout the habitat at several random locations to estimate the population size and density within the entire habitat. The number and size of quadrat samples depends on the type of organisms under study and other factors, including the density of the organism. For example, if sampling daffodils, a 1 m^2 quadrat might be used. With giant redwoods, on the other hand, a larger quadrat of 100 m^2 might be employed. This ensures that enough individuals of the species are counted to get an accurate sample that correlates with the habitat, including areas not sampled.

Figure 45.3 A scientist uses a quadrat to measure population size and density. (credit: NPS Sonoran Desert Network)

For mobile organisms, such as mammals, birds, or fish, scientists use a technique called **mark and recapture**. This method involves marking a sample of captured animals in some way (such as tags, bands, paint, or other body markings), and then releasing them back into the environment to allow them to mix with the rest of the population. Later, researchers collect a new sample, including some individuals that are marked (recaptures) and some individuals that are unmarked (Figure 45.4).

Figure 45.4 Mark and recapture is used to measure the population size of mobile animals such as (a) bighorn sheep, (b) the California condor, and (c) salmon. (credit a: modification of work by Neal Herbert, NPS; credit b: modification of work by Pacific Southwest Region USFWS; credit c: modification of work by Ingrid Taylar)

Using the ratio of marked and unmarked individuals, scientists determine how many individuals are in the sample. From this, calculations are used to estimate the total population size. This method assumes that the larger the population, the lower the percentage of tagged organisms that will be recaptured since they will have mixed with more untagged individuals. For example, if 80 deer are captured, tagged, and released into the forest, and later 100 deer are captured and 20 of them are already marked, we can estimate the population size (*N*) using the following equation:

$$\frac{(\text{number marked first catch x total number of second catch})}{\text{number marked second catch}} = N$$

Using our example, the population size would be estimated at 400.

$$\frac{(80 \text{ x } 100)}{20} = 400$$

Therefore, there are an estimated 400 total individuals in the original population.

There are some limitations to the mark and recapture method. Some animals from the first catch may learn to avoid capture in the second round, thus inflating population estimates. Alternatively, some animals may prefer to be retrapped (especially if a food reward is offered), resulting in an underestimate of population size. Also, some species may be harmed by the marking technique, reducing their survival. A variety of other techniques have been developed, including the electronic tracking of animals tagged with radio transmitters and the use of data from commercial fishing and trapping operations to estimate the size and health of populations and communities.

Species Distribution

In addition to measuring simple density, further information about a population can be obtained by looking at the distribution of the individuals. **Species dispersion patterns** (or distribution patterns) show the spatial relationship between members of a population within a habitat at a particular point in time. In other words, they show whether members of the species live close together or far apart, and what patterns are evident when they are spaced apart.

Individuals in a population can be equally spaced apart, dispersed randomly with no predictable pattern, or clustered in groups. These are known as uniform, random, and clumped dispersion patterns, respectively (Figure 45.5). Uniform dispersion is observed in plants that secrete substances inhibiting the growth of nearby individuals (such as the release of toxic chemicals by the sage plant *Salvia leucophylla,* a phenomenon called allelopathy) and in animals like the penguin that maintain a defined territory. An example of random dispersion occurs with dandelion and other plants that have wind-dispersed seeds that germinate wherever they happen to fall in a favorable environment. A clumped dispersion may be seen in plants that drop their seeds straight to the ground, such as oak trees, or in animals that live in groups (schools of fish or herds of elephants). Clumped dispersions may also be a function of habitat heterogeneity. Thus, the dispersion of the individuals within a population provides more information about how they interact with each other than does a simple density measurement. Just as lower density species might have more difficulty finding a mate, solitary species with a random distribution might have a similar difficulty when compared to social species clumped together in groups.

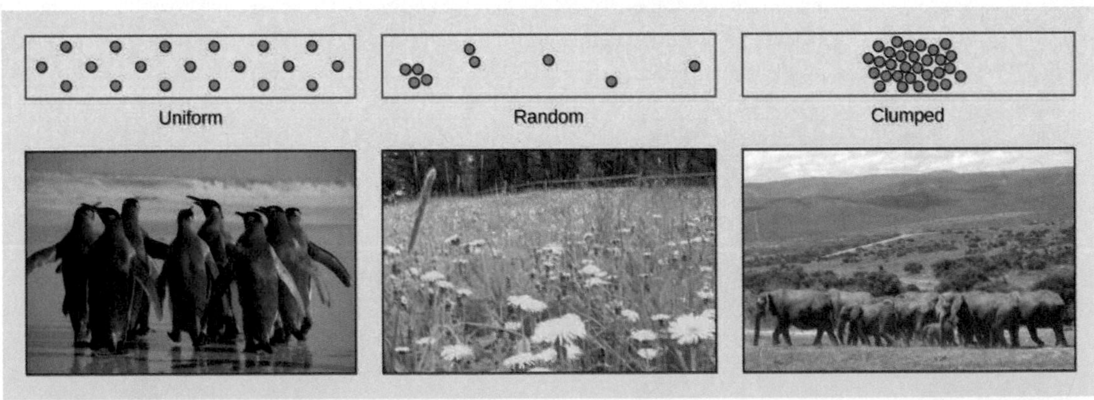

Figure 45.5 Species may have uniform, random, or clumped distribution. Territorial birds such as penguins tend to have uniform distribution. Plants such as dandelions with wind-dispersed seeds tend to be randomly distributed. Animals such as elephants that travel in groups exhibit clumped distribution. (credit a: modification of work by Ben Tubby; credit b: modification of work by Rosendahl; credit c: modification of work by Rebecca Wood)

Demography

While population size and density describe a population at one particular point in time, scientists must use demography to study the dynamics of a population. Demography is the statistical study of population changes over time: birth rates, death rates, and life expectancies. Each of these measures, especially birth rates, may be affected by the population characteristics described above. For example, a large population size results in a higher birth rate because more potentially reproductive individuals are present. In contrast, a large population size can also result in a higher death rate because of competition, disease, and the accumulation of waste. Similarly, a higher population density or a clumped dispersion pattern results in more potential reproductive encounters between individuals, which can increase birth rate. Lastly, a female-biased sex ratio (the ratio of males to females) or **age structure** (the proportion of population members at specific age ranges) composed of many individuals of reproductive age can increase birth rates.

In addition, the demographic characteristics of a population can influence how the population grows or declines over time. If birth and death rates are equal, the population remains stable. However, the population size will increase if birth rates exceed death rates; the population will decrease if birth rates are less than death rates. Life expectancy is another important factor; the length of time individuals remain in the population impacts local resources, reproduction, and the overall health of the population. These demographic characteristics are often displayed in the form of a life table.

Life Tables

Life tables provide important information about the life history of an organism. Life tables divide the population into age groups and often sexes, and show how long a member of that group is likely to live. They are modeled after actuarial tables used by the insurance industry for estimating human life expectancy. Life tables may include the probability of individuals dying before their next birthday (i.e., their **mortality rate**), the percentage of surviving individuals dying at a particular age interval, and their life expectancy at each interval. An example of a life table is shown in Table 45.1 from a study of Dall mountain sheep, a species native to northwestern North America. Notice that the population is divided into age intervals (column A). The mortality rate (per 1000), shown in column D, is based on the number of individuals dying during the age interval (column B) divided by the number of individuals surviving at the beginning of the interval (Column C), multiplied by 1000.

$$\text{mortality rate} = \frac{\text{number of individuals dying}}{\text{number of individuals surviving}} \text{ x } 1000$$

For example, between ages three and four, 12 individuals die out of the 776 that were remaining from the original 1000 sheep. This number is then multiplied by 1000 to get the mortality rate per thousand.

$$\text{mortality rate} = \frac{12}{776} \text{ x } 1000 \approx 15.5$$

As can be seen from the mortality rate data (column D), a high death rate occurred when the sheep were between 6 and 12 months old, and then increased even more from 8 to 12 years old, after which there were few survivors. The data indicate that if a sheep in this population were to survive to age one, it could be expected to live another 7.7 years on average, as shown by the life expectancy numbers in column E.

Life Table of Dall Mountain Sheep[1]

Age interval (years)	Number dying in age interval out of 1000 born	Number surviving at beginning of age interval out of 1000 born	Mortality rate per 1000 alive at beginning of age interval	Life expectancy or mean lifetime remaining to those attaining age interval
0-0.5	54	1000	54.0	7.06
0.5-1	145	946	153.3	--
1-2	12	801	15.0	7.7
2-3	13	789	16.5	6.8

Table 45.1 This life table of *Ovis dalli* shows the number of deaths, number of survivors, mortality rate, and life expectancy at

1Data Adapted from Edward S. Deevey, Jr., "Life Tables for Natural Populations of Animals," *The Quarterly Review of Biology* 22, no. 4 (December 1947):

Age interval (years)	Number dying in age interval out of 1000 born	Number surviving at beginning of age interval out of 1000 born	Mortality rate per 1000 alive at beginning of age interval	Life expectancy or mean lifetime remaining to those attaining age interval
3-4	12	776	15.5	5.9
4-5	30	764	39.3	5.0
5-6	46	734	62.7	4.2
6-7	48	688	69.8	3.4
7-8	69	640	107.8	2.6
8-9	132	571	231.2	1.9
9-10	187	439	426.0	1.3
10-11	156	252	619.0	0.9
11-12	90	96	937.5	0.6
12-13	3	6	500.0	1.2
13-14	3	3	1000	0.7

Table 45.1 This life table of *Ovis dalli* shows the number of deaths, number of survivors, mortality rate, and life expectancy at each age interval for the Dall mountain sheep.

Survivorship Curves

Another tool used by population ecologists is a **survivorship curve**, which is a graph of the number of individuals surviving at each age interval plotted versus time (usually with data compiled from a life table). These curves allow us to compare the life histories of different populations (Figure 45.6). Humans and most primates exhibit a Type I survivorship curve because a high percentage of offspring survive their early and middle years—death occurs predominantly in older individuals. These types of species usually have small numbers of offspring at one time, and they give a high amount of parental care to them to ensure their survival. Birds are an example of an intermediate or Type II survivorship curve because birds die more or less equally at each age interval. These organisms also may have relatively few offspring and provide significant parental care. Trees, marine invertebrates, and most fishes exhibit a Type III survivorship curve because very few of these organisms survive their younger years; however, those that make it to an old age are more likely to survive for a relatively long period of time. Organisms in this category usually have a very large number of offspring, but once they are born, little parental care is provided. Thus these offspring are "on their own" and vulnerable to predation, but their sheer numbers assure the survival of enough individuals to perpetuate the species.

283-314.

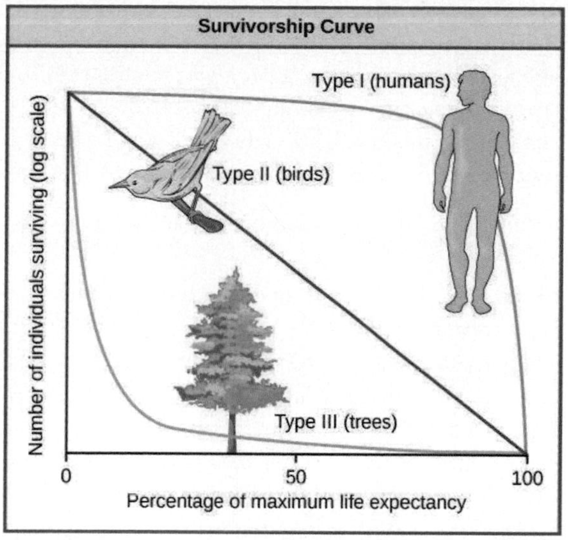

Figure 45.6 Survivorship curves show the distribution of individuals in a population according to age. Humans and most mammals have a Type I survivorship curve because death primarily occurs in the older years. Birds have a Type II survivorship curve, as death at any age is equally probable. Trees have a Type III survivorship curve because very few survive the younger years, but after a certain age, individuals are much more likely to survive.

45.2 Life Histories and Natural Selection

By the end of this section, you will be able to do the following:

- Describe how life history patterns are influenced by natural selection
- Explain different life history patterns and how different reproductive strategies affect species' survival

A species' **life history** describes the series of events over its lifetime, such as how resources are allocated for growth, maintenance, and reproduction. Life history traits affect the life table of an organism. A species' life history is genetically determined and shaped by the environment and natural selection.

Life History Patterns and Energy Budgets

Energy is required by all living organisms for their growth, maintenance, and reproduction; at the same time, energy is often a major limiting factor in determining an organism's survival. Plants, for example, acquire energy from the sun via photosynthesis, but must expend this energy to grow, maintain health, and produce energy-rich seeds to produce the next generation. Animals have the additional burden of using some of their energy reserves to acquire food. Furthermore, some animals must expend energy caring for their offspring. Thus, all species have an **energy budget**: they must balance energy intake with their use of energy for metabolism, reproduction, parental care, and energy storage (such as bears building up body fat for winter hibernation).

Parental Care and Fecundity

Fecundity is the potential reproductive capacity of an individual within a population. In other words, fecundity describes how many offspring could ideally be produced if an individual has as many offspring as possible, repeating the reproductive cycle as soon as possible after the birth of the offspring. In animals, fecundity is inversely related to the amount of parental care given to an individual offspring. Species, such as many marine invertebrates, that produce many offspring usually provide little if any care for the offspring (they would not have the energy or the ability to do so anyway). Most of their energy budget is used to produce many tiny offspring. Animals with this strategy are often self-sufficient at a very early age. This is because of the energy tradeoff these organisms have made to maximize their evolutionary fitness. Because their energy is used for producing offspring instead of parental care, it makes sense that these offspring have some ability to be able to move within their environment and find food and perhaps shelter. Even with these abilities, their small size makes them extremely vulnerable to predation, so the production of many offspring allows enough of them to survive to maintain the species.

Animal species that have few offspring during a reproductive event usually give extensive parental care, devoting much of their energy budget to these activities, sometimes at the expense of their own health. This is the case with many mammals, such as humans, kangaroos, and pandas. The offspring of these species are relatively helpless at birth and need to develop before they achieve self-sufficiency.

Plants with low fecundity produce few energy-rich seeds (such as coconuts and chestnuts) with each having a good chance to germinate into a new organism; plants with high fecundity usually have many small, energy-poor seeds (like orchids) that have a relatively poor chance of surviving. Although it may seem that coconuts and chestnuts have a better chance of surviving, the energy tradeoff of the orchid is also very effective. It is a matter of where the energy is used, for large numbers of seeds or for fewer seeds with more energy.

Early versus Late Reproduction

The timing of reproduction in a life history also affects species survival. Organisms that reproduce at an early age have a greater chance of producing offspring, but this is usually at the expense of their growth and the maintenance of their health. Conversely, organisms that start reproducing later in life often have greater fecundity or are better able to provide parental care, but they risk that they will not survive to reproductive age. Examples of this can be seen in fishes. Small fish, like guppies, use their energy to reproduce rapidly, but never attain the size that would give them defense against some predators. Larger fish, like the bluegill or shark, use their energy to attain a large size, but do so with the risk that they will die before they can reproduce or at least reproduce to their maximum. These different energy strategies and tradeoffs are key to understanding the evolution of each species as it maximizes its fitness and fills its niche. In terms of energy budgeting, some species "blow it all" and use up most of their energy reserves to reproduce early before they die. Other species delay having reproduction to become stronger, more experienced individuals and to make sure that they are strong enough to provide parental care if necessary.

Single versus Multiple Reproductive Events

Some life history traits, such as fecundity, timing of reproduction, and parental care, can be grouped together into general strategies that are used by multiple species. **Semelparity** occurs when a species reproduces only once during its lifetime and then dies. Such species use most of their resource budget during a single reproductive event, sacrificing their health to the point that they do not survive. Examples of semelparity are bamboo, which flowers once and then dies, and the Chinook salmon (Figure 45.7a), which uses most of its energy reserves to migrate from the ocean to its freshwater nesting area, where it reproduces and then dies. Scientists have posited alternate explanations for the evolutionary advantage of the Chinook's post-reproduction death: a programmed suicide caused by a massive release of corticosteroid hormones, presumably so the parents can become food for the offspring, or simple exhaustion caused by the energy demands of reproduction; these are still being debated.

Iteroparity describes species that reproduce repeatedly during their lives. Some animals are able to mate only once per year, but survive multiple mating seasons. The pronghorn antelope is an example of an animal that goes into a seasonal estrus cycle ("heat"): a hormonally induced physiological condition preparing the body for successful mating (Figure 45.7b). Females of these species mate only during the estrus phase of the cycle. A different pattern is observed in primates, including humans and chimpanzees, which may attempt reproduction at any time during their reproductive years, even though their menstrual cycles make pregnancy likely only a few days per month during ovulation (Figure 45.7c).

Figure 45.7 The (a) Chinook salmon mates once and dies. The (b) pronghorn antelope mates during specific times of the year during its reproductive life. Primates, such as humans and (c) chimpanzees, may mate on any day, independent of ovulation. (credit a: modification of work by Roger Tabor, USFWS; credit b: modification of work by Mark Gocke, USDA; credit c: modification of work by "Shiny Things"/Flickr)

LINK TO LEARNING

Play this interactive PBS evolution-based mating game (http://openstax.org/l/mating_game) to learn more about reproductive strategies.

EVOLUTION CONNECTION

Energy Budgets, Reproductive Costs, and Sexual Selection in *Drosophila*

Research into how animals allocate their energy resources for growth, maintenance, and reproduction has used a variety of experimental animal models. Some of this work has been done using the common fruit fly, *Drosophila melanogaster*. Studies have shown that not only does reproduction have a cost as far as how long male fruit flies live, but also fruit flies that have already mated several times have limited sperm remaining for reproduction. Fruit flies maximize their last chances at reproduction by selecting optimal mates.

In a 1981 study, male fruit flies were placed in enclosures with either virgin or inseminated females. The males that mated with virgin females had shorter life spans than those in contact with the same number of inseminated females with which they were unable to mate. This effect occurred regardless of how large (indicative of their age) the males were. Thus, males that did not mate lived longer, allowing them more opportunities to find mates in the future.

More recent studies, performed in 2006, show how males select the female with which they will mate and how this is affected by previous matings (Figure 45.8).[2] Males were allowed to select between smaller and larger females. Findings showed that larger females had greater fecundity, producing twice as many offspring per mating as the smaller females did. Males that had previously mated, and thus had lower supplies of sperm, were termed "resource depleted," while males that had not mated were termed "non-resource-depleted." The study showed that although non-resource-depleted males preferentially mated with larger females, this selection of partners was more pronounced in the resource-depleted males. Thus, males with depleted sperm supplies, which were limited in the number of times that they could mate before they replenished their sperm supply, selected larger, more fecund females, thus maximizing their chances for offspring. This study was one of the first to show that the physiological state of the male affected its mating behavior in a way that clearly maximizes its use of limited reproductive resources.

	Ratio large/small females mated
Non sperm-depleted	8 ± 5
Sperm depleted	15 ± 5

Figure 45.8 Male fruit flies that had previously mated (sperm-depleted) picked larger, more fecund females more often than those that had not mated (non-sperm-depleted). This change in behavior causes an increase in the efficiency of a limited reproductive resource: sperm.

These studies demonstrate two ways in which the energy budget is a factor in reproduction. First, energy expended on mating may reduce an animal's lifespan, but by this time they have already reproduced, so in the context of natural selection this early death is not of much evolutionary importance. Second, when resources such as sperm (and the energy needed to replenish it) are low, an organism's behavior can change to give them the best chance of passing their genes on to the next generation. These changes in behavior, so important to evolution, are studied in a discipline known as behavioral biology, or ethology, at the interface between population biology and psychology.

45.3 Environmental Limits to Population Growth

By the end of this section, you will be able to do the following:

- Explain the characteristics of and differences between exponential and logistic growth patterns
- Give examples of exponential and logistic growth in natural populations
- Describe how natural selection and environmental adaptation led to the evolution of particular life history patterns

Although life histories describe the way many characteristics of a population (such as their age structure) change over time in a general way, population ecologists make use of a variety of methods to model population dynamics mathematically. These more

2Adapted from Phillip G. Byrne and William R. Rice, "Evidence for adaptive male mate choice in the fruit fly *Drosophila melanogaster*," Proc Biol Sci. 273, no. 1589 (2006): 917-922, doi: 10.1098/rspb.2005.3372.

precise models can then be used to accurately describe changes occurring in a population and better predict future changes. Certain long-accepted models are now being modified or even abandoned due to their lack of predictive ability, and scholars strive to create effective new models.

Exponential Growth

Charles Darwin, in his theory of natural selection, was greatly influenced by the English clergyman Thomas Malthus. Malthus published a book in 1798 stating that populations with unlimited natural resources grow very rapidly, and then population growth decreases as resources become depleted. This accelerating pattern of increasing population size is called **exponential growth**.

The best example of exponential growth is seen in bacteria. Bacteria reproduce by prokaryotic fission. This division takes about an hour for many bacterial species. If 1000 bacteria are placed in a large flask with an unlimited supply of nutrients (so the nutrients will not become depleted), after an hour, there is one round of division and each organism divides, resulting in 2000 organisms—an increase of 1000. In another hour, each of the 2000 organisms will double, producing 4000, an increase of 2000 organisms. After the third hour, there should be 8000 bacteria in the flask, an increase of 4000 organisms. The important concept of exponential growth is the accelerating **population growth rate**—the number of organisms added in each reproductive generation—that is, it is increasing at a greater and greater rate. After 1 day and 24 of these cycles, the population would have increased from 1000 to more than 16 billion. When the population size, *N*, is plotted over time, a **J-shaped growth curve** is produced (Figure 45.9).

The bacteria example is not representative of the real world where resources are limited. Furthermore, some bacteria will die during the experiment and thus not reproduce, lowering the growth rate. Therefore, when calculating the growth rate of a population, the **death rate (*D*)** (number organisms that die during a particular time interval) is subtracted from the **birth rate (*B*)** (number organisms that are born during that interval). This is shown in the following formula:

$$\frac{\Delta N \text{ (change in number)}}{\Delta T \text{ (change in time)}} = B \text{ (birth rate)} - D \text{ (death rate)}$$

The birth rate is usually expressed on a per capita (for each individual) basis. Thus, *B* (birth rate) = *bN* (the per capita birth rate "*b*" multiplied by the number of individuals "*N*") and *D* (death rate) = *dN* (the per capita death rate "d" multiplied by the number of individuals "*N*"). Additionally, ecologists are interested in the population at a particular point in time, an infinitely small time interval. For this reason, the terminology of differential calculus is used to obtain the "instantaneous" growth rate, replacing the *change* in number and time with an instant-specific measurement of number and time.

$$\frac{dN}{dT} = bN - dN = (b - d)N$$

Notice that the "*d*" associated with the first term refers to the derivative (as the term is used in calculus) and is different from the death rate, also called "*d*." The difference between birth and death rates is further simplified by substituting the term "*r*" (intrinsic rate of increase) for the relationship between birth and death rates:

$$\frac{dN}{dT} = rN$$

The value "*r*" can be positive, meaning the population is increasing in size; or negative, meaning the population is decreasing in size; or zero, where the population's size is unchanging, a condition known as **zero population growth**. A further refinement of the formula recognizes that different species have inherent differences in their intrinsic rate of increase (often thought of as the potential for reproduction), even under ideal conditions. Obviously, a bacterium can reproduce more rapidly and have a higher intrinsic rate of growth than a human. The maximal growth rate for a species is its **biotic potential, or r_{max}**, thus changing the equation to:

$$\frac{dN}{dT} = r_{\max} N$$

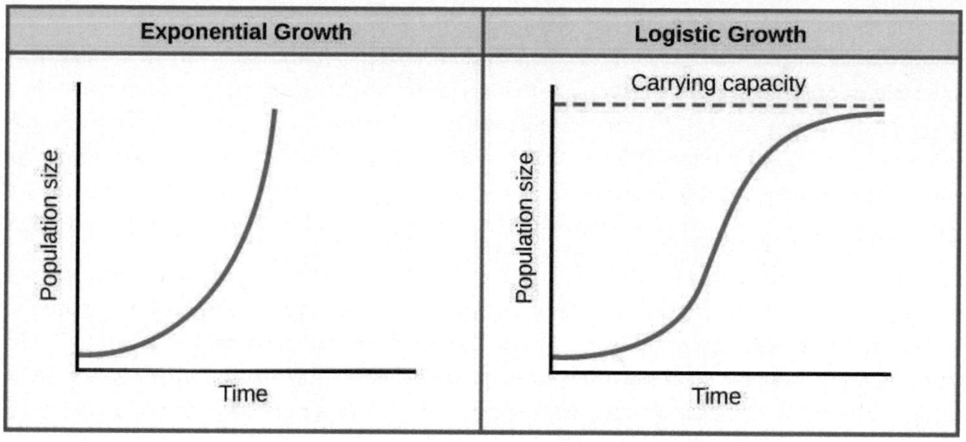

Figure 45.9 When resources are unlimited, populations exhibit exponential growth, resulting in a J-shaped curve. When resources are limited, populations exhibit logistic growth. In logistic growth, population expansion decreases as resources become scarce, and it levels off when the carrying capacity of the environment is reached, resulting in an S-shaped curve.

Logistic Growth

Exponential growth is possible only when infinite natural resources are available; this is not the case in the real world. Charles Darwin recognized this fact in his description of the "struggle for existence," which states that individuals will compete (with members of their own or other species) for limited resources. The successful ones will survive to pass on their own characteristics and traits (which we know now are transferred by genes) to the next generation at a greater rate (natural selection). To model the reality of limited resources, population ecologists developed the **logistic growth** model.

Carrying Capacity and the Logistic Model

In the real world, with its limited resources, exponential growth cannot continue indefinitely. Exponential growth may occur in environments where there are few individuals and plentiful resources, but when the number of individuals gets large enough, resources will be depleted, slowing the growth rate. Eventually, the growth rate will plateau or level off (Figure 45.9). This population size, which represents the maximum population size that a particular environment can support, is called the **carrying capacity, or *K***.

The formula we use to calculate logistic growth adds the carrying capacity as a moderating force in the growth rate. The expression "$K - N$" indicates how many individuals may be added to a population at a given stage, and "$K - N$" divided by "K" is the fraction of the carrying capacity available for further growth. Thus, the exponential growth model is restricted by this factor to generate the logistic growth equation:

$$\frac{dN}{dT} = r_{max}\frac{dN}{dT} = r_{max}N\frac{(K - N)}{K}$$

Notice that when N is very small, $(K-N)/K$ becomes close to K/K or 1, and the right side of the equation reduces to $r_{max}N$, which means the population is growing exponentially and is not influenced by carrying capacity. On the other hand, when N is large, $(K-N)/K$ comes close to zero, which means that population growth will be slowed greatly or even stopped. Thus, population growth is greatly slowed in large populations by the carrying capacity K. This model also allows for the population of a negative population growth, or a population decline. This occurs when the number of individuals in the population exceeds the carrying capacity (because the value of (K-N)/K is negative).

A graph of this equation yields an **S-shaped curve** (Figure 45.9), and it is a more realistic model of population growth than exponential growth. There are three different sections to an S-shaped curve. Initially, growth is exponential because there are few individuals and ample resources available. Then, as resources begin to become limited, the growth rate decreases. Finally, growth levels off at the carrying capacity of the environment, with little change in population size over time.

Role of Intraspecific Competition

The logistic model assumes that every individual within a population will have equal access to resources and, thus, an equal chance for survival. For plants, the amount of water, sunlight, nutrients, and the space to grow are the important resources, whereas in animals, important resources include food, water, shelter, nesting space, and mates.

In the real world, phenotypic variation among individuals within a population means that some individuals will be better adapted to their environment than others. The resulting competition between population members of the same species for resources is termed **intraspecific competition** (intra- = "within"; -specific = "species"). Intraspecific competition for resources may not affect populations that are well below their carrying capacity—resources are plentiful and all individuals can obtain what they need. However, as population size increases, this competition intensifies. In addition, the accumulation of waste products can reduce an environment's carrying capacity.

Examples of Logistic Growth

Yeast, a microscopic fungus used to make bread and alcoholic beverages, exhibits the classical S-shaped curve when grown in a test tube (Figure 45.10a). Its growth levels off as the population depletes the nutrients. In the real world, however, there are variations to this idealized curve. Examples in wild populations include sheep and harbor seals (Figure 45.10b). In both examples, the population size exceeds the carrying capacity for short periods of time and then falls below the carrying capacity afterwards. This fluctuation in population size continues to occur as the population oscillates around its carrying capacity. Still, even with this oscillation, the logistic model is confirmed.

VISUAL CONNECTION

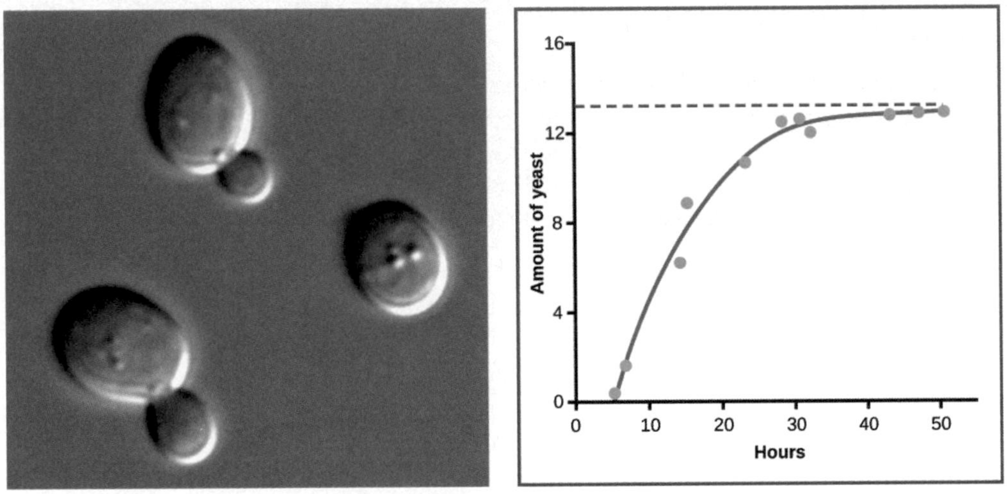

(a)

(b)

Figure 45.10 (a) Yeast grown in ideal conditions in a test tube show a classical S-shaped logistic growth curve, whereas (b) a natural population of seals shows real-world fluctuation.

If the major food source of the seals declines due to pollution or overfishing, which of the following would likely occur?

a. The carrying capacity of seals would decrease, as would the seal population.
b. The carrying capacity of seals would decrease, but the seal population would remain the same.
c. The number of seal deaths would increase but the number of births would also increase, so the population size would remain the same.
d. The carrying capacity of seals would remain the same, but the population of seals would decrease.

45.4 Population Dynamics and Regulation

By the end of this section, you will be able to do the following:

- Give examples of how the carrying capacity of a habitat may change
- Compare and contrast density-dependent growth regulation and density-independent growth regulation, giving examples
- Give examples of exponential and logistic growth in wild animal populations
- Describe how natural selection and environmental adaptation leads to the evolution of particular life-history patterns

The logistic model of population growth, while valid in many natural populations and a useful model, is a simplification of real-world population dynamics. Implicit in the model is that the carrying capacity of the environment does not change, which is not the case. The carrying capacity varies annually: for example, some summers are hot and dry whereas others are cold and wet. In many areas, the carrying capacity during the winter is much lower than it is during the summer. Also, natural events such as earthquakes, volcanoes, and fires can alter an environment and hence its carrying capacity. Additionally, populations do not usually exist in isolation. They engage in **interspecific competition**: that is, they share the environment with other species competing for the same resources. These factors are also important to understanding how a specific population will grow.

Nature regulates population growth in a variety of ways. These are grouped into **density-dependent** factors, in which the density of the population at a given time affects growth rate and mortality, and **density-independent** factors, which influence mortality in a population regardless of population density. Note that in the former, the effect of the factor on the population depends on the density of the population at onset. Conservation biologists want to understand both types because this helps them manage populations and prevent extinction or overpopulation.

Density-Dependent Regulation

Most density-dependent factors are biological in nature (biotic), and include predation, inter- and intraspecific competition, accumulation of waste, and diseases such as those caused by parasites. Usually, the denser a population is, the greater its mortality rate. For example, during intra- and interspecific competition, the reproductive rates of the individuals will usually be lower, reducing their population's rate of growth. In addition, low prey density increases the mortality of its predator because it has more difficulty locating its food source.

An example of density-dependent regulation is shown in Figure 45.11 with results from a study focusing on the giant intestinal roundworm (*Ascaris lumbricoides*), a parasite of humans and other mammals.[3] Denser populations of the parasite exhibited lower fecundity: they contained fewer eggs. One possible explanation for this is that females would be smaller in more dense populations (due to limited resources) and that smaller females would have fewer eggs. This hypothesis was tested and disproved in a 2009 study which showed that female weight had no influence.[4] The actual cause of the density-dependence of fecundity in this organism is still unclear and awaiting further investigation.

3N.A. Croll et al., "The Population Biology and Control of *Ascaris lumbricoides* in a Rural Community in Iran." *Transactions of the Royal Society of Tropical Medicine and Hygiene* 76, no. 2 (1982): 187-197, doi:10.1016/0035-9203(82)90272-3.

4Martin Walker et al., "Density-Dependent Effects on the Weight of Female *Ascaris lumbricoides* Infections of Humans and its Impact on Patterns of Egg Production." *Parasites & Vectors* 2, no. 11 (February 2009), doi:10.1186/1756-3305-2-11.

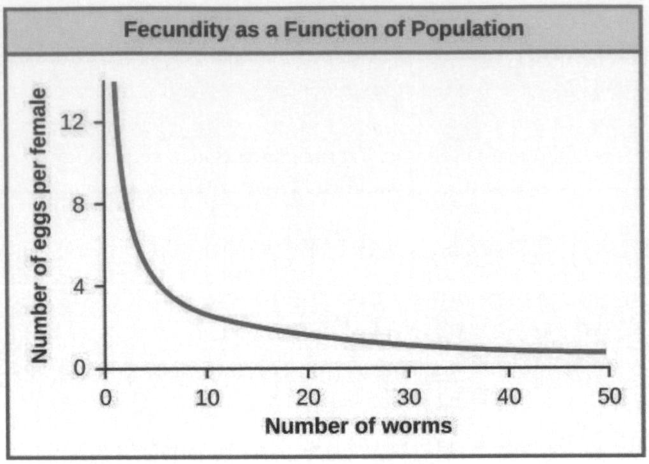

Figure 45.11 In this population of roundworms, fecundity (number of eggs) decreases with population density.[5]

Density-Independent Regulation and Interaction with Density-Dependent Factors

Many factors, typically physical or chemical in nature (abiotic), influence the mortality of a population regardless of its density, including weather, natural disasters, and pollution. An individual deer may be killed in a forest fire regardless of how many deer happen to be in that area. Its chances of survival are the same whether the population density is high or low. The same holds true for cold winter weather.

In real-life situations, population regulation is very complicated and density-dependent and independent factors can interact. A dense population that is reduced in a density-independent manner by some environmental factor(s) will be able to recover differently than a sparse population. For example, a population of deer affected by a harsh winter will recover faster if there are more deer remaining to reproduce.

EVOLUTION CONNECTION

Why Did the Woolly Mammoth Go Extinct?

Figure 45.12 The three photos include: (a) 1916 mural of a mammoth herd from the American Museum of Natural History, (b) the only stuffed mammoth in the world, from the Museum of Zoology located in St. Petersburg, Russia, and (c) a one-month-old baby mammoth, named Lyuba, discovered in Siberia in 2007. (credit a: modification of work by Charles R. Knight; credit b: modification of work by "Tanapon"/Flickr; credit c: modification of work by Matt Howry)

It's easy to get lost in the discussion about why dinosaurs went extinct 65 million years ago. Was it due to a meteor slamming into Earth near the coast of modern-day Mexico, or was it from some long-term weather cycle that is not yet understood? Scientists are continually exploring these and other theories.

Woolly mammoths began to go extinct much more recently, when they shared the Earth with humans who were no different anatomically than humans today (Figure 45.12). Mammoths survived in isolated island populations as recently as 1700 BC. We know a lot about these animals from carcasses found frozen in the ice of Siberia and other regions of the north. Scientists have sequenced at least 50 percent of its genome and believe mammoths are between 98 and 99 percent identical to modern elephants.

It is commonly thought that climate change and human hunting led to their extinction. A 2008 study estimated that climate

5N.A. Croll et al., "The Population Biology and Control of *Ascaris lumbricoides* in a Rural Community in Iran." *Transactions of the Royal Society of Tropical Medicine and Hygiene* 76, no. 2 (1982): 187-197, doi:10.1016/0035-9203(82)90272-3.

change reduced the mammoth's range from 3,000,000 square miles 42,000 years ago to 310,000 square miles 6,000 years ago.[6] It is also well documented that humans hunted these animals. A 2012 study showed that no single factor was exclusively responsible for the extinction of these magnificent creatures.[7] In addition to human hunting, climate change, and reduction of habitat, these scientists demonstrated another important factor in the mammoth's extinction was the migration of humans across the Bering Strait to North America during the last ice age 20,000 years ago.

The maintenance of stable populations was and is very complex, with many interacting factors determining the outcome. It is important to remember that humans are also part of nature. We once contributed to a species' decline using only primitive hunting technology.

Life Histories of *K*-selected and *r*-selected Species

While reproductive strategies play a key role in life histories, they do not account for important factors like limited resources and competition. The regulation of population growth by these factors can be used to introduce a classical concept in population biology, that of *K*-selected versus *r*-selected species.

The concept relates to a species' reproductive strategies, habitat, and behavior, especially in the way that they obtain resources and care for their young. It includes length of life and survivorship factors as well. Population biologists have grouped species into the two large categories—*K*-selected and *r*-selected—although the categories are really two ends of a continuum.

***K*-selected species** are species selected by stable, predictable environments. Populations of *K*-selected species tend to exist close to their carrying capacity (hence the term *K*-selected) where intraspecific competition is high. These species have few, large offspring, a long gestation period, and often give long-term care to their offspring (Table 45.2). While larger in size when born, the offspring are relatively helpless and immature at birth. By the time they reach adulthood, they must develop skills to compete for natural resources. In plants, scientists think of parental care more broadly: how long fruit takes to develop or how long it remains on the plant are determining factors in the time to the next reproductive event. Examples of *K*-selected species are primates (including humans), elephants, and plants such as oak trees (Figure 45.13a).

Oak trees grow very slowly and take, on average, 20 years to produce their first seeds, known as acorns. As many as 50,000 acorns can be produced by an individual tree, but the germination rate is low as many of these rot or are eaten by animals such as squirrels. In some years, oaks may produce an exceptionally large number of acorns, and these years may be on a two- or three-year cycle depending on the species of oak (*r*-selection).

As oak trees grow to a large size and for many years before they begin to produce acorns, they devote a large percentage of their energy budget to growth and maintenance. The tree's height and size allow it to dominate other plants in the competition for sunlight, the oak's primary energy resource. Furthermore, when it does reproduce, the oak produces large, energy-rich seeds that use their energy reserve to become quickly established (*K*-selection).

In contrast, ***r*-selected species** have a large number of small offspring (hence their *r* designation (Table 45.2)). This strategy is often employed in unpredictable or changing environments. Animals that are *r*-selected do not give long-term parental care and the offspring are relatively mature and self-sufficient at birth. Examples of *r*-selected species are marine invertebrates, such as jellyfish, and plants, such as the dandelion (Figure 45.13b). Dandelions have small seeds that are wind dispersed long distances. Many seeds are produced simultaneously to ensure that at least some of them reach a hospitable environment. Seeds that land in inhospitable environments have little chance for survival since their seeds are low in energy content. Note that survival is not necessarily a function of energy stored in the seed itself.

6David Nogués-Bravo et al., "Climate Change, Humans, and the Extinction of the Woolly Mammoth." *PLoS Biol* 6 (April 2008): e79, doi:10.1371/journal.pbio.0060079.

7G.M. MacDonald et al., "Pattern of Extinction of the Woolly Mammoth in Beringia." *Nature Communications* 3, no. 893 (June 2012), doi:10.1038/ncomms1881.

Characteristics of *K*-selected and *r*-selected species

Characteristics of *K*-selected species	Characteristics of *r*-selected species
Mature late	Mature early
Greater longevity	Lower longevity
Increased parental care	Decreased parental care
Increased competition	Decreased competition
Fewer offspring	More offspring
Larger offspring	Smaller offspring

Table 45.2

Figure 45.13 (a) Elephants are considered K-selected species as they live long, mature late, and provide long-term parental care to few offspring. Oak trees produce many offspring that do not receive parental care, but are considered K-selected species based on longevity and late maturation. (b) Dandelions and jellyfish are both considered r-selected species as they mature early, have short lifespans, and produce many offspring that receive no parental care.

Modern Theories of Life History

By the second half of the twentieth century, the concept of K- and r-selected species was used extensively and successfully to study populations. The *r*- and *K*-selection theory, although accepted for decades and used for much groundbreaking research, has now been reconsidered, and many population biologists have abandoned or modified it. Over the years, several studies attempted to confirm the theory, but these attempts have largely failed. Many species were identified that did not follow the theory's predictions. Furthermore, the theory ignored the age-specific mortality of the populations which scientists now know is very important. New **demographic-based models** of life history evolution have been developed which incorporate many ecological concepts included in *r*- and *K*-selection theory as well as population age structure and mortality factors.

45.5 Human Population Growth

By the end of this section, you will be able to do the following:

- Discuss exponential human population growth
- Explain how humans have expanded the carrying capacity of their habitat
- Relate population growth and age structure to the level of economic development in different countries
- Discuss the long-term implications of unchecked human population growth

Population dynamics can be applied to human population growth. Earth's human population is growing rapidly, to the extent that some worry about the ability of the earth's environment to sustain this population. Long-term exponential growth carries the potential risks of famine, disease, and large-scale death.

Although humans have increased the carrying capacity of their environment, the technologies used to achieve this transformation have caused unprecedented changes to Earth's environment, altering ecosystems to the point where some may be in danger of collapse. The depletion of the ozone layer, erosion due to acid rain, and damage from global climate change are caused by human activities. The ultimate effect of these changes on our carrying capacity is unknown. As some point out, it is likely that the negative effects of increasing carrying capacity will outweigh the positive ones—the world's carrying capacity for human beings might actually decrease.

The human population is currently experiencing exponential growth even though human reproduction is far below its biotic potential (Figure 45.14). To reach its biotic potential, all females would have to become pregnant every nine months or so during their reproductive years. Also, resources would have to be such that the environment would support such growth. Neither of these two conditions exists. In spite of this fact, human population is still growing exponentially.

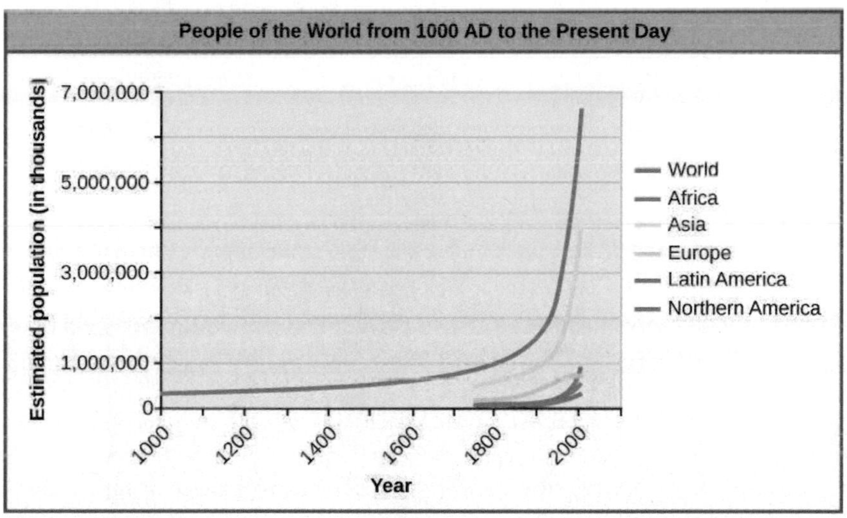

Figure 45.14 Human population growth since 1000 AD is exponential (dark blue line). Notice that while the population in Asia (yellow line), which has many economically underdeveloped countries, is increasing exponentially, the population in Europe (light blue line), where most of the countries are economically developed, is growing much more slowly.

A consequence of exponential human population growth is a reduction in time that it takes to add a particular number of humans to the Earth. Figure 45.15 shows that 123 years were necessary to add 1 billion humans in 1930, but it only took 24 years to add two billion people between 1975 and 1999. As already discussed, our ability to increase our carrying capacity indefinitely my be limited. Without new technological advances, the human growth rate has been predicted to slow in the coming decades. However, the population will still be increasing and the threat of overpopulation remains.

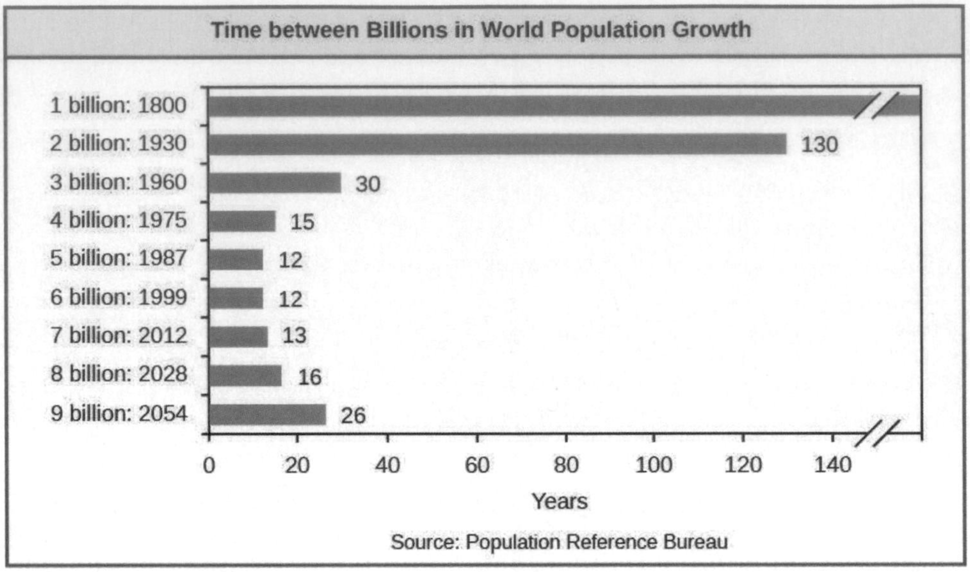

Figure 45.15 The time between the addition of each billion human beings to Earth decreases over time. (credit: modification of work by Ryan T. Cragun)

LINK TO LEARNING

Click through this video (http://openstax.org/l/human_growth) of how human populations have changed over time.

Overcoming Density-Dependent Regulation

Humans are unique in their ability to alter their environment with the conscious purpose of increasing carrying capacity. This ability is a major factor responsible for human population growth and a way of overcoming density-dependent growth regulation. Much of this ability is related to human intelligence, society, and communication. Humans can construct shelter to protect them from the elements and have developed agriculture and domesticated animals to increase their food supplies. In addition, humans use language to communicate this technology to new generations, allowing them to improve upon previous accomplishments.

Other factors in human population growth are migration and public health. Humans originated in Africa, but have since migrated to nearly all inhabitable land on the Earth. Public health, sanitation, and the use of antibiotics and vaccines have decreased the ability of infectious disease to limit human population growth. In the past, diseases such as the bubonic plaque of the fourteenth century killed between 30 and 60 percent of Europe's population and reduced the overall world population by as many as 100 million people. Today, the threat of infectious disease, while not gone, is certainly less severe. According to the Institute for Health Metrics and Evaluation (IHME) in Seattle, global death from infectious disease declined from 15.4 million in 1990 to 10.4 million in 2017. To compare to some of the epidemics of the past, the percentage of the world's population killed between 1993 and 2002 decreased from 0.30 percent of the world's population to 0.14 percent. Thus, infectious disease influence on human population growth is becoming less significant.

Age Structure, Population Growth, and Economic Development

The age structure of a population is an important factor in population dynamics. **Age structure** is the proportion of a population at different age ranges. Age structure allows better prediction of population growth, plus the ability to associate this growth with the level of economic development in the region. Countries with rapid growth have a pyramidal shape in their age structure diagrams, showing a preponderance of younger individuals, many of whom are of reproductive age or will be soon (Figure 45.16). This pattern is most often observed in underdeveloped countries where individuals do not live to old age because of less-than-optimal living conditions. Age structures of areas with slow growth, including developed countries such as the United States, still have a pyramidal structure, but with many fewer young and reproductive-aged individuals and a greater proportion of older individuals. Other developed countries, such as Italy, have zero population growth. The age structure of these populations is more conical, with an even greater percentage of middle-aged and older individuals. The actual growth rates in different countries are shown in Figure 45.17, with the highest rates tending to be in the less economically developed countries of

Africa and Asia.

VISUAL CONNECTION

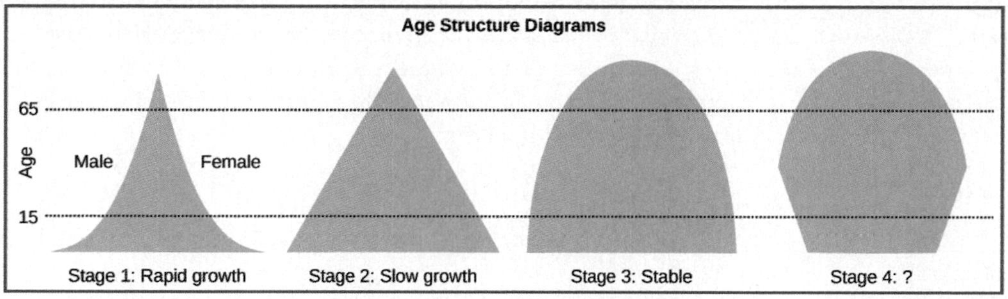

Figure 45.16 Typical age structure diagrams are shown. The rapid growth diagram narrows to a point, indicating that the number of individuals decreases rapidly with age. In the slow growth model, the number of individuals decreases steadily with age. Stable population diagrams are rounded on the top, showing that the number of individuals per age group decreases gradually, and then increases for the older part of the population.

Age structure diagrams for rapidly growing, slow growing, and stable populations are shown in stages 1 through 3. What type of population change do you think stage 4 represents?

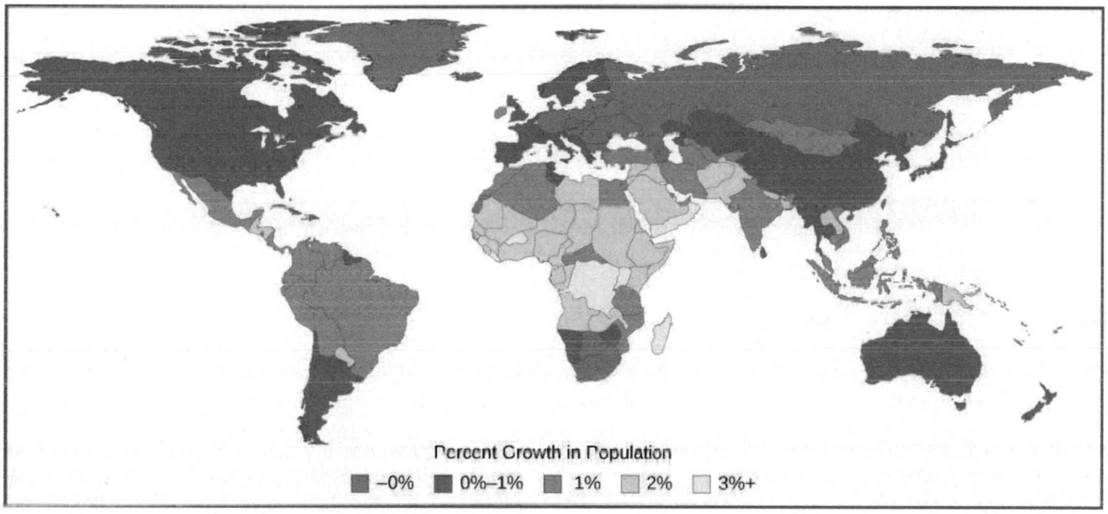

Figure 45.17 The percent growth rate of population in different countries is shown. Notice that the highest growth is occurring in less economically developed countries in Africa and Asia.

Long-Term Consequences of Exponential Human Population Growth

Many dire predictions have been made about the world's population leading to a major crisis called the "population explosion." In the 1968 book *The Population Bomb*, biologist Dr. Paul R. Ehrlich wrote, "The battle to feed all of humanity is over. In the 1970s hundreds of millions of people will starve to death in spite of any crash programs embarked upon now. At this late date nothing can prevent a substantial increase in the world death rate."[8] While many experts view this statement as incorrect based on evidence, the laws of exponential population growth are still in effect, and unchecked human population growth cannot continue indefinitely.

Several nations have instituted policies aimed at influencing population. Efforts to control population growth led to the **one-child policy** in China, which is now being phased out. India also implements national and regional populations to encourage family planning. On the other hand, Japan, Spain, Russia, Iran, and other countries have made efforts to increase population

8Paul R. Erlich, prologue to *The Population Bomb*, (1968; repr., New York: Ballantine, 1970).

growth after birth rates dipped. Such policies are controversial, and the human population continues to grow. At some point the food supply may run out, but the outcomes are difficult to predict. The United Nations estimates that future world population growth may vary from 6 billion (a decrease) to 16 billion people by the year 2100.

Another result of population growth is the endangerment of the natural environment. Many countries have attempted to reduce the human impact on climate change by reducing their emission of the greenhouse gas carbon dioxide. However, these treaties have not been ratified by every country. The role of human activity in causing climate change has become a hotly debated socio-political issue in some countries, including the United States. Thus, we enter the future with considerable uncertainty about our ability to curb human population growth and protect our environment.

LINK TO LEARNING

Visit this website (http://openstax.org/l/populations) and select "Launch movie" for an animation discussing the global impacts of human population growth.

45.6 Community Ecology

By the end of this section, you will be able to do the following:

- Discuss the predator-prey cycle
- Give examples of defenses against predation and herbivory
- Describe the competitive exclusion principle
- Give examples of symbiotic relationships between species
- Describe community structure and succession

Populations rarely, if ever, live in isolation from populations of other species. In most cases, numerous species share a habitat. The interactions between these populations play a major role in regulating population growth and abundance. All populations occupying the same habitat form a community: populations inhabiting a specific area at the same time. The number of species occupying the same habitat and their relative abundance is known as species diversity. Areas with low diversity, such as the glaciers of Antarctica, still contain a wide variety of living things, whereas the diversity of tropical rainforests is so great that it cannot be counted. Ecology is studied at the community level to understand how species interact with each other and compete for the same resources.

Predation and Herbivory

Perhaps the classical example of species interaction is predation: the consumption of prey by its predator. Nature shows on television highlight the drama of one living organism killing another. Populations of predators and prey in a community are not constant over time: in most cases, they vary in cycles that appear to be related. The most often cited example of predator-prey dynamics is seen in the cycling of the lynx (predator) and the snowshoe hare (prey), using nearly 200 year-old trapping data from North American forests (Figure 45.18). This cycle of predator and prey lasts approximately 10 years, with the predator population lagging 1–2 years behind that of the prey population. As the hare numbers increase, there is more food available for the lynx, allowing the lynx population to increase as well. When the lynx population grows to a threshold level, however, they kill so many hares that hare population begins to decline, followed by a decline in the lynx population because of scarcity of food. When the lynx population is low, the hare population size begins to increase due, at least in part, to low predation pressure, starting the cycle anew.

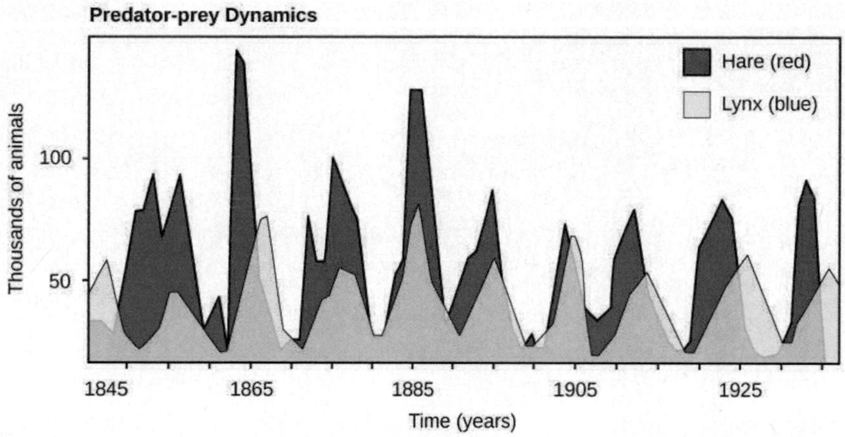

Figure 45.18 The cycling of lynx and snowshoe hare populations in Northern Ontario is an example of predator-prey dynamics.

Some researchers question the idea that predation models entirely control the population cycling of the two species. More recent studies have pointed to undefined density-dependent factors as being important in the cycling, in addition to predation. One possibility is that the cycling is inherent in the hare population due to density-dependent effects such as lower fecundity (maternal stress) caused by crowding when the hare population gets too dense. The hare cycling would then induce the cycling of the lynx because it is the lynxes' major food source. The more we study communities, the more complexities we find, allowing ecologists to derive more accurate and sophisticated models of population dynamics.

Herbivory describes the consumption of plants by insects and other animals, and it is another interspecific relationship that affects populations. Unlike animals, most plants cannot outrun predators or use mimicry to hide from hungry animals. Some plants have developed mechanisms to defend against herbivory. Other species have developed mutualistic relationships; for example, herbivory provides a mechanism of seed distribution that aids in plant reproduction.

Defense Mechanisms against Predation and Herbivory

The study of communities must consider evolutionary forces that act on the members of the various populations contained within it. Species are not static, but slowly changing and adapting to their environment by natural selection and other evolutionary forces. Species have evolved numerous mechanisms to escape predation and herbivory. These defenses may be mechanical, chemical, physical, or behavioral.

Mechanical defenses, such as the presence of thorns on plants or the hard shell on turtles, discourage animal predation and herbivory by causing physical pain to the predator or by physically preventing the predator from being able to eat the prey. Chemical defenses are produced by many animals as well as plants, such as the foxglove which is extremely toxic when eaten. Figure 45.19 shows some organisms' defenses against predation and herbivory.

Figure 45.19 The (a) honey locust tree (*Gleditsia triacanthos*) uses thorns, a mechanical defense, against herbivores, while the (b) Florida red-bellied turtle (*Pseudemys nelsoni)* uses its shell as a mechanical defense against predators. (c) Foxglove (*Digitalis* sp.) uses a chemical defense: toxins produced by the plant can cause nausea, vomiting, hallucinations, convulsions, or death when consumed. (d) The North American millipede (*Narceus americanus*) uses both mechanical and chemical defenses: when threatened, the millipede curls into a defensive ball and produces a noxious substance that irritates eyes and skin. (credit a: modification of work by Huw Williams; credit b: modification of work by "JamieS93"/Flickr; credit c: modification of work by Philip Jägenstedt; credit d: modification of work by Cory Zanker)

Many species use their body shape and coloration to avoid being detected by predators. The tropical walking stick is an insect with the coloration and body shape of a twig which makes it very hard to see when stationary against a background of real twigs (Figure 45.20a). In another example, the chameleon can, within limitations, change its color to match its surroundings (Figure 45.20b). Both of these are examples of **camouflage**, or avoiding detection by blending in with the background.

Figure 45.20 (a) The tropical walking stick and (b) the chameleon use body shape and/or coloration to prevent detection by predators. (credit a: modification of work by Linda Tanner; credit b: modification of work by Frank Vassen)

Some species use coloration as a way of warning predators that they are not good to eat. For example, the cinnabar moth caterpillar, the fire-bellied toad, and many species of beetle have bright colors that warn of a foul taste, the presence of toxic chemicals, and/or the ability to sting or bite, respectively. Predators that ignore this coloration and eat the organisms will experience their unpleasant taste or presence of toxic chemicals and learn not to eat them in the future. This type of defensive mechanism is called **aposematic coloration**, or warning coloration (Figure 45.21).

(a) (b)

Figure 45.21 (a) The strawberry poison dart frog (*Oophaga pumilio*) uses aposematic coloration to warn predators that it is toxic, while the (b) striped skunk (*Mephitis mephitis*) uses aposematic coloration to warn predators of the unpleasant odor it produces. (credit a: modification of work by Jay Iwasaki; credit b: modification of work by Dan Dzurisin)

While some predators learn to avoid eating certain potential prey because of their coloration, other species have evolved mechanisms to mimic this coloration to avoid being eaten, even though they themselves may not be unpleasant to eat or contain toxic chemicals. In **Batesian mimicry**, a harmless species imitates the warning coloration of a harmful one. Assuming they share the same predators, this coloration then protects the harmless ones, even though they do not have the same level of physical or chemical defenses against predation as the organism they mimic. Many insect species mimic the coloration of wasps or bees, which are stinging, venomous insects, thereby discouraging predation (Figure 45.22).

(a) (b)

Figure 45.22 Batesian mimicry occurs when a harmless species mimics the coloration of a harmful species, as is seen with the (a) bumblebee and (b) bee-like robber fly. (credit a, b: modification of work by Cory Zanker)

In **Müllerian mimicry**, multiple species share the same warning coloration, but all of them actually have defenses. Figure 45.23 shows a variety of foul-tasting butterflies with similar coloration. In **Emsleyan/Mertensian mimicry**, a deadly prey mimics a less dangerous one, such as the venomous coral snake mimicking the nonvenomous milk snake. This type of mimicry is extremely rare and more difficult to understand than the previous two types. For this type of mimicry to work, it is essential that eating the milk snake has unpleasant but not fatal consequences. Then, these predators learn not to eat snakes with this coloration, protecting the coral snake as well. If the snake were fatal to the predator, there would be no opportunity for the predator to learn not to eat it, and the benefit for the less toxic species would disappear.

Figure 45.23 Several unpleasant-tasting *Heliconius* butterfly species share a similar color pattern with better-tasting varieties, an example of Müllerian mimicry. (credit: Joron M, Papa R, Beltrán M, Chamberlain N, Mavárez J, et al.)

LINK TO LEARNING

Go to this website (http://openstax.org/l/find_the_mimic) to view stunning examples of mimicry.

Competitive Exclusion Principle

Resources are often limited within a habitat and multiple species may compete to obtain them. All species have an ecological niche in the ecosystem, which describes how they acquire the resources they need and how they interact with other species in the community. The **competitive exclusion principle** states that two species cannot occupy the same niche in a habitat. In other words, different species cannot coexist in a community if they are competing for all the same resources. An example of this principle is shown in Figure 45.24, with two protozoan species, *Paramecium aurelia* and *Paramecium caudatum*. When grown individually in the laboratory, they both thrive. But when they are placed together in the same test tube (habitat), *P. aurelia* outcompetes *P. caudatum* for food, leading to the latter's eventual extinction.

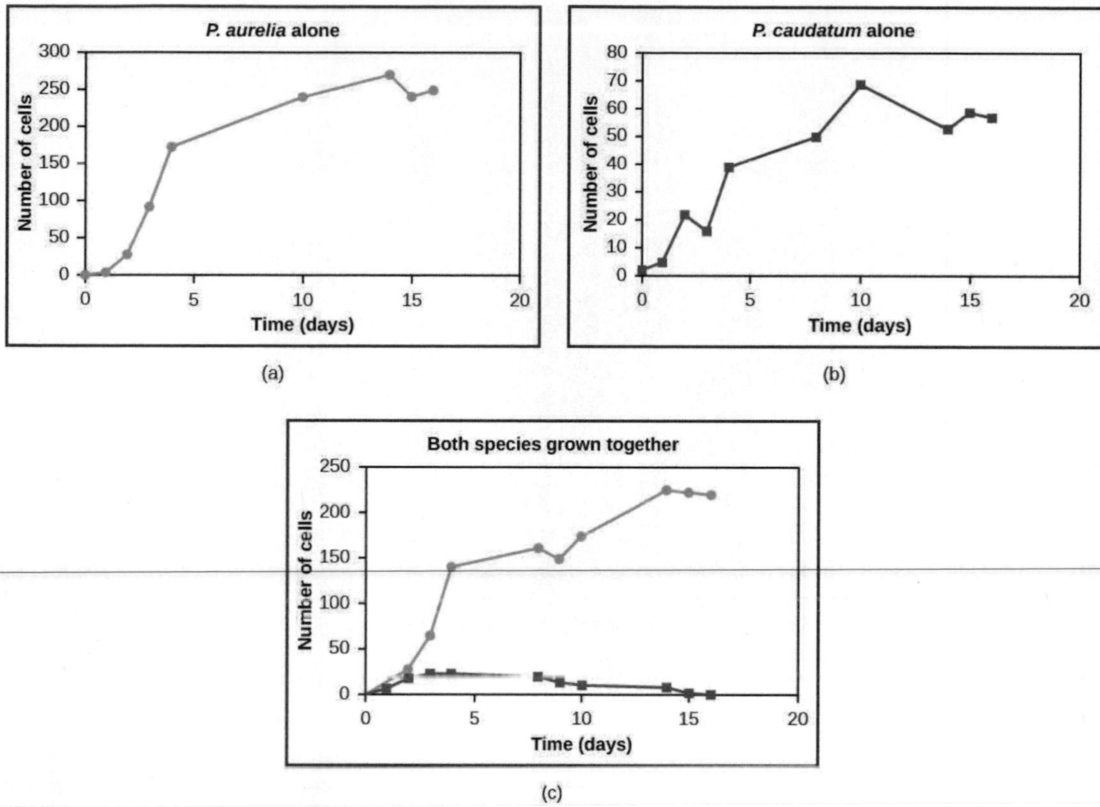

Figure 45.24 *Paramecium aurelia* and *Paramecium caudatum* grow well individually, but when they compete for the same resources, the *P. aurelia* outcompetes the *P. caudatum*.

This exclusion may be avoided if a population evolves to make use of a different resource, a different area of the habitat, or feeds during a different time of day, called resource partitioning. The two organisms are then said to occupy different microniches. These organisms coexist by minimizing direct competition.

Symbiosis

Symbiotic relationships, or **symbioses** (plural), are close interactions between individuals of different species over an extended period of time which impact the abundance and distribution of the associating populations. Most scientists accept this definition, but some restrict the term to only those species that are mutualistic, where both individuals benefit from the interaction. In this discussion, the broader definition will be used.

Commensalism

A **commensal** relationship occurs when one species benefits from the close, prolonged interaction, while the other neither benefits nor is harmed. Birds nesting in trees provide an example of a commensal relationship (Figure 45.25). The tree is not harmed by the presence of the nest among its branches. The nests are light and produce little strain on the structural integrity of the branch, and most of the leaves, which the tree uses to get energy by photosynthesis, are above the nest so they are unaffected. The bird, on the other hand, benefits greatly. If the bird had to nest in the open, its eggs and young would be vulnerable to predators. Another example of a commensal relationship is the pilot fish and the shark. The pilot fish feed on the leftovers of the host's meals, and the host is not affected in any way.

Figure 45.25 The southern masked-weaver bird is starting to make a nest in a tree in Zambezi Valley, Zambia. This is an example of a commensal relationship, in which one species (the bird) benefits, while the other (the tree) neither benefits nor is harmed. (credit: "Hanay"/Wikimedia Commons)

Mutualism

A second type of symbiotic relationship is called **mutualism**, where two species benefit from their interaction. Some scientists believe that these are the only true examples of symbiosis. For example, termites have a mutualistic relationship with protozoa that live in the insect's gut (Figure 45.26a). The termite benefits from the ability of bacterial symbionts within the protozoa to digest cellulose. The termite itself cannot do this, and without the protozoa, it would not be able to obtain energy from its food (cellulose from the wood it chews and eats). The protozoa and the bacterial symbionts benefit by having a protective environment and a constant supply of food from the wood chewing actions of the termite. Lichens have a mutualistic relationship between fungus and photosynthetic algae or bacteria (Figure 45.26b). As these symbionts grow together, the glucose produced by the algae provides nourishment for both organisms, whereas the physical structure of the lichen protects the algae from the elements and makes certain nutrients in the atmosphere more available to the algae.

Figure 45.26 (a) Termites form a mutualistic relationship with symbiotic protozoa in their guts, which allow both organisms to obtain energy from the cellulose the termite consumes. (b) Lichen is a fungus that has symbiotic photosynthetic algae living inside its cells. (credit a: modification of work by Scott Bauer, USDA; credit b: modification of work by Cory Zanker)

Parasitism

A **parasite** is an organism that lives in or on another living organism and derives nutrients from it. In this relationship, the parasite benefits, but the **host** is harmed. The host is usually weakened by the parasite as it siphons resources the host would normally use to maintain itself. The parasite, however, is unlikely to kill the host, especially not quickly, because this would allow no time for the organism to complete its reproductive cycle by spreading to another host.

The reproductive cycles of parasites are often very complex, sometimes requiring more than one host species. A tapeworm is a parasite that causes disease in humans when contaminated, undercooked meat is consumed (Figure 45.27). The tapeworm can live inside the intestine of the host for several years, benefiting from the food the host is eating, and may grow to be over 50 ft long by adding segments. The parasite moves from species to species in a cycle, making two hosts necessary to complete its life cycle.

Another common parasite is *Plasmodium falciparum*, the protozoan cause of malaria, a significant disease in many parts of the world. Living in human liver and red blood cells, the organism reproduces asexually in the gut of blood-feeding mosquitoes to complete its life cycle. Thus malaria is spread from human to human by mosquitoes, one of many arthropod-borne infectious diseases.

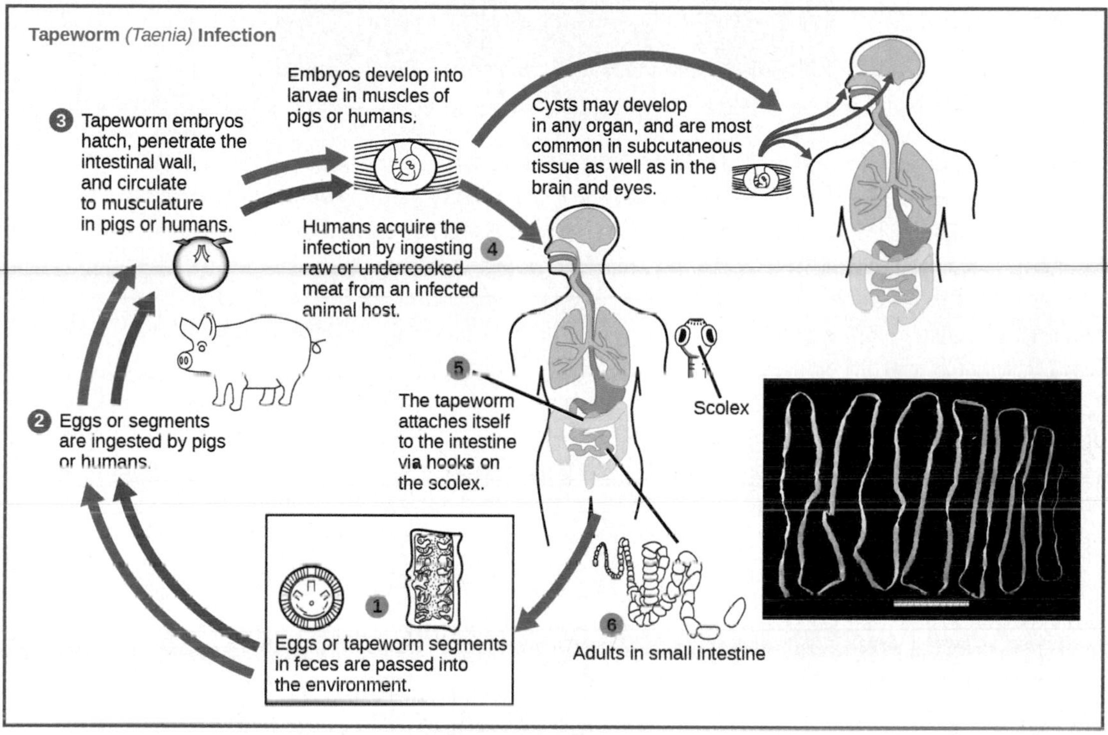

Figure 45.27 This diagram shows the life cycle of a pork tapeworm (*Taenia solium*), a human worm parasite. (credit: modification of work by CDC)

Characteristics of Communities

Communities are complex entities that can be characterized by their structure (the types and numbers of species present) and dynamics (how communities change over time). Understanding community structure and dynamics enables community ecologists to manage ecosystems more effectively.

Foundation Species

Foundation species are considered the "base" or "bedrock" of a community, having the greatest influence on its overall structure. They are usually the primary producers: organisms that bring most of the energy into the community. Kelp, or brown algae, is a foundation species, forming the basis of the kelp forests off the coast of California.

Foundation species may physically modify the environment to produce and maintain habitats that benefit the other organisms that use them. An example is the photosynthetic corals of the coral reef (Figure 45.28). Corals themselves are not photosynthetic, but harbor symbionts within their body tissues (dinoflagellates called zooxanthellae) that perform photosynthesis; this is another example of a mutualism. The exoskeletons of living and dead coral make up most of the reef structure, which protects many other species from waves and ocean currents.

Figure 45.28 Coral is the foundation species of coral reef ecosystems. (credit: Jim E. Maragos, USFWS)

Biodiversity, Species Richness, and Relative Species Abundance

Biodiversity describes a community's biological complexity: it is measured by the number of different species (species richness) in a particular area and their relative abundance (species evenness). The area in question could be a habitat, a biome, or the entire biosphere. **Species richness** is the term that is used to describe the number of species living in a habitat or biome. Species richness varies across the globe (Figure 45.29). One factor in determining species richness is latitude, with the greatest species richness occurring in ecosystems near the equator, which often have warmer temperatures, large amounts of rainfall, and low seasonality. The lowest species richness occurs near the poles, which are much colder, drier, and thus less conducive to life in Geologic time (time since glaciations). The predictability of climate or productivity is also an important factor. Other factors influence species richness as well. For example, the study of **island biogeography** attempts to explain the relatively high species richness found in certain isolated island chains, including the Galápagos Islands that inspired the young Darwin. **Relative species abundance** is the number of individuals in a species relative to the total number of individuals in all species within a habitat, ecosystem, or biome. Foundation species often have the highest relative abundance of species.

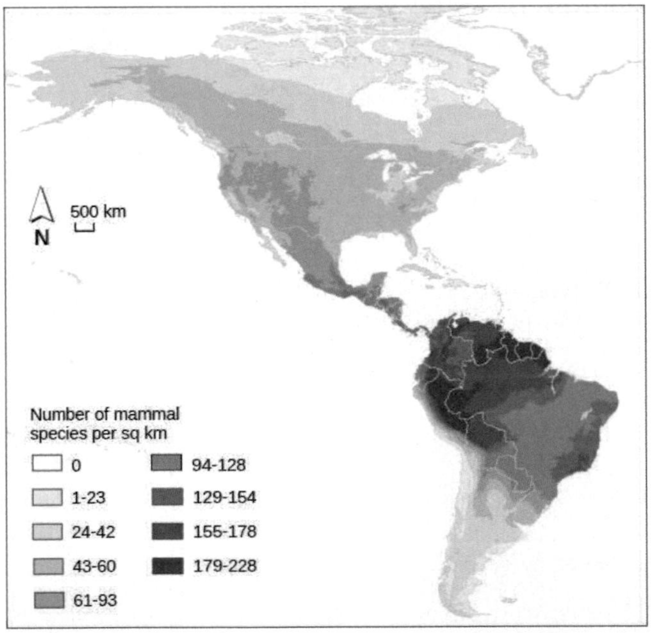

Figure 45.29 The greatest species richness for mammals in North and South America is associated with the equatorial latitudes. (credit: modification of work by NASA, CIESIN, Columbia University)

Keystone Species

A **keystone species** is one whose presence is key to maintaining biodiversity within an ecosystem and to upholding an ecological community's structure. The intertidal sea star, *Pisaster ochraceus*, of the northwestern United States is a keystone species (Figure 45.30). Studies have shown that when this organism is removed from communities, populations of their natural prey (mussels) increase, completely altering the species composition and reducing biodiversity. Another keystone species is the banded tetra, a fish in tropical streams, which supplies nearly all of the phosphorus, a necessary inorganic nutrient, to the rest of the community. If these fish were to become extinct, the community would be greatly affected.

Figure 45.30 The *Pisaster ochraceus* sea star is a keystone species. (credit: Jerry Kirkhart)

Everyday Connection

Invasive Species

Invasive species are nonnative organisms that, when introduced to an area out of their native range, threaten the ecosystem balance of that habitat. Many such species exist in the United States, as shown in Figure 45.31. Whether enjoying a forest hike, taking a summer boat trip, or simply walking down an urban street, you have likely encountered an invasive species.

Figure 45.31 In the United States, invasive species like (a) purple loosestrife (*Lythrum salicaria*) and the (b) zebra mussel (*Dreissena polymorpha*) threaten certain aquatic ecosystems. Some forests are threatened by the spread of (c) common buckthorn (*Rhamnus cathartica*), (d) garlic mustard (*Alliaria petiolata*), and (e) the emerald ash borer (*Agrilus planipennis*). The (f) European starling (*Sturnus vulgaris*) may compete with native bird species for nest holes. (credit a: modification of work by Liz West; credit b: modification of work by M. McCormick, NOAA; credit c: modification of work by E. Dronkert; credit d: modification of work by Dan Davison; credit e: modification of work by USDA; credit f: modification of work by Don DeBold)

One of the many recent proliferations of an invasive species concerns the growth of Asian carp populations. Asian carp were introduced to the United States in the 1970s by fisheries and sewage treatment facilities that used the fish's excellent filter feeding capabilities to clean their ponds of excess plankton. Some of the fish escaped, however, and by the 1980s they had colonized many waterways of the Mississippi River basin, including the Illinois and Missouri Rivers.

Voracious eaters and rapid reproducers, Asian carp may outcompete native species for food, potentially leading to their extinction. For example, black carp are voracious eaters of native mussels and snails, limiting this food source for native fish species. Silver carp eat plankton that native mussels and snails feed on, reducing this food source by a different alteration of the food web. In some areas of the Mississippi River, Asian carp species have become the most predominant, effectively outcompeting native fishes for habitat. In some parts of the Illinois River, Asian carp constitute 95 percent of the community's biomass. Although edible, the fish is bony and not a desired food in the United States. Moreover, their presence threatens the native fish and fisheries of the Great Lakes, which are important to local economies and recreational anglers. Asian carp have even injured humans. The fish, frightened by the sound of approaching motorboats, thrust themselves into the air, often landing in the boat or directly hitting the boaters.

The Great Lakes and their prized salmon and lake trout fisheries are also being threatened by these invasive fish. Asian carp have already colonized rivers and canals that lead into Lake Michigan. One infested waterway of particular importance is the Chicago Sanitary and Ship Channel, the major supply waterway linking the Great Lakes to the Mississippi River. To prevent the Asian carp from leaving the canal, a series of electric barriers have been successfully used to discourage their migration; however, the threat is significant enough that several states and Canada have sued to have the Chicago channel permanently cut off from Lake Michigan. Local and national politicians have weighed in on how to solve the problem, but no one knows whether the Asian carp will ultimately be considered a nuisance, like other invasive species such as the water hyacinth and zebra mussel, or whether it will be the destroyer of the largest freshwater fishery of the world.

The issues associated with Asian carp show how population and community ecology, fisheries management, and politics intersect on issues of vital importance to the human food supply and economy. Socio-political issues like this make extensive use of the sciences of population ecology (the study of members of a particular species occupying a particular area known as a habitat) and community ecology (the study of the interaction of all species within a habitat).

Community Dynamics

Community dynamics are the changes in community structure and composition over time. Sometimes these changes are induced by **environmental disturbances** such as volcanoes, earthquakes, storms, fires, and climate change. Communities with a stable structure are said to be at equilibrium. Following a disturbance, the community may or may not return to the equilibrium state.

Succession describes the sequential appearance and disappearance of species in a community over time. In **primary succession**, newly exposed or newly formed land is colonized by living things; in **secondary succession**, part of an ecosystem is disturbed and remnants of the previous community remain.

Primary Succession and Pioneer Species

Primary succession occurs when new land is formed or rock is exposed: for example, following the eruption of volcanoes, such as those on the Big Island of Hawaii. As lava flows into the ocean, new land is continually being formed. On the Big Island, approximately 32 acres of land is added each year. First, weathering and other natural forces break down the substrate enough for the establishment of certain hearty plants and lichens with few soil requirements, known as **pioneer species** (Figure 45.32). These species help to further break down the mineral rich lava into soil where other, less hardy species will grow and eventually replace the pioneer species. In addition, as these early species grow and die, they add to an ever-growing layer of decomposing organic material and contribute to soil formation. Over time the area will reach an equilibrium state, with a set of organisms quite different from the pioneer species.

Figure 45.32 During primary succession in lava on Maui, Hawaii, succulent plants are the pioneer species. (credit: Forest and Kim Starr)

Secondary succession

A classic example of secondary succession occurs in oak and hickory forests cleared by wildfire (Figure 45.33). Wildfires will burn most vegetation and kill those animals unable to flee the area. Their nutrients, however, are returned to the ground in the form of ash. Thus, even when areas are devoid of life due to severe fires, the area will soon be ready for new life to take hold.

Before the fire, the vegetation was dominated by tall trees with access to the major plant energy resource: sunlight. Their height gave them access to sunlight while also shading the ground and other low-lying species. After the fire, though, these trees are no longer dominant. Thus, the first plants to grow back are usually annual plants followed within a few years by quickly growing and spreading grasses and other pioneer species. Due to, at least in part, changes in the environment brought on by the growth of the grasses and other species, over many years, shrubs will emerge along with small pine, oak, and hickory trees. These organisms are called intermediate species. Eventually, over 150 years, the forest will reach its equilibrium point where species composition is no longer changing and resembles the community before the fire. This equilibrium state is referred to as the **climax community**, which will remain stable until the next disturbance.

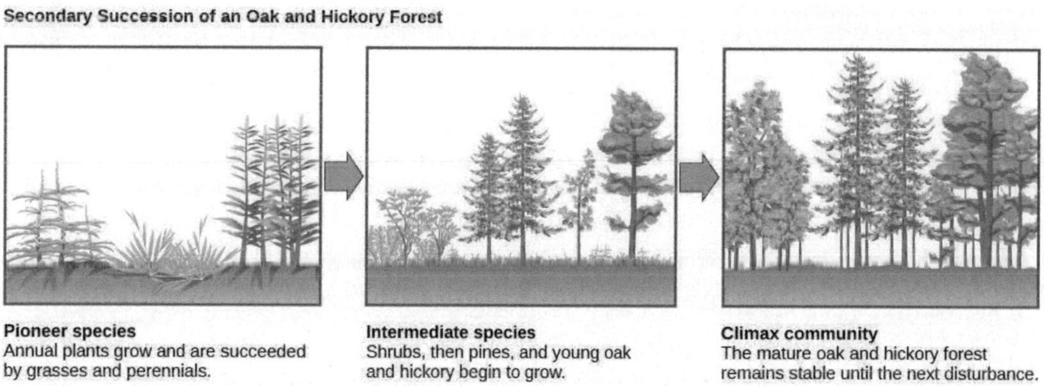

Figure 45.33 Secondary succession is shown in an oak and hickory forest after a forest fire.

45.7 Behavioral Biology: Proximate and Ultimate Causes of Behavior

By the end of this section, you will be able to do the following:

- Compare innate and learned behavior
- Discuss how movement and migration behaviors are a result of natural selection
- Discuss the different ways members of a population communicate with each other
- Give examples of how species use energy for mating displays and other courtship behaviors
- Differentiate between various mating systems
- Describe different ways that species learn

Behavior is the change in activity of an organism in response to a stimulus. **Behavioral biology** is the study of the biological and evolutionary bases for such changes. The idea that behaviors evolved as a result of the pressures of natural selection is not new.

For decades, several types of scientists have studied animal behavior. Biologists do so in the science of **ethology**; psychologists in the science of comparative psychology; and other scientists in the science of neurobiology. The first two, ethology and comparative psychology, are the most consequential for the study of behavioral biology.

One goal of behavioral biology is to distinguish between the **innate behaviors**, which have a strong genetic component and are largely independent of environmental influences, from the **learned behaviors**, which result from environmental conditioning. Innate behavior, or instinct, is important because there is no risk of an incorrect behavior being learned. They are "hard wired" into the system. On the other hand, learned behaviors, although riskier, are flexible, dynamic, and can be altered according to changes in the environment.

Innate Behaviors: Movement and Migration

Innate or instinctual behaviors rely on response to stimuli. The simplest example of this is a **reflex action**, an involuntary and rapid response to stimulus. To test the "knee-jerk" reflex, a doctor taps the patellar tendon below the kneecap with a rubber hammer. The stimulation of the nerves leads to the reflex of extending the leg at the knee. This is similar to the reaction of someone who touches a hot stove and instinctually pulls his or her hand away. Even humans, with our great capacity to learn, still exhibit a variety of innate behaviors.

Kinesis and Taxis

Another activity or movement of innate behavior is **kinesis**, or the undirected movement in response to a stimulus. Orthokinesis is the increased or decreased speed of movement of an organism in response to a stimulus. Woodlice, for example, increase their speed of movement when exposed to high or low temperatures. This movement, although random, increases the probability that the insect spends less time in the unfavorable environment. Another example is klinokinesis, an increase in turning behaviors. It is exhibited by bacteria such as *E. coli* which, in association with orthokinesis, helps the organisms randomly find a more hospitable environment.

A similar, but more directed version of kinesis is **taxis**: the directed movement towards or away from a stimulus. This movement can be in response to light (phototaxis), chemical signals (chemotaxis), or gravity (geotaxis) and can be directed toward (positive) or away (negative) from the source of the stimulus. An example of a positive chemotaxis is exhibited by the unicellular protozoan *Tetrahymena thermophila*. This organism swims using its cilia, at times moving in a straight line, and at other times making turns. The attracting chemotactic agent alters the frequency of turning as the organism moves directly toward the source, following the increasing concentration gradient.

Fixed Action Patterns

A **fixed action pattern** is a series of movements elicited by a stimulus such that even when the stimulus is removed, the pattern goes on to completion. An example of such a behavior occurs in the three-spined stickleback, a small freshwater fish (Figure 45.34). Males of this species develop a red belly during breeding season and show instinctual aggressiveness to other males during this time. In laboratory experiments, researchers exposed such fish to objects that in no way resemble a fish in their shape, but which were painted red on their lower halves. The male sticklebacks responded aggressively to the objects just as if they were real male sticklebacks.

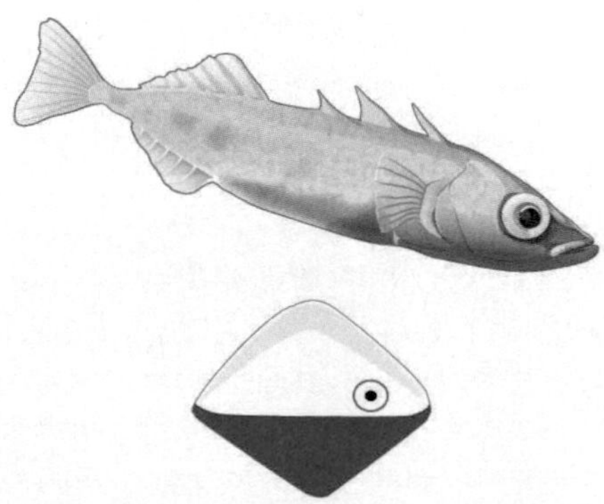

Figure 45.34 Male three-spined stickleback fish exhibit a fixed action pattern. During mating season, the males, which develop a bright red belly, react strongly to red-bottomed objects that in no way resemble fish.

Migration

Migration is the long-range seasonal movement of animals. It is an evolved, adapted response to variation in resource availability, and it is a common phenomenon found in all major groups of animals. Birds fly south for the winter to get to warmer climates with sufficient food, and salmon migrate to their spawning grounds. The popular 2005 documentary *March of the Penguins* followed the 62-mile migration of emperor penguins through Antarctica to bring food back to their breeding site and to their young. Wildebeests (Figure 45.35) migrate over 1800 miles each year in search of new grasslands.

Figure 45.35 Wildebeests migrate in a clockwise fashion over 1800 miles each year in search of rain-ripened grass. (credit: Eric Inafuku)

Although migration is thought of as innate behavior, only some migrating species always migrate (obligate migration). Animals that exhibit facultative migration can choose to migrate or not. Additionally, in some animals, only a portion of the population migrates, whereas the rest does not migrate (incomplete migration). For example, owls that live in the tundra may migrate in years when their food source, small rodents, is relatively scarce, but not migrate during the years when rodents are plentiful.

Foraging

Foraging is the act of searching for and exploiting food resources. Feeding behaviors that maximize energy gain and minimize energy expenditure are called optimal foraging behaviors, and these are favored by natural section. The painted stork, for example, uses its long beak to search the bottom of a freshwater marshland for crabs and other food (Figure 45.36).

Figure 45.36 The painted stork uses its long beak to forage. (credit: J.M. Garg)

Innate Behaviors: Living in Groups

Not all animals live in groups, but even those that live relatively solitary lives, with the exception of those that can reproduce asexually, must mate. Mating usually involves one animal signaling another so as to communicate the desire to mate. There are several types of energy-intensive behaviors or displays associated with mating, called mating rituals. Other behaviors found in populations that live in groups are described in terms of which animal benefits from the behavior. In selfish behavior, only the animal in question benefits; in altruistic behavior, one animal's actions benefit another animal; cooperative behavior describes when both animals benefit. All of these behaviors involve some sort of communication between population members.

Communication within a Species

Animals communicate with each other using stimuli known as **signals**. An example of this is seen in the three-spined stickleback, where the visual signal of a red region in the lower half of a fish signals males to become aggressive and signals females to mate. Other signals are chemical (pheromones), aural (sound), visual (courtship and aggressive displays), or tactile (touch). These types of communication may be instinctual or learned or a combination of both. These are not the same as the communication we associate with language, which has been observed only in humans and perhaps in some species of primates and cetaceans.

A pheromone is a secreted chemical signal used to obtain a response from another individual of the same species. The purpose of pheromones is to elicit a specific behavior from the receiving individual. Pheromones are especially common among social insects, but they are used by many species to attract the opposite sex, to sound alarms, to mark food trails, and to elicit other, more complex behaviors. Even humans are thought to respond to certain pheromones called axillary steroids. These chemicals influence human perception of other people, and in one study were responsible for a group of women synchronizing their menstrual cycles. The role of pheromones in human-to-human communication is not fully understood and continues to be researched.

Songs are an example of an aural signal, one that needs to be heard by the recipient. Perhaps the best known of these are songs of birds, which identify the species and are used to attract mates. Other well-known songs are those of whales, which are of such low frequency that they can travel long distances underwater. Dolphin species communicate with each other (and occasionally even with other species of dolphins) using a wide variety of vocalizations. Male crickets make chirping sounds using a specialized organ to attract a mate, repel other males, and to announce a successful mating.

Courtship displays are a series of ritualized visual behaviors (signals) designed to attract and convince a member of the opposite sex to mate. These displays are ubiquitous in the animal kingdom. Often these displays involve a series of steps, including an initial display by one member followed by a response from the other. If at any point, the display is performed incorrectly or a proper response is not given, the mating ritual is abandoned and the mating attempt will be unsuccessful. The mating display of the common stork is shown in Figure 45.37.

Aggressive displays are also common in the animal kingdom. For example, a dog bares its teeth when it wants another dog to back down. Presumably, these displays communicate not only the willingness of the animal to fight, but also its fighting ability. Although these displays do signal aggression on the part of the sender, it is thought that these displays are actually a mechanism to reduce the amount of actual fighting that occurs between members of the same species: they allow individuals to assess the fighting ability of their opponent and thus decide whether it is "worth the fight." The testing of certain hypotheses using game theory has led to the conclusion that some of these displays may overstate an animal's actual fighting ability and are used to "bluff" the opponent. This type of interaction, even if "dishonest," would be favored by natural selection if it is successful more times than not.

Figure 45.37 This stork's courtship display is designed to attract potential mates. (credit: Linda "jinterwas"/Flickr)

Distraction displays are seen in birds and some fish. They are designed to attract a predator away from the nest. This is an example of an altruistic behavior: it benefits the young more than the individual performing the display, which is putting itself at risk by doing so.

Many animals, especially primates, communicate with other members in the group through touch. Activities such as grooming, touching the shoulder or root of the tail, embracing, lip contact, and greeting ceremonies have all been observed in the Indian langur, an Old World monkey. Similar behaviors are found in other primates, especially in the great apes.

LINK TO LEARNING

The killdeer bird distracts predators from its eggs by faking a broken wing display in this video taken in Boise, Idaho.

Click to view content (https://www.openstax.org/l/killdeer_bird)

Altruistic Behaviors

Behaviors that lower the fitness of the individual but increase the fitness of another individual are termed altruistic. Examples of such behaviors are seen widely across the animal kingdom. Social insects such as worker bees have no ability to reproduce, yet they maintain the queen so she can populate the hive with her offspring. Meerkats keep a sentry standing guard to warn the rest of the colony about intruders, even though the sentry is putting itself at risk. Wolves and wild dogs bring meat to pack members not present during a hunt. Lemurs take care of infants unrelated to them. Although on the surface, these behaviors appear to be altruistic, the truth may not be so simple.

There has been much discussion over why altruistic behaviors exist. Do these behaviors lead to overall evolutionary advantages for their species? Do they help the altruistic individual pass on its own genes? And what about such activities between unrelated individuals? One explanation for altruistic-type behaviors is found in the genetics of natural selection. In the 1976 book, *The Selfish Gene,* scientist Richard Dawkins attempted to explain many seemingly altruistic behaviors from the viewpoint of the gene itself. Although a gene obviously cannot be selfish in the human sense, it may appear that way if the sacrifice of an individual benefits related individuals that share genes that are identical by descent (present in relatives because of common lineage). Mammal parents make this sacrifice to take care of their offspring. Emperor penguins migrate miles in harsh conditions to bring food back for their young. Selfish gene theory has been controversial over the years and is still discussed among scientists in related fields.

Even less-related individuals, those with less genetic identity than that shared by parent and offspring, benefit from seemingly altruistic behavior. The activities of social insects such as bees, wasps, ants, and termites are good examples. Sterile workers in these societies take care of the queen because they are closely related to it, and as the queen has offspring, she is passing on genes from the workers indirectly. Thus, it is of fitness benefit for the worker to maintain the queen without having any direct chance of passing on its genes due to its sterility. The lowering of individual fitness to enhance the reproductive fitness of a relative and thus one's inclusive fitness evolves through **kin selection**. This phenomenon can explain many superficially altruistic behaviors seen in animals. However, these behaviors may not be truly defined as altruism in these cases because the actor is actually increasing its own fitness either directly (through its own offspring) or indirectly (through the inclusive fitness it gains through relatives that share genes with it).

Unrelated individuals may also act altruistically to each other, and this seems to defy the "selfish gene" explanation. An example of this observed in many monkey species where a monkey will present its back to an unrelated monkey to have that individual pick the parasites from its fur. After a certain amount of time, the roles are reversed and the first monkey now grooms the second monkey. Thus, there is reciprocity in the behavior. Both benefit from the interaction and their fitness is raised more than if neither cooperated nor if one cooperated and the other did not cooperate. This behavior is still not necessarily altruism, as the "giving" behavior of the actor is based on the expectation that it will be the "receiver" of the behavior in the future, termed reciprocal altruism. Reciprocal altruism requires that individuals repeatedly encounter each other, often the result of living in the same social group, and that cheaters (those that never "give back") are punished.

Evolutionary game theory, a modification of classical game theory in mathematics, has shown that many of these so-called "altruistic behaviors" are not altruistic at all. The definition of "pure" altruism, based on human behavior, is an action that benefits another without any direct benefit to oneself. Most of the behaviors previously described do not seem to satisfy this definition, and game theorists are good at finding "selfish" components in them. Others have argued that the terms "selfish" and "altruistic" should be dropped completely when discussing animal behavior, as they describe human behavior and may not be directly applicable to instinctual animal activity. What is clear, though, is that heritable behaviors that improve the chances of passing on one's genes or a portion of one's genes are favored by natural selection and will be retained in future generations as long as those behaviors convey a fitness advantage. These instinctual behaviors may then be applied, in special circumstances, to other species, as long as it doesn't lower the animal's fitness.

Finding Sex Partners

Not all animals reproduce sexually, but many that do have the same challenge: they need to find a suitable mate and often have to compete with other individuals to obtain one. Significant energy is spent in the process of locating, attracting, and mating with the sex partner. Two types of selection occur during this process: **intersexual selection**, where individuals of one sex choose mates of the other sex, and **intrasexual selection**, the competition for mates between species members of the same sex. Intersexual selection is often complex because choosing a mate may be based on a variety of visual, aural, tactile, and chemical cues. An example of intersexual selection is when female peacocks choose to mate with the male with the brightest plumage. This type of selection often leads to traits in the chosen sex that do not enhance survival, but are those traits most attractive to the opposite sex (often at the expense of survival). Intrasexual selection involves mating displays and aggressive mating rituals such as rams butting heads—the winner of these battles is the one that is able to mate. Many of these rituals use up considerable energy but result in the selection of the healthiest, strongest, and/or most dominant individuals for mating.

Three general mating systems, all involving innate as opposed to learned behaviors, are seen in animal populations: monogamous, polygynous, and polyandrous.

LINK TO LEARNING

Visit this website (http://openstax.org/l/sex_selection) for informative videos on sexual selection.

In **monogamous** systems, one male and one female are paired for at least one breeding season. In some animals, such as the gray wolf, these associations can last much longer, even a lifetime. Several theories may explain this type of mating system. The "mate-guarding hypothesis" states that males stay with the female to prevent other males from mating with her. This behavior is advantageous in such situations where mates are scarce and difficult to find. Another explanation is the "male-assistance hypothesis," where males that help guard and rear their young will have more and healthier offspring. Monogamy is observed in many bird populations where, in addition to the parental care from the female, the male is also a major provider of parental care for the chicks. A third explanation for the evolutionary advantages of monogamy is the "female-enforcement hypothesis." In this scenario, the female ensures that the male does not have other offspring that might compete with her own, so she actively interferes with the male's signaling to attract other mates.

Polygynous mating refers to one male mating with multiple females. In these situations, the female must be responsible for most of the parental care as the single male is not capable of providing care to that many offspring. In resourced-based polygyny, males compete for territories with the best resources, and then mate with females that enter the territory, drawn to its resource richness. The female benefits by mating with a dominant, genetically fit male; however, it is at the cost of having no male help in caring for the offspring. An example is seen in the yellow-rumped honeyguide, a bird whose males defend beehives because the females feed on their wax. As the females approach, the male defending the nest will mate with them. Harem mating structures are a type of polygynous system where certain males dominate mating while controlling a territory with resources. Harem mating occurs in elephant seals, where the alpha male dominates the mating within the group. A third type of polygyny is a lek system. Here there is a communal courting area where several males perform elaborate displays for females, and the females choose their mate from this group. This behavior is observed in several bird species including the sage grouse and the prairie chicken.

In **polyandrous** mating systems, one female mates with many males. These types of systems are much rarer than monogamous and polygynous mating systems. In pipefishes and seahorses, males receive the eggs from the female, fertilize them, protect them within a pouch, and give birth to the offspring (Figure 45.38). Therefore, the female is able to provide eggs to several males without the burden of carrying the fertilized eggs.

(a) (b)

Figure 45.38 Polyandrous mating, in which one female mates with many males, occurs in the (a) seahorse and the (b) pipefish. (credit a: modification of work by Brian Gratwicke; credit b: modification of work by Stephen Childs)

Simple Learned Behaviors

The majority of the behaviors previously discussed were innate or at least have an innate component (variations on the innate behaviors may be learned). They are inherited and the behaviors do not change in response to signals from the environment. Conversely, learned behaviors, even though they may have instinctive components, allow an organism to adapt to changes in the environment and are modified by previous experiences. Simple learned behaviors include habituation and imprinting—both are important to the maturation process of young animals.

Habituation

Habituation is a simple form of learning in which an animal stops responding to a stimulus after a period of repeated exposure. This is a form of non-associative learning, as the stimulus is not associated with any punishment or reward. Prairie dogs typically sound an alarm call when threatened by a predator, but they become habituated to the sound of human footsteps when no harm is associated with this sound, therefore, they no longer respond to them with an alarm call. In this example, habituation is specific to the sound of human footsteps, as the animals still respond to the sounds of potential predators.

Imprinting

Imprinting is a type of learning that occurs at a particular age or a life stage that is rapid and independent of the species involved. Hatchling ducks recognize the first adult they see, their mother, and make a bond with her. A familiar sight is ducklings walking or swimming after their mothers (Figure 45.39). This is another type of non-associative learning, but is very important in the maturation process of these animals as it encourages them to stay near their mother so they will be protected, greatly increasing their chances of survival. However, if newborn ducks see a human before they see their mother, they will imprint on the human and follow it in just the same manner as they would follow their real mother.

Figure 45.39 The attachment of ducklings to their mother is an example of imprinting. (credit: modification of work by Mark Harkin)

LINK TO LEARNING

The International Crane Foundation has helped raise the world's population of whooping cranes from 21 individuals to about 600. Imprinting hatchlings has been a key to success: biologists wear full crane costumes so the birds never "see" humans. Watch this video to learn more.

Click to view content (https://www.openstax.org/l/whooping_crane)

Conditioned Behavior

Conditioned behaviors are types of associative learning, where a stimulus becomes associated with a consequence. During operant conditioning, the behavioral response is modified by its consequences, with regards to its form, strength, or frequency.

Classical Conditioning

In **classical conditioning**, a response called the conditioned response is associated with a stimulus that it had previously not been associated with, the conditioned stimulus. The response to the original, unconditioned stimulus is called the unconditioned response. The most cited example of classical conditioning is Ivan Pavlov's experiments with dogs (Figure 45.40). In Pavlov's experiments, the unconditioned response was the salivation of dogs in response to the unconditioned stimulus of seeing or smelling their food. The conditioning stimulus that researchers associated with the unconditioned response was the ringing of a bell. During conditioning, every time the animal was given food, the bell was rung. This was repeated during several trials. After some time, the dog learned to associate the ringing of the bell with food and to respond by salivating. After the conditioning period was finished, the dog would respond by salivating when the bell was rung, even when the unconditioned stimulus, the food, was absent. Thus, the ringing of the bell became the conditioned stimulus and the salivation became the conditioned response. Although it is thought by some scientists that the unconditioned and conditioned responses are identical, even Pavlov discovered that the saliva in the conditioned dogs had characteristic differences when compared to the unconditioned dog.

Figure 45.40 In the classic Pavlovian response, the dog becomes conditioned to associate the ringing of the bell with food.

It had been thought by some scientists that this type of conditioning required multiple exposures to the paired stimulus and response, but it is now known that this is not necessary in all cases, and that some conditioning can be learned in a single pairing experiment. Classical conditioning is a major tenet of behaviorism, a branch of psychological philosophy that proposes that all actions, thoughts, and emotions of living things are behaviors that can be treated by behavior modification and changes in the environment.

Operant Conditioning

In **operant conditioning**, the conditioned behavior is gradually modified by its consequences as the animal responds to the stimulus. A major proponent of such conditioning was psychologist B.F. Skinner, the inventor of the Skinner box. Skinner put rats in his boxes that contained a lever that would dispense food to the rat when depressed. While initially the rat would push the lever a few times by accident, it eventually associated pushing the lever with getting the food. This type of learning is an example of operant conditioning. Operant learning is the basis of most animal training. The conditioned behavior is continually modified by positive or negative reinforcement, often a reward such as food or some type of punishment, respectively. In this way, the animal is conditioned to associate a type of behavior with the punishment or reward, and, over time, can be induced to perform behaviors that they would not have done in the wild, such as the "tricks" dolphins perform at marine amusement park shows (Figure 45.41).

Figure 45.41 The training of dolphins by rewarding them with food is an example of positive reinforcement operant conditioning. (credit: Roland Tanglao)

Cognitive Learning

Classical and operant conditioning are inefficient ways for humans and other intelligent animals to learn. Some primates, including humans, are able to learn by imitating the behavior of others and by taking instructions. The development of complex language by humans has made **cognitive learning**, the manipulation of information using the mind, the most prominent method of human learning. In fact, that is how students are learning right now by reading this book. As students read, they can make mental images of objects or organisms and imagine changes to them, or behaviors by them, and anticipate the consequences. In addition to visual processing, cognitive learning is also enhanced by remembering past experiences, touching physical objects, hearing sounds, tasting food, and a variety of other sensory-based inputs. Cognitive learning is so powerful that it can be used to understand conditioning in detail. In the reverse scenario, conditioning cannot help someone learn about cognition.

Classic work on cognitive learning was done by Wolfgang Köhler with chimpanzees. He demonstrated that these animals were capable of abstract thought by showing that they could learn how to solve a puzzle. When a banana was hung in their cage too high for them to reach, and several boxes were placed randomly on the floor, some of the chimps were able to stack the boxes one on top of the other, climb on top of them, and get the banana. This implies that they could visualize the result of stacking the boxes even before they had performed the action. This type of learning is much more powerful and versatile than conditioning.

Cognitive learning is not limited to primates, although they are the most efficient in using it. Maze running experiments done with rats by H.C. Blodgett in the 1920s were the first to show cognitive skills in a simple mammal. The motivation for the animals to work their way through the maze was a piece of food at its end. In these studies, the animals in Group I were run in one trial per day and had food available to them each day on completion of the run (Figure 45.42). Group II rats were not fed in the maze for the first six days and then subsequent runs were done with food for several days after. Group III rats had food available on the third day and every day thereafter. The results were that the control rats, Group I, learned quickly, and figured out how to run the maze in seven days. Group III did not learn much during the three days without food, but rapidly caught up to the control group when given the food reward. Group II learned very slowly for the six days with no reward to motivate them, and they did not begin to catch up to the control group until the day food was given, and then it took two days longer to learn the maze.

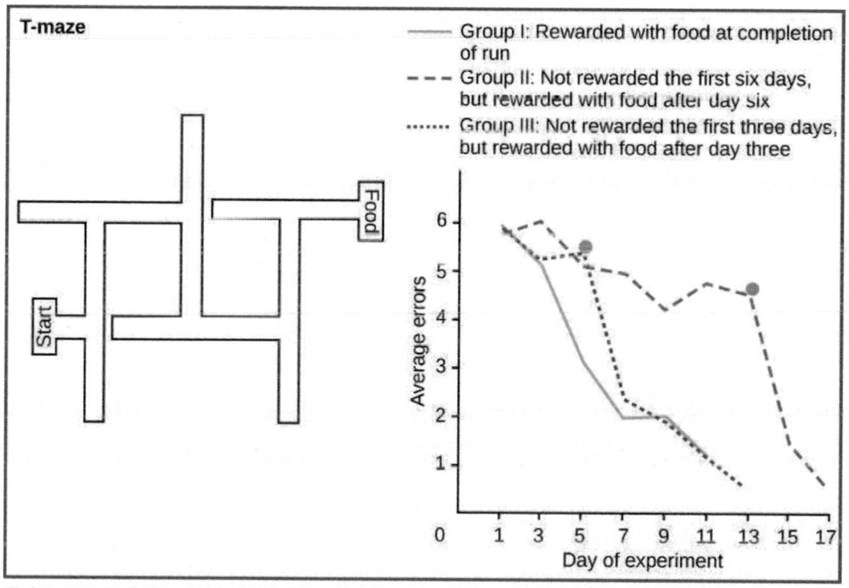

Redrawn after H. C. Blodgett, The effect of the introduction of reward upon the maze performance of rats. Univ. Calif. Publ. Psychol., 1929, 4, No. 8, pages 117 and 120.

Figure 45.42 Group I (the green solid line) found food at the end of each trial, group II (the blue dashed line) did not find food for the first 6 days, and group III (the red dotted line) did not find food during runs on the first three days. Notice that rats given food earlier learned faster and eventually caught up to the control group. The orange dots on the group II and III lines show the days when food rewards were added to the mazes.

It may not be immediately obvious that this type of learning is different than conditioning. Although one might be tempted to believe that the rats simply learned how to find their way through a conditioned series of right and left turns, E.C. Tolman proved a decade later that the rats were making a representation of the maze in their minds, which he called a "cognitive map." This was an early demonstration of the power of cognitive learning and how these abilities were not just limited to humans.

Sociobiology

Sociobiology is an interdisciplinary science originally popularized by social insect researcher E.O. Wilson in the 1970s. Wilson defined the science as "the extension of population biology and evolutionary theory to social organization."[9] The main thrust of sociobiology is that animal and human behavior, including aggressiveness and other social interactions, can be explained almost solely in terms of genetics and natural selection. This science is controversial; noted scientists such as the late Stephen Jay Gould criticized the approach for ignoring the environmental effects on behavior. This is another example of the "nature versus nurture" debate of the role of genetics versus the role of environment in determining an organism's characteristics.

Sociobiology also links genes with behaviors and has been associated with "biological determinism," the belief that all behaviors are hardwired into our genes. No one disputes that certain behaviors can be inherited and that natural selection plays a role retaining them. It is the application of such principles to human behavior that sparks this controversy, which remains active today.

9Edward O. Wilson. *On Human Nature* (1978; repr., Cambridge: Harvard University Press, 2004), xx.

KEY TERMS

age structure proportion of population members at specific age ranges

aggressive display visual display by a species member to discourage other members of the same species or different species

aposematic coloration warning coloration used as a defensive mechanism against predation

Batesian mimicry type of mimicry where a non-harmful species takes on the warning colorations of a harmful one

behavior change in an organism's activities in response to a stimulus

behavioral biology study of the biology and evolution of behavior

biotic potential (r_{max}) maximal potential growth rate of a species

birth rate (*B*) number of births within a population at a specific point in time

camouflage avoid detection by blending in with the background

carrying capacity (*K*) number of individuals of a species that can be supported by the limited resources of a habitat

classical conditioning association of a specific stimulus and response through conditioning

climax community final stage of succession, where a stable community is formed by a characteristic assortment of plant and animal species

cognitive learning knowledge and skills acquired by the manipulation of information in the mind

commensalism relationship between species wherein one species benefits from the close, prolonged interaction, while the other species neither benefits nor is harmed

competitive exclusion principle no two species within a habitat can coexist when they compete for the same resources at the same place and time

conditioned behavior behavior that becomes associated with a specific stimulus through conditioning

courtship display visual display used to attract a mate

death rate (*D*) number of deaths within a population at a specific point in time

demographic-based population model modern model of population dynamics incorporating many features of the *r*- and *K*-selection theory

demography statistical study of changes in populations over time

density-dependent regulation regulation of population that is influenced by population density, such as crowding effects; usually involves biotic factors

density-independent regulation regulation of populations by factors that operate independent of population density, such as forest fires and volcanic eruptions; usually involves abiotic factors

distraction display visual display used to distract predators away from a nesting site

Emsleyan/Mertensian mimicry type of mimicry where a harmful species resembles a less harmful one

energy budget allocation of energy resources for body maintenance, reproduction, and parental care

environmental disturbance change in the environment caused by natural disasters or human activities

ethology biological study of animal behavior

exponential growth accelerating growth pattern seen in species under conditions where resources are not limiting

fecundity potential reproductive capacity of an individual

fixed action pattern series of instinctual behaviors that, once initiated, always goes to completion regardless of changes in the environment

foraging behaviors species use to find food

foundation species species which often forms the major structural portion of the habitat

habituation ability of a species to ignore repeated stimuli that have no consequence

host organism a parasite lives on

imprinting identification of parents by newborns as the first organism they see after birth

innate behavior instinctual behavior that is not altered by changes in the environment

intersexual selection selection of a desirable mate of the opposite sex

interspecific competition competition between species for resources in a shared habitat or environment

intrasexual selection competition between members of the same sex for a mate

intraspecific competition competition between members of the same species

island biogeography study of life on island chains and how their geography interacts with the diversity of species found there

iteroparity life history strategy characterized by multiple reproductive events during the lifetime of a species

J-shaped growth curve shape of an exponential growth curve

***K*-selected species** species suited to stable environments that produce a few, relatively large offspring and provide parental care

keystone species species whose presence is key to maintaining biodiversity in an ecosystem and to upholding an ecological community's structure

kin selection sacrificing one's own life so that one's genes will be passed on to future generations by relatives

kinesis undirected movement of an organism in response to a stimulus

learned behavior behavior that responds to changes in the environment

life history inherited pattern of resource allocation under the influence of natural selection and other evolutionary forces

life table table showing the life expectancy of a population member based on its age

logistic growth leveling off of exponential growth due to limiting resources

mark and recapture technique used to determine population size in mobile organisms

migration long-range seasonal movement of animal species

monogamy mating system whereby one male and one female remain coupled for at least one mating season

mortality rate proportion of population surviving to the beginning of an age interval that die during the age interval

Müllerian mimicry type of mimicry where species share warning coloration and all are harmful to predators

mutualism symbiotic relationship between two species where both species benefit

one-child policy China's policy to limit population growth by limiting urban couples to have only one child or face the penalty of a fine

operant conditioning learned behaviors in response to positive and/or negative reinforcement

parasite organism that uses resources from another species, the host

pioneer species first species to appear in primary and secondary succession

polyandry mating system where one female mates with many males

polygyny mating system where one male mates with many females

population density number of population members divided by the area or volume being measured

population growth rate number of organisms added in each reproductive generation

population size (*N*) number of population members in a habitat at the same time

primary succession succession on land that previously has had no life

quadrat square made of various materials used to determine population size and density in slow moving or stationary organisms

***r*-selected species** species suited to changing environments that produce many offspring and provide little or no parental care

reflex action action in response to direct physical stimulation of a nerve

relative species abundance absolute population size of a particular species relative to the population sizes of other species within the community

S-shaped growth curve shape of a logistic growth curve

secondary succession succession in response to environmental disturbances that move a community away from its equilibrium

semelparity life history strategy characterized by a single reproductive event followed by death

signal method of communication between animals including those obtained by the senses of smell, hearing, sight, or touch

species dispersion pattern (also, species distribution pattern) spatial location of individuals of a given species within a habitat at a particular point in time

species richness number of different species in a community

survivorship curve graph of the number of surviving population members versus the relative age of the member

symbiosis close interaction between individuals of different species over an extended period of time that impacts the abundance and distribution of the associating populations

taxis directed movement in response to a stimulus

zero population growth steady population size where birth rates and death rates are equal

CHAPTER SUMMARY

45.1 Population Demography

Populations are individuals of a species that live in a particular habitat. Ecologists measure characteristics of populations: size, density, dispersion pattern, age structure, and sex ratio. Life tables are useful to calculate life expectancies of individual population members. Survivorship curves show the number of individuals surviving at each age interval plotted versus time.

45.2 Life Histories and Natural Selection

All species have evolved a pattern of living, called a life history strategy, in which they partition energy for growth, maintenance, and reproduction. These patterns evolve through natural selection; they allow species to adapt to their environment to obtain the resources they need to successfully reproduce. There is an inverse relationship between fecundity and parental care. A species may reproduce early in life to ensure surviving to a reproductive age or reproduce later in life to become larger and healthier and better able to give parental care. A species may reproduce once (semelparity) or many times (iteroparity) in its life.

45.3 Environmental Limits to Population Growth

Populations with unlimited resources grow exponentially, with an accelerating growth rate. When resources become limiting, populations follow a logistic growth curve. The population of a species will level off at the carrying capacity of its environment.

45.4 Population Dynamics and Regulation

Populations are regulated by a variety of density-dependent and density-independent factors. Species are divided into two categories based on a variety of features of their life history patterns: *r*-selected species, which have large numbers of offspring, and *K*-selected species, which have few offspring. The *r*- and *K*-selection theory has fallen out of use; however, many of its key features are still used in newer, demographically-based models of population dynamics.

45.5 Human Population Growth

The world's human population is growing at an exponential rate. Humans have increased the world's carrying capacity through migration, agriculture, medical advances, and communication. The age structure of a population allows us to predict population growth. Unchecked human population growth could have dire long-term effects on our environment.

45.6 Community Ecology

Communities include all the different species living in a given area. The variety of these species is called species richness. Many organisms have developed defenses against predation and herbivory, including mechanical defenses, warning coloration, and mimicry, as a result of evolution and the interaction with other members of the community. Two species cannot exist in the same habitat competing directly for the same resources. Species may form symbiotic relationships such as commensalism or mutualism. Community structure is described by its foundation and keystone species. Communities respond to environmental disturbances by succession (the predictable appearance of different types of plant species) until a stable community structure is established.

45.7 Behavioral Biology: Proximate and Ultimate Causes of Behavior

Behaviors are responses to stimuli. They can either be instinctual/innate behaviors, which are not influenced by the environment, or learned behaviors, which are influenced by environmental changes. Instinctual behaviors include mating systems and methods of communication. Learned behaviors include imprinting and habituation, conditioning, and, most powerfully, cognitive learning.

VISUAL CONNECTION QUESTIONS

1. Figure 45.2 As this graph shows, population density typically decreases with increasing body size. Why do you think this is the case?

2. Figure 45.10b If the major food source of the seals declines due to pollution or overfishing, which of the following would likely occur?
    a. The carrying capacity of seals would decrease, as would the seal population.
    b. The carrying capacity of seals would decrease, but the seal population would remain the same.
    c. The number of seal deaths would increase but the number of births would also increase, so the population size would remain the same.
    d. The carrying capacity of seals would remain the same, but the population of seals would decrease.

3. Figure 45.16 Age structure diagrams for rapidly growing, slow growing, and stable populations are shown in stages 1 through 3. What type of population change do you think stage 4 represents?

REVIEW QUESTIONS

4. Which of the following methods will tell an ecologist about both the size and density of a population?
    a. mark and recapture
    b. mark and release
    c. quadrat
    d. life table

5. Which of the following is best at showing the life expectancy of an individual within a population?
    a. quadrat
    b. mark and recapture
    c. survivorship curve
    d. life table

6. Humans have which type of survivorship curve?
    a. Type I
    b. Type II
    c. Type III
    d. Type IV

7. How is a clumped population distribution beneficial for prey animals?
    a. Being a member of a larger group provides protection for each individual from predators.
    b. Prey animals rely on each other to acquire food.
    c. Prey animals live in small family groups to raise young.
    d. Clumped population distributions ensure that at least one member of the population knows how to identify the seasonal migration route.

8. Which of the following is associated with long-term parental care?
    a. few offspring
    b. many offspring
    c. semelparity
    d. fecundity

9. Which of the following is associated with multiple reproductive episodes during a species' lifetime?
    a. semiparity
    b. iteroparity
    c. semelparity
    d. fecundity

10. Which of the following is associated with the reproductive potential of a species?
    a. few offspring
    b. many offspring
    c. semelparity
    d. fecundity

11. Species with limited resources usually exhibit a(n) ________ growth curve.
    a. logistic
    b. logical
    c. experimental
    d. exponential

12. The maximum rate of increased characteristic of a species is called its ________.
    a. limit
    b. carrying capacity
    c. biotic potential
    d. exponential growth pattern

13. The population size of a species capable of being supported by the environment is called its ________.
    a. limit
    b. carrying capacity
    c. biotic potential
    d. logistic growth pattern

14. Species that have many offspring at one time are usually:
    a. *r*-selected
    b. *K*-selected
    c. both *r*- and *K*-selected
    d. not selected

15. A forest fire is an example of ________ regulation.
    a. density-dependent
    b. density-independent
    c. *r*-selected
    d. *K*-selected

16. Primates are examples of:
    a. density-dependent species
    b. density-independent species
    c. *r*-selected species
    d. *K*-selected species

17. Which of the following statements does not support the conclusion that giraffes are *k*-selected species?
    a. Giraffes are approximately 6' tall and weigh 150 lbs at birth.
    b. Wild giraffes begin mating at 6-7 years of age.
    c. Newborn giraffes are capable of coordinated walking within an hour of birth, and running within 24 hours of birth.
    d. Giraffes rarely give birth to twins.

18. Which of the following events would **not** negatively impact Yellowstone's grey wolf carrying capacity?
    a. snow in winter
    b. a beaver damming a river upstream
    c. a forest fire
    d. chronic wasting disease in the deer population

19. A country with zero population growth is likely to be _______.
    a. in Africa
    b. in Asia
    c. economically developed
    d. economically underdeveloped

20. Which type of country has the greatest proportion of young individuals?
    a. economically developed
    b. economically underdeveloped
    c. countries with zero population growth
    d. countries in Europe

21. Which of the following is not a way that humans have increased the carrying capacity of the environment?
    a. agriculture
    b. using large amounts of natural resources
    c. domestication of animals
    d. use of language

22. The first species to live on new land, such as that formed from volcanic lava, are called _______.
    a. climax community
    b. keystone species
    c. foundation species
    d. pioneer species

23. Which type of mimicry involves multiple species with similar warning coloration that are all toxic to predators?
    a. Batesian mimicry
    b. Müllerian mimicry
    c. Emsleyan/Mertensian mimicry
    d. Mertensian mimicry

24. A symbiotic relationship where both of the coexisting species benefit from the interaction is called _______.
    a. commensalism
    b. parasitism
    c. mutualism
    d. communism

25. Which of the following is **not** a mutualistic relationship?
    a. a shark using an aquatic cleaning station
    b. a helminth feeding from its host
    c. a bumblebee collecting pollen from a flower
    d. bacteria living in the gut of humans

26. The ability of rats to learn how to run a maze is an example of _______.
    a. imprinting
    b. classical conditioning
    c. operant conditioning
    d. cognitive learning

27. The training of animals usually involves _______.
    a. imprinting
    b. classical conditioning
    c. operant conditioning
    d. cognitive learning

28. The sacrifice of the life of an individual so that the genes of relatives may be passed on is called _______.
    a. operant learning
    b. kin selection
    c. kinesis
    d. imprinting

29. Why are polyandrous mating systems more rare than polygynous matings?
    a. Only males are capable of multiple rounds of reproduction within a single breeding season.
    b. Only females care for the young.
    c. Females usually experience more intrasexual selection pressure than males.
    d. Females usually devote more energy to offspring production and development.

CRITICAL THINKING QUESTIONS

30. Describe how a researcher would determine the size of a penguin population in Antarctica using the mark and release method.

31. The CDC released the following data in its 2013 Vital Statistics report.

Age interval	Number dying in age interval	Number surviving at beginning of age interval
0-10	756	100,000
11-20	292	99,244
21-30	890	98,953
31-40	1,234	98,164
41-50	2,457	96,811
51-60	5,564	94,352
61-70	10,479	88,788

Table 45.3

Calculate the mortality rate for each age interval, and describe the trends in adult and childhood mortality per 100,000 births in the United States in 2013.

32. Why is long-term parental care not associated with having many offspring during a reproductive episode?
33. Describe the difference in evolutionary pressures experienced by an animal that begins reproducing early and an animal that reproduces late in its lifecycle.
34. Describe the rate of population growth that would be expected at various parts of the S-shaped curve of logistic growth.
35. Describe how the population of a species that survives a mass extinction event would change in size and growth pattern over time beginning immediately after the extinction event.
36. Give an example of how density-dependent and density-independent factors might interact.
37. Describe the age structures in rapidly growing countries, slowly growing countries, and countries with zero population growth.
38. Since the introduction of the Endangered Species Act the number of species on the protected list has more than doubled. Describe how the human population's growth pattern contributes to the rise in endangered species.
39. Describe the competitive exclusion principle and its effects on competing species.
40. Jaguars are a keystone species in the Amazon. Describe how they can be so essential to the ecosystem despite being significantly less abundant than many other species.
41. Describe Pavlov's dog experiments as an example of classical conditioning.
42. Describe the advantage of using an aural or pheromone signal to attract a mate as opposed to a visual signal. How might the population density contribute to the evolution of aural or visual mating rituals?

CHAPTER 46
Ecosystems

Figure 46.1 In the southwestern United States, rainy weather causes an increase in production of pinyon nuts, causing the deer mouse population to explode. Deer mice may carry a virus called *Sin Nombre* (a hantavirus) that causes respiratory disease in humans and has a high fatality rate. In 1992–1993, wet *El Niño* weather caused a *Sin Nombre* epidemic. Navajo healers, who were aware of the link between this disease and weather, predicted the outbreak. (credit "highway": modification of work by Phillip Capper; credit "mouse": modification of work by USFWS)

INTRODUCTION In 1993, an interesting example of ecosystem dynamics occurred when a rare lung disease struck inhabitants of the southwestern United States. This disease had an alarming rate of fatalities, killing more than half of early patients, many of whom were Native Americans. These formerly healthy young adults died from complete respiratory failure. The disease was unknown, and the Centers for Disease Control (CDC), the United States government agency responsible for managing potential epidemics, was brought in to investigate. The scientists could have learned about the disease had they known to talk with the Navajo healers who lived in the area and who had observed the connection between rainfall and mice populations, thereby predicting the 1993 outbreak.

The cause of the disease, determined within a few weeks by the CDC investigators, was the hantavirus known as *Sin Nombre*, the virus with "no name." With insights from traditional Navajo medicine, scientists were able to characterize the disease rapidly and institute effective health measures to prevent its spread. This example illustrates the importance of understanding the complexities of ecosystems and how they respond to changes in the environment.

Chapter Outline

46.1 Ecology of Ecosystems

By the end of this section, you will be able to do the following:

- Describe the basic ecosystem types
- Explain the methods that ecologists use to study ecosystem structure and dynamics
- Identify the different methods of ecosystem modeling
- Differentiate between food chains and food webs and recognize the importance of each

Life in an ecosystem is often about competition for limited resources, a characteristic of the theory of natural selection. Competition in communities (all living things within specific habitats) is observed both within species and among different species. The resources for which organisms compete include organic material, sunlight, and mineral nutrients, which provide the energy for living processes and the matter to make up organisms' physical structures. Other critical factors influencing community dynamics are the components of its physical and geographic environment: a habitat's latitude, amount of rainfall, topography (elevation), and available species. These are all important environmental variables that determine which organisms can exist within a particular area.

An **ecosystem** is a community of living organisms and their interactions with their abiotic (nonliving) environment. Ecosystems can be small, such as the tide pools found near the rocky shores of many oceans, or large, such as the Amazon Rainforest in Brazil (Figure 46.2).

Figure 46.2 A (a) tidal pool ecosystem in Matinicus Island in Maine is a small ecosystem, while the (b) Amazon Rainforest in Brazil is a large ecosystem. (credit a: modification of work by "takomabibelot"/Flickr; credit b: modification of work by Ivan Mlinaric)

There are three broad categories of ecosystems based on their general environment: freshwater, ocean water, and terrestrial. Within these broad categories are individual ecosystem types based on the organisms present and the type of environmental habitat.

Ocean ecosystems are the most common, comprising over 70 percent of the Earth's surface and consisting of three basic types: shallow ocean, deep ocean water, and deep ocean surfaces (the low depth areas of the deep oceans). The shallow ocean ecosystems include extremely biodiverse coral reef ecosystems, and the deep ocean surface is known for its large numbers of plankton and krill (small crustaceans) that support it. These two environments are especially important to aerobic respirators worldwide as the phytoplankton perform 40 percent of all photosynthesis on Earth. Although not as diverse as the other two, deep ocean ecosystems contain a wide variety of marine organisms. Such ecosystems exist even at the bottom of the ocean where light is unable to penetrate through the water.

Freshwater ecosystems are the rarest, occurring on only 1.8 percent of the Earth's surface. Lakes, rivers, streams, and springs comprise these systems. They are quite diverse, and they support a variety of fish, amphibians, reptiles, insects, phytoplankton, fungi, and bacteria.

Terrestrial ecosystems, also known for their diversity, are grouped into large categories called biomes, such as tropical rain forests, savannas, deserts, coniferous forests, deciduous forests, and tundra. Grouping these ecosystems into just a few biome categories obscures the great diversity of the individual ecosystems within them. For example, there is great variation in desert vegetation: the saguaro cacti and other plant life in the Sonoran Desert, in the United States, are relatively abundant compared to the desolate rocky desert of Boa Vista, an island off the coast of Western Africa (Figure 46.3).

(a) (b)

Figure 46.3 Desert ecosystems, like all ecosystems, can vary greatly. The desert in (a) Saguaro National Park, Arizona, has abundant plant life, while the rocky desert of (b) Boa Vista island, Cape Verde, Africa, is devoid of plant life. (credit a: modification of work by Jay Galvin; credit b: modification of work by Ingo Wölbern)

Ecosystems are complex with many interacting parts. They are routinely exposed to various disturbances, or changes in the environment that effect their compositions: yearly variations in rainfall and temperature and the slower processes of plant growth, which may take several years. Many of these disturbances result from natural processes. For example, when lightning causes a forest fire and destroys part of a forest ecosystem, the ground is eventually populated by grasses, then by bushes and shrubs, and later by mature trees, restoring the forest to its former state. The impact of environmental disturbances caused by human activities is as important as the changes wrought by natural processes. Human agricultural practices, air pollution, acid rain, global deforestation, overfishing, eutrophication, oil spills, and waste dumping on land and into the ocean are all issues of concern to conservationists.

Equilibrium is the steady state of an ecosystem where all organisms are in balance with their environment and with each other. In ecology, two parameters are used to measure changes in ecosystems: resistance and resilience. **Resistance** is the ability of an ecosystem to remain at equilibrium in spite of disturbances. **Resilience** is the speed at which an ecosystem recovers equilibrium after being disturbed. Ecosystem resistance and resilience are especially important when considering human impact. The nature of an ecosystem may change to such a degree that it can lose its resilience entirely. This process can lead to the complete destruction or irreversible altering of the ecosystem.

Food Chains and Food Webs

The term "food chain" is sometimes used metaphorically to describe human social situations. Individuals who are considered successful are seen as being at the top of the food chain, consuming all others for their benefit, whereas the less successful are seen as being at the bottom.

The scientific understanding of a food chain is more precise than in its everyday usage. In ecology, a **food chain** is a linear sequence of organisms through which nutrients and energy pass: primary producers, primary consumers, and higher-level consumers are used to describe ecosystem structure and dynamics. There is a single path through the chain. Each organism in a food chain occupies what is called a **trophic level**. Depending on their role as producers or consumers, species or groups of species can be assigned to various trophic levels.

In many ecosystems, the bottom of the food chain consists of photosynthetic organisms (plants and/or phytoplankton), which are called **primary producers**. The organisms that consume the primary producers are herbivores: the **primary consumers**. **Secondary consumers** are usually carnivores that eat the primary consumers. **Tertiary consumers** are carnivores that eat other carnivores. Higher-level consumers feed on the next lower tropic levels, and so on, up to the organisms at the top of the food chain: the **apex consumers**. In the Lake Ontario food chain shown in Figure 46.4, the Chinook salmon is the apex consumer at the top of this food chain.

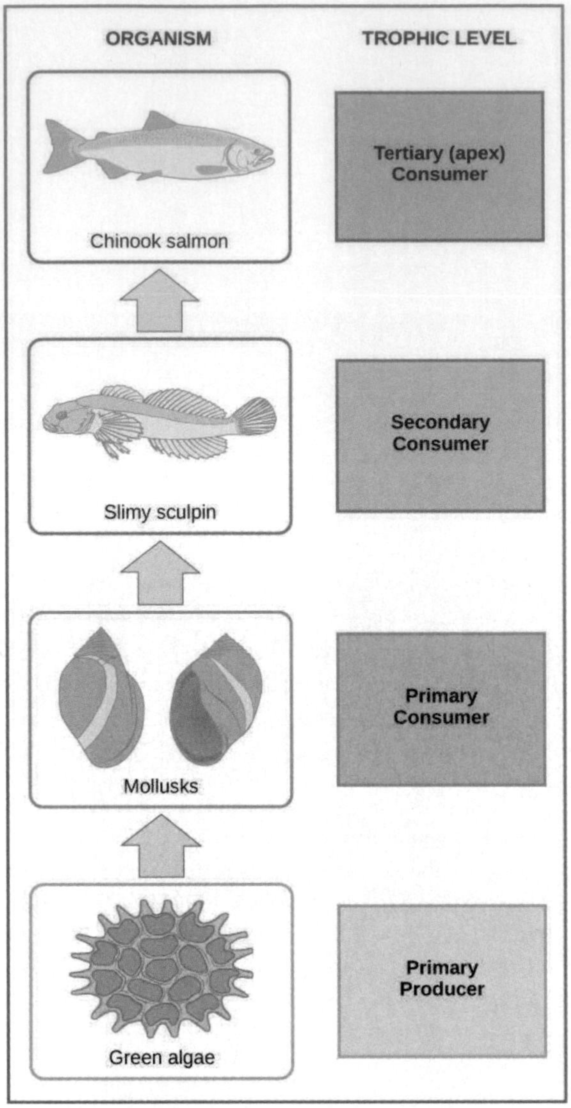

Figure 46.4 These are the trophic levels of a food chain in Lake Ontario at the United States-Canada border. Energy and nutrients flow from photosynthetic green algae at the bottom to the top of the food chain: the Chinook salmon.

One major factor that limits the length of food chains is energy. Energy is lost as heat between each trophic level due to the second law of thermodynamics. Thus, after a limited number of trophic energy transfers, the amount of energy remaining in the food chain may not be great enough to support viable populations at yet a higher trophic level.

The loss of energy between trophic levels is illustrated by the pioneering studies of Howard T. Odum in the Silver Springs, Florida, ecosystem in the 1940s (Figure 46.5). The primary producers generated 20,819 kcal/m^2/yr (kilocalories per square meter per year), the primary consumers generated 3368 kcal/m^2/yr, the secondary consumers generated 383 kcal/m^2/yr, and the tertiary consumers only generated 21 kcal/m^2/yr. Thus, there is little energy remaining for another level of consumers in this ecosystem.

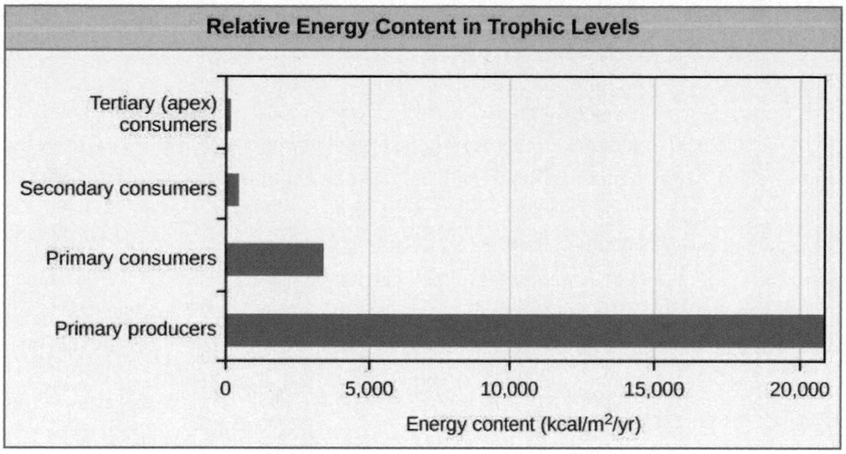

Figure 46.5 The relative energy in trophic levels in a Silver Springs, Florida, ecosystem is shown. Each trophic level has less energy available and supports fewer organisms at the next level.

There is a one problem when using food chains to accurately describe most ecosystems. Even when all organisms are grouped into appropriate trophic levels, some of these organisms can feed on species from more than one trophic level; likewise, some of these organisms can be eaten by species from multiple trophic levels. In other words, the linear model of ecosystems, the food chain, is not completely descriptive of ecosystem structure. A holistic model—which accounts for all the interactions between different species and their complex interconnected relationships with each other and with the environment—is a more accurate and descriptive model for ecosystems. A **food web** is a graphic representation of a holistic, nonlinear web of primary producers, primary consumers, and higher-level consumers used to describe ecosystem structure and dynamics (Figure 46.6).

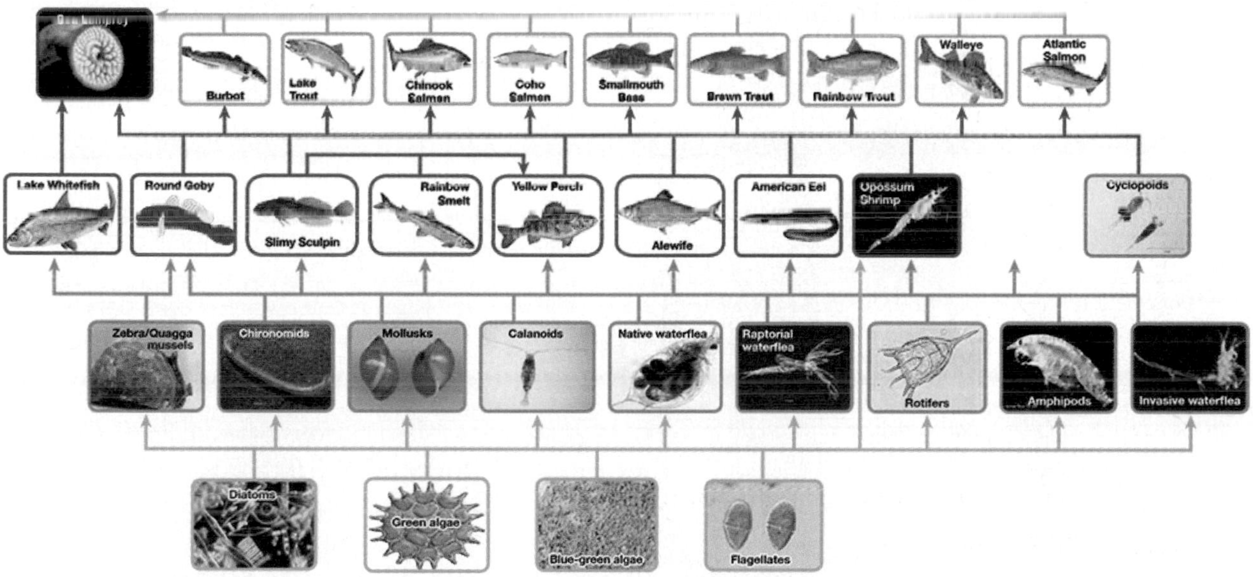

Figure 46.6 This food web shows the interactions between organisms across trophic levels in the Lake Ontario ecosystem. Primary producers are outlined in green, primary consumers in orange, secondary consumers in blue, and tertiary (apex) consumers in purple. Arrows point from an organism that is consumed to the organism that consumes it. Notice how some lines point to more than one trophic level. For example, the opossum shrimp eats both primary producers and primary consumers. (credit: NOAA, GLERL)

A comparison of the two types of structural ecosystem models shows strength in both. Food chains are more flexible for analytical modeling, are easier to follow, and are easier to experiment with, whereas food web models more accurately represent ecosystem structure and dynamics, and data can be directly used as input for simulation modeling.

LINK TO LEARNING

Head to this online interactive simulator (http://openstax.org/l/food_web) to investigate food web function. In the *Interactive Labs* box, under Food Web, click **Step 1**. Read the instructions first, and then click **Step 2** for additional instructions. When you are ready to create a simulation, in the upper-right corner of the *Interactive Labs* box, click **OPEN SIMULATOR**.

Two general types of food webs are often shown interacting within a single ecosystem. A **grazing food web** (such as the Lake Ontario food web in Figure 46.6) has plants or other photosynthetic organisms at its base, followed by herbivores and various carnivores. A **detrital food web** consists of a base of organisms that feed on decaying organic matter (dead organisms), called decomposers or detritivores. These organisms are usually bacteria or fungi that recycle organic material back into the biotic part of the ecosystem as they themselves are consumed by other organisms. As all ecosystems require a method to recycle material from dead organisms, most grazing food webs have an associated detrital food web. For example, in a meadow ecosystem, plants may support a grazing food web of different organisms, primary and other levels of consumers, while at the same time supporting a detrital food web of bacteria, fungi, and detrivorous invertebrates feeding off dead plants and animals.

EVOLUTION CONNECTION

Three-spined Stickleback

It is well established by the theory of natural selection that changes in the environment play a major role in the evolution of species within an ecosystem. However, little is known about how the evolution of species within an ecosystem can alter the ecosystem environment. In 2009, Dr. Luke Harmon, from the University of Idaho, published a paper that for the first time showed that the evolution of organisms into subspecies can have direct effects on their ecosystem environment.[1]

The three-spined stickleback (*Gasterosteus aculeatus*) is a freshwater fish that evolved from a saltwater fish to live in freshwater lakes about 10,000 years ago, which is considered a recent development in evolutionary time (Figure 46.7). Over the last 10,000 years, these freshwater fish then became isolated from each other in different lakes. Depending on which lake population was studied, findings showed that these sticklebacks then either remained as one species or evolved into two species. The divergence of species was made possible by their use of different areas of the pond for feeding called micro niches.

Dr. Harmon and his team created artificial pond microcosms in 250-gallon tanks and added muck from freshwater ponds as a source of zooplankton and other invertebrates to sustain the fish. In different experimental tanks they introduced one species of stickleback from either a single-species or double-species lake.

Over time, the team observed that some of the tanks bloomed with algae while others did not. This puzzled the scientists, and they decided to measure the water's dissolved organic carbon (DOC), which consists of mostly large molecules of decaying organic matter that give pond-water its slightly brownish color. It turned out that the water from the tanks with two-species fish contained larger particles of DOC (and hence darker water) than water with single-species fish. This increase in DOC blocked the sunlight and prevented algal blooming. Conversely, the water from the single-species tank contained smaller DOC particles, allowing more sunlight penetration to fuel the algal blooms.

This change in the environment, which is due to the different feeding habits of the stickleback species in each lake type, probably has a great impact on the survival of other species in these ecosystems, especially other photosynthetic organisms. Thus, the study shows that, at least in these ecosystems, the environment and the evolution of populations have reciprocal effects that may now be factored into simulation models.

1 *Nature* (Vol. 458, April 1, 2009)

Figure 46.7 The three-spined stickleback evolved from a saltwater fish to freshwater fish. (credit: Barrett Paul, USFWS)

Research into Ecosystem Dynamics: Ecosystem Experimentation and Modeling

The study of the changes in ecosystem structure caused by changes in the environment (disturbances) or by internal forces is called **ecosystem dynamics**. Ecosystems are characterized using a variety of research methodologies. Some ecologists study ecosystems using controlled experimental systems, while some study entire ecosystems in their natural state, and others use both approaches.

A **holistic ecosystem model** attempts to quantify the composition, interaction, and dynamics of entire ecosystems; it is the most representative of the ecosystem in its natural state. A food web is an example of a holistic ecosystem model. However, this type of study is limited by time and expense, as well as the fact that it is neither feasible nor ethical to do experiments on large natural ecosystems. It is difficult to quantify all different species in an ecosystem and the dynamics in their habitat, especially when studying large habitats such as the Amazon Rainforest.

For these reasons, scientists study ecosystems under more controlled conditions. Experimental systems usually involve either partitioning a part of a natural ecosystem that can be used for experiments, termed a **mesocosm**, or by recreating an ecosystem entirely in an indoor or outdoor laboratory environment, which is referred to as a **microcosm**. A major limitation to these approaches is that removing individual organisms from their natural ecosystem or altering a natural ecosystem through partitioning may change the dynamics of the ecosystem. These changes are often due to differences in species numbers and diversity and also to environment alterations caused by partitioning (mesocosm) or recreating (microcosm) the natural habitat. Thus, these types of experiments are not totally predictive of changes that would occur in the ecosystem from which they were gathered.

As both of these approaches have their limitations, some ecologists suggest that results from these experimental systems should be used only in conjunction with holistic ecosystem studies to obtain the most representative data about ecosystem structure, function, and dynamics.

Scientists use the data generated by these experimental studies to develop ecosystem models that demonstrate the structure and dynamics of ecosystems. They use three basic types of ecosystem modeling in research and ecosystem management: a conceptual model, an analytical model, and a simulation model. A **conceptual model** is an ecosystem model that consists of flow charts to show interactions of different compartments of the living and nonliving components of the ecosystem. A conceptual model describes ecosystem structure and dynamics and shows how environmental disturbances affect the ecosystem; however, its ability to predict the effects of these disturbances is limited. Analytical and simulation models, in contrast, are mathematical methods of describing ecosystems that are indeed capable of predicting the effects of potential environmental changes without direct experimentation, although with some limitations as to accuracy. An **analytical model** is an ecosystem model that is created using simple mathematical formulas to predict the effects of environmental disturbances on ecosystem structure and dynamics. A **simulation model** is an ecosystem model that is created using complex computer algorithms to holistically model ecosystems and to predict the effects of environmental disturbances on ecosystem structure and dynamics. Ideally, these models are accurate enough to determine which components of the ecosystem are particularly sensitive to disturbances, and they can serve as a guide to ecosystem managers (such as conservation ecologists or fisheries biologists) in the practical maintenance of ecosystem health.

Conceptual Models

Conceptual models are useful for describing ecosystem structure and dynamics and for demonstrating the relationships between different organisms in a community and their environment. Conceptual models are usually depicted graphically as flow charts. The organisms and their resources are grouped into specific compartments with arrows showing the relationship and transfer of energy or nutrients between them. Thus, these diagrams are sometimes called compartment models.

To model the cycling of mineral nutrients, organic and inorganic nutrients are subdivided into those that are bioavailable (ready to be incorporated into biological macromolecules) and those that are not. For example, in a terrestrial ecosystem near a deposit of coal, carbon will be available to the plants of this ecosystem as carbon dioxide gas in a short-term period, not from the carbon-rich coal itself. However, over a longer period, microorganisms capable of digesting coal will incorporate its carbon or release it as natural gas (methane, CH_4), changing this unavailable organic source into an available one. This conversion is greatly accelerated by the combustion of fossil fuels by humans, which releases large amounts of carbon dioxide into the atmosphere. This is thought to be a major factor in the rise of the atmospheric carbon dioxide levels in the industrial age. The carbon dioxide released from burning fossil fuels is produced faster than photosynthetic organisms can use it. This process is intensified by the reduction of photosynthetic trees because of worldwide deforestation. Most scientists agree that high atmospheric carbon dioxide is a major cause of global climate change.

Conceptual models are also used to show the flow of energy through particular ecosystems. Figure 46.8 is based on Howard T. Odum's classical study of the Silver Springs, Florida, holistic ecosystem in the mid-twentieth century.[2] This study shows the energy content and transfer between various ecosystem compartments.

2Howard T. Odum, "Trophic Structure and Productivity of Silver Springs, Florida," *Ecological Monographs* 27, no. 1 (1957): 47–112.

VISUAL CONNECTION

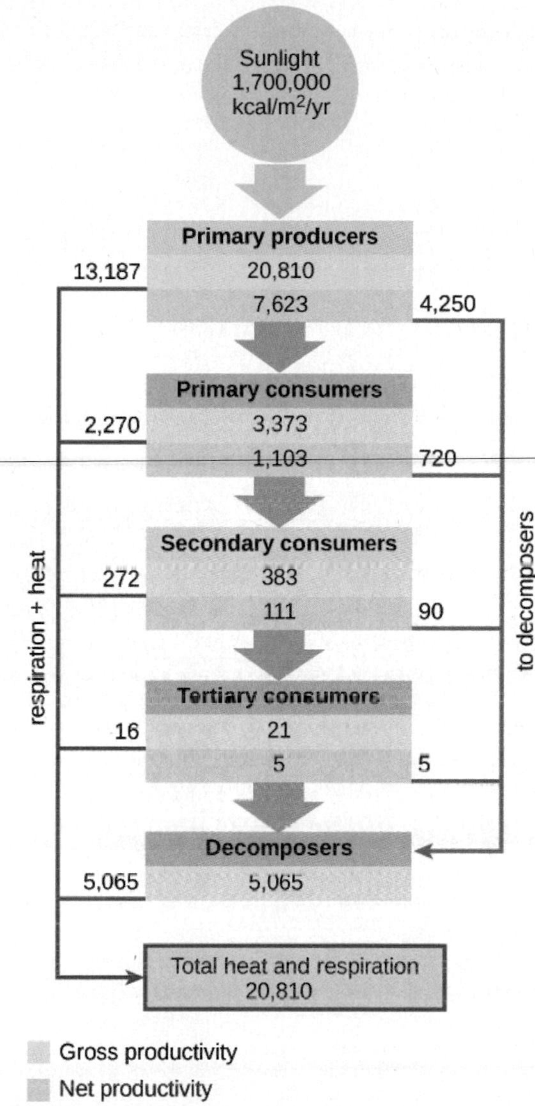

Figure 46.8 This conceptual model shows the flow of energy through a spring ecosystem in Silver Springs, Florida. Notice that the energy decreases with each increase in trophic level.

Why do you think the value for gross productivity of the primary producers is the same as the value for total heat and respiration (20,810 kcal/m^2/yr)?

Analytical and Simulation Models

The major limitation of conceptual models is their inability to predict the consequences of changes in ecosystem species and/or environment. Ecosystems are dynamic entities and subject to a variety of abiotic and biotic disturbances caused by natural forces and/or human activity. Ecosystems altered from their initial equilibrium state can often recover from such disturbances and return to a state of equilibrium. As most ecosystems are subject to periodic disturbances and are often in a state of change, they are usually either moving toward or away from their equilibrium state. There are many of these equilibrium states among the various components of an ecosystem, which affects the ecosystem overall. Furthermore, as humans have the ability to greatly and rapidly alter the species content and habitat of an ecosystem, the need for predictive models that enable understanding of how ecosystems respond to these changes becomes more crucial.

Analytical models often use simple, linear components of ecosystems, such as food chains, and are known to be complex mathematically; therefore, they require a significant amount of mathematical knowledge and expertise. Although analytical models have great potential, their simplification of complex ecosystems is thought to limit their accuracy. Simulation models that use computer programs are better able to deal with the complexities of ecosystem structure.

A recent development in simulation modeling uses supercomputers to create and run individual-based simulations, which accounts for the behavior of individual organisms and their effects on the ecosystem as a whole. These simulations are considered to be the most accurate and predictive of the complex responses of ecosystems to disturbances.

LINK TO LEARNING

Visit The Darwin Project (http://openstax.org/l/Darwin_project) to view a variety of ecosystem models, including simulations that model predator-prey relationships (https://openstax.org/l/Darwin_projct2) to learn more.

46.2 Energy Flow through Ecosystems

By the end of this section, you will be able to do the following:

- Describe how organisms acquire energy in a food web and in associated food chains
- Explain how the efficiency of energy transfers between trophic levels affects ecosystem structure and dynamics
- Discuss trophic levels and how ecological pyramids are used to model them

All living things require energy in one form or another. Energy is required by most complex metabolic pathways (often in the form of adenosine triphosphate, ATP), especially those responsible for building large molecules from smaller compounds, and life itself is an energy-driven process. Living organisms would not be able to assemble macromolecules (proteins, lipids, nucleic acids, and complex carbohydrates) from their monomeric subunits without a constant energy input.

It is important to understand how organisms acquire energy and how that energy is passed from one organism to another through food webs and their constituent food chains. Food webs illustrate how energy flows directionally through ecosystems, including how efficiently organisms acquire it, use it, and how much remains for use by other organisms of the food web.

How Organisms Acquire Energy in a Food Web

Energy is acquired by living things in three ways: photosynthesis, chemosynthesis, and the consumption and digestion of other living or previously living organisms by heterotrophs.

Photosynthetic and chemosynthetic organisms are both grouped into a category known as autotrophs: organisms capable of synthesizing their own food (more specifically, capable of using inorganic carbon as a carbon source). Photosynthetic autotrophs (photoautotrophs) use sunlight as an energy source, whereas chemosynthetic autotrophs (chemoautotrophs) use inorganic molecules as an energy source. Autotrophs are critical for all ecosystems. Without these organisms, energy would not be available to other living organisms and life itself would not be possible.

Photoautotrophs, such as plants, algae, and photosynthetic bacteria, serve as the energy source for a majority of the world's ecosystems. These ecosystems are often described by grazing food webs. Photoautotrophs harness the solar energy of the sun by converting it to chemical energy in the form of ATP (and NADP). The energy stored in ATP is used to synthesize complex organic molecules, such as glucose.

Chemoautotrophs are primarily bacteria that are found in rare ecosystems where sunlight is not available, such as in those associated with dark caves or hydrothermal vents at the bottom of the ocean (Figure 46.9). Many chemoautotrophs in hydrothermal vents use hydrogen sulfide (H_2S), which is released from the vents as a source of chemical energy. This allows chemoautotrophs to synthesize complex organic molecules, such as glucose, for their own energy and in turn supplies energy to the rest of the ecosystem.

Figure 46.9 Swimming shrimp, a few squat lobsters, and hundreds of vent mussels are seen at a hydrothermal vent at the bottom of the ocean. As no sunlight penetrates to this depth, the ecosystem is supported by chemoautotrophic bacteria and organic material that sinks from the ocean's surface. This picture was taken in 2006 at the submerged NW Eifuku volcano off the coast of Japan by the National Oceanic and Atmospheric Administration (NOAA). The summit of this highly active volcano lies 1535 m below the surface.

Productivity within Trophic Levels

Productivity within an ecosystem can be defined as the percentage of energy entering the ecosystem incorporated into biomass in a particular trophic level. **Biomass** is the total mass, in a unit area at the time of measurement, of living or previously living organisms within a trophic level. Ecosystems have characteristic amounts of biomass at each trophic level. For example, in the English Channel ecosystem the primary producers account for a biomass of 4 g/m^2 (grams per square meter), while the primary consumers exhibit a biomass of 21 g/m^2.

The productivity of the primary producers is especially important in any ecosystem because these organisms bring energy to other living organisms by photoautotrophy or chemoautotrophy. The rate at which photosynthetic primary producers incorporate energy from the sun is called **gross primary productivity**. An example of gross primary productivity is shown in the compartment diagram of energy flow within the Silver Springs aquatic ecosystem as shown (Figure 46.8). In this ecosystem, the total energy accumulated by the primary producers (gross primary productivity) was shown to be 20,810 $kcal/m^2/yr$.

Because all organisms need to use some of this energy for their own functions (like respiration and resulting metabolic heat loss) scientists often refer to the net primary productivity of an ecosystem. **Net primary productivity** is the energy that remains in the primary producers after accounting for the organisms' respiration and heat loss. The net productivity is then available to the primary consumers at the next trophic level. In our Silver Springs example, 13,187 of the 20,810 $kcal/m^2/yr$ were used for respiration or were lost as heat, leaving 7,633 $kcal/m^2/yr$ of energy for use by the primary consumers.

Ecological Efficiency: The Transfer of Energy between Trophic Levels

As illustrated in (Figure 46.8), as energy flows from primary producers through the various trophic levels, the ecosystem loses large amounts of energy. The main reason for this loss is the second law of thermodynamics, which states that whenever energy is converted from one form to another, there is a tendency toward disorder (entropy) in the system. In biologic systems, this energy takes the form of metabolic heat, which is lost when the organisms consume other organisms. In the Silver Springs ecosystem example (Figure 46.8), we see that the primary consumers produced 1103 $kcal/m^2/yr$ from the 7618 $kcal/m^2/yr$ of energy available to them from the primary producers. The measurement of energy transfer efficiency between two successive trophic levels is termed the **trophic level transfer efficiency (TLTE)** and is defined by the formula:

$$\text{TLTE} = \frac{\text{production at present trophic level}}{\text{production at previous trophic level}} \times 100$$

In Silver Springs, the TLTE between the first two trophic levels was approximately 14.8 percent. The low efficiency of energy transfer between trophic levels is usually the major factor that limits the length of food chains observed in a food web. The fact is, after four to six energy transfers, there is not enough energy left to support another trophic level. In the Lake Ontario example shown in (Figure 46.6), only three energy transfers occurred between the primary producer, (green algae), and the apex consumer (Chinook salmon).

Ecologists have many different methods of measuring energy transfers within ecosystems. Measurement difficulty depends on the complexity of the ecosystem and how much access scientists have to observe the ecosystem. In other words, some ecosystems are more difficult to study than others, and sometimes the quantification of energy transfers has to be estimated.

Other parameters are important in characterizing energy flow within an ecosystem. **Net production efficiency (NPE)** allows ecologists to quantify how efficiently organisms of a particular trophic level incorporate the energy they receive into biomass; it is calculated using the following formula:

$$\text{NPE} = \frac{\text{net consumer productivity}}{\text{assimilation}} \times 100$$

Net consumer productivity is the energy content available to the organisms of the next trophic level. **Assimilation** is the biomass (energy content generated per unit area) of the present trophic level after accounting for the energy lost due to incomplete ingestion of food, energy used for respiration, and energy lost as waste. Incomplete ingestion refers to the fact that some consumers eat only a part of their food. For example, when a lion kills an antelope, it will eat everything except the hide and bones. The lion is missing the energy-rich bone marrow inside the bone, so the lion does not make use of all the calories its prey could provide.

Thus, NPE measures how efficiently each trophic level uses and incorporates the energy from its food into biomass to fuel the next trophic level. In general, cold-blooded animals (ectotherms), such as invertebrates, fish, amphibians, and reptiles, use less of the energy they obtain for respiration and heat than warm-blooded animals (endotherms), such as birds and mammals. The extra heat generated in endotherms, although an advantage in terms of the activity of these organisms in colder environments, is a major disadvantage in terms of NPE. Therefore, many endotherms have to eat more often than ectotherms to get the energy they need for survival. In general, NPE for ectotherms is an order of magnitude (10x) higher than for endotherms. For example, the NPE for a caterpillar eating leaves has been measured at 18 percent, whereas the NPE for a squirrel eating acorns may be as low as 1.6 percent.

The inefficiency of energy use by warm-blooded animals has broad implications for the world's food supply. It is widely accepted that the meat industry uses large amounts of crops to feed livestock, and because the NPE is low, much of the energy from animal feed is lost. For example, it costs about $0.01 to produce 1000 dietary calories (kcal) of corn or soybeans, but approximately $0.19 to produce a similar number of calories growing cattle for beef consumption. The same energy content of milk from cattle is also costly, at approximately $0.16 per 1000 kcal. Much of this difference is due to the low NPE of cattle. Thus, there has been a growing movement worldwide to promote the consumption of nonmeat and nondairy foods so that less energy is wasted feeding animals for the meat industry.

Modeling Ecosystems Energy Flow: Ecological Pyramids

The structure of ecosystems can be visualized with ecological pyramids, which were first described by the pioneering studies of Charles Elton in the 1920s. **Ecological pyramids** show the relative amounts of various parameters (such as number of organisms, energy, and biomass) across trophic levels.

Pyramids of numbers can be either upright or inverted, depending on the ecosystem. As shown in Figure 46.10, typical grassland during the summer has a base of many plants, and the numbers of organisms decrease at each trophic level. However, during the summer in a temperate forest, the base of the pyramid consists of few trees compared with the number of primary consumers, mostly insects. Because trees are large, they have great photosynthetic capability, and dominate other plants in this ecosystem to obtain sunlight. Even in smaller numbers, primary producers in forests are still capable of supporting other trophic levels.

Another way to visualize ecosystem structure is with pyramids of biomass. This pyramid measures the amount of energy converted into living tissue at the different trophic levels. Using the Silver Springs ecosystem example, this data exhibits an upright biomass pyramid (Figure 46.10), whereas the pyramid from the English Channel example is inverted. The plants (primary producers) of the Silver Springs ecosystem make up a large percentage of the biomass found there. However, the phytoplankton in the English Channel example make up less biomass than the primary consumers, the zooplankton. As with inverted pyramids of numbers, this inverted pyramid is not due to a lack of productivity from the primary producers, but results from the high turnover rate of the phytoplankton. The phytoplankton are consumed rapidly by the primary consumers, thus, minimizing their biomass at any particular point in time. However, phytoplankton reproduce quickly, thus they are able to support the rest of the ecosystem.

Pyramid ecosystem modeling can also be used to show energy flow through the trophic levels. Notice that these numbers are the same as those used in the energy flow compartment diagram in (Figure 46.8). Pyramids of energy are always upright, and an ecosystem without sufficient primary productivity cannot be supported. All types of ecological pyramids are useful for characterizing ecosystem structure. However, in the study of energy flow through the ecosystem, pyramids of energy are the most consistent and representative models of ecosystem structure (Figure 46.10).

VISUAL CONNECTION

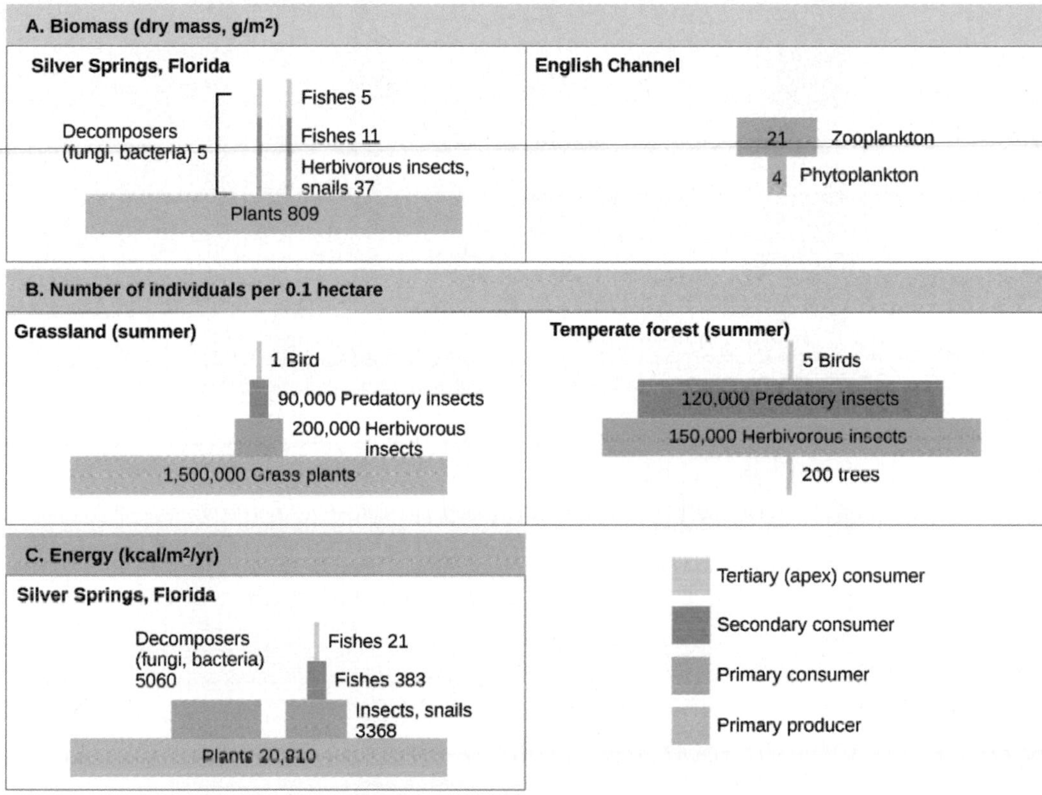

Figure 46.10 Ecological pyramids depict the (a) biomass, (b) number of organisms, and (c) energy in each trophic level.

Pyramids depicting the number of organisms or biomass may be inverted, upright, or even diamond-shaped. Energy pyramids, however, are always upright. Why?

Consequences of Food Webs: Biological Magnification

One of the most important environmental consequences of ecosystem dynamics is biomagnification. **Biomagnification** is the increasing concentration of persistent, toxic substances in organisms at each trophic level, from the primary producers to the apex consumers. Many substances have been shown to bioaccumulate, including the pesticide **d**ichloro**d**iphenyl**t**richloroethane (DDT), which was described in the 1960s bestseller, *Silent Spring*, by Rachel Carson. DDT was a commonly used pesticide before its dangers became known. In some aquatic ecosystems, organisms from each trophic level consumed many organisms of the lower level, which caused DDT to increase in birds (apex consumers) that ate fish. Thus, the birds accumulated sufficient amounts of DDT to cause fragility in their eggshells. This effect increased egg breakage during nesting and was shown to have adverse effects on these bird populations. The use of DDT was banned in the United States in the 1970s.

Other substances that biomagnify are polychlorinated biphenyls (PCBs), which were used in coolant liquids in the United States until their use was banned in 1979, and heavy metals, such as mercury, lead, and cadmium. These substances were best studied in aquatic ecosystems, where fish species at different trophic levels accumulate toxic substances brought through the ecosystem by the primary producers. As illustrated in a study performed by the National Oceanic and Atmospheric Administration (NOAA) in the Saginaw Bay of Lake Huron (Figure 46.11), PCB concentrations increased from the ecosystem's primary producers (phytoplankton) through the different trophic levels of fish species. The apex consumer (walleye) has more than four times the amount of PCBs compared to phytoplankton. Also, based on results from other studies, birds that eat these fish may have PCB levels at least one order of magnitude higher than those found in the lake fish.

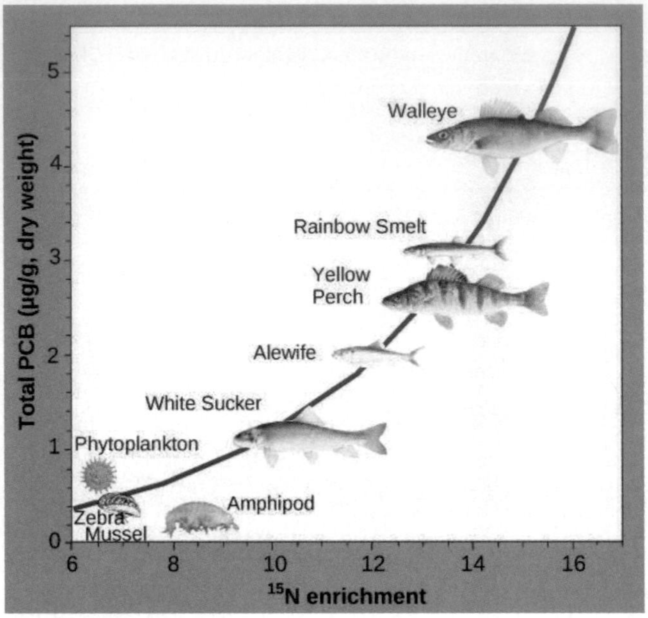

Figure 46.11 This chart shows the PCB concentrations found at the various trophic levels in the Saginaw Bay ecosystem of Lake Huron. Numbers on the x-axis reflect enrichment with heavy isotopes of nitrogen (^{15}N), which is a marker for increasing trophic level. Notice that the fish in the higher trophic levels accumulate more PCBs than those in lower trophic levels. (credit: Patricia Van Hoof, NOAA, GLERL)

Other concerns have been raised by the accumulation of heavy metals, such as mercury and cadmium, in certain types of seafood. The United States Environmental Protection Agency (EPA) recommends that pregnant women and young children should not consume any swordfish, shark, king mackerel, or tilefish because of their high mercury content. These individuals are advised to eat fish low in mercury: salmon, tilapia, shrimp, pollock, and catfish. Biomagnification is a good example of how ecosystem dynamics can affect our everyday lives, even influencing the food we eat.

46.3 Biogeochemical Cycles

By the end of this section, you will be able to do the following:

- Discuss the biogeochemical cycles of water, carbon, nitrogen, phosphorus, and sulfur
- Explain how human activities have impacted these cycles and the potential consequences for Earth

Energy flows directionally through ecosystems, entering as sunlight (or inorganic molecules for chemoautotrophs) and leaving as heat during the many transfers between trophic levels. However, the matter that makes up living organisms is conserved and recycled. The six most common elements associated with organic molecules—carbon, nitrogen, hydrogen, oxygen, phosphorus, and sulfur—take a variety of chemical forms and may exist for long periods in the atmosphere, on land, in water, or beneath the Earth's surface. Geologic processes, such as weathering, erosion, water drainage, and the subduction of the continental plates, all play a role in this recycling of materials. Because geology and chemistry have major roles in the study of this process, the recycling of inorganic matter between living organisms and their environment is called a **biogeochemical cycle**.

Water contains hydrogen and oxygen, which is essential to all living processes. The **hydrosphere** is the area of the Earth where water movement and storage occurs. On or beneath the surface, water occurs in liquid or solid form in rivers, lakes, oceans, groundwater, polar ice caps, and glaciers. And it occurs as water vapor in the atmosphere. Carbon is found in all organic macromolecules and is an important constituent of fossil fuels. Nitrogen is a major component of our nucleic acids and proteins and is critical to human agriculture. Phosphorus, a major component of nucleic acid (along with nitrogen), is one of the main ingredients in artificial fertilizers used in agriculture and their associated environmental impacts on our surface water. Sulfur is critical to the 3-D folding of proteins, such as in disulfide binding.

The cycling of these elements is interconnected. For example, the movement of water is critical for the leaching of nitrogen and phosphate into rivers, lakes, and oceans. Furthermore, the ocean itself is a major reservoir for carbon. Thus, mineral nutrients are cycled, either rapidly or slowly, through the entire biosphere, from one living organism to another, and between the biotic and abiotic world.

LINK TO LEARNING

Head to this website (http://openstax.org/l/biogeochemical) to learn more about biogeochemical cycles.

The Water (Hydrologic) Cycle

Water is the basis of all living processes on Earth. When examining the stores of water on Earth, 97.5 percent of it is non-potable salt water (Figure 46.12). Of the remaining water, 99 percent is locked underground as water or as ice. Thus, less than 1 percent of fresh water is easily accessible from lakes and rivers. Many living things, such as plants, animals, and fungi, are dependent on that small amount of fresh surface water, a lack of which can have massive effects on ecosystem dynamics. To be successful, organisms must adapt to fluctuating water supplies. Humans, of course, have developed technologies to increase water availability, such as digging wells to harvest groundwater, storing rainwater, and using desalination to obtain drinkable water from the ocean.

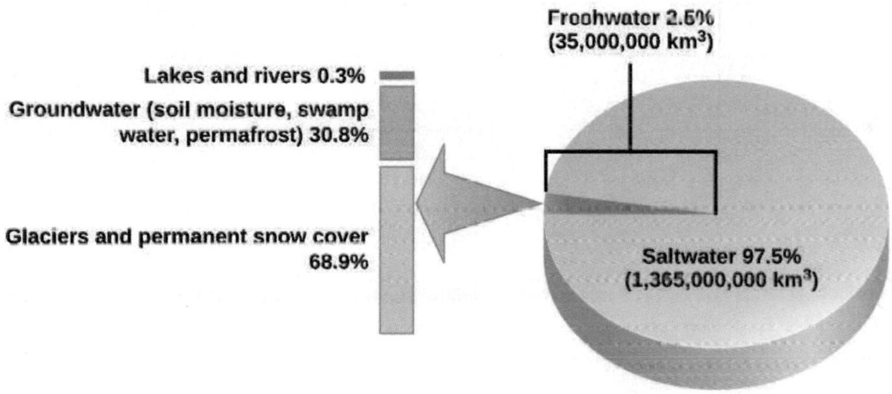

Figure 46.12 Only 2.5 percent of water on Earth is fresh water, and less than 1 percent of fresh water is easily accessible to living things.

Water cycling is extremely important to ecosystem dynamics. Water has a major influence on climate and, thus, on the environments of ecosystems. Most of the water on Earth is stored for long periods in the oceans, underground, and as ice. Figure 46.13 illustrates the average time that an individual water molecule may spend in the Earth's major water reservoirs. **Residence time** is a measure of the average time an individual water molecule stays in a particular reservoir.

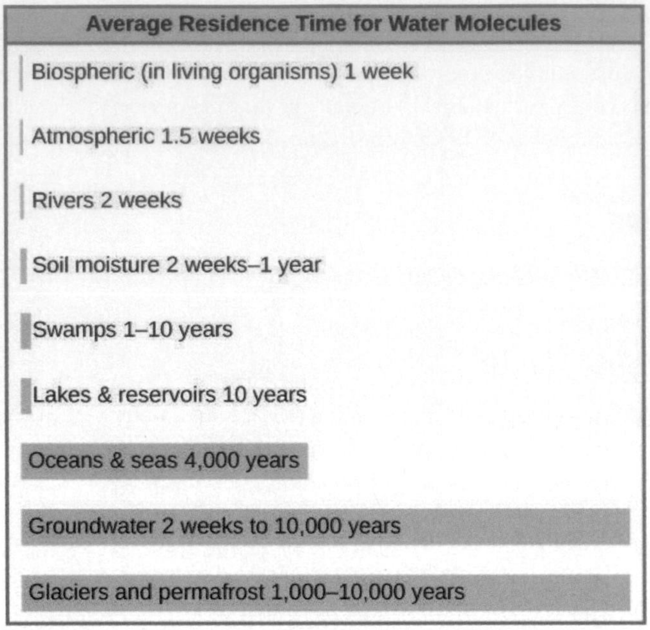

Figure 46.13 This graph shows the average residence time for water molecules in the Earth's water reservoirs.

There are various processes that occur during the cycling of water, shown in Figure 46.14. These processes include the following:

- evaporation/sublimation
- condensation/precipitation
- subsurface water flow
- surface runoff/snowmelt
- streamflow

The water cycle is driven by the sun's energy as it warms the oceans and other surface waters. This leads to the evaporation (water to water vapor) of liquid surface water and the sublimation (ice to water vapor) of frozen water, which deposits large amounts of water vapor into the atmosphere. Over time, this water vapor condenses into clouds as liquid or frozen droplets and is eventually followed by precipitation (rain or snow), which returns water to the Earth's surface. Rain eventually permeates into the ground, where it may evaporate again if it is near the surface, flow beneath the surface, or be stored for long periods. More easily observed is surface runoff: the flow of fresh water either from rain or melting ice. Runoff can then make its way through streams and lakes to the oceans or flow directly to the oceans themselves.

LINK TO LEARNING

Head to this website (http://openstax.org/l/freshwater) to learn more about the world's fresh water supply.

Rain and surface runoff are major ways in which minerals, including carbon, nitrogen, phosphorus, and sulfur, are cycled from land to water. The environmental effects of runoff will be discussed later as these cycles are described.

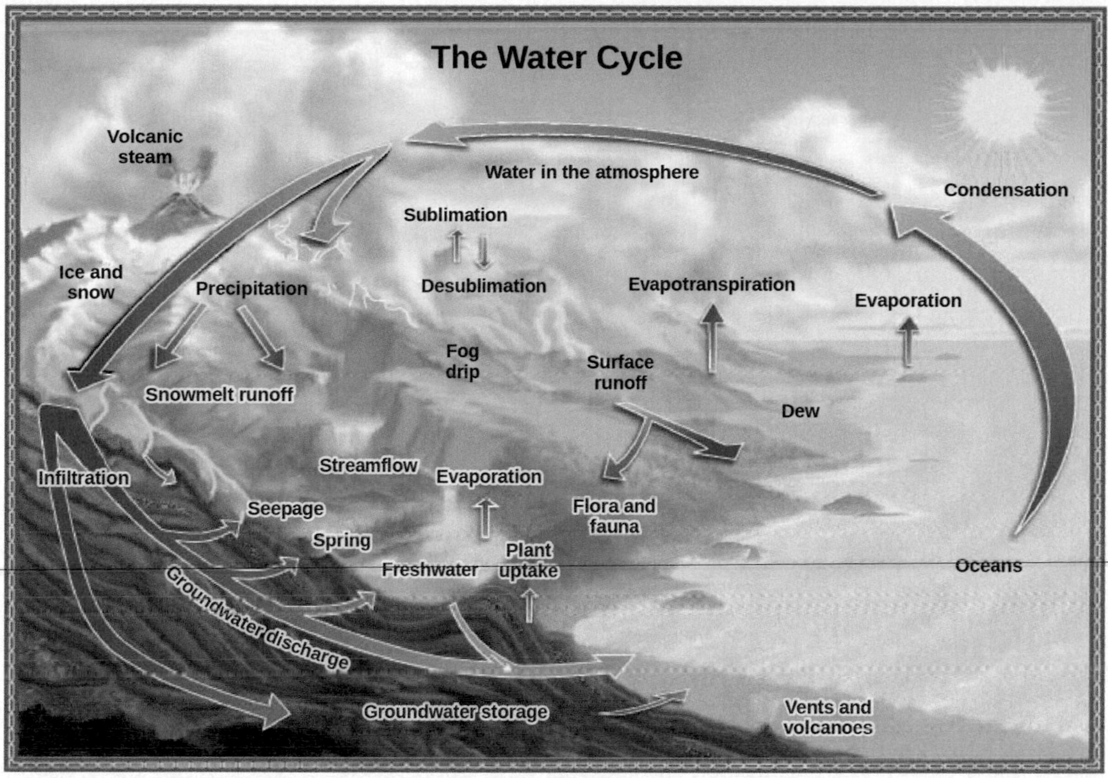

Figure 46.14 Water from the land and oceans enters the atmosphere by evaporation or sublimation, where it condenses into clouds and falls as rain or snow. Precipitated water may enter freshwater bodies or infiltrate the soil. The cycle is complete when surface or groundwater reenters the ocean. (credit: modification of work by John M. Evans and Howard Perlman, USGS)

The Carbon Cycle

Carbon is the second most abundant element in living organisms. Carbon is present in all organic molecules, and its role in the structure of macromolecules is of primary importance to living organisms.

The carbon cycle is most easily studied as two interconnected sub-cycles: one dealing with rapid carbon exchange among living organisms and the other dealing with the long-term cycling of carbon through geologic processes. The entire carbon cycle is shown in Figure 46.15.

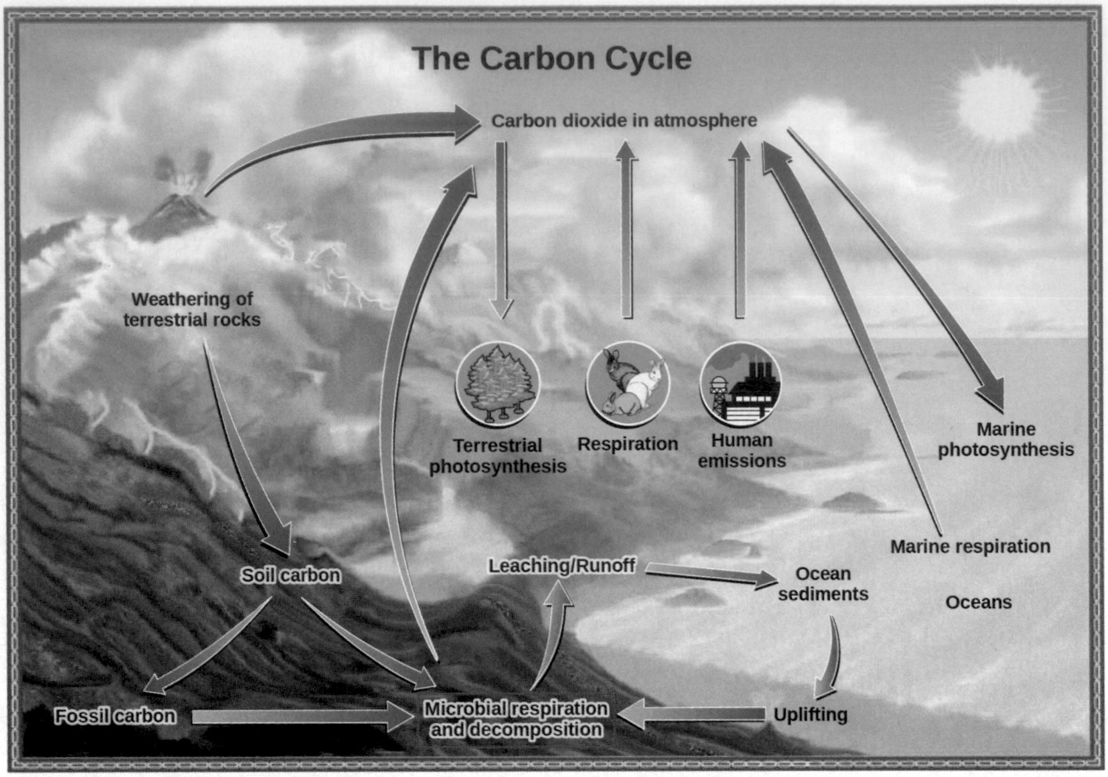

Figure 46.15 Carbon dioxide gas exists in the atmosphere and is dissolved in water. Photosynthesis converts carbon dioxide gas to organic carbon, and respiration cycles the organic carbon back into carbon dioxide gas. Long-term storage of organic carbon occurs when matter from living organisms is buried deep underground and becomes fossilized. Volcanic activity and, more recently, human emissions, bring this stored carbon back into the carbon cycle. (credit: modification of work by John M. Evans and Howard Perlman, USGS)

LINK TO LEARNING

Click this link (http://openstax.org/l/carbon_cycle) to read information about the United States Carbon Cycle Science Program.

The Biological Carbon Cycle

Living organisms are connected in many ways, even between ecosystems. A good example of this connection is the exchange of carbon between autotrophs and heterotrophs within and between ecosystems by way of atmospheric carbon dioxide. Carbon dioxide is the basic building block that most autotrophs use to build multicarbon, high energy compounds, such as glucose. The energy harnessed from the sun is used by these organisms to form the covalent bonds that link carbon atoms together. These chemical bonds thereby store this energy for later use in the process of respiration. Most terrestrial autotrophs obtain their carbon dioxide directly from the atmosphere, while marine autotrophs acquire it in the dissolved form (carbonic acid, $H_2CO_3^-$). However carbon dioxide is acquired, a by-product of the process is oxygen. The photosynthetic organisms are responsible for depositing approximately 21 percent oxygen content of the atmosphere that we observe today.

Heterotrophs and autotrophs are partners in biological carbon exchange (especially the primary consumers, largely herbivores). Heterotrophs acquire the high-energy carbon compounds from the autotrophs by consuming them, and breaking them down by respiration to obtain cellular energy, such as ATP. The most efficient type of respiration, aerobic respiration, requires oxygen obtained from the atmosphere or dissolved in water. Thus, there is a constant exchange of oxygen and carbon dioxide between the autotrophs (which need the carbon) and the heterotrophs (which need the oxygen). Gas exchange through the atmosphere and water is one way that the carbon cycle connects all living organisms on Earth.

The Biogeochemical Carbon Cycle

The movement of carbon through the land, water, and air is complex, and in many cases, it occurs much more slowly geologically than as seen between living organisms. Carbon is stored for long periods in what are known as carbon reservoirs, which include the atmosphere, bodies of liquid water (mostly oceans), ocean sediment, soil, land sediments (including fossil fuels), and the Earth's interior.

As stated, the atmosphere is a major reservoir of carbon in the form of carbon dioxide and is essential to the process of photosynthesis. The level of carbon dioxide in the atmosphere is greatly influenced by the reservoir of carbon in the oceans. The exchange of carbon between the atmosphere and water reservoirs influences how much carbon is found in each location, and each one affects the other reciprocally. Carbon dioxide (CO_2) from the atmosphere dissolves in water and combines with water molecules to form carbonic acid, and then it ionizes to carbonate and bicarbonate ions (Figure 46.16)

Step 1: CO_2 (atmospheric) $\rightleftharpoons$ CO_2 (dissolved)
Step 2: CO_2 (dissolved) + H_2O $\rightleftharpoons$ H_2CO_3 (carbonic acid)
Step 3: H_2CO_3 $\rightleftharpoons$ + H^+ + HCO_3^- (biocarbonate ion)
Step 4: HCO_3^- $\rightleftharpoons$ H^+ + CO_3^{2-} (carbonate ion)

Figure 46.16 Carbon dioxide reacts with water to form bicarbonate and carbonate ions.

The equilibrium coefficients are such that more than 90 percent of the carbon in the ocean is found as bicarbonate ions. Some of these ions combine with seawater calcium to form calcium carbonate ($CaCO_3$), a major component of marine organism shells. These organisms eventually form sediments on the ocean floor. Over geologic time, the calcium carbonate forms limestone, which comprises the largest carbon reservoir on Earth.

On land, carbon is stored in soil as a result of the decomposition of living organisms (by decomposers) or from weathering of terrestrial rock and minerals. This carbon can be leached into the water reservoirs by surface runoff. Deeper underground, on land and at sea, are fossil fuels: the anaerobically decomposed remains of plants that take millions of years to form. Fossil fuels are considered a nonrenewable resource because their use far exceeds their rate of formation. A **nonrenewable resource**, such as fossil fuel, is either regenerated very slowly or not at all. Another way for carbon to enter the atmosphere is from land (including land beneath the surface of the ocean) by the eruption of volcanoes and other geothermal systems. Carbon sediments from the ocean floor are taken deep within the Earth by the process of **subduction**: the movement of one tectonic plate beneath another. Carbon is released as carbon dioxide when a volcano erupts or from volcanic hydrothermal vents.

Humans contribute to atmospheric carbon by the burning of fossil fuels and other materials. Since the Industrial Revolution, humans have significantly increased the release of carbon and carbon compounds, which has in turn affected the climate and overall environment.

Animal husbandry by humans also increases atmospheric carbon. The large numbers of land animals raised to feed the Earth's growing population results in increased carbon dioxide levels in the atmosphere due to farming practices and respiration and methane production. This is another example of how human activity indirectly affects biogeochemical cycles in a significant way. Although much of the debate about the future effects of increasing atmospheric carbon on climate change focuses on fossils fuels, scientists take natural processes, such as volcanoes and respiration, into account as they model and predict the future impact of this increase.

The Nitrogen Cycle

Getting nitrogen into the living world is difficult. Plants and phytoplankton are not equipped to incorporate nitrogen from the atmosphere (which exists as tightly bonded, triple covalent N_2) even though this molecule comprises approximately 78 percent of the atmosphere. Nitrogen enters the living world via free-living and symbiotic bacteria, which incorporate nitrogen into their macromolecules through nitrogen fixation (conversion of N_2). Cyanobacteria live in most aquatic ecosystems where sunlight is present; they play a key role in nitrogen fixation. Cyanobacteria are able to use inorganic sources of nitrogen to "fix" nitrogen. *Rhizobium* bacteria live symbiotically in the root nodules of legumes (such as peas, beans, and peanuts) and provide them with the organic nitrogen they need. (For example, gardeners often grow peas both for their produce and to naturally add nitrogen to the soil. This practice goes back to ancient times, even if the science has only been recently understood.) Free-living bacteria, such as *Azotobacter*, are also important nitrogen fixers.

Organic nitrogen is especially important to the study of ecosystem dynamics since many ecosystem processes, such as primary production and decomposition, are limited by the available supply of nitrogen. As shown in Figure 46.17, the nitrogen that enters living systems by nitrogen fixation is successively converted from organic nitrogen back into nitrogen gas by bacteria. This process occurs in three steps in terrestrial systems: ammonification, nitrification, and denitrification. First, the ammonification process converts nitrogenous waste from living animals or from the remains of dead animals into ammonium

(NH_4^+) by certain bacteria and fungi. Second, the ammonium is converted to nitrites (NO_2^-) by nitrifying bacteria, such as *Nitrosomonas*, through nitrification. Subsequently, nitrites are converted to nitrates (NO_3^-) by similar organisms. Third, the process of denitrification occurs, whereby bacteria, such as *Pseudomonas* and *Clostridium*, convert the nitrates into nitrogen gas, allowing it to reenter the atmosphere.

VISUAL CONNECTION

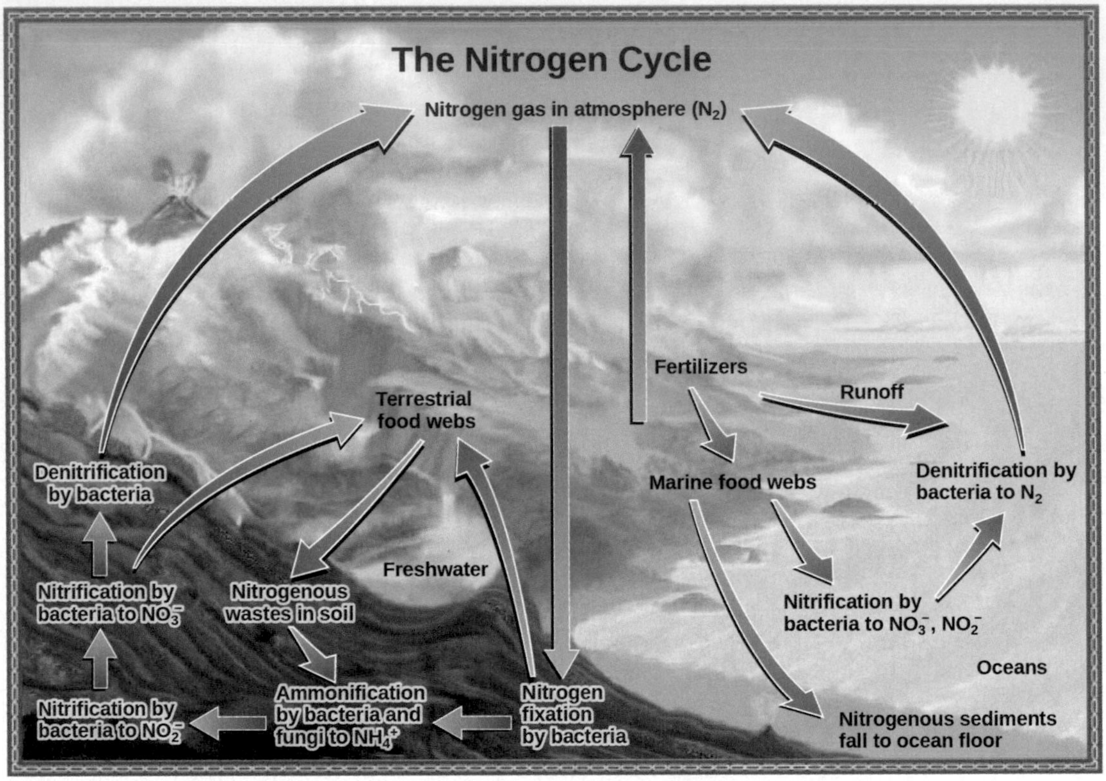

Figure 46.17 Nitrogen enters the living world from the atmosphere via nitrogen-fixing bacteria. This nitrogen and nitrogenous waste from animals is then processed back into gaseous nitrogen by soil bacteria, which also supply terrestrial food webs with the organic nitrogen they need. (credit: modification of work by John M. Evans and Howard Perlman, USGS)

Which of the following statements about the nitrogen cycle is false?

a. Ammonification converts organic nitrogenous matter from living organisms into ammonium (NH_4^+).
b. Denitrification by bacteria converts nitrates (NO_3^-) to nitrogen gas (N_2).
c. Nitrification by bacteria converts nitrates (NO_3^-) to nitrites (NO_2^-).
d. Nitrogen fixing bacteria convert nitrogen gas (N_2) into organic compounds.

Human activity can release nitrogen into the environment by two primary means: the combustion of fossil fuels, which releases different nitrogen oxides, and by the use of artificial fertilizers in agriculture, which are then washed into lakes, streams, and rivers by surface runoff. Atmospheric nitrogen is associated with several effects on Earth's ecosystems including the production of acid rain (as nitric acid, HNO_3) and greenhouse gas (as nitrous oxide, N_2O) potentially causing climate change. A major effect from fertilizer runoff is saltwater and freshwater **eutrophication**, a process whereby nutrient runoff causes the excess growth of microorganisms, depleting dissolved oxygen levels and killing ecosystem fauna.

A similar process occurs in the marine nitrogen cycle, where the ammonification, nitrification, and denitrification processes are performed by marine bacteria. Some of this nitrogen falls to the ocean floor as sediment, which can then be moved to land in geologic time by uplift of the Earth's surface and thereby incorporated into terrestrial rock. Although the movement of nitrogen from rock directly into living systems has been traditionally seen as insignificant compared with nitrogen fixed from the atmosphere, a recent study showed that this process may indeed be significant and should be included in any study of the global nitrogen cycle.[3]

3Scott L. Morford, Benjamin Z. Houlton, and Randy A. Dahlgren, "Increased Forest Ecosystem Carbon and Nitrogen Storage from Nitrogen Rich Bedrock," *Nature* 477, no. 7362 (2011): 78–81.

The Phosphorus Cycle

Phosphorus is an essential nutrient for living processes; it is a major component of nucleic acid and phospholipids, and, as calcium phosphate, makes up the supportive components of our bones. Phosphorus is often the limiting nutrient (necessary for growth) in aquatic ecosystems (Figure 46.18).

Phosphorus occurs in nature as the phosphate ion (PO_4^{3-}). In addition to phosphate runoff as a result of human activity, natural surface runoff occurs when it is leached from phosphate-containing rock by weathering, thus sending phosphates into rivers, lakes, and the ocean. This rock has its origins in the ocean. Phosphate-containing ocean sediments form primarily from the bodies of ocean organisms and from their excretions. However, in remote regions, volcanic ash, aerosols, and mineral dust may also be significant phosphate sources. This sediment then is moved to land over geologic time by the uplifting of areas of the Earth's surface.

Phosphorus is also reciprocally exchanged between phosphate dissolved in the ocean and marine ecosystems. The movement of phosphate from the ocean to the land and through the soil is extremely slow, with the average phosphate ion having an oceanic residence time between 20,000 and 100,000 years.

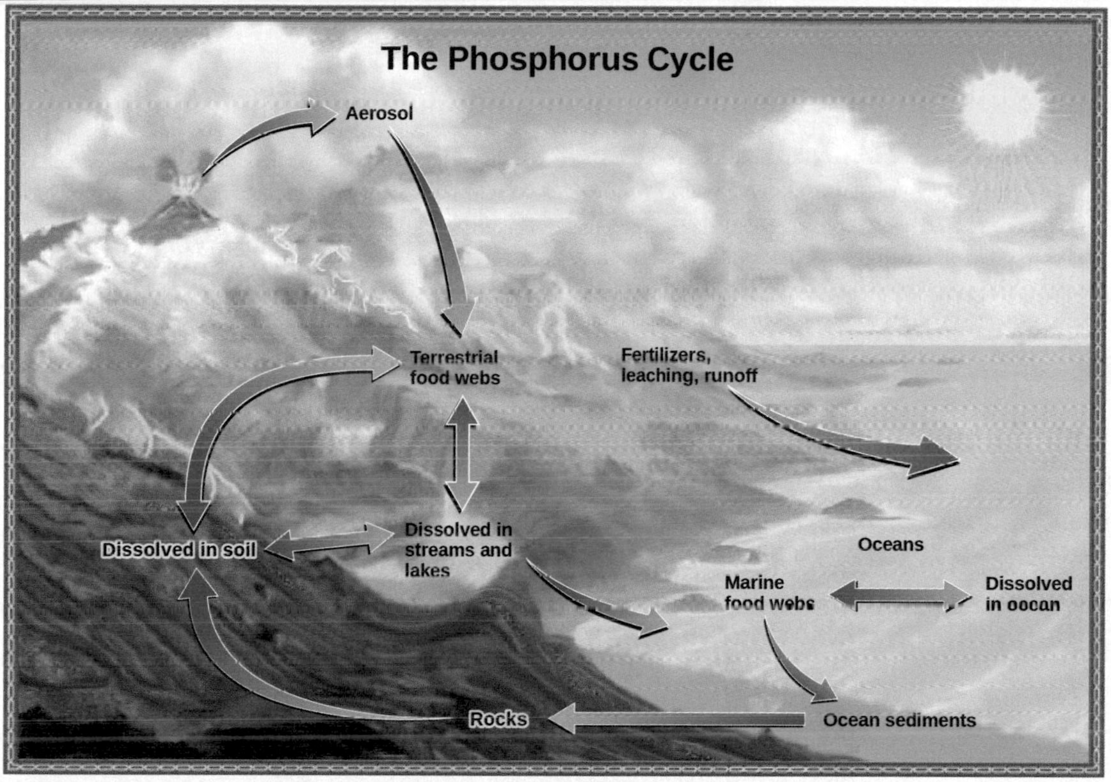

Figure 46.18 In nature, phosphorus exists as the phosphate ion (PO_4^{3-}). Weathering of rocks and volcanic activity releases phosphate into the soil, water, and air, where it becomes available to terrestrial food webs. Phosphate enters the oceans via surface runoff, groundwater flow, and river flow. Phosphate dissolved in ocean water cycles into marine food webs. Some phosphate from the marine food webs falls to the ocean floor, where it forms sediment. (credit: modification of work by John M. Evans and Howard Perlman, USGS)

As discussed in Chapter 44, excess phosphorus and nitrogen that enters these ecosystems from fertilizer runoff and from sewage causes excessive growth of microorganisms and depletes the dissolved oxygen, which leads to the death of many ecosystem fauna, such as shellfish and finfish. This process is responsible for dead zones in lakes and at the mouths of many major rivers (Figure 46.19).

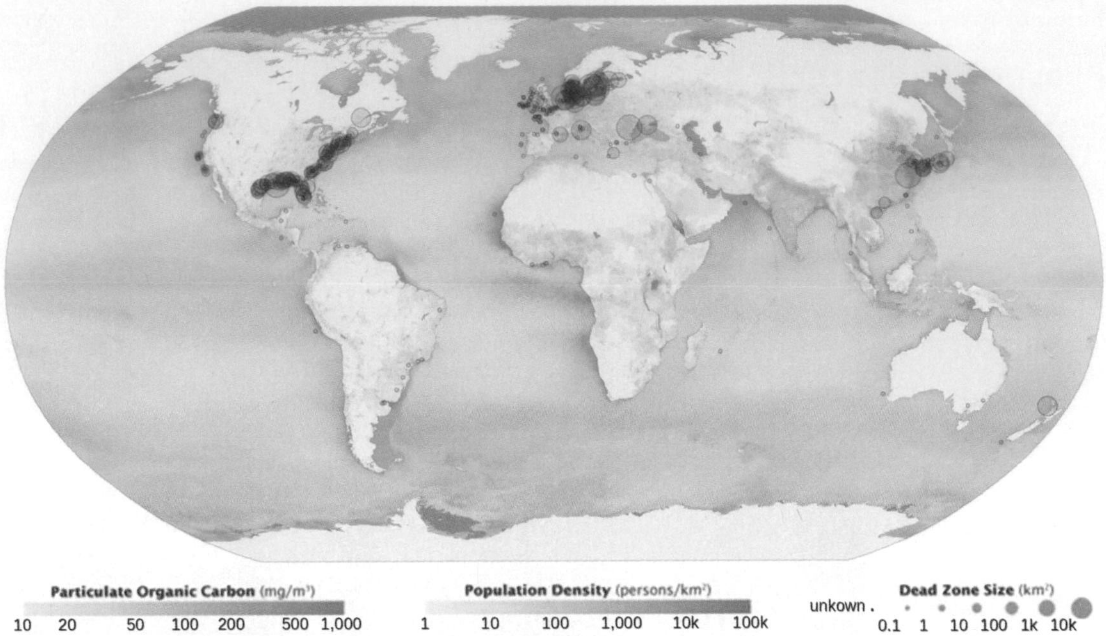

Figure 46.19 Dead zones occur when phosphorus and nitrogen from fertilizers cause excessive growth of microorganisms, which depletes oxygen and kills fauna. Worldwide, large dead zones are found in coastal areas of high population density. (credit: NASA Earth Observatory)

As discussed earlier, a **dead zone** is an area within a freshwater or marine ecosystem where large areas are depleted of their normal flora and fauna; these zones can be caused by eutrophication, oil spills, dumping of toxic chemicals, and other human activities. The number of dead zones has been increasing for several years, and more than 400 of these zones were present as of 2008. One of the worst dead zones is off the coast of the United States in the Gulf of Mexico, where fertilizer runoff from the Mississippi River basin has created a dead zone of over 8463 square miles. Phosphate and nitrate runoff from fertilizers also negatively affect several lake and bay ecosystems including the Chesapeake Bay in the eastern United States.

Everyday Connection

Chesapeake Bay

Figure 46.20 This (a) satellite image shows the Chesapeake Bay, an ecosystem affected by phosphate and nitrate runoff. A (b) member of the Army Corps of Engineers holds a clump of oysters being used as a part of the oyster restoration effort in the bay. (credit a: modification of work by NASA/MODIS; credit b: modification of work by U.S. Army)

The Chesapeake Bay has long been valued as one of the most scenic areas on Earth; it is now in distress and is recognized as a declining ecosystem. In the 1970s, the Chesapeake Bay was one of the first ecosystems to have identified dead zones, which continue to kill many fish and bottom-dwelling species, such as clams, oysters, and worms. Several species have declined in the Chesapeake Bay due to surface water runoff containing excess nutrients from artificial fertilizer used on land. The source of the fertilizers (with high nitrogen and phosphate content) is not limited to agricultural practices. There are many nearby urban areas and more than 150 rivers and streams empty into the bay that are carrying fertilizer runoff from lawns and gardens. Thus, the decline of the Chesapeake Bay is a complex issue and requires the cooperation of industry, agriculture, and everyday homeowners.

Of particular interest to conservationists is the oyster population; it is estimated that more than 200,000 acres of oyster reefs existed in the bay in the 1700s, but that number has now declined to only 36,000 acres. Oyster harvesting was once a major industry for Chesapeake Bay, but it declined 88 percent between 1982 and 2007. This decline was due not only to fertilizer runoff and dead zones but also to overharvesting. Oysters require a certain minimum population density because they must be in close proximity to reproduce. Human activity has altered the oyster population and locations, greatly disrupting the ecosystem.

The restoration of the oyster population in the Chesapeake Bay has been ongoing for several years with mixed success. Not only do many people find oysters good to eat, but they also clean up the bay. Oysters are filter feeders, and as they eat, they clean the water around them. In the 1700s, it was estimated that it took only a few days for the oyster population to filter the entire volume of the bay. Today, with changed water conditions, it is estimated that the present population would take nearly a year to do the same job.

Restoration efforts have been ongoing for several years by nonprofit organizations, such as the Chesapeake Bay Foundation. The restoration goal is to find a way to increase population density so the oysters can reproduce more efficiently. Many disease-resistant varieties (developed at the Virginia Institute of Marine Science for the College of William and Mary) are now available and have been used in the construction of experimental oyster reefs. Efforts to clean and restore the bay by Virginia and Delaware have been hampered because much of the pollution entering the bay comes from other states, which stresses the need for interstate cooperation to gain successful restoration.

The new, hearty oyster strains have also spawned a new and economically viable industry—oyster aquaculture—which not only supplies oysters for food and profit, but also has the added benefit of cleaning the bay.

The Sulfur Cycle

Sulfur is an essential element for the macromolecules of living things. As a part of the amino acid cysteine, it is involved in the formation of disulfide bonds within proteins, which help to determine their 3-D folding patterns, and hence their functions. As shown in Figure 46.21, sulfur cycles between the oceans, land, and atmosphere. Atmospheric sulfur is found in the form of sulfur dioxide (SO_2) and enters the atmosphere in three ways: from the decomposition of organic molecules, from volcanic activity and geothermal vents, and from the burning of fossil fuels by humans.

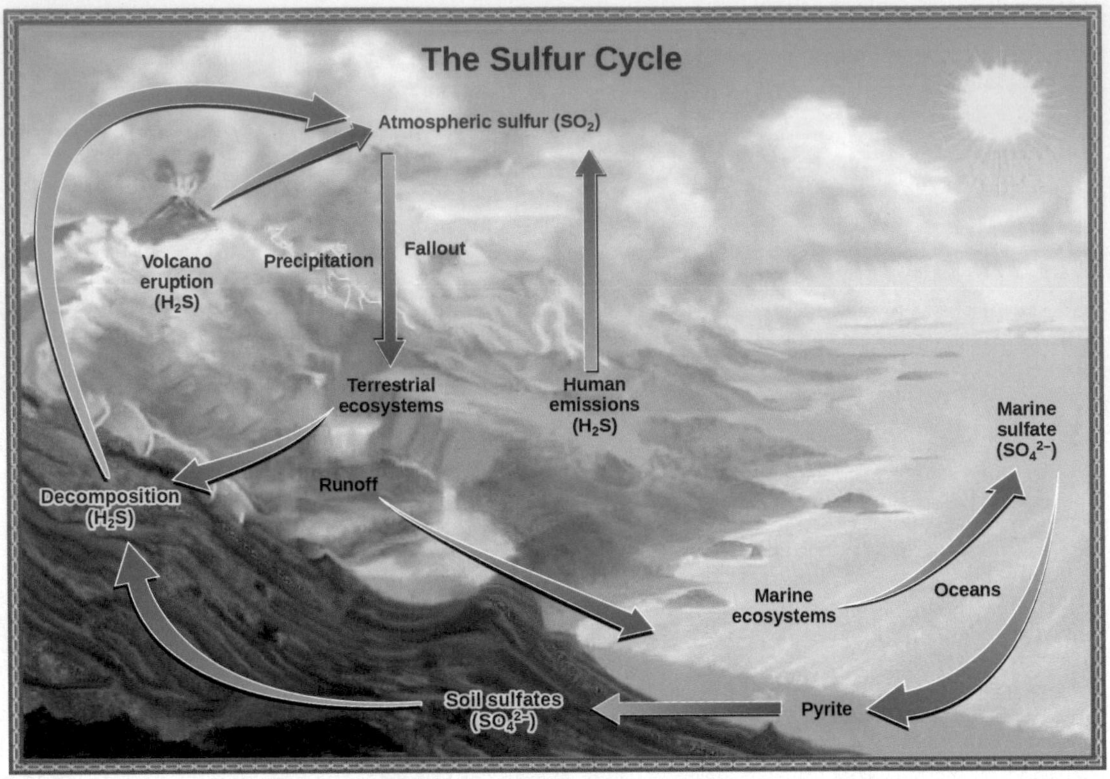

Figure 46.21 Sulfur dioxide from the atmosphere becomes available to terrestrial and marine ecosystems when it is dissolved in precipitation as weak sulfurous acid or when it falls directly to the Earth as fallout. Weathering of rocks also makes sulfates available to terrestrial ecosystems. Decomposition of living organisms returns sulfates to the ocean, soil, and atmosphere. (credit: modification of work by John M. Evans and Howard Perlman, USGS)

On land, sulfur is deposited in four major ways: precipitation, direct fallout from the atmosphere, rock weathering, and geothermal vents (Figure 46.21). Atmospheric sulfur is found in the form of sulfur dioxide (SO_2), and as rain falls through the atmosphere, sulfur is dissolved in the form of weak sulfurous acid (H_2SO_3). Sulfur can also fall directly from the atmosphere in a process called **fallout**. Also, the weathering of sulfur-containing rocks releases sulfur into the soil. These rocks originate from ocean sediments that are moved to land by the geologic uplifting of ocean sediments. Terrestrial ecosystems can then make use of these soil sulfates (SO_4-), and upon the death and decomposition of these organisms, release the sulfur back into the atmosphere as hydrogen sulfide (H_2S) gas.

Figure 46.22 At this sulfur vent in Lassen Volcanic National Park in northeastern California, the yellowish sulfur deposits are visible near the mouth of the vent.

Sulfur enters the ocean via runoff from land, from atmospheric fallout, and from underwater geothermal vents. Some ecosystems (Figure 46.9) rely on chemoautotrophs using sulfur as a biological energy source. This sulfur then supports marine ecosystems in the form of sulfates.

Human activities have played a major role in altering the balance of the global sulfur cycle. The burning of large quantities of fossil fuels, especially from coal, releases larger amounts of hydrogen sulfide gas into the atmosphere. **Acid rain** is caused by rainwater falling to the ground through this sulfur dioxide gas, turning it into weak sulfuric acid. Acid rain damages the natural environment by lowering the pH of lakes, which kills many of the resident fauna; it also affects the man-made environment through the chemical degradation of buildings. For example, many marble monuments, such as the Lincoln Memorial in Washington, DC, have suffered significant damage from acid rain over the years.

LINK TO LEARNING

Click this link (http://openstax.org/l/climate_change) to learn more about global climate change.

KEY TERMS

acid rain corrosive rain caused by rainwater falling to the ground through sulfur dioxide gas, turning it into weak sulfuric acid; can damage structures and ecosystems

analytical model ecosystem model that is created with mathematical formulas to predict the effects of environmental disturbances on ecosystem structure and dynamics

apex consumer organism at the top of the food chain

assimilation biomass consumed and assimilated from the previous trophic level after accounting for the energy lost due to incomplete ingestion of food, energy used for respiration, and energy lost as waste

biogeochemical cycle cycling of mineral nutrients through ecosystems and through the nonliving world

biomagnification increasing concentrations of persistent, toxic substances in organisms at each trophic level, from the primary producers to the apex consumers

biomass total weight, at the time of measurement, of living or previously living organisms in a unit area within a trophic level

chemoautotroph organism capable of synthesizing its own food using energy from inorganic molecules

conceptual model (also, compartment model) ecosystem model that consists of flow charts that show the interactions of different compartments of the living and nonliving components of the ecosystem

dead zone area within an ecosystem in lakes and near the mouths of rivers where large areas of ecosystems are depleted of their normal flora and fauna; these zones can be caused by eutrophication, oil spills, dumping of toxic chemicals, and other human activities

detrital food web type of food web in which the primary consumers consist of decomposers; these are often associated with grazing food webs within the same ecosystem

ecological pyramid (also, Eltonian pyramid) graphical representation of different trophic levels in an ecosystem based of organism numbers, biomass, or energy content

ecosystem community of living organisms and their interactions with their abiotic environment

ecosystem dynamics study of the changes in ecosystem structure caused by changes in the environment or internal forces

equilibrium steady state of an ecosystem where all organisms are in balance with their environment and each other

eutrophication process whereby nutrient runoff causes the excess growth of microorganisms, depleting dissolved oxygen levels and killing ecosystem fauna

fallout direct deposit of solid minerals on land or in the ocean from the atmosphere

food chain linear representation of a chain of primary producers, primary consumers, and higher-level consumers used to describe ecosystem structure and dynamics

food web graphic representation of a holistic, nonlinear web of primary producers, primary consumers, and higher-level consumers used to describe ecosystem structure and dynamics

grazing food web type of food web in which the primary producers are either plants on land or phytoplankton in the water; often associated with a detrital food web within the same ecosystem

gross primary productivity rate at which photosynthetic primary producers incorporate energy from the sun

holistic ecosystem model study that attempts to quantify the composition, interactions, and dynamics of entire ecosystems; often limited by economic and logistical difficulties, depending on the ecosystem

hydrosphere area of the Earth where water movement and storage occurs

mesocosm portion of a natural ecosystem to be used for experiments

microcosm re-creation of natural ecosystems entirely in a laboratory environment to be used for experiments

net consumer productivity energy content available to the organisms of the next trophic level

net primary productivity energy that remains in the primary producers after accounting for the organisms' respiration and heat loss

net production efficiency (NPE) measure of the ability of a trophic level to convert the energy it receives from the previous trophic level into biomass

nonrenewable resource resource, such as fossil fuel, that is either regenerated very slowly or not at all

primary consumer trophic level that obtains its energy from the primary producers of an ecosystem

primary producer trophic level that obtains its energy from sunlight, inorganic chemicals, or dead and/or decaying organic material

residence time measure of the average time an individual water molecule stays in a particular reservoir

resilience (ecological) speed at which an ecosystem recovers equilibrium after being disturbed

resistance (ecological) ability of an ecosystem to remain at equilibrium in spite of disturbances

secondary consumer usually a carnivore that eats primary consumers

simulation model ecosystem model that is created with computer programs to holistically model ecosystems and to predict the effects of environmental disturbances on ecosystem structure and dynamics

subduction movement of one tectonic plate beneath another

tertiary consumer carnivore that eats other carnivores

trophic level position of a species or group of species in a food chain or a food web

trophic level transfer efficiency (TLTE) energy transfer efficiency between two successive trophic levels

CHAPTER SUMMARY

46.1 Ecology of Ecosystems

Ecosystems exist on land, at sea, in the air, and underground. Different ways of modeling ecosystems are necessary to understand how environmental disturbances will affect ecosystem structure and dynamics. Conceptual models are useful to show the general relationships between organisms and the flow of materials or energy between them. Analytical models are used to describe linear food chains, and simulation models work best with holistic food webs.

46.2 Energy Flow through Ecosystems

Organisms in an ecosystem acquire energy in a variety of ways, which is transferred between trophic levels as the energy flows from the bottom to the top of the food web, with energy being lost at each transfer. The efficiency of these transfers is important for understanding the different behaviors and eating habits of warm-blooded versus cold-blooded animals. Modeling of ecosystem energy is best done with ecological pyramids of energy, although other ecological pyramids provide other vital information about ecosystem structure.

46.3 Biogeochemical Cycles

Mineral nutrients are cycled through ecosystems and their environment. Of particular importance are water, carbon, nitrogen, phosphorus, and sulfur. All of these cycles have major impacts on ecosystem structure and function. A variety of human activities, such as pollution, oil spills, and other events have damaged ecosystems, potentially causing global climate change. The health of Earth depends on understanding these cycles and how to protect the environment from irreversible damage.

VISUAL CONNECTION QUESTIONS

1. Figure 46.8 Why do you think the value for gross productivity of the primary producers is the same as the value for total heat and respiration (20,810 kcal/m^2/yr)?

2. Figure 46.10 Pyramids depicting the number of organisms or biomass may be inverted, upright, or even diamond-shaped. Energy pyramids, however, are always upright. Why?

3. Figure 46.17 Which of the following statements about the nitrogen cycle is false?
   a. Ammonification converts organic nitrogenous matter from living organisms into ammonium (NH_4^+).
   b. Denitrification by bacteria converts nitrates (NO_3^-) to nitrogen gas (N_2).
   c. Nitrification by bacteria converts nitrates (NO_3^-) to nitrites (NO_2^-).
   d. Nitrogen fixing bacteria convert nitrogen gas (N_2) into organic compounds.

REVIEW QUESTIONS

4. The ability of an ecosystem to return to its equilibrium state after an environmental disturbance is called ________.
   a. resistance
   b. restoration
   c. reformation
   d. resilience

5. A re-created ecosystem in a laboratory environment is known as a ________.
   a. mesocosm
   b. simulation
   c. microcosm
   d. reproduction

6. Decomposers are associated with which class of food web?
   a. grazing
   b. detrital
   c. inverted
   d. aquatic

7. The primary producers in an ocean grazing food web are usually _______.
   a. plants
   b. animals
   c. fungi
   d. phytoplankton

8. What term describes the use of mathematical equations in the modeling of linear aspects of ecosystems?
   a. analytical modeling
   b. simulation modeling
   c. conceptual modeling
   d. individual-based modeling

9. The position of an organism along a food chain is known as its _______.
   a. locus
   b. location
   c. trophic level
   d. microcosm

10. The loss of an apex consumer would impact which trophic level of a food web?
    a. primary producers
    b. primary consumers
    c. secondary consumers
    d. all of the above

11. A food chain would be a better resource than a food web to answer which question?
    a. How does energy move from an organism in one trophic level to an organism on the next trophic level?
    b. How does energy move within a trophic level?
    c. What preys on grasses?
    d. How is organic matter recycled in a forest?

12. The weight of living organisms in an ecosystem at a particular point in time is called:
    a. energy
    b. production
    c. entropy
    d. biomass

13. Which term describes the process whereby toxic substances increase along trophic levels of an ecosystem?
    a. biomassification
    b. biomagnification
    c. bioentropy
    d. heterotrophy

14. Organisms that can make their own food using inorganic molecules are called:
    a. autotrophs
    b. heterotrophs
    c. photoautotrophs
    d. chemoautotrophs

15. In the English Channel ecosystem, the number of primary producers is smaller than the number of primary consumers because_______.
    a. the apex consumers have a low turnover rate
    b. the primary producers have a low turnover rate
    c. the primary producers have a high turnover rate
    d. the primary consumers have a high turnover rate

16. What law of chemistry determines how much energy can be transferred when it is converted from one form to another?
    a. the first law of thermodynamics
    b. the second law of thermodynamics
    c. the conservation of matter
    d. the conservation of energy

17. The mussels that live at the NW Eifuku volcano are examples of______.
    a. chemoautotrophs
    b. photoautotrophs
    c. apex predators
    d. primary consumers

18. The movement of mineral nutrients through organisms and their environment is called a _______ cycle.
    a. biological
    b. bioaccumulation
    c. biogeochemical
    d. biochemical

19. Carbon is present in the atmosphere as _______.
    a. carbon dioxide
    b. carbonate ion
    c. carbon dust
    d. carbon monoxide

20. The majority of water found on Earth is:
    a. ice
    b. water vapor
    c. fresh water
    d. salt water

21. The average time a molecule spends in its reservoir is known as _______.
    a. residence time
    b. restriction time
    c. resilience time
    d. storage time

22. The process whereby oxygen is depleted by the growth of microorganisms due to excess nutrients in aquatic systems is called _______.
    a. dead zoning
    b. eutrophication
    c. retrofication
    d. depletion
23. The process whereby nitrogen is brought into organic molecules is called _______.
    a. nitrification
    b. denitrification
    c. nitrogen fixation
    d. nitrogen cycling
24. Which of the following approaches would be the most effective way to reduce greenhouse carbon dioxide?
    a. Increase waste deposition into the deep ocean.
    b. Plant more environmentally-suitable plants.
    c. Increase use of fuel sources that do not produce carbon dioxide as a by-product.
    d. Decrease livestock agriculture.
25. How would loss of fungi in a forest effect biogeochemical cycles in the area?
    a. Nitrogen could no longer be fixed into organic molecules.
    b. Phosphorus stores would be released for use by other organisms.
    c. Sulfur release from eroding rocks would cease.
    d. Carbon would accumulate in dead organic matter and waste.

CRITICAL THINKING QUESTIONS

26. Compare and contrast food chains and food webs. What are the strengths of each concept in describing ecosystems?
27. Describe freshwater, ocean, and terrestrial ecosystems.
28. Compare grazing and detrital food webs. Why would they both be present in the same ecosystem?
29. How does the microcosm modeling approach differ from utilizing a holistic model for ecological research?
30. How do conceptual and analytical models of ecosystems compliment each other?
31. Compare the three types of ecological pyramids and how well they describe ecosystem structure. Identify which ones can be inverted and give an example of an inverted pyramid for each.
32. How does the amount of food a warm-blooded animal (endotherm) eats relate to its net production efficiency (NPE)?
33. A study uses an inverted pyramid to demonstrate the relationship between sharks, their aquatic prey, and phytoplankton in an ocean region. What type of pyramid must be used? What does this convey to readers about predation in the area?
34. Describe what a pyramid of numbers would like if an ecologist models the relationship between bird parasites, blue jays, and oak trees in a hectare. Does this match the energy flow pyramid?
35. Describe nitrogen fixation and why it is important to agriculture.
36. What are the factors that cause dead zones? Describe eutrophication, in particular, as a cause.
37. Why are drinking water supplies still a major concern for many countries?
38. Discuss how the human disruption of the carbon cycle has caused ocean acidification.

CHAPTER 47
Conservation Biology and Biodiversity

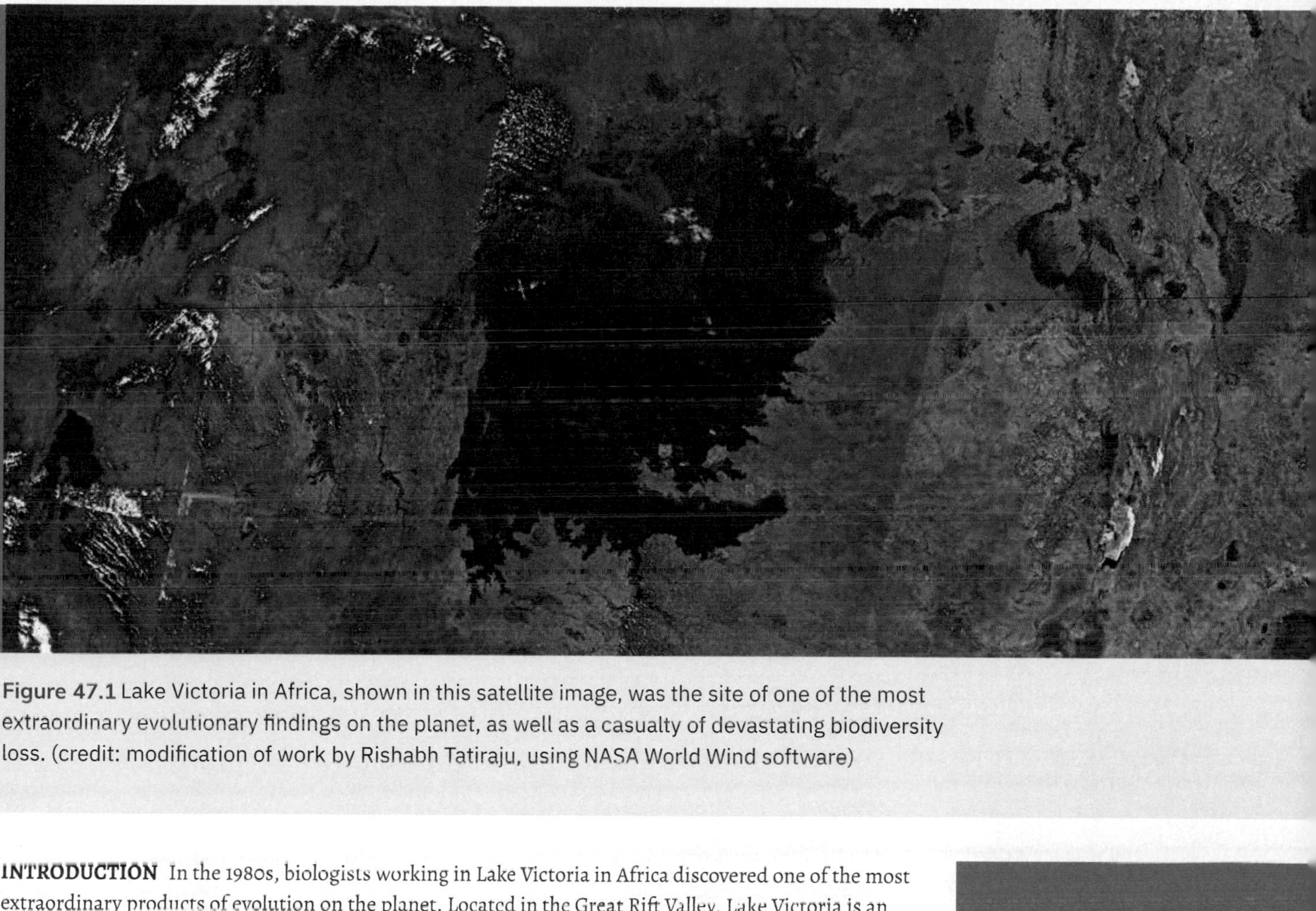

Figure 47.1 Lake Victoria in Africa, shown in this satellite image, was the site of one of the most extraordinary evolutionary findings on the planet, as well as a casualty of devastating biodiversity loss. (credit: modification of work by Rishabh Tatiraju, using NASA World Wind software)

INTRODUCTION In the 1980s, biologists working in Lake Victoria in Africa discovered one of the most extraordinary products of evolution on the planet. Located in the Great Rift Valley, Lake Victoria is an enormous and deep lake about 68,900 km^2 in area (larger than Lake Huron, the second largest of North America's Great Lakes). Biologists were studying species of a family of fish called cichlids. When they sampled for fish in different locations of the lake, the researchers identified over 500 evolved species in total. However, the scientists soon discovered that the invasive Nile Perch was destroying the lake's cichlid population, bringing hundreds of cichlid species to extinction with devastating rapidity.

Chapter Outline

47.1 The Biodiversity Crisis

By the end of this section, you will be able to do the following:

- Define biodiversity in terms of species diversity and abundance
- Describe biodiversity as the equilibrium of naturally fluctuating rates of extinction and speciation
- Identify historical causes of high extinction rates in Earth's history

Traditionally, ecologists have measured **biodiversity**, a general term for the number of species present in the biosphere, by taking into account both the number of species and their relative abundance to each other. Biodiversity can be estimated at a number of levels of organization of living organisms. These estimation indices, which came from *information theory*, are most useful as a first step in quantifying biodiversity between and within ecosystems; they are less useful when the main concern among conservation biologists is simply the loss of biodiversity. However, biologists recognize that measures of biodiversity, in terms of species diversity, may help focus efforts to preserve the biologically or technologically important elements of biodiversity.

The Lake Victoria cichlids provide an example with which we can begin to understand biodiversity. The biologists studying cichlids in the 1980s discovered hundreds of cichlid species representing a variety of specializations to specialized habitat types and specific feeding strategies: such as eating plankton floating in the water, scraping/eating algae from rocks, eating insect larvae from the lake bottom, and eating the eggs of other species of cichlid. The cichlids of Lake Victoria are the product of an complex *adaptive radiation*. An **adaptive radiation** is a rapid (less than three million years in the case of the Lake Victoria cichlids) branching through speciation of a phylogenetic clade into many closely related species. Typically, the species "radiate" into different habitats and niches. The Galápagos Island finches are an example of a modest adaptive radiation with 15 species. The cichlids of Lake Victoria are an example of a spectacular adaptive radiation that formerly included about 500 species.

At the time biologists were making this discovery, some species began to quickly disappear. A culprit in these declines was the Nile perch, a species of large predatory fish that was introduced to Lake Victoria by fisheries to feed the people living around the lake. The Nile perch was introduced in 1963, but its populations did not begin to surge until the 1980s. The perch population grew by consuming cichlids, driving species after species to the point of **extinction** (the disappearance of a species). In fact, there were several factors that played a role in the extinction of perhaps 200 cichlid species in Lake Victoria: the Nile perch, declining lake water quality due to agriculture and land clearing on the shores of Lake Victoria, and increased fishing pressure. Scientists had not even catalogued all of the species present—so many were lost that were never named. The diversity is now a shadow of what it once was.

The cichlids of Lake Victoria are a thumbnail sketch of contemporary rapid species loss that occurs all over Earth that is caused primarily by human activity. Extinction is a natural process of macroevolution that occurs at the rate of about one out of 1 million species becoming extinct per year. The fossil record reveals that there have been five periods of **mass extinction** in history with much higher rates of species loss, and the rate of species loss today is comparable to those periods of mass extinction. However, there is a major difference between the previous mass extinctions and the current extinction we are experiencing: human activity. Specifically, three human activities have a major impact: 1) destruction of habitat, 2) introduction of exotic species, and 3) over-harvesting. Predictions of species loss within the next century, a tiny amount of time on geological timescales, range from 10 percent to 50 percent. Extinctions on this scale have only happened five other times in the history of the planet, and these extinctions were caused by cataclysmic events that changed the course of the history of life in each instance.

Types of Biodiversity

Scientists generally accept that the term biodiversity describes the number and kinds of species and their abundance in a given location or on the planet. Species can be difficult to define, but most biologists still feel comfortable with the concept and are able to identify and count eukaryotic species in most contexts. Biologists have also identified alternate measures of biodiversity, some of which are important for planning how to preserve biodiversity.

Genetic diversity is one of those alternate concepts. **Genetic diversity**, or genetic variation defines the raw material for evolution and adaptation in a species. A species' future potential for adaptation depends on the genetic diversity held in the genomes of the individuals in populations that make up the species. The same is true for higher taxonomic categories. A genus with very different types of species will have more genetic diversity than a genus with species that are genetically similar and have similar ecologies. If there were a choice between one of these genera of species being preserved, the one with the greatest potential for subsequent evolution is the most *genetically diverse* one.

Many genes code for proteins, which in turn carry out the metabolic processes that keep organisms alive and reproducing. Genetic diversity can be measured as **chemical diversity** in that different species produce a variety of chemicals in their cells, both the proteins as well as the products and byproducts of metabolism. This chemical diversity has potential benefit for humans as a source of pharmaceuticals, so it provides one way to measure diversity that is important to human health and welfare.

Humans have generated diversity in domestic animals, plants, and fungi, among many other organisms. This diversity is also suffering losses because of migration, market forces, and increasing globalism in agriculture, especially in densely populated regions such as China, India, and Japan. The human population directly depends on this diversity as a stable food source, and its decline is troubling biologists and agricultural scientists.

It is also useful to define **ecosystem diversity**, meaning the number of different ecosystems on the planet or within a given geographic area (Figure 47.2). Whole ecosystems can disappear even if some of the species might survive by adapting to other ecosystems. The loss of an ecosystem means the loss of interactions between species, the loss of unique features of coadaptation, and the loss of biological productivity that an ecosystem is able to create. An example of a largely extinct ecosystem in North America is the *prairie ecosystem*. Prairies once spanned central North America from the boreal forest in northern Canada down into Mexico. They are now all but gone, replaced by crop fields, pasture lands, and suburban sprawl. Many of the species survive elsewhere, but the hugely productive ecosystem that was responsible for creating the most productive agricultural soils in the United States is now gone. As a consequence, native soils are disappearing or must be maintained and enhanced at great expense.

(a)

(b)

Figure 47.2 The variety of ecosystems on Earth—from (a) coral reef to (b) prairie—enables a great diversity of species to exist. (credit a: modification of work by Jim Maragos, USFWS; credit b: modification of work by Jim Minnerath, USFWS)

Current Species Diversity

Despite considerable effort, knowledge of the species that inhabit the planet is limited and always will be because of a continuing lack of financial resources and political willpower. A recent estimate suggests that the eukaryote species for which science has names, about 1.5 million species, account for less than 20 percent of the total number of eukaryote species present on the planet (8.7 million species, by one estimate). Estimates of numbers of prokaryotic species are largely guesses, but biologists agree that science has only begun to catalog their diversity. Even with what is known, there is no central repository of names or samples of the described species; therefore, there is no way to be sure that the 1.5 million descriptions *is* an accurate accounting. It is a best guess based on the opinions of experts in different taxonomic groups. Given that Earth is losing species at an accelerating pace, science is very much in the place it was with the Lake Victoria cichlids: knowing little about what is being lost. Table 47.1 presents recent estimates of biodiversity in different groups.

Estimates of the Numbers of Described and Predicted Species by Taxonomic Group

	Mora et al. 2011[1]		Chapman 2009[2]		Groombridge & Jenkins 2002[3]	
	Described	Predicted	Described	Predicted	Described	Predicted
Animalia	1,124,516	9,920,000	1,424,153	6,836,330	1,225,500	10,820,000

Table 47.1

	Mora et al. 2011[1]		Chapman 2009[2]		Groombridge & Jenkins 2002[3]	
Chromista	17,892	34,900	25,044	200,500	—	—
Fungi	44,368	616,320	98,998	1,500,000	72,000	1,500,000
Plantae	224,244	314,600	310,129	390,800	270,000	320,000
Protozoa	16,236	72,800	28,871	1,000,000	80,000	600,000
Prokaryotes	—	—	10,307	1,000,000	10,175	—
Total	1,438,769	10,960,000	1,897,502	10,897,630	1,657,675	13,240,000

Table 47.1

There are various initiatives to catalog described species in accessible ways, and the internet is facilitating that effort. Nevertheless, it has been pointed out that at the current rate of new species descriptions (which according to the State of Observed Species Report is 17,000 to 20,000 new species per year), it will take close to 500 years to finish describing life on this planet.[4] Over time, the task becomes both increasingly difficult and increasingly easier as extinction removes species from the planet.

Naming and counting species may seem like an unimportant pursuit given the other needs of humanity, but determining biodiversity it is not simply an accounting of species. Describing a species is a complex process through which biologists determine an organism's unique characteristics and whether or not that organism belongs to any other described species or genus. It allows biologists to find and recognize the species after the initial discovery, and allows them to follow up on questions about its biology. In addition, the unique characteristics of each species make it potentially valuable to humans or other species on which humans depend.

Patterns of Biodiversity

Biodiversity is not evenly distributed on Earth. Lake Victoria contained almost 500 species of cichlids alone, ignoring the other fish families present in the lake. All of these species were found only in Lake Victoria; therefore, the 500 species of cichlids were *endemic*. **Endemic species** are found in only one location. Endemics with highly restricted distributions are particularly vulnerable to extinction. Higher taxonomic levels, such as genera and families, can also be endemic. Lake Michigan contains about 79 species of fish, many of which are found in other lakes in North America. What accounts for the difference in fish diversity in these two lakes? Lake Victoria is an ancient tropical lake, while Lake Michigan is a recently formed temperate lake. Lake Michigan in its present form is only about 7,000 years old, while Lake Victoria in its present form is about 15,000 years old, although its basin is about 400,000 years in age. Biogeographers have suggested these two factors, latitude and age, are two of several hypotheses to explain biodiversity patterns on the planet.

1Mora Camilo et al., "How Many Species Are There on Earth and in the Ocean?" *PLoS Biology* (2011), doi:10.1371/journal.pbio.1001127.

2Arthur D. Chapman, *Numbers of Living Species in Australia and the World*, 2nd ed. (Canberra, AU: Australian Biological Resources Study, 2009). https://www.environment.gov.au/system/files/pages/2ee3f4a1-f130-465b-9c7a-79373680a067/files/nlsaw-2nd-complete.pdf/ (http://openstax.org/l/Aus_diversity) .

3Brian Groombridge and Martin D. Jenkins. *World Atlas of Biodiversity: Earth's Living Resources in the 21st Century*. Berkeley: University of California Press, 2002.

4International Institute for Species Exploration (IISE), *2011 State of Observed Species (SOS)*. Tempe, AZ: IISE, 2011. Accessed May, 20, 2012. http://www.esf.edu/species/ (http://openstax.org/l/observed_species) .

CAREER CONNECTION

Biogeographer

Biogeography is the study of the distribution of the world's species—both in the past and in the present. The work of *biogeographers* is critical to understanding our physical environment, how the environment affects species, and how environmental changes impact the distribution of a species; it has also been critical to developing modern evolutionary theory. Biogeographers need to understand both biology and ecology. They also need to be well-versed in evolutionary studies, soil science, and climatology.

There are three main fields of study under the heading of biogeography: *ecological biogeography, historical biogeography* (called *paleobiogeography*), *and conservation biogeography.* **Ecological biogeography** studies the current factors affecting the distribution of plants and animals. **Historical biogeography**, as the name implies, studies the past distribution of species. **Conservation biogeography**, on the other hand, is focused on the protection and restoration of species based upon known historical and current ecological information. Each of these fields considers both zoogeography and phytogeography—the past and present distribution of animals and plants.

One of the oldest observed patterns in ecology is that species biodiversity in almost every taxonomic group increases as latitude declines. In other words, biodiversity increases closer to the equator (Figure 47.3).

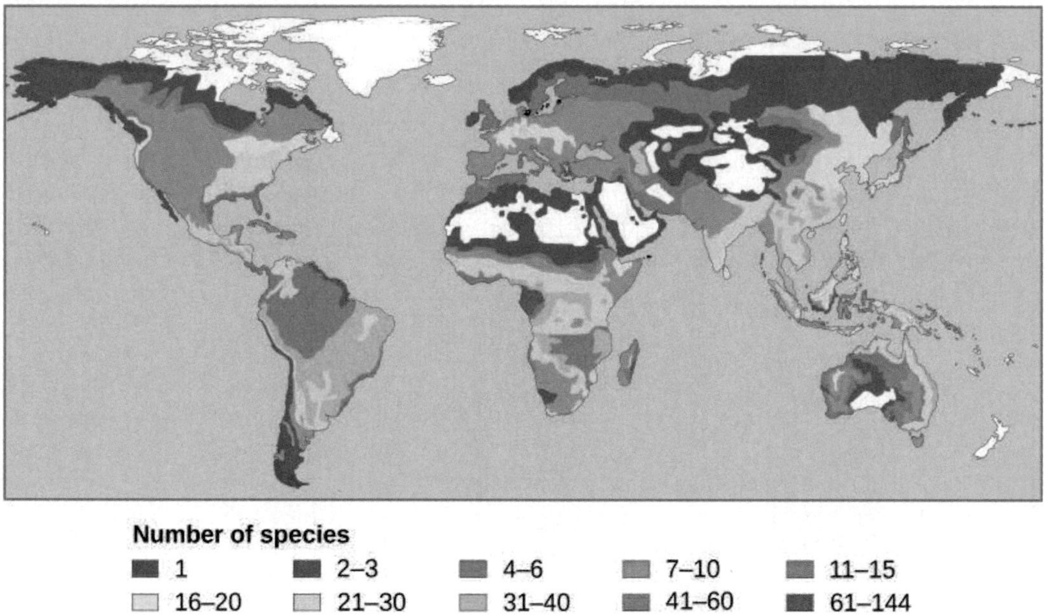

Figure 47.3 This map illustrates the number of amphibian species across the globe and shows the trend toward higher biodiversity at lower latitudes. A similar pattern is observed for most taxonomic groups. The white areas indicate a lack of data in this particular study.

It is not yet clear why biodiversity increases closer to the equator, but scientists have several hypotheses. One factor may be the greater age of the ecosystems in the tropics versus those in temperate regions; the temperate regions were largely devoid of life or were drastically reduced during the last glaciation. The idea is that greater age provides more time for speciation. Another possible explanation is the increased direct energy the tropics receive from the sun versus the decreased intensity of the solar energy that temperate and polar regions receive. Tropical ecosystem complexity may promote speciation by increasing the **heterogeneity**, or *number of ecological niches*, in the tropics relative to higher latitudes. The greater heterogeneity provides more opportunities for coevolution, specialization, and perhaps greater selection pressures leading to population differentiation. However, this hypothesis suffers from some circularity—ecosystems with more species encourage speciation, but how did they get more species to begin with?

The tropics have been perceived as being more stable than temperate regions, which have a pronounced climate and day-length seasonality. The tropics have their own forms of seasonality, such as rainfall, but they are generally assumed to be more stable environments and this stability might promote speciation into highly specialized niches.

Regardless of the mechanisms, it is certainly true that all levels of biodiversity are greatest in the tropics. Additionally, the *rate of endemism* is highest, and there are more biodiversity "hotspots." However, this richness of diversity also means that knowledge of species is unfortunately very low, and there is a high potential for biodiversity loss.

Conservation of Biodiversity

In 1988, British environmentalist Norman Myers developed a conservation concept to identify areas rich in species and at significant risk for species loss: *biodiversity hotspots.* **Biodiversity hotspots** are geographical areas that contain high numbers of endemic species. The purpose of the concept was to identify important locations on the planet for conservation efforts, a kind of conservation triage. By protecting hotspots, governments are able to protect a larger number of species. The original criteria for a hotspot included the presence of 1500 or more endemic plant species and 70 percent of the area disturbed by human activity. There are now 34 biodiversity hotspots (Figure 47.4) containing large numbers of endemic species, which include half of Earth's endemic plants.

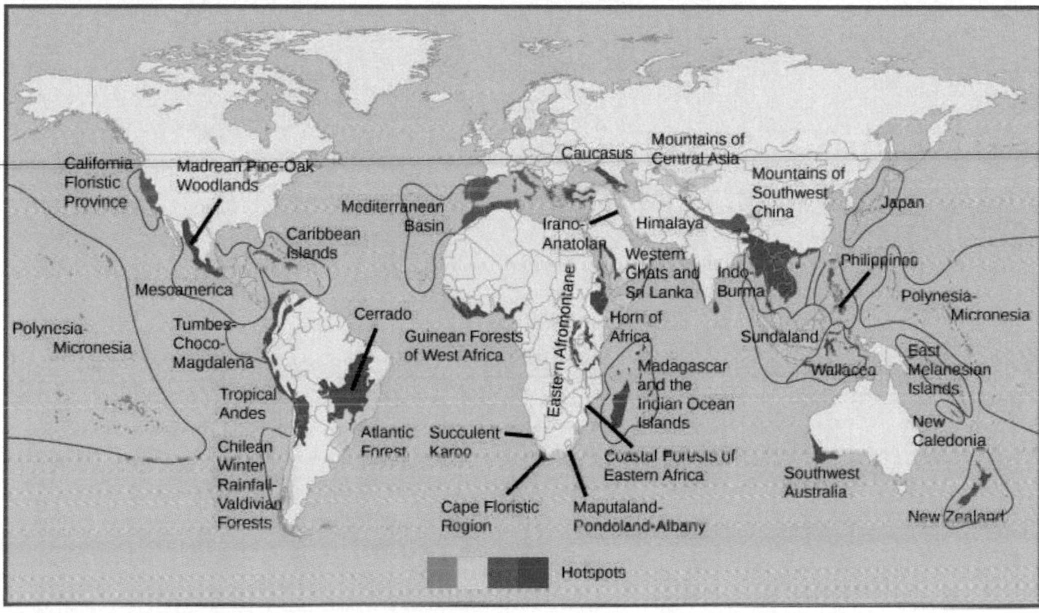

Figure 47.4 Conservation International has identified 34 biodiversity hotspots, which cover only 2.3 percent of the Earth's surface but have endemic to them 42 percent of the terrestrial vertebrate species and 50 percent of the world's plants.

Biodiversity Change through Geological Time

The number of species on the planet, or in any geographical area, is the result of an equilibrium of two evolutionary processes that are continuously ongoing: *speciation* and *extinction*. Both are natural "birth" and "death" processes of macroevolution. When speciation rates begin to outstrip extinction rates, the number of species will increase; likewise, the number of species will decrease when extinction rates begin to overtake speciation rates. Throughout Earth's history, these two processes have fluctuated—sometimes leading to dramatic changes in the number of species on Earth as reflected in the fossil record (Figure 47.5).

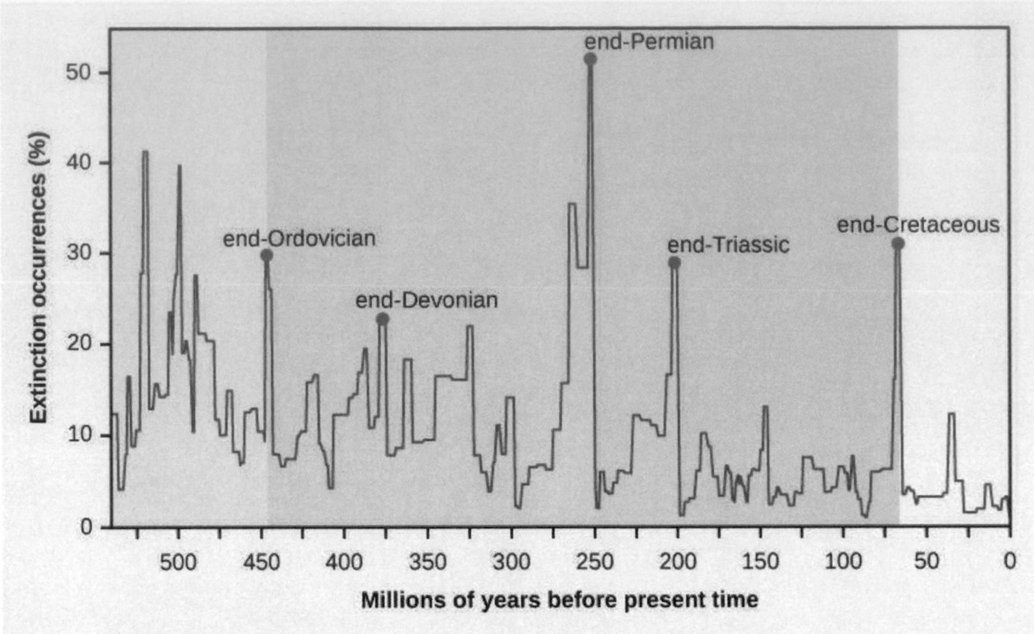

Figure 47.5 Percent extinction occurrences as reflected in the fossil record have fluctuated throughout Earth's history. Sudden and dramatic losses of biodiversity, called *mass extinctions*, have occurred five times.

Paleontologists have identified five strata in the fossil record that appear to show sudden and dramatic (greater than half of all extant species disappearing from the fossil record) losses in biodiversity. These are called **mass extinctions**. There are many lesser, yet still dramatic, extinction events, but the five mass extinctions have attracted the most research. An argument can be made that the five mass extinctions are only the five most *extreme events* in a continuous series of large extinction events throughout the Phanerozoic (since 542 million years ago). In most cases, the hypothesized causes are still controversial; however, the most recent mass extinction event seems clear.

The Five Mass Extinctions

The fossil record of the mass extinctions was the basis for defining periods of geological history, so they typically occur at the transition point between geological periods. The transition in fossils from one period to another reflects the dramatic loss of species and the gradual origin of new species. These transitions can be seen in the rock strata. Table 47.2 provides data on the five mass extinctions.

Mass Extinctions

Geological Period	Mass Extinction Name	Time (millions of years ago)
Ordovician–Silurian	end-Ordovician O–S	450–440
Late Devonian	end-Devonian	375–360
Permian–Triassic	end-Permian	251
Triassic–Jurassic	end-Triassic	205
Cretaceous–Paleogene	end-Cretaceous K–Pg (K–T)	65.5

Table 47.2 This table shows the names and dates for the five mass extinctions in Earth's history.

The **Ordovician-Silurian extinction** event is the first recorded mass extinction and the second largest. During this period, about

85 percent of marine species (few species lived outside the oceans) became extinct. The main hypothesis for its cause is a period of glaciation and then warming. The extinction event actually consists of two extinction events separated by about 1 million years. The first event was caused by cooling, and the second event was due to the subsequent warming. The climate changes affected temperatures and sea levels. Some researchers have suggested that a gamma-ray burst, caused by a nearby supernova, was a possible cause of the Ordovician-Silurian extinction. The gamma-ray burst would have stripped away the Earth's protective ozone layer, allowing intense ultraviolet radiation from the sun to reach the surface of the earth—and *may* account for climate changes observed at the time. The hypothesis is very speculative, and extraterrestrial influences on Earth's history are an active line of research. Recovery of biodiversity after the mass extinction took from 5 to 20 million years, depending on the location.

The **late Devonian extinction** may have occurred over a relatively long period of time. It appears to have mostly affected marine species and not so much the plants or animals inhabiting terrestrial habitats. The causes of this extinction are poorly understood.

The **end-Permian extinction** was the largest in the history of life. Indeed, an argument could be made that Earth became nearly devoid of life during this extinction event. Estimates are that 96 percent of all marine species and 70 percent of all terrestrial species were lost. It was at this time, for example, that the trilobites, a group that survived the Ordovician–Silurian extinction, became extinct. The causes for this mass extinction are not clear, but the leading suspect is extended and widespread volcanic activity that led to a runaway global-warming event. The oceans became largely anoxic, suffocating marine life. Terrestrial tetrapod diversity took 30 million years to recover after the end-Permian extinction. The Permian extinction dramatically altered Earth's biodiversity makeup and the course of evolution.

The causes of the **Triassic–Jurassic** extinction event are not clear, and researchers argue hypotheses including climate change, asteroid impact, and volcanic eruptions. The extinction event occurred just before the breakup of the supercontinent Pangaea, although recent scholarship suggests that the extinctions may have occurred more gradually throughout the Triassic.

The causes of the end-Cretaceous extinction event are the ones that are best understood. It was during this extinction event about 65 million years ago that the majority of the dinosaurs, the dominant vertebrate group for millions of years, disappeared from the planet (with the exception of a theropod clade that gave rise to birds).

The cause of this extinction is now understood to be the result of a cataclysmic impact of a large meteorite, or asteroid, off the coast of what is now the Yucatán Peninsula. This hypothesis, proposed first in 1980, was a radical explanation based on a sharp spike in the levels of *iridium* (which enters our atmosphere from meteors at a fairly constant rate but is otherwise absent on Earth's surface) in the rock stratum that marks the boundary between the Cretaceous and Paleogene periods (Figure 47.6). This boundary marked the disappearance of the dinosaurs in fossils as well as many other taxa. The researchers who discovered the iridium spike interpreted it as a rapid influx of iridium from space to the atmosphere (in the form of a large asteroid) rather than a slowing in the deposition of sediments during that period. It was a radical explanation, but the report of an appropriately aged and sized impact crater in 1991 made the hypothesis more believable. Now an abundance of geological evidence supports the theory. Recovery times for biodiversity after the end-Cretaceous extinction are shorter, in geological time, than for the end-Permian extinction, on the order of 10 million years.

Another possibility, perhaps coincidental with the impact of the Yucatan asteroid, was extensive volcanism that began forming about 66 million years ago, about the same time as the Yucatan asteroid impact, at the end of the Cretaceous. The lava flows covered over 50 percent of what is now India. The release of volcanic gases, particularly sulphur dioxide, during the formation of the traps contributed to climate change, which may have induced the mass extinction.

VISUAL CONNECTION

Figure 47.6 In 1980, Luis and Walter Alvarez, Frank Asaro, and Helen Michels discovered, across the world, a spike in the concentration of iridium within the sedimentary layer at the K–Pg boundary. These researchers hypothesized that this iridium spike was caused by an asteroid impact that resulted in the K–Pg mass extinction. In the photo, the iridium layer is the light band. (credit: USGS)

Scientists measured the relative abundance of fern spores above and below the K–Pg boundary in this rock sample. Which of the following statements most likely represents their findings?

a. An abundance of fern spores from several species was found below the K–Pg boundary, but none was found above.
b. An abundance of fern spores from several species was found above the K–Pg boundary, but none was found below.
c. An abundance of fern spores was found both above and below the K–Pg boundary, but only one species was found below the boundary, and many species were found above the boundary.
d. Many species of fern spores were found both above and below the boundary, but the total number of spores was greater below the boundary.

LINK TO LEARNING

Explore this interactive website (http://openstax.org/l/extinctions) about mass extinctions.

The Pleistocene Extinction

The **Pleistocene Extinction** is one of the lesser extinctions, and a recent one. It is well known that the North American, and to some degree Eurasian, **megafauna**—large vertebrate animals—disappeared toward the end of the last glaciation period. The extinction appears to have happened in a relatively restricted time period of 10,000–12,000 years ago. In North America, the losses were quite dramatic and included the woolly mammoths (with an extant population existing until about 4,000 years ago in isolation on Wrangel Island, Canada), mastodon, giant beavers, giant ground sloths, saber-toothed cats, and the North American camel, just to name a few. In the early 1900s, scientists first suggested the possibility that over-hunting caused the rapid extinction of these large animals. Research into this hypothesis continues today.

In general, the timing of the Pleistocene extinctions correlated with the arrival of paleo-humans, perhaps as long as 40,000 years ago, and not with climate-change events, which is the main competing hypothesis for these extinctions. The extinctions began in Australia about 40,000 to 50,000 years ago, just after the arrival of humans in the area: a marsupial lion, a giant one-ton wombat, and several giant kangaroo species disappeared. In North America, the extinctions of almost all of the large mammals occurred 10,000–12,000 years ago. All that are left are the smaller mammals such as bears, elk, moose, and cougars. Finally, on many remote oceanic islands, the extinctions of many species occurred coincidentally with human arrivals. Not all of the islands had large animals, but when there were large animals, they were often forced into extinction. Madagascar was colonized about 2,000 years ago and the large mammals that lived there became extinct. Eurasia and Africa do not show this pattern, but they also did not experience a recent arrival of hunter-gatherer humans. Rather, humans arrived in Eurasia hundreds of thousands of years ago. This topic remains an area of active research and hypothesizing. It seems clear that even if climate played a role, in most cases human hunting precipitated the extinctions.

Recent Extinctions

The sixth, or **Holocene, mass extinction** appears to have begun earlier than previously believed and is largely due to the disruptive activities of modern *Homo sapiens.* Since the beginning of the Holocene period, there are numerous recent extinctions of individual species that are recorded in human writings. Most of these are coincident with the expansion of the European colonies since the 1500s.

One of the earlier and popularly known examples is the dodo bird. The odd pigeon-like bird lived in the forests of Mauritius (an island in the Indian Ocean) and became extinct around 1662. The dodo was hunted for its meat by sailors and was easy prey because it approached people without fear (the dodo had not evolved with humans). Pigs, rats, and dogs brought to the island by European ships also killed dodo young and eggs.

Steller's sea cow became extinct in 1768; it was related to the manatee and probably once lived along the northwest coast of North America. Steller's sea cow was first discovered by Europeans in 1741 and was overhunted for meat and oil. The last sea cow was killed in 1768. That amounts to just 27 years between the sea cow's first contact with Europeans and extinction of the species!

Since 1900, a variety of species have gone extinct, including the following:

- In 1914, the last living passenger pigeon died in a zoo in Cincinnati, Ohio. This species had once darkened the skies of North America during its migrations, but it was overhunted and suffered from habitat loss that resulted from the clearing of forests for farmland.
- The Carolina parakeet, once common in the eastern United States, died out in 1918. It suffered habitat loss and was hunted to prevent it from eating orchard fruit. (The parakeet ate orchard fruit because its native foods were destroyed to make way for farmland.)
- The Japanese sea lion, which inhabited a broad area around Japan and the coast of Korea, became extinct in the 1950s due to fishermen.
- The Caribbean monk seal was distributed throughout the Caribbean Sea but was driven to extinction via hunting by 1952.

These are only a few of the recorded extinctions in the past 500 years. The International Union for Conservation of Nature (IUCN) keeps a list of extinct and endangered species called the **Red List**. The list is not complete, but it describes 380 extinct species of vertebrates after 1500 AD, 86 of which were driven extinct by overhunting or overfishing.

Estimates of Present-Time Extinction Rates

Estimates of **extinction rates** are hampered by the fact that most extinctions are probably happening without observation. The extinction of a bird or mammal is likely to be noticed by humans, especially if it has been hunted or used in some other way. But there are many organisms that are of less interest to humans (not necessarily of less value) and many that are undescribed.

The **background extinction** rate is estimated to be about one per million species per year (E/MSY). For example, assuming there are about ten million species in existence, the expectation is that ten species would become extinct each year (each year represents ten million species per year).

One contemporary extinction rate estimate uses the extinctions in the written record since the year 1500. For birds alone this method yields an estimate of 26 E/MSY. However, this value may be an underestimate for three reasons. First, many species would not have been described until much later in the time period, so their loss would have gone unnoticed. Second, the number of recently extinct vertebrate species is increasing because extinct species now are being described from skeletal remains. And third, some species are probably already extinct even though conservationists are reluctant to name them as such. Taking these factors into account raises the estimated extinction rate closer to 100 E/MSY. The predicted rate by the end of the century is 1500 E/MSY.

A second approach to estimating present-time extinction rates is to correlate species loss with habitat loss by measuring forest-area loss and understanding species-area relationships. The **species-area relationship** is the rate at which new species are seen when the area surveyed is increased. Studies have shown that the number of species present increases as the size of the island increases. This phenomenon has also been shown to hold true in other island-like habitats as well, such as the mountain-top tepuis of Venezuela, which are surrounded by tropical forest. Turning this relationship around, if the habitat area is reduced, the number of species living there will also decline. Estimates of extinction rates based on habitat loss and species-area relationships have suggested that with about 90 percent habitat loss an expected 50 percent of species would become extinct. Species-area estimates have led to species extinction rate calculations of about 1000 E/MSY and higher. In general, actual observations do not show this amount of loss and suggestions have been made that there is a delay in extinction. Recent work has also called into question the applicability of the species-area relationship when estimating the loss of species. This work argues that the species-area relationship leads to an *overestimate* of extinction rates. A better relationship to use may be the endemics-area relationship. Using this method would bring estimates down to around 500 E/MSY in the coming century. *Note that this value is still 500 times the background rate.*

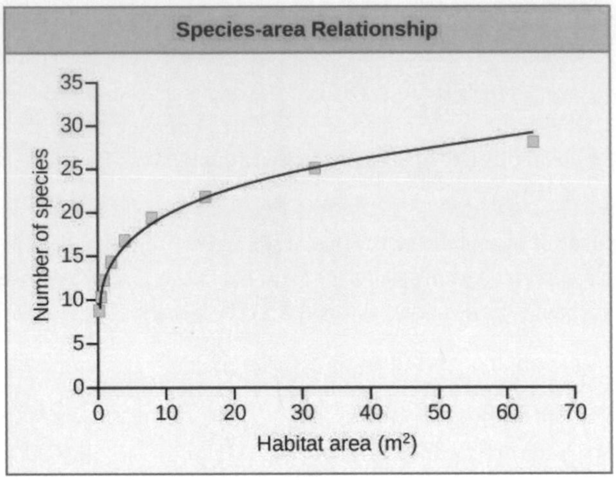

Figure 47.7 Studies have shown that the number of species present increases with the size of the habitat. (credit: modification of work by Adam B. Smith)

LINK TO LEARNING

Check out this interactive exploration (http://openstax.org/l/what_is_missing) of endangered and extinct species, their ecosystems, and the causes of the endangerment or extinction.

47.2 The Importance of Biodiversity to Human Life

By the end of this section, you will be able to do the following:

- Identify chemical diversity benefits to humans
- Identify biodiversity components that support human agriculture
- Describe ecosystem services

It may not be clear why biologists are concerned about biodiversity loss. When biodiversity loss is thought of as the extinction of the passenger pigeon, the dodo bird, and even the woolly mammoth, the loss may appear to be an emotional one. But is the loss practically important for the welfare of the human species? From the perspective of evolution and ecology, the loss of a particular individual species is unimportant (however, we should note that the loss of a keystone species can lead to ecological disaster). Extinction is a normal part of macroevolution. But the *accelerated extinction rate* translates into the loss of tens of thousands of species within our lifetimes, and it is likely to have dramatic effects on human welfare through the collapse of ecosystems and in added costs to maintain food production, clean air and water, and human health.

Agriculture began after early hunter-gatherer societies first settled in one place and heavily modified their immediate environment. This cultural transition has made it difficult for humans to recognize their dependence on undomesticated living things on the planet. Biologists recognize the human species is embedded in ecosystems and is dependent on them, just as every other species on the planet is dependent. Technology smooths out the extremes of existence, but ultimately the human species cannot exist without a supportive ecosystem.

Human Health

Archeological evidence indicates that humans have been using plants for medicinal uses for thousands of years. A Chinese document from approximately 2800 BC is believed to be the the first written account of herbal remedies, and such references occur throughout the global historical record. Contemporary indigenous societies that live close to the land often retain broad knowledge of the medicinal uses of plants growing in their area. Most plants produce **secondary plant compounds**, which are toxins used to protect the plant from insects and other animals that eat them, but some of which also work as medication.

Modern pharmaceutical science also recognizes the importance of these plant compounds. Examples of significant medicines derived from plant compounds include aspirin, codeine, digoxin, atropine, and vincristine (Figure 47.8). Many medicines were once derived from plant extracts but are now synthesized. It is estimated that, at one time, 25 percent of modern drugs contained at least one plant extract. That number has probably decreased to about 10 percent as natural plant ingredients are replaced by synthetic versions. Antibiotics, which are responsible for extraordinary improvements in health and lifespans in developed countries, are compounds largely derived from fungi and bacteria.

Figure 47.8 *Catharanthus roseus*, the Madagascar periwinkle, has various medicinal properties. Among other uses, it is a source of vincristine, a drug used in the treatment of lymphomas. (credit: Forest and Kim Starr)

In recent years, animal venoms and poisons have excited intense research for their medicinal potential. By 2007, the FDA had approved five drugs based on animal toxins to treat diseases such as hypertension, chronic pain, and diabetes. Another five drugs are undergoing clinical trials, and at least six drugs are being used in other countries. Other toxins under investigation come from mammals, snakes, lizards, various amphibians, fish, snails, octopuses, and scorpions.

Aside from representing billions of dollars in profits, these medicines improve people's lives. Pharmaceutical companies are always looking for new compounds synthesized by living organisms that can function as medicines. It is estimated that 1/3 of pharmaceutical research and development is spent on natural compounds and that about 35 percent of new drugs brought to market between 1981 and 2002 were derived from natural compounds. The opportunities for new medications will be reduced in direct proportion to the disappearance of species.

Agricultural Diversity

Since the beginning of human agriculture more than 10,000 years ago, human groups have been breeding and selecting crop varieties. This crop diversity matched the cultural diversity of highly subdivided populations of humans. For example, potatoes were domesticated beginning around 7,000 years ago in the central Andes of Peru and Bolivia. The potatoes grown in that region belong to seven species and the number of varieties likely is in the thousands. Even the Inca capital of Machu Picchu had numerous gardens growing varieties of potatoes. Each variety has been bred to thrive at particular elevations and soil and climate conditions. The diversity is driven by the diverse demands of the topography, the limited movement of people, and the demands created by crop rotation for different varieties that will do well in different fields.

Potatoes are only one example of human-generated diversity. Every plant, animal, and fungus that has been cultivated by humans has been bred from original wild ancestor species into diverse varieties arising from the demands for food value, adaptation to growing conditions, and resistance to pests.

The potato also demonstrates risks of low crop diversity. The tragic Irish potato famine occurred when the single variety grown in Ireland became susceptible to a potato blight, wiping out the entire crop. The loss of the potato crop led to mass famine and the related deaths of over one million people, as well as mass emigration of nearly two million people.

Disease resistance is a chief benefit of crop biodiversity, and lack of diversity in contemporary crop species carries similar risks. Seed companies, which are the source of most crop varieties in developed countries, must continually breed new varieties to keep up with evolving pest organisms. These same seed companies, however, have participated in the decline of the number of varieties available as they focus on selling fewer varieties in more areas of the world.

The ability to create new crop varieties relies on the diversity of varieties available and the accessibility of wild forms related to the crop plant. These wild forms are often the source of new gene variants that can be bred with existing varieties to create varieties with new attributes. Loss of wild species related to a crop will mean the loss of potential in crop improvement. Maintaining the genetic diversity of wild species related to domesticated species ensures our continued food supply.

Since the 1920s, government agriculture departments have maintained seed banks of crop varieties as a way of maintaining crop diversity. This system has flaws because, over time, seed banks are lost through accidents, and there is no way to replace them. In 2008, the **Svalbard Global Seed Vault** (Figure 47.9) began storing seeds from around the world as a backup system to the regional seed banks. If a regional seed bank stores varieties in Svalbard, losses can be replaced from Svalbard. Conditions within the vault are maintained at ideal temperature and humidity for seed survival, but the deep underground location of the vault in the arctic means that failure of the vault's systems will not compromise the climatic conditions inside the vault.

VISUAL CONNECTION

Figure 47.9 The Svalbard Global Seed Vault is a storage facility for seeds of Earth's diverse crops. (credit: Mari Tefre, Svalbard Global Seed Vault)

The Svalbard Global Seed Vault is located on Spitsbergen island in Norway, which has an arctic climate. Why might an arctic climate be good for seed storage?

Crop success is largely dependent on the quality of the soil. Although some agricultural soils are rendered sterile using controversial cultivation and chemical treatments, most contain a huge diversity of organisms that maintain nutrient cycles—breaking down organic matter into nutrient compounds that crops need for growth. These organisms also maintain soil texture that affects water and oxygen dynamics in the soil that are necessary for plant growth. If farmers had to maintain arable soil using alternate means, the cost of food would be much higher than it is now. These kinds of processes are called **ecosystem services**. They occur within ecosystems, such as soil ecosystems, as a result of the diverse metabolic activities of the organisms living there, but they provide benefits to human food production, drinking water availability, and breathable air.

Plant pollination is another key ecosystem service, provided by various species of bees, other insects, and birds. One estimate indicates that honey bee pollination provides the United States a $1.6 billion annual benefit.

Honey bee populations in North America have been suffering large losses caused by a syndrome known as colony collapse disorder, whose cause is unclear. (Evidence suggests the possible culprits may be the invasive varroa mite coupled with the Nosema gut parasite and acute paralysis virus.) Loss of these species would render it very difficult, if not impossible, to grow any of the 150 United States crops requiring pollination, including grapes, oranges, lemons, peppers, most brassica (broccoli and cauliflower), and many berries, melons, and nuts.

Finally, humans compete for their food with crop pests, most of which are insects. Pesticides control these competitors; however, pesticides are costly and lose their effectiveness over time as pest populations adapt and evolve. They also lead to collateral damage by killing non-pest species and risking the health of consumers and agricultural workers. Ecologists believe that the bulk of the work in removing pests is actually done by predators and parasites of those pests, but the impact has not been well studied. A review found that in 74 percent of studies that looked for an effect of landscape complexity on natural enemies of pests, the greater the complexity, the greater the effect of pest-suppressing organisms. An experimental study found that introducing multiple enemies of pea aphids (an important alfalfa pest) increased the yield of alfalfa significantly. This study shows the importance of landscape diversity via the question of whether a diversity of pests is more effective at control than one single pest; the results showed this to be the case. Loss of diversity in pest enemies will inevitably make it more difficult and costly to grow food.

Wild Food Sources

In addition to growing crops and raising animals for food, humans obtain food resources from wild populations, primarily fish populations. In fact, for approximately 1 billion people worldwide, aquatic resources provide the main source of animal protein. But since 1990, global fish production has declined, sometimes dramatically. Unfortunately, and despite considerable effort, few fisheries on the planet are managed for sustainability.

Fishery extinctions rarely lead to complete extinction of the harvested species, but rather to a radical restructuring of the marine ecosystem in which a dominant species is so over-harvested that it becomes a minor player, ecologically. In addition to humans losing the food source, these alterations affect many other species in ways that are difficult or impossible to predict. The collapse of fisheries has dramatic and long-lasting effects on local populations that work in the fishery. In addition, the loss of an inexpensive protein source to populations that cannot afford to replace it will increase the cost of living and limit societies in other ways. In general, the fish taken from fisheries have shifted to smaller species as larger species are fished to extinction. The ultimate outcome could clearly be the loss of aquatic systems as food sources.

LINK TO LEARNING

View a brief video (http://openstax.org/l/declining_fish) discussing declining fish stocks.

Psychological and Moral Value

Finally, it has been clearly shown that humans benefit psychologically from living in a biodiverse world. A chief proponent of this idea is Harvard entomologist E. O. Wilson. He argues that human evolutionary history has adapted us to live in a natural environment and that city environments generate psychological stressors that affect human health and well-being. There is considerable research into the psychological regenerative benefits of natural landscapes that suggests the hypothesis may hold some truth. In addition, there is a moral argument that humans have a responsibility to inflict as little harm as possible on other species.

47.3 Threats to Biodiversity

By the end of this section, you will be able to do the following:

- Identify significant threats to biodiversity
- Explain the effects of habitat loss, the introduction of exotic species, and hunting on biodiversity
- Identify the early and predicted effects of climate change on biodiversity

The core threat to biodiversity on the planet, and therefore a threat to human welfare, is the combination of human population growth and resource exploitation. The human population requires resources to survive and grow, and those resources are being removed unsustainably from the environment. The three greatest proximate threats to biodiversity are habitat loss, overharvesting, and the introduction of exotic species. The first two of these are a direct result of human population growth and resource use. The third results from increased mobility and trade. A fourth major cause of extinction, anthropogenic climate change, has not yet had a large impact, but it is predicted to become significant during this century. Global climate change is also a consequence of human population needs for energy and the use of fossil fuels to meet those needs (Figure 47.10). Environmental issues, such as toxic pollution, have specific targeted effects on species, but they are not generally seen as threats at the magnitude of the others.

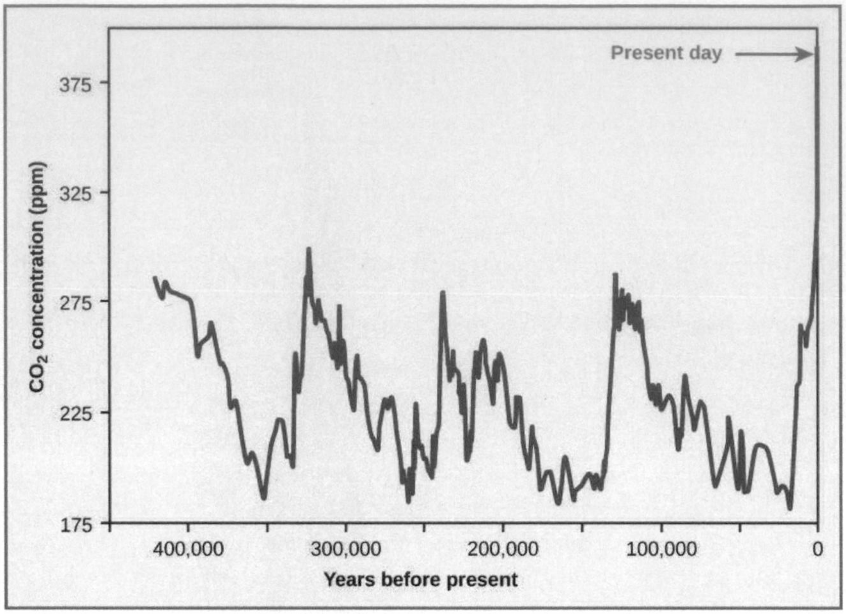

Figure 47.10 Atmospheric carbon dioxide levels fluctuate in a cyclical manner. However, the burning of fossil fuels in recent history has caused a dramatic increase in the levels of carbon dioxide in the Earth's atmosphere, which have now reached levels never before seen in human history. Scientists predict that the addition of this "greenhouse gas" to the atmosphere is resulting in climate change that will significantly impact biodiversity in the coming century.

Habitat Loss

Humans rely on technology to modify their environment and replace certain functions that were once performed by the natural ecosystem. Other species cannot do this. Elimination of their ecosystem—whether it is a forest, a desert, a grassland, a freshwater estuarine, or a marine environment—will kill the individuals belonging to the species. The species will become extinct if we remove the entire habitat within the range of a species. Human destruction of habitats accelerated in the latter half of the twentieth century. Consider the exceptional biodiversity of Sumatra: it is home to one species of orangutan, a species of critically endangered elephant, and the Sumatran tiger, but half of Sumatra's forest is now gone. The neighboring island of Borneo, home to the other species of orangutan, has lost a similar area of forest. Forest loss continues in protected areas of Borneo. All three species of orangutan are now listed as endangered by the International Union for Conservation of Nature (IUCN), but they are simply the most visible of thousands of species that will not survive the disappearance of the forests in Sumatra and Borneo. The forests are removed for timber and to plant palm oil plantations (Figure 47.11). Palm oil is used in many products including food products, cosmetics, and biodiesel in Europe. A five-year estimate of global forest cover loss for the years 2000–2005 was 3.1 percent. In the humid tropics where forest loss is primarily from timber extraction, 272,000 km^2 was lost out of a global total of 11,564,000 km^2 (or 2.4 percent). In the tropics, these losses certainly also represent the extinction of species because of high levels of **endemism**—species unique to a defined geographic location, and found nowhere else.

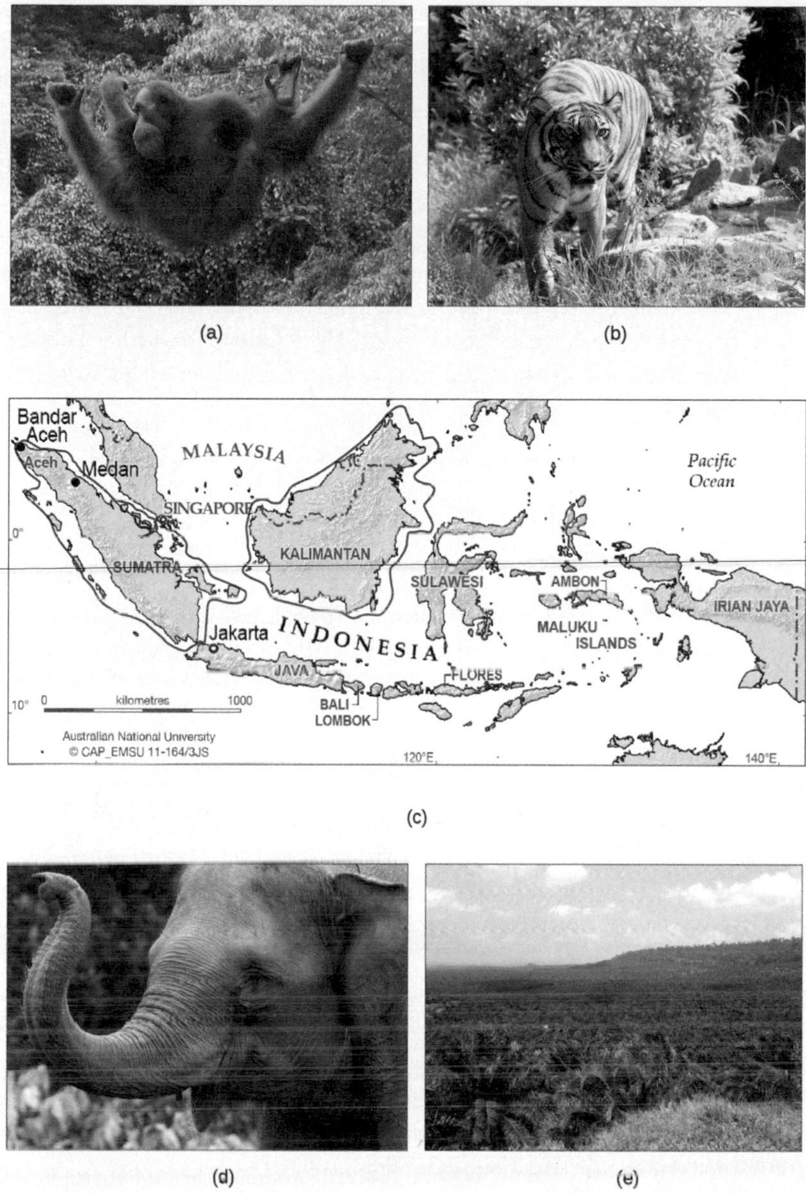

Figure 47.11 (a) One of three species of orangutan, *Pongo pygmaeus,* is found only in the rainforests of Borneo, and an other species of orangutan (*Pongo abelii*) is found only in the rainforests of Sumatra. These animals are examples of the exceptional biodiversity of (c) the islands of Sumatra and Borneo. Other species include the (b) Sumatran tiger (*Panthera tigris sumatrae*) and the (d) Sumatran elephant (*Elephas maximus sumatranus*), both critically endangered species. Rainforest habitat is being removed to make way for (e) oil palm plantations such as this one in Borneo's Sabah Province. (credit a: modification of work by Thorsten Bachner; credit b: modification of work by Dick Mudde; credit c: modification of work by U.S. CIA World Factbook; credit d: modification of work by "Nonprofit Organizations"/Flickr; credit e: modification of work by Dr. Lian Pin Koh)

Everyday Connection

Preventing Habitat Destruction with Wise Wood Choices

Most consumers are not aware that the home improvement products they buy might be contributing to habitat loss and species extinctions. Yet the market for illegally harvested tropical timber is huge, and the wood products often find themselves in building supply stores in the United States. One estimate is that 10 percent of the imported timber stream in the United States, which is the world's largest consumer of wood products, is potentially illegally logged. In 2006, this amounted to $3.6 billion in wood products. Most of the illegal products are imported from countries that act as *intermediaries* and are not the originators of the wood.

How is it possible to determine if a wood product, such as flooring, was harvested sustainably or even legally? The **Forest Stewardship Council** (FSC) certifies sustainably harvested forest products, therefore, looking for their certification on flooring and other hardwood products is one way to ensure that the wood has not been taken illegally from a tropical forest. Certification applies to specific products, not to a producer; some producers' products may not have certification while other products are certified. While there are other industry-backed certifications other than the FSC, these are unreliable due to lack of independence from the industry. Another approach is to buy domestic wood species. While it would be great if there was a list of legal versus illegal wood products, it is not that simple. Logging and forest management laws vary from country to country; what is illegal in one country may be legal in another. Where and how a product is harvested and whether the forest from which it comes is being maintained sustainably all factor into whether a wood product will be certified by the FSC. If you are in doubt, it is always a good idea to ask questions about where a wood product came from and how the supplier knows that it was harvested legally.

Habitat destruction can affect ecosystems other than forests. Rivers and streams are important ecosystems that are frequently modified through land development, damming, channelizing, or water removal. Damming affects the water flow to all parts of a river, which can reduce or eliminate populations that had adapted to the natural flow of the river. For example, an estimated 91 percent of United States rivers have been altered in some way. Modifications include dams, to create energy or store water; levees, to prevent flooding; and dredging or rerouting, to create land that is more suitable for human development. Many fish and amphibian species and numerous freshwater clams in the United States have seen declines caused by river damming and habitat loss.

Overharvesting

Overharvesting is a serious threat to many species, but particularly to aquatic (both marine and freshwater) species. Despite regulation and monitoring, there are recent examples of fishery collapse. The western Atlantic cod fishery is the among the most significant. While it was a hugely productive fishery for 400 years, the introduction of modern factory trawlers in the 1980s caused it to become unsustainable. Fisheries collapse as a result of both economic and political factors. Fisheries are managed as a shared international resource even when the fishing territory lies within an individual country's territorial waters. Common resources are subject to an economic pressure known as the **tragedy of the commons**, in which essentially no fisher has a motivation to exercise restraint in harvesting a fishery when it is not owned by that fisher. Overexploitation is a common outcome. This overexploitation is exacerbated when access to the fishery is open and unregulated and when technology gives fishers the ability to overfish. In a few fisheries, the biological growth of the resource is less than the potential growth of the profits made from fishing if that time and money were invested elsewhere. In these cases—whales are an example—economic forces will always drive toward fishing the population to extinction.

LINK TO LEARNING

Explore a U.S. Fish & Wildlife Service interactive map (http://openstax.org/l/habitat_map) of critical habitat for endangered and threatened species in the United States. To begin, select "Visit the online mapper."

For the most part, fishery extinction is not equivalent to biological extinction—the last fish of a species is rarely fished out of the ocean. At the same time, fishery extinction is still harmful to fish species and their ecosystems. There are some instances in which true extinction is a possibility. Whales have slow-growing populations due to low reproductive rates, and therefore are at risk of complete extinction through hunting. There are some species of sharks with restricted distributions that are at risk of extinction. The groupers are another population of generally slow-growing fishes that, in the Caribbean, includes a number of species that are at risk of extinction from overfishing.

Coral reefs are extremely diverse marine ecosystems that face immediate peril from several processes. Reefs are home to 1/3 of the world's marine fish species—about 4,000 species—despite making up only 1 percent of marine habitat. Most home marine aquaria are stocked with wild-caught organisms, not cultured organisms. Although no species is known to have been driven extinct by the pet trade in marine species, there are studies showing that populations of some species have declined in response to harvesting, indicating that the harvest is not sustainable at those levels. There are concerns about the effect of the pet trade on some terrestrial species such as turtles, amphibians, birds, plants, and even the orangutan.

LINK TO LEARNING

View a brief video (http://openstax.org/l/ocean_matters) discussing the role of marine ecosystems in supporting human welfare and the decline of ocean ecosystems.

Bush meat is the generic term used for wild animals killed for food. Hunting is practiced throughout the world, but hunting practices, particularly in equatorial Africa and parts of Asia, are believed to threaten a number of species with extinction. Traditionally, bush meat in Africa was hunted to feed families directly; however, recent commercialization of the practice now has bush meat available in grocery stores, which has increased harvest rates to the level of unsustainability. Additionally, human population growth has increased the need for protein foods that are not being met from agriculture. Species threatened by the bush meat trade are mostly mammals including many primates living in the Congo basin.

Exotic Species

Exotic species are species that have been intentionally or unintentionally introduced into an ecosystem in which they did not evolve. For example, Kudzu (*Pueraria lobata*), which is native to Japan, was introduced in the United States in 1876. It was later planted for soil conservation. Problematically, it grows too well in the southeastern United States—up to a foot a day. It is now an invasive pest species and covers over 7 million acres in the southeastern United States. If an introduced species is able to survive in its new habitat, that introduction is now reflected in the observed range of the species. Human transportation of people and goods, including the intentional transport of organisms for trade, has dramatically increased the introduction of species into new ecosystems, sometimes at distances that are well beyond the capacity of the species to ever travel itself and outside the range of the species' natural predators.

Most exotic species introductions probably fail because of the low number of individuals introduced or poor adaptation to the ecosystem they enter. Some species, however, possess pre-adaptations that can make them especially successful in a new ecosystem. These exotic species often undergo dramatic population increases in their new habitat and reset the ecological conditions in the new environment, threatening the species that exist there. For this reason, exotic species are also called invasive species. Exotic species can threaten other species through competition for resources, predation, or disease. For example, the Eurasian star thistle, also called spotted knapweed, has invaded and rendered useless some of the open prairies of the western states. However, it is a great nectar-bearing flower for the production of honey and supports numerous pollinating insects, including migrating monarch butterflies in the north central states such as Michigan.

LINK TO LEARNING

Explore an interactive global database (http://openstax.org/l/exotic_invasive) of exotic or invasive species.

Lakes and islands are particularly vulnerable to extinction threats from introduced species. In Lake Victoria, as mentioned earlier, the intentional introduction of the Nile perch was largely responsible for the extinction of about 200 species of endemic cichlids. The accidental introduction of the brown tree snake via aircraft (Figure 47.12) from the Solomon Islands to Guam in 1950 has led to the extinction of three species of birds and three to five species of reptiles endemic to the island. Several other species are still threatened. The brown tree snake is adept at exploiting human transportation as a means to migrate; one was even found on an aircraft arriving in Corpus Christi, Texas. Constant vigilance on the part of airport, military, and commercial aircraft personnel is required to prevent the snake from moving from Guam to other islands in the Pacific, especially Hawaii. Islands do not make up a large area of land on the globe, but they do contain a disproportionate number of endemic species because of their isolation from mainland ancestors.

Figure 47.12 The brown tree snake, *Boiga irregularis*, is an exotic species that has caused numerous extinctions on the island of Guam since its accidental introduction in 1950. (credit: NPS)

It now appears that the global decline in amphibian species recognized in the 1990s is, in some part, caused by the fungus *Batrachochytrium dendrobatidis*, which causes the disease **chytridiomycosis** (Figure 47.13). There is evidence that the fungus is native to Africa and may have been spread throughout the world by transport of a commonly used laboratory and pet species: the African clawed toad (*Xenopus laevis*). It may well be that biologists themselves are responsible for spreading this disease worldwide. The North American bullfrog, *Rana catesbeiana*, which has also been widely introduced as a food animal but which easily escapes captivity, survives most infections of *Batrachochytrium dendrobatidis*, and can act as a reservoir for the disease. It also is a voracious predator in freshwater lakes.

Figure 47.13 This Limosa Harlequin Frog (*Atelopus limosus*), an endangered species from Panama, died from a fungal disease called chytridiomycosis. The red lesions are symptomatic of the disease. (credit: Brian Gratwicke)

Early evidence suggests that another fungal pathogen, *Geomyces destructans*, introduced from Europe is responsible for **white-nose syndrome**, which infects cave-hibernating bats in eastern North America and has spread from a point of origin in western New York State (Figure 47.14). The disease has decimated bat populations and threatens extinction of species already listed as endangered: the Indiana bat, *Myotis sodalis*, and potentially the Virginia big-eared bat, *Corynorhinus townsendii virginianus*. How the fungus was introduced is unclear, but one logical presumption would be that recreational cavers unintentionally brought the fungus on clothes or equipment from Europe.

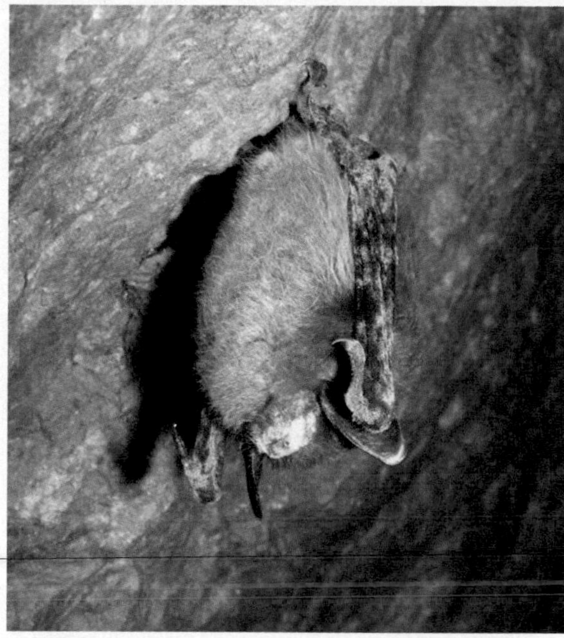

Figure 47.14 This little brown bat in Greeley Mine, Vermont, March 26, 2009, was found to have white-nose syndrome. (credit: Marvin Moriarty, USFWS)

Climate Change

Climate change, and specifically the *anthropogenic* (meaning, caused by humans) warming trend presently escalating, is recognized as a major extinction threat, particularly when combined with other threats such as habitat loss and the expansion of disease organisms. Scientists disagree about the likely magnitude of the effects, with extinction rate estimates ranging from 15 percent to 40 percent of species destined for extinction by 2050. Scientists do agree, however, that climate change will alter regional climates, including rainfall and snowfall patterns, making habitats less hospitable to the species living in them, in particular, the endemic species. The warming trend will shift colder climates toward the north and south poles, forcing species to move with their adapted climate norms while facing habitat gaps along the way. The shifting ranges will impose new competitive regimes on species as they find themselves in contact with other species not present in their historic range. One such unexpected species contact is between polar bears and grizzly bears. Previously, these two distinct species had separate ranges. Now, their ranges are overlapping and there are documented cases of these two species mating and producing viable offspring, which may or may not be viable crossing back to either parental species. Changing climates also throw off species' delicate timed adaptations to seasonal food resources and breeding times. Many contemporary mismatches to shifts in resource availability and timing have already been documented.

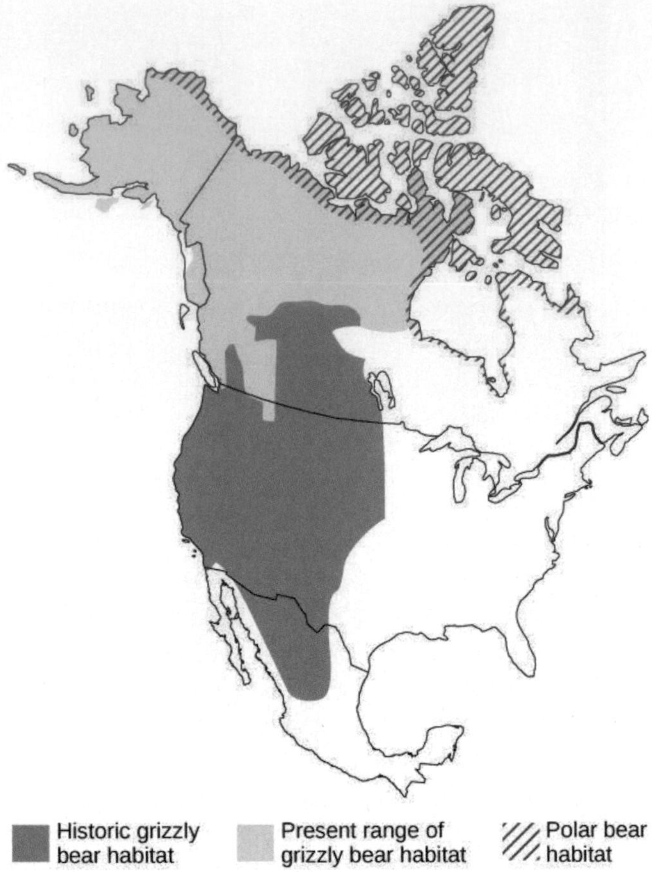

Figure 47.15 Since 2008, grizzly bears (*Ursus arctos horribilis*) have been spotted farther north than their historic range, a possible consequence of climate change. As a result, grizzly bear habitat now overlaps polar bear (*Ursus maritimus*) habitat. The two species of bears, which are capable of mating and producing viable offspring, are considered separate "ecological" species because historically they lived in different habitats and never met. However, in 2006 a hunter shot a wild grizzly-polar bear hybrid known as a grolar bear, the first wild hybrid ever found.

Range shifts are already being observed: for example, some European bird species ranges have moved 91 km northward. The same study suggested that the optimal shift based on warming trends was double that distance, suggesting that the populations are *not* moving quickly enough. Range shifts have also been observed in plants, butterflies, other insects, freshwater fishes, reptiles, and mammals.

Climate gradients will also move up mountains, eventually crowding species higher in altitude and eliminating the habitat for those species adapted to the highest elevations. Some climates will completely disappear. The accelerating rate of warming in the arctic significantly reduces snowfall and the formation of sea ice. Without the ice, species like polar bears cannot successfully hunt seals, which are their only reliable source of food. Sea ice coverage has been decreasing since observations began in the mid-twentieth century, and the rate of decline observed in recent years is far greater than previously predicted.

Finally, global warming will raise ocean levels due to meltwater from glaciers and the greater volume of warmer water. Shorelines will be inundated, reducing island size, which will have an effect on some species, and a number of islands will disappear entirely. Additionally, the gradual melting and subsequent refreezing of the poles, glaciers, and higher elevation mountains—a cycle that has provided freshwater to environments for centuries—will also be jeopardized. This could result in an overabundance of salt water and a shortage of fresh water.

47.4 Preserving Biodiversity

By the end of this section, you will be able to do the following:

- Identify new technologies and methods for describing biodiversity
- Explain the legislative framework for conservation
- Describe principles and challenges of conservation preserve design
- Identify examples of the effects of habitat restoration
- Discuss the role of zoos in biodiversity conservation

Preserving biodiversity is an extraordinary challenge that must be met by greater understanding of biodiversity itself, changes in human behavior and beliefs, and various preservation strategies.

Measuring Biodiversity

The technology of molecular genetics and data processing and storage are maturing to the point where cataloguing the planet's species in an accessible way is now feasible. **DNA barcoding** is one molecular genetic method, which takes advantage of rapid evolution in a mitochondrial gene (cytochrome c oxidase 1) present in eukaryotes, except for plants, to identify species using the sequence of portions of the gene. However, plants may be barcoded using a combination of chloroplast genes. Rapid mass sequencing machines make the molecular genetics portion of the work relatively inexpensive and quick. Computer resources store and make available the large volumes of data. Projects are currently underway to use DNA barcoding to catalog museum specimens, which have already been named and studied, as well as testing the method on less-studied groups. As of mid 2017, close to 200,000 named species had been barcoded. Early studies suggest there are significant numbers of undescribed species that looked too much like sibling species to previously be recognized as different. These now can be identified with DNA barcoding.

Numerous computer databases now provide information about named species and a framework for adding new species. However, as already noted, at the present rate of description of new species, it will take close to 500 years before the complete catalog of life is known. Many, perhaps most, species on the planet do not have that much time.

There is also the problem of understanding which species known to science are threatened and to what degree they are threatened. This task is carried out by the non-profit **IUCN** which, as previously mentioned, maintains the Red List—an online listing of endangered species categorized by taxonomy, type of threat, and other criteria (Figure 47.16). The Red List is supported by scientific research. In 2011, the list contained 61,000 species, all with supporting documentation.

VISUAL CONNECTION

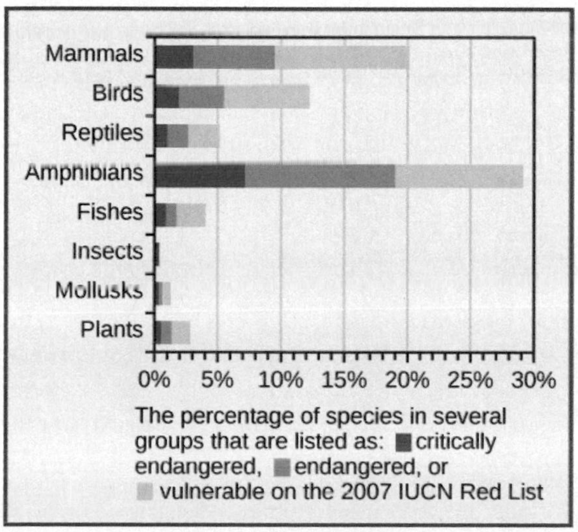

Figure 47.16 This chart shows the percentage of various animal species, by group, on the IUCN Red List as of 2007.

Which of the following statements is not supported by this graph?

a. There are more vulnerable fishes than critically endangered and endangered fishes combined.
b. There are more critically endangered amphibians than vulnerable, endangered and critically endangered reptiles combined.
c. Within each group, there are more critically endangered species than vulnerable species.
d. A greater percentage of bird species are critically endangered than mollusk species.

Changing Human Behavior

Legislation throughout the world has been enacted to protect species. The legislation includes international treaties as well as national and state laws. The Convention on International Trade in Endangered Species of Wild Fauna and Flora **(CITES)** treaty came into force in 1975. The treaty, and the national legislation that supports it, provides a legal framework for preventing approximately 33,000 listed species from being transported across nations' borders, thus protecting them from being caught or killed when international trade is involved. The treaty is limited in its reach because it only deals with international movement of organisms or their parts. It is also limited by various countries' ability or willingness to *enforce* the treaty and supporting legislation. The illegal trade in organisms and their parts is probably a market in the hundreds of millions of dollars. Illegal wildlife trade is monitored by another non-profit: Trade Records Analysis of Flora and Fauna in Commerce **(TRAFFIC)**.

Within many countries there are laws that protect endangered species and regulate hunting and fishing. In the United States, the **Endangered Species Act (ESA)** was enacted in 1973. Species at risk are listed by the Act; the U.S. Fish & Wildlife Service is required by law to develop plans that protect the listed species and bring them back to sustainable numbers. The Act, and others like it in other countries, is a useful tool, but it suffers because it is often difficult to get a species listed, or to get an effective management plan in place once it is listed. Additionally, species may be controversially taken off the list without necessarily having had a change in their situation. More fundamentally, the approach to protecting individual species rather than entire ecosystems is both inefficient and focuses efforts on a few highly visible and often charismatic species, perhaps at the expense of other species that go unprotected. At the same time, the Act has a critical habitat provision outlined in the recovery mechanism that may benefit species other than the one targeted for management.

The Migratory Bird Treaty Act **(MBTA)** is an agreement between the United States and Canada that was signed into law in 1918 in response to declines in North American bird species caused by hunting. The Act now lists over 800 protected species. It makes it illegal to disturb or kill the protected species or distribute their parts (much of the hunting of birds in the past was for their feathers).

As already mentioned, the private non-profit sector plays a large role in the conservation effort both in North America and around the world. The approaches range from species-specific organizations to the broadly focused IUCN and TRAFFIC. The **Nature Conservancy** takes a novel approach. It purchases land and protects it in an attempt to set up preserves for ecosystems.

Although it is focused largely on reducing carbon and related emissions, the Paris Climate Agreement is a significant step toward altering human behavior in a way that should affect biodiversity. If the agreement is successful in its goal of halting global temperature rise, many species negatively affected by climate change may benefit. Assessments of the accord's implementation will not take place until 2023, and measurement of its effects will not be feasible for some time. However, the agreement, signed by over 194 countries, represents the world's most concerted and unified effort to reduce greenhouse gas emissions, embrace alternative energy sources, and ease climate pressure on ecosystems.

Conservation in Preserves

Establishment of wildlife and ecosystem preserves is one of the key tools in conservation efforts. A preserve is an area of land set aside with varying degrees of protection for the organisms that exist within the boundaries of the preserve. Preserves can be effective in the short term for protecting both species and ecosystems, but they face challenges that scientists are still exploring to strengthen their viability as long-term solutions to the preservation of biodiversity *and* the prevention of extinction.

How Much Area to Preserve?

Due to the way protected lands are allocated and the way biodiversity is distributed, it is challenging to determine how much land or marine habitat should be protected. The IUCN World Parks Congress estimated that 11.5 percent of Earth's land surface was covered by preserves of various kinds in 2003. We should note that this area is greater than previous goals; however, it only includes 9 out of 14 recognized major biomes. Similarly, individual animals or types of animals are not equally represented on preserves. For example, high quality preserves include only about 50 percent of threatened amphibian species. To guarantee that all threatened species will be properly protected, either the protected areas must increase in size, or the percentage of high quality preserves must increase, or preserves must be targeted with greater attention to biodiversity protection. Researchers indicate that more attention to the latter solution is required.

Preserve Design

There has been extensive research into optimal preserve designs for maintaining biodiversity. The fundamental principle behind much of the research has been the seminal theoretical work of Robert H. MacArthur and Edward O. Wilson published in 1967 on island biogeography.[5] This work sought to understand the factors affecting biodiversity on islands. The fundamental conclusion was that biodiversity on an island was a function of the origin of species through migration, speciation, and extinction on that island. Islands farther from a mainland are harder to get to, so migration is lower and the equilibrium number of species is lower. Within island populations, evidence suggests that the number of species gradually increases to a level similar to the numbers on the mainland from which the species is suspected to have migrated. In addition, smaller islands are harder to find, so their immigration rates for new species are typically lower. Smaller islands are also less geographically diverse so all things being equal, there are fewer niches to promote speciation. And finally, smaller islands support smaller populations, so the probability of extinction is higher.

As islands get larger, the number of species able to colonize the island and find suitable niches on the island increases, although the effect of island area on species numbers is not a direct correlation. Conservation preserves can be seen as "islands" of habitat within "an ocean" of non-habitat. For a species to persist in a preserve, the preserve must be large enough to support it. The critical size depends, in part, on the home range that is characteristic of the species. A preserve for wolves, which range hundreds of kilometers, must be much larger than a preserve for butterflies, which might range within ten kilometers during its lifetime. But larger preserves have more core area of optimal habitat for individual species, they have more niches to support more species, and they attract more species because they can be found and reached more easily.

Preserves perform better when there are **buffer zones** around them of suboptimal habitat. The buffer allows organisms to exit the boundaries of the preserve without immediate negative consequences from predation or lack of resources. One large preserve is better than the same area of several smaller preserves because there is more core habitat unaffected by edges. For this same reason, preserves in the shape of a square or circle will be better than a preserve with many thin "arms." If preserves must be smaller, then providing **wildlife corridors** between them so that individuals (and their genes) can move between the preserves, for example along rivers and streams, will make the smaller preserves behave more like a large one. All of these factors are taken into consideration when planning the nature of a preserve before the land is set aside.

In addition to the physical, biological, and ecological specifications of a preserve, there are a variety of policy, legislative, and enforcement specifications related to uses of the preserve for functions other than protection of species. These can include anything from timber extraction, mineral extraction, regulated hunting, human habitation, and nondestructive human recreation. Many of these policy decisions are made based on political pressures rather than conservation considerations. In some cases, wildlife protection policies have been so strict that subsistence-living indigenous populations have been forced from ancestral lands that fell within a preserve. In other cases, even if a preserve is designed to protect wildlife, if the protections are not or cannot be enforced, the preserve status will have little meaning in the face of illegal poaching and timber extraction. This is a widespread problem with preserves in areas of the tropics.

Limitations on Preserves

Some of the limitations on preserves as conservation tools are evident from the discussion of preserve design. Political and economic pressures typically make preserves smaller, rather than larger, so setting aside areas that are large enough is difficult. If the area set aside is sufficiently large, there may not be sufficient area to create a buffer around the preserve. In this case, an area on the outer edges of the preserve inevitably becomes a riskier suboptimal habitat for the species in the preserve. Enforcement of protections is also a significant issue in countries without the resources or political will to prevent poaching and illegal resource extraction.

Climate change will create inevitable problems with the *location* of preserves. The species within them may migrate to higher latitudes as the habitat of the preserve becomes less favorable. Scientists are planning for the effects of global warming on future preserves and striving to predict the need for new preserves to accommodate anticipated changes to habitats; however, the end effectiveness is tenuous since these efforts are prediction based.

5Robert H. MacArthur and Edward O. Wilson, E. O., *The Theory of Island Biogeography* (Princeton, N.J.: Princeton University Press, 1967).

Finally, an argument can be made that conservation preserves indicate that humans are growing more separate from nature, and that humans only operate in ways that do damage to biodiversity. Creating preserves may reduce the pressure on humans outside the preserve to be sustainable and non-damaging to biodiversity. On the other hand, properly managed, high quality preserves present opportunities for humans to witness nature in a less damaging way, and preserves may present some financial benefits to local economies. Ultimately, the economic and demographic pressures on biodiversity are unlikely to be mitigated by preserves alone. In order to fully benefit from biodiversity, humans will need to alter activities that damage it.

LINK TO LEARNING

An interactive global data system (http://openstax.org/l/protected_areas) of protected areas can be found at this website. Review data about individual protected areas by location or study statistics on protected areas by country or region.

Habitat Restoration

Habitat restoration holds considerable promise as a mechanism for restoring and maintaining biodiversity. Of course, once a species has become extinct, its restoration is impossible. However, restoration can improve the biodiversity of degraded ecosystems. Reintroducing wolves, a top predator, to Yellowstone National Park in 1995 led to dramatic changes in the ecosystem that increased biodiversity. The wolves (Figure 47.17) function to suppress elk and coyote populations and provide more abundant resources to the guild of carrion eaters. Reducing elk populations has allowed revegetation of riparian areas, which has increased the diversity of species in that habitat. Decreasing the coyote population has increased the populations of species that were previously suppressed by this predator. The number of species of carrion eaters has increased because of the predatory activities of the wolves. In this habitat, the wolf is a **keystone species**, meaning a species that is instrumental in maintaining diversity in an ecosystem. Removing a keystone species from an ecological community may cause a collapse in diversity. The results from the Yellowstone experiment suggest that restoring a keystone species can have the effect of restoring biodiversity in the community. Ecologists have argued for the identification of keystone species where possible and for focusing protection efforts on those species; likewise, it also makes sense to attempt to return them to their ecosystem if they have been removed.

Figure 47.17 (a) The Gibbon wolf pack in Yellowstone National Park, March 1, 2007, represents a keystone species. The reintroduction of wolves into Yellowstone National Park in 1995 led to a change in the grazing behavior of (b) elk. To avoid predation, the elk no longer grazed exposed stream and riverbeds, such as (c) the Lamar Riverbed in Yellowstone. This allowed willow and cottonwood seedlings to grow and recolonized large areas. The seedlings decreased erosion and provided shading to the creek, which improved fish habitat. A new colony of (d) beaver may also have benefited from the habitat change. (credit a: modification of work by Doug Smith, NPS; credit c: modification of work by Jim Peaco, NPS; credit d: modification of work by "Shiny Things"/Flickr)

Other large-scale restoration experiments underway involve dam removal, which is a national movement that is accelerating in importance. In the United States, since the mid-1980s, many aging dams are being considered for removal rather than replacement because of shifting beliefs about the ecological value of free-flowing rivers and because many dams no longer provide the benefit and functions that they did when they were first built. The measured benefits of dam removal include restoration of naturally fluctuating water levels (the purpose of dams is frequently to reduce variation in river flows), which leads to increased fish diversity and improved water quality. In the Pacific Northwest, dam removal projects are expected to increase populations of salmon, which is considered a keystone species because it transports key nutrients to inland ecosystems during its annual spawning migrations. In other regions such as the Atlantic coast, dam removal has allowed the return of spawning *anadromous fish species* (species that are born in fresh water, live most of their lives in salt water, and return to fresh water to spawn). Some of the largest dam removal projects have yet to occur or have happened too recently for the consequences to be measured. The large-scale ecological experiments that these removal projects constitute will provide valuable data for other dam projects slated either for removal or construction.

The Role of Captive Breeding

Zoos have sought to play a role in conservation efforts both through captive breeding programs and education. The transformation of the missions of zoos from collection and exhibition facilities to organizations that are dedicated to conservation is ongoing and gaining strength. In general, it has been recognized that, except in some specific targeted cases, captive breeding programs for endangered species are inefficient and often prone to failure when the species are reintroduced to the wild. However, captive breeding programs have yielded some success stories, such as the American condor reintroduction to the Grand Canyon and the reestablishment of the Whooping Crane along the Midwest flyway.

Unfortunately, zoo facilities are far too limited to contemplate captive breeding programs for the numbers of species that are now at risk. Education is another potential positive impact of zoos on conservation efforts, particularly given the global trend to urbanization and the consequent reduction in contacts between people and wildlife. A number of studies have been performed to look at the effectiveness of zoos on people's attitudes and actions regarding conservation; at present, the results tend to be mixed.

KEY TERMS

adaptive radiation rapid branching through speciation of a phylogenetic tree into many closely related species

biodiversity variety of a biological system, typically conceived as the number of species, but also applying to genes, biochemistry, and ecosystems

biodiversity hotspot concept originated by Norman Myers to describe a geographical region with a large number of endemic species and a large percentage of degraded habitat

bush meat wild-caught animal used as food (typically mammals, birds, and reptiles); usually referring to hunting in the tropics of sub-Saharan Africa, Asia, and the Americas

chemical diversity variety of metabolic compounds in an ecosystem

chytridiomycosis disease of amphibians caused by the fungus *Batrachochytrium dendrobatidis;* thought to be a major cause of the global amphibian decline

DNA barcoding molecular genetic method for identifying a unique genetic sequence to associate with a species

ecosystem diversity variety of ecosystems

endemic species species native to one place

exotic species (also, invasive species) species that has been introduced to an ecosystem in which it did not evolve

extinction disappearance of a species from Earth; local extinction is the disappearance of a species from a region

extinction rate number of species becoming extinct over time, sometimes defined as extinctions per million species–years to make numbers manageable (E/MSY)

genetic diversity variety of genes in a species or other taxonomic group or ecosystem, the term can refer to allelic diversity or genome-wide diversity

heterogeneity number of ecological niches

megafauna large animals

secondary plant compound compound produced as byproducts of plant metabolic processes that is usually toxic, but is sequestered by the plant to defend against herbivores

species-area relationship relationship between area surveyed and number of species encountered; typically measured by incrementally increasing the area of a survey and determining the cumulative numbers of species

tragedy of the commons economic principle that resources held in common will inevitably be overexploited

white-nose syndrome disease of cave-hibernating bats in the eastern United States and Canada associated with the fungus *Geomyces destructans*

CHAPTER SUMMARY

47.1 The Biodiversity Crisis

Biodiversity exists at multiple levels of organization and is measured in different ways depending on the scientific goals of those taking the measurements. These measurements include numbers of species, genetic diversity, chemical diversity, and ecosystem diversity. The number of described species is estimated to be 1.5 million with about 17,000 new species being described each year. Estimates for the total number of species on Earth vary but are on the order of 10 million. Biodiversity is negatively correlated with latitude for most taxa, meaning that biodiversity is higher in the tropics. The mechanism for this pattern is not known with certainty, but several plausible hypotheses have been advanced.

Five mass extinctions with losses of more than 50 percent of extant species are observable in the fossil record. Biodiversity recovery times after mass extinctions vary, but may be as long as 30 million years. Recent extinctions are recorded in written history and are the basis for one method of estimating contemporary extinction rates. The other method uses measures of habitat loss and species-area relationships. Estimates of contemporary extinction rates vary, but some rates are as high as 500 times the background rate, as determined from the fossil record, and are predicted to rise.

47.2 The Importance of Biodiversity to Human Life

Humans use many compounds that were first discovered or derived from living organisms as medicines: secondary plant compounds, animal toxins, and antibiotics produced by bacteria and fungi. More medicines will undoubtedly be discovered in nature. Loss of biodiversity will impact the number of pharmaceuticals available to humans.

Crop diversity is a requirement for food security, and it is being lost. The loss of wild relatives to crops also threatens breeders' abilities to create new varieties. Ecosystems provide ecosystem services that support human agriculture: pollination, nutrient cycling, pest control, and soil development and maintenance. Loss of biodiversity threatens these ecosystem services and risks making food production more expensive or impossible. Wild food sources are mainly aquatic, but few of these resources are being managed for sustainability. Fisheries' ability to provide protein to human populations is threatened when extinction occurs.

Biodiversity may provide important psychological benefits to humans. Additionally, there are moral arguments for the maintenance of biodiversity.

47.3 Threats to Biodiversity

The core threats to biodiversity are human population growth and unsustainable resource use. To date, the most significant causes of extinctions are habitat loss, introduction of exotic species, and overharvesting. Climate change is predicted to be a significant cause of extinctions in the coming century. Habitat loss occurs through deforestation, damming of rivers, and other disruptive human activities. Overharvesting is a threat particularly to aquatic species, while the taking of bush meat in the humid tropics threatens many species in Asia, Africa, and the Americas. Exotic species have been the cause of a number of extinctions and are especially damaging to islands and lakes. Exotic species' introductions are increasing damaging native ecosystems around the world because of the increased mobility of human populations and growing global trade and transportation. Climate change is forcing range changes that may lead to extinction. It is also affecting adaptations to the timing of resource availability that negatively affects species in seasonal environments. The impacts of climate change are greatest in the arctic. Global warming will also raise sea levels, eliminating some islands and reducing the area of all others.

47.4 Preserving Biodiversity

New technological methods such as DNA barcoding and information processing and accessibility are facilitating the cataloging of the planet's biodiversity. There is also a legislative framework for biodiversity protection. International treaties such as CITES regulate the transportation of endangered species across international borders. Legislation within individual countries protecting species and agreements on global warming have had limited success; the Paris Climate accord is currently being implemented as a means to reduce global climate change.

In the United States, the Endangered Species Act protects listed species but is hampered by procedural difficulties and a focus on individual species. The Migratory Bird Act is an agreement between Canada and the United States to protect migratory birds. The non-profit sector is also very active in conservation efforts in a variety of ways.

Conservation preserves are a major tool in biodiversity protection. Presently, 11 percent of Earth's land surface is protected in some way. The science of island biogeography has informed the optimal design of preserves; however, preserves have limitations imposed by political and economic forces. In addition, climate change will limit the effectiveness of preserves in the future. A downside of preserves is that they may lessen the pressure on human societies to function more sustainably outside the preserves.

Habitat restoration has the potential to restore ecosystems to previous biodiversity levels before species become extinct. Examples of restoration include reintroduction of keystone species and removal of dams on rivers. Zoos have attempted to take a more active role in conservation and can have a limited role in captive breeding programs. Zoos also have a useful role in education.

VISUAL CONNECTION QUESTIONS

1. Figure 47.6 Scientists measured the relative abundance of fern spores above and below the K-Pg boundary in this rock sample. Which of the following statements most likely represents their findings?
   a. An abundance of fern spores from several species was found below the K-Pg boundary, but none was found above.
   b. An abundance of fern spores from several species was found above the K-Pg boundary, but none was found below.
   c. An abundance of fern spores was found both above and below the K-Pg boundary, but only one species was found below the boundary , and many species were found above the boundary.
   d. Many species of fern spores were found both above and below the boundary, but the total number of spores was greater below the boundary.

2. Figure 47.9 The Svalbard Global Seed Vault is located on Spitsbergen island in Norway, which has an arctic climate. Why might an arctic climate be good for seed storage?

3. Converting a prairie to a farm field is an example of ________.
   a. overharvesting
   b. habitat loss
   c. exotic species
   d. climate change

4. Figure 47.16 Which of the following statements is not supported by this graph?
   a. There are more vulnerable fishes than critically endangered and endangered fishes combined.
   b. There are more critically endangered amphibians than vulnerable, endangered and critically endangered reptiles combined.
   c. Within each group, there are more critically endangered species than vulnerable species.
   d. A greater percentage of bird species are critically endangered than mollusk species.

REVIEW QUESTIONS

5. With an extinction rate of 100 E/MSY and an estimated 10 million species, how many extinctions are expected to occur in a century?
   a. 100
   b. 10,000
   c. 100,000
   d. 1,000,000

6. An adaptive radiation is_______.
   a. a burst of speciation
   b. a healthy level of UV radiation
   c. a hypothesized cause of a mass extinction
   d. evidence of an asteroid impact

7. The number of currently described species on the planet is about _______.
   a. 17,000
   b. 150,000
   c. 1.5 million
   d. 10 million

8. A mass extinction is defined as _______.
   a. a loss of 95 percent of species
   b. an asteroid impact
   c. a boundary between geological periods
   d. a loss of 50 percent of species

9. A secondary plant compound might be used for which of the following?
   a. a new crop variety
   b. a new drug
   c. a soil nutrient
   d. a pest of a crop pest

10. Pollination is an example of_______.
    a. a possible source of new drugs
    b. chemical diversity
    c. an ecosystem service
    d. crop pest control

11. What is an ecosystem service that performs the same function as a pesticide?
    a. pollination
    b. secondary plant compounds
    c. crop diversity
    d. predators of pests

12. Which two extinction risks may be a direct result of the pet trade?
    a. climate change and exotic species introduction
    b. habitat loss and overharvesting
    c. overharvesting and exotic species introduction
    d. habitat loss and climate change

13. Exotic species are especially threatening to what kind of ecosystem?
    a. deserts
    b. marine ecosystems
    c. islands
    d. tropical forests

14. Certain parrot species cannot be brought to the United States to be sold as pets. What is the name of the legislation that makes this illegal?
    a. Red List
    b. Migratory Bird Act
    c. CITES
    d. Endangered Species Act (ESA)

15. What was the name of the first international agreement on climate change?
    a. Red List
    b. Montreal Protocol
    c. International Union for the Conservation of Nature (IUCN)
    d. Kyoto Protocol

16. About what percentage of land on the planet is set aside as a preserve of some type?
    a. 1 percent
    b. 6 percent
    c. 11 percent
    d. 15 percent

CRITICAL THINKING QUESTIONS

17. Describe the evidence for the cause of the Cretaceous–Paleogene (K–Pg) mass extinction.
18. Describe the two methods used to calculate contemporary extinction rates.
19. Explain how biodiversity loss can impact crop diversity.
20. Describe two types of compounds from living things that are used as medications.
21. Describe the mechanisms by which human population growth and resource use causes increased extinction rates.
22. Explain what extinction threats a frog living on a mountainside in Costa Rica might face.
23. Describe two considerations in conservation preserve design.
24. Describe what happens to an ecosystem when a keystone species is removed.

APPENDIX A

The Periodic Table of Elements

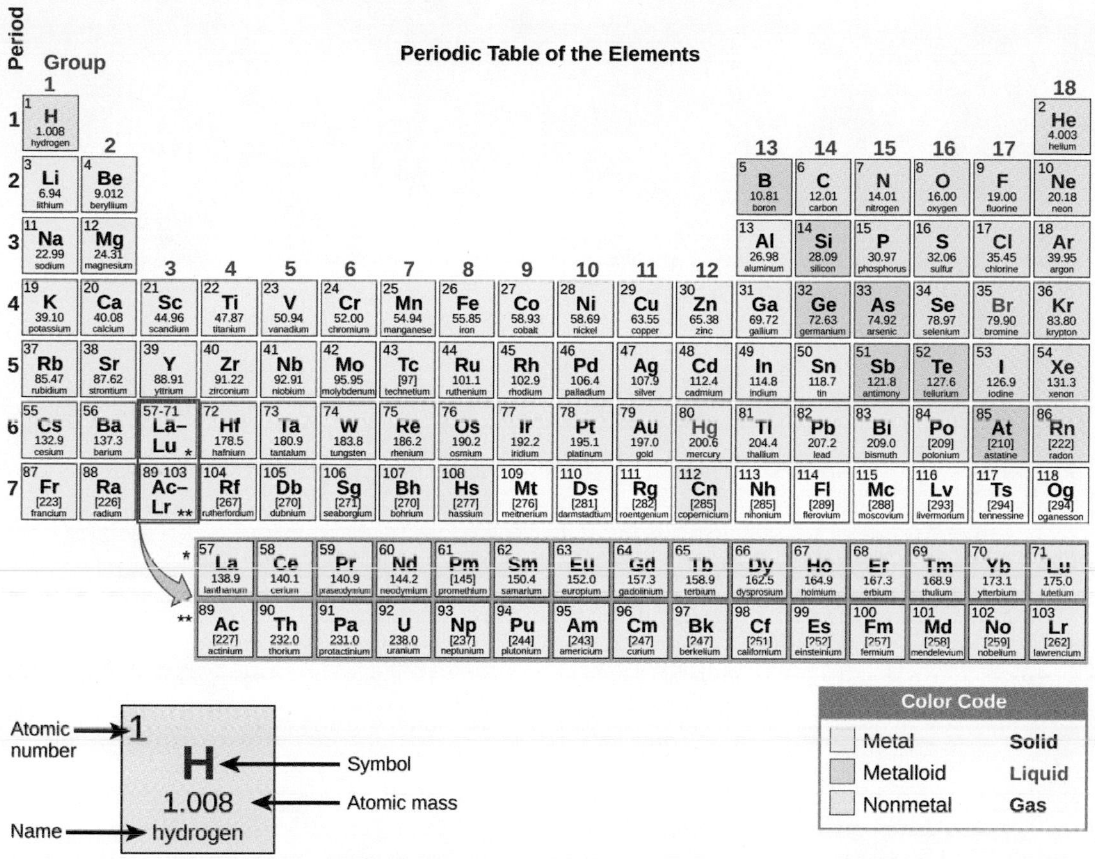

Figure A1

APPENDIX B

Geological Time

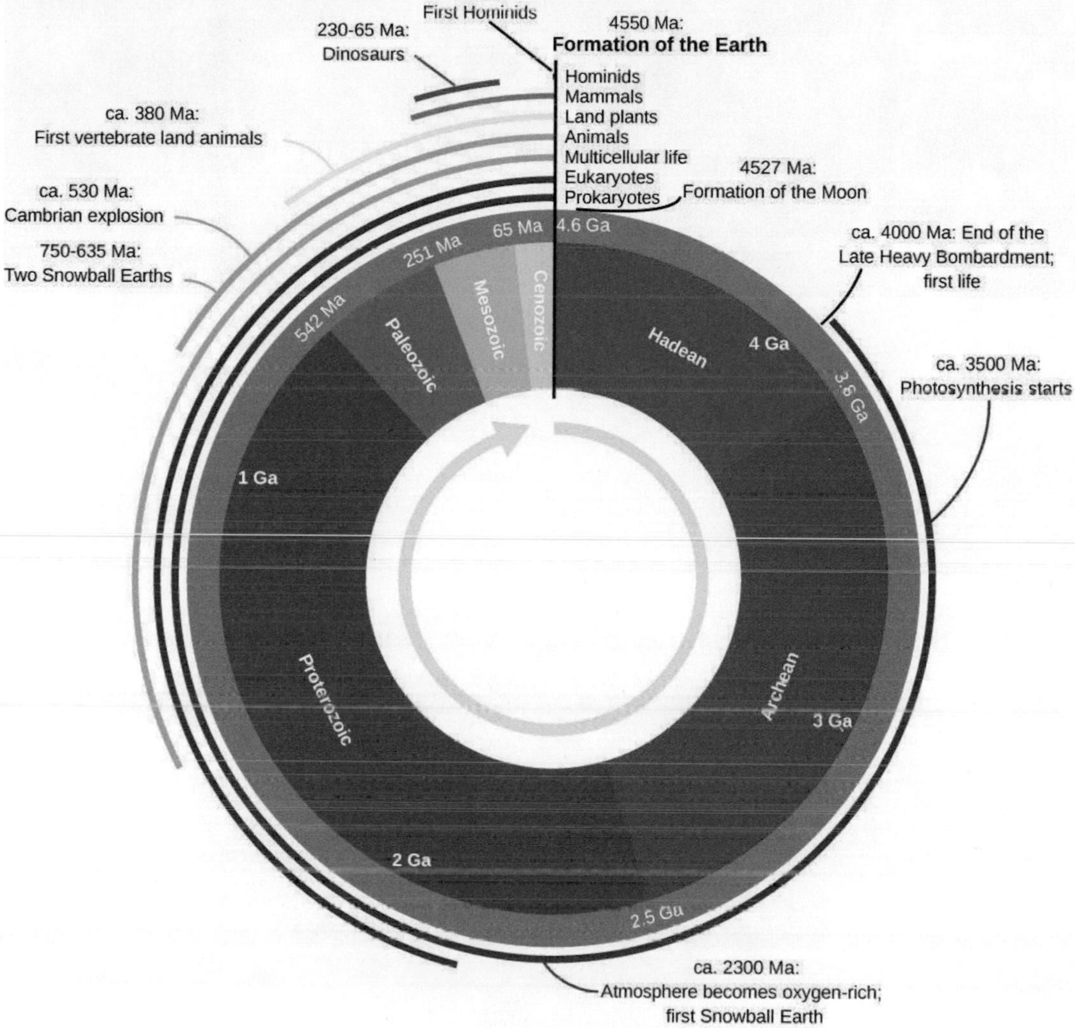

Figure B1 Geological Time Clock

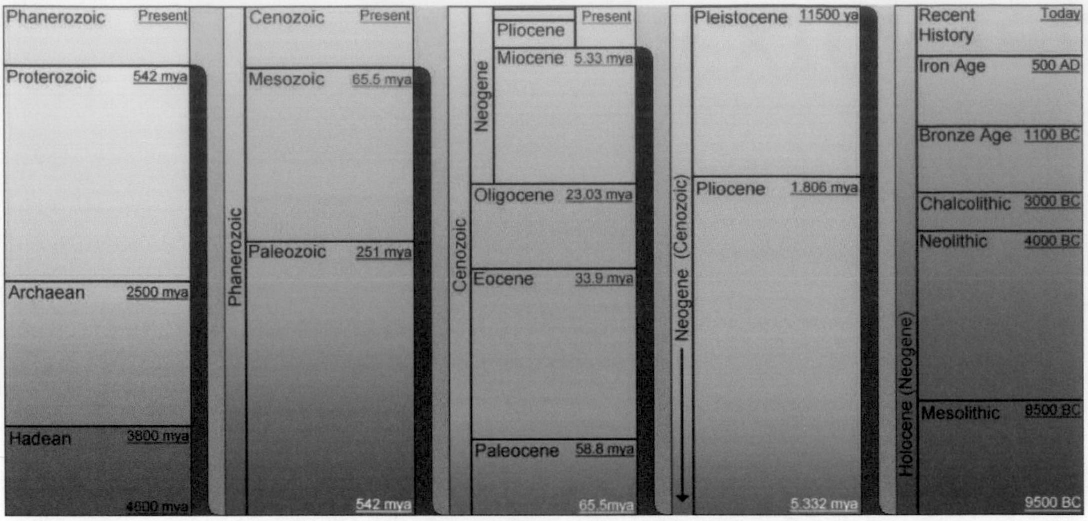

Figure B2 Geological Time Chart
(credit: Richard S. Murphy, Jr.)

APPENDIX C

Measurements and the Metric System

Measurements and the Metric System

Measurement	Unit	Abbreviation	Metric Equivalent	Approximate Standard Equivalent
Length	nanometer	nm	1 nm = 10^{-9} m	1 mm = 0.039 inch 1 cm = 0.394 inch 1 m = 39.37 inches 1 m = 3.28 feet 1 m = 1.093 yards 1 km = 0.621 miles
	micrometer	μm	1 μm = 10^{-6} m	
	millimeter	mm	1 mm = 0.001 m	
	centimeter	cm	1 cm = 0.01 m	
	meter	m	1 m = 100 cm 1 m = 1000 mm	
	kilometer	km	1 km = 1000 m	
Mass	microgram	μg	1 μg = 10^{-6} g	1 g = 0.035 ounce 1 kg = 2.205 pounds
	milligram	mg	1 mg = 10^{-3} g	
	gram	g	1 g = 1000 mg	
	kilogram	kg	1 kg = 1000 g	
Volume	microliter	μl	1 μl = 10^{-6} l	1 ml = 0.034 fluid ounce 1 l = 1.057 quarts 1 kl = 264.172 gallons
	milliliter	ml	1 ml = 10^{-3} l	
	liter	l	1 l = 1000 ml	
	kiloliter	kl	1 kl = 1000 l	
Area	square centimeter	cm^2	1 cm^2 = 100 mm^2	1 cm^2 = 0.155 square inch 1 m^2 = 10.764 square feet 1 m^2 = 1.196 square yards 1 ha = 2.471 acres
	square meter	m^2	1 m^2 = 10,000 cm^2	
	hectare	ha	1 ha = 10,000 m^2	
Temperature	Celsius	°C	—	1 °C = 5/9 × (°F – 32)

Table C1

INDEX

Symbols

A

B

C

D

E

F

G

H

I

J

K

L

N

Q

R

S